S0-BEB-766

Uwe Windhorst • Håkan Johansson (Eds.)

Modern Techniques in Neuroscience Research

Springer

Berlin
Heidelberg
New York
Barcelona
Hong Kong
London
Milan
Paris
Singapore
Tokyo

Uwe Windhorst • Håkan Johansson (Eds.)

Modern Techniques in Neuroscience Research

With 471 Figures, 43 Color Figures, 33 Tables and CD-ROM

Professor Dr. Uwe Windhorst

Center of Musculoskeletat Research
National Institute for Working Life
Petrus Laestadius väg
90713 Umeå, Sweden

Department of Clinical Neurosciences and
Department of Physiology and Biophysics
Faculty of Medicine
The University of Calgary
333 Hospital Drive N.W.
Calgary, Alberta T2N 4N1, Canada

Zentrum Physiologie und Pathophysiologie
Universität Göttingen
Humboldtallee 23
37073 Göttingen, Germany

Professor Dr. Håkan Johansson

Center of Musculoskeletat Research
National Institute for Working Life
Petrus Laestadius väg
90713 Umeå, Sweden

ISBN 3-540-64460-1 Springer-Verlag Berlin Heidelberg New York

Library of Congress Cataolging-in-Publication Data
Modern techniques in neuroscience research / U. Windhorst, H. Johannson (Eds.)
p. cm.
Includes bibliographical referneces and index.
isbn 3-540-64460-1 (hbk. : alk. paper)
1. Neurosciences Laboratory manuals. I. Windhorst, Uwe, 1946-
. II. Johannson, H. (Hakån), 1947- .
RC337.M63 1999
573'.8'072--dc21 99-32065
CIP

Typesetting: medio, Berlin
Cover design: design.& production GmbH, Heidelberg
SPIN: 10574645 27/3136 - 5 4 3 2 1 0 - Printed on acid free paper

Preface

> Nothing tends so much to the advancement of knowledge as the application of a new instrument.
> – *Sir Humphry Davy, 1778–1829*

Neuroscience has become a rapidly expanding endeavor that relies on a number of other sciences, such as mathematics, physics, chemistry, engineering, computer science, general biology and medicine, genetics etc. In fact, many of its recent successes result from the application of ideas and methods borrowed from these fields. Insofar, it is a true interdisciplinary undertaking. This convergence of influences accounts for part of its enormous attractiveness and fascination to students and researchers from diverse walks of life, or science, for that matter. It is probably fair to say that a great number of neuroscience´s most creative and productive proponents have been lured into this field not only by the excitement about the possibility to unmask the secrets of the human mind, but also by the appeal of a vast unknown land needing cultivation and tools to cultivate it.

Danger may arise for any science if it is dominated by methods and techniques of investigation rather than by problems to be solved and concepts to be developed. This might concentrate efforts onto the technically feasible and doable, rather than on the real issues. But, on the other hand, especially the young and growing sciences are heavily dependent on the development and application of methods, often even before a problem relying on these methods may become apparent. In fact, under favorable circumstances, these methods may reveal problems that have not been thought of before. Several examples could be evoked to make this point. On the other hand, the stream of influence is no one-way road. Techniques and findings in neuroscience may have an impact in other sciences and practical applications. All this provides for a fascinating tangle of relationships.

To try and display the variety of techniques used in modern neuroscience research is a tremendous task, just for the sheer vastness and complexity of neuroscience itself and the diversity of methods used. This is well illustrated by other lab manuals in this series, where the description of a single technique, its descendants and ramifications, advantages and limitations, applications and pitfalls may occupy a sizeable volume. Quite the same completeness and richness of detail cannot be achieved in a book attempting at presenting a somewhat comprehensive overview, and quite the same lab-manual protocol style cannot be consistently maintained throughout. However, such an overview may still be worthwhile for two main reasons. First, it may supply just that, an overview for the technically minded ever on the search for applications, that is for those who are appealed by new lands to be explored by tools and the challenge to apply known and to develop new tools. Second, if done well, such a book may present an introduction into neuroscience, from the angle of methods available to tackle questions.

Because neuroscience is an inter-disciplinary endeavor, it is inter-methodological. By this we mean that certain questions can be answered by combining methods borrowed and adapted from different fields. This important idea underlies the organization of this book. The organizing principle is the investigation of levels of the nervous system, from molecular and cellular to whole organisms. We have chosen this principle for two related reasons. First, we would like to show how certain organizational levels can be studied and potentially understood by the application of an array of diverse techniques, thus demonstrating what a particular method can contribute to the understanding of a specific object. Second, we thereby hope to avoid succumbing to the danger of pure technicality, which might arise from ordering the description of techniques according to their inherent principles of coherence (e.g. staining vs. electrophysiological techniques). This might lead to methods living their own life, detached from what they are used for. As a matter of fact, this approach can only be taken *cum grano salis*, for there are techniques that can be employed at various levels (e.g. systems analysis and synergetic approaches).

Thanks are due to a vast number of people. First, we are very grateful to our authors who have taken the time off their busy schedules to contribute one or more chapters to this book. Dr. Rolf Lange at Springer-Verlag has been the enthusiastic initiator of this project, and without his continuing encouragement one or the other editor would probably have given up along the way. Mrs. Anne Clauss at Springer-Verlag has been the ever-reliable manager in Heidelberg, and particularly valuable has been her ability to nudge authors to - please - finally submit their manuscripts when the editors were on the verge of despair. Many thanks are due to supportive staff in the background, particularly the copy editors who, as always, have done an outstanding job in polishing the chapters. Last but not least we would like to thank our wives, Eva and Siggi, for their patience and indulgence in their husbands' obsession with this book.

Uwe Windhorst
Håkan Johansson
Umeå, Juni 1999

Contents

Chapter 16

Optical Recording from Populations of Neurons in Brain Slices

Saurabh R. Sinha and Peter Saggau

Chapter 17

Recording of Electrical Activity of Neuronal Populations

Hakan Johansson, Mikael Bergenheim, Jonas Pedersen and Mats Djupsjöbacka

Chapter 18

Time and Frequency Domain Analysis of Spike Train and Time Series Data

David M. Halliday and Jay R. Rosenberg

Chapter 19

Information-Theoretical Analysis of Sensory Information

Yoav Tock and Gideon F. Inbar

List of Authors

ALEXANDER ADAM
NeuroMuscular Research Center
Boston University
44 Cummington Avenue
Boston, MA 02215
USA

BAGRAT AMIRIKIAN, PH.D.
Brain Sciences Center
Veterans Administration Medical Center
One Veterans Drive
Minneapolis, MN 55417
and
Dept. of Physiology
University of Minnesota
Medical School
6-255 Millard Hall
435 Delaware Street SE
Minneapolis, MN 55455, USA

A. ARIELI
The Grodetsky Center for Research
of Higher Brain Functions
The Weizmann Institute of Science
Rehovot
76100 Israel

KLAUS BALLANYI, DR. RER. NAT.
Universität Göttingen
II. Physiologisches Institut
Abteilung Neuro- und Sinnesphysiologie
Humboldtallee 23
D-37075 Göttingen
Germany

ROBERT W. BANKS, PH.D.
University of Durham
Department of Biological Sciences
Science Laboratories
South Road
Durham DH1 3 LE
UK

WOLFGANG BECKER, DR. -ING.
Sektion Neurophysiologie
Universität Ulm
Albert-Einstein-Str. 47
D-89081 Ulm
Germany

MIKAEL BERGENHEIM, PH.D.
Center of Musculoskeletal Research
National Institute for Working Life
P.O. Box 7654
S-907 13 Umeå
Sweden

STEFAN BLÜML, PH.D.
Clinical Magnetic Resonance Spectroscopy
Unit
Huntington Medical Research Institute
660 S. Fair Oaks Avenue
Pasadena, CA 91105
USA

ANDREW BULLEN
Division of Neuroscience
Baylor College of Medicine
One Baylor Plaza
Houston, TX 77030
and
Department of Physiology
University of Pennsylvania School
of Medicine
Philadelphia
Pennsylvania 19104
USA

KENNETH CAMPBELL, PH.D.
Department of Physiology and
Neuroscience
Wallenberg Neuroscience Center
Lund University
Sölvegatan 17
S-22362 Lund
Sweden

David R. Collins
Center for the Ecological Study of Perception and Action
University of Connecticut
406 Babbidge Road, Box U-20
Storrs, CT 06269-1020
USA

Fabrice Crivello, Ph.D.
Groupe d'Imagerie Neurofonctionnelle
GIP Cyceron
F-14074 Caen Cedex
France

Nicole Datson, Ph.D.
Division of Medical Pharmacology
Leiden/Amsterdam Cente for
Drug Research (LACDR)
P.O. Box 9503
Wassenaarseweg 72
2300 RA Leiden
The Netherlands

Nicholas J. Davey, Ph.D.
Department of Sensorimotor Systems
Division of Neuroscience & Psychological Medicine
Imperial College School of Medicine
Charing Cross Hospital
Fulham Palace Road
London W6 8RF
UK

Scott Day, Ph.D.
Center of Musculoskeletal Research
National Institute for Working Life
P.O. Box 7654
S-90 13 Umeå
Sweden

Jeanette De Jong, Ph.D.
Division of Medical Pharmacology
Leiden/amsterdam Cente for Drug Research (LACDR)
P.O.Box 9503
Waassenaarseweg 72
2300 RA Leiden
The Netherlands

Carlo J. De Luca
NeuroMuscular Research Center
Boston University
44 Cummington Avenue
Boston, MA 02181
USA

Volker Diekmann
Sektion Neurophysiologie
Universität Ulm
Albert-Einstein-Str. 47
D-89081 Ulm
Germany

Mats Djupsjöbacka, Ph.D.
Center of Musculoskeletal Research
National Institute for Working Life
P.O. Box 7654
S-90 13 Umeå
Sweden

Jonathan Dostrovsky, Ph.D.
Dept. of Physiology
University of Toronto
Toronto, ON M5S 1A8
Canada

Addie Ehrenstein, Ph.D.
Department of Psychonomics
Universiteit Utrecht
Heidelberglaan 2 (Room 17.24 C)
NL-3584 CS Utrecht
The Netherlands

Walter H. Ehrenstein, Dr. rer. nat.
Institut fuer Arbeitsphysiologie
an der Universität Dortmund
Ardeystr. 67
D-44139 Dortmund
Germany

Peter H. Ellaway, Ph.D.
Department of Sensorimotor Systems
Division of Neuroscience & Psychological Medicine
Imperial College School of Medicine
Charing Cross Hospital
Fulham Palace Road
London W6 8RF
UK

Sergio N. Erné
Zentralinstitut für Biomedizinische Technik
Universität Ulm
D-89081 Ulm
Germany

Jens Frahm, Dr. rer. nat.
Biomedizinische NMR Forschungs GmbH
am Max-Planck-Institut für biophysikalische Chemie
D-37070 Göttingen
Germany

Peter Fransson
MR Research Center
Department of Clinical Neuroscience
Karolinska Institutet
17176 Stockholm
Sweden

Andrew S. French, Ph.D.
Department of Physiology and Biophysics
Dalhousie University
Halifax, Nova Scotia B3H 4H7
Canada

Rosemary Fricker, Ph.D.
Children's Hospital
Department of Neuroscience
Harvard Medical School
350 Enders Bldg.
320 Longwood Avenue
Boston, MA 02115
USA

Fred H. Gage, Ph.D.
The Salk Institute
Laboratory of Genetics
P.O. Box 85800
San Diego, CA 92186-5800
USA

Björn Gerdle, Ph.D.
Department of Rehabilitation Medicine
Faculty of Health Science
and
Pain and Rehabilitation Centre
University Hospital
S-581 85 Linköping
Sweden

Amiram Grinvald, Ph.D.
The Grodetsky Center for Research of
Higher Brain Functions
The Weizmann Institute of Science
Rehovot
76100 Israel

D. Glaser
The Grodetsky Center for Research of
Higher Brain Functions
The Weizmann Institute of Science
Rehovot
76100 Israel

James D. Guest, M.D., Ph.D.
Department of Surgery and Collaboration
on Repair Discoveries (CORD)
University of British Columbia and
Vancouver Hospital and Health Sciences
Centre
910 West 10th Ave.
Vancouver, BC, V5Z 4E3
Canada
Current Address:
Barrow Neurological Institute
350 West Thomas rd.
Phoenix, Arizona 85013
USA

David M. Halliday, Ph.D.
Division of Neuroscience and Biomedical
Systems
Institute of Biomedical and Life Sciences
West Medical Building
University of Glasgow
Glasgow, G12 8QQ
Scotland
UK

Morten Haugland
Center for Sensory-Motor Interaction (SMI)
Dept. of Medical Informatics and Image
Analysis
Aalborg University
Fredrik Bajers Vej 7D-3
DK-9220 Aalborg
Denmark

Forrest A. Haun, Ph.D.
NeuroDetective, Inc.
1757 Wentz Rd.
Quakertown, Pennsylvania, 18951
USA

Walter Herzog
Human Performance Laboratory
University of Calgary
2500 University Drive N.W.
Calgary, Alberta T2N 1N4
Canada

R. Hildesheim
The Grodetsky Center for Research of
Higher Brain Functions
The Weizmann Institute of Science
Rehovot
76100 Israel

Gideon F. Inbar, Ph.D
Department of Electrical Engineering
Technion - Israel Institute of Technology
Haifa 32000
Israel

Alexei M. Ivanitsky, M.D.
Institute of Higher Nervous Activity and Neurophysiology
Russian Academy of Sciences
5a Butlerov str.
Moscow GSP-7
Russia 117865

George A. Ivanitsky, M.D.
Institute of Higher Nervous Activity and Neurophysiology
Russian Academy of Sciences
5a Butlerov str.
Moscow GSP-7
Russia 117865

Stig Jacobsson
Department of Biomedicine
Division of NBC Defence
Defence Research Establishment (FOA)
and
Department of Pharmacology
Umeå University
S-901 87 Umeå

Håkan Johansson, Ph.D.17
Center of Musculoskeletal Research
National Institute for Working Life
P.O. Box 7654
S-907 13 Umeå
Sweden

Stefan Karlsson
Department of Biomedical Engineering and Informatics
University Hospital
S-901 85 Umeå
Sweden

Amir Karniel
Department of Electrical Engineering
Technion - Israel Institute of Technology
Haifa 32000
Israel

Jan Kehr, Ph.D.
Division of Cell and Molecular Neurochemistry
Department of Neuroscience
Karolinska Institute
S-171 77 Stockholm
Sweden

Bryan E. Kolb, Ph.D.
Department of Psychology
University of Lethbridge
Lethbridge AB T1K 3M4
Canada

Gunnar Krueger
Department of Radiology
Stanford University
1201 Welch Road
Palo Alto, CA 94304
USA

Peter M. Lalley, Ph.D.
University of Wisconsin Madison
Medical Science Center
Department of Physiology
1300 University Avenue
Madison, WI 53706
USA

Daniel H. Lange, D.Sc.
HP Labs Israel
Haifa 32000
Israel

Kenneth A. Lindsay
Department of Mathematics
University of Glasgow
University Gardens
Glasgow G12 8QW
United Kingdom

Milos Ljubisavljević, Ph.D.
Institute for Medical Research
Dr Subotica 4
PO Box 721
11001 Belgrade
Yugoslavia
and
Center of Musculoskeletal Research
National Institute for Working Life
Box 7654
Umeå
Sweden

Gerald E. Loeb, M.D.
Department of Physiology
Queen's University
Kingston, Ontario K7L 3N6
Canada

Peter Lovell, Ph.D.
Department of Cell Biology and Anatomy
Faculty of Medicine
The University of Calgary
3330 Hospital Drive N.W.
Calgary, Alberta T2N 4N1
Canada

Christopher D. Mah, Ph.D.
Department of Physical Medicine and Rehabilitation
Northwestern University
345 E. Superior, # 1406
Chicago, IL 60611
USA

Vasilis Z. Marmarelis, Ph.D.
Biomedical Simulations Resource
University of Southern California
Los Angeles, CA 90089-1451
USA

Bernard Mazoyer Ph.D.
Directeur
Groupe d'Imagerie Neurofonctionnelle
GIP Cyceron
F-14000 Caen Cedex
France

Adonis K. Moschovakis, Ph.D.
University of Crete School of Health Sciences
Department of Basic Sciences
P.O. Box 1393
Iraklion
Crete
Greece

Andrei R. Nikolaev, M.D.
Institute of Higher Nervous Activity and Neurophysiology
Russian Academy of Sciences
5a Butlerov str.
Moscow GSP-7
Russia 117865

Martin Olsson, Ph.D.
Department of Physiology and Neuroscience
Wallenberg Neuroscience Center
Lund University
Sölvegatan 17
S-22362 Lund
Sweden

J.M. Odgen
Department of Mathematics
University of Glasgow
University Gardens
Glasgow G12 8QW
United Kingdom

Jonas Pedersen, Ph.D.
National Institute for Working Life
Center of Musculoskeletal Research
P.O. Box 7654
S-907 13 Umeå
Sweden

M.B. Popovic
Institute for Medical Research
Dr Subotica 4
PO Box 721
11001 Belgrade
Yugoslavia

Jasodhara Ray, Ph.D.
Laboratory of Genetics
Salk Institute for Biological Studies
10010 N. Torrey Pines Road
La Jolla, CA 92037
USA

Edith Ribot-Ciscar, Ph.D.
Laboratoire de Neurobiologie Humaine
UMR CNRS 6562
Université de Provence de St. Jérome
Avenue Escadrille Normandie Niemen
13397 Marseille Cedex 20
France

Ronald Riso
Center for Sensory-Motor Interaction (SMI)
Dept. of Medical Informatics and Image Analysis
Aalborg University
Fredrik Bajers Vej 7D-3
DK-9220 Aalborg
Denmark

Jean-Pierre Roll, Ph.D.
Laboratoire de Neurobiologie Humaine
UMR CNRS 6562
Université de Provence de St. Jérome
Avenue Escadrille Normandie Niemen
13397 Marseille Cedex 20
France

Jay R. Rosenberg, Ph.D.
Division of Neuroscience and Biomedical Systems
Institute of Biomedical and Life Sciences
West Medical Building
University of Glasgow
Glasgow G12 8QQ
Scotland
UK

Brian D. Ross, M.D., D.Phil. (Oxon)
Clinical Magnetic Resonance Spectroscopy Unit
Huntington Medical Research Institute
660 S. Fair Oaks Avenue
Pasadena, CA 91105
USA

Boris V. Safronov, Ph.D.
Physiologisches Institut
Justus-Liebig-Universitaet Giessen
Aulweg 129
D-35392 Giessen
Germany

H. Slovin
The Grodetsky Center for Research of Higher Brain Functions
The Weizmann Institute of Science
Rehovot
76100 Israel

Peter Saggau, Dr. rer. nat.
Division of Neuroscience
Baylor College of Medicine
One Baylor Plaza
Houston, TX 77030
USA

Åke Sellström, Ph.D.
Department of Biomedicine
Division of NBC Defence
Defence Research Establishment (FOA)
S-901 82 Umeå
Sweden

D. Shoham
The Grodetsky Center for Research of Higher Brain Functions
The Weizmann Institute of Science
Rehovot
76100 Israel

A. Shmuel
The Grodetsky Center for Research of Higher Brain Functions
The Weizmann Institute of Science
Rehovot
76100 Israel

E. Shtoyerman
The Grodetsky Center for Research of Higher Brain Functions
The Weizmann Institute of Science
Rehovot
76100 Israel

Saurabh R. Sinha, Ph.D.
Division of Neuroscience
Baylor College of Medicine
One Baylor Plaza
Houston
Texas 77030
USA

Thomas Sinkjaer, Ph.D., Dr. med.
Center for Sensory-Motor Interaction (SMI)
Dept. of Medical Informatics and Image Analysis
Aalborg University
Fredrik Bajers Vej 7D-3
DK-9220 Aalborg
Denmark

Johannes Struijk
Center for Sensory-Motor Interaction (SMI)
Dept. of Medical Informatics and Image Analysis
Aalborg University
Fredrik Bajers Vej 7D-3
DK-9220 Aalborg
Denmark

Naweed I. Syed, Ph.D.
Neuroscience Research Group
Faculty of Medicine
The University of Calgary
3330 Hospital Drive N.W.
Calgary, Alberta T2N 4N1
Canada

Yoav Tock
Department of Electrical Engineering
Technion - Israel Institute of Technology
Haifa 320 00
Israel

Michael T. Turvey, Ph.D.
Department of Psychology
Center for the Ecological Study of Perception and Action
University of Connecticut
406 Babbidge Road, U-20
Storrs, Connecticut 06269-1020
USA

R.E. van Kesteren
Graduate School of Neurosciences Amsterdam
Research Institute Neurosciences
Faculty of Biology
Vrije Universiteit,
De Boelelaan 1087,
1081 HV Amsterdam,
The Netherlands

Jan van Minnen, Ph.D.
Graduate School of Neurosciences Amsterdam
Research Institute Neurosciences
Faculty of Biology
Vrije Universiteit,
De Boelelaan 1087,
1081 HV Amsterdam,
The Netherlands

I. Vanzetta
The Grodetsky Center for Research of
Higher Brain Functions
The Weizmann Institute of Science
Rehovot
76100 Israel

Werner Vogel, Dr. rer. nat.
Physiologisches Institut
Justus-Liebig-Universitaet Giessen
Aulweg 129
D-35392 Giessen
Germany

Erno Vreugdenhil, Ph.D.
Division of Medical Pharmacology
University of Leiden
Sylvius Laboratory
PO BOX 9503
Wassenaarseweg 72
2300 RA Leiden
Belgium

Ian Q. Whishaw, Ph.D.
Department of Psychology
University of Lethbridge
Lethbridge, Alberta T1K 3M4
Canada

Klas Wictorin, M.D. Ph.D.
Department of Physiology & Neuroscience
Wallenberg Neuroscience Ctr
Section Neurobiology
Lund University
Sölvegatan 17
S-223 62 Lund
Sweden

C. Wijnbergen
The Grodetsky Center for Research of
Higher Brain Functions
The Weizmann Institute of Science
Rehovot
76100 Israel

Uwe Windhorst, M.D.
Center of Musculoskeletal Research
National Institute for Working Life
P.O. Box 7654
S-907 13 Umeå
Sweden

Satoshi Yamada, Ph.D.
Advanced Technology R&D Center,
Mitsubishi Electric Corporation
8-1-1, Tsukaguchi-Honmachi,
Amagasaki, Hyogo 661-8661
Japan

Yoshiharu Yamamoto, Ph.D.
Educational Physiology Laboratory
Graduate School of Education
The University of Tokyo
7-3-1 Hongo, Bunkyo-ku
Tokyo 113-0033
Japan

Hasan Zaidi, Ph.D.
Department of Cell Biology and Anatomy
Faculty of Medicine
The University of Calgary
3330 Hospital Drive N.W.
Calgary, Alberta T2N 4N1
Canada

Thomas J. Zwimpfer, M.D., Ph.D.
Departments of Surgery and Zoology and
Collaboration on Repair Discoveries
(CORD)
University of British Columbia and
Vancouver Hospital and Health Sciences
Centre
300-700 West 10th Ave.
Vancouver, BC V5Z 4E5
Canada

Chapter 1

Cytological Staining Methods

R.W. Banks

Introduction

General Introduction

Neurohistology is perhaps traditionally thought of as supplying information only about the spatial or structural aspects of the neuron; however, it is my intention in the present chapter and its companion (Chapter 15 in Sect.II) to present as far as possible a unified view of neurohistology as a set of related problems centred on the relationship between structure and function of nerve cells. Such problems are not unique to the subject, but in the context of neurohistology they are uniquely complex. Neurons, as cells, are unrivalled in diversity of type, and other kinds of cells rarely match any neuron in the complexity of their spatio-temporal properties or in the range of genes expressed. The status of neurohistology as a recognisable discipline is therefore dependent on these properties of nerve cells and nervous tissue, and its history is largely one of the development of methods aimed at overcoming the difficulties presented by them. Of course, recognisable disciplines need not necessarily have sharp boundaries and it is perhaps already apparent that I intend to take a fairly relaxed view as to what constitutes neurohistology. The essential criteria are whether the investigation involves the nervous system and whether it uses microscopy. Beyond those, it is a matter of taste where macroscopically neuroanatomy and neuroimaging give way to neurohistology, and microscopically neurohistology gives way to cellular and molecular biology.

Within the discipline, boundaries must be arbitrary and harder to defend. The division into topics that can be described as cytological (this chapter) and histological *sensu stricto* (Chap. 15) creates such a boundary that is more convenient than real; many of the techniques covered will be applicable in either area. In a similarly cavalier fashion, I shall gather several specific techniques under rather broad and by no means exclusive headings so as to emphasize common purposes of the often disparate methods. It might be argued that the overall purpose is to provide as close as possible a description of neurons and nervous systems in their living state. Clearly neurohistology alone is incapable of reaching that end, but it is essential to its attainment. What is certain is that good neurohistology requires more than the mechanical application of various technical procedures aimed at a static description of the microscopic appearance of the nervous system. I suggest that what is indeed essential is the intelligent and informed combination of structural and functional elements, or at least of the interpretation of structure in functional terms. I hope to demonstrate the truth of this by placing several techniques in the context of specific problems in neuroscience. Any protocols and practical advice given in my chapters will be contained in these case-studies. Equally, if not more, important will be the intervening sections in which the evolutionary development and theoretical

R. W. Banks, University of Durham, Department of Biological Sciences, South Road, Durham, DH1 3LE, UK (phone: +44-191-374-3354; fax: +44-191-374-2417; e-mail: r.w.banks@durham.ac.uk)

backgrounds of various methods are briefly considered in order to highlight their possibilities and limitations.

The Beginnings of Neurohistology

"Often, and not without pleasure, I have observed the structure of the nerves to be composed of very slender vessels of an indescribable fineness, running lengthwise to form the nerve" (Leeuwenhoek, 1717).

Leeuwenhoek's account of his observations on the spinal nerves of cows and sheep, almost certainly the earliest histological description of a part of the vertebrate nervous system, already carries an implicit functional interpretation, for there can be little doubt that his use of the term 'vessels' is a reference to the hydraulic model of neural function proposed by Descartes (1662). His observations are all the more remarkable in view of the necessary limitation of his microtechnique to dissection with fine needles, freehand sections made with a "little knife ... so sharp that it could be used for shaving", and probably air-drying for mechanical stabilisation of tissue. Similar methods remained in virtually exclusive use for the next hundred years or so until Purkinje, who was, significantly, professor of physiology at Wroclaw (Breslau), started hardening tissue in alcohol (spirits of wine), cutting sections with his home-made microtome and staining them with various coloring agents including indigo, tincture of iodine and chrome salts (Phillips, 1987).

These improvements enabled Purkinje to anticipate by two years Schwann's extension of the cell theory to animals by describing nucleated "corpuscles" from a variety of tissues including brain and spinal cord (Hodgson, 1990). But new techniques rarely displace older ones entirely, and it was a combination of serial sectioning and microdissection with needles (teasing) of chromic-acid- or potassium-dichromate-fixed tissue that allowed Deiters (1865) to demonstrate what had eluded Purkinje: the extension of the nerve cell body in dendrites ("protoplasmic processes") of progressively finer divisions, and the continuity of the single axon with the cell body also.

The problem of how to study the contextual relationships of nerve cells and their processes in situ was soon to be spectacularly solved by Golgi (1873) with "la reazione nera", in which the use of silver nitrate was inspired, no doubt, by contemporary experiments in photography. Cajal took those contextual relationships to their classical limits in his magisterial exploitation of Golgi's technique (Cajal, 1995). He espoused Waldeyer's (1891) neuron doctrine in a modified and essentially modern form centred on his concept of the dynamic polarization of the neuron (Cajal, 1906). Yet his insistence on the separate identity of individual neurons had to await half a century and the development of a new technology, electron microscopy, for its confirmation (Palade and Palay, 1954).

Subprotocol 1
Fixation, Sectioning and Embedding

Part 1: General Histological and Cytological Methods

Introduction

"The principles of biological microtechnique may perhaps be reduced to one – the principle that when we make a microscopical preparation of any sort, we ought to try to understand what we are doing..." (Baker, 1958).

The physico-chemical, as well as the spatio-temporal, properties of living nervous tissue are not amenable to much histological work so it is generally necessary to modify them in various ways in order to produce a usable specimen. In this section we shall look at some preparative techniques that are basic to much histological study and that may be conveniently grouped under the heading of fixation, sectioning and embedding. Since they are not specific to neurohistology, the treatment of these techniques will be brief. It is particularly instructive, however, to consider them in the context of their historical development, which, together with that of the various methods of dyeing and staining, is typically a continuing story of progressive problem-solving by eclectic use of technologies derived from contemporary advances in other fields, principally chemistry and physics.

The natural products ethanol, in the form of spirits of wine, and acetic acid, in the form of vinegar, have always been used in the preservation of organic material, but only the first was commonly used in early microtechnique. This is because what was sought was hardening of the tissue, enabling it to be cut into thin sections, and of the two agents only ethanol had the desired effect (Baker, 1958). Hardening by the purely physical method of freezing was also possible and was used by Stilling in 1842 (cited by Cajal, 1995) to prepare sections of brain and spinal cord. With the development of inorganic chemistry in the late 18th and early 19th centuries several substances were found to harden animal tissues sufficiently to allow them to be sectioned, and their particular effects were exploited either as single hardening agents or in various mixtures, many of which continue in use to the present day. The most important are

- mercuric chloride,
- osmium tetroxide,
- chromium trioxide and
- potassium dichromate,

all of which were in use in microtechnique by about 1860. The subsequent rise of organic chemistry led to the introduction of the remaining classical 'hardening agents'

- picric acid (2,4,6-trinitrophenol) and
- formaldehyde (methanal),

the latter as late as 1893 and only after its previous use as a disinfectant (Baker, 1958).

As infiltration and embedding of tissue in solid media became standard practice (see below), the hardening property of these substances lost its relevance and attention could then centre on their role in fixation of the non-aqueous components of the cell. A cell that has been killed or rendered non-viable by chemical action is necessarily artefactual to a greater or lesser extent when compared to the living cell. The amount of artefactual distortion of some feature of interest in the living state can be taken as a measure of the quality of fixation in that respect, whether it be fine structure, enzyme activity, lipid extraction, or whatever. Moreover, in view of the physico-chemical complexity of the cell, it is not surprising that any single substance combines both good and poor fixative qualities when assessed on different criteria. To some extent the deficiencies of one fixative can be counteracted by the complementary benefits of another when used in combination, either sequentially or together. This is necessarily an empirical process, the results of which are in general unpredictable, but it is an approach that has led to the introduction of many important fixatives and fixation procedures.

As an example, we shall follow the development of one of the most widely used procedures, involving a combination of aldehydes with osmium tetroxide, the version in current use in Durham being given in example 2 below. Although osmium tetroxide rapidly destroys enzyme action, Strangeways and Canti (1927) found that it very faithfully preserves the fine structure of the cell as revealed by dark-ground microscopy. Fine-structure preservation is critically important for most electron microscopy because of the very high spatial resolution that it provides, so in the first two or three decades of electron microscopy osmium tetroxide was widely used as the only fixative, typically as a 1 % solution in 0.1 M phosphate or cacodylate buffer at about pH 7.3 (Glauert, 1975). It had the additional advantage of imparting electron density to those components of the specimen that reacted with the osmium tetroxide, and thus increasing image contrast. But the consequent loss of cytochemical information, especially about the localisation of enzyme activity which was preserved by formalin fixation (Holt and Hicks, 1961), prompted Sabatini, Bensch and Barrnett (1963) to assess various aldehydes for their ability to preserve cellular fine structure better than formalin while retaining high levels of enzymic action. Of the nine aldehydes assessed, including formaldehyde and acrolein, the best results were obtained with glutaraldehyde (pentane 1,5-dial, $C_5H_8O_2$), which was used as a 4–6.5 % solution in 0.1 M phosphate or cacodylate buffer at pH 7.2. Its superior performance is usually attributed to its relatively small size, enabling rapid penetration, and its two aldehyde groups, which are thought to allow glutaraldehyde to form stable cross-linkages between various molecules, especially proteins. Moreover, when combined with a second fixation with osmium tetroxide, fine structural preservation was as good as with osmium tetroxide alone even if the blocks had been stored 'for several months' before the second step. In an early modification of the procedure Karnovsky (1965) advocated the inclusion of 4 % formaldehyde in the primary fixative, on the basis that formaldehyde, being much smaller than glutaraldehyde and with only a single aldehyde group, would penetrate tissue more rapidly, stabilizing it sufficiently long for glutaraldehyde to act and thus permit the fixation of larger blocks. Whether or not this is a correct explanation for the action of the aldehyde mixture, the fixative has become probably the most widely used for electron microscopy, though the strength is usually reduced by half, apparently prompted by considerations of the osmotic potential of the fresh solution.

Ever since Leeuwenhoek wielded his "little knife" the importance of sectioning in microtechnique has been clear and, as we have seen, fixation, whether chemical or physical, was initially developed to harden tissue sufficiently for it to be sectioned. Sectioning is necessary not only to make specimens suitably transparent to photons or electrons, but also to reduce the spatial complexity of a specimen to convenient limits. Analysis may be greatly facilitated, and frequently is only made possible at all, by selecting section thickness and orientation appropriate to the scale of spatial structure required of the specimen. The 3-dimensional structure of components larger than the section thickness can then be recovered by reconstruction from serial sections. But in neurohistology, until the discovery of the Golgi method, the complex shapes of complete nerve cells could not easily be traced in sections, and microdissection with needles of the fixed material remained in widespread use throughout much of the latter half of the 19th century. Perhaps because it is incompatible with microdissection, embedding tissue in a medium that could itself be hardened to give mechanical support during sectioning appears to have been adopted relatively late into neurohistology. Embedding, when first used, was just that; the tissue was scarcely, if at all, infiltrated by the medium, but merely surrounded by it in order to retain the relative positions of separate components. Large, gel-forming molecules such as collodion (nitro-cellulose) and gelatine have been used since the earliest days of embedding when, it is no coincidence, both of these substances were also being used in the production of the first photographic emulsions. A low viscosity form of nitro-cellulose ("celloidin") eventually became widely used in neurohistology, particularly when sections greater than about 20 μm in thickness were required. According to Galigher and Kozloff (1964), paraffin wax, a product of the then emergent petroleum industry, was first introduced as a purely embedding medium by Klebs in 1869 but almost immediately (1871) an infiltration method, essentially similar to that in current use, was devised by Born and Strickler. Neurohistologists do not appear to have taken up paraffin embedding immediately, but certainly by the end of the last decade of the 19th century it was being routinely used by them both for thin (2 μm) and serial sections. Biological electron microscopy necessitated the use of new embedding media, opportunely provided by the plastics industry from the 1940s onwards. Glauert (1975) gives a very full account of them: the most widely used are the epoxy resins. Although glutaraldehyde fixation and resin embedding were developed to meet the needs of electron microscopy, the quality of their histological product is such that light microscopy has also benefited, as the following case study will show.

Example 1: The Primary Ending of the Mammalian Muscle Spindle – A Case Study of the Use of 1 μm Thick Serial Sections in Light Microscopy

Materials

Muscle spindles partially exposed by removal of overlying extrafusal muscle fibres for direct observation in the tenuissimus muscle of the anaesthetized cat.

Procedure

Fixation

1. 5% glutaraldehyde in 0.1 M sodium cacodylate buffer pH 7.2 for 5 min. in situ. [Glutaraldehyde is usually obtained as a 25% solution. It polymerizes easily and so should be kept below 4 °C until required.]
2. The same fixative for 4–14 days after excision of portions of muscle each about 10 mm long containing one spindle. [Variation in total fixation time was due to postal despatch between laboratories. There was no obvious difference in the quality of fixation of muscles fixed for different times.]
3. Washed in the buffer for 30 min.
4. 1 % osmium tetroxide, buffered, for 4 hours. [Osmium tetroxide penetrates tissue very slowly, but the tenuissimus muscle is typically less than 1 mm thick and could be adequately fixed in this time. OsO_4 is made up as a 2 % stock solution and kept refrigerated in a sealed bottle. The working strength fixative is made by diluting the stock solution with an equal quantity of 0.2 M sodium cacodylate buffer.]

Dehydration and Embedding

1. Dehydrated in a graded series of ethanol – 70 %, 95 %, 100 % (twice) – for 10 min each at ambient temperature.
2. 50:50 mixture of ethanol and propylene oxide (1,2-epoxy propane) for 15 min. [Propylene oxide is usually included as an intermediate solvent and is analogous to the use of "clearing agents" in paraffin embedding procedures. The refractive index of most clearing agents is similar to that of dehydrated proteins and other cellular components; they were originally used to make fixed tissue transparent, hence the name which has persisted even though they rarely have that function today. For alternative dehydration methods see Glauert (1975).]
3. Propylene oxide for 15 min.
4. 50:50 mixture of propylene oxide and Epon (complete except for the accelerator) left overnight in an unstoppered container in a fume cupboard. [Evaporation of the propylene oxide results in a very well infiltrated block.]

5. Drained excess infiltration medium blotted; transfered to fresh complete Epon.
6. Flat-embedded in an aluminium foil mould; polymerized for 12 hr at 45 °C and 24 hr at 60 °C.

Sectioning and Staining

1. Sections cut manually at 1 µm thickness in groups of 10 on an ultramicrotome with conventional glass knives. [If necessary, the sections can be spread on the water surface using chloroform vapor from a brush held close to them, or by radiant heat from an electrically heated filament. Glass knives need to be replaced regularly; use of a mechanical knife-breaker ensures close similarity of shape in successive knives. Accurate positioning to within a few µm of a new knife with respect to the block face can be achieved by lighting the back of the knife, such that the gap between knife edge and block face appears as a bright line.]
2. Coverslips [50x22 mm is a convenient size] scored with a diamond marker and broken into strips about 3 mm wide were used to collect the sections directly from the water trough of the knife by immersing one end of the strip under the surface of the water (Fig. 1.A). [The sections, either as a ribbon or individually, are easily guided with a toothpick-mounted eyelash onto the strip, which is held in watchmakers' forceps. A simple technique to ensure adequate adhesion of the sections is to draw one face of the strip of coverslip over the tip of the tongue and allow it to dry.]
3. The back of each strip was dried with a soft tissue, leaving the sections free-floating on a small drop of water on the front of the strip.
4. The sections were thoroughly dried onto the strip using a hot plate at about 70° C. [Best done by keeping a glass slide permanently on the hot plate and placing the strips onto the slide (Fig. 1.B).]
5. Stained with toluidine blue (Fig. 2.A) and pyronine (Fig. 2.B) at high pH by placing a drop of the stain on the sections and heating until the stain starts to dry at the edge

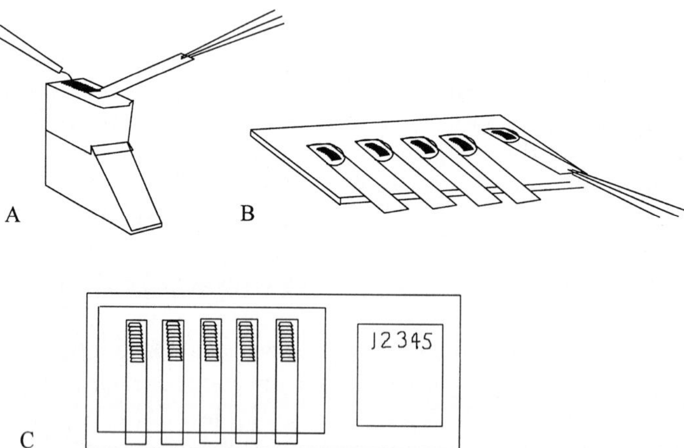

Fig. 1. Stages in the preparation, staining and mounting of serial, 1 µm thick, epoxy resin-embedded sections. A: A sort ribbon of sections is guided onto a strip of glass cut from a coverslip, using an eyelash mounted on a toothpick. The strip of coverslip is held in watchmakers' forceps. B: The back of the coverslip is dried using a soft tissue, leaving the ribbon of sections free-floating on a drop of water on the front of the coverslip, which is then placed on a glass slide on a hot-plate to flatten the sections and dry them. The same arrangement is used to stain the sections as described in the text. C: Several strips are mounted under a single large coverslip and the slide is labelled to indicate the order of the sections.

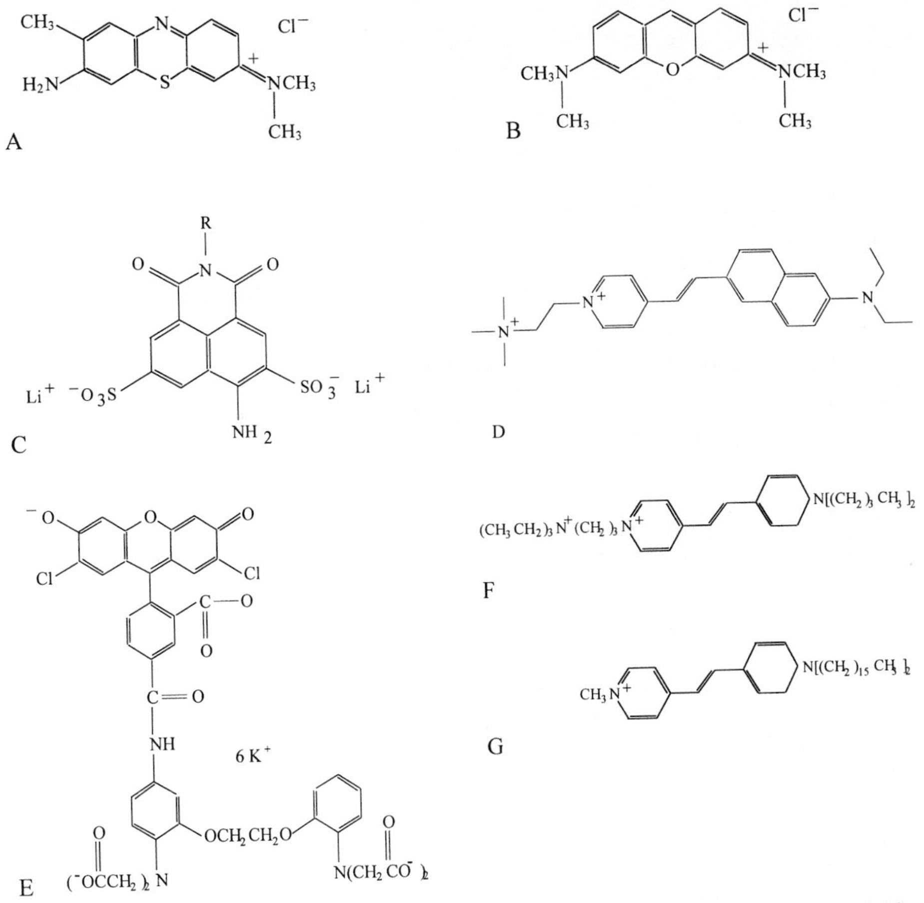

Fig. 2. Structural formulae of various dyes and chromogens mentioned in the text. A: Toluidine blue. B: Pyridine. C: Lucifer yellow. D: JPW1114. E: Calcium Green-1. F: FM1–43. G: DiA.

of the drop. Washed with water and differentiated with 95 % ethanol. [Staining solution is made by dissolving 0.1 g toluidine blue + 0.05 g pyronine + 0.1 g borax (sodium tetraborate) in 60 ml distilled water, and should be filtered periodically.]

6. Dried on the hot plate and mounted using DPX (Distrene-Plasticizer-Xylene). [5 strips each with, say, 10 sections can be conveniently mounted under a single 50x22 mm coverslip (Fig. 1.C). Of course, the strips should be mounted with the sections uppermost.]

Results

The primary ending of a tenuissimus muscle spindle in the cat occupies about 350 µm of the mid portion of the spindle and typically requires some 50 serial, 1 µm, longitudinal sections for its complete examination. The ending is generally considered to comprise the expanded sensory terminals of a single group Ia afferent nerve fibre, together with the system of preterminal branches, both myelinated and unmyelinated, that serve to distribute the terminals among the several intrafusal muscle fibres. There are commonly six intrafusal fibres of three different kinds. Figure 3 shows a selection of micrographs taken with a x100 oil-immersion plan achromat objective (N.A. 1.25); structures consid-

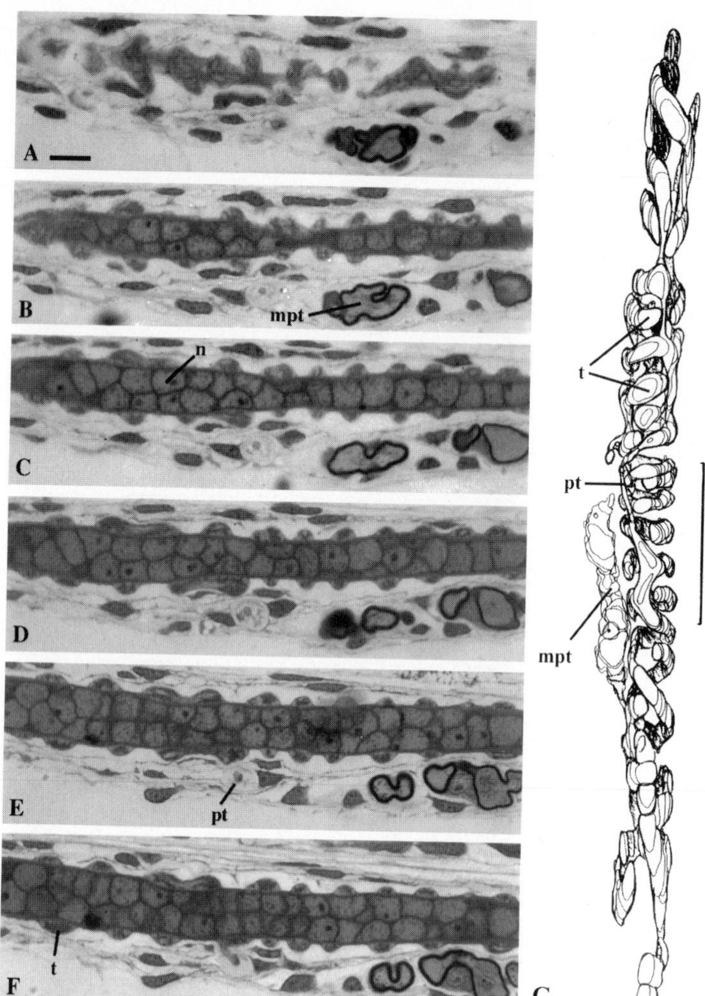

Fig. 3. Examples of results of serial-section analysis using 1 μm epoxy resin-embedded material. A-F: Longitudinal sections taken in the primary sensory region of a mammalian muscle spindle. This spindle contained 5 intrafusal muscle fibres, part of only one of which (a bag_1 fibre) is shown. The sections are serial except that one section has been omitted between A and B, and one between D and E. Scale bar = 10 μm. G: Contour reconstruction of the sensory terminals on the bag fibre shown in A-F. Scale bar = 50 μm. mpt, myelinated preterminal branch; n, nucleus; pt, unmyelinated preterminal branch; t, sensory terminal.

erably less than 0.5 μm in size are easily resolved. Each field of view covers a distance of a little over 100 μm in the long axis of the spindle. The most prominent structure visible in the micrographs is the central portion of one of the intrafusal fibres, specifically the bag_1. In the region of the primary ending, the sarcomeres of the intrafusal fibres are almost entirely replaced by a collection of nuclei (Fig. 3C, n). Projecting from the surface of the fibre are the sensory terminals (Fig. 3F, t). These can be traced between the sections as can portions of the myelinated (Fig. 3B, mpt) and unmyelinated (Fig. 3E, pt) preterminal branches. The dark structures within the terminals are mostly mitochondria. Several accessory, fibroblast-like cells are also visible forming a sheath around the bag_1 fibre. A contour line reconstruction of this part of the sensory ending on the bag_1 fibre, based on these and other intervening sections, is shown in Figure 3G. A 3-dimensional reconstruction of the complete ending was published by Banks in 1986. A similar serial-section analysis was recently used by Banks et al. (1997) in a correlative histophysiological study of multiple encoding sites and pacemaker interactions in the primary ending.

Subprotocol 2 Ultrastructure

Introduction

"The dimensions of the [synaptic] cleft are now known and its detection has led many, perhaps rather hastily, to consider the neuron (discontinuity) versus the reticular controversy (transynaptic cytoplasmic continuity) to be ended." (Gray, 1964).

The advent of the electron microscope removed the barrier to the study of so-called ultrastructure, or spatial organisation, on a finer scale than the resolution of the light microscope. It permitted not only synaptic clefts but also structures one or two orders of magnitude smaller to be made visible in sections of biological material. The effect on microtechnique was, however, more evolutionary than revolutionary except that observation of living cells and tissues is scarcely possible with the electron microscope. It might be supposed that without the possibility of direct comparison with living cells the quality of ultrastructural fixation could only be assessed subjectively, but physical fixation by rapid freezing is entirely feasible (see, for example, Verna, 1983), thus providing an objective standard for chemical methods. Freezing is not generally applicable mainly because of its limitation to very small thicknesses of tissue in order to prevent ice crystal formation (see, for example, Heuser et al., 1979), but it can be important or even essential in some studies and, with sufficient ingenuity, can be applied to relatively inaccessible structures within the brain (Van Harreveld and Fifkova, 1975). Despite the necessity for freezing in some special applications, ultrastructural neurohistology depends overwhelmingly on chemical fixation, the techniques being derived directly from practices and principles originally developed for light microscopy, as has been outlined above. In this section we will look briefly at the role that fixation played in the functional interpretation of synaptic structure. Of primary importance here was the fixation of lipids by OsO_4, so preserving membrane structural integrity. This revealed not only the discontinuity of neurons at the synaptic cleft, but the presence of characteristic round vesicles of 30–50 nm diameter in the presynaptic terminals of synapses with chemically mediated transmission (Gray, 1964). The vesicles were, of course, immediately recognised as being correlated with, or structurally equivalent to, the neurotransmitter quanta. The dynamic nature of vesicle recycling during transmission was clearly established, among others, by Heuser and Reese (1973) who used immersion fixation of frog sartorius muscles, in a Karnovsky-type fixative, after various durations of nerve stimulation and post-stimulation recovery.

Immersion fixation was initially used in ultrastructural studies on the CNS, but it was necessary to cut the tissue finely in order to obtain high quality results, so the spatial relationships of structures greater than about 1 mm in size were lost. Nevertheless, using this technique, Gray (see 1964 review) was able to identify two major types of central synaptic structure and to recognize that they were differentially distributed on the dendrites and somata of the post-synaptic neurons. They were characterised by electron-dense material associated with the post-synaptic membranes that were of greater (type 1) or lesser (type 2) thickness and extent, and their locations led Eccles (1964) to suggest that they might correspond to excitatory and inhibitory synapses, respectively. Despite this and other important advances made using immersion fixation, the advantages of perfusion in maintaining high quality fixation while retaining larger scale structural relationships in the CNS are such that it very soon became the method of first choice (Peters, 1970). At first veronal-acetate-buffered OsO_4 was used (Palay et al., 1962) and subsequently aldehydes, with or without subsequent treatment with OsO_4 (Karlsson and Schultz, 1965; Schultz and Karlsson, 1965; Westrum and Lund, 1966). Immediately, and virtually simultaneously, several authors described the occurrence of flattened presynaptic vesicles in some synapses. Uchizono (1965) was able to correlate round vesicles with Gray type 1 and flattened vesicles with Gray type 2 synapses; utilizing the known interneuronal origins and functional effects of certain synapses in the cerebellar cortex, he further concluded that the first were excitatory and the second inhibitory. The identification was criticised on several grounds, not least that the flattening depended on aldehyde fixation which, if prolonged, would induce even the normally round vesicles to flatten (Lund and Westrum, 1966; Walberg, 1966; Paula-Barbosa, 1975). However, many later observations have substantially confirmed Uchizono's conclusion so that what is perhaps most interesting and instructive in this case is the usefulness of an incidental product of fixation, an artefact that without the functional correlation would otherwise be regarded as undesirable.

Example 2: Synapses of the Cerebellar Cortex

Materials

Cerebellar cortex of the adult rat, anaesthetised with an intraperitoneal injection of sodium pentobarbitone.

Procedure

Fixation

1. Systemic perfusion with a Karnovsky fixative, made up as follows (proportions given for 100 ml)
   - Solution A: 2 g paraformaldehyde dissolved in 40 ml water at 60 °C, 1N NaOH added dropwise (2–6 drops) until the solution clears.
   - Solution B: 10 ml of 25 % glutaraldehyde mixed with 50 ml of 0.2 M sodium cacodylate buffer, pH 7.3.

   Solutions are kept at 4 °C until required, then mixed to give 100 ml complete fixative. Techniques of perfusion vary considerably in their elaboration; the method I have adopted is simple and seemingly reliable: it aims to minimise the time between induction of anaesthesia and effective fixation. A peristaltic pump [Watson-Marlow MHRE 200] is used to provide the driving force [many authors use hydrostatic pressure] and the fixative is introduced immediately the cannula is in place, beginning at a relatively low speed until signs of onset of fixation are evident (limb and tail extension), and progressively increasing the speed over the first few minutes. Fixation is continued for about 10 minutes, consuming about 500 ml fixative for an adult rat. Pressure is not monitored.

   The cannula is fashioned from a 21G hypodermic needle, angled at its mid-point and ground transversely at the tip. A blob of epoxy resin applied to the tip before grinding facilitates introduction of the cannula into the ascending aorta via an incision in the apex of the left ventricle, and the cannula can then be clamped in place using an artery clamp. During surgery and insertion of the cannula, the pump is kept running at a very slow speed to prevent the introduction of air bubbles into the vasculature. As soon as the cannula is clamped in place, the wall of the right atrium is cut and the pump speed is increased to initiate fixation.

2. After perfusion the brain is removed and placed in fresh fixative until required. Blocks or slices are cut sufficiently thin (about 1 mm maximum) to allow penetration of OsO_4. The second fixation with OsO_4, dehydration and embedding are as in example 1 above.

Sectioning

1. 1 μm thick sections for survey and alignment stained with toluidine blue and pyronine as in example 1.
2. Approx. 70–90 nm (silver-pale gold interference color) sections collected on formvar-coated grids and stained with lead citrate and uranyl acetate.

Results

Several different kinds of synaptic association have been described in the cerebellar cortex, most synapses belonging to various kinds of axo-dendritic association. We shall look briefly at three examples, one from the molecular layer and two from the synaptic glomeruli of the granular layer. Synaptically, the molecular layer is dominated by the parallel fibre-dendritic spine synapses between the granule and Purkinje cells. Figure 4A shows the outermost part of the molecular layer with the parallel fibres (pf) cut transversely. Several parallel fibre-dendritic spine synapses (s) may be seen; note that they are usually in close association with glial-cell processes whereas the parallel fibres (the axons of granule cells) are clustered together and lack individual glial-cell sheaths. A similar synapse is shown enlarged in Fig. 4B; it conforms to Gray's type 1, in particular there is a post-synaptic thickening and a sheet of extracellular material lies between the pre- and post-synaptic membranes. The presynaptic vesicles (rv) are round in profile. Figure 4C shows a synaptic glomerulus of the granular layer. This is a complex structure consisting of a central mossy-fibre rosette (mf) surrounded mainly by numerous profiles of granule-cell dendrites (gcd) and Golgi-cell axons. Mossy fibres and Golgi-cell axons both form axo-dendritic synapses with the granule-cell dendrites, but they are of Gray types 1 and 2, respectively. A Golgi cell-granule cell synapse is shown in greater detail in Fig. 4D. The post-synaptic thickening is much less well developed than in a type 1 structure, and there is no obvious extracellular material between the pre- and post-synaptic membranes. Many of the presynaptic vesicles are flattened. As is well known, of course, both parallel and mossy fibres are excitatory, whereas Golgi cells are inhibitory.

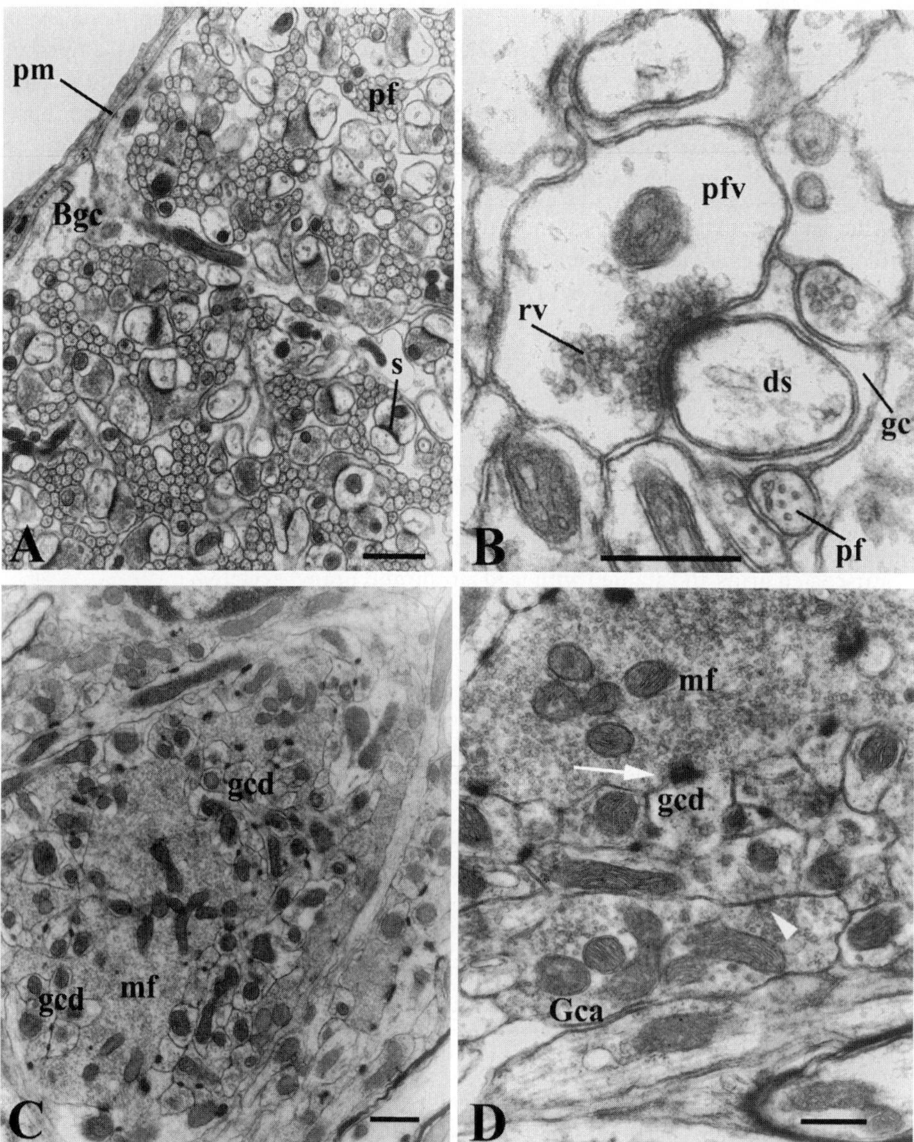

Fig. 4. Examples of electron microscopy of mammalian CNS fixed by perfusion using a mixture of aldehydes. Cerebellar cortex of the rat. A: Outermost part of molecular layer, cut transversely to the parallel fibres. Scale bar = 1 µm. Bgc, Bergmann glial cell process; pf, parallel fibres; pm, pia mater; s, synapse. B: Gray type 1 synapse between a parallel fibre varicosity and a Purkinje cell dendritic spine. Scale bar = 0.5 µm. ds, Purkinje cell dendritic spine; gc, glial cell process; pf, parallel fibre containing microtubules (neurotubules); pfv, presynaptic varicosity of parallel fibre; rv, round vesicles. C: Synaptic glomerulus in the granular layer. Scale bar = 1 µm. mf, mossy fibre rosette filled with round vesicles and forming numerous synaptic contacts with different granule cell dendrites; gcd, granule cell dendrites. D: Gray type 1 (mossy fibre to granule cell dendrites) and type 2 (Golgi cell axon to granule cell dendrites) synapses. Scale bar = 0.5 µm. Gca, Golgi cell axon terminal, with flattened vesicles; gcd, granule cell dendrite; mf, mossy fibre rosette.

Subprotocol 3 The Golgi Method

Part 2: The Differentiation of Single Cells

Introduction

"Golgi is responsible for a method that renders anatomical analysis both a joy and a pleasure." (Cajal, 1995).

The Golgi method is central to neurohistology, almost serving to define the discipline. Here is a technique that by its ability to select single cells, more or less at random, and fill them with a near-black precipitate while leaving the surrounding cells unstained provided a straightforward means to solve the technical problem presented by the complex shapes and interrelationships of cells of the nervous system. These same staining properties render it virtually useless in any other area of histology. We are told that the method was discovered by accident, but insofar as it consisted simply of "prolonged immersion of the pieces [of brain], previously hardened with potassium or ammonium bichromate [sic], in a 0.50 or 1.0 % solution of silver nitrate" (Golgi, 1873), it was probably only a matter of time before someone found it. It is worth recalling that Mueller had introduced potassium dichromate as a hardening agent as recently as 1860 (Baker, 1958).

From its earliest days the method has had its critics, but by acknowledging the criticisms at the outset we can perhaps best appreciate its limitations (and therefore its possibilities). Essentially, the criticisms can be expressed as two questions: i) Does the method provide a representative sample of cells (especially neurons)? ii) When a neuron stains, are all its neurites fully shown? The respective answers – probably not; and perhaps sometimes, but certainly not always – highlight the limitations which, it may be seen, imply that we should be particularly cautious with quantitative results obtained by means of the Golgi method. However, any method that selectively marks individual neurons is liable to suffer the same criticisms and its results will require some sort of complementary control. The Golgi method has continued to be important, even after the introduction of electron microscopy and intracellular staining techniques, due to its particular advantages: economical generation of information on different types of neuron and their interrelationships, and simplicity in execution.

It is not surprising, in view of its importance and long history, that the Golgi method has spawned several variants, though the two principal ones were introduced by Golgi himself. We shall refer to them as the *rapid Golgi* and the *Golgi-Cox* methods. Once again it is instructive to consider briefly how these might have arisen; in the absence of a rational physico-chemical basis for the variants (see below), one suspects them to be due to a process of selection following empirical, if not to say playful, experimentation. Could this be how Golgi came, in the last year or two of the 1870s, to substitute mercuric chloride for silver nitrate after the initial fixation in Mueller's fluid? Mercuric chloride ($HgCl_2$) had only just been popularized as a fixative by Lang, writing in 1878, although it was first used in microtechnique around the middle of the 19th century (Baker, 1958). However, unlike silver nitrate, mercuric chloride led to individual cells being marked by a white precipitate, which needed to be darkened by treatment with alkali. Cox made a relatively minor modification to the method in 1891 by including the mercuric chloride in the primary fixative, and it has remained essentially the same since.

Both the original and, especially, the Golgi-Cox methods suffer from being very prolonged procedures, sometimes up to several months in total. Golgi it was who found that the addition of a small amount of osmium tetroxide to the primary dichromate fixative, originally about 0.33 %, resulted in a great reduction in the amount of time needed for the subsequent silver nitrate exposure. It seems unlikely that Golgi can have predicted this effect of osmium tetroxide, but rather that it was a fortunate side-effect of an attempt to improve the quality of the initial fixation. In any case, in the 1880s Cajal took the new "rapid" Golgi method, played with the fixation a little himself, but more importantly introduced the simple expedient of repeating once or even twice the cycle of chromation and silvering ("double" and "triple" impregnations) in order to improve the extent of the impregnation. Then, only three years after its first use as a fixative (see above), Kopsch (1896) substituted formaldehyde for osmium tetroxide, avoiding the expense of the latter while retaining its effectiveness in speeding the procedure.

During all of these early and very significant developments in the Golgi method, the rational basis for it was hardly understood; which lack, compounded by the method's stochastic nature, was a cause of much of the criticism levelled at it. Today we are more comfortable with stochastic processes and some of the principal steps in the method have become clearer, though a realistic quantitative theory of it is still lacking. The most striking feature is, of course, the confinement of the final reaction product to the interior of individual neurons among similar, unstained cells. It is firstly apparent, therefore, that a barrier to the diffusion of the visible product at the level of the cell membrane is likely to exist throughout the process from fixation to dehydration. The nature of the product, which varies according to the particular method used, has been revealed by X-ray and electron diffraction analyses (Fregerslev et al., 1971a,b; Chan-Palay, 1973; Blackstad et al., 1973). In the chrome-silver variants it is silver chromate (Ag_2CrO_4); analogously, replacement of silver nitrate by mercurous nitrate yields mercurous chromate (Hg_2CrO_4), whereas mercuric nitrate produces mercuric oxide chromate ($Hg_3O_2CrO_4$). With the Golgi-Cox variant the first visible, whitish, product is mercurous chloride (Hg_2Cl_2); according to Stean's (1974) physico-chemical analysis this is converted to mercuric sulphide (HgS) as the final, black, product by alkali treatment. Stean argues that the source of the sulphur is intrinsic and fixative-induced disulphide bonds in protein.

All of the various localised products are characterised by a high degree of insolubility in water and will therefore readily precipitate on completion of the reactions forming them. The crucial question for our understanding of the Golgi method, however, is how the reactants are brought together so as to effect the observed localisation of the reaction product. Examination of electron micrographs of well-impregnated cells (e.g. Blackstad, 1965) shows that the precipitate is microcrystalline and always confined within membrane-bound spaces, usually the cytosol. We shall consider only the chrome-silver technique and note firstly that since the chromate ion (CrO_4^{2-}) is present in trace quantity in a solution of potassium dichromate (Baker, 1958) [CrO_4^{2-}] is presumably limiting, at least during the initial stages of silver impregnation. An important observation reported by Strausfeld (1980) concerns the formation of silver chromate crystals in a block of agarose-chromate gel exposed to silver nitrate solution. A section of such a block seems to show a generally exponential decline in the number of crystals with distance from the exposed surface; moreover the size of the crystals is correlated with their separation, which would seem to imply that local depletion of chromate is responsible for the size limitation. (Superimposed on the overall trend are several Liesegang rings, concentric bands of local variation in the spatial density of crystals, presumably due to the interaction between the rate of advance of Ag^+ and the rate of sequestration of chromate into nascent crystals. Similar "rings", parallel to the free surface of the tissue, can occur as artefacts in samples of brain, see Fig. 5A.) If the probability of nucleation of a crystal is proportional to the local concentration of silver, this distribution may be explained as a consequence of Fick's second law of diffusion. Microcrystals of the type seen in impregnated cells must therefore be nucleated in conditions of relative excess of silver and presumably their formation leads to local depletion of chromate. This would tend to inhibit the subsequent nucleation of silver chromate in adjacent regions; Cajal's double and triple impregnation techniques show that the inhibition can be overcome to some extent by providing more reactants.

The practically simultaneous nucleation of silver chromate throughout a cell, implied by the microcrystalline nature of the reaction product, is borne out by direct observation of the progress of the black reaction (Strausfeld, 1980). The necessity for a relative excess of silver further implies that an earlier event, and one critical for the stochastic nature of the Golgi method, is the accumulation of silver within an individual cytosolic space. At least some of this silver is reduced to the metallic form prior to the earliest appearance of the black reaction, confirming that an excess of silver is present; it may be demonstrated by treatment with ammonium sulphide followed by physical development with hydroquinone and silver nitrate (Strausfeld, 1980). Such development results in an appearance very reminiscent of that produced by the Golgi method itself, with individual cells stained amongst a virtually unstained background. It appears, therefore, that the accumulation of silver within a cytosolic (or more rarely some other membrane-bound) space is a very rapid process, suggesting that a positive-feedback or autocatalytic contribution is present. Although such events are normally confined to single neurons they readily spread between contiguous glial cells, that would be expected to be coupled by gap junctions, suggesting that passive electrical properties could be important. In addition, local removal of free Ag^+ ions by adsorption and reduction to metallic silver might be contributory factors and in any case would tend to inhibit silver accumulation in adjacent spaces.

Example 3: Neurons of the Cerebellar Cortex

The method described here follows closely a rapid-Golgi-aldehyde variant given by Morest (1981). For additional information see also Millhouse (1981) and Scheibel and Scheibel (1978).

Materials

Cerebellum of the adult rat.

Procedure

Fixation

Vascular perfusion using the Karnovsky fixative and method described in example 2 above.

Chromation

Following fixation the brain was removed and cut into blocks about 3 mm in thickness. The blocks were immediately immersed in approximately 25 × their volume of 3 % potassium dichromate and 5 % glutaraldehyde for 7 days at ambient temperature (mean about 20 °C). [The volume restriction is needed to place an appropriate limit on the availability of chromate. pH was not monitored in this process, nor was the solution buffered (see Angulo et al., 1996).]

Silver Impregnation

The blocks were rinsed in 0.75 % silver nitrate, then transferred to approximately 25× their volume of fresh 0.75 % silver nitrate for 6 days at ambient temperature. [Again the solution was not buffered, but see Strausfeld, 1980.]

Dehydration and Embedding

As in example 1 above.

Sectioning Sections were cut at 100 μm thickness using a sledge microtome. The block surface was softened using a heated brass plate immediately before each section was cut. [Alternatively, if electron microscopy is not contemplated, frozen sections could be prepared (Ebbesson and Cheek, 1988).]

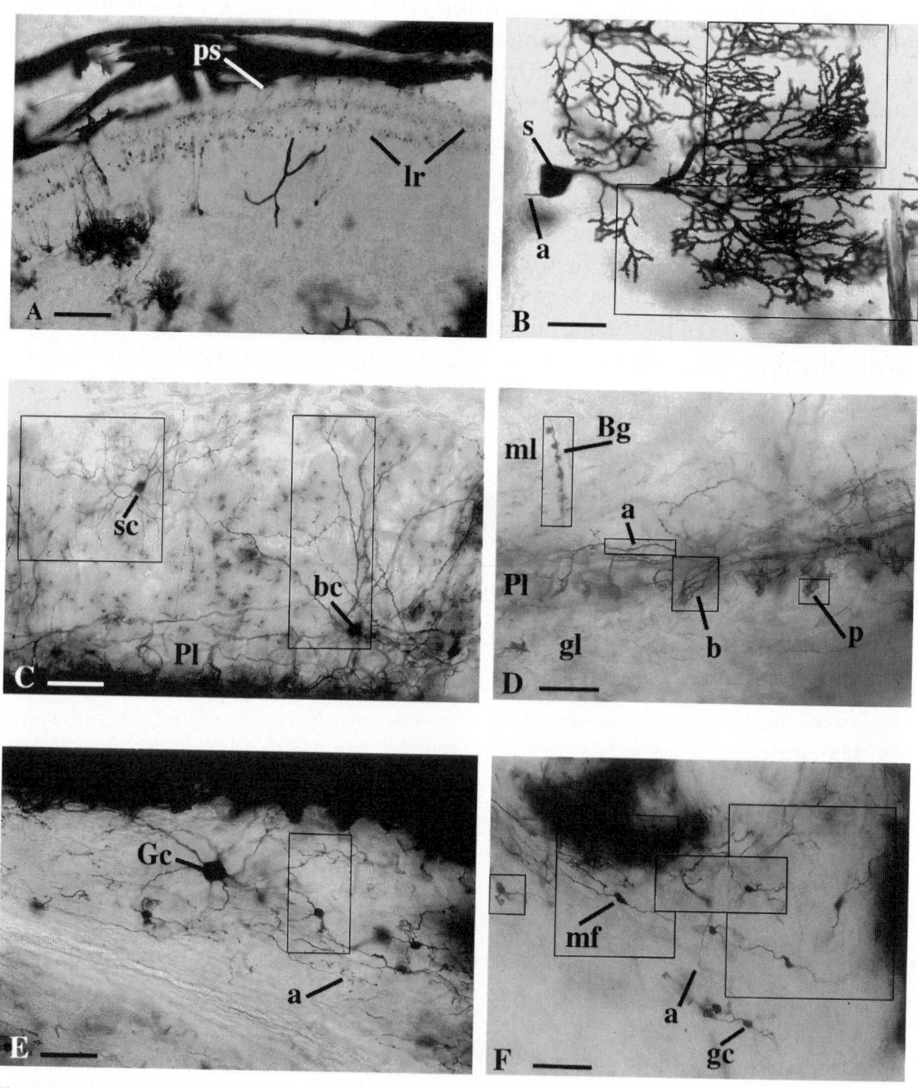

Fig. 5. Examples of mammalian neurons and glial cells stained by a rapid Golgi-aldehyde method. Cerebellar cortex of the rat. A: Full thickness of the cortex, showing two Leisegang's rings parallel to the pial surface. B: Purkinje cell; montage, inset field of view + 2.5 μm with respect to main field. C: Molecular layer cut transversely to the parallel fibres, showing a stellate cell and two basket cells; montage, inset field of view + 5 μm with respect to main field. D: Purkinje layer with adjacent parts of molecular and granular layers, showing several axons of basket cells; montage, inset field of view + 6 μm with respect to main field. E: Granular layer and white matter, showing a Golgi cell and some granule cells; montage, inset field of view + 3.5 μm with respect to main field. F: Granular layer, showing a mossy fibre and several granule cells; montage, inset fields of view (from left) – 9, + 6.5, + 11.5, +6.5 μm with respect to main field. Scale bars = 100 μm (A), 50 μm (B-F). a, axon of Purkinje cell (B), basket cell (D), Golgi cell (E), or granule cell (F); b, basket; bc, basket cell; Bg, Bergmann glial cell process; gc, granule cell; gl, granular layer; lr, Liesegang's rings; mf, mossy fibre rosette; ml, molecular layer; p, pinceau; Pl, Purkinje layer; ps, pial surface; s, soma; sc, stellate cell.

Results

The cerebellar cortex consists of a narrow sheet of neuronal somata, the Purkinje layer, that separates two broad layers: the outermost, finely textured with relatively few neurons, is the molecular layer; the innermost, typified by very large numbers of small neurons, is the granular layer. In the first micrograph (Fig. 5A) the full thickness of the cortex is shown, but very few neurons are stained. A branched blood vessel is prominent near the centre of the field of view, and several cell clusters and individual cells, mainly glial cells, can also be seen. There are large accumulations of silver chromate outside the pial surface (ps), and within the molecular layer are two examples of Liesegang's rings (lr). Note that the rings are parallel to the pial surface and that the outer one is composed of a larger number of smaller clusters of silver chromate crystals than the inner one.

The principal neuron is the Purkinje cell (Fig. 5B); its dendritic tree extends through the full thickness of the molecular layer, but is virtually confined to a single plane orthogonal to the parallel fibres. The soma (s) usually gives rise to a single stem dendrite, which branches repeatedly, and an axon (a) that is directed into the granular layer. Note that only the initial (unmyelinated) segment of the axon is stained. This is quite normal with many variants on the Golgi method. [The often reproduced image from Cajal of a Purkinje cell and its axon is of an immature specimen.] Collectively the somata define the Purkinje layer (Pl in Figs. 5C and D).

Figure 5C shows examples of the two kinds of neuron that occur in the molecular layer: stellate cells (sc), which are found throughout the molecular layer, and basket cells (bc), which occur at the base of the layer adjacent to the Purkinje-cell somata. The characteristic baskets (b, Fig. 5D) appear when side branches of several basket-cell axons surrounding a single Purkinje-cell soma are stained. Each basket continues as the elaborate pinceau (p, Fig. 5D), which encloses the initial segment of the Purkinje-cell axon. Figures 5E and F show the main components of the granular layer. The Golgi cell (Gc) has a radiating dendritic tree that extends into the molecular layer, and a highly branched axon (a, Fig. 5E) confined to the granular layer. Each of the very numerous granule cells (gc) typically has 4 or 5 short dendrites with claw-like branching terminals; their axons (a, Fig. 5F) ascend into the molecular layer to form the parallel fibres. Mossy fibres (mf) are axons arising from outside the cerebellum and as such comprise one of the two afferent systems of the cerebellar cortex, the other being the climbing fibres which were unstained in this material. The swellings at intervals along the mossy fibres are the rosettes that form the central element of the synaptic glomeruli, described in example 2 above. The other principal components of the glomeruli are the Golgi-cell axon and the granule-cell dendrites.

Subprotocol 4 Single-Cell Methods

Introduction

"Because of the multiplicative effect of enzymatic action, the peroxidases are sensitive tracers that may yet have usefulness in marking single cells." (Bennett, 1973).

For all its importance, and despite the doubts surrounding its selectivity and completeness in staining individual neurons, the chief limitation of the Golgi method is the lack of a means of directly relating structure and function in single cells. A complementary frustration accompanied the use of metal microelectrodes. The introduction of the glass capillary microelectrode (micropipette) by Ling and Gerard in 1949 immediately signalled the possibility not only of recording the electrical activity of a single neuron, but of subsequently marking it so as to produce a Golgi-like image of the structure of the cell (Nicholson and Kater, 1973). Early attempts usually involved the formation of an insoluble colored reaction product and were inspired, amongst others, by the Golgi method itself and by techniques of marking the location of the tips of metal microelectrodes, such as the Prussian Blue reaction (cf. Chapter 5). They met with little success mainly because of blockage of the micropipette and failure of the marker to fill the cell, both problems presumably being due to the formation of insoluble salts in the vicinity of the electrode. An alternative approach was the use of colored dyes, such as methyl blue and fast green, in order to obviate the need for a reaction to generate the visible marker, but here the problem was the loss of dye from fine cellular processes during dehydration in preparation for sectioning (Stretton and Kravitz, 1973). The turning point finally came in 1968 with the introduction of *Procion yellow* M4RS (Kravitz et al.) following a systematic survey of over 60 Procion and related dyes (Stretton and Kravitz, 1973).

The Procion dyes had been developed for use in the cotton textile industry, to overcome problems in dying cellulose, about 10 years before their introduction into neurohistology. Each consists of one (Procion M dyes) or two (Procion H dyes) chromogens linked to the reactive group, cyanuric chloride. In the cell the reactive groups form covalent bonds with amino groups of proteins, making the bound dye resistant to loss during subsequent processing. However, in comparison to the rate of this reaction, diffusion of the dye is presumably sufficiently rapid that very fine processes may be filled. Moreover, in this situation (though not, apparently, when covalently linked to cellulose) Procion yellow is fluorescent with a peak emission at about 550 nm. A fluorescent dye was desirable in that as compared with an absorptive dye used in conventional bright-field mode it would provide far higher contrast between the filled cell and its surroundings (Stretton and Kravitz, 1973).

Procion yellow had a spectacular but relatively brief career in neurohistology; once it had demonstrated the feasibility of single-cell marking, new techniques, or rather the new application of established techniques such as enzyme histochemistry, soon followed. One of the most important was that presaged by Bennett in the quotation at the head of this section – histochemical localisation of *horseradish peroxidase*. As with the reactive dyes, horseradish peroxidase could be injected into a previously recorded neuron and allowed to diffuse or be transported throughout the cell (Graybiel and Devor, 1974; Källström and Lindström, 1978; cf. Chap. 5), but in this case the sensitivity of the method is of course due to the marker's enzymic action. Within three years of Bennett's remark the method had been successfully applied by several laboratories, often expanding on results previously obtained by the same authors using Procion yellow, in studies on the spinal cord (Cullheim and Kellerth, 1976; Jankowska, Rastad and Westman, 1976; Light and Durkovic, 1976; Snow, Rose and Brown, 1976), cerebellum (McCrea, Bishop and Kitai, 1976, see example 4 below), neostriatum (Kitai et al., 1976) and leech ganglia (Muller and McMahon, 1976). Among the particular advantages of the use of horseradish peroxidase were that both light and electron microscopy could be applied to the tissue, since the reaction product is osmiophilic, and in comparison with the Golgi method or with injections of Procion yellow or *tritiated glycine* it provided perhaps the most complete filling of the cell yet available (Brown and Fyffe, 1981).

The demise of Procion yellow was due not so much to the adaptation of horseradish peroxidase to single-cell labelling as to the development, only ten years after its first use, of a more highly fluorescent reactive dye – *Lucifer yellow* (Stewart, 1981). The synthesis of Lucifer yellow was based on that of a commercial wool dye, brilliant sulphoflavine, both dyes having similar spectral properties with absorption maxima at 280 and 430 nm, an emission maximum near 540 nm and a quantum yield of 0.25. The chromogen in Lucifer yellow is 4-amino-naphthalimide 3,6-disulphonate, which is N-linked to the reactive group (Fig. 2C). Lucifer yellow VS, where R = m-phenyl-SO_2-CH=CH_2, reacts rapidly with the sulphydryl groups of amino acid residues. Lucifer yellow CH is more commonly used in neurohistology; here R = -NH-CO-NH-NH_2 and the free hydrazide group reacts with aliphatic aldehydes at room temperature (Stewart, 1981). But despite its high fluorescence, Lucifer yellow, like Procion yellow, is liable to fade as well as being undetectable with the electron microscope. A permanent preparation, which is also strongly osmiophilic, can be made by immunoperoxidase staining with a primary antiserum against Lucifer yellow itself (Onn, Pucak and Grace, 1993). This does, however, involve the use of lipid solvents to allow the large antibody molecules free access to the cytoplasm of the labelled cells, so fine structure will be adversely affected. An alternative method of producing an osmiophilic reaction product in high quality fixed material is by the photoconversion of diaminobenzidine (Maranto, 1982). Tissue containing Lucifer yellow-labelled cells is immersed in diaminobenzidine, which is small enough and sufficiently lipid soluble, readily to enter the fixed cells. Exposing the tissue to light of Lucifer yellow's excitation wavelength in the ultraviolet results in the preferential oxidation of the diaminobenzidine in the Lucifer yellow-labelled cells. Whether this is due to the induced fluorescence or some other mechanism does not seem to be known, but diaminobenzidine is known to be photosensitive and to oxidise spontaneously to the colored product on exposure to visible light.

Antibodies, of course, are not the only proteins that can recognize and bind with high affinity to a specific molecule, and such specific binding is widely used in biotechnique. An example is avidin, a glycoprotein isolated from egg-white, which has a very high affinity ($K > 10^{15}\ M^{-1}$) for biotin (vitamin H). Covalent linking of fluorescent dyes, or enzymes, or a recognition marker for a standard immunocytochemical reaction, enables avidin to be used as a highly sensitive detector for biotin in both light and electron microscopy. This in turn allows biotin to be used as a single-cell marker, generally in the form of one of its derivatives such as Nε-biotinyl-L-lysine (biocytin) or N-(2-aminoethyl) biotinamide hydrochloride (biotinamide [Neurobiotin, Vector Labs]) (Horikawa and Armstrong, 1988; Kita and Armstrong, 1991).

It might be supposed that electro- or iontophoresis of horseradish peroxidase, dyes and other intracellular markers would require the microelectrode tip to be located intracellularly also, and indeed overwhelmingly authors have been at pains to ensure stable intracellular recording before, during and preferably after marker injection. Certainly this has provided the most compelling evidence that the marked cell was the one recorded. However, Lynch et al. (1974) reported that single-cell marking was possible using electrophoresis of horseradish peroxidase from extracellular microelectrodes, and Pinault (1994) has produced similar results using biocytin or Neurobiotin as the primary marker, adding electrolytic evidence that the extracellularly recorded cell was the one marked. The mechanism remains obscure, but presumably involves the transfer of some marker molecules directly across the neuronal plasma membrane at the site of maximum field strength, the specificity apparently being due to the electrical relationship of the neuron and electrode. In any case, a non-specific uptake of extracellularly ejected marker seems to be ruled out, since usually just one neuron is marked (Pinault, 1994).

Example 4: Intracellular Marking of the Purkinje Cell with Horseradish Peroxidase

The technique and results for this case study are taken from Bishop and King (1982), where further technical considerations can be found (see also Kitai and Bishop, 1981).

Materials

Cerebellum of the cat.

Procedure

Microelectrode: 0.4–0.9 μm, bevelled and filled with 4–10 % horseradish peroxidase in 0.5 M KCl-Tris buffer pH 7.6; impedance 35–60 MΩ.

Injection: 3–5 per second depolarizing DC pulses of 100–200 ms duration and 3–5 nA intensity for 3–5 minutes.

Fixation: 15 minutes to 30 hours after horseradish peroxidase injection, according to neuronal size and extent of filling required. Vascular perfusion with 0.9 % saline containing 40 mg/kg xylocaine, added as 2 % solution, followed by a Karnovsky-type fixative (1 % paraformaldehyde, 2 % glutaraldehyde and 2.5 % dextrose in sodium phosphate buffer of pH 7.3 and final concentration about 0.1 M).

Sectioning and Processing: 50–60 μm serial frozen (LM) or Vibratome (EM or LM) sections, collected in phosphate buffer; diamino-benzidine reaction; LM-sections mounted on chrome alum-coated slides in gelatine and counterstained [with cresyl violet], EM- or LM-sections dehydrated in acetone and embedded in epoxy resin (Maraglas or Spurr's).

Results

The form of the soma and dendrites of the Purkinje cell present a very similar appearance in both Golgi-stained and HRP-filled neurons, though the finest branches (the dendritic spines) may be more clearly marked using HRP. [In the case of α-motoneurons, at least, the work of Brown and Fyffe (1981) showed that HRP-filling resulted in a much more complete picture of the dendritic structure of the neuron than any method used previously (see above).] However, with HRP filling the axon together with its collateral and terminal branches (Fig. 6) are also filled, whereas with the Golgi method only the unmyelinated initial segment is usually stained (see Fig. 5B). Among other functional implications, this enables the precise organisation of the cortico-nuclear projection to be determined.

Correlation of Techniques

"The picture provided by electron microscopy offers precisely what is missing in that provided by the Golgi methods." (Ramón-Moliner and Ferrari, 1972).

Although what is provided by any particular technique may be fairly clear and fixed, 'precisely what is missing' is likely to depend on the questions being asked. Certainly no

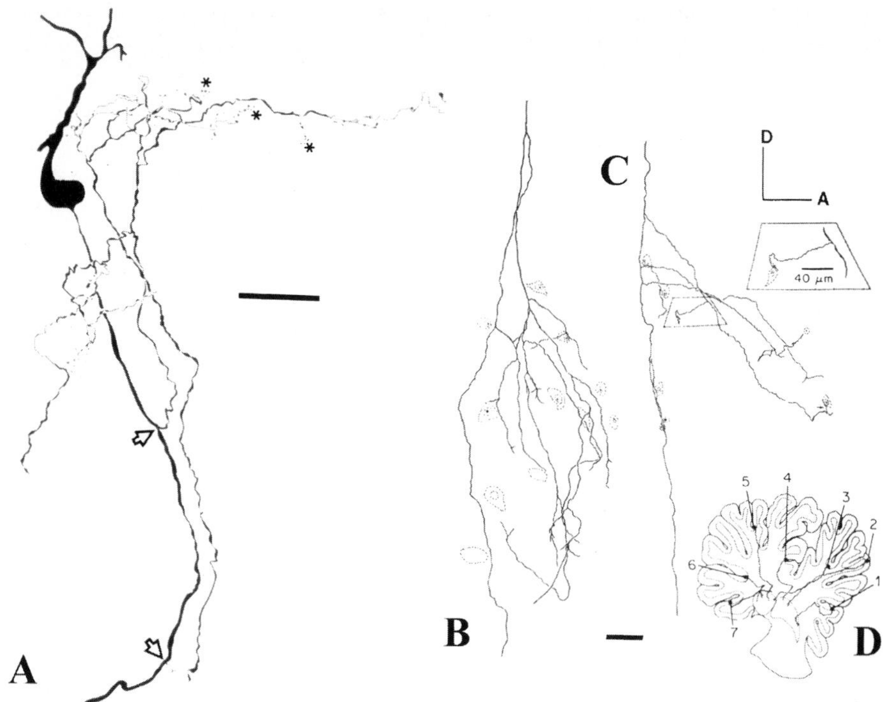

Fig. 6. Examples of staining of single neurons by intracellular iontophoresis of horseradish peroxidase. Purkinje cells of the cerebellar cortex of the cat. A: Reconstruction of the collateral branches of the axon of a Purkinje cell, including the parent axon, soma and a small part of the dendritic tree. Scale bar = 50 μm. B, C: Terminal axonal arborescences of Purkinje cells 1 and 6 of D, respectively. Scale bar = 80 μm. Inset shows a synaptic contact of a branch of C with a deep nuclear cell in more detail. D: Diagram of a sagittal section of the cerebellum showing the sites of foliar origins and deep nuclear terminations of Purkinje cells intracellularly injected with horseradish peroxidase in different experiments. [A: from Bishop and King, 1982, reproduced by kind permission of IBRO; B-D: partly relabelled from Bishop, G. A., McCrea, R. A., Lighthall, J. W. and Kitai, S. T. (1979) An HRP and autoradiographic study of the projection from the cerebellar cortex to the nucleus interpositus anterior and nucleus interpositus posterior of the cat. J. Comp. Neurol. 185, 735–756. Copyright 1979 Wiley-Liss, Inc. Reprinted by permission of Wiley-Liss, Inc., a subsidiary of John Wiley & Sons, Inc.]

single technique can provide a complete description of a cell and its contextual relationships. Sooner or later an investigator will need to make correlations using different methods, preferably applied to the same preparation. In recent years, particularly with the introduction of new fluorescent and other dyes as well as intracellular markers, some of which have been described above, the possibilities for combination have grown enormously. Very often, however, these will involve one or more light microscopical methods with electron microscopy. Thus the variant of the Golgi method given in example 3 was originally developed in order to provide high quality ultrastructural preservation for correlative electron microscopy; even so the silver chromate precipitate was often too dense in the stained cells to allow their fine structure to be seen. Various methods of modifying the precipitate to make it more suitable for both light and electron microscopy were tried, culminating, perhaps, in the photochemical reduction method of Blackstad (1975).

Again, whereas different methods can often be used to answer the same question, they are rarely interchangeable. So intracellular labelling has not entirely replaced the Golgi method, whose value remains its unique ability to stain single cells, of different type, more or less randomly in the same preparation. Indeed, Freund and Somogyi (1983), and Somogyi et al. (1983), have described yet another version – the section-Golgi impregnation procedure – that may be carried out not only after intracellular or retrograde labelling with HRP, but could also be combined with a variety of other histochemical or immunocytochemical techniques. By including gold toning the Golgi-stained material could, moreover, be examined with the electron microscope. As if that were not enough, double or triple staining and even repeat impregnation if the first was not satisfactory were all possible. The random nature of the Golgi method was underlined by the last of these variations, since repeating the impregnation resulted in different cells being stained.

If electrophysiological data are not required, intracellular staining can be carried out after fixation, at least in the case of Lucifer yellow following aldehyde fixation (see Buhl, 1993, for review). We have already seen how an electron-dense reaction product can be deposited at the location of the Lucifer yellow, thus allowing ultrastructural observations to be made. The method may facilitate correlations between neuronal geometry and other properties such as lectin binding (Ojima, 1993) or immunoreactivity, as well as establishing synaptic interconnexions using anterograde degeneration or retrograde marking by a variety of techniques.

The Visualisation of Neuronal Activity

"... the optical approach reported here [ ] will allow the monitoring of calcium dynamics and neural network activity throughout the brain and spinal cord of both normal and mutant lines of zebrafish." (Fetcho and O'Malley, 1995).

Widespread use of tissue slices and of the confocal microscope, together with continued development of the applications of reactive and other dyes, have increased the importance of fluorescent markers in neuroscience. One might also cite technical advances in microelectronics and computing as necessary factors. The markers have made it possible to study various aspects of the activity of living neurons, covering a wide range of the temporal domain, and only a brief summary will be attempted in this final section (for details, see Chaps. 4 and 16). As with many methods or groups of techniques that have become prominent in recent years through such technical advances, the origins of the visual approach to the study of neuronal activity can be traced back a surprisingly long way. I shall mention just one early example: in 1969 Tasaki et al. reported that squid or crab nerves stained with acridine orange fluoresced more intensely during the passage of an action potential. They attributed this to an indirect effect caused by changes in the "physico-chemical properties of the macromolecules around the dye molecules in the nerve membrane". Acridine orange, incidentally, is one of a family of dyes and pharmacologically active substances originally developed in the search for anti-malarial agents. Modern "voltage-sensitive" dyes are more likely to be specifically sought; one such designed for intracellular applications is JPW1114 (Figure 2D), which has a particularly high signal:noise ratio among a group of related fluorescent styryl dyes (Antić and Zečević, 1995).

Many dyes are, of course, sensitive to particular ionic species, including H^+ (pH), Na^+ and Ca^{2+}. Calcium Green-1 [Molecular Probes] (Figure 2E), for example, is a fluorescent dye excited by visible (blue) light, whose fluorescent intensity increases linearly on binding Ca2+. It has been used in studies ranging from the role of Ca^{2+}-mediated regenerative processes in dendritic integration (Schiller et al., 1997), in which the membrane-impermeant fluorophore was injected intracellularly from a micropipette, to the activity of groups of neurons retrogradely filled with a conjugate of Calcium Green and dextran in the zebrafish larva (Fetcho and O'Malley, 1995).

In our last examples we will look briefly at fluorescent dyes with medium to long aliphatic chains that are thought to be incorporated into membranes. Certain styryl dyes, such as FM1-43 [Molecular Probes] (Figure 2F), stain neuromuscular junctions in an activity-dependent manner and have been used to study synaptic vesicle cycling (Betz and Bewick, 1992). Others, such as DiA (also known as 4-Di-16-ASP, [Molecular Probes] Figure 2G), have been used in long-term studies of growth and remodelling in the neuromuscular junction, involving repeated observations on the same junctions at various intervals (Balice-Gordon and Lichtman, 1990).

References

Angulo, A., Fernández, E., Merchán, J. A. and Molina, M. 1996. A reliable method for Golgi staining of retina and brain slices. J. Neurosci. Methods 66, 55–59.

Antić, S. and Zečević, D. 1995. Optical signals from neurons with internally applied voltage-sensitive dyes. J. Neurosci. 15, 1392–1405.

Baker, J. R. 1958. Principles of Biological Microtechnique. Methuen, London.

Balice-Gordon, R. J. and Lichtman, J. W. 1990. In vivo visualization of the growth of pre- and postsynaptic elements of neuromuscular junctions in the mouse. J. Neurosci. 10, 894–908.

Banks, R. W. 1986. Observations on the primary sensory ending of tenuissimus muscle spindles in the cat. Cell Tissue Res. 246, 309–319.

Banks, R. W., Hulliger, M., Scheepstra, K. A. and Otten, E. 1997. Pacemaker activity in a sensory ending with multiple encoding sites: the cat muscle-spindle primary ending. J. Physiol. 498, 177–199.

Bennett, M. V. L. 1973. Permeability and structure of electronic junctions and intercellular movements of tracers. In: Kater, S. B. and Nicholson, C. (eds.) Intracellular Staining in Neurobiology. Springer, Berlin, pp 115–134.

Betz, W. J. and Bewick, G. S. 1992. Optical analysis of synaptic vesicle recycling at the frog neuromuscular junction. Science 255, 200–203.

Bishop, G. A. and King, J. S. 1982. Intracellular horseradish peroxidase injections for tracing neural connections. In: Mesulam, M.-M. (ed.) Tracing Neural Connections with Horseradish Peroxidase. Wiley, Chichester, pp 185–247.

Blackstad, T. W. 1965. Mapping of experimental axon degeneration by electron microscopy of Golgi preparations. Z. Zellforsch. 67, 819–834.

Blackstad, T. W. 1975. Electron microscopy of experimental axonal degeneration in photochemically modified Golgi preparations: a procedure for precise mapping of nervous connections. Brain Res. 95, 191–210.

Blackstad, T. W., Fregerslev, S., Laurberg, S. and Rokkedal, K. 1973. Golgi impregnation with potassium dichromate and mercurous or mercuric nitrate: identification of the precipitate by X-ray and electron diffraction analysis. Histochemie 36, 247–268.

Brown, A. G. and Fyffe, R. E. W. 1981. Direct observations on the contacts made between Ia afferent fibres and α-motoneurones in the cat's lumbosacral spinal cord. J. Physiol. 313, 121–140.

Buhl, E. H. 1993. Intracellular injection in fixed slices in combination with neuroanatomical tracing techniques and electron microscopy to determine multisynaptic pathways in the brain. Microsc. Res. Tech. 24, 15–30.

Cajal, S. Ramon y. 1906. Les structures et les connexions des cellules nerveux. Les Prix Nobel 1904–1906. Norstedt, Stockholm.

Cajal, S. Ramon y. 1995. Histology of the Nervous System of Man and Vertebrates. Translated by Swanson, N. and Swanson, L. W. Oxford, New York.

Chan-Palay, V. 1973. A brief note on the chemical nature of the precipitate within nerve fibers after the rapid Golgi reaction: selected area diffraction in high voltage electron microscopy. Z. Anat. Entwickl.-Gesch. 139, 115–117.

Cox, W. 1891. Imprägnation des centralen Nervensystems mit Quecksilbersalzen. Arch. Mikrosk. Anat. 37, 16–21.

Cullheim, S and Kellerth, J.-O. 1976. Combined light and electron microscopic tracing of neurons, including axons and synaptic terminals, after intracellular injection of horseradish peroxidase. Neurosci. Lett. 2, 301–313.

Deiters, O. F. K. 1865. Untersuchungen über Gehirn und Rückenmark des Menschen und der Säugetiere. Vieweg, Braunschweig.

Descartes, R. 1662. De homine figuris et latinitate donatus a Florentio Schuyl. Francis Moyard and Peter Leff, Leyden.

Ebbesson, S. O. E. and Cheek, M. 1988. The use of cryostat microtomy in a simplified Golgi method for staining vertebrate neurons. Neurosci. Letters 88, 135–138.

Eccles, J. C. 1964. The Physiology of Synapses. Springer, Berlin.

Fetcho, J. R. and O'Malley, D. M. 1995. Visualization of active neural circuitry in the spinal cord of intact zebrafish. J. Neurophysiol. 73, 399–406.

Fregerslev, S., Blackstad, T. W., Fredens, K. and Holm, M. J. 1971a. Golgi potassium dichromate-silver nitrate impregnation. Nature of the precipitate studied by X-ray powder diffraction methods. Histochemie 25, 63–71.

Fregerslev, S., Blackstad, T. W., Fredens, K., Holm, M. J. and Ramón-Moliner, E. 1971b. Golgi impregnation with mercuric chloride: Studies on the precipitate by X-ray powder diffraction and selected area electron diffraction. Histochemie 26, 289–304.

Freund, T. F. and Somogyi, P. 1983. The section-Golgi impregnation procedure. 1. Description of the method and its combination with histochemistry after intracellular iontophoresis or retrograde transport of horseradish peroxidase. Neuroscience 9, 463–474.

Galigher, A. E. and Kozloff, E. N. 1964. Essentials of Practical Microtechnique. Henry Kimpton, London.

Glauert, A. M. 1975. Fixation, Dehydration and Embedding of Biological Specimens. North-Holland, Amsterdam.

Golgi, C. 1873. Sulla struttura della sostanza grigia dell cervello. Gazetta medica italiana Lombarda 33, 244–246 (Translated by M. Santini as 'On the structure of the grey matter of the brain'. In Santini, M. (ed.) 1975, Golgi Centennial Symposium, Raven, New York).

Gray, E. G. 1964. Tissue of the central nervous system. In: Kurtz, S. M. (ed.) Electron Microscopic Anatomy. Academic Press, New York, pp 369–417.

Graybiel, R. and Devor, M. 1974. A microelectrophoretic delivery technique for use with horseradish peroxidase. Brain Res. 68, 167–173.

Heuser, J. E., Reese, T. S., Dennis, M. J., Jan, Y., Jan, L. and Evans, L. 1979. Synaptic vesicle exocytosis captured by quick-freezing and correlated with quantal transmitter release. J. Cell Biol. 81, 275–300.

Heuser, J. E. and Reese, T. S. 1973. Evidence for recycling of synaptic vesicle membrane during transmitter release at the frog neuromuscular junction. J. Cell Biol. 57, 315–344.

Hodgson, E. S. 1990. Long-range perspectives on neurobiology and behavior. Amer. Zool. 30, 403–505.

Holt, E. J. and Hicks, R. M. 1961. Studies on formalin fixation for electron microscopy and cytochemical staining purposes. J. Biophys. Biochem. Cytol. 11, 31–45.

Horikawa, K. and Armstrong, W. E. 1988. A versatile means of intracellular labeling - injection of biocytin and its detection with avidin conjugates. J. Neurosci. Meth. 25, 1–11.

Jankowska, E., Rastad, J. and Westman, J. 1976. Intracellular application of horseradish peroxidase and its light and electron microscopical appearance in spinocervical tract cells. Brain Res. 105, 555–562.

Källström, Y. and Lindström, S. 1978. A simple device for pressure injections of horseradish peroxidase into small central neurones. Brain Res. 156, 102–105.

Karlsson, U. and Schultz, R. C. 1965. Fixation of the central nervous system for electron microscopy by aldehyde perfusion. I. Preservation with aldehyde perfusates versus direct perfusion with osmium tetroxide with special reference to membranes and the extracellular space. J. Ultrastruct. Res. 12, 160–186.

Karnovsky, M. J. 1965. A formaldehyde-glutaraldehyde fixative of high osmolality for use in electron microscopy. J. Cell Biol. 27, 137A–138A.

Kita, H. and Armstrong, W. 1991. A biotin-containing compound N-(2-aminoethyl) biotinamide for intracellular labeling and neuronal tracing studies - comparison with biocytin. J. Neurosci. Meth. 37, 141–150.

Kitai, S. T. and Bishop, G. A. 1981. Horseradish peroxidase: intracellular staining of neurons. In: Heimer, L. and RoBards, M. J. (eds.) Neuroanatomical Tract-Tracing Methods. Plenum, New York, pp 263–277.

Kitai, S. T., Kocsis, J. D., Preston, R. J. and Sugimori, M. 1976. Monosynaptic inputs to caudate neurons identified by intracellular injection of horseradish peroxidase. Brain Res. 109, 601–606.

Kopsch, F. 1896. Erfahrungen über die Verwendung der Formaldehyde bei der Chrom-silber-Imprägnation. Anat. Anz. 11, 727–729.

Kravitz, E. A., Stretton, A. O. W, Alvarez, J. and Furshpan, E. J. 1968. Determination of neuronal geometry using an intracellular dye injection technique. Fed. Proc. 27, 749.

Leeuwenhoek, A. van 1717. Letter to Abraham van Bleiswyk, translated by S. Hoole, 1807 in Anthony van Leeuwenhoek, Selected Works. Philanthropic Society, London.

Light, A. R. and Durkovic, R. G. 1976. Horseradish peroxidase: an improvement in intracellular staining of single, electrophysiologically characterized neurons. Exp. Neurol. 53, 847–853.

Lund, R. D. and Westrum, L. E. 1966. Synaptic vesicle differences after primary formalin fixation. J. Physiol. 185, 7–9P.

Lynch, G., Deadwyler, S. and Gall, C. 1974. Labeling of central nervous system neurons with extracellular recording microelectrodes. Brain Res. 66, 337–341.

Maranto, A. R. 1982. Neuronal mapping: a photooxidation reaction makes Lucifer Yellow useful for electron microscopy. Science. 217, 953–955.

McCrea, R. A., Bishop, G. A. and Kitai, S. T. 1976. Intracellular staining of Purkinje cells and their axons with horseradish peroxidase. Brain Res. 118, 132–136.

Millhouse, O. E. 1981. The Golgi methods. In Heimer, L. and RoBards, M. J. (eds.) Neuroanatomical Tract-Tracing Methods. Plenum, New York, pp 311–344.

Morest, D. K. 1981. The Golgi methods. In Heym, Ch. and Forssmann, W.-G. (eds.) Techniques in Neuroanatomical Research. Springer, Berlin, pp 124–138.

Muller, K. J. and McMahon, U. J. 1976. The shapes of sensory and motor neurons and the distribution of their synapses in ganglia of the leech: a study using intracellular injection of horseradish peroxidase. Proc. Roy. Soc. Lond. B. 194, 481–499.

Nicholson, C. and Kater, S. B. 1973. The development of intracellular staining. In. Kater, S. B. and Nicholson, C. (eds.) Intracellular Staining in Neurobiology. Springer, Berlin, pp 1–19.

Ojima, H. 1993. Dendritic arborization patterns of cortical interneurons labeled with the lectin, Vicia villosa, and injected intracellularly with Lucifer yellow in aldehyde-fixed rat slices. J. Chemical Neuroanat. 6, 311–321.

Onn, S.-P., Pucak, M. L. and Grace, A. A. 1993. Lucifer yellow dye labelling of living nerve cells and subsequent staining with Lucifer yellow antiserum. Neurosci. Protocols. 93-050-17-01-14

Palade, G. E. and Palay, S. L. 1954. Electron microscope observations of interneuronal and neuromuscular synapses. Anat. Rec. 118, 335–336.

Palay, S. L., McGee-Russell, S. M., Gordon, J. and Grillo, M. A. 1962. Fixation of neural tissues for electron microscopy by perfusion with solutions of osmium tetroxide. J. Cell Biol. 12, 385–410.

Paula-Barbosa, M. 1975. The duration of aldehyde fixation as a "flattening factor" of synaptic vesicles. Cell Tiss. Res. 164, 63–72.

Peters, A. 1970. The fixation of central nervous system and the analysis of electron micrographs of the neuropil with special reference to the cerebral cortex. In Nauta, W. H. J. and Ebbesson, S. O. (eds.) Contemporary Research Methods in Neuroanatomy. Springer, Berlin, pp 56–76.

Phillips, C. G. 1987. Purkinje cells and Betz cells. Physiol. Bohemoslov. 30, 217–223.

Pinault, D. 1994. Golgi-like labeling of a single neuron recorded extracellularly. Neurosci. Lett. 170, 255–260.

Ramón-Moliner, E. and Ferrari, J. 1972. Electron microscopy of previously identified cells and processes within the central nervous system. J. Neurocytol. 1, 85–100.

Sabatini, D. D., Bensch, K. and Barrnett, R. J. 1963. Cytochemistry and electron microscopy. The preservation of cellular ultrastructure and enzymatic activity by aldehyde fixation. J. Cell Biol. 17, 19–58.

Scheibel, M. E. and Scheibel, A. R. 1978. The methods of Golgi. In Robertson, R. T. (ed.) Neuroanatomical Research Techniques. Academic, New York, pp 89–114.

Schiller, J., Schiller, Y., Stuart, G. and Sakmann, B. 1997. Calcium action potentials restricted to distal apical dendrites of rat neocortical pyramidal neurons. J. Physiol. 505, 605–616.

Schultz, R. L. and Karlsson, U. 1965. Fixation of the central nervous system for electron microscopy by aldehyde perfusion. II. Effect of osmolarity, pH of perfusates, and fixative concentration. J. Ultrastruct. Res. 12, 187–206.

Snow, P. J., Rose, P. K. and Brown, A. G. 1976. Tracing axons and axon collaterals of spinal neurons using intracellular injection of horseradish peroxidase. Science 191, 312–313.

Somogyi, P., Freund, T. F., Wu, J.-Y. and Smith, A. D. 1983. The section-Golgi impregnation procedure. 2. Immunocytochemical demonstration of glutamate decarboxylase in Golgi-impregnated neurons and in their afferent synaptic boutons in the visual cortex of the cat. Neuroscience 9, 475–490.

Stean, J. P. B. 1974. Some evidence of the nature of the Golgi-Cox deposit and its biochemical origin. Histochemistry 40, 377–383.

Stewart, W. W. 1981. Lucifer dyes – highly fluorescent dyes for biological tracing. Nature 292, 17–21.

Strangeways, T. S. P. and Canti, R. G. 1927. The living cell in vitro as shown by dark-ground illumination and the changes induced in such cells by fixing reagents. Q. Jl. Microsc. Sci. 71, 1–14.
Strausfeld, N. J. 1980. The Golgi method: its application to the insect nervous system and the phenomenon of stochastic impregnation. In Strausfeld, N. J. and Miller, T. A. (eds.) Neuroanatomical Techniques: Insect Nervous System. Springer, New York pp. 131–203.
Stretton, A. O. W. and Kravitz, E. A. 1973. Intracellular dye injection: the selection of Procion yellow and its application in preliminary studies of neuronal geometry in the lobster nervous system. In. Kater, S. B. and Nicholson, C. (eds.) Intracellular Staining in Neurobiology. Springer, Berlin, pp 21–40.
Tasaki, I., Carnay, L., Sandlin, R. and Watanabe, A. 1969. Fluorescence changes during conduction in nerves stained with acridine orange. Science 163, 683–685.
Uchizono, K. 1965. Characteristics of excitatory and inhibitory synapses in the central nervous system of the cat. Nature 207, 642–643.
Van Harreveld, A. and Fifkova, E. 1975. Rapid freezing of deep cerebral structures for electron microscopy. Anat. Rec. 182, 377–386.
Verna, A. 1983. A simple quick-freezing device for ultrastructure preservation: evaluation by freeze-substitution. Biol. Cell 49, 95–98.
Walberg, F. 1966. Elongated vesicles in terminal boutons of the central nervous system, a result of aldehyde fixation. Acta Anat. 65, 224–235.
Waldeyer, W. 1891. Über einige neuere Forschungen im Gebiete der Anatomie des Centralnervensystems. Deutsche medizinische Wochenschrift 17, 1352–1356.
Westrum, L. E. and Lund, R. D. 1966. Formalin perfusion for correlative light- and electron-microscopical studies of the nervous system. J. Cell Sci. 1, 229–238.

Chapter 2

Application of Differential Display and Serial Analysis of Gene Expression in the Nervous System

Erno Vreugdenhil, Jeannette de Jong and Nicole Datson

Introduction

Every biological process in both plant and animal species is associated with changes in gene expression. In the central nervous system (CNS), changes in gene expression are not only causally linked to the development of the CNS but also to complex and as yet not well-understood phenomena such as memory formation, learning and cognition. In addition, changes in gene expression underlie the pathogenesis of many acute and chronic CNS-related disorders such as ischemia, epilepsy, Alzheimer's disease and Parkinson's disease. Thus, insight into and characterization of gene expression profiles, and in particular the changes occurring therein, are crucial for understanding how the brain functions at the molecular level and how malfunction will lead to disease.

Given its complexity, i.e. tens of thousands of genes each expressed at a different level, characterization of gene expression profiles is not a straightforward task. The set of genes expressed and the stochiometry of the resulting messenger RNAs, together called a "transcriptome", determine the phenotype of a cell, tissue and whole organism. The human genome is thought to contain 50,000–100,000 genes of which a subset of approximately 15,000–20,000 genes is expressed in an individual cell. Therefore, gaining insight into gene expression profiles in a particular tissue or cell is a major enterprise, and the identification of a limited set of differentially expressed genes resembles searching for a needle in a haystack.

In the 1980s, several methods aimed at the identification of differentially expressed genes were described, including plus/minus screening and subtractive hybridization methods. Although these methods have proven to be useful in isolating differentially expressed genes, they are technically difficult and labour-intensive, relatively slow and require large amounts of RNA (see e.g. Kavathas et al., 1984; Vreugdenhil et al., 1988).

In the beginning of the 1990s, the sensitivity, speed and accuracy of differential screening techniques were boosted by two major developments: first, polymerase chain reaction techniques were introduced resulting in the possibility to amplify minimal amounts of starting material and making the monitoring of expression of thousands of genes simultaneously possible. Second, the increasing knowledge of DNA sequences of a large number of genes and corresponding transcripts necessitated and resulted in the establishment of nucleotide sequence databases. In addition, different genome projects were initialised to unravel complete nucleotide sequences of several species including several bacterial species, yeast, the nematode, drosophila, mouse and human (McKu-

Correspondence to: Erno Vreugdenhil, Leiden/Amsterdam Center for Drug Research (LACDR), Division of Medical Pharmacology, PO Box 9503, RA Leiden, 2300, The Netherlands (phone: +31-715276230; fax: +31-715276292; e-mail: vreugden@lacdr.leidenuniv.nl)
Jeannette de Jong, Leiden University, Division of Medical Pharmacology, LACDR, P.O. Box 9503, RA Leiden, 2300, The Netherlands
Nicole Datson, Leiden University, Division of Medical Pharmacology, LACDR, P.O. Box 9503, RA Leiden, 2300, The Netherlands

sick, 1997; Rowen et al., 1997; Duboule, 1997; Levy, 1994). At present, the complete genomes of *E. coli* (10^6 bp) and yeast ($2x10^7$ bp) are known, while those of nematode (10^8 bp) and human ($2x10^9$) are partially sequenced; respectively 80% and 5% are known. These known DNA sequences are publicly available and as a consequence application of screening techniques has only to result in a small portion of a particular gene to unambiguously identify it as up- or down-regulated.

The introduction of PCR and the establishment of databases have revolutionized differential screening strategies and resulted in a number of highly sensitive techniques. Here we will discuss two of these: differential display (DD) and serial analysis of gene expression (SAGE).

Differential Display

DD was first described in 1992 by Liang and Pardee (Liang and Pardee, 1992). The tremendous impact of DD is probably best illustrated by the number of approximately 1700 DD-related articles which have been published since its introduction. Many genes linked to numerous CNS-related processes such as neurodegeneration and apoptosis have been identified by DD (Kiryu et al., 1995; Livesey et al., 1997; Tsuda et al., 1997; Imaizumi et al., 1997; Su et al., 1997; Shirvan et al., 1997).

The principle of DD is based on the random amplification and subsequent size separation of cDNA molecules. To this end, total RNA is isolated from a cell or tissue of interest and reverse-transcribed into cDNA. Instead of a single oligodT primer, four different anchored oligodT primers are used (oligodT-MC, oligodT-MG, oligodT-MT and oligodT-MA; M=G/A/C) in four separate cDNA synthesis reactions. Basically, this modified cDNA synthesis divides the original mRNA population into four different cDNA pools. Subsequently, a fraction of each pool of cDNA is randomly amplified using a randomly chosen primer in combination with the same anchored oligodT primer. The PCR conditions, in particular the annealing temperature, are chosen such that approximately 60–100 cDNA fragments are amplified. These cDNA fragments, derived from "stimulated" and "non-stimulated" tissues, are size-separated in parallel on gels. Differentially expressed products are identified by comparing the presence (upregulation) or absence (downregulation) of cDNA fragments in the two situations. This process is repeated with other randomly chosen primers, resulting in the amplification of another portion of the cDNA pool. Finally, differentially expressed cDNA fragments can be excised from gel and further characterized by, e.g., Northern blot analysis, *in situ* hybridization and DNA sequence analysis (see below). The major advantage of this DD approach is its simplicity, its extreme sensitivity and the possibility to identify both up- and downregulated genes in the same experiment. Disadvantages of DD are its labour-intensive character and the generation of many false positives.

Since its introduction, many modifications and improvements of the DD technique have been described. For example, instead of the originally described radioactive DD cDNA fragments, different labels, e.g. fluorescent labels, have been used to monitor DD fragments (Bauer et al., 1993; Ito et al., 1994; Rohrwild et al., 1995; Vreugdenhil et al., 1996b). Consequently, automated DNA sequencers could be used to facilitate the monitoring and analysis of DD fragments. Other efforts have focused on primer design (Liang et al., 1994; Liang et al., 1993; Malhotra et al., 1998). These latter studies have led to the use of extended 20-nucleotide-long primers in more recent reports. Several excellent review articles on the principles of differential display have been published (Liang and Pardee, 1995; Livesey and Hunt, 1996; Vreugdenhil et al., 1996b; Liang and Pardee, 1997).

In this chapter, we will describe in detail two different approaches that worked well in our hands. The first one is based on the use of small oligonucleotides and the use of digoxigenin-labelled primers, in this chapter called "DD-PCR". The second one, called "extended (E)DD-PCR", is based on the use of extended oligonucleotides and fluorescent-labelled primers.

Serial Analysis of Gene Expression (SAGE)

SAGE (Velculescu et al., 1995) is a highly sensitive PCR-based expression profiling method that yields both qualitative and quantitative information on the composition of an mRNA pool or transcriptome. Since its introduction in 1995, a number of SAGE-related articles have been published (Madden et al., 1997; Polyak et al., 1997; Zhang et al., 1997; Velculescu et al., 1997) which have demonstrated the enormous potential of SAGE to detect changes in expression levels of large numbers of genes simultaneously. For example, comparison of expression profiles in colorectal cancer and normal colon epithelium revealed 51 genes with a more than tenfold decrease in expression in primary colorectal cancer cells, while 32 genes were upregulated more than tenfold (Zhang et al., 1997).

SAGE is based on the generation of short, approximately 10-bp transcript-specific sequence tags and ligation of the tags to long strings (concatemers) which are subsequently cloned and sequenced. The frequency of each tag in the concatemers reflects the original stoichiometry of the individual transcripts in the mRNA pool, allowing quantitative assessment of gene expression. By comparing expression profiles derived from different mRNA sources, differentially expressed genes can be identified. An advantage of SAGE is that in a single sequence reaction over 25 tags (cDNAs) can be analyzed, a 25-fold increase in efficiency compared to EST sequencing. Moreover, the short tags are specific enough to uniquely identify each transcript and can be linked to known genes or ESTs present in GenBank, facilitating retrieval of additional sequence of potentially interesting upregulated or downregulated tags and their further characterization. A disadvantage is that in spite of the high efficiency, large numbers of tags need to be sequenced to enable detection of differences in expression of low-abundant transcripts, requiring high-throughput sequencing facilities and robotics. In addition, SAGE is less useful for analysing gene expression in organisms that are relatively underrepresented in GenBank. Another drawback of SAGE is the requirement of a large amount of starting RNA (2.5–5 μg polyA$^+$ RNA). Although SAGE has possible applications in many fields of research, its use is thus restricted to situations in which the amount of starting material is not limited. In addition, use of RNA isolated from complex tissues consisting of heterogeneous cell populations will dilute the relative expression profile of the different cell types and thus perhaps mask relevant changes in expression. These very issues hamper expression profiling in the brain, consisting of many unique, highly specialised, often small substructures, each with their own specific expression profile.

In this chapter we describe a modified SAGE protocol (Protocol B) that requires minimal amounts of starting material, making it extremely suitable for use in neuroscience. Using this protocol we can obtain an expression profile from a single hippocampal punch derived from a 300-μm brain slice, which we estimate to contain at least a factor $5x10^3$ less polyA$^+$ RNA than is required for the original procedure (protocol A).

Table 1. Characteristics of differential display and serial analysis of gene expression

	Differential display	Serial analysis of gene expression
Requirements	Standard molecular biological equipment	Automated DNA sequencer
Quantitativity	No	Yes
False positives	Many	None
End products	cDNA fragments of 100 to 500 bp	10 to 14 bp long tags
Suitable for screening for full-length cDNA	Yes	No
Species preference	Can be applied in every species	For species well represented in GenBank
Starting material	10 ng total RNA	2.5 μg messenger RNA
Detection of low abundant transcripts	Difficult	Difficult
PCR-bias	Yes	No
Labor-intensive	Yes	Yes
Technical approach	Easy	Difficult

Subprotocol 1
Differential Display: Practical Approach

Outline

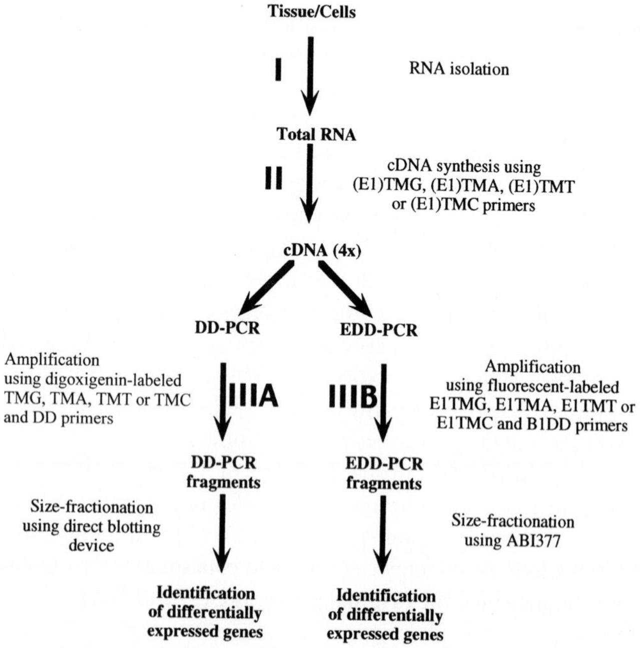

Tissue/Cells

I — RNA isolation

Total RNA

II — cDNA synthesis using (E1)TMG, (E1)TMA, (E1)TMT or (E1)TMC primers

cDNA (4x)

DD-PCR | EDD-PCR

IIIA — Amplification using digoxigenin-labeled TMG, TMA, TMT or TMC and DD primers

IIIB — Amplification using fluorescent-labeled E1TMG, E1TMA, E1TMT or E1TMC and B1DD primers

DD-PCR fragments | EDD-PCR fragments

Size-fractionation using direct blotting device | Size-fractionation using ABI377

Identification of differentially expressed genes | Identification of differentially expressed genes

Fig. 1. Three major steps of differential display will be explained: total RNA has to be isolated (I) which will subsequently be reverse-transcribed into cDNA using DD-specific oligonucleotides (II). This cDNA is used as a template in the polymerase chain reaction (PCR) step by using short oligonucleotides (IIIA) or extended oligonucleotides (**IIIB**).

Materials

Glassware and plasticware

- Sterile glassware (incubated O/N in 180°C oven)
- Sterile Eppendorf tubes (0.5 ml, 1.5. ml and 2.2 ml), Eppendorf tips and sterile falcon tubes (15 ml and 50 ml)
- Filter tips

Note: to avoid false positives it is crucial to eliminate contaminations from the PCR reaction. Though expensive, the use of filter tips can be very useful.

- Sterile plastic gloves and insulated gloves
- Plastic funnel
- Safety glasses
- Quartz cuvettes (0.5 ml or 1.5 ml)

Solutions and buffers

- 4M guanidinium thiocyonate (GTC, for preparation see Chomczynski et al., 1997; Chomczynski and Sacchi, 1987). Several commercially available RNA isolation kits, like TRIzol (GibcoBRL) are based on this method and work very well in our hands.
- Liquid N_2
- Chloroform p/a
- Phenol
- Iso-propanol p/a
- 75% ethanol
- Ethanol absolute
- Glycogen (20mg/ml; purchased from Boehringer). Needed if only a minimal amount of starting material is available.
- DEPC-treated ddH_2O
- 3M NaAc (pH 5.2)
- 0.1 M DTT
- Ethidium bromide (EtBr): 20 mg/ml
- 5x cDNA synthesis buffer
  - -250 mM Tris-HCl
  - -375 mM KCl
  - -15 mM $MgCl_2$
- dNTPs (10mM each)
- 10x polymerase buffer (supplied with the enzyme)
- 25mM $MgCl_2$

Equipment

- 70°C waterbath
- 42°C waterbath
- 25°C waterbath
- Spectrophotometer
- Mortar and pestle
- Thermocycler
- Eppendorf centrifuge
- Gelelectrophoresis equipment
- Direct blotting device (GATC 1500-system)
- Automated DNA sequencer (ABI 310, 373 or 377)

Primers

- DD-PCR
  - T12MG = 5'-DIG -TTT TTT TTT TTT MG -3'
  - T12MA = 5'-DIG -TTT TTT TTT TTT MA -3'

- T12MT = 5'-DIG -TTT TTT TTT TTT MT -3'
- T12MC = 5'-DIG -TTT TTT TTT TTT MC -3'

Note: T12M primers are used for cDNA synthesis *and* PCR

Note: T12M primers are labelled at 5'end.

- DD1=5'-TACAACGAGG-3'
- DD2=5'-TGGATTGGTC-3'
- DD3=5'-CTTTCTACCC-3'
- DD4=5'-TTTTGGCTCC-3'
- DD5=5'-GGAACCAATC-3'
- DD6=5'-AAACTCCGTC-3'
- DD7=5'-TCGATACAGG-3'
- DD8=5'-TGGTAAAGGG-3'
- DD9=5'-TCGGTCATAG-3'
- DD10=5'-GGTACTAAGG-3'
- DD11=5'-TACCTAAGCG-3'
- DD12=5'-CTGCTTGATG-3'
- DD13=5'-GTTTTCGCAG-3'
- DD14=5'-GATCAAGTCC-3'
- DD15=5'-GATCCAGTAC-3'
- DD16=5'-GATCACGTAC-3'
- DD17=5'-GATCTGACAC-3'
- DD18=5'-GATCTCAGAC-3'
- DD19=5'-GATCATAGCC-3'
- DD20=5'-GATCAATCGC-3'

Note: DD primers have been described by Bauer and coworkers (Bauer et al., 1993; Bauer et al., 1994)

- EDD-primers:
  - [FAM]/[HEX]/[DIG]-E1TTTTTTTTTTTTMG
  - [FAM]/[HEX]/[DIG]-E1TTTTTTTTTTTTMA
  - [FAM]/[HEX]/[DIG]-E1TTTTTTTTTTTTMC
  - [FAM]/[HEX]/[DIG]-E1TTTTTTTTTTTTMT
  - E1=5'-CG**G AAT TC**G G-3' ()
  - M=A/G/C

Note: ET12M primers are used for cDNA synthesis *and* PCR

Note: ET12M primers are labelled at 5' end

Note: EDD primers contain *Eco*RI site (shown in bold) to facilitate subcloning

- B1DD primers: B1DD primers are the same as the DD primers, except that they are extended at their 5' end by ten nucleotides called B1.
  E.g. B1DD1=5'-CGT**GGATCC**GTACAACGAGG-3'
  B1DD2=5'-CGT**GGATCC**GTGGATTGGTC-3' etc.

Note: B1 contains a *Bam*HI site (shown in bold) to facilitate subcloning of EDD fragments.

Enzymes

- Reverse Transcriptase: We prefer Superscript II (GibcoBRL). This enzyme is relatively stable and has little batch-to-batch variation.

Note: Upon arrival it is preferable to aliquot the enzyme to avoid reduction in activity due to repetitive freeze/thawing

- RNase-free DNase (5U/µl)
- Taq polymerases

Note: We have tested a series of Taq polymerases and found that Ultma Taq (Perkin Elmer) gave the most consistent data with DD-PCR and Amplitaq Gold (Perkin Elmer) for EDD-PCR.

Procedure

I. RNA Isolation

General remarks

A major factor in the overall success of differential display is the RNA quality. To minimize the chance of RNA degradation:
- Wear gloves
- Use sterile tips and tubes. Incubate all plasticware at 110°C O/N
- Use RNase-free glassware. Incubate glassware at 180°C O/N
- All buffers have to be RNase-free. To this end use diethylpyrocarbonate (DEPC)-treated double-distilled water (add 2.5 ml DEPC to 2.5l double-distilled water, incubate O/N at RT and autoclave).
- Work on ice.

Homogenisation

1. Tissue.
   Freeze tissue (–70°C) and weigh just before processing. Grind frozen tissue with mortar and pestle under liquid N_2 until the tissue is completely transformed into powder. Wear insulated gloves to protect hands from the cold pestle. Use safety glasses and pay attention to the level of liquid N_2 in the mortar. It is crucial that the tissue is submerged in N_2 at this stage since a major source of RNase is derived from damaged cells. Use a funnel to transfer the ground tissue in liquid N_2 from the mortar to a 50ml sterile plastic falcon tube. As soon as all liquid N_2 has evaporated add 1 ml TRIzol per 50–100 mg tissue and shake vigorously until a homogeneous solution is obtained. As soon as all tissue is dissolved completely in TRIzol, RNase activity is completely inhibited by the GTC. At this stage, the solution can be stored O/N at 4^0C. However, we advise processing the RNA immediately until a complete RNase-free situation is created.
2. Cells in monolayer.
   Cells can be lysed directly by adding 1 ml TRIzol to the culture disk (diameter 3.5 cm). Transfer solution to a 2.0 ml Eppendorf tube.
3. Cells in suspension.
   Pellet cells by spinning at 1000g for 5 min. Add 1 ml TRIzol per $0.5–1.0x10^6$ eukaryotic cells and shake vigorously.

Phase separation

1. Incubate TRIzol solution containing the RNA at RT for 5 min.
2. Add 0.2 ml chloroform per ml TRIzol.
3. Vortex for 15 sec. Let stand at RT for 30 sec and vortex again for 15 sec.
4. Transfer solution to 2.0 ml Eppendorf tubes.
5. Spin for 15 min at 15.000*g* at 4°C

RNA precipitation

1. After spinning transfer the colourless upper phase to a new Eppendorf tube.

   Note: avoid transferring material from the interphase. This contains proteins and genomic DNA and thus sincerely affects the quality of the RNA.
2. Add 0.5 ml iso-propanol per ml TRIzol.

   Note: if less than 50 ng total RNA is anticipated, also add 0.5 µl glycogen as a carrier.

3. Vortex and incubate 10 min at RT.
4. Spin for 10 min, 12.000*g* at 4°C.
5. Carefully remove supernatant.
6. Wash the RNA pellet by adding 75% ethanol (1ml ethanol/ml TRIzol).
7. Vortex
8. Spin 5 min, 7.500*g* at 4°C.
9. Carefully remove supernatant.
10. Air-dry pellet for 15 min.

    Note: pellet should not be too dry. Do not use speedvac to dry the pellet as it will be difficult to dissolve.

11. Dissolve pellet in DEPC-treated ddH_2O.

    Note: if all RNA is to be directly used for cDNA synthesis, dissolve pellet in 11 µl ddH_2O.

12. Determine the yield of RNA by measuring the $OD_{260/280}$ (see below)
13. store RNA, add 1/10 volume of 3M NaAc and 2 volumes of absolute ethanol. Aliquot RNA in portions of 2.5 µg and store at –70°C.

Quantitation of RNA

1. Take aliquot from RNA (e.g. one-tenth of the total RNA yield) and add DEPC-dd$H_2$0 to 0.5 ml final volume (if 0.5 ml quartz cuvettes are used) to 1 ml (if 1 ml quartz cuvettes are used).

   Note: at least 2 µg of RNA is required.

2. Determine the $OD_{260/280}$ratio
3. Calculate the yield and quality of the RNA:
   - An OD_{260} of 1 corresponds to an RNA concentration of 40 µg/ml
   - An $OD_{260/280}$of 2.0 indicates a pure RNA sample.

   Note: an $OD_{260/280}$ of less than 2.0 indicates the presence of proteins and/or genomic DNA. In particular the presence of genomic DNA will create false differentially expressed cDNA fragments

DNase treatment

To prevent a possible contamination with genomic DNA, treat RNA samples with RNase-free DNase.

1. Dissolve RNA pellet in 1xDNase buffer.
2. Add 5U DNase.
3. Incubate at 37°C for 15 min.
4. Increase volume with DEPC-treated H_2O, e.g. to 300 µl.
5. Add an equal volume of phenol/chloroform/isoamylalcohol.
6. Vortex vigorously.
7. Spin at 13.000 *g* for 4 min at 4°C
8. Transfer the aqueous upper phase to a clean Eppendorf tube and determine the quality and quantity of the RNA sample by spectrophotometry.
9. Precipitate RNA and store at –70°C.

Quality control of RNA

To check the quality of the RNA run a 1% agarose gel and stain with ethidium bromide (EtBr). Use autoclaved buffers and glassware; clean electrophoresis tank with 0.5M NaOH prior to running gel.

- The ribosomal RNA bands 18S and 28S should be clearly visible.
- If these bands are absent or very faint while most of the EtBr staining is at the bottom of the gel, the RNA is likely degraded and should not be used for DD purposes.

II cDNA Synthesis

General remarks

Four cDNA reactions using four different primers [i.e. (E)T12MA, (E)T12MG, (E)T12MT and (E)T12MC] per RNA are performed. In addition, for each primer a negative control is included. Thus, a total of 8 cDNA reactions per RNA sample will be performed.

Denaturation of RNA

1. Add 2.5 µg total RNA and DEPC-treated ddH_2O. In total this should have a volume of 10 µl.
2. Depending on the type of DD, add 2 µl (25µM) of the appropriate primer.
3. Mix.
4. Incubate at 70°C for 10 min.
5. Place *directly* on ice.

   Note: this incubation at 70°C causes unfolding of tertiary structures which are present in mRNA molecules. Transferring the tube from 70°C to, e.g., RT will result in reannealing and thus in less "full-length" cDNA molecules.

6. Add on ice:
   - 4 µl 5x first strand buffer
   - 2 µl 0.1 M DTT
   - 1 µl dNTPs (10mM)
   - 19 µl total
7. Mix and transfer tube to 25°C waterbath.
8. Incubate for 1 min.
9. Add 1 µl reverse transcriptase (200 units) or 1 µl ddH_2O as a control.
10. Mix and incubate for 10 min.
11. Transfer tube to 42°C waterbath.
12. Incubate for an additional 50 min.
13. Heat-inactivate reverse transcriptase by incubating at 70°C for 10 min.
14. Spin 10 sec to remove condensation from the cap of the tubes.
15. Store cDNA samples at 4°C.

IIIA: DD-PCR

General remark

Use filter tips for all PCR-related pipetting steps.

Protocol

Dilute the 20-µl cDNA sample (see above) to 100 µl with ddH_2O. Four µl of this will be used in the PCR reaction.

Note: When performing DD-PCR for the first time, it is advisable to spend some time optimizing the PCR conditions. For example, a dilution series of cDNA can be performed to determine the optimum template concentration for DD-PCR.

PCR reactions

Note: It is advisable to prepare a pipetting scheme before doing any practical work.

Add the following components:
- 4 µl cDNA
- 4 µl 10x buffer (supplied with the taq polymerase)
- 2µl $MgCl_2$
- 2 µl dNTPs (1 mM)
- 4 µl DIG-labelled T12MN (2.5 µM)
- 4 µl DD-oligo (5 µM)
- 18 µl ddH_2O
- 2 µl Taq polymerase (0.2U/µl)
- 40 µl total

Notes

1. Sometimes $MgCl_2$ is already included in the 10x buffer. In that case do not add $MgCl_2$ and add 20 µl ddH_2O instead of 18 µl ddH_2O.
2. It is advisable to make master mixes of 10x buffer, $MgCl_2$, ddH_2O and of specific primer pairs.
3. If thermocycler is not equipped with a heated lid, use a drop of mineral oil to prevent evaporation.
4. Preheat the PCR sample *except the Taq polymerase* to above 60°C, then add the Taq polymerase. At RT, the conditions for primer annealing are very relaxed. As the enzyme is already slightly active at RT, this will result in too many cDNA fragments. In addition, the reproducibility of the results is less well controlled.

PCR profile

- 3 min at 94°C, 5 min at 37°C and 5 min at 72°C
- Subsequently: 30 sec at 95°C, 2.5 min at 38°C and 45 sec at 72°C; repeat this step 39 times
- 5 min at 72°C

Note: These conditions depend on the type of thermocycler. Our conditions were optimized for a Biometra. It is advisable to vary these conditions when setting up DD.

Gel electrophoresis

We have thus far used a direct blotting device (GATC 1500) for size-separation of DD-PCR generated cDNA fragments. This equipment has been developed specifically to size-separate and detect digoxigenin-labelled DNA molecules which are blotted directly on a nylon membrane during the run of the polyacrylamide (PAA) gel. The DNA fragments on the membrane are visualised by staining the gel with anti-dig antibodies. The advantage of this system is that no radioactivity is required for detecting DNA molecules. In addition, the size-separation range is very long (10 to 800 bp) compared to the classic PAA electrophoresis gels (10 to 300 bp or 150 to 500 bp) and thus less PAA-gels are required. As the protocols are rather specific for this equipment and as it is delivered with an excellent and detailed protocol explaining how to use it, we will here restrict ourselves to some general remarks.

1. To accurately compare cDNA fragments, load PCR samples generated with the same primer pairs next to each other (see Figure 2).
2. A major problem with DD is the occurrence of many false positives (a few of these are indicated by asterisks in Figure 2). To avoid selecting these false positives for further characterization, increase the N number, i.e. do multiple RNA isolations, cDNA syntheses and PCRs independently for each "treatment" and run these next to each other. On gel, select only those cDNA fragments for further analysis that are clearly up- or downregulated during a specific treatment in all cases (see Figure 2). Following this strategy, we have not been confronted with a single false positive.

IIIB: EDD-PCR

Protocol

Twenty µl cDNA is diluted to 2.5 ml with ddH_2O. Of this, 1 µl is used in a PCR amplification. However, when performing EDD-PCR for the first time, we strongly recommend using a dilution series of cDNA as input in the PCR to determine the optimum template concentration.

Note: The primer used in the PCR has to be the same as the one that was used in the cDNA synthesis. E.g., when 5'-[FAM] E1T12MG was used in the cDNA synthesis, 5'-[FAM] E1T12MG should also be used in the PCR.

PCR reactions

In one tube add:

- 1 µl cDNA
- 2 µl 10x PCR-buffer II (supplied with enzyme)

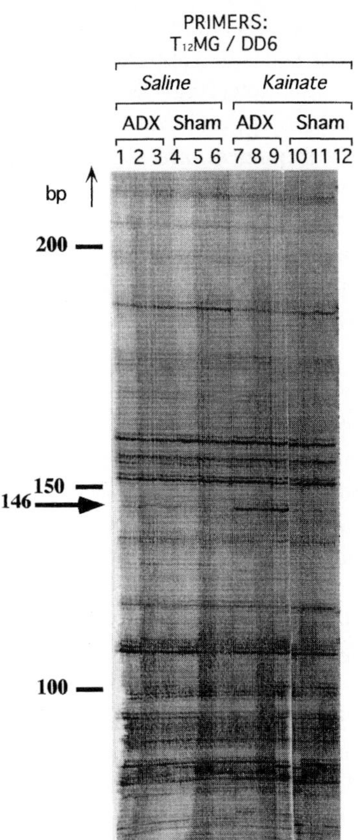

Fig. 2. Differential display of hippocampal RNA derived from animals with different treatments. Animals have been adrenalectomized (1 to 3 and 7 to 9) or sham-operated (4 to 6 and 10 to 12). In addition, kainic acid was administrated (7 to 12) or saline (1 to 6). Subsequently, animals were decapitated and the hippocampus was dissected and processed for DD purposes as described. All RNA samples were individually processed. DD fragments were size-separated on a GATC blotting device. Putative false differentially displayed products are indicated by an asterisk; only fragments differentially expressed in all members of one group (N=3) are selected for further characterization. One such example is indicated by an arrow. The primers used are T12G and DD6. For further details see Vreugdenhil et al., 1996a; Vreugdenhil et al., 1996b

- 1.6 µl MgCl2 (25µM)
- 3.5 µl dNTPs (0.5µM)
- 2 µl Fluorescent, EDD primer (2 µM)
- 2 µl B1DD primer (2 µM)
- 0.4 µl BSA (10 mg/ml)
- 7.1 µl ddH_2O
- 0.4 µl Amplitaq Gold (5U/µl)
- 20 µl total

Notes

1. Make a master mix of all components that are invariable for all PCR incubations. In general these are buffer, $MgCl_2$, dNTPs, BSA, H_2O and Amplitaq Gold. Amplitaq Gold is activated only by incubation at 95°C for 10 min and thus, no hot start is required.
2. Include negative controls ("H_2O" controls) for each primer pair combination. As stated before, EDD-PCR is very sensitive and artefacts are easily generated.

PCR profiles

- 10 min 95°C
- 30 sec 95°C, 2 min 38°C, 2 min 72°C; repeat 4x
- 30 sec 95°C, 1 min 60°C, 1.30 min 72°C; repeat 30x

Note: This "complex" PCR profile is thought to induce the annealing of the B1DD primers to a number of different cDNA molecules in the initial 5 cycles by using low annealing temperatures (38°C). cDNA sequences with moderate homology to the B1DD primer will anneal under these conditions. Subsequent increase of the annealing temperature to 60°C will amplify only those cDNAs which were primed in the initial five cycles and thus false positives caused by a late "random" priming process are prevented.

Gel ectrophoresis of fluorescent-labelled EDD-cDNA fragments

We have analysed the fluorescent EDD-PCR generated cDNA fragments on an automated DNA sequencer. As with the GATC apparatus, the advantage of this system is that no radioactivity is required for detecting DNA molecules and that the size-separation range is very long (10 to 1200 bp) and thus less PAA-gels are required. In addition, the use of an automated DNA sequencer offers specific advantages for the automated analysis of DNA fragments using appropriate software and for digital storage of EDD-data reviewed in Bauer et al., 1993; Vreugdenhil et al., 1996b. As automated DNA sequencers require rather specific and detailed instructions, we feel it is beyond the scope of this chapter to give detailed protocols. For the interested reader we refer to Lewin, 1986; Bauer et al., 1993; Ghosh et al., 1997; Kimpton et al., 1993; Vreugdenhil et al., 1996b.

General remarks

1. The protocols described involve the use of non-radioactive labels. As the kind of label normally does not interfere with enzymatic activity (e.g. reverse transcriptase or taq polymerase), our protocols may be useful for other labels including radioactive ones.
2. (E)DD-PCR fragments of interest can be isolated by:
   a) performing an additional number of four to six cycles and including d(^{32}P-α)ATP. Fragments can subsequently be isolated using standard polyacrylamide gelelectrophoresis.
   b) performing an additional four to six cycles and including dig-labeled primers. Fragments can subsequently be blotted on a membrane in duplo. One half of this membrane can be stained with the anti-dig antibody to localize the fragments of interest; the other halve will be used to excise the piece of membrane corresponding to the fragment of interest using a razor blade. Boil this piece of membrane for 10 min, elute the DNA and perform PCR for 20 to 25 cycles. Identify and characterize fragment with standard agarose gelelectroforesis and cloning procedures. In this procedure do *not*:

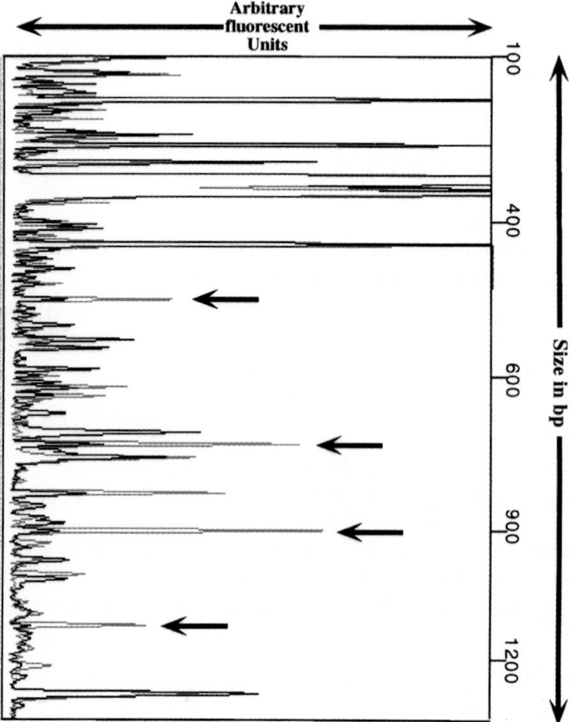

Fig. 3. Differential display using fluorescent-labelled primers and an automated DNA sequencer. EDD-PCR generated cDNA fragments were size-separated on PAA gels using an ABI-377 sequencer. These cDNA fragments were derived from the hippocampus of adrenalectomized and kainic acid-treated animals (red lines) and sham-operated and saline-treated animals (black lines). cDNA fragments were sized using internal size-standards and analysed with Genescan 2.02 software (Applied Biosystems, Perkin Elmer). Note that the scale of size-separation, indicated on top of the figures, ranges from 100 to 1200 bp. For further details see Vreugdenhil et al., 1996b.

- use Hybond N+ as this nylon binds DNA extremely efficiently and thus will not liberate DNA after boiling.
- cross-link DNA after blotting of the membrane-half which is meant for processing the (E)DD-PCR fragment of interest. Again, crosslinking will hamper DNA elution during boiling.

Subprotocol 2
Serial Analysis of Gene Expression (SAGE): Practical Approach

Outline

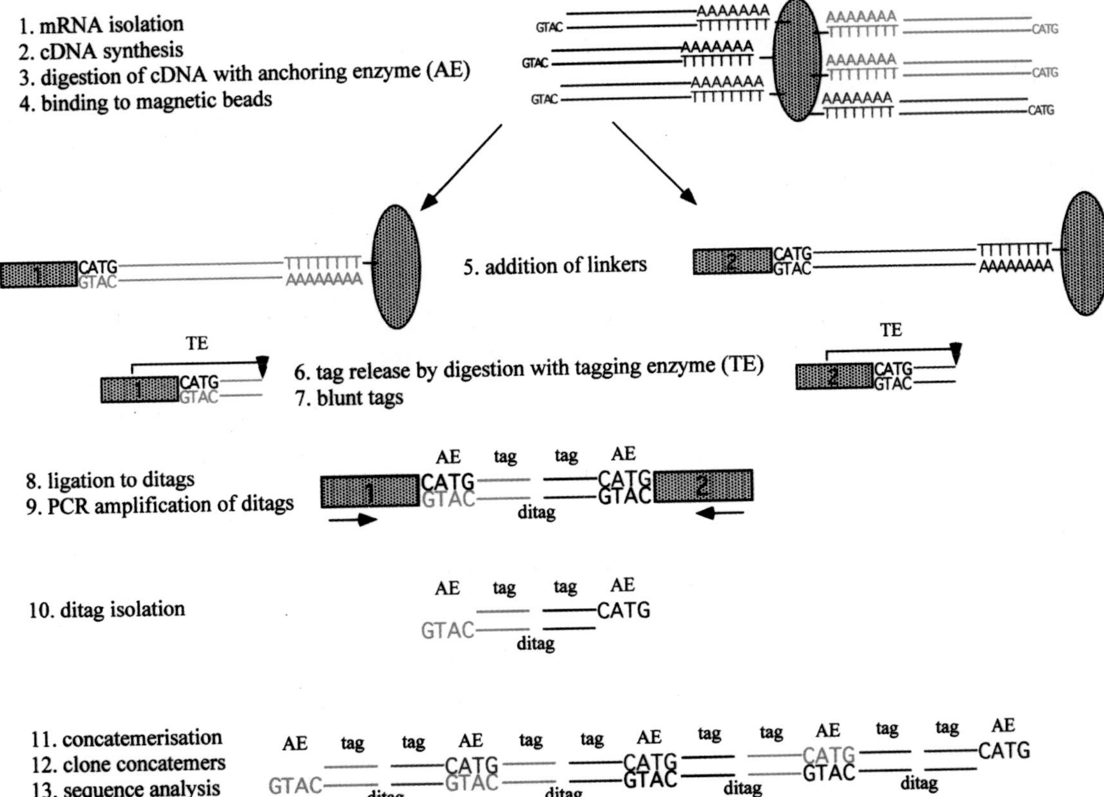

Fig. 4. The various steps comprising the SAGE procedure are outlined in figure 1. PolyA$^+$ RNA or total RNA is isolated from the tissues or cell lines of interest (1) and is reverse-transcribed to cDNA using a biotinylated oligo-d(T) primer (2). After second strand cDNA synthesis, the cDNA is digested with a restriction enzyme with a 6-bp recognition site, called anchoring enzyme (AE) (3), and the 3' ends of the cDNA are captured using streptavidin-coated magnetic beads (4). A linker is ligated to the cDNA (5) which contains a recognition site for a type IIS restriction enzyme, called tagging enzyme (TE). Digestion with the TE, which cuts at a defined distance 3' from the recognition site, results in the release of the linker joined to approximately 10 bp of cDNA, the tag, from the remainder of the bound cDNA (6). The released cDNA tag is subsequently blunted (7) and ligated (8), thus forming ditags of 102 bp. A PCR step is performed on the 102-bp ditags using primers directed against the linker sequences to generate sufficient material for the subsequent steps (9). Since all 102-bp ditags have an equal length, there is no PCR bias favouring amplification of a subset of molecules, ensuring that the stochiometry is not skewed. After sufficient quantities of the 102-bp ditag product have been generated by PCR, the linkers are cleaved off by digestion with the AE, the resulting 23- to 26-bp ditag is isolated from gel (10) and ligated at the AE site to long concatemers of ditags (11). After a size selection, the concatemers are cloned (12) and sequenced (13). The raw sequence data is analysed using the special SAGE software, allowing extraction of individual tags from the concatemers, assessment of tag frequency and comparison with GenBank sequences (14)

Materials

Glassware and plasticware

see DD section

Solutions and buffers

- ddH_2O
- DEPC-treated ddH_2O (see DD section)
- 0.1 M DTT (supplied with SuperScript)
- 5x first strand buffer [250 mM Tris-HCl (pH 8.3), 375 mM KCl, 15 mM $MgCl_2$] (supplied with SuperScript)
- dNTPs (10 mM and 25 mM each) (HT Biotechnology; SB23)
- 5x second strand buffer [100 mM Tris-HCl (pH 6.9), 450 mM KCl, 23 mM $MgCl_2$, 0.75 mM (-NAD+, 50 mM $(NH_4)_2SO_4$]
- 10x restriction buffer (NEBuffer 4, supplied with *Nla*III and *Bsm*FI)
- BSA (10 mg/ml, supplied with *Nla*III and *Bsm*FI)
- LoTE [3 mM Tris-HCL (pH 7.5), 0.2 mM EDTA (pH 7.5)]
- 5x T4 DNA ligase buffer (supplied with ligase)
- 10x PNK buffer (supplied with PNK)
- 10 mM ATP
- Glycogen (Boehringer Mannheim; 20 mg/ml; 901 393)
- 10 M ammonium acetate
- 3 M sodium acetate, pH 5.2
- Ethanol
- 70 % ethanol
- Phenol/chloroform/isoamylalcohol (25:24:1) (PCI)
- PCR buffer II (supplied with AmpliTaq Gold)
- 25 mM $MgCl_2$ (supplied with AmpliTaq Gold)
- DMSO (Sigma; D 8418)
- Polyacrylamide
- TEMED
- 10% ammonium persulphate
- 10 bp ladder (GibcoBrl; 10821-015)
- 100 bp ladder (New England BioLabs; 323-1L)
- Dynabeads M-280 Streptavidin (Dynal; 112.05)
- 2x Bind and Wash buffer [10 mM Tris-HCl (pH 7.5), 1 mM EDTA, 2.0 M NaCl]
- Ethidium bromide (20 mg/ml)

Enzymes

- SuperScript II RNase H^- reverse transcriptase (GibcoBRL; 18064-014)
- DNA polymerase I (GibcoBrl; 10 u/µl; 18010-025)
- T4 DNA ligase (GibcoBrl; 5 u/µl; 15224-041)
- RNase H (Boehringer Mannheim; 1 u/µl; 786 349)
- *Nla*III (New England Biolabs; 125 S)
- Polynucleotide kinase (PNK) (Pharmacia; 27-0736)
- *Bsm*FI (New England Biolabs; 4 u/µl; 572S)
- Klenow (Amersham; E2141 Y)
- AmpliTaq Gold (Perkin Elmer; N808-0247)
- *Sph*I (Promega; 10 u/µl; R6261)

Kits

- mRNA purification kit (Dynal; 610.01) or mRNA DIRECT kit (Dynal; 610.11)
- mRNA Capture Kit (Boehringer Mannheim; 1 787 896)
- QIAquick PCR Purification Kit (Qiagen; 28106)

- Zero Background Cloning Kit (Invitrogen; K2500-01)
- BigDye Primer Kit (Perkin Elmer, cycle sequencing reaction)

Primers

- oligo$(dT)_{18}$ (biotinylated): 5' [biotin] TTTTTTTTTTTTTTTTTT 3'
- primer 1A: 5' TTTGGATTTGCTGGTGCAGTACAACTAGGCTTAATAGGGACATG 3'
- primer 1B: 5' TCCCTATTAAGCCTAGTTGTACTGCACCAGCAAATC [amino mod. C7] 3'
- primer 2A: 5' TTTCTGCTCGAATTCAAGCTTCTAACGATGTACGGGGACATG 3'
- primer 2B: 5' TCCCCGTACATCGTTAGAAGCTTGAATTCGAGC [amino mod. C7] 3'
- primer SAGE 1 (biotinylated): 5' [biotin] GGATTTGCTGGTGCAGTACA 3'
- primer SAGE 2 (biotinylated): 5' [biotin] CTGCTCGAATTCAAGCTTCT 3'

Other components

ELECTROMAX DH10B Cells (GibcoBrl; 18290-015)

Equipment

- Magnetic device (Dynal MPC-E-1 or MPC-E)
- PCR apparatus
- Mini-PROTEAN II vertical electrophoresis system (BioRad)
- Microcentrifuge
- Electroporation cuvettes (BTX; 1-mm gap)
- Gene Pulser II System (BioRad)
- 37°C incubator
- ABI 377 automated sequencer (Perkin Elmer)

Procedure

General remarks

If the amount of RNA available for input in the SAGE procedure is not rate-limiting, it is preferable to use 2.5–5 µg polyA$^+$ RNA. The protocol to follow in this case is protocol A. We also describe an alternative protocol here (protocol B), which has been modified at various steps so that it can be used with very limited amounts of input material (i.e. using this protocol we can obtain an expression profile from a single hippocampal punch from a 300-µm brain slice). The latter protocol is especially relevant for expression profiling in neuronal tissue, in which obtaining large amounts of homogenous tissue for RNA isolation is generally an impossible task due to its complex circuitry and high degree of specialisation.

Steps 1 to 8 of both protocols will be described in detail below, after which they converge. From step 9 onwards there is only a single protocol.

1. mRNA Isolation

Protocol A

There are excellent kits available for isolation of polyA$^+$ RNA. We recommend using the mRNA DIRECT kit (Dynal; 610.11) for isolation of polyA$^+$ RNA from tissue or culture cells or the mRNA purification kit (Dynal; 610.01) for isolation of polyA$^+$ RNA from total RNA. All steps are described extensively in the kit manuals.

Protocol B

1. Isolate total RNA with TRIzol (see DD section)
2. After RNA precipitation resuspend 1–10 µg total RNA in 20 µl lysis buffer (mRNA Capture Kit)

   Note: we have successfully used even smaller amounts of total RNA. However, this requires some modifications in later steps of the SAGE protocol

3. Dilute the biotinylated oligo(dT)20 primer (mRNA Capture Kit) 20x to a final concentration of 5 pmol/µl
4. Add 4 µl of the diluted primer to the RNA in lysis buffer
5. Anneal primer to RNA at 37°C for 5 min
6. Transfer the RNA to a streptavidin-coated PCR tube (mRNA Capture Kit)
7. Incubate at 37°C for 3 min (in this step the µRNA is immobilised to the wall of the tube by streptavidin-biotin binding)
8. Remove the solution from the tube (contains the non-bound RNA fraction: ribosomal RNA, tRNA etc.) and wash tube gently 3x with 50 µl washing solution (mRNA Capture Kit)
9. Remove the washing solution of the final wash step
10. Proceed immediately to cDNA synthesis step

2. cDNA Synthesis

Protocol A

1. Use 2.5–5 µg polyA^{+} RNA in an oligo(dT)-primed cDNA synthesis:
2. Mix together in a 0.5-ml PCR tube (on ice!):
   - 2.5 µl polyA+ RNA (1 µg/µl)
   - 4 µl 5x first strand buffer
   - 2 µl 0.1 M DTT
   - 1 µl 10 mM dNTPs
   - 1 µl oligo(dT)$_{18}$ (biotinylated) (0.5 µg/µl)
   - 1 µl SuperScript II RT (200 u/µl)
   - 8.5 µl DEPC-H_2O
   - total volume: 20 µl
3. Incubate at 42°C for 2 hrs (it is convenient to perform the incubation in a PCR machine)
4. Perform second strand synthesis by adding to the first strand cDNA on ice:
   - (20 µl first strand cDNA)
   - 16 µl 5x second strand buffer
   - 1.6 µl 10 mM dNTPs
   - 2 µl DNA polymerase I (10 u/µl)
   - 1 µl T4 DNA ligase (5 u/µl)
   - 1 µl RNase H (5 u/µl)
   - 38.4 µl ddH_2O
   - total volume: 80 µl
5. Incubate at 16°C for 2 hrs
6. Raise volume to 200 µl with LoTE
7. Add an equal volume phenol/chloroform/isoamylalcohol (25:24:1) (PCI)
8. Vortex
9. Centrifuge in microcentrifuge at 4°C for 5 min at 13.000 rpm
10. Transfer aqueous (top) phase to a new 1.5ml Eppendorf tube
11. Ethanol precipitate by adding:
    - 3 µl glycogen
    - 100 µl 10M ammonium acetate
    - 700 µl ethanol
12. Place at –20°C for 30 min
13. Centrifuge in microcentrifuge at 4°C for 15 min at 13.000 rpm
14. Wash the pellet twice with 500 µl 70 % ethanol by vigorous vortexing

15. Remove 70 % ethanol and allow pellet to air-dry for approximately 15 min
16. Resuspend pellet in 20 µl LoTE

Protocol B

Note: The oligo$(dT)_{20}$ primer (biotinylated; supplied with mRNA Capture Kit) used to capture the $polyA^+$ RNA and immobilise it to the wall of the tube (protocol B, step 1) serves directly as a primer for cDNA synthesis. The synthesised cDNA remains bound to the wall of the tube until it is released in step 6 by digestion with TE.

1. Rinse tube with captured $polyA^+$ RNA once with 50 µl 1x first strand buffer, then remove buffer
2. Replace 1x first strand buffer with (pipet on ice!):
   - 4 µl 5x first strand buffer
   - 2 µl 0.1 M DTT
   - 1 µl 10 mM dNTPs
   - 1 µl SuperScript II RT (200 u/µl)
   - 12 µl DEPC-H_2O
   - Total volume: 20 µl
3. Incubate at 42°C for 2 hrs (it is convenient to perform the incubation in a PCR machine)
4. Remove reaction mixture from tube and rinse once with 50 µl washing solution (mRNA Capture Kit)
5. Remove washing solution and rinse tube once with 50 µl 1x second strand buffer
6. Replace 1x second strand buffer with:
   - 4 µl 5x second strand buffer
   - 0.4 µl 10 mM dNTPs
   - 1 µl DNA polymerase I (10 u/µl)
   - 0.5 µl T4 DNA ligase (5 u/µl)
   - 0.5 µl RNase H (5 u/µl)
   - 13.6µl ddH_2O
   - Total volume: 20 µl
7. Incubate at 16°C for 2 hrs

At this point the double-stranded cDNA from either protocol can be stored at –20°C or directly used in the next step (digestion of cDNA with anchoring enzyme).

Subprotocol 3
Digestion of cDNA with Anchoring Enzyme

If cDNA generated in step 2 of protocol was stored at –20°C, allow it to defrost slowly on ice.

Procedure

Protocol A

1. Digest the double stranded cDNA by adding:
   - (20 µl double stranded cDNA)
   - 20 µl 10x restriction buffer (NEBuffer 4)
   - 2 µl BSA (10 mg/ml)
   - 153 µl LoTE
   - 5 µl *Nla*III (10 u/µl)
   - Total volume: 200 µl

2. Digest at 37°C for 1 hr
3. Heat-inactivate the restriction enzyme by incubating at 65°C for 20 min
4. Extract with an equal volume of PCI and ethanol precipitate as described above (protocol A, step 2)
5. Resuspend washed and air-dried pellet in 20 μl LoTE
6. Proceed to step 4

Protocol B

1. Remove second strand reaction mixture from tube and rinse once with 50 μl washing solution, then remove wash solution
2. Rinse tube once with 50 μl 1x restriction buffer
3. Remove buffer, then add:
   - 2.5 μl 10x restriction buffer (NEBuffer 4)
   - 0.25 μl BSA (10 mg/ml)
   - 20.25 μl LoTE
   - 2 μl *Nla*III (10 u/μl)
   - Total volume: 25 μl
4. Digest at 37°C for 1 hr
5. Heat-inactivate the restriction enzyme by incubating at 65°C for 20 min
6. Proceed to step 5

Subprotocol 4
Binding to Magnetic Beads

This step is only part of protocol A. In protocol B the cDNA is already bound to the streptavidin-coated PCR tubes, so immobilisation using beads is not necessary.

Procedure

Protocol A

1. Resuspend Dynabead M-280 streptavidin slurry by vortexing for 1 min
2. Transfer 2x100-μl portions of Dynabeads to a 1.5-ml Eppendorf tube each
3. Immobilise beads using a magnetic device; remove supernatant
4. Wash beads 3x with 200 μl of 1x Bind and Wash buffer by resuspending, immobilising beads and removing wash solution
5. Divide *Nla*III-digested cDNA into 2 portions of 10 μl; to each portion add 90 μl ddH_2O and 100 μl 2x Bind and Wash buffer. Mix and add to a portion of washed beads
6. Mix well and incubate at room temperature for 30 min (mix every 10 min)
7. Immobilise beads using magnetic device and discard supernatant
8. Wash immobilised beads 3x with 200 μl 1x Bind and Wash buffer, then once with 200 μl LoTE
9. After final wash step, immobilise the beads and remove the LoTE
10. Proceed to step 5

Subprotocol 5
Addition of Linkers

Procedure

Preparation of linkers

The linkers are constructed by allowing two complementary primers to anneal. First it is necessary to phosphorylate the 5' end of primers 1B & 2B.

1. Dilute primers 1B & 2B to a concentration of 350 ng/µl
2. Phosphorylate the 5' end of both primers by adding together:

	Tube 1	Tube 2
primer 1B (350 ng/µl)	36 µl	---
primer 2B (350 ng/µl)	---	36 µl
10x PNK buffer	8 µl	8 µl
10 mM ATP	4 µl	4 µl
PNK (9.3 u/µl)	2 µl	2 µl
LoTE	30 µl	30 µl
Total volume	80 µl	80 µl

3. Incubate at 37°C for 30 min
4. Heat-inactivate PNK at 65°C for 10 min
5. Add 36 µl of primer 1A to the 80 µl of kinased primer 1B. Do the same for primers 2A and 2B (final concentration: 217 ng/µl)
6. Anneal primers by heating to 95°C for 2 min and subsequently incubating at 65°C for 10 min, at 37°C for 10 min and at room temperature for 20 min

Note: Test linkers by ligating 200 ng of each linker followed by electrophoresis on a 12% polyacrylamide gel. In case phosphorylation reaction was successful, the majority (> 70%) of the material should have ligated to form linker dimers (±80 bp).

Protocol A

1. Ligate linker 1 to one of both portions of Dynabeads with bound *Nla*III-digested cDNA, and linker 2 to the other. Add on ice:

	Tube 1	Tube 2
linker 1 (± 200 ng/µl)	10 µl	---
linker 2 (± 200 ng/µl)	---	10 µl
5x T4 DNA ligase buffer	10 µl	10 µl
LoTE	28 µl	28 µl
Total volume	48 µl	48 µl

2. Anneal linkers to *Nla*III-digested cDNA by heating at 50°C for 2 min, then incubating at room temperature for another 15 min
3. Add 2 µl T4 DNA ligase (5u/µl), then ligate linkers at 16°C for 2 hrs
4. After ligation of linkers wash the beads 4x with 200 µl 1x Bind and Wash buffer and 2x with 200 µl 1x restriction buffer
5. After final wash step, immobilise the beads and remove the LoTE

Protocol B

Note: Unlike protocol A, here both linkers are ligated in the same tube. This has a practical reason, since material is bound to the wall of tube and cannot be separated as is possible with Dynabeads.

1. Remove the reaction mixture from *Nla*III digestion from the PCR tube and rinse once with 50 µl washing solution
2. Remove washing solution and tube once with 50 µl 1x T4 DNA ligase buffer, then remove buffer
3. Add to the tube (on ice!):
   - 2.5 µl linker 1 (± 200 ng/µl)
   - 2.5 µl linker 2 (± 200 ng/µl)
   - 5 µl 5x T4 DNA ligase buffer
   - 15 µl LoTE
   - Total volume: 25 µl
4. Anneal linkers to *Nla*III-digested cDNA by heating at 50°C for 2 min, then incubate at room temperature for another 15 min
5. Add 1 µl T4 DNA ligase (5u/µl), then ligate linkers at 16°C for 2 hrs

Subprotocol 6
Tag Release by Digestion with Tagging Enzyme

Procedure

Protocol A

1. Add the following to each tube containing the Dynabeads:
   - 10 µl 10x restriction buffer (NEBuffer 4)
   - 1 µl BSA (10 mg/ml)
   - 88 µl LoTE
   - 1 µl *Bsm*FI (4 u/µl)
   - Total volume: 100 µl
2. Digest at 65°C for 1 hr (mix every 10 min)
3. THIS TIME COLLECT SUPERNATANT! (contains released cDNA tag)
4. Raise volume to 200 µl with LoTE
5. Extract with an equal volume of PCI and ethanol precipitate as described above (protocol A, step 2)
6. Resuspend washed and air-dried pellets in 10 µl LoTE

Protocol B

1. Remove the ligation reaction mixture from the PCR tube and rinse once with 50 µl washing solution
2. Remove washing solution from tube once with 50 µl 1x restriction buffer, then remove buffer
3. Add to tube:
   - 2.5 µl 10x restriction buffer (NEBuffer 4)
   - 0.25 µl BSA (10 mg/ml)
   - 21.25 µl LoTE
   - 1 µl *Bsm*FI (4 u/µl)
   - Total volume: 25 µl
4. Digest at 65°C for 1 hr

5. Add 175 μl LoTE and transfer to a 1.5-ml Eppendorf tube. DO NOT DISCARD since mixture contains the released SAGE tag attached to one of either linkers!
6. Extract with an equal volume of PCI and ethanol precipitate as described above (protocol A, step 2)
7. Resuspend washed and air-dried pellets in 21.5 μl LoTE

Subprotocol 7 Blunting Tags

Procedure

Protocol A

1. Add the following to both tubes containing the released cDNA tags:
   - (10 μl cDNA tags)
   - 10 μl 5x second strand buffer
   - 0.5 μl BSA (10 mg/ml)
   - 1 μl 25 mM dNTPs
   - 3 μl Klenow (1 u/μl)
   - 25.5 μl LoTE
   - Total volume: 50 μl
2. Incubate at 37°C for 30 min
3. Raise volume to 200 μl by addition of 150 μl LoTE
4. Extract with PCI and ethanol precipitate as described above (protocol A, step 2)
5. Resuspend pellet in 4 μl LoTE

Protocol B

1. Add to the precipitated cDNA tags:
   - (21.5μl cDNA tags)
   - 6 μl 5x second strand buffer
   - 0.5 μl BSA (10 mg/ml)
   - 0.5 μl 25 mM dNTPs
   - 1.5 μl Klenow (1 u/μl)
   - Total volume: 30 μl
2. Incubate at 37°C for 30 min
3. Raise volume to 200 μl by addition of 170 μl LoTE
4. Extract with PCI and ethanol precipitate as described above (protocol A, step 2)
5. Resuspend pellet in 4 μl LoTE

Subprotocol 8 Ligation to Ditags

Procedure

Protocol A

1. At this point in the protocol the 2 portions of blunted cDNA tags (derived from the same cDNA synthesis reaction) are combined in order to form ditags. Ligate the blunted tags to ditags by adding:
   - 4 μl blunted ditags ligated to linker 1
   - 4 μl blunted ditags ligated to linker 2

- 2.5 µl 5x T4 DNA ligase buffer
- 1 µl T4 DNA ligase (5 u/µl)
- 1 µl LoTE
- Total volume: 12.5 µl

2. Incubate overnight at 16°C

Protocol B

1. Ligate the blunted tags to ditags by addition of:
   - (4 µl blunted ditags)
   - 1.2 µl 5x T4 DNA ligase buffer
   - 0.8 µl T4 DNA ligase (5 u/µl)
   - Total volume: 6 µl
2. Incubate overnight at 16°C

Subprotocol 9 PCR Amplification of Ditags

Procedure

Protocol A and B

PCR amplification of ligation mix

1. Raise the volume of the ligation mixture to 20 µl by adding LoTE
2. Prepare a 1/100 dilution of the ligation mixture
3. Use 1 µl of the 1/100 dilution as template in a 50 µl PCR reaction:
   - 5 µl 10x PCR buffer II
   - 16 µl 25 mM $MgCl_2$
   - 2 µl dNTPs (25 mM each)
   - 3 µl DMSO
   - 1 µl primer SAGE1 (350 ng/µl)
   - 1 µl primer SAGE2 (350 ng/µl)
   - 20 µl ddH_2O
   - 1 µl AmpliTaq Gold (5 u/µl)
   - Total volume: 49 µl
4. Finally, add 1 µl of the diluted ligation mix as template in the PCR reaction
5. PCR-amplify using the following conditions:
   - 1 cycle of 15 min 95°C
   - 28 cycles of 30 sec 95°C, 1 min 55°C, 1 min 70°C
   - 1 cycle of 5 min 70°C
   - room temperature

After PCR amplification, analyse 10 µl of the PCR mixture by electrophoresis on a 12% polyacrylamide gel. Use a 10-bp ladder as marker. For electrophoresis we recommend the Mini-PROTEAN II vertical electrophoresis system (BioRad).

Purification of large-scale PCR (n = 100)

1. After large-scale amplification of the diluted ligation mix (100x100 µl PCR reaction), pool into a 100-ml glass beaker (pooled volume ± 10 ml)
2. Add 50 ml of PB buffer (QIAquick PCR Purification Kit) to the diluted PCR mixture; mix well
3. Apply 600 µl to each of 16 QIAquick spin columns

4. Centrifuge in microcentrifuge for 45 sec at 13.000 rpm
5. Remove filtrate and repeat another 5x using the same columns, each time applying 600 μl of the diluted PCR mixture
6. After removal of the last filtrate, spin the columns a final time to remove all fluid
7. Wash by applying 0.75 ml buffer PE (QIAquick PCR Purification Kit)
8. Centrifuge in microcentrifuge for 45 sec at 13.000 rpm
9. Discard flow-through and spin another 1 min at full speed to remove all fluid
10. Place the columns in a clean 1.5-ml Eppendorf tube
11. Apply 30 μl of elution buffer (QIAquick PCR Purification Kit) to each column
12. Leave at room temperature for 15 min
13. Elute DNA by centrifuging in microcentrifuge for 1 min at 13.000 rpm
14. Pool the eluted DNA of the 16 columns in a single tube (total volume 480 μl)

Subprotocol 10
Ditag Isolation

Procedure

Protocol A and B

Digestion of ditag with NlaIII to release linkers

1. Add to the 480 μl of purified PCR material the following components:
   - (480 μl purified PCR material)
   - 60 μl 10x restriction buffer
   - 6 μl BSA (10 mg/ml)
   - 34 μl LoTE
   - 20 μl *Nla*III (10 u/μl)
   - Total volume: 600 μl
2. Digest at 37°C for 1 hr (DO NOT HEAT-INACTIVATE ENZYME: THIS WILL CAUSE DENATURATION OF SMALL DITAG OF 22-26 BP)
3. At this point the digested material can be stored overnight at 4°C, or proceed with Dynabead extraction of biotinylated PCR material

Dynabead extraction of biotinylated PCR material

Note: This step enriches for the material released due to removal of the linkers and undigested or partially digested product by biotin-streptavidin capture

1. Resuspend Dynabead M-280 streptavidin slurry by vortexing for 1 min
2. Transfer 3 100-μl portions of Dynabeads to a 1.5-ml Eppendorf tube each
3. Immobilise beads using a magnetic device; remove supernatant
4. Wash beads 3x with 200 μl of 1x Bind & Wash buffer by resuspending, immobilising beads and removing wash solution
5. Divide *Nla*III-digested ditags (volume = 600 μl) into 3 portions of 200 μl; add 200 ml 2x Bind & Wash buffer to each of the 3 portions of washed beads
6. Mix well and incubate at room temperature for 30 min (mix every 10 min)
7. Immobilise beads using magnetic device and transfer supernatant to a new 1.5-ml Eppendorf tube
8. Precipitate the supernatant by addition of:
   - 3 μl glycogen
   - 800 μl ethanol
9. Chill for 15 min in a dry ice/ethanol bath

10. Centrifuge at 4°C in a microcentrifuge for 15 min at 13.000 rpm
11. Wash pellet 1x with 70% ethanol, then allow to air-dry
12. Resuspend in 30 µl LoTE
13. Load entire sample on 8 lanes of a 12 % polyacrylamide gel. Use a 10-bp ladder as marker. (Note: we recommend a loading dye containing Orange G as dye to prevent co-migration with the ditag of 22–26 bp)
14. Run gel at room temperature at 50 V for 2.5 hrs or until Orange G dye is at the bottom of the gel
15. Stain gel with ethidium bromide
16. Excise the ditag band from gel. This band runs at 22–26 bp (see Figure 5) just below the linkers that run at ± 40 bp
17. Divide the excised polyacrylamide fragments between 4 0.5 ml PCR tubes that have been pierced with a 21-gauge needle. Place the PCR tubes in a 1.5-ml Eppendorf tube and spin in a microcentrifuge for 5 min at 13.000 rpm to fragment the polyacrylamide. Remove the 0.5 ml PCR tubes and add 300 µl LoTE to each tube. Vortex and then incubate at 37°C for 15–30 min

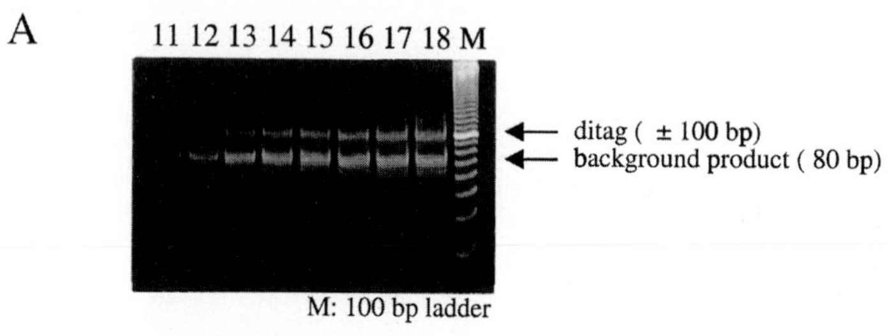

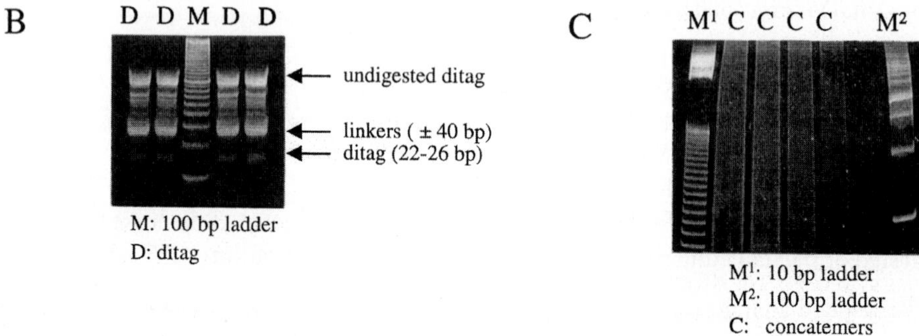

Fig. 5. Typical examples of ditag PCR (A), ditag isolation (B) and concatemerisation (C). A: Ditag PCR products run on a 12% polyacrylamide gel. Optimising the number of PCR cycles is important in the SAGE procedure by increasing the number of PCR cycles by one at a time. The ditag band of approximately 100 bp is clearly visible, as well as the most prominent background product of around 80 bp. At 17–18 cycles in this example the ratio of ditag to background product is optimal. Too many PCR cycles can result in a decrease of the amount of visible ditag due to exhaustion of the PCR reaction and consequent denaturation of formed products. M: marker (100-bp ladder); this marker has a band every 10 bp with the most intense marker band at 100 bp. B: A 12% polyacrylamide gel containing cleaved ditag. The ditag band of 22–26 bp is faintly visible after ethidium bromide staining of the gel. Also visible are the linkers of approximately 40 bp, migrating just above the ditag, and remnants of undigested ditag around 100 bp, as well as some other minor background products. C: Concatemerised ditag run on an 8% polyacrylamide gel, showing a smear ranging from below 100 bp to over 1 kb. The markers are the 10-bp ladder, with bands every 10 bp up to 330 bp and a large fragment high in the gel at 1668 bp, and the 100-bp ladder with a band every 100 bp up to 1200 bp.

18. Transfer the acrylamide suspension to a SpinX column and spin in a microcentrifuge at 13.000 rpm for 5 min
19. Transfer the filtrates each to a new 1.5-ml Eppendorf tube and extract with an equal volume of PCI (protocol A, step 2)
20. Precipitate aqueous phase by addition of:
    100 µl 10 M ammonium acetate
    3 µl glycogen
    900 µl ethanol
21. Chill for 15 min in a dry ice/ethanol bath
22. Centrifuge at 4°C in a microcentrifuge for 15 min at 13.000 rpm
23. Wash pellets 2x with 70% ethanol, then allow to air-dry
24. Resuspend the 4 pellets in a total volume of 7 µl LoTE

Subprotocol 11
Concatemerisation

Procedure

Protocol A and B

1. Ligate the excised and purified ditags to concatemers by adding:
   - (7 µl ditags)
   - 2 µl 5x T4 DNA ligase buffer
   - 1 µl T4 DNA ligase (5 u/µl)
   - Total volume: 10 µl
2. Incubate for 2 hrs at 16°C
3. Load the concatemerised ditags in a single lane of an 8 % polyacrylamide gel. Use a 100-bp ladder as marker. Run the gel at room temperature at 80°C for 3 hrs
4. Stain the gel with ethidium bromide
5. Excise the region of the gel that contains concatemers of 400 bp or larger (see Figure 5C)
6. Purify DNA from the excised gel fragments as described in section 8 for the ditag isolation, except incubate at 65°C instead of 37°C and do not chill in dry ice/ethanol bath prior to centrifugation
7. After precipitation, wash pellet 2x with 70% ethanol, air-dry and resuspend in 4 µl LoTE

Subprotocol 12
Cloning Concatemers

Procedure

Protocol A and B

Ligation of concatemers in SphI-site of pZero

1. Ligate the size-selected concatemers in a vector of choice digested with *Sph* I. We recommend cloning of concatemers in the *Sph* I-site of pZero. For specific details on cloning in this vector we refer to the manual of the Zero Background Cloning Kit:
   - 4 µl size-selected concatemers
   - 1 µl pZero x *Sph*I (10 ng/µl)
   - 2 µl 5x T4 DNA ligase buffer
   - 0.2 µl T4 DNA ligase (5 u/µl)
   - 2.8 µl ddH$_2$O
   - Total volume: 10 µl
2. Incubate overnight at 16°C
3. Raise volume to 200 µl with LoTE
4. Extract with an equal volume of PCI (protocol A, step 2)
5. Precipitate aqueous phase by adding:
   - 3 µl glycogen
   - 1/10th volume 3 M sodium acetate, pH 5.2
   - 2.5 volumes of ethanol
6. Centrifuge in microcentrifuge at 4°C for 15 min at 13.000 rpm
7. Wash pellet 4x (!!!) and air-dry
8. Resuspend in 4 µl LoTE
9. Use 1 µl ligation mix to transform ELECTROMAX DH10B cells

Electroporation

Any transformation procedure can be used, but we recommend electroporation due to its high transformation efficiency in combination with the extremely competent ELECTROMAX DH10B cells (> 10^{10} transformants/µg plasmid).

1. Allow a vial of 100 µl ELECTROMAX DH10B cells to thaw on ice
2. Meanwhile chill the required number of electroporation cuvettes on ice
3. When thawed, proceed immediately with the ELECTROMAX DH10B cells. Aliquot into 35-µl portions (keep on ice!)
4. Add 1 µl ligation mix to a 35-µl portion of cells. Mix by gently tapping the tube
5. Transfer to the chilled cuvette. Pipet cells between the cuvette electrodes and store on ice!
6. Electroporate in a Gene Pulser II using the settings:
   - 2.5 kV
   - 100 Ohm
   - 25 µF
7. Immediately after applying the pulse, add 965 µl SOC medium
8. Incubate at 37°C for 1 hr
9. Plate out in 50- or 100-µl portions on LB plates containing the antibiotic Zeocin

   Note: the details for the plates and SOC medium are described in the manual of the Zero Background Cloning Kit
10. Incubate the plates overnight in a 37°C incubator
11. The following day the plates should contain hundreds of colonies

Subprotocol 13
Sequencing

Procedure

The next steps in the protocol are standard molecular biological techniques, and are therefore not described here in major detail. In brief, the next steps consist of an insert-PCR on bacterial colonies from the generated SAGE library with vector-specific primers to check for insert length. Only PCR products larger than 500 bp are chosen for sequence analysis. In our hands direct sequencing of the PCR products using the BigDye Primer Kit works very well, yielding an average of 20–25 tags per sequence reaction after analysis of the raw data (Figure 6).

Results

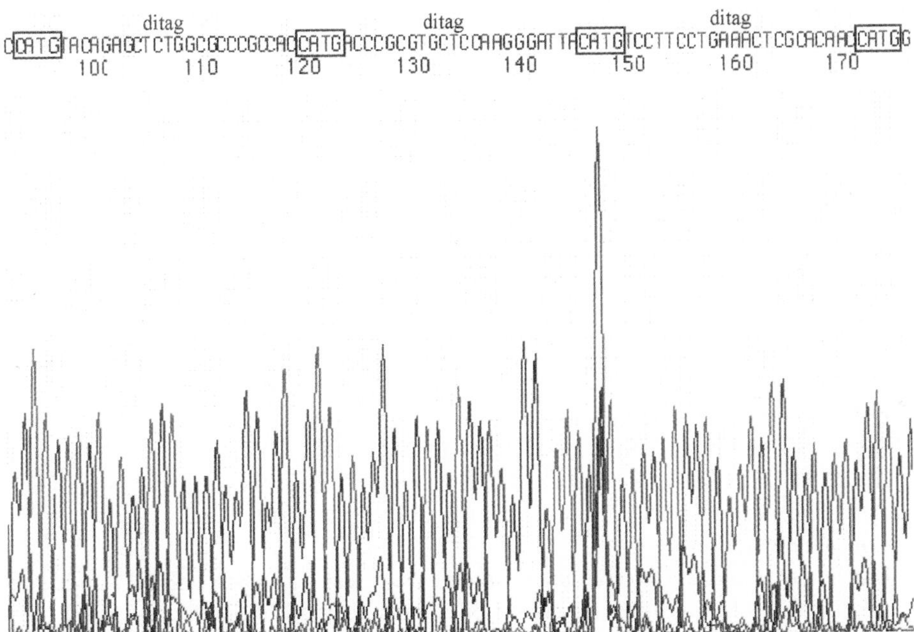

Fig. 6. Example of a part of a sequence of a cloned concatemer, showing the string of ditags separated by the CATG-sequence which is the recognition site for the anchoring enzyme *Nla*III.

Troubleshoot

There are a number of steps that are fairly critical in the SAGE procedure.

- cDNA synthesis
  After synthesis of double stranded cDNA it is advisable to check the quality by performing RT-PCR. If possible, choose primers for a high-abundant transcript (housekeeping gene) and also a low-abundant transcript known to be expressed in the RNA source used. Failure to detect the transcripts by RT-PCR may be indicative of inefficient cDNA synthesis.

- Formation of ditags
  There is no easy way to check that the steps leading to the formation of ditags have worked. The first visible evidence is after PCR amplification of the ditag, when a product of ± 100 bp should be visible, which represents the ditag (22–26 bp) with a linker at both ends (2x ± 40 bp). A background product of ± 80 bp is always visible with an almost equal intensity to the ditag product, and sometimes also a 90-bp product consisting of a single tag (monotag) joined to both linkers. If the 100-bp product is clearly visible (this should be the case using protocol A with 2.5–5 μg poly A^+ RNA as starting material), the PCR is repeated but scaled up to 100 identical 100-μl volume PCR reactions. If there is hardly any 100-bp product visible (this is quite often the case when using protocol B for limited amounts of starting material), the region around 100 bp is excised from the gel, the DNA is extracted and precipitated and subjected to a limited number of re-PCR cycles. In this case it is important to perform a cycle series to determine the optimal number of cycles for re-PCR (Figure 2A).
- PCR amplification of ditags
  In SAGE it is of the utmost importance to prevent PCR contamination, since every experiment yields PCR products of exactly the same length. We recommend a strict separation of handling of pre- and post-PCR, and use of filter tips for pipetting and aliquoting of all PCR reaction components. We standardly take along multiple negative PCR controls. Due to inter-PCR machine variation it is advisable to optimise the PCR conditions for use in your own PCR machine. If the PCR apparatus used does not have a heated lid, add mineral oil on top of the reaction mixture to prevent evaporation.
- Ditag isolation
  After large-scale PCR amplification and purification of the100-bp product the linkers are cleaved off to form the small 22- to 26-bp ditag. Due to its short length this product is easily denatured, which is detrimental because reassociation does not occur easily in the extremely heterogeneous population of molecules. It is therefore important to take precautions to prevent denaturation of the ditag, by performing all centrifugation steps in a chilled microcentrifuge and not heating it above 37°C.
  If you encounter difficulties cleaving off the linkers, there might be a problem with *Nla*III. *Nla*III is fairly unstable at –20°C; therefore it is recommended to aliquot enzyme and store at –80°C; only keep a small quantity at –20°C.
- Concatemerisation
  Ligation of ditags to concatemers should give a smear on an 8 % polyacrylamide gel ranging from the bottom of the gel to well above 1 kb. If this is not the case, some linkers might have been accidentally isolated along with the ditag during the gel separation, giving rise to premature termination of the concatemers. Make sure lanes on gel are not overloaded and that the ditag band of 22–26 bp runs separately from the linker band at 40 bp on the gel.
- Cloning of concatemers
  Be careful not to introduce any traces of DNase in the sample. The presence of even a small proportion of ditags with only one intact sticky end can be detrimental for cloning efficiency. These "faulty" ditags can be added to the concatemers, but do not have another intact sticky end left to add on another ditag. This results in termination of concatemerisation, but also prevents ligation in the cloning vector.

Comments

- Instead of purchasing all components for cDNA synthesis separately, an option is the SuperScript Choice System for cDNA Synthesis (GibcoBrl; 18090-019) which contains all necessary components.
- For all reagents we have mentioned the brands that we standardly use, although other brands might be equally efficient. There is no preference for a certain brand unless stated explicitly.

References

Bauer D, Muller H, Reich J, Riedel H, Ahrenkiel V, Warthoe P Strauss M (1993) Identification of differentially expressed mRNA species by an improved display technique (DDRT-PCR). Nucleic Acids Res 21:4272–4280

Bauer D, Warthoe P, Rohde M Strauss M (1994) Detection and differential display of expressed genes by DDRT-PCR. PCR Methods Appl 4:S97–108

Chomczynski P Sacchi N (1987) Single-step method of RNA isolation by acid guanidinium thiocyanate- phenol-chloroform extraction. Anal Biochem 162:156–159

Chomczynski P, Mackey K, Drews R Wilfinger W (1997) DNAzol: a reagent for the rapid isolation of genomic DNA. Biotechniques 22:550–553

Duboule D (1997) The evolution of genomics [editorial]. Science 278:555

Ghosh S, Karanjawala ZE, Hauser ER, Ally D, Knapp JI, Rayman JB, Musick A, Tannenbaum J, Te C, Shapiro S, Eldridge W, Musick T, Martin C, Smith JR, Carpten JD, Brownstein MJ, Powell JI, Whiten R, Chines P, Nylund SJ, Magnuson VL, Boehnke M Collins FS (1997) Methods for precise sizing, automated binning of alleles, and reduction of error rates in large-scale genotyping using fluorescently labelled dinucleotide markers. FUSION (Finland-U.S. Investigation of NIDDM Genetics) Study Group. Genome Res 7:165–178

Imaizumi K, Tsuda M, Imai Y, Wanaka A, Takagi T Tohyama M (1997) Molecular cloning of a novel polypeptide, DP5, induced during programmed neuronal death. J Biol Chem 272:18842–18848

Ito T, Kito K, Adati N, Mitsui Y, Hagiwara H Sakaki Y (1994) Fluorescent differential display: arbitrarily primed RT-PCR fingerprinting on an automated DNA sequencer. FEBS Lett 351:231–236

Kavathas P, Sukhatme VP, Herzenberg LA Parnes JR (1984) Isolation of the gene encoding the human T-lymphocyte differentiation antigen Leu-2 (T8) by gene transfer and cDNA subtraction. Proc Natl Acad Sci U S A 81:7688–7692

Kimpton CP, Gill P, Walton A, Urquhart A, Millican ES Adams M (1993) Automated DNA profiling employing multiplex amplification of short tandem repeat loci. PCR Methods Appl 3:13–22

Kiryu S, Yao GL, Morita N, Kato H Kiyama H (1995) Nerve injury enhances rat neuronal glutamate transporter expression: identification by differential display PCR. J Neurosci 15:7872–7878

Levy J (1994) Sequencing the yeast genome: an international achievement. Yeast 10:1689–1706

Lewin R (1986) DNA sequencing goes automatic [news]. Science 233:24

Liang P, Pardee AB (1992) Differential display of eukaryotic messenger RNA by means of the polymerase chain reaction [see comments]. Science 257:967–971

Liang P, Averboukh L Pardee AB (1993) Distribution and cloning of eukaryotic mRNAs by means of differential display: refinements and optimization. Nucleic Acids Res 21:3269–3275

Liang P, Zhu W, Zhang X, Guo Z, O'Connell RP, Averboukh L, Wang F Pardee AB (1994) Differential display using one-base anchored oligo-dT primers. Nucleic Acids Res 22:5763–5764

Liang P, Pardee AB (1995) Recent advances in differential display. Curr Opin Immunol 7:274–280

Liang P, Pardee AB (1997) Differential display. A general protocol. Methods Mol Biol 85:3–11

Livesey FJ Hunt SP (1996) Identifying changes in gene expression in the nervous system: mRNA differential display. Trends Neurosci 19:84–88

Livesey FJ, O'Brien JA, Li M, Smith AG, Murphy LJ Hunt SP (1997) A Schwann cell mitogen accompanying regeneration of motor neurons. Nature 390:614–618

Madden SL, Galella EA, Zhu J, Bertelsen AH Beaudry GA (1997) SAGE transcript profiles for p53-dependent growth regulation. Oncogene 15:1079–1085

Malhotra K, Foltz L, Mahoney WC Schueler PA (1998) Interaction and effect of annealing temperature on primers used in differential display RT-PCR. Nucleic Acids Res 26:854–856

McKusick VA (1997) Genomics: structural and functional studies of genomes. Genomics 45:244–249

Polyak K, Xia Y, Zweier JL, Kinzler KW Vogelstein B (1997) A model for p53-induced apoptosis [see comments]. Nature 389:300–305

Rohrwild M, Alpan RS, Liang P Pardee AB (1995) Inosine-containing primers for mRNA differential display. Trends Genet 11:300

Rowen L, Mahairas G Hood L (1997) Sequencing the human genome. Science 278:605–607

Shirvan A, Ziv I, Barzilai A, Djaldeti R, Zilkh-Falb R, Michlin T Melamed E (1997) Induction of mitosis-related genes during dopamine-triggered apoptosis in sympathetic neurons. J Neural Transm Suppl 50:67–78

Su QN, Namikawa K, Toki H Kiyama H (1997) Differential display reveals transcriptional up-regulation of the motor molecules for both anterograde and retrograde axonal transport during nerve regeneration. Eur J Neurosci 9:1542–1547

Tsuda M, Imaizumi K, Katayama T, Kitagawa K, Wanaka A, Tohyama M Takagi T (1997) Expression of zinc transporter gene, ZnT-1, is induced after transient forebrain ischemia in the gerbil. J Neurosci 17:6678–6684

Velculescu VE, Zhang L, Vogelstein B Kinzler KW (1995) Serial analysis of gene expression [see comments]. Science 270:484–487

Velculescu VE, Zhang L, Zhou W, Vogelstein J, Basrai MA, Bassett DE,Jr., Hieter P, Vogelstein B Kinzler KW (1997) Characterization of the yeast transcriptome. Cell 88:243–251

Vreugdenhil E, Jackson JF, Bouwmeester T, Smit AB, van Minnen J, van Heerikhuizen H, Klootwijk J Joosse J (1988) Isolation, characterization, and evolutionary aspects of a cDNA clone encoding multiple neuropeptides involved in the stereotyped egg-laying behavior of the freshwater snail Lymnaea stagnalis. J Neurosci 8:4184–4191

Vreugdenhil E, de Jong J, Busscher JS de Kloet ER (1996a) Kainic acid-induced gene expression in the rat hippocampus is severely affected by adrenalectomy. Neurosci Lett 212:75–78

Vreugdenhil E, de Jong J, Schaaf MJ, Meijer OC, Busscher J, Vuijst C de Kloet ER (1996b) Molecular dissection of corticosteroid action in the rat hippocampus. Application of the differential display techniques. J Mol Neurosci 7:135–146

Zhang L, Zhou W, Velculescu VE, Kern SE, Hruban RH, Hamilton SR, Vogelstein B Kinzler KW (1997) Gene expression profiles in normal and cancer cells. Science 276:1268–1272

Chapter 3

Methods Towards Detection of Protein Synthesis in Dendrites and Axons

J. VAN MINNEN and R.E. VAN KESTEREN

Introduction

General Introduction

The nervous system provides rapid and specific communication between all parts of the body by the action of highly specialized cells, the neurons. Their function is to receive stimuli from the external and internal environment and to coordinate, modify, transmit and translate these stimuli into conscious experiences and coordinated output to target organs. To fulfil this formidable task, neurons are intricately linked to each other and to effector tissues such as, e.g., muscular and glandular tissues. Communication between neurons and the other cells takes place at specialized areas of contact, the synapses. It is estimated that the number of synapses that neurons have can range up into the tens of thousands. This complexity of neurons is also reflected by their morphology. Neurons typically consist of a cell body (soma or perikaryon), multiple dendrites and one axon. Cell bodies may vary in size between four micrometers for granular cells in the cerebellum to over one hundred micrometers for motor neurons in the spinal cord. The most remarkable feature of neurons are their processes, which show a large heterogeneity in their morphology. It ranges from a simple dendrite of a bipolar neuron to the highly complex dendritic tree of Purkinje cells in the cerebellum. Although neurons contain only one axon, its length can vary tremendously: from several micrometers for many neurons in the brain to over one meter for motor neurons in the spinal cord projecting to fingers and toes.

An intriguing question is how a neuron is able to create and maintain its complex morphology and how it provides its far-away regions with proteins that are necessary for synaptic functioning and activity-dependent synaptic plasticity, which occur for instance during learning and memory formation. Until recently, the generally held notion was that all the proteins that are destined for the peripheral domains, *i.e.* the dendrites and axons, are synthesized in the soma and transported to the extrasomal domains. In the last decade an increasing number of studies have challenged this concept and in a number of studies it has been shown that peripheral domains of neurons have the ability to synthesize proteins, independent of the cell body (Crino and Eberwine, 1996; Van Minnen *et al.*, 1997).

The first reports that hinted at the possibility that extrasomal protein synthesis is an inherent capability of neurons go back to the nineteen-sixties, when Edström and co-

J. van Minnen, Research Institute Neurosciences Vrije Universiteit, Graduate School of Neurosciences Amsterdam, Faculty of Biology, De Boelelaan 1087, Amsterdam, 1081 HV, The Netherlands (phone: +31-20-4447107; fax: +31-20-4447123; e-mail: vanminnen@bio.vu.nl)
R.E. van Kesteren, Research Institute Neurosciences Vrije Universiteit, Graduate School of Neurosciences Amsterdam, Faculty of Biology, De Boelelaan 1087, Amsterdam, 1081 HV, The Netherlands (phone: +31-20-4447107; fax: +31-20-4447123)

workers (1962) showed that the Mauthner axon of the goldfish contained considerable amounts of ribonucleic acids and that this axon was capable of incorporating radiolabeled amino acids into proteins (Edström, 1966). Parallel studies on squid giant axons showed that these axons contained ribosomal RNA (rRNA), transfer RNA (tRNA) and a wide variety of messenger RNAs (mRNA; Perrone Capano *et al.*, 1987). Also for mammalian neurons it became apparent that protein synthesis in extrasomal regions is likely to occur. These studies focused mainly on dendrites. The first indication for this possibility came from a study that showed that polysomal complexes were selectively localized beneath postsynaptic sites (Steward, 1983). Later it was shown that the dendritic domain harbors a wide variety of mRNAs, encoding molecules such as structural proteins, growth factors and receptors (Crino and Eberwine, 1996). The important question is whether these mRNAs indeed translate into proteins in these peripheral neuronal domains. The answer is most likely yes, based on the data from a number of studies. Firstly, isolated dendrites and axons synthesize proteins as was concluded from the observation that they incorporate radiolabeled amino-acids into proteins (Davis et al., 1992). Secondly, when a "foreign" mRNA is introduced into neurites, the encoded protein can be detected after an appropriate time of incubation (Van Minnen et al., 1997). Thirdly, isolated neurites synthesize a wide variety of endogenous proteins in a stimulus-dependent way as was detected by metabolic labeling followed by polyacrylamide gel electrophoresis (Bergman et al., 1997).

In the following sections we will describe a number of techniques that have been applied in the studies on mRNAs in dendrites and axons and have created breakthroughs in our way of thinking about how neurons bring about plastic changes in synaptic regions.

These techniques include:

- Subprotocol 1: *In situ* hybridization of *in vitro* cultured neurons.
- Subprotocol 2: *In situ* hybridization at the electron microscopic level.
- Subprotocol 3: Single cell differential mRNA (see also Chapter 2).
- Subprotocol 4: Metabolic labeling of *in vitro* cultured neurites and analysis of protein synthesis.
- Subprotocol 5: mRNA injection into *in vitro* cultured neurons and neurites and analysis of translation.

Subprotocol 1
IN SITU HYBRIDIZATION OF CULTURED NEURONS

Introduction

The aim of *in situ* hybridization (ISH) is to visualize the presence of mRNA molecules in their cellular context. During the ISH procedure high temperatures are used (see below), which imposes specific demands on the fixation procedure of the cells. The mRNA should be fixed in such a way that it cannot diffuse away from its cellular site during the various treatments, yet the fixative should not cross-link the tissue to such an extent that it becomes impermeable to the probe. From our experience, a paraformaldehyde fixative (1–4%), combined with 1% acetic acid (to fix the nucleic acids) gives good preservation of the morphology and a good hybridization signal. Furthermore, the cells have to be permeabilized, since the plasma membrane is not permeable to the nucleic acid probe. We use a solution of 0.5% Nonadet NP 40 which we employ after fixation, but other detergents such as Triton X 100 or Tween 20 may also be used. It may also be necessary to treat the cells with a protease to further increase the accessibility of target RNA sequences to the probes. Proteinase K has been the compound of choice for many years, but since its action is poorly reproducible and may lead to severe damage of the cell's morphology, we use pepsin as an alternative. Since many mRNA species are not abundantly expressed, and thus are not likely to occur in the extrasomal domains in great quantities, the use of RNA probes is recommended, because these probes are considered to be the most sensitive (Dirks, 1996). The probes are labeled with digoxigenin as reporter molecules, which allows an immunocytochemical detection of the hybridized probes. Another often used reporter is biotin, which has a sensitivity equal to the digoxigenin-labeled probes. A disadvantage of the use of biotin may be that some tissues contain biotin, and thus may give false-positive results. A third possibility is the use of radiolabeled (^{35}S) probes, which is still the most sensitive method to demonstrate mRNAs *in situ*. This method is not recommended for the use on cultured neurons because of its low spatial resolution.

If the *in situ* hybridization methods fail to give a positive hybridization signal, other methods such as mRNA amplification can be employed to detect the presence of low-abundant mRNAs in neuritic microdomains (Miyashiro et al., 1994). This method cannot only be employed to detect low levels of previously identified mRNAs, but also allows the characterization of "unknown" mRNAs in particular cellular domains. A disadvantage of this technique is that the information of the cellular context of the signal is lost. In the section "Single Cell Differential mRNA Display" a description of this technique will be given.

Outline

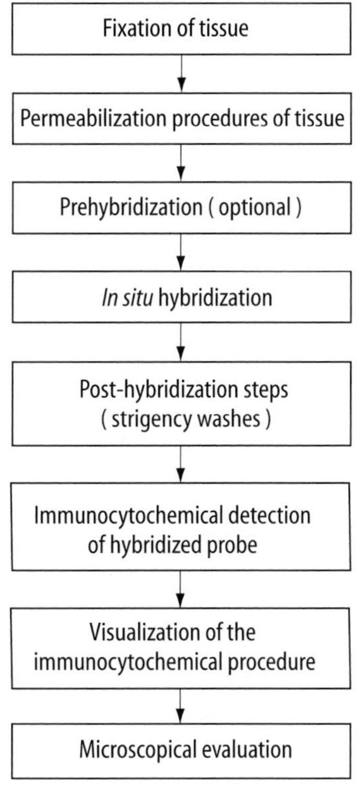

Materials

Note: All materials should be RNase free, so use factory-sterilized plasticware if possible

- Eppendorf tubes
- Filter tips
- Plastic gloves
- 0.22 µm sterile filters
- 10 and 50 ml syringes
- (Inverted) microscope
- Eppendorf centrifuge
- Incubator, temperature range 20–80°C

Solutions and Buffers

Probe Preparation

- T3 or T7 RNA-polymerase (Gibco-BRL)
- NTP mixture: 2.5 mM ATP/CTP/GTP (Gibco-BRL)
- Digoxigenin-UTP labeling mixture (Boehringer Mannheim)
- 5x concentrated transcription buffer (Boehringer Mannheim)
- RNase inhibitor (Promega)
- DNase I (RNase free) (Gibco-BRL)
- ETS buffer: (1 mM EDTA, 0.1% SDS, 10 mM Tris, pH 7.5)
- 1 M 2-(N-morpholine) ethane-sulphonic acid (MES)
- Double distilled water, DEPC-treated
- 5 M NaCl
- 2M NaOH
- 7.5 M ammonium acetate
- Ethanol 100%
- Ethanol 70 %

In Situ Hybridization

- Sodium phosphate buffer: mix 0.5 M Na_2HPO_4 and 0.5 M NaH_2PO_4 until pH reaches 7.0. The solution is DEPC-treated and autoclaved.
- Formamide. In many protocols the use of de-ionized formamide is recommended. We use untreated formamide which in our hands gives excellent results.
- 20x SSC (3.0 M NaCl, 0.3 M sodium citrate, pH 7.0). Filter, DEPC-treat and autoclave.
- Dextran sulphate 10% (w/v), sodium salt
- 50x Denhardt's: 50x = 1% polyvinyl pyrrolidone, 1% bovine serum albumin, 1% Ficoll 400; store at 20°C
- 4 mg/ml acid/alkali cleaved salmon sperm DNA.
- Paraformaldehyde
- Acetic acid
- 2 M HCl
- 0.2 M HCl
- 2 M NaOH
- 1 M Tris HCl, pH 7.4
- Nonadet NP 40
- Hydroxyl ammonium chloride

- Pepsin
- Tris/NaCl Buffer: 0.1 M Tris, 0.15 M NaCl, pH 7.4 (Buffer 1)
- Tris/NaCl/$MgCl_2$ Buffer: 0.1 M Tris, 0.15 M NaCl, 0.05 M $MgCl_2$, pH 9.5 (Buffer 2). Buffers 1 and 2 can be made as 10x stocks and kept at room temperature.
- TBSGT buffer: 0.05 M Tris, 0.15 M NaCl, 0.25% gelatin (w/v), 0.5% Triton X 100 (v/v), pH 7.4.
- Hybridization buffer
- Sheep-anti-digoxigenin, conjugated to alkaline phosphatase (Boehringer Mannheim)
- Alkaline phosphatase substrate: BCIP: 5-bromo-4-chloro-3-indolyl-phosphate (75 mg/ml in 100% dimethyl formamide) and NBT: nitroblue tetrazolium (75 mg/ml in 70% dimethylformamide). Protect BCIP from light.
- Tris-EDTA buffer: 0.01 M Tris, 0.001 M EDTA, pH 7.3
- Levamisole: 1 M in double-distilled water.
- Aquamount (Merck)
- Glycerol: 75% in double-distilled water.

Procedure

Preparation of a Digoxigenin-Labeled Probe

The cDNAs are cloned in pBluescript. For the *in vitro* transcription the plasmids have to be linearized or PCR-amplified. The antisense RNA probes are then synthesized with T3 or T7 RNA polymerase, depending on the orientation of the cDNA insert. After *in vitro* transcription, the DNA template is degraded with RNase-free DNase I and the cRNA is hydrolyzed to obtain fragments of about 200 nucleotides.

- *Linearilization of the plasmid*: The plasmid is cut with a restriction enzyme just downstream of the termination codon of the insert. The DNA is then purified by phenol extraction and ethanol precipitation.
- *PCR-amplification of the insert*: The cDNA insert is PCR-amplified with primers flanking the pBluescript T3 and T7 RNA polymerase promotors and purified by gel-electrophoresis.

(For a comprehensive description of these techniques, see e.g. Maniatis et al. 1982)

Probe Preparation

1. Mix in a sterile Eppendorf tube:
   - 8 µl of 2.5 mM NTP mix
   - 2 µl of UTP/dig UTP mix (final concentration: 0.65 mM UTP, 0.35 mM dig-UTP)
   - 4 µl 5x transcription buffer
   - 1 µl RNase inhibitor
   - 1 µl of T3 or T7 RNA polymerase (depending on orientation of insert)
   - cDNA (100–200 ng PCR product or 1 µg of linearized plasmid
   - make total volume up to 20 µl with sterilized water.
2. Centrifuge briefly and incubate for 2 h. at 37°C.
3. Add 1 µl RNase-free DNase I and incubate 10 min at 37°C.
4. Hydrolyzation of probe: add to tube:
   - 4 µl DEPC-treated water
   - 50 µl ETS buffer
   - 1.7 µl 5 M NaCl
   - 10 µl 2 M NaOH

5. Mix and incubate on ice: 30 min for probes smaller than 1 kB; 60 min for probes longer than 1 kB.
6. Let warm up to RT and add 20 µl 1 M MES
7. Precipitate the probe by adding 28 µl of 7.5 M ammonium acetate and 412 µl of ice-cold ethanol. Allow precipitation to proceed for at least 30 min at –70°C or 2 h. at –20°C.
8. Centrifuge in a table centrifuge at 12.000 RPM for 10 min
9. Discard the supernatant (be careful not to remove the pellet) and wash pellet with ice-cold alcohol 70%
10. Centrifuge at 12.000 RPM, remove supernatant and dry pellet.
11. The yield will be about 2–4 µg RNA, dissolve in 25 µl DEPC-treated water. This gives a concentration of about 80–160 ng/µl.

Quality control of probe: see Chapter 2.

Hybridization Buffer

DNA denatures in 0.1–0.2 M Na^+ at 90–100°C and the maximum rate of renaturation (hybridization) is at 25°C below the melting temperature. Thus, for ISH this would mean that the microscopic preparations have to be hybridized at 65–75°C for prolonged periods of time. These high temperatures will seriously affect the morphology of the cells. A solution for this problem has been found in the use of organic solvents that reduce the thermal stability of double-stranded nucleotides, so that *in situ* hybridization can be performed at lower temperatures. Formamide has been generally applied for this purpose since it reduces the melting temperature of DNA-DNA duplexes in a linear fashion with 0.72°C for each % of formamide. The effect that formamide has on the melting temperature of hybrids can be calculated from the following formula:

For salt concentrations (Na^+) between 0.01 M and 0.2 M, which are generally used for *in situ* hybridizations:

$Tm = 16.6\ {}^{10}\log M + 0.41\ (G+C) + 81.5 - 0.72\ (\%\ formamide)$

Tm = melting temperature

M = salt concentration

G+C = molar percentage of guanine and cytosine of the probe.

Furthermore, the melting temperature and rate of renaturation are also dependent on the fragment length, and especially for shorter oligonucleotides this has to be taken into consideration. The following relation between fragment length and change in melting temperature has been established:

Change in Tm = 600/n

n = number of nucleotides in probe

This leads to the following formula to calculate the melting temperature:

$Tm = 16.6\ {}^{10}\log M + 0.41\ (G+C) + 81.5 - 0.72\ (\%\ formamide) - 600/n$

To calculate the percentage of formamide that is to be used for hybridization the following equation can be used:

$\%\ of\ formamide = ([Tm - 25] - Thyb)/0.72$

Thyb = temperature at which hybridization will be performed.

For example, if the Tm = 82°C, and hybridization will be performed at 37°C,the percentage of formamide is:

$([82 - 25] - 37)/0.72 = 27\%$

RNA forms very stable duplexes with RNA, so when we use an RNA probe for ISH, we use very stringent hybridization conditions. From our experience we have found that a hybridization mixture containing 60% formamide combined with a hybridization temperature of 50°C gives excellent results.

Preparation of the 60% Formamide Containing Hybridization Mixture

1. To a sterile 50 ml polypropylene tube add:
   - 30 ml of 100% deionized formamide
   - 10 ml of 20x SSC
   - 2.5 ml of 0.5 M sodium phosphate buffer, pH 7.0
   - 5 ml of 50x Denhardt's
   - 2.5 ml of 4 mg/ml acid-alkali hydrolysed salmon sperm DNA
   - 5 g dextran sulphate

Protect the hybridization buffer from light and store in the refrigerator.
Acid-alkali hydrolysed salmon sperm DNA is prepared as follows:

1. Transfer one gram salmon sperm DNA to a 50 ml sterile tube. Add 15 ml of DEPC water and allow to soak for 15 min to 2 hours.
2. Add 2.5 ml of 2 M HCl, keep DNA at room temperature. The DNA forms a white precipitate. Shake well until the precipitate sticks together. Gather into a ball with a Pasteur pipette tip over two to three minutes.
3. Add 5.0 ml of 2.0 M NaOH. Shake to resuspend the DNA which should dissolve. Place tube at 50°C for 15 min to increase dissolution.
4. Dilute the mixture to 175 ml with DEPC water, making sure there are no particles.
5. Add 20 ml of 1 M TRIS-HCl, pH 7.4.
6. Titrate with 2 M HCl until the pH of the DNA solution reaches pH 7.0–7.5.

Filter solution through a sterile millipore filter to remove any particles. Measure the OD 260 nm absorbance of the solution. Pipette 20 microliters of DNA solution into 980 microliters of water. The absorbance multiplied by 40 gives the concentration of DNA in micrograms per milliliter. Store in aliquots of 4 mg/ml at –20°C.

In situ Hybridization on Cultured Neurons

The neurons are isolated and cultured according to the methods described in Chapter 10 of this volume.

Since we use RNA probes for *in situ* hybridization, we have to take precautions in order to prevent the degradation of the RNA. Therefore, wear gloves during the experiments, use sterile pipette tips and Eppendorf tubes, and use RNase-free buffers: treat double-distilled water with 0.25% diethylpyrocarbonate (DEPC), incubate O/N at 37°C and autoclave.

All solutions should be filtered using a 0.22 μm filter. Any small particle can knock away the neurons in the culture dish!

1. Fix cells at least for 2h. Add the fixative gently to the culture well, an overly abrupt replacement of the culture medium may result in detachment of the cells from the culture well. We use a 1% paraformaldehyde/1 acetic acid mixture for our invertebrate (molluscan) neurons. For vertebrate neurons a 4% paraformaldehyde fixative is recommended, and up to 5% acetic acid may be added (Dirks, 1996). Acetic acid is an excellent fixative of nucleic acids.
2. Add Nonadet NP 40 to the fixative to a final concentration of 0.5% and incubate for 30 min
3. Wash in buffer 1, 2x10 min.
4. (Optional). Treat cells with 0.1%–0.005 % pepsin in 0.2 M HCl at 37°C for 10 min. This step increases the accessibility of the probe to the target mRNA in the cell and should be optimized for each application. In our experiments we omit the protease treatment entirely, without noticeable loss of signal. Thus, we omit steps 5–8.
5. Fix in 2% paraformaldehyde, 4 min.
6. Treat with 1% hydroxylammoniumchloride in buffer 1, 15 min.

7. Wash in buffer 1, 5 min.
8. Prehybridize cells in hybridization buffer, 1 h at 50°C. This step may not be necessary, but this has to be checked every time, depending on the types of cells that are used.
9. Hybridize cells in hybridization buffer, containing 1 µl probe in 100 µl hybridization buffer. Heat this mixture to 95°C and snap cool on ice immediately prior to hybridization. Incubate at least 3 h to O/N in an incubator at 50°C. Put culture well in an air-tight container with wet tissue to prevent evaporation of the hybridization buffer. Add sufficient hybridization buffer to ensure that the cells are completely covered; we use about 100 µl per culture well.
10. Briefly wash cells twice in 2x SSC, not longer than 2 min per wash
11. Stringent washing of cells: Wash cells 3x20 min at 50°C in a mixture of 2x SSC and 50% formamide. In general, this washing should remove all the non-hybridized probe from the cells. If the background in the cells is still high, a treatment with RNase A may be given (step 12), which will eliminate all the non-hybridized probe.
12. (Optional). Treat cells with RNase A in 2x SSC, 100 ng/µl in buffer 1 for 25 min. If this step is performed, then the stringent washing (step 11) is performed as follows; cells are washed twice before, and twice after the RNase treatment in 50% formamide/2x SSC.
13. Wash in 2x SSC for 5 min.
14. Wash 2x5 min in buffer 1.
15. Wash in TBSGT buffer for 15 min.
16. Incubate in anti-digoxigenin conjugated to alkaline phosphatase, 1:500-diluted in TBSGT buffer, either 2 h at RT, or O/N at 4°C.
17. Wash 2x5 min in buffer 1.
18. Wash 2x5 min in buffer 2.
19. Incubate in substrate for alkaline phosphatase: Add 4.5 µl of NBT and 3.5 µl of BCIP to 1 ml of buffer 2. If the neurons contain endogenous alkaline phosphatase activity, add 10 µl of 1 M levamisole. Add about 0.5 ml of substrate to each culture well. Incubation time may vary from 15 min to O/N, depending on the amount of transcripts present in the cell. Perform incubation in the dark. The alkaline phosphatase substrate used gives a deep-purple reaction product. If a fluorescent substrate is desired, use Fast Red TR/Naphtol (Sigma) in a concentration of 1.0 and 0.4 mg/ml, respectively, in buffer 2. This fluorogen can be viewed using the rhodamine filters of a fluorescence microscope.
20. Stop reaction after visual inspection with Tris-EDTA buffer, 15 min.
21. Rinse cells in double-distilled water
22. Mount cells in aquamount or glycerol
23. View under microscope

Results

We used ISH to study the distribution of neuropeptide-encoding mRNA in primary cultures of neurites of identified *Lymnaea* neurons. We were interested in whether the transcripts were randomly distributed throughout the neurites or whether they were docked at specific microdomains. The ISH experiments showed that the mRNAs are especially abundant in growth cones and varicosities (Fig 1).

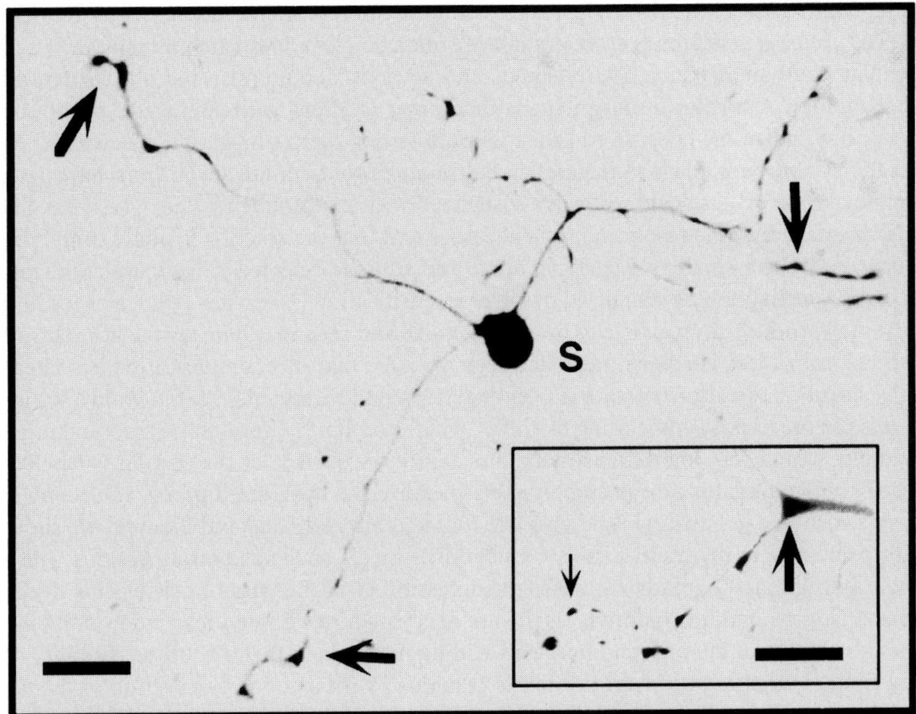

Fig. 1. *In situ* hybridization on a cultured Pedal A neuron using a cRNA probe for Pedal peptide, the endogenous neuropeptide of these neurons. The signal can be observed in the soma (S) and in neurites. Note that not the entire neurites show a hybridization signal, and that the signal is especially prominent in varicosities (*arrows*). This is shown in detail in the inset, where branch points (*large arrow*) and varicosities (*small arrow*) show a strong signal, whereas the connecting neurites are almost devoid of signal. Bar = 100 μm; inset: 20 μm.

Subprotocol 2
In Situ Hybridization at the Electron Microscopic Level

Introduction

To investigate the subcellular localization of specific nucleic acid sequences, many researchers have extended the light microscopic procedures to the ultrastructural level. The availability of non-radioactive probes, with digoxigenin or biotin as reporter molecules, has greatly advanced the field, because these probes allow a precise ultrastructural localization of the target sequences in the section. Moreover, the procedure can be completed in 24h., since no long exposure times are necessary, as is the case for radiolabeled probes. Several strategies for the EM-ISH can be employed, each with its own advantages and disadvantages. We here restrict ourselves to a short description of the various possibilities for EM-ISH; for a comprehensive description of the various techniques the reader is referred to the literature (see *e.g.* Morel, 1993; Morey, 1995).

Pre-embedding ISH techniques are (in theory) superior in sensitivity to post-embedding techniques, since the signal can be observed throughout the entire thickness of the section. However, especially when longer probes are used, these may not penetrate evenly throughout the entire thickness of the section. To increase penetration, freeze/thaw cycles or protease and detergent treatments are often employed, which may result in loss of cellular components such as ribosomes and severely compromise the ultrastructural morphology. A further disadvantage is that larger (>5 nm) gold-conjugates cannot be used to visualize the hybridized probe without strong permeabilization procedures. As an alternative, peroxidase conjugates and ultra-small gold-conjugates (<1 nm) have been employed. Peroxidase conjugates are visualized with diaminobenzidine, which has the disadvantage that the resolution is rather poor and that the reaction product cannot be quantified. Ultra-small gold particles are too small to be detected directly under the microscope and have to be visualized by silver intensification (Danscher 1981), a procedure that may produce silver particles of irregular size and thus may hamper the accuracy of target localization. Furthermore, it has been reported that silver amplification has a negative effect on the ultrastructure (Egger et al. 1994; Macville et al. 1995). A good alternative is the use of post-embedding methods, which combines a good ultrastructural morphology with a reasonable sensitivity. Since only the surface of the section is labeled, large gold-conjugates can be employed to visualize the hybridized probe. Hydrophylic resins such as e.g. Lowicryl K4M and HM20, LR White, LR Gold and Bioacryl are superior compared to the traditional EM embedding media such as Epon or Araldite, since their hydrophilicity greatly increases the accessibility of the target nucleic acids to the probe. Non-embedding methods, as the use of cryosections is often referred to, combine the advantages of the pre- and post-embedding methods in that a good accessibility of the probe to the targets in the section is combined with a good preservation of the ultrastructure. A disadvantage of this method is that it requires expensive equipment and skilled personnel to operate it. Furthermore, the sections are sensitive to denaturing treatments resulting in a loss of morphology and target nucleic acids.

The choice of which method to employ depends on the demands of the researcher and preparation, and should be evaluated from case to case. The advantages and disadvantages of the three methods are summarized in Table 1. In our experience, the non-embedding method yielded the best results in terms of sensitivity and conservation of the ultrastructure. As an additional advantage, the procedure can be easily combined with immunocytochemistry to identify other substances (e.g. neuropeptides) in the same section. We have employed this technique to study the ultrastructural localization of neuropeptide-encoding mRNA in the axonal compartment of molluscan neurons.

Table 1.

Method	Advantages	Disadvantages
Pre-embedding ISH	– Selection of area of interest is easy	– Probe penetration
	– Sensitive	– Non-uniform labeling of sections
	– Good preservation of ultrastructure	– Large gold conjugates do not penetrate
Post-embedding ISH	– Excellent preservation of ultrastructure	– Only surface of section is labeled, probe does not penetrate in section
	– Selection of area of interest can be performed in semithin sections sections	– Resins are toxic
Ultrathin cryosections	– Fast	– Expensive equipment
	– No, or few pretreatments	– Loss of target sequences during denaturing conditions
	– Sensitive	
	– Good preservation of ultrastructure	

Materials

- Cryo-ultramicrotome
- Hot-plate, 37°C
- Antistatic forceps
- Nickel grids, 75 mesh
- Small copper rivets
- Sandpaper (fine)
- Dissection microscope
- Filter paper
- Liquid N_2 storage vessel
- Diamond knife
- Oeses
- Incubator (25°C–80°C)
- µm filters
- 50 ml syringes
- Glass knives
- Electron microscope

Procedure

Fixation of Tissue for Ultrathin Cryosections

General Remarks

- Always use clean glassware or disposables to prepare the fixative
- Always use freshly prepared fixative
- Tissues should be fixed in 2–4% paraformaldehyde and 0.2% glutaraldehyde. Glutaraldehyde is necessary to ensure a good preservation of the ultrastructure. However, higher percentages (>0.5%) will severely hamper the penetration of the probe, because of the cross-linking characteristics of the fixative. Therefore, for each new preparation to be studied, a compromise between preservation of ultrastructure and probe penetration should be reached.

Solutions

- Glutaraldehyde: We use a 25% stock solution (Sigma) that we keep at 4°C. Add the glutaraldehyde to the fixative solution immediately prior to use.
- Paraformaldehyde fixative stock solution: prepare 100 ml of a 10% stock solution:
  1. Add 10 g paraformaldehyde to 75 ml aqua dest. (demi-water).
  2. Place the solution on a stirrer/heater in the fume hood and let it, under constant stirring, reach a temperature of 60°C.
  3. Add a few drops of 1 M NaOH until the solution becomes clear.
  4. Add distilled water to an end volume of 100 ml.
  5. Filter the solution and store in aliquots in the freezer. Each time fixative is required a tube can be thawed and used. Upon thawing the solution will be white. Place the formaldehyde in hot water until it again becomes clear. Do not use the solution if it does not clear.
- 0.2 M and 0.1 M phosphate buffer:
  - A: NaH_2PO_4
  - B: Na_2HPO_4.
  - Add 19 ml of A to 81 ml of B, and, if necessary, adjust pH with A or B to pH 7.4

- 20x SSC stock solution
- Glycine, 0.15% in 0.1 M phosphate buffer
- Liquid N_2
- Double-distilled, filtered (0.22 μm) water
- Formamide
- 70% ethanol
- 60% hybridization buffer
- Gelatin, 2% in 0.1 M phosphate buffer; add sodium azide to a final concentration of 0.02%.
- Bovine serum albumin (BSA) stock solution: 10% in distilled water, set pH to 7.4 with 1 N NaOH
- Phosphate-buffered saline (PBS): 140 mM NaCl, 2.7 mM KCl, 10 mM Na_2HPO_4 and 1.8 mM KH_2PO_4.
- 4% Uranyl acetate (Merck) stock solution in distilled water.
- 2% Uranyl acetate, pH 7: Mix 4% uranyl acetate solution with a 0.15 M solution of oxalic acid in a 1:1 ratio. The pH is set with 1 M NH_4OH. This must be added dropwise since otherwise an insoluble precipitate is formed. Uranyl acetate is sensitive to light and is stored in the dark at 4°C.
- Methylcellulose: Final volume 200 ml; heat 196 ml of distilled water to a temperature of 90°C, then add 4g of methylcellulose (Sigma, 25 centipoises) while stirring. The solution is rapidly cooled on ice while stirring until it has reached a temperature of 10°C. Slow stirring is continued O/N at 4°C and then stopped to let the solution 'mature' for 3 days at 4°C.
- Methylcellulose/uranyl acetate mixture: 180 ml of the above methylcellulose solution is mixed with 20 ml of the 4% uranyl acetate, pH 4.0, solution and gently mixed. Then the solution is centrifuged for 90 min at 12.000 rpm at 4°C. The supernatant is removed and can be stored in the dark for about 3 months at 4°C.
- 2.3 M sucrose
- Sucrose/methylcellulose mixture: Add to 2 volumes of 2.3 M sucrose 1 volume of 2% methylcellulose.
- 1% toluidine blue in distilled water
- Protein A-colloidal gold (5 and 10 nm), commercially available from the University of Utrecht, School of Medicine, Department of Cell Biology, Utrecht, the Netherlands.
- Sheep-anti-digoxigenin (Boehringer Mannheim, Germany)
- Rabbit-anti-sheep (Dakopatts, Glostrup, Denmark)
- Bovine serum albumin (Boehringer Mannheim)
- Cold-water fish gelatin (Aurion, Wageningen, The Netherlands)
- Acetylated bovine serum albumin (BSA-C; Aurion)
- Fixative: Mix together 2 ml of 10% paraformaldehyde stock solution, 5 ml of 0.2 M phosphate buffer and 0.08 ml of 25% glutaraldehyde. Adjust volume to 10 ml with distilled water. Use immediately. (This fixative is used for molluscan tissues, for vertebrate preparations it is recommended to increase the paraformaldehyde concentration to 4%.)

Fixation and Sample Preparation

Small blocks of tissue can be fixed by immersion. For larger tissues, such as rat brain, it is recommended to use a perfusion fixation protocol.

1. Fix small blocks of tissue by immersion O/N at 4°C
2. Wash tissue in 0.1 M phosphate buffer: 2x10 min
3. Wash with 0.15% glycine in 0.1 M phosphate buffer: 2x10 min

Infuse fixed tissue for at least 4h with 2.3. M sucrose in PBS. Rotate the vials containing the tissues during infusion.

Specimen Preparation

Prior to sectioning the tissue blocks have to be mounted on specimen holders; we use copper pins (rivets) for this purpose.

1. Roughen the surface of the pin with sandpaper.
2. Wash pins extensively to remove small pieces of metal
3. Wash pins in 70% ethanol and dry on tissue paper
4. Remove the tissue from the sucrose solution and use a dissection microscope to position the tissue on the pin. Use tissue paper to remove the excess of sucrose.
5. With a clean pair of forceps, plunge the pin into liquid nitrogen.
6. Store samples in liquid nitrogen until use.

Sectioning

1. Make sure that the tissue does not thaw while transferring from the storage vessel to the microtome.
2. The optimal temperature to make ultrathin cryosections is between 100 and 120°C.
3. Sectioning is performed with dry knives
4. Always make a thick (about 500 nm) section for orientation in the tissue and to evaluate the fixation. This section can be stained with a 1% toluidine blue solution.
5. The quality of the ultrathin sections improves when the edges of the specimen are carefully trimmed. This can be done with a razor blade or with the cutting edge of a glass knife.
6. For thin sectioning (40–80 nm, silver to gold sections) the use of a diamond knife is recommended
7. Sections can be removed from the knife by using oeses. These are small wire loops (diameter 2 mm) attached to a 15 cm plastic handle. The loop is filled with 2.3 M sucrose or a sucrose/methylcellulose mixture.
8. After collecting the sections with the oeses, allow them to briefly thaw, and immediately transfer them to the grid.
9. Using the sucrose/methylcellulose mixture to collect the sections has two advantages: it results in a better overall morphology and the grids can be stored for at least three months in the refrigerator.
10. Use formvar- and carbon-coated nickel grids; copper grids will oxidize during the procedure, resulting in a nasty precipitate that will obscure the EM image.

Procedure EM *in situ* Hybridization

Place grids (section side down) on a 2% gelatin solution in a petri dish, first 5 min on a hot plate (temperature 37°C), then 20 min in a 37°C incubator.

All the next steps are performed on droplets. The droplets (5–10 µl for the probe and antibody solutions, 100 µl for the washing steps) are spotted on a piece of parafilm, and the grids are placed section side down on the droplets.

Since the procedure makes use of protein A-gold conjugate, the use of a rabbit-anti-sheep is required as a secondary step, since protein A has high affinity for rabbit IgGs, and low affinity for sheep IgGs, the most common anti-digoxigenin antibody used.

1. 0.15% glycine in PBS, 2x5 min.
2. 2x SSC, 2x5 min.
3. Prehybridize in hybridization buffer (i.e. without probe), 20 min.
4. Add 0.1 µl of probe solution to 10 µl of hybridization mixture.
5. Hybridize grids in a closed container in which wet filter paper is placed to prevent evaporation of hybridization buffer, 3h at 50°C
6. Wash in 2x SSC, 1 min.
7. Wash in a mixture of 50% formamide and 2x SSC, 2x5 min.
8. Wash stringently in 50% formamide/2x SSC, use the closed humidified container, 3x20 min at 50°C
9. Wash in 2x SSC, 2x5 min.
10. Wash in 1% BSA in PBS, 3x5 min.
11. Wash in 1 % fish gelatin in PBS, 10 min.
12. *) Incubate in sheep-anti-digoxigenin, 1:100 dilution in 1% fish gelatin in PBS, 1 hr at RT or O/N, 4°C
13. Wash in 1% fish gelatin in PBS, 5 min.
14. Wash in 1% BSA in PBS, 3x10 min.
15. Wash in 1% fish gelatin in PBS, 5 min.
16. Incubate in rabbit-anti-sheep, 1:100 dilution in 1 % fish gelatin in PBS, 30 min.
17. Wash in 1% fish gelatin in PBS, 5 min.
18. Wash in 1% BSA in PBS, 3x10 min.
19. Incubate in protein A-gold conjugate in 1 % BSA in PBS, dilution according to the manufacturer's instructions, 20 min.
20. Wash in PBS, 3x in 15 min.
21. Fix in 1% glutaraldehyde in PBS, 10 min.
22. Wash in PBS, 5 min.
23. Wash in distilled water, 5x1 min.
24. Contrast in 2% uranyl acetate solution, 5 min.
25. Wash in methylcellulose/uranyl acetate, 2x1 sec.
26. "Embed" sections in methylcellulose/uranyl acetate, 10 min on ice
27. Remove grid from droplet of methylcellulose with oese
28. Remove excess methylcellulose with filter paper, leave grid on the oese and let it dry, 30 min.
29. Transfer grid to grid box

Immunocytochemistry

The EM ISH can be easily combined with immunocytochemistry: After step 22, use the following procedure:

1. Wash in 0.15% glycine in PBS, 3x in 10 min.
2. Wash in 1% fish gelatin and 1% BSA in PBS, 5 min.
3. Incubate in antibody (raised in rabbit) in 1% fish gelatin /1% BSA in PBS, 45 min.
4. Wash in 1% BSA in PBS, 5x1 min.
5. Incubate in protein A-gold conjugate (use a different size of gold than is used for the visualization of the ISH signal) in 1 % BSA in PBS, 20 min.
6. Continue with step 20 of EM ISH protocol

Results

Previously it has been shown that the axonal compartment of the egg-laying hormone (ELH) producing neurons of the mollusc *Lymnaea* contains high amounts of the ELH-encoding mRNA (Van Minnen, 1994). To study the ultrastructural localization of these neuropeptide-encoding mRNAs we used EM-ISH. First, we studied the localization of the transcripts in the cell bodies of the ELH-producing neurons. As expected, the transcripts were found to be mainly associated with the membranes of the rough endoplasmic reticulum (Fig. 2A). Other structures, such as mitochondria, Golgi apparatus and

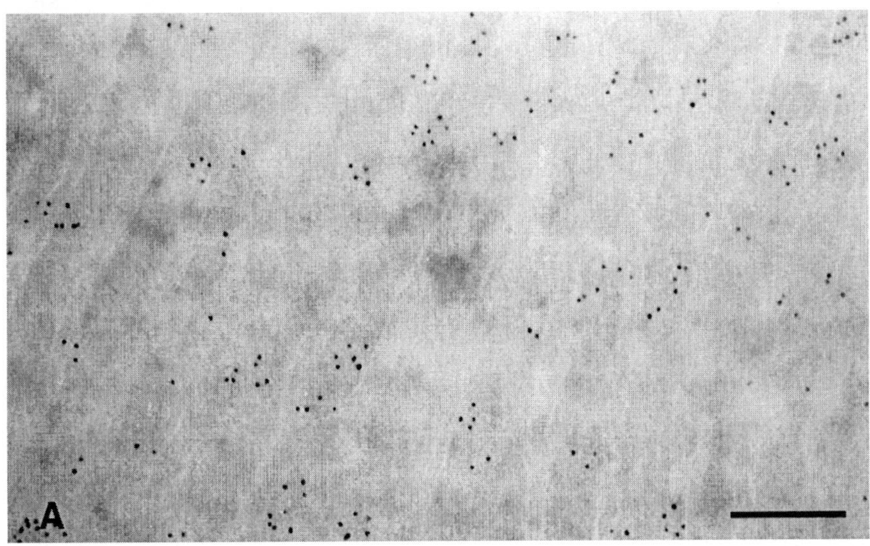

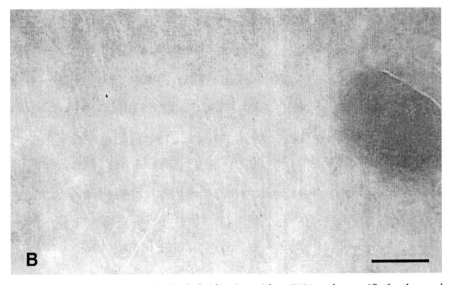

Fig. 2. A: Electron microscopic *in situ* hybridization with a cRNA probe specific for the egg-laying hormone (ELH) encoding mRNA on an ultrathin cryosection of ELH-producing neurons. The RER of these neurons shows a strong hybridization signal, visible as 5 nm gold particles in the immediate vicinity of the membranes of the RER. B: As a negative control we used a cRNA probe specific for one of the molluscan insulin-related peptides (Van Minnen, 1994). The RER of the ELH-producing neurons is devoid of gold particles, demonstrating the specificity of reaction in A. Bar = 100 nm.

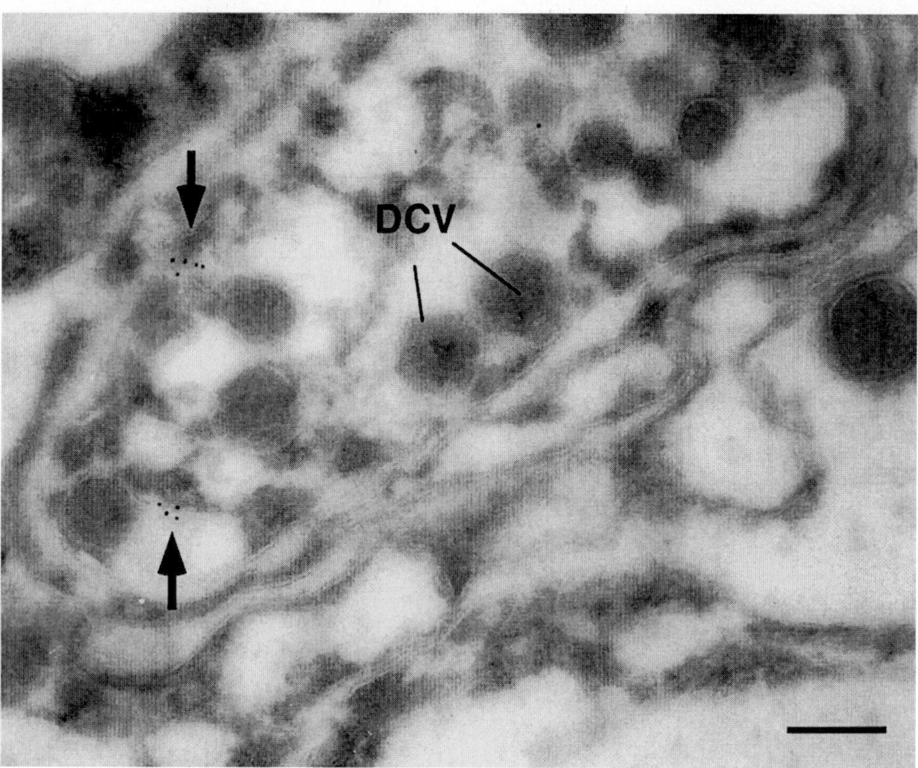

Fig. 3. Electron microscopic *in situ* hybridization with a cRNA probe specific for the egg-laying hormone (ELH) encoding mRNA on an ultrathin cryosection of the axon terminals of the ELH-producing neurons. The hybridization signal (5 nm gold particles; *arrows*) is present in the axoplasm. DCV: ELH-containing dense core vesicles. Bar = 200 nm.

secretory vesicles, were devoid of signal, which can also be concluded from Figure 4, where no colocalization is observed between the ISH signal (small gold particles) and the ELH-immunoreactive material (large gold particles). Next, we studied the localization of the transcripts in the axonal compartments of these neurons. The transcripts appear to be localized in the axoplasm (Fig 3), and not in the ELH-containing dense-core vesicles, as was reported previously (*e.g.* Dirks, 1996).

* In the experiments we performed to investigate the ultrastructural localization of the egg-laying hormone (ELH)-encoding mRNA, we observed that the sheep-anti-digoxigenin antiserum cross-reacted with the ELH-containing dense-core vesicles. To avoid this problem, we used a mouse monoclonal anti-digoxigenin (step 12), and at step 16 of the procedure used a rabbit-anti-mouse, instead of rabbit-anti-sheep. These data results furthermore underline the necessity for appropriate control experiments in performing ISH (see below).

Troubleshoot

Appropriate controls are necessary to assure that the hybridization signals observed in the preparation are indeed the result of the hybridization of the probe to its target sequences in the tissue. There are a number of possible sources of false-positive results. For instance, the probe may bind aspecifically to certain cellular components, without any specific interaction with RNA molecules in the tissue. Furthermore, since an immu-

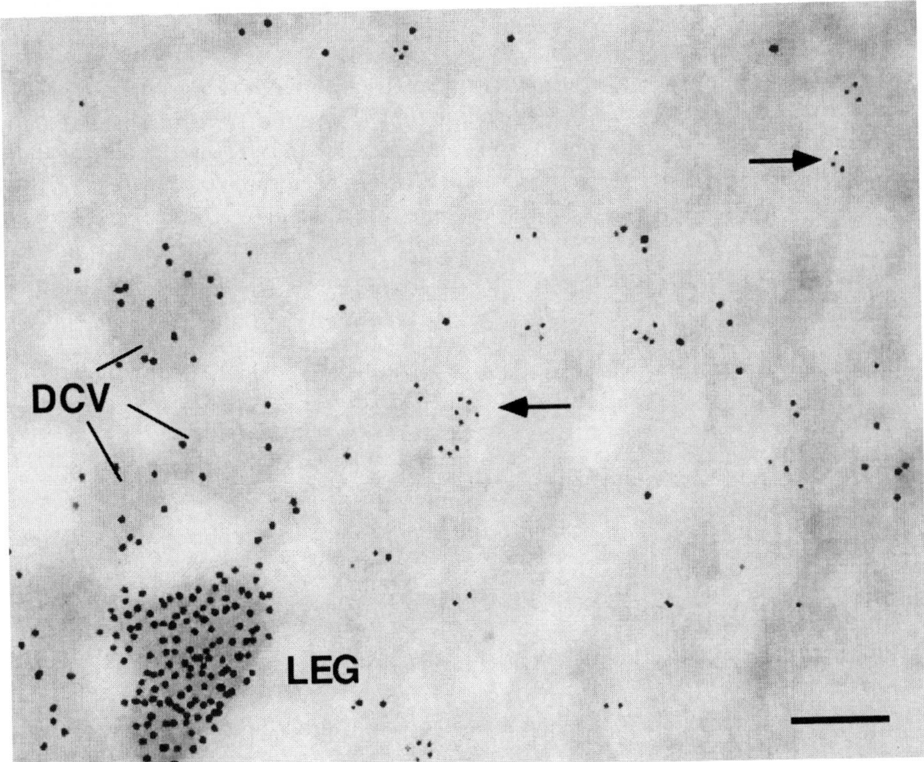

Fig. 4. Electron microscopic *in situ* hybridization combined with immuno-electron microscopy. We used the ELH-specific cRNA probe and an antibody specific for the alpha-peptide encoded on the ELH gene. The hybridization signal is observed on the RER (5 nm gold; *arrows*), whereas the alpha-peptide immunoreactivity (10 nm gold) is present in the dense core vesicles (DCV) and large electron-dense vesicles (LEG; see also Klumperman et al. 1996). Bar = 100 nm

nocytochemical detection system is used, all the well-documented problems of cross-reacting antisera may also be a source of false-positive results. We therefore recommend to carry out the following.

Specificity Control Experiments

1. Perform the entire EM-ISH procedure with omission of the probe; this will tell you whether the immunocytochemical detection system is cross-reacting with tissue constituents.
2. Use the sense probe or an irrelevant probe (e.g. encoding for plasmid sequences) as a negative control to verify whether the hybridization signal observed with the anti-sense probe is due to a specific interaction with the target sequences in the tissue.
3. Remove the endogenous mRNA by treating the cells or tissues with RNase (100 µg/ml in 2 x SSC at 37°C for 60 min).
4. Perform the hybridization procedure in the presence of a 100- to 1000-fold excess of unlabeled antisense probe.

All these control experiments should give negative results.

Another possibility is that the ISH procedure fails to give a hybridization signal. This may be because the target sequences are not or in too low a concentration present in the tissue under investigation. Alternatively, in one of the many steps of the procedure, something may have gone wrong, and below you will find some tips to verify that the procedure has been performed correctly.

First of all, when working with RNA probes, it is important to know that the RNA is intact. This can be easily checked by running the RNA on a 1% ethidium bromide-stained agarose gel. The RNA should be visible as a clear band (see also Chapter 2, quality control of RNA). When the quality of the RNA has been confirmed, we find it very convenient to perform the ISH procedures (both for light and electron microscopy) on nitrocellulose paper. The advantage of using nitrocellulose paper is that one does not have to culture neurons or make sections, and the results are unambiguous: If the procedure is successful, it results in easily identifiable positive dots. Using nitrocellulose paper we can furthermore verify the quality of the probe: Is it (sufficiently) labeled with digoxigenin? The procedure for this test is as follows:

Test Procedure

1. Dilute 1 µl of probe 5, 25, 125 and 625 times and spot 1 µl of probe on nitrocellulose (NC) paper.
2. Cross-link the probe by UV to the NC paper for 10 sec (time will depend on the UV source used).
3. Block the NC paper in a solution containing 0.05% Tween 20 in buffer 1 (see page 61) for 1 min.
4. Continue with step 7–21 of the light microscopic ISH procedure (page 64).

When the 5x and 25x diluted spots show up during the first 5 min of incubation with the substrate, the probe is sufficiently labeled with digoxigenin for the in situ hybridization experiments.

The same procedure can be used to verify whether the immunocytochemical detection system is functioning properly. The only difference is that one must have a probe of which it was previously shown that it is correctly labeled with the reporter molecule (for instance, a different probe used with success in other experiments).

Also the electron microscopic procedure can be verified for proper functioning on NC paper. Since the protein A-gold conjugate is not visible on NC paper, we use a silver intensification step to visualize the protein A-gold. We use a commercially available intensification solution (R-gent, Aurion), which gives intensely stained black dots within ten minutes of incubation.

A commonly observed problem in EM ISH is the aspecific labeling of structures throughout the section. In general this is caused by the aspecific binding of antibody molecules to tissue constituents. Further diluting the antibodies may solve this problem, although care should be taken not to dilute too much in order to prevent loss of specific staining. Furthermore the use of acetylated BSA (BSA-C) in the washing and incubation for the gold conjugate steps may considerably reduce the aspecific background staining, since this blocking agent has an increased negative charge and almost completely inhibits binding of gold conjugates to hydrophobic substrates.

Subprotocol 3
Single-Cell Differential mRNA Display

Introduction

Several methods exist to characterize alterations in expression levels of unknown genes in tissue samples upon a physiological stimulus. Morphological and functional heterogeneity of cell types in many tissues, however, often requires that differences in gene expression are studied at the level of the individual cell, because not all cells may respond to the stimulus, or not all cells may respond in the same way. Thus, it may occur that a significant upregulation of a particular gene in one cell (type) is masked by a downregulation of the same gene in another cell (type). The chances of characterizing important changes in gene expression would significantly increase if one could efficiently study differential gene expression at the single cell level. In the previous parts of this chapter, we described ISH methods that allow the detection of mRNAs derived from previously identified genes. We will now focus on a method that we routinely use to characterize differential expression of unknown genes at the single-cell level.

The limitations of studying differential gene expression at the tissue level hold true especially for the brain, as it is the most complex organ in the body, morphologically, genetically and functionally. Even in a single neuron, gene expression may be differentially regulated in functionally distinct subcellular domains through targeting of specific mRNAs to, for instance, axons and dendrites (see Introduction). We use a single-cell mRNA differential display approach (for a detailed description of differential display, see Liang and Pardee 1992; Livesey and Hunt 1996; and Chap. 2) in order to characterize differential mRNA localization in functionally distinct neuronal domains and to understand the molecular basis of synapse formation between two individual, identified molluscan neurons. The main advantage of differential mRNA display is that it is PCR-based and allows ample amplification of the small amounts of cDNA obtained by reverse transcription of mRNA from individual cells. The main disadvantage is that it is PCR-based, and PCR is known to yield irreproducible results when applied to small amounts of complex templates such as total pools of cDNA with a wide variation in relative abundance. This irreproducibility is due to the random selection of amplification targets in the first rounds of the PCR reaction, and has been termed the "Monte Carlo effect" (Karrer et al. 1995). As this would seriously affect the outcome of a differential mRNA display at the level of individual neurons or subneuronal domains, we have attempted to circumvent Monte Carlo effects by introducing a linear, non-PCR-based amplification of the starting mRNA before proceeding with the differential display PCRs. To this end, we adopted a method that was originally described by Eberwine and colleagues (Mackler et al. 1992; Miyashiro et al. 1994) and makes use of a T7 promoter that is introduced at the 3' end of the cDNA in order to transcribe multiple copies of antisense RNA (aRNA) using T7 RNA polymerase. The T7 promoter sequence is present at the 5' end of the oligo(dT) primer that is used to prime reverse transcription, and will thus become incorporated in each cDNA that is being synthesized. After second strand cDNA synthesis, T7 RNA polymerase-mediated transcription will result in an unbiased 100- to 1000-fold amplification of the cDNA in the form of aRNA molecules. These are then reverse-transcribed into cDNA again using random hexanucleotide primers, and the resulting cDNA is used in a differential display PCR (DD-PCR).

Outline

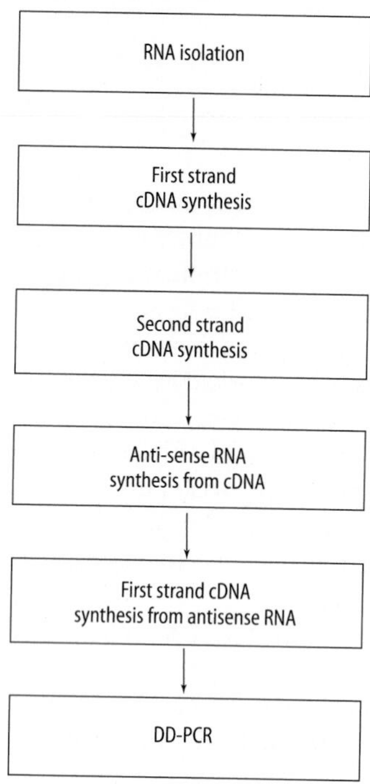

Materials

Equipment

- Water bath
- Thermal cycler (PCR-machine)
- Eppendorf centrifuge

Note: All materials and solutions should be sterile and RNase-free

Plasticware and Filters

- Eppendorf tubes (0.5 ml and 1.5 ml)
- Falcon or Greiner tubes (50 ml)
- Filter tips
- Plastic gloves
- 0.025 μm; MF-Millipore membranes (13 mm Ø)

Solutions and Buffers

- 5x DNase I/proteinase K buffer: [250 mM Tris-HCl (pH 7.5), 10 mM $CaCl_2$, 100 mM $MgCl_2$]
- 5x First strand buffer: [250 mM Tris-HCl (pH 8.3), 375 mM KCl, 15 mM $MgCl_2$]
- 5x Second strand buffer: [94 mM Tris-HCl (pH 6.9), 453 mM KCl, 23 mM $MgCl_2$, 750 mM β-NAD, 50 mM $(NH_4)_2SO_4$]
- 5x T7 RNA polymerase buffer: [200 mM Tris-HCl (pH 8.0), 40 mM $MgCl_2$, 10 mM spermidine-$(HCl)_3$, 125 mM NaCl]
- 10x PCR buffer: [200 mM Tris-HCl (pH 8.4), 500 mM KCl, 20 mM $MgCl_2$]
- TE [10 mM Tris-HCl (pH 7.5)], 1 mM EDTA]
- Gel elution buffer [10 mM Tris-HCl (pH 7.5), 100 mM NaCl, 1 mM EDTA]
- 200 mM EDTA/5% SDS
- DEPC-treated double-distilled H_2O

- 100 mM DTT
- 2 mM DTT
- 3 M sodium acetate (pH 5.2)
- 350 mM KCl
- TE-saturated phenol
- Chloroform p/a
- Isoamylalcohol p/a
- Glycogen (10 mg/ml)
- 100% ethanol
- 70% ethanol
- 10 mM dNTPs (10 mM dATP, 10 mM dCTP, 10 mM dGTP, 10 mM dTTP)
- 200 μM dNTPs (200 μM dATP, 200 μM dCTP, 200 μM dGTP, 200 μM dTTP)
- 10 mM NTPs (10 mM ATP, 10 mM CTP, 10 mM GTP, 10 mM UTP)
- [α-^{32}P]dCTP (3000 Ci/mmol; 10 mCi/ml)
- Formamide loading buffer (95% formamide; 20 mM EDTA; 0.05% Xylene Cyanol FF)
- 40% Polyacrylamide solution (38% Monoacrylamide, 2% Bisacrylamide)
- N,N,N',N'-Tetramethylethylenediamine (TEMED)
- 10% Ammonium Peroxidisulfate (APS)
- Fluorescent ink

Enzymes

- RNasin (Promega; 40 U/μl)
- DNaseI, Amplification Grade (Gibco-BRL; 1 U/μl)
- Proteinase K (Boehringer Mannheim; 20 mg/ml)
- M-MLV Reverse Transcriptase (SuperScript from Gibco-BRL; 200 U/μl)
- *E. coli* DNA polymerase I (Gibco-BRL; 10 U/μl)
- T4 DNA ligase (Gibco-BRL; 1 U/μl)
- RNase H (Gibco-BRL; 2.2 U/μl)
- T7 RNA polymerase (Gibco-BRL; 50 U/μl)
- *Taq* DNA polymerase (Gibco-BRL; 5 U/μl)

Primers

- Oligo(dT)$_{18}$T7 cDNA synthesis primer (10 ng/μl)
  (5'-TAATACGACTCACTATAGGGC TTTTTTTTTTTTTTTTTT-3')
- Random hexanucleotide cDNA synthesis primers (10 ng/μl)
  (5'-NNNNNN-3')
- DD-PCR (dT)-anchor primers (150 ng/μl)
  (5'-TTTTTTTTTTTT(A/C/G)A-3';
  5'-TTTTTTTTTTTT(A/C/G)C-3';
  5'-TTTTTTTTTTTT(A/C/G)G-3';
  5'-TTTTTTTTTTTT(A/C/G)T-3')
- DD-PCR arbitrary decamer primers (17 ng/μl)
  (26 arbitrary decamer primers are used according to Bauer et al. (1993))

Procedure

I. RNA Isolation

1. Collect control and experimental cells (or axons/dendrites) and transfer to 1.5 ml microcentrifuge tubes containing 45 μl 2 mM DTT.
2. Heat-lyse in a 95°C waterbath for 2 min.

3. To each tube add the following:
   - 0.5 µl RNasin (20 U)
   - 15 µl 5x DNase I/proteinase K buffer
   - 3 µl DNase I (3 U)
4. Incubate at 37°C for 30 min.
5. Add the following:
   - 7.5 µl 200 mM EDTA/5% SDS
   - 4 µl Proteinase K
6. Incubate at 37°C for 15 min.
7. Add DEPC-treated H_2O to a final volume of 200 µl.
8. Add 200 µl phenol/chloroform/isoamylalcohol (25:24:1)
9. Vortex for 30 s.
10. Spin in Eppendorf centrifuge at 14,000 *g* for 5 min.
11. Transfer aqueous upper phase to a clean 1.5 ml tube, and add 200 µl chloroform/isoamylalcohol (24:1).
12. Vortex for 30 s.
13. Spin in Eppendorf centrifuge at 14,000 *g* for 2 min.
14. Add 1 µl (10 µg) glycogen, 20 µl 3 M sodium acetate (pH 5.2) and 600 µl 100% ethanol, and precipitate RNA overnight at –20°C.

    Note: Addition of glycogen and overnight precipitation are not necessary in all following precipitations.

15. Spin at 4°C in Eppendorf centrifuge at 14,000 *g* for 30 min.
16. Carefully remove the supernatant and add 200 µl 70% EtOH.
17. Spin briefly in Eppendorf centrifuge.
18. Carefully remove supernatant and dry pellet at room temperature for 5 min.
19. Redissolve pellet as required in the next step.

II. cDNA Synthesis

1. Redissolve RNA pellet in 10 µl DEPC-treated H_2O and add 1 µl (10 ng) oligo$(dT)_{18}$T7.
2. Denature RNA in a 95°C waterbath for 2 min, and quick-chill on ice.
3. Add the following:
   - 4 µl 5x first strand buffer
   - 2 µl 100 mM DTT
   - 1 µl 10 mM dNTPs
   - 0.5 µl RNasin (20 U)
   - 1.5 µl M-MLV Reverse Transcriptase (300 U)
4. Incubate at 37°C for 1h.
5. Add the following:
   - 90 µl DEPC-treated H_2O
   - 32 µl 5x second strand buffer
   - 3 µl 10 mM dNTPs
   - 6 µl 100 mM DTT
   - 6 µl T4 DNA ligase (6 U)
   - 3 µl *E. coli* DNA polymerase I (30 U)
   - 0.6 µl RNase H (1.3 U)
6. Incubate in a 16°C waterbath for 2h.

7. Perform phenol/chloroform/isoamylalcohol extraction, chloroform/isoamylalcohol extraction and ethanol precipitation as described in steps 7–19 of protocol I (RNA isolation).
8. Redissolve ds cDNA pellet in 10 µl H_2O and drop-dialyze against 50 ml RNase-free TE for 4h. (Gently transfer the cDNA solution onto a 0.025 µm Millipore filter floating in a 50 ml conical tube filled with RNase-free TE. After 4h, transfer the drop to a 1.5 ml reaction tube. Rinse spot on filter with 5 µl DEPC-treated H_2O, and transfer to the same tube.)

III. aRNA Synthesis

1. To 15.2 µl of the ds cDNA solution from the previous step, add the following:
   - 5 µl 5x T7 RNA polymerase buffer
   - 1 µl 100 mM DTT
   - 1.3 µl 10 mM NTPs
   - 0.5 µl RNasin (20 U)
   - 2 µl T7 RNA polymerase (100 U)

2. Incubate at 37°C for 4h.
3. Add the following:
   - 17.5 µl H_2O
   - 5 µl 350 mM KCl
   - 2.5 µl DNase I (2.5 U)

4. Incubate at 37°C for 30'.
5. Perform phenol/chloroform/isoamylalcohol extraction, chloroform/isoamylalcohol extraction and EtOH precipitation as described in steps 7–19 of protocol I (RNA isolation).

IV. cDNA Synthesis from aRNA

1. Redissolve aRNA pellet in 10 µl DEPC-treated H_2O and add 1 µl (10 ng) hexanucleotide primers.
2. Denature aRNA in a 95°C waterbath for 2 min and quick-chill on ice.
3. Add the following:
   - 4 µl 5x first strand buffer
   - 2 µl 100 mM DTT
   - 1 µl 10 mM dNTP mixture
   - 0.5 µl RNasin (20 U)
   - 1.5 µl M-MLV reverse transcriptase (300 U)

4. Incubate at 37°C for 1h.
5. Perform phenol/chloroform/isoamylalcohol extraction, chloroform/isoamylalcohol extraction and EtOH precipitation as described in steps 7–19 of protocol I (RNA isolation).
6. Perform drop-dialysis as described in step 8 of protocol II (cDNA synthesis).
7. After transferring the cDNA to a clean 1.5 ml reaction tube, add DEPC-treated H_2O to a final volume of 50 µl.

V. DD-PCR

1. In a 0.5 ml reaction tube mix the following:
   - 2 µl cDNA from the previous step
   - 8.7 µl DEPC-treated H_2O
   - 2 µl one of the (dT)-anchor primers (300 ng)
   - 2 µl one of the arbitrary decamer primers (34 ng)
   - 2 µl 200 µM dNTPs
   - 2 µl 10x PCR buffer
   - 0.3 µl [α-^{32}P]dCTP
   - 1 µl *Taq* DNA polymerase (5 U)

2. When using a thermal cycler that is not equipped with a heated lid, cover the reactions with a drop of mineral oil in order to prevent evaporation.
3. Place the tubes in a preheated thermal cycler at 94°C, and incubate for 3 min before starting the PCR.
4. Cycle 40 times:
   - 25 s at 94°C
   - 50 s at 42°C
   - 25 s at 72°C

5. Add 8 µl formamide loading buffer and denature DNA at 94°C for 5 min.
6. Run 6-µl samples on a denaturing 5% polyacrylamide sequencing gel until the Xylene Cyanol FF has migrated 3/4 of the gel's length.
7. Cover wet gel with plastic foil without fixation and autoradiograph overnight. Use fluorescent ink to mark the corners of the gel.
8. Align processed film and gel accurately using the fluorescent spots as markers, cut the bands of interest out of the gel, and transfer to a clean 1.5 ml reaction tube.
9. Add 500 µl of gel elution buffer to each gel slice and place tubes in a 95°C waterbath for 10 min. Leave tubes at room temperature for 16 h.
10. Transfer 200 µl of the gel elution buffer containing the PCR product of interest to a clean 1.5 ml reaction tube, add 1 µl of glycogen (10 µg), and precipitate the DNA as described in steps 7–19 of protocol I (RNA isolation).
11. Redissolve the DNA pellet in 10 µl DEPC-treated H_2O. The PCR products can now be stored at –20°C. A small sample (*e.g.*, 1–3 µl) can be used for reamplification, using the same primers as in the DD-PCR reaction, and subsequent cloning and sequencing. The number of PCR cycles used for reamplification strongly depends on the recovery of the original PCR product from the polyacrylamide gel. Usually, 20–40 cycles are required to obtain enough material for subsequent cloning steps. For direct cloning of reamplified PCR products, we recommend a commercially available kit like Invitrogen's TOPO TA Cloning Kit.

Results

We have applied single-cell differential mRNA display in the identification of genes that are involved in the process of synapse formation between two identified molluscan neurons (Van Kesteren et al. 1996). Following target cell selection, the formation of synaptic connections between neurons is believed to depend upon changes in gene expression in both the pre- and the postsynaptic cell. Knowing the nature of these changes is important for our understanding not only of *neural development*, but also of *neuronal plasticity*, and *regeneration of the nervous system*. Including an aRNA amplification step in the

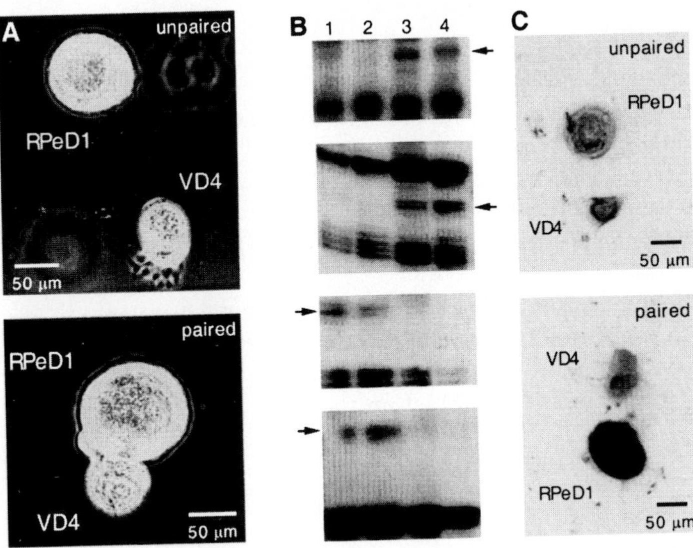

Fig. 5. A: Photographs of identified molluscan neurons RPeD1 and VD4, cultured separately (*upper panel*), and in soma-soma-paired configuration (*lower panel*), in which chemical bidirectional synapses are formed (Feng et al. 1997). **B:** DD-PCR on small pools of individual neurons. Lanes 1 and 2 contain DD-PCR products obtained from 2 independent pools of six unpaired cells, lanes 3 and 4 of six soma-soma-paired cells. Examples of up-regulated (*arrows in the upper two panels*) and downregulated (*arrows in the lower two panels*) gene products are shown. In each panel, at least one gene product is shown of which the expression remains unchanged upon pairing.
C: *In situ* hybridization of unpaired (*upper panel*) and soma-soma-paired (lower panel) cells with one of the upregulated DD-PCR products as a probe, confirming that expression of the corresponding mRNA is indeed induced in RPeD1 upon pairing. The dark staining of the cytoplasm of this cell represents alkaline phosphatase activity, indicating hybridization of the probe with cellular mRNA. (modified from Feng et al. 1997; with permission from the Society for Neuroscience, USA)

protocol significantly enhanced the reproducibility of the differential display patterns obtained. More than 30 up- and downregulated genes were found (for examples, see Figure 5). The identity of these genes is currently being investigated.

Subprotocol 4
Functional Implications of mRNAs in Dendrites and Axons: Metabolic Labeling of Isolated Neurites

Introduction

With the methods that have been described so far, mRNA and rRNA were demonstrated to be present in axons and dendrites (reviewed in Van Minnen, 1994; Steward, 1997). An intriguing question that remains to be answered is whether the mRNAs that reside in the extrasomal domains are indeed translated. The methods described below serve to demonstrate that neurites can indeed sustain protein synthesis, independent of the soma. Several examples of these techniques have appeared in the literature (Crino and Eberwine, 1996, Van Minnen et al., 1997, Bergman et al., 1997).

Materials

Materials and Solutions

- Fixative: 1–4% paraformaldehyde
- Incubation medium: 51.3 mM NaCl, 1.7 mM KCl, 4.0 mM $CaCl_2$, 1.5 mM $MgCl_2$, 0.3 mM glucose, and 1 mM L-glutamine.
- Chloramphenicol (Serva, Heidelberg, Germany)
- Tran-^{35}S label (1175 Ci/mmol; ICN Biochemicals, Irvine, Ca)
- Autoradiographic emulsion (Kodak NTB 2 or Ilford K5, Cheshire, UK)
- Developer (D19, Kodak)
- Fixative for autoradiography: 20% $Na_2S_2O_3$ (sodiumthiosulphate) in water.
- SDS sample buffer: 20% glycerol, 10% 2-mercaptoethanol, 10% SDS.
- Amplify (Ammersham, UK)
- Polyacrylamide gel electrophoresis (PAGE) set-up.

Procedure

Autoradiography

Neurons are cultured for 2 days in conditioned medium, as described in Chapter 10.

1. Separate the neurites from the soma with a sharp microelectrode, remove the soma from the culture dish with an cell-pulling electrode (see Chapter 10)
2. Add chloramphenicol to the culture medium at a final concentration of 0.1 mM to block mitochondrial protein synthesis.
3. Carefully remove the culture medium and wash 6x1 min in incubation medium. Make sure that the neurites are always covered with a small amount of liquid; as soon as they dry out, they will collapse.
4. Label in incubation medium containing 0.5 mCi/ml tran-^{35}S label and 0.1 mM chloramphenicol for 30 min.
5. Wash 6x for 40 min in incubation medium supplemented with 1M unlabeled cystein and methionine and 0.1 mM chloramphenicol.
6. Fix in 1% paraformaldehyde and 1% acetic acid for 2h.
7. Dehydrate through a graded series of ethanol.
8. Dry on air
9. Dip in autoradiographic emulsion
10. Expose 2–7 days; this will depend on the rate of incorporation of the ^{35}S label into proteins.
11. Develop in D19.
12. Rinse briefly in water.
13. Fix in 20% $Na_2S_2O_3$.
14. Wash in water, 20 min.
15. Mount in aquamount or glycerol.

Polyacrylamide Gelelectrophoresis

1. Separate the neurites from the soma and remove soma.
2. Pre-incubate transsected neurites with 0.1 mM chloramphenicol, 30 min.
3. Wash 6x for 1 min in incubation medium.
4. Label in incubation medium containing 0.5 mCi/ml tran-^{35}S label and 0.1 mM chloramphenicol for 2.5h.
5. Wash 6x for 30 min in incubation medium.

6. Harvest neurites in 20 µl SDS sample buffer and transfer to Eppendorf tube.
7. Boil for 5 min.
8. Centrifuge samples, 12,000 rpm, 2 min.
9. Subject supernatant to SDS-PAGE using an 8 % polyacrylamide.
10. Fix gels in 50% methanol and 10% acetic acid for 30 min.
11. Treat gels for 20 min with Amplify.
12. Dry gels.
13. Expose gels to X-ray film (Kodak Biomax MS). Expose from 1 to 3 weeks.
14. Develop film.

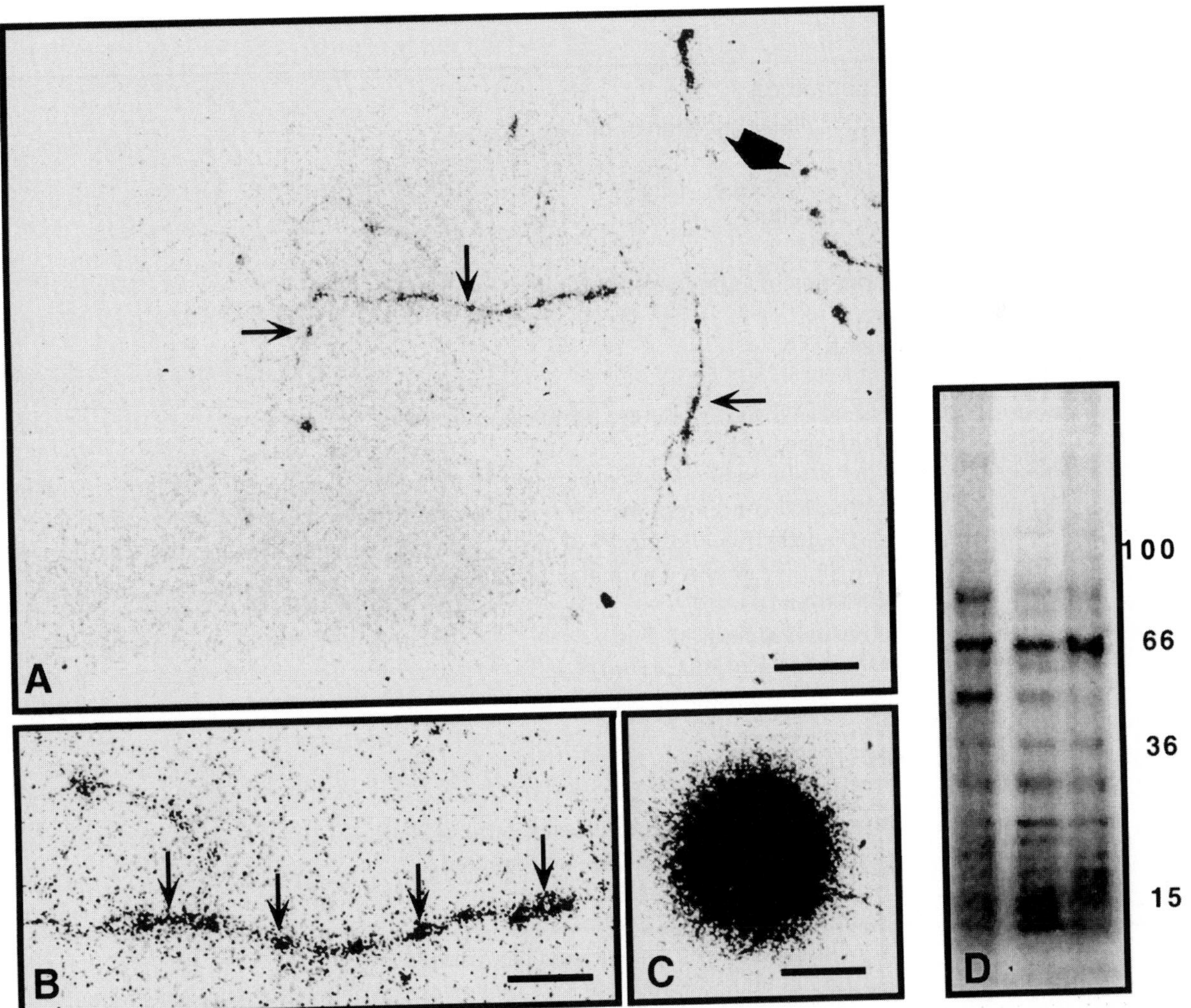

Fig. 6. A: Autoradiograph of isolated neurites of Pedal A cells incubated in tran-^{35}S label and 0.1 mM chloramphenicol for 30 min. The neurites show an autoradiographic signal which indicates that isolated neurites synthesize proteins independently of the soma. Protein synthesis is especially prominent in varicosities (*arrows*) and growth cones (not shown), which are the same subcellular sites where the pedal peptide encoding mRNA was localized (compare Figure 1). The large arrow indicates the site of the soma before it was removed from the dish. Bar is 30 µm. B: Higher magnification of (A) highlighting protein synthesis in the varicosities of the neurites. Bar is 15 µm. C: Autoradiograph of a soma of a Pedal A cell, incubated in the same way as the neurites in (A). The signal is much more intense than that of the neurites, which indicates that most protein synthesis proceeds in the soma. Bar is 30 µm. D: SDS-PAGE using an 8% polyacrylamide gel. Isolated neurites from 10 Pedal A cells were used for each lane. From the autoradiogramm it can be concluded that the isolated neurites produce many proteins ranging in molecular weight from over 100 kD to about 10 kD.

Results

To obtain insight into where in neurites protein synthesis takes place, and the heterogeneity of the proteins produced at these sites, we labeled the isolated neurites with ^{35}S-labeled cysteine and methionine using the tran-^{35}S label. If protein synthesis takes place in the isolated neurites, the radiolabeled amino acids will be incorporated into proteins which can be detected by autoradiographic methods. The results demonstrate that neurites are able to synthesize a large diversity of proteins, and that this process predominantly takes place in varicosities and growth cones (Figure 6).

Subprotocol 5
Intracellular Injection of mRNA

Procedure

Production of Capped mRNA for Intracellular Injections

Materials

- T3 or T7 RNA polymerase (Promega)
- Linearized DNA
- Cap-Scribe buffer, 5 x-concentrated (Boehringer Mannheim). This buffer contains optimized concentrations of ribonucleoside-triphosphates and cap-nucleotide (P^1-5'-(7-methyl)-guanosine P^3 5' guanosine-triphosphate) which ensure that capped mRNA is synthesized with high yield.
- RNaid-kit (Bio 101)
- RNasin (Promega; 40 U/µl)
- DNase I (Gibco-BRL; 1U/µl)

Procedure

1. Mix on ice and make up to a final volume of 20µl:
   - 4 µl Cap Scribe buffer (5 x-concentrated)
   - 0.5 µg of linearized DNA
   - 0.5 µl RNasin.
   - adjust volume to 18 µl with double-distilled water.
   - add 2 µl of the appropriate RNA polymerase to obtain the sense orientation.
2. Incubate at 37°C for 1 h.
3. Add 1 µl of RNase-free DNase I and incubate at 37°C for 10 min.
4. Purify mRNA from the solution using the RNaid kit.
5. Dissolve RNA in 25 µl sterile, RNase-free water.

Preparation of Injection

Materials

- Microelectrodes: coat glass microelectrodes (thin wall, internal diameter 1.5 mm, with filament) with Sigma-coat, and autoclave. Pull on a microelectrode puller to yield a resistance of 10–15 mΩ when filled with saturated solution of K_2SO_4.
- Microelectrode puller (Kopf Instruments, USA).
- Microloader pipette tips (Eppendorf), autoclave before use.
- Micromanipulator (Narashige 110 203)
- Inverted microscope with phase contrast optics.
- Pressure injector (Narashige).

Solutions

(For solutions not mentioned here the reader is referred to the section "*In Situ* Hybridization on Cultured Neurons".)
- Ethylenediamine
- Primary antiserum to the protein encoded by the injected mRNA.
- Fluorescent secondary antibody

For isolation procedure and culture conditions of neurons, the reader is referred to Chapter 12.

Procedure

1. Isolate neurons from CNS with a long segment of their original axon intact.
2. Culture in conditioned medium for 18–48h, until sufficient new neurites have sprouted from the original axon
3. Transect axon about 1 cell diameter away from the cell body with a sharp glass electrode
4. Backfill a glass electrode with mRNA (concentration 20–50 ng/µl) by means of an Eppendorf microloader pipette tip. Only the tip is filled with the mRNA solution (use about 0.3 µl per electrode).
5. Impale axon and inject mRNA into neurite by applying 20 psi pulses of 1–15 s duration. The injected mRNA is in general visible as a clear white droplet under phase contrast illumination.
6. Incubate. The time of incubation may depend on the cell type used and the species of mRNA used. In our experiments, using an mRNA encoding the egg-laying prohormone of *Lymnaea*, already 2h. after injection the translated protein could be demonstrated.
7. Fix and permeabilize cells; follow steps 1 through 3 of ISH procedure (page 63)
8. Wash in TBSGT (page 61), 2x5 min.
9. Incubate in primary antiserum, 2h. at RT, or O/N at 4°C.
10. Wash in TBSGT, 2x10 min.
11. Incubate in secondary antiserum, 1–2h. at RT. For our experiments we generally use a FITC-conjugated secondary antiserum. Alternatively, also enzyme-conjugated antibodies can be employed. For the procedure of an alkaline-phosphatase-conjugate secondary antibody the reader is referred to the section "*In Situ* Hybridization of Cultured Neurons".
12. Wash in TBSGT buffer, 1x5 min.
13. Wash in water, 2x10 min.
14. Mount in 75% glycerol in H_2O, to which 0.5% ethylenediamine is added to slow down fading of the FITC.
15. Observe through an inverted microscope with epifluorescent illumination.

Results

To investigate whether isolated neurites are able to translate mRNA into proteins, we injected mRNA encoding the egg-laying hormone (ELH) prohormone into neurites of Pedal A cells. We had first established that these cells do not express the ELH gene. We

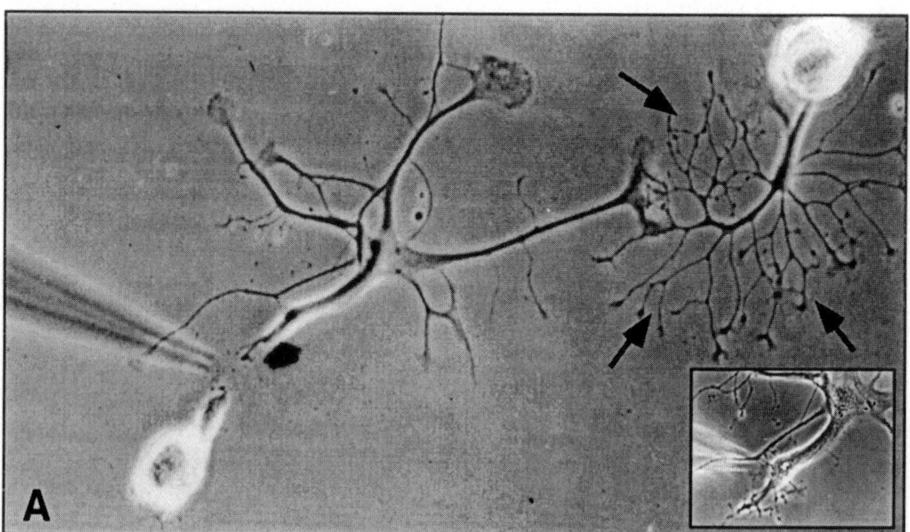

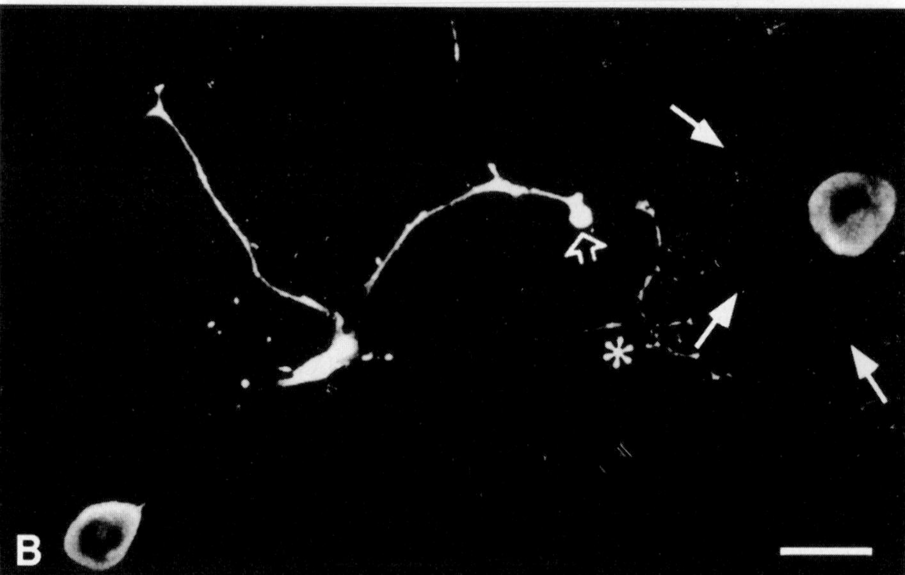

Fig. 7. Translation of ELH mRNA in the isolated axon of a Pedal A cell. A: The axon of a Pedal A cell was transsected using a sharp glass pipette (*large arrow*) and the mRNA was injected into the axon (A, inset). Note that following severance from the cell body and mRNA injection the neurites and growth cones of the truncated axon continued to extend (e.g., see growth cone at open arrow in B). The Pedal A cell in the top left corner was not injected and served as a control for the immunocytochemical reaction. B: Four hours post injection the cells were fixed and processed for immunocytochemistry to detect ELH immunoreactivity, using a FITC-conjugated secondary antiserum. ELH-immunoreactive material was detected in neurites (*asterisk*) and growth cones (*open arrow*). Note that the neurites of the uninjected control cell (indicated by *arrows* in A and B) were completely devoid of ELH immunoreactivity. The faint fluorescent signal of the somata of both pedal A cells is due to autofluorescence of the cell body. Bar = 50 µm.

assayed for translation of the injected mRNA by means of immunocytochemistry using an antiserum directed to ELH. It appeared that isolated neurites readily translate the injected mRNA into an immuno-detectable protein. The translated product was preferentially localized in growth cones and varicosities (Figure 7).

References

Bauer D, Mueller H, Reich J, Riedel H, Ahrenkiel V, Warthoe P, Strauss M (1993) Identification of differentially expressed mRNA species by an improved display technique (DDRT-PCR). Nucleic Acids Research 21: 4272–4280.

Bergman JJ, Syed NI, Kesteren ER van, Smit AB, Geraerts WPM, Van Minnen J (1997) Modulation of local protein synthesis in neurites of identified *Lymnaea* neurons. Soc Neurosci Abstracts 23: 596.

Crino PB, Eberwine J (1996) Molecular characterization of the dendritic growth cone: regulated mRNA transport and local protein synthesis. Neuron 17:1173–1187.

Danscher G (1981) Light and electron microscopic localization of silver in biological tissue. Histochemistry 71: 177–86.

Davis L, Dou P, DeWit M, Kater SB (1992) Protein synthesis within neural growth cones. J Neurosci 12:4867–4877.

Dirks RW (1996) RNA molecules lighting up under the microscope. Histochemistry Cell Biol 106: 151–166.

Edström JE, Eichner D, Edström A (1962) The ribonucleic acid extracted from isolated Mauthner neurons. Biochim. Biophys. Acta 61:178–184.

Edström A (1966) Amino Acid Incorporation in isolated Mauthner nerve fibre components. J Neurochem 13: 315–321.

Egger D, Troxler M, Bienz K (1994) Light and electron microscopic in situ hybridization: Non-radioactive labeling and detection, double hybridization, and combined hybridization-immunocytochemistry. J Histochem Cytochem 42: 815–822.

Feng ZP, Klumperman J, Lukowiak K, Syed NI. (1997) *In vitro* synaptogenesis between the somata of identified *Lymnaea* neurons requires protein synthesis but not extrinsic growth factors or substrate adhesion molecules. J Neurosci 17: 7839–7849.

Karrer EE, Lincoln JE, Hogenhout S, Bennett AB, Bostock RM, Martineau B, Lucas WJ, Gilchrist DG, Alexander D (1995) *In situ* isolation of mRNA from individual plant cells: creation of cell-specific cDNA libraries. Proc Natl Acad Sci USA 92:3814–3818.

Klumperman J, Spijker S, Van Minnen J, Sharpbaker H, Smit AB, Geraerts WPM (1996) Cell type-specific sorting of neuropeptides – a mechanism to modulate peptide composition of large dense-core vesicles. J Neurosci 16: 7930–7940.

Liang P, Pardee AB (1992) Differential display of eukaryotic messenger RNA by means of the polymerase chain reaction. Science 257:967–971.

Livesey FJ, Hunt SP (1996) Identifying changes in gene expression in the nervous system: mRNA differential display. Trends Neurosci 19: 84–88.

Mackler SA, Brooks BP, Eberwine JH (1992) Stimulus-induced coordinate changes in mRNA abundance in single postsynaptic hippocampal CA1 neurons. Neuron 9: 539–548.

Macville MV, Wiesmeijer KC, Dirks RW, Fransen JA, Raap AK (1995) Saponin pre-treatment in pre-embedding electron microscopic in situ hybridization for detection of specific RNA sequences in cultured cells: a methodological study. J Histochem Cytochem 43: 1005–1018.

Maniatis T, Fritsch EG, Sambrook J (1982) Molecular cloning: a Laboratory Manual. Cold Spring Harbor: Cold Spring Harbor Laboratory

Miyashiro K, Dichter M, Eberwine J (1994) On the nature and differential distribution of mRNAs in hippocampal neurites: Implications for neuronal functioning. Proc Natl Acad Sci USA 91: 10800–10804.

Morel G (1993) [Ultrastructural in situ hybridization. Pathol Biol (Paris) 41:187–193.

Morey AL (1995) Non-isotopic in situ hybridization at the ultrastructural level. J Pathol 176: 113–121.

Perrone Capano C, Giuditta A, Castigli E, Kaplan BB (1987) Occurrence and sequence complexity of polyadenylated RNA in squid axoplasm. J Neurochem 49: 698–704.

Steward O (1983) Polyribosomes at the base of dendritic spines of central nervous system neurons: Their possible role in synapse construction and modification. Cold Spring Harbour Symposia on Quantitative Biology 48:745–759.

Steward O (1997) mRNA localization in neurons: A multipurpose mechanism? Neuron 18: 9–12.
Van Kesteren RE, Feng Z-P, Bulloch AGM, Syed NI, Geraerts WPM (1996) Identification of genes involved in synapse formation between identified molluscan neurons using single cell mRNA differential display. Soc Neurosci Abstracts 22: 1948.
Van Minnen J (1994) Axonal localization of neuropeptide-encoding mRNA in identified neurons of the snail *Lymnaea stagnalis*. Cell Tissue Res 276:155–161.
Van Minnen J, Bergman JJ, Van Kesteren ER, Smit AB, Geraerts WP, Lukowiak K, Hasan SU, Syed NI. (1997) De novo protein synthesis in isolated axons of identified neurons. Neuroscience 80: 1–7.

Chapter 4

Optical Recording from Individual Neurons in Culture

Andrew Bullen and Peter Saggau

Introduction

Methods for optically recording dynamic processes in single living neurons must be considered in light of two fundamental questions: *what to record* and *how to record* it. Specifically, deciding *what to record* involves determining a parameter of interest (e.g., membrane potential or ion concentration), the nature of the information required (e.g., qualitative or quantitative) and the optical indicator best suited to making these measurements. Likewise, deciding *how to record* these signals involves consideration of recording methodologies (i.e., photometry or imaging), experimental procedures (e.g., loading and staining protocols) and data processing techniques (i.e., signal processing and analysis). Irrespective of which combination of methods are chosen it is important to understand the essential factors that contribute to obtaining high quality optical signals. By fully understanding the fundamental limits of this recording methodology, the novice investigator should be able to maximize signal quality and effectively solve any technical problems that might arise.

This chapter considers both instrumentation and experimental factors and their implications for making both qualitative and quantitative optical recordings from individual neurons in culture. In particular, appropriate methods for making fast recordings of various physiological parameters, with subcellular resolution, are documented. Two classes of indicators, voltage-sensitive dyes and calcium indicators, are used to illustrate the principles underlying successful optical recording. These principles can easily be extrapolated to other indicator types.

The scope of this chapter is limited to the consideration of methods for making physiological recordings from individual neurons or small group of cells in culture. Therefore, we do not consider methods used to examine fine structural details or localize cellular markers. Nor do we consider the cell culture methods necessary to produce neurons suitable for optical recording (instead, see Chapter 10). However, several cell culture properties that should be optimized to facilitate this kind of recording are documented. Optical recording methods for use in more complex tissues such as brain slices or in vivo preparations are considered in other chapters in this volume. In particular, the reader is referred to Sinha and Saggau (Chapter 16) and Grinvald et al. (Chapter 34).

There are a number of preliminary steps that must be undertaken prior to conducting an experiment employing optical recording methods. The remainder of this section examines several such issues. These include:

- Desirable properties of cell cultures for optical recording
- Methods for visualizing single neurons

A. Bullen, Baylor College of Medicine, present address: University of Conneticut Health Center, Center for Biomedical Imaging Technology, Farmington, CT, USA
Correspondence to: P. Saggau, Baylor College of Medicine, Division of Neuroscience, One Baylor Plaza, Houston, TX, 77030, USA (phone: +01-713-798-5082; fax: +01-713-798-3904; e-mail: psaggau@bcm.tmc.edu)

- Choice of indicator(s) (e.g. parameter, type of signal)
- Choice of recording technique (e.g. photometry or imaging).
- Choice of instrumentation (including light sources, optics and detectors).

Desirable Properties of Cell Cultures for Optical Recording

Neuronal cell cultures can take many forms depending on the cell type and culture conditions employed. Unfortunately, the conditions that favor good optical recordings are quite stringent and sometimes require modified culture methods. This is especially true in cases where indicators are bath-applied or subcellular resolution is required. In these cases the density of neurons or neuronal processes (i.e., axons and dendrites) should be sufficiently low to allow clear identification of the signal source. This is normally achieved by producing a single layer of cells at low density (e.g., ~200 cells per mm^2). Additionally, the number of non-neuronal cells in the culture (e.g., glia) must be kept to a minimum as they are also stained by optical indicators and can easily generate non-specific optical signals. Finally, cells should be cultured on a substrate optically compatible with high numerical aperture objective lenses and the contrast-enhancement techniques documented in the next section (i.e., < 150 μm thickness). In particular, while some plastic substrates have the preferred surface on which to grow some types of neurons, they must also be thin enough to accommodate the working distance of the objective lens. In addition, these substrates should be non-polarizing; otherwise they can disrupt image formation in cases where polarization-dependent contrast enhancement techniques are used (see below).

Methods for Visualizing Single Neurons

Single neurons in culture possess inherently low visual contrast. This is especially apparent with traditional brightfield illumination. In cell culture, phase contrast microscopy is commonly used to identify and assess the viability of these neurons. Typically, healthy neurons are very phase-bright, while unhealthy cells and glia are phase-dim. However, transillumination with phase contrast optics does not necessarily provide the high spatial resolution or the perception of depth appropriate to distinguish between cells or amongst small cellular structures. Hence other illumination methods are often used to select cells or parts of cells for optical recording. Some examples of these other methods are:

- Normarski or Differential Interference Contrast (DIC)
- Hoffman Modulation Contrast (HMC)
- Varel Contrast or variable relief contrast (VC)

While HMC is relatively inexpensive, easy to implement and insensitive to birefringence, its optical sectioning capabilities are typically less than DIC. However, DIC performs poorly with highly birefringent structures such as myelinated axons. Moreover, as DIC uses polarized light, care needs to be taken with plastic dishes or coverslips because their anisotropic nature can seriously degrade image quality. Varel relief contrast is a recent innovation and currently is available from only one vendor. Like HMC, VC is relatively low cost and easy to implement. However, it is more difficult to combine with epifluorescence optics. For more detailed information on these methods see Spector et. al. (1998).

Choice of Indicator

One advantage of optical recording techniques is the variety of different physiological parameters (e.g., ion concentration, membrane potential and second messengers) that can be examined. However, for each of these parameters there are a vast array of seemingly similar indicators to choose from. For instance, within each indicator class, a range of dyes are available based on a number of properties. These properties can include:

- Mode of application (bulk loading vs. single cell);
- Spectral properties (UV vs. visible excitation);
- Absolute binding affinities (for ions or membranes);
- Special properties (e.g., near-membrane dyes or low-leakage or dextran conjugates).

Quantitative vs. Qualitative Indicators

Prior to undertaking studies using optical indicators a number of important choices must be made regarding the indicator to be employed. These decisions are typically based on the underlying experimental objectives and the likely signal characteristics. For instance, an investigator must decide whether a qualitative result is sufficient or a quantitative result is required and therefore ratiometric methods should be employed. To aid the reader in this decision making process a simple flowchart is presented below (Figure 1.).

Other more advanced considerations (*e.g., sensitivity, brightness, and photostability)* that are important in choosing between these seemingly similar indicators are discussed in a later section ("Criteria for Comparing Indicators"). Two excellent sources of information about currently available optical indicators are Haugland (1996) and Johnson (1998).

Fluorescence vs. Absorbance

A variety of optical parameters can be measured and related to a physiological variable of interest. For instance, some indicators require that fluorescence be monitored while others are best assayed with absorbance methods. In some instances, both fluorescence and absorbance can be measured from the same indicator. Still other indicators are luminescent and require no excitation at all.

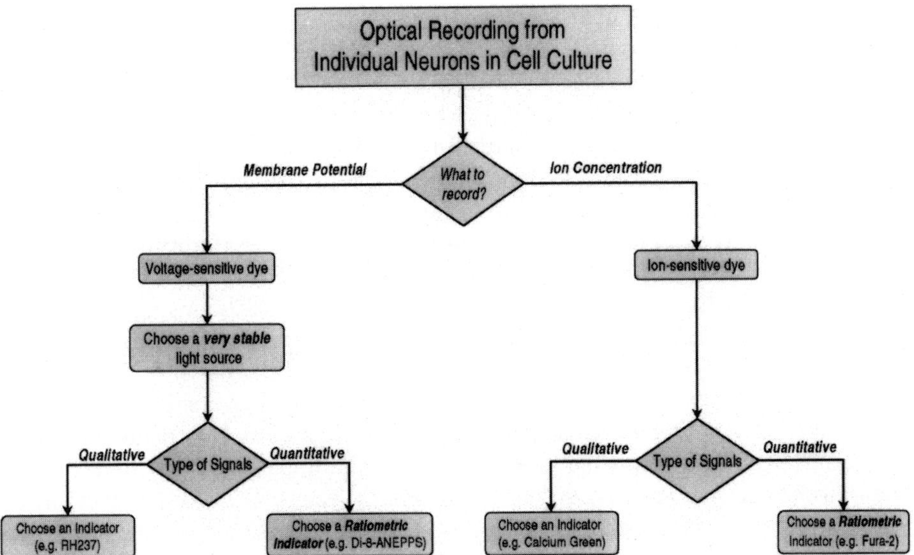

Fig. 1. Decision making scheme to determine "What to record".

In general, when optical signals are assayed from small structures or when a low concentration of indicator molecules is used, fluorescence methods are preferred. This is especially true in experiments utilizing cell cultures where fluorescence methods are almost always favored over other possibilities. In this case, fluorescent signals are usually superior to those recorded with absorbance approaches because of the limited surface area and/or volume of subcellular structures from which the signal is generated. Furthermore, the signals from fluorescent dyes ($\Delta F/F$) are typically larger than corresponding signals ($\Delta I/I$) from absorbance indicators and thus the instrumentation requirements are less rigorous. In light of these facts, subsequent discussions are limited to fluorescent indicators.

The choice of optical indicator will, in part, dictate which light sources, optical configuration and detector(s) are employed. Different classes of optical indicators exhibit different characteristics and require optimization of different parameters. In this chapter we have chosen to examine two fundamentally different kinds of fluorescent optical indicators: *voltage-sensitive dyes* and *Ca indicators* (see below). Each of these indicators presents a distinctly different set of instrumentation requirements and methodological problems.

Voltage-sensitive Dyes

Optical indicators that are sensitive to membrane potential are commonly referred to as *voltage-sensitive indicators* or *voltage-sensitive dyes (VSDs).* Two classes of VSDs are recognized based on their response time to changes in membrane potential. They are referred to as either *slow-response VSDs* or *fast-response VSDs.* Slow-response VSDs are also called *redistribution dyes.* Typically these membrane-permanent dyes partition between extracellular and intracellular compartments in a manner dependent on membrane potential. In general, redistribution dyes exhibit high sensitivity to membrane potential ($\Delta F/F > 10^{-2}$ per 100 mV). However, the slow response time of these dyes (usually in the second range) caused by their electrodiffusion across the membrane, strongly limits their usefulness for measuring neural activity. Fast-response VSDs are typically amphipathic and *membrane-impermanent* dyes that attach to membranes and change either their orientation or parts of their structure in response to a change in the electrical field. Such structural changes cause these molecules to exhibit a change in the *fluorescence.* In general, fast-response VSDs show quite small changes in fluorescence with respect to a change in membrane potential (10^{-4}-10^{-2} per 100mV). However, these indicators have fast response times (usually in the μsec range) which make them attractive for measuring neural activity. Except for some dyes that exhibit voltage-dependent spectral shifts, the absolute calibration of fast-response VSD signals is difficult. Dyes exhibiting voltage-dependent spectral shifts are commonly designated *electrochromic* and allow ratiometric measurements and therefore absolute calibration.

Deciding between different VSDs within the same class is a difficult task. In addition to differences in absolute voltage-sensitivity and brightness, most VSDs exhibit species and cell type differences in voltage sensitivities and membrane affinity (Ross and Reichardt, 1979). Often the best voltage-sensitive dye for a particular application is determined empirically. Some of the most widely used fluorescent VSDs for single cell studies are documented in Table 1.

Ca Indicators

Most modern *calcium-sensitive dyes* (CaSD) are *tetracarboxylic dyes* which were derived from the *calcium buffer* BAPTA. In fact, a large family of fluorescent calcium indicators has been created by conjugating BAPTA with different fluorophores. Members of this family, which includes Fura-2 and Fluo-3, are highly selective for calcium over other cations. Two types of measurements are typically made with these modern calcium indicators: qualitative or quantitative. Different indicator molecules are required for each type of measurement. Qualitative measurements reflect changes in calcium levels without

Table 1. Fluorescent Voltage-sensitive Dyes Used in Single Cell Studies

VSD	Site of Application	Structure	Signal Size (per 100 mV)	Relative Membrane Affinity	Refs.
RH 237	Extracellular	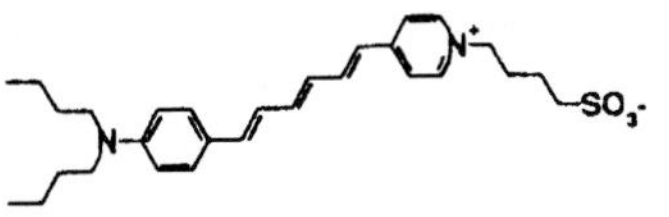	2 %	Low	1, 2
RH 421	Extracellular	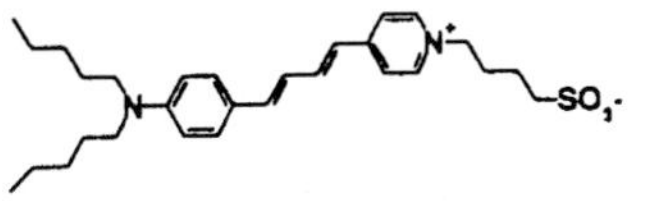	5 %	Moderate	3
di-4-ANEPPS	Extracellular	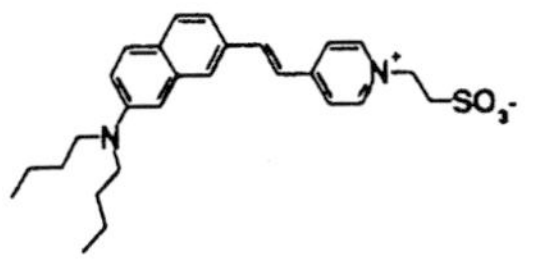	10 %	Moderate	4
di-8-ANEPPS	Extracellular	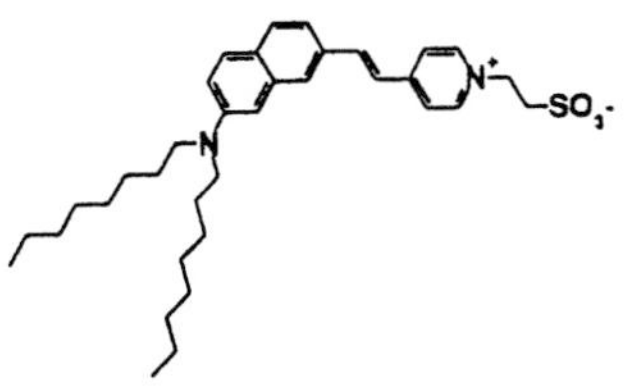	20 %	High	5, 6
di-2-ANEPEQ	Intracellular	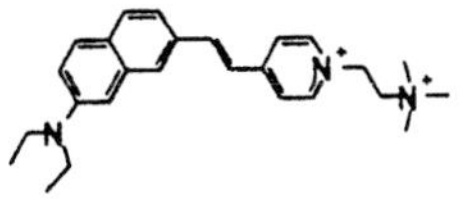	3 %	Low	7,8

Pine (1991a) (2) Chien and Pine (1991b), (3) Meyer et al. (1998), (4) Kleinfield *et al.* (1994), (5) Rohr and Salzberg (1994), (6) Bullen and Saggau (1997) (7) Antic and Zecevic (1995), (8) Zecevic (1996).

reference to resting levels or the absolute size of these changes. This kind of measurement is normally depicted as the change in fluorescence normalized by the overall mean fluorescence (ΔF/F). In contrast, quantitative measurements are made ratiometrically and give an estimate of absolute calcium changes. Ratiometric measurements are particularly useful because they inherently eliminate distortions caused by photobleaching, variations in probe loading and retention, and by instrumentation factors such as long-term illumination instability. However, this kind of measurement typically requires a post-experiment calibration. Commonly, indicators employed for ratiometric determinations undergo calcium-dependent spectral shifts while qualitative indicators simply change their brightness in proportion to bound calcium. CaSDs are also distinguished by their *binding affinity* and *relative sensitivity*. The *binding affinity* indicates their sensitivity for Ca ions and is described by the dissociation constant (K_d). In contrast, the *relative sensitivity* indicates the magnitude of their fluorescence change to fluctuations in calcium concentration. This parameter is normally given as a ratio of calcium-bound and calcium-free levels. When compared to VSDs, calcium indicators typically produce a much larger change in ΔF/F (i.e., 10^{-2}-10^{0}) and consequently are less affected by source noise (i.e., variations in illumination intensity from the light source that are directly reflected in the resulting fluorescence).

Some fluorescent CaSDs commonly used in single cell studies are documented in the adjacent table (Table 2.).

Table 2. Fluorescent Calcium-sensitive Dyes Used in Single Cell Studies

CaSD	Ratiometric	Structure	Ca Affinity (K_d)	Relative Sensitivity	Refs.
Indo-1	Emission	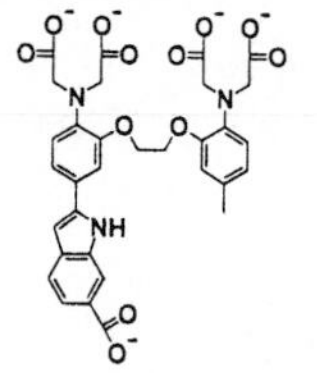	Std. - 230 nM 1EF - 33 μM	R_{max}/R_{min} 20 20	1, 2
Fura-2	Excitation	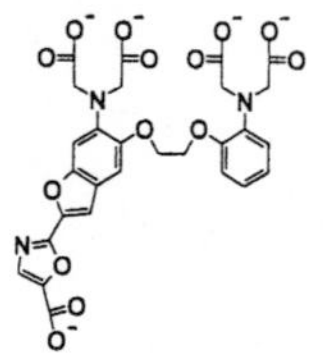	Std. - 145 nM 2FF - 35 μM	R_{max}/R_{min} 45 45	1, 2
Fluo-3	No	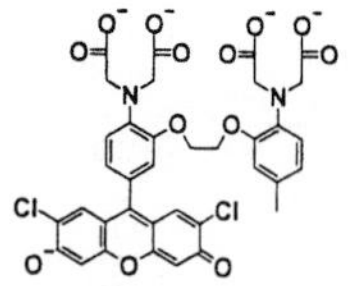	Std. - 390 nM 3FF - 41 μ	F_{Ca}/F_{Free} 200 120	2, 3
Calcium Green	No	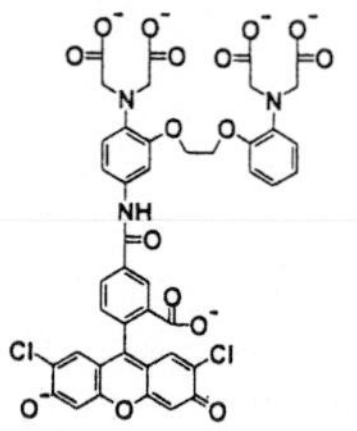	1N - 19 nM 2N - 550 nM 5N - 14μM	F_{Ca}/F_{Free} 14 100 38	2, 4
Oregon Green	No	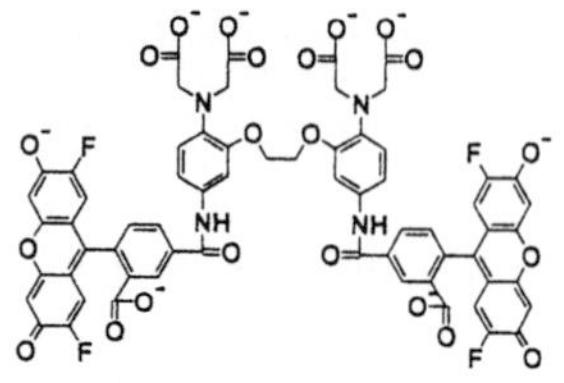	1–170 nM 2 - 580 nM 5 - 20μM	F_{Ca}/F_{Free} 14 100 44	2
Calcium Orange	No	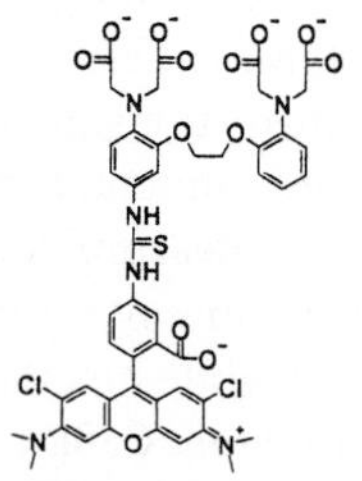	1N - 185 nM 5N - 20μM	F_{Ca}/F_{Free} 3 5	2, 4

Molecular structures shown refer to Indo-1(Std), Fura-2(Std), Fluo-3(Std), CaGn-1N, OrGn-488-BAPTA-2 and CaOr-5N. (1) Grynkiewicz et al. (1985), (2) Haugland (1996) (3) Minta et al. (1989), (4) Eberhard and Erne (1991).

Dual Indicator Studies

In principle, measurements with two or more optical indicators could be made simultaneously from the same tissue using two or more dyes. However, simultaneous recording of two or more physiological signals from the same point in space and time is considerably more difficult than examining a single parameter alone (for details see Bullen and Saggau, 1998 and Morris, 1992).

Choice of recording technique

A number of recording techniques are available to obtain optical measurements from single neurons. Typically, an investigator must choose between some form of photometry and imaging. In pure photometry, a continuous measurement is made from the whole field or a subset of the field defined by a fixed aperture. While in imaging approaches, the signal is included in a series of images recorded at equally spaced intervals in time. Different variations of these basic techniques are possible and some of the more common types are depicted schematically in Figure 2.

Each of these approaches possesses various advantages and disadvantages depending on the experimental objectives. A comparison of these approaches and their compatibility with various types of optical indicators is documented in the table below (Table 3).

High-speed, random-access, laser scanning microscopy

In its simplest form, photometry involves a single measurement from a predefined area. As shown in Figure 2, this area can encompass a whole cell, parts of a single cell or

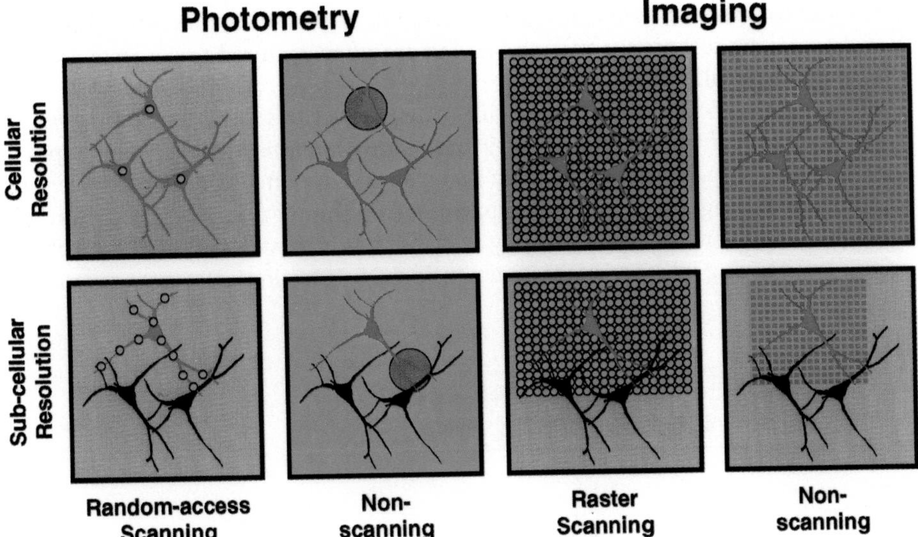

Fig. 2. Different types of Optical Recording Schemes Used in Single Cell Studies. Note: the cell(s) stained with optical indicator are always shown in red.

Table 3. Comparison of Optical Recording Methods for Use with Single Neurons

	Photometry		Imaging	
	Non-scanning (e.g., PMT or single photodiode)	Scanning (e.g., random access scanning microscopy)	Non-scanning (e.g., imaging detector)	Scanning (e.g., confocal microscopy)
Spatial Resolution	None	High	High	High
Temporal Resolution	High	High	Low to moderate	Low
Compatible with CaSDs	Yes	Yes	Yes	No
Compatible with VSDs	Yes	Yes	No	No

parts of multiple cells. The obvious advantage of this approach is recording speed which can be very high. However, simple photometry only provides information from one site or area at a time. An alternative form of photometry with spatial resolution is "scanning photometry." In this approach optical recordings are made from multiple interlaced recording sites with a scanning light source. One implementation of scanning photometry is "high-speed, random-access, laser scanning microscopy". In this composite approach image capture and optical recording function are performed separately to gain temporal bandwidth and/or spatial resolution. Through the use of a very fast scanning scheme based on acousto-optical deflection this method repeatedly samples a series of predetermined scanning sites, with high digitizing resolution and at rates compatible with the fastest physiological events. Thus, this approach is able to optimize both spatial and temporal resolution. For more details about this approach see Bullen et al. (1997).

Imaging applications can also be divided into scanning and non-scanning classes. Non-scanning approaches typically use a scientific grade video camera (i.e., cooled CCD) or a lower spatial resolution photodiode array. The spatial resolution of these systems is dependent on the number of pixels per dimension which can be quite high (i.e., 1024). However, the drawback of many of these systems is their poor temporal resolution. Moreover, the full range of spatial resolution cannot always be utilized because spatial averaging techniques such as binning are often required to generate a useful physiological signal.

An alternative to camera-based imaging systems is various forms of scanning *microscopy*; in particular, *confocal microscopy* and *multi-photon microscopy*. Many types of confocal microscopes are commercially available and are often considered for optical recording from single neurons. In comparison, two-photon microscopy is a relatively recent innovation and there are still only a few systems in existence. Both these technologies allow imaging in complex three-dimensional preparations. However, several fea-

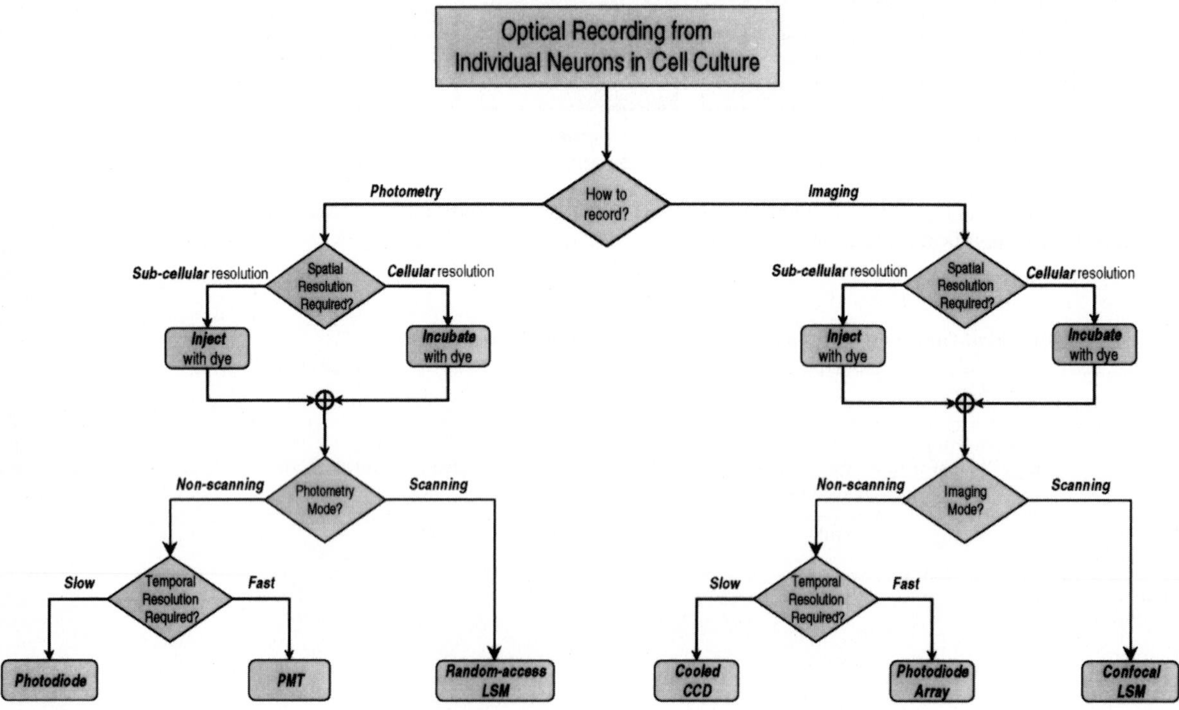

Fig. 3. Decision making scheme to determine "How to record".

tures of these instruments limit their suitability for optical recording from single neurons. Most obviously, these instruments are almost exclusively imaging devices with limited overall frame rates and therefore they are not very useful for recording fast physiological signals. Moreover, commercial versions of these instruments commonly employ only 8-bit digitizing resolution, which is sometimes sufficient for CaSDs but is generally insufficient for use with VSDs. Finally, the considerable cost of these instruments is a significant issue for the novice investigator. Furthermore, because cultured neurons are typically a monolayer of cells, there is little benefit to be gained by adding an optical sectioning capability or from sub-micron spatial resolution. To aid the reader in deciding between this array of recording possibilities the following flowchart was derived to simplify the process (Figure 3).

Instrumentation

Correct instrumentation choices are important. In particular, it is critical to optimally illuminate the preparation to ensure success. Furthermore, collecting the maximum amount of emission light available ensures the highest possible signal quality. These instrumentation choices include consideration of:

- Light sources
- Optics
- Photodetectors

Light Sources

Important factors in determining the best light sources include:

- Brightness
- Spectral distribution
- Intensity stability

The most common light sources for optical imaging are:

- Tungsten halogen bulbs
- High pressure gas discharge bulbs
- Lasers

Tungsten halogen bulbs exhibit the highest intensity stability ($\Delta I/I = 10^{-5}$-10^{-4}). The spectrum of tungsten bulbs can be regarded as mostly white light, with a weak emission in the UV. *High-pressure gas discharge bulbs* such as Hg- or Xe-burners are light sources with much higher intensity but their amplitude noise is somewhat larger (i.e., 10^{-4}-10^{-3}). While Hg-burners emit an inhomogeneous spectrum with many peaks down to the UV, the spectrum of Xe-burners is quite homogeneous. If Hg-burners are chosen for fluorescent applications, care should be exercised to ensure that the absorbance spectra of the fluorophores and the associated excitation filter correspond to a known Hg emission line (e.g., 365, 405, 436, and 546 nm). *Lasers* are an increasingly attractive light source. In particular, their ability to generate a diffraction-limited illumination spot makes them a perfect high intensity light source for various scanning applications. Unfortunately, the relative noise of lasers is commonly quite high, usually in the 10^{-3} –10^{-2} range. However, some recently introduced lasers with improved cavity designs have performed much better in this regard (i.e. 10^{-5}-10^{-4}). In addition, the coherence of laser light can give rise to interference-based speckles that adds to the total noise, thus making the use of lasers for absorbance measurements difficult. Lasers inherently emit monochromatic light but only at a discrete number of wavelengths (or lines). Thus, if a laser source is to be used, indicators must be chosen that match the available laser lines.

Optics

Optical considerations include:

- Type of microscope (upright or inverted)
- Objective lens and condenser

- Filters
- Dichroic mirrors

I. Type of microscope: The type of microscope that is typically considered for optical recording from single neurons could in principle be either upright or inverted. However, inverted microscopes are commonly chosen because they allow the use of high magnification, high numerical aperture objective lenses.

Likewise in fluorescence applications, an epifluorescence configuration is normally chosen over transfluorescence. There are several logical reasons for this choice. Firstly, with epifluorescence illumination, the excitation light and emitted fluorescence travel in opposite directions and are easily separated by a dichroic beam splitter. Secondly, the objective also serves as the condenser, assuring perfect alignment and maximal illumination and collection efficiencies. Epifluorescence illumination in combination with an inverted microscope also allows easy access with micropipettes. Finally, epifluorescence illumination is most easily combined with other transmitted light techniques such as phase contrast or DIC. A typical epifluorescence-recording configuration demonstrating these aspects is shown in Figure 4.

II. Objective lens and condenser: The objective lens is perhaps the most critical component in any microscope. It governs, among other things, the resolution, magnification and light gathering capabilities of the total system. In choosing an appropriate objective for optical recording from single neuron in culture the following features are critical:
- Magnification
- Numerical aperture
- Working distance
- Optical corrections

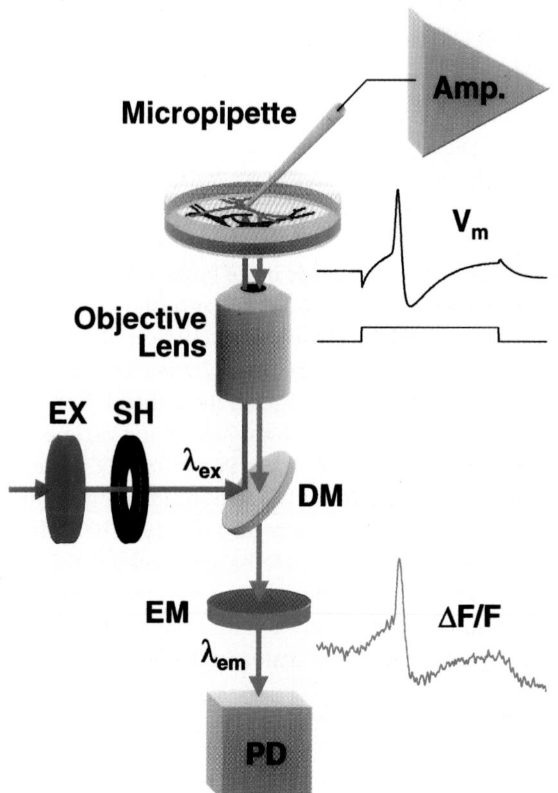

Fig. 4. Typical Optical Recording Configuration with Epifluorescence Illumination as Implemented on an Inverted Microscope.
EX = excitation filter,
SH = shutter,
λ_{ex} = excitation wavelength,
DM = dichroic mirror,
EM = emission filter,
λ_{em} = emission wavelength,
PD = photodetector,
Amp. = amplifier

Magnification: The magnification required to accurately measure fluorescence from a single neuron or parts of individual cells with spatial resolution corresponding to the structures of interest is typically 40x. Other commonly used objective lenses include 63x or 100x. It is important to remember that any additional magnification secondary to the objective, while increasing image size, does not add resolution or light gathering capability and therefore should be considered "empty magnification".

Numerical aperture: In addition to its resolution, the numerical aperture (N.A.) of an objective lens determines its collection efficiency and hence the image brightness. Image brightness is proportional to the *brightness of illumination* and the *light-gathering power of the objective.* Both parameters are determined by the square of the N.A. of the lens and therefor in an epifluorescence configuration there is a 4th power relationship between image brightness and N.A. As a consequence, a relatively small change in numerical aperture can significantly change image brightness and thus signal strength in optical recording.

Working distance: Working distance (W.D.). is an important parameter listed in the specification of all lenses and refers to the free space between an *objective lens* and the specimen. It becomes a major issue with upright microscopes when imaging approaches are combined with micropipette applications. W.D. specifications are normally given by manufacturers as the distance above a normal thickness coverslip (#1). W.D. is inversely proportional to numerical aperture and thus the resolving power and collection efficiency of high N.A. lens normally come with a W.D. penalty.

Optical corrections: High quality objective lenses are commonly corrected for spherical and chromatic lens aberrations. While many of these corrections are important in image formation, they are in large part inconsequential for optical recording where other factors are limiting and most manufacturers have developed objectives specialized for fluorescence recording. However, it is important for an investigator to know the limitations of the various objective lens choices and understand the specifications given for each lens type.

Another important factor in choosing an objective lens is whether it is designed to act as a fixed tube-length or infinity-corrected lens. A fixed tube-length objective lens directly projects a real image within the microscope, normally at 160 mm from the back focal plane. In contrast, an infinity-corrected lens projects this image towards infinity. The advantage of an infinity-corrected lens arises from the reduced number of subsequent relay lenses required in the optical path of a typical compound microscope. Fewer lenses in any given light path yield fewer internal reflections and higher relative transmission. This is especially true in cases such as DIC microscopy or epifluorescence microscopy where additional prisms or filters are inserted into the optical path.

Typically all these parameters are best met with high power, high N.A., immersion lenses specialized for fluorescence recording. Each microscope vendor offers one or more objective lenses that are specialized for recording fluorescence. Currently we use the Zeiss Fluar 100x (N.A. 1.3) objective but many other compatible lenses are available from Nikon, Leica, Olympus, etc. The condenser is of no practical importance for epifluorescence microscopy. However, it plays an important role in image formation with transillumination and for this reason, the condenser selected should be closely matched to the numerical aperture of the objective lens. This ensures that the resolving power of the objective is not limited by that of the condenser.

III. Filters and dichroic mirrors. The spectral properties of filters (i.e. excitation and emission) and the dichroic mirror(s) determine:

- The strength and appropriateness of excitation light.
- The relative separation between excitation and emission light.
- The collection efficiency for emission light.

Together these factors are strong determinants of signal strength and therefore signal quality. Under ideal conditions the excitation and emission filters (also known as exciter and emitter) chosen should be centered on the dye's respective absorption and emission peaks. To maximize the signal strength it is advantageous to employ filters that also have wide bandwidths. However, this strategy requires a sufficient separation between the excitation and emission spectra of the dye; otherwise optical cross-talk can occur between channels. A commonly used alternative is to employ a relatively narrowband excitation filter, because source light is plentiful and can be easily increased. Single laser lines are also an attractive excitation source for the same reason. The extra spectral separation achieved in these cases can then be used by the emission filter which should be as wide as possible to ensure all emission light is collected. In most cases the spectral properties of fluorescent indicators used in optical recording from single neurons are well characterized and standard filter sets are commonly available. A selection of these are documented in the following table (Table4.). In special cases where novel filter designs are required, several manufacturers, notably Chroma or Omega (both of Brattleboro, VT), are proficient in producing customized optical elements.

Note: In some cases the signal size ($\Delta F/F$) of selected VSDs can be made larger by employing excitation and emission filters that do not correspond to the spectral maximum of the dye. This phenomenon arises from the voltage-dependent spectral shifts that occur in addition to the absolute amplitude changes caused by depolarization. For example, while the excitation and emission peaks of di-8-ANEPPS are 476nm and 570nm, respectively, Rohr and Salzberg (1994) found the best signal-to-noise ratio with this dye could be achieved with a 530(25) exciter, a DCLP560 dichroic mirror and a OG570 emission filter.

Table 4. Filters, Dichroic Mirrors and Laser Lines for Fluorescent Dyes Commonly Used in Single Cell Studies.

Dyes		Excitation (nm)			Dichroic Mirror	Emission (nm)	
		Peak	Filters (FWHM)	Laser Lines		Peak	Filters (FWHM)
Calcium Sensitives Dyes	Indo-1	346	350 (20)	351, 354, 355	DCLP379, DCLP455	401, 475	400 (40), 480 (40)
	Fura-2	340, 380, (363)	340 (20), 380 (20)	334, 364	DCLP430	512	510 (60)
	Fluo-3	503	490 (20)	488	DCLP505	525	525 (30)
	Calcium Green	506	490 (20)	488	DCLP505	530	530 (40)
	Oregon Green	496	490 (20)	488	DCLP505	524	530 (40)
	Calcium Orange	549	540 (20)	532	DCLP560	575	580 (30)
Voltage-Sensitive Dyes	RH 237	506	500 (40)	488, 514	DCLP560	687	OG610
	RH 421	493	500 (40)	488	DCLP550	638	OG590
	di-4-ANEPPS	476	470 (40)	476, 488, 514	DCLP570	605	OG570
	di-8-ANEPPS	476	470 (40)	476, 488, 514	DCLP565	600	OG570 or 540 (60) & 600 (60)
	di-2-ANEPEQ	497	500 (40)	488, 514	DCLP575	640	OG570

Photodetectors

A substantial role in the overall performance of an optical recording system is played by the photodetector employed, and this component should be selected carefully. Detection devices can range from a single element photodetector used in the simplest forms of photometry to high resolution, scientific-grade cooled CCD cameras used in advanced imaging applications. In deciding between different detector types the main parameters of concern are:

- Sensitivity
- Quantum efficiency (QE)
- Dark noise
- Dynamic range
- Spectral response
- Cost

In imaging detectors, *readout speed* and *digitizing resolution* are also important considerations.

I. Non-imaging detectors The most prevalent single element photodetectors are *photodiodes* and *photomultipliers*. Such detectors have no spatial resolution but are still important for many imaging applications. For instance, in scanning microscopy, where an image can be produced with point illumination by sequentially scanning a whole preparation or parts thereof. Semiconductor *photodiodes* are an attractive photodetector option due to their high dynamic range, high quantum efficiency (>90%) and low cost. Despite their somewhat higher complexity and cost, *photomultiplier tubes* (PMT) are particularly attractive for scanning applications because they possess both sensitivity and response speed. These devices are an integral combination of a vacuum photodiode and a multistage photocurrent amplifier that makes use of multiple amplification stages to generate secondary photoelectrons. While the internal gain of a photomultiplier can be quite high (the gain increases exponentially with the number of stages), the true quantum efficiency is much less when compared to a semiconductor photodiode (about 10%, meaning that only every 10th photon generates a photoelectron). In addition, this internal gain also amplifies the dark noise. In comparing photodiodes and PMTs, it is somewhat difficult to make a global statement about which detector is better because it depends on the amount of light to be detected and the bandwidth required. Above a certain light level, the photodiode will always outperform the PMT and will only be constrained by shot noise limitations. However, below this level, the dark noise of the photodiode and its accompanying electronics will become dominant and the PMT will produce a better signal-to-noise ratio. This general principle is confounded by bandwidth considerations. The response speed of the photodiode at low light levels is limited by the time constant imposed by the large feedback resistors (i.e., GΩ range) required in the current-to-voltage conversion process. This is not a problem for the typical PMT because of the current amplification in each internal gain stage. Thus, in the case where both sensitivity (i.e., low light levels) and speed (i.e., scanning applications) are required, the PMT is probably superior to the photodiode.

II. Imaging detectors Various types of *video cameras* are the most common imaging detectors for optical recording. However, due to the requirements imposed by both the optical indicators and the speed of the signals to be measured, only very few imaging detectors are really suited to record neural activity. A normal video camera provides 30 frames/second and has a maximal intensity resolution of ~0.5% which is insufficient for optical imaging of neural activity with VSDs that requires a detector that supplies 10^3 frames/second and the ability to resolve signals that are 10^{-4} of the static light intensity. One example of a scientific grade imaging camera that is often used in neurobiology is a frame-transfer cooled CCD camera. This device possesses high sensitivity, low noise but only modest frame rates. A more appropriate imaging detector for recording fast neuronal activity is the low resolution *photodiode matrix (PDM)*. This photodetector can be regarded as an *array of single photodiodes*. Consequently, the favorable quantum efficiency and high sensitivity of photodiodes also applies to the photodiode matrix. The array sizes in these devices vary from 5x5 to 128x128 elements, with 10x10 or 12x12 being the most common. For very fast applications true *parallel access* can be used with arrays of up to 32x32 elements. Each photodiode is connected to its individual *current-to-voltage converter*, which can allow for gap-free recording (i.e., no shift or readout time delay). As was the case with a single photodiode, the bandwidth of such imaging detectors depends largely on the amount of light and the required signal-to-noise ratio. For a more detailed discussion of these issues see section "Data Acquisition and Digitization Issues".

The following table (Table5.) summarizes the relative merits of each photodetector.

Table 5. Comparison of Photodetector Properties

	Sensitivity	Quantum efficiency	Dark noise	Dynamic range	Readout speed	Spatial resolution	Cost
Photodiode	+	+++	-	+++	N / A	1	$
Photomultiplier	+++	+	--	++	N / A	1	$$
Photodiode matrix	+	+++	-	+++	+++	16x16	$$$
CCD (ccoled, frame-transfer)	++	++	---	++	+	512x512	$$$

Note: Quantum efficiency is defined as the ratio of photons detected over total number of incident photons. Sensitivity is defined as a measure of the minimum amount of detectable light.

Outline

The important steps in undertaking optical recording from single neurons can be divided into three parts:

- Instrumental design and construction
- Experimental design and implementation
- Signal analysis and presentation

Most of the instrumental considerations were addressed in the previous section and outlined in the flowcharts in Figure1 and 3. The following sections document important considerations for the remaining two areas.

Materials

Cells: Previously prepared and plated at an appropriate density (i.e. 200–300 cells per mm^2).

Solutions: Physiological saline(s) and drugs depending on experimental objectives.

Optical Indicators: Stock solutions ready for each specific application. Optical indicators are available form a number of sources with the best being Molecular Probes (Eugene, OR). See Haugland (1996).

Optical Recording System: Including microscope, light source, filters, epifluorescence optics and detector(s).

Auxiliary Electrophysiological Equipment: Such as stimulators, amplifiers, perfusion system and micromanipulators (cf. Chapter 5).

Data Acquisition System: Including computer, A/D and D/A plug in boards and data storage devices (cf. Chapter 45).

Procedure

This section addresses a number of methodological issues or experimental techniques important for successful optical recording. No single *procedure* is presented because experiments using optical recording methods can be quite heterogeneous in nature. Instead, those elements that are common to all experiments are considered. Many of these elements are quite mundane but often represent the difference between successful experiments and unnecessary frustration. These considerations include:
- Mixture and storage of optical indicators
- Loading and staining protocols
- Experimental design issues
- Calibration procedures
- Signal processing methods

Mixture and Storage of Optical Indicators

Solubility and storage of VSDs

There is no universal way to solubilize and store these indicators. In most instances the optimal conditions are empirically determined for each dye. Because of their amphipathic nature most VSDs are not inherently water-soluble and external agents are sometimes required to solubilize them. In addition, other agents are sometimes required to aid in partitioning these dyes into membranes. Various solvents or combination of solvents and other external agents have been proposed to achieve these tasks.

These include:
- Ethanol (EtOH)
- Methanol (MeOH)
- Dimethyl Sulfoxide (DMSO)
- Dimethyl Formamide (DMF)
- Pluronic F-127
- Bile salts (e.g. sodium cholate)
- Stained vesicles

A few examples of ways to solubilize and store VSDs commonly used in single cell studies are given below.

Example 1: di-8-ANEPPS in DMSO/ F-127 (after Rohr and Salzberg, 1994; Bullen et al., 1997). One vial (5 mg) of di-8-ANEPPS (#D-3167; Molecular Probes, Eugene, OR) is dissolved with 625 µl of a Pluronic F-127/ DMSO solution (25% and 75% w/w resp.) for a final concentration of 8 mg/ml or 13 mM. Aliquots of 12.5 µl (i.e., single experiment size) are made. These aliquots are stored desiccated and protected from light at 4 °C.

Example 2: RH421 in bile salts (after Meyer et al., 1997). RH421 (#S-1108; Molecular Probes) is solubilized (20 mg/ml) in the bile salt sodium cholate (10 mM in water; Sigma, C1254) at a molar ratio of about 2 to 1 to produce a 300–400x stock solution that can be added directly to the physiological saline bathing cells. Staining times of 3–5 minutes are normally sufficient to produce good signals. Store at 4 °C and protect from light.

Example 3: di-2-ANEPEQ in water (after Antic and Zecevic, 1995).A stock solution of di-2-ANEPEQ (also known as JPW1114: #D-6923; Molecular Probes) is made in water (3 mg/ml). Prior to microinjection this solution is filtered (0.22 μm pore size). This stock solution can be stored for several months at 4 °C.

Note: In many cases increased temperature and sonication are also required to get these indicators into solution. In general, the stock solutions of VSDs can be stored at 4°C without any loss of function or brightness.

Solubility and storage of CaSDs

The ion-sensitive indicators typically exist in two forms: free salt and acetoxymethyl (AM) esters. The requirements for solubility and storage in each case are different.
Free Salt: The salt forms of most CaSDs are water-soluble and stable for long periods at –20 °C, whether stored as a solid or in solution. Typically, these salts are used for microinjection or dialysis and therefore are prepared as concentrated stock solutions in pure (Ca-free) water. There are no special precautions required to make these solutions; however, they are easiest to deal with when made and stored as concentrated aliquots (50–100x). These aliquots are best stored desiccated at –20 °C.

Note: Some investigators mix these dyes with the internal solution of the patch pipette and store them together frozen. However, our experience indicates that dye stored in this manner will degrade faster over time.

Example 1: Oregon Green 488 BAPTA-1, Hexapotassium salt. One 500 μg vial of Oregon Green 488 BAPTA-1 (#O-6806; Molecular Probes) is dissolved in 90 μl of pure, distilled, deionized water for a stock concentration of ~5 mM. This mixture is then briefly centrifuged and sonicated to ensure complete mixing. Single experiment size aliquots are then made and stored desiccated at –20 °C.
AM ester: AM esters are normally supplied pre-aliquoted and should be reconstituted using high quality DMSO. Some AM esters also require the inclusion of a dispersing agent such as Pluronic F-127 (1–20 % w/v) to achieve complete solubility. Whether or not Pluronic F-127 is used, it is advisable to make these stock solutions at the highest possible concentration (i.e., 1–5 mM). This increases stability and minimizes the amount of solvent finally present in the bathing medium. These stock solutions should then be stored well sealed, frozen and desiccated. In fact, it is advisable to use these stock solutions immediately; otherwise the solvent will readily take up moisture, leading to decomposition of the dye.

Example 2: Calcium Orange AM for bath application. A 50 μg vial of Calcium Orange AM (#C-3015; Molecular Probes) is dissolved in a solution of DMSO/Pluronic F-127(10% w/v) for a stock concentration of 4 mM. This mixture is then briefly centrifuged and sonicated to ensure complete mixing. The stock solution is then tightly sealed and keep on ice until used (up to 2–3 hours only).

Loading/Staining Protocols

There are a variety of potential methods for loading/staining with optical indicators. These methods fall into two main categories:
- Bulk loading
- Single cell loading

In bulk loading studies all the cells present are loaded or stained indiscriminately. Examples of bulk loading procedures include:
- Bath incubation
- AM ester loading
- Electroporation
- Cationic liposome delivery
- Hypoosmotic shock

The most popular method for introducing calcium dyes into cells is via their acetoxymethyl esters (AM). AM esters work by shielding the strongly negatively charged parts of the dye molecule (see Table 2) and hence make them membrane-permanent. Once inside the cell, nonspecific esterases cleave these esters back to their calcium-sensitive form and thus trap the dye intracellularly. In single cell studies loading is normally achieved via microinjection or dialysis through a patch pipette although localized electroporation is also an option.

We will consider the three methods most commonly used for optical recording purposes. They are:
- Bath application
- Microinjection
- Dialysis (through a patch pipette).

In each case the optimal staining/loading conditions vary depending on the indicator used. In many instances the best conditions are determined empirically; however, a few representative examples are documented here as a guide.

Example 1: Bath Application of extracellular VSDs, di-8-ANEPPS (after Bullen et al., 1997). A 12.5 μl aliquot of di-8-ANEPPS stock solution is gently heated to melt it. To this, 1 ml of physiological Ringers is added, giving a concentration of 163 μM. This solution is then sonicated briefly (20–30 seconds). Prior to staining, cells are washed once with PBS. Cells are then incubated at dye concentration between 75 μM and 163 μM. Generally ten minutes of staining is sufficient. Excess dye can be removed by a further PBS wash although is not always necessary.

Note: Avoid staining with or using VSDs in the presence of serum or large protein concentration as this can act as a sink for the dye and disrupt cell staining or even destain cells.

Example 2: Injection of Intracellular VSDs di-2-ANEPEQ into snail neurons (after Zecevic, 1996). A nearly saturated and prefiltered stock solution (3 mg/ml) of di-2-ANEPEQ is injected directly into *Helix aspera* neurons using repetitive, short pressure pulses (5–60 p.s.i., 1–50 ms) through a micropipette (resistance = 2–10 MΩ). Cells are then incubated at 15 °C for 12 hours to allow complete diffusion of the dye throughout the cell.

Example 3: Dialysis of the Intracellular VSDs, di-2-ANEPEQ, into cultured mammalian neurons (after Bullen and Saggau: Unpublished Observations). An aliquot of stock solution (5 mM) is added directly to the internal solution of the patch pipette each day for a final di-2-ANEPEQ concentration of 100–500 μM. The internal solution of the patch pipette is (in mM): KCl 140, $MgCl_2$ 1, NaATP 5, NaGTP 0.25, EGTA 10, HEPES 10, pH 7.4. Seal formation and dialysis into the cell are conducted using standard methods. Diffusion of the dye away from the soma occurs at approximately 1 μm per minute for distances less than 150 μm.

Example 4: Bath Application of CaSDs Fluo-3 AM in cultured rat cortical neurons (after Murphy et al., 1992). Fluo-3 AM is dissolved in DMSO at 5 mg/ml and further diluted into Hank's balanced salt solution, in the presence of 0.25% pluronic F-127, for a working concentration of 10 μg/ml. Cells are incubated with this solution for 1 hour at room temperature. These cells are then washed twice in Hank's prior to use.

Note: Avoid trying to use sharp microelectrodes for dye injection with cultured mammalian CNS neurons as this procedure has an extremely low rate of successful penetration. Dialysis through a patch pipette is a much more efficient method.

Example 5: Oregon Green 488 BAPTA-1 for dialysis via a patch pipette into cultured hippocampal neurons (after Bullen and Saggau: Unpublished Observations). An aliquot (5–15 μL) of stock solution (5 mM) is added directly to the internal solution (1 ml) of the patch pipette each day for a final Oregon Green 488 BAPTA-1 concentration of 25–75 μM. The internal solution of the patch pipette is (in mM): KCl 140, $MgCl_2$ 1, NaATP 5, NaGTP 0.25, HEPES 10, pH 7.4. Seal formation and dialysis into the cell are conducted using standard methods. Allow 10 to 20 minutes for the dye to equilibrate inside the cell before commencing any experimental manipulations.

Note: In experiments with calcium indicators don't include any additional calcium buffer (e.g., EGTA or BAPTA) in the internal pipette solution.

Note: Avoid using large concentrations of calcium indicator (i.e., greater than 100 μM) inside cells as this can result in significant buffering and distortion of calcium transients.

Note: When using CaSDs with patch pipettes it is important to avoid mixing of the internal solution (where [Ca] is nominally zero) and the bath saline (with millimolar calcium). For this reason it is important to apply positive pressure to the pipette when initially entering the bath. Additionally, it is wise to puff out the solution at the pipette tip immediately before seal formation.

Experimental Design Issues

There are numerous considerations critical to the design and execution of experiments with optical indicators. Many of these factors are prerequisites to obtaining useful data and avoiding artifactual results. These considerations can be divided into factors that are general to all experiments and those specific to experiments with optical indicators.

General Design Considerations

Determining the authenticity of any effect arising from an experimental manipulation or drug application requires several basic criteria to be satisfied. These include:

- Measurement baseline: Was a steady baseline accomplished before an experimental manipulation or pharmacological agent was applied?
- Repeatability: Was the experimental effect observed repeatable?
- Reversibility: Was the experimental effect reversible upon removal of the manipulation or drug?
- Graded response: Could the response be graded with stimulus strength?
- Pharmacology: Can the response be blocked or potentiated with appropriate pharmacological agents?

Specific Design Considerations

Design considerations specific to the use of optical indicators usually address aspects of dye application, signal optimization and integration with complementary electrophysiological techniques.

- Dye application: Specific criteria are needed to establish whether the indicator concentration and/or sensitivity was constant throughout the experiment. Non-uniform dye concentration or sensitivity can arise due to incomplete dialysis from a patch pi-

pette or internalization of voltage-sensitive dye. The response to standard or a control stimulus can be used to confirm a constant responsiveness.
- Signal optimization: In some cases, the overall signal is composed of specific and nonspecific components and therefore procedures should be in place to distinguish between these components. One example of the nonspecific fluorescence is that due to cell autofluorescence. This intrinsic fluorescence can be emitted from a specimen independent of any extrinsic fluorescent molecules and is commonly a problem when illuminating biological preparation in the near UV. The solution to this problem is to measure cell fluorescence in the absence of the optical indicator and subtract this value from the resting fluorescence in the presence of the dye. This value is often measured before staining or from an equivalent unstained site. Another important experimental issue is whether or not signal averaging or digital oversampling is required to detect the signals of interest. In cases where the signals are small and averaging is required, enough similar traces must be collected to allow use of this procedure. Finally, if illumination intensity is large, consideration often needs to be given to whether a bleaching correction is required. Bleaching corrections are commonly made with control traces that are collected under experimental conditions but in the absence of the stimulus or experimental manipulation.
- Integration: The integration of optical and electrophysiological techniques often requires specific procedural changes. For instance, VSDs dissolved with solvents, especially DMSO/F-127, can inhibit seal formation between a patch pipette and the cell membrane and it is sometimes necessary to form this seal before staining the cell.

Calibration Procedures

The calibration of optical signals is necessary if the goal of an experiment is to determine a quantitative result or measure an absolute change in the parameter of interest. Likewise, if a comparison between optical signals from different experiments or between points within the same experiment is required, these signals should also be calibrated. While it may appear to be advantageous if all signals were just recorded in a quantitative manner, this approach is not always possible because other factors such as recording bandwidth are often comprised in the process. The calibration of optical signals can be achieved in one of three ways:
- Single wavelength measurements
- Ratiometric measurements
- Hybrid measurements

Without doubt, ratiometric measurements give the most reliable results. This type of measurement is possible with indicators that show a spectral shift in either their excitation or emission spectrum that is dependent on the variable of interest. These spectral shifts allow the comparison of two wavelengths where the fluorescence intensities are changing in opposite directions or between one wavelength and a spectral isosbestic point (i.e. point insensitive to the parameter of interest). In addition to providing a quantitative result, ratiometric measurements reduce or eliminate systematic variations in fluorescence due to:
- Indicator concentration
- Excitation pathlength
- Excitation intensity
- Detector efficiency.

Furthermore, ratiometric methods are important in eliminating a variety of artifacts and nonsystematic factors. These include:

- Photobleaching
- Indicator leakage over time
- Non-uniform indicator distribution
- Variable cell thickness.

In some cases ratiometric measurements are also more sensitive because the changes in fluorescence at each wavelength are usually of opposite sign and therefore the magnitude of the change in the ratio is greater than the change in either wavelength alone.

Under some experimental conditions it is impractical to perform true ratio measurements. An alternative is to perform hybrid measurements (Lev-Ram et al., 1992). In a hybrid protocol, quantitative measurements are combined with qualitative estimations performed at a different instant in time. For example, an initial baseline could be determined quantitatively with a ratiometric measurement. Subsequently, fast changes in the same parameter are followed qualitatively at a single wavelength but with much higher measurement frequency. However, it is important to note that this approach assumes all other variables (especially indicator concentration) remain constant during recording of the single wavelength measurements.

Both ratiometric and non-ratiometric methods have been used with calcium-sensitive and voltage-sensitive dyes. Examples of each type of calibration procedure are documented below. Table6 outlines schematically how these measurements are made in each case. In addition, some general guidelines common to both types of indicator are outlined below.

VSD Calibration

Fluorescent voltage-sensitive dyes are generally considered "linear voltmeters without scale." Specifically, they provide information about voltage changes but the absolute amplitude of this signal can vary due to differences in dye staining and variations in local sensitivity. Hence these indicators are most commonly used *uncalibrated and absolute comparisons between points and across preparations are not attempted.* However, calibrated measurements are possible in some circumstances and the success of these procedures is easily verified with concurrent electrical measurements. This type of measurement includes those made at a:

- Single wavelength
- Two excitation wavelengths based on an excitation spectral shift
- Two emission wavelengths based on an emission spectral shift

I. Single wavelength measurements for comparisons between points in same preparation: Fromherz and others (Fromherz and Vetter, 1992; Fromherz and Muller, 1994) have devised a way to compare the relative magnitude of voltage signals from different points in the same preparation. Briefly, these authors choose to examine the ratio of fluorescence changes to voltage changes of opposite sign. The rationale behind this approach is that differences in local sensitivity and the fraction of fluorescing molecules would cancel out and reflect only the ratio of the underlying voltages. Thus:

$$\frac{\Delta F_2}{\Delta F_1} = \frac{\Delta V_2}{\Delta V_1}$$

where ΔF refers to the change in fluorescence and ΔV the change in membrane potential. The subscripts 1 and 2 correspond to separate locations. Theoretically, if an electrical measurement is also made at one of these points it would be possible to calculate absolute ΔV at the other point.

Note: The sensitivity and accuracy of this approach remains to be proven and whether it is an advance over traditional data display methods (i.e., ΔF/F) requires empirical evaluation.

Table 6. Calibration Methods used in Single Studies.

Indicator		Example	Epifluorescence configuration	Equation	Refs.
VSDs	Single wavelength	RH421	532 nm	$V_{m1} \approx \left(\frac{V_{m2}}{F_2}\right) \cdot (F_1)$	1, 2
	Excitation ratio	Di-8-ANEPPS	440 nm, 530 nm	$V_m = CF \times R'$	3, 4, 5
	Emission ratio	Di-8-ANEPPS	488 nm, DCLP570	$V_m = CF \times R'$	6, 7
CaSDs	Single wavelength	Fluo-3	488 nm	$\left[Ca^{2+}\right] = K_d \cdot \frac{F - F_{min}}{F_{max} - F}$	8, 9
	Excitation ratio	Fura-2	340 nm, 380 nm	$\left[Ca^{2+}\right] = K_d^* \cdot \frac{R - R_{min}}{R_{max} - R}$	9, 10
	Emission ratio	Indo-1	350 nm, DCLP455	$\left[Ca^{2+}\right] = K_d^* \cdot \frac{R - R_{min}}{R_{max} - R}$	9, 10

(1) Fromherz and Vetter (1992), (2) Fromherz and Muller (1994), (3) Montana et al., (1989), (4) Bedlack et al. (1994), (5) Zhang et al. (1998), (6) Beach et al. (1996) (7) Bullen and Saggau (1997), (8) Minta et al, (1989), (9) Haugland (1996), (10) Grynkiewicz et al., (1985).

II. Ratiometric methods using ***two excitation wavelengths*** **and based on an** ***excitation spectral shift*** **for measuring absolute changes in membrane potential** (after Montana et al., 1989): In addition to undergoing voltage-dependent changes in the amplitude of the emission spectrum, some VSDs also exhibit voltage-dependent spectral shifts. Loew and colleagues have used the excitation spectral shift of di-8-ANEPPS as the basis for ratiometric VSD measurements. By alternatively exciting this dye on the wings of its absorption spectrum (440 and 530 nm) and measuring wide-band fluorescence (>570 nm) they have derived a ratiometric parameter that is linear with membrane potential over the physiological range. These authors have extended this approach to include single cell imaging (Bedlack et al., 1994). By interlacing images captured at each excitation wavelength they generated a ratiometric map of membrane potential throughout a whole cell. Recently, they have extended this approach to include a more accurate calibration procedure by employing patch clamp techniques for absolute determination of membrane potential (Zhang et al, 1998). The disadvantage of this excitation ratio formation procedure is the requirement to interlace two images and/or switch excitation filters, which is time-consuming and limits the overall temporal bandwidth to less than that required to capture fast events such as action potentials.

The equation for converting normalized ratio data into an absolute membrane potential value (in mV) is:

$$V_m = CF \times R'$$

Where C.F. is the conversion factor between the ratio value and membrane potential and R` is the normalized ratio (typically normalized to R at 0 mV).

III. Ratiometric methods using ***two emission wavelengths*** **and based on an** ***emission spectral shift*** **for measuring absolute changes in membrane potential** (after Bullen and Saggau, 1999; Beach et al., 1996): An alternative method to make ratiometric determinations of membrane potential with this kind of indicator employs a single excitation wavelength and undertakes simultaneous measurements at dual emission wavelengths. This method is based on the voltage-dependent shift in the emission spectra of the voltage-sensitive dye, di-8-ANEPPS. Typically, fluorescence measurements are made at two emission wavelengths using a secondary dichroic beamsplitter (e.g., DCLP570) or prism and dual photodetectors (<570 and >570 nm). The signal at each wavelength changes in opposite directions and the ratio of these signals is linearly related to membrane potential. One implementation of this scheme employs a high-speed, random-access, laser-scanning microscope (Bullen et al., 1997) with dual photodetectors for simultaneous detection at two emission wavelengths (Bullen and Saggau, 1999). In this approach, measurements are made with a discrete laser line (476 or 488 nm) and hence acquisition speed can be very high because there is no requirement to switch excitation filters. Furthermore, because this excitation wavelength coincides with a voltage-insensitive point in the excitation spectra, the excitation spectral shift is removed as a confounding influence. Concurrent current clamp measurements can be used to calibrate this method. The formula to convert ratio values into absolute membrane potentials is the same as described in the previous section.

CaSD Calibration

Three forms of calibrated measurements are possible with this type of indicator. They are:
- Single wavelength
- Ratiometric: based on an excitation shift or an emission shift
- Hybrid measurements

I. Single wavelength measurements: The calibration equation for a single wavelength can be written in terms of the fluorescence values:

$$\left[Ca^{2+}\right] = K_d \cdot \frac{F - F_{\text{min}}}{F_{\text{max}} - F}$$

where
K_d = dissociation constant determined *in vitro*.
F = measured fluorescence.
F_{max} = maximal fluorescence intensity in saturating calcium.
F_{min} = minimum fluorescence intensity in zero calcium or saturation with a quenching agent (e.g., Mn^{2+}).

This kind of measurement can be undertaken with any kind of calcium indicator (e.g., Calcium Green) but is susceptible to variations in path length, dye concentration etc.

II. Ratiometric measurements When indicators are employed that shift their fluorescence spectra upon binding to calcium (e.g., the excitation spectrum of Fura-2 or the emission spectrum of Indo-1), measurements are commonly made at two distinct wavelengths (λ_1, λ_2) to obtain a ratio ($R=F_{\lambda 1}/F_{\lambda 2}$). The calibration equation for dual wavelength indicators is:

$$\left[Ca^{2+}\right] = K_d^* \cdot \frac{R - R_{\text{min}}}{R_{\text{max}} - R}$$

where
K_d^* = K_d (F_{max}/F_{min}),
R_{min} = ratio in zero calcium or following saturation with a quenching agent (e.g., Mn^{2+}).
R_{max} = ratio in a saturating concentration of calcium.

R_{min} and R_{max} are most accurate when they are obtained under conditions that approximate the experimental milieu (i.e., in the cell). This kind of approach is considered much more accurate than those described for a single wavelength measurement (see previous section) and overcomes discrepancies in path length, dye concentration etc.

III. Hybrid measurements: An alternative to ratiometric measurement is to perform hybrid measurements. In this case, the *resting calcium concentration* is first determined with a ratiometric measurement. Subsequently, *fast changes in the calcium concentration*, $\Delta[Ca^{2+}]$, are optically followed at a single wavelength but with much higher measurement frequency. The calibration equation for this kind of measurement (Lev-Ram et al., 1992) is:

$$\Delta\left[Ca^{2+}\right] = \left(K_d + \left[Ca^{2+}\right]\right) \cdot \frac{\frac{\Delta F}{F}}{\frac{\Delta F_{\text{max}}}{F}}$$

where $\Delta F/F$ is the fractional change in fluorescence and $\Delta F_{max}/F$ = $(F_{max}$-$F)/F$ is the maximal fractional change from resting to saturating calcium concentrations and $[Ca^{2+}]$ is the resting calcium concentration measured ratiometrically at an earlier time. This kind of approach could, for example, employ Fura-2 as both a ratiometric indicator and a single wavelength dye. Normally this indicator requires mechanical filter switching, which is an inherently slow process, to alternate excitation wavelengths. However, by alternately employing dual and single wavelength measurements the hybrid approach overcomes this limitation and allows faster estimations of calcium changes. It is important to note that this approach requires an invariant indicator concentration during recording (i.e., no bleaching or change in dye concentration from extrusion or dialysis).

General guidelines for the calibration of optical indicators

An important step in converting optical measurements to the physiological parameter of interest is the *post-experiment calibration*. While the conversion factors for both voltage and calcium indicator calibration can be determined in solution or various simplified preparations (i.e., vesicles), these conditions often do not approximate the true intracellular milieu. Factors that are not well reproduced in these situations include:

- Temperature
- pH
- Ionic strength
- Interaction of dyes with proteins or membranes.

Moreover, the interactions of some CaSD with intracellular proteins have been shown to change the apparent K_d (Kurebayashi et al., 1993). In short, in situ calibrations are typically better than equivalent in vitro procedures and should be adopted whenever possible.

These calibrations are typically achieved by chemical clamping of the cell with pore-forming ionophores. For example, ratios of membrane potential can be calibrated with valinomycin. Specifically, a series of valinomycin-mediated K^+ diffusion potentials are used to step through the range of interesting membrane potential while measuring fluorescence ratios. For a detailed description of this procedure see Loew (1994). Likewise, ratiometric calcium measurements can be calibrated in situ with the ionophores such as ionomycin or calcimycin (or its analog 4-bromo-A-23187). However, care should be taken when using these compounds because they possess a relatively high level of autofluorescence, especially in the UV. For a detailed description of these procedures see the chapter by Kao (in Nuccitelli, 1994). This chapter also documents many of the underlying assumptions and practical limitations of these calibration procedures.

Note: Always subtract off any autofluorescence or other offsets before forming a ratio or calculating $\Delta F/F$.

Signal Processing Methods

Even after the brightest indicators have been chosen and instrumentation considerations optimized, some optical signals are weak and/or noisy. In other instances, the sensitivity of the probes being used can be poor or the underlying physiological events are quantal in nature. In each case, extra care must be taken to extract the signal from background noise. Noise can be:

- Systematic
- Random

In some cases systematic noise can be measured and removed by subtraction or division. In contrast, random noise is harder to separate from the underlying signal. Signal averaging is one way to cancel out the effects of truly random noise. However, averaging is not always possible (i.e., non-stationary events). Within a single sweep, only those components of random noise that are spectrally separate from the signal can be removed (i.e., by filtering).

To overcome noise problems in optical recording experiments a number of signal processing and noise reduction techniques are available. These include:

- Source noise ratio formation
- Digital filtering
- Signal averaging

Source noise ratio formation

Systematic noise present in experimental records can be of two types: additive or multiplicative. Typically, additive noise can be removed by subtraction while multiplicative noise is best corrected by ratio formation. Multiplicative noise generated by variations in source intensity is the most common noise source in optical recording experiments. This is particularly apparent in cases where the relative changes in fluorescence are equal to or less than the fluctuations arising from the light source. In this case, it is difficult to resolve the signal from the noise. This is a particularly prevalent problem with laser sources where such intensity variations can be as much as 5 % peak-to-peak. However, these variations can be measured and removed from the signal by ratio formation. In practice, ratio formation between the signal and a reference measurement is the most efficient way to remove source noise variations. The advantage of this scheme over subtraction is that there is no requirement to match amplitudes between signal and reference. The effectiveness of this procedure is documented in Bullen et al. (1997) and is demonstrated with real signals in Figure 5.

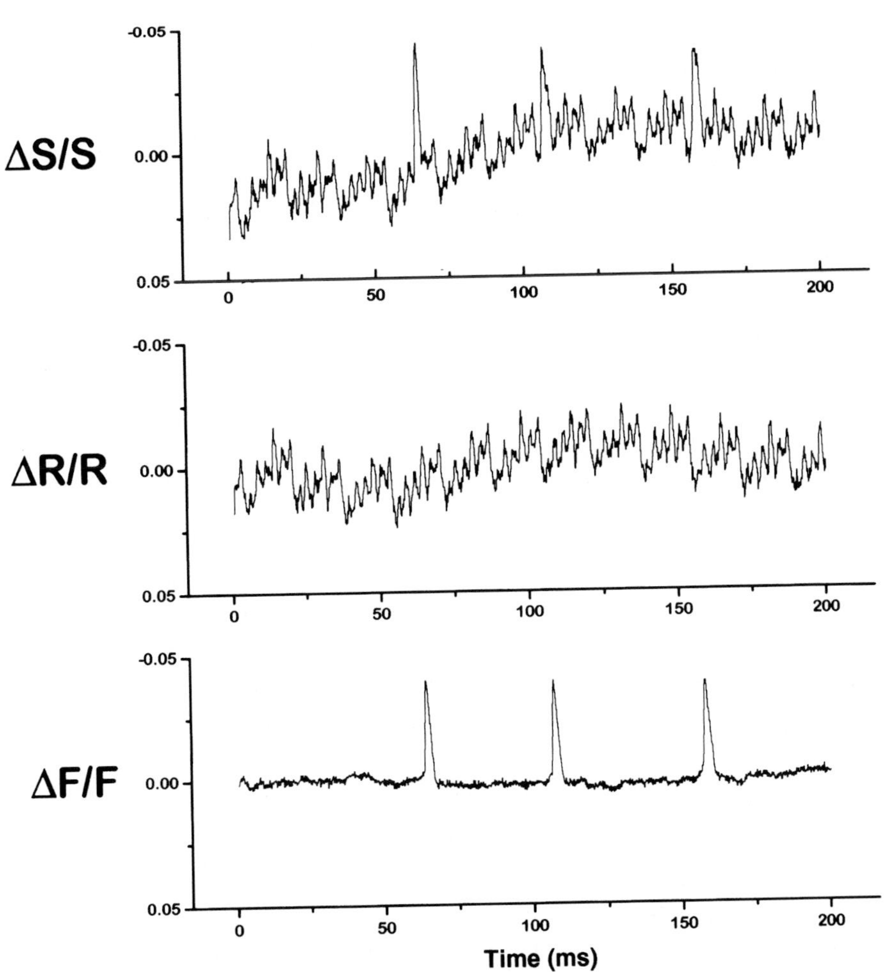

Fig. 5. Demonstration of Source Noise Removal following Ratio Formation with a Reference Signal. Top trace (ΔS/S) is the raw signal that includes both the underlying signal and the contaminant source noise. Middle trace (ΔR/R) is a reference signal containing the source noise sampled directly from the excitation source which in this case was a laser. The bottom trace (ΔF/F) is the processed signal after source noise has been removed. Note how the source noise from laser intensity fluctuations is completely removed by this procedure.

Another way to remove source noise variations is to form ratios between two emission wavelengths. In this case the source noise variations are present in both wavelengths as common mode signals and effectively cancel out in the ratio formation process.

Digital Filtering

Digital filtering is an important signal conditioning tool used to reduce the contribution of random noise or unwanted signals (cf. Chapter 45). This is especially true for non-stationary signals or for signals that cannot be averaged. The principle behind digital filtering is that frequencies of interest can be separated based on whether they represent signal or noise. There are four main types of digital filters:
- Low-pass
- High-pass
- Band-pass
- Band-stop

Low-pass filters are important in restricting the bandwidth of experimental records to that containing useful signal components while removing high-frequency components. *High-pass* filtering is also referred to as *A/C coupling* and is useful in removing the DC component of a signal and thus highlighting only that part that is changing. *Band-stop* (or *notch*) filters reject discrete frequency bands and are particularly useful in removing AC-line noise from experimental records.

Other special filters exist that conserve high frequency components while still acting as a *low-pass* filter. One example is the Savitzky-Golay filter method which essentially performs a local polynomial regression to determine the smoothed value for each data point. This method is superior to other filtering methods because it tends to preserve features of the data such as peak height and width, which are usually 'washed out' by adjacent averaging or *low-pass* filtering.

Different implementations of these digital filters are commonly found in scientific plotting and graphing packages (such as Origin or SigmaPlot) or in specialized mathematical environments (such as Matlab or Mathematica).

Note: Care must be taken to avoid filtering frequencies that contain important signal components. Moreover, it should be recognized that some filtering methods can introduce small phase shifts into the data. However, modern finite impulse response (FIR) digital filters can be used in opposite directions to counteract this problem. Finally, care should be taken to ensure the sampling theorem is not violated (see "Data Acquisition and Digitization Issues").

Note: It is advisable to use analog filters (active or passive) at all data acquisition and processing steps (i.e., from the detector onwards). This typically reduces the accumulation of unwanted noise at each step and reduces the need for subsequent digital filtering.

Signal Averaging

Signal averaging refers to the grouping in space or time of areas or repeated trials to reduce the contribution of random noise (cf. Chapter 45). If the noise is truly random, then signal averaging will reduce it by a factor of $1/\sqrt{N}$, where N is the number of trials. This kind of averaging requires stationary events and involves no loss in frequency components occurring, provided the signals being averaged are strongly time-locked. If signal averaging procedures are adopted, care should be taken not to introduce any temporal jitter (i.e., biological or instrumental) into the process as this can result in a low-pass filter effect. In cases where this is a problem, events such as action potentials can be aligned by their peaks and averaged together to improve the overall signal quality.

The effectiveness of signal averaging is demonstrated in Figure6. In this example, drawn from experiments examining the linearity of ratiometric VSD data, the relative noise decreases in direct proportion to the number of trials averaged.

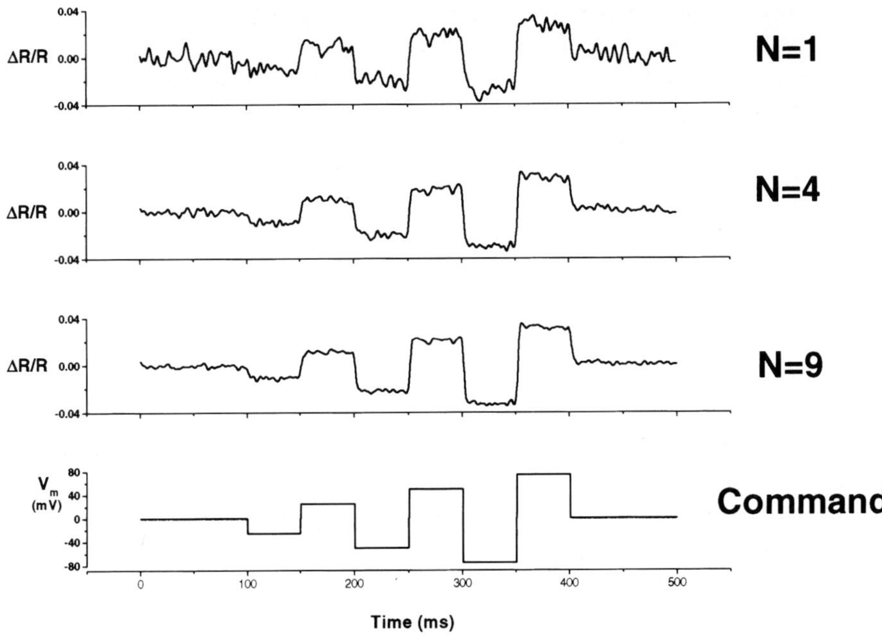

Fig. 6. Example of Noise Reduction by Signal Averaging. The top three traces represent ongoing average of ratiometric signals recorded from the same scanning point under identical conditions during an optical calibration experiment. The number of traces averaged is given to the right of each trace. The bottom trace is the voltage command waveform used to elicit these signals. Note the degree of improvement (i.e., the reduction in noise) with the increasing number of traces averaged. Experiment conducted with the voltage-sensitive dye di-8-ANEPPS in cultured hippocampal neurons.

One disadvantage with temporal averaging is that the overall measurement frequency is typically reduced because of the time required to capture enough traces for averaging. This restricts the number of points that can be sampled within a fixed time frame. This can be a problem in cases where the response time course to a drug or an experimental manipulation is important.

Results

This section documents a number of considerations important in the presentation and display of experimental results. These include:
- Data presentation
- Important properties of experimental records
- Types of experiments and representative records

Data Presentation

The results of optical recordings from single neurons or parts of neurons can be presented in a number of ways. These include:
- One-dimensional records: Signals recorded from a single site or data extracted from individual points or regions of interest (ROI) in an image and displayed as a one-dimensional trace against time.

- Pseudo-color images: A series of images colored in a way to distinguish activity levels or changes in ion concentration.
- Live video: A reproduction, sometimes with altered timing, derived directly from the images captured during the experiment.

While video records and pseudo-color images provide a graphically pleasing and qualitative means of displaying data, it is often difficult to show time courses and/or quantitative changes with this format. Furthermore, it is often difficult to reconcile these images with other parameters measured concurrently (i.e., currents and voltages) that are inherently one-dimensional.

Important Properties of Experimental Records

An unfortunate trend in modern papers is presentation of heavily reduced data and in many cases original records are omitted completely. This is especially true in studies employing optical recording techniques where many recordings are often reduced to a single pseudo-color image. However, for others to judge the quality of the underlying data it is important to present examples of original records. In examining this kind of data it is important to consider a number of different questions. For instance, do the records presented show:

- Detection sensitivity: Were the recording method and indicator(s) used sensitive enough to make useful measurements?
- Signal-to-noise ratio: Is the signal-to-noise ratio exhibited sufficient to make experimentally useful conclusions and could it be further optimized?
- Spatiotemporal resolution: Did the method used posses sufficient spatiotemporal resolution to answer the experimental questions posed?
- Fidelity: Are the records presented an accurate reflection of the underlying physiological events or was some disruption or alteration caused by the recording methodology itself?

Types of Experiments and Representative Records

A number of experimental records are presented here to demonstrate the types of applications possible in experiments with single neurons. For instance, the ratiometric signals presented earlier (see Figure 6) were derived from an experiment examining the linearity of the VSD di-8-ANEPPS. This record was made from a scanning site immediately adjacent to a patch pipette and demonstrates the strong similarity in terms of response time and amplitude between the optical signals and voltage-clamp command waveform. This type of calibration can subsequently be used to quantify similar optical measurements made under more physiological conditions.

In a separate example using the same VSD, the pattern of postsynaptic potential integration and conduction in the dendrites of cultured hippocampal neurons were examined (see Figure 7.). Several facets of this experiment further illustrate the usefulness of this kind of optical recording. In particular, these recordings were made from very small cellular structures in a noninvasive manner. Moreover, several measurements were made simultaneously from different recording sites (2 μm diameter) and in way not easily possible with other experimental methods. Furthermore, these signals were obtained at rates (i.e., 2 kHz) sufficient to adequately sample the physiological events of interest.

In a similar experiment, the spatial differences in calcium signals generated from adjacent points in the same cell following a series of action potentials were examined

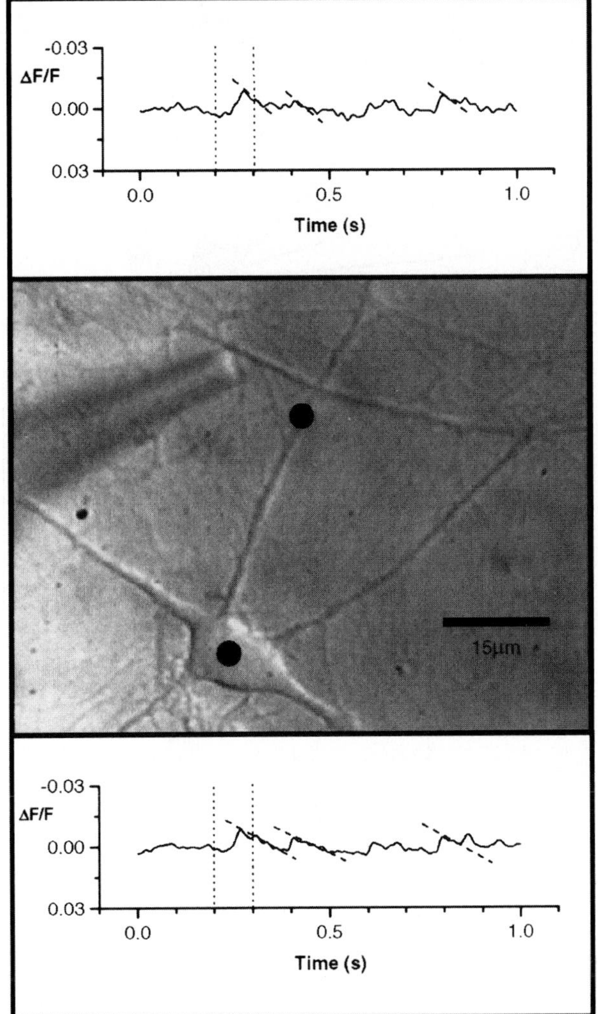

Fig. 7. Representative Voltage-sensitive dye recordings. Optical recordings of postsynaptic potentials in the dendrites of hippocampal neurons made with di-8-ANEPPS at two independent scanning sites. Focal stimulation of presynaptic terminals undertaken with hyperosmotic saline occurs between the two dashed lines.

(Figure8.). This experiment demonstrates spatial differences in the magnitude and kinetics of calcium transients examined at different points within the same cell at high temporal resolution.

Troubleshoot

This section documents a number of problems commonly encountered in optical recording experiments. In particular the following issues are examined:
- Signal quality issues
- Prevention of photodynamic damage
- Loading and staining problems

Signal Quality Issues

The most important factor in all recording techniques is the magnitude of *signal-to-noise ratio* (S/N) that can be achieved. This is especially true for some optical indicators (especially VSDs) where the relative change in fluorescence can be very small. Many factors can influence signal quality. These include:

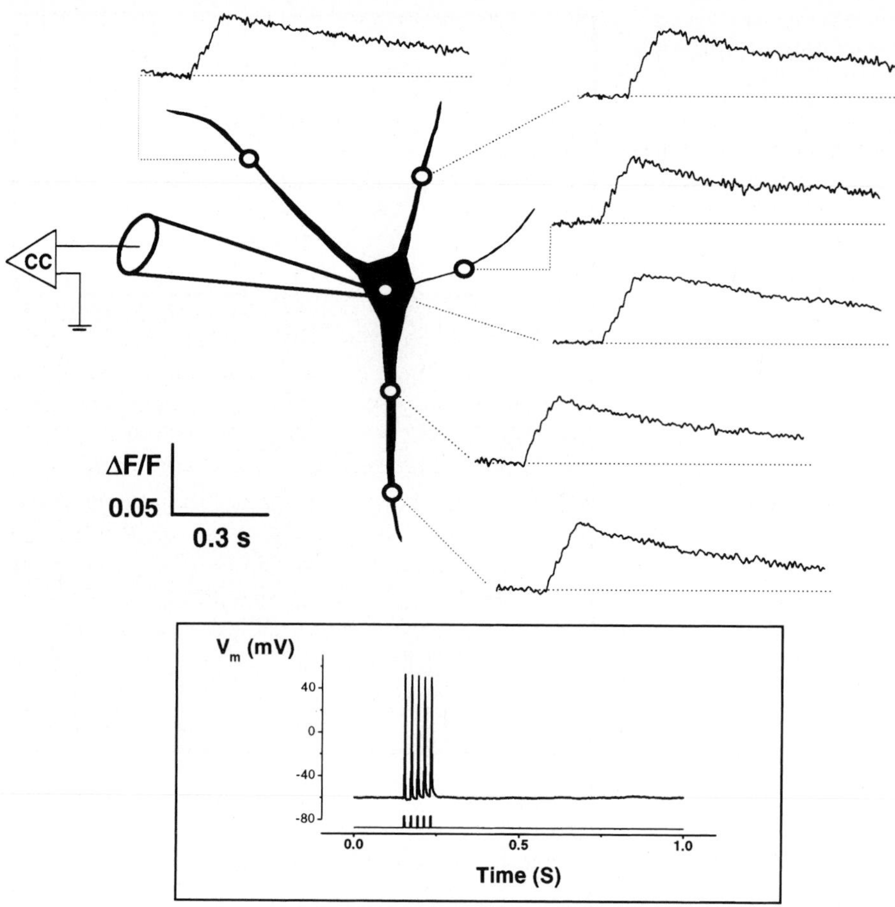

Fig. 8. Representative Calcium-sensitive dye recordings. Optical recordings of a calcium transient elicited by somatic current injection (see inset) in a hippocampal neuron and measured with Oregon Green BAPTA-1 (100μM). Measurements made with a Random-access, laser-scanning microscope (after Bullen et al., 1997). The relative position of each scanning site (2μm) is indicated on an outline drawing of the cell.

- Excitation intensity: The fluorescence arising from an optical indicator is directly proportional to the excitation intensity and hence the strongest light source possible should be employed. However, care should be taken to avoid bleaching and phototoxicity associated with excessive illumination intensity.
- Indicator concentration: Emission light intensity is also proportional to the indicator concentration and hence the maximal amount of probe possible should be utilized. In this case, care should be exercised to avoid probe concentrations that have pharmacological or buffering effects and thereby change the underlying physiology (see section "Loading and Staining Problems"). Moreover, at very high indicator concentrations emission light intensity begins to decline due to "quenching."
- Excitation volume: In scanning microscopy, the signal from a large excitation volume (i.e., large scanning spot) will be greater than from a smaller one because more fluorophores are excited. Hence there is a tradeoff between the spatial resolution achieved with a small scanning spot and the consequent reduction in excitation volume and signal strength.
- Indicator sensitivity: The best signal quality is likely to arise from optical indicators with large sensitivities (i.e., large ΔF/F, F_{Ca}/F_{Free} or R_{max}/R_{min}) and therefore in cases

where there is a choice between indicators, the candidate with the largest sensitivity should be preferred.

- Indicator brightness: When a choice between indicators exists it is always advisable to choose the brightest one. Guidelines on comparing absolute indicator brightness are included in a later section ("Criteria for Comparing Indicators").
- Maximum collection efficiency (N.A. and Filters): It is important to remember that collection efficiency is determined by the product of many factors. These include the numerical aperture, the absolute transmission of the objective lens, the relative throughput of the microscope and the transmission of the dichroic mirror and related filters. A total collection efficiency of 10% would not be uncommon and in a worst case scenario this value could be much lower. Thus, in cases where there is a choice between objective lenses it is always best to choose the lens with the highest N.A.. This serves to maximize the illumination intensity and to optimize the collection of emission light. Similarly, care should be taken in the selection of dichroic mirror and emission filters such that large portions of the signal are not excluded. Other potential sources of light loss include: the use of DIC optics without the analyzer removed from the emission pathway, dirty optics or optical misalignment. Just the DIC analyzer alone will reduce emission light intensity by another 50%.
- Maximum detection efficiency (Q.E.): In all cases, but especially when light levels are low, it is advantageous to choose a detector with the highest possible quantum efficiency. This ensures that all photons captured are converted into a useful signal (see Table 5).
- Recording bandwidth: Recording bandwidth should be just enough to capture signals of interest. Inclusion of temporal bandwidth in excess of that needed to record the signal will only contribute noise. However, if extra temporal bandwidth exists, it can be used to improve signal quality by *digital oversampling*. Oversampling is a form of temporal smoothing that averages several closely spaced points where only one was previously sampled. This procedure serves to reduce random noise with the improvement in the signal-to-noise being proportional to $1/\sqrt{N}$, where N is the number of extra samples.

Signal Quality Checklist

✓ Maximum-tolerated Excitation Intensity

✓ Maximum Probe Concentration

✓ Most Sensitive Indicator

✓ Brightest Indicator available

✓ Maximum Collection Efficiency (N. A. & Filters)

✓ Maximum Detection Efficiency (Q. E.)

✓ Minimum Badwidth (or max. oversampling)

Prevention of Photodynamic Damage

Excessive illumination intensity can sometimes lead to photodamage or phototoxicity. In many cases this phenomenon arises from the production of free radicals, such as singlet oxygen, which are a byproduct of fluorescent excitation. Among other things, these free radicals can severely disrupt membrane integrity. This is a particularly prevalent problem for VSDs that reside in the plasma membrane. There are various methods for reducing photodamage from fluorescent indicators. Some of these methods attempt to

Table 7. Antioxidant Formulations for Use with Fluorescent Indicators.

	Mol. Wt.	Solubility	Max. Soluble concentration	Best Storage Option	Stock Concentration (1, 5)	Final Concentration
Ascorbic Acid (A4403) (2)	176.1	Water	10 mg/ml	–20°C	11 mM or 2 mg/ml	110 μM
Citric Acid (C5920) (2)	294.1	Water	100 mg/ml	Room Temp or –20°C	27 mM or 8 mg/ml	270 μM
Glutathione (3) (G6013) (2)	307.3	Water	50 mg/ml	Use Fresh	Use Fresh	32 μM or 1 mg/100 ml
Pyruvate (P5280) (2)	110	Water	100 mg/ml	–20°C	25 mM or 2.8 mg/ml	250 μM
Tocopherol (4) (T1539) (2)	430.7	100 % EtOH	25 mg/ml	–20°C	46 mM or 20 mg/ml	46 μM
Tocopherol Acetate (4) (T1157) (2)	472.8	100 % EtOH	25 mg/ml	–20°C	42 mM or 20 mg/ml	42 μM

Notes: (1) Make fresh each day as some components are not stable more than 3 hours. (2) Sigma Catalog Numbers included for each reagent. (3) Reduced form of Glutathione. (4) May also require warming to get into solution. (5) Either 100x stock in physiological saline or 1000x stock in ethanol.

minimize photodamage by the addition of large quantities of antioxidant to the bathing solution, others employ a lesser amount of strategically placed antioxidant (i.e., in membranes) while still others act by removing oxygen from the solution altogether. Some examples of these approaches are:

- ACE et al.: The simplest strategy to reduce the effect(s) of free radicals in solution is to add large quantities of antioxidant to the bathing medium. Various formulations of vitamins (A, C and E; ACE) and other similar agents (e.g., Trolox; Fluka) have been suggested. However, the reported success of these formulations is relatively modest, probably due to the lack of immediate protection at the site of injury (i.e., membrane). Representative agents of this sort, together with solubility and storage information, are documented in Table7.
- Astaxanthin: A more direct approach employs the natural carotenoid, Astaxanthin. Initially this agent was utilized as an antioxidant for use in combination with a novel VSD based on fluorescence-resonance energy transfer principles (Gonzalez and Tsien, 1997). The effectiveness of this approach should in principle be greater than antioxidants in solution because of its physical proximity in the membrane and its potent ability to quench reactive oxygen species. Furthermore it has been shown that relatively large concentrations of this agent can be used without fear of toxic effects. This, together with its relative water solubility when compared to other carotenoids such as β-carotene, make it an extremely attractive antioxidant agent. However, preliminary reports describing the use of Astaxanthin with other VSDs have revealed only modest effectiveness. For information about the solubility, storage and application of Astaxanthin, see Cooney et al. (1993).
- Glucose Oxidase/Catalase: The combination of glucose oxidase and catalase is a potent oxygen removal scheme and is particularly effective when used with the VSD di-8-ANEPPS (Obaid and Salzberg, 1997). Glucose oxidase takes oxygen from solution to form H_2O_2 which is subsequently converted to water by catalase. The high catalytic activity of these enzymes means that relatively small amounts (G.O. 40U/ml and Cat. 800U/ml) are needed to de-oxygenate a solution. Glucose Oxidase (G-6125) and Catalase (C-40) are both available from Sigma (St. Louis, MO).
- Oxyrase: Oxyrase is a biocatalytic oxygen reducing system. This commercial preparation also works by removing oxygen from the bathing solution surrounding cells. Oxyrase is prepared from *E. coli* and is a crude preparation that uses lactic acid, formic acid or succinic acid added to the bathing solution as a hydrogen donor. Prelim-

inary experiments with this formulation and fluorescent dyes indicate it is an effective antioxidant in many instances. Oxyrase is available from Oxyrase Inc. (Mansfield, OH).

Comparison of antioxidant formulations

There is no universal solution to the problem of photodamage. In many cases the best method for a particular application must be determined empirically. However, in general terms some of the above methods are more effective than others. In our experience with VSDs the Glucose Oxidase/Catalase combination is the best method for use with single cells in short-term experiments.

Loading and Staining Problems

VSDs

There are three types of problems commonly associated with VSD staining. They are:

- Poor staining/Low membrane affinity: Some VSDs (e.g., RH414) have a relatively modest membrane binding affinity for some cell types while others stain cells at a particularly slow rate (e.g., di-8-ANEPPS). The binding affinity of these dyes is due in part to their structure but membrane composition may also play a role. Furthermore, there are significant species and cell type differences (Ross and Reichardt, 1979). There are three potential solutions to this problem. Firstly, a series of suitable dyes should be screened prior to beginning a new series of experiments to find those which exhibit the highest membrane affinity and largest signal amplitude. Secondly, a wide range of staining time and conditions should be considered to determine the optimal staining conditions. Finally, there are agents (i.e., 0.05% Pluronic F-127) that can be added to the dye stock solution to facilitate dye insertion into the membrane.
- Overstaining: Using excessive amounts of VSDs can have various deleterious effects. In particular, nonspecific fluorescence arising from unbound dye or dye bound in a nonspecific orientation can seriously degrade signal size and quality. Furthermore, some VSDs have been shown to exhibit pharmacological effects when used at high concentrations and thus it is recommended that the minimal dye concentration compatible with good signals should be employed.
- Dye internalization: Because VSDs reside in the outer leaflet of the plasma membrane there is always the possibility that some dye molecules can cross into the inner leaflet or even directly onto membranes of intracellular organelles. Various membrane recycling processes can also serve to transport VSDs into cells. Such internalization of VSDs typically results in decreased signal size because these dye molecules have either no sensitivity to membrane potential or a directly opposite sensitivity to that of normally oriented dye molecules. The solution to this problem is to use dyes that are less likely to be internalized (e.g. di-8-ANEPPS). Additionally, incubation at greater than room temperature apparently increases the likelihood of dye being internalized into the cell, so this should be avoided.

CaSDs

There are four types of problems commonly associated with loading cells with CaSDs. Two of these problems are specific to AM ester loading techniques. They are:

- Compartmentalization: Under ideal conditions, fluorescent indicators loaded with this technique are uniformly distributed in the cytosol and absent from other cellular compartments. However, AM esters are capable of accumulating in any membrane-enclosed compartment within the cell. In addition, some polyanionic forms of these indicators can be sequestered within various organelles by active transport processes. This aberrant compartmentalization is normally more pronounced at higher loading temperatures and can be avoided by reducing loading temperature. The use of indicators conjugated to dextran can also retard compartmentalization and sequestration.

- Incomplete ester hydrolysis: Low or slow rates of de-esterification can result in a significant proportion of intracellular dye being partially de-esterified and hence insensitive to calcium but still somewhat fluorescent. This can result in an serious underestimation of the true cytosolic calcium concentration. Additionally, incomplete ester hydrolysis can also promote compartmentalization. Fluorescence quenching by Mn^{2+}, which only binds to the de-esterified form, is one way to quantify this effect. One means to avoid the confounding effects of fluorescence from AM esters is to select indicators whose esters are non-fluorescent. For instance, Calcium Green and Oregon Green 488 BAPTA are basically non-fluorescent as AM esters. In contrast, AM esters of Fura-2 and Calcium Orange still posses some basal fluorescence.

Two additional problems occur with both AM ester loading and micro-injection or dialysis of free salt forms of CaSDs. They are:

- Overstaining: Whether AM esters or simple salts are used to load single neurons, some care must be taken to not overload the cells as this can cause unwanted buffering effects. This buffering can affect the resting calcium concentration, the size and kinetics of calcium transients and disrupt various cellular processes dependent on calcium. In short, the consequences of using incorrect indicator concentrations vary from poor signal quality to distorted physiology. One way to demonstrate that the signals recorded are free of any buffering effects is to conduct experiments across a range of indicator concentrations.
- Extrusion: Various anionic indicators tends to leak out or are actively extruded by some cell types. In some cases various pharmacological tools can be used to block this problem (i.e., probenecid, sulfinpyrazone and verapamil). Another solution is to use indicators designed to be resistant to these phenomena. In particular, dextran-conjugated dyes are typically resistant to extrusion and leakage. Recently, Texas Fluorescence Labs (Austin, TX) have developed a number of such calcium dye variants that are leak-resistant (i.e., Fura PE3, Indo PE3 and Fluo LR).

Comments

This section examines various issues of secondary importance that should be considered in the design of optical recording experiments. These include:

- Indicator binding kinetics
- Criteria for comparing indicators
- Data acquisition and digitization issues

Indicator Binding Kinetics

Ion-sensitive indicators bind to free ions and form a complex in a process described by simple bimolecular binding kinetics. Binding of free indicator [X] and free calcium $[Ca^{2+}]$, and unbinding of bound calcium [CaX] are determined by the *on-rate* and *off-rate* (k_+ and k_-, respectively).

$$[X]+\left[Ca^{2+}\right]\underset{k_-}{\overset{k_+}{\leftrightarrow}}[CaX]$$

There are several important consequences that arise from this simple binding scheme. Firstly, these indicators exhibit a classical sigmoidal-binding curve and thus the relationship between calcium concentration and fluorescence is not linear. However, a straight line can be used to approximate this relationship over the steepest part of the

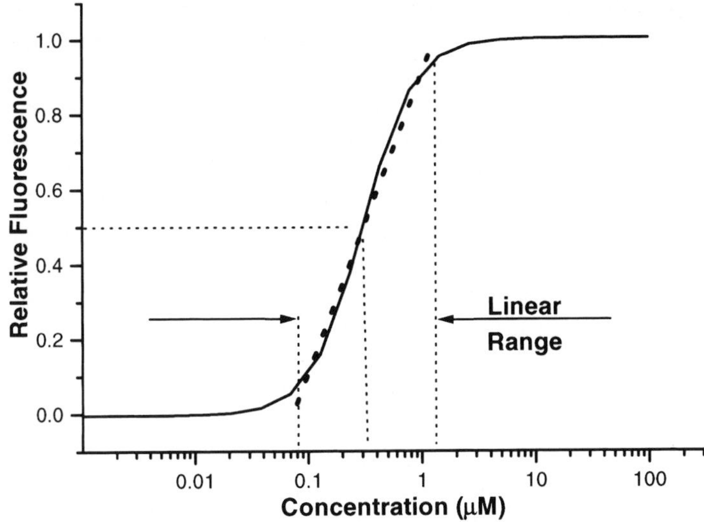

Fig. 9. Typical Binding Curve for a Calcium-sensitive dye. Binding curve of an idealized calcium indicator with a K_d = 350nM. Note how the linear range of this indicator (dashed line) extends from about 0.1x to 10x the K_d.

curve (see Figure9). An important consequence of this simplifying assumption is that the linear range of such an indicator only encompasses calcium concentrations between 0.1x and 10x the K_d (see the example in Figure 9). Outside this range, the fluorescence produced is not related to the underlying calcium concentration in the same linear fashion. To overcome this limitation each kind of calcium indicator is commonly available in a range of affinities (see Table2). In choosing between calcium indicators of different affinities, it is safest to match high end of this linear range with the maximum concentration expected.

Another limitation arising from the kinetics of these indicators is the relatively slow off-rate, especially in high-affinity versions of these molecules. A functional consequence of this rate is that the decline in fluorescence measured during a typical calcium transient probably reflects the rate of dye unbinding rather than the rate of calcium removal. Furthermore, it is unlikely that fast repetitive events will be adequately tracked with these probes.

Another important consideration in experiments examining the diffusion of calcium within cells is the that the Ca-dye complex can diffuse with faster kinetics than calcium alone, which is subject to cytosolic buffering. Thus, it is possible to get the incorrect impression of actual calcium movement.

Note: The use of dextran-conjugated dyes significantly decreases the magnitude of dye diffusion and reduces the likelihood of artifactual calcium movement.

Criteria for Comparing Indicators

The extensive array of available indicators often requires the investigator to choose between many seemingly similar dyes. Objective criteria for comparing different dyes typically includes consideration of their:

- Sensitivity
- Brightness
- Photostability

Sensitivity The sensitivity of an optical indicator refers to the magnitude of its change in fluorescence relative to the change in the parameter it is being used to measure. Typically, a sensitive indicator will undergo a large change in fluorescence for even a small change in ion concentration or voltage. However, while some dyes (e.g., Fluo-3) exhibit very large sensitivities, they are not necessarily suitable for all applications because they lack brightness.

Brightness Brightness refers to the strength of the fluorescence generated by an optical indicator. The fluorescence output from a given dye depends on the efficiency with which it absorbs and emits photons and its ability to undergo repeated excitation/emission cycles. Absorption efficiency is normally quantified using the molar extinction coefficient (ε) which is a value typically determined at the peak of the excitation spectrum. Fluorescence is normally characterized by its quantum efficiency (QE), or the ratio of photons emitted per photons absorbed. The quantum efficiency is normally a measure of total emission over the entire emission spectrum. Brightness or fluorescent intensity per dye molecule is proportional to the product of ε and QE and is the most useful way to compare the potential signal of two similar indicators.

Photostability Under intense illumination, the irreversible destruction or photobleaching of particular fluorophores becomes a limiting factor in some experiments. While there is little objective information to compare the photostability of different indicators, some recently developed fluorophores (i.e., Oregon Green) have been engineered for improved photostability over their predecessors (i.e., Fluorescein).

Data Acquisition and Digitization Issues

In order to process and/or store analog optical signals in a computer, they have to be digitized (cf. Chapter 45). The interface between the analog and digital world is commonly a device called an analog-to-digital converter (A/D converter). In general, the selection of an A/D converter depends on two parameters:
- Digitizing resolution
- Speed of conversion

Resolution The functional digitizing resolution required in an experiment depends on the relative intensity change in the signal. For instance, a typical VSD signal ($\Delta F/F$) is 1 % per 100 mV and an experimentally useful resolution would be 2 mV in membrane potential (i.e., 1/50 of 1%). The corresponding digitizing resolution needed under these conditions would be 5000 digitizing steps (50x100 = 5000). This number corresponds to 13 bits of digitizing resolution (i.e., 8192 steps) because 12 bit digitization (i.e., 4096) exhibits insufficient sensitivity. Failure to employ sufficient digitizing resolution will lead to the appearance of digitizing noise (i.e., quantitized steps) in experimental records and/or the failure to adequately resolve the signals of interest.

Speed of Conversion The speed of an A/D converter should be high enough to adequately reproduce all interesting frequency components in the signal. The sampling rate (f_{sample}) of an A/D converter depends on the *temporal bandwidth* of the signals (Δf) and the number (N) of the channels sampled simultaneously:

According to the sampling theorem this is the minimal sampling rate that can adequately reproduce the signal. Ideally, an oversampling factor of 2–5 should be employed. Typically, the frequency bandwidth for electrophysiological signals is between 500 Hz and 5 kHz and thus the minimal f_{sample} is 10,000 times the number of channels sampled in parallel. This factor becomes a serious consideration when a large number of points or channels are recorded and digitized serially (e.g., an image). Moreover, maximum digitization rates typically scale inversely with digitizing resolution. Thus, under circumstances where signal sizes are small and 14- or 16-bit digitization is required, the throughput of an A/D converter can severely limit the bandwidth of optical recordings.

Acknowledgement: This work was supported in part by grants from NSF BIR-95211685 (A.B.), NSF IBN-9723871 (P.S.) and NIH NS33147 (P.S.). We also gratefully acknowledge the excellent technical assistance provided by Dr. S.S. Patel in various aspects of this project.

References

A Practical guide to the study of calcium in living cells. Methods in Cell Biology: Vol. 40. R. Nuccitelli (Ed.) Academic Press, San Diego. 1994.

Ebner, T.J. and G. Chen. (1995) Use of voltage-sensitive dyes and optical recordings in the central nervous system. Prog. Neurobiol. 46: 463–506.

Fluorescent and luminescent probes for biological activity. W.T. Matson, (ed.) Academic Press. (1993)

Haugland, R.P. Handbook of fluorescent probes and research chemicals. Molecular Probes. Eugene, OR. (1996).

Light Microscopy and Cell Structure. Vol. 2. Cells. A laboratory manual. D. Spector, R.D. Goldman and L. A. Leinwand. Cold Spring Harbor Press (1998).

Loew, L.M. (1994) Voltage-sensitive dyes and imaging neuronal activity. Neuroprotocols 5:72–79.

Optical methods in cell physiology. P. DeWeer and B.M. Salzberg, (Editors). Wiley-Interscience. New York (1986).

Antic, S. and Zecevic, D. (1995) Optical signals from neurons with internally applied voltage-sensitive dyes. J. Neurosci, 15: 1392–1405.

Beach, J.M., McGahren, E.D., Xia, J. and B.R. Duling. (1996) Ratiometric measurement of endothelial depolarization in arterioles with a potential-sensitive dye. Am. J. Physiol. 39:H2216–2227.

Bedlack, R.S., Wei, M-d., Fox, S.H., Gross, E. and L.M. Loew. (1994). Distinct electric potentials in soma and neurite membranes. Neuron. 13:1187–1193.

Bullen, A., Patel, S. S., and Saggau, P. (1997). High-speed, random-access fluorescence microscopy 1. High- resolution optical recording with voltage-sensitive dyes and ion indicators. Biophys. J. 73(1): 477–491.

Bullen, A. and Saggau, P. (1999). High-Speed, random-access fluorescence microscopy. II Fast quantitative measurements with voltage-sensitive dyes. Biophys. J. 76(4):2272–2287.

Bullen, A. and Saggau, P. (1998) Indicators and Optical Configuration for Simultaneous High-resolution recording of membrane potential and Intracellular Calcium using Laser Scanning Microscopy. Eur. J. Physiol. 436:827–846.

Chien, C.-B. and J. Pine. (1991a) Voltage-sensitive dye recordings of action potentials and synaptic potentials from sympathetic microcultures. Biophys. J. 60: 697–711.

Chien, C.-B. and J. Pine. (1991b) An apparatus for recording synaptic potentials from neuronal cultures using voltage-sensitive fluorescent dyes. J. Neurosci. Methods. 38: 93–105

Cooney, R.V., Kappock, T.J., Pung, A. and J.S. Bertram (1993). Solubilization, cellular uptake, and activity of β-carotene and other carotenoids as inhibitors of neoplastic transformation in cultured cells. Methods in Enzymology 214:55–68.

Eberhard, E. and P. Erne (1991) calcium binding to fluorescent calcium indicators: Calcium green, calcium orange and calcium crimson. Biochem. Biophys. Res. Com. 180(1):209–215.

Fromherz, P. and C.O. Muller (1994) Cable properties of a straight neurite of a leech neuron probed by a voltage-sensitive dye. P.N.A.S 91:4604–08.

Fromherz, P. and T. Vetter (1992) Cable properties of arborized Retzius cells of the Leech in Culture as probed by a voltage-sensitive dye. P.N.A.S 89:2041–45.

Gonzalez, J.E. and R.Y. Tsien. 1997. Improved indicators of cell membrane potential that use fluorescence resonance energy transfer. Chemistry and Biology 4:269–277.

Grynkiewicz, G., Poenie, M. and R.Y. Tsien (1985) A new generation of calcium indicators with greatly fluorescence properties. J. Biol. Chem. 260(6): 3440–3450.

Haugland, R.P. (1996). Handbook of fluorescent probes and research chemicals. Molecular Probes. Eugene, OR.
Johnson, I. (1998) Fluorescent probes for living cells. Histochemical J. 30: 123–140.
Kao, J.P.Y., Harootunian, A.T. and R.Y. Tsien (1989) Photochemically generated cytosolic calcium pulses and their detection by Fluo-3. J. Biol. Chem. 264(14): 8179–8184.
Kleinfeld, D., Delaney, K.R., Fee, M.S., Flores, J.A, Tank, D.W. and A. Gelperin (1994) Dynamics of propagating waves in the olfactory network of a terrestrial mollusk: An electrical and optical study. J. Neurophysiol. 72(3):1402–19.
Kurebayashi, N. Harkins, A.B. and S.M. Baylor (1993). Use of fura red as an intracellular indicator in frog skeletal muscle fibers. Biophysical. J. 64:1934–1960.
Lev-Ram, V., Miyakawa, H., Lasser-Ross, N. and W.N. Ross (1992). Calcium transients in cerebellar purkinje neurons evoked by intracellular stimulation. J. Neurophysiol. 68(4):1167–1177.
Loew, L.M. (1993). Potentiometric membrane dyes. In: Fluorescent and luminescent probes for biological activity. W.T. Matson, (ed.) Academic Press, London. pp 150–160.
Loew, L.M. (1994) Voltage-sensitive dyes and imaging neuronal activity. Neuroprotocols 5:72–79.
Meyer, E. Muller, C.O. and P. Fromherz (1997). Cable properties of dendrites in hippocampal neurons of the rat mapped by a voltage-sensitive dye. Eur. J. Neuroscience 9:778–785.
Minta, A. Kao, J.P.Y., and R.Y. Tsien (1989) Fluorescent indicators for cytosolic calcium based on rhodamine and fluorescein chromophores. J. Biol. Chem. 264(14): 8171–8178.
Montana, V., Farkas, D.L. and L.M. Loew. (1989). Dual-wavelength ratiometric fluorescence measurements of membrane potential. Biochemistry. 28:4536–4539.
Morris, S.J. (1992) Simultaneous multiple detection of fluorescent molecules. In: B. Herman and J.J. Lemaster (eds.) Optical Microscopy: New Technologies and Applications. Academic Press.
Murphy, T.H., Blatter, L.A., Wier, W.G. and J.M. Baraban (1992). Spontaneous synchronous synaptic calcium transients in cultured cortical neurons. J. Neuroscience. 12(12):4834–45.
Obaid, A.L. and B.M. Salzberg (1997) Optical studies of an enteric plexus: Recording the spatiotemporal patterns of activity of an intact network during electrical stimulation and pharmacological interventions. Soc. Neuroscience Abstract no. 816.1. 23(2):2097.
Rohr, S. and B.M. Salzberg. (1994) Multiple site optical recording of transmembrane voltage (MSORTV) in patterned growth heart cell cultures: Assessing electrical behavior, with microsecond resolution, on a cellular and subcellular scale. Biophys. J. 67: 1301–1315.
Ross, W.N. and Reichardt, L.F. (1979). Species-specific effects on the optical signals of voltage-sensitive dyes, J. Mem. Biol. 48: 343–356.
Spector, D., Goldman, R.D and L. A. Leinwand (1998). Light Microscopy and Cell Structure. Cells. A laboratory manual. Vol. 2. Cold Spring Harbor Press.
Zecevic, D. (1996). Multiple spike-initiation zones in single neurons revealed by voltage-sensitive dyes. Nature, 381(6580): 322–325.
Zhang, J., Davidson, R.M., Wei, M-d. and L.M. Loew. (1998). Membrane electric properties by combined patch clamp and fluorescence ratio imaging in single neurons. Biophysical J. 74:48–53.

Chapter 5

Electrical Activity of Individual Neurons In Situ: Extra- and Intracellular Recording

Peter M. Lalley, Adonis K. Moschovakis and Uwe Windhorst

Introduction

Microelectrode recording of electrical activity provides a means to measure the discharge patterns of nerve cells with high spatial and temporal resolution and with minimal damage to nervous tissue. For these reasons it has long been the principal method for analyzing the behavior and function of neurons and neural networks. An additional and extremely useful application of microelectrode technology is the ability to inject tracers directly into neurons through an intracellular microelectrode in order to label the cells and identify their location, morphology, and synaptic contacts with other neurons and effectors.

The first investigations of single neuron activity were carried out with microelectrodes for extracellular recording, which led to the identification of previously uncharacterized cell types and synaptic circuits (e.g., Lorente de Nó, 1938; Renshaw, 1946; see Eccles (1964) and McLennan (1970) for additional examples). Shortly thereafter, micropipettes for intracellular recording were developed (Ling and Gerard 1949). These were used to measure membrane potentials and to uncover voltage- and time-dependent properties that determine neuron excitability. Intracellular recording also revealed the nature and functional significance of excitatory and inhibitory postsynaptic potentials and helped to identify the underlying membrane conductance mechanisms (Combs, Eccles and Fatt 1955a–c). Subsequent adaptation of voltage-clamp technology for use with microelectrodes (Brennecke and Lindemann 1974a,b; Wilson and Goldner 1975; Adams and Gage 1979; Finkel and Redman 1984) permitted measurement of membrane currents and estimates of conductances *in vitro* (Adams et al. 1982a,b; Johnston et al. 1980) and *in vivo* (Dunn and Wilson 1977; Finkel and Redman 1983a; Richter et al. 1996). Further advances in neurophysiological investigation came with the development of techniques which enabled investigators to record the electrical activity of neurons and then label them by intracellular injections of fluorescent dyes (Thomas and Wilson 1966; Stretton and Kravitz 1968).

Microelectrode recording of electrophysiological properties and labeling of neurons continue to be important tools for analyzing the behavior and function of single nerve

Peter M. Lalley, University of Wisconsin Medical Science Center, Dept. Physiology, 1300 University Avenue, Madison, WI, 53706, USA (phone: +01-608-263-4697; fax: +01-608-265-5512; e-mail: pmlalley@facstaff.wisc.edu)
Adonis K. Moschovakis, University of Crete School of Health Sciences, Dept. of Basic Sciences, P.O. Box 1393, Iraklion, Crete, Greece (phone: +30-81-394509; fax: +30-81-394530; e-mail: moschov@med.uch.gr)
Uwe Windhorst, National Institute for Working Life, Dept. of Musculoskeletal Research, P.O.Box 7654, Umea, 90713, Sweden (phone: +46-90-7867453; fax: +46-90-7865027; e-mail: uwe.windhorst@niwl.se), University of Calgary, Faculty of Medicine, Dept. Of Clinical Neuroscience and Dept. Of Physiology and Biophysics, 1403 - 29th St. N.W., Calgary, Alberta, T2N 2T9, Canada,
Universität Göttingen, Zentrum Physiologie und Pathophysiologie, Umboldtallee 23, Göttingen, 37073, Germany (e-mail: uwe@neuro-physiol.med.uni-goettingen.de)

cells and neural networks. In this chapter, we address problems and describe methods of recording from neurons in the *in situ* preparation, since applications to *in vitro* preparations are more straightforward and are addressed in other chapters. Moreover, we hope to stimulate renewed interest in research that relates the activity patterns of individual neurons to the properties of networks in the intact nervous system and ultimately to the behavior of the intact organism. Ideally, there will be an expanding, fruitful merger of *in vivo* microelectrode methodology with cellular and molecular biology to provide new perspectives on how the nervous system functions.

Subprotocol 1
General Arrangement and Preparation for Electrophysiological Recording and Data Acquisition

Overview of Equipment and Arrangement for Electrophysiological Experimentation

In this section we present an inventory of equipment used in recording, measuring and storing electrophysiological data for *in situ* research on mammals. Obviously, requirements will vary according to the goals of the research. Figure 1 illustrates basic equipment used to record electrophysiological signals with a microelectrode. Although the schematic diagram illustrates an *in vitro* setup, the same type of arrangement is used for electrophysiological investigation in the intact animal. The figure is followed by a list of essential or useful electrophysiological devices and ancillary equipment.

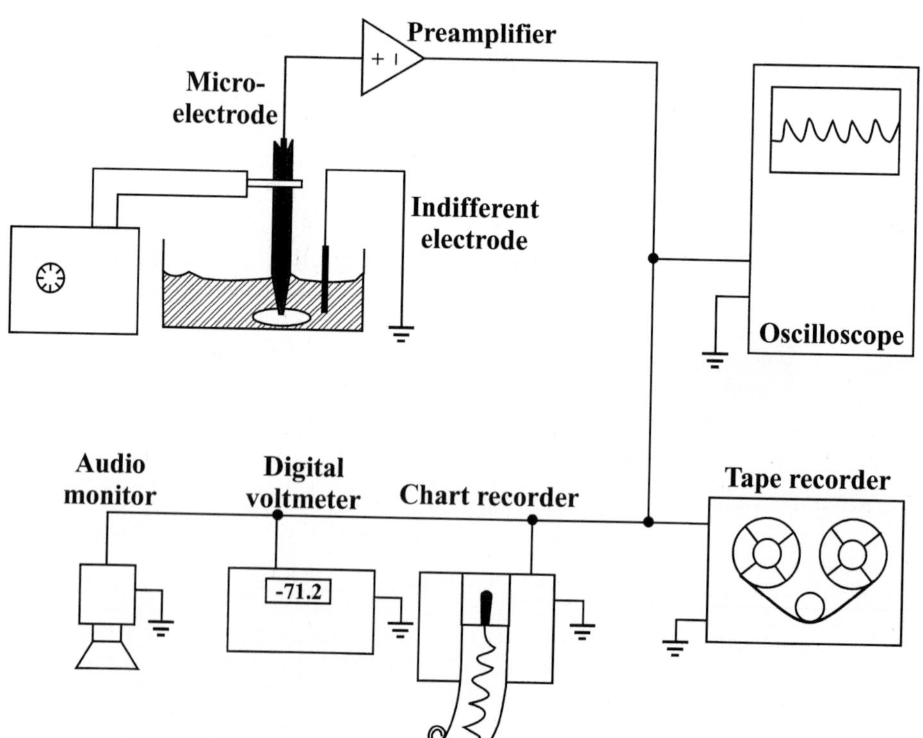

Fig. 1. Essential instruments for recordi

Materials

- Anesthetic agents such as urethane, a-chloralose or pentobarbital
- Instruments for microsurgery: dental drill, cautery, forceps, scissors, scalpels, hemostats, bone cutters, bone wax, gel foam, etc.
- Heating pad and infrared lamps
- Telethermometer to monitor body (core) temperature
- Device to continuously measure blood pressure
- Ventilatory gas monitor (expired CO_2, inspired O_2)
- Mechanical ventilator for artificial respiration
- Surgical microscope
- Antivibration, shock-absorbing table
- Stereotaxic head-holder and spinal fixation frame
- Micromanipulators to hold microelectrodes, nerve electrodes, pressure feet
- Voltage/constant current pulse generators with stimulus isolation units
- Horizontal pipette puller with programmable heat and pull parameters (intracellular electrodes) or vertical puller (extracellular recording)
- Motor-driven microdrive to advance and withdraw microelectrodes in nervous tissue
- High-gain AC preamplifiers with bandpass and 50 or 60 Hz notch filters for extracellular recording
- DC amplifier for intracellular recording (e.g., Axoclamp 2B)
- DC-offset controllers for use with tape recorder and oscilloscope

  Note: Voltage offset controllers are used to ensure that DC signals presented to the inputs of tape recorders and oscillographs are in range for proper storage.

- Multi-channel oscilloscope
- Multi-channel paper oscillograph recorder, DC-10 kHz frequency response
- Audio-amplifier

  Note: Audio amplifiers are extremely useful for detecting action potentials. Often, one hears cell discharges before the cell is visible on the oscilloscope and chart recorder. In addition, they allow one to hear synaptic noise during intracellular recording. Much valuable information can be obtained from the sound of subthreshold postsynaptic potentials, such as their temporal occurrence with respect to other recorded activities and the condition of the cell after penetration.

- Neuronal spike discriminator and rate meter to measure action potential frequency
- Digital tape recorder
- Faraday cage

  Note: Faraday cages may be necessary to shield out extraneous noise from laboratory lights and power supplies when recording bioelectric signals that require high amplification, e.g., extracellular spike potentials in the microvolt to low millivolt range. However, careful grounding of the animal and equipment will often reduce noise to acceptable levels without the need for a Faraday cage. A very effective means of grounding is to connect the chassis grounds of all electrical equipment to a common grounding post, which is in turn connected to the ground input of a DC- or AC-preamplifier. Another useful method to further reduce noise is to enclose the area around the preamplifier head stage with aluminum foil which is connected by a cable to the common ground post.

- Microprocessor and computer interface (A/D converter(s)) with software for data acquisition and analysis

Note: Many varieties of microprocessors, interfaces and software packages are available, such as those supplied by Axon Instruments, Inc., and Cambridge Electronics Design, U.K. Such systems are not only extremely valuable for retrospective analysis, they can be used with their scrolling programs to survey on-line data, thereby reducing the need for continuous paper oscillograph registration.

- Back-up device(s)

Procedure

Preparation

Here we describe general procedures for recording nerve signals from intact animals. Specific aspects related to recording bioelectric potentials will be discussed where appropriate in PARTs 2–4. Figure 2A illustrates how the preparation might appear after surgi-

A

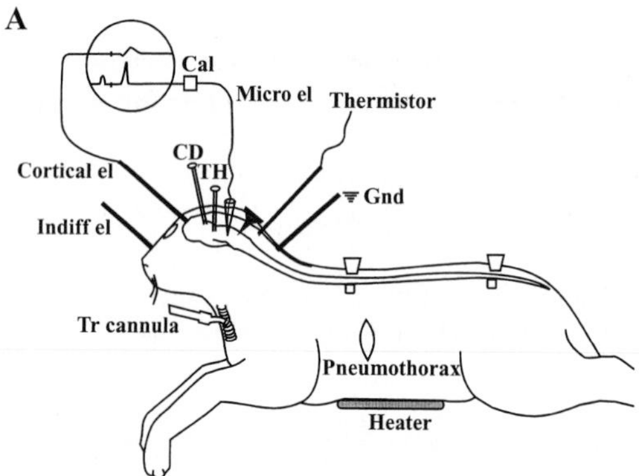

B

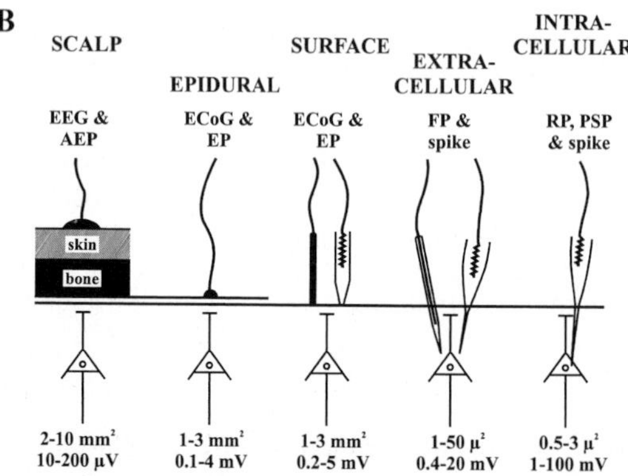

Fig. 2. Surgical preparation and arrangement of recording and stimulation electrodes for intracellular recording from neurons in the brain in a small animal. Upper circle is an oscilloscope screen showing extracellular recording of evoked cortical surface response; lower trace, intracellular recording from a subcortical neuron, preceded by a calibration (*Cal*) pulse. Pick up amplifiers interposed between the cortical recording electrode (*Cortical el*), the micrelectrode (*Micro el*) and the oscilloscope have been omitted. *Indiff el*: indifferent electrode; *CD* and *Th* are stimulating electrodes in the caudate nucleus and thalamus. *Gnd* is the ground wire. *Tr*, tracheal cannula. (Redrawn and modified from Willis and Grossman 1973). **B:** Electrode arrangements for different types of recording: *EEG & AEP*: recording electroencephalogram and auditory evoked potentials. *ECoG & EP*: electrocorticogram and epidural potentials. *FP & spike*: field potentials and unitary action potentials. *RP, PSP & spike*: resting potential, postsynaptic potential and transmembrane action potential. Numbers below the schematized neurons give the recording areas of the electrodes (ranges) and the amplitudes of the detected signals. (Redrawn and modified from Purves 1981)

cal preparation and placement of electrodes for recording. The lower panel (B) summarizes different types of electrodes and their use in detecting bioelectrical potentials.

Choice of Preparation: Anesthetized or Unanesthetized Decerebrate?

Whether to utilize anesthetized or unanesthetized, decerebrate preparations will primarily depend on the goals of the research and the effects of anesthesia on the neurons to be analyzed. Most general anesthetics have the advantages of increasing recording stability by eliminating unwanted locomotor activity and maintaining stable patterns of spontaneous discharge. However, suppression of excitatory as well as inhibitory synaptic influences on neurons often limit their usefulness (Zimmerman et al. 1994). Urinary bladder reflexes, for example, are greatly depressed by anesthetic doses of barbiturates and necessitate the use of either decerebrate unanesthetized preparations, or α-chloralose (deGroat and Ryall, 1968). Decerebration at the midcollicular level of the brain stem (see Zheng et al. 1991) eliminates the problem of anesthesia, but such preparations take longer and are more difficult to prepare.

Pre-Surgical Medication

Animals should be premedicated with atropine to prevent salivary secretions. Depending on species and route of administration, effective doses range from 0.1 to 0.3 mg/kg. The methylbromide salt of atropine is preferred since, unlike atropine sulfate, it does not cross the blood-CNS-barrier. Premedication with dexamethasone (0.3 mg/kg) is useful in minimizing edema of central nervous tissue produced by craniotomy or laminectomy. When surgical procedures are extensive, it is often beneficial to slowly infuse lactated Ringer or Glucose-Ringer solutions intravenously during recording (Richter et al. 1991).

Surgical Methods

Craniotomy, laminectomy and opening of the dura mater are performed not only to expose central nervous tissue for microelectrode exploration, but also to allow drainage of cerebrospinal fluid for stability of microelectrode recording and visualization of tissue surfaces for proper microelectrode insertion. They should be as wide as possible for these purposes and to minimize herniation of nervous tissue. It is often necessary to further stabilize recording conditions by performing bilateral thoracotomy. This minimizes movement of brain and spinal tissue associated with spontaneous respiration. A wide thoracotomy is made after paralysis with i.v. gallamine triethiodide (4–10 mg/kg initially, followed by 4–10 mg/hr) or pancuronium bromide (0.1–0.2 mg/kg, then 0.07–0.1 mg/hr) and mechanical ventilation. To prevent lung collapse after thoracotomy, positive pressure is applied to the expiratory outflow by immersing a tube connected to the output port of the ventilator in 1–2 cm H_2O.

Other procedures which aid stability of recording include fixing the margins of the cut dura mater to surrounding bone or muscle with fine ligatures or cyanoacrylate glue, infiltration of space between bone and tissue with 3% agar in Ringer, removal of arachnoid membrane from all exposed tissue and positioning of a pressure foot over the site of microelectrode insertion. Pial membrane must be also be removed over the recording site to prevent tissue dimpling at the site of insertion and damage to the microelectrode tip (Richter et al. 1996).

After completion of surgery all exposed tissue should be covered with either agar solution, paraffin oil, a mixture of vaseline and paraffin oil, or covered with moist gauze. Covering the tissue will prevent drying, which can result in contraction of muscle and rapid deterioration of tissue.

Preparation of Nervous Tissue for Microelectrode Insertion and Recording

Irrespective of the type of electrode utilized, and whether the experiment is designed for extra- or intracellular recording, it is important to properly prepare nervous tissue for insertion of microelectrodes. As mentioned previously, this may necessitate removal of dural membrane and sometimes arachnoid membrane if glass micropipettes or metal microelectrodes with fine, fragile tips are utilized. In addition, removal of patches of pia over spots where microelectrodes for intracellular recording are inserted is usually necessary. Otherwise, tissue will dimple and eventually recoil as the microelectrode is advanced, resulting in damage to surface nervous tissue, injury to neurons along the penetration track and an inaccurate estimate of the depth of the electrode in tissue. In removing membranes, care must be taken to avoid surface bleeding which impairs insertion of microelectrodes and damages the tips.

It is also important to prevent excessive accumulation of cerebrospinal fluid (CSF) on the surface. Surface CSF increases capacitance, reducing the response time of the microelectrode. Often, CSF builds up because plasma and tissue pCO_2 are high as a result of less-than-adequate ventilation. This can be corrected by increasing the rate of mechanical ventilation. Drains made from cotton wool placed at borders of the surgical field are also helpful.

Grounding the Preparation

In the *in situ* experiment, the preparation is grounded by placing a coiled silver or platinum wire (sufficient for AC recording) or a chlorided silver wire (DC recording) into low-resistance contact with moistened non-nervous tissue. Sintered silver chloride discs that are connected to wire are now commercially available and are ideal for animal grounding. The Ag, AgCl or Pt wire is encased in a Ringer-soaked cotton wool pledget and either sown into place against muscle, or inserted into the mouth of the animal. To maintain proper grounding, It is important that the cotton wool is kept saturated with electrolyte solution.

Seeking Cells

In this section we review some general procedures that are used to seek and record intra-or extracellularly in the brain or spinal cord.

Amplifier Gain and Filtering

When searching for cells with electrodes meant for intracellular recording, 10x amplifier gain is normally used. For extracellular recording, AC amplifier gains of 1000x to 5000x are used to detect single unit potentials or evoked field potentials. Bandpass filtering at 100–3000 Hz is frequently used for extracellular recording.

Manipulators for Advancing Microelectrodes During the Cell Search

A variety of high-precision manipulators are available for positioning and advancing microelectrodes. It is convenient in the *in situ* experiment to mount a motor- or piston-driven micromanipulator on an X-Y-Z or polar coordinate manipulator such as the Narishige Canberra-type. This greatly facilitates placement of the driver and microelectrode over the area of insertion. The microelectrode is then advanced into the tissue by the microdrive. For extracellular recording, oil-driven microdrives are acceptable, but are generally less desirable for intracellular recording because they cannot be driven at high step velocity to penetrate the neuron. In addition, stability is reduced by temperature-related expansion and contraction of oil in the drive system. Motor- or piezoelectric drivers are preferred and are available from a number of commercial sources. Such devices should be capable of advancing the microelectrode by remote control in small or large steps or continuously, through a range of velocities and accelerations and with minimum backlash and lateral displacement. Small steps at high velocity greatly facilitate penetration of cell membranes and minimize cell damage. A disadvantage of some motor drivers is high frequency hum which may be problematic for example in auditory neurophysiology experiments.

Tip Clearing as an Aid to Detection and Penetration of Nerve Cells

Tissue accumulated on the tip and shank of an electrode increases resistance and, therefore, noise, making it more difficult to detect cells. In addition, tissue accumulation will result in damage to the cell membrane during impalement for intracellular recording.

To keep microelectrode tips clear, brief pulses, about 1 s duration, of positive current (100 nA) are passed through the electrode manually via a toggle switch or with a trigger circuit as the electrode is advanced in steps of a few μm. The positive current also aids in penetration of the cell membrane. Many newer amplifiers for intracellular recording come equipped with built-in clearing current units. Alternatively, one can pass current via the stimulus input from an external stimulator triggered manually with a spring-loaded toggle switch or electronic trigger.

Field Potentials as Aids in Localizing Cells

Neurons that are clustered in pools can often be quickly localized by recording their electrically-evoked antidromic or paucisynaptic field potentials. The shape and amplitude of field potentials evoked by electrical pulses are used to guide the microelectrode to its final destination. In the sacral spinal cord, for example, pools of sacral parasympathetic preganglionic neurons responding to ventral root stimulation (De Groat and Ryall, 1968) produce a surface-positive field potential which, as the microelectrode is advanced, grows in amplitude, then reverses polarity in the region where pools of preganglionic neurons are located. Unitary action potentials within a pool are detected by gradually reducing the pulse intensity, until a single neuron response is detected by its tendency to drop out and reappear around threshold stimulus intensity. As additional examples, field potential recording has been used to localize vagal preganglionic neurons (Porter 1963) and inspiratory R_β neurons (Bachman et al. 1984) in the medulla that respond to cervical vagus nerve stimulation.

Growth in Amplitude and Change in Configuration of Extracellular Unitary Action Potentials

Extracellular unitary action potentials occurring spontaneously or evoked by stimulation are less than 1 mV in amplitude and tend to be negative or triphasic (negative-positive-negative) when the electrode is relatively distant from the cell. The spikes often show a deflection or notch on the rising phase which distinguishes them from axons, although the latter can also exhibit such deflections if the electrode tip makes axon contact. When the electrode is very close to the cell, extracellular potentials can exceed 5 mV and are negative-positive, as illustrated in Fig. 3. Methods for recording such responses are presented below in parts 2 and 3.

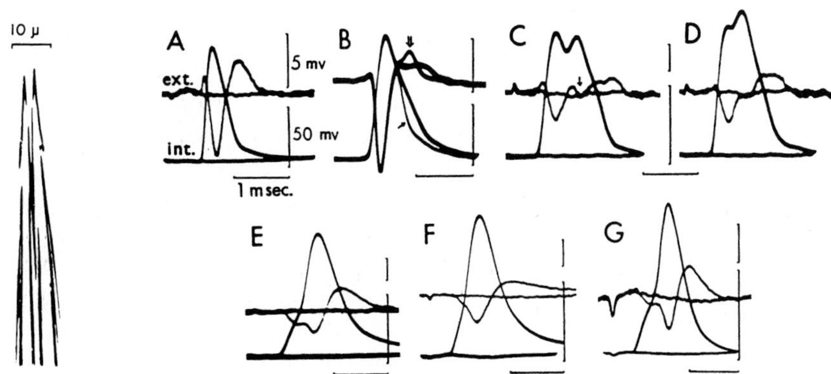

Fig. 3. Electrical potential changes recorded simultaneously, inside and outside spinal motoneurons. *Left panel*, photomicrograph of two microelectrodes connected in parallel for extra- and intracellular recording. *Right panel*, extra- (*ext.*) and intracellular (*int.*) action potentials recorded from motoneurons of the lumbar spinal cord in response to single pulses applied at just-threshold intensity to the ventral roots. Flat traces indicate lack of responsiveness to stimulation at just below threshold intensities. **A-G**, recordings presumed to be from the motor axon (**A**), from the initial segment (**B-D**), and from the soma and proximal dendrites (**E-G**). In **C** and **D**, inflections near the peaks of the *int.* action potentials reflect the delay of the action poptential as it propagates from the initial segment to soma and dendrites (*IS-SD* break). Arrow above the *ext.* trace in **C** denotes the *IS-SD* break recorded extracellularly. (Modified from Terzuolo and Araki, 1961).

Subprotocol 2
Extracellular Recording

Introduction

Recording the electrical activity of individual nerve, glial and muscle cells is important in research and clinical settings. Although one might want to record transmembrane potentials in every instance (see Part 3), this is technically difficult, often not feasible and, indeed, frequently not necessary. Researchers have therefore employed methods based on extracellular electrodes. Variations of these techniques have been developed to provide estimates of intracellular events as well as, in most cases, sequences of action potentials. In the latter application, it is most often the event of an action potential occurring at a particular time instant that is of importance, and not its amplitude or shape. In fact, in most recording arrangements, the amplitude of externally recorded action potentials is appreciably less than intracellularly recorded ones, and this attenuation depends on a number of factors, such as placement of electrodes relative to the membrane surface, electrical resistance and geometry of the external medium between them.

There are a huge variety of extracellular recording methods and electrode fabrication techniques, which cannot be reviewed here in any detail. Moreover, some procedural issues are addressed in other parts. Therefore, in this part we will instead present overviews with references to more specific literature.

Procedure

Overview : Recordings from Nerve and Muscle Fibers

In this section, we discuss the principles of recording from nerve or muscle fibers using fairly gross external electrodes.

Gross External Electrodes

Assume a nerve or muscle fiber is in contact with two external electrodes, as illustrated in Fig. 4. When an action potential is generated and propagates over the fiber (assumed uniform) from left to right, at any particular moment in time the membrane potential will be reversed from negative interior to positive interior over a small area of membrane whose extent depends linearly on conduction velocity. This is indicated by the stippled areas in Fig. 4. In all cases shown, the stippled areas progressively shift from left to right as time advances (indicated by increasing numbers).

- Monophasic recording (left column). When the right electrode is placed in contact with the crushed (depolarized) right end of the nerve (indicated by the dashed area), there is a small "injury" potential difference between the two electrodes, as indicated by the rightward deviation of the potentiometer's needle in panel 1. This corresponds to the resting state because the excitation has not yet reached the left electrode. When this occurs (panel 2), the potentiometer's needle swings to the left as the underlying area becomes depolarized. With the excited area moved between the electrodes (panel 3), the needle returns to its initial position. This does not change when the excitation reaches the dead region of the nerve and dissipates (panel 4). Similar results are obtained when fiber conduction between the two electrodes is blocked by, e.g., local anesthetics or damage, or when the right electrode is moved to a remote place (e.g., musculature or other tissue), yielding an *indifferent electrode. In conclusion, placing one electrode in a neutral, inactive place leads to a monophasic recording as shown on*

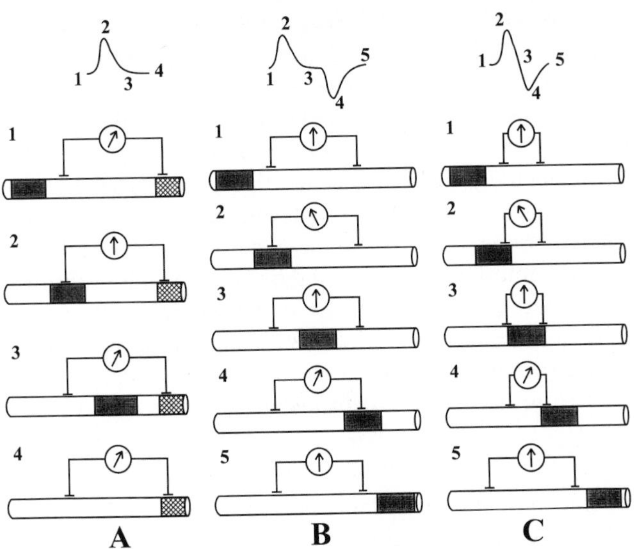

Fig. 4. Monophasic and biphasic recording of extracellular action potentials in nerve fibers. The curves at the top of the figure display the shapes of action potentials recorded under the conditions specific for that column. In the left column, the right end of the nerve is dead, i.e., the nerve is permanently depolarized in the region indicated by the cross-hatching. The middle column shows an intact nerve with widely spaced electrodes and the right column an intact nerve with closely spaced electrodes. The shaded areas represent the region of the nerve where the membrane potential is momentarily reversed. The action potential propagates from left to right. The numbers next to the curves indicate values of instantaneous membrane potentials which correspond to the location of the action potential relative to the recording electrodes, as shown underneath. The direction of current flow is indicated by the positions of the *arrows* (needles) in the potentiometers. (Redrawn and modified from Brinley 1980)

top of the column. Essentially this is also the situation when an active (different) microelectrode is advanced close to a neuron in nervous tissue and the inactive (indifferent) electrode is placed in some tissue (e.g., muscle or skin) remote from the recording site.

- Biphasic recording with widely spaced electrodes (middle column). When both electrodes are placed on viable fiber segments at a distance large in comparison to the active section, the reversed membrane potential will successively elicit opposite (biphasic) excursions of the potentiometer needle as shown on top of the middle column.
- Biphasic recording with closely spaced electrodes (right column). Spacing the electrodes closer in relation to the active zone's extent (right column) leads to an overlap of negative- and positive-going potential changes as shown on top of the right column.

With appropriate variations, these principles apply to recordings made under the following circumstances:

- Recordings from fibers in peripheral nerves, spinal roots or central tracts split down to thin filaments in acute animal preparations and humans. Splitting of the filaments is usually done manually with fine clockmaker or specially designed forceps. The filaments are then usually placed on two metal (silver, tungsten, platinum) hook electrodes, one of which can also be made an indifferent electrode by various means (setting equivalent to that in Fig. 4, left column; see also Chapter 17). Often, the filaments contain still other active fibers, so that the recording contains an *interference pattern*

of superposed action potential trains from several fibers (*multi-unit activity*); the experimenter's skill and experience are then required to *functionally isolate* a particular fiber by more splitting and spike separation devices or algorithms.
- Recordings from intact nerves or nerves dissected free (neurograms) and from muscles (electromyograms; see Chapter 26). Such recordings almost always yield *interference patterns with multi-unit activity*, and unless that is the objective, recording of single-unit activity will require specific data processing techniques. In other applications, *compound multi-unit activity* is recorded from these multi-stranded structures in response to electrical or mechanical or other stimulation which yields *evoked potentials*. Such investigations are frequent in neurology to determine whether pathways are interrupted or compound conduction velocities are normal or reduced.
- Recordings from intact nerves (neurograms) and from muscles (electromyograms) in chronic animal preparations or humans with chronically implanted nerve electrodes (see below).

Suction Electrodes

Often, it is technically not feasible to mount long filaments on hook electrodes, for instance when nerve or spinal root stumps are short. Using negative pressure, these stumps can then be sucked into a glass pipette or other fine tube filled with an electrolyte to make electrical contact (Snodderly 1973). With the suction electrode, extracellular recordings can easily be made from very small hearts (e.g., of insects), intestinal nerves, eye neurons, and muscles can be stimulated by an electrical pulse to the neurons innervating them. It is also possible to record from moving tissue using a plastic tube instead of glass. Suction electrodes can be self-manufactured, but are also available commercially (A-M Systems).

Chronically Implanted Nerve Electrodes

A number of electrode systems have been developed that are implanted chronically in animals or humans to record from various nervous structures, ranging from peripheral nerves to spinal roots, tracts and nuclei. Some of these systems will be dealt with in other chapters (Chapters 28, 29), and only a brief overview is given here.
- A variety of implantable nerve cuffs (see Loeb et al. 1996; and Chapter 29) have been developed for chronic recording, stimulation and pharmacological modulation of nerves. Recordings can also be obtained from longitudinally implanted intrafascicular electrodes (LIFEs) to be inserted into nerves. Metallized polymer fibers are said to be particularly suitable for long-term implantation into nerves (see McNaughton and Horch 1996; also further references), because they are highly flexible, small, biocompatible, and suitable as leadwires for multi-unit recording and as stimulating electrodes.
- Wires implanted in spinal roots and/or tracts. Free-floating fine wires have been implanted chronically into dorsal root ganglia (e.g., Loeb et al. 1977), dorsal roots (e.g., Prochazka et al. 1977) or ventral roots (e.g., Hoffer et al. 1981) of freely moving cats to record from randomly sampled single units. Arrays of wires are now routinely being implanted into central structures, such as the hippocampus of rats (e.g., Wilson and McNaughton 1993), to record simultaneously from several single neurons.

Overview : Recording with Microelectrodes

Single-unit recordings are now routinely performed with fine microelectrodes, made of several types of metal or carbon-fiber electrodes or glass micropipettes filled with electrolyte. Using microelectrodes requires a careful preparation of the nervous tissue to be penetrated as described below.

Selection and Preparation of Microelectrodes

Extracellular recording poses fewer problems than does intracellular recording (see Part 3). Therefore, the selection of the microelectrode is not that critical. Any small-tip electrode with sufficiently high resistance and low noise will do. Microelectrodes can be adapted to various uses, and can be combined into arrays that can be put to several uses, e.g. to record from a neuron while drugs are simultaneously applied nearby (see Chapter 7). Here we provide a brief overview of the several ways to record extracellularly the activity of neurons. There are many ways to modify them to suit particular needs. The enormous variety of microelectrodes prohibits a detailed comprehensive presentation of methods to fabricate them.

Micropipettes

By comparison to extracellular ones, intracellular microelectrodes must meet more stringent performance criteria. Accordingly, this section is limited to a brief summary of the desired properties of electrodes while a more extensive treatment is deferred until Subprotocol 3. Desired properties are:

- Inertness to tissue;
- Low noise;
- Minimal, unchanging tip resistance;
- Absence of tip potentials;
- Selectivity;
- Ability to contain dyes or markers.

For extracellular recording, vertical pipette pullers are suitable for preparing micropipettes. Since the intraelectrode electrolyte solution may diffuse out of the electrode, it should be compatible with the extracellular milieu; thus, for extracellular recording, NaCl is used more often, usually in a filtered solution of 2–4 M. The micropipette tip may be broken back with a glass rod under a microscope to yield low electrode resistances (2–10 MΩ).

Metal Electrodes

The main advantage of metal microelectrodes is their stability and flexibility. They therefore lend themselves particularly well to chronic recordings, for example from the monkey cortex, where they may be driven through the intact dura mater. They usually give more stable recordings than micropipettes and are less noisy. But they are equally suited for recordings from acute preparations and for electrical stimulation of nervous tissue. Metal electrodes may be combined with micropipettes used for iontophoretic application or pressure ejection of drugs (for references see Tsai et al. 1997).

The general properties desired of metal microelectrodes are:

- Ruggedness, mechanical stability and flexibility;
- Long, stable recordings;
- Low noise, high signal-to-noise ratio;
- Selectivity (good isolation of spikes of one cell from those of surrounding cells);
- Low bias (cells of different size should ideally be sampled with equal frequency).

There are scores of recipes for the fabrication of metal electrodes because many laboratories have tried to optimize them to specific uses (Snodderly 1973; Loeb et al. 1995). The following is a brief overview from a selection of papers where details are discussed and additional references can be found.

- Platinum electrodes insulated with glass. Their tips can be coated with platinum black to increase the effective tip area and lower the resistance. The electrodes can be strengthened by adding iridium to yield platinum/iridium electrodes. The most serious disadvantage of platinum electrodes is that they have very fine, fragile tips (Snodderly 1973).
- Tungsten electrodes are very stiff. When their tip diameter is small they are better for isolating cells and picking up high-frequency signals (Snodderly 1973), but they are

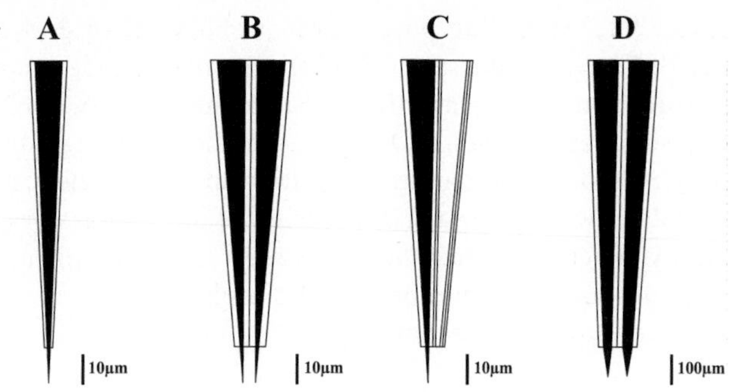

Fig. 5. Single and multiple tungsten microelectrodes insulated with glass. **A**: Single tungsten-in-glass electrode. **B**: Bipolar tungsten-in-glass electrode. **C**: Tungsten electrode/micropipette assembly, in the pipette is an inner microtube. **D**: Coarser bipolar tungsten electrode suitable for stimulation. (Redrawn and modified from Li et al. 1995)

noisy. The insulation can be made of lacquer or glass. Information about how to collapse glass capillaries onto a tungsten wire, to manipulate the insulation and the shape of the electrode tip and to recycle single and double barrel electrodes can be found in Li et al. (1995).
- Stainless steel electrodes are useful not only for recording but also for marking the electrode tip position. This is achieved by passing current through the tip and thereby depositing iron in its environment. The deposit is made visible as a Prussian blue spot by perfusing the animal with potassium ferrocyanide at the end of the experiment (Snodderly 1973). Stainless-steel *microwire* can be threaded into a glass capillary which is collapsed onto the wire to form glass-coated wire microelectrodes using a commercial electrode puller. Such microelectrodes can be combined with a usual micropipette in a *recording-injection system* (Tsai et al. 1997).
- Parylene-C coated iridium wires are said to be well suited for:
  - Penetrating hard connective tissue such as dura mater;
  - Recording from very small neurons;
  - Making multiple electrolytic lesions to mark recording sites along a track (see below);
  - Arrangements in multiple-electrode arrays;
  - Chronic microstimulation without electrochemical damage (Loeb et al. 1995).

Various types of metal electrodes are also commercially available (see Suppliers List at end of this chapter).

Carbon-Fiber Electrodes

Since the late seventies, carbon-fiber electrodes have been used extensively (for references see Kuras and Gutmaniene 1995). Their advantages are:
- Have much lower impedance than NaCl-filled micropipettes;
- Can be used in multi-barrel arrays with some of the barrels devoted to local iontophoresis and some (those containing a carbon fiber) to recording;
- Can be used to mark the extracellular recording site;
- Can also be used to scan the chemical environment of the cell via voltammetry (Stamford et al. 1993).

Their main disadvantage is that low-frequency noise hinders their use for recording of slower potentials, but silver-plating of the tips reduces the noise (Millar and Williams 1988), and so does treatment with pulsed electric current (Kuras and Gutmaniene 1995).

Overview: Marking of Electrode Tracks

In many studies, it is important to be able to correlate neurophysiological recordings with neuroanatomy, i.e. to localize the neurons recorded from. This problem is nontrivial, especially in chronic preparations where multiple electrode penetrations are made over many weeks or months. Various methods have been used to mark electrode tracks.

Electrolytic Lesions

Electrolytic lesions are the most common way of marking an electrode track, either at several places along the way or at the end. To make these, small constant currents (5–10 mA) need to be applied for several seconds, say 20 s. These lesions can be seen in unstained histological sections, even after several months when the tissue is stained for markers of glial reactions such as cytochrome oxidase. On the other hand, lesions can cause extensive tissue damage and are often difficult to separate and identify in closely spaced multiple tracks.

Stains

As mentioned above, *stainless steel electrodes* can be made to deposit iron by passing current through them, and the iron can then be visualized as a Prussian blue spot by potassium ferrocyanide in the perfusate.

Other methods involve stains such as

- Thionin (Nissl stain) (Powell and Mountcastle 1959);
- Cytochrome oxidase (Livingstone and Hubel 1984);
- Immunological staining using an antibody against glial fibrillary acidic protein (Benevento and McCleary 1992).

Again, one problem with these techniques is that closely spaced lesions from adjacent tracks are difficult to identify. Furthermore, they are less effective when the time between penetration and perfusion is long (DiCarlo et al. 1996).

Dyes

To circumvent such problems, a recently proposed marking technique involves coating the electrode with a fluorescent dye. During penetration, this slowly diffuses off the electrode and can later be viewed in histological sections under fluorescent optics (DiCarlo et al. 1996). The advantages are:

- Tissue sections need not be processed or stained;
- There is no detectable tissue damage;
- Adjacent tracks can be uniquely identified by using different dyes with different absorption/emission spectra or various mixtures of such dyes;
- Dye coating can be used with any extracellular electrode arrangement.

Overview: Attempts at Assessing Intracellular Events with Extracellular Electrodes

In addition to intracellular recording with microelectrodes (see Part 3), various methods have been used to assess transmembrane potentials with extracellular electrodes.

Sucrose Gap

One method of recording transmembrane potentials from elongated structures such as axons is the sucrose gap technique (see Fig. 6). The basic idea rests on two facts. Firstly, the usual way of recording action potentials by a pair of extracellular electrodes, as shown in Fig. 4, attenuates potential size because the low-resistance extracellular space effectively shunts the potential drop along the fiber. Thus, increasing resistance between electrodes by some kind of barrier counteracts this drop (see Purves 1981). Secondly, equalizing the extracellular and intracellular potentials by depolarization enables the recording of transmembrane potentials. Thus, using a strongly depolarizing extracellular fluid, such as highly concentrated K_2SO_4, on one side of the barrier essentially renders the electrode in this place intracellular.

The barrier can be created by any high-resistance medium, such as air, paraffin (mineral) oil or a solution of non-polarized substance. A solution of highly purified sucrose

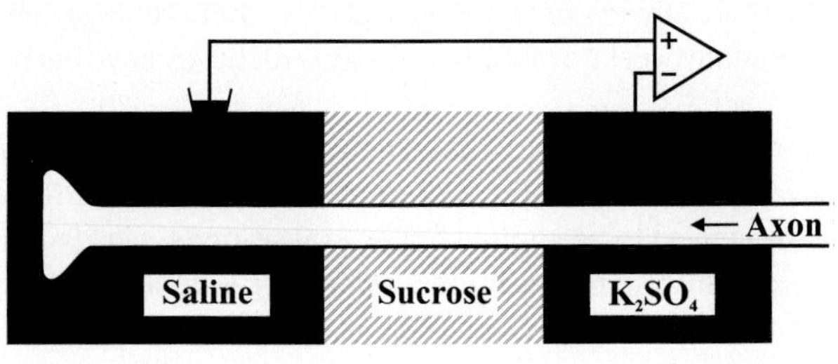

Fig. 6. Sucrose-gap recording in which an extracellular electrode can be utilized to record membrane potential. The concentrated K_2SO_4 solution depolarizes the neuron so as to make this part of the fiber isopotential with the extracellular fluid. (Redrawn and modified from Pittman 1986)

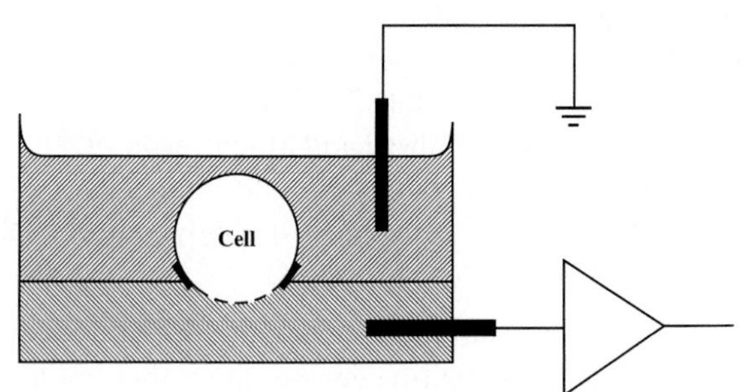

Fig. 7. Kostyuk´s method of "intracellular" recording. A large cell is sealed over a hole and the protruding membrane ruptured to provide electrical access to the intracellular compartment. No transmembrane electrode is needed. (Redrawn and modified from Purves 1981)

in de-ionized water penetrates between cells and can thus create very high extracellular resistance (Purves 1981).

Kostyuk's Method

A method applicable to large cells (e.g., snail neurons, invertebrate oocytes) isolated *in vitro* is illustrated in Fig. 7. It can be used for fast voltage-clamping and allows one to change the intracellular medium.

Estimation of Postsynaptic Potentials by Spike Trains

Since intracellular recordings are impossible to achieve in freely moving animals and in humans (at least under normal circumstances), methods have been developed to indirectly assess postsynaptic potentials (EPSPs, IPSPs) from extracellular recordings. The idea is that in a discharging neuron, the probability of postsynaptic discharge is in part due to membrane potential changes such as EPSPs and IPSPs. That is, an EPSP will transiently increase the firing probability and an IPSP will decrease it. But the precise relationship between firing probabilities and potential changes is not straightforward because of the nonlinearity of the encoder transforming potential changes into discharge (see, e.g., Fetz and Gustafsson 1983; Midroni and Ashby 1989). These issues are dealt with in Chapter 19.

Results

To illustrate the use of extracellular recordings, we selected two examples from the vast body of literature. The first is drawn from an early investigation of the mammalian (cat) retina (Kuffler 1953), the second from a study investigating the effects of the red nucleus (ruber) on the responses of Renshaw cells in the spinal cord (Henatsch et al. 1986).

Cat Retinal Ganglion Cells

Retinal ganglion cells are characterized by their "receptive" field, i.e., the area on the retina whose illumination affects their discharge. The size of this field varies with the size and intensity of the spot of light employed and with the level of ambient background illumination, shrinking with increasing illumination and expanding after dark adaptation (e.g., Kuffler 1953). This field is usually not homogeneous in that it can be divided into subdivisions.

Let's consider a specific example. In Fig. 8 (from Kuffler 1953), a ganglion cell in the cat's retina was recorded with a glass-coated platinum-iridium wire electrode. The experimental arrangement is depicted in the lower right inset. The electrode tip was just above the recorded ganglion cell in the middle. A light spot of 0.2 mm diameter and an intensity of 3–5 times threshold shined into the central region *b*, evoked an intense discharge as shown in panel b. Although persistent, this response adapted over longer periods of illumination (indicated by the lower trace in *b*; see legend). When the stimulus was moved to locations *a*, *b* or *d* in the inset (still within a central area of ca. 0.5 mm diameter), the respective discharges were much weaker (panels *a*, *b* and *d*, respectively) and adapted much faster. Thus, even within such a small central area which could evoke a discharge, the retinal ganglion cell excitability decreased with distance from the center.

Such topographical dependence was even more pronounced when more extensive areas were explored (not shown). Moving the stimulus beyond the central area would cause the cell to fire not during illumination, but after its end. This is called an off-response. Mixed on- and off-responses were obtained in response to illumination of intermediate region. Other cells displayed the opposite pattern, i.e., an off-response from

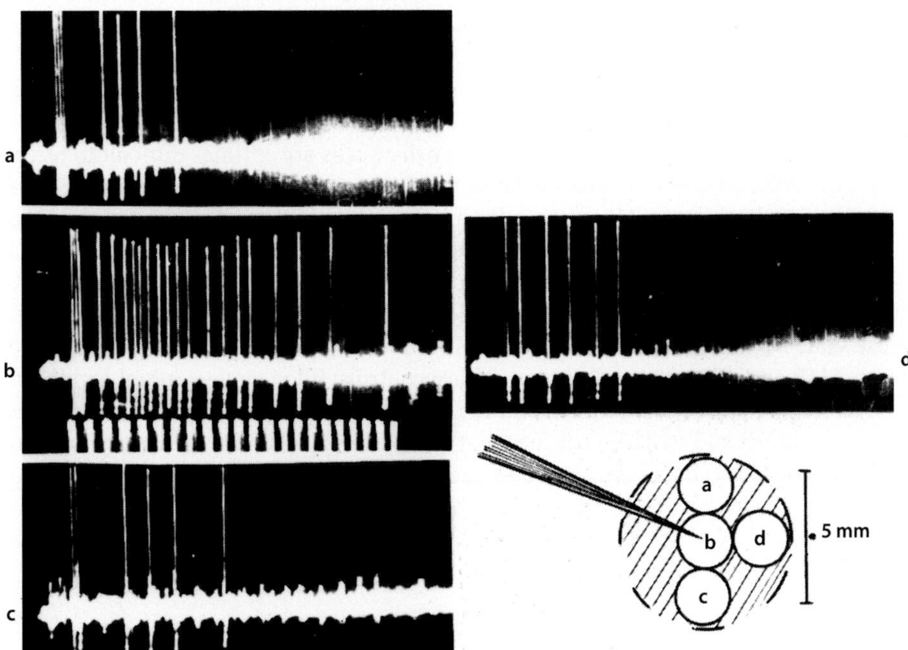

Fig. 8. Center portion of receptive field. Ganglion cell activity caused by circular light spot 0.2 mm in diameter, 3–5 times threshold. Background illumination was about 30 meter candles. Positions of light spot indicated in diagram. In *b* an "on" discharge persists for duration of flash. Intensity modulation at 20/s. Movement of spot to positions *a*, *c*, amd *d* causes lower frequency discharge which is not maintained for duration of light stimulus. Movement of spot beyond shaded area fails to set up impulses. Potentials 0.5 mV. (Reproduced, with permission, from Kuffler 1953).

their receptive field center and an on-response from the surround. Using two small spots of light to study interactions between the different areas within a receptive field indicated that off-areas inhibited and on-areas excited a particular ganglion cell. Different mixtures of these influences accounted for a multitude of retinal ganglion cell discharge patterns (Kuffler 1953). Such studies contributed significantly to our present knowledge of signal processing in the mammalian retina.

Spinal Renshaw Cells

Many spinal interneuronal systems receive modulating inputs from supraspinal motor structures; so do Renshaw cells which receive excitatory recurrent axon collaterals from motoneurons and in turn inhibit various spinal neurons (see, e.g., Windhorst 1990, 1996). The investigation, from which the following results are reproduced (cf. Fig. 9), was undertaken to study the effects of electrical stimulation of the mesencephalic red nucleus on the excitatory coupling between spinal motoneurons and Renshaw cells (Henatsch et al. 1986). In chloralose-urethane-anesthetized cats, monosynaptic reflexes (*MR*) were recorded with hook electrodes from lumbosacral ventral roots in response to electrical (test) stimulation of selected hindlimb muscle nerves, and the discharge of Renshaw cells recurrently excited by the reflex-activated motoneurons was simultaneously recorded with NaCl (3 M) filled micropipettes (cf. Fig. 9C). A coaxial bipolar steel electrode was stereotaxically positioned in the contralateral red nucleus to deliver conditioning stimulation (the electrode tip position was later verified by electrolytic lesion and histological techniques; see above). Conditioning stimulation consisted of trains of alternating current pulses which differed in strengths (between 30 and 250 μA peak to peak), rate (between 500 and 1000 Hz) and duration (5–25 ms) and time relation to the test stimuli, the onset of the train usually preceding the test stimuli by several milliseconds.

Typical examples of the results obtained are displayed in Fig. 9. Fig. 9C shows individual *MR* and *RC* responses without (upper two traces) and with rubral conditioning (lower two traces). Because of their small size, Renshaw cells are at times difficult to record from; this explains the low signal-to-noise ratio. The number of *RC* spikes in the burst (*Nb*) decreased, while the *MR* amplitude (*AMR*) slightly increased upon stimulation of the red nucleus. For comparison, both *Nb* and *AMR* were measured for 20 test stimuli at each stimulus strength, while stimulus strength varied widely within the range for activation of group *Ia* fibers. Figure 9A is a plot of *Nb* versus the average *AMR* measured at the same *Nb* and stimulus condition. Rubral conditioning displaced the relation between *AMR* and *Nb* to the right (open circles: controls; filled circles: rubral conditioning). Similar changes were found in another two cells from two different experiments as shown in Fig. 9B and D. This indicates that the excitatory coupling between motoneurons and Renshaw cells is subject to modulation, in this case inhibitory, from the red nucleus (for more details see Henatsch et al. 1986).

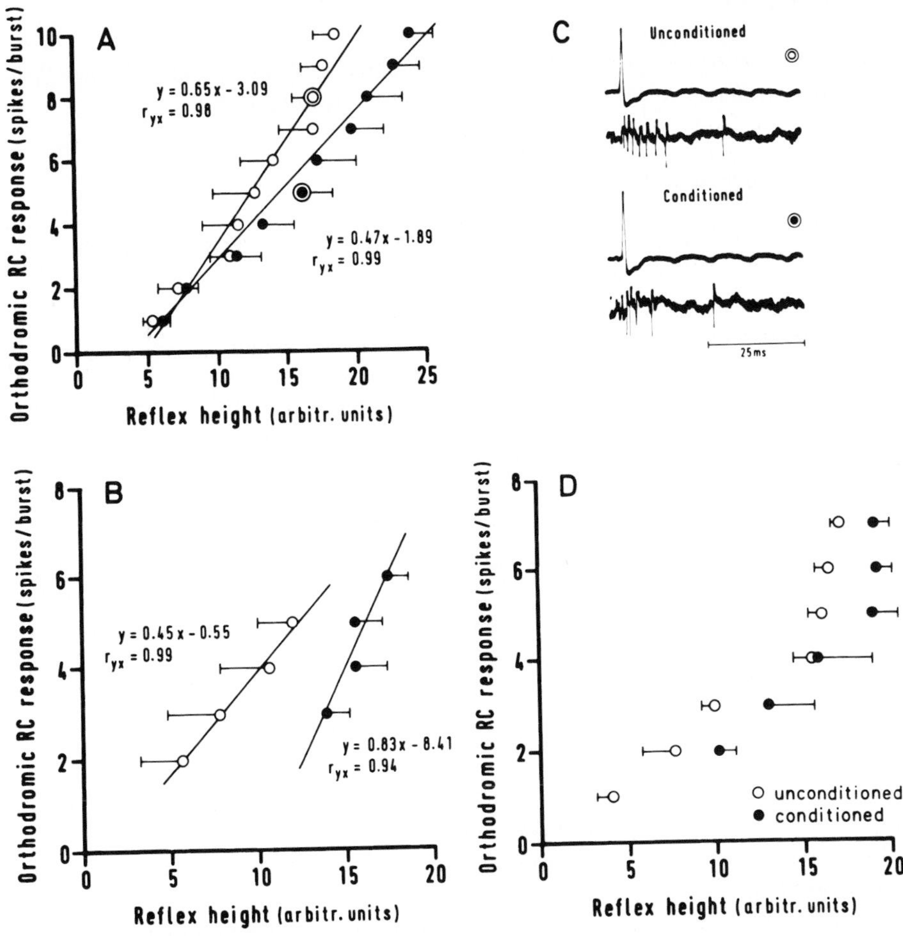

Fig. 9. "Linkage characteristics" of the Renshaw cell-monosynaptic reflex relation with and without nucleus ruber conditioning. Results from three Renshaw cells (*RC*) from different cats. Orthodromic test shock trials were delivered in groups of systematically varied strengths. Monosynaptic reflexes (*MR*) and orthodromic responses of a linked *RC* were recorded simultaneously. Ordinates: Number of spikes per orthodromic *RC* burst. Abscissae: Averages of *MR* amplitude for all reflexes occurring at the same given number of *RC* spikes (with *SD* = standard deviation, drawn in only one direction). From the paired data graphical "characteristics" of the functional linkage between the reflex-producing motoneuron (*MN*) pool and the *RC* receiving recurrent collaterals from that pool were obtained. Regression lines were drawn in samples **A** and **B**, but not in sample **D** due to the nonlinear course of this relation. The regression equations are given, with *x* denoting *MR* amplitude and *y RC* spike number. R_{yx} denotes the correlation coefficient. *Open circles*: unconditioned controls. *Filled circles*: with conditioning ruber stimulation. In **A**, the rubral stimulation effects a reduction in slope; in **B**, the slope is increased, but the whole curve is shifted towards higher values of reflex amplitude; in **D**, most of the points are roughly parallel-shifted to higher reflex amplitudes. In all cases, the overall effect is a displacement of the curve in an inhibitory direction (for the *RC*), indicating a reduction of the *MN-RC* linkage. *Inset* **C** (related to **A**): Two original records of orthodromic unconditioned (upper two traces) and conditioned (lower two traces) *RC* responses (lower traces) at about equal *MR* amplitudes (upper traces). The corresponding points in the two curves in **A** are encircled. (Reproduced, with permission, from Henatsch et al. 1986).

Applications

Extracellular recording, particularly when combined with other methods, can help solve several problems but is not devoid of certain interpretative difficulties.

Neuron Characterization and Identification

One goal of extracellular recording is to identify and characterize neurons. In this context, several parameters are of use.

- Location
  In invertebrate ganglia, individual neurons can be reproducibly identified from one animal to the next. In the *in situ* vertebrate *CNS*, this is not possible for individual cells, but certain classes of cells, such as the pyramidal cells of the cortex, can often be identified with the help of additional criteria such as their responsiveness to the electrical stimulation of particular fiber bundles (such as the cerebral peduncles in the case above) or the depth of the electrode tip.
- Response profiles
  The major impact of extracellular recording has been the description of the discharge patterns of neurons as they relate to sensory inputs and motor acts. This work has generated most of the presently available knowledge about the signal traffic through sensory-motor and other systems.

Functional Connections

Extracellular recording can be used to infer functional connections between neurons. This pertains to both inputs and outputs of a neuron.

The *afferent connections* of a neuron can be studied in several ways.

- Stimulation of afferent pathways
  One very common method is to electrically stimulate afferent pathways (e.g., peripheral nerves or central tracts or nuclei) and record from the postsynaptic neuron, as illustrated in Fig. 10A. The latencies of postsynaptic discharge may give some indica-

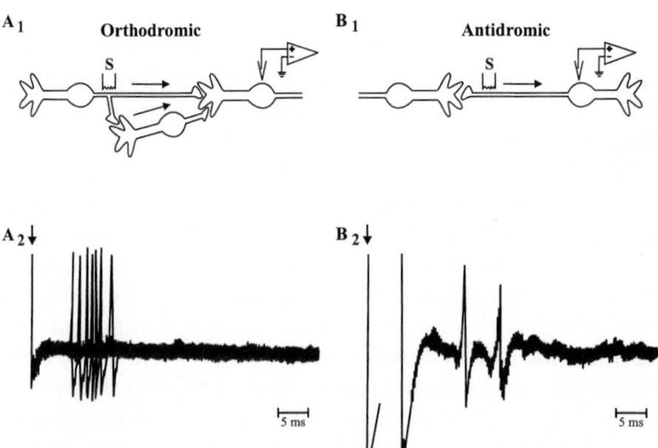

Fig. 10. Ortho- and antidromic potentials. **A:** Stimulation of afferent pathways to a neuron (**A1**) with an electrode *S* may elicit action potentials in the postsynaptic neuron, usually at variable latencies, because the intervening synapses render the coupling loose. This can be seen in the five superimposed oscilloscope traces in **A2**. **B:** Stimulation of an axon of a neuron (**B1**) with an electrode *S* results in an antidromic potential which occurs at a fixed latency because it does not cross any synapses. Five superimposed oscilloscope sweeps (**B2**) illustrate that antidromic potentials followed two stimuli (*arrows*) with invariant latency. (Redrawn and modified from Pittman 1986)

tion of the number of synapses interposed between the stimulated and the responding neurons.

- Cross-correlation of the discharges of presynaptic and postsynaptic cells
  Another way is to use two extracellular electrodes to simultaneously record from a presynaptic and a postsynaptic neuron (not illustrated). Using cross-correlation techniques (see Chapter 18), connections between the neurons can be indirectly inferred. This method requires that both neurons produce discharges at rather modest discharge rates. If they do not do so on their own, discharges can be evoked by iontophoretic application of excitatory substances, such as excitatory amino acids (cf. Chapter 7).

The *efferent projection* of a cell can be studied as follows. A cell's axon is electrically stimulated in some anatomical structure and the evoked action potential is recorded from the soma, as illustrated in Fig. 10B. The evoked action potentials should 1) have constant latencies, 2) follow repetitive stimuli at high rate, and 3) collide with orthodromic action potentials, as illustrated in Fig. 11. Combined with distance measurements, this method may yield estimates of the cell's conduction velocity.

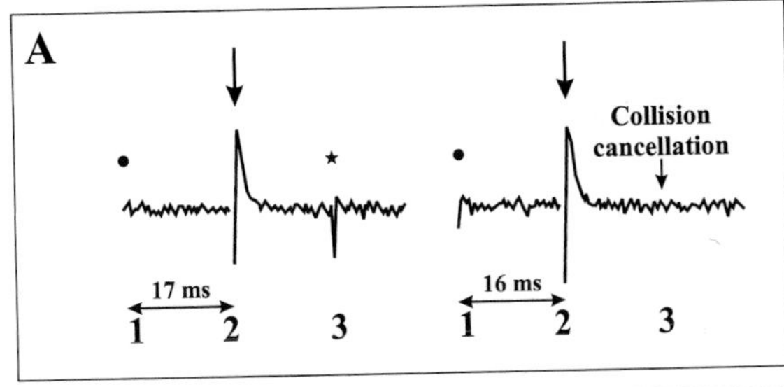

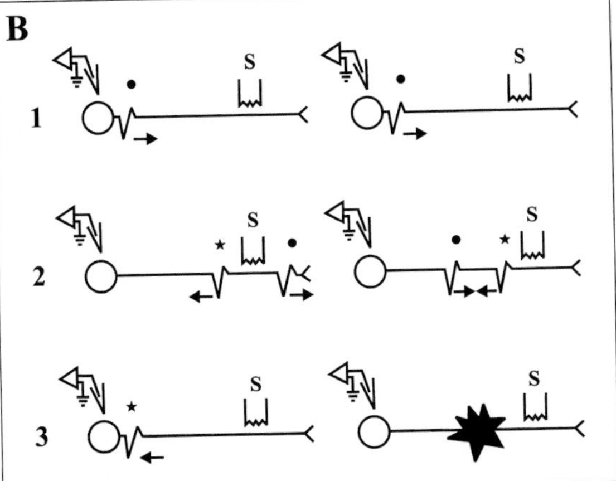

Fig. 11. Collision test for antidromic invasion. **A:** Oscilloscope traces illustrating extracellular voltage records of a spontaneous action potential (•) followed 17 ms later by stimulus artifact (Ø). An antidromic potential (*) followed after a 16 ms latency. On the right is seen a similar sequence of events; however, the stimulus follows the spontaneous potential by 16 ms, which is within the antidromic latency and the antidromic potential is therefore not seen due to collision. The mechanisms involved in this are explained in **B** at the times indicated by the numbers (*S* denotes the stimulating electrode). 1, The spontaneous action potential (•) invades the axon in an orthodromic direction. 2, On the left, the stimulus elicits an antidromic potential (*) *after* the orthodromic potential (•) has passed the point of stimulation. On the *right*, the stimulus is delivered *before* the orthodromic potential has reached the stimulation site. 3, On the *left*, the antidromic potential invades the soma where it is recorded by the electrode; on the right, the antidromic and orthodromic potentials have collided and therefore no antidromic potential invades the cell body. (Redrawn and modified from Pittman 1986)

Sampling Bias

Despite its relative ease and the importance of the results it provides, extracellular recording poses certain problems. One is the *sampling problem.* Since large neurons produce more current than small ones, the larger the cells the easier it is for extracellular electrodes to pick up the currents produced by them. This leads to a sampling bias that favors bigger cells at the expense of smaller ones. Also, unless there is a way of activating all cells, it is only spontaneously active neurons that are sampled. Nor is the number and identity of the spontaneously responsive neurons unrelated to the type and dose of the anesthesic employed (Towe 1973; Towe and Harding 1970). These two factors might distort the proportions of different kinds of cells in a sample. This is particularly disturbing when their estimation is one of the motives for performing the study. Finally, extracellular recording yields much less information about signal processing within a cell than does intracellular recording. For instance, slower membrane potential changes and postsynaptic potentials are not seen, and can only be estimated using very indirect statistical methods (see above).

Subprotocol 3 Intracellular Recording with Sharp Electrodes

Introduction

The principal advantage of intracellular recording is that it allows measurements of events and phenomena that are not possible with extracellular recording, including subthreshold events such as excitatory and inhibitory postsynaptic potentials (EPSPs and IPSPs), and postsynaptic currents (EPSCs and IPSCs), neuron input resistance and reversal potentials for PSPs and PSCs. Moreover, the contributions of cytoplasmic and extracellular ions to PSPs and PSCs can be estimated by ion substitution (Eccles 1964), and the roles of second messengers can be assessed by injecting intracellular protein activators and inhibitors from the recording electrode (Richter et al. 1997). The disadvantage lies in the difficulty and lower probability of obtaining and maintaining stable, high-quality recording.

In this part of the chapter, we describe how micropipettes are selected and prepared for intracellular recording, how cells are penetrated, and how membrane potentials and membrane currents are recorded.

Procedure

Preparation of Micropipettes

Desired Micropipette Properties

Micropipettes that are filled with electrolyte solution should have the following properties:

- They should produce minimal damage during tissue insertion, cell search and cell penetration. They should, therefore, have a needle-like tip that will pierce the cell membrane with minimal damage, and a profile which, when the electrode is inserted to its deepest depth, minimizes tissue displacement and damage at the surface.
- Noise level should be minimal to improve the signal-to-noise ratio of the recording. The noise-related problems inherent to micropipettes and recording systems as well

as suggestions of how to solve them are presented in detail elsewhere (Purves 1981; Halliwell et al. 1994; Benndorf 1995). The noise produced by a micropipette filled with electrolyte solution will be related, among other things, to the composition of the glass (see below), the tip diameter, the type of electrolyte solution, and additives for intracellular injection. Also, RC noise will increase with depth of penetration in tissue (Benndorf 1995). Tip resistance can be reduced by bumping or beveling. RC resistance is reducible by coating the micropipette with Sylgard up to about 50 μm from the tip.
- Electrode resistance should remain steady throughout the period of recording. To ensure that such is the case, microelectrode resistance is monitored by injecting current pulses (see below).
- Tip potentials should be eliminated or minimized. Tip potentials are liquid junction potentials that develop at the tip of the micropipette. The DC offset of the intracellular amplifier should be used to compensate for these potentials.
- The microelectrode should be capable of passing relatively large currents with minimal noise and rectification.
- They should have rapid settling times after passaging current pulses.

Current-passing capability and rapid settling are particularly critical for discontinuous single-electrode voltage clamp (dSEVC) experiments. A rapid and simple method of coating sharp microelectrodes with Dricoat and Vaseline to enhance these properties for dSEVC experiments has been described (Juusola et al. 1997).

Selection of Micropipette Glass

Glass capillaries for fabrication of micropipettes are commonly constructed from borosilicate, aluminosilicate or quartz. Micropipettes made from quartz have low noise properties and are extremely tough and durable, allowing the micropipette to be driven through dura membrane without damaging the tip. A pipette puller that utilizes a laser beam to generate sufficient energy to soften the glass for pulling is necessary, and the tips cannot be readily broken back or beveled to reduce tip resistance. Borosilicate glass is most commonly utilized, being harder, more durable and having better electrical resistivity than soft glass. It is easy to work with and produces micropipetes of varying length, shape and tip profile with the proper pipette puller. Aluminosilicate glass is harder and more durable than borosilicate, and has a lower coefficient of thermal expansion and reduced electrical conductivity. In addition, aluminosilicate glass thins out as it is drawn to a tip, allowing very fine tips to be formed at moderately short tapers.

Glass capillaries are also selected on the basis of wall thickness, inner and outer diameters, and internal structures such as capillary filaments or septa that facilitate filling with electrolyte solution. Our bias is in the direction of relatively thick-walled borosilicate glass, e.g. 1.5 mm outside diameter x0.86 mm inside diameter (Clark Electromedical Instruments, UK). Micropipettes of varying length, profile and final tip resistance can be fabricated with relative ease from such capillaries. They are well suited for producing sharp microelectrodes that are long and slender, yet durable and resistant to bowing and thus suitable for intracellular recording deep inside the brain.

Choice of Pipette Pullers

For intracellular recording from all but the largest cells, the best results are obtained with horizontal pullers that pull pipettes in two stages and can be programmed to adjust heating and pulling parameters. These devices offer greater flexibility in producing micropipettes of optimal length, profile and tip diameter than vertical pullers. Examples of high-quality programmable horizontal horrizontal pullers are the Brown-Flaming type and the Zeiss type (DM Zeiss, Germany).

Choice of Micropipette Filling Solutions

Aqueous solutions of potassium salts are generally utilized, such as:
- Potassium chloride (3M): KCl produces the lowest tip resistance for a given tip diameter and is therefore the most frequently used electrolyte in voltage clamp experi-

ments. Moreover, during filling, diffusion to the tip of the pipette is relatively rapid with a minimum of bubble formation. Its principal limitation is that dialysis of the cytoplasm with KCl will reverse chloride-dependent IPSPs.

- Potassium acetate (3–5M): With K-acetate there is of course no chloride-dependent IPSP reversal, but electrode resistance is higher, so there is a longer settling time after current pulses, and electrodes tend to rectify.
- Potassium citrate (3M): K-citrate is relatively viscous, therefore micropipettes tend to fill somewhat less readily. However, current passing capability is high and settling time after a current pulse is fast.
- Potassium methylsulfate (2M): Current-carrying capability is excellent, settling time is fast and rectification is relatively low. Tip resistance remains relatively stable during long recording periods. According to one report (Zhang et al. 1994), K-methylsulfate is the least disruptive to intracellular structures and Ca^{2+} homeostasis. Afterhyperpolarizing potentials and currents are minimally affected. K-methylsulfate would, therefore, be an obvious choice for voltage clamp experiments where membrane potential must be clamped to negative potentials with anionic current.
- BAPTA (5mM or 10mM) can be added to any of the above solutions to buffer Ca^{2+} that might have entered the cell or have been released following microelectrode penetration. BAPTA seems to improve membrane sealing and, therefore, improve recording quality.

Filling Micropipettes

Micropipette solutions should be first filtered through standard, medium-retention filter paper and stored refrigerated when not used immediately. On the day of experimentation, solutions are injected from a syringe through a 0.2 μm membrane filter and fine-gauge needle into the micropipette, with the latter held in a vertical position. The solution should be at the tip of the needle before injection. It is important that there be no bubbles trapped in the solution near the tip of the recording micropipette after injection, since they will introduce noise and impede current passage. Bubbles may be removed with cat whiskers that have been pre-soaked in 70 % ethanol. The whisker is inserted gently, as close to the tip as possible, and twirled. In addition, storing filled micropipettes vertically and tip down in a humidified chamber for several minutes allows bubbles to migrate from the tip to the surface of the filling solution.

Coupling the Micropipette to the Recording Amplifier

Connections between the micropipette and the amplifier are made with non-polarizable electrodes, generally silver/silver chloride wires. For best results, these should be prepared just before use, but this is not an absolute requirement. If prepared days before use, they must be stored in darkness to prevent oxidation of AgCl. Ag/AgCl wires are prepared by connecting a length of silver wire to the positive pole of a 9V DC source (e.g., battery) and immersing the other end to a desired length in aqueous 1N HCl solution. A second silver wire of larger diameter is connected to the negative pole and positioned in the solution. This should result in the formation of white AgCl within a few seconds on the micropipette wire, and formation of oxygen bubbles on the other wire. The wire is inserted into the micropipette, chlorided to a length greater than the column of electrolyte. The wire is secured in place in the micropipette with wax or fast-drying cement, and attached to the headstage of the amplifier via a short BNC cable (150mm or less) and crocodile clip.

Wire-attached, sintered AgCl pellets small enough to fit inside micropipettes are commercially available. Small chambers equipped with Ag/AgCl pellets and an O-ring to hold the micropipette in place can also be purchased (e.g., World Precision Instruments, USA). The chambers are filled with an electrolyte solution identical to the micropipette solution, and are directly connected by a metal post to the headstage of the amplifier. Sintered AgCl pellets or discs are also available that are light-stable and capable of carrying more current than Ag/AgCl wire (Hallowell 1994).

The disadvantage of chambers is that they bring the headstage close to the preparation, which may be a disadvantage in restricted spaces.

Sintered AgCl pellets or Ag/AgCl wires are also used to ground the animal through the recording system (see Section 1.1.5).

Recording Intracellularly From Neurons

Methods for preparing nervous tissue for locating and finding neurons with intracellular micropipettes were described previously (Section 1.1.4). Here we describe methods that are used after the pipettes are inserted in the brain or spinal cord.

Approaching the Neuron

Neurons are generally searched for with the amplifier in the Bridge Mode (see below, 3.3.1). For successful impalement, the tip of the micropipette must be free of tissue debris. Dirty tips will greatly increase electrode resistance (R_E) and, therefore, noise as the microelectrode is advanced. It is often advisable to frequently check R_E, either by passing negative-going constant current pulses (50–100 μs duration) through the pipette and observing the voltage deflections, or by reading R_E directly from a monitor built into the amplifier. Tips should be cleared with clearing current before attempting to impale a neuron.

Cell-searching and impalement are best done with a high-performance, motor-driven microdrive. The micropipette is advanced at low velocity in the continuous mode to the general region where pools of sought-after neurons are located, then advanced in single steps for cell impalement. Step sizes may range from 1 to 5 μm, however there is no formula for determining the step-length when approaching a neuron. The optimal length will be determined empirically as experience is gained with the species and age of the animal, the size of neurons, the arborization of neurites and the condition of the tissue, among other things.

Impaling the Neuron

Two methods are most often used to impale neurons *in situ*. The first consists of manually applying a sudden "jolt" of positive clearing current (about 100 nA) through a toggle switch simultaneously with a single forward step of the micropipette. Some amplifiers come equipped with current pulsers to aid impalement. Our experience is that manual application through a spring-loaded toggle switch is more effective. The second method consists of over-compensating the electrode capacitance to produce ringing for a very brief period as the micropipette is advanced through a single step. Many amplifiers come equipped with such "buzzing" circuits.

Successful penetration will be signaled by a steady, negative shift of membrane potential, −50 mV or more negative when the cell is not discharging or generating PSPs. Sealing of membrane around the pipette can sometimes be aided by very small backward or forward steps (0.5 μm). Other signs of successful impalement include action potentials with rapid rates of depolarization that overshoot and terminate in well-developed afterhyperpolarizations, and well-defined PSPs.

Recording Modes

After cell impalement, neurons are usually recorded from in either of three modes: *Bridge Mode, Discontinuous Single Electrode Current-Clamp (dSECC), Discontinuous Single Electrode Voltage-Clamp (dSEVC)*. Common to all three recording modes is the necessity to compensate for the various sorts of capacitance that develop along the microelectrode in tissue and in the recording system. This is accomplished through the utilization of capacitance neutralization circuits.

It should be noted that two-microelectrode voltage-clamping with a fast Field Effect Transistor-controlled switching circuit has also been very effectively utilized to analyze currents in spinal motoneurons *in situ*. The methods will not be described here, but the interested reader can refer to Schwindt and Crill (1980) for details.

Recording in Bridge Mode

In the bridge mode, an operational amplifier monitors the microelectrode voltage continuously. Moreover, continuous current can be injected. The bridge circuit allows voltage differences arising from current flow through the microelectrode (IR_E) to be balanced, or nulled out so that only bioelectric potentials remain. The value of R_E at the null point can then be read from the dial setting.

The advantage of the bridge mode in recording membrane potential is that there is no circuit noise associated with switching between current-inject- and potential-record-modes, as in dSECC recording. However, R_E often changes with time and with passage of current and should therefore be frequently checked and balanced. Many investigators do this by injecting small negative-going constant current pulses, duration 50–100 ms at fixed intervals (e.g., at the beginning of each sweep of the oscilloscope).

1. With the micropipette in the extracellular fluid, the response to a current pulse should approximate a square wave, with fast transients at the start and finish of the pulse. The fast transients are minimized by increasing capacity compensation through a potentiometer to produce the most rapid decay without producing overshoots, undershoots or oscillations.
2. Capacitance should be adjusted frequently as a cell is searched for, including soon before cell penetration. Step-by-step procedures and illustrations for proper bridge-balancing and capacitance compensation are included with the *Instructions for Use* manuals supplied with most commercially available microelectrode amplifiers.
3. When a neuron is adequately impaled and before the bridge is balanced, the voltage change that is due to a current pulse (ΔE) equals $IR_E + IR_N$. Between the capacitance transients, ΔE will reach a peak and decay in two phases, fast and slow, related to the time constants of the glass and cell membrane, respectively. The fast components are eliminated by balancing the bridge, leaving a slow voltage component resulting from current passage across the neuronal membrane (IR_N). If applied current is constant and known, the neuronal input resistance (R_N) can be calculated by measuring the slow voltage component.
4. Tip potentials will sometimes develop during intracellular recording, which can result in overestimation of membrane potentials. This will be evident from negative values of potential recorded when the microelectrode is withdrawn from the cell. Therefore, a useful countermeasure is to subtract from the intracellular measurements the value of the membrane potential that is measured after the microelectrode is backed several hundred microns away from the cell.

Recording in Discontinuous Current Clamp Mode

In discontinuous single electrode current clamp (dSECC) mode, amplifier-, sample-and-hold- and switching circuits enable the microelectrode to cycle at high frequency (kHz) between passing current and picking up membrane potential. The arrangement allows the recording of membrane potential independently of IR_E, provided R_E is not exceedingly great. Cycling rates must be sufficiently high, generally greater than ten cycles per membrane time constant ($R_m \times C_m$), to allow C_m to smooth the membrane voltage response to an injected current pulse. An advantage of dSECC is that small changes of R_E will not contaminate electrotonic potentials during measurements of R_N. Recording is, however, noisier than in Bridge Mode because of switching.

Protocol – same as recording in bridge mode

Recording in Discontinuous Voltage Clamp Mode

Over twenty years ago, Dean and Wilson (1976) showed that discontinuous single electrode voltage clamp mode (dSEVC) was useful for recording membrane currents of spinal motoneurons in the intact preparation, and more recently the method has been utilized to record currents in other types of neurons *in situ*, including those of the medullary respiratory network (Richter et al. 1996). The theory of recording from neurons in dSEVC mode has been described in detail (Brennecke and Lindemann 1974a,b; Wilson and Goldner 1975; Johnston and Brown 1983; Finkel and Redman 1984) and will be only briefly described here. Concise and informative descriptions of voltage-clamp methodology are given in Pellmar (1986) and Halliwell et al. (1994).

dSEVC is an extremely powerful technique with a unique advantage of enabling the recording and measurement of voltage- and time-dependent currents with a single electrode and with high temporal resolution. The main disadvantages of dSEVC lie in the limitations imposed on the method by cell size and geometry and by the electrical properties of the micropipette. In addition, inherent in the relative complexity of dSEVC is the greater possibility of error in measurements of membrane currents resulting from improper circuit adjustments.

The dSEVC operating system includes a command voltage source, buffer and feedback amplifiers, sample-and-hold circuits and a switching circuit which alternates at high frequency between current-injecting and voltage-recording modes. Membrane potential is held at a pre-determined level while membrane currents flowing at that potential are measured. A feedback circuit compares the cell membrane potential (V_m), sampled at a time when IR_E has decayed to zero, with the command holding potential (V_H). Current is injected to compensate for any difference between V_m and V_H produced by spontaneous synaptic potentials (PSPs) or by injected voltage steps or ramps. The injected current is measured and recorded, which is equal and opposite to the current flowing across the cell membrane (I_m) at a corresponding value of V_m.

Protocol

1. The first step in dSEVC recording is to ensure that the micropipette is functioning properly. The micropipette must not rectify appreciably, as evidenced by equal capability in passing positive and negative current, and V_E must settle quickly to baseline after a current pulse. Gentle bumping of the micropipette tip under microscopic control after filling will increase current-carrying capability (Richter et al. 1996). Fast settling time is necessary to allow high-frequency sampling of V_m but not V_E. Electrode settling time is decreased by keeping electrode capacitance to a minimum. This can be accomplished by using steeply-tapering microelectrodes (although this will result in greater tissue displacement and damage when recording from deep layers), minimizing the depth of electrolyte in the microelectrode and on the surface of nervous tissue, pre-coating the micropipette to reduce capacitance (Juusula et al. 1997), placing a layer of mineral oil over the tissue surface (which coats the microelectrode), and keeping cable length between the microelectrode and the amplifier head-stage to a minimum. In addition, microelectrodes can be shielded with an insulated metal coating which is connected to the case of the unity-gain head stage (Finkel and Redman 1983b). The disadvantage of the latter approach is that it increases high-frequency noise.
   Recently, amplifiers have become commercially available (npi electronic GMBH, Germany) that greatly decrease settling time and increase the capability to record V_m responses with high resistance microelectrodes through the use of a method called supercharging (Strickholm 1995a,b). Supercharging can be used in both dSECC and dSEVC modes to reduce settling times of sharp microelectrodes (40–80 MΩ) to 2–3 ms. This permits switching frequencies of 30–40 kHz with a duty cycle (time in the current-injecting mode) of 25% (Richter et al. 1996).
2. The properties of the microelectrode should be tested in the ECF prior to impalement of a neuron. Offset voltage is applied to bring the voltage to ground level. Current

pulses are injected while monitoring the sampled voltage in the bridge mode. The bridge is balanced to eliminate the fast voltage transients produced by the micropipette, then the amplifier is switched to dSECC and capacitance is adjusted to reproduce the sampled voltage deflection recorded in balanced Bridge Mode. The injected current should not change the sampled voltage appreciably. Furthermore, the microelectrode should carry depolarizing and hyperpolarizing currents equally well. Duty cycle is also selected at this point. Short duty cycles are better to allow complete settling, but the larger currents required to bring V_m to the desired level may exceed the current-passing capability of the microelectrode. Duty cycles of 25–33% seem to be most often used.

3. After impalement, bridge balance and capacitance compensation are readjusted. The unsampled electrode potentials are monitored on an oscilloscope in dSECC mode to ensure complete settling and correct capacitance adjustment. Capacity compensation is adjusted to allow the fastest switching frequency that permits full decay of injected current pulses without ringing. Switching frequency should be as high as necessary to capture fast changes of V_m without imparting a switching artifact associated with insufficient decay of IR_E.
4. The amplifier is then switched to dSEVC mode with the holding potential (V_H) set to a desired pre-selected level. The open-loop clamp gain (nA/mV) of the constant current source is then increased from zero. The gain refers to the number of nanoamperes that the output current will change for each mV of difference between the command potential and the membrane potential. High gain is desired to minimize clamp error (ε), the difference between command voltage and desired membrane potential. Large values of ε will result in errors in measuring I_m. If the gain is too low, currents will be underestimated. On the other hand, gains that are too high will introduce current over- and undershoot, or oscillations in I_m and V_m recordings. It is therefore important to monitor the actual V_m as well as the command potential.

Problems

The Problem of Cell Geometry and Space Clamping during dSEVC

Most neurons of the CNS are not electrically compact. Currents generated at more remote dendritic locations may not be voltage-clamped owing to decrementation of current injected in the soma. Thus, synaptic currents arising in remote regions will not be accurately represented by soma recordings. In addition, spontaneous action potentials may not be adequately clamped and reversal potentials may be over-estimated. For a more detailed discussion of inadequate space clamping and its consequences, see Johnston and Brown (1983).

Results

Results Obtained with dSEVC Recording and Supercharging in Deep Layers of the Brain Stem

An amplifier equipped with supercharging capability has been used to record spontaneous fluctuations of I_m in respiratory neurons located in deep layers of the medulla by Richter and colleagues (Richter et al. 1996). As illustrated in Fig. 12, the supercharging circuit enables recording of spontaneous, respiratory-rhythmic fluctuations of membrane potential (dSECC) and current (dSEVC) using uncoated sharp microelectrodes with resistances in the range of 40–80 MΩ. This has enabled high-resolution recording of fast spontaneous postsynaptic currents (PSCs), postsynaptic potentials (PSPs) and

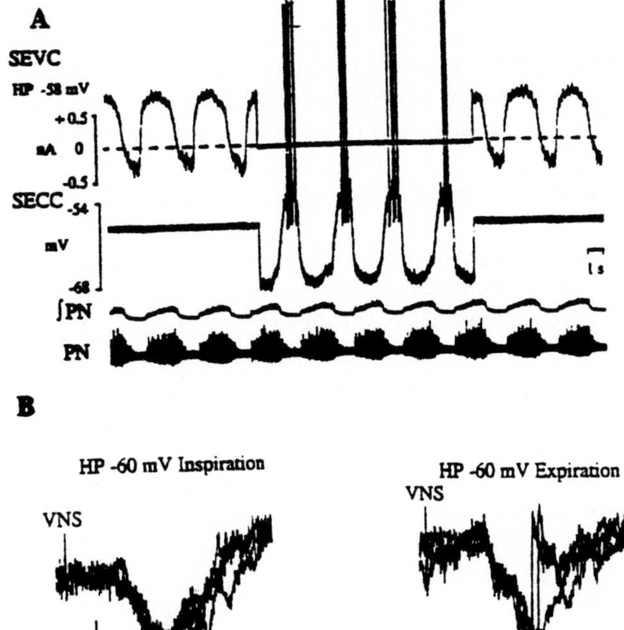

Fig. 12. Recording of membrane potentials and currents from an expiratory neuron in the caudal medulla oblongata of an anesthetized cat, using a supercharging microelectrode amplifier. The neuron was more than 2 mm deep in brain stem tissue (modified from Richter et al. 1996). **A:** Respiratory fluctuations of membrane current (I_m) in *dSEVC* mode (nA, *left* and *right*; holding potential, −58 mV) and membrane potential (V_m) in SECC mode (mV, *middle*). Lower traces are extracellular action potential discharges of the phrenic nerve (*PN*) recorded with bipolar hook electrodes, and the time integral of *PN* discharge frequency (*PN*). During the silent phases of *PN*, I_m in the expiratory neuron is inward and V_m depolarizes to threshold for action potential discharge. During the *PN* discharge, I_m is outward and V_m hyperpolarizes. **B:** Superimposed excitatory postsynaptic currents (*EPSCs*) evoked by stimulating the ipsilateral cervical vagus nerves with single shocks during the *PN* discharge phase (Inspiration) and silent period (Expiration).

stimulus-evoked PSCs and PSPs, as well as the slower excitatory and inhibitory respiratory drive currents and potentials produced by temporal summation of fast PSCs and PSPs.

Intracellular Iontophoresis During dSECC and dSEVC

During intracellular recording, agents that alter intracellular second messengers or membrane ion channels can be injected into cells. Using this approach, Richter and colleagues analyzed the contributions of cAMP-dependent Protein Kinase A, Protein Kinase C and Ca^{2+} currents to pattern generation in respiratory neurons (Haji et al. 1996; Pierrefiche et al. 1996; Richter et al. 1996; Lalley et al. 1997). Charged compounds were injected by current pulses, then their effects on respiratory-rhythmic fluctuations of I_m, V_m and I_{Ca} were evaluated. The advantage of pulsed-current iontophoresis is that the amount of substance ejected from the electrode will be proportional to the applied charge, hence dose-response relationships can be established. Moreover, the cells are not rapidly dialyzed with the chemical so there is adequate time for control recording. Of course, compounds that are negatively charged will be continuously expelled during voltage clamp with negative current.

Intracellular Stimulation

Here we discuss how intracellular stimulation is used in bridge and dSECC modes to investigate basic electrophysiological properties in neurons. We also briefly describe voltage-pulse protocols in dSEVC mode which are used to identify various types of membrane currents.

Functional Properties of Neurons Detected by Intracellular Stimulation in Bridge and dSECC Modes

Intracellular stimulation in either Bridge or dSECC modes has been very useful in identifying basic electrophysiological properties of neurons, such as membrane input resistance, threshold, membrane time constants, electrotonic length, total cell capacitance, action potential after-hyperpolarization and discharge properties. For more detailed descriptions of the methods of analysis, see Burke and ten Bruggencate (1971), Barrett and Crill (1974), Gustafson and Pinter (1985), Zengel et al. (1985), Sasaki (1990), Viana et al. (1993), Engelhart et al. (1989, 1995), Liu et al. (1995) .

- Membrane input resistance *(R_N)* is most often determined by injecting hyperpolarizing current (duration 50 ms or greater) and measuring the voltage drop produced by the current pulse.
- Threshold: and Membrane time constant (τ)

There are several approaches for deriving threshold:

1. A rather precise method of estimating threshold is to measure *the rheobase,* or minimal current required to evoke an action potential by a long depolarizing current pulse. Carried out rigorously, a strength-duration relationship is established by injecting depolarizing current pulses of varying duration and determining at each duration what the threshold intensity will be. This provides not only an estimate of threshold but also the membrane time constant, which reflects cell size, degree of dendritic arborization, membrane capacitance and resistivity (Liu et al,1995). The relationship between current intensity (I) and duration of current (t) at threshold is an exponential function:

$$I_{rh} / I = 1 - e^{-t/\tau} \tag{1}$$

where I_{rh} is the rheobasic current, and τ is the membrane time constant. I is graphically plotted as multiples of I_{rh} (ordinate, logarithmic scale) vs. time (abscissa, linear scale). Where $I=3/2\ I_{rh}$ on the plot, $t=\tau$. In a homogeneous population of neurons, estimates of τ might be made to follow the course of growth-, age- or disease-related neurological (Engelhardt et al. 1989) and toxicological processes (Liu et al, 1995).

2. A strength-latency relationship may also be derived by measuring latency for impulse initiation following current pulses of long duration and different strengths, where the above mathematical relationship of equation (1) also applies (Frank and Fuortes 1956).
3. Rheobase can be simply defined as the minimum current in a long current pulse (e.g., 50 ms) that evokes a direct spike 100 % of the time (Liu et al. 1995).

To estimate the respective contributions of the soma *and dendritic compartments to* τ, Burke and ten Bruggencate (1971) measured the time course of decay of voltage transients evoked by constant current pulses applied in the soma. The time-related change of the voltage transient is given by:

$$V(t) = V_{ss} + \sum_{n=0}^{\infty} C_n \exp(-t/\tau) \tag{2}$$

where V_{ss} is the steady state voltage at the soma during the current pulse and C_n is a weighting coefficient. Equation (2) can be differentiated to give:

$$dV/dt = \sum_{n=0}^{\infty} -(C_n / \tau_n)\exp(-t/\tau_n) \tag{3}$$

Values of dV/dt are obtained by measuring differences of V between consecutive pairs of t-values on greatly expanded records of the voltage transient. A semilogarithmic plot of dV/dt. (ordinate, logarithmic scale) vs. time (abscissa, linear scale) yields a curve which is nonlinear at early values of t. At later values of t the data fit a straight line which is back-extrapolated to time zero. The reciprocal of the slope is taken as τ_m, the membrane time constant (τ_0 in equations 2 and 3). Values of dV/dt from the extrapolated line are subtracted from those on the curvilinear region for corresponding values of t and antilog values of the differences are re-plotted. Extrapolation of the linear portion yields 1/slope, the first "equalizing" time constant, τ_1. This process (peeling of exponentials) is repeated until all curvilinear components are eliminated, yielding values of $\tau_2, \tau_3 \ldots \tau_n$. These equalizing time constants reflect spread and distribution of the voltage transient from the soma into primary, secondary, tertiary...etc. branches of the dendritic tree.

Equations (2) and (3) apply to voltage transients that are free of initial "overshooting" and "undershooting". Otherwise, corrections for these non-linearities are required, as discussed in Burke and ten Bruggencate (1971), Zengel et al. (1985) and Engelhardt et al. (1989).

- Electrotonic length (L):
  The electrotonic length (L) of the soma and dendrites, a passive cable property of the cell that determines how voltage distributes between the soma and dendrites as a function of length and branching of dendrites, is given by (Rall 1969; Burke and ten Bruggencate 1971):

$$L = \frac{\pi}{\sqrt{\tau_m / \tau_1 - 1}} \tag{4}$$

- Calculations of L have also been useful for comparing functional architectural features among different types of neurons (Burke and ten Bruggencate 1971) and evaluating how age, growth and toxicological processes affect functional architecture. In addition, estimates of L obtained by measuring stimulus-evoked PSPs and PSCs have been used in combination with staining of soma and dendrites (see Part 4 of this chapter) to approximate electrotonic distances of synapses from the soma (Johnston and Brown, 1983).
- Total cell capacitance (C_{cell}) is obtained from the relationship:

$$C_{\text{cell}} = \tau L / \left[R_N \tanh(L)\right] \tag{5}$$

  and can be used to estimate neuron surface area by dividing C_m, the area-specific capacitance ($2\,\mu F/cm^2$; range, 1–4; Rall 1977), into the value derived for C_{cell} from Eq. 5.
- Action potential afterhyperpolarization (AHP) is thought to be related to activation of calcium-dependent potassium channels and can be a major factor in limiting interspike intervals in neurons (Baldissera and Gustaffson, 1971; Viana et al. 1993). Amplitude and duration of the AHP are measured by injecting suprathreshold current pulses into a neuron (Ito and Oshima, 1965), or by stimulating its axon to evoke an antidromic action potential (Sasaki, 1991).

Pulse Protocols for Voltage Clamp Experiments

Voltage pulse protocols are utilized to separate and identify currents generated in whole cells into constituent components associated with a type of channel. Such protocols take advantage of the fact that whole cell currents can be activated or inactivated within different ranges of membrane potential. Only a few examples of various types of membrane currents and pulse protocols that are used to activate them will be given here.

- Leak currents

  In order to estimate the magnitude of a specific type of membrane current, non-specific (ohmic) leak currents must be subtracted out. When depolarizing steps are utilized to activate and detect specific currents, a correction for leak current is made by adding hyperpolarizing command pulses which are subtracted from the activating pulse. When very large depolarizing steps are required, smaller negative pulses can be scaled up by computer. See Cachelin et al. (1994) for a more detailed description.
- Potassium currents

  Several types of potassium currents have been described, including delayed rectifier (K_v), rapidly-inactivating A-type (K_A), muscarinic receptor-activated (M) current, inwardly-rectifying (K_{ir}), calcium-dependent (K_{Ca}) and ATP-sensitive (K_{ATP}) currents (Hille 1992). The A-type potassium current is inactive at normal resting potential. This current is de-inactivated in some types of neurons by holding potentials or hyperpolarizing pre-pulses negative to −60 mV. Application of long depolarizing pulses then activates the rapidly-decaying A-type current (Connor and Stevens 1971; Adams et al. 1982b). The non-inactivating M-current that has a voltage threshold at −60mV has been visualized in sympathetic neurons and hippocampal neurons by holding membrane potential under voltage clamp at a relatively depolarized level (e.g., −30 mV) to activate I_M, and applying steps to −60 mV for 0.5–1 s. The result, as shown in Fig. 13, is a slow inward relaxation seen on the current trace, representing I_M (Adams et al. 1982a; Halliwell and Adams 1982). In cat lumbar motoneurons voltage-clamped *in situ,* Crill and Schwindt (1983) identified fast and slow types of potassium current, I_{KF} and I_{KS}, after blocking sodium currents with tetrodotoxin. The fast-type of I_K seems to be associated with rapid repolarization of the action potential, whereas I_{KS} is probably a calcium-dependent current (Fig. 14).
- Sodium currents

  Fast, persistent, and "resurgent" sodium currents have been identified during voltage clamp in mammalian neurons. The fast I_{Na} responsible for the sodium spike has been measured in lumbar motoneurons *in situ* in conventional fashion, i.e., by applying

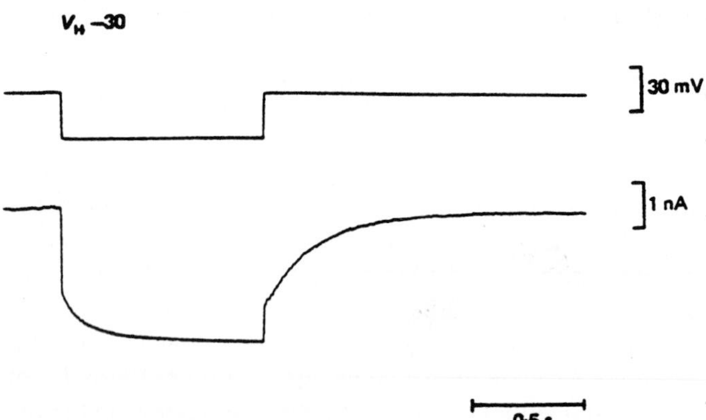

Fig. 13. Clamp currents recorded from a bullfrog sympathetic neuron during a 30 mV hyperpolarizing voltage-command from a holding potential (V_H) of −30 mV. Upper trace, voltage; lower trace, current. The slow inward relaxation following the instantaneous current step reflects the deactivation of I_M. (Modified from Adams et al. 1982a).

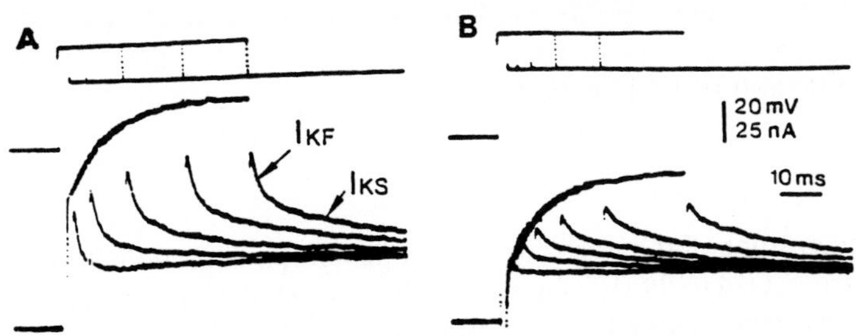

Fig. 14. Outward potassium currents in a voltage-clamped cat lumbar motoneuron. **A** and **B**: Superimposed traces of imposed voltage-clamp steps (*upper*) and outward currents (*lower*) before (**A**) and after (**B**) extracellular iontophoresis of the potassium channel blocker tetraethylammonium (*TEA*). Voltage clamp steps consist of prepulses of variable duration and amplitude sufficient to activate I_{KF} and I_{KS}, followed by down-steps to the same potential at various times after the prepulse. This protocol separates I_{KF} and I_{KS}. *TEA* greatly diminished I_{KF} but had no effect on I_{KS}. (Reproduced from Crill and Schwindt 1983).

depolarizing steps from negative holding potentials (Crill and Schwindt 1983). The persistent I_{Na}, which seems to contribute to repetitive discharge properties, has been visualized by blocking I_K with TEA and I_{Ca} with cobalt, holding membrane potential 20 mV positive to resting potential to inactivate fast sodium spikes, and applying a 20 mV depolarizing step (Stafstrom et al. 1985). TTX-sensitive resurgent sodium currents have been recorded by applying repolarizing voltages (–60 to –20 mV) after long depolarizations to +30 mV to produce maximal inactivation of I_{Na} (Raman and Bean 1997). Resurgent sodium currents may be responsible for the complex spikes recorded from cerebellar Purkinje neurons.

- Calcium currents
  At least six types of voltage-gated calcium channels have been identified electrophysiologically and pharmacologically: T, L, N, P, Q and R. The T-type Ca^{2+} current is a transient calcium current that is activated by small depolarizations from negative holding potentials (low-voltage-activated or LVA currents). The current begins to activate at –70 mV and is maximal between –50 and –20 mV. L-, N-, P- and Q-type calcium currents require stronger membrane depolarization (high-voltage-activated; HVA) to reach threshold (positive to –20 mV) and are maximally activated at +10 to +20 mV (Carbone and Lux 1986; Fox et al. 1987).

Subprotocol 4
Intracellular Recording and Tracer Injection

Introduction

The desire to know the position of the electrode tip and the morphological identity of the neurons recorded from it led to the development of the intracellular tracer injection technique. Early attempts relied on methyl blue and fast green (Thomas and Wilson 1965, 1966). Ever since their introduction (Stretton and Kravitz 1968), fluorescent dyes have been extensively used in particular in *in vitro* preparations. They have not been widely used in *in vivo* experiments largely due to the fact that they tend to fade following prolonged exposure to light and the fact that they do not travel well in axons. Several important results have been obtained with the help of intracellular injections of cobalt in invertebrates (Pitman et al. 1972), but this tracer has not been used very often in vertebrates, largely due to the fact that it does not stain long axonal segments. Since its development by several labs working independently (Czarkowska et al. 1976; Jankowska et al. 1976; McCrea et al. 1976; Snow et al. 1976), the intracellular horseradish peroxidase (HRP) technique has been used extensively to revolutionize the way we see and think of the brain. Here, we describe procedures to be used with the two tracers that have been most extensively used in mammals, namely HRP and biocytin.

Outline

- *Surgery*: Placement of head bolt, electrode(s), eye coil(s), recording chamber(s), etc.
- *Preparation of pipettes:* Pulling, filling and beveling
- *Cell impalement*
- *Tracer injection*
- *Animal perfusion:* Transcardial with saline followed by fixative
- *Brain section*
- *Histochemistry* for HRP or biocytin
- *Evaluation of sections*

Materials

Equipment for Intracellular Tracer Injection in Anesthetized Paralyzed Animals

Most of the basic equipment needed to surgically prepare the animal for recording (e.g., implantation of chamber(s), head holder(s), electrode(s), eye coil(s), etc.) has been listed in Part 1. The following is needed for experiments in both acute and alert chronic preparations.

Tracer solution: 10% solution of HRP in 0.5 M KCl and 100 mM Tris Buffer (pH = 7.4). The solution should be filtered through paper with pores smaller than 0.45 mm in diameter. Millipore (P.O. Box 255, Bedford MA 01730) makes appropriate filters and its Swinny 13 mm stainless holder has a reasonably small dead space.

Equipment for Intra-axonal Tracer Injection in Alert Behaving Animals

In addition, the following items are needed for the intra-axonal experiments:
- Primate chair (for monkeys) or cloth bag (for rabbits and cats) with appropriate head holding devices

- Light-weight, hydraulic microelectrode holder
- Computer-driven apparatus to control the behavior of trained animals

Equipment for Perfusion-fixation of Injected Animals

- Surgical instruments
- Two bottles (one for saline and one for the fixative) for gravity perfusion, connected through tubing and a 3-way stopcock to the perfusion cannula
- Phosphate-buffered saline, pH 7.4 (PBS; 0.02 M phosphate buffer and 0.9 % NaCl)
- Fixative: 1 % paraformaldehyde, 1 % glutaraldehyde and 5% sucrose in 50 mM phosphate buffer (pH = 7.4) works well for HRP material.

To prepare 1 liter of this fixative, dissolve 10 g of paraformaldehyde in 400 ml of distilled water, heat to 60 °C, add a few drops of NaOH (40%) while stirring until the solution clears, filter (using the Buchner funnel and the vacuum) and cool. Then, add 40 ml of 25% glutaraldehyde, 500 ml of 0.1 M phosphate buffer and bring to 1 liter with distilled water. A fixative of 4% paraformaldehyde in 0.1 M phosphate buffer (pH: 7.4) works well for biocytin material and a fixative of 2.5 % glutaraldehyde, 0.5% paraformaldehyde, 0.2% picric acid in 0.1 M phosphate buffer is to be preferred if the material is processed for electron microscopy.

Histology Equipment

- Freezing microtome or vibratome
- Rotating table
- Section holder
- Glassware
- Chrome-alum coated slides

To prepare the slides, dip them in a 1:1 solution of acetic acid (glacial) and alcohol (95%) and then into consecutive baths of distilled water until it sheets off evenly (usually three baths are adequate). Then dip in the chrome-alum solution and leave in an oven preheated to 37–40 °C for 2–3 hours to dry. Store covered at room temperature. To prepare the chrome-alum solution, add 0.3 g chromium potassium sulfate and 3 g gelatin in 500 ml of hot distilled water. Filter (three times).

Solutions for Histochemistry

- Phosphate buffer: To prepare 0.4 M phosphate buffer (pH=7.4), dissolve 21.324 g KH_2PO_4 and 90.690 g Na_2HPO_4 into distilled water and bring to 2 liters. Solutions of this molarity can be stored at room temperature. The 0.1 M solutions used in the protocols should be kept in the refrigerator. To prepare 1 liter of 0.1 M phosphate buffer, add 250 ml of 0.4 M phosphate buffer and bring up to 1 liter with distilled water.
- Tris buffered saline (TBS), 0.05 M, pH: 7.4: To prepare 1 liter, add 9 g NaCl and 6.61 g Trizma HCl and 0.97 g Trizma Base to distilled water and bring up to 1 liter.
- 0.05 % Diaminobenzidine tetrahydrochloride (DAB): To make 200 ml add 100 mg DAB (kept always in the freezer) to 200 ml 0.1 M phosphate buffer in a small beaker. To rapidly dissolve the DAB place the beaker in an ultrasonic cleaner (e.g., the UBATH by WPI). Filter with Watman paper #1 using the Buchner funnel and the vacuum.
- PBS-T: PBS: Prepared as in *4.2.3.*, plus 1% Triton X100.

Procedure

Preparation of Micropipettes for Intracellular Injection of Tracers

1. Clean the glass (in a solution made of 200 g potassium dichromate dissolved in 1 liter of distilled water contained in a Pyrex piece of glassware to which is then added 750 ml concentrated sulfuric acid).

2. Pull the micropipettes so that the length of the shaft and the conformation of the tip fit the needs of the experiment (size of the neurons of interest and their distance from the point of entry into the brain). This will also determine the choice of glass to be used. For example, micropipettes which need to travel long distances (more than 4–5 mm) should have fairly long shafts (more than 15 mm), and to avoid bending they should be made by pulling 2–4 mm OD glass (Corning Glass).
3. Prepare tracer solution (10% solution of Boehringer HRP or 5% solution of Sigma biocytin in 0.5 M KCl and 0.1 M Tris Buffer, pH =7.4).
4. Fill micropipettes. The MicroFil nonmetallic syringe needle (WPI) is adequate for filling micropipettes made from omega-dot or triangular tubing. Pipettes without filament can be pressure back-filled with smaller micropipettes also pulled from 0.81 mm OD glass (50% wall thickness; Frederick & Dimmock Inc.).
5. Store filled micropipettes in humid container.
6. Bevel tip diameter to less than 1 mm (impedance: 40–80 MΩ). We obtain consistent results with the Sutter Instruments BV-10, but other methods also work well.

Intrasomatic Tracer Injection

1. Penetrate soma as described above. This is signaled by a 10–80 mV DC shift and the appearance of 10–50 mV action potentials and/or synaptic responses time-locked to the search electrical stimulus.
2. Wait until recording stabilizes, injecting small amounts of hyperpolarizing current (nA range) if necessary.
3. Document the antidromic and/or synaptic phenomena.
4. Inject HRP for 3–30 min by means of depolarizing current pulses of 10–30 nA, 1 per second 60% duty cycle while monitoring the evoked phenomena.
5. Let animal survive for 4–10 hours after the last injection, while insuring normal blood pressure and proper ventilation and anesthesia.

Intraaxonal Tracer Injection

1. Penetrate fiber by gently tapping on the microelectrode holder. This is signaled by a 10–80 mV DC shift and the appearance of 5–50 mV action potentials on the oscilloscope screen.
2. Document the behavioral relevance of the fiber's discharge pattern.
3. Inject HRP for 7–18 min by means of depolarizing current pulses of 7–15 nA, 500 ms, 1 Hz, while monitoring the resting membrane potential and periodically inspect the unit's discharge pattern during brief interruptions of the current injection.
4. Return the animal to its housing quarters and let it survive for 30–50 hours after the last injection.

Terminal Anesthesia, Transcardial Perfusion, Tissue Fixation, Brain Blocking and Sectioning for Immuno-Histochemical Detection of Injected Tracer

1. Anesthetize animals deeply with sodium pentobarbital and wait for total absence of responses to strong stimuli.
2. Expose the femoral vein and inject heparin (5000 units, i.v.).
3. Wait for about half an hour.
4. Open the thoracic cavity by cutting along the sternum and enlarge the opening with a retractor. Ligate the descending aorta (when not interested in spinal neurons) and

make a small incomplete incision through the wall of the apex of the heart. Insert the perfusion cannula in the left ventricle and the ascending aorta and secure it near its exit.

5. Perfuse with 1 liter of PBS, pH: 7.4 (for a cat or a rhesus monkey; about half a liter for a squirrel monkey) to wash out the blood. At the start of the perfusion inject 1 ml of sodium nitrite or nitroprusside (a few crystals will do) in PBS and cut an incision in the right atrium. Wait until the venous return becomes clear (10 min or less).
6. Continue with 2 liters of fixative (for a cat or a rhesus monkey; about one liter for a squirrel monkey). The initial 200 ml of this should be at a relatively high pressure (through a large volume glass syringe or by increasing the pressure in the bottle containing the fixative by about 1 Atm with the help of a manometer).
7. Complete the perfusion with about 1 liter of fixative containing 15% sucrose (half a liter is adequate for a 1 kg animal such as the squirrel monkey).
8. Block the brain *in situ* in the stereotaxic coordinates that are most appropriate for the plane of sectioning (frontal, parasagittal or horizontal), the need to identify electrode tracks and the location of the injected cells and their terminal fields.
9. Dissect out the blocks of the brain and/or the spinal cord.
10. Store tissue in a reasonably large volume of cold 20% sucrose in 0.1 M phosphate buffer (pH 7.4) for cryo-protection. Refrigerate until the tissue sinks to the bottom of the vial before histological processing.
11. Place tissue on a pedestal made of frozen sucrose solution on the stage of the microtome (the stage of the sliding microtome we use employs a Peltier device to make the frozen pedestal and freeze the tissue). Freeze the tissue and cut 60 mm (biocytin) or 80 mm (HRP) thick serial sections.
12. Collect sections over ice in 0.1 phosphate buffer (HRP) or PBS (biocytin).

HRP Procedure (Adams 1981)

There are several ways to react the tissue in order to visualize the intracellularly injected HRP (e.g., Itoh et al. 1979; Bishop and King 1982). One of the most commonly used is Adams' modification of the DAB reaction (Adams 1977; Adams 1981).

1. Place in 0.05% DAB, 5% $CoCl_2$ and 0.2% nickel ammonium sulfate in 0.1 M phosphate buffer for 20–30 min, in the dark.
2. Add 1 ml of 0.3% H_2O_2 per 100 ml of DAB solution and incubate sections for an additional 30 min.
3. Rinse in phosphate buffer 3x10 min. Mount and let dry.
4. Counterstain (e.g., with cresyl violet).
5. Dehydrate, clear and cover slip.

Biocytin Procedure (Horikawa and Armstrong 1988)

1. Rinse 4x20 min in PBS-T.
2. Incubate in a 1:200 solution of ABC kit (Vectastain Std., series elite, ref. PK6000, Vector Labs.). To prepare, mix 1000 ml sol. A and 1000 ml sol. B in 200 ml PBS-T).
3. Leave overnight on moving plate and room temperature.
4. Rinse 3x10 min in TBS.
5. Blot trays thoroughly between solutions.
6. Preincubate in 0.2 % nickel ammonium sulfate in TBS, 10 min. Do not filter.
7. Preincubate in 0.05% DAB + 0.2 % nickel ammonium sulfate in TBS, 10 min.
8. Incubate in 0.05% DAB + 0.2 % nickel ammonium sulfate + 0.006% H_2O_2 in TBS, 5–20 min.

9. Rinse 3x10min in TBS.
10. Mount and let dry.
11. Counterstain (e.g., with cresyl violet). Dehydrate, clear and cover-slip.

Results

Because they combine information about the synaptic responses or the discharge pattern of neurons with detailed information about the dendritic appearance and the terminal fields of their axons, intracellular tracer injections have been used extensively to obtain a detailed picture of the microcircuitry of the brain. In this section we describe two classes of cells which were discovered and/or studied in detail with the help of these techniques. Our selection was guided by the ease with which their salient morphological features can be related to important functional properties of the mammalian central nervous system.

Functionally Identified Cells of the Visual Cortex Intrasomatically Injected with HRP in the Anesthetized Cat

Figure 15 illustrates the pattern of axonal terminations of a pyramidal complex cell of layer 5 of the visual cortex that was intracellularly injected with HRP in the anesthetized paralyzed cat and reconstructed from serial sections by Gilbert and Wiesel (1983). As shown here, its axon projected fairly heavily to layer 6 and extended for more than 4mm away from the cell body down the medial bank of area 17. Axonal projections such as these can account for the unique receptive field properties of layer 6 neurons which show summation with bar length up to very large values (Gilbert and Wiesel 1983). The fact that layer 5 projections are necessary for the response properties of layer 6 cells was confirmed when it was shown (Bolz et al. 1989) that responses of layer 6 cells to long bars decreased while their responses to short bars remained unchanged following the inactivation (with the injection of GABA through a micropipette) of layer 5 cells (within a radius consistent with the length of the axonal arbor of Fig. 15). On a more organismic scale, this pattern of projections is typical of mechanisms that are needed to account for perceptual phenomena which require long range interactions between widely separated parts of the visual field (Spillman and Werner 1996). An example of such a phenomenon is provided by the Craik-O'Brien-Cornsweet illusion. The central region of the gray scale pattern illustrated near the top of the inset of Fig. 15 looks darker than the side regions, despite the luminance profile illustrated near the bottom of the inset (the illusion disappears if one uses thin strips to cover the discontinuities of the gray scale pattern indicated on the luminance profile).

Contacts Established by Physiologically Identified Ia Fibers Intraaxonally Injected with HRP onto Functionally Identified Ankle Extensor α-Motoneurons Intrasomatically Injected with HRP in the Anesthetized Cat

The monosynaptic excitatory connection between muscle spindle primary afferents and α-motoneurons underlies much of the myotatic reflex and is deservedly one of the best studied in the central nervous system. However, the fact that the composite excitatory postsynaptic potential (EPSP) can be largely accounted for by the density and dendritic location of group Ia synapses onto α-motoneurons had to await definitive demonstration until labeling of both the presynaptic and the postsynaptic structures be-

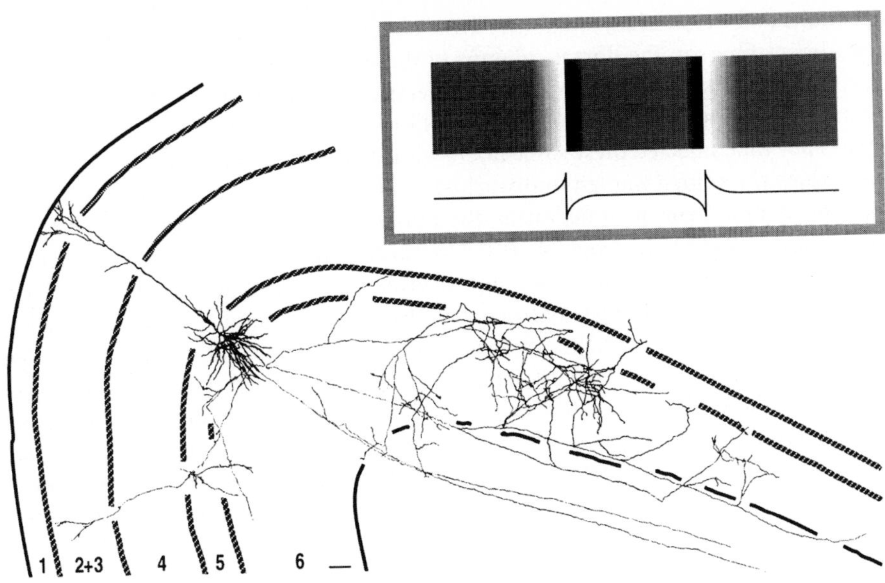

Fig. 15. Camera lucida drawing of a pyramidal neuron located in layer 5 of the striate cortex of the cat (modified from Gilbert and Wiesel 1983, with permission). The cell had a standard complex receptive field and showed no end-inhibition. Scale bar is equal to 100mm. The inset illustrates a grey-scale pattern that gives rise to the Craik-O'Brien-Cornsweet illusion (*top*) and its luminance profile (*bottom*).

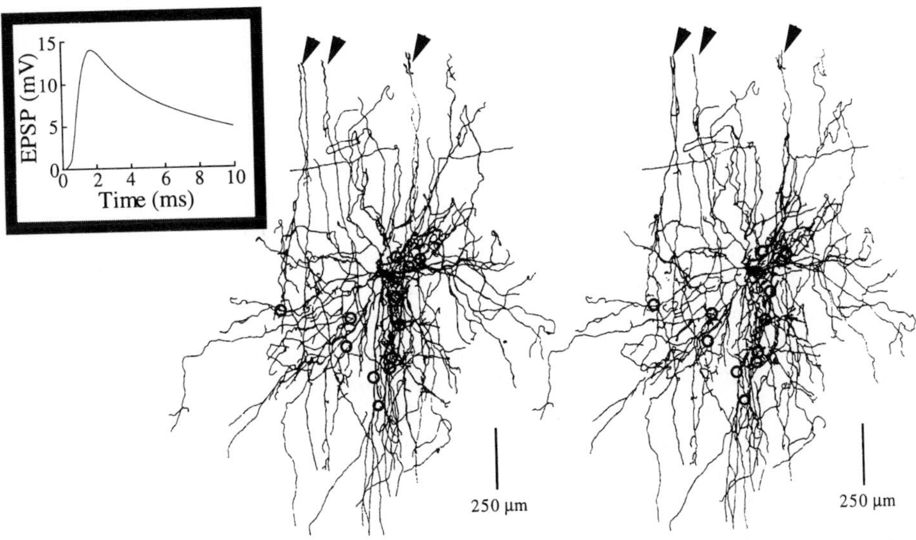

Fig. 16. Stereo drawing of the contacts between a single soleus group Ia afferent fiber and a single type *S* (-motoneuron (modified from Burke and Glenn 1996, with permission). *Open circles* show the locations of regions of synaptic contact. The *inset* illustrates a computer simulation of the composite EPSP evoked by a compartmental model of a type *S* (-motoneuron in response to the temporally dispersed activation of 300 synapses (modified from Segev et al. 1990, with permission).

came feasible (Segev et al. 1990; Burke and Glenn 1996). Figure 16 is an example of relevant data presented as a stereo drawing of the contacts (open circles) between collaterals (arrowheads) arising from a single soleus Ia afferent fiber and a single type S α-motoneuron (Burke and Glenn 1996). To obtain this data, Burke and Glenn (1996) impaled primary afferent fibers in the dorsal columns of the spinal cord of the cat and identified them from their short latency responses to electrical stimulation of individual muscle nerves and from their responses to low-amplitude high-frequency sinusoidal muscle stretch. They then injected these same fibers with HRP. Following a period of 4–6 hours during which the animal was left undisturbed to allow diffusion of HRP into intraspinal collaterals, α-motoneurons innervating the muscle of origin of the injected Ia fibers were impaled and also injected with HRP. After perfusion, sectioning of the spinal cord and histochemical treatment of the sections to visualize the HRP, both the afferent fibers and the motoneurons were serially reconstructed and the position of each contact between them was plotted in Euclidean space. Then, compartmental models of injected motoneurons were constructed with the help of methods detailed elsewhere (Fleshman et al. 1988; Burke et al. 1994). After taking into consideration the number of Ia afferent fibers innervating each cell in the relevant motoneuron pools, the number of synapses between them and the conductance transients they evoke, the synaptic responses of model motoneurons (Fig. 16, inset) reproduced the amplitude and time course characteristic of experimentally observed composite group Ia EPSPs (Segev et al. 1990).

Comments and Troubleshooting

Slides and Coverslips

Chrome-alum coated slides are excellent for the purposes of L/M studies. In general, the coating and the size of the slides and the coverslips and the make of the embedding material depend on the morphological analysis to be undertaken. Wilson and Groves (1979) describe an excellent recipe for cases when the L/M study is followed by an E/M one.

Preparation of Micropipettes and Choice of Tracer

The desired properties of micropipettes used in intracellular experiments and how to attain them are dealt with in Procedure 3.1. To a large extent, they consist in steps which ensure excellent electrical properties while minimizing cell damage. When recording is followed by tracer injection, one must additionally ensure that the tracer is readily ejected through the same pipette. Cleaning of the pipette before filling, and appropriate beveling of their tips as well as avoiding blood vessels and fiber tracts or careful passage through them if unavoidable can prove helpful in this regard.

Biocytin is a smaller molecule than HRP, which can account for the longer sections of axons visualized when it is employed. On the other hand, a smaller percentage of cells is recovered when biocytin is used as a tracer and it does not seem to fill axonal branches as completely as HRP.

Survival, Anesthetic Agents and Length of Transport

Longer survival periods generally permit more complete fills of the axons which can be followed for longer distances. On the other hand, excessively long survival periods result in degeneration of filled neurons and longer distances between the two retraction bulbs that form at the cut end of axons after intraaxonal tracer injections. The choice of anesthetic (and tracer) can also be important in this regard. Pentobarbital can inhibit the uptake (Turner 1977) and axonal transport of HRP (Mesulam and Mufson 1980). If the animal must remain anesthetized after the injection, it may be better to use urethane instead (Rogers et al. 1980).

Deviations from the Tracer Injection Protocols

Biocytin can stain single neurons even when applied juxtacellularly through glass micropipettes of tip diameter roughly equal to 1 mm and filled with a 2% solution of biocytin in 0.5 M KCl (Pinault 1996). It is critical to be able to modulate the firing of well-isolated and identified neurons with depolarizing current pulses of low intensity (less than 10 nA) through the micropipette. Responding neurons have been successfully labeled with depolarizing current pulses (intensity: 0.5–3 nA; duration: 200 ms) applied for 10–20 min at a rate of 2.5 Hz (50% duty cycle) under continuous monitoring of cell firing. Single cells can also be stained with biocytin in tissue fixed prior to cell impalement (Li et al. 1990). In such cases, micropipettes are often visually guided towards cells of interest that were previously labeled with a different tracer. Also, micropipettes carry an aqueous solution of lucifer yellow (0.5%) in addition to the biocytin (5%). Dye is injected with hyperpolarizing current (0.5–3 nA) for 10–30 min and until distal processes brightly fluoresce for lucifer yellow.

HRP Procedure

The solution used for the heavy metal intensification of the DAB reaction product is unstable and should be prepared just before use. Also, the timing of the DAB reaction is critical and progress must be monitored to ensure that it does not go too far.

CAUTION: DAB may be carcinogenic. Use appropriate care in handling. Gloves and mask should be used while handling DAB in powder form. Mark the glassware you use for the DAB procedure and dispose of DAB solutions in containers containing chlorine.

Background and its Reduction

Red blood cells, vessels and certain brain stem nuclei display peroxidase activity. This is often desirable as cells of interest turn gray-brown and may stand out, thus obviating the need for counterstaining. If this background peroxidase activity is undesirable, it can be eliminated by pretreatment for 1 hour with a 0.1% freshly prepared solution of phenylhydrazine (pH: 7.0) at 37°C (Straus 1979). A simpler method is to take sections through a series of ethanol solutions (50%x10 min followed by 70%x15 min and then by 50%x10 min) prior to the DAB reaction (Metz et al. 1989).

Summary and Conclusions

Microelectrode recording and staining of neurons have provided considerable information about the individual behaviors, network functions, morphology and locations of neurons in the CNS. In this chapter, we have deliberately focused on protocols developed for use in intact animals, recognizing that CNS control of many functions such as cognition, blood pressure and heart rate, motor coordination and locomotion, respiration and micturition, to name but a few, are best analyzed in the *in situ preparation.* In other chapters of this book, modern electrophysiological and optical techniques that reveal functional and structural properties of neurons and channels in the *in vitro* preparation are discussed. Hopefully, many of the techniques and information from such research will be utilized for studies of neuron and network behavior in the intact animal, including studies of motor and higher-order nervous function in humans (Section III of this book).

Supplier List

We do not endorse any of the companies listed below.

General equipment needed to conduct the experiment to the stage of microelectrode recording can be purchased from Harvard Apparatus, Columbus Instruments, Fine Science Tools, Inc. More companies can be found on website: http://weber.u.washington.edu/~chudler/comm.html#stereo

Product	Manufacturers
Vibration table	Harvard Apparatus, Newport Technical Manufacturing Corp.
Feline stereotaxic system and spinal frame	Medical Systems Corp.
Stereotaxic instruments	David Kopf Instruments Narishige Scientific Instruments
Micromanipulators	
hydraulic	Narishige Scientific Instruments
motorized	Newport Scientific Precision Instruments Sutter Instruments
piezoelectric	Burleigh Sutter Instruments
pneumatic	Technical Products International
Microscopes, surgical	Jena Leitz Wild Zeiss
Amplifiers	
Extracellular	A-M Systems BAK Frederick Haer & Co. Grass Instruments npi electronic GMBH
Intracellular	
Bridge	A-M Systems Dagan Corporation
dSECC, dSEVC	Axon Instruments, Inc. Dagan Corporation npi electronic GMBH
Two electrode voltage and current clamp	Medical Systems Corp.
Current/voltage pulse generators	A.M.P.I. A-M Systems Coulbourn Digitimer, Ltd. Grass Instruments Harvard Apparatus World Precision Instruments

Product	Manufacturers
Pipette glass	A-M Systems, Inc. Clark Electromedical Instruments Frederick Haer & Co. Medical Systems Corp. Sutter Instruments World Precision Instruments
Pipette pullers	DMZ (Zeiss) Medical Systems Corp Narishige Scientific Instruments Sutter Instruments
Pipette coaters	Ala Scientific Instruments Dow Corning Sylgard 184 Medical Systems Corp Narishige Scientific Instruments Q-dope
Metal microelectrodes	A-M Systems, Inc Frederick Haer & Co. Micro Probe Inc. Transidyn General World Precision Instruments
Carbon fiber electrode	Frederick Haer & Co. Medical Systems Corp.
Suction electrodes	A-M Systems, Inc.
Microelectrode holders	Alpha Omega Engin Axon Instruments, Inc World Precision Instruments E.W. Wright
Oscilloscopes	Gould Instrument Systems Hitachi Tektronix
Audio amplifiers	Grass Instruments Harvard Apparatus
Tape recorders (VCR-based)	Bio-Logic InstruTech Neurodata Racal Vetter
Chart recorders	Astro-Med Columbus Instruments Gould Instrument Systems Grass Instruments Harvard Apparatus
Spike processors	Alpha Omega Engin Digitimer, Ltd. Frederick Haer & Co.
Computer interfaces	Astro-Med Axon Instruments, Inc. Cambridge Electronic Design, Ltd. Coulbourn Instruments Dagan Corporation Data Translation Digitimer, Ltd. Gould Instrument Systems Harvard Apparatus Instrutech Corp. National Instruments

References

Adams JC (1977) Technical considerations on the use of horseradish peroxidase as a neuronal marker. Neuroscience 2:141–145

Adams JC (1981) Heavy metal intensification of DAB-based HRP reaction product. J Histochem Cytochem 29:775

Adams DJ, Gage PW (1979) Ionic currents in response to depolarization in an Aplysia neurone. J Physiol (Lond) 289:115–141

Adams PR, Brown DA, Constanti A (1982a) M-currents and other potassium currents in bullfrog sympathetic neurones. J Physiol (Lond) 330:537–572

Adams PR, Constanti A, Brown DA, Clark RB (1982b) Intracellular Ca^{2+} activates a fast voltage sensitive K^{+} current in vertebrate sympathetic neurones. Nature 296:746–749

Backman SB, Anders K, Ballantyne D, Röhrig N, Camerer H, Mifflin S, Jordan D, Dickhaus H, Spyer KM, Richter DW (1984) Evidence for a monosynaptic connection between slowly adapting pulmonary stretch receptor afferents and inspiratory beta neurones. Pflügers Arch 402:129–136

Baldissera F, Gustafsson B (1971) Regulation of repetitive firing in motoneurones by the afterhyperpolarization conductance. Brain Res. 30: 431–434

Barrett JN, Crill WE (1974) Specific membrane properties of cat motoneurones. J Physiol (Lond) 239:301–324

Benevento LA, McClearly LB (1992) An immunocytochemical method for marking microelectrode tracks following single-unit recordings in long surviving, awake monkeys. J Neurosci Methods 41:199–204

Benndorf K (1995) Low-noise recording. In: Sakmann B, Neher E (eds) Single-channel recording, 2nd Ed. Plenum Press, New York, pp 129–145

Bishop G, King J (1982) Intracellular horseradish peroxidase injections for tracing neural connections. In: Mesulam M (ed) Tracing neural connections with horseradish peroxidase. John Wiley & Son, Chichester, pp 185–247

Bolz J, Gilbert CD, Wiesel TN (1989) Pharmacological analysis of cortical circuitry. Trends Neurosci 12: 292–296

Brennecke R, Lindemann B (1974a) Theory of a membrane-voltage clamp with discontinuous feedback through a pulsed current clamp. Rev Sci Instr 45:184–188

Brennecke R, Lindemann B (1974b) Design of a fast voltage clamp for biological membranes, using discontinuous feedback. Rev Sci Instr 45:656–661

Brinley FJ Jr (1980) Excitation and conduction in nerve fibers. In: Mountcastle VB (ed) Medical physiology, vol 1, 14th Ed. The C.V. Mosby Company, St. Louis Toronto London, pp 46–81

Burke RE, Fyffe REW, Moschovakis AK (1994) Electrotonic architecture of cat gamma motoneurons. J Neurophysiol 72:2302–2316

Burke RE, Glenn LL (1996) Horseradish peroxidase study of the spatial and electrotonic distribution of group Ia synapses on type-identified ankle extensor motoneurons in the cat. J Comp Neurol 372:465–485

Burke RE, ten Bruggencate G (1971) Electrotonic characteristics of alpha motoneurones of varying size. J Physiol (Lond) 212:1–20

Cachelin AB, Dempster J, Gray PTA (1994) Computers. In: Ogden D (ed) Microelectrode techniques. The Plymouth workshop handbook, 2nd Ed. Cambridge, The Company of Biologists Limited, Cambridge, pp 209–254

Carbone E, Lux HD (1987) Kinetics and selectivity of a low-voltage-activated calcium current in chick and rat sensory neurones. J Physiol (Lond) 386:547–570

Coombs JS, Eccles JC, Fatt P (1955a) The specific ion conductances and ionic movements across the motoneuronal membrane that produce the inhibitory postsynaptic potential. J Physiol (Lond) 130:326–373

Coombs JS, Eccles JC, Fatt P (1955b) Excitatory synaptic action in motoneurones. J Physiol (Lond) 130:374–395

Coombs JS, Eccles JC, Fatt P (1955c) The inhibitory suppression of reflex discharges from motoneurones. J Physiol (Lond) 130:396–413

Connor JA, Stevens CS (1971). Voltage clamp studies of a transient outward membrane current in gastropod neural somata. J. Physiol. 213: 21–30

Crill WE, Schwindt PC (1983) Active currents in mammalian central neurons. Trends Neurosci 6:236–240

Czarkowska J, Jankowska, Sybirska E (1976) Axonal projections of spinal interneurones excited by group I afferents in the cat, revealed by intracellular staining with horseradish peroxidase. Brain Res 118:115–118

DeGroat WC, Ryall RW (1968) The identification and characteristics of sacral parasympathetic preganglionic neurones. J Physiol (Lond) 196:563–577
DiCarlo JJ, Lane JW, Hsiao SS, Johnson KO (1996) Marking microelectrode penetrations with fluorescent dyes. J Neurosci Methods 64:75–81
Dunn PF, Wilson WA (1977) Development of the single microelectrode current and voltage clamp for central nervous neurons. Electroenceph Clin Neurophysiol 43:752–756
Eccles JC (1964) The physiology of synapses. Springer-Verlag, New York
Engelhardt JK, Morales FR, Castillo PE, Pedroarena C, Pose I, Chase MH (1995) Experimental analysis of the method of "peeling" exponentials for measuring passive electrical properties of mammalian motoneurons. Brain Res 675:241–248
Engelhardt JK, Morales FR, Yamuy J, Chase MH (1989) Cable properties of spinal motoneurons in adult and aged cats. J Neurophysiol 61:194–201
Fetz EE, Gustafsson B (1983) Relation between shapes of post-synaptic potentials and changes in firing probability of cat motoneurones. J Physiol (Lond) 341:387–410
Finkel AS, Redman SJ (1983a) The synaptic current evoked in cat spinal motoneurones by impulses in single group Ia axons. J Physiol (Lond) 342:615–632
Finkel AS, Redman SJ (1983b) A shielded microelectrode suitable for single-electrode voltage clamping of neurons in the CNS. J Neurosci Methods 9:23–29
Finkel AS, Redman SJ (1984) Theory and operation of a single microelectrode voltage clamp. J Neurosci Meth 11:101–127
Fleshman JW, Segev I, Burke RE (1988) Electrotonic architecture of type-identified alpha-motoneurons in the cat spinal cord. J Neurophysiol 60: 60–85
Fox AP, Nowycky MC, Tsien RW (1987) Kinetic and pharmacological properties distinguishing three types of calcium currents in chick sensory neurones. J Physiol (Lond) 394:149–172
Frank K, Fuortes MGF (1956) Stimulation of spinal motoneurones with intracellular electrodes. J Physiol (Lond) 134:451–470
Gilbert CD, Wiesel TN (1983) Clustered intrinsic connections in cat visual cortex. J Neurosci. 3:1116–1133
Gustafsson B., Pinter MJ (1985) JFactors determining the variation of the afterhyperpolarization duration in cat lumbar α–motoneurones. Brain Res. 326: 392–395
Haji A, Pierrefiche O, Lalley PM, Richter DW (1996) Protein Kinase C pathways modulate respiratory pattern generation in the cat. J Physiol (Lond) 494:297–306
Halliwell J, Whitaker M, Ogden D (1994) Using microelectrodes. In: Ogden D (ed) Microelectrode techniques, The Plymouth workshop handbook, 2nd Ed. The Company of Biologists Limited, Cambridge, pp 1–15
Halliwell JV, Adams PR (1982) Voltage-clamp analysis of muscarinic excitation in hippocampal neurons. Brain Res 250:71–92
Halliwell JV, Plant TD, Robbins J, Standen NB (1994) Voltage clamp techniques. In: Ogden D (ed). Microelectrode techniques. The Plymouth workshop handbook, 2nd Ed. The Company of Biologists Limited, Cambridge, pp 17–35
Henatsch H-D, Meyer-Lohmann J, Windhorst U, Schmidt J (1986) Differential effects of stimulation of the cat's red nucleus on lumbar alpha motoneurones and their Renshaw cells. Exp Brain Res 62:161–174
Hille B (1992) Ionic channels of excitable membranes, 2nd Ed. Sinauer Associates, Inc., Sunderland, MA
Hoffer JA, O'Donovan MJ, Pratt CA, Loeb GE (1981) Discharge patterns of hindlimb motoneurons during normal cat locomotion. Science 213:466–467
Horikawa K, Armstrong WE (1988) A versatile means of intracellular labeling: injection of biocytin and its detection with avidin conjugates. J Neurosci Meth 25:1–11
Itto M, Oshima T (1965) Electrical behavior of the motoneurone membrane during intracellularly applied current steps. J. Physiol. 180:607–635.
Itoh K, Konishi A, Nomura S, Mizuno N, Nakamura Y, Sugimoto T (1979) Application of coupled oxidation reaction to electron microscopic demonstration of horseradish peroxidase: cobalt-glucose oxidase method. Brain Res 175:341–346
Jankowska E, Rastad J, Westman J (1976) Intracellular application of horseradish peroxidase and its light and electron microscopical appearance in spinocervical tract cells. Brain Res 105:557–562
Johnston D, Brown TH (1983) Interpretation of voltage clamp measurements in hippocampal neurons. J Neurophysiol 50:464–486
Johnston D, Hablitz JJ, Wilson WA (1980) Voltage-clamp discloses slow inward current in hippocampal burst-firing neurones. Nature 286:391–393

Juusola M, Seyfarth EA, French AS (1997) Rapid coating of glass-capillary microelectrodes for single-electrode voltage-clamp. J Neurosci Methods 71:199–204

Kuffler SW (1953) Discharge patterns and functional organization of mammalian retina. J Neurophysiol 16:37–68

Kuras A, Gutmaniene N (1995) Preparation of carbon-fibre microelectrode for extracellular recording of synaptic potentials. J Neurosci Methods 62:207–212

Lalley PM, Pierrefiche O, Bischoff AM, Richter DW (1997) cAMP-dependent protein kinase modulates expiratory neurons. J Neurophysiol 77:1119–1131

Li C-Y, Xu X-Z, Tigwell D (1995) A simple and comprehensive method for the construction, repair and recycling of single and double tungsten microelectrodes. J Neurosci Methods 57:217–220

Li D, Seeley PJ, Bliss TVP, Raisman G (1990) Intracellular injection of biocytin into fixed tissue and its detection with avidin-HRP. Neurosci Lett 38S:81

Ling G, Gerard RW (1949) The normal membrane potential of frog sartorius fibers. J Cell Comp Physiol 34:383–396

Liu R-H, Yamuy J, Xi M-C, Morales FR, Chase MH (1995) Changes in the electrophysiological properties of cat spinal motoneurons following intramuscular injection of adriamycin compared with changes in the properties of motoneurons in aged cats. J Neurophysiol 74:1972–1981

Livingstone MS, Hubel DH (1984) Anatomy and physiology of a color system in the primate visual cortex. J Neurosci 4:309–356

Loeb GE, Bak MJ, Duysens J (1977) Long-term unit recording from somatosensory neurons in the spinal ganglia of the freely walking cat. Science 197:1192–1194

Loeb GE, Peck RA, Martyniuk J (1995) Toward the ultimate metal microelectrode. J Neurosci Methods 63:175–183

Lorente de Nó R (1938) Limits of variation of the synaptic delay of motoneurons. J Neurophysiol 1:187–194

McCrea RA, Bishop GA, Kitai ST (1976) Intracellular staining of Purkinje cells and their axons with horseradish peroxidase. Brain Res 118:132–136

McLennan, H (1970) Synaptic transmission, 2nd Ed. W.B. Saunders Co, Philadelphia

McNaughton TG, Horch KW (1996) Metallized polymer fibers as leadwires and intrafascicular microelectrodes. J Neurosci Methods 70:103–110

Midroni G, Ashby P (1989) How synaptic noise may affect cross-correlations. J Neurosci Methods 27:1–12

Mesulam M-M, Mufson EJ (1980) The rapid anterograde transport of horseradish peroxidase. Neuroscience 5:1277–1286

Metz CB, Schneider SP, Fyffe REW (1989) Selective suppression of endogenous peroxidase activity: application for enhancing appearance of HRP-labeled neurons in vitro. J Neurosci Meth 26:181–188

Pellmar T (1986) Single-electrode voltage clamp in mammalian electrophysiology. In: HM Geller (Ed) Electrophysiological Techniques in Pharmacology. Vol 3. Modern methods in Pharmacology. Alan R. Liss, Inc. New York. pp. 91–102

Pierrefiche O, Haji A, Richter DW (1996) *In vivo* analysis of voltage-dependent Ca^{2+} currents contributing to respiratory bursting. Soc Neurosci Abstr 22:1746.

Pinault D (1996) A novel single-cell staining procedure performed in vivo under electrophysiological control: morpho-functional features of juxtacellularly labeled thalamic cells and other central neurons with biocytin or neurobiotin. J Neurosci Meth 65:113–136

Pitman RM, Tweedle CD, Cohen MJ (1972) Branching of central neurons: Intracellular cobalt injection for light and electron microscopy. Science 176:412–414

Pittman QJ (1986) How to listen to neurons? Discussions Neurosci 3, No 2

Porter R (1963) Unit responses evoked in the medulla oblongata by vagus nerve stimulation. J Physiol (Lond) 168:717–735

Powell TPS, Mountcastle VB (1959) The cytoarchitecture of the postcentral gyrus of the monkey *Macaca mulatta*. Bull Johns Hopkins Hosp 105:108–131

Prochazka A, Westerman RA, Ziccone SP (1977) Ia afferent activity during a variety of voluntary movements in the cat. J Physiol (Lond) 268:423–48

Purves RD (1981) Microelectrode methods for intracellular recording and ionophoresis. Academic Press, London San Diego New York Boston Sydney Tokyo Toronto

Rall W (1969) Time constants and electrotonic length of membrane cylinders and neurons. Biophys J 9:1483–1508

Rall W (1977) Core conductor theory and cable properties of neurons. In: Kandel ER (ed) Handbook of physiology, Sect 1: The nervous system, vol 1, part 1: Cellular biology of neurons. Am Physiol Soc, Bethesda, MD, p 39–97

Raman IM, Bean BP (1997) Resurgent sodium current and action potential formation in dissociated cerebellar Purkinje neurons. J Neurosci 17:4517–4526
Renshaw B (1946) Central effects of cenripetal impulses in axons of spinal ventral roots. J Neurophysiol 9:191–204
Richter DW, Lalley PM, Pierrefiche O, Haji A, Bischoff AM, Wilken B, Hanefeld F (1997) Intracellular signal pathways controlling respiratory neurons. Resp Physiol 110:113–123
Richter DW, Bischoff AM, Anders K, Bellingham M, Windhorst U (1991) Response of the medullary respiratory network of the cat to hypoxia. J Physiol (Lond) 443:231–256
Richter DW, Pierrefiche O, Lalley PM, Polder HR (1996) Voltage-clamp analysis of neurons within deep layers of the brain. J Neurosci Methods 67:121–131
Rogers RC, Butcher LL, Novin D (1980) Effects of urethane and pentobarbital anesthesia on the demonstration of retrograde and anterograde transport of horseradish peroxidase. Brain Res 187:197–200
Sasaki M (1990) Membrane properties of external urethral and external anal sphincter motoneurones in the cat. J. Physiol. 440: 345–366
Schwindt P, Crill W (1980) Role of a persistent inward current in motoneuron bursting during spinal seizures. J Neurophysiol 43:1296–1318
Segev I, Fleshman JWJ, Burke RE (1990) Computer simulation of group Ia EPSPs using morphologically realistic models of cat a-motoneurons. J Neurophysiol 64:648–660
Snodderly DM Jr (1973) Extracellular single unit recording. In: Thompson RF, Patterson MM (eds) Bioelectric recording techniques. Part A. Cellular processes and brain potentials. Academic Press, New York, pp 137–163
Snow PJ, Rose PK, Brown AG (1976) Tracing axons and axon collaterals of spinal neurons uing intracellular injection of horseradish peroxidase. Science 191:312–313
Spillman L, Werner JS (1996) Long-range interactions in visual perception. Trends Neurosci 19:428–434
Stafstrom CE, Schwindt PC, Chubb MC, Crill WE (1985) Properties of persistent sodium conductance and calcium conductance of layer V neurons from cat sensorimotor cortex in vitro. J Neurophysiol 53:153–170
Stamford JA, Palij P, Davidson C, Jorm ChM, Millar J (1993) Simultaneous "real-time" electrochemical and electrophysiological recording in brain slices with a single carbon fibre microelectrode. J Neurosci Methods 50:279–290
Straus W (1979) Peroxidase procedures. Technical problems encountered during their application. J Histochem Cytochem 27:1349–1351
Stretton AOW, Kravitz EA (1968) Neuronal geometry; determination with a technique of intracellular dye injection. Science 162:132–134
Strickholm A (1995a) A supercharger for single electrode voltage and current clamping. J. Neurosci. Meth. 61: 47–52
Strickholm A (1995b) A single electrode voltage, current- and patch-amplifier with complete stable series resistance compensation. J. Neurosci. Meth. 61: 53–66
Terzuolo CA, Araki T (1961) An analysis of intra- versus extracellular potential changes associated with activity of single spinal motoneurons. Ann NY Acad Sci 94:547–558
Thomas RC, Wilson VJ (1965) Precise localization of Renshaw cells with a new marking technique. Nature 206:211–213
Thomas RC, Wilson VJ (1966) Marking single neurons by staining with intra-cellular recording microelectrodes. Science 151:1538–1539
Towe AL (1973) Sampling single neuron activity. In: Thompson RF, Patterson MM (eds) Bioelectric recording techniques. Part A. Cellular processes and brain potentials. Academic Press, New York, pp 79–93
Towe AL, Harding GW (1970) Extracellular microelectrode sampling bias. Exp Neurol 29:366–381
Tsai ML, Chai CY, Yen C-T (1997) A simple method for the construction of a recording-injection microelectrode with glass-insulated microwire. J Neurosci Methods 72:1–4
Turner PT (1977) Effect of pentobarbital on uptake of horseradish peroxidase by rabbit cortical synapses. Exp Neurol 54:24–32
Viana F, Bayliss DA, Berger AJ (1993) Multiple potassium conductances and their role in action potential repolarization and repetitive firing behavior of neonatal rat hypoglossal motoneurons. J Neurophysiol 69:2150–2163
Willis WD, Grossman RG (1973). Techniques in neuroanatomy and neurophysiology. In: Medical neurobiology. Appendix 1. The C.V. Mosby Co, St. Louis, pp 384–394
Wilson CJ, Groves PM (1979) A simple and rapid section embedding technique for sequential light and electron microscopic examination of individually stained central neurons. J Neurosci Meth 1:383–391

Wilson MA, McNaughton BL (1993) Dynamics of the hippocampal ensemble code for space. Science 261:1055–1058
Wilson WA, Goldner MM (1975) Voltage clamping with a single microelectrode. J Neurobiol 6:411–422
Windhorst U (1990) Activation of Renshaw cells. Prog Neurobiol 35: 135–179
Windhorst U (1996) On the role of recurrent inhibitory feedback in motor control. Prog Neurobiol 49: 517–587
Zengel JE, Reid SA, Sypert GW, Munson JB (1985) Membrane electrical properties and prediction of motor-unit type of medial gastrocnemius motoneurons in the cat. J Neurophysiol 53:1323–1344
Zhang L, Weiner JL, Valianta TA, Velumian AA, Watson PL, Jahromi SS, Schertzer S, Pennefather P, Carlen PL (1994) Whole-cell recording of the Ca^{2+}-dependent slow afterhyperpolarization in hippocampal neurones: effects of internally applied anions. Pflügers Archiv-Europ J Physiol 426:247–253
Zheng Y, Barillot JC, Bianchi AL (1991) Patterns of membrane potentials and distributions of the medullary respiratory neurons in the decerebrate rat. Brain Res 546:261–270
Zimmerman SA, Jones MV, Harison NL (1994) Potentiation of γ-aminobutyric $acid_A$ receptor Cl^- current correlates with *in vivo* anesthetic potency. J Pharmacol Exper Ther 270:987–991

Chapter 6

Electrical Activity of Individual Neurons: Patch-Clamp Techniques

Boris V. Safronov and Werner Vogel

Introduction

During the last two decades the patch-clamp technique has become one of the major tools of modern electrophysiology. Originally used for measurements of single-channel currents (Neher and Sakmann, 1976; developed by Hamill et al., 1981), it has turned out to be a powerful method in studying cell excitability, functions and pharmacology of ionic channels as well as mechanisms of their regulation by different metabolic factors. Several recording configurations of the patch-clamp technique enable investigation of macroscopic currents of entire cells as well as elementary single-channel currents in microscopic membrane pieces (patches). An important advantage of the method is the possibility to make recordings under conditions where voltages and solutions at both sides of the membrane are controlled and can be manipulated during the experiment.

Most patch-clamp measurements are performed in a voltage-clamp mode, where the membrane potential is held at a given level, while the current through ionic channels is displayed. In addition, the patch electrodes are successfully used in current-clamp studies, predominantly with entire cells, for recording changes in membrane potential in response to injection of command current pulses. The current-clamp mode allows to monitor different forms of cell activity, for example, action potentials, excitatory and inhibitory post-synaptic potentials as well as changes in membrane potentials due to activation of electrogenic membrane transporters (e.g. Na^+-K^+ pump). In general, the voltage-clamp experiments are performed to describe the biophysical properties of the channels underlying ionic currents in biological membranes, whereas current-clamp recordings provide important information about the channel functions in the cell.

The amplitudes of ionic currents studied by means of the patch-clamp technique range in most cases from 0.1 pA to 10 nA (10^{-13} to 10^{-8} A). Such small currents are usually measured by recording a voltage drop across a large resistor. Designed for this purpose, the basic recording circuitry of the patch-clamp amplifier (Fig. 1A) consists of a standard current-to-voltage (I-V) converter with a large feedback resistor (R_F) and a differential amplifier (DA). The I-V converter forces the pipette potential (V_P) to follow the applied command potential (V_C). The recorded pipette current (I_P) flows through R_F producing a voltage drop of I_PR_F. The voltage at the output of the I-V converter represents the sum of V_C and I_PR_F. The differential amplifier circuit employed for subtraction of V_C gives an output signal $V_{OUT} = I_PR_F$. Typical R_F values used for single-channel recordings are in the range of 10 to 100 GΩ, whereas those for recordings of macroscopic currents are smaller by about two orders of magnitude.

Boris V. Safronov, Justus-Liebig-Universität Giessen, Physiologisches Institut, Aulweg 129, Giessen, 35392, Germany (phone: +49-641-99-47268; fax: +49-641-99-47269; e-mail: boris.safronov@physiologie.med.uni-giessen.de)
Werner Vogel, Justus-Liebig-Universität Giessen, Physiologisches Institut, Aulweg 129, Giessen, 35392, Germany (phone: +49-641-99-47260; fax: +49-641-99-47269; e-mail: werner.vogel@physiologie.med.uni-giessen.de)

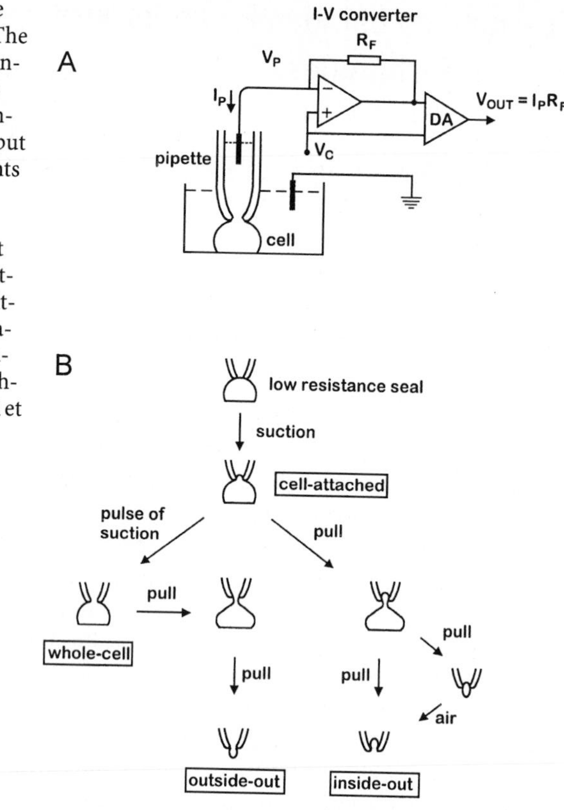

Fig. 1. General scheme of the patch-clamp technique. A: The basic recording circuitry consists of a current-to-voltage (I-V) converter and differential amplifier (DA). The output signal $V_{OUT} = I_PR_F$ represents the voltage drop across the feedback resistor (R_F) produced by the pipette current (I_P). V_C is the command voltage and V_P is the pipette voltage. B: Schematic representation of four major configurations of the patch-clamp technique. Modified from Hamill et al. (1981).

Electrodes used in the patch-clamp experiments are saline-filled glass pipettes with a tip diameter of about 1 µm. The rim of the pipette tip is smoothed in a special fire-polishing procedure. The contact with the cell is established by sealing the pipette tip onto the cytoplasmatic membrane. The four major recording configurations (or modes) of the patch-clamp technique are: "cell-attached", "whole-cell", "outside-out" and "inside-out" (Fig. 1B). Each of these modes has its own advantages and limitations and is used to solve specific problems.

The *cell-attached* recording mode is the first step necessary for establishing any other patch-clamp configuration. In order to form the cell-attached mode, a pipette tip is placed on the surface of the cell, forming a low resistance contact (seal) with its membrane. Slight suction applied to the upper end of the pipette results in formation of a tight seal with a resistance of 1 to 100 GΩ. Such a seal with a resistance in the range of gigaohms is called "giga-seal". Formation of a giga-seal is extremely important for reduction of noise during single-channel recordings. In the cell-attached mode, recordings are made from the membrane area under the pipette, while the interior of the cell remains intact. The potential at the cytoplasmatic surface of the patch is equal to the resting potential of the cell (V_R) measured with respect to zero potential of the reference electrode placed in the bath. If the potential of the pipette electrode is also set to zero, the patch membrane experiences the voltage drop equal to the resting potential. Applying the command potential V_C to the pipette electrode would result in a voltage drop of V_R-V_C on the patch membrane. There are two major advantages of the cell-attached configuration. (1) The channel activity is recorded from the intact membrane which did not lose important cytosolic factors. For this reason, cell-attached patches are widely used for investigation of ion channels with non-modified gating kinetics as well as in experiments studying channel modulation by diffusible second messengers. (2) In many preparations, the cell-attached mode enables recording from only a small number of ionic channels.

A disadvantage of cell-attached recording is the unknown resting potential of the cell. This problem is usually solved in two different ways. (1) V_R can be directly measured at the end of an experiment when the patch membrane is broken by applying a negative pressure to the pipette (see below). Such a measurement is correct if V_R did not change considerably during the experiment. (2) In the experiment V_R of the cell can be artificially set (depolarized) to 0 mV by perfusing the bath with a solution containing a high K^+ concentration (about 150 mM).

Disrupture of the cell-attached patch by applying a pulse of suction to the pipette, without damaging the seal between the pipette rim and the membrane, results in formation of the *whole-cell* recording mode. In this mode the pipette solution diffuses into the cell. Since the internal volume of the recording pipette is much larger than that of the cell, the pipette solution will completely substitute the intracellular solution. Depending on size and geometry of the cell and the patch pipette, this diffusion can last from several seconds to 1–2 minutes. In the whole-cell configuration the patch electrode has a low resistance access to the interior of the cell and therefore membrane potential in the cell is equal to that of the patch electrode (unless large currents are recorded, see below). Thus the pipette could be used either for measurement of the membrane potential or for clamping it at a given level. The whole-cell mode is the most frequently used configuration of the patch-clamp technique employed for recording macroscopic currents in the membrane of the entire cell. In contrast to the classical sharp electrode which impales the cell, the patch pipette seals onto its membrane producing much less damage and therefore enables recordings with much smaller leakage currents.

An excised *outside-out* patch is obtained from the whole-cell recording configuration by slowly pulling the pipette away from the cell until a small patch of the membrane reseals on the tip of the pipette. Under these conditions, the internal surface of the cell membrane is exposed to the pipette solution, whereas its external surface is in contact with the bath solution. The potential of the pipette electrode is applied to the internal side of the membrane and the external surface of the membrane is held at zero potential of the bath electrode. The outside-out configuration is normally used for recording the channel activity with a possibility of changing external solutions. The membrane area of an outside-out patch is usually larger than that of a cell-attached or an inside-out patch obtained using pipettes with the same diameters of tip opening. In many preparations outside-out patches contain more than one ionic channel. In this case recordings from a single channel are possible only in cell-attached or inside-out modes.

An excised *inside-out* membrane patch can be directly obtained from the cell-attached mode by pulling the pipette away from the cell. In some cases this directly results in formation of the inside-out patch. More frequently, however, the cell membrane reseals on the tip of the pipette forming a vesicle. The outer membrane of the vesicle can be easily destroyed by shortly exposing the tip of the pipette to air. In the inside-out mode, the pipette solution and the pipette potential are applied to the external (extracellular) surface of the patch membrane, while the cytoplasmic side of the membrane is exposed to the bath solution and has the potential of the reference electrode. The inside-out patch configuration gives the possibility of solution exchange at the cytoplasmic surface of the membrane.

The choice of the solution for filling the pipette is determined by the recording configuration to be established. For recordings in the whole-cell and outside-out modes the pipette should contain an internal solution (see Table 1), whereas those used for cell-attached and inside-out recordings are filled with an external solution.

The patch-clamp technique is often used for estimation of the channel densities in biological membranes. This is usually done by dividing the number of active channels in the patch by its area. Since the membrane patch at the tip of the pipette is at the border of resolution of light microscope, the patch area is usually estimated using the pipette resistance (R_P), a parameter which can be easily measured in electrophysiological experiments. As seen in Fig. 1B, the patch area depends on the diameter of the pipette tip. The diameter of the pipette tip and its geometry also determine the R_P. A relationship between the patch area a (μm^2) and R_P can be described using an empirical equation (Sakmann and Neher, 1995):

$$a = 12.6(1/R_P + 0.018),$$

where R_p (M Ω) is measured for the pipettes filled with normal physiological saline. In spite of some differences in the patch geometry mentioned above, this equation provides a good estimate of the area of cell-attached, inside-out as well as outside-out membrane patches.

Materials

Set-up

A general scheme of a standard patch-clamp set-up is shown in Fig. 2. The recording part of the set-up includes a patch-clamp amplifier, analog-to-digital and digital-to-analog (AD-DA) converter, and a personal computer for generation of command pulses and signal recording. Although a built-in filter is available in the patch-clamp amplifier, an additional low-pass 8-pole Bessel filter is desirable. For convenience the set-up is usually equipped with an independent pulse generator and an oscilloscope. The data can also be stored using a conventional DAT (digital audio tape) recorder.

The following devices and materials are necessary for any patch-clamp laboratory:

- Pneumatic anti-vibration table with a massive baseplate (e.g. Physik Instrumente, Waldbronn, Germany or Newport, Irvine, CA) designed to reduce vibrations normally present in any building. The microscope, preparation plate and micromanipulator with the headstage are mounted on the baseplate of the anti-vibration table.

  Note: Special precautions should be taken when organizing the set-up in order to provide the high mechanical stability required in patch-clamp experiments. A drift between the patch pipette and the preparation is minimized by mounting the prepa-

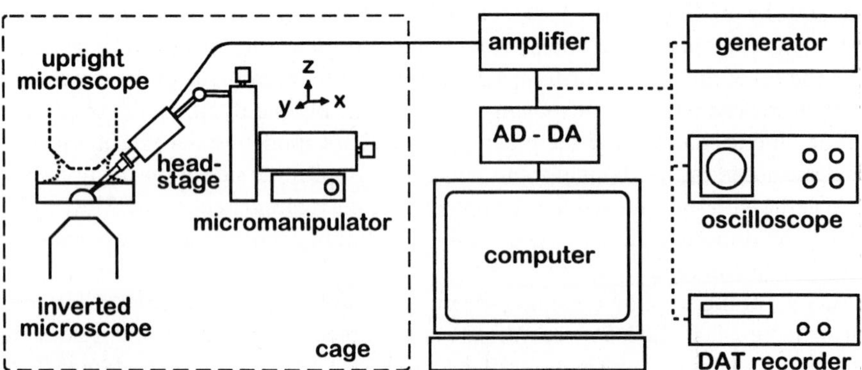

Fig. 2. General scheme of a standard patch-clamp set-up. An inverted microscope is used in experiments with isolated and cultured cells, whereas an upright microscope with a water immersion objective (shown by dotted lines) is installed in a set-up for studying the cells in tissue slices.

ration table and the micromanipulator on a common tower separated from the microscope.

- Inverted microscope (e.g. Axiovert 135, Carl Zeiss, Jena) is used for recordings from isolated or cultured cells (cf. Chapter 10). It allows visualization of the cells from the bottom and provides free access for the patch pipette from the top. An upright microscope (e.g. Axioscop FS, Carl Zeiss, Jena) with a water immersion objective (x40–60) is employed in experiments with tissue slices (cf. Chapter 9). The working distance of this objective must be sufficiently long (about 1.5 mm) to enable positioning of the patch pipette between slice and objective. In most cases an overall magnification of the microscopes of x400 is sufficient. For convenience the microscope is often connected with a video camera and the approach of the pipette to the preparation is monitored using a standard TV set.
- A stable motor-driven micromanipulator providing movement in three axes (e.g. HS6, Märzhäuser, Wetzlar, Germany).
- A patch-clamp amplifier used for both whole-cell and excised patch recordings (e.g. Axoclamp 200B, Axon Instruments, Foster City, CA). The amplifier usually includes a pipette holder and a headstage pre-amplifier.
- AD-DA converter (e.g. Axon Instruments, Foster City, CA).
- A personal computer with software (e.g. pClamp, Axon Instruments, Foster City, CA).
- Faraday cage shielding the patch-clamp recordings from surrounding electrical noise (e.g. Grittmann , Heidelberg, Germany). The cage should stand on the floor so that it does not touch the anti-vibration table. This helps to avoid transmission of mechanical vibration to the baseplate of the table. The cage is also often used for mounting the perfusion systems.

Fabrication of Pipettes

The following equipment and materials are used for fabrication of the patch pipettes.

- Pipette puller (e.g. Sutter Instruments Co., Novato, CA). This is a horizontal microprocessor-controlled device pulling the pipettes in several steps.
- A microforge is used for fire-polishing the tips of the patch pipettes. Usually, this is a self-made device consisting of a compound high-magnification inverted microscope with a long-distance objective, a hand-driven manipulator and a heating element (Fig. 3B). Manipulator and heating element are installed on the stage of the microscope. The heating element includes a thin platinum wire (diameter of about 100 µm) connected to a power supply. In order to prevent evaporation of the metal onto the pipette tip during polishing, a part of the wire should be covered with glass. This can be easily performed by melting a pipette to the heated wire. An air stream can be directed to the wire and electrode in order to restrict the heating to the vicinity of the heating filament. In most cases, however, the pipettes polished without air stream have good geometry of the tip and provide stable seals. The objective of the microscope is protected from dust and overheating by appropriately positioning a standard cover glass.
- Pipette glass (e.g. GC 150-7.5, Clark, Reading, United Kingdom). This is a thick-walled glass capillary with an external diameter of 1.5 mm and an internal diameter of 0.86 mm. It contains an internal glass filament that aids the filling of the pipette with electrolyte solution. Pipette glass without internal filament can also be used. In this case, pipette filling is performed in two stages (see below). There are many different types of glass that are successfully used for fabrication of patch-clamp electrodes (for more information see Corey and Stevens, 1983; Rae and Levis, 1992). In

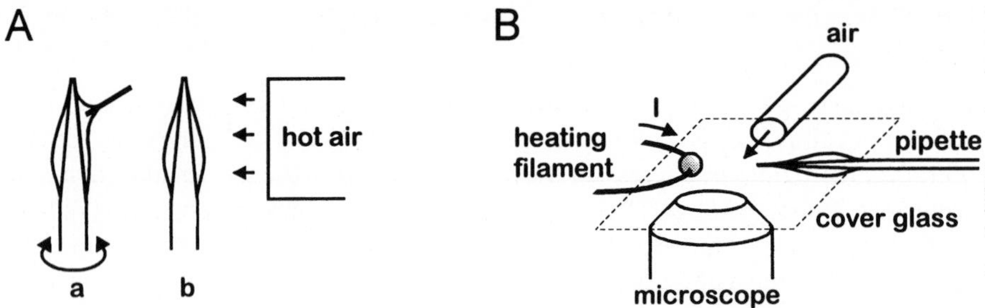

Fig. 3. Coating and polishing the patch pipettes. A: The pipette is coated by applying Sylgard with a fine needle (a). Sylgard is cured by moving the pipette tip into a jet of hot air (b). B: Microforge for fire-polishing the pipette. Under control with an inverted microscope, a heated wire covered with a glass ball is briefly approached to the pipette tip, smoothing the rim.

general, thick-walled glass tubing is suitable for fabrication of electrodes for single-channel recording whereas the pipettes for whole-cell experiments are often made from thin-walled glass.

- Sylgard resin (Sylgard 184, Dow-Corning, Midland, MI) is used for coating the pipettes. The coating is performed to reduce capacity and electrical noise of the recording pipette.
- Equipment for fast curing of the Sylgard. It includes a low-magnification microscope for visual control and a hot air jet or heated wire coil for rapid curing.

Solutions

Solutions applied to the extracellular and cytoplasmic surfaces of the membrane will be referred to as external and internal solutions, respectively. Compositions of typical external and internal solutions used in patch-clamp experiments with mammalian neurons are given in Table 1.

Oxygenated external-1 solution is used in experiments with brain slices. External-2 solution buffered with HEPES-NaOH is suitable for experiments with freshly isolated and cultured cells as well as for recordings from membrane patches.

Table 1. Solutions

	External-1	External-2	Internal
NaCl	115 mM	141 mM	5 mM
KCl	5.6 mM	5.6 mM	155 mM
$CaCl_2$	2 mM	2 mM	–
$MgCl_2$	1 mM	1 mM	1 mM
glucose	11 mM	11 mM	–
EGTA	–	–	3 mM
NaH_2PO_4	1 mM	–	–
Buffer	25 mM $NaHCO_3$, pH 7.4 is set when bubbled with 95%–5% mixture of O_2-CO_2	10 mM HEPES, pH 7.3 with NaOH	10 mM HEPES, pH 7.3 with KOH

Compositions of these solutions can be modified in a wide range depending on the type of ionic current to be investigated. There are, however, several requirements for all solutions used in patch-clamp experiments.

- Both internal and external solutions must be filtered, for example, using 0.2 μm syringe filter. Contamination of the pipette tip or cell membrane in non-filtered solution can prevent formation of seal.
- At least 10 mM Cl^- must be present in each solution to provide a stability of chlorided silver electrodes. If a full substitution of Cl^- is required, electrodes should be connected to the solutions via an agar bridge.
- Internal solution should contain low concentrations ($<10^{-7}$ M) of Ca^{2+} ions. EGTA or BAPTA (fast Ca^{2+} buffer) are widely used for chelating the intracellular Ca^{2+}.

Electrodes and Grounding

A chlorided silver wire directly immersed into the bath solution provides a good reference electrode. If contamination of the bath solution by Ag^+ ions should be avoided, a chlorided silver wire is placed in a small saline-filled pot electrically connected to the bath via an agar bridge. The solution for filling agar bridges should have an ionic composition similar to that of the bath solution (e.g. 150 mM NaCl).

Since the currents recorded in single-channel experiments are in the range of 0.1–20 pA, special care should be taken in grounding the reference electrode and all devices (cf. Chapter 45). Both quality and pattern of grounding play an important role in reducing the line-frequency (50–60 Hz) interference. The bath electrode is connected to a high-quality ground terminal of the headstage. Objects like table, microscope, micromanipulator, Faraday cage etc. are grounded using low-resistance cables to a common point which is connected to the ground terminal on the amplifier case. In a patch-clamp amplifier, the ground terminal of the headstage and ground terminal on the amplifier case are internally connected through a headstage cable.

Note: It is important to avoid any ground loops picking up interference signal.

Preparations

In general, the choice of preparation is determined by the type of problem to be solved. In this section we would like to mention the kinds of cell preparations suitable for patch-clamp investigation. The most important criterion for establishing the giga-seal contact is a clean surface of the cell membrane. Excised outside-out and inside-out patches can be obtained only from cells attached to the bottom of the recording chamber. The most widely used preparations are: enzymatically dissociated cells, cells in culture and tissue slices (cf. Chapter 9 and 10). The cells are usually dissociated after treatment of tissue pieces with proteolytic enzymes such as trypsin, collagenase or protease. The membrane of enzymatically isolated and cultured cells is often clean and is directly accessible for the patch electrode. Several detailed protocols for dissociation and culturing the cells suitable for patch-clamping are given elsewhere (Trube, 1983; Bottenstein and Sato, 1985; Standen and Stanfield, 1992).

The brain slice technique enables the investigation of intact neurons that were not subjected to enzymatic treatment and did not lose synaptic connections with other neurons. Since this method has become more and more popular during the last years, it is worthwhile to give a short description here. The slices are prepared according to a procedure proposed by Edwards et al. (1989). After decapitation of the animal, the brain or the spinal cord are quickly removed and placed in an ice-cold oxygenated external-1 solution (see Table 1). The region of interest is cut out and glued onto the glass stage fixed in the chamber of the vibrating tissue slicer. Small pieces of the brain tissue or the spinal cord that cannot be directly glued should be first embedded in agar. The block of agar containing the tissue can be glued onto the glass stage. The cut slices with a thickness of 200–300 µm are incubated for 30–60 min at 37 °C and then stored at room temperature until use. Although some healthy cells can be found on the surface of the slice, the majority of them is covered by connective tissue. There are several possibilities to obtain access to these cells with a patch pipette. For example, the tissue covering the neuron can be removed by blowing and sucking the saline by means of a separate large-diameter or broken patch pipette (Edwards et al., 1989). An alternative possibility is the "blow and seal" technique (Stuart et al., 1993). Positive pressure is applied to the recording patch pipette so that the solution stream pushes connective tissue away from the pipette tip as it advances through the slice. After touching the cell membrane, the positive pressure is released and the applied suction leads to formation of a tight seal.

Procedure

Fabrication of the Patch Pipette

1. Pulling
   The invention of programmed pullers has considerably simplified the procedure of pipette manufacturing. One should just select the corresponding pulling protocol, fix the glass capillary and pull two complementary pipettes. It is more difficult, however, to program different pulling protocols, since the variation of several parameters influences the geometry of the pipette tip and the diameter of its opening. Detailed instructions on programming the pulling protocols can usually be found in the manuals supplied with the device.
   One should keep in mind that the geometry of the pipette is the major factor determining its resistance. Modern microprocessor-controlled devices are able to pull the pipette in several steps. The first step thins the glass capillary over 7–10 mm and the following steps separate two electrodes and determine the size of the tip openings. A standard patch pipette is pulled in 2–6 steps. In general, increasing the number of steps makes it possible to pull the pipettes with a larger angle of the tip cone and therefore with smaller resistance at a given diameter of the tip opening. Microprocessor-controlled pullers can store the programs for several pulling protocols. Another advantage of programmable pullers is the high reproducibility of the patch electrodes.

   Note: The pipettes used for whole-cell recordings usually have a resistance of 1–5 MΩ, whereas those for study of single channels are in the range of 5–30 MΩ.

2. Coating with Sylgard
   In order to reduce the capacity of the recording pipette, its tip (region adjacent to the opening) is coated with an insulator. This leads to a reduction of both the background noise and the capacity currents masking the recorded signal. The coating is especially important for single-channel recordings from cell-attached, inside-out and outside-out patches. It is not necessary in whole-cell experiments where recorded currents are large and the membrane capacity of the cell is considerably larger than that of the patch pipette. In general, the coating procedure is very critical for obtaining a giga-seal contact.
   Sylgard is a two-component mixture widely used for pipette coating. Resin and catalyst are mixed and allowed to precure for 2–3 hours at room temperature. Precured

Sylgard filled in syringes can be stored in a freezer (–18 °C) for several months. Sylgard is applied by a fine needle to the pipette tip under microscope (x10–20) observation, while the pipette is rotated (Fig. 3Aa). Although the coating of the pipette as close to the tip as possible is desired, care should be taken to avoid covering the pipette tip opening. Since only clean glass seals to the membrane, the last 10–50 µm adjacent to the tip must remain uncoated. The total length of the coated shank in most cases is 5–7 mm. Immediately after application, Sylgard should be cured until full solidification, for example, by positioning the pipette tip into the stream of hot air (Fig. 3Ab) or into a heated wire coil.

Note: If during an experiment a giga-seal has not been achieved with coated pipettes, uncoated ones should be tested. If coating is the reason for the failure of getting a good seal, try to increase the thickness of the Sylgard by prolonging its precuring and/or increase the length of uncoated region at the pipette tip.
At this stage, the pipettes cannot provide stable seals, since (a) they have sharp and often uneven tip rims and (b) in spite of all precautions taken during the coating procedure, uncured Sylgard penetrates to the pipette tip, covering it with a very thin film. The rim is smoothed and the Sylgard film is burned off by fire-polishing the pipette tip.

3. Polishing the Tip
The pipette is polished under visual control using a microforge (Fig. 3B). The tip of the pipette is positioned 10–20 µm away from the filament heated to a dull red glow. The whole procedure lasts a few seconds. It can be observed as the glass wall at the pipette tip becomes thicker and the rim is smoothed.

Note: Polishing determines the final diameter of the pipette tip opening. By stronger polishing it is possible to considerably reduce the diameter of tip opening. Polishing is always done after the pipette coating, just before the experiment. The polished patch pipette should be protected from dust.

4. Filling
A pipette made from a glass capillary tubing with inner filament can simply be back-filled using a standard syringe with a fine-diameter needle. A pipette made from a glass capillary without inner filament should be filled in two steps. The first one is the filling of the pipette tip. The tip is immersed into a small beaker containing pipette solution and then a negative pressure is applied to the pipette back opening by means of a 10-ml syringe connected via silicone tubing. Depending on the diameter of the tip opening, the filling of the tip lasts from several seconds to 1–2 minutes. The suction must be released before the pipette tip is withdrawn from the solution. The second step is a standard backfilling of the pipette with a syringe. Air bubbles remaining in the pipette tip after filling can be removed by gently flicking the pipette shank with a fingernail. The filled part of the pipette should not exceed 5–10 mm. Larger filling can increase the pipette capacity and electrical noise.

Note: The solution used for pipette filling is determined by the kind of patch-clamp configuration to be established. The pipettes for cell-attached and inside-out patches are filled with an external solution, whereas those for whole-cell and outside-out recordings should contain an internal one. A filter (0.2 µm) must be inserted between the needle and the syringe used for backfilling the pipette and filling the beaker with a pipette solution. If the pipette tip is filled in a beaker, it is important to apply a slightly positive pressure at the moment when the pipette tip crosses the air-solution and solution-air borders. This helps to avoid a contamination of the tip by dust unavoidably appearing on the surface of solution. EGTA-containing solutions can interact with a metal needle of the syringe used for pipette filling. In this case, a pulled plastic capillary can be used instead of a metal needle. Another possibility would be to discard an appropriate amount of solution from the syringe just before electrode filling.

Patch-Clamp Experiment

5. The filled pipette is mounted in a pipette holder and fixed by a screw cap. It is important to control whether a rubber ring fixing the pipette fits tightly; if not, the system will lose pressure and formation of the seal will be compromised. Pressure is applied to the system via a silicone tubing, connected to a special outlet on the holder, either simply by mouth or by means of a U-shaped water manometer. In order to prevent contamination of the pipette tip, positive pressure should be applied to the system, so that solution streams out of the pipette tip while crossing the air-solution border and moving the pipette towards the cell.
   Two procedures have to be performed before the pipette tip has touched the cell membrane: measurement of the pipette resistance and compensation of offset potentials. The amplifier is in the voltage-clamp mode and the holding potential (V_H) is set to 0 mV. Rectangular 50-ms depolarizing command pulses to V_C of 1–2 mV are generated once per second (Fig. 4A) to visualize changes in the leakage current (Fig. 4B). As the solution in the pipette represents a linear ohmic resistance, the current response also has a rectangular shape (transient currents due to pipette capacity are negligible) with an amplitude of ΔI.
6. The pipette resistance (R_P) is determined by dividing ($V_C - V_H$) by ΔI. In the experiment shown in Fig. 4B the R_P value is about 5 MΩ.
7. An offset potential between the pipette and the bath electrode should be nulled using the offset compensation circuit of the amplifier. This is important, since uncompensated offset potentials would be added to V_H.
8. Formation of the giga-Seal.
   As soon as the pipette touches the cell membrane (usually indicated by some reduction in (ΔI), the positive pressure is released and a slight negative pressure (suction) is applied. As the membrane seals to the pipette tip, the resistance of the contact in-

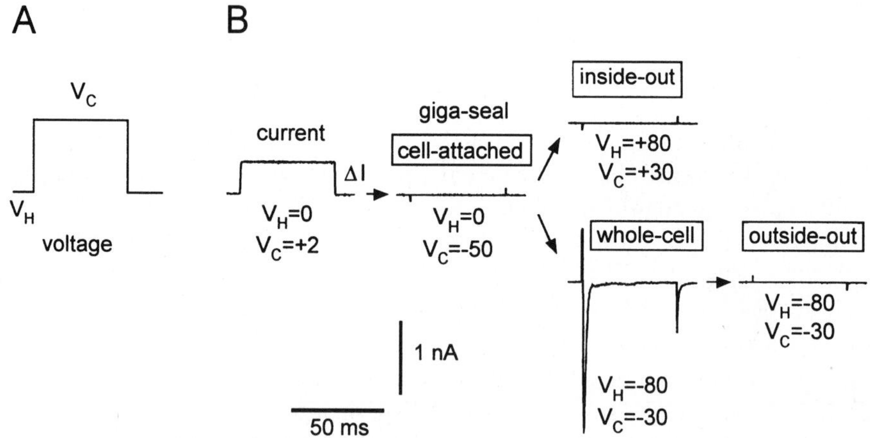

Fig. 4. Changes of pipette currents accompanying formation of four major patch-clamp modes. A: Voltage pulse applied to the electrode. Holding potential (V_H) and command potential (V_C) are set in accordance with the recording mode to be established. B: Current response of a 5 MΩ pipette to a 2 mV voltage pulse recorded before touching the cell (left). The current response is dramatically reduced after formation of the giga-seal. The following recordings show currents monitored in cell-attached, whole-cell, outside-out and inside-out modes, respectively. Typical V_H and V_C values used for the recordings are indicated near the corresponding traces.

creases, leading to a dramatic reduction of (ΔI (Fig. 4B, giga-seal). Although sometimes a giga-seal can occur spontaneously, in most cases some suction is needed for its formation. The resistance of the giga-seal is usually in the range of 1 and 100 GΩ. Formation of the giga-seal can last from 1 s to 1–2 min and is always accompanied by a considerable reduction in background noise.

Note: A patch pipette can be used only once. If the giga-seal contact has not been established with the cell membrane, another pipette should be used and steps 1–8 have to be repeated. If a seal has not been achieved, one can try (1) to check whether the pipette holder is losing pressure, (2) to use uncoated pipettes, (3) to prepare new solutions, (4) to make a new preparation or (5) to change the glass ball covering the fire-polishing filament.
Steps 1–8 are common for all patch-clamp protocols. The following five steps (9–13) show how different configurations of the patch-clamp technique are obtained.

Cell-Attached Mode

9. Basically, the cell-attached configuration has already been established at the end of step 8 (Fig. 4B, cell-attached). In addition, fast capacity currents should be electronically compensated using a corresponding circuit of the patch-clamp amplifier. The gain is increased to that used for single-channel recordings.

Note: A V_H of 0 mV is a good choice for the cell-attached recording. Assuming a membrane resting potential of V_R=–80 mV, a voltage drop on the patch membrane in this case would be V_R–V_H=–80 mV. In order to depolarize the patch membrane, for example, to –30 mV, a command pulse of V_C=–50 mV should be applied. Exact measurement of V_R is performed in current-clamp mode at the end of the experiment after the patch membrane has been destroyed by suction. If for any reason the measurement of V_R at the end of the experiment is not possible, a membrane potential can be reported in respect to unknown V_R.

Whole-Cell Mode

10. After establishing the giga-seal (step 8), V_H and V_C are prepared for the whole-cell mode. V_H is usually set to –80 mV, corresponding to the expected resting potential of the cell, and a command pulse steps to V_C=–30 mV. The fast component of the transient current arising from charging of the pipette capacity is cancelled. Breaking the patch membrane by applying a brief pulse of suction results in formation of whole-cell mode which is accompanied by the appearance of large slow transients and increase in leakage current and electrical noise (Fig. 4B, whole-cell). All these changes result from an increase in the area of clamped membrane after transition from the cell-attached to the whole-cell configuration. At the given pulse protocol, the successful establishment of the whole-cell mode with a typical neuron is also indicated by the appearance of voltage-activated Na^+ currents.
At this moment the slow transients should be compensated using appropriate circuitries of the patch-clamp amplifier. During compensation, the cell can be stimulated with small hyperpolarizing pulses that do not activate fast conductances.

Note

Two main problems appear during whole-cell recording.

a) Voltage-Clamp Error due to series Resistance.
In the whole-cell mode, the membrane potential is expected to be equal to that generated on the output of stimulation circuitry of the patch-clamp amplifier. Unfortunately, this is the case only when small ionic currents are recorded. With larger currents, however, the real potential on the membrane (V_M) differs from the command potential, V_C, applied to the pipette. The equivalent electrical circuit is shown in Fig. 5. A potential V_C applied to the patch electrode will drop on the resistance of the cell membrane (R_M) and on the series resistance of the electrode (R_S). R_S is not equal

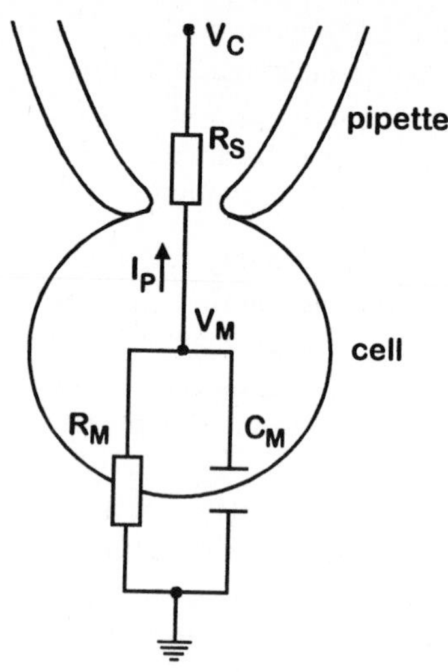

Fig. 5. Voltage error produced by resistance in series (R_S). Equivalent electrical circuit of the whole-cell recording. The membrane potential (V_M) is not equal to the given command potential (V_C) due to voltage drop on R_S. R_M and C_M are resistance and capacity of the cell membrane, respectively.

to the pipette resistance R_P measured in bath solution before touching the cell. In a real experiment, remnants of the broken membrane and cytoplasmatic elements sucked into the pipette tip will increase the value of R_P by a factor of 2–5. According to Ohm's law for the scheme of Fig. 5, $V_M = V_C + I_P R_S$, where I_P is the current flowing through the pipette and the cell membrane. Thus, the voltage drop on the series resistance equal to $I_P R_S$ represents the difference between V_C and V_M or, in other words, an error in the clamped potential. Assuming an amplitude of Na^+ current of 3 nA, $R_S = 3R_P$ and $R_P = 2$ MΩ, the error would be 18 mV. In this case, instead of the –40 mV applied, the cell membrane will see –22 mV. Such an error changes the shape of the current-voltage relationships for Na^+ currents, shifting it to more hyperpolarized potentials and steepening its activation. Additionally, a large R_S slows the charging of the cell membrane, so that it cannot follow fast changes in V_C. There are several possibilities to reduce the error produced by the resistance in series.

- One can use low-resistance pipettes ($R_P \approx 1$ MΩ) or reduce the amplitude of the currents recorded. The latter can be done either by partially blocking the currents or by reducing the concentration of the major carried ion in external or internal solutions. The amplitude of Na^+ current can also be reduced by using more depolarized holding potentials inactivating channels.
- If for some reason the reduction in R_S or current amplitude is undesirable, electronic series resistance compensation should be done. The corresponding circuitry is incorporated in standard patch-clamp amplifiers.

In any case, it is advisable first to estimate the value of the error due to series resistance by simply multiplying the maximal current by a triple R_P. In most experiments the series resistance can remain uncompensated if the error is smaller than 2–3 mV. Larger errors should be compensated in one of the ways mentioned above.

b) Space-Clamp Problem.

The next problem appears when the whole-cell currents are recorded from large cells with long processes, for example, from neurons in brain slices. The pipette is usually placed on the cell soma and therefore the potential is adequately clamped only in the adjacent somatic membrane, whereas it remains uncontrolled in remote membrane regions such as dendrites and axon. Ionic currents recorded under such conditions represent a mixture of responses from both well- and poorly clamped regions. Kinetics and activation properties of these currents can considerably differ from those recorded in adequately clamped cells.

There is no possibility to improve the clamp conditions in cells with a complicated geometry. A reliable description of ionic currents by means of the patch-clamp technique should be performed with small round cells (10–30 μm) without long processes or with isolated membrane patches (see below).

11. Current-Clamp Recordings.
The whole-cell mode is also used for current-clamp recordings where the membrane current is clamped while a change in membrane potential is displayed. The current-clamp mode is used for measurements of the membrane resting potential as well as for recording postsynaptic potentials and action potentials. Switching between voltage- and current-clamp modes is simply performed by using the corresponding buttons of the patch-clamp amplifier.

Outside-Out Patch

12. The outside-out patch is obtained from the whole-cell recording configuration (step 10, before compensation of transients has been done). V_H is set to –80 mV and the command pulse steps to V_C=–30 mV. The pipette is slowly pulled away from the cell until the connection is lost. The formation of the outside-out patch is accompanied by a progressive reduction of Na^+ currents, the disappearance of the slow component of capacity currents and a dramatic reduction of background noise. Ionic currents in an outside-out patch are seen only when the amplification factor is increased to that used for single-channel recordings. The fast component of the transient current should be compensated.

Note: If the attempt to pull an outside-out patch was not successful, try to pull the next patch more slowly. It is advisable to withdraw the pipette tip several millimeters away from the cell to be sure that the patch has lost its connection to the cell membrane.

Inside-Out Patch

13. After establishing a giga-seal (step 8), V_H and V_C should be prepared for an inside-out patch. In the inside-out mode the pipette potential is applied to the extracellular surface of the cell membrane, whereas its intacellular surface is at the potential of the bath electrode. Therefore, the membrane will experience a physiological voltage gradient if V_H is set to +80 mV. A command pulse V_C can be set to +30 mV. Thereafter, the pipette is carefully withdrawn until the contact with the cell has been lost. There are two ways of obtaining an inside-out patch configuration.

a) In some cases, withdrawal of the pipette directly results in formation of the inside-out patch (Fig. 1B).
b) In most cases, however, a vesicle is formed (Fig. 1B). The external membrane of the vesicle should be destroyed by a short exposure of the pipette tip to air. Formation of the inside-out patch is followed by compensation of the fast transients corresponding to charging of the pipette capacity. The amplification factor should be appropriately increased to enable single-channel recordings.

Some problems may arise if inside-out patches are studied in the same bath where the preparation, i.e. the cultured cells or a brain slice, is located. In these experiments the bath is usually perfused with an internal solution containing a high concentration of K^+ ions which depolarizes the membrane of the cells to 0 mV. As a consequence of long exposure to internal solution, many cells either die or become unable to seal to the pipette. In these cases, the culture dish or brain slice has to be discarded after each attempt to record from an inside-out patch, even if the patch did not contain channels of interest or had not survived the whole experimental procedure. This may be quite disappointing in experiments where:

- the number of tissue slices or culture dishes is limited
- several inside-out patches from the same cell should be obtained

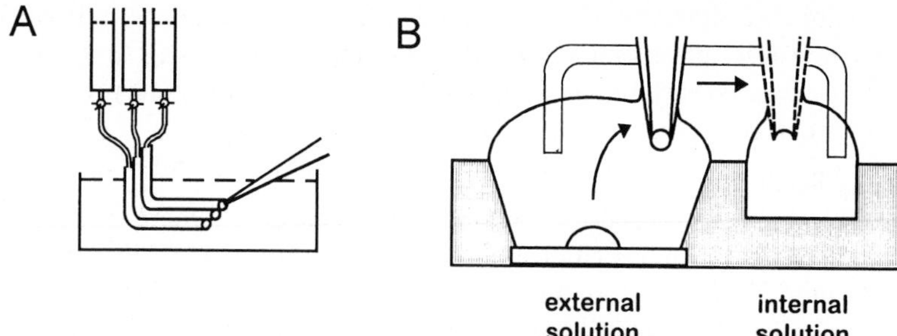

Fig. 6. Methods for studying inside-out patches. A: Multi-barrel perfusing system. The pipette tip with the patch is inserted into the corresponding barrel. B: Experimental chamber with the main and the small additional baths containing external and internal solutions, respectively. The jump from the main into the additional chamber results in formation of an inside-out patch. Modified from Safronov and Vogel (1995).

- preparing the cell for sealing requires some additional time-consuming procedures like cleaning from connective tissue in a brain slice.

Multi-Barrel Perfusion System

Investigation of inside-out patches without destroying the preparation can be performed by using a multi-barrel perfusion system (Yellen, 1982). Barrels made from glass capillaries are connected via silicone tubings with syringes containing different solutions (Fig. 6A). Solutions are slowly moved by gravitational force. The velocity of the solution flow is adjusted so that the barrel solution at the outlet is not diluted by the bath solution. The tip of the pipette with an inside-out patch is inserted into the barrel with an appropriate solution. The solution at the cytoplasmic side of an inside-out patch can be changed by moving the pipette tip into another barrel. A multi-barrel perfusing system can be used for studying both inside- and outside-out membrane patches. This system has two disadvantages. (1) Installation of the multi-barrel system needs some additional space in the recording bath. The system is not always convenient, for example, in experiments with brain slices where much space in the bath is occupied by a water immersion objective and an additional increase of the bath volume is not desired, since it will decrease the intensity of slice perfusion with external solution. (2) The bath solution is contaminated by internal solutions flowing out of the barrels.

Additional Bath

The inside-out patches can also be studied in a small additional bath (Fig. 6B, Safronov and Vogel, 1995). This bath, containing internal solution, is located 0.2–0.5 mm away from the main bath and is electrically connected via a standard agar bridge. It is important to fill both baths in such a way that solutions form high menisci. The principle of the inside-out patch formation is very similar to that of Fig. 1B. The pipette is sealed to the cell membrane, and a vesicle is formed by slowly withdrawing the pipette. The external membrane of the vesicle is destroyed by exposure to air during a fast movement (jump) into the additional bath. The jump is visually controlled under the dissecting microscope. The major steps of this procedure are:

a) The microscope is focused on the border between the chambers. To make focusing easier one can draw a point by a water-resistant pen or make a fine scratch on or near the border;
b) The microscope is moved 300–400 μm upwards;
c) The pipette is slowly moved towards the border between the chambers, and its tip is positioned in focus just near the border. The stability of both menisci is controlled;
d) The pipette is accelerated so that maximal velocity is reached at the moment the pipette tip leaves the main bath.

In general, patch exposure to air during the jump is shorter than during the standard procedure when the pipette tip is withdrawn and inserted into the same bath.

This method allowed us to obtain up to twenty inside-out patches from the same neuron. The following observations were made in our laboratory during work with an additional bath. (1) The success rate of the jump is high (about 90 %) when performed with a vesicle leaving the main chamber. It can, however, be lower when an inside-out patch has already been formed in the main bath. (2) If the external membrane of the vesicle has not been destroyed during the jump, it should be repeated, if necessary, several times. (3) Jumps could be successfully performed using both coated and uncoated pipettes. (4) The additional bath can be perfused with different solutions during single-channel recording. For this purpose two cannulae, for supplying and sucking, should be added to the bath. (5) In some cases the success rate of jump is higher if the pipette potential is reduced from +80 to about +60–+40 mV. A lower potential appears to be less stressful for the membrane. After the jump, the pipette potential is set again to +80 mV.

Investigation of inside-out patches in an additional bath has the following advantages:

- Small volumes (200 µl) of internal solution are needed to fill an additional bath. It can therefore be very helpful in experiments where expensive substances are used,
- Baths are separated and therefore the main bath is not contaminated by an intracellular solution.

Note: The material of which the experimental chamber is made plays a crucial role in determining the shape and stability of menisci. It is important to choose hydrophobic material to avoid merging of solutions during the jump. Teflon is a good candidate for the chamber, enabling formation of stable menisci. Unfortunately, it is a relatively soft material and fabrication of the chamber with a narrow border between the baths demands much skill. An excellent alternative to Teflon is delrin (polyoxymethylene, Du Pont, France). Compared to Teflon, delrin is mechanically harder and therefore easier to handle in manufacturing, but it possesses lower hydrophobicity, and the menisci are less stable in a chamber made of delrin. This disadvantage can however be compensated by covering the surface of a delrin chamber with a very thin film of vaseline. For this purpose, vaseline is first spead over to the surface and is then wiped away with a dry tissue until it becomes hardly visible. A very thin film of remaining vaseline is still sufficient to provide stable menisci. Once covered with vaseline, the chamber can be used for several months. The chamber is cleaned at the end of the experiment with distilled water.

Results

Here we would like to show an example of patch-clamp recording of voltage-activated Na^+ current. In spite of careful compensation of capacity currents made during the experiment, a small part of them always remains uncompensated. Unfortunately, remaining transients are very large in comparison with single-channel currents and, in some cases, they may completely mask early openings of Na^+ channels at the beginning of a depolarizing pulse. In addition, a non-specific leakage current flowing through the seal between pipette tip and membrane is always present in each recording. The simplest way to correct for both capacity and leakage currents is their subtraction by using a divided pulse protocol.

An example of recordings from a large outside-out patch containing 10–20 Na^+ channels is shown in Fig. 7. A depolarizing voltage pulse from –80 to –20 mV (Δ_1=60 mV) activates Na^+ channel current which is masked by a capacity transient. The amplitude of the leakage current in this recording is comparable with that of the Na^+ current investigated. A recording used for the correction was obtained by stepping up the voltage from

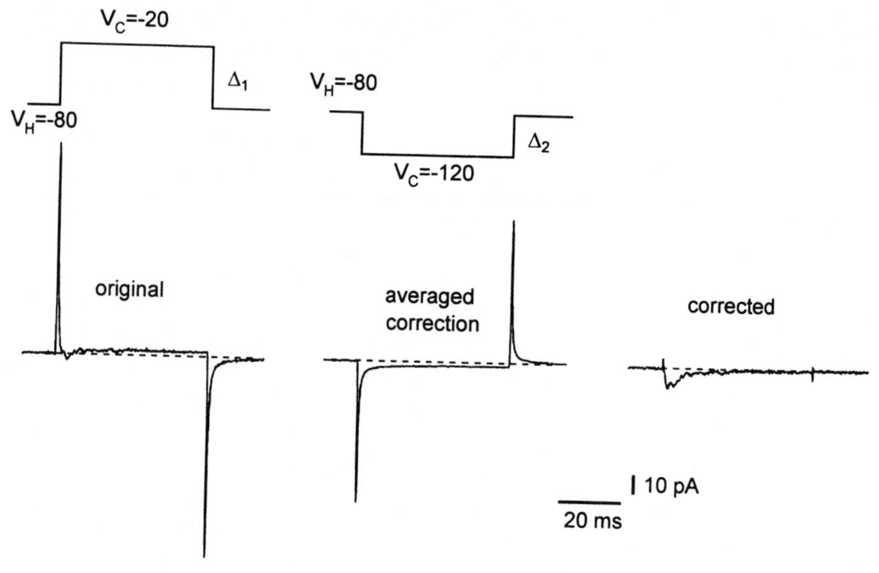

Fig. 7. Subtraction of capacity and leakage currents from original recording. *Left*, original recording of Na^+ current in a large outside-out membrane patch. *Middle*, correction trace obtained by averaging 20 single recordings. Pulse protocols are shown above the traces. *Right*, corrected Na^+ current obtained by subtracting averaged correction trace multiplied by a factor of –1.5 from original recording.

–80 to –120 mV (Δ_2=–40 mV). This negative voltage pulse did not activate ionic channels and the membrane response consisted of capacity and leakage components only. It is important to repeat the correction pulse 10 to 20 times using an averaged trace for current correction. In order to subtract capacity and leakage currents from the original recording, the averaged correction trace multiplied by a factor of $k=-\Delta_1/\Delta_2=1.5$ is added. The corrected recording is shown on the right.

- Corrections for capacity and leakage currents are made in nearly all experiments studying the properties of voltage-gated channels activated by a voltage step.
- It is important to make sure that the voltage pulse used for correction recording does not activate ionic, e.g. inward-rectifier, currents.
- Since the averaged correction trace is often multiplied by large factors, it is important to minimize its noise by increasing the number of averaged recordings.
- Overly small correction pulses of –10 to –20 mV will require high multiplication factors *k*, especially if membrane responses to large depolarizations (+40–+60 mV) are investigated. This would introduce additional noise to the corrected recording.
- Overly strong hyperpolarizing correction pulses can damage the membrane and lead to cell death.

Comments

Filtering

In order to reduce background noise, the recorded signal should be filtered. Overly strong filtering, however, will cut off useful information about channel gating. For recording of fast processes like activation of voltage-gated Na^+ currents, a low-pass filter frequency of 3–5 kHz is usually used. A low-pass filter frequency of 1 kHz is adequate, for example, for recording slower delayed-rectifier K^+ currents. In any particular case, filter frequency is determined by the kinetics of the process to be investigated.

Note: The sample rate of data acquisition depends on the frequency of filtering. As a rule, the frequency of digitization should be at least two times higher than that of filtering (cf. Chapter 45).

Liquid Junction Potentials

Liquid junction potentials always appear at the border between two solutions of different ionic composition, i.e., at the pipette tip opening where pipette and bath solutions contact each other. They are produced by electric charge accumulation which accompanies diffusion of ions with different mobilities along their concentration gradients between two solutions. The liquid junction potential is added to the offset potential of a chlorided silver electrode, and they are compensated prior to pipette sealing (see step 7) by applying a voltage of opposite sign to the pipette. After formation of the seal, however, the ions can no longer diffuse between pipette and bath solutions and the liquid junction potential disappears. Under these conditions the voltage applied for its compensation will be directly added to the potential clamped and therefore the real membrane potential will differ from that shown on the display of the patch-clamp amplifier.

Depending on the ionic composition of solutions, the liquid junction potentials vary from a few to 12 mV. They are relatively small (around 3 mV) for solutions containing physiological NaCl and KCl concentrations, becoming much larger when low mobility ions are used. Junction potentials can be calculated using the Henderson equation (Barry and Lynch, 1991) or may be determined experimentally (Neher, 1992). In most cases, a correction for liquid junction potentials is made for data presentation by adding the calculated or measured value to the membrane potential recorded in the experiment. If for any reason such a correction has not been done, an estimate of junction potentials can be given in the methods section.

Note: It is very important to keep the pipette holder clean and dry throughout the experiment. The pipette holder and bath electrodes should be regularly cleaned with methanol. The set-up must be protected from dust.

Applications

Some Other Applications of the Patch-Clamp Technique

In addition to the four major modes of the patch-clamp technique described above, there are several other modifications of this method developed for studying more specific problems of membrane excitability.

Perforated Patch

In some experiments the properties of ionic currents are modified as important cytoplasmic factors diffuse from the cell into the pipette after establishing whole-cell configuration. One possibility to prevent this diffusion is recording whole-cell currents through a perforated membrane. The idea of the method is illustrated in Fig. 8A. A standard cell-attached recording mode is established using a pipette filled with a solution containing a channel-forming substance. Voltage-independent channels incorporated into the membrane of a cell-attached patch ("perforated patch") provide a low-resistance access to the cell interior. These narrow channels are permeable for small monovalent cations and anions only, allowing their free diffusion between the pipette and the cell, as in a standard whole-cell mode. In contrast, the large organic molecules remain impermeable and therefore cannot diffuse away from the cell. A perforated patch enables voltage-clamp recordings from a whole cell under conditions where intracellular Na^+, K^+ and Cl^- concentrations are determined by the pipette solution but the major cytosolic factors remain intact.

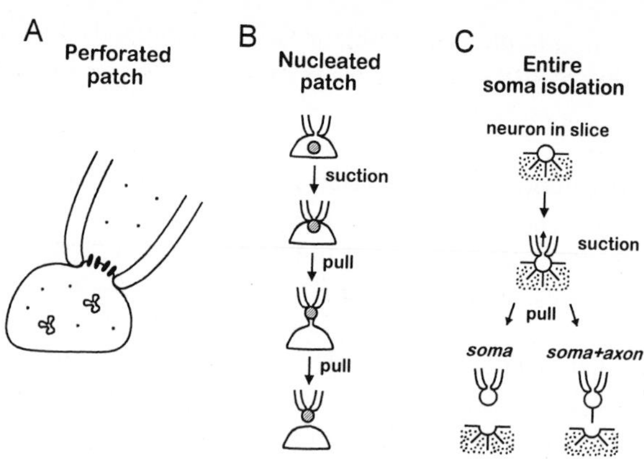

Fig. 8. Some modifications of the patch-clamp technique. A: Perforated patch. Redrawn and modified from Marty and Neher (1995). B: Nucleated patch. C: The method of entire soma isolation. Modified from Safronov, Wolff and Vogel (1997).

Commonly used channel-forming substances are ATP (Lindau and Fernandez, 1986), the antibiotic nystatin (Horn and Marty, 1988; Korn et al., 1991) and the antibiotic amphotericin B (Rae et al., 1991).

There are several problems to be kept in mind when dealing with perforated patches.

- Resistance of the perforated patch membrane R_{PP} is added to the series resistance R_S. In general, R_{PP} is 5 to 10 times larger than R_S and therefore a voltage error due to series resistance will be increased accordingly.
- Nystatin inhibits formation of the seal. Therefore, the very tip of the pipette in these experiments is filled with nystatin-free solution (see filling of the pipette tips, step 4), whereas the rest of the pipette is back-filled with a nystatin-containing one. The seal should be formed before nystatin molecules diffuse into the pipette tip.

Additional information about recordings using perforated patches can be found in more specialized literature (Marty and Neher, 1995).

Giant Membrane Patch

Giant patches are used for investigation of electrogenic membrane transporters, recordings of macroscopic ionic currents or single-channel recordings of channels expressed at low densities. Giant patches are mostly studied in inside-out configuration. They are obtained using pipettes with a large tip diameter of 12–40 μm. These pipettes are fabricated from standard patch pipettes by appropriately cutting their tips by means of the technique described elsewhere (Hilgemann, 1995). Because of difficulties in obtaining a stable seal with a large-diameter pipette tip, seal formation is usually enhanced by treatment (or coating) of the pipette tip with a special hydrocarbon mixture. Under these conditions giant inside-out patches with a seal resistance of 1–10 GΩ can be routinely obtained from membranes of many cell types.

Nucleated Patch

A giant outside-out patch can also be obtained in the form of a nucleated membrane patch (Sather et al., 1992). The nucleated patch represents a membrane patch covering the nucleus after its extraction from the parent cell. The major procedure of obtaining the nucleated patch is similar to that described for the outside-out patch. The principal difference is that the patch excision from the cell is accompanied by an application of suction through the patch electrode. Applied suction attracts the cell nucleus to the pipette tip so that the nucleus is extracted from the cell as the pipette is withdrawn (Fig. 8B). The patch membrane reseals after extraction of the nucleus. Pulling the nucleated patch usually takes 1–2 min. These patches have a spherical shape with a diameter of 5–8 μm. Advantages of nucleated patches in comparison with standard outside-out patches are:

- Ionic currents in nucleated patches are much larger and therefore current kinetics can be studied without averaging the recordings obtained by repetitive patch stimulation;
- Nucleated patches survive substantially longer after their excision due probably to the membrane support provided by the nucleus.

Entire Soma Isolation

One further modification of the patch-clamp technique is the method of entire soma isolation (Safronov, Wolff and Vogel, 1997). The scheme of this method is shown in Fig. 8C. It can be applied to relatively small neurons with a soma diameter of about 10 µm studied in brain or spinal cord slices. At the beginning a standard whole-cell configuration with a neuron in the slice is established and Na^+ currents from an intact cell are recorded. Recording of Na^+ current at this stage is important for further identification of the isolated structure. Thereafter, slight suction is applied through the patch pipette, and it is withdrawn in a way similar to that described for the formation of nucleated patches. In most cases withdrawal of the pipette results in the isolation of the entire soma (*soma*) with no processes attached. Na^+ currents recorded from the *soma* are considerably smaller than those of the intact neuron, reflecting the axonal location of the majority of Na^+ channels. In other cases, the result of isolation is an entire soma with an adjacent process. The process can be identified by recording Na^+ currents in the isolated structure. If the amplitude of the current is as large as in the intact neuron before isolation, the process is identified as the axon and the isolated structure is *soma+axon*. If the Na^+ current is as small as in a typical *soma*, the process can be identified as a dendrite and the isolated structure is called *soma+dendrite*. The isolated *somata* and *soma+axon* structures have proven to be useful in studying the spatial distribution of ionic channels in the neuronal membrane as well as the functional roles of the soma and axon in action potential generation (Safronov et al., 1997). The major advantages of the entire soma isolation method are listed below.

- The isolated structures are generally in good physiological shape and are able to maintain the resting potentials recorded before isolation. The *soma+axon* complexes are also able to generate full action potentials.
- By comparing the currents in an intact neuron with those in an isolated *soma* and *soma+axon* complex, the spatial distribution of the channels between soma, axon and dendrites of small neurons can be studied.
- The method can be used to investigate the contributions of the soma and axon to the generation of action potentials in the neuron.
- The currents (usually 50–300 pA) or their kinetics can be studied in an isolated *soma* under conditions of minimum voltage error due to resistance in series or insufficient space clamp.

Acknowledgement: We would like to thank Dr. M. E. Bräu for stimulating discussions and critically reading the manuscript, and B. Agari and E. Sturmfels for technical assistance. Support from the Deutsche Forschungsgemeinschaft (Vo188/16, 19) is gratefully acknowledged.

References

Barry PH, Lynch JW (1991) Liquid junction potentials and small cell effects in patch-clamp analysis [published erratum appears in J Membr Biol 1992 Feb;125(3):286]. Journal of Membrane Biology 121:101–117

Bottenstein JE, Sato G (1985) Cell cultures in the neurosciences. Plenum Press, New York

Corey DP, Stevens CF (1983) Science and technology of patch-recording electrodes. In: Sakmann B, Neher E (eds) Single-channel recording. Plenum Press, New York, pp 53–68

Edwards FA, Konnerth A, Sakmann B, Takahashi T (1989) A thin slice preparation for patch clamp recordings from neurones of the mammalian central nervous system. Pflugers Archiv 414:600–612

Hamill OP, Marty A, Neher E, Sakmann B, Sigworth FJ (1981) Improved patch-clamp techniques for high-resolution current recording from cells and cell-free membrane patches. Pflugers Archiv 391:85–100

Hilgemann DW (1995) The giant membrane patch. In: Sakmann B, Neher E (eds) Single-channel recording. Plenum Press, New York, pp 307–327

Horn R, Marty A (1988) Muscarinic activation of ionic currents measured by a new whole-cell recording method. Journal of General Physiology 92:145–159

Korn SJ, Marty A, Conner JA, Horn R (1991) Perforated patch recording. Methods in Neurosciences 4:264–273

Lindau M, Fernandez JM (1986) IgE-mediated degranulation of mast cells does not require opening of ion channels. Nature 319:150–153

Marty A, Neher E (1995) Tight-seal whole-cell recording. In: Sakmann B, Neher E (eds) Single-channel recording. Plenum Press, New York, pp 31–52

Neher E (1992) Correction for liquid junction potentials in patch clamp experiments. Methods in Enzymology 207:123–131

Neher E, Sakmann B (1976) Single-channel currents recorded from membrane of denervated frog muscle fibres. Nature 260:799–802

Rae J, Cooper K, Gates P, Watsky M (1991) Low access resistance perforated patch recordings using amphotericin B. Journal of Neuroscience Methods 37:15–26

Rae JL, Levis RA (1992) Glass technology for patch clamp electrodes. Methods in Enzymology 207:66–92

Safronov BV, Vogel W (1995) Single voltage-activated Na^+ and K^+ channels in the somata of rat motoneurones. Journal of Physiology 487:91–106

Safronov BV, Wolff M, Vogel W (1997) Functional distribution of three types of Na^+ channel on soma and processes of dorsal horn neurones of rat spinal cord. Journal of Physiology 503:371–385

Sakmann B, Neher E (1995) Geometric parameters of pipettes and membrane patches. In: Sakmann B, Neher E (eds) Single-channel recording. Plenum Press, New York, pp 637–650

Sather W, Dieudonne S, MacDonald JF, Ascher P (1992) Activation and desensitization of N-methyl-D-aspartate receptors in nucleated outside-out patches from mouse neurones. Journal of Physiology 450:643–672

Standen NB, Stanfield PR (1992) Patch clamp methods for single channel and whole cell recording. In: Stamford JA (ed) Monitoring neuronal activity. Oxford University Press, New York, pp 59–83

Stuart GJ, Dodt HU, Sakmann B (1993) Patch-clamp recordings from the soma and dendrites of neurons in brain slices using infrared video microscopy. Pflugers Archiv 423:511–518

Trube G (1983) Enzymatic dispersion of heart and other tissues. In: Sakmann B, Neher E (eds) Single-channel recording. Plenum Press, New York, pp 69–76

Yellen G (1982) Single Ca^{2+}-activated nonselective cation channels in neuroblastoma. Nature 296:357–359

Chapter 7

Microiontophoresis and Pressure Ejection

PETER M. LALLEY

Introduction

The ability to introduce minute amounts of drugs and chemicals by iontophoresis or pressure ejection from fine-tipped micropipettes into the microenvironment of nerve cells offers several advantages over other modes of administration: (1) Direct application circumvents diffusional barriers and limits enzymatic breakdown that might otherwise prevent substances from reaching the site of action. (2) The actions and effects of the substances can be confined to a single neuron. (3) Cell receptors for neurotransmitters and neuromodulators can be identified pharmacologically. (4) The functional significance of neurotransmitters can be assessed by comparing their effects through local application with stimulus-evoked responses, and observing how each are affected by receptor antagonists or by agents which block neurotransmitter breakdown or uptake. (5) Local application of membrane-permeable second messenger-selective agents can be used to identify intracellular signal transduction mechanisms.

The iontophoretic technique for applying neuroactive substances directly to neurons in the CNS was first utilized over 60 years ago. In 1936, Suh et al. identified cholinoceptive pressor regions in the brain stem by iontophoresing acetylcholine intracisternally. Later, Nastuk (1953) and Del Castillo and Katz (1955, 1957a,b) reintroduced the method to investigate the actions of acetylcholine at the neuromuscular junction. Curtis and Eccles (1958a,b) utilized multibarrel microiontophoresis to study the actions of pharmacological agents on spinal Renshaw cells. Since then, the technique has been used in numerous investigations to study the actions of various neurotransmitter candidates and neuroactive chemicals at synapses throughout the CNS (Krnjevic´ and Phillis 1963; Curtis 1964; Krnjevic 1964; McLennan and York 1967; Phillis et al. 1967; Salmoiraghi and Stefanis 1967; Curtis and Crawford 1969; Bradley and Candy 1970; Diamond et al. 1973; Hill and Simmonds 1973; Krnjevic 1974; Bloom 1974; Simmonds 1974; Lalley 1994; Lalley et al. 1994, 1995, 1997; Parker and Newland 1995; Young et al. 1995; Bond and Lodge 1995; Wang et al. 1995; Haji et al. 1996; Schmid et al. 1996; Heppenstall and Fleetwood-Walker 1997; Remmers et al. 1997; Zhang and Mifflin 1997). The technique of pressure-injecting substances from micropipettes was introduced by Reyniers in 1933 and improved by Chambers and Kopac (1950; see Keynes 1964). Krnjevic and Phillis (1963) applied glutamate by pressure-ejection onto single cerebral cortical neurons. Later, McCaman et al (1977), Sakai et al. (1979) and Palmer et al. (1980) published detailed accounts for micropressure-injecting known volumes of substances intracellularly and extracelluarly in the CNS. The method has proven to be very effective in measuring the actions and effects of poorly-charged substances on neurons (Dufy et al. 1979; Siggins and French 1979; Palmer and Hoffer 1980; Palmer et al. 1980; Sorensesn et al.

Peter M. Lalley, University of Wisconsin Medical Science Center, Department of Physiology, 1300 University Avenue, Madison, WI, 53706, USA (phone: +01-608-263-4697; fax: +01-608-265-5512; e-mail: pmlalley@facstaff.wisc.edu.)

1981; Palmer 1980, 1982; Palmer et al 1986). Intracellular pressure-injection has been used to stain neurons (Sakai et al. 1978), to demonstrate regulation of acetylcholine content by acetylcholinesterase (Tauc et al. 1974), track the transmembrane flux of calcium with fluorescent dyes (Chang et al. 1974), and study neurotransmitter synthesis and axonal transport (Schwartz 1974; Thompson et al. 1976). The technique has been utilized in awake animals for extracellular (Suvorov et al. 1996) and intracellular (Szente et al. 1990) application of neurochemicals.

In this chapter, methods for applying substances by microiontophoresis or micropressure-ejection to neurons in the CNS will be described. Emphasis will be given to methods used in the *in vivo* preparation. The techniques have also been used to study drug actions in brain and spinal cord slices and other types of *in vitro* preparation (Andrade and Nicoll 1987; Nicoll et al. 1990).

The reader is encouraged to read the more detailed discussions of microiontophoresis and micropressure-ejection in Curtis (1964), Bloom (1974), Simmonds (1974), Hicks (1984) and Palmer et al. (1986).

Subprotocol 1 Microiontophoresis

General Description

Microiontophoresis involves the controlled ejection of charged substances from glass micropipettes into the extracellular milieu of nerve cells or into a cell's cytoplasm. Between ejections, the substance is retained by applying current opposite in polarity to the ejecting current. For intracellular microiontophoresis, a single micropipette can be used to record membrane potential and inject substances with current in bridge mode (Chapter 5). Alternatively, theta glass tubing (which has a dividing septum) may be used. One side of the divided micropipette is filled with a solution of the chemical to be ejected and is fitted with a silver wire connected to a current pump, the other side contains the recording solution and is connected to the recording amplifier.

For extracellular microiontophoresis, it is common to eject several test substances near a neuron from a multibarrel array. A number of glass tubes are fused together and pulled to give an array of micropipettes, each containing a solution of a neuroactive substance. The tip of the array is bumped with a glass rod under microscopic control, such that each barrel has a final diameter of about 1 µm. One of the barrels is typically filled with 2–4 M NaCl for extracellular recording. Figure 1A illustrates a 5-barrel configuration. The center barrel would be used for recording while the other 4 barrels would be used to eject neuroactive chemicals. The parallel microelectrode array which is seen in Figure 1B was designed by Curtis (1968), modified by Oliver (1971) and is used in experiments involving intracellular recording and extracellular microelectrophoresis. A micropipette for intracellular recording is glued to a multibarrel pipette assembly for microiontophoresis. The arrangement allows one to monitor the effects of neuroactive chemicals on membrane potential and membrane currents. Substances can also be injected into target neurons to identify transduction mechanisms that couple membrane receptor activation to excitability (Lalley et al, 1997; Richter et al, 1997). The array in Figure 1C is a photomicrograph of a coaxial assembly from Remmers et al (1997) that was designed by Sonnhof (1973) for intracellular recording and extracellular microiontophoresis

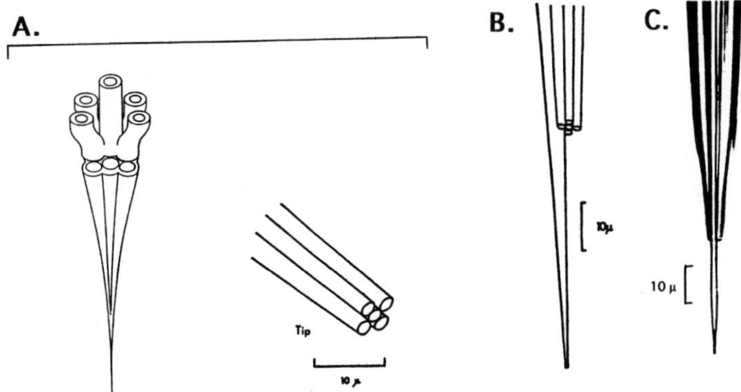

Fig. 1. Multibarrel arrays for microiontophoresis and recording from nerve cells. **A**: Drawing of five-barrel array for extracellular recording and local application of neurochemicals. *Left*, overall configuration. *Right*, tips at high magnification. Center barrel is typically filled with 2–4M NaCl solution and used for recording. (Modified from McLennan 1970). **B**: Drawing of parallel micropipette assembly for intracellular recording and extracellular chemical application. Five-barrel array is glued in parallel to intracellular recording pipette. **C**: Photomicrograph of coaxial array for intracellular recording and extracellular microiontophoresis. A long, slender micropipette for intracellular recording is inserted down the center barrel of a multibarrel pipette assembly. Surrounding barrels are used for iontophoresis. (Modified from Remmers et al. 1997)

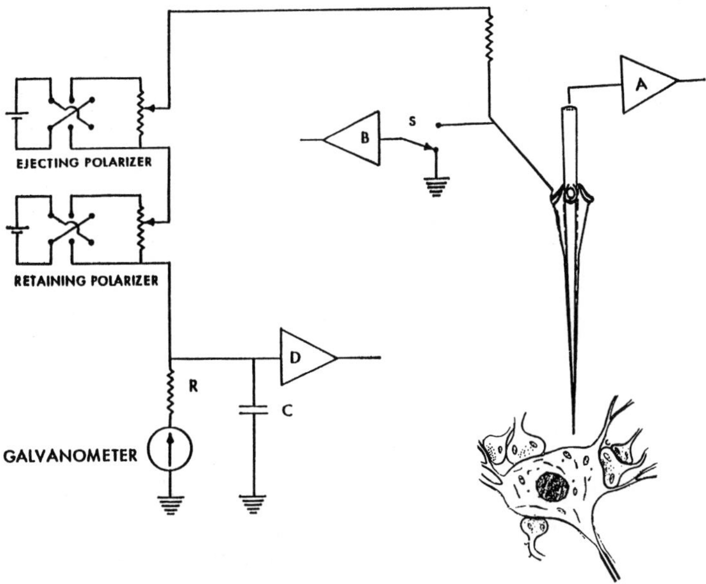

Fig. 2. Simplified electrical circuit for microiontophoresis. Amplifiers D and B and the galvanometer are used to monitor ejecting and retaining currents drawn from the polarizers. Amplifier A monitors bioelectric potentials. (Modified from Curtis 1964).

A simplified electrical circuit for microiontophoresis and extracellular recording is illustrated in Figure 2. See Curtis (1964) for a detailed explanation. One of the barrels is shown connected to ejecting and retaining "polarizers", which in former times were 90V and 45V batteries. More sophisticated constant-current pumps are now used to eject and retain substances and to measure the respective currents.

Materials

Basic Equipment Needs

The essential equipment items for iontophoresis experiments include:
- Equipment required for extra- or intracellular recording (see Chapter 5)
- Vertical pipette puller capable of generating sufficient current and pull for preparation of multibarrel micropipettes
- Heavy-duty heating coils for use with the pipette puller
- Programmable current pump for microiontophoresing neuroactive substances
- Window discriminator-ratemeter. Action potentials are detected in a voltage window. Each spike within the window triggers a TTL pulse which is passed to a ratemeter and to other counting and processing devices
- Computer interface, microprocessor, and software programs for constructing histograms and for various types of data analysis
- Pulse generator and stimulus isolation unit for evoking action potentials in response to stimulation of peripheral sensory or motor nerves or nerve tracts.

Choosing a Microiontophoresis Current Pump

Circuits for constant-current microiontophoresis can be constructed from published designs (for example, Walker et al, 1995). Alternatively, excellent multichannel current generators are commercially available (see equipment list at the end of this chapter). The microiontophoresis pump should have the following features:
- Several independently programmable channels to control ejecting and retaining currents
- Low noise properties
- Numerical display of currents
- Timers that switch on and switch off currents, so that substances can be pulsed or applied continuously
- Inputs and circuits that enable computer-control of currents
- Event pulses to chart recorders, oscilloscopes and computers signaling the occurrence and magnitude of retaining and ejecting currents
- A monitor for each channel to signal blocking of micropipette currents
- Capability to measure micropipette resistance
- A grounding switch for each channel
- Small headstage that can be positioned close to the preparation. Headstage should be equipped with a grounding post.

Procedure

Procedure 1: Preparation of Microelectrode Assemblies

Glass capillary tubing containing filaments or septa are used to facilitate filling of micropipettes. Tubing can be purchased already fused into 2–7 barrel arrays. Their advantage is that they need only be cut at appropriate lengths and pulled with a vertical pipette puller to the desired length and configuration before filling with drug solutions. The disadvantage is that care must be taken in filling to avoid spillover during filling from one barrel to another.

The alternative is to fabricate multibarrel arrays from single tubing and bend the tubes away from each other at the filling ends. The method described here is a modified version of the method described by Crossman et al. (1974).

Protocol 1

1. Capillary tubes are bundled together and held securely in place about 15 mm from each end with cuffs made from heat-shrinkable tubing.
2. The array is fixed firmly in the chucks of a vertical pipette puller. Heat is applied through the coil, with the magnetic pull set to zero, until the array is just soft enough to begin lengthening under the influence of gravity. Manual pressure is applied to allow only a small amount of lengthening, the array is rotated quickly through 270 degrees and the heating current is switched off to allow the glass to cool. The position of the coil is adjusted to the mid-region of the twisted, tapered glass column. Heating current and magnet pull are adjusted to desired levels and the array is pulled, yielding two usable multibarrel pipettes.
3. A metal hook is inserted into the open end of a barrel while applying heat to the shank from a Bunsen burner in order to bend the barrel by about 30 degrees. All barrels are bent in this fashion except the recording barrel.
4. Melted dental wax is applied between the bent and unbent barrels to form a crown. The crown and the cuff of heat-shrinkable tubing keep the barrels securely together. The micropipette array is then ready for filling and bumping to the proper tip diameter.
5. The recording barrel is connected to a manipulator and microdrive that advances the assembly into tissue.

Procedures for fabricating parallel and coaxial multibarrel arrays for intracellular recording and microiontophoresis have been described in detail (Curtis 1968; Oliver 1971; Sonnhof 1973; Lalley et al. 1994; Remmers 1997). The parallel electrode is simpler to fabricate and coupling resistances and capacitances are lower. The recording pipette must be strongly bonded to the electrophoresis assembly or they will separate in tissue. The coaxial type of electrode is much more difficult to fabricate; however, the recording and iontophoresing pipettes cannot separate.

The parallel array utilized by Lalley et al. (1994) is produced according to the following procedures and specifications:

Protocol 2

1. The distance between the recording tip and those of the microiontophoresis assembly is kept at 40 μm.
2. Single micropipettes for intracellular recording are pulled on a horizontal puller to yield the desired length, profile and tip. The recording barrel is bent with heat applied from a heating coil at an angle of 15–20 degrees, 10 mm from the tip.
3. The shaft of the recording barrel below the bend is positioned under microscopic control (400x amplification) in the crease between two barrels of the multi-barrel iontophoresis assembly.
4. Light-sensitive dental cement (3M Corporation, USA) is applied to the crease before bringing the two components together.
5. After correct positioning, the connection is made secure by applying 900 nm light from a curing gun for 30 seconds, then a second layer is applied to enclose the barrels and light-cured.

6. Cranioplast cement is applied at the bend in the recording barrel to make a small collar that further prevents separation.

The final product produces no more displacement of tissue than a standard 5-barrel array for extracellular microiontophoresis. This makes it highly suitable for recording responses of neurons deep in tissue (Richter et al. 1996).

Parallel array microelectrodes have been utilized for extracellular recording and microiontophoresis by Crossman et al. (1974). The tip of the recording pipette protrudes beyond the multibarrel assembly by 5–15 μm. The assemblies have excellent recording properties, with large signal-to-noise ratios, and are reported to be quite useful for recording responses of small cells in subcortical regions.

Carbon Fiber Electrodes

Low-noise recording electrodes for extracellular recording and microiontophoresis have been constructed by placing a carbon filament in the recording barrel of a multibarrel array and etching the fiber to produce a fine tip that is very suitable for single unit recording (Armstrong-James and Millar 1979; Fu and Lorden 1996). Multibarrel micropipettes with carbon fibers in two barrels have been used to record neuron discharges and to measure concentrations of iontophoresed catecholamines and 5-hydroxytryptamine (5-HT) (Armstrong-James et al. 1981; Crepsi et al. 1984).

Preparing Solutions

Procedures and precautions used in filling micropipettes for microiontophoresis are similar to those described for preparation of recording micropipettes in Chapter 5. The solutions should be filtered through 0.2 μm filter membranes to remove debris. The tips must be free of bubbles, otherwise the micropipettes will fail to iontophorese adequately and current passed through the filling solution will generate noise which will be picked up by the recording pipette. It is advisable to fill the micropipettes at least 10–15 minutes before use and store them in vertical position in a humidified chamber. The barrels should be checked microscopically for bubbles and precipitate immediately before use. Bubbles can sometimes be displaced by twirling a clean cat whisker in the barrels.

Solutions should be adjusted to a pH that will facilitate ejection of charged substances by iontophoresis or electro-osmosis without altering the substance's pharmacological properties. In solutions of pH 4 or less the neuronal actions of monoamines – which can excite or depress neurons depending on the subtype of catecholamine- or 5-HT receptor activated – are biased in favor of excitation (Frederickson et al. 1971). Responses related to pH are generally avoided when the pH of solutions are kept between 4.5 and 8.

Electrical Coupling to the Current Source

Clean silver wire should be used to apply current to the pipette solution. Chloriding is not necessary, and, in fact, may cause blockage due to flaking off of silver chloride, which clogs micropipette tips. To prevent contact and transfer of current between silver wires, it is convenient to solder a 60–70 mm length of silver wire to an 80 mm length of thin, flexible insulated cable and connect the latter to the iontophoresis current pump. The insulated cable can be fitted to an amphenol pin for connection to the headstage of the current pump. After the wires for iontophoresis and recording are inserted into the micropipette barrels, they are separated and held securely in place by carefully pouring a small amount of melted dental wax or paraffin over the pipette orifice.

Micropipette Tip Size

Micropipettes fill more readily when they are filled before bumping; however, they can be successfully filled even after bumping. Bumping is performed under microscopic control, with a glass rod mounted on a micromanipulator, to yield individual tip diameters of approximately 1 μm. This tip size generally results in low tip potentials. Larger tip diameters increase the likelihood of uncontrolled diffusion of substances.

Procedure 2: Recording and Microiontophoresis

1. Neurons are searched for using procedures described in Chapter 5.
2. The current pump is turned on when the micropipette tips are in the tissue. Normally, retaining currents of 5–10 nA are used. Ejecting currents are set to relatively low levels, such as 5 nA, and increased gradually to 60–80 nA. This procedure is used to test the current-passing properties of each micropipette. If blocking or unacceptable noise occurs, the assembly should be withdrawn and discarded.
3. It is advisable to first evaluate the effectiveness of test substances and ejecting currents on cells other than the target neurons. Such tests may be impractical for intracellular recording and extracellular microiontophoresis since damage to the recording tip might occur, but they should be performed whenever possible. Any unidentified neuron that fires spontaneously without an injury discharge can be tested for this purpose.
4. Microelectrophoresis pipettes need to be warmed up by passage of ejecting current. Sometimes, passage of even large ejecting currents will initially elicit no response since retaining currents have evacuated the test substance from the solution near the tip of the micropipette. This dead space will have to be recharged by repeated passage of ejecting current. A useful procedure is to apply ejecting currents of increasing intensity at fixed intervals with a 50% duty cycle, up to 100 nA. Once a response is elicited, tests should be repeated several times to insure that stable responses are evoked, indicating that the concentration of substance has reached equilibrium at the tip of the micropipette.
5. The multibarrel assembly should now be ready for tests on the intended neurons. Target neurons should be selected which have stable recording properties. Tests should not be made on neurons which exhibit injury discharges or action potential inactivation. In tests made with extracellular recording, action potentials should be of negative polarity and be at least several hundred µV greater than background noise. It is highly desirable to record the neuron in isolation from other neurons. If this is not possible, window discrimination may be used to count the spike frequency of the test neuron. In that case, it is important that the test substances are affecting only the target neuron. Cells tested during intracellular recording should exhibit stable membrane potential properties as described in Chapter 5.
6. Precautions to prevent current artifacts should be taken. Currents can evoke responses independent of the substance ejected, especially when the microiontophoresis assembly is close to a neuron. Current artifacts should be suspected if action potential frequency or membrane potential suddenly changes, coincident with the time course of the applied current. Generally, positive current inhibits discharges and hyperpolarizes membrane potential whereas negative current has opposite effects. During extracellular recording, current artifacts rarely occur if action potentials are negative in polarity, indicating that the assembly is sufficiently distant to prevent current spread to the cell membrane. A procedure to minimize current artifacts involves current-balancing. One micropipette of the assembly is filled with 165 mM NaCl and connected by silver wire to the balancing channel of the microelectrophoresis unit. The channel delivers current equal and opposite in polarity to the sum of currents delivered through all other barrels. A second procedure is to test for current artifacts by passing current through a barrel containing only 165 mM NaCl.
7. The adequacy of retaining currents should be determined. Normally, 5–10 nA currents will be adequate; however, the effectiveness of retaining currents should be questioned if responses to neuroactive substances are unexpectedly persistent or if responses change with a gradual time course when the intensity of the retaining current is varied. Currents greater than 20 nA are generally avoided since they produce dead space at the micropipette tips.

Overview 1: Measuring and Interpreting Responses

Spike Frequency Analysis

Action potential frequency can be measured as either instantaneous frequency, or as the moving average after processing by a leaky integrator circuit (Eldridge 1971). Moving averages can provide information about how neuroactive substances affect the pattern as well as the frequency of cell discharge. Another approach is to collect the spikes into time bins by computer. Histograms can be constructed which illustrate spike frequency or numbers of spikes vs. time. In addition, interspike interval histograms (ISIHs) or post-stimulus histograms (PSTHs) can be constructed. ISIHs can reveal whether neurochemicals preferentially depress either high- or low-frequency components in a train of action potentials (Hoffer et al. 1971). Figure 3 illustrates the use of moving averages and ISHSs to measure the depression of spike frequency in medullary inspiratory neurons by DAMGO, a μ-opioid receptor agonist (Lalley et al. 1997).

PSTHs are useful for analyzing drug effects on stimulus-evoked monosynaptic spike discharges (Bloom 1974).

Membrane Potential and Neuron Input Resistance

Intracellular recording reveals how chemicals affect subthreshold events such as resting membrane potential and membrane current, PSPs and PSCs, membrane conductances and input resistance, R_N, and discharge properties. Measurements of R_N will often indi-

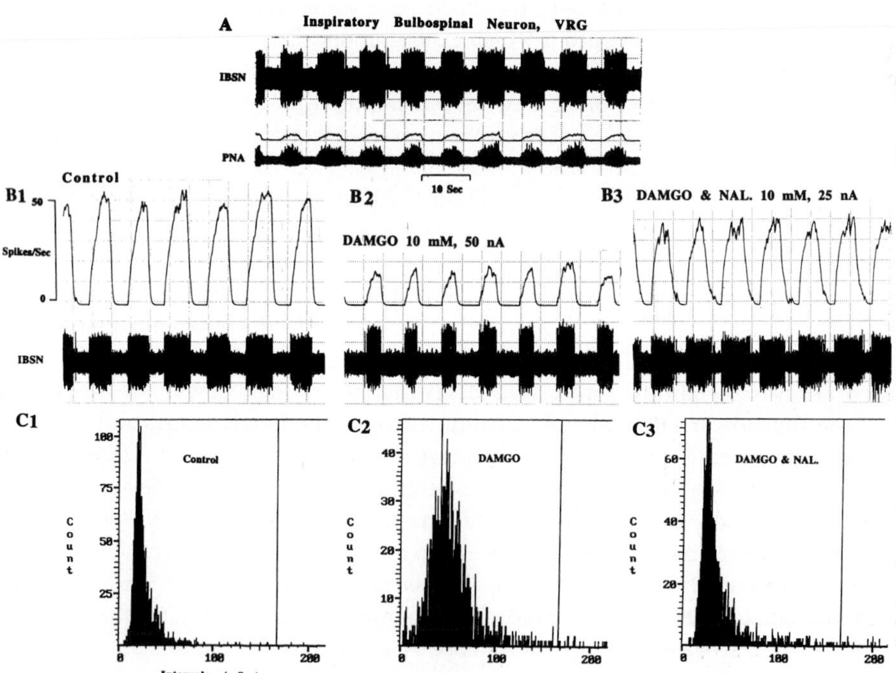

Fig. 3. Inhibition of action potential discharge frequency and burst duration in a medullary bulbospinal inspiratory neuron (IBSN), as revealed by ratemeter records and interspike interval histograms during microiontophoresis of DAMGO, a μ-opioid receptor agonist. **A:** Records from top to bottom are: action potentials recorded extracellularly with a 5-barrel microiontophoresis assembly, ratemeter records (moving averages) of a population of inspiratory motor axons of the phrenic nerve innervating the diaphragm and the electroneurograms of phrenic discharge activity (PNA). **B:** Ratemeter records (moving averages) of IBSN action potential frequency and microelectrode recordings. **C:** Interspike interval histograms of IBSN discharge.

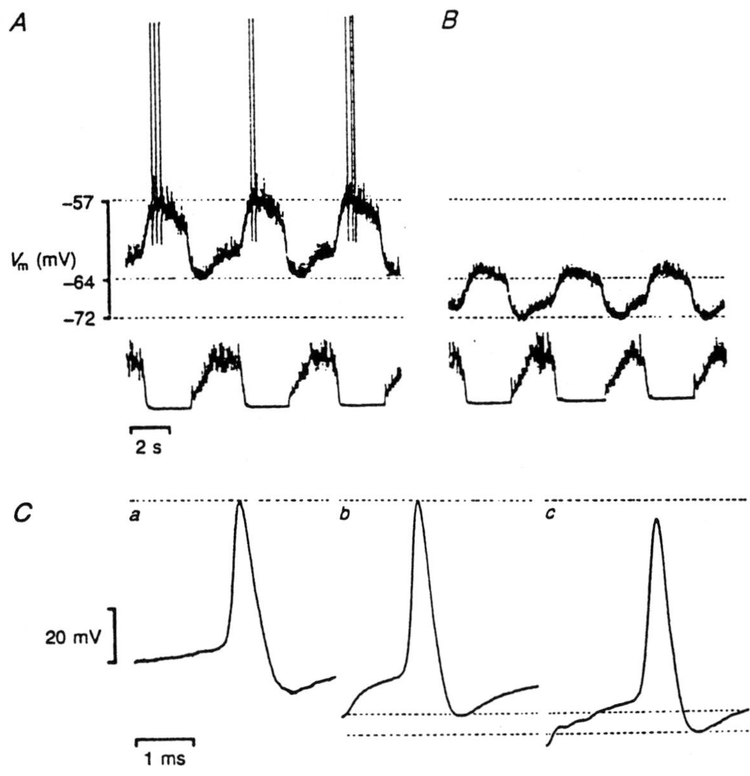

Fig. 4. Depression of cell discharge, hyperpolarization of membrane potential and decreased neuronal input resistance produced during microiontophoresis of the 5-HT-1A receptor agonist 8-OHDPAT. **A** and **B** show membrane potential (*top*; V_m) of a medullary expiratory neuron and ratemeter records (moving average) of phrenic nerve activity (*bottom*). **A**: Control records. **B**: 8-OH-DPAT microiontophoresis hyperpolarized membrane potential and abolished action potentials. Ca: Averages of 20 spontaneously occurring action potentials recorded under control conditions. Cb: Average of 20 action potentials evoked by depolarizing constant current pulses recorded in balanced bridge mode in the absence of 8-OHDPAT application. Cc: Average of action potentials evoked in the presence of 8-OHDPAT. Membrane potential is hyperpolarized, amplitude of depolarizing electrotonic potential is decreased, signifying increased membrane conductance resulting from postsynaptic inhibition.

cate whether the effects of an iontophoresed neurochemical are direct on the target neuron, or affected through actions on a presynaptic neuron. Figure4 from Lalley et al. (1994) illustrates how R_N measurements revealed that inhibition mediated by 8-OHDPAT, a 5HT-1A receptor agonist, was direct on a medullary respiratory neuron.

A mechanism of postsynaptic inhibition is consistent with data showing that 8-OH-DPAT hyperpolarizes the vast majority of CNS neurons having 5-HT-1A receptors, and does so in association with increased membrane permeability to potassium ions (Zifa and Fillion 1992). Had the depression of neuron discharges been associated with reduced waves of depolarization and increased input resistance, presynaptic inhibition would have been implicated.

Dose-Response Relationships

In studies of classical dose-response relationships, the concentrations of drugs are known and it is assumed that drugs are uniformly distributed around a finite number of receptors, such that the response will vary in proportion to the number of agonist-activated receptors. Such conditions may exist *in vitro* when drugs in a bath solution are applied to thin tissue slices, hence sigmoidal dose-response curves can be constructed for analyzing drug-receptor interactions (Gero 1971). In microiontophoresis experiments, the quantity of drug around the target neuron is generally unknown. This is because: (1) The amount of drug released from the micropipette cannot be calculated from the ejecting current (2); it is not known how the concentration of drug at the site of action is related to the distance from its release site.

Amount of Drug Released by Current

Microiontophoresis puts into action Faraday's laws of electrolysis, which state that the mass of a charged substance produced at an electrode in an aqueous solution by electrolysis is directly proportional to the quantity of electricity that has passed through the solution and the equivalent weight of the substance. Mathematically, Faraday's laws can be expressed as:

$$M = n(IT / ZF) \tag{1}$$

where M is the number of moles of the charged substance, I is the applied current (amperes), T is the duration of current application (seconds), Z is the valence state of the substance, F is Faraday's constant and n is the transport number, a proportionality factor which depends on the chemical properties of the charged substance and solvent medium.

In microelectrophoresis experiments, M is the number of moles of substance expelled at the micropipette tip by current. This means that a current applied to a wire inserted in a micropipette will repel charged molecules of similar polarity from the micropipette solution into the surrounding liquid medium, and retain molecules if the current and charge on the molecule are of opposite polarity. For example, positive current passed through a micropipette solution containing acetylcholine chloride (ACh^+Cl^-) in distilled water will expel ACh^+ and retain Cl^-. Negative current will retain ACh^+ and expel Cl^-. Substances are mobilized in aqueous solution not only by electrolysis, but also by electro-osmosis, in which the charge mobilizes the substance by acting on molecules of water of hydration that surround it. According to Curtis (1964), the contribution of electro-osmosis to the total ejection of highly charged molecules is relatively small, whereas it is the principal means of expelling poorly-charged substances. For example, hydrochloride, methochloride or methobromide salts of bicuculline, a convulsant alkaloid which blocks $GABA_A$ receptors, are dissolved in a solution of 165 mM NaCl, acidified to pH 3 and ejected by electro-osmosis with positive current.

The movement of a substance in solution is affected by the tip potential (TP) and by the zeta potential, ζ, an electrokinetic potential set up at the glass-solution interface in the interior of the micropipette tip. The total TP, to which ζ is contributory, is normally negative in sign. TP is higher when micropipette tips are finer (electrode resistance is high) and the solutions inside and outside the micropipette are dissimilar, and lower when the filling solutions are acidic.

Critical to determining the quantity M of a neuroactive substance expelled by a micropipette is an accurate determination of n. The transport number is influenced by temperature, ζ, electrode tip diameter, composition of the pipette class, the solvent, the various molecules within the pipette solution and the properties of the medium into which substances are iontophoresed. The transport number of a substance varies significantly between micropipettes even when glass composition is identical, tip sizes are similar and the test substance is expelled into identical solutions such as Ringer's solution. Values of **n** for relatively few chemicals have been measured by ejecting radioactive compounds into distilled water, Ringer's or physiological saline solutions; however, there are large differences between values of **n** obtained in such solutions and in brain tissue (Hoffer et al. 1971). Other factors that affect n are:

- Changes in the charge at micropipette tips resulting from passage of current
- Migration of chemicals away from the tip during application of retaining currents
- Current transfer between adjacent micropipettes

A further complication is that n will change during testing if the chemical properties of nervous tissue are altered, as might occur from multiple tissue penetrations, edema, changes of pH, etc.

Concentration-Distance Relationship

Estimates have been made of the concentration of a drug at a target neuron by applying concentration-distance equations developed from the principles of diffusion. Nicholson and Phillips (1981) evaluated the effect of distance and duration of iontophoretic application on drug concentration in the vicinity of cerebellar Purkinje neurons from the equation:

$$C(d,t) = (Q\lambda^2 / 4\pi D\alpha d) \times \mathrm{erfc}(d\lambda / 2\overline{A}(Dt)) \quad (2)$$

where C is the concentration of the drug at a distance d, and at time t of drug application, Q is the amount of drug expelled at microelectrode tip (mols/s), λ is the tortuosity factor in the medium from source to target, α is the volume fraction, erfc is the error function and D is the diffusion constant ($cm^{-2}s^{-1}$). In the experiments of Nicholson and Phillips the test substances were measured directly in the tissue, so that C, d and t were known and α, λ could be estimated. However, in most studies, measurements of C are not possible or practical. Perhaps the only term in equation (2) which can normally be determined with certainty is **d** when compound electrodes are used to record responses from impaled neurons. Otherwise it is doubtful that accurate estimation of the actual concentration can be made, for several reasons.

First, in equation (2), Q varies with transport number and ejecting current:

$$Q = nI / F \quad (3)$$

and values of n, for reasons presented above, are not trustworthy.

Second, λ and α will vary in different regions of the nervous system and with experimental conditions.

Third, D will be influenced by:

- Enzymatic breakdown, neuronal and glial uptake of microiontophoresed chemicals.
- Barriers created by neurites, glial cells, blood vessels and tissue debris resulting from microelectrode insertion. Such barriers can block access to a neuron or cause non-uniform distribution of the ejected substance.

Estimating Relative Potency

Since concentrations are not usually measured directly and cannot be estimated with accuracy, an alternative approach is to measure the relative potencies of drugs. For accurate estimates of relative potency, the following prerequisites must be satisfied:

- Comparisons must be restricted to neurons of the same functional type, having similar electrophysiological and morphological (size, dendritic arborization) properties. Test substances must be compared for relative effectiveness on the same neuron.

- Test substances must be ejected and responses must be compared under identical recording conditions.
- At each current intensity and for each substance, the duration of drug ejection must be long enough to achieve a steady response level.
- Recovery to control levels must be allowed between drug ejections.
- Responses should be normalized for the analysis of dose-response properties.

Two examples are given here to indicate how relative potencies of drugs are determined and antagonist efficacy is evaluated.

Example 1

Curtis et al. (1971) analyzed the dose-related inhibitory effects of the amino acid glycine and the effectiveness of strychnine as an antagonist on spinal interneurons in the following manner:
- A steady level of cell discharge was evoked with microiontophoretic application of DL-homocysteic acid, an excitatory glutamate-like amino acid.
- Glycine was applied during the steady state discharge with different ejecting currents to effect graded degrees of depression of spike frequency.
- The sequence was repeated two more times during which strychnine was applied with 5 nA and 10 nA ejecting currents. Glycine and strychnine were applied sufficiently long to obtain steady levels of response.
- Results were plotted graphically as percent of maximum inhibition of discharge frequency (ordinate) vs. logarithm of glycine-ejecting current (nA). As shown in Figure 5A, the curves thus derived were sigmoidal. Strychnine produced current-related shifts to the right, consistent with competitive antagonism at glycine receptors. Curtis et al. (1971) pointed out that difficulties associated with such studies of dose-response relationships can arise by virtue of non-uniform drug distribution, although such problems were not encountered in their study.

Example 2

Hill and Simmonds (1973) and Simmonds (1974) used similar microiontophoresis procedures, but utilized a different method of graphical analysis. In their investigation:
- Steady firing of cortical neurons was evoked with glutamate iontophoresis.
- Graded levels of discharge depression were evoked by GABA applied with different ejecting currents.
- As shown in Figure 5B, plots of % inhibition of firing vs. log of ejection time yielded sigmoidal curves which shifted in parallel with different intensities of ejecting currents. The index of relative potency was chosen as T50, the time taken to achieve 50 % inhibition of neuronal firing (indicated by the dashed horizontal line).

Time-response curves for GABA were also displaced to the right along the time axis by microiontophoretic application of $GABA_A$ receptor antagonists. The displaced curves were parallel to the control curve when responses were plotted against linear time rather than log time. Relative effectiveness of antagonists was assessed by comparing values of T50.

Critical to the accuracy of such procedures is the requirement of restricting comparisons to neurons of similar type. This eliminates dissimilarities in responsiveness which are linked to different intrinsic drug sensitivities among different types of neurons.

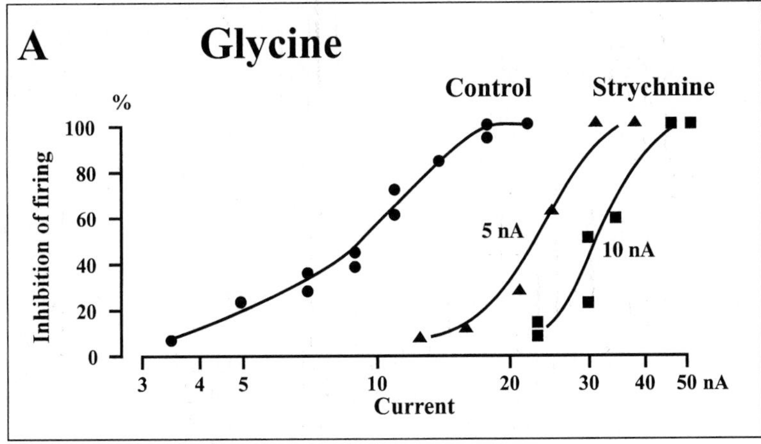

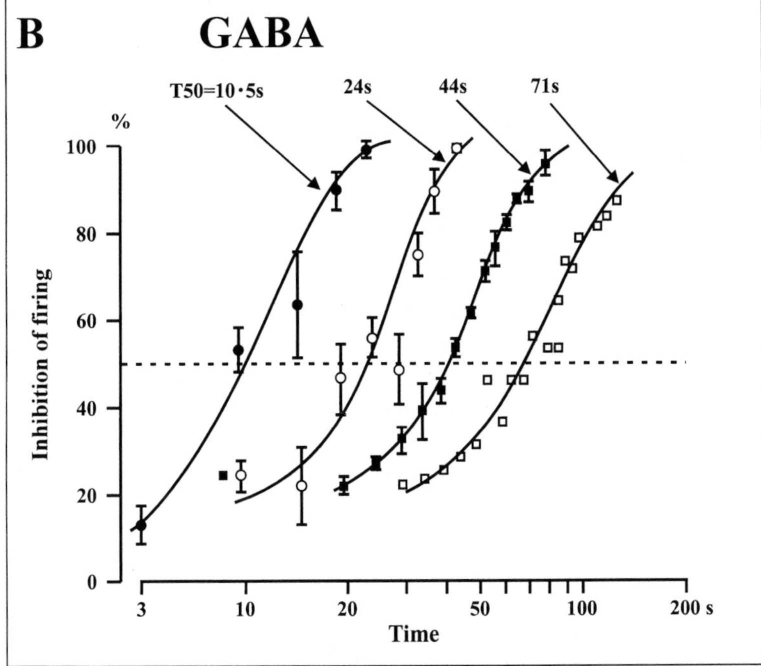

Fig. 5. Graphical analysis of the relative potencies of receptor agonists and antagonism by a competitive receptor antagonist applied by microiontophoresis. **A:** Plots of inhibition of action potential frequency in cat spinal interneurons by glycine and antagonism by strychnine. Glycine was tested on a steady level of action potential firing produced by iontophoresis of DL-homocysteic acid. Ordinate, percent of maximal inhibition of action potential frequency by different glycine-ejecting currents. Abscissa, logarithm of ejecting current. Control curve (glycine only) is on the *left*. *Middle*, concurrent ejection of glycine and strychnine, 5nA. *Right*, ejection of glycine and strychnine, 10nA. Note parallel shifts produced by strychnine, indicative of competitive antagonism at glycine receptors. (Modified from Curtis et al., 1971). **B:** Plots of inhibition of cat cortical neuron discharges during microiontophoresis of γ-aminobutyric acid (GABA) with four different currents: 20nA (•), 10 nA (o), 5nA (■) and 2nA (❑). Steady firing of the neuron was induced by microiontophoresis of L-glutamate, 20nA. Vertical bars represent standard errors. Values of T50 are times required to produce 50% inhibition of evoked firing by GABA. (Reproduced from Hill and Simmonds 1973).

Potential Sources of Error

Problems of reproducibility and interpretation will arise if sources of artifact are not eliminated. Those associated with pH and current effects have already been discussed. Other potential sources of error include undesired local anesthetic and other types of non-selective drug actions, network-related changes in cellular behavior and presynaptic effects.

- Local Anesthetic and Other Non-Selective Drug Actions
  Certain drugs including muscarinic-, adrenergic- and serotonin-receptor blocking drugs have local anesthetic effects at concentrations greater than those responsible for their primary effects. The hallmark of local anesthetic activity is depression of action potential amplitude and lengthening of duration. At high concentrations, drugs may also non-selectively depress spike discharges. Such effects can usually be eliminated by reducing the strength of ejecting currents.
  Other examples of non-selective effects include enhancement of responses to depressant substances such as GABA by barbiturates and other general anesthetics (Nicoll et al. 1990), and blockade of receptors for other neurotransmitters by "selective" receptor antagonists, for example blockade of β–adrenoceptors by Ketanserine, a 5-HT-2A receptor blocker. In the former situation, it may be necessary to perform tests on unanesthetized decerebrate preparations. In situations such as the latter, it is important that the test substance be not only selective for a subtype of receptor, but also be free of actions on receptors for other neurotransmitters, neuromodulators and hormones.
- Changes Related to Altered Network Properties
  Changes in the responsiveness of neurons to microiontophoresed substances may occur during the course of testing because the intensity of spontaneous synaptic input to neurons changes. Synaptic input may vary with anesthetic state, blood pressure, alveolar and arterial oxygen and carbon dioxide, etc. It is therefore imperative to monitor and minimize variables that are known to affect the network behavior of test neurons. It is also helpful to record some index of network behavior, such as discharges of a phrenic nerve when testing brain stem or spinal respiratory neurons, or blood pressure when testing sympathetic vasomotor neurons.
- Presynaptic Effects
  Drug actions on presynaptic neurons should be suspected if effects on nerve cell properties are opposite to the known actions of the drug. For example, presynaptic depression would be suggested if cell discharges increase during application of a depressant drug. Presynaptic actions should also be suspected if receptor agonists and antagonists have similar effects on the target neuron (Wang et al. 1995).

Cell Structure and Interpreting Functional Significance

Responsiveness to a neuroactive substance will be influenced by the location of receptors. Hence, cell size and geometry should be taken into account when interpreting the functional significance of endogenous neurochemicals based on responses to iontophoresis. For example:

- Small cells are usually more sensitive than large cells to microiontophoretically applied drugs because, for the same dose, a relatively greater percentage of the cell surface comes into contact with the drug. The greater sensitivity might lead to an erroneous conclusion that excitability in small cells is more effectively modulated under physiological circumstances by an endogenous neurohumoral substance.

- Greater numbers of receptors for neurotransmitters and neuromodulators may be located at distant dendritic synapses. A weak or absent response to iontophoresis may mean that the neurochemical has not reached remote synapses where its endogenous counterpart has physiologically important actions.

Subprotocol 2
Micropressure Ejection

General Uses and Description

Pressure ejection is used to deliver uncharged or poorly charged substances to the vicinity of neurons. The technique is useful for *in vivo* and *in vitro* studies. In the *in vitro* slice preparation, separate recording and pressure-ejection pipettes are positioned close to the target neuron. In the *in situ* preparation, micropressure ejection has been used in two general ways:

1. Single micropipettes with relatively large tips (10 µm or greater) have been used to inject nanoliter volumes for the purpose of altering the excitability of groups of neurons in a small area. For example, nanoliter volumes of solutions containing GABA- and glycine receptor antagonists have been ejected into the Pre-Boetzinger Complex of the medulla oblongata, a critical respiratory rhythm-generating area, to evaluate the significance of inhibitory synaptic transmission on the respiratory rhythm in the adult mammal *in vivo* (Pierrefiche et al., in press, 1998).
2. Volumes less than 1 nl of neuroactive drugs and neuromodulators have been applied to single neurons from multibarrel assemblies during extracellular or intracellular recording (Palmer et al. 1986; Szente et al. 1990).

Volumes of neuroactive substances are ejected from micropipettes connected by soft catheter tubing and high-pressure tubing to a source of compressed gas, usually nitrogen. A switch- or TTL pulse-controlled solonoid valve is used to deliver pulses of desired pressure and duration, either as single pulses are as programmed pulses.

Materials

Equipment and Supplies

Equipment for electrophysiological recording and analysis is the same as described in Chapter 5 and in Part 1 of this chapter. Additional equipment and supplies include:

- A source of inert gas, such as a cylinder of pressurized nitrogen.
- Electronically-controlled solonoid pressure valves to regulate degree and duration of pressure
- A source of trigger pulses for the pressure valves (TTL pulse generator)
- Surgical microscope equipped with reticule to measure displacement of solutions from micropipettes
- High-pressure tubing and soft catheter tubing to connect pipettes to output ports of pressure valve.

Capillary Glass

Single or multibarrel pipettes containing filaments are used to aid filling. Multibarrel assemblies are commercially available. In most cases the barrels are fused all along the length of the assembly, hence it is not possible to attach tubing to each barrel to independently apply various substances. However, assemblies can be fabricated as described in above. Procedures for assembling multi-barrel pipette assemblies are also described by Palmer et al. (1986).

Solutions

Drugs should be dissolved in mock CSF or Ringer's solution, preferably in low micromolar concentrations and adjusted to pH 7.4.

Procedure

Test procedures are the same as those described for microiontophoresis. The performance of the pressure-ejecting system should be checked before insertion of the assembly in tissue. Connections between micropipette and catheter tubing (3–5 inch lengths) should be secure. Once in the tissue, performance of the pressure system should be checked again. Drug volumes ejected with various pressures and pulse durations should be selected and preliminary tests of drug effectiveness on non-target neurons should be made whenever feasible.

Methods of Analysis

Methods of analysis are identical to those described for microiontophoresis. The procedures described by Curtis et al (1971) and Hill and Simmonds (1973) are very suitable for comparing relative potencies of drugs and the effectiveness of antagonists.

Calculating Volume and Quantity Ejected

The general procedure is to microscopically measure with a reticule the length (L) of fluid ejected under pressure from a pipette of known internal radius (r). Volume (V) can then be calculated from $V = \pi r^2 L$. From the known concentration of the pipette solution, the amount of substance ejected can be calculated; however, the concentration of drug at the target site will not be known (see above: Concentration-Distance Relationships).

Volume Ejected vs. Micropipette Resistance

Micropipettes with tip resistances between 1.0 and 1.4 MΩ resistance will usually eject uniform volumes for a given pressure and duration over long test periods. Pipettes with finer tips tend to plug in brain or spinal cord tissue whereas larger tips (resistance less than 1.0 MΩ) produce variable results, and larger volumes are ejected that are more likely to produce volume-related response artifacts (Sakai et al. 1979; Palmer et al. 1986).

Artifacts

pH, Local Anesthetic and Other Nonspecific Effects

Response artifacts similar to those that might be encountered with microiontophoretic application can be produced by micropressure ejection. It is therefore important to:

- Adjust pH as close as possible to pH 7.4. Solutions of pH less than 5.5 and greater than 8 should not be used.
- Drug concentrations in pipette solutions should be dilute, preferably in the low micromolar range.

Volume Artifacts

Solution volume can produce injury discharges and abrupt changes of membrane potential. Larger volumes can also move cells away from the recording micropipette. Such artifacts should be suspected when changes of neuron discharge, membrane potential, or neuronal network behavior occur simultaneously with the volume ejection. Artifacts can be tested for by ejecting drug-free solution from a barrel of the assembly.

Solvent Artifacts

Solvents other than water, such as ethanol or dimethylsulfoxide, may alter neuron behavior. Hence, they should be diluted as much as possible with Ringer's solution or mock CSF. Solutions of high osmolarity should also be avoided, since they might affect neuron responses through redistribution of cytoplasmic and extracellular water (Curtis 1964).

Comments

The methods described in this chapter have turned out to be very useful for delivering drugs and chemicals directly to the vicinity of target neurons. Studies using these methods have provided important information about the identity and functional characteristics of neurotransmitters and neuromodulators. Cellular mechanisms of drug action have been elucidated, and signal transduction processes that link membrane receptor activation to ion channel functions have been uncovered. Methods for local drug application will continue in the future to be very useful for evaluating the direct effects of chemicals on neurons in the CNS. Modern techniques in molecular biology and genetics have led to the discovery of a large number of potentially important neuromodulators whose functional significance and intracellular transduction mechanisms await identification through the use of microiontophoresis and micropressure-ejection.

Suppliers

Programmable Multi-Channel Microintophoresis Pumps

Micro-Iontophoresis Current Generator 6400	Dagan Corporation, USA
MVCS Iontophoretic Series	npi electronic GMBH, FRG
Neurophore BH-2	Medical Systems Corp, USA

Electronically-controlled Pressure Valve Systems

Picospritzer	General Valve Corporation, USA
PPM-2	Medical Systems, USA
Pneumatic Picopump PV830	World Precision Instruments, USA
Hydraulic NanoPump, A1400	World Precision Instruments, USA
Nanoliter Injector A203	World Precision Instruments, USA
PDES-2/4	npi elecronic GMBH, FRG

References

Andrade R, Nicoll RA (1987) Pharmacologically distinct actions of serotonin on single pyramidal neurones of the rat hippocampus recorded in vitro. J Physiol (Lond) 394:99–124

Armstrong-James M, Millar J (1979) Carbon fibre microelectrodes. J Neurosci Methods 1: 279–287

Armstrong-James M, Fox K, Kruk ZI, Millar J (1981) Quantitative iontophoresis of catecholamines using multibarrel carbon fibre microelectrodes. J Neurosci Methods 4: 385–406

Bloom FE (1974) To spritz or not to spritz: The doubtful value of aimless iontophoresis. Life Sci 14: 1819–1834

Bond A, Lodge D (1995) Pharmacology of metabotropic glutamate receptor-mediated enhancement of responses to excitatory and inhibitory amino acids on rat spinal neurones in vivo. Neuropharmacol 34: 1015–1023

Bradley PB, Candy JM (1970) Iontophoretic release of acetylcholine, noradrenaline, 5-hydroxytryptamine and D-lysergic acid diethylamide from micropipettes. Br J Pharmacol 40: 194–201

Chang JJ, Gelperin A, Johnson FH (1974) Intracellularly injected alquorin detects transmembrane calcium flux during action potentials in an identified neuron from the terrestrial slug. Brain Res 77: 431–432

Crepsi F, Paret J, Keane, PE, Morre M (1984) An improved differential pulse voltametry technique allows the simultaneous analysis of dopaminergic and serotonergic activities in vivo with a single carbon-fibre electrode. Neurosci Lett 52:159–164

Crossman AR, Walker RJ, Woodruff GN (1974) Problems associated with iontophoretic studies in the caudate nucleus and substantia nigra. Neuropharmacol 13:547–552

Curtis DR, Crawford JM (1969) Central synaptic transmission – microelectrophoretic studies. Annu Rev Pharmacol 9:209–240

Curtis DR, Eccles RM (1958a) The effect of diffusional barriers upon the pharmacology of cells within the central nervous system. J Physiol (Lond) 141: 446–463

Curtis DR, Eccles RM (1958b) The excitation of Renshaw cells by pharmacological agents applied electrophoretically. J Physiol (Lond) 141: 435–445

Curtis DR, Duggan AW, Johnston GAR (1971) The specificity of strychnine as a glycine antagonist in the mammalian spinal cord. Exp Brain Res 12:547–565

Curtis DR (1964) Microelectrophoresis. In: Nastuk WL (ed) Physical Techniques in Biological Research, Volume V. Electrophysiological Methods, Part A., Chapter 4. Academic Press: New York, pp 144–190

Curtis DR (1968) A method for assembly of "parallel" micro-pipettes. Electroenceph Clin Neurophysiol 24:587–589

Del Castillo J, Katz B (1955) On the localization of acetylcholine receptors. J Physiol (Lond) 128:157–181

Del Castilo J, Katz B (1957a) The identity of "intrinsic" and "extrinsic" acetylcholine receptors in the motor end-plate. Proc R Soc Lond Ser B 146:357–361

Del Castilo J, Katz B (1957b) Interaction at end-plate receptors between different choline derivatives. Proc R Soc Lond Ser B 146: 369–381

Diamond J, Roper, S, Yasargil GM (1973) The membrane effects and sensitivity to strychnine of neural inhibition of the Mauthner cell, and its inhibition by glycine and GABA. J Physiol (Lond) 232:87–111

Dufy B, Vincent J-D, Fleury H, Pasquier P, Gourdji D, Tixler-Vidal A (1979) Membrane effects of thyrotropin-releasing hormone and estrogen shown by intracellular recording from pituitary cells. Science 204:509–511

Eldridge FL (1971) Relationship between phrenic nerve activity and ventilation. Am J Physiol 221: 535–543

Eisenstadt M, Goldman JE, Kandel ER, Koike H, Koester J, Schwartz, J (1973) Intrasomatic injection of radioactive precursors for studying transmitter synthesis in identified neurons of *Aplysia*. Proc Nat Acad Sci 70:3371–3375

Frederickson RCA, Jordan LM, Phillis JW (1971) The action of noradrenaline on cortical neurons: effects of pH. Brain Res 35:556–560

Fu J, Lorden JF (1996) An easily constructed carbon fiber recording and micriontophoresis assembly J Neurosci Methods 68:247–251

Gero A (1971) Intimate study of drug action III: Mechanisms of molecular drug action. In: DiPalma JR (ed) Drill's Pharmacology in Medicine, Fourth Edition, Chapter 5. McGraw-Hill Book Company: New York, pp 67–98

Haji A, Furuichi S,Takeda R (1996) Effects of iontophoretically applied acetylcholine on membrane potential and synaptic activity of bulbar respiratory neurones in decerebrate cats. Neuropharmacol 35:195–203

Heppenstal PA, Fleetwood-Walker SM (1997) The glycine site of the NMDA receptor contributes to neurokinin 1 receptor agonist facilitation of NMDA receptor agonist-evoked activity in rat dorsal horn neurons. Brain Res 744:235–245

Hicks TP (1984) The history and development of microiontophoresis in experimental neurobiology. Prog Neurobiol 22:185–240

Hill RG, Simmonds MA (1973) A method for comparing the potencies of (-aminobutyric acid antagonists on single neurones using micro-iontophoretic techniques. Br J Pharmacol 48:1–11

Hoffer BJ, Neff NH, Siggins GR (1971) Microiontophoretic release of norepinephrine from micropipettes. Neuropharmacol 10:175–180

Hoffer BJ, Siggins G.R, Bloom FE (1971) Studies on norepinephrine-containing afferents to Purkinje cells of rat cerebellum. II. Sensitivity of Purkinje cells to norepinephrine and related substances administered by microiontophoresis. Brain Res 25:523–534

Keynes RD (1964) Addendum: Microinjection, to: Microelectrophoresis by DR Curtis. In: Nastuk WL (ed) Physical Techniques in Biological Research, Volume V, Electrophysiological Methods, Part A, Chapter 4. Academic Press: New York, pp 183–188

Krnjevic K, Phillis JW (1963) Iontophoretic studies of neurones in the mammalian cerebral cortex. J Physiol (Lond) 165:274–304

Krnjevic K (1974) Chemical nature of synaptic transmission in vertebrates. Physiol Rev 54:418–540

Krnjevic K (1964) Micro-iontophoretic studies on cortical neurones. Int Rev Neurobiol 7: 41–98

Lalley PM (1994) The excitability and rhythm of medullary respiratory neurons in the cat are altered by the serotonin receptor agonist 5-methoxy-N,N-dimethyl-tryptamine. Brain Res 648:87–98

Lalley PM, Ballanyi K, Hoch B, Richter, DW (1997) Elevated cAMP levels reverse opioid- or prostaglandin-mediated depression of neonatal rat respiratory neurons. Soc Neurosci Abstr 27:222

Lalley PM, Bischoff AM, Richter DW (1994) 5HT-1A receptor-mediated modulation of medullary expiratory neurones in the cat. J Physiol (Lond) 476:117–130

Lalley PM, Bischoff AM, Schwarzacher SW, Richter DW (1995) 5HT-2 receptor- controlled modulation of medullary respiratory neurones. J Physiol (Lond) 487:653–661

Lalley PM, Pierrefiche O, Bischoff AM, Richter DW (1997) cAMP-dependent protein kinase modulates expiratory neurons in vivo. J Neurophysiol 77:1119–1131

McCaman RE, McKenna DG, Ono JK (1977) A pressure ejection system for intracellular and extracellular ejections of picoliter volumes. Brain Res 136:141–147

McLennan, York DH (1967) The action of dopamine on neurones of the caudate nucleus. J Physiol (Lond) 189: 393–402

Nastuk WL (1953) Membrane potential changes at a single muscle endplate produced by transitory application of acetylcholine with an electrically controlled microjet. Fedn Proc 12: 102P

Nicoll RA, Malenka RC, Kauer JA (1990) Functional comparison of neurotransmitter receptor subtypes in mammalian central nervous system. Physiol Rev 70:513–565

Nicholson C, Phillips JM (1981) Ion Diffusion modified by tortuosity and volume fraction in the extracellular microenvironment of the rat cerebellum. J Physiol (Lond) 321:225–257

Oliver AP (1971) A simple rapid method for preparing parallel micropipette electrodes. Electroenceph Clin Neurophysiol 31:284–286

Palmer MR, Hoffer BJ (1980) Catecholamine modulation of enkephalin-induced electrophysiolical responses in cerebral cortex. J Pharmacol Exper Ther 213:205–215

Palmer MR (1982) Micro pressure-ejection: A complimentary technique to microiontophoresis for neuropharmacological studies in the mammalian central nervous system. Electrophysiol Tech 9:123–139

Palmer MR, Fossom LH, Gerhardt GA (1986) Micro pressure-ejection techniques for mammalian neuropharmacological investigations. In: Geller HM (ed), Electrophysiological Techniques in Pharmacology. Alan R. Liss, Inc: New York, pp 169–187

Palmer MR, Wuerthele, SM, Hoffer BJ (1980) Physical and physiological characteristics of micropressure ejection of drugs from multibarreled pipettes. Neuropharmacol 19:931–938

Parker D, Newland PL (1995) Cholinergic synaptic transmission between proprioceptive afferents and a hind leg motor neuron in the locust. J Neurophysiol 73:586–594

Phillis JW, Tebecis AK, York DH (1967) A study of cholinoceptive cells in the lateral geniculate nucleus. J Physiol (Lond) 192: 695–713

Pierrefiche O, Foutz AS, Champagnat J, Denavit-Saubie M (1994) NMDA and non-NMDA receptors may play distinct roles in timing mechanisms and transmission in the feline respiratory network. J Physiol (Lond) 474: 509–523

Remmers JE, Schultz SA, Wallace J, Takeda R, Haji A (1997) A modified coaxial compound micropipette for extracellular iontophoresis and intracellular recording: fabrication, performance and theory. Jap J Pharmacol 75:161–169

Richter DW, Lalley PM, Pierrefiche O, Haji A, Bischoff AM, Wilken B, Hanefeld F (1997) Intracellular signal pathways controlling respiratory neurons. Resp Physiol 110: 113–123

Sakai M, Sakai H, Woody CD (1978). Intracellular staining of cortical neurons by pressure microinjection of horseradish peroxidase and recovery by core biopsy. Exp Neurol 58:138–144

Sakai M, Swartz BE, Woody CD (1979) Controlled micro-release of pharmacological agents: Measurements of volumes ejected in vitro through fine-tipped glass microelectrodes by pressure. Neuropharmacol 18:209–213

Salmoiraghi GC, Stefanis CN (1967) A critique of iontophoretic studies of central nervous system neurons. Int Rev Neurobiol 10:1–30

Schmid K, Foutz AS, Denavit-Saubie M (1996) Inhibitions mediated by glycine and GABA receptors shape the discharge pattern of bulbar respiratory neurons. Brain Res 710:150–160

Schwartz JH (1974) Synthesis, axonal transport and release of acetylcholine by identified neurons of *Aplysia.* Soc Gen Physiol Ser 28: 239

Siggins GR, French ED (1979) Central neurons are depressed by iontophoretic and micro-pressure applications of ethanol and tetrahydropapveroline. Drug Alcohol Depend 4:239–243

Simmonds MA (1974) Quantitative evaluation of responses to microiontophoretically applied drugs. Neuropharmacol 13:401–415

Sonnhof U (1973) A multi-barreled coaxial electrode for iontophoresis and intracellular recording with a gold shield of the central pipette for capacitance neutralization. Pfluegers Arch - Europ J Physiol 341:351–358

Sorensen S, Dunwiddie T, McClearn G, Freedman R, Hoffer B (1981) Ethanol-induced depression in cerebellar and hippocampal neurons of mice selectively bred for differences in ethanol sensitivity: An electrophysiological study. Pharmacol Biochem Behav 14: 227–234

Suh TH, Wang CH, Lim RKS (1936) The effect of intracisternal application of acetylcholine and the localization of the pressor centre and tract. Chin J Physiol 10:61–78

Suvorov NF, Mikhailov AV, Voilokova NL, Il'ina E.V (1996) Universal method for microelectrode and neurochemical investigations of subcortical brain structures of awake cats. Neurosci Behav Physiol 26:251–255

Szente MB, Baranyi A,Woody CD (1990) Effects of protein kinase C inhibitor H-7on membrane properties and synaptic responses of neocortical neurons of awake cats. Brain Res 506: 281–286

Tauc L, Hoffman A, Tsuji S, Hinzen DH, Faille L (1974) Transmission abolished in a cholinergic synapse after injection of acetylcholinesterase into the presynaptic neuron Nature 250: 496–498

Thompson EB, Schwartz JH, Kandel ER (1976). A radioautographic analysis in the light and electron microscope of identified Aplysia neurons and their processes after intrasomatic injection of L-^{3}H-Fucose. Brain Res 112:251–281

Walker T, Dillman N, Weiss ML (1995) A constant current source for extracellular microiontophoresis. J Neurosci Methods 63:127–136

Wang Y, Jones JF, Ramage AG, Jordan D (1995) Effects of 5-HT and 5-HT1A receptor agonists and antagonists on dorsal vagal preganglionic neurones in anesthetized rats: an ionophoretic study. Br J Pharmacol 116: 2291–2297

Young MR, Fleetwood-Walker SM , Mitchell R, Dickinson T (1995) The involvement of metabotropic glutamate receptors and their intracellular signaling pathways in sustained nociceptive transmission in rat dorsal horn neurons. Neuropharmacol 34:1033–1041

Zhang J, Mifflin SW (1997) Influences of excitatory amino acid receptor agonists on nucleus of the solitary tract neurons receiving aortic depressor nerve inputs. J Pharmacol Exp Ther 282: 639–647

Zifa E, Fillion G (1992) 5-Hydroxytryptamine receptors. Pharmacol Rev 44:401–458

Chapter 8

An Introduction to the Principles of Neuronal Modelling

K. A. Lindsay,* J. M. Ogden,* D. M. Halliday,** J. R. Rosenberg**

Introduction

Neuronal modelling is the process by which a biological neuron is represented by a mathematical structure that incorporates its biophysical and geometrical characteristics. This structure is referred to as the mathematical model or the model of the neuron. The behavior of this representation may serve a number of purposes: for example, it may be used as the basis for estimating the biophysical parameters of real neurons or it may be used to define the computational and information processing properties of a neuron. Neuronal modelling requires not only an understanding of mathematical and computational techniques, but also an understanding of the what the process of modelling entails. A general treatment of models, however, would necessarily lead to the examination of a number of philosophical questions. Here we simply discuss some aspects of modelling that in our experience have proved to be useful in the construction and application of models. These topics are not usually considered in the neurophysiological modelling literature, but an understanding of the basic assumptions of modelling, and the presumed relation between model and reality is essential for constructive work in computational neuroscience.

The main themes of this chapter concern:

- the mathematical formulation of a model of a dendritic tree based on elementary conservation laws, leading to a new generalisation of the Rall equivalent cylinder;
- the numerical treatment of this model using traditional finite difference schemes, highlighting some previously unrecognised shortcomings of these schemes;
- the use of the finite difference representation to generate fully equivalent cables for passive dendrites of arbitrary geometry;
- procedures for generating stochastic spike trains of known statistical characteristics as well as an arbitrarily large number of stochastic spike trains with any desired correlation structure (e.g. weakly to strongly correlated);
- the introduction of a generalised model of a neuron based on the selection of designated points as opposed to compartments, and for which the Rall compartmental model, and those related to it, appear as special cases;
- an introduction to the spectral methodology for solving partial differential equations with applications to simple and branched dendritic structures, including a comparison of the numerical predictions based on the spectral technique with the results derived from the analytic solutions.

Much contemporary modelling work, leading to an improved understanding of the importance of dendrites and their functional role in shaping the behavior of neurons, developed from the pioneering studies of Wilfrid Rall (Segev, Rinzel and Shepherd, 1995). An important aspect of this work has been directed toward the estimation of membrane parameters and how these parameters might be influenced by neuron geometry (Segev *et al.*, 1995; Rall, Burke, Holmes, Jack, Redman and Segev, 1992). This work has received a number of

*Corresponding author: K. A. Lindsay, Department of Mathematics, University of Glasgow, Glasgow G12 8QQ

**Division of Neuroscience and Biomedical Systems University of Glasgow, Glasgow G12 8QQ

excellent and extensive reviews and is not reviewed here (Segev *et al.*, 1995). There is also an extensive literature on the derivation of neuronal models exhibiting particular types of spike train output behavior such as bursting or periodic spike trains. The reader is referred to specific reviews of this material (e.g. Rinzel and Ermentrout, 1989; Getting, 1989; Cohen, Rossignol and Grillner, 1988). The discussion of this chapter will concentrate on the fundamental issues of modelling that are important for any application. As always, linear models of neuronal behavior play a prominent role in any discussion of neuronal behavior since they form the baseline against which nonlinear behavior is measured (Rall *et al.*, 1992). In addition, these models have distinctive features, such as exact solutions and equivalent cables, each providing insights into neuronal function that may not be apparent in nonlinear models (e.g. Evans and Kember, 1998; Evans, Kember and Major, 1992; Evans, Major and Kember, 1995; Kember, 1995; Major, 1993; Major, Evans and Jack, 1993a; Major, Evans and Jack, 1993b; Major and Evans, 1994; Ogden, Rosenberg and Whitehead, 1999; Whitehead and Rosenberg, 1993). The numerical techniques described in this chapter, and applied to linear dendritic models, extend naturally to the nonlinear case simply because the nonlinearities in dendritic modelling occur through the dependence of current inputs on membrane potential, while the differential operator remains unchanged[1]. Fully nonlinear problems have nonlinear differential operators.

It is assumed that readers are familiar with elementary linear algebra and basic calculus. Indeed, readers competent in linear algebra and knowledgeable in programming and the numerical solution of systems of linear equations should, on the basis of this chapter, be able to construct their own neuronal representation of complex branching dendritic structures in a straightforward algorithmic way and also obtain solutions. In addition, an understanding of the material of this chapter should prove invaluable in making effective use of the high quality computer packages (Deschutter, 1992; Hines and Carnevale, 1997) for neuronal modelling now freely available over the web. These packages are comprehensive and cover both the linear and non-linear behavior of neurons.

A Philosophy of Modelling

Modelling is the process by which a complex phenomenon or concrete object is replaced by another entity called "the model" whose environment and operational characteristics are defined by prescribed rules, and whose behavior is taken to represent that of the concrete object or phenomenon (Regnier, 1966). The rules describing the behavior of the model are almost invariably couched in the language of mathematics. Indeed, the term "modelling" is often used interchangeably with "Mathematical Modelling", although the underlying philosophy and procedures of modelling are not inherently mathematical.

Note that experimental data or empirical descriptions of the concrete object that result from experimental measurement are not by themselves logically coherent. The work of the theoretician is to create, beyond the empirical descriptions of the data, the theoretical representation that constitutes the model, and at the same time provides a logical structure for the data. The model may also be seen as a substitute for the real phenomenon that translates difficult questions concerning this phenomenon into easier questions concerning the model. In this sense the model facilitates one's intuition concerning the real phenomenon. The primary purpose of modelling then becomes that of describing reality in terms of models that are inherently coherent, and thereby order and predict the behavior of the real entity based on that of the model. In order to achieve this aim, the interaction between the real entity and its model is quantified by experiment. The greater the variety of experimental conditions in which the predictions of the model are in acceptable agreement with

[1] Nonlinearities in neuronal modelling merely add technical difficulties, but do not change the fundamental and conceptual issues of modelling.

reality, the stronger the conviction that the predictions of the model and the experience of reality will be in agreement in uncharted territory, although this remains to be tested. In effect, the model is used to "interpolate" reality much in the same way as a finite or discrete set of observations are used to represent a continuous process. Over the domain of experiment where the model is valid, reality is understood by the successful agreement of experimentation and model predictions. Implicit in this strategy of experimentation and model construction is the belief that reality consents to organise itself according to the same logical structure as that of the model (Regnier, 1966, 1974).

Without doubt the success of contemporary science and technology is founded on the quality with which the existing models and paradigms explain observations. Some models are so effective that the boundary between model characteristics and reality becomes sufficiently blurred that the user perceives the model to be reality. For example, our experience suggests that a surprisingly large fraction of those versed in mechanics actually believe that the motion of an automobile is controlled by Newton's Laws, all be it with the embellishments afforded by air resistance and the like!

The Modelling Cycle

Modelling is often perceived as a vague or woolly process shrouded in some kind of mystique. In fact, modelling is a cyclical process with well defined stages that must be pursued in a prescribed order, irrespective of the complexity of the entity to be modelled. The initial stage of model development must specify aims and quantifiable criteria against which the predictions of the model are to be tested. Once done, the next stage of the modelling procedure requires the specification of model parameters and variables, including assumptions and the limitations that are incumbent on the model as a result of these assumptions. For example, a discretised model of a dendritic tree based on cylinders *cannot* be associated with a unique dendritic tree, but rather is associated with a continuum of trees that are geometrically/electrically "close" — a "fuzzy" dendrite. Since a number of different trees are represented by the same model, decisions on discretisation are not independent of the specification of aims and objective criteria against which the model is to be validated.

Model variables and parameters, once defined with their associated limiting assumptions, are now connected by rules/principles that have proved to be successful in similar applications. These either express conservation of model properties — conservation laws — (e.g. conservation of electrical charge) or specify relationships between model variables — constitutive laws — (e.g. Ohm's law connecting current through a resistive medium and potential difference). The electrical behavior of a dendritic tree is fundamentally controlled by the conservation of charge. This is identified as a conservation law because its validity is independent of circumstance — the principle of conservation of charge is universally valid. By contrast, constitutive laws define the properties of the dendritic material by postulating relations connecting measurable variables such as current and potential difference. For example, dendritic current flow is typically assumed to obey Ohm's law, that is, the dendritic current is taken to be a linear function of potential gradient[2]. Similarly, membrane leakage currents obey Ohms law.

Once the conservation and constitutive laws are satisfied, the model is defined and the implications of these laws can now be pursued using mathematical methods. However, answers must always be interpreted in the light of the modelling assumptions and limitations that are implicit in the construction of the model. In particular, manifestations of the model should not be confused with properties of the real entity. For example, it is known that passive multi-cylinder models of dendritic trees can be transformed into equivalent cables which may contain disconnected sections (Ogden *et al.*, 1999; Whitehead and Rosenberg,

[2]Not all dendritic current flow is Ohmic. For example, Hodgkin-Huxley currents are nonlinear functions of voltage.

1993). Are we therefore to believe that real dendritic trees have regions that are physically disconnected from the soma? Of course not — this would be an inappropriate interpretation of the mathematical result. Rather, the model suggests that there are configurations of inputs to the dendritic tree that exert no net electrical effect at the soma. Essentially the model solution must always be appraised in the light of the modelling assumptions and limitations, and not taken at face value.

Finally, the predictions of the model must always be compared with real measurement whenever possible. If model predictions are at odds with measurement, the modelling cycle should be re-entered entailing possible adjustments to the choice of variables and parameters, revision of the aims and objective criteria through which the model is to be accredited and a potential re-appraisal of the underlying conservation and constitutive laws on which the model is based.

Formulation of Dendritic Model

Within the context of neuronal modelling, the term "cable theory" refers to the collection of linear and non-linear models that have been used to describe the time dependent electrical activity in arbitrarily branched dendritic trees, axons and branched axonal terminals with varying degrees of biophysical realism. A historical overview of cable theory can be found in Rall (1977). The general one dimensional non-linear cable equation for a cylindrical dendritic limb of arbitrary, but "small", taper is now formulated, the derivation being based on the conservation of electrostatic charge together with a variety of assumptions concerning the electrical properties of the biological material from which neurons are built.

A dendrite is typically modelled as an intracellular Ohmic medium enclosed by a thin highly resistive membrane which is itself immersed in a perfectly conducting fluid. The configuration is illustrated in Fig. 1. The cross-sectional dimensions of the intracellular region are sufficiently small compared with its length so that the electrostatic potential is effectively constant over the dendritic cross-section to a first approximation, and exactly constant in the idealised model. As a result, the corresponding current flow within the intracellular region is predominantly axial with negligible flow perpendicular to the axial direction, and in the idealised model is entirely axial. Importantly, this means that axial length along the dendritic core and axial length along the dendritic membrane surface are identical to the level of resolution afforded by the one dimensional model unlike a full three dimensional model where surface arc length must be distinguished from axial distance. This is the criterion for a "small" taper. In a full three dimensional model, surface current densities are applied over the membrane surface, which may be parameterised in terms of the axial coordinate x, but is regarded as locally conical and not locally cylindrical. This departure from locally cylindrical geometry now inevitably generates significant non-axial currents[3].

The mathematical formulation of the cable equation presented in this article will ignore potential variations and current flow across the dendritic cross-section. This model of current flow in dendrites may therefore be regarded as but the leading term in the asymptotic expansion of the full dendritic potential. In fact, a simplistic nondimensionalisation of the full three dimensional cable equation suggests that the fine structure in the membrane potential arising from non-axial current flow is of the order of the ratio of the cross-sectional area of the dendrite to its length squared. This may be as small as 10^{-6}. Formal analyses of three dimensional current flow in the intracellular region for dendrites of right circular cross-section has been given by Rall (1969), for striated muscle fibers by Falk and

[3]The distinction between dx, the differential of axial length, and ds, the differential of axial length along the dendritic membrane surface is essential in, for example, the theory of elasticity.

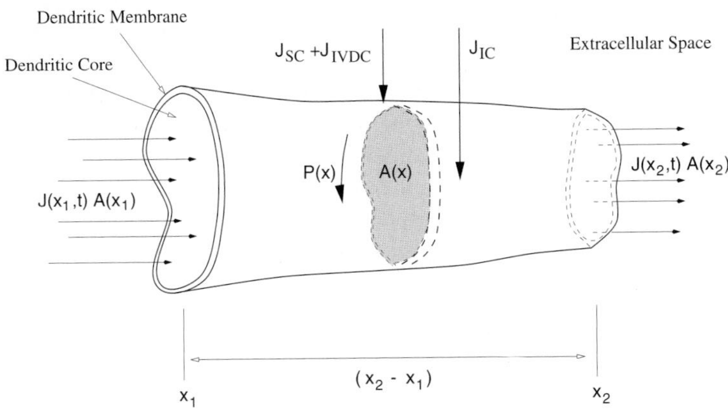

Fig. 1. Diagrammatic representation of a dendritic segment of length $(x_2 - x_1)$ illustrating axial current flow $J(x_1, t)A(x_1)$ into the segment and axial current flow $J(x_2, t)A(x_2)$ out of the segment where $J(x, t)$ is the axial current density (rate of flow of charge per unit area) and $A(x)$ is the cross sectional area. The perimeter of the segment is $P(x)$. Note that both A and P are generally functions of the axial coordinate x. Here J_{IC} represents injected current (exogenous), J_{SC} represents synaptic current and J_{IVDC} represents the sum of intrinsic and voltage dependent currents. Both J_{IC} and J_{IVDC} are given by constitutive formulae. Coordinate x measures length along the dendrite increasing away from the soma.

Fatt (1964) and considered in general by Eisenberg & Johnson (1970). Their conclusions reinforced the validity of the one dimensional model of a dendrite.

Model of a Dendritic Cylinder

The construction of the model dendrite requires a mathematical representation of the intracellular dendritic core, the dendritic membrane and the surrounding extracellular medium together with descriptions of the soma of the cell, synaptic inputs onto dendrites and boundary conditions at dendritic terminals. At internal branch points, the membrane potential is continuous and current is conserved. The objective of the modelling procedure is to provide a quantitative description of the behavior of the membrane potential at the soma (and by default, elsewhere) as it is shaped by the time and spatial characteristics of injected currents as well as synaptic inputs. The model therefore provides the basis for estimating membrane parameters and describes how different distributions of synaptic inputs and different dendritic geometries affect the potential at the soma or elsewhere. The behavior of the potential at the soma will determine the temporal properties of the output spike train from the neuron[4]. The mathematical representation of each component part of the model is now presented.

Definitions and Conventions

Let coordinate x be axial distance along a dendrite limb of length L and make the conventional decision that $x = 0$ always denotes the end of the limb closest to the soma so that $x \in [0, L]$. Charge can be associated with the highly resistive dendritic membrane (a capacitance effect). Redistribution of this charge through the membrane itself and the axial diffusion of this charge along the resistive intracellular core of the dendrite establishes a

[4]Much theoretical work has been devoted to modelling spike train outputs from neurons by an integrate-to-threshold-and-fire model where the threshold crossing times of the membrane potential depend on the characteristics of the simulated spike train inputs (see Tuckwell, 1988a; Holden, 1976)

transmembrane potential difference $V(x,t)$ between the dendrite's core and the extracellular medium (see Fig. 1). Note that the assumption that the extracellular material is perfectly conducting means that it is at a uniform potential, which may be arbitrarily fixed at zero, and that charge distributions in the extracellular region are equilibrated instantly. Associated with the potential $V(x,t)$ is the axial current density $J(x,t)$, measured per unit area of the cross-section of the dendritic core. The fundamental relationship between V and J is phenomenological and defines the electrical properties of the dendritic core material. In practice, the core is widely assumed to be an ohmic conductor so that

$$J(x,t) = -g_{\mathrm{A}} \frac{\partial V(x,t)}{\partial x} \tag{8.1}$$

where g_{A} is the axial conductivity of the intracellular material. By convention, J is the current density in the direction of *increasing* x and therefore $J(L,t)$ is current flow out of the dendritic core at $x = L$ whereas $J(0,t)$ is current flow from the soma or a contingent dendritic limb into the dendritic core at $x = 0$. By implication the current flow into the soma from a contingent dendrite is $-J(0,t)$. In general, J is a vector although in this one dimensional application, it behaves like a signed algebraic quantity. The consistent application of these conventions is essential in more complex dendritic structures, particularly when synaptic inputs and injected currents are included in the model.

For readers who are more familiar with the formulation of Ohms law in circuit theory, it is useful to digress for a moment and demonstrate that equation (8.1) is a more fundamental description of Ohms law embodying $V = IR$ as a special case. Consider an ohmic conductor formed into a slab of thickness d and cross-sectional area A (see Fig. 2). If current I flows through this slab when potential difference $V = V_d - V_0$ is applied across its faces as illustrated in Fig. 2, then the corresponding current density is $J = I/A$ and the associated potential gradient satisfies

$$J = \frac{I}{A} = -g\frac{dV}{dx}, \qquad x \in (0,d), \qquad\qquad V(0) = V_0$$

where g (assumed constant) is the electrical conductivity of the material from which the slab is constructed. This equation has solution $V(x) = -Ix/gA + V_0$ and the requirement that $V(d) = V_d$ leads to the conclusion that $V = (V_0 - V_d) = Id/gA = IR$ where R is the ohmic resistance of the slab. Hence the slab behaves like an ohmic conductor of resistance[5] $R = d/gA = \varrho d/A$ where ϱ (Ω m) is the resistivity corresponding to conductivity g.

Derivation of the Cable Equation

The intracellular core of a dendrite is enclosed by a thin layer or membrane of highly resistive tissue. Although this membrane is largely impermeable, it contains a distribution of ion channels which can allow charge to move between the intracellular and extracellular regions. Following the convention of Getting (1989), the cable equation will incorporate three distinct types of current density[6], namely, injected current $-J_{\mathrm{IC}}$, intrinsic and voltage-dependent currents $-J_{\mathrm{IVDC}}$ and synaptic current $-J_{\mathrm{SC}}$, all measured per *unit area* of the dendritic membrane. In reality, intrinsic or pumping currents arise from active transport processes, while the voltage-dependent currents (of which Hodgkin-Huxley type current

[5]If the slab has cross-sectional area $A(x)$, then the resistance has the general form $R = \varrho \int_0^d \frac{ds}{A(s)}$. If it is unreasonable to model current flow using formula (8.1), it is common to postulate that $J = J(V - V_0)$ in which case conductance is defined to be $\partial J/\partial V$. Note that such conductances are dimensionally different from that arising in (8.1) but are dimensionally consistent.

[6]The positive direction of radial current flow is in the direction of increasing radial coordinate, that is, outwards. Therefore current inputs from the domain exterior to the dendritic core are conventionally negative.

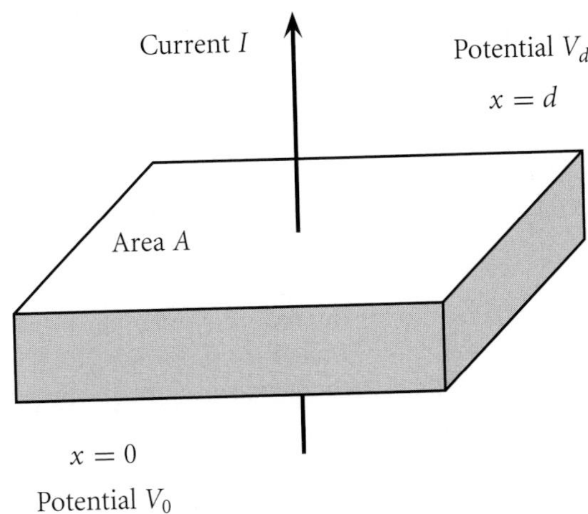

Fig. 2. The diagram illustrates a slab of ohmic material with thickness d and surface area A through which a uniform current I flows under a constant potential difference $V = V_0 - V_d$.

is an example) are due to various ion channels. If J_{IVDC} is a nonlinear function of membrane potential (i.e. non-ohmic) or synaptic currents are present, the dendritic membrane is said to be *active* otherwise it is *passive*. Linear cable theory deals with arbitrary dendritic geometry and passive membranes.

The equation describing the evolution of the membrane potential $V(x, t)$ is constructed from a consideration of charge conservation in the section of dendritic limb $x_1 \leqslant x \leqslant x_2$. Suppose that this limb has cross-sectional area $A(x)$ with associated perimeter $P(x)$, then the net current *into* the dendritic section is

$$A(x_1)J(x_1, t) - A(x_2)J(x_2, t) - \int_{x_1}^{x_2} J_{\mathrm{in}}(x, t)P(x)\,dx \tag{8.2}$$

where

$$J_{\mathrm{in}}(x, t) = J_{\mathrm{IC}}(x, t) + J_{\mathrm{IVDC}}(x, t) + J_{\mathrm{SC}}(x, t)\,. \tag{8.3}$$

By definition, J_{in} is just the effective current density applied per unit area of the dendritic membrane, being the sum of the injected current J_{IC}, the synaptic current J_{SC} and the intrinsic and voltage dependent currents J_{IVDC} (all measured per unit area of the dendritic membrane). Notice again that equations (8.2) and (8.3) make no distinction between axial coordinate x and axial length s along the dendritic surface, despite the presence of dendritic taper. To distinguish between x and s contradicts a premise of the model since it is tantamount to assuming that current flow in the dendritic core is not axial. During the time interval $[t_1, t_2]$, this net influx of current increases the charge stored in this dendritic segment by

$$\int_{t_1}^{t_2} \left[A(x_1)J(x_1, t) - A(x_2)J(x_2, t) - \int_{x_1}^{x_2} J_{\mathrm{in}}(x, t)P(x)\,dx \right] dt\,. \tag{8.4}$$

That additional charge is associated with the dendritic membrane which is assumed to have capacitance C_{M} per unit area. By definition, the change in stored charge over the time interval $[t_1, t_2]$ is

$$\int_{x_1}^{x_2} C_{\mathrm{M}}\,P(x)V(x, t_2)\,dx - \int_{x_1}^{x_2} C_{\mathrm{M}}\,P(x)V(x, t_1)\,dx\,. \tag{8.5}$$

In the absence of externally injected charge during the time interval $[t_1, t_2]$, conservation of charge requires that expressions (8.4) and (8.5) are equal, that is,

$$\int_{x_1}^{x_2} C_M P(x)\Big[V(x,t_2) - V(x,t_1)\Big]\,dx = \int_{t_1}^{t_2}\Big[A(x_1)J(x_1,t) - A(x_2)J(x_2,t) - \int_{x_1}^{x_2} J_{\rm in}(x,t)P(x)\,dx\Big]\,dt\,. \tag{8.6}$$

This fundamental equation describes the temporal and spatial evolution of the membrane potential $V(x,t)$. It can be manipulated into a more familiar form provided $V(x,t)$ is a sufficiently differentiable function of space and time. On dividing equation (8.6) by (t_2-t_1) and taking limits as $t_1, t_2 \to t$, it follows by elementary calculus that

$$\int_{x_1}^{x_2} C_M P(x)\frac{\partial V(x,t)}{\partial t}\,dx = A(x_1)J(x_1,t) - A(x_2)J(x_2,t) - \int_{x_1}^{x_2} J_{\rm in}(x,t)P(x)\,dx\,. \tag{8.7}$$

A similar operation applied to the spatial components of equation (8.7) reveals that $V(x,t)$ and $J(x,t)$ satisfy the partial differential equation

$$C_M\frac{\partial V(x,t)}{\partial t} = -\frac{1}{P(x)}\frac{\partial(A(x)J(x,t))}{\partial x} - \Big(J_{\rm IC}(x,t) + J_{\rm IVDC}(x,t) + J_{\rm SC}(x,t)\Big)\,. \tag{8.8}$$

where $J_{\rm in}$ has been replaced by its definition (8.3). Equation (8.8) is a consequence of charge conservation *only* and makes no assumptions whatsoever about the constitutive properties of the dendrite. In practice, it is almost universally assumed that the dendritic core is well modelled as an Ohmic conductor in which case V and J are connected through Ohms Law (8.1). In this case, the membrane potential V is seen to satisfy the partial differential equation (Cable Equation)

$$C_M\frac{\partial V(x,t)}{\partial t} = \frac{1}{P(x)}\frac{\partial}{\partial x}\left(g_A A(x)\frac{\partial V}{\partial x}\right) - \Big(J_{\rm IC}(x,t) + J_{\rm IVDC}(x,t) + J_{\rm SC}(x,t)\Big)\,. \tag{8.9}$$

To complete the specification of the problem, it is necessary to quantify the input current densities $J_{\rm IC}$, $J_{\rm IVDC}$ and $J_{\rm SC}$, the initial membrane potential and the conditions to be applied at the soma, terminal boundaries and at dendritic branches.

Specification of Currents

As already seen in expression (8.3), the input current density $J_{\rm in}(x,t)$ consists of three philosophically different contributions. The first component $J_{\rm IC}$ is exogenous and describes current injected into the dendritic core from outside. The second and third components of $J_{\rm in}$ are constitutive in nature and quantify respectively the properties of the ion channels within the dendritic membrane in terms of their reaction to membrane potential and the presence of synaptic inputs from other neurons.

Typically $J_{\rm IVDC}$, the second component of $J_{\rm in}$, is a constitutive function of membrane potential V with the property that $J_{\rm IVDC} = 0$ when $V = E_L$, the resting membrane potential. We assume the existence of E_L, the potential at which pumping and ion channel currents are balanced. Without loss of generality, the existence of the resting potential indicates that $J_{\rm IVDC} = C(V) - C(E_L)$ where C is a constitutive function of membrane potential. Assuming that C is a suitably differentiable function of V, then the mean value theorem states that

$$J_{\rm IVDC} = \frac{dC(V^*)}{dV}(V - E_L) = g(V^*)(V - E_L) \tag{8.10}$$

where $V^* = V^*(V, E_L)$ is a potential between V and E_L and $g(V^*)$ may be considered to be a nonlinear conductance. Similarly, the mean value theorem applied to $g(V^*)$ yields

$$g(V^*) = g(E_L) + \frac{dg(V^{**})}{dV^*}(V^* - E_L) = g_M + g_{NL}(V) \tag{8.11}$$

where $g_M = g(E_L)$ is the passive membrane conductance, $V^{**} = V^{**}(V^*, E_L)$ is a potential between V^* and E_L (and consequently between V and E_L) and g_{NL} defines a voltage dependent conductance from equation (8.11). Note that g_{NL} is implicitly a function of V and is zero when $V = E_L$. Assembling results (8.10, 8.11) together gives

$$J_{IVDC} = g(E_L)(V - E_L) + g_{NL}(V - E_L) = g_M(V - E_L) + g_{NL}(V - E_L)\,. \tag{8.12}$$

Therefore for small deviations in potential from the resting potential, J_{IVDC} will be dominated by the linear term $g_M(V - E_L)$ and the dendrite is said to be passive. As the membrane potential increasingly deviates from its resting value, g_{NL} may assume larger and larger values and may eventually dominate the membrane behavior. Once g_{NL} is considered to be significantly different from zero, the membrane is said to be active. The form of g_{NL} is usually determined experimentally[7].

The third component of J_{in} is due to synaptic current input to the dendritic membrane. Synaptic inputs temporarily open ion channels across the dendritic membrane allowing movement of charge in sympathy with the prevailing potential difference. The opening and closing of these channels is commonly modelled by a time varying conductivity, while the actual current flow is assumed to be ohmic. Suppose that the channels are located at sites[8] $x = x_k$, $k = 1, 2, \ldots, N$, then the synaptic current density $J_{SC}(x, t)$ has general form

$$J_{SC}(x, t) = \sum_{j=1}^{\infty} \sum_{k=1}^{N} \sum_{\alpha} g_{syn}^{\alpha}(t - t_{kj}^{\alpha})(V(x_k, t) - E_{\alpha})\delta(x - x_k) \tag{8.13}$$

where $t_{k1}^{\alpha}, t_{k2}^{\alpha}, \ldots$ are the times (stochastic) at which synapse k associated with ionic species α becomes active while $g_{syn}^{\alpha}(t)$ models the conductance and E_{α} is the reversal potential associated with this species and $\delta(x - x_k)$ is the Dirac delta function at $x = x_k$. The function g is typically modelled by the so-called "alpha" function

$$g_{syn}^{\alpha}(t) = \begin{cases} 0 & t \leqslant 0 \\ G_{\alpha} \dfrac{t}{T_{\alpha}} e^{(1-t/T_{\alpha})} & t > 0 \end{cases} \tag{8.14}$$

where G_{α} is the maximum conductance and T_{α} is the time constant associated with species α (Getting, 1989).

In conclusion, the cable equation for a dendritic limb has final form

$$C_M \frac{\partial V(x, t)}{\partial t} + g_M(V - E_L) = \frac{1}{P(x)} \frac{\partial}{\partial x}\left(g_A A(x) \frac{\partial V}{\partial x}\right) - J_{extra} \tag{8.15}$$

where J_{extra} is a collective current density which includes exogenous currents, synaptic currents and the nonlinear components of the intrinsic and voltage-dependent currents, that is,

$$J_{extra} = J_{IC}(x, t) + J_{SC}(x, t) + g_{NL}(V - E_L) \tag{8.16}$$

in which g_{NL} is a voltage dependent conductance and J_{SC}, J_{IC} have their defined meaning. In the absence of exogenous input current, synaptic current activity and nonlinear membrane current flow, $J_{extra} = 0$.

[7]For example, the Hodgkin-Huxley equations provide a particular form for a nonlinear conductance.

[8]Channels associated with different ionic species may be located at the same axial coordinate x_k but clearly not on the same area of the membrane. The synaptic current at coordinate x_k is just the sum of the synaptic currents due to all the species at x_k.

Initial and Boundary Conditions

To obtain specific solutions to the cable equation (8.15), it is necessary to supply $V(x,0)$, the initial value of the membrane potential, and the boundary conditions to be satisfied at dendritic terminals, at the soma and at dendritic branch points (where appropriate).

Consider a dendritic terminal at membrane potential $V(L,t)$ leaking charge to the extracellular environment at potential V_{ex} (already taken to be zero) and into which current $-I_{\text{end}}$ is injected. Since the axial coordinate x is conventionally chosen to increase away from the soma, then the net *outflow* of current at this terminal is $A(L)J(L,t) - I_{\text{end}}$. This outflow is due to potential difference $V(L,t) - V_{\text{ex}}$ between the dendritic tip and the extracellular region. Assuming that the flow is ohmic then the terminal boundary condition has form

$$A(L)J(L,t) - I_{\text{end}} = g_{\text{L}}\, A(L)\Big(V(L,t) - V_{\text{ex}}\Big) \tag{8.17}$$

where g_{L} is the leakage conductance at the dendritic terminal. When J is replaced by expression (8.1), the membrane potential $V(L,t)$ is seen to satisfy the Robin condition

$$g_{\text{A}} \frac{\partial V(L,t)}{\partial x} + g_{\text{L}}\Big(V(L,t) - V_{\text{ex}}\Big) = -\frac{I_{\text{end}}}{A(L)} \,. \tag{8.18}$$

Several common boundary conditions are now considered in detail.

Current injection condition When $g_{\text{L}} = 0$, boundary condition (8.18) reduces to the current injection condition

$$A(L)\, g_{\text{A}} \frac{\partial V(L,t)}{\partial x} = -I_{\text{end}} \,.$$

A *sealed end* is the special case of this condition in which $I_{\text{end}} = 0$, that is, there is no injected current at $x = L$ and the region between the dendrite and the extracellular material is perfectly insulating. The sealed end boundary condition is therefore

$$\frac{\partial V(L,t)}{\partial x} = 0 \,. \tag{8.19}$$

This condition is often taken as the natural terminal condition for dendritic tips.

General voltage condition A general voltage condition may be applied to a dendritic terminal whenever the region between the dendritic terminal and the extracellular material is perfectly conducting. Essentially g_{L} is infinite in equation (8.18) so that

$$V(l,t) = V_{\text{ex}}(t) \,. \tag{8.20}$$

A dendritic tip is said to be a *cut end* whenever $V_{\text{ex}}(t) = 0$.

Dendritic branch points At a dendritic branch point, the membrane potential on each limb must be continuous and the sum of currents into the branch point must total zero. Suppose that current $-I_{\text{BP}}$ is injected into a branch point, then the net flow of charge *into* the branch point is

$$-I_{\text{BP}} + A^{(p)}(L^{(p)})J^{(p)}(L^{(p)},t) - \sum_{k=1}^{n} A^{(c_k)}(0)J^{(c_k)}(0,t)$$

where $\varphi^{(p)}$ is the value of φ on the parent limb and $\varphi^{(c_k)}$ $(1 \leqslant k \leqslant n)$ is the value of φ on the kth child dendrite. Voltage continuity and current conservation at the branch point now requires that

$$V^{(p)}(L^{(p)},t) = V^{(c_1)}(0,t) = \cdots = V^{(c_n)}(0,t) \,, \tag{8.21}$$

$$-I_{\text{BP}} - A^{(p)}(L^{(p)})g_{\text{A}}^{(p)} \frac{\partial V^{(p)}(L^{(p)},t)}{\partial x} + \sum_{k=1}^{n} A^{(c_k)}(0) g^{(c_k)} \frac{\partial V^{(c_k)}(0,t)}{\partial x} = 0 \,, \tag{8.22}$$

where current densities $J^{(p)}$ and $J^{(c_k)}$ have been replaced by potential gradients from Ohms law.

Somatic boundary condition

Somatic and terminal boundary conditions differ in the respect that charge can reside on the soma membrane surface, that is, the soma exhibits capacitance by virtue of its lumped geometrical construction. Suppose that m dendrites are connected to the soma and that $-I_S$ is the current entering the soma across its membrane surface, then the total rate of supply of charge to the soma is

$$-I_S - \sum_{i=1}^{m} A_i(0) J_i(0,t) = -A_S J_{\text{soma}} + \sum_{i=1}^{m} A_i(0) g_A \frac{\partial V_i(0,t)}{\partial x},$$

where $A_i(0)$ is the cross-sectional area of limb i, and it is assumed without significant loss of generalisation that g_A is the same for each dendrite. Continuity of potential at the soma-to-tree connection requires that $V_S(t) = V_1(0,t) = V_2(0,t) = \cdots = V_m(0,t)$. Since the rate of increase in somal charge is just $C_S\, dV_S(t)/dt$, where C_S is the capacitance of the soma, then conservation of charge requires that $V_S(t)$ satisfies the ordinary differential equation

$$C_S \frac{dV_S(t)}{dt} = -I_S + \sum_{i=1}^{m} A_i(0) g_A \frac{\partial V_i(0,t)}{\partial x}. \tag{8.23}$$

In common with dendritic membranes, the input current I_S is the sum of exogenous current injection I_{SIC}, possible synaptic current activity I_{SSC} (which may be instrumental in discharging the soma) and intrinsic voltage-dependent currents I_{SIVDC} modelled as the sum of an ohmic leakage current $A_S\, g_S\, (V_S(t) - E_L)$ and a nonlinear voltage dependent current $A_S\, g_{\text{SNL}}(V_S)(V_S - E_L)$ where A_S is the surface area of the soma. The synaptic and intrinsic voltage-dependent currents in this case model charge movement across the somal membrane due to the presence of ion channels. After minimal algebra, the somal boundary condition has final form

$$C_S \frac{dV_S(t)}{dt} + A_S\, g_S\, (V_S(t) - E_L) = -I_{\text{extra}} + \sum_{i=1}^{m} A_i(0)\, g_A \frac{\partial V_i(0,t)}{\partial x}. \tag{8.24}$$

Here I_{extra} is the sum of the exogenous current I_{SIC} input, the synaptic current activity I_{SSC} and the nonlinear voltage dependent currents $A_S\, g_{\text{SNL}}\, (V_S)(V_S - E_L)$ across the somal membrane. Note that g_A has dimension $\Omega^{-1}\text{m}^{-1}$ but that g_S has dimension $\Omega^{-1}\text{m}^{-2}$.

Cable Equation for Uniform Dendrites

For non-tapering dendrites, the cross sectional area $A(x)$ and the perimeter $P(x)$ are constant. The simplified form of the cable equation (8.15) is now

$$\frac{C_M}{g_M}\frac{\partial V}{\partial t} + (V - E_L) = \frac{g_A A}{g_M P}\frac{\partial^2 V}{\partial x^2} - \frac{J_{\text{extra}}}{g_M} \tag{8.25}$$

where J_{extra} is defined by equation (8.16). Let time τ and length λ be defined by

$$\tau = \frac{C_M}{g_M}, \qquad \lambda^2 = \frac{g_A A}{g_M P} \tag{8.26}$$

then equation (8.25) now becomes

$$\tau \frac{\partial V}{\partial t} + (V - E_L) = \lambda^2 \frac{\partial^2 V}{\partial x^2} - \frac{J_{\text{extra}}}{g_M}. \tag{8.27}$$

The homogeneous form of this equation corresponds to the usual cable equation, for example, Rall (1989). This dimensional equation may be non-dimensionalised by rescaling time and length by τ and λ respectively using the coordinate transformations $t^* = t/\tau$ and $x^* = x/\lambda$. In this process, the length of the dendrite is rescaled to the non-dimensional length $l = L/\lambda$ and the current density J_{extra} is rescaled to a non-dimensional linear current density $J^* = \lambda P J_{\text{extra}}$. By convention, τ is the time constant and λ is the length constant of a dendritic limb. Typically τ is assumed to be constant throughout a dendritic tree (determined by electrical parameters only) while λ varies from limb to limb. The resulting nondimensional cable equation is

$$\frac{\partial V}{\partial t} + (V - E_{\text{L}}) = \frac{\partial^2 V}{\partial x^2} - \frac{J(x,t)}{\lambda g_{\text{M}} P} = \frac{\partial^2 V}{\partial x^2} - \frac{J(x,t)}{\sqrt{g_{\text{A}} g_{\text{M}} A P}}, \qquad x \in (0, l) \tag{8.28}$$

where x, t and J are nondimensional in this equation although the superscript * has been suppressed, and will be suppressed in all future calculations for representational convenience. Non-dimensional length measurements are commonly referred to as *electrotonic* units and J in equation (8.28) is electrotonic current density of input currents. The total axial current I_{A} in a dendritic limb is now

$$I_{\text{A}} = A J_{\text{A}} = -A \frac{g_{\text{A}}}{\lambda} \frac{\partial V}{\partial x} = -\sqrt{g_{\text{M}} g_{\text{A}} A P}\, \frac{\partial V}{\partial x}\,. \tag{8.29}$$

Since $(g_{\text{M}} g_{\text{A}} A P)^{1/2}$ has dimension Ω^{-1} (dimension of conductance), it will henceforth be called the *g-value* of the uniform dendrite. The final non-dimensional forms of the cable equation and axial dendritic current are respectively

$$\frac{\partial V}{\partial t} + (V - E_{\text{L}}) = \frac{\partial^2 V}{\partial x^2} - \frac{J}{g}, \qquad I_{\text{A}} = -g \frac{\partial V}{\partial x}, \qquad x \in (0, l) \tag{8.30}$$

where g, the g-value of the cable, is given by the expression

$$g = \sqrt{g_{\text{A}} g_{\text{M}} A P}\,. \tag{8.31}$$

For example, a uniform dendrite with circular cross-section of diameter d has electrotonic scaling factor $\lambda = \sqrt{g_{\text{A}} d / 4 g_{\text{M}}}$ in equation (8.26). The expression for axial current is

$$I_{\text{A}} = -\frac{\pi}{2} \sqrt{g_{\text{M}} g_{\text{A}}}\, d^{3/2} \frac{\partial V}{\partial x} = -\frac{\pi}{2}\, c\, \sqrt{g_{\text{M}} g_{\text{A}}}\, \frac{\partial V}{\partial x} \tag{8.32}$$

where $c = d^{3/2}$ is usually called the c-value of the cable. As a general remark, the ratio of g-value to c-value (i.e. g/c) for any limb of a dendrite is universally constant whenever dendritic electrical properties are uniform everywhere.

Rall Equivalent Cylinder

The considerations of the previous sections lead naturally to the development of the *Rall equivalent cylinder* (Rall, 1962a). Consider a dendritic branch point comprising a parent dendrite and N child dendrites all of which have constant values for g_{M}, g_{A} and the same time constant τ. Suppose further that each child dendrite has uniform cross section (not all necessarily equal), identical electrotonic length l (each limb will likely have a different physical length) and identical terminal boundary condition. The membrane potential for branch k $(1 \leqslant k \leqslant N)$ satisfies the cable equation

$$\frac{\partial V_k}{\partial t} + (V_k - E_{\text{L}}) = \frac{\partial^2 V_k}{\partial x^2} - \frac{J_k}{g^{(k)}}, \qquad x \in (0, l)\,, \tag{8.33}$$

where it is now assumed that J_k *is independent of membrane potential* (i.e. the dendrite is passive so that J_k is only injected current). Continuity of potential among the parent and child branches requires that

$$V_1(0,t) = V_2(0,t) = \cdots = V_N(0,t) = V^{(p)}(l^{(p)},t) \tag{8.34}$$

in which $V^{(p)}(l^{(p)},t)$ is the parent membrane potential at the branch point. Equation (8.29) gives the form for the total axial current in each limb. Conservation of current requires that the total currents *into* the branch point sum to zero and therefore

$$\sum_{k=1}^{N} g^{(k)} \frac{\partial V_k(0,t)}{\partial x} - g^{(p)} \frac{\partial V^{(p)}(l^{(p)},t)}{\partial x} = 0\,. \tag{8.35}$$

Let the potential function $\psi(x,t)$ and conductance G_S be defined by the formulae

$$\psi(x,t) = \frac{1}{G_S} \sum_{k=1}^{N} g^{(k)} V_k(x,t)\,, \qquad G_S = \sum_{k=1}^{N} g^{(k)}\,. \tag{8.36}$$

Clearly $\psi(x,t)$ is just a weighted sum of the membrane potentials in the child limbs. Furthermore, the continuity of potential at the branch point embodied in equation (8.34) ensures that $\psi(0,t) = V^{(p)}(l^{(p)},t)$. The superposition property of the cable equation also guarantees that ψ satisfies the cable equation

$$\frac{\partial \psi}{\partial t} + (\psi - E_L) = \frac{\partial^2 \psi}{\partial x^2} - \frac{C(x,t)}{G_S}\,, \qquad C(x,t) = \sum_{k=1}^{N} J_k\,, \qquad x \in (0,l)\,. \tag{8.37}$$

Thus the current $C(x,t)$ is simply a weighted sum of input current densities on the child limbs. The superposition principle indicates that the terminal boundary condition for the ψ potential is determined by the terminal boundary conditions for the child potentials. In the special cases in which each child limb is cut, then so is ψ, while if each child limb is sealed, then ψ is likewise sealed.

To complete the construction of the Rall equivalent cylinder, it remains to show that ψ conserves current at the branch point. This can be achieved with a suitable choice for the electrical and geometrical properties associated with the ψ cable. By differentiating ψ with respect to x and then using the current balance condition of (8.35), it is now immediately obvious that $I_A^{(p)}$, the axial current in the parent branch, satisfies

$$-I_A^{(p)} = g^{(p)} \frac{\partial V^{(p)}(l^{(p)},t)}{\partial x} = G_S \frac{\partial \psi(0,t)}{\partial x}\,. \tag{8.38}$$

Expression (8.29) defines the axial current in the ψ cable to be

$$I_A^{(\psi)} = -g^{(\psi)} \frac{\partial \psi(x,t)}{\partial x} = -\sqrt{g_M^{(\psi)} g_A^{(\psi)}}\,(AP)_\psi^{1/2} \frac{\partial \psi(x,t)}{\partial x}$$

at coordinate x and so the current at $x = 0$ on the ψ cable is

$$I_A^{(\psi)} = g^{(\psi)} \frac{\partial \psi(0,t)}{\partial x} = \frac{g^{(\psi)}}{G_S} I_A^{(p)}$$

when $\partial \psi(0,t)/\partial x$ is replaced from (8.38). Therefore, to conserve current between the parent and the ψ cable, it is necessary that $I_A^{(\psi)} = I_A^{(p)}$ and this is ensured by the choice

$$g^{(\psi)} = G_S = \sum_{k=1}^{N} g^{(k)} \quad \equiv \quad \sqrt{g_M^{(\psi)} g_A^{(\psi)}}\,(AP)_\psi^{1/2} = \sum_{k=1}^{N} \sqrt{g_M^{(k)} g_A^{(k)}}\,(AP)_k^{1/2}\,. \tag{8.39}$$

This is the condition to be satisfied for a generalised Rall cable. The definition mixes electrical and geometrical characteristics of the child dendrites. Several simplifications of the generalised Rall cylinder are possible. For example, if the electrical properties of the dendrites are all equal and equal to that of the parent then condition (8.39) now becomes the purely geometrical condition

$$(AP)_{\psi}^{1/2} = \sum_{k=1}^{N} (AP)_{k}^{1/2} . \tag{8.40}$$

Finally, if dendrites are assumed to have circular cross-section then $A = \pi d^2/4$, $P = \pi d$ so that $(AP)^{1/2} = \pi d^{3/2}/2$. In this situation condition (8.40) simplifies to the well known Rall three-halves rule

$$d_{\psi}^{3/2} = \sum_{k=1}^{N} d_{k}^{3/2} . \tag{8.41}$$

If the parent dendrite also has diameter d_{ψ}, then the new cable extends seamlessly the length of the parent dendrite by l and therefore opens up the possibility of a further Rall reduction.

The notion of full equivalence is discussed in detail in a later section, but full equivalence must allow for preservation of information and an essential condition for full equivalence is that electrotonic length is preserved. The Rall equivalent cylinder is necessarily incomplete since the original branched structure consisted of N child limbs, each of length l, by contrast with the equivalent Rall cylinder which has length l, the length of only one branch. In fact, the form of ψ indicates that the Rall equivalent cylinder is an average of the individual child cables. The characteristics of these individual cables cannot be retrieved from the Rall equivalent cylinder. Clearly, a truly equivalent structure enables the properties of the original child dendrites to be reconstructed from it. In this sense, the Rall equivalent cylinder is incomplete, but can be made complete by the inclusion of a further $N-1$ disconnected cables formed from the $N-1$ pair-wise differences of the potentials on the child limbs. Each of these potentials describes a cable disjoint from every other cable, with one end cut and the other end satisfying the terminal condition of the original child branches.

The Discrete Tree Equations

One popular way to solve the model equations for a dendritic tree is to use finite difference methods to reformulate them as a discrete system of ordinary differential equations to be integrated with respect to time, the dependent variables in these equations being the membrane potential at a predetermined set of points (or nodes) distributed uniformly over the dendritic tree (Mascagni, 1989 and references therein). Since the underlying differential operator in the cable equation (8.15) is linear, then the discretised equations have matrix representation

$$\frac{dV}{dt} = AV + B \tag{8.42}$$

where V is a vector of membrane potentials, B is a vector of electrotonic input currents and A is a square matrix embodying the electrical and geometrical features of the tree including the connections between limbs, the terminal boundary conditions and a soma condition. For this reason, A is often called the *structure* matrix of the dendrite. If the dendrite is passive, this matrix will be independent of time. The formulation of the dendritic structure matrix[9] presented here follows the procedure described in detail by Ogden *et al.* (1999). At

[9]For other matrix representations see Tuckwell (1988a), Perkel and Mulloney (1978a, 1978b), Perkel, Mulloney and Budelli (1981) and Mascagni (1989).

this stage, it should be noted that equation (8.42) may be nonlinear since B is the sum of the exogenous currents and the nonlinear components of the intrinsic voltage dependent currents.

The continuous dendrite is first associated with a continuous "model dendrite" of multi-cylinder construction in a way that recognises that individual limb lengths can, in reality, be measured only to a prescribed experimental accuracy. Through a variety of steps involving the assignment of nodes, the representation of spatial derivatives, the treatment of branch points, points of discontinuity in cable cross-sectional area, terminal boundary conditions and soma boundary condition, the equations and boundary conditions implicit in the specification of the model dendrite are manipulated into the finite set of ordinary differential equations (8.42). Each component of the discretisation procedure is now described in detail.

Canonical Cable and Electrotonic Length

A dendrite is modelled by a family of interacting cable equations, one for each limb, such that membrane current is everywhere conserved and membrane potential is continuous at dendritic branch points and the soma. Each cable equation describes how the effects of membrane capacitance (as measured by $C_M \,\partial V/\partial t$), membrane resistance (as measured by $g_M(V - E_L)$) and the axial diffusion of membrane potential (as measured by $(1/P(x))(\partial/\partial x)(g_A A(x)\partial V/\partial x)$) combine to determine the spatial-temporal behavior of the membrane potential on that limb. The first step in the construction of the model dendrite requires a transformation of each cable equation to a canonical form. This canonical form is formulated so as to give equal importance to each of the three interacting processes that shape the membrane potential. This procedure also serves to simplify and nondimensionalise the cable equation so as to make it more amenable to analytical and numerical[10] methods.

Recall that the membrane potential $V(x,t)$ in a limb of a dendrite of cross-sectional area $A(x)$ and perimeter $P(x)$ satisfies the partial differential equation

$$C_M \frac{\partial V(x,t)}{\partial t} + g_M(V - E_L) = \frac{1}{P(x)} \frac{\partial}{\partial x}\left(g_A\, A(x) \frac{\partial V}{\partial x} \right) - J_{\text{extra}} \,, \tag{8.43}$$

$$J_{\text{extra}} = J_{\text{IC}}(x,t) + J_{\text{SC}}(x,t) + g_{\text{NL}}(V)(V - E_L) \,. \tag{8.44}$$

Classical analysis of dendritic structure usually begins by partitioning the physical dendrite into uniform cylinders (see Rall, 1962a). This discretisation process is based on subjective criteria[11] to delimit cylinder boundaries. The subsequent rescaling of length from $x \to z$ and time from $t \to t^* = t/\tau$ (i.e. non-dimensionalisation) cannot be achieved until these cylinders have been defined. The rescaled length of *each cylinder* is defined as its *electrotonic length*, while its associated non-dimensionalised cable equation has form

$$\frac{\partial V}{\partial t^*} + (V - E_L) = \frac{\partial^2 V}{\partial z^2} - \frac{J^*}{g} \tag{8.45}$$

where J^* is the electrotonic current corresponding to J_{NL} defined in equation (8.44) and g is the g-value of the cable. The consequence of this non-dimensionalisation procedure is that the coefficients of the membrane potential and its spatial and temporal derivatives are all unity, and therefore equation (8.45) is the canonical form for a uniform cylinder, since each term carries equal weight. Indeed, the original scaling was motivated by this very requirement; it is the simplified from of equation (8.45) that recommends its acceptance

[10] Strictly speaking, numerical methods can only be applied to nondimensionalised equations since they deal with arithmetical quantities that are implicitly nondimensional.

[11] A new cylinder is started after a diameter change of at least 0.2 μm (e.g. Segev *et al.* 1989).

as a canonical form for the equation. However, the restriction to uniform cylinders implies that equation (8.45) is likely to be a special case of a more general canonical form.

Ideally a non-dimensionalisation procedure is required that does not depend on the *a priori* definition of cylinders, but which reduces to equation (8.45) for a uniform cylinder. The more general canonical equation for non-uniform dendrites carries with it a generalised definition of electrotonic length, which is not subjective, but gives the well recognised electrotonic length for uniform cylinders and also leads to an objective procedure for discretising the dendritic tree.

This aim is achieved by rescaling dendritic geometry in a nonlinear way. Let nondimensional axial coordinate z and time t^* be defined by the transformations

$$z = \int_0^x \sqrt{\frac{g_M P(s)}{g_A A(s)}}\, ds\,, \qquad t^* = t/\tau\,, \quad \tau = C_M/g_M\,. \tag{8.46}$$

These nondimensional coordinates lead to the generalised canonical form of the cable equation for non-uniform dendrites. With these coordinate transformations,

$$\frac{\partial V}{\partial t} = \frac{g_M}{C_M}\frac{\partial V}{\partial t^*}\,, \qquad \frac{\partial V}{\partial x} = \frac{g_M P(x)}{C_M A(x)}\frac{\partial V}{\partial z}\,.$$

After further straightforward analysis, the corresponding generalised nondimensionalised form of the cable equation (8.43) and (8.44) may be shown to be

$$\frac{\partial V}{\partial t^*} + (V - E_L) = \frac{1}{g(z)}\frac{\partial}{\partial z}\left(g(z)\frac{\partial V}{\partial z}\right) - \frac{J}{g(z)}\,, \tag{8.47}$$

in which

$$g(z) = \sqrt{g_A g_M A(x) P(x)}\,, \qquad z = \int_0^x \sqrt{\frac{g_M P(s)}{g_A A(s)}}\, ds\,. \tag{8.48}$$

In equation (8.47), J is electrotonic linear current density and is related to J_{NL} by the requirement that for arbitrary $0 < a < b < L$,

$$\int_a^b P(x)\, J_{\text{extra}}\, dx = \int_{z(a)}^{z(b)} J\, dz \quad \equiv \quad J_{\text{extra}} = J\sqrt{\frac{g_M}{g_A A(x) P(x)}}\,. \tag{8.49}$$

Equation (8.47) is the generalised canonical equation of a dendrite. For uniform dendritic cylinders, $A(x)$ and $P(x)$ are constant functions of x. In this event, $g(z)$ is also constant and the canonical equation (8.47) simply reduces to the traditional form (8.45). Furthermore, z is simply a constant multiple of x, that is, the electrotonic length of a dendritic cylinder is a fixed multiple of its physical length. The familiar Rall definition of electrotonic length applies this result to each of the uniform cylinders comprising a dendritic limb. However, electrotonic length l and physical length L of a dendrite are generally connected through the nonlinear relationship

$$l = \int_0^L \sqrt{\frac{g_M P(s)}{g_A A(s)}}\, ds\,, \tag{8.50}$$

which becomes linear only when the integrand is constant.

Uniformly Tapered Dendrite

It is of practical interest to determine the electrotonic length of a uniformly tapered dendrite whose physical length is L, and whose left hand and right hand cross-sectional area and perimeter are $A_{\rm L}$, $P_{\rm L}$ and $A_{\rm R}$, $P_{\rm R}$ respectively. It is shown in the section on generalised compartmental models (p. 270) that

$$A(x) = (1-\lambda)^2 A_{\rm L}\,,\quad \lambda \in [0, L/H]\,,\quad P(x) = (1-\lambda)P_{\rm L}\,,\quad x = \lambda H$$

where H is the theoretical length of the dendrite if it were tapered to extinction. Assuming that $g_{\rm A}$ and $g_{\rm M}$ are constant, then from equation (8.50), the electrotonic length of this dendrite is

$$\begin{aligned} l &= \int_0^L \sqrt{\frac{g_{\rm M}}{g_{\rm A}}}\sqrt{\frac{(1-\lambda)P_{\rm L}}{(1-\lambda)^2 A_{\rm L}}}\,dx\,, \qquad x = \lambda H \\ &= H\sqrt{\frac{g_{\rm M}}{g_{\rm A}}}\sqrt{\frac{P_{\rm L}}{A_{\rm L}}}\int_0^{\lambda_{\rm R}} (1-\lambda)^{-1/2}\,d\lambda \\ &= 2H\sqrt{\frac{g_{\rm M}}{g_{\rm A}}}\sqrt{\frac{P_{\rm L}}{A_{\rm L}}}\left[1-\sqrt{(1-\lambda_{\rm R})}\right] \\ &= \sqrt{\frac{g_{\rm M}}{g_{\rm A}}}\sqrt{\frac{P_{\rm L}}{A_{\rm L}}}\frac{2H\lambda_{\rm R}}{1+\sqrt{(1-\lambda_{\rm R})}} \\ &= \sqrt{\frac{g_{\rm M}}{g_{\rm A}}}\frac{P_{\rm L}}{\sqrt{A_{\rm L}}}\frac{2L}{\sqrt{P_{\rm L}}+\sqrt{P_{\rm R}}}\,. \end{aligned}$$

For a tapering dendrite, it is obvious that $P(x)/\sqrt{A(x)} = P_{\rm L}/\sqrt{A_{\rm L}}$ for all x. In effect, the constancy of the ratio $P(x)/\sqrt{A(x)}$ for all x is a test for uniform taper. The actual value of this ratio depends on the geometry of the dendritic cross-sectional area. For circular dendrites, the constant is $2\sqrt{\pi}$. Since the circle encloses maximum area for a given perimeter (isoperimetric inequality of calculus of variations), then $P_{\rm L}/\sqrt{A_{\rm L}} \geqslant 2\sqrt{\pi}$ under all circumstances. Therefore, for dendrites of uniform taper,

$$l = \sqrt{\frac{g_{\rm M}}{g_{\rm A}}}\frac{P_{\rm L}}{\sqrt{A_{\rm L}}}\frac{2L}{\sqrt{P_{\rm L}}+\sqrt{P_{\rm R}}} \geqslant \sqrt{\frac{\pi g_{\rm M}}{g_{\rm A}}}\frac{4L}{\sqrt{P_{\rm L}}+\sqrt{P_{\rm R}}}$$

with equality if and only if the dendrite has circular cross-section.

Construction of the Model Dendrite

The non-dimensionalisation procedure described in the previous subsections, when applied to a dendritic tree with n limbs of physical lengths $L_1, \ldots, L_n$, generates a morphologically equivalent structure with limbs of electrotonic length $l_1, \ldots, l_n$ respectively. Although the notion of largest common denominator is not generally sensible for arbitrary lengths $l_1, \ldots, l_n$, it can be replaced by the idea that, given a set of n arbitrarily positive constants $\varepsilon_1, \varepsilon_2, \ldots, \varepsilon_n$, there exist non-negative integers $m_1, \ldots, m_n$ and a length l such that

$$|l_k - m_k\, l| \leqslant \varepsilon_k\,, \qquad 1 \leqslant k \leqslant n\,. \tag{8.51}$$

Notionally, $\varepsilon_1, \varepsilon_2, \ldots, \varepsilon_n$ are the non-dimensional measurement errors inherent in the determination of the electrotonic lengths $l_1, \ldots, l_n$ respectively and represent the scaled errors in $L_1, \ldots, L_n$. They may also be interpreted as a user-specified acceptable error. Once this error is specified, the discretisation procedure is automatic. Although the choice of l and the accompanying integers $m_1, \ldots, m_n$ is not unique, for a prescribed choice of $\varepsilon_1, \varepsilon_2, \ldots, \varepsilon_n$, there is a largest value of l, say l_{max}, that is unique by definition. This largest value will henceforth be called the *quantum length* and be unambiguously denoted by l. Even for the quantum length, the values of $m_1, \ldots, m_n$ may still be non-unique but now acceptable choices for any integer m_k $(1 \leqslant k \leqslant n)$ differ by at most one. For the purpose of dendritic modelling, the morphology of the original electrotonic dendrite is now approximated by a new "model dendrite" formed by replacing limb k (exact length l_k) by a limb of length $m_k\, l$. Once a quantum length and a set of integers $m_1, \ldots, m_n$ are specified, there are infinitely many electrotonic dendrites that are represented by the same model dendrite. The habitant of these "equivalent morphologies" can be thought of as a fuzzy zone around the model dendrite, the size of this zone being dependent on the values of $\varepsilon_1, \varepsilon_2, \ldots, \varepsilon_n$. Conspicuous by its absence in this discussion is the thickness of the original dendrite; it has no direct bearing on the choice of model dendrite, although it indirectly controls the specification of the ε values by influencing the choice of l, the quantum length. Clearly there is a relationship between l, the maximum length of model dendrite over which no changes in cross-sectional area and perimeter are permitted, and the sensitivity with which changes in dendrite thickness can be specified through the predetermined choice of $\varepsilon_1, \varepsilon_2, \ldots, \varepsilon_n$. Therefore (8.50) and (8.51) together with the choice of an acceptable error completely define the discretisation procedure.

Once the quantum length is chosen and the model dendrite constructed, a series of nodes, uniformly spaced distance h apart, are now distributed throughout the dendrite so that one node lies on each cylinder boundary and there is at least one (internal) node within each dendritic cylinder. Of course, smaller values of h provide superior refinement of the membrane potential but at the cost of reduced computational speed and increased memory storage. The nodes, once placed, are numbered individually so that node z_0 corresponds to the tree-to-soma connection point. In a branched structure, it is impossible to guarantee consecutive numbering of nodes on all cylinders since there are nodes that lie on more than one cylinder, but it is possible to guarantee that node numbers increase away from the soma. The enumeration scheme illustrated in figure 3 is implemented in this article because it simplifies the matrix representation of a dendritic tree and the construction of the equivalent cable. A key point in the enumeration algorithm is that nodes at which the membrane potential is known, for example, *clamped ends* (membrane potential externally fixed) or *cut ends* (membrane potential is zero) are left un-numbered. For a dendritic tree with N terminals, it is evident that consecutive numbering of nodes is broken on exactly $(N-1)$ occasions.

Finite Difference Formulae

As a preamble to the discretisation of the cable equations and the discussion of related issues, Taylor's theorem is used to derive some familiar and less familiar finite difference formulae for a function of a single variable. For sufficiently differentiable functions f of the single variable x, Taylor's theorem states that

$$f(x+h) = f(x) + hf'(x) + \frac{h^2}{2} f''(x) + \frac{h^3}{6} f'''(x) + O(h^4) \tag{8.52}$$

provided h is suitably small. Similarly, it follows from Taylor's theorem that

$$\begin{aligned}
f(x-h) &= f(x) - hf'(x) + \frac{h^2}{2} f''(x) - \frac{h^3}{6} f'''(x) + O(h^4)\,, \\
f(x+2h) &= f(x) + 2hf'(x) + 2h^2 f''(x) + \frac{4h^3}{3} f'''(x) + O(h^4)\,, \\
f(x-2h) &= f(x) - 2hf'(x) + 2h^2 f''(x) - \frac{4h^3}{3} f'''(x) + O(h^4)\,.
\end{aligned}$$

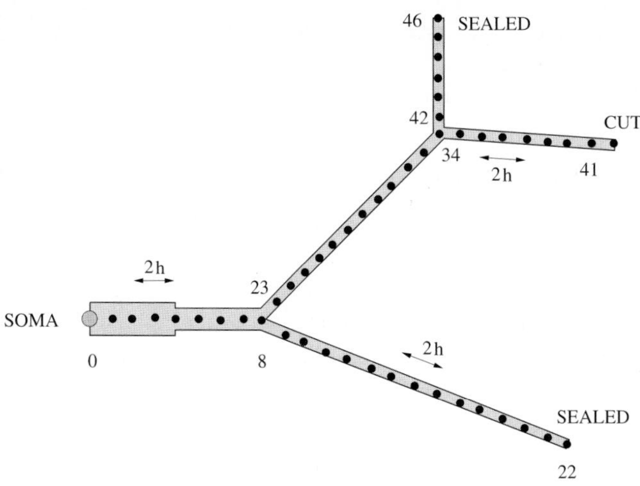

Fig. 3. Discretisation of a dendritic tree. All nodes are spaced h apart in electrotonic space. Node numbering starts from z_0 at the soma.

Based on these expansions, it is a matter of elementary algebra to demonstrate that $f'(x)$ and $f''(x)$ can be approximated to second order in h by the familiar central difference formulae

$$f'(x) = \frac{f(x+h) - f(x-h)}{2h} + O(h^2)\,, \tag{8.53}$$

$$f''(x) = \frac{f(x+h) - 2f(x) + f(x-h)}{h^2} + O(h^2)\,. \tag{8.54}$$

Equally important, but less familiar, are the right and left hand finite difference formulae

$$f'(x) = \frac{3f(x) + f(x-2h) - 4f(x-h)}{2h} + O(h^2)\,, \tag{8.55}$$

$$f'(x) = \frac{4f(x+h) - 3f(x) - f(x+2h)}{2h} + O(h^2)\,. \tag{8.56}$$

Results (8.55) and (8.56) are used in the finite difference treatment of boundary conditions involving gradients, for example, sealed ends, branch points etc.

The Discrete Formulation

Suppose that the dendritic tree has n nodes denoted symbolically by $z_0, z_1, \ldots, z_{n-1}$ where z_0 is chosen to be the soma. The soma is treated separately from the tree itself. For tree nodes (i.e. $j > 0$), let v_j denote the membrane potential at node z_j, let g_j denote the g-value of the dendritic segment (a mathematical cylinder of length h) which has z_j as its distal end and let J_j be the electrotonic current density applied to the dendrite at node z_j. The construction of the discrete equations requires the consideration of three different classes of nodes:

Terminal nodes Given a distribution of nodes on a model dendritic tree, a node is classified as a terminal node if it is either the node at the soma-to-tree connection or is the last *enumerated* node on any pathway from the soma to a dendritic tip. For leaky or sealed ends, the terminal node will be located at the dendritic tip but for a cut or clamped end, the terminal node will be the node connected to the dendritic tip, that is, the penultimate node on that limb.

Shared nodes Shared nodes correspond to branch points and points of discontinuity in cable geometry.

Internal nodes Any node that is not a shared node or a terminal node is, by definition, an internal node.

For any node z_j, let r_j^{-1} be defined as the sum of the g-values of all the segments of the finite difference representation of the dendrite that contain z_j. For example, r_0^{-1} is the sum of the g-values of all the segments containing the soma z_0 while if z_j is a branch point at which N child dendrites meet a parent dendrite, then $r_j^{-1} = g_j + \sum_{r=1}^{N} g_{k_r}$ where g_{k_r} is the g-value of the rth child dendrite. If node z_j is a dendritic terminal then $r_j^{-1} = g_j$; if z_j is an internal node then $r_j^{-1} = 2g_j$ and, finally, if z_j is a point of discontinuous cable geometry then $r_j^{-1} = g_j + g_{j+1}$.

Finite difference formulae are used to approximate the first and second spatial derivatives of membrane potential. Let z_i, z_j and z_k be three nodes, physically in sequence along a dendritic limb but not necessarily sequentially numbered and such that $i < j < k$, then it follows from finite difference formulae (8.53) and (8.54) that

$$\frac{\partial v(z_j,t)}{\partial x} = \frac{v_k - v_i}{2h} + O(h^2) \tag{8.57}$$

$$\frac{\partial^2 v(z_j,t)}{\partial x^2} = \frac{v_k - 2v_j + v_i}{h^2} + O(h^2)\,. \tag{8.58}$$

One sided gradient calculation

Suppose that z_j is a shared node, the somal node or a node located at a tip of a dendritic tree. The computation of the potential gradient at z_j is now non-trivial since z_j is either flanked by a single node (soma or dendritic tip) or alternatively, a different model equation applies to each of its flanking nodes (shared node). This impasse is often resolved using an idea borrowed from the numerical treatment of boundary value problems for ordinary differential equations, namely the notion of extending the differential equation to z_j itself by creating a suitably positioned fictitious node z_i or z_k. The potential at the fictitious node is then manipulated to fix the gradient value at z_j. To be specific, suppose that z_i and z_j are physically contingent nodes and that z_k is a fictitious node more distal from the soma than z_j. A rearrangement of the finite difference formulae for the potential gradient and cable equation at z_j yields

$$v_k = h^2 \frac{dv_j}{dt} - v_i + (h^2+2)v_j + O(h^4)\,, \qquad v_k = v_i + 2h\,\frac{\partial v}{\partial x}\bigg|_{\text{node } j} + O(h^3)\,.$$

Taken together, these two equations enable the fictitious potential v_k at z_k to be eliminated to obtain

$$\frac{2}{h}\frac{\partial v(z_j^{(-)},t)}{\partial x} = \frac{dv_j}{dt} + \beta v_j - 2\alpha v_i + O(h) \tag{8.59}$$

where the symbolism $z_j^{(-)}$ indicates that the derivative has been computed by introducing a fictitious node to the right of z_j. Equation (8.59) therefore provides a formula for the potential gradient at z_j in terms of known potentials less distal from the soma. Similarly, the formula

$$\frac{2}{h}\frac{\partial v(z_j^{(+)},t)}{\partial x} = -\frac{dv_j}{dt} + 2\alpha v_k - \beta v_j + O(h) \tag{8.60}$$

expresses the potential gradient at z_j in terms of potentials more distal from the soma. Results (8.59) and (8.60) are needed in the discussion of terminal and shared nodes but are not required for internal nodes that are now examined.

Internal nodes

Suppose that z_j is an internal node flanked by nodes z_i and z_k with z_k most distal from the soma. The cable equation at z_j, namely,

$$\frac{\partial V(z_j,t)}{\partial t} + V(z_j,t) = \frac{\partial^2 V(z_j,t)}{\partial x^2} - \frac{J_j}{g_j}$$

is replaced by a finite difference scheme centred on z_j. The resultant equation is

$$\frac{dv_j}{dt} = \alpha v_i - \beta v_j + \alpha v_k - \frac{J_j}{g_j} \tag{8.61}$$

to $O(h^2)$ where $\alpha = 1/h^2$ and $\beta = (1+2\alpha)$. If V and J are the vector of membrane potentials and electrotonic input currents respectively then, in the notation of linear algebra, this equation has form

$$I_j \frac{dV}{dt} = A_j V - 2R_j J$$

where I_j is the jth row of the $n \times n$ identity matrix I_n, R_j is the jth row of the $n \times n$ diagonal matrix R where $R_{j,j} = r_j$ and A_j is the jth row of the $n \times n$ matrix A whose entries in the jth row are all zero except $a_{j,i} = a_{j,k} = \alpha$ and $a_{j,j} = -\beta$.

Terminal nodes

Terminal nodes are more difficult to handle than internal nodes since one of z_k or z_i is either fictitious, as happens with the somal node or at a dendritic tip, or one is not a variable of the model as happens with cut and voltage clamped ends. The treatment of terminal nodes reduces to three different possibilities.

Cut end boundary condition: If z_j is a terminal node adjacent to a cut end then z_j essentially behaves like an internal node with $v_k = 0$. It follows immediately from (8.61) that

$$\frac{dv_j}{dt} = \alpha v_i - \beta v_j - \frac{J_j}{g_j} \,. \tag{8.62}$$

In the notation of linear algebra, equation (8.62) is expressed in the form

$$I_j \frac{dV}{dt} = A_j V - 2R_j J \,,$$

where I_j is the jth row of the $n \times n$ identity matrix I_n, R_j is the jth row of the $n \times n$ diagonal matrix R where $R_{j,j} = r_j$ and A_j is the jth row of the $n \times n$ matrix A whose entries in the jth row are all zero except $a_{j,j} = -\beta$ and $a_{j,i} = \alpha$. The voltage clamped boundary condition is formulated in a similar way.

Current injected boundary condition: Suppose that z_j is a dendritic terminal into which current $I_j^{(i)}$ is injected but which leaks charge to the extracellular region at a rate proportional to the potential difference between the potential v_j of the dendritic terminal and the potential of the extracellular material, taken to be zero (Ohmic leakage). If g_L is the lumped leakage conductance of the dendritic tip, then the current injected boundary condition is

$$g_j \frac{\partial v(z_j^{(-)},t)}{\partial x} + g_L v_j = -I_j^{(i)} \,. \tag{8.63}$$

When formula (8.59) is used to replace the right handed gradient in (8.63) and it is recognised that $i = j-1$, the final leakage boundary condition at node z_j takes the form

$$\frac{dv_j}{dt} = 2\alpha v_{j-1} - \beta v_j - \frac{g_L v_j + I_j^{(i)}}{g_j h} \tag{8.64}$$

to $O(h)$. In the notation of linear algebra, equation (8.64) has matrix representation

$$I_j \frac{dV}{dt} = A_j V - 2R_j J$$

where I_j is again the jth row of the $n \times n$ identity matrix I_n, R_j is the jth row of the $n \times n$ diagonal matrix R where $R_{j,j} = r_j$, the jth entry of the vector J is $I_j^{(i)}/(2h)$ and A_j is the jth row of the $n \times n$ matrix A whose entries in the jth row are all zero except $a_{j,j-1} = 2\alpha$ and $a_{j,j} = -\beta - (2g_{\text{L}}/g_j h)$.

In particular, a sealed terminal corresponds to $g_{\text{L}} = 0$ and $I_j^{(i)} = 0$.

Tree-to-soma condition: Finally, the soma (or local origin when transforming a sub-tree) behaves like a terminal node in the sense that the soma must contribute a boundary condition in order to complete the system of nodal equations. Suppose that N dendritic limbs emanate from the soma and that limb r, $(1 \leqslant r \leqslant N)$, starts with z_0 and has second node z_{k_r}. If total electrotonic current J_0 is applied to the soma, then the somal potential $v_{\text{S}} = v_0$ satisfies

$$g\left(\varepsilon \frac{dv_{\text{S}}}{dt} + v_{\text{S}}\right) = -J_0 + \sum_{r=1}^{N} g_{k_r} \frac{\partial v_{k_r}(z_0^{(+)}, t)}{\partial x}, \tag{8.65}$$

in which g, the total somal conductance, and ε are defined by the expressions

$$g = A_{\text{S}}\, g_{\text{S}} \qquad \varepsilon = \frac{g_{\text{M}} C_{\text{S}}}{g_{\text{S}} C_{\text{M}}}.$$

The finite difference formulation of the somal condition (8.65) is derived by using formula (8.60) to replace potential gradients in equation (8.65). The result is

$$g\left(\varepsilon \frac{dv_0}{dt} + v_0\right) = -J_0 + \frac{h}{2} \sum_{r=1}^{N} g_{k_r} \left[-\frac{dv_0}{dt} + 2\alpha v_{k_r} - \beta v_0\right]$$

to $O(h^2)$. After some algebraic manipulation, it can be shown that the somal condition (8.65) finally simplifies to

$$\frac{dv_0}{dt} = -\frac{2g + \beta h g_0}{2\varepsilon g + h g_0} v_0 + \frac{2\alpha h}{2\varepsilon g + h g_0} \sum_{r=1}^{N} g_{k_r} v_{k_r} - \frac{2J_0}{2\varepsilon g + h g_0}. \tag{8.66}$$

In the notation of linear algebra, the somal condition (8.66) has representation

$$I_0 \frac{dV}{dt} = A_0 V - 2R_0 J$$

where I_0 is the 0th row of the $n \times n$ identity matrix I_n, R_0 is the 0th row of the $n \times n$ diagonal matrix R where $R_{0,0} = 1/(2\varepsilon g + h g_0)$ and A_0 is the 0th row of the $n \times n$ matrix A whose entries in the 0th row are all zero except $a_{0,k_r} = 2\alpha h g_{k_r}/(2\varepsilon g + h g_0)$ and $a_{0,0} = -(2g + \beta h g_0)/(2\varepsilon g + h g_0)$.

Shared nodes

Shared nodes occur at branch points and at point of discontinuity in limb geometry (change in cross-sectional area of limb). Both are characterised by the need to conserve axial current.

Discontinuous cable geometry: Suppose that z_j is a node at which limb geometry changes discontinuously. If electrotonic current J_j is injected at z_j, then conservation of current requires that

$$-g_j \frac{\partial v(z_j^{(-)}, t)}{\partial x} - J_j + g_{j+1} \frac{\partial v(z_j^{(+)}, t)}{\partial x} = 0\,.$$

Each partial derivative is replaced by its finite difference approximation (8.59) and (8.60) respectively. After some elementary algebra, it follows that

$$\frac{dv_j}{dt} = -\beta v_j + \frac{2\alpha g_j}{g_j + g_{j+1}} v_i + \frac{2\alpha g_{j+1}}{g_j + g_{j+1}} v_{j+1} - \frac{2J_j}{h(g_j + g_{j+1})} \tag{8.67}$$

to $O(h)$. The definition of r_j allows equation (8.67) to be expressed in the more elegant form

$$\frac{dv_j}{dt} = -\beta v_j + 2\alpha r_j g_j v_i + 2\alpha r_j g_{j+1} v_{j+1} - 2r_j \frac{J_j}{h} \tag{8.68}$$

with matrix representation

$$I_j \frac{dV}{dt} = A_j V - 2R_j J$$

where I_j is the jth row of the $n \times n$ identity matrix I_n, the vector J has j^{th} component J_j/h, R_j is the jth row of the $n \times n$ diagonal matrix R where $R_{j,j} = r_j$ and A_j is the jth row of the $n \times n$ matrix A whose entries in the jth row are all zero except $a_{j,i} = 2\alpha r_j g_j$, $a_{j,j+1} = 2\alpha r_j g_{j+1}$ and $a_{j,j} = -\beta$.

Branch point condition: Suppose that z_j is a branch point at which a parent branch with penultimate node $(j-1)$ meets N child branches each beginning with z_j but such that the second node on child r is z_{k_r}, $(1 \leqslant r \leqslant N)$. If electrotonic current J_j is injected at z_j, then conservation of current at this branch point requires that

$$-g_j \frac{\partial v(z_j^{(-)}, t)}{\partial x} - J_j + \sum_{r=1}^{N} g_{k_r} \frac{\partial v_{\text{child}-r}(z_j^{(+)}, t)}{\partial x} = 0\,.$$

The partial derivatives are now replaced by their respective expressions using results (8.59) and (8.60) to obtain

$$-g_j \left[\frac{dv_j}{dt} + \beta v_j - 2\alpha v_{j-1} \right] + \sum_{r=1}^{N} g_{k_r} \left[-\frac{dv_j}{dt} + 2\alpha v_{k_r} - \beta v_j \right] - \frac{2J_j}{h} = 0$$

to $O(h)$. Bearing in mind that r_j is the sum of the g-values of all limbs impinging on z_j, this condition simplifies to

$$\frac{dv_j}{dt} = -\beta v_j + 2\alpha r_j g_j v_{j-1} + 2\alpha r_j \sum_{r=1}^{N} g_{k_r} v_{k_r} - 2r_j \frac{J_j}{h}\,. \tag{8.69}$$

In the parlance of linear algebra, equation (8.69) has representation

$$I_j \frac{dV}{dt} = A_j V - 2R_j J$$

where I_j is the jth row of the $n \times n$ identity matrix I_n, the vector J has j^{th} component J_j/h, R_j is the jth row of the $n \times n$ diagonal matrix R where $R_{j,j} = r_j$ and A_j is the jth row of the $n \times n$ matrix A whose entries in the jth row are all zero except $a_{j,j-1} = 2\alpha r_j g_j$, $a_{j,k_1} = 2\alpha r_j g_{k_1}, \ldots, a_{j,k_N} = 2\alpha r_j g_{k_N}$ and $a_{j,j} = -\beta$.

Matrix Representation of a Dendritic Tree

The previous analysis indicates how the discrete formulation of the dendritic tree produces one equation for each node. These equations can be assembled to form the $n \times n$ system of differential equations

$$\frac{dV}{dt} = AV - 2RJ\,. \tag{8.70}$$

In these equations, A is commonly called the *tree matrix* or *structure matrix* of the dendrite and is determined entirely by the dendritic geometry and terminal boundary conditions. The tree matrix is independent of tree potentials although it may be dependent on time if synaptic inputs are active. On the other hand, the input currents J may be dependent on both time and potential. The enumeration scheme, together with the use of second order central differences to approximate derivatives, ensures that A is largely tri-diagonal although it has unavoidable off-tri-diagonal entries. Branch points always generate off-tri-diagonal entries since nodes can be numbered consecutively only on one cylinder. Some obvious properties of A are now described.

(a) A branch point consisting of one parent and N child branches has $(N-1)$ pairs of off-tri-diagonal elements; a binary branch point has a pair of off-tri-diagonal entries. The examples in Figs 4 and 5b illustrate this structure. Unbranched structures (cables) always have completely tri-diagonal tree matrices (Figure 5a). Indeed, several unbranched cables may be placed together in the same matrix as illustrated in Figure 5c.

(b) Although A is not a symmetric matrix, it is guaranteed to be structurally symmetric in the sense that $a_{ij} \neq 0 \iff a_{ji} \neq 0$. This property of A rises from the reflexive nature of connectivity; if node i is connected to node j then node j is connected to node i. In particular, A has zero entries in its sub and super diagonals corresponding to connected nodes that are not numbered sequentially.

Tree matrix example

The representation of dendritic structures using second order central differences is ideally suited to numerical methods. However, it is instructive to demonstrate the above ideas with reference to a nontrivial dendritic tree whose technical details are algebraically feasibility. A Y-junction with a parent limb of length l and g-value g_P connected to a soma at one end and connected to two sealed limbs of length l and g-values g_L and g_R at its other end, provides the simplest branched structure with a disconnected section. The configuration and designation of nodes are illustrated in Fig 4.

The tree matrix for this simple but non-trivial tree may be expressed in the algebraically convenient form

$$A = \begin{bmatrix} -\beta_S & \alpha k^2 & 0 & 0 & 0 & 0 & 0 \\ \alpha & -\beta & \alpha & 0 & 0 & 0 & 0 \\ 0 & \alpha p^2 & -\beta & \alpha q^2 & 0 & \alpha r^2 & 0 \\ 0 & 0 & \alpha & -\beta & \alpha & 0 & 0 \\ 0 & 0 & 0 & 2\alpha & -\beta & 0 & 0 \\ 0 & 0 & \alpha & 0 & 0 & -\beta & \alpha \\ 0 & 0 & 0 & 0 & 0 & 2\alpha & -\beta \end{bmatrix} \tag{8.71}$$

where $g_s = g_P + g_R + g_L = r_2^{-1}$ and

$$k^2 = \frac{2hg_P}{2\varepsilon g + hg_P}\,, \quad p^2 = \frac{2g_P}{g_S}\,, \quad q^2 = \frac{2g_R}{g_S}\,, \quad r^2 = \frac{2g_L}{g_S}\,. \tag{8.72}$$

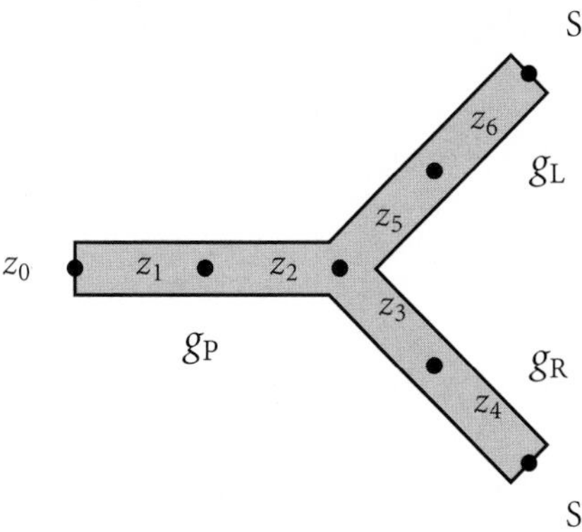

Fig. 4. The figure represents a dendritic tree consisting of a parent limb of length l and g-value g_P connected to a soma at z_0 and branching into two sealed limbs each of length l and with g-values g_R and g_L at z_2. Each limb is described by three nodes giving a total of seven nodes over the entire tree.

Nodes z_4 and z_6 are sealed terminals, z_0 is the soma-to-tree connection and z_2 is a binary branch point. Nodes z_1, z_3 and z_5 are all internal. This tree will be analysed in detailed later. In particular, subsequent analysis will use the result $p^2 + q^2 + r^2 = 2$.

Figures 5a–c illustrate other tree matrices schematically. Note the similarity between the tree matrices in Figures 5b and 5c. The off-tri-diagonal elements indicate that two unbranched segments are connected at the junction.

Formal Solution of Matrix Equations

This section seeks to establish two non-trivial properties of the tree matrix A, namely

(i) that there exist an $n \times n$ diagonal matrix S such that $S^{-1}AS$ is symmetric, that is, A is similar to a symmetric matrix;
(ii) that the eigenvalues of A are always real and negative.

Although, the first of these properties appears theoretical, it is a key ingredient of the argument used to establish the second result, and also leads eventually to the concept of the equivalent cable. The second of these results essentially states that this model of a dendritic tree predicts exponential dissipation of input currents without the oscillations that occur in weakly damped systems. In this respect, the dendritic model is consistent with observed dendritic behavior. Such a property of the model needs to be justified and cannot simply be taken for granted. Perhaps the best illustration of this idea is in mechanics/thermodynamics where it is possible to construct environments in which heat flows from cold bodies to hot bodies without the expenditure of energy. This process is entirely consistent with the undisputed principles of conservation of mass, conservation of energy, conservation of linear momentum and conservation of angular momentum, but has never been observed to occur spontaneously in reality nor is it ever likely to be observed in the future because it implies the existence of machines that convert heat into energy with perfect efficiency. The concept of entropy is introduced with the specific purpose of prohibiting spontaneous heat flow from cold to hot bodies. The motto is that although the model appears to encapsulate all the reasonable features of the real phenomenon, it can never be assumed that the model does not also admit undesirable effects that are not enjoyed by the real phenomenon.

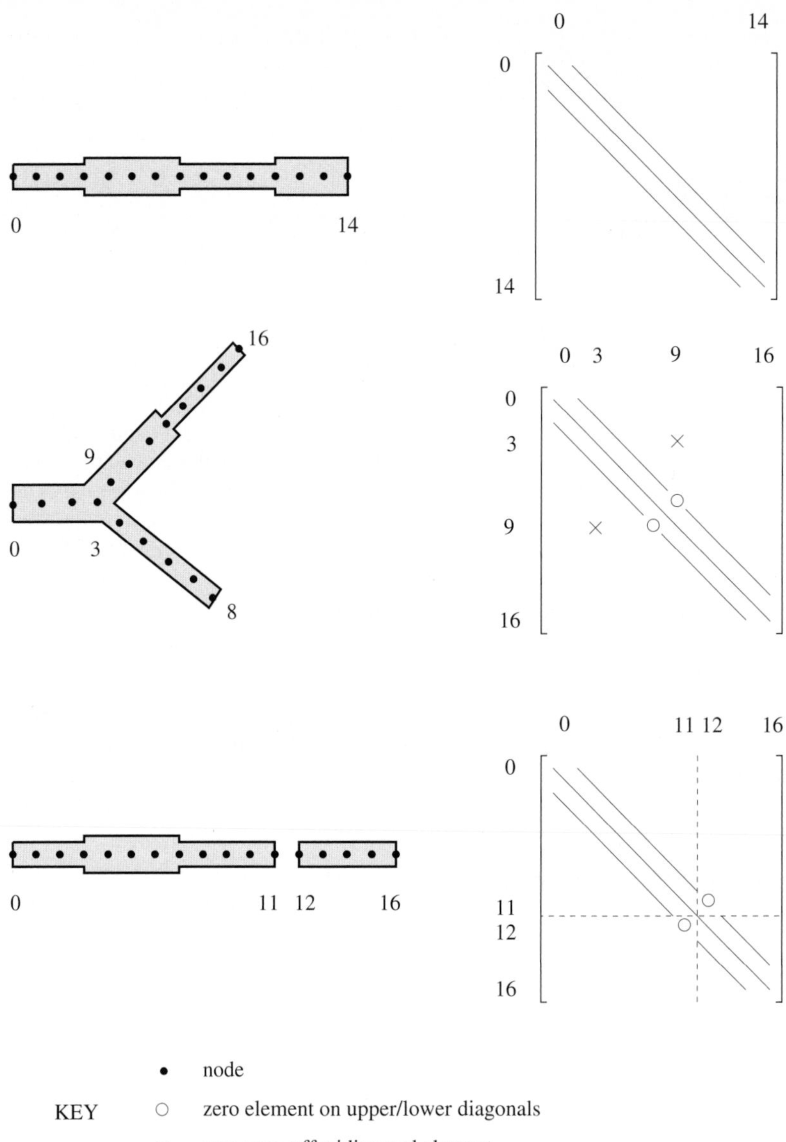

Fig. 5. Branched trees, unbranched cables and their tree matrix representation. In (a) any single unbranched structure has a tree matrix that is tri-diagonal; in (b) a singly branched tree has an almost tri-diagonal tree matrix; in (c) placing two unbranched cables are in the same matrix representation gives a tri-diagonal matrix, but the elements representing each cable are now separated by two zero elements along the diagonal.

Structure of the Tree Matrix

The structural symmetry of the tree matrix A already alluded to in the previous sections is now described in detail. It will be seen that a tree described by n nodes has a tree matrix comprising n non-zero diagonal entries and $2(n-1)$ non-zero off-diagonal entries distributed symmetrically; a total of $3n-2$ non-zero entries. While the number of diagonal entries is self-evident, the successful structural analysis of A is critically dependent on the distribution and number of its off-diagonal entries. Two contrasting discussions are presented here, both reaching the same conclusion but using different perspectives.

Local argument

Any tree node can be classified as either an internal node (a node connected to a node on either side) or a node corresponding to the soma, a branch point or a dendritic terminal. Internal nodes generate two off-diagonal entries arising from their connection to their two neighbours, while nodes corresponding to dendritic terminals generate a single off-diagonal entry in A since such nodes have only one neighbour. With this information in mind, the idea is to start at the tips of a dendritic tree and proceed towards the soma, counting the surplus/deficit in off-diagonal entries with respect to two entries per node. Suppose that N terminal branches and a parent limb meet at a branch point then one off-diagonal entry is generated for each cylinder contingent on the branch point — $(N + 1)$ off-diagonal entries in total. Thus the branch point generates a surplus of $(N - 1)$ off-diagonal entries to compensate for the deficit of N off-diagonal entries, one from each terminal limb. Thus there is a deficit of exactly one off-diagonal entry in A arising from the complete description of that branch point. In effect, the penultimate node of the parent limb of that branch point now behaves like a terminal node. This argument may be repeated until the soma is reached; each repetition revealing a deficit of exactly one off-diagonal element in A. However, the counting at the soma has presupposed that the soma is connected to a parent limb, which it is not, and therefore has over counted off-diagonal entries by one. Thus there is a deficit of two off-diagonal entries in A, that is, the tree matrix A consists of exactly $(n - 1)$ pairs of non-zero off-diagonal entries.

Global argument

Any tree with N terminals can be divided into a set of N paths, each consisting of consecutively numbered nodes. There is one path, the *soma path*, where numbering starts from "0" and ends on a dendritic terminal, plus $N - 1$ additional paths each starting with a node connected to a branch point and ending on a dendritic terminal (or a node connected to a terminal in the case of a voltage boundary condition). Observe that these paths are uniquely determined by the numbering scheme and have the property that every tree node lies on exactly one path. Each path must contribute a tri-diagonal portion to the tree matrix, plus additional off-tri-diagonal elements arising from the connection of that path to the rest of the tree.

Now consider any path starting with node p and connected to another path at the branch point described by node j then $p > j + 1$. Tree connectivity ensures that elements a_{jp} and a_{pj} of the matrix A are non-zero and off-tri-diagonal, while elements $a_{p(p-1)} = a_{(p-1)p} = 0$ since node $p - 1$, which must be a dendritic terminal, and therefore cannot be connected to node p. Since the soma path does not connect to another path, each non-somal path contributes a pair of non-zero off-tri-diagonal entries and a pair of zero elements on the sub- and super-diagonals of A, in total $N - 1$ pairs of non-zero off-tri-diagonal elements and $N - 1$ pairs of zero elements on the sub- and super-diagonals of A.

If the tree is represented by n nodes then there are $n - 2N$ nodes that lie within paths. Each such element, j say, must connect to nodes $j - 1$ and $j + 1$ and so contributes two off-diagonal elements to the jth row of A, giving a total of $2(n - 2N)$ off-diagonal elements. The off-tri-diagonal elements due to connections between paths have already been determined as $2(N - 1)$. Since the starting node, p say, of each path (including the soma path) must also be connected to node $p + 1$, there are a further N off-diagonal elements, one each in row p. Finally, each of the N terminal nodes must connect to just one other node, yielding a total of N off-diagonal elements. In total, then, there are $2(n - 2N) + 2(N - 1) + N + N = 2(n - 1)$ off-diagonal elements.

Symmetrising the Tree Matrix

The symmetrising argument consists of three observations. The first is that the structural symmetry of A guarantees that it has an even number of non-zero off-diagonal entries, while the properties of the finite difference algorithm guarantees that these entries are all positive.

Now let S be a nonsingular $n \times n$ real diagonal matrix with entry s_i in the ith row and column, then the matrix $S^{-1}AS$ has $(i,j)^{\text{th}}$ entry $(s_j/s_i)a_{i,j}$ and $(j,i)^{\text{th}}$ entry $(s_i/s_j)a_{j,i}$. Self evidently, the diagonal entries of $S^{-1}AS$ are just the diagonal entries of A. For each pair of non-zero off-diagonal entries $a_{i,j}$ and $a_{j,i}$, the values of s_i and s_j $(j \neq i)$ are chosen to satisfy

$$\frac{s_i}{s_j}a_{j,i} = \frac{s_j}{s_i}a_{i,j} \quad \rightarrow \quad \frac{s_j}{s_i} = \sqrt{\frac{a_{j,i}}{a_{j,i}}} \tag{8.73}$$

where it should be recognised that the argument of this square root is always positive. Since each row of A must contain at least one off-diagonal entry and there are exactly $(n-1)$ different pairs of off-diagonal elements in A, then equations (8.73) are a set of $(n-1)$ simultaneous equations in the n unknowns $s_0, \ldots, s_{n-1}$. Their general solution involves an arbitrary constant which may be taken to be s_0 without loss of generality. Typically s_0 is set to unity. Consequently there exists a diagonal matrix S such that $S^{-1}AS$ is a symmetric matrix, that is, the tree matrix A is similar to a symmetric matrix.

Indeed, the non-zero off-diagonal entry of $S^{-1}AS$ in the (i,j) location is $\sqrt{a_{ij}a_{ji}}$ while the symmetrising matrix has diagonal entry $s_i = \sqrt{r_0/r_i}$ in the ith row and column. Therefore, there is no practical need to construct A in a numerical procedure since $S^{-1}AS$ and S can be constructed directly.

Example of the symmetrisation process

As an example of the symmetrisation procedure, the tree matrix (8.71) may be shown to have symmetric form

$$S^{-1}AS = \begin{bmatrix} -\beta_S & \alpha k & 0 & 0 & 0 & 0 & 0 \\ \alpha k & -\beta & \alpha p & 0 & 0 & 0 & 0 \\ 0 & \alpha p & -\beta & \alpha q & 0 & \alpha r & 0 \\ 0 & 0 & \alpha q & -\beta & \sqrt{2}\alpha & 0 & 0 \\ 0 & 0 & 0 & \sqrt{2}\alpha & -\beta & 0 & 0 \\ 0 & 0 & \alpha r & 0 & 0 & -\beta & \sqrt{2}\alpha \\ 0 & 0 & 0 & 0 & 0 & \sqrt{2}\alpha & -\beta \end{bmatrix} \tag{8.74}$$

where S is the diagonal matrix

$$S = \text{diag}\left(1, \frac{1}{k}, \frac{p}{k}, \frac{p}{kq}, \frac{p\sqrt{2}}{kq}, \frac{p}{kr}, \frac{\sqrt{2}p}{kr}\right) . \tag{8.75}$$

Eigenvalues and Eigenvectors of Matrices

A non-zero vector X is an *eigenvector* of a square matrix A provided there is a scalar μ, called an *eigenvalue*, such that

$$AX = \mu X .$$

Since $(A - \mu I)X = 0$ with $X \neq 0$ then the eigenvalues of A are just the solutions of the polynomial equation $\det(A - \mu I) = 0$. Let S be a nonsingular square matrix of the same type as A then

$$\det(S^{-1}AS - \mu I) = \det\left(S^{-1}(A - \mu I)S\right) = \det S^{-1} \det(A - \mu I) \det S = \det(A - \mu I) . \tag{8.76}$$

Hence μ is an eigenvalue of A if and only if it is an eigenvalue of $S^{-1}AS$. When A is a real matrix, the polynomial equation $\det(A - \mu I) = 0$ is real and so its solutions can be either real or complex conjugate pairs. Hence real matrices need not have real eigenvalues. However, it is well known that real symmetric matrix have real eigenvalues. To appreciate this fact, let A be a real symmetric matrix with eigenvalue μ and corresponding eigenvector X so that $AX = \mu X$. Bearing in mind that A is a real matrix, the complex conjugate of $AX = \mu X$ yields $A\bar{X} = \bar{\mu}\bar{X}$ where $\bar{z}$ is the complex conjugate of z. The transpose of $A\bar{X} = \bar{\mu}\bar{X}$ gives

$$\bar{\mu}\bar{X}^{\mathrm{T}} = (A\bar{X})^{\mathrm{T}} = \bar{X}^{\mathrm{T}}A^{\mathrm{T}} = \bar{X}^{\mathrm{T}}A\,,$$

and when this equation is post multiplied by X, the result is

$$\bar{\mu}\bar{X}^{\mathrm{T}}X = \bar{X}^{\mathrm{T}}AX = \bar{X}^{\mathrm{T}}(\mu X) \quad \rightarrow \quad (\bar{\mu} - \mu)\bar{X}^{\mathrm{T}}X = 0\,.$$

Since $X \neq 0$ then $\bar{\mu} = \mu$ and so μ is real. This is a general property of real symmetric matrices.

Recall now that every tree matrix A has an associated non-singular diagonal matrix S such that $S^{-1}AS$ is symmetric. Result (8.76) indicates that A and $S^{-1}AS$ have the same eigenvalues while the result on real symmetric matrices indicates that the eigenvalues of $S^{-1}AS$ are real. Hence the eigenvalues of A are real although A is not symmetric. Furthermore, if $AX = \mu X$ then for all $0 \leqslant i < n$,

$$\sum_{j=0}^{n-1} a_{ij}x_j = \mu x_i \quad \rightarrow \quad (a_{ii} - \mu)x_i = -\sum_{j=1, j\neq i}^{n-1} a_{ij}x_j$$

and if x_i is the largest component of X then

$$|a_{ii} - \mu| \leq \sum_{j=1, j\neq i}^{n-1} |a_{ij}|\frac{|x_j|}{|x_i|} \leqslant \sum_{j=1, j\neq i}^{n-1} |a_{ij}|\,.$$

This simple result is commonly known as Gershgorin's "circle theorem". It reveals an important property of tree matrices.

Let A be a tree matrix. It is an algebraic fact that tree matrices are diagonally dominated, that is, the modulus of the each main diagonal entry exceeds the sum of the moduli of all the off-diagonal entries in the row containing that entry. The comparisons are

Type of node	Diagonal entry	Sum of moduli of off-diagonal entries
Soma	$-\dfrac{2g + h\beta g_0}{2\varepsilon g + hg_0}$	$\dfrac{2h\alpha g_0}{2\varepsilon g + hg_0}$
Internal nodes and branch points	$-(1 + 2\alpha)$	2α
Terminal nodes	$-(1 + 2\alpha)$	α

Since the diagonal entries of tree matrices A are negative without exception then their family of Gershgorin circles is contained entirely within the left half-plane. No circle contains the origin and so $\mu = 0$ is *not* an eigenvalue of A. But A is guaranteed to have real eigenvalues and so all the eigenvalues of A are real and negative. This is an important result since it will shortly be clear that this model of a dendritic tree predicts the attenuation of current input as is observed in real physical dendrites.

Solution of the Discretised Cable Equations

It has already been demonstrated that the tree matrix A has real eigenvalues, all of which are negative. It can be shown that there is a real nonsingular matrix P (the columns of P are some ordering of the eigenvectors of A) such that $P^{-1}AP = D$ where D is a real diagonal matrix containing the eigenvalues of A. Alternatively, $A = PDP^{-1}$. Using this representation of A, the original system of differential equations (8.70) becomes

$$\frac{dV}{dt} = PDP^{-1}V - 2RJ \,. \tag{8.77}$$

Let $Y = P^{-1}V$ then Y satisfies

$$\frac{dY}{dt} = DY - 2P^{-1}RJ \,. \tag{8.78}$$

Given any constant square matrix M, it can be proved formally that $d(e^{Mt})/dt = Me^{Mt}$. This idea, when applied to equation (8.78) yields

$$\frac{d(e^{-Dt}Y)}{dt} = -2e^{-Dt}P^{-1}RJ$$

which now integrates to

$$Y(t) = e^{Dt}Y(0) - 2\int_0^t e^{D(t-s)}P^{-1}RJ(s)\,ds \,. \tag{8.79}$$

In conclusion $V = PY$ satisfies

$$V(t) = Pe^{Dt}Y(0) - 2\int_0^t Pe^{D(t-s)}P^{-1}RJ(s)\,ds = Pe^{Dt}Y(0) - 2\int_0^t e^{A(t-s)}RJ(s)\,ds \,. \tag{8.80}$$

Since $D = \text{diag}\,(\mu_0, \mu_1, \ldots, \mu_{n-1})$ is a diagonal matrix then it is self evident that e^{Dt} is the diagonal matrix diag $(e^{\mu_0 t}, e^{\mu_1 t}, \ldots, e^{\mu_{n-1} t})$. The corresponding time constants for this representation of the dendritic tree are therefore $(\mu_k)^{-1}$, $0 \leqslant k < n$. Of course, the real dendritic tree has a countably infinite set of time constants. It is only the leading eigenvalues (the least negative in this case) that are adequately approximated by the finite difference representation of the tree. A finer discretisation of the dendritic tree can improve the accuracy of the matrix determination of the leading time constants both numerically and in number but still only captures a finite number of eigenvalues by contrast with the analytical analysis of a dendritic tree. Of course, it is only the leading eigenvalues that are important practically since the others decay too quickly to have any serious impact on the solution over long periods of time .

In practice, the finite difference representation of a dendritic tree provides neither a practical algorithm for finding tree time constants nor a sensible way to integrate the tree equations with respect to time, particularly if accuracy is required. The reason lies in the algebraic (powers of h or reciprocal powers of n) convergence of finite difference schemes, made worse by the reduced accuracy inherent in the treatment of branch points, points of discontinuity and terminal boundary conditions. The strength of the finite difference representation of a dendritic tree lies in the ability of the matrix A to capture the connectivity and geometrical structure of the passive dendrite, leading in turn to the construction of the equivalent cable representation of that dendrite, conditioned on the discretisation of the original dendrite. The following example illustrates the numerical deterioration as a direct result of the formulation of boundary conditions at external nodes.

Comparison of Gradient Boundary Conditions

To appreciate the difference between the algebraic and analytical implementation of boundary conditions involving gradients, it is sufficient to consider the finite difference solution of an initial boundary value problem for the one dimensional diffusion equation with a gradient boundary condition. All the essential features of a fully blown solution of a dendritic tree are embodied, in microcosm, in this simple problem. To be specific, suppose that $v(x,t)$ is the solution of the initial boundary value problem

$$\frac{\partial v}{\partial t} = \frac{\partial^2 v}{\partial x^2} + g(t,x) , \quad (t,x) \in (0,\infty) \times (0,\pi/2) \tag{8.81}$$

with initial condition $v(x,0) = v_0(x)$ and boundary conditions

$$v(0,t) = 0 , \qquad \frac{\partial v(\pi/2,t)}{\partial x} = 0 .$$

Let $[0,\pi/2]$ be subdivided by $(n+1)$ uniformly spaced nodes $x_0, x_1, \ldots, x_n$ where $x_k = kh$ and $h = \pi/(2n)$ and suppose that $v_k(t) = v(x_k,t)$ and $g_k(t) = g(x_k,t)$. Clearly $v_0(t) = 0$ for all time since this is just the boundary condition on $x = 0$. At the internal node x_k, $0 < k < n$, the second spatial derivative appearing in equation (8.81) may be written

$$\left.\frac{\partial^2 v}{\partial^2 x}\right|_{x=x_k} = \frac{v_{k+1} - 2v_k + v_{k-1}}{h^2} + O(h^2)$$

so that the differential equation at the internal nodes is approximated to order h^2 by the system of $(n-1)$ ordinary differential equations

$$\frac{dv_k}{dt} = \frac{v_{k+1} - 2v_k + v_{k-1}}{h^2} + g_k , \qquad v_k(0) \text{ given} , \qquad 0 < k < n . \tag{8.82}$$

The specification of the problem is completed by implementing the boundary condition at $x = \pi/2$. There are two quite different ways to do this.

Algebraic treatment of gradient boundary condition

One way to enforce the zero gradient condition at $x = x_n = \pi/2$ is to recognise from (8.55) that

$$\left.\frac{\partial v}{\partial x}\right|_{x=x_n} = \frac{3v_n + v_{n-2} - 4v_{n-1}}{2h} + O(h^2) = 0 .$$

Hence the zero gradient boundary condition is satisfied to $O(h^2)$ by requiring that

$$v_n = \frac{4v_{n-1} - v_{n-2}}{3} .$$

The disadvantage of this approach is that it apparently destroys the tri-diagonal structure of the finite difference scheme when applied to branch points in a dendritic tree. In fact, this is untrue — it's just that an extra step (involving the solution of a system of linear equations) is required to calculate the membrane potentials at all dendritic branch points and terminals.

Analytical treatment of gradient boundary condition

The analytical treatment of the gradient boundary condition at $x_n = \pi/2$ is based on a fictitious node $x_{n+1} = x_n + h$ exterior to the region of solution. As already described in this section, the fictitious potential v_{n+1} at this node satisfies

$$\begin{aligned} \frac{dv_n}{dt} &= \frac{v_{n+1} - 2v_n + v_{n-1}}{h^2} + g_n , \\ \left.\frac{\partial v}{\partial x}\right|_{x=x_n} &= \frac{v_{n+1} - v_{n-1}}{2h} + O(h^2) . \end{aligned}$$

When v_{n+1} is eliminated between these two equations and the gradient of v is set to zero at $x = \pi/2$, it follows easily that v_n satisfies the ordinary differential equation

$$\frac{dv_n}{dt} = \frac{2(v_{n-1} - v_n)}{h^2} + g_n ,$$

the equation being accurate to $O(h)$. The initial condition $v_n(0)$ is calculated from the initial data. The crucial point here is that the tri-diagonal form of the equations is preserved but at the cost of a boundary condition whose accuracy is an order of magnitude less than that at internal nodes. In practice, this inaccurate description of the boundary condition infects the whole numerical scheme.

Illustration These theoretical considerations are all fine and well but there is no substitute for a concrete application to appreciate the numerical properties of the two different implementations of the gradient condition.

It is demonstrated easily that the exact solution of equation (8.81) for initial condition $v_0(x) = \sin x$ and input $g(t,x) = 2bte^{\alpha t} \sin x$ is

$$v(t,x) = \begin{cases} (1+bt^2)e^{\alpha t} \sin x\,, & \alpha = -1 \\ e^{-t} \sin x + \dfrac{2b \sin x}{(1+\alpha)^2} \left((1+\alpha)te^{(1+\alpha)t} + 1 - e^{(1+\alpha)t} \right) . & \alpha \neq -1 \end{cases}$$

The exact solutions at $t = 4, \alpha = -1$ and $t = 4, \alpha = 1$ are used to check the numerical accuracy of the approximate solutions derived using the finite difference algorithm based on the central difference formulae (8.53). The two competing schemes are integrated to high temporal accuracy so that the resulting errors are due purely to spatial truncation of the second derivative while the variation between both sets of errors (algebraic and analytic) is due entirely to the different implementations of the boundary conditions; otherwise both systems of equations are identical.

Table 8.1. Comparison of errors in the finite difference solution of $u_t = u_{xx} + g$ for algebraic and analytical forms of the gradient boundary condition when $\alpha = -1$.

x value at $t = 4$	Errors for 10 nodes ($h \approx 0.1571$) Algebraic bdry cond	Analytical bdry cond	Errors for 20 nodes ($h \approx 0.0785$) Algebraic bdry cond	Analytical bdry cond
0.000	0.00000	0.00000	0.00000	0.00000
0.157	0.00065	0.00128	0.00024	0.00032
0.314	0.00126	0.00253	0.00047	0.00063
0.471	0.00182	0.00372	0.00069	0.00093
0.628	0.00231	0.00482	0.00089	0.00120
0.785	0.00268	0.00579	0.00105	0.00145
0.942	0.00293	0.00663	0.00119	0.00166
1.100	0.00304	0.00730	0.00129	0.00182
1.257	0.00300	0.00779	0.00135	0.00195
1.414	0.00280	0.00809	0.00136	0.00202
1.571	0.00243	0.00820	0.00132	0.00205

By way of contrast, the same calculations are done for $\alpha = 1$. In this case, the driving force $g(t,x)$ grows exponentially in time.

Table 8.2. Comparison of errors in the finite difference solution of $u_t = u_{xx} + g$ for algebraic and analytical forms of the gradient boundary condition when $\alpha = 1$.

x value at $t = 4$	Errors for 10 nodes ($h \approx 0.1571$) Algebraic bdry cond	Analytical bdry cond	Errors for 20 nodes ($h \approx 0.0785$) Algebraic bdry cond	Analytical bdry cond
0.000	0.00000	0.00000	0.00000	0.00000
0.157	0.17010	0.26361	0.05429	0.06590
0.314	0.33059	0.52072	0.10656	0.13018
0.471	0.47190	0.76502	0.15484	0.19125
0.628	0.58459	0.99047	0.19719	0.24761
0.785	0.65939	1.19155	0.23177	0.29788
0.942	0.68720	1.36327	0.25681	0.34081
1.100	0.65908	1.50144	0.27068	0.37535
1.257	0.56626	1.60262	0.27185	0.40065
1.414	0.39998	1.66436	0.25891	0.41608
1.571	0.15144	1.68509	0.23060	0.42127

In all circumstances, the algebraic form of the boundary condition is superior to the analytical form, although this superiority diminishes as spatial resolution improves. Recall that each algorithm is identical except in the treatment of the gradient boundary condition. It is clear that even for fine spatial resolution, the reduced error of the analytical treatment rapidly infects the entire solution. Of course, the largest errors are associated with estimates of the slowest time constants of the cable equation $u_t + u = u_{xx} + g$. Note that this equation is formally identical to $v_t = v_{xx} + G$ in which $v = e^t u$ and $G = e^t g$.

Generating Independent and Correlated Stochastic Spike Trains

To investigate the behavior of model neurons under conditions that approach that of real neurons, it is necessary to generate large numbers of spike train inputs in which the statistical characteristics of the individual spike trains can be specified and the correlation between the spike trains set at any desired strength. Large numbers of inputs are required to match the features of real neurons which receive, for example, in the case of a cortical pyramidal cell, as many as 10^4–10^5 synaptic inputs (Bernander, Koch and Usher, 1994). Large scale synaptic background activity is known to influence the spatial-temporal integration of the effects of synaptic inputs within individual neurons (Bernander, Douglas, Martin and Koch, 1991; Bernander *et al.*, 1994; Murthy and Fetz 1994; Rapp, Yarom and Segev, 1992). The ability to generate spike trains with known statistical properties will provide a necessary tool for investigating different hypotheses on how local signal processing may occur within individual neurons. The structure of the correlation between spike trains is also an important factor in determining how features of particular signals may be extracted by individual neurons or by populations of neurons (Halliday, 1998a).

Exponentially Distributed Inter-spike Intervals

Real neuronal spike trains are frequently characterised by the distribution of their inter-spike intervals which may be correlated and may also vary widely (i.e not necessarily periodic). However, many naturally occurring spike trains are realistically modelled by a point process in which the successive intervals are totally random, i.e. a Poisson process (Holden, 1976). The interval lengths of a Poisson process are characterised by, and can be generated from an exponential distribution. To appreciate this result, let $N(u, v)$ denote the number of spikes in $(u, v]$, then the Poisson process with mean spike rate M per unit time and history $\mathscr{H}_t$ satisfies

$$\begin{aligned} \text{Prob}\,(N(t, t+\tau) = 1 \mid \mathscr{H}_t) &= \tau/M + o(\tau) \\ \text{Prob}\,(N(t, t+\tau) > 1 \mid \mathscr{H}_t) &= o(\tau) \end{aligned} \tag{8.83}$$

for positive τ and all times t. Let $F(t) = \text{Prob}\,(T \leqslant t)$ where T is elapsed time since the last spike, then

$$F(t+\tau) = F(t) + \Big(1 - F(t)\Big)\frac{\tau}{M} + o(\tau)\,, \qquad F(0) = 0\,.$$

Consequently,

$$\frac{F(t+\tau) - F(t)}{\tau} = \frac{1 - F(t)}{M} + \frac{o(\tau)}{\tau}$$

and by taking limits of this equation with respect to τ as $\tau \to 0^+$, it follows that F satisfies the differential equation $\dot{F} = (1 - F)/M$ with initial condition $F(0) = 0$. It can be shown easily that $F(t)$ and its related probability density function $f(t)$ are given by

$$F(t) = 1 - e^{-t/M}\,, \qquad f(t) = \frac{dF}{dt} = \frac{1}{M}e^{-t/M}\,. \tag{8.84}$$

Furthermore, realisations of T can be obtained from $T = T(F)$, the inverse mapping of $F = F(t)$, by treating F (or equivalently, $1-F$) as a uniformly distributed random variable in $(0, 1)$. Thus for any choice of M, deviates T are constructed from uniform deviates U by the formula

$$T = -M\log(1 - F) = -M\log U\,, \qquad U \in U(0, 1)\,. \tag{8.85}$$

Correlated point processes may be described by an analysis similar to that given for the Poisson process by providing an appropriate form for equations (8.83). If spike times occur at instantaneous rate $M(s)$ at elapsed time s after the occurrence of the previous spike, then the related point process has cumulative density F and probability density f where

$$F(t) = 1 - \exp\Big(-\textstyle\int_0^t M^{-1}(s)\,ds\Big)\,, \qquad f(t) = M^{-1}(t)\exp\Big(-\textstyle\int_0^t M^{-1}(s)\,ds\Big)\,.$$

Moreover, $M(s)$ itself can have a stochastic origin in which case the entire process is said to be doubly stochastic (see Cox and Isham, 1980).

Normally Distributed Inter-spike Intervals

Weakly periodic spike trains in which the inter-spike intervals tend to be clustered about some mean value are often modelled by the normal distribution. Normally distributed inter-spike intervals are characterised by their mean and variance by contrast with inter-spike intervals generated as Poisson deviates; the latter being completely characterised by a single parameter M, their mean spike rate.

Normal deviates are usually generated using a variant of the Box-Muller (Box and Muller, 1958) algorithm commonly called Polar-Marsagalia (Marsagalia and Bray, 1964). The algorithm consists of two stages.

Stage I

If u_1 and u_2 are two independent uniform random deviates in $(0, 1)$ then

$$v_1 = 2u_1 - 1\,, \qquad v_2 = 2u_2 - 1 \tag{8.86}$$

are two independent uniformly distributed deviates in $(-1, 1)$. The pair (v_1, v_2) is accepted as seed for the second stage of the algorithm provided $w = v_1^2 + v_2^2 < 1$, otherwise it is rejected and a new pair constructed by two further drawings from the uniform random number generator $U(0, 1)$. Clearly the pair (v_1, v_2) is accepted with probability $\pi/4$ ($\approx$ 78.6 %). However, this apparently inefficient use of the uniform random number generator is generously rewarded in the second stage of the algorithm in that no computationally expensive trigonometrical calculations are needed by contrast with the Box-Muller method which requires the computation of sine and cosine functions.

Stage II

Given a pair (v_1, v_2) of random deviates such that $0 < w = v_1^2 + v_2^2 < 1$, then

$$x_1 = \mu + \sigma v_1 \sqrt{-2\log(w)/w}\,, \qquad x_2 = \mu + \sigma v_2 \sqrt{-2\log(w)/w}\,. \tag{8.87}$$

are a pair of independent normal deviates with mean μ and standard deviation σ. Code for this algorithm is contained in the example of Appendix I.

To appreciate this result, it is enough to recognise that the Jacobian of the mapping $(v_1, v_2) \to (x_1, x_2)$ defined by equations (8.87) satisfies

$$\left|\frac{\partial(v_1, v_2)}{\partial(x_1, x_2)}\right| = \left(\frac{1}{\sigma\sqrt{2}}\exp\left[-\frac{(x_1-\mu)^2}{2\sigma^2}\right]\right)\left(\frac{1}{\sigma\sqrt{2}}\exp\left[-\frac{(x_2-\mu)^2}{2\sigma^2}\right]\right). \tag{8.88}$$

Since (v_1, v_2) is uniformly distributed in the unit circle, it has joint probability density function $1/\pi$ and therefore (x_1, x_2) has joint probability density function $f(x_1, x_2)$ where

$$f(x_1, x_2) = \left(\frac{1}{\sigma\sqrt{2\pi}}\exp\left[-\frac{(x_1-\mu)^2}{2\sigma^2}\right]\right)\left(\frac{1}{\sigma\sqrt{2\pi}}\exp\left[-\frac{(x_2-\mu)^2}{2\sigma^2}\right]\right). \tag{8.89}$$

Since $f(x_1, x_2)$ is separable in (x_1, x_2) and has sample space R^2, then x_1 and x_2 are two independent normally distributed deviates with mean μ and standard deviation σ.

Uniform Deviates

The construction of exponential and normal deviates relies fundamentally on the ability to generate uniform random numbers drawn from $(0, 1)$. The vast majority of such generators rely on congruence relations of the form

$$X_{n+1} = aX_n + b \qquad (\text{mod } c)$$

where a, c are suitably chosen *positive* integers and b is a non-negative integer. The generator is "seeded" by a choice of the starting integer X_0 (the seed) but thereafter the algorithm generates a series of integers ranging from 0 to $c - 1$. The rational number $U_n = X_n/c$ is then a good approximation of a uniform deviate in $[0, 1]$.

Modern computers have a voracious appetite for uniform deviates as simulations become ever more ambitious. Primitive random number generators such as `rand( )` in the C programming language are useful for limited simulations only. For serious simulation work, professional uniform random number generators are best but it's often convenient to have access to a portable but high quality uniform random number generator. The generator advocated by Wichmann and Hill (1982) is implemented by the pseudo-code:

- choose three integers i, j and k, for example, by using a primitive random number generator such as `srand` and `rand`;

- recompute i, j and k according to the formulae

  $$i = \mathrm{mod}\,(171i, 30269) \quad j = \mathrm{mod}\,(172j, 30307) \quad k = \mathrm{mod}\,(170k, 30323)\ .$$

- now compute the rational number

  $$z = i/30269 + j/30307 + k/30323$$

  then mod $(z, 1.0)$ may be regarded as a pseudo-random number in $(0, 1)$.

This surprisingly simple algorithm (coded in the example in Appendix I) has a portfolio of approximately 2.7×10^{13} deviates and relies for its effectiveness on the fact that 30269, 30307 and 30323 are three large consecutive prime numbers. Recent research on random number generators has focussed on "lagged-Fibonacci" generators defined by the recurrence sequence

$$X_n = X_{n-r}\ (\text{binary operation})\ X_{n-s}$$

where $0 < s < r$ are the lagging parameters (integers) with $n \geqslant r$ and the binary operation referred to is addition/subtraction (mod c) (see Kloeden, Platen and Schurz, 1994; Knuth, 1997). Such recursive sequences can generate very long periods. For example, with $r = 17$, $s = 5$ and the operation of addition (mod 2^{31}), the period of the lagged-Fibonacci generator is $(2^{17} - 1)2^{31}$ (approximately 2.8×10^{14}) if at least one starting seed (17 in total) is *odd* — say $X_0 = 1$.

Correlated Spike Trains

The algorithms described so far in this section have concentrated on the generation of single spike trains of known statistical characteristics, although clearly they can be extended to the generation of arbitrary numbers of independent spike trains. However, to understand the effects of correlated spike trains on the behavior of neurons, it is necessary to be able to generate correlated spike trains whose strength of correlation can be controlled. Halliday (1998a) introduces and describes a novel procedure for generating correlated spike trains using integrate-to-threshold-and-fire type encoders.

The design of these encoders is based on a leaky integrator circuit with an incorporated threshold. Spike times are taken as the times of threshold crossing. In its general form, the integrate-to-threshold-and-fire encoder can be expressed in the differential form

$$\tau dv = G(n + y)\,dt - v\,dt \tag{8.90}$$

where v is the output of the encoder[12], n is a noise process, G (fixed at unity) is the gain of the encoder, τ (2.5×10^{-2} sec.) is its time constant and

$$y(t) = A \sum_i \left(H(t - t_i) - H(t - t_i - a)\right)\ . \tag{8.91}$$

The function $y(t)$, referred to as the correlating signal, is expressed in terms of the Heaviside function H and describes a train of pulses of amplitude A and duration a initiated at times t_i which may be generated deterministically or stochastically. Encoders of this type form the basic building block for generating temporally correlated spike trains. For a given noise process, the strength of correlation is governed largely by the product $A\,a$. In practice, when generating large numbers of inputs to a dendritic tree, some of which are correlated, many encoders are required. A subset of correlated inputs is generated by feeding a selected group of these encoders with the common signal y.

[12]Although not explicit, encoders based on leaky integrators incorporate exponentially decaying memory. Other neurophysiological applications may require different memory functions for the encoder. For example, encoders based on a Weibull density $\varkappa t^{\varkappa-1}$ lead to super-exponential decay typified by $\exp(-t^{\varkappa})$.

There are two cases to consider depending on the nature of the noise input to each encoder. A pseudo-white noise input stochastically fixes the level of the white noise but maintains that level for a fixed interval of time. Thus pseudo-white noise is a piecewise differentiable function. By contrast, white noise is nowhere differentiable. For a given choice of mean and variance, the behavior of encoders is markedly different for the two types of inputs.

Pseudo-white noise input

A pseudo-white noise process is one in which Gaussian noise is generated at discrete points in time and held uniform over the time interval between these points. White noise produced in this way essentially behaves like a function of bounded variation. For pseudo-white noise, equation (8.90) may be represented meaningfully in the more familiar form

$$\tau \frac{dv}{dt} = G(n+y) - v = Gx - v \,, \qquad v(0) = 0 \,, \tag{8.92}$$

and can be integrated numerically using the conventional backward Euler algorithm

$$v_{k+1} = \frac{\tau v_k + G x_{k+1} h}{\tau + h} \,, \qquad x_k = n_k + y_k \,. \tag{8.93}$$

In this scheme, n_k is the noise, y_k is the correlating signal and v_k is the solution of (8.92) at time t_k. The projected solution at the next time step $t_{k+1} = t_k + h$ is v_{k+1}, where h (10^{-3} sec., Halliday, 1998a) is the duration of the time step. After each time step, the value of v_{k+1} at time t_{k+1} is compared with the constant threshold v_{th} (set at unity). If v_{k+1} exceeds the threshold, an output spike is generated at time

$$t_k + \frac{v_{\text{th}} - v_k}{v_{k+1} - v_k} \tag{8.94}$$

and the encoder value is reset to $v_{k+1} - v_{\text{th}}$ at time t_{k+1}. Of course, this new initial value depends on h, the step length, but since the purpose of the encoder is to fire spikes, subtleties of this nature are ignored since they have a negligible effect on the statistics of these spike trains. Fundamentally, this encoder presupposes that v is a differentiable function of t except possibly at the points $t_1, t_2, \ldots$. In this sense, the operating characteristics of the encoder are inextricably linked to the choice of h used in the numerical integration of the algorithm. The efficacy of this class of encoder lies in its ease of implementation and the fact that it generates spike trains with good numerical efficiency and realistic neurobiological properties.

In summary, correlated dendritic inputs may be created from a family of identical encoders whose individual members supply input at a selected location (synapse) on the model dendrite. Each encoder receives two types of input; one that is a pseudo-white noise process that is independent of all the other encoders and a second type that is a correlating signal common to some encoders, but not necessarily all encoders. In the absence of the correlating signal, each encoder will generate a spike train whose characteristics are determined by the properties of the pseudo-white noise input. Table 8.3, taken from Halliday (1998a), gives some guidance as to the selection of parameters for the pseudo-white noise component of encoder input and the frequency and coefficient of variation of the corresponding spike train output.

By means of a framework of common inputs to an ensemble of encoders, it is possible to produce temporally correlated outputs from these encoders. These outputs can be used to provide correlated inputs at selected synapses on a dendrite. Encoders that are fed by that signal and are close to threshold, can be selectively triggered by adjusting the strength of the correlating signal. Figure 6 illustrates a family of n encoders fed with the common input shown as a sequence of pulses. Each pulse increments the independent inputs causing some encoders to fire synchronously. By adjusting the properties of the

Table 8.3. An example of some choices for the parameters of the pseudo-white noise input and the corresponding properties of the output spike train with encoder gain $G = 1$, encoder threshold $\nu_{th} = 1$ and encoder time constant $\tau = 0.025$.

Pseudo-white noise input	Mean	1.020	1.015	1.269	0.892
	Std. Dev.	0.065	0.150	0.307	6.200
Output spike train	Spikes/sec	10	10	25	32
	CoV	0.1	0.2	0.1	1.0

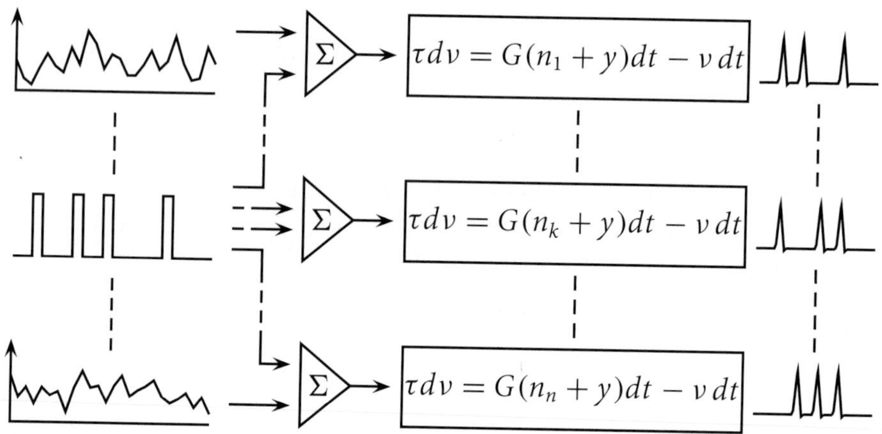

Fig. 6. Diagrammatic representation of n encoders, each receiving noise that is uncorrelated with itself and with that of the other encoders. In this example, each encoder also receives a sequence of pulses common to all encoders.

common and independent components comprising the total input to each encoder, and the number of encoders that are to receive common inputs of a prescribed class, correlated spike trains with a desired correlation structure can be generated. Further details on choices of parameters are given in Halliday (1998a).

White noise input To appreciate the difference between pseudo-white noise and white noise, it is convenient to recognise that equation (8.90) has symbolic solution

$$\nu(t) = \nu(0)e^{-t/\tau} + \frac{G}{\tau}\int_0^t n(s)e^{-(t-s)/\tau}\, ds + \frac{G}{\tau}\int_0^t y(s)e^{-(t-s)/\tau}\, ds\,. \qquad (8.95)$$

When n and y are functions of bounded variation, the integrals in (8.95) may be interpreted in the sense of Riemann integration. Specifically, the value of the integral is independent of the limiting procedure. On the other hand, when n is white noise, i.e. is not of bounded variation, then the value of the first integral depends critically on the limiting procedure. For example, the limiting procedure based on the midpoint of intervals defines the *Stratonovich* integral, while that based on the left hand endpoint of intervals defines the *Ito* integral (see Kloeden and Platen, 1995). The latter is often preferred simply because it is not predictive. When n is white noise, ν is continuous everywhere and differentiable nowhere. Consequently, equation (8.92) makes no sense and the backward Euler scheme (8.93) is invalid because it implicitly assumes that ν is a differentiable function.

Suppose that the encoder is driven by white noise of mean a and standard deviation b, then $n\,dt = a\,dt + b\,dW$ where dW is the differential of the Weiner process $W(t)$, defined as a continuous standard Gaussian process with independent increments such that

$$W(0) = 0 \qquad \text{with probability one}$$

$$E[W(t)] = 0, \qquad \text{Var}\,[W(t) - W(s)] = t - s\,, \quad 0 \leqslant s \leqslant t$$

where $E[X]$ and Var $[X]$ denote respectively the expected value and variance of the random variable X. Thus $v(t)$, the state of the encoder, satisfies the stochastic differential equation

$$\tau dv = G(a\,dt + b\,dW + y\,dt) - v\,dt \tag{8.96}$$

and may be determined by the numerical integration of (8.96) using the stochastic backward Euler algorithm

$$v_{k+1} = \frac{\tau v_k + Gh(a + y_{k+1}) + bGdW_k}{\tau + h}\,. \tag{8.97}$$

Table 8.4 provides a comparison[13] of the output spike rate and coefficient of variation when the input to the encoder is a white noise process as opposed to the pseudo-white noise process in Table 8.3. The specification of the noise is now independent of the step size h. Consequently the characteristics of the output spike train are fashioned by the mean and variance of the white noise input by contrast with the three parameters required for pseudo-white noise.

Table 8.4. An example of some choices for the parameters of the white noise input and the corresponding properties of the output spike train when encoder gain $G = 1$, encoder threshold $v_{\text{th}} = 1$ and encoder time constant $\tau = 0.025$. This table, when compared to Table 8.3, demonstrates the marked difference between the effects of pseudo-white and white noise of given mean and variance.

White noise input	Mean	1.020	1.015	1.269	0.892
	Std. Dev.	0.065	0.150	0.307	6.200
Output spike train	Spikes/sec	21	33	62	790
	CoV	0.6	0.8	1.2	5.4

In any event, the solution (8.96) is a continuous random variable in time. Without evaluating $v(t)$, some of its important statistical characteristics can be extracted.

Features of the Encoder Response to White Noise Input

It has already been observed that the stochastic equation $\tau dv = Gx(t)\,dt - v\,dt$ has symbolic solution (see 8.95)

$$v(t) = v(0)e^{-t/\tau} + \frac{G}{\tau}\int_0^t x(s)e^{-(t-s)/\tau}\,ds \tag{8.98}$$

where $x = n + y$. Suppose that $v(0)$, the initial value of v and $x(s)$, the random input at time s are uncorrelated random variables for all times s. By taking the expected value of equation (8.98), it follows immediately that $E[v(t)]$, $E[v(0)]$ and $E[x(s)]$ satisfy

$$E[v(t)] = E[v(0)]e^{-t/\tau} + \frac{G}{\tau}\int_0^t E[x(s)]e^{-(t-s)/\tau}\,ds\,. \tag{8.99}$$

[13] Appendix I gives a C program to generate pseudo-white noise and white noise driven spike trains. Incorporated in the program is code to generate uniform and normal deviates.

The linearity of expression (8.98) now ensures that

$$\nu(t) - E[\nu(t)] = \Big(\nu(0) - E[\nu(0)]\Big)e^{-t/\tau} + \frac{G}{\tau}\int_0^t \Big(x(s) - E[x(s)]\Big)e^{-(t-s)/\tau}\,ds\,. \quad (8.100)$$

The variance of ν therefore satisfies

$$\begin{aligned}\mathrm{Var}\,[\nu(t)] &= \mathrm{Var}\,[\nu(0)]e^{-2t/\tau} \\ &\quad + \frac{2G}{\tau}\int_0^t E\Big[\Big(x(s) - E[x(s)]\Big)\Big(\nu(0) - E[\nu(0)]\Big)\Big]e^{-(2t-s)/\tau}\,ds \\ &\quad + \frac{G^2}{\tau^2}\int_0^t\int_0^t E\Big[\Big(x(s) - E[x(s)]\Big)\Big(x(u) - E[x(u)]\Big)\Big]e^{-(2t-s-u)/\tau}\,ds\,du\,.\end{aligned} \quad (8.101)$$

Since $\nu(0)$ and $x(s)$ are uncorrelated

$$E\Big[\Big(x(s) - E[x(s)]\Big)\Big(\nu(0) - E[\nu(0)]\Big)\Big] = 0\,, \qquad \forall s > 0\,. \quad (8.102)$$

If the covariance between $x(s)$ and $x(u)$ is denoted by $\mathrm{Cov}\,[x(s), x(u)]$ and defined by

$$\mathrm{Cov}\,[x(s), x(u)] = E\Big[\Big(x(s) - E[x(s)]\Big)\Big(x(u) - E[x(u)]\Big)\Big]\,,$$

then in view of result (8.102), equation (8.101) simplifies to

$$\mathrm{Var}\,[\nu(t)] = \mathrm{Var}\,[\nu(0)]e^{-2t/\tau} + \frac{G^2}{\tau^2}\int_0^t\int_0^t \mathrm{Cov}\,[x(s), x(u)]e^{-(2t-s-u)/\tau}\,ds\,du\,. \quad (8.103)$$

In practice, $x(s)$ is the sum of a white noise input of mean a (constant) and standard deviation b (constant) and correlating signal $y(s)$. Hence

$$E[x(s)] = a + E[y(s)]\,, \qquad \mathrm{Cov}\,[x(s), x(u)] = b^2\delta(s-u) + \mathrm{Cov}\,[y(s), y(u)]\,. \quad (8.104)$$

Thus equations (8.99) and (8.103) yield expressions for the mean and variance of the random variable ν at any time t. Therefore given sufficient information to specify $E[x(s)]$ and $\mathrm{Cov}\,[x(s), x(u)]$ in equation (8.104), then the first two moments of the encoder output at any time prior to firing are determined. Since Gaussian deviates are completely specified by their mean and variance, one perception of $\nu(t)$ is that it is approximately a normal deviate whose mean and variance at time t are determined by the solution of equations (8.99) and (8.103). For example, in the absence of the correlating signal, the mean and variance of $\nu(t)$ are respectively

$$\begin{aligned} E[\nu(t)] &= Ga(1 - e^{-t/\tau}) + E[\nu(0)]e^{-t/\tau}\,, \\ \mathrm{Var}\,[\nu(t)] &= \frac{G^2b^2}{2\tau}(1 - e^{-2t/\tau}) + \mathrm{Var}\,[\nu(0)]e^{-2t/\tau}\,. \end{aligned} \quad (8.105)$$

Examples

The following examples illustrate (1) the procedure used to estimate the strength of correlation of a sample of weakly correlated spike trains, and (2) demonstrate the powerful effect that a weakly correlated small percentage of the total synaptic input to a model neuron has on the timing of output spikes from the neuron.

Example 1 The sample of weakly correlated spike trains used in the first example was generated by the procedure introduced in Halliday (1998a), described above, and illustrated in Fig. 6. The independent noise inputs to 100 encoders were adjusted so that each encoder generated a spike train with a mean rate centred on 12 Hz. The correlating signal, $y(t)$, common to the encoders consisted of a pulse sequence with a mean rate centred on 25 Hz. Since the dominant frequency component of the common input to the sample of encoders is centred on 25 Hz, on theoretical grounds, one would expect that the coherence between any pair of spike trains generated by the encoders would have a peak centred about the frequency of this common input (see Rosenberg, Halliday, Breeze and Conway, 1998). A sample of 100 spike trains with the characteristics determined by the combined noise and correlating inputs to the encoders, each of 100 s, was generated.

Fig. 7a shows the estimated coherence between an arbitrarly selected pair of spike trains generated by the encoders. Clearly a sample of duration 100 s is not sufficient to detect any correlation between these signals. However, the pooled coherence (refer to Chapter 18, and Amjad, Halliday, Rosenberg and Conway, 1997), shown in Fig 7b, estimated from twenty pairs of encoder outputs gives small but significant values centred on the known dominant frequency of the correlating signal. The absence of a detectable coherence from individual pairs of spike trains (Fig. 7a), coupled with the small peak value of the pooled coherence (0.009 in Fig.7b), indicates that the strength of correlation between the encoder generated spike trains is extremely weak. Details of the choice of parameters for generating the correlated spike trains of any desired strength of correlation and their analysis by a pooled coherence estimate are given in Halliday (1998a).

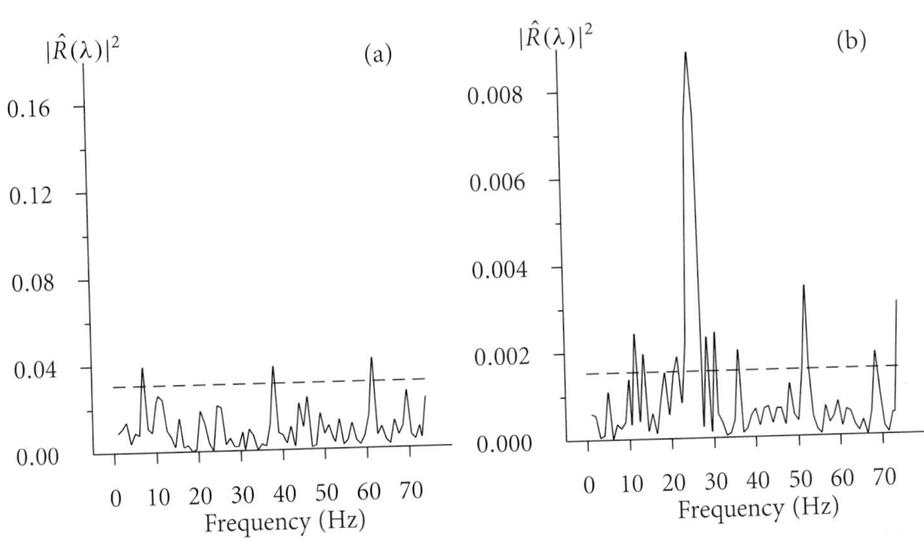

Fig. 7. (a) Estimated coherence $|\hat{R}(\lambda)|^2$ between two arbitrarily selected encoder generated spike trains from a sample of 100 weakly correlated processes, and (b) Pooled coherence estimated from 20 pairs of spike trains from the sample of 100. The mean and standard deviation of the independent pseudo-white noise input $\mu = 1.247$ and $\sigma = 0.306$, whereas the correlating signal $y(t)$ had amplitude $A = 0.425$ with $a = 0.002$ and period 40 ms. The dashed horizontal line represents the upper level of an approximate 95 % confidence interval assuming that the two processes are independent.

Example 2

The second example demonstrates that, although the correlation between spike trains may be weak, the effect that this correlation has on the timing of spike outputs may, nevertheless, be profound. A two cell model consisting of identical compartmental models of motoneurons that share a percentage of their synaptic input is used to illustrate the effect of weakly correlated signals on the timing of output spikes from these neurons. The coherence between the output spike trains from the two neurons is used to provide an indirect measure of the effect of correlated inputs on the timing of spike outputs. It has been demonstrated both theoretically (Rosenberg *et al.*, 1998) and in practice (Farmer, Bremner,

Halliday, Rosenberg and Stephens, 1993) that the coherence between two spike trains from neurons known to receive common inputs reflects the frequency content of the common inputs. Since the two model neurons are identical, the effect of correlating the input spike trains on the timing of output spikes from one neuron will mirror those from the other. The coherence between the output spike trains can then be used to assess how changes in the correlation between common inputs will effect the timing of output spikes, through the identification of the frequency content of the common inputs.

Two cases are examined when 5 % of the total synaptic input is common to the two model neurons. In the first case the common inputs are uncorrelated, whereas in the second they are weakly correlated.

Each model neuron receives 996 inputs, distributed uniformly over the cell, giving rise to a total synaptic input of 31 872 EPSPs/sec. In the absence of common inputs, each cell will discharge at approximately 12 spikes/s. When 5 % of the inputs to the cells are made common, where each common input is driven by a 25 Hz signal, the overall rate of synaptic input to each cell is adjusted to remain at 31 872 EPSPs/sec. The autospectrum of the output from one model neuron, shown in Fig. 8a, has a dominant peak at approximately 12 Hz corresponding to the mean rate of discharge of the cell.

When the common inputs are uncorrelated, the estimated coherence between the two output spike trains (Fig. 8b), based on a 100 s sample, is not significant, suggesting that an uncorrelated 5 % common input may not influence the timing of output spikes from these neurons. In the second case the 25 Hz common inputs are weakly correlated at a peak value equal to that of the pooled coherence shown in Fig. 7b. The coherence between the output spike trains, shown in Fig. 8d, now has a significant peak centred about 25 Hz — the frequency of the common inputs; the mean rate of the neuron, however, remains unchanged as indicated by its autospectrum (Fig. 8c). Although the peak value of the pooled coherence for the common inputs is only 0.009 — indicating weakly correlated common inputs — the coherence between the output spike trains is approximately 20 times greater. Simply by weakly correlating 5 % of the synaptic input to a neuron a significant effect on the timing of its output spikes is produced. The effects of weakly correlated synaptic inputs has been examined more fully in Halliday (1998b).

Equivalent Cable Construction

The idea of "equivalent structure" is intuitively understood but often not precisely defined. The object of this section is to give a definition of equivalence in the context of neuronal modelling that is both consistent with common usage and also is mathematically precise.

Recall that any model of a concrete object, in this case a dendritic tree, is an abstract object whose description is taken by us to be a description of the concrete object. Importantly, the abstract object is entirely determined by its definition and that the concrete object, by contrast, is never susceptible to an exhaustive description. One says that the abstract object is a model of the concrete object when the definition of the former is taken for a representation of the latter.

In the context of neuronal modelling, the concrete structure is a dendritic tree and soma, whereas the abstract object is simply a collection of connected cylinders. Clearly these two objects can never be truly equivalent in the dictionary sense of equivalence. Even as precise geometry of the original dendrite is approached in the limit, the electrical properties may still not be matched between the dendrite and its model representation.

Various degrees of equivalence can be associated with the preservation of a range of features of the concrete object in the abstract object. Examples of such features are total dendritic length (as opposed to maximum soma to tip length), total dendritic membrane area, the values of electrical parameters, etc. Therefore equivalence is about levels of information preservation between the concrete and abstract objects while mathematical equivalence is

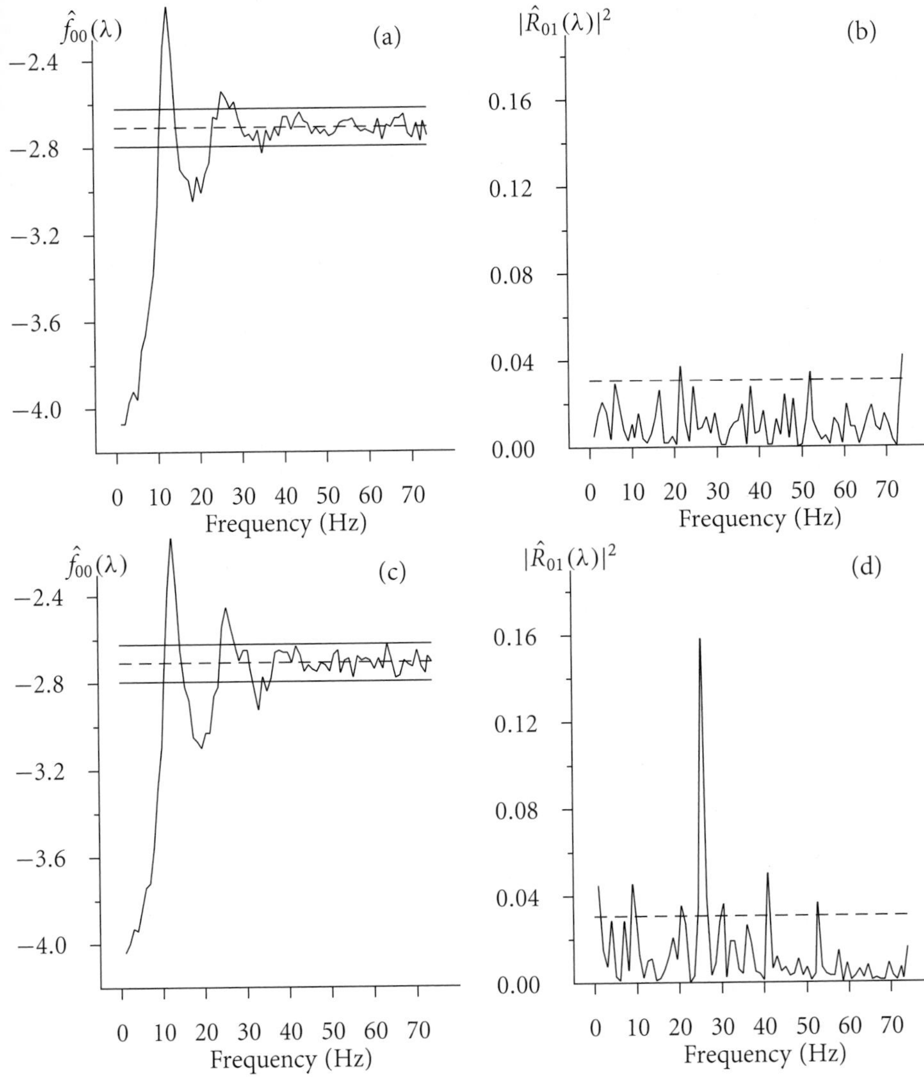

Fig. 8. (a) Estimated autospectrum, $\hat{f}_{00}(\lambda)$, of a spike train generated by one of two model neurons when their common inputs are uncorrelated while (c) is the same autospectrum for weakly correlated common inputs. (b) Estimated coherence, $|\hat{R}_{01}(\lambda)|^2$, between the output spike trains of the two neurons for uncorrelated common inputs to the neurons while (d) is the same coherence for weakly correlated common inputs. The horizontal dashed lines in (b) and (d) represent the upper level of an approximate 95 % confidence interval under the assumption that the two processes are independent. The dashed and solid horizontal lines respectively in (a) and (c) represent the $P/2\pi$ level where P is the mean rate of the process and the approximate 95 % confidence interval under the assumption that the spike train was generated by a Poisson process.

about precise information preservation between respective models. For example, an arbitrarily branched multi-cylinder model can be replaced by an unbranched multi-cylinder model that is absolutely equivalent to it in the mathematical (and dictionary) sense.

A Brief History of Equivalent Models

All cable models are inspired by the success of Rall's original equivalent cylinder (Rall, 1962a, 1962b) which gave insight into the role of passive dendrites in neuron function, and

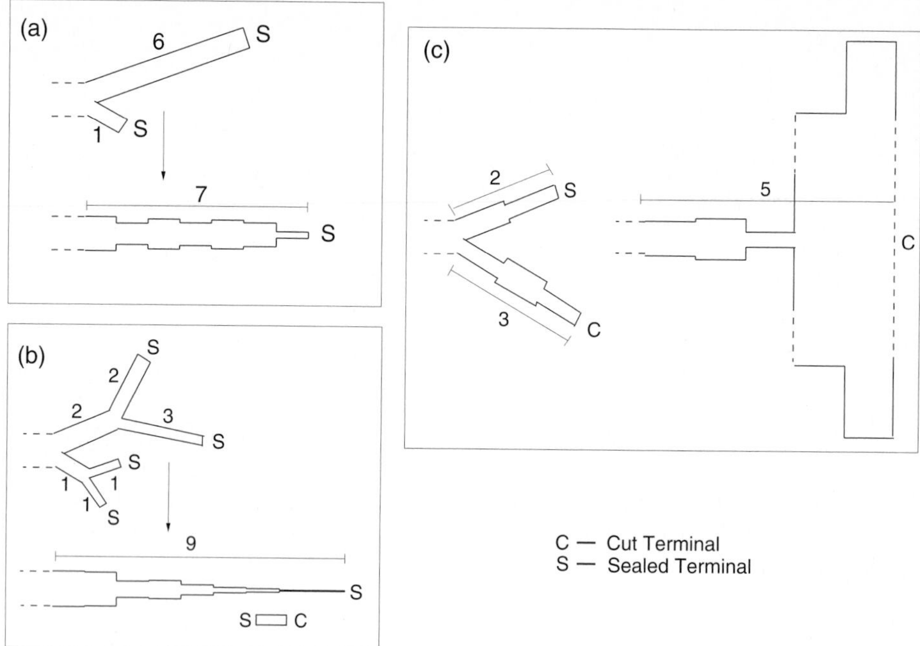

Fig. 9. Diagrammatic representation of some equivalent cables. (a) A simple non-degenerate Y-junction with one short limb and one long limb, and both terminals sealed. (b) A tree with two orders of branching and all terminals sealed. The Equivalent cable is generated after three Y-junctions are transformed. (c) A Y-junction with one sealed and one cut terminal. The cut terminal induces the large jump in diameters indicated by dotted lines.

allowed the estimation of specific electrical parameters. The restricted tree geometries for which this result is valid have prompted several efforts to extend Rall's result.

Rall (1962a) extended the cylinder model to include tapering tree geometry by introducing the idea of fractional orders of branching. However, the concept is clearly mathematical and divorced from physiological reality. On the other hand, Burke (1997), Clements and Redman (1989) and Fleshman, Segev and Burke (1988) have derived empirical equivalent cables which do not impose restrictions on tree geometry and have proven more successful in use. For these cables, equivalence is measured by their ability to approximate somatic voltage transients when a current pulse is injected at the soma. Analysis of such transients enables improved estimates of passive electrical parameters such as axial resistivity, membrane resistivity, and effective electrotonic length of the dendritic tree to be made (Rall *et al.*, 1992; Burke, Fyffe and Moschovakis, 1994). These cables have also been used to dynamically reduce sections of complex tress to improve the efficiency of computer modelling of neurons (e.g., see Manor, Gonczarowski and Segev, 1991).

The two most familiar models are Rall's Equivalent Cylinder and the Lambda Cable. Significantly, however, none of these previous models are fully equivalent (in a mathematical sense) to the original representation of the dendritic structure, simply because each fails to preserve total dendritic electrotonic length. Configurations of inputs on the original representation cannot be reconstructed from those on the Rall Equivalent Cylinder or the Lambda Cable. Different configurations of inputs on the original representation can give rise to the same configuration of inputs on these equivalent models, which therefore do not contain the information necessary for the construction of a unique configuration of inputs on the original representation that give the same effect at the soma. Only by preserving total dendritic electrotonic length can a unique relationship between tree and cable be established.

Overview of Equivalent Cables

Two model representations are said to be mathematically equivalent or just *equivalent* if any configuration of inputs on one structure can be associated uniquely with a configuration of inputs on the second structure and vice versa, such that the responses at the soma in both cases are identical. In mathematical terms, this association is called a *mapping*. The uniqueness requirement ensures that the mapping is injective while the second requirement guarantees that the mapping is surjective. Mappings that are both injective and surjective are called bijective mappings. Under this definition, model properties describing geometrical structure, boundary conditions and electrical activity will be preserved; all characterisable phenomena on one model are reproducible exactly in any equivalent model.

Any dendritic tree model formed from multiple uniform segments, each described by the linear cable equation, and each with electrotonic length a multiple of some quantum length l, may be transformed to its equivalent cable provided (1) the membrane time constant $\tau = C_M/g_M$ is a universal constant over the entire tree, and (2) each terminal satisfies either a current injection or a cut end boundary condition. The soma, which is the point with respect to which the cable is generated (the origin), doesn't influence the equivalent cable structure and may take any boundary condition.

The equivalent cable preserves total electrotonic length and may consist of many distinct sections, only one of which (called the *connected section*) is attached to the soma of the original tree. The remaining sections are all *disconnected sections*, are not attached to the soma, and therefore define electrical activity over the tree that will not influence the soma.

Basic branching structure

The basic geometrical unit of dendritic construction is the simple Y-junction comprising two daughter branches arising from a single parent branch.

The equivalent cable for a Y-junction contains a connected section plus at most one disconnected section. A Y-junction is classified as *degenerate* if its equivalent cable contains a disconnected section, otherwise it is *non-degenerate*. This degeneracy can be associated with repeated eigenvalues in the tree matrix representation.

Any multi-branched dendritic structure can be constructed from Y-junctions. The geometrical complexity of any dendrite can therefore be associated with the number of basic geometric units, i.e. Y-junctions, required to build the dendrite. An equivalent model can be generated by collapsing any terminal Y-junction. This new model is equivalent to the original dendrite but will have reduced geometrical complexity since one branch point has been removed. However, this reduction is achieved at the expense of a more complex input structure. Fig. 10 illustrates the reduction in geometrical complexity and the corresponding increase in the complexity of the input current structure. By continuing the process of selectively reducing terminal Y-junctions, a hierarchy of equivalent models can be generated. The process terminates with a final unbranched equivalent model or equivalent cable.

Examples of an Equivalent Cable

Fig. 9 illustrates three simple dendritic trees and their equivalent cables.

It is instructive to examine in detail the properties of the equivalent representation of a simple Y-junction with limbs of equal length (Fig. 10), partly because of its native simplicity and partly because such junctions turn out to be commonplace in the reduction of general dendritic trees to their equivalent cable.

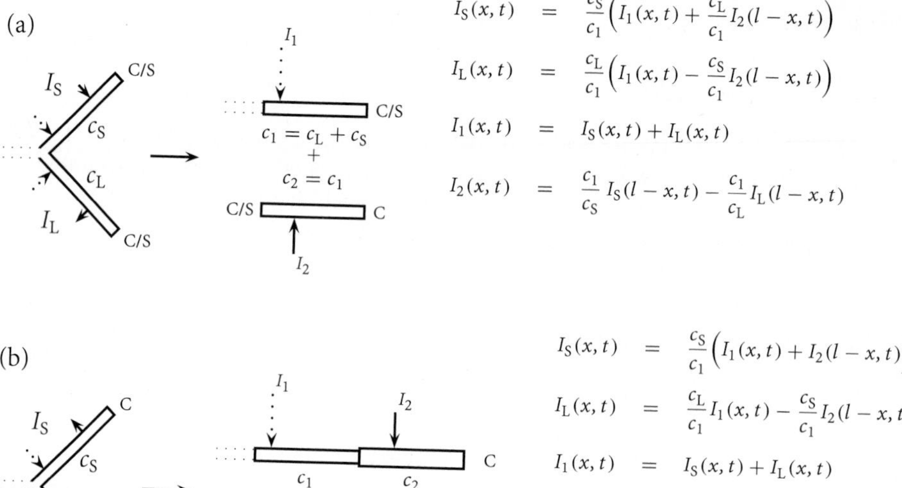

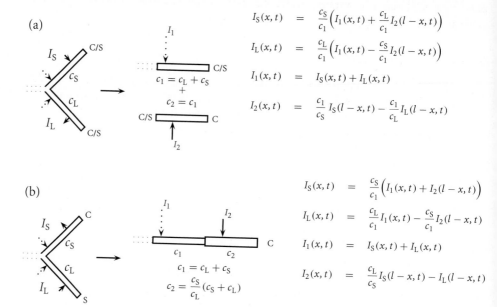

Fig. 10. Construction of an equivalent cable for simple uniform Y-junctions with limbs of equal length. The notation "C/S" refers to cut/sealed boundary conditions respectively. In (a) both boundary conditions are the same whereas in (b) they are different. The equations in (a) and (b) give the bijective mappings between the tree and equivalent cylinder. Arrows directed into limbs are excitatory while those directed away from limbs are inhibitory.

In all circumstances, the equivalent cable consists of an unbranched piecewise uniform cable with sections of known diameter and with total length that of the original dendrite, together with a bijective mapping between the dendritic tree and the unbranched equivalent cable. In the example of figure (10ab), the equivalent cable consists of two sections, one of which is connected to the parent dendrite while the other may either be connected to this section or be entirely disconnected from it depending on the nature of the terminal boundary conditions.

Let $V_S(x,t)$ and $V_L(x,t)$ be the membrane potentials in the respective limbs of the Y-junction illustrated in figure 10 and let $I_S(x,t)$ and $I_L(x,t)$ be the corresponding current inputs. Continuity of membrane potential and conservation of current at the junction demand respectively that $V_S(x,t)$ and $V_L(x,t)$ satisfy the boundary conditions

$$
\begin{aligned}
& V_S(0,t) = V_L(0,t) = V_P(L,t)\,, && \text{(Continuity of potential)} \\
& c_S \frac{\partial V_S(0,t)}{\partial x} + c_L \frac{\partial V_L(0,t)}{\partial x} = c_P \frac{\partial V_P(L,t)}{\partial x} && \text{(Conservation of current)}
\end{aligned}
\tag{8.106}
$$

where $V_P(x,t)$ is the potential in the parent branch and L is its length. By an analysis similar to that in the section on the Rall equivalent cylinder (p. 224), it can be demonstrated that the potentials

$$
\left.
\begin{aligned}
\psi_1(x,t) &= \frac{c_S V_S(x,t) + c_L V_L(x,t)}{c_S + c_L} \\
\psi_2(x,t) &= V_S(x,t) - V_L(x,t)
\end{aligned}
\right\}
\qquad 0 < x < l\,, \quad t > 0\,,
\tag{8.107}
$$

are cable solutions. The continuity of potential in boundary condition (8.106a) implies that $\psi_1(0,t) = V_P(L,t)$ whereas conservation of current in (8.106b) yields

$$(c_S + c_L)\frac{\partial \psi_1(0,t)}{\partial x} = c_S \frac{\partial V_S(0,t)}{\partial x} + c_L \frac{\partial V_L(0,t)}{\partial x} = c_P \frac{\partial V_P(L,t)}{\partial x} .$$

Hence $\psi_1(x,t)$ is the potential in a cable that connects to the parent branch in the sense that it preserves both continuity of membrane potential and conservation of current provided this cable is chosen to have $c_1 = c_S + c_L$. Of course, this is just the familiar Rall result in another guise. Thus $\psi_1(x,t)$, $0 \leqslant x \leqslant l$, is the potential of Rall's equivalent cylinder for this Y-junction. Moreover, it should be observed that $\psi_2(x,t)$ is also a solution of a cable equation satisfying the cut boundary condition $\psi_2(0,t) = 0$ in view of the continuity of V_S and V_L at $x = 0$. The previous remarks are completely general for this dendrite but still insufficient to specify the equivalent cable. To obtain the fully equivalent cable, it is necessary to incorporates the terminal boundary conditions into the construction process. Two possibilities exist here: either both tips of the Y-junction satisfy different terminal boundary conditions as in figure 10b or they share the same terminal boundary condition as in figure 10a.

Whenever the dendritic terminals are both cut or both sealed, ψ_1 and ψ_2 are respectively both cut or both sealed at $x = l$. Hence ψ_1 and ψ_2 are two complete and independent cable solutions and therefore represent two separate cables. However, only ψ_1 is connected to the parent branch because it alone satisfies continuity of potential and conservation of current with the parent branch. In this case, ψ_2 is a disconnected section. Clearly the equivalent cable requires both the connected and disconnected sections to resolve potentials everywhere on the original Y-junction.

Alternatively when the dendritic terminals satisfy different boundary conditions, ψ_1 and ψ_2 cannot individually terminate at $x = l$, but $\psi_1(x,t)$ (at $x = l$) can be connected to $\psi_2(l - x, t)$ (at $x = 0$) in a way that preserves continuity of potential and conservation of current. Consequently the equivalent cable has no disconnected section in this instance (Fig. 10b) and terminates on a cut end.

This analytical construction serves to illustrate all the properties of equivalent cables and the associated mapping. In fact, the entire process can be performed numerically using a matrix representation of the dendritic structure. The following section provides a methodology for constructing equivalent cables using this representation while at the same time providing a framework for the numerical integration of the cable equations expressed within a finite difference scheme.

Equivalent Cable Construction

The construction of equivalent cables relies on the fact that tree matrices corresponding to dendrites with combinations of cut and sealed terminals can be transformed into tri-diagonal matrices that have a natural interpretation as an unbranched dendrite (or *equivalent cable*). The procedure consists of three stages, the first of which has already been described in the section on symmetrising the tree matrix (p. 239) and involves the construction of the *symmetric tree matrix* $S^{-1}AS$ where S is a real diagonal matrix. The second and third stages involve respectively the generation of the *symmetric cable matrix* C from $S^{-1}AS$ and its de-symmetrisation into the matrix E which has an interpretation as a cable, called the *equivalent cable*. Householder reflections play an important role in the construction of the equivalent cable and are now discussed.

Householder reflections

Given any vector V with components v_i, the entries h_{ij} of the *Householder reflection* matrix H (see Golub and Van Loan, 1990) are defined by

$$h_{ij} = \delta_{ij} - \frac{2v_i v_j}{v_r v_r} , \tag{8.108}$$

where δ_{ij} is Kronecker's delta and a repetition of indices implies summation. By construction, H is both symmetric and idempotent. The former is obvious since $h_{ij} = h_{ji}$ while the latter follows from the calculation

$$\begin{aligned}
h_{ik}h_{kj} &= \left(\delta_{ik} - \frac{2v_i v_k}{v_r v_r}\right)\left(\delta_{kj} - \frac{2v_k v_j}{v_s v_s}\right) \\
&= \delta_{ik}\delta_{kj} - 2\delta_{ik}\frac{v_k v_j}{v_s v_s} - 2\delta_{kj}\frac{v_i v_k}{v_r v_r} + 4\frac{v_i v_k}{v_r v_r}\frac{v_k v_j}{v_s v_s} \\
&= \delta_{ij} - 4\frac{v_i v_j}{v_r v_r} + 4\frac{v_i v_j}{v_r v_r} \\
&= \delta_{ij}\,.
\end{aligned}$$

Thus H is a symmetric orthogonal matrix and is also its own inverse, that is, $H^{\mathrm{T}} = H^{-1}$.

Householder reflections are traditionally used as the first stage in the reduction of a matrix to its Jordan canonical form. By skilful choices of the vector V, it is possible to construct a sequence of Householder reflections that reduce any matrix to upper Hessenberg form[14]. In particular, Householder reflections reduce symmetric matrices to tri-diagonal form (see Golub and Van Loan, 1990). However, when the Householder algorithm is applied in the traditional way to $S^{-1}AS$, the resulting tri-diagonal form has no immediate interpretation as a cable or unbranched dendrite, and the method appears to fail. The difficulty stems from the fact that conventional applications of Householder's algorithm always start with the last column and bottom row of a matrix and progressively sweep through the matrix structure finally arriving at its top left hand corner. The resulting tri-diagonal matrix has no direct interpretation as a cable.

Instead another strategy is required. In overview, it can be shown that repeated pre-and-post multiplication of $S^{-1}AS$ by a series of suitably chosen Householder matrices eventually reduces $S^{-1}AS$ to a symmetric cable matrix C which may then be associated with a de-symmetrised cable matrix E. Each pre- and post-multiplication by a Householder matrix is called a *Householder operation* and has the property that it zeroes a single pair of elements in the reduction of $S^{-1}AS$ to C. In this respect the algorithm is similar to a Given's rotation (Golub and van Loan, 1990), except that the latter destroys the matrix structure essential for cable formation whereas the former sequentially generates elements of the symmetrised cable matrix as off-tri-diagonal entries are progressively cleared from successive rows of the partially tri-diagonalised symmetric tree matrix.

Let A be a $n \times n$ symmetric matrix A. The Householder reduction algorithm that preserves cable structure consists of two complementary operations now described in detail. Suppose that p and q are respectively the *minimum* row index and the *minimum* column index in row p such that $a_{p,q}$ and $a_{q,p}$ $(q > p + 1)$ is a non-zero off-tri-diagonal element. If p and q do not exist then A is tri-diagonal, otherwise choose v_i in matrix (8.108) by the formula

$$\begin{aligned}
&v_i = \sqrt{1-\alpha}\,\delta_{i,(p+1)} - \sqrt{1+\alpha}\,\delta_{i,q}\,, \\
&\alpha = \frac{a_{p,(p+1)}}{\sqrt{a^2_{p,(p+1)} + a^2_{p,q}}}\,, \qquad \beta = \frac{a_{p,q}}{\sqrt{a^2_{p,(p+1)} + a^2_{p,q}}}\,.
\end{aligned} \tag{8.109}$$

[14] An upper Hessenberg matrix has all its entries below the principal sub-diagonal zero and is the required starting form for a QZ or QL reduction to Jordan canonical form.

The Householder reflection defined by this choice of V is denoted by $H^{(p,q)}$ and has block matrix form

$$H^{(p,q)} = \begin{bmatrix} I_p & 0 & 0 & 0 & 0 \\ 0 & \alpha & 0 & \beta & 0 \\ 0 & 0 & I_{q-p-2} & 0 & 0 \\ 0 & \beta & 0 & -\alpha & 0 \\ 0 & 0 & 0 & 0 & I_{n-q} \end{bmatrix} \tag{8.110}$$

where I_j denotes the $j \times j$ identity matrix and $\alpha^2 + \beta^2 = 1$. It can be demonstrated that the $(p,q)^{\text{th}}$ and $(q,p)^{\text{th}}$ entries of $H^{(p,q)}AH^{(p,q)}$ are zero; the effect of the Householder operation defined by $H^{(p,q)}$ is to distribute the $(p,q)^{\text{th}}$ and $(q,p)^{\text{th}}$ elements of A around the $(p,p+1)^{\text{th}}$ and $(p+1,p)^{\text{th}}$ entries of $H^{(p,q)}AH^{(p,q)}$ and other off-tri-diagonal elements lying below row p in the new matrix $H^{(p,q)}AH^{(p,q)}$. Repeated application of this idea with a carefully chosen sequence of Householder operations enables all the off-tri-diagonal elements $S^{-1}AS$ to be "chased" out of the original matrix. In conclusion, there is a series of Householder reflections $H_1, H_2, \ldots, H_k$ such that

$$T = (H_k H_{k-1} \cdots H_2 H_1)(S^{-1}AS)(H_1 H_2 \cdots H_{k-1} H_k) \tag{8.111}$$

is a symmetric tri-diagonal matrix. Unfortunately pairs of negative entries can arise on the sub/super diagonals of T so T itself is not the symmetric cable matrix C. However, T can be transformed into C by a sequence of elementary matrix operations.

Suppose that $t_{p,(p+1)}$ and $t_{(p+1),p}$ are a pair of negative elements in the sub/super diagonal of T and that all the off-tri-diagonal elements in the first $(p-1)$ rows of T are zero, then the algebraic sign of the elements in the $(p,p+1)^{\text{th}}$ and $(p+1,p)^{\text{th}}$ entries of $R^{(p)}TR^{(p)}$ are reversed by the action of the symmetric orthogonal matrix

$$R^{(p)} = \begin{bmatrix} I_p & 0 & 0 \\ 0 & -1 & 0 \\ 0 & 0 & I_{n-p-1} \end{bmatrix} . \tag{8.112}$$

In fact, $R^{(p)}$ changes the algebraic sign of all the entries in the pth row and column of T except the (p,p) entry (which actually has its algebraic sign changed twice). Thus there is a sequence of reflections $R_1, R_2, \ldots, R_m$ such that T is transformed into the symmetric cable matrix C

$$C = (R_m R_{m-1} \cdots R_2 R_1) T (R_1 R_2 \cdots R_{m-1} R_m) \tag{8.113}$$

whose sub and super diagonals contain non-negative elements only. In conclusion, there is a series of Householder reflections and $H_1, \ldots, H_k$ and a series of correction matrices $R_1, \ldots, R_m$ such that

$$C = Q^{-1}(S^{-1}AS)Q\,, \qquad Q = R_m \cdots R_1 H_k \cdots H_1\,, \qquad Q^T Q = QQ^T = I\,.$$

Fig. 11 illustrates the procedure schematically for a general Y-junction symmetric tree matrix.

Extracting the equivalent cable

The matrix C is now regarded as the symmetrised form of a tree matrix, E, corresponding to an unbranched tree or *equivalent cable*. It therefore remains to extract E from C together with the symmetrising $n \times n$ diagonal matrix $X = \text{diag}\,(x_0, \ldots, x_{n-1})$ whose first entry is $x_0 = 1$ without loss of generality, and for which $C = X^{-1}EX$. Clearly

$$C = X^{-1}EX \qquad \Longleftrightarrow \qquad c_{ij} = \frac{x_j e_{ij}}{x_i}\,.$$

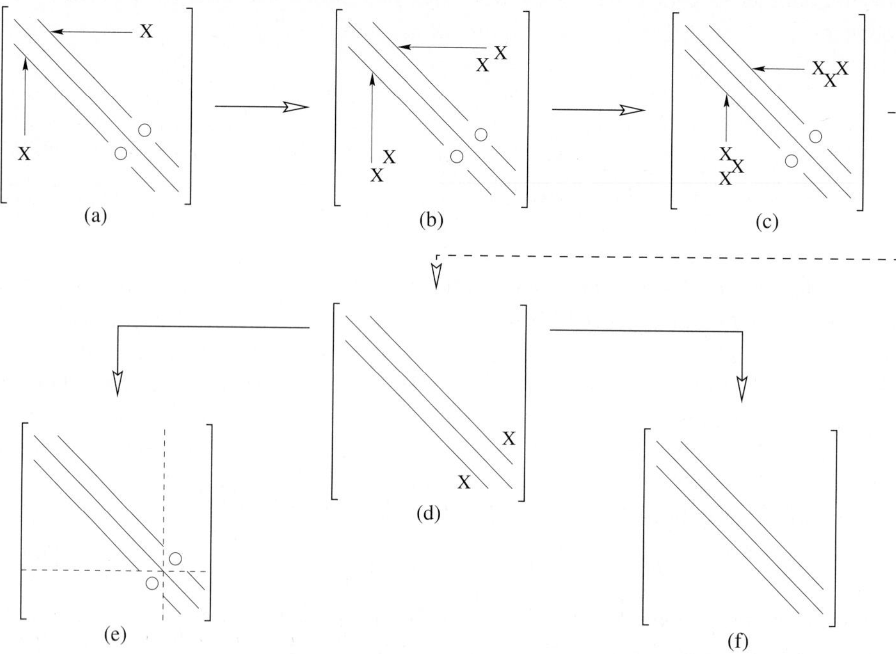

Fig. 11. Schematic of the Householder tri-diagonalisation procedure applied to the tree matrix for a general Y-junction. (a) The symmetric tree has one pair of off-tri-diagonal elements. (b) zeroing this element produces new off-tri-diagonal element further towards the lower-right corner of the matrix. (c)–(d) repeat until tri-diagonality is achieved. The resulting cable matrix will represent either (e) two sections (one connected, one disconnected) or (f) one section (connected).

The connection between X and E mirrors the connection between the original tree matrix A and its symmetrising matrix S. By construction, E and C have the same main diagonal. The sub and super diagonal entries of C and E satisfy

$$c_{i,i+1} = \frac{x_{i+1} e_{i+1,i}}{x_i}\,, \qquad c_{i+1,i} = \frac{x_i e_{i,i+1}}{x_{i+1}}\,, \qquad 0 \leqslant i < n-1\,.$$

From the symmetry of C it now follows that

$$e_{i+1,i} = \frac{c^2_{i,i+1}}{e_{i,i+1}}\,, \qquad 0 \leqslant i < n-1\,. \tag{8.114}$$

$$x_{i+1} = x_i \sqrt{\frac{e_{i,i+1}}{e_{i+1,i}}}\,, \qquad 0 \leqslant i < n-1\,. \tag{8.115}$$

Since the diagonal entries of E and C are identical, the task is therefore to construct the sub and super diagonals of E from C. The technical construction of the E (equivalent cable) is best understood through an appreciation of the overall strategy. The procedure by which C is de-symmetrised to get E may be usefully decomposed into three distinct phases, namely stage I, starting a cable; stage II, building a cable body and stage III, terminating a cable and, if necessary, restarting a new cable. Cables terminate either naturally because the de-symmetrisation process exhausts all the entries of C or prematurely because C has a non-trivial block tri-diagonal structure[15] that imposes a non-trivial block tri-diagonal structure on E. Of course, each tri-diagonal block of E corresponds to a cable, but only the block of E occupying the top left hand corner is connected to the soma (somal cable): each

[15]The matrix C is block tri-diagonal whenever zero elements occur in its sub/super diagonals.

remaining block can be identified as a cable that is disconnected from the soma. Therefore configurations of inputs on the model dendrite that map to inputs on this disconnected cable have no impact on the somal potential.

Each block tri-diagonal matrix of E defines a distinct cable. The terminal boundary conditions satisfied by this cable are embedded in the first and last rows of the block, while all other rows are identified with internal nodes. In particular, the finite difference representation of derivatives guarantees that the sum of the off-diagonal entries of these non-boundary rows is 2α. Sealed terminals correspond to a boundary row whose off-diagonal entry is also 2α while boundary rows corresponding to cut terminals have α as off-diagonal entry and require the cable to be extended by one internodal distance to the actual cut terminal. Once a starting boundary condition is identified for any cable, the body of the cable and the nature of its other terminal boundary are determined algorithmically by alternate use of result (8.114) and the fact that the sum of all off-diagonal entries in each non-boundary row is 2α.

Stage I The value of e_{01}, the off-diagonal entry in the first row of E is found by recognising that the first section of the equivalent cable is the "Rall" sum of the limbs contingent on the soma. Recall from (8.66) that the somal condition for the original tree is

$$\frac{dv_0}{dt} = -\frac{2g + \beta h g_0}{2\varepsilon g + h g_0} v_0 + \frac{2\alpha h}{2\varepsilon g + h g_0} \sum_{r=1}^{N} g_{k_r} v_{k_r} - \frac{2J_0}{2\varepsilon g + h g_0} .$$

Therefore the first row of E contains the pair of entries

$$e_{00} = -\frac{2g + \beta h g_0}{2\varepsilon g + h g_0} , \qquad e_{01} = \frac{2\alpha h}{2\varepsilon g + h g_0} \sum_{r=1}^{N} g_{k_r} = \frac{2\alpha h g_0}{2\varepsilon g + h g_0} .$$

Stage II Once e_{01} is known then equation (8.114) asserts that $e_{10} = c_{01}^2 / e_{01}$ and consequently $e_{12} = 2\alpha - e_{10}$. This procedure is repeated, that is, $e_{i+1,i}$ is calculated from $e_{i,i+1}$ and then $e_{i+1,i+2}$ is calculated from $e_{i+1,i}$ according to the prescriptions

$$e_{i+1,i} = \frac{c_{i,i+1}^2}{e_{i,i+1}} , \qquad e_{i+1,i+2} = 2\alpha - e_{i+1,i} \tag{8.116}$$

provided $e_{i,i+1} \neq 0$. Once started, this algorithmic procedure generates the body of a cable and concludes the second phase of cable construction.

Stage III Sooner or later the algorithm described in Stage II fails because either $i + 1 = n - 1$ (i.e. E is completely determined) or $e_{i,i+1} e_{i+1,i} = 0$ so that the equivalent cable now contains disconnected sections. In the latter case, $e_{i,i+1} = e_{i+1,i} = 0$. In both cases, the nature of the boundary condition at cable termination is determined from the numerical value of $e_{i,i-1}$, the last non zero off-diagonal entry of E to be determined prior to disconnection/termination. If $e_{i,i-1} = 2\alpha$ then the cable ends on a sealed end otherwise $e_{i,i-1} = \alpha$ and the cable ends on a cut terminal after one further internodal distance. There are no other possible values for $e_{i,i-1}$.

Self evidently, the diagonal entries of the symmetrising matrix X are calculated directly from equation (8.115) until a natural or premature cable termination. Whenever a cable terminates prematurely and C still has elements to be de-symmetrised, the de-symmetrisation procedure must be restarted at either a sealed or cut terminal, and the remaining elements of X initialised with $x_{i+1} = 1$. The difficulty stems from the absence of a somal boundary condition, and is resolved by inspecting the properties of the mapping between the original dendrite and the symmetrised cable matrix C. If a cable is to be initiated at node $(i+1)$ then it will restart with a sealed end if the membrane potential at node z_{i+1} features in the $(i+1)^{\text{th}}$ mapping vector, otherwise it will restart with a cut end.

Diameters of equivalent cable sections

The description of the equivalent cable is completed by the specification of the g-value of each cable section. If G_i is the g-value of a cable section that has node Z_i at its distal end then the finite difference representation of the cable equation at node Z_i is

$$\frac{dV_i}{dt} = -\beta V_i + \frac{2\alpha G_i}{G_i + G_{i+1}} V_{i-1} + \frac{2\alpha G_{i+1}}{G_i + G_{i+1}} V_{i+1} = -\beta V_i + e_{i,i-1} V_{i-1} + e_{i,i+1} V_{i+1} \,.$$

By inspection, it is clear that

$$e_{i,i-1} + e_{i,i+1} = 2\alpha \,, \qquad G_{i+1} = G_i \, \frac{e_{i,i+1}}{e_{i,i-1}} \,.$$

The first result has already been used in the construction of E while the last result determines the g-values of all the sections of the component cables of the equivalent cable from the g-value of their first section. For the somal cable, the g-value of the first section is simply the sum of the g-values of the limbs connected to the soma. For a disconnected section, the g-value of its first section may be arbitrarily fixed at unity without loss of generality.

The electrical mapping

The pathway from original dendrite to equivalent cable is now seen to consist of three distinct steps, namely, symmeterisation of the original dendritic representation, reduction of that symmetrised representation to symmetric cable form and finally, de-symmeterisation of the symmetric cable to get the equivalent cable. In the formalism of matrix algebra, these operations are described respectively by the similarity transformations

$$C = H^{-1}(S^{-1}AS)H \,, \qquad C = X^{-1}EX$$

so that the original tree matrix A and its equivalent form E are now connected by the similarity transformation

$$E = M^{-1}AM \,, \qquad M = SHX^{-1} \,. \tag{8.117}$$

Furthermore, the original cable equation now becomes

$$\frac{d(MV)}{dt} = E(MV) - 2MRJ \qquad \equiv \qquad \frac{dV_{\text{E}}}{dt} = EV_{\text{E}} - 2MRJ$$

where V_{E} is seen to be the membrane potentials on the equivalent cable. Thus the matrix M may be interpreted as the electrical mapping between the original tree with membrane potentials V and its equivalent cable with membrane potentials $V_{\text{E}} = MV$. The matrix M will be called the Electro-Geometric-Projection matrix (EGP).

Computational Considerations

The tree and cable matrices can be efficiently stored since they are sparse and nearly tri-diagonal. It has already been explained that a tree matrix based on n numbered nodes contains $3n-2$ non-zero entries. The elements of the main diagonal can be stored as two elements, one associated with the soma node and one corresponding to the all the remaining nodes. The $2n-2$ off-diagonal elements can be stored as triplets whose first two elements are respectively the row and column of the element while the last is the element value.

The symmetrising matrix S is stored as a vector of length n while the symmetric tree matrix $S^{-1}AS$ is stored as $(n-1)$ triplets. The form for S is constructed directly from A as described in equations (8.73) and defines the form for $S^{-1}AS$.

The simple structure of the Householder reflection $H^{(p,q)}$ guarantees that only rows $p+1$ and q and columns $p+1$ and q of the current intermediate matrix are modified by this Householder operation. Moreover, it is only elements in rows/columns p or greater that are actually altered. Therefore, the Householder operation $H_{(p,q)}$ modifies a maximum of $2(n-p-1)$ elements. The Householder operations by which the symmetric tree matrix is manipulated into the symmetric cable matrix maintain a high level of sparsity in intermediate matrices. The temporary off-tri-diagonal elements are small in number and may be stored in triplet form.

The electrical mapping between tree and its equivalent cable can be stored in terms of the sequence of individual Householder reflections, each of which may be stored as a triplet. For example, $H_{(p,q)}$ is the triplet (p, q, β) (α and the structure of $H_{(p,q)}$, follow from equation 8.109 and 8.110). Of course, to appreciate the connection between dendritic tree and equivalent cable, the EGP matrix is required.

The structure of the equivalent cable is stored with similar efficiency to the original dendritic tree as is the symmetrising matrix X.

In conclusion, the passage from dendritic tree to equivalent cable can be achieved with high speed and efficient memory utilisation in view of the sparse nature of dendritic structure matrices. However, the electrical mapping, being an association between points on the dendritic tree and its equivalent cable, requires calculations on full matrices for a complete specification and therefore is inevitably slow and memory intensive.

Full Example of the Householder Procedure

It has already been observed that the diagonal matrix S defined by equation (8.75), when applied to the tree matrix A defined in equation (8.71), gives the symmetric tree matrix

$$S^{-1}AS = \begin{bmatrix} -\beta_S & \alpha k & 0 & 0 & 0 & 0 & 0 \\ \alpha k & -\beta & \alpha p & 0 & 0 & 0 & 0 \\ 0 & \alpha p & -\beta & \alpha q & 0 & \alpha r & 0 \\ 0 & 0 & \alpha q & -\beta & \sqrt{2}\alpha & 0 & 0 \\ 0 & 0 & 0 & \sqrt{2}\alpha & -\beta & 0 & 0 \\ 0 & 0 & \alpha r & 0 & 0 & -\beta & \sqrt{2}\alpha \\ 0 & 0 & 0 & 0 & 0 & \sqrt{2}\alpha & -\beta \end{bmatrix}.$$

The Householder reduction of $S^{-1}AS$ to tri-diagonal form is now illustrated for this matrix and is achieved by two Householder operations.

Step I The first Householder operation is designed to zero the entry αr in the third row and sixth column of $S^{-1}AS$. With this intention in mind, let the symmetrised tree matrix $S^{-1}AS$ and the Householder reflection H_1 have block diagonal forms

$$S^{-1}AS = \begin{bmatrix} T & U \\ U^T & B \end{bmatrix}, \qquad H_1 = \begin{bmatrix} I_3 & 0_{34} \\ 0_{43} & Q \end{bmatrix}$$

in which the forms for T (a 3×3 matrix), U (a 3×4 matrix) and B (a 4×4 matrix) are evident from the expression for $S^{-1}AS$. Specifically

$$Q = \begin{bmatrix} \gamma & 0 & \delta & 0 \\ 0 & 1 & 0 & 0 \\ \delta & 0 & -\gamma & 0 \\ 0 & 0 & 0 & 1 \end{bmatrix}, \qquad \begin{aligned} \gamma &= \frac{q}{\sqrt{r^2+q^2}} \\ \delta &= \frac{r}{\sqrt{r^2+q^2}} \end{aligned} .$$

Using matrix block multiplication, it follows that

$$H_1(S^{-1}AS)H_1 = \begin{bmatrix} T & UQ \\ (UQ)^{\mathrm{T}} & QBQ \end{bmatrix}$$

in which UQ and QBQ are respectively 3×4 and 4×4 matrices. Let $w = \sqrt{r^2+q^2}$ then it is simply a matter of matrix algebra to verify that

$$H_1(S^{-1}AS)H_1 = \begin{bmatrix} -\beta_S & \alpha k & 0 & 0 & 0 & 0 & 0 \\ \alpha k & -\beta & \alpha p & 0 & 0 & 0 & 0 \\ 0 & \alpha p & -\beta & \alpha w & 0 & 0 & 0 \\ 0 & 0 & \alpha w & -\beta & \sqrt{2}\alpha\gamma & 0 & \sqrt{2}\alpha\delta \\ 0 & 0 & 0 & \sqrt{2}\alpha\gamma & -\beta & \sqrt{2}\alpha\delta & 0 \\ 0 & 0 & 0 & 0 & \sqrt{2}\alpha\delta & -\beta & -\sqrt{2}\alpha\gamma \\ 0 & 0 & 0 & \sqrt{2}\alpha\delta & 0 & -\sqrt{2}\alpha\gamma & -\beta \end{bmatrix} . \tag{8.118}$$

Step II The second Householder operation is designed to zero the entry $\sqrt{2}\alpha\delta$ in the fourth row and seventh column of $H_1(S^{-1}AS)H_1$. In this case, let $H_1(S^{-1}AS)H_1$ and the Householder reflection H_2 have block diagonal forms

$$H_1(S^{-1}AS)H_1 = \begin{bmatrix} T & U \\ U^{\mathrm{T}} & B \end{bmatrix}, \qquad H_2 = \begin{bmatrix} I_4 & 0_{43} \\ 0_{34} & Q \end{bmatrix}$$

in which the forms for T (a 4×4 matrix), U (a 4×3 matrix) and B (a 3×3 matrix) are evident from expression (8.118) for $H_1(S^{-1}AS)H_1$. Specifically

$$Q = \begin{bmatrix} \gamma & 0 & \delta \\ 0 & 1 & 0 \\ \delta & 0 & -\gamma \end{bmatrix} .$$

Let $H = H_2 H_1$, then it is again a matter of algebra to demonstrate that

$$H(S^{-1}AS)H = \begin{bmatrix} -\beta_S & \alpha k & 0 & 0 & 0 & 0 & 0 \\ \alpha k & -\beta & \alpha p & 0 & 0 & 0 & 0 \\ 0 & \alpha p & -\beta & \alpha w & 0 & 0 & 0 \\ 0 & 0 & \alpha w & -\beta & \sqrt{2}\alpha & 0 & 0 \\ 0 & 0 & 0 & \sqrt{2}\alpha & -\beta & 0 & 0 \\ 0 & 0 & 0 & 0 & 0 & -\beta & \sqrt{2}\alpha \\ 0 & 0 & 0 & 0 & 0 & \sqrt{2}\alpha & -\beta \end{bmatrix} \tag{8.119}$$

where the final Householder operations are embodied in the orthogonal matrix

$$H = H_2 H_1 = \begin{bmatrix} 1 & 0 & 0 & 0 & 0 & 0 & 0 \\ 0 & 1 & 0 & 0 & 0 & 0 & 0 \\ 0 & 0 & 1 & 0 & 0 & 0 & 0 \\ 0 & 0 & 0 & \gamma & 0 & \delta & 0 \\ 0 & 0 & 0 & 0 & \gamma & 0 & \delta \\ 0 & 0 & 0 & \delta & 0 & -\gamma & 0 \\ 0 & 0 & 0 & 0 & \delta & 0 & -\gamma \end{bmatrix}. \tag{8.120}$$

Fully equivalent cable

The equivalent cable matrix E is obtained from (8.119) by the de-symmetrisation algorithm described previously. The presence of a pair of zero elements in the (5, 6) and (6, 5) entries of the symmetrised cable matrix $H(S^{-1}AS)H$ indicates that the equivalent cable in this instance has a connected section of length $2l$ and a disconnected section of length l. The de-symmetrisation procedure begins by recognising that the first row of the de-symmetrised matrix E has second entry αk^2. Thereafter, the process is mechanical until the fifth row of E is complete. At this point

$$E = \begin{bmatrix} -\beta_S & \alpha k^2 & 0 & 0 & 0 & 0 & 0 \\ \alpha & -\beta & \alpha & 0 & 0 & 0 & 0 \\ 0 & \alpha p^2 & -\beta & \alpha w^2 & 0 & 0 & 0 \\ 0 & 0 & \alpha & -\beta & \alpha & 0 & 0 \\ 0 & 0 & 0 & 2\alpha & -\beta & 0 & 0 \\ 0 & 0 & 0 & 0 & 0 & \cdots & \cdots \\ 0 & 0 & 0 & 0 & 0 & \cdots & \cdots \end{bmatrix}, \tag{8.121}$$

while the de-symmetrising diagonal matrix is

$$X = \text{diag}\left(1, \frac{1}{k}, \frac{p}{k}, \frac{p}{kw}, \frac{p\sqrt{2}}{kw}, \ldots, \ldots\right). \tag{8.122}$$

Furthermore, the form of the fifth row of E indicates that the connected section of the fully equivalent cable ends on a sealed terminal. The g-values of the first four sections of the equivalent cable are

$$g_1 = g_2 = g_P\,, \qquad g_3 = g_4 = g_2\left(\frac{w^2}{p^2}\right) = g_L + g_R\,. \tag{8.123}$$

After four sections, the equivalent cable must be restarted. The restarting condition is determined by properties of the sixth row of H displayed in (8.120). The terminal nodes z_4 and z_6 of the original dendritic tree are not present in the electrical mapping associated with the sixth row of H and therefore the equivalent cable must be restarted on a cut terminal. The completed components of E in (8.121), X in (8.122) and the cable g-values in (8.123) are respectively

$$\begin{bmatrix} -\beta & \alpha \\ 2\alpha & -\beta \end{bmatrix}, \qquad \left(\ldots, 1, \frac{1}{\sqrt{2}}\right), \qquad g_5 = g_6 = g_R + g_L\,.$$

It is now evident that the disconnected section of the equivalent cable begins on a cut terminal, ends on a sealed terminal and is of uniform thickness. Bringing together the symmetrising matrix S given in (8.75), the Householder operations H given in (8.120) and the de-symmetrising matrix X given in (8.122), the complete electrical mapping from dendritic tree to equivalent cable is now seen to be

$$M = SHX^{-1} = \begin{bmatrix} 1 & 0 & 0 & 0 & 0 & 0 & 0 \\ 0 & 1 & 0 & 0 & 0 & 0 & 0 \\ 0 & 0 & 1 & 0 & 0 & 0 & 0 \\ 0 & 0 & 0 & 1 & 0 & \xi & 0 \\ 0 & 0 & 0 & 0 & 1 & 0 & \xi \\ 0 & 0 & 0 & 1 & 0 & \eta & 0 \\ 0 & 0 & 0 & 0 & 1 & 0 & \eta \end{bmatrix}, \qquad \begin{aligned} \xi &= \frac{rp}{wkq} \\ -\eta &= \frac{qp}{wkr} \end{aligned}\,.$$

Generalised Compartmental Models

Rall's equivalent cylinder was used as a mathematical model to simplify the "exploration of the physiological implications of dendritic branching" (see Rall, 1964). However, Rall recognised both the limited utility of this model when applied to the spatiotemporal analysis of branching and tapering dendritic structures with complex patterns of synaptic activity, and the difficulty of solving the partial differential equations (cable equations) characterising these complex structures. To simplify the analytical and computational problems associated with the direct application of the cable equation to dendritic systems, Rall introduced a compartmental model of a neuron whose underlying mathematical formulation was expressed in terms of ordinary differential equations as opposed to the partial differential equations that appear in neuronal cable theory (Rall, 1964). This procedure has the advantage that well understood numerical algorithms are readily available to solve these systems of ordinary differential equations.

The Rall compartments are a collection of contiguous sections of the neuron, each one of which is considered to be spatially uniform. The compartments themselves were modelled by the usual equivalent circuit for the electrical behavior of a nerve membrane and their interaction with neighbouring compartments is governed by Kirchhoff's circuit laws. The practical implementation of the Rall compartmental model associates a distinct point (typically its midpoint) with each compartment, whose primary role is then to determine the biophysical characteristics of the equivalent circuit at that point. The Rall compartmental model of a neuron is often described as a series of isopotential regions coupled by resistances to its immediate neighbours (see Rall *et al.*, 1992; Perkel and Mulloney, 1978a).

Since the equations governing the Rall compartmental model arise from a consideration of electrical circuits, these equations are obliged to have a matrix representation that is tri-diagonal except for rows pertaining to dendritic branch points. It must be emphasised that this tri-diagonal structure is inherent in Rall's compartmental model and should not be confused with numerical schemes to integrate the cable equation specification of the dendrite. Any tri-diagonal structure possessed by the latter arises through the choice of numerical scheme (e.g. second order central differences) and is not an obligatory feature of the model. For example, fourth order central differences schemes give penta-diagonal matrices and spectral methods give full matrices. In summary, there is a subtle but important distinction between the structure of Rall's compartmental model, which is exact, and numerical schemes that have a similar mathematical structure but are approximate.

In Rall's compartmental model of a neuron, statements about the interactions between compartments (distinct physical regions of a dendrite) become equivalent to statements describing the interaction between points representing these regions. The equivalence between compartment and point provides the motivation for a more general description of compartmental models through the mathematical notion of duality: compartments and points are defined as dual elements so that results for compartments may be regarded as results for points and vice-versa. However, although both sets of results are mathematically similar, the procedures for formulating a description based on compartments is different from that used for points. In the latter, points form a set of designated sites on the dendritic tree, and the mathematical model is formulated for the membrane potential at these points under the modelling assumption that current can flow across the dendritic membrane only at these designated points. It will be seen that Rall's compartmental model is a special case of this class of compartmental model.

The question now arises as to the connection, if any, between a compartmental representation of a neuron and one based on interconnected limbs described by cable equations. Rall partly answers this question by demonstrating that the limit of his compartmental model is the cable equation. The inference of this result is that sufficiently refined compartmental models give neuronal behavior that is close to that predicted from analytical solutions of the cable equations. Specific comparisons (see Segev, Fleshman, Miller and Bunow, 1985) do indeed bare out this presumption. However, it is important to recognise that each compartmentalisation of a neuron is a different model, and that comparisons between compartmental and cable models have necessarily been limited to comparisons for finite times. To prove that the cable and compartmental models are identical, it is necessary to show that, for each location on a real dendrite, the maximum difference in the membrane potential calculated (without numerical error) for the two models can be made smaller than some closeness criterion for all time, and not just for a finite time. That is, the convergence between models is uniform in time. If this can be proved, then agreement between solutions to dendritic models based on compartments and based on the cable equation is entirely expected and confirms Rall's insight that compartmental models do indeed capture the behavior of neurons (as defined by the cable equation model).

The development of the generalised model brings with it several benefits to compartmental modelling that are not available for the special models. Once a desired spatial resolution for dendritic potentials is specified, the values for axial resistance connecting compartments and the membrane capacitance for each compartment follow from the application of several simple rules, which for tapering dendrites give exact expressions. Taking the general model as a reference, errors arising in the choice of compartment parameters for particular models may be estimated. In particular compartmental models, synaptic inputs are assigned to the compartments in which they happen to fall, irrespective of their position within this compartment. The general model provides a means for partitioning synaptic input between neighbouring compartments in a way that is natural and consistent with dendritic physiology.

Formulation of a General Compartmental Model

Building on the notion of duality, a compartmental model based on points (as opposed to compartments) is now developed. Let z_{j-1}, z_j and z_{j+1} be three physically sequential points on a dendritic limb at which charge flow between the intracellular and extracellular media is possible. Recall that current flow across the membrane is restricted to these points only. The usual ladder network (Fig. 12) used in the formulation of Rall's compartmental model is a useful aid in motivating the development of the general compartmental model. Each rung of the ladder now models the membrane properties of the dendrite in the neighbourhood of the designated point z_j. Each rung is constructed by the usual equivalent circuit in which a capacitor is connected in parallel with individual batteries and resistors as illustrated in figure 12. The backbone of the ladder describes the resistive (axoplasmic) coupling between adjacent points.

The general compartmental equations are now constructed using Kirchhoff's circuit laws and the properties of standard circuit components. The first law requires conservation of current at z_j, that is

$$I_j^{(\mathrm{m})} = I_{j-1,j} - I_{j,j+1} \, . \tag{8.124}$$

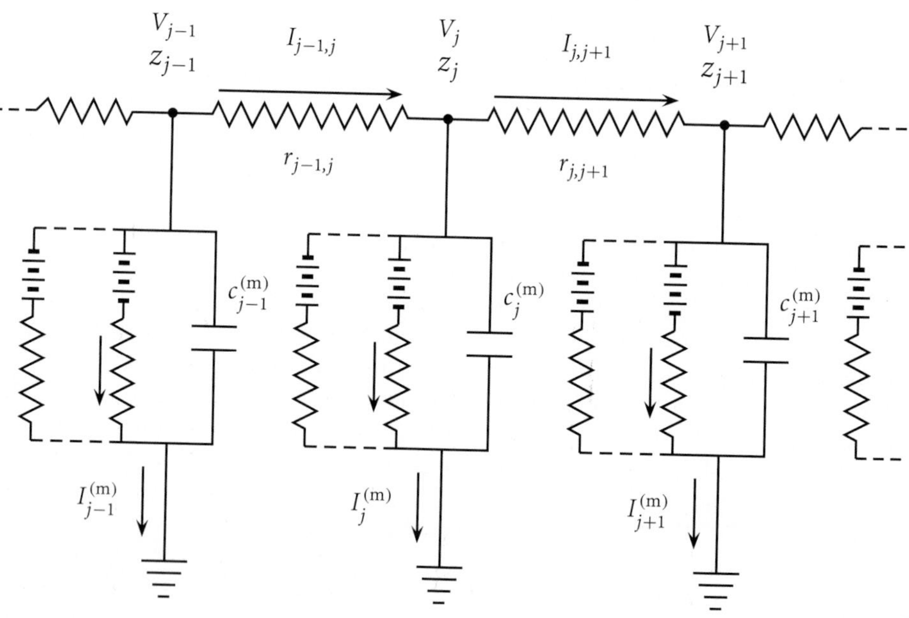

Fig. 12. A diagrammatic representation of the ladder network used to represent the general compartmental model. Membrane current flowing between intracellular and extracellular media in the neighbourhood of designated point z_j is modelled by the usual equivalent electrical circuit consisting of a capacitor in parallel with a battery and one or more resistances representing ionic currents. Axial currents $I_{j-1,j}$ and $I_{j,j+1}$ are ohmic and governed by the resistances $r_{j-1,j}$ between z_{j-1} and z_j and by $r_{j,j+1}$ between z_j and z_{j+1}.

The current flow from z_{j-1} to z_j is based on Ohms law, and in the first instance, assumes no current leakage across the membrane connecting these designated nodes. In order to model realistic synaptic activity at any location on the dendrite, the model will be extended to include discrete current inputs between z_{j-1} and z_j. This extension will allow, synaptic input to be partitioned in a natural way between nodes.

Let $I_{j-1,j}(t)$ be the total axial current flowing along a dendrite limb between nodes z_j and z_{j-1} in the absence of current input. Suppose also that the dendritic limb has cross-sectional area $A(x)$ where x is an axial coordinate along the limb then $I_{j-1,j}$ satisfies

$$I_{j-1,j}(t) = -g_A A(x) \frac{\partial V(t,x)}{\partial x}$$

where $V(t,x)$ is the dendritic membrane potential at x. The potential difference between points z_{j-1} and z_j is therefore

$$V(t,z_j) - V(t,z_{j-1}) = V_j - V_{j-1} = \int_{z_{j-1}}^{z_j} \frac{\partial V}{\partial x}\,dx = -\frac{I_{j-1,j}(t)}{g_A}\int_{z_{j-1}}^{z_j} \frac{ds}{A(s)}$$

In conclusion,

$$I_{j-1,j} = g_A \frac{V_{j-1} - V_j}{r_{j-1,j}}, \qquad r_{j-1,j} = \int_{z_{j-1}}^{z_j} \frac{ds}{A(s)}. \tag{8.125}$$

The characteristics of the equivalent circuit indicate that

$$I_j^{(m)} = I_j^{(\text{ionic})} + c_j^{(m)} \frac{dV_j}{dt} \tag{8.126}$$

where $c_j^{(m)}$ is a lumped capacitance at z_j (see Hines and Carnevale, 1997), that is,

$$c_j^{(m)} = C_M \int_{(z_{j-1}+z_j)/2}^{(z_j+z_{j+1})/2} P(x)\,dx. \tag{8.127}$$

Technically, this integral should be taken over the dendritic surface and not along the dendritic axis. However, if surface and axial measures of length are significantly different[16], that is, the dendritic taper is severe, then non-axial current flow is almost certainly important and the validity of dendritic compartment models themselves is doubtful. Thus the equation governing the evolution of the membrane potential at z_j is obtained by combining the component equations (8.124–8.126) and is

$$c_j^{(m)} \frac{dV_j}{dt} + I_j^{(\text{ionic})} = g_A \frac{V_{j-1} - V_j}{r_{j-1,j}} + g_A \frac{V_{j+1} - V_j}{r_{j,j+1}}. \tag{8.128}$$

This equation is formally identical to those for the Rall compartmental model (see Segev, Fleshman and Burke, 1989) of a neuron. Indeed, all compartmental models of neurons will necessarily take the form of equation (8.128) for suitable choices of $c_j^{(m)}$ and $r_{j-1,j}$ etc. with different choices giving different models. Equations for terminal boundaries, branch points and the tree-to-soma connection are treated in a similar way. A branch point is modelled by the membrane potential at the junction but the associated region of dendrite is star-like and not cable-like. For example, Rall's compartmental model corresponds to the choice

$$r_{j-1,j} = g_A \frac{(z_j - z_{j-1})}{2}\left(\frac{1}{A_{j-1}} + \frac{1}{A_j}\right), \tag{8.129}$$

$$c_j^{(m)} = C_M \frac{(z_{j+1} - z_{j-1})}{2} P\left(\frac{z_{j-1} + 2z_j + z_{j+1}}{4}\right). \tag{8.130}$$

[16]Recall that $ds = dx\sqrt{1 + (dy/dx)^2}$ for a plane curve so that ds and dx differ to second order in gradient/taper.

Formulae (8.129) and (8.130) are immediately recognisable as numerical quadratures for $r_{j-1,j}$ in equation (8.125) and c_j in equation (8.127). Rall's expression for $r_{j-1,j}$ is based on the trapezoidal rule

$$\int_{z_{j-1}}^{z_j} f(x)\,dx = \frac{(z_j - z_{j-1})}{2}\Big(f(z_{j-1}) + f(z_j)\Big) - \frac{(z_j - z_{j-1})^3}{12} f''(\xi_j) \tag{8.131}$$

while that for c_j uses the midpoint rule

$$\int_{z_{j-1}}^{z_j} f(x)\,dx = (z_j - z_{j-1}) f\Big(\frac{z_{j-1} + z_j}{2}\Big) + \frac{(z_j - z_{j-1})^2}{3} f''(\eta_j)\,. \tag{8.132}$$

Rall's compartmental model is now seen to be an approximation of a more general compartmental model based on points and not isopotential segments of a dendrite. In particular, the difference between the Rall compartmental model and the general compartmental model may be attributed to errors in approximating quadratures for $r_{j-1,j}$ and $c_j^{(\mathrm{m})}$. Another popular model of dendritic structure assumes contiguous cylinders of uniform cross-section so that $A(x)$ and $P(x)$ are piecewise constant functions of x. In this event, the quadratures for $r_{j-1,j}$ and $c_j^{(\mathrm{m})}$ may be evaluated exactly.

Finally, it should be noted that if information regarding the geometric structure of a dendrite is available at selected locations, then the quadratures for $r_{j-1,j}$ and $c_j^{(\mathrm{m})}$ may be estimated numerically using the trapezoidal rule. If geometrical data is available at uniformly spaced nodes then Simpson's rule can be used for improved numerical accuracy.

Uniformly Tapering Dendrites

The general compartmental model of a dendrite raises the possibility of quantifying the errors incurred in representing tapering dendritic geometry by a stepped sequence of uniform contiguous cylinders. Prior to this discussion, some elegant and exact results for inter-nodal resistances are developed for tapering dendrites. The concept of taper, although primitive, involves an element of subtlety that is not immediately apparent. Let $\mathscr{A}_{\mathrm{L}}$ be the cross-sectional area of a tapering dendritic section and let point $(a, b, 0)$ be interior to $\mathscr{A}_{\mathrm{L}}$. A section of a uniformly tapering dendrite may be viewed as a frustrum[17] of the cone formed by a pencil of lines drawn from the point $V(a, b, H)$ $(H > 0)$ to $\delta\mathscr{A}_{\mathrm{L}}$, the boundary of $\mathscr{A}_{\mathrm{L}}$. Without loss of generality, suppose that $\delta\mathscr{A}_{\mathrm{L}}$ has length P_{L} and is the parametric curve $\boldsymbol{x} = (x_0(u), y_0(u), 0)$ where $u \in \mathscr{I}$, an interval of the real line. Since any point on the line joining $V(a, b, H)$ to the point with coordinates $(x_0(u), y_0(u), 0)$ on $\delta\mathscr{A}_{\mathrm{L}}$ has position

$$\boldsymbol{x} = \lambda(a, b, H) + (1 - \lambda)(x_0(u), y_0(u), 0)\,, \qquad \lambda \in [0, 1]$$

then the surface of the cone with vertex $V(a, b, H)$ has parametric equations

$$\begin{aligned} x &= a\lambda + (1 - \lambda)x_0(u)\,, \\ y &= b\lambda + (1 - \lambda)y_0(u)\,, \qquad (u, \lambda) \in \mathscr{I} \times [0, 1]\,. \\ z &= H\lambda\,, \end{aligned} \tag{8.133}$$

It follows directly from Green's theorem that the cross-sectional area of this cone exposed by the plane λ constant, is given by the line integral

$$A(\lambda) = \frac{1}{2}\oint (x\,dy - y\,dx)$$

[17] Subtle tapered dendritic geometries can be constructed by repositioning the vertex of the generating cone after each section is generated.

$$= \frac{1}{2}\oint \Big[(a\lambda + (1-\lambda)x_0)(1-\lambda)dy_0 - (b\lambda + (1-\lambda)y_0)(1-\lambda)dx_0\Big]$$
$$= \frac{1-\lambda}{2}\oint (a\lambda\, dy_0 - b\, dx_0) + \frac{(1-\lambda)^2}{2}\oint (x_0\, dy_0 - y_0\, dx_0)$$

where the curve of integration is the perimeter of the cone when λ is constant. Since

$$\oint dy_0 = 0\,, \qquad \oint dx_0 = 0\,, \qquad \frac{1}{2}\oint (x_0\, dy_0 - y_0\, dx_0) = A_{\mathrm{L}}$$

then it follows immediately that

$$A(\lambda) = (1-\lambda)^2 A_{\mathrm{L}}\,, \qquad \lambda \in [0,1]\,, \tag{8.134}$$

for all tapered dendrites. Similarly $P(\lambda)$, the perimeter of $A(\lambda)$, has value

$$\begin{aligned} P(\lambda) = \oint ds &= \int_{\mathcal{I}} \sqrt{\left(\frac{\partial x}{\partial u}\right)^2 + \left(\frac{\partial y}{\partial u}\right)^2}\, du \\ &= \int_{\mathcal{I}} (1-\lambda)\sqrt{\left(\frac{\partial x_0}{\partial u}\right)^2 + \left(\frac{\partial y_0}{\partial u}\right)^2}\, du \\ &= (1-\lambda)P_{\mathrm{L}}\,. \end{aligned}$$

Suppose that a tapering section of a dendrite is modelled by the frustrum of this cone defined by $\lambda \in [0, \lambda_1]$, $(\lambda_1 < 1)$. If the length of the section (height of the frustrum) is L then $L = H\lambda_1$ and the cross-sectional areas of the left and right hand faces of the frustrum are A_{L} and $A_{\mathrm{R}} = (1-\lambda_1)^2 A_{\mathrm{L}}$ respectively from formula (8.134). Since $z = H\lambda$ then

$$\int_0^L \frac{dz}{A(z)} = \int_0^{\lambda_1} \frac{H\, d\lambda}{(1-\lambda)^2 A_{\mathrm{L}}} = \frac{H\lambda_1}{(1-\lambda_1)\, A_{\mathrm{L}}} = \frac{L}{\sqrt{A_{\mathrm{L}} A_{\mathrm{R}}}} \tag{8.135}$$

By contrast, Rall's compartmental model replaces this integral by its trapezoidal estimate. Using the well known result that the arithmetic mean of two positive numbers is never less than their geometric mean, it follows that

$$\frac{L}{2}\left[\frac{1}{A_{\mathrm{L}}} + \frac{1}{A_{\mathrm{R}}}\right] \geqslant \frac{L}{\sqrt{A_{\mathrm{L}} A_{\mathrm{R}}}}$$

where the left hand side of this inequality denotes the Rall approximation and the right hand side is the exact area. Thus the classical association between dendritic axial resistance and geometry tends to overestimate axial resistance for dendrites with a pure taper. The suggestion is therefore that the expression (8.135) for dendritic axial resistance captures more accurately the electrical properties of the dendrite and enjoys the advantage that it is exact for dendrites with a pure taper. In particular, the procedures by which dendritic limbs are partitioned into cylinders are redundant for the purpose of assigning axial resistance. Similarly, the lumped capacitance associated with each node of a tapering dendrite is

$$\int_0^L C_{\mathrm{M}} P(x)\, dx = C_{\mathrm{M}}\, H \int_0^{\lambda_1} (1-\lambda)P_{\mathrm{L}}\, d\lambda = C_{\mathrm{M}} \frac{(P_{\mathrm{L}} + P_{\mathrm{R}})L}{2}\,.$$

The general compartmental model also provides insight as to how real dendritic cross-sectional areas might taper to zero. To interpret the integral expression for $r_{j-1,j}$ at a dendritic tip z_j where $A(z_j) = 0$, the function $A^{-1}(x)$ must have an integrable singularity. This condition requires that $A(x) = O\big((z_j - x)^k\big)$ where $0 < k < 1$ which is consistent with bull-nosed shaped dendritic terminals, and is inconsistent with dendritic terminals that end on a taper. The clear suggestion is that dendritic limbs may be well modelled by tapering sections except near terminals where a bull-nose shaped cross-section should be matched to the tapered section to achieve termination.

Discrete Internodal Input

Discrete internodal current inputs such as might arise in the consideration of synaptic activity are now discussed for the general compartmental model. The usual approach is simply to assign a synaptic input to the compartment on which it naturally falls. This procedure ignores the exact location of the synaptic input and the effects that this input may have on neighbouring compartments. The generalised model suggests a procedure for partitioning the effects of synaptic activity between compartments.

Suppose that a synapse is active at site z_s between the nodes z_j and z_{j+1} and let $V^{(s)}$ be the membrane potential at z_s. If $I_{j,j+1}$ is the current leaving z_j in the direction of z_{j+1} then the balance of currents at the site of the synaptic input requires that $I_{j,j+1}+g_s(t)(V_k^{(s)}-E_\alpha)$ is the current entering z_{j+1}. The potentials V_j, $V^{(s)}$ and V_{j+1} are therefore connected by the equations

$$V^{(s)} - V_j = \frac{I_{j,j+1}}{g_A}\int_{z_j}^{z^{(s)}} \frac{ds}{A(s)}\,, \tag{8.136}$$

$$V_{j+1} - V^{(s)} = \frac{I_{j,j+1} + g_s(t)(V^{(s)} - E_\alpha)}{g_A}\int_{z^{(s)}}^{z_{j+1}} \frac{ds}{A(s)}\,. \tag{8.137}$$

By elimination of $V^{(s)}$ between these equations, $I_{j,j+1}$ is determined in terms V_j and V_{j+1}. In this way, the effect of the synaptic activity at z_s is incorporated into the system of compartmental equations for the membrane potential at the nodes z_j and z_{j+1}. This idea for one synapse can be extended to many synapses and gives rise to a system of linear equations comparable to (8.136, 8.137). However, since synaptic activity occurs stochastically and evolves in time, the system matrix itself is dynamic and the solution process is excessively time consuming.

An approximate way to partition synaptic activity in a way that is both numerically efficient and responsive to variations in location of synaptic inputs is based on the assumption that synaptic currents are small enough to ensure that the potential distribution between nodes is not significantly different from that based on zero internodal current input. In this event,

$$V(z) = V_j - \frac{I_{j,j+1}(t)}{g_A}\int_{z_j}^{z} \frac{ds}{A(s)}\,, \qquad V_{j+1} - V_j = -\frac{I_{j,j+1}(t)}{g_A}\int_{z_j}^{z_{j+1}} \frac{ds}{A(s)}\,.$$

Eliminating the current $I_{j,j+1}$ between the equations yields

$$V(z) - E_\alpha = \frac{\left(V_j - E_\alpha\right)\displaystyle\int_{z}^{z_{j+1}} \frac{ds}{A(s)} + \left(V_{j+1} - E_\alpha\right)\displaystyle\int_{z_j}^{z} \frac{ds}{A(s)}}{\displaystyle\int_{z_j}^{z_{j+1}} \frac{ds}{A(s)}}\,. \tag{8.138}$$

Suppose now that a synapse is active at z_s, then the related input current is modelled by $g_s(t)(V(z_s) - E_\alpha)$. In view of formula (8.138) for $V(z)$, it is clear that synaptic input at $z = z_s$ may be approximately redistributed as $f_s g_s(t)$ at z_j and $(1 - f_s)g_s(t)$ at z_{j+1} where

$$f_s = \frac{\displaystyle\int_{z_s}^{z_{j+1}} \frac{ds}{A(s)}}{\displaystyle\int_{z_j}^{z_{j+1}} \frac{ds}{A(s)}}\,.$$

This approximate result is derivable from the general procedure outlined at the start of this subsection by expanding the solutions for $I_{j,j+1}$ to order $O(g_s^2)$.

For uniformly tapering dendrites, it can be shown that

$$f_s = \frac{A_R^{-1/2} - A_S^{-1/2}}{A_R^{-1/2} - A_L^{-1/2}} = \frac{P_R^{-1} - P_S^{-1}}{P_R^{-1} - P_L^{-1}} .$$

Time Integration

By restricting attention to the formulation of compartmental models of neurons, this section has so far avoided any confusion that may arise between compartmental models and apparently similar numerical schemes used to treat partial differential equation models of dendrites. The discussion has indicated that compartmental models of neuronal behavior are formulated typically as a system of ordinary differential equations for the dendritic membrane potential together with a set of initial conditions. Although these equations have many linear terms, they are generally nonlinear due to the presence of intrinsic voltage dependent currents.

All subsequent discussion is directed towards the numerical solution of the compartmental equations. The presence of transient solutions in the model equations presents the primary difficulty in their numerical integration. Although these transients are short lived, they impose a global limitation on the size of time step for which numerical schemes based on forward integration are feasible. Such algorithms may require absurdly small integration time steps at *all times* despite the fact that the true solution may be very well behaved once the transients have decayed. Differential equations with two widely different time scales (transient time and total observation time, for example) are said to be stiff. By way of illustration, consider the numerical solution of the differential equation

$$\tau \frac{dy}{dt} = -(y-1) , \qquad y(0) = A > 1 \tag{8.139}$$

using the standard Euler scheme

$$y_{n+1} = y_n - \frac{h}{\tau}(y_n - 1) , \qquad y_0 = A$$

with fixed time step h and where $y_n = y(nh)$. By inspection, the exact solution is $y(t) = 1 + (A-1)e^{-t/\tau}$ and the numerical solution is $y_n = 1 + (A-1)\big(1 - h/\tau\big)^n$. For a given τ, the behavior of the numerical solution can be classified into the three distinct regions $h \in (0, \tau)$, $h \in (\tau, 2\tau)$ and $h \in (2\tau, \infty)$.

Stable solution: When $0 < h < \tau$ then $0 < (1 - h/\tau) < 1$ and the numerical scheme and analytical solution are in agreement. This is the only situation in which the forward scheme reflects accurately the analytical solution.

Oscillatory and bounded: When $\tau < h < 2\tau$ then $-1 < (1 - h/\tau) < 0$ and the numerical scheme oscillates boundedly, although $y_n \to 1$ as $n \to \infty$, that is, the limit of the numerical and analytical schemes are identical. However, this is only time at which the analytical and numerical solutions are uniformly close. At finite times, the analytical and numerical solutions differ markedly.

Oscillatory and unbounded: When $h > 2\tau$ then $(1 - h/\tau) < -1$ and the numerical scheme oscillates unboundedly. The numerical and analytical schemes are nowhere identical other that at the initial point. Clearly the numerical scheme is unstable in this instance.

In particular, these result apply to the numerical solution at all times since any re-adjustment of h after time T is equivalent to a new initial value problem with starting value

$y(T)$ at time $t = T$. This simple example illustrates the archetypal numerical behavior of stiff equations.

Consider now the numerical scheme

$$y_{n+1} = y_n - \frac{h}{\tau}(1-\alpha)(y_n - 1) - \frac{h}{\tau}\alpha(y_{n+1} - 1) \tag{8.140}$$

for the solution of equation (8.139) in which $\alpha \in [0, 1]$. Clearly $\alpha = 0$ is the (forward) Euler scheme just discussed while $\alpha = 1$ is a fully backward Euler scheme to solve (8.139). Again, it is straightforward to verify that scheme (8.140) has solution

$$y_n = 1 + (A-1)\left(\frac{\tau + h\alpha - h}{\tau + h\alpha}\right)^n .$$

By inspection, this scheme may oscillate unboundedly if $\alpha < 1/2$ and may oscillate boundedly if $1/2 \le \alpha < 1$. However, if $\alpha = 1$, that is, the scheme is fully backward then

$$y_n = 1 + (A-1)\left(\frac{\tau}{\tau + h}\right)^n$$

and the scheme is unconditionally stable irrespective of the choice of h. The key feature of algorithm (8.140) is that errors in y_n are *reduced* in y_{n+1} because of the contraction property of the multiplier $\tau/(\tau+h)$. It is the unconditional stability of backward integration schemes that render them most suitable for the integration of dendritic compartmental equations. Suppose that n steps of size h take the solution to time t then $t = nh$ and

$$y_n = y_n(t) = 1 + (A-1)\left(\frac{\tau}{\tau + h}\right)^n = 1 + (A-1)\left(1 + \frac{t}{\tau n}\right)^{-n}$$

is now the estimate of $y(t)$ using the numerical scheme. Using the standard result that $(1 + x/n)^{-n} \to e^{-x}$ as $n \to \infty$, it follows that $y_n(t) \to y(t)$, the exact solution at time t, as $n \to \infty$.

In traditional numerical work involving the integration of stiff equations, it is common practice to use commercial software libraries such as NAG or IMSL simply because they provide high quality adaptive bootstrapping schemes (variable order), the best known of which is undoubtedly due to Gear (1971). However, it would be misleading to suggest that the compartmental equations arising in dendritic modelling can be classified as traditional for two straightforward reasons. Firstly, the equations are sparse since compartments interact with nearest neighbours only and secondly, the synaptic activity on a dendrite is stochastic and therefore integration must proceed in small time steps to appreciate the statistics of this activity. It is primarily for these two reasons that numerical methods have evolved in tandem with the dendritic compartmental models themselves (see Mascagni, 1989; Hines and Carnevale, 1997), whilst commercial implementations of stiff integrators have been effectively sidelined. Suppose that the compartmentalised equations for a dendrite have form

$$\frac{dV}{dt} = AV + F(V), \qquad V(0) = V_0 , \tag{8.141}$$

where $V = \big(v_0(t), \ldots, v_n(t)\big)^{\mathrm{T}}$ are the membrane potentials at nodes $z_0, \ldots, z_n$ respectively and F is an $(n+1)$-vector whose components are *non-linear* functions of the components of V, that is, $v_0, \ldots, v_n$ (typically arising from intrinsic and voltage dependent currents). In particular, A is an $(n+1) \times (n+1)$ matrix whose entries may be dependent on time but are independent of $v_0, \ldots, v_n$. Equation (8.141) is typically integrated using the backward integration scheme

$$V_{k+1} = V_k + h\Big(A_{k+1}V_{k+1} + F(V_{k+1})\Big)$$

which may be re-arranged to form

$$(I - hA_{k+1})V_{k+1} = V_k + hF(V_{k+1}) \,. \tag{8.142}$$

Compartmental models enjoy the property that A_{k+1} has real and negative eigenvalues so that $(I - hA_{k+1})$ has real eigenvalues all exceeding unity. Thus the inverse of $(I - hA_{k+1})$ exists and is a contraction mapping, being a generalisation of the scalar contraction $\tau/(\tau + h)$. The algorithm (8.142) is now implemented in the form

$$V_{k+1} = (I - hA_{k+1})^{-1}\big(V_k + hF(V_{k+1})\big) \,. \tag{8.143}$$

Given V_k, values of V_{k+1} are first predicted and then corrected while the contraction property of $(I - hA_{k+1})^{-1}$ ensures numerical stability. In practice, the tri-diagonal predominance is such that no matrix inverse is formally computed in the execution of the iterative scheme (8.143).

The Spectral Methodology

There are a number of neurophysiology packages (see De Schutter, 1992) that can be used to investigate dendritic trees modelled by equation (8.15). These packages almost universally resolve the spatial dependence of the tree potential using the method of finite differences. However, the ease with which finite difference schemes can be programmed for dendritic trees should now be counterbalanced against the poor numerical resolution inherent in such algorithms and the need to embrace the much larger and more complex dendritic structures required in contemporary neurophysiology. Nowadays finite difference schemes in applied mathematics have been largely superseded by finite element and spectral algorithms for intensive calculations. The latter method is particularly suitable for the description of dendritic trees. Furthermore, spectral techniques usually enjoy exponential convergence to the analytical solution unlike finite differences methods which converge only algebraically. Canuto, Hussaini, Quarteroni and Zang (1988) make this comparison for $V(x,t)$, the solution of the initial boundary value problem

$$\frac{\partial V}{\partial t} = \frac{\partial^2 V}{\partial x^2} \,, \qquad (x,t) \in (0,1) \times (0,\infty) \tag{8.144}$$

with boundary and initial data

$$V(0,t) = V(1,t) = 0 \,, \qquad V(x,0) = \sin \pi x \,. \tag{8.145}$$

The exact solution is $V(x,t) = e^{-\pi^2 t} \sin \pi x$ while the spectral solution based on Chebyshev polynomials can be shown to be

$$V(x,t) = 2 \sum_{k=1}^{\infty} \sin(k\pi/2) J_k(\pi) e^{-\pi^2 t} T_k(x) \tag{8.146}$$

where $T_k(x)$ is the Chebyshev polynomial of order k (to be defined shortly) and $J_k(x)$ is the Bessel function of the first kind of order k. It can be demonstrated that the truncated series (8.146) converges to $V(x,t)$ exponentially. Canuto *et al.* (1988) quote maximum errors for the Chebyshev collocation algorithm varying from 4.58×10^{-4} with 8 polynomials to 2.09×10^{-11} with 16 polynomials. Second order finite differences with 16 degrees of freedom achieves an accuracy of 0.135.

Typically g_L/g_A, the ratio of membrane leakage conductance to core conductance, and a/l^2, the ratio of dendritic radius to length squared are both order one. Hence the operational conditions of the cable equation (8.15) are very similar to those quoted here for the heat equation (8.144). In practice, the vast majority of C and Fortran compilers in the marketplace cannot take advantage of the exponential convergence inherent in a spectral algorithm although there is now increasing pressure to remedy this situation. In the current environment, we believe that spectral algorithms in dendritic modelling are probably best deployed to reduce the number of variables needed to capture the essence of each limb of a dendritic tree thereby freeing resources to be used either to treat larger dendrites or alternatively to do faster computations on a given dendrite.

Mathematical Preliminaries

The family of Chebyshev polynomials $T_0(z), T_1(z), \ldots T_n(z) \ldots$ is defined by the identity

$$T_n(\cos\vartheta) = \cos(n\vartheta)\,, \qquad n = 0, 1, \ldots\,. \tag{8.147}$$

For example, $T_0(z) = 1$, $T_1(z) = z$ and $T_2(z) = 2z^2 - 1$ are the first three Chebyshev polynomials. By replacing ϑ by $\pi - \vartheta$ in (8.147), and by substituting $\vartheta = 0$ and $\vartheta = \pi$ in definition (8.147), it follows that

$$\left.\begin{array}{ll} T_n(-z) = (-1)^n T_n(z)\,, & z \in [-1, 1]\,, \\ T_n(1) = 1\,, & T_n(-1) = (-1)^n\,, \end{array}\right\} \quad n = 0, 1, \ldots\,. \tag{8.148}$$

Furthermore $T_n'(-1)$ and $T_n'(1)$ can be found by differentiating definition (8.147) with respect to ϑ and then recognising that

$$T_n'(-1) = \lim_{\vartheta\to\pi} \frac{n \sin n\vartheta}{\sin\vartheta} = (-1)^n n^2\,, \qquad T_n'(1) = \lim_{\vartheta\to 0} \frac{n \sin n\vartheta}{\sin\vartheta} = n^2\,. \tag{8.149}$$

Results (8.148) and (8.149) are used in the treatment of boundary conditions involving potential and currents respectively. From definition (8.147), it can also be demonstrated in a straightforward way that Chebyshev polynomials possess the properties

$$2T_n(z)T_m(z) = T_{m+n}(z) + T_{|n-m|}(z)\,, \tag{8.150}$$

$$2T_n(z) = \frac{T_{n+1}'(z)}{n+1} - \frac{T_{n-1}'(z)}{n-1}\,, \qquad n > 1\,, \tag{8.151}$$

together with the orthogonality condition

$$\int_{-1}^{1} \frac{T_n(z)T_m(z)}{\sqrt{1-z^2}}\,dz = \begin{cases} 0 & n \neq m \\ \pi & m = n = 0 \\ \pi/2 & n = m \geqslant 1 \end{cases}\,. \tag{8.152}$$

A large class of important functions have Chebyshev series expansions. The situation is analogous to Fourier series, and just as with Fourier series, the orthogonality condition is used to calculate the coefficients of a Chebyshev series expansion from its parent function. Suppose that f is a function defined in $[-1, 1]$ and is square integrable with respect to the weight function $w(z) = (1 - z^2)^{-1/2}$, that is

$$\int_{-1}^{1} \frac{f^2(z)\,dz}{\sqrt{1-z^2}} < \infty$$

then f has a Chebyshev series expansion

$$f(z) = \sum_{k=0}^{\infty} \hat{f}_k T_k(z)\,, \tag{8.153}$$

whose coefficients $\hat{f}_k$ are formed from (8.153) by multiplying both sides of this equation by $T_n(z)/\sqrt{1-z^2}$ and then integrating over $[-1, 1]$. It follows immediately that

$$\hat{f}_0 = \frac{1}{\pi}\int_{-1}^{1} \frac{f(z)}{\sqrt{1-z^2}}\,dz\,, \qquad \hat{f}_n = \frac{2}{\pi}\int_{-1}^{1} \frac{f(z)T_n(z)}{\sqrt{1-z^2}}\,dz \qquad n \geqslant 1\,. \tag{8.154}$$

The non-periodic nature of Chebyshev series makes then an ideal way to represent solutions of partial differential equations whose boundary conditions are non-periodic.

Suppose now that the representation (8.153) involves only $(N+1)$ polynomials so that $\hat{f}_{N+1} = \hat{f}_{N+2} = \cdots = 0$. In this case $f(z)$ has Chebyshev series *approximation*

$$f(z) = \sum_{k=0}^{N} f_k T_k(z)\,. \tag{8.155}$$

Superficially it would appear that the series (8.155) is just a truncated version of (8.153) in which f_k is an approximation of $\hat{f}_k$, the coefficient in the corresponding infinite expansion. In fact, the truncated series (8.155) is more accurately viewed as a compromise description of $f(z)$ in which the coefficients $f_0 \cdots f_N$ also contain the aliased effects of the residual series $\hat{f}_{N+1}T_{N+1} + \cdots$ in expansion (8.153). For the finite Chebyshev expansion, the expression (8.154) for the coefficient $\hat{f}_n$ is replaced by

$$f_n = \frac{1}{c_n}\int_{-1}^{1} \frac{f(z)T_n(z)}{\sqrt{1-z^2}}\,dz\,, \qquad c_n = \begin{cases} \pi & n = 0, N \\ \pi/2 & 0 < n < N\,. \end{cases}$$

Discrete Chebyshev Transform

The computation of $f_0, f_1, \ldots, f_N$ from $f(z)$ is now addressed. Given a set of $N+1$ points $(z_0, F_0), (z_1, F_1), \ldots, (z_N, F_N)$ where $F_k = f(z_k)$, then the coefficients $f_0, f_1, \ldots, f_N$ may be evaluated as the solution of the system of $N+1$ linear equations

$$F_k = \sum_{j=0}^{N} f_j T_j(z_k)\,, \qquad k = 0, 1, \ldots, N\,.$$

The dissection $z_k = \cos(k\pi/N)$ $(0 \leqslant k \leqslant N)$, given by the nodes of a Gauß-Lobatto quadrature of order N (see Davis and Rabinowitz, 1983)), are often favoured and leads to the discrete Chebyshev transform. This states that if $z_0, z_1, \ldots, z_N$ are points in $[-1, 1]$ defined by $z_k = \cos(\pi k/N)$ and if $F_k = f(z_k)$ then

$$F_j = \sum_{k=0}^{N} f_k \cos\frac{\pi jk}{N}\,, \qquad j = 0, 1, \ldots, N \tag{8.156}$$

$$f_k = \frac{2}{Nc_k}\sum_{j=0}^{N} \frac{1}{c_j} F_j \cos\frac{\pi jk}{N}\,, \qquad k = 0, 1, \ldots, N \tag{8.157}$$

where

$$c_0 = c_N = 2\,, \qquad c_1 = c_2 = \cdots = c_{N-1} = 1\,. \tag{8.158}$$

Of course, other families of orthogonal polynomials possess similar discrete transforms. However, the discrete Chebyshev transform is singled out by the fact that, with minor modification, it can be recast in the framework of the Fast Fourier Transforms (FFT) so that the transition from coefficient (spectral space) to value (physical space) and vice-versa is performed with high precision and speed. This is one of the major ingredients of the Chebyshev magic.

Function Computation

The discrete Chebyshev transform, when implemented by means of the fast Fourier transform, provides an efficient way to move from coefficient space to value space provided function values at $z_0, z_1, \ldots, z_N$ are needed. However, it is often necessary to compute function values at points lying between the Gauß-Lobatto nodes. These can be computed easily from (8.153) by first determining the angle $\vartheta \in [0, \pi]$ corresponding to $z = \cos\vartheta$ and then recognising that

$$f(z) = f(\cos\vartheta) = \sum_{k=0}^{N} f_k \cos k\vartheta \,. \tag{8.159}$$

Alternatively, if such function values are required frequently (e.g the treatment of spatially distributed synaptic inputs), another more sophisticated (and also more efficient) algorithm is appropriate.

Let $y_0, y_1, \ldots, y_{N+2}$ be a sequence defined by the iteration

$$y_k = 2zy_{k+1} - y_{k+2} + f_k \,, \quad k = 0, 1, \ldots, N \qquad y_{N+1} = y_{N+2} = 0 \,, \tag{8.160}$$

where f_k is the coefficient of $T_k(z)$ in the Chebyshev series approximation of $f(z)$. From property (8.150), it follows that $2zT_k(z) = T_{k+1}(z) + T_{k-1}(z)$. This result, together with the definition of f_k in terms of y_k, when used in expression (8.155) for $f(z)$, yields

$$\begin{aligned} f(z) &= \sum_{k=0}^{N} f_k T_k(z) = \sum_{k=0}^{N} \Big(y_k - 2zy_{k+1} + y_{k+2}\Big) T_k(z) \\ &= \sum_{k=0}^{N} y_k T_k(z) - 2zy_1 - \sum_{k=1}^{N} y_{k+1}\Big(2zT_k(z)\Big) + \sum_{k=0}^{N} y_{k+2} T_k(z) \\ &= \sum_{k=0}^{N} y_k T_k(z) - 2zy_1 - \sum_{k=1}^{N} y_{k+1}\Big(T_{k+1}(z) + T_{k-1}(z)\Big) + \sum_{k=0}^{N} y_{k+2} T_k(z) \\ &= \sum_{k=0}^{N} y_k T_k(z) - 2zy_1 - \sum_{k=2}^{N+2} y_k T_k(z) - \sum_{k=0}^{N-1} y_{k+2} T_k(z) + \sum_{k=0}^{N} y_{k+2} T_k(z) \\ &= y_0 - zy_1 \end{aligned}$$

This iterative computation of $f(z)$ is based on Clenshaw's algorithm (see Clenshaw, 1962). Although more difficult to implement than formula (8.159) for $f(z)$, it *never* requires the determination of $T_k(z)$ nor does it require the evaluation of the cosine function. Furthermore, the iterative scheme (8.160) is stable provided $|z| \leqslant 1$ (always true!). Thus if f needs to be evaluated frequently at points that are not Gauß-Lobatto nodes, then iterative scheme (8.160) is recommended and leads to the final result $f(z) = y_0 - zy_1$.

Spectral Differentiation

Suppose that f is a differentiable function of z in $(-1, 1)$ then $f'(z)$ can be estimated from the representation (8.155). Indeed, it is obvious from property (8.151) that

$$\frac{df}{dz} = \sum_{k=0}^{N} f_k \frac{dT_k(z)}{dz} = \sum_{k=0}^{N} f_k^{(1)} T_k(z) \,, \tag{8.161}$$

where the coefficients $f_k^{(1)}$ are generated from f_k by the iterative formula

$$\begin{aligned} f_k^{(1)} &= f_{k+2}^{(1)} + 2(k+1) f_{k+1} \qquad 1 \leqslant k \leqslant N-1 \,, \\ 2f_0^{(1)} &= f_2^{(1)} + 2f_1 \,, \qquad f_N^{(1)} = f_{N+1}^{(1)} = 0 \,. \end{aligned} \tag{8.162}$$

This result is central to the numerical solution of partial differential equations using Chebyshev polynomials. Its justification relies on that fact that $T_k(z)$ can be replaced by derivatives of Chebyshev polynomials from property (8.151). Using this idea together with the trivial observation that $T_0'(z) = 0$, the manipulation of formula (8.161) gives

$$\begin{aligned}\sum_{k=1}^{N} 2f_k T_k'(z) &= 2f_0^{(1)} T_0(z) + 2f_1^{(1)} T_1(z) + \sum_{k=2}^{N} f_k^{(1)} \left(\frac{T_{k+1}'(z)}{k+1} - \frac{T_{k-1}'(z)}{k-1} \right) \\ &= 2f_0^{(1)} T_1'(z) + \frac{f_1^{(1)}}{2} T_2'(z) + \sum_{k=3}^{N+1} f_{k-1}^{(1)} \frac{T_k'(z)}{k} - \sum_{k=1}^{N-1} f_{k+1}^{(1)} \frac{T_k'(z)}{k} \\ &= f_0^{(1)} T_1'(z) + \sum_{k=1}^{N} \left(f_{k-1}^{(1)} - f_{k+1}^{(1)} \right) \frac{T_k'(z)}{k}\end{aligned}$$

bearing in mind that $f_N^{(1)} = f_{N+1}^{(1)} = 0$. Equating the coefficients of $T_k'(z)$ on both sides of this identity yields the result

$$f_{k-1}^{(1)} - f_{k+1}^{(1)} = 2kf_k\,, \quad 2 \leqslant k \leqslant N\,, \qquad\qquad 2f_0^{(1)} - f_2^{(1)} = 2f_1\,.$$

The iterative relationship (8.161) is a practical way to calculate $f_0^{(1)}, \ldots, f_{N-1}^{(1)}$, the coefficients of the Chebyshev series for $f'(z)$. Furthermore, the algorithm can be repeated again to compute $f_0^{(2)}, \ldots, f_{N-2}^{(2)}$, the Chebyshev coefficients of $f''(z)$.

Boundary Conditions

Of course, once the Chebyshev coefficients of $f'(z)$ are known then $f'(z)$ can be evaluated for every $z \in [-1, 1]$ by summation of the series, and can be computed very efficiently at the Chebyshev nodes using the inverse discrete Chebyshev transform. The endpoints $z = 1$ and $z = -1$ are particularly important since $f(1)$, $f(-1)$, $f'(1)$ and $f'(-1)$ are ubiquitous ingredients of boundary conditions. It is a piece of tedious algebra to verify that

$$f'(1) = \frac{2N^2+1}{6} F_0 + \sum_{j=1}^{N-1} \frac{F_j(-1)^j}{\sin^2(\pi j/2N)} + \frac{1}{2} F_N\,, \tag{8.163}$$

$$f'(-1) = -\frac{1}{2} F_0 - \sum_{j=1}^{N-1} \frac{F_j(-1)^j}{\cos^2(\pi j/2N)} - \frac{2N^2+1}{6} F_N \tag{8.164}$$

in which $F_k = f(z_k)$. Thus boundary conditions involving derivatives can be used to extract information concerning boundary values of the function f itself. For example, suppose that a cable is sealed at $z = 1$ so that the potential $V(z, t)$ has zero gradient at $z = 1$. In view of identity (8.163), the sealed end condition is arranged by ensuring that $V_0(t)$, the potential at $z = 1$, satisfies

$$\frac{\partial V(1, t)}{\partial z} = \frac{2N^2+1}{6} V_0(t) + \sum_{j=1}^{N-1} \frac{V_j(t)(-1)^j}{\sin^2(\pi j/2N)} + \frac{1}{2} V_N(t) = 0$$

from which it follows that

$$V_0(t) = -\frac{6}{2N^2+1} \sum_{j=1}^{N-1} \frac{V_j(t)(-1)^j}{\sin^2(\pi j/2N)} - \frac{3}{2N^2+1} V_N(t)\,. \tag{8.165}$$

Representation of Synaptic Input

It has already been recognised that synaptic current inputs into a dendrite are commonly described by a time dependent conductance whose spatial behavior is delta-like. For example, the activation of a synaptic input at time τ and location $x = X$ on a dendrite of length L involving species S_α is modelled by the current density $g_{\text{syn}}(t-\tau)(V(X,t)-E_\alpha)\delta(x-X)$ in the cable equation for that limb where typically

$$g_{\text{syn}}(t) = \begin{cases} 0 & t \leqslant 0 \\ G^{(\alpha)}\dfrac{t}{T}e^{(1-t/T)} & t > 0\,. \end{cases}$$

Suppose that

$$g_{\text{syn}}(t)\delta(x-X)(V(x,t)-E_S) = \hat{g}(z,t) = g_{\text{syn}}(t)\sum_{k=0}^{N} g_k T_k(z) \tag{8.166}$$

then the coefficients $g_0, g_1, \ldots, g_N$ are given by the formulae

$$g_k = \frac{1}{c_k}\int_{-1}^{1}\frac{\hat{g}(z)}{\sqrt{1-z^2}}\,dz\,, \qquad c_0 = c_N = \pi, \quad c_k = \frac{\pi}{2}, \quad (0 < k < N)\,. \tag{8.167}$$

However, the computation of $\hat{g}(z)$ requires some formal recognition of the properties of delta functions. In all subsequent analyses the mapping $z = -1 + 2x/L$ will be used to associate point x on a dendritic segment of length L (i. e. $x \in [0, L]$) with point $z \in [-1, 1]$. Under this mapping, $X \in [0, L]$ on the real dendrite maps to $Z \in [-1, 1]$ where $Z = -1 + 2X/L$. Thus

$$\delta(x-X) = \delta\left(\frac{L(1+z)}{2} - \frac{L(1+Z)}{2}\right) = \delta\left(\frac{L(z-Z)}{2}\right) = \frac{2}{L}\delta(z-Z)\,.$$

Suppose that a synaptic input becomes active at time τ then it is clear from (8.166) that

$$\hat{g}(z) = \frac{2g_{\text{syn}}(t-\tau)}{L}\delta(z-Z)(V(z,t)-E_S) \tag{8.168}$$

where $V(z,t)$ is used unambiguously to mean $V(x(z),t)$. It now follows from (8.167) that

$$\begin{aligned} g_k &= \frac{2g_{\text{syn}}(t-\tau)}{Lc_k}\,\frac{\Big(V(Z,t)-E_\alpha\Big)T_n(Z)}{\sqrt{1-Z^2}} \\ &= \frac{2g_{\text{syn}}(t-\tau)}{Lc_k}\,\frac{\Big(V(\cos\Theta,t)-E_\alpha\Big)\cos k\Theta}{\sin\Theta}\,, \end{aligned} \tag{8.169}$$

where $Z = \cos\Theta$. These expressions therefore give the coefficients of the Chebyshev expansion of a synaptic input occurring at $x = X$.

Solution Procedure

The cable equation (8.15) may be represented in the general format

$$\frac{\partial V}{\partial t} = \mathscr{L}(V) \tag{8.170}$$

where $\mathscr{L}$ is an operator involving t and x but only spatial derivatives. The dendrite $x \in [0, L]$ is first mapped into $[-1, 1]$ by the formula $z = -1+2x/L$. Similarly each occurrence of the differential operator $\partial/\partial x$ in $\mathscr{L}$ is replaced by $(2/L)\partial/\partial z$. Suppose now that the modified potential $V(z,t)$ is approximated by the discrete Chebyshev series

$$V(z,t) = \sum_{k=0}^{N} v_k(t)T_k(z)\,, \qquad (z,t) \in [-1,1] \times (0,\infty) \tag{8.171}$$

then

$$\frac{\partial V(z)}{\partial t} = \sum_{k=0}^{N} \frac{\partial v_k}{\partial t} T_k(z)\,. \tag{8.172}$$

Two distinct solution strategies are now evident; either solve for $v_0(t), \ldots, v_N(t)$, the coefficients of the spectral representation of $V(z,t)$ or solve for $V(z_0,t), \ldots, V(z_N,t)$, the values of the potential at the Chebyshev nodes $z_0, \ldots, z_N$. The former is commonly called a “spectral” or “pure spectral” approach because it seeks to find the coefficients of the spectral series for V rather than function values. The latter is often called a “pseudo-spectral” or “collocation” approach because it seeks to solve for function values at the nodes of a dissection. Briefly, the former is often harder to implement and arguably less physical in its representation of the boundary conditions. On the other hand it executes faster and is more powerful in the respect that it can deal more effectively with diverging solutions. However, the well behaved nature of dendritic potentials allied with the number and complexity of the boundary conditions for the interconnected limbs of a dendritic tree suggests that the collocation approach is probably the better strategy in this instance.

Spectral Algorithm

For completeness, the basic construction of the spectral algorithm is now described for equation (8.170). The potential and its time derivative are first represented in the form described in equations (8.171) and (8.172). The latter equation is now multiplied by $(1-z^2)^{-1/2}\,T_j(z)$ and integrated over $[-1,1]$ to obtain

$$\frac{dv_j}{dt} = \frac{2}{\pi c_j}\int_{-1}^{1} \frac{\partial V}{\partial t}\frac{T_j(z)}{\sqrt{1-z^2}}\,dz = \frac{2}{\pi c_j}\int_{-1}^{1} \frac{\mathscr{L}(V)T_j(z)}{\sqrt{1-z^2}}\,dz\,, \quad j = 0,1,\ldots,N\,. \tag{8.173}$$

Each integer value of j from $j = 0$ to $j = N-2$ generates an ordinary differential equation connecting the coefficients $v_0, v_1, \ldots, v_N$. However, the equations in (8.173) arising from $j = N-1$ and $j = N$ misrepresent $\dot{v}_{N-1}$ and $\dot{v}_N$ due to missing contributions from Chebyshev polynomials of order higher than N. These two equations are therefore replaced by boundary conditions. In fact, two boundary conditions are used to extract algebraic expressions for v_{N-1} and v_N in terms of the other spectral coefficients $v_0, v_1, \ldots, v_{N-2}$.

Collocation Algorithm

The collocation solution of equation (8.170) requires that

$$\frac{\partial V(z_j,t)}{\partial t} + J(z_j,t) = \mathscr{L}(V)|_{z=z_j}\,. \tag{8.174}$$

be true arithmetically at all internal points $z_1, \ldots, z_{N-1}$ of the dissection $(-1,1)$. Boundary conditions are solved algebraically for $V(z_0,t)$ and $V(z_N,t)$. It should be stated here that theoretical analyses leading to the formulation of equations (8.173) and (8.174) often present these derivation in terms of trial functions (Chebyshev polynomials in this case) and test functions (delta functions here). Of course, such analyses are essential in an appreciation of the errors inherent in the numerical procedures but have negligible impact on the practical implementation of the algorithms.

Irrespective of implementation methodology (pure spectral or pseudo spectral), boundary conditions are applied *first* before the computation of $\mathscr{L}(V)$ is performed. Furthermore, the treatment of the boundary conditions is nearly always stabilizing in the sense that computational errors in v_{N-1}, v_N or alternatively $V(z_0, t)$, $V(z_N, t)$ are multiplied by a factor less than unity and therefore are controlled.

Two passive non-tapering prototype dendritic trees are now used to illustrate the method. Each dendritic configuration has a closed form analytical solution that will be extracted and compared with that arising from the numerical solution of the underlying partial differential equations. Of course, the numerical strategy is equally implementable for non-uniform active dendrites whereas the analytical methodology breaks down for tapered dendrites[18].

Spectral and Exact Solution of an Unbranched Tree

Consider a simple dendrite consisting of a uniform single limb of radius a and length l attached to a soma of area A_S. The dendrite is assumed to be passive, and therefore described by the linear cable equation. In the absence of active synaptic input current J_S, synaptic current activity is *simulated* by the time dependent alpha function input current $J_{in}(x, t) = G(t/T)e^{(1-t/T)}\delta(x - fl)$ acting at location $x = fl$, $f \in (0, 1)$ and turned on at $t = 0$. The configuration is illustrated in figure 13.

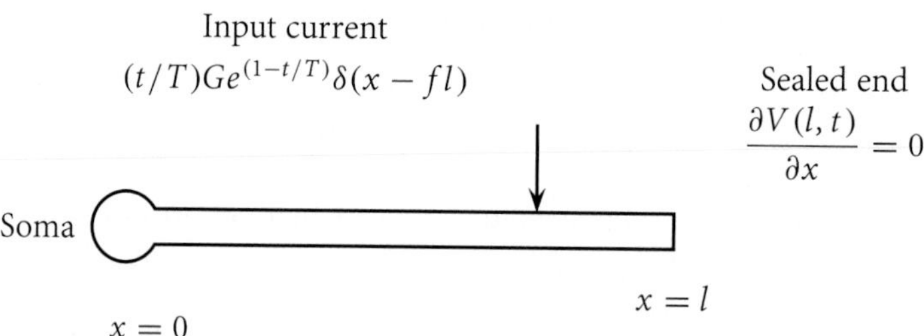

Fig. 13. A passive dendritic limb with sealed end and attached soma is stimulated by an injected "alpha" function pulse of current at $x = fl$, where $(0 < f < 1)$.

The deviation from the resting potential of the passive dendrite in figure 13 is modelled by the partial differential equation

$$\tau \frac{\partial V}{\partial t} + V + \frac{J_{in}}{g_L} = \frac{a g_A}{2 g_L} \frac{\partial^2 V}{\partial x^2}, \qquad x \in (0, l) . \tag{8.175}$$

Equation (8.175) is solved with boundary conditions

$$\frac{\partial V(l,t)}{\partial x} = 0 , \qquad \tau_S \frac{\partial V(0,t)}{\partial t} + V(0,t) = \frac{g_A}{g_S} \frac{\pi a^2}{A_S} \frac{\partial V(0,t)}{\partial x} \tag{8.176}$$

and initial condition

$$V(x, 0) = 0 . \tag{8.177}$$

In order to simultaneously simplify the representation of the problem for analytical purposes and also develop a formulation of the problem suitable for numerical solution using

[18]This is perhaps an overstatement of reality for uniformly tapered dendrites. In fact, the trigonometric functions appearing in the analytical solution need to be replaced by Bessel functions. The entire operation becomes more intractable and requires sophisticated mathematics.

spectral methods, it is convenient to non-dimensionalise time, length and voltage in equations (8.175, 8.176, 8.177) by the coordinate transformations $t^* = t/\tau$, $z = -1 + 2x/l$ and $\varphi = V/V_0$ where $V_0 = 2G\sigma e/lg_L$ and $\sigma = \tau/T$. The corresponding non-dimensionalised equations, boundary conditions and initial condition now become

$$\frac{\partial \varphi}{\partial t} + \varphi + i(z,t) = \beta \frac{\partial^2 \varphi}{\partial z^2}, \qquad z \in (-1, 1), \tag{8.178}$$

$$\frac{\partial \varphi(1,t)}{\partial z} = 0, \tag{8.179}$$

$$\chi \left(\varepsilon \frac{\partial \varphi(-1,t)}{\partial t} + \varphi(-1,t) \right) = \frac{\partial \varphi(-1,t)}{\partial z} \tag{8.180}$$

$$\varphi(z,0) = 0 \tag{8.181}$$

in which superscript stars have been suppressed but it is understood that all variables are non-dimensional. Although $i(z,t) = te^{-\sigma t}\delta(z - (2f - 1))$ in this application, the analysis is pursued generally without explicit reference to this particularly simple form. The non-dimensional parameters σ, β, χ and ε are defined in terms of the biophysical parameters of the problem by

$$\sigma = \frac{\tau}{T}, \qquad \beta = \frac{2a}{l^2} \frac{g_A}{g_L}, \qquad \chi = \frac{l}{2} \frac{g_S}{g_A} \frac{A_S}{\pi a^2}, \qquad \varepsilon = \frac{\tau_S}{\tau}.$$

Parameter σ is the ratio of the time constant of the dendritic membrane to that of the input current while ε is the ratio of the somal and dendritic time constants. In the final specification of the problem it will be assumed that $\varepsilon = 1$ but, for the time being, the analysis proceeds on the basis that ε is an arbitrary positive constant.

Finite Transforms

For real λ let φ_λ and i_λ be defined by

$$\varphi_\lambda = \int_{-1}^{1} \varphi(z,t) \cos \lambda(1-z)\, dz \qquad i_\lambda = \int_{-1}^{1} i(z,t) \cos \lambda(1-z)\, dz\,. \tag{8.182}$$

Equation (8.178) is now multiplied by $\cos \lambda(1-z)$ and the resulting equation integrated over $(-1, 1)$. Bearing in mind that φ satisfies the boundary condition (8.179), after two integrations by parts, it follows easily that

$$\frac{d\varphi_\lambda}{dt} + (1 + \beta\lambda^2)\varphi_\lambda + i_\lambda = -\beta \cos 2\lambda \frac{\partial \varphi(-1,t)}{\partial z} + \beta\lambda \sin 2\lambda\, \varphi(-1,t)\,. \tag{8.183}$$

The somal boundary condition (8.180) is now used to remove the potential gradient at $x = 0$. After some straightforward algebra, equation (8.183) can be recast in the form

$$\frac{d\varphi_\lambda}{dt} + (1 + \beta\lambda^2)\varphi_\lambda + \varepsilon\beta\chi \cos 2\lambda \left[\frac{d\varphi(-1,t)}{dt} + \frac{\chi \cos 2\lambda - \lambda \sin 2\lambda}{\varepsilon\chi \cos 2\lambda} \varphi(-1,t) \right] = -i_\lambda\,. \tag{8.184}$$

The key idea is now to choose λ (as yet unspecified) to satisfy the transcendental equation

$$1 + \beta\lambda^2 = \frac{\chi \cos 2\lambda - \lambda \sin 2\lambda}{\varepsilon\chi \cos 2\lambda} \qquad \equiv \qquad \tan 2\lambda = \chi \left(\frac{1-\varepsilon}{\lambda} - \beta\varepsilon\lambda \right). \tag{8.185}$$

With this choice of λ, the function

$$\psi_\lambda(t) = \varphi_\lambda + \beta\varepsilon\chi \cos(2\lambda)\, \varphi(-1,t) \tag{8.186}$$

satisfies the ordinary differential equation

$$\frac{d\psi_\lambda}{dt} + (1+\beta\lambda^2)\psi_\lambda = -i_\lambda$$

with general solution

$$\psi_\lambda(t) = \psi_\lambda(0)e^{-(1+\beta\lambda^2)t} - \int_0^t i_\lambda(s)e^{-(1+\beta\lambda^2)(t-s)}\,ds\,.$$

In view of the initial condition (8.181), $\psi_\lambda(0) = 0$ in this application so that

$$\psi_\lambda(t) = -\int_0^t i_\lambda(s)e^{-(1+\beta\lambda^2)(t-s)}\,ds\,. \tag{8.187}$$

The primary purpose of this analysis is to provide closed form solutions for the potential in some trial dendrites so that direct comparisons can be made with solutions extracted by the numerical procedures described in the previous sections. With this aim in mind, it makes sense to ease the technical complications arising in the construction of the exact solution by assuming now that $\varepsilon = 1$. Thus λ satisfies the transcendental equation

$$\tan 2\lambda + 2\gamma\lambda = 0\,, \qquad 2\gamma = \beta\chi = \frac{g_S}{g_L}\frac{A_S}{\pi a l}\,. \tag{8.188}$$

In order to appreciate the role of the ψ's in the formation of the analytical cable solution, it is first necessary to develop an orthogonality condition.

Orthogonality of Eigenfunctions

Let λ and η be two distinct solutions of the transcendental equation (8.188) then it is verified easily that

$$\int_{-1}^{1} \sin\lambda(1-z)\sin\eta(1-z)\,dz = \begin{cases} 0 & \lambda \neq \eta \\ 1+\gamma\cos^2 2\lambda & \lambda = \eta\,. \end{cases} \tag{8.189}$$

Thus $\sin\lambda(1-z)$ and $\sin\eta(1-z)$ are orthogonal functions over $[-1,1]$ whenever λ and η are distinct solutions of (8.188). In fact, the eigenfunctions $\cos\lambda(1-z)$ corresponding to each solution λ of (8.188) also form a complete space so that $\varphi(z,t)$ has representation

$$\varphi(z,t) = A_0 + \sum_\lambda A_\lambda \cos\lambda(1-z)\,. \tag{8.190}$$

The task is now to compute A_0 and A_λ. In order to determine A_0, observe first that

$$\int_{-1}^{1} \varphi(z,t)\,dz = 2A_0 + \sum_\lambda A_\lambda \int_{-1}^{1} \cos\lambda(1-z)\,dz = 2A_0 + \sum_\lambda A_\lambda \frac{\sin 2\lambda}{\lambda}\,.$$

Assuming that the series for $\varphi(z,t)$ converges to $\varphi(-1,t)$ when $z=-1$ (as indeed it does) then

$$\varphi_S(t) = A_0 + \sum_\lambda A_\lambda \cos 2\lambda\,.$$

The previous two results are now added together to obtain

$$\psi_0 = A_0(2+\beta\chi) + \sum_\lambda A_\lambda\left(\frac{\sin 2\lambda}{\lambda} + \beta\chi\cos 2\lambda\right) = 2(1+\gamma)A_0\,. \tag{8.191}$$

In conclusion,

$$A_0 = \frac{\psi_0}{2(1+\gamma)} = -\frac{1}{2(1+\gamma)} \int_0^t i_\lambda(s) e^{-(t-s)}\, ds\,. \tag{8.192}$$

Furthermore, the orthogonality property (8.189) leads to the derivation

$$\int_{-1}^{1} \frac{\partial \varphi}{\partial z} \sin\lambda(1-z)\, dz = \sum_\alpha A_\alpha \int_{-1}^{1} \sin\alpha(1-z)\sin\lambda(1-z)\, dz = (1+\gamma\cos^2\lambda)A_\lambda\,.$$

Using integration by parts, it follows immediately that

$$\int_{-1}^{1} \frac{\partial \varphi}{\partial z} \sin\lambda(1-z)\, dz = \varphi_\lambda - \frac{\sin 2\lambda}{\lambda}\varphi_S(t) = \psi_\lambda$$

which in turn leads to the conclusion that

$$A_\lambda = \frac{\psi_\lambda}{1+\gamma\cos^2\lambda}\,. \tag{8.193}$$

Hence the coefficients A_λ in the series expansion of $\varphi(z,t)$ are determined from ψ_λ and consequently the deviation of the dendritic membrane potential from its resting state has closed form expression

$$\varphi(z,t) = \frac{\psi_0(t)}{2(1+\gamma)} + \sum_\lambda \frac{\psi_\lambda(t)\cos\lambda(1-z)}{1+\gamma\cos^2 2\lambda} \tag{8.194}$$

where λ are the solutions of the transcendental equation $\tan 2\lambda + 2\gamma\lambda = 0$.

Convergence of Analytical Solution

There are two quite separate issues underlying the construction of the series solution (8.194) for $\varphi(z,t)$. Arguably the more important of these concerns the question of completeness of the space of eigenfunctions; is every solution representable as a weighted sum of eigenfunctions (c.f. Fourier series)? Allied to this issue are complementary questions relating to the quality of series convergence. On the other hand, boundary and initial conditions are satisfied through the choice of λ's and the time dependence of the ψ's. In fact the eigenvalue problem here is complete, although not all apparently well posed problems enjoy this property. The following cautionary example illustrates this point. The functions $\chi(z,t) = \sin\lambda(1-z)$ demonstrably satisfy the orthogonality property

$$\int_{-1}^{1} \sin\lambda(1-z)\sin\eta(1-z)\, dz = \begin{cases} 0 & \lambda \neq \eta \\ 1-\gamma\cos^2 2\lambda & \lambda = \eta \end{cases}$$

where λ and η are solutions of $\tan 2\lambda = 2\gamma\lambda$ with $\gamma > 0$. In this instance, it can be shown by counterexample that the corresponding eigenspace formed by $\chi_\lambda(z,t)$ is incomplete when $0 < \gamma < 1$ but is partially complete (complete for a restricted class of functions) when $\gamma = 1$ and complete when $\gamma > 1$. It is unclear whether or not these deficiencies can be remedied when $\gamma \leqslant 1$ as they appear to originate from the fact that the equation $\tan x = \gamma x$ has only the trivial solution $x = 0$ on the principle branch of the tangent function in this case. When $\gamma > 1$, $\tan x = \gamma x$ also has a non-trivial solution on the principle branch of the tangent function.

To test the quality of convergence, the previous discussion can be used to deduce the identity

$$1+z = \frac{1}{1+\gamma} + 2\sum_\lambda \left(\frac{\sin\lambda}{\lambda}\right)^2 \frac{\cos\lambda(1-z)}{1+\gamma\cos^2 2\lambda}\,, \qquad z \in [-1,1]\,, \tag{8.195}$$

Table 8.5. The table illustrates the accuracy that can be expected from an exact representation of the solution to the cable equation for various numbers of eigenvalues. In fact, 5 000 eigenvalues is only marginally superior to 500.

Gamma	Error at $z=-1.0$	Error at $z=-0.5$	Error at $z=0.0$	Error at $z=0.5$	Error at $z=1.0$
		Accuracy based on 10 eigenfunctions			
$\gamma=0.1$	1.83×10^{-3}	2.78×10^{-3}	6.22×10^{-4}	4.96×10^{-3}	4.07×10^{-2}
$\gamma=1.0$	7.69×10^{-5}	2.22×10^{-3}	3.14×10^{-4}	4.90×10^{-3}	4.05×10^{-2}
$\gamma=10.0$	6.49×10^{-6}	2.18×10^{-3}	2.77×10^{-4}	4.90×10^{-3}	4.05×10^{-2}
		Accuracy based on 50 eigenfunctions			
$\gamma=0.1$	1.60×10^{-5}	8.91×10^{-5}	5.89×10^{-6}	2.11×10^{-4}	8.11×10^{-3}
$\gamma=1.0$	6.25×10^{-7}	8.78×10^{-5}	2.65×10^{-6}	2.11×10^{-4}	8.11×10^{-3}
$\gamma=10.0$	5.27×10^{-8}	8.77×10^{-5}	2.32×10^{-6}	2.11×10^{-4}	8.11×10^{-3}
		Accuracy based on 500 eigenfunctions			
$\gamma=0.1$	1.63×10^{-8}	9.99×10^{-9}	6.19×10^{-9}	1.24×10^{-8}	8.11×10^{-4}
$\gamma=1.0$	1.82×10^{-10}	7.49×10^{-10}	3.71×10^{-9}	9.37×10^{-9}	8.11×10^{-4}
$\gamma=10.0$	1.25×10^{-10}	9.16×10^{-10}	2.22×10^{-9}	1.04×10^{-8}	8.11×10^{-4}

where $\tan 2\lambda + 2\gamma\lambda = 0$ and $\gamma > 0$ is arbitrary. Table 8.5 provide some indication of the expected error in representing $1+z$ by a truncated series of eigenfunctions.

In this example, 50 eigenfunctions provide adequate accuracy except near $z = 1$. Series using more eigenfunctions clearly provide better accuracy although, in fact, hardly any improvement in accuracy is achieved with more than 500 eigenfunctions primarily due to losses in arithmetical significance and the poor convergence characteristics of the truncated series. A critical inspection of the convergence properties of the orthogonal series reveals that the largest inaccuracies occur near $z = 1$ and that the convergence error there is roughly inversely proportional to the number of polynomials used in the expansion. The non-uniform convergence of series solutions suggests the likelihood of serious practical difficulties when analytical solutions are used to study the characteristics of large, heavily branched passive dendritic structures. Of course, the analytical solution will always produce numbers, but their relevance must be appraised carefully.

Solution for Alpha Current Injection

The specific potential for a dendrite excited by an injected alpha function current input at $x = fl$ is now considered. It has already been shown that this current input is described by $i(z,t) = te^{-\sigma t}\delta(z-(2f-1))$ so that $i_\lambda = \cos 2\lambda(1-f)te^{-\sigma t}$. Thus

$$\psi_\lambda = -\cos 2\lambda(1-f)\int_0^t s\,e^{-\sigma s}e^{-(1+\beta\lambda^2)(t-s)}\,.$$

After some further integration by parts, it can be shown that

$$\psi_\lambda = \frac{\cos 2\lambda(1-f)}{(\sigma-1-\beta\lambda^2)^2}\left[(\sigma-1-\beta\lambda^2)te^{-\sigma t}+e^{-\sigma t}-e^{-(1+\beta\lambda^2)t}\right]. \tag{8.196}$$

Numerical Solution

Let $\varphi_0(t), \varphi_1(t), \ldots, \varphi_N(t)$ denote the potential in the cable at nodes $z_0, z_1, \ldots, z_N$ and let $i_0, i_1, \ldots, i_N$ be a distribution of currents at the nodes that are equivalent to the input current $i(z,t) = te^{-\sigma t}\delta(z - (2f - 1))$, modulo the precision available (only $N + 1$ Chebyshev polynomials are in use). It follows directly from (8.161) that if

$$i(z,t) = \sum_{j=0}^{N} \hat{i}_j T_j(z)$$

then

$$\hat{i}_j = \frac{1}{c_j} \frac{te^{-\sigma t} \cos 2j \cos^{-1}(2\sqrt{f})}{\sqrt{f(1-f)}}\,, \qquad c_j = \begin{cases} \pi & j = 0, N \\ \pi/2 & 0 < j < N\,. \end{cases}$$

The values $\varphi_0, \ldots, \varphi_N$ are now determined as follows. First choose φ_0 by the prescription

$$\varphi_0(t) = -\frac{6}{2N^2+1} \sum_{j=1}^{N-1} \frac{\varphi_j(t)(-1)^j}{\sin^2(\pi j/2N)} - \frac{3}{2N^2+1}\varphi_N(t)\,, \tag{8.197}$$

then $\varphi(z,t)$ satisfies the sealed condition at $z = 1$. Now $\varphi_1, \ldots, \varphi_{N-1}$ are obtained as the solutions of the ordinary differential equations

$$\frac{d\varphi_k}{dt} = -\varphi_k + \chi^{-1}\varphi''(z_k,t) - i_k(t)\,, \qquad k = 1, 2, \ldots, N-1\,. \tag{8.198}$$

The second spatial derivative in (8.198) is computed by first using the Chebeyshev cosine transform to get the Chebyshev coefficients of $\varphi(z,t)$ followed by two differentiations in spectral space followed lastly by the inverse Chebyshev cosine transform back into physical space. Finally, φ_N is determined from the somal boundary condition

$$\frac{d\varphi_N}{dt} = -\varphi_N - \chi^{-1}\left(\frac{\varphi_0}{2} + \sum_{j=1}^{N-1} \frac{\varphi_j(t)(-1)^j}{\cos^2(\pi j/2N)} + \frac{2N^2+1}{6}\varphi_N\right).$$

Fig. 14 illustrates the agreement between the exact and numerical solutions (based on 8 and 16 polynomials) for the potential at the soma in this unbranched dendrite. Differences between the exact and numerical solutions are entirely due to the ability of any numerical solution to treat delta-like inputs at points other than nodes.

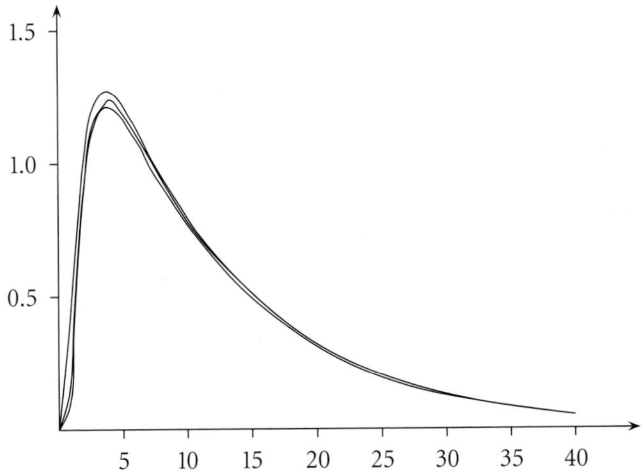

Fig. 14. Comparison of true potential and spectral estimate of the potential at the soma. The spectral estimate is based on 8 (peaks highest) and 16 (lowest curve) polynomials.

Spectral and Exact Solution of a Branched Tree

Major *et al.* (1993a) describe another analytical treatment of this problem and its various generalisations using separation of variables. This approach may present the following problems:

1. Fourier series cannot be freely evaluated and differentiated at endpoints;
2. Spatially and temporally distributed inputs can be difficult to incorporate into these schemes — the method of separation of variables intrinsically applies to homogeneous equations.

These difficulties can often be overcome once an orthogonality condition is established. However the transform methodology illustrated here is quite general and automatically allows non-homogeneous terms to be included in the membrane potential in a natural way.

Consider a dendrite consisting of a uniform limb of radius a_1 and length l_1 attached at one end to a soma of area A_S and branched at its other end into two uniform dendrites of radii a_2 and a_3 and lengths l_2 and l_3 respectively. The branched dendrite is assumed to be passive, and therefore described by the linear cable equation. In the absence of active synaptic input current J_S, synaptic current activity is *simulated* at $t = 0$ by the time dependent alpha function input currents $J^{(k)}(x,t) = G^{(k)}(t/T^{(k)})e^{(1-t/T^{(k)})}\delta(x - f_k l_k)$ acting at location $x = f_k l_k$, $f_k \in (0,1)$ on limb k. Figure 15 illustrates the configuration.

The deviation from the resting potential of the passive dendrite in figure 15 is modelled by the partial differential equations

$$\tau \frac{\partial V^{(k)}}{\partial t} + V^{(k)} + \frac{J_{\text{in}}^{(k)}}{g_M} = \frac{a_k g_A}{2 g_M} \frac{\partial^2 V^{(k)}}{\partial x^2}, \qquad x \in (0, l_k), \qquad k = 1, 2, 3. \tag{8.199}$$

The crucial assumption in this analysis is that each limb has an identical time constant τ and is consistent with the assumption that the passive electrical properties of dendritic material are site independent. In mathematically terms, this uniformity admits a "separable solution" to the complete set of cable equations. If time constants vary from limb to limb, an exact

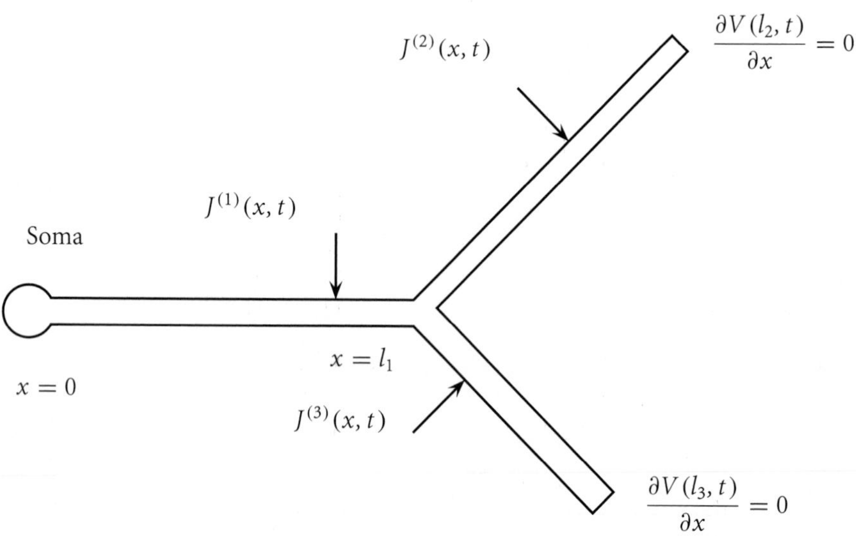

Fig. 15. A passive branched dendrite with sealed ends is stimulated on limb k by injected "alpha" function pulses of current $J^{(k)}$ at $x = f_k l_k$, where $(0 < f_k < 1)$.

mathematical solution can be found by Fourier series methods, but it is not a separable solution and it does not lead to an elementary transcendental equation for the modes of decay. Here the statement of the cable equations assumes *a priori* that axial conductance g_A and membrane conductance g_M are universally constant. This assumption is made for algebraic convenience and may be relaxed. Equation (8.199) is solved with boundary conditions

$$\begin{aligned} \frac{\partial V^{(2)}(l_2,t)}{\partial x} = \frac{\partial V^{(3)}(l_3,t)}{\partial x} &= 0\,, \\ V^{(1)}(l_1,t) = V^{(2)}(0,t) &= V^{(3)}(0,t)\,, \\ \pi a_2^2 g_A \frac{\partial V^{(2)}(0,t)}{\partial x} + \pi a_3^2 g_A \frac{\partial V^{(3)}(0,t)}{\partial x} &= \pi a_1^2 g_A \frac{\partial V^{(1)}(l_1,t)}{\partial x}\,, \\ \tau_S \frac{\partial V^{(1)}(0,t)}{\partial t} + V^{(1)}(0,t) &= \frac{g_A}{g_S}\frac{\pi a_1^2}{A_S}\frac{\partial V^{(1)}(0,t)}{\partial x} \end{aligned} \tag{8.200}$$

and initial condition

$$V^{(1)}(x,0) = V^{(2)}(x,0) = V^{(3)}(x,0) = 0\,. \tag{8.201}$$

As in illustration I, each limb is mapped into $[-1,1]$ so that the analytical representation of the problem is also a suitable basis for the development of the numerical solution using spectral methods. Let biophysical parameters χ, γ_k, α_k and σ_k be defined by

$$\begin{aligned} \chi &= \frac{g_S}{g_A}\frac{A_S l_1}{2\pi a_1^2}\,, & \gamma_k &= \left(\frac{l_k^2 g_M}{2a_k g_A}\right)^{1/2}, & & k = 1,2,3\,. \\ \alpha_k &= \frac{2e\sigma_k J^{(k)}}{g_M l_k}\,, & \sigma_k &= \frac{\tau}{T^{(k)}}\,, \end{aligned} \tag{8.202}$$

Parameter σ_k is the ratio of the time constant of the dendritic membrane to that of the input current on limb k. When time and length in equations (8.199) are non-dimensionalised by the coordinate transformations $t^* = t/\tau$ and $z = -1 + 2x/l_k$ respectively, the cable equations (8.199) become

$$\frac{\partial V^{(k)}}{\partial t} + V^{(k)} + i^{(k)}(z,t) = \frac{1}{\gamma_k^2}\frac{\partial^2 V^{(k)}}{\partial z^2}\,, \qquad z \in (-1,1)\,, \qquad k = 1,2,3\,, \tag{8.203}$$

in which superscript stars have been suppressed but is understood that all variables are non-dimensional and where

$$i^{(k)}(z,t) = \alpha_k t e^{-\sigma_k t}\delta(z - (2f_k - 1))\,, \qquad k = 1,2,3\,. \tag{8.204}$$

As in illustration I, the analytical development uses the specific expression for $i^{(k)}$ in the final derivation of the potentials but is otherwise completely general. The boundary and initial conditions accompanying equations (8.203) have non-dimensional form

$$\frac{\partial V^{(2)}(1,t)}{\partial z} = \frac{\partial V^{(3)}(1,t)}{\partial z} = 0\,, \tag{8.205}$$

$$V^{(1)}(1,t) = V^{(2)}(-1,t) = V^{(3)}(-1,t)\,, \tag{8.206}$$

$$\frac{A_2}{\gamma_2^2}\frac{\partial V^{(2)}(-1,t)}{\partial z} + \frac{A_3}{\gamma_3^2}\frac{\partial V^{(3)}(-1,t)}{\partial z} = \frac{A_1}{\gamma_1^2}\frac{\partial V^{(1)}(1,t)}{\partial z}\,, \tag{8.207}$$

$$A_S\frac{2g_S}{g_M}\left(\varepsilon\frac{\partial V^{(1)}(-1,t)}{\partial t} + V^{(1)}(-1,t)\right) = \frac{A_1}{\gamma_1^2}\frac{\partial V^{(1)}(-1,t)}{\partial z}\,, \tag{8.208}$$

where A_k $(k = 1, 2, 3)$ is the curved surface area of dendrite k. In condition (8.208), ε is again the ratio of the somal to dendritic time constant. For the time being, the analysis proceeds on the basis that ε is an arbitrary positive constant although it will eventual take the value unity.

Finite Transforms

For arbitrary real λ let $V_\lambda^{(1)}$, $V_\lambda^{(2)}$ and $V_\lambda^{(3)}$ be defined by

$$\begin{aligned} V_\lambda^{(1)} &= \int_{-1}^{1} V^{(1)}(z,t)\Big(\cos\lambda\gamma_1(1-z) + \mu_1 \sin\lambda\gamma_1(1-z)\Big)\, dz\,, \\ V_\lambda^{(2)} &= \int_{-1}^{1} V^{(2)}(z,t)\frac{\cos\lambda\gamma_2(1-z)}{\cos 2\lambda\gamma_2}\, dz\,, \\ V_\lambda^{(3)} &= \int_{-1}^{1} V^{(3)}(z,t)\frac{\cos\lambda\gamma_3(1-z)}{\cos 2\lambda\gamma_3}\, dz\,, \end{aligned} \tag{8.209}$$

in which μ_1 is an arbitrary parameter and let $i_\lambda^{(1)}$, $i_\lambda^{(2)}$ and $i_\lambda^{(3)}$ be similarly defined by

$$\begin{aligned} i_\lambda^{(1)} &= \int_{-1}^{1} i^{(1)}(z,t)\Big(\cos\lambda\gamma_1(1-z) + \mu_1 \sin\lambda\gamma_1(1-z)\Big)\, dz\,, \\ i_\lambda^{(2)} &= \int_{-1}^{1} i^{(2)}(z,t)\frac{\cos\lambda\gamma_2(1-z)}{\cos 2\lambda\gamma_2}\, dz\,, \qquad i_\lambda^{(3)} = \int_{-1}^{1} i^{(3)}(z,t)\frac{\cos\lambda\gamma_3(1-z)}{\cos 2\lambda\gamma_3}\, dz\,. \end{aligned} \tag{8.210}$$

The cable equations for limbs two and three are now multiplied by $\cos\lambda\gamma_2(1-z)/\cos 2\lambda\gamma_2$ and $\cos\lambda\gamma_3(1-z)/\cos 2\lambda\gamma_3$ respectively and then integrated over $(-1, 1)$. Bearing in mind that $V^{(2)}$ and $V^{(3)}$ satisfy the boundary conditions (8.205), it can be shown that

$$\frac{dV_\lambda^{(2)}}{dt} + (1+\lambda^2)V_\lambda^{(2)} + i_\lambda^{(2)} = -\frac{1}{\gamma_2^2}\left[\frac{\partial V^{(2)}(-1,t)}{\partial z} - \lambda\gamma_2 \tan 2\lambda\gamma_2\, V^{(2)}(-1,t)\right], \tag{8.211}$$

$$\frac{dV_\lambda^{(3)}}{dt} + (1+\lambda^2)V_\lambda^{(3)} + i_\lambda^{(3)} = -\frac{1}{\gamma_3^2}\left[\frac{\partial V^{(3)}(-1,t)}{\partial z} - \lambda\gamma_3 \tan 2\lambda\gamma_3\, V^{(3)}(-1,t)\right]. \tag{8.212}$$

Similarly, the cable equation for $k = 1$ is multiplied by $\cos\lambda\gamma_1(1-z) + \mu_1 \sin\lambda\gamma_1(1-z)$ and the resulting partial differential equation integrated over $(-1, 1)$. In this instance, the result is

$$\begin{aligned} \frac{dV_\lambda^{(1)}}{dt} + (1+\lambda^2)V_\lambda^{(1)} + i_\lambda^{(1)} &= -\frac{1}{\gamma_1^2}\Bigg[(\cos 2\lambda\gamma_1 + \mu_1 \sin 2\lambda\gamma_1)\frac{\partial V^{(1)}(-1,t)}{\partial z} \\ &\quad -\frac{\partial V^{(1)}(1,t)}{\partial z} - \lambda\gamma_1\mu_1 V^{(1)}(1,t) - \lambda\gamma_1(\sin 2\lambda\gamma_1 - \mu_1\cos 2\lambda\gamma_1)V^{(1)}(-1,t)\Bigg]. \end{aligned} \tag{8.213}$$

Let ξ_λ and i_λ be defined by the formulae

$$\begin{aligned} \xi_\lambda(t) &= A_1 V_\lambda^{(1)} + A_2 V_\lambda^{(2)} + A_3 V_\lambda^{(3)} \\ i_\lambda(t) &= A_1\, i_\lambda^{(1)} + A_2\, i_\lambda^{(2)} + A_3\, i_\lambda^{(3)}\,. \end{aligned} \tag{8.214}$$

In view of the current balance condition (8.207), it is straightforward algebra to verify that

$$\frac{d\xi_\lambda}{dt} + (1+\lambda^2)\xi_\lambda + i_\lambda = -\frac{A_1}{\gamma_1^2}\Big(\cos 2\lambda\gamma_1 + \mu_1 \sin 2\lambda\gamma_1\Big)\frac{\partial V^{(1)}(-1,t)}{\partial z}$$

$$+\frac{A_1\lambda}{\gamma_1}\Big(\sin 2\lambda\gamma_1 - \mu_1\cos 2\lambda\gamma_1\Big)V^{(1)}(-1,t) \qquad (8.215)$$
$$+\lambda\left[\frac{A_1\mu_1}{\gamma_1}+\frac{A_2}{\gamma_2}\tan 2\lambda\gamma_2+\frac{A_3}{\gamma_3}\tan 2\lambda\gamma_3\right]V^{(1)}(1,t)\,.$$

It is now beneficial to choose the arbitrary parameter μ_1 such that

$$\frac{A_1\mu_1}{\gamma_1}+\frac{A_2}{\gamma_2}\tan 2\lambda\gamma_2+\frac{A_3}{\gamma_3}\tan 2\lambda\gamma_3=0\,. \qquad (8.216)$$

Since $A_k\gamma_k^{-1} = 2(2g_A/g_M)^{1/2}c_k$ where $c_k = d_k^{3/2}$, then whenever the electrical properties of the dendrite are uniform, the computation of μ_1 is dependent only on the c-values of each limb. To digress further, if limbs two and three have the same electrotonic length, that is, $\gamma_2 = \gamma_3$ then clearly

$$\mu_1 = -\frac{c_2+c_3}{c_1}\tan 2\lambda\gamma_2 \qquad (8.217)$$

and if, in addition, the parent limb satisfies $c_1 = c_2 + c_3$ so that the child limbs can be collapsed and connected seamlessly to the parent (the Rall cylinder) then

$$\mu_1 = -\tan 2\lambda\gamma_2 \qquad (8.218)$$

Assuming now that μ_1 is chosen according to the prescription of formula (8.216), then clearly ξ_λ satisfies the simplified equation

$$\begin{aligned}\frac{d\xi_\lambda}{dt}+(1+\lambda^2)\xi_\lambda+i_\lambda \quad &= \quad -\frac{A_1}{\gamma_1^2}\Big(\cos 2\lambda\gamma_1+\mu_1\sin 2\lambda\gamma_1\Big)\frac{\partial V^{(1)}(-1,t)}{\partial z}\\ &\quad +\frac{A_1\lambda}{\gamma_1}\Big(\sin 2\lambda\gamma_1-\mu_1\cos 2\lambda\gamma_1\Big)V^{(1)}(-1,t)\,.\end{aligned} \qquad (8.219)$$

The somal boundary condition (8.208) is now used to remove the potential gradient at $z = -1$. After some straightforward algebra, the right hand side of equation (8.219) can be recast in the form

$$-\beta_\lambda\left[\frac{dV^{(1)}(-1,t)}{dt}+\left(\frac{1}{\varepsilon}-\frac{A_1\lambda}{2\gamma_1 A_S}\frac{C_M}{C_S}\frac{\tan 2\lambda\gamma_1-\mu_1}{1+\mu_1\tan 2\lambda\gamma_1}\right)V^{(1)}(-1,t)\right] \qquad (8.220)$$

where

$$\beta_\lambda = 2A_S\frac{C_S}{C_M}(\cos 2\lambda\gamma_1+\mu_1\sin 2\lambda\gamma_1)\,. \qquad (8.221)$$

Just as in illustration I, the key idea is now to choose λ (as yet unspecified) to satisfy the transcendental equation

$$1+\lambda^2 = \frac{1}{\varepsilon}-\frac{A_1\lambda}{2\gamma_1 A_S}\frac{C_M}{C_S}\frac{\tan 2\lambda\gamma_1-\mu_1}{1+\mu_1\tan 2\lambda\gamma_1}\,. \qquad (8.222)$$

Again, in the particular case of a branched network with a Rall equivalent cylinder representation, recall from (8.218) that $\mu_1 = -\tan 2\lambda\gamma_2$. Consequently it is a matter of simple trigonometry to recognise that, in the case of a Rall Y-junction, λ is a solution of

$$1+\lambda^2 = \frac{1}{\varepsilon}-\frac{A_1\lambda}{2\gamma_1 A_S}\frac{C_M}{C_S}\tan 2\lambda(\gamma_1+\gamma_2)\,. \qquad (8.223)$$

Thus the secular condition for λ reduces to that of a soma with a single unbranched dendrite of electrotonic length $(\gamma_1 + \gamma_2)$. In effect, this a just a restatement of the fact that the branched structure is now equivalent to a Rall cylinder of electrotonic length $(\gamma_1 + \gamma_2)$ connected to the soma.

Now let ψ_λ be defined by the expression

$$\psi_\lambda(t) = \xi_\lambda(t) + \beta_\lambda V^{(1)}(-1, t) \tag{8.224}$$

and return to the original analysis. When λ is determined by the secular condition (8.222), ψ_λ satisfies the ordinary differential equation

$$\frac{d\psi_\lambda}{dt} + (1 + \lambda^2)\psi_\lambda = -i_\lambda$$

with general solution

$$\psi_\lambda(t) = \psi_\lambda(0)e^{-(1+\lambda^2)t} - \int_0^t i_\lambda(s)e^{-(1+\lambda^2)(t-s)}\,ds\,. \tag{8.225}$$

In view of the initial condition (8.201), $\psi_\lambda(0) = 0$ in this application and so

$$\psi_\lambda(t) = -\int_0^t i_\lambda(s)e^{-(1+\lambda^2)(t-s)}\,ds\,. \tag{8.226}$$

Since the primary purpose of this analysis is to obtain exact solutions for comparison with numerical solutions, it is now convenient to set $\varepsilon = 1$ and $C_M = C_S$. It this case, λ satisfies the transcendental equation

$$2\lambda\gamma_1\frac{A_S}{A_1} + \frac{\tan 2\lambda\gamma_1 - \mu_1}{1 + \mu_1\tan 2\lambda\gamma_1} = 0 \tag{8.227}$$

in which μ_1 is specified by formula (8.216). At first sight, it would appear that the equation for λ involves the tangent function and all the incumbent book keeping that its asymptotes necessitate. In fact, the form of the secular condition (8.227) has an elegant structure that does not involve asymptotes. Suppose $f(\lambda; p, q)$ is defined by the formula

$$\begin{aligned} f(\lambda; p, q) &= \eta\Big(2A_S\gamma_1\lambda\cos 2\lambda\vartheta + A_1\sin 2\lambda\vartheta\Big) \\ \eta &= \frac{A_1}{\gamma_1} + p\frac{A_2}{\gamma_2} + q\frac{A_3}{\gamma_3}\,, \qquad \vartheta = \gamma_1 + p\gamma_2 + q\gamma_3 \end{aligned} \tag{8.228}$$

then equation (8.227) can be rewritten

$$f(\lambda; 1, 1) + f(\lambda; 1, -1) + f(\lambda; -1, 1) + f(\lambda; -1, -1) = \sum f(\lambda; p, q) = 0 \tag{8.229}$$

where the sum is taken over all possible combinations of the integers p and q satisfying $|p| = |q| = 1$.

Orthogonality of Eigenfunctions

As in illustration I, the complete solution of the problem ultimately requires the derivation of an orthogonality condition. Here the process is complicated by the fact that the eigenfunctions contain components in each limb of the branched dendrite. Let λ and η be two distinct solutions of the transcendental equation (8.222) then it is straightforward but tedious to verify that

$$\int_{-1}^{1}\frac{\sin\lambda\gamma(1-z)}{\cos 2\lambda\gamma}\,\frac{\sin\eta\gamma(1-z)}{\cos 2\eta\gamma}\,dz = \begin{cases} \dfrac{\eta\tan 2\lambda\gamma - \lambda\tan 2\eta\gamma}{\gamma(\lambda^2-\eta^2)} & \lambda \neq \eta \\[2ex] \sec^2 2\lambda\gamma - \dfrac{\tan 2\lambda\gamma}{2\lambda\gamma} & \lambda = \eta\,. \end{cases} \tag{8.230}$$

Similarly it can be shown that

$$\int_{-1}^{1}\Big(\sin\lambda\gamma(1-z)-\mu^{(\lambda)}\cos\lambda\gamma(1-z)\Big)\Big(\sin\eta\gamma(1-z)-\mu^{(\eta)}\cos\eta\gamma(1-z\Big)\,dz \tag{8.231}$$

has value

$$\begin{cases}\dfrac{\cos 2\lambda\gamma\cos 2\eta\gamma}{\gamma(\lambda^2-\eta^2)}\Big[\eta(\tan 2\gamma\lambda-\mu^{(\lambda)})(1+\mu^{(\eta)}\tan 2\gamma\eta) \\ \qquad -\lambda(\tan 2\gamma\eta-\mu^{(\eta)})(1+\mu^{(\lambda)}\tan 2\gamma\lambda)\Big]+\dfrac{\eta\mu^{(\lambda)}-\lambda\mu^{(\eta)}}{\gamma(\lambda^2-\eta^2)} & \lambda\neq\eta \\ \dfrac{\sin 2\lambda\gamma\cos 2\lambda\gamma}{2\gamma\lambda}\Big(\mu^{(\lambda)}-\tan 2\gamma\lambda\Big)^2+\Big(1+\mu^{(\lambda)2}-\dfrac{\tan 2\lambda\gamma}{2\lambda\gamma}\Big) & \lambda=\eta\,.\end{cases} \tag{8.232}$$

The key idea is now to combine results (8.230), (8.231) and (8.232) by calculating

$$\begin{aligned}&A_1\int_{-1}^{1}\Big(\sin\lambda\gamma_1(1-z)-\mu_1^{(\lambda)}\cos\lambda\gamma_1(1-z)\Big)\Big(\sin\eta\gamma_1(1-z)-\mu_1^{(\eta)}\cos\eta\gamma_1(1-z\Big)\,dz \\ &+A_2\int_{-1}^{1}\frac{\sin\lambda\gamma_2(1-z)}{\cos 2\lambda\gamma_2}\frac{\sin\eta\gamma_2(1-z)}{\cos 2\eta\gamma_2}\,dz+A_3\int_{-1}^{1}\frac{\sin\lambda\gamma_3(1-z)}{\cos 2\lambda\gamma_3}\frac{\sin\eta\gamma_3(1-z)}{\cos 2\eta\gamma_3}\,dz\end{aligned} \tag{8.233}$$

when $\lambda=\eta$ and when $\lambda\neq\eta$. The calculations are again straightforward but tedious. It can be shown that expression (8.233) evaluates to

$$\begin{cases}A_1\dfrac{\cos 2\lambda\gamma_1\cos 2\eta\gamma_1}{\gamma_1(\lambda^2-\eta^2)}\Big[\eta(\tan 2\lambda\gamma_1-\mu^{(\lambda)})(1+\mu^{(\eta)}\tan 2\eta\gamma_1) \\ \qquad -\lambda(\tan 2\eta\gamma_1-\mu^{(\eta)})(1+\mu^{(\lambda)}\tan 2\lambda\gamma_1)\Big] & \lambda\neq\eta\,, \\ A_1\Big(1+\mu_1^{(\lambda)2}\Big)+A_2\sec^2 2\lambda\gamma_2+A_3\sec^2 2\lambda\gamma_3 \\ \qquad +A_S\cos^2 2\lambda\gamma_1\Big(1+\mu_1^{(\lambda)}\tan 2\lambda\gamma_1\Big)^2 & \lambda=\eta\,.\end{cases} \tag{8.234}$$

Whenever $\lambda\neq\eta$, the secular condition (8.227) ensures that expression (8.233) is identically zero. Define the vector $\boldsymbol{v}(z)$ and the diagonal 3×3 matrix D by

$$\begin{aligned}\boldsymbol{v}_\lambda(z)&=\Big(\sin\lambda\gamma_1(1-z)-\mu_1^{(\lambda)}\cos\lambda\gamma_1(1-z),\frac{\sin\lambda\gamma_2(1-z)}{\cos 2\lambda\gamma_2},\frac{\sin\lambda\gamma_3(1-z)}{\cos 2\lambda\gamma_3}\Big) \\ D&=\operatorname{diag}(A_1,A_2,A_3)\end{aligned} \tag{8.235}$$

then it now follows from (8.227), (8.233) and (8.234) that

$$\langle\boldsymbol{v}_\lambda,\boldsymbol{v}_\eta\rangle=\int_{-1}^{1}\boldsymbol{v}_\lambda(z)^{\mathrm{T}}D\boldsymbol{v}_\eta(z)\,dz=\begin{cases}0 & \lambda\neq\eta\,, \\ A(\lambda) & \lambda=\eta\end{cases} \tag{8.236}$$

where

$$\begin{aligned}A(\lambda)=A_1\Big(1+\mu_1^{(\lambda)2}\Big)&+A_2\sec^2 2\lambda\gamma_2+A_3\sec^2 2\lambda\gamma_3 \\ &+A_S\cos^2 2\lambda\gamma_1\Big(1+\mu_1^{(\lambda)}\tan 2\lambda\gamma_1\Big)^2\,.\end{aligned} \tag{8.237}$$

Thus the vectors $\boldsymbol{v}_\lambda(z)$ are mutually orthogonal with respect to the inner product $\langle\boldsymbol{u},\boldsymbol{v}\rangle$ defined in (8.236). This important result is now used to construct the full analytical solution to the initial boundary value problem.

Solution Representation

Within this eigenspace, the potentials $V^{(1)}(z,t)$, $V^{(2)}(z,t)$ and $V^{(3)}(z,t)$ have representation

$$V^{(1)}(z,t) = V_0(t) + \sum_\lambda V_\lambda(t)\Big(\cos\lambda\gamma_1(1-z) + \mu_1^{(\lambda)}\sin\lambda\gamma_1(1-z)\Big) \quad (8.238)$$

$$V^{(2)}(z,t) = V_0(t) + \sum_\lambda V_\lambda(t)\frac{\cos\lambda\gamma_2(1-z)}{\cos 2\lambda\gamma_2} \quad (8.239)$$

$$V^{(3)}(z,t) = V_0(t) + \sum_\lambda V_\lambda(t)\frac{\cos\lambda\gamma_3(1-z)}{\cos 2\lambda\gamma_3} \quad (8.240)$$

where λ are the solutions of secular equation (8.227). The forms for $V^{(1)}(z,t)$, $V^{(2)}(z,t)$ and $V^{(3)}(z,t)$ satisfy continuity of membrane potential at the branch point by construction and also preserve current balance through the choice of $\mu_1^{(\lambda)}$. The task is now to compute $V_0(t)$ and $V_\lambda(t)$. Each equation in (8.238) is differentiated partially with respect to z and the result used to show that $V_\lambda(t)$ satisfies

$$\left\langle\left(\frac{A_1}{\gamma_1}\frac{\partial V^{(1)}}{\partial z}, \frac{A_2}{\gamma_2}\frac{\partial V^{(2)}}{\partial z}, \frac{A_3}{\gamma_3}\frac{\partial V^{(3)}}{\partial z}\right), \boldsymbol{v}_\eta\right\rangle = \sum_\lambda \lambda V_\lambda(t)\langle \boldsymbol{v}_\lambda, \boldsymbol{v}_\eta\rangle = \eta A_\eta V_\eta(t) \quad (8.241)$$

where $\langle \boldsymbol{u}, \boldsymbol{v}\rangle$ is defined in (8.236) and A_η is given by expression (8.237). Every term in the summation appearing in equation (8.241) is eliminated using the orthogonality condition (8.236) except the term that arises when $\lambda = \eta$. The left hand side of equation (8.241) is now integrated by parts to deduce that

$$\begin{aligned}
\eta A_\eta V_\eta(t) &= \left[\left\langle\left(\frac{A_1}{\gamma_1}V^{(1)}, \frac{A_2}{\gamma_2}V^{(2)}, \frac{A_3}{\gamma_3}V^{(3)}\right), \boldsymbol{v}_\eta\right\rangle\right]_{z=-1}^{z=1} \\
&\quad - \left\langle\left(\frac{A_1}{\gamma_1}V^{(1)}, \frac{A_2}{\gamma_2}V^{(2)}, \frac{A_3}{\gamma_3}V^{(3)}\right), \frac{d\boldsymbol{v}_\eta}{dz}\right\rangle \\
&= -V^{(1)}(1,t)\Big(\mu_1^{(\eta)}\frac{A_1}{\gamma_1} + \frac{A_2}{\gamma_2}\tan 2\eta\gamma_2 + \frac{A_3}{\gamma_3}\tan 2\eta\gamma_3\Big) \\
&\quad - \frac{A_1}{\gamma_1}V^{(1)}(-1,t)\Big(\sin 2\eta\gamma_1 - \mu_1^{(\eta)}\cos 2\eta\gamma_1\Big) \\
&\quad + \eta\Big(A_1V_\eta^{(1)} + A_2V_\eta^{(2)} + A_3V_\eta^{(3)}\Big)\,.
\end{aligned} \quad (8.242)$$

The definition of $\mu_1^{(\eta)}$ ensures that the first of these brackets is zero. The third bracket is immediately identified as $\xi_\eta(t)$ from definition (8.214) and hence it now follows from (8.242) that

$$\begin{aligned}
\eta A_\eta V_\eta(t) &= \eta\xi_\eta(t) - \frac{A_1}{\gamma_1}V_S(t)\Big(\sin 2\eta\gamma_1 - \mu_1^{(\eta)}\cos 2\eta\gamma_1\Big) \\
&= \eta\xi_\eta(t) - \frac{A_1}{\gamma_1}V_S(t)\Big(-2\eta\gamma_1\frac{A_S}{A_1}\Big)\Big(\cos 2\eta\gamma_1 + \mu_1^{(\eta)}\sin 2\eta\gamma_1\Big) \\
&= \eta\xi_\eta(t) + 2\eta A_S V_S(t)\Big(\cos 2\eta\gamma_1 + \mu_1^{(\eta)}\sin 2\eta\gamma_1\Big) \\
&= \eta\xi_\eta(t) + \eta\beta_\eta V_S(t)\,,
\end{aligned} \quad (8.243)$$

where β_η is defined previously in (8.221) and $C_M = C_S$ in this instance. In view of the definition of $\psi_\eta(t)$ in (8.225) and its value in (8.226), this analysis leads directly to the crucial result

$$V_\eta(t) = \frac{\psi_\eta(t)}{A_\eta} = -\frac{1}{A_\eta}\int_0^t i_\lambda(s)e^{-(1+\eta^2)(t-s)}\,ds\,. \quad (8.244)$$

In remains to find $V_0(t)$. Assuming that the series for $V^{(1)}(z,t)$ converges to $V^{(1)}(-1,t)$ when $z=-1$ (as indeed it does) then

$$V_S(t) = V^{(1)}(-1,t) = V_0(t) + \sum_{\lambda} V_\lambda(t)\Big(\cos 2\lambda\gamma_1 + \mu_1^{(\lambda)} \sin 2\lambda\gamma_1 \Big) . \tag{8.245}$$

It follows directly from the expressions for $V^{(1)}(z,t)$, $V^{(2)}(z,t)$ and $V^{(3)}(z,t)$ in (8.238) that

$$\int_{-1}^{1} \Big(A_1 V^{(1)}(z,t) + A_2 V^{(2)}(z,t) + A_3 V^{(3)}(z,t)\Big)\, dz = 2(A_1+A_2+A_3)V_0(t)$$
$$+\sum_{\lambda} \frac{V_\lambda(t)}{\lambda}\left[\frac{A_1\mu_1^{(\lambda)}}{\gamma_1} + \frac{A_2}{\gamma_2}\tan 2\lambda\gamma_2 + \frac{A_3}{\gamma_3}\tan 2\lambda\gamma_3 + \frac{A_1}{\gamma_1}\Big(\sin 2\lambda\gamma_1 - \mu_1^{(\lambda)} \cos 2\lambda\gamma_1\Big)\right] . \tag{8.246}$$

Taking account of the definition of $\mu_1^{(\lambda)}$ and the secular equation (8.227) satisfied by the eigenvalues, it follows from (8.246) that

$$\xi_0(t) = \int_{-1}^{1} \Big(A_1 V^{(1)}(z,t) + A_2 V^{(2)}(z,t) + A_3 V^{(3)}(z,t)\Big)\, dz = 2(A_1+A_2+A_3)V_0(t)$$
$$-2A_S \sum_{\lambda} V_\lambda(t)\Big(\cos 2\lambda\gamma_1 + \mu_1^{(\lambda)} \sin 2\lambda\gamma_1 \Big) . \tag{8.247}$$

Equations (8.245) and (8.247) are now combined so as to eliminate $V_\lambda(t)$. The final result is

$$\xi_0(t) + 2A_S = 2(A_S + A_1 + A_2 + A_3)V_0(t) = \psi_0(t) \tag{8.248}$$

so that

$$V_0(t) = -\frac{1}{2(A_S + A_1 + A_2 + A_3)} \int_0^t i_0(s) e^{-(t-s)}\, ds . \tag{8.249}$$

Hence the coefficients $V_0(t)$ and V_λ in the series expansion of $V^{(1)}(z,t)$, $V^{(2)}(z,t)$ and $V^{(3)}(z,t)$ are generally determined from $\psi_0(t)$ and $\psi_\lambda(t)$ and in this instance are specified by formulae (8.249) and (8.244) in which λ is a solution of the secular equation (8.227).

Specific Solution

The specific potential for a dendritic structure excited by injected alpha function current inputs on limbs 1, 2 and 3 at $x_1 = f_1 l_1$, $x_2 = f_2 l_2$ and $x_3 = f_3 l_3$ is now considered. It has already been shown in formula (8.204) that

$$i^{(1)}(z,t) = \alpha_1 t e^{-\sigma_1 t} \delta(z - (2f_1 - 1)) , \tag{8.250}$$
$$i^{(2)}(z,t) = \alpha_2 t e^{-\sigma_2 t} \delta(z - (2f_2 - 1)) , \tag{8.251}$$
$$i^{(3)}(z,t) = \alpha_3 t e^{-\sigma_3 t} \delta(z - (2f_3 - 1)) , \tag{8.252}$$

where the meanings of α_k and σ_k are defined in (8.202). The transformed current densities $i_1^{(\lambda)}$, $i_2^{(\lambda)}$ and $i_3^{(\lambda)}$ are computed from expressions (8.210) and yield

$$i_\lambda^{(1)} = \alpha_1 t e^{-\sigma_1 t}\Big(\cos 2\lambda\gamma_1(1-f_1) + \mu_1^{(\lambda)} \sin 2\lambda\gamma_1(1-f_1)\Big) ,$$
$$i_\lambda^{(2)} = \alpha_2 t e^{-\sigma_2 t} \frac{\cos 2\lambda\gamma_2(1-f_2)}{\cos 2\lambda\gamma_2} , \qquad i_\lambda^{(3)} = \alpha_3 t e^{-\sigma_3 t} \frac{\cos 2\lambda\gamma_3(1-f_3)}{\cos 2\lambda\gamma_3} . \tag{8.253}$$

The current density $i_\lambda(s)$ is now formed from $i_\lambda^{(1)}$, $i_\lambda^{(2)}$ and $i_\lambda^{(3)}$ according to recipe (8.214) and this in turn enables $\psi_\lambda(t)$ to be computed from its integral value in formula (8.226). Hence all the coefficients of the series expansions of the membrane potential in limb are now determined, that is, the analytical solution is found.

Numerical Solution

Unlike the unbranched dentrite, the implementation of boundary conditions at dendritic terminals and the soma, and continuity conditions at the branch point, needs detailed explanation. Let $\varphi_0^{(k)}(t), \varphi_1^{(k)}(t), \ldots, \varphi_N^{(k)}(t)$ denote the potential at nodes $z_0, z_1, \ldots, z_N$ on branch k of the dendritic structure and let $i_0^{(k)}, i_1^{(k)}, \ldots, i_N^{(k)}$ be a distribution of currents at these nodes that is equivalent to the input current $i^{(k)}(z,t) = \alpha_k t e^{-\sigma_k t}\delta(z-(2f_k-1))$, modulo the precision available (only $N+1$ Chebyshev polynomials are in use). It follows directly from (8.161) and the formula for $i^{(k)}(z,t)$ that if

$$i^{(k)}(z,t) = \sum_{j=0}^{N} \hat{i}_j^{(k)} T_j(z)$$

then

$$\hat{i}_j^{(k)} = \frac{1}{c_j} \frac{\alpha_k t e^{-\sigma_k t} \cos 2j \cos^{-1}(2\sqrt{f_k})}{\sqrt{f_k(1-f_k)}}, \qquad c_j = \begin{cases} \pi & j = 0, N \\ \pi/2 & 0 < j < N\,. \end{cases}$$

The values $\varphi_0^{(k)}, \ldots, \varphi_N^{(k)}$ are now determined so as to satisfy terminal boundary conditions, the soma condition and the continuity conditions at the branch point. Continuity of membrane potential at the branch point requires that

$$V_B(t) = \varphi_0^{(1)}(t) = \varphi_N^{(2)}(t) = \varphi_N^{(3)}(t)\,. \tag{8.254}$$

The sealed conditions at $z = 1$ on limbs 2 and 3 are satisfied by the requirements

$$\varphi_0^{(2)}(t) + \frac{\omega_N}{2}\varphi_N^{(2)}(t) = \varphi_0^{(2)}(t) + \frac{\omega_N}{2}V_B = -\omega_N \sum_{j=1}^{N-1} \frac{\varphi_j^{(2)}(t)(-1)^j}{\sin^2(\pi j/2N)} \tag{8.255}$$

$$\varphi_0^{(3)}(t) + \frac{\omega_N}{2}\varphi_N^{(3)}(t) = \varphi_0^{(3)}(t) + \frac{\omega_N}{2}V_B = -\omega_N \sum_{j=1}^{N-1} \frac{\varphi_j^{(3)}(t)(-1)^j}{\sin^2(\pi j/2N)} \tag{8.256}$$

where

$$\omega_N = \frac{6}{2N^2+1}\,. \tag{8.257}$$

Conservation of current at the branch point is expressed in the condition (8.207) and requires that

$$\begin{aligned} &\frac{A_2}{\gamma_2^2}\left[\frac{\varphi_0^{(2)}(t)}{2} + \sum_{j=1}^{N-1}\frac{\varphi_j^{(2)}(t)(-1)^j}{\cos^2(\pi j/2N)} + \frac{\varphi_N^{(2)}(t)}{\omega_N}\right] + \frac{A_3}{\gamma_3^2}\left[\frac{\varphi_0^{(3)}(t)}{2} + \sum_{j=1}^{N-1}\frac{\varphi_j^{(3)}(t)(-1)^j}{\cos^2(\pi j/2N)}\right. \\ &\qquad \left. + \frac{\varphi_N^{(3)}(t)}{\omega_N}\right] = \frac{A_1}{\gamma_1^2}\left[\frac{\varphi_0^{(1)}(t)}{\omega_N} + \sum_{j=1}^{N-1}\frac{\varphi_j^{(1)}(t)(-1)^j}{\sin^2(\pi j/2N)} + \frac{\varphi_N^{(1)}(t)}{2}\right], \end{aligned} \tag{8.258}$$

The second spatial derivative in (8.198) is computed by first using the Chebyshev cosine transform to get the Chebyshev coefficients of $\varphi(z,t)$ followed by two differentiations in spectral space followed lastly by the inverse Chebyshev cosine transform back into physical space. Finally, $V_S = \varphi_N^{(1)}(t)$ is determined from the somal boundary condition

$$\frac{dV_S}{dt} = -V_S - \chi^{-1}\left(\frac{V_B}{2} + \sum_{j=1}^{N-1}\frac{\varphi_j^{(1)}(t)(-1)^j}{\cos^2(\pi j/2N)} + \frac{2N^2+1}{6}V_S\right)\,.$$

A comparison between the analytical and numerical solutions for the branched dendrite is given in figure 16.

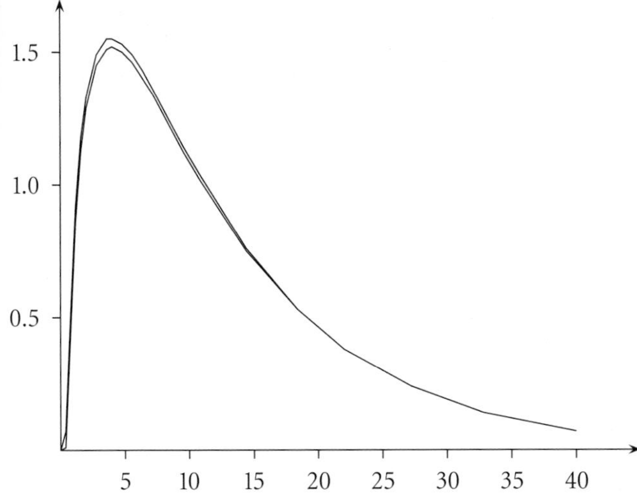

Fig. 16. Comparison of true potential and spectral estimate of the potential at the soma. The spectral estimate based on 16 polynomials is almost indistinguishable from the exact solution while that based on 8 polynomials has a slightly lower peak.

References

Amjad AM, Rosenberg JR, Halliday DM, Conway BA (1997) An extended difference of coherence test for comparing and combining several independent coherence estimates: theory and application to the study of motor units and physiological tremor. J NeuroSci Meth 73: 69–79

Bernander Ö, Douglas RJ, Martin KAC, Koch C (1991) Synaptic background activity influences spatio-temporal integration in single pyramidal cells. Proc Natl Acad Sci USA 88: 115692–11573

Bernander Ö, Koch C, Usher M (1994) The effect of synchronised inputs at the single neuron level. Neural Comput 6: 622–641

Box BD, Muller M (1958) A note on the generation of random normal variables. Ann Math Stat 29: 610–611

Burke RE (1997) Equivalent cable representations of dendritic trees: variations on a theme. Soc Neurosci abstr 23: 654

Burke RE, Fyffe REW, Moschovakis AK (1994) Electrotonic architecture of cat gamma-motoneurons. J Neurophysiol 72(5): 2302–2316

Canuto C, Hussaini MY, Quarteroni A, Zang TA (1988) Spectral methods in fluid mechanics. Springer series in computational physics. Springer, Berlin Heidelberg New York

Clements JD, Redman SJ (1989) Cable properties of cat spinal motoneurones measured by combining voltage clamp, current clamp and intra-cellular staining. J Physiol Lond 409: 63–87

Clenshaw CW (1962) Mathematical tables. Vol 5 National Physical Laboratory, London HM Stationery Office

Cohen AH, Rossignol S, Grillner S (1988) Neural control of rhythmic movements in vertebrates. John Wiley and Sons, New York

Cox DR, Isham V (1980) Point processes. Monographs in applied probability and statistics, Chapman and Hall, New York

Davis PJ, Rabinowitz P (1983) Methods of numerical integration. (2nd edition) Academic Press, Harcourt Brace Jovanovich Publishers, San Diego, New York, London, Sydney, Tokyo, Toronto

De Schutter E (1992) A consumer guide to neuronal modelling software. TINS 15(11): 462–464

Eisenberg RS, Johnson EA (1970) Three dimensional electric field problems in physiology. Prog Biophys Mol Biol 20: 1–65

Evans JD, Kember GC (1998) Analytical solutions to a tapering multi-cylinder somatic shunt cable model for passive neurons. Math. Biosci. 149(2): 137–165

Evans JD, Kember GC, Major G (1992) Techniques for obtaining analytical solutions to the multi-cylinder somatic shunt cable model for passive neurons. Biophys J 63: 350–365

Evans JD, Major G, Kember GC (1995) Techniques for the application of the analytical solution to the multicylinder somatic shunt cable model for passive neurons. Math Biosci 125(1): 1–50

Falk G and Fatt P (1964) Linear electrical properties of striated muscle fibres observed with intracellular electrodes. Proc Roy Soc Lond B 160: 69–123

Farmer SF, Bremner ER, Haliday DM, Rosenberg JR, Stephens JA (1993) The frequency content of common synaptic inputs to motoneurons studied during voluntary isometric contractions in man. J Physiol 470: 127–155

Fleshman JW, Segev I, Burke RE (1988) Electrotonic architecture of type-identified α-motoneurons in the cat spinal cord. J Neurophysiol 60(1): 60–85

Gear CW (1971) Numerical initial value problems in ordinary differential equations. Prentice Hall, Englewood Cliffs, NJ

Getting PA Reconstruction of small neural networks. In: Koch C, Segev I (eds) Methods in neuronal modelling: from synapses to networks. 1th edn. MIT press, Cambridge, MA, pp 171–194

Golub GH, Van Loan CF (1990) Matrix computations. (2nd edition) John Hopkins University Press, Baltimore, Maryland USA

Halliday DM (1998a) Generation and characterisation of correlated spike trains. Comp Biol Med 28: 143–152

Halliday DM (1998b) Weak stochastic temporal correlation of large scale synaptic input is a major determinant of neuronal bandwidth. Neural Comput (in press)

Hines ML, Carnevale NT (1997) The NEURON simulation environment. Neural Comp 9(6): 1179–1209

Holden AV (1976) Models of the stochastic activity of neurons. Lecture notes in biomathematics. 12 Springer, Berlin Heidelberg New York

Kember GC, Evans JD (1995) Analytical solutions to a multicylinder somatic shunt cable model for passive neurons with spines. IMA J Math Applied in Medicine and Biology 12(2): 137–157

Kloeden PE, Platen E (1995) Numerical solution of stochastic differential equations. Springer, Berlin Heidelberg New York

Kloeden PE, Platen E, Schurz H (1994) Numerical solution of SDE through computer experiments. Springer, Berlin Heidelberg New York

Knuth DE (1997) The art of computer programming: Vol II Seminumerical algorithms. Addison Wesley, Reading MA, Harlow England, Don Mills Ontario, Amsterdam, Tokyo

Major G (1993) Solutions for transients in arbitrarily branching cables: III Voltage clamp problems. Biophys J 65: 469–491

Major G, Evans D (1994) Solutions for transients in arbitrarily branching cables: IV Non-uniform electrical parameters. Biophys J 66: 615–634

Major G, Evans D, Jack JJB (1993a) Solutions for transients in arbitrarily branching cables: I Voltage recording with a somatic shunt. Biophys J 65: 423–449

Major G, Evans D, Jack JJB (1993b) Solutions for transients in arbitrarily branching cables: II Voltage clamp theory. Biophys J 65: 450–468

Manor Y, Gonczarowski J, Segev I (1991) Propogation of action potentials along complex axonal trees: model and implementation. Biophys J 60: 1411–1423

Marsagalia G, Bray TA (1964) A convenient method for generating normal variables. SIAM Review 6: 260–264

Mascagni MV (1989) Numerical methods for neuronal modelling. In: Koch C, Segev I (eds) Methods in neuronal modelling: from synapses to networks MIT press, Cambridge, MA, pp 255–282

Murthy VN, Fetz EE (1994) Effects of input synchrony on the firing rate of a 3-conductance cortical neuron model. Neural Comp 6: 1111–1126

Ogden JM, Rosenberg JR, Whitehead RR (1999) The Lanczos procedure for generating equivalent cables. In: Poznanski RR (ed) Modelling in the neurosciences: from ion channels to neural networks. Harwood Academic, pp 177–229

Perkel DH, Mulloney B (1978a) Calibrating compartmental models of neurons. Amer J Physiol 235: R93–R98

Perkel DH, Mulloney B (1978b) Electrotonic properties of neurons: steady-state compartmental model. J Neurophysiol 41: 621–639

Perkel DH, Mulloney B, Budelli RW (1981) Quantative methods for predicting neuronal behaviour. Neuroscience 6: 823–837

Rall W (1962) Electrophysiology of a dendritic model. Biophys J 2(2): 145–167

Rall W (1962) Theory of physiological properties of dendrites. Ann NY Acad Sci 96: 1071–1092

Rall W (1964) Theoretical significance of dendritic trees for neuronal input-output relations. In: Neural theory and modelling. (ed) Reiss R Stanford University Press, Stanford, California pp 73–97

Rall W (1969) Distributions of potential in cylindrical coordinates for a membrane cylinder. Biophys J 9: 1509–1541

Rall W (1977) Core conductor theory and cable properties of neurons. In: Kandel ER, Brookhardt JM, Mountcastle VB (eds) Handbook of Physiology: the Nervous System, Vol I, Williams and Wilkinson, Baltimore, Maryland, pp 39–98

Rall W (1989) Cable theory. In: Koch C, Segev I (eds) Methods in neuronal modelling: from synapses to networks. 1th edn. MIT press, Cambridge, MA, pp 9–62

Rall W, Burke RE, Holmes WR, Jack JJB, Redman SJ, Segev I (1992) Matching dendritic neuron models to experimental data. Physiol Rev suppl 72(4): S159–S186

Rapp M, Yarom Y, Segev I (1992) The impact of parallel fibre background activity on the cable properties of cerebellar Purkinje cells. Neural Comp 4: 518–533

Regnier A (1966) Les infortunes de la raison. Collection science ouverte aux Éditions du Seuil, Paris

Regnier A (1974) La crise du langage scientifique. Éditions anthropos, Paris

Rinzel J, Ermentrout GB (1989) Analysis of neural excitability and oscillations. In: Koch C, Segev I (eds) Methods in neuronal modelling: from synapses to networks. MIT press, Cambridge, MA, pp 135–170

Rosenberg JR, Halliday DM, Breeze P, Conway BA (1998) Identification of patterns of neuronal connectivity-partial spectra, partial coherence, and neuronal interactions. J NeuroSci Meth 83: 57–72

Segev I, Fleshman JW, Burke RE (1989) Compartmental models of complex neurons. In: Methods in neuronal modelling. Koch C, Segev I (eds),1th edn. MIT Press, Cambridge, MA pp 63–96

Segev I, Fleshman JW, Miller JP, Bunow B (1985) Modelling the electrical properties of anatomically complex neurons using a network analysis program: passive membrane. Biol Cybern 53: 27–40

Segev I, Rinzel J, Shepherd G (eds) (1995) The theoretical foundations of dendritic function: selected papers of Wilfrid Rall with commentaries. MIT press, Cambridge, MA

Tuckwell HC (1988a) Introduction to theoretical neurobiology Vol I — linear cable theory and dendritic structure. Cambridge University Press, Cambridge

Tuckwell HC (1988b) Introduction to theoretical neurobiology Vol II — nonlinear and stochastic theories. Cambridge University Press, Cambridge

Whitehead RR, Rosenberg JR (1993) On trees as equivalent cables. Proc Roy Soc Lond B 252: 103–108

Wichmann BA, Hill ID (1982) Appl Statistics 31(2): 188–190

Notation and definitions

$a \in \mathscr{A}$	a is a member of set $\mathscr{A}$
$[a, b]$	closed interval $a \leqslant x \leqslant b$
(a, b)	open interval $a < x < b$
$(a, b]$	semi-open interval $a < x \leqslant b$
$O(f)$	expression divided by f remains bounded as $f \to 0$
$o(f)$	expression divided by f tends to zero as $f \to 0$
$\mathscr{A} \times \mathscr{B}$	set of pairs (a, b) where $a \in \mathscr{A}$ and $b \in \mathscr{B}$
$V(x, t)$	transmembrane potential at time t and position x
$J(x, t)$	axial current density (current per unit area of the dendritic cross-section) at time t and position x
g_A	axial conductivity of intercellular material
J_{IC}	injected current per unit length of dendrite
J_{IVDC}	intrinsic voltage-dependent current per unit area of dendritic membrane
J_{SC}	synaptic current per unit area of dendritic membrane
$A(x)$	dendritic cross-sectional area at position x
$P(x)$	dendritic perimeter at position x
A_s	surface area of soma
$-J_{soma}$	current density into soma across membrane surface
ϱ	resistivity (Ohm·m)
$\delta(x - a)$	Dirac delta function $\delta(x - a) = 0$ if $x \neq a$, and $\int_{-\infty}^{\infty} \delta(x - a)dx = 1$, $\int_{-\infty}^{\infty} \delta(x - a)f(x)dx = f(a)$
C_s	capacitance per unit area of soma
$V_s(t)$	transmembrane potential of soma
E_L	resting membrane potential
E_α	equilibrium potential for ionic species α
g_S	conductance per unit area of soma ($\text{Ohm}^{-1}\text{m}^{-2}$)
J_{SI}	current injected into soma
g_M	passive membrane conductance
g_{NL}	non-linear (active) membrane conductance
g_L	leakage conductance
τ	time constant of a dendritic limb $\tau = C_M/g_M$
λ	length constant of a dendritic limb $\lambda^2 = g_A A/g_M P$
L	length of a dendritic limb
l	nondimensional or electronic length of a dendritic limb, $l = \int_0^L \sqrt{\frac{g_M P(s)}{g_A A(s)}}\, ds$
g-value	$g = \sqrt{g_M g_A A P}$
d	diameter of dendrite
c-value	$c = d^{3/2}$
z_i	designated node on a dendritic tree
v_i	membrane potential at node z_i
g_i	g-value of a dendritic cylinder
r_j^{-1}	Sum of the g-values of all the segments of the finite difference representation that contain z_j
ε	soma dendritic conductance ratio $\varepsilon = g_M C_S/g_S C_M = \tau_S/\tau$
A^{-1}	inverse of matrix A
A^T	transpose of matrix A
$\bar{A}$	complex conjugate of matrix A

$A^{(p,q)}$	$(p, q)^{\text{th}}$ entry of matrix A
$H(t)$	Heaviside unit step function $H(t) = 1$ if $t > 0$ or 0 if $t < 0$
$\langle \boldsymbol{u}, \boldsymbol{v} \rangle$	inner product of vectors $\boldsymbol{u}$ and $\boldsymbol{v}$
$\mathrm{E}[x]$	expectation of random variable x
$\mathrm{Var}[x]$	variance of random variable x
$\mathrm{Cov}[x(s), x(u)]$	covariance of the values of the random variable x at s and u,
δ_{ij}	Kronecker delta defined by $\delta_{ij} = 1$ if $i = j$ and 0 otherwise

Appendix I

This appendix gives the C computer program that was used to determine the properties of an encoder in the absence of correlated inputs. The last two functions in the program generate normal deviates and uniform random numbers in (0, 1) respectively and are implementations of the algorithms described in the sections on pages 247 and 248.

```
#include <stdio.h>
#include <stdlib.h>
#include <math.h>

/*
**  Builds SPIKE TRAINS using a integrate-to-threshold-and-fire
**  methodology for the ENCODER \tau dz_t=G(A dt+B dW_t)-z_t dt.
*/

#define   ZMAX    1.0     /* Threshold at which ENCODER fires */
#define   GAIN    1.0     /* Gain of the ENCODER */
#define     X0    0.0     /* Starting state for ENCODER */
#define      H    0.001   /* Step size for numerical integration */
#define    TAU    0.025   /* ENCODER time constant */
#define      A    1.269   /* Mean white noise input */
#define      B    0.307   /* STD DEV of white noise input */
#define   NSIM    10000   /* Number of simulations to be done */
#define   FLAG    0       /* Set FLAG=0 for pseudo-white noise
                             Set FLAG=1 for white noise */

int main( void )
{
    int i, nspike=0;
    double spike_time[NSIM], mu, sd, tnow=0.0, zold=X0, znow=X0,
           tmp, sigma, coeff01, coeff02, coeff11, coeff12,
           coeff13, normal(double,double);

/*  Step 1. - Initialise counters, times and ENCODER state */
    sigma = sqrt(H);
    coeff01 = coeff11 = TAU/(H+TAU);
    coeff02 = GAIN*H/(H+TAU);
    coeff12 = GAIN*H*A/(H+TAU);
    coeff13 = GAIN*B/(H+TAU);

/*  Step 2. - Simulate ENCODER operation for pseudo-white noise */
    do {
        if ( znow < ZMAX ) {          /* ENCODER below threshold */
            zold = znow;
            znow = coeff01*znow+coeff02*normal(A,B);
            tnow += H;
        } else if ( znow == ZMAX ) { /* ENCODER on threshold */
            spike_time[nspike] = tnow;
            tnow = znow = zold = 0.0;
            nspike++;
        } else {                      /* ENCODER above threshold */
            tmp = H*(znow-ZMAX)/(znow-zold);
```

```
            spike_time[nspike] = tnow-tmp;
            nspike++;
            znow = fmod(znow,ZMAX);
            zold = znow;
            tnow = tmp;
        }
    } while ( nspike<NSIM && FLAG==0 );

/*  Step 3. - Simulate ENCODER operation for white noise */
    do {
        if ( znow < ZMAX ) {          /* ENCODER below threshold */
            zold = znow;
            znow = coeff11*znow+coeff12+coeff13*normal(0.0,sigma);
            tnow += H;
        } else if ( znow == ZMAX ) { /* ENCODER on threshold */
            spike_time[nspike] = tnow;
            tnow = znow = zold = 0.0;
            nspike++;
        } else {                      /* ENCODER above threshold */
            tmp = H*(znow-ZMAX)/(znow-zold);
            spike_time[nspike] = tnow-tmp;
            nspike++;
            znow = fmod(znow,ZMAX);
            zold = znow;
            tnow = tmp;
        }
    } while ( nspike<NSIM && FLAG==1 );

/*  Step 4. - Compute MEAN (mu) and STD DEV (sd) of spike train */
    for ( mu=0.0,i=0 ; i<NSIM ; i++ ) mu += spike_time[i];
    mu /=  ((double) NSIM);
    for ( sd=0.0,tmp=0.0,i=0 ; i<NSIM ; i++ ) {
        sigma = spike_time[i]-mu;
        tmp += sigma;
        sd += sigma*sigma;
    }
    sigma = ((double) NSIM);
    sd = (sd-tmp*tmp/sigma)/(sigma-1.0);
    sd = sqrt(sd);
    printf("\n    Mean spike rate %6.1lf",1.0/mu);
    printf("\n CoV of spike_train %6.1lf",sd/mu);
    exit(0);
}


 /************************************************************
          Function returns Gaussian deviate.
  ************************************************************/
double normal( double mean, double sigma)
{
    static int start=1;
    static double g1, g2;
```

```
    double v1, v2, w, ran( int);

    if ( start ) {
        do {
            v1 = 2.0*ran(1)-1.0;
            v2 = 2.0*ran(1)-1.0;
            w = v1*v1+v2*v2;
        } while ( w==0.0 || w>=1.0 );
        w = log(w)/w;
        w = sqrt(-w-w);
        g1 = v1*w;
        g2 = v2*w;
        start = !start;
        return (mean+sigma*g1);
    } else {
        start = !start;
        return (mean+sigma*g2);
    }
}

 /*********************************************************
      Function returns primitive uniform random number using
      algorithm AS183 by Wichmann and Hill Appl. Stat. (1982)
  *********************************************************/
double ran( int n)
{
    static int start=1;
    void srand( unsigned int);
    static unsigned long int ix, iy, iz;
    double temp;

    if ( start ) {
        srand( ((unsigned int) abs(n)) );
        ix = rand( );
        iy = rand( );
        iz = rand( );
        start = 0;
    }

/*  1st item of modular arithmetic  */
    ix = (171*ix)%30269;
/*  2nd item of modular arithmetic  */
    iy = (172*iy)%30307;
/*  3rd item of modular arithmetic  */
    iz = (170*iz)%30323;
/*  Generate random number in (0,1)  */
    temp = ((double) ix)/30269.0+((double) iy)/30307.0
           +((double) iz)/30323.0;
    return fmod(temp,1.0);
}
```

Chapter 9

In Vitro Preparations

Klaus Ballanyi

Introduction

Many recent neuroscience techniques, such as patch-clamp (Chapter 6) or confocal optical (Chapter 4 and 14) recordings, can only be applied to very few structures in the intact brain. To use these powerful tools for analysis of neuronal and glial functions, the brain region of interest must be isolated and kept *in vitro*. Besides cell cultures, acute brain slices and (perfused) *en bloc* preparations are the most commonly used *in vitro* nervous tissues as they enable investigation of neuronal properties *in situ*. This chapter describes procedures for preparation as well as specific properties of these isolated tissues. It also summarises factors pivotal to establishing and maintaining "physiological" conditions *in vitro*. These procedures vary considerably for different brain regions and also depend on species and developmental stage of the animal. In addition, a number of factors, associated with the *in vitro* situation, can perturb cellular properties, and therefore hamper functional analysis. For these reasons, a simple schematic "lab manual" cannot be provided for *in vitro* preparations. Features of *in vitro* brain structures are exemplified with cellular data from hippocampus and respiratory or other medullary regions.

In Vitro Models

In vitro investigations on mammalian nervous functions have been traditionally performed on brain slices from adult rodents. Mice are used in an increasing number of reports as they allow for analysis of functional consequences of gene manipulation. Most of the work on transgenic mice is done on juvenile, newborn or even fetal animals, since "knock out" of certain genes often causes death shortly after or even before birth. Juvenile rodents are also preferred for neurophysiological studies, as patch clamping in combination with imaging in superficial cells is easier to perform (Edwards et al., 1988; Konnerth, 1990; Eilers et al., 1995). In contrast to the use of rodents for *in vitro* studies, most *in vivo* analyses, e.g. on the visual or the respiratory system, are done in adult cats.

Accordingly, species differences and maturation processes have to be taken into account when choosing an appropriate *in vitro* model to study a specific brain function (Aitken et al., 1995). For example, a subpopulation of neurons of the dorsal lateral geniculate nucleus generates rhythmic high frequency bursts of action potentials in brain slices from cats, but not from guinea-pigs (McCormick and Pape, 1990). Furthermore, functional properties of the respiratory network, isolated in brainstem preparations

Klaus Ballanyi, Universität Göttingen, II. Physiologisches Institut, Humboldtallee 23, Göttingen, 37073, Germany (phone: +49-551-399632; fax: +49-551-399676; e-mail: kb@neuro-physiol.med.uni-goettingen.de)

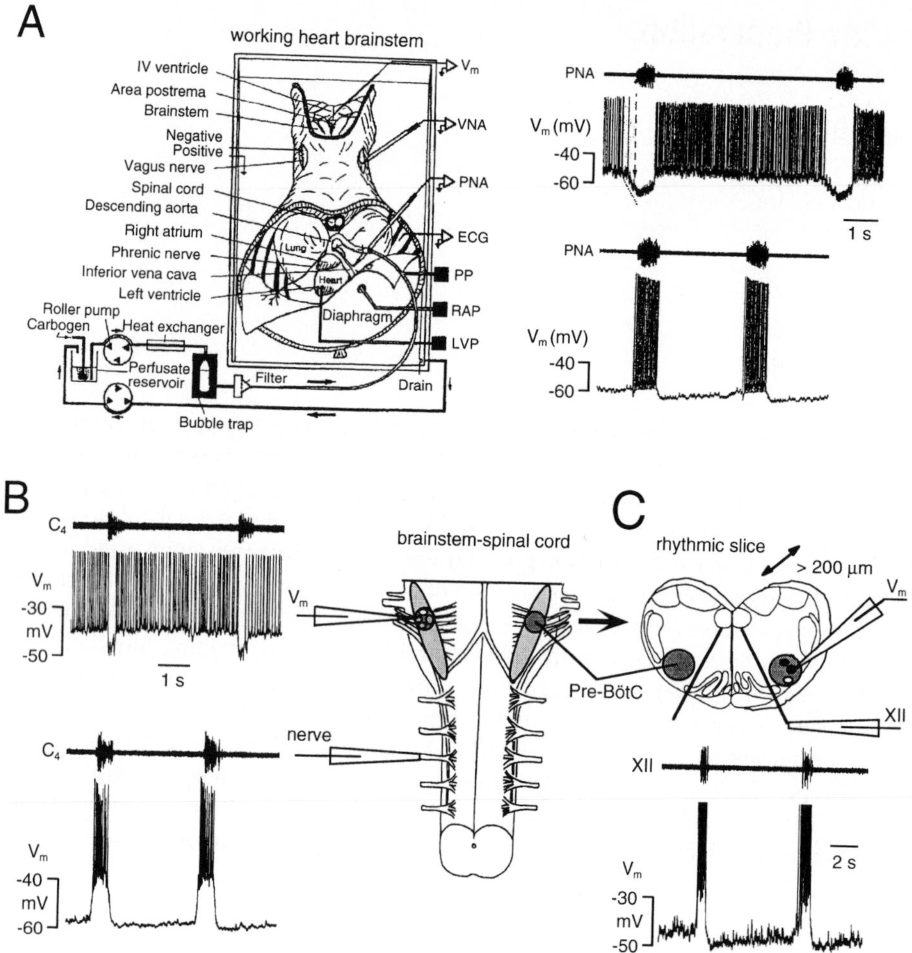

Fig. 1. *In vitro* isolation of the medullary respiratory network. A, in the working heart-brainstem preparation of the mature mouse adequate oxygenation is mediated by perfusing carbogenated artificial cerebrospinal fluid (ACF) -dextran solution at 31 °C via the descending aorta. Phrenic (PNA) and central vagus nerve activities (VNA), electrocardiogram (ECG), perfusion (PP), right atrial (RAP) and left ventricular (LVP) pressures and membrane potential (V_m) are monitored. Viability is > 5 h and set-up time 40–45 min. V_m recordings stem from neurons of the ventral respiratory group (VRG), which are inhibited (upper traces) or excited (lower traces) during inspiration-related PNA. (Reproduced with permission from Paton, 1996.) Similar inspiratory spinal nerve activities and V_m fluctuations of VRG neurons can be recorded for up to 12 h after isolation of the brainstem-spinal cord from neonatal rats, superfused at 25–27 °C with ACF containing 30 mM D-glucose. C, after further reduction of the *en bloc* medulla preparation to a transverse slice containing the Pre-Bötzinger Complex (Pre-BötC) of the VRG as the respiratory center, inspiratory activity can be recorded from hypoglossal (XII) nerve rootlets. In slices with a rostro-caudal thickness of less than 250 μm, rhythmic activity is only revealed in inspiratory Pre-BötC neurons (for details, see Smith et al., 1995; Ballanyi et al., 1999).

from perinatal rodents of apparently related species, can differ profoundly. For example, regular respiration-related nerve activity can be recorded in *en bloc* brainstem preparations from neonatal rats or laboratory mice (Fig. 1; Suzue, 1984; Brockhaus and Ballanyi, 1998; Ballanyi et al., 1999), whereas no rhythmic activity is revealed in the corresponding preparations from newborn "spiny" mice (Greer et al., 1996). As discussed in the review article by Ballanyi et al., 1999, such differences are related to the fact that spiny mice (just as guinea-pigs) are born in a considerably more mature state, since they have a gestation period of 40 days instead of about 22 and 20 days for laboratory rats and mice, respectively. It is likely that isolated brain tissue from precocious newborn animals has a lower tolerance to insufficient *in vitro* supply of oxygen or glucose than that of immature newborns.

The sensitivity of neuronal structures to neuroactive substances can also rapidly change within the perinatal period. For example, prostaglandins (Wolfe and Horrocks, 1994) stimulate the isolated respiratory network in embryonic E18–19 rats, whereas they cause long-lasting block of *in vitro* respiratory rhythm at embryonic days E20–21 and also within the first 2 postnatal days (Fig. 2; Meyer et al., 1999). As a further example, cortical layer formation during the postnatal period is associated with profound

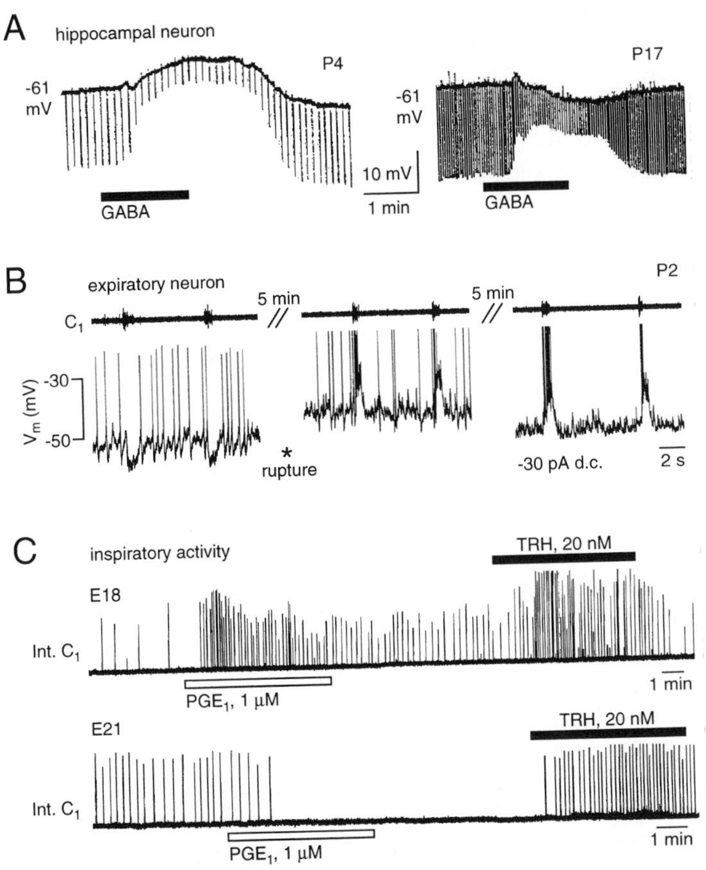

Fig. 2. Developmental differences of responses to neuromodulators. A, γ-aminobutyric acid (GABA) evokes membrane depolarisation and a decrease in membrane resistance (measured by regular injection of hyperpolarising current pulses via the microelectrode) in a hippocampal pyramidal neuron from a 4-day-old (P4) rat, whereas hyperpolarisation and a fall in resistance are observed in a cell from a P17 rat. (Reproduced with permission from Cherubini et al., 1991.) B, a gramicidin-perforated patch recording with a high Cl^- containing patch electrode reveals an inspiration-related hyperpolarisation in a medullary expiratory neuron of the brainstem-spinal cord preparation from neonatal rats. After rupture of the patch, the inspiratory hyperpolarisation reverses into an excitatory depolarisation due to dialysis of the cell with Cl^-. (Reproduced with permission from Brockhaus and Ballanyi, 1998.) C, bath-application of prostaglandin E_1 (PGE_1) results in frequency stimulation of (integrated) inspiration-related spinal (C_1) nerve activity in a brainstem-spinal cord preparation from an embryonic day 18 (E18) rat, whereas the drug abolishes rhythmic activity in a preparation from an E21 rat. At both embryonic ages, thyrotropin releasing hormone (TRH) exerts a strong stimulatory action. (From T. Meyer, B. Hoch and K. Ballanyi, in preparation).

changes of neuronal properties. γ-amino butyric acid (GABA) and glycine exert an excitatory action on neonatal neocortical and hippocampal neurons, whereas they act as "classical" inhibitory neurotransmitters after postnatal maturation of these structures (Fig. 2; Cherubini et al., 1991). In contrast, the medullary respiratory network is rather mature at birth, since it must generate a robust rhythm to provide sufficient oxygen supply to the organism. Accordingly, GABA- and glycinergic Cl^--dependent IPSPs of respiratory interneurons are hyperpolarising, and therefore inhibitory in neonatal animals (Fig. 2; Brockhaus and Ballanyi, 1998). These last examples show that the maturity of distinct brain regions can differ considerably at a given age.

En bloc Preparations

In some cold-blooded animals, the entire brain can be kept isolated to study complex neuronal functions *ex vivo*. For more than 20 years, the isolated central nervous system of the lamprey has been used to investigate, for example, the neural mechanisms related to fictive swimming (Grillner et al., 1991). Due to the small dimensions of the spinal cord, rhythmic neuronal processes such as oscillations of intracellular Ca^{2+} can be visualised during locomotor activity with high spatial resolution (Bacskai et al., 1995). A further established *en bloc* preparation is the isolated turtle brain (Hounsgaard and Nicholson, 1990). This preparation has the merit of allowing for the analysis of interactions of extended neuronal networks (e.g. cerebellum and brainstem) under defined *in vitro* conditions without occurrence of noise or vibration associated with heartbeat and respiration. Besides other applications, the isolated turtle brain is used by several groups to study the cellular mechanisms of the anoxia tolerance of cold-blooded vertebrates (Lutz and Nilsson, 1997). Due to its extreme tolerance to anoxia, it is understandable that the isolated turtle brain remains functional for up to several days *in vitro* without internal perfusion (Hounsgaard and Nicholson, 1990).

Similar to cold-blooded animals such as the turtle, the neonatal mammalian brain is highly tolerant to anoxia. Accordingly, non-perfused *en bloc* preparations of diverse brain regions of newborns remain viable for periods of more than 10 hours. For example, the intact hippocampal formation of neonatal or young rats can be kept alive for an extended period with excellent morphological preservation (Khalilov et al., 1997). Field or patch-clamp recordings, intracellular Ca^{2+} measurements, and 3-D reconstruction of labelled cells can be performed routinely and network-driven activities can be studied between connected intact structures such as the septum and the hippocampus (Khalilov et al., 1997). Similarly, the respiratory network and diverse reflex pathways remain functionally intact in non-perfused brainstem-spinal cord and spinal cord-hindlimb preparations from neonatal rats (Suzue, 1984; Onimaru et al., 1998; Smith et al., 1988). With the exception of elevated (30 mM) concentration of glucose in the superfusate to provide sufficient supply of substrate (Suzue, 1984; Ballanyi et al., 1996b), particular procedures or solutions usually must not be applied to maintain vitality in submerged experimental chambers. As an example of neuronal behaviour in the *en bloc* medulla of newborn rats, different classes of neurons of the ventral respiratory group exert a characteristic pattern of membrane potential fluctuations (Fig. 1; Brockhaus and Ballanyi, 1998; Onimaru et al., 1995; Ballanyi et al., 1999). These membrane oscillations remain almost unaltered after further reduction of the *en bloc* preparation to an inspiratory active transverse brainstem slice, containing the pre-Bötzinger complex as the respiratory center (Smith et al., 1991). However, for brain structures like the neocortex or hippocampus, which are particular immature at birth, it must be considered that membrane properties in *en bloc* preparations of newborn mammals can considerably differ from those in the corresponding structure isolated from the adult brain.

In the mature brain, membrane properties and complex functions of neuronal networks can be studied in whole brain and brainstem-cerebellum *en bloc* preparations from adult rodents (Llinas and Mühlethaler, 1988). In contrast to the brain of cold-blooded vertebrates or newborn mammals, the isolated mature brain tissues require arterial perfusion with artificial cerebrospinal fluid (Fig. 1; Llinas and Mühletaler, 1988; Paton, 1996). Depending on the brain area of interest and also on the species (Ballanyi et al., 1992; Schäfer et al., 1993; Morawietz et al., 1995), the perfusate should contain oxygen carriers like perfluoro-tributylamine (FC-43) to provide sufficient substrate supply. For the same purpose, the flow rate of the perfusate should preferentially be close to 20 ml/min (Paton, 1996). In some perfused preparations, however, flow rates of more than 2–3 ml/min cannot be applied without occurrence of oedema and subsequent loss of neuronal function (Ballanyi et al., 1992; Schäfer et al., 1993; Morawietz et al., 1995). It was also reported that addition of oncotic substances such as polyvinylpyrrolidone K30 or dextrane as well as antibiotics such as penicillin and/or streptomycin improve viability of the preparations (Llinas and Mühlethaler, 1988; Morawietz et al., 1995; Paton, 1996). Since the surgical procedures and also the *in vitro* superfusion (and perfusion) techniques vary considerably for *en bloc* brain preparations, it is beyond the scope of the present study to describe these manoeuvres in more detail. It should also be taken into account that these preparations are, at present, only used by a fairly limited number of groups, since the vast majority of research on isolated nervous tissue *in situ* is done on brain slices.

Nevertheless, *en bloc* preparations turned out to be powerful tools for the study of cellular mechanisms underlying integrative or autonomous brain functions. As described above, the respiratory network remains functionally active in isolated medulla preparations. In particular in the recently developed arterially perfused "working heart-brainstem preparation" (Paton, 1996), a variety of reflexes involved in the control of cardio-respiratory functions are preserved. In this preparation, intracellular recordings can be obtained routinely from respiratory neurons (Fig. 1; Paton, 1996). Developmental changes in the organisation of the respiratory network can be analysed by comparing data from the non-perfused medulla preparations of perinatal rodents with those obtained in the perfused preparations from mature animals (Ballanyi et al., 1992). Such studies showed that central chemosensitivity is not only functional in the isolated medulla of adult rats, but also in the brainstem-spinal cord of neonatal pups (Fig. 3; Voipio and Ballanyi, 1997; Ballanyi et al., unpublished observations). It was also demonstrated in these preparations that the respiratory network of newborn rats exerts an extreme tolerance to anoxia, whereas oxygen depletion results in rapid block of respiratory rhythm and major perturbance of ion homeostasis in the perfused brainstem of adult rats (Fig. 4; Ballanyi et al., 1992; Brockhaus et al., 1993; Morawietz et al., 1995; Ballanyi et al., 1996b; Richter and Ballanyi, 1996). In the context of such studies on perturbation of neuronal functions by metabolic disturbances, perfused preparations have the merit of allowing to discriminate between the effects of hypoxia, hypoglycemia and ischemia by variation of the composition of the perfusate (Schäfer et al., 1993; Morawietz et al., 1995). Furthermore, it can be tested to which extent novel drugs can be used therapeutically, since the blood brain barrier appears to be intact in arterially perfused brain tissues.

Brain Slices

Besides cell cultures, brain slices are the most commonly used *in vitro* preparations for analysis of brain function in mammals. The procedures for preparation of slices described below are mainly based on previous reports (Edwards et al., 1989; Konnerth, 1990; Aitken et al., 1995). Organotypic slice cultures, which are constituted by a quasi-monolayer of cells, retaining their characteristic dendritic, axonal and synaptic mor-

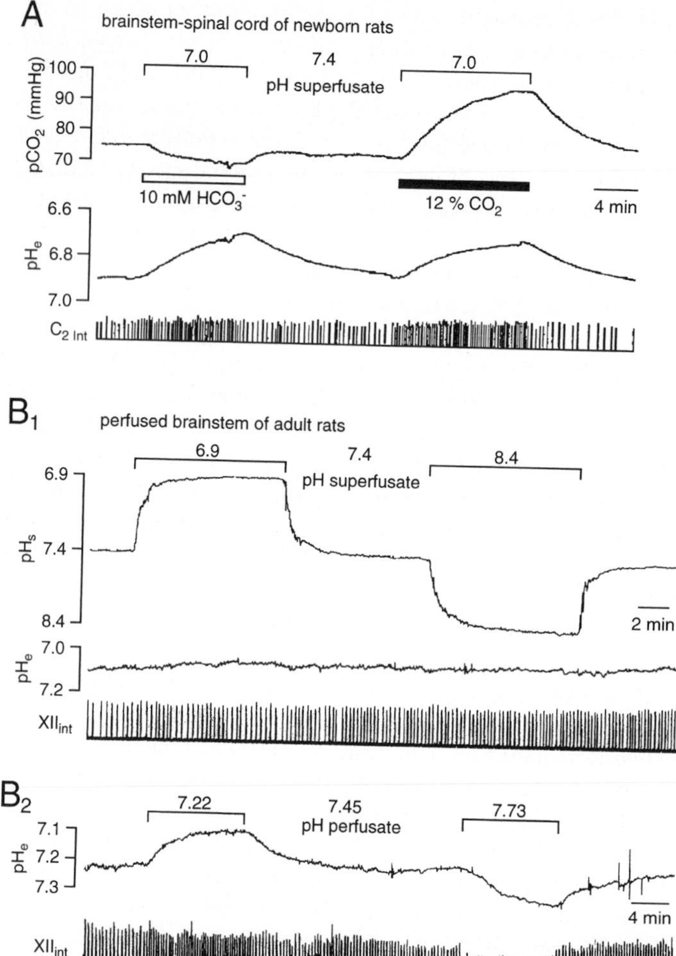

Fig. 3. Preservation of central chemosensitivity in *en bloc* medulla preparations. A, in the brainstem-spinal cord preparation of newborn rats lowering of the pH of the superfusate from 7.4 to 7.0 by either decreasing HCO_3^- concentration from 25 to 10 mM or by elevating the CO_2 content from 5 to nominally 12 % results in a comparable fall of interstitial pH (pH_e) in the region of the ventral respiratory group (VRG) at a recording depth of 300 µm. Both types of extracellular acidosis lead to a reversible increase in (integrated) inspiration-related phrenic (C_2) nerve activity, despite opposing changes of partial pressure of tissue CO_2 (pCO_2) in the VRG. (Reproduced with permission from Voipio and Ballanyi, 1997.) B, in the arterially perfused brainstem of adult rats, a change of the pH of the superfusate by ±0.5 pH units neither affects pH_e in the VRG (recording depth 1 mm), nor integrated inspiratory hypoglossal (XII) nerve activity. In contrast, a fall in pH of the perfusion fluid leads to frequency stimulation of *in vitro* respiratory rhythm, whereas elevation of perfusate pH causes reversible suppression of rhythmic activity. (K. Ballanyi, J. Voipio, S. Kuwana, G. Morawietz and D.W. Richter, unpublished). These results indicate that central chemosensitivity is closely associated with H^+-sensitive structures of VRG neurons.

phology, as well as their connectivity and physiological and pathophysiological phenotypes, are described elsewhere (Gähwiler and Knöpfel, 1990). A particularly interesting aspect of organotypic slices is the possibility of co-culturing brain regions that are functionally connected, but too distant to be included in one acute slice. Some experience with cell culture is of advantage to establish organotypic slices, whereas acute slice techniques are usually easier to handle and might altogether be somewhat cheaper.

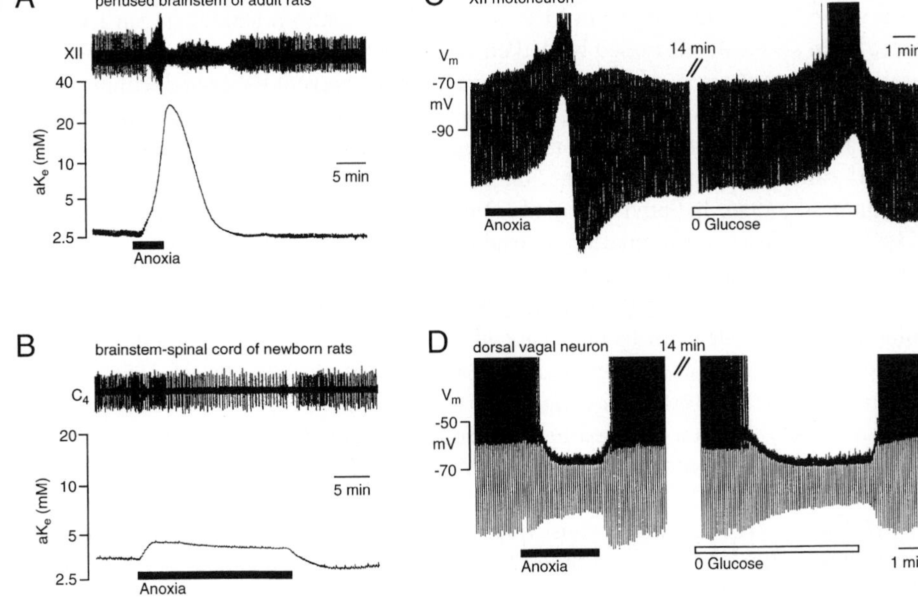

Fig. 4. Age-dependent and regional differences in the response of medullary neuronal structures to energy depletion. A, in adult rats anoxia evoked by arterial perfusion of hypoxic, N_2-gassed solution leads to reversible block of inspiratory hypoglossal (XII) nerve activity, accompanied by a prominent rise of extracellular K^+ (aK_e) in the ventral respiratory group (VRG). This K^+ elevation indicates depolarisation of respiratory and /or XII neurons as judged by the tonic XII activity at the peak of the anoxic aK_e rise. B, anoxia evoked by superfusion of the brainstem-spinal cord preparation of neonatal rats with hypoxic solution only results in a minor perturbance of aK_e and slowing of inspiratory phrenic (C_4) nerve activity. (A,B reproduced with permission from Ballanyi et al., 1992.) C, in a hypoglossal (XII) motoneuron of a 400 μm submerged slice preparation from a 3-week-old-rat both anoxia and glucose depletion evoke progressive depolarisation, decrease of membrane resistance and concomitant spike discharge. (From K. Ballanyi and J. Doutheil, unpublished). D, these types of metabolic disturbance elicit stable hyperpolarisation mediated by ATP-sensitive K^+ channels that are accompanied by a fall in resistance and block of spontaneous spike discharge in a neuron of the neighboring dorsal vagal nucleus of the same type of brainstem slice. (Reproduced with permission from Ballanyi et al., 1996a.)

Anesthesia

It is established that anesthetics modify neuronal functions (Aitken et al., 1995). Since anesthetics differ in their effects on cellular properties, in particular on ionotropic neurotransmitter receptors (Gage and Hamill, 1981), the appropriate drug should be chosen according to the function under investigation. It is generally accepted that anesthesia with ether or isoflurane produces only minor interference, since these substances are rapidly washed-out *in vitro*. Nevertheless, in a number of studies animals are killed either by exposure to CO_2 or by decapitation to avoid direct effects of anesthetics.

Solutions

Artificial cerebrospinal fluid (ACF) of quasi-physiological ionic content (in mM) 118 NaCl, 3 KCl, 1.5 $CaCl_2$, 1 $MgCl_2$, 1 NaH_2PO_4, 25 $NaHCO_3$ (for glucose concentration, see below) should be used for rinsing during isolation and cutting of brain tissue, for storage of slices, and for recording in the experimental chamber. To reduce cytotoxic effects of Ca^{2+} influx during isolation and slicing, the tissue should be exposed to ice-cold ACF, with a Ca^{2+} concentration reduced to 0.5 mM. This solution should be sufficient for most brain slices. As discussed by Aitken et al. (1995), it might be beneficial for some brain tissues to use special pre-incubation buffers to prevent damage during preparation and early phase of storage of slices. For this purpose, solutions are often used in which a major proportion of NaCl is replaced with sucrose. However, lack of extracellular Na^+ can perturb neuronal function due to the Na^+ dependence of pH regulation (Trapp et al., 1996; Ballanyi and Kaila, 1998). Reduction of $[Ca^{2+}]$ and elevation of $[Mg^{2+}]$ in the ACF, in combination with addition of ketamine or other glutamatergic antagonists, is thought to protect slices from glutamate-mediated toxicity during slicing and/or storage of slices (Aitken et al., 1995). To avoid cell swelling, addition of a high molecular weight dextran to increase colloid osmotic pressure is suggested by some groups (Fig. 1; Llinas and Mühlethaler, 1988; Aitken et al., 1995; Paton, 1996). Finally, substances such as ascorbic acid might help to reduce formation of free radicals. $[K^+]$ of the superfusate is often elevated to increase the excitability of reduced neuronal networks *in vitro*. A K^+ concentration of 6 mM is physiological for studies on peripheral nervous tissues.

pH: The CO_2/HCO_3^- system constitutes the predominant pH buffer of the interstitial fluid. When adjusting the pH of the ACF by gassing with carbogen (95 % O_2, 5 % CO_2), the temperature dependence of pH must be taken into account. For *en bloc* preparations and thick slices, it should be considered that on-going metabolic activity produces a considerable tissue pH gradient. Accordingly, a low extracellular pH in the vicinity of respiratory neurons in *en bloc* medulla preparations (Fig. 3) can be partly counteracted by a superfusate with a pH of 7.8 (Voipio and Ballanyi, 1997; Ballanyi et al., 1999). Isolated (cultured) cells are often studied in CO_2/HCO_3^- free solutions pH-buffered with substances like Hepes. Due to differences in pK values and/or buffering power with respect to the physiological buffer, and also because of the HCO_3^- dependence of pH regulation, such buffers might disturb neuronal function in slices (Trapp et al., 1996; Ballanyi and Kaila, 1998).

Glucose: For studies carried out at physiological temperature thick slices containing neurons with a high metabolic rate or *en bloc* preparations require a superfusate concentration of D-glucose of 10–30 mM (Suzue, 1984; Ballanyi et al., 1996b). Such glucose levels can evoke hyperglycemic cell damage *in vivo*. However, steady energy consumption in slices produces a gradient of interstitial glucose (Lowry et al., 1998). Thus, glucose levels in the center of brain slices might well be close to the physiological range of 2–5 mM in the arterial blood or even lower (Lowry et al., 1998). It should also be considered that glucose can affect neuronal membrane properties. In medullary dorsal vagal neurons, lowering of glucose elicits opening of ATP-sensitive K^+ (K_{ATP}) channels, resulting in a block of tonic spike discharge due to membrane hyperpolarisation (Fig. 4; Ballanyi et al., 1996a; Karschin et al., 1998).

Isolation of Brain Tissue

Surgical procedures for isolation of a particular brain region can modify cellular properties. For example, ischemia induced by circulatory arrest after killing of the animal might not only lead to release of neurotransmitters with excitotoxic effects like glutamate, but also to post-mortem synthesis of prostaglandins from released arachidonic acid (Wolfe and Horrocks, 1994). Consequently, it is difficult to estimate basal levels of transmitters or neuroactive substances in acute slices of different brain regions. Interestingly, brain regions difficult to dissect like pineal gland or hypothalamus have the highest prostaglandin levels (Wolfe and Horrocks, 1994). To minimise ischemia-related release of endogenous substances and changes in gene expression, dissection should be done quickly (ideally between 1–2 min). As soon as the skull is opened and also later, the exposed brain (in particular from animals older than 1 week) should be rinsed with ice-cold ACF. After isolation, the brain block should be kept at 0–4 °C to reduce metabolism-related ischemic damage and to improve the texture for slicing. The tissue can be kept under these conditions for at least 3–60 min, thus allowing for consecutive slicing of different brain regions.

Cutting

For studies in which visualisation of recorded cells is not required, slices with a thickness of 300–500 μm containing a functional neuronal circuitry can be obtained using "McIllwain"-type tissue choppers (Aitken et al., 1995). However, the use of a vibratome or vibroslice is advisable if electrophysiological and/or optical recordings from superficial neurons or glia under microscopic control are to be performed (Figs. 5,6; Edwards et al., 1989; Konnerth, 1990). A rather clear-cut slice surface can be obtained with low-cost vibratomes like the Campden LE127ZL (Campden, UK). Cutting-related debris, in particular of dendritic structures or of interneurons in the vicinity of the slice surface, is apparently reduced using vibratomes from FTB (Vibracut, FTB Weinheim, Germany) or Leica (VT 1000S; Leica Bensheim, Germany).

To improve mechanical stability during cutting, a larger block of cooled brain tissue is first manually cut. One surface of the block, cut parallel to the prospective orientation of the slice, is positioned on the stage of the vibratome which is covered immediately before fixation with a thin layer of cyanoacrylate glue. The slicing chamber, preferentially containing a solid base of frozen low Ca^{2+} ACF, is then immediately filled with ice-cold solution of the same composition. For some types of brain tissues from neonatal animals (e.g. respiratory active brain slices), cutting in ice-cold ACF is not obligatory and slicing at room temperature might even result in improved cellular responses (K. Ballanyi, unpublished observations). Tissues that are too small to be glued directly to the stage (i.e. the spinal cord of embryonic rodents) can be first embedded in 2 % agar dissolved in ACF. The heated agar must be cooled to below 40 °C before embedding the brain tissue. Solidification of the agar is improved by careful rinsing with ice-cold ACF. First, one or more slices need to be cut to obtain a flat surface of the tissue block, and ultimately flat slices of the brain region of interest. Slicing can be done under visual control with a dissecting microscope or a magnifying lens. When using a vibratome with a cutting blade that moves on the horizontal plane, a frequency of about 5–10 Hz is optimal for most brain tissues. In contrast, depending on the texture of tissue (e.g. white matter *vs.* grey matter), the speed of cutting should be adjusted to 10–20 mm/min to avoid pushing of the tissue block. Also depending on the texture of the tissue, the angle of the blade and the amplitude of the vibration might be critical for the quality of the obtained slice. If the tissue block contains brain surface structures, the arachnoid (and pia mater) should be removed prior to slicing.

Storage

Immediately after cutting, the slices are transferred with a cut and fire-polished Pasteur pipette to oxygenated ACF. The slices are positioned on a plastic petri dish whose bottom has been replaced by a fine cotton mesh. For humidification of the gas phase above the slices ("interface"-like conditions), the top of the petri dish is fixed to the beaker in which the solution is bubbled from the bottom (Fig. 5; Edwards et al., 1989). Several

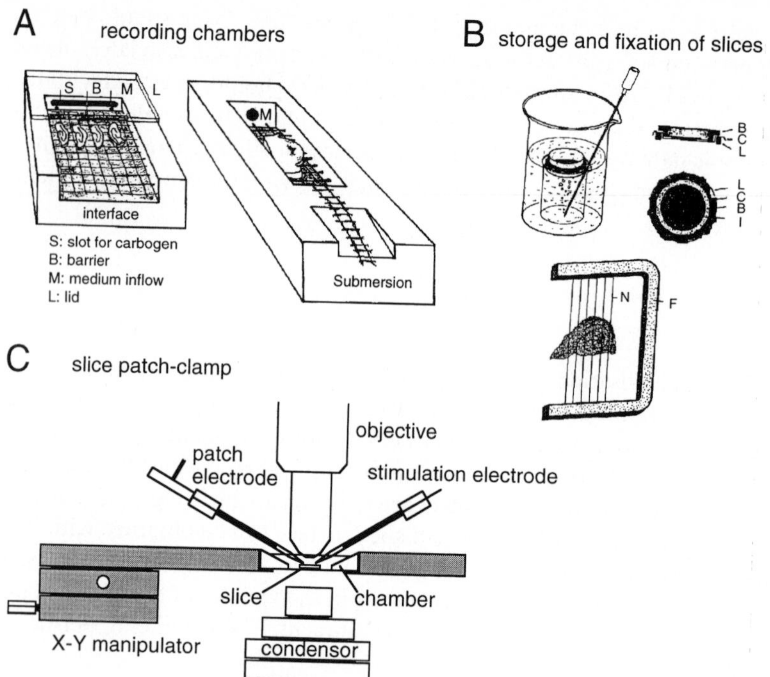

Fig. 5. Accessories for maintenance, mechanical fixation and recording of brain slices. A, in an "interface"-type recording chamber (left) slices rest on a net with the fluid level high enough to keep the surface moist and low enough to avoid mechanical disturbance. Warmed and moistened carbogen is blown over the slices from a slot (S) and directed to their surface by a lid (L; wet paper weighted by a glass slide). The artificial cerebrospinal fluid (ACF) also enters in the back (B). Another barrier may be used at the front end; the fluid level is adjusted by the amount of draining material at the front end, from which the fluid drips. In "submersion"-type chambers (right) slices are either stabilised with a grid (see B) or weights. They can also be fixed with insect needles pinned into a Sylgard bottom layer of the chamber. (Reproduced with permission from Haas and Büsselberg, 1992). B, a slice-holding chamber is placed on top of a 50 ml beaker inserted into a 250 ml beaker. These are filled with ACF to the level of the top of the holding chamber. A bubbler inserted through the spout of the inner beaker oxygenates the slices. The whole assembly is placed in a water bath (at 25–37 °C, depending on the slice type) and covered to prevent evaporation. A suitable holding chamber is made by breaking the top (L) and bottom (B) out of a small (10x35 mm) plastic petri dish, forming two rings. When inverted, the ring formed by L fits tightly onto the lip of B. A piece of fine cotton mesh (C) is stretched over B and can be clamped in place by L. Grids for fixation of slices consist of nylon threads (N) glued to a platinum frame (F). (Reproduced with permission from Edwards et al., 1989). C, schematic diagram of the experimental set-up for patch-clamping visually identified neurons in slices. The slice is fixed on the glass at the bottom of the chamber, which is mounted on the moveable stage of an upright microscope equipped with a long-distance, water immersion objective. (Reproduced with permission from Konnerth, 1990).

groups reported that "interface" pre-incubation and storage for up to several hours improve neuronal responses in brain slices (Haas and Büsselberg, 1992; Aitken et al., 1995). To create an interface between solution and humidified gas, the slices are placed on small pieces of lens paper or directly on a sheet of filter paper, covering the top of a beaker. The beaker, which is filled up to the rim with ACF, is kept in a closed system providing a saturated atmosphere when the fluid is sufficiently bubbled (Fig. 5). At a temperature of 20–37 °C, slices of most brain regions remain vital for periods of about 10 h, although especially cells close to the surface might die within 3–4 h.

Recording Chambers

The requirements for the construction of chambers for recording from brain slices are normally met with very simple designs and components (Haas and Büsselberg, 1992). These systems must allow for (i) adequate supply of the structure under study with fluid of the desired temperature, pH and O_2 content; (ii) mechanical stability of the slices in particular when superfusates are exchanged; (iii) easy access to the tissue by recording and/or stimulating electrodes.

Either "interface" or "submersion" chambers have been traditionally used. In both systems, ACF is administered to the slices either by gravity or a roller pump and removed by suction or dripping over the rim. During transport of the medium to the experimental chamber, oxygen content and pH can be kept constant by using Tygon (Kronlab, Sinsheim, Germany) tubings which have a sufficiently low permeability for O_2 and CO_2. In a variety of studies, the superfusate is re-circulated and thus passes several times through the experimental chamber. Such an approach might be justified while applying expensive drugs. If possible, re-perfusion should be avoided, as substances released from the slices, such as prostaglandins, can exert effects even in picomolar concentrations (Haas and Büsselberg, 1992; Wolfe and Horrocks, 1994; Meyer et al., 1999).

In interface chambers, slices are positioned on a nylon mesh sometimes covered by a piece of lens or filter paper. Fluid level is adjusted to the upper surface of the slice, thus stabilising the tissue by surface tension. Drying out is prevented by establishing a humid O_2/CO_2 atmosphere over the slices. A major drawback of interface chambers is the slow kinetics of responses to bath-applied drugs. Improved kinetics of drug effects can be achieved by local application via pressure or ionophoresis. Advantages of interface chambers are superior visibility of nervous structures like cortical layers or fiber tracts and large amplitude of field potentials due to less electrical shunt (Aitken et al., 1995). The use of a dissection microscope enables precise positioning of recording and stimulating electrodes in complex experimental arrangements. Even better optical resolution can be achieved with organotypic slice cultures with a thickness of only 1–3 cell layers (Gähwiler and Knöpfel, 1990).

The study of superficial neurons and glial cells in submerged slices using upright or inverted microscopes with high magnification (x40 or x63) water immersion objectives was established several years ago (Figs. 5,6; Edwards et al., 1989; Konnerth, 1990). Application of infrared, confocal or two-photon microscopical techniques allows the visualisation not only of the somatic region of the cell under study, but also of subcellular structures such as dendrites or axons (MacVicar, 1984; Eilers et al., 1995; Dodt and Zieglgansberger, 1995). These optical methods applied to brain slices enable monitoring of dynamic changes of cellular ions like Ca^{2+}, pH, or intracellular processes like mitochondrial depolarisation in addition to simultaneous analysis of membrane electroresponsiveness (Eilers et al., 1995; Trapp et al., 1996; Ballanyi and Kaila, 1998). This powerful experimental approach is used in an increasing number of studies and also most *en bloc* preparations are kept submerged. Accordingly, submersion chambers currently prevail over interface chambers. In submersion chambers, slices are mechanically stabilised with a grid (Fig. 5; Edwards et al., 1989) or small weights. Alternatively, submerged slices (preferentially positioned on a mesh to allow for subfusion of ACF) can be fixed with insect needles pinned into a Sylgard layer at the bottom of the chamber (Fig. 5; Haas and Büsselberg, 1992; see also Khalilov et al., 1997). As a major advantage with regard to interface systems, submersion chambers provide better and faster perfusion of the slice (or *en bloc* preparation) in pharmacological experiments.

Determinants of *Ex Vivo* Brain Function

Dimension

Analysis of complex synaptic interactions between different types of pyramidal cells and interneurons (Fig. 6), and also of synaptic plasticity associated with long-term potentiation or depression, was established more than 2 decades ago in acute hippocampal slices with a thickness of 200–500 μm (Fowler, 1988; Edwards et al., 1989; Aitken et al., 1995). Similarly, the characteristic features of most other neuronal structures can be preserved in brain slices of these dimensions. Thicker slices from adult animals can normally not be used due to insufficient supply with oxygen and glucose of the central cellular layers. Surprisingly, almost physiological connectivity and synaptic plasticity are

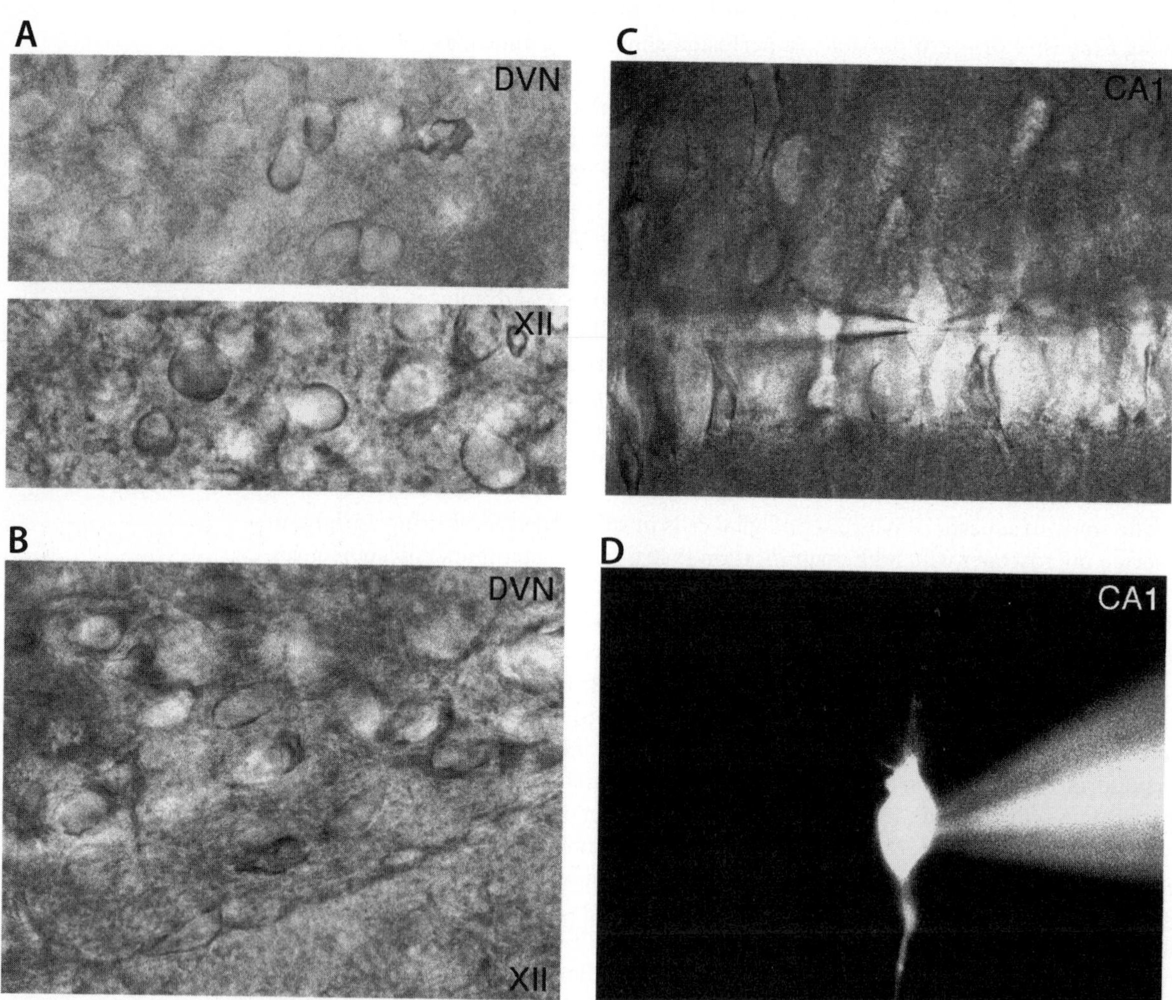

Fig. 6. Visualisation of neuronal structures in submerged brain slices. A, somata of both dorsal vagal (DVN upper panel) and hypoglossal neurons (XII lower panel) are revealed close to the surface of a 150 μm thin brainstem slice from a 4-day-old rat as visualised according to the techniques described in Fig. 5C. B, dorsal vagal neurons (upper part) are also clearly visible in a slice from a 13-day-old rat, whereas no somata are detected in the hypoglossal nucleus (lower right part). C, pyramidal cells and interneurons are visible in the CA1 soma region of the hippocampal formation from a 10-day-old rat. A pH-sensitive microelectrode (left) and a patch electrode (right) are positioned next to the soma of a pyramidal neuron. D, fluorescence image of the cell in C after filling with lucifer yellow. (From M. Lückermann and K. Ballanyi, unpublished.)

also retained in quasi-one-layered organotypic (hippocampal) slice cultures (Gähwiler and Knöpfel, 1990; Thompson et al., 1992).

Transverse brainstem slices from perinatal rodents need a rostro-caudal thickness of more than 200 μm to produce respiratory activity. To evoke regular rhythmic activity in such medullary slices, it is necessary to increase neuronal excitability by elevating $[K^+]$ of the ACF to 5–9 mM (Smith et al., 1991). In contrast, a stable rhythm is revealed using 3 mM K^+-containing ACF in slices with a thickness of more than 600 μm (Smith et al., 1991). It is thought that thinner slices are not rhythmic in physiological $[K^+]$ due to cutting-related removal of tonic excitatory drive from more caudal and/or rostral neurons to the respiratory network (Smith et al., 1995; Ballanyi et al., 1999). Two further examples point out the relevance of the dimension in *in vitro* preparations. In the brainstem-spinal cord preparation of newborn rats, disturbance of cranial inspiratory nerve activity, evoked by block of $GABA_A$ receptors with bicuculline, is suppressed after transection at the spinomedullary junction (Fig. 8; Brockhaus et al., 1998). This shows that bicuculline-induced seizure-like spinal phrenic activity does not originate in the respiratory network of this preparation, but is rather transferred from disinhibited spinal networks to the medullary neurons. Furthermore, expiratory neurons are not found in brainstem slices less than 250 μm thick, although the respiratory rhythm is almost identical to that in thicker slices (compare Fig. 1; Smith et al., 1995; Ballanyi et al., 1999). In brainstem slices as well as in the *en bloc* medulla, a subpopulation of inspiratory neurons, presumably responsible for generation of respiratory rhythm (Smith et al., 1991; Smith et al., 1995; Onimaru et al., 1995; Ballanyi et al., 1999), is capable of endogenous bursting. Intrinsic rhythmic bursting was also revealed, for example, in thalamic slices (McCormick and Pape, 1990). This shows that not only synaptic connectivity and plasticity, but also intrinsic membrane conductances mediating complex neuronal responses are retained in isolated brain tissues.

Metabolism

In contrast to neonatal pups, no rhythmic activity can be evoked in non-perfused *en bloc* medulla preparations of more mature rats containing the entire rostro-caudal extension of the respiratory network. As in other brain structures, this is related to the rapid decrease in the tolerance to anoxia within the first postnatal week (Fig. 4; Ballanyi et al., 1992; Ballanyi et al., 1996b; Richter and Ballanyi, 1996; Lutz and Nilsson, 1997; Ballanyi et al., 1999). The age-dependent loss of capability to tolerate insufficient oxygen supply is inversely related to a postnatal increase in utilisation of aerobic metabolism, which is secondary to a profound increase in metabolic rate. In addition, metabolic rate can vary considerably even in distinct structures of one brain slice. For example, a much steeper profile of tissue partial pressure of oxygen (pO_2) is revealed in the gray matter of the hypoglossal nucleus than in the white matter of the pyramidal tract (Fig. 7; Jiang et al., 1991).

Stimulation of anaerobic metabolism ("Pasteur effect") is sufficient to provide long-term maintenance of physiological intracellular ATP levels in many *in vitro* preparations from newborn mammals (Ballanyi et al., 1992; Ballanyi et al., 1996b; Lutz and Nilsson, 1997). Accordingly, "normal" responses can be measured in respiratory neurons (Brockhaus and Ballanyi, 1998) of neonatal brain tissue thicker than 1.5 mm despite the occurrence of an anoxic core (Fig. 7; Brockhaus et al., 1993; Ballanyi et al., 1999). The effects of oxygen depletion on neuronal conductances like K_{ATP} or O_2-sensitive K^+ channels (Fig. 4; Ballanyi et al., 1996a) should, however, not be neglected. Due to increased postnatal resting oxygen consumption, anoxia is revealed in more superficial layers of slices from adult brains compared with slices of similar dimensions from newborns

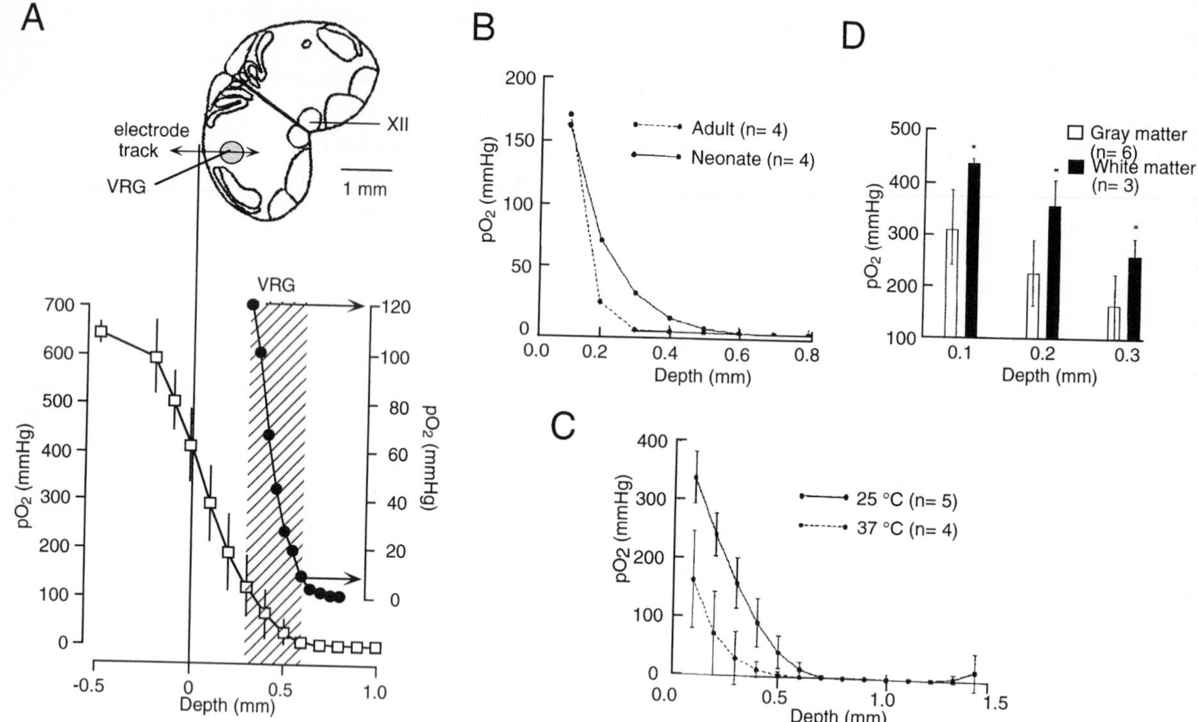

Fig. 7. Age- and temperature-dependent as well as regional differences in tissue profiles of partial pressure of tissue oxygen (pO_2). A, in the brainstem-spinal cord preparation of newborn rats pO_2 profiles were measured with an O_2-sensitive microelectrode. In the schematic section of the medulla at the rostrocaudal level of the pre-Bötzinger complex (see Smith et al., 1991 for details), the ventral respiratory group (VRG) is indicated by the filled circle. The 0µm position in the diagram corresponds to the ventral surface of the medulla. The magnified inset in the right part of A shows that pO_2 in the VRG ranges from a hyperoxia (120 mmHg) to normoxia (7 mmHg). (Reproduced with permission from Brockhaus et al., 1993.) B, anoxic regions are found with O_2 microelectrodes in the region of the hypoglossal nucleus (XII in the scheme of A) in adult brainstem slices of 600µm and in neonatal slices slices of 1500µm thickness. C, in neonatal brainstem slices (1500µm), significantly different pO_2 profiles are observed between 37 and 25 °C. D, independent of recording depth, baseline levels of pO_2 differ (*$p < 0.05$) in gray (XII nucleus) and white matter (pyramidal tract) of 400µm thick brainstem slices. Values given as means ± SD. (B,C,D reproduced with permission from Jiang et al., 1991.)

(Fig. 7; Jiang et al., 1991; Brockhaus et al., 1993; Ballanyi et al., 1996a). Thus, the thickness of slices from the mature brain should (at reduced temperature) not exceed 500µm, especially for *in vitro* studies on metabolically active brain structures like neocortex or hippocampus.

Due to the steep pO_2 profile in individual slices or *en bloc* preparations, it should be noted that neurons close to the surface might be subjected to hyperoxia ($pO_2 > 90$ Torr), whereas neurons in the center of slices might rather be hypoxic (Fig. 7). This somehow corresponds to the *in vivo* situation, where brain cells located in the vicinity of the arterial capillaries are exposed to a considerably higher pO_2 than those close to the venous aspects (Grote et al., 1981). On-going metabolic activity evokes accumulation of CO_2 in brain slices, leading to formation of H^+ (Voipio and Ballanyi, 1997). Due to the resulting pH gradient, extracellular pH differs from that of the superfusate or perfusate (Fig. 3) and is also different for neurons located at different depths within the tissue (Brockhaus et al., 1993; Morawietz et al., 1995; Trapp et al., 1996; Voipio and Ballanyi, 1997). In the center of thicker slices or *en bloc* preparations, extracellular K^+ is also elevated by up to several mM (Brockhaus et al., 1993; Ballanyi et al., 1999). It is likely that neurons in slices are also exposed to a gradient of glucose (Lowry et al., 1998). Depending on the thickness of the preparation and also on the metabolic rate of the brain region, it might therefore be necessary to supply the preparations with elevated (up to 30mM) glucose (Suzue, 1984; Edwards et al., 1989; Ballanyi et al., 1996b).

Not only the occurrence of metabolism-related tissue gradients for glucose, O_2, CO_2, pH and K^+ needs to be considered. Under identical conditions, there might also be striking differences in the tolerance or the response to any of these substrates among different neuronal populations. For example, both anoxia and glucose depletion produce a major depolarisation and functional impairment in mature hypoglossal motoneurons, whereas the cells of the neighboring dorsal vagal nucleus respond to such metabolic insults with sustained hyperpolarisations mediated by K_{ATP} channels (Fig. 4; Ballanyi et al., 1996a; Karschin et al., 1998). As a further example, the tissue pH gradient might affect H^+-sensitive K^+ channels, and therefore membrane behaviour of neurons at different depths in slices (see also Fig. 3; Trapp et al., 1996; Ballanyi and Kaila, 1998).

The above considerations are not only valid for measurements in traditional "thick" slices or *en bloc* preparations. Also for recording from superficial cells under visual control, it should be noted that one surface of the submerged "thin" slices is attached to the glass bottom of the recording chamber. This might lead to insufficient substrate supply of the cells at this interface. The resulting tissue gradients for metabolic products and ions might, in some cases, influence the behaviour of the recorded cells (Trapp et al., 1996; Ballanyi and Kaila, 1998).

Temperature

Due to the steep temperature dependence of neuronal processes, such as spike duration or synaptic transmission, an *in vitro* temperature of 37 °C would be desirable for comparison with *in vivo* data. However, at physiological temperature metabolic rate and demand for metabolic substrates are greatly increased in brain slices. This might result in expansion of the hypoxic/anoxic core especially in thick slices (Fig. 7), leading to potentiation of tissue gradients of ions. Physiological temperature may also lead to accumulation of metabolism-related neuroactive substances such as adenosine or prostaglandins, modulating cellular properties. To minimise such effects, *in vitro* temperatures of 25–30 °C have been typically chosen when thick slices or *en bloc* preparations have been used. At 37 °C, slices from mature brains should typically not have a thickness of more than 400μm (Fig. 7). As an example for the relation between temperature and neuronal properties, the frequency of inspiratory activity in the brainstem-spinal cord from newborn rats increases considerably upon elevation of the *in vitro* temperature from 27 to 37 °C. However, the *in vitro* rhythm is stable for up to more than 10h at reduced temperature, whereas respiratory activity vanishes after only a few hours at physiological temperature (Suzue, 1984; Ballanyi et al., 1999).

Neuromodulation

Neuronal properties might change *in vitro* within several hours or even immediately after isolation of a brain tissue (Aitken et al., 1995). For example, frequency of respiratory rhythm varies by about 5–15 inspiratory bursts/min possibly due to slight differences during isolation of the brainstem-spinal cord from newborn rats. Neuromodulators such as serotonin, acetylcholine or thyrotropin releasing hormone exert a strong stimulatory effect on respiratory frequency in prenatal preparations with a low burst rate, whereas these substances are ineffective in postnatal preparations showing a higher resting frequency (Meyer et al., 1999). It has been hypothesised that a low bursting frequency is caused by a decrease of cAMP in respiratory neurons involved in rhythm generation (Ballanyi et al., 1997; Ballanyi et al., 1999). Consistent with that view, prostaglandins are assumed to depress *in vitro* respiratory frequency in neonatal rats (Fig. 2) due to impairment of adenylyl cyclase, thus resulting in a fall of cAMP (Ballanyi et al., 1997; Meyer et al., 1999). It might well be that low respiratory frequencies in some postnatal brainstem-spinal cord preparations are related to formation of prostaglandins during the isolation procedure (Wolfe and Horrocks, 1994). This is in agreement with findings that brain cAMP levels change profoundly upon decapitation or ischemia (Aitken et al., 1995). Since several other neuroactive substances also modify cellular cAMP levels, it is likely that their wash-out (or wash-in) contributes to the variability in the burst rate of *in vitro* respiratory activity.

As an example for metabolism-related neuromodulation, it was suggested that extracellular adenosine, constituted due to anoxia in the core of the *en bloc* medulla, might act as an endogenous anticonvulsant on the isolated respiratory network of neonatal rats (Brockhaus et al., 1998). This assumption is based upon the observation that bicuculline does not elicit seizure-like discharge in this preparation, whereas the drug evokes epileptiform discharge in 600–750μm thick respiratory active medulla slices with no anoxic core (Shao and Feldman, 1997; compare Fig. 7). Related to the latter findings, it was demonstrated that exogenous adenosine suppresses bicuculline-induced rhythmic seizures in the spinal aspect of the brainstem-spinal cord preparation of newborn rats (Fig. 8; Brockhaus et al., 1998) as well as in organotypic hippocampal slice cultures (Fig. 8; Thompson et al., 1992). That metabolically produced endogenous adenosine modifies neuronal properties in hippocampal slices is indicated by the observation that attenuation of antidromic spike afterpotentials in response to reduction of the flow rate of the superfusate is antagonised by theophylline, a blocker of A_1 adenosine receptors (Fig. 8; Fowler, 1988).

Such examples of the influence of *in vitro* conditions on endogenous neuromodulators may also help to explain inconsistencies in the effects of some drugs or experimental procedures in some preparations. For example, in a proportion of brainstem slices block of aerobic metabolism leads to K_{ATP} channel activation in almost every dorsal vagal neuron tested, whereas in brainstem slices from other rats (even of the same litter) no activation is observed. This difference does not seem to be due to a "bad" preparation, since other typical electrophysiological features of these cells are indiscernible (Karschin et al., 1998). More likely, the complex interactions between a number of distinct intracellular constituents determining the activity state of these metabolism-regulated channels differ in individual preparations (Ballanyi and Kulik, 1998; Karschin et al., 1998).

Intracellular Dialysis

In the context of inconsistencies of drug effects, it should be noted that prolonged storage of slices might also result in changes not only in the amount or effectiveness of extracellular neuromodulators, but also of intracellular processes like redox state (Karschin et al., 1998). Spontaneous activation of K_{ATP} channels in dorsal vagal neurons occurs within several minutes after establishing a whole-cell patch-clamp recording in medullary slices kept *in vitro* for several hours (Ballanyi and Kulik, 1998). In contrast, resting current remains stable in the same type of neurons recorded with patch electrodes only a few hours after preparation of slices or during microelectrode recordings from "old" slices (Ballanyi et al., 1996a; Karschin et al., 1998).

This latter example shows that neuronal properties are affected not only by processes associated with long-term storage of slices, but also by dialysis of the intracellular mi-

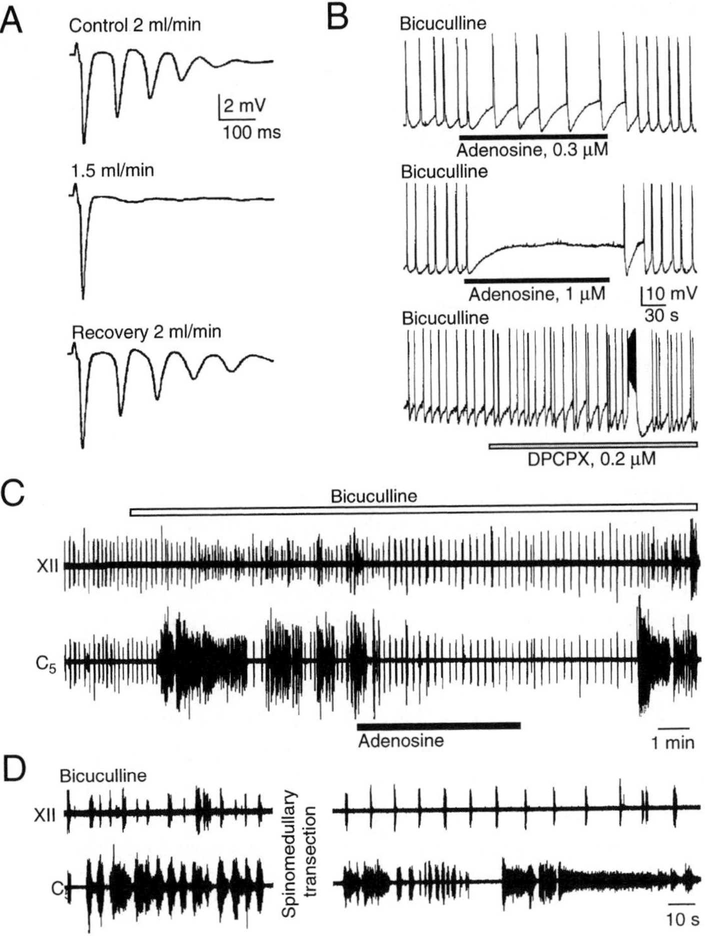

Fig. 8. Modulation of *in vitro* neuronal properties by adenosine. A, in hippocampal slices superfused with low Ca^{2+}, high Mg^{2+} solution intended to suppress synaptic activity, reduction of the flow rate leads to a reversible block of extracellularly recorded afterpotentials of CA1 pyramidal neurons evoked by antidromic alveolar stimulation, whereas the initial spike remains unaffected. Not illustrated here, the suppression of afterpotentials is reversed by theophylline, a blocker of A_1 adenosine receptors. (Reproduced with permission from Fowler, 1988.) B, bath-application of 1 μM adenosine (middle panel) effectively suppresses bicuculline (40 μM)-induced rhythmic seizure-like activity in CA3 pyramidal neurons of organotypic hippocampal slices, whereas 0.3 μM of the drug evokes slowing of rhythmic bursting (upper panel). 0.2 μM of the A_1 receptor antagonist DPCPX leads to an increase in the frequency of bicuculline-induced bursting, indicating an inhibitory effect of endogenous adenosine. (Reproduced with permission from Thompson et al., 1992.) C, seizure-like perturbance of inspiratory hypoglossal (XII) and phrenic (C_5) activity in the brainstem-spinal cord preparation from newborn rats, evoked by 50 μM bicuculline, is reversibly suppressed by 500 μM adenosine. C, the bicuculline-induced perturbance of inspiratory cranial XII activity disappears after transection of the brainstem-spinal cord at the spinomedullary junction, whereas seizure-like activity, but not respiratory rhythm, persists in the spinal aspect. (B,C from J. Brockhaus and K. Ballanyi, unpublished.)

lieu during whole-cell recording. Indeed, it has been shown that wash-out of intracellular constituents is responsible for run-down of several ion conductances. In many recent electrophysiological studies, whole-cell recordings from superficial cells are combined with optical measurements of intracellular ion changes (Figs. 5,6; Konnerth, 1990; Eilers et al., 1995; Trapp et al., 1996; Ballanyi and Kaila, 1998; Ballanyi and Kulik, 1998). Although such optical measurements provide a powerful tool for functional analysis of neuronal processes, it must be considered that dyes like fura-2 or BCECF, used for recording of intracellular Ca^{2+} or pH, respectively, affect the normal intracellular buffering capacity for these ions (Eilers et al., 1995; Trapp et al., 1996; Ballanyi and Kulik, 1998). Furthermore, it is known that radical formation during excitation of such dyes might impair cellular functions (Eilers et al., 1995; Trapp et al., 1996).

Vulnerability of Superficial Neurons

In vitro analysis of superficial neuronal structures or properties under visual control is limited by the age of the animal for most brain regions. In slices from rodents younger than 2 weeks, neurons of most brain aspects can be easily visualised, despite some region-specific differences (e.g. brainstem *vs.* hippocampus; Fig. 6). In preparations from older animals, visualisation is in many cases hampered by formation of myelin blobs, secondary to impairment of glial structures in the course of slicing (Edwards et al., 1989; Konnerth, 1990). This is, however, not a generalised rule: Purkinje neurons of cerebellar slices from 4-week-old mice, for example, can well be patch-clamped under visual control. In contrast, in brainstem slices from rats older than 10 days, superficial hypoglossal motoneurons are somehow "sick", as judged by the round shape of the soma and a large nucleus. Consequently, whole-cell recording is not possible from these superficial cells, whereas deeper hypoglossal motoneurons in thick slices (from the same tissue block) are viable for whole-cell and microelectrode recording (Fig. 4; K. Ballanyi, unpublished observations). Furthermore, superficial hypoglossal neurons look healthy and are patchable in brainstem slices from young rats (Fig. 6; Lips and Keller, 1998). It has been proposed that some neuronal populations are particularly sensitive to transient perturbation of intracellular Ca^{2+} during slicing (Lips and Keller, 1998). In agreement with this view, whole-cell recording is possible from fairly mature hypoglossal neurons in deeper layers of thin slices. For these measurements, the neurons need to be visualised by infrared microscopic techniques (MacVicar, 1984; Dodt and Zieglgansberger, 1995).

Conclusions

This chapter presented some evidence for the suitability of *in vitro* preparations for analysis of functional properties of nervous tissues *ex vivo*. Synaptic connectivity and plasticity as well as intrinsic membrane characteristics are retained in slice or *en bloc* preparations of most brain regions. However, brain slices might lack cellular elements that are involved in complex neuronal behaviour in the intact animal. Besides the appropriate choice of species and age of the animal to be used as an *in vitro* model, it should be taken into account that cellular properties might change in the course of *in vitro* isolation or during long-term storage of brain tissue. It must also be considered that the physiological concentrations of neuroactive substances might change in isolated preparations, and that tissue gradients for substrates and ions might develop. Therefore, recordings should preferentially be performed at identical depths in nervous tissue isolated from animals of a narrow "time window" of age. Finally, the thickness of non-perfused *in vitro* preparations should be chosen according to the metabolic rate of the brain region under study, which depends, among other factors, on the maturation stage.

Acknowledgement: The author thanks Drs. Paola Pedarzani and Mark Lückermann for valuable comments on the manuscript and Lucia Secchia-Ballanyi for steady encouragement and patience. The study was supported by the Hermann und Lilly Schilling-Stiftung, W. Sander-Stiftung and the Deutsche Forschungsgemeinschaft.

References

Aitken PG, Breese GR, Dudek FF, Edwards F, Espanol MT, Larkman PM, Lipton P, Newman GC, Nowak Jr. TS, Panizzon KL, Raley-Susman KM, Reid KH, Rice ME, Sarvey JM, Schoepp DD, Segal M, Taylor CP, Teyler TJ, Voulalas PJ (1995) Preparative methods for brain slices: a discussion. J Neurosci Methods 59: 139–149

Bacskai BJ, Wallen P Lev-Ram V, Grillner S, Tsien RY (1995) Activity-related calcium dynamics in lamprey motoneurons as revealed by video-rate confocal microscopy. Neuron 14: 19–28

Ballanyi K, Onimaru H, Hommo I (1999) Respiratory network function in the isolated brainstem-spinal cord preparation of newborn rats. Progress Neurobiol (in press)

Ballanyi K, Kaila K (1998) Activity-evoked changes in intracellular pH. In: pH and Brain Function. Eds. Kaila K, Ransom BR. Wiley-Liss, Inc., pp. 283–300

Ballanyi K, Kulik A (1998) Intracellular Ca^{2+} during metabolic activation of K_{ATP} channels in spontaneously active dorsal vagal neurons in medullary slices. Eur J Neurosci 10: 2574–2585

Ballanyi K, Doutheil J, Brockhaus J (1996a) Membrane potentials and microenvironment of rat dorsal vagal cells *in vitro* during energy depletion. J Physiol 495: 769–784

Ballanyi K, Völker A, Richter DW (1996b) Functional relevance of anaerobic metabolism in the isolated respiratory network of newborn rats. Eur J Physiol 432: 741–748

Ballanyi K, Hoch B, Lalley PM, Richter DW (1997) cAMP-dependent reversal of opioid-and prostaglandin-mediated depression of the isolated respiratory network in newborn rats. J Physiol 504: 127–134

Ballanyi K, Kuwana S, Völker A, Morawietz G, Richter DW (1992) Developmental changes in the hypoxia tolerance of the in vitro respiratory network of rats. Neurosci Lett 148: 141–144

Brockhaus J, Ballanyi K (1998) Synaptic inhibition in the isolated respiratory network of neonatal rats. Eur J Neurosci 10: 3823–3839

Brockhaus J, Nikouline V, Ballanyi K (1998) Adenosine mediated suppression of seizure-like activity in the respiratory active brainstem-spinal cord of neonatal rats. In: Göttingen Neurobiology Report. Eds. Elsner N, Wehner R. Stuttgart, New York: Thieme, p. 270

Brockhaus J, Ballanyi K, Smith JC, Richter DW (1993) Microenvironment of respiratory neurons in the in vitro brainstem-spinal cord of neonatal rats. J Physiol 462: 421–445

Cherubini E, Gaiarsa JL, Ben-Ari Y (1991) GABA: an excitatory transmitter in early postnatal life. Trends Neurosci 12: 515–519

Dodt HU, Zieglgansberger W (1995) Infrared videomicroscopy: a new look at neuronal structure and function. Trends Neurosci 17: 453–458

Edwards FA, Konnerth A, Sakmann B, Takahashi T (1989) A thin slice preparation for patch clamp recordings from neurones of the mammalian central nervous system. Eur J Physiol 414: 600–612

Eilers J, Schneggenburger R, Konnerth A (1995) Patch clamp and calcium imaging in brain slices. In: Sakmann B, Neher, E (eds) Single Channel Recording. Plenum Press, New York, pp. 213–229

Fowler JC (1988) Modulation of neuronal excitability by endogenous adenosine in the absence of synaptic transmission. Brain Res 463: 368–373

Gähwiler BH, Knöpfel T (1990) Cultures of brain slices. In: Jahnsen H (ed) Preparations of vertebrate central nervous system *in vitro*. Wiley & Sons, Chichester, New York, Brisbane, Toronto, Singapore, pp. 77–100

Gage PW, Hamill OP (1981) Effects of anesthetics on ion channels in synapses. Int Rev Physiol 25: 1–45

Greer JJ, Carter JE, Allan DW (1996) Respiratory rhythm generation in a precocial rodent in vitro preparation Respir Physiol 103: 105–112

Grillner S, Wallen P, Brodin L (1991) Neuronal network generating motor behaviour in lamprey: circuitry, transmitters, membrane properties, and simulation. Ann Rev Neurosci 14: 169–199

Grote J, Zimmer K, Schubert R (1981) Effects of severe arterial hypocapnia on regional blood flow regulation in the brain cortex of cats. Eur J Physiol 391:195–199

Haas HL, Büsselberg D (1992) Recording chambers-slices. In: Kettenmann H, Grantyn R (eds) Practical Electrophysiological Methods. Wiley-Liss, New York, Chichester, Brisbane, Toronto, Singapore, pp.16–19

Hounsgaard J, Nicholson C (1990) The isolated turtle brain and the physiology of neuronal circuits. In: Jahnsen H (ed) Preparations of vertebrate central nervous system *in vitro*. Wiley & Sons, Chichester, New York, Brisbane, Toronto, Singapore, pp. 155–182

Jiang C, Agulian S, Haddad GG (1991) O_2 tension in adult and neonatal brain slices under several experimental conditions. Brain Res 568: 159–164

Karschin A, Brockhaus J, Ballanyi K (1998) K_{ATP} channel formation by the sulphonylurea receptors SUR1 with Kir6.2 subunits in rat dorsal vagal neurons *in situ*. J Physiol 509: 339–346

Khalilov I, Esclapez M, Medina I, Aggoun D, Lamsa K, Leinekugel X, Khazipov R, Ben-Ari Y (1997) A novel in vitro preparation: the intact hippocampal formation. Neuron 19: 743–749.

Konnerth A (1990) Patch-clamping in slices of mammalian CNS. Trends Neurosci 13: 321–323

Lips MB, Keller BU (1998) Endogenous calcium buffering in motoneurones of the nucleus hypoglossal from mouse. J Physiol 511: 105–117

Llinas R, Mühletaler M (1988) An electrophysiological study of the *in vitro* perfused brain stem-cerebellum of adult guinea pigs. J Physiol 404: 215–240.

Lowry JP, O'Neill RD, Boutelle MG, Fillenz M (1998) Continuous monitoring of extracellular glucose concentrations in the striatum of freely moving rats with an implanted glucose biosensor. J Neurochem 70: 391–396

Lutz PL, Nilsson GE (1997) The brain without oxygen. Springer, New York, Berlin

McVicar BA (1984) Infrared video microscopy to visualize neurons in the in vitro brain slice preparation. J Neurosci Methods 12: 133–139

McCormick DA, Pape HC (1990) Properties of a hyperpolarization-activated cation current and its role in rhythmic oscillations in thalamic relay neurones. J Physiol 431: 291–318

Meyer T, Hoch B, Ballanyi K (1999) Endogenous frequency depression of the isolated respiratory network of fetal rats. Eur J Physiol 437: P36–1

Morawietz G, Ballanyi K, Kuwana S, Richter DW (1995) Oxygen supply and ion homeostasis of the respiratory network in the in vitro perfused brainstem of adult rats. Exp Brain Res 106: 265–274

Onimaru H, Arata A, Homma I (1995) Intrinsic burst generation of preinspiratory neurons in the medulla of brainstem-spinal cord preparations isolated from newborn rats. Exp Brain Res 106: 57–68

Paton JFR (1996) The ventral medullary respiratory network of the mature mouse studied in a working heart-brainstem preparation. J Physiol 493: 819–831

Richter DW, Ballanyi K (1996) Response of the medullary respiratory network to hypoxia: a comparative analysis of neonatal and adult mammals. In: Tissue oxygen deprivation: from molecular to integrated function. Eds. Haddad, GG, Lister G. Dekker Inc., New York, Basel, Hong Kong, pp. 751–777

Schäfer T, Morin-Surun MP, Denavit-Subie M (1993) Oxygen supply and respiratory-like activity in the isolated perfused brainstem of the adult guinea pig. Brain Res 618: 246–250

Shao YM, Feldman JL (1997) Respiratory rhythm generation and synaptic inhibition of expiratory neurons in pre-Bötzinger complex: differential roles of glycinergic and GABAergic transmission. J Neurophysiol 77: 1853–1860

Smith JC, Feldman JL, Schmidt BJ (1988) Neural mechanisms generating locomotion studied in mammalian brainstem-spinal cord in vitro. FASEB J 2: 2283–2288

Smith, J.C., Funk, G.D., Johnson, S.M. and Feldman, J.L. (1995) Cellular and synaptic mechanisms generating respiratory rhythm: insights from in vitro and computational studies. In: Trouth CO, Millis RM, Kiwull-Schöne HF, Schläfke ME (eds) Ventral brainstem mechanisms and control of respiration and blood pressure. New York, Basel, Hongkong, M. Dekker, Inc., pp.463–496.

Smith JC, Ellenberger HH, Ballanyi K, Richter DW, Feldman JL (1991) Pre-Bötzinger complex: a brain region that may generate respiratory rhythm in mammals. Science 254: 726–729

Suzue T (1984) Respiratory rhythm generation in the in vitro brain stem-spinal cord preparation of the neonatal rat. J Physiol 354: 173–183

Thompson SM, Haas HL, Gähwiler BH (1992) Comparison of the actions of adenosine at pre- and postsynaptic receptors in the rat hippocampus *in vitro*. J Physiol 451: 347–363

Trapp S, Lückermann M, Brooks PA, Ballanyi K (1996) Acidosis of rat dorsal vagal neurons in situ during spontaneous and evoked activity. J Physiol 496: 695–710

Voipio J, Ballanyi K (1997) Interstitial P_{CO2} and pH, and their role as chemostimulants in the isolated respiratory network of neonatal rats. J Physiol 499: 527–542

Wolfe LS, Horrocks LA (1994) Eicosanoids. In: Siegel GJ, Agranoff BW, Albers RW, Molinoff PB (eds) Basic neurochemistry. Raven Press, New York, pp. 475–492

Chapter 10

Culturing CNS Neurons: A Practical Approach to Cultured Embryonic Chick Neurons

Åke Sellström and Stig Jacobsson

Introduction

The need for an accurate, i.e. true and reproducible, in-vitro model of the neuron is obvious and therefore not further commented on. The prospect of cultured neuronal cells to meet this need has generated a wealth of experience and several methods to choose from. Trying to condense this experience in a few lines, the following is said:

The choice of methodology for neuronal cultures is normally a trade-off between simplicity, reproducibility and biological accuracy.

Tumoral cell lines, for instance, have to yield in biological accuracy to the primary cultures, which on the other hand are less reproducible and more cumbersome to handle. By biological accuracy we mean that the model to a high degree reflects the desired *in situ* properties. Frequently this is addressed as a question of differentiation.

The starting material for primary neuronal cultures are post-mitotic but undifferentiated neurons, whereas cell lines generally are dividing tumor cells lacking the property of contact inhibition. Both cell lines and primary cultures need to undergo differentiation during the culturing period in order to achieve better accuracy. Primary neuronal cultures to some extent differentiate with time in culture. Accordingly, primary cultures made from various regions of the nervous system are becoming more and more common, since they reflect properties specific to their region of origin (Table 1). Primary

Table 1. The Use of Primary Cultures 1997

Origin of Neural Tissue	Percent of Articles
Cortex	39
Cerebellum	14
Hippocampus	16
Striatum	8
Hypothalamus	3
Spinal Cord	3
Retina	3
Ganglion	13
Other	1

The table shows the distribution of articles recovered when primary neuronal cultures were searched in 1997. A total of 114 articles was classified.

Correspondence to: Åke Sellström, Defence Research Establishment (FOA), Division of NBC Defence, Department of Biomedicine, Umeå, SE-901 82, Sweden (phone: +46-90-10-67-29; fax: +46-90-10-68-03; e-mail: Sellstrom@ume.foa.se)
Stig Jacobsson, Department of Pharmacology and Clinical Neuroscience, Umeå University, Umeå, SE-90187, Sweden

cultures and cell lines may also be manipulated so as to induce observable changes, which are referred to as differentiated, i.e. some property expected from the differentiated neuron may be potentiated by this manipulation.

For a defined and "isolated" task, where the property to be studied is documented as properly expressed in a cell line, this may be the model of choice. For a more explorative task or for studies where proper expression is important, the primary culture is more likely to serve one's purposes. Please bear in mind that a proper expression always has to be confirmed one way or the other.

This contribution is devoted to the description and comments on primary neuronal cultures, which according to the authors is the generally preferred method. The maintenance of cell lines holds little magic. Because discussions on establishing a cell line and/or attempting to differentiate the established cell line would open up a totally new area, it is not within the scope of this article.

A third way to establish neuronal cultures exists, i.e. the organotypical culture (Stoppini et al. 1991). These cultures have the advantage of maintaining a brain-like environment, which should favor a reasonable accuracy of the neurons present. However, while this preparation may provide a "better" environment for its neuronal cells, it also carries a number of non-neuronal elements just like the brain itself.

Developmental trends within the area of primary cultures include the use of various regions in order to obtain better-defined cellular models. Depending on the species and the region used, different ages of the embryonal tissue are recommended. Frequently, 8-day-old embryonic rat cerebellum is used for cultures of glutamatergic granular neurons (Kingsbury et al. 1985; Wroblewski et al. 1985) and 14-day-old rat cerebral cortex for GABAergic interneurons (Schousboe and Hertz 1987). Although systematic studies of this issue have not been performed, we may conclude that the subpopulation of neurons to survive and differentiate in the culture should have reached the stage of a post-mitotic neuron. Recently, adult tissues have also been used as starting material for neuronal cell cultures (Brewer 1997). This and other developmental trends, as well as the use of defined media, are dealt with below.

The use of embryonic material when establishing primary neuronal cultures adds an extra dimension of complication to this method. The exact and reproducible timing of embryonic age may be difficult and some researchers may find the abortion of fetuses stressful. For these reasons, our own laboratory is presently using the embryonic chick as starting material. The protocol given below is, accordingly, described for primary neuronal chick cultures but can also be applied to other preparations.

Chick telencephalic neurons are most readily dissected and cultured if removed from embryonic day 7.5–8 (Hamburger-Hamilton Stage 34). The embryonic brain is extremely fragile and soft and the use of isotonic dissection solution is essential for optimal preservation of the morphology. HEPES-buffered solutions are preferable when working for extended periods of time in atmospheric CO_2 concentrations since solutions buffered with bicarbonate tend to rapidly become alkaline. Removal of divalent cations facilitates dissociation and decreases a potential excitotoxic challenge. Therefore, many dissection solutions lack magnesium and calcium. If the desired structures are difficult to uncover and the dissection time-consuming, it may, however, be advisable to add calcium and magnesium to the dissection solution to preserve tissue integrity.

The dissociation of neural tissue into a suspension of single cells can be accomplished by either mechanical dissociation or enzymatic treatment followed by mechanical dissociation. Simple mechanical dissociation is recommended for studies in which minimum disturbance of the membrane receptor proteins is desired. Both fetal and neonatal brain tissue can be easily dissociated mechanically by trituration with a glass Pasteur pipette. For some applications the use of enzymes is necessary; this subject is discussed in Comments.

Neuronal attachment and, hence, the long-term survival of primary neuronal cultures are affected by the culture substrate. The most common substrates for neurons are poly-L-lysine or poly-D-lysine but other substrates are also successfully used, e.g. collagen, laminin, polyethyleneimine and polyornithine.

Traditionally, neuronal cells have been cultured in medium containing serum. Fetal serum is most commonly used, since serum derived from young, i.e. fast-growing individuals, is regarded as the most growth promoting. To restrict the proliferation of non-neuronal cells, some authors use anti-mitotic agents such as cytosine arabinoside or fluorodeoxyuridine after one to two days in culture (for more information see Comments).

The composition of the serum must be regarded as undefined. Moreover, its quality may vary between suppliers, as well as between different batches from a single supplier. As an attempt at a controlled and optimized culture condition, chemically defined serum-free medium is frequently introduced. This can be done following initial plating in serum-supplemented medium which, generally, does not require the addition of anti-mitotic inhibitors to the cultures. In addition, the medium may be defined to prevent unrestricted proliferation of non-neuronal cells (see Comments). The use of a chemically defined, serum-free medium is recommended for studies in which better control of the factors present in the nutrient medium is desired .

The density, i.e. number of cells per mm^2, at which a culture is initiated is critical. It is critical for both growth characteristics and the composition of the final culture. The seeding density should therefore be maintained as reproducibly as possible. It is claimed that neuronal cells will only grow in culture in serum-containing medium at a density above 300 cells/mm^2. Using available commercial serum-free medium (e.g. Neurobasal), much lower seeding densities have been reported. Neurobasal is a commercially available serum-free base medium optimized for long-term survival of hippocampal neurons plated at low density (Brewer 1995). Recently, adult hippocampal and cortical neurons from rats have been regenerated and supported by this serum-free medium for more than three weeks (Brewer 1997). We recommend the use of serum-free media when culturing embryonic chick telencephalic neurons. More information on this issue may be found in Comments.

Outline

- Incubation of eggs
- Preparation of growth substrates and culture media
- Dissection of tissue
- Preparation of cell suspension
- Plating and feeding of cells
- Use of cells

Materials

A laminar flow hood (LAF-hood) equipped with a High Efficiency Particulate Air (HEPA)-filter, a CO_2-incubator with controlled temperature and humidity, and a good quality phase-contrast microscope are necessary basic equipment. The LAF-hood should be equipped with a throttled gas burner and technical air suction. It is also important to have access to centrifuges, pH-meter and sterilization equipment (e.g. an autoclave).

Other necessary utensils are

- Disposable bottle top filter (500 ml funnel)
- Culture medium bottles
- Disposable plastic pipettes (5 and 10 ml capacity)
- Long-form glass Pasteur pipettes
- Disposable plastic conical culture tubes (15 and 50 ml capacity)
- Nylon mesh cell strainers (40 or 70 μm)
- Disposable plastic syringes (10 and 20 ml capacity)
- Syringe filter units (0.22 μm)
- Large and fine serrated curved forceps and fine sharp watchmaker's forceps
- Pipettes (5 μl to 1 ml capacities) with tips
- A multichannel pipette (10 channels) with syringes (when using multiwell clusters)
- Disposable 60-mm tissue culture dishes
- Disposable tissue culture flasks or multiwell clusters (depending on the experiment)
- Hemocytometer (Burcher chamber)
- Conical Eppendorf tubes (1.5 ml)

Note that all non-presterilized instruments or labware should be sterilized (autoclaved, flame-sterilized, or scrubbed with 70% ethanol or disinfection solution)

Culture media and reagents

- Sterilized double-distilled water
- Phosphate-buffered saline (PBS)
- Poly-D-lysine hydrobromide (m.w. 70,000–150,000)
- Sodium tetraborate (Borax)
- Boric acid
- Calcium- and magnesium-free Hank's balanced salt solution (CMF-HBSS)
- HEPES (free acid)
- Sodium pyruvate
- Dulbecco's modified Eagle medium (DMEM) with GLUTAMAX I
- Fetal calf serum (FCS) or fetal bovine serum (FBS)
- Neurobasal medium
- B-27 serum-free supplement (50x)
- Penicillin and streptomycin (PEST, 10 000 IU/ml)
- L-glutamine (100x)
- Trypan blue solution (0.4%)

B-27, FCS, L-glutamine and PEST should be stored at –20 °C, and to avoid repeated thawing and freezing, aliquots of appropriate size should be prepared in advance and frozen. This concentration of L-glutamine, stored in 1.5 ml Eppendorf tubes, will precipitate when thawed, but re-dissolving is quick and does not compromise the cell culture.

Procedure

Incubate fertile White Leghorn eggs in force-draft, humidified egg incubator (e.g., Agroswede, Malmö, Sweden) at 37.8 °C for eight days (or the required time). The eggs should be placed with the blunt end upwards and turned automatically every two hours. Note that the use of animal tissue for experimental work may require legal approval.

Preparing the Growth Substrata

1. Prepare and sterilize borate buffer (2.37 g borax and 1.55 g boric acid dissolved in 500 ml double-distilled water, pH 8.4).
2. Dissolve 50 μg/ml poly-D-lysine in borate buffer.

3. Apply enough of the poly-D-lysine solution to cover the surface of the culture dish or flask.
4. Incubate at room temperature for one hour or overnight.
5. Rinse thoroughly (2–3 washes) with sterile, double-distilled water.
6. The dishes are allowed to dry and then exposed to UV light for 10 min (optional). Although it is preferable to prepare the dishes immediately before use, they can be stored at +4 °C for one week.

Preparing Dissociation Solutions and Culture Media

- 0.15 M PBS buffer:
  - 9.0 g NaCl
  - 0.73 g Na_2HPO_4 x 7 H_2O
  - 0.21 g KH_2PO_4
  - add 500 ml double-distilled water, adjust pH to 7.4, sterilize and store at +4 °C.
- Dissociation buffer:
  - 1.19 g HEPES
  - 0.055 g sodium pyruvate
  - 5 ml (or 100 IU/ml) PEST
  - add CMF-HBSS to 500 ml, adjust pH to 7.4, sterilize by filtration and store at +4 °C.
- DMEM + 20% FCS (100 ml):
  - 79 ml DMEM with glutaMAX I
  - 20 ml FCS
  - 1 ml or 100 IU/ml PEST
- DMEM + 5% FCS (100 ml):
  - 94 ml DMEM with glutaMAX I
  - 5 ml FCS
  - 1 ml (or 100 IU/ml) PEST
- B27-supplemented Neurobasal (100 ml):
  - 96.75 ml Neurobasal medium
  - 2 ml B-27 serum-free supplement
  - 1 ml (or 100 IU/ml) PEST
  - 0.25 ml 200 mM L-glutamine

All culture media are mixed in sterile culture bottles and stored at +4 °C for no more than 5 days. Note that all culture media should be pre-warmed to +37 °C before feeding.

Dissection of Embryonic Chick Telencephali

1. Swab a small metallic or plastic box with 70% EtOH or disinfection solution containing 70% isopropanol.
2. Take out the egg, holding the blunt end upward. The airspace is located in the blunt end of the egg and the embryo is most easily observed and removed from the egg through the membrane lining the airspace.
3. Swab the egg with alcohol or disinfection solution and put it in the box with the blunt end uppermost.
4. When all the incubated eggs have been disinfected and allowed to dry, move to the sterile hood.

5. Fill three 60mm plastic petri dishes with about 10ml cold dissociation buffer (depending on the number of embryos collected).
6. To remove the embryos from eggs, crack the shell in a line around the airspace by tapping with the back side of a forceps and remove this portion of the shell and the membrane lining the airspace.
7. Carefully remove the embryo with a pair of curved forceps by hooking them around the neck of the embryo. Be careful not to bring the yolk sac or its membranes together with the embryo.
8. Place embryo in the petri dish containing dissociation buffer. When all embryos have been collected, transfer them one at a time to another petri dish with buffer for dissection.
9. Place the embryo on its ventral side and fix it by introducing a pair of sharp fine forceps dorsally through the large mesencephalon.
10. With a pair of fine serrated curved forceps grab the telencephalon and pinch out the two hemispheres.
11. Transfer the telencephali with the forceps to the last petri dish with fresh dissociation buffer.
12. When the entire telecephalon has been dissected out and transferred to the last petri dish, carefully remove all meninges, membranes and vessels with a pair of sharp forceps.
13. Add 2ml dissociation buffer to a 15ml conical tube. Carefully transfer the telencephali with a Pasteur pipette to the tube.

Dissociation of Tissue to Single Cell Suspension

1. Disaggregate the telecephalon into a single cell suspension by trituration using siliconized, sterile, long-form glass Pasteur pipettes. This is a very critical step and will be developed only through practice. Firmly press the tip of the pipette to the bottom of the conical tube. Slowly draw up the tissue pieces and at a similar pace expel the solution back into the tube. This represents one pass. Complete 10 to 20 more passes. Care should be taken to minimize air introduction (air bubbling), which will lyse cells.
2. Fire-polish another pipette and reduce the diameter by about one-half. Continue with another 10 to 20 passes at a higher rate and force, but still treating the tissue with some tenderness.
3. An apparently trivial point is important here. A thick-walled pipette teat is essential to provide the right degree of force for trituration, and the pipette must not be chipped or cracked. The number of passes necessary for dissociation to single cell suspension must be dependent on whether enzymatic digestion is used or not and will be learned by experience. Try not to disrupt all tissue pieces of visual size, since they mostly consist of membranes and unwanted cells.
4. Dilute the single cell suspension with 2 volumes (4ml) DMEM + 20% FCS to restore divalent cations and nutrition, and let the non-dispersed tissue settle for three minutes.
5. With a Pasteur pipette, carefully transfer the supernatant to a fresh 15ml conical tube and centrifuge for 1min at 200x g.
6. Discard the supernatant from the centrifuge tube and gently add 5ml DMEM + 20% FCS to the loose pellet. Gentle flicking of the bottom of the tube is often sufficient to resuspend the pellet. In this step the cell suspension is cleared of debris and cytotoxic substances released from dead cells.
7. Place a 40μm cell strainer on top of a 50ml plastic conical culture tube and sieve the cell suspension through the mesh. With a 15ml plastic pipette, wash additional 15 to

45 ml DMEM + 20% FCS through the cell strainer mesh. In this step aggregated cells are removed from the suspension.

8. Remove 50 μl of the cell suspension and mix this gently with an equal volume of trypan blue solution. Count the number of viable (unstained, bright) cells using a hemocytometer. The yield of the cells varies depending on the age of the embryo and on the ratio between the number of embryos used and medium volume added. A good cell preparation should have approximately 10% dead cells, i.e. stained by trypan blue.

Plating and Feeding of Cells

1. The plating density used is based on the experimental requirements. Generally, a high plating density increases cell survival but also increases the amount of non-neuronal cells. When comparing results obtained in different sizes of dishes and flasks, it is recommended to measure the density as number of cells per area unit rather than number of cells per volume unit. We suggest a plating density between 500 and 3000 cells/mm^2.
2. The volume of medium used also depends on the culture dish or flask used. For 35, 60, and 100 mm dishes, 2, 3, and 10 ml medium, respectively, are added. T-25, T-75, and T-160 culture flasks should be filled with 5, 15, and 30 ml, respectively. To each well on cluster plates we add 50–100 μl to 96-well, 0.5 ml to 24-well, 1 ml to 12-well, and 2 ml to 6-well plates.
3. Use a pipette (or multichannel pipette) with sterile tips to add an appropriate amount of cell suspension to the pre-coated culture vessels. Incubate the culture in a humidified 5% CO_2 37 °C incubator.
4. Within 24 hours, change the medium to DMEM + 5% FCS.
5. After additional 48 hours of incubation, replace the serum-containing medium with the serum-free B27/Neurobasal medium.
6. The cultures should then be fed on every second or third day (normally, Monday, Wednesday and Friday).
7. Since we want to obtain a chemically-defined serum-free condition, it is important to remove all medium when changing to lower serum concentration or serum-free medium. To successfully do so, we use a Pasteur pipette connected to a technical air suction device to remove all medium from the culture. Here it is very important to work fast when replacing the medium and not allow the cells to dry out. At the same time it is important not to add medium directly on the cell surface, since this will ultimately detach the cells.

Results

Quantitative and Qualitative Analysis of Neuronal Cultures

The most important and obvious way to confirm the status of a primary neuronal culture is to examine it using a high quality phase-contrast microscope. One day after plating, all of the viable neurons should be attached to the substratum. It is very difficult at this stage to judge whether the growing cells are indeed neurons. After three to four days in culture, cell bodies should look smooth and phase-bright, and many neurites should have initiated and extended several cell body diameters. At this stage the specialist may be able to assess the cellular composition of the culture, but the task is not easy.

In order to ensure that a primary neuronal cell culture has successfully been established, the use of specific markers is called for. Staining techniques used in neuroanatomy are not recommended, since these normally will not provide a conclusive result. Instead immunocytochemistry using specific antigenes is recommended. Commonly used markers for neurons are neuron-specific enolase (NSE), microtubulus-associated protein (MAP) or neuron-specific antigens such as transmitter-producing enzymes, e.g. glutamic acid decarboxylase (GAD) or tyrosine hydroxylase (TH). Ready-to-use antibodies against several neuronal markers are commercially available.

Using markers to confirm the presence of neurons may not be sufficient. Negative markers, i.e. markers for other cells, may be necessary for control; normally the presence of glial cells should be looked for. The astrocyte-specific antigen glial fibrillary acidic protein (GFAP) is the most commonly used marker for this purpose. The most accurate estimate of the cellular population contained in a culture will be obtained using double-labelling techniques, i.e. using a neuron-specific and a glial-specific antibody in parallel in the same preparation.

When studying characteristics of the differentiated neuron, such as the presence of transmitter synthesis or receptors, nothing should be taken for granted. In fact, the usefulness of an individual culture model for every specific purpose has to be controlled and confirmed.

Troubleshoot

- Rapid pH shift in the medium could be due to incorrect CO_2 tension in the incubator, overly tight caps on tissue culture flasks or insufficient bicarbonate buffering in medium. It could also be due to bacterial, yeast, or fungal contamination. Precipitate in the medium with a change in pH indicates bacterial or fungal contamination, and the culture should be discarded.
- Destroyed (rough) cells and phase-dark cell bodies with large amounts of cellular debris in a one-day culture indicate that the dissociation process was too efficient or too harsh. NB cultures grown in serum-free media often appear "dirty" partly due to increased cell death, but also due to the lack of macrophage activity.
- Poor attachment and stunted neurite outgrowth indicate that the substratum or the medium supplementation is not optimal. The growth substrate (poly-D-lysine) might be of bad quality and fresh batches should be obtained. Bad cell attachment could also indicate mycoplasma contamination.
- Unexpected death of the culture might be due to low CO_2 tension or temperature fluctuations. These parameters should be monitored, and it is also important to minimize opening and closing of incubator doors. Another possible cause is the build-up of toxic metabolites in the medium if it is not changed regularly.

Comments

Disintegrating the Tissue into a Cell Suspension

The dissociation of neural tissue into single cells can be accomplished by either mechanical dissociation or enzymatic treatment followed by mechanical dissociation. Simple mechanical dissociation is recommended for studies in which little disturbance of the membrane receptor proteins is desired. Both fetal and neonatal brain tissue can be easily dissociated mechanically by trituration with a siliconized, sterile, long-form glass Pasteur pipette. It is important to gently triturate the tissue up and down with the Pasteur pipette at a steady pace and without introducing air bubbles, since this will lyse cells. The more developed brain tissue (i.e. tissue from older individuals) is harder to dissociate to a single cell suspension by simple mechanical dissociation without handling the tissue too roughly, and thereby increasing the number of dead cells obtained in the cell suspension. Therefore, it is in many cases preferable to use enzymatic digestion before the mechanical treatment. Trypsin is a very potent protease that works well to dissociate brain tissue, but if digestion conditions are not carefully controlled, trypsin can reduce neuronal viability. Trypsin lots can be very variable, even when purchased from the same supplier. When using trypsin digestion, a working concentration of 0.2% is sufficient but can be reduced by adding ethylenediaminetetraacetic acid (EDTA). EDTA is a calcium chelator and reduces cell-to-cell adhesion, thereby enhancing trypsin digestion. Trypsin-EDTA solutions (0.05% trypsin, 0.53 mM EDTA) are commercially available from Life Technologies (Cat. No. 35400).

Other enzymes can be used, such as proteinase K, dispase, collagenase and papain. The enzyme concentrations used during the dissociation step may vary among the different species/ages used (e.g. how "hard" the brain tissues are) or the incubation time and temperature.

Serum-free Culturing Media

In order to mimic the in-vivo conditions, it has been common to culture cells and tissue in medium containing many different biological fluids, such as serum, tissue extracts and ascites fluids. Since these tissue fluids often have unknown origin and vary in composition, there have been attempts to replace them with solutions of salts, amino acids, vitamins and radical scavengers. For most cell types, though, it has been necessary to keep small amounts of serum in the medium to reach a satisfactory cell growth. In serum-supplemented medium, serum from human, horse, fetal calf and serum from newborn calf is commonly used. Serum derived from young, i.e. fast-growing individuals, is regarded as more growth promoting and therefore fetal serum is the most commonly used. The quality of serum may vary among suppliers, as well as among different batches from a single supplier (often due to differences in the animal stock or drain conditions); it is therefore recommended to buy a significant portion of the same batch. The suppliers usually screen their products for toxicity, minimizing the risk for serum-derived cell toxicity.

A major disadvantage of serum-supplementation when culturing neurons is the unrestricted proliferation of non-neuronal cells that can overgrow the neurons and reduce the viability of neurons during long-term culturing. In a culture containing mixed cell types, it is difficult to determine whether a certain agent acts on the other cell types and indirectly influences the neurons. It is important to be careful during the dissections and dissociation procedure to remove meninges, vessels and other membranes from the tissue, since these structures bring large quantities of fibroblasts and satellite cells to the culture. Even if the dissection/dissociation procedure is well conducted, a small number of non-neuronal proliferating cells can overgrow neurons within a short period of time when culturing in serum-containing medium. To restrict the proliferation of non-neuronal cells, the concentration of serum can be reduced by replacing the old culture media with a fresh medium containing lower concentrations of serum. It is also possible to treat the culture with anti-mitotic agents such as cytosine arabinoside (araC) or fluorodeoxyuridine after one to two days in culture. Note that anti-mitotic agents only kill fast-dividing cells such as fibroblasts and not more slowly dividing glial cells. Bear also in mind the toxicity of these agents, and use them at the lowest possible concentrations (below 10 μM). Finally, cultures can be grown in a chemically defined serum-free medium after initial plating in serum-supplemented medium. This procedure generally does not require the addition of anti-mitotic inhibitors to the cultures.

Another important disadvantage of the use of serum-supplemented medium is the limited control of components present in the medium. The aim of serum removal from culture medium is to get a chemically defined medium. The use of a chemically defined, serum-free medium is recommended for studies in which better control of the factors present in the nutrient medium is desired. Studies for neuronal development, plasticity, electrophysiology, gene expression, pharmacology, and neurotoxicity may benefit from serum-free culture medium. It should be noted that in some serum-free media, the serum is replaced with another biological component, such as pituitary extracts or ascites fluid. This makes the medium serum-free, but not chemically defined. To use cell and tissue cultures as a model for the behavior of cells and tissue in an organism is of course a defective method, since the cells have been deprived of their natural surrounding and become dedifferentiated. On the other hand, there is no need to believe that serum-free culturing makes it more difficult to maintain a differentiated state. On the contrary, many cell types differentiate during serum-free conditions. Serum gives selective growth stimulation of proliferating cell types (e.g. fibroblasts). Chemically defined serum-free media prevent unrestricted proliferation of non-neuronal cells, and longer survival of neuron populations in the culture may be achieved.

Seeding the Cells

Brain neurons will only grow in culture in serum-containing medium above a density of 300 cells/mm^2, which precludes the study of individual neurons without a feeder layer of glial cells. Neurobasal is a commercially available serum-free base medium optimized for long-term survival of hippocampal neurons plated at low density. It has been shown to be superior to DMEM for growth of hippocampal neurons at densities below 640 cells/mm^2 (Brewer et al. 1993). Factors contributing to this better performance include reductions in osmolarity, glutamine and cysteine, and elimination of toxic ferrous sulphate. Neurobasal also contains amino acids and a vitamin missing from DMEM: alanine, aspargine, proline, and vitamin B12. Neurobasal supplemented with the commercially available B-27 serum-free supplement produces more than 30-day survival of hippocampal neurons at densities down to 80 cells/mm^2 and also supports growth of neurons from other brain regions such as striatum, substantia nigra, septum, cerebral cortex, cerebellum, and dentate gyrus (Brewer, 1995). The B-27 serum-free supplement contains optimized concentrations of insulin, transferrin, progesterone, putrescine, and selenium along with the thyroid hormone T3 and fatty acids. B-27 also contains antioxidants such as vitamin E, glutathione, selenium, catalase, and superoxide dismutase. For studies of free radicals, a B-27 version lacking antioxidants is also commercially available (Life Technologies, Cat. No. 10889). Even the regeneration of adult hippocampal and cortical neurons from rats is supported by the B27/Neurobasal serum-free medium, and adult neurons have been cultured for more than three weeks (Brewer 1997). We have successfully used B27/Neurobasal when culturing embryonic chick telencephalic neurons.

Applications

Choice of Tissue-Culturing Dishes

The choice of culturing dishes or flasks depends of course on the subsequent application of the cultured cells.

- Immunocytochemical experiments are very common with tissue-cultured cells. Today there are many different suppliers of slide chambers developed for this purpose. Cell cultures in 35 or 60 mm plastic dishes are also useful.
- Survival studies or toxicological studies involving morphological evaluation are very well performed in 35 or 60 mm plastic dishes.
- In studies where large amounts of cells are required as homogenate for subsequent analysis, e.g. Western or Northern blots, radioligand binding studies, or other biochemical studies. Culturing should be conducted in 25 cm^2 (T-25), 75 cm^2 (T-75) or 160 cm^2 (T-160) culture flasks.
- For pharmacological and toxicological studies in neuronal culture systems where untreated sister cultures are needed as controls; flat-bottomed multiwell culture plates (cluster plates) are perhaps most commonly used. For studies that employ spectrophotometric or fluorescence plate readers, it is typical to use 24-, 48-, or 96-well plates. To monitor the cultures morphologically, 24-well culture plates are preferred, since it is difficult to get a good phase-contrast view in the 96-well plates.

Suppliers

Fertilized White Leghorn eggs can be obtained from any farmer with a trimmed and fit rooster. CMF-HBSS (Cat. No.14175), HEPES (Cat. No. 11344), Sodium pyruvate (Cat. No. 11840), DMEM with glutaMAX I (Cat. No.21885), FBS (Cat. No. 10106), Neurobasal medium (Cat. No.21103), B-27 serum-free supplement (Cat. No.17504), PEST (Cat. No.15140), and L-glutamine (Cat. No.25030) can be purchased from Life Technologies Ltd., Gaithersburg, MD, USA.

Poly-D-lysine hydrobromide (Cat. No.P0899), BORAX (Cat. No. B9876), Boric Acid (Cat. No. B0257), Trypan blue solution (Cat. No. T8154), and other cell culture reagents can be obtained from Sigma, St. Louis, MO, USA.

Cell culture vessels, sterile filters, and other disposable plastic labware can be obtained from several sources such as Corning Costar, Acton, MA, USA; Nalge Nunc International, Rochester, NY, USA; and Becton Dickinson, Franklin Lakes, New Jersey, USA.

References

Brewer GJ (1995) Serum-free B27/Neurobasal medium supports differentiated growth of neurons from the striatum, substantia nigra, septum, cerebral cortex, cerebellum, and dentate gyrus. J Neurosci Res 42: 674–683

Brewer GJ (1997) Isolation and culture of adult rat hippocampal neurons. J Neurosci Meth 71: 143–155

Brewer GJ, Torricelli EK, Evege EK, Price PJ (1993) Optimized survival of hippocampal neurons in B27-supplemented Neurobasal, a new serum-free medium combination. J Neurosci Res 35: 567–576

Kingsbury A, Gallo V, Woodham P, Balazs R (1985) Survival, morphology and adhesion properties of cerebellar interneurons cultured in chemically defined and serum-supplemented medium. Brain Res 349: 17–25

Schousboe A, Hertz L (1987) Primary cultures of GABAergic and glutamatergic neurons as model systems to study neurotransmitter functions. II. Developmental aspects. In Vernadakis A (ed): Model systems of development and aging of the nervous system, Publ. Martinus Nijhoff, Dordrecht Boston, pp 33–42
Stoppini L, Buchs P-A, Muller D (1991) A simple method for organotypic cultures of nervous tissue. J Neurosci Meth 37: 173–182
Wroblewski JT, Nicoletti F, Costa E (1985) Different coupling of excitatory amino acid receptors with Ca^{2+} channels in primary cultures of cerebellar granule cells. Neuropharmacology 9: 919–921

Abbreviations

CMF-HBSS	calcium- and magnesium-free Hank's balanced salt solution
DMEM	Dulbecco's modified Eagle medium
EDTA	ethylenediaminetetraacetic acid
FBS	fetal bovine serum
FCS	fetal calf serum
GAD	glutamic acid decarboxylase
GFAP	glial fibrillary acidic protein
HBSS	Hank´s balanced salt solution
MAP	microtubulus-associated protein
NSE	neuron-specific enolase
PBS	phosphate-buffered saline
PEST	penicillin and streptomycin
TH	tyrosine hydroxylase

Chapter 11

Neural Stem Cell Isolation, Characterization and Transplantation

Jasodhara Ray and Fred H. Gage

Introduction

During development, nerve cells in the mammalian central nervous system (CNS) are generated by the proliferation of multipotent stem or progenitor cells that migrate, find their site of final destination and ultimately terminally differentiate. However, the mechanisms for development of diversified cell types in the CNS and the environmental stimuli involved in these processes are not well understood. In analogy with the hematopoietic system, it has been proposed that during CNS development, a self-renewing population of stem cells gives rise to a more restrictive population of progenitor cells. However, the existence of these elusive progenitor cells has not been proven due to the unavailability of specific phenotypic markers. In the adult brain only a small number of stem cells exist and they differentiate into neurons at specific neurogenic sites at a slow rate (Morshead et al., 1994; Kuhn et al., 1996).

Attempts have been made for decades to isolate and culture stem or progenitor cells (will be referred to as stem cells in this chapter) that give rise to neuroblasts and glioblasts, which upon differentiation, generate mature neurons, astrocytes and oligodendrocytes, i.e., the cells that are the major building blocks of the central nervous system (CNS). To obtain long-term proliferative cultures, cells from embryonic brains were immortalized by using oncogenic transgenes like v-myc or SV40 large T antigen (Cepko, 1988a; Lendahl and McKay, 1990; Whittemore and Snyder, 1996). The immortalization process arrests cells at specific stages of development and halts their terminal differentiation (Cepko, 1988a; Lendahl and McKay, 1990). Clonal cultures of cells representing a specific stage in development were isolated and used for *in vitro* and *in vivo* studies (Gage et al., 1995a; Whittmore and Snyder; 1996; Ray et al., 1997; Fisher, 1997). Although immortalized cells offer a number of advantages, they do not always represent their primary counterparts. Isolation and long-term culturing of primary stem/progenitor cells have been advanced by the findings that the mitogenic growth factors epidermal growth factor (EGF) and basic fibroblast growth factor (FGF-2) have proliferative effects on these cells (Weiss et al., 1996a; Ray et al., 1997; McKay, 1997).

Isolated and cultured stem cells not only provide an important source of cells for *in vitro* studies to address issues related to fate choice and differentiation, but they are also an important source of CNS cells that can be used in transplantation studies (Ray et al., 1997, 1998). To explore the possible use of these cells for therapy, embryonic or adult brain-derived stem cells have been grafted in normal CNS to determine their *in vivo* survival and potential for fate choice, differentiation and integration (Hammang et al., 1994; Gage et al., 1995b; Suhonen et al., 1996; Ray et al., 1997). Adult rat hippocampus-derived FGF-2 responsive cells grafted to homo- or heterotypic neurogenic sites (hip-

Correspondence to: Jasodhara Ray, Salk Institute for Biological Studies, Laboratory of Genetics, 10010N. Torrey Pines Road, La Jolla, CA, 92037, USA (phone: +01-858-453-4100 ext. 1006; fax: +01 858-597-0824; e-mail: jray@salk.edu)

pocampus or olfactory system) differentiated into neurons (Gage et al., 1995b; Suhonen et al., 1996), whereas when they were grafted to non-neurogenic site (cerebellum) no neuronal differentiation was observed (Suhonen et al., 1996). In contrast, EGF-responsive embryonic mouse stem cells grafted in intact neonatal mouse cortex and spinal cord showed poor survival and no neuronal differentiation (Hammang et al., 1994). Similar results were obtained with EGF-responsive stem cells from embryonic rats engrafted into lesioned adult rat brains (Svendsen et al., 1996). The usefulness of stem cells as a vehicle to deliver trophic factors in the brain has been explored in two recent studies. EGF-responsive stem cells were cultured from transgenic mice in which GFAP promoter drives the expression of human nerve growth factor (NGF). Cells grafted in adult rat striatum survived up to 3 weeks (latest time point examined), did not migrate from the graft site and differentiated into astrocytes (Carpenter et al., 1997; Kordower et al., 1997). The secreted bioactive NGF induced hypertrophy and sprouting of the endogenous cholinergic neurons within the intact rodent brain (Carpenter et al., 1997) and prevented the degeneration of striatal neurons in a rodent model of Huntington's disease (Kordower et al., 1997). These studies show promise that neural stem cells can be exploited for cell replacement and gene therapy purposes to repair CNS injuries or disorders.

Neural tissues are composed of both neuronal and nonneuronal cells as well as connective and vascular tissues. When removed from the brain and put in culture, the cells lose the physiological connections, anchorages and the humoral environments. Strategies to isolate and culture stem cells from CNS involved: i) isolation of cells from an intertwined network of thousands of adhesive contacts without causing damage to the cells; ii) separating stem cells from other brain cells and connective tissue debris; and iii) providing the appropriate environmental conditions, including specific nutrients and growth factors, required for the survival and the proliferation of stem cells. Neural cultures are generally maintained at pH 7.2–7.6 and at the appropriate osmolarity. The most commonly used methods for the isolation and culture of stem cells use serum-free culture medium supplemented with various hormones and nutrients (Bottenstein and Sato, 1979) and mitogenic growth factors EGF or FGF-2. EGF has been used to culture subependymal/forebrain stem cells as neurospheres from embryonic and adult mouse (Reynolds et al., 1992; Reynolds and Weiss, 1992). However, EGF failed to induce proliferation of stem cells isolated from adult mouse spinal cord which required a combination of EGF and FGF-2 for proliferation (Weiss et al., 1996b). In contrast, FGF-2 alone has been successfully used to establish monolayer cultures of stem cells from different embryonic and adult brain regions and the spinal cord (Ray et al., 1993, 1994; Gage et al., 1995b; Palmer et al., 1995; Minger et al., 1996; Shihabuddin et al., 1997) and to generate neurospheres from lateral ventricle/forebrain of adult mouse (Gritti et al., 1996).

The choice of a culture method depends on the specific issues being addressed by the researchers. Methods for culturing neural stem cells have been described previously (Ray et al., 1995). This chapter describes the strategies and the latest techniques for culturing stem cells by the two most commonly used methods, and discusses the generation of clonal populations and the characterization of the cells and their grafting in adult rat brain.

Outline

Dissect out embryonic or adult brain regions, digest the tissues by enzymatic digestion to release cells from connective tissues, partially purify stem cells from debris and other cells by percoll density gradient centrifugation (optional), and plate cells (Fig. 1).

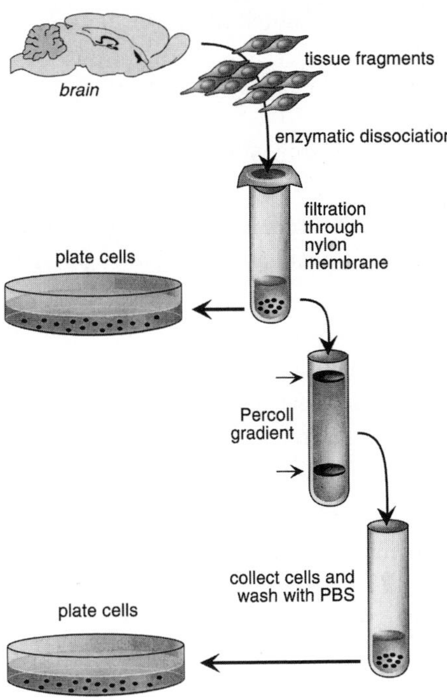

Fig. 1. Schematic diagram of the protocol for isolation of stem cells from adult brain. Tissue pieces from brains are subjected to enzymatic digestion, debris removed and the cell suspension is plated. Alternatively, stem cells can be partially purified by density gradient centrifugation and then plated.

Materials

Chemicals

- Phosphate-buffered saline (PBS; 1 x and 10X; Gibco)
- Dulbecco's Ca^{++}, Mg^{++} free PBS (D-PBS; Gibco)
- Ca^{++}, Mg^{++} free Hanks' balanced salt solution (HBBS; Irvine Scientific)
- Papain, DNase I, trypsin (Worthington Biochemicals)
- Neutral protease (Dispase, grade II, Boehringer Mannheim)
- ATV trypsin (Trypsin/EDTA solution; Irvine Scientific)
- Poly-D-lysine (pDL; Sigma); Poly-L-lysine (pLL; Sigma), Poly-L-ornithine hydrobromide (PORN), Mw 30–70,000 (Sigma); Laminin (Gibco)
- Dulbecco's modified essential medium (DMEM): Ham's F12 (1:1) medium (Irvine Scientific) containing 3.1 g/liter glucose and 1 mM L-Glutamine
- Fungizone (Irvine scientific), Penicillin-streptamycin (100X; Gibco)
- N2 supplements (100X; Gibco)
- Percoll (Amersham; Sigma)
- FGF-2 (Collaborative research, Almone Laboratories; R&D systems, Gibco)
- EGF (Gibco)
- Fetal bovine serum (FBS; Sigma, Gemini)
- Dimethylsulfoxide (DMSO, Sigma)
- Neomycine analog G418 (Gibco)
- Bromodeoxyuridine (BrdU; Sigma)
- Antibodies against lineage specific proteins and BrdU
- 4',6-diamidino-2-phenylindole (DAPI; Sigma)
- 1,4-di-diazobicyclo-[2.2.2]-octane (Sigma)
- Polyvinyl alcohol (Sigma)
- Acepromazine maleate, Ketamine, Rompun (all from Henry Schein Veterinary)

Supply

- Tissue culture dishes and Labtek or Nunc Chamber slides (Fisher)
- Cryovials (Nalgene)
- Sterile filters (0.45 μm and 0.22 μm; Nalgene)
- Betadine (J. A. Webster, Inc.)
- Nylon mesh (pore size 15 μm, Nitex, TETKO, Inc.)

Equipment

- Laminar flow hood; CO_2 Incubator (Forma Scientific)
- Freezing chambers (Nalgene) and liquid nitrogen tank

Solutions and Medium

1. Anesthetic solution
   - 7.5 ml Ketamine (10 mg/ml)
   - 0.75 ml Acepromazine maleate (100 mg/ml)
   - 1.9 ml Rompun (20 mg/ml)
   - 9.85 ml Saline
   - Inject 0.25 ml/160 g rat
2. Artificial cerebrospinal fluid (aCSF)
   - 124 mM NaCl
   - 5 mM KCl
   - 1.3 mM $MgCl_2$
   - 2 mM $CaCl_2$
   - 26 mM $NaHCO_3$
   - 10 mM D-glucose
   - 1 x penicillin-streptomycin
   - pH 7.35, Total osmolality ~280 mOsm
3. Papain-Protease-DNase (PPD) solution
   - HBBS supplemented with 12.4 mM $MgSO_4$
   - 0.01% papain
   - 0.1% neutral protease
   - 0.01% DNase I
   - Filter sterilize through 0.22 μm filter
   - Store aliquots at –20°C
4. Trypsin-hyaluronidase-kynurenic acid solution
   - 124 mM NaCl
   - 5 mM KCl
   - 3.2 mM $MgCl_2$
   - 0.1 mM $CaCl_2$
   - 26 mM $NaHCO_3$
   - 10 mM D-glucose
   - 1 x penicillin-streptomycin
   - 1.33 mg/ml trypsin
   - 0.67 mg/ml hyaluronidase
   - 0.2 mg/ml kynurenic acid
   - pH 7.35, Total osmolality ~280 mOsm
5. N2 supplement culture medium (N2 medium)
   - DMEM:F12
   - 2.5 μg/ml Fungizone, 1 x penicillin-streptomycin
   - N2 supplement (1 x)
   - Store at 4 °C for <1 month
   - Prewarm to 37 °C before use

6. 4% Paraformaldehyde Solution
   - Distilled water at 50–60°C
   - 4% paraformaldehyde
   - 3–5 pellets of NaOH
   - 0.2 M $NaPO_4$, pH 7.2
   - Dissolve, cool, filter
7. TCS (Tissue Collecting Solution)/Cryoprotectant
   - 25% Glycerin
   - 30% Ethylene Glycol
   - 0.1 M $NaPO_4$, pH 7.2
   - Store at room temperature for <3 months.
8. PAV-DABCO solution
   - 25% glycerol
   - 10% polyvinyl alcohol
   - 2.5% 1,4-diazobicyclo-[2.2.2]-octane
   - 100 mM Tris HCl, pH 8.5
   - Store in aliquots at –20°C

Procedure

Dissection of Embryonic and Adult CNS Regions

The methods for tissue dissection from embryonic and adult hippocampus and spinal cord of rat are briefly described here. A more detailed protocol can be found in Chapter 14.

Embryonic CNS

1. Deeply anesthetize timed pregnant rats by interperitoneal injection of the anesthesia cocktail: ketamine (44 mg/kg), acepromazine (4.0 mg/kg) and rompun (0.75 mg/kg).
2. Remove uterine horns by cesarean section, place on ice, take out embryos from their individual sacs and place them in cold D-PBS.
3. Measure the crown-rump length to verify the embryonic age.

Hippocampus

1. Remove brain from each embryo and place in D-PBS. Using forceps to stabilize the brain, peel back the cortex on one side from the midline and lay out flat. Hold the tissue pieces with a sharp forceps and cut out the hippocampus, lying just underneath the cortex, with a pair of erridectomy scissors.
2. Place brains under microscope. Transfer the tissue to D-PBS.
3. Take out the hippocampus from the contralateral side. Remove any meningeal membranes or blood vessels still attached to the tissue with sharp forceps. Pool hippocampi of 15–20 embryos.

Spinal Cord

1. Place the embryo on its side in sterile D-PBS under a microscope and make an initial cut lateral to the spinal canal with a pair of erridectomy scissors. Make subsequent cuts on the opposite side, and at the level of the cervical and lower lumbar regions.
2. Remove spinal cord from the embryo and remove residual connective tissue. Pool spinal cords from 10–15 embryos.

Adult CNS Hippocampus

1. Anesthetize adult female Fischer 344 rats (3–6 months old), kill by decapitation, remove brain and place them in cold D-PBS in a petri dish.
2. Place the brain under a microscope, separate the right and the left cortex above thalamus by bisecting the corpus collosum and rolling back the posterior margin of each

cortical hemisphere. Separate the fornix and subiculum along the internal and external edge of each hippocampal formation and remove hippocampus.

Adult CNS Spinal Cord

1. Place anesthetized rat on its side and make a cut lateral to the spinal canal. Make cuts on the opposite side and gently remove the spinal cord from the animal. Remove any connective tissues attached to the cord. Microdissect sacral, thoracic, lumber and cervical areas.

Establishment of Primary Cultures from Embryonic CNS Regions

Culture of Stem Rat Cells as Monolayers with FGF-2 (Ray et al., 1993)

1. Transfer the dissected tissues to a 15 ml centrifuge tube containing ~5 ml D-PBS, resuspend the tissues by gently tapping the tube and wash 3–4 times by centrifugation at 1000 g for 3 min.
2. Resuspend tissue in 1 ml DMEM:F12 medium, break the tissue pieces by tituration with a fire-polished Pasteur pipet and then make a single cell suspension by repeated tituration with a large to medium bore (1.5–1.0 mm) Pasteur pipet (about 20 x). Avoid forming bubbles by not blowing out the last bit of the medium.
3. Wash tissues 1–2 x with DMEM:F12 medium by centrifugation to remove debris and make a final single cell suspension by tituration with a large to medium bore Pasteur pipet (5–10 x).
4. Make an appropriate dilution of an aliquot of the cell suspension and count the number of cells using a hemocytometer. Adjust the cell density of the suspension and plate 1–2 X 10^4 cells/cm^2 onto PORN or PORN/laminin coated dishes.
5. Change medium every 3–4 days or longer depending on the confluency. If the cell density is low, replace half of the medium with fresh medium but increase FGF-2 concentration to make it 20 ng/ml.

Culture of Stem Mouse Cells with EGF (Reynolds et al., 1992; Reynolds and Weiss, 1996)

1. Wash dissected tissues in DMEME:F12 medium, mechanically dissociate the tissue with a medium bore Pasteur pipet as described before.
2. Make a cell suspension, count cells and plate onto PORN coated glass cover slips in 24 well culture dishes at a density of 2500 cells/cm^2. Culture cells in N2 medium containing 20 ng/ml EGF.
3. After 10–14 days, replace medium with fresh medium and repeat the medium change every 2–4 days.

Establishment of Stem Cell Cultures from Adult CNS Regions

Enzymatic Digestion

The two most commonly used methods for the digestion of connective tissues to release the cells are described below.

Papain-protease-DNase (PPD) Digestion (Ray et al., 1995)

1. Transfer tissue pieces to a 10 cm petri dish and cut them into small pieces (1–2 mm^3).
2. Wash dissected tissues with 5 ml HBSS or D-PBS 3 x and remove the final wash by aspiration.
3. Resuspend tissue pieces in 5 ml PPD solution and incubate for 15 min in a 37 °C waterbath with occasional gentle shaking to keep the tissue pieces resuspended. Titurate with a 5 ml pipet to break up the large chunk of tissues and incubate for additional 15 min in the 37 °C waterbath with occasional shaking. Titurate with a 5 ml pipet until the cell suspension is free of visible tissue pieces (suspension will turn milky).

4. Remove the remaining tissue pieces by filtration through a nylon mesh. Centrifuge the filtered cell suspension at 1000g for 3min and remove PPD solution by gentle aspiration. Cell suspension is not tightly packed, so do not disturb the cell pellet.
5. Resuspend cells in DMEM:F12 containing 10% FBS (10 ml/g of starting tissue weight) and wash the cell pellet two to five times by centrifugation. Aspirate the medium and resuspend cells in N2 medium containing 10% FBS (1–2ml). The cell suspension still contains small tissue pieces, myelin and red blood cells. Cells can be plated without further purification or can be separated from contaminating cells and debris by percoll density gradient centrifugation (see subprotocol).

Trypsin-Hyaluronidase-Kynurenic Acid Digestion (Gritti et al., 1996; Weiss et al., 1996b)
1. Place the brain/spinal cord of adult mice in 95% O_2/5% CO_2 oxygenated artificial cerebrospinal fluid (aCSF). Cut the tissues into small pieces (~1–2 mm^3) and transfer to spinner flasks (Bellco Glass) containing the enzyme mixture (trypsin, hyaluronidase and kynurenic acid).
2. Aerate the tissue suspension with 5% O_2/5% CO_2 and incubate at 32–35 °C for 90min with constant stirring.
3. Transfer the tissue to DMEM:F12 medium containing 0.7 mg/ml ovamucoid and dissociate cells by mechanical tituration with a fire-polished Pasteur pipet. Centrifuge dissociated cell suspension at 1000g for 5min, wash pellet once in the same buffer and then plate cells without further purification or separate them from contaminating cells and debris by Percoll density gradient centrifugation (see subprotocol).

Culturing of Rat Cells as Monolayers (Ray et al., 1995; Gage et al., 1995b).
1. Resuspend cells in N2 medium containing 10% FBS and plate at least $1 x 10^4$ cells/cm^2 in 10cm uncoated tissue culture plates.
2. On the next day, change medium with serum-free N2 medium containing 20 ng/ml FGF-2.
3. Feed cultures every 3–4 days and if the cell density is low, exchange half the medium with fresh medium. Add double amounts of FGF-2 to increase the concentration to 20 ng/ml.

Culturing of Mouse Cells as Neurospheres (Reynolds and Weiss, 1992; Gritti et al., 1996; Weiss et al., 1996b)
1. Plate cells (25–1000 cells/cm^2) in uncoated 6 well plates in serum-free N2 medium containing EGF, FGF-2 or both at concentrations of 20 ng/ml. Cells can be maintained and passaged as neurospheres or can be grown as monolayer cultures (J. Ray, unpublished results).
2. To grow mouse stem cells as monolayers, allow spheres to grow until they are big and attach to the substratum.
3. Passage monolayer cultures by trypsinization onto uncoated tissue culture dishes (see subprotocol).

Isolation of Clonal Cultures

The main reason for isolating clonal cultures is to determine whether the stem cells are multipotent and can give rise to both neurons and glia. Clonal cultures can be generated in two ways:

Limiting Dilution

1. Plate cells from bulk cultures at clonal density (1–2 cells/well; 96 well plate or 1 cell/7 cm^2 in a 35mm petri dish) onto PORN/laminin coated dishes in serum-free N2 medium containing appropriate growth factor.

2. To monitor a particular cell, mark its position on the dish by scratching the bottom of the plate. Stem cells migrate in the dish, so always make sure that the same cell is monitored.
3. Feed cells every 4–5 days with medium containing EGF or FGF-2 (20 ng/ml) and supplemented with 50% conditioned medium collected from a high density stem cell culture (see subprotocol). Since cells present at low density will not be effectively able to condition their own medium, factors present in the conditioned medium collected from high density culture will support the survival/proliferation of stem cells plated at clonal density.
4. Monitor the cells until the density of the clones reaches a critical mass (>100 cells/colony), then either passage cells and expand or characterize the clonal populations by immunocytochemistry.

Genetic Marking of Cells to Establish Clonal Cultures

1. Infect a bulk population of stem cells with the limiting dilutions of the retroviral vector of choice expressing a marker gene like green fluorescent protein (GFP), *E. coli* LacZ gene or alkaline phosphatase (Suhonen et al., 1997; see comments).
2. Plate 1% of the cells in the presence of the minimum amount of G418 needed to select for stable transfectants. Usually start with 40 μg/ml G418 and increase the concentration slowly to 100 μg/ml. To increase cell survival during the selection process, increase the concentration of G418 gradually and use medium supplemented with 50% conditioned medium collected from a high density stem cell culture.
3. Feed cells every 3–4 days until clusters of proliferative cells appear. Selection can be stopped when stably transfected cultures are established but select cells periodically to remove cells that have lost the selectable marker gene.
4. Passage individual clones with agarose/trypsin (see subprotocol).
5. Determine the clonality of cultures by Southern blot analysis (Sambrook et al., 1989). Briefly, prepare genomic DNA by lysis of cells and digest with appropriate restriction enzymes that cut once within the vector or once in each viral long terminal repeat. Resolve digested DNA on agarose gels, transfer to nylon membranes and probe with ^{32}P-labeled *neo* or transgene-specific probes.

Immunocytochemical Analysis of Stem Cells (Peterson et al., 1996)

All staining procedures are carried out at room temperature except where indicated and washing steps are done for 10 min each.

1. Grow rat or mouse stem cells in PORN/laminin coated glass chamber slides until 50–70% confluent.
2. Fix cells for 10 min in 4% paraformaldehyde and wash twice in 100 mM Tris-buffered saline (TBS). Fixed cells can be stored in TBS at 4 °C for about a week or can be processed immediately for immunocytochemistry.
3. Preincubate cells for at least 1 h in TBS containing 10% donkey serum and 0.25% Triton X-100 (blocking buffer).
4. Incubate with pooled primary antibody (polyclonal and monoclonal) diluted in TBS containing 0.25% Triton X-100 (TBS$^+$). If the antibody recognizes cell surface molecules then exclude Triton from the incubation buffer.
5. After 24–48 h at 4 °C, wash cells 3x with the blocking buffer and incubate for 2 h in the dark with species-specific secondary antibody conjugated to the desired fluorophores like fluorescein isothiocyanate (FITC), Texas Red Cy-5 or Cy-3 diluted in TBS$^+$.
6. Wash cells twice in TBS, incubate in TBS containing DAPI (10 ng/ml) for 1 min and then coverslip in PAV-DABCO solution. Analyze cell phenotype by confocal or fluorescence microscopy.

7. If necessary, the signal for specific antigens can be amplified by using biotin-streptavidin amplification. After the primary antibody and wash step, incubate cells in biotinylated donkey anti-species antibody diluted in TBS^+ for 2h, wash twice in the same buffer and then incubate in streptavidin conjugated to the desired fluorophore.
8. To detect both cell surface and nuclear or cytoplasmic antigens in the same cell, preincubate cells in TBS containing 10% donkey serum and then incubate with antibodies against cell surface antigens for a minimum of 2h at room temperature or overnight at 4 °C. Wash cells three times in TBS, postfix in 4% paraformaldehyde for 5min, wash three times in TBS and then preincubate in blocking buffer. Proceed with the antibody staining protocol as described before.

Differentiation of Stem Cells

1. Plate cells in PORN/laminin coated glass chamber slides at a density of $1X10^5/cm^2$ (high density) or $2.5–3X10^3/cm^2$ (low density) and grow for 24h in serum-free N2 medium containing FGF-2.
2. Replace the medium with fresh medium containing differentiating agents like serum (0.5, 2 or 10%), retinoic acid (1µM), forskolin (5µM), brain-derived neurotrophic factor (BDNF; 20 ng/ml), neurotrophin 3 (NT-3; 40 ng/ml), ciliary neurotrophic factor (CNTF; 10–20 ng/ml), leukemia inhibitory factor (LIF; 10 ng/ml) and thyroid hormone T3 (3 ng/ml). To density arrest, plate cells in the absence or the presence of a low concentration of FGF (1 ng/ml) (Vicario-Abejon et al., 1995; Johe et al., 1996; Palmer et al., 1997).
3. Change medium every 2 days and allow the cells to differentiate for 6 days. Fix cells with paraformaldehyde and analyze by immunocytochemistry.

Stereotaxic Implantation of Stem Cells into the Adult Brain

The procedure for the preparation of animals for the implantation of stem cells is similar to that described for the implantation of fetal cells in Chapter 14. In addition, detailed protocols related to the grafting procedures, perfusion of the animals, sectioning of the brain tissues and their characterization by using histochemical and immunocytochemical methods can be found in Suhonen et al. (1997). Methods are briefly described below.

Preparation of Cells for Grafting

1. Passage stem cell cultures 3–5 days prior to grafting from a 70–80% confluent culture at 1:1 or 1:2 split ratio.
2. Cultures should be 50–60% confluent by the start of BrdU treatment. Two days prior to grafting change medium with fresh medium containing appropriate growth factor(s) and 5µM BrdU (stock 5 mM in water).
3. Repeat the process the next day.
4. Detach cells from the flasks by using ATV trypsin, transfer to a 15ml centrifuge tube with D-PBS. Centrifuge at 1000g for 3min, wash cells with D-PBS twice and resuspend in D-PBS by using medium-to-small bore Pasteur pipets.
5. Count the cell number in a hemocytometer.
6. Remove the appropriate number of cells to a 0.5ml Eppendorf tube, spin for 1min in a microcentrifuge and resuspend cells at a concentration of $5X10^4$-$1X10^5$ cells/µl in D-PBS containing 20 ng/ml of FGF-2 for rat or EGF or both the factors and heparin for mouse.

Grafting of Stem Cells in Adult Rat Brain

1. Determine the coordinates of the injection sites from either the adult rat brain atlas (Paxinos and Watson, 1986) or by injecting a small amount of dye at a particular coordinate to determine its location (should be done as a separate experiment prior to grafting experiments. See Suhonen et al., 1997).
2. Anesthetize the recipient animals with an intramuscular injection of anesthesia solution.
3. Shave the skull, sterilize the skin with Betadine and place the anesthetized rat in the stereotaxic frame.
4. Make a single incision in the skin from a point midline between the eyes to a point between the ears using a No. 10 scalpel blade. Separate the skin flaps, clean and dry the skull surface of blood and connective tissue with swabs.
5. Identify Bregma. Drill a 1.0 mm wide hole over the desired point into the cranium. Cut the dura with the point of 26 gauge needle.
6. Gently resuspend the cells by tapping the tube containing the cell suspension. Draw up the required volume (1–3 µl; $5X10^4$-$1X10^5$ cells/µl) of cell suspension into the syringe mounted to the electrode manipulator on the stereotaxic frame, avoiding air bubbles.
7. Lower the syringe to the specific vertical distance below dura assuming a stereotaxic frame of reference measured from Bregma for the anterior-posterior and medial-lateral coordinates, and from dura for the vertical coordinates. Slowly inject required volume of solution at a rate of 1 µl per min or slower (2–3 µl/min).
8. When the injection is completed, raise the needle 1 mm and then leave the syringe needle in place for an additional 2 min to minimize cell diffusion up the needle track and then gently withdraw over a 1–2 min period.
9. If there is more than one injection site, repeat injection steps at the new sites.
10. Remove the rat from the stereotaxic frame, clean the skull, sprinkle antibiotic powder, put the skin flaps together, close the skin incision with wound clips and transfer the animal to a recovery cage.

Perfusion of Adult Rat

A detailed protocol can be found in Suhonen et al. (1997).

1. At appropriate time points after surgery anesthetize the rats.
2. Perfuse the animal transcardially with ice-cold 0.9% saline (50 ml/rat) using a perfusion pump at a flow rate of ~1000 ml/h followed by fixation with 4% paraformaldehyde (250 ml/rat or 150 ml for head perfusion only). Add 0.1% glutaraldehyde if electron microscopy is planned or if certain antibodies require glutaraldehyde fixation. In our experience, the presence of a low concentration of glutaraldehyde does not interfere with the antigenicity of the tissue.
3. Remove brains, postfix overnight at 4 °C on a shaker table. Transfer the brains to 30% sucrose (in 0.1 M $NaPO_4$, pH 7.2), and keep at 4 °C for at least 3 days or until the brains sink before sectioning. If fresh tissue is used for the experiments then omit the sucrose step.

Sectioning of the Rat Brains

1. To cut brain sections in a freezing sliding microtome, trim the brain to the minimum size containing the desired target areas.
2. Mount the trimmed-down brain on the chuck of a freezing sledge microtome containing OTC compound and then freeze with powdered dry ice for 15 minutes. OTC compound facilitates the sectioning by holding the tissue firmly to the chuck.

3. Cut the sections and transfer the slices from the knife into tissue culture wells (24 or 96 wells) containing TCS. Sections can be stored in a –20 °C freezer indefinitely.

Coating of Tissue Culture Plates

All procedures should be done under sterile conditions (in a laminar flow hood) using sterile solutions. For cell attachment and growth as a monolayer, cultures are grown on tissue culture dishes coated with PORN or PORN/laminin, pDL or pLL, PORN or PORN/laminin (Ray et al., 1995)

1. Make a 10 mg/ml stock solution of PORN in sterile water and filter sterilize by passing through 0.22 μm filter. Store aliquots at –20 °C.
2. Add enough PORN (10 μg/ml in water) to entirely cover the surface of the dish and incubate at room temperature for 24 h.
3. Wash 2–3 times with sterile water, store plates in water in sealed plastic bags at –20 °C for future use.
4. Alternatively, air dry PORN coated plates at room temperature in a laminar flow hood. Before plating the cells, wash plates 2–3 times with water followed by a wash with the culture medium. If coating with laminin, use PBS instead of water for washing.
5. Make a stock solution of laminin (5 mg/ml mouse or rat laminin) in PBS. Store in small aliquots at –80 °C. Do not freeze-thaw laminin more than once or twice.
6. Add enough laminin (5 μg/ml in PBS) to PORN coated plates to cover the surface and incubate at 37 °C for 24 h.
7. Store the plates sealed in plastic bags at –20 °C. The plates can be stored for 1–2 months. Wash once in PBS or medium before plating cells.

pDL or pLL (Juurlink, 1992)

1. Make a stock solution of pDL or pLL (1.0 mg/ml) in water or in 0.1 M boric acid-NaOH buffer, pH 8.4, and filter sterilize. Store in small aliquots at –20 °C.
2. Dilute the stock in water or appropriate buffer to make a solution of desired concentration (10–50 μg/ml). Cover the surface of the plates with enough pDL or pLL solution, incubate at 37 °C for 2–24 h.
3. Wash plates with sterile water 3–4 times and air dry. Plates can be used for several weeks.

Percoll Gradient Purification of Stem Cells

Stem cells can be partially purified from contaminating debris and other cell types by using Percoll density gradients.

1. Dilute stock Percoll solution 9:1 (vol/vol) with 10x PBS.
2. For discontinuous density gradient, form layers containing 50, 40, 30, 20 and 10% Percoll. Layer cell suspension (obtained after enzymatic digestion and filtration) over step gradient (Maric et al., 1997).
3. Centrifuge at 400 g for 20–30 min at room temperature.
4. Collect the layer between 40–50% gradient.
5. Dilute 2–5 fold in cold PBS (containing antibiotics and fungizone) and wash cells at least 3 times by centrifugation at 1000 g for 3 min. The cell pellet is very small; to avoid losing the cell pellet leave behind ~1 ml wash liquid in between aspiration and washes.
6. Resuspend the cell pellet in 1 ml of the appropriate plating medium, count cells and then plate ~1.3–4 X 10^4 cells/cm^2 (1–3 X 10^6 cells/75 cm^2 flask). If a smaller number of cells is obtained, plate in 35 or 60 mm plates.

7. Cells can also be separated from debris by continuous density gradient centrifugation. Mix cells with Percoll (1:1), centrifuge and then collect the liquid in between myelin layer at the top and the red blood cell layer at the bottom. Proceed from step 5.

Passaging and Re-culturing of Neural Stem Cells

Monolayer Cultures

1. Add 1.0–1.5ml ATV trypsin/10cm plate or T 75 flask (add less for smaller dishes) prewarmed to 37 °C. Swirl plate/flask to distribute the liquid evenly.
2. Let sit for 1min. Hit the side of the plate gently to dislodge cells.
3. Transfer cells to a 15ml sterile centrifuge tube by using PBS. Wash the plate once with PBS and transfer to the same tube. Centrifuge at 1000g for 3min.
4. Remove supernatant slowly so as not to disturb the cell pellet. Resuspend cells in 1 ml serum-free N2 medium, titurate with a medium bore fire-polished Pasteur pipet.
5. Plate portions of the cells (split-ratio will depend on initial cell density and the growth rate of cells) on PORN/laminin coated (rat cells) or uncoated (mouse cells) plates in the same medium containing FGF-2 (rat cells) or EGF, FGF-2 and heparin (mouse cells).
6. If necessary freeze cells in liquid nitrogen for long-term storage.

Passaging of Neurospheres

1. Collect the culture medium containing the spheres in 15ml sterile centrifuge tubes and centrifuge at 1000g for 3min. Remove supernatant slowly without disturbing the cell pellet.
2. Resuspend cells in 1ml serum-free N2 medium containing EGF and make a single cell suspension by titration (10–20x) with a medium bore Pasteur pipet.
3. Plate cells onto uncoated plates or freeze cells in liquid nitrogen for long-term storage.

Freezing of Cells

1. Suspend cells in serum-free N2 medium containing 10% DMSO and appropriate growth factors.
2. Aliquot 1ml in each freezing vial.
3. Put the vials in freezing chambers and place the chamber in –70 °C freezer to allow the cells to freeze slowly.
4. On the next day transfer the vials to a box kept in liquid nitrogen.

Re-culturing of Frozen Cells

1. Remove vials from liquid nitrogen and thaw cells quickly by constant shaking of the tube in 37°C waterbath.
2. Transfer cells to a sterile 15ml centrifuge tube with DMEM:F12 medium, centrifuge at 1000 g for 3min. Remove supernatant.
3. Wash cells once in the same medium. Resuspend cells in 1ml N2 medium, make a single cell suspension by titurating with a medium bore Pasteur pipet.
4. Plate on PORN/laminin coated (for rat cells) or uncoated (for mouse cells) plates in serum-free N2 medium containing appropriate growth factors.

Agarose/trypsin Method for Passaging Cells

1. Pick clones from plates that have relatively big (>100 cells/clone) and well-separated colonies. Mark a clone by marking it on the back of the dish.
2. Melt 3% agarose solution (made in PBS) in a microwave oven, cool to ~45–50 °C and mix 1 ml agarose with 2ml ATV trypsin (warmed to 37 °C).
3. Remove culture medium and immediately add agarose/trypsin mixture to the dish containing the colonies, swirl the plate to gently spread over the cells and allow to solidify for 2–3min.
4. With a sterile Pasteur pipet gently cut around the clones and lift the agarose plugs (with cells attached to them) and transfer them to individual wells of a 24 well plate

containing serum-free N2 medium supplemented with 50% conditioned medium and G418. Gently wash the area of the plug twice with the medium (~100μl) and transfer it to the well containing the cells.
5. Change medium every 3–4 days and allow the cells to grow until desired confluency is reached and cells are ready to be passaged.

Preparation of Condition Medium for Use in Culturing

1. Collect the conditioned medium from a high density stem cell culture after at least 24 h incubation.
2. Centrifuge at 1000g for 5min and freeze aliquots. The conditioned medium can also be filter sterilized to prevent accidental contamination with residual cells.

Analyses of Brain Sections

The fate and the nature of the grafted stem cells *in vivo* are characterized by immunocytochemistry, in situ hybridization and electron microscopy. Only the immunocytochemistry method is described here.
1. To detect BrdU, pretreat free-floating CNS sections for 2h in 50% formamide/2XSSC at 65°C followed by a 30-min incubation in 2M HCl at 37 °C.
2. Analyze the pretreated sections by immunocytochemistry (as described above) for the expression of specific marker proteins followed by confocal or fluorescence microscopy. Co-localization of BrdU and a specific marker protein will identify the *in vivo* fate of the grafted cells.

Cell numbers in brain sections are quantified by using unbiased stereology. The method has been described in detail previously (Sterio, 1984; Peterson et al., 1994; West and Slomianka, 1998) and only an outline is given here.
1. Sample a systematic series of tissue sections completely containing the structure to be quantified in a uniform, random manner using the optical dissector principle so that all cells within the tissue have equal probability of being sampled.
2. Count cells directly within the three-dimensional unbiased counting frame according to the sampling criteria for the optical dissector.
3. Calculate results either directly from the known subsample of the total structure (the optical fractionator procedure) or by combining the numerical density achieved from the optical dissector counts with the estimation of total structure volume determined using the Cavalieri procedure (the N_V-V_{Ref} procedure).

Results

Growth Properties and Morphology of Rat and Mouse CNS-derived Stem Cells

When rat stem cells are plated as monolayers at a high density (2–5X10^4 cells/cm^2), proliferating cells can be seen within 3–5 days (depending on the initial plating density; Fig.2*a*). At least half the cells in culture have a stem cell morphology of small phase bright cell bodies and two or more long processes (Fig.2 b, c). The cultures also contain cells of different morphologies, including some flat and some long phase dark elongated cells (Fig.2 *b*, *c*). Stem cells seem to be generated from flat cells (Fig.2 c). The flat cells do not express any stem/progenitor cell marker proteins (nestin, vimentin, O4 and A2B5) and may represent true stem cell populations (J. Ray, unpublished observations).

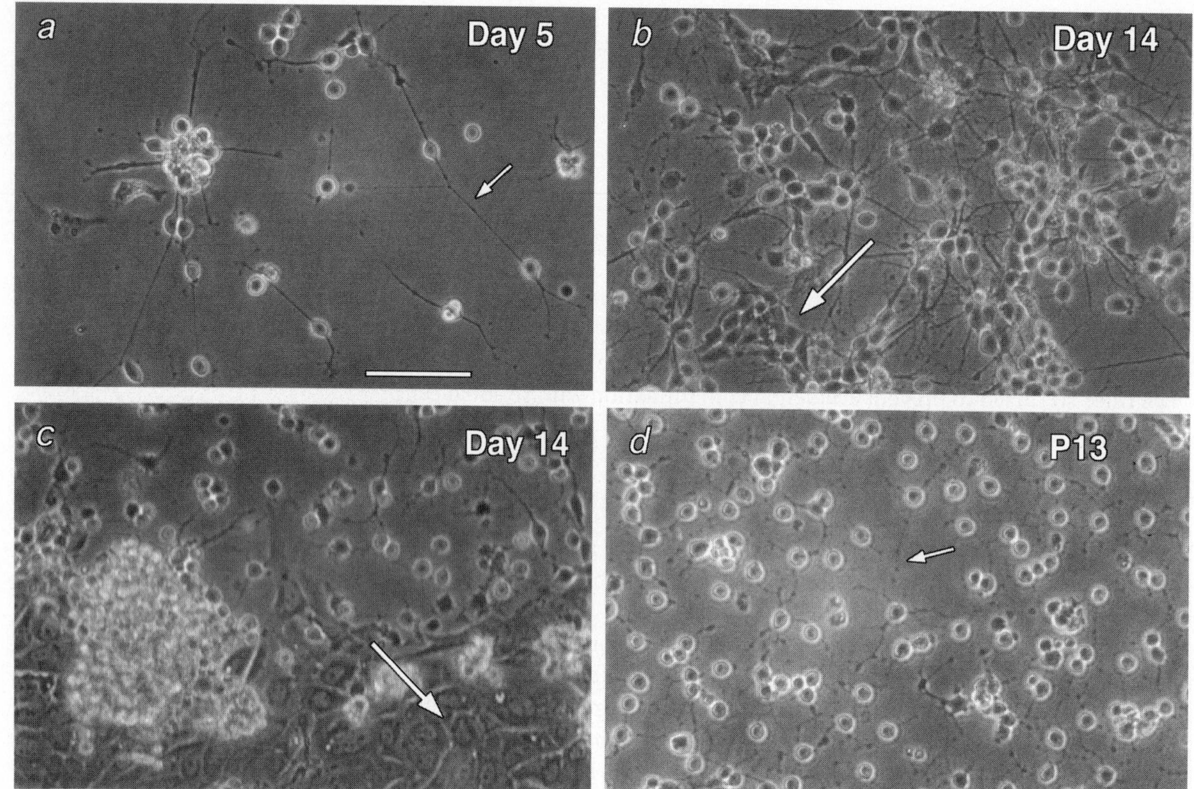

Fig. 2. Morphology of adult rat neural stem cells cultured in serum-free N2 medium containing FGF-2. (*a*) Proliferating cells can be seen by 3–5 days in vitro (DIV). Stem cells have small phase bright cell bodies and two or more long processes (small arrows in *a*, *d*). (*b*) By 14 DIV, a large number of stem cells are present. (*c*) The cultures also contain flat cells (indicated by long arrows) which do not stain for any stem or precursor cell markers. Small phase bright cells seem to generate on top of these flat cells. (*d*) Mostly stem cells are present in the passaged cultures. Scale bar: 100μm

Mouse stem cell proliferation and neurosphere formation are detected 3–5 days after plating (Fig. 3 *a*) and they increase in size with time in culture (Fig. 3 *b*). When some of the neurospheres reach a critical mass, they attach to the substratum and start to generate streams of cells. The cells grow out of the spheres (Fig. 3 *c*), and on subsequent passaging cells will grow as monolayers (Fig. 3 *d*). Some spheres do not attach and grow as spheres even after passaging. Although it is not possible to determine the morphology of cells within the neurospheres, the morphology of cells grown as monolayers is different from rat cells. They have bigger, more elongated cell bodies and smaller processes than rat stem cells.

Troubleshoot

- Purity of the cultures to a large extent depends on the clean dissection of tissues. Contaminating connective tissues can increase the non-neuronal cell population that will eventually overtake the cultures.
- Enzymatic digestion of tissues with PPD should not be done for more than 40 min. Longer enzymatic digestion will lower the yield of stem cells.
- Embryonic rat cells plated and cultured in uncoated tissue culture dishes in serum-free medium containing EGF generated neurospheres, but they grew poorly and

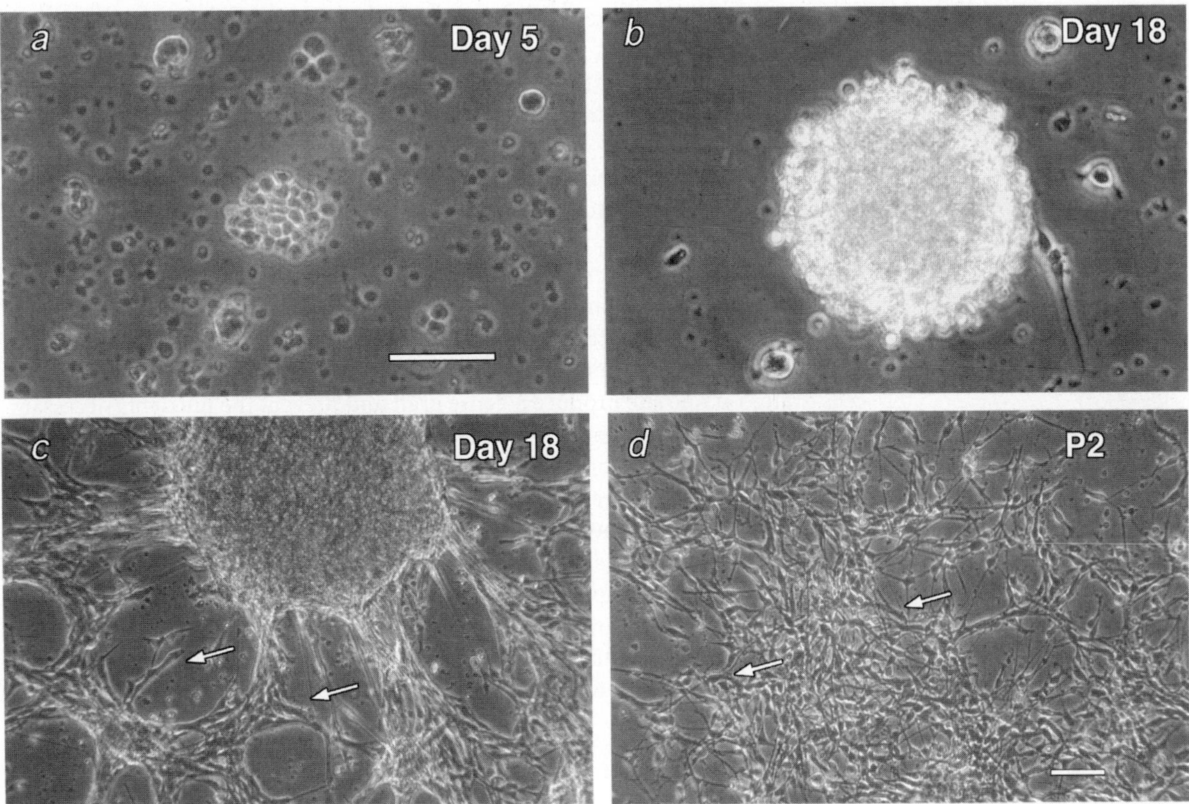

Fig. 3. Morphology of adult mouse neural stem cells cultured in serum-free N2 medium containing EGF, FGF-2 and heparin. (*a*) Neurospheres are visible by 5 DIV. With time in culture neurospheres increase in size (*b*) and some of the spheres attach to the substratum (*c*). Cells stream out of the spheres and grow as monolayer. (*d*) Upon passage attached cells grow as monolayers. Mouse stem cells have more elongated cell bodies (arrows in *c*, *d*) and smaller processes than rat stem cells. Scale bar: 100 µm

could not be expanded for more than 4 weeks (Svendsen et al., 1997; J. Ray, unpublished observation).

- Survival of rat stem cells grown as monolayer cultures is density dependent, and when plated at <1000 cells/cm^2 they do not survive even in the presence of FGF-2 J. Ray, unpublished observation). However, when mouse stem cells are grown as neurospheres, where cells remain in close contact, they grow even when plated at a low density, albeit at a slow rate. For example, adult mouse striatal stem cells plated in EGF or FGF-2 at 200 cells/cm^2 began to divide by 5 days *in vitro* (DIV) and generated spherical clusters of cells by 21 DIV (Gritti et al., 1996). No significant difference was found between the number of spheres per plate generated in response to EGF or FGF-2.
- The culturing conditions and requirements of factors for the growth of rat, mouse, monkey and human CNS-derived stem cells are different (Ray et al., 1995; Svendsen et al., 1997; Sah et al., 1997; J. Ray, unpublished observation). The best condition to grow adult rat stem cells, irrespective of CNS regions, is to culture them as monolayers in serum-free N2 medium containing FGF-2 (20 ng/ml) (Gage et al., 1995b). Conditions for the establishment of stem cell cultures from mouse CNS regions are quite different from those for rat. The plating conditions used for rat cells do not generate stem cell populations from mouse CNS (J. Ray, unpublished observation)
- Do not trypsinize cells for more than 2 min. Longer incubation will result in extensive cell death. Stem cells detach easily and shorter trypsin treatment will enrich pas-

saged cultures with stem-like cells while leaving behind flat cells and more differentiated glia and neurons.

- For fixing cells, use fresh or frozen (–20 °C) aliquots of paraformaldehyde. For perfusion, prepare paraformaldehyde just before use, keep on ice or in the cold room until use.
- To achieve maximum cell survival after grafting, prepare cells just before grafting, keep them at room temperature during the grafting process and resuspend once every 30–40 minutes.
  - Graft cells within 3 h after the preparation of the cell suspension. If grafting a large number of animals, prepare the cells in batches to minimize the time they are kept in suspension before grafting.
  - The optimal concentration of cell suspension is important as a large number of cells grafted at one site produce a dense graft and prevent the nutrients necessary for graft survival from reaching the center of the graft. As a result cells start to die and yellow necrotic centers are formed. A concentration of $5 X 10^4$–$1 X 10^5$ cells/μl and a deposit of 1–2 μl are used in the authors' laboratory. If it is necessary to implant more cells, make multiple deposits.
  - Use 26 gauge or smaller size needles (that are not too narrow) to avoid shearing cells as this will result in poor survival of the grafted cells. To avoid clogging, regularly rinse the syringe and the needle with ethanol and saline solution.
  - To check the viability of the cells used for grafting, replate the extra cells and analyze how many cells survive after 1–2 days.
- For immunostaining it is best to use 40–50 μm free-floating brain sections although thinner sections can result in better resolution. Depending on the desired thickness cut the sections on a freezing sledge microtome (30–100 μm) or on the cryostat (1–50 μm).

Comments

Some of the parameters important to consider for *in vitro* culturing of stem cells and their analyses are discussed below.

Age and Brain Regions

Although stem cells can be cultured from both embryonic and adult brain from different species, it is easier to establish cultures from embryonic brains. Development of brain regions and the cascade of expression of various receptors for growth factors that are essential for their survival and proliferation take place during discrete developmental periods. For culturing EGF or FGF-2-responsive stem cells choose the optimum embryonic age at which the receptors for these mitogenic growth factors are expressed abundantly. Use freshly dissected tissues whenever possible as this will generate better yield of stem cells.

Proteolytic Digestion

No comparison has been made of the two most commonly used enzymatic digestion methods described here to extricate live cells from the three-dimensional adult CNS network and it is not possible to say which is the better method. In the authors' laboratory the PPD digestion method is routinely used to culture stem cells from various adult brain regions as well as from different species and has yielded large numbers of viable cells. A number of other proteases alone or in combination have also been successfully used to isolate stem cells from adult brain (Reynolds and Weiss, 1992; Weiss et al., 1996). Comparison of the viability of cells isolated from embryonic brains by various protocols involving mechanical or enzymatic (collagenase, trypsin or papain) methods showed that papain optimizes cell viability compared to other methods (Maric et al., 1997). A study that compared the ability of different proteases (trypsin/proteinase K, protease XXIII, papain, collagenase and dispase) to isolate viable neurons from 6-week-old rats showed that, although almost equal numbers of cells could be isolated with different enzymes, the viability of cells isolated with papain alone was better (Brewer, 1997).

Medium and Supplements

Most methods described for the culture of neural stem cells use DMEM or DMEM:F12 medium, N2 supplement and EGF, FGF-2 or both. Neurobasal medium (Gibco) containing B27 supplements and EGF has also been used to culture embryonic (E18) rat brain stem cells (Svendsen et al., 1995) and adult rat hippocampal neurons (Brewer, 1997). B27 supplement includes a range of hormones, anti-oxidants and retinal acetate (Brewer et al., 1993) in addition to the basic formulation of N2 (Bottenstein and Sato, 1979). However, the survival rather than the proliferation of stem cells following their initial plating is strongly increased in B27 medium containing EGF (Svendsen et al., 1995). To generate proliferating stem cells, cultures established in B27 medium can be transferred to N2 medium containing EGF without any detrimental effect.

Substratum

Composition of the substratum is important for the adhesion, survival, proliferation and differentiation of cells. To grow cells as monolayer cultures they are plated on uncoated plastics in the presence of serum for 2–24h. Factors in serum provide components for cell attachment. Alternatively, plates can be coated with agents like polymers of basic amino acids PORN or pLL or pDL and cells will attach on the basis of charge. Laminin, a cell adhesion molecule, can be used in addition to PORN as substratum. However, neural stem cells can also be cultured in uncoated tissue culture dishes where they form neurospheres and remain as suspension cultures. Some of the neurospheres loosely attach to the plates and grow as monolayers. Substratum has also been reported to influence the fate choice of stem cells (Stemple and Anderson, 1992). For example, subclones of neural crest cells generate only astrocytes when plated on fibronectin but neurons when plated on fibronectin/pDL.

Choice of EGF vs. FGF-2

Although both FGF-2, EGF or their combinations have been successfully used to culture stem cells from mouse and rat, the cells from different species respond differently to these growth factors. Direct comparison of growth rates and survival of stem cell neurospheres isolated from embryonic rat or mouse striatum and cultured in EGF showed that, although mouse cells can be expanded over 50 days, rat cells died between days 21 and 28 (Svendsen et al., 1997). FGF-2 alone did not lead to an expansion of rat striatal cells under the conditions used in the experiments. Although combinations of FGF-2 and EGF acted synergistically on the growth of rat cells, they did not prevent their senescence and death. In summary, EGF, FGF-2 or their combinations generate neurospheres from mouse and rat brains but the presence of FGF-2 limits the expansion capability of rat cell neurospheres. In contrast, rat stem cells cultured as monolayers in the presence of FGF-2 can be expanded and maintained in culture for a long period of time (Ray et al., 1993; Ray and Gage, 1994; Gage et al., 1995b; Shihabuddin et al., 1997). This difference may reflect the difference between two culture conditions, or EGF and FGF-2 may be recruiting cells that differ in their "stemness." Future studies will address these issues.

Neurospheres vs. Monolayer Cultures

It must be noted that no study at this point has compared the properties of stem cells cultured as monolayers or neurospheres with FGF-2 or EGF alone or with a combination of the factors. As a result it is not possible to say whether cells isolated and cultured by different methods using the two growth factors recruit the same type of cells and expand them in culture or if they are different cells. It is possible that the same cells are recruited but that they exhibit different properties in EGF and FGF-2. For example, EGF-responsive stem cells can be subsequently expanded in FGF-2 (Vescovi et al., 1993) but FGF-2-responsive cells cannot be subsequently cultured in EGF (J. Ray, unpublished observation).

There are several disadvantages for culturing cells as neurospheres. The morphology of cells in the spheres cannot be determined visually and due to lack of penetration of the antibodies, they will immunostain cells present on or close to the surface of the spheres. Consequently, the nature of cells present within the spheres cannot be determined. If the spheres are big, cells situated well within the spheres would die due to lack of nutrients needed for their survival and hence cell yield will reduce with time and passage. Neurosphere cultures cannot be used for some biochemical assays like ligand binding studies to determine the number of specific receptor binding sites present on the cells, rate of cell division and for proliferation assay by ^{3}H-thymidine incorporation to determine the mitogenic effects of growth factors.

Retroviral Vectors

Retroviral vectors are derived from Moloney murine leukemia virus (MoMLV) and contain the gene of interest (transgene) and a selectable marker like neomycin-resistant gene, *neo* (Verma and Somia, 1997). Retroviral vectors can infect only dividing cells and integrate randomly in the cellular genome. As a result, all the progeny of a single infected cell will inherit a unique and identifiable integration site (Cepko, 1988b) and the clonality of a population of cell can be determined by restriction enzyme digest followed by Southern blotting (Sambrook et al., 1989).

Choice of Methods for Sectioning of Brain

Thick cryostat or freezing sliding microtome-cut sections are used for immunohistochemical analyses and thin cryostat sections for in situ hybridization. Freezing sliding microtome is used to cut thicker (40–50 μm) sections from fixed brains and it is difficult to get sections <25 μm thick. The thicker sections are well suited for analysis by confocal microscopy where the requisition of multiple focal planes produces a data set from which three-dimensional information about the cytoarchitecture can be obtained. Another advantage of the microtome is that multiple brains can be sectioned simultaneously. The cryostat can be used to cut thinner sections (10–30 μm) from fresh or fixed brains and mounted to glass slides directly after cutting. Thin cryostat sections cut from fresh brain are used for in situ hybridization. Brain sections cut in the vibratome are thicker (>50 μm) and are generally used for electron microscopy. One disadvantage of vibratome cut sections is that they are susceptible to compression artifacts.

Applications

The ability to generate long-term cultures of neural stem cells from normal or transgenic animals, to obtain clonal cultures and to transfer transgene into these cells has opened up the possibility of studying how environmental stimuli influence the fate choice and differentiation potentials of these cells both*in vivo* and *in vitro*. In addition, stem cells can be genetically modified to express specific growth factors and neurotransmitters. Stem cells, normal or genetically modified, can be used in animal models of neurodegenerative diseases with the ultimate aim of providing an expandable source of well-characterized cells for cell replacement therapy or gene therapy in humans.

Acknowledgement: We thank M. L. Gage for her helpful critique of the manuscript. The work in our laboratory was supported by grants from American Paralysis Association, Hollfelder Foundation, STTR fund from NIA (R42 AG12576-03) and by Contract NO1-NS-6-2348 from NIH. The content of this publication does not necessarily reflect the views or policies of the Department of Human and Health Services; neither does mention of trade names, commercial products, or organizations imply endorsement by the U. S. Government.

References

Brewer GJ, Torricelli JR, Evege EK, Price PJ (1993) Optimized survival of hippocampal neurons in B27-supplemented Neurobasal, a new serum-free combination. J Neurosci Res 35:567–765

Brewer GJ (1997) Isolation and culture of adult rat hippocampal neurons. J Neurosci Methods 71:143–155

Bottenstein JE, Sato G (1979) Growth of rat neurobalstoma cell line in serum-free supplemented medium. Proc Natl Acad Sci USA 76:514–517

Carpenter MK, Winkler C, Fricker R, Emerich DF, Wong SC, Greco C, Chen E-Y, Chu Y, Kordower JH, Messing A, Björklund A, Hammang JP (1997) Generation, and transplantation of EGF-responsive neural stem cells derived from GFAP-hNGF transgenic mice. Exp Neurol 148:187–204

Cepko CL (1988a) Immortalization of neural cells via retrovirus-mediated oncogene transduction. Trends Neurosci. 11:6–8.

Cepko CL (1988b) Retroviral vectors and their applications in neurobiology. Neuron 1:345–353

Fisher LJ (1997) Neural precursor cells: applications for the study and repair of the central nervous system. Neurobiol Dis 4:1–22.

Gage FH, Ray J, Fisher LJ (1995a) Isolation, characterization and use of stem cells from the CNS. Ann Rev Neurosci 18:159–192

Gage FH, Coates PW, Palmer TD, Kuhn HG, Fisher LJ, Suhonen JO, Peterson DA, Suhr ST, Ray J (1995b) Survival and differentiation of adult neuronal progenitor cells transplanted to the adult brain. Proc Natl Acad Sci USA 92:11879–11883.

Gritti A, Parati EA, Cova L, Frolichsthal P, Galli R, Wanke E, Faravelli L, Morassutti DJ, Roisen F, Nickel DD, Vescovi AL (1996) Mutipotential stem cells from the adult mouse brain proliferate and self-renew in response to basic fibroblast growth factor. J Neurosci 16:1091–1100.

Hammang JP, Reynolds BA, Weiss S, Messing A, Duncan ID (1994) Transplantation of epidermal growth factor-responsive neural stem cell progeny into the murine central nervous system. Methods in Neurosci 21:281–293

Johe KK, Hazel TG, Muller T, Dugich-Djordjevic MM, McKay RD (1996) Single factor direct the differentiation of stem cells from the fetal and adult central nervous system. Genes Develop 10:3129–40.

Juurlink BH (1992) Chick spinal somatic motoneurons in culture. In: Federoff S and Richardson A (eds) Protocols for neural cell culture, Humana Press, Totowa, NJ. pp 39–51.

Kordower JH, Chen E-Y, Winkler C, Fricker R, Charles V, Messing A, Mufson EJ, Wong SC, Rosenstein JM, Björklund A, Emerich DF, Hammang JP, Carpenter MK (1997) Grafts of EGF-responsive neural stem cells derived from GFAP-hNGF transgenic mice: trophic and tropic effects in a rodent model of Huntington's disease. J Comp Neurol 387:96–113

Kuhn HG, Dickinson-Anson H, Gage FH (1996) Neurogenesis in the dentate gyrus of the adult rat: age-related decrease of neuronal progenitor proliferation. J Neurosci 16:2027–2033.

Lendahl U, McKay RDG (1990) The use of cell lines in neurobiology. Trends Neurosci 13:132–137.

Maric O, Maric I, Ma W, Lahojuji F, Somogyi R, Wen X, Sieghart W, Fritschy J-M, Barker JL (1997) Anatomical gradients in proliferation and differentiation of embryonic rat CNS accessed by buoyant density fractionation: alpha 3, beta 3 and gamma 3 $GABA_A$ receptor subunit co-expression by post-mitotic neocortical neurons correlates directly with cell buoyancy. Eur J Neurosci 9:507–522.

McKay RD (1997) Stem cells in the central nervous system. Science 276:66–71.

Minger SL, Fisher LJ, Ray J, Gage FH (1996) Long-term survival of transplanted basal forebrain neurons following *in vitro* propagation with basic fibroblast growth factor. Exp Neurol 141:12–24.

Morshead CM, Reynolds BA, Craig CG, McBurney MW, Staines WA, Morassutti D, Weiss S, van der Kooy D (1994) Neural stem cells in the adult mammalian forebrain: A relatively quiescent subpopulation of subependymal cells. Neuron 13:1071–1082.

Palmer TD, Ray J, Gage FH (1995) FGF-2-Responsive neuronal progenitors reside in proliferative and quiescent regions of the adult rodent brain. Mol Cell Neurosci 6:474–486

Palmer TD, Takahashi J, Gage FH (1997) The adult rat hippocampus contains primordial neural stem cells. Mol Cell Neurosci 8:389–404.

Paxinos G, Watson C (1986) The rat brain in stereotaxic coordinates. Academic Press, San Diego, CA.

Peterson DA, Lucidi-Phillipi CA, Eagle KL, Gage FH (1994) Perforant path damage results in progressive neuronal death and somal atrophy in layer II of entorhinal cortex and functional impairment with increasing postdamage age. J Neurosci 14:6872–6885

Peterson DA, Lucidi-Phillipi CA, Murphy D, Ray J, Gage FH (1996) FGF-2 protects layer II entrohinal glutamatergic neurons from axotomy-induced death. J Neurosci 16:886–898

Ray J, Peterson DA, Schinstine M, Gage FH (1993) Proliferation, differentiation, and long-term culture of primary hippocampal neurons. Proc Natl Acad Sci USA 90:3602–3606.

Ray J, Gage FH (1994) Spinal cord neuroblasts proliferate in response to basic fibroblast growth factor. J Neurosci 14:3548–3564.

Ray J, Raymon HK, Gage FH (1995) Generation and culturing of precursor cells and neuroblasts from embryonic and adult central nervous system. In: Vogt PK, Verma IM (eds) Oncogene techniques. Methods in Enzymology, vol 254. Academic Press, San Diego, pp 20–37

Ray J, Palmer TD, Suhonen JO, Takahasi J, Gage FH (1997) Neurogenesis in the adult brain: Lessons learned from the studies of progenitor cells from embryonic and adult central nervous system In: Gage FH and Christen Y (eds) Research and Perspective in Neurosciences. Isolation, characterization and utilization of CNS stem cells. Fondation IPSEN, Springer, Heidelberg, pp 129–149

Ray J, Palmer TD, Shihabuddin LS, Gage FH (1998) The use of neural progenitor cells for therapy in the CNS disorders. In: Tuszynski MH, Kordower J H and Bankiewicz K (eds) CNS regeneration: basic science and clinical applications Academic Press, San Diego. (in press)

Reynolds BA, Tetzlaff W, Weiss S (1992) A multipotent progenitor cell produces neurons and astrocytes. J Neurosci 12:4565–4574.

Reynolds BA, Weiss S (1992) Generation of neurons and astrocytes from isolated cells of the adult mammalian central nervous system. Science 255:1707–1710.

Reynolds BA, Weiss S (1996) Clonal and population analyses demonstrate that an EGF-responsive mammalian embryonic CNS precursor is a stem cell. Develop Biol 175:1–13.

Sambrook J, Fritsch EF, Maniatis T (1989) Molecular cloning: a laboratory manual. Cold Spring Harbor Laboratory, Cold Spring Harbor, NY.

Sah DWY, Ray J, Gage FH (1997) Bipotent progenitor cell lines from the human CNS respond differently to external cues. Nature Biotech 15:574–580.

Shihabuddin LS, Ray J, Gage FH (1997) FGF-2 alone is sufficient to isolate progenitors found in the adult mammalian spinal cord. Exp Neurol 148:577–586

Stemple DL and Anderson DJ (1992) Isolation of a stem cell for neurons and glia from the mammalian neural crest. Cell 71:973–985

Sterio DC (1984) The unbiased estimation of number and size of arbitrary particles using the dissector. J Microsc 134:127–136

Suhonen JO, Peterson DA, Ray J, Gage FH (1996) Differentiation of adult-derived hippocampal progenitor cells into olfactory bulb neurons. Nature 382:624–627

Suhonen JO, Ray J, Blömer U, Gage FH (1997). *Ex vivo* and *in vivo* gene delivery to the brain. Curr Prot Hum Gene Supplement 11:13.3.1–13.3.24

Svendsen CN, Fawcett JW, Bentlag C, Dunnett SB (1995) Increased survival of rat EGF-generated CNS precursor cells using B27 supplemented medium. Exp Brain Res 102:407–414.

Svendsen CN, Clarke DJ, Rosser AE, Dunnett SB (1996) Survival and differentiation of rat and human epidermal growth factor-responsive precursor cells following grafting into the lesioned adult ventral nervous system. Exp Neurol 137:376–388

Svendsen CN, Skepper J, Rosser AE, ter Borg MG, Tyres P, Ryken T (1997) Restricted growth potential of rat neural precursors as compared to mouse. Dev Brain Res 99:253–258

Verma IM, Somia N (1997) Gene therapy – promises, problems and prospects. Nature 389:239–42.

Vescovi AL, Reynolds BA, Fraser DD, Weiss S (1993) bFGF regulates the proliferative fate of unipotent (neuronal) and bipotent (neuronal/astroglial) EGF-generated CNS progenitor cells. Neuron 11:951–966.

Vicario-Abejon C, Johe KK, Hazel TG, Collazo D, McKay RD (1995) Function of basic fibroblast growth factor and neurotrophins in the differentiation of hippocampal neurons. Neuron 15:105–114

Weiss, S, Reynolds BA, Vescovi AL, Morshead C, Craig CG, van der Kooy D (1996a) Is there a neural stem cell in the mammalian forebrain? TINS 19:387–393

Weiss S, Dunne C, Hewson J, Wohl C, Wheatley M, Peterson A C, Reynold B A. (1996b) Multipotent CNS stem cells are present in the adult mammalian spinal cord and ventricular neuroaxis. J Neurosci 16:7599–7609

West JM, Slomianka L (1998) Total number of neurons in the layers of the human entorhinal cortex. Hippocampus 8:69–82

Whittemore SR, Snyder EY (1996) Physiological relevance and functional potential of central nervous-system-derived cell lines. Mol Neurobiol 12:13–38.

Suppliers

Company: Almone Laboratories, Shatner Center 3P.O. Box 4287, Jerusalem, 91042, Israel (phone: 972-2-652-8002)

Company: Amersham Corp., 2636S. Clearbrook Dr., Arlington Hts., IL, 60005, USA (phone: 800-323-9750)

Company: Boehringer Mannheim, P.O. Box 50414, Indianapolis, IN, 46250, USA (phone: 800-262-1640)

Company: Collaborative Research Inc., Two Oak Pk., Bedford, MA, 01730, USA (phone: 800-343-2035)

Company: Fischer Scientific, 2761 Walnut Ave., Tustin, CA, 92081, USA (phone: 800-766-7000)

Company: Forma Scientific, P.O. Box 649, Marietta, OH, 45750, USA (phone: 800-848-3080)

Company: Gemini Bioproducts, 5115-M Douglas Fir Road, Calabasas, CA, 91302-1441, USA (phone: 800-543-6464)

Company: Gibco BRL (Life Technologies), P.O. Box 6009, Garthersburg, MD, 20877, USA (phone: 800-638-8992)

Company: Henry Schein Veterinary, 5 Harbor Park Drive, Port Washington, New York, NY, 11050, USA (phone: 800-872-4346)

Company: Irvine Scientific, 2511 Daimler St., Santa Ana, CA, 92905-5588, USA (phone: 714-261-7800)

Company: J. A. Webster, Inc., 86 Leaminister Rd., Sterling, MA, 05164-7911, USA (phone: 800-225-7911)

Company: Nalgene Brand Products, 75 Panorama Creek Dr., P.O. Box 20365, Rochester, NY, 14602-0365, USA (phone: 716-264-3942)

Company: R&D Systems, 614 McKinley Place N. E., 6 Minneapolis, MN, 55413, USA (phone: 800-343-7475)

Company: Sigma Chemical Co., P.O. Box 14508, St. Louis, MO, 63178, USA (phone: 800-325-3010)

Company: Tetko, 525 Monterey Pass Rd., Monterey Park, CA, 91754, USA (phone: 800-283-8182)

Company: Worthington Biochemicals, Halls Mill Rd., Freehold, NJ, 07728, USA (phone: 800-445-9603)

Abbreviations

FGF-2	Basic fibroblast growth factor
EGF	Epidermal growth factor
PORN	Polyornithine
pDL	Poly-D-lysine
pLL	Poly-L-lysine
PBS	Phosphate-buffered saline
DMEM	Dulbecco's modified essential medium
BrdU	Bromodeoxyuridine
DAPI	4',6-diamidino-2-phenylindole
DAVCO	1,4-di-diazobicyclo-[2.2.2]-octane
BDNF	Brain derived neurotrophic factor
NT-3	Neurotrophin 3
CNTF	Ciliary neurotrophic factor
LIF	Leukemia inhibitory factor

Glossary

Stem cells — A self-sustaining population of cells that give rise to all the cells of the CNS. Retroviral vectors: Most retroviral vectors are derived from Moloney murine leukemia virus (MoMLV) in which the native viral protein sequences are replaced with recombinant sequences like the gene of interest and selectable marker genes. The transgene is expressed from long terminal repeats (LTR) promoter or from an internal promoter.

Chapter 12

In Vitro Reconstruction of Neuronal Circuits: A Simple Model System Approach

Naweed I. Syed, Hasan Zaidi and Peter Lovell

Introduction

In neurobiology, a variety of *in vitro* techniques have evolved historically from our efforts to understand how the nervous system controls various animal behaviors. The reductionist approach, once "ill favored" by most neuroethologists, is now gaining ground. This renewed faith in reduced preparations stems from the fact that most motor programs and their underlying neuronal circuits have proven too complex to understand – at both – the whole animal and at an organ system's level. For instance, not only has the identification and characterization of neuronal circuits proven challenging, but the cellular and synaptic mechanisms have also remained elusive. This lack of fundamental knowledge regarding the cellular basis of most animal behaviors, owes its existence to the complexity of an intact mammalian brain and the intricate nature of synaptic connections between its neurons. Many reduced preparations, on the other hand, offer attractive alternatives because such intricacies can be reduced to a manageable level. However, the data obtained from such preparations should be treated with caution. In our view, an *in vitro* isolated neuron, or a reconstructed network for that matter, provides insights into its own inherent behavior. These data should not, therefore, be considered "sufficient" to explain how an entire behavioral repertoire is organized in the intact animals. Having stated that, here we demonstrate that by utilizing *in vitro* cell culture techniques, a three-cell network underlying respiratory behavior in the fresh water mollusc *Lymnaea* can be reconstituted in cell culture. This *in vitro* reconstructed circuit generated rhythmical motor patterns that were similar to those seen *in vivo*. The *in vitro* cell culture approach has since allowed us to address fundamental questions regarding the neuronal basis of rhythmicity – at a resolution, that is unattainable both in the intact and semi-intact animals.

Networks of neurons often termed as central pattern generators (CPG) control a variety of rhythmic behaviors such as locomotion, respiration feeding etc. (Delcomyn, 1980; Harris-Warrick and Johnson, 1989; Kirstan, 1980; Pearson, 1995; Selverston, 1980). In most preparations studied to date, the CPG neurons appear sufficient to generate rhythmical motor outputs in the absence of peripheral feedback. The identification and characterization of the CPG neurons (intrinsic membrane and synaptic properties) is therefore considered essential for understanding how the nervous system controls various rhythmic behaviors. Our research program is directed towards understanding how a network of central pattern generating (CPG) neurons controls respiratory behavior in the fresh water mollusk *Lymnaea stagnalis*. We utilize this invertebrate preparation because its respiratory behavior is simple, well characterized (Syed et al., 1991), and the underlying circuit has also been identified both *in vivo* and in the semi-

Correspondence to: Naweed I. Syed, University of Calgary, Faculty of Medicine, Department of Cell Biology and Anatomy, Calgary, Alberta, T2N 4N1, Canada (phone: (403) 220-5479; fax: (403) 270-8928; e-mail: nisyed@acs.ucalgary.ca)

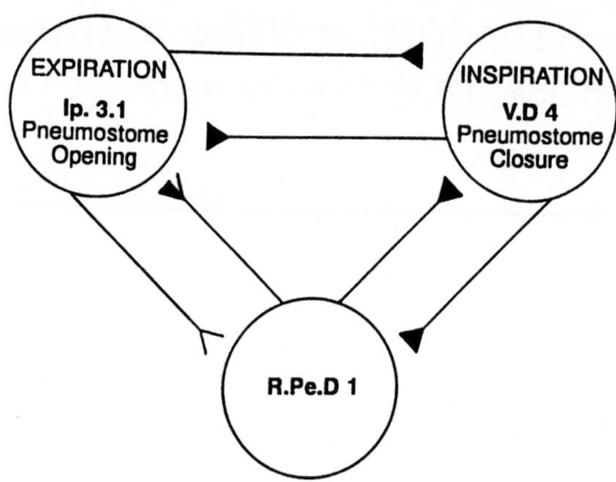

Fig. 1. Schematic diagram depicting synaptic connections between the respiratory central pattern generating neurons. RPeD1 (right pedal dorsal 1); IP3I (input 3 interneuron); VD4 (visceral dorsal 4). *Open* and *closed symbols* represent excitatory and inhibitory synaptic connections, respectively. *Half open/closed symbol* represents mixed excitatory and inhibitory synaptic connection.

intact animals (Syed and Winlow, 1991; Syed et al., 1991). Remarkably, the neural circuit mediating pulmonary ventilation is comprised of a small number of identifiable motor neurons and interneurons (Syed and Winlow, 1991). The CPG consists of three interneurons, namely: right pedal dorsal 1 (RPeD1 – initiates the respiratory rhythm), input 3 interneuron (IP3I – opens pneumostome – expiration) and visceral dorsal 4 (VD4 – closes pneumostome – inspiration). Reciprocal inhibitory connections exist between IP3I and VD4 (the "half center") and between RPeD1 and VD4. RPeD1 induces a biphasic response in IP3I (inhibition followed by excitation). IP3I in turn excites RPeD1. The synaptic connections between respiratory neurons are shown in Figure 1.

In vivo, studies have indicated that these three interneurons alone produce the respiratory rhythm (Syed and Winlow, 1991). Whether this simple circuit provides the necessary and sufficient neural substrate for respiratory rhythmogenesis could not be determined *in vivo* (see below). Similarly, some other questions that could not be addressed either in *in vivo* or *in situ* preparations include:

- Are IP3I and VD4 endogenous or conditional bursters?
- Are synaptic connections between the CPG neurons chemical and monosynaptic?
- Are IP3I and VD4 sufficient to generate respiratory rhythm?
- Is RPeD1 necessary for rhythm generation?
- Is rhythm generation a function of intrinsic membrane or network properties?

Outline

Answers to the above questions are critical for our understanding of both the cellular and synaptic mechanisms underlying rhythmicity in this system; however as mentioned above, this is difficult to achieve either in an intact, semi-intact or an isolated brain preparation. An alternative approach would be to attempt to re-construct the entire respiratory circuit in cell culture. However, for an *in vitro* approach to work, the isolated cells must:

1. Survive the isolation procedure.
2. Adhere well to any given substrate.
3. Maintain their intrinsic membrane properties.
4. Regenerate their axonal processes.
5. Re-establish their appropriate synaptic connections.
6. Synaptically interconnected neurons must generate rhythmical pattern, which is similar to that seen *in vivo*.

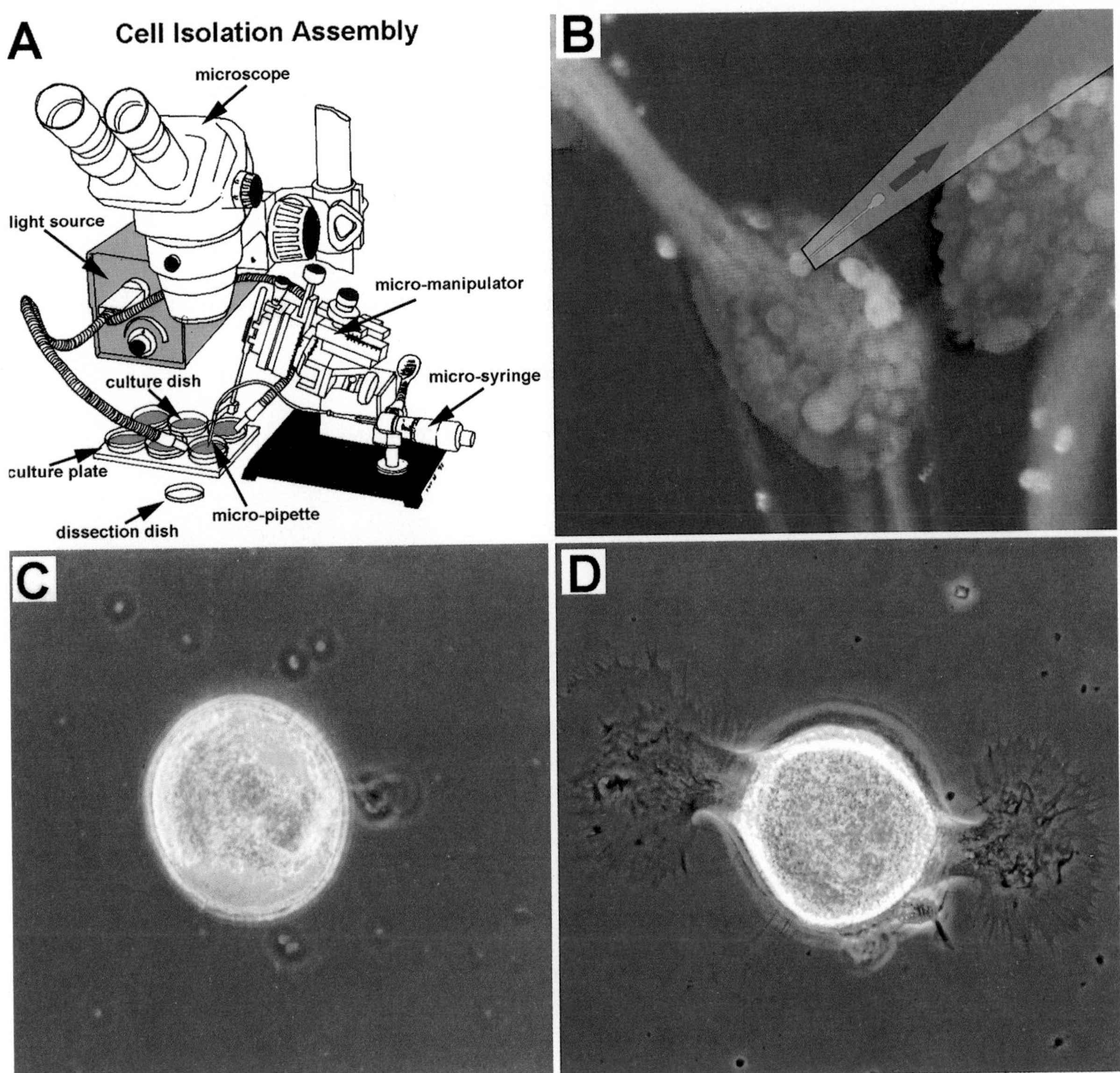

Fig. 2. Experimental set-up for *in vitro* isolation of identified neurons. A) Cell isolation assembly. B) A glass micro-pipette is positioned on top of VD4 somata – located in the visceral ganglia. C) Individually isolated somata of an identified *Lymnaea* neuron. D) In the presence of brain conditioned media, growth cones emerged from the somata within a few hours of cell plating.

To address some of the above issues, we first developed techniques to culture *Lymnaea* neurons. Individually isolated neurons were then used to reconstruct the entire respiratory CPG in a culture dish. The schematic diagram depicting the cell isolation procedure is given in Figure 2.

Materials

Cell Isolation Assembly

To assemble cell isolation apparatus, the following items are put together as depicted in Figure 2A.

- Dissection microscope – with moveable boom stand
- MM-33 Micromanipulator – with magnetic base
- Light source – either fiber optics or portable
- Gillimont syringe

Laminar air flow cabinet – Tissue culture hood

A clean air bench or tissue culture hood is required to prepare solutions and also to perform sterile dissection.

Note: it is critical to purchase a vibration free tissue culture hood.

Incubator/ desiccator

Required for both short- and long-term maintenance of cultured cells and also to prepare brain conditioned media (CM).

Electrophysiology and photomicrography

- Inverted microscope
- DC preamplifiers (2x)
- Chart recorder
- Oscilloscope
- Stimulator
- Micromanipulators (2x) mounted onto the microscope
- 35mm camera
- Pressure injection system – for the application of various compounds
- Peristaltic pump – for the superfusion of chemicals
- Electrode puller – required to prepare sharp electrodes and cell extraction pipettes
- Microforge – required to fire polish cell isolation pipettes

General supply items for cell culture (sterile)

- Pipettes (5–25ml)
- Syringes (1–60cc)
- Needles
- Falcon cell culture dishes (3001 and 3008)
- Filters (Millipore 0.22µm and 500ml)
- 100–1000µL Eppendorf pipette with sterile tips

Chemicals

Chemicals

- NaCl, KCl, $CaCl_2$, $MgCl_2$, HEPES (for normal saline and defined media)
- Leibovitz's L-15 Medium (without inorganic salts and L-Glutamine – Special order, GIBCO, USA – formula # 82-5254EL)
- D-glucose
- Gentamycin – required for antibiotic saline and the DM
- Poly L- Lysine – required to coat cell culture dishes

  Note: that Poly L-Lysine comes in a wide range of molecular weights; it is therefore very important to obtain the "right kind" (see below)

- Trypsin (Sigma Type III, catalog # T-8253) – required for tissue softening prior to cell extraction
- Trypsin inhibitor (Type III – soybean, catalog # T-9003) – required to stop the enzymatic activity

- Tris buffer
- Listerine (mouth-wash) – is a very effective antibacterial agent, which also serves well as an anesthetic

  Note: Use only on the intact animals, the CNS must never be directly exposed to this agent
- Ethanol (70%) – required for general sterilization of bench top and surgical equipment. 70% ethanol is the most optimal concentration of alcohol to use – higher concentrations may cause bacteria to encapsulate and survive until a more favorable environment is available. Lower concentrations, on the other hand, are not as effective
- High Grade super Q water – required for all solutions

Solutions

Normal Lymnaea saline

Prepare 4X-concentration stock solution by mixing:
- 160.0 ml 1 M NaCl
- 6.8 ml 1 M KCl
- 16.4 ml 1 M $CaCl_2$
- 6.0 ml 1 M $MgCl_2$
- 40.0 ml 1 M HEPES

Use super Q water to bring the total volume to 1 L and adjust the final pH to 7.9. To prepare normal saline (NS) dilute 4x stock saline (as is needed) to yield final concentrations of:
- NaCl 40.0 mM
- KCl 1.7 mM
- $CaCl_2$ 4.1 mM
- $MgCl_2$ 1.5 mM
- HEPES 10.0 mM

Antibiotic saline (ABS)

Fill autoclavable bottles with 500 ml of normal saline and autoclave. Add 10 ml of Gentamycin stock solution (7500 µg/ml – filtered through a Millipore filter) to 500 ml of sterile saline – this will yield a final Gentamycin concentration of 150 µg/ml.

Defined Media (DM)

DM is made from powdered medium supplied by Grand Island Biological Company (Grand Island, NY, USA -GIBCO – a special order). It's formula # is 82-5154EL Leibovitz's L-15 Medium, w/o inorganic salts and L-glutamine. Each packet (27 grams) makes 5 L of non-diluted stock solution of L-15 medium. Store at 2–8 degrees Celsius until reconstituted.

Defined Medium – Stock Solution

1. Use highest purity water source available
2. Add 950ml of high purity water in a glass container
3. While stirring, empty the contents of 1 packet L-15 powder
4. Once dissolved, adjust pH to 7.4 with either HCl or NaOH
5. Bring total volume to 1L with SQ water, check pH and adjust as necessary
6. In a laminar flow hood, filter (Millipore Sterivex-GV 0.22 µM filter unit) DM into autoclaved bottles – 200 ml/bottle
7. Freeze at –20 degrees Celsius

Normal DM

1. Thaw 1x200 ml aliquot of non-diluted L-15 stock medium
2. Sonicate for 15 minutes at room temperature

3. Dilute to 50% by adding:
   - 60 mg L-glutamine
   - 21.62 mg D+ glucose
   - 100 ml 4x saline stock solution
   - 98.93 ml distilled water (highest purity)
   - 1.325 ml Gentamycin stock solution
4. Filter through Nalgene 0.45 μM filter unit and store at 4 degrees Celsius

Brain conditioned medium

Conditioned media (CM) refers to defined media in which *Lymnaea* central ring ganglia were incubated for a period of time. During this incubation, growth factors are released from the brain directly into the media. These growth factors (identities not yet fully determined), are required for neurite outgrowth.

In addition to its growth promoting capabilities, CM is also a perfect habitat for bacterial and fungal growth. A strict aseptic approach is therefore essential for CM preparation.

1. Isolated central ring ganglia must be washed in a series of ABS dishes (5 washes, 6 brains/dish, 15 minutes for each wash).
2. Under sterile conditions, transfer 2 brains/ml of defined media into presterilized and Sigmacote treated (Sigma SL-2) 60 mm x 15 mm Pyrex or Kimax glass petri dish. Do not exceed the limit of 10 ml of media/ dish – it is ideal to use 5–7ml/ dish.
3. Incubate in a humidifier (80–90% humidity) for 72 hours.
4. On day 3, remove brains and filter CM through low protein binding syringe filter (Millipore, 0.22 μM). The filtered CM should be frozen in cryovials or polypropylene tubes at = –20 degrees Celsius. Thaw CM at room temperature prior to use.
   To avoid protein loss over numerous media transfers, CM can also be prepared in various other forms.
   - "**SAM**" (Substrate Adsorbed Material) – incubate antibiotic treated ganglia (4-brains/2 ml DM) directly into poly -L-lysine coated falcon 3001 tissue culture dishes. Most trophic factors released from the central ring ganglia will adhere to poly-L-lysine substrate. After 72 hours of incubation, remove both brains and the supernatant and fill dishes with DM. The growth factors attached to the bottom of the poly-L-lysine dish are sufficient to promote robust neurite outgrowth.
   - "Super SAM" – incubate brains for 72 hours in DM as described for SAM. Remove ring ganglia after 72 hours, but leave the supernatant in the dish. This dish now contains both substrate bound and diffusible trophic molecules, which also promote robust growth (for further details see Wong et al., 1981). The central ring ganglia can also be re-harvested to prepare additional CM.

Poly-L-lysine dishes

Poly-L-lysine provides a good substrate for neuronal adhesion. Either plastic or glass coverslip- attached dishes can be treated in the following manner.

- Day 1)
  Prepare 0.1% poly-L-lysine in Tris buffer (for 20 dishes, make up 40 mg poly-L-lysine in 40 ml buffer), filter (22μm Millipore filter) and store in siliconized glass container. Use within 4 weeks. To prepare dishes, dispense 2 ml poly-L-lysine solution into each 35-mm falcon culture dish in the tissue culture hood. Leave overnight at room temperature. Make sure that all dishes are covered with their lids.
- Day 2)
  In a sterile hood, remove poly L-lysine solution and immediately wash each dish.
  - 3x with sterile water (15 minutes/wash)
  - 1x with sterile saline (leave for 20 minutes)
  - 3x with sterile water (15 minutes/wash)

- Air-dry inside the hood
- Wrap dishes in parafilm and store in a humid incubator for at least 3 days prior to use

Note: Selecting an appropriate substrate is one of the most important steps towards successful cell culture. Lack of an appropriate substrate will prevent neuronal adhesion, whereas excessive poly-L-lysine will kill the cells. Similarly, neuronal growth patterns and the extent of total neurite outgrowth are also substrate contingent. For instance, on various different substrates, an identified *Lymnaea* neuron exhibits growth patterns that are unique to each substrate. Figure 3 shows a typical example of an identified *Lymnaea* neuron that was cultured either on (A) poly-L-lysine, (B) laminin, (C) fibronectin and (D) concavallin A (Con A). Most extensive neurite outgrowth is observed on both poly-L-lysine and Con A coated dishes, whereas fewer neurites extend on laminin and fibronectin. On a Con A substrate, neurites are generally thin, and they fasciculate with one another to form larger bundles. Moreover, neurons grown on a Con A substrate almost exclusively establish electrical synapses, whereas poly-L-lysine substrate promotes appropriate chemical synapses. In summary, an appropriate substrate is not only necessary for neurite outgrowth but also for specific synapse formation.

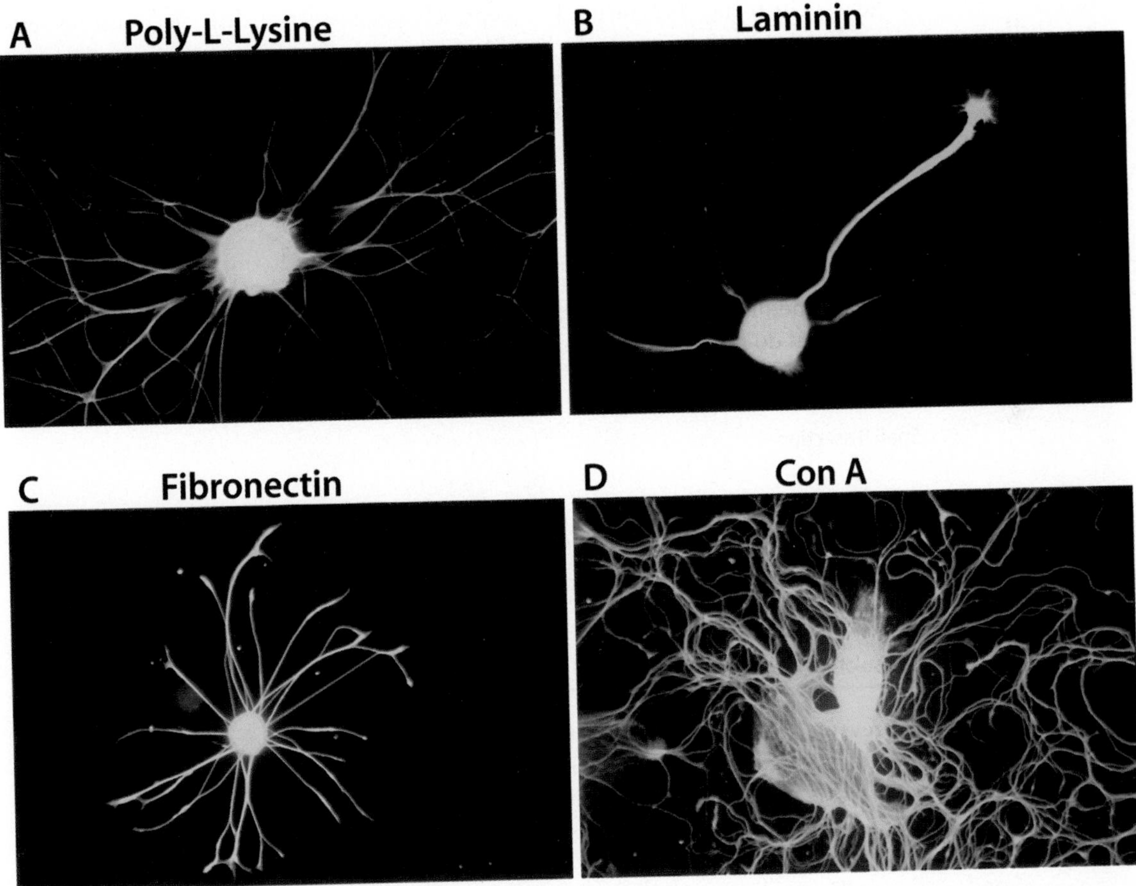

Fig. 3. Growth patterns from identified Lymnaea neurons are substrate-type dependent. **A)** An isolated neuron exhibited extensive outgrowth on the Ply L-Lysine substrate. **B)** Fewer neurites extend on a Laminin substrate. **C)** Fibronectin-stimulated neurite outgrowth was similar to that of Poly L-Lysine, however the neurite length was much less extensive. **D)** Most extensive outgrowth was observed on a Con a substrate. These neurites were however, thinner and more extensively branched as compared with any other substrate. All neurons were double-stained with actin (red) and tubulin (green) antibodies.

Procedure

Sterile Environment – a Must Requirement for Cell Culture.

Most cell culture procedures are carried out in a laminar flow hood. Once properly sterilized, it will prevent air-born microorganisms from infecting the culture media.

Hood Sterilization

The following procedures should be performed prior to every experiment and following any time after during which the hood is turned off.

1. To start the airflow, turn on the hood blower
2. Spray all surfaces inside the hood with 70% ethanol
3. Soak a paper towel with ethanol and wipe all surfaces inside the hood (ethanol sprayed from a bottle may damage the filters)
4. Wait for hood to dry and re-spray as necessary

   Note: From here on, always spray hands with ethanol before working in the hood

5. Soak all instruments to be used (forceps, scissors, dissection dish, etc.) with 70% ethanol and place them inside the hood to dry
6. Since most solutions used in cell culture are sterile (including antibiotic saline, defined media, conditioned media, 1 molar glucose) all bottles should be opened only in a properly sterilized hood

Hints to prevent accidental contamination

- Do not eat or drink near the hood (especially beverages or food made from yeast!).
- Keep the hood uncluttered.
- Any item that is temporarily removed from hood (e.g.: forceps) must be sprayed with ethanol before returning. Always spray hands with ethanol before working inside the hood, but do not introduce ethanol to the culture dishes.

Snail Dissection

Place de-shelled animals into 10–25% Listerine solution for 10 min. Snails are subsequently pinned down in a dissection dish containing ABS saline and their CNS removed under sterile conditions (see Syed et al., 1990 and Ridgway et al., 1991 for details).

Cell Isolation Pipettes

To prevent neurons from sticking to the pipette glass, it should be coated with either Sigmacote or serum. Sigmacote (Sigma catalog # SL-2 is a special silicone – heptane solution) forms a tight, microscopically thin film on glass and prevents cells from attaching to the pipette.

Step 1

1. Use 1.5 mm diameter glass capillary tubes – **with no filament.**
2. In a fume hood, fill capillary tubes (via capillary action) with Sigmacote and turn them upside down allowing the Sigmacote to flow out from the other end – repeat and let dry overnight. Do not use tissue culture hood for any of the toxic chemicals, as it poses serious health hazards!

To prepare cell isolation pipettes from Sigmacote-treated capillaries. Step 2

3. Draw capillaries on an electrode puller as if you were to prepare a sharp, intracellular electrode. This will produce an electrode with long shaft and a fine tip.
4. A diamond knife pencil is used to cut the electrode shaft half way along its length. The distance from the cut end will determine the final diameter of cell isolation pipette.
5. Snap the sharp tip off by applying gentle pressure at the tip. The electrode should break cleanly.
6. Fire- polish the electrode tip on a microforge. This should remove any sharp edges that could be detrimental to neuronal viability. Use microforge graticule to determine the tip diameter. It is important to **note** that the electrode tip should always be slightly larger than that of the somata, for which it is being used.
7. Remove electrode from the microforge and fire-polish its opening (opposite end from the sharp tip) on a Bunsen burner. This prevents the electrode from damaging the tubing and clogging the pipette tip.

Cell Isolation

Neuronal isolation refers to the process of removing individual cell bodies from the central nervous system and placing them under appropriate culture conditions. This is a very delicate and highly sophisticated procedure, with a rather steep initial learning curve. With some practice, patience, persistence, and perseverance, it becomes relatively easy to remove individually identified neurons. The following protocol outlines the basic steps.

Note: All procedures should be carried out aseptically, in a sterile culture hood

1. Place desired number of central ring ganglia in a falcon 3001 plastic culture dish containing ABS. These are subsequently transferred between 3 falcon dishes, each containing around 3 ml ABS. Leave in each wash for 10–15 minutes. When transferring between dishes, hold connective tissue attached to the ganglia with forceps – **do not** damage the ganglia. To prevent cross-contamination between dishes, keep lids closed all the time except during changes.
2. During antibiotic treatment, prepare 2 falcon dishes for enzyme treatment. Fill each dish with 3 ml DM and either 6 mg trypsin or 6 mg trypsin inhibitor (i.e. 2 mg/ml to yield a final concentration of 0.2% by volume). Label dishes appropriately to prevent confusion and possible catastrophe.
3. Following washes, transfer brains (12 / dish) into a trypsin/DM dish and leave at room temperature (18–20 °C) for 20–25 minutes. Swirl the dish every 5 minutes to ensure even and full enzymatic degradation of the connective tissue surrounding the ganglia. Following enzyme treatment, transfer brains to the trypsin inhibitor/DM dish and leave for 15 minutes. Again, swirl the dish every 5 minutes to prevent further enzymatic digestion.

   Note: Enzymatic treatment is also one of the most important steps in cell culture. Not only is it critical to select the right type/s of enzymes but also to determine the precise timing and the temperatures. If an enzyme bottle is left out at room temperature for longer periods of time, its activity will decline over time. This would need to be compensated for subsequent usage, by either increasing, a) the timing or b) the enzyme concentration itself. It is therefore important to store enzyme bottles in freezer, immediately after their use. The choice of any given enzyme is based on various tissue parameters. For instance, if the central ring ganglia have an extensive connective tissue sheath, then stronger enzymes (such as protease) are usually required for longer periods of time. Generally, collagenase-dispase is considered best, because it contains two enzymes – the collagenase which works best on collagen-based connective tissue; whereas dispase (protease) digests the sheath surrounding the ganglia. The choice for any given enzyme should however be determined "experimentally".

4. In a sterile beaker, prepare "high osmolarity DM" by adding 750 μl of 1M glucose solution to 20 ml DM. This solution will raise normal osmolarity of the DM from 130–145 to 180 –195 mOsm. A high osmolarity DM will cause neuronal shrinkage, thus making them "tougher" to withstand the extraction process.
5. Transfer brains to a dissection dish containing high osmolarity DM and pin down the central ring ganglia.
6. To prevent evaporative loss of solvent, keep dissection dish covered at all the times except during cell isolation and surgery. This is very important, because solvent loss will result in higher osmolarity and subsequent cell damage.
7. Attach a Sigmacote-treated glass pipette of suitable size to the tubing, and sterilize with ethanol for at least 5 minutes. Rinse micro-syringe, tubing and the pipette thoroughly with ABS (10 ml) prior to filling them with high osmolarity DM. It is vital that there are no bubbles either in the tubing or the micro-syringe itself, as it will render cell extraction difficult.
8. Add appropriate amount of medium (DM or CM 2.5–3.0ml) to cell culture dishes and place them in circular cubbyholes, made in a homemade plastic dish holder (Fig. 2A). The cell culture dishes will either be plain plastic (3001) or have poly-L-lysine coated glass cover slips attached to their bottom. The latter is preferable for neurite outgrowth since glass provides for a better optical clarity as compared to plastic. Attaching glass cover slips to the bottom of a plastic dish is laborious; this approach does however offer several advantages. For instance, all tissue culture dishes used for antibiotic treatment can be recycled to attach glass cover slips. Briefly, a circular hole is drilled at the bottom of a 3001 dish and a glass cover slip is glued via non-toxic material. These dishes are washed with super Q water and UV treated. It is important to use non-polished and non-coated glass cover slips made from a German glass (Bellco Glass, Inc. Biological Glassware and Equipment, USA). These dishes provide the best optical resolution for both the phase contrast and the Nomarski optics microscopy.
9. Use fine forceps to remove the outer connective tissue sheath surrounding each ganglion. To prevent excessive hand movement, try to rest forearms and possibly wrists on the hood base plate and anchor fingers to both sides of the dissecting dish. You may notice that excessive coffee and subsequent de- sheathing do not go "hand in hand"!
10. Remove (with fine forceps) the inner connective sheath surrounding each ganglion. To prevent neuronal damage, always pinch at the ganglion away from the desired neurons and tear gently. Avoid touching the cell bodies with forceps, as this may damage the somata.
11. Gently squeeze the commissure on either side of the desired ganglion with a pair of fine forceps. This will axotomize the axons of most neurons and will thus facilitate cell extraction.
12. Maneuver (via the micromanipulator) pipette tip on top of the cell body and apply gentle suction pressure through the micro-syringe. The cell body should be gently sucked up into the pipette. At this point, it would look somewhat like a "balloon on a string", the somata being the balloon and the axon a string. Continue to apply gentle suction (Fig. 2A) until the axon snaps and the cell moves up in the pipette.
13. Once the cell floats into the pipette and is visible within a few mm of its opening, move both microscope (using moveable boom stand arm) and the light source to

locate the center of a tissue culture dish (See Figure 2A). Carefully manipulate the pipette tip towards the bottom and gently flush out the cell. If done correctly, the undamaged cell should settle slowly at the bottom of the culture dish.

Note: All cell transfers between dissecting and the culture dishes are made by moving the cell pulling assembly back and forth – the culture dishes should never be disturbed.

14. Repeat steps 10–13 to obtain desired number of cells. It is important that plated cells should never be disturbed physically. Neurons should always be plated at some distance from one another (5–10 soma diameter apart), except in instances where chemical synapses are desired. For this purpose, and to achieve strong chemical synapses, cells must be plated in close proximity to one another.
15. Leave culture dishes undisturbed (overnight) to allow for cell attachment, growth, and/or synapse formation.
16. Remove forceps, scissors, micro-syringe and tubing etc. from hood, rinse with 70% ethanol and store.

Results

Lymnaea Neurons Survive Cell Isolation Procedure and Exhibit Robust Neurite Outgrowth in Culture.

Within a few hours of their isolation, the identified *Lymnaea* neurons assume a spherical shape as the axon stumps resorb into their somata (Fig. 2C). Most neurons exhibit neurite outgrowth within a few hours of platting on a poly-L-lysine coated dish containing CM. Typically, the primary growth cones emanating directly from the somata are large (almost the somata size, see Fig. 2D) and they exhibit cytoskeletal features that resemble other growth cones from both vertebrate and invertebrate neurons. Specifically, both actin and tubulin are predominantly located at the filopodia and lamelipodia, respectively (Fig. 4A and B). After 12–24 hrs the cultured neurons exhibited extensive outgrowth (Fig. 4C).

Respiratory CPG Neurons IP3I and VD4 Are Conditional Bursters.

To ask whether individually cultured *Lymnaea* neurons maintain their intrinsic properties *in vitro* and to test if IP3I and VD4 were endogenous or conditional bursters, we made direct intracellular recordings from single cells. As shown in Figure 5a,b, both IP3I and VD4 were quiescent, firing bursts of action potentials only after their depolarization. RPeD1, on the other hand, was tonically active (Fig. 5c). Consistent with our *in vivo* data, this experiment not only resolved that both IP3I and VD4 are conditional bursters, but also showed that neurons did indeed maintain their intrinsic membrane properties.

Paired Respiratory CPG Neurons Re-Establish Appropriate Synaptic Connections in Cell Culture.

To test whether paired respiratory CPG neurons establish appropriate synaptic connections in cell culture, and if synaptically connected cells IP3I and VD4 were sufficient to generate respiratory rhythm, cells were cultured in pairs. Direct intracellular recordings were made to test for the presence of synapses. All cell types exhibited extensive out-

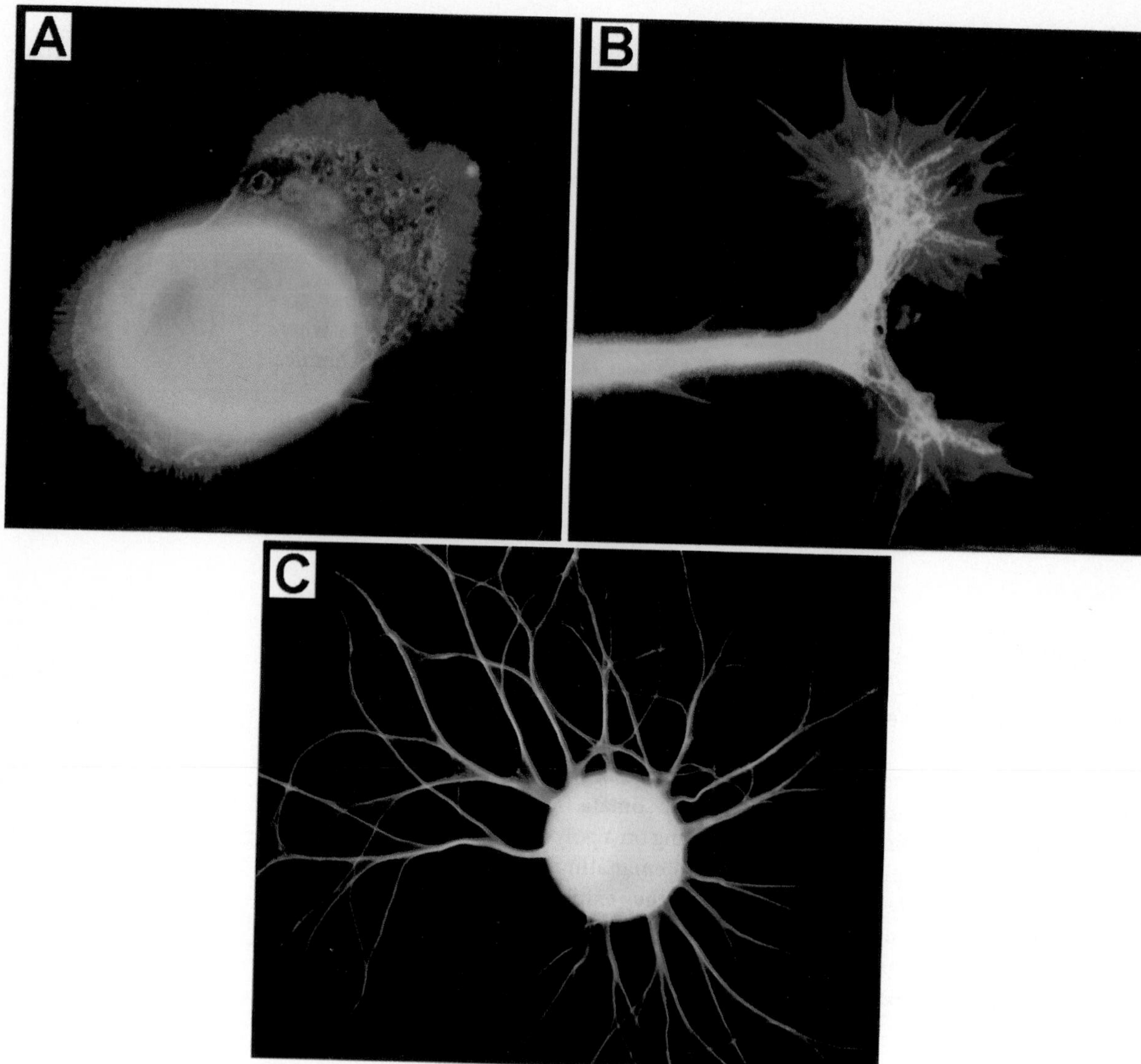

Fig. 4. Actin and microtubular organization at various stages of neurite outgrowth. A) During early sprouting, actin (*red*) is most prominent at the peripheral domain. Note that at this time point (1–2 hrs of culture), microtubules (*green*) are not well organized. B) Actin dominates the peripheral domain of the growth cones, whereas microtubules are now well organized within the neurite (*green/yellow* – 12–18 hrs *in vitro*). C) After 24 hrs in cell culture, microtubules make up the core of cytoskeletal elements.

growth and established appropriate synaptic connections which were similar to those seen *in vivo* (Fig. 6). However, stimulation of either IP3I or VD4 failed to initiate bursting activity in its partner cell. Specifically, both IP3I and VD4 established reciprocal inhibitory synapses with one another (Fig. 6e,f). RPeD1 excited IP3I via a dual inhibitory-excitatory response (Fig. 6c), whereas RPeD1 was excited by IP3I (Fig. 6d). A mutually inhibitory synapse developed between RPeD1 and VD4 (Fig. 6a,b). These data provided unequivocal evidence that synaptic connections seen *in vivo* are indeed direct and chemical. Moreover, they demonstrated that any given cell pair is not sufficient to generate respiratory rhythm.

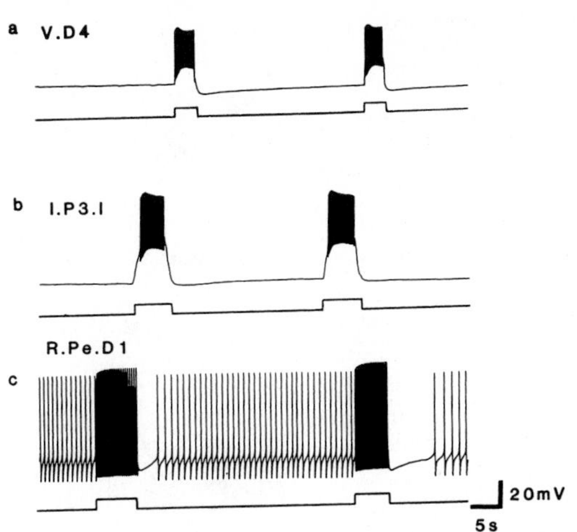

Fig. 5. Activity patterns of individually isolated respiratory CPG neurons. Intracellular recordings were made from individually cultured respiratory neurons VD4, IP3I and RPeD1. Both VD4 (**a**) and IP3I (**b**) were found to be quiescent. However, when stimulated via current injections, both cells fired a burst of spikes. RPeD1, on the other hand, was tonically active, increasing its fire frequency upon current injection (**c**). Figure modified from Syed et al., 1990.

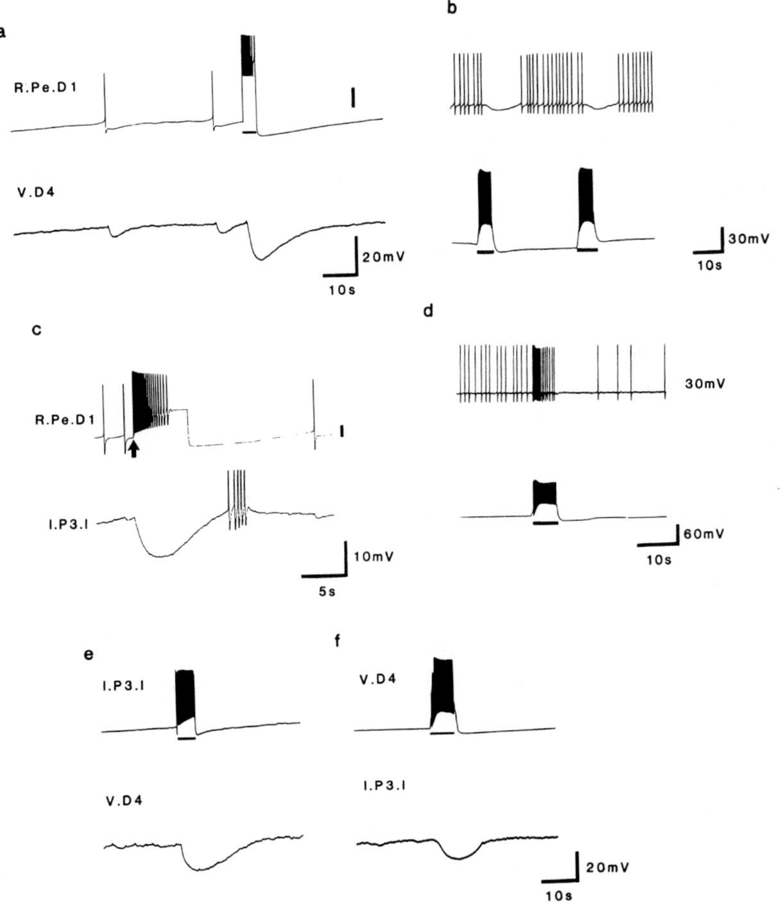

Fig. 6. Paired respiratory CPG neurons established appropriate synaptic connections in cell culture. Cells were paired in the presence of brain conditioned media and allowed to extend neurites. 12–24hrs after their isolation, neurons extend elaborate processes. Following neuritic overlap, simultaneous recordings were made from both neurons. Intracellular recordings from RPeD1 – VD4 pair revealed mutual inhibitory synapses between the cells (**a,b**). RPeD1 stimulation (*at arrow*) excited IP3I via a dual excitatory-inhibitory synapse (**c**), whereas IP3I excited RPeD1. (**d**) IP3I – VD4 pair showed mutual inhibitory synapses (**e,f**). Figure modified from Syed et al., 1990.

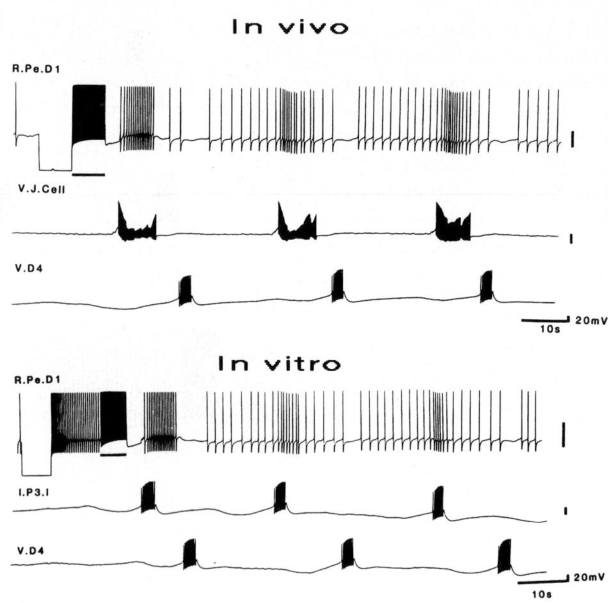

Fig. 7. *In vitro* reconstructed CPG generated respiratory rhythm similar to that seen *in vivo*. CPG neurons RPeD1, IP3I and VD4 were cultured together and simultaneous intracellular recordings were made. RPeD1 stimulation generated rhythmical bursting activity in previously quiescent neurons IP3I and VD4. This rhythmical activity was similar to that seen *in vivo* (*top panel*). It is important to note that in the *in vivo* experiments (*upper panel*) an indirect evidence for IP3I activity was obtained from one of its postsynaptic neurons (VJ). Figure modified from Syed et al., 1990.

RPeD1, IP3I and VD4 Are Sufficient to Generate Respiratory Rhythm in an *in vitro* Reconstructed Network

To test whether respiratory rhythm generation in this circuit was a function of network properties, RPeD1, IP3I and VD4 were cultured together. Following their neuritic overlap, simultaneous intracellular recordings were made, and RPeD1 was stimulated via direct current injection. A bust of action potentials in RPeD1 activated IP3I via a biphasic response (excitation followed by inhibition). IP3I in turn excited RPeD1. Combined activity in both RPeD1 and IP3I produces a burst of spikes in VD4, which in turn inhibited both neurons. However, the entire network continued to generate rhythmical bursting activity that lasted for several inspiratory and expiratory cycles. This rhythmical pattern that was recorded *in vitro* (Fig. 7b) resembled that seen *in vivo* (Fig. 7a). Together, these data provided direct evidence that indeed three CPG neurons are sufficient to generate respiratory rhythm and that rhythm generation in this circuit is a function of network properties.

Utilizing whole cell patchclamp techniques, we are currently determining the involvement of intrinsic membrane properties in respiratory rhythmogenesis (Barnes et al., 1994).

Reconstruction of Synapses between the Somata of Identified *Lymnaea* Neurons

As shown above, the respiratory rhythmogenesis in *Lymnaea* is a function of network synaptic properties; however, even in *in vitro* preparation, synapses formed between cultured neurons are inaccessible for direct electrophysiological analysis. To gain direct access to both the somata and its synaptic sites, we developed synapses between the cell bodies of the respiratory CPG neurons. Specifically, both RPeD1 and VD4 somata were isolated and juxtaposed in cell culture. Within 18–24 hours, synapses developed between the cell bodies in the absence of neurites (Fig. 8). Both morphologically and electrophysiologically these synapses were similar to those seen *in vivo* (Feng et al., 1997).

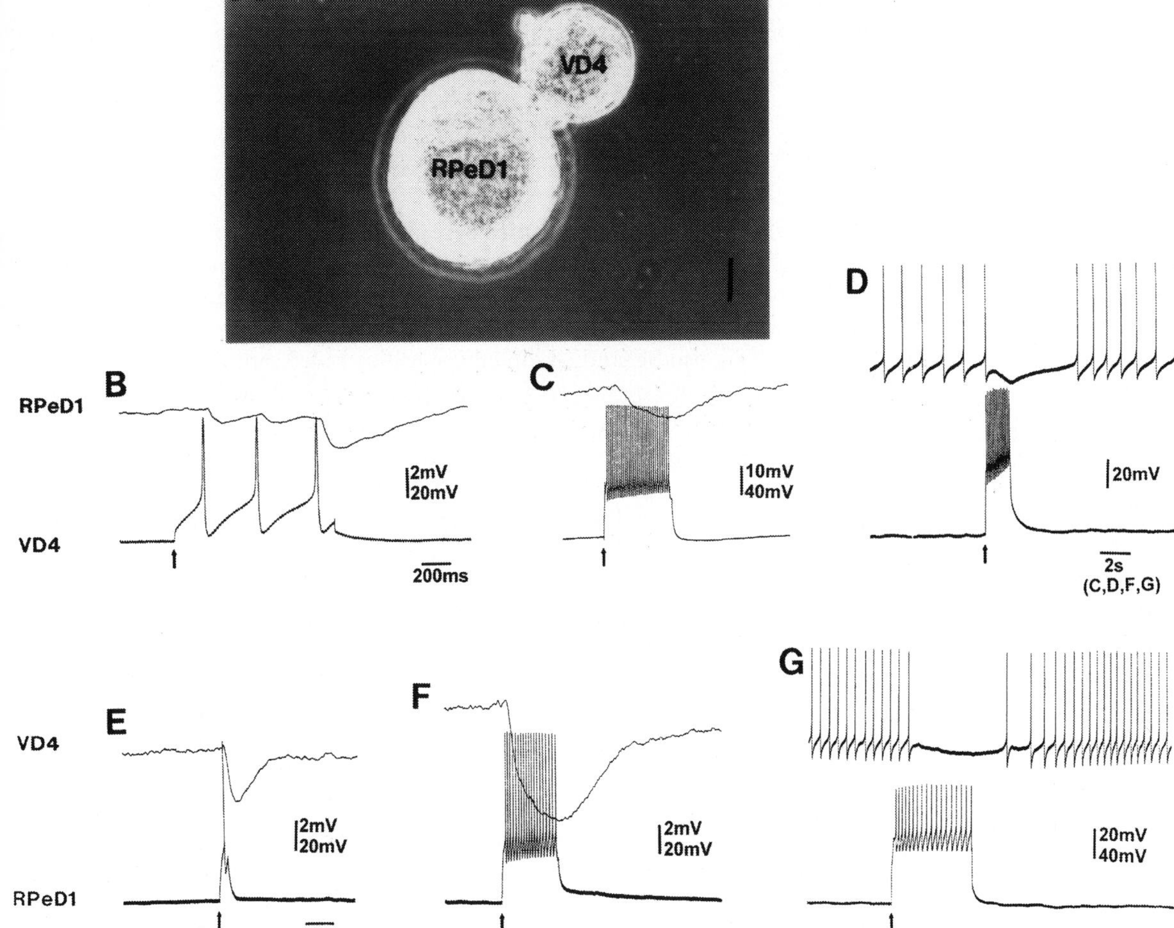

Fig. 8. Specific synapses between the respiratory CPG neurons reform in a soma-soma configuration. **A**) Identified respiratory CPG neurons RPeD1 and VD4 were isolated from their respective ganglia and somata juxtaposed *in vitro*. Mutual inhibitory synapses, similar to those seen *in vivo*, developed between the cell bodies in the absence of neurite outgrowth. Specifically, VD4 stimulation produced 1:1 inhibitory postsynaptic potentials in RPeD1 (**B**). A burst of spikes in VD4 (*at arrow*) produced a compound inhibitory postsynaptic potential in RPeD1 (**C**). Similarly VD4 stimulation (*at arrow*) inhibited the spiking activity in a spontaneously active RPeD1 (**D**). A similar inhibitory synapse was observed between RPeD1 and VD4 (**E-G**). Figure from Feng et al., 1997.

However, an added advantage of this preparation is that both ion channel and the synaptic mechanisms underlying respiratory rhythmogenesis can now be analyzed directly.

Does the *in vitro* Data Apply *in vivo*?

The propensity of adult molluscan neurons to regenerate and reconnect with their target cells is retained in *in situ* organ culture (see Moffette, 1995). We have also demonstrated that when transplanted either into the host ganglia of an intact animal or in an organ culture, a single implanted neuron also regrows its processes and relocates its synaptic partners with a remarkable degree of accuracy. Moreover, the transplanted neuron can also restore a behavioral deficit by incorporating itself into the host's respiratory circuitry (Syed et al., 1992b).

Taken together, the studies outlined above demonstrate the effectiveness of an *in vitro* cell culture approach in elucidating both the cellular and synaptic mechanisms underlying rhythm generation in this circuit. Moreover, these *in vitro* studies can also be revisited *in vivo* to validate data obtained from cell culture experiments.

Conclusions

The *in vitro* cell culture approach adopted here offers a wonderful opportunity to elucidate ionic, cellular and synaptic mechanisms underlying rhythm generation in our model. Moreover, since synapses form *in vitro* in a relatively short time period, both the cellular and molecular mechanisms underlying specificity of target cell selection and synaptogenesis can also be studied directly. Specifically, we can now begin to unravel the mechanisms by which networks of pattern generating neurons are put together during early development and regeneration. This preparation can also be used to study mechanisms of axonal pathfinding, growth cone behavior, regeneration and plasticity. Moreover, utilizing this approach, we were also able to build neuronal circuits from cells that were derived from two different molluscan species (Syed et al., 1992a). Finally, we recently extended our *in vitro* cell culture approach to successfully isolate neurons from select regions of a vertebrate brain (Turner et al., 1995). These neurons not only survived well in primary cell culture but also exhibited neurite outgrowth. A variety of cell culture approaches (see Bulloch and Syed, 1992; Ridgway et al., 1991; Feng et al., 1997; Jacklet et al., 1996; Saver et al., 1998; Turner et al., 1995) have since allowed us insights into cellular and synaptic properties of neurons in both invertebrate and vertebrate species, at a resolution that is unattainable *in vivo*.

Acknowledgement: The authors acknowledge excellent technical support by Mr. Wali Zaidi. This work was supported by MRC (Canada). NIS is an AHFMR senior scholar and PL and HZ were supported by a graduate and a summer studentship, respectively.

References

Barnes, S., Syed, N.I., Bulloch, A.G.M. and Lukowiak, K. (1994). Multichannel modulation by dopamine in an interneuron of the respiratory central pattern generator of *Lymnaea*. J. Exp. Biol. 189: 37–54.

Bulloch, A. G. M. and Syed, N.I. (1992). *In vitro* reconstruction of neural circuits. TINS 15:442–427.

Delcomyn, F. (1980). Neural basis of rhythmic behaviour in animals. Science 210:492–498.

Feng, Z-P., Klumperman, J., Lukowiak, K. and Syed, N.I. (1997). *In vitro* synaptogenesis between the somata of identified *Lymnaea* neurons requires protein synthesis but not extrinsic growth factors or substrate adhesion molecules. J. Neurosci., 17 (20): 7839–7849.

Getting, P.A. (1989). Emerging principles governing the operation of neural networks. An Rev Neurosci. 12: 185–204.

Harris-Warrick, R. M. and Johnson, B. R. (1989). Motor pattern networks: Flexible foundations for rhythmic pattern production. In: Perspectives in Neural Systems and Behaviour (ed. T. J. Rew and D. B. Kelly) pp 51–71, New York: Alan R-Liss Inc.

Jacket, J., Barnes, S., Bulloch, A.G.M., Lukowiak, K. and Syed, N.I. (1996). Rhythmic activities of isolated and cultured pacemaker neurons and photoreceptors of the *Aplysia* retina in culture. J. Neurobiol. 31: 16–38.

Kristan, W. B., Jr. (1980). Generation of rhythmic motor patterns. In: Information Processing in the Nervous System (ed. H. M. Pinsker and W. D. Willis, Jr.) pp.241–261, New York: Raven press.

Moffett, S. B. (1995). Neural regeneration in gastropod molluscs. Pro. in Neurobiol. 46:289–330.

Pearson, K. G. (1993). Common principles of motor control in vertebrates and invertebrates. An Rev Neurosci. 16: 265–297.
Ridgway, R.L., Syed, N.I., Lukowiak, K. and Bulloch, A.G.M. (1991). Nerve growth factor (NGF) induces sprouting of specific neurons of the snail, *Lymnaea stagnalis*. J. Neurobiol. 22(4):377–390.
Saver, M.A., Wilkens, J.L., Syed, N.I. (1998) *In situ* and *in vitro* identification and characterization of cardiac ganglion neurons in the crab, *Carcinus Maenas*. J. Neurophysiol. (In press).
Selverston, A. I. (1980). Are central pattern generators understandable? Behav. Brain Sci. 3: 535–571.
Syed, N. I., Bulloch, A. G. M. and Lukowiak, K. (1990). *In vitro* reconstruction of the respiratory central pattern generator of the mollusk *Lymnaea*. Science 250: 282–285.
Syed, N.I., Harrison, D. and Winlow, W. (1991). Respiratory behavior in the pond snail *Lymnaea stagnalis*. I. Behavioral analysis and relevant motor neurons. J. Comp. Physiol. 169:541–555.
Syed, N.I., Lukowiak, K. and Bulloch, A.G.M. (1992a). Specific *in vitro* synaptogenesis between identified *Lymnaea* and *Helisoma* neurons. NeuroReports 3:793–796.
Syed, N.I., Ridgway, R.L., Lukowiak, K. and Bulloch, A.G.M. (1992b). Transplantation and functional integration of an identified interneuron that controls respiratory behavior in *Lymnaea*. Neuron 8:767–774.
Syed, N. I. and Winlow, W. (1991). Respiratory behaviour in the pond snail *Lymnaea* stagnalis. II. Neural elements of the central pattern generator (CPG). J. Comp. Physiol. A 169: 557–568.
Turner, R.W., Borj, L.L. and Syed, N. I. (1995). A technique for the primary dissociation of neurons from restricted regions of the vertebrate CNS. J. Neurosci. Meth. 56: 57–70.
Wong, R. G., Martil El, Kater S. B. (1983). Conditioning factor(s) produced by several molluscan species promote neurite outgrowth in cell culture. J. Exp. Biol. 105: 389–393.

Chapter 13

Grafting of Peripheral Nerves and Schwann Cells into the CNS to Support Axon Regeneration

Thomas J. Zwimpfer and James D. Guest

Introduction

Prerequisites for functional recovery following axonal interruption (i.e. axotomy) in the central (CNS) or peripheral nervous system (PNS) are: 1) Survival of the injured neuron; 2) Axon regrowth of sufficient length to reach its target; 3) Axon guidance and pathfinding such that the appropriate connections are reformed and; 4) Formation and maintenance of functional synapses. Neurons that survive axotomy extend their axons only a short distance (approximately 1mm) within the CNS of adult mammals and higher vertebrates. This is in contrast to the robust growth of injured axons within the PNS of vertebrates and in some regions of the CNS of lower vertebrates (Gaze and Keating 1970; Sharma et al. 1993).

Neuroanatomic tracing techniques have demonstrated that many types of neurons in the brain and spinal cord of adult rodents can extend axons into PN grafts (Richardson et al. 1980; David and Aguayo 1981; Aguayo 1985; Vidal-Sanz et al. 1987; Cheng et al. 1996; Kobyashi et al. 1997) or Schwann cell (SC)-seeded implants (Guénard et al. 1993; Xu et al. 1995a,b; Xu et al. 1997; Guest et al. 1997b; see Bunge and Kleitman 1997 for reviews) for distances long enough to reach their normal targets. These include neurons in the retina (e.g. retinal ganglion cells; Vidal-Sanz et al. 1987), brain (e.g. motor and somato-sensory cortex, visual cortex, olfactory bulb, basal ganglia, thalamus, hippocampus, deep cerebellar nuclei; reviewed in Aguayo 1985), brainstem, spinal cord (e.g. motoneurons; Carlstedt et al. 1986), and descending and ascending spinal tracts (e.g. rubrospinal, reticulospinal, vestibulospinal, raphespinal, coerulospinal and primary sensory axons in the dorsal columns; Guénard et al. 1993; Oudega et al. 1994; Xu et al. 1995a; Cheng et al. 1996; Kobyashi et al. 1997; Guest et al. 1998).

Implantation of PN grafts or SC implants into the CNS promotes the lengthy regrowth of injured CNS axons and may be a future treatment for human brain and spinal cord injuries. Presently, these strategies provide the opportunity to investigate the capacity of regenerating CNS axons for axon pathfinding, target recognition and formation of persistent and functional synaptic connections. Substitution of the optic nerve (ON) by a PN graft, following ON transection in the adult rodent, has demonstrated that injured retinal ganglion cell (RGC) axons can grow for sufficient distances to reach their

Corresponding author for Peripheral Nerve Grafting Protocol: Thomas J. Zwimpfer, University of British Columbia and Vancouver Hospital and Health Sciences Centre, Departments of Surgery (Neurosurgery) and Zoology and Collaboration on Repair Discoveries (CORD), 300 - 700 West 10th Avenue, Vancouver, British Columbia, V5Z 4E5, Canada (phone: 604-873-4766; fax: 604-873-6422; e-mail: zwimpfer @interchange.ubc.ca)
Corresponding author for Schwann Cell Implant protocol: James D. Guest, Department of Neurosurgery, Miami Project to Cure Paralysis, University of Miami School of Medicine, P.O. Box: 016960(R-48.), Miami, Florida, 33101, USA (phone: +01-305-243-6001; fax: +01-305-243-6017; e-mail: jimguest@ compuserve.com)

normal target, the superior colliculus (Vidal-Sanz et al. 1987). Following insertion of the free end of the PN graft into the superior colliculus, RGC axons can leave the PN graft, grow into and arborize within the superior colliculus, and form persistent and functional synaptic connections in the normal retinorecipient layers of the superior colliculus (Vidal-Sanz et al. 1987; Carter et al. 1989; Keirstead et al. 1989). Use of PN grafts to guide regenerating RGC axons into non-retinal targets (e.g. cerebellar cortex and inferior colliculus) revealed that injured adult CNS axons can grow into and form persistent synaptic connections in novel targets (Zwimpfer et al. 1992). Therefore, in addition to promoting lengthy axon growth, PN grafts and SC implants could also be used to guide regenerating CNS axons into close proximity of their normal synaptic targets to minimize the formation of anomalous synaptic connections in the injured CNS.

There are several major limitations to the use of these permissive conduits to restore anatomical and functional connections in the injured CNS. Simply providing a PN environment for injured CNS axons may not be sufficient for long axon growth into these conduits. For example, there is little axon growth into PN grafts from injured RGCs (Richardson et al. 1982) or rubrospinal neurons (Kobyashi et al. 1997) when axons are injured a long distance from the neuronal cell body. Following axotomy close to the cell body, most CNS axons are capable of lengthy growth within permissive substrates (e.g. PN grafts). This axon regeneration is correlated with a neuronal cell body response that involves increased expression of regeneration-associated genes (RAGs) such as GAP-43 (Tetzlaff et al. 1991; Schaden et al. 1994), Tα1-tubulin (Tetzlaff et al. 1991), and c-jun (Jenkins et al. 1993). Infusion of neurotrophic factors or transplantation of small PN segments in close proximity to the neuronal cell body increases RAG expression and enhances extension of injured CNS axons into PN grafts (Ng et al. 1995; Oudega and Hagg 1996; Kobyashi et al. 1997) or SC implants (Xu et al. 1995b).

Secondly, only a small proportion of regenerating CNS axons are able to leave the PN graft or SC implant and re-enter the CNS. This is possibly due to scar formation at the PNS-CNS interface, which includes reactive astrocytes, fibroblasts (FBs) and microglia. Reactive astrocytes appear to inhibit axon growth by inducing a physiological stop signal within growing axons (Luizzi and Tedeschi 1992). The molecules that mediate this signal are not well characterized but reactive astrocytes within a glial scar do express higher levels of putative inhibitory molecules such as tenascin and chondroitin sulphate proteoglycans (Bovolenta et al. 1993; Pindzola et al. 1993).

Finally, axons that are successful in re-entering the CNS only grow short distances (usually 1 to 2 mm) within the adult mammalian CNS. Neutralization of molecules in the CNS that are putative inhibitors of axon growth is an additional strategy that appears to facilitate the growth of a small proportion of transected CNS axons (Schnell and Schwab 1990, 1993; Bregman et al. 1995; Keirstead et al. 1995). Recent reports suggest that transplantation of a different type of glial cells called "olfactory ensheathing cells (OECs)" into the injured spinal cord (Ramon-Cueto and Nieto-Sampedro 1994; Li et al. 1997; Ramon-Cueto et al. 1998) or brain (Smale et al. 1996) may further facilitate axon growth. A discussion of OECs and their potential use in facilitating CNS regeneration is beyond the scope of this chapter (see the following reviews for additional information on OECs: Doucette 1995; Ramon-Cueto and Valverde 1995).

The SC membrane, its basal lamina and/or factors released by SCs appear to be necessary for the promotion of axon growth in a PN environment. For example, PN grafts in which SCs were killed by repeated freeze-thawing failed to support axon growth (Smith and Stevenson 1988) while injured PNS and CNS axons did grow through tubes containing only cultured SCs (see review by Bunge and Kleitman 1997). SCs produce molecules that have been shown to promote axon growth. These include extracellular matrix molecules (e.g. laminin, fibronectin, collagen I and IV), cell adhesion molecules (e.g. L1, N-cadherin, and N-CAM; see Carbonetto and David 1993 for review), other cell surface molecules (e.g. p75 NGF receptor) and a wide variety of trophic factors (e.g. NGF, BDNF, NT-3, PDGF and GDNF). Loss of axon contact due to axotomy results in reactive changes in SCs that include increased mitotic activity (maximum at 3 to 4 days after axotomy; Bradley and Asbury 1970) and an enhanced production of trophic factors (Heumann et al. 1987; Meyer et al. 1992).

Peripheral Nerve Grafts

Because of its long length, relatively large cross-sectional area and superficial location, the most common donor nerve used in animal studies is the sciatic nerve or one of its branches, especially the common peroneal nerve. Other donor nerves, such as intercostal nerves, have also been used especially in studies that assess recovery of lower limb function following injury (Cheng et al. 1996). PN grafts can be harvested immediately after nerve transection (fresh) or can be left *in situ* and allowed to undergo Wallerian degeneration for a period of time prior to harvest (predegenerated). Some studies have demonstrated a mild to moderate increase in the rate and extent of axon growth within predegenerated as opposed to fresh PN grafts (Zhao and Kerns 1994; Oudega et al. 1994).

Advantages

- Readily available and inexpensive.
- Use of the common peroneal nerve results in very low morbidity to the donor animal. The resulting absence in ankle dorsiflexion does not significantly interfere with walking or ability to reach food. The area of sensory loss appears to be small as the rate of self-mutilation of the foot (i.e. autotomy) is extremely low. Both common peroneal nerves can be harvested from the same animal without obvious detriment to the animal's health or life span. The entire sciatic nerve has been used routinely but does result in increased gait difficulty and autotomy.
- Use of autologous grafts (i.e. harvested from and transplanted into the same animal) negates the need for immunosuppression or the use of an immune-compromised host, both of which result in increased costs as well as morbidity and mortality.

Disadvantages

- Limited number (two) and length (4cm) of graft material if autologous peroneal nerves are used. Up to 18 intercostal nerves have been harvested from one donor animal (Cheng et al. 1996) but each of these are much smaller in length and cross-sectional area, and harvesting requires extensive and prolonged surgery.
- Use of the entire sciatic nerve increases the risk of self-mutilation (i.e. autotomy) of the foot and interferes with functional assessment of the lower limb.
- Difficult to modify the cellular (i.e. SCs) or non-cellular components (i.e. level of trophic factors) of the graft.

Schwann Cell Implants

The cultivation of purified Schwann cell (SC) populations provides the opportunity to construct grafts containing novel combinations of components (cellular and acellular) for CNS grafting. SC implants have been used in models of neural injury in the brain, spinal cord and PN (Blakemore and Crang 1985; Kromer and Cornbrooks 1985; Guénard et al. 1992; Montero-Menei et al. 1992; Neuberger et al. 1992; Kim et al. 1994; Levi and Bunge 1994; Levi et al.. 1994; Li and Raisman 1994; Xu et al. 1995a,b, 1997; Honmou et al. 1996; Stichel et al. 1996; Levi et al. 1997). Implants may consist simply of an unpurified SC suspension or SCs may be purified and/or combined with natural biomaterials (collagen, Matrigel etc.) or with artificial materials such as guidance channels. Another method of SC transplantation in the spinal cord is the creation of SC "trails" in which stereotactically implanted SCs form a well-defined pathway in the spinal cord distal to the transection site (Menei et al. 1997). This chapter will focus on SC grafts within guidance channels, which are tubular structures comprised of biocompatible materials such as silicone or PAN/PVC.

When SCs are mixed with Matrigel (a basal lamina mixture) and suspended within a PAN/PVC channel, the cell suspension contracts and undergoes re-organization into a solid cable of uniformly dispersed, longitudinally oriented cell processes (Guénard et al. 1992). PAN/PVC channels have an inner membrane with uniform pores (MW cutoff of 50,000 Kd) that results in a controlled environment by restricting movement of both cellular (i.e. inflammatory cells) and acellular (i.e. proteins) components. As a result, molecules released by SCs can be concentrated within channels (and potentially adverse molecules excluded), which may account for the influence of channel pore size on axon regeneration (Aebischer et al. 1989). In addition, trophic molecules can be delivered into and concentrated within channels over fixed time periods. For example, channels were adapted for delivery of trophic factors (NT-3 and BDNF) by osmotic minipumps (Xu et al. 1995b) or slow release from a biological (i.e. fibrin) glue apposed to the channel (Guest et al. 1997a). Finally, axonal tracers can be readily placed into grafts in these channels to label regenerating axons with low risk of spillage and non-specific labeling.

Advantages

Implants of cultured SCs offer certain advantages over PN grafts.

- Purification of the cellular population prior to grafting.
- Expansion of population to massive numbers.
- Storage of frozen SCs to be used when needed.
- SCs can be genetically altered (i.e. produce larger quantities of trophic factors; Menei et al. 1997).
- Induction into a specific state of activation (e.g., treatment with forskolin causes increased cAMP levels secondary to adenylate cyclase activation which increases SC sensitivity to protein growth factors (Stewart et al. 1991).
- The components of implants (both cellular and acellular) can be altered to produce novel combinations of constituents that do not occur naturally. For example, Guénard et al. (1994) studied the impact of varying the proportion of cultured astrocyte and Schwann cell combinations on PN regeneration within guidance channels.
- Represents a potential strategy for autologous SC transplantation into the injured human nervous system. Small donor segments can be expanded to large purified SC populations.

Disadvantages

- Very labor-intensive.
- Expensive in terms of facilities and materials required.
- The large number of protocol steps increases the chance of error, culture infection or toxicity to cells and requires extensive planning and coordination for experiments.
- Need for host immunosuppression or use of an immune-compromised host when the host and donor are different animals. However, SCs that are pooled from a number of animals of a specific in-bred strain (i.e. Fischer rats) and transplanted into an animal of the same strain have not shown significant evidence of rejection in an extensive series of experiments (Xu et al. 1995a,b, 1997).
- *In vitro* manipulation and expansion of SC populations is associated with theoretical risks of genetic instability and immortalization (i.e. tumor formation), particularly if the cells are cultured in serum-containing conditions (Rawson et al. 1991).

Organization of the Chapter

Protocols will be presented in two parts.
Part I: Harvest and Implantation of Peripheral Nerve (PN) Grafts into the CNS.
Part II: Preparation and Implantation of Schwann Cell Guidance Channels.

Subprotocol 1
Harvest and Implantation of PN Grafts into the CNS

Outline

1. Harvesting an autologous common peroneal nerve graft (Fig. 1).
2. Substitution of the optic nerve by a PN graft (Figs. 2, 3).
3. Direct implantation of a PN graft into the CNS (Fig. 4).
4. Labeling of CNS neurons that extend axons into PN grafts (Fig. 5).

Materials

Specialized Materials

- 10-0 suture (Ethicon).
- Instruments:
  - Jeweler's forceps (# 4 or 5; Dumont Medical).
  - Fine-needle driver or 45°-angled forceps (i.e. ss/45; Dumont Medical).
  - Spring microscissors.
  - Fine-tipped bone rongeurs (i.e. Model 16000-14; Fine Science Tools, North Vancouver, B.C., Canada).
- Absorbable gelatin sponge (Gelfoam; Upjohn, Kalamazoo, Mich., USA) for hemostasis.
- Mydriatic eye drops (i.e. 1% Cyclopentolate HCL, Alcon).
- Rat head holder (Model # 320) and Universal stand (Model # 310; David Kopf, Tujunga, Calif., USA).
- Bone wax (hemostasis for bone bleeding).
- Electric bone drill (Dremel, Racin, Wis., USA).
- Antibiotic ointment (i.e. polysporin).
- Hypodermic needles (26G or smaller) to open pia.
- Glass micropipettes
- 10-μl Hamilton syringe (Model # 710; Hamilton, Reno, Nev., USA).

Procedure

All procedures in this section are done under anesthetic administered by intraperitoneal injections (i.e. Ketamine and Xylazine) and can be done routinely in hamsters and rats. These procedures are feasible in mice but are technically more difficult, especially PN grafting of the optic nerve.

Harvesting an Autologous Common Peroneal Nerve Graft (Figure 1)

Nerve Exposure
The animal is positioned prone and the leg fully extended, held by tape applied to the foot. A straight skin incision is made along the long axis of the leg, extending from the

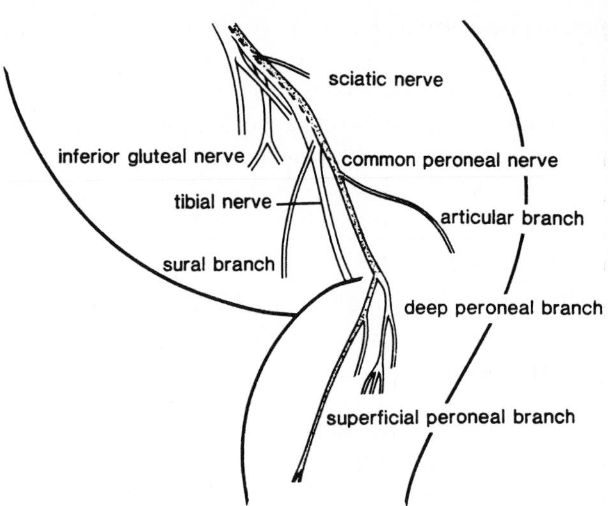

Fig. 1. Major peripheral nerves in the rodent's leg. *Dotted region* outlines the portion of the common peroneal nerve that can be harvested as a graft. *Hatched line* indicates the portion of the epineurium of the sciatic nerve that must be incised to separate the peroneal and tibial components (modified from Vidal-Sanz 1990; see also Greene 1963).

hip to just below the knee. The gluteus and hamstring muscles are incised at their insertion into the femur and retracted medially with a suture to expose the underlying sciatic nerve. The entire sciatic nerve can be harvested, but to decrease postoperative leg weakness and lessen the risk of self-mutilation of the foot (autotomy), the common peroneal component can be harvested with sparing the tibial nerve. Animals that autotomize will usually need to be sacrificed.

Transection and Mobilization of the Distal Common Peroneal Nerve Graft
The dissection begins distally by cutting the two main branches of the common peroneal nerve (the deep and superficial peroneal nerves) at the point where they enter the calf muscles just below the knee (Fig. 1). It is best to pick up the common peroneal nerve at the distal end using a fine forceps (i.e. jeweler's forceps, #4 or 5; Dumont Medical) held in the non-dominant hand, while the dissection and mobilization of the peroneal nerve is done using a fine-tipped microscissors held in the dominant hand. It is important not to place excessive stretch on the nerve and to prevent it from drying out during the dissection by periodically wetting it with saline.

The common peroneal nerve passes under a large vein in the distal thigh, and to prevent bleeding, it is best to spare this vessel and mobilize the peroneal nerve by passing its distal end under the vein from distal to proximal. If bleeding from this vein does occur, it can be controlled with a small piece of absorbable gelatin sponge (Gelfoam; Upjohn, Kalamazoo, Mich., USA) and local pressure. The distal end of the common peronal nerve is picked up with a jeweler's forceps and the nerve is mobilized to where it joins the tibial nerve to form the sciatic nerve. A small articular branch of the proximal common peroneal nerve will need to be cut (Fig. 1).

Division of the Epineurial Sheath and Intraneural Mobilization of the Peroneal Nerve
The two components of the sciatic nerve, the common peroneal and tibial nerves, can be separated up to the level of the sciatic notch by incising their common epineurial sheath. This sheath is incised with microscissors at the distal end of the sciatic nerve and the rest of the sheath cut longitudinally (hatched line in Fig. 1) and superficial to the underlying sciatic nerve up to the level of the sciatic notch. This is made easier by placing the peroneal nerve on gentle downward and lateral stretch to facilitate separation of the peroneal and tibial components. Once the sheath is filleted open, the tibial component can be easily mobilized from the peroneal portion by blunt dissection using the tip of the microscissors. This intraneural dissection is more painful and the animal may re-

quire some additional anesthetic at this point. However, some involuntary muscle contraction is normal during this part of the dissection. The common peroneal nerve is transected at the level of the sciatic notch, just distal to the inferior gluteal nerve.

This dissection should provide a PN segment of 4 cm in length in an adult rat and 3.5 cm in the adult hamster. The PN graft should be immediately placed in saline and remain there until the grafting portion of the procedure. No problems have been experienced with delays of several hours between harvesting and implanting the PN graft as long as the graft is prevented from drying out.

Wound Closure and Post-Operative Care
Using a running 4-0 silk suture, the gluteal and hamstring muscles are reapproximated to the lateral portion of the quadriceps muscle at their point of insertion into the femur. The skin is closed using a running 2-0 or larger non-absorbable suture (i.e. silk) or with metallic clips, but clips need to be removed in 5 to 7 days. Use of a skin suture smaller than 2-0 will likely result in early dehiscence of the wound due to chewing of the suture by this or other rats in the cage. However, 4 to 5 animals are routinely kept in one cage without problems. On awakening, animals are able to walk on the operated leg although there is permanent weakness of ankle dorsiflexion.

Use of Predegenerated versus Fresh PN Grafts
A PN graft can be harvested immediately after nerve transection (fresh) or the nerve can be cut at its proximal end and left within the animal to undergo Wallerian degeneration and harvested later (predegenerated). Some studies have demonstrated a mild to moderate increase in the rate and extent of axon growth within predegenerated as opposed to fresh PN grafts (Zhao and Kerns 1994; Oudega et al. 1994). This effect is likely due to the axotomy-induced increase in SC mitotic activity and trophic factor production (reviewed in the Introduction). The most common period of predegeneration used is 7 days, but a positive effect on axon growth can be seen with predegeneration times as short as 3 days and as long as 35 days (Lewin-Kowalik et al. 1992). In some studies, predegenerated PN grafts either did not result in enhanced axon regrowth compared to fresh grafts (Ellis and McCaffrey 1985) or the increased regeneration was only evident at early time points with no difference at longer survival times (Lewin-Kowalik et al. 1990). Fresh PN grafts do support excellent growth of injured PNS and CNS axons without the need for an additional anesthetic and surgical procedure.

Substitution of the Optic Nerve by a PN Graft

The protocol presented here was originally described by Vidal-Sanz et al. (1987).

Optic Nerve Transection within the Orbit
Funduscopy to Check Retinal Vasculature. After induction of anesthesia but prior to surgery, the blood supply of the retina should be checked by funduscopy. This is facilitated by applying several drops of a mydriatic solution (1% Cyclopentolate HCL; Alcon) to the eye to dilate the pupil. A moistened plastic cover slip or glass microscope slide is placed on the cornea and the retina viewed using the operating microscope.

Head Immobilization. Microsurgery of the orbit requires the animal's head to be rigidly held but rotation of the head provided to maximize exposure. A standard stereotactic frame can be used but no rotation is available. A head holder that clamps around the animal's nose and provides unlimited head rotation as well as neck flexion and extension is preferable (Model #320 and 310; David Kopf, Tujunga, Calif., USA). Surgery within the orbit is greatly facilitated through the use of gentle suction and pieces of mois-

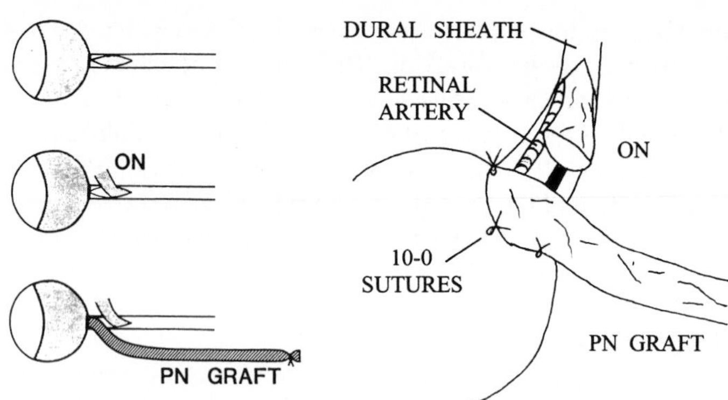

Fig. 2. Substitution of the optic nerve by a PN graft. **Left.** *Upper*: Longitudinal incision of the dura of the intraorbital portion of the optic nerve (ON). *Middle*: Transection of the ON immediately behind the globe. *Lower*: Apposition of the free end of a PN (common peroneal nerve) graft to the ocular stump of the transected ON. **Right.** Diagram of the view through the operating microscope. One end of the PN graft overlies the ocular stump of the transected ON. The PN graft is attached by three 10-0 sutures placed through the epineurium of the PN and the sclera, where the sclera is continuous with the dural sheath around the ocular stump of the ON. The retinal artery and vein lie on the inferior dural sheath, just outside and underneath the ON. To prevent retinal ischemia, these retinal vessels must be spared when the ON is transected. (Adapted from Vidal-Sanz 1990).

tened cotton that can be used as wicks to absorb the venous oozing that can obstruct vision.

Intraorbital Exposure and Transection of the Optic Nerve. A midline scalp incision is made from the nasal region to the insertion of the occipital muscles. The scalp and periosteum on the operative side are mobilized laterally. The orbit is entered by incising the periosteum at the supraorbital ridge and extending the opening down through the fascia of the temporalis muscle. Periosteum and muscle are retracted laterally by a 4-0 suture. This may result in avulsion of a small vein that enters the frontal bone through a small foramen just below the supraorbital margin. The bleeding from the foramen frequently stops spontaneously or can be controlled by packing the foramen with bone wax. The lacrimal gland is mobilized anteriorly and covered with moistened cotton. This will expose the levator palpebrae and superior rectus muscles overlying the superior aspect of the globe. These muscles are cut 5 mm from their site of insertion into the globe. A 6-0 suture is placed through these muscles and used to gently rotate the eye laterally and down to expose the optic nerve (ON) immediately behind the globe. Excessive or prolonged traction on these muscles (by the 6-0 suture) can result in kinking and occlusion of the retinal artery and produce retinal ischemia or infarction. The dura of the ON is incised longitudinally immediately behind the globe (Fig. 2, left) and the ON is picked up with a jeweler's forceps and bluntly mobilized away from the retinal artery and vein, which lie outside and underneath the ON on the inferior dural sheath (Fig. 2, right). The ON is transected at its point of exit from the globe, ensuring no damage to the retinal artery and vein.

Development of cataracts within the lens of the operated eye is not uncommon and the animal does not need to be sacrificed. Cataracts do not affect retinal ganglion cell survival or axon regeneration, but they will make intravitreal injection of tracers more difficult (see section D).

Attachment of the PN Graft to the ON Stump

Placement of 10-0 Sutures in the PN Graft. The 6-0 suture through the levator palpebrae and superior rectus muscles should be relaxed while the PN graft is prepared and the three 10-0 sutures are placed through the epineurium of the distal end of the common peroneal graft. It is best to place the entire PN graft over the exposed frontal bone and cover it with moistened cotton, leaving only the distal end exposed. The distal end of the PN graft is used as it retains an intact epineurium. Three monofilament 10-0 sutures (Ethicon) are placed from outside to inside the nerve through the epineurium of the distal end, equidistant from each other (i.e. at 12, 4 and 8 o'clock when the distal end is viewed end-on). The small needle can be held by a microsurgical needle-driver or with a blunt ended forceps with a 45°-angled tip (i.e. SS/45; Dumont Medical). To minimize the length of time that the globe is rotated during grafting to the ON, it is best to place all three sutures through the PN graft before placing tension on the 6-0 suture to re-expose the ON stump.

Placement of 10-0 Sutures around the ON Stump. Gentle traction is placed on the 6-0 suture to rotate the eye laterally and re-expose the ON stump. Each 10-0 suture is then passed through the sclera (from inside to outside) at the point where the dural sheath of the ON becomes the sclera of the globe (Fig. 2, right). The sutures should be placed in the following order: The 8 o'clock suture in the PN graft (when looking at the PN graft end-on) should be placed at the 10 o'clock position around the ON stump, the 4 o'clock PN graft suture placed at 2 o'clock around the ON, and the 12 o'clock PN graft suture placed at 6 o'clock. The 10-0 sutures are **not** put through the ON itself as the ON has no true epineurium. Care must be taken to not injure or kink the retinal vessels lying on the inferior dural sheath. All three sutures should be passed through the sclera before tying a knot, otherwise the PN graft will be pulled down onto the back of the globe and hinder placement of the remaining scleral sutures. Instrument-tied knots are made by creating a single loop around the tip of the needle driver with the portion of the suture on the needle side and grasping the free end (i.e. non-needle end) of the suture and pulling it through the loop and pulling it tight. Each knot should have a minimum of three throws. This will result in the end of the PN graft being apposed to and covering the ON stump of the globe (Fig. 2).

Placement of the PN Graft in a Channel Drilled in the Skull. The remaining portion of the PN graft is placed over the surface of the brain in a 5-mm-wide channel that is drilled in the frontal and parietal bones with an electric drill (Dremel; Racine, Wis., USA; Fig. 3, top). Compared to PN grafts left over the intact skull, grafts exposed to the meninges on the surface of the brain survive better, likely through improved vascularization of the graft. The free end of the PN graft is buried in the cervical musculature and marked with a 4-0 suture to facilitate its later identification. The eye is rotated back to its usual position, the temporalis fascia is closed with 4-0 silk and the scalp is closed with a running 2-0 silk. The retina is again examined by funduscopy as described above. Animals in which the retinal arteries are narrowed or appear empty and that do not return to normal within a few minutes are discarded. Antibiotic ointment (i.e. Polysporin) is applied to both eyes to prevent corneal exposure.

Direct Implantation of a PN Graft into the CNS

A PN graft can be inserted into the brain or spinal cord to either:

- Induce local CNS neurons to extend axons out of the CNS and into the graft or;
- To guide regenerating axons (PNS or CNS) that are already within a PN graft into the vicinity of specific CNS regions to facilitate re-innervation and synapse formation.

The technical aspects of graft insertion are similar for both. Insertion of a PN graft into the ventral horn of the cervical spinal cord, to induce re-growth of motoneuron axons, will be presented as an example. The free end of a PN graft that has been attached to the back of the eye is usually inserted into regions of the CNS that normally receive RGC input (i.e. superior colliculus) 6 to 10 weeks after grafting to the eye. See Vidal-Sanz et al. (1987), Carter et al. (1989) and Zwimpfer et al. (1992) for details of PN graft insertion into the brain.

Implantation of a PN Graft into the Spinal Cord in a Model of Nerve Root Avulsion
Exposure of the Spinal Cord and Root Avulsion. The animal is placed prone and its head secured in a head holder as described above. A midline incision is made from the occiput to the C7 spinous process and the cervical musculature and paraspinal muscles are mobilized away from the spine bilaterally and retracted with sutures. The left C4 and C5 lamina and adjacent facet joints are removed piecemeal with a fine-tipped bone rongeur, and the dura is opened over the left lateral aspect of the spinal cord. The left C5 and C6 ventral and dorsal roots are avulsed with a fine-tipped 90°-angled glass micropipette (see above for details of preparation) such that the rootlets are disrupted at their point of entry into the spinal cord. C5 and C6 roots are chosen for avulsion to preserve most of the sensation to the paw (i.e. C7, C8 and T1) in order to decrease the chance of self-mutilation (autotomy) of the upper limb.

Implantation of the PN Graft into the Spinal Cord. In both this example and when the PN graft is inserted to guide regenerating axons that are already within the graft into the CNS, it is preferable to tease the end of the graft into 2 or 3 smaller fascicles and insert each fascicle at a separate site rather than implant the entire end of the graft at one site. This increases the area of the CNS exposed to the PN graft.

The end of the PN graft is desheathed of its epineurium for a distance of 1 cm and the exposed nerve is picked up with two jeweler's forceps and gently teased into 2 or 3 fascicles of similar diameter (Fig. 4). The pia of the brain and spinal cord is quite tough and should be opened separately prior to graft insertion. A nick is made in the pia with a small hypodermic needle (i.e. size 26G or smaller), just large enough to accommodate a single fascicle. The end of a fascicle is freshly cut and inserted at a depth of 1mm into the ventrolateral aspect of the spinal cord, with a sharp-tipped 90°-angled glass micropipette (Fig. 4). The depth of insertion in other regions of the CNS will depend upon the location of the target neurons. The fascicles and PN graft must not be under tension and the fascicles can be held in the CNS by covering them with Gelfoam or with fibrin glue. To minimize the chance of having the fascicles pulled out of the CNS, the epineurium of the graft close to the CNS should be anchored to the surrounding paraspinal muscles with a 10-0 suture. The free end of the PN graft is marked with a 4-0 silk suture and left under the cervical musculature overlying the occiput.

Preparation of 90°-angled Glass Micropipettes Used for Both Spinal Nerve Root Avulsion and PN Graft Insertion into the CNS.

The center of a glass micropipette (diameter 1.5 to 2mm) is held by hand in a gas flame and the pipette is pulled parallel to its long axis to result in gradual thinning and elongation. Once the center of the pipette is thin and while it is still hot, the two ends are pulled perpendicular to the long axis of the tube to produce a short vertical section and two 90°-angle bends. After allowing the tube to cool, the vertical portion is broken in the center (the thinnest part) to produce two glass micropipettes with a 90°-angled tip. This 90° angle facilitates access to regions of the CNS that are deep in the surgical wound, such as the ventrolateral aspect of the spinal cord.

Labeling of CNS Neurons that Extend Axons into PN Grafts

Use of a Retrogradely Transported Axonal Tracer (Figs. 3, 4)
Application of the Tracer to the Free End of a PN Graft. The cell bodies of CNS neurons that extend axons into a PN graft can be labeled by retrograde transport of axonal tracers that are applied to the free end of the PN graft. Tracers such as Horseradish Peroxi-

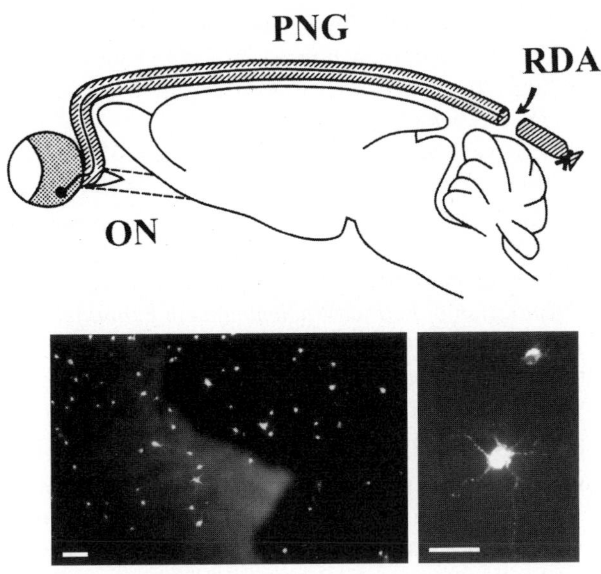

Fig. 3. Retrograde labeling of regenerated retinal ganglion cell (RGC) axons. **Upper:** Application of Rhodamine-Dextran Amine (RDA) to the cut end of a PN graft (PNG), 2 months after the PNG was apposed to the ocular stump of the transected optic nerve (ON). Retrogradely transported RDA will label RGC cell bodies that regenerated axons to the end of the PN graft. **Lower:** Fluorescence photomicrographs of wholemount preparations of retina. Many RDA-labeled RGC cell bodies are evident at low power (*left*; Bar = 100 µm.) At higher power (*right*; Bar = 50 µm), RGC cell bodies are characterized by a nucleus and/or labeled dendrites.

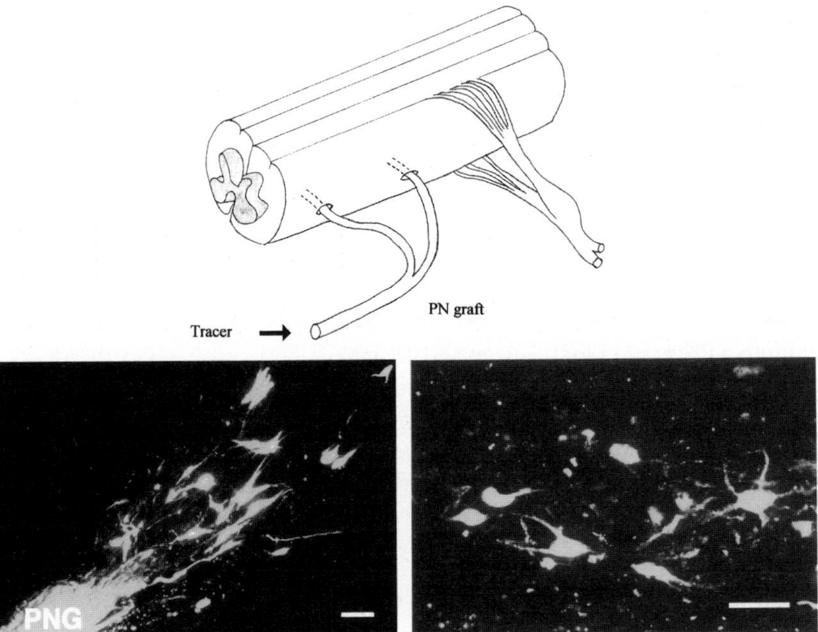

Fig. 4. Retrograde labeling of regenerated ventral horn neurons. **Upper:** Application of the tracer Rhodamine-Dextran Amine (RDA) to the cut end of a PN graft, 2 months after avulsion of the ipsilateral 5th and 6th cervical ventral and dorsal roots. The end of the PN graft was teased into two fascicles and each fascicle inserted at a depth of 1 mm into the ventrolateral aspect of the spinal cord. **Lower:** Fluorescence micrographs of RDA-labeled ventral horn neurons that regenerated axons into the PN graft (PNG). Bars = 100 µm.

dase (HRP; Sigma, St. Louis, Mo., USA) and RDA (Dextran, tetramethylrodamine; "Fluoro Ruby"; Molecular Probes, Cat. # D-1817) are primarily taken up by injured axons. They are applied to the proximal stump of a freshly cut PN graft either by a small piece of Gelfoam soaked in the tracer or as a small crystal of the tracer (Figs. 3, 4). Fluorogold (Fluorochrome, Englewood, Col., USA), however, is a fluorescent tracer that is readily taken up both by injured and intact axons (Koliatsos et al. 1994) and can be directly injected into an otherwise intact graft by pressure injection through a fine-tipped glass micropipette.

Tracers such as HRP have been estimated to be retrogradely transported at a rate of 50 to over 100 mm per day (LaVail, 1975) and therefore survival times of 3 to 5 days usually result in adequate neuronal labeling. The tracer should be applied to the graft at least 1 cm from the site of entry of the graft into the CNS to minimize the chance of non-specific neuronal labeling by diffusion of the tracer.

Preparation of Retinal Wholemounts to Examine Retrogradely Labeled Retinal Ganglion Cells (RGCs) from a PN Graft (Fig. 3). Animals are sacrificed 3 to 5 days after application of the tracer to the cut end of a PN graft that had been transplanted to the back of the eye after intraorbital ON transection. Saline and then 4% paraformaldehyde (PF) is perfused intracardially and followed by removal of the experimental eye.

Note: The fixative glutaraldehyde must **not** be used as the retina is very difficult to dissect away from the outer layers of the eye. After incising the cornea, the lens is removed and then four equally spaced radial cuts are made through the full thickness of the eye cup from the sclero-corneal junction towards but not reaching the optic cup. The retina is detached from the eye cup, placed in 4% PF for 1 hour and then washed several times in 0.1 M phosphate buffer (a more detailed description of the preparation of retinal wholemounts can be found in "The Wholemount Handbook" by Stone 1981). The retina is laid vitreous side up on a microscope slide, covered with a coverslip and then viewed under light or fluorescence microscopy, depending on the type of tracer used. If HRP is used as the tracer, the retina must be reacted with an appropriate substrate (i.e. Tetramethylbenzidine; see Mesulam 1982 for details of HRP histochemistry) to produce an HRP reaction product that is visible under light and electron microscopy.

Retrogradely labeled RGC cell bodies are characterized by a nucleus and/or labeled dendrites and can be easily counted (Fig. 3). Following intraorbital ON transection, approximately 0.5 to 3% of the total population of RGCs within each retina of the adult Sprague-Dawley rat (estimated at 110,000; Perry et al. 1983) regenerates axons for at least 2.5 cm into a PN graft (Vidal-Sanz et al. 1987).

Preparation and Examination of the Spinal Cord for Neurons that Regenerated Axons into a PN Graft (Fig. 4). Three to 5 days after tracer application, the animal is perfused intracardially with saline followed by 4% PF, and the spinal cord is dissected out being careful not avulse the PN graft from the cord. After overnight cryoprotection in an 18% sucrose solution, transverse or longitudinal sections of the spinal cord are cut on a cryostat and viewed under fluorescence microscopy for labeled neurons and axons (Fig. 4).

Anterograde Tracing of CNS Axons Regenerating within a PN Graft and into the CNS (Fig. 5)

Tracers injected into CNS nuclei can be transported anterogradely along both intact axons and previously transected axons that have regenerated into a PN graft inserted into the CNS (Cheng et al. 1996). Anterograde tracers include HRP, the plant lectin *Phaseolus vulgaris* Leukoagglutinin (PHA-L; Gerfen and Sawchenko 1984), biotinylated dextran amine (BDA) and some fluorescent tracers such as RDA and rhodamine-B-isothiocyanate (RITC; Thanos et al. 1987).

Anterograde labeling of RGCs does not require injection into the brain but involves injection of tracer into the vitreous body in the posterior chamber of the eye. HRP is commonly used for these purposes as it is endocytosed by mechanically intact RGCs following intravitreal injection and is transported anterogradely to label axons and terminals of both intact and regenerating RGCs (Fig. 5) (Vidal-Sanz et al. 1987; Carter et al. 1989; Zwimpfer et al. 1992).

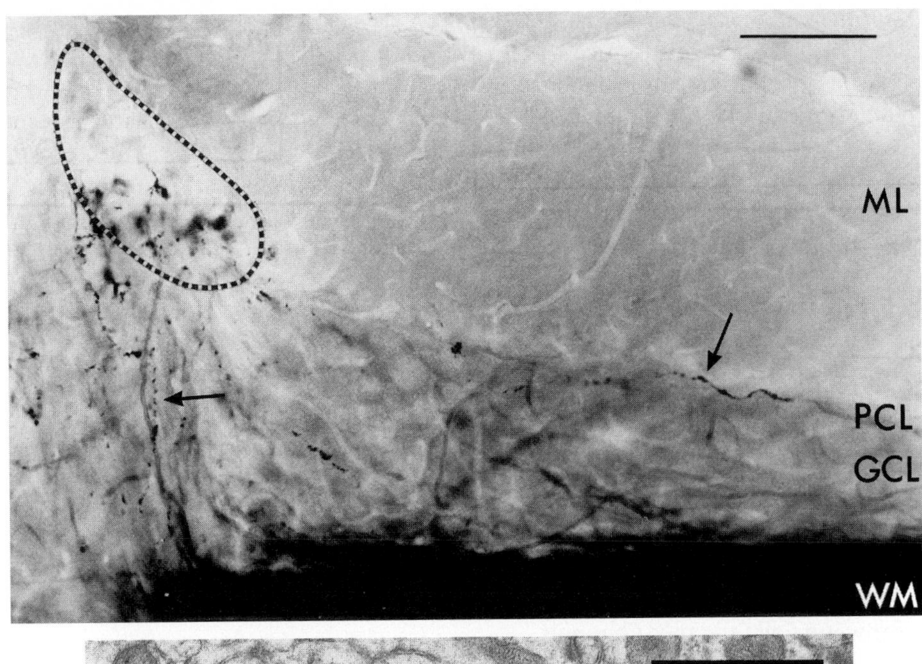

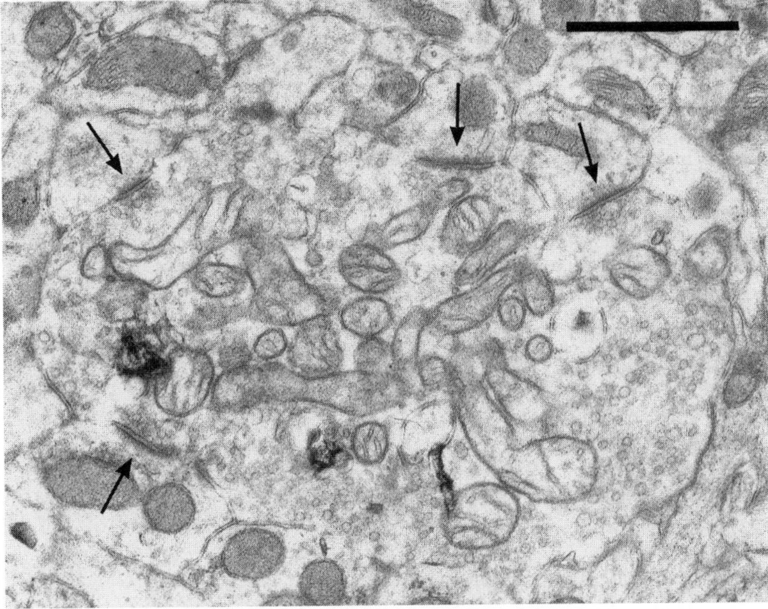

Fig. 5. Anterograde labeling of regenerated RGC axons and axon terminals. The tracer, horseradish peroxidase (HRP), was injected into the vitreous body of the operated eye 3 months after substitution of the optic nerve by a PN graft and insertion of the free end of the PN graft into the cerebellum. HRP is anterogradely transported and labels regenerated RGC axons and axon terminals. **Upper:** Light micrograph. Linearly arranged HRP reaction product (*small black dots*) indicates the RGC axons (*arrows*) that have grown from the end of a PN graft (*outlined by the broken line*) and into the cerebellum. At this plane of focus, HRP-labeled axons were only seen in the ganglion cell layer (GCL) of the cerebellum (ML= molecular layer; PCL= Purkinje cell layer; WM= white matter). Bar = 100 μm. **Lower:** Electron micrograph of a regenerated RGC axon terminal within the GCL. The terminal contains HRP reaction product (*black deposits*) and forms synapses (*arrows*) with dendrites of cerebellar granule cells. Bar = 1 μm. (Used with permission from Zwimpfer et al. 1992).

Intravitreal Injection of Axonal Tracers. A few drops of a mydriatic solution (1% Cyclopentolate Hydrochloride, Alcon) are placed onto the left eye to facilitate visualization of the intraocular contents by funduscopy (see above for details). A small hole is placed into the sclera just adjacent to the corneo-scleral junction using a 26G needle. Five ml of a 30% solution of HRP (Boehringer-Mannheim, Ingelheim, Germany) is injected into the vitreous using a 10-ml Hamilton syringe (Model # 710; Hamilton, Reno, Nev., USA). The scleral opening is closed using gelfoam or bipolar cautery. Survival times as short as 2 days are adequate to label axons and their terminals both at light and electron microscopy.

Examination of HRP-labeled Regenerating RGC Axons within a PN Graft and the Brain. Details of the histological preparation of the PN graft and brain regions into which the distal end of the PN graft was inserted (i.e. superior colliculus or cerebellum) are outlined in Vidal-Sanz et al. (1987), Carter et al. (1989) and Zwimpfer et al. (1992). The HRP reaction product within regenerated RGC axons and axon terminals can be identified under both light microscopy (Fig. 5 top) and EM (Fig. 5 bottom).

Subprotocol 2 Schwann Cell Guidance Channels

Outline

The steps involved in preparation and implantation of a Schwann cell (SC) guidance channel are summarized in Fig. 6.

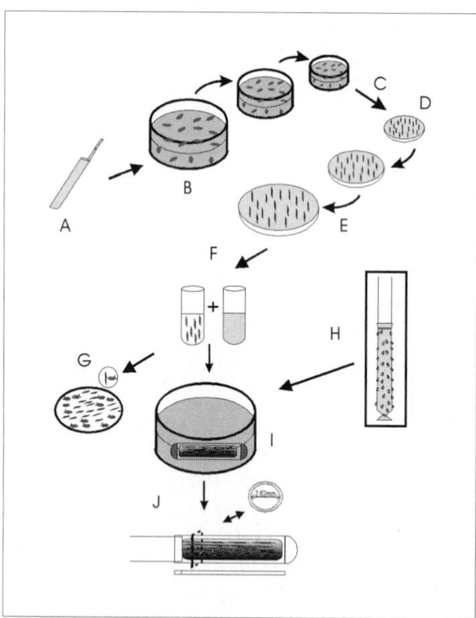

Fig. 6. Basic methods of SC graft preparation within guidance channels. **A.** Peripheral nerves are dissected into their component fascicles stripped of their epineurium and cut into segments 2–3 mm in length and placed in medium. **B.** Explants are allowed to attach to the base of a plastic dish and then moved to new dishes after the fibroblasts (FB) outgrowth becomes confluent, thereby increasing the SC/FB ratio. **C.** Explants are enzymatically dissociated into their component cells and **(D)** plated as monolayers. **E.** The cellular population is expanded and purified by serial plating and growing to confluence. **F.** The expanded cells are dissociated, harvested and mixed with Matrigel to achieve the desired final concentration. **G.** Cell purity is established immunohistochemically. **H.** The guidance channel is cleaned by pressure flushing. **I.** The SC/Matrigel mixture is loaded into a channel, which is kept in medium overnight to allow graft organization/syneresis to occur prior to graft implantation. **J.** Implantation of a closed-ended SC-seeded channel.

Materials

Specialized Materials

- Semipermeable 60:40 polyacrylonitrile/polyvinylchloride (PAN/PVC) copolymer channels (2.6mm i.d., 3.0mm o.d.; 50 kDa MW cutoff) and PAN/PVC glue were obtained as a gift from Dr. Patrick Aebischer (Centre Hospitalier Universitaire Vaudois, Lausanne, Switzerland). Channels are fabricated by conventional spinning techniques as outlined in Cabasso (1980) and Aebischer et al. (1991). Channels are available through Dr. Aebischer or via agreement with Cytotherapeutics, Rhode Island, USA.
- Semipermeable PAN/PVC copolymer channels (2.6mm i.d., 3.0mm o.d.; 50 kDa MW cutoff) and PAN/PVC glue were obtained as a gift from Dr. Patrick Aebischer (see Acknowledgments).
- Mitogen-recombinant heregulin B1 (rHRGß$1_{177-241}$ 10nM)-Genentech-South San Francisco, Calif., USA; cholera toxin (100ng/ml- Sigma, St. Louis, Mo., USA), and forskolin (1µM; Sigma) were also required.
- Matrigel (Collaborative Research, Bedford, Mass., USA)
- Anti-S-100 antibody (1:100; Dakopatts, Glostrup, Denmark). Hoechst dye (5µl, Hoechst 33342; Sigma).
- Neuronal tracers:
  - Retrograde: Fast Blue (Sigma, St. Louis, Mo., USA).
  - Anterograde: PHA-L (*Phaseolus vulgaris* leucoagglutinin; Vector Lab, Burlingame, Calif., USA),
  - RDA (Rhodamine dextran amine, Dextran, tetramethylrodamine; "Fluoro Ruby"; Molecular Probes, Cat. # D-1817)
  - BDA- Biotinylated dextran amine (Molecular Probes, Eugene, Ore., USA)].
- Precision current source for iontophoresis (Trankinetics, CS3).

Procedure

Methods will be divided into three *sections*:
1. Cellular graft construction: Harvesting and *in vitro* expansion of Schwann cells.
2. Preparation of SC grafts within guidance channels and implantation into rats.
3. Anatomical methods to document axonal regeneration.

Cellular Graft Construction: Harvesting and *In Vitro* Expansion of Schwann Cells

SC cultivation is a complex topic and this protocol presents only one of several possible methods. Recent papers describe serum-free conditions that may be superior for SC expansion (Li et al. 1996). However, it is important to point out that our *in vivo* observations were obtained using the cell culture techniques described here. Significant alterations of *in vitro* conditions may alter the *in vivo* function of cells. The derivation of purified cultures of SCs from peripheral nerves involves the following key steps (Fig. 6):
1. Dissection and subdivision of nerve fascicles into tissue units whose viability can be supported by diffusion, (e.g. small explants).
2. Increasing SC/FB (FB = Fibroblast) ratio within explants, i.e. SC purification. Serial explantation is one method to achieve this.
3. Enzymatic dissociation of nerve explants into component cells.
4. Plating of cells onto suitable substrates to obtain purified cellular monolayers.
5. Expansion and further purification of the cellular population by SC mitogenesis and passaging.

6. Harvesting of cells with minimal loss and maximal viability.
7. Characterization of dissociated cells prior to implantation.

Dissection of Nerve Fascicles from Human or Rodent Nerves and Subdivision into Tissue Units Whose Viability Can Be Supported by Diffusion

When using human donor nerve components, consideration must be given to the potential for infectious disease transmission and appropriate precautions maintained. In addition to peripheral nerves, cauda equina roots are an excellent tissue from which to derive SCs. Nerves are harvested within 30 min of aortic clamping and stored in RPMI (GIBCO Laboratories, Grand Island, N.Y., USA) at 4 °C for no more than 24 h before enculturation. Each nerve or cauda equina root is prepared for culture within a sterile laminar flow hood and under a dissecting microscope, according to the protocol of Morrissey et al. (1991). The nerves are washed three times in Liebovitz's L15 medium (GIBCO), the epineurium is stripped off, and individual fascicles are isolated. L15 is advantageous for this tedious work because it maintains a stable pH in room air (versus DMEM). The fascicles are then cut into 2–3 mm segments, placed in 100-mm plastic tissue culture dishes, allowed to attach by using a small initial volume of medium and kept in a humidified atmosphere with 5% CO. The medium is replaced three times per week with DMEM (Dulbecco's Modified Eagle's Medium, GIBCO) with 10% fetal calf serum (FCS; Hyclone Laboratories, Logan, Utah, USA) and penicillin/streptomycin (50 U/ml and 50 µg/ml, respectively). Other methods have been described to increase the SC/FB ratio during this pre-dissociation phase (Casella et al. 1996).

Troubleshooting

It is important to ensure that explants actually contact the plastic surface so that FB outgrowth can occur. If there is too much medium, the explants will tend to tumble and not adhere.

Serial Explantation - Increase in SC/FB Ratio within Explants

Individual explants are transplanted to new dishes after a circumferential monolayer of predominantly FBs is formed, usually at 7–14 days. After three to five such explantations, the nerve explants are largely depleted of FBs and they may be enzymatically dissociated into predominantly SCs according to the protocol of Pleasure et al. (1986).

Enzymatic Dissociation of Nerve Explants into Component Cells

Explants are pooled and placed in 1–2 ml of an enzyme cocktail consisting of 1.25 U/ml dispase (Boehringer Mannheim Biochemicals, Germany), 0.05% collagenase (Worthington Biochemicals, Freehold, N.J., USA) and 15% FCS in DMEM. The explants are left in the enzyme solution overnight and gently triturated the following morning with a flame-narrowed bent-tip glass pipette until individual explants can no longer be recognized. The cells are then centrifuged and washed in L15 and 10% FCS three times and plated onto 100-mm tissue culture dishes coated with 200 g/ml poly-L-lysine (PLL; Sigma, St. Louis, Mo., USA) or another suitable substrate (see D below).

Troubleshooting

Gentle trituration is important as many cells can be irretrievably damaged during this step. A poor cell yield can make expansion a very lengthy process.

Plating of Cells to Obtain Purified Cellular Monolayers

Schwann cells can be cultivated on a number of substrates including: plastic, PLL, collagen and laminin. Some substrates have been shown to support cell division to a greater or lesser extent (see Casella et al. 1996 and Li et al. 1996), and substrate choice is particularly important when culturing human SCs where substrate adherence decreases at higher passage numbers. SCs grow very well on collagen but are dissociated from it with difficulty. They grow well on PLL and are easily dissociated by routine trypsinization. PLL is a good choice for rodent SC culture.

Troubleshooting

Plating density is important. If it is too low, FBs from the dissociated explants will proliferate rapidly and dominate the culture. In addition, the SCs appear less healthy at low density, possibly due to autocrine effects of secreted factors that may be involved. A good initial density is 2×10^6 cells/100mm dish.

Expansion and Further Purification of the Cellular Population

SC expansion is essential to graft preparation as FBs proliferate readily in serum-containing media and can rapidly become the predominant cell type. Thus, specific steps must be taken to support SC division, while FBs are repressed. A number of molecules can be employed *in vitro* to increase the rate of SC mitogenesis. These include: pituitary extract (glial growth factor), PDGF, forskolin, cholera toxin (Levi et al. 1995; Morrissey et al. 1995; Casella et al. 1996; see reviews by Scarpini et al. 1993; Bunge, Fernandez-Valle 1995). Forskolin, an activator of adenylate cyclase, appears to decrease FB division while facilitating the action of some mitogens (Rutkowski et al. 1995). To obtain expanded populations of human SCs, the dissociated cells were exposed to DMEM plus 10% FCS and a triple mitogen cocktail of recombinant heregulin B1 (rHRGß$1_{177-241}$ 10nM), cholera toxin (100ng/ml), and forskolin (1 μM). Serum-free media that can both promote SCs and suppress FB division have been described (Li et al. 1996).

Once the SCs become confluent, they are removed from the PLL-coated culture plates by rinsing twice with Ca^{2+}/Mg^{2+}-free Hank's Balanced Salt Solution (HBSS; GIBCO) followed by trypsin (0.05%) and EDTA (0.02%; Sigma) in HBSS for 5–10min at 37 °C with gentle shaking. The cells are collected and rinsed twice in L15 and 10% FCS before re-plating onto PLL-coated dishes with the above D10-mitogen cocktail. Human SCs in the second or third passage were utilized for grafting, while rodent SCs are often employed at later passages.

Troubleshooting

Limits of expansion for human SCs. After approximately five passages human SCs may appear senescent, cease division and lose substrate adherence. Similar changes may occur with cultured rodent cells but only at much higher passage numbers. These changes may not be seen if defined media are employed (Li et al. 1996).

Schwann cell culture infections. We have never successfully eradicated infection in an SC culture and therefore infected plates are immediately discarded. Both bacterial and fungal contamination may be seen. Attempts to salvage infected cultures increase the risk of complete culture system loss. Infections can be minimized by strict attention to aseptic techniques and frequently cleaning incubators and hoods. We employ media containing both penicillin and streptomycin (see section A).

Harvesting of Cells with Minimal Cell Loss and Maximal Viability

Rigorous conditions for dissociation from the culture substrate (i.e. enzyme concentration and duration of exposure and shaking) can adversely affect cell yield, which is a critical variable in constructing grafts. Viability (checked with Trypan blue exclusion) should exceed 90%.

Troubleshooting

Cell damage may lead to DNA release which can cause cell cohesion and clumping, which makes graft preparation very difficult. The inclusion of a small concentration of DNAase (0.1%; Sigma) may be helpful.

Characterization of Dissociated Cells prior to Implantation

Interpretation of experiments employing cellular transplants requires an accurate understanding of the composition of the transplant population. Visual microscopic assessment is unreliable. Cell-specific immunomarkers such as S-100 for SCs are available. The following method can aid in the discrimination of SCs from FBs.

On the day the guidance channels are filled with the human or nude rat SCs, a small fraction is seeded onto Aclar (Allied Fiber and Plastics, Pottsville, Pa., USA) mini-dishes coated with ammoniated collagen (Kleitman et al. 1998). The next day the cells are immunostained for S-100 (1:100 dilution), an antigen present in SCs but not FBs (Scarpini et al. 1986), to assess the purity of the SC preparation. The cultures are then mounted on slides with a drop of Citifluor (London, UK) containing 5 μM Hoechst dye. The Hoechst dye labels all nuclei in the culture under fluorescence microscopy. SCs are both S-100-positive and Hoechst-positive; FBs (polymorphic flat cells) are Hoechst-positive but S-100-negative. SC purity should range from 90% to 99%.

Preparation of SC Grafts within Guidance Channels and Implantation into Rats (Fig. 6)

Steps:

1. Channel cleaning
2. Forming grafts within channels
3. Implantation
4. Postoperative care

Channels provide an environment within which cellular grafts can organize. There are many types of channels available and their structural characteristics can significantly affect the grafting outcome (reviewed in: Aebischer et al. 1988, 1989, 1990; Valentini 1995). At a minimum they must be biocompatible and nontoxic. The channels used in our studies are created from biocompatible synthetic polymers. They have an inner smooth membrane which contains pores of a regular diameter. These channels were extensively evaluated in animal PNS studies (Guénard et al. 1992; Levi et al. 1994, 1997) prior to their initial employment in the spinal cord by Xu et al. (1995a). We utilize Matrigel to uniformly disperse SCs within the channel, to enhance cable formation, and as a source of extracellular matrix molecules such as laminin and collagen. However, guidance channels with only Matrigel and no SCs supported less regeneration of PNS axons than channels with only saline (Valentini et al. 1987).

Channel Cleaning

Prior to segmentation into 10-mm lengths, the 6–8 cm PAN/PVC channel tubes are cleaned and sterilized as described by Aebischer et al. (1988). This involves pressure flushing of the tube with 0.1 N HCL followed by graded ETOH solutions (95%, 70% and 50%) and finally sterile saline. We employ an infusion pump for flushing with the goal of observing fluid beading at the channel pores (see Fig. 6H).

Troubleshooting

It is very important to clean such tubes thoroughly; however, if an excessive pressure (i.e. rate of infusion) is used, the dimensions of the tube can be distorted by expansion and the inner membrane may be damaged.

Forming Grafts within Channels

SCs were collected by centrifugation after trypsinization as described above and washed in L15. A sample was counted using a hemocytometer. The cell count determines the volume of graft material that can be created. Channel volume is calculated based on length in mm x πr^2 (r = channel radius). We employed final densities of 120×10^6 cells/ml based on previous experiments in the PNS. At this density, 2–3 confluent 100-mm plates are required to fill each 10-mm length of 2.6-mm diameter channel. The cells were gently and thoroughly re-suspended in an ice-cooled (4 °C) 70/30 (v/v) solution of L15/Matrigel and drawn into the prepared channels by controlled suction.

Troubleshooting
This is a crucial step. Firstly, all components to be employed, including plastic dishes, pipettes, cells, media and Matrigel, must be chilled on ice. Matrigel, particularly, must be slowly thawed to 4 °C and not allowed to warm suddenly; if this occurs, it will gel prematurely. Secondly, care must be taken that the cell/Matrigel solution does not undergo excessive evaporation within the laminar flow hood as this could alter the cell density. Thirdly, it is easy to accidentally introduce air bubbles into the Matrigel, which will become persisting inhomogeneities within the graft. In our experience this is the most common cause of poor cable formation. Finally, large numbers of cells may be lost if the graft is placed into media prior to gelation of the Matrigel. To prevent this, the channels are capped at both ends. We employ PAN/PVC copolymer glue for this purpose, but other materials such as fibrin glue have also been used. Grafts are maintained in DMEM/10% FCS with mitogens and penicillin/streptomycin (50 U/ml and 50 (g/ml, respectively) overnight at 37 °C.

It may be important to maintain the grafts in a serum-containing environment prior to implantation to avoid serum withdrawal effects on cellular viability.

Implantation (Fig. 7)
Rats are handled and maintained according to NIH Guidelines for the Care and Use of Laboratory Animals. We employ athymic female nude rats (Harlan Hsd: RH-rnu/rnu; Harlan) weighing 165–185 g for human SC transplantation in order to obviate immune rejection. The rats are anesthetized using a combination of oxygen (0.9 L/min), nitrous oxide (0.4 L/min) and halothane (1.5%). Preoperatively, cefazolin (15mg/100g body weight; Eli Lilly, Indianapolis, Ind., USA) is administered IM to reduce infection. All surgery is performed under sterile conditions, with the rat placed prone and its head held with ear bars.

A T7 - T10 laminectomy is performed and the T8 - T11 cord segments are exposed through a longitudinal dural incision. The T9 - T11 cord segments are removed using a freshly broken razor blade as a scalpel to transect the spinal cord. An effort is made to leave the ventral dura intact as it supports the graft and helps to minimize separation of the cord stumps. Temporary traction is applied to the animal's tail to increase the gap between the cord stumps and facilitate channel implantation.

The distal cord stump is often softer and it is best to first direct the channel over this stump for approximately 1 mm before the proximal stump is lifted and inserted into the channel. Interface apposition is optimal if the preformed graft is slightly compressed by the distal and proximal cord stumps following the release of tail traction. The remaining dura is pulled to wrap the sides of the channel, and a piece of hydrocephalus shunt film (Durafilm; Codman Surlef, Randolph, Mass., USA) or silastic sheet is positioned over the channel and interfaces. This film makes it much easier to access the channel if subsequent tracing procedures are planned and to recover the graft and cord following perfusion. The muscle and skin are then closed in layers.

Troubleshooting
Anesthesia. Close attention must be paid to the anesthetic depth and adequacy of ventilation in paraplegic rats. Mortality is clearly associated with the duration of surgery. The animals are unusually vulnerable to intraoperative hypoventilation and sudden apnea. Spinal deformities and fixation procedures may also reduce the animal's capacity to ventilate. It is important to minimize all forms of retraction that limit respiratory excur-

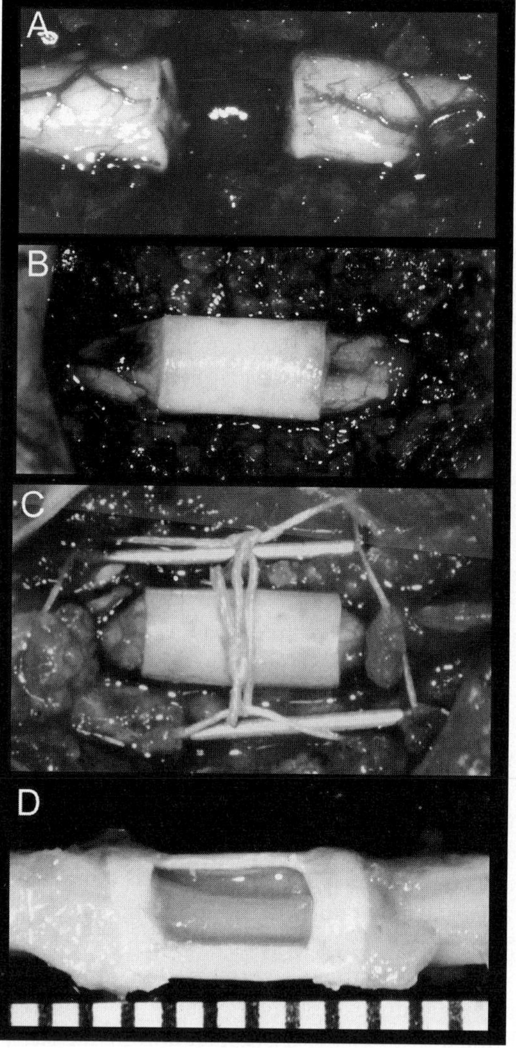

Fig. 7. Spinal cord transection and SC channel implantation. **A.** A 5-mm segment of thoracic cord (T9 - T11) is resected. (Rostral = Left). **B.** An SC-seeded channel is placed between the cord stumps. **C.** Illustrates one technique employed to prevent excess vertebral movement at the grafting site (see De Medicinelli and Wyatt 1993). **D.** The appearance of a perfused human SC cable 35 days following grafting. (Demarcations below the channel = 1 mm).

sions and to lighten anesthesia if necessary. Thermal regulation is grossly abnormal following spinal cord transection and we employ continuous temperature monitoring with feedback-regulated heating pads. Inhalational anesthesia gives a greater degree of flexibility in anesthetizing these fragile animals.

Surgical Technique. If the cord transection is not made with a very sharp instrument, the stumps may swell dramatically. As a result, the stump may be compressed by the channel and undergo secondary necrosis. Decreased ischemia of cord stumps may be one mechanism by which methylprednisolone improves axon regeneration into the SC guidance channels (Chen et al. 1996).

Both the dorsal vein and ventral artery may bleed quite profusely, but local gentle pressure with small pledgets of thrombin-soaked Gelfoam (Upjohn, Kalamazoo, Mich., USA) will control this bleeding. It is important that a hematoma not develop between the spinal cord stump and the graft.

Spinal Deformities. We often noted marked postoperative deformities following cord transection and channel grafting. In some cases the kyphosis and scoliosis were sufficient to disrupt the graft. To prevent this, we have employed previously described spinal column-stabilizing procedures (Fig. 7C; De Medinaceli and Wyatt 1993; Cheng and Olson 1995). These procedures are effective but increase the duration and technical difficulty.

Postoperative Care
Immediately after implantation, rats are returned to their cages and food and water are made easily accessible. A subcutaneous injection of 5 ml lactated Ringer's solution (Kendall McGaw Lab., Irvine, Calif., USA) is administered to compensate for blood loss. Cefazolin (15 mg/100 g) is injected IM twice daily as prophylaxis against urinary tract infections until spontaneous bladder emptying recovers. Bladder expression is performed three times a day for the first week and twice daily thereafter until spontaneous emptying recovers. The rats are quarantined in rooms containing HEPA filters, handled in sterile laminar flow hoods, and fed autoclaved food and water ad libitum.

Troubleshooting
Bladder care is demanding in spinal-transected rodents. They are prone to urinary tract infections, calculi and bladder rupture. We have found that surgical cystostomy can be life-saving and is well tolerated in animals that develop severe urinary obstruction or bladder rupture.

Anatomical Methods to Document Axonal Regeneration

Numerous anatomical methods may be employed to document axonal regeneration following grafting into the injured spinal cord (Fig. 8). In ascending order of complexity, we have employed:

1. Light microscopy (LM)
2. Immunohistochemistry (IHC)
3. Electron Microscopy (EM)
4. Retrograde and Anterograde Neuroanatomical Tracing

The first three methods are well established, widely available and will not be further discussed except to provide some examples and describe some of the sources of difficulty we have experienced. Application of these methods to channel/SC spinal cord grafts has been described (Xu et al. 1995a,b, 1997; Chen et al. 1996; Guest et al. 1997a,b, 1998).

Based on LM, IHC and EM, we have observed that the resulting cables within channels differed from both normal peripheral nerves and CNS tracts in the following ways:

1. The vast majority of axons and fascicles were found at the periphery of the graft (Fig. 9B,C).
2. CNS axons became fasciculated within the SC graft but were disorganized at the spinal cord-SC graft interface (Fig. 10B).
3. CNS axons were myelinated with peripheral myelin (Fig. 9A,C).

Troubleshooting

1. SC/Matrigel grafts are easily damaged or even lost during tissue processing for histology, particularly with sagittal sections for frozen sectioning and immunostaining. Care must be taken to preserve the interface regions to allow meaningful interpretation of results. To achieve this, we pre-embedded the spinal cord stumps, grafts and channels in gelatin using the method described by Oudega et al. (1994).
2. The graft/host interface is often partially disrupted by cysts or frank scar. Because of these inhomogeneities and because a relatively small number of fibers regenerate into and beyond SC grafts, it is important to assess all of the tissue, particularly when working with sagittal sections.

Neuroanatomical Tracing

Figs. 8B, C, 11 and 12). We employ the following methods of axon tracing:

1. Retrograde tracing with Fast Blue, Fluorogold and other Fluoro-labels.
2. Anterograde tracing with PHA-L, RDA, BDA.

HRP, which is widely employed in both retrograde and anterograde tracing, is discussed in the section on peripheral nerve grafting.

Retrograde Tracing

Method: See the following references for details (Gerfen and Sawchenko 1984; Xu et al. 1995a,b; Guest et al. 1998).

The channel is an excellent site for the delivery of retrograde tracers into the graft. A window is cut in the dorsal channel wall to visualize the graft. It can be difficult to estimate the exact region of the graft-host interface since this is often a three-dimensional concave surface (see Fig. 10A). Therefore, we attempt to inject the tracer more than 3 mm from the proximal cord stump. This minimizes the possibility of non-specific retrograde labeling of axons in the proximal stump by diffusion of the tracer. Tracers vary in their diffusion characteristics (Richmond et al. 1994). In addition, appropriate control animals should be included, such as identically grafted animals whose graft cables are severed proximally prior to injection of a retrograde tracer, particularly when using a tracer for the first time.

The scar tissue around PAN/PVC channels prevents accidental leakage of the tracer into the subarachnoid space, while the CSF spaces above and below the graft are not

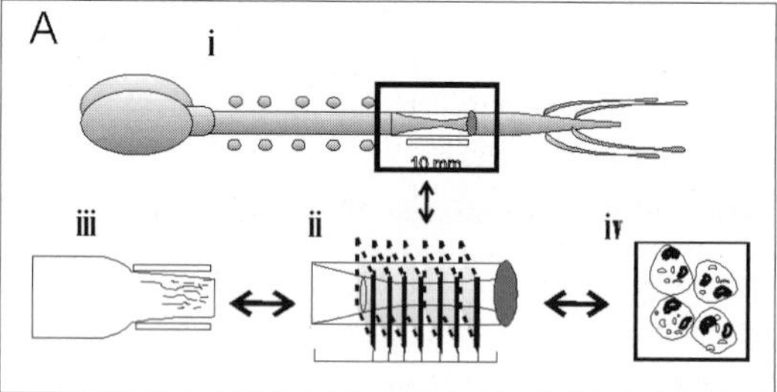

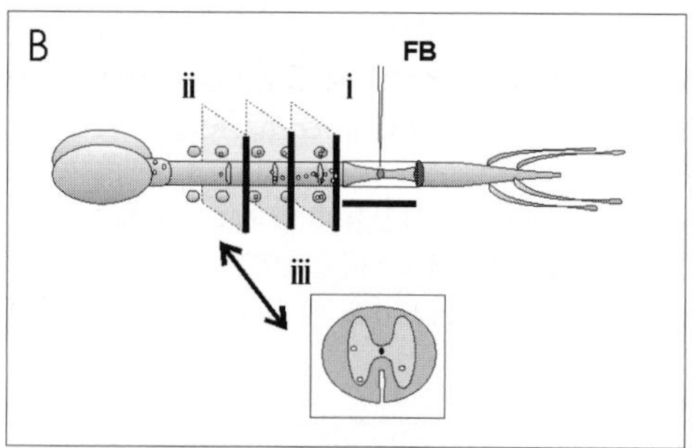

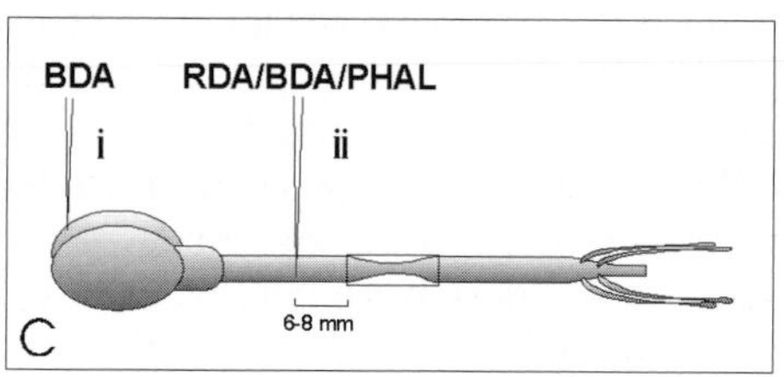

Fig. 8. Methods of histological assessment. **A. i)** Schematic illustration of the rat neural axis with a graft and channel. **ii)** Axial sectioning of a perfused cord and graft for **(iv)** immunohistochemistry, EM and light microscopy. **iii)** The host-graft interface is sectioned sagittally. **B.** Retrograde tracing. **i)** Fast Blue (FB) is injected into the cable of a distally closed-ended channel at least 3mm from the rostral host-graft interface. **ii)** The entire rostral neural axis is sectioned axially and assessed for FB-filled neuronal somata **(iii)**. **C.** Anterograde tracing methods. **i)** Corticospinal (CST) tract tracing with BDA. **ii)** Injection of the anterograde tracers RDA, BDA or PHA-L 6–8mm rostral to the channel to label descending tracts.

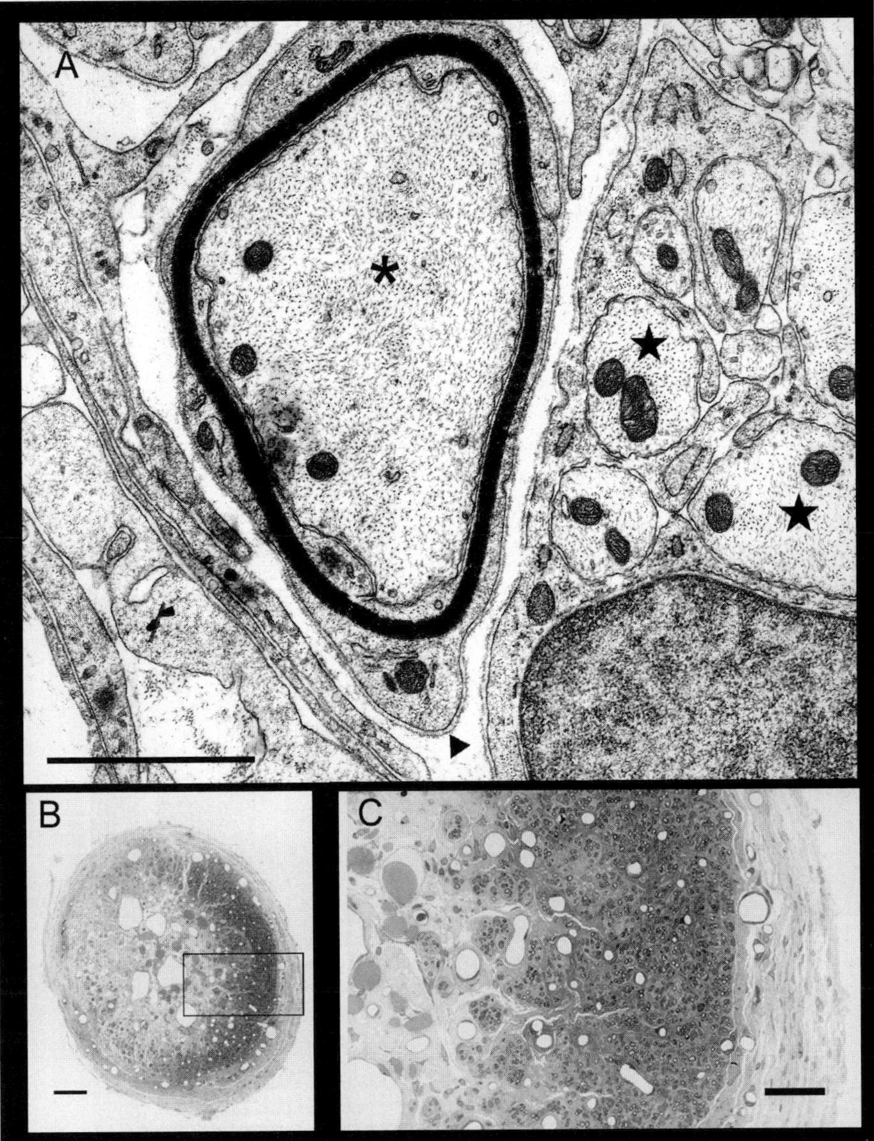

Fig. 9. A. Electron microscopy of human Schwann cell graft. * = axon with peripheral myelin and characteristic basal lamina on the surface membrane (*arrowhead*). Ensheathed non-myelinated axons (*star*). Bar = 1 μm. **B.** Light microscopy. Axial section through a human Schwann cell graft 35 days following implantation. Bar = 100 μm. **C.** High-power photomicrograph of area outline by the box in B. Myelinated axons are found in the periphery of the graft. Bar = 50 μm.

continuous with the channel and graft. Therefore, accidental contamination of the CSF with the tracer substance is unlikely.

Troubleshooting

The graft and cord are easily injured during tracer injections, particularly with animal movement during respiration or tracer injections. To limit movements, temporary mechanical fixation of spinous processes both above and below the site of injection is used. We also apply 0.1% xylocaine to the surface of the spinal cord.

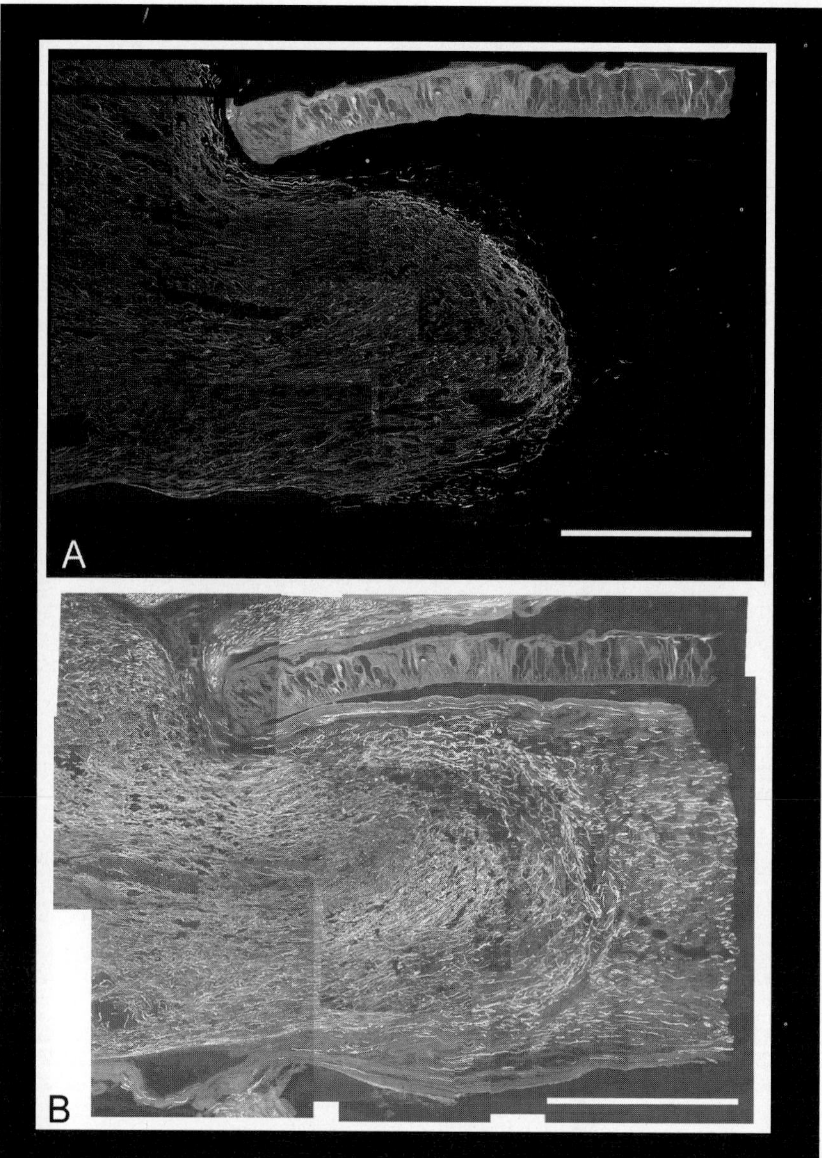

Fig. 10. Immunohistochemistry of a sagittal section of the proximal spinal cord/graft interface (Left = proximal cord). **A.** GFAP staining demarcates the concave shape of the proximal spinal cord stump within the guidance channel. Bars = 1 mm. **B.** Neurofilament staining labels axons within the cord and proximal SC graft.

Assessment of Retrograde Tracing. Axial and sagittal sections of the brainstem and proximal spinal cord reveal retrogradely labeled neuronal cell bodies (Fig. 11B,C).

Anterograde Tracing

This technique allows visualization of regenerated axons and axon terminals. This permits the study of the behavior of the regenerating axon in key regions such as the host/graft interface (see Guest et al. 1997a).

From the Brain. Anterograde tracer injections into the brain are reproducible, generally faster and technically simpler than spinal cord or graft injections. We require ap-

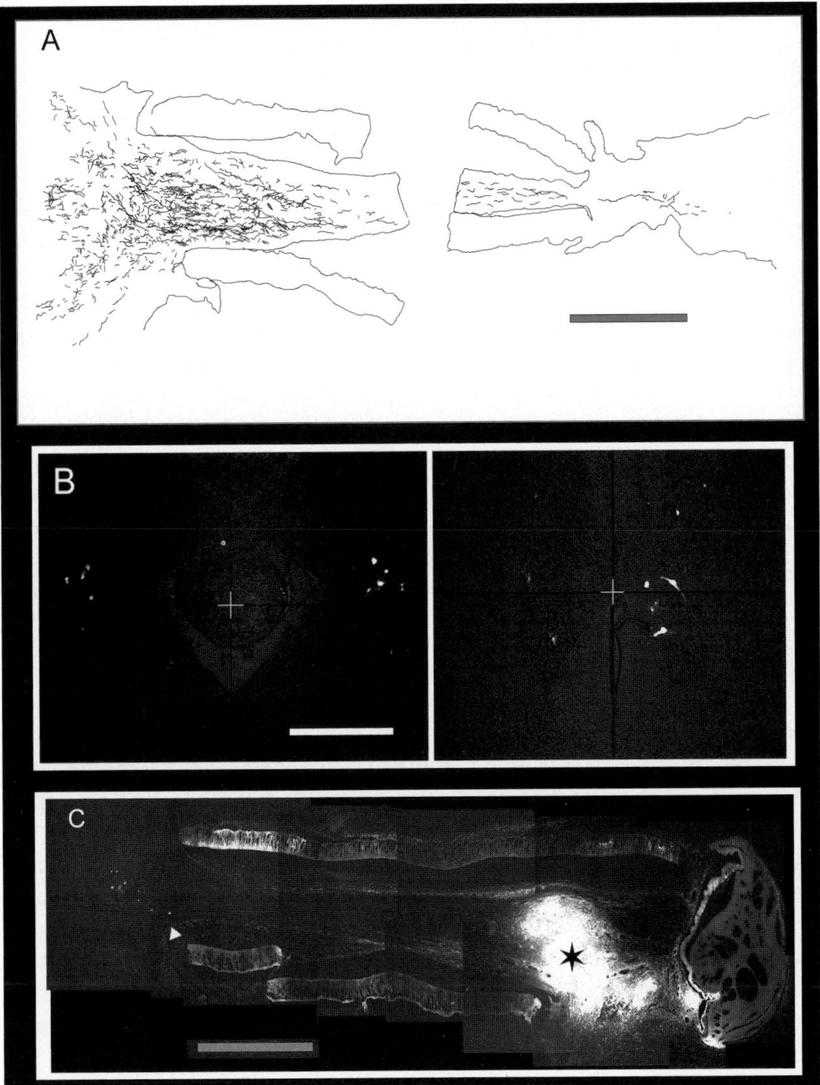

Fig. 11. A. Digitized camera lucida composite of PHA-L anterograde tracing. The tracer was electrophoretically deposited into the spinal grey matter 6–8 mm proximal to the graft. Labeled fibers extend into the proximal graft. A few fibers extend into the distal spinal cord (*far right*) for up to 2.6 mm. Bar = 2.5 mm. See Fig. 12B for corresponding ligth micrograph. **B.** FB-retrogradely labeled neurons in the lateral vestibular nucleus in the brainstem (*left*) and the spinal grey matter (*right*). FB was injected into the distal end of the SC graft (see Fig. 3C). Bar = 2 mm. **C.** Photomontage of a sagittally sectioned human Schwann cell graft and proximal spinal cord (*extreme left*). FB (*) was deposited into the distal end of the graft. The PAN/PVC channel and distal glue cap can be seen. There are some retrogradely labeled neuronal somata (*white arrowhead*) in the spinal cord just proximal to the cord/graft interface. Bar = 2 mm.

proximately 12–14 days for optimum anterograde transport of RDA and BDA from the motor cortex to the distal thoracic spinal cord (see Guest et al. 1997a,b; Xu et al. 1997 for specific details). Other methods of anterograde labeling of the corticospinal tract have been described (Schnell and Schwab 1990, 1993; Li and Raisman 1994).

From the Spinal Cord. Injection of tracers directly into the spinal cord may be performed to either label intrinsic spinal cord neurons (from the grey matter) and spinal cord tracts (i.e. fibers of passage) in the white matter. There is extensive regeneration of intrinsic spinal cord neurons (predominantly propriospinal and motoneurons) into SC grafts (Figs. 11 A, 12A, B; Xu et al. 1995a; Guest et al. 1997b).

Either pressure injections or iontophoretic delivery of tracers may be employed and it is important that the animal's spinal column be immobilized as described in the section on retrograde tracing. Pressure injections are a reliable method but create a region of tissue injury. The size of injury depends on the volume and rate of injection, tracer toxicity, and animal movement during injection. Iontophoresis requires precisely fashioned needles, longer injection times and specialized equipment but causes minimal tissue trauma. To optimize the balance between the proportion of fibers labeled and the injurious effects of tracer injection, we have combined small initial pressure injections with subsequent iontophoresis ("pressure-iontophoresis"; Guest et al. 1997a,b). Large injection sites, with minimal trauma, were obtained with RDA, BDA or PHA-L.

Troubleshooting

Anterograde tracing from the rostral thoracic spinal cord to assess regeneration of propriospinal neurons is technically difficult in the rat as there is a large vein that runs from the cervicothoracic fat pad into the epidural venous system at T4 or T5. If this vein is injured, severe hemorrhage can occur (De Medinaceli 1986).

Acknowledgement:

Peripheral Nerve Grafts

PN grafting studies were performed at the Montreal General Hospital Research Centre (McGill University) and the University of British Columbia (Collaboration on Repair Discoveries and the Department of Surgery). TJZ was supported by the Medical Research Council of Canada (Fellowship and Clinician-Scientist Award), the Vancouver Hospital and Health Sciences Centre, B.C. Health Research Foundation and the Neuroscience Network of Canada. Dr. M. Vidal-Sanz first described substitution of the optic nerve by a peripheral nerve graft. Special thanks to Drs. M. Vidal-Sanz and D. Carter for technical instruction. The following individuals provided surgical and technical assistance in the studies from which the figures in this chapter were taken: M. David, J. Liu, S. Shinn, K. Stilwell, C. Tarazi and W. Wilcox.

Schwann Cell Implants

Human SC work was performed at the Miami Project to Cure Paralysis, University of Miami School of Medicine. The methods of purification and expansion of human SCs were developed under the direction of Drs. R.P. (deceased) and M.B. Bunge with primary involvement of Drs. A. Levi, N. Kleitman, T. Morissey, P. Wood and G. Casella. PAN/PVC guidance channels were generously provided by Dr. Patrick Aebischer (Centre Hospitalier Universitaire Vaudois, Lausanne, Switzerland). Drs. A. Levi and V. Guénard performed the initial experiments employing SC grafts within PAN/PVC channels in the PNS. Dr. X.M. Xu carried out the first studies with rodent SC grafts in the CNS. Human peripheral nerves were graciously provided by L. Olsen and the University of Miami transplant team. Technical assistance of the following individuals is acknowledged: K. Akong, M. Bates, J.P. Brunschwig, R. Camarena, D. Hesse, I. Margitich, B. Puckett, A. Rao, D. Santiago and C. Vargas. Funding was provided by NIH grants NS28059, NS09923, the Hollfelder foundation, the Miami Project and the American Association of Neurological Surgeons Research Foundation.

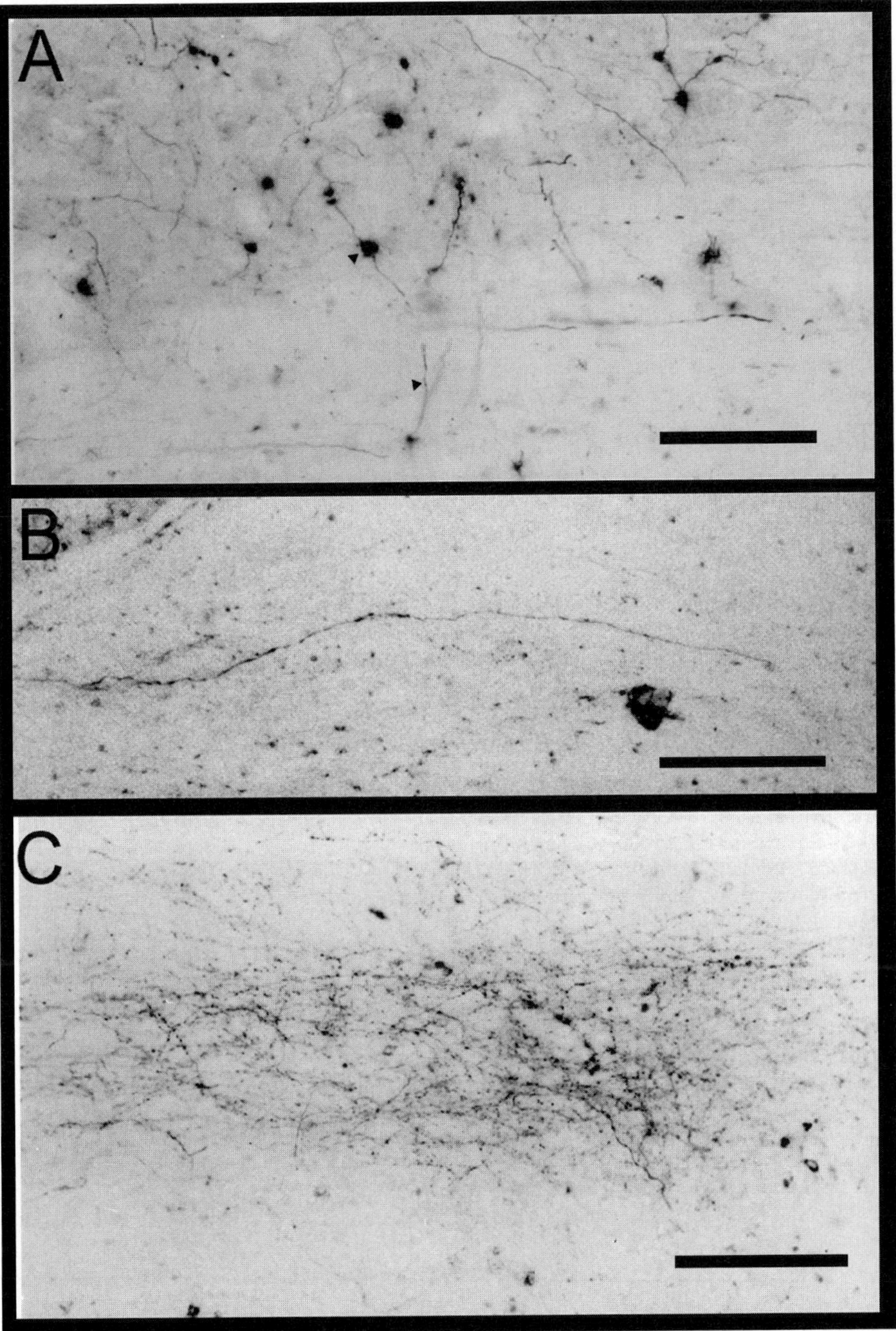

Fig. 12. Anterograde tracing. **A.** PHA-L labeling in the spinal grey matter proximal to a human Schwann cell graft. Labeled somata extend processes that run proximally and distally in the white matter, suggesting that they are propriospinal neurons. Bar = 100 µm. **B.** A PHA-L-labeled fiber extends into the spinal cord distal to a human Schwann cell graft. Bar = 150 µm. PHA-L was injected into the spinal grey matter 6–8 mm proximal to the graft. See Fig. 11A for corresponding camera lucida. **C.** Distal termination of BDA-labeled corticospinal (CST) axons within the stump of the cord proximal to a human Schwann cell graft + aFGF-impregnated fibrin glue. BDA was injected into the motor cortex to anterogradely label the CST tract. No CST axons entered the SC graft. Bar = 150 µm.

References

Aebischer P, Guénard V, Winn SR, Valentini RF, Galletti PM (1988) Blind-ended semipermeable guidance channels support peripheral nerve regeneration in the absence of a distal nerve stump. Brain Res 454:179–187.

Aebischer P, Guénard V, Brace S (1989) Peripheral nerve regeneration through blind-ended semipermeable guidance channels: effect of the molecular weight cutoff. J Neurosci 9:3590–3595.

Aebischer P, Guénard V, Valentini RF (1990) The morphology of regenerating peripheral nerves is modulated by the surface microgeometry of polymeric guidance channels. Brain Res 531:211–218.

Aebischer P, Wahlberg L, Tresco PA, Winn SR (1991) Macroencapsulation of dopamine-secreting cells by coextrusion with an organic solution. Biomaterials 12:50–56.

Aguayo A.J 1985. Axonal regeneration from injured neurons in the adult mammalian central nervous system. In Synaptic Plasticity (C.W. Cotman Ed.), pp.457–483. Guilford Press, London.

Blakemore W, Crang A (1985) The use of cultured autologous Schwann cells to remyelinate areas of persistent demyelination in the central nervous system. J Neurological Sci 70: 207–223.

Bovolenta P, Wandosell F, Nieto-Sampedro F 1993. Neurite outgrowth inhibitors associated with glial cells and glial cell lines. Neuroreport 13:345–348.

Bradley W, Asbury AK (1970) Duration of synthesis phase in neurilemma cells in mouse sciatic nerve during degeneration. Exp Neurol 26:275–282.

Bregman BS, Kunkel-Bagden E, Schnell L, Dal HN, Gao D, Schwab ME (1995) Recovery from spinal cord injury mediated by antibodies to neurite growth inhibitors. Nature 378:498–501.

Bunge MB Kleitman N (1997) Schwann cells as facilitators of axonal regeneration in CNS fiber tracts. Chapter 31:319–333 in Cell Biology and Pathology of Myelin eds. Juurlink et al.., Plenum Press, NY

Bunge RP, Fernandez-Valle (1995) Basic biology of the Schwann cell. In: Neuroglia. Kettenmann H, Ransom BR (eds.), Oxford Univ. Press, NY, pp 44–57.

Cabasso F (1980) Hollow fiber membranes, In: Encyclopedia of Chemical Technology, Kirk- Othner (ed.), New York; Jon Wiley and Sons, 12: 492–517.

Carbonetto S, David S (1993) Adhesive molecules of the cell surface and extracellular matrix in neural regeneration. In Neuroregeneration (A. Gorio, Ed.), pp. 77–100. Raven Press, New York.

Carlstedt T, Linda H, Cullheim S, Risling M (1986) Reinnervation of hind limb muscles after ventral root avulsion and implantation in the lumbar spinal cord of the adult rat. Acta Physiol Scand 128:645–646

Carter DA, Bray GM, Aguayo AJ (1989) Regenerated retinal ganglion cell axons can form well-differentiated synapses in the superior colliculus of adult hamsters. J Neurosci 9:4042–4050.

Casella GTB, Bunge RP, Wood PM (1996) Improved method for harvesting human Schwann cells from mature peripheral nerve and expansion in vitro. Glia 17:327–338.

Chen A, Xu XM, Kleitman N, Bunge MB (1996) Methylprednisolone administration improves axonal regeneration into Schwann cell grafts in thoracic rat spinal cord. Exp Neurol 138:261–276.

Cheng H, Olson L (1995) A new surgical technique that allows proximodistal regeneration of 5-HT fibers after complete transection of the rat spinal cord. Exp Neurol 136:149–161.

Cheng H, Cao Y and Olson L (1996) Spinal cord repair in adult paraplegic rats: Partial restoration of hind limb function. Science 273:510–513.

David S, Aguayo AJ (1981) Axonal elongation in PNS "bridges" after CNS injury in adult rats. Science 214:931–933.

De Medinaceli, L (1986). Research Note: An anatomical landmark for procedures on rat thoracic spinal cord. Exp Neurol 91:404–8. 91: 404–408.

De Medinaceli L, Wyatt R (1993) A method for shortening of the rat spine and its neurologic consequences. J. Neural Transpl and Plast. 4(1): 39–52.

Doucette R (1995) Olfactory ensheathing cells: potential for glial cell transplantation into areas of CNS injury. Histol and Histopath 10:503–507.

Ellis JC, McCaffery TV (1985) Nerve grafting. Functional results after primary vs delayed repair. Arch Otolaryngol 111:781–785.

Gaze RM, Keating MJ (1970) The restoration of the ipsilateral visual projection following regeneration of the optic nerve in the frog. Brain Res. 21:207–216.

Gerfen CR and Sawchenko PE (1984) An anterograde neuroanatomical tracing method that shows the detailed morphology of neurons, their axons and terminals: immunohistochemical localization of an axonally transported plant lectin *Phaseolus vulgaris* Leukoagglutinin (PHA-L). Brain Res 290:219–238.

Greene EC (1963) Anatomy of the rat. Translations of the American Philosophical Society. Vol. 27, Hafner Publishing, New York.

Guénard V, Xu Xao Ming, Bunge MB (1993) The use of Schwann cell transplantation to foster central nervous system repair. Sem In Neurosci 5:401–411.

Guénard V, Aebischer P, Bunge R (1994) The astrocyte inhibition of peripheral nerve regeneration is reversed by Schwann cells. Exp Neurol 126:44–60.

Guénard V, Kleitman N, Morrissey TK, Bunge RP, Aebischer P (1992) Syngeneic Schwann cells derived from adult nerves seeded in semipermeable guidance channels enhance peripheral nerve regeneration. J Neurosci 12:3310–3320.

Guest JD, Hesse D, Schnell L, Schwab ME, Bunge MB, Bunge RP (1997a) Influence of IN-1 antibody and acidic FGF-Fibrin glue on the response of injured corticospinal tract axons to human Schwann cell grafts. J Neurosci Res 50:888–905.

Guest JD, Rao A, Olson L, Bunge MB, Bunge RP (1997b) The ability of human Schwann cell grafts to promote regeneration in the transected nude rat spinal cord. Exp Neurol 148(2):502–522.

Guest J, Aebischer P, Akong, K, Bunge M, Bunge R (1998) Human Schwann cell transplants in transected nude rat spinal cord: Graft survival, axonal regeneration, and myelination. Submitted to J Neuroscience.

Heumann R, Korching S, Bandtlow C, Thoenen H (1987) Changes of nerve growth factor synthesis in nonneuronal cells in response to sciatic nerve transection. J Cell Biol 104:1623–1631.

Honmou O, Felts P, Waxman S, Kocsis J (1996) Restoration of normal conduction properties in demyelinated spinal cord axons in the adult rat by transplantation of exogenous Schwann cells. J Neurosci 16:3199–3208. 16(10): 3199–3208.

Jenkins R, Tetzlaff, Hunt SP (1993). Differential expression of immediate early genes in rubrospinal neurons following axotomy in rat. Eur J Neurosci 5:203–209.

Keirstead HS, Dyer JK, Sholomenko GN, McGraw J, Delaney KR, Steeves JD (1995). Axonal regeneration and physiological activity following transection and immunological disruption of myelin within the hatchling chick spinal cord. J Neurosci 15:6963–6974.

Keirstead SA, Rasminsky M, Fukuda Y, Carter DA, Aguayo AJ, Vidal-Sanz M (1989) Electrophysiologic responses in hamster superior colliculus evoked by regenerating retinal axons. Science 246:255–257.

Kim D, Connolly S, Kline D, Voorhies R, Smith A, Powell M, Yoes T, et al. (1994) Labeled Schwann Cell Transplants Versus Sural Nerve Grafts in Nerve Repair. J Neurosurg 80: 254–260.

Kleitman N, Wood P, Bunge R (1998) Tissue culture methods for the study of myelination. Neuronal Cell Culture, 2nd ed. G. Banker and K. Goslin. Boston, MA, MIT Press (In press).

Kobayashi NR, Fan D, Giehl KM, Bedard AM, Wiegand SJ, Tetzlaff W (1997) BDNF and NT-4/5 prevent atrophy of rat rubrospinal neurons after cervical axotomy, stimulate GAP-43 and Tα1-Tubulin mRNA expression, and promote axonal regeneration. J Neurosci 17:9583–9595.

Koliatsos VE, Price WI, Pardo CA, Price DL (1994) Ventral root avulsion: An experimental model of death of adult motor neurons. J Comp Neurol 342:35–44.

Kromer, L, C Cornbrooks (1985) Transplants of Schwann cell cultures promote axonal regeneration in the adult mammalian brain. Proc. Natl. Acad. Sci. USA. 82:6330–6334.

LaVail, JH (1975) Retrograde cell degeneration and retrograde transport techniques. In: The use of axonal transport for studies of neuronal connectivity. Eds: WM Cowan and M Cuenod), pp. 217–248, Elsevier, Amsterdam.

Levi A, Bunge R (1994) Studies of Myelin Formation after Transplantation of Human Schwann Cells into the Severe Combined Immunodeficient Mouse. Experimental Neurology 130: 41–52.

Levi A, Guénard V, Aebischer P, Bunge R (1994) The functional characteristics of Schwann cells cultured from human peripheral nerve after transplantation into a gap within the rat sciatic nerve. J Neurosci 14:1309–1319

Levi ADO, Sliwkowski MX, Lofgren J, Hefti F, Bunge RP (1995) The influence of heregulins on human Schwann cell proliferation. J Neurosci 15:1329–1340.

Levi A, Sonntag V, Dickman C, Mather J, Li R, Cordoba S, Bichard B, et al. (1997) The role of cultured Schwann cell grafts in the repair of gaps within the peripheral nervous system of primates. Exp Neurol 143:25–36. 143:25–36.

Lewin-Kowalik J, Sieron AL, Krause M, Kwiek S (1990) Predegenerated peripheal nerve grafts facilitate neurite outgrowth from the hippocampus. Brain Res Bull 25:669–673.

Lewin-Kowalik J, Sieron AL, Krause M, Barski JJ, Gorka D (1992) Time-dependent regenerative influence of predegenerated nerve grafts on hippocampus. Brain Res Bull 29:831–835.

Li R-h, Slikowski M, Lo J, Mather J (1996) Establishment of Schwann cell lines from normal adult and embryonic rat dorsal root ganglia. J Neurosci Meth 67:57–69.

Li Y, Raisman G (1994) Schwann cells induce sprouting in motor and sensory axons in the adult spinal cord. J Neurosci 14:4050–4053.

Li Y, Field PR and Raisman G (1997) Repair of adult rat corticospinal tract by transplants of olfactory ensheathing cells. Science 277:2000–2002.

Liuzzi FJ, Tedeschi (1992) Axo-glial interactions at the dorsal root transitional zone regulate neurofilament protein synthesis in axotomized sensory neurons. J Neurosci 12:4783–4792.

Menei P, Montero-Menei C, Whittemore S, Bunge R, Bunge M (1997) Schwann cells genetically modified to secrete human BDNF promote enhanced axonal regrowth across adult rat spinal cord. Eur J of Neurosci 10: 607–621

Mesulam, M.M. (1982) Tracing neural connections with horseradish peroxidase. J. Wiley and Sons, New York, pp.12–20.

Meyer M, Matuoka I, Wetmore C, Olsen L and Thoenen H (1992) Enhanced synthesis of brain-derived neurotrophic factor in the lesioned peripheral nerve: different mechanisms are responsible for the regulation of BDNF and NGF mRNA, J Cell Biol 199:143–153.

Montero-Menei, C, A Pouplard-Barthelaix, M Gumpel, A Baron-Van Evercooren (1992) Pure Schwann cell suspension grafts promote regeneration of the lesioned septo-hippocampal cholinergic pathway. Brain Research 570:198–208.

Morrissey TKM, Kleitman N, Bunge RP (1991) Isolation and functional characterization of Schwann cells derived from adult peripheral nerve. J Neurosci 11(8):2433–2442.

Morrissey TK, Levi ADO, Nuijens A, Sliwkowski MX, Bunge RP (1995) Axon-induced mitogenesis of human Schwann cells involves heregulin and p185 erbB2. Proc Natl Acad Sci 92:1431–1435.

Neuberger T, Cornbrooks C, K LF (1992) Effects of delayed transplantation of cultured Schwann cells on axonal regeneration from central nervous system cholinergic neurons. J Comp Neurol 315:16–33.

Ng TF, So K, Chung SK (1995) Influence of peripheral nerve grafts on the expression of GAP-43 in regenerating retinal ganglion cells in adult hamsters. J. Neurocyt. 24:487–496.

Oudega M, Varon S and Hagg T (1994) Regeneration of adult rat sensory axons into intraspinal nerve grafts: promoting effects of conditioning lesion and graft predegeneration. Exp Neurol 129:194–206.

Oudega M and Hagg T (1996) Nerve growth factor promotes regeneration of sensory axons into adult rat spinal cord. Exp Neurol 140:218–229.

Perry VH, Henderson Z, Linden R (1983) Postnatal changes in retinal ganglion cell and optic axon populations in the pigmented rat. J Comp Neurol 219:356–368.

Pindzola RR, Doller C, Silver J (1993) Putative inhibitory extracellular matrix molecules at the dorsal root entry zone during development and after root and sciatic nerve lesions. Dev Biol 156: 34–48.

Pleasure D, Kreider B, Sobue G, Ross AH, Koprowski H, Sonnenfeld KH, Rubenstein AE (1986) Schwann-like cells cultured from human dermal fibromas. Ann NY Acad Sci 486:227–240.

Ramon-Cueto A and Nieto-Sampedro M (1994) Regeneration into the spinal cord of transected dorsal root axons is promoted by ensheathing glia transplants. Exp Neurol 127:323–244.

Ramon-Cueto A and Valverde F (1995) Olfactory bulb ensheathing glia: A unique cell type with axonal growth-promoting properties. Glia 14:163–173

Ramon-Cueto, A, G Plant, J Avila, M Bunge (1998). Long-distance axonal regeneration in the transected adult rat spinal cord is promoted by olfactory ensheathing glia transplants. J Neurosci 18(10): 3803–3815.

Rawson, C, S Shirahata, T Natsuno, D Barnes (1991). Oncogene transformation frequency of senscent SFME is increased by c-myc. Oncogene 6: 487–489.

Richardson, P.M., U.M. McGuiness, and A.J. Aguayo. 1980. Axons from CNS neurons regenerate into CNS grafts. Nature Lond. 284:264–265.

Richardson PM, Issa VMK, Shemie S (1982) Regeneration and retrograde degeneration of axons in the rat optic nerve. J Neurocytol 11:949–966.

Richmond FJR, Creasy JL, Kitamura S, Smits E (1994) Efficacy of seven retrograde tracers, compared in multiple-labelling studies of feline motoneurons. J Neursoci Meth 53:35–46.

Rutkowski JC, Kirk M, Lerner G, Tennekoon (1995) Purification and expansion of human Schwann cells in vitro. Nature Medicine 1:80–83.

Scarpini EG, Meola G, Baron P, Beretta S, Velicogna M, Scarlato G (1986) S-100 protein and laminin: Immunocytochemical markers for human Schwann cells *in vitro*. Exp Neurol 93:77–83.

Scarpini E, Baron P, Scarlato G, Eds (1993) Human Schwann Cells in Culture. Neurogeneration. New York, Raven Press.

Schaden H, Stuermer CAO, Bahr M (1994) GAP-43 immunoreactivity and axon regeneration in retinal ganglion cells of the rat. J Neurobiol 25:1570–1578.

Schnell L, Schwab M (1990) Axonal regeneration in the rat spinal cord produced by an antibody against myelin-associated neurite growth inhibitors. Nature 343:269–272

Schnell L, Schwab ME (1993) Sprouting and regeneration of lesioned corticospinal tract fibers in the adult rat spinal cord. Eur. J. Neurosci. 5:1156–1171.

Sharma SC, Jadhao AG, Prasada Rao PD (1993) Regeneration of supraspinal projection neurons in the adult goldfish. Brain Res. 620:221–228.

Smale KA, Doucette R and Kawaja MD (1996) Implantation of olfactory ensheathing cells in the adult rat brain following fimbria-fornix transection. Exp Neurol 137:225–233.

Smith GV, Stenvenson JA (1988) Peripheral nerve grafts lacking viable Schwann cells fail to support central nervous system axonal regeneration. Exp Brain Res. 69:299–306.

Stewart H, Eccleston P, Jessen K, Mirsky (1991) Interaction between cAMP elevation, identified growth factors, and serum components in regulating Schwann cell growth. J Neurosci Res 30: 346–352.

Stichel C Lips K, Wunderlich G, Muller H (1996) Reconstruction of transected postcommissural fornix in adult rat by Schwann cell suspension grafts. Exp. Neurol 140: 21–36.

Stone J. (1981) The wholemount handbook: a guide to the preparation and analysis of retinal wholemounts. Maitland, Sydney.

Thanos S, Vidal-Sanz M, Aguayo AJ (1987) The use of rhodamine-B-isothiocyanate (RITC) as an anterograde and retrograde tracer in the adult visual system. Brain Res 406:317–321.

Tetzlaff W, Alexander SW, Miller FD, Bisby MA (1991) Response of facial and rubrospinal neurons to axotomy: changes in mRNA expression for cytoskeletal proteins and GAP-43. J. Neurosci. 11:2528–2544.

Valentini R (1995) Nerve Guidance Channels. The Biomedical Engineering Handbook. B. JD, CRC Press: 1985–1996.

Valentini RF, AP, Winn SR, Galletti PM (1987) Collagen- and laminin- containing gels impede peripheral nerve regeneration through semipermeable nerve guidance channels. Exp Neurol 98:350–356.

Vidal-Sanz M, Bray GM, Villegas-Perez MP, Thanos S, Aguayo AJ (1987) Axonal regeneration and synapse formation in the superior colliculus by retinal ganglion cells in the adult rat. J. Neurosci. 7:2894–2909.

Vidal-Sanz M (1990) Regenerative responses of injured adult rat retinal ganglion cells: Axonal elongation, synapse formation and persistence of connections. PhD thesis, McGill University.

Xu XM, Guénard V, Kleitman N, Bunge MB (1995a) Axonal regeneration into Schwann cell-seeded guidance channels grafted into transected adult rat spinal cord. J Comp Neurol 351:145–160.

Xu XM, Guénard V, Kleitman N, Aebischer P, Bunge MB (1995b) A combination of BDNF and NT-3 promotes supraspinal axonal regeneration into Schwann cell grafts in adult rat thoracic spinal cord. Exp Neurol 134:261–272.

Xu, X, Chen A, Guénard V, Kleitman N, Bunge M (1997) Bridging Schwann cell transplants promote axonal regeneration from both the rostral and caudal stumps of transected adult rat spinal cord. J Neurocytol 26:1–16.

Zhao Q, Kerns JM (1994) Effects of predegeneration on nerve regeneration through silicone Y-chambers Brain Res 633:97–104.

Zwimpfer TJ, Aguayo AJ, Bray GM (1992) Synapse formation and preferential distribution in the granule cell layer by regenerating retinal ganglion cell axons guided to the cerebellum of adult hamsters. J Neurosci 12:1144–1159.

Chapter 14

Cell and Tissue Transplantation in the Rodent CNS

Klas Wictorin, Martin Olsson, Kenneth Campbell and Rosemary Fricker

Introduction

Transplantation of cells/tissue to the central nervous system (CNS) is a widely used experimental tool for studies of developmental and regenerative processes (for reviews, see in Dunnett and Björklund (eds) (1994), and Gaiano and Fishell (1998)). Transplantation is also of major interest with regard to neurodegenerative diseases as well as to several other CNS disorders, since experiments in related animal models have shown that grafts can ameliorate lesion-induced deficits. Depending on the characteristics of the disease model and of the implanted cells/tissue, such graft-derived functional effects have been shown to be mediated through different mechanisms, such as partial reconstruction of damaged circuitries, diffuse release of substances, trophic support and stimulation of regenerative processes (Dunnett and Björklund 1994). Indeed, transplantation is today also applied in clinical trials in patients with, for example, Parkinson´s disease or Huntington´s chorea (Lindvall 1994, Kopyov et al 1998, and in Freeman and Widner (eds) 1998).

This chapter focuses on the use of transplantation to the CNS in experimental rodent studies, and presents protocols for preparing cells/tissue from different regions of the embryonic CNS, for implantation into either adult, neonatal or embryonic recipients. Apart from directly dissected, primary embryonic or fetal tissue, other forms of tissue or cells, such as isolated and propagated stem cells (see Chapter 11) or genetically modified cells from the CNS or from other regions are also being used for grafting into the CNS, for instance in applications of *ex vivo* gene transfer (Kawaja et al, 1992) . Furthermore, Chapter 13 in this volume describes the use of the transplantation technique in studies of axonal growth.

The first experiment with neural transplantation into the brain was reported in 1890 by Thompson, who attempted to move large pieces of neocortical tissue between adult cats and dogs. However, from what is known today about the importance of the age of the donor and of species-differences, this approach most probably gave poor results. Among the first reports on successful grafting into the CNS were those of Ranson (1909), who implanted spinal ganglia into the cerebral cortex of developing rats, and Dunn (1917), who reported survival of implanted neonatal cortex into the cortex of ne-

Klas Wictorin, Lund University, Dept Physiological Sciences, Wallenberg Neuroscience Center, Sölvegatan 17, Lund, 223 62, Sweden (phone: +46-46-2220564; fax: +46-46-2220561)
Martin Olsson, Lund University, Dept Physiological Sciences, Wallenberg Neuroscience Center, Sölvegatan 17, Lund, 223 62, Sweden (phone: +46-46-2220564; fax: +46-46-2220561)
Kenneth Campbell, Lund University, Dept Physiological and Sciences, Wallenberg Neuroscience Center, Sölvegatan 17, Lund, 223 62, Sweden (phone: +46-46-2220564; fax: +46-46-2220561; e-mail: kenny@ biogen.wblab.lu.se)
Rosemary Fricker, Harvard Medical School, Children´s Hospital, Dept Neuroscience, 350 Enders Bldg., 320 Longwood Avenue, Boston, MA, 02115, USA (phone: +01-617-355-721-3526; fax: +01-617-355-3636; e-mail: fricker_r@hub.tch.harvard.edu)

onatal recipients. Furthermore, LeGros Clark (1940) was the first to describe successful grafting of embryonic CNS tissue into neonatal recipients (cortex into cortex). Altogether, these and other early investigators documented the general feasibility of neural grafting, and made early observations on, for instance, the neurotrophic effects of grafted CNS tissue, the importance of the age of the donor tissue and the recipient, and the effects of different implantation sites (see Björklund and Stenevi 1985, for review on the history of neural transplantation). The recent history of this field started around thirty years ago, when new autoradiographic and histochemical methods were applied, and thus Das and colleagues (Das and Altman 1971, 1972) studied the survival of fetal neural tissue implanted into neonatal brain, Olson and coworkers (Olson and Malmfors 1970, Olson and Seiger 1972) used the anterior eye chamber as a site to study the development and growth of neural tissue implants, and Björklund and colleagues (Björklund and Stenevi 1971, Björklund et al 1971) investigated sprouting of central monoaminergic fibers, and growth of these fibers into intracerebral grafts of smooth muscle. Since then, the field of transplantation into the CNS has become a gradually more and more intense area of modern neurobiological research.

It is clearly beyond the scope of this methodological chapter to give a full account of all different aspects of the field of transplantation into the CNS. For instance, the transplantation approach is widely used for developmental studies also in species other than rodents (for a recent review, see Gaiano and Fishell 1998). The references given in the protocols below are merely illustrations and present some selected examples of the application of the transplantation technique. For each of the tissue types and different animal models described, there exists a large and constantly growing body of literature. For more extensive reviews, the readers are advised to consult, for instance, Dunnett and Richards (eds) (1990) and Dunnett and Björklund (eds) (1994).

Outline

The basic steps involved in the protocols of this chapter are schematically presented in the flow chart below:

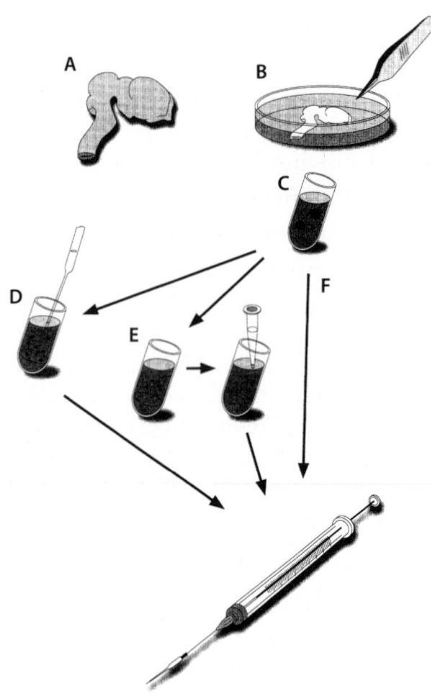

Fig. 1. Flowchart illustrating the major steps in the transplantation protocol. After retrieving embryos from the pregnant rodent (A), the regions of interest are dissected out (B). The dissected pieces are collected in a small tube (C) and can be further prepared either as a dispersed cell suspension (D) or a single cell suspension (E), or used directly in the form of solid tissue pieces (F) for transplantation from a microsyringe.

Subprotocol 1
Dissection of Embryonic/Fetal CNS Tissue

Materials

- Dissection bench or hood

  Note: If a ventilated hood is available, that is a preferable place to work in. Otherwise, as clean as possible conditions should be used, when dissecting on a normal laboratory bench. Infections following transplantation of embryonic tissue into rodents are however very rare, even when dissecting outside of a hood.

- Instruments for microsurgery, i.e., fine forceps and spring (iridectomy) scissors (curved and straight). Larger instruments, scissors and forceps, for removal of the embryos from the pregnant rodent. Instruments are sterilized prior to use.
- Stereoscopic dissecting microscope, with transillumination or incident light.

  Note: When using incident light, placing a black piece of paper or plastic under the dish improves visibility and contrast.

- Pregnant rodents
- Dissection medium.

  Note: Different investigators use different media. In many previous studies a basic medium comprising 0.6% D-glucose in sterile 0.9% saline has been used (see, e.g., Brundin et al 1985a). Most investigators now use different types of tissue culture media, such as calcium- and magnesium-free Hanks´ balanced salt solution (HBSS). Also Dulbecco´s modified eagle medium (DMEM) is often used (Nikkhah et al, 1994a, b), and has been shown to give better cell survival, as compared to HBSS, following implantation in the *in oculo* grafting model (Björklund et al, 1997). Also Watts et al (1998) have recently studied the influence of the dissection medium for ventral mesencephalic (vm) tissue, and suggest that using DMEM results in better graft survival and TH (tyrosine hydroxylase) cell counts.

- Anesthetics.

  Note: The pregnant rodent is either killed or deeply (terminally) anesthetized just prior to retrieval of the embryos, using for instance a large dose of barbiturates. The choice of anesthetics and the procedure should be adapted to local regulations and recommendations.

Procedure

Tissue directly dissected from the CNS will survive transplantation only when obtained from embryonic/fetal or neonatal donors, with optimal survival often coinciding with the period when the neurons are undergoing their last cell divisions, but before they have established extensive axonal projections (Björklund and Stenevi 1984).

Ordering and Staging of Pregnant Rodents

1. Pregnant rats or mice should be obtained preferably with a defined time of mating (date of plug positivity). After overnight mating, the day with a positive vaginal plug

Table 1. Correlation between gestational age in days for fresh rat embryos/fetuses, and crown-rump length (CRL) in mm (modified from Dunnett and Björklund, 1992).

Embryonic age (E), days (rat)	Crown-to-rump length (CRL), mm
12	8
13	9
14	10–11
15	12–14
16	15–16
17	17–19
18	21–23
19	24–25

is defined as embryonic day (E) 0. The age of the embryo can also be assessed, through the known correlation between the embryonic age and the crown-to-rump length (CRL; see Table 1). For this, the embryo is placed on a glass slide and measured using a millimeter scale. With training, it is also possible to palpate the pregnant rodents (under light anesthesia, using, e.g., ether), and to thereby determine the approximate age of the embryos. As a reference point for rats, at around E13 the embryos feel like distinct spheres, with a diameter approximating the CRL. For details, see Dunnett and Björklund (1992). From E18 and onwards the rat embryo is instead referred to as a fetus.

2. The optimal age of the embryonic tissue varies with the region of interest and the particular questions raised in a certain experiment. See below for examples of useful embryonic ages for the different CNS regions described. The gestation period for rodents varies somewhat from strain to strain, but is for rats 21–22 days, and for mice 19–20 days. The protocols given here refer primarily to the dissection of rat embryos, unless otherwise stated. The readers should consult specific literature for information on optimal developmental stages for dissection of corresponding regions from the mouse brain. Alvarez-Bolado and Swanson (1996) summarize a useful starting point for comparisons as follows: "during the first week of gestation rat development is about one to one and a half day behind that in mouse; during the second week it is about two days behind; and by birth it is about three to four days behind."

Retrieval of the Embryos and CNS

1. After deep or terminal anesthesia, the abdomen is shaved and swabbed using 70% ethanol. Using sterile instruments, the abdominal wall is opened (laparotomy). With another set of sterile forceps and scissors, the uterine horns are removed, and placed into a sterile tube containing dissection medium (see above), whereafter the pregnant rodent is killed by overdose or decapitation. The embryos are then collected one by one from the uterine horns, using small scissors and forceps, and put into a small sterile petri dish with medium, placed under the dissecting microscope.
2. The embryos are decapitated at a caudal level (unless a specific part of the spinal cord is wanted), and the heads moved into a new culture dish. The brains and upper spinal cords are then removed, by first making a cut with microscissors (or a fine scalpel blade) under the brain, at the level of the eyes, and then by using two fine forceps to tear off the tissues surrounding the CNS, without damaging any of the underlying tissue. Alternatively, a midline incision can be made in the developing skull and the pieces of the developing cartilage peeled gently away.

3. The CNS tissue is then transferred into a new dish with fresh medium for further dissection of the specific region. At this stage the tissue may be passed through several dishes of fresh medium to rinse away any extraneous blood and loose tissue fragments.

Dissection of the Embryonic/Fetal CNS

The following section describes ways to dissect some of the most frequently used CNS regions. Figure 2 schematically depicts the location of some of these regions in a developing rodent brain.

For further reading, detailed atlases and reviews on rodent CNS development are available (Altman and Bayer 1995; Bayer and Altman 1995; Alvarez-Bolado and Swanson 1996). The following protocols are recommendations from the literature and from the authors of this chapter. However, with emerging knowledge on the developmental characteristics of particular CNS regions, adjustments of the detailed dissections may be necessary to optimize the procedure for a particular experiment. Using different developmental time-points could lead to the inclusion of different cell types in the dissected pieces and a different mix of precursors and more mature, differentiated cells. Information on the dissection of different CNS regions is also available in Seiger (1985), Dunnett and Björklund (1992, 1997).

For any dissection, the following general steps are recommended:

1. The CNS should be held gently with forceps, contacting regions only at some distance away from the actual region of interest, which should itself never be directly pinched or squeezed. The selected region is cut out using a fine pair of microdissection scissors.
2. Also, when transferring a small, dissected piece into a tube or another dish, it should not be grasped directly with forceps, but rather lifted from underneath using, e.g., the scissors or a pair of curved forceps, and the surface tension of the dissection medium.
3. The dissected pieces can be kept at room temperature in dissection medium during the dissection procedure.

Ganglionic Eminences

The ganglionic eminences are located in the floor of the telencephalic vesicle (Fig. 3) and are known to give rise to the corpus striatum (including both the striatum, or caudate-putamen, and the pallidum; Smart and Sturrock 1979). Neurogenesis in the rat caudate-

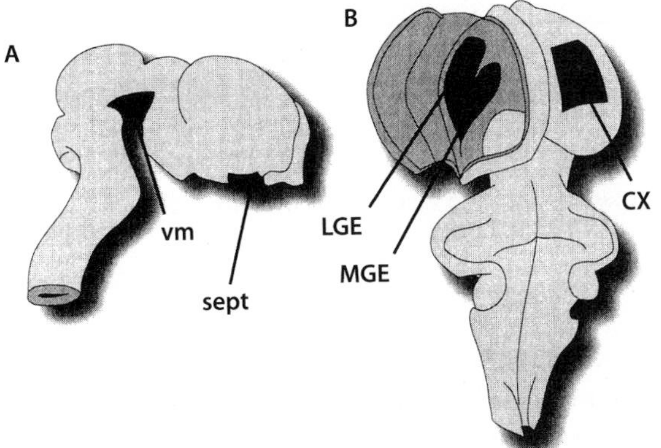

Fig. 2. Overview of the embryonic rat CNS at approximately E14 showing the locations of some of the most often dissected regions. cx, cortex; LGE, lateral ganglionic eminence; MGE, medial ganglionic eminence; sept, septum; vm, ventral mesencephalon.

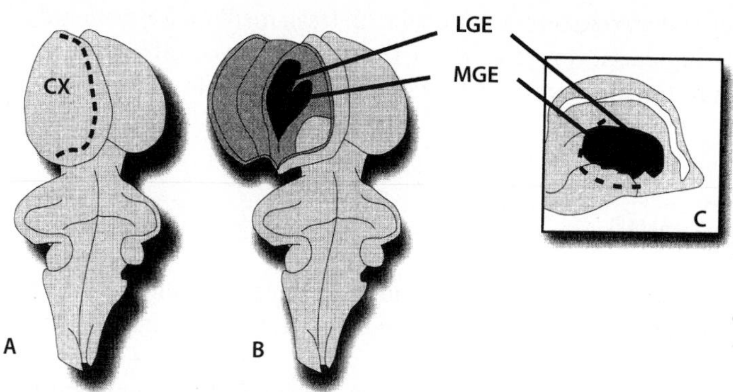

Fig. 3. Dissection of the ganglionic eminences, from an E14 rat embryo. After making a paramedial cut in the cortex (A) and folding the cortical piece laterally, the lateral ganglionic eminence (LGE) and the medial ganglionic eminence (MGE) are dissected out (B, C), either as a whole piece or each piece separately. See the text for details. (Modified from Wictorin et al 1989)

putamen starts around E12 and continues until postnatal day (P) 3, with a peak around E15 (Bayer 1984, Marchand and Lajoie 1986). The two ganglionic eminences, the lateral (LGE) and the medial (MGE), have been extensively used in transplantation experiments, mainly in the animal model of Huntington´s disease (HD; see, e.g., Wictorin 1992 and Björklund et al 1994, for reviews). Recently, several studies have addressed the relative contributions of these two elevations to the corpus striatum and to other related areas. Thus, it has been shown that the LGE is the major source of striatal neurons, generating both the projection neurons (see, e.g., Pakzaban et al 1993), as well as a population of the striatal interneurons colocalizing GABA and somatostatin (Olsson et al 1998b). The MGE gives rise to neurons of the pallidum and basal forebrain, and also to the cholinergic interneurons of the striatum (Olsson et al 1998b). These findings have led to the use of different dissection protocols for these structures, including either only LGE or both LGE and MGE, depending on exactly which regions or cell types are desired. The exact dissection procedure for LGE only or for LGE and MGE together is also influenced by the age of the embryos. The most often used ages for rat are between E12 and E15, and for mouse between E11 and E14. Further details on the dissection of the ganglionic eminences can be found in Olsson et al (1995).

The **protocol** below presents standard procedures for either the combined dissection of both LGE and MGE or only one of the structures separately. The combined dissection is schematically outlined in Fig 3, which shows an E14–15 rat embryo.

1. With the brain placed on its ventral surface, a longitudinal cut is made along the medial ridge of the cortex of the telencephalic vesicle, and the cortical sheet is folded laterally to expose the underlying ganglionic eminences.
2. To retrieve both LGE and MGE, a heart-shaped piece is cut from a caudal angle, as shown in Fig 3. To dissect only LGE (or only MGE), a first cut can be made through the sulcus between the two eminences, followed by a horizontal cut to get hold of the desired piece. Superficial cuts will include mostly the germinal zones, whereas deeper cuts will include also differentiating cells.

Septum

The developing cholinergic cells of the septum and diagonal band nuclei are born mostly between E12 and 16 in the rat (Semba and Fibiger 1988). Tissue from these regions have been used, for instance, in models of *dementia* and *aging*, and good survival and

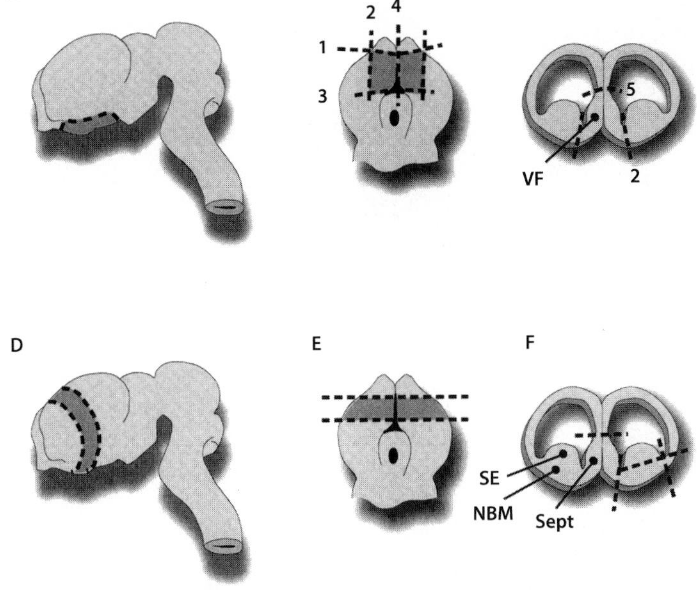

Fig. 4. Dissection of the septum from an E15 rat embryo. This region can be dissected either using a standard dissection (A-C) with a ventral approach or using a differential dissection (D-F) where both the septal and nucleus basalis pieces are dissected after first making a coronal slab through the forebrain. For details, see the text. NBM, nucleus basalis of Meynert; SE, striatal eminence; Sept, septum; VF, ventral forebrain. (Modified from Dunnett and Björklund 1992)

functional effects have been reported when using tissue from E14–15 rat (Nilsson et al 1988). As shown in Fig 4 and described below, these regions can be dissected using either a standard procedure to include both the septum and variable portions of the diagonal band, or using a differential dissection (Dunnett et al 1986; Nilsson et al 1988; Dunnett and Björklund 1992).

1. Standard dissection (Fig 4A-C): With the brain placed on its dorsal surface, coronal cuts are made from the ventral side, one just behind the rudimentary olfactory bulbs (1 in Fig 4B) and one just rostral to the hypothalamus (3 in Fig 4B). This is followed by sagittal cuts bilaterally, just medial to the ganglionic eminences (or striatal eminence; 2 in Fig 4B and C), and a final cut dorsal at the fusion between the septum and the cortical anlage (5 in Fig 4C).
2. Differential dissection (Fig 4D-F): A coronal slice is first made through the forebrain by performing cuts just behind the developing olfactory bulbs and just rostral to the hypothalamus, as indicated in Fig 4D in a lateral view and in E from the ventral side. Then the septal piece is dissected from the ventral medial parts of the coronal slab (Fig 4F), and the region including the developing nucleus basalis of Meynert (NBM) dissected from the ventral lateral portions after removal of the overlying striatal eminence (SE).

Ventral Mesencephalon

The embryonic ventral mesencephalon (vm) is often dissected to obtain the developing dopaminergic neurons of this region for use in animal models of Parkinson´s disease (see, e.g., Brundin et al 1994). The dopaminergic neurons of the vm in rats are born between E 13 and 15 (Bayer and Altman 1995) and it is from these stages that donor tissue

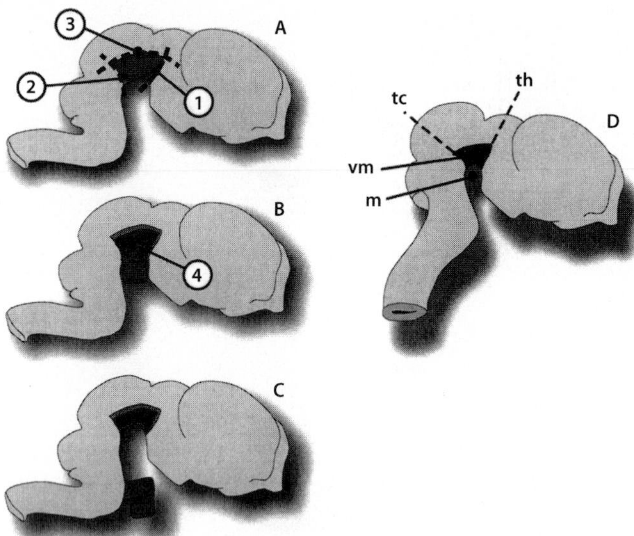

Fig. 5. Dissection of the ventral mesencephalon (vm) from an E14 rat embryo depicted in a lateral view. (A, B, C) shows the stepwise dissection procedure involving the different cuts, as described in detail in the text. (D) illustrates the angles of the first two cuts, with imaginary lines continued to intersect the dorsal surface of the tectum (tc) and the thalamus (th). m, mesencephalic flexure. (Modified from Dunnett and Björklund 1992)

for grafting has been successfully obtained. The dissection of the vm is described in Fig. 5, and further details and references can also be found in Dunnett and Björklund (1997).

1. With the embryo placed on its side, coronal cuts are made at the levels of the mes-diencephalic junction and at a caudal level of the mesencephalic flexure, whereafter the entire ventral mesencephalon is dissected bilaterally. In more detail, the first and most rostral coronal cut (1 in Fig5A) is made at the rostral end of the mesencephalic flexure. Then, a second coronal cut is placed at the caudal end of the mesencephalon (2 in Fig5). The positions of these coronal cuts can also be defined using landmarks on the overlying thalamic and tectal surfaces, as indicated by imaginary lines in Fig5D. Finally, two sagittal cuts (3 in Fig5A and 4 in Fig5B) are made to obtain a piece shaped as shown in Fig. 5C.
2. Special care should be taken to remove all the meningeal coverings. This can be done after completing the anatomical dissection by lifting the mesenchymal layer off under the dissection microscope.

Hippocampus

Hippocampal tissue has been used extensively in transplantation experiments, for instance, in models of ischemia (Tönder et al 1989; Hodges et al 1994). Hippocampal neurons are born relatively late in embryonic development, with dentate granule cells added to a large degree also postnatally (Bayer and Altman 1995). Indeed, it has recently been shown that neurogenesis occurs in the adult dentate gyrus of rodents throughout life (Gage et al 1998). To obtain grafts rich in pyramidal neurons, tissue from E18 rat has been successfully used, as well as from developmental stages well into the postnatal period (Sunde and Zimmer 1983). At those later time-points it is also possible to make

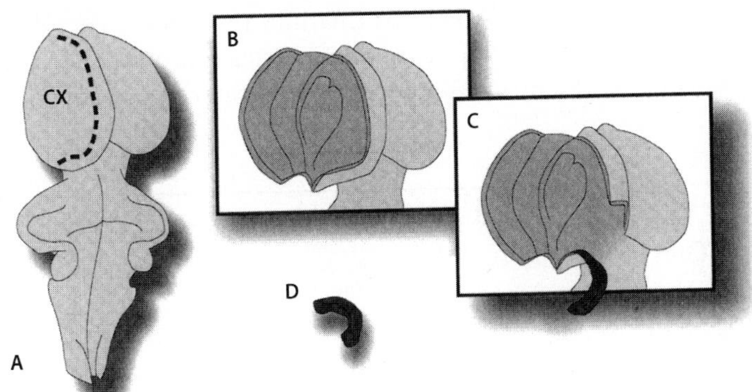

Fig. 6. Dissection of the hippocampal anlage from an E18 rat embryo. After a paramedial cut in the cortex following the borders of the hippocampus caudally (A), the cortex is folded laterally (B) and the hippocampal piece is dissected out (C, D), as described in detail in the text. (Modified from Dunnett and Björklund 1992)

more selective dissections of certain hippocampal subregions. Figure6 describes a dissection from an E18 rat embryo.

1. To reach the hippocampal formation a paramedial longitudinal cut is made through the entire length of the cortical anlage (Fig6A). In the caudal portion, the cut is made along and just lateral to the hippocampal anlage. After folding the developing cortex laterally (Fig6B), a cut is made through the cingulate cortex and underlying fimbria. The hippocampus is then gently pulled and folded backwards to free it from the underlying thalamus (Fig6C). Finally, a cut is made at the lower end of the hippocampal piece, and the obtained piece (Fig6D) is trimmed from excess neocortical tissue and the meninges removed.

Neocortex

Parietal neocortex can be obtained as a longitudinal strip of the full depth of the cortical anlage. Neocortical tissue has been dissected and transplanted into both newborn rats and into adult rats in models of, for example, cortical infarcts (see, e.g., Castro et al 1988 and Grabowski et al 1992, and Kolb and Fantie 1994 for review). The neocortex develops relatively late, and good survival of grafts has been reported when using tissue from mid-gestation and from several days into the postnatal period.

1. First, a cut is made longitudinally through the neocortex along the medial portion of the telencephalic vesicle. The sheet of neocortical tissue is then folded laterally, and a large strip is collected by making rostral and caudal coronal cuts, as well as a longitudinal cut lateral to the ganglionic eminences. The exact size and area dissected out could of course vary, depending on whether a particular cortical subregion is desired.
2. The meningeal coverings can be removed either before cutting the piece out or just after, using a pair of fine forceps.

Other Regions

Transplantation experiments have also been made using tissue from several other regions of the developing CNS as donor tissue, such as spinal cord (see, e.g., Bregman 1994), cerebellar anlage (see, e.g., Sotelo and Alvarado-Mallart 1987), retina (see, e.g.,

Lund et al 1987), hypothalamus (see, e.g., Wood and Charlton 1994), locus coeruleus (see, e.g., Björklund et al 1986) and dorsal raphe (see, e.g., Foster et al 1988). See also Seiger (1984) and Dunnett and Björklund (1992) for dissection protocols.

Subprotocol 2
Preparation of Tissue/Cells

Materials

- Incubator (for 37 °C), or water bath
- Fire-polished Pasteur pipettes
- Micropipettes (Eppendorf; for 1ml and 200μl)
- Haematocytometer
- Chemicals, e.g., trypsin, DNAse

Procedure

Different protocols for preparing the dissected tissue for transplantation are described in the literature, such as
- Dissociated cell suspensions – but with some remaining tissue pieces or aggregates,
- Single cell suspensions, and
- Whole tissue pieces or slabs.

Examples from experiments with transplantation of tissue pieces can be found in, for example, Stenevi et al (1985), where transplantation into surgically prepared cavities is described. Also, for instance, Sotelo and his colleagues (Sotelo and Alvarado-Mallart 1987) have implanted pieces of cerebellar anlage into cerebellar deficient mice. One reason for using whole tissue pieces could be that a relatively intact organization of the dissected piece is desired, with a certain orientation. For the use of fragmented non-dissociated tissue for transplantation, see Björklund and Dunnett (1992).

This chapter focuses on the use of cell suspensions. A major advantage of the suspension method is that it allows for precise stereotaxic injections of the cells using, for instance, a Hamilton syringe and/or a glass capillary, also into deep sites in the brain (Björklund et al 1980, 1983). The suspension method also makes possible a direct assessment of the viability and the numbers of cells that are injected into each site (see below, and, e.g., Brundin et al 1985a; Nikkhah et al 1994a, Barker et al 1995).

Whether the method with a dissociated cell suspension or the use of a single cell suspension or the technique with whole tissue pieces is preferable, in terms of cell and graft survival and functional effects' should be determined for each specific application. The methods given below represent selected protocols from our own laboratory which have proven useful in certain experimental paradigms. With new emerging data, these procedures will continuously be subject to modifications.

Preparing the Cell Suspension

To achieve a suspension of cells, the dissected tissue pieces are generally both enzymatically and mechanically dissociated. The procedure will affect the viability of the suspended cells and of the graft following transplantation, and much effort is therefore currently put into optimizing the different steps in the preparation of the cells (see, e.g., Barker et al 1995). It is, for example, known that when embryonic dopaminergic neurons are grafted into the adult brain, only about 5–20% of the grafted neurons survive (Sauer and Brundin 1991), and in that model, the addition of lazaroids, which inhibit lipid peroxidation, has been found to increase the survival of the grafted cells (Nakao et al 1994). Recently, caspase inhibitors were found to reduce apoptosis and increase survival of dopaminergic neurons grafted into hemiparkinsonian rats (Schierle et al 1999). With regard to the enzymatic digestion, Björklund et al (1986) described that good graft survival with noradrenergic tissue from the locus coeruleus was only obtained when not using trypsin in the protocol. Furthermore, Olsson et al (1997) found that omitting the trypsin treatment (i.e., leaving the cell surface molecules intact) resulted in a more specific homotypic integration of grafted ganglionic eminence cells after injections into the lateral ventricles of rat embryos. Apart from optimizing the procedure for preparing the cell suspensions, attempts are now also being made to expand mesencephalic precursors in vitro, prior to implantation, using trophic factors and specific culturing methods (Studer et al 1998).

1. After dissecting a certain embryonic region, as described above, the pieces are collected in a small Eppendorf tube or vial in dissection medium.
2. The pieces are then incubated in 0.1% trypsin (Worthington, N.J., USA)/0.5% deoxyribonuclease (DNase, Sigma) in dissection medium at 37 °C for 20 min.
3. The tissue pieces are rinsed to remove the trypsin by gently replacing the medium four times.
4. The final volume is brought to an equivalent of, for instance, 10 μl per tissue piece. Thus, add medium with DNase to the test tube, with a total volume depending on the desired concentration of cells, for instance, 5–6 μl per dissected VM, or around 5 μl for each ganglionic eminence (LGE+MGE).
5. The tissue is then mechanically dissociated into a milky cell suspension through repeated trituration with fire-polished Pasteur pipettes of decreasing diameters (usually, two different diameters are used). Care should be taken to avoid air bubbles. The resulting suspension normally contains individual cells mixed with small aggregates or clumps of tissue. Excessive pipetting should be avoided, and usually a total of 10–15 gentle strokes of the pipette results in a good balance between dissociation and trauma to the cells (Björklund and Dunnett 1992). The concentration (and viability, see below) of the suspension can then be evaluated through the Trypan blue dye exclusion method or using acridine orange/ethidium bromide, and cell counting in a haematocytometer (Brundin et al 1985a).

A Suspension of Single Cells

In order to be able to do so-called microtransplantation, i.e., inject smaller volumes of cells, and use a small-diameter glass capillary to inject the cells, a modification of the traditional cell suspension technique has been developed (Nikkhah et al 1994a, b). Using this technique a homogeneous single cell suspension can be obtained, which also allows for a more exact assessment of the density of the cells in the suspension, and hence reproducibility in the number of grafted cells. Nikkhah et al. (1994a, b) have used the method for VM-grafts, and this protocol is based on those experiments.

1. The tissue pieces are collected, incubated with trypsin and washed, as described under 2.1.1–3 above. The volume of the pieces and medium is adjusted to be appropriate for trituration (e.g., 250 or 500 μl), depending on the amount of dissected tissue.
2. The pieces are then mechanically dissociated through repeated trituration, starting with a 1-ml Eppendorf pipette, to get a milky cell suspension. Then the suspension is

triturated repeatedly with a 200-µl Eppendorf pipette to obtain a homogeneous single cell suspension. Then follows assessment in a haematocytometer of the concentration of the suspension and, with a known total volume, also the total number of cells.
3. The tube is then centrifuged at 600 rpm for 5 mins.
4. The pellet is then resuspended using a few triturations with a 200-µl Eppendorf Pipetteman in a pre-determined final volume of medium. This volume is chosen depending on the total number of cells to achieve an appropriate final concentration. For example, Nikkhah et al (1994a) used a final volume of 100 µl for 25 VM-pieces, with a final concentration of 100.000–150.000 cells/µl.

Viability and Storage of the Cells

The viability of a newly made cell suspension is normally very good, with, for instance, 90–95% living cells for VM-tissue, as determined through the Trypan blue dye exclusion method or using acridine orange/ethidium bromide, and cell counting in a haematocytometer (Brundin et al 1985a). The suspension can be kept at room temperature and remain viable for several hours during the transplantation session, although viability decreases over time and also depends on the type of tissue and the age of the dissected embryos (Brundin et al 1985a). In most experiments, we keep the cell suspension on ice or in the refrigerator during the transplantation session.

In order to be able to store the tissue for longer times, prior to grafting successful attempts have been made to hibernate, for example, ventral mesencephalic or ganglionic eminence tissue for several days (Sauer and Brundin 1991, Frodl et al 1995, Grasbon-Frodl et al 1996, Nikkhah et al 1995). Also, freezing in liquid nitrogen has been investigated, although with marked reductions in graft survival and function (Sauer et al, 1992).

Subprotocol 3
Transplantation into Adults

Materials

- Anesthetics

  Note: Different laboratories have their own preferences, also depending on the local regulations. See Björklund and Dunnett (1992) for discussion.

- Stereotaxic frame

  Note: In our laboratory, we use frames from Kopf Instruments, USA.

- Dental drill
- Microsurgical instruments

  Note: Standard scalpel blades, fine forceps and syringes.

- Operation microscope
- Suture thread, clips

Procedure

This section focuses on the implantation of cell suspensions into the adult CNS parenchyma. Other approaches described in literature include transplantation into the ventricles (see, e.g., Freed 1985), and implantation of solid grafts into intracerebral transplantation cavities (Stenevi et al 1985).

Anesthesia and General Surgical Aspects

1. The rats should be anesthetized according to local laboratory guidelines.
2. The rat is placed in a stereotaxic frame, which allows for an exact and reproducible placement of the transplants.
3. When the animal is deeply anesthetized, the fur over the head is shaved and swabbed with 70% ethanol. Using a scalpel blade the skin is opened longitudinally, and the overlying muscle and connective tissue removed to clear the bone of the skull.
4. Coordinates are calculated from either the bregma or the lambda. There exists a large number of lesion models and different implantation sites described in the literature, with published stereotaxic coordinates and other specific parameters for implantation into, for instance, the striatum (see, e.g., Brundin et al 1985a), substantia nigra (see, e.g., Nikkhah et al 1994b), hippocampus (see, e.g., Bengzon et al 1990), and thalamus (see, e.g., Peschanski and Isacson, 1988). For planning of experiments, the readers are also advised to study reviews in, for instance, Björklund and Dunnett (eds; 1994). Information for determining new stereotaxic coordinates is also given in rat brain atlases, like Paxinos and Watson (1986, 1997) and Swanson (1992). For mouse brain, see, e.g., Franklin and Paxinos (1997).
5. When coordinates have been determined, a small hole is drilled through the bone, but leaving the dura intact.

The Injection Instrument

A fine microsyringe, e.g., a 2- or 10-μl Hamilton syringe (Hamilton Bonaduz, Switzerland), is used to deliver the cells into the implantation site. The cells can be injected directly from the metal cannula (stainless steel needle) of the syringe, with a diameter appropriate to allow passage of the suspension. Alternatively, to reduce the diameter of the tip and hence minimize tissue damage and backflow of the suspension after injection, for instance, Nikkhah et al (1994a) have used a fine glass capillary with an outer diameter of 50–70μm, attached to the tip of the needle. This is schematically depicted in Fig.7. Using a standard pipette puller, a long-shanked pipette or capillary is thus pulled from a glass capillary (i.d. 0.55 mm; o.d. 1.0mm). The glass capillary is then fitted onto the metal cannula (o.d. 0.5mm) of the syringe using a cuff of polyethylene tubing (i.d. 0.58mm, o.d. 0.965mm) as an adapter. The polyethylene tubing is first heated slightly (with hot air from a hair-dryer) and pulled to achieve a conically shaped cuff with gradually smaller diameter which can be mounted tightly onto the tip of the metal cannula. Thereafter, the glass capillary can be tightly fitted onto the cuff (Nikkhah et al 1994a).

When using a glass capillary the optimal diameter depends on the type of suspension that is used (Nikkhah et al 1994a). For the homogeneous single cell suspensions, capillaries with very fine diameters can be used, but aggregates and very dense suspensions may require a larger diameter to avoid clogging. Wider instruments should be employed when injecting whole tissue pieces.

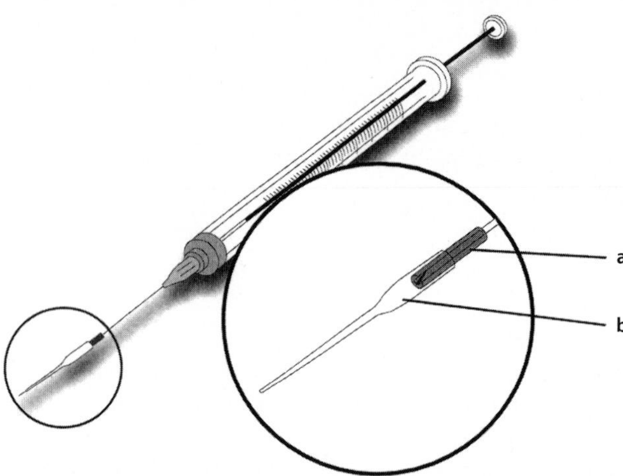

Fig. 7. Schematic drawing of a microsyringe with an attached glass capillary used for transplantation of cell suspensions. The inset shows the glass capillary (b) mounted onto the metal cannulus of the syringe using a piece of polyethylene tubing. See text for details.

Injection of the Cells

1. We usually keep the cell suspension on ice or in the refrigerator during the transplantation session. The suspension should be gently resuspended using a pipette when visible sedimentation of the suspension has occurred, normally just before loading new cells into the syringe for a new injection.
2. The needle of the syringe or the glass capillary is filled by back suction. When using a glass capillary connected to the Hamilton syringe, the whole system of syringe and capillary is first filled with medium through injection into the top of the syringe after removing the injection wire. Care should be taken to avoid air bubbles in the syringe or capillary and to make sure that there is no leakage where the glass capillary is attached using the cuff (see above). An additional advantage of using the glass capillary is that the cells can be inspected through the microscope as they pass through the capillary during the actual injection.
3. The vertical coordinate is, in most cases, calculated from the dural surface (see Fig. 8). Approximately 1 μl can be injected per minute, divided over small pulses and with a 2-min interval before slowly retracting the syringe.
4. The skin is sutured or clips applied.

On the Need for Immunosuppression

The CNS is considered as an immunologically privileged site, denoting a prolonged survival rate of an incompatible graft compared with rates obtained in other sites of the body (see, e.g., Widner and Brundin 1988, Sloan et al 1990, Widner 1995 for reviews). There are several factors influencing the immune response following transplantation, such as the degree of donor-recipient disparity (*allogeneic* = grafts between different strains of animals within a species; *concordant xenogeneic* = grafts between closely related species; *discordant xenogeneic* = grafts between distantly related species), origin

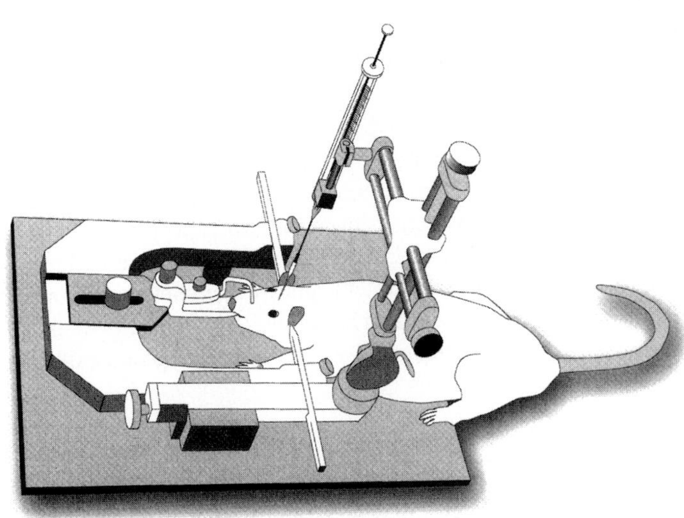

Fig. 8. Schematic illustration of a stereotaxic frame with a microsyringe for transplantation into an adult rat. See text for details.

of the vasculature, content of antigen-presenting cells, and locally acting immunosuppressive factors. Within the brain there are also heterogeneous survival rates depending on the donor tissue preparation and on the implantation sites. Cell suspensions of allogeneic, and occasionally xenogeneic grafts placed into, e.g., the striatum have high survival rates, whereas similar grafts placed into the hippocampus, or grafts of solid pieces placed in the ventricles, are rejected unless immunosuppression is given (Widner 1995).

Most studies referred to in this chapter describe the use of syngeneic grafts (i.e., transplantation within the same inbred strain), where there is usually no need for immunosuppressive treatment. For xenografts with mouse tissue implanted into rats, cyclosporin A has been shown to increase graft survival (Brundin et al, 1985b). In our own laboratory, the protocol consists of daily intraperitoneal (ip) injections of cyclosporin A at 10mg /kg (Sandimmun; Sandoz, Basel, Switzerland) – dissolved in Cremophor and saline at 10 mg/ml. We also give antibiotics (e.g., Borgal; Hoechst, Munich, Germany) in the drinking water during the first ten days of the immunosuppression. In a meta-analysis of the relative effectiveness of systemic immunosuppression for neural tissue xenografts, Pakzaban and Isacson (1994) found that any kind of immunosuppression improves the overall survival rate to approximately 75%, as compared to below 30% without any treatment. A combination of several immunosuppressive drugs (cyclosproin A, azathioprine and prednisolone) is very effective, as demonstrated by Pedersen et al (1995).

There are several different effective immunosuppressive treatments, although none without complications and expenses. Apart from the protocol with daily i.p. injections of cyclosporin, there are several other single-drug protocols which will allow for unimpaired graft survival for up to 10–15 weeks, such as daily injections of methylprednisolone at 30mg /kg (Duan et al. 1996). See also Czech et al. (1999) for recent recommendations, where it is furthermore shown that the single-drug protocols are improved by addition of daily prednisolone oral galvage at 20 mg/kg or daily methylprednisolone 10 mg/kg i.p. injections for 2 weeks. Mouse recipients are best treated with anti-lymphocyte sera (ALS) or monoclonal antibodies, since regular drugs need to be given at high dosages (e.g., 50–80 mg/kg for cyclosporin) and the toxicity is high. Antibody treatments are usually well tolerated, can be given i.p. but need to be repeated every 3–

5 days and started already a few days ahead of implantation. This can occasionally result in transplantation tolerance (Wood et al. 1996). Rodent antibody treatment is comparatively expensive.

As in any transplantation situation, immunological tolerance, with spontaneous indefinite graft survival, can be achieved provided the recipient animals are neonatal (Brent 1990). As reviewed by Lund and Banerjee (1992), grafts of mouse retina implanted into neonatal rat brains can survive for as long as two years, whereas the same tissue is rapidly rejected when placed into an adult recipient. It seems that the transition between survival and rejection occurs when the recipient rats are aged between 8 and 12 days postnatal at the time of implantation (Lund and Banerjee 1992). In our own experiments, we usually implant mouse cells into neonatal rats at P1–2, with good results.

Notes on Perfusion and Further Processing of the Tissue

In some transplantation studies, the grafts are assessed already in the living animals through, for instance, microdialysis to determine transmitter release (see, e.g., Campbell et al 1993), behavioral testing (see, e.g., Dunnett 1994) or using different imaging techniques (see, e.g., Fricker et al 1997). For histological analysis, the animals are sacrificed and prepared for subsequent analysis using specific protocols depending on the desired analysis. For detailed anatomical studies, the sacrifice of the animals is often precluded by injections of neuroanatomical tracers. Some of the most widely used techniques to analyze the grafts involve *immunocytochemistry* and *in situ hybridization*. See Chapter 15 for details on modern histological techniques.

For standard immunohistochemical analysis in our laboratory, the animals are deeply anesthetized using a high dose of, for instance, pentobarbital. The brains are then fixed through transcardial perfusion, first with a brief (approximately 1 min) prerinse of saline or phosphate-buffered saline (PBS), followed by 250–300ml of 4% paraformaldehyde (in PB) and a postfixation for a few hours or overnight, depending on the subsequent analysis. After immersion in 20% buffered sucrose solution, the brains are then cut on a freezing microtome for further analysis.

Subprotocol 4
Transplantation into Neonates

Materials

- Neonatal transplantation stand or frame
- Wet ice for hypothermic anesthesia
- Dry ice
- 70% ethanol
- Fine suture thread (e.g., 7.0)
- See also Part 3

Procedure

Transplantation into neonates follows in many ways the same procedure as for adult recipients, but there are also some major differences. Most conducted experiments, and the protocol given here, refer to early postnatal rats, although also later postnatal recipients and different species have been used by others (Lund and Yee 1992).

General Aspects and Anesthesia

As described above, the gestational period for rats is around 21–22 days, and timed pregnancies are preferable in order to be able to plan the day for surgery ahead. A quiet environment should be arranged for the pregnant rat, both before and after delivery. Stressed animals may give premature birth and also kill and eat their pups, also after some days into the postnatal period. The pregnant rats should be housed in individual cages. The day of birth is designated as postnatal day 0 (P0). In our laboratory, we use mostly rats from P1 to P3.

1. The time the pups spend away from the mother should be minimized by preparing everything for surgery beforehand. The pups can however spend at least a few hours on their own.
2. A few pups are moved at a time from the mother into the surgery room. Gloves are always used when handling the pups. The pups are kept under a warm lamp on a clean and soft surface, avoiding overheating and dehydration.
3. Each pup is anesthetized through hypothermia by submersion under wet ice for 2–5 min (depending on species and age) until no reflexes or movements are seen.

Transplantation

Due to their small size, it is difficult to accurately mount the pups and do precise surgery. A number of different set-ups for operations on neonates are found in literature. For instance, Lund and Yee (1992) describe a transplantation stand with a fiber-optic light source illuminating the animal from below. With such transillumination, it is possible to see major venous sinuses and to use those as landmarks for implantation either stereotaxically or under visual guidance. The pup can be fixed in a certain position using, for example, modeling clay. The more recently developed hypothermic miniaturized stereotaxic frame (Cunningham and McKay 1993), which is also used in our labo-

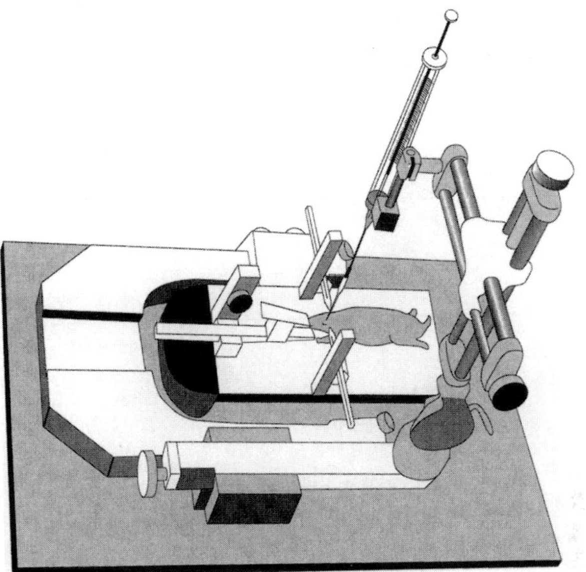

Fig. 9. Schematic illustration of the Cunningham miniaturized stereotaxic instrument (Stoelting) mounted onto a regular stereotaxic frame for transplantation into neonatal rats. See Cunningham and McKay (1993) and text for details.

ratory, allows for fixation of the animal with small earbars, adjustment of the angle of the earbars relative to the nose, and exact stereotaxic surgery (see Fig. 9). The device can be used by itself or mounted onto a conventional adult frame. This frame has an adjoining well, where dry ice is kept in 70% ethanol to maintain hypothermia during surgery.

1. The skin is lifted from the skull, and a gentle incision is made through the skin with a scalpel blade or using a pair of microscissors. It is important not to cut through into the bone and cartilage of the forming skull and especially to avoid damaging the sagittal sinus. The pup should be kept hypothermic during all surgery.
2. The pup is mounted onto the frame or stand. When using ear plugs to mount the pup, the skin should be pulled laterally to reveal the developing external auditory meati, under which the ear bars are gently placed.
3. In our laboratory, the system with a small diameter glass capillary attached to a Hamilton syringe is used for injections of cell suspensions into the neonates, as described above. Other investigators describe the implantation of larger pieces or intact tissue, from, e.g., the retina or neocortex, also by direct placement into a certain accessible site (see, e.g., Lund and Yee 1992). It is important to minimize damage to the dura and underlying parenchyma when transplanting to deeper sites in the brain.
4. The transplantation syringe can be mounted in the standard stereotaxic frame and coordinates obtained by using the large frame as for adult surgery (see Fig. 9).
5. The procedure for loading the syringe is identical to that used for adult surgery. Inject maximum 2 µl per site, preferably 1 µl in small regions, such as the hippocampus. It is better to spread the volume over several sites.
6. For injections to mouse recipients nanoliter volumes may be preferred. For this, it is recommended to mount a glass capillary directly to a nanoinjector which can then be mounted either on to the adult stereotaxic frame or micromanipulator with its own stereotaxic frame. Mice may either be mounted approximately in the rat neonatal frame using earbars to lightly press the external meati, otherwise immobilizing the head using modeling clay.
7. After completion of surgery the midline incision is sutured using fine thread, taking care not to tear the skin. Excess suture should be trimmed close to the knot to avoid mothers attempting to remove the sutures. Pups are then cleaned of all blood and warmed slowly under a lamp. A small amount of pressure applied to the tail can be used to stimulate breathing if necessary. Pups should be fully awake and active before returning to the mother.

Subprotocol 5
Transplantation into Embryos

Materials

- Glass capillaries

  Note: For free-hand embryonic injections, we use a glass microcapillary connected to a 10µl Hamilton to minimize the trauma while penetrating the uterine wall. The tip of the capillary is beveled to allow easy penetration into the tissue. When injecting into embryos at late gestational stages (E17.5-birth), the glass microcapillary should not be as long to provide more durability while penetrating the thicker tissue.

- Ultrasound machine

Note: Ultrasound backscatter microscopy (UBM) is a high frequency (40–100 MHz) ultrasound imaging method resulting in high spatial resolution (90 µm at 40 MHz) over a limited penetration depth (7–10 mm at 40 MHz; Turnbull et al 1995a, b). 8 mm x 8 mm UBM images (512x512 pixels) are produced by a mechanically scanned, focused 40-MHz transducer at image frame rates of 4 or 8 images per second. Images are captured either in direct digital format into an IBM-compatible PC computer, or onto a VCR from the video output of the scanner. A commercial UBM system (Ultrasound Biomicroscope Model 840; Humphrey Instruments, San Leandro, Calif., USA) has also been used to guide injections into mouse embryos. This scanner also operates at 40 MHz, producing 5 mm x 5 mm images at a frame rate of 8 images per second, with image quality comparable to the prototype UBM described above. To facilitate *in utero* injections, the mechanical probe of the UBM is mounted on a motorized 3-axis positioning stage, and fine XYZ positioning of the UBM image plane is maintained during the injection procedure using a joystick controller (Newport-Klinger, Irvine, Calif., USA).

- Micromanipulator

Note: UBM-injections are performed as described below, using a 3-axis micromanipulator (Narishige, Tokyo, Japan) to position the injection needle. These injection needles are made from glass micropipettes which are pulled to produce a long taper, broken under a dissection microscope at an inner diameter between 30 and 50 µm, and sharpened to produce a bevel angle between 20° and 25°.

- Microsyringe pump

Note: An oil-filled manual microsyringe pump (Stoelting, Wood Dale, Ill., USA) with a 25-µl Hamilton syringe is used to draw cell suspensions into the injection microcapillary and to inject a precise volume of cell suspension into each embryo.

Procedure

In developmental studies, transplantation of cells into the embryonic brain, *in utero*, has proven to be a very useful approach (Campbell et al 1995; Brüstle et al 1995; Fishell 1995; Olsson et al 1997, 1998a). Using free-hand injections, the cells can be placed into the ventricles of the developing embryo. With a more refined technique, ultrasound-guided injections can be made into the parenchyma of the recipient embryos (Olsson et al 1997) or into the cavity of the neural tube (Liu et al 1998).

In utero transplantation has been used in experiments with transplantation of forebrain precursors, and here comparisons can be made with related experiments in neonatal or adult graft recipients. Transplantation experiments offer a unique tool to investigate the degree to which neural precursors at certain developmental stages are committed to generate specific neural phenotypes, as well as the role the local environment plays in this process (i.e., developmental potential). In this context, heterotopic transplantation of midgestation embryonic telencephalic precursors into adult hosts has revealed a rather restricted developmental potential; the transplanted cells differentiate largely into phenotypes appropriate to the site of cellular origin rather than those typical of their new location (see, e.g., Olsson et al 1995). These experiments are, however, limited by the fact that neurogenesis is largely a prenatal event and many of the instructive cues normally involved in the specification of progenitors may be lacking in the adult and neonatal hosts. In fact, transplantation of midgestation striatal precursors into the forebrain ventricle of similar stage embryonic hosts, *in utero*, has clearly demonstrated a greater developmental potential of these progenitors than was previously observed after transplantation into adult or neonatal hosts (Campbell et al 1995, Brüstle et al 1995; Fishell 1995, Olsson et al 1997, 1998a).

Cells injected into the embryonic forebrain ventricle do not, however, incorporate equally well into all regions of the developing brain and this model is therefore likely to test the degree of positional specification present in neural progenitors more accurately than their true potential to respecify after placement in ectopic regions. To more fully test the commitment of neural progenitors, an ultrasound-guided *in utero* grafting paradigm was developed which allows for intraparenchymal injections into discrete developing regions of E12.5–13.5 mouse embryos (Olsson et al 1997), using high resolution ultrasound backscatter microscopy (Turnbull et al 1995b). This ultrasound-guided embryonic injection technique also allows for injections into the neural tube cavity of embryos soon after closure of the cephalic folds (E9–9.5; Liu et al 1998).

Free-hand Intraventricular Injections

In our laboratory, we have successfully performed intraventricular *in utero* injections into rat embryos ranging from E15 to E20. At the later stages of development, the embryos fill out the uterus and the forebrain becomes more clearly visible and also more accessible.

1. To minimize the actual surgery time, the cell suspension and all arrangements around the transplantation procedure should be prepared in advance, prior to anesthetizing the pregnant rodent.
2. With the animal placed on its back, the abdomen is swabbed using 70% ethanol and a midline laparotomy performed. Cover with sterile tissue/paper.
3. Sterility should be retained throughout the procedure and one uterine horn moved at a time. Then one embryo at a time is oriented such that the telencephalic vesicles and calvarian sutures become evident by transilluminating the uterine sac using fiber optics. Thus, with an optimized orientation of the embryo, the forebrain ventricles become clearly visible.
4. Injections are performed free-hand using the 10-µl Hamilton syringe equipped with the glass micropipette and with a second experimenter assisting in holding the embryo in place. The injections are completed within seconds and the needle can be withdrawn following a delay of ~ 5–10 s. To optimize embryonic survival, care should be taken to lose as little amniotic fluid as possible during injections. After transplantation, the uterine horns are placed back into the abdomen, and the mother can be sutured and left to give birth.

Ultrasound-guided Injections into Parenchymal Targets or into the Early Embryonic Neural Tube Cavity

1. Timed pregnant mice (e.g., CD1 from Charles River, N.Y., USA) with embryos at a gestational age of E9.5–13.5 are anesthetized. The abdomen is wet-shaved and a 2-cm midline laparotomy is performed. Each uterine horn is carefully taken out individually and one side is chosen for injection. The uterine horns are repositioned in the abdomen with the embryo closest to the ovary of the side to be injected left outside the abdomen. The pregnant mouse is then placed in the lower level of a two-level holding stage. Petri dishes are modified by punching a 25-mm-diameter hole in the bottom of each dish and covering the hole with a thin rubber membrane (see Fig. 10). The petri dish is mounted with 2 pins to the top level of the holding stage over the mouse's abdomen, and the part of the uterus containing the first embryo is gently

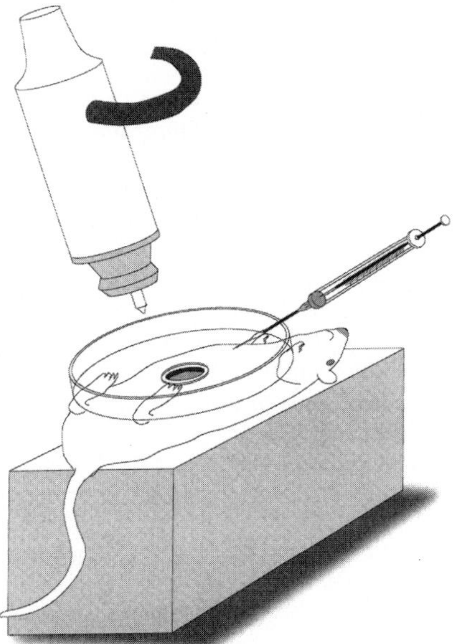

Fig. 10. Schematic overview of the transplantation into rat embryos. Procedures for either free-hand or ultrasound-guided injections of cells are described in the text.

pulled through a slit in the rubber membrane (Olsson et al 1997; see Liu et al 1998 for complete description). Fluid coupling between the tissue and transducer is maintained by filling the petri dish over the mouse with sterile PBS containing calcium- and magnesium-chloride. The tip of the microcapillary can be identified as a bright dot on the screen displaying the ultrasound image, and by maximizing the UBM signal (image brightness) one ensures that the needle tip is located in the image plane being monitored.

2. Using UBM-guidance, embryos are positioned so as to yield a coronal or horizontal ultrasound image for intraparenchymal injections while the injections into the early neural tube cavity are most easily performed while imaging the embryo in the sagittal plane. After positive identification of the parenchymal target (we have successfully injected into the LGE, MGE, the cerebellar anlage, tectum or tegmentum) or the neural tube cavity, the glass microcapillary is inserted through the uterine wall and 1–2 µl cell suspension is injected into the chosen site. The backscatter signal from the cell suspensions are sufficiently high so that injected suspensions can be visualized distributing in the parenchymal target or filling up the ventricular space. After injecting one embryo, the next embryo is gently pulled through the slit in the rubber membrane and positioned for injection, while the first embryo is placed back into the abdominal cavity. In this way, it is possible to inject all embryos on one side of the uterine horn (4–10) in 30–60 minutes. The position of each injected embryo along the uterine horn is registered for identification at sacrifice. After injections, the uterine horn is placed back into the abdomen and the mother is sutured.

Acknowledgement: We thank Bengt Mattsson for providing the illustrations.

References

Altman J, Bayer SA (1995) Atlas of prenatal rat brain development. CRC Press Inc, Ann Arbor.

Alvarez-Bolado G, Swanson LW (1996) Developmental brain maps: structure of the embryonic rat brain. Elsevier, Amsterdam.

Barker RA, Fricker RA, Abrous DN, Fawcett J, Dunnett SB (1995) A comparative study of preparation techniques for improving the viability of nigral grafts using vital stains, in vitro cultures, and in vivo grafts. Cell Transplantation 4(2):173–200.

Bayer SA (1984) Neurogenesis in the rat neostriatum. Int J Dev Neurosci 2:163–175

Bayer SA, Altman J (1995) Neurogenesis and neuronal migration. In The rat nervous system. Second Edition (ed Paxinos G) pp 1041–1078, Academic Press, San Diego.

Bengzon J, Kokaia M, Brundin P, Lindvall O (1990) Seizure suppression in kindling epilepsy by intrahippocampal locus coeruleus grafts: evidence for an alpha-2-adrenoreceptor mediated mechanism. Exp Brain Research 81:433–437.

Björklund A, Katzman R, Stenevi U, West KA (1971) Development and growth of axonal sprouts from noradrenaline and 5-hydroxytryptamine neurones in the rat spinal cord. Brain Res 31:21–33.

Björklund A, Stenevi U (1971) Growth of central catecholamine neurons into smooth muscle grafts in the rat mesencephalon. Brain Res 31:1–20.

Björklund A, Schmidt RH, Stenevi U (1980) Functional reinnervation of the neostraitum in the adult rat by use of intraparenchymal grafting of dissociated cell suspensions from the substantia nigra. Cell Tiss Res 212:39–45.

Björklund A, Stenevi U, Schmidt RH, Dunnett SB, Gage FH (1983) Intracerebral grafting of neuronal cell suspensions. I. Introduction and general methods of preparation. Acta Physiol Scand 522:1–7

Björklund A, Stenevi U (1984) Intracerebral neural implants: neuronal replacement and reconstruction of damaged circuitries. Ann Rev Neurosci 7:229–308.

Björklund A, Stenevi U (1985) Intracerebral neural grafting: a historical perspective. In Neural grafting in the mammalian CNS (eds Björklund Am Stenevi U) pp 3–14. Elsevier, Amsterdam.

Björklund A, Nornes H, Gage FH (1986) Cell suspension grafts of noradrenergic locus coeruleus neurons in rat hippocampus and spinal cord: reinnervation and transmitter turnover. Neuroscience 18: 685–698.

Björklund A, Dunnett SB (1992) Neural transplantation in adult rats. In Neural transplantation. A practical approach. (eds Dunnett SD, Björklund A) pp57–78, IRL Press, at Oxford University Press, Oxford.

Björklund A, Campbell K, Sirinathsinghji DJS, Fricker RA, Dunnett SB (1994) Functional capacity of striatal transplants in the rat Huntington model. In Functional neural transplantation (eds Dunnett SB, Björklund A) pp 157–195, Raven Press, New York.

Björklund L, Spenger C, Strömberg I (1997) Tirilazad mesylate increases dopaminergic neuronal survival in the in oculo grafting model. Exp Neurol 148(1):324–333.

Bregman BS (1994) Recovery of function after spinal cord injury: transplantation strategy. In Functional neural transplantation (eds Dunnett SB, Björklund A) pp 489–530. Raven Press, New York.

Brent L (1990) Immunologically privileged sites. In: Pathophysiology of the Blood-Brain Barrier (eds Johansson BB, Owman C, Widner H) pp 383–402, Elsevier, Amsterdam.

Brundin P, Isacson O, Björklund A (1985a) Monitoring of cell viability of embryonic tissue and its use as a criterion for intracerebral graft survival. Brain Res 331:251–259.

Brundin P, Nilsson OG, Gage FH, Björklund A (1985b) Cyclosporin A increases the survival of cross-species intrastriatal grafts of embryonic dopamine neurons. Exp Brain Res 60: 204–208.

Brundin P, Duan W-M, Sauer H (1994) Functional effects of mesencephalic dopamine neurons and adrenal chromaffin cells grafted to the rodent striatum. In Functional neural transplantation (eds Dunnett SB, Björklund A) pp 9–46. Raven Press, New York.

Brüstle O, Maskos U and McKay RDG (1995) Host-guided midration allows targeted introduction of neurons into the embryonic brain. Neuron 15, 1275–1285.

Campbell K, Kalén P, Lundberg C, Wictorin K, Mandel RJ, Björklund A (1993) Characterization of GABA release from intrastriatal striatal transplants: Dependence on host-derived afferents. Neuroscience 53:403–415.

Campbell K, Olsson M, Björklund A (1995) Regional incorporation and site -specific differentiation of striatal precursors transplanted to the embryonic forebrain ventricle. Neuron 15, 1259–1273.

Castro AJ, Tönder N, Sunde NA, Zimmer J (1988) Fetal neocortical transplants grafted to the cerebral cortex of newborn rats receive afferents from the basal forebrain, locus coeruleus and midline raphe. Exp Brain Res 69:613–622.

Czech KA, Larsson L, Wahlgren L, Bennett W Korsgren O, Widner H (1999) Short-term combination immunosuppressive treatment reduces host responses to porcine mesencephalic xenografts. Transplantation, in prep.

Cunningham MG, McKay RD (1993) A hypothermic miniaturized stereotaxic instrument for surgery in newborn rats. J Neurosci Meth 47 (1–2):105–114.

Das GD, Altman J (1971) Transplanted precursors of nerve cells: their fate in the cerebellum of young rats. Science 173:637–638.

Das GD, Altman J (1972) Studies on the transplantation of developing neural tissue in the mammalian brain. I. Transplantation of cerebellar slabs into the cerebellum of neonate rats. Brain Res 98:233–249.

Duan W-M, Brundin P, Grasbon-Frodl E, Widner H (1996) Methylprednisolone prevents rejection of intrastriatal grafts of xenogeneic embryonic neural tissue in adult rats. Brain Res 712:199–212.

Dunn EM (1917) Primary and secondary findings in a series of attempts to transplant cerebral cortex in the albino rat. J Comp Neurol 27:565–582.

Dunnett SB, Whishaw IQ, Bunch ST, Fine A (1986) Acetylcholine-rich neuronal grafts in the forebrain of rats: effects of environmental enrichment, neonatal noradrenaline depletion, host transplantation site and regional source of embryonic donor cells on graft size and acetylcholineesterase-positive fiber outgrowth. Brain Res 378:357–373.

Dunnett SB, Richards S-J (eds) (1990) Neural transplantation. From molecular basis to clinical applications. Elsevier, Amsterdam.

Dunnett SD, Björklund A (1992) Staging and dissection of rat embryos. In Neural transplantation. A practical approach. (eds Dunnett SD, Björklund A) pp 1–19, IRL Press, at Oxford University Press, Oxford.

Dunnett SB (1994) Strategies for testing learning and memory abilities in transplanted rats. In Functional neural transplantation (eds Dunnett SB, Björklund A) pp 217–251, Raven Press, New York.

Dunnett SD, Björklund A (1994) Mechanisms of function of neural grafts in the injured brain. In Functional neural transplantation (eds Dunnett SB, Björklund A) pp 531–567, Raven Press, New York.

Dunnett SD, Björklund A (eds) (1994) Functional neural transplantation. Raven Press, New York.

Dunnett SD, Björklund A (1997) Basic neural transplantation techniques. I. Dissociated cell suspension grafts of embryonic ventral mesencephalon in the adult rat brain. Brain Res Protoc 1(1):91–9.

Fishell G (1995) Striatal precursors adopt cortical identities in response to local cues. Development 121, 803–812.

Foster GA, Schultzberg M, Gage FH, Björklund A, Hökfelt T, Nornes H, Cuello AC, Verhofstad AAJ, Visser TJ (1988) Transmitter expression and morphological development of embryonic medullary and mesencephalic raphé neurons after transplantation to the adult rat central nervous system. I. Grafts to the spinal cord. Exp Brain Res 70:242–255.

Franklin KBJ, Paxinos G (1997) The mouse brain in stereotaxic coordinates. Academic press, San Diego.

Freed WJ (1985) Transplantation of tissues to the cerebral ventricles: methodological details and rate of graft survival. In Neural grafting in the mammalian CNS (eds Björklund A, Stenevi U) pp 31–49. Elsevier, Amsterdam.

Freeman TB, Widner H (eds) (1998) Cell transplantation for neurological disorders. Toward reconstruction of the human central nervous system. Humana Press, Totowa, New Jersey.

Fricker RA, Torres EM, Hume SP, Myers R, Opacka-Juffrey J, Ashworth S, Brooks DJ, Dunnett SB (1997) The effects of donor stage on the survival and function of embryonic striatal grafts in the adult rat brain. II. Correlation between positron emission tomography and reaching behaviour. Neuroscience 79:711–721.

Frodl EM, Sauer H, Lindvall O, Brundin P (1995) Effects of hibernation or cryopreservation on the survival and integration of striatal grafts placed in the ibotenate-lesioned rat caudate-putamen. Cell Transplant 4:571–577

Gage FH, Kempermann G, Palmer TD, Peterson DA, Ray J (1998) Multipotent progenitor cells in the adult dentate gyrus. J Neurobiol, 36(2):249–266.

Gaiano N, Fishell G (1998) Transplantation as a tool to study progenitors within the vertebrate nervous system. J Neurobiol, 36(2):152–161.

Grabowski M, Brundin P, Johansson BB (1992) Vascularization of fetal neocortical grafts implanted in brain infarcts in spontaneously hypertensive rats. Neuroscience 51:673–682.

Grasbon-Frodl EM, Nakao N, Brundin P (1996) Lazaroids improve the survival of embryonic mesencephalic donor tissue stored at 4 °C and subsequently used for cultures or intracerebral transplantation. Brain Res Bull 39:341–347.

Hodges H, Sinden JD, Meldrum BS, Gray JA (1994) Cerebral transplantation in models of ischemia. In Functional neural transplantation (eds Dunnett SB, Björklund A) pp 347–386. Raven Press, New York

Kawaja MD, Fisher LJ, Shinstine M, Jinnah HA, Ray J, Chen LS, Gage FH (1992) Grafting genetically modified cells within the rat central nervous system: methodological considerations. In Neural transplantation. A practical approach. (eds Dunnett SD, Björklund A) IRL Press, at Oxford University Press, Oxford.

Kolb B, Fantie BD (1994) Cortical graft function in adult and neonatal rats. In Functional neural transplantation (eds Dunnett SB, Björklund A) pp 415–436. Raven Press, New York.

Kopyov OV, Jacques S, Kurth M, Philpott L, Lee A, Patterson M, Duma C, Lieberman, A, Eagle KS (1998) Fetal transplantation for Huntington´s disease: Clinical studies. In: Cell transplantation for neurological disorders. Toward reconstruction of the human central nervous system (eds Freeman TB, Widner H) pp 95–134, Humana Press, Totowa, New Jersey.

LeGros Clark WE (1940) Neuronal differentiation in implanted foetal cortical tissue. J Neurol Psychiat 3:263–284.

Lindvall O (1994) Neural transplantation in Parkinson´s disease. In Functional neural transplantation (eds Dunnett SB, Björklund A) pp 103–138. Raven Press, New York.

Liu A, Joyner AL, Turnbull DH (1998) Alteration of limb and brain patterning in early mouse embryos by ultrasound-guided injection of Shh-expressing cells. Mechanisms of Development 75, 107–115.

Lund RD, Rao K, Hankin M, Kunz HW, Gill III TJ (1987) Transplantation of retina and visual cortex to rat brains of different ages. Maturation, connection patterns, and immunological consequences. In Cell and tissue transplantation into the adult brain (eds Azmitia AC, Björklund A) pp 227–241, Ann NY Acad Sci 495, New York.

Lund RD, Banarjee R (1992) Immunological considerations in neural transplantation. In Neural transplantation. A practical approach. (eds Dunnett SD, Björklund A) pp 161–176, IRL Press, at Oxford University Press, Oxford.

Lund RD, Yee KT (1992) Intracerebral transplantation to immature hosts. In Neural transplantation. A practical approach. (eds Dunnett SD, Björklund A) pp 79–91, IRL Press, at Oxford University Press, Oxford.

Marchand R, Lajoie R (1986) Histogenesis of the striatopallidal system in the rat. Neuroscience 17:573–590.

Nakao N, Frodl EM, Duan W-M, Widner H, Brundin P (1994) Lazaroids improve the survival of grafted rat embryonic dopamine neurons. Proc Natl Acad Sci USA 91:12408–12412.

Nikkhah G, Olsson M, Eberhard J, Bentlage C, Cunningham MG, Björklund A (1994a) A microtransplantation approach for cell suspension grafting in the rat Parkinson model. A detailed account of the methodology. Neuroscience 63:57–72.

Nikkhah G, Bentlage C, Cunningham MG, Björklund A (1994b) Intranigral fetal dopamine grafts induce behavioural compensation in the rat Parkinson model. The Journal of Neuroscience 14:3449–3461.

Nikkhah G, Eberhard J, Olsson M, Björklund A (1995) Preservation of fetal ventral mesencephalic cells by cool storage: In vitro viability and TH-positive neuron survival after microtransplantation to the striatum. Brain Research 687:22–34.

Nilsson OG, Clarke DJ, Brundin P, Björklund A (1988) Comparison of growth and reinnervation properties of cholinergic neurons grafted to the deafferented hippocampus. J Comp Neurol 268: 204–222.

Olson L, Malmfors T (1970) Growth characteristics of adrenergic nerves in the adult rat. Fluorescence histochemical and 3H-noradrenaline uptake studies using tissue transplantations to the anterior chamber of the eye. Acta Physiol Scand Suppl 348:1–112.

Olson L, Seiger Å (1972) Early prenatal ontogeny of central monoamine neurons in the rat: fluorescence histochemical observations. Z Anat Entwickl 137:301–316.

Olsson M, Campbell K, Wictorin K, Björklund A (1995) Projection neurons in fetal striatal transplants are predominantly derived from the lateral ganglionic eminence. Neuroscience 69, 1169–1182.

Olsson M, Campbell K, Turnbull D (1997) Specification of mouse telencephalic and mid-hindbrain progenitors following heterotopic ultrasound-guided transplantation. Neuron 19, 761–772.

Olsson M, Bjerregaard C, Winkler C, Gates M, Campbell K, Björklund A (1998a) Incorporation of mouse neural progenitors transplanted into the rat embryonic forebrain is developmentally regulated and dependent on regional and adhesive properties. The European Journal of Neuroscience 10, 71–85.

Olsson M, Björklund A, Campbell K (1998b) Early specification of striatal projection neurons and interneuronal subtypes in the lateral and medial ganglionic eminence. Neuroscience 84:867–876.

Pakzaban P, Deacon TW, Burns LH, Isacson O (1993) Increased proportion of acetylcholinesterase-rich zones and improved morphological integration in host striatum of fetal grafts derived from the lateral but not medial ganglionic eminence. Exp Brain Research 97:13–22.

Pakzaban P, Isacson O (1994) Neural xenotransplantation: reconstruction of neuronal circuitry across species barriers. Neuroscience 62 (4):989–1001.

Paxinos G, Watson C (1986) The rat brain in stereotaxic coordinates. Second edition. Academic press, Australia.

Paxinos G, Watson C (1997) The rat brain in stereotaxic coordinates. Compact third edition. Academic press, San Diego.

Pedersen EB, Poulsen FR, Zimmer J, Finsen B (1995) Prevention of mouse-rat brain xenograft rejection by a combination therapy of cyclosporin A, prednisolone and azathioprine. Exp Brain Research 106 (2):181–186

Peschanski M, Isacson O (1988) Fetal homotypic transplants in the excitotoxically neuron depleted thalamus. I. Light microscopy. J Comp Neurol 274:449–463.

Ranson SW (1909) Transplantation of the spinal ganglion into the brain. Quart Bull Northwest Univ Med School 11:176–178.

Sauer H, Brundin P (1991) Effects of cool storage on survival and function of intrastriatal ventral mesencephalic grafts. Restor Neurol Neurosci 2:123–135.

Sauer H, Frodl EM, Kupsch A, ten Bruggencate G, Oertel WH (1992) Cryopreservation, survival and function of intrastriatal fetal mesencephalic grafts in a rat model of Prkinson´s disease. Exp Brain Res 90:54–62.

Schierle GS, Hansson O, Leist M, Nicotera P, Widner H, Brundin P (1999) Caspase inhibition reduces apoptosis and increases survival of nigral transplants. Nature Medicine 5(1):97–100.

Seiger Å (1985) Preparation of immature central nervous system regions for transplantation. In Neural grafting in the mammalian CNS (eds Björklund Am Stenevi U) pp 71–77. Elsevier, Amsterdam.

Semba K, Fibiger HC (1988) Time of origin of cholinergic neurons in the rat basal forebrain. J Comp Neurol 269:87–95.

Sloan DJ, Baker BJ, Puklavec M, Charlton HM (1990) The effect of site of transplantation and histocompatibility differences on the survival of neural tissue transplanted to the CNS of defined inbred rat strains. Prog Brain Res 82: 141–152.

Smart IHM, Sturrock RR (1979) Ontogeny of the neostriatum. In: The neostriatum (Divac I, Öberg RGE eds) pp 127–146, Pergamon Press, Oxford.

Sotelo C, Alvarado-Mallart RM (1987) Reconstruction of defective cerebellar circuitry in adult Purkinje cell degeneration mutant mice by Purkinje cell replacement through transplantation of solid embryonic implants. Neuroscience 20:1–22.

Stenevi U, Kromer LF, Gage FH, Björklund A (1985) Solid neural grafts in intracerebral transplantation cavities. In Neural grafting in the mammalian CNS (eds Björklund A, Stenevi U) pp 41–49. Elsevier, Amsterdam.

Studer L, Tabar V, McKay RDG (1998) Transplantation of expanded mesencephalic precursors leads to recovery in parkinsonian rats. Nature Neurosci, 1(4):290–295.

Sunde NA, Zimmer J (1983) Cellular histochemical and connective organization of the hippocampus and fascia dentata transplanted to different regions of immature and adult rat brains. Dev Brain Res:165–191.

Swanson LW (1992) Brain maps: structure of the rat brain

Thompson WG (1890) Successful brain grafting. NY Med J 51:701–702.

Turnbull DH, Starkosi BG, Harasiewicz KA, Semple JL, From L, Gupta AK, Sauder DN, Foster FS (1995a) A 40–100MHz B-scan ultrasound backscatter microscope for skin imaging. Ultrasound Med Biol 21, 79–88.

Turnbull DH, Bloomfield T, Baldwin HS, Foster FS, Joyner AL (1995b) Ultrasound backscatter microscope analysis of early mouse embryonic brain development. Proc. Natl. Acad. Sci. USA 92, 2239–2243.

Tönder N, Sörensen T, Zimmer J, Jörgensen MB, Johansson FF, Diemer NH (1989) Neural grafting to ischemic lesions of the adult rat hippocampus. Exp Brain Res 74:512–526.

Watts C, Caldwell MA, Dunnett SB (1998) The development of intracerebral cell-suspension implants is influenced by the grafting medium. Cell Transplant 7:573–583.

Wictorin K, Ouimet CC, Björklund A (1989) Intrinsic organization and connectivity of intrastriatal striatal transplants in rats as revealed by DARPP-32 immunohistochemistry: specificity of connections with the lesioned host brain. Eur J Neurosci 1:690–701.

Wictorin K (1992) Anatomy and connectivity of intrastriatal striatal transplants. Prog Neurobiol 38:611–639.

Widner H, Brundin P (1988) Immunological aspects of grafting in the mammalian central nervous system. A review and speculative synthesis. Brain Res Rev 13:287–324.

Widner H (1995). Transplantation of neuronal and non-neuronal cells into the brain. In Immune response in the nervous system (ed Rothwell NJ), pp 189–217, Bios Scientific Publishers.

Wood MJA, Charlton HM (1994) Hypothalamic grafts and neuroendocrine function. In Functional neural transplantation (eds Dunnett SB, Björklund A) pp 451–466. Raven Press, New York.

Wood MJA, Sloan DJ, Wood KJ, Charlton HM (1996) Indefinite survival of neural xenografts induced with anti-CD4 monoclonal antibodies. Neuroscience 70 (3): 775–789.

Chapter 15

Histological Staining Methods

R. W. Banks

Introduction

In this chapter we shall examine the development and application of histological methods at the tissue and system level of neural organisation. The principal techniques – fixation, dyeing, metal impregnation and histochemistry – have already been encountered in Chapter 1, where the emphasis was at the cytological level. The two chapters are intended to be considered together, thus the specific techniques covered in each one are applicable at both cellular and tissue levels. Once again, the methods to be described here will, whenever possible, be placed in the context of particular functional problems. There are, however, no hard and fast rules about which is the most appropriate method to apply to a given problem; this will depend on many factors not all under the investigator's control, including one's experience, material resources and availability of imaging and data-recording facilities. Of course, it should be borne in mind that different methods will almost certainly generate somewhat different data while addressing the same problem. This point will be illustrated in example 4 below with a case study from my own work on the innervation of muscle spindles.

Subprotocol 1 Architectonics

Introduction

"[Architectonic] studies assume, with justification, that there are diversities of structure in the cortex which are reliable indices of functional differentiation. It is a far step from this assumption, however, to the conclusion that every recognizable difference in cortical structure represents a functional differentiation." (Lashley and Clark, 1946).

At the tissue level of neural organisation one of the principal concerns of neurohistologists has been the characterisation and delimitation of recognisable regions of the nervous system, such as cortical areas and specific nuclei. Classically this has relied on scrutiny of the size, shape and number of nerve cells, particularly of their somata and proximal dendrites (cytoarchitectonics), and on the density and disposition of myelinated axons (myeloarchitectonics). Although it is essentially microanatomical in its approach, architectonics has always been based on the premise of the existence of at least partial congruity between structurally and functionally differentiated regions of the

R. W. Banks, University of Durham, Department of Biological Sciences, South Road, Durham, DH1 3LE, UK (phone: +44-191-374-3354; fax: +44-191-374-2417; e-mail: r.w.banks@durham.ac.uk)

nervous system. Histological and physiological studies should therefore complement and inform each other at the tissue level just as at the cellular level.

We may trace the origin of architectonics to the work of Meynert (1867–1868), who gave a detailed description of the horizontal layering of the cerebral cortex and recognised that the basic pattern was subject to regional variation. But this was more than ten years before the techniques were introduced that were to transform the subject: Weigert's (1882) acid fuchsin method for myelin, and Nissl's (1894) basic aniline dye method for nerve-cell somata. Maps and atlases followed in the wake of the new techniques: though they may have been considered by their authors to be provisional (for example, Vogt and Vogt, 1919), their utility as topographical standards has ensured their frequent and typically uncritical citation (for example, that of Brodmann, 1908). The greatest problem with the early maps was the large element of subjectivity required in their construction, as was forcefully demonstrated by Lashley and Clark (1946). Nevertheless, it was clear that the topographical variations described by the maps did frequently coincide with important functional divisions, and highly influential maps continued to be produced by the same subjective means, such as that by Rexed (1954) of the cat's spinal cord. With automatic image analysis now made possible and ever-increasingly powerful by the microcomputer, a more objective approach to architectonics has been developed (Zilles, Schleicher and Kretschmann, 1978) and the resulting maps are now appearing in recent atlases (e.g. Paxinos and Watson, 1998). Objective and unbiased measurement is crucial to architectonics, but consideration of quantitative methods is beyond the scope of this chapter. For a recent review of stereological techniques, including several references to neuroanatomy and neurohistology, see Mayhew (1991).

Current histological techniques required for architectonic analysis are the direct descendants of those developed by Weigert and Nissl. We shall look first at the Nissl method. The introduction of methylene blue (Fig. 1A) as a vital stain of nerve cells by Ehrlich in 1886 was quickly followed by its adoption by Nissl (1894) for use on sections of alcohol-hardened brain. It is perhaps ironic that, whereas Nissl objected to speculation about the chemical nature of the chromophilic material and instead concentrated on its detailed structural arrangement, naming several different types (Clarke and O'Malley, 1996), today the chemical nature is taken for granted (the RNA content of ribosomes) and we glibly and indiscriminately speak only of Nissl granules. The method itself continues to be of great importance for, among others, the construction of architectonic maps as reference standards, the location of electrode placements or experimental lesions, and as counterstains for Golgi and intracellular staining techniques (see Chap. 1). Nissl staining can be produced using a variety of basic dyes that range in colour from red, through purple and violet, to blue. Windle, Rhines and Rankin (1943) reviewed several of these dyes and recommended a veronal-acetate buffered solution (pH 3.65) of thionine (Fig. 1B). Cresyl violet was introduced as a Nissl stain by Powers and Clark (1955), who also recommended an acetate buffer (at pH 3.5). It is used in example 1 below as the unbuffered acetate (Fig. 1C).

In his original myelin stain Weigert used the dye acid fuchsin, a sulphonated triarylmethane compound derived from rosaniline (Fig. 1D), but the method is better known today in one of its variants that use haematoxylin, such as Weigert-Pal. Weigert could only obtain staining after prolonged fixation of the tissue with potassium dichromate, which thus acts as a mordant for the dye as well as a fixative for lipids (Baker, 1958). It is supposed that the dye forms coordination compounds or ionic bonds with the chromium. The process was completed by differentiation using alkaline ethanol. In example 1 we use the dye chromoxane cyanine R (solochrome cyanine, Fig. 1E), which was introduced as a myelin stain by Page (1965). Like acid fuchsin, it is a triarylmethane compound; to stain myelin it is used at pH 1.5 with a mordant containing ferric ions with which it forms several coordination compounds, the most stable being $[Fe_2(dye)]^{2-}$ (Kiernan, 1984a,b).

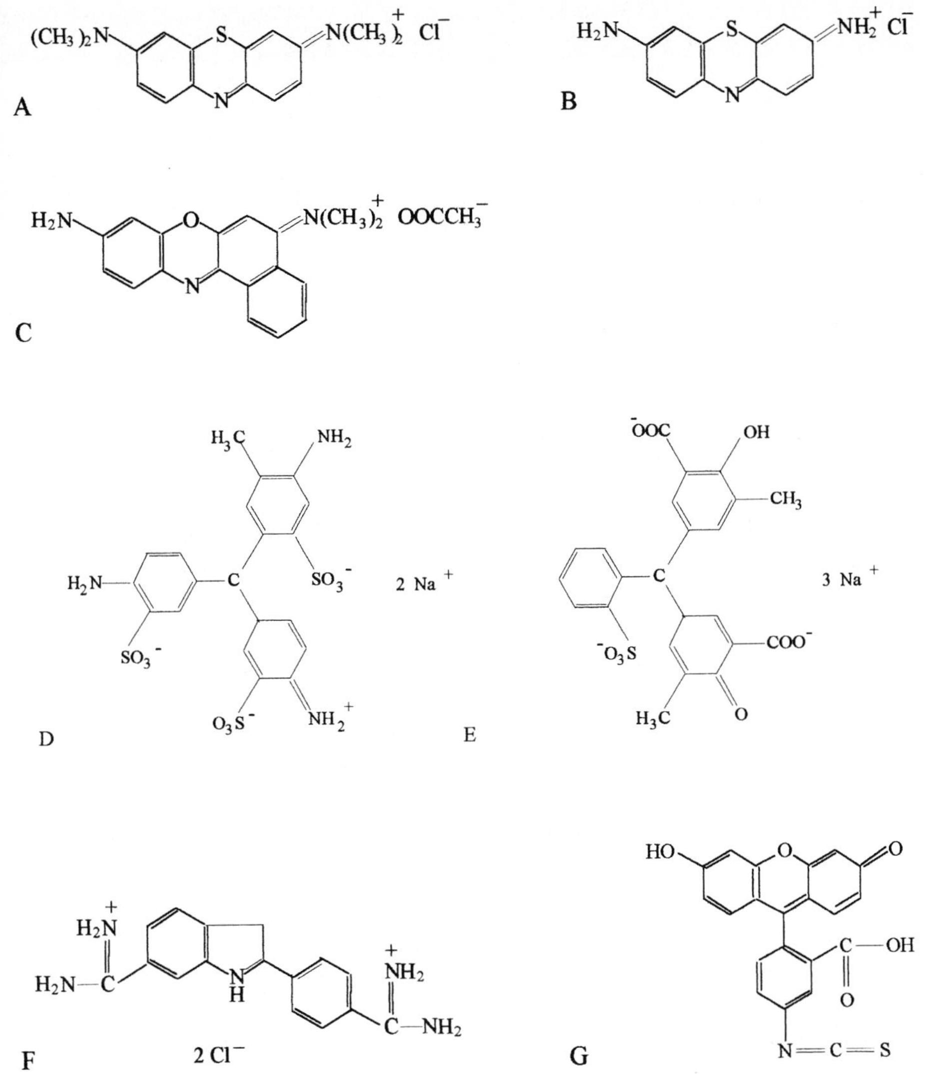

Fig. 1. Structural formulae of various dyes and chromogens mentioned in the text. A: Methylene blue. B: Thionine. C: Cresyl violet acetate. D: Acid fuchsin. E: Chromoxane cyanine R (solochrome cyanine). F: 4',6-diamidino-2-phenylindol (DAPI). G: Fluorescein isothiocyanate (FITC).

Example 1: Nissl granules and myelin

Materials

Cerebrum of the adult rat.

Procedure

Fixation

1. Systemic perfusion as in example 2, Chapter 1, but using formal-saline (approx. 4% formaldehyde in 0.85% saline).
2. Brain removed and stored in fresh formal-saline.
3. Coronal slices cut manually with a single-edged razorblade, about 2 mm thick.

Dehydration and embedding

1. Dehydrated through a graded series (70%, 95%, absolute) of ethanol, 3 changes of 4 hrs in each. 3 changes of 4 hrs each, or until clear, in Histo-Clear (National Diagnostics).
2. Embedded by immersion for 3 changes of 8 hrs each in molten paraffin wax (about 55 °C).

Sectioning and staining

1. Sections cut at 10 μm.
2. Wax removed with Histo-Clear, and sections hydrated through a graded series of ethanol [absolute, 95%, 70%]
3. Staining [note that in this example the different staining methods were carried out on separate sections]

   Nissl granules
   a) Nissl's original method was highly regressive, which is to say that the tissue was grossly overstained and then differentiated, and a regressive technique continues to be a feature of most modern methods.
   b) stained for 20 s with 0.1% cresyl violet acetate [The time required for adequate staining can be very variable, depending on the source of the dye, the age and pH of the solution, the age of the tissue and probably other factors.]
   c) dehydrated through a graded series of ethanol [70%, 95%, absolute], cleared with Histo-Clear
   d) differentiated with 95% ethanol to reduce background staining
   e) dehydration with absolute ethanol completed and sections again cleared with Histo-Clear. Step c. may be repeated, if necessary.

   myelin (Kiernan, 1984b)
   a) stained for 15–20 min with chromoxane cyanine R [0.21 M aqueous ferric chloride (5.6% w/v $FeCl_3.6H_2O$), 20 ml; chromoxane cyanine R, 1.0 g; conc. (95–98% w/w) H_2SO_4, 2.5 ml; water to 500 ml]
   b) washed in tap water to remove excess dye
   c) differentiated in ferric chloride (5.6%)
   d) washed in tap water (5 min)
   e) [may be counterstained for Nissl granules with 0.5% neutral red]
   f) dehydrated through a graded series of ethanol [70%, 95%, absolute], cleared with Histo-Clear

4. Sections mounted with DPX.

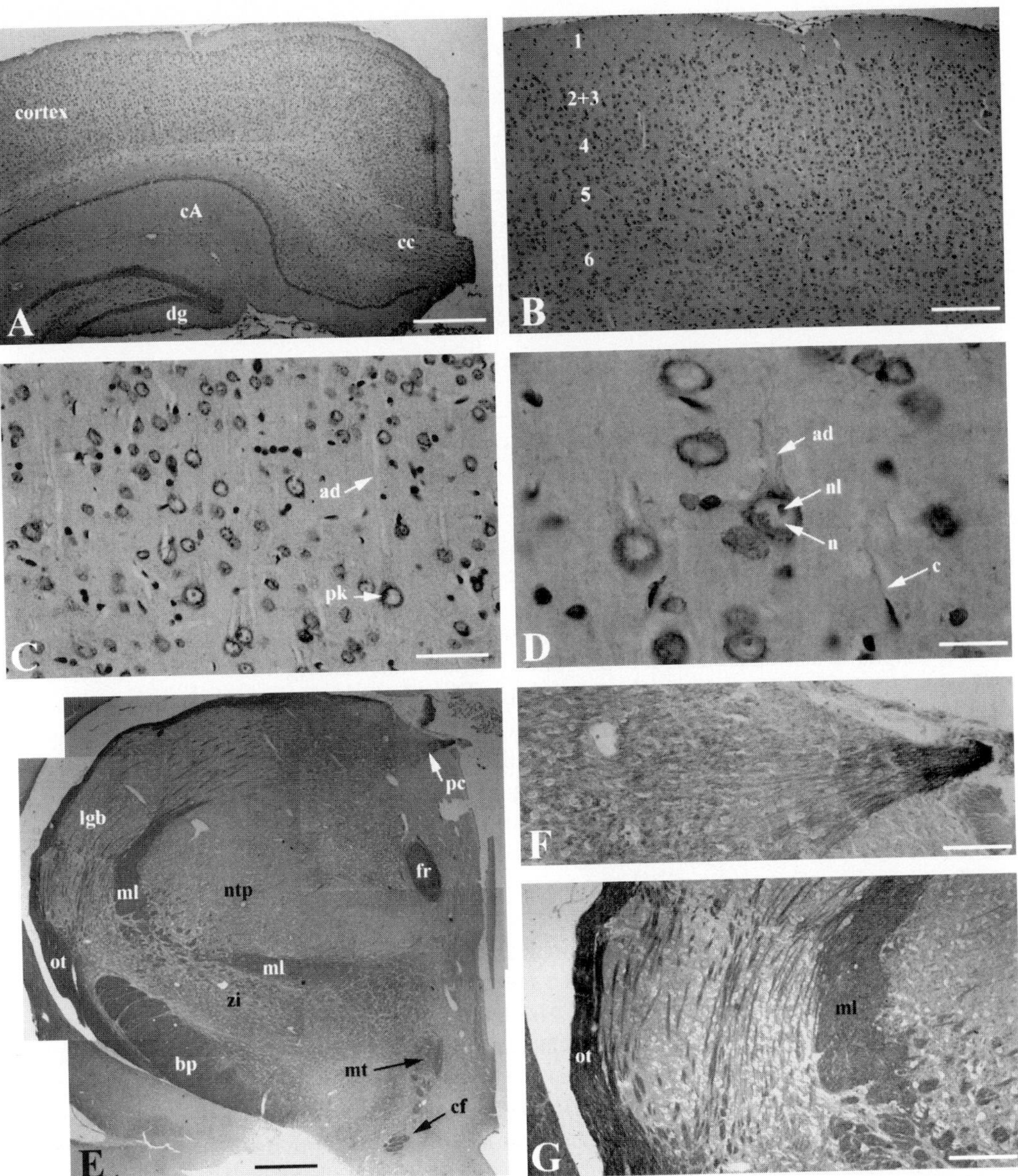

Fig. 2. Coronal sections of rat brain to show examples of staining for Nissl granules using cresyl violet acetate (A – D) and for myelin using chromoxane cyanine R (E – F). A, the medio-dorsal cerebrum, showing the cortex and passing through the corpus callosum (cc) and the cornu Ammonis (cA) and dentate gyrus (dg) of the hippocampal formation. The compact layers of neuronal perikarya making up the stratum pyramidale of cA and the stratum granulosum of dg are particularly clear. Scale bar = 500 µm. B, the full depth of the dorso-lateral cortex from the same section as A. Horizontal layers may be discerned on the basis of perikaryal size and distribution. The layers are numbered conventionally, layer 1 being the outermost, adjacent to the free surface of the cortex which retains a partial covering of arachnoid mater. Scale bar = 200 µm. C, detail of cortical layer 5. The largest neurons, one of whose perikaryon (pk) and apical dendrite (ad) are indicated, are deep pyramidal cells. Scale bar = 50 µm. D shows a field of view containing a few deep pyramidal cells. The Nissl granules are confined to the perikarya and proximal dendrites, the euchromatic nuclei (n) of the neurons appearing unstained, though the nucleoli (nl) are intensely stained. The endothelial cells of capillaries (c) have heterochromatic nuclei that are also intensely stained, as are those of most glial cells. Scale bar = 20 µm. E, a section through the posterior nucleus of the thalamus (ntp), showing several major myelinated fibre tracts: the basis pedunculi (bp), the column of the fornix (cf), the fasciculus retroflexus (fr), the medial lemniscus (ml), the mamillothalamic tract (mt), the optic tract (ot) and the posterior commissure (pc). The optic tract supplies small bundles of fibres to the lateral geniculate body (lgb); numerous similarly small bundles of fibres, mainly forming reciprocal connexions between the thalamus and cerebral cortex, are scattered throughout the zona incerta (zi). Scale bar = 500 µm. F, detail of the posterior commissure, showing radiating bundles of myelinated fibres. Scale bar = 100 µm. G, detail of the lateral geniculate body. Scale bar = 200 µm.

Results

The perikarya and proximal dendrites of neurons are filled with purplish Nissl granules after staining with cresyl violet acetate. The euchromatic nuclei of neurons are virtually unstained with the exception of the usually single and prominent nucleolus in each one, which is intensely stained. Heterochromatic nuclei of glial, ependymal and endothelial cells are heavily stained with an intensity roughly proportional to the degree of condensation of chromatin. After chromoxane cyanine staining, myelin appears intensely blue against a greyish background of neuropil. Necrotic cells and erythrocytes are especially heavily stained, so for the best results good quality perfusion fixation is necessary.

Subprotocol 2 Hodology

Introduction

"The transection of axons in vertebrates leads to marked morphological changes, peripheral as well as central to the lesion." Grant and Walberg, 1974.

Whereas the Golgi method could provide information on local circuits of neurons, other techniques were needed to trace the routes taken by axons in making long-distance connexions. Classically, these relied on the morphological changes referred to in the quotation from Grant and Walberg; more recently methods involving the use of various tracers have been developed, but these ultimately depend on the same cellular processes that underlie the morphological changes. Waller's demonstration of the degeneration of myelin distal to a nerve section provided the basis for the earliest tract-tracing studies: by Türck, whose observations were made on pathological human tissue, and by Gudden, who used experimental lesions in animals (Clarke and O'Malley, 1996). As early as 1886, Marchi provided a specific stain for degenerating myelin using a mixture of potassium dichromate and osmium tetroxide (Clarke and O'Malley, 1996), which allowed many of the major myelinated pathways to be traced; but it was to be another 60 years before primary changes in the axons themselves and their terminals were intensively studied, following Glees's (1946) modification of the Bielschowsky reduced silver method (see below). Glees's method, which stained both normal and degenerating fibres, was soon complemented by another Bielschowsky variant in which the argyrophilia of the normal fibres was suppressed by pretreatment involving phosphomolybdic acid (Nauta and Gygax, 1951). Interpretation of the results obtained with silver impregnation was always difficult, partly due to the need to assess the quality of impregnation and partly because of a long debate concerning the identity of the terminal structures that were being stained. These types of problem are not unique to silver methods, however, and eventually could be controlled to large extent when the increased resolution provided by the electron microscope became widely available. Broadly speaking, in the non-suppressive techniques argyrophilia was found to be associated with neurofilaments, so those terminals in which neurofilaments were sparse or lacking did not stain (Heimer and Ekholm, 1967; Walberg, 1971). The addition of the suppressive pretreatment abolished this neurofilament-associated argyrophilia and replaced it with one based on membranous structures including mitochondria; it is supposed that the phosphomolybdic acid preferentially occupies the argyrophilic sites of the neurofilaments (Lund and Westrum, 1966), but the physico-chemical basis for the specific staining of degenerating membranous structures remains unclear. Despite increased knowledge of the rational basis of the reduced silver methods, and the consequently greater control that can be exercised over their variable elements, they are no longer widely used in hodology and technical protocols will not be given here: the interested reader is referred to the review by de Olmos, Ebbesson and Heimer (1981). The characteristic ultrastructural changes that take place during synaptic bouton degeneration (Colonnier and Gray, 1962; Grant and Walberg, 1974) have been used to analyse fine details of connectivity, often in combination with one or more light microscopical techniques (see, for example, Somogyi, Hodgson and Smith, 1979, and for a general consideration of ultrastructural studies in neurohistology see Chapter 1).

With the cytological detail afforded by his new method Nissl was able to show that, in addition to the anterograde Wallerian degeneration, changes retrograde to the lesion also occurred, characterised by the temporary dissolution of the chromophilic granules. This chromatolytic reaction for a while provided another important means of tract-tracing (Brodal, 1940). Furthermore, if the neurons themselves died as a result of the injury they could be specifically stained by a Nauta method (Grant and Aldskogius, 1967). Neurons in juvenile animals were particularly susceptible, which was generally seen as being due to a continuing dependence on peripherally derived trophic factors. Indeed, the trophic interdependence of neuron and target, as well as the growth and maintenance of the axon itself, were widely supposed to depend on the movement of substances, in particular proteins, both anterogradely and retrogradely along axons. We may trace the origin of the modern family of tract-tracing methods involving the uptake of various kinds of markers to two studies whose primary purpose was to elucidate aspects of the axonal transport process: firstly Droz and Leblond (1963), who used autoradiography to follow the anterograde movement of proteins after giving a pulse-label of radioactive amino acid; and secondly Kristensson's and Olsson's (1971) enzyme histochemical demonstration of the uptake and retrograde transport by motorneurons of an exogenous protein, horseradish peroxidase, after its injection into a muscle. Indeed, just as the use of Procion yellow as an intracellular label catalysed the development of a range of techniques for marking physiologically identified single cells *in vivo* (Chapter 1), so these proofs of the existence of the anterograde and retrograde components of axonal transport may be seen as the catalysts for the modern methods of tract tracing.

Perhaps the greatest advantage of the new methods is their very variety; not only can each act as a check on the others, but much more importantly they may be used in combination on the same material to reveal information about collateral pathways or topographic organisation in a highly efficient manner (Steward, 1981). Furthermore, they can also be used in conjunction with various other techniques such as immunohistochemistry and *in situ* hybridization for correlative studies on neural connexions, transmitters and receptors (Skirboll et al., 1989; Chronwall et al., 1989). However, it should not be supposed that problems of the types associated with silver impregnation were immediately and automatically solved. The most serious of these, and one which applies, in principle, to many hodological methods, is that of the so-called fibres of passage (Graybiel, 1975). Thus studies involving the identification of degenerating axons and terminals must always take into account that, no matter how precise and localised the site of the experimental lesion, axons not arising from the cells killed by the lesion are likely, and frequently certain, to be damaged also. When the purpose is not to kill cells, but to induce them to take up one or other marker substances, the iontophoretic or pressure injection is still likely to cause some damage, and there is the possibility that some markers may be taken up not only by cell bodies or by axonal terminals, but also by normal axons along their length. For similar reasons it is normally desirable that the marker should not pass easily from one neuron to another, at least not in an uncontrolled way.

Autoradiography, the first modern method of tract tracing, largely overcame the fibres of passage problem and provided a specifically anterograde technique (Cowan et al., 1972), presumably because amino acids and sugars are taken up in significant amounts only by the somata, and it is in this region of the neuron that proteins are synthesized. Cowan et al. (1972) noted that the newly synthesized proteins were transported at two very different rates: i) >100 mm/day, with accumulation mainly in presynaptic terminals, and ii) 1–5 mm/day, the protein in this case tending to remain in the axon. They pointed out that by selecting the survival time after injection of the radioactive tracer, information could be obtained both on the detailed patterns of synaptic connectivity (by electron microscopy) and on the pathway followed by the axons to their destinations (by light microscopy). Different levels of terminal and axonal labelling also result from selection of different amino acids, and, if desired, transneuronal marking can be obtained using proline or fucose. For a consideration of the choice of marker substance and other technical matters, see Edwards and Hendrickson (1981). Autoradiography therefore provided a very powerful hodological method (Droz, 1975) whose principal drawback is the length of time needed to make the radiographic exposure.

The horseradish peroxidase method relies on the enzymatic cleavage of the substrate hydrogen peroxide and its reduction to water by the coupled oxidation of a suitable hydrogen donor. The method has undergone many changes and improvements since it was first introduced for retrograde tract tracing in the central nervous system by LaVail and LaVail (1972), who used diaminobenzidine (DAB) at pH 7.2–7.4 as the hydrogen donor. Among these changes we may cite: i) the reduction in pH of the incubation medium using DAB to 4.9–5.3 [it must be remembered that the pH optimum for a histochemical reaction may be quite different from the *in vitro* optimum of the enzyme, due to fixation, dissociation constant of the hydrogen donor, temperature and other factors] (Malmgren and Olsson, 1978); ii) a further increase in sensitivity by the use of tetramethyl benzidine (TMB) as donor at pH 3.3, which, as a result, showed that horseradish peroxidase could also be taken up by cell bodies and transported anterogradely (Mesulam, 1978) [this pH dissociates antigen:antibody complexes, so the TMB reaction cannot be used for immunohistochemistry]; iii) enhancement of horseradish peroxidase uptake using a membrane solubilizer (dimethyl sulphoxide; Keefer, 1978) or a lectin-horseradish peroxidase conjugate (Gonatas et al., 1979); iv) combination of enhanced uptake with TMB as hydrogen donor, resulting in improved tracing of anterograde (West and Black, 1979) and transganglionic (Brushart and Mesulam, 1980) projections; and v) the development of a section-on-the-slide incubation procedure (Sickles and Oblak, 1983). For a monograph on the use of horseradish peroxidase in tract tracing see Mesulam (1982).

The use of fluorescent dyes in hodology also has its origin in the studies of protein uptake and retrograde transport by Kristensson, whose original tracer (1970) was albumen stained with Evans blue; indeed, it was the necessity to demonstrate that protein and not just the dye was being transported that subsequently led Kristensson and Olsson (1971) to try horseradish peroxidase. Somewhat surprisingly, therefore, it was not until 1977 that Kuypers et al. first introduced fluorescent dyes as axonal tracers in their own right, explicitly to provide the possibility of double labelling of neurons that project to more than one site, by a simpler means than combining horseradish peroxidase and autoradiography. Initially Kuypers et al. (1977) combined Evans blue (fluoresces red) with a mixture of primuline (cytoplasmic granules fluoresce golden) and 4',6-diamidino-2-phenylindol (DAPI, Fig. 1F, light blue fluorescence), then shortly after (1979) they also introduced bisbenzimide (yellow-green) and propidium iodide (orange). In 1982 Aschoff and Hollander examined the properties of nine fluorescent compounds as retrograde tracers in comparison with horseradish peroxidase and concluded that all four diamidino compounds tested (DAPI, Fast blue [trans-1-(5-amidino-2-(6-amidino-2-indolyl)-ethylene-dihydrochloride)], Granular blue [2-(4-(2-(4-amidinophenoxy)ethoxy) phenyl)indol-6-carboxamidino-dihydrochloride], and True blue [trans-1,2-bis(5-amidino-2-benzofuranylethylene)-hydrochloride] gave similar results to horseradish peroxidase; whereas bisbenzimide and Nuclear yellow (benzimidazols) gave false positive results by anterograde transport, and propidium iodide, Evans blue and primuline all stained fewer cells than the enzyme.

With the development of epifluorescence and subsequently confocal microscopy, fluorescent methods have become progressively more important in microtechnique in general, using the dyes both directly, as in the early tract-tracing methods just described, and indirectly by covalent bonding to antibodies or avidin. With indirect fluorescence in tract tracing the primary tracer might be biotin or one of its derivatives (see Chapter 1), or a lectin such as *Phaseolus vulgaris* leucoagglutinin (PHA-L; Gerfen, Sawchenko and Carlsen, 1989). The affinities of particular lectins for specific combinations of saccharide groups may also be exploited to differentiate neuronal subtypes that express different glycoproteins (e.g. Mori, 1986).

Example 2: The striatonigral projection in the rat (from Gerfen,1985)

In this case study we shall consider only the simultaneous dual fluorescent labelling used by Gerfen to map the spatial distribution of the striatonigral pathway, by retrograde transport of Fast blue and diamidino yellow. In addition, Gerfen used anterograde transport of *Phaseolus vulgaris* leucoagglutinin demonstrated by the avidin-biotin immunoperoxidase technique, and a mixture of [^{3}H]proline and [^{3}H]leucine demonstrated by autoradiography; and he also combined retrograde fluorescence tracing using Fast blue with an indirect immunofluorescent technique to demonstrate the presence of various neuropeptides in the striatum. For further details of these methods see Gerfen (1985).

Materials

Cerebrum of the adult rat.

Procedure

Dye injection

30 nl each of Fast blue and diamidino yellow injected into different sites in the same substantia nigra using a 1-µl Hamilton syringe. [Fast blue stains the cytoplasm and diamidino yellow the nuclei of the retrogradely labelled neurons. A survival time of 22 h limits the diffusion of diamidino yellow into neuronal cytoplasm or surrounding glial cells.]

Terminal anaesthesia and fixation

1. Chloral hydrate overdose
2. Perfusion (see Chapter 1, example 2) successively with:
   a) 100 ml 0.9% saline, 250 ml 4% formaldehyde (as paraformaldehyde) in 0.1 M sodium acetate buffer pH 6.5
   b) 350 ml 4% formaldehyde in 0.1 M sodium borate buffer pH 9.5.
3. Brains removed and fixed overnight in the last perfusate plus 20% sucrose.

Sectioning and microscopy

1. Frozen 30-µm sections
2. Epifluorescence – excitation 330–380 nm, barrier 420 nm.

Results

Dual labelling of separate sites within the same substantia nigra illustrate the topography of the striatonigral projection in that components of the rostrocaudal and mediolateral dimensions of the corpus striatum are transformed into the mediolateral dimension of the substantia nigra (Fig. 3). The topographic component is, however, somewhat approximate or blurred, since relatively well separated injection sites result in overlapping zones of stained cells in the corpus striatum (Fig. 3B, B').

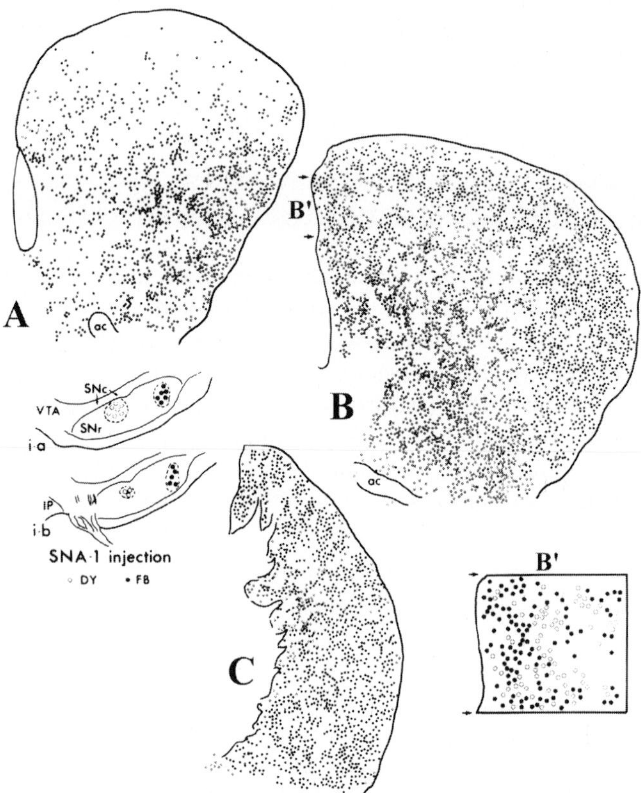

Fig. 3. Diagrammatic representation of sample results from an experiment to demonstrate the organisation of the striatonigral projection in the rat, by means of retrograde transport of the fluorescent dyes diamidino yellow (DY) and fast blue (FB) injected into separate sites in the substantia nigra, pars reticulata (SNr). The injection sites are shown in two sections through the substantia nigra, i.a and i.b, respectively rostrad and caudad. A – C, coronal sections through the corpus striatum, rostrad to caudad, the symbols marking the positions of perikarya labelled with one or other dye. Double-labelled cells (5 %, evenly distributed) are not shown. FB-labelled cells predominate at levels B and C, especially in more lateral locations, where they occur virtually exclusively. DY-labelled cells predominate at level A, but are restricted to progressively more medial locations in levels B and C. They are always intermingled with some FB-labelled cells, as shown in the detail of B enlarged as B' (DY-labelled cells, open circles; FB-labelled cells, filled circles. In the original figure, filled and open circles are also used in A – C; as reproduced here they may still be distinguishable with a hand lens). The medio-lateral axis of the substantia nigra thus maps approximately into the rostro-caudal and medio-lateral axes of the corpus striatum. ac, anterior commissure; IP, interpeduncular nucleus; SNc, substantia nigra, pars compacta; VTA, ventral tegmental nucleus of Tsai. [Partly relabelled from Gerfen, C. R. (1985). The neostriatal mosaic. I. Compartmental organization of projections from the striatum to the substantia nigra in the rat. J. Comp. Neurol. 236, 454–476. Copyright © 1985 Wiley-Liss. Reprinted by permission of Wiley-Liss, a subsidiary of John Wiley & Sons]

Subprotocol 3
Histochemical Methods: Neurochemistry and Functional Neurohistology, Including the Molecular Biology of Neurons

Introduction

"[..] Voss [..] invented the term histotopochemistry specifically to signify the prime concern of histochemistry with localisation. Although this was, and still is, an excellent term, it was criticised as cumbersome and lacking in euphony. Thus it never passed into general use." Pearse, 1980.

The huge technology of histochemistry serves to apply known physicochemical principles to the localisation of specific substances or biochemical processes in cells and tissues. The scope and generality of histochemistry ensure that it has a major role to play in neuroscience, but not one that is more or less specific to the discipline in the way that most of the other techniques that we have considered so far undoubtedly are. In this short review of neurohistology, therefore, it will be necessary to be highly selective in the treatment given to histochemistry, and to restrict the coverage to a brief consideration of the main subdivisions as they have been applied to the nervous system. In particular, my usual historical approach to the development and significance of the technology will have to be abandoned, since clearly in the case of histochemistry it does not involve neurohistology or neurochemistry sufficiently closely. The interested reader may consult Pearse (1980) to rectify this omission.

We have, of course, already encountered methods based on histochemical principles, both in the present chapter and in Chapter 1, including those employing horseradish peroxidase and antibodies. In essence, these methods, and most of the following ones, share as a unifying principle the exploitation of the specific affinity of a macromolecule for a particular ligand, or class of ligands, to locate either the ligand (the antibody methods we have met) or the macromolecule (the horseradish peroxidase methods). In addition we should note the highly sensitive demonstration of the monoamine neurotransmitters by their reaction with aldehydes or glyoxylic acid to form various intensely fluorescent heterocyclic compounds (Axelsson et al., 1973; Moore, 1981), and the use of the method to localise the aminergic pathways of the brain (Björklund, 1983).

Each of the following subdivisions – enzyme and receptor histochemistry, immunohistochemistry, and *in situ* hybridization and gene expression – represents a large family of methods that usually include both structural (light microscopical) and ultrastructural (electron microscopical) variants. There will be space only for an outline treatment of the common basis of the family. Equally, it will be impossible to attempt anything like a comprehensive coverage of the applications of each subdivision to neuroscience; instead, general references and a few specific examples will be provided as entries to the literature. We shall take one case study as example 3, the distribution of zebrin I in the cerebellar cortex as demonstrated by immunohistochemistry, in a little more detail.

Enzyme and Receptor Histochemistry

Under this heading we will consider those intrinsic proteins that may be located by their affinity for a specific substrate or ligand. In neuroscience the main classes of interest are the enzymes involved in the synthesis or degradation of neurotransmitters and the various receptors for those transmitters. The use of radio-labelled inhibitors or agonists, such as [^{3}H]ouabain and [^{3}H]naloxone, provides a direct means of mapping the distribution of enzymes and receptors that has the additional advantage of being potentially quantitative (Kuhar, 1983; Geary and Wooten, 1989). The method has been extensively applied in the case of receptors, but much less so for enzymes, where instead the catalytic action of the enzyme itself is usually employed to generate a visible reaction product at, or close to, the site of the enzyme (Kiernan, 1990). The product must therefore be insoluble and, if it is not formed directly by the enzymic action, so must its precursors. Consequently it is often necessary to use an experimental substrate that is an analogue of the natural one. For example, in the demonstration of acetylcholinesterase activity the substrate most commonly used is acetylthiocholine (Koelle and Friedenwald, 1949), the primary reaction product is the insoluble crystalline copper-thiocholine iodide, and the final coloured product is brown amorphous cuprous sulphide formed by reaction with hydrogen, sodium, or ammonium sulphide. Karnovsky and Roots (1964) introduced a direct colouring method by including ferricyanide in the incubation medium. The ferricyanide is reduced to ferrocyanide by the sulphydryl group of the thiocholine liberated from the ester by the enzyme, and the insoluble Hatchett's brown (copper ferrocyanide) is immediately precipitated. Acetylcholine and its analogue are also hydrolysed by non-specific (or "pseudo-") cholinesterase, which is primarily responsible for the hydrolysis of higher acyl esters of choline. It is therefore necessary to inhibit its action using ethopropazine hydrochloride, for example, if the activity of acetylcholinesterase alone is required. For a review of studies on acetylcholinesterase activity in the central nervous system see Butcher (1983).

Cytochrome oxidase is another enzyme that has come to prominence in recent work on the nervous system, following the fortuitous discovery that its activity is especially marked in a columnar system of blobs in Brodmann's area 17 (see above) of the primate visual cortex (Horton and Hubel, 1981). The neurons within the blobs possess receptive field properties that are quite different from those of the cytochrome-oxidase-poor cells of the surrounding ("interblob") areas (Livingstone and Hubel, 1984). However, unlike the distribution of acetylcholinesterase, the reason for the regional differences in the activity of cytochrome oxidase is unknown and remains obscure, though it is presumably linked in some way to local metabolic requirements. This is because cytochrome oxidase is a mitochondrial marker, the terminal enzyme of the electron transport chain, catalysing the oxidation of cytochrome c. In a modern version of the histochemical method to demonstrate cytochrome oxidase activity in the nervous system, diaminobenzidine is the hydrogen donor, and the method is sufficiently sensitive to permit a brief fixation of the tissue with phosphate-buffered formaldehyde prior to freezing and cryostat sectioning (Silverman and Tootell, 1987).

Functional Neurohistology - [^{14}C]-2-Deoxyglucose Autoradiography

I have stressed as a basic principle the necessity for good histology to be always informed by function (and vice versa), but in the method of [^{14}C]-2-deoxyglucose autoradiography the complementary approaches are especially close in that a histological preparation is generated as the end product of a particular neural activity. Insofar as it involves the affinity of enzymes for specific substrates, it is included here under the general heading of enzyme histochemistry. The method was devised by Sokoloff et al. (1977); it exploits the inability of mature neurons to store glycogen and their consequent extreme dependency on a continuously available supply of blood glucose for their energy requirements. 2-deoxyglucose is transported into neurons by the same high affinity mechanism as glucose, in a competitive manner, and is phosphorylated by hexokinase, again in competition with glucose. However, 2-deoxyglucose-6-phosphate does not act as a substrate for phosphohexoisomerase and therefore it tends to accumulate in the neuron, with a half-life in the grey matter of 7.7 h. Since the uptake of glucose is directly linked to its consumption by the neuron, and this in turn is linked to the neuron's metabolic activity, differential stimulation of groups of neurons leads to the differential accumulation of 2-deoxyglucose after its systemic administration (Hand, 1981). The method has been used, for example, to demonstrate the arrangement of orientation columns in the visual cortex (Hubel, Wiesel and Stryker, 1978).

Immunohistochemistry

Two features of the structure of the immunoglobulin molecule are at the heart of the family of immunohistochemical methods: the extreme variability in the amino-acid sequence of the binding sites of immunoglobulins secreted by different plasma cells; and the bivalent property, or the presence of two similar antigen-binding sites per molecule. The first of these means that immunohistochemistry is virtually limitless in its applications, and in studies on the nervous system, as elsewhere, it is now usually the method of first choice and often the only method. Radiolabelling of the molecule combined with microdissection techniques has been used, for example, to produce maps of the distribution of various neuropeptides by radioimmunoassay (Palkovits and Brownstein, 1985), each peptide being recognised by a specific antibody. The enormous size of the molecule means that it can also be labelled with relatively small molecules, by covalent linkage, without affecting its antigenic affinity, provided that the binding site itself is not involved. This type of labelling is by far the most widely used in immunohistochemistry, one of the earliest methods being the spontaneous linkage of fluorescein isothiocyanate (FITC, Fig. 1G), mainly to the ε-amino group of lysine residues, when the pH is in the range 9–10. The fluorescent antibody may then be used to detect its specific antigen directly, as was originally done, but the direct method is now rarely used because of its comparative lack of sensitivity and the need to generate a labelled antibody for each antigen of interest.

Sensitivity may be improved by increasing the number of labelled molecules eventually deposited next to each molecule of antigen. There is now a variety of such indirect methods to choose from, for which the secondary detection systems and often the primary antisera are commercially available. In the indirect methods the primary antibody, either polyclonal and typically raised in a rabbit, or monoclonal, is unlabelled. In addition to recognizing the specific antigen of interest, the primary antibody itself acts as an antigen for, say, an anti-rabbit antibody, typically raised in a goat by immunization with rabbit immunoglobulin. The goat anti-rabbit antibody may be labelled to be used as the secondary detection system itself, or may be unlabelled to be used as a means of linking the secondary detection system to the primary antibody, so providing yet further sensitivity increase. It is this step that exploits the bivalent nature of the immunoglobulin molecule. Commonly used secondary detection systems that may be linked in this way are: (i) fluorescently labelled rabbit immunoglobulin; (ii) the peroxidase-antiperoxidase (PAP) complex consisting of horseradish peroxidase bound antigenically to rabbit anti-peroxidase immunoglobulin, which is then demonstrated by the peroxidase reaction; and (iii) the avidin-biotin complex (ABC, Vectastain) in which the secondary rabbit immunoglobulin, covalently labelled with biotin, or biotinylated, is coupled via avidin (see Chapter 1) to biotinylated horseradish peroxidase. For a concise summary of the advantages and disadvantages of the various immunohistochemical methods and the controls required for each, see Kiernan (1990).

Example 3: Zebrin I in cerebellar cortex (from Hawkes and Leclerc, 1987)

In this case study we examine the distribution of an antigen, identified as a polypeptide (120,000 MW), then unknown but specific to the Purkinje cells of the cerebellar cortex and recognized by a monoclonal antibody, designated mabQ113. The antigen was called zebrin I on account of its characteristic occurrence in parasagittal bands within the cortex. For details of the generation of the primary antibody see Hawkes and Leclerc (1987).

Materials

Cerebellum of the adult rat.

Procedure

Anaesthesia and fixation

1. Anaesthesia – sodium pentobarbital
2. Fixation
   - Perfusion with 4% formaldehyde (from paraformaldehyde), 0.2% glutaraldehyde in 0.1 M phosphate buffer at pH 7.4
   - Overnight in the fixative without glutaraldehyde
   - Cerebellum removed from cranium and stored in phosphate buffer with sodium azide for 2–4 days prior to sectioning [longer storage reduces antigenicity].

Sectioning

50-μm Vibratome or freezing-microtome sections.

Incubation with primary antibody

Overnight at room temperature with constant agitation in mabQ113, diluted 1/32 with 10% normal horse serum, 0.1 M phosphate buffer at pH 7.4 and 0.15 M NaCl.

Secondary antibody

1. 15-min wash in three changes of phosphate buffer.
2. 2 hours in 1/100 rabbit antimouse PAP complex (Dako).

Peroxidase reaction

15-min incubation with 4-chloro-1-naphthol as hydrogen donor (see above).

Results

Peroxidase reaction product, and presumably therefore the antigen, was found in the somata and dendrites of subsets of Purkinje cells that were arranged in some 7 parasagittal bands symmetrically disposed about the midline (Fig. 4). All the remaining Purkinje cells were unstained. The details of the bands were highly reproducible in different individuals and their parasagittal arrangement was subsequently found to correspond in a precisely congruent, complementary, or overlapping fashion with several other structural, functional and biochemical divisions of the cerebellar cortex and its connexions (Hawkes and Gravel, 1991).

In situ hybridization and gene expression

The histological demonstration and localisation of gene expression by *in situ* hybridization requires but a brief mention in the present chapter. In principle the technique is

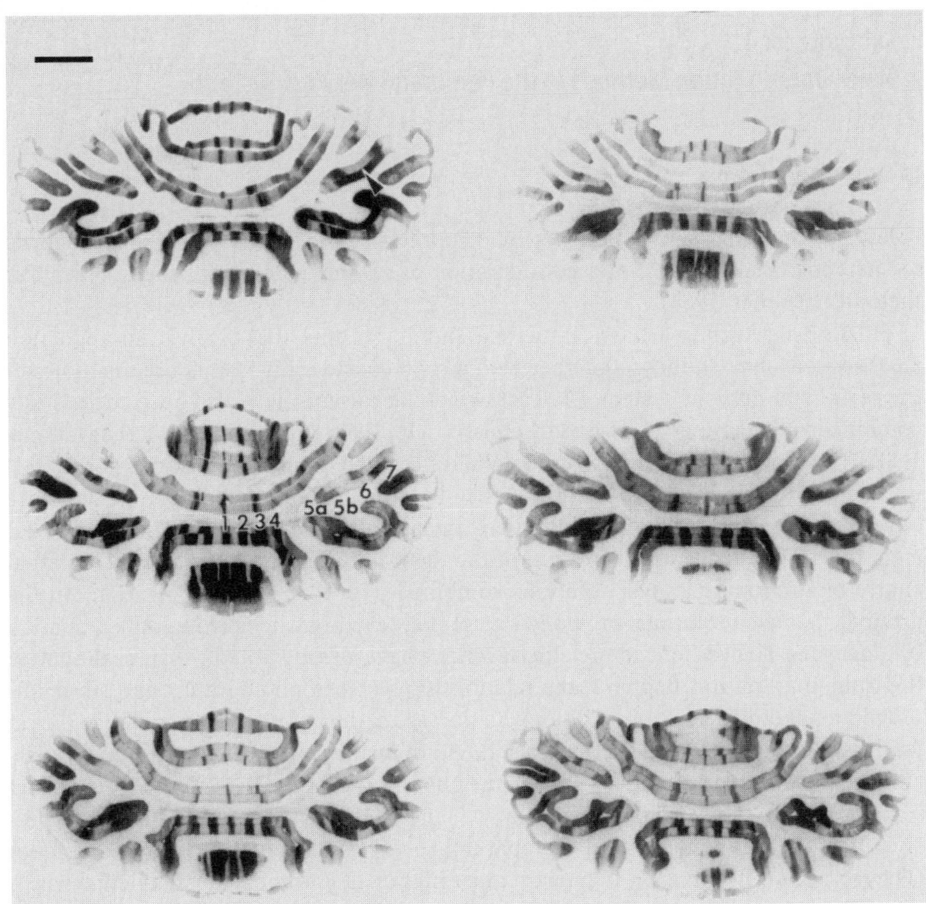

Fig. 4. Localisation of the protein zebrin I in parasagittal bands of Purkinje cells in the rat cerebellum by an immunoperoxidase method. Specific recognition of zebrin I was by means of a monoclonal antibody, mabQ113. Corresponding sections taken at the same level through different cerebella demonstrate the reproducible pattern of zebrin I-containing bands, numbered 1–7. Scale bar = 2 mm. [from Hawkes, R. and Leclerc, N. 1987. Antigenic map of the rat cerebellar cortex: the distribution of parasagittal bands as revealed by monoclonal anti-Purkinje cell antibody mabQ113. J. Comp. Neurol. 256, 29–41. Copyright © 1987 Wiley-Liss. Reprinted by permission of Wiley-Liss, a subsidiary of John Wiley & Sons]

straightforward, relying as it does on the specific affinity of complementary oligomeric strands of nucleotides. It was devised by Gall and Pardue in 1969 and is mainly used to infer the expression of a particular gene by detecting the corresponding mRNA with a radiolabelled, or occasionally fluorescently labelled, antisense probe (Uhl, 1988). Increasingly, synthetic oligonucleotide cDNA probes are being used, which, among other things, have the advantage that they can be directed to a nucleotide sequence within a specific exon. To give a single example of its application in neuroscience, it was one of the methods used to confirm the identity of zebrin II, a protein expressed in a subset of Purkinje cells, with aldolase C (Ahn et al., 1994). In this case the probe was a complementary single-stranded RNA (riboprobe), synthesized from templates derived from a cDNA library prepared from whole-cerebellum mRNA. Clones expressing zebrin II were identified by immunochemistry.

Subprotocol 4
Silver-Impregnation Methods in the Peripheral Nervous System

Introduction

"Max Bielschowsky was a pathologist who appreciated the advantages of formaldehyde for nervous tissue, and from a consideration of its chemistry proposed a new silver method" (Holmes, 1968).

Finally, we return to a classical method and its variants, that of silver impregnation and reduced silver staining, together with a brief consideration of its use in studies of the peripheral nervous system. (At least two of the variants have also been successfully combined with cholinesterase histochemistry – Ip, 1967; Diaz and Pécot-Dechavassine, 1987.) The reduced silver methods have an undeserved reputation for capriciousness – undeserved because it seems that the difficulty in their use is due to the critical nature of various steps, rather than any inherently uncontrollable events or irrationality of design. Indeed we may trace the origin of the method to Bielschowsky's (1902) rational choice of ammoniacal silver solutions combined with formaldehyde fixation, on the grounds that such solutions provided a test for aldehyde groups through the reduction by aldehydes of silver ions to metallic silver. We have already noted, when dealing with the Golgi method in Chapter 1, the relationship between photography and silver impregnation methods in neurohistology; in the case of the reduced silver methods the relationship is particularly close (Holmes, 1968). In further rational, though undoubtedly empirical, steps, Bielschowsky included the photographic processes of unreduced silver removal using sodium thiosulphate ("hypo") and gold toning. Argyrophilia is a property of most proteins, reducible silver apparently being associated with histidine residues (Peters, 1955a). In the axon it appears to be mainly due to the neurofilaments, where metallic silver is thought to be preferentially deposited in the form of nuclei, each of a very few atoms, under the conditions of low (10^{-5}–10^{-3} M) Ag^+ concentration and mild alkalinity of the impregnation (Peters, 1955c). It must be supposed that similar nuclei are formed in association with other proteinaceous structures, perhaps in smaller quantity or with lesser affinity, but for whatever reason they are more easily removed under the alkaline conditions of development. This differentiation is affected by the particular fixation procedure employed (FitzGerald, 1963).

After the impregnation stage the essential chemical basis of the method is in fact quite well understood (Samuel, 1953; Peters, 1955b), proceeding by a reduction reaction similar to photographic development. A commonly used developer is hydroquinone, which, catalysed by the nuclei of metallic silver, reduces Ag^+ with the formation of *p*-benzoquinone, resulting in the deposition of more metallic silver on the nuclei:

$$2\,Ag^{+} + \underset{\text{hydroquinone}}{HO{-}C_6H_4{-}OH} + 2\,OH^{-} \rightarrow 2\,Ag\downarrow + \underset{\text{p-benzoquinone}}{O{=}C_6H_4{=}O} + 2\,H_2O$$

In most variants of the method the source of silver is that remaining in the aqueous phase of the tissue after its immersion in silver nitrate, as well as that released from non-axonal proteins during development. Alternatively a photographic-type physical developer may be used, containing a known amount of silver nitrate (Cruz et al., 1984; Novotny and Gommert-Novotny, 1988; Schweizer and Kaupenjohann, 1988). When the nuclei, or silver particles, become sufficiently large (in the range 3–70 nm as revealed by electron microscopy), they produce enough light-scattering that the neurofilaments appear dark within the axon.

The great advantage of the reduced silver methods is their ability, in well-stained preparations, to display the complete innervation of a region of tissue or of an entire small sense organ such as a muscle spindle. In the majority of published variants the parameters have been optimized for sectioned material, but it can be much more useful in some cases to stain the tissue *en bloc* and then tease out the structures of interest. This is the approach adopted here in our case study of the innervation of the mammalian muscle spindle. In my own work, I have also tackled the problem of spindle innervation by the much more laborious method of reconstruction from serial, 1-μm, plastic sections stained with toluidine blue and pyronine (see Chapter 1). As mentioned in the Introduction, these different approaches to the same problem provide somewhat different information. Thus, with serial sections, intrafusal muscle fibres of all three types (the so-called bag_1, bag_2 and chain fibres) may be easily traced individually from end to end and all myelinated axons may also be traced with complete confidence (provided critical sections are not missing!), but the finest unmyelinated axons, whose diameter may be less than the resolving power of the light microscope, cannot always be so traced. Conversely, with a teased, silver-impregnated preparation, the chain fibres cannot usually be traced individually along their entire length, but both myelinated and unmyelinated axons can be followed without difficulty, probably also including some of smaller diameter than the theoretical resolution of the microscope, due to the light-scattering by the colloidal silver particles.

Example 4: The innervation of the muscle spindle

Materials

Tenuissimus muscle (m. caudocruralis) of the adult cat.

Procedure

The method is that of Barker and Ip (1963)

Fixation
5 days in freshly prepared chloral hydrate, 1 g; 95% ethanol, 45 ml; distilled water, 50 ml; conc. nitric acid, 1 ml.

Wash
24h in running tap water.

Alkalinization
24h in 95% ethanol, 25 ml; ammonia [s.g. 0.88, about 35% w/w], 1 drop from a Pasteur pipette. [This follows the original method; one could be more systematic and determine empirically the best value of pH, though the results do not suggest that this is necessary].

Silver impregnation
Surplus fluid blotted or drained, muscle incubated for 4 days in 1.5% w/v silver nitrate at 37 °C. [The time here is critical for success, and the value given differs from that in the original paper, since those authors studied larger muscles. After encountering problems of inadequate staining some years ago, I carried out a study (unpublished) involving systematic variation of all the most important steps in the method. It was clear that

by far the most critical was the silver impregnation and especially its duration, though this could be offset to some extent by varying the concentration of silver nitrate. Too short a duration resulted in a rather general staining, particularly of any and every cell nucleus, thus obscuring the staining of the axons, and incidentally confirming a similar observation made by Palmgren (1960). Too long a duration, similarly counterintuitive perhaps, resulted in whole areas of the tissue, including axons, being unstained while small regions were intensely stained. These observations demonstrate that the process of differential silver impregnation described above is a dynamic one, and does not involve a stable end-point. When dealing with a new preparation, therefore, it may be necessary to adopt an iterative approach by first guessing the correct time, then adjusting

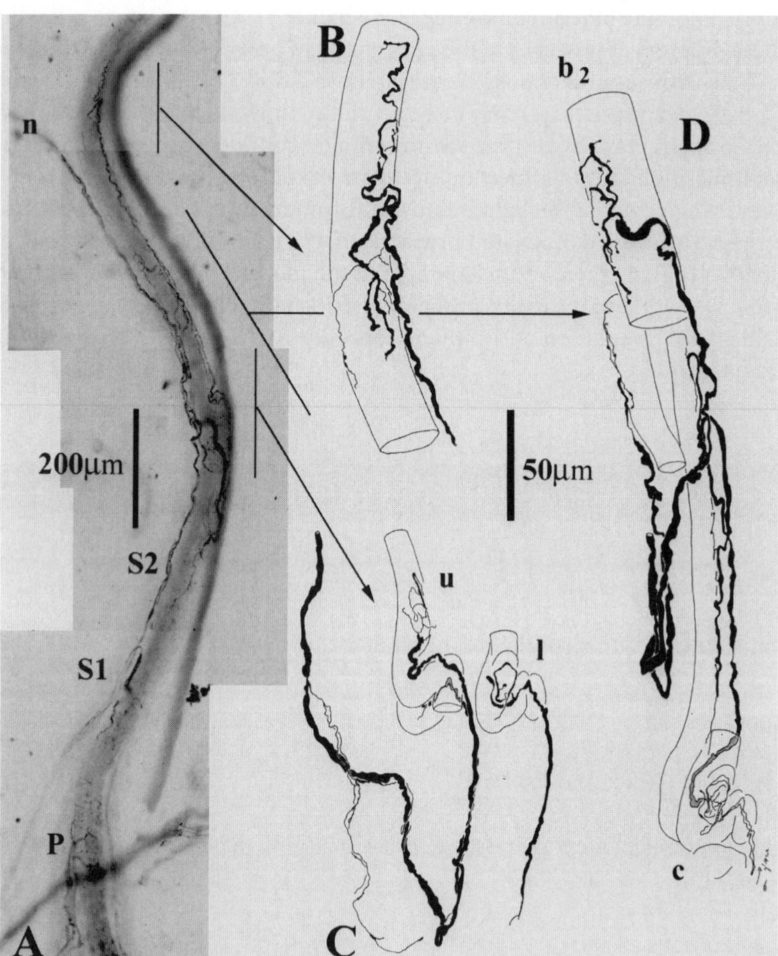

Fig. 5. Analysis of the motor innervation supplying the bag_2 and chain intrafusal muscle fibres in one pole of a muscle spindle from the tenuissimus muscle of a cat, stained by a reduced silver method prior to isolation of whole spindles by microdissection (teasing). A, montage of the equatorial region containing the primary sensory ending (P), the juxta-equatorial region containing two secondary sensory endings (S1, S2), and the polar region containing the motor innervation of the spindle, partly supplied by the small nerve marked n. B – D, camera lucida drawings of the preterminal and terminal branches of the three motor axons that severally innervated the bag_2 fibre alone (B), two chain fibres together (C) and the bag_2 (b_2) and three chain (c) fibres together (D), the locations of the individually traced regions being indicated on the montage. In C the two chain fibres are shown in their relative positions on the left, with the upper fibre (u) obscuring the lower fibre. The lower fibre (l) is repeated separately on the right.

it as indicated by the resulting quality of impregnation. For any given preparation, it has been my experience since those systematic experiments that such an optimized time will almost always give excellent or at least satisfactory results.]

Reduction (development)

2 days in freshly prepared hydroquinone, 2g; 25% formic acid, 100ml. [This stage does appear to reach a stable end-point and is therefore not critical, provided it is long enough.]

Clearing

1. Rinsed in distilled water
2. Cleared and stored for several days in glycerol prior to teasing.

Permanent mount

Teased spindles and intramuscular nerves mounted in glycerol, using a circular cover-slip ringed with pitch.

Results

In a critically impregnated preparation the axons and their terminals are dark brown or almost black against a background that ranges from pale yellow (straw-coloured) to light brown. It may or may not be possible to distinguish nuclei in the background, but sarcomeres are usually visible. Myelin is normally unstained, but the locations of nodes of Ranvier are normally clear due to the associated constriction in the axon. At high resolution it is often possible to see the fibrillar or filamentous nature of the reduced silver deposits; the stain does not, therefore, fill the entire axon and the expanded sensory terminals in particular may not be visible in their entirety. For further details see Banks (1994).

Acknowledgement: It is a pleasure to thank Christine Richardson and Paul Sidney for their invaluable technical help in the preparation of specimens and figures for this chapter and for Chapter1.

References

Ahn, A., Dziennis, S., Hawkes, R. and Herrup, K. 1994. The cloning of zebrin II reveals its identity with aldolase C. Development 120, 2081–2090.

Axelsson, S., Björklund, A., Falck, A., Lindvall, O. and Svensson, L.-A. 1973. Glyoxylic acid condensation: a new florescence method for the histochemical demonstration of biogenic monoamines. Acta Physio. Scand. 87, 57–62.

Baker, J.R. 1958. Principles of Biological Microtechnique. Methuen, London.

Banks, R.W. 1994. Intrafusual motor innervation: A quantitative histological analysis of tenuissimus spindels in the cat. J. Anat. 185, 151–172.

Barker, D. and Ip, M. C. 1963. A silver method for demonstrating the innervation of mammalian muscle in teased preparations. J. Physiol. 169, 73–74P

Bielschowsky, M. 1902. Die Silberimprägnation der Axencylinder. Neurol. Zentralbl. 21, 579–584.

Björklund, A. 1983. Fluorescence histochemistry of biogenic amines. In Björklund, A. and Hökfelt, T. (eds.) Handbook of Chemical Neuroanatomy. Vol. 1. Methods in Chemical Neuroanatomy. Elsevier, Amsterdam. pp. 50–121.

Brodal, A. 1940. Modification of Gudden method for study of cerebral localization. Arch. Neurol. 43, 46–58.

Brodmann, K. 1908. Beiträge zur histologischen Lokalisation der Grosshirnrinde. VI. Mitteilung: Die Cortexgliederung des Menschen. J. Psychol. Neurol., Leipzig. 10, 231–246.

Brushart, T. M. and Mesulam, M.-M. 1980. Transganglionic demonstration of central sensory projections from skin and muscle with HRP-lectin conjugates. Neurosci. Lett. 17, 1–6.

Butcher, L. L. 1983. Acetylcholinesterase histochemistry. In Björklund, A. and Hökfelt, T. (eds.) Handbook of Chemical Neuroanatomy. Vol. 1. Methods in Chemical Neuroanatomy. Elsevier, Amsterdam. pp.1–49.

Clarke, E. and O'Malley, C. D. O. 1996. The Human Brain and Spinal Cord. 2e Norman Publishing, San Francisco

Colonnier, M and Gray, E. G. 1962. Degeneration in the cerebral cortex. In Breese, S. S. (ed) Electron Microscopy, Fifth International Congress for Electron Microscopy. Academic Press, New York. vol. 1, p U-3

Cowan, W. M., Gottlieb, D. I., Hendrickson, A. E., Price, J. L. and Woolsey, T. A. 1972. The autoradiographic demonstration of axonal connections in the central nervous system. Brain Res. 37, 21–51.

Chronwall, B. M., Lewis, M. E., Schwaber, J. S. and O'Donohue, T. L. 1989. In situ hybridization combined with retrograde fluorescent tract tracing. In Heimer, L. and Záborszky, L. (eds.) Neuroanatomical Tract-Tracing Methods 2: Recent Progress. Plenum, New York. pp 265–297.

Cruz, M. C., Jeanmonod, D., Meier, K. and Van der Loos, H. 1984. A silver and gold technique for axons and axon-bundles in formalin-fixed central and peripheral nervous tissue. J. Neurosci. Methods 10, 1–8

de Olmos, J. S., Ebbesson, S. O. E. and Heimer, L. 1981. Silver methods for the impregnation of degenerating axoplasm. In Heimer, L. and RoBards, M. J. (eds) Neuroanatomical Tract-Tracing Methods. Plenum, New York. pp 117–170.

Diaz, J. and Pécot-Dechavassine, M. 1987. An improved combined cholinesterase stain and silver impregnation method for quantitative analysis of innervation patterns in frog muscle. Stain Technol. 62, 161–166.

Droz, B. 1975. Autoradiography as a tool for visualizing neurons and neuronal processes. In Cowan, W. M. and Cuénod, M. (eds) The Use of Axonal Transport for Studies of Neuronal Connectivity. Elsevier, Amsterdam. pp 127–154.

Droz, B. and Leblond, C. P. 1963. Axonal migration of proteins in the central nervous system and peripheral nerves as shown by radioautography. J. Comp. Neurol. 121, 325–346.

Edwards, S. B. and Hendrickson, A. 1981. The autoradiographic tracing of axonal connections in the central nervous system. In Heimer, L. and RoBards, M. J. (eds) Neuroanatomical Tract-Tracing Methods. Plenum, New York. pp 171–205.

Ehrlich, P. 1886. Über die Methylenblaureaction der lebenden Nervensubstanz. Dt. Med. Wschr. 12, 49–52.

FitzGerald, M. J. T. 1963. A general-purpose silver technique for peripheral nerve fibers in frozen sections. Stain Technol. 38, 321–327.

Gall, J. G. and Pardue, M. L. 1969. Formation and detection of RNA-DNA hybrid molecules in cytological preparations. Proc. Natl. Acad. Sci. USA 63, 378–383.

Geary, W. A. and Wooten, G. F. 1989. Receptor autoradiography. In Heimer, L. and Záborszky, L. (eds.) Neuroanatomical Tract-Tracing Methods 2: Recent Progress. Plenum, New York. pp 311–330.

Gerfen, C. R. 1985. The neostriatal mosaic. I. Compartmental organization of projections from the striatum to the substantia nigra in the rat. J. Comp. Neurol. 236, 454–476.

Gerfen, C. R., Sawchenko, P. E. and Carlsen, J. 1989. The PHA-L anterograde axonal tracing method. In Heimer, L. and Záborszky, L. (eds.) Neuroanatomical Tract-Tracing Methods 2: Recent Progress. Plenum, New York. pp 19–47

Glees, P. 1946.Terminal degeneration within the central nervous system as studied by a new silver method. J. Neuropath. Exp. Neurol. 5, 54–59.

Gonatas, N. K., Harper, C., Mizutani, T. and Gonatas, J. 1979. Superior sensitivity of conjugates of horseradish peroxidase with wheat germ agglutinin for studies of retrograde axonal transport. J. Histochem. Cytochem. 27, 728–734.

Grant, G. and Aldskogius, H. 1967. Silver impregnation of degenerating dendrites, cells and axons central to axonal transection 1. A Nauta study on the hypoglossal nerve in kittens. Exp. Brain Res. 3, 150–162.

Grant, G. and Walberg, F. 1974. The light and electron microscopical appearance of anterograde and retrograde neuronal degeneration. In Fuxe, K., Olson, L. and Zotterman, Y. (eds) Dynamics of Degeneration and Growth in Neurons. Pergamon, Oxford. pp 5–18.

Graybiel, A. M. 1975. Wallerian degeneration and anterograde tracer methods. In Cowan, W. M. and Cuénod, M. (eds) The Use of Axonal Transport for Studies of Neuronal Connectivity. Elsevier, Amsterdam. pp 173–216.

Hand, P. J. 1981. The 2-deoxyglucose method. In Heimer, L. and RoBards, M. J. (eds) Neuroanatomical Tract-Tracing Methods. Plenum, New York. pp 511–538.

Hawkes, R. and Gravel, C. 1991. The modular cerebellum. Prog. Neurobiol. 36, 309–327.

Hawkes, R. and Leclerc, N. 1987. Antigenic map of the rat cerebellar cortex: the distribution of parasagittal bands as revealed by monoclonal anti-Purkinje cell antibody mabQ113. J. Comp. Neurol. 256, 29–41

Heimer, L. and Ekholm, R. 1967. Neuronal argyrophilia in early degenerative states: a light and electron-microscopic study of the Glees and Nauta techniques. Experientia 23, 237–239.

Holmes, W. 1968. Empiricism - silver methods and the nerve axon. In McGee-Russell, S. M. and Ross, K. F. A. (eds.) Cell Structure and its Interpretation: essays presented to John Randal Baker FRS. Edward Arnold, London. pp. 95–102.

Horton, J. C. and Hubel, D. H. 1981. A regular patchy distribution of cytochrome-oxidase staining in primary visual cortex of the macaque monkey. Nature 292, 762–764.

Hubel, D. H., Wiesel, T. N. and Stryker, M. P. 1978. Anatomical demonstration of orientation columns in Macaque monkey. J. Comp. Neurol. 177, 361–380.

Ip, M. C. 1967. A combined method for demonstrating the cholinesterase activity and the nervous structure of mammalian peripheral motor endings in teased preparations. J. Physiol. 192, 801–803.

Karnovsky, M. J. and Roots, L. 1964. A "direct colouring" thiocholine method for cholinesterases. J. Histochem. Cytochem. 12, 219–221.

Keefer, D. A. 1978. Horseradish peroxidase as a retrogradely-transported, detailed dendritic marker. Brain Res. 140, 15–32.

Kiernan, J. A. 1984a. Chromoxane cyanine R. I. Physical and chemical properties of the dye and some of its complexes. J. Microsc. 134, 13–23.

Kiernan, J. A. 1984b. Chromoxane cyanine R. II. Staining of animal tissues by the dye and its iron complexes. J. Microsc. 134, 25–39.

Kiernan, J. A. 1990. Histological and Histochemical Methods: Theory and Practice. 2nd ed. Pergamon, Oxford.

Koelle, G. B. and Friedenwald, J.S. 1949. A histochemical method for localizing cholinesterase activity. Proc. Soc. Exp. Biol. N.Y. 70, 617–622.

Kristensson, K. 1970. Transport of fluorescent protein tracer in peripheral nerves. Acta Neuropath. (Berl.) 16, 293–300.

Kristensson, K. and Olsson, Y. 1971. Retrograde axonal transport of protein. Brain Res. 29, 363–365.

Kuhar, M. J. 1983. Autoradiographic localization of drug and neurotransmitter receptors. In Björklund, A. and Hökfelt, T. (eds.) Handbook of Chemical Neuroanatomy. Vol. 1. Methods in Chemical Neuroanatomy. Elsevier, Amsterdam. pp.398–415.

Kuypers, H. G. J. M., Catsman-Berrevoets, C. E. and Padt, R. E. 1977. Retrograde axonal transport of fluorescent substances in the rat's forebrain. Neurosci. Lett. 6, 127–135.

Kuypers, H. G. J. M., Bentivoglio, M., van der Kooy, D. and Catsman-Berrevoets, C. E. 1979. Retrograde transport of bisbenzimide and propidium iodide through axons to their parent cell bodies. Neurosci. Lett. 12, 1–7.

Lashley, K. S. and Clark, G. 1946. The cytoarchitecture of the cerebral cortex of Ateles: a critical examination of architectonic studies. J. Comp. Neurol. 85, 223–305.

LaVail, J. H. and LaVail, M. M. 1972. Retrograde axonal transport in the central nervous system. Science 176, 1416–1417.

Livingstone, M. S. and Hubel, D. H. 1984. Anatomy and physiology of a color system in the primate visual cortex. J. Neurosci. 4, 309–356.

Lund, R. D. and Westrum, L. E. 1966. Neurofibrils and the Nauta method. Science 151, 1397–1399.

Malmgren, L. and Olsson, Y. 1978. A sensitive method for histochemical demonstration of horseradish peroxidase in neurons following retrograde axonal transport. Brain Res. 148, 279–294.

Mayhew, T. M. 1991. The new stereological methods for interpreting functional morphology from slices of cells and organs. Exp. Physiol. 76, 639–665.

Mesulam, M.-M. 1978. Tetramethyl benzidine for horseradish peroxidase neurohistochemistry: a non-carcinogenic blue reaction-product with superior sensitivity for visualizing neural afferents and efferents. J. Histochem. Cytochem. 26, 106–117.

Mesulam, M.-M. (ed.) 1982. Tracing Neural Connections with Horseradish Peroxidase. Wiley, Chichester.

Meynert, T. 1867–1868. Der Bau der Gross-Hirnrinde und seine örtlichen Verschiedenheiten, nebst einem pathologisch-anatomischen Corollarium. Vjschr. Psychiat., Vienna. 1, 77–93, 198–217; 2, 88–113.

Moore, R. Y. 1981. Fluorescence histochemical methods: neurotransmitter histochemistry. In Heimer, L. and RoBards, M. J. (eds) Neuroanatomical Tract-Tracing Methods. Plenum, New York. pp 441–482.

Mori, K. 1986. Lectin *Ulex europaeus* agglutinin I specifically labels a subset of primary afferent fibers which project selectively to the superficial dorsal horn of the spinal cord. Brain Res. 365, 404–408.

Nauta, W. J. H. and Gygax, P. A. 1951. Silver impregnation of degenerating axon terminals in the central nervous system (1) technic (2) chemical notes. Stain Technol. 26, 5–11.

Nissl, F. 1894. Über die sogenannten Granula der Nervenzellen. Neurol. Zbl. 13, 676–685, 781–789, 810–814.

Novotny, G. E. K. and Gommert-Novotny, E. 1988. Silver impregnation of peripheral and central axons. Stain Technol. 63, 1–14.

Page, K. M. 1965. A stain for myelin using solochrome cyanin. J. Med. Lab. Technol. 22, 224–225.

Palkovits, M. and Brownstein, M. J. 1985. Distribution of neuropeptides in the central nervous system using biochemical micromethods. In Björklund, A. and Hökfelt, T. (eds.) Handbook of Chemical Neuroanatomy. Vol. 4. GABA and Neuropeptides in the CNS. Elsevier, Amsterdam. pp.1–71.

Palmgren, A. 1960. Specific silver staining of nerve fibres. I. Technique for vertebrates. Acta Zool. 41, 239–265

Paxinos, G. and Watson, C. 1998. The Rat Brain in Stereotaxic Coordinates. 4e. Academic Press, New York.

Pearse, A. G. E. 1980. Histochemistry: Theoretical and Applied. 4th edition, volume 1. Preparative and Optical Technology. Churchill Livingstone, Edinburgh.

Peters, A. 1955a. Experiments on the mechanism of silver staining. I. Impregnation. Quart. J. Microsc. Sci. 96, 84–102.

Peters, A. 1955b. Experiments on the mechanism of silver staining. II. Development. Quart. J. Microsc. Sci. 96, 103–115.

Peters, A. 1955c. Experiments on the mechanism of silver staining. III. Electron microscope studies. Quart. J. Microsc. Sci. 96, 317–322.

Powers, M. and Clark, G. 1955. An evaluation of cresyl echt violet as a Nissl stain. Stain Technol. 30, 83–92.

Rexed, B. 1954. A cytoarchitectonic atlas of the spinal cord in the cat. J. Comp. Neurol. 100, 297–379.

Samuel, E. P. 1953. The mechanism of silver staining. J. Anat. 87, 278–287.

Schweizer, H. and Kaupenjohann, H. 1988. A silver impregnation method for motor and sensory nerves and their endings in formalin-fixed mammalian muscles. J. Neurosci. Methods 25, 45–48

Sickles, D. W. and Oblak, T. G. 1983. A horseradish peroxidase labeling technique for correlation of motoneuron metabolic activity with muscle fiber types. J. Neurosci. Methods 7, 195–201.

Silverman, M. S. and Tootell, R. B. H. 1987. Modified technique for cytochrome oxidase histochemistry: increased staining intensity and compatibility with 2-deoxyglucose autoradiography. J. Neurosci. Methods 19, 1–10.

Skirboll, L. R., Thor, K., Helke, C., Hökfelt, T., Robertson, B. and Long, R. 1989. Use of retrograde fluorescent tracers in combination with immunohistochemical methods. In Heimer, L. and Záborszky, L. (eds.) Neuroanatomical Tract-Tracing Methods 2: Recent Progress. Plenum, New York. pp 5–18.

Somogyi, P., Hodgson, A. J. and Smith, A. D. 1979. An approach to tracing neuron networks in the cerebral cortex and basal ganglia. Combination of Golgi staining, retrograde transport of horseradish peroxidase and anterograde degeneration of synaptic boutons in the same material. Neuroscience 4, 1805–1852.

Steward, O. 1981. Horseradish peroxidase and fluorescent substances and their combination with other techniques. In Heimer, L. and RoBards, M. J. (eds) Neuroanatomical Tract-Tracing Methods. Plenum, New York. pp 279–310.

Uhl, G. R. 1988. An approach to in situ hybridization using oligonucleotide cDNA probes. In van Leeuwen, F. W., Buijs, R. M., Pool, C. W. and Pach, O. (eds.) Molecular Neuroanatomy. Elsevier, Amsterdam. pp 25–41.

Vogt, O. and Vogt, C. 1919. Allgemeinere Ergebnisse unserer Hirnforschung. J. Psychol. Neurol., Leipzig. 25, 273–462.

Walberg, F. 1971. Does silver impregnate normal and degenerating boutons? A study based on light and electron microscopical observations of the inferior olive. Brain Res. 31, 47–65.

Weigert, C. 1882. Über eine neue Untersuchungsmethode des Centralnervensystems. Z. med. Wiss. 20, 753–757, 772–774.

West, J. R. and Black, A. C. 1979. Enhancing the anterograde movement of HRP to label sparse neuronal projections. Neurosci. Lett. 12, 35–40.

Windle, W. F., Rhines, R. and Rankin, J. 1943. A Nissl method using buffered solutions of thionin. Stain Technol. 18, 77–86.

Zilles, K., Schleicher, A. and Kretschmann, H.-J. 1978. A quantitative approach to cytoarchitectonics. I. The areal pattern of the cortex of Tupaia belangeri. Anat. Embryol. 153, 195–212.

Chapter 16

Optical Recording from Populations of Neurons in Brain Slices

SAURABH R. SINHA and PETER SAGGAU

Introduction

Optical recording involves the use of molecular indicators whose optical properties (absorbance or fluorescence) change with parameters of cellular activity. Available indicators include those sensitive to membrane potential (V_m), to calcium concentration and to concentrations of other ions (H^+, Na^+, K^+, Mg^{2+}, Zn^{2+} and Cl^-). Optical recording provides several advantages over conventional approaches. In the case of electrical activity, even heroic efforts involving large arrays of microelectrodes (e.g., Breckenridge et al. 1995) cannot give the spatial resolution that can be obtained relatively easily with optical approaches with significantly less damage to the preparation. Optical techniques also allow for recording from sites inaccessible to conventional approaches. In the case of ionic concentrations, they also provide higher temporal resolution than conventional methods such as ion-sensitive electrodes. Application of optical recording techniques to single neurons in culture is discussed in Chapter 4 (Bullen and Saggau). In the present chapter we describe techniques for optically recording from populations of neurons in mammalian brain slices. These techniques are illustrated with examples from our laboratory of recording membrane potential and intracellular calcium concentration ($[Ca^{2+}]_i$) from rodent hippocampal slices. First, we briefly discuss some of the basic issues involved including optical indicators, loading techniques, optical filters and photodetectors. Intrinsic optical properties of brain tissue are also discussed but only in so far as they affect measurement of signals from optical indicators. Chapter 34 (Grinvald et al.) deals with how intrinsic optical properties can be used to record activity in intact brain structures.

Optical Indicators

In this section we briefly discuss the characteristics of some commonly used optical indicators with emphasis on voltage-sensitive and calcium-sensitive dyes, the two types of optical indicators used in the protocols described in this chapter. Additional details on these indicators are given in chapter 4 (Bullen and Saggau) of this book; furthermore, detailed information on most commercially available optical indicators can be found in the Molecular Probes catalog (Haugland 1996).

Voltage-sensitive Dyes

Voltage-sensitive dyes (VSDs) are molecules whose optical properties reflect either the membrane potential or changes in V_m (for review see Cohen and Lesher 1986; Ebner and Chen 1995). VSDs can be subdivided into two classes based on the speed of their response. "Slow" VSDs are membrane permeable and distribute themselves across the

Correspondence to: Peter Saggau, Baylor College of Medicine, Division of Neuroscience, One Baylor Plaza, Houston, TX, 77030, USA (phone: +01-713-798-5082; fax: +01-713-798-3946; e-mail: psaggau@bcm.tmc.edu)

Table 1. Dyes and Optical Filters/Dichroic Beam Splitters

Indicator	Type	K_d (nM)	λ_{ex} (nm)	λ_{em} (nm)	EX Filter (nm)	DB (nm)	EM Filter (nm)
RH-155	Abs	--	638	--	650/40	--	--
RH-414	Fluor	--	531	714	535/45	605	610LP
Calcium Orange-AM	Fluor	328	549	575	535/45	565	570LP
Fura-2-AM	Fluor	224	335,362	512	360,380/10	450	510/40
Furaptra-AM	Fluor	50,000	330,370	511	360,380/10	450	510/40

cell membrane according to the Nernst equation. These indicators can be useful for measuring slow changes in V_m but are generally not of use for measuring fast changes such as those encountered during neuronal activity. "Fast" VSDs are amphipathic molecules that insert into the cell membrane and exhibit potential-dependent changes in either absorbance (A) or fluorescence (F). Their response times are in the range of microseconds; the response of VSDs to change in membrane potential is linear in the physiological range (Ross et al. 1977), i.e., the change in absorbance (ΔA) or fluorescence (ΔF) is proportional to the change in membrane potential (ΔV_m). Several non-fluorescent (e.g., RH-155 and RH-492) and fluorescent (e.g., RH-414 and RH-795) fast VSDs have been used in brain slices. Absorbance measurements have the chief advantage that for a given intensity of incident light, the intensity of light reaching the detector will be larger than it would be for a fluorescence measurement; also, the setup required for an absorbance measurement is marginally simpler than one for fluorescence (see Fig. 1). On the other hand, fluorescence measurements generally provide slightly larger fractional changes in the signals compared to absorbance measurements, $\Delta F/F \sim 10^{-3}$ versus $\Delta A/A \sim 10^{-4}$. Another fact about VSDs to keep in mind is the same VSD may behave very differently in different preparations; this is most likely related to differences in the composition of the plasma membranes. Thus, it is usually necessary to screen several VSDs whenever using them in a new preparation. Some of the VSDs commonly used in our lab are listed in Table 1 along with the characteristics of the optical filters used with them.

Due to their amphipathic nature, VSDs are very difficult to load into individual cells. Significant success with intracellular loading of VSDs has been reported only in invertebrate preparations with large cells that can be kept viable for long periods of time after injection (>12 h). This is necessary to allow for pressure injection of large quantities of dye and for subsequent diffusion of dye into the processes (Zecevic 1996). Thus, although some initial success has recently been reported with loading VSDs into individual neurons in brain slices (Antic et al. 1997), VSDs are generally bath-applied for optical recording from both single neurons in culture and from populations of neurons in brain slices. There are some initial reports of selective loading of VSDs into certain subpopulations of neurons by retrograde transfer of dye by cut axons in some preparations (Wenner et al. 1996).

Calcium-sensitive Dyes

Calcium-sensitive dyes (CaSDs) are calcium chelators whose optical properties change upon binding Ca^{2+}. Absorbance measurements are rarely used with CaSDs today. Most modern CaSDs were derived from the calcium chelator BAPTA by conjugation with various fluorophores (Tsien 1980; Grynkiewicz et al. 1985; Minta et al. 1989). These highly fluorescent compounds bind only one Ca^{2+}, with high selectivity over Mg^{2+} in the physiological range, and with relative insensitivity to H^+. Upon binding Ca^{2+}, either the efficiency with which they absorb photons (extinction coefficient, ε) or the efficiency with which they convert an absorbed photon to an emitted one (quantum efficiency) is altered. For several of these indicators, there is an actual wavelength shift in their absorption (Fura-2, Furaptra) or emission (Indo-1) spectra; this allows for ratiometric calibration of the signals. However, this is generally not useful when recording from groups of neurons, as the value will actually be an average of the concentrations in several cells and is relatively meaningless unless it is assumed that all the loaded cells are in the same state of activity.

Unlike VSDs which respond linearly to changes in membrane potential, physiological changes in $[Ca^{2+}]_i$ can be such that the response of the CaSD may not be linear. Ca^{2+} binding to CaSD is a first order chemical reaction with a dissociation constant, K_d, an on-rate constant, k_{on}, and an off-rate constant, k_{off}. For any given CaSD, the change in fluorescence is approximately linear to the change in $[Ca^{2+}]_i$ over the range $0.1K_d$ to $10K_d$. A nonlinear response will be observed both below and above this range. Thus, the range of expected $[Ca^{2+}]_i$ is an important consideration in selecting a CaSD for a particular application. For example, if one is interested in detecting changes in calcium concentrations near the resting level (~100 nM in most neurons), a CaSD with a K_d in the range of hundreds of nanomolar is most appropriate, e.g., Fura-2 or Calcium Orange (see Table 1). However, if one is instead interested in relative changes in the peak levels of calcium concentration in synaptic terminals following a single action potential or a train, an indicator such as Furaptra with a K_d of tens of micromolar is more appropriate. Also because the CaSD signal depends on actual binding and unbinding of Ca^{2+} to the calcium indicator, the time course of the signal depends on the kinetics of this reaction as well as the actual change in $[Ca^{2+}]$i. The time constant of the rise in the CaSD signal (τ_{on}) for a step rise in $[Ca^{2+}]_i$ is:

$$\tau_{on} = \frac{1}{k_{on} \cdot ([Ca^{2+}] + [CaSD]) + k_{off}}$$

For a step decline, the time constant of the fall in the CaSD signal (τ_{off}) is:

$$\tau_{off} = \frac{1}{k_{off}}$$

Most of the modern indicators have a similar k_{on}, being close to diffusion-limited; therefore, the kinetics of their responses are determined by their K_d ($= k_{off}/k_{on}$). The higher the K_d, the faster the response and the more accurately the time course of the CaSD signal reflects the time course of $[Ca^{2+}]_i$. Indicators with a high K_d also have the advantage that they are less likely to be saturated and are less likely to significantly contribute to the calcium buffering properties of the neuron. For a more detailed discussion of these issues, see Sinha et al. (1997). The K_d values of the indicators we use routinely are given in Table 1.

In order to record $[Ca^{2+}]_i$, these indicators must be loaded into the cells of interest. If the only available means for loading these indicators were to inject them using microelectrodes, recording from any significant number of cells would not be possible. Loading of CaSDs was greatly simplified by the advent of acetoxymethyl (AM) esters of these indicators (Tsien 1981): these esters are membrane permeable but become trapped inside cells when cleaved by intracellular esterases. Thus, preparations can be loaded by simply bathing slices in the AM ester of the CaSD. This type of loading is very useful when non-selective loading of a large number of cells is desired instead of selective loading of a single cell. Local application of AM esters to selected anatomical structures such as axon tracts in some preparations allows for the selective loading of subcellular structures such as presynaptic terminals (Regehr and Tank 1991; Wu and Saggau 1994) or dendrites (Wu and Saggau 1994). The specific CaSDs that we regularly use in our laboratory are listed in Table 1.

Other Ion-sensitive Dyes

As mentioned previously, optical indicators that are sensitive to ions other than Ca^{2+} are commercially available (e.g., H^+, Na^+, K^+), many in the AM ester form which allows for loading into populations of neurons. However, we have not used any of these indicators in our laboratory for recording transients in brain slices, nor are we aware of other investigators who have done so. The main issues involved in using these indicators to record from populations of neurons should be the same as those for CaSDs. Several of these indicators have been used in cell cultures and in individually loaded neurons in brain slices. pH-indicators are available with various spectral characteristics and sensitivities (pK_a); examples include BCECF, SNAFL and SNARF. Na^+-sensitive dyes that are amenable to loading into populations of neurons include SBFI and Sodium Green; PBFI is a K^+-sensitive dye that is also available as an AM ester. A general problem with currently available potassium and sodium indicators is relatively poor selectivity between the two ions. For more details on these and other ion-sensitive indicators (e.g., Cl^- or Zn^{2+}), the reader is referred to the Molecular Probes catalog (Haugland 1996). This is also an excellent source for information about new optical indicators, which are constantly being developed.

Microscopes

Several types of microscopes can be used in optical recording applications using brain slices including conventional microscopes, scanning microscopes, confocal microscopes and various combinations of the above (for a more detailed discussion of the various types of microscopes, see chapter 4, Bullen and Saggau). In this chapter, our discussion will be limited to conventional microscopes. A schematic of a basic setup for absorbance measurements is shown in Fig. 1A. It should be noted that only water-immersion objectives with their small working distances (e.g., ~1.7 mm for 40×, 0.9NA, Zeiss) can be used for absorbance measurements. Air objectives cannot be employed with submerged brain slices, as ripples at the air-liquid interface will cause changes in the amount of light reaching the detectors. These changes can be much larger than the actual optical indicator signal (recall that fractional changes in the absorbance of an optical indicator are typically well below 0.1%). The main problem with small working distance is that micropipette access is severely limited. Even using an inverted microscope would not avoid this problem, since in the case of an absorbance measurement, a water immersion condenser would be necessary.

For fluorescence measurements, an inverted microscope is preferable for several reasons: (1) the top of the recording chamber is not hindered by the presence of an objective lens, allowing easy access for micropipettes; and (2) both air and oil immersion objectives can be used. Fig. 1B is a schematic of the basic inverted epifluorescence microscope setup that we use in our laboratory. However, upright microscopes have their own advantages; for example, with an upright microscope, the side of the preparation to which one has micropipette access is also the side from which most of the optical signals are emerging. This is an important consideration when recording from single cells in a brain slice. Fig. 1C shows a schematic of an upright epifluorescence microscope. As discussed above, the only option is a water immersion objective for such a setup.

The performance characteristics of a conventional microscope are largely determined by its objective lens. The two main characteristics of an objective are its magnification (*mag*) and its numerical aperture (*NA*). The magnification determines the size of the area seen by the objective; in conjunction with the number of individual elements in the photodetector, this determines the spatial resolution of the optical signal. The *NA* describes the light gathering capability of the objective: the higher the *NA*, the larger the fraction of light emanating from the preparation that is collected by the objective (for

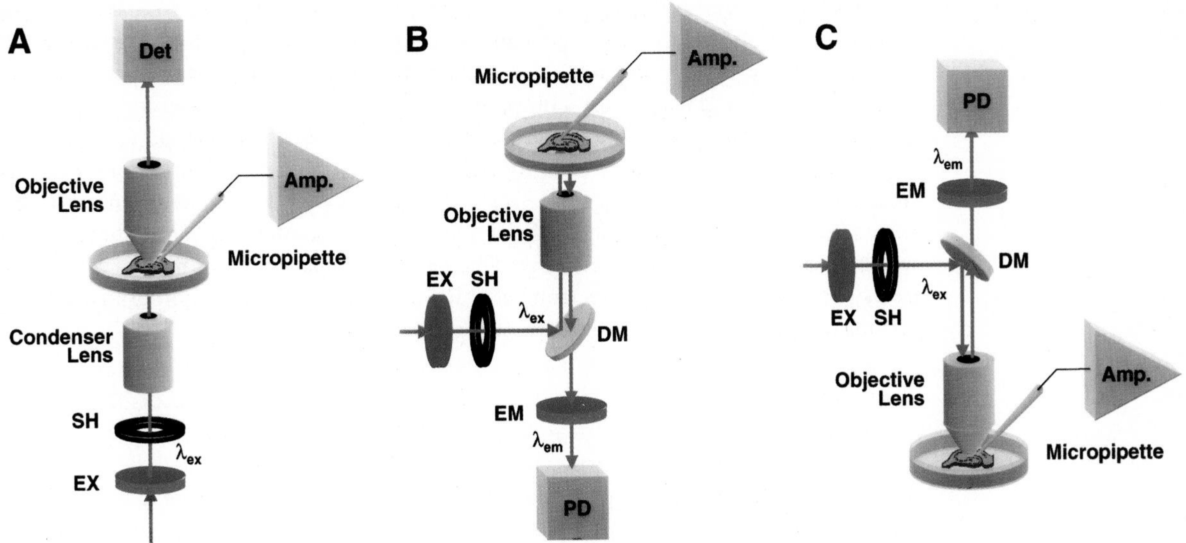

Fig. 1. Schematics of typical optical recording setup. **A.** Microscope for absorbance measurement. Light is limited by an excitation filter (EX) to the appropriate wavelength (λ_i), an electromechanical shutter (SH) to control the time of illumination and is directed onto the preparation via a condenser lens. Transmitted light is collected by the objective lens (water-immersion) and quantified by a photodetector (Det). **B.** Inverted microscope for epifluorescence measurement. Excitation light (λ_{ex}) is directed onto the preparation by a dichroic beamsplitter (DB) and the objective lens (air or oil-immersion). Fluorescence emitted from the preparation (λ_{em}) is collected by the same lens, passes the DB, is filtered by the emission filter (EM) and quantified by DET. **C.** Upright microscope for epifluorescence measurement. Setup is analogous to that in B, except that it is now in an upright configuration, utilizing a water immersion lens.

an oil immersion objective, the upper limit for *NA* is 1.4). The magnification also affects how much light is collected by the objective because it determines the size of the area from which the objective collects light and, thus, the number of fluorescent molecules. For an epifluorescence microscope, the image intensity is proportional to:

$$\frac{NA^4}{mag^2}$$

This relationship can be used to determine the characteristics needed for an objective, the most common scenario being that one has an objective lens of a given *mag* and *NA* that collects sufficient light for a given application and now a different *mag* is desired. The above equation can be used to determine the *NA* the new lens should have to collect the same amount of light. The objective lenses that we commonly use are listed in Table 2.

Table 2. Objective Lenses

Objective	NA	Type	WD (mm)	$NA^4/mag^2 \times 10^{-4}$
5x	0.25	Air	5	1.56
10x	0.5	Air	0.88	6.25
25x	0.8	Oil/water	0.13	6.55
40x	0.75	Water	1.7	1.98
40x	0.9	Oil/water	0.13	4.10
50x	0.9	Oil	0.3	2.62
63x	0.9	Water	1.7	1.65

Other integral parts of an optical imaging setup include the light source, shutter, optical filters and photodetectors. Filters and detectors are discussed below. The two types of light sources we commonly use are a tungsten halogen lamp and a xenon burner (see chapter 4, Bullen and Saggau for a more detailed discussion of light sources). The chief advantage of the tungsten lamp is that it can be operated with a battery-powered source providing very high amplitude stability (RMS noise ~10^{-5}–10^{-4}). The reason for using the noisier xenon burner (RMS noise ~10^{-3}–10^{-4}) is that it is ~5× brighter than the tungsten lamp; the relative brightness of the xenon burner is even higher at short wavelengths, i.e., near the ultraviolet range. Due to noise considerations, mercury burners are usually not advisable. An electromechanical shutter is necessary in order to limit the periods of light exposure to only the actual recording periods.

Optical Filters

Optical filters are an integral part of any optical recording setup. A brief description of the more commonly used types of filters is given here; the exact specifications of the filters used with various indicators are given in Table 1. A bandpass filter is characterized by its center wavelength (the wavelength at which it has maximal transmission) and its full-width at half-maximal transmission (the range of wavelengths over which the filter transmits >50% of its maximal transmission). These numbers are usually expressed as center/FWHM (e.g., 535/25 nm). A longpass filter transmits all visible wavelengths above its characteristic wavelength (the wavelength at which it transmits 50% of its maximum). A dichroic beamsplitter is a hybrid of an optical filter and mirror which reflects all light below its characteristic wavelength and transmits all light above it.

Fig. 2 shows the relationship between the filters and the excitation and emission spectra of an optical indicator. In a typical optical recording setup equipped for epifluorescence (Fig. 1B,C), the excitation filter has bandpass properties and the emission filter is either bandpass or longpass. The dichroic beamsplitter has a characteristic wavelength in between the emission and excitation filter so that it can separate the excitation and emission light.

Photodetectors

Several types of photodetectors are available today, differing in sensitivity, spatio-temporal resolution and intensity resolution. We will first discuss characteristics of photodetectors in general and then specifically describe the types of optical detectors we use in our laboratory. Additional information on this topic is given in Chapter 4 of this book (Bullen and Saggau).

General Considerations About Photodetectors

The sensitivity of a photodetector is proportional to the number of electrons that are generated for each photon that is absorbed by the detector (quantum efficiency, QE). Silicon photodiodes have the highest quantum efficiencies with QE > 0.8 and, therefore, are the detectors of choice with respect to sensitivity.

Spatial resolutions of available photodetectors cover a large range, from single photodiodes with no spatial resolution, to photodiode matrices (PDMs) with relatively low spatial resolution, to charge-coupled device (CCD) cameras with high spatial resolution. While high QE values also apply to PDMs, most CCD cameras have significantly lower values (QE ~ 0.5) due to their complex layout. There are two main tradeoffs that come with increasing spatial resolution. First, the higher the number of individual ele-

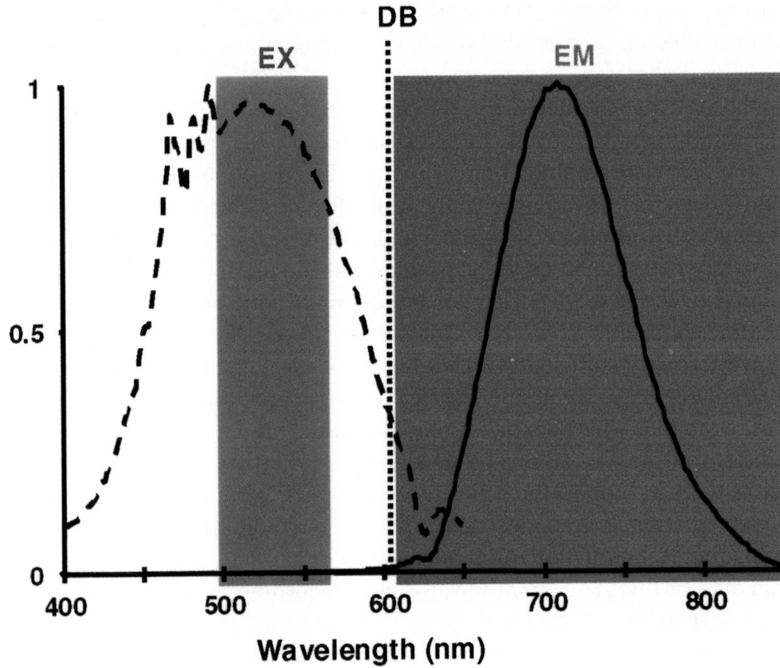

Fig. 2. Selection of optical filters and dichroic mirror. Excitation (*dashed line*) and emission (*solid line*) spectra of fluorescent indicator (specifically the VSD RH-414) are shown along with typical characteristics of the excitation filter (EX), dichroic beamsplitter (DB) and emission filter (EM).

ments in the detector, the lower the amount of light that reaches any given element. Second, the larger the number of elements, the more data that must be gathered, thus reducing the speed of data collection. This is a more significant problem for CCD cameras because most are serial readout devices and data from all the elements must pass through a single amplifier.

The temporal resolution of a photodetector refers to the speed with which a complete frame of data can be collected; the intensity resolution refers to the smallest fractional change in signal (typically voltage, thus $\Delta V/V$) that can be recorded. Both of these resolutions have to be matched by the characteristics of the analog-to-digital (A/D) converter used: the conversion rate for temporal resolution and the number of bits for the intensity resolution. For CCD cameras, the A/D converter is typically an integral part of the device, while this is not the case for single photodiodes and PDMs. With respect to conversion rate, the minimum sampling frequency necessary to record a signal of bandwidth Δf is

$$f_{sample} = 2 \cdot \Delta f \cdot N,$$

where N is the number of elements in the detector. For an m-bit A/D converter, the intensity resolution is:

$$\frac{\Delta V}{V} = 2^{-m}.$$

Thus, intensity resolution for a 16-bit converter is $\sim 1.5 \times 10^{-5}$; for a 12-bit converter, it is $\sim 2.5 \times 10^{-4}$. For a given price range, there is typically a tradeoff between the number of bits and the conversion rate an A/D converter provides. This requires one to critically balance the need for intensity resolution versus temporal and spatial resolution as the spatial resolution affects the conversion rate necessary to obtain a desired frame rate.

Specific Photodetectors

The simplest and least expensive detector is the single photodiode, which provides very high temporal resolution and sensitivity but no spatial resolution. A single photodiode detector with the appropriate amplifiers can be built for less than $50.00 in components; furthermore, such detectors are commercially available. A schematic of the amplifier that we use for single photodiodes is shown in Fig. 3. The response time constant of this photodiode is mainly determined by the capacitance of the photodiode itself and the feedback resistance of the first stage (R_f); for typical values of components shown in Fig. 3, response time constants range from <100 μsec to 1 msec. For single photodiodes ($N=1$), A/D converters with sufficiently high intensity resolution and high digitization rates are easily available (e.g., 16-bit and 10 kHz). A single photodiode thus makes an ideal detector for situations where spatial resolution is not an issue; our laboratory uses single photodiodes routinely to record CaSD signals from selectively loaded pre- or postsynaptic structures in hippocampal slices. Also, because of its low cost, flexibility and ease of implementation, it is an ideal detector for pilot studies involving new preparations, dyes, loading techniques or equipment.

The two main options that are currently available for recording spatially resolved signals are photodiode matrices (PDMs) and cooled CCD cameras. The latter have the advantage of high spatial resolution (several hundred pixels in each dimension). They also have the capability for temporal integration on the detector chip and the flexibility to vary their spatial resolution by integration of pixels (binning). This integration feature makes them especially useful in low light applications. The main disadvantage of CCD

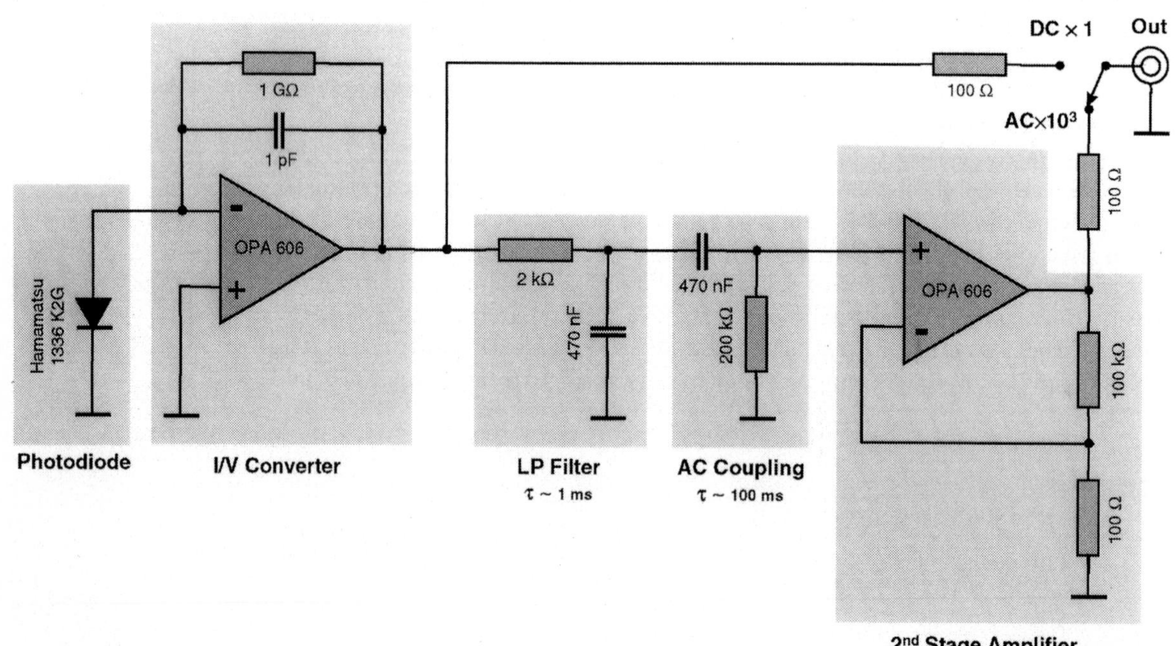

Fig. 3. Amplifier for single photodiode. The amplifier consists of an I-V converter, the output of which is directly available (DC x 1) and also forms the input to the second stage of the amplifier. The trans-impedance amplification of the first stage is due to the feedback resistor R_f; its time constant is $R_f \cdot C_f$. The second stage is preceded by a low-pass filter (LP filter) and provides an additional AC-amplification (AC x 1000).

Table 3. Photodetectors

Photodetector	Sensitivity	Spatial Res.	Temporal Res.	Intensity Res.	Ease of Use	Cost
Photodiode	+++	+	++++	++++	+++	+
PDM	+++	+++	+++	+++	+	+++
CCD camera	++	++++	+/++	+	++	+++

cameras relates to their usually low temporal resolution and sensitivity. Due to their serial readout scheme, their temporal resolution only approaches a few hundred frames/sec, even with substantial spatial integration (Lasser-Ross et al. 1991). In practice, their temporal resolution is even lower due to temporal integration, which is often necessitated by their low sensitivity to light. However, back-thinned CCD cameras with QEs approaching that of photodiodes have recently become commercially available. Also, newly developed CCD cameras with multiple readout amplifiers may overcome the speed problem. Another limitation is that most currently available CCDs only provide 12-bit intensity resolution; therefore, they can at best measure a fractional change of $\sim 2.5\times10^{-4}$, which is not sufficient to record the fractional changes observed with VSD signals but is sufficient to record some CaSD signals. For example, CCD cameras have been used to record transients in individual mossy fiber terminals in hippocampal slices in which mossy fibers were selectively loaded with CaSD (Regehr and Tank 1991; Wu and Saggau, unpublished observation).

The detector that we use for most applications requiring spatial resolution is the photodiode matrix. PDMs allow for a reasonable compromise between the high spatial resolution of CCDs and high intensity and temporal resolution of single photodiodes; furthermore, their QE is similar to that of single photodiodes. PDMs are available in various configurations from 2×2 to 128×128 (Centronic, Hamamatsu, Fuji). Each element of the PDM requires its own amplifier analogous to the single photodiode amplifier shown in Fig. 3. For the 10×10 PDM that we use (MD-100, Centronic), this means that 100 individual amplifiers are necessary. In such a case, physical placement of these amplifiers near the PDM becomes a challenge. We have overcome this problem by building an amplifier consisting of stacks of circuit boards on which the various stages of 100 individual amplifiers are located in parallel. The building of such an amplifier is likely to be outside the reach of many users; PDMs with amplifiers are now commercially available from several sources (e.g., Hamamatsu, OptImaging).

In addition to the physical requirements placed on the amplifier for the PDM, additional requirements on the design are placed by the need to digitize a large number of channels at a high sampling rate with sufficient intensity resolution. A/D converters with sufficiently high sampling frequencies to record from large arrays and with high intensity resolution are available; however, they are very expensive. Instead, we use a relatively straightforward procedure to increase the virtual intensity resolution of a fast (400 kHz sampling frequency) 12-bit A/D converter. This procedure relies on the use of AC coupling and subsequent amplification to measure the optical signal which consists of a large static component (static fluorescence, F) and a much smaller component that changes with the parameter being measured (ΔF). The static fluorescence from the brain slice is first measured with the amplifier in DC coupling mode with an amplification of 1; the fluorescence transient during the activity to be recorded is measured in AC coupling mode with a gain of 10^3. This additional gain of 10^3 effectively extends a 12-bit into a 22-bit A/D converter ($10^3 \approx 2^{10}$). A schematic of the amplifier we use for our PDM

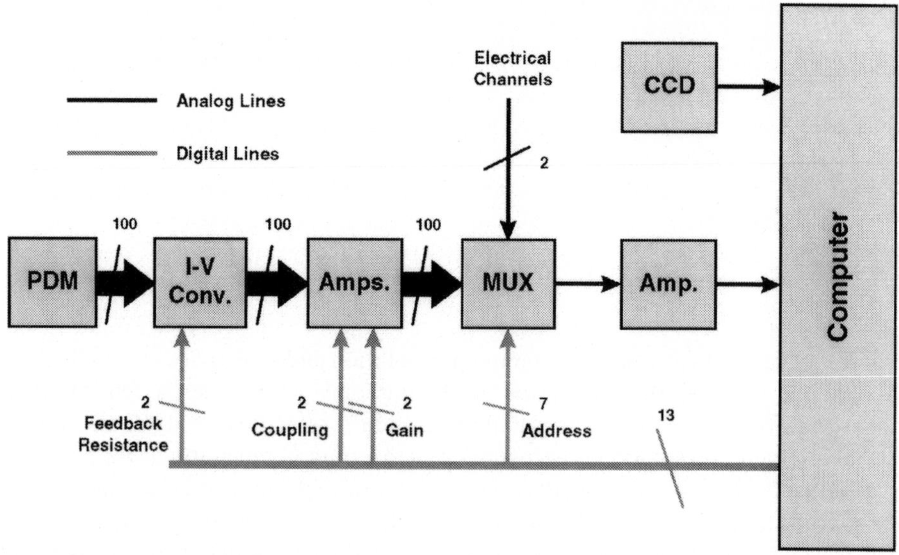

Fig. 4. Amplifier for PDM. The amplifier consists of 100 two-stage amplifiers similar to the single photodiode amplifier shown in Fig. 3. The outputs of these this amplifiers are multiplexed (MUX) together with two additional external channels, one of which is used for the microelectrode signal. A single channel amplifier is located after the MUX for additional gain; its output is digitized by the computer. The signal from a CCD camera, which is used to obtain a background image of the recording area, is also fed into the computer. Digital lines from the computer control various amplifier characteristics such as feedback resistance of the I-V converter, AC- versus DC-coupling, gain of second stage, and the MUX address.

is illustrated in Fig. 4. A disadvantage of using AC coupling is the need to increase the duration of light exposure for the preparation. When an AC-coupled amplifier is first exposed to light (opening of shutter), a voltage transient is observed, which settles back to baseline with its characteristic AC time constant (100 msec for our design). Therefore, after opening the shutter, one needs to wait ~5–10 time constants (>500 msec) before starting to record. With DC coupling, this delay only needs to be long enough for mechanical artifacts from the shutter opening to decay before starting to record (typically <50 msec).

Tissue Properties

Autofluorescence refers to the fluorescence intrinsic to the tissue. It is mainly due to intracellular aromatic molecules such as FAD and tryptophan. While there have not been any systematic studies of autofluorescence, several basic principles are well established.

- Autofluorescence depends on the wavelength of the excitation light: it is highest for low excitation wavelengths (UV to blue); for higher excitation wavelengths such as green light (500–540 nm), it is almost negligible. Thus, autofluorescence is a significant issue when using indicators excited at short wavelengths such as Fura-2 or Furaptra, but is not a factor when using RH-414 or Calcium Orange, which are excited by green light.
- The level of autofluorescence is highly variable; it varies depending on the type of tissue (even within different regions of the same brain slice), the animal species being used and the age of the animal.

Thus, regardless of the indicator being used, it is important to measure the level of autofluorescence and to correct for it. Ideally, one would like to measure the autofluorescence at the exact location where one is recording; however, this is usually not practical. A reasonable compromise is to measure the autofluorescence in the corresponding region in a separate brain slice from the same animal or, in the case of selective loading of pre- or postsynaptic structures, to measure the autofluorescence in a corresponding region which has not been stained.

Another tissue property to keep in mind when recording optically from brain slices is that most tissues scatter light extensively. The main effect of light scattering on optical recording from brain slices is that it places a limitation on the spatial resolution possible regardless of the level of magnification or the number of elements in the detector. As with autofluorescence, the amount of scattering is highly variable; unfortunately, unlike autofluorescence, it is difficult to correct for light scattering. Scattering increases with the thickness of the tissue and varies inversely with the wavelength of light, i.e., higher scattering for shorter wavelengths. Therefore, multi-photon excitation of fluorescent indicators in the infrared has been demonstrated to significantly decrease this effect, allowing for optical recording from non-superficial structures in light scattering preparations (for review see Denk and Svoboda 1997).

Outline

As must be apparent from the above discussion, many issues need to be considered before attempting to record optical signals from a brain slice preparation. Each application must be customized with respect to the indicator used, loading technique, microscope, objective, optical filters and photodetectors. Fig. 5 shows a very basic decision tree that should be used when starting a new application. The first decision is to select a class of optical indicators based on what is to be measured: VSD for membrane potential, CaSD for calcium concentration or another ion-sensitive indicator for concentration of other ions. Then the specific indicator must be selected. For VSDs it is largely a matter of selecting an absorbance versus fluorescence indicator, the desired spectral characteristics (based on available optical filters, light sources and tissue properties such as autofluorescence), and trial and error due to the variability in dye loading from preparation to preparation. For ion-sensitive dyes, one must also consider affinity and kinetics of the dye needed for the particular application as well as the spectral characteristics. Furthermore, for ion-sensitive dyes one must choose between selective loading of specific structures or bath-application; whereas, for the VSDs, only bath-application is commonly available, although see section "Voltage-sensitive Dyes". The next step

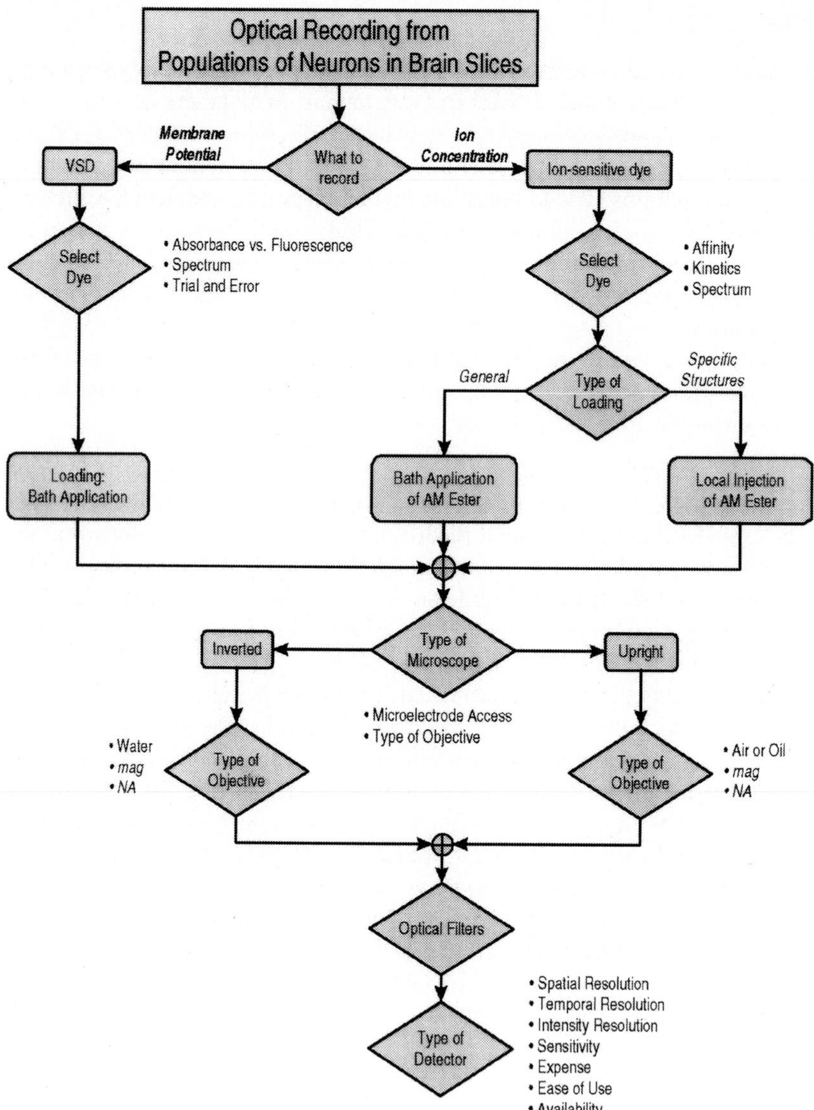

Fig. 5. Outline of experiment planning. The decisions involved in planning an optical recording experiment are illustrated. Next to the decision boxes (*diamonds*) are the various issues that should be considered in making that selection. See text for details.

is to select the type of microscope (upright or inverted) and objective. The relative advantages and disadvantages of different microscope configurations have already been discussed. For the objective lens, the main factors will be the type (air or oil for inverted and water for upright), the desired *mag* and *NA*, and the working distance (see Table 2). Optical filters should be selected based on the characteristics of the indicator. Lastly, the type of detector must be selected. All of the issues that were discussed previously need to be considered thoroughly before making a decision. In addition, we highly recommend that a single photodiode be used first to test the feasibility of the planned experiments, including the quality of indicator loading, size of signals, intensity of light reaching the detector (see Troubleshooting section for more details). Below, we give some specific examples of optical recording procedures that we have implemented after going through a decision process similar to that discussed here.

Materials

- Inverted microscope with epi-fluorescence capability (e.g., IM35 or Axiovert10, Zeiss); see section "Results" for the specific microscope objectives used.
- Light source: tungsten halogen lamp (12 V, 100 W, e.g., Xenophot HLX, Osram) or xenon burner (e.g., XBO 75W/2, Osram).
- Power supply for light source with RMS noise <0.01%: (e.g., ATE 75-8, Kepco, Flushing, NY) or battery-powered constant current source.
- Computer-controlled shutter (e.g., Uniblitz, Vincent, Rochester, NY).
- Photodetector: single photodiode, photodiode array or cooled CCD camera.
- Dyes and appropriate optical filters (see Table 1).
- Bath solution (in mM): NaCl 124, KCl 5, $CaCl_2$ 2, $MgCl_2$ 1.2, $NaHCO_3$ 26, D-glucose 10, saturated with 95% O_2/5% CO_2. All solutions should be bubbled constantly.
- DMSO (Sigma, St. Louis, MO), Pluronic acid (Molecular Probes, Eugene, OR).
- VSD staining chamber: small volume chamber with provision for staining solution access to both sides of the brain slice and for bubbling with 95% O_2/5% CO_2 mixture. For example, a 5 cc syringe with the plunger removed and with the other end sealed with heat or silicone, containing a mesh (plastic or stretched nylon pantyhose) placed over a plastic ring of the appropriate size. For bubbling, a 20 to 25 gauge needle can be used.
- CaSD staining chamber: 35 mm plastic Petri dish with small hole in cover that is sealed with dental wax after passing 20 to 25 gauge needle. The needle should be angled such that the 95% O_2/5% CO_2 mixture blows over the surface of the staining solution rather than in it. This is necessary because the CaSD staining solution contains pluronic acid, a detergent, and blowing gas directly into the solution leads to the formation of very large bubbles and foam.
- Recording chamber: a submersion chamber with a glass cover slip (#0) for the bottom. Chamber should have provisions for circulating solution, for holding the brain slice in place, and, if desired, for adjusting the temperature of the circulating solution (31–32 °C for the experiments shown here).

Procedure

Preparation of Hippocampal Slices

The procedure used in our laboratory to prepare transverse hippocampal slices is described briefly here; however, any procedure that gives viable slices may be used. The appropriate staining solution should be made before cutting brain slices.

1. Anesthetize the animals with methoxyflurane and quickly decapitate with a guillotine.
2. Remove the brain and place in ice-cold bath solution and allow to cool for ~3–5 min.
3. Dissect the hippocampi free from the remainder of the brain.
4. Mount hippocampi on stage of a vibrating tissue slicer (e.g., Vibratome 1000, TPI) and cut 400 μm slices from the middle 1/3 of the hippocampus.
5. Store slices at room temperature in bath solution for at least 1 h before staining, except when staining with bath-application of CaSD, in which case slices may be transferred to the staining solution immediately.

Dye Loading

VSD Loading

1. Prepare stock solution of RH-414: 4 mM in distilled water. This solution can be stored in the freezer for up to several months.

2. On day of experiment, add 10–20 µl of RH-414 stock solution to 4 ml of bathing solution to obtain staining solution (bathing solution with 25–50 µM RH-414). Place staining solution in VSD staining chamber with bubbling needle.
3. Transfer one slice to staining chamber. Stain for 15 min; bubble solution at a rate that will just not move the slices.
4. Wash slice in 20 ml of bath solution for 15–30 min before transferring to recording chamber on stage of microscope.

CaSD Loading

Bath-application

1. Prepare solution of 75% DMSO and 25% pluronic acid by weight. Gentle heating will be required to get pluronic acid into solution. Solution may be stored at room temperature and used for up to 1 week. It may be necessary to gently reheat the solution on subsequent days. After heating this solution, allow it to cool to room temperature before using.
2. Add 10 µl of DMSO/pluronic acid solution to 50 µg container of Calcium Orange-AM. Let solution stand for 30 min at room temperature.
3. Immediately after cutting hippocampal slices, add 1 ml of bathing solution to container. Shake vigorously to mix.
4. Sonicate for 5–10 min at highest setting.
5. Transfer above solution and 4–5 brain slices to the CaSD staining chamber.
6. Loosely seal with Parafilm and start blowing O_2/CO_2 mixture over staining solution at the highest flow rate that does not cause the slices to move.
7. Stain for 3–3.5 h at 30 °C.
8. Unseal container, add 2 ml bath solution and reseal until slices are needed. Slices may be kept in this holding solution for several hours.
9. Wash one slice in 20 ml of bath solution for 15–30 min before transferring to recording chamber on stage of microscope.

Selective Loading

1. Prepare 75% DMSO/25% pluronic acid solution as described in step 1 of section on bath-application of CaSD.
2. Dissolve 5 µl DMSO/pluronic acid solution to 50 µg container of the CaSD-AM (Fura-2-AM or Furaptra-AM). Let solution stand for 30 min at room temperature.
3. Add 50 µl bath solution to container. Shake vigorously to mix.
4. Sonicate for 5–10 min at highest setting.
5. Prepare micropipette (~2 µm tip diameter).
6. Load tip of micropipette with small amount of dye solution by gently applying suction to back of micropipette with a syringe connected to plastic tubing.
7. At least one hour after cutting brain slices, transfer one slice to recording chamber on stage of microscope.
8. Pressure-inject (~10 p.s.i., 20 msec, 5–10 pulses) a small amount of the dye solution (<<1 µl) into the appropriate axonal tract, 0.5–1 mm away from the recording site (e.g., Schaffer collaterals for loading presynaptic terminals of area CA3 pyramidal cells, see Fig. 6, or alveus for loading postsynaptic structures of area CA1 pyramidal cells).

   Note: best results are obtained when the direction of fluid flow in the recording chamber is such that any excess dye at the injection site is carried away from the recording site.

9. Recording may commence ~1 h after injection when fluorescence emerges in the recording area.

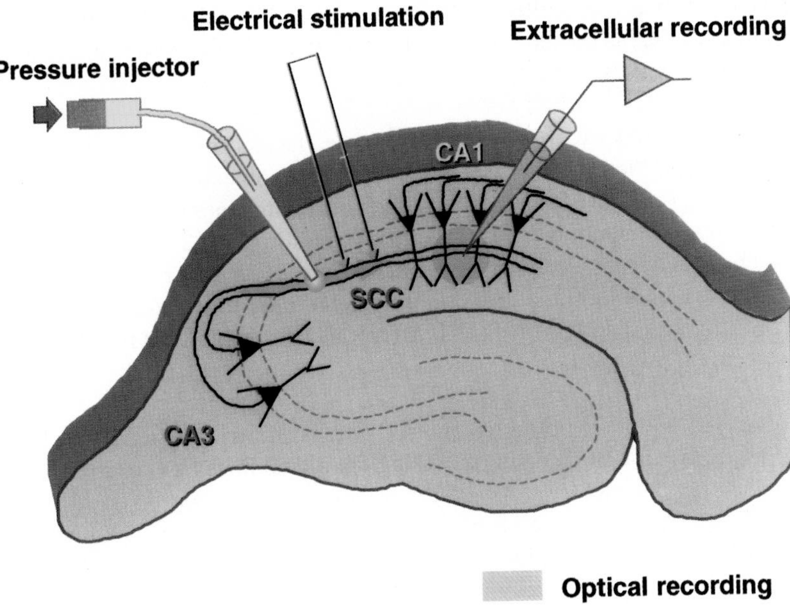

Fig. 6. Technique for selective loading of CaSD. A small amount of CaSD (~ 1 mM of AM ester in DMSO/pluronic solution) is pressure-injected into *str. radiatum* of area CA1, which is the location of CA3 pyramidal cell axons (Schaffer collateral/commissural pathway, SCC). Indicator is taken up, trapped intracellularly by esterases and diffuses within the axons, filling presynaptic terminals at the recording site. Approximate locations of stimulation and recording electrodes are shown.

Optical Recording Procedure

Below we describe the specifics of the procedures for recording evoked or spontaneous signals from slices loaded using the various protocols described above. Regardless of the specifics of the experiments, there are several things that should be done in any optical recording experiment. First, the autofluorescence of the tissue should be measured; as discussed previously, this is especially important when using indicators that are excited by short wavelengths of light. In the case of slices where specific structures are loaded by injection of CaSD into axonal tracts, autofluorescence may be measured within the same slice in a region analogous to the recording area that has not been loaded or in the recording area itself prior to loading. In the case where the brain slices are loaded by bath-application of the indicator, autofluorescence should be measured in an analogous area in another brain slice from the same animal. It is important that the illumination intensity be the same during the measurement of the autofluorescence and during the actual recording period.

Another measurement that should be made during each experiment is a bleaching trace. During exposure to light, molecules of optical indicator are bleached, i.e., lose their fluorescence. It is not clear if these molecules are actually destroyed or if they are put into a state where they are temporarily unable to fluoresce. Regardless of the exact mechanism, bleaching is a mono-exponential process for which it is easy to compensate. The easiest bleaching correction procedure requires the recording of a trace in a period without activity – neither evoked nor spontaneous. The parameters for collecting the bleaching trace (digitization rate, duration, time when shutter is opened) should be exactly the same as for collecting the data trace. Another possibility is that, because bleaching is an exponential process, if the shutter is opened for a sufficiently long period of time before the actual data collection is started, then bleaching will have nearly reached a steady state during the data collection. This latter technique is useful when recording spontaneous activity.

Evoked Activity with Single Photodiode

For recording evoked activity with a single photodiode, the recording procedure is relatively straightforward. We typically use an oil-immersion objective (Achroplan 50×, *NA* 0.9, Zeiss). For digitization of data and providing the digital lines used to control the shutter and electrical stimulator, we use a multipurpose I/O card (12-bit, 50 kHz, DAS-50, Keithley). The signal from a conventional microelectrode used to record the extracellular field potential and the signal from the single photodiode are both digitized by this card. Fig. 7 illustrates the steps involved in a typical experiment. First the appropriate autofluorescence is measured.

Evoked Activity with PDM

For recording activity with the photodiode matrix, we typically use a 10×, *NA* 0.5 objective (Zeiss). For digitization of data and providing the digital lines used to control the PDM amplifier, shutter and electrical stimulator, we use a multipurpose I/O card (Flash 12, Strawberry Tree). The A/D converter on this card is 8 channels, 12-bit, 400 kHz with 256k sample on-board memory; it also has 8 TTL I/O lines and a 2-channel D/A converter. The field potential recorded by a conventional microelectrode is fed into the PDM amplifier and is multiplexed with all the optical channels as shown in Fig. 4. The steps involved in a typical experiment are illustrated in Fig. 8. For the reasons discussed

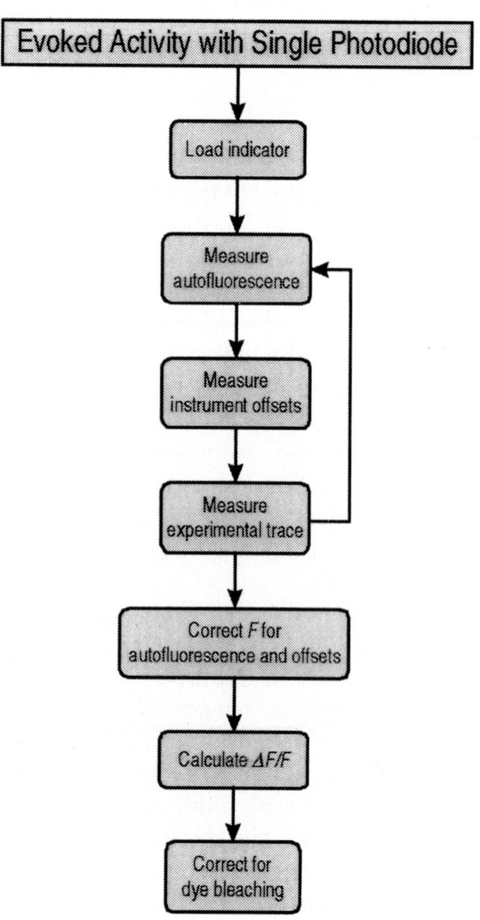

Fig. 7. Procedure for recording evoked activity with single PD. See text for details.

previously, separate AC and DC measurements were necessary. Because the optical indicators we typically used with the PDM were those excited by longer wavelengths, it was unnecessary to perform an autofluorescence correction as the autofluorescence was much lower than the level of offset in the amplifier.

Spontaneous Activity with PDM

The same hardware and basic procedures are used to record spontaneous activity with the PDM as for evoked activity. The main additional challenge in recording spontaneous activity is the lack of an event on which to trigger data collection. Continuous recording, which is often used to record spontaneous activity with microelectrodes, is not an option due to the large amount of data involved (>100 channels). We have employed two procedures for recording spontaneous activity with PDMs. The first procedure uses software which relies on luck to capture spontaneous activity (Colom and Saggau 1994; Sinha et al. 1995): data epochs of a defined length are collected and displayed; when the user sees an event of interest, data collection is stopped and the last epoch is saved. This procedure is sufficient for recording relatively frequent events, i.e., those occurring at frequencies faster than 0.2 Hz; however, with this technique it is very difficult to reliably record more infrequent events.

A slight modification of this recording procedure allows for reliable recording of events occurring at frequencies $< 0.05\,\text{Hz}$. A flow chart of this modified procedure is

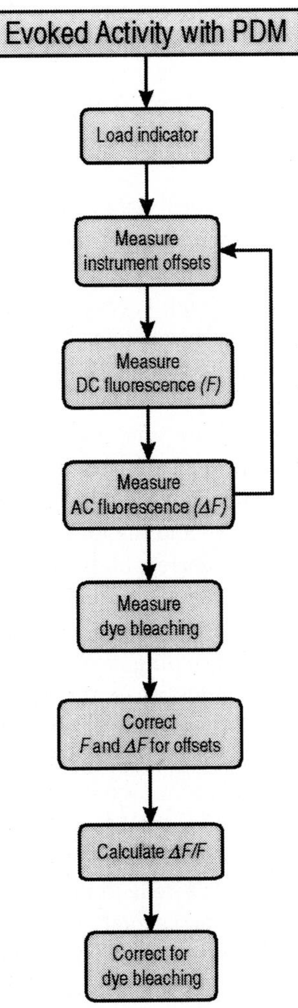

Fig. 8. Procedure for recording evoked activity with PDM. See text for details.

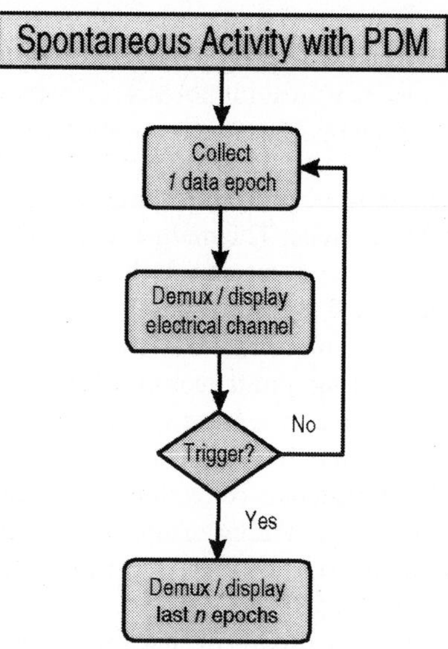

Fig. 9. Procedure for recording spontaneous activity with PDM. See text for details.

shown in Fig. 9. An epoch of data (typically 0.5–1 sec) is collected, and only the electrical channel is demultiplexed and displayed. This process is repeated until the user observes spontaneous activity and triggers the computer; the last *n* epochs collected previous to the trigger (typically 2–3) are then demultiplexed, displayed and available for further manipulation and storage.

Processing of Optical Data

The first step in processing the optical data is to correct for any autofluorescence and instrumentation offsets. In the case of a single photodiode, these quantities are simply subtracted from all points in the optical trace; in the case of the PDM, they are subtracted from the DC fluorescence, which represents the static fluorescence. All optical signals are displayed as change in fluorescence divided by resting fluorescence ($\Delta F/F$). This corrects for variations in dye concentration, in illumination intensity and in the sensitivity of the individual detector elements. For a single photodiode, resting fluorescence (F) is simply the average fluorescence in the portion of the data trace before any activity is observed, i.e., before the application of any stimuli; change in fluorescence (ΔF) is obtained by subtracting F from each data point in the trace. For the PDM, the DC fluorescence is the resting fluorescence (F) and the AC trace is the change in fluorescence (ΔF). For the CaSD an increase in $\Delta F/F$ corresponds to an increase in $[Ca^{2+}]_i$. For the VSD a decrease in $\Delta F/F$ corresponds to depolarization; therefore, all VSD traces are inverted so that a depolarization corresponds to an upward deflection.

For both types of detectors, bleaching correction is performed by first calculating $\Delta F/F$ for both the data and bleaching traces and then subtracting, point by point or in the case of the PDM, element by element, the bleaching trace from the data trace.

For quantification of optical signals, we use two different measurements. For both the VSD and CaSD, the amplitude of the signal can be used. For the VSD alone, another quantity, the mean window amplitude (MWA; Albowitz and Kuhnt 1995) is also a useful measure. The MWA is simply the mean value of the VSD signal during a defined time window. Unlike the amplitude, MWA reflects changes in duration of activity as well as its maximal amplitude. Also, for small signals, it has the additional advantage of reducing the effect of noise: the contribution of noise components with periods smaller than the period of time over which the MWA is calculated will be reduced in calculating this time-averaged value. It is important to note that calculating the MWA for a CaSD signal is relatively meaningless as the duration of the CaSD signal is affected by the kinetics of the indicator in addition to the duration of the calcium concentration transient. A quantity that is useful for CaSD signals is the amplitude of the first derivative of the signal: if the CaSD signal is proportional to $[Ca^{2+}]_i$ and no significant calcium release from internal stores occurs, then the derivative is proportional to the calcium influx. Furthermore, if a dye with fast kinetics is used, i.e., one with a low affinity for Ca^{2+}, then the duration of this derivative is approximately the duration of the calcium influx (for more details, see Sinha et al. 1997).

Results

Using the techniques described above, we have conducted a number of investigations in the hippocampal slice looking at different aspects of synaptic transmission, modulation of transmitter release, plasticity and epileptiform activity in the hippocampal slice (Wu and Saggau 1994a,b, 1995, 1997; Qian and Saggau 1997a,b; Colom and Saggau 1994; Sinha et al. 1995, 1997). Below we present examples taken from these investigations that illustrate the various techniques. For more examples and details, the reader should refer to the individual articles.

Evoked Signals with Selective Loading of CaSD

Fig. 10A illustrates an experiment where the CaSD Fura-2 was selectively loaded into presynaptic terminals of area CA3 pyramidal cells by pressure injecting into the Schaffer collaterals (see Fig. 6). Signals were recorded in *stratum radiatum* of area CA1. Selective recording is insured by the fact that only the axons of the CA3 pyramidal cells travel from the site of injection to the site of recording; some interneurons and glia may also have processes which span this region, but they are vastly outnumbered by the axons. The selectivity of the loading can be illustrated by blocking all postsynaptic activity with the glutamate receptor antagonists CNQX and D-APV; these antagonists completely block the postsynaptic response but do not affect the CaSD signal, confirming its presynaptic origin.

The dependence of the CaSD signal on the K_d (and k_{on} and k_{off}) is illustrated in Fig. 10B,C, which shows the same calcium concentration transient measured with a high affinity CaSD (Fura-2) and a low affinity CaSD (Furaptra). Both indicators are measuring the calcium transient evoked in CA3-CA1 presynaptic terminals by a single action potential. The faster kinetics of the low affinity indicator, Furaptra, are apparent. A combined modeling and experimental study (Sinha et al. 1997) demonstrated that while the high affinity indicator, Fura-2, may be locally saturated by the high local $[Ca^{2+}]_i$ near the plasma membrane after a single action potential, its overall amplitude is still proportional to this local concentration. Thus, Fura-2 can be used to investigate presynaptic calcium transients evoked by single action potentials. However, for calcium transients evoked by multiple action potentials, it is more prudent to use a lower affinity indicator such as Furaptra.

Our laboratory has used this technique extensively to investigate which calcium channels are involved in transmitter release at the CA3-CA1 synapse, how presynaptic calcium influx is altered during synaptic plasticity, and the role of presynaptic calcium in modulation of transmitter release (for review, see Wu and Saggau 1997). We have also used this technique to specifically load other structures in the hippocampal slice; furthermore, others have used a similar technique (using local perfusion with CaSD rather than injection) to specifically load structures in hippocampal and cerebellar slices. The

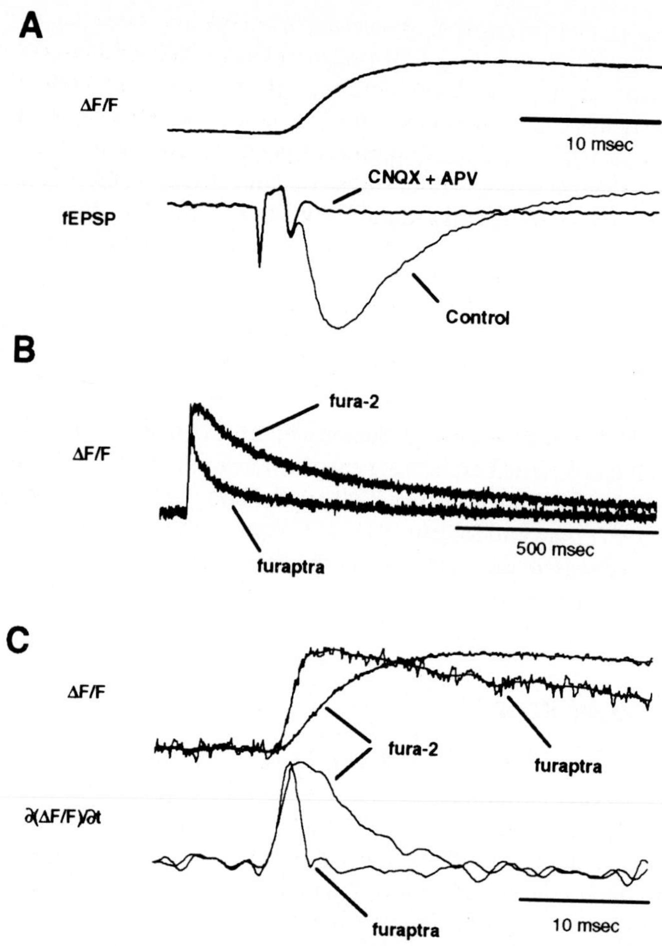

Fig. 10. CaSD transients from selectively loaded presynaptic terminals recorded with a single photodiode. Presynaptic axons and terminals of CA3 pyramidal cells were loaded with the CaSD Fura-2, using the technique described in Fig. 6. **A.** Overlaid fEPSPs and CaSD transients evoked by a single stimulus in area CA1 under control conditions and after the application of glutamate receptor antagonists (CNQX, 10 µM, + D-APV, 25 µM) to block postsynaptic activity. Blocking postsynaptic activity does not alter the CaSD signal, indicating a presynaptic origin.. **B.** Comparison of presynaptic calcium transients measured with a low affinity (Furaptra) and a high affinity (Fura-2) CaSD. CaSDs were loaded as described in A. Normalized transients evoked by a single electrical stimulus are shown. **C.** The same transients on a faster time scale (*top*) and the first derivative of the calcium transients (calcium influx, *bottom*). The low affinity indicator allows for the observation of more rapid transients with less distortion than the high affinity indicator.

postsynaptic CA1 pyramidal cells can be selectively loaded by injecting CaSD into the alveus (Regehr and Tank 1991; Wu and Saggau 1994a). Presynaptic mossy fiber terminals in area CA3 can be loaded by injecting CaSD into the hilus (Regehr and Tank 1994; Qian and Saggau, unpublished observation). Also, parallel fiber presynaptic terminals in the cerebellum can be selectively loaded using this technique (Mintz et al. 1995).

Evoked Signals with Bath-applied CaSD

Fig. 11 shows an experiment where the CaSD Calcium Orange was bath-applied to a hippocampal slice, and calcium transients evoked by a single electrical stimulus applied to the Schaffer collaterals were recorded in area CA1. Because the indicator was bath-applied, it was non-selectively loaded into the various pre- and postsynaptic neuronal structures. This can be illustrated by applying the ionotropic glutamate receptor antagonists CNQX and D-APV to block the postsynaptic response without altering the presynaptic activity. The relative contributions of the various cellular structures to the overall CaSD signal depends on two main factors: (1) the size of the change in $[Ca^{2+}]_i$ in those structures and (2) the volume of the structures:

$$CaSD = \sum(\Delta[Ca^{2+}]_i \times Volume_i)$$

The second factor is due to the fact that the number of indicator molecules in a particular structure is roughly proportional to its volume. For a more detailed discussion of the contribution of different cellular elements to the CaSD signal see Sinha et al (1995).

Fig. 11C illustrates how this technique can be used to obtain information about the spatio-temporal characteristics of neuronal activity. CaSD signals are first seen in the dendritic regions, the location of the presynaptic terminals, closest to the site of stimu-

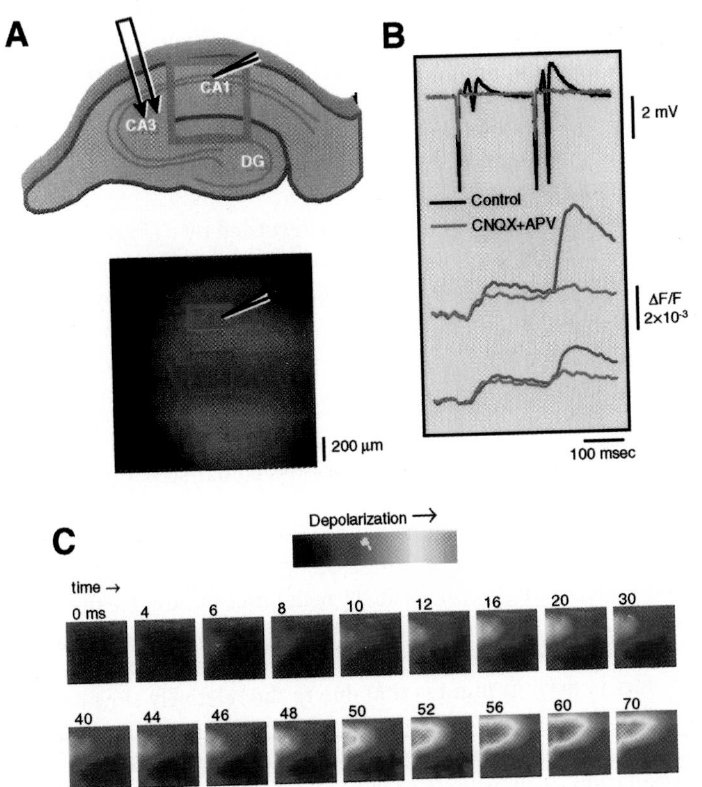

Fig. 11. Evoked CaSD transients recorded with a PDM. The CaSD Calcium Orange was bath-applied as described in section "CaSD Loading". Signals evoked by stimulating the Schaffer collaterals were recorded in area CA1 with a 10×10 PDM as described in Fig. 8 **A.** Schematic of brain slice showing recording area and positions of recording and stimulation electrodes (*top*); CCD image of recording area indicating groups of PDM elements (framed) from which data is shown in B and C (*bottom*). Green frame covers *str. pyramidale*; blue frame covers *str. radiatum*. **B.** Evoked activity in response to pair of electrical stimuli recorded under control conditions and in the presence of CNQX (10 µM) + D-APV (25 µM). Field electrode recordings (*top*) and sample CaSD signals are shown (*bottom*). **C.** Time lapse illustrating the spatio-temporal nature of the CaSD signals recorded under control conditions. Time 0 corresponds to the first stimulus; 60 msec corresponds to the second stimulus.

lation (to the left of the region shown in the figure). They then spreads across the brain slice away from the site of stimulation and into the cell body layer, *stratum pyramidale*, the location of the postsynaptic cell bodies. While there is significant loading of CaSD into glial cells (Albowitz et al. 1997), changes in $[Ca^{2+}]_i$ in glial cells tend to occur on much slower time scales, thus contributing little to the transients shown here.

Spontaneous Epileptiform Activity Recorded with Bath-applied VSD

Optical recording techniques are extremely well suited for investigations of the spatio-temporal characteristics of complex activity in neuronal populations, such as spontaneous epileptiform activity. We have previously used this technique to extensively investigate different aspects of spontaneous interictal epileptiform activity in the hippocampal slice (Colom and Saggau 1994; Sinha et al. 1995, 1996). Fig. 12 shows a sample experiment where synchronized activity in the hippocampal interneuronal network was recorded using the VSD RH-414 and a PDM. This activity occurs spontaneously in the presence of the K^+ channel antagonist 4-aminopyridine (4-AP, 100 μM) and can be isolated from spontaneous interictal epileptiform activity by the application of the ionotropic glutamate receptor antagonist CNQX and D-APV (Michelson and Wong 1994; Sinha et al. 1996). This activity is mediated by $GABA_A$ receptor-mediated depolarizing responses among interneurons. Using this technique has allowed us to investigate the spatio-temporal characteristics of this activity and to compare it to spontaneous interictal epileptiform activity (Sinha et al. 1996). We are currently in the process of further characterizing this activity and investigating the specific types of interneurons involved in generating it and the mechanisms responsible.

As discussed above for $[Ca^{2+}]_i$ and CaSDs, the activity recorded by a single element of the PDM when using a VSD actually represents a weighted average of the membrane potential transients in all the structures located in the recording area covered by that detector element. For VSDs, which remain on the surface of cells, this weight is the surface area of the structures:

$$VSD = \sum(\Delta V_{m,i} \times \text{SurfaceArea}_i)$$

For a more detailed discussion of the contribution of various cellular structures to the VSD signal, see Sinha et al. (1995).

Because the VSD signals represent an average of the membrane potential in multiple structures, the possibility exists that if some of these structures are hyperpolarized while others are depolarized, their activities may cancel each other out to give the false impression of the absence of activity. This is not as big a concern with CaSD signals because few physiological manipulations cause a rapid decrease in $[Ca^{2+}]_i$ from resting levels. Another important fact to keep in mind is that due to the relatively slow nature of the activity being recorded, the AC coupling constant of the PDM (~100 msec) may actually affect the signals. This problem could be avoided by using an A/D converter with sufficient intensity resolution to allow for recording only in DC-coupled mode, as with the single photodiode. Note: the sensitivity of this technique is not sufficient to record spontaneous activity in individual neurons; the activity must be occurring in a population of neurons.

Troubleshoot

Optical recording is a technically demanding endeavor and a systematic approach is essential. This is as true for the initial planning and implementation as it is for troubleshooting of the techniques.

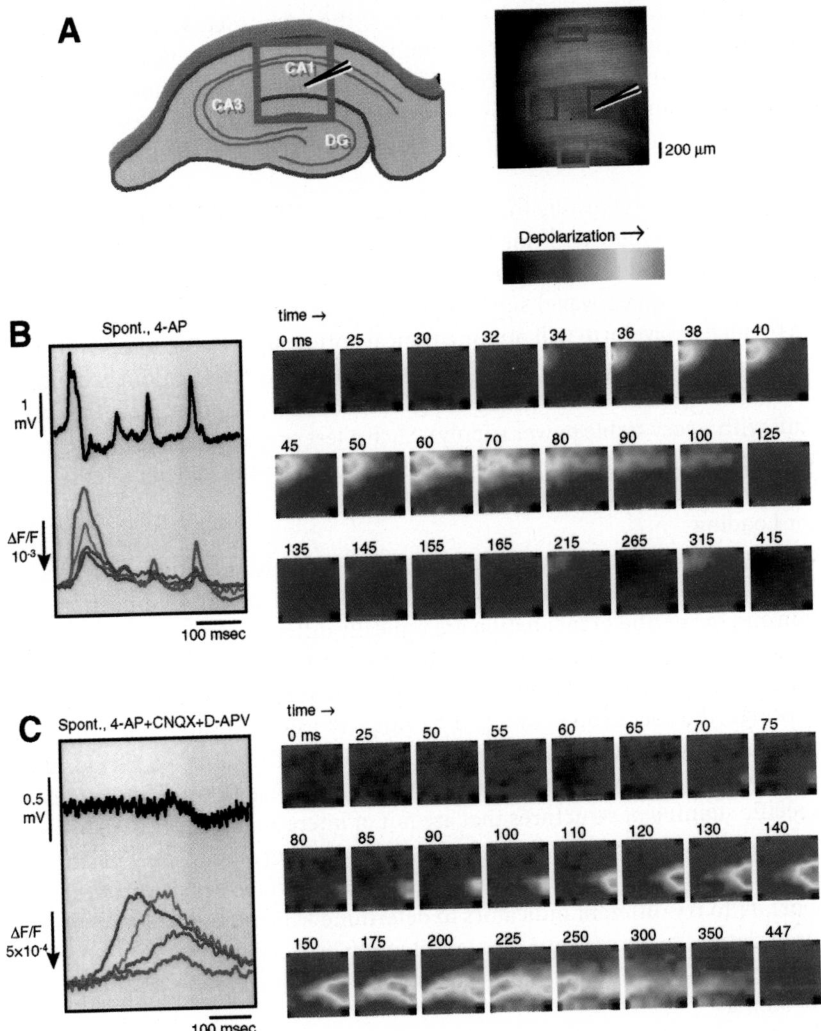

Fig. 12. VSD transients associated with spontaneous epileptiform activity recorded with a PDM. The VSD RH-414 was loaded by bath-application as described in section "VSD Loading"; spontaneous synchronized activity was recorded with a 10×10 PDM as described in Fig. 9. **A.** Schematic of brain slice showing recording area and position of recording electrode (*left*); CCD image of recording area indicating groups of PDM elements (framed) from which data is shown in B and C (*right*). **B.** Spontaneous epileptiform activity recorded in 100 µM 4-AP. Field electrode recording and sample VSD signals are shown (*left*) along with a time lapse illustrating the spatio-temporal nature of the activity (*right*; time 0 corresponds to the beginning of the traces shown on the left). The activity originated in area CA3 and spread across the CA1; it was confined mainly to *str. pyramidale* and proximal dendritic regions. **C.** Spontaneous synchronized interneuronal activity recorded in 4-AP + 10 µM CNQX + 25 µM D-APV. This activity originated in the subicular side of area CA1 and spread towards CA3; it was largest in the distal apical dendritic region.

A step that can greatly aid in troubleshooting is to start with a single photodiode as the photodetector. In addition to the advantages discussed previously, the output of a single photodiode with its amplifier can be directly monitored with an oscilloscope. This removes the added complexity of the data acquisition hardware, which is only necessary for later processing, i.e., signal averaging, and storage of data for a single photodiode (with CCD cameras or PDMs it is much more difficult or even impossible to monitor the data without the data collection hardware). Thus, we highly recommend that the first step should be to fine tune all other aspects of the experimental design (indicator loading, light source noise, optical filter selection, data collection hardware) using a single photodiode. Once this has been accomplished, then the desired photodetector can replace the single photodiode. The only troubleshooting that should then remain will involve the new photodetector and/or its interface with the data acquisition hardware.

- Test Equipment
  A relatively simple way of testing the photodetector and data collection hardware and procedures is to use a light-emitting diode (LED) powered by a function generator as a test preparation. Most function generators have the capability to output a small AC component (e.g., a square wave) superimposed on a larger DC component. An LED powered by such a waveform will mimic a typical optical signal: a small ΔF with a relatively large F. This method can easily generate a signal with a fractional change on the order of 10^{-2} (generation of smaller fractional changes requires a custom-made test circuit with a very stable power supply). Such a test preparation allows one to test the entire detector and data acquisition hardware without the added complications that a biological preparation introduces.
- Indicator Loading
  There are several aspects of indicator loading that need to be optimized and checked in order to obtain the best signals for a given experimental design. The optimal concentration and the period of incubation are typically different for different indicators and preparations. The numbers given in the protocols in this chapter are those we have found to be optimal for the specific applications and to provide a good starting range; however, the user should always try a range of values. It should be noted that using the highest indicator concentration for the longest incubation period might provide the highest fluorescence but not necessarily the best signal. This can be due to nonspecific staining of structures that are not of interest and, in the case of membrane-permeable AM esters, loading of subcellular structures (e.g., endoplasmic reticulum or mitochondria). As mentioned in section "Voltage-sensitive Dyes", it is also important to try different indicators to determine which is best suited for a particular preparation. In the case of AM esters, it might be worthwhile to try brain slices from younger animals to obtain improved loading.
- Origin of Signals
  Another aspect of indicator loading that is very important when recording from populations of neurons in brain slices is the location of the loaded indicator and, thus, where the optical signals are originating. For indicator that has been bath-applied this is mainly a matter of determining whether the signals are from neuronal or glial structures. The time course of signals can be useful in this respect: glial signals tend to be much slower and/or smaller than neuronal signals (Prince et al. 1973; Dani et al. 1992) and generally require more intense stimuli (Porter and McCarthy 1996). Also, for fluorescent indicators, confocal microscopy can be used to determine the indicator location (Albowitz et al. 1997). In the case where the attempt is made to load specific structures in a slice, one can use pharmacological tools (e.g., blocking postsynaptic activity to determine if signals are originating for pre- or postsynaptic structures) or selective activation of a group of cells.
- Artifactual Signals
  Once an optical signal has been obtained, it is important to determine whether it is an actual indicator signal or an artifact. The two main sources of artifacts that may mimic an indicator signal are movement and intrinsic signals. Because of the small size of optical indicator signals, very small movements in the preparation can lead to changes in the light seen by the photodetector that are on the same order of magnitude as the optical signal or even larger. In addition, as discussed in Chapter 34 (Grinvald et al.), the intrinsic optical properties of neuronal tissue change with activity. In

order to determine if the recorded signals are actually from the indicator, several approaches can be taken. One is to determine the spectral properties of the signal, as movement artifacts and intrinsic optical signals are relatively wavelength-independent, whereas optical indicator signals are not. Furthermore, with some optical indicators, the direction of the voltage- or calcium-dependent change of the optical signal will actually vary with the wavelength (e.g., VSD RH-155 or CaSD Fura-2); artifactual signals will not do this. Another way of distinguishing between actual signals and artifacts is to compare recordings from stained and unstained preparations.

- Indicator Toxicity
  Although optical recording techniques are generally much less invasive than conventional approaches, the presence of an indicator may still affect the preparation. One obvious effect is the possibility of indicator toxicity; while this was a more significant problem with earlier indicators, it is still an important factor to keep in mind. Actually testing for toxicity can be as simple as monitoring the effect of the indicator on electrical responses such as field potentials. If the indicator is noted to significantly deteriorate the health of the preparations, several options are available such as reducing the indicator concentration if possible, or trying another indicator. Also, because the toxicity of most indicators, specially the VSDs, is related to substances produced by intense illumination in the presence of oxygen, reducing the amount of light exposure will typically reduce toxicity. Unfortunately, the use of antioxidants or even removal of oxygen from the bathing solution is not a viable option for mammalian brain slices.
- Buffering by Indicators
  Aside from toxicity, optical indicators can also alter properties of the preparation. This is specially true for ion-sensitive indicators, which act as buffers for the ion that they bind and thus can alter the concentration and dynamics of that ion. In general, the higher the concentration of the indicator and the higher its affinity for the ion, the more likely it is to have such an effect. This is very important to keep in mind when interpreting optical signals from CaSDs. The extent to which this is a factor can be determined by varying either the concentration of the indicator (difficult to do when loading by bath-application) or to use an indicator with a lower affinity (e.g., Furaptra instead of Fura-2).
- Signal-to-Noise Ratio
  Because of the small size of most optical signals, obtaining a sufficient signal-to-noise ratio (*SNR*) is frequently a problem. Noise in an optical recording system can be classified into one of three categories: dark noise, source noise and shot noise. Dark noise is present even in the absence of light; it is the noise that is intrinsic to recording apparatus and is independent of the amount of light:

  $$N_{dark} = \text{const.}$$

  Source noise is due to fluctuations in the intensity of the light emanating from the light source. It can be the result of noise in the power supply, noise intrinsic to the light source, or noise that is somehow introduced in the light path (e.g., motion artifact). Source noise is proportional to the amount of light:

  $$N_{source} \propto I$$

  Shot noise is due to the quantal nature of light. It is proportional to the square root of the light intensity:

  $$N_{shot} \propto \sqrt{I}$$

  The first step in determining how to improve *SNR* is to identify the predominant source of the noise. This can done by determining how noise varies with light intensity: invariant (N_{dark}), proportional (N_{source}) or exponential (N_{shot}). Because the optical signal is proportional to light intensity, the effect of varying the light intensity on *SNR* for these various types of noise are:

$$SNR_{dark} \propto I$$

$$SNR_{source} = \text{const.}$$

$$SNR_{shot} \propto \sqrt{I}$$

Thus if the limiting noise is either N_{dark} or N_{shot}, increasing the intensity of illumination will improve *SNR*. However, if N_{source} is the limiting noise, then only decreasing N_{source} (i.e., using a more stable light source and/or power supply) will improve *SNR*. A simple approach to improving *SNR* is to average signals. This can refer to taking the average of a number of trials (temporal averaging) or, in the case of spatially-resolving photodetectors, the average of several photodetector elements (spatial averaging). In both cases, the improvement in *SNR* obtained by averaging is:

$$SNR \propto \sqrt{n}$$

where n is the number of trials or elements that are averaged.

Comments

Optical recording techniques are powerful tools in cellular physiology. They provide the means for studies that would not be feasible with more conventional approaches, such as using micropipettes. In addition, they often offer an elegant approach to problems that may be very difficult to address with other techniques. In this chapter we have attempted to cover the basic issues involved in optically recording from populations of neurons in brain slices. Optical recording is a very broad and constantly growing field; even the variety of specific examples that we have shown from our own work can only provide a limited picture of the vast array of specific techniques and applications . However, the basic issues we have discussed are generally applicable and should provide a good starting point for anyone interested in using optical recording techniques to record from populations of neurons. Because of the large body of literature available in this field and because of the extremely rapid rate with which new techniques, indicators and hardware are being developed, we highly recommend that the next step after reading this article should be to scour the literature for examples that closely match the specific application. Chances are someone has at least tried it.

Acknowledgement: This work was supported in part by grants from NIMH MH 10491 (S.R.S.), NSF IBN-9723871 µP.S.) and NIH NS33147 (P.S.). We are grateful to L.G. Wu and Dr. J. Qian for their contributions to some of the reviewed techniques and studies. We also gratefully acknowledge the excellent technical assistance provided by Dr. S.S. Patel in various aspects of this project.

References

Albowitz B, Konig P, Kuhnt U (1997) Spatio-temporal distribution of intracellular calcium transients during epileptiform activity in guinea pig hippocampal slices. J Neurophysiol 77:491–501.

Antic S, Major G, Chen WR, Wuskel J, Loew L, Zecevic D (1997) Fast voltage-sensitive dye recording of membrane potential changes at multiple sites on an individual nerve cell in rat cortical slice. Biol Bull 193:261.

Breckenridge LJ, Wilson RJA, Connolly P, Curtis ASG, Dow JAT, Blackshaw SE, Wilkinson CDW (1995) Advantages of using microfabricated extracellular electrodes for *in vitro* neuronal recordings. J Neurosci Res 42:266–276

Colom LV, Saggau P (1994) Spontaneous interictal-like activity originates in multiple areas of the CA2-CA3 region of hippocampal slices. J Neurophysiol 71:1574–1585

Dani JW, Chernjavsky A, Smith SJ (1992) Neuronal activity triggers calcium waves in hippocampal astrocyte networks. Neuron 8:429–440

Denk, W. and Svoboda, K. (1997) Photon upmanship: Why multiphoton imaging is more than a gimmick. Neuron 18:351.

Grynkiewicz G, Poenie M, Tsien RY (1985) A new generation of Ca^{2+} indicators with greatly improved fluorescence properties. J Biol Chem 260:3440–3450

Haugland, RP (1996) Handbook of fluorescent probes and research chemicals, Molecular Probes, Eugene, OR

Michelson HB, Wong RKS (1994) Synchronization of inhibitory neurones in the guinea pig hippocampus in vitro. J Physiol (Lond) 477:35–45

Minta A, Kao JPY, Tsien RY (1989) Fluorescent indicators for cytosolic calcium based on rhodamine and fluorescein chromophores. J Biol Chem 265:8171–8178

Mintz IM, Sabatini BL, Regehr WG (1995) Calcium control of transmitter release at a cerebellar synapse. Neuron 15:675–688

Porter JT, McCarthy KD (1996) Hippocampal astrocytes in situ respond to glutamate released from synaptic terminals. J Neurosci 16:5073–5081

Prince DA, Lux HD, Neher E (1973) Measurement of extracellular potassium activity in cat cortex. Brain Res 50:489–495

Qian JQ, Colmers WF, Saggau P (1997) Inhibition of synaptic transmission by neuropeptide Y in rat hippocampal area CA1: modulation of presynaptic Ca^{2+} entry. J Neurosci 17:8169–8177

Qian JQ, Saggau P (1997) Presynaptic inhibition of synaptic transmission in the rat hippocampus by activation of muscarinic receptors: involvement of presynaptic calcium influx. Brit J Pharm 122:511–519

Regehr WG, Delaney KR, Tank WD (1994) The role of presynaptic calcium in short-term enhancement at the hippocampal mossy fiber synapse. J Neurosci 14:523–537

Regehr WG, Tank WD (1991) Selective Fura-2 loading of presynaptic terminals and nerve cell processes by local perfusion in brain slice. J Neurosci Methods 37:111–119

Ross WN, Salzberg BM, Cohen LB, Grinvald A, Davila HV, Waggoner AS, Wang CH (1977) Changes in absorption, fluorescence, dichroism, and birefringence in stained giant axons: optical measurement of membrane potential. J Membrane Biol 33:141–183

Sinha SR, Patel S, Saggau P (1995) Simultaneous optical recording of evoked and spontaneous transients of membrane potential and intracellular calcium concentration with high spatio-temporal resolution. J Neurosci Meth 60:49–60

Sinha SR, Saggau P (1996) Spontaneous synchronized activity in guinea pig hippocampal slices induced by 4-AP in the presence of ionotropic glutamate receptor antagonists. Soc Neurosci Abstr 22:823.17

Sinha SR, Wu L-G, Saggau P (1997) Presynaptic calcium dynamics and transmitter release evoked by single action potentials at mammalian central synapses. Biophys J 72:637–651

Tsien RY (1980) New calcium indicators and buffers with high selectivity against magnesium and protons: design, synthesis, and properties of prototype structures. Biochemistry 19:2396–2404

Tsien RY (1981) A non-disruptive technique for loading calcium buffers and indicators into cells. Nature 290:527–528

Wenner P, Tsau Y, Cohen LB, O'Donovan JH, Dan Y (1996) Voltage-sensitive dye recording using retrogradely transported dye in the chicken spinal cord: staining and signal characteristics. J Neurosci Methods 70:111–120

Wu L-G, Saggau P (1994a) Presynaptic calcium is increased during normal synaptic transmission and paired-pulse facilitation, but not in long-term potentiation in area CA1 of hippocampus. J Neurosci 14:645–654

Wu L-G, Saggau P (1994b) Pharmacological identification of two types of presynaptic voltage-dependent calcium channels at CA3-CA1 synapses of the hippocampus. J Neurosci 14:5613–5622
Wu L-G, Saggau P (1994c) Adenosine inhibits evoked synaptic transmission primarily by reducing presynaptic calcium influx in area CA1 of hippocampus. Neuron 12:1139–1148
Wu L-G, Saggau P (1995a) Block of multiple presynaptic channel types by ω-conotoxin-MVIIC at hippocampal CA3 to CA1 synapses. J Neurophysiol 73:1965–1972
Wu L-G, Saggau P (1995b) $GABA_B$ receptor-mediated presynaptic inhibition in guinea-pig hippocampus is caused by reduction of presynaptic Ca^{2+} influx. J Physiol (Lond) 485:649–657
Wu L-G, Saggau P (1997) Presynaptic inhibition of elicited neurotransmitter release. TINS 20:204–212
Zecevic D (1996) Multiple spike-initiation zones in single neurons revealed by voltage-sensitive dyes. Nature 381:322–325

Suppliers

Company: Molecular Probes, Eugene, OR, USA
Company: Centronic, Newbury Park, CA, USA
Company: Chroma Technology Corporation, Brattleboro, VT, USA
Company: Fuji, Stamfort, CT, USA
Company: Hamamatsu, Bridgewater, NY, USA
Company: Kepco, Flushing, NY, USA
Company: Neuroplex, OptImaging, Fairfield, CT, USA
Company: Vincent, Rochester, NY, USA
Company: Zeiss, Rochester, NY, USA

Abbreviations

$[Ca^{2+}]_i$	intracellular calcium concentration
CaSD	calcium-sensitive dye
CCD	charge-coupled device (camera)
DB	dichroic beamsplitter
EM	emission filter
EX	excitation filter
PDA	photodiode matrix amplifier
PDM	photodiode matrix
SH	shutter
V_m	membrane potential
VSD	voltage-sensitive dye

Chapter 17

Recording of Electrical Activity of Neuronal Populations

H. JOHANSSON, M. BERGENHEIM, J. PEDERSEN and M. DJUPSJÖBACKA

Introduction

The problem of how information is encoded and transmitted within the nervous system has challenged neurophysiologists for years. Many attempts to find and describe some general principles for such neural coding have been made, and it is probably fair to say that the opinion about these general principles has changed lately. Today, it seems clear that most information within both the central and the peripheral nervous system is encoded and transmitted by large populations of neurons. For the sensory nervous system there is a continuously growing body of evidence indicating that even simpler peripheral stimuli are transmitted to the central nervous system by large populations of sensory neurons.

Due to this, neurophysiologists have shifted their interest from the activity of single neurons to the multi-unit activity in such populations. Due to technical and practical difficulties, population activity was for a long time assessed indirectly by using sequential recordings of single neurons, and the population was artificially constructed afterwards by pooling these sequentially recorded single neurons. However, the neurons within a neural population must be simultaneously recorded if their population characteristics are to be investigated, and indeed, recent experiments indicate that information is lost if the neural populations are recorded sequentially and pooled (Johansson et al. 1995b). This fact has led to a rapid development of techniques enabling the recording of several neurons simultaneously both in the central and peripheral nervous system.

So why is there a problem with pooled sequential recordings? Well, there are several easily conceived problems associated with such recordings of neural activity. One of the more important problems is that it is difficult to reproduce a stimulus perfectly. The reason for this is that even if a series of stimuli might seem identical in terms of some qualities measured by the experimenter (e.g., changes in muscle length), it does certainly not mean that identical effects are produced at the receptor level. For all types of neural activity the receptor responses produced by a given stimulus are likely to change over time due to variations in a number of experimental conditions, e.g., temperature, circulation, depth of anesthesia. Therefore, there is little doubt that the ideal way to study population coding is indeed to record the neurons within the population simultaneously (for elaborate discussion see Deadwyler and Hampson 1997, Johansson et al. 1995b).

This chapter will briefly review some methods for recording multi-unit activity on both the central and the peripheral level in *in vivo* animal experiments. These are, however, not the only areas where multi-unit recording techniques have been developed. As an example, there are several methods for such recordings of neurons in cell culture preparations. The interested reader is referred to the paper by Breckenridge and co-workers (1995) describing a 64 electrode array, or the paper by Stoppini and colleagues

---

H. Johansson, National Institute for Working Life, Department of Musculoskeletal Research, Umeå, Sweden (phone: +46-90-785-50-60; fax: +46-90-786-50-27; e-mail: hakan.johansson@niwl.se)

(1997), where an array of 30 biocompatible microelectrodes is described. Both of these electrode arrays are designed for extracellular recordings.

Naturally, this chapter cannot deal with all the developed techniques and it will not be a step-by-step guide to multi-unit recordings. It will, however, give some references to techniques used, and some recent examples will be discussed more elaborately. Furthermore, an example of analysis of population coding will be presented together with some examples of results attained using this analytic method.

It should perhaps be made clear already at this early stage that for a good multi-unit recording setup, a combination of a multi-unit electrode and proper software for spike separation is desirable. Below, these issues will be discussed separately.

Subprotocol 1 Multi-Unit Recording

Procedure

Multi-unit Recording on the Peripheral and Dorsal Root Level

As concerns multi-unit recordings on the peripheral level and the level of the dorsal root, there are not a lot of techniques presented in the literature. We would like, however, to mention a couple of principally different setups. Quite early, an interesting type of multi-electrode was developed in order to record from regenerating severed nerve fibers (Marks 1965). In this method, the severed nerve was made to regenerate through an array of gold tubes, and each of these gold tubes served as an electrode. These techniques could only be used for severed nerve fibers, and naturally, the time between insertion of the electrodes and the time for recording was quite long. Methods were needed for multi-unit recordings in acute experiments. In early attempts to perform such multi-unit recordings, a setup of several single conventional electrodes was used. However, since such electrodes needed a lot of space (partly because they used one reference electrode each), the number of axons that could be recorded from simultaneously was quite limited.

As a response to this demand Djupsjöbacka and coworkers (1994) presented a multi-channel hook electrode for acute cat experiments some years ago. This electrode is shown in Fig. 1, and consists of 12 silver wire hooks which are fixed on a common 'holder' made of a 3-mm-thick black PVC plastic plate. The holder is shaped in a semi-circular fashion to fit optimally in the limited space between the entrance of the L_7-L_6 spinal roots into the spinal canal, and the pelvic bone. In this holder 12 notches are milled radially. The notches have a width of 0.45 mm and a depth of 1.0 mm. In each of these notches a hook is mounted tightly by pressing it down into the notch. The hooks are made of 0.50-mm solid silver wire, which is chlorided in order to obtain a stable half-cell potential and a low impedance between the hook and the nerve filament (Geddes et al. 1969). The chloriding is performed after the hooks are mounted on the holder. Each of the 12 hooks constitutes a separate channel. Each channel on the electrode has an individually shielded cable with two conductors – the positive input wire is soldered to the corresponding Ag/AgCl hook, and the negative input wire is soldered to an Ag/AgCl reference needle (see Fig. 1B). Since it would be impractical to have a unique reference needle for every channel, all reference wires are connected to a single reference needle. The shield of the cables is connected to ground. The cables have a cross-sectional area of 0.02 mm^2 consisting of 10 copper fibers. This makes a thin and flexible bundle of cables. In

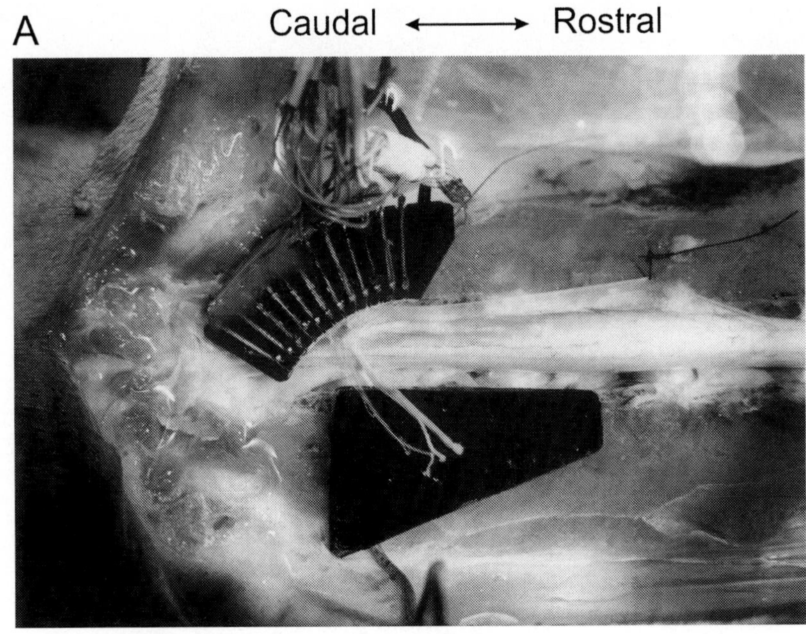

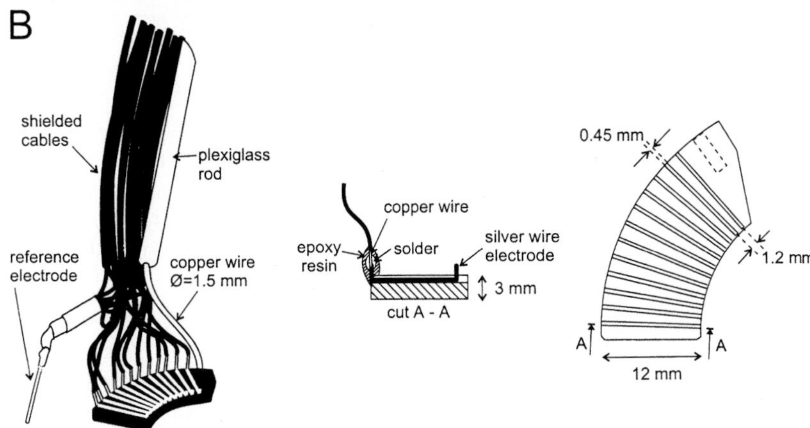

Fig. 1. A: Photograph of a multichannel hook electrode in place. This photo is taken during the recording from the L7 dorsal root in anesthetized cat. **B**: Schematic drawing of the same electrode.

order to avoid short-circuiting of the channels due to contact with the tissue covering the pelvic bone (see Fig 1A), the back of the electrode, where the hooks leave the holder and are soldered to the cables, is embedded in epoxy (see Fig. 1B). In order to make this process easy we found it important to design the electrode with maximum distance between the upward pointing part of the hook, where the dorsal root filaments are wound to the silver wire, and the back part of the holder.

Today, a somewhat different electrode is used in our lab (see Fig. 2). This electrode is a modified version of the one described above. For some experiments, it has some clear-cut advantages. Firstly, the silver-wire hooks are flexible, and can be bent in any direction. Secondly, they are not fixed to the PVC plate, but only held by it. This means that the wires can be pulled backwards and forwards enabling a control of the length of each electrode separately. For some particular experiments such a flexible setup might have major advantages.

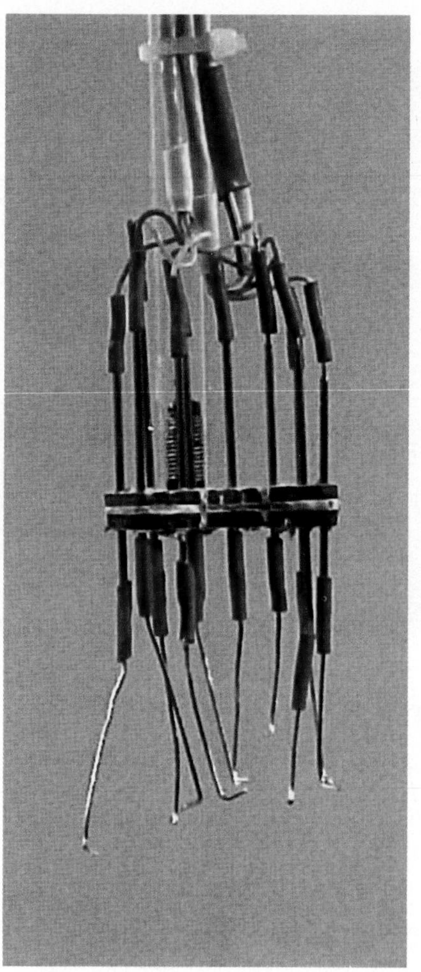

Fig. 2. Photograph of a more recent version of the multichannel hook electrode illustrated in fig 1. Here, the silver wires are not fixated to the PVC plate but only held in place by it. Also, the silver wires are flexible and can be bent in any desired direction.

The quality of recording is the same for these two types of electrodes. Figure 3 gives an example of such recordings. In this particular experiment, the responses of 12 muscle spindle afferents (MSAs) from two cat hind limb muscles (the gastrocnemius and soleus muscles: GS; and the posterior biceps and semitendinosus muscles: PBSt) were simultaneously recorded via the multi-channel hook electrode from 12 L_7 dorsal root filaments. Both receptor-bearing muscles were subjected to sinusoidal stretching (GS: 1 Hz; PBSt: 0.9 Hz; peak-to-peak amplitude: 2 mm) around a mean length of 2 mm below maximal physiological length. The 12 uppermost traces in the figure show the amplified signals from each channel sampled at 20 kHz. The two lowermost traces show the sinusoidal stretching of the muscles.

The electrodes mentioned above are easiest to use when the cat is rigidly immobilized. There are, however, certain preparations where such immobilization is not possible. One such preparation is the decerebrated cat, actively locomoting on a treadmill. In this setup the muscle tissue, the spinal cord and also the dorsal roots themselves are constantly moving in relation to the frame where the electrode is fixed. Due to this movement, the use of the types of electrodes described above might be difficult. Taylor and coworkers (personal communication) solved this problem by fixating the electrode to the muscle tissue surrounding the recording site. With this method of fixating the electrode, the authors have recorded up to 8 single muscle spindle afferents simultaneously (Taylor et al. 1998) in the cat locomoting on a treadmill. This technique has not

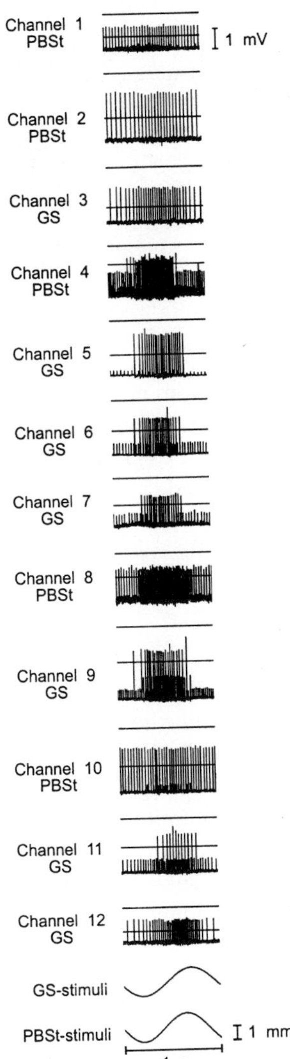

Fig. 3. Example of a recording using the multichannel hook described in Figs. 1 and 2. This figure illustrates a simultaneous recording of 12 nerve filaments. The responses of 12 muscle spindle afferents from the gastrocnemius-soleus (GS) muscles (7 afferents) and the posterior biceps-semitendinosus (PBSt) muscles (5 afferents) to sinusoidal stretches are shown. The two lowermost traces show the sinusoidal stretching of the muscles (1 Hz for the GS muscles and 0.9 Hz for the PBSt muscles).

yet been described in a full-length paper, but the interested reader can contact P.H. Ellaway (author of Chapter 33) for more detailed information.

The recording of multi-unit activity becomes even more complicated if one wants the cat to be awake and freely moving. To this end, a different approach based on microwire electrodes has been developed, and it is actively used today in the lab of A. Prochazka and coworkers in Edmonton, Canada. Here, chronically implanted so-called floating microwire electrodes are used, and several single afferents on the dorsal root level can be recorded simultaneously. The advantage with this kind of setup is that neural activity can be studied in awake, behaving cats rather than in anesthetized ones. This set-up is illustrated in Fig. 4 and as can be seen in the figure, microwires are floating on the neural tissue, stuck and/or sutured to the dura mater. The cable shield is fixed and stabilized with a dental acrylic cap in the L6 spinous process. For references concerning this method see Prochazka et al. (1993) and Loeb et al. 1977.

In conclusion, in spite of the fact that simultaneous recordings are necessary for the investigation of population coding, there are not a lot of techniques described allowing such recordings on the peripheral level. This is certainly a pity since investigation of population coding on the peripheral level has the advantage of focusing on what infor-

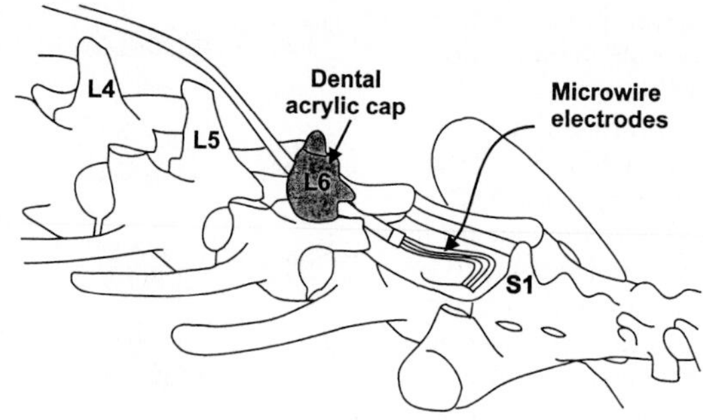

Fig. 4. The chronically implanted floating microwire electrode. As can be seen in the figure, the microwires are floating freely around the dorsal root and the cable shield is fixated with a dental acrylic cap to the L6 spinous process. (Redrawn from Prochazka et al. 1983, with permission)

mation is *available* to the CNS, but does not necessarily make any assumptions about the unknown decoding in the CNS. Since many of the analytical methods applied on the CNS level cannot avoid making such assumptions, analysis on the peripheral level should perhaps sometimes be preferred.

Multi-unit Recording on the Central Level

Recording of multiple neurons on the central level demands a technically different approach to that on the peripheral level, and over the years, there have been several attempts to make such recordings. As on the peripheral level, the earlier attempts were simply a setup of several conventional microelectrodes (see, e.g., Verzeano 1956). Following these early trials, techniques using parallel micropipettes were developed (see, e.g., Terzuolo and Araki 1961). In fact, quite a few attempts using several individual microelectrodes were made, and one of the problems that arose was how to move and control these electrodes individually. To solve this problem, Humphrey (1970) built a microelectrode drive enabling the individual advance of five microelectrodes (separation of recording points, about 3 mm).

In the early eighties, Krüger and Bach (1980, 1981) put forward a setup consisting of an array of 30 rigidly coupled microelectrodes. Even though the individual control of each microelectrode was not possible for this construction, successful recordings of 18 single units from the monkey striate cortex were obtained.

A different approach is the technique based on microetching (for review, see Prickard 1979a,b). Being rather large (often several millimeters), these electrodes were mainly intended for surface potential recordings. As an example of such an electrode, one might mention the 20–30 electrode device developed by Hanna and Johnson (1968).

More recently, techniques using multiple thin-shaft probes have been developed. (see, e.g., Eckhorn 1991, 1992; Krüger 1983; Reitboeck 1983). These recording techniques are particularly valuable if the multiple probes can be advanced separately. Furthermore, it is crucial that these probes cause minimal damage to the tissue, and that a flexible geometrical arrangement of the probes is possible.

An example of a manipulator with some of these advantages is the "Reitboeck Manipulator" (Reitboeck 1983). This manipulator has been used in several laboratories performing recordings from chronically prepared cats (see, e.g., Eckhorn 1991, 1992) and monkeys (see, e.g., Mountcastle et al. 1987).

However, tissue penetrations with such probes (including fiber and wire microelectrodes) using standard drives were limited to a depth that was below the buckling length of the probes.

To solve this problem, a new insertion method was developed by Eckhorn and Thomas (see Fig. 5). In this method, stretched rubber tubes are used for guiding the thin-

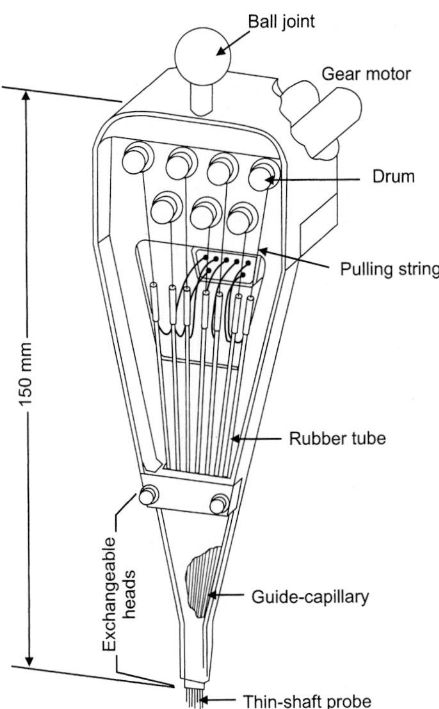

Fig. 5. A rubber tube drive with 7 independently movable thin-shafted probes. The guide capillaries are made of stainless steel. The gear motor is computer controlled and winds up the pulling strings on each drum separately. Hereby, each probe can be guided individually. (Redrawn from Eckhorn and Thomas 1993 with permission)

shaft probes. The authors claim that this new method has the following *advantages* compared with the traditional Reitboeck manipulator:

1. Several times higher mechanical driving forces are available and can be chosen over a broad range.
2. Probe positioning errors are smaller and do not accumulate, independent of the number of single movements and their directions.
3. "Microphonic potentials", picked up by high impedance probes during mechanical vibrations, are small or even absent in the rubber tube drive; this enables continuous recordings while the probes scan the tissue at low velocity.
4. Probes can be driven simultaneously and independently at different velocities in two directions.
5. The probe insertion device can be made small and lightweight.
6. Many types of thin probes can be handled, including fiber and wire electrodes with shaft diameters down to some tens of micrometers.
7. A broad range of commercially available micropositioners can be used for probe movement.

In this method, a flexible stretched rubber tube is fixated to a metal tube (capillary) that is mounted on the main support at the insertion device. The probe is inside the tube and the capillary guides it in the desired direction. A micropositioner exerts a pulling force. This micropositioner acts at the upper end of the rubber tube and stretches it to the extent required. A clamp (crimped silver-plated copper capillary) connects the end of the probe and the rubber tube with the support pin or wire of the micropositioner. Positioning of the probe's tip is made by translatory movement of positioners support.

This setup has been used for parallel independent insertion of 7 fiber microelectrodes through the intact dura into the brain of chronically prepared cats and monkeys (Eckhorn et al. 1993, Eckhorn and Obermueller 1993).

By using DC-micromotors with gear reduction (1/4096), the electrodes can be separately positioned. This is made possible by winding a string (Teflon-coated thread,

200 μm) around a drum (3.5 mm) fixed at the gear's axis. A position sensor (10 imp/rev) connected to the motors enables a resolution of 0.27 μm and allows respective computer control of the electrode's position.

Quite recently, these manipulators have been successfully tested for two years in a series of multiple microelectrode recordings from visual cortical areas of cats and monkeys. In these experiments the intact dura was penetrated singly by the electrodes (Eckhorn et al. 1993, Eckhorn and Obermueller 1993). Low mechanical interference between the microelectrodes enabled stable extracellular recordings of single-cell action potentials with several fiber microelectrodes (3–7 MΩ at 1 kHz), even when one electrode was slowly driven (at 0.5–10 μm/s). This low mechanical interference was due to the smooth movement in the rubber tube drive, the thin shafts of the fiber electrodes and their anti-adhesive coating.

The authors claim that the manipulator is small, light, and relatively cheap. Furthermore, they state that probes for both stimulation and recording of neuronal and muscular signals can be inserted, as well as thin injection pipettes and acupuncture needles.

Equipment for the Reitboeck method of multiple microelectrode recordings is commercially available from the U. Thomas Recording company (see SUPPLIERS below). The company manufactures and distributes several multi-channel recording systems and multi-fiber microelectrodes together with filters, fiber-electrode manipulators, tip pullers, tip grinding machines and positioning systems.

Another manufacturer of multi-microelectrode drive systems is Fredrick-Haer (FHC) (see SUPPLIERS below). The system advances 4 or 8 microelectrodes independently in 0.25-μm steps, and the electrodes can be arranged in different matrix configurations (i.e., 1*4, 2*2, 2*4, 1*8). The company also ensures compatibility with the major producers of relevant software, e.g., DataWave Technologies (Longmont, Colorado, USA). FHC can also provide customized solutions for needles and/or microelectrode drivers.

Quite an interesting device which combines multiple single-unit recordings with microiontophoresis has recently been described by Haidarliu and coworkers (1995). This metal-cored, multi-barrelled electrode consists of a 9-cm shaft and has a diameter of 0.6 mm at 7 cm from the tip. The authors claim that this electrode is suitable for several units in deep regions of the brain of various animals, including monkeys. Furthermore they suggest a setup with one combined electrode of the type described above, and three tungsten-in-glass electrodes in combination with four spike sorters (see below), enabling the recording of up to 12 units simultaneously. A pleasant aspect of this electrode is that the pharmacological environment of the recorded cells can be continuously modified during the recording. The combined electrode can be prepared using standard tungsten rods, glass capillaries, and standard equipment for pulling micropipettes (for details concerning the preparation of this electrode, see Haidarliu et al. 1995).

The microdrive system for this setup consists of a microdriving terminal and a compact remotely controlled microelectrode drive system allowing the independent advance and measuring of four microelectrodes. The microdriving terminal was originally developed by Abeles and Vaadia, Hebrew University, Jerusalem, and the remotely controlled drive system was designed by Ribaupierre, University of Lausanne. The system was tested on guinea pigs, in whom isolated single unit recordings could be obtained for several hours.

Very recently, Nicolelis and coworkers (1997) presented a method enabling the recording of large populations of neurons from the cortical and subcortical levels in behaving rats. In this method, 48 microwires were implanted in the brain stem, thalamus and somatosensory cortex of rats. Quite impressively, 86 % of the implanted wires gave single neuron activity, and with the use of a Many Neuron Acquisition Processor (MNAP, see SPIKE SEPARATION TECHNIQUES below) an average of 2.3 single spike trains was discriminated per microwire.

According to the authors, the optimal spacing of the microwires for cortical and subcortical implants was 100–250 μm, and with this setup they could make simultaneous recordings from neurons in the trigeminal ganglion, principal and spinal nuclei of the trigeminal brain stem complex, the ventral posterior medial nucleus of the thalamus, and the primary somatosensory cortex.

Using a blunt version of the wires, the ensemble recordings remained constant for several hours, and furthermore, the implanted microwires could even be used a couple of weeks after surgery, as they still yielded recordable single neuron spike trains.

The different configurations of the microelectrode matrices were designed by NB Laboratories (see SUPPLIERS below), and for details concerning the various matrices, see Nicolelis et al. (1997).

Spike Separation Techniques

Today, there are a lot of commercially available spike separation programs. Some years ago, however, this was not the case, and a substantial amount of work was put into the development of such techniques. In this section, we will briefly present some examples of this work (reviewed by Schmidt 1984a, b).

One early spike discriminator that deserves mention is the software-based algorithm for extracellular multi-neuron recordings described by Salganicoff et al. (1988), and the low-cost single board solution designed by Kreiter and coworkers (1989). A neural network solution was presented by Jansen (1990), who used a backpropagation network to reconstruct individual spike trains from extracellular multi-neuron recordings. In a similar method, a three-layer connectionist network was used for classification of multi-unit recordings from intrafascicular electrodes implanted in cats (Mirfakhraei and Horch 1994). Recently, Oghalai et al. (1994) described an adaptive spike discriminator, based on the ART2 algorithm (Carpenter and Grossberg 1987), for classification of multi-unit spike trains recorded on a single channel. Even more recently, Öhberg and coworkers (1996) described a software-based method for real-time spike discrimination which can be used for classification of single spikes in several channels with simultaneous multi-neuron recordings. An example of a multiple-unit recording where this spike separating method was used is shown in Fig. 6.

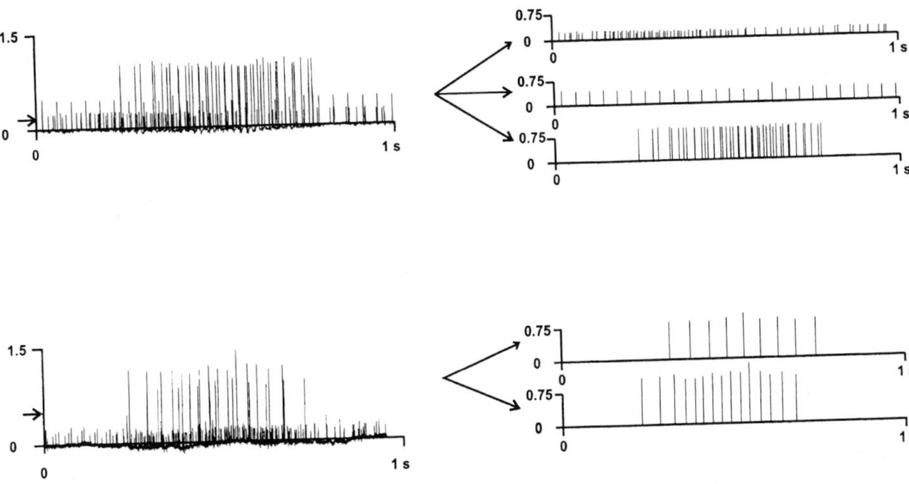

Fig. 6. Classification of different spikes from multineuron filament recordings. In the uppermost diagram three individual spike trains can be identified, and in the lowermost diagram two spike trains have been discriminated.

A so-called Many Neuron Acquisition Processor (MNAP, from Spectrum Scientific, see SUPPLIERS below), has recently been used and described by Nikolelis and coworkers (1997). This is a 96 channel system that allows discrimination of up to 4 individual spike trains from each channel. Both the MNAP itself and data storage transmitted from the MNAP to a PC by a fast-speed MXI bus can be controlled by a single microcomputer. The MNAP handles data sampling, waveform discrimination, and it can also generate real-time signals for synchronization of external devices. It has an output board which allows visual and audio monitoring of the analog signals. This processor is described in further detail by Nicolelis et al. (1997), where examples of recordings can be seen.

There is also some equipment commercially available for data acquisition which includes spike discriminators using linear clustering methods (e.g., Spike2, CED, see SUPPLIERS below).

Subprotocol 2
Examples of Analysis and Results

If recording multi-unit activity simultaneously is a difficult issue, then the analysis of such activity is even more complex. There are probably as many methods for analysis as there are scientists working on the problem. A recent review by Deadwyler and Hampson (1997) discusses quite a few of these analytical approaches. It provides a good review of methods using pattern detection and population vector calculations, and methods based on dimensional analysis, such as principal component analysis and linear- and canonical discriminant analysis.

There are however some problems associated with the analysis of ensemble firing patterns. The most common approach so far has been to look for similarities between the representations of the ensemble response to the stimulus and the stimulus itself, and to use the degree of similarity as a measure of how well a stimulus is represented by afferent neural activity. Examples of findings in such studies might be that the sum (Crago et al. 1982) or the mean (Hulliger et al. 1995) of the discharge rates of a population of sequentially recorded Golgi tendon organ afferents show an almost linear relationship with total muscle force. This line of investigation assumes that there have to be similarities between the representations of the stimulus in the ensemble response and the stimulus itself, or in other words, that the decoding mechanisms in the CNS would require afferent neural codes that are isomorphic representations of stimuli (e.g., could be represented with isometric curves). However, it is by no means obvious that these assumptions are correct.

Procedure

A relatively new method developed in our lab is based on *Principal Component Analysis* (PCA) combined with algorithms for the calculation of *stimulus separation* (discrimination). This method has several advantages compared to other analytic methods (see above). Firstly, the method allows the quantification of the ability of ensembles of afferents to discriminate between stimuli, thus enabling a direct comparison of the encoding capacity of differently composed ensembles of afferents. The possibility to make these comparisons is essential for the investigation of ensemble coding. Secondly, the method can be used to evaluate to which extent each afferent in an ensemble contributes to the discrimination of stimuli. Thirdly, this method focuses on stimulus discrimination and compares the relative discriminative ability of different ensembles of receptor afferents, and therefore, makes few assumptions about the unknown decoding within the CNS. Finally, the method is based on simultaneous recordings of afferents. Since most of the current knowledge about ensemble coding is based on single sequential recordings, and the questionable assumption is made that the actual ensemble behavior may be inferred from them, the new method offers an important advantage.

The ability of the ensemble of simultaneously recorded muscle spindle afferents (MSAs) to discriminate between stimuli was assessed by calculating the capacity of the ensembles to separate five different test stimuli. This method has been thoroughly described elsewhere (for details see Johansson et al. 1995a, b, Bergenheim et al. 1995, 1996). Therefore, the following section is restricted to a brief summary of the basic features of the method.

Multivariate Analysis

The responses of ensembles of MSAs to the sinusoidal stretches were arranged in a data table (see Fig. 7 A). Each row of such a table contains all the data from one test stimulus (i.e., one stretch sequence). Each afferent is represented by 500 columns in the table, and the variables representing the different afferents follow each other successively in every row (see Fig. 7A).

As the first step, a PCA of the data table was performed. The rows of the data table can be represented as points in a p-dimensional space (p = number of variables or columns in the table). We computed the first three principal components for the data set; in this way the p-dimensional system was reduced to three dimensions. In our experiments three principal components always described more than 90% of the variance of the data set. The significance of the computed principal components was checked with cross-validation. Our PCA analysis was performed with a program package (SIMCA, Umetri, Sweden) run on a PC-486 computer.

Quantification of Stimulus Separation

After the PCA, each repetition of a stimulus was represented as a point in a three-dimensional system spanned by the first three principal components. The three objects for one particular sinusoidal amplitude were denoted an object group (see Fig. 7B). An algorithm was designed to estimate the average separation of the stimuli (objects) in p-space. The algorithm computes the distances between all the different object groups in the hyperplane, and it also takes into account the spreading of the objects within the object groups (see Fig. 7C). Thus an algorithm was used where maximal separation would be obtained for maximal distances between object groups combined with minimal spreading within the object groups. Thus, a relative measure of the degree of separation for a certain population of afferents could be obtained (for a more elaborate description, see Bergenheim et al. 1995, 1996; Johansson et al. 1995a,b).

Results

An example of results from an experiment is illustrated in Fig. 8 which shows the average calculated stimulus separation for all combinations of 1, 2, 3, 4 etc. afferents in an experiment with 16 simultaneously recorded afferents (11 primary MSAs, 1 secondary MSA and 4 Golgi tendon organ afferents) responding to 5 stimuli consisting of sinusoidal stretches with different amplitudes. The filled circles illustrate the average stimulus sep-

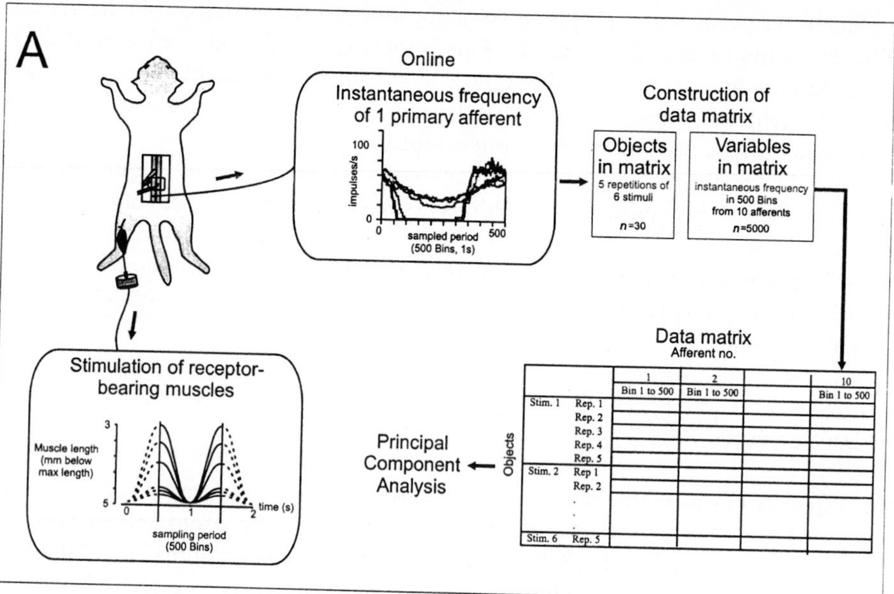

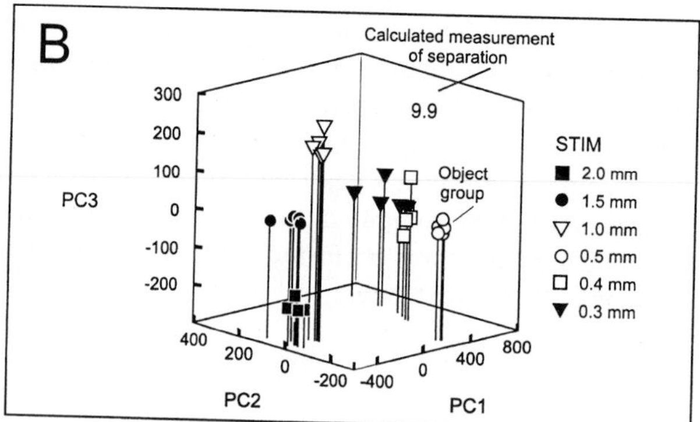

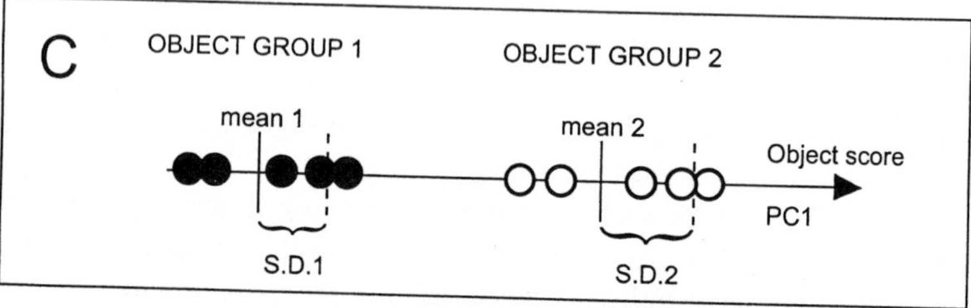

Fig. 7. Experimental setup, recording and analysis. **A**: Preparation, stimulation, recording and organization of the data matrix. The activity of several afferents is recorded simultaneously with one of the multichannel hooks decribed previously (see Figs. 1 and 2). Each set of afferents is tested with a number of sinusoidal stretches (i.e., 1 Hz, peak-to-peak amplitude: 0.3, 0.4, 0.5, 1.0, 1.5, and 2.0 mm). For each test, one second of the stimuli ("sampling period" in figure) was sampled, and the instantaneous frequency during each 2 ms was calculated and stored as variables in a data matrix. Each test stimulation is represented as a row (object) in this data matrix. Each afferent is represented by 500 columns, and the variables representing different afferents follow each other successively in each row. Hereby, each object will contain the instantaneous frequency responses from all the simultaneously recorded afferents. **B**: Example of an object score plot for three principal components. **C**: Factors used for quantification of stimulus separation. Mean 1 represents the mean object score for the 5 objects within object group 1, for the first principal component. S.D.1 is the standard deviation for these scores. Mean 2 and S.D.2 are the same values for object group 2.

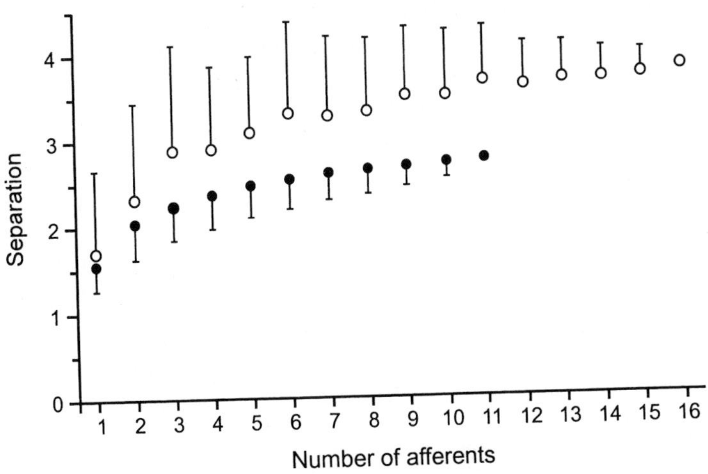

Fig. 8. Example of results. The average calculated stimulus separation for a population of 11 primary muscle spindle afferents (filled circles) and a mixed population of 16 afferents (11 primary muscle spindle afferents, 4 Golgi tendon organ afferents, and 1 secondary muscle spindle afferent open circles) between 5 different stimuli, for all combinations of 1, 2, 3 etc. afferents. Horizontal bars indicate the standard deviation. (Redrawn from Bergenheim et al. 1996 with permission)

aration for all ensemble combinations constructed from the 11 primary MSAs. The open circles show the average stimulus separation for ensembles constructed from the total population of simultaneously recorded afferents (i.e., 16 afferents), consisting of all three types of afferents. As can be seen in the figure, mixed ensembles of primary MSAs, secondary MSAs and Golgi tendon organ afferents as well as the ensembles of only primary MSA increased the separation of stimuli with increased ensemble size; maximal separation was obtained with a limited number of afferents, and the variability in discriminative ability decreased with increasing ensemble size. Furthermore, the results clearly demonstrate that mixed ensembles of primary MSAs, secondary MSAs and Golgi tendon organ afferents discriminated better between different muscle stretches than populations of only one type of afferent. These results corroborate the theoretical prediction that at least two factors are likely to be of importance for the encoding ability of an ensemble: (1) the individual receptors in an ensemble should not respond identically to the same stimulus, and (2) the sensitivity tuning of the individual receptors in the ensemble should be overlapping. The variation in response profiles or sensitivity tuning among the individual afferents of an ensemble will certainly be greater for mixed populations of primary MSAs, secondary MSAs and Golgi tendon organ afferents than for populations of only primary MSAs.

Comments

It is probably safe to conclude that the general opinion in modern neuroscience is that: (1) information in the nervous system seems to be coded in populations of neurons, and (2) for studies of this population coding, as many neurons as possible within the population should be recorded simultaneously.

This is the reason why there has been such a rapid development of multi-unit recording techniques in recent years, and today, using a combination of electrodes and spike separation methods, we have the possibility to record large numbers of neurons simultaneously on both the central and the peripheral level of the nervous system.

Acknowledgement: This study was supported by grants from The Swedish Council for Work Life Research and Inga Britt and Arne Lundbergs Forskningsstiftelse.

References

Bergenheim M, Johansson H and Pedersen J (1995) The role of the (-system for improving information transmission in populations of Ia afferents. Neurosci Res 23:207–215.

Bergenheim M, Johansson H, Pedersen J, Öhberg F and Sjölander P (1996) Ensemble coding of muscle stretches in afferent populations containing different types of muscle afferents. Brain Res 734:157–166.

Breckenridge LJ, Wilson RJA, Connolly P, Curtis ASG, Dow JAT, Blackshaw SE and Wilkinson CDW (1995) Advantages of using microfabricated extracellular electrodes for in vitro neuronal recording. J Neurosci Res 42:266–276.

Carpenter GA and Grossberg S (1987) ART 2: self-organization of stable category recognition codes for analog input patterns. Applied Optics 26:4919–4930.

Crago PE, Houk JC and Rymer WZ (1982) Sampling of total muscle force by tendon organs. J Neurophysiol 47:1069–83.

Deadwyler SA and Hampson RE (1997) The significance of neural ensemble codes during behavior and cognition. Annu Rev Neurosci 20:217–244.

Djupsjöbacka M, Johansson H, Bergenheim M and Sandström U (1994) A multichannel hook electrode for simultaneous recording of up to 12 nerve filaments. J Neurosci Methods 52:69–72.

Eckhorn R (1991) Stimulus-evoked synchronizations in the visual cortex: linking of local features into glabal figures? In Neural Cooperativity, Springer Series in Synergetics. Krüger J pp 184–224. Springer, Heidelberg.

Eckhorn R (1992) Connections between local and global principles of visual processing may be uncovered with parallel recordings of single-cell- and group-signals. In Proc Reisensburg Symposium.Aertsen A and and Braitenberg V pp 385–420. Springer, Heidelberg.

Eckhorn R and Obermueller A (1993) Single neurons are differently involved in stimulus-specific oscillations in cat visual cortex. Exp Brain Res 95:177–182.

Eckhorn R and Thomas U (1993) A new method for the insertion of multiple microprobes into neural an muscular tissue, including fiber electrodes, fine wires, needles and microsensors. J Neuosci Methods 49:175–179.

Eckhorn R, Frien A, Bauer R, Woelbern T and Kehr H (1993) High frequency (69–90 Hz) oscillations in primary visual cortex of awake monkeys. Neuroreport 4:243–246.

Geddes LA, Baker LE and Moore AG (1969) Optimum electrolytic chloriding of silver electrodes. Med Biol Eng 7:49–56.

Hanna GR and Johnson RN (1968) A rapid and simple method for the fabrication of arrays of recording electrodes. Electroencephalogt Clin Neurophysiol 25:284–286

Haidarliu S, Shulz D and Ahissar E (1995) A multi-electrode array for combined microiontophoresis and multiple single-unit recordings. J Neurosci Methods 56:25–31.

Hulliger M, Sjölander P, Windhorst UR and Otten E (1995) Force coding by populations of cat golgi tendon organ afferents: The role of muscle length and motor unit pool activation strategies. In Alpha and Gamma Motor Systems. Taylor A, Gladden MH and Durbaba R pp 302–308. Plenum Press, New York and London.

Humphrey DR (1970) A chronically implantable multiple microelectrode system with independent control of electrode positions. Electroencephalogr Clin neurophysiol 29:616–620

Jansen RF (1990) The reconstruction of individual spike trains from extracellular multineuron recordings using a neural network emulation program. J Neurosci Methods 35:203–213.

Johansson H, Bergenheim M, Djupsjöbacka M and Sjölander P (1995a) Analysis of encoding of stimulus separation in ensembles of muscle afferents. In Alpha and Gamma Motor Systems. Taylor A, Gladden M and Durbaba R pp 287–293. Plenum Press, New York, London.

Johansson H, Bergenheim M, Djupsjöbacka M and Sjölander P (1995b) A method for analysis of encoding of stimulus separation in ensembles of afferents. J Neurosci Methods 63:67–74.

Kreiter AK, Aertsen MHJ and Gerstein GL (1989) A low-cost single-board solution for real-time, unsupervised waveform classification of multineuron recordings. J Neurosci Methods 30:59–69.

Krüger J and Bach M (1980) A 30 fold multi-microelectrode for simultaneous single unit recording. Pfluegers Arch 384:R33

Krüger J and Bach M (1981) Simultaneous recording with 30 microelectrodes in monkey visual cortex. Exp Brain Res 41:191–194

Krüger J (1983) Simultaneous individual recordings from many cerebral neurones: techniques and results. Rev Physiol Biochem Pharmacol 98:177–233.

Loeb GE, Bak MJ and Duysens J (1977) Long-term unit recording from somatosensory neurons in the spinal ganglia of the freely walking cat. Science 197:1192–1194

Marks WB (1965) Some methods for simultaneous multiunit recordings. In: Nye PW (ed) Proc symp information processing in sight sensory systems. California Institute of Technology, Pasadena, pp 200–206

Mirfakhraei K and Horch K (1994) Classification of action potentials in multi-unit intrafascicular recordings using neural network pattern recognition techniques. IEEE Trans Biomed Eng 41:89–91.

Mountcastle VB, Poggio GF, Reitboeck HJ, Georgopoulos AP, Johnson KO, Longerbeam MB, Steinmetz MA, Phillips JR, Brandt DK and Habbel CG (1987) A system for multiple microelectrode recording from the neocortex of waking monkeys. Proc Soc Neurosci 629.

Nicolelis MA, Ghazanfar AA, Faggin BM, Votaw S and Oliveira LMO (1997) Reconstructing the engram: simultaneous, multisite, many single neuron recordings. Neuron 18:529–537.

Öhberg F, Johansson H, Bergenheim M, Pedersen J and Djupsjöbacka M (1996) A neural network approach to real-time spike discrimination during simultaneous recording from several multiunit nerve filaments. J Neurosci Methods 64:181–187.

Oghalai JS, Street WN and Rhode WS (1994) A neural network-based spike discriminator. J Neurosci Methods 54:9–22.

Prickard RS (1979a) A review of printed circuit microelectrodes and their production. J Neurosci Meth 1:301–318

Prickard RS (1979b) Printed circuit microelectrodes Trends Neurosci 2:259–269

Prochazka A (1983) Chronique techniques for studying neurophysiology of movement in cats. In: Lemon R (ed) Methods for neuronal recording in conscious animals. IBRO handbook Ser.: Methods neurosci. Wiley, New York, pp 113–128

Reitboeck HJP (1983) A 19-channel matrix drive with individually controllable fiber microelectrodes for neurophysiological applications. IEEE Transactions on Systems, Man and Cybernetics 13:676–682.

Salganicoff M, Sarna M, Sax L and Gerstein GL (1988) Unsupervised waveform classification for multi-neuron recordings: a real-time, software-based system. I. Algorithms and implementation. J Neurosci Methods 25:181–187.

Schmidt EM (1984a) Computer separation of multi-unit neuroelectric data: a review. J Neurosci Methods 12:95–111.

Schmidt EM (1984b) Instruments for sorting neuroelectric data: a review. J Neurosci Methods 12:1–24.

Stoppini L, Duport S and Correges P (1997) A new extracellular multirecording system for electrophysiological studies: application to hippocampal organotypic cultures. J Neurosci Methods 72:23–33.

Taylor A, Ellaway PH, Durbaba R and Rawlinson S (1998) Multiple single unit spindle recording during active and passive locomotor movements in the decerebrate cat. Abstract for: Peripheral and spinal mechanisms in the neural control of movement. Tucson. Arizona, Nov 4–6

Terzuolo CA and Araki T (1961) An analysis of intra- versus extracellular potential changes associated with activity of single spinal motoneurons. Ann NY Acad Sci. 94:547–558

Verzeano M (1956) Activity of cerebral neurons in the transition from wakefulness to sleep. Science 124:366–367

Suppliers

Company: U. Thomas Recording, Marburger Strasse 30, Marburg, 35043, Germany (phone: +49-6421-46442; fax: +49-6421-46404; home page: http://home.t-online.de/home/ThomasRecording/TRecMain.htm)

Company: Fredrick-Haer Inc. (FHC), 9 Main Street, Bowdoinham, ME, 04008, USA (phone: +01-207-666-8190; fax: +01-207-666-8292; e-mail: fhcinc@fh-co.com; home page: www.fc-co.com)

Company: Cambridge Electronic Design Limited (CED), Science Park, Milton Road, Cambridge, CB4 4FE, England (phone: +44-1223-420186; fax: +44-1223-420488; e-mail: info@ced.co.uk ; home page: www.ced.co.uk)

Company: NB laboratories (NBLABS), Dennison, Texas, USA

Company: Many Neuron Acquisition Processor (MNAP), Spectrum Scientific, Dallas, USA

Abbreviations

CNS	Central Nervous System
GS	Gastrocnemius and Soleus
MSA	Muscle spindle afferent
PBSt	Posterior Biceps and Semitendinosus
PC	Personal Computer
PCA	Principal Component Analysis

Chapter 18

Time and Frequency Domain Analysis of Spike Train and Time Series Data

David M. Halliday and Jay R. Rosenberg

Introduction

The concept of a spike triggered average will be familiar to many neurophysiologists. The first application in neurophysiology by Mendell and Henneman (1968, 1971) was used to examine the magnitude of monosynaptic excitatory postsynaptic potentials (EPSP) from muscle spindle Ia afferents onto homonymous motoneurons, which provided a major piece of evidence in the development of the size principle for motoneuron recruitment (see Henneman and Mendell, 1981). The technique has gained widespread acceptance, and been widely used to investigate the strength of synaptic connections in the mammalian central nervous system (e.g. Watt et al., 1976; Stauffer et al., 1976; Kirkwood and Sears, 1980; Cope et al., 1987), leading to new insights and an increased understanding of basic neurophysiological mechanisms.

The basic principle in the above studies is that averaging of an intracellular recording from a motoneuron triggered by action potentials from a single intact afferent will reveal a waveform which is taken as an estimate of the postsynaptic potential (PSP) for that input. Averaging is required due to the presence of unrelated activity within the cell, which can be regarded as a noise component. Spike triggered averaging can detect weak effects (Cope et al., 1987), averages involving 10^5 or more triggers are commonly used.

A similar procedure is often used to assess the coupling between two simultaneously recorded sequences of action potentials. In this case it is the timing of spikes in one spike train which is averaged with respect to the timing of spikes in a second spike train. This leads to a histogram based measure, frequently referred to as the cross-correlation histogram, which shows the relative timing of spikes in one spike train with respect to a second spike train. First used by Griffith and Horn (1963) to study functional coupling between cells in cat visual cortex, this method has subsequently been widely accepted and used in other areas of neurophysiology (e.g. Sears and Stagg, 1976; Kirkwood and Sears, 1978; Datta and Stephens, 1990).

Spike triggered averaging and cross-correlation histogram analysis can be considered as the detection of a *correlation* between two signals. Both are affected by the presence of noise, particularly when studying weak interactions. In most cases the signals which are being studied contain noise, and can therefore be considered as stochastic processes. The study of stochastic signals, and the detection of correlated activity in the presence of noise are major research areas in engineering and statistics, an extensive literature exists on the related questions of characterizing stochastic signals, and the estimation of correlated activity in the presence of noise (e.g. Brillinger and Tukey, 1984). In this chapter we are concerned with the question: Given two stochastic signals (which

Correspondence to: David M. Halliday, University of Glasgow, Division of Neuroscience and Biomedical Systems, Institute of Biomedical and Life Sciences, West Medical Building, Glasgow, G12 8QQ, UK (phone: +44-141-330-4759; fax: +44-141-330-4100; e-mail: gpaa34@udcf.gla.ac.uk)

can be either a sequence of spike times, or a sampled waveform) then how can these signals, and any correlation between them, be characterized? The *mean* and *variance* of a regularly sampled waveform are useful measures which characterize the distribution of amplitude values. *Auto-spectra* provide a more informative picture of the data, auto-spectral estimates can be interpreted as statistical parameters related to the variance of the signal at discrete frequencies. Similar comments apply to spike train data, which can be treated as a sequence of *point events*, and the distribution of intervals between successive spikes subjected to a similar analysis. The question of correlation between two signals can be considered as an investigation of the joint distribution of two stochastic processes, which leads to covariance analysis and cross-spectral analysis. These concepts will be expanded upon in the following sections, the object of these introductory remarks is to place the problem of assessing the correlation between neurophysiological signals in the domain of engineering and statistics, which allows the extensive methods which have been developed in these fields to be applied to the problem.

An important aspect of studying correlation in the presence of noise is the ability to place some error bounds on parameter estimates. This is often missing in the application of spike triggered averaging and cross-correlation techniques. The use of statistics should form an essential part of any analysis, both for dealing with error and uncertainty, and for testing hypotheses about the correlation structure between signals.

The object of this chapter is to present a framework within which the correlation between spike train data and/or sampled waveform data can be studied, where both time domain and frequency domain measures are used in a complementary fashion to maximise the insight into and inferences from experimental data. In this framework, the spike triggered average and cross-correlation histogram are closely related to cumulant density functions, a time domain parameter estimated from an inverse Fourier transform of a cross-spectrum. A key feature of the framework is the unified aspect for dealing with both spike train, sampled waveform and mixed spike train/waveform data. This is achieved using a Fourier based set of estimation procedures. In this context, a key result is that the large sample statistical properties of the finite Fourier transform of a stochastic signal are simpler than those of the process itself (Brillinger, 1974; 1983) and are the same for both types of data (Brillinger, 1972).

We consider first the analysis of spike train data. Part 1 defines time domain parameters which can be used to assess the correlation between spike trains, describes estimation procedures, based on the cross-correlation histogram, and Example 1 illustrates a sample analysis of two motor unit spike trains. The majority of the chapter is then concerned with a Fourier based framework for dealing with both spike train and waveform data. Part 2 discusses the Fourier transform of both data types, and defines and gives procedures for estimating second order spectra. Part 3 defines parameters which can be used to characterize the correlation between pairs of signals, application of these to different data sets is illustrated in Examples 7–10. Part 4 discusses how to extend the framework to deal with the interactions between several simultaneously recorded signals. Part 5 describes an approach for summarizing the correlation structure between several independent pairs of signals. Part 6 outlines an alternative approach to the analysis of spike train data, based on maximum likelihood methods. In the Concluding Remarks we discuss the limitations of the techniques presented.

In this chapter, the statistical presentation is kept to a minimum, with only the basic definitions and estimation procedures presented. The techniques, however, are presented in sufficient detail to allow the interested reader to undertake the analyses themselves. This chapter summarizes several years of interdisciplinary work, further details can be found in the references cited below, and in the following publications and references cited therein (Rosenberg et al., 1982, 1989, 1998; Halliday et al., 1992, 1995a; Amjad et al., 1997).

PART 1: Time Domain Analysis of Neuronal Spike Train Data

Stochastic Point Process Parameters for Time Domain Analysis of Neuronal Spike Train Data - Definitions

In this section we define a number of time domain parameters, estimates of which can be used to characterize interactions between spike trains. Estimation procedures and the setting of *confidence limits* are described in the next section.

In dealing with neuronal spike train data, the quantities normally available for analysis are the times of occurrence of each spike. The duration of action potentials is short compared with the spacing between, thus the sequences of spike times are normally considered as a series of point events. In addition, the interval between events is not deterministic but contains random fluctuations. We can therefore consider neuronal spike trains (or any other sequence of events which meet these requirements) as realizations of stochastic point processes. A *stochastic point process* may be defined formally as a random non-negative integer-valued measure (Brillinger 1978). In practice this represents the ordered times of occurrence of spikes (or events) in terms of a multiple of the sampling interval, dt, which should be chosen sufficiently small such that at most one event occurs in any interval. A point process which satisfies this condition is known as *orderly*.

We further assume that the point process data is *weakly stationary*, i.e. parameters which characterize the data do not change with time, and that widely spaced differential increments are effectively independent. This latter is known as a *mixing condition*. Discussion of these assumptions can be found in Cox and Isham (1980), Cox and Lewis (1972) and Daley and Vere-Jones (1988), and in relation to neuronal spike trains in Conway et al. (1993). The assumption of orderliness is important, since it allows certain point process parameters to be interpreted in terms of expected values or as probabilities (Cox and Lewis, 1972; Srinivasan, 1974; Brillinger, 1975), which provides a useful guide to the interpretation of these parameters.

For a point process, denoted by N_1, the *counting variate* $N_1(t)$ counts the number of events in the interval $(0,t]$. An important elementary function of point processes is the differential increment. The *differential increment* for process N_1 is denoted by $dN_1(t)$, and defined as $dN_1(t)= N_1(t, t + dt]$. It can be considered as a counting variate, which counts the number of events in a small interval of duration dt starting at time t. For an orderly point process, $dN_1(t)$ will take on the value 0 or 1 depending on the occurrence of a spike in the sampling interval dt.

Two simultaneously recorded spike trains can be considered as a realization of a bivariate point process. Let (N_0, N_1) be such a realization of a stationary bivariate point process with differential increments at time t given by $\{dN_0(t), dN_1(t)\} = \{N_0(t+dt], N_1(t+dt]\}$. Stationarity of (N_0, N_1) implies that the distribution of the differential increments in the intervals $(t, t+dt]$ and $(t+\tau, t+\tau+dt]$ is independent of τ. The following point process parameters may be defined in terms of the differential increments (Brillinger 1975, 1976; Rosenberg et al., 1982, 1989, Conway et al., 1993).

The *mean intensity*, P_1, of the point process N_1 is defined as

$$E\{dN_1(t)\} = P_1\, dt \tag{1}$$

where E{ } denotes the averaging operator or mathematical expectation of a random variable. The assumption of orderliness allows expression (1) to be interpreted in a probabilistic manner as

$$\text{Prob}\{N_1 \text{ event in } (t, t+dt]\} \tag{2}$$

The mean intensity, P_0, of the point process N_0 is defined in a similar manner. The second order cross-product density at lag u, $P_{10}(u)$, between the two point processes N_0 and N_1 is defined as

$$E\{dN_1(t+u)\,dN_0(t)\} = P_{10}(u)\,du\,dt \tag{3}$$

This expression can be interpreted as

$$\text{Prob}\{N_1 \text{ event in } (t+u, t+u+du] \;\;\&\;\; N_0 \text{ event in } (t, t+dt]\} \tag{4}$$

The second order product-density functions, $P_{00}(u)$ and $P_{11}(u)$, are defined as in (3) by equating N_0 and N_1. A *conditional mean intensity* can be defined as

$$m_{10}(u) = P_{10}(u)/P_0 \tag{5}$$

which can be interpreted as a conditional probability as

$$\text{Prob}\{N_1 \text{ event in } (t+u, t+u+du] \;\text{ given an } N_0 \text{ event at } t\} \tag{6}$$

From the mixing condition, the differential increments $dN_1(t+u)$ and $dN_0(t)$ become independent as u becomes large. Therefore we can write

$$\lim_{|u|\to\infty} P_{10}(u) = P_1 P_0 \tag{7}$$

This leads to the definition of the second order *cross-covariance function*, also called the second order *cumulant density function*, $q_{10}(u)$, as

$$q_{10}(u) = P_{10}(u) - P_1 P_0 \tag{8}$$

The two *auto-covariance functions*, $q_{00}(u)$ and $q_{11}(u)$, are defined similarly. This function will tend to zero as $|u| \to \infty$, and has the interpretation

$$\text{cov}\{dN_1(t+u), dN_0(t)\} = q_{10}(u)\,du\,dt \tag{9}$$

where cov{ } denotes covariance. In the case of the individual processes we must write

$$\begin{aligned} \text{cov}\{dN_0(t+u), dN_0(t)\} &= (\delta(u) + q_{00}(u))\,du\,dt \\ \text{cov}\{dN_1(t+u), dN_1(t)\} &= (\delta(u) + q_{11}(u))\,du\,dt \end{aligned} \tag{10}$$

where $\delta(\cdot)$ is a *Dirac delta function* which has to be included to take into account the behavior of the covariance density at u=0.

Estimation Procedures and Confidence Limits for Time Domain Point Process Parameters

The mean intensity, P_1, for a sample of duration R of the point process N_1 can be estimated as

$$\hat{P}_1 = \frac{N_1(\text{R})}{\text{R}} \tag{11}$$

For example, if N_1 is a sample spike train of 60 seconds duration with a sampling interval of 1 millisecond, containing 500 events, then R=60,000 and $\hat{P}_1$ = 500/60000. To distinguish between a parameter and its estimate, we use the notation $\hat{P}_1$ to denote an estimate of the parameter P_1.

Estimates of the second order product densities defined above can be constructed by the following procedure. If we denote the set of spike times for N_0 as $\{r_i\,;\, i=1,\ldots N_0(\text{R})\}$

and the set of spike times for N_1 as $\{s_j; j=1,\ldots N_1(\mathrm{R})\}$, we can construct a counting variate $J_{10}^{\mathrm{R}}(u)$ such that (Griffith and Horn, 1963; Cox 1965)

$$J_{10}^{\mathrm{R}}(u) = \#\left\{(r_i, s_j) \text{ such that } \left(u - \frac{b}{2}\right) < (s_j - r_i) < \left(u + \frac{b}{2}\right)\right\} \tag{12}$$

where $\#\{A\}$ indicates the number of events in set A. The variate $J_{10}^{\mathrm{R}}(u)$ counts the number of occurrences of N_1 events falling in a bin of width b, whose midpoint is u time units away from an N_0 event. In the neurophysiological literature, the variate $J_{10}^{\mathrm{R}}(u)$ is often called the *cross-correlation histogram*. The expected value of this variate is (Cox 1965; Cox and Lewis, 1972)

$$E\left\{J_{10}^{\mathrm{R}}(u)\right\} \approx b\,\mathrm{R}\,P_{10}(u) \tag{13}$$

This equation illustrates the relationship between the cross-correlation histogram and the second order product density, and leads to the following approximately unbiased estimates for $P_{10}(u)$ and $m_{10}(u)$

$$\hat{P}_{10}(u) = J_{10}^{\mathrm{R}}(u)/b\mathrm{R} \tag{14}$$

$$\hat{m}_{10}(u) = J_{10}^{\mathrm{R}}(u)/bN_0(\mathrm{R}) \tag{15}$$

where $\hat{m}_{10}(u)$ denotes an estimate of $m_{10}(u)$. The cumulant density, $q_{10}(u)$, can be estimated using equation (8) as

$$\hat{q}_{10}(u) = \hat{P}_{10}(u) - \hat{P}_1\hat{P}_0 = J_{10}^{\mathrm{R}}(u)/b\mathrm{R} - \hat{P}_1\hat{P}_0 \tag{16}$$

The large sample properties of the above estimates of $P_{10}(u)$ and $m_{10}(u)$ are considered in Brillinger (1976), where it is shown that the estimates (14) and (15) of the product density and cross-intensity are approximately *Poisson random variables*, which can be approximated by the two normal distributions $\mathrm{N}\{P_{10}(u), P_{10}(u)/b\,\mathrm{R}\}$ and $\mathrm{N}\{m_{10}(u), m_{10}(u)/b\,P_1\}$, respectively. $\mathrm{N}\{A, B\}$ refers to a normal distribution with mean A and variance B. In both cases, the variance of the estimate depends on the value of the parameter being estimated. In such cases it is usual to apply a variance stabilising transform (Kendall and Stuart, 1966; Jenkins and Watts, 1968). Brillinger (1976) proposes a square root transformation giving

$$\hat{P}_{10}(u)^{1/2} \approx \mathrm{N}\left\{P_{10}(u)^{1/2}, (4b\mathrm{R})^{-1}\right\} \tag{17}$$

$$\hat{m}_{10}(u)^{1/2} \approx \mathrm{N}\left\{m_{10}(u)^{1/2}, (4bN_0(\mathrm{R}))^{-1}\right\} \tag{18}$$

These parameter estimates have constant variance, which allows us to set *confidence limits* to test the hypothesis of independent (uncorrelated) spike trains. These confidence limits can be set in the following manner. Given an estimate $\hat{z}$ of a parameter z, which is approximately normallv distributed with variance $\mathrm{var}\{\hat{z}\}$, then 95% confidence limits can be set at $\pm 1.96\left(\mathrm{var}\{\hat{z}\}\right)^{1/2}$. The asymptotic values for $(\hat{P}_{10}(u))^{1/2}$ and $(\hat{m}_{10}(u))^{1/2}$ for large u can be estimated using equations (7) and (5), and indicate the expected values for two independent spike trains. Therefore we have the following asymptotic distribution and 95% confidence limits for $(\hat{P}_{10}(u))^{1/2}$

$$\left(\hat{P}_1\hat{P}_0\right)^{1/2} \pm 1.96(4b\mathrm{R})^{-1/2} \tag{19}$$

and for $\left(\hat{m}_{10}(u)\right)^{1/2}$

$$\left(\hat{P}_1\right)^{1/2} \pm 1.96\left(4bN_0(\mathrm{R})\right)^{-1/2} \tag{20}$$

Estimated values lying inside the upper and lower confidence limits can be interpreted as evidence of uncorrelated spike trains.

The asymptotic distribution of the estimated cumulant density, $\hat{q}_{10}(u)$, is discussed in Rigas (1983), and can be approximated by

$$\mathrm{var}\{\hat{q}_{10}(u)\} \approx \frac{2\pi}{\mathrm{R}} \int_{-\pi/b}^{\pi/b} f_{11}(\lambda)\, f_{00}(\lambda)\, d\lambda \tag{21}$$

where $f_{11}(\lambda)$ and $f_{00}(\lambda)$ are the auto-spectra of processes N_1 and N_0, respectively. The auto-spectra of spike train data will be discussed below, however, under the assumptions of Poisson spike trains the variance in (21) can be approximated by $(P_1\,P_0/\mathrm{R}b)$. For two independent spike trains, an asymptotic value and upper and lower confidence limit for the estimate of $\hat{q}_{10}(u)$ given in (16) can be set at

$$0 \pm 1.96\left(\hat{P}_1\,\hat{P}_0/\mathrm{R}b\right)^{1/2} \tag{22}$$

It is worth pointing out that once the cross-correlation histogram, $\mathrm{J}_{10}^{\mathrm{R}}(u)$, has been obtained, the product density, cross-intensity and cumulant density can be estimated very easily, and confidence limits which depend only on the quantities b, R, $N_0(\mathrm{R})$ and $N_1(\mathrm{R})$, and not on the characteristics of the spike trains (other than the mean rates), can easily be determined.

Results

Example 1: Time Domain Point Process Parameter Estimates

We illustrate the application of the above point process parameters with a sample analysis of a pair of motor unit spike trains, recorded from the middle finger portion of the extensor digitorum communis (EDC) muscle of a normal healthy subject during a maintained postural contraction. The two spike trains contain 1293 and 919 spikes, the record duration is 100 seconds, sampled at 1 ms intervals. The first order statistics are $N_0(\mathrm{R}) = 1293$, $N_1(\mathrm{R}) = 919$, $\mathrm{R} = 100{,}000$, $\hat{P}_0 = 0.01293$, $\hat{P}_1 = 0.00919$, and in addition $(\hat{P}_0\hat{P}_1)^{1/2} = 0.0109$ and $\hat{P}_1^{\,1/2} = 0.096$. Shown in Fig. 1 are estimates of (A) the cross-correlation histogram, $\mathrm{J}_{10}^{\mathrm{R}}(u)$, (B) the square root of the product density, $(\hat{P}_{10}(u))^{1/2}$, (C) the square root of the cross-intensity function, $\hat{m}_{10}(u)^{1/2}$, and (D) the cumulant density function, $\hat{q}_{10}(u)$. All parameters have a bin width b=1.0, giving estimates with the same resolution as the sampled spike trains (1 ms). The relationship between the cross-correlation histogram and the other three parameter estimates, made explicit by equations (13) to (16), is clearly demonstrated in this figure. The three parameter estimates all have the same basic shape determined from the cross-correlation histogram. The main difference is in the asymptotic value of the three estimates, which reflect the different probability descriptions given in equations (4), (6) and (9). The asymptotic values and upper and lower confidence limits are 0.0109±0.0031 for $\hat{P}_{10}(u)^{1/2}$, 0.096±0.027 for $\hat{m}_{10}(u)^{1/2}$, and $0 \pm 6.7\times10^{-5}$ for $\hat{q}_{10}(u)$. The main feature is the large peak at lag u=+5 ms, which exceeds the upper confidence limit in all three parameter estimates, indicating correlated motor unit activity. Also present in the cross-correlation histogram in Fig. 1A are smaller oscillatory features on either side of the central peak, these features are often referred to as sidebands and taken to reflect the presence of common rhythmic inputs to the two motoneurons whose motor unit activity is being studied (Moore et al., 1970). The confidence limits in the three parameter estimates indicate that these features are only of marginal significance at the 5% level. These features will be further discussed

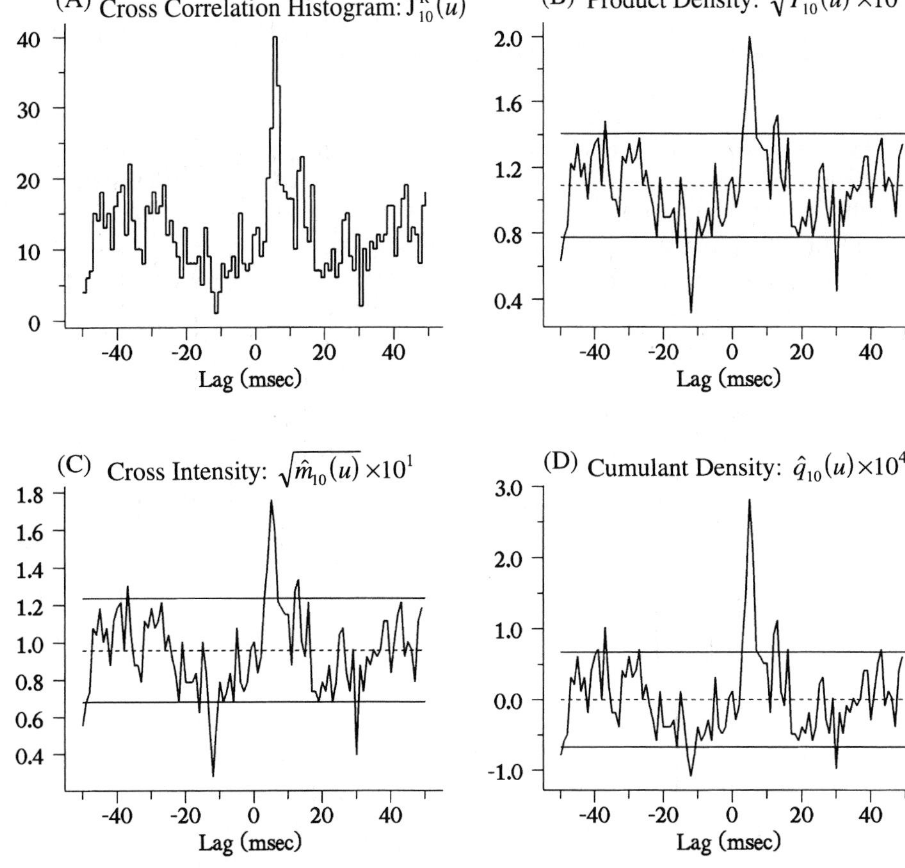

Fig. 1. Time domain analysis of the correlation between two motor unit spike trains. Estimates of (A) Cross-correlation histogram, $J^R_{10}(u)$, which has the units of counts, (B) square root of the product density, $(\hat{P}_{10}(u))^{1/2}\times10^2$, (C) square root of the cross-intensity function, $(\hat{m}_{10}(u))^{1/2}\times10^1$, and (D) cumulant density function, $\hat{q}_{10}(u)\times10^4$. Estimates have a bin width of b=1.0, and N_0(R)= 1293, N_1(R)=919, R=100,000. The dashed horizontal lines in B, C and D are the estimated asymptotic values, the solid horizontal lines are the estimated upper and lower 95% confidence limits under the assumption of independence.

below, with respect to frequency domain analysis, which is illustrated in Fig. 3. For uncorrelated spike trains, the three parameter estimates, $\hat{P}_{10}(u)^{1/2}$, $\hat{m}_{10}(u)^{1/2}$, and $\hat{q}_{10}(u)$, will fluctuate about their respective asymptotic values, the expected range of fluctuations can be estimated from the upper and lower confidence limits. Any significant departure outside these values, as in the peak at u=+5 ms in Fig. 1, can be taken to indicate a significant dependency between the spike trains at that particular lag value. Figure 1 illustrates the advantages of including confidence limits by focusing quickly on significant features in parameter estimates. The two parameter estimates $\hat{P}_{10}(u)^{1/2}$ and $\hat{m}_{10}(u)^{1/2}$ contain information about firing rates in their asymptotic distributions related to the probability descriptions in previous section. However, unless this is of specific interest, the presence of a statistically significant correlation can be inferred from any of the three estimates. In such situations, only one of the parameter estimates needs to be constructed, and the use of cumulant density estimates has the added advantage of being easily incorporated into the unified Fourier based framework presented below for dealing with spike train and/or time series data.

Part 2: Frequency Domain Analysis

The Finite Fourier Transform of Point Process and Time Series Data

All frequency domain analyses described below are based on parameter estimates formed from arithmetic combinations of *finite Fourier transforms*. The finite Fourier transform is therefore central to this analysis. This section defines and discusses how to estimate the finite Fourier transform of spike train and regularly sampled waveform data. Spike trains are assumed to be realizations of stochastic point processes, and waveform data are assumed to be realizations of time series. Point process data are assumed to meet the assumptions of orderliness discussed above, and time series data are assumed to be zero mean. Both types of data are further assumed to satisfy the two conditions of weak stationarity and a mixing condition. Weak stationarity implies that parameters which characterize a stretch of data do not change with time. The mixing condition implies that point process differential increments and/or time series sample values which are widely separated in time are independent. These assumptions are discussed in Halliday et al. (1995a). Both time series and point process data can be considered as belonging to the class of *stationary interval functions* considered in Brillinger (1972). The finite Fourier transform of a segment of point process N_1 containing T differential increments is defined as (Brillinger, 1972; Rosenberg et al., 1989)

$$d_{N_1}^{\mathrm{T}}(\lambda)=\int_0^{\mathrm{T}} e^{-i\lambda t}\, dN_1(t) \tag{23}$$

The integral in equation (23) can be thought of in a heuristic sense as comparing the periodicities of the sinusoids and cosinusoids of the complex Fourier exponential with the spacing between the spikes of process N_1, which allows information about periodic components in the spike train to be extracted. Modern spectral analysis methods invariably use a *fast Fourier transform* (FFT) algorithm to compute the finite Fourier transform of a sequence of data, at an equispaced set of Fourier frequencies. This requires equally spaced samples as input, which is achieved through the use of the differential increments to represent the point process N_1 in (23). Therefore, to estimate the quantity $d_{N_1}^{\mathrm{T}}(\lambda_k)$, using an FFT of length T we can approximate the integral in equation (23) by a discrete summation as (Brillinger, 1972; Rosenberg et al., 1982, 1989, Halliday et al., 1992)

$$d_{N_1}^{\mathrm{T}}(\lambda_j)\approx \sum_{t=0}^{\mathrm{T}-1} e^{-i\lambda_j t}\, dN_1(t)=\sum_{t=0}^{\mathrm{T}-1} e^{-i\lambda_j t}\left[N_1(t+\Delta t)-N_1(t)\right] \tag{24}$$

where τ_n are the times of occurrence of the N_1 events in the interval (0,T]. Since the point process is assumed orderly, the differential increments will have the value 0 or 1. The use of differential increments in equation (24) is therefore equivalent to representing the point process N_1 as a regularly sampled 0–1 time series. The Fourier frequencies, λ_j, are given by $\lambda_j = 2\pi j / T$, for j=0,…T/2. This defines a range for λ_j of $(0, \pi)$, where π corresponds to the *Nyquist frequency* (cf. Chapter 45). The Fourier frequencies can be expressed in Hz as $j/\mathrm{T}\Delta t$, where T is the number of points in the finite Fourier transform, and Δt the sampling interval (seconds). The quantity $1/\mathrm{T}\Delta t$ represents the fundamental Fourier frequency in cycles/s, i.e. the lowest frequency which can be resolved of 1 complete cycle of duration T, and represents the minimum spectral resolution of any parameters formed from $d_{N_1}^{\mathrm{T}}(\lambda_j)$.

In a similar fashion to (23), the Fourier transform of a segment of length T from a time series, x, is written as (Brillinger 1972, 1974)

$$d_x^{\mathrm{T}}(\lambda) = \int_0^{\mathrm{T}} e^{-i\lambda t}\, x(t)\, dt \tag{25}$$

Since $x(t)$ is a regularly sampled waveform, the sampled values can be readily evaluated by an FFT algorithm

$$d_x^{\mathrm{T}}(\lambda_j) \approx \sum_{t=0}^{\mathrm{T}-1} e^{-i\lambda_j t}\, x_t \tag{26}$$

where x_t are the sample values of $x(t)$ at time t. The Fourier frequencies, λ_j, are defined as for (24). Equation (26) performs a Fourier decomposition of the segment of $x(t)$ into constituent frequency components, which highlights distinct periodic components in the data (Brillinger, 1983).

Efficient FFT routines can be found in Bloomfield (1976, Chapter 4), Sorensen et al. (1987), and in the compendium of numerical methods by Press et al. (1989) which are available in a number of programming languages.

Definition and Estimation of Second Order Spectra

In this section we discuss the construction of estimates of second order spectra based on complex products of the finite Fourier transforms of point process and time series data discussed above. Following Halliday et al. (1995a) we use the term *hybrid* to characterize a parameter that depends on a time series and a point process. The auto-spectrum of time series x is denoted by $f_{xx}(\lambda)$, and of point process N_1 by $f_{11}(\lambda)$. The hybrid cross-spectrum between the two processes is denoted by $f_{x1}(\lambda)$. The asymptotic distribution of the finite Fourier transform for a broad variety of stationary processes, called "stationary interval functions" which include stationary time series and stationary point process data, is discussed in Brillinger (1972), where it is shown that for $\mathrm{T} \to \infty$, the asymptotic distribution of $d_{N_1}^{\mathrm{T}}(\lambda)$ and $d_x^{\mathrm{T}}(\lambda)$ is a complex normal. In the case of a times series x, this leads to consideration of the following statistic as an *estimate of the auto-spectrum*

$$\frac{1}{2\pi \mathrm{T}} \left| d_x^{\mathrm{T}}(\lambda) \right|^2 \tag{27}$$

This quantity, which is often referred to as the *periodogram*, $I_{xx}^{\mathrm{T}}(\lambda)$, was first proposed by Schuster (1898) to search for hidden periodicities in a series, x. The periodogram is not a consistent estimate of the spectrum $f_{xx}(\lambda)$, and requires further smoothing. One method to achieve this is averaging periodograms based on disjoint sections of data. Such an approach leads to

$$\hat{f}_{xx}(\lambda_j) = \frac{1}{2\pi \mathrm{LT}} \sum_{l=1}^{\mathrm{L}} \left| d_x^{\mathrm{T}}(\lambda_j, l) \right|^2 \tag{28}$$

as an estimate of the auto-spectrum of series x, at the Fourier frequencies λ_j defined above. Using this procedure involves splitting the complete record of R samples into L non-overlapping disjoint sections of length T, the quantity $d_x^{\mathrm{T}}(\lambda_j, l)$ refers to the finite Fourier transform of the lth segment (l=1,…,L). An estimate of the auto-spectrum of the point process N_1, $\hat{f}_{11}(\lambda)$, is obtained by replacing $d_x^{\mathrm{T}}(\lambda_j, l)$ with $d_{N_1}^{\mathrm{T}}(\lambda_j, l)$ in equation (28). Using this approach, the *hybrid cross-spectrum* between N_1 and x, $\hat{f}_{x1}(\lambda)$ can be estimated as

$$\hat{f}_{x1}(\lambda_j) = \frac{1}{2\pi \mathrm{LT}} \sum_{l=1}^{\mathrm{L}} d_x^{\mathrm{T}}(\lambda_j, l)\, \overline{d_{N_1}^{\mathrm{T}}(\lambda_j, l)} \tag{29}$$

Cross-spectra between two time series and between two point processes can be estimated in a similar manner by substitution of the appropriate finite Fourier transforms into equation (29). The approach of smoothing periodograms is a widely used method of spectral estimation (Bartlett, 1948; Brillinger 1972; 1981; Rosenberg et al., 1989; Halliday et al., 1995a).

For large T and $\lambda \neq 0$, the estimated cross-spectrum $\hat{f}_{x1}(\lambda_j)$ may be seen to have the same form as a complex covariance parameter, $\mathrm{cov}\{A,B\} = E\{(A-E\{A\})\overline{(B-E\{B\})}\}$, and can be interpreted as the covariance between the components of processes N_1 and x at each Fourier frequency λ_j. The estimated auto-spectra, $\hat{f}_{11}(\lambda_j)$ and $\hat{f}_{xx}(\lambda_j)$, have the same form as a variance parameter which provides a measure of the variance (or power) at each Fourier frequency λ_j of the process x (Tukey, 1961). Other methods of estimating spectra of spike train data are discussed in Halliday et al. (1992).

Second order spectra can also be defined in terms of the Fourier transform of the appropriate auto- or cross-covariance (cumulant density) functions. For example, the spectrum of x can be defined as (Jenkins and Watts, 1968; Brillinger, 1981)

$$f_{xx}(\lambda) = \frac{1}{2\pi} \int_{-\infty}^{\infty} q_{xx}(u)\, e^{-i\lambda u}\, du \tag{30}$$

where $q_{xx}(u)$ is the auto-covariance or cumulant density of process x. In the point process case the expression becomes (Bartlett, 1963)

$$f_{11}(\lambda) = \frac{P_1}{2\pi} + \frac{1}{2\pi} \int_{-\infty}^{\infty} q_{11}(u)\, e^{-i\lambda u}\, du \tag{31}$$

where the additional term arises from the inclusion of the Dirac delta function in the definition of the auto-covariance (10). For u large $q_{11}(u)$ will tend to zero, which gives the asymptotic distribution of $f_{11}(\lambda)$ as $\lim_{\lambda \to \infty} f_{11}(\lambda) = P_1 / 2\pi$. This approximation is used above in the derivation of (22) from (21) with respect to the estimation of the variance of point process cumulant density functions. For the hybrid cross-spectrum, $f_{x1}(\lambda)$, the appropriate expression is

$$f_{x1}(\lambda) = \frac{1}{2\pi} \int_{-\infty}^{\infty} q_{x1}(u)\, e^{-i\lambda u}\, du \tag{32}$$

Other cross-spectra can be defined in a similar fashion. Equations (30) to (32) show the close relationship between cumulant density (covariance) functions and spectra. These relationships also show that it is possible to estimate spectra indirectly through the Fourier transform of covariance functions. Such an approach formed the basis of the first major practical digital time series analysis (Blackman and Tukey, 1958), before the advent of the FFT. Modern methods of spectral estimation generally use direct procedures based on FFT algorithms to estimate the finite Fourier transforms, as described above.

For estimates of the auto-spectrum, $\hat{f}_{xx}(\lambda)$, obtained via (28), it can be shown that the variance can be approximated by $\mathrm{var}\{\hat{f}_{xx}(\lambda)\} \approx \mathrm{L}^{-1}(f_{xx}(\lambda))^2$ (Brillinger 1972, 1981; Bloomfield, 1976), where L is the number of disjoint sections used to estimate the spectrum. This expression contains the value of the actual spectrum at a particular frequency and will therefore change with changing frequency. The appropriate variance stabilising transform is the natural log, giving $\mathrm{var}\{\ln(\hat{f}_{xx}(\lambda))\} \approx \mathrm{L}^{-1}$. It is customary practice to plot spectra on a $\log_{10}$ scale, thus

$$\mathrm{var}\left\{\log_{10}\left(\hat{f}_{xx}(\lambda)\right)\right\} = \left(\log_{10}(\mathrm{e})\right)^2 \mathrm{L}^{-1} \tag{33}$$

with a resulting estimate and 95% confidence limits at frequency λ of

$$\log_{10}\left(\hat{f}_{xx}(\lambda)\right) \pm 0.851\, \mathrm{L}^{-1/2} \tag{34}$$

An alternative method of indicating the confidence limit, which is independent of the value of the spectrum, is to plot a scale bar of magnitude $\left(1.7\,\mathrm{L}^{-1/2}\right)$ as a guide to interpret any distinct features in the estimated spectrum. In the point process case, where the spectra have an asymptotic value, the three lines

$$\log_{10}\left(\frac{\hat{\mathrm{P}}_1}{2\pi}\right),\quad \log_{10}\left(\frac{\hat{\mathrm{P}}_1}{2\pi}\right)\pm 0.851\,\mathrm{L}^{-1/2} \tag{35}$$

can be used as a guide to interpret features in the estimate of $\hat{f}_{11}(\lambda)$.

Results: Auto-Spectra

In Fig. 2 are shown examples of auto-spectral estimates of both point process and time series signals. All these estimates have been constructed from data sampled at 1 ms ($\Delta t = 10^{-3}$ s), and a segment length of 1024 points (T=1024) in the finite Fourier transforms (24) and (26), giving a spectral resolution of 0.977 Hz.

Example 2: Motor Unit Spectrum

Figures 2A and 2B are from the same data set as analysed in Fig. 1. This data consists of two individual motor unit discharges recorded from the middle finger portion of the (EDC) muscle in a human subject, and the tremor recorded simultaneously from the distal phalanx of the finger using an accelerometer, while the subject maintained the unrestrained middle finger extended in a horizontal position (see Conway et al., 1995b and Halliday et al., 1995a, for further details of the experimental protocol). The duration of the data set is 100 seconds ($R=10^5$), giving L=97 for this data set. The spike train whose spectral estimate is illustrated in Fig2A, N_0, has 1293 spikes ($N_0(R)=1293$), a mean rate of 12.9 spikes/s, and the coefficient of variation (c.o.v.) is 0.17. For the tremor signal the mean RMS value is 6.91 cm/s^2. The dominant feature of the log plot of the motor unit spectrum is the large peak around 13 Hz, corresponding to the mean rate of firing, illustrating that the mean firing rate is the dominant rhythmic component in this motor unit discharge. There is a less well-defined peak centered around 26 Hz, this frequency matches the expected first harmonic of the spectral peak corresponding to the mean discharge rate. At higher frequencies the estimate lies almost entirely inside the expected upper and lower 95% confidence intervals shown as the solid horizontal lines. These intervals are constructed under the assumption of a random (Poisson) spike train, and at higher frequencies this motor unit discharge behaves as a random spike train. This behavior is a consequence of the mixing condition discussed above, where differential increments widely separated in time tend to become independent (a characteristic of Poisson spike trains). At lower frequencies, however, the spectrum of this spike train exhibits significant departure from that expected for a Poisson spike train. For spike trains whose dominant spectral component reflects the mean firing rate, the appearance of harmonic components in their spectral estimates is related to the c.o.v. A more regular discharge generally has a spectral estimate which contains more harmonic components.

Example 3: Tremor Spectrum

The log plot of the estimated tremor spectrum, $\hat{f}_{xx}(\lambda)$, in Fig. 2B has a peak around 21 Hz, this is the dominant component of physiological tremor recorded from the unrestrained finger, and is due in part to the natural resonance of the extended finger. For a

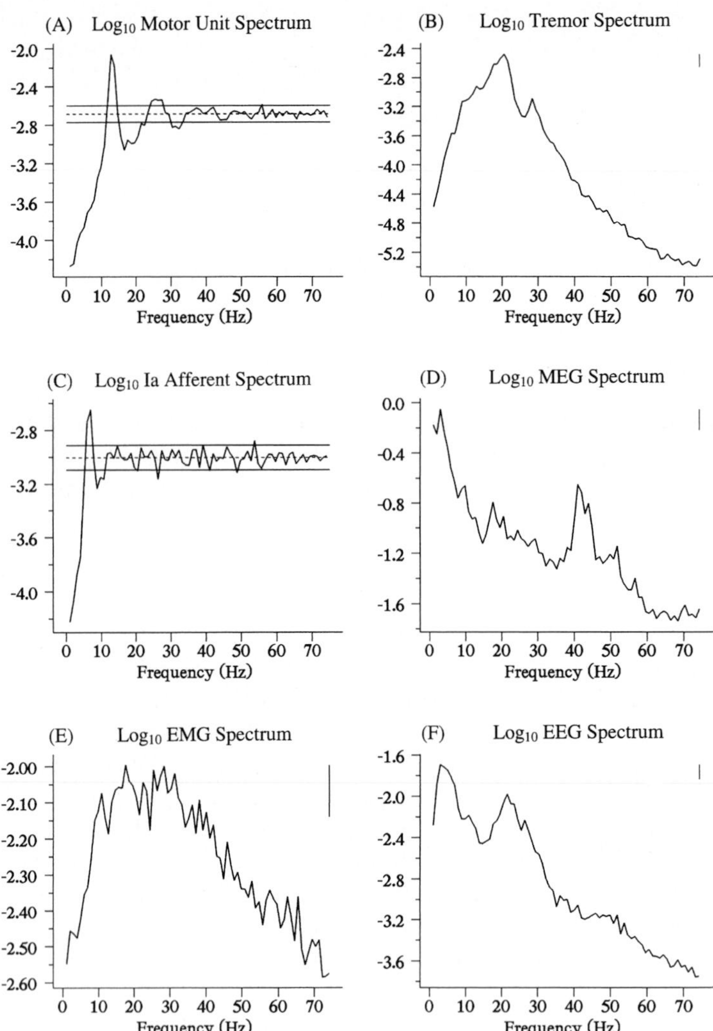

Fig. 2. Examples of spectral estimates. Log plots of estimated power spectra for (A) motor unit spike train during position holding of the extended middle finger, and (B) simultaneous tremor acceleration signal, (C) Ia afferent spike train, (D) Magnetoencephalogram recorded over the sensorimotor cortex, (E) rectified surface EMG from wrist extensors during maintained wrist extension, and (F) simultaneous EEG recorded over the sensorimotor cortex. Dashed horizontal lines in (A, C) represent the asymptotic value of each estimate, solid horizontal lines give the estimated upper and lower 95% confidence limits, based on the assumption of a Poisson spike train. Solid vertical lines at the top right in (B, D, E, F) give the estimated magnitude of a 95% confidence interval for each spectral estimate. For all data the sampling interval Δt=1 ms, and for all estimates T=1024 points.

more detailed discussion of the spectrum of physiological tremor see Stiles and Randall, (1967) and Halliday et al. (1995a). There is another smaller peak around 28 Hz. The confidence interval for the spectral estimate can be used as a guide to help establish if this smaller peak is significant. The confidence interval is shown as the solid vertical line in the top right of Fig. 2B, of magnitude 0.173 dB. This is the same as the spacing between the upper and lower confidence limits (solid lines) in Fig. 2A, estimated from the same number of segments (see equations (34) and (35) above). Comparison of the local fluctuation around 28 Hz with the scale bar suggests that this localised peak in the spectrum does reflect a distinct rhythmic component as opposed to chance fluctuations in the spectral estimate, which we would expect to be of a magnitude smaller than this scale bar. The correlation between the single motor unit discharge in Fig. 2A and this tremor signal is examined below in Fig. 4.

Example 4: Ia Afferent Spectrum

In Fig. 2C is shown another point-process spectral estimate, this is for a single Ia afferent discharge recorded using the microneurography technique (Vallbo and Hagbarth, 1968) from a human subject during low force voluntary isometric contraction of the fourth finger portion of the EDC muscle (for further details see Halliday et al., 1995b). The data is 89 seconds in duration (R=89,000; L=86), and contains 553 spikes. The mean rate is 6.2 spikes/s, and the c.o.v. is 0.21. The log plot of this spectral estimate contains a clear peak around 6 Hz, reflecting the mean firing rate of the Ia.

Example 5: MEG Spectrum

The spectral estimate in Fig. 2D is for human cortical activity, in this case recorded as the *magnetoencephalogram* (MEG; cf. Chapter 37) over the sensorimotor cortex during a maintained contraction of a contralateral intrinsic hand muscle (for details, see Conway et al., 1995a). This spectral estimate, which is for a record of 110 seconds (R=110,000; L=107), has most power present at lower frequencies, with decreasing power at higher frequencies. Comparison of local fluctuations in this estimate with the 95% confidence interval (top right) indicates distinct rhythmic components centered about 18 and 42 Hz, the latter is particularly distinct. The functional significance of these rhythms is discussed in Conway et al. (1995a).

Example 6: EMG and EEG Spectra

The two spectral estimates illustrated in Fig. 2E and F are for a simultaneously recorded surface *electromyogram* (EMG; cf. Chapter 26) from the wrist extensors and a bipolar *electroencephalogram* (EEG; cf. Chapter 35) from over the contralateral sensorimotor cortex in a human subject during maintained wrist extension (for details see Halliday et al., 1998). These spectra are estimated from a total of 138 seconds of data. The log plot of the EMG spectrum is for the rectified surface EMG signal, and exhibits a broad peak from around 10 to 40 Hz. The reason for analysing rectified EMG is discussed in Example 9 (in connection with Fig. 5). The EEG spectral estimate has a concentration of power at low frequencies (the decrease at the lowest frequencies reflects the high-pass filtering of 3 Hz associated with the instrumentation), and a clearly defined peak centered around 22 Hz. The 95% confidence intervals have the same magnitude for both estimates. The correlation between these two signals is examined below in Fig. 5. It is interesting to compare this EEG spectral estimate to the MEG spectral estimate (Fig. 2D), recorded from a similar location in a different subject (see discussion in Halliday et al., 1998).

The above six spectral estimates illustrate analysis of a broad range of signal types, which are typical of those encountered in neurophysiological experiments. The next part discusses how the correlation between pairs of such signals can be investigated. The starting point for these analyses are estimates of the auto-spectra of individual signals, and cross-spectra between pairs of signals.

Part 3: Correlation Between Signals

Within the present framework the dependence between two signals can be characterized by parameters which assess the correlation between the signals. In the frequency domain it is customary to consider the magnitude squared of the correlation between the Fourier transforms of the two signals under consideration. For the bivariate point processes (N_0, N_1) this leads to (Brillinger 1975; Rosenberg et al., 1989)

$$\lim_{T\to\infty} \left|\mathrm{corr}\left\{d_{N_1}^T(\lambda), d_{N_0}^T(\lambda)\right\}\right|^2 \tag{36}$$

as a measure of the correlation between processes N_0 and N_1. This quantity is called the *coherence function* (Wiener, 1930), denoted by $|R_{10}(\lambda)|^2$, estimates of which provide a measure of the strength of correlation between N_0 and N_1 as a function of frequency. The definition of the correlation $\mathrm{corr}\{d_{N_1}^T(\lambda), d_{N_0}^T(\lambda)\}$ between the Fourier transforms of the two point processes N_0 and N_1 in terms of variance and covariance, given by the expression: $\mathrm{corr}\left\{d_{N_1}^T(\lambda), d_{N_0}^T(\lambda)\right\} = \mathrm{cov}\left\{d_{N_1}^T(\lambda), d_{N_0}^T(\lambda)\right\} \Big/ \sqrt{var\left\{d_{N_1}^T(\lambda)\right\} var\left\{d_{N_0}^T(\lambda)\right\}}$, leads to the alternative definition for the coherence function between point processes N_0 and N_1 as

$$|R_{10}(\lambda)|^2 = \frac{|f_{10}(\lambda)|^2}{f_{11}(\lambda) f_{00}(\lambda)} \tag{37}$$

Coherence functions provide a normative measure of linear association between two processes on a scale from 0 to 1, with 0 occurring in the case of independent processes (Brillinger, 1975; Rosenberg et al., 1989). Expression (37) leads to an estimation procedure by substitution of the appropriate spectral estimates to give

$$|\hat{R}_{10}(\lambda)|^2 = \frac{|\hat{f}_{10}(\lambda)|^2}{\hat{f}_{11}(\lambda) \hat{f}_{00}(\lambda)} \tag{38}$$

where $|\hat{R}_{10}(\lambda)|^2$ denotes an estimate of $|R_{10}(\lambda)|^2$. A similar procedure can be used to estimate the coherence between a time series x and point process N_1

$$|\hat{R}_{x1}(\lambda)|^2 = \frac{|\hat{f}_{x1}(\lambda)|^2}{\hat{f}_{xx}(\lambda) \hat{f}_{11}(\lambda)} \tag{39}$$

and the coherence between two time series x and y, $|R_{xy}(\lambda)|^2$, can be estimated in a similar manner. Estimates of the necessary second order spectra can be constructed using the method of disjoint sections outlined above in Part 2. Coherence estimates obtained in this way all have the same large sample properties for any combination of point process and/or time series data (Halliday, 1995a). A confidence interval at the 100α% point which is based on the assumption of independence, i.e. $|R_{x1}(\lambda)|^2=0$, is given by the value $1-(1-\alpha)^{1/(L-1)}$, where L is the number of disjoint sections used to estimate the second order spectra (Bloomfield, 1976; Brillinger, 1981). Therefore an upper 95% confidence limit can be set at the constant level

$$1-(0.05)^{1/(L-1)} \tag{40}$$

and estimated values of coherence below this level can be taken as evidence for a lack of correlation between the two processes at a particular frequency. The setting of confidence limits about the estimated values of coherence when significant correlation is present is discussed in Halliday et al. (1995a).

Coherence estimates assess the magnitude of correlation between two signals in the frequency domain. Information relating to timing can be obtained by examining the phase difference between the two signals. For point process N_1 time and series x, the *phase spectrum*, $\Phi_{x1}(\lambda)$, is defined as the argument of the cross-spectrum

$$\Phi_{x1}(\lambda) = \arg\{f_{x1}(\lambda)\} \tag{41}$$

This function can be estimated by direct substitution of the estimated cross-spectrum, equation (29) as

$$\hat{\Phi}_{x1}(\lambda) = \arg\{\hat{f}_{x1}(\lambda)\} \tag{42}$$

Phase spectra between other combinations of point-process and/or time series data can be defined and estimated in similar fashion. Phase estimates are only valid when there is significant correlation between the two signals. In practice $|\hat{R}_{x1}(\lambda)|^2$ can be used to indicate the regions where $\hat{\Phi}_{x1}(\lambda)$ has a valid interpretation. Phase estimates can be interpreted as the phase difference between harmonics of N_1 and x at frequency λ. The arctan function can be used to obtain the argument of the cross-spectrum, resulting in a phase estimate over the range $[-\pi/2, \pi/2]$ radians. However, the signs of the real and imaginary parts of $\hat{f}_{x1}(\lambda)$ can be used to determine in which quadrant the arctangent falls, so extending the range to $[-\pi, \pi]$ radians.

Phase estimates can often be interpreted according to different theoretical models. A useful model is the phase curve for two signals which are correlated with a fixed time delay, where the theoretical phase curve is a straight line, passing through the origin (0 radians at 0 frequency) with slope equal to the delay, and a positive slope for a phase lead, and negative slope for a phase lag (see Jenkins and Watts, 1968). In situations where there is significant correlation over a wide range of frequencies and a delay between two signals, it is reasonable to extend the phase estimate outside the range $[-\pi, \pi]$ radians, which avoids discontinuities in phase estimates. Such a phase estimate is often referred to as an *unconstrained phase estimate*. The representation of phase estimates is discussed in Brillinger (1981). Different theoretical phase curves for other forms of correlation structure are discussed in Jenkins and Watts (1968). Details for the construction of confidence limits about estimated phase values can be found in Halliday et al. (1995a). In situations where the correlation structure between two signals is dominated by a delay it is possible to estimate this delay from the phase curve, such an approach, based on weighted least squares regression, is described in the Appendix in Rosenberg et al. (1989). This method has the advantage of providing an estimate of the standard error for the estimated delay.

As discussed above, correlation analysis in neurophysiology has traditionally been performed in the time domain. We next discuss how correlation between point-process and/or time series data can be characterized as a function of time. Within the present Fourier based analytical, the correlation between two signals as a function of time can be estimated using *cumulant density functions*. These can be defined in terms of the *inverse Fourier transform of the cross-spectrum* (Jenkins and Watts, 1968; Brillinger, 1974). The second order hybrid cumulant density function between processes N_1 and x, $q_{x1}(u)$ is defined as

$$q_{x1}(u) = \int_{-\pi}^{\pi} f_{x1}(\lambda)\, e^{i\lambda u}\, d\lambda \tag{43}$$

This expression and equation (32) illustrate the equivalence between time and frequency domain analysis, cumulant densities and spectra form a Fourier transform pair, cf.

(30) and (32). The above hybrid cumulant can be estimated by the following expression

$$\hat{q}_{x1}(u) = \frac{2\pi}{T} \sum_{|j| \le T/2b} \hat{f}_{x1}(\lambda_j) e^{i\lambda_j u} \tag{44}$$

where $\lambda_j = 2\pi j/T$ are the Fourier frequencies, and b is the desired time domain bin width ($b \geq 1.0$), a value of $b=1.0$ results in a time domain estimate with the same temporal resolution as the sampling rate of the two signals. Equation (44) can be implemented using a real valued inverse FFT algorithm (e.g. Sorensen et al., 1987). The point process cumulant density function $q_{10}(u)$ and the time series cumulant density function $q_{xy}(u)$ can be defined and estimated in terms of the appropriate cross-spectra, $f_{10}(\lambda)$ and $f_{xy}(\lambda)$, respectively, using equations (43) and (44).

Cumulant density functions can be interpreted as statistical parameters which provide a measure of linear dependence between two signals (Brillinger, 1972; Rosenblatt, 1983; Mendel, 1991). If the two signals are independent, the value of the cumulant is zero. Cumulant densities can assume either positive or negative values. Unlike coherence estimates, they are not bounded measure of association, therefore there is no upper limit indicating a perfect linear relationship. In the case of hybrid data, $\hat{q}_{x1}(u)$ has an interpretation similar to a spike triggered average (Rigas, 1983; Halliday et al., 1995a). The relationship between the point-process cumulant density and the cross-correlation histogram was discussed in Part 1. It is possible to define and estimate cumulant density functions directly in the time domain, however, estimation via the frequency domain using equation (44) provides a unified framework for dealing with different data types, and is necessary for the construction of confidence limits for cumulant density estimates.

The variance of the hybrid cumulant density estimate (44) based on the assumption of independent processes can be approximated by (Rigas, 1983)

$$\operatorname{var}\{\hat{q}_{x1}(u)\} \approx \frac{2\pi}{R} \int_{-\pi/b}^{\pi/b} f_{xx}(\lambda) f_{11}(\lambda)\, d\lambda \tag{45}$$

where R is the record length, and b is the bin width of the estimate used in equation (44). This expression can be estimated using a discrete summation and substituting estimates of the spectra $f_{xx}(\lambda)$ and $f_{11}(\lambda)$ giving

$$\operatorname{var}\{\hat{q}_{x1}(u)\} \approx \left(\frac{2\pi}{R}\right)\left(\frac{2\pi}{T}\right) \sum_{j=1}^{(T/2-1)/b} 2\, \hat{f}_{xx}(\lambda_j) \hat{f}_{11}(\lambda_j) \tag{46}$$

where $\lambda_j = 2\pi j/T$, R is the record length, b the bin width, and T is the segment length used in the estimation of the finite Fourier transforms (24) and (26). Under the assumption of independent processes, the asymptotic value and upper and lower 95% confidence limits for the estimated cumulant (44) are given by

$$0 \pm 1.96 \left[\left(\frac{2\pi}{R}\right)\left(\frac{2\pi}{T}\right) \sum_{j=1}^{(T/2-1)/b} 2\, \hat{f}_{xx}(\lambda_j) \hat{f}_{11}(\lambda_j) \right]^{1/2} \tag{47}$$

Values of the estimated cumulant $\hat{q}_{x1}(u)$ lying inside the upper and lower 95% confidence limits can be taken as evidence of no linear correlation between point process N_1 and time series x at a particular value of lag u. Equations (45) to (47) are valid for cumulant density estimates constructed from other combinations of time series and/or point process data. For the point process cumulant density $\hat{q}_{10}(u)$, the assumption of Poisson spike trains allows the asymptotic values of the two point process spectra

$\lim_{\lambda \to \infty} f_{00}(\lambda) = P_0 / 2\pi$, and $\lim_{\lambda \to \infty} f_{11}(\lambda) = P_1 / 2\pi$ to be used to approximate the values of the spectra in (47), resulting in the simplified expression (22). This simplification allows confidence intervals for point process cumulants to be estimated without having to estimate the spectra. For the stochastic spike train discharges encountered in neurophysiological experiments this approximation results in estimated confidence intervals which plotted on a graph are almost indistinguishable from those obtained by integration of the spectra.

Equation (43) defines cumulant densities in terms of the inverse Fourier transform of the cross-spectrum, which leads to the estimation procedure in equation (44). It is also possible to define and estimate cumulant densities directly in the time domain. In the case of point process data this was done in Part 1. The interpretation of cumulant densities does not, in general, depend on the method of estimation.

For point process data, the derivation of cumulant densities directly in the time domain in terms of the correlation between differential increments is discussed in Brillinger (1975) and Rosenberg et al. (1989). For two point processes N_0 and N_1 the correlation between differential increments of the two point processes leads to the expression

$$\operatorname{corr}\{dN_1(t+u), dN_0(t)\} = q_{10}(u) \frac{\sqrt{du\,dt}}{\sqrt{P_1 P_0}} \tag{48}$$

The term relating to the sampling interval in the differential increments, $\sqrt{du\,dt}$, illustrates the unbounded nature of cumulant density estimates. Other point process time domain measures of association, based on product density functions, are given in Brillinger (1975).

In the case of hybrid data, the spike triggered average between point process N_1 and time series x, which we denote by $\mu_{x1}(u)$, can be estimated as (Rosenberg et al., 1982; Rigas, 1983)

$$\hat{\mu}_{x1}(u) = \frac{1}{\mathrm{R}} \sum_{i=1}^{N_1(\mathrm{R})} x(\tau_i + u) \tag{49}$$

where R is the record length and τ_i are the times of the events in process N_1, (i= 1,...,N_1(R)). This estimate of the spike triggered average can be used to estimate the hybrid cumulant $\hat{q}_{x1}(u)$ as (Rosenberg et al., 1982; Rigas, 1983)

$$\hat{q}_{x1}(u) = \hat{\mu}_{x1}(u) - \hat{P}_1 \hat{\mu}_x \tag{50}$$

where $\hat{P}_1$ is the estimated mean rate of process N_1, see (11), and $\hat{\mu}_x$ is the mean of process x. This equation illustrates the close relationship between the spike triggered average and the hybrid cumulant density; indeed, for a zero mean time series (which is a usual assumption for time series analysis) the two parameters are the same.

For zero mean time series data, $x(t)$ and $y(t)$, the cumulant density can be estimated for small lag u, as the *cross-covariance function* (Parzen, 1961; Bloomfield, 1976; Brillinger, 1981)

$$\hat{q}_{xy}(u) = \frac{1}{\mathrm{R}} \sum_{t=1}^{\mathrm{R}-|u|} x_{t+u}\, y_t \tag{51}$$

The use of a Fourier based estimation framework has advantages over the more traditional approach of estimating time domain parameters directly in the time domain. Firstly it provides a unified estimation procedure, both for constructing parameter estimates and for constructing confidence intervals. Equations (44) and (46) are valid for any pairwise combination of point process and/or time series data, thus the same software routines can be used for all data types. The same is not true for direct time estimation in the time domain, as illustrated by the differences in equations (12) for point process data, (49) for hybrid data and (51) for time series data. As mentioned above the method of estimation does not affect the interpretation of parameter estimates. Cumulant densities can be interpreted as statistical parameters which provide a measure of dependence between point process and/or time series data. In the case of point process data the cumulant density is closely related to the more traditionally used cross-correlation histogram, and for a zero mean time series, the hybrid cumulant density is the same as the more traditionally used spike triggered average. The present report presents these parameters in the context of a unified Fourier based framework which can assess the correlation between signals in both the time domain and frequency domain – an approach which can offer extra insight into complex neural systems. The time and frequency domain parameters should be viewed as complementary to each other. Equations (32) and (43) illustrate the mathematical equivalence of cumulant density functions and second order spectra (via the Fourier transform). However, mathematical equivalence does not necessarily result in equivalence of representation, a point often made in the writings of J.W. Tukey (Tukey, 1980; see also Fig15, Rosenberg et al., 1989), and illustrated below in Fig. 3. Therefore both time and frequency domain parameters should routinely be used for data analysis.

The above Fourier based methods all involve an assumption of linearity, extension of these methods to higher-order analyses is described in Halliday et al. (1995a).

Results : Time and Frequency Domain Correlation Analyses

In this section we illustrate application of the parameters defined above to characterize the correlation between pairs of signals. This is done for four data sets which illustrate the analysis of different combinations of data: point process data, hybrid data (point process and time series) and time series data.

Example 7: Motor Unit – Motor Unit Coherence and Phase

The first example considers the same motor unit pair whose time domain analysis is illustrated in Fig. 1. The estimated cumulant density for this data set is illustrated in Fig. 1D, this is the direct time domain estimate, based on equation (16). The Fourier based estimate (equation (44), b=1.0, T=1024, L=97) is graphically indistinguishable from this. The upper and lower 95% confidence intervals in Fig. 1D are based on the simplified expression (22) and have the values $\pm 6.76\times10^{-5}$, those estimated by integration of the cumulant (47) have the values $\pm 6.73\times10^{-5}$. For the type of spike trains illustrated in this report, the simplified estimate for point process cumulant density confidence intervals provides almost identical values to those obtained by integration of the spectra. The simplified expression (22) will be useful for spike trains whose spectra only deviate from the asymptotic value for Poisson spike trains over a limited range of frequencies compared with the Nyquist frequency (500 Hz for this data). The estimated coherence, $|\hat{R}_{10}(\lambda)|^2$, and phase, $\hat{\Phi}_{10}(\lambda)$ (T=1024), for this data are shown in Fig. 3. The coherence between the two motor units, N_0 and N_1, has two distinct bands, from 1–6 Hz and 20–30 Hz. These frequency bands do not coincide with the peak in the spectrum of motor unit 0 (fig 2A) or motor unit 1 (not shown), thus we can conclude that the motor unit firing rate does not contribute to the coherence between this pair of motor units. The coherence estimate reflects periodic components present in the common inputs which are responsible for the synchronized motor unit firing (see Farmer et al., 1993 for

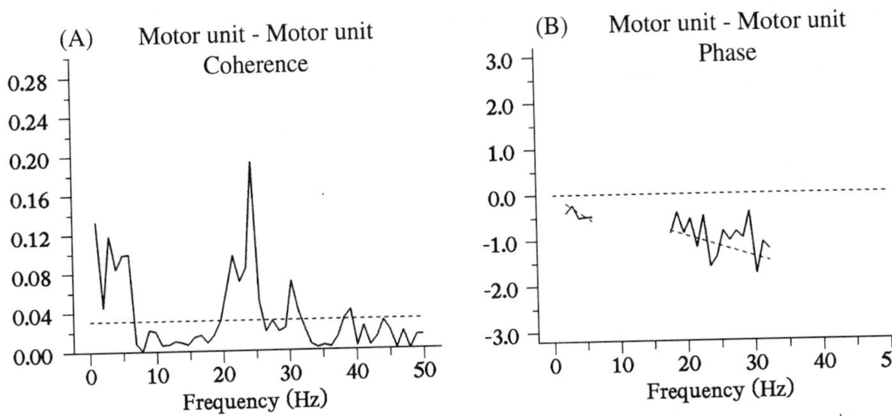

Fig. 3. Frequency domain analysis of motor unit correlation. Estimate of (A) coherence, $|\hat{R}_{10}(\lambda)|^2$, and (B) phase, $\hat{\Phi}_{10}(\lambda)$, between same motor unit data illustrated in Fig. 1. The horizontal dashed line in (A) gives the estimated upper 95% confidence limit based on the assumption of independence. The phase estimate in (B) is plotted where the coherence (A) is significant. The dotted lines through each section of the phase estimate are the theoretical phase curves for delays of 15.5 ms in the 1–6 Hz section, and 7.3 ms in the 18–32 Hz section (see text for details).

further discussion). The peak in the cumulant density estimate (Fig. 1D) indicates synchronized motor unit activity, the small sidebands which occur at around 40 ms on either side of the central peak further suggest a periodic component around 25 Hz is present, as indicated by the coherence estimate. However, comparison of Figs. 1D and 3A illustrates that different parameters can emphasize different features of the data – the cumulant estimate has sidebands which are only marginally significant, whereas the coherence estimate has a clear peak around 20–30 Hz – and argues in favor of both time and frequency domain analysis. The phase estimate is shown in Fig. 3B (solid lines), plotted in radians over the two regions where there is significant correlation between the motor units, estimated by inspection of the coherence estimate as 1–6 Hz and 18–32 Hz. In the time domain, the two motor units have a correlation structure which is dominated by a delay, as indicated by the peak at +5 ms in the cumulant (Fig. 1D). The phase estimate between the two motor units also reflects this delay, and the regression method described in the Appendix in Rosenberg et al. (1989) was used to estimate the slope of the phase curve in each section. These give delays of 15.5±3.6 ms for the delay in the 1–6 Hz section, and 7.3±1.2 ms in the 18–32 Hz section. The delay in the 18–32 Hz region corresponds with the latency estimated from the time to the peak in the cumulant, whereas the delay in the lower frequency range is longer. The cumulant estimate in Fig. 1D does have a smaller peak centered around +13 ms, which is consistent with the delay estimated from the phase over 1–6 Hz. To summarize, this example has examined the correlation between a pair of motor unit discharges, which exhibit a tendency for correlated firing dominated by a delay between the firing times and involving two distinct rhythmic components. The central peak in the cumulant density indicates synchronized firing, the maximum occurs at a latency of +5 ms, and small sidebands centered around 40 ms on either side suggest the presence of a rhythmicity around 25 Hz in the correlation. The frequency domain analysis reveals that the dominant component in the motor unit firing is the rhythmic component associated with the mean firing rate, that the motor units are coupled over two frequency bands, 1–6 and 18–32 Hz which do not correspond with the firing rate, and that different delays are associated with the coupling in each frequency band.

Example 8: Motor Unit – Tremor Coherence, Phase and Cumulant

The second example considers hybrid data, and examines the relationship between one of the motor unit discharges from the previous example and a simultaneously recorded tremor signal. The spectrum of the tremor signal is discussed above in Section 2.2, and is shown in Fig. 2B. Figure 4 illustrates the estimated coherence, $|\hat{R}_{x0}(\lambda)|^2$, phase, $\hat{\Phi}_{x0}(\lambda)$ (T=1024), and cumulant density, $\hat{q}_{x0}(u)$ (b=1.0), between the motor unit, N_0, and the tremor, x. The coherence estimate (Fig. 4A) has significant values over a broad range of frequencies, with evidence of two distinct bands which have maximum values around 6 Hz and 22 Hz. These correspond to the same frequency bands as the motor unit correlation, however, the magnitude of this coherence estimate is greater than that between the two motor units (see Conway et al., 1995b for further discussion). The phase estimate, Fig. 4B, in this example plotted in unrestrained form (see Part 3), has values which are constantly decreasing, however, the slope is not constant over the broad frequency range where the coherence is significant. Thus the pure delay model is not appropriate to explain this phase curve. The dominant feature in the hybrid cumulant den-

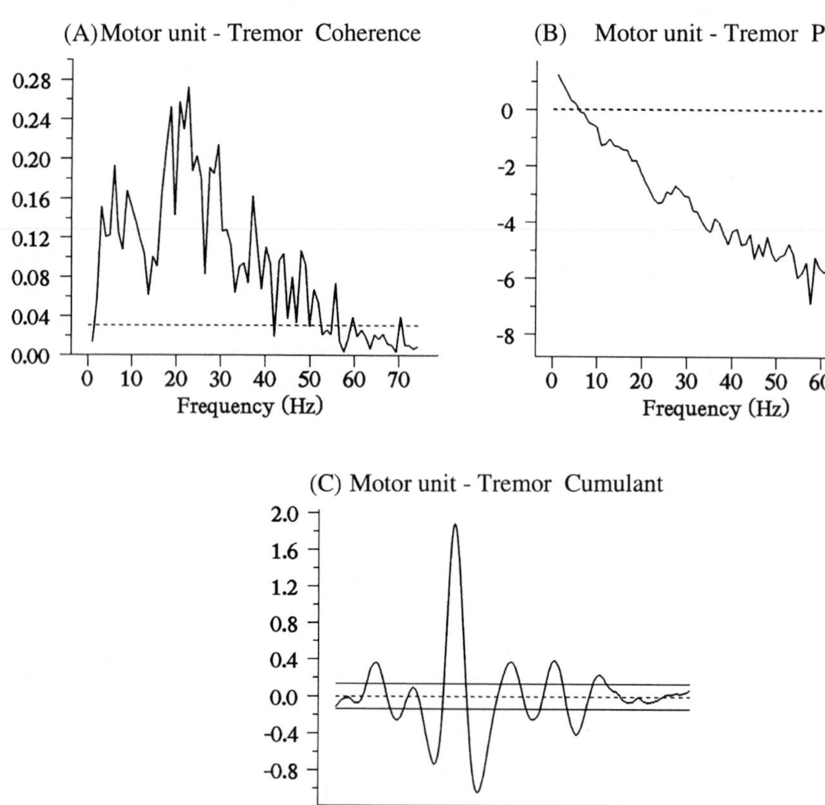

Fig. 4. Correlation between a single motor unit and a tremor acceleration signal. Estimate of (A) coherence, $|\hat{R}_{x0}(\lambda)|^2$, (B) phase, $\hat{\Phi}_{x0}(\lambda)$, and (C) cumulant density, $\hat{q}_{x0}(u)\times 10^4$, between one of the motor units illustrated in Figs. 1, 3 and a simultaneously recorded tremor acceleration signal. The horizontal dashed line in (A) gives the estimated upper 95% confidence limit based on the assumption of independence. The phase estimate in (B) is plotted in unrestrained form. The horizontal lines in (C) are the asymptotic value (dashed line at zero) and estimated upper and lower 95% confidence limits, based on the assumption of independence. The motor unit spectrum is illustrated in Fig. 2A, the tremor spectrum is shown in Fig. 2B.

sity estimate (Fig. 4C) is the large peak just after time zero, with a maximum around +17 ms. As discussed above, this cumulant can be interpreted as a spike triggered average, and provides an estimate of the acceleration response to motor unit impulses. It has a general form like a damped oscillation. A regression analysis on the part of the phase curve where the slope does pass through zero, from 11 Hz to 47 Hz, gives a delay of 17.9±0.7 ms, which matches well the peak value in the cumulant. An alternative interpretation of a similar data set, based on linear systems analysis, is described in Halliday et al. (1995a).

Example 9: EMG - EEG Correlation

The third example studies the correlation between two time series, namely the EEG and surface EMG signals whose auto-spectra are illustrated in Fig. 2E and 2F (see Part 1 and Halliday et al., 1998 for a description of the experimental protocol). Figure 5 illustrates the coherence, $|\hat{R}_{yx}(\lambda)|^2$, phase, $\hat{\Phi}_{yx}(\lambda)$ (T=1024) and cumulant density, $\hat{q}_{yx}(u)$ (b= 1.0), estimates between the EEG signal, x, and rectified surface EMG signal, y. The justification for using rectified surface EMG (without any smoothing) is that it reduces the components in the finite Fourier transform (26) which are due to the shape of individual muscle action potentials, while retaining information related to their spacing. This is further discussed in Halliday et al. (1995a).

In constructing the coherence and phase estimates shown in Fig. 5, the auto- and cross-spectral estimates have been further smoothed using a *Hanning filter*. This involves a weighted average of adjacent values in the frequency domain. If we denote the cross-spectral estimate between x and y with Hanning as $\hat{f}'_{yx}(\lambda_j)$, at frequency λ_j, this is obtained from the original estimate, based on (29), $\hat{f}_{yx}(\lambda_j)$ as

$$\hat{f}'_{yx}(\lambda_j) = \frac{1}{4}\hat{f}_{yx}(\lambda_{j-1}) + \frac{1}{2}\hat{f}_{yx}(\lambda_j) + \frac{1}{4}\hat{f}_{yx}(\lambda_{j+1}) \tag{52}$$

This procedure applies a moving average with weights $^1/_4$, $^1/_2$, $^1/_4$ to obtain the smoothed spectral estimates, which are then used to form coherence and phase estimates as described in Part 2. The use of smoothing invalidates the expressions for confidence limits given above. Confidence limits indicate the expected degree of variability in spectral estimates, any additional smoothing will reduce the variability, resulting in smaller confidence limits. Hanning can be considered as a specific case of a generalised weighting scheme which can be written as

$$\hat{f}'_{yx}(\lambda_j) = \sum_{k=-m}^{+m} w_k \hat{f}_{yx}\left(\lambda_j + \frac{2\pi k}{\mathrm{T}}\right),$$

where T is the segment length in the finite Fourier transform, and w_k; k= 0, ±1, ±2,..., $\pm m$ are the weights. It is customary for the weights to satisfy the condition: $\sum w_k = 1$. The correction to the variance of the spectral estimate is given by the factor (Brillinger, 1981)

$$\sum_{k=-m}^{+m} w_k^2 \tag{53}$$

The variance of the log transform of the auto-spectral estimate constructed from L disjoint sections with Hanning is then given by

$$\mathrm{var}\left\{\log_{10}\left(\hat{f}_{xx}(\lambda)\right)\right\} = \left(\log_{10}(\mathrm{e})\right)^2 \mathrm{L}^{-1} \sum w_k^2 .$$

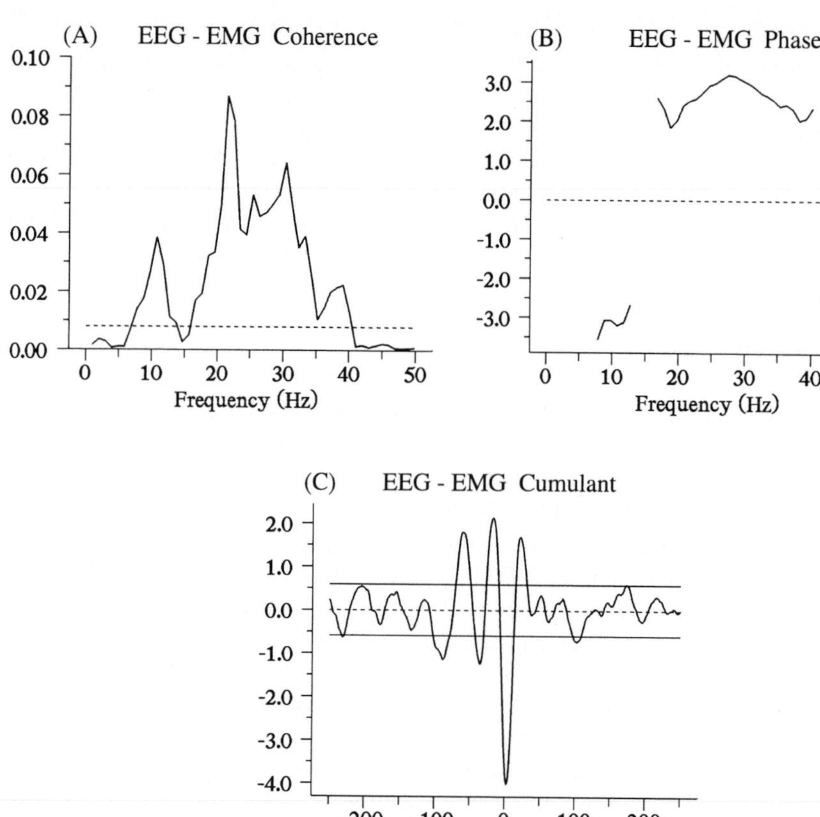

Fig. 5. Correlation between EEG and EMG during a maintained contraction. Estimate of (A) coherence, $|\hat{R}_{yx}(\lambda)|^2$, (B) phase, $\hat{\Phi}_{yx}(\lambda)$, and (C) cumulant density, $\hat{q}_{yx}(u)\times 10^4$, between a bipolar EEG recorded over the sensorimotor cortex and a surface EMG from the wrist portion of the extensor digitorum muscle. Surface EMG was full wave rectified (without time constant) before processing. Coherence and phase estimates are further smoothed with a Hanning window before plotting. The horizontal dashed line in (A) gives the estimated upper 95% confidence limit based on the assumption of independence. The phase estimate in (B) has two sections, plotted where the coherence estimate is significant, 8–12 Hz and 16–40 Hz. The horizontal lines in (C) are the asymptotic value (dashed line at zero) and estimated upper and lower 95% confidence limits, based on the assumption of independence. The EMG spectrum is illustrated in Fig. 2E, the EEG spectrum is shown in Fig. 2F.

For Hanning $\sum w_k^2 = 0.375$, which results in 95% confidence intervals for auto-spectral estimates of $\pm 0.521\,\mathrm{L}^{-1/2}$. For coherence estimates the correction for further smoothing results in the expression

$$1-(0.05)^{1/\left((\mathrm{L}-1)\sum w_k^2\right)} \tag{54}$$

for the upper 95% confidence limit based on the assumption of independent processes. The process of applying Hanning to spectral estimates based on (29) results in reduced variability, however, this is achieved at a cost of increased spectral bandwidth, fine structure in spectra is smoothed out by Hanning. Although the spectra are defined at the same Fourier frequencies, $\lambda_j = 2\pi j / T$, the effective spectral bandwidth of parameter estimates will be increased by the additional smoothing. In the present situation where weak correlation exists over a range of frequencies, application of Hanning results in smoother coherence and phase estimates, which better define the correlation structure between EEG and EMG. The cumulant density is estimated from the original cross-spectral estimate, $\hat{f}_{yx}(\lambda)$, without the use of Hanning.

The coherence estimate (Fig. 5A) is significant over the range 15 to 40 Hz, with a smaller peak at 10 Hz. The maximum value of the coherence is around 0.08, which indicates weak coupling. The phase curve has two distinct sections, plotted where the coherence is significant, from 8 to 12 Hz and from 16 to 40 Hz. The cumulant has an oscillatory structure, with three clear positive peaks separate by 40 ms. This corresponds to a frequency of 25 Hz, agreeing with the coherence estimate. The cumulant has a prominent dip around time zero (minimum at +2 ms), this can be interpreted as indicating synchronous activity between the two signals. The negative value of the cumulant around time zero indicates the signals are out of phase. The phase curve (Fig. 5B) fluctuates around $\pm\pi$ radians, providing further evidence in favor of this interpretation. However, the phase section from 20 to 28 Hz has a constant slope that passes through the origin when extrapolated. A weighted regression analysis (see Appendix in Rosenberg et al., 1989) on this section gives a phase lead of 18.5±0.35 ms. In the estimate $\hat{\Phi}_{yx}(\lambda)$, the EEG, x, is the reference signal, this leads to an alternative interpretation of EMG leading EEG by around 18.5 ms over the frequency range 20–28 Hz. This latency matches the peak in the cumulant at −19 ms. This latter interpretation only explains the timing relationship between EEG and EMG over part of the frequency range at which they are correlated. A more detailed discussion of the coupling between cortical activity and motor unit firing in humans can be found in Conway et al. (1995a) and Halliday et al. (1998). This example illustrates the problems associated with interpretation of a complex correlation structure between two signals.

Example 10: EMG - EMG Correlation

The fourth example considers the interaction between two surface EMGs recorded from different muscles. The study of activation and control of co-contracting muscle groups involved in a common motor task is often referred to as the study of muscle synergy (for a review see Hepp-Raymond et al. 1996). One mechanism thought to be responsible for such muscle synergy is the presence of shared drive to the different motoneuron pools (Gibbs et al. 1995). This process can be studied experimentally by examining the cross-correlation between EMG signals recorded from the different muscles. The presence of a peak around time zero in the cross-correlation histogram is taken to reflect the presence of a common excitatory drive to both motoneuron pools (Gibbs et al. 1995). This example considers the question of muscle synergy in terms of common frequency components present in surface EMG records recorded from Abductor Digiti Minimi (ADM) and the Extensor Digitorum muscle during a postural task involving maintained wrist extension with fingers spread apart (Conway et al. 1998). The analysis of the two surface EMG signals is illustrated in Fig. 6, the data consists of four 61 second records combined to give a total of 264 seconds of data (R=264,000; T=1024; L=256). Both EMG signals were full wave rectified, without any time constant, before analysis. The log plots of the two spectral estimates, $\hat{f}_{xx}(\lambda)$ and $\hat{f}_{yy}(\lambda)$, are shown in Fig. 6A and 6B. The ADM spectrum (Fig. 6A) has a broad peak from 12 to 25Hz, the wrist extensor EMG spectrum is dominated by a sharp peak at 10 Hz. The coherence estimate (Fig. 6C) exhibits significant correlation in the range 18–26 Hz, with maximum values around 23 Hz. The phase estimate (Fig. 6D) over this frequency range exhibits a phase lead, with constant slope, for which the weighted regression scheme (Rosenberg et al., 1989) gives a time lead of 8.5±0.98 ms. The estimated cumulant (Fig. 6E) has a peak centered about −10 ms, and additional peaks around 45 ms on either side of this central peak. Both coherence and cumulant density estimates indicate a rhythmic correlation structure between the two

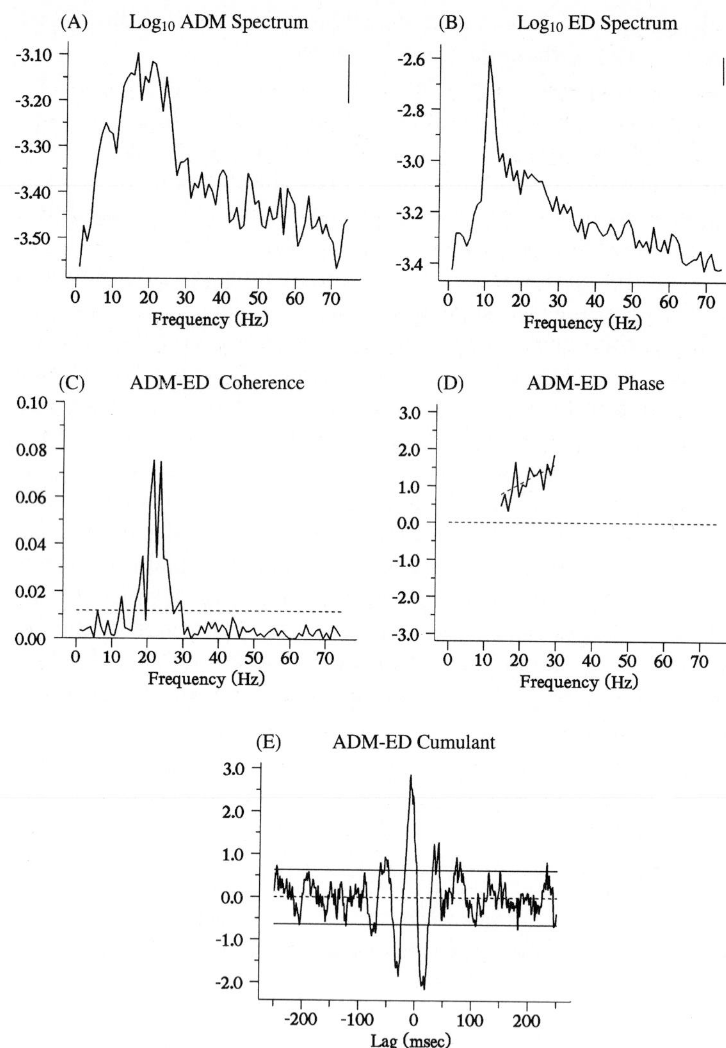

Fig. 6. Correlation between two different EMG signals during a maintained postural task. Log plot of estimated power spectra of surface EMG recorded from (A) abductor digiti minimi (ADM), and (B) wrist portion of the extensor digitorum (ED) muscle. Estimated (C) coherence, $|\hat{R}_{yx}(\lambda)|^2$, (D) phase, $\hat{\Phi}_{yx}(\lambda)$, and (E) cumulant density, $\hat{q}_{yx}(u)\times 10^4$, between the two EMG signals. Surface EMGs were full wave rectified (without time constant) before processing. Solid vertical lines at the top right in (A, B) give the estimated magnitude of a 95% confidence interval for these spectral estimates. The horizontal dashed line in (C) gives the estimated upper 95% confidence limit based on the assumption of independence. The dotted line through the phase estimate is the theoretical phase curve for a time lead of 8.5 ms. The horizontal lines in (E) are the asymptotic value (dashed line at zero) and estimated upper and lower 95% confidence limits, based on the assumption of independence.

EMG signals during the maintained contraction, suggesting that muscle synergy is in part generated by common rhythmic synaptic drive to different motor pools (Conway et al., 1998). Gibbs et al. (1995) studied correlation between EMG signals by applying a constant threshold to each EMG signal, and using a cross-correlation histogram analysis to characterize the correlation between the two sequences of spike trains generated by this thresholding of the EMG signals. This approach requires the choice of a suitable threshold. The analysis in Fig. 6 illustrates an alternative approach, which treats the rectified surface EMGs as time series.

Part 4: Multivariate Analysis

The above methods can be extended to examine the correlation structure between several simultaneously recorded signals. This is called a multivariate analysis, and is equivalent to*multivariate regression analysis*, except that parameters are estimated at each frequency of interest.

Two related questions which can be addressed by such analysis are 1) whether the correlation between two signals results from the common (linear) influence of a third signal, and 2) whether a third signal is capable of predicting the correlation between two signals. Both these questions can be addressed by estimating the *partial coherence, partial phase* and *partial cumulant density* which characterize the correlation between the two original signals after removing the common linear effects of the third (predictor) signal from each. This multivariate analysis can be performed on any combination of point process and/or time series data (Halliday et al., 1995a).

Partial Spectra

The starting point for the multivariate analysis are the estimates of second order spectra described in Part 2, with the requirement of three simultaneously recorded signals and spectral estimates which have been estimated with the same segment length, T. For example, to estimate the partial correlation between point process N_0 and time series x, with time series y as predictor, we start by defining partial spectra, estimates of which are then used to construct estimates of the other partial parameters. The *partial cross-spectrum* between N_0 and x, with y as predictor, is defined as (Brillinger, 1981)

$$f_{x0/y}(\lambda) = f_{x0}(\lambda) - \frac{f_{xy}(\lambda)\, f_{y0}(\lambda)}{f_{yy}(\lambda)} \tag{55}$$

The *partial auto-spectra*, $f_{xx/y}(\lambda)$, is defined as

$$f_{xx/y}(\lambda) = f_{xx}(\lambda) - \frac{f_{xy}(\lambda)\, f_{yx}(\lambda)}{f_{yy}(\lambda)} = f_{xx}(\lambda) - \frac{\left|f_{xy}(\lambda)\right|^2}{f_{yy}(\lambda)} \tag{56}$$

The other partial auto-spectrum, $f_{00/y}(\lambda)$, is defined in a similar manner. These partial spectra can be used to estimate the first order *partial coherence* between N_0 and x, with y as predictor, denoted by $\left|R_{x0/y}(\lambda)\right|^2$, as

$$\left|R_{x0/y}(\lambda)\right|^2 = \frac{\left|f_{x0/y}(\lambda)\right|^2}{f_{xx/y}(\lambda)\, f_{00/y}(\lambda)} \tag{57}$$

This equation has a similar form to that for the ordinary coherence function, (37). The corresponding first order *partial phase* is defined as

$$\Phi_{x0/y}(\lambda) = \arg\left\{f_{x0/y}(\lambda)\right\} \tag{58}$$

This function provides information about the timing relation of any residual coupling between N_0 and x after the removal of the common effects of process y. Partial coherence functions, like ordinary coherence functions, are bounded measures of association, with values between 0 and 1 (Brillinger, 1975, 1981; Rosenberg et al., 1989; Halliday et al., 1995a).

The above partial parameters can be estimated by substitution of estimates of the appropriate spectra, for example, the first order partial cross-spectrum in (55) can be estimated by:

$$\hat{f}_{x0/y}(\lambda)=\hat{f}_{x0}(\lambda)-\frac{\hat{f}_{xy}(\lambda)\,\hat{f}_{y0}(\lambda)}{\hat{f}_{yy}(\lambda)} \tag{59}$$

with the necessary second order spectra obtained from (29). The partial coherence and phase can then be estimated using direct substitution as in Part 2. The setting of confidence limits for estimates of the partial coherence (57), based on the assumption of independence, is similar to that for ordinary coherence functions, with a correction for the number of predictors used. For the case of 1 predictor, as in (57), the upper 95% confidence limit is estimated as the constant value $1-(0.05)^{1/(L-2)}$ (Halliday et al., 1995a). The setting of confidence limits about the estimated partial coherence involves the same procedures as ordinary coherence estimates, see Halliday et al. (1995a) for details.

The most convenient manner to estimate *partial cumulant density functions* is to use the inverse Fourier transform of the appropriate partial cross-spectrum. For the above three processes, this is denoted as $q_{x0/y}(u)$, which can be estimated as

$$\hat{q}_{x0/y}(u)=\frac{2\pi}{\mathrm{T}}\sum_{|j|\le \mathrm{T}/2b}\hat{f}_{x0/y}(\lambda_j)\,e^{i\lambda_j u} \tag{60}$$

where $\lambda_j=2\pi j/\mathrm{T}$ are the Fourier frequencies, and b the bin width ($b\geq1.0$). Expression (60) is similar to (44). This function provides a measure of any residual dependency between processes N_0 and x, as a function of time, after removal of any common linear influence of process y. Expressions (45) to (47) can be used to determine a confidence limit for this estimate, under the assumptions of independence (see discussion in Halliday et al., 1995a). Equation (22) can be used for partial cumulant density estimates which involve the correlation between two point processes.

An alternative definition of the first order partial coherence, $|R_{x0/y}(\lambda)|^2$, as the magnitude squared of the correlation between the finite Fourier transforms of N_0 and x, after removal of the effects of process y from each, may be written (suppressing the dependencies on λ) as (Brillinger, 1975, 1981; Rosenberg et al., 1989; Halliday et al., 1995a)

$$|R_{x0/y}|^2=\lim_{\mathrm{T}\to\infty}\left|\operatorname{corr}\left\{d_x^{\mathrm{T}}-\left(\frac{f_{xy}}{f_{yy}}\right)d_y^{\mathrm{T}},\, d_{N_0}^{\mathrm{T}}-\left(\frac{f_{0y}}{f_{yy}}\right)d_y^{\mathrm{T}}\right\}\right|^2 \tag{61}$$

Expansion of this expression in a similar manner to (36) leads to equation (57). The two terms (f_{xy}/f_{yy}) and (f_{0y}/f_{yy}) represent the regression coefficients which give the *optimum linear prediction* of d_x^{T} and $d_{N_0}^{\mathrm{T}}$, respectively, in terms of d_y^{T}. Estimates of $|R_{x0/y}(\lambda)|^2$ test the hypothesis that the coupling between N_0 and x can be predicted by process y, in which case the parameter will have the value zero.

The partial coherence defined in (57) and (61) is a first order partial coherence, which examines the correlation between two signals after removing the common effects of a single predictor. This framework can be extended to define and estimate partial coherence functions of any order. Full details, including estimation procedures and the setting of confidence limits can be found in Halliday et al. (1995a).

Results

Example 11: Partial Coherence and Cumulant

As an example of partial parameters we consider the motor unit correlation in Fig. 1, with the inclusion of the simultaneous recording of finger tremor. The hypothesis we wish to test is whether the tremor signal is a useful predictor of motor unit synchronization. This represents a multivariate analysis of the correlation between motor units N_0 and N_1 with the tremor, x, as predictor, leading to consideration of the partial coherence $|R_{10/x}(\lambda)|^2$. The spectrum of one motor unit and the tremor are shown in Fig. 2A and 2B. The correlation between the two motor units is illustrated in Fig. 3, and between one motor unit and the tremor in Fig. 4. Shown in Fig. 7 are the partial coherence estimate $|\hat{R}_{10/x}(\lambda)|^2$ and partial cumulant density estimate $\hat{q}_{10/x}(u)$. The partial coherence estimate (Fig. 7A), when compared with the ordinary coherence estimate (Fig. 3A), has almost no significant features, apart from peaks at 1 Hz and 24 Hz, indicating that the above hypothesis (that physiological tremor can predict motor unit correlation) is largely correct. The partial cumulant (Fig. 7B) has a greatly reduced central peak compared to the ordinary cumulant (Fig. 1D) and no clear sidebands. The partial phase estimate is not illustrated since it is only valid at 1 Hz and 24 Hz. This example illustrates the usefulness of a multivariate framework in testing hypotheses relating to the dependency between different signals, since we can now state conclusively that, for this data set, physiological tremor is a good predictor of motor unit synchronization.

Multiple Coherence

A second question which can be answered within a multivariate framework is the assessment of the dependence of one signal upon two or more different signals. Such a question leads to consideration of *multiple coherence functions*. For example, the multiple coherence function which assesses the strength of dependence of a time series x on

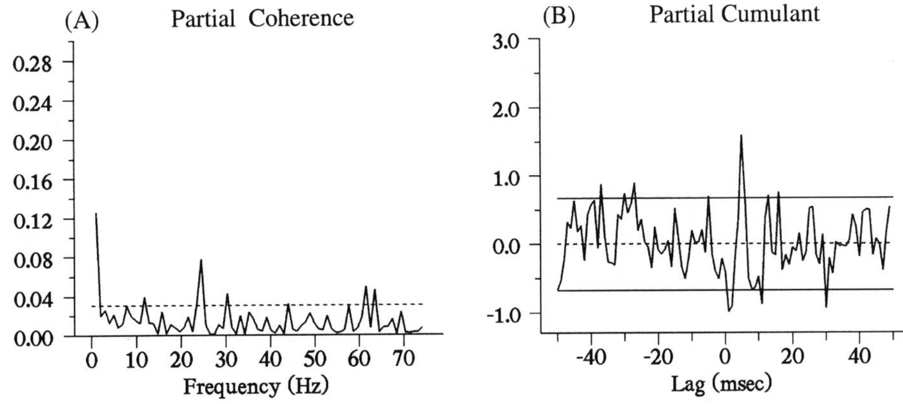

Fig. 7. Partial correlation analysis. Estimated (A) partial coherence, $|\hat{R}_{10/x}(\lambda)|^2$, and (B) partial cumulant density, $\hat{q}_{10/x}(u)\times10^4$, between two motor units using a tremor signal as predictor. The horizontal dashed line in (A) gives the estimated upper 95% confidence limit based on the assumption of independence. The horizontal lines in (B) are the asymptotic value (dashed line at zero) and estimated upper and lower 95% confidence limits, based on the assumption of independence. The partial coherence should be compared with the ordinary coherence estimate in Fig. 3A, and the partial cumulant estimate should be compared with the ordinary cumulant estimate in Fig. 1D for the same motor unit pair.

two point processes N_0 and N_1 is denoted by $|R_{x\cdot 10}(\lambda)|^2$. This can be defined in terms of ordinary coherence and partial coherence functions between the three signals as

$$|R_{x\cdot 10}(\lambda)|^2 = |R_{x0}(\lambda)|^2 + |R_{x1/0}(\lambda)|^2 \left[1 - |R_{x0}(\lambda)|^2\right] \tag{62}$$

This function can be estimated by substituting estimates of coherence and partial coherence functions into the above equation. An example of the application of multiple coherence analysis can be found in Halliday et al. (1995a). Multiple coherence functions, like other coherence functions, are bounded measures of association with values between 0 and 1. They can also be defined for more than two predictors, the derivation and estimation procedures for such higher order multiple coherence functions is given in Halliday et al. (1995a), along with procedures for the setting of a confidence limit, based on the assumption of independence.

Comments

Multivariate parameters greatly extend the range of questions which can be addressed. In many experimental protocols designed to examine the relationship between two variables, it may not be possible (or desirable) to control other variables which influence the relationship between the two variables. In such cases, and where it is possible to record the other variables, partial parameters can be used to characterize the interdependence between all the variables, and to distinguish common effects from a direct relationship.

The derivation and estimation of partial parameters directly in the time domain depends on the types of data and requires more complex procedures. However, the Fourier based framework allows partial cumulant estimates to be constructed using identical methods to those for construction of ordinary cumulant density functions (see equations (44) and (60)). Partial cumulant density estimates can be compared with the original cumulant to provide a description of any residual coupling as a function of time.

Partial coherence estimates can only be interpreted conclusively when the partial coherence estimate exhibits a complete reduction of the coupling present in the ordinary coherence estimate (as in Fig. 7). Care should be exercised in the interpretation of a partial coherence estimate which exhibits only a part reduction in magnitude when compared with the original ordinary coherence estimate. A part reduction in magnitude can have several explanations:

- the predictor signal can only predict part of the correlation, other (unobserved) effects could exert a common influence on the two signals,
- there could be a direct causal relationship between the two signals which is independent of the effects of the predictor signal, or
- the predictor could influence the two variables in a non-linear manner (see Appendix in Rosenberg et al., 1998).

Rosenberg et al. (1989, 1998) discuss in detail the use of partial parameters to identify patterns of neuronal connectivity.

Part 5: Extended Coherence Analysis – Pooled Spectra and Pooled Coherence

The previous section dealt with a description of the correlation between many simultaneously observed signals, or dependent data. This section deals with independent data, and describes a technique which can be used to characterize a number of independent coherence estimates, the aim of which is to obtain a single measure of correlation which is representative of the correlation between signals across a number of independent pairs of signals. The derivation below is valid for point process and/or time series data which satisfy the assumptions of weak stationarity, and the mixing condition discussed above. It is further assumed that all the original coherence estimates have been obtained from independent data sets, for example from repeat trials, or from observations across a number of subjects.

In the following definition it is assumed we are considering k independent pairs of processes, with each pair denoted by $(a_i, b_i: i=1,\ldots,k)$, and that L_i is the number of disjoint sections used to estimate the second order spectra for the ith pair of processes. The processes a and b can represent any combination of point process and/or time series data. The *pooled coherence estimate* which summarizes the correlation across the k pairs of processes is obtained by a weighted average of the individual spectra as (Amjad et al., 1997)

$$\frac{\left|\sum_{i=1}^{k}\hat{f}_{a_ib_i}(\lambda)\ \mathrm{L}_i\right|^2}{\left(\sum_{i=1}^{k}\hat{f}_{a_ia_i}(\lambda)\ \mathrm{L}_i\right)\left(\sum_{i=1}^{k}\hat{f}_{b_ib_i}(\lambda)\ \mathrm{L}_i\right)} \tag{63}$$

In equation (63) $\hat{f}_{a_ib_i}(\lambda)$ denotes an estimate of the second order spectrum $f_{a_ib_i}(\lambda)$, estimated from L_i disjoint sections, according to (29). The above derivation requires that all second order spectral estimates have been estimated with the same number of points in the finite Fourier transforms (24) and (26). Pooled coherence estimates, like ordinary coherence estimates, have values between 0 and 1. The upper 95% confidence limit for the estimate (63), based on the assumption of independence between the k pairs of processes, is (Amjad et al., 1997)

$$1-(0.05)^{1/(\Sigma \mathrm{L}_i-1)} \tag{64}$$

where $\Sigma\mathrm{L}_i$ is the total number of segments in the pooled coherence estimate. Estimated values of pooled coherence below this level at a particular frequency, λ, can be interpreted as evidence that, on average, no coupling occurs between the k pairs (a_i, b_i) at that frequency.

It is also possible to use the individual terms in expression (63) to obtain estimates of pooled spectra, which requires a correction factor of

$$\left(\sum_{i=1}^{k}\mathrm{L}_i\right)^{-1}$$

to obtain the correct value for the two pooled auto-spectral estimates and the pooled cross-spectral estimate. Thus, the *complex valued pooled cross-spectrum* can be estimated as

$$\frac{\sum_{i=1}^{k}\hat{f}_{a_ib_i}(\lambda)\ \mathrm{L}_i}{\left(\sum_{i=1}^{k}\mathrm{L}_i\right)} \tag{65}$$

This can be used to obtain a *pooled phase estimate*, and a *pooled cumulant density estimate* (via an inverse Fourier transform) using methods similar to those described in Part 3. Pooled cumulant density functions provide a single time domain measure of association which can be used to summarize the correlation between many different pairs of processes. For further details see Amjad et al. (1997) and Halliday et al. (1995a).

Example 12: Pooled Coherence, Phase and Cumulant

Figure 8 illustrates an example of the application of pooled coherence to a large data set consisting of 190 individual records of motor unit pairs recorded from the third finger portion of EDC in a total of 13 healthy adult subjects during maintained postural contractions (see Conway et al., 1995b, Halliday et al., 1995a for details of experimental protocol). The average record duration is 89 seconds, range: 20 to 180 seconds. The pooled coherence, pooled phase and pooled cumulant estimates are shown in Fig. 8, constructed from a total of 16,384 segments (T=1024), equivalent to 279.6 minutes of data. Before estimating the pooled parameters, the original motor unit data underwent a temporal alignment procedure, such that the peak in individual cumulant density estimates always occurred at time zero, see Amjad et al. (1997) for further discussion of this procedure. The interpretation of these parameter estimates is similar to the ordinary coherence, phase, and cumulant density estimates shown previously, except that now we are dealing with the population behavior. The coherence estimate has two clearly defined bands, a low frequency band from 1–10 Hz, and a higher frequency band centered around 25 Hz. The magnitude of the estimate is very small, with maxima of 0.035 and

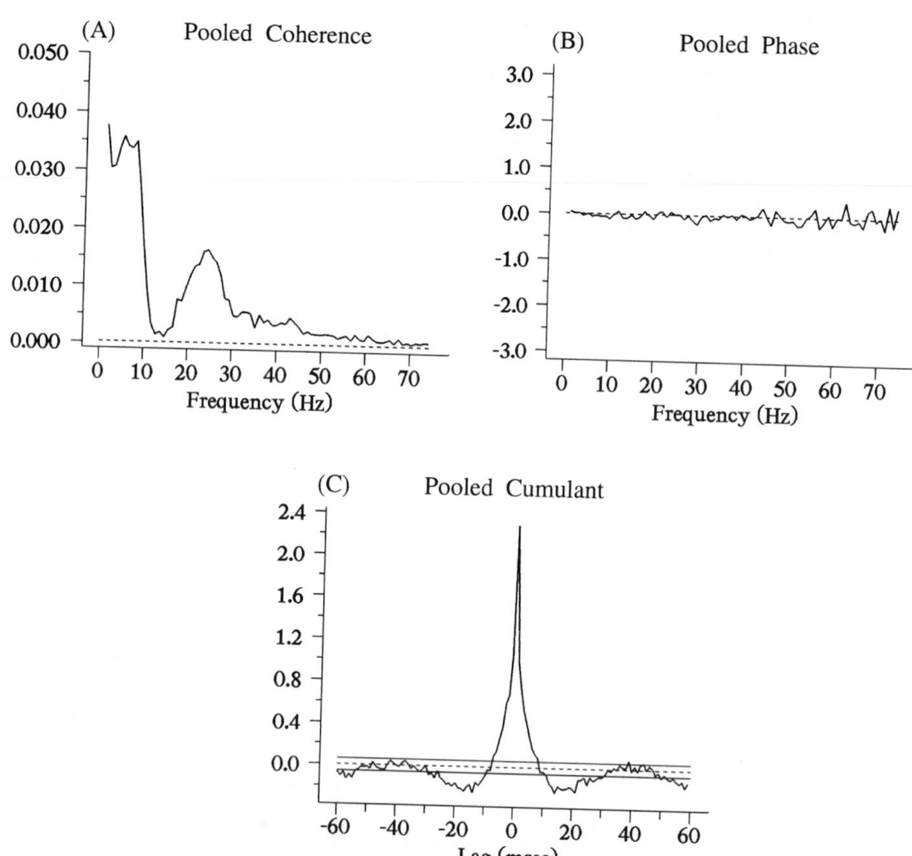

Fig. 8. Extended coherence analysis. Estimate of (A) pooled coherence, $|\hat{R}_{10}(\lambda)|^2$, (B) pooled phase, $\hat{\Phi}_{10}(\lambda)$, and (C) pooled cumulant density, $\hat{q}_{10}(u)\times 10^4$, for a population consisting of 190 separate records from motor unit pairs in EDC. The horizontal dashed line in (A) gives the estimated upper 95% confidence limit based on the assumption of independence. The horizontal lines in (C) are the asymptotic value (dashed line at zero) and estimated upper and lower 95% confidence limits, based on the assumption of independence.

0.016 in these frequency bands. The large quantity of data used to construct this estimate results in greatly reduced standard errors, which allows the weak coupling present to be more accurately specified than is possible for a single example. The estimated pooled phase is constant at zero radians, this reflects the results of the temporal alignment process. The estimated pooled cumulant has a clearly defined time course, with the central peak and sidebands well defined.

Comments

Pooled coherence analysis is useful to summarize a large data set, as in the above example. The more traditional approach is to select and present a "typical" example from the data set. However, the presentation of selected examples from a larger data set can often lead to misleading conclusions, by emphasizing features not typical of the population as a whole (see discussion in Fetz, 1992).

The framework for pooled coherence in Amjad et al. (1997) also includes a statistical test to determine if the coherence estimates in the pooled estimate can be considered to have the same magnitude at each frequency. This test provides a rigorous means of examining task-dependency in a set of coherence estimates. Amjad et al. (1997) illustrate the application of this test to investigate the relationship between a single motor unit and physiological tremor during altered inertial loading. In situations where the test for equal coherence estimates is violated, as in the above data (not shown), pooled coherence can still provide a single representative measure which summarizes the coherence structure within a larger data set.

Part 6: A Maximum Likelihood Approach to Neuronal Interactions

The Fourier based methods described in the previous sections provide a framework for analysis of spike train and/or time series data. However, they are non-parametric methods, since they do not provide estimates of parameters which have a direct neurophysiological interpretation. In this section we describe an alternative *parametric time domain* approach to analysis of neuronal interactions based on a conceptual neuron model which describes the relationship between input and output spike trains. Parameters of the model are estimated in the time domain using likelihood methods. Such an approach is commonly used in statistics to provide a model based description of data. Brillinger (1988a,b) describes a maximum likelihood approach to the analysis of neuronal interactions based on an *"integrate to threshold and fire"* model (cf. Chapter 21). This is a threshold based model which incorporates the linear summation of effects due to pre-synaptic input spikes with a recovery process, which, among other things, represents intrinsic properties of the neuron after firing an output spike. When the additive effects of the summation and recovery processes exceed threshold, the neuron will fire an action potential. An expression for the probability or likelihood of the observed output spike train is constructed in terms of a threshold-crossing probability. The arguments of this probability function depend on summation, recovery, and threshold functions. Maximum likelihood is used to estimate the parameters characterizing these functions. Breeze et al. (1994) and Emhemmed (1995) give several examples of the application of maximum likelihood to both model generated and experimental data.

The first step in the likelihood method is the construction of a *probability model for the output spike train*. We assume orderly spike train data (see Part 1), which allows spike trains to be represented as a 0–1 time series, and a standard binomial probability model to be set up for the output spike train, except that the probability of a spike occurring at time t is not constant, but will depend on t. If we let N_t denote a spike train and Δ a small time interval, then at time t

$$N_t = \begin{cases} 1 & \text{if there is a spike in } (t, t+\Delta) \\ 0 & \text{otherwise} \end{cases} \tag{66}$$

for $t = 0, \pm\Delta, \pm 2\Delta, \ldots$. If H_t represents the history (or times of occurrence of spikes) of N_t up to and including t, then the conditional probability of a spike occurring at time t may be written as

$$P_t = \text{Prob}\{N_t = 1 \mid H_t\} \tag{67}$$

and the likelihood, $l(N_t, \underline{\vartheta})$, of observing a particular spike train N_t is given as (Brillinger 1988a,b; Emhemmed, 1995)

$$l(N_t, \underline{\vartheta}) = \prod_t P_t^{N_t} (1 - P_t)^{1-N_t} \tag{68}$$

where ϑ represents the set of parameters to be estimated.

Using likelihood procedures requires a model for P_t in terms of parameters that are thought to influence N_t. Following Brillinger (1988a,b) we construct a model which consists of a summation function, which describes the effects of individual input spikes, a recovery function, which accounts for refractoriness and spontaneous firing, and a random threshold function. When the combined action of the summation and recovery functions exceeds threshold, the neuron discharges an action potential and is reset to a resting level. If we denote the *summation function* as $a(u)$, the effects of an input spike train, $\underline{M(t)}$, can be modelled as a linear summation over the time of occurrence of all input spikes since the last output spike as

$$\int_0^{\gamma(t)} a(u)\, dM(t-u) = \sum_{t=0}^{\gamma_t - 1} a_u\, M_{t-u} \tag{69}$$

where $\gamma(t)$ denotes the time elapsed since the previous output spike. This equation is a linear summation of a_u over the times of occurrence of spikes, M_t, during the interval, γ_t, since the last output spike. This approach may be extended to include non-linear terms (Brillinger, 1988b), continuous (time series) inputs (Brillinger, 1988b), or a combination of time series and spike train inputs (Emhemmed, 1995).

The *recovery function* is modelled as a polynomial of the time elapsed since the last output spike, which can be written as

$$\sum_{\upsilon=1}^{k} \theta_\nu \gamma_t^\nu \tag{70}$$

where γ_t is the time since the last output spike, and θ_ν, $\nu=1,\ldots k$, are the parameters to be estimated. The recovery function on its own can be used to model the spontaneous discharge of a neuron.

The threshold is assumed to be either constant, θ_0, or to decay exponentially from a constant value following an output spike, which requires two extra parameters: a magnitude μ, and a time constant λ. It includes a noise term, $\varepsilon(t)$, to account for contributions from unobserved inputs which also influence the neuron. The *threshold* can be written as

$$\theta_0 + \mu e^{-\lambda t} + \varepsilon(t) \tag{71}$$

An output spike will occur when the value of the summation function and the recovery function exceed threshold.

A linear function which compares the sum of the summation and recovery functions with the threshold function, referred to as the *linear predictor*, and denoted by Z_t, can now be constructed by combining (69), (70) and (71) to give

$$Z_t = \sum_{t=0}^{\gamma_t - 1} a_u M_{t-u} + \sum_{v=1}^{k} \theta_v \gamma_t^v - (\theta_0 + \mu e^{-\lambda \gamma_t}) \tag{72}$$

A natural way to meet the requirement that the probability model for the occurrence of a spike remains between zero and one is to apply a transformation to Z_t. The function used to transform Z_t is referred to as a *link function* (McCullagh and Nelder, 1992). Brillinger (1988ab) proposes the standard cumulative normal, $\Theta(\cdot)$, as a link function, other suitable link functions are discussed in (Emhemmed, 1995). Using the standard cumulative normal the conditional probability, P_t, becomes

$$\begin{aligned} P_t &= \Theta(Z_t) \\ &= \Theta\left(\sum_{t=1}^{\gamma_t - 1} a_u M_{t-u} + \sum_{v=1}^{k} \theta_v \gamma_t^v - \left(\theta_0 + e^{-\lambda \gamma_t}\right) \right) \end{aligned} \tag{73}$$

and the corresponding likelihood function (68) is

$$l(N_t, \underline{\vartheta}) = \prod_t \Theta(Z_t)^{N_t} (1 - \Theta(Z_t))^{1-N_t} \tag{74}$$

The set of parameters to be estimated is $\underline{\vartheta} = (\{a_u\}, \{\theta_v\}, \theta_0, \lambda, \mu)$. Maximum likelihood estimates these parameters to maximise the value of the likelihood function. This procedure may be carried out using the *statistical package GENSTAT*, which also provides standard errors for all of the estimated parameters. Emhemmed (1995) describes setting up a GENSTAT program for this analysis. An alternative implementation using the *statistical package GLIM* to investigate the relationship between three interconnected neurons is described in Brillinger (1988a).

An important aspect is assessing the *goodness of fit* of the likelihood model based on the binomial distribution. Brillinger (1988a,b) discusses a procedure for assessing the goodness of fit by comparing the estimated probability of occurrence of a spike in the modelled spike train with the theoretical probability. This procedure is based on a visual comparison between the estimated probability of occurrence of a spike in the modelled spike train with the theoretical probability when both are plotted against selected values of the linear predictor, Z_t.

Results

We illustrate the application of the likelihood method with two examples, based on analysis of data derived from a *conductance based neuron model* (Getting, 1989, Halliday, 1995), where transmembrane ionic currents are assumed to flow through channels with a linear instantaneous current voltage relationship obeying Ohm's law (Hille, 1984). The present simulations are based on point neuron models, where the intracellular membrane potential for each cell is given by the equation (Getting, 1989)

$$C_m \frac{dV_m}{dt} = -I_{leak}(V_m) - \sum_{j=1}^{n} I_{syn}^j(V_m, t) - \sum_{i=1}^{k} I_{ahp}^i(V_m, t) - I_{ext}(t) \tag{75}$$

where V_m represents the membrane potential at time t and C_m is the cell capacitance. $I_{leak}(V_m)$ is the passive leakage current, $I^j_{syn}(V_m,t)$ is the current due to the jth pre-synaptic spike, with the summation over the total number of pre-synaptic spikes, denoted by n. The afterhyperpolarization (AHP) current due to the ith post-synaptic spike is $I^{i}_{ahp}(V_m,t)$, with the summation over the total number of post-synaptic spikes, denoted by k. $I_{ext}(t)$ is a time dependent external current applied to the cell which is used to simulate a population of unobserved inputs responsible for spontaneous background firing. In practice this is achieved by using a non-zero mean normal distribution to simulate synaptic noise (Lüscher, 1990).

The cell leakage current is estimated as $I_{leak}(V_m) = (V_m - V_r)/R_m$, where R_m is the cell input resistance. The synaptic current due to a single pre-synaptic spike at time t=0 is estimated as $I_{syn}(V_m,t) = g_{syn}(t)\,(V_m - V_{syn})$, where $g_{syn}(t)$ is a time dependent conductance change associated with the opening of ionic channels following neurotransmitter release, and V_{syn} is the equilibrium potential for this ionic current. The AHP current due to a single post-synaptic spike at time t=0 is estimated as $I_{ahp}(V_m,t) = g_{ahp}(t)\,(V_m - V_{ahp})$, where $g_{ahp}(t)$ is a time dependent conductance change, and V_{ahp} is the equilibrium potential. Expressions for $g_{syn}(t)$ and $g_{ahp}(t)$ are given below. Each pre-synaptic input spike activates one extra term in the summation over n in equation 75, which lasts for the duration of the $g_{syn}(t)$ for that input. Similarly, each post-synaptic spike activates one extra term in the summation over k in equation 75, which lasts for the duration of the $g_{ahp}(t)$ for that cell.

The voltage V_m is compared with a threshold voltage, V_{th}, at each time step to determine if an action potential has occurred. A time varying threshold is incorporated into the simulation, this allows point neuron simulations to duplicate a wide range of repetitive firing characteristics (Getting, 1989). The threshold is specified by three variables, the asymptotic level, θ_∞, the level to which the threshold is elevated after each output spike, θ_0, and the decay time constant with which the threshold decays to the asymptotic level, τ_θ.

The selection of simulation parameters is done in the same order, and at each stage parameters are selected so that the behavior of the simulation matches experimental observations for the type of cell being simulated. First passive parameters are selected, then cell membrane/input resistance, R_m, and time constant τ_m, are chosen, where $\tau_m = R_m C_m$. This determines the cell capacitance, C_m. The cell resting potential, V_r, and threshold parameters, θ_∞, θ_0 and τ_θ, are chosen. These determine the rheobase current required for repetitive firing of the cell. The time course of the AHP can be adjusted under constant current stimulus by altering the conductance function $g_{ahp}(t)$. The characteristics of a single excitatory post synaptic potential (EPSP), or a single inhibitory post synaptic potential (IPSP) from rest can be adjusted by altering the conductance $g_{syn}(t)$, and the equilibrium potential V_{syn}. The resulting EPSP or IPSP can be characterized by rise time, half width and magnitude. EPSP and IPSP conductances are modelled by an alpha function: $g_{syn}(t) = A\,(t/\tau_a)\exp(-t/\tau_a)$, (Rall, 1967) requiring the choice of a scaling factor, A, and a time constant, τ_a. Once these have been determined, the firing rates for pre-synaptic inputs have to be chosen. Selecting an appropriate mean firing rate for the input, along with any applied external current, $I_{ext}(t)$, determines the mean output firing rate of the simulation, and can be adjusted to give the desired output rate for each cell.

Example 13: Motoneuron

The first example is based on a class of cells which have been widely studied, namely motoneurons. The first simulated data set was derived from a simulation using passive parameters within the range of values quoted in Rall (1977) for experimental studies on spinal motoneurons, with R_m=5 MΩ, τ_m=5 ms, C_m= 1 µF, and a resting potential of V_r=

−70 mV. The AHP conductance used a simplified version of the three-term model proposed by Baldissera and Gustafsson (1974) for observed AHP time courses in cat lumbar motoneurons, based on an exponential conductance function $g_{ahp}(t) = A \exp(-t/\tau_a)$, with τ_a=14 ms, and A=1.0e-08, with V_{ahp}=−75 mV. The threshold parameters were θ_∞= −65mV, θ_0=−55 mV, and τ_θ=20 ms. In this example, no synaptic noise was applied, the output discharge was entirely due to the single pre-synaptic input, which was activated by a random, or Poisson, spike train with a mean firing rate of 50 spikes/s. The EPSP conductance values were τ_a=2 ms, and A=1.1e^{-08}, with V_{syn}=0.0 mV. The EPSP parameters for a single conductance activated from rest are a rise time (10%–90%) of T_r=3.1 ms, a half width of T_{hw}=9.8 ms, and a magnitude of 10.62 mV. These values are outside the upper limits of 2.1 ms, 7.7 ms and 0.54 mV reported for the same parameters measured using spike triggered averaging of single fibre Ia connections to cat spinal motoneurons (Cope et al., 1987), but were necessary, however, to obtain repetitive firing with only a single pre-synaptic input. The EPSP magnitude and duration during repetitive firing are reduced due to the shunting action of the large AHP conductance. The simulation was used to generate 60 seconds of data at 1 ms sampling intervals with these parameters, in total 2416 output spikes were obtained, a mean rate of 40.3 spikes/s.

Illustrated in Fig. 9A is the estimated cumulant density, $\hat{q}_{10}(u)$, see equation (16), between the input and output spike trains. This estimate suggests individual inputs have an excitatory effect on the output discharge, the duration of which is about 4 ms. There is a subsequent dip in the estimated cumulant which is outside the lower 95% confidence limit, however, this feature can be interpreted to reflect the mapping to the cumulant density of structure in the auto-correlation of the output discharge (Moore et al., 1970). In contrast, the likelihood approach separates out these two effects. The estimated summation function, Fig. 9C, which we denote by $\hat{a}(u)$, has significant values for lags up to 7 ms, suggesting that the duration of the excitatory effect for a single input lasts about 7 ms. The recovery function used a third order polynomial, k=3 in equation (70), however, using a threshold with an exponential term results in values of θ_2 and θ_3 which were not significant, and can be neglected, resulting in a first order recovery function with constant slope. The difference between the estimated recovery and threshold functions (Fig. 9E) indicates that the probability of an output spike is small up to about 15 ms after an output spike, after which they converge more rapidly, indicating an increase in firing probability. The features of the summation, recovery, and threshold function correspond well with the structure of the simulated motoneuron, in which the half width of a single excitatory postsynaptic potential was 9.8 ms, and the mean rate of the output spike train was 40 spikes/s.

Example 14: Invertebrate Neuron

In the second example the simulated neuron is based on studies of small networks responsible for rhythmic pattern generation in invertebrates (Getting, 1989). The simulation was set up with parameters R_m=12.5 MΩ, τ_m=50 ms, giving C_m=4 μF, and a resting potential of V_r=−60 mV. The threshold parameters were θ_∞=−45mV, θ_0=−35 mV, and τ_θ=10 ms. Synaptic noise was simulated by an applied $I_{ext}(t)$ with a mean value of 2.9 nA, and a standard deviation of 1.7 nA. In this example an inhibitory input was applied, with conductance parameters of τ_a=1 ms, A=2.2e^{-07}, and V_{syn}=−80.0 mV. The IPSP parameters for a single conductance activated from rest were T_r=5.7 ms, T_{hw}=39.9 ms, and a magnitude of −1.0 mV. The input firing rate was set at 50 spikes/s with a random discharge, and in 60 seconds the simulation produced 1400 output spikes, 23.3 spikes/s. The estimated cumulant density, Fig. 9B, suggests an inhibitory effect lasting about 8 ms fol-

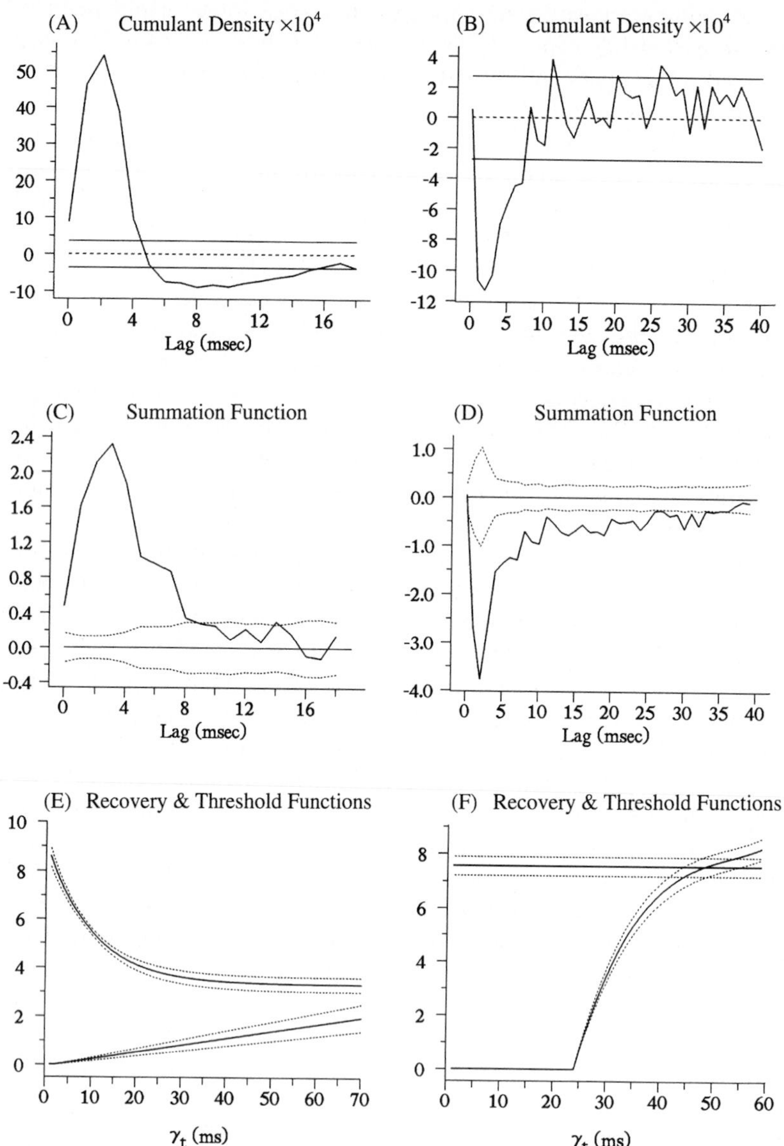

Fig. 9. (A) Estimated cumulant density, $\hat{q}_{10}(u)\times 10^4$, between a random excitatory spike train input and the output discharge for a simulated neurone. (C) Estimated summation function, $\hat{a}(u)$, and (E) estimated threshold function (upper traces) and recovery function (lower traces) for the same data set as in (A). (B) Estimated cumulant density, $\hat{q}_{10}(u)\times 10^4$, between a random inhibitory spike train input and the output discharge for a simulated neurone. (D) Estimated summation function, $\hat{a}(u)$, and (F) estimated threshold function (upper traces) and recovery function (lower traces) for the same data set as in (B). Horizontal lines in (A, B) indicate the asymptotic value (dashed line at zero) and estimated upper and lower 95% confidence limits, based on the assumption of independence. Dotted lines in (C, D) indicate ±1.96 standard error limits plotted around zero. Dotted lines in (E, F) represent ±1.96 standard error limits for threshold and recovery function estimates.

lowing each input spike, whereas the summation function (Fig. 9D) suggests that the inhibitory effect lasts for more than 30 ms, which corresponds more closely with the 40 ms half width of a single inhibitory post synaptic potential. This example uses a constant threshold in the likelihood model, the recovery function is fixed at zero for lags up to

25 ms, which is the minimum interspike interval for the output discharge, after this the recovery function, modelled by a third order polynomial, rises quickly, indicating a rapid increase in the probability of firing. In the output discharge there are only 43 intervals which exceed 60 ms.

Comments

These two examples illustrate the application of likelihood methods to describe the input/output relationship for neuronal data. The estimated parameters have a direct neurophysiological interpretation, the summation function matches closely the time course of individual EPSP and IPSP functions. The estimated recovery and threshold functions give insight into intrinsic properties and refractory behavior after firing. This is in contrast to the non-parametric cross-correlation based estimates where effects due to single inputs and intrinsic effects contribute to the time course of the estimated cumulant density. It is worth pointing out that in both these examples random (Poisson) input sequences were used in an attempt to minimise this effect.

The likelihood analysis is based on an "integrate to threshold and fire" model which incorporates a random threshold, and where output spike times are determined by a threshold-crossing process. An important assumption in likelihood analysis is that the underlying model is valid. In this context, the biophysically motivated conductance based model (Getting, 1989) used to generate the data for the two examples is similar in form to the conceptual likelihood model. This suggests that the likelihood model will be useful to investigate a variety of neuronal spike train data. Examples of the application to experimental data are given in Brillinger (1988a,b) and Emhemmed (1995). The likelihood method is flexible, and allows for arbitrary numbers of neurons and neuronal mechanisms, which can be incorporated by the addition of extra terms in the linear predictor, equation (72).

Likelihood analysis is computationally more intensive that the Fourier based methods presented above. However, if the aim of an analysis is to obtain estimates of parameters of models which underlie neuronal processes, the likelihood approach may be more appropriate. An initial analysis using Fourier based techniques can provide guidance in the types of model which should be considered.

Concluding Remarks

In this chapter we have presented a framework for the analysis of neurophysiological data, which includes both time and frequency domain parameters. The framework relies on Fourier based estimation methods, which provides a unified framework for the analysis of both data types encountered in neuroscience (spike train and/or waveform data). Time domain parameters are closely related to the more traditionally used cross-correlation histogram and spike triggered averaging methodologies. Two extensions of the Fourier based framework have also been described, which extend the range of questions which can be addressed. A multivariate framework which can deal with the relationship between several simultaneously recorded signals was described in Part 4. The extended coherence analysis described in Part 5 can be used to summarize the correlation structure within a large number of data sets, and to explore questions of task dependency. Both these extensions use Fourier estimation methods, however, equivalent time domain parameters can be obtained by an inverse Fourier transform. In Part 3 we stress the complementary nature of time and frequency domain parameters for characterizing the correlation structure between neural signals. Part 6 outlines an alternative parametric time domain model based approach to characterizing neuronal spike train data.

Throughout the chapter we have stressed the importance of using confidence limits on parameter estimates. In most cases the expressions are easy to compute. The use of confidence limits are an essential part of any statistical analysis, both for dealing with uncertainty in parameter estimates, and for testing hypothesis. Their use is illustrated in the examples presented in this chapter, for example in distinguishing between distinct rhythmic components and chance fluctuations in spectral estimates (see Fig. 2), and in testing the hypothesis that the estimated correlation between two signals exceeds that expected by chance (Figs. 1, 3, 4, 5, 6, 8). The duration of the individual data sets in these examples ranges from 89,000 points to 264,000 points. The inferences which can be made from analysis of these data are a result, in part, of the large numbers of points in these data sets. Analysis of data sets which consist of a few thousand points, or less, without the use of confidence limits may lead to misinterpretation of parameter estimates, particularly in situations involving weak correlation, since the uncertainty in parameter estimates may have a similar order of magnitude as the parameter being estimated. In addition, over-interpretation of apparent fine details in parameter estimates may also give misleading results without careful use of confidence limits for parameter estimates. The setting of confidence limits about estimated values of coherence is described in Halliday et al. (1995a, Section VI). For time domain parameter estimates the sampling distribution for parameter estimates is more complex, and are only valid under the restricted condition of independent processes (see Amjad et al. 1997). In applying these techniques, it is important to distinguish between a parameter and its estimate, all parameter estimates have error and uncertainty, partly due to the problem of estimation associated with finite quantities of data. The use of raw cross-correlation histograms and raw spike triggered averages should therefore be avoided if possible. It seems appropriate to quote from one of the first studies to apply the cross-correlation technique to spike train data (Griffith and Horn, 1963): "... it is essential to have some idea of what deviations may be regarded as significant ...".

The above comments are not intended to discourage the potential user. Interpretation of parameter estimates cannot be done according to a set of pre-defined rules. Confidence limits only provide a guide to interpretation, the user should also be guided by their knowledge of the system under study. Nonetheless, these techniques do provide a comprehensive framework for analysis of neurophysiological data (and other types of signals which meet the assumptions of weak stationarity, mixing condition, and orderliness for point process data). The multivariate methods in Part 4 are particularly suited to take advantage of experimental developments involving multi-electrode recording techniques (cf. Chapter 17). Correlation analysis has underpinned many developments in neuroscience, and can continue to contribute to many studies which address questions related to tracing signal pathways, to the relationship between cortical activity, electromyographic activity and motor output, to studying the relationship between distant neural groups, and to issues related to information processing in neural circuits, and other dynamic aspects of neural behavior. The experimental data presented in this chapter are all from normal subjects, we conclude by commenting that these methods are equally applicable to clinical studies.

Analysis Software

A software archive is available to perform some of the above analyses. Details of this archive can be obtained by sending an e-mail request to: gpaa34@udcf.gla.ac.uk.

Acknowledgement: Supported by grants from The Wellcome Trust (036928; 048128), and the Joint Research Council/HCI Cognitive Science Initiative. We would like to thank Peter Breeze and Yousef Emhemmed for their help in preparing the section on maximum likelihood.

References

Amjad AM, Halliday DM, Rosenberg JR, Conway BA (1997) An extended difference of coherence test for comparing and combining independent estimates: theory and application to the study of motor units and physiological tremor. J. Neurosci. Meth. 73: 69–79

Baldissera F, Gustafsson B (1974) Afterhyperpolarization conductance time course in lumbar motoneurones of the cat. Acta Physiol Scand 91: 512–527

Bartlett MS (1948) Smoothing periodograms from time series with continuous spectra. Nature 161: 686–687

Bartlett MS (1963) The spectral analysis of point processes. J Roy Statist Soc 25: 264–281

Blackman RB, Tukey JW (1958) The measurement of power spectra from the point of view of communications engineering. Bell Sys Tech J 37: 183–282, 485–569 (Reprinted Dover Press, New-York, 1959)

Bloomfield P(1976) Fourier Analysis of Time Series: An Introduction. Wiley, New York.

Breeze P, Emhemmed YM, Halliday DM, Rosenberg JR (1994) Likelihood analysis of a model for neuronal input-output data. J Physiol 479P: 112P

Brillinger DR (1972) The spectral analysis of stationary interval functions. In: LeCam LM, Neyman J, Scott E (eds) Proceedings 6th Berkeley Symposium Mathematics Statistics Probability, Univ California Press, Berkeley, pp 483–513

Brillinger DR (1974) Fourier analysis of stationary processes. Proc IEEE 62: 1628–1643

Brillinger DR (1975) Identification of point process systems. Ann Probability 3: 909–929

Brillinger DR (1976) Estimation of second-order intensities of a bivariate stationary point process. J Roy Statist Soc B38: 60–66

Brillinger DR (1978) Comparative aspects of the study of ordinary time series and of point processes. In: Krishnaiah PR (ed) Developments in Statistics, vol 1. Academic Press, New York, pp 33–133

Brillinger DR (1981) Time Series – Data Analysis and Theory, 2nd edn. Holden Day, San Francisco

Brillinger DR (1983) The finite Fourier transform of a stationary process. In: Brillinger DR, Krishnaiah PR (eds) Handbook of Statistics, Elsevier, pp 21–37

Brillinger DR (1988a) Maximum likelihood analysis of spike trains of interacting nerve cells. Biol Cybernet 59: 198–200

Brillinger DR (1988b). The maximum likelihood approach to the identification of neuronal interactions. Ann Biomed Eng 16: 3–16

Brillinger DR, Tukey JW (1984) Spectrum analysis in the presence of noise: Some issues and examples. In: The collected works of John W Tukey, Volume II, Time series: 1965–1984. Wadsworth , Belmont, CA, pp 1001–1141

Conway BA, Halliday DM, Rosenberg JR (1993) Detection of weak synaptic interactions between single Ia-afferents and motor-unit spike trains in the decerebrate cat. J Physiol 471: 379–409

Conway BA, Halliday DM, Farmer, SF, Shahani U, Maas P, Weir AI, Rosenberg JR (1995a) Synchronization between motor cortex and spinal motoneuronal pool during the performance of a maintained motor task in man. J Physiol 489: 917–924

Conway BA, Farmer, SF, Halliday DM, Rosenberg JR (1995b) On the relation between motor unit discharge and physiological tremor. In: Taylor A, Gladden MH, Durbaba R (eds) Alpha and Gamma Motor Systems. Plenum Press, New York, pp 596–598

Conway BA, Halliday DM, Bray K, Cameron M, McLelland D, Mulcahy E, Farmer SF, Rosenberg JR (1998) Inter-muscle coherence during co-contraction of finger and wrist muscles in man. J Physiol 509: 175P

Cope TC, Fetz EE, Matsumura M (1987) Cross-correlation assessment of synaptic strength of single Ia fibre connections with triceps surae motoneurones in cats. J Physiol 390: 161–188

Cox DR (1965) On the estimation of the intensity function of a stationary point process. J Roy Statist Soc B27: 332–337

Cox DR, Isham V (1980) Point Processes. Chapman and Hall, London

Cox DR, Lewis PAW (1972) Multivariate point processes. In: LeCam LM, Neyman J, Scott E (eds) Proceedings 6th Berkeley Symposium Mathematics Statistics Probability, vol 3. University of California Press, Berkeley pp 401–488

Daley DJ, Vere-Jones D (1988) An introduction to the Theory of Point Processes. Springer, New York

Datta AK, Stephens JA (1990) Short-term synchronization of motor unit activity during voluntary contractions in man. J Physiol 422: 397–419

Emhemmed YM (1995) Maximum likelihood analysis of neuronal spike trains. PhD Thesis, University of Glasgow, 246pp

Fetz EE (1992) Are movement parameters recognizably coded in the activity of single neurons. Behavioral Brain Sci, 15: 679–690

Farmer SF, Bremner FD, Halliday DM, Rosenberg JR, Stephens JA (1993) The frequency content of common synaptic inputs to motoneurones studied during voluntary isometric contraction in man. J Physiol 470: 127–155

Getting PA (1989) Reconstruction of small networks. In: Koch C, Segev I (eds) Methods in neuronal modeling: From synapses to networks, MIT Press, pp 171–194

Gibbs J, Harrison LM, Stephens JA (1995) Organization of inputs to motoneuron pools in man. J Physiol, 485: 245–256

Griffith JS, Horn G (1963) Functional coupling between cells in the visual cortex of the unrestrained cat. Nature, 199: 873, 893–895

Halliday DM, Murray-Smith DJ, Rosenberg JR (1992) A frequency domain identification approach to the study of neuromuscular systems – a combined experimental and modelling study. Trans Inst MC, 14: 79–90

Halliday DM (1995) Effects of electronic spread of EPSPs on synaptic transmission in motoneurones – A simulation Study. In: Taylor A, Gladden MH, Durbaba R (eds) Alpha and Gamma Motor Systems. Plenum Press, New York, pp 337–339

Halliday DM, Rosenberg JR, Amjad AM, Breeze P, Conway BA, Farmer SF (1995a) A framework for the analysis of mixed time series/point process data – Theory and application to the study of physiological tremor, single motor unit discharges and electromyograms. Prog Biophys molec Biol, 64: 237–278

Halliday DM, Kakuda N, Wessberg J, Vallbo ÅB, Conway BA, Rosenberg JR (1995b) Correlation between Ia afferent discharges, EMG and torque during steady isometric contractions of human finger muscles. In: Taylor A, Gladden MH, Durbaba R (eds) Alpha and Gamma Motor Systems. Plenum Press, New York, pp 547–549

Halliday DM, Conway BA, Farmer SF, Rosenberg JR (1998) Using electroencephalography to study functional coupling between cortical activity and electromyograms during voluntary contractions in humans. Neurosci Lett, 241: 5–8

Hille B (1984) Ionic channels of excitable membranes. Sinauer

Henneman E, Mendell LM (1981) Functional organization of motoneuron pool and its inputs. In: Brookhart JM, Mountcastle VB (eds) Handbook of Physiology, Section 1, Vol 2, Part 1, The nervous system: Motor control. American Physiological Society, Bethesda, MD, pp 423–507

Hepp-Raymond M-C, Huesler EJ, Maier MA (1996) Precision grip in humans: Temporal and spatial synergies. In: Wing AM, Haggard P, Flanagan JR (eds) Hand and Brain, the neurophysiology and psychology of hand movements. Academic Press, London, pp 37–68

Jenkins GM, Watts DG (1968) Spectral analysis and its applications. Holden-Day.

Kendall DG, Stuart A (1966) The advanced theory of statistics, vol 1. Griffin, London. Kirkwood PA, Sears TA (1978) The synaptic connections to intercostal motoneurones revealed by the average common excitation potential. J Physiol, 275: 103–134

Kirkwood PA, Sears, TA (1980) The measurement of synaptic connections in the mammalian central nervous system by means of spike triggered averaging. In: Desmedt JE (ed) Progress in Clinical Neurophysiology, vol 8, Spinal and supraspinal mechanisms of voluntary motor control and locomotion. Basel, S Karger, pp 44–71

Lüscher H-R (1990) Transmission failure and its relief in the spinal monosynaptic reflex arc. In: Binder MD, Mendell LM (eds) The segmental motor system, Oxford University Press, pp 328–348 McCullagh P, Nelder JA (1992) General Linear Models (2nd edition). Monographs on statistics and Applied Probability 37. Chapman Hall, London

Mendel J (1991) Tutorial on higher-order statistics (spectra) in signal processing and system theory: theoretical results and some applications. Proc IEEE, 79: 278–305

Mendell LM, Henneman E (1968) Terminals of single Ia fibres: Distribution within a pool of 300 homonymous motoneurones. Science 160: 96–98

Mendell LM, Henneman E (1971) Terminals of single Ia fibres: location, density and distribution within a pool of 300 homonymous motor neurons. J Neurophysiol, 34: 171–187

Moore GP, Segundo JP, Perkel DH, Levitan H (1970) Statistical signs of synaptic interaction in neurones. Biophys J 10: 876–900

Parzen E (1961) Mathematical considerations in the estimation of spectra. Technometrics, 3: 167–190 Press WH, Flannery BP, Teukolsky SA, Vetterling WT (1989) Numerical recipes. Cambridge University Press

Rall W (1967) Distinguishing theoretical synaptic potentials computed for different soma-dendritic distributions of synaptic inputs. J Neurophysiol 30: 1138–1168

Rall W (1977) Core conductor theory and cable properties of neurones. In: Kandel ER, Brookhart JM, Mountcastle VB (eds) Handbook of physiology: The nervous system, vol 1, part 1. Williams and Wilkins, Maryland, pp 39–97

Rigas A (1983) Point Processes and Time Series Analysis: Theory and Applications to Complex Physiological Systems. PhD Thesis, University of Glasgow, 330pp

Rosenberg JR, Murray-Smith DJ, Rigas A (1982) An introduction to the application of system identification techniques to elements of the neuromuscular system. Trans Inst MC 4: 187–201

Rosenberg JR, Amjad AM, Breeze P, Brillinger DR, Halliday DM (1989) The Fourier approach to the identification of functional coupling between neuronal spike trains. Prog Biophys molec Biol, 53: 1–31

Rosenberg JR, Halliday DM, Breeze P, Conway BA (1998) Identification of patterns of neuronal activity – partial spectra, partial coherence, and neuronal interactions. J Neurosci Meth 83:57–72

Rosenblatt M (1983) Cumulants and Cumulant Spectra. In: Brillinger DR, Krishnaiah PR (eds) Handbook of Statistics vol 3. North Holland, New York, pp 369–382

Schuster A (1898) On the investigation of hidden periodicities with application to a supposed 26 day period of meteorological phenomenon. Terr Mag, 3: 13–41

Sears TA, Stagg D (1976) Short-term synchronization of intercostal motoneurone activity. J Physiol 263: 357–387

Sorensen HV, Jones DL, Heideman MT, Burrus CS (1987) Real valued Fast Fourier Transform Algorithms. Proc IEEE ASSP, 35: 849–863; Corrections p 1353

Srinivasan SK (1974) Stochastic Point Processes and their Applications. Monograph No 34. Griffin, London

Stauffer EK, Watt DGD, Taylor A, Reinking RM, Stuart DG (1976) Analysis of muscle receptor connections by spike triggered averaging. 2. Spindle group II afferents. J Neurophysiol, 39: 1393–1402

Stiles RN, Randall JE (1967) Mechanical factors in human tremor frequency. J App Physiol 23: 324–330

Tukey JW (1961) Discussion, emphasizing the connection between analysis of variance and spectrum analysis. Technometrics, 3: 191–219

Tukey JW (1980) Can we predict where "time series" should go next? In: Brillinger DR, Tiao GC(eds) Directions in Time Series IMS, Hayward, California, pp 1–31

Vallbo ÅB, Hagbarth K-E (1968) Activity from skin mechanoreceptors recorded percutaneously in awake human subjects. Exp Neurol 21: 270–289

Watt DGD, Stauffer EK, Taylor A, Reinking RM, Stuart DG (1976) Analysis of muscle receptor connections by spike triggered averaging. I. Spindle primary and tendon organ afferents. J Neurophysiol 39: 1375–1392

Wiener N (1930) Generalized harmonic analysis. Acta Math 55: 117–258

Chapter 19

Information-Theoretical Analysis of Sensory Information

Yoav Tock and Gideon F. Inbar

Introduction

It is clear that, in order to survive, organisms must exchange information between the body and its environment as well as between different body parts. They do so via various systems employing different means, one being the nervous system. At the nervous system's periphery, *sensory neurons* transduce a physico-chemical analog signal into a train of discrete action potentials (spike trains), which can be transmitted over long distances. A similar process also occurs within the central nervous system as most neurons need to transduce their synaptic potentials, i.e., analog signals, into spike trains.

Sensory systems can be analyzed from many different viewpoints. For example, consider the muscle spindle which is generally taken to be a muscle length receptor (see, e.g., Prochazka 1996). In a *linear systems* approach (see Chapter 21), the frequency transfer function of the sensor, relating length input to spike train output, may be estimated and analyzed using various methods (e.g., Matthews and Stein, 1969; Rosenthal et al., 1970; Hulliger et al., 1977; Poppele, 1981; Kröller et al., 1985). From a *control* viewpoint, the muscle spindle has been modeled and examined with respect to its role in the stretch reflex feedback loop (see, for example, Houk, 1963; McRuer et al., 1968; Inbar, 1972; Koehler and Windhorst, 1981). Taking a different approach, Inbar and Milgram (1975) looked specifically at the encoding process by assuming a nonlinear model for the neural transduction in the receptor, and then explored the decoding problem, i.e., how the analog sensory input signal could be retrieved from the spike train. In addition they explored the impact of sensory transmission via parallel channels on the quality of the decoded signal (Milgram and Inbar, 1976).

Different approaches use different criteria, according to the assumed goal of the system. Therefore, the quality of the signal provided by the muscle spindle may be different according to the methodology used and the question the research is trying to answer. Having this in mind, we here take a communication viewpoint in analyzing sensory systems. In this approach, the world outside the nervous system is the source of information, and the nervous system is the target. We have said above that, in order to convey this information, sensory signals are transduced into spike trains. This transduction between different types of signal corresponds to an encoding process. The sensor does not try to encode *all* the information present in the world outside the nervous system, but is selective, and presumably encodes just the parts that are relevant to the organism. In addition, noise inherent in the system or impinging on it may interfere with the encoding process and distort the information carried by the encoded signal. The basic question then is how much information is transferred between the sensory world and the

Yoav Tock, Technion – Israel Institute of Technology, Department of Electrical Engineering, Haifa, Israel (phone: 972-4-8294633; fax: 972-4-8323041; e-mail: ytock@tx.technion.ac.il)
Gideon F. Inbar, Technion - Israel Institute of Technology, Department of Electrical Engineering, Haifa, Israel (phone: 972-4-8294718; fax: 972-4-8323041; e-mail: inbar@ee.technion.ac.il)

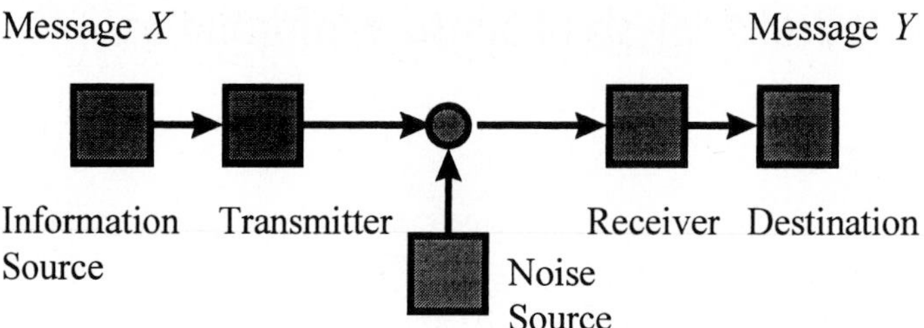

Fig. 1. A block diagram presentation of the basic communication problem.

nervous system. Many researchers use the term information in an ill-defined, informal, qualitative manner. In this chapter, we take a rigid quantitative approach and describe a method that quantifies information transmission between the sensory signal impinging upon the sensor and the resulting spike train.

In so doing, let us take the organism's point of view, or the view of its central nervous system. The organism faces this problem: Looking at the spike train, it is trying to extract information about the stimulus (sensory signal), in face of noise interfering with information transmission (see also Inbar and Milgram, 1975).

This problem has much in common with the basic problem tackled by *information theory* which has been developed to deal with problems in technical communication systems. Information theory was first formulated by Shannon in 1948, in a paper called "A Mathematical Theory of Communication" (Shannon, 1948; reprinted in Shannon and Weaver, 1949). The basic problem, graphically presented in Fig. 1, is as follows.

A statistical *information source*, X, emits characters (defined signals) chosen at random from a predefined *alphabet*. For example, if the source randomly emits just ones and zeroes, it is referred to as a *binary source*. The message X is encoded and transmitted over a noisy channel. At the other end, a receiver decodes the message. The recipient looks at the channel output Y and tries to gain information about the input X. That is, looking at the received message Y, he wants to know what was the transmitted message X. If the channel is noiseless, this is an easy job, since Y will always equal X. But, in most practical cases, the channel will be noisy, and Y will often be different from X. In some extreme cases, the noise will be large, or the communication channel will be broken, so that the output Y will not help us to decide which letter was transmitted. Between those two extremes, we want to quantify the amount of "information" that the signal Y carries about signal X.

In our context, studying sensory systems, X is a signal such as light intensity, muscle length, air pressure, etc., and Y is a spike train. In order to use information theory, we must assume that the system under discussion can be characterized statistically. In the natural setting, the sensory signal X has the statistical structure of the world, if indeed we can find such a structure. In the laboratory, we can control its statistics, and measure the resulting spike train.

Information theory deals with the statistical aspects of information. This theoretical framework provides us with powerful tools to analyze and design communication systems, and to deal with many analogous problems.

Ever since its publication, there have been efforts to apply the tools of information theory to the analysis of the nervous system. Only four years after its publication, MacKay and McCulloch (1952) tried to estimate the information transmission capacity of a single spiking neuron. Two decades later, Eckhorn and Pöpel (1974, 1975) applied information theory to the afferent visual system of the cat. Their work has laid the foundation to much recent work (for a review see: Hertz, 1995). The method we are going to describe is due to the seminal work of Bialek and his colleagues (Bialek et al. 1991), which has recently appeared in a book (Rieke et al, 1997). This method relies on reconstruction of the stimulus from the spike train. This reconstruction is inevitably less than perfect, resulting in a reconstruction error. The reconstruction error is used to estimate the noise in the system, and this estimated noise is used to calculate a lower bound for the transmitted information rate.

The goals of this chapter are twofold. The first goal is to present the method. The second goal is to make the theoretical grounds for this method accessible to researchers without an extensive engineering or mathematical background. Since information theory and random signal theory can be highly mathematical, we have chosen an intuitive rather then rigorous mathematical approach, as outlined in the next section.

Outline

The main goal of this chapter is to describe a method for quantifying information transmission, in sensory receptors, between sensory signals and the resulting spike train. The description of the method is divided into several building blocks.

- We start by presenting a few concepts related to the *neural code*. These culminate in a simple method of reading the neural code, that is, reconstructing the stimulus from the spike train. Reconstructing the stimulus is not a goal in itself, but merely a building block of the main method.
- Then we explain the *basic terms of information theory* to be used in describing the method. Information theory relies on probability theory, therefore we present in passing the needed terms from *probability theory*. This section deals only with the information carried by discrete events. In order to generalize the results to continuous time signals,
- the next section presents a few fundamentals of *random signal theory*. Random signal theoretical results are used in
- the following section to generalize the basic results of information theory to *random time signals*. The first four sections prepare the theoretical ground for
- the next section, in which we present the main method. The main product this section is a *lower bound to the information rate*.
- Then, we present the *upper bound to the information rate*, and cite experimental results from sensory systems. These results elucidate the type of knowledge obtained by applying this methodology to a single sensory neuron.
- Finally, we bring together the results of an *experiment on the cat muscle spindle*, and a *simulation model* of the same system, in order to demonstrate this technique and in order to provide a clear benchmark.

The Neural Code

In the following section we are going to review two dominant viewpoints in the analysis of the neural code. We do this in order to explain the rationale behind a simple method of reconstructing the input stimulus from the spike train. This reconstruction scheme is a building block in the information transmission analysis method.

The Neural Code – Traditional Approach

When we come to investigate the neural code, we adopt a probabilistic approach. It is well known that the neuron is a noisy device. If we present the neuron with the same stimulus $s_1(t)$, $0<t<T$, over and over again, it responds with a different spike train $\{t_i\}_{i=1}^{N}$

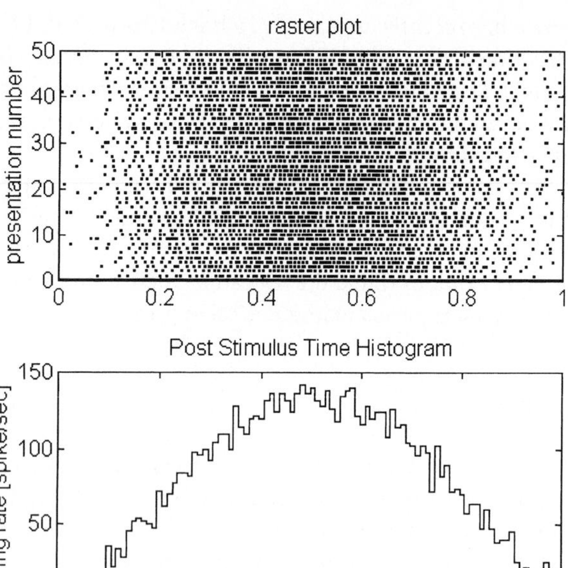

Fig. 2. Variability of neural response and construction of the PSTH. The top figure is a raster plot, showing the response of the neuron to 50 presentations of the same stimulus. Each dot represents the occurrence of a spike. The data is from a computer model of a noisy neuron. The input waveform was the bell-shaped cosine window: $s(t)$=1-cos($2\pi t$). The bottom figure is the PSTH, the average number of spikes in a time bin, normalized by the bin size (10 ms in this case). We see that the firing rate tends to approximate the input stimulus.

each time. In this notation, t_i stands for the spike occurrence time in the interval $(0,T)$. If we repeat this simple experiment many times, we will be able to estimate the probability of finding a specific spike train $\{t_1,t_2,\ldots,t_N\}$ in response to that given stimulus – $s_1(t)$. This is the *conditional distribution* – $P[\{t_i\}|s(t)=s_1(t)]$. We can repeat this experiment, but with many different input waveforms $s(t)$. The input stimulus is then chosen from $P[s(t)]$, the probability distribution that defines an ensemble of waveforms. Finally we arrive at the *stimulus-conditional distribution* – $P[\{t_i\}|s(t)]$, which defines the probability of a spike train $\{t_i\}$ given some input stimuli $s(t)$. Estimating this distribution is difficult because very large amounts of data are needed. Generally, we can estimate just the first few moments – mean, variance, and so on.

Take for example the *Post Stimulus Time Histogram (PSTH)* seen in Fig. 2. The PSTH is the average response to a stimulus, averaged over many presentations of the same stimulus. Spike occurrence time is determined with resolution $\Delta\tau$ (bin width). All we have to do is count the number of spikes we find in a $\Delta\tau$ time bin, and normalize by the number of presentations and the size of the time bin. Done in this way, the result is a time-dependent firing rate $r(t)$, which can be interpreted as the probability per unit time of firing a spike.

It can be shown that the firing rate is $r(t)=E[\{t_i\}|s(t)]$, the first moment (expectation E) of the distribution $P[\{t_i\}|s(t)]$. In an analogous manner, we can formally define the variance and covariance of this distribution. It can be shown that higher order statistics like the auto-correlation and the interspike interval histogram are related to the second moment of this distribution.

Taking the Organism's Point of View

As defined above, the firing rate $r(t)$ is derived from ensembles of spike trains created by repeated applications of the same stimulus. These trains are all drawn from the stimulus-conditional distribution $P[\{t_i\}|s(t)]$. As a measure defined over this ensemble, the firing rate $r(t)$ cannot be determined by observing a single occurrence of a spike train, only by observing an ensemble of responses. An ensemble of responses can be collected in two possible ways. The first possibility is by observing the system for a long time, over repeated presentations of the same stimulus, as in Fig. 2. This possibility is clearly meaningless from a behavioral point of view. The second possibility is by observing many identical neurons responding simultaneously to the same stimulus. This possibility is also behaviorally meaningless because there are rarely ever two identical neurons, and most importantly, in many organisms, especially insects, there are not enough neurons responding to the same stimulus in order to form this population average.

Let us take the organism's point of view, or the view of its central nervous system. The question is this: Can the organism base its behavioral decisions on the firing rate, or any other statistic drawn from $P[\{t_i\}|s(t)]$? The answer is no. The organism does not "know" $P[\{t_i\}|s(t)]$, simply because it does not know what the stimulus $s(t)$ was. The organism does not look at the stimulus and extract information about the spike train. The organism faces the opposite problem: looking at the spike train, it is trying to extract information about the stimulus (see also Inbar and Milgram, 1975). In other words, the distribution $P[\{t_i\}|s(t)]$ defines the encoding process, whereas the organism's nervous system is concerned with the decoding process.

Response-Conditional Ensemble

The organism's point of view corresponds to the *response-conditional distribution* $P[s(t)|\{t_i\}]$. Given a spike train $\{t_i\}$, what is the probability of an input $s(t)$? What we would like to do is to *characterize the neural code from this point of view*. Again, it is difficult to estimate this distribution, because of the amount of data needed. What we can do is to try and estimate the first few moments. The first step is to estimate the mean of the distribution. This mean has a special meaning: Given a specific spike train $\{t_1,t_2,\ldots,t_N\}$, what is the average stimulus that triggers it?

We can perform this procedure for many possible spike trains, starting from the simple ones. For example, we can find the average stimulus trajectory that triggers (and, hence, precedes) one spike, or the average stimulus trajectory that triggers two spikes at a certain interspike interval (and, hence, precedes the second spike), and so on. Typical examples of such a characterization are depicted in Fig. 3. Note that such average stimulus trajectories deviate from zero only over a finite time span because, like in every physical system with finite memory (relaxation time constant), the effect of an input on a system output subsides with time.

This characterization can be interpreted as a "dictionary". With this dictionary we can, in principle, look at one spike train, and come up with a reasonable estimate of the stimulus that caused it. This dictionary enables us to "read" the neural code.

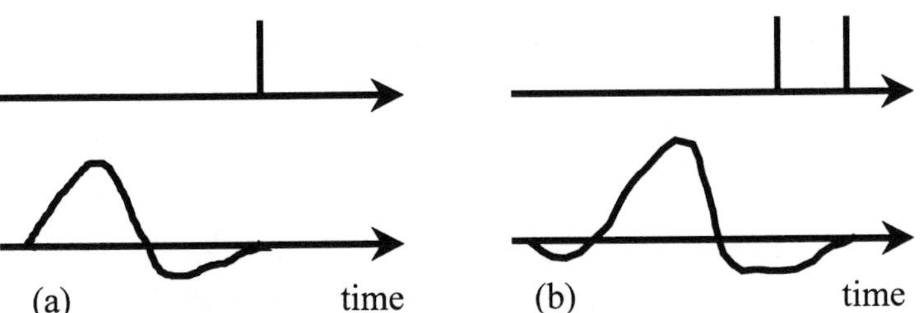

Fig. 3. Sketch of (a) the average stimulus (lower trace) that triggers one spike (upper trace); (b) the average stimulus that triggers two spikes with a certain gap between them.

Reading the Code

The fact that we can build a dictionary based on the response-conditional distribution leads to a simple method to *hypothetically reconstruct the input stimulus.* (It is important to realize that estimating the input is only a mathematical tool to be used below. It does not mean that we assume the neuron "wants" to, or is "designed" to do so.)

Consider first a spike train in which the spikes are temporally very far apart. Then we could use the following procedure to reconstruct the stimuli.

- Estimate the average stimulus that triggers a spike (like the one in Fig. 3a). This waveform is called the *estimation kernel.* As noted above, it has a limited duration.
- Place this kernel after every spike, and add them all up. This procedure is illustrated in the lower two traces of Fig. 4, where the estimation kernel (lower trace) is appended to each spike. Since the spikes are farther apart than the duration of the estimation kernel (indicated by vertical dashed lines), the kernels do not overlap, in this case. The upper trace in Fig. 4 shows the original stimulus generating the spike train, which differs from the reconstructed stimulus in two aspects: It has a slightly different time course (see below) and precedes the reconstructed stimulus by the *estimation delay* because the estimation kernels follow the spikes rather than precede them. The estimation kernels follow the spikes because of causality, only after a spike arrives we can reconstruct the stimulus that triggered it.

The described reconstruction method corresponds to a mathematical operation called *convolution* (see Chapter 21), where the spike train is convolved with the kernel, and can be written as follows:

$$s_{est}(t) = \sum_{i=1}^{N} K_1(t - t_i)$$

where $K_1(t)$ is the estimation kernel, and $\{t_i\}_{i=1}^{N}$ are the spike occurrence times. In practice, of course, spikes are not always far away from each other, so the kernels $K_1(t)$ superimpose when appended to the spikes.

Formally, the best would be if one arrived at a kernel that minimized the *Mean Square Error* (*MSE*), which is:

$$MSE = \frac{1}{T}\int_0^T \left| s(t - \tau_{delay}) - s_{est}(t) \right|^2 dt$$

As seen in Fig. 4, causal estimation of the stimuli from the spike train causes a delay – τ_{delay}. This delay is determined by the width of the kernel $K_1(t)$. It is a standard optimi-

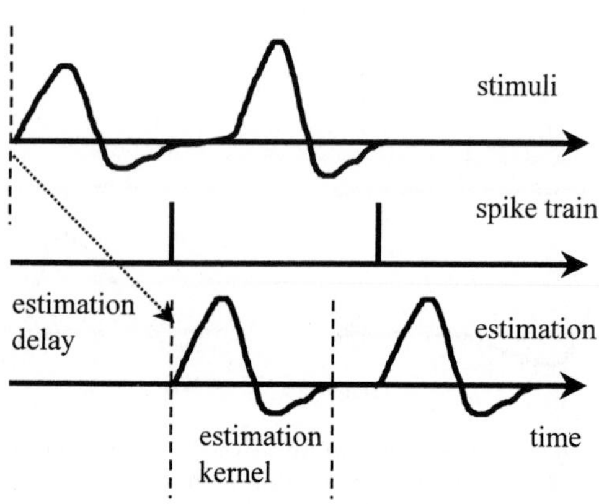

Fig. 4. A sketch of the estimation procedure. Estimating the stimuli from the spike train by placing a kernel in front of every spike and adding them all up. In this example, the estimation kernel is the average stimulus that triggers a spike.

zation problem to find the shape and width of the kernel that minimize the *MSE* criterion. If we start with narrow kernels and progress to wider ones, we find that at some stage the estimation quality does not improve (i.e., the error does not decrease). In experiments conducted on the fly visual system, the estimation quality was found to saturate after a delay of about 50 ms (Rieke et al, 1997). In practice, one should find the appropriate delay in each case.

The linear decoding scheme described above, which uses only the first-order estimation kernel, has achieved good experimental results. Bialek et. al. (1991) showed that the use of higher order decoding techniques, involving averages such as in Fig. 3b, does not improve the decoding quality by much. This is intriguing, because we know that neurons exhibit very nonlinear input-output behaviors. How is it possible, then, to describe the neural code with a linear decoding scheme? Part of the answer lies in the fact that decoding is a very different problem than encoding. Apparently, even with non-linear encoding, a linear decoding scheme may work (for a mathematical analysis of this phenomenon, see Bialek et. al. 1993).

Basics of Information Theory

In order to present the basics of information theory, we have to review and define a few basic terms from probability theory. For a detailed presentation of probability theory, see Papoulis (1991). For a lucid, in-depth presentation of information theory, see the textbook by Cover and Thomas (1990).

Self-Information

The basic entity in probability is a *Probability Mass Function (PMF)*. This function defines discrete events and the probability of those events happening. For example, the fair die. It has six faces, therefore six possible events, all equally likely. The *PMF* of the fair die looks like this:

$$P_{\text{fair die}}[X=x]=\begin{cases} 1/6 & x=1 \\ 1/6 & x=2 \\ 1/6 & x=3 \\ 1/6 & x=4 \\ 1/6 & x=5 \\ 1/6 & x=6 \end{cases}$$

Self-information is the information that a single event gives us. Mathematically it is defined as follows:

$$I_s=\log_2\left(\frac{1}{P[X=x]}\right)\ \text{[bit]}$$

The self-information of an event is the dual logarithm of the inverse of the probability of that event. When the logarithm is base 2, this quantity is measured in bits. The single-event self-information of the fair die is ~2.58 bit. Note that the probability of the event is in the denominator of the logarithm's argument. This means that low-probability events have high self-information. Take for example a PMF that describes the probability of hearing the firing alarm on any single day:

$$P_{\text{fire alarm}}[X=x]=\begin{cases} 0.001 & x=\text{fire} \\ 0.999 & x=\text{no fire} \end{cases}$$

Using the expression above, we find the self-information these two possible events carry: $I_s(\text{fire}) \approx 10$ [bits] and $I_s(\text{no fire}) \approx 0.001$ [bits]. The (hopefully) rare event of "fire" gives us a lot of information, and the frequent event of "no fire" gives us very little information. To get an intuitive feel about this, reflect on your situation right now – you probably don't pay much attention to the fact that the fire alarm *does not* ring, but if it would start ringing now, it would give you a lot of information – you know you have to do something – call the fire department, or run for your life... Therefore, at least in a statistical sense, a rare event carries more information.

Entropy

Self-information quantifies the information carried by a single event based on its probability of occurrence. If we want to know how much information is carried by the *PMF* itself, that is, to weigh together the self-information carried by all the events defined by the *PMF*, we use the *mean self-information*:

$$H(X) = E\left[\log_2\left(\frac{1}{p(x)}\right)\right] = \sum_x p(x)\log_2\left(\frac{1}{p(x)}\right)$$

This is the mathematical definition of entropy. Entropy can be interpreted in several ways. One interpretation, in line with the definition of self-information, relates to the available information in a *PMF*. The other interpretation relates to the degree of uncertainty or randomness in a *PMF*. For example, consider the entropies of the fair coin and fire alarm *PMF*s:

$$H_{\text{fair coin}} = \tfrac{1}{2}\cdot 1 + \tfrac{1}{2}\cdot 1 = 1 \text{ [bit]}$$

$$H_{\text{fire alarm}} \cong 0.001\cdot 10 + 0.999\cdot 0.001 \cong 0.011 \text{ [bit]}$$

The outcomes of the flips of a fair coin are much more random than the outcomes of the fire alarm, and this is reflected by a higher entropy. The higher the entropy, the more random, uncertain and unpredictable is a *PMF*.

Suppose we can set the probability of each face of a die. How would you distribute the probabilities among the faces of the die so that it has maximal entropy among all possible dice? The die with maximum entropy is the fair die with all faces equally likely. This idea of *maximum entropy* is very important and we will use it later. In the discrete case, when random variables take discrete values, the uniform distribution (e.g., fair die, fair coin), in which all states are equally likely, is the distribution that maximizes the entropy, among all distributions with the same number of states.

When all possible states are equally likely, the entropy is just the logarithm of the number of possible states, for example:

$$H_{\text{fair dice}} = \sum_{i=1}^{6} \tfrac{1}{6}\log_2\left(\frac{1}{1/6}\right) = \log_2(6) \cong 2.58 \text{ [bit]}$$

Therefore, 2.58 [bit] is the maximum amount of entropy that can be achieved by any *PMF* that has six states.

Entropy of the Gaussian Distribution

A continuous *Random Variable (RV)*, x, can take any real value $x \in (-\infty, +\infty)$. The function that describes continuous *RV*s is the *Probability Density Function (PDF)*, $P(x)$. With this function we can determine the probability of finding the *RV* within an interval (a,b):

$$\text{Prob}\left(a<x<b\right)=\int_a^b P(x)dx$$

Take for example the familiar Gaussian *PDF*, with mean *M* and variance σ^2:

$$P(x)=\frac{1}{\sqrt{2\pi\sigma^2}}\exp\left[-\frac{(x-M)^2}{2\sigma^2}\right]$$

In the continuous case, entropy is defined in the following way:

$$H(X)=\int p(x)\log_2\left(\frac{1}{p(x)}\right)dx$$

and is called *Differential Entropy*.[1] Note that in the transition from discrete *RVs* to continuous *RVs* the sum is replaced by an integral. Using this expression, we find the entropy of the Gaussian *PDF*:

$$H_g=\frac{1}{2}\log_2\left(2\pi e\sigma^2\right)$$

The entropy of the Gaussian distribution depends only on the variance σ^2. The variance is, of course, a measure of variability. As in the discrete case, we see that again entropy is a measure of variability and uncertainty. Note also that the entropy of the Gaussian *PDF* is maximal among all probability density functions that have the same variance.

Entropy of Spike Trains

Let us put the knowledge about entropy to use and estimate the entropy of spike trains. Suppose we sample a spiking neuron and determine the spike arrival time with resolution $\Delta\tau$. We measure a mean firing rate of $\bar{r}$ spikes per second. The time resolution must be fine enough so that there would be no more then one spike in a $\Delta\tau$ bin. We model the spike train as a binary string, where a 1' denotes a spike, and a 0' denotes no-spike, as depicted in Fig. 5. If we look at an interval T with resolution $\Delta\tau$, we get $N=T/\Delta\tau$ bins. When the resolution is fine enough, the probability of seeing a 1' in any bin becomes relative to the rate and the resolution: $p=\bar{r}\cdot\Delta\tau$. Since we know the average rate, we know that a typical T second interval will contain $N_1=\bar{r}\cdot T$ ones. The number of zeroes is of course $N_0=N-N_1$.
From combinatorics we find that there are $N!/(N_1!N_0!)$ possible binary strings with N_1 ones and N_0 zeroes. If we make the assumption that all these possible strings are equally

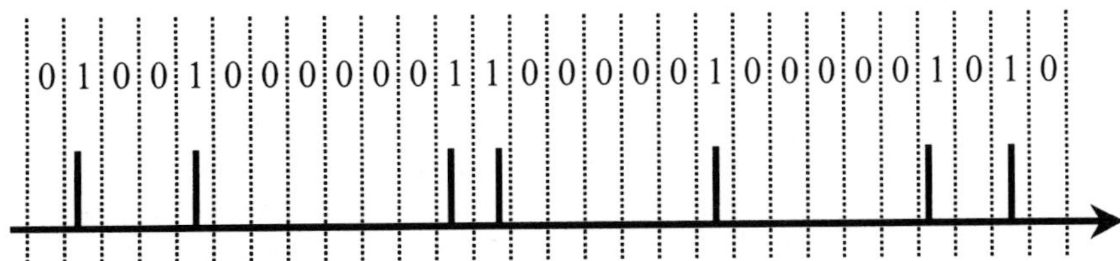

Fig. 5. Modeling the spike train as a binary string.

1 Mathematically, differential entropy is a little different from the entropy of a discrete *RV*, but we will not elaborate on this here. For details see Cover and Thomas (1990).

likely, then the entropy is $\log_2$ (number of states):

$$H = \log_2\left(\frac{N!}{N_1!N_0!}\right) \text{ [bit]}$$

This is the entropy of a spike train of T second duration. As the interval T increases, the entropy increases with it. Dividing the entropy by T, we have the *entropy rate*, the amount of entropy per unit time. Using the relations mentioned above and dividing by T, we can express the entropy rate with $\bar{r}$ and $\Delta\tau$ only:

$$H/T \underset{\bar{r}\Delta\tau<<1}{\approx} \bar{r}\log_2\left(\frac{e}{\bar{r}\Delta\tau}\right) \text{ [bit / sec]}$$

where e is the base of the natural logarithm. Take for example a typical result where $\bar{r} \approx 40\ \text{s}^{-1}$ and $\Delta\tau \approx 1$ ms; using the above formula we get an entropy rate of 243.46 bits per second, or 6.08 bits per spike. Note that because we assumed that all possible spike trains with rate $\bar{r}$ are equally likely, we actually got the maximum entropy of a spike train with rate $\bar{r}$. This expression is a practical tool that will allow us later to assess experimental results.

Mutual Information

Now let us return to the original question: How much "information" does signal Y carry about signal X? (See Fig. 1). Before we answer that question, we have to define the *conditional entropy*. In the discrete case, the mathematical definition is:

$$H(X|Y) = E_{xy}\left[\log_2\left(\tfrac{1}{p(x|y)}\right)\right] = \sum_y p(y) \sum_x p(x|y)\ \log_2\left(\tfrac{1}{p(x|y)}\right)$$

and in the continuous case it becomes:

$$H(X|Y) = \int p(y) \int p(x|y)\ \log_2\left(\tfrac{1}{p(x|y)}\right) dx\, dy$$

The conditional entropy is the uncertainty about the input X after observing the output Y. If we have a perfect channel, without noise, then Y tells us exactly what X was, and there is no uncertainty, resulting in: $H(X|Y)=0$. If, on the other hand, the channel is disconnected or very noisy, then observing Y is useless, and gives us no information about the input. In that case the uncertainty about X is maximal, yielding $H(X|Y)=H(X)$.

Mutual information is the information we gain about X after observing Y, defined as:

$$I(X;Y) = H(X) - H(X|Y)$$

Information gain can be interpreted as a reduction in uncertainty. Before we observe the output, all we have is $H(X)$, the uncertainty about the input. After we observe the output, we are left with $H(X|Y)$, the uncertainty about the input X after observing the output Y. The difference in uncertainty is the information we gained by looking at the output. Therefore, in a perfect channel, after looking at the output, the reduction in uncertainty is maximal and equals the input entropy $H(X)$. In a disconnected channel, looking at the output gives no information, and there is no reduction in uncertainty, yielding $I(X;Y)=0$. Note that mutual information is symmetric with respect to its arguments: , and is always non-negative: $I(X;Y)\geq 0$.

Mutual Information of the Gaussian Channel

Let us calculate the mutual information between input and output in a simple channel. In this channel, depicted in Fig. 6, the output equals the input multiplied by some gain, and there is noise added to the channel output. Hence, the channel output is $y=gs+n$, where s is the input, g is the gain, and n is the noise. In this channel, the input and the-noise are Gaussian *RVs*, with zero mean and variance $\langle s^2\rangle$, $\langle n^2\rangle$, respectively: $s \sim N(0,\langle s^2\rangle)$, $n \sim N(0,\langle n^2\rangle)$. Because the input and output can take any real value $s, y \in (-\infty,+\infty)$, we say that they are drawn from an infinite alphabet.

It is sometimes convenient to transform this channel model by referring the noise to the input, that is, model the channel as if the noise is added to the input and not to the output, as seen in Fig. 7. After referring noise to the input, the channel output will equal $y= g\ (s+n_{eff})$. The noise, when it is referred to the input, is called *effective noise*, and equals: $n_{eff}=n/g$. Since the noise n is Gaussian, the effective noise is also Gaussian with zero mean and variance $\langle n_{eff}^2\rangle = \langle n^2\rangle / g^2$; i.e., $n_{eff} \sim N(0,\langle n^2\rangle / g^2)$.

It can be shown that the mutual information between input and output is:

$$I(x;y)=\frac{1}{2}\log_2\left(1+\frac{\langle s^2\rangle}{\langle n_{eff}^2\rangle}\right)=\frac{1}{2}\log_2(1+SNR)$$

$$SNR=\frac{\langle s^2\rangle}{\langle n_{eff}^2\rangle}$$

where *SNR* is the *signal-to-noise* ratio.

The mutual information thus depends on the *SNR*, i.e., the ratio between the signal variance and effective noise variance. When the signal variance is large relative to the effective noise variance, the output contains little noise, and the mutual information is large. When the *SNR* is very small, the output consists almost entirely of noise, and the mutual information is small, approaching zero as the noise grows.

Random Continuous Time Signals

In the previous sections, we have discussed discrete events, not continuous time signals. In order to extend the information-theoretical discussion to continuous time signals, we briefly present a few fundamentals of *random signal theory*. For a detailed presentation see Candy (1988) or Porat (1994).

It is well known that a real continuous time signal $f(t)$, within a time window of duration T, can be represented as a *Fourier series* (see Chapter 18; also Kwakernaak and

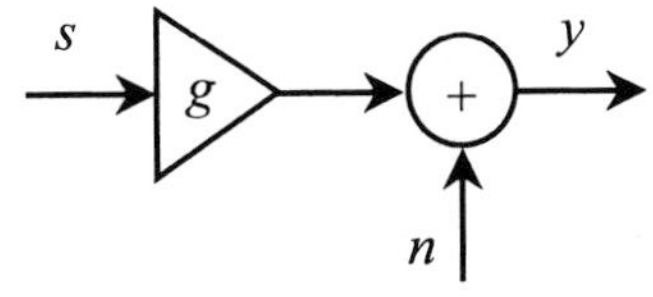

Fig. 6. The additive Gaussian channel. The channel is represented by the amplifier (open triangle) with input signal s, and noise n added to the output.

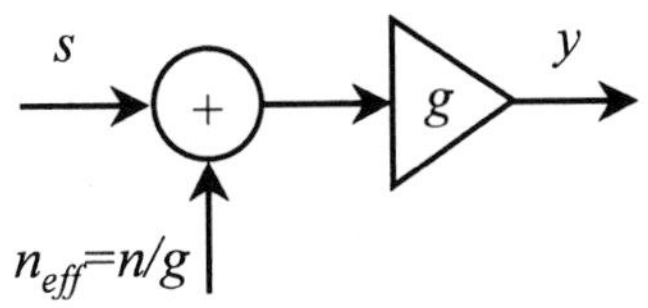

Fig. 7. The equivalent Gaussian channel after referring noise to the input.

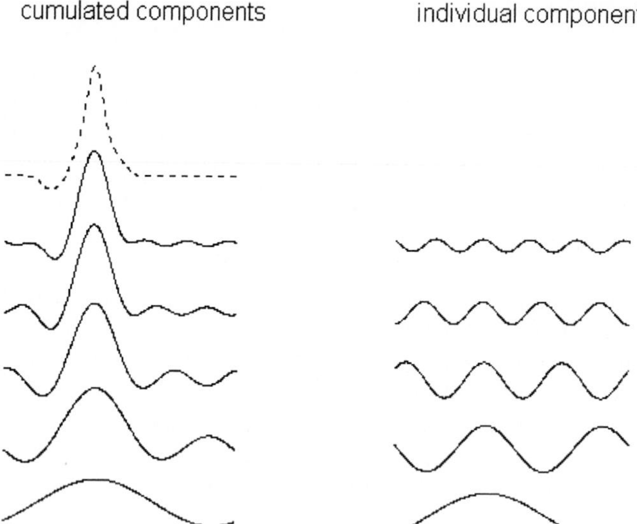

Fig. 8. The composition of an action potential-like waveform from harmonic components. The dashed left top waveform is the final waveform. Below it are successive approximations of this waveform. On the right are the harmonic components (sine and cosine together) that add into each approximation. Note that the amplitude of the components diminishes as frequency grows.

Sivan, 1991). In this representation, we compose the signal from harmonics (sine and cosine waves) at frequencies that are integer multiples of the fundamental frequency $2\pi/T$. Each harmonic is multiplied by a coefficient, and they are all summed together to form the signal:

$$f(t) = f_0 + \sum_{k=1}^{\infty} a_k \cos(w_k t) + \sum_{k=1}^{\infty} b_k \sin(w_k t)$$

where $w_k = 2\pi \cdot k/T$ is the frequency of the kth harmonic. Figure 8 shows the cumulative composition of an action potential-like waveform from its harmonics.

The set of Fourier coefficients $\{a_k, b_k\}$ completely defines the waveform. When we choose the Fourier coefficients randomly, we get a random time signal. In principle, we can choose the coefficients any way we want. However, we are looking for the time signal equivalent of the Gaussian random variable. If we choose the Fourier coefficients $\{a_k, b_k\}$ independently from a Gaussian distribution $a_k, b_k \sim N[0, \sigma_k^2]$, we get a Gaussian random signal, the equivalent of the Gaussian random variable.

The Power Spectrum

The variances of the Fourier coefficients $\{a_k, b_k\}$ at every frequency w_k completely define an ensemble of random Gaussian signals. Note that the variance of the sine and cosine coefficients is equal. The variance of the coefficient can be thought of as the "power" at the respective frequency. We can think of a function that gives the coefficient's variance at every frequency: $\sigma^2(w_k) = \langle a_k^2 \rangle = \langle b_k^2 \rangle$. This is a *spectral representation* of the ensemble of random signals. Since the spectral resolution $\Delta f = 1/T$ depends on the interval size T, it is convenient to look at the power within a spectral band . This is the *power spectral density*, or *power spectrum*:

$$S(w_k) = \frac{1}{\Delta f} \sigma^2(w_k) = T\sigma^2(w_k)$$

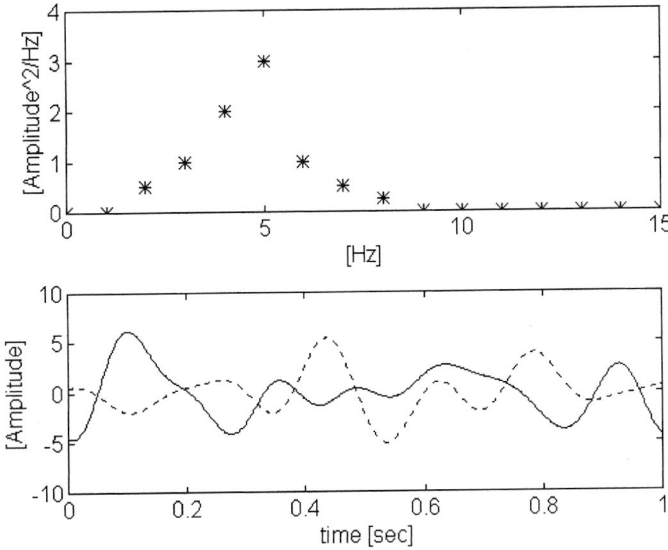

Fig. 9. The power spectrum (top). The time interval here is T=1 second; therefore the frequency resolution is 1 Hz. In the top plot, we see the variance of the Fourier coefficients, divided by the frequency resolution. There is power at only 7 frequencies, with a dominant frequency of 5 Hz. This power spectrum completely defines an ensemble of random Gaussian signals. In the bottom plot, we see two sample functions drawn from that ensemble.

As the interval size T grows, this function tends towards a continuous function of frequency. In Fig. 9, we see an example of the power spectrum (upper plot), along with two sample functions drawn from the ensemble of random Gaussian signals it defines (lower plot). When the power at every frequency is equal, the result is a *flat spectrum*, also called *white Gaussian noise*. Otherwise, a non-flat spectrum, as exemplified in Fig. 9, is called *colored Gaussian noise*.

Information Transmission with Continuous Time Signals

Now we can analyze the mutual information in a channel with continuous time signals. As usual, we start with a simple case, the case of the Gaussian channel; but this time, the input, output and noise are time signals.

Parallel Gaussian Channels

In analogy to the Gaussian channel presented in Fig. 6, in Fig. 10 we have a channel in which the input and the noise are Gaussian random signals, with power spectrum $S(w)$ and $N(w)$, respectively. The function $G(w)$ is a linear transfer function, which defines the gain in every frequency.

Because the Fourier coefficients of Gaussian random signals for the different frequencies are independent, it is easier to analyze this channel in the frequency domain. In the frequency domain, each Fourier coefficient is represented by a channel of its own. In Fig. 11, we see that this results in an array of parallel independent channels. Each one of these channels is similar to the Gaussian channel we already analyzed (see Figs. 6 and 7).

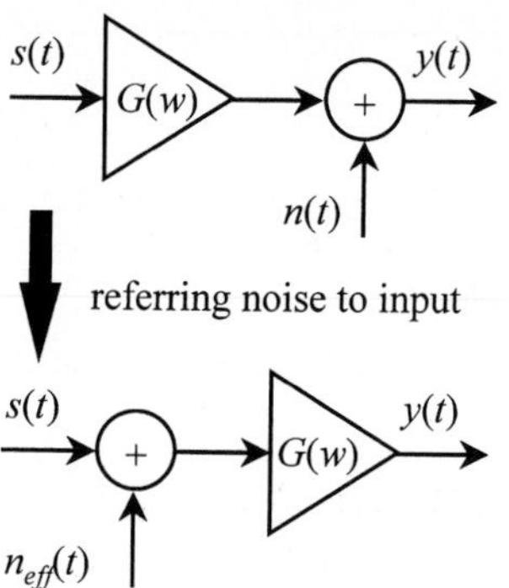

Fig. 10. The Gaussian channel, time signals case. Two equivalent forms, before and after referring noise to the input.

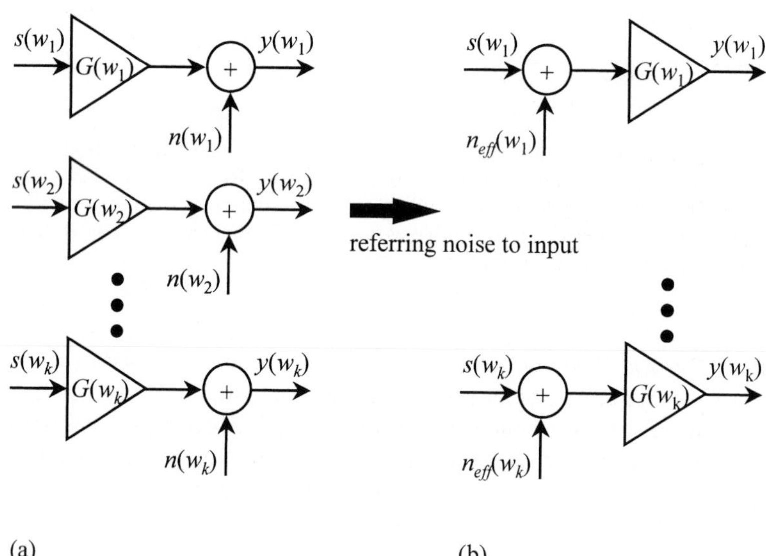

Fig. 11. A parallel array of independent Gaussian channels. Before (a) and after (b) referring noise to the input. $s(w_k)$, $n(w_k)$, $n_{eff}(w_k)$ and $y(w_k)$ are the Fourier coefficients of the time signals $s(t)$, $n(t)$, $n_{eff}(t)$ and $y(t)$, respectively. This is a slightly different notation then before, where we used $\{a_k, b_k\}$ to write the Fourier coefficients of the signal. The new notation originates from the complex Fourier series expansion (see Chapter 18, also Kwakernaak and Sivan, 1991), but for our purpose, is conceptually the same. The main idea is that each coefficient is represented by a channel of its own, and since they are all independent, these channels can be analyzed independently.

Suppose we are looking at an interval of T seconds on such a parallel channel. Each Fourier coefficient is associated with a separate channel, similar to the one we have already analyzed. Therefore, we can calculate the mutual information carried by each coefficient:

$$I_k = \tfrac{1}{2}\log_2\left(1 + SNR(w_k)\right) \quad \text{[bit]}.$$

Since all the Fourier coefficients are independent, we can sum up the individual contributions to find the total mutual information:

$$I = \tfrac{1}{2}\sum_k \log_2\left(1 + SNR(w_k)\right) \quad \text{[bit]}.$$

Information Rate

In the previous equation we found the mutual information in an interval of T seconds. If the channel operates continuously and T grows, mutual information grows with it. If we divide mutual information by T, we get the *information rate*. When T grows to infinity, the sum becomes an integral and we have:

$$R_{\text{info}} = \left\{ \frac{I}{T} \right\}_{T \to \infty} = \frac{1}{2} \int_{-\infty}^{\infty} \log_2 (1 + SNR(w)) \frac{dw}{2\pi}$$

where $SNR(w) = S(w)/N_{eff}(w)$. This result, called *Shannon's Formula*, is one of the most important in information theory.

Let's see what this expression can tell us. In Fig. 12, the operation of a possible Gaussian channel is sketched, with the noise referred to the input. In this case, the effective noise, with power spectrum $N_{eff}(w)$, is a property of the channel. Generally, the power spectrum of the effective noise will be low in some frequency regions, and high in other regions, as exemplified by the solid curve in the lower plot of Fig. 12. The regions where the effective noise is low are regions that can be used to transfer information with high fidelity. Therefore, $N_{eff}(w)$ is sometimes called a *tuning curve*, because it reveals the frequencies the channel is "tuned to". The input stimuli, with power spectrum $S(w)$, can be controlled in the experiment. In Fig. 12 (lower plot, dashed line), it is assumed to be flat over a considerable frequency range and then to fall off at high frequencies. The $SNR(w)$ (upper plot in Fig. 12) shows at which frequencies information is transferred with high fidelity and at which frequencies it is transferred with low fidelity. Note that the SNR is a combined feature of the channel and the input stimuli, changing for different stimuli. The effective noise, on the other hand, is a property of the channel alone (in the case of the Gaussian channel discussed here). It is clear that in studying sensory information transmission it is of great interest to find out $N_{eff}(w)$ and $SNR(w)$ for each sensory system within the physiological range of $S(w)$.

Noise Whitening

Suppose we have a Gaussian channel with a given effective noise curve as exemplified by the heavy solid line in the $N_{eff}(w)$ plot of Fig. 13 (lower plot). We now ask what is the input power spectrum $S(w)$ that maximizes the information rate R_{info}?

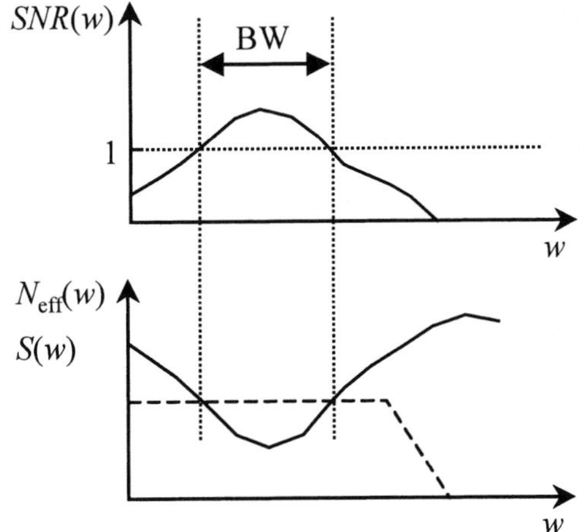

Fig. 12. In the lower plot we see the presumed input power spectrum (dashed line) and the effective noise power spectrum (solid line). In the top plot is shown the SNR. At frequencies where the SNR is high, information is transferred with good fidelity. The width of the region where $SNR > 1$ can be thought of as the bandwidth (BW) of the channel (indicated by vertical dotted lines).

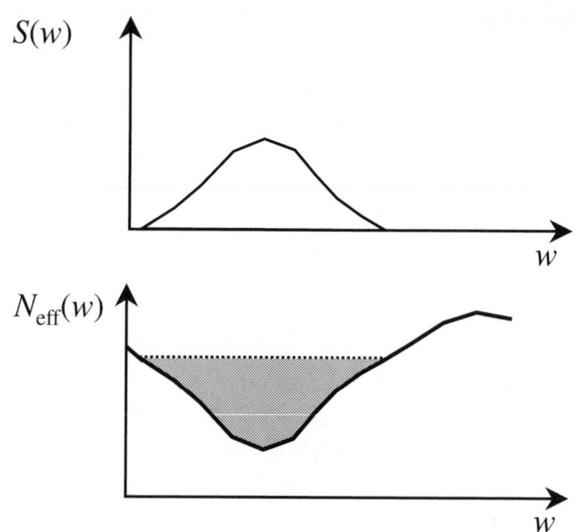

Fig. 13. Noise whitening. The heavy solid curve in the bottom plot is the effective noise power spectrum. The shaded region that "water-fills" this curve corresponds to the inverted input power spectrum that maximizes the information rate. The input power spectrum is shown in the top plot. The area of the shaded region is the total power of the input signal. The output power spectrum equals the noise power spectrum in those frequency regions where the input power spectrum is zero, and the horizontal dotted line (lower plot) in regions where the input power spectrum is non-zero.

If we had no power limitation, we could just use a very powerful input signal, and in general, achieve any information rate. This is not very practical, however, nor interesting. In general, the total power is limited, being equal to the area under the $S(w)$ graph. The objective then is to find that $S(w)$ curve which – with a defined area – maximizes R_{info}.

The input power spectrum that maximizes R_{info} is the one that complements $N_{eff}(w)$ in a *water filling analogy*: "pour" the input power spectrum into the trough of the $N_{eff}(w)$ curve, until the allotted area (power limitation) is used up. This is shown as the shaded area in Fig. 13 (lower plot), along with the resulting inverted input power spectrum on an independent scale (upper plot).

When, with the above optimization, we look at the output of the channel, we see a power spectrum that is flat, or "white", in the frequency regions where the input is non-zero. This is why this technique is also called "noise whitening". This means that the output looks very noisy. If we were just looking at the output, we could not determine if it is really a noisy channel, or a channel with an optimal encoding scheme. In order to tell the difference, one has to look at the input and the output, and measure the mutual information. This has a correlate in neurons. Neurons have sometimes been referred to as noisy, just because their output spike train looks noisy and irregular. From the noise whitening concept we understand that this is an invalid conclusion. What we should do is look at both the input and output of a neuron, and try to measure the mutual information. We describe how to do this in the following section.

Information Transmission – The Method

It is difficult to measure entropy or mutual information directly. So we take a standard engineering technique, i.e., to bound the desired parameter between an upper and lower bound. We do that because it is sometimes much easier to calculate these bounds than to estimate the parameter itself. The desired parameter is guaranteed to be between these bounds, and if the lower and upper bounds are close enough, this is almost as good as estimating the parameter directly. In this section we are going to describe a method that provides a *lower bound* on information rate. This method is due to Bialek et. al. (1991), and appears in a book by the same group of researchers (in Chapter 3 of Rieke et. al., 1997). The *upper bound* will be presented in the next section.

Outline of the Method

We assume a laboratory setup, in which we provide the input stimuli to the neural system under study, and measure the resulting spike train. The input stimulus must be a Gaussian random signal for this method to work properly. The method, graphically presented in Fig. 14, consists of four main stages:

- Reconstruct the input stimuli from the spike train, using the simple estimation method described earlier, aiming at minimizing the mean square error between the known input and the reconstructed one;
- Use the estimation error to estimate the effective noise in the channel;
- Estimate the power spectrum of the input stimuli and the effective noise, and use them to calculate the *SNR*;
- Use the *SNR* to calculate a lower bound to the information rate.

In the following, we explain how this method works, and why it provides a lower bound to the information rate.

Mutual Information

Suppose we observe an interval of T seconds of the experimental input-output data. What we actually want to find is the mutual information between the spike train and the input stimuli, or, in other words, the information the spike train provides about the input. According to the definition, mutual information is:

$$I[\{t_i\} \rightarrow s(t)] = H(s(t)) - H(s(t)|\{t_i\})$$

From the experimental setup, we know the entropy of the input signal, $H[s(t)]$, because we determine its statistics. If we overestimate the conditional entropy $H[s(t)|\{t_i\}]$, we will get a lower bound on the mutual information:

$$I[\{t_i\} \rightarrow s(t)] \geq H(s(t)) - H_{OverEst}(s(t)|\{t_i\})$$

Overestimating the Conditional Entropy

The question then is how to overestimate the conditional entropy $H[s(t)|\{t_i\}]$. This mathematical stage is beyond the scope of this chapter. We recommend looking up the

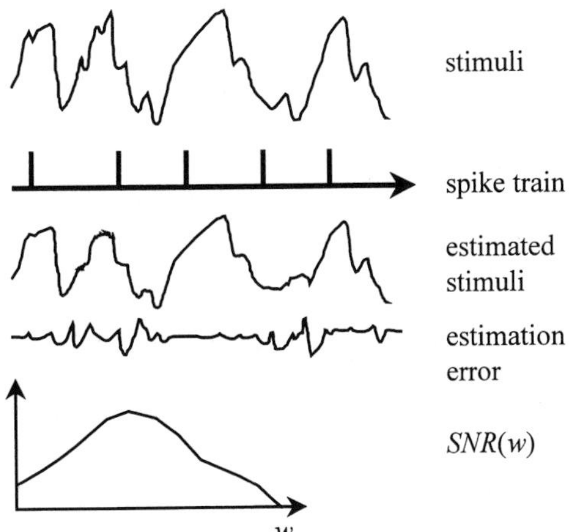

Fig. 14. Outline of the method. Stimulate the system with a Gaussian random signal and record the spike train (upper two traces). Estimate (reconstruct) the stimuli from the spike train (third trace from top). Find the estimation error (fourth trace). Use the estimation error to estimate the effective noise in the system. Find the *SNR* (lower trace). Use the *SNR* to calculate a lower bound to R_{info}.

details from the original text (in Chapter 3 of Rieke et. al., 1997). Let us just mention two of the main ideas that allow us to solve this problem.

First, we use the fact that the entropy of the Gaussian distribution is maximal among all distributions with the same variance. Therefore, if we assume $P[s(t)|\{t_i\}]$ is Gaussian, we overestimate the conditional entropy. Second, we know that Gaussian random signals are completely characterized by the power spectrum, the analogue of the variance in the case of the Gaussian random variable. Thus, if we overestimate the power spectrum, we will again overestimate the conditional entropy. Simplifying things greatly, we can say that by overestimating the power spectrum of the noise in the system, we also overestimate the conditional entropy, and as a result, underestimate the mutual information.

The practical way to do this is to construct an estimator that takes the spike train $\{t_i\}$ as input, and returns an estimate of the input $s(t)$. We then use the estimator output $s_{est}(t)$ to find the estimation error: $n_{est}(t)=[s(t)-s_{est}(t)]$. The estimator should minimize the mean square error criterion:

$$MSE = \frac{1}{T}\int_0^T |n_{est}(t)|^2\, dt$$

In principle, any type of estimator can be used. In practice, the simple linear estimator we presented before is enough in many cases. Rieke et. al. (1997) show how to use higher order estimators, but we ignore this issue here.

As we said before, it can be shown that the power spectrum of the estimation errors, $N_{est}(w)$, is bigger than the power spectrum of the noise in the system (when we assume that the system is Gaussian).

A Lower Bound on Information Rate

Until now, we have handled the noise as seen at the output of the system. In analogy to what we did in our discussion on parallel Gaussian channels, we would like to refer the noise $N_{est}(w)$ to the input and find the effective noise spectra $N_{eff}(w)$. Using frequency-domain methods, we refer the noise to the input, and separate random and systematic errors.

The experiment is divided into M segments of T_0 seconds length. We transform the signal and the estimate in each segment to the frequency domain using the Fourier transform: $s^i(t) \to \tilde{s}^i(w)$, $s^i_{est}(t) \to \tilde{s}^i_{est}(w)$, where $1 \le i \le M$ is the segment number. For each signal and for each frequency w_k, we can define a row vector of M complex Fourier coefficients:

$$\tilde{\mathbf{s}}(w_k) = [\tilde{s}^1(w_k), \tilde{s}^2(w_k), \ldots, \tilde{s}^M(w_k)]$$

$$\tilde{\mathbf{s}}_{est}(w_k) = [\tilde{s}^1_{est}(w_k), \tilde{s}^2_{est}(w_k), \ldots, \tilde{s}^M_{est}(w_k)]$$

If we "plot" the Fourier components of the estimate against the corresponding components of the input, we get a scatter plot with M points, which should resemble a straight line with scatter:

$$\tilde{\mathbf{s}}_{est}(w_k) = g(w_k)[\tilde{\mathbf{s}}(w_k) + \tilde{\mathbf{n}}_{eff}(w_k)]$$

The slope of the best fit, $g(w_k)$, corrects for systematic errors. It can be found by:

$$g(w_k) = \frac{\tilde{\mathbf{s}}_{est}(w_k) \cdot \tilde{\mathbf{s}}(w_k)^*}{\tilde{\mathbf{s}}(w_k) \cdot \tilde{\mathbf{s}}(w_k)^*}$$

where s^* is the complex conjugate transpose of vector s. Finally, the scatter along the x-axis, $\tilde{\mathbf{n}}_{eff}(w)$, is the effective noise:

$$\tilde{\mathbf{n}}_{eff}(w_k) = \tilde{\mathbf{s}}_{est}(w_k)/g(w_k) - \tilde{\mathbf{s}}(w_k)$$

These are the Fourier coefficients of the effective noise. Their variance allows us to estimate the effective noise power spectrum:

$$N_{eff}(w_k) = \left\langle \left| \tilde{n}_{eff}(w_k) \right|^2 \right\rangle T_0$$

where the average $\langle \cdot \rangle$ is taken over the collection of M segments. Since we know the input power spectrum $S(w)$, we can find the SNR:

$$SNR(w_k) = \frac{\left\langle \left| \tilde{s}(w_k) \right|^2 \right\rangle T_0}{\left\langle \left| \tilde{n}_{eff}(w_k) \right|^2 \right\rangle T_0} = \frac{S(w_k)}{N_{eff}(w_k)}$$

Thus, because the input is Gaussian, and because we modeled the channel as Gaussian, we can place a lower bound on the information rate using the formula:

$$R_{\text{info}} \geq \tfrac{1}{2} \int_{-\infty}^{\infty} \log_2 \left(1 + SNR(w)\right) \frac{dw}{2\pi}$$

Summary – the Practical Procedure

- Stimulate the neural system with an input stimulus $s(t)$ that is a Gaussian random signal with a predetermined power spectrum.
- Measure the resulting spike train $\{t_i\}$
- Estimate the input stimulus from the spike train: $\{t_i\} \rightarrow s_{est}(t)$
- Using frequency-domain methods, refer the noise to the input and compute the effective noise in the estimate:

$$\tilde{n}_{eff}(w) = \tilde{s}_{est}(w) \big/ g(w) - \tilde{s}(w)$$

- Find the power spectrum of the input and effective noise:

$$S(w_k) = \left\langle \left| \tilde{s}(w_k) \right|^2 \right\rangle T_0, \; N_{eff}(w_k) = \left\langle \left| \tilde{n}_{eff}(w_k) \right|^2 \right\rangle T_0$$

- Compute the signal-to-noise ratio:

$$SNR(w) = \frac{S(w)}{N_{eff}(w)}$$

- Use the SNR to calculate a lower bound to the information rate:

$$R_{\text{info}} \geq R_{\text{info}}^{\text{LB}} = \tfrac{1}{2} \int_{-\infty}^{\infty} \log_2 \left(1 + SNR(w)\right) \frac{dw}{2\pi}$$

Results

Cricket Mechanoreceptor

In a series of experiments, Warland and his colleagues (Warland et. al. 1992) investigated the information-transmission properties of cricket sensory hairs. These sensory hairs are mechanoreceptors, and one type of them, filiform hairs, are mainly sensitive to air displacements. Each one of the sensory hairs grows out of a sensory neuron, which in turn sends an axon to a ganglion.

In these experiments the investigators grabbed one sensory hair, and delivered random displacement stimuli to it, while recording the resulting spike train from its axon. The input stimulus was Gaussian noise with flat spectrum in the range of 25–525 Hz. The results were a signal-to-noise ratio of $SNR \sim 1$ over a bandwidth spanning ~300 Hz (see Rieke et al 1997; Fig. 3.19). After integration, this resulted in a lower bound to information rate of 294 ± 6 bits/s, equivalent to 3.2 ± 0.07 bits per spike.

This result can be interpreted in a few ways. One interpretation is that an information rate of ~300 bit/s, observed over a 1-s interval, implies the output spike train can identify one out of $2^{300} \sim 10^{90}$ possible input signals every second. Another interpretation is that the spike train can tell just between two signals (a positive or negative deflection of the hair), but this information is updated every ~3 ms. The second interpretation is in line with the fact that information is transferred with low *SNR* over a wide bandwidth. From this analysis, it seems that these sensory hairs are specialized for transferring large amounts of information about rapidly changing stimuli.

Upper Bound to Information Rate and Coding Efficiency

In this section we present the upper bound on the information rate and the concept of *coding efficiency*. We survey results achieved by applying the above method to various neural systems, and present the impact of using input stimuli with a naturally shaped power spectrum. All the results cited below pertain to a single channel, that is, a single sensory neuron.

Upper Bound to Information Rate

The ~300 bits/s result from the cricket filiform hair is a lower bound to the information rate. The true information rate is higher than that. In order to know if the lower bound is tight, we must develop an upper bound. Spike-train entropy sets a natural limit to information transmission. This is true because the spike-train entropy over a time window of T seconds is always greater than the mutual information between input and output:

$$I[\{t_i\} \to s(t)] = H(\{t_i\}) - H(\{t_i\}|s(t)) \leq H(\{t_i\})$$

Note: $H(\{t_i\}|s(t)) \geq 0$

Estimating the spike-train entropy $H(\{t_i\})$ directly can be difficult, but we have shown earlier (see section "Entropy of Trains") that we can easily approximate the spike-train entropy rate, H/T, depending on the average firing rate and timing resolution only:

$$R_{\text{info}}^{\text{UB}} = H/T \approx \bar{r} \log_2\left(\frac{e}{\bar{r}\Delta\tau}\right)$$

In fact, the entropy rate according to this formula is maximal among all spike trains with the same average firing rate and timing resolution. Thus, $R_{\text{info}}^{\text{UB}}$ is an upper bound[2] to the information rate. When we divide the upper bound by the lower bound, we get an estimate of *coding efficiency*:

$$CE = \frac{R_{\text{info}}^{\text{LB}}}{R_{\text{info}}^{\text{UB}}} = \frac{\left(\frac{1}{2}\int \log_2(1+SNR(w)\frac{dw}{2\pi}\right)}{(H/T)}$$

This number represents the percentage of the output entropy used to transfer information about the input. The rest of the output entropy is due to noise in the system. Note that the true R_{info} is always bigger than the lower bound, and that the true $H(\{t_i\})/T$ is always lower than the upper bound. This means that we actually underestimate the true coding efficiency.

Results

Frog Sacculus

Rieke and his colleagues (Rieke et al. 1993) investigated the information-transmission properties of the frog sacculus. Frogs sense vibrations of the ground in order to detect potential predators. The frog sacculus is a sensory organ that utilizes hair cells similar to those in the human cochlea in order to sense these vibrations. It is possible to investigate this system by recording from the afferent nerve while shaking the whole frog.

In these experiments the displacement input stimulus was Gaussian noise with an approximately flat spectrum in the range of 30–1000 Hz. Using the same method as described before, the results revealed a signal-to-noise ratio of 3 to 4 between 40 and 70 Hz (see Rieke et al. 1997; figure 3.21). After integration, this resulted in a lower bound to information rate of 155 ± 3 bits/s, equivalent to ~3 bits per spike. Coding efficiencies were in the range of 0.5–0.6. Coding efficiency of 0.5 means that half the spike train entropy is used to transfer information about the input. This is an important result, especially in light of earlier views of the neuron as a very noisy device.

Natural Stimuli – Frog Calls

From the "water filling" analogy we learned that for a given channel, there is an input signal ensemble that maximizes information transmission in the channel, subject to a total power limitation. The "water filling" analogy tells us how to find this distribution in the case of the Gaussian channel. Which signal distribution maximizes information transmission of a sensory neuron?

Until now, the input stimulus in the reviewed experiments was Gaussian noise with flat power spectrum. Using flat spectra has its advantages, mainly because there is power at all frequencies. However, flat Gaussian noise certainly does not represent the stimuli occurring in nature. The power spectrum of naturally occurring signals is highly structured. The question arises as to whether sensory systems take advantage of this structure in producing a more efficient representation of the sensory world.

One step in answering this question is to use stimuli that have naturalistic power spectra, and to compare the resulting information rate and coding efficiency with that of flat Gaussian noise. Rieke, Bodnar and Bialek (1995) did just that. They investigated the bullfrog auditory system. The natural stimuli of this system consist mainly of frog calls. Spectral analysis of frog calls reveals that it is made up of 20 nearly harmonic bands, with a fundamental frequency near 100 Hz. So, Rieke et al. (1995) used a synthetic approximation of this power spectrum as stimuli: Gaussian noise with a power spectrum similar to the natural power spectrum. The other type of stimulus was Gaussian noise with a flat power spectrum (see Rieke et al. 1997; Fig. 3.25).

2 Note that one undesired feature of this upper bound is that it increases as the timing resolution increases (time t in $\Delta\tau$ decreases). This upper bound can be improved. For a discussion on the upper bound to neural information transmission, see Tock and Inbar (1999).

The results were dramatic. The *SNR* of the naturally shaped spectrum was higher than the *SNR* of the flat spectrum at almost every frequency (see Rieke et al 1997; Fig. 3.25). After integration, they found a lower bound on information transmission. When the stimulus was a flat spectrum, $R_{\text{info}}^{\text{LB}} \sim 46$ bits/s, increasing to $R_{\text{info}}^{\text{LB}} \sim 133$ bits/s with naturally shaped spectra. The coding efficiency of the stimulus with flat power spectrum was ~0.2, increasing to ~0.9 for the stimulus with naturally shaped power spectrum. This result is very important. A 90% coding efficiency means that the neuron is very close to the fundamental limit of information rate, the entropy rate of the output. This remarkable efficiency is achieved with naturally shaped power spectrum, which implies that the neuron uses the structure of naturally occurring signals in order to achieve a higher information rate and coding efficiency. This result emphasizes the importance of using natural signals in the investigation of even the most peripheral neural system.

The Muscle Spindle: Experimental and Simulation Results

Experimental Results

The muscle spindle (*MS*) is a mechanoreceptor which is sensitive to mechanical events in the muscle (Hulliger, 1984). Two types of muscle spindle afferent (*MSA*) nerve fibers convey sensory information to the spinal cord. Primary *MSA*s, or group Ia afferents, carry information about muscle length and the rate of change of muscle length (velocity). Secondary *MSA*s, or group II afferents, carry information mainly about muscle length. The *MS* responsiveness to mechanical events is controlled by the central nervous system, via the gamma system. In the experiment described below[3], a random displacement stimulus was delivered to the tendon of an isolated muscle of the cat hindlimb, while recording from afferent Ia fibers from the dorsal roots. During the experiment the cat was anaesthetized and the ventral roots were cut. The input stimulus was Gaussian noise with an approximately flat spectrum in the range of 0.5–20 Hz. The experiment was analyzed using the same method as described above. Figure 15 shows a segment of the input stimulus (top plot, dashed line), the reconstructed input signal (top plot, solid line), and the recorded spike train (bottom plot). Notice how the reconstruction signal follows the high-frequency changes of the input signal, but fails to record the low-frequency changes.

As can be seen in Fig. 15, the input signal does not have a zero mean. To overcome this issue we added a bias term to the linear reconstruction expression:

$$s_{est}(t) = Bias + \sum_{i=1}^{N} K_1(t - t_i)$$

and the optimization routine finds both $K_1(t)$ and the *Bias* that minimize the *MSE*. These are shown in Fig. 16.

The power spectrum of the input and effective noise, along with the resulting *SNR*, are shown in Fig. 17. One can see that the *SNR*>1 between the frequencies of 5 and 20 Hz, and that the maximal *SNR* is about ~8. The *SNR* is high for high frequencies and low for low frequencies within the range where the input spectrum is non-zero. This means that more information is being transferred about high frequencies of the input signal than about low frequencies. The *SNR* curve graphically represents the observation made in

3 The experiment was carried out during the summer of 1998, in collaboration with Profs. H. Johansson and U. Windhorst, The Department of Musculoskeletal Research, National Institute for Working Life, Umeå, Sweden. The experiment is part of an ongoing research project on information transmission in the sensorimotor system. The results mentioned here are preliminary, and are brought up only in order to elucidate the analysis method.

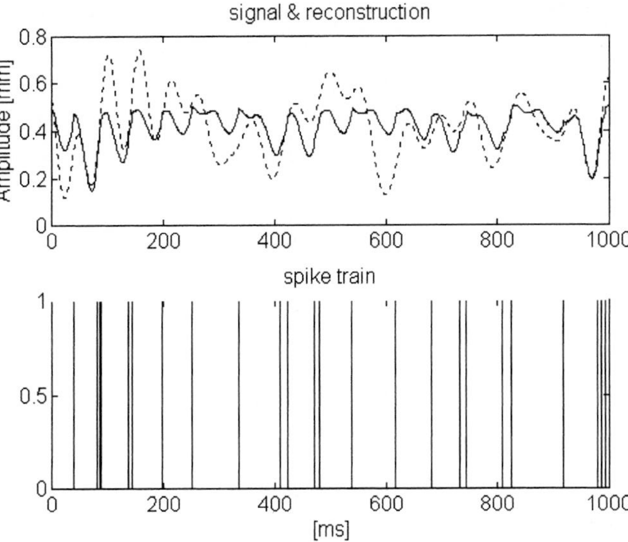

Fig. 15. A segment of the input signal (top plot, dashed line) and the reconstructed input signal (top plot, solid line), along with the recorded spike train (bottom plot) of the *MS* experiment. One can see that the reconstruction signal follows the high-frequency changes but not the low-frequency changes of the input signal.

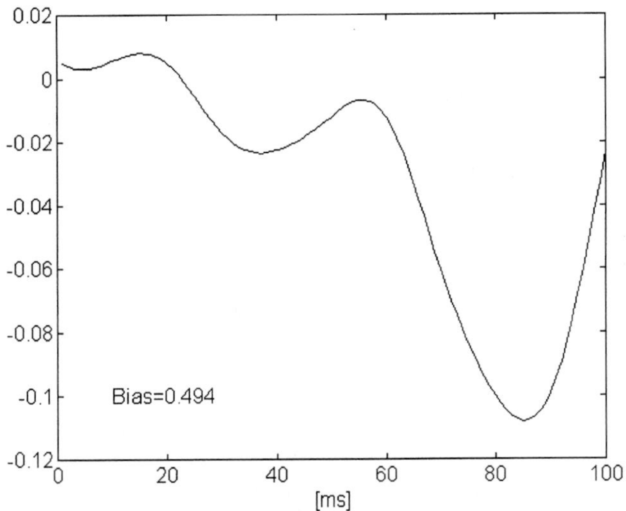

Fig. 16. The reconstruction kernel $K_1(t)$ and the bias term of the MS experiment. Intuitively, this waveform can be interpreted as the average stimulus that triggers a spike (in arbitrary units of stimulus strength).

Fig. 15, that the reconstruction quality for high frequencies is better than for low frequencies.

After integration this results in a lower bound to the information rate of 39.48 ± 0.75 [bits/s]. With an average spike frequency of 28.61 [spikes/s], this corresponds to 1.38 ± 0.026 [bits/spike]. The coding efficiency was 0.21.

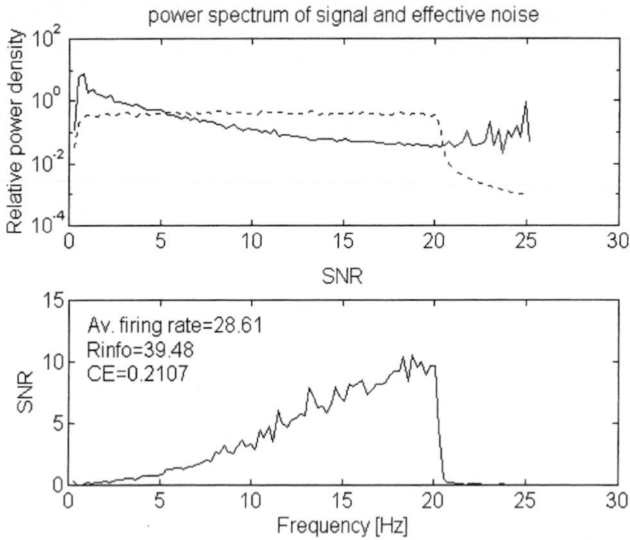

Fig. 17. The power spectral density of the input (upper plot, dashed line) and effective noise (upper plot, solid line), along with the signal-to-noise ratio (bottom plot) of the *MS* experiment. According to the *SNR* curve the bandwidth is between 5–20 Hz. At this range the *SNR* increases with frequency, and the best *SNR* is achieved at ~18 Hz. This is consistent with Fig. 15, where we see that the reconstruction quality is better at high frequencies.

Simulation Results

In order to provide insight into the neural encoding process and experimental results, simulations were conducted. These simulations may corroborate the experimental results and demonstrate how structural properties of the muscle spindle are related to its response characteristics, as revealed by this analysis method.

The SSIPFM Neuron Model

The Single-Sign Integrated Pulse Frequency Modulator (*SSIPFM*) is a simple model of the neuron encoder. In this model the input signal $s(t)$ is integrated and a spike is produced whenever the integrator state crosses a threshold, *a*. After spiking, the integrator is reset to zero. Mathematically, spike times $\{t_i\}$ are determined according to:

$$\int_{t_{i-1}}^{t_i} s(t)dt = a$$

Noise in the encoding process can be modeled by introducing a threshold that fluctuates with time: $a(t)$. For simplicity, the process $a(t)$ was modeled as white Gaussian noise. The mean threshold $E(a)$ depends on the *DC* level to maintain a fixed average firing rate: $\bar{r} \approx DC/E(a)$ [spikes/s]. Hence, the "noisiness" was quantified by the ratio of the standard deviation to the mean: $STD(a)/E(a) = 5\%$. The spectral content of the input signal and the output mean firing rate were similar to the muscle spindle experimental situation described above. The input was Gaussian random noise with a flat power spectrum in the range of [0.5–20] Hz. The noise was clipped to ±3 standard deviations, and *DC* was added to achieve 80% modulation depth. The encoder threshold was set to achieve an average spike frequency of ~30 spikes/s, similar to the experiment.

The power spectrum of the input and the effective noise, along with the resulting *SNR*, are shown in Fig. 18. The calculated information rate is ~47.87 [bits/s], with a coding efficiency of ~0.25.

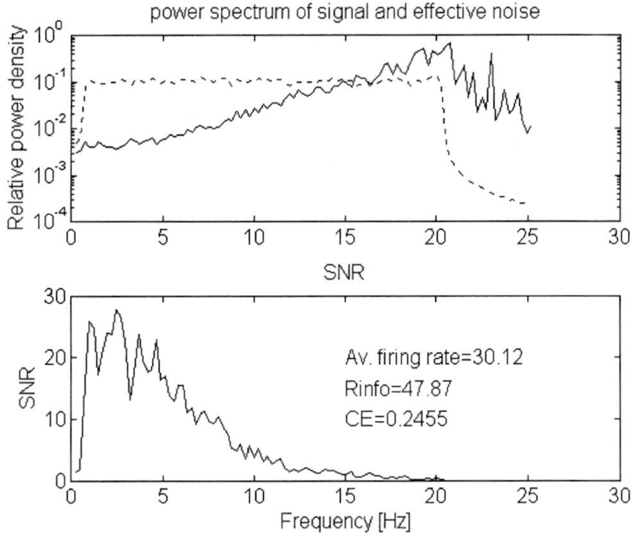

Fig. 18. The power spectral density of the input (upper plot, dashed line) and the effective noise (upper plot, solid line) along with the signal-to-noise ratio (bottom plot) of the *SSIPFM* encoder. The *SNR* decreases with frequency and is maximal at low frequencies, which means that reconstruction quality is better for low frequencies. This is inconsistent with the experimental result in Fig. 17.

The shape of the *SNR* shows that the *SSIPFM* encoder transfers more information about the low frequencies of the input signal than the high frequencies. A comparison between Figs. 17 and 18 reveals that this is in opposite to experimental results from the muscle spindle.

The Muscle Spindle Model

The fact that results from the *SSIPFM* model are inconsistent with experimental results is not surprising. The *SSIPFM* encoder is a reasonable model for the spike initiation site. However, the displacement stimulus of the muscle does not go directly to that site. Instead, it goes through mechanical filtering by the intrafusal fibers that constitute the muscle spindle. After experimenting with various models, a simple non-linear model was found to account for this effect. This model is based on the mechanical properties of the spindle, in line with the model proposed by McRuer et. al. (1968) and the model used by Milgram and Inbar (1976). It is clear therefore that different filtering and different N_{eff}, *SNR*, and R_{info} can be expected not only for the secondary (type II) ending, but for different muscle spindles and under different gamma system activity levels.

To approximate the response of primary (Ia) afferents, we add a velocity component to the displacement signal, and feed the positive part of this composite signal into a noisy *SSIPFM* encoder:

$$s(t) \rightarrow \left[s(t)+s'(t)\right]^{+} \rightarrow \mathrm{SSIPFM}$$

where $s'(t)$ is the velocity, and $y=[x]^{+}$ means: if $x<0$ then $y=0$, else $y=x$. The input signal was the same as before, and the firing rate was predetermined to be ~30 [spikes/s].

The effective noise power spectrum and the *SNR*, shown in Fig. 19, are very similar to those received in the animal experiment (see Fig. 17). The lower bound to the information rate was ~35.8 [bit/s], with coding efficiency of ~0.19, very close to those received in the animal experiment.

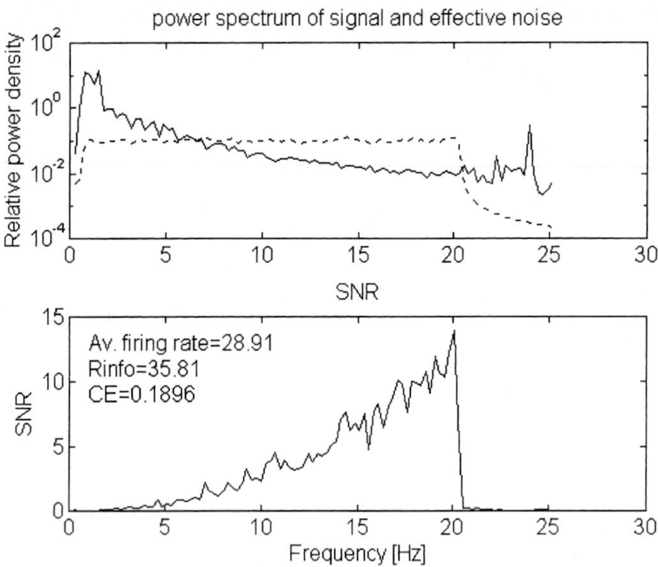

Fig. 19. The power spectral density of the input (upper plot, dashed line) and the effective noise (upper plot, solid line), along with the signal-to-noise ratio (bottom plot) of the muscle spindle model. Note the resemblance to the experimental results in Fig. 17.

Conclusions

From the results presented above, it is evident that different sensory systems transfer information in different frequency bands and with different *SNR* curves. While cricket sensory hairs utilize a wide bandwidth (~300 Hz) and low *SNR* (~1), in the frog sacculus we find a narrower bandwidth (~30 Hz) and a moderate *SNR* (~3–4). In the muscle spindle we find an even higher *SNR* (~8) over a narrower bandwidth (~15 Hz). Other results may be obtained in the *MS* case with different input signals. The differences between various sensory systems and modalities stem from the unique structure and function of the systems surveyed. The simulation model demonstrated that these results can be related to structural and functional aspects of the sensor, and enhance our understanding of it. With further analysis, it may be possible to correlate between this type of results and the sensor's function within the framework of the organism.

Coding efficiencies in the experiments mentioned above range from 20% to 90%. The gap between the lower bound and the upper bound could be attributed to three factors: (1) the upper bound ($R_{\text{info}}^{\text{UB}}$) is higher than the spike-train entropy $H(\{t_i\})$; (2) the spike-train entropy is higher than the true information rate (R_{info}); and (3) the true information rate is higher than its lower bound ($R_{\text{info}}^{\text{LB}}$). Ideally, if the upper bound is close to the spike-train entropy, and the lower bound is close to the true information rate, then the coding efficiency represents the percentage of the output entropy used to transfer information about the input. The rest of the entropy could be attributed to noise in the system. However, two problems may decrease the coding efficiency. One arises if the linear decoding scheme fails to extract most of the information in the signal. In this case it may be necessary to employ higher order estimation techniques. The second problem arises if the upper bound severely overestimates the spike-train entropy. In this case, the use of more sophisticated entropy estimation techniques may help (see for example: Farach et. al., 1995).

Used carefully, the method described in this paper may bring to light the information transmission properties of a sensory system. This is a powerful tool that may give us much insight into the function of the system.

Acknowledgement: We would like to thank Prof. U. Windhorst for his comments and helpful suggestions during the preparation of this chapter. We would like to thank Prof. H. Johansson, head of the Department of Musculoskeletal Research, National Institute for Working Life, Umeå, Sweden, for the collaboration, and Dr. M. Bergenheim, Dr. J. Pedersen, Mr. F. Hellström and Mr. J. Thunberg for carrying out the animal experiment and providing us with the data.

References

Bialek W, DeWeese M, Rieke F, Warland D (1993). Bits and brains: information flow in the nervous system. Physica A, 200: 581–593.

Bialek W, Rieke F, de Ruyter van Ssteveninck R, Warland D (1991). Reading a neural code. Science 252: 1854–1857.

Candy JV (1988). Signal processing: The modern approach. McGraw-Hill, New York.

Cover TM, Thomas JA (1990). Elements of information theory. John Wiley & Sons, New York

Eckhorn R, Pöpel B (1974). Rigorous and extended application of information theory to the afferent visual system of the cat. I. Basic concepts. Kybernetik 16: 191–200.

Eckhorn R, Pöpel B (1975). Rigorous and extended application of information theory to the afferent visual system of the cat. II. Experimental results. Biol Cybern 17: 7–17.

Farach M, Noordewier M, Savari S, Shepp L, Wyner A, Ziv J (1995). On the entropy of DNA: Algorithms and measurements based on memory and rapid convergence. Proceedings of the 1995 Symposium on Discrete Algorithms, pp. 48–57.

Hertz J (1995). Sensory coding and information theory. In: Arbib MA (ed), The handbook of brain theory and neural networks. MIT Press, Cambridge, MA.

Houk J C (1963). A mathematical model of the stretch reflex in human muscle systems. MS thesis, Massacchusetts Institute of Technology, Cambridge.

Hulliger M (1984). The mammalian muscle spindle and its central control. Rev Physiol, Biochem Pharmacol 101: 1–110.

Hulliger M, Matthews PBC, Noth J (1977) Static and dynamic fusimotor stimulation on the response of Ia fibres to low frequency sinusoidal stretching of widely ranging amplitudes. J Physiol (Lond) 291:233–249

Inbar G F (1972). Muscle spindles in muscle control. Kybernetik, 11, 119–147.

Inbar G F, Milgram P (1975). Estimation of intracellular potentials from evoked neural pulse trains. IEEE Trans BioMed Engin, BME-22: 379–383.

Koehler W, Windhorst U (1981) Frequency response characteristics of a multi-loop representation of the segmental muscle stretch reflex. Biol Cybern 40:59–70

Kröller J, Grüsser OJ, Weiss L R (1985). The response of primary muscle spindle endings to random muscle stretch: a quantitative analysis. Exp Brain Res 61: 1–10.

Kwakernaak H, Sivan R (1991) Modern signals and systems. Prentice-Hall International, Inc.

MacKay D, McCulloch WS (1952). The limiting information capacity of a neuronal link. Bull Math Biophys 14: 127–135.

McRuer D T, Magdaleno R E, Moore G P (1968). A neuromuscular actuation system model. IEEE Trans Man-Machine Systems MMS-9: 61–71.

Matthews PBC, Stein RB (1969) The sensitivity of muscle spindle afferents to small sinusoidal changes in length. J Physiol (Lond) 200:723–743.

Milgram P, Inbar GF (1976). Distortion suppression in neuromascular information transmission due to interchannel dispertion in muscle spindle firing thresholds. IEEE Trans BioMed Engin BME-23: 1–15.

Papoulis A (1991). Probability, random variables and stochastic processes. 3nd edition. McGraw-Hill, New York

Poppele RE (1981). An analysis of muscle spindle behavior using randomly applied stretches. Neuroscience 6: 1157–1165.

Porat B (1994) Digital processing of random signals: theory and methods. Prentice-Hall, Inc, New Jersey.
Prochazka A (1996) Proprioceptive feedback and movement regulation. In Rowell L, Shepard J (eds) Integration of motor, circulatory, respiratory and metabolic control during exercise, pp 89–127. American handbook of physiology. Sect A. Neural control of movement. Oxford University Press, New York
Rieke F, Bodnar D, Bialek W (1995). Naturalistic stimuli increase the rate and efficiency of information transmission by primary auditory neurons. Proc Royal Soc Lond Series B 262: 259–265.
Rieke F, Warland D, Bialek W (1993). Coding efficiency and information rates in sensory neurons. Europhysics Letters, 22: 151–156.
Rieke F, Warland D, de Ruyter van Stevenink R, Bialek W (1997). Spikes: Exploring the neural code. The MIT Press, Cambridge, MA.
Rosenthal NP, McKean TA, Roberts WJ, Terzuolo CA (1970) Frequency analysis of stretch reflex and its main subsystems in triceps surae muscles of the cat. J Neurophysiol 33: 713–749.
Shannon CE (1948). A mathematical theory of communication. Bell System Technical Journal, 27: 379–423, 623–656.
Shannon CE, Weaver W (1949). The mathematical theory of communication. University of Illinois Press, Urbana.
Tock Y, Inbar Gf (1999). On the upper bound to neural information transmission rate. EE Pub. 1220, Technion – IIT, Israel.
Warland D, Landolfa M, Miller J P, Bialek W (1992). Reading between the spikes in the cercal filiform hair receptors of the cricket. In: Eeckman F (ed) Analysis and modeling of neural systems, pp. 327–333. Kluwer Academic, Boston.

Chapter 20

Information-Theoretical Analysis of Small Neuronal Networks

SATOSHI YAMADA

Introduction

Determining the synaptic connection structure of neuronal networks is critical for understanding the organization and function of the nervous system. Multi-neuron recordings such as multi-channel optical recordings (Cohen and Lesher 1986, Nakashima et al. 1992) and multi-unit extracellular recordings (Gerstein et al. 1983, Novak and Wheeler 1986, Wilson and McNaughton 1993) have been developed and shown to be useful in understanding the structure and function of neuronal networks. However, they do not provide any direct information about synaptic connectivity. Cross-correlation analysis of two action potential trains (Perkel et al. 1967a, 1967b; Gerstein and Aertsen 1985; Surmeier and Weinberg 1985; Melssen and Epping 1987; Palm et al. 1988; Aertsen et al. 1989; Yang and Shamma 1990) or three action potential trains (Perkel et al. 1975) has been widely used for this purpose. However, they pose some difficulties.

- It is sometimes difficult to distinguish a significant cross-correlation from noise
- They do not provide quantitative estimation of synaptic strengths.
- They do not allow to infer the connection structure.

The cross-correlation analysis using information theory will tackle the above difficulties. Several researchers used information theory to detect neuronal responses or interactions (Eckhorn and Pöpel 1974; Tsukada et al. 1975; Eckhorn et al. 1976; Fagen 1978; Nakahama et al. 1983; Fuller and Looft 1984; Optican and Richmond 1987; Reinke and Diekmann 1987; Williams et al. 1987; Optican et al. 1991). However, their methods were not systematic analysis methods and lacked procedures to deduce the connection structure of networks consisting of n neurons (n-neuron networks). We developed analysis methods to deduce the connection structure of n-neuron networks (Yamada et al. 1993, 1996) using information theory. In this chapter, the cross-correlation method using information theory will be reviewed. First, several quantities used in the analysis will be defined. Second, the overview of the analysis will be described. Third, for demonstration, the methods developed will be applied to data obtained by the simulation of neural network models, and examples of the results will be presented. Finally, the merit and deficiencies of this method will be discussed.

Theory

Relation between Cross-Correlation Analysis and Information Theory

One purpose of the cross-correlation analysis is to detect correlations between action potential (spike) trains of neurons as exemplified in Fig. 1b. From information theory, we may define a measure called "mutual information" (see below) that represents the

Satoshi Yamada, Advanced Technology R&D Center, Mitsubishi Electric Corporation, 8-1-1, Tsukaguchi-Honmachi, Amagasaki, Hyogo, 661-8661, Japan (phone: 81-6-6497-7066; fax: 81-6-6497-7289; e-mail: yamada@bio.crl.melco.co.jp)

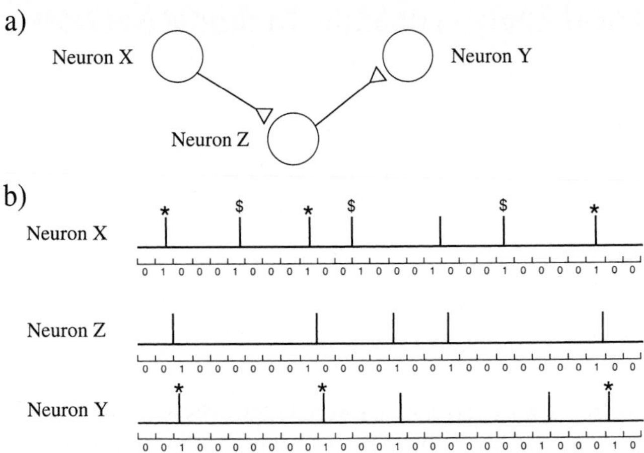

Fig. 1. A three-neuron network and an example of action potential trains of the neurons. a) The connection structure of the network, b) Example of action potential trains of the neurons and the corresponding sequences of 0's and 1's.

common information carried by sets of neurons and is quantitatively related to the correlation between these neurons and the strengths of their synaptic connections.

Another purpose of the present approach is to deduce the connection structure. For this purpose, it is necessary to distinguish between direct connections and indirect interactions. To do so, the contribution of other neurons must be estimated. A suitable measure is what is called "conditional mutual information" (see below), which represents the mutual information independent of other neurons and can thus estimate other neurons' contribution.

To exemplify these notions, consider a network of three neurons labelled *X*, *Y* and *Z*, as depicted in Figure 1a. In this network, there are excitatory connections from *X* to *Z* $X \to Z$ and from *Z* to *Y* $Z \to Y$, but *X* and *Y* have no direct connection. Nonetheless, the cross-correlation between *X* and *Y* will show significant values because of *Z*'s mediation. In Figure 1b, this is suggested by the following facts. Action potentials of *X* are often followed, at a time interval t^0_{XY}, by spikes of *Y*, as indicated by "*", and for these marked spike pairs, *X*'s spikes are followed by spikes of *Z* at an interval t^0_{ZX}, which in turn precede those of *Y* by t^0_{ZY}. However, most often spikes are not produced by *Y* when *Z* is silent after *X*'s firing (in Figure 1b, *X*'s firings without *Z*'s firings are indicated by "$"). This does not imply that spikes of *X* may not rarely be followed by spikes of *Z* without eliciting spikes of *Y*, and that spikes in *Y* may not rarely be preceded by spikes in *Z* without the latter being preceded by spikes in *X*. After all, we are dealing with statistical connections and signals.

If the correlation between *X* and *Y* in the case of active *Z* and the correlation in the case of silent *Z* are calculated separately, the correlation independent of *Z* may be calculated. The conditional mutual information does exactly that, namely calculate such values (see equation (8) in the following section) and thus estimate the correlation independent of *Z*.

In this information-theoretical approach, spiking neurons are thought of as producing event processes with two states, either active (1: firing an action potential) or at rest (0: not firing). That is, action potential trains are converted into sequences of 1's and 0's by dividing time into discrete time bins of duration Δ and counting spikes into them, as illustrated in Figure 1b below the respective spike trains. From these sequences, probabilities and the following information-theoretical quantities are calculated.

Mutual Information

If **X** is a point process with a discrete and finite state space, the Shannon entropy is given by:

$$H(\mathbf{X}) = -\sum_{s(\mathbf{X})} p(s(\mathbf{X}))\log[p(s(\mathbf{X}))], \tag{1}$$

where *s(X)* and *p(s(X))* denote the state of **X** and the probability of *s(X)*, respectively (Shannon 1948). Similarly, the joint Shannon entropy between processes **X** and **Y** is given by

$$H(\mathbf{X},\mathbf{Y}) = -\sum_{s(\mathbf{X}),s(\mathbf{Y})} p(s(\mathbf{X}),s(\mathbf{Y}))\log[p(s(\mathbf{X}),s(\mathbf{Y}))], \tag{2}$$

where *p(s(X),s(Y))* denotes the joint probability.

Two-point mutual information (2pMI), representing the information carried in common by two processes **X** and **Y**, is given by

$$I(\mathbf{X}:\mathbf{Y}) = H(\mathbf{X}) + H(\mathbf{Y}) - H(\mathbf{X},\mathbf{Y}). \tag{3}$$

Three-point mutual information (3pMI) is defined equivalently (Ikeda et al. 1989)

$$I(\mathbf{X}:\mathbf{Y}:\mathbf{Z}) = H(\mathbf{X}) + H(\mathbf{Y}) + H(\mathbf{Z}) - H(\mathbf{X},\mathbf{Y}) - H(\mathbf{Y},\mathbf{Z}) - H(\mathbf{Z},\mathbf{X}) + H(\mathbf{X},\mathbf{Y},\mathbf{Z}). \tag{4}$$

As the expansion of the above definitions, n-point mutual information(npMI) is defined as (Yamada et al. 1996):

$$I(\mathbf{N}_1:\mathbf{N}_2:\cdots:\mathbf{N}_n) = \sum_{m=1}^{n} \sum_{k_1,\cdots,k_m \in \{1:\cdots:n\}} (-1)^{m+1} H(\mathbf{N}_{k_1},\cdots,\mathbf{N}_{k_m}), \tag{5}$$

where the inner summation is the sum of all combinations of m neurons drawn from a collection of n neurons.

I(X:Y:Z) is related to *I(X:Y)* as follows:

$$I(\mathbf{X}:\mathbf{Y}:\mathbf{Z}) = I(\mathbf{X}:\mathbf{Y}) - I(\mathbf{X}:\mathbf{Y}|\mathbf{Z}), \tag{6}$$

where *I(X:Y|Z)* is the two-point conditional mutual information (2pCMI) between **X** and **Y** if the state of **Z** is given.

Similarly, npMI is also given by 2pMI and the conditional mutual information,

$$I(\mathbf{N}_1:\mathbf{N}_2:\cdots:\mathbf{N}_n) = I(\mathbf{N}_i:\mathbf{N}_j) + \sum_{m=1}^{n-2} \sum_{k_1,\cdots,k_m \in \{1:\cdots:n\}, k_1,\cdots,k_m \neq i,j} (-1)^m I(\mathbf{N}_i:\mathbf{N}_j|\mathbf{N}_{k_1},\cdots,\mathbf{N}_{k_m}) \tag{7}$$

where $I(\mathbf{N}_i:\mathbf{N}_j|\mathbf{N}_{k_1},\cdots,\mathbf{N}_{k_m})$ is the two-point joint conditional mutual information (2pJCMI).

The mutual information and the conditional mutual information are described by probabilities as follows:

$$I(\mathbf{X}:\mathbf{Y}) = \sum_{s(\mathbf{X}),s(\mathbf{Y}) \in \{0;1\}} p(s(\mathbf{X}),s(\mathbf{Y}))\log\frac{p(s(\mathbf{Y})|s(\mathbf{X}))}{p(s(\mathbf{Y}))}$$

$$I(\mathbf{X}:\mathbf{Y}|\mathbf{Z}) = \sum_{s(\mathbf{X}),s(\mathbf{Y}),s(\mathbf{Z}) \in \{0;1\}} p(s(\mathbf{X}),s(\mathbf{Y}),s(\mathbf{Z}))\log\frac{p(s(\mathbf{X}),s(\mathbf{Y})|s(\mathbf{Z}))}{p(s(\mathbf{X})|s(\mathbf{Z}))p(s(\mathbf{Y})|s(\mathbf{Z}))} \tag{8}$$

$$I(\mathbf{N}_i:\mathbf{N}_j|\mathbf{N}_{k_1},\cdots,\mathbf{N}_{k_m}) = \sum_{s(\mathbf{N}_i),s(\mathbf{N}_j),s(\mathbf{N}_{k_1}),\cdots,s(\mathbf{N}_{k_m}) \in \{0;1\}} p(s(\mathbf{N}_i),s(\mathbf{N}_j),s(\mathbf{N}_{k_1}),\cdots,s(\mathbf{N}_{k_m})) \times$$

$$\times \log\frac{p(s(\mathbf{N}_i),s(\mathbf{N}_j)|s(\mathbf{N}_{k_1}),\cdots,s(\mathbf{N}_{k_m}))}{p(s(\mathbf{N}_i)|s(\mathbf{N}_{k_1}),\cdots,s(\mathbf{N}_{k_m}))p(s(\mathbf{N}_j)|s(\mathbf{N}_{k_1}),\cdots,s(\mathbf{N}_{k_m}))}$$

Correlations between action potential trains must be considered at relative time differences. For example, equation (5) becomes

$$I(N_1(t_{11}):N_2(t_{12}):\cdots:N_n(t_{1n})) = \sum_{m=1}^{n} \sum_{k_1,\cdots,k_m \in \{1:\cdots:n\}} (-1)^{m+1} H(N_{k_1}(t_{1k_1}),\cdots,N_{k_m}(t_{1k_m})), \quad (9)$$

where $t_{11},\cdots,t_{1n}$ ($t_{11} \equiv 0$) are relative time differences of $N_1,\cdots,N_n$ from N_1, respectively. All the above equations must be changed in a similar way.

The npMI expresses the amount of information shared among n processes (McGill 1955). If any neuron or any group of neurons has no connection with the remaining neurons, then npMI is equal to zero (see Appendix of Yamada et al. 1996 for the proof). If a significant npMI peak is found (statistical significance will be discussed later), it indicates that all neurons are interconnected, either directly or indirectly via other neurons. The time differences obtained at the peaks are candidates of time differences for effective connections.

Signed Channel Capacity

Consider a presynaptic neuron X coupled to a postsynaptic neuron Y. 2pMI is dependent on the firing probability of presynaptic neuron X, $p(s(X))$. To estimate the connectivity quantitatively, the channel capacity is used. The channel capacity is the maximum of the mutual information with respect to $p(s(X))$ (Shannon 1948) and thus independent of $p(s(X))$. The channel capacity is calculated (Blahut 1972) as:

$$C(X:Y) = \max_{p(s(X))} I(X:Y) \quad (10)$$

The channel capacity and 2pMI are non-negative for both excitatory and inhibitory interactions. To distinguish excitatory from inhibitory interactions, we introduce a signed channel capacity (SCC) and a signed two-point mutual information (SMI) (Yamada et al. 1993). Because inhibitory synapses reduce the firing probability of post-synaptic neurons after the firing of pre-synaptic neurons, $p(s(Y)=1|s(X)=1)$, the $p(s(Y)=1|s(X)=1)$ will be smaller than the spontaneous firing probability, $p(s(Y)=1|s(X)=0)$. Let SCC and SMI be negative (inhibitory) if the following inequality is satisfied (see Figure 2-d and 2-e)

$$p(s(Y)=1|s(X)=1) < p(s(Y)=1|s(X)=0). \quad (11)$$

Statistical Significance

In a cross-correlational analysis, we can deduce a model with a minimal number of "effective connections" that could replicate the observed feature of action potential trains (Aertsen et al. 1989, Yamada et al. 1996). The corresponding neuron pair is considered to have an "effective connection" if 2pMI, 2pCMI or 2pJCMI is larger than the following bound.

Consider the bounds for $I(X:Y(t_{XY}))$. For a given significance level α (usually α=1% or 5% are used), bounds b are defined as follows (Palm et al. 1988, Yamada et al. 1996):

$$p(|I(X:Y(t_{XY}))| > b) = \alpha \quad (12)$$

where $p(|I(X:Y(t_{XY}))| > b)$ denotes the probability that $I(X:Y(t_{XY}))$ surpasses bounds b. Equation (12) means that the probability that $I(X:Y(t_{XY}))$ surpasses bounds b is only α when X and Y are independent (X and Y have no synaptic connection). So if $I(X:Y(t_{XY}))$

surpasses b, the X-Y connection is considered to be statistically significant with a significance level α.

The bounds b for $I(\boldsymbol{X}{:}\boldsymbol{Y}(t_{XY}))$ are the maximum and minimum of $I(\boldsymbol{X}{:}\boldsymbol{Y}(t_{XY}))$ with the probability of 1-α when neurons X and Y are independent. In order to calculate bounds for $I(\boldsymbol{X}{:}\boldsymbol{Y}(t_{XY}))$, it is necessary to calculate bounds for the coincident number of Y's action potentials ($b(C_y)$). The expectation number of coincident Y's action potentials at a certain time difference is $N_x p_y$, where N_x and p_y denote the action potential number of neuron X and the firing probability of neuron Y, respectively. The coincident number of action potentials is a random variable, and its distribution is approximated by the Gaussian distribution if $N_x p_y$ is sufficiently large. In the Gaussian distribution, the standard deviation is calculated by $\sigma = \sqrt{N_x\, p_y (1 - p_y)}$. Using the standard deviation, the upper and lower bounds of the coincident number of action potentials can be calculated : $b(C_y) = N_x p_y \pm \omega\sigma$ (ω=1.96 for α=5%, ω=2.58 for α=1%, from the Gaussian distribution). Then the upper and lower bounds for probabilities are estimated by using $b(C_y)$, and finally bounds for $I(\boldsymbol{X}{:}\boldsymbol{Y}(t_{xy}))$ are calculated.

Similarly, bounds for npMI, 2pCMI and 2pJCMI are calculated. The upper and lower bounds for the coincident action potential number and probabilities used in the calculation are calculated similarly. The bounds for npMI, 2pCMI and 2pJCMI are the maximum and minimum among values calculated by the possible combinations of the upper and lower bounds for probabilities.

Role of Conditional Mutual Information

In the correlation analysis of action potential trains, it is most important to distinguish direct connections from indirect interactions. The conditional mutual information can estimate the contribution of other neurons. In a two-neuron network, a statistically significant 2pMI peak indicates an "effective" connection between the two neurons. In a three-neuron network, a significant 2pMI peak, however, does not indicate a direct "effective" connection because there is a possibility of indirect interaction via another neuron.

For example, consider a network of three neurons (X, Y and Z) (Figure 1-a). In this network, there are connections of $X \to Z$ and $Z \to Y$ at t^0_{ZX} and t^0_{ZY}, respectively. X and Y do not have a direct connection, but $I(\mathbf{X}{:}\mathbf{Y}(t^0_{XY}))$ $t^0_{XY} = t^0_{ZY} - t^0_{ZX}$ shows significant values because X and Y have correlated action potentials dependent on Z. However, the elevation of Y's firing probability is not caused by X, but by Z. So X and Y are statistically independent when the state of Z is given,

$$p(s(\boldsymbol{X}(t^0_{ZX})), s(\boldsymbol{Y}(t^0_{ZY}))|s(\boldsymbol{Z})) = p(s(\boldsymbol{X}(t^0_{ZX}))|s(\boldsymbol{Z})) \times p(\boldsymbol{Y}(t^0_{ZY}))|s(\boldsymbol{Z}))$$

(which means that correlated Y's firing probability does not show further elevation dependent on X) and then

$$I(\boldsymbol{X}(t^0_{ZX}){:}\boldsymbol{Y}(t^0_{ZY})|\boldsymbol{Z}) = 0$$

(In practice, $I(X{:}Y|Z)$=0 means that $I(X{:}Y|Z)$ is less than its bound.) Therefore, an effective connection between X and Y is confirmed when 2pCMI is not 0 or refused when 2pCMI=0. Under rather unlikely conditions, 2pCMI=0 if the corresponding connection exists (see "Comments" and Yamada et al. 1996 for details).

Since 2pMI and 2pCMI for a 3pMI peak play a crucial role in determining an effective connection, they should be calculated as accurately as possible. Their calculation should be done over the 3pMI peak region rather than over a square region artificially determined by Δ. The 2pMI and 2pCMI over the 3pMI peak should be calculated by using probabilities recalculated over the whole 3pMI peak region.

In an n-neuron network, $I(N_i{:}N_j|N_{k_1},\cdots,N_{k_m})$ is used to distinguish direct connections from indirect interactions.

Estimation of Effective Connections

Consider a three-neuron network. A significant 3pMI peak means that every neuron in the network is connected with at least one other neuron. Three time differences, t^0_{ZX}, t^0_{ZY} and t^0_{XY} $(=t^0_{ZY}-t^0_{ZX})$, are obtained from the peak position. They are the candidates of the time differences of effective connections. Recalculated 2pMI and 2pCMI over the peak region is used to determine the effective connections.

First, consider the case where all three time differences are non-zero. The neurons are rearranged and renamed A, B, and C in the order of the time differences. (e.g., if $t^0_{ZX}=-10$, $t^0_{ZY}=10$, then $X\to A, Z\to B, Y\to C$). With these time differences, C cannot affect the correlation between A and B. Thus, whether A and B have an effective connection or not is estimated by $I(A{:}B)$. In contrast, because of the possible indirect interactions the statistical effectiveness of A-C and B-C is estimated by $I(A{:}C|B)$ and $I(B{:}C|A)$, respectively. If either $I(A{:}B)$, $I(A{:}C|B)$ or $I(B{:}C|A)$ is zero, the corresponding connection is statistically refused, *i.e.*, not effective. Namely, the following connections are considered to be effective:

$$I(\boldsymbol{A}{:}\boldsymbol{B})=0\Rightarrow A\to C, B\to C$$
$$I(\boldsymbol{A}{:}\boldsymbol{C}|\boldsymbol{B})=0\Rightarrow A\to B, B\to C$$
$$I(\boldsymbol{B}{:}\boldsymbol{C}|\boldsymbol{A})=0\Rightarrow A\to B, A\to C$$

If all of them are non-zero, all three connections are considered to be statistically effective. If two or three of them are zero, some of them must be zeroed by the unlikely special conditions (see "Comments" and Yamada et al. 1996). If the time difference from A to B is zero, the same procedure is applicable, but the direction of the A-B connection cannot be determined. If the time difference from B to C is zero, the connection structure cannot be deduced except for the case of no effective connection between B and C. If all the time differences are zero, the connection structure cannot be determined.

Consider an n-neuron network. The interaction among the n neurons can be detected by a significant peak of npMI in the $t_{12},\cdots,t_{1n}$ hyperplane. Time differences, $t^0_{12},\cdots,t^0_{1n}$ and $t^0_{ij}(=t^0_{1j}-t^0_{1i})$ $(i,j=2,\cdots,n,\ i\neq j)$, are obtained from the significant peak position. The n neurons are rearranged and renamed $N_1,N_2,\cdots,N_n$ in the order of the time differences. Because N_m $(m>j>i)$ cannot affect the correlation from N_i to N_j, but N_p $(p<j)$ has the possibility to affect it, the effectiveness of the N_i-N_j connection can be estimated by $(I(N_i(t^0_{1i}){:}N_j(t^0_{1j})|N_{k_1}(t^0_{1k_1}),\cdots,N_{k_{j-2}}(t^0_{1k_{j-2}}))$ $(k_1,\cdots,k_{j-2}<j,\ k_1,\cdots,k_{j-2}\neq i)$. When i=1 and j=2, $I(N_1{:}N_2(t^0_{12}))$ is used. Similarly to the analysis of the three-neuron network, if 2pMI=0 or 2pJCMI=0, the corresponding connection is not effective. If one or more time differences $(t^0_{1k_1},\cdots,t^0_{1k_{j-2}})$ are equal to t^0_{1j}, the statistical effectiveness of the N_i-N_j connection cannot be estimated, because in the case of $t^0_{1l}=t^0_{1j}$ the direction between N_j and N_l cannot be determined. But if all such connections are ineffective, i.e., $I(N_l{:}N_j|N_{k_1},\cdots,N_{k_{j-2}})=0$, the effectiveness of the N_i-N_j connection can be estimated.

Outline

The procedure to deduce the connection structure of an n-neuron network by using the information-theoretical analysis of action potential trains is as follows:

1. Get action potential train data that indicate the occurrence times of action potentials.

2. Convert an action potential train of each neuron into the sequence of 1's and 0's in discrete time bins with an interval Δ.
3. Calculate cross-correlation histograms.
4. Calculate probabilities required to calculate *n*pMI.
5. Calculate *n*pMI and its bounds.
6. Find statistically significant *n*pMI peaks.
7. Determine peak region of significant *n*pMI peaks.
8. Recalculate 2pMI and 2pJCMI over the significant peak regions and their bounds.
9. Determine statistically effective connections.

Procedures and Results

Analysis of Two-Neuron Network

First, the analysis of a two-neuron network is described. Consider the network of neurons *X* and *Y*. Figure 2-a shows two action potential trains (top traces). These data are first converted to the sequence of 1's and 0's in discrete time bins with an interval of Δ.

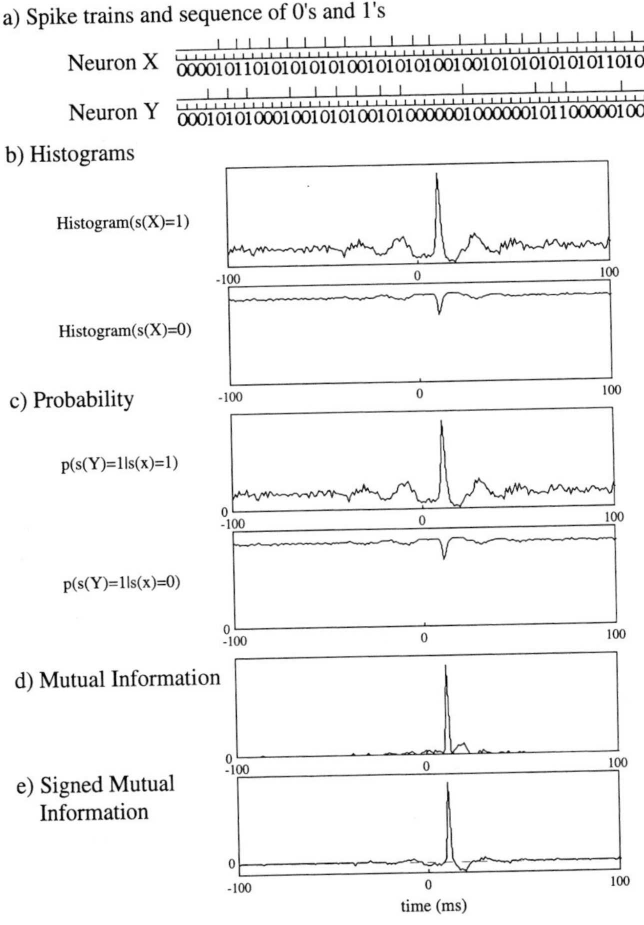

Fig. 2. Explanation of information-theoretical analysis of a two-neuron network. a) Action potential trains of two neurons, *X* and *Y*, and the corresponding sequences of 1's and 0's. b) Cross-correlation histograms for s(*X*)=1 and s(*X*)=0. c) Conditional probabilities, $p(s(Y)=1|s(X)=1)$ and $p(s(Y)=1|s(X)=0)$. d) 2pMI. e) SMI.

The size of the time bin should be smaller than the minimum action potential intervals occurring in any train such that a single time bin contains at most one action potential. The smaller the time bin used, the larger the amounts of data required to find a significant correlation. So the time bin should be as large as possible as long as it does not exceed the minimum action potential interval. The time axis is devided into many blocks with the time bin determined above. If an action potential exists in a time bin, the symbol of the time bin is 1, otherwise 0 (Figure 2-a).

To calculate 2pMI, the following probabilities at each time difference have to be calculated.

$$p(s(X)=0),$$
$$p(s(X)=1),$$
$$p(s(Y,t_{xy})=0),$$
$$p(s(Y,t_{xy})=1),$$
$$p(s(Y,t_{xy})=0|s(X)=0),$$
$$p(s(Y,t_{xy})=1|s(X)=0),$$
$$p(s(Y,t_{xy})=0|s(X)=1),$$
$$p(s(Y,t_{xy})=1|s(X)=1),$$

where t_{xy} is the time difference between neurons X and Y. In this case, neuron X is temporarily set to be the presynaptic neuron. There are equations among the above probabilities, i.e.,

$$p(s(X)=0)=1-p(s(X)=1)$$
$$p(s(Y,t_{xy})=0)=1-p(s(Y,t_{xy})=1)$$
$$p(s(Y,t_{xy})=0|s(X)=0)=1-p(s(Y,t_{xy})=1|s(X)=0)$$
$$p(s(Y,t_{xy})=0|s(X)=1)=1-p(s(Y,t_{xy})=1|s(X)=1)$$
$$p(s(Y,t_{xy})=1)=p(s(Y,t_{xy})=1|s(X)=0)\times p(s(X)=0)+p(s(Y,txy)=1|s(X)=1)\times p(s(X)=1)$$

Three probabilities ($p(s(X)=1)$, $p(s(Y,t_{xy})=1|s(X)=0)$, and $p(s(Y,t_{xy})=1|s(X)=1)$) at each time difference should be estimated. The $p(s(X)=1)$ is easily estimated from the ratio of 1's in the sequence. The $p(s(Y,t_{xy})=1|s(X)=0)$ and $p(s(Y,t_{xy})=1|s(X)=1)$ are estimated from the cross-correlation histogram (Figure 2-b). Figure 2-c shows the conditional probabilities. Using these probabilities at each time difference, 2pMI at each time difference is calculated (Figure 2-d) by using equation (8). To distinguish inhibitory interactions from excitatory ones, SMI is calculated (see inequality (11)) (Figure 2-e). Dotted horizontal lines in Figure 2-e represent bounds for SMI. In this example, an excitatory peak at a time difference of 10 ms is found to be significant.

SMI amplifies the significant changes as opposed to the ordinary cross-correlation methods, *e.g.*, scaled cross-coincidence histogram (SCCH) (Melssen and Epping 1987) and "surprise" (SUR) (Aertsen et al. 1989). As shown in Figure 3, SMI shows bigger correlation peaks than SCCH and SUR for both excitatory (Figure 3-a) and inhibitory (Figure 3-b) interactions. This amplification is due to the nonlinear function of SMI as shown in Figure 4.

The SCC gives good measures for synaptic strengths. To obtain quantitative estimates of synaptic strength, the channel capacity over the significant peak should be calculated. Figure 5 shows the dependency of recalculated SCC over the significant peak and SCCH at the peak on synaptic strengths. As shown in Figure 5, the SCC gives a quan-

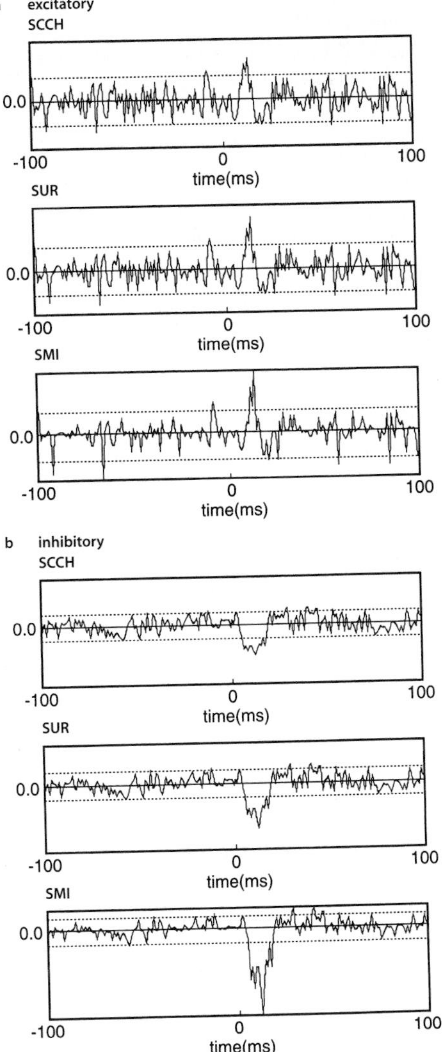

Fig. 3. Figure 3. Comparison of cross-correlation graphs. a) Cross-correlation graphs between neurons X_1 and Y by SCCH, SUR, and SMI. Neuron X_1 has an excitatory synapse with Y. b) Cross-correlation graphs between neurons X_2 and Y. Neuron X_2 has an inhibitory synapse with Y. The two horizontal dashed lines indicate upper and lower bounds (α=5%). The action potential trains were simulated by the Hodgkin-Huxley neuron model (reprinted from Yamada. et al. 1993)

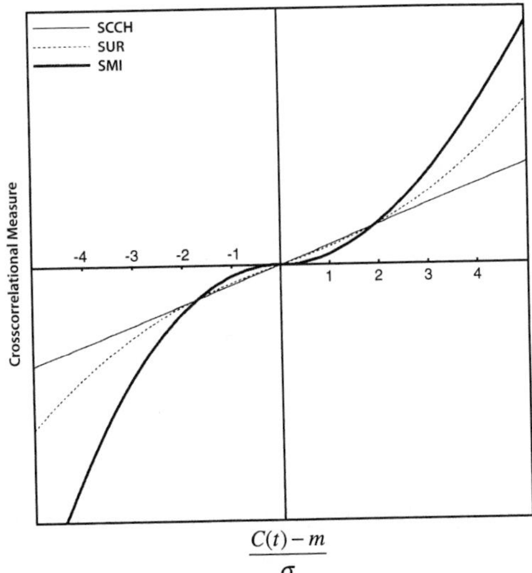

Fig. 4. Function of SMI, SCCH and SUR with respect to C(t), where C(t) denotes the coincident number of action potentials at time differences t. The x-axis represents the standardized coincident counts, (C(t)–m)/σ, where m and σ denote the mean and standard deviation of the distribution of the coincident number, respectively, and the y-axis represents the normalized cross-correlation measures. The normalization was done by using the values at the upper bound. Bold line: SMI, thin line: SCCH, dashed line: SUR (reprinted from Yamada et al. 1993).

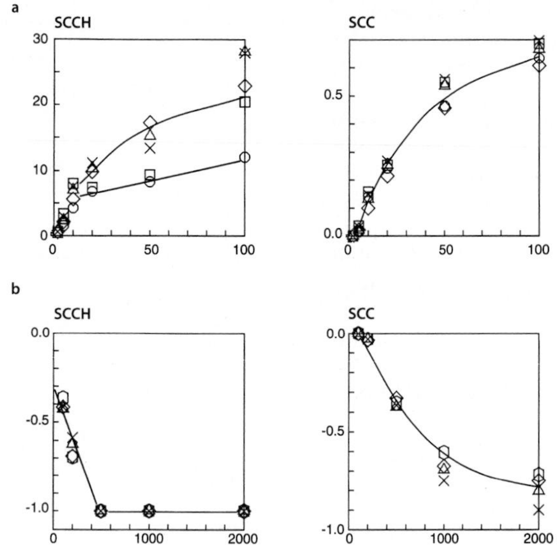

Fig. 5. Synaptic strength dependency of SCCH and SCC. a) Excitatory interaction between X_1 and Y. Recalculated SCC over the peak region or peak height of SCCH is plotted against synaptic strengths w, varying the firing probability of X_1. The action potential number of X_1 in each simulation is 2000. b) Inhibitory interaction between X_2 and Y. The action potential number of X_2 in each simulation is 2000. The firing frequency of X_1 (a) and X_2 (b); ○: 50/s, □: 25/s, X: 14.2/s, Δ: 10/s, ◇: 5/s, ⬡: 2.5/s (reprinted from Yamada et al. 1993).

titative measure independent of firing probability of the presynaptic neuron. In contrast, SCCH is dependent on firing probability.

Analysis of a Three-Neuron Network

In this section, the analysis method to deduce the connection structure of a three-neuron network is described. Consider the network of X, Y and Z. The conversion of action potential trains into the sequence of 1's and 0's is done in a way similar to the method described in the previous section. Let Z temporarily be a presynaptic neuron. In the case of a three-neuron network, the following probabilities at each time difference should be estimated from the cross-correlation histogram.

$$p(s(Z)=1),$$
$$p(s(X,t_{zx})=1|s(Z)=0),$$
$$p(s(X,t_{zx})=1|s(Z)=1),$$
$$p(s(Y,t_{zy})=1|s(Z)=0),$$
$$p(s(Y,t_{zy})=1|s(Z)=1),$$
$$p(s(X,t_{zx})=1,s(Y,t_{zy})=1|s(Z)=0),$$
$$p(s(X,t_{zx})=1,s(Y,t_{zy})=1|s(Z)=1).$$

Other probabilities are calculated by using equations of probabilities, e.g.,

$$p(s(X,t_{zx})=1,s(Y,t_{zy})=0|s(Z)=1)$$
$$=p(s(X,t_{zx})=1|s(Z)=1)-p(s(X,t_{zx})=1,s(Y,t_{zy})=1|s(Z)=1)$$

$$p(s(X,t_{zx})=1,s(Y,t_{zy})=1)=p(s(X,t_{zx})=1,s(Y,t_{zy})=1|s(Z)=0)\times p(s(Z)=0)+$$
$$+p(s(X,t_{zx})=1,s(Y,t_{zy})=1|s(Z)=1)\times p(s(Z)=1)$$

3pMI at each time difference is calculated according to equations (6) and (8).

Figure6-a shows the 3pMI graph of action potential trains produced by the simulation of a three-neuron network model. Three significant 3pMI peaks, (t_{zx},t_{zy})=(17,12),

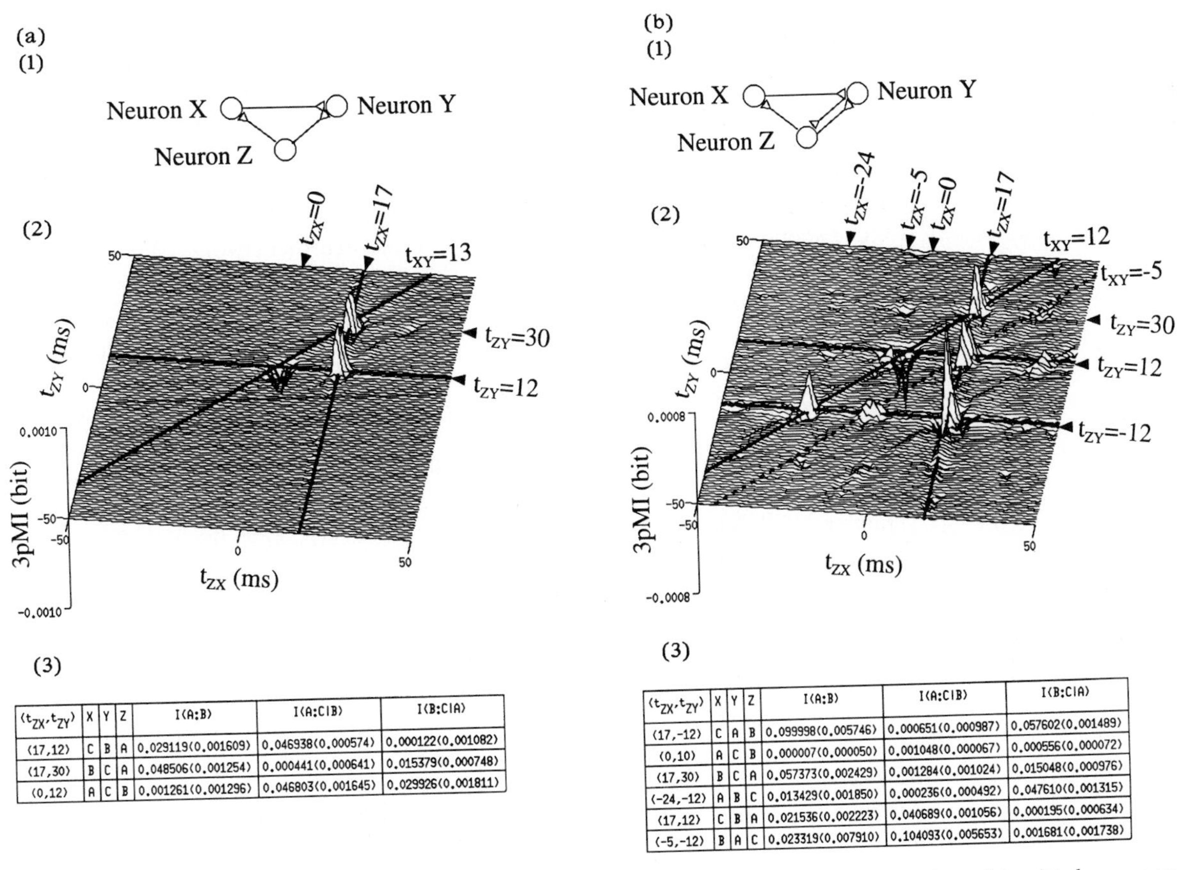

(t_{ZX},t_{ZY})	X	Y	Z	I(A:B)	I(A:C|B)	I(B:C|A)
(17,12)	C	B	A	0.029119(0.001609)	0.046938(0.000574)	0.000122(0.001082)
(17,30)	B	C	A	0.048506(0.001254)	0.000441(0.000641)	0.015379(0.000748)
(0,12)	A	C	B	0.001261(0.001296)	0.046803(0.001645)	0.029926(0.001811)

(t_{ZX},t_{ZY})	X	Y	Z	I(A:B)	I(A:C|B)	I(B:C|A)
(17,-12)	C	A	B	0.099998(0.005746)	0.000651(0.000987)	0.057602(0.001489)
(0,10)	A	C	B	0.000007(0.000050)	0.001048(0.000067)	0.000556(0.000072)
(17,30)	B	C	A	0.057373(0.002429)	0.001284(0.001024)	0.015048(0.000976)
(-24,-12)	A	B	C	0.013429(0.001850)	0.000236(0.000492)	0.047610(0.001315)
(17,12)	C	B	A	0.021536(0.002223)	0.040689(0.001056)	0.000195(0.000634)
(-5,-12)	B	A	C	0.023319(0.007910)	0.104093(0.005653)	0.001681(0.001738)

Fig. 6. 3pMI analysis of action potential trains obtained by the simulation of three-neuron network models. **a)** A three-neuron network model with Z→X, Z→Y and X→Y connections. To have spontaneous action potentials, each neuron had a pulse input with random intervals that had a normal distribution. m(*X*) (mean of the pulse input interval of *X*)=95ms, m(*Y*)=90 ms, m(*Z*)= 60 ms. The simulation was done for 200s. The total number of action potentials; *X*:4000, *Y*: 5400, *Z*:3150. (1) The connection structure of a three-neuron network. (2) 3pMI analysis of action potential trains. 3pMI has two positive peaks at (17,12) and (17,30), and one negative peak at (0,12). Arrowheads indicate the position of the significant 3pMI peaks. Bold line represents the time difference of direct connections determined by the analysis. (3) Recalculated I(***A***:***B***), I(***A***:***C***|***B***) and I(***B***:***C***|***A***) and their bounds (α=1%) over the peak region. Bound values are shown in parentheses. **b)** A three-neuron network model with Z→X, Z→Y, Z←Y and X→Y connections. m(*X*)=105 ms, m(*Y*)=120 ms, m(*Z*)=90 ms. The total number of action potentials; *X*:4300, *Y*:5150, *Z*:4850. (1) The connection structure of a three-neuron network. (2) 3pMI analysis of action potential trains. 3pMI has five major peaks, four positive peaks at (17,–12), (17,12), (17,30) and (–24,–12), and one negative peak at (0,10), and one secondary peak (–5,–12) is also statistically significant. Dashed line represents the time difference of the indirect interaction. (3) Recalculated I(***A***:***B***), I(***A***:***C***|***B***) and I(***B***:***C***|***A***) and their bounds (α=1%) over the peak region (reprinted from Yamada et al. 1996, with kind permission from Elsevier Science).

(17,30) and (0,12), are found in the graph. These time differences and corresponding time differences of t_{xy}, e.g., t_{xy}=–5 at (17,12), are candidates of effective connections. The recalculated 2pMI and 2pCMI over the peak region are used to estimate the effectiveness of each connection. For example, consider the analysis of the peak at (17,12). The neurons are rearranged and renamed *A*, *B* and *C* in the order of the time difference. In this case, $Z \rightarrow A$, $Y \rightarrow B$, $X \rightarrow C$. As described in the section on "Estimation of Effective Connections", the statistical effectiveness of the *A*-*B* connection is estimated by *I(A:B)*. In contrast, the statistical effectiveness of *A*-*C* and *B*-*C* connections must be estimated by *I(A:C|B)* and *I(B:C|A)*, respectively. If *I(A:B)*, *I(A:C|B)* or *I(B:C|A)* are less than their bounds, the corresponding connections are not considered to be effective. Recalculated 2pMI and 2pCMI are shown in Figure 6-a(3). Because *I(B:C|A)* is less than its bound, the *B-C* connection is not effective. Since others are larger than their bounds, the *A-B* and *A-C* connections correspond to effective ones. So t_{zy}=12, and t_{zx}=17 are considered to be time differences of effective connections.

A similar analysis is applicable to the other peaks. From the peak of (17,30), t_{zx}=17 and t_{xy}=13 are effective connections (see 2pMI and 2pCMI at (17,30) of Figure 6-a(3)). From the peak of (0,12), t_{xy}=12 and t_{zy}=12 are effective connections. Results obtained in three significant 3pMI peaks are consistent with each other. The t_{zy}=12, t_{zx}=17 and t_{xy}= 12 correspond to effective connections in this three-neuron network.

Figure 6-b shows another example of the analysis of a three-neuron network. In this case, there are five major statistically significant peaks: four positive peaks at (17,–12), (17,30), (–24,–12), (17,12), and one negative peak at (0,10). Similar analysis deduces the connection structure. Recalculated 2pMI and 2pCMI indicate that $Z \rightarrow X$ (t_{zx}=17), $Z \leftarrow Y$ (t_{zy}=–12), $Z \rightarrow Y$ (t_{zy}=12), and $X \rightarrow Y$ (t_{xy}=12) connections are effective (Figure 6-b(3)).

There are some small peaks in Figure 6-b. For example, the positive peak at (–5,–12) indicates apparently direct, effective $Z \leftarrow Y$ (t_{zy}=–12) and $X \leftarrow Y$ (t_{xy}=–7) connections (Figure 6-b(3)). But the latter is shown to be indirect at (17,12) on the line of t_{xy}=–5 (the dashed line in Figure 6-b).

In a cross-correlation method, an indirect interaction can be determined by estimating other neurons' contribution, while no direct evidence for a direct connection can be obtained. Therefore, a correlation that is not determined to be indirect at all the peaks at the corresponding time difference must be considered to be direct, while one that is determined to be indirect at one or more peaks must be considered to be indirect. Because the time difference of t_{xy}=–5 is determined to be indirect at (17,12), the *X-Y* correlation of t_{xy}=–5 is considered to be indirect. The peak at (–5,–12) is produced by the combination of direct connection of t_{zy}=–12 and the indirect interaction of t_{xy}=–7; such a peak is called a "secondary peak".

Expansion to *n*-Neuron Network Analysis and Example of a Four-Neuron Network Analysis

The analysis of an *n*-neuron network can be done in a similar procedure. You should convert action potential trains into sequences of 0's and 1's, calculate cross-correlation histograms and probablilities necessary for *n*pMI calculation, calculate *n*pMI at each time difference and its bounds, find significant *n*pMI peaks, recalculate 2pMI and 2pJCMI over each significant *n*pMI peak region, and then estimate the statistical effectiveness of each connection.

Figure 7 shows the procedure of the analysis of a four-neuron network. First, calculate 4pMI at each time difference, and find out the statistically significant peaks. In this case, there are 14 statistically significant peaks, 4 positive and 10 negative peaks (the upper table of Figure 7(2)). For every peak, neurons are rearranged according to time differences, and then 2pMI or 2pJCMI and their bounds are calculated, and finally effective connections are determined. For example, at the peak of (t_{wx},t_{wy},t_{wz})=(–11,–22,10) neurons are rearranged to *A*, *B*, *C* and *D* (*C* as *W*, *B* as *X*, *A* as *Y* and *D* as *Z*). $I(A{:}B)$, $I(A{:}C|B)$, $I(B{:}C|A)$, $I(A{:}D|B,C)$, $I(B{:}D|A,C)$ and $I(C{:}D|A,B)$ are used to determine the effective connections. As shown in the lower table of Figure 7(2), $X \leftarrow Y$ (t_{xy}=–11), $W \leftarrow X$ (t_{wx}=–11), $W \rightarrow Z$ (t_{wz}=10) are considered to be effective. The results for all significant peaks are consistent, and show that $W \leftarrow X$ (t_{wx}=–11), $W \rightarrow Y$ (t_{wy}=11), $W \rightarrow Z$ (t_{wz}=11), $X \leftarrow Y$ (t_{xy}=–11), $X \leftarrow Z$ (t_{xz}=–11), $Y \rightarrow Z$ (t_{yz}=11) and $Y \leftarrow Z$ (t_{yz}=–11) are considered to have effective connections.

We assume that all neurons in an *n*-neuron network have connections with others. If one or more neurons or a group of neurons have no connection with the remaining

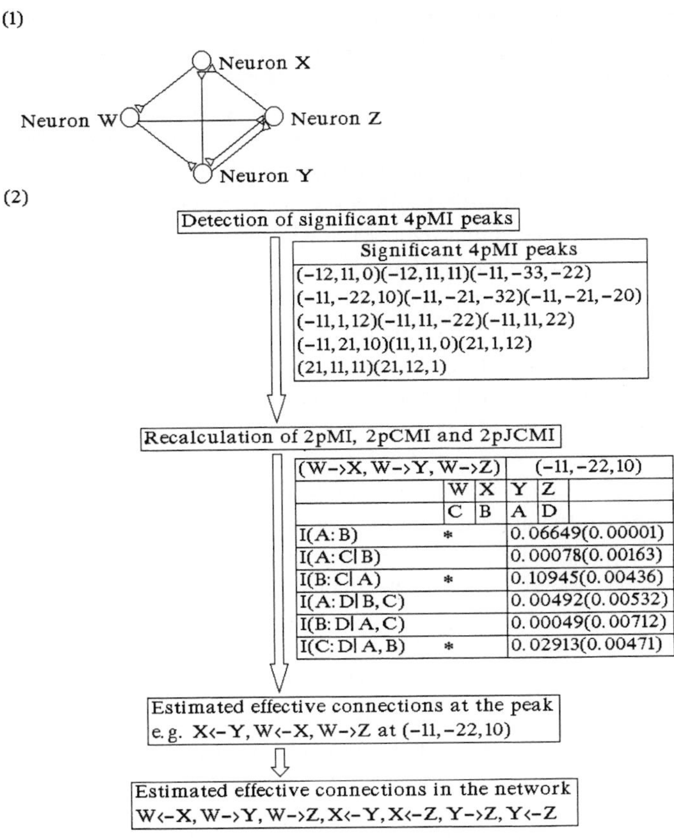

Fig. 7. Analysis of a four-neuron network (W, X, Y, and Z). m(W)=73ms, m(X)=69ms, m(Y)=62ms, m(Z)=56ms. The simulation was done for 3100 s. The total number of action potentials; W:89390, X:104390, Y:106950, Z:108160. (1) The connection structure of the four-neuron network. (2) Analysis method of action potential trains of the four-neuron network. Detect significant 4pMI peaks, recalculate 2pMI and 2pJCMI over the peak region, estimate the effectiveness of each connection structure of the network. The upper table shows the significant 4pMI peak positions, and the lower table shows recalculated 2pMI and 2pJCMI and their bounds for one of the significant 4pMI peaks at (−11,−22,10). The asterisk in the lower table shows the effective connections (reprinted from Yamada et al. 1996, with kind permission from Elsevier Science)

ones, there is no significant npMI peak and no analysis can be done. In such a case, a combination of neurons that produces significant npMI peaks and has a maximal neuron number should be found out and analyzed.

Comments

As described in the previous sections, the information-theoretical analysis has the following advantages:

- It can infer the connection structure of n-neuron networks.
- It amplifies correlational measures beyond the bound of noise by its nonlinear function as shown in Figure 4.
- The recalculated SCC allows a fairly effective estimation of synaptic strengths.

The networks used in Figures 6 and 7 contained only excitatory synapses. The information-theoretical analysis is also applicable to networks with inhibitory synapses. For example, a neural network containing an inhibitory synapse (a neural network of three neurons containing excitatory $X \rightarrow Z$ and inhibitory $Z \rightarrow Y$ synapses) was also analyzed, resulting in accurate prediction of the connection structure.

However, there are some deficiencies of the information-theoretical analysis.

- *Possibility of missing a connection.* The main point in information-theoretical analysis is that a connection is not considered to be effective if the corresponding 2pMI or 2pJCMI is zero. However, under unlikely conditions, 2pMI or 2pJCMI is equal to zero even if the connection of the corresponding pair exists (see Yamada et al. 1996 for the detailed conditions). One example is that $I(A{:}C|B)=0$ when $p(s(B)=1|s(A)=1)=1$ and $p(s(B)=1|s(A)=0)=0$. This condition means that B always fires after A's firing and has no spontaneous firing. Because the activity of B is not an independent variable but completely depends on A, no statistical method can determine whether $A \rightarrow C$ connection or $B \rightarrow C$ connection is effective.
- *Requirement of large amounts of data.* To detect the correlation among n neurons, large amounts of data are required. For example, about 1000 and 50,000 action potentials were needed to detect the significant 3pMI peaks in Figure6 and the signficant 4pMI peaks in Figure7, respectively. The minimum size of data to detect the correlation does not depend on the analysis method but on characteristics of interactions, e.g., synaptic strengths, firing probabilities *etc.* As shown in Figures 3 and 4, the information-theoretical analysis has the advantage of detecting smaller correlations than a scatter diagram (Perkel et al. 1975), SCCH (Melssen and Epping 1987), SUR (Aertsen et al. 1989) and the method by Yang and Shamma (1990).
- *Possibility of incomplete network reconstruction.* The method of analysing n-neuron networks is constructed under the condition that the activity of all neurons in the network is recorded. If the activity of some neurons is not recorded, a deduced connection structure is no longer conclusive: for example, any estimated direct connections could be indirect interactions depending on an unrecorded neuron. No statistical method can avoid this disadvantage. Even if an optimal statistical method exists, it can only determine a minimum effective connection structure among the recorded neurons, which can be obtained by the information-theoretical analysis.
- *Applicability to simple networks only.* The current information-theoretical analysis is expected to be powerful in deducing the connection structure of a simple nervous system, because it assumes simple synaptic interactions. The information-theoretical analysis will detect significant correlations, but it is unlikely to be easy to infer a connection structure of a complex nervous system like the mammalian brain where an enormous number of neurons is cooperatively working to perform sophisticated tasks. For example, it is difficult to analyze the complex dynamic interactions among cortical neurons proposed by Softky and Koch (1992).

As discussed above, it is difficult to infer the complete connection structure of real neuronal networks consisting of n neurons. However, the information-theoretical analysis can detect the interactions between neuron pairs and estimate indirect correlations among neuron triplets. Results of such analyses will be the starting point to infer the connection structure of real neuronal networks.

References

Aertsen AMHJ, Gerstein GL, Habib MK, Palm G (1989) Dynamics of neuronal firing correlation: modulation of "effective connectivity". J Neurophysiol 61: 900–917

Blahut RE (1972) Computations of channel capacity and rate-distortion functions. IEEE Trans Inform Theory IT-18: 460–473

Cohen LB, Lesher S (1986) Optical monitoring of membrane potential: methods of multisite optical measurements. Soc Gen Physiol Ser 40: 71–99

Eckhorn R, Grüsser OJ, Kröller J, Pellnitz K, Pöpel B (1976) Efficiency of different neuronal codes: information transfer calculations for three different neuronal systems. Biol Cybern 22: 49–60

Eckhorn R, Pöpel B (1974) Rigorous and extended application of information theory to the afferent visual system of the cat. I. Basic concepts. Kybernetik 16: 191–200

Fagen RM (1978) Information measures: statistical confidence limits and inference. J Theor Biol 73: 61–79

Fuller MS, Looft FJ (1984) An information-theoretic analysis of cutaneous receptor responses. IEEE Trans Biomed Eng BME-31: 377–383

Gerstein GL, Bloom MJ, Espinosa IE, Evanczuk S, Turner MR (1983) Design of a laboratory for multineuron studies. IEEE Trans Syst Man Cybern SMC-13: 668–676

Gerstein GL, Aertsen AMHJ (1985) Representation of cooperative firing activity among simultaneous recorded neurons J Neurophysiol 54: 1513–1528

Ikeda K, Otsuka K, Matsumoto K (1989) Maxwell-Bloch turbulence. Prog Theor Phys Suppl 99: 295–324

McGill WJ (1955) Multivariate information transmission. IRE Trans Inf Theory 1: 93–111

Melssen WJ, Epping WJM (1987) Detection and estimation of neural connectivity based on crosscorrelation analysis. Biol Cybern 57: 403–414

Nakahama H, Yamamoto M, Aya K, Shima K, Fujii H (1983) Markov dependency based on Shannon's entropy and its application to neural spike trains. IEEE Trans Syst Man Cybern SMC-13: 692–701

Nakashima M, Yamada S, Shiono S, Maeda M, Satoh F (1992) 448-detector optical recording system: development and application to *Aplysia* gill-withdrawal reflex. IEEE Trans Biomed Eng BME-39: 26–36

Novak JL, Wheeler BC (1986) Recording from the *Aplysia* abdominal ganglion with a planar microelectrode array. IEEE Trans Biomed Eng BME-33: 196–202

Optican LM, Richmond BJ (1987) Temporal encoding of two-dimensional patterns by single units primate inferior temporal cortex. III. Information theoretic analysis. J Neurophysiol 57: 162–178

Optican LM, Gawne TJ, Richmond BJ, Joseph PJ (1991) Unbiased measures of transmitted information and channel capacity from multivariate neuronal data. Biol Cybern 65: 305–310

Palm G, Aertsen AMHJ, Gerstein GL (1988) On the significance of correlations among neuronal spike trains. Biol Cybern 59: 1–11

Perkel DH, Gerstein GL, Moore GP (1967a) Neuronal spike trains and stochastic point processes I. The single spike train. Biophys J 7: 391–418

Perkel DH, Gerstein GL, Moore GP (1967b) Neuronal spike trains and stochastic point processes II. Simultaneous spike trains. Biophys J 7: 419–440

Perkel DH, Gerstein GL, Smith MS, Tatton WG (1975) Nerve-impulse patterns: a quantitative display technique for three neurons. Brain Res 100: 271–296

Reinke W, Diekmann V (1987) Uncertainty analysis of human EEG spectra: a multivariate information theoretical method for the analysis of brain activity. Biol Cybern 57: 379–387

Shannon CE (1948) A mathematical theory of communication. Bell Syst Techn J 27: 379–423

Softky WR, Koch C (1992) Cortical cells should fire regularly, but do not. Neural Comp 4: 643–646

Surmeier DJ, Weinberg RJ (1985) The relationship between crosscorrelation measures and underlying synaptic events. Brain Res 331: 180–184

Tsukada M, Ishii N, Sato R (1975) Temporal pattern discrimination of impulse sequences in the computer-simulated nerve cells. Biol Cybern 17: 19–28

Williams WJ, Shevrin H, Marshall RE (1987) Information modeling and analysis of event related poteintials. IEEE Trans Biomed Eng BME-34: 928–937

Wilson MA, McNaughton BL (1993) Dynamics of the hippocampal ensemble code for space. Science 261: 1055–1058

Yamada S, Nakashima M, Matsumoto K, Shiono S (1993) Information theoretic analysis of action potential trains: I. Analysis of correlation between two neurons. Biol Cybern 68: 215–220

Yamada S, Matsumoto K, Nakashima M, Shiono S (1996) Information theoretic analysis of action potential trains: II. Analysis of correlation among *n* neurons to deduce connection structure. J Neurosci Methods 66: 35–45

Yang X, Shamma SA (1990) Identification of connectivity in neural networks. Biophys J 57: 987–999

Abbreviations

SMI	signed two-point mutual information
SCC	signed channel capacity
SCCH	scaled cross-coincidence histograms
SUR	surprise
2pMI	two-point mutual information
2pCMI	two-point conditional mutual information
2pJCMI	two-point joint conditional mutual information
3pMI	three-point mutual information
*n*pMI	*n*-point mutual information

Chapter 21

Linear Systems Description

AMIR KARNIEL and GIDEON F. INBAR

Introduction

The systems approach is a widely used practice in modeling artificial as well as natural phenomena. Each process or sub-process is viewed as an input-output system, as described graphically in Fig.1.

This approach is used extensively in engineering, for example in modeling electronic and mechanical systems and in chemical process description. In this chapter we describe this approach and its application to biological systems in general and the nervous system in particular. The systems approach can be used as a modeling tool to comprehend the function of the system and to produce a hypothetical model which can be tested in experiments. It is useful in describing and characterising experimental results, at times by relating the anatomical and physiological properties to the measured variables (see for example the muscle spindle transfer function, Houk 1963). Mathematical modeling of part of the neurological system can be used to study that and other parts by simulation. (See McRuer et al. 1968 for an example of combination of models to the motor neurons, the muscle and the muscle spindle in a closed loop). The systems approach modeling is also useful for building interfaces to engineering systems in order to develop measurement devices or artificial organs such as hearing aids, pacemakers and artificial limbs.

Linear systems are highly popular models due to their simplicity and since they are very convenient for mathematical analysis. Beyond the above technical advantage, many systems can be modeled as linear systems at least for a limited range of operation.

Let us begin with a short description of the terminology of this field. Figure 1 describes the general notion of an input-output system in a block diagram. The input is u and the output is y. They usually describe a physical quantity as potential, current force or position. In this chapter their value may be scalar real numbers, or vectors of real

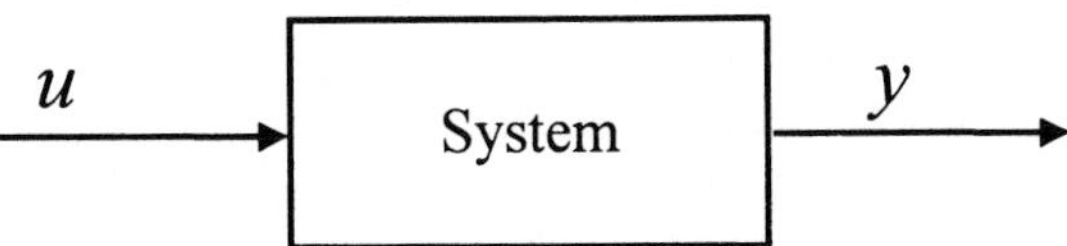

Fig. 1. Input-Output system. u is the input and y is the output. They can be scalars or vectors that represent physical values such as potential, current, force, or position. They can also be functions of time, that is trajectories. The output is generally a function of the input and possibly of the state of the system.

Amir Karniel, Technion – Israel Institute of Technology, Department of Electrical Engineering, Haifa, Israel (phone: 972-4-8294662; fax: 972-4-8323041; e-mail: karniel@tx.technion.ac.il)

Gideon F. Inbar, Technion – Israel Institute of Technology, Department of Electrical Engineering, Haifa, Israel (phone: 972-4-8294718; fax: 972-4-8323041; e-mail: inbar@ee.technion.ac.il)

numbers in the case of multiple inputs and outputs. The inputs and outputs can also be functions of time, that is trajectories, either discrete or continuous. We will concentrate on deterministic mapping systems, which means that for a specific input there is a specific singular output, i.e., the output is a function of the input, $y=f(u)$. Let us add two important qualifiers for a system:

- A system is *time-invariant*, roughly, if the system properties do not change with time.
- A system is *linear* if it satisfies the property of superposition, that is, for any couple of inputs and outputs $y^1=f(u^1)$ and $y^2=f(u^2)$, the equation $ay^1+by^2=f(au^1+bu^2)$ will be satisfied for any couple of scalars a and b.

A system that satisfies both of these properties is naturally called a linear time-invariant (LTI) system. All the systems in this chapter are LTI unless otherwise mentioned.

In this chapter we introduce linear systems, static and dynamic, then we move to a detailed description of each step in describing, modeling and analyzing linear systems, as is outlined in the next section.

Outline

In this section we combine a short outline of this chapter with a description of the biological system modeler work. The main stages are illustrated in Fig. 2.

The first step in biological modeling by the systems approach is to choose or define the inputs and outputs. This can be done by inspection of the anatomical structure of the system and incorporating prior knowledge about the physiological function of the modeled system. Such inspection can lead to an electrical or mechanical model, or sometimes directly to an equation that describes relations between inputs and outputs. For example, by looking at the physiological structure of a small region in the retina, one can choose the input to be the light intensity and the output to be the firing rate of a related axon. Then one can suggest a simple linear model such as $f = a \cdot I$, where f is the firing rate, I is the light intensity, and a is a constant that is sometimes called a "gain" . Another suggestion may include a more sophisticated electrical model of the nerve cell, which finally produces a differential equation that relates the output to the inputs. Part 1 deals with the first suggestion, i.e., static relations, where linear artificial neural networks are described and an example for an associative memory is given. The major part of the chapter is about dynamic models, that is, where the inputs and outputs are functions of time and the output may be a function of previous events and not only of the current input. Part 2 introduces dynamic linear systems and Part 3 deals with electrical and mechanical models, and how to derive the differential equations from the graphic description of the systems model; these procedures are based on Kirchhoff's and Newton's Laws. This part contains various examples for modeling the nervous system, synapses and muscles. Once we have a model, that is, a set of equations that describes the biological system, we can check the behavior of the model in various cases in order to produce hypotheses that can be later checked on real data from the biological system. Laplace and Z transforms are powerful tools to analyze and manipulate linear systems and they are the subjects of Part 4.

A model usually contains some parameters, for example, the parameter a in the simple model of the retina above. One of the objectives of the modeler is to estimate the values of these parameters. This estimation is based on measurements of the system's input and output. An estimation method for linear systems is described in Part 5. Part 6 describes how to integrate linear models of subsystems in a block diagram in order to get a model of the complete system; this method is used extensively in control theory, and therefore examples from the field of motor control and of temperature regulation are given. Models of artificial means can also be incorporated, such as measurement devices, artificial organs or functional neuromuscular stimulation for the paralyzed patient. Recently it has become a fashion to discuss nonlinear models and chaos, which seems to appear in many natural systems. This observation is correct; however, in many cases, the powerful linear system description tools can still be used in order to describe and analyze nonlinear systems. This is the subject of Part 7. The most common tool to handle nonlinear systems is linearization, which is finding a linear model that is similar to the nonlinear system in some area of interest. Other options are linear time-varying or parameter-varying models such as the Hodgkin and Huxley membrane model; two other options are pre- or post-processing of the linear system which are the terms that are used in the field of neural computation. One should note that the work of the modeler usually consists of a few iterations of improving the model, designing new experiments to obtain new data, estimating the parameters and analyzing the results, as illustrated graphically in Fig. 2.

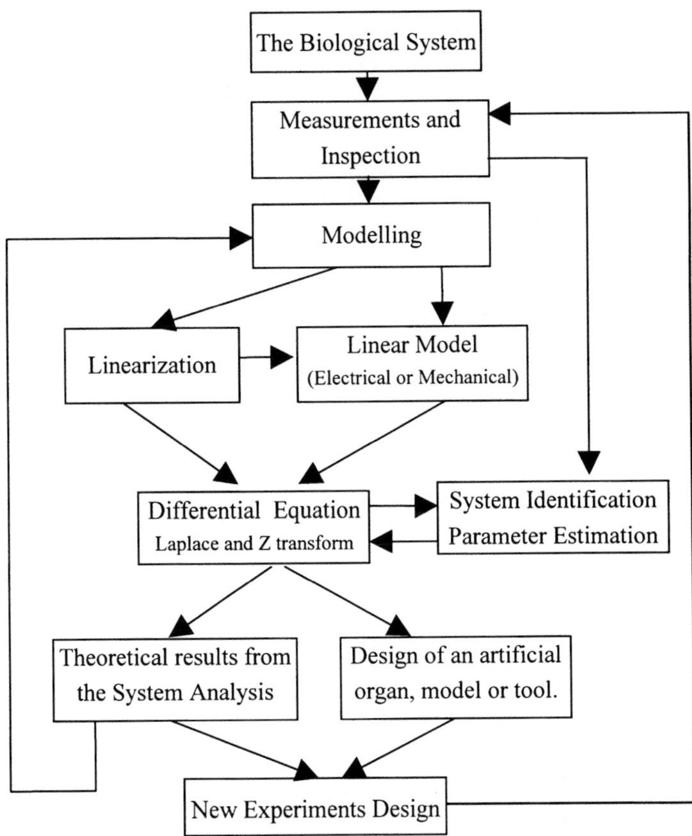

Fig. 2. Outline of linear systems description: The first step is to choose or define the inputs and the outputs and to acquire data about the system from measurements and inspection of the physical (i.e., anatomical) structure of the system. Then one can move directly to system identification by trying to fit a linear model to the data, or first draw a physical or electrical model and then estimate its parameters. The given system may be linear or nonlinear; in the case of a linear system we can describe it with a mechanical or an electrical equivalent model and then write the differential equations. In the case of a nonlinear system we can use a linear approximation and then continue as if we had a linear system. The linear time-invariant difference equation can be transformed to the Laplace domain to get the transfer function. These functions can be used for various goals, such as system identification, artificial control and modeling in order to anticipate the system behaviour, and analyzing its properties. Finally, one can use the results of this procedure to design a new experiment and go back to the measurement step.

Part 1: Static Linear Systems

In a static system, the output depends on the input only and does not depend on time. The simplest linear input-output static system is the system $y = a \cdot u$ where a is a constant. If we wish to extend this system to multiple inputs and outputs, we can use vector notation and write $Y = A \cdot U$ where Y and U are the output and input vectors and A is the transfer matrix. We restrict our description to homogeneous systems, i.e., those characterized by zero in the input producing zero in the output. However, it is easy to move to the general case by introducing a new input that is constant and then, for one dimension, the relation would be $y = a_0 + a_1 \cdot u$.

A basic element in many neural network models and in artificial neural networks is such a linear relation as illustrated in Fig. 3. The inputs u can model the activity of neurons that influence the modeled neuron; the constants w_i represent the synaptic strength or position; and the output can represent the neuron potential or firing rate.

In the case of multiple outputs, that is, multiple neurons, we can construct an artificial neural network (ANN), as illustrated in Fig. 4. The relation between the output and the input is $y_j = \sum_i w_{ij} u_i$. For m outputs and n inputs one can write the input-output relation as $Y = W^T \cdot U$ using the following matrixes notation:

$$Y = \begin{bmatrix} y_1 \\ y_2 \\ \vdots \\ y_m \end{bmatrix} \quad U = \begin{bmatrix} u_1 \\ u_2 \\ \vdots \\ u_n \end{bmatrix} \quad W = \begin{bmatrix} w_{11} & w_{12} & \cdots & w_{1m} \\ w_{21} & w_{22} & \cdots & w_{am} \\ \vdots & \vdots & \ddots & \vdots \\ w_{n1} & w_{n2} & \cdots & w_{nm} \end{bmatrix} \quad W^T = \begin{bmatrix} w_{11} & w_{21} & \cdots & w_{n1} \\ w_{12} & w_{22} & \cdots & w_{n2} \\ \vdots & \vdots & \ddots & \vdots \\ w_{1m} & w_{am} & \cdots & w_{mn} \end{bmatrix}$$

Each single weight represents the relation or association between a specific pair of input and output. Therefore this network is sometimes called "associative network". By implementing the rule of Hebb (Hebb 1949), i.e., adding strength to connections between neurons that act simultaneously, one can construct a basic model for associative memory.

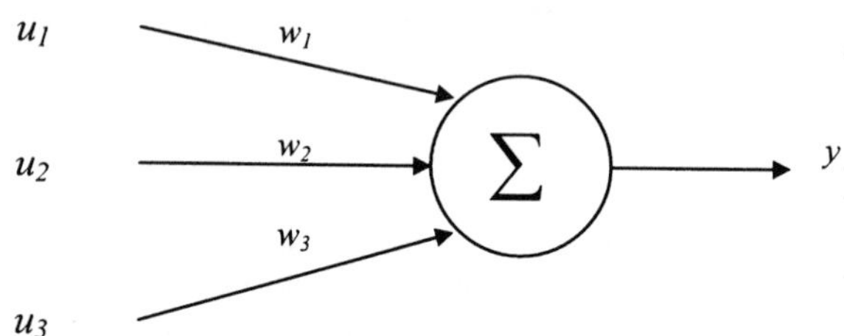

Fig. 3. Static linear system: y= 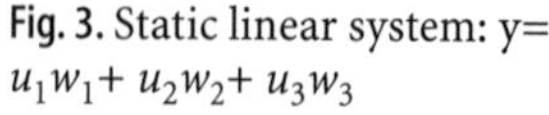 $u_1 w_1 + u_2 w_2 + u_3 w_3$

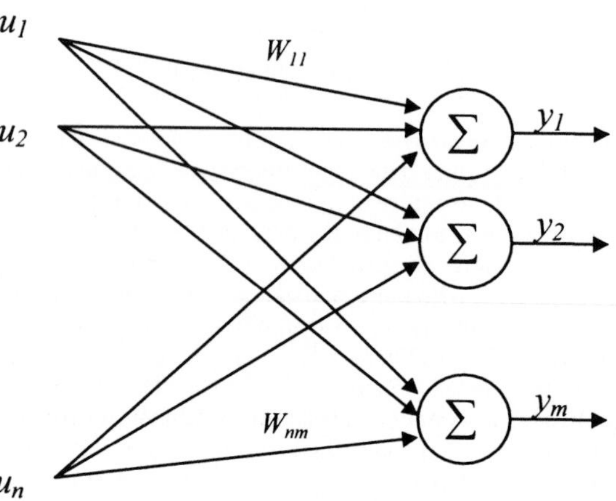

Fig. 4. An associative linear artificial neural network

The technical definition of memory is a device that can store and recall information, the input is called an address and the output is the data. Associative memory can recall the data in face of an inaccurate address, as long as the corrupted address is close enough to the true address. This property is very similar to the human brain memory. The next example describes a simple implementation of associative memory with the linear neural network.

Example 1: Linear Neural Network as an Associative Memory

This example demonstrates the use of a linear neural network as an associative memory. The input can be referred to as the address and the output as the data. For example we can think of the problem of face recognition, where the input would represent the face (maybe a vector of bitmap from a camera, or better a vector of features) and the output would be an identity number of the person. The structure of the network described in Figure4 suggests that $y_j = \sum_i w_{ij} \cdot u_i$. Let us choose the domain of the inputs and outputs to be the binary range {-1,+1} and denote the items that we wish to insert into the memory with superscript l=1,2...,L where L is the number of items (these items are sometimes referred to as the "learning examples"). According to the rule of Hebb, the weights represent the correlation between two neurons, and in this case the correlation between the output and the input. This can be done mathematically by the equation $w_{ij} = \varepsilon \cdot \sum_{l=1}^{L} x_i^l \cdot y_i^l$ where the constant ε is chosen to be $1/n$.

Let us look at a simple numerical example of two memory items.

$$u^1 = [-1,1,1,-1] \quad y^1 = [1]$$

$$u^2 = [1,1,1,1] \quad y^2 = [-1]$$

In this case the indexes are in the range $i \in \{1,2,3,4\}$ $j = 1$ $l \in \{1,2\}$ and the weights matrix will be:

$$W = \left\{ \varepsilon \cdot \sum_{i=1}^{L} u_k^1 y_j^1 \right\} = \frac{1}{4} \begin{bmatrix} -1 \cdot 1 + 1 \cdot -1 \\ 1 \cdot 1 + 1 \cdot -1 \\ 1 \cdot 1 + 1 \cdot -1 \\ -1 \cdot 1 + 1 \cdot -1 \end{bmatrix} = \begin{bmatrix} -1/2 \\ 0 \\ 0 \\ -1/2 \end{bmatrix}$$

One can verify that the output of each memory item is correct. The next question is related to the generalization capability, that is, what will be the result of a new input vector that was not learned?

Associative networks should produce the nearest stored item. Let us check it for the following vector: $u = [-1,-1,-1,-1]$. For this vector, the output is $y = \sum_i w_i \cdot u_i = 1/2 + 1/2 = 1$, which is the result of the first item above, and one can see that the new item is closer to the first item (only two bits were inverse, compared with four bits in the second item).

This associative memory has many drawbacks, the stored data should be binary orthogonal vectors, there are a lot of connections and the capacity is low. There are other nonlinear associative memories, but they lack the simplicity and the mathematical tractability of the linear model and they are beyond the scope of this chapter. For more information about this and other ANN architectures see Chapter 25 and Fausett (1994)

The architecture described above and illustrated in Fig. 4 is the most general static linear neural network, since adding more layers of neutrons will not change the capability of this architecture. Note that this is not the case for nonlinear neural networks, where adding layers can enhance the capabilities of the architecture.

Part 2: Dynamic Linear Systems

Most of this chapter will deal with dynamic linear systems. In this section we will explain the notion of dynamic linear systems and the common methods to describe them. In the following sections specific procedures and examples will be described.

In dynamic linear systems, the inputs and outputs are functions of time, i.e., trajectories. Five common methods to describe such a system are described below:

- *Electronic or mechanical diagram.* One common way to describe a system is through graphic description of its physical elements and their connection. This method provides comprehensive description of the system's structure and an easy way to get qualitative understanding of the system by an expert. However, it is not always easy to predict the exact behavior of the system by examining the graphic diagram; therefore it is usually numerically simulated or transformed to a set of equations that can be mathematically analyzed and manipulated.
- *Differential or difference equation.* A system can be described by a set of differential equations relating the input and the output (or difference equation in the discrete case), the output will be the solution of the equation. A simple way to represent a linear dynamic system as a differential equation is the following standard differential equation: $y=\sum_i w_i \cdot u_i = y(t) = w_1 u(t) + w_2 \dot{u}(t) + \cdots + w_{N+1}\dot{y}(t) + w_{N+2}\ddot{y}(t)\cdots$ where a dot over a variable represents the time derivative, i.e.,

  $$\dot{u}(t) = \frac{\partial u(t)}{\partial t}, \quad \ddot{u}(t) = \frac{\partial^2 u(t)}{\partial t^2}$$

  etc. or in the discrete time: $y(t) = w_1 u(t) + w_2 u(t-1) + \cdots + w_{N+1} y(t-1) + w_{N+2} y(t-2)\cdots$ where t is a natural number, i.e., $t \in N$.
- *State space description.* The notion of state of the system helps us in separating the dynamic part of the system. One introduces a set of new variables that represent the state of the system in a way that the output is a static function of the state and possibly the inputs. The behavior of the variables is dominated by its own differential equation. The state variables may have a physical interpretation, such as the potential of a capacitor. In the linear case these equations are linear and can be written with matrix notation as follows:

  $$\dot{x} = Ax + Bu$$
  $$y = Cx + Du$$

  where x is the state, u is the input, and y is the output.
- *Impulse response.* As noted above, the basic property of linear systems is the superposition property; that is, the response of a linear system to the sum of two inputs is equal to the sum of the system's responses to each input. Thus, if we had a simple input, such that any other input could be produced as a linear sum of that input, and if we knew the system's response to that simple input, we could calculate its response to any other input. Such an input function is the impulse (also known as the delta function, $\delta(t)$), and the system's response to the impulse is called "impulse response". So if we know the impulse response, we practically know everything about the system. That is one of the major beauties of linear systems. Let us first describe the impulse function and then see how to calculate any response when the impulse response is given. Strictly speaking, an impulse is an abstract mathematical concept. To understand this concept, imagine a rectangular pulse lasting from $t=0$ to $t=\Delta t$ (duration Δt) and amplitude $1/\Delta t$, so that its area equals 1 (unit pulse). Now let Δt approach zero. In the limit, the pulse will be of infinitely short duration and infinitely large amplitude and is called unit impulse (because of its unit area) or delta function, $\delta(t)$, which

is a singular function. Impulses of other areas are obtained by appropriate multiplication with a factor. Unless specified otherwise, impulses are assumed to occur at zero time. For example, $\delta(t-t_o)$ is an impulse that occur at $t=t_o$.
The most important property of the impulse is that, for any regular function that is continuous at $t=0$, $\int_{-\infty}^{\infty} \delta(t) \cdot \phi(t) = \phi(0)$. That is, the impulse "highlights" the function at $t=0$. This function can therefore be thought of as being composed of sequences of such highlights generated by integrals whose integrands are products of and successive, infinitely closely spaced delta functions. Any linear system's response $y(t)$ to an input $u(t)$ therefore is the superposition of the system's impulse response $h(t)$ to all the successive function highlights. This superposition is called a *convolution*. The convolution of h and u is defined as $y(t) = h(t) * u(t) = \int_{-\infty}^{\infty} h(t-\tau) \cdot u(\tau) \cdot d\tau$.
In the discrete case, the delta function is much simpler, its value is one at $t=0$ and zero otherwise. The convolution in the discrete case is defined as $y(t) = h(t) * u(t) = \sum_{m=-\infty}^{\infty} h(t-m) \cdot u(m)$ where t and m are natural numbers. More information about the delta function and the convolution integral can be found in most of the advanced linear systems textbooks, such as Kwakernaak and Sivan (1991) and Lathi (1974), which also contains a graphic view of the convolution integral.
Comments: (i) Many other functions can be used as inputs to a linear system and produce all the information about the system, actually any function that contains all the frequencies. In many cases the step function and the step response are used, and sometimes a random noise signal is used. (ii) The impulse response is a very useful mathematical tool. However, in experiments of the biological system, it is usually not recommended to try to introduce an impulse. In fact, it is not possible to introduce a pure impulse but even an approximated high-energy impulse can cause damage to the system. The biological system is seldom linear in all the frequencies and an impulse can activate nonlinear modes of the system, therefore it is recommended to test and model the system only in its linear regions.

- *Transfer function in the Laplace or Z transform domain*. Given the impulse response and an arbitrary input, one can calculate the output as mentioned in the previous method, but the calculation involves evaluating the convolution integral, which may be a hard task. The main idea of the transforms is to move to another space where the convolution becomes a simple multiplication. The impulse response is transformed to a function that is called the transfer function, and the output in the Laplace domain is the Laplace domain input multiplied by the transfer function. In this way one can combine subsystems to a complex system in a block diagram, as will be described later in this chapter.

In the next parts we further describe, explain and demonstrate how to use these mathematical description tools.

Coherence

Before we start to build models and fit them to our data, we need to validate our assumption that we do have a linear system. A practical method to check the linearity of an unknown system is by calculating the coherence function between the input and the output signals. The coherence function, Γ, is defined as follows:

$$\Gamma(z) = \begin{cases} \dfrac{S_{xy} \cdot S_{yx}}{S_{xx} \cdot S_{yy}} & S_{xx} \cdot S_{yy} \neq 0 \\ 0 & S_{xx} \cdot S_{yy} = 0 \end{cases}$$

where S_{uv} is the cross spectrum of the signals u and v. Most of the mathematical software, such as MATLAB, has the toolboxes and command to calculate the coherence function. (See Chapter 18 for more information about the coherence function.) For an LTI system without noise, the value of the coherence function is one. Therefore, if we find small values of the coherence function, we cannot be sure whether the system is not linear or very noisy. In both of these cases there is no point in estimating parameters for an LTI model. If the coherence function is close to one at the frequencies of interest, one can go on to estimate the parameters, build an LTI model and expect small errors. See Cadzo and Solomon (1987) for a thorough description of linear modeling and the coherence function, and Inbar (1996) for an example of typical coherence values in EMG measurements to estimate mechanical transfer function.

Note: One should notice the estimation procedure in order to get an accurate value of the coherence function, see Benignus (1969) for an estimation procedure.

Part 3: Physical Components of Linear Systems

Dynamic linear systems can be described schematically with simple basic physical components, electrical or mechanical. This method is natural for the description and design of physical systems, and therefore was widely used and developed by mechanical and electrical engineers. The advantage of this description is in the graphic description that is more comprehensible then differential equations due to the correspondence between the graphic elements and the modeled system. There are many simulation programs, such as SPICE, which provide graphic description as well as numerical simulation of such models (see Conant 1993, and Nilsson and Riedel 1996). In this section we will introduce the basic elements of electrical and mechanical systems, and the procedure to get the differential equation from the graphic description, which is based on the well-known laws of Newton and Kirchhoff. For more information about circuit theory and about modeling dynamic systems, see for example Charles and Kuh (1969) and Dorny (1993).

All the examples are from the field of neuromuscular system modeling.

Electrical Models

The basic elements of electrical models are resistor (R), capacitor (C), inductor (L) and sources of potential (V) or current (I). In some cases it is convenient to use conductance (g) instead of resistor, but the relation is simple, they are just the inverse of one another, i.e., g=1/R. From a graphic scheme of the electrical model one can extract the flow of current and the potential at each place in the model. The following Procedure 3.1 and Fig. 5 describe the methods and steps required in order to extract the differential equations out of the schematic description.

Procedure 3.1: Writing the Differential Equation of a Linear Electrical Circuit

1. Using an arrow, mark the current flow direction in each branch of the model.
2. For each node, write the Kirchhoff current law, stating that the sum of incoming currents equals zero: $\sum I = 0$
3. Replace each current by its value according to Fig. 5.
4. Solve or simplify the set of equations.

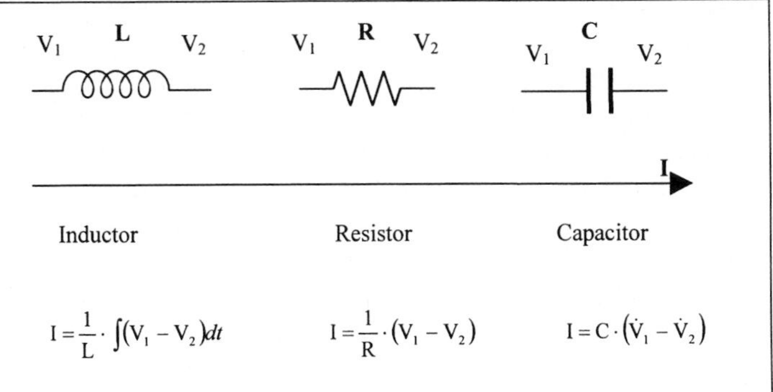

Fig. 5. The basic components of linear electrical circuits.

Example 3.1: Linear Neuron Model

The simplest dynamic linear model of a nerve cell is depicted in Fig. 6. The dendritic tree is represented by the resistors R_1, R_2 and R_3, which transmit the currents generated by the input voltages V_1, V_2 and V_3 to the cell body. The potentials V_1, V_2, and V_3 are due to synapses from other neurons. The currents generated in the dendritic tree are integrated by the capacitor C, representing the cell body membrane capacity. This integrator is "leaky", as represented by the membrane resistance R_4, this model therefore being referred to as "leaky integrator". Let us write the differential equation of this model according to Procedure 3.1.

1. Mark all the currents with an arrow that is going out of the point V_c
2. Due to the Kirchhoff current law we can write:

$$I_{R_1} + I_{R_2} + I_{R_3} + I_{R_4} + I_C = 0$$

3. Now replace the currents by their values according to Fig. 5:

$$\frac{V_c - V_1}{R_1} + \frac{V_c - V_2}{R_2} + \frac{V_c - V_3}{R_3} + \frac{V_c}{R_4} + C\frac{dV_c}{dt} = 0$$

4. Finally, some simplification can be made to get a standard first-order differential equation:

$$\left(\frac{1}{R_1} + \frac{1}{R_2} + \frac{1}{R_3} + \frac{1}{R_4}\right)V_c - \left(\frac{V_1}{R_1} + \frac{V_2}{R_2} + \frac{V_3}{R_3}\right) + C\frac{dV_c}{dt} = 0$$

This first-order equation can be solved analytically or numerically, and can also be transformed to the Laplace domain for further systems modeling and integrating, as will be described in the next sections. At this stage, let us consider some more examples for the execution of Procedure 3.1.

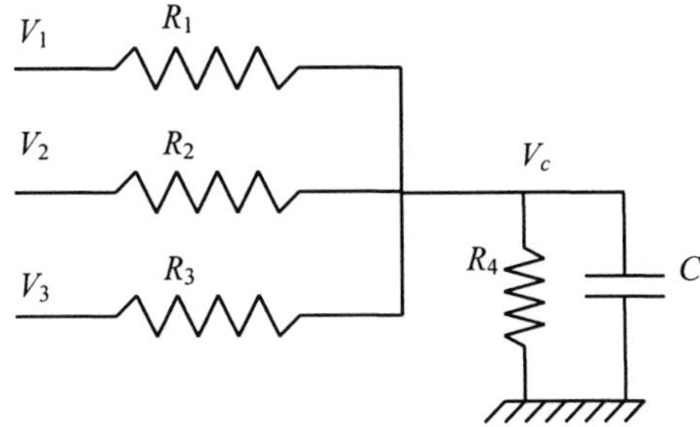

Fig. 6. A leaky integrator model for the nerve

Example 3.2: Linear Membrane Model

The membrane of nerves and muscles, when the potential is near the resting potential, can be modeled as the linear electric circuit in Fig. 7. C_m is the membrane capacity, and each branch represents a current of a single ion type. The potential sources V_K, V_{Na}, and V_{Cl} represent the Nernst potentials of potassium, sodium and chloride, respectively. The resistors R_K, R_{Na}, and R_{Cl} represent the resistance of the membrane to currents of potassium, sodium and chloride, respectively. The resistance is the macroscopic manifestation of the microscopic state of channels within the cell membrane. Note the arrows beside each voltage variable $(V_m, V_{Cl}, V_{Na}, V_K)$. Each voltage variable denotes the potential difference between the arrow's head and its tail. For example, the membrane potential is defined as $V_m = V_{in} - V_{out}$. The direction of the potential sources (the longer line is positive) represents the typical value of the ion potential, as chloride and potassium have negative Nernst potentials and sodium has a positive Nernst potential. In writing the equations of such a model we regard only the arrow's direction; for a biologically plausible model the given data or the results of the calculations are expected to be compatible with the sources' directions as represented by the long and short lines.

Let us write the differential equation of this model according to Procedure 3.1:

$$I_c + I_{Na} + I_K + I_{CL} = 0$$

$$C_m \cdot \frac{\partial V_m}{\partial t} + \frac{(V_m - V_{Na})}{R_{Na}} + \frac{(V_m - V_K)}{R_K} + \frac{(V_m - V_{Cl})}{R_{Cl}} = 0$$

Let us check what happens in the steady state, i.e., when V_m does not change. In this case the time derivative of V_m is equal to zero and one can write the value of the membrane potential as follows:

$$V_m = \left(\frac{V_{Na}}{R_{Na}} + \frac{V_K}{R_K} + \frac{V_{Cl}}{R_{Cl}} \right) \cdot \left(\frac{1}{R_{Na}} + \frac{1}{R_K} + \frac{1}{R_{Cl}} \right)^{-1}$$

Notice that this is a weighted average of the Nernst potentials of the ions, where the weights depend on the membrane resistivity.

This result is similar to the Goldman equation (see Plonsey and Barr, Capter 3):

$$V_m = +\frac{K \cdot T}{q} \cdot \ln \left(\frac{P_K \cdot [K]_e + P_{Na} \cdot [Na]_e + P_{Cl} \cdot [CL]_i}{P_K \cdot [K]_i + P_{Na} \cdot [Na]_i + P_{Cl} \cdot [CL]_e} \right)$$

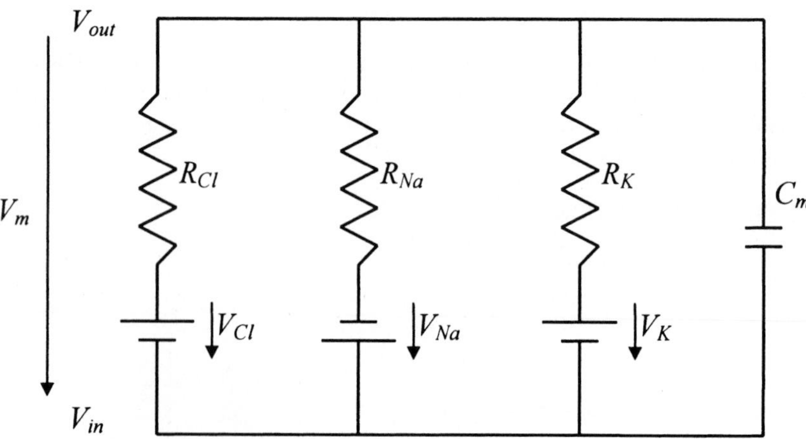

Fig. 7. The subthreshold membrane model

where P_k, P_{Na}, P_{Cl} are the permeability of potassium, sodium and chloride, respectively, [X] stands for the concentration of ion X in the fluid, index i stands for internal fluid and e stands for external fluid. KT/q is a constant equal to 26mV at room temperature.

In the extreme case, such as when the membrane is permeable to one ion only, the results are the same, i.e., the membrane potential equals the ions' Nernst potential. However, these two equations are not identical since their development is based on different sets of assumptions.

Example 3.3: Model of the Postsynaptic Membrane

Many chemical synapses in the nervous system can be described as follows:
- The presynaptic neuron releases neurotransmitters that operate on special sites in the postsynaptic membrane that open channels to specific ions. These ions flow according to the electrodiffusion forces and change the postsynaptic potential.
- The following electrical model (Fig. 8) describes the changes in the postsynaptic potential as a result of the change in the number of opened channels. The potential V_s represents the Nernst potential of the ion, and Δg_srepresents the change in the membrane admittance as a result of one channel that has been opened.

Let us write the equation of this model according to Procedure 3.1. Note that this is a static model (without capacitors or inductors), and remember that the admittance is the inverse of the resistance.

The number of open channels, that is, the number of close switches in Fig. 8, is denoted by n, each close switch adding one more branch to the circuit. Therefore, with I_{g_0} flowing through g_o and I_g flowing through Δg_s:

$$I_{g_0} + \sum_{i=1}^{n} I_g = 0$$

$$g_0 \cdot V_{Post} + n \cdot \Delta g_s \cdot (V_{Post} - V_s) = 0$$

$$V_{post} = \frac{n \cdot \Delta g_s \cdot V_s}{g_0 + n \cdot \Delta g_s}$$

This is a linear model with respect to its currents and potential; however, notice that the relation between the number n of channels and the postsynaptic voltage is nonlinear.

Note: One can add an exponential relation between the presynaptic potential and the number of channels opened. The result is that the relation between post- and presynaptic potential is a sigmoidal function which is very common in ANN models.

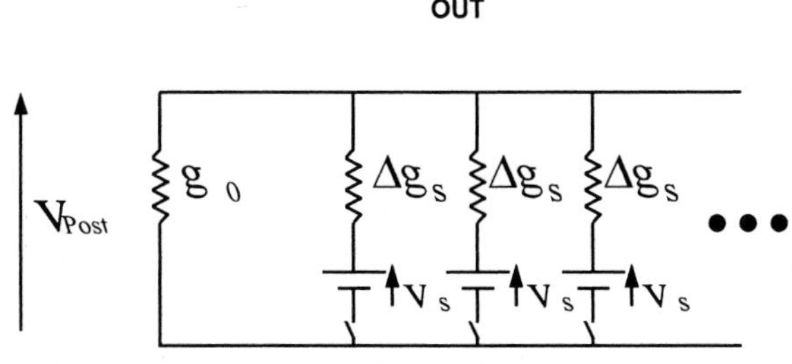

Fig. 8. A model of the postsynaptic membrane

Example 3.4: Cable Model of the Passive Nerve Fiber

This example introduces a widely used model for propagation of potentials within axons and dendrites. This model is valid for the same regime as in Example 3.2, i.e., in the subthreshold region where the membrane can be viewed as a linear system.

In this model fibers are idealized as having a cylindrical geometry as described in Fig. 9 (see Plonsey and Barr 1988, Chapter 6, for a thorough description).

Let us look at the continuous fiber, as if it were constructed from small segments. With slight abuse of the term *dx*, we can refer to the length of each segment as *dx*, and then when *dx* approaches zero, it becomes the integral and differential operator. The electrical model is described in Fig. 10, with the following definitions:

- $r_i \cdot dx$ is the resistance of the internal liquid of the fiber segment to axial current;
- $r_e \cdot dx$ is the resistance of the extracellular liquid of the segment;
- r_m / dx is the resistance of the segment to current through the membrane;
- $C_m dx$ is the capacity of the membrane segment;
- $i_m dx$ is the current through the membrane in one segment of the fiber;
- $i_p dx$ is the current due to an external electrode in one segment of the fiber; it is positive for current entering the extracellular space via polarising electrodes.
- I_i is the axial current in the fiber,
- I_e is the current outside the fiber.

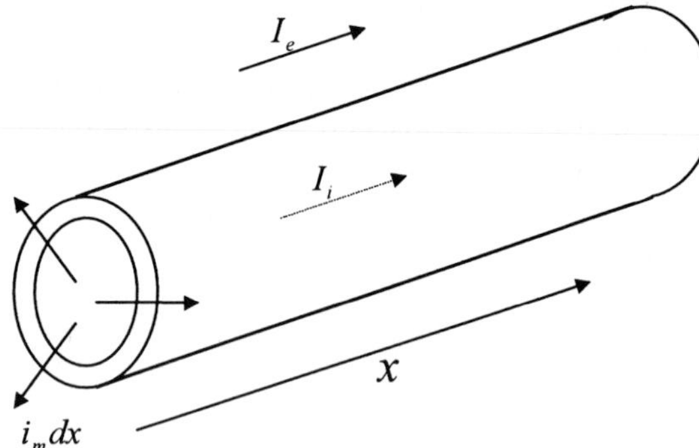

Fig. 9. The nerve fiber as a cylindrical fiber and the currents related to it.

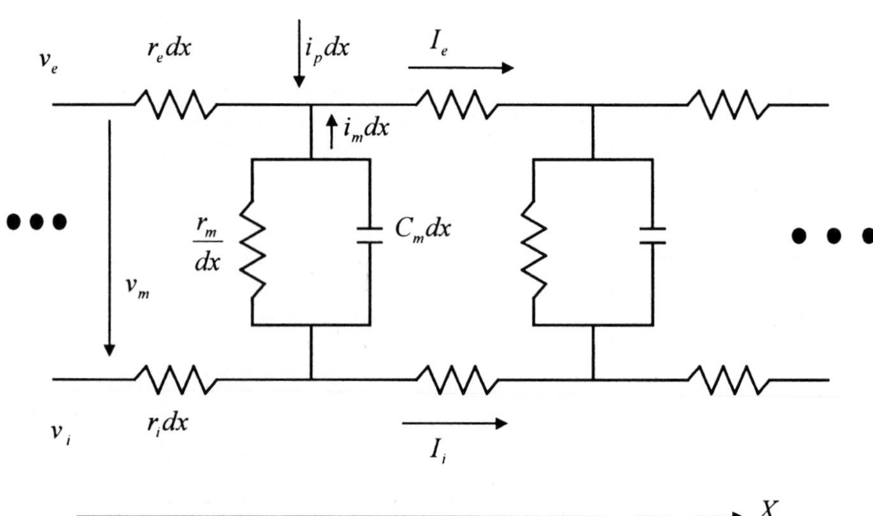

Fig. 10. The electrical linear model of the passive nerve fiber.

Now we can use Procedure 3.1 to write the current law and find out the differential equation that governs the membrane potential in the passive fiber.

From Kirchhoff's current law, we can write the following three equations:

$$i_p dx + i_m dx - dI_e = 0$$
$$i_m dx + dI = 0_i$$
$$i_m = \frac{V_m}{r_m} + C_m \frac{dV_m}{dt}$$

From Ohm's law (or from the definition of the resistive element) we have:

$$\frac{dV_e}{dx} = -I_e r_e$$
$$\frac{dV_i}{dx} = -I_i r_i$$

Let us recall the definition of the membrane potential, $V_m \equiv V_i - V_e$, apply differentiation with respect to x to both sides, and use the above relations.

$$\frac{dV_m}{dx} = \frac{dV_i}{dx} - \frac{dV_e}{dx} = -I_i r_i + I_e r_e$$

With a second differentiation and the three current law equations above we can write the following differential equation for the membrane potential:

$$\frac{d^2 V_m}{dx^2} = -\frac{dI_i}{dx} r_i + \frac{dI_e}{dx} r_e = i_m r_i + \left(i_m + i_p\right) r_e = r_e i_p + \left(r_i + r_e\right) C_m \frac{dV_m}{dt} + \left(r_i + r_e\right) \frac{V_m}{r_m}$$

Let us introduce some useful notations, which make that equation more compact:

$$\tau \equiv r_m \cdot c_m \qquad \lambda^2 \equiv \frac{r_m}{r_i + r_e} \qquad D \equiv \frac{\lambda^2}{\tau} \qquad q(x,t) \equiv -D \cdot r_e \cdot i_p$$

$$D \cdot \frac{d^2 v_m}{dx^2} - \frac{dv_m}{dt} - \frac{v_m}{\tau} = -q(x,t)$$

The last equation is known as the cable differential equation, which can be solved analytically or numerically.

If we introduce an impulse in $q(x,t)$, with some mathematical manipulation, we get the following impulse response:

$$V_h(x,t) = \frac{1}{2\sqrt{\pi \cdot D \cdot t}} \cdot e^{-\frac{x^2}{4 \cdot D \cdot t}} \cdot e^{-\frac{t}{\tau}}$$

See Fig. 11 for an illustration of this impulse response function.

With the impulse response, one can calculate the response of the system to any given input, $q(x,t)$, by the convolution operator, which in this two-dimensional case is the following integral:

$$V_m(x,t) = (V_h ** q)(x,t) = \int_{\eta=0}^{t} \int_{\xi=-\infty}^{\infty} \frac{q(\xi,\eta)}{2\sqrt{\pi \cdot D \cdot (t-\eta)}} \cdot e^{-\frac{(x-\xi)^2}{4 \cdot D \cdot (t-\eta)}} \cdot e^{-\frac{(t-\eta)}{\tau}} \cdot d\eta \cdot d\xi$$

This integral can be solved analytically for some simple cases and numerically for practically any input function.

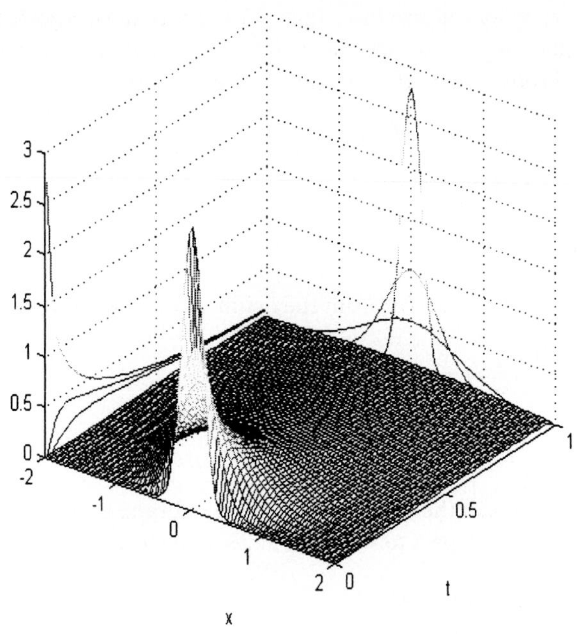

Fig. 11. Two-dimensional impulse response of the passive fiber. X is in units of $[1/\sqrt{D}]$ and t is in units of $[1/\tau]$. The sections on the sides of the picture are at times 0.01, 0.07 and 0.21 $[1/\tau]$, and in distances 0.03, 0.5 and 0.8 $[1/\sqrt{D}]$.

Results

The passive fiber model above was used extensively to check the conduction velocity of nervous fibers with and without myelin, to calculate the optimal distance between the Ranvier nodes, to analyze the propagation of postsynapses potential in the dentritic tree and the propagation of action potentials in the axon. For further examples see Plonsey and Barr (1988) and Stein (1980).

Mechanical Models

There is great interest in modeling muscle and joint dynamics. There are two main reasons for this. One is that muscle is the main motor output of the nervous system and therefore an important window into the operation of the nervous system. Another reason lies in building prostheses and artificial limbs and in external excitation of muscles in paralyzed patients, which is called "functional neuromuscular stimulation" (FNS) (see, for example, Allin and Inbar 1986). All these fields require the construction of a model for the system. Below are two examples for muscle and joint modeling with mechanical elements.

The basic elements of mechanical models are:

- Spring (K), also known as elastic element;
- Damper (B), also called friction element;
- Mass (M);
- Force or tension generator (F, P or T).

The position (X) can be fixed to one location or be free to change according to the forces acting on it. From a graphic scheme one can extract the position velocities and forces at each place in the model. The following Procedure 3.2 and Fig. 12 describe the methods and steps required in order to extract the differential equations from the schematic description.

Procedure 3.2: Writing the Differential Equation of a Linear Mechanical System

1. Using an arrow, mark the force direction at each branch.
2. For each node, apply Newton's second law, stating that the forces applied to a mass equal the acceleration multiplied by the mass: $\sum F = M \cdot \ddot{x}$
3. Replace each force with its value according to Fig. 12.
4. Solve or simplify the equations.

Example 3.5: Second-order Mechanical Muscle Model

Figure 13 depicts a linear lumped model, approximating muscle behavior for a small signal (see McRuer et al. 1968). In this model,

- P represents the internal force in the muscle that is the result of the neural excitation;
- K and B are the elastic and viscous elements that represent the passive mechanical properties of the muscle tissue;
- M is the mass of the muscles and the joint.

According to Procedure 3.2:

1. Mark all the force directions to the right.
2. Of interest in this model is the position of the mass, so write Newton's law for that point: $F_p + F_K + F_B = M \cdot \ddot{x}$
3. Replace the forces according to Fig. 12: $-P + K \cdot (0 - x) + B \cdot (0 - \dot{x}) = M \cdot \ddot{x}$
4. The simplification stage is trivial here and leads to the following second-order equation: $P + K \cdot x + B \cdot \dot{x} + M \cdot \ddot{x} = 0$

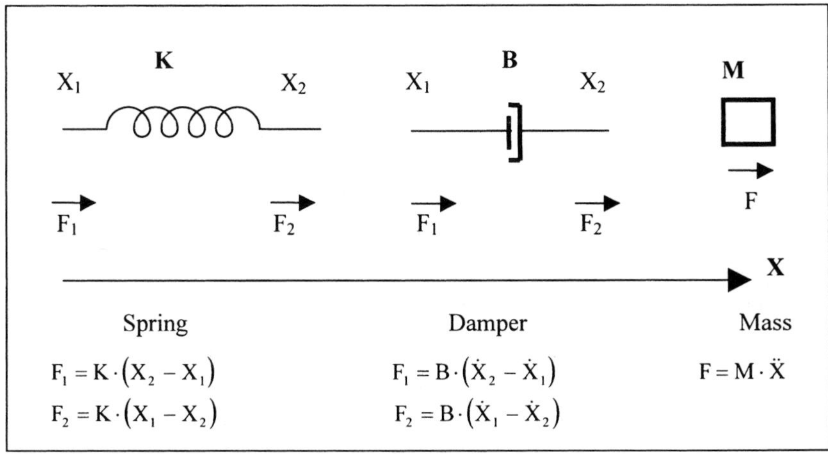

Fig. 12. The basic components of a linear mechanical system.

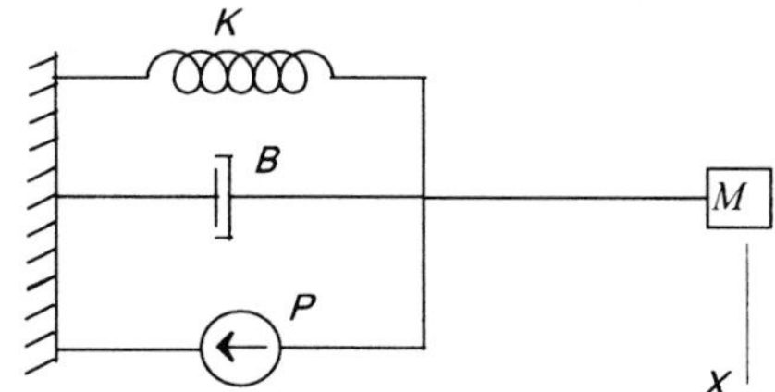

Fig. 13. Second-order mechanical model of muscles and joint

Example 3.6: A More Complex Mechanical Muscle Model

This is a slightly more complicated mechanical model of the muscle. This model is a linearized version of the Hill model, which will be discussed later in this chapter. The model represents one muscle, so it has to be combined with other muscles that act on a specific joint and with the joint mass in order to get a complete model for specific movements. In this model (see Fig. 14),

- P represents the internal force in the muscle resulting from neural excitation;
- B is the viscous element that represents the relation between force and velocity in the muscle;
- K_s is a serial elastic element that represents the mechanical property of the tendon;
- K_p and B_p represent the mechanical properties of other tissues around the muscle and the joint;
- F is the force between the joint and the muscle.

Let us write the equations according to Procedure 3.2.

There are two points of interest, X and X_1, the latter being the connection point of B, P and K_s. Both points are not associated with a mass, leading to the following two equations:

$$F - B_p\dot{x} - k_p x + k_s(x_1 - x) = 0$$

$$-P - B\dot{x}_1 + k_s(x - x_1) = 0$$

Extracting x_1 from the first equation and inserting it into the second leads to

$$x_1 = \frac{-F + B_p\dot{x} + k_p x}{k_s} + x$$

$$-P - B\left(\frac{-\dot{F} + B_p\ddot{x} + k_p\dot{x}}{k_s} + \dot{x}\right) + k_s\left(\frac{F - B_p\dot{x} - k_p x}{k_s}\right) = 0$$

$$-Pk_s - B\left(-\dot{F} + B_p\ddot{x} + k_p\dot{x} + k_s\dot{x}\right) + k_s\left(F - B_p\dot{x} - k_p x\right) = 0$$

This final equation is a third-order differential equation that can be solved numerically or transformed to the Laplace domain for further manipulation or incorporation in a larger model as will be done in the next section.

The reader has surely noticed the similarity between Procedures 3.1 and 3.2. One can transform a mechanical model to an electrical model and vice versa according to Table 1. This transform can be useful if one is an expert in one kind of scheme, or if one has good simulation software for a specific kind of modeling scheme.

This equivalence underscores the major advantage of linear systems modeling. Linear modeling of any kind of system, mechanical, electrical, hydraulic, chemical or other,

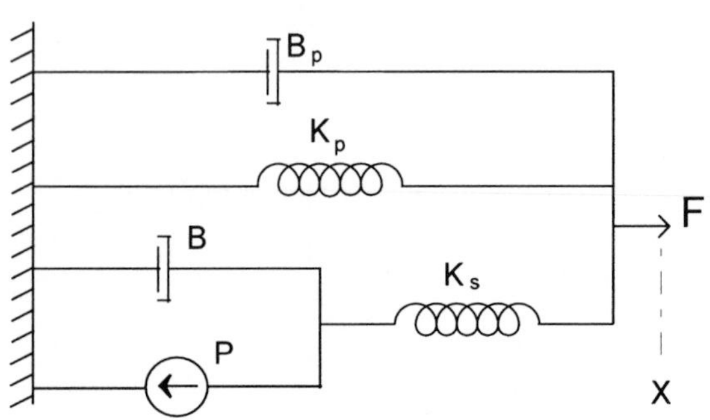

Fig. 14. A more complex linear mechanical muscle model

Table 1. The equivalence of electrical and mechanical components

Mechanical	F	$\dot{X}$	B	K	M
Electrical	I	V	1/R	1/L	C

is always reduced to a differential equation that has standard solutions and can be transformed to the Laplace domain and treated with the same tools. This contrasts with non-linear models that are usually unique to a specific system and therefore require a special theory to be built for each case.

Part 4: Laplace and Z Transform

Linear systems can be described and analyzed in the frequency domain. For linear signal analyses the Fourier transform is very popular, and for linear systems description the Laplace and Z transforms are used for continuous and discrete description, respectively. The main advantage of this description is in finding the transfer function of the system, which is the Laplace transform of the impulse response, as mentioned briefly in Part 2. In the Laplace domain, complex operations simplify, such that e.g. differentiation reduces to multiplication.

Let us begin with the definition of the transform and an introductory example.

The Laplace transform of the continuous signal *x(t)* and the Z transform of the discrete signal are the following:

$$X(s)=\int_{-\infty}^{\infty}x(t)\cdot e^{-s\cdot t}dt \qquad X(z)=\sum_{n=-\infty}^{\infty}x(n)\cdot z^{-n}$$

The domain of the variables *s* and *z* consists of all the complex numbers for which the integral (or sum) above converges. For example, let us look at a system that executes an integration of its input, that is, the system $y(t)=\int_{-\infty}^{t}u(t)dt$. The impulse response of this system is a step function and the Laplace transform of the impulse response is 1/*s*. Therefore, in the Laplace domain the relation between the output and the input is $Y(s)=U(s)/s$, which is a much simpler relation than the integration above. Figure 15 describes this idea graphically.

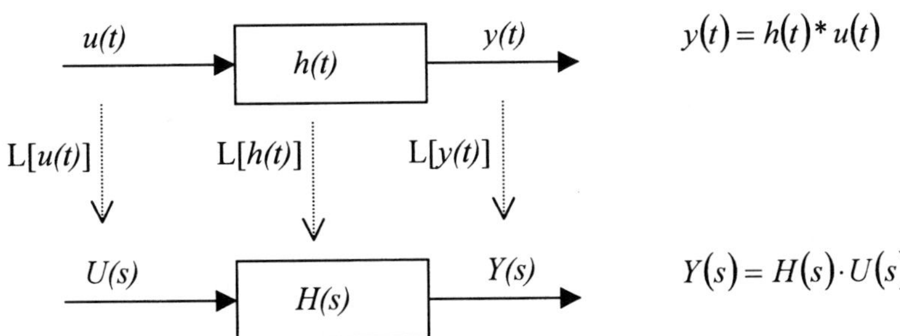

Fig. 15. Linear input-output system: In the time domain, top, *h(t)* is the impulse response and * denotes the convolution operation. In the Laplace domain, bottom, *H(s)* is the transfer function.

Procedure 4.1: Determining Laplace and Z Transforms

The calculation of the transform of a specific function requires some practice. However, for practical purposes, we can just look up a table of Laplace or Z transforms. A short table is given below (see Table 2) and detailed tables can be found in Kwakernaak and Sivan (1991). Another practical approach is to use numerical or symbolic software such as MATLAB that has built-in functions to these transforms.

The following subprotocols introduce some standard properties of these transforms that are used for systems description and analysis. The comprehensive theory and practice of these transforms is beyond the scope of this chapter, and the interested reader is referred to Kwakernaak and Sivan (1991).

Procedure 4.2: Transfer Function from Differential Equation

This simple procedure is based on the following property of the Laplace transform:

$$L\left\{\frac{df(t)}{dt}\right\} = s \cdot L\{f(t)\}$$

This property is correct when one assumes zero initial conditions, i.e., *f(0)=0*.

Therefore the transform of the following general differential equation will be:

$$y(t) = w_1 u(t) + w_2 \dot{u}(t) + \cdots + w_N u^{(N-1)}(t) + w_{N+1}\dot{y}(t) + w_{N+2}\ddot{y}(t) \cdots + w_{N+M} y^{(M)}(t)$$
$$Y(s) = w_1 U(s) + w_2 sU(s) + \cdots + w_N s^{N-1} U(s) + w_{N+1} sY(s) + w_{N+2} s^2 Y(s) + \cdots + w_{N+M} s^M Y(s)$$

Table 2. Laplace and Z transforms: Some useful functions and properties.

Continous Time $f(t)$	Laplace Transform $F(s)$	Discrete Time $f(n)$	Z Transform $F(z)$
$\delta(t)$	1	$\Delta(n)$	1
$f(t) = \begin{cases} 1 & t \geq 0 \\ 0 & t < 0 \end{cases}$	$\frac{1}{s}$	$f(n) = \begin{cases} 1 & n \geq 0 \\ 0 & n < 0 \end{cases}$	$\frac{z}{z-1}$
$e^{-a \cdot t}$	$\frac{1}{s+a}$	a^n	$\frac{z}{z-a}$
$e^{-a \cdot t} \sin(w \cdot t)$	$\frac{w}{(s+a)^2 + w^2}$	$a^n \sin(\Omega \cdot n)$	$\frac{a \cdot \sin(\Omega) \cdot z}{z^2 - 2a\cos(\Omega)z + a^2}$
$a \cdot f(t) + b \cdot g(t)$	$a \cdot F(s) + b \cdot G(s)$	$a \cdot f(n) + b \cdot g(n)$	$a \cdot F(z) + b \cdot G(z)$
$\frac{d}{dt} f(t)$	$s \cdot F(s) - f(0)$	$f(n+1)$	$z \cdot F(z)$
$f(t) * g(t)$	$F(s) \cdot G(s)$	$f(n) * g(n)$	$F(z) \cdot F(z)$

From the last expression one can directly extract the transfer function of the system:

$$\frac{Y(s)}{U(s)}=\frac{w_1+w_2s+\cdots+w_Ns^{N-1}}{1+w_{N+1}s+w_{N+2}s^2+\cdots+w_{N+M}s^M}$$

A similar procedure holds for the Z transform, that is, the difference equation

$$y(n)=\sum_{i=0}^{N}a_i\cdot x(n-i)-\sum_{j=1}^{M}b_j\cdot y(n-j)$$

will be transform to the following algebraic equation

$$Y(z)=\sum_{i=0}^{N}a_i\cdot z^{-i}\cdot X(z)-\sum_{j=1}^{M}b_j\cdot z^{-j}\cdot Y(z)$$

Comment: Rational Transfer Functions

The notion of transfer function is illustrated in Fig. 15. We just saw how to transform a differential equation to a transfer function where the numerator and denominator were polynomial functions. The transfer function can therefore be described as follows:

$$H(s)=\frac{OUT(s)}{IN(s)}=k\cdot\frac{\Pi(s-z_i)}{\Pi(s-p_j)}$$

where k is called the "gain", the z_i are called "zeros", and the p_i are called "poles".

This formalization is easy to analyze. There is a vast literature about the influence of the location of the poles and zeros on system behavior, and there are many names for all kinds of such systems. If there are only poles, the system is called "auto-recursive" (AR); if there are only zeros, the system is called "moving average" (MA); and the general case is called "auto-recursive moving average (ARMA) system". The case of zeros only is also called "finite impulse response" (FIR) because the influence of the impulse is gone after a short period, while adding poles produces an infinite impulse response (IIR).

Another useful property of the Laplace transform enables the calculation of the steady state of the system in the Laplace domain:

$$\lim_{t\to\infty}f(t)=\lim_{s\to 0}s\cdot F(s)$$

All these properties have their equivalent in the Z transform domain for discrete signals and systems.

Procedure 4.3: Discretization

Since we usually use a computer and discrete measurements, it is very useful to have a discrete model. Nevertheless, the physical and biological world is a continuous world. Discretization is the procedure of converting a continuous model to a discrete one. There are different procedures for discretization just as there are many ways of numerical integration. The simplest method, the Euler's forward method, is to move to the Z transform by replacing each s by $(z-1)/T$. This method is demonstrated in Example 4.2 (see Santina et al. 1996 for more details about discretization methods).

Procedure 4.4: Check for Stability

A *system* is stable if a bounded input produces a bounded output. In the Laplace domain, there is a simple procedure to check the stability of a system.

1. Write the transfer function as a rational function.
2. Find the roots of the denominator, that is, the poles.
3. If the real part of all the poles is negative, then the system is stable. Otherwise the system is not stable.

Example 4.1: Stability and Step Response

This example serves to practice the linear systems description and the transform tools. We will show how to find the conditions for a second-order system to be stable, find an expression for its response to a step input and find the time to reach the maximum in a response to a step input.

For this purpose, let us first highlight some pertinent properties of first- and second-order systems. A first-order system is most common in biological modeling. Most of the examples in this chapter are of a first-order system starting with the leaky integrator of Example 3.1. The transfer function of a typical first-order system is $1/(s+a)$, and from Table 2, one can see that the impulse response of such a system is an exponential decay. The step response of this system is illustrated on the left side of Fig. 16. This function is characterised by its time constant $\tau=1/a$, which is the time when the response has reached about 2/3 of its final value.

Second-order systems can also produce oscillatory behavior. The muscle and joint model in Example 3.5 is a second-order system. A description of the step response of a typical second-order system is illustrated on the right side of Fig. 16. In order to describe the parameters of such a system, let us write the transfer function of a second-order system in a standard form:

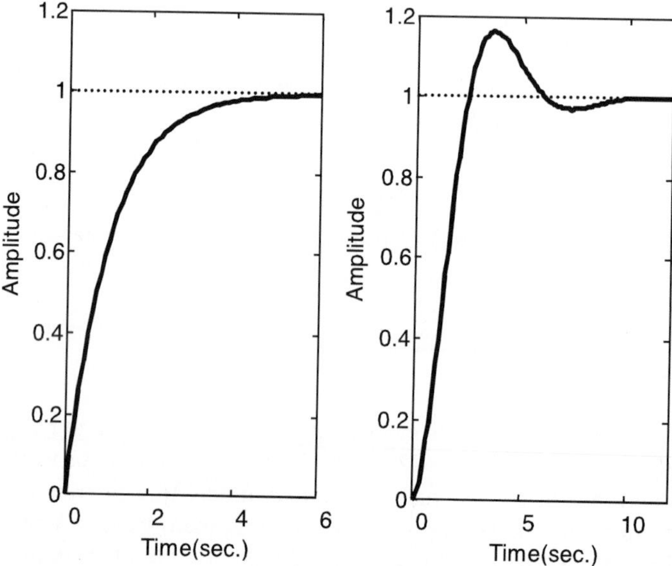

Fig. 16. Step response of a first-order system (left) and of a second-order system (right). The transfer functions of these systems are $1/(1+s)$ and $1/(s^2+s+1)$, respectively.

$$H(s)=\frac{w_n^2}{s^2+2\cdot\xi\cdot w_n\cdot s+w_n^2}$$

This system has been analyzed extensively, and there are expressions for every possible feature of it. For example, the overshoot, which is the ratio of the response that is beyond the steady state response to the latter, is the following function of the parameter

$$\xi: \text{O.S.}=e^{-\frac{\pi\cdot\xi}{\sqrt{1-\xi^2}}}.$$

The frequency of the oscillations is

$$f_d=\frac{w_n}{2\pi}\cdot\sqrt{1-\xi^2},$$

the time constant is

$$\tau=\frac{1}{\xi\cdot w_n}$$

and the settling time, which is the time to reach a region of 2% of the final value and stay there, is

$$t_s\cong\frac{4}{\xi\cdot w_n}.$$

Let us now check the stability of the above second-order system according to Procedure 4.4. For that purpose, we have to find the location of the poles, that is, the roots of the denominator of the transfer function, which are:

$$s_{1,2}=-\xi\cdot w_n\pm jw_n\cdot\sqrt{1-\xi^2}$$

For stability, both poles must have a negative real part. Therefore the requirement for stability is: $\xi\cdot w_n>0$. Notice that the second-order system has an oscillatory behavior when the poles have imaginary parts.

Let us now move to the calculation of the step response. We know that in the Laplace domain the output is the input multiplied by the transfer function. According to Table 2, the Laplace transform of the step function is 1/s, and the output will be the second-order systems transfer function divided by *s:*

$$Y(s)=\frac{w_n^2}{s\cdot\left(s^2+2\cdot\xi\cdot w_n\cdot s+w_n^2\right)}$$

The inverse transform of this function can be calculated using Table 2 and some mathematical manipulations, by looking at an extended Laplace transforms table or by using mathematical software. The resulting function is the response to a step function, which is:

$$y(t)=1-\frac{e^{-\xi\cdot w_n\cdot t}}{\sqrt{1-\xi^2}}\cdot\sin\left[w_n\cdot t\cdot\sqrt{1-\xi^2}+\tan^{-1}\left(\frac{\sqrt{1-\xi^2}}{\xi}\right)\right]$$

The maximum is the first place where the tangent is flat, that is,

$$\frac{dy(t)}{dt}=0.$$

The time of this event is:

$$t_p = \frac{\pi}{w_n \cdot \sqrt{1-\xi^2}}$$

Example 4.2: Second-order Mechanical Muscle Model

The second-order mechanical muscle model was introduced in Example 3.5 (see Fig. 13). With Procedure 4.1 we can transform the differential equation to the following transfer function:

$$\frac{X(s)}{P(s)} = \frac{-1}{M \cdot s^2 + B \cdot s + K}$$

One can derive a similar relation for external force and its relation to the position, or any other desired relationship. With Procedure 4.3, we can transform the above function to the Z transform domain:

$$\frac{X(z)}{P(z)} = \left. \frac{-1}{M \cdot s^2 + B \cdot s + K} \right|_{s=\frac{z-1}{T}} =$$

$$\frac{-T^2}{M \cdot z^2 + (B \cdot T - 2 \cdot M) \cdot z + M - B \cdot T + K \cdot T^2}$$

From the Z transform we can move to discrete time by applying Procedure 4.1 inversely:

$$X(n) = \frac{-T^2}{M} \cdot P(n-2) - \frac{(B \cdot T - 2 \cdot M)}{M} \cdot X(n-1) - \frac{\left(M - B \cdot T + K \cdot T^2\right)}{M} \cdot X(n-2)$$

This difference equation may be useful for parameter estimation and handling of sampled data from this system, as will be demonstrated in Example 5.2. The last difference equation can be formalized as follows: $X(n) = w_1 \cdot P(n-2) + w_2 \cdot X(n-1) + w_3 \cdot X(n-2)$, where w_i are the parameters. If the sample time T is given, it is equivalent to know w_i or M,B,K. Therefore, for each time step, the system can be viewed as a simple static system with three inputs and one output, as described graphically in Fig. 3.

Example 4.3: Complex Mechanical Muscle Model

This example demonstrates how to extract the transfer function and the steady-state behavior of the Hill-type mechanical muscle model that was introduced in Example 3.6. From the graphic model in Fig. 14, we derived the following differential equation that can be further simplified:

$$-Pk_s - B\left(-\dot{F} + B_p\ddot{x} + k_p\dot{x} + k_s\dot{x}\right) + k_s\left(F - B_p\dot{x} - k_p x\right) = 0$$

From this equation we can move to the Laplace domain as described above in Procedure 4.1:

$$(k_s + Bs)F - Pk_s - \left(BB_p s^2 + Bk_p s + Bk_s s + k_s B_p s + k_s k_p\right)X = 0$$

Now we can extract any transfer function we need according to the input and output we define. In an isometric experiment when the length is held constant, we can subtract the constant force due to the constant length and find the transfer function between the force, F, and the neural excitation to the muscle, which is related to P. (In the linear model we assume a linear relation between the firing rate of the motor neuron and the hypothetical internal force P.)

$$\frac{F}{P} = \frac{k_s}{(k_s + Bs)}$$

This is a first-order system and one can investigate its behavior according to the description in Example 4.1 above. For example, the force response to a step in the input will look like the left graph in Fig. 16.

The same manipulation can be done in order to investigate an isotonic experiment, where the force is held constant.

The steady-state behavior of the model can be calculated by practically replacing each s with a zero, which brings us to the following relation:

$$\frac{F - P}{k_p} = x$$

One should notice that the viscous elements have no role in the steady-state behavior, since they produce forces only at times of change in position.

A good example for the application of these mathematical procedures to the physiological neuromuscular system is the study of eye movements. The time constant of the mechanical model of the system is of the order of seconds while we know that the eye can move from one position to the other in tenths of a second. Therefore, a simple step in neural excitation will not satisfy the observations of eye movement. An alternative hypothesis is that the neural excitation signal contains an initial pulse added to the step to accelerate the movement. This was found to be consistent both with the model and the measurements. This control strategy of pulse plus step neural excitation is also used in limb movement control models.

Comment: Notice that when a step neural excitation is mentioned, we relate to the firing rate and not to the nerve cell potential, that is, a unit step at time zero means that the cell starts to fire an action potential once a second from time zero and thereafter.

Part 5: System Identification and Parameter Estimation

In many disciplines of science and technology we frequently face data from an unknown system and our aim is to find a model of this system. A parametric model belongs to a family of models characterized by a finite number of parameters. The modeler's task is first to choose a proper model family and then to estimate the parameter values. In this section we will first describe the estimation problem in general and then concentrate on linear models.

The general problem of parameter estimation can be formalized as follows: Let $\Theta(u,a)$ be a family of parametric functions, that is, for each parameter vector a_0, $y = \Theta(u,a_0)$ is a static input/output function or a transfer function in the Laplace domain, where u is the input and y is the output. Suppose that we have an unknown system $F(u)$ that is assumed to belong to the above function family, that is, $F(u)=\Theta(u,a_0)$ for a specific but unknown parameter vector a_0. As a result of an experiment on this unknown system, we collect a group of measurements of input/output pairs $\{u_i,y_i\}$ that naturally satisfy $y_i=F(u_i)$. (In the presence of measurement noise or uncertainty in the generating function, that is, if we are not positive about the assumption that the unknown system belongs to the selected family of parametric functions, we can relax the requirements from the data to $|y_i-F(u_i)|<n$, where n represents the noise or the uncertainty in the fitness of the model to the system). The problem is to find the vector of parameters a that will best fit the measurement pairs according to a given criterion. If one uses the least-squares criterion, the problem is to solve the following minimization:

$$\hat{a}=\underset{a}{\arg\min}\sum_i\left(y_i-\Theta(x_i,a)\right)^2$$

There are many methods to solve this problem and to formalize parametric groups of functions (see for example Sjoberg et al. 1995). In this chapter we concentrate on the linear group of functions, implying that the function can be transformed to the Laplace domain, resulting in a transfer function. In the discrete case the same can be done with a difference equation and the Z transform.

Procedure 5.1: Estimation Scheme

The basic way to estimate the parameters of a linear model is the following:
The linear model is $y=\sum_i w_i\cdot u_i$ or in matrix notation $y=W^TU$. The real system may not be linear and the data we have may be noisy; however, the optimal linear model according to the least-squares criterion is the following:

$$W_{OPT}=\Phi^{-1}\cdot P$$

$$P\equiv E\left[Y\cdot U\right]\qquad\Phi\equiv E\left[U\cdot U^T\right]$$

where E stands for expectation. For the origin and proof, see any textbook on linear parameter estimation (e.g., Porat 1994).

In practice we estimate the expectation as a numerical average over the measurements, that is,

$$P=\frac{1}{N}\sum_{l=1}^{N}Y^l\cdot U^l\qquad\Phi=\frac{1}{N}\sum_{l=1}^{N}U^l\cdot U^{lT}$$

Example 5.1: Two Inputs-One Output System

This is a simple synthetic numerical example to demonstrate the use of Procedure 5.1. Assume that we have a static system with two inputs and one output and we wish to find an optimal linear model for this system. The model will be $y=w_1\cdot u_1+w_2\cdot u_2$, the input vector will be $U=[u_1,u_2]^T$ and the parameter vector will be $W=[w_1,w_2]^T$. In order to estimate the parameters of the model, an experiment was conducted and the following four measurements were obtained:

u_1	u_2	y
1	1	3.1
1	−1	1.2
−1	1	−0.8
−1	−1	−2.9

Let us calculate the estimation of P and Φ according to procedure 5.1.

$$P \equiv E[Y \cdot U] = E\begin{bmatrix} y \cdot u_1 \\ y \cdot u_2 \end{bmatrix} = \frac{1}{4}\begin{bmatrix} 3.1+1.2+0.8+2.9 \\ 3.1-1.2-0.8+2.9 \end{bmatrix} = \begin{bmatrix} 2 \\ 1 \end{bmatrix}$$

$$\Phi \equiv E\left[U \cdot U^T\right] = E\begin{bmatrix} u_1 \cdot u_1 & u_1 \cdot u_2 \\ u_2 \cdot u_1 & u_2 \cdot u_2 \end{bmatrix} = \cdots = \begin{bmatrix} 1 & 0 \\ 0 & 1 \end{bmatrix}$$

Now we can calculate the optimal parameters:

$$W_{OPT} = \Phi^{-1} \cdot P = \begin{bmatrix} 2 \\ 1 \end{bmatrix}$$

and with these parameters we can calculate the model's output to the measurement data and check the fitness of the model to the data: One can see that the model outputs are

u	u_2	y_m
1	1	3
1	-1	1
-1	1	-1
-1	-1	-3

similar to the actual data.

Comment: In practice, the measurements may be random and there may be more noise, therefore more examples are needed in order to get a good estimation of the parameters.

Example 5.2: Estimation of Muscle Model

In Example 4.2we found the following relation for the linear muscle model that was introduced in Example 3.5:

$$X(n) = w_1 \cdot P(n-2) + w_2 \cdot X(n-1) + w_3 \cdot X(n-2)$$

In this example we demonstrate the parameter estimation procedure for this dynamic model. We can combine our input components $X(n-2)$, $X(n-1)$, $P(n-2)$ to form an input vector U, and denote the output vector, which in this case has just one element, $X(n)$, by the letter Y. Now we can use the optimal solution of procedure 5.1.

Let us illustrate this estimation scheme by a simulation example. A random sequence of P was chosen (normally distributed noise with standard deviation (STD) equal to one, and zero mean). The length X was calculated according to the model with the following nominal value of the parameters: M=5, B=3, K=2, T=0.1, that is, W_1=–0.002, W_2=1.94, W_3=–0.944.

Figure17 shows the results of the simulation. The first graph is the random input P, the second is the calculated X. An additional random noise was added to simulate measurement noise or uncertainties in the model (normally distributed noise with STD=0.01 and zero mean); this sequence appears in the third graph. Then the optimal parameters were calculated according to Procedure 5.1, and the results were W_1= –0.0016, W_2= 1.744, W_3= –0.739, which is close to the nominal parameters, as expected. Finally, the output of the estimated model for each time step was calculated, and it appears in the fourth graph being similar to the second graph, which is the actual model output.

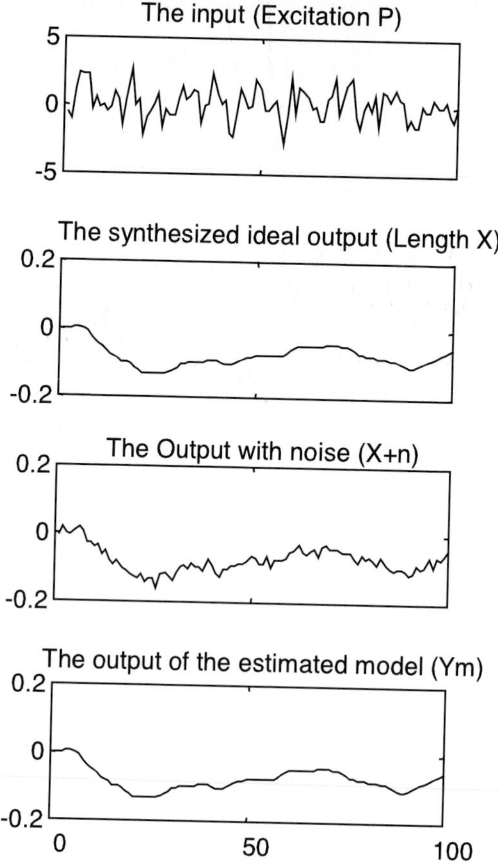

Fig. 17. An example of parameter estimation of the linear (ARMA) muscle model. The simulation is discrete, therefore the abscissa consists of time steps. The length of each time step in this simulation is 0.1 second.

Comments: The above example is a straightforward implementation of Procedure 5.1. There are various pitfalls that are listed below:

- This example regards the discrete data as a set of independent examples of a static model and the optimal model is checked for each couple of input-output independently. In practice, the error combines from one time step to the other, since the model may use its own output to estimate the next time step, and not the real system outputs. This problem can be severe when the system has some unstable poles, then the error might grow very fast.
- The calculated model, that is, the estimated parameters, should be checked on a new data set and not only on the data that was used for parameter estimation. This check is called "generalization check" and can assist in avoiding over-fitting of the data. We discuss this method of validation in the following subsection.
- One should remember that the biological system is generally a time-varying system. For example, muscles can change their properties due to fatigue. Therefore the duration of the experiment must be short in order to justify the assumption that the system is time-invariant.
- Finally, we must mention here that the simple optimal parameter calculation in Procedure 5.1 is not always stable numerically. There are many improvements and practical methods that can be found in modern numerical software, such as MATLAB (see Ljung 1986).

Problem: Choosing the Right Model Order

In the last example the structure of the model was known and the only problem was estimating the parameters, but in most biological cases the model is unknown.

In order to estimate the parameters, we first need to establish the order of the model, which means, for example in the ARMA case, choosing N and M in the following discrete model:

$$y(n) = \sum_{i=0}^{N} a_i \cdot x(n-i) - \sum_{j=1}^{M} b_j \cdot y(n-j)$$

At first glance, one might suppose that the more parameters the model has, the better it will fit the actual system, but this is not the case (see Paiss and Inbar 1987 for an extensive treatment of the model order selection problem for the case of surface electromyography). Too many parameters are not only a computational burden but they may cause errors in the model. One can be wrong by either choosing too many or too few parameters. See Fig. 25 for a description of the pitfalls in choosing the wrong number of parameters.

Many approaches have been suggested for choosing the proper order. For linear models a commonly used approach is the Akaike information criterion (AIC) which is based on a discrepancy measure. For the ARMA model it will take the following form:

$$N_T^{-1} \cdot AIC(N,M) = \hat{\sigma}^2 + \frac{2 \cdot (N+M+1)}{N_T}$$

where N and M are the model size, see the discrete ARMA model above, N_T is the total number of samples and $\hat{\sigma}^2$ is the estimation of the error.

Since the first term, the estimation of the error, $\hat{\sigma}^2$, monotonically decreases with increasing model size and the second term increases, one can find an optimal model size by finding the minimal value of the *AIC*. Another method to choose the order of the model is validation. This method is commonly used in pattern recognition and classification where part of the data is kept from the learning phase (in our case this will be the fitting phase), and then the model is chosen for its generalization capabilities checked on the kept data. For more information about parameter estimation and systems identification see Porat (1994), Sjoberg et al. (1995) and Ljung (1986).

Part 6: Modeling The Nervous System Control

We have seen that linear systems can be described in the Laplace domain by their transfer functions. These transfer functions make it easy to analyze complex systems that include many modules. This section describes how to use block diagrams in modeling the nervous system.

Procedure 6.1: Block Diagrams

There are only two basic elements in linear block diagrams: summer and transfer function block. The summation element is usually symbolized by a circle and a sign at each one of its inputs that determines whether is should be added or subtracted. The transfer function is represented as a block with the transfer function or an impulse response function in it. A special case of transfer function is a mere gain that is sometimes symbolized by a triangle. Each block is connected to other blocks by arrows that symbolize the direction of information flow.

The procedure of writing the transfer function between two points in a block diagram is as follows:

1. Name each arrow that has no name with a unique variable.
2. Write an equation for each variable. For example, if the variable is y and the input(s) to the block before it is u, then write the following:
   For an output of a summation element, write $y = \sum_i u_i$.
   For an output of a transfer function H, write $y=Hu$ (in the Laplace domain).
3. Simplify the set of equations in order to get a transfer function between the input and the output or any other relation needed.

Example 6.1: Feedback Control

Feedback control is based on using the outcome of the process or the controlled system, which is usually called the "Plant", in order to control it, in other words, using the error between the desired output, y_d, and the actual output, y, in order to reduce it. This scheme is widely used to describe the nervous system control of the musculoskeletal system.

The analogy of the feedback scheme, Fig. 18, to motor control is the following: The plant corresponds to the muscles, the bones and the dynamics of the environment; the feedback corresponds to the output of the sensory systems, and the controller corresponds to the nervous system. Let us follow Procedure 6.1 in order to find the transfer function of the complete system in Fig. 18. (The Laplace variable s is omitted for simplicity.)

1. Let us call the output of the feedback x_1, and the output of the sum x_2.
2. There are four elements and therefore four equations:

$$Y = P \cdot u \quad u = C \cdot x_2 \quad x_1 = F \cdot Y \quad x_2 = Y_d - x_1$$

3. From the above equations one can extract the following transfer function:

$$\frac{Y}{Y_d} = \frac{P \cdot C}{P \cdot C \cdot F + 1}$$

One major advantage of the feedback control scheme is the reduced sensitivity to changes in the parameters of the plant, and to changes in the environment. The sensitivity of system H to changes in the parameter k is defined as follows:

$$S_H^k \equiv \left| \frac{\partial H}{\partial k} \cdot \frac{k}{H} \right|$$

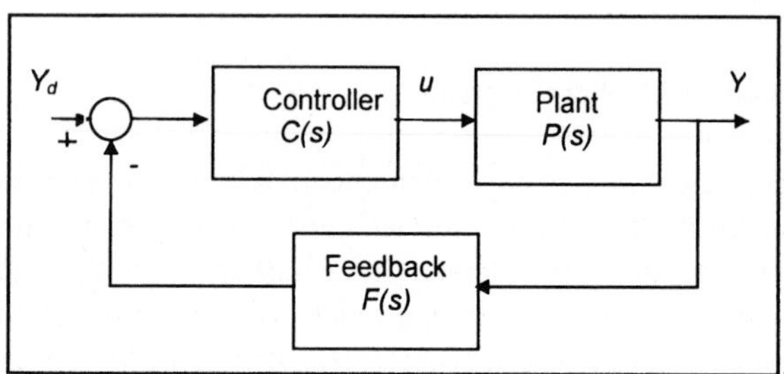

Fig. 18. Feedback control

When the value of the sensitivity function is zero, the system is insensitive to changes in the parameter. Let us look at the system without feedback, where H is the transfer function, and k is a gain parameter. The system in an open loop is $H = k \cdot P$. The sensitivity of the system would be:

$$S_H^k \equiv \left| \frac{\partial H}{\partial k} \cdot \frac{k}{H} \right| = P \cdot \frac{k}{k \cdot P} = 1$$

However, in closed-loop mode, the system is $H = \dfrac{k \cdot P}{k \cdot P \cdot F + 1}$ and the sensitivity will be:

$$S_H^k \equiv \left| \frac{\partial H}{\partial k} \cdot \frac{k}{H} \right| = \frac{P \cdot (k \cdot P \cdot F + 1) - k \cdot (P)^2 \cdot F}{(k \cdot P \cdot F + 1)^2} \cdot \frac{k \cdot P \cdot F + 1}{P} = \frac{1}{k \cdot P \cdot F + 1} < 1$$

So when the loop gain, k, is high, the sensitivity to changes is low.

There is a vast literature on the stability of such systems and on methods to choose a controller when the specifications of the desired performances are given (see for example Kwakernaak and Sivan 1991 and section III in Levine 1996).

Comments: The first problem in using this simple feedback scheme to model biological systems occurs when one tries to measure the loop gain. In biological systems, one often finds very low loop gains in the order of one. Therefore reduced sensitivity to changes in the parameters frequently does not occur in biological systems. Another problem results from delays in biological systems that can cause instability and oscillation. See Karniel and Inbar (in press) for a review of these and other problems in biological motor control.

Example 6.2: Multiple Feedback Loops

The importance of the loop gain in reducing the sensitivity to parameter changes was mentioned in the previous example. The loop gain can also be a major factor in determining the stability of the system. In order to measure the loop gain, one should open the loop, introduce a test input at one end and measure the output at the other. However, in biological systems there are typically multiple parallel feedbacks (see, e.g., Milgram and Inbar 1976; Windhorst 1996). For example, in the temperature regulation system, there are sensors in the skin, in the core of the body and in the hypothalamus, and they all influence the temperature regulation mechanisms (see Brown and Brengelmann 1970). In movement control, there are feedback loops from sensors in the muscles, joints and skin (i.e., muscle spindles, Golgi tendon organs, pressure transducers etc.), and furthermore there are many sensors of each type operating in parallel. The primary advantage of such multiple loops, and of any redundancy, is robustness. That is, if one subsystem fails, there are other options to operate the system. There is a great danger in trying to estimate the loop gain in such a system because there may be loops that we

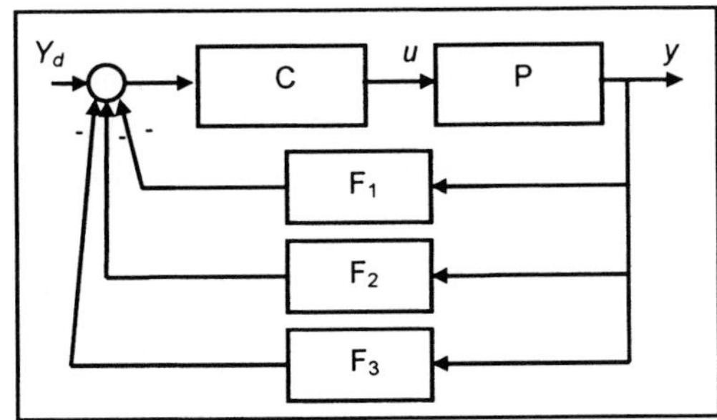

Fig. 19. Multiple feedback loops

cannot open or are unaware of. In such a case we may underestimate the loop gain. For example, if we open the first two loops in Fig. 19 and leave F_3 connected, the transfer function from Y_d to Y will be $P \cdot C / (P \cdot C \cdot F_3 + 1)$ instead of $P \cdot C$ when there is no additional loop. So one should be aware of these multiple loops.

Part 7: Modeling Nonlinear Systems with Linear Systems Description Tools

Many physical and virtually all biological systems are not linear, and many are time-varying systems. Still, we may wish to use the powerful linear systems description tools that were described in this chapter. In this section we will broaden our scope to illustrate the application of linear systems description tools to nonlinear systems. Linearization is a method to find a linear system that is similar to the modeled nonlinear system, at least for a small-signal region and a short time. We will show how nonlinear systems can be described as time-varying or parameter-varying systems, and then we will discuss nonlinear systems as a linear sum of nonlinear functions or as a nonlinear function of a linear system.

Linearization

Linearization is the procedure of finding a linear system that is similar to the modeled nonlinear system in some domain near a point that is called the "working point". Graphically one can imagine the linearized model as a tangent of the nonlinear function. Mathematically the linearization represents the first two terms in the Taylor expansion of the function.

Procedure 7.1: Static Systems Linearization

Static systems linearization simply means to take the first two components in the Taylor expansion. For a single input and single output system, this means taking the following linear estimation $F_L(u)$ of the nonlinear system $F(u)$ near the point $\hat{u}$ which is called the *working point.*

$$F_L(u) = F(\hat{u}) + \left.\frac{dF}{du}\right|_{u=\hat{u}} \cdot (u - \hat{u})$$

This estimation is good for smooth functions, near the working point, and it may be very poor for distant points. The same estimation can be implemented for multiple input static systems.

Procedure 7.2: Dynamic Systems Linearization

Dynamic systems can be represented as a set of differential equations in the state space. If we assume continuity of the function, we can write the first two parts of the Taylor's expansion of $\dot{X} = F(X)$ near the point of interest $\hat{X}$, which is:

$$\dot{x} = f_i(x_1, \ldots, x_n) = f_i(\hat{x}_1, \ldots, \hat{x}_n) + \left.\frac{\partial f_i}{\partial x_1}\right|_{X=\hat{X}} (X_1 - \hat{X}_1) + \ldots + \left.\frac{\partial f_i}{\partial xn}\right|_{X=\hat{X}} (X_n - \hat{X}_n)$$
$$+ O\left(\|X - \hat{X}\|^2\right)$$

The Jacobean matrix is:

$$A = \begin{bmatrix} \frac{\partial f_1}{\partial x_1} & \cdots & \frac{\partial f_1}{\partial x_n} \\ \vdots & & \vdots \\ \frac{\partial f_n}{\partial x_1} & \cdots & \frac{\partial f_n}{\partial x_n} \end{bmatrix}_{X=\hat{X}}$$

And by defining a new variable $Z = X - \hat{X}$, we can get the following linear state function:

$\dot{z} = A \cdot z$

The solution of this differential equation is an exponential function. It is interesting to know whether the solution decays, that is, whether the system is stable. We can answer this question by checking the eigenvalues λ_i of the Jacobean matrix A: The system is stable if and only if the real part of all the eigenvalues λ_i is negative. The eigenvalues can be calculated by finding the roots of the equation $|A - \lambda \cdot I| = 0$, or by just writing the proper command in MATLAB or in any other mathematical software.

Example 7.1: Relation between Force and Length of the Muscle

Striated muscles consist of actin and myosin filaments that slide one over the other. As a result of this infrastructure, there is an optimal length, L_o, at which the muscle can produce maximum force. Therefore, the relation between the length of a muscle and its force is nonlinear and can be approximated by the following nonlinear equation:

$$F = F_{\max} - (L_0 - x)^2$$

Suppose that we want a linear model of this muscle near a working point $\hat{x} = L_0 / 2$; we can then linearize the above relation according to Procedure 7.1 as follows:

$$F_L(x) = F(\hat{x}) + \left.\frac{dF}{dX}\right|_{x=\hat{x}} \cdot (x - \hat{x}) = F_{\max} - \left(L_0 - \frac{L_0}{2}\right)^2 + 2 \cdot \frac{L_0}{2} \cdot \left(x - \frac{L_0}{2}\right) =$$

$$= F_{\max} - \frac{3 \cdot L_0^{\,2}}{4} + L_0 \cdot x$$

Comments

1. The model in Example 3.5 is actually a linearized model of the muscle (see McRuer et al. 1968).
2. The membrane model in Example 3.2 can be seen as linearization of the membrane properties near equilibrium points.
3. One should notice that the linearized model is close to the modeled function only for small perturbations near the working point.

Nonlinear Systems as Linear Time- or Parameter-varying Systems

In Procedures 3.1 and 3.2 we considered electrical and mechanical models of linear systems. These procedures can be used to describe nonlinear systems if one allows change in the values of the elements as a function of time or of other values in the system. We give here two examples for such a modeling approach: the membrane electric model and the muscle mechanical model.

Example 7.2: Hodgkin-Huxley Model

This model was introduced in Example 3.2, where the values of the resistive elements were fixed. However, the membrane is not linear and resistance to each ion current is not a constant (see Hodgkin and Huxley (1952) for a comprehensive description of this nonlinear model).

This model is illustrated in Fig. 20, where the arrows on the resistors mean that the resistance is not a constant. The membrane resistance to sodium and its resistance to potassium change as a function of the membrane potential and as a function of time.

The equations that describe this model are the following set of nonlinear differential equations:

$$I_m = C_m \frac{\partial V_m}{\partial t} + g_{Na} \cdot (V_m - V_{Na}) + g_K \cdot (V_m - V_K) + g_{cl} \cdot (V_m - V_{cl})$$

$$g_{Na} = \bar{g}_{Na} \cdot m^3 \cdot h$$

$$g_K = \bar{g}_K \cdot n^4$$

$$g_{cl} = \bar{g}_{cl} = const$$

$$\dot{n}(t) = \alpha_n \cdot (1-n) - \beta_n \cdot n$$

$$\dot{m}(t) = \alpha_m \cdot (1-m) - \beta_m \cdot m$$

$$\dot{h}(t) = \alpha_h \cdot (1-h) - \beta_h \cdot h$$

The first equation can be written according to Procedure 3.1, it is the differential equation of the electrical model in Fig. 20, which is similar to the equation in Example 3.2, where resistance, R, was used instead of conductance, g. We will not go into the details of the change in membrane resistivity, but one can notice in the equations above that this change is also described by linear systems description tools, in fact by a simple first-order differential equation.

Example 7.3: Hill-Type Muscle Model

Let us consider the Hill-type mechanical muscle model in Fig. 21. This model is taken from Zangemeister et al. (1981), with minor changes (see Karniel and Inbar 1997). Note that this model combines a mechanical description with a block diagram in the Laplace domain. It is similar to the model in Example 3.6, but shows three differences: The parallel spring was omitted, a first-order filter was added to describe the excitation-con-

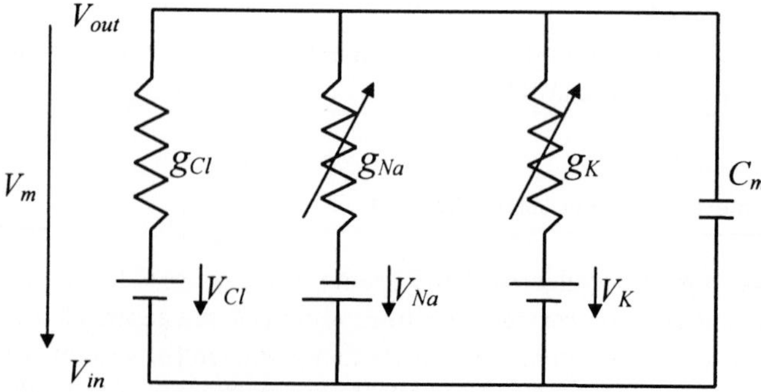

Fig. 20. The Hodgkin-Huxley electrical model of the membrane.

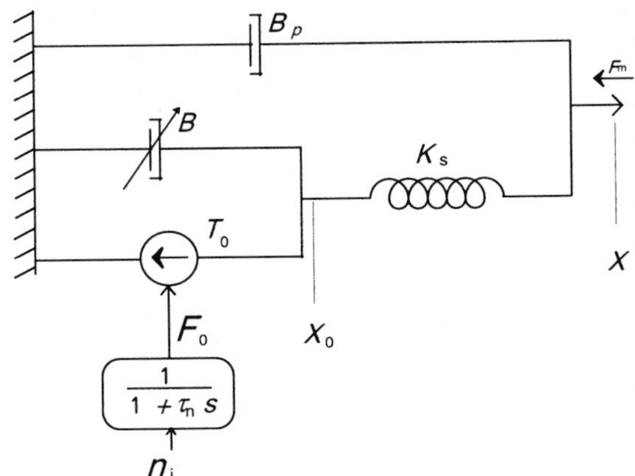

Fig. 21. Mechanical muscle model. n_i is the neural input. The first-order filter represents the excitation-contraction coupling. T_o is the hypothetical force in the muscle. B represents the relation between force and velocity from Hill's model. The other elements represent the mechanical properties of the tendon and other connective tissues around the joint.

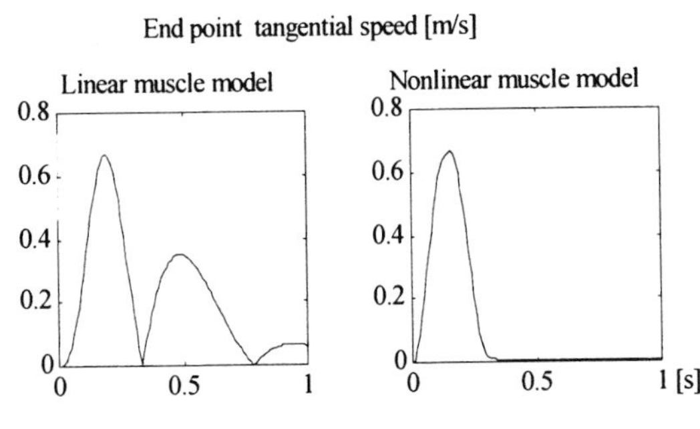

Fig. 22. A comparison of the speed profile of the end point of a two-degrees-of-freedom anthropomorphic arm with a linear muscle model (left) and with a nonlinear muscle model (right) in response to typical rectangular pulse activation of the muscles. Only the nonlinear muscle model yields a bell-shaped speed profile with a smooth stop (for more details, see Karniel and Inbar 1997)

traction coupling and the recruitment of the motor units, and the main difference is the viscous element which is not a constant.

Following are the differential equations of this mechanical model:

$$\dot{F}_0 = \frac{1}{\tau_n} \cdot (n_i - F_0)$$

$$T_0 = F_0 \cdot F_{\max}$$

$$\dot{X}_0 = \frac{(K_s \cdot (X - X_0) - T_0)}{B}$$

$$F_m = B_P \cdot \dot{X} + K_s \cdot (X - X_0)$$

This model was derived from the Hill model where the value of the viscous element, B, depends on the internal force and on the contraction velocity:

$$B = \begin{cases} (a \cdot T_0)/(b + v) & v \geq 0 \\ a' \cdot T_0 & v < 0 \end{cases}$$

The value of B was taken as a constant in several models, for the sake of simplicity, in order to get a linear model of the muscle. This linear model is under-damped and therefore overshoots, and oscillations are most likely to appear in the controlled movement. This problem is avoided by the use of the more realistic nonlinear model, as demonstrated in Fig. 22 for a very basic movement, the reaching movement.

This example demonstrates how nonlinearity might be exploited advantageously by nature. However, in order to simulate and analyze this phenomenon, we exploited the linear systems description tools.

Comment Another useful model is the second-order muscle model which was introduced in Example 3.5. In this model, in order to get a more realistic behavior, the stiffness must change as a function of the activation, the length and the velocity of the muscle (see for example Inbar 1996).

Pre-processing or Post-processing

In this subsection we describe two simple ways of combining linear systems in nonlinear modeling. The first one is to describe a linear combination of nonlinear fixed functions, which is called a pre-processing, since the inputs are processed prior to their entrance to the linear model, see Fig. 23. The second way is to describe a nonlinear function of a linear combination, which is called "post-processing" since the output of the linear model is nonlinearly processed, see Fig. 24. Both models can take advantage of the linear parameter estimation tools in order to estimate the parameters of the linear part of the model.

Example 7.3: Artificial Neural Networks

The growing field of neural computation is based on combinations of linear and nonlinear elements. The perceptron which is the basic threshold element of neural networks is built as in Fig. 24, where the function $F(x)$ is a step function or any other sigmoidal function.

It is well known that any function can be approximated with the Taylor expansion as a polynomial function. So one can choose the functions $F_i(u_i)$ in Fig. 23 to be: 1, u, u^2, u^3, etc. and then this model can estimate any continuous function. The field of neural computation contains numerous examples for these kinds of models, see Chapter 25.

Comment: Overfit and Underfit

The problem of choosing the order of the model raised in the previous section on linear systems is a major problem in the field of neural computation. Both too many or too few parameters should be avoided. *Under-fit* is the situation where the model is less complex

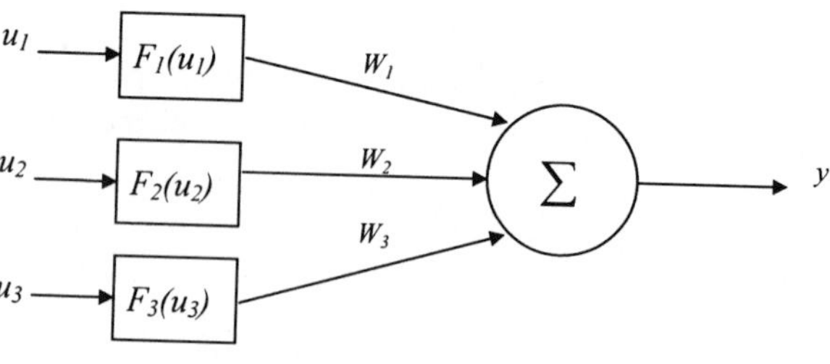

Fig. 23. Linear combination of nonlinear functions

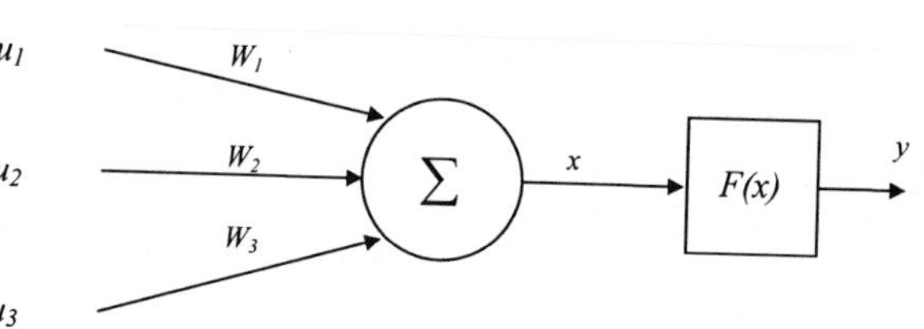

Fig. 24. Nonlinear function of a linear sum

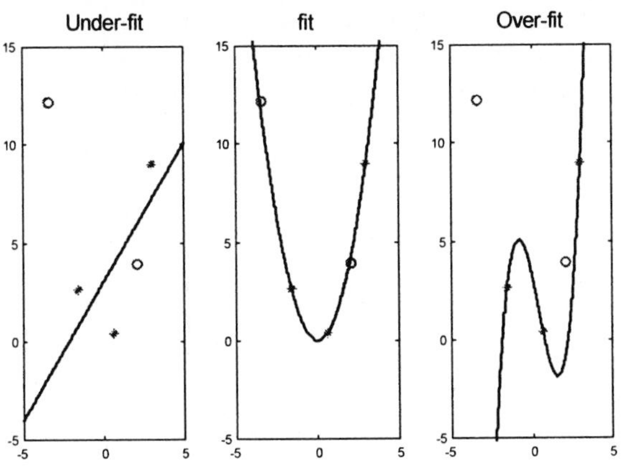

Fig. 25. Fitting a model to data. In this illustration, the three stars are the data taken from an underlying unknown function. On the left, a linear function was fitted to the data. In the middle, a quadratic function was fitted; and on the right, a third-order polynomial function was fitted. After the fitting was completed, two more examples were taken from the same underlying function (the two circles). One can see that the left model is too simple, i.e., under-fits the data, and the right model is too complex, i.e., over-fits the data, but unfortunately does not fit the underlying system.

than the actual system. In this case, the model is unable to fit the data. See Fig. 25 on the left.

Over-fit is the situation where the model is more complex than the actual system. In this case, the model will fit the observations. However, if there is noise or insufficient observations (that is, the number of independent observations is smaller then the number of parameters), then the model will not fit the actual system, and in the validation process it may fail to predict the outcome of the system. (In the validation process we check the generalization, that is, the ability of the identified model to deal with examples that were not used for the fitting). See Fig. 25 on the right.

Example 7.5: Single-sign Integrated Pulse Frequency Modulation

The transformation of graded membrane potentials into sequences of action potentials in nerve cells is often modeled by single-sign integrated pulse frequency modulation (IPFM), as illustrated in Fig. 26. The input (membrane potential) is integrated (the block $1/s$), and when the value of the integrator reaches a threshold A, the pulse shaper (P.S.), which can be anatomically related to the axon hillock, produces an action potential that resets the integrator and is the output of the system. This model has a linear part which is the integration, and a nonlinear part which is the threshold.

This model can also be combined with the model in Example 3.1 in order to account for multiple inputs from different synapses that influence the value of the integrator.

Let us analyze this nonlinear model and perform linearization in order to find a similar linear model.

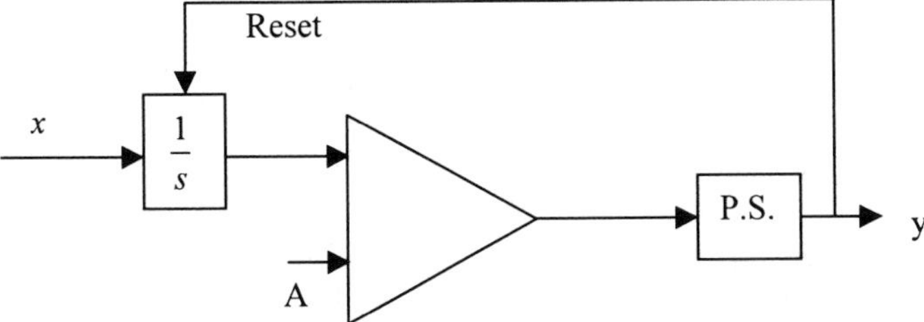

Fig. 26. The integrated pulse frequency modulation (IPFM) model for neural coding

The output is a series of pulses, and we are interested only in the time of each pulse. We also know that the integrator is reset exactly when a new pulse is generated, therefore we know that the integral of the input between two pulses is equal to the threshold, that is:

$$A = \int_{t_k}^{t_{k+1}} x(t) \cdot dt = X(t_{k+1}) - X(t_k) \qquad \text{where } x(t) \equiv \frac{dX(t)}{dt}$$

Let us write the first elements of the Taylor expansion for $X(t_{k+1})$ near t_k:

$$X(t_{k+1}) \cong X(t_k) + x(t_k) \cdot (t_{k+1} - t_k)$$

Let us define the frequency of the output as one over the interval between two pulses and combine the above equations to get

$$A = X(t_{k+1}) - X(t_k) \cong x(t_k) \cdot (t_{k+1} - t_k) = \frac{x(t_k)}{f}$$

Therefore we can conclude that the relation between the frequency of the output and the input signal is

$$f(t) \cong \frac{1}{A} \cdot x(t)$$

We have arrived at the simplest linear system there is. We can take this IPFM as a model of a piece of the retina, the input x as the intensity of light and the output as the firing rate of the optic nerve, and therefore we can close this chapter with the very first example that opened it.

Conclusions

Linear systems description is a very powerful tool used extensively in all branches of science and technology. Biological systems are generally not linear, and a purely linear model is thus seldom satisfactory. Nevertheless, linear systems description tools have important advantages due to their simplicity, analyzability and tractability. They can also be used in one or more of the following model types: locally linear, short-term linear, linear time-varying, linear parameter-varying, linear combinations of nonlinear functions and nonlinear functions of linear combinations. Therefore linear systems description tools are not expected to become obsolete in the near future.

References

Allin J, Inbar GF (1986) FNS Parameter Selection and Upper Limb Characterization. IEEE transactions on biomedical engineering 33:809–817

Benignus VA (1969) Estimation of the coherence spectrum and its confidence interval using the fast fourier transform, IEEE Trans. Audio Electroacoustics, 17:145–150 and correction in 18:320.

Brown AC, Brengelmann GL (1970) The interaction of peripheral and central inputs in the temperature regulation system. In JD Hardy, AP Gagge, JAJ Stolwijk (Eds) Physiological and Behavioral Temperature Regulation, Chapter 47, pp 684–702.

Cadzo JA and Solomon OM (1987) Linear Modeling and the Coherence Function. IEEE Trans. On Acoustic, Speech, and Signal Processing ASSP-35:19–28

Conant R (1993) Engineering Circuit Analysis with Pspice and Probe. McGraw-Hill, Inc, New York.

Charles AD, Kuh ES (1969) Basic Circuit Theory, International student edition, McGraw-Hill KogaKusha, Ltd.

Dorny CN (1993) Understanding Dynamic Systems. Prentice-Hall, Inc.
Fausett Laurene (1994) Fundamentals of Neural Networks. Prentice Hall International, Inc.
Hebb DO (1949) The organization of behaviour, New York: Wiley.
Hodgkin AL, Huxley AF (1952) A Quantitative description of membrane current and its application to conduction and excitation in nerve. J. Physiol. 117:500–544
Houk JC (1963) A mathematical model of the stretch reflex in human muscle systems. M.S. Thesis, Massachusetts Institute of Technology, Cambridge.
Inbar GF (1996) Estimation of Human Elbow Joint Mechanical Transfer Function During Steady State and During Cyclical Movements. In:Gath I & Inbar GF (eds.) Advances in Processing and Pattern Analysis of Biological Signals. Plenum Press.
Karniel A, Inbar GF (1997) A Model for Learning Human Reaching Movements. Biological Cybernetics 77:173–183.
Karniel A, Inbar GF (in press) Human Motor Control: Learning to control a Time-Varying Non-Linear Many-to-One System. Accepted for publication in IEEE Transactions on system, man, and cybernetics Part C.
Kwakernaak H. and R. Sivan (1991) Modern signals and systems. Prentice-Hall International, Inc.
Lathi BP (1974) Signals, Systems, and Controls. Intext Educational Publishers, New York.
Levine WS Ed.(1996) The control handbook. CRC press.
Ljung L (1986) System Identification Toolbox: The manual. The Mathworks Inc. 1st Ed., (4th ed. 1994) Natick, MA.
McRuer DT, Magdaleno RE, Moore GP (1968) A Neuromuscular Actuation System Model. IEEE Trans. On man-machine systems 9:61–71
Milgram P, Inbar GF. (1976) Distortion Suppression in Neuromuscular Information Transmission Due to Interchannel Dispersion in Muscle Spindle Firing Thresholds. IEEE transactions on biomedical engineering 23:1–15
Nilsson JW, Riedel SA (1996) Using Computer Tools for Electric Circuits. Fith Edition Addison-Wesley, Inc.
Paiss O, Inbar GF (1987) Autoregressive Modeling of Surface EMG and Its Spectrum with Application to Fatigue. IEEE transactions on biomedical engineering 34:761–770
Plonsey R. and R.C. Barr (1988) Bioelectricity A quantitative approach. Plenum Press, New York and London.
Porat B. (1994) Digital processing of random signals: theory and methods. Prentice-Hall, Inc, New Jersey.
Santina SM, Stubberud AR, Hostetter GH (1996) Discrete-Time Equivalents to Continuous-Time Systems. In Levine WS (Ed) The Control Handbook , CRC press Inc. pp:265–279.
Sjoberg J, Zhang Q, Ljung L, Benveniste A, Delyon B, Glorennec P, Hjalmarsson H, and Juditsky A (1995) Nonlinear Black-box Modeling in System Identification: a Unified Overview. Automatica, 31:1691–1724.
Stein RB (1980) Nerve and Muscle membranes, cells, and systems. Plenum Press, New York.
Windhorst U (1996) Chapter 1. Regulatory principles in physiology. In: Greger R, Windhorst U (eds) Comprehensive human physiology. From cellular mechanisms to integration, pp. 21–42. Springer-Verlag, Berlin Heidelberg
Zangemeister WH, Lehman S, Stark L (1981) Simulation of Head Movement Trajectories: Model and Fit to Main Sequence. Biol. Cybern. 41:19–32

Chapter 22

Nonlinear Analysis of Neuronal Systems

ANDREW S. FRENCH and VASILIS Z. MARMARELIS

Introduction

Linear analysis provides powerful tools for understanding the behavior of neuronal components and systems, but all real physical components are nonlinear under some conditions, and many neuronal components demonstrate such strongly nonlinear behavior that linear analysis provides only a rough approximation to reality. A common example is the production of action potentials by a neuron, where the output signal has no simple linear relationship to the input. The development of general methods for analyzing nonlinear systems and interpreting the results of analysis has been much slower than for linear systems, due to the increased theoretical and computational complexities involved. However, significant progress has been made, and the analysis of neuronal systems has played an important part in these developments. Here we describe the implementation of the major methods that are now available, and their application to neuronal problems.

A linear, time-invariant system with one input and one output can be represented by the convolution integral (see Chapter 21, this volume):

$$y(t) = k_0 + \int_0^\infty k_1(\tau)x(t - \tau)d\tau \tag{1}$$

which means that the output, $y(t)$, is obtained by integrating the input past, $x(t)$, multiplied by a memory function, $k_1(\tau)$, over all of previous time. A constant value, k_0, can be added for completeness. System analysis consists of obtaining k_0 and k_1. This idea can easily be extended to include nonlinear combinations of the input in the form:

$$y(t) = k_0 + \int_0^\infty k_1(\tau)x(t-\tau)d\tau + \int_0^\infty\int_0^\infty k_2(\tau_1,\tau_2)x(t-\tau_1)x(t-\tau_2)d\tau_1 d\tau_2 + \ldots \tag{2}$$

where k_2, k_3, etc. are higher order memory functions. This is usually called a Volterra series after its creator (Volterra 1930), and the analysis task is again to obtain values for the memory functions, which are generally called the kernels of the system. The kernels cannot be obtained for a general nonlinear system with an arbitrary input unless the series is finite. However, Wiener (1958) proposed a method that could be used for a wide range of nonlinear systems, based on the properties of Gaussian White Noise (GWN). Briefly, Wiener showed that if GWN was used as the input and the expansion was suitably modified, a series of orthogonal terms could be produced that allowed each kernel

Correspondence to: Andrew S. French, Dalhousie University, Department of Physiology and Biophysics, Halifax, Nova Scotia, B3H4H7 Canada (phone: +902-494-1302; fax: +902-494-2050; e-mail: andrew.french@dal.ca)
Vasilis Z. Marmarelis, University of Southern California, Biomedical Simulations Resource, Los Angeles, California, USA (phone: +01-213-740-0841; fax: +01-213-740-0343; e-mail: vzm@bmsrs.usc.edu)

to be estimated independently. The new series has new kernels that are now called the Wiener kernels.

Writing before the general availability of digital computers, Wiener proposed a measurement scheme based on electronic analogs of Laguerre functions, but as computers became available Lee and Schetzen (1965) proposed measuring the kernels by stimulation with GWN and performing a series of multi-dimensional cross-correlations between the input and output signals to obtain the kernels. This method was used by Stark (1969) to measure the second order behavior of the pupillary reflex in the human eye, in one of the earliest applications of general nonlinear systems analysis to a neuronal system, and then by Marmarelis and Naka (1972) to examine neuronal responses in the catfish retina. At about the same time, French and Butz (1973) showed that the cross-correlations could be carried out more efficiently in the frequency domain, and this approach was used to measure nonlinear responses in an insect mechanosensory neuron (French and Wong 1977). A thorough treatment of the white noise approach to physiological system modeling, including nonlinear analysis and synthesis methods, was produced by Marmarelis and Marmarelis (1978), and a series of workshops organized by the Biomedical Simulations Resource (BMSR) at the University of Southern California in Los Angeles have dealt with many of the subsequent advances in the field. The proceedings of these workshops provide a useful introduction to the interested reader (Marmarelis 1987; 1989a; 1994).

Nonlinear analysis inevitably involves significant computational effort, so the rapidly increasing power and speed of digital computers has been a major factor in bringing nonlinear analysis into the laboratory. However, this has also been accompanied by important developments in methods for measuring and interpreting the kernels. These include the fast orthogonal (Korenberg 1988), parallel cascade (Palm 1979; Korenberg 1991) and Laguerre expansion (Marmarelis 1993) methods of analysis, and various block-structured nonlinear models for synthesis (Hunter and Korenberg 1986; Korenberg and Hunter 1986; Marmarelis and Orme 1993; Marmarelis and Zhao 1997). The general principles of the most commonly used methods of analysis and synthesis will be described below, together with guidance on implementing these methods in the neuroscience laboratory.

Outline

The general procedure for analyzing a nonlinear neuronal system is shown in Fig. 1. The input consists of a random signal that normally approximates GWN, although a Poisson distributed impulse train may also be used for systems where the input normally consists of action potentials. Nonlinear systems analysis produces estimates of the system

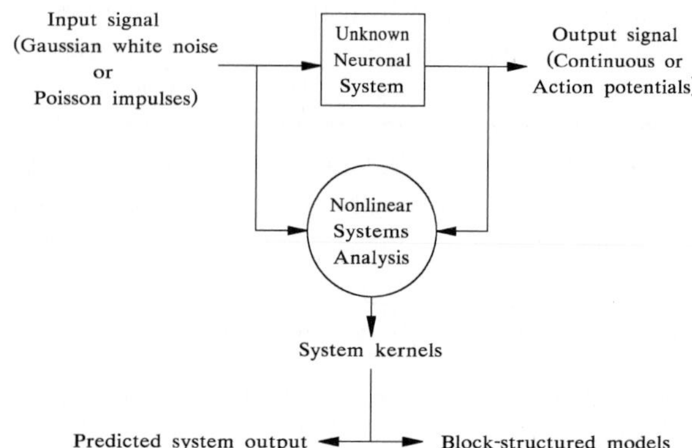

Fig. 1. The general scheme to be followed for characterizing an unknown nonlinear neuronal system.

kernels, analogous to the impulse response or frequency response that is obtained from linear systems analysis. The kernels may be used to predict the output to an unknown input, and may also be used to create functional models of the system. The most popular of these are the block-structured models.

Procedure

The procedure for performing nonlinear systems analysis follows directly from Fig. 1.

1. Create a suitable random or pseudo-random input signal.
2. Filter the input signal to restrict its bandwidth.
3. Apply the input signal to the unknown system for an appropriate period of time.
4. Filter the output signal (optional – depending on the input).
5. Sample the input and output signals (normally by A/D converters) and store digital values.
6. Run the nonlinear systems analysis software to estimate the system kernels.
7. Predict the system output using a section of data that was not used for kernel estimation.
8. Measure the mean-square error between experimental and predicted output.
9. Construct functional model(s) of the unknown system based on the kernel estimates.

Creating and Sampling the Input Signal

Although perfect GWN is impossible to produce from real electronic components, a variety of band-limited substitutes have been used satisfactorily, and their use for linear and nonlinear systems analysis has been discussed before (Marmarelis and Marmarelis 1978). Noise sources are available commercially, but can also be constructed quite easily in the laboratory. Probably the most commonly used method is based on feedback binary shift registers generating maximum-length pseudo-random sequences, or m-sequences. These can be created in software and transmitted to the experiment via a digital to analog converter, or a simple logic output. They can also be constructed independently of the computer from logic integrated circuits. Both methods have been used extensively.

Whatever the origin of the noise, it must be restricted to a known bandwidth before being used in the experiment to avoid aliasing during the sampling process. Normally, a low-pass filter is used, with its upper frequency limit determined by the desired sampling rate. It is advisable to use a sharp cut-off low-pass filter to give useful noise power up to the Nyquist frequency (half the sampling frequency), so eight- or nine-pole filters are commonly used. If the noise is being created by a computer, then it can be filtered digitally within the computer, and the rate of digital to analog conversion must be at least twice as fast as the highest frequency component.

The duration of the experiment depends on several factors, including any unwanted noise in the preparation that will degrade the measurement, and on the final level of accuracy required. As a rough guide, all of the methods described here will work well with about ten thousand pairs of input-output data, and often with less, so an experiment with an upper bandwidth of 500 Hz would require a sampling rate of 1 kHz and a duration of 10 s, while a bandwidth of 50 Hz would require sampling at 100 Hz and for 100 s, and so on.

Sampling the Output Signal

For linear systems analysis, decisions about sampling rate and filter frequency are straightforward, once the bandwidth to be investigated has been chosen. For nonlinear systems the problem is more difficult, because the system output bandwidth can be much wider than the input bandwidth. In general, the highest frequency component in the output will be n-times the highest frequency component in the input, where n is the highest order nonlinearity. Therefore, two choices are available: (1) Sample the input and output signals several times faster than the Nyquist frequency of the input signal. For example, use a filter that cuts off at 100 Hz in the input signal and sample at 1 kHz (5 times the Nyquist frequency). (2) Filter the output to remove any components above the Nyquist frequency. The problem with the second method is that the filter on the output becomes part of the system that is being identified, so it is normal to filter both input and output signals with identical filters, and special matched electronic filters are available for this purpose. Digital filtering of both input and output signals is also possible, but the initial sampling rate must again be high enough to avoid any aliasing.

Action Potentials

The input and/or output signals of a neuronal system are often in the form of action potentials, and this creates special problems in generating appropriate stimuli, in sampling and in filtering signals. An input signal consisting of action potentials can clearly not be GWN, but alternative analysis using a Poisson impulse train to produce system kernels is available (Krausz 1975) and has been used extensively for analyzing hippocampal systems (Berger et al. 1994).

If the input is continuous and the output is a sequence of action potentials, then several possibilities exist, all based on treating the action potentials as Dirac delta functions, whose time of occurrence is the only significant parameter: (1) Cross-correlation can be performed very efficiently with delta functions, so it is possible to modify the software to use the action potentials directly (Bryant and Segundo 1976; deBoer and Kuyper, 1968). However, this does not filter the output signal, which may still contain frequency components above the Nyquist frequency of the input sampling rate. (2) A perfect filter that passes all frequency components below the Nyquist frequency and nothing beyond has an impulse response given by the function: $\sin(2\pi f_N t)/2\pi f_N t$, where f_N is the Nyquist frequency and t is time. Therefore, replacing each action potential with this function and then sampling at $2f_N$ produces a perfectly filtered and sampled output that can be used immediately for analysis. This operation must be done digitally after recording the data, because the function extends both forwards and backwards in time, but can be performed very efficiently by the French-Holden Algorithm (French and Holden 1971; Peterka et al. 1978). (3) A simple method is to produce a sampled output signal that has value zero everywhere except at those times coinciding with the peak of action potentials, where the value is unity. This is essentially the same as the French-Holden method, but assuming that each action potential falls exactly on a sampling point. Although it is an approximation, the method has been used successfully in several studies (Sakuranaga et al. 1987; French and Korenberg 1991).

Analysis Methods

All of the methods in common use are based on the original formulations by Volterra and Wiener, but there is no general agreement regarding the best way to estimate the kernels. All of the descriptions here will assume that the system being investigated has a single input that can be driven with a wide range of input signals, including random noise, and has a single output. It is also assumed that the system is stationary, *i.e.*, that its behavior does not change with time. Extension of some methods to systems with multiple inputs, multiple outputs, non-stationary behavior etc. is possible, but beyond the present scope. Articles in the BMSR workshop volumes (Marmarelis 1987; 1989a; 1994) provide a good starting point for readers interested in pursuing these more complex problems.

Analysis method 1: Cross-correlation

Assuming that the input signal, $x(t)$, is GWN, the output of the system, $y(t)$, is given by:

$$y(t) = \sum_{n=0}^{\infty} G_n[h_n, x(t)] \tag{3}$$

where G_n are the Wiener functionals and h_n are the Wiener kernels. The first four functionals of the Wiener series are:

$$G_0 = h_0 \tag{4}$$

$$G_1 = \int_0^{\infty} h_1(\tau)x(t-\tau)d\tau \tag{5}$$

$$G_2 = \int_0^{\infty}\int_0^{\infty} h_2(\tau_1,\tau_2)\,x(t-\tau_1)x(t-\tau_2)d\tau_1 d\tau_2 - P\int_0^{\infty} h_2(\tau,\tau)\,d\tau \tag{6}$$

$$\begin{aligned} G_3 = &\int_0^{\infty}\int_0^{\infty}\int_0^{\infty} h_3(\tau_1,\tau_2,\tau_3)x(t-\tau_1)x(t-\tau_2)x(t-\tau_3)d\tau_1 d\tau_2 d\tau_3 \\ &- 3P\int_0^{\infty} h_3(\tau_1,\tau_2,\tau_2)x(t-\tau_1)d\tau_1 d\tau_2 \end{aligned} \tag{7}$$

where P represents the power level of the input signal. Lee and Schetzen (1965) were able to show that the kernels can be obtained from the cross-correlations:

$$h_m(\tau_1,\tau_2,\ldots,\tau_m) = \frac{1}{m!P^m} E[y(t)x(t-\tau_1)x(t-\tau_2)\ldots x(t-\tau_m)] \tag{8}$$

where E[] represents the expected value. For the evaluation of the kernel at diagonal points where $\tau_i=\tau_j$ the term $y(t)$ must be replaced by the *m-th* response residual:

$$y_m(t) = y(t) - \sum_{n=0}^{m-1} G_n(t) \tag{9}$$

More complete descriptions of the method are given by Lee and Schetzen (1965) and by Marmarelis and Marmarelis (1978).

The Fourier transform method (French and Butz 1973) is a frequency domain version of the cross-correlation method, in which the input and output signals, $x(t)$ and $y(t)$, are first Fourier transformed and then used to compute frequency response functions by multiplication of complex conjugates. This method is faster than time domain correlation but improved computing speed has reduced this advantage. The frequency domain method may still be useful for analyzing systems that have traditionally been characterized by linear transfer functions, and it has the advantage of being able to compensate for non-white input signals. A complete practical description of the method (French 1977) and its application to a sensory receptor (French and Wong 1977) are available. An alternative frequency domain method that has been used in several interesting applications uses an input consisting of sums of sinusoids to obtain the kernel estimates (Victor and Shapley 1980).

As noted above, the output may consist of action potentials that can be treated as Dirac delta functions of time, $\delta(t)$, simplifying the cross correlation operations. Methods for measuring the kernels based on cross-correlation with delta functions have been described by Bryant and Segundo (1976) and by Krausz (1976).

Analysis method 2: The fast orthogonal method

The fast orthogonal method (Korenberg 1988) is based on an earlier orthogonal method (Korenberg et al. 1988) of estimating the Volterra kernels (Eq. 3). Both methods aim to minimize the mean-square error, e, between the experimental system output, $y(n)$, and the output, $y_s(n)$, predicted by a Volterra model:

$$e = \overline{\left(y(n) - y_s(n)\right)^2} \tag{10}$$

where the bar indicates a time average over the length of the experimental record. Note that we have changed from a continuous time variable, t, to a discrete time variable, n. Since algorithms must eventually be written in discrete format, this representation will be used extensively. If the Volterra model is limited to a finite order, such as second order with kernels k_0, k_1, k_2, then simultaneous equations can be written to solve for the kernels by linear regression. Of course, the number of equations to be solved could be very large. For example, the kernels k_0, k_1, k_2, above, each with a memory length of M, would require the inversion of a square matrix having $1+M+M(M+1)/2$ columns, so that a memory length of 50 or 100 lags would produce thousands of columns. The orthogonal method avoids the matrix inversion by rewriting the Volterra series as a sum of functions, W_i, that are orthogonal over the actual data record:

$$y_s(n) = \sum_{i=0}^{L} g_i W_i(n) \tag{11}$$

The W_i(n) are constructed via Gram-Schmidt orthogonalization as linear combinations of the functions: 1, $x(n-j_1)$, $x(n-j_1)x(n-j_2)$; $j_1=0,\ldots,M-1$, $j_2=j_1,\ldots M-1$. Since the $W_i(n)$ are mutually orthogonal, the weights, g_i, can be estimated independently of each other, avoiding the solution of simultaneous equations. The mean-square error is minimized when each weight is given by:

$$g_i = \frac{\overline{y(n)W_i(n)}}{\overline{W_i^2(n)}} \tag{12}$$

It is then straightforward to convert the weights into the kernel values.

The fast orthogonal algorithm avoids the creation of the actual orthogonal functions by observing that they are only needed as parts of time averages (Eq. 12). Starting from the zero order terms, each time average is calculated recursively from previous time averages. The only step-by-step averages over each value of n required are the input mean, input autocorrelation, output mean, and cross-correlations between input and output. The increased efficiency provided by the fast orthogonal algorithm is impressive, with time savings exceeding a factor of 10^6 under some conditions. Because of the orthogonalization procedure used, the method does not require that the input signal is white, or has special autocorrelation properties although the model must be complete to avoid estimation bias in the kernels. However, the use of GWN is strongly recommended, because the kernels produced by the method then closely approximate the Wiener kernels, which also provide a minimum mean-square error between the experimental and fitted outputs of the system.

Analysis method 3: The parallel cascade method

Palm (1979) showed that any nonlinear system can be approximated to arbitrary accuracy by a parallel set of LNL cascades, where LNL indicates a linear system, followed by

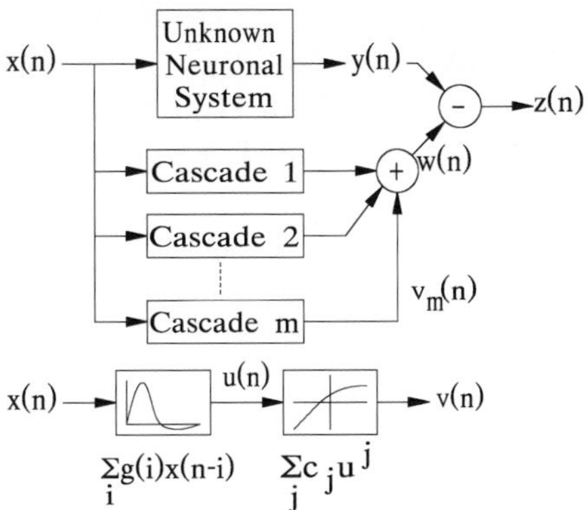

Fig. 2. Schematic of the parallel cascade method. The block diagram at the bottom shows the internal structure of each cascade.

a zero-memory nonlinearity, followed by a second linear system, and suggested that a general system of nonlinear identification might be based on this approach. Korenberg (1991) showed that a parallel set of LN cascades can represent any system that can also be represented as a finite Volterra series with finite memory, and developed efficient methods for fitting an unknown system with a set of parallel LN cascades and using them to calculate the Volterra kernels.

Figure 2 shows the method schematically. The unknown system is stimulated with an appropriate input, and both input, $x(n)$, and output, $y(n)$, are recorded. The linear part of the first cascade is chosen randomly from one of the first, second or higher order cross-correlations between the input and the output, and then the static nonlinearity, in the form of a polynomial, is fitted by a procedure that minimizes the mean-square error between the cascade output, $w(n)$, and the real output, $y(n)$. Then $w(n)$ is subtracted from $y(n)$ to give a residual, $z(n)$. Now the procedure is repeated but using the residual, $z(n)$, as the output that is to be fitted, and so on. Thus the output of the real system is modeled as the sum of the outputs of the cascaded pathways. As each LN cascade is added, the mean-square error of the residual, $z(n)$, is computed after the *m-th* cascade as:

$$e_m = \frac{\overline{z_m^2(n)}}{\overline{y^2(n)}} \tag{13}$$

where the bars indicate mean values. Processing stops when the error or the number of cascades reach some pre-determined value. The Volterra kernels are obtained by summing the linear systems of all cascades multiplied by the respective coefficients in the polynomial expansions. A nice practical feature of the method is the ease of displaying the kernels, or parts of them, as they are built up from each cascade. Again, the input signal, $x(n)$, is arbitrary, but the use of GWN produces kernels that are good approximations to the Wiener kernels. While the fitted set of cascades does not provide a unique description of the unknown system, the kernels are still unique. Korenberg (1991) points out that the method is particularly suited to higher order kernels and many lag values because it avoids the need for high order cross-correlations. He also provides a thorough description of the method.

Analysis method 4: Laguerre functions

Wiener (1958) first proposed that the kernels could be represented by Laguerre expansions so that the identification problem is to estimate the unknown coefficients of the Laguerre functions, and a digital alternative to Wiener's continuous method was suggested by Ogura (1985). Watanabe and Stark (1975) were the first to implement Wiener's suggestions, and implementation methods were discussed by Korenberg and Hunter (1990).

Marmarelis (1993) introduced an efficient implementation using least-squares fitting, instead of covariance time-averaging, and showed that this method can be effective with non-white inputs. The method considers the expansion of the Volterra kernels on the Laguerre orthonormal basis $\{b_j(\tau)\}$, $j=0,1,\ldots,L$:

$$k_1(\tau)=\sum_{j=0}^{L} c_1(j)b_j(\tau) \tag{14}$$

$$k_2(\tau_1,\tau_2)=\sum_{j_1=0}^{L}\sum_{j_2=0}^{L} c_2(j_1,j_2)b_{j_1}(\tau_1)b_{j_2}(\tau_2) \tag{15}$$

etc. Then the Volterra series (2) becomes:

$$y(n)=k_0+\sum_{j=0}^{L} c_1(j)v_j(n)+\sum_{j_1=0}^{L}\sum_{j_2=0}^{L} c_2(j_1,j_2)v_{j_1}(n)v_{j_2}(n)+\ldots \tag{16}$$

where,

$$v_j(n)=\sum_{\tau=0}^{\infty} b_j(\tau)x(n-\tau) \tag{17}$$

The system identification problem now reduces to estimating the Laguerre expansion coefficients c_1, c_2, etc. This method is very efficient when the system kernels can be represented with a relatively small number of Laguerre functions. Several recent applications have shown that no more than 12 Laguerre functions are adequate for this purpose, leading to significant computational savings. For instance, in a study of renal autoregulation 8 Laguerre functions were found to be adequate to represent the system kernels. Since time domain representation of the kernels required about 80 lag values in this case, the computational savings were on the order of 100 and 1000 for a second order and third order model, respectively. In addition to the computational savings, the estimation accuracy is greater because of the reduction in the number of estimated parameters. Finally, the Laguerre expansion technique (LET) suggests an equivalent block-structured model form for the Volterra class of systems that is composed of the Laguerre filter bank receiving the system input and feeding a multi-input static nonlinearity that generates the system output. This block-structured model form has led to compact model representations for nonlinear dynamic systems (see below).

Critical in the application of LET is the selection of the Laguerre parameter α, which determines the rate of exponential decay of the functions (Marmarelis 1993) and the number L of required Laguerre functions for a given physiological system. In the current version of the algorithm available from BMSR, α is selected automatically on the basis of the parameter values for L and M. The latter are determined by successive trials.

Synthesis Methods

One purpose of performing nonlinear systems analysis is to obtain a functional model of the unknown system. Having estimated the kernels, the Volterra or Wiener series immediately provide such a model, which may be useful in its own right. For example, Berger et al. (1994) have used Volterra series models of different parts of the hippocampal slice as modules to construct a more complete model of the entire system. However, many investigators use the kernel values to construct alternative models whose parameters are more likely to be related to known physical components of the system, thus combining the kernel values with information gained by other approaches, including areas such as neuroanatomy, or neuropharmacology.

Synthesis method 1: Reconstruction of the output signal

This can be achieved by simply inserting the estimated kernel values into the appropriate model. For a second order Volterra model in which the zero, first and second order kernels had already been estimated, we would calculate:

$$y_s(n) = k_0 + \sum_{i=0}^{M} k_1(i)x(n-i) + \sum_{i=0}^{M}\sum_{j=0}^{M} k_2(i,j)x(n-i)x(n-j) \tag{18}$$

where M is the number of lag values for which the kernels have been estimated. Then the predicted and actual output can be compared visually. At this stage it is also useful to measure the percentage mean-square error between the estimated output, y_s, and the real output, y:

$$e = 100\frac{\overline{(y(n) - y_s(n))^2}}{\overline{y^2(n)} - \overline{y(n)}^2} \tag{19}$$

where the bar again indicates a time average over all of the available n. Note that this series prediction and error estimation should be done with data that was not used in the kernel estimation. For example, one could take twenty thousand sample pairs, use ten thousand for measuring the kernels, and the remaining ten thousand for the prediction and error calculation.

Synthesis method 2: Block-structured models

Cascades of linear systems with static nonlinearities, sometimes called LN cascades, were introduced in the description of the parallel cascade method above. Cascade models have also been widely used to model overall system behavior. Marmarelis and Naka (1972) used the Wiener cascade (LN) to model the horizontal-bipolar-ganglion cell system in the catfish retina, while French and Wong (1977) used the same model for an insect mechanoreceptor. Marmarelis and Marmarelis (1978) and Hunter and Korenberg (1986) reviewed the use of three cascade structures (Fig. 3) and provided step-by-step procedures for obtaining the parameters of the cascades from the estimated kernels. These three cascades are important because it is possible to derive unique solutions for their best-fitting values from the kernel estimates (except for any linear scaling factors within the cascades). It should also be noted that the Wiener (LN) and Hammerstein (NL) models are both special cases of the LNL model. While other cascade models with more L or N boxes, including multi-path cascades, have been used, they cannot provide unique representations of a system's behavior. A simple test is available to determine if the LNL cascade can be applied to a nonlinear system. If a kernel of order n is integrated along $(n-1)$ lag axes, then the resulting functions should be proportional to the first order kernel (Chen et al. 1986; Korenberg 1973; Korenberg and Hunter 1986).

An alternative, parallel block-structured model which has a unique solution is the method of neuronal modes (Marmarelis 1989b; Marmarelis and Orme 1993) illustrated in Fig. 4. Each mode is a linear filter (Eq. 1) and the multi-input static nonlinearity is given by:

$$y(n) = f(u_1(n), u_2(n), \ldots u_m(n)) \tag{20}$$

where f is a static function that combines the m values that are produced by the filters at each time n. Analysis consists of determining the number of significant modes, their

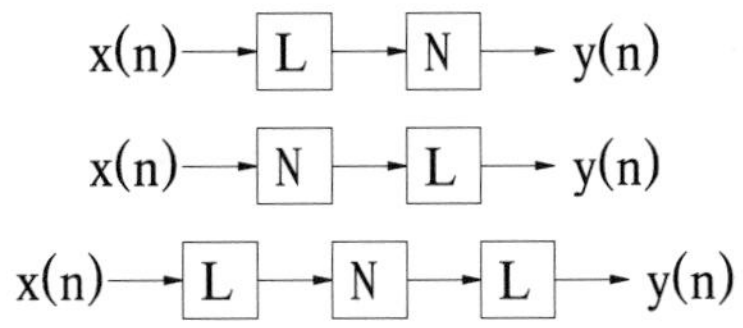

x(n) → L → N → y(n)
x(n) → N → L → y(n)
x(n) → L → N → L → y(n)

Fig. 3. Popular cascade block-structured models. From the top are shown the Wiener, Hammerstein and LNL models.

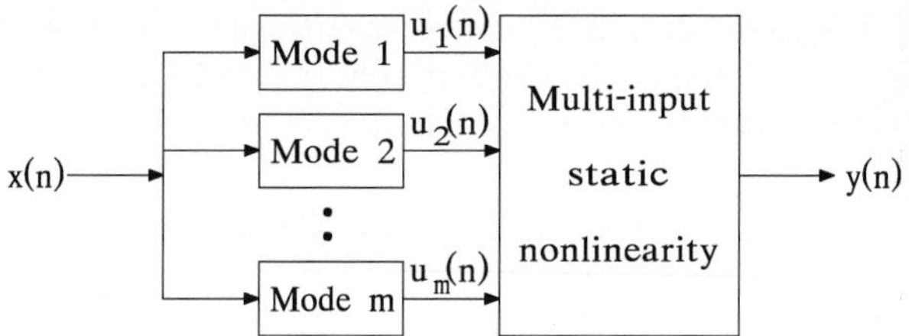

Fig. 4. The neuronal mode model. Each mode is a dynamic linear filter that takes the system input, $x(n)$, and produces an output, $u_i(n)$. The mode outputs are combined by the static nonlinearity to give the model output, $y(n)$.

impulse responses and the form of the function *f*. The method that has been used to perform the analysis follows directly from the Laguerre expansion approach, described above. The estimated Laguerre expansion coefficients of the first two Volterra kernels are used to form a symmetric coefficient matrix **C**. Eigen decomposition of the matrix **C** allows selection of the most significant or "principal" modes of the system from inspection of the relative absolute values of the eigenvalues. The eigenvectors corresponding to the significant eigenvalues are then used to construct the impulse responses, $g_i(n)$, of the selected principal modes. Although the Laguerre coefficients are more computationally efficient, the same results can be obtained via eigen decomposition of a matrix based on the first and second order kernels themselves (Marmarelis, 1997).

Finally, the multi-input static nonlinear function is obtained analytically or graphically. The input signal, $x(n)$, is convolved with each of the principal modes to produce the multiple input signals, $u_i(n)$, to the static nonlinearity. For analytical solutions, some mathematical form of the nonlinearity is postulated, such as a polynomial function, and its unknown parameters are estimated by least-squares fitting between the predicted and actual system outputs. Graphical evaluation is feasible when only two modes are significant. Then a surface can be produced by averaging all of the output values that correspond to each two-dimensional bin in the plane defined by u_1, u_2. When the output of the system is action potentials, the graphical representation can take the form of a trigger region, defined as the locus of points in u_1, u_2 that correspond to an output action potential over the data record. Due to noise and stochastic fluctuations of the threshold, the boundary of the trigger region cannot be sharply defined, but rather takes the form of a "trigger zone" over which the probability of firing changes gradually from 0 to 1. The form of the function defining this probability of firing can be estimated from the data by counting the number of times that a given combination of values of u_1, u_2 leads to an output action potential (Marmarelis and Orme 1993). A complete analysis of action potential encoding using the method of neuronal modes is given in French and Marmarelis (1995).

While the parameters of block-structured models can be derived from fitted kernel values, it is also possible to fit their parameters directly from the input-output data. For example, the linear block of an LN model could be described by a simple function with a small number of parameters, or even as a series of lag values that describe the impulse response. Similarly, the static nonlinear block could be described by an algebraic function, such as a polynomial. Then, if there are many more input-output data pairs available from an experiment with a sufficiently rich input signal, some nonlinear fitting procedure, such as the Levenberg-Marquardt method (Press et al. 1990), can be used to estimate the unknown parameters of the block-structured model. This approach has been used to fit LNL cascades and more complex cascades (French and Patrick 1994; Juusola et al. 1995; Weckström et al. 1995).

Results

Plotting and publishing results

First order kernels and system input-output signals can be plotted by normal graphical methods. Second order or higher order kernels are multi-dimensional and require more sophisticated plotting. Both perspective plots and contour maps of three-dimensional objects are in common use. The former tend to be easier to interpret, and the latter allow easier quantitative measurements to be taken. Probably the best approach is to present the data in both ways if possible. Alternatively, single slices through a second order kernel can be shown as normal two-dimensional plots. Higher order kernels must be shown as slices, producing either two-dimensional or three-dimensional plots. Routines for creating perspective and contour plots are widely available, and form a part of the BMSR LYSIS package (see below).

It is important to calculate the units, and to indicate the dimensions of the axes carefully. One method for second order kernels is to indicate the maximum and minimum values in the legend.

It is common practice to smooth kernels, particularly second and higher order kernels, before plotting, since this can make them easier to interpret. While this may be a useful practice, it should always be made clear how the smoothing was done. Press et al. (1990, pp. 514–516) discuss some methods of data smoothing, as well as the various justifications for smoothing. A simple smoothing technique is to replace each point in a kernel with the weighted average of the point and its nearest neighbors. The averaging must be modified slightly at the edges of the kernel to either extrapolate or ignore values beyond the edge. One advantage of the Laguerre expansion method of kernel estimation is the smoothness of the estimated kernels, since higher order terms in the Laguerre expansion are often insignificant and therefore are not included in the estimate.

Obtaining Software, Information and Advice

BMSR provides a free package of programs called LYSIS (Greek for solution) that includes all of the primary components needed to carry out kernel estimation using various methods, noise generation, plotting of results and synthesis of the fitted system (e.g., principal mode analysis). Further information can be obtained from BMSR, University of Southern California, Los Angeles, CA 90089-1451, USA, or from their web page at: http://www.usc.edu/dept/biomed/BMSR.

A commercial software package called KERNEL is available from ASF Software, Halifax, Canada. KERNEL performs linear and nonlinear systems analysis, together with kernel displays and system prediction. Further information is available by e-mail from ASF.Software@iname.com or from their web page at: http://www.rockwood.ns.ca/asf-software.

Investigators interested in using other methods must either write the programs themselves, based on the original publications, or seek help from the authors of the original publications. The latter route is clearly most feasible, because of the significant effort that is usually required to produce a reliable program, and because the authors are usually happy to collaborate in new applications of their work, as well as being in the best position to advise on questions of accuracy and interpretation.

Examples

It is beyond the scope of the present article to show examples of all the methods of analysis and synthesis that have been applied to nonlinear neuronal systems. Interested

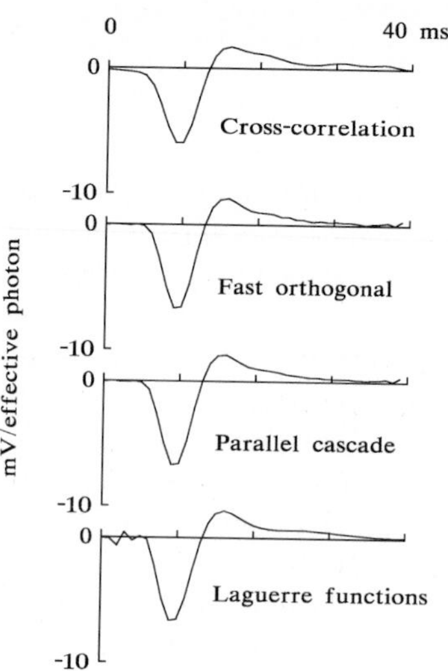

Fig. 5. First order kernels for the response measured in a large monopolar cell of a fly compound eye during random light stimulation, using the four different methods described in the text. The original data set contained 8,192 input-output pairs. The parallel cascade calculation used 1000 cascades, with second order polynomial functions for the nonlinear components. The Laguerre function calculation used the first 10 Laguerre functions.

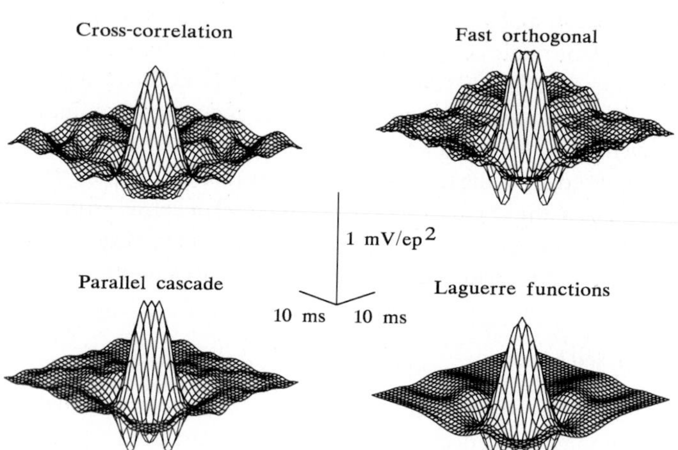

Fig. 6. Second order kernels obtained during the same analyses as Fig. 5. All the kernels were identically smoothed by averaging each point with its nearest neighbors. Total excursions from minimum to maximum values for the four kernels (in units of mV/effective photon^2) were: 0.99 (cross-correlation), 1.30 (fast orthogonal), 1.36 (parallel cascade) and 1.34 (Laguerre functions).

readers should consult some of the original journal articles that have been cited above. Here we show examples of first order kernels (Fig. 5) and second order kernels (Fig. 6) that were estimated from experimental data using the four analysis methods described here. The experiment was an intracellular recording of the membrane potential in a large monopolar cell of the fly compound eye, during stimulation with a randomly fluctuating light (Juusola et al. 1995). Figure 7 shows the actual recording during a short period of the experiment, together with the output predicted by the first order kernel alone, and the output predicted by the first and second order kernels combined. These examples show that all of the methods produce reliable results, and that the addition of the second order kernel produced a significant improvement in the predicted system output.

For these examples we used 40 time lags to produce the kernels. Therefore, the total numbers of free parameters in the kernels were: one for the zero order, 40 for the first order, and 820 for the second order (note that the second order kernel is symmetric

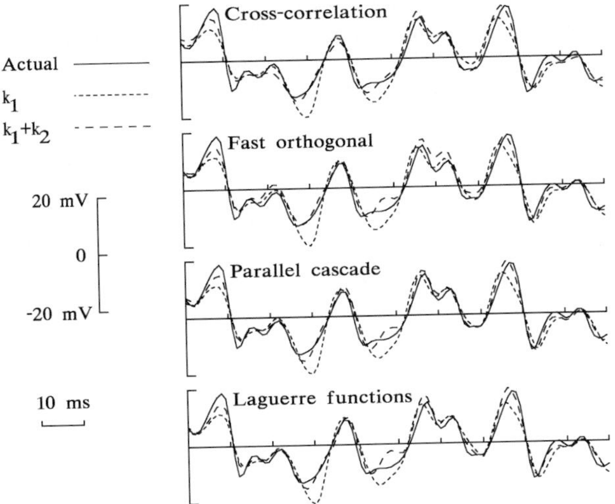

Fig. 7. Actual and predicted membrane potential records in the large monopolar cell during a 100 ms period. Predictions are shown for each of the four methods, using the first order kernels alone (k_1), and the first order plus second order kernels (k_1+k_2).

around the diagonal) for a total of 861 parameters. However, the number of parameters required by the Laguerre method was actually smaller, because the kernels were produced from 10 Laguerre functions, corresponding to 66 free parameters. The initial stage of the parallel cascade method actually involves more parameters, because each cascade contains the number of lags, plus the polynomial coefficients. However, the final number is the same as the cross-correlation and fast orthogonal methods when the kernels are calculated at the end of the process.

Acknowledgement: We thank Matti Weckström for permission to use experimental data gathered in his laboratory in preparing Figures 5–7. Our work is supported by the Medical Research Council of Canada and by grant number RR-01861 from the National Center for Research Resources at the National Institutes of Health of the USA.

References

Berger TW, Harty TP, Choi C, Xie X, Barrionuevo G, Sclabassi RJ (1994) Experimental basis for an input/output model of the hippocampal formation. In: Marmarelis VZ (ed) Advanced Methods of Physiological System Modeling, Plenum Press, New York, N.Y., 3: 29–53

Bryant HL, Segundo JP (1976) Spike initiation by transmembrane current: a white noise analysis. J Physiol 260:279–314

Chen HW, Ishi N, Suzumura N (1986) Structural classification of nonlinear systems by input and output measurements. Int J Syst Sci 17:741–774

deBoer E, Kuyper P (1968) Triggered correlation. IEEE Trans Biomed Eng 15:169–179

French AS, Holden AV (1971) Alias-free sampling of neuronal spike trains. Kybernetik 8:165–171

French AS, Butz EG (1973) Measuring the Wiener kernels of a non-linear system using the fast Fourier transform algorithm. Int J Ctrl 17:529–539

French AS (1977) Practical nonlinear system analysis by Wiener kernel estimation in the frequency domain. Biol Cybern 24:111–119

French AS, Wong RKS (1977) Nonlinear analysis of sensory transduction in an insect mechanoreceptor. Biol Cybern 26:231–240

French AS, Korenberg MJ (1991) Dissection of a nonlinear cascade model for sensory transduction. Ann Biomed Eng 19:473–484

French AS, Patrick SK (1994) A nonlinear model of step responses in the cockroach tactile spine neuron. Biol Cybern 70:435–441

French AS, Marmarelis VZ (1995) Nonlinear neuronal mode analysis of action potential encoding in the cockroach tactile spine neuron. Biol Cybern 73:425–430

Hunter IW, Korenberg MJ (1986) The identification of nonlinear biological systems: Wiener and Hammerstein cascade models. Biol Cybern 55:135–144

Juusola M, Weckström M, Uusitalo RO, Korenberg MJ, French AS (1995) Nonlinear models of the first synapse in the light-adapted fly retina. J Neurophysiol 74:2538–2547

Korenberg MJ (1973) Identification of biological cascades of linear and static nonlinear systems. Proc 16th Midwest Symp Circuit Theory 18.2:1–9

Korenberg MJ, Hunter IW (1986) The Identification of nonlinear biological systems: LNL cascade models. Biol Cybern 55:125–134

Korenberg MJ (1988) Identifying nonlinear difference equation and functional expansion representations: the fast orthogonal algorithm. Ann Biomed Eng 16:123–142

Korenberg MJ, Bruder SB, Mcilroy PJ (1988) Exact orthogonal kernel estimation from finite data records: extending Wiener's identification of nonlinear systems. Ann Biomed Eng 16:201–214

Korenberg MJ, Hunter IW (1990) The identification of nonlinear biological systems: Wiener kernel approaches. Ann Biomed Eng 18:629–654

Korenberg MJ (1991) Parallel cascade identification and kernel estimation for nonlinear systems. Ann Biomed Eng 19:429–455

Krausz HI (1975) Identification of nonlinear systems using random impulse train inputs. Biol Cybern 19:217–230

Lee YW, Schetzen M (1965) Measuring the Wiener kernels of a nonlinear system by cross-correlation. Int J Ctrl 2:237–254

Marmarelis PZ, Naka K-I (1972) White noise analysis of a neuron chain: an application of the Wiener theory. Science 175:1276–78

Marmarelis PZ, Marmarelis V Z (1978) Analysis of Physiological Systems: The White-Noise Approach. Plenum Press, New York, N.Y.

Marmarelis VZ (1987) Advanced Methods of Physiological System Modeling vol 1. Biomedical Simulations Resource, Los Angeles, California

Marmarelis VZ (1989a) Advanced Methods of Physiological System Modeling vol 2. Plenum Press, New York, N.Y.

Marmarelis VZ (1989b) Signal transformation and coding in neural systems. IEEE Trans Biomed Eng 36:15–24

Marmarelis VZ (1993) Identification of nonlinear biological systems using Laguerre expansions of kernels. Ann Biomed Eng 21:573–589

Marmarelis VZ, Orme ME (1993) Modeling of neural systems by use of neuronal modes. IEEE Trans Biomed Eng 41:1149–1158

Marmarelis VZ (1994) Advanced Methods of Physiological System Modeling vol 3. Plenum Press, New York, N.Y.

Marmarelis VZ (1997) Modeling methodology for nonlinear physiological systems. Ann Biomed Eng 25:239–251

Marmarelis VZ, Zhao X (1997) Volterra models and three-layer perceptrons. IEEE Trans Neural Networks 8:1421–1433

Ogura H (1985) Estimation of Wiener kernels of a nonlinear system and a fast algorithm using digital Laguerre filters. Proc 15th NIBB Conf. Pp. 14–62, Okazaki, Japan

Palm G (1979) On representation and approximation of nonlinear systems. Biol Cybern 34:49–52

Peterka RJ, Sanderson AC, O'Leary DP (1978) Practical considerations in implementing the French-Holden algorithm for sampling neuronal spike trains. IEEE Trans Biomed Eng 25:192–195

Press WH, Flannery B P, Teukolsky S A, Vetterling W T (1990) numerical recipes in c. the art of scientific computing. Cambridge University Press, Cambridge

Sakuranaga M, Ando Y-I, Naka K-I (1987) Dynamics of ganglion cell response in the catfish and frog retinas. J Gen Physiol 90:229–259

Stark L (1969) The pupillary control system: its non-linear adaptive and stochastic engineering design characteristics. Automatica 5:655–676

Victor JD, Shapley R (1980) A method of nonlinear analysis in the frequency domain. Biophys J 29:459–484

Volterra V (1930) Theory of Functions and Integral and Integro-differential Equations. Dover Publications Inc., New York, N.Y.

Watanabe A, Stark L (1975) Kernels method for nonlinear analysis: Identification of a biological control system. Math Biosci 27:99–108

Weckström M, Juusola M, Uusitalo RO, French AS (1995) Fast-acting compressive and facilitatory non linearities in light-adapted fly photoreceptors. Ann Biomed Eng 23:70–77

Wiener N (1958) Nonlinear Problems in Random Theory. The MIT Press, Cambridge, Massachusetts

Chapter 23

Dynamical Stability Analyses of Coordination Patterns

David R. Collins and M.T. Turvey

Introduction

The synchronizing of rhythmically moving limbs and limb segments is one of the most fundamental achievements of biological movement systems. It has attracted considerable scientific attention because it is a primary expression of how biology (a) organizes movements in space and time, (b) resolves issues of movement efficiency, and (c) meets the competing challenges of movement stability and flexibility. In broad theoretical terms, a multisegmental rhythmic pattern can be viewed as one of the nervous system's original forms of a collective organization in which many component parts, distributed across the body, act coherently to produce a global action.

Quantitative mathematical models of synchronized limbs based on physicochemical principles and neurobiological facts have proven difficult to formulate and validate. The intractability arises, in large part, from the richness and magnitude of the interactions among neural, muscular, metabolic, and mechanical processes. The methods identified in the present chapter represent an alternative approach to this complexity. They are part and parcel of efforts to develop a qualitative dynamical model that incorporates, in broad strokes, the essential features of synchronized rhythmic behavior.

The chapter is divided into three parts pertaining to particular cases of a quantifying or collective variable for a rhythmic coordination. Part 1 examines characterizing a potential function which captures the dynamics of a collective variable which is expected to be stationary on the time scale of typical experiments. Part 2 examines correlation and recurrence methods for nonstationary collective variables. Part 3 examines the composition of the dynamics at the level of a stationary collective variable through determination of how many dynamically active degrees of freedom are required to produce the dynamics. Finally, future directions are discussed in a Postscript.

Part 1: Stationary Methods

Stationary Models

The measures to be discussed in this chapter are particularly motivated by the theory of synergetics (e.g., Haken, 1983, 1996; Kelso, 1995) and discussions of degrees of freedom in biological movement systems (e.g., Bernstein, 1967; Turvey, 1990). A typical strategy of synergetics is to seek a quantifying or *collective variable*, which characterizes the

Correspondence to: David R. Collins, University of Connecticut, Department of Psychology, Center for the Ecological Study of Perception and Action, 406 Babbidge Road, Box U-20, Storrs, CT, 06269-1020, USA (phone: +01-860-486-2212; fax: +01-860-486-2760; e-mail: drc93001@uconnvm.uconn.edu)
M.T. Turvey, University of Connecticut, Department of Psychology, Center for the Ecological Study of Perception and Action, 406 Babbidge Road, Box U-20, Storrs, CT, 06269-1020, USA

qualitative changes of a phenomenon of interest as a consequence of continuous changes in one or more variables referred to as *control parameters.*

A traditional exemplary phenomenon is a *phase transition* of a substance freezing from a liquid to a solid, which depends on the control variables of temperature or pressure. At standard air pressure (1.0 atmospheres), water freezes to ice when the temperature decreases to 0 degrees Celcius (273 Kelvin). If the temperature is fixed at 0 degrees C, water will freeze when the pressure is increased to 1.0 atmospheres. Different substances will exhibit different relations to temperature and pressure; for example, alcohol freezes at much lower temperatures than water.

An example of a phase transition in biological coordination is the gait transition of a trot to a gallop exhibited by most quadrupeds, in which the four legs change their relative timing. While the trot-to-gallop transition may be intuitively clear, a more experimentally accessible phase transition occurs, for increasing movement speed, from antiphase to inphase coordination of two fingers, each rotating about a single joint (Kelso, 1984). A collective variable which captures this phase transition is the *relative phase* ϕ (Haken, Kelso & Bunz, 1985), defined as the difference between the phase progression of the two fingers treated as limit cycle oscillators from the theory of nonlinear dynamics (e.g., Strogatz, 1994). The *phase* is the (time-varying) angle defined by the arctangent of the angular velocity divided by the angular position. *Inphase* is identified with synchronous synchopation of the muscles of the two fingers, while *antiphase* is identified with antisynchronous synchopation.

Once a collective variable is identified, a *potential function* can be defined which captures the preferences of the system of interest. Perhaps the most familiar potential is that due to gravity. Gravity pulls objects down, while a supporting surface such as the ground may partly or fully prevent downward motion. The analysis is simpler if inertia is negligible, for example, for objects immersed in a thick fluid. Together, gravity, the ground and the fluid serve to define a potential function. For example, if a ball is placed on a hilltop it will stay at rest, but as soon as it gets a slight push, it will roll in the direction of the push until it rests stably in a valley bottom, with no overshoot if inertia is neglected.

An example for the coordination of two limbs or limb segments is the potential defined for relative phase (Haken, Kelso & Bunz, 1985). Kelso (1984) had observed that inphase and antiphase were both stable at slow speeds, but antiphase lost its stability at higher speeds. The presence of two stable states indicates the need for two valleys or wells in the potential, while the loss of antiphase stability indicates that the corresponding well disappears for higher speeds. Also, at slower speeds the inphase coordination pattern is more stable than antiphase coordination pattern (exhibiting lower standard deviation), suggesting that the corresponding inphase well is deeper with steeper walls. These observations are captured in the following potential V,

$$V = -a\cos\phi - b\cos 2\phi \qquad \bmod 2\pi \tag{1}$$

which is a function of the relative phase ϕ and the parameters a and b. The b term provides for two identical wells, while the a term indicates the degree to which the inphase well is more stable. The ratio b/a is a control parameter which determines the relative stability of antiphase to inphase. Further, when b/a decreases to 0.25, the equation undergoes a bifurcation in which the antiphase well disappears. The right hand term, mod 2π, indicates that V is to be treated modularly within a particular interval of 2π (such as $-\pi/2 \le \phi < 3\pi/2$); that is, 2π is added or subtracted from ϕ until it is within the chosen interval. Because Equation 1 is periodic with a period of 2π, the modularity simply indicates that a single interval will be chosen for convenience.

An important addition to Equation 1 is a term which accounts for differences between the two oscillators. The differences have been characterized in terms of the characteristic frequencies of each oscillator, derived by assuming that a complicated biological mass distribution can be approximated as a physical compound pendulum. An example is a person's hand holding a rod like a ski pole. The hand can be approximated as a hollow cylinder surrounding the handle, with mass equal to a small percent of the total body mass. The handle, rod and possibly weights attached at the bottom are cylinders with easily estimated mass, height and radius. As long as the hand grasps the handle firmly during a coordination, the pieces can be treated as composing a rigid compound pendulum. If a person holding the rod lets the hand freely oscillate following gravity (pretending that there is no friction), then the frequency of the oscillation is in principle the same as if all the mass were focused at a characteristic length from the point of rotation (e.g., the center of the wrist), from which a characteristic frequency can be derived (e.g., standard physics texts such as Symon, 1971).

When the frequencies of two oscillators are different, the one with the higher frequency tends to lead by a stable amount. The lead or lag has been termed a *detuning* effect in connection to, for example, a detuned guitar string which is slightly longer or shorter than preferred. The characterization of the detuning that is typically used is the arithmetic difference between two oscillators, although this fails to account for the experimental data when carefully examined (as does the ratio of the frequencies; Sternad, Collins & Turvey, 1995; Collins, Sternad & Turvey, 1996). The standard introduction of the detuning δ into Equation 1 is by addition of a linear term, $\delta\,\phi$. This term provides a convenient account for the direction of phase transitions, but it will be important for the stochastic analyses to be discussed that the potential remain periodic. A periodic version of the detuning is included in the following form of the potential (see Collins, Park & Turvey, 1998),

$$V = -\delta/2\sin 2\phi - a\cos\phi - b\cos 2\phi \qquad \bmod 2\pi \tag{2}$$

The left panel of Figure 1 shows a simple example of Equation 2 for which an attractive well exists near both $\phi = 0$ and $\phi = \pi$, but the well for ϕ near 0 is deeper and hence more stable. A small shift to the right of both wells from 0 and π is due to a small positive δ. It is worth noting that Equation 2 provides a broadly generic model of coordination which is sufficient to capture many robust phenomena: phase transitions between multiple states, greater stability of a particular state, and the shifting of states due to nonidentical components.

The deterministic changes of ϕ can be derived as an *equation of motion* from the potential as the negative derivative, $\dot{\phi} = -\partial V/\partial\phi$ (the overdot represents a derivative with

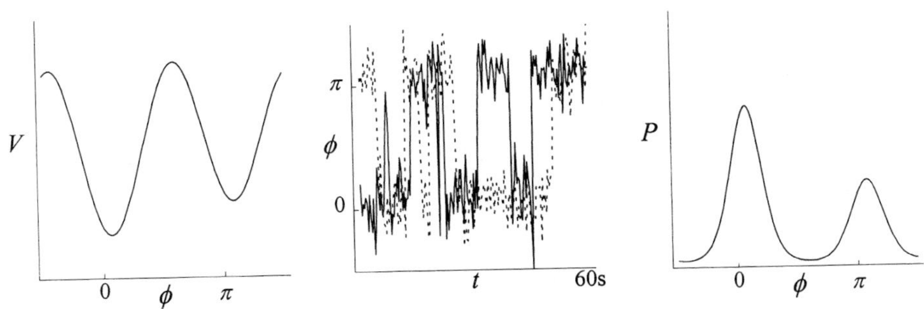

Fig. 1. *Left panel*: A potential function $V(\phi)$ (Equation 2) exhibiting modes shifted to the right of $\phi = 0$ and $\phi = \pi$ due to a positive δ. *Center panel*: Two time series simulated from the approximate equation of motion (Equation 15) initialized at $\phi = 0$ (solid trajectory) or $\phi = \pi$ (dashed trajectory). *Right panel*: The stationary probability distribution $P(\phi)$ (Equation 5). The parameter settings $a = 0.39$, $b = 1.57$, $\delta = 1.57$, and $Q = 2.4$ were used in the three panels.

respect to time; e.g., velocity is the time derivative of position). However, it is clear from experimental observations that some form of uncertainty is involved in almost any biological process and its measurement. The uncertainty can be approximated as a *gaussian white noise process*. That the noise is gaussian or normally distributed means that it follows a bell curve, such that small deviations from the mean are most likely, but occasionally large deviations occur. The characterization of the noise as white means that each value is uncorrelated with subsequent and previous values. While other types of noise are possible, gaussian white noise is the most amenable to analytic treatment and is the best starting point unless different detailed properties are known to be important. However, characterization of the type of noise can be an interesting question in itself, and is involved in the nonstationary analyses of Part 2.

The presence of biological noise can be included in deriving an equation of motion by adding a gaussian white noise process, ξ_t (with zero mean and unit variance), which is scaled by a noise strength which is typically included as the square root of Q (Schöner, Haken & Kelso, 1986). Thus, a stochastic differential equation of motion is derived from Equation 2,

$$\dot{\phi} = \delta \cos 2\phi - a \sin \phi - 2b \sin 2\phi + \sqrt{Q}\xi_t \qquad \mod 2\pi \tag{3}$$

Equation 3 describes, for example, how a ball will move on the potential landscape of Equation 2 when it is subject to noise. The center panel of Figure 1 shows a time series initialized at $\phi = 0$ (solid trajectory) and $\phi = \pi$ (dashed trajectory) and simulated from a discrete approximation of Equation 3 (see Equation 16 and discussion below) using the same parameters as for V in the left panel.

The stochastic differential Equation 3 has an equivalent form in probability theory as a Fokker-Planck equation which relates a *probability distribution* P to a potential and gaussian white noise. The general Fokker-Planck equation is the partial differential equation,

$$\dot{P} = \nabla \cdot (P \nabla V) + \nabla^2 (QP/2) \tag{4}$$

Here, ∇, read as del, is used for various multidimensional partial derivatives with respect to the components of ϕ if ϕ is a vector (a multidimensional variable; e.g., Arfken, 1985) and Q may be a matrix function of a vector ϕ. Equation 4 may only be analytically solved for special cases, but several approximation techniques are available (e.g., Risken, 1996). For this chapter, it is sufficient to let ϕ and Q be scalars (single dimensional) and let Q be independent of ϕ. The special case of interest is the time-invariant or *stationary* solution, using the periodicity of V to set the boundary conditions. The more standard form of the detuning introduces a drift to the probability which would prevent a stationary solution (the slant to the potential introduced by the linear detuning predicts that a ball will run downhill forever to positive or negative infinity for negative or positive detuning, respectively). In contrast, the periodic version in Equation 2 provides for a stationary probability distribution (the left and right hand sides of a 2π interval meet, so if the ball leaves from one side, it is returned at the same height on the other side), which has the form,

$$P = Ne^{-2V/Q} \qquad \mod 2\pi \tag{5}$$

in which N is a normalization constant that ensures that the integral over a 2π interval is unity. Because Equation 5 depends on the ratio of the potential to the noise, it will be convenient to refer to the ratio $2V/Q$ as the relative potential. It is worth emphasizing that the occurrence of V in Equation 5 implies the full form of Equation 2; that is, a, b and δ occur in Equation 5.

Returning to the example of a ball on an underwater landscape perturbed by water, Equation 5 indicates the average behavior of the ball over all time, whereas Equation 3 generates the behavior of the ball in time. This can be illustrated by applying the algorithm to be discussed below for time series data of the ball's position. An example of P is shown in the right panel of Figure 1 using the same parameters as used for the left and center panels. Comparing V and P in Figure 1, it can be observed that P is roughly V flipped over; this is qualitatively true because the exponential of the negative relative potential in Equation 5 acts as an inverse.

Because the stationary distribution eliminates, in principle, the temporal development of a system, it is useful to have complementary measures by which to recover the time scale of an interesting phenomenon. One such measure is the *relaxation time*, which may be experimentally accessible as the time taken to recover a mean state following a perturbation (Scholz, Kelso & Schöner, 1987) or from a spectral analysis of a time series (see Fuchs & Kelso, 1994; Gardiner, 1985; Schöner et al., 1986). The relaxation time refers to the fast time-scale fluctuations of the collective variable about an attractive state. Another time scale measure is the *mean first passage time*, which refers to the mean (average) time taken from initialization to a phase transition that changes the collective variable from the region of one attractor to another (or more generally an exit from a specified region). The contrast can be seen for the example of a ball by the time scale of fluctuations in a well versus the fluctuations that move a ball between wells (for the relaxation time versus the mean first passage time, respectively).

Note: The term phase transition typically denotes a change from one stable state to another. However, in this chapter, phase transition will be used more generally to include any loss of stability because the definition of the mean first passage time T_2 depends only on a passage beyond the stable region for a particular state.

For a scalar ϕ, the relaxation time T_1 within a well is approximately

$$T_1 = \left(\left. \frac{d^2V}{d\phi^2} \right|_{\min} \right)^{-1} \tag{6}$$

and the mean first passage time between wells is approximately

$$T_2 = \frac{\pi}{\sqrt{\left| \left. \frac{d^2V}{d\phi^2} \right|_{max} \left. \frac{d^2V}{d\phi^2} \right|_{min} \right|}} e^{2(V_{max} - V_{min})/Q} \tag{7}$$

where *min* and *max* subscripts indicate that the term is to be evaluated at the value of ϕ for which the potential is minimum or maximum (well bottom or peak top), respectively (Gardiner, 1985; Gilmore, 1981). The time scales can also be approximated in the case of a vector or multidimensional ϕ, but particularly for T_2 the approximation is more difficult and depends more sensitively on details of the potential, and will not be discussed further.

A third and longest time scale is the *equilibration time* or the expected time for the system to evolve from the initial to final distribution. This time will depend roughly on how different the initial and final distributions are and how strong the noise is relative to the potential, but in principle the equilibration time is the limit of infinite time for Equation 6. For practical purposes, it should be much larger than T_2, as will be discussed in the next subsection.

Assumptions behind the Collection of Stationary Data Measures

As mentioned above, it is possible to estimate a stationary probability distribution *P* from experimental time series data. The assumptions behind the algorithm (see below) are:

1. The observation time is long enough.
2. Enough phase transitions occur in the data set.
3. The collective variable is bounded.
4. A good initial distribution can be composed of an equal number of trials per mode.

If these four assumptions are satisfied, then it is reasonable to interpret the estimate of *P* as stationary and compare it to theory as discussed below. Further, *P* may provide a better estimate of moments (mean and standard deviation) than direct calculation from time series if phase transitions occur on some trials, even if *P* is unlikely to be stationary because the observation time is too short (Assumption 1 fails).

Observation time will be defined by the trial duration times the number of trials (from one or more participants) composing the data set from which a particular *P* is calculated. This simple definition implies an equivalent trade-off between duration and number of trials that may not be accurate, but is a good starting assumption pending detailed analyses that suggest preferences for longer or more trials. However, several considerations suggest that it is advantageous to combine at least a few trials, rather than estimate *P* from a single, very long trial. Long trials (e.g., more than a minute or two) may require modifications of computer software to introduce or alter the use of memory buffers to augment computer RAM (Random Access Memory; basically the computer's working memory). A more stringent restriction is imposed by endurance and patience, particularly for human participants. Boredom and fatigue may limit trials to very few minutes for intense coordination tasks. Likewise, typical experimental sessions are limited to about an hour. Because more conditions across trials means less time per condition, a session limit also reduces trial duration.

Given these trial duration limits, the data set for one condition of a single participant may not yield a sufficient observation time from which to estimate a stationary *P*. An option may be multiple sessions, typically composed of a different order of the same trials. Another option is to ignore individual differences and combine data from a number of participants, effectively multiplying the observation time by the number of participants. It may be advantageous to normalize individual data using a characterization of individual differences if available, but such a characterization implies a detailed understanding of the coordination processes which is more likely to remain the goal rather than the starting point for coordination research in the near future. Lacking a means of normalization, one possible technique is to estimate model parameters (see below) for a data set composed from many participants, then examine how well partial data sets compare to the full set. For example, Collins and Turvey (1997) determined model parameters using 10 (or 12) participants, then plotted the goodness-of-fit parameter r^2 as a function of number of participants included in a partial data set. The plots suggested an asymptote near the number of subjects examined, a result which was supported through the use of simulated data.

The second assumption involved in estimating a stationary *P* is that "enough" phase transitions occur in the data set. For multimodal data, a phase transition can be defined by a departure of the collective variable from the basin of attraction of the initial attractor (i.e., for the two modes of inphase and antiphase, the collective variable of relative phase can be parsed into two halves, one characterized as the inphase basin, the other as the antiphase basin). For periodic unimodal data, a phase transition can be identified when the collective variable passes into an adjacent period and is remapped into the central period. For examples such as posture, it may be useful to interpret a stumble or a fall as a phase transition leaving the attractive (and stable) region. The expected time T_2 for a phase transition (Equation 9) applies to each of these cases.

Phase transitions as well as local fluctuations (with time scale given as the relaxation time T_1, Equation 8) are important as the mechanism of achieving a stationary P which differs from the initial distribution. Local fluctuations serve to spread the distribution about the mean state on a time scale T_1 perhaps on the order of a few seconds, whereas phase transitions shift probability between mean state regions, with a time scale T_2 that may be on the order of typical trial times, or in considerable excess. Following a discussion of boundedness, a suggestion for satisfying these assumptions will be offered.

The third assumption for estimating a stationary P is that the collective variable is appropriately *bounded*, given the observation time. For relative phase, it is appropriate to apply periodic boundary conditions such that every cycle is equivalent, and restrict the measured P to a convenient period. For a measure of postural coordination, such as the center-of-posture (*COP*), a stationary distribution may not be exhibited for trials with durations of several minutes or less. However, it is likely that given a long enough observation time, the *COP* will fill the region of stability beyond which one stumbles or falls over. If posture is only observed for a minute or two, the *COP* may, for example, remain primarily in the forward left region of the bounded space. Individual trials longer than a few minutes may be confounded by fatigue effects, and combining data from separate trials and separate participants requires some normalization to a center or mean of the *COP* which may not be known *a priori*. Given these constraints, the nonstationary methods in Part 2 should be pursued.

The fourth assumption is more of a suggestion for satisfying the previous assumptions with a relatively short observation time. Two extreme *initial distributions*, the delta function distribution and the uniform distribution, which are frequently assumed in solving partial differential equations such as Equation 4 are not very amenable to experimentation, at least with human participants. Instead, a compromise distribution is suggested below which consists of combining the data of an equal number of trials initialized per mode.

The *delta function distribution* is a fully deterministic initial condition, for which it is assumed that the initial probability is entirely focused at one exact value of the collective variable. For example, it might be assumed that initially $\phi = 0$, exactly.

Note: The delta function strains the definition of a function (and in fact led to the mathematics of generalized functions) because it requires that a finite probability area be contained at a single value. This contradictory "area" consists of an infinite height and a zero width.

The exactness constraint can be relaxed by assuming a small random (gaussian distributed) error on the initial condition. Most individual experimental trials essentially assume such a relaxed initial distribution in that a trial is begun when the participant is performing the task close enough to the expected starting value. While this initial condition is easy to implement experimentally, it does not lend itself well to an estimation of a stationary distribution because the system would have to be observed for perhaps an extraordinarily long time in order for it to visit every possible state with the expected long-term probability. Recalling the ball moving on the V of Figure 1 (left panel), a ball started near the bottom of either well would have to be watched long enough to move around for a while in that well, then switch in one direction to the other well, move around in that well for a while, eventually move back, and overall move between both wells in both directions several times. It may be useful to change metaphors to a pillar of thick fluid (representing probability mass) started at a particular value on V, say the bottom of a well. Once the pillar is released, the fluid will collapse into the well. Changing the role of the noise from randomly moving the ball to randomly shaking the potential back and forth (which implies a nontraditional but feasible addition of a noise term to the potential rather than the equation of motion), some of the fluid will eventually move into the other well. After some time (which would depend on the viscosity or thickness of the fluid, the steepness of the wells and the rate of shaking), the fluid will fill both wells to some equitable value. Then this fluid can be frozen and flipped over to suggest the stationary probability distribution.

The metaphorical connection between a fluid and probability mass is also helpful for building intuitions about the alternate extreme of a uniform distribution. A *uniform distribution* is defined by equal probability of any value of the collective variable, which is reflected by a flat fluid of constant (uniform) height. When released, the flat fluid (initial distribution) will conform to the shape of the wells (potential). The noise, again interpreted as shaking the potential, will redistribute the fluid (probability mass) based on the relative depths of the wells and the heights of the interposing peaks. Eventually the long-term stationary distribution will be achieved. The previous metaphor must be adjusted such that multiple balls are started with equal space between each ball. Then the balls must all be watched to observe how they eventually distribute their numbers and time spent over regions of the potential.

The balls provide the better metaphor for individual trials, and serve to suggest an initial distribution in between the delta function and uniform distribution. Given standard constraints on trial duration and number of trials, it is suggested that for each experimental condition (e.g., each movement speed) an equal number of trials be prepared for each expected mode. For example, if the coordination exhibits two modes, in-phase and antiphase, then one relatively long trial (or two shorter trials) should be started in each mode. If only one mode is expected, for example, if posture is stable about a single fixed mean, then only one trial per condition is needed. This *compromise distribution* would be mathematically achieved (with appropriate qualifications) by combining as many delta function distributions (each with an added gaussian random component about the expected mean) as there are modes.

One advantage of the compromise distribution (equal trials per mode) is that it counterbalances potential biases of initialization within a particular mode. We have unpublished observations that suggest a tendency to return to the initial phase following a transition despite typical instructions to permit transitions to occur. Unless it becomes clear how to incorporate this bias into the model, it is hoped that it can be averaged out by combining an equal number of trials per initial condition. The other aspect of initial condition bias follows from the dependency of T_2 on the minimum from which a phase transition occurs: on average, it will take longer to observe phase transitions from the more (or most) stable mode. Accordingly, the observation time would have to be extended if trials were only initialized in the more (most) stable mode. Without attempting a detailed analysis, the required observation time for the compromise distribution should depend roughly on the average T_2, rather than the longest T_2. The compromise distribution also avoids initially knowing or assuming which is the more (most) stable mode.

If satisfied, the preceding assumptions allow for the following algorithm. Checking the assumptions for calculating P should reveal with what degree of accuracy an experimental data set is likely to be replicable. That is, if P is obviously choppy rather than smooth (e.g., there are several peaks of similar height near the presumed mean), then the chosen bin number is too ambitious for the data set and a more moderate bin number (and correspondingly lower degree of expected accuracy) should be examined. Conversely, if P is particulary smooth, a greater and less conservative bin number may be appropriate. Understanding how well P is estimated will be important to interpreting the calculation of moments in the next section, and even more so in the last section of Part I, which discusses how to estimate parameters from P.

Procedure for Estimating a Stationary Probability Distribution

1. Collect discretely sampled time series of the collective variable
2. Combine data for an equal number of trials initialized in each mode
3. Combine data from multiple trials
4. Combine data from multiple participants
5. Divide the range of ϕ into n bins of width $2\pi/n$
6. Count the occurrences of ϕ in each bin
7. Divide each count by the total number of data points

A good choice of sampling rate should depend on the time scale of interesting experimental features. Behavioral tremor is unlikely to exceed 20 Hz, while changes in movement direction or speed often occur on time scales faster than 1 Hz. Thus, a sampling rate between 5 and 15 Hz is likely to be sufficient for coordination studies. However, higher sampling rates may be appropriate for initial variables, particularly if derivatives or coordinate changes will be used in deriving a collective variable. This is the case, for example, in determining a continuous relative phase because it is calculated from either a Cartesian or angular position variable as the phase angle (a polar coordinate) in the phase space of position and velocity (first derivative). It may be desirable to downsample from a high sampling rate (e.g., maintain every ninth data point to downsample from 90 to 10 Hz) once the collective variable is derived.

While it is suggested that data be combined from an equal number of trials initialized in each mode, the combination of data from multiple trials and/or participants is more an option than a suggestion which depends on achieving sufficient phase transitions. The combination of trials is really an averaging process, because the distribution implicit for each trial will be combined, then normalized. Thus, standard concerns with averaging data should be kept in mind, particularly when parameters are estimated below. If enough data can be gathered per condition for each participant, then it is probably worth estimating P for each condition and each participant. Then it becomes possible to run an ANOVA or MANOVA using the parameters as separate data points or separate dependent measures. If the data set is thought to be insufficient to estimate P for each participant, one must be satisfied with P estimated from the total data set. As discussed below, the nonlinear regression analysis used to estimate parameters should provide parameter uncertainties which can be used as a t-test to determine if each parameter is statistically significant.

The statistical significance in either of these cases must be considered in the context of the number of bins chosen in Step 5 of the algorithm, as well as the number of data points determined by the sampling rate and the number of trials. Pending a better statistical understanding of how these choices effect the estimate of P, a working criterion is that the resulting P look smooth enough. Very many data points are likely to support a refined P with a large number of *bins*, each representing a small width of the collective variable. Fewer data points are likely to result in a less smooth P at the same number of bins, so a smaller number of bins (each of larger width) should be used to achieve a good degree of smoothness. For example, Experiment 1 of Collins and Turvey (1997) used 120 $\times$ 120 $\times$ 2 bins for P estimated from 622,080 data points (= 36 sec $\times$ 90 Hz $\times$ 2 trials $\times$ 8 conditions $\times$ 12 participants) for three ϕ variables, but Experiment 3 reduced to 72 $\times$ 72 bins for P estimated from 64,800 data points (= 9 sec $\times$ 90 hz $\times$ 4 trials $\times$ 4 conditions $\times$ 5 participants) for two ϕ variables.

Note: The 90 Hz was most likely excessive, and should have been downsampled by 9 or 10 Hz. This is particularly true because, as mentioned before, ϕ involves a smoothed derivative and a coordinate transform, so it is unlikely to be accurate at the original sampling rate.

Once the data set is composed and the bin number chosen, it helps to have a computer program to count the data points occurring in each bin. The resulting counts per bin are divided by the total data points to normalize the distribution so it can be interpreted as a probability distribution (whereas it is a frequency histogram prior to normalizing). With one further normalization step, this resulting P is directly comparable to the form of Equation 5, as discussed below.

Moments of Stationary Probability Distributions

Once P has been estimated, the more traditional measures of mean and standard deviation can be recovered based on moments of P. The *zero probability moment* p_m (where m is the mode) is an additional measure that serves to quantify the probability lost, due to phase transitions, from one mode in favor of other modes, and is needed to calculate other moments. Each moment must be determined locally, or for each basin of attraction, when more than one mode occurs in the data. For the example of inphase and antiphase coordination, ϕ can be divided into basins of attraction, which for inphase is the region $(-\pi/2 \leq \phi < \pi/2)$ and for antiphase is the region $(\pi/2 \leq \phi < 3\pi/2)$. For calculating moments from the data-estimate of P, these regions imply dividing the bins into halves. Thus, if the total bin number is 120, the first 60 bins (1–60) constitute the inphase basin, while the last 60 bins (61–120) constitute the antiphase basin. Letting B denote bin number, B_i denote the initial bin of the basin (e.g., $B_i = 61$ for antiphase) and B_f denote the final bin of the basin (e.g., $B_f = 120$ for antiphase), p_m is defined for each mode as the sum Σ of probabilities for each of these bins,

$$p_m = \sum_{B=B_i}^{B_f} P(\phi_B) \tag{8}$$

This measure can be interpreted in terms of the behavior of two coordination balls, one initialized inphase at $\phi = 0$ and the other initialized antiphase at $\phi = \pi$. Say the first ball remains in the inphase basin for a full 50 sec of observation. The second ball leaves the antiphase basin after 30 sec, remaining for the final 20 sec in the inphase basin. Combining the data for the two balls, the fraction of the time spent inphase is 70/100 s or $p_0 = 0.7$, while the fraction of the time spent antiphase is 30/100 sec or $p_\pi = 0.3$. Further, it can be said that the time spent inphase relative to antiphase is 7/3 ($= p_0/p_\pi$).

The *local mean* or average $<\phi_m>$ for each mode is defined as sum of the probability of each bin weighted by the value ϕ_B at the center of the bin, divided by p_m for the mode,

$$\langle \phi_m \rangle = \frac{1}{p_m} \sum_{B=B_i}^{B_f} \phi_B P(\phi_B) \tag{9}$$

Recalling that variance can be determined by subtracting the square of the mean from the second moment, the *local standard deviation* SD_m for each mode can be defined,

$$SD_m = \sqrt{\left[\frac{1}{p_m} \sum_{B=B_i}^{B_f} \phi_B^2 P(\phi_B) \right] - \langle \phi_m \rangle^2} \tag{10}$$

The advantage of calculating local moments of P whenever phase transitions occur rather than directly calculating means and standard deviations from time series of the collective variable(s) is suggested by the example for p_0 and p_π. A mean of about $3\pi/5$ rad [= (30 s*π rad + 20 s*0 rad) /50 s] would be calculated directly from observing the second ball, whereas the statistical median and mode would both be near π rad. None of these measures accurately characterizes the bimodality of the time series for the second ball, while for the first ball all three measures of central tendency would yield about 0 rad. Further, a standard deviation calculated directly from the second ball's time series would suggest much wilder fluctuations that are actually present apart from the large but rapid fluctuation that carries the ball to the inphase basin. Following the algorithm of the preceding section and calculating the moments using Equations 8–10 would provide much more representative measures of the combined behavior of the two balls.

However, in practice, a third approach is very often used, which is to throw away any time series exhibiting a phase transition, such as for ball two. Ball two would be repeatedly initialized in the antiphase basin, and time series would be recorded until one was found for which the ball remained in the antiphase basin throughout. It is hopefully clear that this is a procedure very biased toward confirming two very different mean states, which would suggest a very misleading view of the balls' behavior, particularly when the standard deviations are interpreted. Rather than proposing to throw away phase transition data, this section provides a method of making good use of this data by calculating the relative probability of modes, while preserving the intent of calculating mean and standard deviation.

Finally, note that the moments can be calculated from P whether it can be considered stationary or not. If some but not enough phase transitions occur, then p_m will not be very accurate, but $<\phi_m>$ and SD_m will still be more accurately and representatively calculated from P rather than directly from the time series because these measures account for the phase transitions. P will provide no benefit if no phase transitions occur, but should still provide good estimates for $<\phi_m>$ and SD_m. When no phase transitions occur, p_m will simply be the fraction of trials initialized in that mode to the total trials used to calculate P, which should be the same for all modes – this is just a restatement that an equal number of trials were initialized in each mode.

Time Evolution of Probability Distributions

Whereas the preceding paragraphs focus on a stationary P, understanding the time evolution to P is also an important and useful endeavor. At present, it will suffice to examine two estimates of time scale that complement the stationary P, but future studies may be able to examine the time – evolving P directly, and make use of the full form of the Fokker-Planck Equation (Equation 4).

The two time scale estimates discussed above were the relaxation time T_1 and the mean first passage time T_2. T_1 has been measured from perturbation experiments in which one of two fingers was pushed at peak velocity to disturb its coordination with the second finger. T_1 was measured as the time from perturbation to recovery of the prior $<\phi_m>$ (Scholz, Kelso, & Schöner, 1987). However, trials were discounted in determining the average T_1 if a phase transition occurred rather than a recovery from the perturbation. As discussed previously, discarding data is not to be encouraged and will bias the estimate of T_1 to be smaller, but neither is an alternative clear. It has also been suggested that T_1 can be measured from a peak of a Fourier frequency transformation of the time series (Gardiner, 1985; Schöner et al., 1986; Fuchs & Kelso, 1994), but we have not yet seen this technique implemented in the coordination literature. The idea is that the width measured across the peak at the height which is half the full peak height is a measure of the relaxation time as the response of the frequency spectra slightly displaced from a principle frequency peak.

T_2 has been measured from simulated time series by taking the average of the times from initialization to a departure from the basin of attraction (Collins, Park, & Turvey, 1998). However, some creative technique is required for incorporating the information provided by trials in which no phase transition occurred unless the experimental design allows trials to run until a transition occurs. One possibility may be to use the fraction of trials not exhibiting a transition to the total trials as an estimate of the total probability of the tail of the T_2 distribution (the integral of T_2 for all times greater than the experimental limit). Another difficulty with estimating T_2 is that the results for simulations of data analogous to 20 participants did not agree very well with predictions. Because real experimental data is unlikely to provide better results, and collecting appropriate data from significantly more than 20 participants would be a very large undertaking, application of T_2 may provide a rough guideline rather than an accurate complement to the stationary distribution.

The authors are currently investigating T_1, T_2 and possible uses of the time-evolving P, but currently it seems that experimental measures of time scale are not yet sufficiently reliable, and are better used as guidelines rather than careful measures.

Estimating Model Parameters from Data

Of interest in this subsection is how to connect models and data. The strategies of synergetics can be very powerful methods of identifying a clear collective variable and developing a model of its preferences and development in terms of a potential function V. When applied to the flexible behavior of biological systems, the presence of variability may at first seem a barrier to interpretation using a deterministic potential, but the variability plays a critical role in synergetic theory in the form of stochastic dynamics. By practically approximating the wide range of sources of variability as gaussian white noise, the techniques of probability and statistics have been applied to examine the probability distribution P determined by the potential and noise through Equation 4, the Fokker-Planck Equation. A prediction for a stationary P is offered in Equation 5, but specific predictions depend on the noise strength and the parameters of the potential. That is, until specific numbers for each parameter (e.g., $Q = 1.2$, $a = 3.7$, etc.) are entered into Equation 5, only generic features of the model can be studied; the model is still too general. A very specific model can be achieved by comparison to data, once the data is put into the right format, which for Equation 5 means estimating P from the data as developed in the previous subsection. The comparison is done through a nonlinear regression analysis that determines the parameters which serve to best fit the model to the data.

Whereas ANOVA techniques can be used to determine what experimental variables influence the data, regression techniques can determine how much and what type of influence the variables may have, in the context of a particular model. For example, an ANOVA-based experiment may determine if height categories (short, medium and tall) and weight categories (light, medium and heavy) influence grip strength. Given statistically significant effects, the order may be determined by comparing means. Regression could work on the categories, or on height and weight treated as continuous variables using their magnitudes. For a linear regression, height and weight can be tested as separate independent variables IV in a model for the dependent variable DV of grip strength,

$$DV = a + bIV \tag{11}$$

where the intercept a is the value of the DV when the IV is zero and the slope b is how quickly the DV increases with respect to the IV. Given a data set of DV values for a variety of IV values, the best fitting a and b can be determined, as well as a 95% confidence interval about these estimates. The confidence interval provides a t-test such that if zero is included, the parameter is not significant. Also, the goodness-of-fit r^2 and the significance level of the full model can be determined.

A multiple regression model generalizes the linear regression model to estimate the slopes and significance of multiple *IV*s and simple nonlinear terms (e.g., the square of the height or the product of height and weight) expressed as additional *IV*s. The multiple regression can be written for *n* linear *IV*s,

$$DV = a + \sum_{i=1}^{n} b_i IV_i \tag{12}$$

For nonlinear models that cannot be reduced to a linear form, multiple regression will be insufficient. For some models, there are other ways of getting at the parameters that best characterize the data. A very important example is determining mean and standard deviation because these two parameters fully specify a gaussian or normal distribution, which takes the familiar form of a bell-curve. Another way of getting these parameters is to follow the algorithm above to estimate *P*, then apply Equations 8–10 to get the mean μ and standard deviation σ, assuming a single mode. Alternately, a nonlinear regression analysis can be used to estimate μ and σ from the explicit form of the normal distribution,

$$P = \frac{1}{\sqrt{2\pi}\sigma} \exp\left[-(\phi-\mu)^2 / \left(2\sigma^2\right)\right] \tag{13}$$

Equation 13 is included not to suggest this is the best way to estimate μ and σ, but rather to provide a connection with the stationary probability distribution Equation 5. For the potential Equation 2, Equation 5 becomes,

$$P = N \exp\left[-2(-\delta/2\sin 2\phi - a\cos\phi - b\cos 2\phi)/Q\right] \tag{14}$$

However, the form of Equation 14 does not provide for estimates of δ, a, b and Q separately, and is better reduced using the relative parameters $\delta' = \delta/Q$, $a' = 2a/Q$ and $b' = 2b/Q$,

$$P = N \exp\left[-\delta' \sin 2\phi - a' \cos\phi - b' \cos 2\phi\right] \tag{14a}$$

For comparison to Equation 14a (or an analogous equation using a different *V* in Equation 7), each value of *P* estimated from data should be divided by the width of the bin, e.g., divide *P* by $2\pi/60$ for a 2π interval with 60 bins. Normalizing by the width ensures that the area under the curve (the sum of width times heights) will be approximately 1.0 as expected for probability; further, the best-fitting curve can subsequently be plotted over the *P* values because the height units will agree. Using the *P* values as the dependent variable for a nonlinear regression on the values of ϕ provides estimates of δ', a' and b'. Similar to a multiple regression, estimates of the 95% confidence interval about each parameter can be used as a *t*-test of significance at the 0.05 probability level. Further, a goodness-of-fit r^2 can be defined and estimated with the nonlinear regression, but the significance of the total model cannot be tested in general, whereas it can be for linear and multiple regression.

As noted, the stationary *P* is in principle time-independent and serves only to establish the strength of parameters of a potential relative to the noise strength. An additional measure, such as the mean relaxation time T_1 or the mean first passage time T_2, is required to disambiguate the potential and the noise. Recalling that we consider noise to be a place-holder for the wide range of processes yet to be understood (perceptual and motor variability, contribution of muscular and neural subsystems, high-level brain or cognitive influences, and external air flow and floor vibrations, just to name a few), disambiguating the noise from the potential can suggest just how much "slop" is still left for posterity.

There will be as many equations as there are relative parameters (e.g., $2a/Q$) estimated for the relative potential ($2V/Q$). The one additional equation needed to separate the relative parameters from Q is formed by entering the time scale estimate T_1 or T_2 into the appropriate Equation 6 or 7. For example, for the potential Equation 2 in Equation 5, three relative parameters will be estimated, $2\delta/Q$, $2a/Q$ and $2b/Q$. Equation 6 or 7 provides a fourth equation with which to solve for four unknown parameters, δ, a, b, and Q.

As a cautionary note, rather than consider the magnitudes of the parameter estimates as a final result, they are probably better used as preliminary data to compare similar models of increasing generality (to determine if all the terms of a more general model are needed, e.g., Collins & Turvey, 1997) or to examine the effects of an experimental manipulation (e.g., increasing movement speed, varying detuning, and their interaction, Collins, Park & Turvey, 1998). Considering the numeric values as preliminary returns the emphasis to the experimental manipulations, rather than details (e.g., number and duration of trials; number of participants) which are likely to change the numeric values but should not change the interpretation of the manipulations.

Once determined, the parameter estimates can be used to generate time series simulations, from which the stationary P and time scale measure T_1 or T_2 can be estimated to examine how well the assumptions of the methods are satisfied. Time series can be generated using a discrete approximation of Equation 4 (see Kloeden & Platen 1992 for a detailed discussion of simulating stochastic differential equations), recalling that in general $\dot{\phi} = -\partial V/\partial\phi + \sqrt{Q}\xi_t$

$$\phi_{i+1} = \phi_i + (\delta\cos 2\phi_i - a\sin\phi_i - 2b\sin 2\phi_i)\Delta t + \sqrt{Q}\Delta W_i \qquad \bmod 2\pi \quad (15)$$

where ΔW_i is a Wiener process with 0 mean and variance equal to the time step Δt, produced using a gaussian pseudo-random number generator (Press et al., 1992). The values for the time step and the number of time series can be set to those used in the experiment from which the parameter estimates were derived, or they can be varied to predict the effects of smaller or larger sampling rate (varying Δt) or number of trials and participants (varying the number of time series). Also, if the time scale estimates were not used to disambiguate V from Q, then several values of Q can be selected and used in the simulations to qualitatively compare the results to the experimental time series (e.g., Collins & Turvey, 1997).

Part 2: Nonstationary Analyses

Our focus to this point has been on methods that bear most directly on rhythmic intersegmental coordination of the kind that typifies locomotion. It has been the case, however, that a major inspiration for the development and/or application of stochastic and nonlinear analyses has been the nonrhythmical coordination among body segments required simply to stand still.

The *center-of-pressure* or *COP* is the vertical ground reaction vector equal and opposite to a weighted average of all downward forces acting between the feet and the ground. It is related to, but not identical with, the center-of-gravity vector. Two independent muscular subsystems seem to govern the *COP*: plantar flexion/dorsi flexion at the ankle and adduction/abduction at the hip (Winter, Price, Frank, Powell & Zabjek, 1996). These subsystems are responsible, respectively, for motions of the *COP* in the *anteroposterior* (*AP*) and *mediolateral* (*ML*) directions. Figure2 shows a typical *COP* "record in time" expressed as a trajectory or trail in *AP* and *ML* coordinates. It portrays apparently erratic behavior over both relatively shorter and relatively longer time scales. We will consider a method (*average mutual information*) for evaluating the proposed independence of fluctuations in the *AP* and *ML* directions and we will consider

Fig. 2. A typical center-of-pressure *COP* trajectory or trail for quiet stance, shown in anteroposterior *AP* and mediolateral *ML* coordinates.

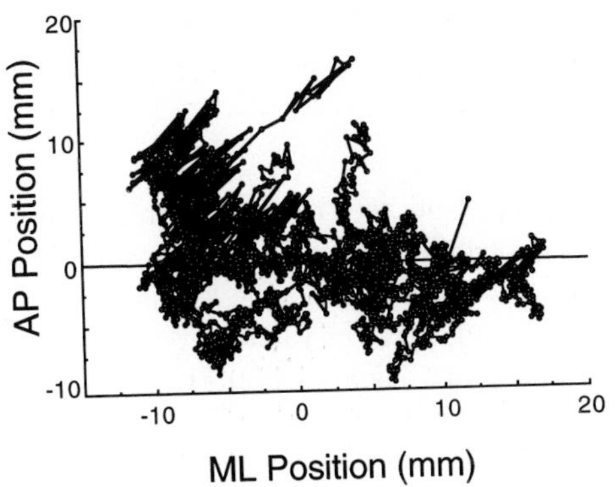

methods (*Hurst's rescaled range analysis* and *fractional Brownian motion*) that suggest memory processes in *COP* records potentially indicative of neural control mechanisms.

Average Mutual Information

Mutual information (cf. Chapter 19, 20) is a measure (in bits) that indicates how much information of one time series at time *t* can be learned from the measurement of another time series at time *t* (Gallager, 1968). More specifically, in respect to records in time of *COP* fluctuations in the *AP* and *ML* directions, the quantification takes the form of a series of *AP* measurements $ap_1, ap_2, \ldots, ap_m, \ldots, ap_M$, and a series of *ML* measurements $ml_1, ml_2, \ldots, ml_n, \ldots, ml_M$, and the question of "how much can be learned about the measurement ap_m from the measurement ml_n?" The answer is:

$$\log_2 \{[P_{AP\,ML}(ap_m, ml_n)]/[P_{AP}(ap_m)\,P_{ML}(ml_n)]\}.$$

If ap_m and ml_n are independent, then the mutual information measure should not differ significantly from 1, meaning that zero bits of information are shared. Mutual information is like other variables defined on the space of measurements (ap_m, ml_n); it is distributed over that space and its moments can be interpreted in the standard way (Abarbanel, 1996). The *average mutual information*, or mean value over all measures, is given by:

$$I_{AP\,ML} = \sum_{ap_m, ml_n} P_{AP\,ML}(ap_m, ml_n) \log_2 \left[\frac{P_{AP\,ML}(ap_m, ml_n)}{P_{AP}(ap_m) P_{ML}(ml_n)} \right] \tag{16}$$

Analysis of experimental data shows that the average mutual information of *AP* and *ML* fluctuations is effectively zero. One of the important properties of this measure of nonlinear correlation is its invariance under smooth changes of coordinate system (Fraser & Swinney, 1986). Among other things, this property means that the average mutual information is robust against noise (in the form of altered locations of points and changed details of local distributions of points) introduced by contaminated measurements (e.g., of ap_m and ml_n; Abarbanel, 1996).

Rescaled Range Analysis

Turning to the time series of *COP*, we can take advantage of a remarkable fact about records in time of very many natural phenomena. Although they are seemingly unstructured at short and long time scales (such as the water level of major rivers over many centuries), they exhibit power law behavior.

The *rescaled range analysis* characterizes a time series such as the *COP* by a *Hurst exponent H*, computable from,

$$[R(\Delta t)/S(\Delta t)] \propto \Delta t^H \tag{17}$$

In the preceding equation, referred to as Hurst's Empirical Law (e.g., Feder, 1988), R is the difference between the maximum and minimum *COP* values within a given time interval or lag of Δt. Obviously, the range R must be determined for very many time windows of ever increasing size. The values of Δt set the temporal resolution. For any given interval, S is the standard deviation of the *COP* values in that interval which rescales the range R and renders R/S dimensionless (thereby permitting comparisons of observed ranges of *COP* under different postural restraints). Equation 17, therefore, is an expression of how the rescaled range depends on the temporal resolution at which it is measured.

The exponent H may range in value from 0 to 1 with $H = 0.5$ signifying that the differences between consecutive values are uncorrelated. When H is other than 0.5 it signifies correlated or memory-like processes (Feder, 1988; Liebovitch, 1998). If $0.5 < H \leq 1$, then past and future increments are positively correlated. They are said to be *persistent*. For *COP* it would mean that is has a tendency to continue moving in its current direction. With respect to the *COP* trail in Figure 2, persistence would entail a tendency, but not an obligation, to avoid self-intersection (Mandelbrot, 1982). If $0 \leq H < 0.5$, then past and future increments are negatively correlated. They are said to be *antipersistent*. For *COP*, it would mean that it has a tendency to turn back constantly toward the point it came from.

Fractional Brownian Motion

The rescaled range analysis and the notions of persistence and antipersistence are modeled by fractional Brownian motion. Keeping the focus on *COP*, one can think of its behavior under the independent *AP* and *ML* subsystems as that of a particle in random walk (J. Collins & DeLuca, 1993). The two subsystems "kick" the *COP qua* particle, causing it to hop or jump, with the succession of jumps composing the *COP*'s random walk. The classic example of a random walk model is that of Brownian motion: the mean square displacement of the particle along a line (a one-dimensional random walk) is proportional to the time interval over which it is measured. Thus,

$$\langle \Delta x^2 \rangle = 2D\Delta t \tag{18}$$

where the brackets denote averaging, Δt is the given time interval, and the parameter D is the *diffusion coefficient*. D is a measure of the average level of stochastic activity in the random walk, and is clearly related to both the frequency and the amplitudes of the hops or jumps composing the walk. For a Brownian motion over Δt, greater values of D are obtained when the mean square displacement is greater, which can result from higher frequency and/or greater amplitudes of the jumps.

In pure Brownian motion there is no correlation between successive jumps. For random walks that exhibit scaling regions over which the jumps at successive time intervals are correlated, the appropriate theory is *fractional Brownian motion*, a generalization of Equation (18) introduced by Mandelbrot and van Ness (1968; see also Feder, 1988). The generalization involves the scaling law

$$\langle \Delta x^2 \rangle \approx \Delta t^{2H} \tag{19}$$

The method for preparing and analyzing *COP* data in the form of Equation (19) has been termed the *stabilogram-diffusion method* (see J. Collins & De Luca, 1993). For each experimental trial, the squared displacements between all pairs of data points separated by a given time interval Δt are calculated and then averaged over the number of intervals of size Δt in that trial. The foregoing procedure is repeated for multiple magnitudes of Δt (e.g., from 10 ms to 10 s in steps of 10 ms). In general, for a given Δt that spans m data intervals

$$\left\langle \Delta x^2 \right\rangle_{\Delta t} = \left[\sum_{i=1}^{N-m} (\Delta x_i)^2 \right] \Big/ (N-m) \qquad (20)$$

where x is the displacement and N is the total number of data points in the trial. To determine H, (Δx^2) is plotted in double logarithmic coordinates against Δt.

Application of the method to *COP* data suggests that the H value, the slope of the function in logarithmic coordinates, may differ for short and long time scales. Specifically, the implication is that *COP* increments partition into persistence over small time intervals ($0.5 < H \leq 1.0$) and antipersistence over long time intervals ($0 \leq H < 0.5$; e.g., J. Collins & De Luca, 1993; Riley, Mitra, Stoffregen, & Turvey, 1997; Riley, Wong, Mitra & Turvey, 1997). This partitioning suggests open and closed loop controls, respectively (J. Collins & De Luca, 1993). Some have also suggested that the time scale partitioning signifies exploratory processes (persistence is information gathering) and performatory processes (antipersistence is adjusting on the basis of obtained information), respectively (Riley, Mitra et al., 1997).

An alternative to dichotomizing the time scales has been examined by analyzing the data as an *Ornstein-Ulenbeck process*, for which the correlation function (see, e.g., Gardiner, 1985) smoothly varies from the apparent short-term, open-loop regime, to an apparent long-term, closed-loop regime (Newell, Slobounov, Slobounova, & Molenaar, 1997). This Ornstein-Ulenbeck process can be expressed in the context of Part 1 as an equation of motion derived by adding noise to the negative derivative of a quadratic potential,

$$V = -ar^2 \qquad (21)$$

in which a reflects the steepness of the potential well corresponding to the degree of perceptual/biomechanical control.

Another application of the preceding correlation procedures is in respect to the stride interval fluctuations of human gait (Hausdorff et al., 1996; Liebovitch & Todorov, 1996). Long-range correlations, indicative of fractal dynamics and $1/f$ scaling (cf. Chapter 24), were found for participants walking at freely selected paces of fast, normal or slow. The long-range correlations persisted through at least 1000 strides. They disappeared when participants were paced with a metronome set to the average of each type of gait (fast, normal or slow), suggesting that supraspinal influence may be able to negate a natural tendency toward long-range correlations (Hausdorff et al., 1996).

Recurrence Quantification Analysis

From the point of view of many algorithms, the data collected on biological movement systems is often far from ideal. Most notably, the data tend to be nonstationary. Within the time course of observation, the system descriptors (e.g., means, standard deviations) are inconstant due to nonlinear interactions (both feedforward and feedback), influences of externally originating noise, and dynamical state changes. *Recurrence quantification analysis* (RQA) is a direct response to these biological idiosyncracies. It is a nonlinear and multi-dimensional technique that does not require data stationarity and, further, imposes no rigid constraints on the statistical distribution or size of the data (Eckmann, Kamphorst, & Ruelle, 1987). Its contribution as a tool for studying movement is the variety of measures it provides of a given time series whose stationarity is suspect. It yields, for example, quantification of deterministic structure and nonstationarity (Webber, 1991; Webber & Zbilut, 1994, 1996).

RQA examines the local recurrence (or neighborliness, or time correlations) of data points in a reconstructed phase space (see Part 3). Temporally separated data points that are (spatial) neighbors in the reconstructed space reflect recurrence. Over time, as the dynamics unfold, data points return to the same region of phase space (that is, they recur).

The basic measurement strategy consists of taking a sphere of radius r centered on a point $x(i)$ in the reconstructed space and counting the number of points that are within the distance r from $x(i)$. For each $i = 1, \ldots, N$ (where N is the number of data points), one measures the distance between data points $x(j)$, where $j = 1, \ldots, N$, and $x(i)$. That is, for each data point in the embedded series, the distance between it and every other data point is calculated. Points are recurrent wherever the distance between them is less than or equal to r. The degree (and nature) of recurrence in a time series is represented graphically through the *recurrence plot* (Eckmann et al., 1987).

Figure 3 displays a recurrence plot for a time series of fluctuations in a person's center of pressure during quiet standing (Riley, Balasubramaniam & Turvey, 1999). Each darkened pixel on the recurrence plot represents a recurrent point. To obtain such a point, take a value of i along the abscissa, calculate for each value of j along the ordinate the distance between $x(i)$ and $x(j)$ and darken a pixel at (i, j) everytime this distance is less than or equal to r. Recurrence plots such as Figure 3 emerge as one continues this process for all i. For $i = j$, a point is being compared with itself yielding a distance of zero and, in consequence, the central diagonal line in the recurrence plot; and for a fixed r, the two triangular halves of the recurrence plot will be reflections of one another (Webber, 1991; Webber & Zbilut, 1994). Several input parameters are required for producing a recurrence plot such as Figure 3 and conducting the subsequent RQA. One has to choose, for example, a time lag, an embedding dimension (cf. Chapter 24), and r. The art of implementing and interpreting recurrence plots and RQAs lies in the selection of these and other input parameters.

The qualitative features of recurrence plots, large-scale typologies and small-scale textures, are important signs to a behavior's defining characteristics (Eckmann et

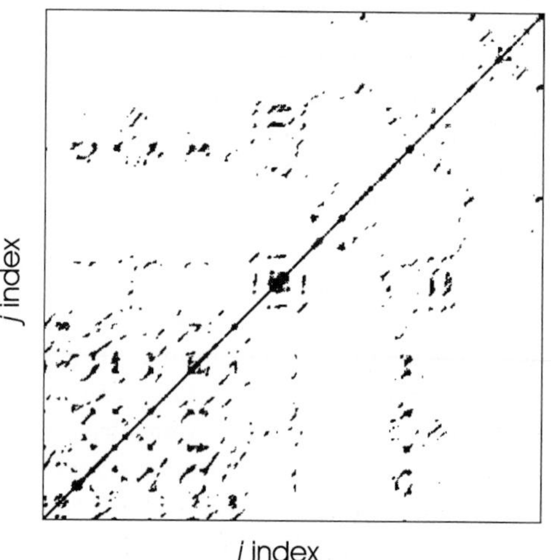

Fig. 3. A recurrence plot for a *COP* trajectory during quiet stance. The indices i and j refer to the position of a value in a time series. Points below the diagonal refer to values which recur within a specified radius at a time index j later than time index i in the time series, points above the diagonal refer to the same set of recurrent values with indices reversed, and points on the diagonal trivially indicate that the indices $i = j$ refer to the same value.

al.,1987; Webber & Zbilut,1994). To begin with, *homogeneity* is a large-scale typology – a homogeneous distribution of recurrent points throughout the plot yielding an essentially uniform pattern (which is clearly not a feature of Figure 3). It signals white noise or the absence of dynamical structure. It might also signal deterministic chaos. *Drift* is a large-scale typology that refers to the tendency of a recurrence plot to become less dense, or more pale, with increasing distance from the main diagonal. A uniformly progressive drift may reflect nonstationarity in the form of a gradual trend in the data. Abrupt drift – a rapid change in density with increasing distance from the diagonal – may reflect a sudden change in level in the time series. Drift is evident in Figure 3. A third large-scale typology is *long diagonal lines parallel to the main diagonal* indicating the presence of *rhythmic structure*. Figure 3 does not suggest a strong periodic component in the depicted series.

One type of small-scale texture is in the form of *single, isolated recurrent points*. These indicate stochastic behavior (several such points may be seen in Figure 3). Another type is in the form of *short line segments*. If they are diagonal and parallel to the main diagonal they indicate determinism – strings of vector patterns in the time series repeating themselves multiple times during the period of observation. In terms of attractor dynamics (cf. Chapter 24), the system is revisiting the same region of the attractor at different times. A random-number generator would not be expected to show these diagonal lines, but a deterministic system (a sine wave, a chaotic attractor, etc.) would. Several of these upward diagonal line segments may be seen in Figure 3. If the segments are diagonal but perpendicular to the main diagonal, this means that vector sequences at different locations in the series are mirror images of one another. This is expected for simple oscillations such as sine waves; if the wave is bisected at a peak or trough, the two resulting halves are mirror images. Such texture is present, although not widespread, in Figure 3. If the line segments are horizontal or vertical, this means that isolated vector sequences match closely with a repeated string of vectors further along the dynamic (separated in time). This texture is also present in Figure 3.

A third small-scale texture is the *checkerboard texture*, which is the grouping of line segments into small regions of the plot. This reflects the system visiting different areas of the attractor, and switching back and forth between them. The checkerboard texture is not present in Figure 3.

Two additional qualitative features may be identified (Webber & Zbilut, 1994). First, *bands of white space* (no recurrent points) indicate transient activity or a sudden change in level, and may reflect a change of underlying state. Large bands of white space are seen in Figure 3 – for example, from about one-half to about three-quarters of the length (from bottom to top) of the j-axis, the plot shows few recurrent points in the upper triangular half. Second, there may be a *sudden change in the density of recurrent points* as one moves along the i-axis of the plot; this is indicative of a change of dynamical regimes, and may be observed after a brief period of transient behavior (e.g., it may be seen after a band of white space). This feature is not seen in Figure 3.

While visual inspection of recurrence plots may be revealing, a significant advance in this analysis was the quantification of recurrence plots (RQA) provided by Webber and Zbilut (1994). RQA allows, for example, comparison across experimental conditions using standard statistical techniques such as analysis of variance (ANOVA). Five measures may be obtained through *RQA*: *percent recurrence* (%RECUR), *percent determinism* (%DET), the *ratio* of these two quantities, *entropy*, and *trend*. All of these variables are computed based on recurrent points in one triangular half only of the recurrence plot. The measures are briefly explained below; see Webber and Zbilut (1994) for more details.

- *%RECUR* is the number of recurrent points in the plot expressed as a percentage of the number of possibly recurrent points (i.e., of the total number of i-j distance comparisons, with the exception of when $i = j$ – the main diagonal). It is thus a measure of the extent to which the recurrence plot is covered by recurrent points, or, equivalently, the percentage of points within a distance r of one another. The level of %RECUR obtained for a given time series will depend to an extent upon the specified value of r.

- *%DET* is the percentage of recurrent points which fall on the upward diagonal line segments. The observed level of %DET will depend upon the specified definition of the number of points forming a line segment; this is usually set to two adjacent recurrent points with no intervening white space (larger, more conservative values may be chosen). This measure reflects the degree of determinism observed because, as stated above, upward diagonal line segments indicate that the system is re-visiting the same region of the attractor (or, the same region of the reconstructed space) repeatedly; that is, the dynamics are reliable or repeat themselves, which is, essentially, what "determinism" means.
- *Entropy* is computed as the Shannon entropy (cf. Chapter 19, 20) of a histogram of line segment lengths (a simple frequency histogram in which the number of observed upward diagonal lines of different lengths are counted). It is computed only with respect to the upward diagonal lines that indicate determinism. It is therefore a measure of the complexity of the deterministic structure of the time series (see Webber & Zbilut, 1994).

Two quantities, ratio (%DET/%RECUR) and trend, address the nonstationarity in recurrence plots.

- *Ratio* may be useful in the detection of changes in physiological state (Webber & Zbilut, 1994). During transitions between states, %RECUR usually decreases while %DET usually changes very little. This quantity is more useful to this end if RQA is performed over a moving window (e.g., over repeated epochs of overlapping or non-overlapping data windows). In fact, all of the measures discussed here may be obtained in this fashion if changes in underlying state are of interest.
- *Trend* is a quantification of drift, the decline in density (the increasing paleness) of recurrence plots away from the main diagonal, and is computed as the slope of the line of best fit drawn for %RECUR as a function of distance from the main diagonal. Non-zero trend indicates drift in the system, while zero (or very close to zero) values indicate stationarity. Trend is expressed in units of %RECUR per1000 data points, and, since trend is a quantification of the increasing paleness of recurrence plots away from the diagonal, trend values will usually be negatively signed (i.e., if %RECUR decreases with increasing distance from the diagonal, the regression line will have a negative slope).

In this overview of RQA, we have limited ourselves to the qualitative and quantitative aspects in order that the reader might readily see what the method offers in respect to seemingly recalcitrant data sets. The technical issues of parameter selections, and the use of shuffled data sets to facilitate inferences from the qualitative and quantitative measures, are not dealt with here. The reader is referred to Webber and Zbilut (1996), and Riley et al. (1999) for details about these matters.

Part 3: Phase Space Reconstruction

With respect to the equations for interlimb coordination defined above, the pattern ϕ is a collective variable that arises from underlying subsystems and, in turn, directs or enslaves them. This special circularity of complex systems (see Haken, 1983, 1996; Kugler & Turvey, 1987, 1988) raises the key question: How many independent subsystems are directed by a collective variable and its dynamics? Or, equivalently, what is the minimal number of active degrees of freedom (*ADFs*) needed to produce a collective variable and its dynamics? The method described in this subsection, referred to as *phase-space reconstruction* (cf. also Chapter 24), provides an estimate of that number.

The keys to the method are (a) the understanding that measurement of a single scalar suffices to reveal a nonlinear system's dynamics, because all variables are generically connected in a nonlinear process, and (b) the embedding theorem (Takens, 1981; Casdagli, Sauer, & Yorke, 1991; Eckmann & Ruelle, 1985). This latter theorem asserts that a measured scalar time series such as $x(t)$ can be embedded in a reconstructed phase space of vectors $y(t)$ that is related to the original phase space by smooth, differentiable transformations, with the result that several fundamental invariants of the original dynamics are preserved in the reconstructed phase space (see Abarbanel, 1996, p.18). Accordingly, the systematic study of the geometry of the dynamics in the reconstructed phase space provides valuable information about the original, unknown dynamics of the system.

There are three possibilities for the original dynamics. First, the system may be very high-dimensional, such that no low-dimensional model of it will absorb any significant aspect of its behavior. Second, there may be a few variables that are responsible for most of the system's behavioral variations, with some additional, and relatively low-amplitude, high-dimensional aspects (e.g., measurement noise). Third, the system may be purely low-dimensional, meaning that a set of first-order, autonomous, ordinary differential equations written over a few variables can capture its behavior, which, of course, may still look quite irregular and seemingly high-dimensional.

Average Mutual Information Revisited

The vectors $y(t)$ referred to above are constructed with the coordinates $[x(t), x(t + T\tau), x(t + 2T\tau)\ldots]$, where T is some integer multiple of the sampling period, and τ is an appropriate time delay. The parameter τ is chosen by inspecting the nonlinear correlation function defined above, namely, the *average mutual information*, adjusted to measures taken at two different times from a single time series (Abarbanel, 1996; Abarbanel, Brown, Sidorowich, & Tsimring, 1993). In the present context, the average mutual information function provides an estimate, over a set of measurements, of the amount of information (in bits) that can be learned about $x(t + T\tau)$ from knowing the value of $x(t + [T-1]\tau)$.Because the coordinates of the embedding space are composed of time lagged values of the measurement series $x(t)$, the smaller the average mutual information at a particular time lag, the more effective (in eliciting new information about the original dynamics) is the addition of a measurement at that time lag as a coordinate of $y(t)$. In practice, the value of the time delay at which the first minimum of the average mutual information function occurs is chosen as τ(see Fraser & Swinney, 1986).

False Nearest Neighbors

It is fundamentally important for the method to recognize that for the scalar measurement series $x(t)$, obtained over multiple cycles, points on the line close to a given point may be false neighbors. They are in the proximity of the given point not because they are dynamically related to the given point, but because of the severity of the projection (that is, representing the dynamics of the system by one coordinate). The process of *unfolding* the attractor, or the geometric entity in the appropriately dimensioned (and, in this case, reconstructed) phase space on which the dynamics evolve, is, accordingly, that of adding time-delayed values of $x(t)$ as coordinates of $y(t)$ until false-neighborliness due to projection is completely eliminated.

The quantitative aspect of unfolding the attractor in the above manner involves the determination of global *false nearest neighbors* (*FNN*) in any given embedding space. The essence of the procedure lies in taking sets of points nearest to every data point from a series $y(t)$ embedded in d-space, calculating the Euclidean distances in that space, and finding out whether these distances change substantially when the same calculations are made on $y(t)$ embedded in $d+1$-space. If the changes in distance exceed an adopted threshold, then embedding in $d+1$-space is taken to have removed false neighbors that remained in d-space embedding. For time series in which all regions of the attractor are well represented, percent *FNN* calculations have been shown to be robust against changes in threshold settings, and the issue of optimum values for these settings has itself received extensive study in the context of a variety of physical time series (Abarbanel, 1996).

When enough coordinates have been added to $y(t)$, such that all *FNN* have been eliminated, adding further coordinates does not reveal any more about the dynamics of the system. The number of dimensions d_E at which this occurs is called the *embedding dimension*. There are two points worth noting about d_E. First, for a given dynamical system, the d_E that is calculated can change as a function of which projection of its full dynamics is captured by the measurement series. Thus, two different experimental variables taken from the same system may yield different values of d_E. However, as discussed in the next section, important invariants of the original system can be extracted by studying *how* the system evolves on the attractor as unfolded in d_E-space. Second, if the measurement series is a projection of a system that contains very high-dimensional dynamics (or is contaminated by measurement noise), unfolding the attractor in the manner above may never succeed in removing all *FNN*s. Noise is formally infinite-dimensional. No matter how many dimensions are added to $y(t)$, noisy $x(t)$ will always wish to be unfolded in a larger dimension. For finite-precision numerical calculations on finite amounts of data, however, d_E analysis will fail to remove all *FNN*s only when the level of noise is significant as compared to the size of the attractor. Thus, the results of d_E analysis can indicate whether a behavior contains low-dimensional, deterministic dynamics or high-dimensional dynamics. The latter might recommend stochastic methods for modelling, provided, of course, that the source of the noise is not of system-extrinsic, experimental origin.

Active Degrees of Freedom

As just noted, the value of the embedding dimension d_E that is obtained during phase space reconstruction can change depending upon the choice of measurement variable. In general, if the dimension of the true phase space of the system's attractor is d_A, then unfolding a measurement projection in an embedding space of dimension $d_E > 2d_A$ suffices to undo all intersections due to projection (Abarbanel, 1996, p. 19). (Whereas d_A may be either a fractal or integer dimension, d_E is always an integer dimension.) However, since the dynamical system underlying the measurement remains the same no matter what one chooses to measure of its behavior, the number of *active degrees of freedom ADF* (i.e., the number of first-order, autonomous, ordinary differential equations) of the system is invariant over changes in the measurement projection. The geometric interpretation of this invariant is the number of dimensions required to describe the *local* evolution of the orbit of the system around the attractor. The critical intuition is that this local dimension can be, and often is, less than the dimension of the space in which the attractor itself lives. Consider as an example an oscillator that has two irrationally related frequencies in its motion. The orbit of this system lies on a two-torus, a geometric object closely resembling a doughnut (cf. Chapter 24). While the number of dynamical degrees of freedom of the system is clearly two, the attractor itself would require $d_E = 3$ to unfold fully.

Abarbanel's (1996) method for determining the number of *ADF*s, that is, the local dimension d_L analysis, begins in a working (reconstructed) space of dimension greater than d_E to ensure that the attractor is fully unfolded (i.e., all neighbors are true neighbors). Then, for any point in this space, the procedure tries to find a sub-space of dimension $d_L \geq d_E$ in which accurate local neighborhood-to-neighborhood maps of the data can be constructed. Neighborhoods of several sizes are specified by taking sets of N neighbors of a given point $y(t)$, and then a local rule is abstracted for how these points evolve in one time-step into the same N points in the neighborhood of $y(t+1)$. The success rate of the rule is measured as percent bad predictions (henceforth, %bad). The target of the analysis is to determine a value of d_L at which %bad becomes independent of d_L and of the number of neighbors N. This value d_L quantifies *ADF*.

Besides indicating the number of variables that should be used in attempts to model the dynamics of a system, d_L analysis also provides indications of local, low-amplitude (and/or fast time-scale) noise that may not be detectable in the attractor-level analysis of d_E. We noted above that d_E analysis fails to remove all *FNN*s only when the level of noise at the scale of the attractor is high. In d_L analysis, however, the value of %bad at which it becomes independent of d_L and N can serve as an indicator of noise at finer space-time scales.

The phase-space reconstruction method has been used to address the assumption (see above) that the attractor for an individual oscillating limb is a limit cycle (Mitra, Riley, & Turvey, 1997). Figure 4 (upper panels) shows the orbits in the phase plane of velocity and displacement to be thick bands. The variability evident in Figure 4 raises the contrasting possibilities that (a) the limit cycles are noisy or (b) motions from higher dimensional phase spaces are being projected (inappropriately) onto the phase plane. The phase-space reconstruction method confirms possibility (b). It yielded 3 *ADF*s for the equivalent of a large oscillating limb and 4 *ADF*s for the equivalent of a small oscillating limb, respectively, as can be seen in Figure 4 (lower panels).

Phase-space reconstruction yields the number of the *ADF*s but not the *identity of the ADFs*. Identification must be achieved by other scientific means. Experiments directed at the neural mechanisms responsible for controlling the postures and oscillations of limbs suggest a minimum of three neural control variables for single-joint rhythmic movement – the r, c, and μ commands (Feldman, 1980; Feldman & Levin, 1995). The r and c commands determine, respectively, joint position and joint compliance with the μ command adjusting the r and c commands (precisely, the muscle thresholds that they specify) to movement velocity. This close match between the proposed minimal number of neural control variables and *ADF* revealed by phase-space reconstruction is of potential significance. It is suggestive of future points of convergence between conventional neuroscience and the tools of nonlinear dynamics (Mitra et al., 1997, 1998).

Postscript

We hope that at this stage a diligent reader will not be left with a question of which technique is right for his or her coordination time series data. Instead, we would have the reader asking which should be applied first, and perhaps answering with a random number generator. The analyses address disparate levels of temporal and dynamical structure and may all be required before an appreciable understanding of any one coordination system can be begun. Further, many more techniques will be needed before coordination is "explained".

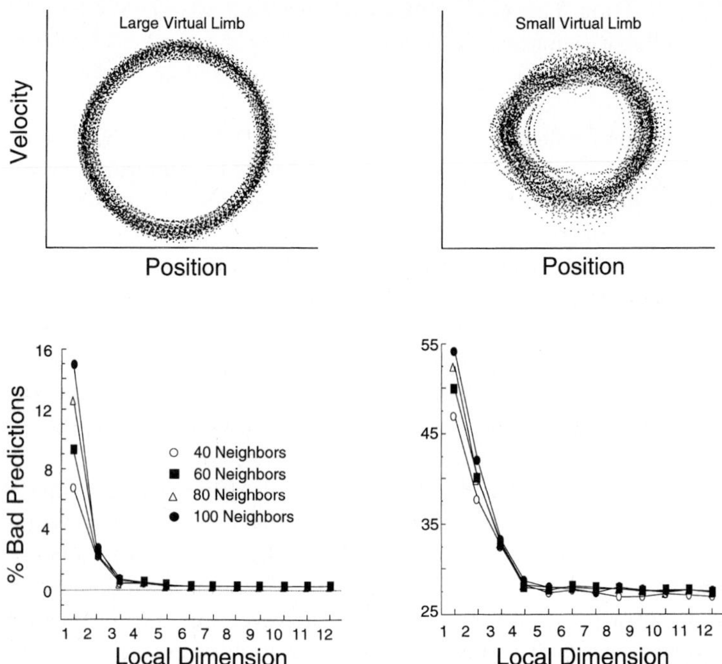

Fig. 4. *Upper panels*: Trajectories or trails in the phase space of position and velocity for the oscillatory coordination of a person holding a long or short pendulum in the hand (effecting a large or small virtual limb in the left and right panels, respectively). *Bottom panels*: The nearly zero %Bad predictions for a local dimension of 3 or greater in the left panel indicates that 3 (deterministic) active degrees of freedom *ADF*s suffice to capture the dynamics for the above trajectory of the large virtual limb. In contrast to the small virtual limb trajectory in the right panel, the %Bad predictions asymptote around 27% for a local dimension of 4, indicating that 4 *ADF*s plus an additive noise are required to capture the dynamics.

The first part addressed the relatively long time scale at which a coordination system may be considered stationary, which is more likely to be several or many one minute trials rather than just one. At this level, coordination can be viewed across several possible modes of performance, such as inphase and antiphase coordination of two oscillators. This first part examined the quantification of a potential function model using connections between stochastic theory and data put into the amenable format of a probability distribution. Parts 2 and 3 do not have this explicit connection between a model and data, which is advantageous in that no model is required, but is also more limiting because no model is suggested either.

In addition to not requiring a specific model, the methods discussed in Part 2 do not require the data to be stationary. Rather, a quantification of the nonstationarity is sought in terms of correlation and recurrence over time steps. Depending on the researchers' perspective, these nonstationary techniques may be considered preliminary to developing a long-term stationary model across a number of trials, or the techniques may be considered the refinement of a stationary model, or may be the goal in itself, particularly if stationarity is not of interest or is not accessible. The latter may be the case, for example, in posture, for which a stationary analysis might unreasonably require the consideration of several hours of continuous quite stance.

The technique of phase space reconstruction in the third part requires a single time series to be considered stationary, yet investigates the deeper dynamical structure behind the time series. In particular, the culmination is the estimation of the number of active degrees of freedom; how many first-order, autonomous ordinary differential equations are needed to generate the dynamics, in principle in a manner which is invariant over the particular time series measured. This technique may make the estimation of a simple, single dimensional potential function seem trivial, except that the analysis only offers how many equations are needed, not which ones. Still, if used conservatively, this quantification may provide a very useful constraint on theorizing.

Acknowledgement: The writing of this chapter was supported by NSF Grants SBR 97-28970 and SBR 97-09678. We would like to thank Michael Riley for help with Figures 2–4, and his conceptual contribution to Part 2. Additionally, we acknowledge the conceptual contributions of Ramesh Balasubramaniam to Part 2 and Suvobrata Mitra to Part 3. MT Turvey is also at Haskins Laboratories, New Haven, CT, USA.

References

Abarbanel HDI (1996) Analysis of observed chaotic data. New York: Springer-Verlag

Abarbanel HDI, Brown R, Sidorowich JJ (SID) & Tsimring LSh (1993) The analysis of observed chaotic data in physical systems. Reviews of Modern Physics, 65:1331–1392

Arfken G (1985) Mathematical methods for physicists. London: Academic Press Limited

Bernstein NA (1967) The control and regulation of movement. London: Pergamon Press

Casdagli M, Sauer T & Yorke JA (1991) Embeddology. Journal of Statistical Physics, 65:579–616

Collins DR, Park H & Turvey MT (1998) Relative coordination reconsidered: A stochastic account. Motor Control, 2:228–240

Collins DR, Sternad D & Turvey MT (1996) An experimental note on defining frequency competition in intersegmental coordination dynamics. Journal of Motor Behavior 28:299–303

Collins DR & Turvey MT (1997) A stochastic analysis of superposed rhythmic synergies. Human Movement Science 16:33–80

Collins JJ & DeLuca CJ (1993) Open-loop and closed-loop control of posture: A random walk analysis of center-of-pressure trajectories. Experimental Brain Research, 95, 308–318

Eckmann J-P, Kamphorst SO & Ruelle D (1987) Recurrence plots of dynamical systems. Europhysics Letters, 4:973–977

Eckmann JP & Ruelle D (1985) Ergodic theory of chaos and strange attractors. Reviews of Modern Physics, 57:617

Feder J (1988) Fractals. New York: Plenum Press

Feldman A (1980) Superposition of motor programs. I. Rhythmic forearm movements in man. Neuroscience, 5:81–90

Feldman A & Levin MF (1995) The origin and use of positional frames of reference in motor control. Behavioral and Brain Sciences, 18:723–786

Fraser AM & Swinney HL (1986) Independent coordinates for strange attractors from mutual information. Physical Review A, 33:1134–1140

Fuchs A & Kelso JAS (1994) A theoretical note on models of interlimb coordination. Journal of Experimental Psychology: Human Perception and Performance 20: 1088–1097

Gallager RG (1968) Information theory and reliable communication. New York: John Wiley and Sons

Gardiner CW (1985) Handbook of Stochastic methods for physics, chemistry, and the natural sciences (vol. 13, 2nd Ed.) Berlin: Springer-Verlag

Gilmore R (1981) Catastrophe theory for scientists and engineers. New York: Wiley

Haken H (1983) Synergetics: An introduction. Berlin: Springer Verlag

Haken H (1996) Principles of Brain Functioning. Berlin: Springer

Haken H, Kelso JAS & Bunz H (1985) A theoretical model of phase transitions in human hand movements. Biological Cybernetics 51: 347–356

Hausdorff JM, Purdon PL, Peng C-K, Ladin Z, Wei JY, Goldberger AL (1996) Fractal dynamics of human gait: stability of long-range correlations in stride interval fluctuations. Journal of Applied Physiology, 80:1448–1457

Kelso JAS (1984) Phase transitions and critical behavior in human bimanual coordination. Americal Journal of Physiology, 246, R1000-R1004

Kelso JAS (1995) Dynamics Patterns: The Self-Organization of Brain and Behavior. Cambridge, MA: The MIT Press

Kloeden PE & Platen E (1992) Numerical solutions of stochastic differential equations. Berlin: Springer-Verlag
Kugler PN & Turvey MT (1987) Information, natural law, and the self-assembly of rhythmic movement. Hillsdale, NJ: Erlbaum
Kugler PN & Turvey MT (1988) Self-organization, flow-fields, and information. Human Movement Science, 7:97–130
Liebovitch LS (1998) Fractals and chaos simplified for the life sciences. New York: Oxford University Press
Liebovitch LS & Todorov AT (1996) Invited editorial on "Fractal dynamics of human gait: Stability of long-range correlations in stride interval fluctuations". Journal of Applied Physiology, 80:1446–1447
Mandelbrot BB (1982) The Fractal Geometry of Nature. San Francisco: Freeman
Mandelbrot BB & van Ness JW (1968) Fractional Brownian motions, fractional noises, and applications. SIAM Review, 10, 422–437
Mitra S, Riley MA & Turvey MT (1997) Chaos in human rhythmic movement. Journal of Motor Behavior, 29:195–198
Newell KM, Slobounov SM, Slobounova ES & Molenaar PCM (1997) Stochastic processes in postural center-of-pressure profiles. Experimental Brain Research, 113:158–164
Press WH, Teukolsky SA, Vetterling WT & Flannery BP (1992) Numerical recipes in C: The art of scientific computing, 2nd Ed. Cambridge: University Press
Riley MA, Balasubramaniam R & Turvey MT (1999) Recurrence quantification analysis of postural fluctuations. Gait and Posture, 9:65–78
Riley MA, Mitra S, Stoffregen TA & Turvey MT (1997) Influences of body lean and vision on unperturbed postural sway. Motor Control, 1, 229–246
Riley MA, Wong S, Mitra S & Turvey MT (1997) Common effects of touch and vision on postural parameters. Experimental Brain Research, 117, 165–170
Risken H (1996) The Fokker-Planck equation; methods of solution and applications. Berlin: Springer-Verlag
Scholz JP, Kelso JAS & Schöner GS (1987) Nonequilibrium phase transitions in coordinated biological motion: Critical slowing down and switching time. Physics Letters A, 123:390–394
Schöner G, Haken H & Kelso JAS (1986) A stochastic theory of phase transitions in human hand movements. Biological Cybernetics, 53: 247–257
Schöner G, Jiang WY & Kelso JAS (1990) A synergetic theory of quadrupedal gaits and gait transitions. Journal of Theoretical Biology, 142: 359–391
Sternad D, Collins DR, & Turvey MT (1995) The detuning factor in the dynamics of interlimb rhythmic coordination. Biological Cybernetics 67: 27–35.
Strogatz SH (1994) Nonlinear dynamics and chaos: with applications to physics, biology, chemistry and engineering. Reading, MA: Addison-Wesley
Symon KR (1971) Mechanics 3rd Ed. Philippines: Addison-Wesley
Takens F (1981) Detecting strange attractors in turbulence. In: Dynamical systems and turbulence, DA Rand & L-S Young (eds) New York: Springer-Verlag, pp 366–381
Turvey MT (1990) Coordination. American Psychologist, 45:938–953
Webber CL Jr (1991) Rhythmogenesis of deterministic breathing patterns. In: Haken H & Koepchen H-P (eds) Rhythms in physiological systems. Berlin: Springer-Verlag, pp 171–191
Webber CL Jr & Zbilut JP (1994) Dynamical assessment of physiological systems and states using recurrence plot strategies. Journal of Applied Physiology, 76:965–973
Webber CL Jr & Zbilut JP (1996) Assessing deterministic structures in physiological systems using recurrence plot strategies. In Khoo MCH (ed) Bioengineering approaches to pulmonary physiology and medicine. New York: Plenum Press pp137–148
Winter DA, Price F, Frank JS, Powell C & Zabjek KF (1996) Unified theory regarding A/P and M/L balance in quiet stance. Journal of Neurophysiology, 75, 2334–2343

Abbreviations

ϕ	relative phase
V	potential function
P	probability distribution
T_1	relaxation time
T_2	mean first passage time

p_m	zero probability moment
$<\phi_m>$	local mean
SD_m	local standard deviation
COP	center-of-pressure
AP	anteroposterior
ML	mediolateral
H	Hurst exponent
D	diffusion coefficient
RQA	recurrence quantification analysis
FNN	false nearest neighbors
d_E	embedding dimension
d_A	the dimension of the attractors' true phase space
d_L	the local dimension, which serves to quantify ADF
ADF	active degrees of freedom

Glossary

a variable which captures the essense of a phenomenon	**collective variable**
variables which dictate changes in a corresponding collective variable	**control variable**
an abrupt change in properties of a substance (such as water to ice) or relationship (such as from antiphase to inphase coordination)	**phase transition**
the difference between the phase progression of two limit cycle oscillators	**relative phase ϕ**
the angle defined by the arctangent of the ratio of velocity and position	**phase**
relative phase of 0 rad (0 deg)	**inphase**
relative phase of π rad (180 deg)	**antiphase**
a function for which minima correspond to attractive states of the associated dynamics, and maxima correspond to repelling states	**potential function V**
the quantification of a positive (lead) or negative (lag) relative phase, particularly due to differences between the oscillators	**detuning**
defined as the negative derivative of a potential function, specifies the changes of the (collective) variable	**equation of motion**
a noise process with gaussian or normally distributed values for which correlations are zero at all time scales	**gaussian white noise process**
a distribution (or density) function that specifies the likelihood or probability of states of a (collective) variable	**probability distribution P**
time-invariant or unchanging with respect to time	**stationary**
time taken to recover a mean state following a perturbation	**relaxation time T_1**
the average or mean time from initialization until a departure from a specified region occurs (e.g., a phase transition)	**mean first passage time T_2**
expected time for a process to evolve from an initial distribution to a final, stationary distribution	**equilibration time**
the sum of trial durations composing a data set from which a probability distribution is calculated	**observation time**
restricted to a particular region of a collective variable	**bounded**
the initial or beginning condition for a probability distribution	**initial distribution**
a deterministic distribution with zero variance from an exact state	**delta function distribution**
equal probability of any state in a bounded region	**uniform distribution**

compromise distribution	the combination of an equal number of trials initialized in likely mode (e.g., an equal number of inphase and antiphase trials)
bins	segments of a small fixed width of a collective variable
zero probability moment p_m	the sum or integral of probability in the region surrounding a particular mode
local mean $<\phi_m>$	the sum of probability weighted by the central value for each bin in the region surrounding a particular mode
local standard deviation SD_m	the square root of the difference between (a) the sum of the probability weighted by the squared central value for each bin, and (b) the square of the local mean
center-of-pressure COP	the vertical ground reaction vector equal and opposite to a weighted average of all downward forces acting between the feet and the ground
anteroposterior AP	the plane of forward and backward postural sway
mediolateral ML	the plane of left and right postural sway
average mutual information	the mean over all time steps of the amount of information that can be learned about one time series from another
rescaled range analysis	calculation of H from the relation of the time step to the range divided by the standard deviation of a time series
Hurst exponent H	an exponent quantifying the correlation of steps in a time series
persistence	positive correlations of past and future increments, $0.5 < H \leq 1$
antipersistence	negative correlations of past and future increments, $0 \leq H < 0.5$
Brownian motion	a random walk with uncorrelated steps
fractional Brownian motion	a random walk exhibiting correlated steps
diffusion coefficient D	a measure of the average level of stochastic activity in a random walk
stabilogram-diffusion method	calculation of H from the scaling law relating the time step to the mean of the squared displacements
recurrence quantification analysis RQA	a technique designed to test and quantify nonstationarity in time series with no rigid constraints on distribution or size of the data
recurrence plot	a plot generated using position in the time series as both horizontal and vertical axes, for which a point is plotted whenever the time series value for each axis falls within a particular radius (all points on the diagonal are trivially recurrent)
phase-space reconstruction	a method of investigating properties which are invariant over smooth coordinate transformations in order to study the original dynamics from which a single time series is measured
false nearest neighbors	FNN: points which are in proximity due to the phase space being of insufficient dimension rather than due to the dynamics
embedding dimension d_E	the number of dimensions beyond which nothing further is revealed about the dynamics
active degrees of freedom ADF	the number of first-order, autonomous, ordinary differential equations which is invariant over changes in measurement projection

Chapter 24

Detection of Chaos and Fractals from Experimental Time Series

YOSHIHARU YAMAMOTO

Introduction

The neurosciences deal with the various levels of nervous system from a single neuron, a highly nonlinear element, to the whole brain, a highly complex system. Consequently, when observing neural activities and/or the end-organ responses, one can easily find signals exhibiting irregular, complex, and seemingly random dynamics. Until recently, the common strategy to study these dynamics has been to use various types of stochastic models to deal with the randomness (Tuckwell 1989). The theory of nonlinear dynamics, especially that of (low-dimensional) *chaos* and *fractals*, is changing this traditional strategy.

Chaos theory has opened the possibility that nonlinear systems with a few degrees of freedom, while deterministic in principle, can create output signals that appear complex and mimic stochastic signals from the point of view of conventional time series analysis (Sauer 1997). The concept of fractal has also indicated that many complex structures in the natural world can be described by a simple but nontrivial extension of the classical random-walk model (Mandelbrot 1982).

In this chapter, I will introduce some of the typical numerical techniques for the detection of chaos and fractals from experimental signal time series, inevitably containing measurement and/or intrinsic noise, after giving brief theoretical backgrounds for the methodologies. Although these methods will be shown to be useful in detecting hidden nonlinear structures in real world data, one has to keep in mind that no single technique is sufficiently powerful to "judge" chaos and/or fractals with 100% confidence (Casdagli 1991). It can therefore be said, and will be emphasized later, that the critical application is a key to success in gaining deeper insights into the underlying structures in one's data.

Part 1: Theoretical Backgrounds

Aperiodicity

A unique feature of signals with seemingly irregular dynamics is the absence of *periodicity*, or being *aperiodic*. Traditionally, therefore, spectral analysis is first applied to a given time series to search for hidden periodicities that might explain the source of signal variability (cf. Chapter 18). For example, in the simplest case, the presence of a harmonic or sinusoidal oscillator is suspected (Fig. 1A) when a power spectrum $P_{xx}(\omega)$, giv-

Yoshiharu Yamamoto, The University of Tokyo, Graduate School of Education, Educational Physiology Laboratory, 7-3-1 Hongo, Bunkyo-ku, Tokyo, 113-0033, Japan (phone: +81-3-5841-3971; fax: +81-3-5689-8069; e-mail: yamamoto@educhan.p.u-tokyo.ac.jp)

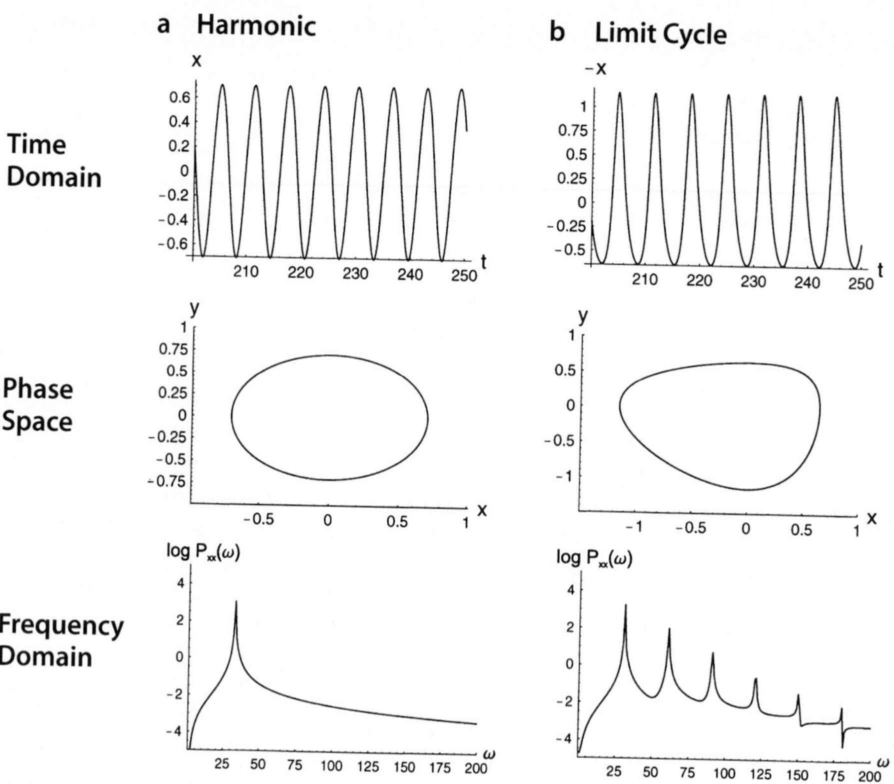

Fig. 1. A simple harmonic motion (A) and periodic dynamics of two-dimensional dynamical system (B): $\dot{x}=y(1-x)$, $\dot{y}=-x(1-y)$. From top to bottom, the panels show a time series of x, the dynamics in a phase space (x, y), and the power spectra ($P_{xx}(\omega)$) as a function of frequency (ω).

en by the squared norm of the Fourier transform of a time series x(t) as

$$P_{xx}(\omega)=\left\|\int_0^\infty e^{i\omega t}x(t)dt\right\|^2, \tag{1}$$

has a single sharp peak.

The harmonic motion represented by the peak is the solution of a linear differential equation $\ddot{x}=-x$, and this equation can simply be reduced to a system of first-order (linear) differential equations as $\dot{x}=y,\ \dot{y}=-x$. Generally, a system of first-order differential equations in the form of

$$\dot{x}_i=F_i(x_1,\ldots,x_n) \tag{2}$$

is called a *dynamical system*. The forms of *trajectories* or *orbits* of the dynamical system in *phase space* ($x_1,\ldots,x_n$), e.g., (x, y) in the two-dimensional case mentioned above, characterize the dynamics of the solution of the original differential equation. For example, for the harmonic motion, the orbit is an ellipsoid, indicating that the sustained periodic oscillation is sinusoidal (Fig. 1A). In a nonlinear two-dimensional dynamical system, such as $\dot{x}=y(1-x),\ \dot{y}=-x(1-y)$, the time series is still highly periodic although the final trajectory, called an *attractor*, is not ellipsoidal and the spectrum contains higher harmonics (Fig. 1B). This type of periodic oscillations is called a *limit cycle*.

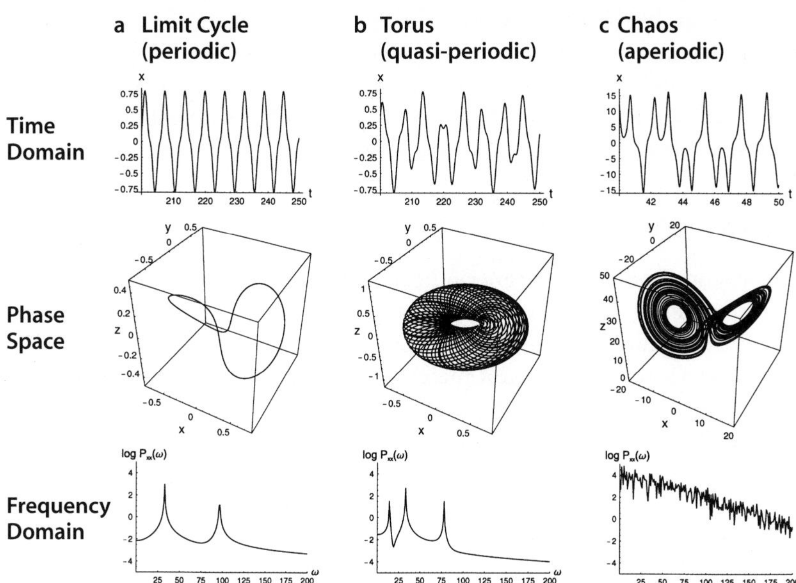

Fig. 2. A limit cycle (A) and a torus (B) of nonlinearly coupled harmonic oscillators (Eq. (3)), and chaotic dynamics of the Lorenz equations (C; Eq. (4)). From top to bottom, the panels show a time series of x, the dynamics in a phase space (x, y, z), and the power spectrum ($P_{XX}(\omega)$) as a function of frequency (ω).

The situation is much different in dynamical systems in third- or higher-dimensional phase space. Fig. 2A and 2B show the dynamics of two nonlinearly-coupled harmonic oscillators with frequencies of ω_1 and ω_2, of which the solution is given by

$$
\begin{aligned}
x(t) &= (0.5 + 0.3\cos\omega_2 t)\cos\omega_1 t, \\
y(t) &= (0.5 + 0.3\cos\omega_2 t)\sin\omega_1 t, \\
z(t) &= 0.3\sin\omega_2 t.
\end{aligned}
\tag{3}
$$

Note that the system is a simple extension of the solution in Fig. 1A: $x(t) = \cos\omega_0 t$, $y(t) = \sin\omega_0 t, \omega_0 = 1$. When the frequency ratio ω_2/ω_1 is a rational number (Fig. 2A), the attractor is a periodic limit cycle with two distinct peaks in the power spectrum. In the case where ω_2/ω_1 is an irrational number (Fig. 2B), however, the time series does not seem to be periodic and the orbit is finally trapped into a three-dimensional surface of a *torus*. This is an example of *quasi-periodic* motion, but the spectrum still has a few distinct peaks.

Nonlinear chaotic systems exhibit dynamics very different from limit cycles and tori. For example, when looking at a time series generated by the famous Lorenz equations (Lorenz 1963):

$$
\begin{aligned}
\dot{x} &= 10(y - x), \\
\dot{y} &= 28x - y - xz, \\
\dot{z} &= -\frac{8}{3}z + xy,
\end{aligned}
\tag{4}
$$

the phase space trajectory has a very complicated geometrical structure (Fig. 2C). Although the orbit never re-visits the same point again, implying it is aperiodic, the trajectory does not fill the entire phase space, implying it is bounded, and is constrained in a *strange or chaotic attractor*. Consequently, the power spectrum has a broadband characteristic which mimics stochastic signals (Fig. 2C).

The above examples clearly show that spectral analysis is in fact useful for searching for hidden periodicities in time series data. They also indicate that, even if spectral analysis fails, it is still possible that the time series contains a nonlinear structure. Thus, methodologies for detecting nonlinear structures, especially those for chaotic motions, have attracted much attention in the last few decades.

Sensitive Dependence on Initial Conditions

The reason why chaotic systems show aperiodic dynamics is that phase space trajectories that have nearly identical initial states will separate from each other at an exponentially increasing rate captured by the so-called *Lyapunov exponent*. This is defined as follows. Consider two (usually the nearest) neighboring points in phase space at time 0 and at time t, the points' distances in the i-th direction being $\|\delta x_i(0)\|$ and $\|\delta x_i(t)\|$, respectively (Fig. 3A). The Lyapunov exponent is then defined by the average growth rate λ_i of the initial distance,

$$\frac{\|\delta x_i(t)\|}{\|\delta x_i(0)\|} = 2^{\lambda_i t} (t \to \infty) \text{ or}$$
$$\lambda_i = \lim_{t\to\infty} \frac{1}{t} \log_2 \frac{\|\delta x_i(t)\|}{\|\delta x_i(0)\|} \quad (5)$$

Chaotic systems are characterized by having at least one positive λ_i. This indicates that any neighboring points with infinitesimal differences at the initial state abruptly separate from each other in the i-th direction. In other words, even if the initial states are close, the final states are much different. This phenomenon is sometimes called *sensitive dependence on initial conditions*. Although the exponential separation causes chaotic systems to exhibit much of the same long-term behavior as stochastic systems, the positive Lyapunov exponent is only observed for chaotic systems. The Lyapunov exponents for stochastic signals are zero, indicating that $\|\delta x_i(0)\|$ and $\|\delta x_i(t)\|$ remain the same independent of time.

Relation to Information Theory

The Lyapunov exponent opens a relation to *information theory* as follows (cf. Chapters 19, 20). If P(t) denotes the probability of observing a distance of $\|\delta x_i(t)\|$ at time t, Eq. (5) can be rewritten as

$$\lambda_i = \lim_{t\to\infty} \frac{1}{t} (\log_2 P(t) - \log_2 P(0)) \quad (6)$$

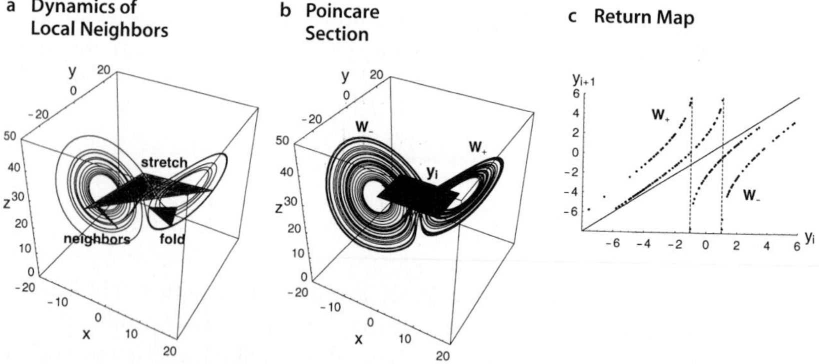

Fig. 3. (A) Stretching-and-folding dynamics, (B) the Poincaré section, and (C) a return map of the Lorenz attractor (Eq. (4)). The return map was constructed by using y of Eq. (4).

Thus, the magnitude of the Lyapunov exponent quantifies the rate at which system processes create or destroy information in bits per unit of time (e.g., bit/s). In chaotic systems with the positive Lyapunov exponents, these processes create information by increasing the degree of uncertainty about the initial conditions with increasing time steps, resulting in the overall information production rate $K = \sum_{\lambda_i > 0} \lambda_i$, which is called the *Kolmogorov entropy*.

Boundedness, Fractal Geometry

Another unique feature of strange attractors is their *stretching-and-folding* dynamics (Fig. 3A). By definition (Eq. (5)), every Lyapunov exponent λ_i corresponds to a coordinate of phase space. If all the Lyapunov exponents are positive, the trajectories neither converge onto a certain attractor nor do they exhibit stationary dynamics. Thus, chaotic systems should have at least one negative Lyapunov exponent so that, in the corresponding direction, the distance between trajectories decreases with time. The consequence of this is that, in addition to the exponential stretching of a neighboring region due to the positive Lyapunov exponent, the region is folded back into a small subspace (Fig. 3A).

In contrast to stochastic dynamics, the stretching-and-folding dynamics of strange attractors prevents the orbits from filling the entire local subspace. Rather, an iteration of the same deterministic stretching-and-folding operation keeps the trajectory confined in a unique *self-similar* geometrical structure called fractal. (The term fractal has two different meanings in nonlinear dynamics. The details will be introduced later in this chapter.) This unique geometry can be characterized by the so-called *fractal dimension* d_F. The geometry in the local subspace of a limit cycle (Fig. 2A) and a torus (Fig. 2B) is respectively a line (one-dimensional) and a plane (two-dimensional). In case of space-filling stochastic dynamics, the geometry is a cube which has the same dimensional value as the three-dimensional Euclidian phase space. By contrast, the fractal geometry is characterized by a non-integer or fractal dimension. For instance, the calculated dimension of the Lorenz attractor (Fig. 2C) is 2.06 (Moon 1992), indicating that the dynamics is more complex than that of a torus but is far from filling the three-dimensional phase space.

Determinism

Even if aperiodic chaotic motions mimic stochastic signals in some respects, they possess a hidden order generating the complex and seemingly irregular behavior. The order behind chaos can be in vestigated and visualized by examining the so-called *(Poincaré return map* (Fig. 3C). The return map of an attractor in n-dimensional phase space is a sequence of stroboscopic projections on the (n−1)-dimensional plane called the *Poincaré section* (Fig. 3B). (Note that this is where a discrete map, not autonomous (i.e., without input forcings) continuous systems, of lower than three dimensions exhibits chaotic behavior.) That is, the return map maps the repeated crossings of the trajectory through the Poincaré section. For example, when looking at the y-component of the Lorenz attractor's return map (though not "one-to-one"; Fig. 3C), there is indeed a fairly regular pattern indicating that some types of *determinism* are present in the dynamics. And each crossing of the orbit across the plane $z = 27$ (Fig. 3B) is governed by a switching between two N-shaped maps from y_i to y_{i+1}. In low-dimensional chaotic systems, it is sometimes, though not always, possible to find this type of clear deterministic map in the Poincaré section.

Embedding Theorem

In the theoretical cases discussed above, the number of phase space coordinates or the number of differential equations describing a given dynamical system is known *a priori*, i.e., n in Eq. (2) is known. Once this information is given, the dynamics and/or geometry of attractors in n-dimensional phase space can be used to search for potentially chaotic motions. For experimental time series data, however, n is hardly known in advance, and in most cases, the number of variables available for measurements is only one. It is thus necessary to "reconstruct" the geometry in phase space from a single record of time series measurement.

Geometric methods for reconstructing attractors from a time series of measurements began with the pioneering work of Packard et al. (1980) and Takens (1980). In these methodologies, the common strategy is to construct a vector time series $X_M(t)$ in an M-dimensional Euclidean space by taking (*embedding*) delayed samples of an experimental time series x(t) as coordinates, such that

$$X_M(t) = (x(t), x(t+L), \ldots, x(t+(M-1)\cdot L)) \tag{7}$$

where L is a fixed lag (see Fig. 4 with M = 3). From sampled data for x(t) (x_i), the discrete version $X_M(i)$ is frequently extracted. In this case, the lag L is chosen as an integer multiple of the sampling interval. The integer M is called an *embedding dimension*. Takens (1980) showed that, given the "sufficient" condition of $M > 2\ d_F$, the geometry of $X_M(t)$ captured the topological characteristics of the original attractor necessary for applying the above tools for chaos detection. Furthermore, Packard et al. (1980) showed that the Lyapunov exponents calculated from the reconstructed attractors were the same as those of the originals. Fig. 4B shows an example of the embedding with M = 3, using the x-variable of the Lorenz attractor (Fig. 2C). (Note that the condition $M > 2\ d_F$ is not always necessary; see Abarbanel (1997).) It can be seen that the geometrical structure of $X_M(t)$ well preserves that of the original attractor in the three-dimensional phase space (Fig. 2C).

The embedding theorem works perfectly if we have an infinite amount of infinitely accurate data. In practice, however, the length of experimental data is limited and the accuracy is far from being infinite. This is where the choice of the lag L and/or the embedding dimension M becomes important for the actual data analyses (Abarbanel 1997).

The theory behind the optimal choice of lag L can intuitively be understood by looking at the embeddings of the x-variable of the Lorenz attractor with too small (Fig. 4A)

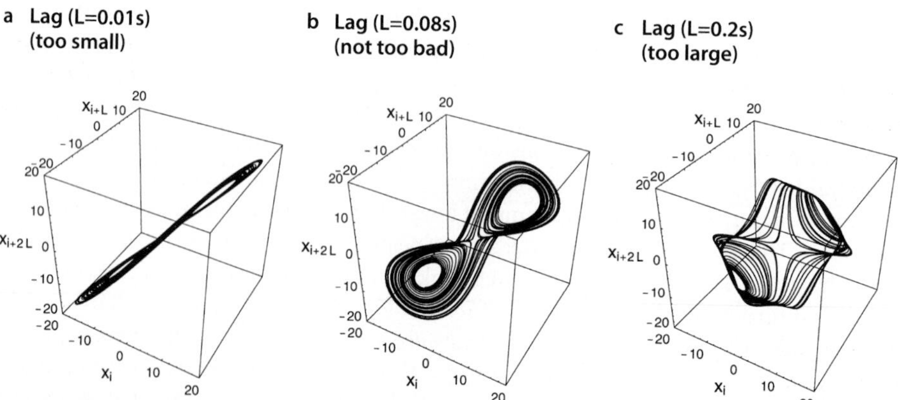

Fig. 4. Time series embeddings of the x-variable of the Lorenz attractor (Eq. (4) with the embedding dimension of M = 3. The lag L in Eq. (7) is set to 0.01 s (A), 0.08 s (B), and 0.2 s (C).

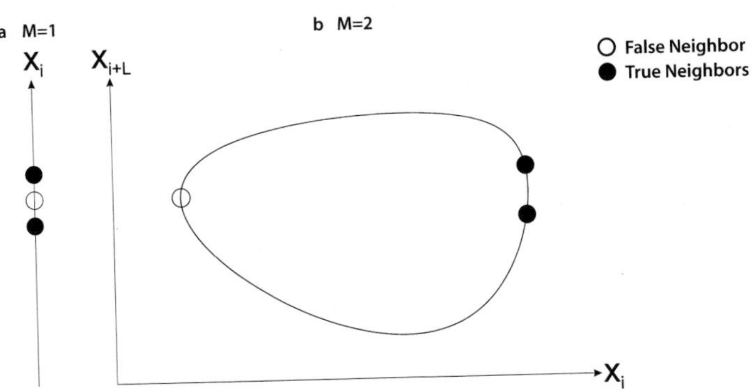

Fig. 5. Schematic representation of the concept of false (nearest) neighbors. The embedding of a two-dimensional limit cycle (B) with too small an embedding dimension (A; M = 1) results in the false neighbor which is not a neighboring point in the original phase space.

and too large an L (Fig. 4C). The use of too small a lag results in strong correlations among the components of $X_M(t)$ and the local geometry of embedding much like a line, i.e., $d_F = 1$ (Fig. 4A). (Again, with an infinite amount of infinitely accurate data, this embedding preserves the geometry of the original attractor.) On the other hand, choosing too large a lag gives rise to another spurious correlation by which self-crossing of the orbit contaminates the local structure of the attractor (Fig. 4C). Therefore, the lag L is typically chosen so that the components of $X_M(t)$ are uncorrelated or independent of each other. For this purpose, two measures are used to determine the optimal lag, namely, the auto-correlation time, i.e., the time at which the auto-correlation of x(t) drops to e^{-1} of the value at time zero, and/or the time at which the first local minimum of mutual information content, a measure of dependence of the coordinates, is observed (Fraser and Swinney 1986). For low-dimensional systems, "visualization" as in Fig. 4 is still useful.

According to the embedding theorem, the choice of M requires *a priori* knowledge of d_F of the original attractor, which is unrealistic for experimental data. When M is chosen arbitrarily and happens to be too small as compared to d_F of the original attractor, this may sometimes result in so-called *false nearest neighbors* (FNN) in the reconstructed phase space (Kennel et al. 1992). The FNN is a point of a data set that comprises the (local) nearest neighbors solely because one is viewing the orbit in too small an embedding space which introduces self-crossing of the orbit. In the simplest case, for example, embedding a two-dimensional limit cycle with M = 1 results in the FNN (Fig. 5). In this case, by adding sufficient coordinates to the embedding space (i.e., increasing M), the FNN would not be the nearest neighbor anymore, whereas a true neighbor would still be (see Fig. 5 for the simplest case). Kennel et al. (1992) recommended to increase M in a stepwise fashion until a number of the FNN dropped substantially to zero. The same strategy is usually followed when measures for chaos such as the Lyapunov exponent and fractal dimension are calculated while scanning M from low to sufficiently high values. If the embedding dimension M becomes adequate, values for the measures are expected to converge onto "true" values.

Part 2: Procedure and Results

Introduction

In what follows, analytical procedures for the detection of chaos and fractals from time series data are introduced with tips and cautions in interpreting the results of the analyses. For illustration, two sample data sets will be analyzed: outputs of an integrate-and-fire neuron model driven by chaotic dynamics (Sauer 1997), and spontaneous fluctuations in the synchronized firing intervals of human sinus node cells, sometimes called *heart rate variability* (HRV) (Malliani et al. 1991; Saul 1990). The former model has mainly been prepared to show that the methodologies actually work for a known low-dimensional chaotic system. However, to keep neuroscientific applications in perspective, analyses of chaos in interspike interval (ISI) data are also presented. The latter type of experimental data have also been proposed to contain chaotic (Babloyantz and Destexhe 1988; Goldberger 1991; Yamamoto et al. 1993) as well as fractal dynamics (Goldberger 1991; Yamamoto and Hughson 1994). The HRV data also serve as an example of high-dimensional systems.

Tools for the detection of chaos and/or fractals generally involve many numerical calculations. The details of these calculations cannot be given here. Various relevant software packages are available either commercially or non-commercially. To collect the packages, the information can be obtained, for example, from http://amath-www.colorado.edu:80/appm/faculty/jdm/faq-[5].html. For the analyses in this chapter, I am using a freeware package called Time Series Statistical Analyses System (TSAS).

The TSAS is an assembly of modules that has been developed by our research group for the study of nonlinear dynamics in (neuro-)physiological time series data. Each module performs tasks such as filtering data, calculating spectra and making the summary reports, analyzing data in a reconstructed phase space for nonlinear dynamics, and drawing simple character-based graphs. The package is supplied with generic C source codes with part of the algorithms ported from the famous Numerical Recipes (Press et al. 1988), thus easy to use (and debug!) on various platforms like PC and workstations. The binary and source codes are available from ftp://psas.p.u-tokyo.ac.jp. The names and the usages of the TSAS modules below are written in a `typewriter style`.

Chaos in Interspike Interval Data from a Neuron Model

The most convincing evidence for the existence of chaos in neurophysiological data has been found in the response of a single neuron and of mathematical neuronal models to periodic current stimulations (Holden 1986). In these studies, it has been shown clearly that the periodic inputs to the neuron sometimes result in aperiodic responses in the interspike interval data. While the accuracy of the actual electrophysiological data is inevitably limited, very similar responses have also been obtained for the Hodgkin-Huxley equations, a nonlinear dynamical system with four variables.

Attention has recently been paid to the opposite question: When a neuron is forced by inputs generated by chaotic motions, can the dynamics be reflected or probed in the output ISI data? Due to nonlinear input-output relationships of the actual neurons and neuronal models, this question is generally difficult to answer. Sauer (1997), however, recently showed that, with the minimal nonlinearity in an integrate-and-fire neuron given by

$$\int_{T_i}^{T_{i+1}} x(t)dt = \Theta, \tag{8}$$

where Θ is a positive number representing the firing threshold, a sequence of interspike intervals $I_i = T_i - T_{i-1}$ could be used to reconstruct attractor-generated input data.

For example, Fig. 6 illustrates the ISI data generated by feeding the integrate-and-fire model with the x-variable of the three dynamical systems studied in Fig. 2. Each input

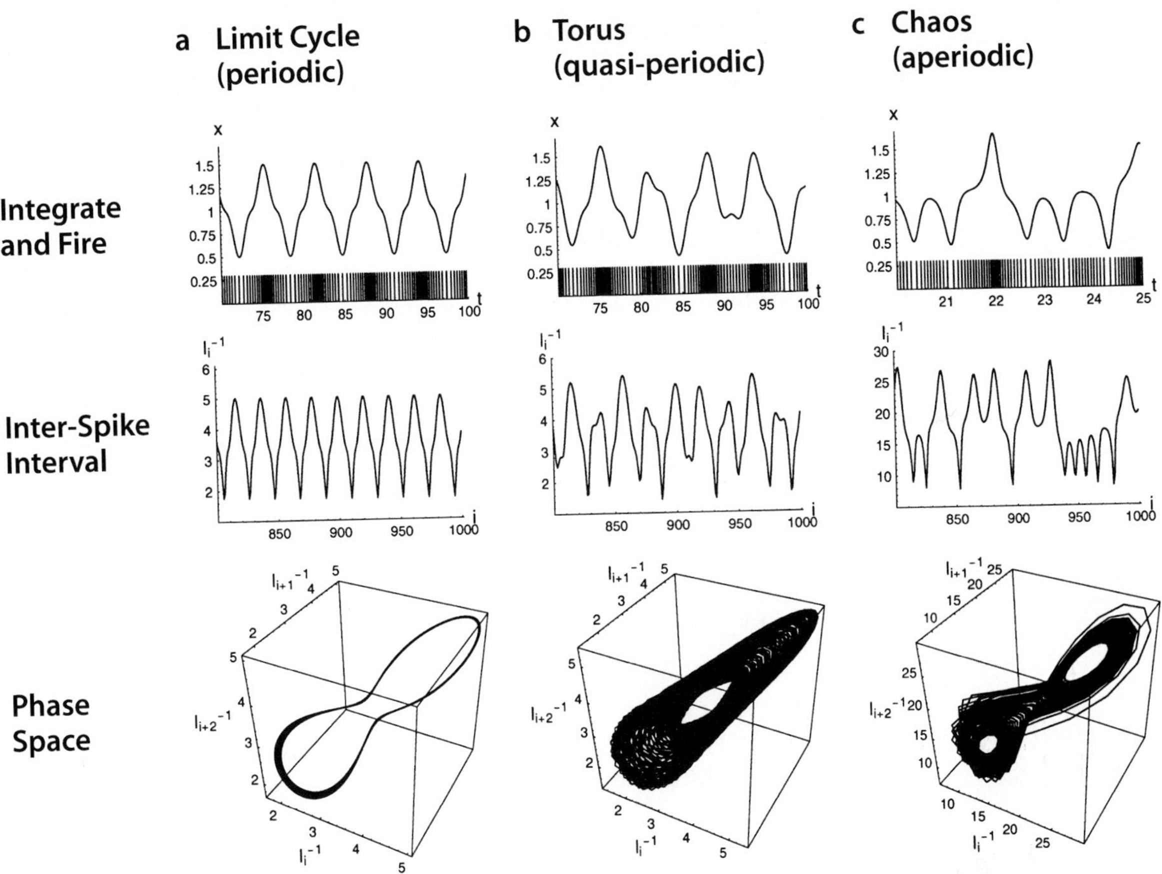

Fig. 6. Generation of interspike intervals (ISIs) by an integrate-and-fire model (Eq. (8)) driven by a limit cycle (A), a torus (B), and chaotic dynamics of the Lorenz equations (C; Eq. (4). From top to bottom, the panels show a time series of x (the same as those in Fig. 2) and the resultant spikes, the dynamics of ISI data (I_i^{-1}; thus transformed to the instantaneous spike frequencies), and the trajectories in the reconstructed space.

data set was transformed to have a unit mean and a standard deviation of 0.3, and the threshold Θ was set so that the number of ISIs for each data set was about 1,500 from 9,000 original data points. It can be seen that the sequences of ISIs and the embeddings in the three-dimensional reconstructed phase space for the limit cycle (Fig. 6A), the torus (Fig. 6B), and the Lorenz equations (Fig. 6C) well preserve the characteristic patterns of the underlying continuous dynamics (Fig. 2). However, due mainly to the discreteness of the ISI data, the local structure of the reconstructed attractors may be destroyed, especially in the chaotic case (Fig. 6C). This possibility can be further examined quantitatively, using the tools of chaos theory.

Embedding of ISI Data

The analyses of ISI data starts by embedding ISIs into a reconstructed phase space by a TSAS module: `embed␣/ed:M␣/lg:L`. The program feeds a sequence of ISI data (in a line-by-line text format) from the standard input and prints the M-dimensional vectors with the lag L to the standard output. The lag of one ISI is thought to be acceptable when looking at the reconstructed attractors in Fig. 6.

Assessment of Fractal Geometry and Attractor Boundedness: Correlation Dimension

The fractal geometry or the boundedness of attractors is examined most frequently by calculating the so-called *correlation dimension* ν (Grassberger and Procaccia 1983). In this method, the self-similar property of an attractor is probed by the scaling behavior of a *correlation integral:*

$$C_M(r) = (1/N^2)\sum_{i \neq j} H\left(r - \left\|X_M(i) - X_M(j)\right\|\right) \tag{9}$$

where H(·) is 1 for positive and 0 for negative arguments. The $C_M(r)$ is a (normalized) number of data points in the reconstructed phase space within a sphere of radius r. For unbounded, stochastic signals, the correlation integral for an M-dimensional space is expected to be scaled as $C_M(r) \propto r^M$ because the trajectory uniformly fills the M-sphere of radius r. For bounded signals, however, there is a finite scaling exponent such that $C_M(r) \propto r^{\nu}(\nu < M)$. If ν has a non-integer value, the phase space is expected to have fractal geometry. [Strictly speaking, ν is not exactly equal to d_F (Hentschel and Procaccia 1983). However, the great advantage of the correlation dimension algorithm is that it is computationally very efficient.]

A TSAS module `cordim` is then used to calculate $C_M(r)$ as well as the slope of the linear part of log r vs. log $C_M(r)$ as an estimate for ν (Fig. 7). For each M, this can be done

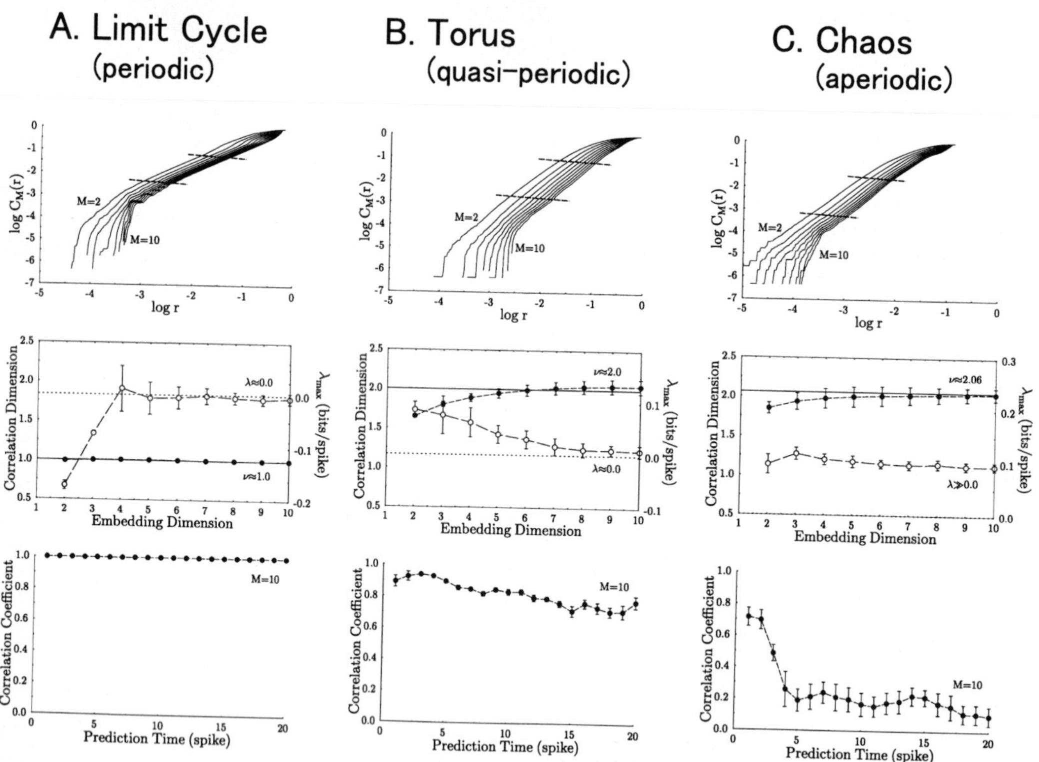

Fig. 7. Results of calculating the correlation dimension (ν), maximal Lyapunov exponent (λ_{max}), and nonlinear predictability for the ISI data generated (in Fig. 6) by a limit cycle (A), a torus (B), and the Lorenz attractor (Eq. (4). The top panels show the scaling relationship of correlation integral $C_M(r)$ with different values of the embedding dimension M. The dashed lines indicate regions used to calculate the slope (i.e., ν). In the middle panels, mean ν (closed circles) and λ_{max} (open circles) and the standard deviations (vertical lines) for 20 data sets are presented. The horizontal solid and dotted lines represent respectively ν and λ_{max} for the corresponding continuous systems. The bottom panels show mean and standard deviations of the correlation coefficients between the observed and the predicted values as a function of prediction time (see text for details).

by: embed␣/ed:M␣/lg:1␣<filename␣|␣cordim␣/sl:1 (filename; an ISI filename), on a DOS or Unix shell. By default, a linear scaling region is selected according to the method by Holzfuss and Mayer-Kress (1986). With sufficiently higher M (see the embedding theorem), the average ν for the ISI data generated by Eq. (8) approached the value for the continuous counterparts, i.e., 1.0 for the limit cycle (Fig. 7A), 2.0 for the torus (Fig. 7B), and 2.06 for the Lorenz attractor (Fig. 7C). The difference between the values for the latter two cases was, however, dubious, and it was difficult to conclude that the ν for the ISI data driven by the Lorenz attractor (Fig. 6C) had a non-integer value (i.e., having a fractal geometry). Generally, for most experimental data, the correlation dimension is not sensitive enough to search for a non-integer or fractal dimension d_F that might be generated by chaotic motions. What can be interpreted safely from the results of the analysis of correlation dimension (Fig. 7) is that all the ISI data are neither stochastic nor unbounded because the estimates for ν are much smaller than M. While the interpretation sounds trivial, especially for the low-dimensional dynamics, it is still significant because the power spectrum of the ISI data from the Lorenz chaos, as shown in Fig. 2C, may mimic stochastic signals.

Estimation of the Lyapunov Exponent

The TSAS has a module called lypexp to calculate the maximal or the largest Lyapunov exponent (λ_{max}); this module can be used to search for a sensitive dependence on initial conditions by the positive λ_{max}. The original algorithm was proposed by Wolf et al. (1985), and realizes a step-by-step piecewise calculation of Eq. (5). [There is another method for determining a complete set of the λ_i; see Echmann and Ruelle (1985).]

The module, invoked by: embed␣/ed:M␣/lg:1␣<filename␣|␣lypexp␣/sl:1, was used to calculate λ_{max} for the ISI data (Fig. 7). Consequently, the ISIs generated by the Lorenz attractor were found to have positive λ_{max} ($\approx$ 0.1 bits of information per spike) and the value was much greater than that relating to either the limit cycle or the torus. It is thus concluded that the embedding with an integrate-and-fire neuron preserves a sensitive dependence on initial conditions of the input chaotic system. It should be noted, however, that the λ_{max} for the torus also seemed to be positive at some embedding dimensions (Fig. 7B). Frequently, this type of "gray zone" result is obtained, especially in cases of experimental data, because λ_{max} (and ν) is reported by a single number whereas the value *per se* contains some amount of error.

Method of Nonlinear Prediction

Another useful and intuitive approach for detection of chaos in a time series is the *nonlinear prediction* (or forecasting) technique (Sugihara and May 1990; Casdagli 1991). This method utilizes *both* of two rather conflicting characteristics of chaotic systems: the short-term prediction is possible due to their deterministic nature while the long-term forecasting is impossible because the phase space trajectories that initially have nearly identical states separate from each other at an exponentially fast rate.

A TSAS module called nlpred implements an algorithm by Sugihara and May (1990) for the nonlinear prediction. In brief, using the embedded vectors $X_M(t)$, a minimal neighborhood of each vector is first chosen from the second half of the data such that the vector is contained within the smallest (in diameter) simplex formed from its M+1 closest neighbors from the second half of the data set. To obtain the predicted value, the computation is made for where the original vector has moved within the range of this simplex after a certain number of time steps or prediction times, giving exponential

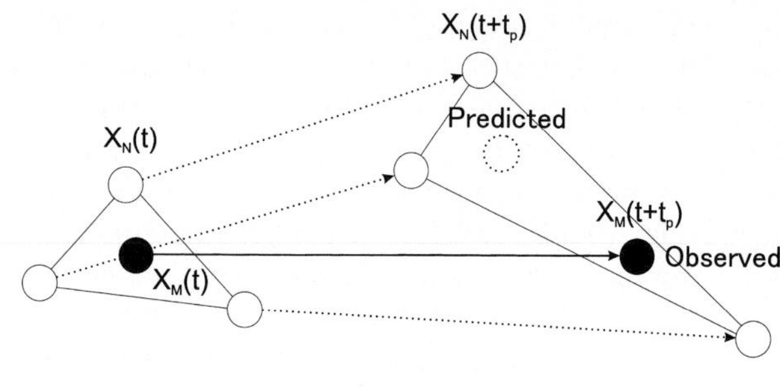

Fig. 8. Schematic representation of the method of nonlinear prediction. For the embedded vectors from the actual data ($X_M(t)$, the neighboring points $X_N(t)$ are first chosen and evolved for the prediction time t_p to obtain the predicted values. The performance of the prediction is evaluated by calculating a correlation coefficient between the observed and the predicted values.

weight to its original distances from the relevant neighbors. The performance of the prediction is evaluated by calculating a correlation coefficient (ρ) between the observed values (after the same prediction time) and the predicted values (Fig. 8).

In case of stochastic signals, the ρ should be low, usually close to zero, and independent of prediction time. In case of deterministic but non-chaotic systems such as periodic or quasi-periodic motions, the ρ should be high and independent of prediction time. In case of deterministic chaotic signals, the ρ should be initially high with an abrupt drop as the prediction time increases. This is due to the exponential growth of the variance in differences between the observed and the predicted values with the index directly related to the Kolmogorov entropy or a sum of positive Lyapunov exponents (Wales 1991).

When looking at the results for the ISI data, calculated after taking the first difference of the ISI sequences to diminish signals associated with simple cyclic variation (Sugihara and May 1990), only the ISI by the Lorenz attractor exhibited a typical pattern for the existence of chaos (Fig. 7C), ρ initially being high with the abrupt drop, while the predictability for the ISI by the torus remained relatively high independent of prediction time (Fig. 7B). This analysis was done by issuing: embed␣/ed:10␣/lg:1␣<filename␣| nlpred␣/sl:1 (filename; the first differenced ISI filename). The first differenced ISI time series was prepared in advance by taking differences in the components of two-dimensional embedded vectors.

In summary, the analyses above confirm the recent findings by Sauer (1997) that, with the model integrate-and-fire neuron (Eq. (8)), the ISI data can be used to study the underlying dynamics and/or to detect chaotic motions in its input.

Correlation Dimension and Nonlinear Prediction of Heart Rate Variability

It has been hypothesized that beat-to-beat fluctuations of human heartbeat intervals (HRV) are chaotic (Babloyantz and Destexhe 1988; Goldberger 1991; Yamamoto et al. 1993). As the normal human HRV is largely controlled by the balance between parasympathetic and sympathetic nervous system activity imposed upon the spontaneous discharge frequency of the sinoatrial node (Dexter et al. 1991; Malliani et al. 1991; Saul 1990), this hypothesis has the intriguing implication that the possibly chaotic dynamics in HRV could reflect such behavior in the autonomic centers in the brain.

As an example of experimental data, HRV from resting humans will be used here to try to detect chaos and/or fractals. The logic used in the analyses of ISI above is considered also to be relevant in analyses of HRV, i.e., "interbeat intervals".

As in the previous section, twenty recordings of human HRV for 8,500 beats were analyzed by cordim and nlpred with the embedding dimension M scanning from 2 to 20 (Fig. 9). [To obtain the HRV data, the surface electrocardiogram was sampled on a real time basis with 1-ms accuracy and the intervals between successive QRS spikes were re-

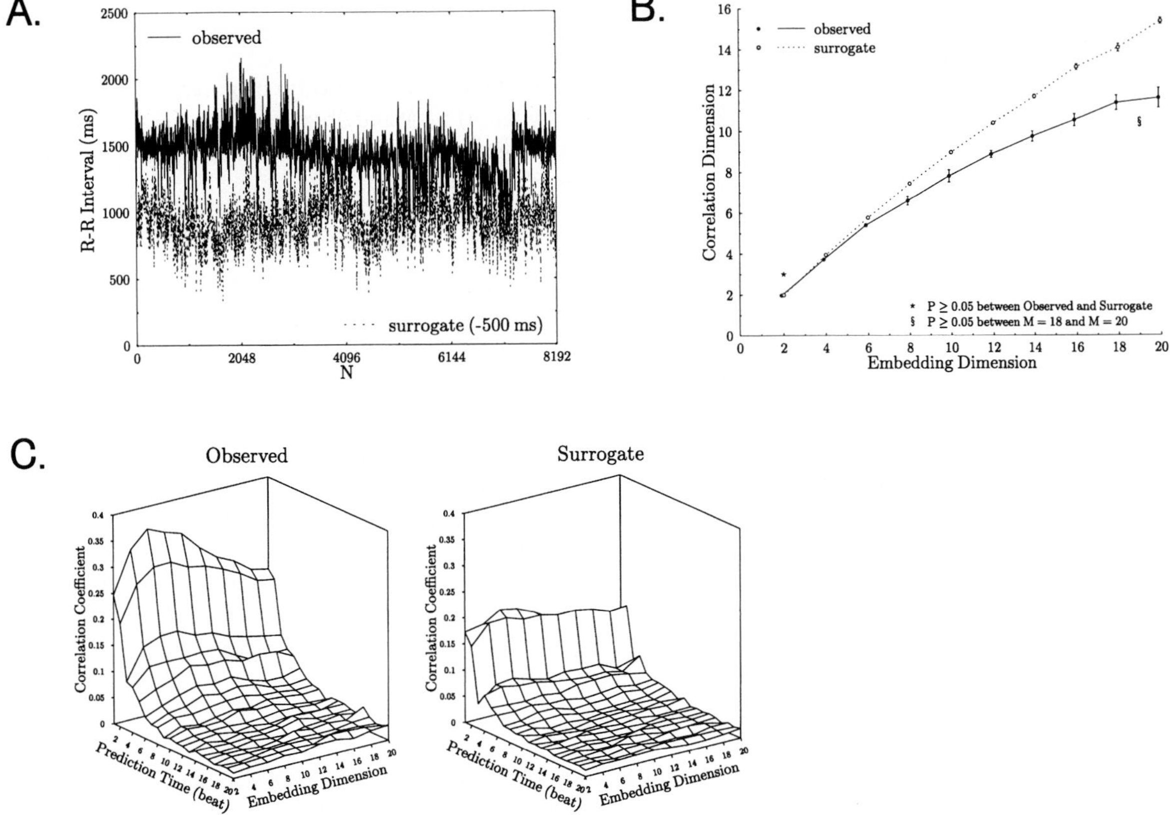

Fig. 9. (A) An example of long-term heart rate variability (HRV; i.e., R-R Interval) in a resting human and its stochastic surrogate (shifted 500 ms down). (B) Averages for estimated correlation dimensions of long-term HRV with the increment of embedding dimension. Vertical bars express standard deviations for 20 recordings. (C) Average correlation coefficients between the observed and the nonlinearly predicted δ HRV as functions of prediction time and embedding dimension. Data from Yamamoto and Hughson (1994).

corded. The experimental details can be found in Yamamoto and Hughson (1994).] The lag L was determined as the first local minimum of mutual information content (Fraser and Swinney 1986) of the original HRV and the first differenced HRV (Δ HRV) for the analyses of correlation dimension and nonlinear predictability, respectively.

The ν for the observed HRV increased with increasing M although it had a tendency to reach a plateau at higher M without significant differences between the values for M = 18 and M = 20 (Fig. 9B). At the highest M of 20, an estimate for ν was 11.6 ± 0.5. The nonlinear predictability of the HRV indicated the existence of chaos because the initial abrupt drops in correlation coefficients between the observed and the predicted values at lower prediction time were observed with the optimal embedding (M = 6; Fig. 9C). Thus, a tentative conclusion would be that the resting human HRV contains high-dimensional chaotic motion. This is probably the most simple or naive interpretation of the results. Life is not so easy, however, especially in the high-dimensional case.

Power Spectral Analysis

There is considerable confusion in the literature over two distinct definitions of the term fractal in the context of time series analysis. It has come to describe both power-law behavior in the Fourier spectrum of the time series, and strange attractors in the (usually reconstructed) phase space of chaotic time series. The human HRV has long been known to be fractal in the former sense (Kobayashi and Musha 1982; Yamamoto and Hughson 1994).

The study of stochastic processes with power-law spectra originated in the classical work of *fractional Brownian motion* (fBm) by Mandelbrot and Van Ness (1968). After appropriate setting of initial conditions, the fBm is defined as satisfying the following relationship:

$$x(ht) \overset{d}{=} h^H x(t), \tag{10}$$

where $\overset{d}{=}$ implies that both hands of $\overset{d}{=}$ have the same statistical distribution function. This relationship (10), given a self-similar or fractal signal (Mandelbrot 1982), implies that the distribution remains unchanged by the factor h^H ($h > 0$) even after one changes the time scale to read it. The constant $0 < H < 1$, called *Hurst exponent*, introduces a general power-law scaling. In case of an ordinary Brownian motion, an additive process of Gaussian white noise, the value of H is 0.5. This corresponds to a well-known `law' that h-times additions of Gaussian random variables result in an increase in the standard deviation by the factor of $\sqrt{h} = h^{0.5}$. The extension of H to values other than 0.5, however, does not permit simple (linear) additions of random variables.

The fBm also shows power-law behavior in the Fourier spectrum: there is a negatively linear relationship between the log of spectral power versus log of frequency plane (Fig. 10). The inverse of the slopes of the log-log plots, called a *spectral exponent* β, is related to H (Yamamoto and Hughson 1993) by:

$$H = (\beta - 1)/2. \tag{11}$$

The spectrum of fBm, which is different from that of white noise or the low-pass filtered version, characteristically increases toward low frequencies (Fig. 10), indicating the existence of a *long-range correlation*.

What happens if one embeds fBm in a phase space? The answer is, following Mandelbrot (1985), when the embedding dimension M is greater than the so-called *fractal dimension of trail* $d_T = 1/H$, this stochastic signal does not fill the entire phase space as in the case of deterministic dynamics. Osborne and Provenzale (1989) also showed that the correlation dimension ν of fBm had a finite value approximating d_T. Furthermore, Provenzale et al. (1991) reported that the estimate for the Kolmogorov entropy had a positive finite value for fBm. Considering this, the observed $\nu = 11.6 \pm 0.5$ for the human HRV may simply indicate that it contains random fractal time series (i.e., fBm), rather than chaotic motions, of which the spectral exponent β is about 1.2 by Eq. (11) and $d_T = 1/H$. As the reported value of β for human HRV was within this range (Yamamoto and Hughson 1994), this expectation is very likely to be right.

This possibility is even more important when considering the ubiquitousness of power-law, fBm-type variability in neurophysiological data from firing patterns of a single neuron (Teich 1989) to collected neuronal activities in the brain (Yamamoto et al. 1986; Inouye et al. 1994). Thus, a method to distinguish chaos from the random fractal time series is indeed necessary. (The actual situation is much more complicated because there is indeed a high-dimensional chaos which is very difficult to distinguish from power-law, fractal noise.)

Method of Surrogate Data

In the early 1990s, the method of *surrogate data* was proposed (Theiler et al. 1992) as a means to study possible chaotic dynamics and discriminate them from stochastic noise. In this method, stochastic surrogate data are generated that have the same power spec-

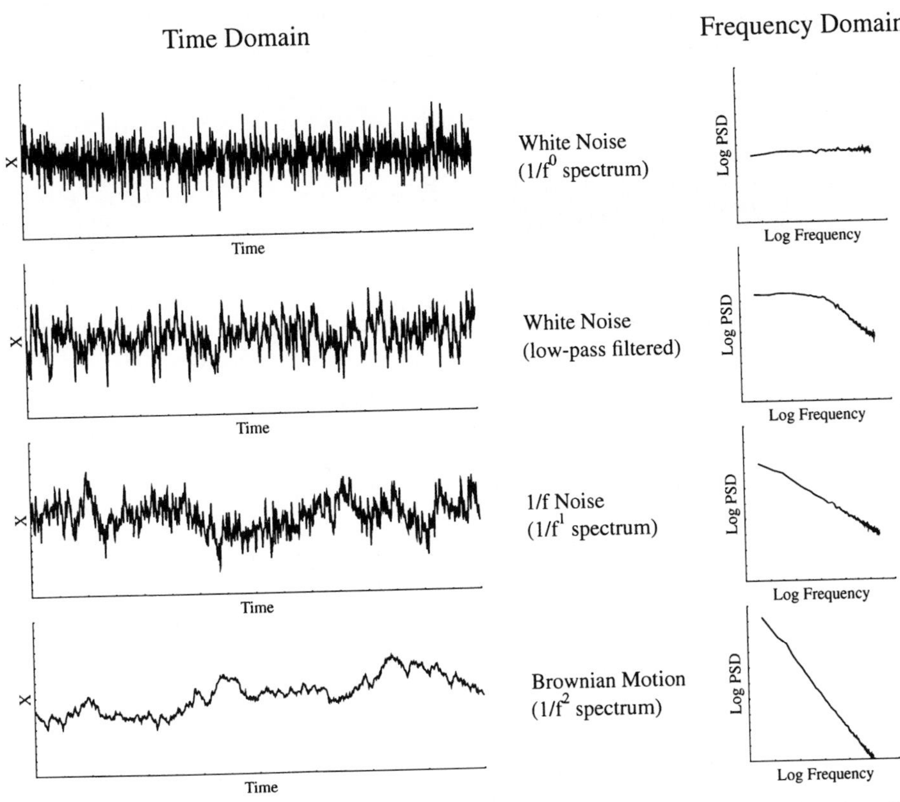

Fig. 10. Computer-generated fractional Brownian motion with different values of the spectral exponent β (left) and the power-law ($1/f^{\beta}$-type) spectra in log-log axes (right). For comparison of the low-frequency behavior in the spectra, low-pass filtered white noise and its spectrum are shown in the second panels.

tra as the original, but have random phase relationships among the Fourier components. If any numerical procedures for studying chaotic dynamics produce the same results for the surrogates as for the original data, we cannot reject a null hypothesis that the observed dynamics is generated by a linear stochastic model rather than representing deterministic chaos. (This is because the surrogate data generated as such can be regarded as an output of a linear, e.g., autoregressive, model.) While measures for chaos such as the correlation dimension and the (largest) Lyapunov exponent are usually given by a single number, the repeated generation of surrogate data provides a confidence interval for the null effect range, enabling hypothesis testing.

A TSAS module `gnoise` feeds spectral data from the standard input and prints the *isospectral surrogate* data to the standard output. To generate the surrogate data sets from the original data, the `gnoise` should be used after calculating power spectra of the originals using other modules called `filter` and `ftspec`. In TSAS, this can be done by: `filter␣<␣filename␣|␣ftspec␣/sl:4␣|␣gnoise␣/sl:4` (`filename`; a filename for the original data).

In Fig. 9, the results of analyses for correlation dimension and nonlinear prediction using the surrogate HRV are also presented. [The isospectral surrogate used here is only valid if a distribution of x(t) is well approximated by the Gaussian distribution (true for HRV data). Otherwise, a different type of surrogate data should be prepared; see Kaplan and Glass (1993).] For all M > 2, estimates of ν for the observed HRV were significantly smaller than those for the surrogates which increased almost linearly with increasing M (Fig. 9B). As for nonlinear predictability, the ρ for the surrogates was not different for

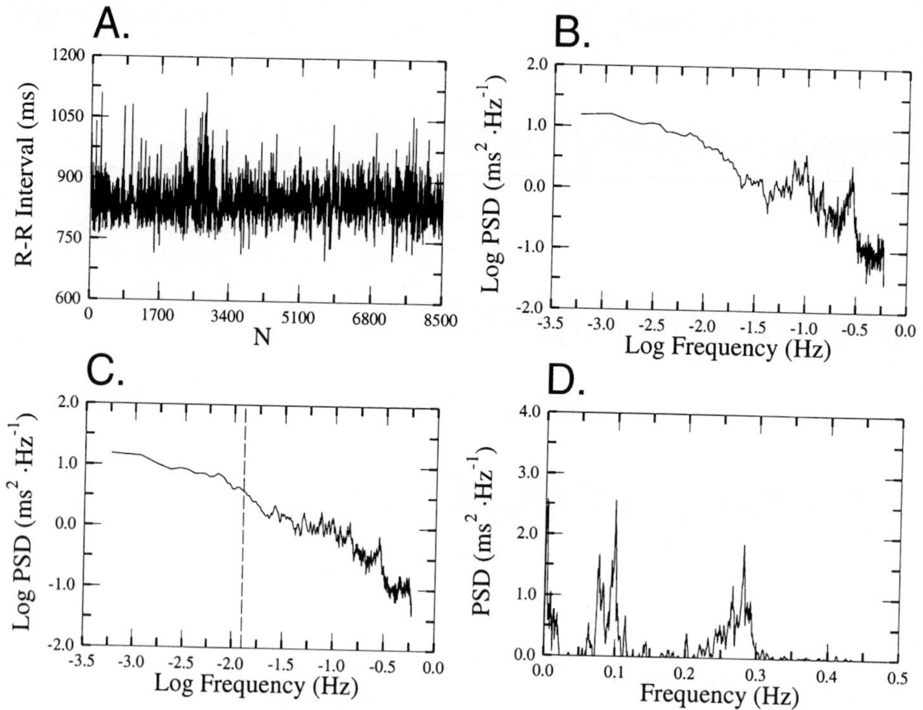

Fig. 11. An example of human R-R interval sequence (HRV; A), the total (B), the fractal (C), and the harmonic (i.e., total-fractal; D) power spectra. The length of data was 8,500 heartbeats. The dashed line in C indicates the critical frequency above which a linear fit was performed to calculate the spectral exponent β. Modified from Yamamoto and Hughson (1994).

the various levels of M and decreased gradually with the increment of prediction time of <10 beats (Fig. 9C). By contrast, for the observed HRV, the ρ at <4 beats was significantly (P<0.01) higher than that for the surrogates especially at the optimal embedding dimension of M = 6 (Fig. 9C). Note that the power spectra of both the original and the surrogate HRV (Fig. 9A) show the power-law dependence. (See Fig. 11C for the original data. The surrogates have the same power spectra by definition.) Thus, the conclusion would be that the normal human HRV is something more than a pure fractional Brownian motion.

One may be disappointed by this final conclusion. One should remember, however, that even a quasi-periodic motion exhibits a finite ν and a positive Lyapunov exponent when calculated from finite ISI data without measurement errors (Fig. 7B). In addition, the calculation of ν (and probably λ_{max}) has been reported to be very sensitive to the limited length of data (Ruelle 1990). (This is why human HRV data is used as an example. The stable recording of neurophysiological data extending 10^4 spikes may sometimes be difficult.) Thus, solely on grounds of smaller ν and more susceptible nonlinear predictability for the observed data than the surrogates, a definitive conclusion about the existence of chaotic dynamics in the human HRV is considered to be difficult.

Coarse Graining Spectral Analysis

As a matter of fact, human HRV is known to have well-defined periodic oscillations due to the feedback regulation of blood pressure and the external forcing by respiration (Malliani et al. 1991; Saul 1990), and these periodic motions are indeed "something more" than a pure fractional Brownian motion. To extract random fractal components from such data containing a mixture of periodic and random fractal signals, coarse graining spectral analysis (CGSA) has been proposed by Yamamoto and Hughson (1993).

CGSA is based on the property of self-similarity of a random fractal process given by relationship (10). Practically, the time series x(t) is often obtained as the discrete form x(i). In this case, with h = 2, for example, the new time series x(2i) can be constructed by taking every other point from x(i). Hence, this procedure is called *coarse graining*. A direct consequence of this is that for a random fractal process, the Fourier spectrum of the coarse-grained signal has a form identical to that of the original signal (Barnsley et al. 1988). Therefore, for the fractal process, the cross-power spectrum between the original and the coarse-grained data is similar to the auto-power spectrum of the original time series by the factor h^H. Conversely, coarse graining a simple harmonic or periodic signal results in total loss of power. Therefore, it is possible by CGSA to evaluate the contribution of the fractal process in a given time series quantitatively as the ratio of the gain of the cross-spectrum between x(i) and x(hi) to the auto-power spectrum of x(i).

Fig. 11 shows an example of the application of CGSA to the human HRV. In TSAS, this was done by: `filter␣<␣filename␣|␣ftspec␣/sl:4␣/cg` (`filename`; a file containing beat-to-beat HRV). The original data are shown in Fig. 11A. When standard spectral analysis was applied (Fig. 11B), a broadband spectrum with inverse power-law scaling was observed as shown previously. The existence of periodic components was suggested by several small, but distinct peaks. However, when CGSA was used to calculate the fractal power spectrum (Fig. 11C), these peaks disappeared. Actually, subtracting panel C from panel B resulted in a textbook example of a harmonic spectrum (Malliani et al. 1991; Saul 1990) with high and low frequency peaks (Fig. 11D).

The analyses of 20 recordings of HRV revealed that the contribution of a random fractal process in the human HRV accounted for >85% of the total variance, and the average β was 1.08 (Yamamoto and Hughson 1994). As the remaining spectral components strongly indicated the existence of periodic motions (Fig. 11D), there seemed to be very little room for deterministic but aperiodic dynamics (i.e., chaos) to be observed in these data sets. The positive results obtained by tools for detecting chaos might possibly be due to the massive existence of fBm which mimics high-dimensional (in terms of d_T) dynamics in the phase space. The detection of random fractal components in a given time series is thus important in itself as this type of signal has frequently been reported to fool the algorithms used to study chaotic dynamics (Tsonis and Elsner 1992) and has recently been shown to be of functional significance in information transmission in a mathematical neuron model (Nozaki and Yamamoto 1998).

Concluding Remarks

In this chapter, several techniques for the detection of chaos and/or fractals from experimental time series have been introduced. Of course, the ultimate question is: "Is my irregular or complex system chaotic and/or fractal?" The above examples have shown that, if the underlying dynamics seems to be generated by a low-dimensional attractor, the answer can be given by using these methodologies.

If these tools suggest the existence of a high-dimensional nonlinear dynamical system, however, one should be very cautious in interpreting the result. Especially, as emphasized by Casdagli (1991), "when only one technique is used to analyze a time series, the results are expected to be at best incomplete, and at worst misleading". The use of various techniques, even those with stochastic models, is recommended because finding $\nu \gg 10$ for noisy experimental data is substantially the same as saying that your system is stochastic.

References

Abarbanel HDI (1997) Tools for the analysis of chaotic data. Fields Inst Comm 11:1–16
Babloyantz A, Destexhe A (1988) Is the normal heart a periodic oscillator? Biol Cybern 58:203–211
Barnsley MF, Devaney RL, Mandelbrot BB, Peitgen HO, Saupe D, Voss RF (1988) The Science of Fractal Images. Springer-Verlag, New York
Casdagli M (1991) Chaos and deterministic *versus* stochastic non-linear modelling. J R Statist Soc B 54:303–328
Dexter F, Rudy Y, Levy MN, Bruce E (1991) Mathematical model of cellular basis for the respiratory sinus arrhythmia. J Theor Biol 150:157–176
Eckmann JP, Ruelle D (1985) Ergodic theory of chaos and strange attractors. Rev Mod Phys 57:617–656
Fraser AM, Swinney HL (1986) Independent coordinates for strange attractors from mutual information. Phys Rev A 33:1134–1140
Goldberger AL (1991) Is the normal heartbeat chaotic or homeostatic? News Physiol Sci 6:87–91
Grassberger P, Procaccia I (1983) Measuring the strangeness of strange attractors. Physica D 9:189–208
Hentschel HGE, Procaccia I (1983) The infinite number of generalized dimensions of fractals and strange attractors. Physica D 8:435–444
Holden AV (1986) Chaos. Manchester University Press, Manchester
Holzfuss J, Mayer-Kress G (1986) An approach to error-estimation in the application of dimension algorithms. In: Mayer-Kress G (ed) Dimensions and Entropies in Chaotic Systems. Quantification of Complex Behavior. Springer-Verlag, Berlin Heidelberg, pp 114–122
Inouye T, Ukai S, Shinosaki K, Iyama A, Matsumoto Y, Toi S (1994) Changes in the fractal dimension of alpha envelope from wakefulness to drowsiness in the human electroencephalogram. Neurosci Lett 174:105–108
Kaplan DT, Glass L (1993) Coarse-grained embeddings of time series: random walks, Gaussian random processes, and deterministic chaos. Physica D 64:431–454
Kennel MB, Brown R, Abarbanel HDI (1992) Determining embedding dimension for phase-space reconstruction using a geometrical construction. Phys Rev A 45:3403–3411
Kobayashi M, Musha T (1982) 1/f Fluctuation of heartbeat period. IEEE Trans Biomed Eng 29:456–457
Lorenz EN (1963) Deterministic nonperiodic flow. J Atmos Sci 20:130–141
Malliani A, Pagani M, Lombardi F, Cerutti S (1991) Cardiovascular neural regulation explored in the frequency domain. Circulation 84:482–492
Mandelbrot BB (1982) The Fractal Geometry of Nature. W. H. Freeman & Company, New York
Mandelbrot BB (1985) Self-affine fractals and fractal dimension. Physica Scripta 32:257–260
Mandelbrot BB, Van Ness JW (1968) Fractional Brownian motions, fractional noises and applications. SIAM Rev 10:422–436
Moon FC (1992) Chaotic and Fractal Dynamics. An Introduction for Applied Scientists and Engineers. John Wiley & Sons, New York
Nozaki D, Yamamoto Y (1998) Enhancement of stochastic resonance in a FitzHugh-Nagumo neuronal model driven by colored noise. Phys Lett A 243:281–287
Osborne AR, Provenzale A (1989) Finite correlation dimension for stochastic systems with power-law spectra. Physica D 35:357–381
Packard NH, Crutchfield JP, Farmer JD, Shaw RS (1980) Geometry from a time series. Phys Rev Lett 45:712–716
Press WH, Flannery BP, Teukolsky SA, Vetterling WT (1988) Numerical Recipes in C. The Art of Scientific Computing. Cambridge University Press, Cambridge, U.K.
Provenzale A, Osborne AR, Soj R (1991) Convergence of the K_2 entropy for random noises with power law spectra. Physica D 47:361–372
Ruelle D (1990) Deterministic chaos: the science and the fiction. Proc R Soc Lond A 427:241–248
Sauer T (1997) Reconstruction of integrate-and-fire dynamics. Fields Inst Comm 11:63–75
Saul JP (1990) Beat-to-beat variations of heart rate reflect modulation of cardiac autonomic outflow. News Physiol Sci 5:32–37
Sugihara G, May RM (1990) Nonlinear forecasting as a way of distinguishing chaos from measurement error in time series. Nature 344:734–741
Takens F (1980) Detecting strange attractors in turbulence. In: Rand DA, Young LS (eds) Lecture Notes in Mathematics. Dynamical Systems and Turbulence. Springer-Verlag, Berlin Heidelberg New York, pp 366–381

Teich MC (1989) Fractal character of the auditory neural spike train. IEEE Trans Biomed Eng 36:150–160

Theiler J, Eubank S, Longtin A, Galdrikian B, Farmer JD (1992) Testing for nonlinearity in time series: the method of surrogate data. Physica D 58:77–94

Tsonis AA, Elsner JB (1992) Nonlinear prediction as a way of distinguishing chaos from random fractal sequences. Nature 358:217–220

Tuckwell HC (1989) Stochastic Processes in the Neurosciences. Society for Industrial and Applied Mathematics, Philadelphia

Wales DJ (1991) Calculating the rate of loss of information from chaotic time series by forecasting. Nature 350:485–488

Wolf A, Swift JB, Swinney HL, Vastano JA (1985) Determining Lyapunov exponents from a time series. Physica D 16:285–317

Yamamoto M, Nakahama H, Shima K, Kodama T, Mushiake H (1986) Markov-dependency and spectral analyses on spike-counts in mesencephalic reticular neurons during sleep and attentive states. Brain Res 366:279–289

Yamamoto Y, Hughson RL (1993) Extracting fractal components from time series. Physica D 68:250–264

Yamamoto Y, Hughson RL (1994) On the fractal nature of heart rate variability in humans: effects of data length and β-adrenergic blockade. Am J Physiol 266 (Regulatory Integrative Comp Physiol 35): R40-R49

Yamamoto Y, Hughson RL, Sutton JR, Houston CS, Cymerman A, Fallen EL, Kamath MV (1993) Operation Everest II: an indication of deterministic chaos in human heart rate variability at simulated extreme altitude. Biol Cybern 69:205–212

Chapter 25

Neural Networks and Modeling of Neuronal Networks

BAGRAT AMIRIKIAN

Introduction

The past decades have seen an explosive growth in accumulation of experimental data in neuroscience research. The detailed anatomical and physiological data alone, however, are not enough to understand how the nervous system works. It is the recognition of this fact that makes modeling studies a significant part of mainstream research in neuroscience. The combination of theoretical methods, including mathematical analyses and computer simulations, together with modern experimental techniques has led to the emergence of a new discipline of *computational neuroscience* with the ultimate goal of explaining how neural signals represent and process information in the brain. Modeling of neuronal networks is a powerful tool that enables accomplishment of this goal by understanding how specific parts of the nervous system perform certain operations (for instance, learning specific motor skills, computing the direction of reaching movement, decoding spatial information, etc.) and is complementary to traditional techniques in neuroscience research.

Another field of modern science, often referred to as *neural computation*, is concerned with learning and computing in networks of artificial, neuron-like units. Though closely related to computational neuroscience, the field of neural computation differs fundamentally in its goal. One of the motivations for studying network computations is the fact that in many tasks the human brain outperforms even the fastest supercomputers available today. Inspired by the knowledge from neuroscience, the artificial neural networks realize an alternative computational paradigm to the classical one introduced by von Neumann. Its main concern is what the artificial networks can do to learn and implement a particular task. Thus, utilizing the idea of parallel and distributed processing, which is widely believed to be the way the brain operates, the field of neural computation does not try to be biologically realistic. What is the best hardware that solves the task? That is the question. In contrast, the field of computational neuroscience is interested in how the task is solved by the nervous system, i.e., how the biological hardware solves the task. Despite the difference in goals, the importance of interplay between modeling biologically plausible and artificial neural networks should not be underestimated. The ideas and concepts developed in one field drive the other, and vice versa.

The relationship between theory and experiment plays a particularly crucial role and creates a wide spectrum of approaches in the modeling of neuronal networks (Koch and Segev, 1989; Abeles, 1991; Marder and Abbott, 1995; Marder et al., 1997). Some models are heavily based on the anatomical and electrophysiological properties of the actual bi-

Bagrat Amirikian, Veterans Administration Medical Center, Brain Sciences Center, One Veterans Drive, Minneapolis, MN, 55417, USA (phone: +01-612-725-2282; fax: +01-612-725-2291; e-mail: amiri001@maroon.tc.umn.edu), University of Minnesota Medical School, Department of Physiology, 6-255 Millard Hall, 435 Delaware Street SE, Minneapolis, MN, 55455, USA

ological structures involved. Studies along this line usually proceed from the detailed description of single cells to the behavior of the network. This approach is most useful when accurate experimental data at the spatial and biophysical levels are available, the function of a neuronal network is already known, and the network itself is relatively small. Such models can determine whether existing data are sufficient to explain observed network behavior, and intend to pinpoint drawbacks and missing components in the model. The alternative to this *data-driven/bottom-up approach* is a *theory-driven/top-down approach*. Here, the emphasis shifts to descriptions of higher level functions such as a perceptual ability. Based on the theoretical analysis, an algorithm that performs the desired function is developed first and then embedded into the simplified network while imposing known biological constraints. This kind of approach tends to be more loosely bound to particular experimental data. However, by sacrificing specificity, the theory-driven approach attempts to address fundamental and puzzling questions, and can help in formulating and testing what kind of computational algorithms the brain is using in different tasks. In the long run, this approach is expected to suggest new experiments and research directions. Whereas the data-driven and theory-driven approaches represent two opposite extremes, there are varieties of other approaches that combine different proportions of the "abstract" and "realistic" components of the modeling and fill in the gap between these two extremes.

It is important to realize that any modeling, by definition, accepts *intentional simplification* of a real system. The reason for this is not merely the limit of computational power available today. Suppose that one would be able to explicitly simulate on a computer the nervous system at the level of every single ionic channel. This would not advance, even for an iota, our understanding of how the brain functions. The simplifying models are necessary to find out which properties of the system are crucial for a particular phenomenon and which are not. This could be achieved by incorporating into a model only those features that are most relevant to the phenomenon under investigation and omitting all other details that a modeler believes are less important. Such an approach raises a fundamental question appropriate for any phenomenological modeling in general and modeling of neuronal networks in particular. Namely, what is the relationship between the complexity of a model, i.e., its realism, and the credibility of the model, i.e., its predictive power? In other words, given two models that equally well explain the same data set, which model is preferable, the simple one or the more complex? *Occam's razor*, the *principle of parsimony*, suggests a criterion for selecting the credible model. Though its validity has never been proven in full generality, the parsimony principle has been incorporated into the methodology of science a long time ago and can be formulated as follows. If two explanations conform equally well to past observations, the simpler of the two has a better chance to predict future observations. The application of this general principle has been extremely useful in many areas of science. I strongly believe that the principle of parsimony is also appropriate for neuronal network modeling and should be exercised in order to achieve a reasonable compromise between tractability and realism. Based on this principle, the idealized strategy for developing *biologically plausible models* can be formulated as follows. Design the simplest neural network that incorporates a set of experimental data relevant to a phenomenon under investigation. If the network operation accurately simulates the phenomenon, then it is an appropriate model for studying that phenomenon. Biologically plausible models must not contain all the known features of the target system, they need to include only those features that are necessary to accurately simulate the phenomena under study. The model should be made more complex only when it contradicts new experimental observations related to the phenomena of interest.

The goal of this chapter is not to review all kinds of models currently used in neuroscience research. The significant theoretical work built around biological neural networks is so huge that to cover it all is perhaps impossible. Rather, I will concentrate on major concepts and procedures relevant to the modeling of large-scale neuronal networks thus conveying knowledge to a reader on "how to go on". For readers who would be further interested in the theory-driven approaches I recommend the monograph by Hertz et al. (1991) that focuses not so much on the level of detailed modeling as on the level of algorithms and representations. In contrast, the textbook by Anderson (1995) approaches to neural networks from a broad neuroscience perspective, with an emphasis on the biology behind the assumptions of the models, as well as on what the models might be used for. Finally, the link between the theoretical studies and experimental approaches is the main concern of the book by Koch and Segev (1989). It has an excellent collection of papers for those interested in biophysical mechanisms for computation in neurons.

Network Architecture and Operation

Units and Connections

Usually the first step in modeling a neuronal network is the definition of the network architecture. Depending on the complexity of a target system and the desired level of realism, each unit of the model network may simulate either a single neuron or a set of similar neurons with coherent functions and properties. The communication of activity from one cell to another is modeled by means of a connection between a corresponding pair of model units. The entire set of units and the pattern of connections between them define the architecture of the network. The realistic design of architecture requires knowledge of the underlying neural structure from neuroanatomical studies. When such data are not available, which is often the case, an *educated guess* based on other indirect electrophysiological studies could be useful.

Types of Units

Each unit can be classified as input, output, or hidden depending on the role that it plays in the operation of the network. *Input units* receive external signals that may represent sensory signals, signals from other networks, or some events in the external environment of an organism. The stimulation of input units may then change the activities of *hidden units* that they are connected to. This perturbation may further propagate across the whole network through connections to other hidden units and reach, ultimately, the *output units*. The role of the latter is to provide an input to another network or simulate events at the behavioral level, for example, a motor action. Depending on the network architecture and its interpretation the input, hidden, and output units may partially or completely overlap.

Types of Architecture

There are two prevailing network architectures that have been used in the theory and modeling of neural networks. Figure 1A shows an example of layered *feed-forward architecture*. The role of the input units is to feed external signals to the rest of the network. All connections are feed-forward. There are no connections between units in the same layer. Units in the intermediate layers are considered as hidden units, whereas the last layer represents the output units. These types of architecture are also known as *perceptrons* (Rosenblatt, 1962).

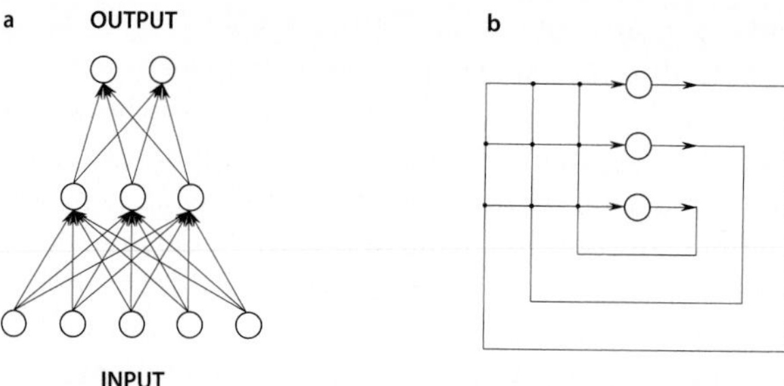

Fig. 1. Network architecture. Circles represent units whereas lines show connections between them. The arrow indicates the direction of a connection. *A*: Feed-forward network with one hidden layer. *B*: Fully recurrent network with three units.

Networks that are not strictly feed-forward, but include feedback connections are called *recurrent networks.* An example of a fully recurrent network is presented in Fig. 1B. In this architecture each cell receives inputs from, and sends its output to all other cells in the network. Unlike the feed-forward architecture, there is no explicit distinction between input, hidden and output units. Rather, in the framework of this architecture, each unit may play a single (input, hidden, or output), dual (input-hidden or hidden-output), or even triple (input-hidden-output) role.

Dynamical Rules

The specification of architecture, model neurons, and connections between them are necessary but not sufficient for the complete definition of network operation. The missing component that must also be provided is the dynamical rule that stipulates how and when the state of each neuron is updated. Once the network is fully specified its operation is to transform the input signals into output ones.

Model Neurons, Connections and Network Dynamics

McCulloch-Pitts Model

One of the first attempts to understand how the brain works can be traced back to Aristotle, the ancient Greek philosopher who lived more than 2000 years ago. However, the first mathematical models of a neuron, the elementary processing unit of the brain, were proposed relatively recently. The model suggested in the 1940s by Warren S. McCulloch, a neurophysiologist, and Walter Pitts, a mathematician, played a particularly critical role, and is often considered as the ultimate ancestor of all artificial neural networks. In the framework of their model (McCulloch and Pitts, 1943), the neuron is a simple *binary threshold element* operating in a *discrete time scale*, t=0, 1, 2,..., spending one time unit per processing step. The basic idea is that each neuron computes a weighted sum of activities of other units that have synaptic connections to it. The neuron then updates its output to either active or inactive state according to whether the sum is above or below a certain threshold. Formally, if at time t the activity of the jth neuron is $V_j(t)$, then the output of neuron i one time step later, $V_i(t+1)$, is given by

$$V_i(t+1) = H\left(\sum_j w_{ij} V_j(t) - \theta_i\right). \tag{1}$$

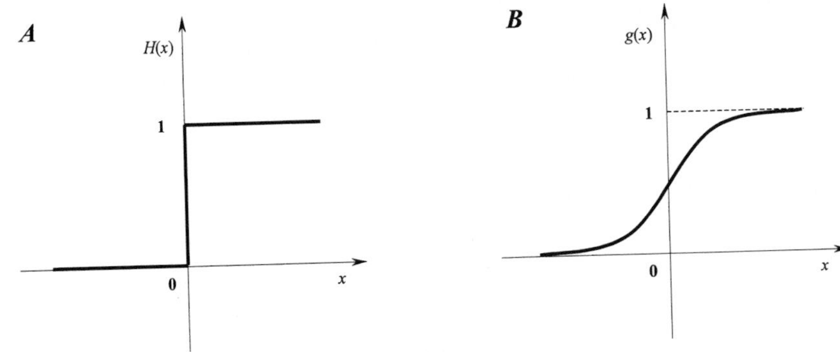

Fig. 2. Activation functions. *A*: The unit (Heaviside) step function. *B*: Sigmoid function.

Here the weight w_{ij} stands for the *strength of the synapse* connecting neuron *j* to neuron *i*. A positive w_{ij} corresponds to an *excitatory synapse* whereas a negative one to an *inhibitory synapse*. If there is no synaptic connection between neuron *j* and *i*, w_{ij} is set to zero. The parameter θ_i is a threshold value of *i*th neuron. The function is the unit (Heaviside) step function:

$$H(x)=\begin{cases}1 & \text{if } x\geq 0;\\ 0 & \text{otherwise}\end{cases}, \tag{2}$$

see Fig. 2A. From Eqs. 1 and 2 it follows that the variable V_i can be either 1 or 0, thus representing the state of neuron *i* as active (V_i=1) or inactive (V_i=0).

Despite the simplicity of a single McCulloch-Pitts neuron, a network assembled from these units is computationally very powerful. McCulloch and Pitts showed that synchronous operation of a sufficiently large number of such units, when the connections between them are suitably chosen, could in principle perform any desired computation.

Hopfield (1982) suggested a similar binary threshold unit model of a neuron. However, the dynamical rule of network operation is different. Unlike the McCulloch-Pitts model, each Hopfield neuron updates its state asynchronously, that is at a random time, independently of any other neuron. Another difference is the interpretation of the binary states. According to Hopfield, the active state corresponds to "firing at maximum rate" whereas inactive to "not firing". In contrast, McCulloch and Pitts interpret the active state as the event of firing of a single spike rather than a prolonged firing.

Leaky Integrator Model

While discrete-time, binary-threshold models are still widely used in the theory of neural computation, they no longer play an active role in the field of computational neuroscience. Most of the accumulated experimental data demonstrate that average firing rate, and not individual spike times, correlate with external stimuli or behavioral variables ("rate code"). These observations justified the modeling of biological neurons as units that respond to their inputs in a continuous fashion by gradually changing their average firing rate. A simple way to incorporate this feature into the binary unit models discussed above is to substitute the step function $H(x)$ in Eq. 1 by a continuous nonlinear function $g(x)$, usually called the *activation function*. Because the firing rate of real neurons is bounded, the activation function usually has a sigmoid form, see Fig. 2B. These types of units are often called *graded response neurons* as opposed to binary units. The commonly used activation functions include $g(x)=1/(1+\exp(-x))$ and $g(x)=\tanh(x)$. The former varies between 0 and 1, whereas the latter between −1 and +1. Mathematical convenience rather than biological reasoning usually dictates the choice of a particular function. In any case, the lower and upper bounds of $g(x)$ may be offset and re-scaled to bring them into correspondence with the lower and upper bounds of the firing range of a given cell.

The graded response neurons, like binary-threshold units, may operate in a discrete time scale updating their states either synchronously or asynchronously. However, there is also a new possibility for operation in a *continuous time scale* (Cohen and Grossberg, 1983; Hopfield, 1984). Namely, all neurons continuously and simultaneously change their states according to dynamical rules given by a set of differential equations. The best-known example of such a model is Hopfield's (1984) network of graded response neurons. The basic idea is to represent each neuron in terms of its highly simplified equivalent electrical circuit. Then the dynamical rules that govern the time evolution of the network of such neurons are given by a set of resistance-capacitance charging equations, which can be transformed to the following system of coupled differential equations:

$$\begin{aligned} \tau_i \frac{du_i}{dt} &= -u_i(t) + \sum_j w_{ij} V_j(t) + u_i^0 \\ V_i(t) &= g(u_i(t)) \end{aligned} \quad . \tag{3}$$

Here, $u_i(t)$ is an internal state variable conceptually representing the cell's membrane potential, whereas $V_i(t)$ corresponds to the output activity of the cell in terms of the average firing rate. The time constant $\tau_i = R_i C_i$ depends on the resistance, R_i, and capacitance, C_i, properties of the cell membrane, and defines the time scale of the network dynamics. The constant u_i^0 is included for generality and represents a fixed external current to neuron i expressed in units of the potential. Neurons obeying dynamical rules described by Eq. 3 are known as *leaky-integrator neurons*. Such a name serves to emphasize the opposite contribution of the first, $-u_i(t)$, and the second, $\sum_j w_{ij} V_j(t)$, term on the right side of Eq. 3. If only the latter would be present, then the neuron would simply integrate its inputs:

$$u_i(t) = u_i(0) + \frac{1}{\tau_i} \int_0^t \sum_j w_{ij} V_j(x) dx. \tag{4}$$

The term $-u_i(t)$ provides a "leakage" of the potential, thus opposing the integration. At steady state, when the increase of the potential is compensated by its leakage, $u_i(t)$ ceases to change so $du_i/dt=0$ for all i. The solution of Eq. 3 in this case gives the steady state output activity of cell i:

$$V_i = g\left(\sum_j w_{ij} V_j + u_i^0 \right). \tag{5}$$

Equation 5 shows that at steady state the relationship between the firing rate of the leaky integrator neuron and its net input is similar to the corresponding relationship for the McCulloch-Pitts neuron (cf. Eq. 1).

Integrate-and-Fire Model

Recent experimental investigations led to a view of neural coding that is quite distinct from the classical one based exclusively on average firing rates. They provide supporting evidence for the "temporal coding" based on the precise timing of single spikes fired by a group of neurons in the same or different cortical areas (Fetz, 1997; Gerstner et al., 1997; and references in them). This controversy concerning neural coding represents one of the hottest topics in current neuroscience research. The adequate neuron model for investigating this issue would be the one that produces action potentials rather than continuously varying firing rate.

The simplest model that generates action potentials can be obtained by coupling the leaky integrator neuron to a firing threshold. The basic idea is to divide the operation of the neuron into two qualitatively distinct modes. First, the model neuron builds up its potential starting from a specific value u_i^{rst}, called the reset potential, by temporally integrating its inputs. This mode is described by an ordinary differential equation similar to Eq. 3:

$$\tau_i \frac{du_i}{dt} = -u_i(t) + U_i(t). \tag{6}$$

Here, $U_i(t)=R_iI_i(t)$, where $I_i(t)$ is the total synaptic current charging the spike emitting part of the cell, soma. Second, once the soma potential $u_i(t)$ reaches a specific threshold value u_i^{thr}, the cell instantaneously fires a spike and resets its potential to u_i^{rst}. After an absolute refractory period, during which the cell cannot fire spikes, the neuron restarts its operation in the first mode. Thus the outcome of the model is the alteration of the prolonged period of integration and instantaneous firing. Note that the action potentials have no structure in this model. Therefore the output train of cell i is completely described by the sequence of times $\{t_i^k, k=1,2,...\}$ at which spikes occur. This model, known in the literature as *integrate-and-fire neuron*, was introduced by Lapicque (1907), a neurophysiologist who first employed it in the calculation of firing times.

In the framework of integrate-and-fire models, there are several approaches to estimate the effective synaptic current $I_i(t)$ charging the soma. The dynamics of $I_i(t)$ depends on a set of synaptic time constants $\{\tau_{ij}^{syn}\}$. Each τ_{ij}^{syn} characterizes the temporal variation of the synaptic conductance of neuron i invoked by arriving spikes fired by neuron i. In the approximation $\tau^{syn} \ll \tau$, i.e., when the characteristic time of the synaptic current changes is much shorter than that of the charging of the soma, the effective total current $I_i(t)$ is represented by a sum of elementary contributions made at the time of arrival of individual spikes (Frolov and Medvedev, 1986; Amit and Tsodyks, 1991):

$$I_i(t) = \tau_i \sum_j w_{ij} \sum_k \delta(t - t_j^k - \Delta_{ij}), \tag{7}$$

where Δ_{ij} is the delay in the arriving time at synapse i of spikes fired by neuron j, $\delta(x)$ is the Dirac delta function. The strength of synaptic connection, w_{ij}, is expressed in units of the current.

Conductance-Based Models

A more realistic approach in representing the effective charging current $I_i(t)$ utilizes a conductance-based model that accounts for a variety of transmembrane ionic currents. In the framework of this approach, Eq. 6 is usually given in the following equivalent form:

$$C_i \frac{du_i}{dt} + I_i^{ion}(t) = 0. \tag{8}$$

Here, $I_i^{ion}(t)$ designates the net transmembrane ionic current, including the leakage. It is assumed that all ionic current flow occurs through membrane channels and the instantaneous voltage-current relationship obeys Ohm's law. The ionic current through channels of a particular type is then given by a linear expression:

$$I_i^{chn}(t) = g^{chn}\left(u_i(t) - E^{chn}\right), \tag{9}$$

whereas the net ionic current $I_i^{ion}(t)$ is a simple sum of the currents through different types of channels: $I_i^{ion}(t) = \sum_{chn} I_i^{chn}(t)$. Here, g^{chn} is the conductance associated with a specific type of channel. The sign of the expression in Eq. 9, which indicates whether the current is outward or inward, depends on whether the membrane potential $u_i(t)$ is above or below the channel reversal potential, E^{chn}. It is usually assumed that E^{chn} does not explicitly depend on time or potential. The known ion channels can be divided into three distinct types: passive or leak, synaptic, and active. Depending on the type of the channel, the corresponding conductance g^{chn} may have a mathematical description that ranges from very simple to very complex. For example, the passive channels are represented by a constant (time- and voltage-independent) conductance. Other channels, such as those located at synapses, change their conductance to certain ions when the appropriate chemical agents (e.g., neurotransmitters or second messengers) bind to their receptors. Because the release of chemical agents is triggered by a presynaptic action potential, the conductance of the synaptic channels is modeled as a time-dependent but voltage-independent function that has a sharp peak at the spike arrival time. The active channels, which are the most complex from a modeling viewpoint, have conductances that are both voltage- and time-dependent. The model neuron that incorporates these types of nonlinear channels may produce membrane responses that mimic not only a subthreshold mode but also the generation of action potentials. Unlike the integrate-and-fire model, in which spikes are unstructured and discontinuous in time, here the spike generation occurs in a continuous-time fashion. Therefore, the model neuron of this type, often referred to as a *biophysical model*, may produce action potentials that have a shape similar to those observed in experiments. The best-known example of such a model, which played a crucial role for the development of biophysics of nerve cells, is Hodgkin and Huxley's (1952) description of initiation and propagation of action potentials in the squid giant axon.

Compartmental Approach and Realistic Modeling

The models we have considered so far are *single-point models* that disregard the underlying spatial structure of the neuron. The application of cable theory to nerve axons (Hodgkin and Rushton, 1946) and dendrites (Rall, 1959), as well as the introduction of a compartmental approach (Rall, 1964), made it possible to develop increasingly realistic models of a single neuron (cf. Chapter 8). Advanced biophysical models of this kind, which are trying to incorporate as much morphological and physiological data as possible, represent a neuron as a set of electrically coupled isopotential compartments. The basic assumption is that the continuously distributed system can be divided into small segments, called compartments. The geometry of compartments is modeled as an ellipsoid (soma) or cylinder (dendrites, axon and its branches) of various sizes. Electrically, each compartment is considered as isopotential and modeled as a resistance-capacitance pair. Adjacent compartments are connected by series resistances. It is assumed that nonuniformity in physical properties of the neuron (i.e., geometry, specific electric characteristics) and differences in potential occur between compartments rather than within them. From a modeling perspective, however, one must be aware that detailed biophysical models incorporate a vast number of adjustable parameters. While these models are adequate for studying in detail the behavior of a single neuron or a small neural circuitry consisting of a few cells, their application to a large-scale network may be inappropriate. Such a network would be so complex that it would be impractical to simulate and analyze its behavior. By choosing a simpler model of a single neuron one may reach a reasonable (in the context of phenomenon of interest) compromise between tractability and realism.

Learning and Generalization

Basics

Any model, including a neural network model, has a set of *adjustable parameters*. Their values are usually determined by bringing the performance of the model into correspondence with a particular set of experimental data related to a phenomenon under study. The performance of a neural network, that is, the relation between the input signals and the activities of output units produced by the network, depends on several factors such as the network architecture, the number of model neurons, the strengths of synaptic connections between them, etc. Traditionally, in the field of neural networks, the adjustable parameters are associated with the strengths of synaptic connections, $\mathbf{w} = \{w_{ij}\}$, whereas other parameters are kept constant. This approach is consonant to numerous experimental observations which indicate that during a relatively short time period, a key mechanism by which biological neuronal networks change their behavior is the modification of the conductivity of preexisting synapses (i.e., modification of strengths of synaptic connections) rather than the variation of the number of neurons or formation of new connections between them (i.e., modification of the network size and architecture, respectively).

A fixed set of connection weights $\mathbf{w}$ corresponds to a specific input-to-output transformation task implemented by the network. As the values of the weights change, the same signals acting on the input units generate different activities of the output units. Therefore, by varying the w_{ij}'s one can implement different transformation tasks. This also means that the network *memorizes* the task that it implements in the set of synaptic connections $\mathbf{w}$. A key question in the theory of neural computation is concerned with the problem of *learning*: "How do we choose the connection weights so the network implements a specific task of interest?" The process of a systematic adjustment of the connection weights, the goal of which is to find a solution to this problem, is called training or learning and is described by a corresponding *learning algorithm.*

Generalization by Induction

The most common approach to network learning, which dates back to the pioneering work on perceptrons (Rosenblatt, 1962), takes the following form. The known examples of a particular transformation task to be learned are divided into two subsets: the training set and the testing set. The former is used to train the network to produce an appropriate output (in terms of the transformation task under consideration) for each example in the set by applying a specific learning algorithm. It is expected that after training, the network would generate correct (or nearly correct) responses to all examples in the training set. Next, one would like to check whether the network indeed has learned the transformation task or whether it has simply memorized examples in the training set. For that purpose, examples from the testing set are presented to the network. If the responses to the novel examples of the same task are correct, then it is said that a *generalization* has taken place. If, however, the number of correct responses is at a chance level, then there is no generalization.

It is the ability of neural networks to generalize by induction that originated much of the excitement about them. As was illustrated in numerous papers, the networks, indeed, could be astonishingly successful at generalization. However, this is not always the case. The key question is then: "Which properties of the networks and the tasks they are trained on determine the success of generalization?" A theoretical framework for generalization is usually formulated in terms of the probability that the trained network generates a correct output for a novel input as a function of the number of examples in the training set (Denker et al., 1987; Carnevali and Pattarnello, 1987; Anshelevich et al., 1989). In the framework of this consideration some general analytical results can be obtained. They provide an estimate of either the average probability (over all possible sets of connection weights that are consistent with the training examples; e.g., Schwartz et al., 1990; Levin et al., 1990) or the worst-case bounds of the probability (among those sets of connection weights that are consistent with the training examples; e.g., Blumer et al., 1989; Baum and Haussler, 1989). These kinds of theoretical analyses are important especially in the field of neural computation as they might provide general guidelines and tips for designing the network architecture optimal not only for generalization, but also for training time, cost of computations, etc. For example, it is widely believed that by limiting the number of units and the number of adjustable connections (thus embedding into the network as much information about the learned task as possible) one can reduce not only computational costs and possibly training time, but also improve generalization ability. It is important to realize, however, that these are merely general recommendations, not strict rules that guarantee a success in training a network on examples of a specific task. In fact, Amirikian and Nishimura (1994) demonstrated that appropriate rules for selecting the optimal architecture can be drastically different depending on what particular task has to be learned. Specifically, the authors addressed the question of how the generalization ability depends on the network size. It turned out that as the number of units in the network considered in Amirikian and Nishimura (1994) increases, the generalization of some tasks worsens whereas generalization of others improves. Therefore, the answer to the question of which network, small or large, is good for generalization will depend on the particular task to be learned.

Type of Learning Paradigms

Two general classes of learning paradigms are commonly distinguished: supervised and unsupervised. *Supervised learning* requires knowledge of correct answers to all examples in the training set. In this approach, which is also known as *learning with a teacher*, a direct comparison of the produced outputs with known answers provides a feedback to the network about any errors. The comparison is done in terms of an *error function* $E(\mathbf{w})$ (sometimes called *cost function* or *objective function*) that tells us how well the network performs on examples from the training set. Although there are many functions suitable for that purpose, there is one that is most commonly used. It is a simple quadratic function of the differences between the produced outputs and corresponding correct answers:

$$E(\mathbf{w}) = \frac{1}{2} \sum_{\mu} \sum_{i} \left(O_i^{\mu}(\mathbf{w}) - A_i^{\mu} \right)^2 \tag{10}$$

Here $O_i^{\mu}(\mathbf{w})$ is the value of the output variable i when example μ is presented to the network with a set of synaptic connections $\mathbf{w}$, and A_i^{μ} is the known correct value of the same variable. It is important to notice that, in the context of computational neuroscience, the output variables O_i^{μ} and their correct values A_i^{μ} given in the answers could be given at the neuronal level (i.e., in terms of activities of the output units) or behavioral level (e.g., direction of motor action). The learning algorithm is an iterative procedure that adjusts synaptic connections based on the feedback error $E(\mathbf{w})$. Its ultimate goal is to find such a set of connections $\mathbf{w}$ that minimizes the error function. Thus, once a particular form of the $E(\mathbf{w})$ is chosen, the issue of supervised learning merely reduces to the solving of an optimization problem.

In practice, information supplied by the correct answers may not always be complete, that is, not all of the correct values of the output variables might be known. In the extreme case there is just a single bit of information specifying whether the output is right or wrong. Because in this case the feedback is only evaluative, it is often called *reinforcement learning* as opposed to *instructed learning*, which specifies what the correct output is.

Sometimes knowledge of correct answers to specific examples is not available at all and the learning goal is not explicitly defined. In such cases *unsupervised learning* algorithms, which do not require any feedback from the environment about the performance, could be useful. In the course of unsupervised learning the network is expected to find out by itself correlations, features, or regularities in the input data and represent them in the output in some appropriate way. In unsupervised learning, changes in synaptic connections are influenced only by local events (i.e., strength of a connection and activities of a pair of units that it links) whereas in supervised learning, due to the global character of the feedback error, the learning algorithm is affected by remote events (i.e., activities of the output units or behavioral consequences of that activity). There are two other points that are noteworthy. First, unsupervised learning can be useful only when examples in the training set are *redundant*. For example, if a network is required to learn similarities between presented examples and categorize them accordingly, the examples must indeed contain similar (though not the same) cases, i.e., be redundant. And second, unsupervised learning may be useful even when supervised learning would be possible. In particular, supervised learning, because it responds to remote events, sometimes could be extremely slow, while unsupervised learning that is, in contrast, affected by local events might be faster.

Supervised Learning

In supervised learning, the error function $E(\mathbf{w})$ can be regarded as a surface in a multidimensional space of connection weights w with a landscape consisting of some wells and hills. Figure 3 illustrates this idea schematically. The goal of a supervised learning algorithm is then to find the deepest well on the *error landscape*. In contrast to all other wells that are called *local minima*, the deepest well is called a *global minimum*. As the number of connection weights in networks is usually large, the search for the global minimum, as a rule, is a hard computational problem. The point is that due to the highly nonlinear character of neural networks, the error landscape usually consists of many local minima. Though there are numerous optimization algorithms that find minima, detecting the absolutely best solution among them could be extremely time-consuming. In many applications, however, the practical difference between a very good solution (i.e., a very deep local minimum) and the absolutely best solution (i.e., the global minimum) is small. Therefore, a trade-off goodness of solution against computational costs makes practical sense.

Concerning the algorithm that finds minima, it could be either a general optimization algorithm or one that is specially dedicated to a specific architecture. *Simulated annealing* (Kirkpatrick et al., 1983) is an example of a general algorithm that is commonly used in supervised learning. It is based on ideas taken from statistical physics. In the framework of this approach, a fixed set of synaptic connections $\mathbf{w}$ is treated as the

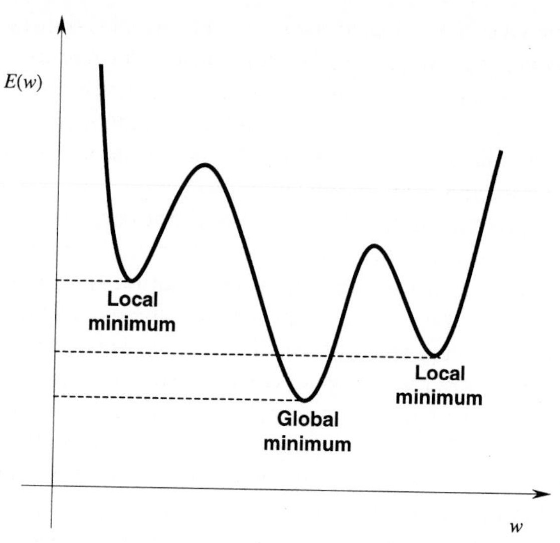

Fig. 3. Schematic illustration of an error landscape. In this case, among three minima only one is global.

"state" of the system while the error function $E(\mathbf{w})$ is treated as the "energy" of the system in that state. In these terms, our goal is to find the "ground state" of the system, that is, the lowest energy state. Search for the ground state is realized by means of an iterative procedure that adjusts synaptic weights. A typical simulated *annealing protocol* can be described as follows. Suppose the network is in a state $\mathbf{w}_1$ with energy $E_1=E(\mathbf{w}_1)$. During each cycle of iteration, a new state $\mathbf{w}_2$ of the system is picked up at random and its energy $E_2=E(\mathbf{w}_2)$ is computed. The energy change between these two states is given by $\Delta E=E_2-E_1$. If the new state has lower energy, i.e., when $\Delta E<0$, then the system unconditionally moves to the new state $\mathbf{w}_2$. If, however, the new state has higher energy, i.e., when $\Delta E>0$, then the system moves to the new state with the probability $\exp(-\Delta E/T)$ and remains in its previous state with the probability $1-\exp(-\Delta E/T)$. (In other words, the algorithm accepts with some probability even those changes in the synaptic connections that worsen performance of the network.) Here the parameter T has a meaning of "temperature". The probability of transition from one state to another is chosen such that the system eventually (after many cycles of iteration) reaches its equilibrium and obeys the Boltzmann distribution at temperature T. The key idea of this method, however, is not to keep the system at a constant temperature but rather gradually cool the system down. The simulated annealing procedure is initialized at a sufficiently high temperature, at which states with higher energies are easily accepted. As the temperature falls, the system state will be more and more likely to be in the lower energy states. If the cooling is slow enough for equilibrium to be established at each temperature, the ground state is reached in the limit of $T=0$. In practical application of simulated annealing, a major challenge is to determine the best *annealing schedule*. If the cooling rate is too fast, the system may be trapped and then freeze out in one of its local minima with relatively higher energy states. If, on the other hand, the annealing schedule is too slow, vast computational resources can be wasted. To achieve reasonable results in practice, a good deal of experimentation should be involved. An exponential schedule $T_k=T_0\exp(-\alpha k)$, where T_0 is an initial temperature, T_k is a temperature on the kth cycle of iteration, and α is a positive scaling constant, could be a good starting point. Another practical issue is how to choose a new state of the system. Although many different schemes are possible, the most commonly used and perhaps the simplest one is to select at random a connection weight and assign to it some new value.

Another well-known and widely used supervised learning algorithm is *back-propagation*. As often happens in science, it was invented independently several times (Werbos, 1974; Parker, 1985; Le Cun 1985). However, the publication by Rumelhart et al. (1986) in the journal *Nature* had perhaps the strongest impact on the field of neural computation. Back-propagation belongs to a broader family of optimization algorithms known as *gradient descent* and is specifically dedicated to networks with feed-forward architecture consisting of graded-response neurons with differentiable activation functions. Later, the back-propagation approach was extended also to recurrent networks (e.g., Pineda, 1987).

The idea of gradient descent is conceptually much simpler than that of simulated annealing. The straightforward way to find a minimum on the error landscape is to check very many points $\mathbf{w}$ in the multi-dimensional space of connection weights and pick up the one that has the lowest value of the error function $E(\mathbf{w})$. In practice, this method is rarely used because the number of points in the space is usually astronomically large and the power of existing computers is far from being enough to solve the problem by this brute-force approach. Instead of exploring the weight space globally, a smarter algorithm would explore it locally, for example, by making small steps in several different directions and selecting the one that goes downhill on the error landscape. Iterative application of this procedure will eventually bring us to one of the nearest (with respect to the starting point) local minima in the error landscape. The gradient descent algorithm is even more clever because it says in what specific direction to move that will make the largest reduction in error at each iteration cycle. The gradient of the error function $E(\mathbf{w})$, often denoted $\nabla E(\mathbf{w})$, is the vector in the multi-dimensional weight space that points in the direction of the steepest ascent on the error landscape at the point $\mathbf{w}$. Correspondingly, the opposite vector $-\nabla E(\mathbf{w})$ points in the direction of the steepest descent. At the point of a minimum or maximum $\nabla E(\mathbf{w}) = 0$. The standard gradient-descent algorithm corresponds to sliding downhill along the direction of the largest decrease of the error function and suggests a simple update rule for the connection weights at the k th iteration cycle:

$$\mathbf{w}^{k+1} = \mathbf{w}^{k} - \eta \nabla E(\mathbf{w}^{k}), \tag{11}$$

Here η is a positive parameter often called *learning rate*. When a minimum is reached, $\nabla E(\mathbf{w}) = 0$ and connection weights $\mathbf{w}$ cease to change. The learning rate η together with the magnitude of $\nabla E(\mathbf{w})$ determines the length of the step along the gradient direction. The choice of η is critical to the operation of the method. If it is too small, the downhill sliding on the error landscape may be unacceptably slow. If it is too large, however, the minimum could be overshot.

Gradient-descent methods may be useful only when the cost of calculation of the gradient vector at each iteration cycle is relatively low. Utilizing architectural constraints of feed-forward networks, back-propagation scheme, as was suggested in Rumelhart et al. (1986), provides an elegant and cost-effective way to calculate all the components of the vector $\nabla E(\mathbf{w})$. The weights $\mathbf{w}$ are updated in a sequential procedure that starts from the connections feeding the top output layer and proceeds down, layer by layer, until the bottom input layer is reached. Thus, while the connections of the network propagate signals forwards (see Fig. 1A), the corrections of the weights caused by the output error $E(\mathbf{w})$ propagate backwards. Hence the name – "error back-propagation" or simply "back-propagation". Although the rediscovery of back-propagation in 1986 provoked an explosion in neural network studies, the algorithm itself is not the ultimate solution to neural network training. Perhaps one of the most serious difficulties is its speed. Despite the fact that calculation of the gradient is relatively fast, in practical applications the rate of convergence toward a minimum is exceedingly slow. Furthermore, like any other gradient-descent algorithm, the standard back-propagation gets stuck in one of the local minima. Therefore, a problem is usually solved many times from random starting weights until a satisfactory solution is found. These and some other difficulties of gradient-descent methods, in general, and back-propagation, in particular, are considered in Anderson (1995). It is worth noticing, however, that back-propagation has been extensively studied in the past decade, and many variations and modifications of the standard algorithm have been suggested to cope with some of these difficulties (see, e.g., Hertz et al., 1991).

Unsupervised Learning

In unsupervised learning there is no teacher. The information available to a synaptic connection during the training phase is local and very limited. Specifically, it knows only its own state, i.e., the strength of connection w_{ij}, and the states of the two neurons that it links, i.e., their activities V_i and V_j. Therefore, the unsupervised learning algorithm, which modifies the strength of synaptic connections, can be expressed, in a very general form, as

$$\frac{dw_{ij}}{dt} = f\left(w_{ij}, V_i, V_j\right), \tag{12}$$

where f is an arbitrary function. In a typical unsupervised learning protocol, an example is drawn from the training set and presented to the network. The synaptic connections are then adjusted according to the learning algorithm given by Eq. 12. This procedure is repeated until all examples from the training set are shown. In practice, the time t in Eq. 12 is a discrete rather than continuous variable and corresponds to an interval in the training phase when an individual example is presented. Accordingly, Eq. 12 takes a form that specifies the changes Δw_{ij}, often called *learning rules*, that occur during that interval.

The choice of learning algorithm f is usually based on intuitively plausible suggestions. Consider a simple case in which the synaptic connection is changing only as a function of its own state:

$$\frac{dw_{ij}}{dt} = -\alpha w_{ij}. \tag{13}$$

Solution of Eq. 13 is given by $w_{ij}(t)=w_0\exp(-\alpha t)$, where w_0 is a value of the synaptic weight at time t=0, and α is a positive parameter. This, in fact, represents the case of an exponential decay of memory in the absence of adjacent neuronal activity or, in other words, a *forgetting*. The rate of this process is determined by the parameter α.

In order to learn something useful, the exponential decay must be overridden by some supportive signal that may come from either of the neurons participating in the synaptic connection. A simple intuition, based on the life experience of "strengthening by use", suggests reinforcing the synapse when the pre-synaptic neuron is active. The main drawback of this approach, however, is its lack of selectivity: the change in the connection will occur whatever the activity of the post-synaptic neuron is. Therefore another intuitive idea, "strengthening by coincidence", seems more appropriate. The first person who explicitly phrased that learning involves the activities of both connected neurons was Hebb (1949). In particular, he hypothesized that concurrent firing in the pre- and post-synaptic neurons strengthens a synaptic connection. This idea, known as *Hebbian learning*, can be formalized by adding a second term in Eq. 13 proportional to the product of the activities of the involved neurons:

$$\frac{dw_{ij}}{dt} = -\alpha w_{ij} + \eta V_i V_j. \tag{14}$$

Here π is a positive parameter that controls, as previously, the learning rate. It is worth noting that most of the unsupervised learning algorithms are based on Hebbian learning or use its various modifications.

So far we have considered networks in which multiple output units can be active together. In the context of some tasks, however, only one output unit should be active at a time. The units compete for being the one to fire, and are therefore often called *winner-take-all units*, whereas the learning algorithm that ensures such a behavior is called *competitive learning* and takes the following form:

$$\frac{dw_{ij}}{dt} = \eta V_i \left(V_j - w_{ij} \right). \tag{15}$$

Interestingly, the competitive learning algorithm given by Eq. 15 can be obtained from Hebbian learning with decay by a formal substitution $\alpha = \eta V_i$ (cf. Eq. 14). At first glance, it is not obvious why Eq. 15 corresponds to a competitive process. The underlying idea is that if the ith unit is a winner, then its activity $V_i \approx 1$. If, however, it is a loser, $V_i \approx 0$. Therefore, the learning algorithm given by Eq. 15 changes connections to the winner and does not change connections to losers.

Finally, unlike supervised learning algorithms, unsupervised learning does not perform any explicit optimization. In some cases, however, there is a well-defined quantity (e.g., *information content* or variance of the output) that is being maximized indirectly while a learning algorithm specified in the form of Eq. 12 is applied (for some examples see Hertz et al., 1991).

Acknowledgement: I would like to thank Apostolos Georgopoulos for critically reading the manuscript and for suggestions. Without Dr. Georgopoulos' help and encouragement this work would have been impossible. This work was supported by the American Legion Brain Sciences Chair and the Department of Veterans' Affairs.

References

Abeles M (1991) Corticonics: Neural circuits of the cerebral cortex. Cambridge University Press, Cambridge New York Port Chester Melbourne Sydney

Amirikian B, Nishimura H (1994) What size network is good for generalization of a specific task of interest? Neural Networks 7:321–329

Amit DJ, Tsodyks MV (1991) Quantitative study of attractor neural network retrieving at low spike rates I: Substrate – spikes, rates and neuronal gain. Netw Comput Neural Syst 2:259–273

Anderson JA (1995) An introduction to neural networks. MIT Press, Cambridge London

Anshelevich VV, Amirikian BR, Lukashin AV, Frank-Kamenetskii MD (1989) On the ability of neural networks to perform generalization by induction. Biol Cybern 61:125–128

Baum EB, Haussler D (1989) What size net gives valid generalization? Neural Comp 1:151–160

Blumer A, Ehrenfeucht A, Haussler D, Warmuth M (1989) Learnability and the Vapnik-Chervonenkis dimension. J ACM 36:929–965

Carnevali P, Patarnello S (1987) Exhaustive thermodynamical analysis of boolean learning networks. Europhys Lett 4:1199–1204

Cohen MA, Grossberg S (1983) Absolute stability of global pattern formation and parallel memory storage by competitive neural networks. IEEE Trans Syst Man Cyber SMC-13:815–826

Denker J, Schwartz D, Wittner B, Solla S, Howard R, Jackel L, Hopfield J (1987) Large automatic learning, rule extraction, and generalization. Complex Syst 1:877–922

Fetz EE (1997) Temporal coding in neural populations? Science 278:1901–1902

Frolov AA, Medvedev AV (1986) Substantiation of the "point approximation" for describing the total electrical activity of the brain with use of a simulation model. Biophys 31:332–337

Gerstner W, Kreiter AK, Markram H, Herz AVM (1997) Neural codes: Firing rates and beyond. Proc Natl Acad Sci USA 94:12740–12741

Hebb DO (1949) The organization of behavior. Wiley, New York
Hertz J, Krogh A, Palmer RG (1991) Introduction to the theory of neural computation. Addison-Wesley, Redwood City
Hodgkin AL, Huxley AF (1952) A quantitative description of membrane current and its application to conduction and excitation in nerve J Physiol Lond 117:500–544
Hodgkin AL, Rushton WAH (1946) The electrical constants of crustacean nerve fiber. Proc Roy Soc Lond B 133:444–479
Hopfield JJ (1982) Neural networks and physical systems with emergent collective computational abilities. Proc Natl Acad Sci USA 79:2554–2558
Hopfield JJ (1984) Neurons with graded response have collective computational properties like those of two-state neurons. Proc Natl Acad Sci USA 81:3088–3092
Kirkpatrick S, Gelatt CD, Vecchi MP (1983) Optimization by simulated annealing. Science 220:671–680
Koch C, Segev I (1989) Methods in neuronal modeling: From synapses to networks. MIT Press, Cambridge London
Lapicque L (1907) Recherches quantitatifs sur l'excitation electrique des nerfs traitee comme une polarizsation. J Physiol Pathol Gen Paris 9:620–635
Le Cun Y (1985) Une procédure d'apprentissage pour réseau à seuil assymétrique. In: Cognitiva 85: A la frontière de l'intelligence artificielle des sciences de la connaissance des neurosciences. CESTA, Paris, pp 599–604
Levin E, Tishby N, Solla S (1990) A statistical approach to learning and generalization in layered neural networks. Proc IEEE 78:1568–1574
Marder E, Abbott LF (1995) Theory in motion. Curr Opin Neurobiol 5:832–840
Marder E, Kopell N, Sigvardt K (1997) How computation aids in understanding biological networks. In: Stein PSG, Grillner S, Selverston AI, Stuart DG (eds) Neurons, networks, and motor behavior. MIT Press, Cambridge London, pp 139–149
McCulloch WS, Pitts W (1943) A logical calculus of ideas immanent in nervous activity. Bull Math Biophys 5:115–133
Parker DB (1985) Learning logic. Technical report TR-47, Center for computational research in economics and magnetic science, MIT, Cambridge
Pineda FJ (1987) Generalization of back-propagation to recurrent neural networks. Phys Rev Lett 59:2229–2232
Rall W (1959) Branching dendritic trees and motoneuron membrane resistivity. Exp Neurol 2:503–532
Rall W (1964) Theoretical significance of dendritic tree for input-output relation. In: Reiss RF (ed) Neural theory and modeling. Stanford University Press, Stanford, pp 73–97
Rosenblatt F (1962) Principles of neurodynamics. Spartan, New York
Rumelhart DE, Hinton GE, Williams RJ (1986) Learning representations by back-propagating errors. Nature 323:533–536
Schwartz DB, Samalam VK, Solla SA, Denker JS (1990) Exhaustive learning. Neural Comp 2:371–382
Werbos P (1974) Beyond regression: new tools for prediction and analysis in the behavioral sciences. Ph.D. thesis, Harvard University

Chapter 26

Acquisition, Processing and Analysis of the Surface Electromyogram

Björn Gerdle, Stefan Karlsson, Scott Day and Mats Djupsjöbacka

Introduction

During muscle activation but prior to contraction and the production of force, small electrical currents are generated by the exchange of ions across muscle fiber membranes. The electric signal generated during muscle activation, often referred to as the myoelectric signal, can be measured through electrodes (conductive elements) applied to the skin surface or inserted into the muscle (cf. Chapter 27). The signal represents the electrical activation of the mechanical system of the muscle fibers and thus the activity preceding the mechanical events. An example of one indwelling recording technique is outlined in Chapter 27. For a complementary overview of other indwelling techniques the reader is referred to (Sanders and Stålberg 1996; Stålberg 1980; Yu and Murray 1984)

The current chapter is restricted to techniques involved with surface EMG recordings. What does the surface electromyogram (EMG) represent? Where does it originate? How does it relate to peripheral motor drive and muscle force (torque)? Why do we record this signal? The aim of this chapter is to address these questions both conceptually and pragmatically. Furthermore, the main emphasis is on providing an overview of the most common acquisition and processing techniques used in surface EMG studies. A comprehensive account of all the methodologies used in surface EMG recordings is beyond the scope of this chapter; however, additional references are provided when concepts, techniques or results are omitted or insufficiently described.

The surface electromyogram is a bioelectric signal that represents, in some filtered form, the aggregate activity of the motor units within a critical distance from the recording electrode. Therefore, the EMG signal also reflects the level of peripheral motor drive to a muscle, but in a manner which is not well understood. Since the level of peripheral motor drive has a direct influence on the force produced by a muscle, the EMG signal reflects muscle force, but again in a poorly defined manner. So if the EMG signal is not capable of accurately predicting the level of peripheral motor drive or muscle force, then what is the motivation for recording the signal? For a start, the direct measurement of the ensemble activity of a motor pool (i.e., peripheral motor drive) is not technically feasible in human studies, and the situation is only slightly better in chronic animal studies. Furthermore, measurement of force or joint torque produced during muscle contraction is only possible for a limited number of applications with the use of

Correspondence to: Björn Gerdle, Department of Rehabilitation Medicine, Faculty of Health Sciences, Linköping, 581 85, Sweden (PHONE: +4613221574; FAX: +4613224465; E-MAIL: Bjorn.Gerdle@rehab.inr.liu.se)

Stefan Karlsson, University Hospital, Department of Biomedical Engineering and Informatics, Umeå, 901 85, Sweden

Scott Day, National Institute for Working Life, Department of Musculoskeletal Research, P.O. Box 7654, Umeå, 90713, Sweden

Mats Djupsjöbacka, National Institute for Working Life, Department of Musculoskeletal Research, P.O. Box 7654, Umeå, 90713, Sweden

functionally restrictive equipment. For virtually all applications, surface EMG measurement is a non-invasive technique capable of providing information on the onset time, duration and relative intensity of muscle activation. By combining surface EMG measurement with other techniques such as peripheral nerve stimulation, it is possible to identify alterations in motor nerve conduction. This is just one example of the growing number of clinical applications where surface EMG is a useful diagnostic tool. In addition, information may be extracted from the raw surface signal through various manipulations; for example, by computing frequency variables of the signal the progression of muscle fatigue can be monitored. Finally, for basic researchers, the primary motivation is to investigate the properties of the surface EMG signal in relation to various physiological variables.

Part 1: Muscle Anatomy and Physiology

Motor Unit Anatomy

Anatomically, each skeletal motor unit (MU) is composed of a motoneuron, motor axon and all of the muscle fibers that it innervates (for a comprehensive review see Burke 1981; Close 1972). Any individual macroscopic skeletal muscle is composed of sets of such MUs. Contraction in skeletal muscles is controlled by the nervous system through the discrete activation of individual members of the MU pool and by varying their activation rates. As such, MUs comprise the functional units of muscle contraction. The number of MUs varies considerably both between muscles for the same subject and across subjects for the same muscle (McComas et al. 1993). The majority of data on MU counts originates from cat research; however, a limited number of human data exist from histological and electrophysiological studies. Table 1 provides estimates of MU counts for various human muscles (see also McComas et al. 1993).

The motor axon, originating from the motoneuron soma in the ventral horn of the spinal cord, begins numerous divisions as it approaches its peripheral target muscle.

Table 1. Estimates of MU counts in various human skeletal muscles. Note that there are inaccuracies with both techniques so the values should be interpreted as rough estimates. (The histological data were taken from Feinstein et al., 1955 and the electrophysiological data were taken from McComas et al., 1993)

Muscle	Technique	Observations	Mean # of Units	SD
Biceps Brachii	electrophysiological	64	111	46
Brachioradialis	histological	2	333	
Extensor digitorum brevis	electrophysiological	58	132	54
External rectus	histological	1	2 970	
First dorsal interosseus	histological	1	119	
First lumbrical	histological	2	96	
Hypothenar	electrophysiological	36	409	183
Medial gastrocnemius	histological	1	579	
Platysma	histological	1	1 096	
Thenar (median nerve)	electrophysiological	64	240	92
Tibialis anterior	electrophysiological	13	252	109
Tibialis anterior	histological	1	445	
Vastus medialis	electrophysiological	24	224	112

Normally, each branch of the motor axon innervates a separate muscle fiber (myocyte) at the motor endplate, a specialized region located midway between the fiber's tendon origin and insertion. The constituent muscle fibers of an MU are distributed over considerable areas and intermingled with the fibers from other MUs. Bodine-Fowler and coworkers (1990) observed, in various hindlimb muscles of the cat, that the absolute cross-sectional areas of MUs varied across muscles, ranging from 16 to 47 mm^2. Although the range is quite wide, the absolute cross-sectional area was typically found to vary in relation to the particular muscle's innervation ratio (i.e., number of muscle fibers comprising a motor unit). That is, the average MU area is often smaller for muscles with low innervation ratios and high for those with high innervation ratios. In most human limb muscles, MU territories are estimated to extend over a diameter of 5 to 10 mm (Buchthal and Schmalbruch 1980). At first glimpse, these data would suggest that human muscles have a lower cross-sectional area than cat muscles; however, the human estimates were obtained electrophysiologically with needle electrodes and are likely to underestimate the total area due to the intrinsic nature of the technique. Hence, the results from human limb muscles are generally consistent with the animal data.

Motor unit fibers do not typically extend over the entire length of the muscle (Smits et al. 1994). Rather, for most limb muscles in humans, the ratio between fiber and muscle belly length is less than 0.5 (Wickiewicz et al. 1983). This data would suggest that the territories of individual MUs are also dispersed throughout the length of the muscle. The axon terminals and motor endplates are normally found in the middle of muscle fibers. Buchthal and colleagues (1955) estimated from electrophysiological evidence in the biceps brachii that an average MU innervation zone is longitudinally spread over a distance roughly equal to 10% of the fiber length. Globally, the number of distinct innervation zones, and the spatial distribution of each within a muscle, has been the subject of some debate. Earlier reports by Coers (1959) and Christensen (1959) indicate that most muscles have a single spatially restricted innervation zone. However, other electrophysiologic reports suggest that the innervation zones are quite diffusely arranged (Eccles and O'Connor 1939; Masuda et al. 1985). Furthermore, other investigators have suggested that in some human muscles compartmentalization is a likely design feature (Windhorst et al. 1989). In support of this concept, ter Haar Romeny et al. (1984) located three distinct groups of MUs in the long head of the biceps brachii. As will become clear at a later stage, the number and distribution of innervation zones influence the shape of the motor unit action potentials (MUAPs) and the aggregate EMG interference pattern. Thus, knowledge about the location of the innervation zone(s) is important when selecting the location of surface electrodes.

Mechanical Properties of the Motor Unit

Skeletal MUs have been classified in terms of their structural, biochemical and physiologic properties (for a comprehensive review see Burke 1981; Close 1972). MUs are commonly classified in terms of their contraction characteristics as either slow twitch (type S) or fast twitch. MUs with fast twitch properties are further subdivided according to their resistance to fatigue as either resistant (FR), intermediate (FI; not as commonly used), or fast fatiguable (FF). There are also two common muscle fiber classification schemes that are based upon fiber histochemical properties (Enoka 1995). In practice, the two muscle fiber schemes are comparable to the MU classification scheme. Type S MUs are usually comprised of type I or slow oxidative (SO) muscle fibers, type FR MUs normally contain type IIa or fast oxidative-glycolytic (FOG) fibers, and type FF MUs are most commonly composed of type IIb or fast glycolytic (FG) fibers. Regardless of the classification scheme, the muscle fibers of individual MUs show considerable variability

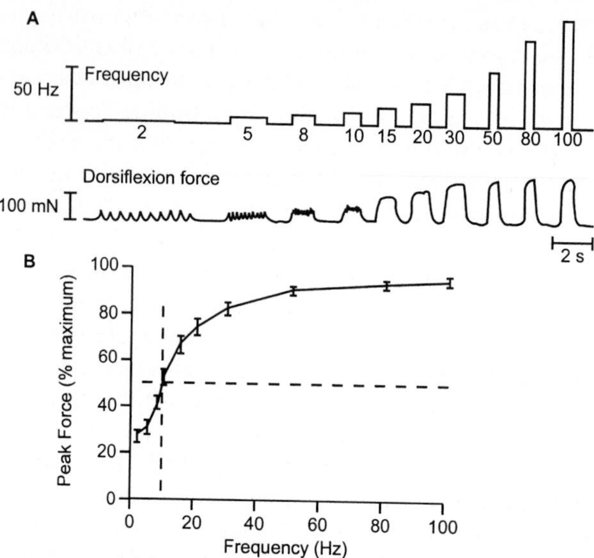

Fig. 1. Relation between firing frequency and mechanical force output of toe-extensor motor units in humans. A) Periodic stimulation at increasing frequencies (upper panel) was introduced via a tungsten micro-electrode producing the force profiles in the second panel. B) Percentage of peak force as a function of increasing stimulation rates, averaged across 13 MUs. Note the distinctive sigmoidal shape of the force – frequency relationship (see also text). [From Enoka, 1993. By Permission.] (This figure was produced from the data of Drs. Vaughn Macefield, Andrew Fuglevand and Brenda Bigland-Ritchie, taken from an unpublished manuscript).

of mechanical properties within each class (Burke and Tsairis 1973). Hence, it may be more accurate to view MUs as a continuum, from those with the slowest twitch and highest fatigue resistant properties to those with the fastest twitch and lowest fatigue resistant properties.

For MUs of each fiber type, the force increases non-linearly with increasing firing rates in an approximately sigmoidal manner (Botterman et al. 1986; Burke et al. 1976; Kernell et al. 1983). Moreover, as the firing rate is progressively increased from 1 s^{-1} the contraction changes from single twitches (lower arm of sigmoid) to unfused tetanus (middle portion featuring approximate linear increases) and finally to fused tetanus (upper arm, force saturation) (Fig. 1).

Generally, the maximum tetanic force produced by an MU at any constant frequency is proportional to the product of its cross-sectional area (cm^2) and the muscle fiber type-specific tension ($kg * cm^{-2}$; Close 1972). The specific tension is a value calculated from the MU's maximum tetanic force divided by the product of the number of fibers and their average cross-sectional area. Within the same fiber type, the specific tension is relatively consistent between muscles and preparations (Close 1972). For example, in the cat, MUs with fast twitch properties (FR or FF) have specific tensions ranging between 2.2 and 3.5 kg^*cm^{-2}, while MUs with slow twitch properties (S) yield values of approximately 0.6 kg^*cm^{-2} (Burke 1981). In mixed muscles, the average size (cross-sectional area) and number of fibers are generally lowest for type S motor units, increasing for both FR and FF units, with the latter revealing the highest values (Bodine et al. 1987; Burke 1981). Nevertheless, there is also a considerable amount of variability in these values for MUs within the same fiber type (Burke 1981). Also, there are reports that there exist sex differences; females having fast twitch (type 2) muscle fibers that are smaller than the slow twitch (type 1) muscle fibers (Gerdle et al. 1988; Mannion et al. 1997; Simoneau and Bouchard 1989; Simoneau et al. 1985). Hence, for a given muscle, the range of tetanic forces produced by the pool of MUs and the overlap in forces between MU types are considerable. For example, measuring twitch force in motor units of the extensor digitorum brevis, Sica and McComas (1971) found that twitch tension varied from 20 to 140 mN, whereas, in the medial gastrocnemius, Garnett and coworkers (1979) observed an extensive range between 15 and 2000 mN.

Electrophysiology of Muscle Fiber Depolarization and Action Potential Propagation

Under most conditions, the activation of a motoneuron results in the activation of all of the muscle fibers that it innervates. Depolarization of the motoneuron axon terminals at the neuro-muscular junction results in the release of acetylcholine (Ach). For a single terminal axon branch, the released Ach binds to motor endplate receptors on the subsynaptic membrane of the muscle fiber, opening channels permeable to cations, in particular Na^+ and K^+. At rest, the intracellular (sarcoplasmatic) space is negatively charged with respect to the extracellular space. However, after channel activation (i.e., opening), the intracellular space becomes more positively charged due to the movement of Na^+ into and K^+ out of the cell. The depolarization causes voltage-gated channels in the endplate surroundings to open, allowing Na^+ to enter the cell and elicit an action potential (AP). This AP then propagates in both directions to the musculo-tendinous junctions. Repolarization of the membrane (i.e., action potential termination) follows quickly in the wake of the depolarization, with the outward movement of K^+ through another set of ion-specific voltage-gated channels. Hence, only one small fiber region on each side of the motor endplate is depolarized at any given moment (see Fig. 2). The localized region of depolarization is referred to as the current sink, while the two neighboring membrane segments comprise two current sources (Rosenfalck 1969). Together, the two current sources and the current sink represent a tripole in a point-source model (Stein and Oguztöreli 1978). The flow of current from each source is in opposite directions parallel to the muscle fiber membrane.

Figure 2 schematically shows the charge distributions inside and outside the muscle fiber during one instant of AP propagation. Because the surrounding tissues are ionic solutions, currents also travel in a radial direction through the extracellular milieu. It is these currents which can be detected as voltage changes at a distance from the muscle fiber (Fuglevand et al. 1992). Figure 2 also depicts the radially directed current fields generated in a point-source model at a given instant in time. If, in such a static, single-instant figure, one tries to imagine how AP propagation occurs, a monopolar electrode can be envisaged as moving along the current field in a direction opposite to AP propagation. As the electrode approaches the leading (positive) edge of the tripole current field, a positive potential is detected. With further movement of the electrode, the amplitude progressively increases, then decreases back to zero at the isopotential region. In a similar manner, movement across the region overlying the current sink results in the measurement of a negative voltage. Crossing the second isopotential region results in the reversal to a positive potential. As in the previous two cases, the magnitude of the potential increases and subsequently decreases to zero. Due to the electrochemical properties of muscle fibers, the first two phases are relatively large and brief, while the trailing phase is small and prolonged. For a more detailed description see the review by Dumitru and DeLisa (1991).

The volume surrounding the muscle fiber provides a medium with a much higher impedance for radial current flow than for current flow along the muscle fibers (Buchthal et al. 1957a; Lindström and Petersen 1983). This impedance is responsible for the spatial dispersion of the current fields and the decline in potential magnitude at increasing distances from the current source (see Fig. 2). Moreover, as the distance increases

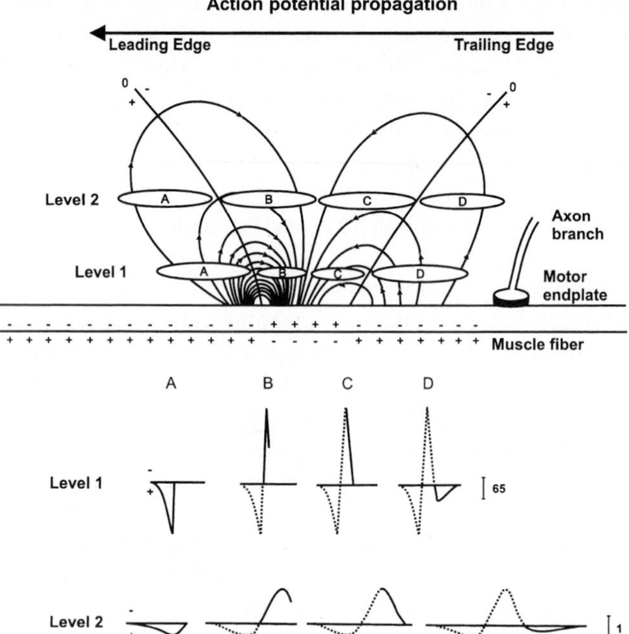

Fig. 2. Schematic diagram of the current fields generated by the propagation of an action potential along a muscle fiber, and the potential differences that would be obtained over difference regions of the current field. Note the charge distribution along the muscle fiber in relation to the current fields, and the changes in potential (A-D) occurring at different current field regions. Also note the reduction in potential magnitude and the increase in potential duration as the distance between the fiber and the measurement location is increased. See text for explanation [Redrawn from AAEM Minimonograph #10: Volume conduction, Dumitru and DeLisa (1988), Muscle & Nerve, Copyright © (1998 Muscle & Nerve). Reprinted by permission of John Wiley & Sons, Inc.]

between the muscle fiber and the recording electrode, there is a steep decline in the amplitude and a modest increase in the duration of the recorded AP (Fuglevand et al. 1992). This phenomenon is commonly referred to as tissue filtering. With increasing electrode-muscle fiber distances the AP will be progressively less distinguishable from the background noise. Intuitively and in practice, the smallest phase will become indistinguishable first. Incidentally, subcutaneous epidermis and adipose layers have greater tissue impedance than muscle, and thus are more effective tissue filters of the bioelectric signal (Basmajian and DeLuca 1985).

Conduction Velocity

The conduction velocity (CV) and amplitude of the muscle fiber AP increase in essentially a linear manner with muscle fiber radius (Håkansson 1956, 1957). These parameters also appear to be dependent on the muscle fiber type (Hanson 1974). Fast twitch fibers are on average larger in radius and have higher conduction velocities than slow twitch muscle fibers. The AP duration is inversely proportional to the CV (Buchthal et al. 1957a; Lindström and Petersen 1983). That is, on average, fast twitch muscle fibers have shorter AP duration. Several other factors linked to fiber type might be associated with differences in conduction velocity. Fast twitch fibers possess a higher ATPase activity, a higher rate of Ca^{2+} release and uptake, shorter potential rise and twitch contraction times and a more developed T-system and sarcoplasmatic reticulum than slow twitch fibers (for references see Gerdle et al. 1991).Using needle electrodes in human biceps brachii, Stålberg (1966) and Buchthal et al. (1955) obtained average and standard deviation CV estimates of 3.69 ± 0.71 and 4.02 ± 0.6 m s^{-1} for groups of muscle fibers. This compares well with the range of conduction velocities obtained for MUs with extracellular electrodes in the mixed (i.e., slow and fast twitch fibers) biceps brachii muscle in humans using a variety of indirect analytic techniques (3.2–5.3, Lynn 1979; 3.23–5.72, Gydikov et al. 1984).

The Composite Motor Unit Action Potential

Normally, activation of the motoneuron results in near-synchronous activation of all muscle fibers that it innervates. It is commonly accepted that the electrical fields generated by the ensemble of muscle fibers comprising an MU summate linearly to produce the composite MUAP. Formally this could be described as the MUAP being a weighted linear sum of all single fiber APs. For example, in the biceps brachii of healthy subjects, Buchthal and colleagues (1954, 1957) observed that the APs recorded are typically biphasic or triphasic with an average duration of 8.7 ms. Since, with the same concentric electrode recording arrangement, the average duration of single-fiber APs is less than 2.5 ms, the longer duration of MUAPs is due to temporal dispersion, arising from the spatial distribution of the muscle fibers and endplates (Buchthal and Schmalbruch 1980; Gootzen et al. 1991). There are also slight differences in the activation timing of an MU's individual muscle fibers, as the length of the terminal axon branches and resultant conduction delay vary between fibers. However, the effect of this temporal dispersion is quite negligible compared to the spatial dispersion of muscle fibers.

Discharge Properties of Motor Units

The MUs of skeletal muscles that are active during intermittent or sustained contractions normally discharge a sequence of APs. This sequence of MUAPs is most commonly characterized either by the firing rate (i.e., number of events over a selected time period) or by a series of inter-pulse intervals (i.e., list of periods between successive MUAPs). Furthermore, the instantaneous firing rate is obtained by inverting the inter-pulse interval (i.e., 1/ inter-pulse interval). The discharge properties are best described as stochastic. Instead of discharging APs with a constant inter-pulse interval (IPI), the timing of successive APs fluctuates as a result of the asynchronous arrival of both excitatory and inhibitory synaptic inputs to the motoneurons from descending, peripheral and segmental sources (Stålberg and Theile 1973). Figure 3 schematically illustrates the difference between a sequence of regularly (Figure 3A) and irregularly occurring events (Figure 3B) for the same mean firing rate. In addition, Figure 3C shows a train of APs recorded from the cat soleus muscle during the activation of a single MU. This exemplifies the variability that is incorporated in a sequence of APs discharged by motoneurons at a given firing rate under conditions of voluntary contraction.

Effect of Electrode Orientation

If a bipolar electrode is located at sufficient distance from an MU's innervation zone, then the recorded AP shapes will depend only on the relative spatial distribution of fibers (see above). This is likely the case for selected electrode locations on a muscle with a small number of highly confined innervation zones. Furthermore, with such a record-

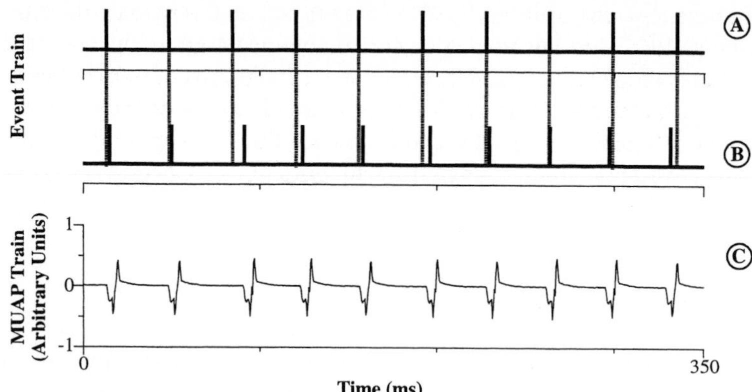

Fig. 3. Trains of rectangular pulses at a mean rate of 28/s, (A) without and (B) with variability in the duration of successive inter-pulse intervals. The variability in inter-pulse intervals is highlighted by the shaded line drawn from panel A to B. The timing of events in panel B is directly reflected in the timing of the MUAP train in panel C. The AP train was recorded from the cat soleus muscle during the isolated stimulation of a small group of α-motor axons. The timing of successive MUAPs reflects the irregularity of MU firing that is observed during voluntary activation. [Data from Day, 1997]

ing arrangement, the MUAPs should exhibit simple triphasic features (Winter 1990). Alternatively, for muscles with widely distributed and non-uniform innervation zones, or for small muscles where the electrode occupies a relatively substantial proportion of the surface area, the bipolar electrode will inevitably be very close to the motor endplate regions of a number of MUs. Moreover, for some MUs the electrode may straddle the motor endplate region, or come within detection distance of the bi-directional propagating APs (Fuglevand et al. 1992). In these cases, the detected AP would be expected to reveal quite complex features, which are specific for the given electrode-MU territory arrangement. On aggregate, a wide dispersion of electrode-MU orientations will result in a diverse ensemble of MUAP shapes, with a considerable number of quite complex waveforms (Loeb and Gans 1986). Although most of the observations are qualitative rather than quantitative, they indicate that attention should be paid when applying surface electrodes (see below). Similar to individual muscle fiber potentials, the magnitude and duration of individual MUs is critically dependent on the distance between the electrode and the MU territory. This is a reoccurring theme, but one which is important to obtain a more complete understanding of what the EMG signal represents.

Voluntary Activation of a Muscle's Motor Unit Pool

Denny-Brown and Pennybaker (1938) introduced the principle of orderly recruitment of MUs according to the force that they produce. Henneman (1957) extended this principle, suggesting that during graded isometric contractions of increasing force, MUs are recruited in an orderly manner that is related to the size of their motoneurons. Although recruitment according to size may be thought of as a general rule, there is indirect evidence that it may not hold for all types of contractions (Glendinning and Enoka 1994; Powers and Rymer 1988). For example, with rapid contractions it is quite possible that the largest MUs are activated at the onset of contraction.

The pattern of MU discharge upon recruitment has been described as either phasic or tonic, depending on whether short bursts (e.g., during gait) or sustained trains of motoneuron APs are generated (e.g., during sustained isometric or isokinetic contractions). For the latter, most studies indicate that the minimum firing or firing rate is between 5 and 7 s^{-1} (De Luca et al. 1982a; Kernell and Sjöholm 1975), although occasionally rates as low as 3 s^{-1} have been observed (Broman et al. 1985b). The highest initial rate of discharge varies among studies, but commonly ranges between 12 (Clamann 1970) and 26 s^{-1} (Kukulka and Clamann 1981); nevertheless, initial rates as high as 35 s^{-1} have been reported (Erim et al. 1996). For a detailed review, the reader is referred to Enoka (1995).

Following stable recruitment of an MU, further increments in synaptic excitation of a motoneuron typically result in an upward modulation of firing rates. This phenomenon is commonly referred to as rate coding or rate modulation. The extent of this modulation is dependent on the temporal association of the recruitment of an MU and increases in contraction intensity (Milner-Brown et al. 1973). That is, MUs recruited at high force levels will demonstrate smaller rate modulation increases than those MUs recruited at low force levels for submaximal contractions (Person and Kudina 1972) (Broman et al. 1985b). Consistent with this observation, lower-threshold MUs exhibit higher maximal firing rates during isometric contractions at submaximal intensities (De Luca et al. 1982b; Monster and Chan 1977). However, for contractions approaching the maximum voluntary level, the firing rates of MUs with quite diverse thresholds tend to converge. Furthermore, higher-threshold MUs may surpass their lower-threshold counterparts in maximum firing rates during maximal contractions (data from Erim et al. 1996; Kernell and Sjöholm 1975). Typically, maximum firing rates appear to vary between muscles and are dependent on the range of contraction intensities investigated. Values ranging from 20 to 35 are most commonly reported from limb muscles in human studies (De Luca et al. 1982b; Erim et al. 1996; Kukulka and Clamann 1981).

Contributions of Recruitment and Rate Modulation

Earlier it was thought that the recruitment of MUs was the primary regulator of muscle force production (Kukulka and Clamann 1981). However, the work of Milner-Brown and colleagues (1973) demonstrated convincingly that, at least for the first dorsal interosseous muscle of the hand, rate modulation plays a more important role in force regulation. For this muscle and other small muscles of the hand, recruitment of MUs occurs over the first 50% of force production in combination with rate modulation, while the final 50% of voluntary force is achieved by rate modulation alone (Kukulka and Clamann 1981; De Luca et al. 1982b). This is in contrast to other muscles, especially large limb muscles, which demonstrate recruitment over the first 70 to 88% of muscle force, with a small fraction of force production left to rate modulation alone (Kukulka and Clamann 1981) (De Luca et al. 1982b).

Motor Pool Activation Statistics

Most reports in the literature support the postulate that MUs discharge independently of one another. Cross-correlation techniques (see Chapter 18) can be used to assess the discharge properties of one MU with respect to another (Milner-Brown et al. 1973). While concomitantly active MUs often reveal increases in the level of synchronization above a value that would be expected with two randomly occurring processes (Milner-Brown et al. 1973; Nordstrom et al. 1992), under normal conditions such synchronization tends to be weak and restricted to relatively short periods (Kirkwood and Sears 1978; Datta and Stephens 1980). It can therefore be generalized that MUs discharge in an asynchronous manner during voluntary contractions.

Interactions Between MUAPs in Generating the EMG Interference Pattern

MUs and their associated innervation zones are spatially distributed within the muscle's geometric territory. The size and shape of each contributing MUAP will therefore depend on its orientation with respect to the recording electrode. Aside from the impact of muscle fiber endplate distribution on MUAP properties (see above), the magnitude and duration of each MUAP is a function of the distance between the electrode(s) and MU territory. Larger distances result in MUAPs that are of lower magnitude and of prolonged duration due to tissue filtering (Basmajian and De Luca 1985; Lindström and Petersen 1983). Therefore, during muscle contraction, each MUAP contributes a different quantity of signal to the EMG. In addition to the spatial dispersion of MUs within a muscle's territory, there is also dispersion in the timing of individual MUAP events. This temporal dispersion is responsible for the interference nature of the EMG signal under conditions of voluntary activation. Examples of the EMG interference pattern recorded during a static contraction and during a sequence of dynamic contractions are illustrated in Figs. 4A and B. To gain an appreciation of the nature of the composite EMG signal, it is necessary to understand how coinciding MUAPs interact. In a recent study, Day (1997) observed that individual MUAPs summate almost algebraically to produce the EMG interference pattern. To illustrate the effect of algebraic summation, let us consider the interaction between two MUAPs, represented as simple sinusoids with the same amplitude and duration. If the two MUAPs are in phase (i.e., begin at exactly the same time), algebraic summation will produce a signal twice the size of each MUAP. In contrast, if one MUAP is phase-shifted 180° with respect to the other, the negative phase of the first MUAP will coincide with the positive phase of the second. Algebraic summation of the negative and positive phases will result in an effective loss of signal. This process is often referred to as signal cancellation.

With asynchronous activation the phase relation between overlapping MUAPs is by definition random. Since the negative and positive phases of each MUAP have an equivalent area, there is an equal chance of signal addition as signal cancellation. Hence, the amount of signal cancellation is determined by the probability of MUAP overlap. In turn, the probability of MUAP overlap is determined by the number of active MUs, and

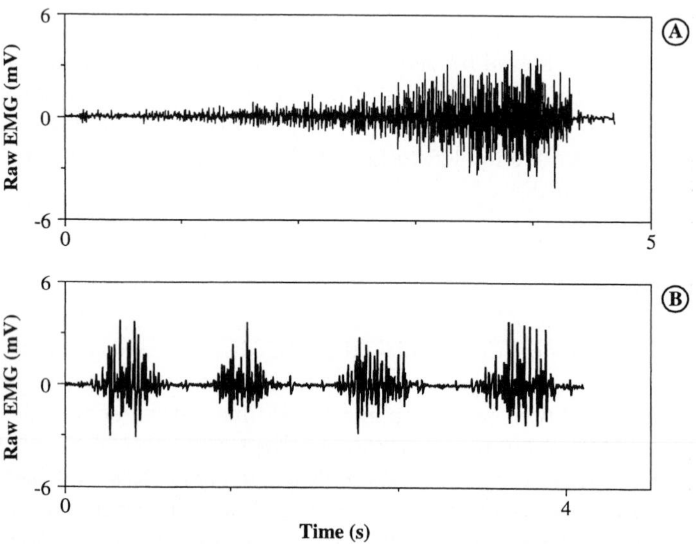

Fig. 4. EMG interference patterns from a static contraction (A) and a sequence of dynamic contractions (B).

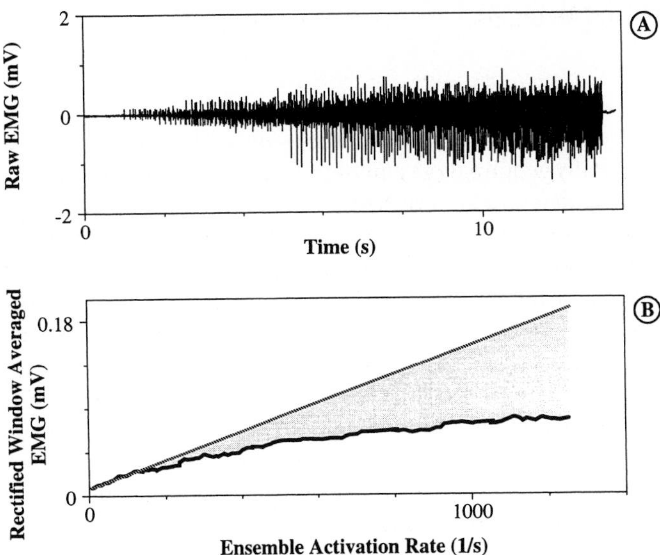

Fig. 5. Raw (A) and quantitative (B) estimates of the surface EMG obtained from the cat soleus muscle during an experimental simulation of muscle activation. A pool of 40 MUs were activated independently and asynchronously with a combined strategy of recruitment and rate modulation, mimicking one such strategy observed under conditions of voluntary activation. Note that the raw surface EMG signal (A) has qualitative features very similar to those in Figure4A. Also note that the rectified, window-averaged signal (AEMG) exhibits marked saturation with increasing levels of ensemble activation (EAR). A tangential line is drawn from the initial slope of the AEMG-EAR relation to highlight the presence of signal cancellation, due to the temporal overlap of signals with positive and negative potentials. [Data taken from Day,1997]

their individual firing rates and MUAP durations (Day 1997). The relative impact of signal cancellation at increasing levels of isometric, asynchronous activation is illustrated in Fig. 5. The raw surface EMG signal, plotted in panel A, was rectified and window-averaged over 100 ms epochs and plotted in panel B against the measured level of motor activation. The level of motor activation is expressed as the ensemble activation rate, which is the simple product of the number of active MUs and the mean MU firing rate. Instead of increasing linearly, the measure of averaged rectified EMG (i.e., absolute area of the signal) reveals appreciable saturation. Furthermore, the shaded area above the line is a qualitative estimate of the amount of signal cancellation with increasing ensemble activation rates. This observation contradicts the common notion that the magnitude of the EMG signal increases linearly with increasing motor drive from the central nervous system. For a more detailed and quantified description of the process of signal cancellation the reader is referred to the paper by Day (1997).

Relationship between EMG and Mechanical Output: The Activation-Contraction Delay

The onset of EMG activity in the muscle precedes the output of mechanical force by around 10 to 100 ms (for references see Zhou 1996). This delay is due to the process of excitation-contraction coupling (for reviews see Rios et al. 1992; and Stephenson et al. 1995), which can vary between MUs of the same or different fiber types. Complicating matters further, the activation-contraction time delay can vary over the duration of the contraction for the same MU. Moreover, temperature changes in the muscle can also alter this delay (Loeb and Gans 1986). The delay should be given due consideration when relating the EMG signal to mechanical measures (such as force, joint torque, joint angle etc.).

Part 2: Signal Acquisition and Materials

General

As a result of the popularity of personal computers and low-cost add-in components, it is now easier than ever to digitally acquire large amounts of data (also see Chapter 45). For example, large quantities of data can be recorded on any mass storage media (e.g., large hard disks, magnetic optic disks, compact disks etc.). Later, data can be quickly retrieved for viewing, and, more importantly, for performing post-run data analysis and signal processing. There are a large number of computer-based numerical methods that can be used to reduce noise, compensate for instrumental artifacts, perform statistical tests, optimize measurement strategies, diagnose measurement difficulties, and decompose complex signals into their component parts. These techniques can often make difficult measurements easier by extracting more information from the available data. Many of these techniques are based on laborious mathematical procedures that were not in practical use before the advent of computerized instrumentation. Generally, measurement systems are composed of three basic components: sensor, processor, and reproducer. Every individual component in any system designed for reliable reproducibility must possess amplitude and phase linearity and also adequate bandwidth.

Most biomedical measurements begin with a sensor (electrode) that converts a measurable physiological quantity, such as ion current in the body, to an electrical signal as shown in Fig. 6. The amplifier is sometimes called signal conditioner that converts

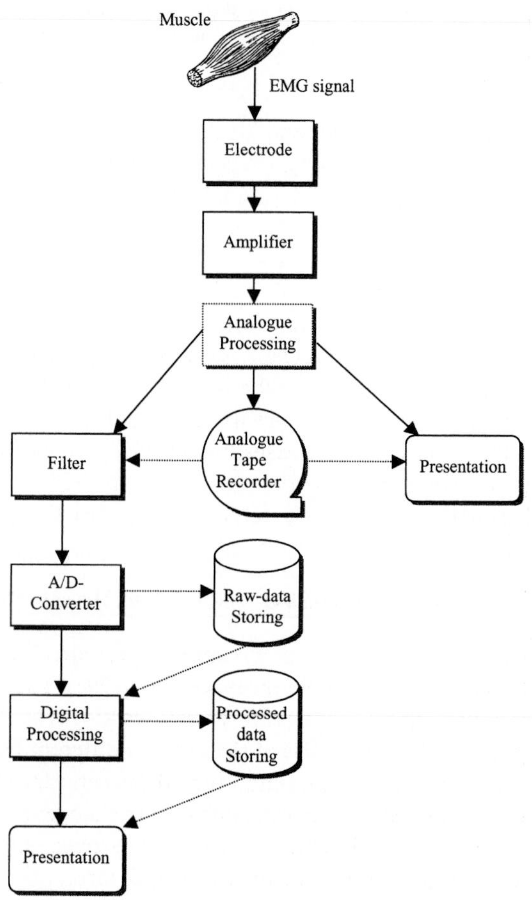

Fig. 6. A schematic drawing of the building blocks of a possible surface EMG acquisition systems.

the measured signal to an appropriate level for an analogue storage (e.g., tape recorder) and/or signal processing. Besides amplification, signal conditioning can include filtering, isolation etc., while signal processing can include analogue-to-digital (A/D) conversion, linearization, parameter calculation and more. The signal could in some applications be temporarily stored on, for example, tape for later digitizing and digital processing. Alternatively, the analogue processing may manipulate the signal to provide the desired type of display, for example, envelope detection of the measured signal (see Part 4: Signal Processing below). Before digitizing, the signal must be passed through a bandpass filter with appropriate bandwidth. This step is necessary to eliminate noise from the signal but also to prevent the generation of spurious signals (aliasing) due to the sampling process (see A/D Converter). If the signal is digitized and processed while the physiological event is occurring, the technique is called on-line processing. An alternative method is to replay the tape-recorded analogue signal or raw-data stored digital signal. Finally, the conditioned, digitized and processed signal can be stored as well as displayed.

Surface Electrodes

In the case of surface EMG, the surface electrode is an essential component of the measurement system, since it has a tremendous impact on the quality of the measured signal. For instance, the signal-to-noise ratio (SNR) is determined almost exclusively by the electrodes, as long as adequate amplifiers and A/D-converters are used. In order to measure the potential difference from biologically generated current fields, it is necessary to provide some interface between the body and the electronic measuring equipment. The current in the body is carried by ions, whereas electrons are the carriers in the electrodes and its lead wires (i.e., cables). Thus, the electrodes serve as a transducer to change an ionic current into an electronic current.

Two kinds of surface electrodes can be distinguished: 1) dry electrodes in direct contact with the skin, and 2) floating electrodes using an electrolytic gel as a chemical interface between the skin and the metallic part of the electrode. Dry electrodes are mainly used in applications where geometry or size of the electrodes does not allow gel. Bar-electrodes and multi-electrodes are typical examples of electrodes where gel is not in use. The electrodes may either be resistively or capacitively coupled and are often made of noble metals. Most popular is a resistively coupled silver electrode.

With floating electrodes, oxidative and reductive chemical reactions take place in the contact region of the metal surface and the gel. The extent of these reactions in the gel progressively declines with increasing distance from the metal surface, which produces a direct current (DC) potential difference known as the half-cell potential (Geddes 1972). The magnitude of this potential is determined by the metal involved, its area, the concentration of its ions in electrolyte, and temperature, as well as other second-order factors such as perspiration accumulation etc. When pairs of electrodes are used, it is far better to use the same material for each electrode, because the half-cell potentials are approximately equal. That means that the net DC potential seen at the input of the amplifier connected to the electrodes is relatively small. This minimizes possible saturation effects as in the case of surface EMG application when using high-gain direct-coupled amplifiers.

The electrode-skin interface generates a similar potential that is mainly caused by a large increase in impedance from the outermost skin layer, consisting of dead skin material. This potential, however, can be minimized by removing this skin layer or at least a part of it, by dry shaving or rubbing with sandpaper and then cleaning with alcohol and ether solution (4:1). The contact impedance is typically reduced by at least a factor of ten by proper preparation (Merletti and Migliorini 1998); moreover, this also reduces the noise generated in the metal-skin junction.

There is also an alternate current (AC) overpotential that is generated by factors such as fluctuations in the impedance between the electrolyte gel and the electrode. One popular method to reduce the effect of this overpotential is to use silver-silver chloride electrodes (Ag-AgCl). This electrode consists of a silver metal surface with an additional thin layer of AgCl composite. The silver chloride layer allows current to pass more freely across the electrolyte-electrode junction, and therefore minimizes the influence of the overpotential. These electrodes are used in nearly 80% of surface EMG applications (Duchene and Goubel 1993). In addition, the Ag-AgCl electrodes introduce less electric noise than the equivalent metallic Ag electrodes.

Electrodes may be constructed as either passive or active units. With passive units, a dry or floating electrode is connected to signal condition electronics at a later stage. Passive units are relatively inexpensive and are of low mass and size, making fixation of the electrode(s) less problematic. In contrast, active units have signal conditioning electronics that are built into the electrode (e.g., small pre-amplifier). As such, active electrodes need an auxiliary source of power. With active electrodes the input impedance is greatly increased, which, in turn, reduces the impact of the impedance between the electrode and skin. Therefore, less environmental noise (i.e., 50 or 60 Hz power line noise) is introduced into the electrode's lead wires.

With modern EMG amplifiers of high input impedance, the electrode-skin impedance becomes less critical. However, one drawback with high input impedance is that power-line noise and movement artifacts are introduced in the lead wires, because of the small inter-wire capacitance. Another method to reduce these effects is to add a pre-amplification stage close to the passive electrodes. The lead wires should be as short as possible, and the pre-amplifier should have high input and low output impedance. This method also offers the possibility to transfer the signal to the main amplifier by various telemetric techniques. Telemetry is beneficial for many applications, such as gait studies, where a freer range of movement is required.

Electrode Configurations

There are many possible configurations of surface electrodes that can be used to record the EMG signal. However, there are three basic electrode configurations:

Monopolar Configuration

The monopolar configuration is used when recording with a single electrode placed on the skin above the muscle with respect to a reference electrode. The reference electrode must be located on an electrically neutral tissue away from the active electrode. This method is used because of its simplicity but the SNR and spatial resolution of the recorded EMG are inferior to EMG records obtained using differential amplification (e.g., bipolar configuration).

Bipolar Configuration

The EMG signal is most commonly recorded using a bipolar electrode configuration. Two detection surfaces (i.e., electrodes), typically spaced 10–20 mm apart in the muscle fiber direction, are used to detect two potentials on the skin above the muscle, each with respect to a reference electrode. The exact location, spacing, and orientation of those two leads are critical because they dictate which gradients of local potential the electrodes will detect (see Electrode Placement below). Bipolar electrode arrangements are used with a differential amplifier, which functions to suppress signals common to both electrodes. In essence, differential amplification subtracts the potential at one electrode from that at the other electrode and then amplifies the difference. Correlated signals common to both sites, such as distant AC signals from power cords and electrical devices, but also EMG signals from more distant muscles, will be suppressed, whereas signals from the muscle tissue close to the electrodes will not be correlated and therefore amplified. Consequently, bipolar recordings offer the advantage of increased spatial resolution and improved SNR. Moreover, DC components such as half-cell potentials at the electrode-electrolyte junction will be detected with similar amplitude in both electrodes and therefore also be suppressed. The selection of the reference electrode will also be less critical as to choice of material, since the effect of the half-cell potential will be eliminated by the bipolar configuration. Unfortunately, bipolar configurations band-pass-filter the recorded signal. The filtering effect produced is dependent on the inter-electrode distance: decreasing the inter-electrode distance shifts the EMG bandwidth to higher frequencies and lowers the amplitude of the signal. For a more detailed exposition see Lindström and Magnusson (1977).

Multipolar Configuration

Recent electrode developments, i.e., the advent of multipolar or multielectrode systems, allow analysis of a spatial two-dimensional EMG signal topography, as well as determination of muscle fiber orientation, conduction velocity, and motor-point localization (for review see Ramaekers et al. 1993).

Functional Electrode Classification

There are many different types of electrodes available commercially. An inventory of commonly used surface electrodes has been performed by SENIAM (Surface EMG for the Non-Invasive Assessment of Muscles, a concerted action, funded by the European Commissions' Biomed 2 Research Program) (Hermens and Freriks 1998). Unfortunately, it is common to use the term electrode when defining both the properties of the conductive detection sites, and the number of detection sites the electrode comprises. Individually, electrodes may be described in terms of their composite material, shape, size, surface structure, mechanical electrode construction etc. On aggregate, electrodes are often described by the number and arrangement of the detection's sites, and, if applicable, any additional signal conditioning equipment. Functionally, electrodes may be classified by the nature of the information sought and the muscle examined. One functional classification scheme distinguishes between:
- Electrodes used to assess the global activity from a muscle;
- Electrodes used to isolate single MU activity.

Bipolar electrodes (i.e., configurations), differing in electrode sizes, shapes and inter-electrode distances, primarily occupy the first class; whereas the second class consists of more complex multi-electrode systems, consisting of one- or two-dimensional arrays of electrodes.

Electrode Placement

Before electrodes can be placed, the skin needs to be properly prepared by dry shaving, rubbing with sandpaper and cleaning with an alcohol and ether solution (4:1). The recording electrodes, regardless of whether monopolar, bipolar or multipolar configurations are used, should be placed on the longitudinal midline of the muscle between the zone of muscle fiber innervation (motor point) and the tendon. For optimal results, using bipolar configuration, the two detection surfaces of the electrodes should be orientated in the direction of the muscle fibers (i.e., parallel to the muscle fibers). As much as 50% less amplitude will be recorded if the electrodes are perpendicular rather then parallel to the muscle fibers (Vigreux et al. 1979). Also, the frequency contents of the recorded EMG signal will be effected since the spectrum reflects underlying physiological changes in MUAP trains (Hogrel et al. 1998; Karmen and Caldwell 1996). Determining the muscle fiber orientation may be difficult in practice, but every attempt should be made to do so. Moreover, the center-to-center electrode distance (10–20 mm) should be fixed between recordings, so that quantitative comparisons can be made.

A reference electrode is necessary for providing a common reference to the input(s) of the amplifiers and should be placed close to the detection point(s), preferably over an electrically neutral tissue (Zipp 1982). However, in the bipolar configuration, placing the ground electrode over an active point will not generally affect the recorded signal, unless the amplifier's common mode rejection ratio is poor. If a common ground electrode is used for several electrode detection points, the ground electrode should be placed as far away as possible from these points on an electrically neutral tissue (e.g., the wrist). If the surface EMG signal is recorded on shoulder-, neck- or back-muscles and is contaminated with electrocardiogram (ECG) artifacts, the processus spinosus of C7 is a suitable alternative reference site (Hermens and Freriks 1998). The size of the reference electrode should be relatively large to minimize electrode-skin impedance. Another way to minimize impedance is to use a so-called lip clip electrode. This electrode is 'clipped' to the lip of the subject so that the electrode surface is in contact with the oral mucosa. This arrangement will give a low impedance and may reduce noise in the recording (Turker et al. 1988).

The electrode-skin contact must be fixed to limit the risk of movements of the electrodes over the skin as well as minimize risk of pulling cables (Hermens and Freriks 1998). Inadequate fixation can cause common mode disturbance in amplifiers and movement artifacts, which will effectively reduce the SNR of the recorded EMG.

Amplifier and Filters

The electrode-detected amplitude of the surface EMG signals can range from 0 to 6 mV (peak to peak) or 0 to 1.5 mV (root-mean-square value: RMS) (Basmajian and DeLuca 1985). Before processing, the signal must be amplified to appropriate levels to optimize the resolution of the recording or digitizing equipment (see A/D Converter below). In addition, the EMG signal should be bandpass-filtered to remove movement artifacts and other noise. Typical bandpass frequency ranges are from between 10 and 20 Hz (high-pass filtering) to between 500 and 1000 Hz (low-pass filtering). High-pass filtering is necessary because movement artifacts contain mostly low-frequency components (i.e., less than 20 Hz), while low-pass filtering is desirable to remove high-frequency noise not originating from the EMG signal and to avoid aliasing (see below).

High-quality EMG amplifiers have adjustable gains of typically between 100 and 10,000 to maximize the SNR of the measured EMG signal during each recording. The gain should be set as high as possible without clipping peaks or distorting the recorded EMG signal. The input impedance should be high in the TΩ range, in parallel with 5 pF capacitance (Basmajian and DeLuca 1985), to minimize the effects of unbalance or extremely high electrode impedance. The input range should be linear in the detected EMG signal range, e.g., ±5–10 mV, while the output range should typically in the ±5 V range, depending on the recording equipment range. The common-mode rejection ratio (CMMR) for a differential amplifier should be higher then 100 dB. The input noise should be less than 4 μV (RMS). The noise is measured by short-circuiting the detection surfaces to the reference and then measuring the root mean squared value of the output noise with respect to the gain.

A/D Converter

Perhaps the most convenient means of recording the signal is via a digital computer. In this case, the signal must be converted from its natural analogue form to a digital representation. In order to capture the full range of frequency components in the EMG signal, digital sampling should be conducted at an adequate sampling rate (see Chapter 45). According to the Nyquist sampling theorem, the minimum sampling rate should be twice the signal bandwidth (Oppenheim and Schafer 1989). That is, if the EMG signal contains frequency components up to 500 Hz, the required sampling rate must exceed 1 kHz. Generally, it is desirable to sample at a sampling rate higher than the minimum: 2 kHz is commonly accepted as a sufficient rate for most surface EMG applications. If the sampled signal (including noise) contains frequency components that are higher than half the sampling rate, aliasing (folding) will occur, such that high-frequency components are reflected onto the lower frequencies of biological relevance. For example, if the signal bandwidth includes frequency components up to 600 Hz and the sampling rate is 1,000 Hz, then the highest frequency component (i.e., 600 Hz) will be represented by the 400 Hz component in the digitized signal. To avoid aliasing problems when digitizing the signal, an analogue low-pass filter, also called anti-aliasing filter, must be used. In addition, the signal amplitude should be set by the amplifier to an appropriate level to minimize the quantization error. A quantization error is due to the difference between the original signal value and the digitized value. In other words, adjusting the gain as large as possible without clipping the peaks of the EMG signal will optimize the SNR with respect to the quantization error.

Storage

The digitized surface EMG signal can be stored as raw-data and/or as processed data. The choice as to how to store data depends on the need of real-time presentation of calculated parameters, the storage capacity and post-processing requirements of the recording system. The reader is referred to Chapter 45 for more detailed information regarding signal acquisition.

Part 3: Registration Procedures

The International Society of Electrophysiology and Kinesiology (ISEK) has recently set up a group with the goal of giving recommendations about how the registration should be done and reported. Important work within this field is also currently performed within SENIAM, see Appendix (this chapter).

Registrations of EMG

1. Locate appropriate places for electrode placements. Generally the surface electrodes should be placed between the innervation zone and the tendon on the longitudinal midline of the muscle. The reader is strongly recommended to consult the report from SENIAM, the state of the art on sensors and sensor placement procedures for surface ElectroMyoGraphy: a proposal for sensor placement procedures (Eds: Hermens and Freriks 1998), for professional advice on how to place the electrodes with respect to a certain muscle.
2. The selected skin area must then be prepared:
   - dry shaved,
   - rubbed with alcohol and ether (4:1) solution and/or slight skin abrasion using sandpaper.

3. If floating electrodes are used, apply gel above the electrode material of the surface electrodes. Avoid air bubbles in the gel.
4. Attach the surface electrodes with 10–20 mm center-to-center distance along the muscle fibers of the muscle under investigation. Use the same distance for all electrode pairs and for all recording sessions in a study.
5. Attach the ground electrode outside the area of investigation, preferable over bony parts.
6. Connect the electrodes with the amplifier and data acquisition equipment.
7. Check skin impedance with the purpose of achieving balance between the electrodes. This can be done either using a special impedance meter or, if present, using the common-mode test of the amplifier. This has to be done by measuring the impedance between each electrode and the reference.
8. It can be most valuable to take a photograph of the attachment of the electrodes together with some anatomical landmarks for documentation of exact placement in the particular measurement.
9. Make static test contractions and adjust the gain (must be individually adjusted) to ensure that good electrode-skin contact exists and that RMS noise levels are as low as possible.
10. Register the background noise in the recording while the muscle of interest is at rest to obtain data for noise level compensation of amplitude variables. This is especially important for measurements of low-level contractions.
11. Make appropriate test contractions to enable normalization of amplitude variables.
12. Perform the measurements.

Part 4: Signal Processing

Signal Analysis of Surface EMG

The purpose of processing the EMG signal is to extract useful information. The EMG signal recorded from the skin surface is a composite of both underlying physiological excitation of the skeletal muscle and the acquisition system.

However, it must be noted that the number of active muscle fibers, muscle fiber length and diameter, electrode type, location relative to motor points, distance (tissue thickness) and orientation towards active muscle fibers are of great importance as to the amplitude and frequency spectrum of the signal. Moreover, temperature and long-term changes in the electrode-skin interface contribute to the recorded signal properties.

At any time instant, the surface EMG signal, detected by the electrode, consists of two classes of linear summations: 1) APs from individual muscle fibers innervated by a single active motoneuron, and 2) all active MUs.

The APs of all muscle fibers in the same MU almost fire simultaneously. The shape of one individual MUAP detected on the skin is therefore to some extent affected by motor endplate spatial arrangement, but also by different depths of fibers within the muscle and different distances from the detection electrode (Karmen and Caldwell 1996). Each fiber AP contributes to the amplitude and frequency contents of the characteristics of the MUAP trains. Fibers close to the electrode make a greater contribution to the amplitude and frequency contents (see below and Fig. 7a). Individual MUAPs and MUAPs evoked by electrical stimulation (compound APs or M-wave; see Part 7: Special Applications below) could be regarded as quasi-deterministic. However, when the number of irregularly firing MUs increases so that the MUAPs overlap, the myoelectric signal is well described as a Gaussian stochastic process with a zero mean (Basmajian and DeLu-

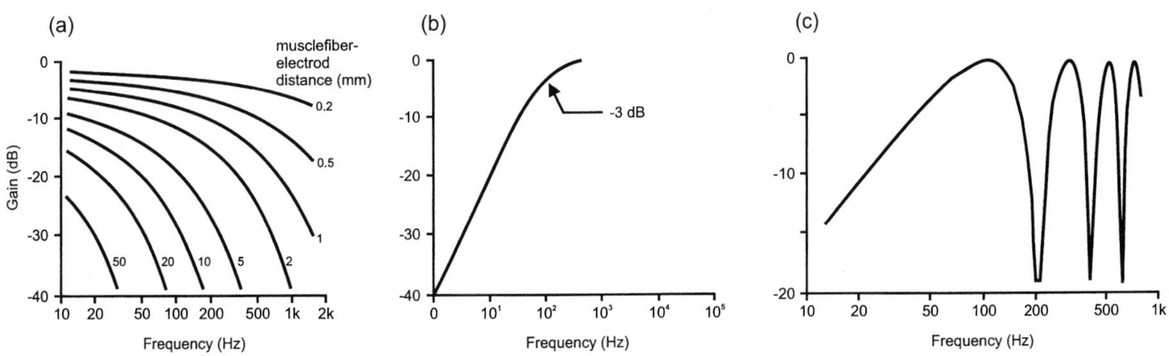

Fig. 7. Filtering effects caused by a) tissue, b) electrode-skin, c) bipolar electrode configuration. [From a, c) Lindstöm and Magnusson (Lindström and Magnusson 1977) By Permission (©1999 IEEE) , b) Basmajian and DeLuca (Basmajian and DeLuca 1985) By Permission from Lippincott Williams & Wilkins]

ca 1985). Stochastic signals can be stationary or nonstationary and are best characterized in statistical terms. A non-stationary EMG signal contains frequency components varying with amplitude throughout the frequency spectrum.

Note: In practice, there exist a less stringent condition that is called wide-sense stationary or quasi-stationary, if mean is constant (i.e. independent of time) and its autocorrelation depends only on the time indices difference.

The frequency spectrum ranges from DC to approximately 500 Hz, with the dominant energy being in the 50–150 Hz range. However, because of a series of filtering effects caused by body tissue, skin-electrolyte-electrode interface, electrode configuration and amplifiers, the measured EMG signal bandwidth will be effected. Muscle fibers, subcutaneous fat and skin are anisotropic and act as a spatial filter with low-pass filtering properties, where an increasing distance between the muscle fiber and the electrode increases the filtering effect (Fig. 7 a) (Lindström and Magnusson 1977). This means that higher frequency components are attenuated more when the electrode-to-muscle fiber distance increases. The transmission between skin end electrode detection surface acts as a high-pass filter (Fig. 7 b) (Basmajian and DeLuca 1985). If bipolar electrode configurations in a differential amplifier arrangement are used, a smaller distance between the electrodes shifts the bandwidth to higher frequencies and lowers amplitude of the signal (Fig. 7 c) (Lindström and Magnusson 1977). Finally the amplifier typically limits the EMG signal spectrum to 10–500 Hz. The amplitude of the surface EMG signal can typically range from 0 to 6 mV (peak-to-peak) or 0 to 1.5 mV (RMS) (Basmajian and DeLuca 1985).

There are numerous variables used in the literature to describe the EMG signal. However, here we only present the most commonly used. These classical surface EMG variables can be divided into two main groups; amplitude and frequency variables.

It is important to note that extraction of some variables can be performed with hardware prior to A/D conversion but also digitally after sampling. Digital processing of the EMG signals can be performed during the recording for real-time presentation or as post-processing of digitally stored raw EMG signals. Moreover, also note that the variables can be presented/calculated during non-overlapping time intervals or segments of the EMG signal or continuously as a running variable, e.g., as a moving average. Calculating variables for segments gives the advantage of data reduction.

Amplitude Variables

The raw (unprocessed) EMG signal has long been analyzed by visual inspection on the oscilloscope or on a paper printout (cf. Fig. 4). This simple approach is often used in kinesiological studies to identify the on-off or active-inactive transitions of muscle contractions.

To quantify the amplitude, simple averaging will only yield zero, therefore more sophisticated analysis methods are needed for correct interpretation of the time-varying muscle activities. Integrated EMG (IEMG) was one of the earliest methods (Inman et al. 1952). These authors used a linear envelope detector to follow the fully rectified EMG signal. Note that the term integration is not correctly used since it has a well-defined mathematical meaning (see below). A more correct term is "envelope detector". Methods that have been used are rectification (half or full), linear envelope, integration, root-mean-square (for reviews see Duchene and Goubel, 1993 and Merletti and Lo Conte, 1995):

Rectification Since averaging of the EMG signal gives zero, the simplest method is rectification that translates the bipolar EMG signal to single polarity. The rectification could be half or full. Half rectification removes the negative values whereas full rectification inverts the negative values and therefore preserves the signal energy.

Average Rectified Value (ARV) The rectified signal still varies randomly, but by averaging (smoothing or low-pass filtering) the rectified signal, its fluctuations can be suppressed. The ARV is the area (or integral) between the rectified signal and the time axis computed during a time interval τ and divided by τ. The technique to move a time window τ over the rectified signal is called moving average.

Integrated EMG (IEMG) The integrated EMG is calculated during a predefined time window, τ. This method is a subset of the ARV method where only a division by τ is removed. Since the rectified values are always positive and not normalized to the time window, the integrated value will increase with increasing time window. Note that this is not the linear envelope detector described above.

Root-Mean-Square (RMS) The RMS value measures the power of the signal (more correct the root thereof) and therefore provides more information than the previously described methods, since it has a clearer physical meaning. It is recommended by Basmajian and DeLuca (1985) above the other methods and is preferred in most applications. The RMS is the area between the squared signal and the time axis computed during a time interval τ and divided by τ and the square root thereof. Note that the RMS value and the square thereof give an estimate of the standard deviation and variance, respectively, for the surface EMG signal. This variable could also be presented as a running RMS value by letting the window τ glide over the analyzed signal.

Frequency Variables

The first step towards the computation of the spectral (frequency) variables is the estimation of the power spectral density (PSD) function of the signal (see Chapter 18). The most widely used method for estimating the frequency spectrum of the surface EMG signal is the Fourier transform, but also parametric (auto-regressive) methods have been used. The Fourier transform decomposes a waveform (signal) into sinusoids of different frequency sinusoids. It identifies or distinguishes the different frequency sinusoids and their respective amplitudes. The sum of all sinusoids with their corresponding amplitudes gives the original waveform. One major explanation for the popularity of the Fourier technique is the development of the fast Fourier transform (FFT) algorithm by Cooley and Tukey (1965), which drastically reduces the number of computations.

During a sustained voluntary contraction, the EMG signal is non-stationary simply due to the AP summation of irregular discharges of active MUs. However, during relatively low-level (20–30% of maximal voluntary contraction: MVC) and short contractions, the EMG signal may be assumed to be wide-sense stationary, i.e., locally stationary or quasi-stationary. For higher levels of MVC the local stationarity holds for shorter segments. This leads to the concept of time-dependent spectra, which, in effect, amounts to subdividing the signal into segments, during which the wide-sense stationarity hypothesis holds, and then computing the spectra for each segment. Commonly used segment durations range from 0.5 to 2 s, where the shortest segment is used for highly non-stationary signals during intense contractions (Merletti et al. 1992). During dynamic contractions, spectral analysis must be handled with great care, because there is a change in the number of active MU, in the geometric relation between the active muscle fibers and the electrode, in the geometric relation and the innervation zone, in the muscle fiber length etc. All these factors greatly increase the non-stationarity of the EMG signal. Thus, Fourier and other classical signal processing methods may not be appropriate for the analyses. Therefore, time-frequency methods which do not require stationarity of the signal have been introduced in EMG analysis.

Several methods have been used to characterize the frequency content of the EMG signal, such as peak amplitude, spectral bandwidth, partitioning the spectrum into frequency bands or calculation of statistical measures such as the mean or median frequency of the power density spectrum. Sometimes also higher-order spectral moments like skewness and kurtosis are studied (Merletti et al. 1995).

Peak Amplitude (PA)

The peak amplitude of the PSD is defined as the dominant frequency component of the EMG signal. However, due to the stochastic behavior of the EMG signal the peak must be estimated. The idealized version of a typical EMG signal spectrum is shown in Fig. 8, where the peak amplitude is designated PA.

Spectral Bandwidth

The spectral bandwidth of the PSD is defined at the 3 dB (log-scale) or the 0.5 (linear-scale) amplitude of a normalized spectrum (Fig. 8).

Median Frequency (MDF)

The median frequency of the PSD is defined as the frequency that divides the PSD into two areas with equal energy (Fig. 8).

Mean Frequency (MNF)

The mean frequency or the weighted average frequency of the PSD is statistically also commonly termed normalised first spectral moment (Fig. 8). In the literature it has earlier been also labeled fmean and MPF.

Zero Crossing (ZC) and Turns Counting (TC)

In contrast to the other frequency variables, zero crossing and turns counting are not based on the PSD, but are defined as half the number of times the raw EMG signal crosses the zero level or reaches a peak, respectively, within a given time interval. This measure is used as a frequency variable in the assessment of muscle fatigue (Hägg 1991). For contractions of low force level the relationship between ZC or TC and the number of MUAPs is linear (Lindström and Magnusson 1977). However, when the contraction level increases and therefore also the number of active MUs increases, the EMG signal takes the shape of random, band-limited Gaussian noise, and the linear relationship does not longer exist.

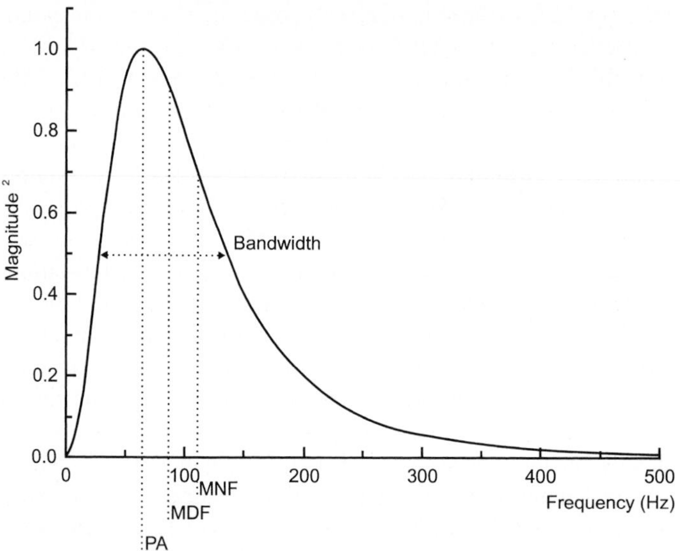

Fig. 8. Schematic drawing of the frequency spectra of a surface EMG signal.

Normalization of EMG Variables

The recorded EMG amplitude is never absolute, which is mainly due to the fact that the impedance between the active muscle fibers and the electrodes varies and is unknown. The impedance also varies between measurements within the same subject (as soon as electrodes are removed and replaced) and naturally also between subjects.

Thus, when comparing amplitude variables between measurements, some kind of normalization has to be made. Normalizing the signal amplitude with respect to force/torque for each muscle is a commonly used technique; typically a maximal contraction or a submaximal contraction at a known force level is used. However, this may not be possible in some applications, thus several other methods of normalization have been proposed (for further reading see Winkel et al., 1995); see also Part 5: Results of the present chapter).

Also, when the measurements involve low-level contractions, it is advisable to compensate the amplitude variables for the measurement noise. This can be done by recording EMG when the muscle under investigation is totally relaxed. In the case of the RMS variable, the square of the variable should be reduced by the square of the noise RMS, and the root of the difference is the compensated value.

During studies of muscle fatigue, regression techniques have often been used to monitor the variations in time of both amplitude and frequency variables. Commonly, the values are normalized using the intercept of the regression (Fig. 9). For a review of such techniques see Merletti et al. (1995). These authors also presented a new index for the quantification of fatigue (the area ratio index) (Merletti et al. 1991). Note that normalization by dividing the values of the variables with an initial (single) estimate is not advisable since all data will then be dependent on the reliability of the initial value.

Figure 9 shows a number of normalized EMG signal variables during a 20 s isometric contraction of the human tibialis anterior muscle: A) stimulated at 40 Hz and B) voluntary contraction at 80% of MVC. Note the much smaller fluctuation of the variables during the stimulated contractions. The fluctuation in the voluntarily performed contractions is due to the irregular firing of MUs.

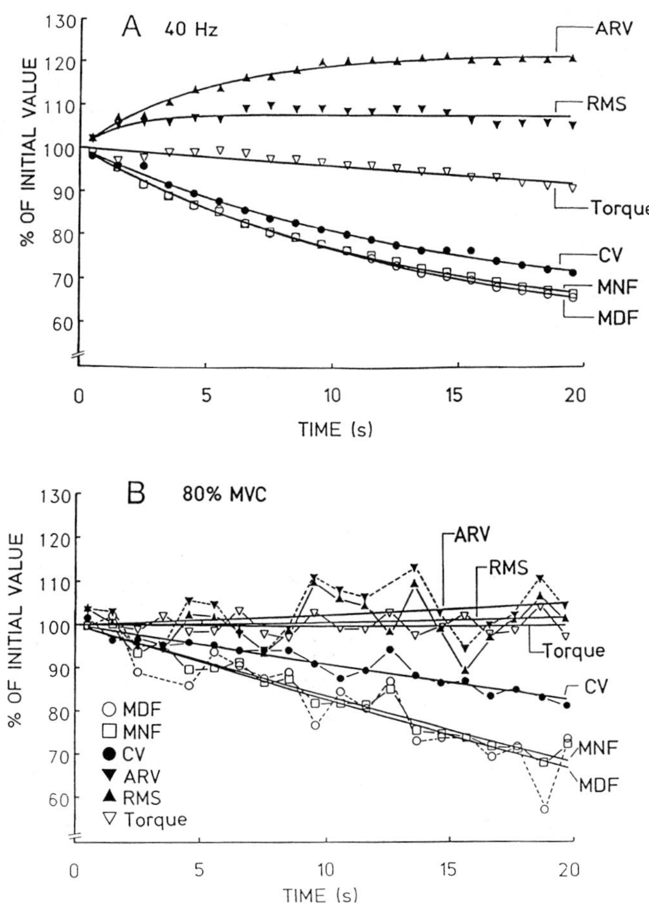

Fig. 9. Normalized values of ARV, RMS, MNF, MDF, torque and CV (average muscle fiber conduction velocity; AMFCV) during a 40Hz electrically evoked contraction (A) and a voluntary contraction at 80% of MVC (B) of the tibialis anterior muscle in the same healthy subject. [From Merletti et al., 1992. By Permission.]

Aspects of Reproducibility

Since surface EMG suffers from significant variability, reproducibility (reliability) should not be taken for granted. However, EMG registrations during static contractions are generally considered as reproducible (see also Aarås et al. 1996; Bilodeau et al. 1994; Ng and Richardson 1996; Viitasalo and Komi 1975; Viitasalo et al. 1980; Yang and Winter 1983 for references). Good reproducibility of frequency variables has been reported for static contractions recorded during-the-day and between-days (Daanen et al. 1990; Viitasalo and Komi 1975).

Interpretating EMG obtained with dynamic contractions might be more difficult – especially for frequency spectrum variables – because the movement per se introduces additional factors that might affect spectrum characteristics, for instance, changes in the geometrical relations between the active MUs and the electrodes and changes in muscle fiber length. From static contractions it is known that such factors have the potential of significantly influencing the frequency variables (Duchene and Goubel 1993; Potvin 1997; Öberg 1992). Also, during dynamic contractions the number of active MUs and their firing rate changes rapidly throughout the range of motion, which implies non-stationary spectra. In theory several seconds (0.5–2s) of stationary EMG are needed to obtain stable estimates (Hägg 1992; Merletti and Knaflitz 1992). Hence, the problems with non-stationarity of the signal increase the risk of erroneous interpretations.

Sleivert and Wenger (1994) reported very similar intra-class correlation coefficients (ICCs) for integrated EMG during static and dynamic contractions.

Finucane et al. (1998) reported RMS values for high submaximal concentric and eccentric isokinetic contractions to be reliable. Öberg (1992) recommended multiple measurements or regression analysis to minimize the effect of random variation of the frequency-spectrum variables. A rather limited number of studies exist concerning the reproducibility and validity of surface EMG variables in the frequency domain during dynamic contractions, some of them to the affirmative (Fugl-Meyer et al. 1985; Potvin and Bent 1997; Shankar et al. 1989; Elert et al. 1998). Potvin (1997) reported similar MNF for concentric and eccentric contraction.

Part 5: Results

It is beyond the scope of this chapter to give an extensive review of the use of surface EMG. Instead, this section displays some examples from research and clinical applications.

Muscle Tension

Patients with chronic pain conditions are clinically often described as having increased muscle tension. Contrary to such clinical descriptions no increased RMS levels are observed when these subjects are investigated at rest (for references see Svensson et al. 1998). However, the clinical concept of muscle tension is heterogeneous and complicated especially in relation to muscle pain (Simons and Mense 1998). As will be exemplified below, subjects with chronic muscle pain (myalgia) are unable to relax between bursts of activity.

Veiersted and coworkers (1990) investigated surface EMG activity of the trapezius muscle during stereotyped work (operating a chocolate packing machine). They determined the pattern of spontaneous short periods of low muscle activity (gaps; at least 0.2s, during which the muscle activity is below 0.5% MVC). Workers with complaints had both a significantly higher median static muscle load and significantly lower number of gaps than workers without complaints (Veiersted et al. 1990). In a prospective study of new female employees of a chocolate manufacturing plant the same group showed that employees that developed trapezius myalgia had a lower frequency of EMG gaps at the start of employment than the employees without myalgia (Veiersted et al. 1993).

Svebak and coworkers (1993) found that non-intended muscle tension (i.e., the gastrocnemius muscle activity during joy-stick operations of the hand) was positively and significantly related to the proportion of slow twitch fibers (Svebak et al. 1993).

Non-fatiguing Contractions

Static Contractions

The surface EMG signal can be used to provide information about the force contribution of muscles and muscle groups. In rare cases the resultant muscular moment is due to one muscle, which enables one to investigate the force-EMG relationship in vivo (DeLuca 1997). In clinical settings, the EMG is commonly used in bio-feedback applications or in studies of static work. Such bio-feedback applications have been used, for instance, in patients with chronic musculo-skeletal pain and in hemiplegic patients (Kasman et al. 1998; Schleenbaker and Mainous 1993). The relationship between biomechanical output (force or torque) and the signal amplitude (RMS) is then determined for a short (few seconds) and non-fatiguing, gradually increasing contraction (ramp). Then the mathematical relationship can be determined between these two variables. RMS will increase with increasing force output (linearly or according to some other mathematical function). No unique mathematical relationship can be expected (DeLuca 1997). In the clinical situation (e.g., bio-feedback), the registered RMS level then is "translated" to biomechanical output and often signals are used to alert the subject that a certain level of contraction has been reached. For research purposes, it has recently been concluded that such translations might be associated with low validity (Mathiassen et al. 1995). The major part of the studies reported concern the upper part of the trapezius muscle. In ergonomic studies of static work, the levels of contraction throughout the working day/period are calculated using the amplitude probability distribution (APDF) (Jonsson 1978, 1982). The static level is defined as the contraction (activity) level, above which the muscle activity is found for 90 % of the recording time. The median and peak level are defined as the contraction levels, above which the muscle activity is found for 50 and 10 % of the time, respectively (Veiersted 1995). In such studies the mathematical relationships between output and signal amplitude are very often determined only at low submaximal force levels (for instance, 0–30 % MVC or 0–50% MVC). Very recently Mathiassen and coworkers reviewed the literature and discussed different normalization procedures (Mathiassen et al. 1995), which are necessary when the electrode setup is changed.

At present, there is no clear picture of the relation between force/torque and frequency parameters of the EMG (Duchene and Goubel 1993). Hence, increases in the lower part of the force/torque range, decreases and constancy of the frequency parameters as a function of increasing biomechanical output have been reported (for references see Gerdle and Karlsson 1994; Gerdle et al. 1993).

Fatiguing Contractions

Note: Muscle fatigue is defined as "any-exercise-induced reduction in the capacity to generate force or power output" (Vollestad 1997).

During a sustained static contraction, the frequency content of the EMG is compressed towards lower frequencies, and simultaneously the RMS increases during prolonged static contractions (Fig. 10).

As evident from Fig. 10, the degree of decrease throughout the test period appears to correlate with the force/torque level even after normalization for differences in endur-

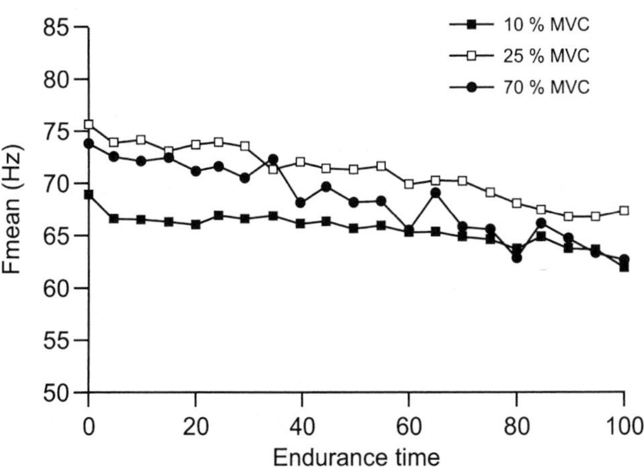

Fig. 10. Mean values of the mean frequency of vastus lateralis (Hz) throughout the normalized endurance time (%, every 5th % are shown) at three different torque levels (10, 25 and 70 % MVC) in clinically healthy men. [From Gerdle and Karlsson (1994). By Permission].

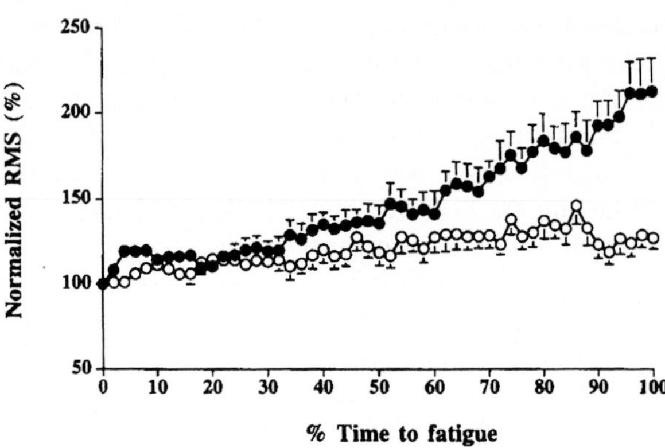

Fig. 11. The relationship between relative time to fatigue and relative RMS during sustained contractions at 25% MVC (filled circles) and 70% MVC (open circles). [From Crenshaw et al. (1997). By Permission].

ance time. By contrast, the observed increases in RMS are inversely correlated with the force/torque level (at least at sub-maximal contractions) (Fig. 11).

The precise mechanisms behind the decrease in MNF or MDF during sustained contractions are at present unclear (Duchene and Goubel 1993). However, it is generally agreed that a decrease in AMFCV is an important factor behind the decrease in MNF/MDF.

There has recently been an increased interest in how muscle morphological factors influence the surface EMG, especially the frequency parameters. When different muscles are compared, it becomes apparent that muscles with a relatively low proportion of slow twitch (type I) fibers undergo a faster spectral shift than muscles with a high proportion of such fibers (for references see Duchene and Goubel 1993; Komi and Tesch 1979; Tesch et al. 1983). However, the picture is less clear when only one muscle in different subjects with different proportions of slow fibers is investigated, even though significant correlations have been described; more studies are needed (Gerdle et al. 1997; Kupa et al. 1995; Mannion et al. 1998; Moritani et al. 1985; Pedrinelli et al. 1998).

Dynamic Contractions Using Dynamometers

Figure 12 shows a series of dynamic isokinetic contractions with simultaneously registered raw EMG (top). The raw EMG activity has a pattern similar to the torque variable. Between the contractions, no EMG activity is generally observed (the subjects are instructed to completely relax during this part of the contraction cycle).

No significant effect of the angular velocity (range: 0.57–3.13 rad s^{-1}) was found either on the signal amplitude (RMS) or the mean frequency of the EMG of the knee extensors (Gerdle et al. 1988).

Fatiguing Isokinetic Contractions

Such studies have been conducted in clinically healthy subjects or patients, most commonly using protocols consisting of series of 50 or more maximal-effort, isokinetic contractions. The output (i.e., peak torque, work or power) usually decreases in a more or less linear way during the initial part of the test (typically the initial 30–60 contractions) (for references see Elert 1991; Johansson 1987; Komi and Tesch 1979). Thereafter no significant change occurs. The frequency parameters (at least of the prime movers) show a pattern very similar to the biomechanical output; i.e., an initial decrease followed by a constant level (Fig. 13).

The degree of frequency shift appears to be most prominent in the prime movers and less prominent in more postural muscles. But differences are also observed, which cannot be easily explained by the type of muscle (prime mover or postural); instead factors such as amount of subcutaneous tissue, different degrees of movement between the electrodes and the active muscle fibers, different proportions of fiber types and different amounts of cross-talk might explain such differences.

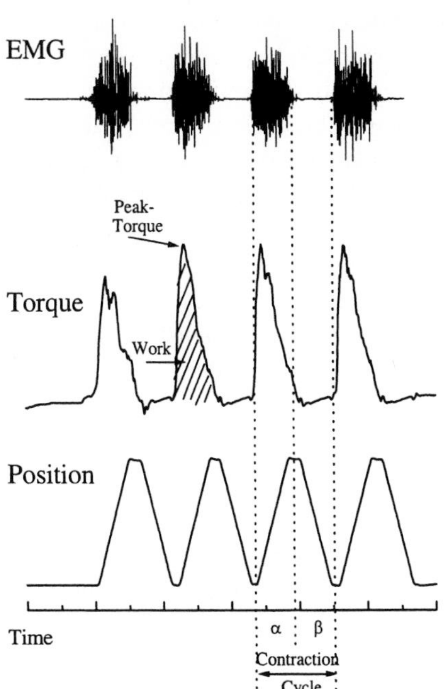

Fig. 12. Schematic drawing of the different phases of the contraction cycle during isokinetic contractions. The upper trace shows the raw EMG, and the middle and bottom traces show torque and position, respectively. The contraction cycle is divided into an active (labeled α) and a passive part (labeled β). A position window is placed within each phase (α and β) of the contraction cycle and used as a basis for calculation of the biomechanical and electromyographic variables [Reprinted from Journal of Electromyography and Kinesiology, volume 4, Karlsson, Erlandsson and Gerdle, A personal computer-based system for real-time analysis of surface EMG signals during static and dynamic contractions, pages 170–180, no. 3, Copyright © 1998, with permission from Elsevier Science].

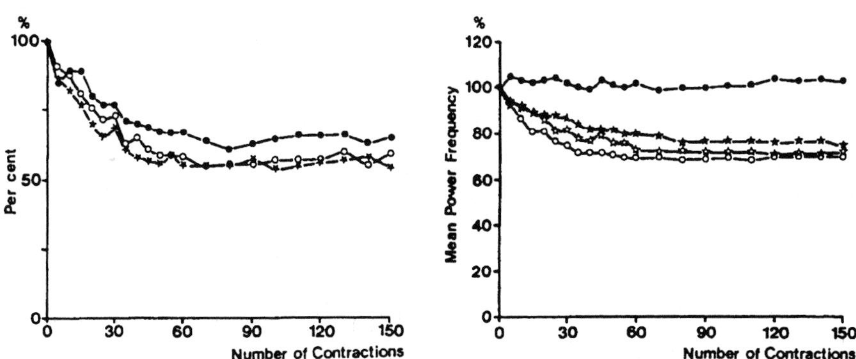

Fig. 13. The relationships between biomechanical output [left graph: peak torque (filled circles), work (filled stars) and power (open circles)] and mean frequency of the shoulder flexors [right graph: m. trapezius (filled stars), m. deltoid (open circles), m. infraspinatus (open stars) and mm. biceps brachii (filled circles)], respectively, and the number of maximum isokinetic shoulder forward flexions. [From Eur J Appl Physiol, Muscular fatigue during repeated isokinetic shoulder forward flexions in young females. Gerdle, Elert & Henriksson-Larsén. 58:666–673, figures 2 and 3, 1989, copyright 1999, Springer-Verlag GmbH & Co.KG, by permission].

No consistent pattern is observed for RMS during isokinetic contractions. The following patterns have been described:

- Increases during the initial, approximately 20 contractions, followed by a constant level or a slow decrease throughout the test to the initial level;
- Decreases during the initial 20–30 contractions, followed by a constant level;
- Constant throughout the test.

The reasons for the heterogeneous picture are not clear at present.

Recently a variable describing the relative ability to relax in between active, maximal-effort, isokinetic contractions has been presented. The ratio between the RMS of the passive and the active part of the contraction cycle can be calculated for the terminal 50 contractions (i.e., mean value) (signal amplitude ratio, SAR); a high ratio indicates a high level of activity during the passive part of the contraction cycle, and vice versa. Increased SAR levels of the shoulder flexors have been observed in groups of patients with different chronic pain conditions (i.e., work-related trapezius myalgia, fibromyalgia, whiplash-associated disorders) (Elert et al. 1992; Fredin et al. 1997). The similarity with the silent-gap results are obvious (see above). However, Carlson et al. (1996), investigating ambulatory EMG in the upper trapezius region, were unable to find any differences between patients with and without muscle pain in less standardized situations (Carlson et al. 1996).

Free Dynamic Contractions

This is the typical situation when analyzing gait or running, the EMG signals then usually being transmitted via a telemetric system. These analyses are associated with several problems when determining the EMG activity of each muscle. To facilitate the interpretation of the results, the raw EMG is mostly rectified and smoothed, which is exemplified from a study of running in Fig. 14.

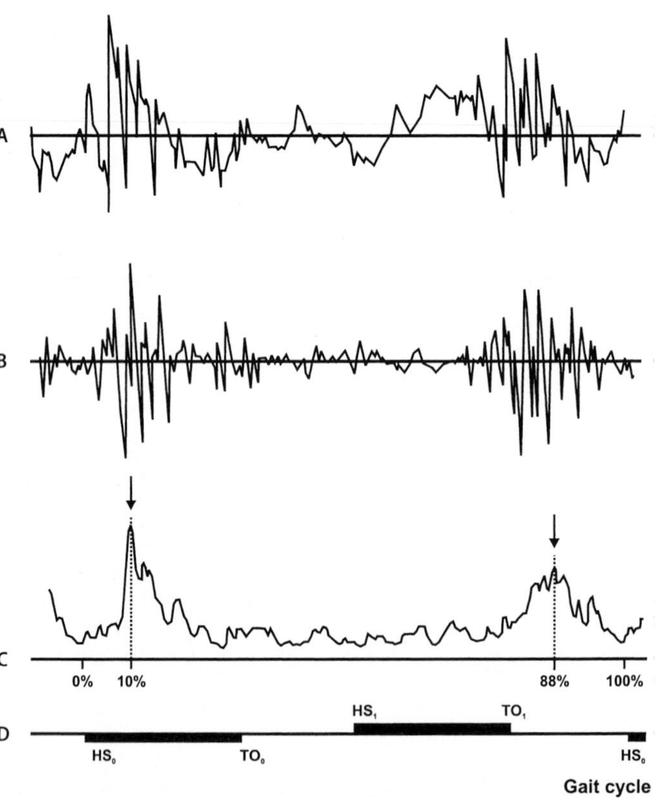

Fig. 14. EMG signal collected from the biceps femoris muscle during running. Part a shows raw EMG with low-frequency movement artifacts. Part b shows the EMG from part a but high-pass filtered (50 Hz). Part c shows the EMG of part b after rectification and smoothing. Each arrow indicates the highest point during EMG activation (EMG peak). Part d shows different parts of the running cycle. [Reprinted from Journal of Electromyography and Kinesiology, volume 6, Guidetti, Rivellini and Figura, EMG patterns during running: Intra- and Inter-individual variability, pages 37–48, no. 1, Copyright © 1998, with permission from Elsevier Science].

Some authors describe the EMG activity of each muscle in terms of time of onset, while others describe it as time of peak activity. No standard method is used in the literature for determining the onset time (Hodges and Bui 1996), and there is always the problem with cross-talk from neighboring muscles. It is also necessary to handle the problem of movement artifacts, which generally is done by applying some high-pass filter (see Fig. 14). Moreover, it is important to pay attention to cross-talk (see below under section Troubleshooting). DeLuca and Merletti (1988) reported that a substantial part of the EMG activity in leg muscles may be due to cross-talk (DeLuca and Merletti 1988). Both intra- and inter-individual variability of the EMG patterns of different muscles have been investigated in gait and running (Arsenault et al. 1986; Basmajian and DeLuca 1985; Guidetti et al. 1996).

Part 6: Troubleshooting

Noise, Artifacts and Cross-talk

Many of the problems with surface EMG are related to noise. In a measurement, the signal is defined as that component of a variable which contains information about the object quantity, whereas noise is defined as a component unrelated to the object quantity. There are many sources of noise in physical measurements, such as electric power fluctuations, air currents, stray radiation from nearby electrical apparatus, etc. The quality of a signal is often expressed quantitatively as the signal-to-noise ratio (SNR), which is the ratio of the true signal amplitude to the standard deviation of the noise. One of the fundamental problems in signal measurement is distinguishing the noise from the signal.

Typical noise sources during EMG signal measurements are:

- *Electrode terminal noise* (4–6 μV_{rms}) because of the electrolyte-electrode and the electrode-skin interfaces. It can be minimized by use of Ag-AgCl electrodes and careful skin preparation, respectively.
- *Amplifier input noise* ($<4\ \mu V_{rms}$) is a quality parameter of the amplifier. The spectrum of this noise ranges from DC to several kHz, and it can only be reduced by using a high-quality amplifier.
- *Movement artifacts* are caused by relative movements between electrolyte-skin and electrode and can also be reduced by the choice of electrode (Ag-AgCl, design etc.) and careful skin preparation. Using short wires and/or active electrodes can reduce artifacts in wires. High-pass filtering with a cut-off frequency of about 10 Hz can also reduce this kind of errors (see Fig. 14).
- *Ambient noise* caused by physical or chemical events outside the object and measuring system. Typical examples of noise are cross-talk (see below), electromagnetic inferences in tissue and cables, power-line frequency etc. The main source of interference noise is the 50 Hz (or 60 Hz) power source radiation. The amplitude of the noise signal may have a magnitude that is greater than the EMG signal. One method to reduce this dominant source of noise is to use a notch-filter at this frequency. However, in practice this filter also removes adjacent frequency components, and since the dominant energy of the EMG signal is located in the 20–100 Hz frequency range, this is a serious problem. Therefore, using notch-filters in EMG signal measurements is not advisable.
- *Quantization noise or quantization error* is caused by the analogue-to-digital conversion. The error is the difference between the true analogue value and the digitized value. This noise can be minimized by utilizing as much as possible of the input range of the A/D converter.

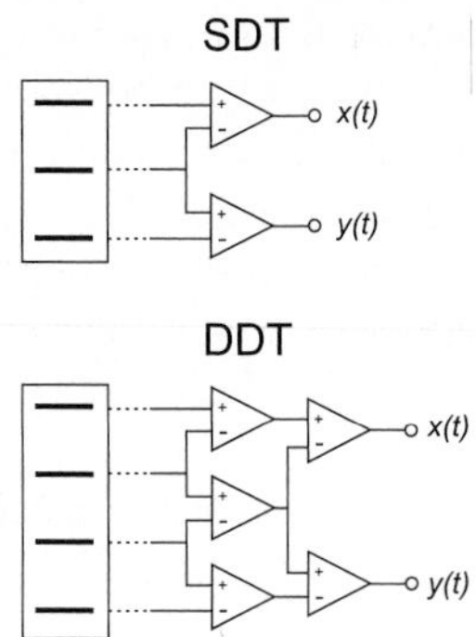

Fig. 15. The single differential technique (SDT, upper panel) and the double differential technique (DDT, lower panel).

- *Cross-talk.* It is important to recognize that the bipolar surface EMG is not always a selective representation of the electrical activity of the muscle/s directly underlying the recording electrodes. Although signal sources close to the electrode will dominate the EMG, more distant sources from other muscles will often contaminate the EMG to some degree. This phenomenon is referred to as cross-talk.
  The first step towards minimizing cross-talk is to place the electrodes on the midline of the muscle belly so that the distance to neighboring muscles is maximized (DeLuca 1997).
  Another possibility is to reduce the inter-electrode distance, which reduces the pick-up volume and thus limits the influence from distant signal sources. Also, instead of using single differential recordings, the double differential technique (see Fig. 15) can be used, since this arrangement acts as a "near-pass" filter. Koh and Grabiner (1993) investigated the effect of this technique on the cross-talk from the triceps surae muscles to the surface EMG recorded from the tibialis anterior muscle. Electrical stimulation of the nerve to the triceps surae, at intensities which generated an EMG of equal magnitude as maximal effort, resulted in cross-talk in the tibialis anterior EMG corresponding to 10–12% of maximal voluntary effort for the single differential recording. Using the double differential technique the cross-talk was reduced to about 5%. For both recording types, two different inter-electrode distances were studied, 8 and 20 mm. The smaller distance reduced cross-talk somewhat, but it also resulted in a substantially increased inter-test variability in EMG amplitude. Similar data, supporting the double differential technique, has also been documented for recordings from the lateral and medial hamstring muscles when stimulating the quadriceps (Koh and Grabiner 1992). Thus, in situations when surface EMG is recorded while neighboring muscles are highly active, using the double differential technique, while avoiding too small inter-electrode distance, will reduce cross-talk without sacrificing the reliability of the measurement.

Other Problems

In practice the following common problems arise:

- Insufficient contact between surface electrodes and skin;
- Inappropriate amount of gel;
- Too many bubbles in the contact gel or too dry gel;
- Pressure against the surface electrodes. Note that it is important that, during the muscle contraction, the electrodes not press against each other;
- Non-optimal gain. Too low gain will result in a too small signal. Too high gain will result in clipping effects;
- Insufficient application of the grounding electrode can result in noise on all registration channels;
- Insufficient contact between cables and electrodes;

- Movement of cables due to insufficient fixation;
- Cross-talk;
- Strong electric fields in the building (for instance, X-ray equipment, mobile phones);
- Perspiration during strenuous work (for instance, fatigue studies).

Part 7: Special Applications

Average Muscle Fiber Conduction Velocity

Background

The average muscle fiber conduction velocity (AMFCV) can be estimated using surface EMG. Measurements can be made either during voluntary or electrically evoked static contractions. The muscle fiber AP originates at the innervation zone and travels bidirectionally along the muscle fibers. The AMFCV is estimated by measuring the time delay τ between two surface EMG signals, recorded at two locations parallel to the muscle fiber direction. Usually τ is derived from the location of the peak of the cross-correlation function of the signals. The AMFCV is then obtained by dividing the distance between the EMG recording sites by τ.

Acquisition

The two spatially shifted EMG signals x(t) and y(t) necessary for AMFCV measurement can be obtained using:
- The *single differential technique* (SDT, Fig. 15, upper panel), which uses a three-contact electrode and generates two single differentiated EMG signals.
- The *double differential technique* (DDT, Fig. 15, lower panel), which uses a four-contact surface electrode and generates two double differentiated EMG signals (Broman et al. 1985a).

However, the two EMG signals x(t) and y(t) generated by the SDT method may contain non-delayed activity. The origin of this phenomenon is not clear, but it may arise from signals from distant sources, modified by the inhomogeneous and anisotropic volume conductor properties of the tissues. Thus, the SDT method may give rise to overestimation of AMFCV, since the non-delayed activity will "push" the peak of the cross-correlation function between the two signals towards lower values and hence bias the AMFCV estimates towards too high values.

The DDT method, introduced by Broman et al. (1985a), reduces the risk of overestimating AMFCV, since the double differentiation acts as a "near-pass" filter, suppressing distal signal sources. Thus, this method should be used if possible.

Materials

The materials needed for AMFCV are:
- Special-purpose electrode with three to four sites mounted on a fixture so that inter-site distances are constant. The most commonly used electrode consists of silver wire bars, 1 mm in diameter and 5 to 10 mm long, mounted on a fixture with 10 mm between the bars (Broman et al. 1985a; Fiorito et al. 1994; Hägg 1993).
- Filters for bandpass filtering at 40–250 Hz.
- Standard EMG differential amplifiers
- Data storage equipment

Procedures

Electrode Positioning
It is recommended to use the double differential technique (DDT); a four-contact surface electrode generates two double differentiated EMG signals. The possibility to detect the propagation of muscle fiber APs with surface electrodes, and hence to enable AMFCV measurements, critically depends on the location of the electrode relative to the in-

nervation zone and the muscle fiber architecture (Roy et al. 1986). The optimal electrode position is between the muscle tendon and the first motor point, i.e., innervation zone, and parallel to the fiber direction. Thus, there must be a sufficiently large portion of the muscle, accessible with surface electrodes, where the muscle fiber APs travel in parallel and in the same direction. Masuda and Sadoyama (1987) have investigated the possibility for AMFCV measurement for 26 muscles, for 19 of these muscles they could locate recording sites showing propagation of muscle fiber APs.

Processing

The AMFCV is computed by dividing the distance between the recording sites by the time delay τ between the two EMG signals x(t) and y(t). The two EMG signals x(t) and y(t) should be bandpass-filtered. The high-frequency limit is usually set between 160 to 250 Hz, and the low-frequency limit should be set rather high, at 40 to 80 Hz. This is done to eliminate frequency components with wavelengths close to, or longer than, the lengths of the muscle, which may give rise to reflections and other distortions that will interfere with the AMFCV estimation (Broman et al. 1985a).

The most commonly used technique for measuring the time delay τ is by computing the cross-correlation coefficient function $R_{xy}(\theta)$ of the signals x(t) and y(t). This function describes the similarity between the signals as they are displaced in time in relation to each other, a correlation coefficient of one signifies perfect similarity. Thus, the sought time delay τ is the time shift for the peak in the correlogram. When the time delay τ is computed using the cross-correlation technique, it is generally held that the correlation coefficient $R(\tau)$ should be greater than 0.8 for AMFCV estimates to be valid (Arendt-Nielsen and Zwarts 1989). This is no guarantee for a valid estimate, though. At present there seems to be no definite method for evaluating the validity.

To obtain a relevant resolution of the AMFCV estimate, τ should be computed with a resolution of about 10 μs (assuming an inter-electrode distance of 10 mm). This can be obtained by interpolating the cross-correlogram to achieve the relevant time resolution (see, e.g., Sollie et al. 1985).

It is desirable to compute the AMFCV and monitor the cross-correlation coefficient function on-line. This enables the investigator to scan the muscle to obtain the best electrode location and thus the most reliable AMFCV estimate. Several systems to this end have been described (e.g., Hägg and Gloria 1994; Fiorito et al. 1994).

Results

The AMFCV is a variable that may give information on several physiological aspects of muscle function, such as the development of fatigue, metabolic activity, fiber-type composition and MU recruitment strategy. It can also be used in the diagnosis of several muscle pathologies. The physiological range of AMFCV measured with surface EMG technique is about 2.5 to 5.5 m/s (for review see Arendt-Nielsen and Zwarts 1989). For example, during a voluntary isometric contraction of the biceps brachii muscle at 30% of maximal voluntary contraction (MVC), Krogh-Lund and Jørgensen (1993) found that the AMFCV decreased from an initial value of 4.6±0.56 m/s to 3.9±0.69 m/s at exhaustion (data from eight subjects), for the brachioradialis muscle the corresponding figures were 4.3±0.52 m/s to 2.9±0.76 m/s (data from 10 subjects).

Electrically Evoked Surface EMG

Background

Muscle contraction can be artificially elicited through electrical stimulation of motor nerves in the peripheral nervous system or motor pathways in the central nervous system. Magnetic stimulation can achieve the same result and is discussed in Chapter 33. Similar to voluntary contraction, current fields are generated during evoked muscle activation, which can be measured with surface or indwelling electrodes. However, unlike voluntary activation, the evoked EMG does not resemble a stochastic signal. Rather, the evoked compound action potential or M-wave has a closer resemblance to an MUAP. This is no coincidence, as M-waves and MUAPs are generated by the synchronous activation of groups of muscle fibers: either all the muscle fibers of a single MU (i.e., MUAP), or all the muscle fibers of a group of MUs (i.e., M-wave).

In the human central nervous system, electrically evoked muscle activation can be achieved by stimulating the motor cortex, descending pyramidal tracts, or select areas in the spinal cord. With centrally evoked contractions, large stimulation intensities must be generated to activate neurons through the skull and vertebrae. Since high stimulus intensities also cause appreciable discomfort, studies using centrally evoked contractions are conducted less frequently. Instead, it is more common to electrically evoke contractions through stimulating cutaneous peripheral motor nerves, muscle nerve branches, or the muscle itself. A discussion of the techniques involved in direct muscle stimulation can be found in a paper by Enoka (1988). Cutaneous nerves are often selected since the stimulation intensity required to evoke muscle contraction with surface electrodes is limited; and therefore, the subject's discomfort is minimized. Another consideration when selecting a site for electrical stimulation is the distance between the electrode and the muscle. Generally, as the distance decreases the selectivity of stimulation increases (i.e., fewer muscles will be activated as the distance between the stimulus and recording sites decreases). To ensure selective muscle activation, it is possible to stimulate the motor nerve at the point of muscle entry or even branches of the nerve further distally. Nevertheless, the risk with this technique is that the stimulus will only activate a select number of muscle nerve branches and subsequently only a fraction of the MU pool.

Without the highly conductive properties of biological tissues, nerve stimulation would not be feasible. Actually, the currents induced by stimulation and those generated during muscle contraction propagate through the surrounding tissue in roughly the same manner. It is the size of the current sources (i.e., stimulating electrode vs. muscle fiber) and the current density that are disproportionately higher in the case of surface stimulation. When discussing current fields, it is common to describe the magnitude or strength of the field in terms of current density. In simplistic terms the current density may be defined as the quantity of current within a predefined area. The units are in amperes/meters^2 (i.e., A/m^2), although for electrical stimulation the quantities of current density are more commonly expressed in mA/cm^2.

During the movement of current through a resistive substrate, i.e., biological tissues in the current context, a potential difference (i.e., voltage difference) is generated along the current path. This creates an area of relative depolarization and a region of relative hyperpolarization. If a sufficient potential gradient per unit time is generated along the length of the target nerve, the area of relative depolarization will activate one or more axons. The resultant AP initiated in an axon will propagate peripherally and centrally. The number of α-motor axons activated will depend on the potential difference along the nerve. Generally, at threshold levels of stimulation, myelinated axons with the largest diameter are activated first. Since α-motor axons (i.e., MU axons) are among the largest in diameter, they are recruited at lower levels of stimulation. MUs producing the largest force tend to have the largest axon diameters (Bawa et al. 1984), although the relation between axon diameter and force is relatively inconsistent (i.e., cross-correlation coefficient is less than 1). Another important mitigating factor determining the recruitment order of α-motor axons is the distance from the current source. Again, increasing the distance from the current source results in a reduction of current density (i.e., stimulus strength) at the site where the nerve is to be activated. Hence, there will be a bias towards recruiting the axons located closest to the stimulation site. The extent of this bias is determined by the topographical dispersion of α–motor axon pool within the nerve and between nerve branches. When all of a muscle's motor axons are located in one nerve bundle, the influence of the recruitment bias is rather low; however, the bias becomes increasingly important as the target muscle is approached and the nerve begins to branch. Moreover, the recruitment bias is especially prevalent during motor-point stimulation directly over the target muscle. In summary, during artificial electrical stimulation, recruitment generally proceeds from the larger to the smaller MUs, and the amount of overlap in the recruitment order increases as the stimulation site approaches the target muscle.

The largest-diameter, myelinated axons are those innervating muscle spindles (Ia afferents), so these are activated at the lowest stimulus strengths. As a result, axons of primary muscle spindle afferents will also be activated at stimulus intensities sufficient to activate α-motor axons. This volley of APs in primary muscle spindle axons travels to the spinal cord. Since these afferents synapse directly onto α-motor neurons (i.e., motor units), an additional but delayed motor or muscle reflex can be evoked. The EMG response elicited through this monosynaptic reflex pathway is called the Hoffmann or H-reflex. With sufficiently high stimulus intensities the H-reflex may be present in the EMG record, but delayed with respect to the M-wave response due to the increased conduction distance and synaptic delay associated with this reflex pathway.

Acquisition and Materials

There are considerable variations in the equipment and techniques employed to electrically elicit muscle contraction. In humans, evoked activation of a nerve is most commonly achieved by applying a stimulus to the skin overlying the nerve with surface electrodes. Some studies have used indwelling electrodes to stimulate the nerve tissues more directly; however, there are a number of developmental and safety issues that need to be resolved before indwelling stimulation can be used more widely. Further information on prospective indwelling stimulation techniques and their applications can be found in papers by Grill and coworkers (1996) and Loeb and colleagues (Cameron et al. 1997; Loeb and Peck 1996).

Stimulators

It is necessary to highlight the dangers of using improper stimulation equipment with human subjects. Only stimulators that are optically and/or electromagnetically isolated should be used on humans. Isolation implies that the output stage of the stimulator is not connected to any other equipment, and its power source does not share a common ground. Connecting the stimulator to other equipment will negate the isolation! Furthermore, most other equipment (e.g., computers, oscilloscopes etc.) connected to the wall socket has a common grounding point. If your stimulator shares a common ground, then a malfunction with any of this equipment could cause a short-circuiting of the ground. The end result would be the transmission of an extremely high current through the stimulator and into your subject. In addition, connecting the stimulator to a common ground can have a negative impact on signal quality.

Stimulators used in electrophysiological applications, including evoked EMG studies, have a number of standard features. One important feature is its mode of operation. That is, the pulses generated by the stimulator can either be voltage-controlled or current-controlled. Voltage control implies that the potential difference during the stimulus between the cathode and the anode can be selected and maintained at a relatively stable value; whereas current control implies that the current travelling between the cathode and anode can be set and maintained. If the impedance of the tissues separating the cathode and anode is isotropic (i.e., having consistent properties), then the current would be constant under voltage control and vice versa. However, the impedance of skin, muscle and connective tissues is rather inconsistent (i.e., anisotropic) and may vary over time with activation history. Also, the impedance between the electrode and the skin can change over time. As a result, it is important to choose the stimulator mode of operation, based upon the requirements of your experimental design. Current-control stimulation is less sensitive to the variations in impedance, and hence, provides greater control over the number of axons that are activated. Unfortunately, the discharge of current from the tissues following each stimulus pulse is prolonged with this mode of stimulator output. As a result, the presence of stimulus artifacts can be more problematic for the experimenter. In contrast, voltage-controlled stimulation is more susceptible to changes in tissue and electrode-skin impedance, but produces shorter discharge transients after each stimulus pulse. Thus, it is more difficult to maintain the same level of activation (i.e., same subset of axons), but the presence of stimulus artifacts is less problematic.

Another important stimulation feature is the shape and duration of the output waveforms. Most stimulators output rectangular waveforms (pulses) that are either monophasic or biphasic. Stimulation with monophasic pulses is adequate for applications where the duration of exposure is limited; however, for applications using chronic stimulation, monophasic stimulation should be avoided. Monophasic stimulation produces a net flow of charge in one direction (i.e., a direct current), altering the chemical composition of tissues (Loeb and Gans 1986). With sufficient cumulative exposure, this unbalanced current can lead to irreversible damage of the underlying tissues. To reduce the impact of stimulation on the underlying tissues, charge-balanced biphasic pulses should be used. Biphasic stimulation also has the advantage of reducing the amount of stimulus current that reaches the recording site and contaminates the EMG signal (i.e., M-wave). This contaminating signal is commonly referred to as the stimulus artifact.

Consideration should also be given to the duration of the stimulus pulses. To reduce the impact of stimulation on tissue integrity, each phase of the stimulus pulse should not exceed 0.5 ms, although even shorter pulse durations are recommended (e.g., 0.1 or 0.2 ms). The pulse duration has an impact on the intensity of stimulation needed to activate nerve axons. The relationship between pulse duration and threshold stimulus intensity is hyperbolic: increasing pulse duration reduces the threshold stimulus intensity. The functional impact of this hyperbolic relationship is that a level of nerve activation can be maintained for a larger range of stimulation intensities by using short stimulus pulses. Stated differently, it is easier to grade the level of activation (i.e., recruitment of axons) of the nerve with increasing stimulus intensities by using pulses of shorter duration. For a more thorough discussion on the effect of pulse duration on stimulation the reader is referred to Loeb and Gans (1986) and Tehovnik (1996).

There are a number of methods to minimize the amount of stimulus artifact in the M-wave record. Firstly, balanced biphasic stimulation should be used with bipolar electrode arrangements (see below). Also, the orientation of the bipolar stimulating electrodes should be approximately perpendicular to the orientation of the bipolar recording (EMG) electrodes. Slight modifications in the position of the EMG electrodes around this perpendicular orientation can have a large effect on the level of recorded stimulus artifact. When feasible, use voltage-controlled stimulation. As previously discussed, voltage-controlled stimulation is less effective in maintaining a fixed level of activation; however, in many instances such as supra-maximal nerve stimulation (i.e., activation of the entire motor pool) the level of activation is of secondary importance. There are other more sophisticated techniques to suppress stimulus artifacts that are nicely described by Merletti and coworkers (1992).

Surface Electrodes: Apparatus and Techniques

Electrodes

Surface stimulation is typically conducted using either a bipolar or a monopolar electrode configuration. In the bipolar configuration, two small stimulation electrodes of roughly equal size (see below) are secured to the skin overlying the nerve, so that the orientation of the electrodes is parallel to the path of the underlying nerve. A parallel orientation of the electrode and nerve will minimize the level of stimulus current or voltage needed to activate the nerve. One electrode is connected to the stimulator's cathode output port (-ve) and the other to the anode port (+ve). When using biphasic stimulation pulses, nerve activation occurs at lower stimulus intensities under the site of the cathodal electrode (Dreyer et al. 1993; Winkler and Stålberg 1988). At higher stimulus intensities nerve activation will also occur near the anodal electrode site. As a general rule, the cathode should be placed closer to the EMG recording site than the anode. Using this configuration will ensure that only the APs initiated under the cathode site will reach the target muscle. There is a risk that double discharges, initiated under the anode and cathode from a single stimulus pulse, will reach the muscle at high stimulus intensities, when the anode is placed closer to the target muscle. Double discharges would produce a contaminated and very unstable M-wave recording. For details on the proposed mechanisms refer to the paper by Dreyer and coworkers (1993).

In the monopolar configuration, a relatively small stimulating electrode is placed over the target nerve, while a relatively large reference or dispersive electrode is placed distal to both the stimulation electrode and the target muscle. Since lower stimulus intensities are required to activate peripheral nerves, the stimulation electrode should be connected to the stimulator's cathodal output port. If the stimulator has a ground port, then the reference electrode should be connected there. Alternatively, if the stimulator has only anodal and cathodal ports, the reference (dispersal electrode in this case) should be connected to the anodal port. In most cases the effect is the same, as long as the stimulus pulses are monophasic. However, before configuring the monopolar arrangement refer to the operations manual for your stimulator. The function of the reference electrode is to create a controlled path for the flow of stimulus current. The reference electrode should be quite large, because the current emanating from the stimulation electrode spreads out rapidly with increasing distance. While, in reality, the reference electrode reduces the amount of "stray" current, the majority of this current dissipates in the surrounding tissues. If the stimulus site is too close to the target muscle, the amount of stimulus artifact that is present in the M-wave record may be unacceptably high. In this case, it may be more advisable to use the bipolar stimulation technique. Due to the shorter inter-electrode distance and the smaller electrode size with the bipolar configuration, the field of current is restricted to a smaller area. Therefore, the use of a bipolar electrode configuration will minimize the quantity of stimulus artifact in the evoked EMG signal when the distance between the stimulation and recording sites is small. Furthermore, it is recommended to use bipolar stimulation with charge-balanced biphasic pulses as a first choice (see above).

Electrode Materials

The choice of electrode material is at least as important as the selection of electrode size. The materials comprising the stimulation electrode should be both highly conductive and stable to electrolytic corrosion. Some investigators and practitioners use disposable, self-adhesive electrodes that are used for recording EMG. The conductive substrate in these electrodes is typically Ag/AgCl. While such electrodes have good conductive properties, the stability of silver (Ag) to electrolytic corrosion during stimulation is unacceptably poor. Therefore, the authors advise against using these electrodes for applications where the duration or intensity of the stimulation is more than incidental. Other electrode materials that are susceptible to electrolytic corrosion and should be avoided include gold, copper and nickel (Loeb and Gans 1986). The authors are unaware of any disposable electrodes having materials that are acceptable for high-intensity or chronic stimulation. Materials that are less susceptible to electrolytic corrosion, in a general order of increasing stability and cost, include stainless steel, tungsten, platinum and irid-

Table 2. Stimulation protocols and electrode properties from some electrophysiological studies.

Authors	Muscle	Stim. Location	Stimulation Properties				
			Mono - Bipolar	Type	Size	Inter-electrode Distance	Pulse Duration
Cioni et al., 1985	Tibialis Anterior	Peroneal nerve: below head of the fibula	Monopolar	Ag/AgCl recording electrodes	-ve –10 mm	Ref. Location: medial tibia	0.2 ms
Dreyer et al., 1993	Thenar (8 cm distal to stim. electrode	Median nerve: ventral wrist	Monopolar	Disk electrode *	-ve - 1 cm diam., ref. - 2 cm diam.	Ref. Location: dorsal radius	0.1, 0.5 ms
Matre et al., 1998	Soleus	Tibial nerve: popliteal fossa	Monopolar	Ball electrode *	2.5 cm diam	Ref Location: anterior thigh	1 ms
Merletti et al.,	Vastus Medialis	Motor point: Vastus medialis	Monopolar	Sponge electrode	-ve: 3x4 cm, +ve: 8x12 cm	Ref. Location: dorsal thigh	0.1 ms
Passeo et al., 1994	First dorsal interosseous	Ulnar nerve: Above the elbow joint	Bipolar	Disk electrodes			0.2 ms
Schmid et al., 1990	Biceps & Abd. Digiti Minimi	Cervical motor roots: C7	Bipolar		Area 1 cm^2	6 cm	
Winker & Stålberg, 1988	Sensory evoked potential records	Median nerve: ventral wrist	Bipolar	Felt pad electrode	8 mm diam.	1.2 cm	0.015 - 0.2 ms

* Denotes unknown electrode material

ium (Tehovnik 1996). Refer to the book by Loeb and Gans (1986) for a more detailed description. To reduce the impedance between the electrode and skin, thereby reducing the necessary stimulus intensity, it is common to introduce a wetted material on the tip of the stimulating electrode. This may take the form of a cloth wrapped around the tip, or as a separate sponge that is connected to the tip. Table 2 provides examples, taken directly from the literature, of the types of electrodes that are used. This table demonstrates that there is no standard regarding the conductive materials that are used. The best advice is to avoid certain materials, as mentioned previously, and to experiment with the acceptable materials that are locally or commercially available. Finally, to reduce the additional noise between the stimulator's output stage and the electrodes, the lead cables should be i) unshielded, ii) of relatively short length and iii) twisted together for at least a portion of their length (i.e., where it does not interfere with electrode attachment).

Electrode Sizes

The stimulus intensity necessary to generate a maximal M-wave response (i.e., complete muscular recruitment) is dependent on the distance between the electrodes and the target nerve. The relationship between the threshold stimulus (e.g., current) and the electrode-target distance is non-linear; whereby the threshold current increases exponentially with distance (Tehovnik 1996). The value of the exponent is dependent on the particular impedance of the tissues, but can be approximated as the square (i.e., power of 2). For any level of stimulus current, the density of the current is determined by electrode size. That is, current density equals the current divided by the electrode area. In addition to the duration of the stimulus pulse, the current density is also a very important concern regarding tissue damage. To limit or avoid tissue damage, the current density should be limited by using relatively large stimulus electrodes. Unfortunately, empirical evidence is lacking on what size of electrodes is adequate for various surface stimulus intensities. Perhaps the best way to decide is to use a size that does not cause undue subject discomfort. Table 2 provides a list of electrode sizes that are used for various stimulation protocols using different target muscles. Within this table, there are also examples of typical sizes of reference electrodes that are used in monopolar stimulation protocols. As a final recommendation, increase the size of the electrode as the subcutaneous depth of the target nerve increases.

Electrode Placement: Peripheral Nerve Stimulation
It is clear that the stimulating electrode(s) should be placed over the target nerve, so the distance between the electrode and the nerve is minimized. In the monopolar configuration, the distance between the stimulating and reference electrode is not of particular importance, only the general area of reference electrode placement is of concern. The main issue then is what the distance between anodal and cathodal electrodes should be in the bipolar configuration. As with the chosen electrode materials and electrode sizes, there is no standardized inter-electrode distance that is used for a given target nerve and muscle. Table 2 provides some information on the inter-electrode distances used with different stimulation protocols. A general recommendation is to increase the inter-electrode distance as the subcutaneous depth of the nerve increases.

Procedures

Recording evoked EMG involves the same procedures as during static or dynamic recordings. However, aside from stimulation considerations there are a few differences that exist. For a description of stimulation procedures for centrally evoked contractions see Rothwell et al. (1987) and Schmidt et al. (1990).

1. Read the evoked contraction section above.
2. After locating the target nerve, place the stimulating electrode(s) over the site. Reduce the stimulus intensity to zero, select a low frequency (1–2 Hz) and then turn on the stimulator's output. Increase the stimulus intensity until muscle twitches are observed or until the subject can feel the stimulus on the skin. Systematically move the electrode(s) position until the strongest twitch response is elicited. If no response is elicited (i.e., except skin sensations), consult an anatomical reference text to ensure that the electrode placement corresponds to the area where the target nerve runs most superficially. In addition, decrease the impedance of the electrode with the skin by attaching a wetted cloth or sponge. Reposition the electrode and start from the beginning. If a response is still not observed, increase the stimulus intensity, but do not exceed 20 mA or 100 V. If no response is elicited at these intensities, then select another target nerve.
3. Mark the electrode position on the skin, or fix the electrode to the optimal stimulating spot.
4. Follow points 1 to 4 of the general recording procedures outlined above for static and dynamic contractions.
5. Place the EMG reference electrode, so that the EMG recording electrodes are between the stimulation electrodes and the EMG reference.
6. Follow point 6 in the procedures above, and set up an oscillograph, oscilloscope or a computer-based on-line graphics package.
7. Begin to stimulate and slowly rotate the recording electrodes about a central point. While slowly rotating the recording electrodes, view the measured signal. Fix the electrodes to the position where the stimulus artifact is the smallest. Check the electrode-skin impedance as detailed in point 7 for the general procedures.
8. Adjust stimulation intensity to achieve the desired level of activation.
9. See point 9 above.

Results

The shape of the recorded M-wave during maximal evoked nerve activation reflects the synchronous activation of the entire MU pool of a muscle. In most cases the shape of the recorded M-wave is strikingly similar to that of a single MUAP. Moreover, M-wave shape during an isometric contraction is determined, to a great extent, by the relative orientation of the electrode to the constituent muscle fibers and their endplate distribution in the muscle. If the muscle geometry is consistent prior to the arrival of each nerve stimulus (i.e., for isometric contractions), the M-wave responses from successive stimuli may be directly compared. Making comparisons of M-wave responses where muscle geometry is not consistent between stimuli should be avoided, unless this is the variable of interest (i.e., you are interested in the effect of changing muscle geometry on M-wave shape). Changes in muscle length will alter the orientation of the recording electrode and muscle fibers, which will, in turn, change the shape of the M-wave. Figure 16A illustrates the effect of transient changes in M-wave shape, which are concomitant with the development of muscle force and changes in muscle fiber length. In the current context, the M-wave becomes increasingly temporally compressed with the development of force. Also, there is a slight reduction in the M-wave's peak-to-peak amplitude with force development. The problem is that these changes in M-wave shape are inconsistent, varying between electrode recording positions and muscles. Muscle geometry can be assumed constant, when a sufficiently low or high rate of nerve stimulation is employed. The rate is sufficiently low when mechanical force twitches are discrete events that do not overlap. Normally, for most muscles and conditions, discrete twitches are observed for stimulation rates up to 3 s^{-1}. A sufficient rate of stimulation, for the upper bound, is achieved when a fused tetanus is observed. In practice, a stimulation rate of 30 s^{-1} is sufficient for most limb muscles under normal conditions. Figure 16 also demonstrates the consistency of successive evoked M-waves at stimulation rates of approximately 2 (Fig. 16B) and 30/s (Fig. 16C), where the muscle length was held relatively constant at the onset of each successive stimuli. Other sub-tetanic stimulation rates between these values can be used as long as the limitations in doing so are realized and precautions taken. Moreover, it is imperative that the intervals between successive stimuli are constant. This is achieved by maintaining a constant stimulation rate over time (i.e., instantaneous stimulation rate). In addition, the M-waves at the beginning of the contraction, during force development to a steady state level, should be discarded. For a constant stimulation rate, force development takes on the order of 300 ms for low rates, to 800 ms for high rates. For dynamic contractions, a similar but more extreme alteration of muscle geometry is observed, compared with the development of force under isometric conditions. Hence, it is even more difficult to compare the M-wave responses from successive stimuli during contractions evoked under dynamic conditions.

Merletti and coworkers investigated the repeatability of electrically evoked EMG signals in the human vastus medialis muscle and reported that parameters of spectral variables were more repeatable than those of amplitude variables and conduction velocity (Merletti et al. 1998). However, during voluntary contractions other investigators have found that conduction velocity was the most repeatable variable during voluntary contractions of the biceps brachii (Linssen et al. 1993).

Applications

With evoked activation of a muscle, the experimenter controls both the number of MUs activated and the relative timing of their activation. Generally, there is ample opportunity to investigate whether or not the evoked surface EMG signal has any utility in or-

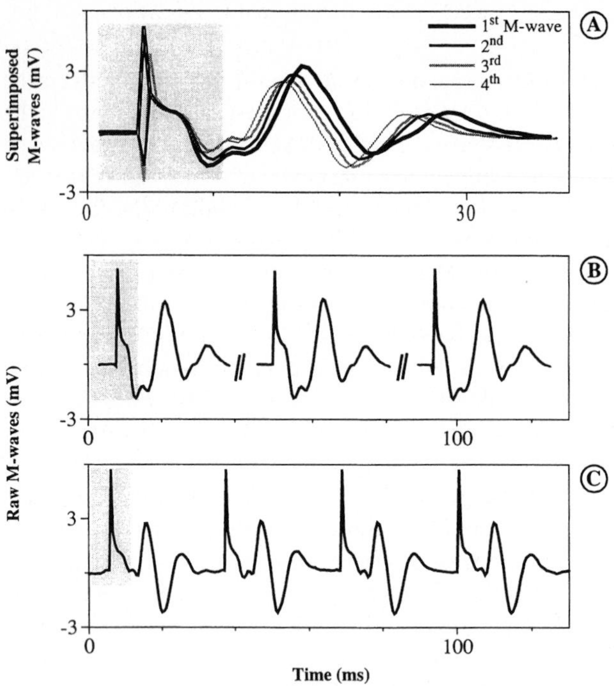

Fig. 16. Evoked surface M-wave responses recorded from one subject's tibialis anterior muscle during stimulation of the common peroneal nerve. A) Superimposed traces of 4 successive M-waves recorded during the initial development of muscle force. B) Interrupted and C) continuous time series records of successive M-waves at a stimulation rate of 2/s (B) and 30/s (C). Note that in each panel the waveform (shaded area) preceding the M-wave response represents the stimulus artifact produced by monopolar stimulation. The reference electrode was located dorsally on the lower thigh. The M-waves of panels B and C are much more consistent in shape than those of panel A.

thotic, therapeutic or diagnostic applications. Currently, the limited number of practical studies conducted is almost exclusively restricted to diagnostic applications. Potentially, a large amount of information can be gained from the M-wave response and in part used for clinical purposes:

- Conduction velocity of the motor axons.
  First, the delay between the stimulus and the onset of the M-wave can be used to calculate the conduction velocity and thereby to assess the condition of motor axons (for review see Iyer 1993).
- Integrity of neuromuscular transmission.
  By assessing changes in temporal (i.e., shape) and spectral (i.e., frequency spectrum) properties of successive M-wave responses during repetitive stimulation, it is possible to detect failure in transmission at the neuromuscular junction. In addition to extreme muscle fatigue, there are a number of conditions where the integrity of the pre- and/or post-synaptic region of the neuromuscular junction is compromised. These include Eaton-Lambert syndrome, botulism and tic paralysis, as well as certain other neuronopathies and neuropathies (Eng et al. 1984). It is quite possible that, in the future, the surface M-wave recordings will be used as a diagnostic screening tool to identify the presence of conditions affecting neuromuscular transmission.
- Speed of muscular AP propagation.
  The temporal and spectral (i.e., frequency domain) properties of the M-wave also give an indication of the muscular AP conduction velocity. For example, during the onset and progression of muscle fatigue, the conduction velocity of MUAPs declines,

resulting in an M-wave that is prolonged in the time domain and spectrally compressed in the frequency domain. For a more comprehensive discussion on alterations in M-wave shape with progressive muscle fatigue see the review by Merletti and coworkers (1992).

- Upper and lower motoneuron disorders.
  Cioni and colleagues (1985) investigated whether the surface EMG signal, recorded from voluntary and evoked isometric contractions, could be used to distinguish between normal subjects and those with upper and lower motoneuron disorders. They used the ratio between the RMS values obtained under evoked conditions and those obtained under voluntary conditions. The normal subjects presented an average ratio of 3.22 ± 0.67, whereas the hemiparetic patients (CNS origin) presented ratios between 5.16 and 9.33on the affected side and 2.59 and 4.10 on the unaffected side. In addition, polyneuropathic patients (i.e., disorders of the peripheral nervous system) presented ratios consistent with the normal subjects, although the RMS values obtained under voluntary and evoked conditions were proportionally reduced. These results suggest that the differences originated from the inability of hemiparetic patients to voluntarily activate the muscle to its full potential.

With further investigation, there is little doubt that M-wave studies can be used in other diagnostic applications. There are other types of analysis that can be employed; however, an overview of these techniques is beyond the scope of this article (see Merletti et al. 1990; Eng et al. 1984; Merletti et al. 1992; Nakashima et al. 1989).

Acknowledgement: Monica Edström, Majken Rahm, Barbo Larsson, Jessica Elert, Tomas Bäcklund and Christian Karlberg are gratefully acknowledged for valuable comments especially concerning the troubleshooting part. The present chapter was supported by grants from the Swedish Council for Work Life Research.

References

Basmajian J. V. and DeLuca C. J., Muscles alive: Their functions revealed by electromyography, 5th ed., Williams & Wilkins, Baltimore, 1985.

DeLuca C. J., The use of surface electromyography in biomechanics, J. Appl. Biomechanics, 13:135–163, 1997.

Duchêne J. and Gouble F., Surface electromyogram during voluntary contraction: Processing tools and relation to physiological events, Crit. Rev. in Biomed. Eng. 21(4):313–397, 1993.

Hermens H.J, Hägg G. and Freriks B. (Eds.) European Applications on Surface ElectroMyoGraphy, proceedings of the second general SENIAM workshop Stockholm, Sweden, June 1997. 1997, Roessingh Research and Development b.v.

Hermens H.J. and Freriks B. (Eds.) The state of the art on sensors and sensor placement procedures for surface electromyograpghy: A proposal for sensor placement procedures, Deliverable of the SENIAM project, 1998, Roessingh Research and Development b.v.

Hägg G.M. Interpretation of EMG spectral alterations and alteration indexes at sustained contraction. J Appl Physiol 1992, 73:1211–1217.

Kasman G.S., Cram J.R. and Wolf S.L. Clinical Applications in surface Electromyography – Chronic Musculoskeletal Pain. 1997, Aspen Publishers, Gaithersburg. pp 1–415

Lindström L. and Petersén I. Power spectrum analysis of EMG signals and its applications. In Computer-aided electromyography. Desmedt J.E (Ed) 1983, pp 1–51. S. Karger AG, Basel.

Loeb G.E. and Gans C. Electromyography for experimentalists. 1986, The University of Chicago Press, Chicago, London.

Aarås A, Veieröd MB, Larsen S, Örtengren R, Ro O (1996) Reproducibility and stability of normalised EMG measurements on musculus trapezius. Ergonomics 39: 171–185

Arendt-Nielsen L, Zwarts M (1989) Measurement of muscle fiber conduction velocity in humans: Techniques and applications. Journal of clinical neurophysiology 6: 173–190

Arsenault AB, Winter DA, Marteniuk RG (1986) Is there a "normal" profile of EMG activity in gait? Med Biol Eng Comput 24: 337–343

Basmajian JV, De Luca CJ (1985) In Muscles Alive. Their Function Revealed by Electromyography. Williams & Wilkens, Baltimore.

Basmajian J, DeLuca CJ (1985) Muscles alive: Their function revealed by electromyography. Wiliams & Wilkins, Baltimore.

Bawa P, Binder MD, Ruenzel P, Henneman E (1984) Recruitment order of motoneurons in stretch reflexes is highly correlated with their axonal conduction velocity. Journal of Neurophysiology 52: 410–420

Bilodeau M, Arsenault AB, Gravel D, Bourbannais D (1994) EMG power spectrum of elbow extensors: A reliability study. Electromyogr clin neurophysiol 34: 149–158

Bodine SC, Roy RR, Eldred E, Edgerton VR (1987) Maximal force as a function of anatomical features of motor units in the cat tibialis anterior. Journal of Neurophysiology 57: 1730–1745

Bodine-Fowler S, Garfinkel A, Roy RR, Edgerton VR (1990) Spatial distribution of muscle fibers within the territory of a motor unit. Muscle & Nerve 13: 1133–1145

Botterman BR, Iwamoto GA, Gonyea WJ (1986) Gradation of isometric tension by different activation rates in motor units of cat flexor carpi radialis muscle. Journal of Neurophysiology 56: 494–506

Broman H, Bilotto G, De Luca CJ (1985a) A note on the noninvasive estimation of muscle fiber conduction velocity. IEEE Trans Biomed Eng 32: 341–344

Broman H, Bilotto G, De Luca CJ (1985b) Myoelectric signal conduction velocity and spectral parameters: influence of force and time. Journal of Applied Physiology 58: 1428–1437

Buchthal F, Guld C, Rosenfalck P (1954) Action potential parameters in normal human muscle and their dependence on physical variables. Acta Physiologica Scandinavica 32: 200–218

Buchthal F, Guld C, Rosenfalck P (1955a) Propagation velocity in electrically activated muscle fibres in man. Acta Physiologica Scandinavica 34: 75–89

Buchthal F, Guld C, Rosenfalck P (1955b) Innervation zone and propagation velocity in human muscle. Acta Physiologica Scandinavica 35: 174–190

Buchthal F, Guld C, Rosenfalck P (1957a) Volume conduction of the spike of the motor unit potential investigated with a new type of multielectrode. Acta Physiologica Scandinavica 38: 331–354

Buchthal F, Guld C, Rosenfalck P (1957b) Multi-electrode study of a territory of a motor unit. Acta Physiologica Scandinavica 39: 83–104

Buchthal F, Schmalbruch H (1980) Motor unit of mammalian muscle. Physiological Reviews 60: 90–142

Burke RE (1981) Motor units: Anatomy, physiology, and functional organization. In Handbook of Physiology – The Nervous System II. Brooks VB pp 345–422. American Physiological Society, Bethesda.

Burke RE, Rudomin P, Zajac FE 3d (1976) The effect of activation history on tension production by individual muscle units. Brain Research 109: 515–529

Burke RE, Tsairis P (1973) Anatomy and innervation ratios in motor units of cat gastrocnemius. Journal of Physiology (London) 234: 749–765

Cameron T, Loeb GE, Peck RA, Schulman JH, Strojnik P, Troyk PR (1997) Micromodular implants to provide electrical stimulation of paralyzed muscles and limbs. IEEE Transactions on Biomedical Engineering 44: 781–790

Carlson CR, Wynn KT, Edwards J, Okekson JP, Nitz AJ, Workman DE, Cassisi J (1996) Ambulatory electromyogram acativity in the upper trapezius region. Patients with muscle pain vs. pain-fre control subjects. Spine 21: 595–599

Christensen E (1959) Topography of terminal motor innervation in striated muscles from stillborn infants. American Journal of Physical Medicine 38: 65–78

Cioni R, Paradiso C, Battistini N, Starita A, Navona C, Denoth F (1985) Automatic analysis of surface EMG (preliminary findings in healthy subjects and in patients with neurogenic motor diseases). Electroencephalography and Clinical Neurophysiology 61: 243–246

Clamann HP (1970) Activity of single motor units during isometric tension. Neurology 20: 254–260

Close RI (1972) Dynamic properties of mammalian skeletal muscles. Physiological Reviews 52: 129–197

Coers C (1959) Structural organization of the motor nerve endings in mammalian muscle spindles and other striated muscle fibers. American Journal of Physical Medicine 38: 166–175

Cooley JW, Tukey JW (1965) An algorithm for the machine computation of complex. Fourier series. Mathematics of Computation 19: 297–301

Crenshaw A, Karlsson S, Gerdle B, Fridén J (1997) Differential responses in intramuscular pressure and EMG fatigue indicators during low versus high level static contractions to fatigue. Acta Physiol Scand 160: 353–362

Daanen HAM, Mazure M, Holewijin M, Van der Velde EA (1990) Reproducibility of the mean power frequency of the surface electromyogram. Eur J Appl Physiol 61: 274–277

Datta AK, Stephens JA (1980) Short-term synchronization of motor unit firing in human first dorsal interosseous muscle. Journal of Physiology (London) 308: 19–20

Day SJ (1997) The Properties of Electromyogram and Force in Experimental and Computer Simulations of Isometric Muscle Contractions: Data from an Acute Cat Preparation. Dissertation, University of Calgary, Calgary.

De Luca CJ, LeFever RS, McCue MP, Xenakis AP (1982a) Control scheme governing concurrently active human motor units during voluntary contractions. Journal of Physiology (London) 329: 129–142

De Luca CJ, LeFever RS, McCue MP, Xenakis AP (1982b) Behaviour of human motor units in different muscles during linearly varying contractions. Journal of Physiology (London) 329: 113–128

DeLuca CJ (1997) The use of surface electromyography in biomechanics. J Appl Biomechanics 13: 135–163

DeLuca CJ, Merletti R (1988) Surface EMG crosstalk among muscles of the leg. Electroencephalography and Clinical Neurophysiology 69: 568–575

Denny-Brown D, Pennybacker JB (1938) Fibrillation and fasciculation in voluntary muscle. Brain 61: 311–334

Dreyer SJ, Dumitru D, King JC (1993) Anodal block V anodal stimulation. Fact or fiction. American Journal of Physical Medicine and Rehabilitation 72: 10–18

Duchene J, Goubel F (1993) Surface electromyogram during voluntary contraction: Processing tools and relation to physiological events. Critical Reviews in Biomedical Engineering 21: 313–397

Dumitru D, DeLisa JA (1991) Aaem minimonograph #10: Volume conduction. Muscle & Nerve 14: 605–624

Eccles JC, O'Connor WJ (1939) Responses which nerve impulses evoke in mammalian striated muscles. Journal of Physiology (London) 97: 44–102

Elert J (1991) The pattern of activation and relaxation during fatiguing isokinetic contractions in subjectswith and without muscle pain. Medical dissertation, Umeå. pp 1–46.

Elert J, Karlsson S, Gerdle B (1998) One-year reproducibility and stability of the signal amplitude ratio and other variables of the EMG: test-retest of a shoulder forward flexion test in female workers with neck and shoulder problems. Clin Physiol 18: 529–538

Elert J, Rantapää-Dahlqvist S, Henriksson-Larsén K, Lorentzon R, Gerdle B (1992) Muscle perforamnce, electromyography and fibre type composition in fibromyalgia and work-related myalgia. Scand J Rheumatol 21: 28–34

Eng GD, Becker MJ, Muldoon SM (1984) Electrodiagnostic tests in the detection of malignant hyperthermia. Muscle & Nerve 7: 618–625

Enoka RM (1988) Muscle strength and its development. New perspectives. Sports Medicine 6: 146–168

Enoka RM (1995) Morphological features and activation patterns of motor units. Journal of Clinical Neurophysiology 12: 538–559

Erim Z, De Luca CJ, Mineo K, Aoki T (1996) Rank-ordered regulation of motor units. Muscle & Nerve 19: 563–573

Finucane SDG, Rafeei T, Kues J, Lamb RL, Mayhew TP (1998) Reproducibility of electromyographic recordings of submaximal concentric and eccentric muscle contractions in humans. Electromyography and Motor Control - Electroencephalography and Clinical Neurophysiology 109: 4 p290–4

Fiorito A, Rao S, Merletti R (1994) Analogue and digital instruments for non-invasive estimation of muscle fibre conduction velocity. Med Biol Eng Comput 32: 521–529

Fredin Y, Elert J, Britschgi N, Vaher A, Gerdle B (1997) A decreased ability to relax between repetitive muscle contractions in patients with chronic symptoms after whiplash trauma of the neck. J Musculoskel Pain 5: 55–70

Fugl-Meyer AR, Gerdle B, Eriksson B-E, Jonsson B (1985) Isokinetic plantar flexion endurance. Scand J Rehabil Med 20: 89–92

Fuglevand AJ, Winter DA, Patla AE, Stashuk D (1992) Detection of motor unit action potentials with surface electrodes: influence of electrode size and spacing. Biological Cybernetics 67: 143–153

Garnett RA, O'Donovan MJ, Stephens JA, Taylor A (1979) Motor unit organization of human medial gastrocnemius. Journal of Physiology (London) 287: 33–43

Geddes LA (1972) Electrodes and the measurement of bioelectric events. John Wiley & Sons, London. Gerdle B, Edström M, Rahm M (1993) Fatigue in the shoulder muscles during static work at two different torque levels. Clin Physiol 13: 469–482

Gerdle B, Elert J, Henriksson-Larsén K (1989) Muscular fatigue during repeated isokinetic shoulder forward flexions in young females. Eur J Appl Physiol 58: 666–673

Gerdle B, Henriksson-Larsén K, Lorentzon R, Wretling M-L (1991) Dependence of the mean power frequency of the electromyogram on muscle force and fibre type. Acta Physiol Scand 142: 457–465

Gerdle B, Karlsson S (1994) The mean frequency of the EMG of the knee extensors is torque dependent both in the unfatigued and the fatigued states. Clinical Physiology 14: 419–432

Gerdle B, Karlsson S, Crenshaw AG, Fridén J (1997) The relationship between EMG and muscle morphology throughout sustained static knee extension at two submaximal force levels. Acta Physiol Scand 160: 341–351

Gerdle B, Wretling M-L, Henriksson- Larsén K (1988) Do the fibre-type proportion and the angular velocity influence the mean power frequency of the electromyogram? Acta Physiol Scand 134: 341–346

Glendinning DS, Enoka RM (1994) Motor unit behavior in Parkinson's disease. Physical Therapy 74: 61–70

Gootzen THJM, Stegeman DF, Van Oosterom A (1991) Finite limb dimensions and finite muscle length in a model for the generation of electromyographic signals. Electroencephalography and Clinical Neurophysiology 81: 152–162

Grill WM, Mortimer JT (1996) Non-invasive measurement of the input-output properties of peripheral nerve stimulating electrodes. Journal of Neuroscience Methods 65: 43–50

Guidetti L, Rivellini G, Figura F (1996) EMG patterns during running: Intra- and inter-individual variability. J electromyogr Kinesiol 6: 37–48

Gydikov A, Kostov K, Kossev A, Kosarov D (1984) Estimation of the spreading velocity and the parameters of the muscle potentials by averaging of the summated electromyogram. Electromyography and Clinical Neurophysiology 24: 191–212

Hanson J (1974) The effects of repetitive stimulation on the action potential and the twitch of rat muscle. Acta Physiologica Scandinavica 90: 387–400

Henneman E (1957) Relation between size of neurons and their susceptibility to discharge. Science 126: 1345–1347

Hodges PW, Bui BH (1996) A comparison of computer-based methods for the determnation of onset ofmuscle contraction using electromyography. Electroencephalography and clinical neurophysiology 101: 511–519

Hogrel JY, Duchene J, Marini JF (1998) Variability of some SEMG parameter estimates with electrode location. Journal of Electromyography and Kinesiology 8: 305–315

Håkansson CH (1956) Conduction velocity and amplitude of the action potential as related to circumference in the isolated fibre of frog muscle. Acta Physiologica Scandinavica 37: 14–34

Håkansson CH (1957) Action potentials recorded intra- and extra-cellularly from the isolated frog muscle fibre in ringer's solution and in air. Acta Physiologica Scandinavica 39: 291–312

Hägg GM (1991) Zero crossing rate as an index of electromyographic spectral alterations and its applications to ergonomics. Arbetsmiljöinstitutet, Göteborg. pp 1–37.

Hägg GM (1992) Interpretation of EMG spectral alterations and alteration indexes at sustained contraction. J Appl Physiol 73: 1211–1217

Hägg GM (1993) Action potential velocity measurements in the upper trapezius muscle. Journal of Electromyography and Kinesiology 3: 231–235

Hägg GM, Gloria R (1994) Surface EMG muscular conduction velocity measurement system implemented on a standard personal computer without A/D convertor. Med Biol Eng Comput 32: 691–694

Inman VT, Ralston HJ, Saunders JBCM, Feinstein B, Wright EW (1952) Relation of human electromyogram to muscular tension. Electroencephalogr Clin Neurophysiol 4: 187–194

Iyer VG (1993) Understanding nerve conduction and electromyographic studies. Hand Clinics 9: 273–287

Johansson C (1987) Elite sprinters, ice hockey players, orienteers and marathon runners. Isokinetic leg muscle performance in relation to muscle structure and training. Medical Disseration, Umeå. pp 1–31.

Jonsson B (1978) Kinesiology – with special reference to electromyographic kinesiology. In Contemp. Clin. Neurophysiol. Cobb WA, van Duijn H pp 417–428. Elsevier, Amsterdam.

Jonsson B (1982) Measurement and evaluation of local muscular strain in the shoulder during constrained work. J Hum Ergol (Tokyo) 11: 73–88

Karlsson S, Erlandsson B, Gerdle B (1994) A personal computer-based system for real-time analysis of surface EMG signals during static and dynamic contractions. J Electromyogr Kinesiol 4: 170–180

Karmen G, Caldwell GE (1996) Physiology and interpretation of the electromyogram. Journal of Clinical Neurophysiology 13: 366–384

Kasman GS, Cram JR, Wolf SL (1998) Clinical applications in surface electromyography – chronic musculoskeletal pain. Aspen Publishers, Inc, Gaithersburg

Kernell D, Eerbeek O, Verhey BA (1983) Relation between isometric force and stimulus rate in cat's hindlimb motor units of different twitch contraction time. Experimental Brain Research 50: 220–227

Kernell D, Sjöholm H (1975) Recruitment and firing rate modulation of motor unit tension in a small muscle of the cat's foot. Brain Research 98: 57–72

Kirkwood PA, Sears TA (1978) The synaptic connexions to intercostal motoneurones as revealed by the average common excitation potential. Journal of Physiology (London) 275: 103–134

Koh TJ, Grabiner MD (1992) Cross talk in surface electromyograms of human hamstring muscles. Journal of Orthopaedic Research 10: 701–709

Koh TJ, Grabiner MD (1993) Evaluation of methods to minimize cross talk in surface electromyography. Journal of Biomechanics 26 Suppl 1: 151–157

Komi PA, Tesch P (1979) EMG frequency spectrum, muscle structure and fatigue during dynamic contractions in man. Eur J Appl Physiol 42: 41–50

Krogh-Lund C, Jørgensen K (1993) Myo-electric fatigue manifestations revisited: power spectrum, conduction velocity, and amplitude of human elbow flexor muscles during isolated and repetitive endurance contractions at 30% maximal voluntary contraction. Eur J Appl Physiol 66: 161–173

Kukulka CG, Clamann HP (1981) Comparison of the recruitment and discharge properties of motor units in human brachial biceps and adductor pollicis during isometric contractions. Brain Research 219: 45–55

Kupa EJ, Roy SH, Kandarian SC, DeLuca CJ (1995) Effects of muscle fiber type and size on EMG median frequency and conduction velocity. J Appl Physiol 79: 23–32

Lindström LH, Magnusson RI (1977) Interpretation of myoelectric power spectra: A model and its applications. Proceedings of the IEEE 65: 653–662

Lindström L, Petersen I (1983) Power spectrum analysis of EMG signals and its applications. In Computor-Aided Electromyography. Desmedt JE pp 1–51. Karger, Basel.

Lindström LH, Magnusson RI (1977) Interpretation of myoelectric power spectra: a model and its applications. Proceedings of the IEEE 65: 653–662

Linssen WHJP, Stegeman DF, Joosten EMG, van't Hof MA, Binkhorst RA, Notermans SLH (1993) variability and interrelationships of suface EMG parameters during local muscle fatigue. Muscle Nerve 16: 849–856

Loeb GE, Gans C (1986) In Electromyography for Experimentalists. University of Chicago Press, Chicago.

Loeb GE, Peck RA (1996) Cuff electrodes for chronic stimulation and recording of peripheral nerve activity. Journal of Neuroscience Methods 64: 95–103

Lynn PA (1979) Direct on-line estimation of muscle fiber conduction velocity by surface electromyograhy. IEEE Transactions on Biomedical Engineering BME-26: 564–571

Mannion AF, Dumas GA, Cooper RG, Espinosa FJ, Faris AW, Stevenson JM (1997) Muscle fibre size and type distributation in thoracic and lumbar regions of erector spinae in healthy subjects without low back pain: normal values and sex differences. J Anat 190: 505–513

Mannion AF, Dumas GA, Stevenson JM, Cooper RG (1998) The influence of muscle fiber size and type distribution on electromyographic measures of back muscle fatigability. Spine 23: 576–584

Masuda T, Miyano H, Sadoyama T (1985) The position of innervation zones in the biceps brachii investigated by surface electromyography. IEEE Transactions on Biomedical Engineering BME-32: 36–42

Masuda T, Sadoyama T (1987) Skeletal muscles from which the propagation of motor unit action potentials is detectable with a surface electrode array. Electroencephalography and clinical neurophysiology 67: 421–427

Mathiassen SE, Winkel J, Hägg GM (1995) Normalization of surface EMG amplitude from the upper trapezius muscle in ergonomic studies – a review. J Electromyogr Kinesiol 5: 197–226

Matre DA, Sinkjær T, Svensson P, Arendt-Nielsen L (1998) Experimental muscle pain increases the human stretch reflex. Pain 75: 331–339

McComas AJ, Galea V, de Bruin H (1993) Motor unit populations in healthy and diseased muscles. Physical Therapy 73: 868–877

Merletti R, Lo Conte LR, Orizio C (1991) Indices of muscle fatigue. Journal of Electromyography and Kinesiology 1: 20–33

Merletti R, Fiorito A, Lo Conte MR, Cisari C (1998) Repeatability of electrically evoked EMG signals in the human vastus medialis muscle. Muscle & Nerve 21: 184–193

Merletti R, Gulisashvili A, Lo Conte LR (1995) Estimation of shape characteristics of surface muscle signal spectra from time domain data. IEEE Trans Biomed Eng 42: 769–776

Merletti R, Knaflitz M, De Luca CJ (1990) Myoelectric manifestations of fatigue in voluntary and electrically elicited contractions. Journal of Applied Physiology 69: 1810–1820

Merletti R, Knaflitz M, DeLuca CJ (1992) Electrically evoked myoelectric signals. Critical Reviews in Biomedical Engineering 19: 293–340

Merletti R, Lo Conte LR (1995) Advances in processing of surface myoelectric signals: Part 1. Med & Biol Eng & Comput 33: 362–372

Merletti R, Migliorini M (1998) Surface EMG electrode noise and contact impedance. Proceedings of the third general SENIAM workshop

Milner-Brown HS, Stein RB, Yemm R (1973) Changes in firing rate of human motor units during linearly changing voluntary contractions. Journal of Physiology (London) 230: 371–390

Monster AW, Chan H (1977) Isometric force production by motor units of extensor digitorum communis muscle in man. Journal of Neurophysiology 40: 1432–1443

Moritani T, Gaffney FD, Carmichael T, Hargis J (1985) Interrelationships among muscle fiber types, electromyogram and blood pressure during fatiguing isometric contraction. In Biomechanics, IXA. International series on Biomechanics. Winter DA, Norman RW, Wells RP, Hayes KC, Patla AE pp 287–292.

Nakashima K, Azumi T, Ohta M, Hamasaki N, Takahashi K (1989) Electromyographic responses in leg muscles after electrical stimulation in myelopathy patients with tonic seizures. Electromyography and Clinical Neurophysiology 29: 203–211

Ng JK-F, Richardson CA (1996) Reliability of electromyographic power spectral analysis of back muscle endurance in healthy subjects. Arch phys med rehabil 77: 259–264

Nordstrom MA, Fuglevand AJ, Enoka RM (1992) Estimating the strength of common input to human motoneurons from the cross-correlogram. Journal of Physiology (London) 453: 547–574

Oppenheim AV, Schafer RW (1989) In: Discrete-time signal processing. Prentice Hall.

Passero S, Paradiso C, Giannini F, Cioni R, Burgalassi L, Battistini N (1994) Diagnosis of thoracic outlet syndrome. Relative value of electrophysiological studies [see comments]. Acta Neurologica Scandinavica 90: 179–185

Pedrinelli R, Marino L, Dell'Omo G, Siciliano G, Rossi B (1998) Altered surface myoelectric signals in peripheral vascular disease: correlations with muscle fiber composition. Muscle & Nerve 21: 201–210

Person RS, Kudina LP (1972) Discharge frequency and discharge pattern of human motor units during voluntary contraction of muscle. Electroencephalography and Clinical Neurophysiology 32: 471–483

Potvin JR, Bent LR (1997) A validation of techniques using surface EMG signals from dynamic contractions to quantify muscle fatigue during repetitive tasks. J Electromyogr Kinesiol 7: 131–139

Powers RK, Rymer WZ (1988) Effects of acute dorsal spinal hemisection on motoneuron discharge in the medial gastrocnemius of the decerebrate cat. Journal of Neurophysiology 59: 1540–1556

Ramaekers VT, Disselhorst-Klug C, Schneider J, Silny J, Forst J, Forst R, Kotlarek F, Rau G (1993) Clinical application of a noninvasive multi-electrode array EMG for the recording of single motor unit activity. Neuropediatrics 24: 134–138

Rios E, Pizarro G, Stefani E (1992) Charge movement and the nature of signal transduction in skeletal muscle excitation-contraction coupling. Annual Review of Physiology 54: 109–133

Rosenfalck P (1969) Intra- and extracellular potential fields of active nerve and muscle fibres. A physico-mathematical analysis of different models. Thrombosis et Diathesis Haemorrhagica Supplementum 321: 1–168

Rothwell JC, Thompson PD, Day BL, Dick JP, Kachi T, Cowan JM, Marsden CD (1987) Motor cortex stimulation in intact man. 1. General characteristics of EMG responses in different muscles. Brain 110: 1173–90

Roy SH, De Luca CJ, Schneider J (1986) Effects of electrode locaiton on myoelectric conduciton velocity and median frequency estimates. J Appl Physiol 61: 1510–1517

Sanders DB, Stålberg EV (1996) AAEM minimonograph #25: single-fiber electromyography. Muscle & Nerve 19: 1069–1083

Schleenbaker RE, Mainous AG (1993) Electromyographic biofeedbackk for neuromuscular reeducation in the hemiplegic stroke patient: A meta-analysis. Arch Phys Med Rehabil 74: 1301–1304

Schmid UD, Walker G, Hess CW, Schmid J (1990) Magnetic and electrical stimulation of cervical motor roots: technique, site and mechanisms of excitation. Journal of Neurology, Neurosurgery and Psychiatry 53: 770–777

Shankar S, Gander RE, Brandell BR (1989) Changes in the myoelectric signal (MES) power spectra during dynamic contractions. Electroencephalography and clinical Neurophysiology 73: 142–150

Sica RE, McComas AJ (1971) Fast and slow twitch units in a human muscle. Journal of Neurology, Neurosurgery and Psychiatry 34: 113–120

Simoneau JA, Lortie G, Boulay MR, Thibault MC, Theriault G, Bouchard C (1985) Skeletal muscle histochemical and biochemical characteristics in sedentary male and female subjects. Can J physiol pharmacol 63: 30–35

Simoneau J-A, Bouchard C (1989) Human variation in skeletal muscle fiber-type proportion and enzyme activities. Am J physiol 257: 567–572

Simons DG, Mense S (1998) Understanding and measurement of muscle t one as related to clinical muscle pain. Pain 75: 1–17

Sleivert GG, Wenger HA (1994) Reliability of measuring isometric and isokinetic peak torque, rate of torque development, integrated electromyography, and tibial nerve conduction velocity. Arch Phys Med Rehabil 75: 1315–1521

Smits E, Rose PK, Gordon T, Richmond FJ (1994) Organization of single motor units in feline sartorius. Journal of Neurophysiology 72: 1885–1896

Sollie G, Hermens HJ, Boon KL, Wallings-De Jonge W, Zilvold G (1985) The measurement of the conduction velocity of muscle fibres with surface EMG according to the cross-correlation method. Electromyogr clin neurophysiol 25: 193–204

Stålberg E (1966) Propagation velocity in human muscle fibers in situ. Acta Physiologica Scandinavica Supplementum 287: 1–112

Stålberg E (1980) Some electrophysiological methods for the study of human muscle. Journal of Biomedical Engineering 2: 290–298

Stålberg E, Theile B (1973) Discharge pattern of motoneurones in humans. In New Developments in Electromyography and Clinical Neurophysiology. Desmedt J pp 234–241. Karger, Basel.

Stein RB, Oguztöreli MN (1978) The radial decline of nerve impulses in a restricted cylindrical extracellular space. Biological Cybernetics 28: 159–165

Stephenson DG, Lamb GD, Stephenson GM, Fryer MW (1995) Mechanisms of excitation-contraction coupling relevant to skeletal muscle fatigue. Advances in Experimental Medicine and Biology 384: 45–56

Svebak S, Braathen ET, Sejersted OM, Bowim B, Fauske S, Laberg JC (1993) Electromyographic activation and proportion of fast versus slow twitch muscle fibers: A genetic disposition for psychogenic muscle tension? Int J Psychophysiol 15: 43–49

Svensson P, Graven-Nielsen T, Matre D, Arendt-Nielsen L (1998) Experimental muscle pain does not cause long-lasting increases in resting electromyographic activity. Muscle & Nerve 21: 1382–1389

Tehovnik EJ (1996) Electrical stimulation of neural tissue to evoke behavioral responses. Journal of Neuroscience Methods 65: 1–17

ter Haar Romeny BM, Denier van der Gon JJ, Gielen CCAM (1984) Relation between location of a motor unit in the human biceps brachii and its critical firing levels for different tasks. Experimental Neurology 85: 631–650

Tesch PA, Komi PV, Jacobs I, Karlsson J, Viitasalo JT (1983) Influence of lactate accumulation of EMG frequency spectrum during repeated concentric contractions. Acta Physiol Scand 119: 61–67

Turker KS, Miles TS, Le HT (1988) The lip-clip: a simple, low-impedance ground electrode for use in human electrophysiology. Brain Research Bulletin 21: 139–141

Veiersted KB (1995) Medical Dissertation. National Institute of Occupational Health and University of Oslo, Oslo. pp 1–77.

Veiersted KB, Westgaard RH, Andersen P (1993) Electromyographic evaluation of muscular work pattern as a predictor of trapezius myalgi. Scand J Work Environ Health 19: 284–290

Veiersted K, Westgaard R, Andersen P (1990) Pattern of muscle activity during stereotyped work and its relation to muscle pain. Int Arch Occup Environ Health 62: 31–41

Vigreux B, Cnockaert JC, Pertuzon E (1979) Factors influencing quantified surface EMGs. European Journal of Applied Physiology and Occupational Physiology 41: 119–129

Viitasalo JHT, Komi PV (1975) Signal characteristics of EMG with special reference to reproducibility of measurements. Acta Physiol Scand 93: 531–539

Viitasalo JT, Saukkonen S, Komi PV (1980) Reproducibility of measurements of selected neuromuscular performance variables in man. Electromyogr clin neurophysiol 20: 487–501

Vollestad NK (1997) Measurement of human muscle fatigue. J Neurosci Methods 74: 219–227
Wickiewicz TL, Roy RR, Powell PL, Edgerton VR (1983) Muscle architecture of the human lower limb. Clinical Orthopaedics and Related Research 275–283
Windhorst U, Hamm TM, Stuart DG (1989) On the function of muscle and reflex partitioning. Behavioral and Brain Sciences 12: 629–681
Winkel J, Mathiassen SE, Hägg GM (1995) Normalization of upper trapezius EMG amplitude in ergonomic studies. Journal Of Electromyography And Kinesiology 5: 197–226
Winkler T, Stålberg E (1988) Surface anodal stimulation of human peripheral nerves. Experimental Brain Research 73: 481–488
Winter DA (1990) In Biomechanics and Motor Control of Human Movement. John Wiley & Sons, Inc., New York.
Yang JF, Winter DA (1983) Electromyography reliability in maximal and submaximal isometric contractions. Arch phys med rehabil 64: 417–420
Yu YL, Murray NM (1984) A comparison of concentric needle electromyography, quantitative EMG and single fibre EMG in the diagnosis of neuromuscular diseases. Electroencephalography and Clinical Neurophysiology 58: 220–225
Zhou S (1996) Acute effect of repeated maximal isometric contraction on electromechanical delay of knee extensor muscle. J Electromyogr Kinesiol 6: 117–127
Zipp P (1982) Recommendations for the standardization of lead positions in surface electromyography. J Appl Physiol 50: 41–54
Öberg T (1992) Trapezius muscle fatigue and electromyographic frequency analysis. Medical disseration. Linköping University, Linköping.

Abbreviations

τ	denotes a time difference
A/D	analogue to digital
Ach	acetylcholine
AMFCV	average muscle fiber conduction velocity
AP	action potential
ARV	average rectified value
CMMR	common-mode rejection ratio
CNS	central nervous system
CV	conduction velocity of the muscle fiber
DDT	double differential technique
EMG	electromyogram
FF	fast twitch fiber or motor unit, fatiguable
FFT	fast Fourier transform
FI	fast twitch fiber or motor unit, intermediary
FR	fast twitch fiber or motor unit, fatigue resistant
H-reflex	Hoffman reflex
IEMG	integrated electromyogram
IPI	inter-pulse interval
MDF	median frequency
ME	myoelectric
MNF	mean frequency
MU	motor unit
MUAP	motor unit action potential
M-wave	evoked compound action potential
PSD	power spectral density
RMS	root-mean-square
S	slow twitch fiber or motor unit
SDT	single differential technique
SNR	signal-to-noise ratio
TC	turns counting
ZC	zero crossing

Appendix

Reprinted from Merletti, Wallinga, Hermens & Freriks, 1997, Guidelines for reporting SEMG data, In: The state of the art on sensors and sensor placement procedures for surface electromyography: a proposal for sensor placement procedures, deliverable of the SENIAM project (Eds: Hermens & Freriks), Roessingh Research and Development, By Permission.

Guidelines for Reporting SEMG data

Merletti R., Wallinga W., Hermens H.J., Freriks B.

Starting point for reporting on experiments in which Surface EMG (SEMG) has been used is that the reporting should be such that an unambiguous insight in the methodology and techniques used is obtained in such a way that reproduction of the results is possible. To do so, the following details should be included in the report:

Electrodes
- material of which the electrodes are made (Ag/AgCl, Au etc.)
- shape of the electrodes (circular, square, rectangular)
- size of the electrodes: diameter in mm for circular electrodes, length x width in mm for square / rectangular electrodes

Electrode Placement Procedure
- description of the skin preparation method (e.g., skin abrasion, shaving of hair, cleaning method and means)
- whether gel or paste was used
- electrode location on the muscle with respect to tendons, motor point and other muscles; if possible: describe the electrode location as a position on a leadline between 2 anatomical landmarks
- electrode orientation with respect to the direction of muscle fibers
- inter electrode distance (IED): center to center distance in mm

SEMG detection equipment
- manufacturer and type of the SEMG (pre)amplifier
- detection mode (monopolar, differential, double differential etc.)
- actual gain and amplitude range in volts
- signal-to-noise ratio (S/N) in dB
- input impedance in mOhm
- Common Mode Rejection Ratio (CMRR)
- filters applied on the raw SEMG:
  - type (Butterworth, Chebyshev etc.)
  - kind (lowpass, bandpass etc.)
  - bandwidth and/or low and high pass cut-off frequencies in Hz
  - order and/or slopes of the cut-offs in dB/octave or dB/decade

Rectification method (if applicable)
Specify whether full or half-wave rectification was carried out.

Sampling SEMG signal into the computer
- manufacturer and type of the A/D board
- sampling frequency in Hz
- number of bits
- input amplitude range used in volts

SEMG amplitude processing method

- There are several methods of EMG processing.
  Smoothing the rectified signal with a low-pass filter of a given time constant (10–250 ms) is often described as "smoothing with a low-pass filter with a time constant of x ms". This process can be described as "linear envelope detection" by giving the time constant value and/or the cut-off frequency and the order of the low-pass filter used. Designating the SEMG resulting from this procedure as the "integrated EMG" (IEMG) is incorrect (see below).
- The mean value of the rectified EMG over a time interval T is defined as "Average Rectified Value" (ARV) or "Mean Amplitude Value" (MAV) and is computed as the integral of the rectified EMG over the time interval T divided by T. T should be reported in sec.
- Another method of providing amplitude information is the "Root Mean Square" (RMS) defined as the square root of the mean square value. Just as the ARV, this quantity is defined for a specific time interval T which must be indicated in sec. Smoothed, low-pass filtered, average rectified or RMS values are voltages and are measured in volts (V).
- Integrated EMG (IEMG) is sometimes reported. In this case the signal is integrated (not filtered!) over a time interval T. IEMG is therefore the area under a voltage curve and is measured in Vs.

SEMG frequency domain processing method

The Power Density Spectra presentation of the SEMG should include:

- the type of window used prior to taking the Fourier Transform (rectangular, Hamming etc.)
- the length of the time epoch used for each spectral estimate
- the algorithm used (e.g., FFT)
- whether zero padding was applied and the resultant frequency resolution
- the equation used to calculate the Median Frequency (MDF), Mean Frequency (MNF), moments etc.

Other processing techniques, especially novel techniques, must be accompanied by a full scientific description.

SEMG normalization method

When normalizing an SEMG recording, the amplitude parameter obtained from the recording is divided by the same amplitude parameter obtained from a Reference Contraction (RC), e.g., a Maximum Voluntary Contraction (MVC). The following information should be provided about the RC:

- how were the subjects trained to obtain the RC
- what was the joint angle and/or muscle length during the RC
- what were the conditions and angles of adjoining joints during the RC (e.g., for studies on elbow flexion, the condition of the wrist and shoulder joints should be provided)
- the rate of rise of force
- the velocity of shortening or lengthening
- the ranges of joint angle or muscle length in non-isometric contractions
- the load applied in non-isometric contractions

SEMG processing method for estimation of Muscle Fiber Conduction Velocity (MFCV)

Estimates of muscle fiber conduction velocity (MFCV) should include:

- report about the electrodes used (shape, size, material etc… (see above))
- the interelectrode distance in mm

- type of signals used to estimate MFCV (two single or two double diff. signals, multiple signals from a linear array etc.)
- the algorithm used for delay estimation (delay between reference points such as zero crossings, cross-correlation in the time domain, estimates in the frequency domain etc.)
- the delay resolution obtained

Chapter 27

Decomposition and Analysis of Intramuscular Electromyographic Signals

CARLO J. DE LUCA and ALEXANDER ADAM

Introduction

The clinical community has long shown interest in the concept of extracting as many *motor unit action potentials* (MUAPs) as possible from an intramuscular *electromyographic* (EMG) signal. Adrian and Bronk (1929) developed the first concentric needle electrode to identify both shape and firing rate of the MUAPs. Subsequent manual approaches of graphically measuring and quantifying the EMG signal evolved into computer-based techniques directed at identifying individual action potentials and discharge times by shape discrimination. The Precision Decomposition technique described in this chapter recovers all the usable information available in the EMG signal. The information can be conveniently grouped into two categories: morphology and control properties. Morphology describes the parameters of the MUAP shape such as the peak-to-peak amplitude, the time duration, the number of phases, and the area. These parameters are provided by the recovered Concentric and Macro MUAP. The morphology of the MUAP describes features that are related to the anatomical and physiological properties of the muscle fibers. These are the parameters which the clinician is accustomed to evaluating during a standard clinical EMG examination. The control properties of the motor units dictate the firing characteristics of the motor units. Therefore, the firing characteristics provide a description of how the motor units are controlled by the central nervous system and to some extent the peripheral nervous system. Clinically, they quantify upper motoneuron diseases.

The technique of Precision Decomposition has been under development by our group since the late 1970s. The first public description of it was in the form of an abstract published in the Abstracts of the Society for Neuroscience (LeFever and De Luca, 1978). The signal processing concepts which underlie the approach appeared in the IEEE Transaction of Biomedical Engineering (LeFever et al. 1982a, b). A more pragmatic description of the algorithms and workings of the technique was provided by Mambrito and De Luca (1984). This paper also described a generic foolproof method of measuring the accuracy of any decomposition technique. Stashuk and De Luca (1989) have provided an update on useful modifications and applications of the technique while De Luca (1993) recently provided a comprehensive, hands-on account of the methodology.

Principle

The term decomposition has been commonly used to describe the process whereby individual MUAPs are identified and uniquely classified from a set of superimposed *motor*

Correspondence to: Carlo J. De Luca, Boston University, NeuroMuscular Research Center, Department of Biomedical Engineering and Department of Neurology, Deerfield Street, Boston, MA, 02215, USA (phone: +01-617-353-9757; fax: +01-617-353-5737; e-mail: cjd@bu.edu)
Alexander Adam, Boston University, NeuroMuscular Research Center, Department of Biomedical Engineering, Deerfield Street, Boston, MA, 02215, USA

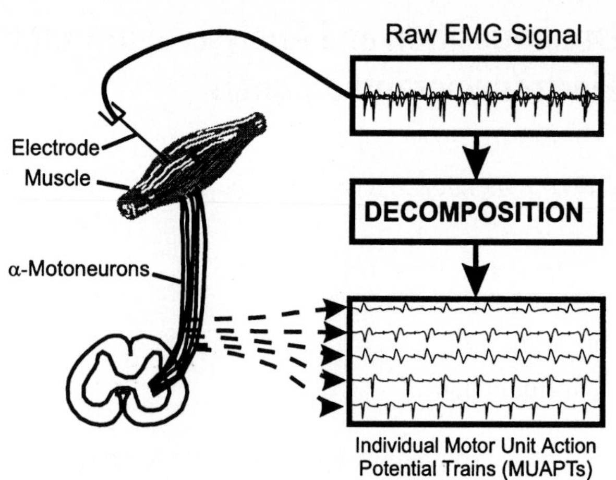

Fig. 1. Pictorial outline of the decomposition of the EMG signal into its constituent MUAPs (From De Luca et al. 1982a).

unit action potential trains (MUAPTs) belonging to concurrently active motor units. The concept of decomposition is depicted in Fig. 1. It involves the breaking down of the interference EMG signal that is recorded when more than one motor unit is active in the vicinity of the detection electrode. Identification refers to the categorization of the times of occurrences of the MUAP as well as the description of its morphological characteristics. From the above description it is apparent that the process of decomposing an EMG signal may range from a trivial task when only two MUAPTs with distinctly different MUAP shapes are present to a theoretical impossibility when many MUAPTs with nearly similar and unstable MUAP shapes are present.

A completely decomposed EMG signal provides all the information available in the signal. The timing information provides a complete description of the inter-firing interval (IFI), firing rate and synchronization characteristics; the availability of all the MUAPs which are discharged by a specific motor unit enables a more consistent expression of the shape by averaging the shape over a set of discharges.

Field of Application

The comprehensive, more accurate and more reliable information provided by decomposition finds applicability in both clinical and research environments. It is a new tool which enables us to explore the workings of the nervous system in normal and dysfunctional modalities. In the field of neurology, the ability to measure the behavior of firing rates and synchronization of motor unit discharges holds the promise of more analytically classifying dysfunctions of CNS origin. Consider the potential advantages of diagnosing a CNS abnormality by inserting a needle into a muscle with no direct assault to the CNS. The ability to obtain more reliable representations of MUAP shapes by averaging over several correctly identified MUAPs of an individual motor unit provides a more accurate basis for diagnostics based on morphological measurements. Furthermore, the capability of storing and measuring numerous MUAPs makes more convenient the laborious process of obtaining normative data. In fact, it enables individual laboratories to obtain their own normative data, thus allowing them to develop improvements in methodologies and approaches for measuring the characteristics of the MUAPs.

In the field of neurophysiology, the decomposition technique allows researchers to study the behavior of several concurrently active motor units and determine their characteristics beyond those relating to individual motor units and to discharge-to-discharge occurrences. It is now feasible to search for information transmission within the nervous system beyond individual neuron-to-neuron interaction. We can now explore more comprehensively and more effectively the orchestration of neuronal activation within and among muscles. It will be possible to execute these studies in the cooperative human performing voluntary contractions and without destroying the environment of the system under investigation. Furthermore, any enhancement of knowledge obtained from fundamental studies on the normal CNS can augment the clinical armament for performing diagnoses.

Scope

- Accuracy ranges from 85 to 100% depending on the complexity of the signal and the number of motor units.
- Able to decompose EMG signals detected up to 100% maximal contraction, nominally up to 80%; high threshold motor unit behavior may be studied.
- Resolves the occurrence of two or more superimposed MUAPs.
- Able to decompose up to 11 concurrently active motor units.
- Decomposition time = k(Accuracy)(Number of Motor Units)(Complexity of signal)

Unique Aspects of the Precision Decomposition

- Uses a special needle or wire electrode which has four detection surfaces, each 50 μm in diameter.
- The quadrifilar electrode detects Micro Signals. These are less selective than Single Fiber signals, but more selective than Concentric Signals.
- The Micro Signals are recorded with a bandwidth of 1 kHz–10 kHz.
- Use of quadrifilar wire electrodes allows for minor anisometric and long contractions
- May be operated in automatic or operator-assisted modes. Automatic mode is faster but less accurate.

Characteristics of the Precision Decomposition

- Morphological information of Concentric and Macro MUAPs when using needle electrode
- Control properties of individual motor units:
  - Dot Plot shows the duration of all IFIs of each motor unit. This plot may be used to study the behavior of the motor unit firings.
  - Bar Plot shows the location of all the MUAPs of each motor unit. This plot may be used to determine the recruitment and derecruitment threshold of a motor unit.
  - Firing Rate Plot shows the behavior of the firing rates of concurrent motor units.
- Groups of Motor Units:
  - Cross-correlation Plot shows the amount and relative latency of the cross-correlation (cf. Chapter 18) among the firing rates of concurrently active motor units.
  - Synchronization Plot shows the amount of synchronization between any pair of MUAPTs that have been decomposed.

Limitations

- With needle electrodes, use is generally limited to isometric contractions.
- With wire electrode, use includes slow dynamic contractions.
- Works only on EMG signals recorded during force rates less than 40% MVC/s.
- Can be challenging if the signal is complex and/or signal quality is low.

Outline

The decomposition procedure consists of the following components (See Fig. 2):

1. Signal detection
2. Needle positioning and needle movement
3. Signal acquisition and sampling
4. The decomposition algorithms
5. Data reconstruction
6. Data analysis and presentation.

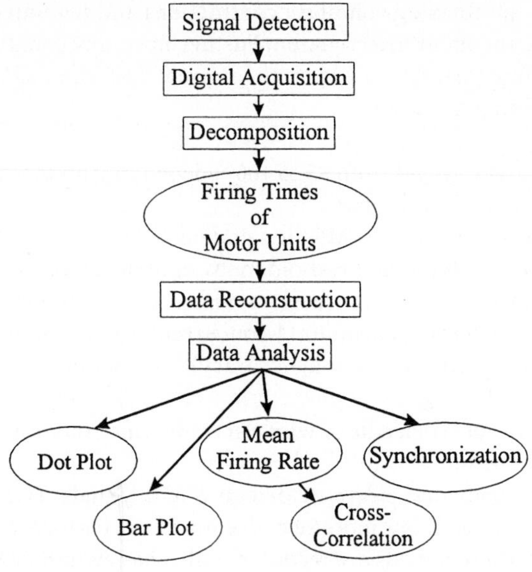

Fig. 2. Flowchart of the decomposition process.

Materials

- Quadrifilar needle electrode or quadrifilar fine wire electrode
- Needle front end: Buffer amplifiers and differential amplifiers
- PC with high-speed data acquisition board
- Sampling and compression software
- Decomposition software
- Analysis software

Procedure

Signal Detection

An important feature of the signal detection segment of the Precision Decomposition consists of having a *special electrode* that simultaneously detects three channels of an EMG signal from the same proximity of muscle tissue. This is a fundamentally important aspect of our approach. The three channels of signals are necessary to reduce ambiguities in the decision making processes which are employed by the computer algorithms to distinguish among different MUAPTs in the EMG signal. This task is accomplished by a specially designed quadrifilar electrode. Two types of quadrifilar electrodes are used: one is a needle electrode and the other is a wire electrode.

Electrode Choice

The needle electrode has the following *advantages*:
- It may be used to obtain the morphological characteristics of the action potentials.
- It may be used for obtaining Macro EMG signals.
- It may be precisely located in the muscle.
- It may be manipulated to locate it in regions which provide high quality signals.
- It is useful for clinical studies.

It has the following *drawbacks*:
- It cannot be used for dynamic contractions.

- It is painful when recording from deeply located muscles.
- It cannot be left in the muscle for prolonged periods; typically 1 hour.

The wire electrode has the following *advantages*:
- It may be used with slow dynamic contractions.
- It is not painful.
- It may be kept in the muscles for prolonged periods (several hours).

It has the following *drawbacks*:
- It cannot be relocated, thus compromising the ability to explore locations which yield signals that are more decomposable.
- It may migrate during a contraction.

The Needle Electrode

See Fig. 3A for details of the needle electrode and means for connecting it to the amplifiers. The salient feature of this electrode is its four detection surfaces located in a side port of the needle cannula. The wires are either 50 μm or 75 μm in diameter (depending on the fiber density of the particular muscle being studied) and are correspondingly spaced either 150 μm or 200 μm apart on the corners of a square configuration. These configurations were determined by empirical tests which found these dimensions of the detection surfaces to yield the signal quality required by the decomposition algorithms. The detection surfaces are connected in bipolar configurations through differential amplifiers to provide three differential outputs, each carrying an EMG signal. We found that the electrode with the smaller detection surfaces is more selective and therefore better suited for use in muscles with higher muscle fiber density and for contractions at

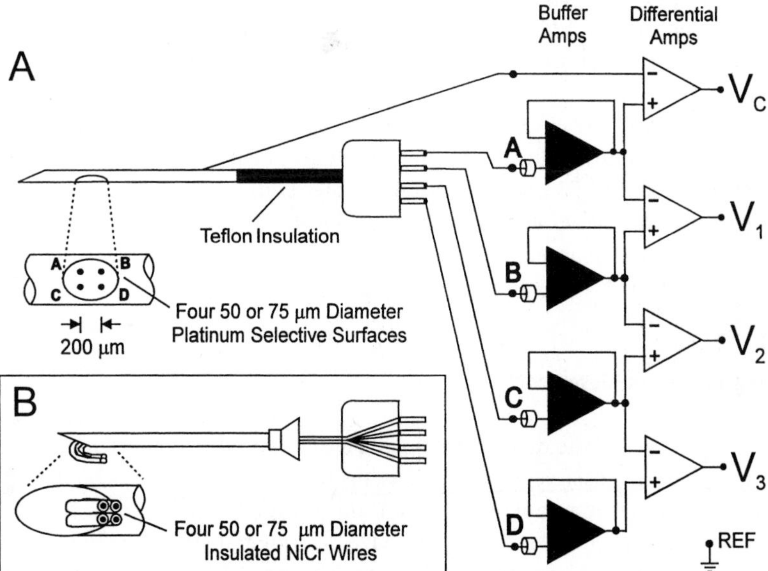

Fig. 3. A schematic representation of the electrode set-up for use in the Precision Decomposition technique. A) Quadrifilar needle electrode configured to detect 3 differential channels of EMG signals (V1 to V3) from the selective surfaces of the needle side port as well as an additional channel from the needle cannula (VC). Another needle configuration (not shown here) allows signals to be recorded from a concentric surface at the tip of the needle. B) Instead of the needle electrode, a quadrifilar wire electrode can be connected to the amplifiers. The wires are inserted into the muscle with a disposable hypodermic needle which is removed after the electrode is positioned. The distal end of the wires form a barb which helps to secure the electrode in the muscle. (Adapted from Mambrito and De Luca 1984)

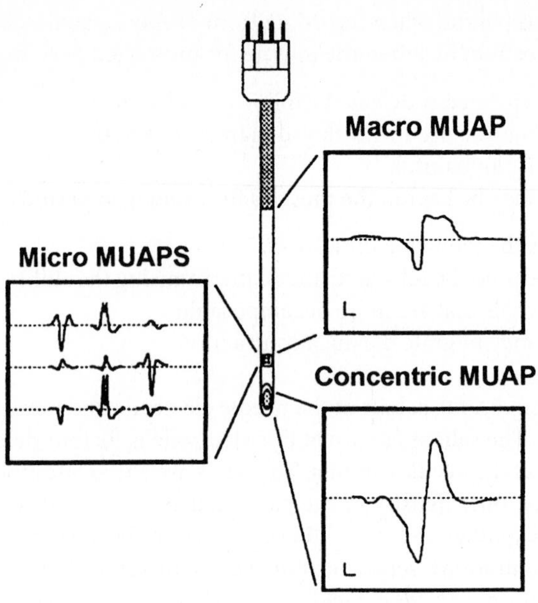

Fig. 4. Examples of the Micro, Macro and Concentric signals detected by the quadrifilar needle electrode. When using the quadrifilar wire electrode only the Micro signals are recorded. (From De Luca 1993)

higher force levels. The quadrifilar electrode is also able to detect either a concentric needle EMG signal, or via another configuration the Macro EMG signal. Examples of these arrangements are shown in Fig. 4. The concentric needle EMG is detected by a plug located at the tip of the needle. The detection surface of the plug has an area of 2.7 mm^2, the standard dimension used in concentric needles. Similarly, the Macro detection surface is standard size. Thus, depending on the configuration, a quadrifilar needle electrode detects three channels of EMG signals from the side port and either a concentric EMG signal or a Macro EMG signal for a total of four channels of EMG signals.

The Wire Electrode

See Fig. 3B for details of the wire electrode and means for connecting it to the amplifiers. The electrode consists of four Nichrome or platinum wires coated with nylon. The diameter of the wire is 50 μm or 75 μm, depending on the required detection selectivity, the smaller wire configuration being more selective. The wire is cut in cross-section, exposing the minimal amount of area. The distal 1 mm of the wire bundle is curved forming a barb which assists in securing the electrode into the muscle.

Needle Positioning and Needle Movement

The highly selective nature of the quadrifilar needle electrode, which makes identification of several motor units possible, makes collecting stable EMG signals a challenging task. Small movements that may normally have little or no effect on signals detected with concentric or macro needle electrodes can have detrimental effects on the signals detected from the quadrifilar electrode. For this reason we have developed several techniques for inserting and positioning the quadrifilar electrode that minimize its movement and maximize signal quality.

Needle Insertion

The most stable EMG signals are those that are obtained from a needle that is inserted oblique to the direction of the muscle fibers, typically at an angle of 30 degrees. If inserted more parallel to the muscle fibers, there tends to be more sliding of the fibers with respect to the electrode. At angles steeper than 30 degrees, there tends to be more needle movement due to shear forces on the cannula. To minimize needle movement, one has to find the angle where the shearing forces are just enough to act as an anchor on any fibers that may be moving. Our experience has shown this angle to be about 30 degrees. If possible, position the detection surfaces of the quadrifilar electrode so that they are near a motor point. In this region one finds MUAPs that propagate in opposite directions. When measured with the quadrifilar electrode, such MUAPs have opposite polarity. The difference in their MUAP shapes makes them easier to distinguish and decompose the signal.

Needle Rotation

Due to its unique design, the quadrifilar needle electrode requires a slightly different positioning technique than a standard concentric needle electrode. The detection surface on a concentric electrode is relatively large and located on the tip. Rotation of the electrode causes small changes in signal quality compared to inward and outward translations. This type of electrode is typically positioned by moving it inward and outward, making many needle tracks. The detection surfaces on the quadrifilar electrode are much smaller and located on a side port, rendering the electrode more sensitive to rotation. When positioning the quadrifilar electrode, it is helpful to rotate it back and forth as it is slowly inserted or withdrawn. This enables the experimenter to sample the maximum area, and minimizes the number of needle tracks.

Audio Feedback

Sound is a useful tool when searching for MUAPs. As the detection surfaces of the quadrifilar electrode are positioned closer to active muscle fibers, the signals tend to increase in amplitude and frequency. This is reflected by louder and sharper popping and crackling sounds when the signals are fed into an audio device. By using this audio feedback, the experimenter can concentrate on how he is positioning the electrode rather than constantly turning his head to look at a screen or oscilloscope. Consequently, better signals can be found faster.

Contraction Force

It may take several minutes to locate a position where the motor unit shapes are distinct. If the subject maintains a force level that is too high during positioning, the muscle will fatigue too quickly; if the force is too low, the needle will not give an accurate representation of how it will anchor itself in the muscle. When positioning the electrode, have the subject maintain approximately 10 % MVC and try to find two or three distinct motor unit shapes. Once a suitable position has been found, have the subject slowly increase the force level to the desired amount while monitoring the signal. Generally two or three more motor units will be observed. If the signal does not remain stable when the force is increased, try another position.

Stability

Perhaps the most difficult part of needle positioning is finding a signal that remains stable while the force level changes. Even during isometric contractions, there are slight movements of the muscle fibers relative to each other, the fascia, the skin, and the electrode. Unfortunately, these movements can cause signal instability. There are, however, several ways to deal with this. It is sometimes useful to watch how the needle moves and see how the signals change during a contraction. The needle can then be positioned such that the movement will improve signal quality. If this is not successful, hold the needle and keep it from moving. This tends to work better, but it can also create problems. Remember, one wants to keep the detection surfaces in a fixed position relative to the local muscle fibers. Thus, it is sometimes necessary for the needle to move with the rest of the muscle so that the detection surfaces can remain stable relative to the local fibers. Holding the needle may prevent this from occurring. In general, the most trouble-free signals are obtained from electrodes that are well secured in the muscle and are not held.

Signal Acquisition and Sampling

Amplification and Filtering

Any attempt to record an EMG signal should always amplify the signal as much as possible, without distorting the signal, prior to digitizing the signal. In doing so, the sampling resolution of the digitized signal is maximized because the full sampling range is used. The recommended procedure is to adjust the gains of the amplifiers to the value where the EMG signal is as large as possible without any clipping at its peaks. The signals detected from the fine wires at the tip of the needle or from the wire electrodes are band-limited to 1 kHz to 10 kHz. This is a unique feature of our decomposition procedure. It purposefully distorts the shapes of the action potentials, thus rendering them unfamiliar in appearance to the investigator, but particularly useful to the decomposition algorithms. The shapes become shorter and sharper in appearance; the tails of the MUAPs are considerably reduced in length, decreasing the chance of superposition among the different MUAPs. It is recommended that the concentric needle EMG signal be recorded with a bandwidth of 10 Hz to 10 kHz and the Macro EMG signal with a bandwidth of 10 Hz to 1 kHz.

Sampling Threshold

When the signal in any of the channels surpasses a preset threshold value, the signals are sampled and digitized. The threshold has a default value which has been found to be appropriate for a typical signal. By judiciously setting the threshold value, it is possible to improve the signal-to-noise ratio of the sampled signal that will be analyzed. This operation requires practice and is only recommended when the user has become familiar with the system. Efficiency in data storage is achieved by storing only those parts of the signal where there is activity above the threshold value. A complete time reference is provided by storing the amount of time between stored signal epochs represented by the number of skipped samples. Thus, no information pertinent to the decomposition process is lost. An example of the time-compressed data is presented in Fig. 5 (lower panel)

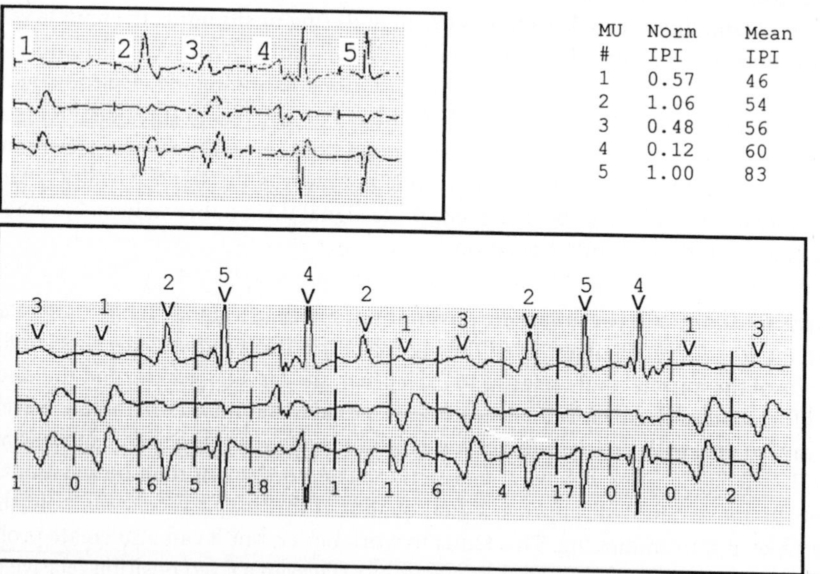

Fig. 5. TOP LEFT: Three-channel representation of motor unit (MU) templates from five different MUs. TOP RIGHT: Average IPI and normalized IPI for the MUs identified on the left. BOTTOM: Compressed three-channel EMG signal with assigned MU numbers. Vertical bars indicate skipped time intervals. Numbers below bars represent the skipped time in ms. Height of the bars represents the save threshold, i.e., minimum signal amplitude which is not skipped during the digitization process.

which presents a view of the screen during an operator-assisted decomposition. Note the vertical bars indicating the skipped time and the number of milliseconds skipped which appears directly below the skipped interval.

We have called the signals detected at the side port detection surfaces, Micro EMG signals. They are detected from a few (approximately 3 or 4) fibers per motor unit. Thus the Micro EMG signals are not as selective as those from a single fiber electrode, but are much more selective than those detected from a concentric electrode. The Micro EMG signals are sampled at a rate of 50 kHz. This sampling rate is well above the *Nyquist rate* (see Chapter 45). It is necessary to provide the required resolution of the MUAPs so that sufficiently accurate alignments can be made when comparing MUAPs in the decomposition algorithm. This is an important feature of the Precision Decomposition because the decision space is in the time domain. The concentric needle and macro EMG signals can be sampled at a lower rate, typically 2 kHz, because they have narrower frequency bandwidths.

The Decomposition Algorithms

The decomposition routines are complex rule-based algorithms which have evolved over a period of two decades and contain information know-how for dealing with the peculiarities encountered in real EMG data. These algorithms identify action potentials using *template matching* and *probability of firing statistics*, resolve superpositions, and allocate the action potentials to motor units. User-interactive editing algorithms are used to check the accuracy and make modifications according to well-established rules. For details see De Luca (1993).

Data Reconstruction

With the time record of the MUAP firings of each MUAPT in the EMG signal established by the decomposition of the Micro EMG signals, it is possible to extract the Concentric and Macro EMG MUAPs from the corresponding EMG signals. This is accomplished by *waveform averaging* or in physiological terminology, *spike-triggered averaging* (cf. Chapter 18). That is, each time a MUAP of a particular motor unit is present, select the corresponding time interval from the concentric or Macro EMG signal and save it. Then average the waves in all the time intervals. In this fashion, that part of the waveform (in the time intervals) which belongs to the MUAP will add across the time intervals and that part not associated with the MUAP will tend to cancel out because the positive and negative phases of the other action potentials in the time intervals will overlap considerably. This averaging procedure works best if the number of time intervals is large, the "noise" signals are small in number and low in amplitude, and there is little synchronization among the motor units.

The other necessary factor for the trigger-averaging to work is that the MUAP being recovered must be present in both the Micro signals and the Concentric or Macro signal being considered. The geometry of the quadrifilar electrode is organized with this requirement in mind. An example of recovered Concentric MUAP and Macro MUAP is presented in Fig. 4.

Data Analysis and Presentation

When an action potential has been identified as belonging to a specific motor unit, the algorithms seek the greatest value of the amplitude of the action potential and store the time of its occurrence. In so doing, a time series of all the discharges of each motor unit

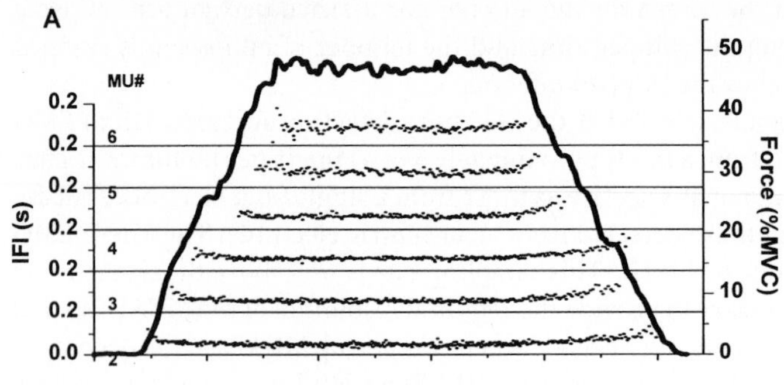

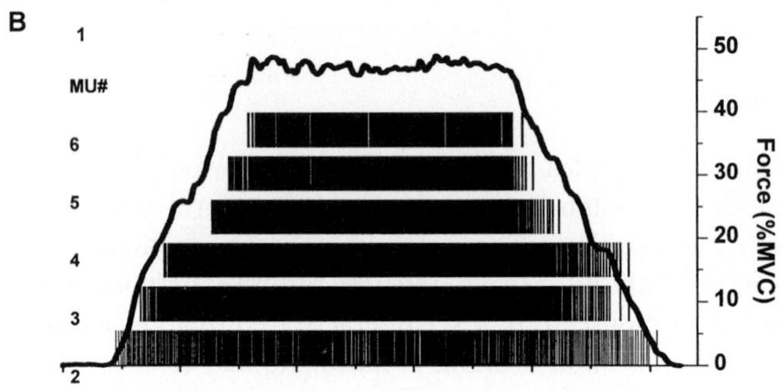

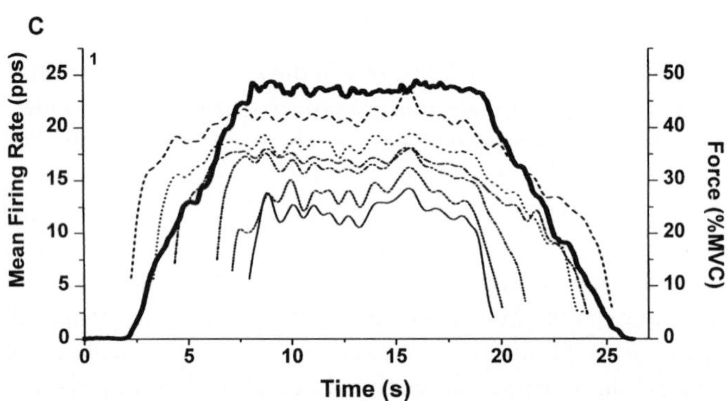

Fig. 6. Plots A to C are derived from the same decomposed EMG signal record. Contraction time is measured on the horizontal axis and contraction force (solid line), normalized to the maximum voluntary contraction (MVC), is measured on the right vertical axis. A) Dot Plot: Each inter-firing interval (IFI) of a MUAPT is plotted sequentially on the vertical scale. B) Bar Plot: A bar is placed in the location of each motor unit action potential (MUAP). C) Firing Rate Plot: The mean number of pulses per second of each motor unit is plotted as a function of time.

is obtained. The time intervals between discharges are plotted as a function of contraction time as shown in Fig. 6A. Such a Dot Plot is used in the user-interactive editing procedure to check for decomposition errors. To illustrate the recruitment order of motor units, the firing time data are plotted as Bar Plots, an example of which is provided in Fig. 6B. The individual discharges are useful for investigating motor unit characteristics such as synchronization and other discharge-to-discharge relationships such as reflex responses, but provide little useful information concerning the control aspects of the motor unit firings. For this purpose it is more useful to study the behavior of the firing rates of motor units that provide a more mechanically relevant relationship. The firing rates may be obtained in a variety of ways. We prefer to *low-pass filter* each motor unit's firing time impulse train with a *Hanning window* to produce a continuous-time, mean firing rate signal. For most applications we prefer a Hanning window width of 400 ms. However, the amount of smoothing will depend on the information to be extracted from the firing rates. Figure 6C shows the mean firing rate signals obtained from the motor unit discharges using a smoothing window of 400 ms.

Results

Applications for neuroscience focus on the behavior characteristics of the firings of concurrently active motor units. Consequently, we will describe results relevant to this issue. For clinically relevant provisions of the Precision Decomposition technique, please refer to De Luca (1993).

Firing Rate Decay

The first observation directly resulting from Precision Decomposition analysis was the firing rate decay (De Luca and Forrest, 1973; De Luca, 1985; De Luca et al., 1996). We reported that during isotonic and isometric contractions, the firing rate of the motor units decreased as a function of time (Fig. 7A). As the firing rate decreased, we never saw a new motor unit being recruited during the first 20 s of a contraction. We first suggested (De Luca, 1979) and later interpreted (De Luca et al., 1996) the firing rate decrease during sustained voluntary contractions to be a manifestation of two phenomena: a) The intrinsic property of the motoneuron to exhibit a firing rate decay over time when stimulated with a DC current which was first described in an animal preparation by Kernell (1965). b) A reduced need to fire a motor unit due to the increase in amplitude and duration of its force twitch upon repeated discharge, commonly referred to as twitch potentiation.

Common Drive

The second observation was the phenomenon of common drive (De Luca et al., 1982a, b). We found that the firing rates of motor units fluctuated in unison with essentially no time delay between them. This was seen by cross-correlating the firing rates of pairs of concurrently active motor units (Fig 7B). We saw this behavior in all muscles tested, ranging from small distal muscles to large proximal muscles. Even motor units belonging to different motoneuron pools exhibited common firing rate fluctuations when controlled as one functional unit; this we observed during antagonist muscle co-activation (De Luca and Mambrito, 1987). The existence of the common drive has been verified by independent investigators (Miles, 1987; Stashuk and de Bruin, 1988; Guiheneuc, 1992; Iyer et al., 1994; Semmler et al., 1997). It indicates that the CNS has evolved a relatively simple strategy for controlling motor units. Additionally we found that the irregular nature of the firing rates and the common drive phenomenon imply that muscles cannot produce smooth constant forces. We verified this fact by cross-correlating the firing rates and the force output of the muscle and found a significant correlation with a latency due to the mechanical delay in force buildup of the muscle fibers and force transmission through the muscle and tendon tissue (Fig. 7C).

Synchronization

The existence of a high degree of cross-correlation between the firing rates of motor units does not imply that the individual firings of the motor units are synchronized. By synchronization it is meant that motor units fire at a fixed time latency with respect to each other. Synchronization occurs in two modalities: short-term and long-term (Fig. 8). A study of motor unit pairs detected during isometric isotonic contractions in six muscles

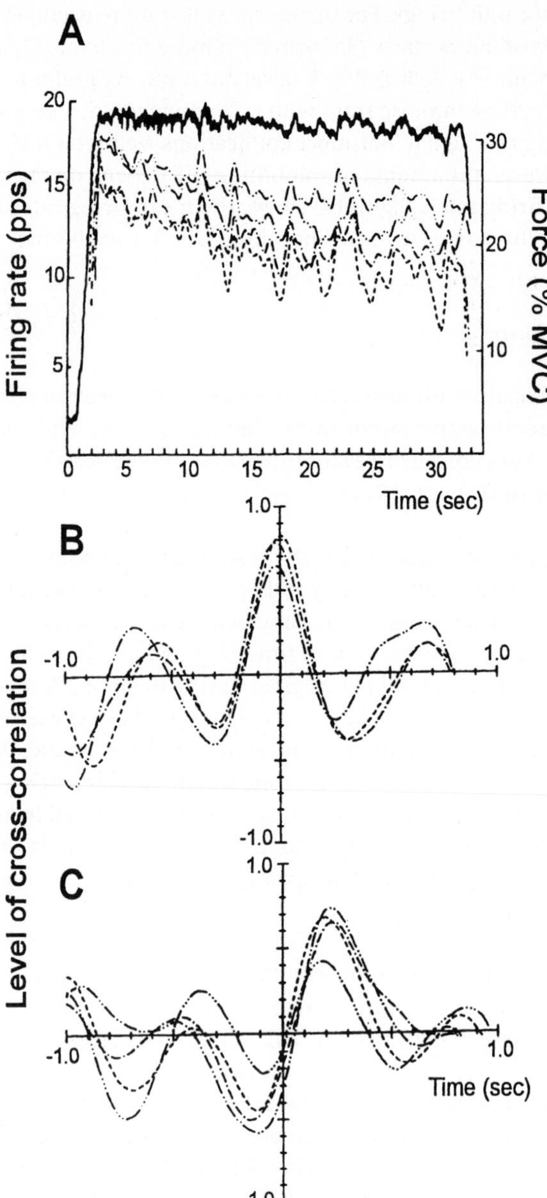

Fig. 7. A) Firing rate records of four concurrently active motor units (dashed lines) are shown superimposed on the force output (solid line) recorded during an isometric constant-force contraction of the deltoid muscle. The force level is given as a percentage of MVC on the right. B) The cross-correlations of the mean firing rates of a motor unit with those of the other units. Note that the peaks occur at zero time. C) The cross-correlations of the firing rates of all four motor units with the force output of the muscle. Peaks occurring at positive time lags indicate that the firing rate leads the force as is expected due to the time required to build up the force in the muscle after the fibers have been activated. (From De Luca et al. 1982b)

revealed that an average of 8 % of the firings were short-term synchronized and only 1 % long-term synchronized (De Luca et al., 1993). Short-term synchronized firings occurred at sporadic intervals and in bursts of typically one or two consecutive firings which had no apparent effect on the force produced by the muscle. We concluded that synchronization of motor unit firings is an epi-phenomenon with no physiological design of its own.

Onion Skin

The third observation was the onion skin phenomenon. Along with Person and Kudina (1972) as well as Tanji and Kato (1973), we (De Luca and Forrest, 1973) were among the first to report that during isometric contractions lasting less than 20 s, the earlier recruited motor units always fired at greater average rates than later recruited motor units. (When the firing rates are plotted as a function of time, the hierarchical values of the motor unit firing rates form overlapping layers resembling the structure of the skin

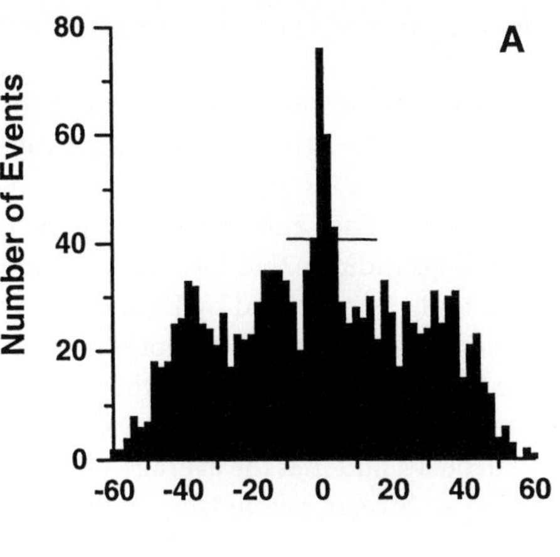

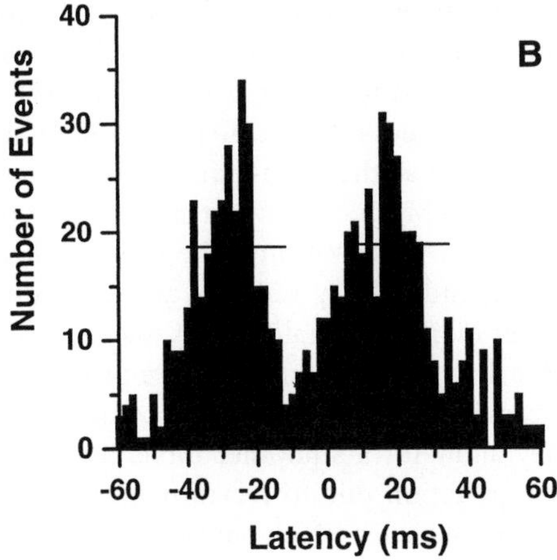

Fig. 8. Synchronization plots: The amount or synchronization between a pair of motor units is studied by calculating a cross-interval histogram. A cross-interval histogram accumulates the occurrences of the time interval between the firing of one motor unit and the first subsequent and previous firing that occur in the companion motor unit. A) Example of a cross-interval histogram which displayed short-term synchronization. B) Example of long-term synchronization.

of an onion. See Fig. 6C). Subsequently (De Luca et al., 1982a), we documented this phenomenon in detail. Independent verifications followed by Hoffer et al. (1987) and Stashuk and de Bruin (1988). Thus, the later recruited, more glycolytic, faster-twitch motor units which require a greater firing rate than the earlier recruited, more oxidative, slower-twitch motor units to fuse would be less likely to tetanize. Even during high level contractions in the neighborhood of 80 to 100 % MVC, the firing rates of the high threshold motor units are in the range of 20–30 pulses per second (pps), an amount likely to be insufficient for complete tetanization. This finding ran counter to the previously held belief that higher threshold motor units would be expected to fire faster so as to produce more force. The onion skin phenomenon begs the question as to why motor unit control developed so as to not maximize the force-generating capacity of a muscle. After all, if the purpose of a muscle was to generate force, it was reasonable to speculate that the motor unit control would be organized to make the most of available mechanical capacity within the muscle. Why should muscles evolve to have an apparent Reserve Capacity not commonly accessible during voluntary contractions? This an intriguing question. One possible explanation is that the higher threshold motor units, which are faster fatiguing, would become exhausted quickly if they fired fast. A control system so organized would not provide sustained contractions at high force levels which would be necessary to cope with life-threatening situations and ensure the survival of the species. It appears that the motor unit control developed to maximize a combination of contraction force and contraction time rather than only the contraction force. The available reserve capacity for generating force over brief periods of time may explain the occurrence of exceptional feats of strength that are reported to occur during life-threatening situations.

Two corollary observations were also made for the behavior of the firing rates: a) The later recruited motor units had greater initial firing rates as previously indicated by Clamann (1970). b) The firing rates of all units converged to a near common value during maximal contractions (De Luca and Erim, 1994; Erim et al., 1996).

All the above findings indicate that the control signals (net excitation) act on the motoneuron pool as a unit. As proposed by Henneman and colleagues (Henneman et al., 1965a,b) the individual properties of the motoneurons determine the recruitment hierarchy in response to the net excitation. To that enlightening observation we now add that the firing rate of the individual motor units responds to the net excitation communally and simultaneously, and that the average value of the firing rates is also hierarchically organized with an inverse relationship to the recruitment threshold.

Diversification

The fourth observation was the diversification of the control properties (De Luca et al., 1982a). The motor units of smaller, distal muscles such as the first dorsal interosseous tend to be recruited in the force range up to 50 % MVC and have mean firing rates that reach relatively high values (approx. 40 pps) at 80 % MVC. Whereas those from larger, proximal muscles such as the deltoid and the trapezius recruit their motor units in a force range up to 80 % MVC and have firing rates that reach relatively lower values (approx. 30 pps). A similar observation in the adductor pollicis and biceps brachii muscles was reported independently by Kukulka and Clamann (1981). The reduced dynamic range of the larger more proximal muscles may be due to the increased recurrent inhibition of the Renshaw system which is more prominent in these muscles as shown by Rossi and Mazzachio (1991). These diverse control properties are useful in at least two ways. First, they allow for a smoother force. Smaller muscles have less motor units, therefore, force gradation due to recruitment would be coarser throughout the full range than in larger muscles which have many more (an order of magnitude or more) motor units. Second, the larger more proximal muscles tend to be more postural and are required to produce sustained contractions more often. The lower firing rates in these muscles delay the progression of fatigue.

Exercise

We found that long-term exercise appears to induce modifications in the motor unit control properties (Adam et al., 1988). Comparing the motor unit control parameters of the first dorsal interosseous muscles of the dominant and non-dominant hands performing isometric, isotonic contractions at the same MVC level, we found that motor units of the dominant side had lower firing rates for the same level of contraction, and a larger number of motor units were recruited at lower force levels. This finding is consistent with the previously known fact that the dominant hand has slower twitch muscle fibers, probably due to the life-long preferential use. The twitches of slower fibers fuse at lower firing rates allowing for a reduced excitation and decreased firing rates in the dominant hand without a reduction in force output.

Aging

We have recently reported that aging causes alterations in the motor unit control properties (Erim et al., in press). In our study in the first dorsal interosseous muscle of elderly subjects above 65 years of age, we found that the firing rate and the recruitment threshold of motor units became modified in the same manner as that induced by exercise. This observation was not surprising because it is well known that aged muscles contain a greater percentage of slow twitch Type I fibers, as is the case in exercised muscles, although the cause for the increased percentage of Type I fibers appears to be different. When we studied the common drive in the elderly, we saw that in approximately one-half of them the cross-correlation between pairs of motor units was severely re-

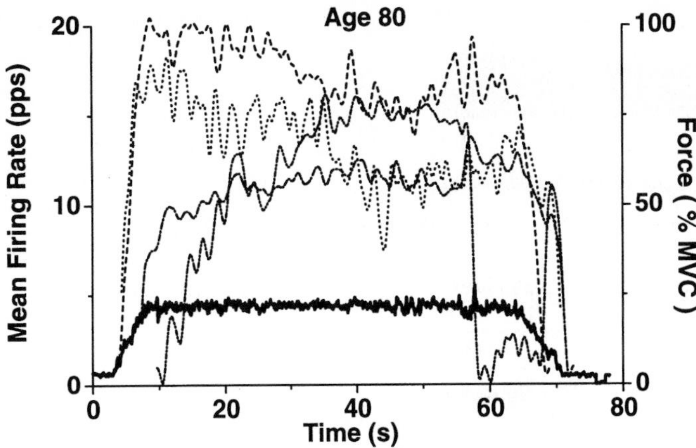

Fig. 9. Example of cross-over of motor unit firing rates in an elderly subject. See legend to Figure 6C for details of plot. (From Erim et al., in press)

duced and in some cases apparently nonexistent. Also, in the elderly, the onion skin phenomenon was disrupted (Fig. 9). When plotted as a function of contraction time, the firing rates of numerous motor units crossed over those of earlier recruited ones and the behavior of the firing rates was not orderly in a hierarchical sense; some decreased while others increased during an isometric, isotonic contraction. We surmise that this dissociation among the firing rates of motor units leads to an inefficient force generation scheme.

Motor Unit Substitution

All the above observations were made on relatively short-duration (less than 20 s) isometric, isotonic contractions. They may not fully describe the behavior of the control properties during sustained contractions of limb muscles or postural muscles which are commonly required to contract for prolonged periods of time. Recently, we have seen cases where the onion skin property is disturbed during short-term (20 s or less) contractions of normal healthy trapezius muscle and during long-duration (150 s or more) contractions in normal healthy first dorsal interosseous muscles. We suspect that the cross-over of the firing rates is due to at least two factors which cause the firing rates of earlier recruited motor units to decrease below the value of the newly recruited motor units: a) The Renshaw recurrent inhibition of earlier recruited motor units which is more dominant in proximal muscles such as the trapezius, hence the disturbed onion skin during short-duration contractions. b) The motoneuron adaptation process reported by Kernell (1965) which decreases the firing rates of motor units during sustained activation, causing the discharge rates of earlier recruited motor units to decrease below that of later recruited motor units.

While studying long-duration contractions in the range of 5 min to 60 min with our colleague Westgaard, we observed definite examples of motor unit substitution (Westgaard and De Luca, 1999). These are cases where a motor unit stopped firing during a sustained contraction when the activity level decreased slightly, and in response to a subsequent slight increase in the force output, a new motor unit was recruited in place of the one that was derecruited (Fig. 10). We believe this phenomenon is the result of adaptation of the recruitment threshold of active motoneurons. The recruitment threshold of a motor unit which had been active for some time would have become greater than the recruitment threshold of the next one in the hierarchy. In this fashion the next motor unit becomes recruited in response to an increase in the net excitation to the motoneuron pool.

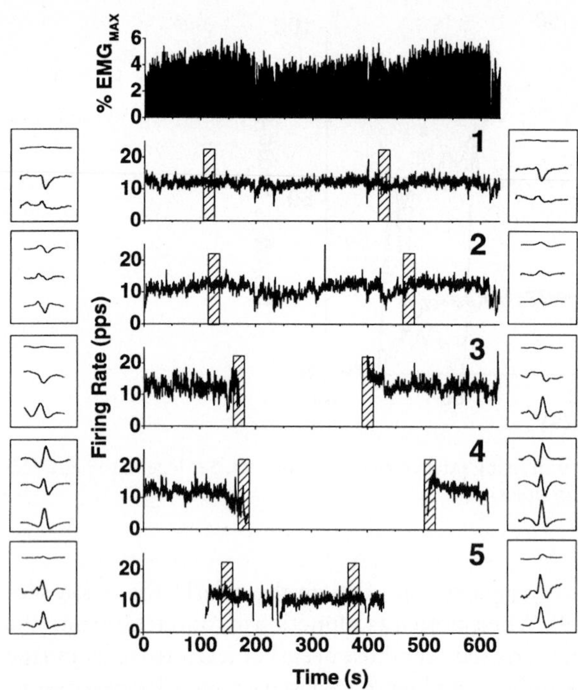

Fig. 10. Motor unit substitution during sustained contraction of the trapezius muscle. The top panel shows the root-mean-square (RMS) amplitude of the surface EMG signal normalized to the value measured at the MVC. The plots below represent the firing rates of five motor units identified by Precision Decomposition from intramuscular EMG signals obtained with a quadrifilar wire electrode. Motor units #1 and #2 fire continuously, while motor unit #3 ceases firing as the surface EMG signal decreases (t = 170 s) and becomes active again as the amplitude of the EMG signal increases (t = 400 s). Motor unit #4 behaves in a similar fashion. The novel observation is the fact that motor unit #5, the highest threshold unit in this group, fires when the lower threshold motor units are silent (200 s < t < 400 s). We refer to this phenomenon as motor unit substitution. Boxes to the left and right of the firing rate plots contain the characteristic Micro EMG shapes for each of the MUAPTs. Motor unit action potential shapes are averages obtained at time intervals indicated by shaded bars in each of the firing rate plots. (Modified from Westgaard and De Luca 1999)

Troubleshoot

Difficult Superpositions

During an automatic or operator-assisted decomposition, superpositions of several MUAPs almost always create problems. In the operator-assisted mode the program stops and asks the operator for help; in the automatic mode it tends to skip the superimposed waves. In either case, the superpositions need to be resolved by the operator if 100% accuracy is desired. The following steps may be helpful when trying to resolve the more difficult superpositions.

Normalized IFI

When subtracting templates from a superposition, remove the most obvious components first. As each template is removed, components of other templates may become easier to recognize. Use the normalized IFI information to determine which motor units have the greatest probability of firing. Look ahead in the signal to determine which waves, other than those that may be contained in the superposition, fulfill this criterion. This tactic may give hints as to which waveforms are hidden in the superposition. Keep in mind that rapid force changes can cause peculiar firing patterns; thus IFI information may not always be useful.

Sometimes templates may not be cleanly subtracted, leaving a residual that clutters the rest of the superposition. This can be due to peak misalignments which are caused by the overlapping of several shapes. Try subtracting the template from a peak on a different channel or try subtracting templates in a different order. This approach may simplify the resolution of the superposition. If all else fails, skip the superposition and go back to it later. When more information is known about the signal, it may be easier to resolve.

Error Detection

The Dot plot, a plot of the IFIs as a function of contraction time for every motor unit, is useful to detect incorrect motor unit allocation. An example of this plot for a decomposed signal record containing errors is presented in Figure 11A. If the firing of a motor unit has been missed, the amplitude of the IFI will be twice as great as that of the average firing interval. An example of this case may be seen in motor unit #2, where at approximately 11 s an abnormally great IFI occurs. The missed detection of a firing is likely accompanied by an incorrect allocation to another motor unit. Therefore, in the time vicinity of the skipped firing of one motor unit there will be one or two short IFIs in another unit. This occurrence is noted in motor unit #4 at roughly the same time. Using an

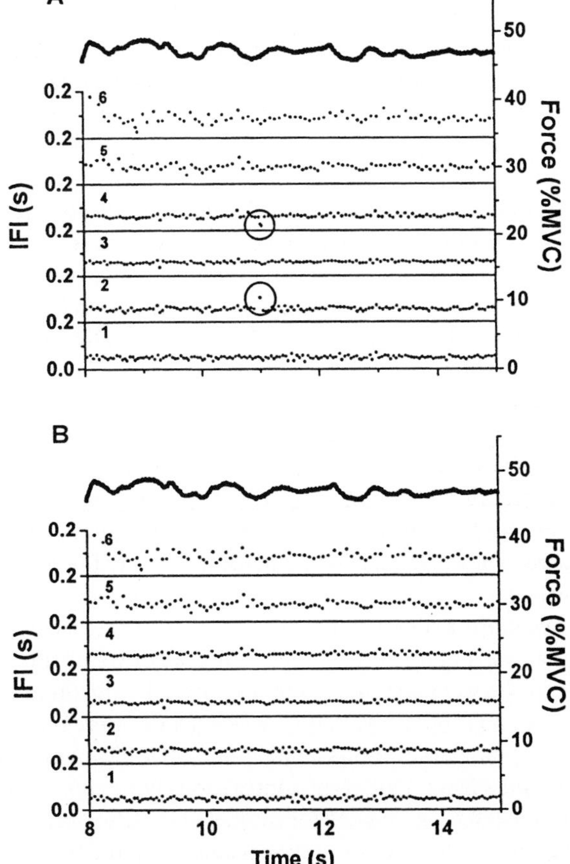

Fig. 11. A) An example of the decomposed IFIs of six concurrently active motor units containing two classification errors. Circles indicate the occurrence of one large and two small IFIs, which are due to the misqualification of motor unit #4 for motor unit #2. The use of the Dot Plot in identifying discordant motor unit firings becomes apparent. B) The same record of data after the editing procedure.

editor program which displays the original EMG record as well as the assigned motor unit templates and IFI statistics, the classification of motor units is adjusted. The new Dot Plot (Fig. 11B) shows the error removed from the IFI plots.

Comments

Validation of Decomposition Technique

A basic fundamental question arises when decomposing a signal that has an unobservable source, such as the EMG signal, into its constituent units (MUAPs). That is – how does one know that the decomposed sequence of motor unit discharges represents the true and unique solution? Therefore, it is essential to assess the accuracy of any EMG signal decomposition system and to validate the results obtained using such a technique. Furthermore, the decomposition technique may be highly interactive and during decomposition many decisions may be made by the operator. Thus, it is also necessary to assess the consistency of the results produced by different operators.

Testing for Consistency

The issue of the consistency is the simpler of the two, and it has been addressed by LeFever et al. (1982 b). Briefly, the following test was performed. Two highly trained operators (each with at least 400 hours experience in decomposing EMG signals) and a third, less experienced, operator (16 hours of EMG signal decomposition) were required to independently decompose the same EMG signal record which was considered `difficult' (i.e., at the limit of the decomposition technique capabilities according to the two experienced operators). The EMG signal selected contained 5 MUAPTs which the skilled operators believed had been reliably detected. Both skilled operators were 100% in agreement for the detection of a total of 479 MUAPs from 5 motor units. The results of the untrained operator decomposition contained a total of 12 discrepancies with respect to the two trained operators. Since the original, the consistency has been tested in a similar fashion on many other occasions.

Testing for Accuracy

The issue of the accuracy is much more complicated. It is impossible to measure the decomposition accuracy in an absolute sense with real EMG signal, since occurrence times of all the MUAPs and precise definitions of all MUAP waveforms in the EMG signal are unknown a priori or a posteriori. This limitation has been circumvented in two ways.

First, the accuracy was tested on synthetically generated EMG signal. For details on the procedure to generate a synthetic EMG signal and execution of the test, refer to LeFever et al. (1982 a, b). Briefly, the synthetic EMG was constructed by linearly superimposing 8 mathematically generated MUAPTs along with Gaussian noise. The standard deviation of the zero mean Gaussian noise was 40% of the peak amplitude of the smallest MUAP waveform. A skilled operator was able to decompose the record with an accuracy of 99.8%, incurring one error in a total of 435 classifications. This particular record is now used as a benchmark to identify the performance criterion of new operators.

Secondly, a direct test of the accuracy of the decomposition technique on real EMG signal was obtained in the following way (Mambrito and De Luca 1984). Two needle electrodes were inserted in the same muscle (tibialis anterior) about 1 cm apart. The two sets of EMG signals from the two electrodes were recorded simultaneously and decomposed. Some motor units contributed MUAPTs in both sets of signals. A comparison of the results from 3 different contractions with two "common" MUAPTs per contraction showed 100% agreement for a total 1415 detections of the "common" MUAPs. In this case, an undetected error in the results from the "common" MUAP detections could occur only if a simultaneous error of the same kind (wrong classification of a MUAP or missed detection) is made in the decomposition of the two records. The chances of such an event are incalculably small. Thus, the consistency of the decomposition data of the same units from two different electrodes provides an indirect measure of the accuracy in real data decomposition. This test has been repeated numerous times with similar results.

References

Adam A, De Luca CJ, Erim Z (1988) Hand dominance and motor unit firing behavior. J Neurophysiol 80:1373–1382

Adrian ED and Bronk DW (1929) Motor nerve fibers. Part II. The frequency of discharge in reflex and voluntary contractions. J Physiol 67:19–151

Clamann HP (1970) Activity of single motor units during isometric tension. Neurology 20:254–260

De Luca C (1979) Physiology and mathematics of myoelectric signals. IEEE Trans Biomed Engin BME-26:315–325

De Luca CJ (1985) Control properties of motor units. J Exp Biol 115:125–136

De Luca CJ, Erim Z (1994) Common Drive of Motor Units in Regulation of Muscle Force. Trends Neurosci 17:299–305

De Luca CJ, Roy AM, Erim Z (1993) Synchronization of motor-unit firings in several human muscles. J Neurophysiol 70:2010–2023

De Luca CJ, Foley PJ, Erim Z (1996) Control Properties of Motor Units in Constant-Force Isometric Contractions. J Neurophysiol 76:1503–1516

De Luca CJ (1993) Precision decomposition of EMG signals. Methods Clin Neurophysiol 4:1–28

De Luca CJ, Mambrito B (1987) Voluntary control of motor units in human antagonist muscles: Coactivation and reciprocal activation. J Neurophysiol 58:525–542

De Luca CJ, Forrest WJ (1973) Some properties of motor unit action potential trains recorded during constant force isometric contractions in man. Kybernetik 12:160–168

De Luca CJ, LeFever RS, McCue MP, Xenakis AP (1982a) Behavior of human motor units in different muscles during linearly-varying contractions. J Physiol (Lond) 329:113–128

De Luca CJ, LeFever RS, McCue MP, Xenakis AP (1982b) Control scheme governing concurrently active motor units during voluntary contractions. J Physiol 329:129–142

Erim Z, Beg MF, Burke DT, De Luca CJ (in press) Effects of aging on motor unit firing behavior. J Neurophysiol 23:1833

Erim Z, De Luca C, Mineo K, Aoki T (1996) Rank-Ordered regulation of motor units. Muscle & Nerve 19:563–573

Guiheneuc P (1992) Le Recruitment de Unités Motrices: Méthodologie, Physiologie et Pathologie. In: Cadilhac J, Dapres G (Eds.) EMG: Actualités en Electromyographie, pp 35–39. Sauramps Medical; Montpellier

Henneman E, Somjen G, Carpenter DO (1965a) Excitability and inhibitability of motoneurons of different sizes. J Neurophysiol 28:599–620

Henneman E, Somjen G, Carpenter DO (1965b) Functional significance of cell size in spinal motoneurons. J Neurophysiol 28:560–580

Hoffer JA, Sugano N, Loeb GE, Marks WB, O'Donovan MJ, Pratt CA (1987) Cat hindlimb motoneurons during locomotion. II. Normal activity patterns. J Neurophysiol 57:530–552

Iyer MB, Christakos CN, Ghez C (1994) Coherent modulations of human motor unit discharges during quasi-sinusoidal isometric muscle contractions. Neurosci Lett 170:94–98

Kernell D (1965) The adaptation and the relation between discharge frequency and current strength of cat lumbosacral motoneurones stimulated by long-lasting injected currents. Acta Physiol Scand 65:65–73

Kukulka CG, Clamann PH (1981) Comparison of the recruitment and discharge properties of motor units in human brachial biceps and adductor pollicis during isometric contractions. Brain Res 219:45–55

LeFever, RS and De Luca, C J (1978) Decomposition of action potential trains. Proceedings of 8th Annual Meeting of the Society for Neuroscience 229

LeFever RS, De Luca CJ (1982a) A procedure for decomposing the myoelectric signal into its constituent action potentials. Part I. Technique, theory and implementation. IEEE Trans Biomed Engin BME-29: 149–157

LeFever RS, Xenakis AP, De Luca CJ (1982b) A procedure for decomposing the myoelectric signal into its constituent action potentials. Part II. Execution and test for accuracy. IEEE Trans Biomed Engin BME-29: 158–164.

Mambrito B, De Luca CJ (1984) A technique for the detection, decomposition and analysis of the EMG signal. EEG Clin Neurophysiol 58: 175–188.

Miles TS (1987) The cortical control of motor neurons: some principles of operation. Medical Hypotheses 23:43–50

Person RS, Kudina LP (1972) Discharge frequency and discharge pattern of human motor units during voluntary contractions in man. EEG Clin Neurophysiol 32:371–483

Rossi A, Mazzachio R (1991) Presence of homonymous recurrent inhibition in motoneurons supplying different lower limb muscles in humans. Exp Brain Res 84:367–373

Semmler JG, Nordstrom MA, Wallace CJ (1997) Relationship between motor unit short-term synchronization and common drive in human first dorsal interosseous muscle. Brain Res 767:314–320

Stashuk D, De Bruin H (1988) Automatic decomposition of selective needle-detected myoelectric signals. IEEE Trans Biomed Engin BME-35:1–10

Stashuk D, De Luca CJ (1989) Update on the decomposition and analysis of EMG signals. In: Desmedt JE (ed) Computer-aided electromyography and expert systems, pp 39–53. Elsevier: Amsterdam

Tanji J, Kato M (1973) Firing rate of individual motor units in voluntary contraction of abductor digiti minimi muscle in man. Exp Neurol 40:771–783

Westgaard RH, De Luca CJ (1999) Motor Unit Substitution in Long-Duration Contractions of the Human Trapezius Muscle. J Neurophysiol 82:501–504

Abbreviations

EMG	Electromyographic
IFI	Inter-Firing Interval
MVC	Maximum Voluntary Contraction
MUAP	Motor Unit Action Potential
MUAPT	Motor Unit Action Potential Train

Chapter 28

Relating Muscle Activity to Movement in Animals

Gerald E. Loeb

Introduction

Electromyography (EMG) provides a direct indication of the level of activity in individual muscles and indirect data about the net synaptic drive to pools of spinal motor neurons. It is usually desirable to combine such data with a description of the motor behavior that was produced by the muscle activity. A subjective description may be useful to identify when the animal was engaged in a typical example of the behavior to be studied. An objective, quantitative measure of the behavior may be necessary to understand the biomechanical requirements of a task and to interpret the EMG recordings in the light of an hypothesized strategy for motor control. This chapter summarizes the core methodology for obtaining synchronized EMG and video data from intact, behaving animals and points out opportunities and requirements to extend their analysis to quantitative biomechanics. These methods and many variations on them for different preparations and paradigms are described in greater detail by Loeb and Gans (1986).

Outline

The numbered steps indicated with solid boxes and lines in Fig. 1 represent the core methodology:

- methods for the design, construction and surgical implantation of bipolar EMG electrodes;
- techniques for recording EMG and other analog data so that they can be synchronized with the corresponding fields of video data;
- procedures for identifying specific body movements (kinematics) and relating them to muscle activity.

The lettered steps indicated with dashed lines in Fig. 1 represent optional procedures required for kinetic analysis, the quantitative study of the relationship between forces and movements. Kinetic analysis itself can be divided into forward kinetic analysis (left side of Fig. 1), in which EMG data are used to estimate the forces and torques generated by muscles, and inverse dynamic analysis (right side of Fig. 1), in which the laws of physical motion are applied to the kinematic data to compute the net torques that the muscles must have produced at each joint to create the observed movement. While a detailed consideration of kinetic methodologies is outside the scope of this chapter, it is important to consider the additional demands that such analyses place on the core experimental procedures.

Gerald E. Loeb, University of Southern California, Department of Biomedical Engineering, Los Angeles, CA 90089-1451, USA (phone: 213-740-7237; fax: 213-740-0343; e-mail: g.loeb@bmsrs.usc.edu)

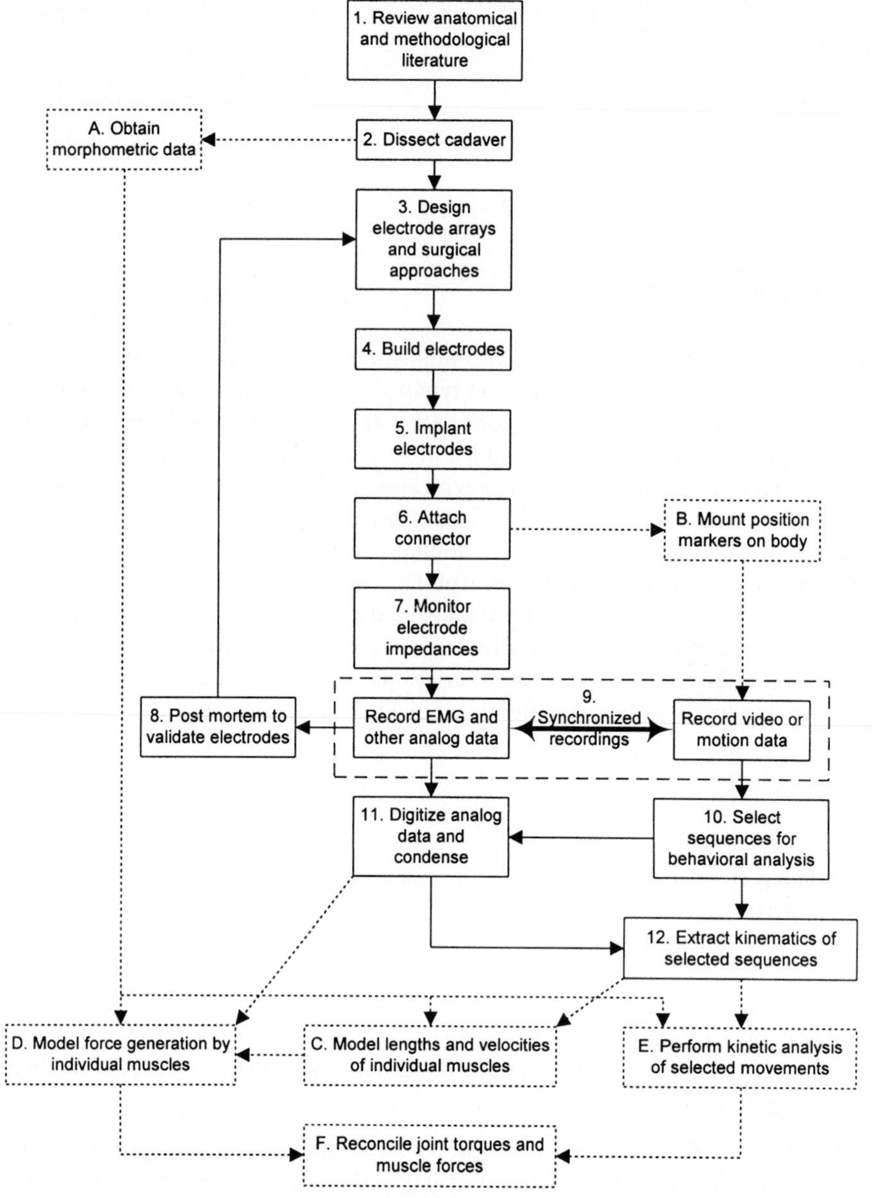

Fig. 1. Protocol for EMG and kinesiology studies in animals. Solid boxes and lines indicate basic steps for subjective understanding of the relationship between muscle activity and behavioral performance; dotted boxes and lines indicate optional steps required for quantitative kinetic analysis.

Materials

EMG Electrode Materials

- Multistranded stainless steel wire with Teflon insulation (Cooner Wire AS631)
- Dacron-reinforced silicone sheeting (Specialty Silicone Fabricators)
- Silicone elastomer adhesive (Dow Corning Medical Adhesive Type A)
- Nonresorbable sutures (e.g. Ethicon Ethibond X936H)

Recording Accessories

- Ribbon cable connectors (e.g. 3M Scotchflex or similar)
- Reflective markers (3M #7610 Retro Reflective tape)
- Differential preamplifiers (e.g. Bak Electronics model MDA-2). Features to look for include high common-mode rejection ratio, adjustable gain (range 200–5000x) and an appropriate range of high and low pass filters (50–200Hz and 1–5kHz, respectively).
- Bin integrator (e.g. Bak Electronics model PSI-1). This is required only if it is not practical or desirable to digitize the complete EMG waveform with sufficient temporal resolution to avoid aliasing (sampling rate at least twice the bandwidth).
- Impedance meter (e.g. Bak Electronics model IMP-1). Features to look for include a low-current AC test signal (typically less than 1 μA @ 1 kHz), floating differential input (or battery operated) and direct read-out of impedance magnitude (range 1–100 kilo-ohms for EMG use).

Video Recording Equipment

- SMPTE timecode generator/reader (e.g. Horita model TRG-50PC). Features to look for include simultaneous generation of visible video characters ("burn-in"), digital vertical interval timecode (VITC; inserted into video signal from camera in the blank scan lines during vertical flyback between fields), audio frequency code (can be recorded on audio edge tracks of video and analog tape recorders), and remote electronic control and transfer of timecode data (usually via serial port to a personal computer controlling the experiment or performing the analysis).
- Videocassette recorder with still-field playback (consumer electronics grade). Features to look for include playback of single fields rather than frames (to avoid image blurring from interlacing of successive fields), multiple heads to provide stable images without "tear" lines associated with a field boundary, and smooth and stable switching between forward and reverse during slow motion or single field "jog" playback.
- solid-state video camera with electronic shutter (consumer electronics grade). Look for sufficient light sensitivity to produce an acceptable picture when shuttered at 1 ms or less per field to prevent smearing the image. Note that most modern CCD cameras are sufficiently sensitive in the near infrared that arrays of IR-LEDs may be used to augment or replace visible light.
- direct motion-analysis recording equipment (optional). There are several different systems that directly compute the trajectories of specific markers in three-dimensional space, usually by computer analysis of images from two or more specialized video cameras. These are used mostly for human motion studies and often require relatively large reflective markers or LEDs that are unsuitable for use on animals. These systems often provide for digitization of multiple sources of analog data such as EMG, but some means of synchronization to an additional, conventional video recording is usually necessary and may be difficult to achieve. Systems that permit computer-assisted post hoc analysis of conventional videotape are most suitable (e.g. Peak Performance Technologies Inc.).

Procedure

Electrode Strategy

1. Review literature: A great deal of trouble can be saved by searching thoroughly for all prior references to the neuromuscular architecture, functional anatomy and electromyography of the target structure. This work is often quite old but thorough. Many muscles have complex patterns of compartmentalized histochemistry, innervation and function that must be understood in order to design experiments that will account for behaviors accurately, completely and reproducibly.

2. Dissect cadaver: Even when working with a familiar preparation, it is highly advisable to perform at least one dissection to identify surgical landmarks, safe incision lines and all muscles that could participate functionally or present a source of EMG cross-talk. Make detailed annotated drawings showing the locations of electrodes, fixation sites and routing of leads.
3. Design electrodes: During the cadaver dissection, determine the number of recording channels and the designs of the electrodes. Make templates for epimysial electrodes by cutting plastic transparency sheets to the desired size for each site and marking the desired electrode contact positions with indelible marker (use solid lines to indicate contacts facing up and dashed lines to indicate contacts facing down on two-sided patches). The main consideration is to sample from as much of the cross-sectional area of each muscle as possible while not picking up cross-talk from adjacent muscles. Always use bipolar electrodes spaced parallel to the fascicles. Muscles with separate compartments may require more than one recording channel at least until it is clear that those compartments are recruited similarly. If feasible, recordings should be made from all adjacent muscles whether they are relevant to the behavior or not, in order to identify any possible sources of cross-talk.
4. Build implant hardware:

Epimysial Electrodes

Small or thin muscles in superficial layers are best recorded by epimysial patch electrodes, in which bipolar contacts are affixed to the surface of a sheet of silicone. The silicone patch is sutured at its corners to one fascial surface of the muscle. It positions the contacts with the appropriate alignment and spacing while acting as a nonconductive barrier to reduce cross-talk from the adjacent muscle. Two or more channels can be positioned on one or both sides of the patch to record from multiple compartments or muscles. Blank patches can be used to prevent cross-talk from spreading through the instrumented muscle from an adjacent, uninstrumented muscle. As shown in Fig. 2A, dimension x is typically 3–5 mm depending on the size of the muscle.

Note: Monkeys seem to be highly reactive to the silicone patch material; use intramuscular electrodes instead, as described below.

a) Make electrode contacts by stripping a segment of insulation x units long from near the end of each end of a wire that is about three times the length between the muscle and the connector sites. This can be done by cutting carefully through the Teflon sleeve without nicking the wires and sliding the sleeve towards but not completely over the end of the wire.
b) Cut the patch material to the appropriate size (at least 3x units square).
c) Thread the electrode contacts through the patch material, using a hypodermic needle to create holes and feeding the wire ends through the end of the needle to pass through the patch.
d) Center the exposed portions of the wires on the muscle-facing side of the patch and cut off the distal protruding ends on the other side of the patch to a length of about 3mm.
e) Apply and cure silicone adhesive in small blobs over the four points where the wire leads emerge on the back of the patch and where they run over the edge of the patch.
f) Sterilize by autoclave.

Intramuscular Electrodes

Thick and/or deep muscles are best recorded by intramuscular hook electrodes, in which bipolar contacts are attached to a suture that is used to drag them into the middle of the muscle belly, parallel to the fascicles. The springy, multistranded wire con-

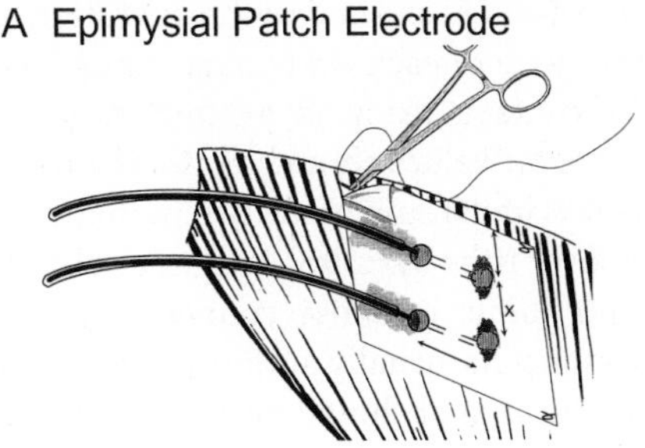

Fig. 2. Designs for chronically implanted, bipolar EMG electrodes for sampling whole muscle activity in animals. A. Epimysial patch electrode, suitable to minimize cross-talk in thin and/or superficial muscles. B. Intramuscular wire electrodes, suitable for large and/or deep muscles.

tacts are tied to the suture with the required bipolar spacing and bent backwards so that they will spring outward and lodge in the muscle. The leading end of the suture will be looped through the fascia and tied loosely to the tail of the suture. As shown in Fig. 2B, dimension x is typically 3–5 mm depending on the size of the muscle.

a) Strip x mm of Teflon insulation from both ends of a length of wire, being careful not to nick the wire strands. This may be done by stretching the Teflon sleeve over the end of the wire, cutting it off, shrinking it back over the wire with the heat from a soldering iron and cutting off the protruding strands to the desired length.
b) Tie the two wire ends side by side with the appropriate spacing in the middle of the suture as shown in Fig. 2B.
c) Bend the exposed wire contacts backwards along the suture.
d) Sterilize by autoclave.

Surgical Implantation

5. Implant electrodes: Expose the recording sites to be reached via each incision and attach the electrodes to the muscles in the appropriate manner. Attach one electrode wire with a large exposed contact to fascia to act as a ground. Tunnel or drag leads subcutaneously to exit at connector site. Leave some excess lead length coiled in subcutaneous pockets and strain relieve leads by suturing to fascia as required to deal with motion or growth.
6. Attach connectors: When multiple electrodes are implanted, each wire loop signifying one bipolar channel can be identified by tying in a glass "ishi" bead of a different color. Multipin connectors are best mounted either rigidly to the skull using bone screws and dental methacrylate adhesive or flexibly through loose skin via heavy sutures to underlying bone or strong fascial layers (Hoffer et al., 1987). The stainless steel EMG leads must be tinned carefully after dressing their bare ends with acid flux (10% HCl). After soldering the leads to the connector pins, cover carefully with silicone adhesive to prevent moisture from shunting between the leads and pins. See also Loeb et al. (1995) for an alternative to soldering.

Validate Electrodes

7. Monitor impedances: At the end of the operation and daily thereafter, measure the electrical impedance of each individual contact vs. the ground electrode and each bipolar pair of contacts using an AC impedance meter. For the designs described here, the monopolar values should be in the range 1–10 kΩ and the bipolar values should be 70–90% of the sum of the two monopolar contacts. Impedances normally change slowly over time by ±30%. Larger or abrupt increases signify broken leads, defective solder joints or loose connector pins; decreases signify fluid shunts around connector pins, usually at the back where the leads are soldered.
8. Post mortem: The animal should be sacrificed as soon after the last recording session as feasible. Each electrode site should be surgically explored to determine that electrodes remained intact and correctly positioned. If there was any question about the accuracy of the lead identification and/or the integrity of the connections, electrical continuity can be determined by probing the individual electrode contacts and corresponding connector pins directly with a simple ohmmeter.

Record Data

9. Synchronized analog and video data: It is usually necessary to record multichannel analog data such as EMG and high resolution motion information such as video on separate media. Digital computers are just starting to have sufficient speed and storage capacity to digitize both together, but any substantial recording period will generate huge files that are cumbersome to handle. Synchronizing two types of media is best accomplished with a common timecode in formats suitable for each medium. SMPTE timecode comes in formats that are synchronized to North American (NTSC, 30 frames/s) or European (PAL, 25 frames/s) video and that can be written as visible images or digitally encoded bits in the video stream as well as an audio-frequency, pulse-width-modulated carrier (8kHz bandwidth) that can be recorded on audio edge tracks of video or analog tape recordings. It is helpful if the timecode generator/reader can be controlled by a personal computer via serial port. This permits the computer to digitize analog data into files that begin at a known timecode (see Fig. 3). If a general timecode is not feasible, multimedia records can be synchronized by introducing easily identifiable, unique events on all media, such as a simultaneous light flash and audible click or electrical pulse.

Analyze Data

10. Select behaviors: Behavioral data from animals are usually inconsistent. Perhaps the best reason for making a video recording even if only analog or direct motion analysis data are required is to screen the recorded behaviors for appropriate instances of the desired behavior. A videotape system capable of slow-motion and still-field playback allows the start and stop timecodes of the relevant data segments to be identified from the "burn-in" characters on the screen.
11. Analyze EMG: Using the timecode boundaries from the previous step, create and view as multiple traces the EMGs and any other transducer data, making sure there is no electrical noise, artefact, cross-talk or other questions about the quality of the records (Fig. 3). It is important to perform this step on the data at the earliest stage of its analysis, because aggressive data reduction techniques such as smoothing and averaging of multiple records tend to obscure such problems. The usual analysis of EMG includes extraction of the envelope of modulation (Bak and Loeb, 1979). This can be accomplished digitally if the amplified EMG was digitized at a high enough frequency to avoid aliasing (sample rate at least twice the bandwidth). The speed and capacity of the digitization system can be greatly reduced by preprocessing the EMG signal by full-wave rectification and integration into discrete bins whose duration corresponds to the sampling interval of the digitizer.
12. Correlate with movement: It is usually practical to perform a detailed kinematic analysis (such as extracting limb trajectories from the position of markers in suc-

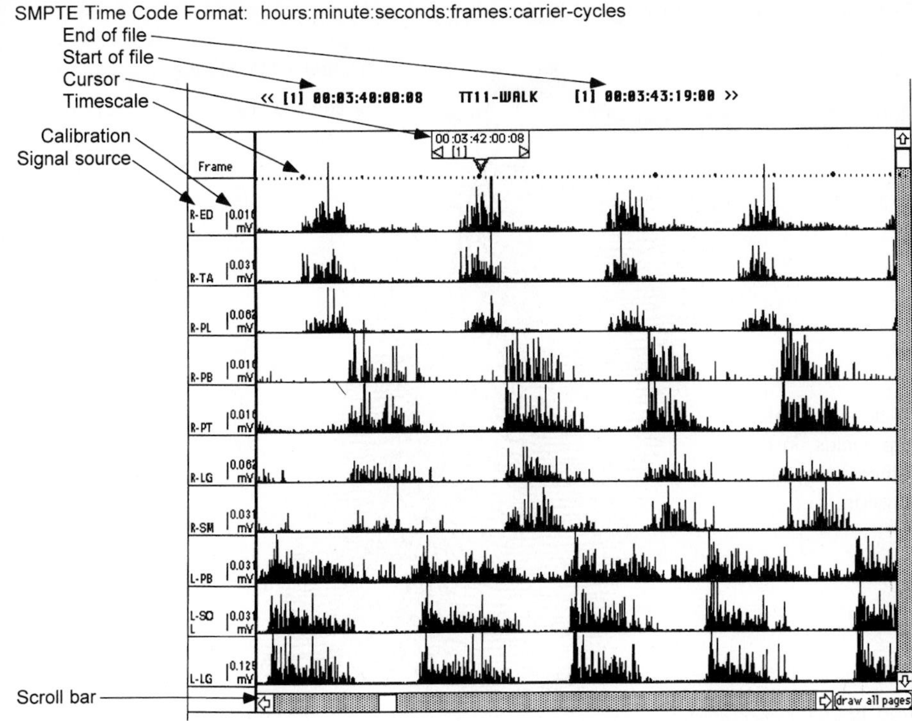

Fig. 3. Custom graphic display of a digitized multichannel EMG file synchronized to SMPTE timecode on a simultaneously recorded videotape.

cessive video fields) on only a subset of the EMG data to be analyzed. The multitrace EMG record should be scanned visually or analyzed digitally to identify the timecodes corresponding to particular inflection points (e.g. onsets or peaks of muscle activity) around which to concentrate such kinematic analysis. Multiple records of the same behavior but with slightly differing durations can be averaged by interpolating the records between inflection points so as to provide the same number of bins for each record (Fig. 4).

Subprotocol for Kinetic Analysis

Morphometry

A. Inverse dynamic analysis requires data on the length, mass, rotational inertia and center of mass for each body segment. Forward kinetic analysis requires data on the mass, fascicle length (at optimal sarcomere length), series elastic length (tendon + aponeurosis) and moment arms of each muscle. Pinnation angle has negligible effects unless greater than 25° (Scott and Winter, 1991).

Skeletal Markers

B. Both forward and inverse analyses depend on being able to compute accurately the position and intersegmental angles of all skeletal segments from the position of whatever visible markers are employed. This requires quantitative information about the relationship between the markers and the centers of rotation of the joints. If the skin surface tends to slip with respect to the underlying bones, it may be necessary to attach percutaneous markers directly to underlying bone or to use trigonometry

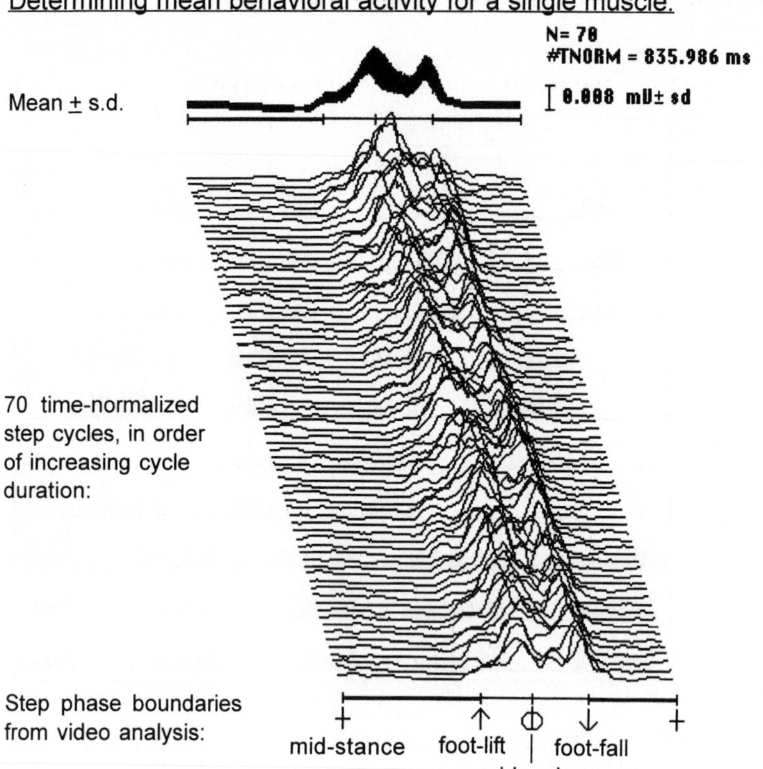

Fig. 4. Rastered EMG activity from multiple cycles of locomotion that were time-normalized based on step cycle inflections identified from videotape (symbols at bottom).

to compute the position of some joints from the distance between flanking markers and the lengths of the skeletal segments to which they are attached.

Musculoskeletal Kinematics

C. Muscle lengths and velocities are computed from the joint angles between the skeletal segments and the morphometric data. Rather than computing moment arms from the attachment points of muscles and tendons on bone, it is usually preferable to "map" muscle length changes over the entire range of joint angles (e.g. Young et al. 1993), particularly for joints with multiple degrees of freedom, tendons with complexly tethered paths and/or muscles crossing multiple joints. This map permits joint angles to be converted into a "path length" for each muscle, which usually will need to be decomposed into a fascicle length (normalized to optimal sarcomere length) and a non-contractile series-elastic component consisting of tendon + aponeurosis. In muscles with relatively long series-elastic components, significant length changes may occur in the contractile elements as a result of stretching and shortening of the series-elastic elements under changing muscle forces.

Active Force Models

D. EMG data can be converted into a reasonable representation of the activation level of the contractile apparatus. The "envelope" of EMG amplitude must be delayed to account for the rise-time and fall-time of such activation, which will depend on the types of fibers contributing to the EMG activity. It is necessary to scale the EMG amplitude to represent some fraction of the total cross-sectional area of the muscle. A recorded sample of a behavior for which the muscle is expected to be nearly maximally activated is particularly useful. The total muscle force can be estimated from activation, length and velocity of the contractile machinery (Brown et al., 1996; for the most current model, see http://brain.phgy.queensu.ca/muscle_model/). This force must then be extrapolated to joint torque, which requires data about the moment arm(s) of the muscle in the appropriate axes of each joint at each joint angle.

Inverse Dynamics

E. The kinematic data about skeletal position must be differentiated twice to produce angular acceleration, from which the torques required to overcome inertia are computed according to Newton's laws of motion (e.g. Hoy and Zernicke, 1985; Hoy et al., 1985). Noise in the kinematic data for individual marker positions is particularly troublesome at this stage. Other terms that must be considered are the effects of gravity, external forces at contact points, internal constraints such as elastic ligaments and boney restrictions on range of motion, and intersegmental coriolis effects (which arise from the non-torque forces conveyed across joints as a result of compression and tension in the cartilage and ligaments that constrain their motion to rotation). The Working Model software packages available from Knowledge Revolution Inc. are useful for modeling 2D and 3D systems with idealized segments and linkages.

Kinetic Validation

F. In principle, it should be possible to add up the torques computed from the forward kinetic analysis of each muscle and reconcile them with the net torques computed from the inverse dynamic analysis of the movement. In practice, this is rarely feasible because it requires complete data and highly accurate models for all muscles crossing all joints in the system under analysis. Nevertheless, it is useful to examine all available EMG data at least qualitatively to determine whether it accounts reasonably completely for the biomechanical requirements of the task under study.

Results

Figures 3–4 show typical EMG recordings at various stages of analysis. The graphs were made using custom software that operates on spreadsheet data on a Macintosh computer. Similar graphical analyses can now be accomplished with a variety of commercial data acquisition and analysis packages. Figure 3 shows an analysis (at the level of step 10 in Fig. 1) based on multichannel EMG data that was digitized with a sample interval of 3.3 ms after full-wave rectification and integration into 3.3 ms bins synchronized with the digitizer (Bak and Loeb, 1979). The calibration bars shown next to the trace identifiers reflect a calibration that accounts for both the amplifier gains and the bin duration; i.e. doubling the bin duration doubles the amplitude of the integrated EMG, so the voltage label associated with the calibration bar is halved. A given amplitude of bin-integrated EMG corresponds to a peak-to-peak value about ten times higher in the raw signal.

The starting SMPTE timecode of the digitized file was known and the sampling interval was synchronized with the 8 kHz SMPTE timecode carrier. This makes it possible to place a cursor on the data traces indicating the exact corresponding timecode. A time trace was added to the top of the record showing the even seconds (large dots) and individual video frames and 10th frames (small and large ticks). This makes it possible to select video fields for still analysis to extract the kinematics of the corresponding limb movements.

A more advanced analysis is shown in Fig. 4. A long record consisting of multiple step cycles with similar but not identical durations was divided into discrete step cycle phases based on the video analysis of limb position (key at bottom shows inflection points identified on the video). For each EMG channel, the EMG data were lightly smoothed digitally and interpolated in time for each phase so as to provide the same number of samples in each step cycle. The time-normalized step cycles were then plotted in a raster format in order of actual step cycle duration, after rejecting any step cycles with sudden changes in duration indicative of excessive acceleration or deceleration. The trace at the top shows the range of EMG values throughout the step cycle as mean ± the standard deviation of the time-normalized samples.

References

Bak MJ, Loeb GE (1979) A pulsed integrator for EMG analysis. Electroencephalogr Clin Neurophysiol 47:738–741.

Brown IE, Scott SH, Loeb GE (1996) Mechanics of feline soleus: II. Design and validation of a mathematical model. J Muscle Res Cell Motility 17:219–232.

Hoffer JA, Loeb GE, Marks WB, O'Donovan MJ, Pratt CA, Sugano N (1987) Cat hindlimb motoneurons during locomotion: I. Destination, axonal conduction velocity and recruitment threshold. J Neurophysiol 57:510–529.

Hoy MG, Zernicke RF (1985) Modulation of limb dynamics in the swing phase of locomotion. J Biomech 18:49–60.

Hoy MG, Zernicke RF, Smith JL (1985) Contrasting roles of inertial and muscle moments at knee and ankle during paw-shake response. J Neurophysiol 54:1282–1294.

Loeb GE, Gans C (1986) Electromyography for experimentalists. Chicago: University Chicago Press.

Loeb GE, Peck RA, Smith DW (1995) Microminiature molding techniques for cochlear electrode arrays. J Neurosci Meth 63:85–92.

Scott SH, Winter DA (1991) A comparison of three muscle pennation assumptions and their effect on isometric and isotonic force. J Biomech 24:163–167.

Young RP, Scott SH, Loeb GE (1993) The distal hindlimb musculature of the cat; Multiaxis moment arms of the ankle joint. Exp Brain Res 96:141–151.

Suppliers

Company: Bak Electronics, Inc., P.O. Box 623, 6411 Ridge Rd., Mt. Airy, Maryland, 21771, USA (phone: +01-301-607-8300; fax: +01-301-607-9018)

Company: Cooner Wire, 9265 Owensmouth, Chatsworth, California, 91311, USA (phone: +01-818-882-8311)

Company: Horita, P.O. Box 3993, Mission Viejo, California, 92690, USA (phone: +01-949-489-0240; fax: +01-949-489-0242)

Company: Knowledge Revolution, 66 Bovet Road, Suite 200, San Mateo, California, 94402, USA (phone: +01-415-574-7777; fax: +01-415-574-7541)

Company: Peak Performance Technologies Inc., 7388 S. Revere Parkway, Suite 601, Englewood, Colorado, 80112-9765, USA (phone: +01-303-799-8686; fax: +01-303-799-8690)

Company: Specialty Silicone Fabricators, Inc., 3077 Rollie Gates Dr., Paso Robles, California, 93446, USA (phone: +01-805-239-4284; fax: +01-805-239-4146)

Chapter 29

Long-term Cuff Electrode Recordings from Peripheral Nerves in Animals and Humans

Thomas Sinkjær, Morten Haugland, Johannes Struijk and Ronald Riso

Introduction

A peripheral nerve contains thousands of nerve fibers, each of them transmitting information, either from the periphery to the central nervous system or from the central nervous system to the periphery. The efferent fibers transmit information to actuators, mainly muscles, whereas afferent fibers transmit sensory information about the state of organs and events, such as muscle length, touch, skin temperature, joint angles, nociception, and several other modalities of sensory information.

Most of the peripheral nerves contain both afferent and efferent fibers, and the peripheral nerve can thus be seen as a bi-directional information channel. Since nerve fibers are excitable by electric stimuli, this gives the possibility to investigate and influence the neuromuscular system by activating muscles using electrical stimulation of nerves.

Hoffer et al. (1974) and Stein et al. (1975) introduced cuff electrodes for long-term recording of the Electro-Neurogram (ENG) from peripheral nerves. The cuffs were used to obtain higher signal amplitudes than would be possible without cuffs, at least in chronic recordings, and to decrease the pick-up of noise, especially from active muscles.

Nerve-recording cuffs have several favorable characteristics, such as demonstrated safety and recording stability. Also, since the cuffed nerve encloses a large number of afferent fibers, a nerve cuff can potentially be used to sample large areas of skin or many muscle afferents with only a single device. The effect of spatial summation, however, can also be a disadvantage since information about the location of the active receptors is imprecise and constrained only by the combined receptive field territories of all the recorded fibers. Despite the limitation of selectivity, nerve-recording cuffs are presently the most viable technology for chronic peripheral nerve interfaces.

There have been many studies about the neuromuscular system's (patho)physiology, in particular chronic studies in freely moving animals. Examples can be found in Hanson et al. (1987), who recorded the "central respiratory drive" from the phrenic nerve in fetal sheep; Hoffer et al. (1981; 1987) used cuffs to estimate conduction velocities to assess the order of motor unit recruitment in walking cats; Marshall and Tatton (1990) used cuffs to record from joint receptors in the cat; Milner et al. (1991) recorded cutaneous afferent activity from the median nerve of a monkey during grasping and lifting; Little (1986) and Sinkjær and Hoffer (1990) recorded peripheral nerve traffic during spinal reflexes in normal and reduced cat models; Loeb et al. (1987) and Loeb and Peck (1996) used cuff electrodes to record motor activity during locomotion; Palmer et al. (1985) obtained cutaneous receptor activity from median nerves in cats; and Stein et al. (1981) used cuffs to classify sensory patterns during locomotion. Another set of studies relates to monitoring the state of the nerve, especially regeneration after induced dam-

Correspondence to: Thomas Sinkjær, Aalborg University, Center for Sensory-Motor Interaction (SMI), Fredrik Bajers Vej 7D-3, Aalborg, 9220, Denmark (phone: +45-96-35-88-28; fax: +45-98-15-40-08; e-mail: ts@smi.auc.dk)

age. Examples of studies of axotomized nerves and regeneration of nerve fibers were presented by Davis et al. (1978), Gordon et al. (1980; 1991), Hoffer et al. (1979), Krarup and Loeb (1987), and Krarup et al. (1988; 1989).

Several studies pertain to the use of sensory signals as feedback information to control neuro-prosthetic devices. Studies in animals comprise those by Haugland (1994a; b), Haugland and Hoffer (1994), Hoffer and Sinkjær (1986), Nikolic et al. (1994), and Popovic et al. (1993). To develop a *functional electrical stimulation* (FES) device to manage apnea in humans, Sahin et al. (1997) used the spiral cuff to record respiratory output in cats from the hypoglossal and phrenic nerves. In addition, Woodbury and Woodbury (1991) stimulated and recorded with cuff electrodes around the vagus nerve in rats to study electro-stimulation as an anti-convulsive treatment, and Jezernik et al. (1999) recorded from bladder afferents with the intention to use the recorded signals in future systems for bladder control.

In humans, cuff electrodes for sensory recordings have been used in hemiplegic subjects for the control of dropfoot stimulation (Andreasen et al., 1996; Haugland and Sinkjær, 1995; Upshaw and Sinkjær, 1998), and in tetraplegic SCI subjects for the control of hand-grasp (Haugland et al., 1999; Riso et al., 1995; Riso and Slot, 1996; Sinkjær et al., 1994; Sinkjær et al., 1993; Slot et al., 1997).

The nerve-cuff electrode still has an unrivaled position as a tool for recording signals from peripheral nerves in chronic experiments and as a means to provide sensory information from natural sensors to be used in neural prostheses. Other kinds of electrodes, such as intra-fascicular electrodes, may provide advantages with respect to selectivity, but until now, their practical application has been rather limited. This chapter describes the characteristics of the afferent signals obtained with cuff electrodes in long-term implants. This is done to evaluate their suitability to monitor the neural traffic in peripheral nerves and particularly their suitability to perform as sensors in neural prosthetic devices. For an extensive review on the application of cuff electrodes in animals see Hoffer (1990), on the properties of the signals in long-term implants see Struijk et al. (1999), and on their application in automatic adjustment in neural prosthetic devices see Haugland and Sinkjær (1999) and Riso (1998).

Procedure

The Cuff Electrode

Design and Recording Characteristics

Two designs of chronic nerve-recording cuffs for implantation around peripheral nerves are shown in Fig. 1. Both devices contain three electrodes attached to the interior wall of a silicone tube that encircles the nerve. One electrode is placed at the cuff center and there is one at each cuff end. The cuff shown in Fig. 1 (top) (Kallesøe et al., 1996) evolved from a design described by Stein and his colleagues that was used extensively in animal research (Stein et al., 1975; Hoffer, 1990).

In the Hoffer-style cuff, a longitudinal slit is provided along the length of the cylinder to allow the cuff to be installed around the nerve. A key design feature is the method of cuff closure that should not allow any gap to compromise the electrical insulating action of the cuff structure. Cuffs with a longitudinal slit have been employed with good results in chronic studies involving volunteer human subjects (Haugland and Sinkjær, 1995; Haugland et al., 1999) and were used to record the human neural data to be presented below. The design of the self-coiling cuff shown in Fig. 1B in part uses fabrication techniques first described by Naples et al. (1988). After installation, the silastic sheeting comprising the self-coiling cuff wraps with several overlapping turns around the nerve.

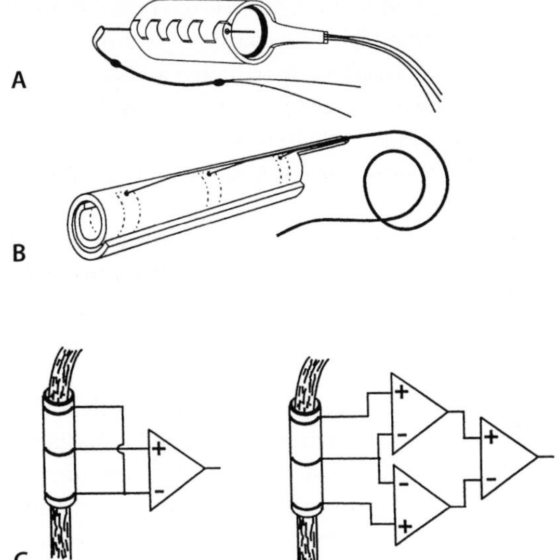

Fig. 1. Two designs for chronically implanted peripheral nerve-recording cuffs. Each cuff contains three separate and equally spaced electrodes. Electrodes are formed from strips of platinum foil welded to stainless steel lead wires. Typically, such cuffs are 2 to 3 cm in overall length. Hoffer and his associates have developed the cuff design shown at the top of the figure (A). It contains a longitudinal slit that allows the cuff to be pulled open while the nerve is placed into it. The slit is then held closed by insertion of the metal pin shown. The second cuff shown (B) is a self-coiling design which has the main advantage of being self-sizing, allowing it to fit closely around the nerve. The principle for recording neural signals with an implanted tripolar nerve-cuff electrode is shown in (C). See text for details

Several factors are key to the operation of neural recording cuffs. The silastic tube provides electrical insulation that constrains the weak neural currents to travel within the tube. The longer the cuff, the less current can leak through the tissue surrounding the cuff, and thus the recorded signals have higher amplitude with increases in cuff length. It has recently been shown in a study by Thomsen et. al. (1996) that the optimal cuff length for recording from large myelinated afferents is more than 6 cm. This contradicts earlier studies (e.g., Stein et. al. 1975; Marks and Loeb, 1976) which reported the optimal length to be below 4 cm. Nevertheless, it would be difficult to find space to implant such a long (6 cm) cuff in many applications, and, moreover, a 4-cm-long cuff seems to provide about 95% of the maximum signal strength that could be recorded with the longer cuff anyway.

Another characteristic of cuff electrodes is that they favor the registration of activity from larger-diameter fibers versus smaller-diameter fibers. This occurs because the action potentials of larger fibers produce greater action currents. Finally, the signal-to-noise ratio is more favorable when recording from small-diameter nerves than from larger nerves, and this is because smaller (narrower) nerve cuffs produce a more restrictive internal path for the flow of the discharge currents.

Effect of Electrode Number and Connectivity Scheme

There are a few options with regard to the scheme used when the electrodes are connected to the amplifier. The signal is recorded by shorting the cuff-end electrodes and amplifying the difference between the center electrode and the two end electrodes. The advantage of using a tripolar cuff instead of a monopolar or bipolar cuff is the reduction of artifacts from the surroundings (Stein et. al., 1975). It has been proposed that the reason for the artifact reduction in tripolar recording is that there can be no potential gradient across the cuff in the longitudinal direction when the end electrodes are shorted (Stein et al., 1975). More recently, it was shown that the linearization of external fields within the cuff (because of the insulating wall) has an effect on reducing artifacts (Struijk and Thomsen, 1995). If the impedance of the end electrodes is equal and the cuff is symmetrical, the shorted end electrodes give the same potential as the center electrode with respect to the external field. When these potentials are subtracted in a differential amplifier, the artifacts are reduced (see below).

More recent studies, however, have indicated that if the outer cuff electrodes are not shunted together, but instead are used as separate inputs to two differential amplifiers, it is possible to achieve an enhanced ability to reject extraneous signals such as EMG or the artifacts that result when FES stimulus pulses are applied in the vicinity of the recording cuff (Pflaum and Riso, 1996).

The end electrodes in a tripolar cuff take up some space and make the cuff effectively shorter than it would be with a monopolar recording. In an innovative design, Thomsen et al. (1999) took advantage of the larger signal amplitude obtained with monopolar recording for registering the desired neural signal. At the same time, however, a bipolar recording was made of the extraneous artifacts. The artifact signal could then be subtracted from the desired signal to enhance the signal-to-noise ratio. This approach would be especially advantageous when the use of normal-length cuffs (20–30 mm) is prohibited by anatomical constraints, but when a short monopolar cuff could be accommodated. For the case of a nerve trunk containing multiple fascicles, efforts have been made to obtain spatial selectivity in the recorded signals. To achieve this, cuffs have been fashioned that contain up to twelve electrode elements with independent lead wires. Put simplistically, depending on the positions of the electrode elements and the underlying fascicles within the nerve trunk, a given fascicle may be preferentially registered by the electrode elements nearest to it. Initial results from testing this concept have not been encouraging, however, and only modest selectivity has been demonstrated. Moreover, since the electrodes are positioned on the outside of the nerve trunk, the registration from superficial fibers is favored over fibers that are located more deeply within the nerve.

In an effort to enhance the spatial resolution of the recorded activity and to obtain better recording access to the fascicles lying deeper in a trunk nerve, cuffs have been applied with electrodes mounted radially, so that they tend to insinuate themselves into the nerve (i.e., in between the fascicles), as reported by Tyler and Durand (1996) in their work to develop a cuff for selective stimulation of nerve fascicles. These designs and others seek to incorporate mechanical effects that try to remodel the morphology of the whole nerve to facilitate a spatial isolation of the nerve's component fascicles.

Alternatively, selective recordings can be obtained by placing a separate cuff around each nerve that is to be independently monitored. This solution is difficult to implement in practice, however, because of the bulkiness of the cuffs and lead wires, and because of the need for multiple surgical access sites during the cuff implantations.

Nerve Signal Amplification and Processing

Cuff impedance usually ranges between 1 and 2 kOhm when a 1 kHz sinusoidal test signal is used. Because the signal amplitude is small, an ultra-low-noise differential preamplifier must be used (cf. Sect IV). Preferably, the amplifier noise should be below $4\,\text{nV}/\sqrt{\text{Hz}}$ (corresponding to $0.25\,\mu V_{RMS}$ at a 4 kHz bandwidth), which is the theoretical thermal noise of a 1 kOhm resistor at a temperature of 37 °C. This is a value that can be difficult to attain in practical situations, implying that amplifier noise will often be the dominating noise source.

Effective neural recording with cuff electrodes requires an amplifier with a high common-mode rejection ratio (CMRR) preferably higher than 120 dB (see below), which can be difficult to achieve at the frequencies of interest (typically 1–5 kHz). One of the

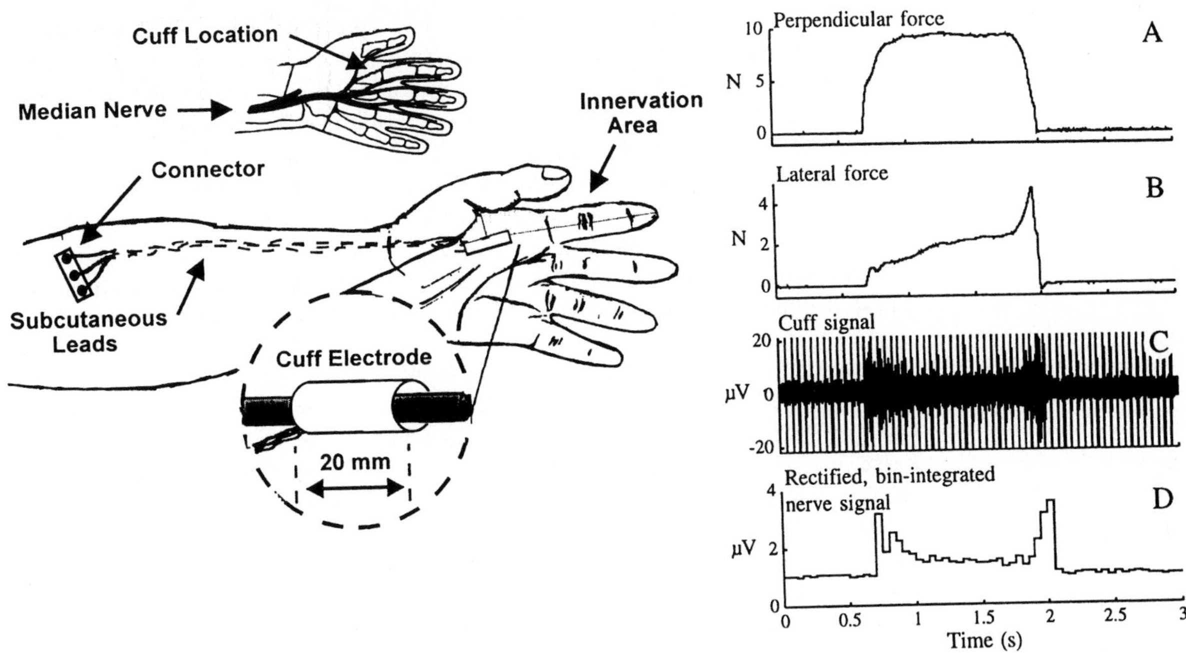

Fig. 2. Left: Location of a nerve cuff used to record cutaneous activity from the index finger as indicated. Three lead wires from the cuff are routed subcutaneously to a site on the volar forearm where they exit and can be accessed via a small connector. The cuff is located on a branch of the palmar digital nerve that in turn derives from the median nerve. Right: Signals recorded from nerve-recording cuff while stimulating nearby muscles electrically. Force is applied to the skin within the innervation area of the nerve by a hand-held force probe. A: Force applied perpendicularly on the skin. B: Lateral force applied along the skin. In this case the lateral force was increased so that the probe slipped across the skin at the end of the trial. C: Signal recorded from the nerve. D: The same signal after rectification and bin-integration of the artifact-free periods. Notice that bin-integration introduces a delay of half a bin-width, on average. Modified from Fig. 2 in Riso (1998) and Fig. 3 in Haugland and Sinkjær (1999).

drawbacks of using a cuff electrode is the low signal amplitude, which is typically below 5 μV when recording the activity from skin receptors in human applications (Sinkjær et al., 1994; Haugland et al., 1995; 1999).

The nerve cuff and the amplifier configuration reject a large fraction of the external noise. When nearby muscles are stimulated, however, the cuff will still pick up some stimulation artifact. This is illustrated by the example data shown in Fig. 2. The left figure shows details regarding the implantation of the tripolar nerve cuff used in this investigation, and the right figure shows the recorded nerve and mechanical signals. These data were recorded from a digital nerve in a human subject while the flexor pollicis longus muscle (FPL) was stimulated at a fixed rate of 20 Hz. Stimulation artifacts appear in the cuff signal as large spikes that saturate the amplifier in both the positive and negative direction. There was no attempt to blank the artifacts by shorting or disconnecting the input to the amplifier as is often done when recording EMG during stimulation (e.g., Knaflitz and Merletti, 1988; also cf. Sect IV). If this is done to an amplifier with high gain, it can easily cause even larger artifacts due to switching noise and changes in source impedance. Instead, an amplifier was used that did not block ("hang") when driven into saturation. In this way the artifacts could be removed from the signal after amplification by simply ignoring the periods when the artifacts were present.

If the stimulated muscles are very close to the cuff, evoked EMG responses may also be picked up. These responses can be selectively suppressed by high-pass filtering (see below) because they contain frequencies mainly below 1 kHz, and the nerve signal (depending on the exact dimensions of the cuff and the types of nerve fiber in the nerve) usually contains frequencies above 1 kHz. The information contained in the cuff-recorded nerve signal appears to be stored in the amplitude of the signal rather than in the frequency content. A simple way of measuring amplitude is to rectify and integrate the signal. If this is done in bins containing all the valid data between two stimulation artifacts, a signal can be obtained as illustrated in the lower trace (D) in Fig. 2. The method of high-pass filtering and bin-integration of nerve-cuff signals has been described in Haugland and Hoffer (1994). More sophisticated signal analysis such as the use of higher order statistics (Upshaw and Sinkjær, 1998; Sinkjær et al., 1998) has also been used and shown to be able to extract more reliable information from the nerve signal.

Nerve Damage in Long-Term Implants

The issue of nerve damage is important, especially with regard to applications of nerve-cuff electrodes in human subjects. Obviously, the presence of the nerve cuff induces changes in the tissues. Connective tissue covers the cuff electrodes, and the shape of the nerve changes so as to completely fill the cuff (see, for example, Larsen et al., 1998). According to Hoffer (1990), the cuff should be at least 20% larger than the diameter of the nerve to prevent compression neuropathy caused by post-surgical edema (Davis et al., 1978; Strain and Olson, 1975). Compression neuropathy affects the larger fibers most severely (Gillespie and Stein, 1983; Sunderland, 1978). Naples et al. (1988) tried to minimize compression neuropathy by development of the self-spiraling cuff.

In a nine-month implant in the hindlimb of one cat (lateral gastrocnemius-soleus nerve), Stein et al. (1977) observed a small reduction in the number of large fibers as compared with the same nerve in the contralateral leg. No associated changes in electrophysiological recordings were reported. Recently, a large histological study with a total of 40 rabbits including experimental, sham-operated, and control animals was performed in our laboratories by Larsen et al. (1998). Animals were assigned to a long-term implant group (16 months) and a short-term implant group (14 days). The cuff electrodes were placed on the tibial nerve, just distal to the knee. Traction of the wires on the electrode was minimized by having the wires exit the electrode distally and routing them, via a loop just proximal to the knee, to a subcutaneous connector placed in the groin. Each animal was isolated in a cage of 1×1 m^2 floor area for 10 days after the operation. Thereafter the animals were housed in groups of five per cage (2.1×1.3 m^2).

At the conclusion of the studies, transverse nerve sections were made at levels proximal to the cuff, at the mid-cuff level, and distal to the cuff. Sections from the nerve in the contralateral leg served as additional control material. Nerve fiber diameter histograms could thus be obtained at various places with respect to the nerve cuff.

The histological study (see Larsen et al., 1998, for details) showed that there was a statistically significant loss of myelinated fibers inside the cuff of 27% ($p=0.002$) and distal to the cuff of 24% ($p=0.01$) 14 days post-surgery, whereas the numbers of fibers proximal to the cuff and in the contralateral leg nerve were unchanged (all numbers compared with the 14-day control group). The sham-operated animals showed no significant loss of fibers. Sixteen months post-surgery there was no longer any significant decrease in the total number of myelinated fibers. No change was found in the number of unmyelinated fibers at either two weeks or sixteen months after implantation (Fig. 3).

The initial loss (after 14 days) appeared to be non-specific for fiber size, except for sparing of the smallest fibers. After 16 months, although the total number of fibers was the same as in the control group (Fig. 3 top), there appeared to have been a shift from the largest fiber group (the largest 20% of fiber diameter) towards smaller fibers, probably indicating that the fibers regenerated, but not to their original diameter (Fig. 3 bottom). This effect was most pronounced distal to the cuff, but also visible inside the cuff. Proximally and contralaterally there was no change. The results demonstrate that implanted cuff electrodes may cause initial loss of myelinated fibers, which subsequently

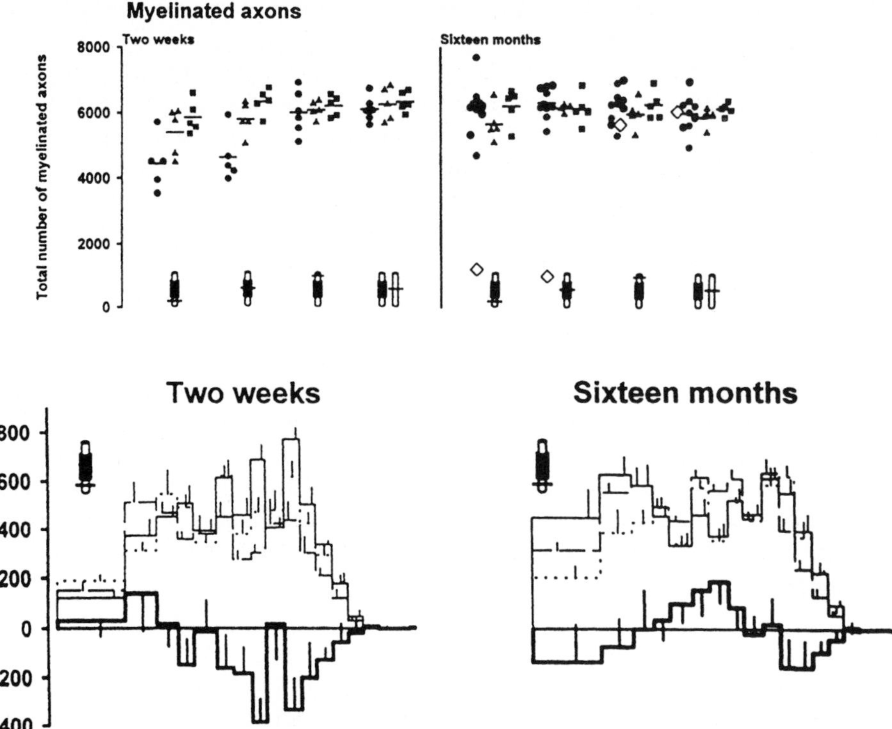

Fig. 3. **Top** panels show total number of myelinated axons at two weeks and sixteen months. Circles represent treated animals. The open diamonds indicate result from one outlier animal. Triangles refer to sham-operated animals, and squares to control animals. The cross-section level is shown at the bottom of the top figure with icons. The horizontal lines indicate the group mean. (Modified from Fig. 5 in Larsen et al. 1998.) **Bottom** panels show the averaged absolute size distribution of myelinated axons from the cross-section level distal to the cuff electrode, at two weeks and sixteen months. Dashed lines show treated animals, dotted lines sham-operated animals, and solid lines control animals. Vertical bars show SEM. The thick solid lines show the difference between treated and control distributions, i.e., the distribution of the loss of myelinated axons. The ordinate shows the absolute (total) number of axons in each class; abscissa is logarithmic. (Modified from Fig. 7 in Larsen et al. 1998).

regenerate. This implies that long-term implantation is possible without a significant loss of nerve fibers.

Neural Recordings with Long-Term Implants

Cuff electrodes are relatively stable in long-term recordings (Stein et al., 1977), but the stability has never been quantified. In the 16-month rabbit implants (Larsen et al., 1998), we did not see any deformation of the electrodes, nor migration despite large changes in the electrode impedance shortly after implantation (Thomsen 1998; Struijk et al., 1999). This initial decrease in impedance can be attributed to the initial fluid accumulation resulting from an increase in vascular permeability in response to the surgical procedure (Grill and Mortimer, 1994).

The initial amplitude of the recorded nerve decreases within the first days after implant (Thomsen 1998, Thomsen et al., 1999), and a strong linear relationship exists between the nerve signal amplitude and the electrode impedance. This implies that dividing the recorded potentials by the impedance gives a better estimate of the true bio-electrical source strength, which was shown to be relatively constant over time. Care should be taken, however, since the measured impedance is not just the "tissue" impedance between the central contact and the two end-contacts (Struijk et al., 1999).

The initial loss of nerve fibers seems to be reflected also in the amplitude and the median frequency (cf. Chapter 26) of the electrically evoked signal (Struijk et al., 1999). The changes in median frequency can be attributed to the regeneration of smaller myelinated fibers which shifts the relative contribution of fiber activity towards those regenerated fibers (Fig. 3, bottom left). The regenerated fibers have a lower frequency content than larger fibers, because of their lower conduction velocity (cf. Chap. 26). Clinical neurophysiological measurements from long-term implants in four human subjects over more than 10 implantation years have shown only minor changes in the fastest conduction velocity (CV) and in the absolute amplitude of the nerve signal over the time of implantation (Haugland and Sinkjær, 1995; Upshaw and Sinkjær, 1998; Haugland et al., 1999). There was an exception in that in one subject a decrease in amplitude was observed within the first six months of implantation. After this time, the amplitude recovered and was constant thereafter throughout the two-year implantation time (Slot et al., 1997). No change was observed in the maximal CV. We speculate that this patient had an initial loss of nerve fibers and that the fibers regenerated as seen in the animal studies. This will not affect the maximal CV as long as some of the largest fibers are not damaged, but it should decrease the median frequency. It is therefore recommended to calculate the median frequency as described by Struijk et al. (1999) if reversible nerve damage is suspected to have taken place.

Application of Nerve-Cuff Signals to Control Neural Prosthesis

Despite substantial progress over nearly three decades of development, many challenges remain to provide better functionality from Functional Electrical Stimulation (FES) systems. Chief among these is the replacement of sensory information lost due to spinal cord or brain injury. The following is a description of how signals extracted from long-term peripheral nerve recordings can be applied to improve motor functions in impaired human subjects.

The responses of particular classes of afferents are described first, after which applications using those responses to control neural prostheses are presented before introducing another class of sensory afferent. Cutaneous afferents are described first, with an emphasis on their possible use in FES systems in the upper and lower extremities. This is followed by a presentation of the characteristics of cuff-recorded activity from muscle afferents. The section concludes with a discussion of how mechanoreceptors within the bladder may be used to monitor bladder fullness for application to neuroprostheses that assist in the management of bladder dysfunction.

Cuff-Recorded Responses from Cutaneous Afferents

Large cutaneous afferents provide the substrate for tactile sensation and play an important role in the motor control of grasp function. In the routine task of picking up a drinking glass, for example, tactile afferents signal key events such as the initial contact between the fingers and the glass, and the detection of slippage. Analysis of normal grasp control has revealed a tight regulation between the grip force and the manipulative force (e.g., lifting or pulling force). When an object is lifted, both the grip force and the lifting force increase in parallel, but the grip force is regulated to be modestly greater than the lifting force at each instant so that the object does not slip. A particular ratio of the grip and lifting forces is established, and this ratio is adapted for the current frictional conditions present at the skin-object interface. The system operates with a "safety margin" in that the grasp force is usually between 30 and 50% greater than the minimum force needed to prevent the object from slipping. With this scheme, the muscular effort can be relaxed when it is not needed and fatigue is minimized. The mechanoreceptive afferents in the digital skin continuously monitor the normal (grip) skin forces and the shear-directed (load) forces, and studies have shown that the automatic grip force regulation is totally disrupted if the digital skin is anesthetized (Westling and Johansson, 1984; Johansson and Westling, 1987; Johansson et al., 1992a;b;c).

The possibility that a similar control strategy could be employed for neuro-prosthetic (FES) hand grasp systems is compelling and has resulted in a series of ongoing studies which are described below.

A chronic neural recording cuff was implanted on a digital nerve in a quadriplegic volunteer who was also fitted with a set of intra-muscular stimulation electrodes so that lateral grasp function could be produced. The particular site for the recording cuff allowed registration of the cutaneous afferents that come from the lateral border of the index finger (see Fig. 2 – left), which is the skin area that is contacted during lateral grasp activities (also called *"Key Grip")*. The studies have shown that slippage events during object manipulation can be detected in the nerve-cuff recordings and that these events can be used to upgrade the muscle commands with a sufficiently short delay, so that object slippage can be arrested (Haugland et al., 1999). Lately this has been shown to be true in functional motor tasks such as eating. Moreover, in a comparative study against normal individuals who performed the same object restraint task, the response latencies of the grip force upgrades for the FES "artificial reflex controller" were not appreciably different (Lickel et al., 1996; Haugland et al., 1999).

Feedback for Lower-Extremity FES Systems

Motion of the hip, knee and ankle, as well as foot-floor contact and limb loading, comprise information that might be useful in providing cognitive feedback to users of lower extremity FES systems, as well as for fully integrated FES control systems. Using a vibrotactile sensory substitution approach, Phillips (1988) demonstrated the importance of limb loading information for FES users in maintaining balance. Erzin and his colleagues (1996) employed an electro-cutaneous feedback display and reported that users achieved a more optimal cadence.

The potential application of "natural" cutaneous sensors to provide information regarding foot-floor contact has been recognized, and a few efforts have already been commenced in this area. Two individuals with "dropfoot", as a consequence of cerebral palsy in one and multiple sclerosis in the other, were each implanted with a chronic neural recording cuff, applied to the sural nerve in one subject and the tibial nerve in the other subject (Sinkjær et al., 1994; Haugland and Sinkjær, 1995; Upshaw and Sinkjær, 1998).

As shown in Fig. 4, the receptive field territory of the sural nerve is such that the implanted recording cuff can register skin contact at the lateral border of the foot and heel. The status of contact of the foot with the floor was extracted from the nerve signal and was used in place of the usual shoe-insole-heel switch to toggle the muscle stimulation on and off in synchrony with the swing and stance phases of the gait.

Cuff-Recorded Responses from Muscle Afferents

Normal motor function relies heavily on signals from muscle afferents for information about the static and dynamic forces acting about the joints. It might be useful for future FES systems if control signals could be derived from recordings of this afferent activity, and initial successes in this area by Yoshida and Horch (1996) lend support to this effort. Using the rabbit ankle as a model for the human ankle, Riso et al. (1999) characterized the responses evoked in a pair of complementary mixed nerves (the tibial and peroneal components of the sciatic nerve) that carry muscle afferents from the main ankle extensor and ankle flexor muscles. Simultaneous recordings were obtained using tripolar cuffs installed around the two nerves. For the initial study, a servo apparatus was used to rotate the ankle with ramp and hold movements that alternated from flexion to extension. Figure 5 shows the basic responses to this passive motion. Dorsiflexion (see "**2**") movement stretches the extensor muscles, and after a position threshold "**3**" is exceeded, evokes vigorous activity in the tibial nerve. At the start of the plateau (hold)

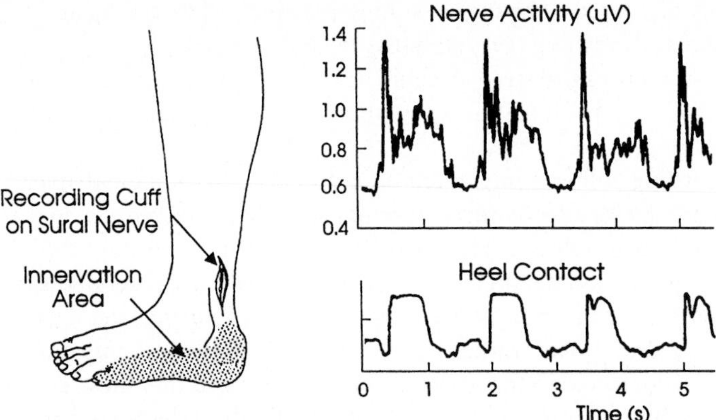

Fig. 4. Showing how natural sensory signals recorded from the sural nerve can be used to control an FES footdrop orthosis. The control signal was derived by rectifying and then integrating the activity (at right) recorded with a nerve cuff applied chronically around the sural nerve in a patient with multiple sclerosis. Contact of the foot with the floor was determined by applying a threshold to the processed signal. The patient wore shoes during the collection of the data. (Modified from Haugland and Sinkjær, 1995).

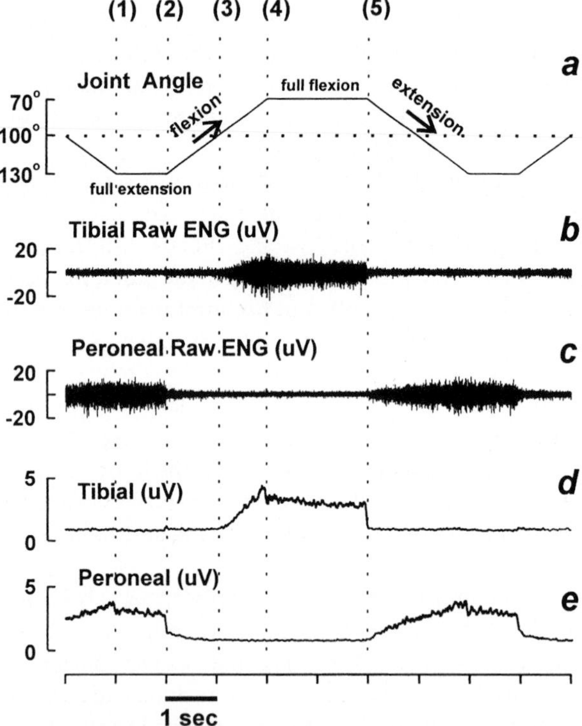

Fig. 5. Nerve activity recorded from the tibial and peroneal nerves in a rabbit during passive ankle movement over a range of 60° from near-maximum flexion to near-maximum extension. The nerve activities before processing appear in b and c, while d and e show the respective activities after rectification and integration using a moving window averager. The data shown in b-e represent the mean results of superimposing twenty responses. (For further details, see text). Figure modified from figure8 in Riso (1998)

phase ("4"), there is a reduction in the tibial activity because the dynamic response from the spindle primary afferents is curtailed, likely leaving only the spindle secondary (static) afferents discharging and possibly some contribution from the Golgi tendon organ afferents. At "5" the motion reverses direction toward extension with two consequences: The activity halts abruptly in the tibial recording (since the spindles are being shortened), and in a complementary manner, an increase is seen in the afferent activity recorded from the peroneal nerve.

The most rudimentary analysis of the dual nerve recordings reveals the direction of the movement of the joint. Moreover, if the rate of the motion is increased, then the dynamic response is also increased. Calibration of the dynamic response, however, has been difficult to achieve because it depends strongly on the initial stretch (i.e., related to the starting position) of the responding muscles (Jensen et al., 1998; Riso et al., 1999). A final consideration with regard to the intended application for FES concerns the effect of contractions within the muscles whose afferents are being recorded, since it is well known that contractions unload the spindle receptors and interrupt their activity. At the same time, however, the Golgi tendon receptors will be activated by the contraction. The activity recorded by the cuff will represent a summation of these opposing effects, and studies have begun to quantify these effects. One way of circumventing the complications of muscle contractions would be to not apply FES to some muscles around each joint so as to receive passive movement signals from them. In situations where FES is given only to the muscle(s) on one side of a joint at any given time (i.e., co-contraction of antagonistic pairs is avoided), the muscles opposite to the FES muscle will be passively stretched. In either case, stimulus artifacts from nearby stimulated muscles will still contaminate the cuff recordings, but there will be "clear" periods between the artifacts when the neural activity can be studied.

Feedback for FES-Assisted Micturition and Continence

Delayed or incomplete bladder voiding poses serious health risks to individuals with spinal cord injury (SCI). This includes not only bladder and kidney infections from stale urine and reflux back to the kidneys, but also a condition known as autonomic dysreflexia in which excessive bladder pressure can upset the autonomic regulation of the blood pressure.

If the person with spinal injury could be alerted that the bladder was approaching full capacity, then voiding could be performed in a timely manner. Working with an anesthetized cat preparation, Häbler et al. (1993) showed that single afferents recorded from the S2 sacral root discharged in relation to the state of fullness of the bladder. With the purpose of applying this finding in neuroprostheses for artificial bladder control and micturition in man, Jezernik et al. (1997, 1999) have performed acute studies in pigs and demonstrated that cuff recordings from the pelvic nerve and sacral roots also show increased neural activity when the bladder pressure rises. Fig. 6 shows the bladder pressure and pelvic nerve signal during spontaneous bladder contractions in an anesthezied pig. It follows from the figure that the nerve signal reflects the rhythmic changes in the bladder pressure and to a lesser extent the absolute increase in the bladder volume.

The findings demonstrated in Fig. 6 suggest that patients with a hyper-reflexsive bladder (SCI, some incontinent patients) could, when needed, receive a closed-loop

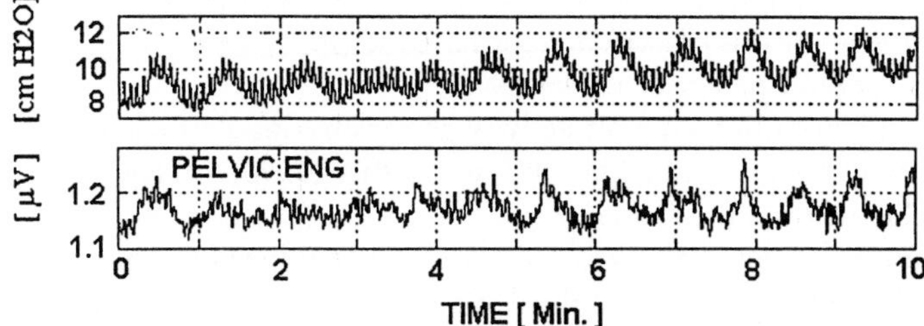

Fig. 6. Spontaneous bladder contractions during slow bladder filling at low bladder volume. Top figure shows the bladder pressure and bottom figure the rectified and filtered nerve signal from pelvic nerve. The contractions are clearly reflected in the nerve signal. The amplitude of the recorded pelvic nerve changes was high in comparison to the rather small pressure fluctuations (only 2 cm H_2O each). Modified from Fig. 10 in Jezernik et al., 1999.

controlled FES implant that uses this recorded sensory input to detect and inhibit the unwanted bladder contraction. Present solutions, i.e., suppression of reflex contractions by drugs and emptying the bladder by catheterization, need to be done many times during the day and night and are often unpleasant for the patient. In some cases drugs do not work, then surgical intervention is needed, by which the detrusor is deafferented by cutting the sacral dorsal roots to prevent reflex contractions. The bladder can then be emptied by the use of a sacral root stimulator that has electrodes on the sacral ventral roots. Dorsal rhizotomy increases the bladder capacity, but reflex erection in male patients is lost. To avoid cutting sacral dorsal roots, one could detect fast pressure rises and detrusor activation with nerve-cuff recordings from bladder nerves. The controller could then take appropriate actions as, e.g., inhibit detrusor contractions by stimulating pudendal or penile nerves, or block efferent or afferent pelvic nerve transmission to prevent reflex detrusor contractions (Jezernik et al., 1999; Rijkhoff et al., 1998). In this way continence could be reestablished, low-pressure voiding achieved, functional bladder capacity increased; and besides medical status improvement, patients would become more independent and could socialize more easily.

Results

The application of cuff electrodes to peripheral nerves to register the activity of cutaneous and muscle afferents was described with emphasis on electrode properties, nerve damage, and electrode design. Future developments of cuff electrodes will probably focus on the fabrication methods, such as the use of thin film electrodes, addition of electronics on the cuff, improvement of signal-to-noise ratio, and cuffs for fascicle-selective recordings.

Recent studies performed with chronically implanted nerve cuffs in disabled human subjects were presented as examples of the potential value of nerve-cuff recordings to furnish sensory information for feedback control in fully implantable FES systems. This included the development of an artificial grasp reflex based on cutaneous activity recorded from a digital nerve and the use of a neurally derived signal from the foot skin for controlling the stimulation of the dorsiflexor muscles in a dropfoot stimulator. Plans to exploit neurally derived signals for feedback control with bladder neuroprostheses were presented.

New findings from studies in rabbits were presented regarding the muscle afferent responses that can be recorded from peripheral sensory nerves when the ankle flexor and extensor muscles are stretched by passive joint motion. The reciprocal behavior of the afferent activity of the tibial and peroneal nerves served as a model for antagonistic muscle pairs acting at various joints in the human such as the elbow, wrist or knee. An initial focus, however, is to apply this muscle afferent-based sensor system to the case of FES-assisted standing in paraplegia (Jensen et. al. 1998).

Strategies for using cuff-recorded afferent activity as sensory inputs for cognitive feedback systems have only been described briefly, but see Riso (1998) for suggestions given for possible developments in the future.

The role of neuro-motor prostheses for increasing the quality of life of disabled individuals is becoming more evident each day. As the demand increases to develop systems capable of providing expanded functionality, so too does the need to develop adequate sensors for control. The use of natural sensors represents an innovation. Future research will show whether cuffs and other types of electrodes can be used to reliably extract signals from the large number of other receptors in the body to improve and expand on the use of natural sensors in neuro-prosthetic systems.

Acknowledgement: This work was supported by The Danish National Research Foundation.

References

Andreasen LNS, Jensen W. (1996) Characterization of the calcaneal and sural ENG during standing - An experimental study. Master thesis, report nr. S10-M11, Aalborg University, Denmark

Davis LA, Gordon T, Hoffer JA, Jhamandas J, Stein RB. (1978) Compound action potentials recorded from mammalian peripheral nerves following ligation or resuturing. J. Physiol. (Lond) 285:543–559

Erzin R, Bajd T., Kralj A., Savrim R., Benko H. (1996) Influence of sensory biofeedback on FES assisted walking, Elektroteh. Vestn., 63:53

Gillespie MJ, Stein RB. (1983) The relationship between axon diameter, myelin thickness and conduction velocity during atrophy of mammalian peripheral nerves. Brain Res., 259:41–56

Gordon T, Gillespie J, Orozco R. Davis L. (1991) Axotomy-induced changes in rabbit hindlimb nerves and the effects of chronic electrical stimulation. J. Neurosci., 11:2157–2169

Gordon T, Hoffer JA, Jhamandas J, Stein RB. (1980) Long-term effects of axotomy on neural activity during cat locomotion. J. Physiol. (Lond) 303:243–263

Grill WM, Mortimer JT (1994) Electrical properties of implant encapsulation tissue. Annals of Biomed. Eng. 22:23–33

Hanson MA, Moore PJ, Nijhuis JG. (1987) Chronic recording from the phrenic nerve in fetal sheep in utero. J. Physiol. (Lond) 394:4P

Haugland, M., Hoffer, J.A. (1994a) Artifact-free sensory nerve signals obtained from cuff electrodes during functional electrical stimulation of nearby muscles. IEEE Transactions on Rehabilitation Engineering, 2:37–39

Haugland MK, Hoffer JA, Sinkjær T. (1994) Skin contact force information in sensory nerve signals recorded by implanted cuff electrodes. IEEE Trans. Rehab. Eng., 2:18–28

Haugland MK, Hoffer JA. (1994b) Slip information provided by nerve cuff signals: Application in closed-loop control of functional electrical stimulation. IEEE Trans. Rehab. Eng. 2:29–36

Haugland MK, Lickel A, Haase J, Sinkjær T. (1999) Control of FES thumb force using slip information obtained from the cutaneous electro-neurogram in quadriplegic man. IEEE Trans. Rehab. Eng., Vol. 7.

Haugland M, Lickel A, Riso R, Adamczyk MM, Keith M, Jensen IL, Haase J, Sinkjær T. (1995) Restoration of lateral hand grasp using natural sensors. Proc. of the 5th Vienna Int. Workshop on FES, Vienna, pp. 339–342

Haugland MK, Sinkjær T. (1995) Cutaneous whole nerve recordings used for correction of footdrop in hemiplegic man. IEEE Trans. Rehab. Eng., 3:307–317

Haugland MK, Sinkjær T. (1999) Control with natural sensors. Invited chapter to section VIII. Synthesis of Posture and movement in Neural Prostheses. Book Editors: Jack Winters and Pat Crago. In Press

Häbler HJ, Jänig W, Koltzenburg M. (1993) Myelinated primary afferents of the sacral spinal cord responding to slow filling and distention of the cat urinary bladder, J. Physiol. (Lond) 463:449

Hoffer JA. (1990) Techniques to record spinal cord, peripheral nerve and muscle activity in freely moving animals. In: Neurophysiological Techniques: Applications to Neural Systems. Neuromethods 15, A. A. Boulton, G.B. Baker and C.H. Vanderwolf, Eds. Humana Press, Clifton, N.J., pp. 65–145

Hoffer JA, Loeb GE, Pratt CA. (1981) Single unit conduction velocities from averaged nerve cuff electrode records in freely moving cats. J. of Neurosc. Meth., 4:211–225

Hoffer JA, Loeb GE, Marks WB, O'Donovan MJ, Pratt CA, Sugano N. (1987) Cat hindlimb motoneurons during locomotion. I. Destination, axonal conduction velocity and recruitment threshold. J. Neurophysiol., 57:510–529

Hoffer JA, Marks WB, Rymer WZ. (1974) Nerve fiber activity during normal movements, Soc. Neurosci. Abstr., 4:300

Hoffer JA, Sinkjær T. (1986) A natural "force sensor" suitable for closed-loop control of functional neuromuscular stimulation. Proc. 2nd Vienna Int. Workshop on Functional Electrostimulation, pp.47–50

Hoffer JA, Stein RB, Gordon T. (1979) Differential atrophy of sensory and motor fibers following section of cat peripheral nerves. Brain Res., 178:347–361

Jensen W., Riso R.R. and. Sinkjær T. (1998) Position information in whole nerve cuff recordings of muscle spindle afferents in a rabbit model of normal and paraplegic standing. Proceedings of the IEEE/EMBS Annual Meeting, Hong Kong, Nov.

Jezernik S, Wen JG, Rijkhoff NJM, Djurhuus JC, Sinkjær T. (1999) Analysis of nerve cuff electrode recordings from preganglionic pelvic nerve and sacral roots in pigs. J. Urology, Submitted

Jezernik S, Wen JG, Rijkhoff NJM, Haugland M., Djurhuus JC, Sinkjær T. (1997) Whole nerve cuff recordings from nerves innervating the urinary bladder, Second Annual IFESS Conference / Fifth Triennial Neural Prostheses Conference, Vancouver, Canada, Proceedings pp. 45–46, August

Johansson RS, Häger C, Backström L. (1992c) Somatosensory control of precision grip during unpredictable pulling loads: III. Impairments during digital anesthesia, Exp. Brain Res., 89:204

Johansson RS, Häger C, Riso RR, (1992b) Somatosensory control of precision grip during unpredictable pulling loads. II. Changes in load force rate, Exp. Brain Res., 89:192

Johansson RS, Riso RR, Häger C, Backström C. (1992a) Somatosensory control of precision grip during unpredictable pulling loads: 1. Changes in load force amplitude, Exp. Brain Res., 89:204

Johansson RS, Westling G. (1987) Signals in tactile afferents from the fingers eliciting adaptive motor responses during precision grip, Exp. Brain Res., 67:141

Kallesøe JA, Hoffer JA, Strange K, Valenzuela I. (1996) Implantable cuff having improved closure: United States Patent No.5,487,756, awarded January 30

Knaflitz M, Merletti R. (1988) Suppression of stimulation artifacts from myoelectric-evoked potential recordings. IEEE Transactions on Biomedical Engineering, 35(9):758–763

Krarup C, Loeb GE. (1987) Conduction studies in peripheral cat nerve using implanted electrodes: I. Methods and findings in control. Muscle & Nerve, 11:922–932

Krarup C, Loeb GE, Pezeshkpour GH. (1988) Conduction studies in peripheral cat nerve using implanted electrodes: II The effects of prolonged constriction on regeneration of crushed nerve fibers. Muscle & Nerve, 11:933–944

Krarup C, Loeb GE, Pezeshkpour GH. (1989) Conduction studies in peripheral cat nerve using implanted electrodes: III The effects of prolonged constriction on the distal nerve segment. Muscle & Nerve, 12:915–928

Larsen JO, Thomsen M, Haugland M, Sinkjær T. (1998) Degeneration and regeneration in rabbit peripheral nerve with long-term nerve cuff electrode implant. A stereological study of myelinated and unmyelinated axons. Acta Neuropathologica, 96:365–378

Lickel A, Haugland MK, Sinkjær T. (1996) Comparison of catch responses between a tetraplegic patient using an FES system and healthy subjects, Proc. 8th Annual International Conference of IEEE/EMBS

Little JW. (1986) Serial recording of reflexes after feline spinal cord transection. Exp. Neurol., 93:510–521

Loeb GE, Marks WB, Hoffer JA. (1987) Cat hindlimb motoneurons during locomotion. IV. Participation in cutaneous reflexes. J. Neurophysiol., 57:563–573

Loeb GE, Peck RA, (1996) Cuff electrodes for chronic stimulation and recording of peripheral nerve activity. J. Neurosc. Meth., 64:95–103

Marks WB, Loeb GE. (1976) Action currents, internodal potentials and extracellular records of myelinated mammalian nerve fibres derived from node potentials. Biophys. J., 16:655–668

Marshall KW, Tatton WG. (1990) Joint receptors modulate short and long latency muscle responses in the awake cat. Exp. Brain Res., 83:137–150

Milner TE, Dugas C, Picard N, Smith AM. (1991) Cutaneous afferent activity in the median nerve during grasping in the primate. Brain Res., 548:228–241

Naples GG, Mortimer JT, Schemer A, Sweeney JD. (1988) A spiral cuff electrode for peripheral nerve stimulation, IEEE Trans. Biomed. Eng., 35:905

Nicolic ZM, Popovic DB, Stein RB, Kenwell Z. (1994) Instrumentation for ENG and EMG recordings in FES systems. IEEE Trans. Biomed. Eng., 41:703–706

Palmer CI, Marks WB, Bak MJ. (1985) The responses of cat motor cortical units to electrical cutaneous stimulation during locomotion and during lifting, falling and landing. Exp. Brain Res., 58:102–116

Pflaum C, Riso RR. (1996) Performance of alternative amplifier configurations for tripolar nerve cuff recorded ENG, Proc. ^{18}th Annual meeting IEEE/Engr. In Med. Biol. Soc., Amsterdam

Phillips CA (1988) Sensory feedback control of upper and lower extremity; motor prostheses, CRC Crit. Rev. Biomed. Eng., 16:105

Popovic DB, Stein RB, Jovanovic KL, Rongching D, Kostov A, Armstrong WW. (1993) Sensory nerve recording for closed-loop control to restore motor functions. IEEE Trans. Biomed. Eng., 40:1024–1031

Rijkhoff NJM, Wijkstra H, van Kerrebroeck PEV, Debruyne FMJ. (1998) Selective detrusor activation by sacral ventral nerve root stimulation: First results of intraoperative testing in humans during implantation of a Finetech-Brindley system. World Journal of Urology, 16:337–341

Riso RR. (1998) Perspectives on the role of natural sensors for cognitive feedback in neuromotor prostheses. Automedica, 16:329–353

Riso RR. Slot PJ. (1996) Characterization of the ENG activity from a digital nerve for feedback control in grasp neuroprostheses, In: Neuroprosthethics from basic research to clinical applications, Pedotti A, Ferrarin M., Quintern J., Riener R., Eds, Springer, pp. 354–358

Riso RR, Mosallie FK, Jensen W, Sinkjær T. (1999) Nerve Cuff recordings of muscle afferent activity from tibial and peroneal nerves in rabbit during passive ankle motion. IEEE Trans. on Rehab. Eng. Provisionally accepted

Riso RR, Slot P, Haugland M, Sinkjær T. (1995) Characterization of cutaneous nerve responses for control of neuromotor prostheses, Proc. 5th Vienna Intl. Workshop on Functional Electrical Stimulation, p. 335

Sahin M, Haxhiu MA, Durand DM, Dreshaj IA. (1997) Spiral nerve cuff electrode for recording of respiratory output. J. Appl. Physiol., 83:317

Sinkjær T, Hansen M, Upshaw B, Haugland M, Kostov A. (1998) Processing sensory nerve signals meant for control of paralyzed muscles. NORSIG '98 IEEE Nordic Signal Processing Symposium. 8th–11th June, Vigsø Holiday Resort, Denmark, 17–24.

Sinkjær T, Haugland MK, Haase J. (1994) Natural neural sensing and artificial muscle control in man. Exp. Brain Res., 98:542

Sinkjær T, Haugland M, Haase J. (1993) Neural cuff electrode recordings as a replacement of lost sensory feedback in paraplegic patients. Neurobionics, 267–277

Sinkjær T, Hoffer JA. (1990) Factors determining segmental reflex action in normal and decerebrate cats. J. Neurophysiol., 64:1625–1635

Slot P, Selmar P, Rasmussen A, Sinkjær T. (1997) Effect of long-term implanted nerve cuff electrodes on the electrophysiological properties of human sensory nerves. J. Artificial Organs, 21:207–209

Stein RB, Charles D, Davis L, Jhamandas J, Mannard A, Nichols TR. (1975) Principles underlying new methods for chronic neural recording. Canad. J. Neurol. Sci., 2:235–244

Stein RB, Gordon T, Oguztöreli, Lee RG. (1981) Classifying sensory patterns and their effects on locomotion and tremor. Can. J. Physiol. Pharmacol., 59:645–655

Stein RB, Nichols TR, Jhamandas J, Davis L, Charles D. (1977) Stable long-term recordings from cat peripheral nerves. Brain Res., 128:21–38

Strain RE, Olson WH. (1975) Selective damage of large diameter peripheral nerve fibers by compression: An application of Laplace's law. Exp. Neurol., 47:68–80

Struijk, J.J., Thomsen, M., Larsen, J.O., Sinkjær, T. (1999) The use of cuff electrodes in long-term recordings of natural sensory information from peripheral nerves. IEEE Engineering in Medicine and Biology Magazine. May/June 1999

Struijk JJ, Thomsen M. (1998) Tripolar nerve cuff recording: Stimulus artifact, EMG and the recorded nerve signal. 17th Annual International Conference IEEE Engineering in Medicine and Biology Society, September, Montreal, Quebec, Canada. Only available on CD-ROM

Thomsen M. (1998) Characterisation and optimisation of whole nerve cuff recording cuff electrodes. Ph.D.-thesis, Aalborg University, Denmark

Thomsen M, Struijk JJ, Sinkjær T. (1996) Artifact reduction with monopolar nerve cuff recording electrodes. 18th Annual Int. Conference of the IEEE Engineering in Medicine and Biology Society, October-November, Amsterdam, The Netherlands

Thomsen M, Struijk JJ, Sinkjær T. (1999) Nerve cuff recording with a combined mono-and bi-polar electrode, IEEE Trans. Rehab. Eng., Submitted

Tyler DJ, Durand DM. (1996) Selective stimulation with a chronic slowly penetrating interfascicular nerve electrode. Proceedings of the 18th Annual Meeting of the IEEE/EMBS , Amsterdam, paper #582

Upshaw, B., Sinkjær, T. (1998) Digital signal processing algorithms for the detection of afferent nerve activity recorded from cuff electrodes. IEEE Transactions on Rehabilitation Engineering, 6:172–181

Upshaw B, Sinkjær T. (1997) Natural vs. artificial sensors applied in peroneal nerve stimulation. Journal of Artificial Organs, 21(3):227–231

Westling G, Johansson RS. (1984) Factors influencing the force control during precision grip, Exp. Br. Res., 53:277

Woodbury JW, Woodbury DM. (1991) Vagal stimulation reduces the severity of maximal electroshock seizures in intact rats: Use of a cuff electrode for stimulating and recording. Pace, 14:94–107

Yoshida K, Horch K. (1996) Closed – loop control of ankle position using muscle afferent feedback with functional neuromuscular stimulation, IEEE Trans. Biomed. Eng., 43(2):167

Chapter 30

Microneurography in Humans

Mikael Bergenheim, Jean-Pierre Roll and Edith Ribot-Ciscar

Introduction

The microneurographic technique allows the researcher to observe and monitor the continuous ongoing neural activity of peripheral nerves in awake human subjects. It enables the study of both multiunit and single neuronal activities in practically all kinds of nerve fibers regardless of size or myelinization. The scientist has the possibility to correlate neuronal activity with peripheral stimuli, vegetative and motoric activity, as well as subjective experience. The technique is a relatively new one (first publication by Hagbarth and Vallbo in 1967), but has already attracted neuroscientists working in many different areas of the nervous system, in both normal and pathological conditions. During the 30 or so years the technique has been available, it has enabled major scientific advances in many of these areas.

The aim of this chapter is first of all to provide a practical tool for scientists wishing to use the microneurographic technique. Furthermore, it will more elaborately discuss some examples of results attained using microneurography, illustrating the role of muscle spindles in kinesthesia (see below; RESULTS). However, the role of other afferents in kinesthesia has also been investigated. As an example of this, microneurographic recordings have revealed that cutaneous afferents might provide information about joint movement, and that there are some slow adapting (SA II) receptors that seem to provide accurate information about joint position (Edin 1992). Furthermore, joint afferents have been shown to respond with a slowly adapting discharge to extreme angular displacements in one or more axes of passive joint rotation (Macefield et al. 1990). Concerning the sense of touch, the properties of the Aα fibers from receptors in the glabrous skin (FA I, FA II, SA I and SA II) of the human hand were thoroughly described by Vallbo and collaborators (see, e.g., Johansson 1976). In the field of pain, the functional properties of unmyelinated C fibers in human skin nerves were studied by Torebjörk and Hallin (1976), and amongst other features, their polymodality and receptive field sizes were illustrated.

The microneurographic technique has also been used to explore efferent activity. Such work has, for example, described the reflex influence of muscle afferents on the firing rates of α motoneurons (Macefield et al. 1993). Furthermore, Aniss and coworkers (1990) demonstrated selective reflex activation of fusimotor neurons in response to non-painful stimuli. Direct recordings from fusimotor neurons performed in our lab have revealed that the activity in these neurons can be modulated under various condi-

Correspondence to: Mikael Bergenheim, National Institute for Working Life, Division of Physiology, Box 7654, Umea, 907 13, Sweden (phone: +46 90 7866389; +33 0491 288296; fax: 46 90 7865027; +33 04 91 28 86 69; e-mail: mikael.bergenheim@niwl.se; lnh@newsup.univ-mrs.fr)
Jean-Pierre Roll, UMR 6562 Université de Provence / CNRS, Laboratoire de Neurobiologie Humaine, Marseille, France
Edith Ribot-Ciscar, UMR 6562 Université de Provence / CNRS, Laboratoire de Neurobiologie Humaine, Marseille, France

tions; e.g., cognitive factors such as listening to maneuver instructions or performing mental computation, behavioral activities such as laughing or talking, and environmental disturbances such as somebody entering the room or a short unexpected auditory stimulus (Ribot et al. 1986).

On the vegetative level, important work concerning the central control of sympathetic outflow to both skin (Wallin et al. 1976; Hallin and Torebjörk 1974) and muscle (Sundlof and Wallin 1978) has been performed quite early in the history of microneurography.

Briefly mentioned above are some examples of physiological studies in which microneurography has been an important tool. However, the technique has also been used in pathophysiology. Here one might mention the studies of Parkinsonian rigidity (Burke et al. 1977), and spasticity (Szumski et al. 1974, Hagbarth et al. 1973).

The micro-electrode normally used to record neural activity in microneurography can also be used for stimulation. In this way perceptual responses to microstimulation of single joint, muscle (Macefield et al. 1990), skin (Ochoa and Torebjörk 1983, Macefield et al. 1990) and C nociceptor afferents (Ochoa and Torebjörk 1989) of the human hand were explored.

The aim of the present chapter is not to review the extensive work performed using the microneurographic technique, as this has been done several times previously (for recent summary of references see Gandevia and Hales 1997), but rather to provide a practical description of how to perform and employ the technique. Naturally, this paper cannot give justice to all the different techniques used in all the laboratories performing microneurography, and the core of the chapter will be the methods used in our lab. However, in case of technical or methodological discrepancies between our and other laboratories, we will address these differences as objectively as possible.

Finally, it should be mentioned that, although the microneurography technique might seem relatively simple (and in theory it is), in practice it demands both a lot of patience and quite a bit of luck.

Materials

The Tungsten Electrode

The electrodes most commonly used are insulated tungsten micro-electrodes. These electrodes have a conical recording surface of variable size which enables a recording in three dimensions from many surrounding fibers in the nerve fascicle (see Fig. 1). Since the tungsten electrode must be sterile, and since the properties of the electrode might change after usage, it cannot be used repeatedly. Properties:

- Impedance: 500 kΩ to 1MΩ tested at 1kHz
- Tip diameter: approx. 5–8 μm
- Length: 31 mm

Many laboratories prefer to produce their electrodes themselves. However, there are electrodes that are commercially available (see SUPPLIERS) and need only to be slightly modified before usage. This commercially available electrode is ordered without connector in order to diminish its weight and facilitate its "free floating" in the tissue (see PROCEDURE). Therefore, before the electrode can be used, a teflon-covered silver wire must be soldered to its end. This might be done in the following manner (see Fig. 2):

1. The teflon cover is removed from approx. one cm at the end of the wire (see Fig. 2A), and this peeled section of the wire is then tightly wrapped around the uninsulated end of the electrode (5–7 turns) and soldered in place (see Fig. 2B).
2. The soldered part and any remaining uninsultaed part of the electrode and wire are then insulated using a heated plastic tube (approx. 1cm) which is tightly squeezed in place (see Fig. 2C).

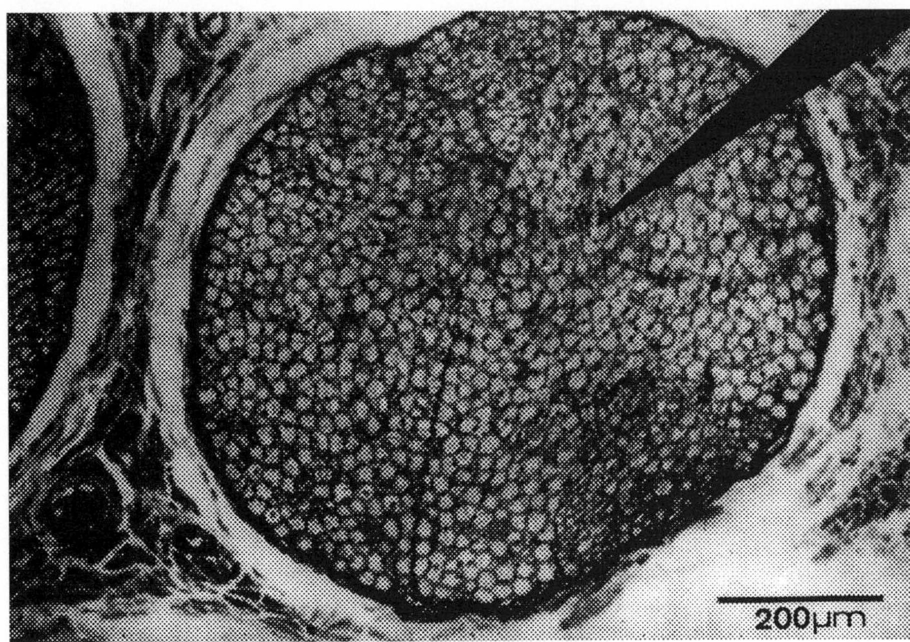

Fig. 1. Microphotograph of a transection of a peripheral nerve. Mounted to the photo is the black silhouette of the tip of a microelectrode.

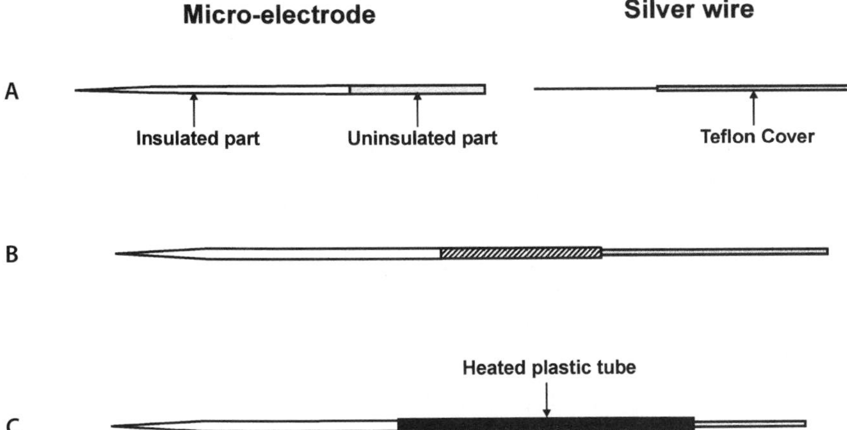

Fig. 2. The modification of the commercially available microelectrode. **A:** The microelectrode and the peeled silver wire. **B:** The silver wire is wrapped around the uninsulated part of the electrode **C:** The uninsulated area of the electrode and the peeled section of the silver wire are covered by a heated plastic tube. The tube is squeezed tightly around the electrode.

3. In order to keep the electrodes sterile, they are placed in a hermetically closed box containing formaldehyde tablets. Before using the electrodes, they are carefully rinsed in alcohol.

The Concentric Needle Electrode

Recently, a concentric needle electrode has been described (Hallin 1990, Hallin and Wu 1998). This electrode differs from the normal tungsten one in that it has a unidimensionally oriented and smaller recording surface. It consists of a commercially available hypodermic needle and a small central core wire of either tungsten or platinum-iridium. The electrode is filled with Araldite. The authors claim that this electrode "in several respects reflects the intraneural biology of the peripheral nerve more faithfully than the tungsten electrode", and ascribe this to the properties of the recording surface. For this reason, the electrode seems to be particularly useful for the recording of C fibers (Hallin 1990). The authors further suggest that the electrode has some advantages to the traditional tungsten electrode, e.g., the electrical and mechanical properties of the electrode are stable, enabling repeated use, and multi-channel recordings can be achieved by adding more recording surfaces to the electrode (Hallin and Wu 1998).

Properties:

- Shaft diameter: 200–250 μm
- Wire diameter: 10–30μm (tungsten wire), or 20–30μm (platinum-iridium wire)
- Impedance: 500–700 kΩ (thin wire, 10μm), 200–300 kΩ (thicker wire, 30 μm) tested at 1 kHz

For details concerning the manufacturing of the electrode, see Hallin and Wu (1998).

Recording and Analysis

A standard electrophysiological recording system can be used. The bandwidth of the recording should be maintained between 300 and 3000 Hz to obtain optimal signal-to-noise ratios.

There are several options as to the choice of the reference for the recording. In our lab, we use a damp band (fastened around the thigh of the subject, see Fig. 4) as ground, and at the same time as reference. In other laboratories, an electrode similar to the registration electrode but with a slightly larger uninsulated part is inserted a couple of centimeters away from the recording electrode and used as a reference (see, e.g., Gandevia and Hales 1997).

In the case of multiunit recordings or an insufficient signal-to-noise-ratio, individual spike trains can be discriminated. Over the years, several different methods have been used (for review, see Schmidt 1984a, b). For example, a software-based algorithm for real-time sorting of extracellular multi-neuron recordings was described by Salganicoff and coworkers (1988), and a low-cost single-board solution for real-time classification of multi-neuron recordings was produced by Kreiter and colleagues (1989). A backpropagation neural network solution was presented by Jansen (1990), and a similar method was later described by Mirfakhraei and Horch (1994). Oghalai and coworkers (1994) described an adaptive spike discriminator, based on the ART2 algorithm, for classification of multi-unit spike trains recorded on a single channel. More recently, a multi-channel, real-time, unsupervised spike discriminator based on SOM algorithms was presented by Öhberg et al. (1996). There are also several commercial systems available for data aquisition (e.g., Discovery; DataWave Technologies, Longmont, CO, USA or Spike2; CED, Cambridge, England) which include spike discriminators using linear clustering methods. For various reasons it might be of interest to analyze the morphology of the recorded action potentials (e.g., to look for the occurrence of conduction blocks). As the methods for this analysis have been extensively described elsewhere (Gandevia and Hales 1997, Inglis et al. 1996), the current chapter will not discuss this issue.

Subjects

Most commonly, the subjects are paid volunteering students, with a general interest in the experimental approach. After their health history has been ascertained, they are carefully selected on the basis of the following criteria:

- *Corpulence.* Subjects with an average corpulence are chosen. Adipose or muscular subjects are excluded since the nerve is often too deeply located in these subjects, and thus, not easily palpated. On the other hand, in subjects that are too thin, the nerve might be too superficial. In these subjects, locating and approaching the nerve is certainly easy; however, since the electrode cannot be planted deeply enough, it is difficult to stabilize its position after it has been released by the experimenter.
- *Sex.* Male subjects are generally preferred as it is a common observation that the nerve often is easier to palpate and locate in these subjects.
- *Ability to stay calm.* This is important, since the subject is required to stay seated in a chair, more or less immobile during a full experiment. This is one of the reasons why an experiment should not last longer than four hours. However, in some laboratories the subject lies down during the experiment.
- *Reaction* to the experimental environment and protocol, and in particular to the sting of the electrode. Naturally, the subjects that are too sensitive are excluded. It has also been observed that subjects that are too impressed by the laboratory equipment or the experimental procedure in general are not suitable.

Once a subject has been selected on the basis of the criteria presented above, he is asked to sign an experimental consent in accordance with the Declaration of Helsinki (and any additional consents required in the country where the experiment is performed), and an appointment for the future experiment is made.

If more than one experiment is to be performed on a particular subject, the interval between the different experiments should not be less than 1 month.

The Investigated Nerves

In theory, microneurography can be performed on any peripheral nerve. However, for practical reasons, the technique has been most commonly applied to a limited number of nerves of which a few will be briefly mentioned below (see Fig. 3):

- *Lateral peroneal nerve.* This nerve has the advantage of being superficially located in man and thus, the path of the nerve is often easily palpated. For a right-handed ex-

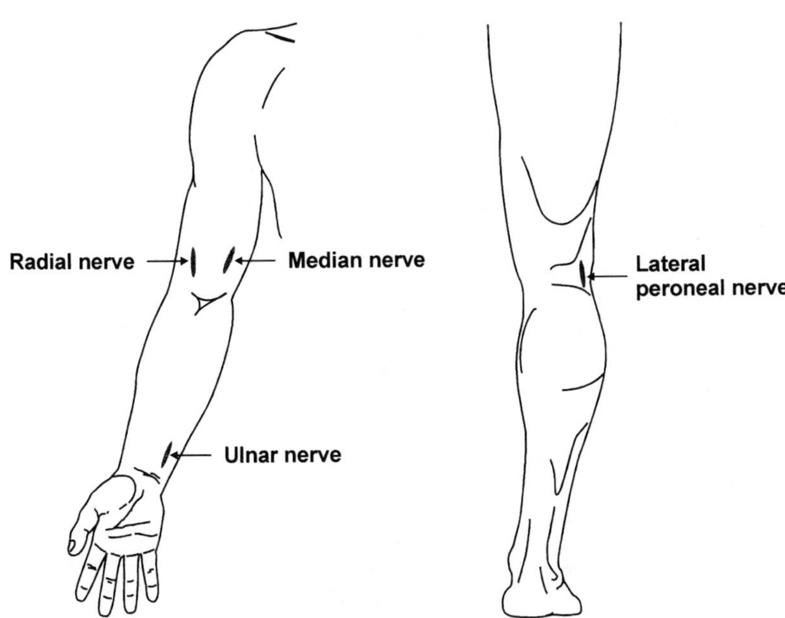

Fig. 3. Anatomical location of the most frequently investigated nerves. The shaded areas indicate the sites most suitable for application of the electrode.

perimenter, it is preferable to work on the left leg of the subject. The electrode is inserted at the level of the head of the fibula or at the beginning of the politeal fossa of the left leg.

- *Median nerve.* The median nerve can be recorded both from the wrist and just proximal to the elbow (see, e.g., Hallin 1990)
- *Radial nerve.* The radial nerve is often used in microneurographic experiments, and can be accessed 7–8 cm proximal to the elbow (see, e.g., Rothwell et al. 1990).
- *Ulnar nerve.* This nerve is most easily reached 5–10 cm proximal to the wrist (see, e.g., Rothwell et al. 1990).

Procedure

As mentioned previously, the aim of this chapter is to provide a practical guide to microneurography. In the following section we will describe and illustrate the technique as it is performed in our lab when recording from the lateral peroneal nerve. Although many of the remarks below are general and valid for recording of other nerves as well, it is inevitable that some comments will be specific and applicable to our experimental set-up only.

The experiments should be performed in an electrically shielded room with a pleasant atmosphere for the subjects as concerns, e.g., light, sound and color. Furthermore it might be recommended to keep as much as possible of the laboratory equipment hidden. Before starting the experiment, the subject is comfortably seated in an armchair (see Fig. 4). The legs are positioned in cushioned grooves so that a standardized relaxed position can be maintained without any muscular activity occurring. The knee joint is at an angle of about 120–130 degrees. The right foot lies on a stationary plate, and the left foot is fixed on a rotating pedal, the axis of which is centered in front of the ankle.

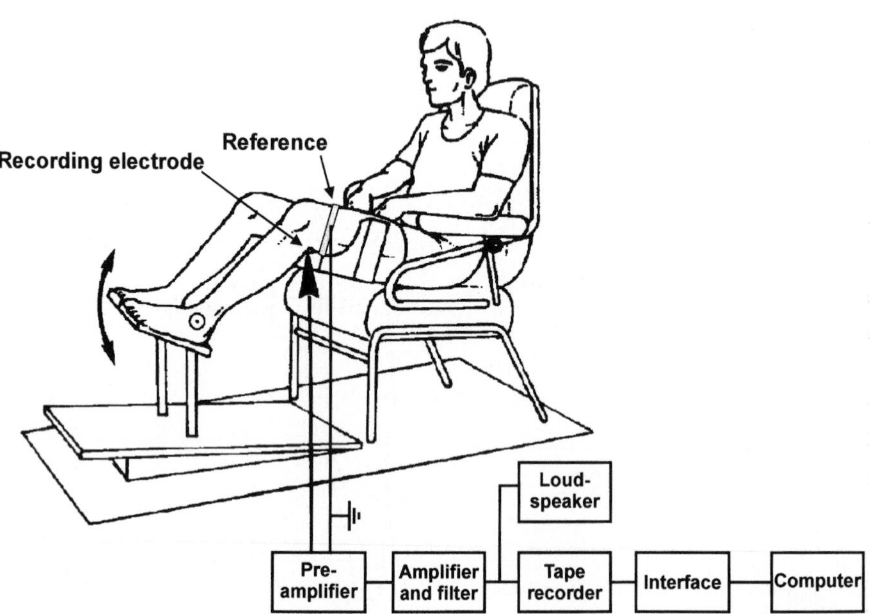

Fig. 4. Schematic illustration of the experimental set-up, and the flow-chart of the sampling equipment. In our lab we use Grass P5 series pre-amplifiers and amplifiers, a Luxman graphic equalizer as filter, a 1401 PLUS interface, a Biologic DTR-1802 tape recorder, and a PC586 computer.

The rotating pedal is connected to an electromechanical motor allowing movements whose parameters (e.g., velocity, amplitude) can be controlled. The amplified signal from the micro-electrode is connected to an oscilloscope and a loudspeaker (see Fig. 4), enabling continuous monitoring of the nerve activity throughout all the different stages of the experiment. In order to record movement data (i.e., dorsal or plantar flexions), either in passive or active conditions, the subject is equipped with a system involving two low-mass hinged bars, one attached to the leg of the subject, and the other to the foot by means of rubber bands. Any movement occurring at the hinge between the rods, which is centered in front of the external maleolus, is recorded using a linear potentiometer. For hygienic reasons, but also to soften the skin, the skin around the point of insertion is thoroughly cleaned using antiseptic soap. Standard surface EMG electrodes (for details see Chapter 26) are placed on the muscles of interest, depending on the experimental protocol. Before the recordings start, the subject should be trained to perform any voluntary activity demanded (if any) by the experimental protocol, without producing a "disturbing" gross muscular activity. It is of major importance to avoid such activities during microneurographic recordings for at least two reasons. Firstly, they might cause displacement of the electrode, resulting in loss of the unit or, in the worst case, damage of the nerve. Secondly, even minor EMG activity in the muscles near the electrode might lead to a contamination of the recorded nerve signals by motor unit activity.

Progress of the Experiment

1. *Palpating the nerve.* The first phase of the work consists of precisely locating the path of the nerve and marking it with a pen on the surface of the skin. If the nerve appears to be too deep or too superficial, its depth can be changed by a light extension/flexion of the knee.
2. *Inserting the electrode.* The microelectrode is then manually inserted percutaneously by alternate light pushes and releases (allowing the elastic skin to relieve the electrode slightly during the release phase), such that it advances slowly through the skin. When the electrode has passed through the skin, it is advisable to release it for a minute and observe the reaction of the subject. During all these steps, it might happen that, despite all the precautions, the subject is affected by the experiment or the laboratory environment and faints. For this reason one must pay careful attention to the subject's state. Signs like scratching the head, repetitive yawning, sensations of heat, and perspiration are all warning signals. If these signs appear, the experiment should be stopped immediately.
3. *Approaching the nerve through the subcutaneous tissue.* The nerve is approached using the same pushing-releasing technique as described above. The time required for this progression of the electrode varies greatly from one experiment to the other. Sometimes the nerve is reached in just a few minutes, but usually it takes about one hour. In order to guide the electrode tip into the nerve, many groups deliver weak electrical pulses through the electrode (e.g., 1–6 V, 0.2 ms duration), eliciting paresthesia or local muscle contraction in the fascicle territory. However, the use of this technique is not without problems. On the psychological level, the fact that electrical stimulation will be used in the experiment appears frightening to some subjects, and should therefore be avoided. Also, it is our experience that electrical stimulation not performed in a perfect way might cause powerful motoric activity, i.e., "starting reactions," and may thus jeopardize both the stability of the recording and the safety of the subject. In either case, while approaching the nerve, the subject is asked to give continuous information concerning any sensation he/she may perceive. This infor-

mation can give the experimenter important clues as to whether he/she should keep or change the direction of the penetration. For example, a painful sensation reported as "deep and/or diffuse" could indicate that the electrode is pressing against a muscular tendon. If this occurs, or indeed, as soon as the subject feels pain of any sort, the electrode should be released or maybe even drawn back slightly. After this, the pain usually disappears. On the other hand, the subject may perceive a tactile sensation, induced by mechanical stimulation of the nerve, projecting to an area that he can easily locate. The description of such an area might confirm to the experimenter that he/she is headed in the right direction. Here it is important to emphasize that the subject must be instructed not to point out the area of projection (as this might induce gross muscular activity), but rather to name it.

4. *Reaching the nerve.* When the nerve is reached, multiunit activity is usually heard from the audioamplifier. It is now time to find the right type of unit within the nerve. A precise identification of a unit can only be attained after the unit has been isolated and stabilized and while the electrode is "floating free". The procedure of this precise identification is presented as a separate subprotocol (see IDENTIFICATION OF THE FIBERS). However, before attempting to stabilize the recording and let the electrode float free, a *preliminary* identification should be made. For this purpose, it is useful to continuously stroke the skin of the area innervated by the nerve, or to stimulate the innervated muscles by pressing the tendons, muscle bellies, or by slightly stretching the muscle passively. Such stimulations allow the experimenter to roughly identify the modality and localization of any found afferents and their receptors. A detailed flow-chart of this preliminary identification is presented in Fig. 5. Already at the time when the nerve is reached and entered, the spontaneous behavior of the fiber might indicate its receptor origin. Most fibers discharge at very high frequency at

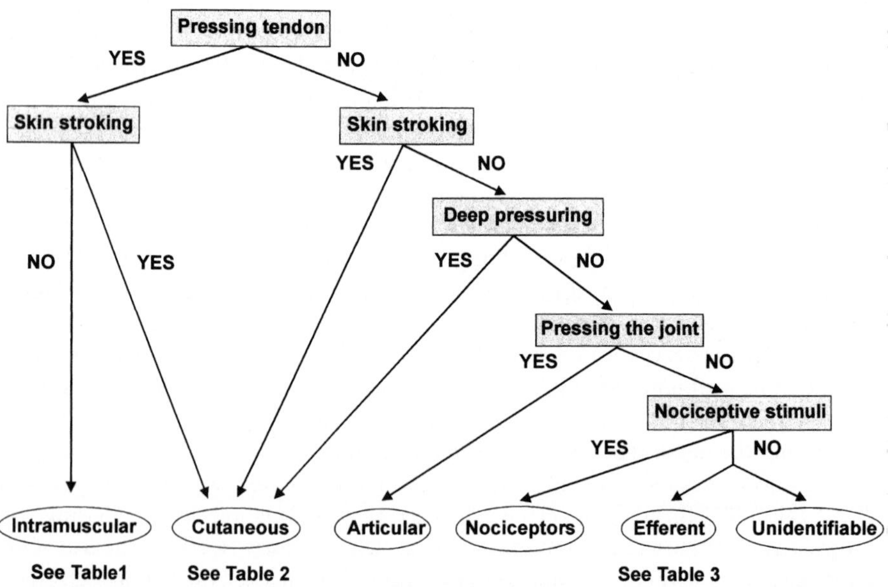

Fig. 5. Schematic flow-chart describing a possible method for the preliminary identification of a recorded nerve fiber. The different types of stimulation are given in the gray boxes. If the fiber responds to this stimulation, the procedure should be continued in the direction indicated by the YES arrow. If there is no response to the stimulus, the NO direction should be followed. The identities of the fibers are given in the circles at the bottom of the figure. For a precise identification of the intramuscular and cutaneous receptors and the efferent fibers, see IDENTIFICATION OF THE FIBERS and Tables 1, 2, and 3, respectively.

the moment when the nerve is reached. If this response drops to complete silence within a second or two, it is most probably a cutaneous afferent. If, on the other hand, the response does not drop to complete silence, but a sustained spontaneous activity remains, it is most likely an afferent from an intramuscular receptor. If the resting spontaneous activity is in regular bursts, it might be a vegetative efferent.

5. *Stabilizing the recording.* Finally, when a signal from a single neuron has been picked up by the electrode, this signal must be isolated and stabilized. This can be achieved by moving the electrode in minute steps guided by the auditory feedback produced by a continuous stimulation of the afferent. The major problem of this phase is that when the experimenter releases the electrode to let it float free in the tissue, the elasticity of the skin generally pulls out the electrode slightly, whereby the signal is easily lost. Even though it might be tempting to use some sort of micro-manipulator at this stage, they are not useful due to the variable mechanical resistance and elasticity of the skin. Furthermore, any attempts to fixate the electrode in such a way should be considered as dangerous for the nerve in case of sudden movements, or involuntary motor activities of the subject. Although at this stage, when the recording has been stabilized, one has already an idea of the modality of the unit, it is necessary to more precisely identify the registered fiber.

Identification of the Fibers

In contrast to animal experiments, measurements of the conduction velocity of nerve fibers are not easily obtained in man. Therefore, the fibers must be classified on the basis of qualitative differences, and those that are difficult to classify should be abandoned. This qualitative classification is not without problems, and in fact this is the reason why microneurographic papers often state that the afferents were "probably identified as...". Some of the problems associated with this qualitative classification are discussed in a recent paper by Burke (1997). For a more elaborate discussion concerning the classification, see, e.g., Edin and Vallbo (1990). It should be pointed out that it is not realistic to perform all the tests described in the different tables on every afferent as it is not always easy to keep a stable recording for a sufficient time. It is up to the experimenter to adopt an identification protocol suitable and sufficient for the current experiment. It is, however, of major importance that the precise criteria used for fiber identification are clearly described in any resulting scientific reports.

Musculo-tendinous Afferents

A differentiation between the musculo-tendinous afferents requires a number of physiological tests presented in Table 1. However, some general comments deserve mentioning. There are three types of musculo-tendinous afferents: the primary muscle spindle afferent (primary MSA) innervating the primary muscle spindle endings, the secondary muscle spindle afferent (secondary MSA) innervating the secondary muscle spindle endings, and the Golgi tendon organ afferent (GTO afferent) innervating the Golgi tendon organs. All types of afferents are generally activated by pressing the tendon of the receptor-bearing muscle. Nevertheless, the force necessary to activate the primary and secondary MSAs is much lower than for the GTO afferents. Moreover, the GTO afferents are highly sensitive to even low-intensity active muscle contractions. Their response, in terms of instantaneous frequency, is highly regular during maintained low-level voluntary contraction, and a stepwise increase in instantaneous frequency can be seen during a progressive increase of the level of contraction, most probably due to a progressive recruitment of new motor units.

In most cases it is possible to locate the receptor whose afferent is being registered. This can be done by pressing the muscle belly or tendon with a relatively small and blunt

Table 1. Possible tests and qualities used to identify the intramuscular receptors

Unit Type	Ia	II	Ib
Regularity of Spontaneous Discharge	-	+	+
Response To Ramp Stretch	+	+	±
Initial Burst	+	-	-
High Dynamic Index	+	-	-
Prompt Silencing During Passive Shortening	+	-	-
Response To Twitch Test	-	-	+
Response To Isometric Contraction	±	±	+
Rapid Relaxation Burst	+	±	-
Stretch Sensitization	+	±	-
Driving By Tendon Vibration (60–80 Hz)	+	-	-

Twitch test: an isometric contraction elicited by percutaneous electrical stimulation of the muscle nerve.

Rapid Relaxation Burst: A slowly increasing voluntary contraction followed by a sudden relaxation.

Stretch sensitization: Repeated rapid stretches are performed. Thereafter the muscle is kept in either a long or a short position for a few seconds. Following this hold phase, a slow ramp stretch is performed and the response to this slow stretch is noted. For a positive stretch sensitization, the response to the slow stretch after the muscle has been kept in a short position should be enhanced as compared to that after the muscle has been kept in the longer position.

device (such as the upper end of a pen), and finding the location where the biggest response is produced.

Cutaneous Afferents

Some physiological and morphological differences used to identify the different types of cutaneous receptors in glabrous skin are presented in Table 2 (for details see Vallbo and Johansson 1984, Johansson and Vallbo 1983). The tactile units are of four main types; the slowly adapting (SA) and the fast adapting (FA) subdivided into two groups depending on the receptive field properties (type I and type II). In general, these afferents show no spontaneous activity. As with the GTO afferents, these afferents are briefly activated when the electrode enters the nerve but fall silent within a few seconds. Thus, for stabilizing the recording and locating the receptor field, it is necessary to continu-

Table 2. Possible tests and qualities used to identify the cutaneous receptors

Receptor Type	FA I	SA I	FA II	SA II
Spontaneous Activity	0	0	0	frequent
Receptive Field Borders	distinct	distinct	diffuse	diffuse
Receptive Field Width	small	small	large	large
Adaption To Maintained Indentation	quick	slow	quick	slow
Response To Remote Mechanical Stimulation	no	no	yes	yes
Location	superficial	superficial	deep	deep
Drop In Vibration Response	sharp	progressive	sharp	progressive
Sensitivity To Edge Contours	yes	yes	no	no
Sensitivity to Joint Movements	sometimes	sometimes	yes	yes

Drop in vibration response: Response drops either sharply or progressively above one-to-one driving.

ously rub the innervated skin area. For a preliminary identification, the rapidly adapting afferents are found using superficial tactile stimulation (gentle stroking of the skin, see "Skin Stroking" in Fig. 5). The slowly adapting receptors are activated by more deeply applied static or dynamic pressures (see "Deep Pressuring" in Fig. 5).

Articular Afferents

Identifying articular afferents may be somewhat difficult. Even so, an afferent can be identified as innervating a joint receptor on the basis of the following criteria:
- It does not respond to light stroking of the skin;
- It does not respond to pressure of the adjacent muscle;
- It responds to maintained pressure applied directly over the joint capsule but not over adjacent non-articular bone.

Nociceptive Afferents

The polymodal C fibers can be activated by squeezing or applying pressure to the skin, by giving needle pricks in the receptive area or by applying warm or hot stimuli. Such stimuli might be derived from, e.g., a radiant heat source or from a commercially available temperature stimulator (Somedic AB, Stockholm).

Efferents

α, γ and sympathetic efferent activity has been recorded using the microneurographic technique. Whereas the sympathetic efferents are quite easy to identify, it is more complicated to differentiate between an α and a γ motoneuron. Some physiological differences between these motoneurons are presented in Table 3 (for details see Ribot et al. 1986).

Multiunit sympathetic activity appears in bursts with an easily recognizable temporal pattern. Delius and coworkers (1972) have described how this activity can be identified:
- The bursts appear in relaxed subjects with no accompanying EMG activity;
- They appear in irregular sequences and in synchrony with the cardiac pulse and therefore produce a very typical temporal pattern;
- The bursts have a lower mean frequency content than the afferent mechanoreceptive discharges.

Single sympathetic units have also been recorded. For example, Hallin and Torebjörk (1974) recorded such activity from intact skin nerves. They presented the following criteria as signs of a unit being sympathetic (for details see Hallin and Torebjörk 1974):
- Correlation with sympathetic mass discharges. The single isolated unit activity often appears together with bursts of mass discharges such as described above;

Table 3. Possible tests and qualities used to separate the α and γ efferents

Efferent Type	Alpha	Gamma
Spontaneous Activity At Rest	-	+
Activity During Isometric Voluntary Contraction	+	+
Tonic Vibration Reflex	+	-
Voluntary Stopping Of Fiber Activity	+	±
Activated By Clenching Of The Fists	-	+
Active During Mental Computation	-	+
Activated By Unexpected Sounds	-	+
Response To Changes in Emotional State	-	+
Response To Environment Changes	-	+
Response To Pinna Twisting	-	+

- Non-defined receptive fields. These fibers can be activated by stimuli anywhere on the body.
- Latency of reflex responses. The reflex latencies from different types of stimuli are long (exceeding 0.5 s).
- The activity is not blocked by Lidocaine (1%) applied just distal to the recording electrode.
- Correlation with "Galvanic skin response".

Results

The results presented in this section will illustrate how the microneurographic technique has been used to study the neurosensory mechanisms underlying kinesthetic perception (proprioception) in man. Today it seems clear that proprioception is a compound sense, where simultaneous activity in a number of different types of peripheral receptors (e.g., joint, skin and muscle receptors), possibly along with central signals derived from the motor command itself (i.e., efferent copy, see, e.g., Gandevia and Burke 1992), contribute to the sensation of position and movement.

As to the feedback contribution to proprioception, the information from intramuscular receptors seems to be of major importance. This was illustrated by the fact that mechanical vibration of the muscle tendons of a limb produced the illusion that the limb was moving. In fact, vibration always produces illusions of movements consistent with perceived lengthening of the vibrated muscle or muscle group (see, e.g., Goodwin et al. 1972 ; Roll and Vedel 1982). At this time, the general interpretation of these kinesthetic illusions was that the vibration activated preferably musculotendinous receptors, properties which had previously been described only in cat. With the development of the microneurographic technique it became possible to study the exact afferent messages produced in the situations where illusions were perceived. More precisely, the following questions could be addressed : (1) Which of the musculotendinous receptors (primary or secondary MSAs or GTO afferents) were activated by the vibrations ? (2) Which were the laws relating stimulus parameters and afferent activities ? (3) Finally, with a knowledge of the exact organization of the afferent messages during the actual production of specific movement, one could construct a vibration pattern generating an afferent message as similar as possible to the "real" message. In such a case, what would be the link between the perceived illusion during the vibration and the originally performed movement ?

Using the microneurographic technique (see Fig. 6A), it was found that the primary MSAs were highly sensitive to mechanical tendon vibrations (0.25–0.5 mm peak-to-peak, applied to resting muscle; see, e.g., Burke et al. 1976, Roll and Vedel 1982, Roll et al. 1989). On the other hand, the secondary MSAs and the GTO afferents showed only a modest sensitivity to the vibrations (see Fig. 6B).

For the primary MSAs, the existence of a one-to-one stimulus response linkage within the 1 to 100 Hz frequency range was demonstrated (see Fig. 6B). This meant that by modulating the vibration frequency, it was possible to induce a proportional change in the primary MSA discharge frequencies (see Fig. 6A).

Thus microneurography in man enabled the demonstration that a vibration stimulus applied to the tendon activates quite selectively the primary MSA channel. The strict one-to-one linkage between the vibration frequency and the afferent response frequency makes the tendon vibration a qualified experimental tool suitable for studying perceptual aspects of the proprioceptive information transmitted by these afferents. Certain characteristics of the afferent messages during a natural movement were revealed in these studies (see Fig. 6A, uppermost diagram, and Fig. 7A):

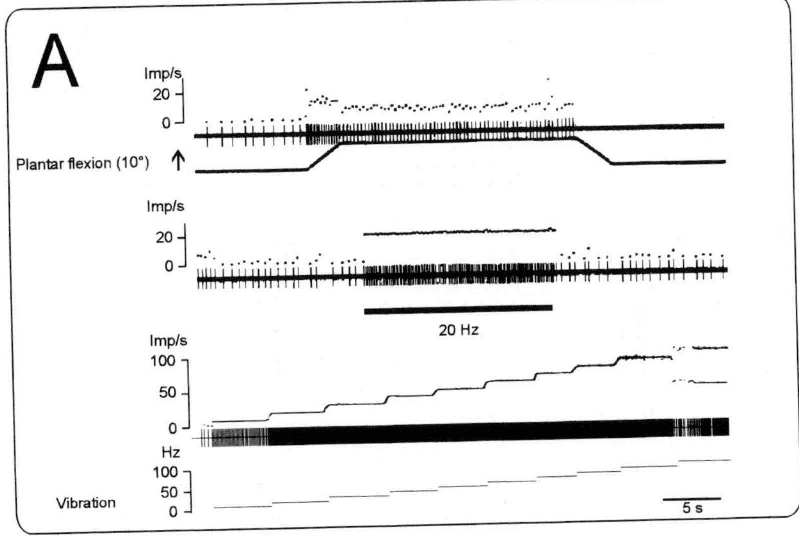

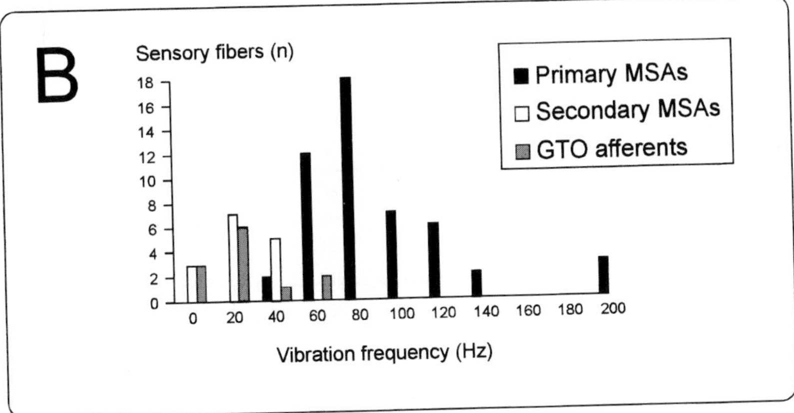

Fig. 6. **A:** Response of a primary muscle spindle afferent (MSA) from the tibialis anterior muscle (TA) to passive ankle movement (uppermost diagram), and to 20 Hz vibration (middle diagram). In the lowermost diagram, the response of a primary MSA to a modulated vibration pattern is shown. Each of these diagrams are (from top to bottom): the instantaneous frequency curve, the afferent spike train, and finally the ankle position or vibration pattern. **B:** Comparison of the upper limit of one-to-one driving by vibration of 50 primary MSAs (filled columns), 15 secondary MSAs (open columns), and 12 Golgi tendon organ afferents (gray columns).

- The primary MSA is abruptly activated at the beginning of the movement;
- Its discharge frequency continuously increases during a movement of constant velocity;
- At the end of the movement, it returns to a stable frequency depending on the attained position;
- The muscle in which the activity is increased is in fact the muscle that is passively stretched, i.e., the antagonist of the prime mover.

These characteristics were then used for the construction and functional usage of vibration patterns, and the qualities of the illusions were studied. As an example, a vibration pattern applied to the extensor muscle and mimicking the general characteristics mentioned above produced the illusion of a flexion at constant velocity (see Fig. 7B).

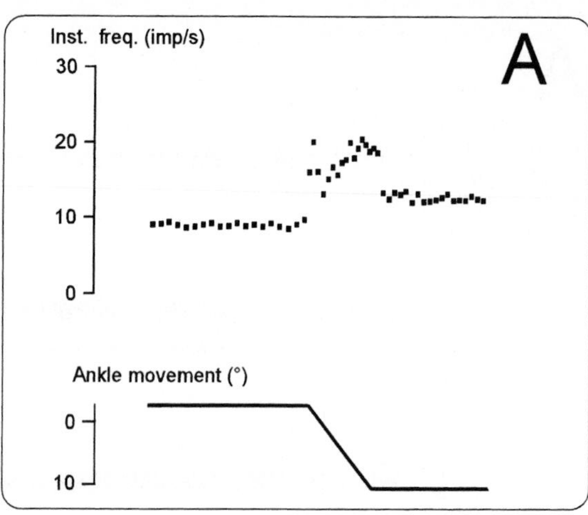

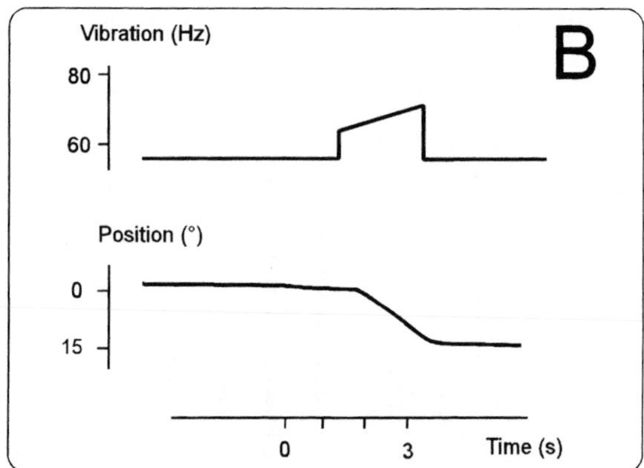

Fig. 7. A: Reponse of a primary muscle spindle afferent (MSA) from the tibialis anterior muscle (TA) to a 10° passive plantar flexion of the ankle joint. In the uppermost diagram, the instantaneous frequency of the afferent is shown. The position of the ankle is illustrated in the lowermost diagram. **B:** Illusion produced by vibration of the TA tendon. The frequency of the vibration can be seen in the uppermost diagram (amplitude; 0.25–0.5 mm peak-to-peak). The illusion, as reproduced by the non-vibrated contralateral ankle, is illustrated in the lowermost diagram (plantar flexion in the downward direction).

In fact, by further mimicking the primary MSA responses attained with microneurography, it was possible to use more complex vibration patterns, using several vibrators applied, for example, to the four muscle groups of the wrist (see Fig. 8A). By varying the vibration frequency, the duration of each stimulus and the vibrator onsets, and by applying the vibrators either simultaneously or successively (see Fig. 8B), more complex kinesthetic illusions could be induced, involving geometrical shapes such as rectilinear figure drawing (see Fig. 8C) (Roll and Gilhodes 1995).

Even though the microneurographic technique has traditionally been used to study more basic motor and sensory neurophysiological functions, the results mentioned above illustrate that the technique can also be used to approach more integrative brain functions, revealing cognitive information processing like that used for memorization and recognition of motor forms.

Comments

Health Risks

The risks of nerve damage must be considered acceptably low. As an example, Hagbarth (1979) reports to have seen some remaining signs of local neuropathy, such as cutaneous hyperesthesia or partial muscle paresis, in only 3 of about 1000 experiments. In

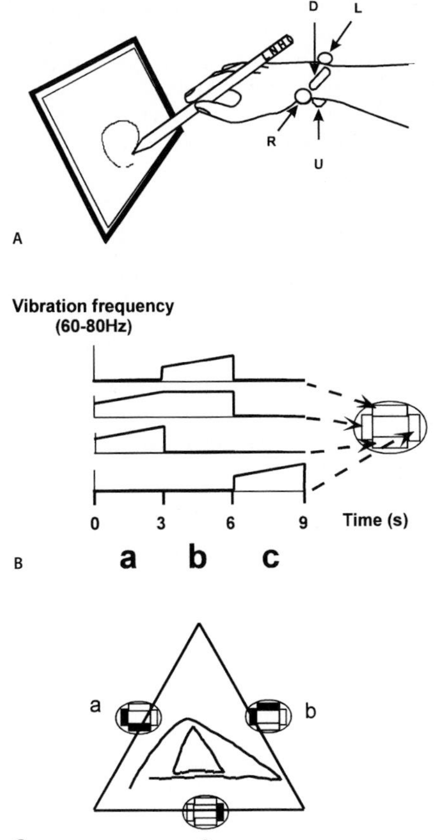

Fig. 8. Illusions produced by complex vibration patterns. **A:** Experimental set-up. The positions of the four vibrators (R, D, L, and U) are indicated. The subject holds a pen connected to a digitizing board. **B:** The pattern of vibration of the four vibrators evoking the illusion of triangle drawing. **C:** Example of illusion produced by the vibration pattern illustrated in B. The figure also shows at which phase of the illusion the vibrators were active (active vibrator in black). The letters *a*, *b*, and *c* bring out the temporal correlation between the vibration pattern illustrated in **B**, and the illusion shown in **C**.

these cases, the symptoms lasted for 2 to 6 months. More recently, the risks of long-term effects of microneurography as investigated in both man and animals were reviewed by Gandevia and Hales (1997).

Acknowledgement: This work was supported by The Swedish Council for Work Life Research, The Swedish Foundation for International Cooperation in Research and Higher Education, CNRS, and INSERM grants.

References

Aniss AM, Diener HC, Hore J, Burke D, Gandevia SC (1990) Reflex activation of muscle spindles in human pretibial muscles during standing. J Neurophysiol 64:671–679

Burke D, Hagbarth KE, Löfstedt L, Wallin BG (1976) The responses of human muscle spindle endings to vibration of non-contracting muscles. J Physiol (London) 261: 673–693

Burke D, Hagbarth K-E, Wallin BG (1977) Reflex mechanisms in Parkinsonian rigidity. Scand J Rehabil Med 9:15–23

Burke D (1997) Unit identification, sampling bias and technical issues in microneurographic recordings from muscle spindle afferents. J Neurosci Methods 74:137–144

Delius W, Hagbarth K-E, Hongell A, Wallin G (1972) General characteristics of sympathetic activity in human muscle nerves. Acta Physiol Scand 84:65–81

Edin BB, Vallbo AB (1990) Classification of human muscle stretch receptor afferents; a Bayesian approach. J Neurophysiol 63:1314–1322

Edin BB (1992) Quantitative analysis of strain sensitivity in human mechanoreceptors from hairy skin. J Neurophysiol. 67:1105–1113

Gandevia SC, Hales J P (1997) The methodology and scope of human microneurography. J Neurosci Methods 74:123–136

Gandevia SC, Burke D (1992) Does the nervous system depend on kinesthetic information to control natural limb movements? Behav Brain Sci. 15: 614–632

Goodwin GM, McCloskey DI, Matthews PCB (1972) The contribution of muscle afferents to kinesthesia shown by vibration-induced illusions of movement and by the effect of paralysing joint afferents. Brain 95:705–748

Hagbarth KE, Vallbo AB (1967) Mechanoreceptor activity recorded percutaneously with semimicroelectrodes in human peripheral nerves. Acta Physiol Scand. 69:121–122

Hagbarth K-E, Wallin G, Löfstedt L (1973) Muscle spindle responses to stretch in normal and spastic subjects. Scand J Rehabil Med 5:156–159

Hagbarth K-E (1979) Extereoceptive, proprioceptive, and sympathetic activity recorded with microelectrodes from human peripheral nerves. Mayo Clin Proc 54:353–365

Hallin RG (1990) Microneurography in relation to intraneural topography: somatotopic organisation of median nerve fascicles in humans. J Neurol Neurosurg Psychiat 53:736–744

Hallin RG, Torebjörk HE (1974) Single unit sympathetic activity in human skin nerves during rest and various manoeuvres. Acta Physiol Scand 92:303–317

Hallin RG, Wu G (1998) Protocol for microneurography with concentric needle electrodes. Brain Res Prot 2:120–132

Inglis JT, Leeper JB, Burke D, Gandevia SC (1996) Morphology of action potentials recorded from human nerves using microneurography. Exp Brain Res 110:308–314

Jansen RF (1990) The reconstruction of individual spike trains from extracellular multineuron recordings using a neural network emulation program. J Neurosci Methods 35:203–213

Johansson RS (1976) Skin mechanoreceptors in human hand: receptive field characteristics. In: Zotterman Y (ed) Sensory functions of the skin in primates: With special reference to man. Pergamon Press, Oxford, pp 475–487

Johansson RS, Vallbo AB (1983) Tactile sensory coding in the glabrous skin of the human hand. Trends Neurosci 6:27–32

Kreiter AK, Aertsen MHJ, Gerstein GL (1989) A low-cost single-board solution for real-time, unsupervised waveform classification of multineuron recordings. J Neurosci Methods 30:59–69

Macefield VG, Gandevia SC, Burke D (1990) Perceptual responses to microstimulation of single afferents innervating joints, muscles and skin of the human hand. J Physiol (Lond) 429:113–129

Macefield VG, Gandevia SC, Bigland-Ritchie B, Gorman RB, Burke D (1993) The firing rates of human motoneurones voluntarily activated in the absence of muscle afferent feedback. J Physiol (Lond) 471:429–443

Mirfakhraei K, Horch K (1994) Classification of action potentials in multi-unit intrafasicular recordings using neural network pattern recognition techniques. IEEE Transactions Biomed Engin 41:89–91

Ochoa JL, Torebjörk JR (1983) Sensations evoked by intraneural microstimulation of single mechanoreceptor units innervating the human hand. J Physiol (Lond) 342:633–654

Ochoa JL, Torebjörk JR (1989) Sensations evoked by intraneural microstimulation of C nociceptor fibers in human skin nerves. J Physiol (Lond) 415:583–599

Oghalai JS, Street WN, Rhode WS (1994) A neural network-based spike discriminator. J Neurosci Methods 54:9–22

Ribot E, Roll JP, Vedel JP (1986) Efferent discharges recorded from single skeletomotor and fusimotor fibres in man. J Physiol (London) 375:251–268

Roll JP, Vedel JP (1982) Kinesthetic role of muscle afferents in man, studied by tendon vibration and microneurography. Exp Brain Res 47:177–190

Roll JP, Vedel JP, Ribot E (1989) Alteration of proprioceptive messages induced by tendon vibration in man: microneurographic study. Exp Brain Res 76:213–222

Roll JP, Gilhodes JC (1995) Proprioceptive sensory codes mediating movement trajectory perception: human hand vibration-induced drawing illusions. Can J Physiol Pharmacol 73:295–304

Rothwell JC, Gandevia SC, Burke D (1990) Activation of fusimotor neurones by motor cortical stimulation in human subjects. J Physiol (Lond) 431°:743–756

Salganicoff M, Sarna M, Sax L, Gerstein GL (1988) Unsupervised waveform classification for multi-neuron recordings: a real-time, software-based system. Algorithms and implementation. J Neurosci Methods 25:181–187

Schmidt EM (1984a) Instruments for sorting neuroelectric data: a review. J Neurosci Methods 12:1–24

Schmidt EM (1984b) Computer separation of multi-unit neuroelectric data: a review. J Neurosci Methods 12:95–111

Sundlof G, Wallin BG (1978) Human muscle nerve sympathetic activity at rest: Relationship to blood pressure and age. J Physiol (Lond) 274:621–637

Szumski AJ, Burg D, Struppler A, Velho F (1974) Activity of muscle spindles during muscle twitch and clonus in normal and spastic human subjects. Electroencephal Clin Neurophysiol 37:589–597

Torebjörk, HE, Hallin, RG (1976) Skin receptors supplied by unmyelinated (C) fibres in man. In: Zotterman Y (eds) Sensory functions of the skin in primates: With special reference to man. Pergamon Press, Oxford, pp 475–487

Vallbo AB, Johansson RS (1984) Properties of cutaneous mechanoreceptors in the human hand related to touch sensation. Human Neurobiol 3:3–14

Wallin BG, Konig U. (1976) Changes of skin nerve sympathetic activity during induction of general anesthesia with thiopentone in man. Brain Res 103:157–160

Öhberg F, Johansson H, Bergenheim M, Pedersen J. Djupsjöbacka M (1995) A neural network approach to real-time spike discrimination during simultaneous recording from several multi-unit nerve filaments. J Neurosc Methods 64:181–187

Suppliers

Commercially available electrode

Company: Frederick Haer Inc., 9, Main Street, Bowdoinham, ME, 04008, USA (phone: +01-207-666-8190 ; fax: +01-207-666-8292; e-mail: fhcinc@fh-co.com)

Abbreviations

FA	Fast adapting receptor
RA	Rapidly adapting receptor
PC	Pacinian corpuscle
SA I	Slowly adapting type I receptor
SA II	Slowly adapting type II receptor
MSA	Muscle spindle afferent
GTO	Golgi tendon organ

Chapter 31

Biomechanical Analysis of Human and Animal Movement

W. HERZOG

Introduction

Biomechanical analyses of human and animal movements are performed for two basic reasons. The first reason is to analyze the internal and external forces acting on the human or animal system, and to determine the effects of these forces on musculoskeletal tissues, such as tendons, ligaments, articular cartilage, muscle and bone. These studies are typically centered around one of two questions: (1) what is the normal response of musculoskeletal tissues to everyday loading conditions; and (2) what are the loading conditions that may produce acute failure (ligament tears or breaking of bones) or long-term, chronic tissue failure, for example, the slow and continuous breakdown of articular cartilage in osteoarthritis? This first set of investigations may be called the classic biomechanical investigations. The second principal reason for analyzing human and animal movement has been to determine the mechanisms underlying movement control. Analyzing task-specific movement patterns in normal and perturbed situations has allowed to formulate frameworks of movement control. Similarly, the direct measurement of single and multiple muscle forces during animal locomotion has provided a tremendous understanding of the coordination of synergistic muscles during prescribed movements, such as locomotion.

In this chapter, I am not particularly concerned about the aims of specific studies that are introduced; rather I will attempt to focus on the techniques and methods that were used for the measurement and calculation of the desired quantities. The chapter is divided into two parts. The first part deals with the "external" biomechanics, and the second part with the "internal" biomechanics of the system of interest. External biomechanics is concerned with the mechanical analysis of all forces and movements acting on the outside of the human or animal body. Such external forces include the weight of the system and any contact forces of the system with its environment. Internal biomechanics is concerned with the mechanical analysis of all forces acting on musculoskeletal tissues (tendons, ligaments, articular cartilage, muscle and bone) and the effects that these forces produce on those tissues.

The part on "internal" biomechanics is further divided into experimental and theoretical methods and approaches. This distinction was not made for the part on "external" biomechanics, because externally most parameters of interest can be measured experimentally, therefore rendering theoretical approaches unnecessary. In contrast, internally most parameters of interest cannot (and have not been) measured experimentally; therefore, sophisticated theoretical approaches have been used in an attempt to estimate the internal mechanics during human and animal movement.

W. Herzog, Faculty of Kinesiology, University of Calgary, Calgary, Alberta, T2N IN4, Canada

Space limitations do not allow a complete and detailed description of all biomechanical tools used in the analysis of movements. Therefore, I have selected specific methods that will be described using exemplary studies from the scientific published literature. I would like to apologize in advance to all those whose studies could easily have been mentioned in the context of this chapter, but were not. The selected methods are biased towards those that I have used personally, and thus feel competent to describe. The selected literature examples are those that, in my opinion, illustrate a particular point best.

Part 1: External Biomechanics

The external mechanics of a system are defined by the forces acting externally on the system of interest and the movements of the system or parts of it. In the analysis of human movement, the external forces typically consist of the weight and the contact forces of the subject with the environment. Movements of the body are typically measured using high-speed video; however, techniques such as high-speed cine-film, goniometry, strobe-photography, and accelerometry have been used. For conciseness, I will restrict my comments to the most frequently used tools for analyzing external movements; these include ground reaction force measurements using (commercially available) *force platforms* and movement measurements using *high-speed video*. Furthermore, I will also discuss external muscle activity measurements using *surface electromyography*. Although surface electromyography (cf. Chapter 26, 28) does not measure a mechanical quantity, it provides an estimate of the activation of skeletal muscles, and so provides insight into aspects of muscular force production and coordination during movement.

External Force Measurements Using Force Platforms

A force platform is a device that typically has a hard top surface with force sensors underneath it. The principle of a force platform may be explained using the example of a bathroom scale. A bathroom scale measures the vertical ground reaction force of a person who steps onto it. If you stand absolutely quietly on a bathroom scale (i.e., your body is in static equilibrium), the reaction force of the bathroom scale is equal to the person's body weight. If you possess a good bathroom scale, you may observe that the measurement is never quite constant, but the force measured varies slightly about a mean value. These variations in force correspond to the accelerations and decelerations of blood that is pumped during every heart beat and causes small vertical oscillations (and thus accelerations) of the center of mass of the whole body. Similarly, if you do not stand still on the bathroom scale, the force dial on the scale will also not be still. For example, if you let yourself drop from an upright standing position into a squat position, the bathroom scale will initially show a decrease in force (because your center of mass is accelerated downward, which can only occur if the weight force is larger than the vertical ground reaction force) and then an increase in force (because your center of mass is stopped from dropping, which corresponds to an upward acceleration of the center of mass that can only occur if the vertical ground reaction force is larger than the body weight).

Early "force platforms" used in scientific investigations were not so different from ordinary bathroom scales. Marey (1873) built a "force platform" that consisted of spirals of Indian rubber tubes installed in a wooden frame. Elftman (1938) used a force platform moving vertically up and down on four springs. The vertical displacements of the top plate were recorded optically to determine the vertical ground reaction forces during walking.

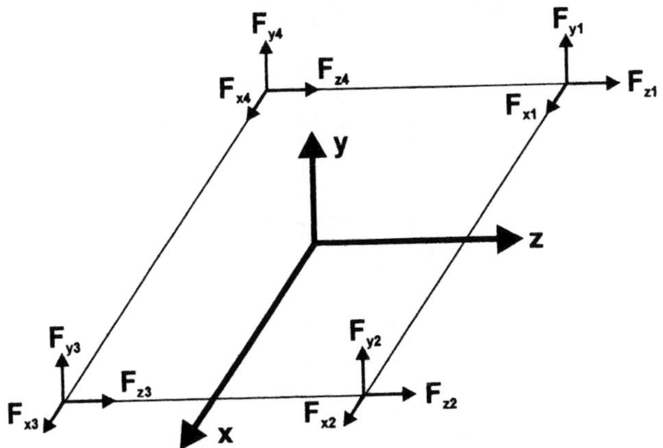

Fig. 1. Schematic diagram of the measuring plane of a force platform showing the force components that are measured at each corner, and showing the platform embedded x-y-z reference frame at the centre of the platform.

Today's force platforms are considerably more sophisticated than those described above. Commercially available platforms typically consist of a rectangular steel plate supported by force-sensors in each of the four corners. The force sensors used today are either piezo-electric or strain-gauge based sensors.

Typically, each of the four force sensors of a platform is designed to measure forces in three mutually perpendicular directions (Fig. 1). The force directions are along the longitudinal and the transverse axis of the force platform in the horizontal plane (x- and z-direction according to the guidelines of the International Society of Biomechanics), and perpendicular to the force platform, vertically upwards (the y-direction) (Fig. 1).

The resultant force acting from the platform on a subject (for example, a person walking across the platform) is

$$F_x = F_x1 + F_x2 + F_x3 + F_x4$$
$$F_y = F_y1 + F_y2 + F_y3 + F_y4$$
$$F_z = F_z1 + F_z2 + F_z3 + F_z4$$

where F_x, F_y and F_z are the resultant components in each of the three mutually perpendicular directions defined above, and F_{xi}, F_{yi} and F_{zi} represent the corresponding force components measured at corner i where i = 1, 2, 3, 4. Finally the resultant ground reaction force (vector), **F** is defined as

$$\mathbf{F} = F_x\hat{\mathbf{i}} + F_y\hat{\mathbf{j}} + F_z\hat{\mathbf{k}}$$

where $\hat{\mathbf{i}}$, $\hat{\mathbf{j}}$ and $\hat{\mathbf{k}}$ are unit vectors in the x, y and z direction, respectively.

From the force components measured at each of the four corners of the force platform, the corresponding moments can be derived about any point of interest, for example, the point of application of the resultant force. Therefore, a modern force platform allows for a full, three-dimensional description of the external forces and moments acting on the human (or animal) body while in contact with the force platform.

Figure 2 shows the average ground reaction forces in the anterior-posterior (x) and vertical (y) direction of human and cat walking. Note the similarity in shape of the two components of the ground reaction force for both bi-pedal and quadro-pedal locomotion.

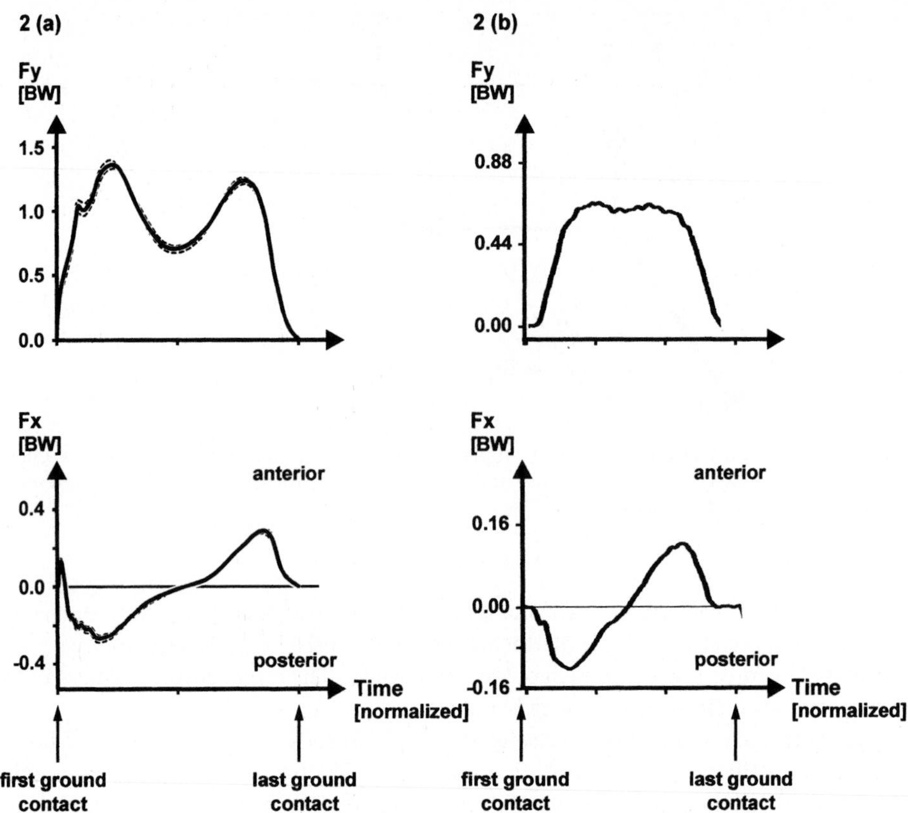

Fig. 2. Vertical (Fy) and anterior-posterior (Fx) ground reaction forces for human (a) and cat (b) locomotion.

External Movement Measurements Using High-Speed Video

Only 100 years ago, movement quantification was technically extremely difficult. The best technique to observe movement was observation with the human eye. Many professions still rely on movement quantification by direct observation. These professions include coaches, medical doctors, teachers and many other health, fitness, and movement professionals.

Early movement analysis techniques in biomechanics relied on multiple exposures of a single photograph using a regular photographic camera. Multiple exposure of a single photograph was achieved in one of two ways: by chrono-photography or by stroboscopy.

Chrono-photography refers to techniques in which the light entering the photographic lens is cyclically interrupted. In practice, this was achieved by placing a disc with a small opening in front of the camera and rotating the disc at a given frequency. The rotating disc acted like an "external" shutter allowing light to enter through the continuously open shutter of the photographic camera at regular intervals. Knowing the frequency of rotation, the inter-exposure interval times could be calculated, thereby providing a time scale for the movements captured using this multiple exposure method.

Stroboscopy refers to techniques in which light enters through the continuously open shutter of a photographic camera through multiple flash exposures. Stroboscopic measurements are made in an otherwise completely dark environment. The shutter of the camera is continuously open, and the movement of the subject is captured by multiple flash exposures that typically occur at periodic intervals. Knowing the frequency of the flash, and thus the time period between successive exposures, a time scale is provided on the multiple exposure photograph.

With the popularization of cine-cameras, specialized high-speed cameras shooting at frame rates of up to 10,000 Hz have been developed and used in the analysis of human and animal movement. However, because of the cost of cine-film and the long processing time, biomechanical movement analyses are performed virtually exclusively using high-speed video. Cine-cameras are only used in special applications in which either high temporal or high spatial resolution is required.

Video cameras have a shutter mechanism that allows light to enter through the camera lens. Light-sensitive cells detect the incoming light and emit an electrical signal proportional to the amount of light. The cells are scanned and the electrical signals are stored on-line to a computer (or a videotape).

Analysis of movements captured on a videotape is made by digitizing the video records frame by frame. However, many video movement analysis systems can be bought with semi-automatic or automatic data analysis capabilities. For automated systems, subjects wear markers at the anatomical points to be digitized, for example, the joints of the lower limbs (cf. Chap. 32). These markers can be identified automatically (i.e., by contrast) using computer software. Once all markers have been correctly identified frame by frame, the resulting coordinates of the markers are directly fed to computer software that typically calculates the displacements, velocities and accelerations (linear and angular) for all points and segments of interest, as well as for the entire body.

Limitations

Because of the discrete size of the light-sensitive cells in video cameras, the spatial resolution of video movement analysis is restricted to finite values. For most commercially available systems, the error in the three-dimensional coordinate determination of a marker (point) is about 1/600 of the diagonal image size. That means, for a frame size of about 600 mm (diagonally), the expected error (under optimal laboratory conditions) is about 1 mm. For many practical applications in biomechanics, such an error can be (and has been) tolerated. However, in Part 2 on internal biomechanics, I will discuss an example (and the corresponding technique) that requires a spatial resolution of <20 μm for movement analysis. Most commercially available video systems have maximal temporal resolutions of about 5 ms (i.e., 200 Hz). Again, this temporal resolution is sufficient for most (but not all) biomechanical analyses of human and animal movements.

Surface Electromyography

Surface electromyography is a cheap, non-invasive, simple approach to estimate times and magnitudes of muscular activation, and to obtain a quantitative picture of muscle coordination during human movement. Many textbooks have been devoted exclusively to the measurement, analysis, and interpretation of electromyographical (EMG) signals (Basmajian and de Luca, 1985; Loeb and Gans, 1986). Furthermore, EMG is covered elsewhere in this book; therefore just a brief and basic account of this methodology will be given here.

When recording an EMG signal from a single muscle fiber, one measures changes in electrical potential across the muscle fiber membrane. At rest, the potential of a muscle fiber is approximately −90 mV inside vs. outside that is set to zero for reference (see Fig. 3). With sufficient stimulation, the potential inside the cell rises temporarily to about +40 mV. This change in potential, representing a fiber action potential, can be recorded. In an intact muscle, a fiber is never stimulated by itself but always together with all the other fibers that make up a motor unit. The EMG signal recorded from the depolarization of a motor unit is called a motor unit action potential (MUAP). In general, the

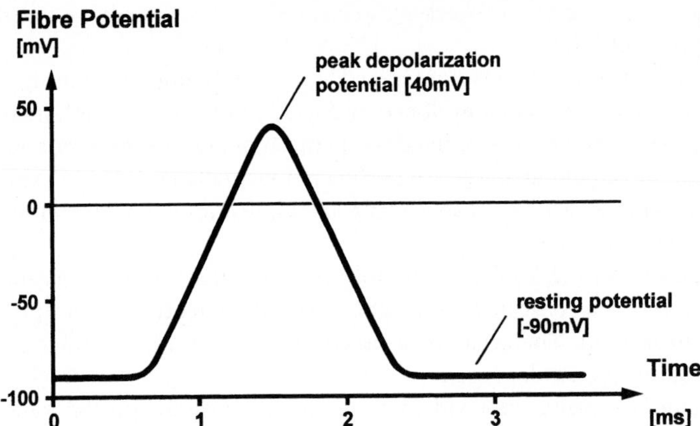

Fig. 3. Schematic illustration of an action potential in a skeletal muscle fibre.

actual EMG signal obtained from voluntarily contracting muscles is a record of many motor units firing at different mean rates. Since the placement of the recording electrode on a muscle determines its geometric relation to the motor units, recordings obtained using different electrode placements on the same muscle will usually be different.

The resting membrane potential in a muscle fiber is caused by the concentrations of ions inside and outside the fiber. The key ions are potassium (K^+) and sodium (Na^+). There are two basic mechanisms that must be accounted for when explaining the resting membrane potential. For illustrating the first mechanism, let us concentrate on the K^+ ions exclusively. The ratio of K^+ ions inside versus outside the muscle cell is about 30:1 (Wilkie, 1968). K^+ ions diffuse out of the muscle fiber, following the concentration gradient, but when they leave the cell, the electrical voltage (potential difference) across the cell membrane increases and prevents further loss of K^+ ions. The K^+ ions reach an equilibrium state when the two opposing forces acting on them – the concentration gradient and the electrical voltage of the cell – are equal. At equilibrium, K^+ ions neither gain nor lose energy when crossing the cell membrane.

The membrane potential can have only one value, which may be determined for the K^+ ions. However, the corresponding membrane potential for the Na^+ ions cannot be equal to that of the K^+ ions, since the ratio of Na^+ inside versus outside the cell is about 1:7.7. Therefore, a second mechanism must influence the resting potential of a muscle fiber. This second mechanism is associated with the fact that the muscle cell membrane is not freely permeable to all ions. For example, Na^+ ions cannot easily permeate the cell membrane, and if they do enter the cell, they are actively (i.e., at the expense of metabolic energy) pumped out again by the so-called sodium pump. The facts that the muscle cell membrane cannot be freely crossed by some ions, and that electrical and osmotic equilibrium conditions influence ion traffic through the cell membrane, must be considered when determining the resting concentration of ions in muscle fibers.

The resting potential of a muscle fiber is about –90 mV inside the cell, and depends on the selective permeability of the cell membrane. When an action potential of a motor neuron reaches the presynaptic terminal, a series of chemical reactions takes place, culminating in the release of acetylcholine (ACh). Acetylcholine diffuses across the synaptic cleft, binds to receptor molecules of the muscle fiber membrane, and causes a change in membrane permeability. Most importantly, permeability to Na^+ ions increases, and if the depolarization of the membrane, caused by Na^+ ion diffusion, exceeds a critical threshold, an action potential propagates along the muscle fiber. If such an action potential were measured using an electrode inside the muscle fiber, it would go from about (90 mV (resting potential) to about +40 mV (peak depolarization potential), and back again to the resting value (Fig. 3). The membrane potential and active state of the contractile machinery of the muscle fiber are tightly linked. A depolarization of the cell membrane exceeding a critical threshold will cause contraction of the muscle fiber.

Electromyographic signals are typically recorded by electrodes measuring differences in voltage (potential) between two points. EMG electrodes may be grouped broadly into four categories. The first two categories describe where the electrodes are placed: inside the muscle (indwelling electrodes) or on the surface of the muscle (surface electrodes). The remaining two categories describe the electrode configuration: mono-polar or multi- (typically bi-) polar. In the context of this chapter, we will only consider surface electromyography.

Surface Electrodes

Surface electrodes are placed on the skin overlying the muscle of interest. They come in a variety of types. The most frequently used surface electrodes are commercially available silver-silver chloride electrodes.

Before recording with surface electrodes, the electrical impedance of the skin must be decreased by shaving the area of electrode placement and by applying rubbing alcohol or abrasive pastes to remove dead cells and oils. The recording electrodes are then attached using electrode gel, and slight pressure is applied to improve contact between the electrode and the skin. Electrode gels are commercially available, and pressure may be applied by fixing the electrodes with adhesive tapes or elastic bands.

Surface electrodes are simple to use and non-invasive. Surface EMG recordings are typically used to obtain a general picture of the electrical activity of an entire muscle or muscle group. However, in humans, surface electrodes can only be used for recording from superficial, relatively large muscles.

Typically, EMG recordings are obtained using a *bi-polar electrode configuration*. A bi-polar configuration implies that there are two electrical contacts that are used to measure an electrical potential relative to a common ground electrode. The two potentials measured by the electrode pair are sent to a differential amplifier that amplifies the voltage difference between the two electrodes. In a *mono-polar configuration*, the difference in voltage is recorded from a single recording electrode relative to the ground electrode. The mono-polar electrode configuration has the disadvantage that it detects any electrical signal in its vicinity – not only the signal originating from the muscle of interest. In a bi-polar electrode configuration, two signals are recorded relative to the ground electrode. Each of the two signals contains all the electrical activity in the vicinity of the recording electrodes, but subtracting the two signals from each other, using differential amplification, will cancel most of the unwanted electrical activity from outside the muscle, since it appears as a similar signal at both recording electrodes (see below). The electrical signal originating from muscular activity is received by each recording electrode as an essentially different signal, and thus, is enhanced in the differential amplification process.

EMG Signal Processing

Figure 4a shows an unprocessed EMG signal obtained from human rectus femoris muscle during an isometric contraction at a level of 70% of the maximal voluntary knee extension force. The record was obtained using bi-polar surface electrodes. Raw EMG signals resemble noise signals with a distribution around the zero point. Any interpretation of the raw EMG signal with respect to the force production of the muscle, its relative activation or its fatigue state is difficult. Therefore, EMG signals are typically processed before they are used for assessing the contractile state of a muscle. Processing of EMG signals may be done in the time or frequency domain (see below).

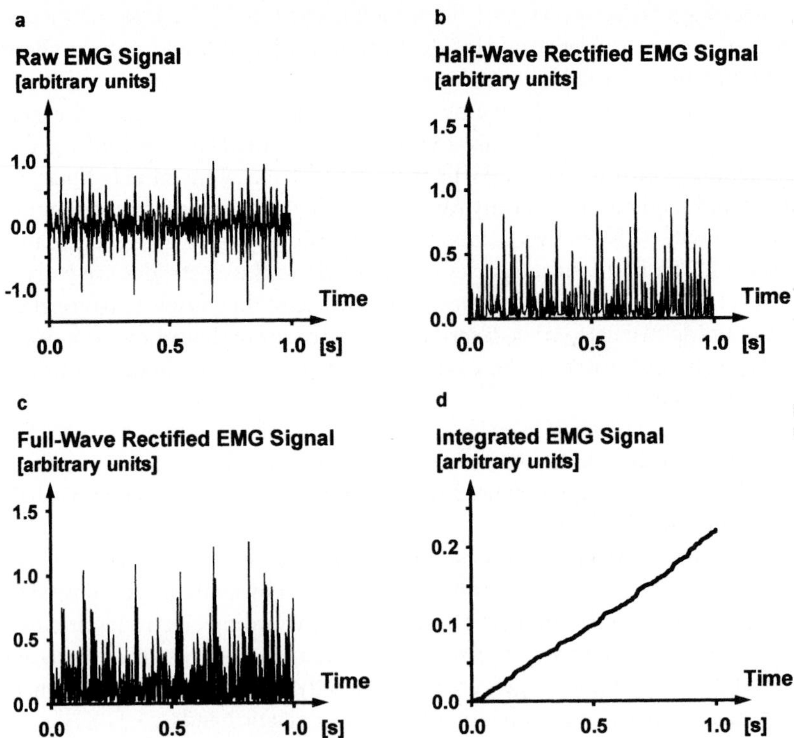

Fig. 4. Raw EMG signal (a), half-wave (b), and full-wave rectified (c) EMG signal, and integrated (d) EMG signal from human rectus femoris contracting isometrically at about 70% of its maximal voluntary force (see text for further details).

EMG Signal Processing in the Time Domain; Rectification

Any type of averaging of the EMG signal in the time domain will give a value of close to zero, independent of the number of motor units contributing to the signal and their mean firing rates. Therefore, before performing any type of averaging procedure, it is necessary to rectify the EMG signal. Rectification is the name of the process by which either all negative values of the EMG signal are eliminated from the analysis procedure (half-wave rectification, Fig. 4b) or only the absolute magnitudes of the signal are considered (full-wave rectification, Fig. 4c). Typically, full-wave rectification is preferred since it retains the entire signal. Often the rectified signal is used for further signal processing.

Smoothing

The rectified signal contains the high-frequency content of the raw signal. Often it is desirable to eliminate the high-frequency content of the EMG records in order to better relate the EMG signal to contractile features of the muscle. Elimination of the high-frequency content of the EMG signal is accomplished using any type of low-pass filtering approach. For details of filtering procedures, see Basmajian and De Luca (1985) and below.

Integration

The integrated EMG (IEMG) has been related to muscular force more often than any other form of the processed EMG (e.g., Bigland and Lippold, 1954; Bouisset and Goubel, 1971). Integration of the EMG signal is performed on the rectified form of the signal and refers to the mathematical integration of the EMG records with respect to time (Fig. 4d). Integration of an EMG signal is equivalent to calculating the area under the rectified

EMG-time curve, and is defined as:

$$\text{IEMG} = \sum_{t}^{t+T} \left|\text{EMG}(t)\right| \cdot dt$$

Where $|\text{EMG}(t)|$ represents the rectified EMG signal.Since integration of the EMG signal gives steadily increasing values over time, integrations are typically performed either over a sufficiently small time period, T, or the integrator is reset to zero when the integrated EMG value reaches a specified limit.

Root Mean Square

The root mean square (RMS) value of the EMG signal is an excellent indicator of the magnitude of the signal and is frequently used in studying muscular fatigue. RMS values are calculated by summing the squared values of the raw EMG signal, determining the mean of the sum, and taking the square root of the mean so obtained:

$$\text{RMS} = \left(\frac{1}{T} \sum_{t}^{t+T} \text{EMG}^2(t)\,dt \right)^{1/2}$$

EMG Signal Analysis in the Frequency Domain

The study of EMG signals in the frequency domain has received much attention because of the loss of the high-frequency content of the signal with increasing muscular fatigue (e.g., Bigland-Ritchie et al., 1981; Hagberg et al., 1981; Komi and Tesch, 1979; Lindström et al., 1977; Petrofsky et al., 1982). Power density spectra of the EMG signal can be obtained readily by using the Fast Fourier Transformation techniques for stationary, time-independent EMG signals, or wavelet analysis for non-stationary, time-varying signals (Fig. 5).

The most frequently used parameters for analyzing the power density spectrum of EMG signals are the mean and the median frequencies, although the bandwidth and the peak power frequency have also been used to describe the spectrum (Basmajian and de Luca, 1985). The mean frequency is defined as the sum of the first moments of frequency divided by the area under the power-frequency curve. The median frequency is defined as the frequency that divides the power of the EMG spectrum into two equal areas. The peak power frequency is defined as the frequency at which the power of the signal is greatest; and the bandwidth of the frequency spectrum is typically defined as the range

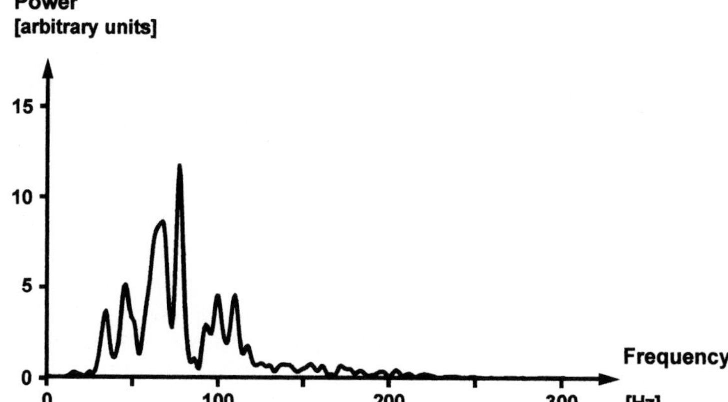

Fig. 5. Frequency spectrum of an EMG signal obtained from human rectus femoris during an isometric contraction at a 70% maximal voluntary effort.

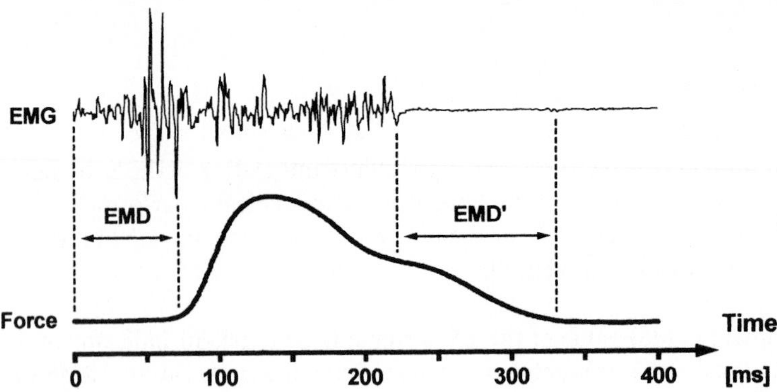

Fig. 6. Electromechanical delay in the cat soleus during walking at a nominal speed of 1.2 m/s. See text for further explanation.

of frequencies associated with a power value of 0.5 or higher, when the peak power is defined as 1.0.

Electromechanical Delay

Electromechanical delay refers to the time delay that is observed between the onset of EMG activity and the first sign of muscular force production. Typically, the electromechanical delay is assumed to be constant. However, this assumption is not correct as the duration of the EMG activity is rarely the same as the corresponding muscle force. For example, the EMG and force signal from the cat soleus, shown for the stance phase of walking at a speed of 1.2 m/s, are different (Fig. 6). The electromechanical delay, as defined above, is approximately 72 ms; however, the EMG activity ends approximately 109 ms before the force drops to zero. Therefore, if one wants to relate EMG activity to muscular force, as many people have attempted to do (Guimaraes et al., 1994a; Guimaraes et al., 1994b; Hof and Van den Berg, 1981a; Hof and Van den Berg, 1981b; Hof and Van den Berg, 1981c; Hof and Van den Berg, 1981d; Lippold, 1952; Milner-Brown and Stein, 1975; Moritani and deVries, 1978; Savelberg and Herzog, 1997; Sherif et al., 1983; van den Bogert et al., 1988; van Ruijven and Weijs, 1990), it is important to account for a variable electro-mechanical delay within a given activation cycle.

Part 2: Internal Biomechanics

The field of biomechanics, with its force and movement measurement systems, became popular in the late 1960s and early 1970s. At that time, much of the research focus was on measurement of the external biomechanics using methods described in Part 1. However, as early as in the late 1970s, there appeared experimental studies that were aimed exclusively at quantifying the internal mechanics of biological systems. In the 1980s and up until now, the focus of biomechanics has shifted from the purely descriptive studies of the external mechanics of human and animal movement to studies aimed at understanding the mechanics and control of movement. These studies relied heavily on measuring the internal forces acting on the musculoskeletal system during movement. Here I would like to discuss two of the many approaches used in this area of research; the first is concerned with the measurement of *in vivo* muscular forces, the second with determining joint contact pressures.

The topic of muscle force measurements has been chosen for a variety of reasons. Muscle forces are the only active forces. They can be controlled by conscious effort, and muscle forces during human and animal movement are much larger than any other external forces. Therefore, they are very important forces to consider when analyzing the loading on the human and animal musculoskeletal system. However, aside from the mechanical importance of muscle forces, knowledge of muscle forces also provides tremendous insight into the mechanisms of movement control.

The topic of joint contact pressure measurements has been chosen because joint degenerative diseases, such as osteoarthritis, are some of the most prevalent and least understood diseases in the industrialized world. Osteoarthritis is thought to be caused by mechanical factors, thus making mechanical investigations particularly relevant to understanding this disease. Also, dynamic *in vivo* measurements of joint contact pressures have not been possible to date because of technical limitations. Therefore, this area of research offers exciting avenues for experimental approaches and theoretical studies.

Muscle Force Measurements

Walmsley et al. (1978) pioneered the field of multiple *in vivo* muscle force measurements. These investigators measured the forces in soleus and medial gastrocnemius muscles of freely moving cats. They used a so-called buckle-type force transducer that has been used in a number of studies with some modification to its size and shape depending on the target muscles and the preference of design of the research group (Fig. 7).

The basic design of all buckle-type transducers is the same. They consist of one or two stainless-steel base elements. These base elements have a shape such that the target tendon can be woven through the element with a slight deflection from its natural path (Fig. 7). If the muscle produces force and pulls on the tendon, the tendon tends to align along its natural path, thereby deforming the stainless-steel base element. The deformation of the base element is measured using (typically two) strain-gauges. Therefore, one obtains a strain-gauge signal that is proportional to the muscular force. The exact relationship between the strain-gauge signal and the muscle force is obtained in a variety of ways in a terminal experiment (e.g., Walmsley et al., 1978). A detailed description of tendon-force transducer design and implantation, as well as the corresponding techniques

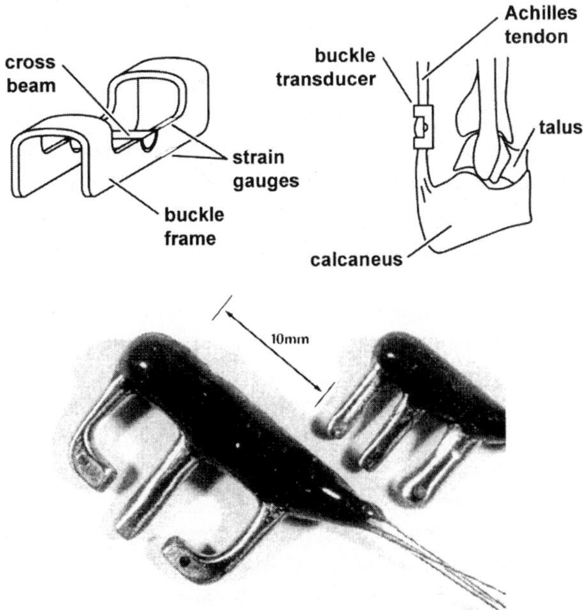

Fig. 7. Various designs of buckle-type force transducers. The two bottom transducers are photographs of transducers used to measure forces in the cat gastrocnemius (large transducer) and soleus (small transducer) during unrestrained movements.

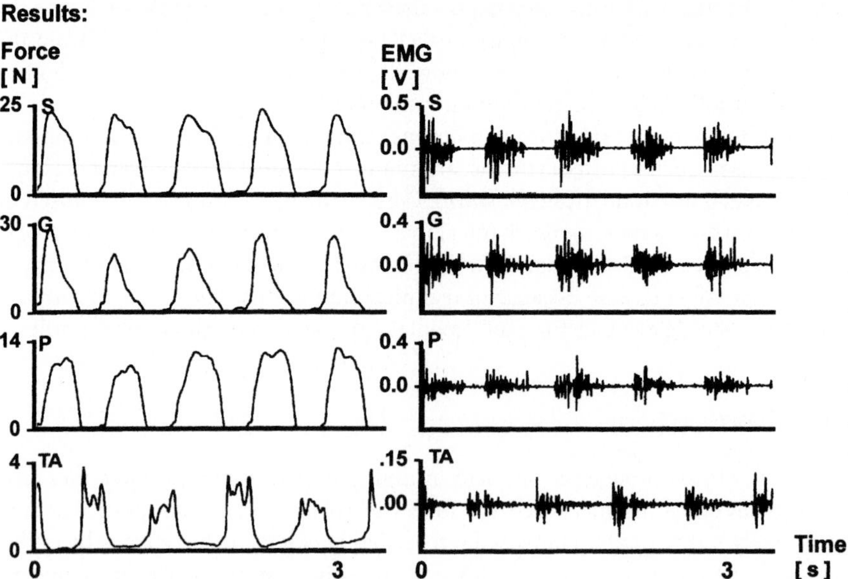

Fig. 8. Forces and EMG records obtained from cat soleus (S), gastrocnemius (G), plantaris (P), and tibialis anterior (TA) during walking at a nominal speed of approximately 0.8 m/s. Five full step cycles are shown. All forces and EMGs were obtained from the same hindlimb in a single animal.

to measure muscle forces using cable or telemetry transition during unrestrained animal locomotion may be found elsewhere (Herzog et al., 1993a; Herzog et al., 1995). Figure 8 shows an example of raw muscle force and EMG data collected for a series of step cycles in the cat gastrocnemius, soleus, plantaris and tibialis anterior.

The advantages of buckle-type transducers for muscle force measurements are that they are cheap and easy to build, they are easy to implant in animals, and they are very robust. Furthermore, by design, they measure the forces transmitted by the entire tendon, and buckle-type force transducers can be calibrated accurately and in a straightforward manner.

The disadvantage of using buckle-type transducers is that they are large relative to the tendon, therefore multiple tendon force measurements at the same joint may be difficult because of space limitations. Also, buckle-type transducers can only be used in tendons that do not run along bones or other structures that might produce forces on the transducer. Although buckle-type transducers have been used for *in vivo* force measurements in the cat patellar tendon (Hoffer et al., 1987) and the extensor digitorum longus (Abraham and Loeb, 1985), these tendons/muscles are good examples of where buckle-type transducers should not be used because they get invariably impinged between tendon and bone. These impingements cause artifactual deformation of the transducer, producing signals that are not directly related to muscular force production.

Because of some of the disadvantages of buckle-type force transducers, and because of the desire of investigators to measure muscle forces from muscles whose tendons run along bones, new force transducers have been developed in the past decade. One family of such transducers is generically referred to as implantable force transducers or IFTs. The name for these transducers comes from the fact that they are surgically implanted into the target tendon.

One of the first IFT designs was proposed by Xu et al. (1992). These investigators used an Omega-shaped base element (see Fig. 9) that was implanted into the mid-section of the target tendon, causing a deflection of the tendon around the IFT element. Upon

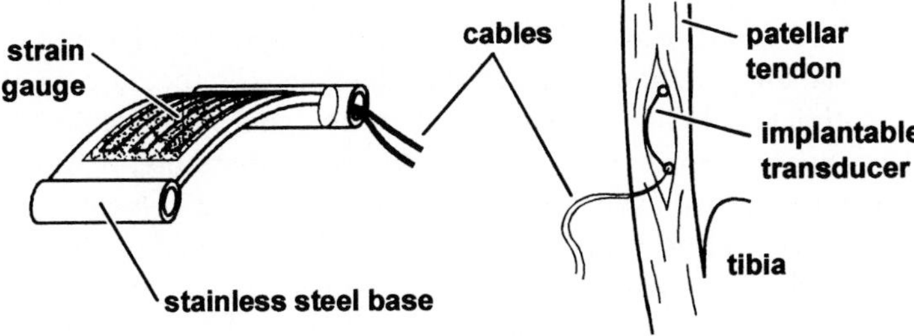

Fig. 9. Schematic drawing of an implantable force transducer used in measurements of cat patellar tendon forces. See text for additional details.

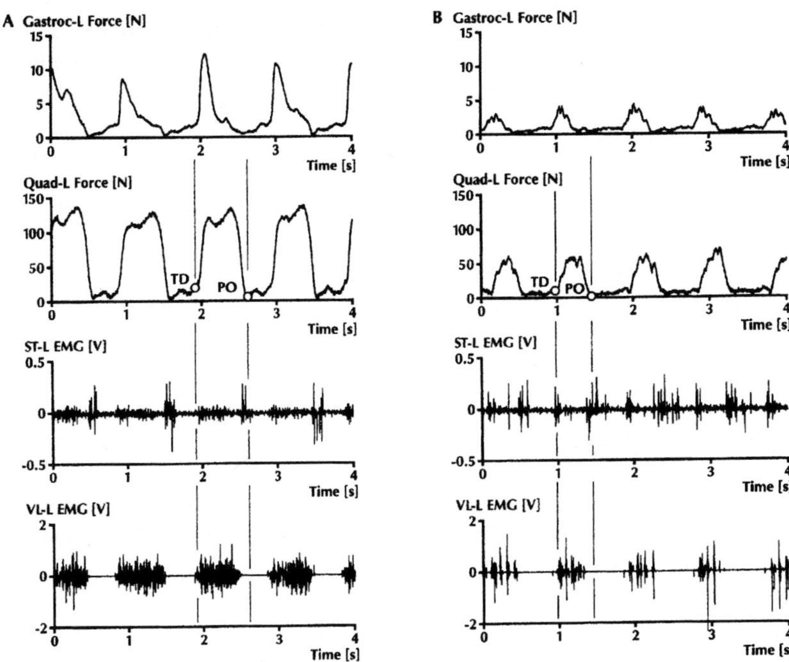

Fig. 10. Forces in gastrocnemius (gastroc) and quadriceps (quad) (patellar tendon), as well as EMGs records from semitendinosus (ST) and vastus lateralis (VL) obtained during cat walking at about 0.4–0.7 m/s. A: before transection of the anterior cruciate ligament; B: after transection of the anterior cruciate ligament. TD = Touchdown of paw; PO = Paw off at the end of the stance phase. Measurements were made in the same hindlimb of one animal approximately two weeks apart.

muscular force production, the deflected tendon fibers tend to align along their natural undeflected path, thereby pressing on the IFT element and causing compression. The corresponding deformation of the Omega-shaped element is measured via strain-gauges, and, as for the buckle-type transducers, is proportional to muscle force.

The advantage of IFTs is that they can be used for muscle force measurements in tight spaces. For example, Fig. 10 shows force measurements from the cat patellar tendon before and after surgical removal of the anterior cruciate ligament (an important stabilizing ligament in the knee). For completeness, the corresponding electromyographic activities of the knee extensors are also shown for walking at a speed of about 0.4–0.7 m/s.

Measurements of muscle forces using IFTs have been rare. Implantable force transducers are relatively hard to build (because of gluing strain-gauges on a small bent surface), and proper surgical implantation requires considerable skill (because small movements of the IFT within the tendon dramatically alter the signal output of the IFT relative to the produced muscular force; Herzog et al., 1996a). Furthermore, IFTs only measure part of the force transmitted by the tendon, therefore tendon alignment in the final calibration procedure is of utmost importance. Minute deviations of the tendon from its *in vivo* path can easily cause errors of 50% or more (Herzog et al., 1996a). Finally, calibration curves of IFTs have been shown (for the cat patellar tendon) to depend on joint angle and speed of contraction, and they show a distinct hysteresis between loading and unloading (Herzog et al., 1996b).

In summary, although IFTs can be used for force measurements in some muscles for which force measurements cannot be obtained with buckle-type transducers, IFTs are hard to build, implant and use. In contrast, buckle-type transducers are easy to build, implant and use.

Before leaving the topic of muscle force measurements, it should be pointed out that many other methods have been tried to measure muscle forces. Many of these methods have relied on measuring tendon strains associated with muscle force production, for example, by using strain transducers based on liquid metals in silastic tubing that deform with the tendon (Brown et al., 1986), or by using transducers based on the "Hall" effect (Arms et al., 1983). However, these types of transducer have invariably been found to be inaccurate, and therefore have not been used widely in the systematic determination of *in vivo* muscle forces.

Joint Contact Pressure Measurements

Human and animal diarthrodial joints allow for movement and are the transmission sites of forces from one segment to the next. At the contact interface, bones are covered with a thin layer of articular cartilage (0.3–1.0 mm in the cat knee and 1–3 mm in the human knee). Without articular cartilage, contact at joints would be made directly from bone onto bone, i.e., two hard surfaces with a high coefficient of friction. The loss of cartilage from the articular surfaces, such as in arthritis, causes stress concentrations, high friction, pain and disability.

Despite being very thin, articular cartilage distributes the forces that are transmitted across a joint over a large contact area. When joint contact forces increase, so does the articular contact area, therefore average contact pressure does not increase nearly as fast as the contact forces (Herzog et al., 1998). Despite the importance of articular cartilage in force transmission across joints, little is known about the *in vivo* pressure distributions in diarthrodial joints.

Many attempts have been made in the past decade to measure pressure distributions in diarthrodial joints *in vitro* and *in situ* (*in vivo* measurements of true pressure distributions have not been made to date). Here I would like to discuss one approach that has probably been used more often than any other and has provided the most insight into joint contact pressure distributions under a variety of different conditions. This approach involves the use of Fuji pressure-sensitive film. Fuji pressure-sensitive film consists of two sheets: a sheet containing liquid, dye-filled micro-capsules of various diameters which rupture at specified stress states, and a thin sheet containing a developer substance which provides a recording of the dye from ruptured capsules.

Fuji pressure-sensitive film has been used in many *in vitro* studies; however, just a handful of *in situ* studies have been performed. *In vitro*, in this context, refers to isolated joints from cadaver specimens, and *in situ* refers to experiments in alive but deeply an-

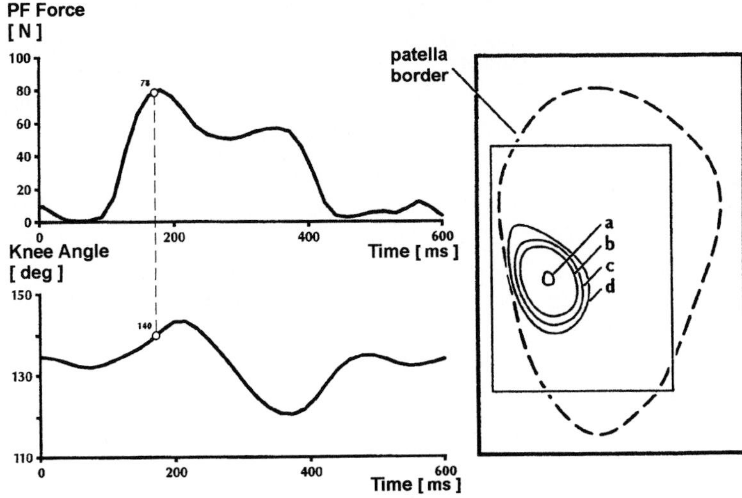

Fig. 11. Patellar tendon force (PF), knee angle, and contact pressure distribution on the retropatellar surface, quantified using iso-pressure lines. a = 14 MPa, b = 10 MPa, c = 7 MPa, d = 5 MPa.

esthetized animals. Haut et al. (1995) measured the peak pressure distribution in rabbit patello-femoral joints during impact loading of the joint during loading conditions that might occur during car accidents. These investigators showed that contact pressures exceeding approximately 30 MPa cause full-depth fissures in articular cartilage which leads to secondary osteoarthritis (Haut et al., 1995).

Normal average joint contact pressures have been found to be about 5–14 MPa in the patello-femoral joint of walking cats (Fig. 11). However, the peak pressures that can be produced by supramaximal stimulation of the cat knee extensor muscles exceed 40 MPa (Herzog et al., 1998).

Joint contact pressure distributions may change substantially as a function of articular cartilage adaptation/degeneration. In a study aimed at quantifying the possible changes in joint contact pressure distribution in the early stages of osteoarthritis, we transected the anterior cruciate ligament in cats, a procedure known to lead to osteoarthritis (Herzog et al., 1993b). Joint contact pressure measurements were made in the experimental and contralateral control patello-femoral joints 16 weeks following intervention. It was found that, for a given contact force across the patello-femoral joint, contact areas were increased (about 20%) and peak pressures were decreased (about 50%) in the experimental compared to the contralateral joint (Fig. 12).

Although Fuji pressure-sensitive film has been used on many occasions to determine the pressure distribution in diarthrodial joints, Fuji film has many disadvantages that must be overcome in the near future to make better measurements of the joint contact mechanics. The sensitivity and accuracy of Fuji pressure-sensitive film is limited. The accuracy is about ±10% (Fukubayashi and Kurosawa, 1980; Hale and Brown, 1992; Liggins et al., 1995; Singerman et al., 1987). Furthermore, Fuji film is about 100 times harder than normal articular cartilage, and when assembled for measurement is about 0.3 mm thick. Therefore, its introduction into joints causes another error of 10%–20%, depending on the joint geometry and cartilage mechanical properties (Wu et al., 1998). Together, the low resolution of Fuji film and its thickness and hardness cause average errors in pressure measurements of up to 20–30%. However, about half of this error may be accounted for by the use of theoretically based correction algorithms (Wu et al., 1998).

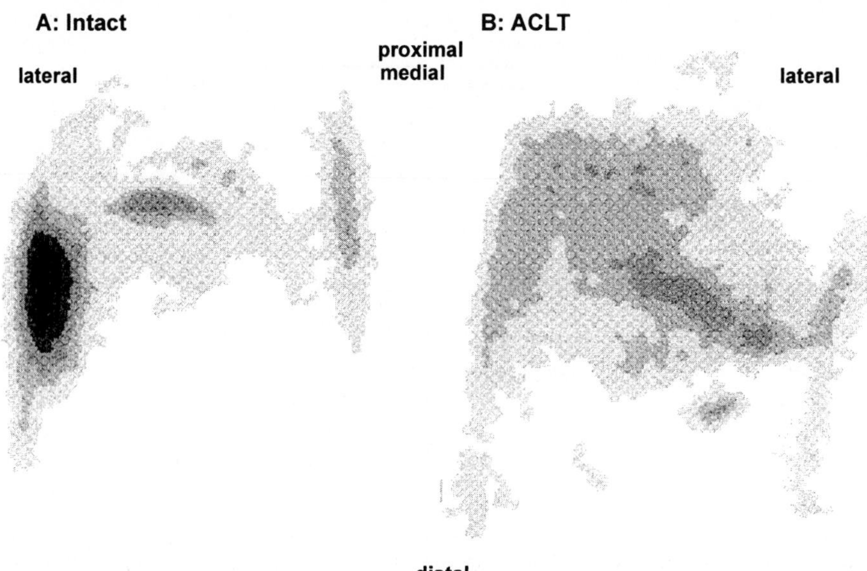

Fig. 12. Processed contact pressures on the cat patellofemoral joint for a given contact force. Left: intact healthy joint; right: anterior cruciate ligament-transected joint 16 weeks following intervention. Note, the larger contact area and the smaller contact pressures (less dark) in the transected compared to the intact knee.

Aside from its accuracy, Fuji pressure-sensitive film has the disadvantage that it can only record a single static pressure distribution that reflects the peak pressures occurring during a measurement. This limitation has two consequences; first, no dynamic pressure measurements are possible, and second, the peak pressures obtained may not all occur at the same instant in time during a measurement. However, the time-history of local peak pressure occurrences cannot be determined from the final record. Future devices aimed at measuring contact pressures should be built for dynamic pressure distribution measurements, and ideally, should be thinner and softer than Fuji film to reduce the error associated with introducing the pressure measurement device into a joint.

Movement Measurements

In order to fully characterize the mechanics of human and animal movement, it may not be sufficient to just measure the external movement of the entire body or segments of the body. Internal movements of the musculoskeletal system should be of prime importance when attempting to answer questions about soft tissue strains (and the corresponding forces) or joint mechanics.

Here I would like to discuss some of the techniques that have been used for analyzing bone movements during locomotion. Typically, bones in animals and humans are covered by large muscles and skin. These tissues move relative to the underlying bone, therefore analysis of the external movement of a limb segment gives only an inaccurate account of the underlying movement of the bone.

There are two basic techniques to analyze bone movement during dynamic tasks; invasive techniques that require attachment of some marker system on the target bone, and non-invasive techniques. Non-invasive techniques are typically based on X-ray technology, for example, video fluoroscopy. Video fluoroscopy is based on X-ray exposure of the target limb and video recording of X-ray pictures (typically at 30 Hz) of the moving limb. Using video fluoroscopy, a "continuous" record of bone movement can be obtained. Digitization and analysis of video fluoroscopy can be made using the same approaches as are used for normal high-speed movement recording. Digital video fluoroscopy systems based on frame grabber technology are now commercially available; therefore, the fluoroscopy pictures can be stored online to a computer for offline analysis of bone movement. Using digital fluoroscopy, Tashman et al. (1995; 1997) demonstrated convincingly that transection of the anterior cruciate ligament in dogs caused a large anterior displacement of the tibia relative to the femur at paw contact of walking in the anterior cruciate ligament-deficient but not in the intact knee.

Invasive techniques involve the attachment of a marker system to the target bones. For example, Korvick et al. (1994) performed a similar study to that described above by Tashman et al. (1997), i.e., they were interested in the relative movement of the tibia to the femur in walking dogs before and after anterior cruciate ligament transection. The approach of Korvick et al. (1994) consisted of rigidly attaching an external metal frame to the target bones during data acquisition. From the kinematics of the external frames, the relative movements of the "rigidly" connected bones could be derived. Although simple conceptually, this particular approach proved difficult to implement. Korvick et al. (1994) mentioned problems of loosening of the external frame from the bones, and further reported an attrition of three out of five experimental animals because of lameness. It did not become clear whether this lameness was associated with the invasiveness of the approach or the lack of surgical expertise. Nevertheless, the partial results of Korvick et al. (1994) were confirmed by the findings of Tashman et al. (1995; 1997), i.e., that there was a substantial increase in anterior translation of the tibia relative to the femur at the instant of paw contact during walking following transection of the anterior cruciate ligament.

Rather than using rigidly attached external metal frames to measure the movements of bones, researchers have used bone pins that are rigidly fixed to bones. The pin movements were filmed or recorded by video during the activity of interest, pin movements were digitized, and then, through appropriate mathematical translations based on rigid body kinematics, the three-dimensional movements of the bones could be determined. Using bone pins, rather than entire external fixation frames, is less invasive and has been used in human and animal experimentation (Lafortune et al., 1992; Reinschmidt et al., 1997).

The disadvantage of the above-mentioned invasive and non-invasive procedures is that they have a limited resolution either in time and/or space. For example, the video-based movement analysis procedures are associated with an error of 1 point in 600, as mentioned before. That means that, for a frame size of 600 mm, the best-case error is about 1 mm. Although such an error may be acceptable for many applications, it is not acceptable for all experimental questions. Similarly, the temporal resolution of video-based systems is typically limited to 30, 60 Hz, or for high-speed systems, to 180 or 200 Hz.

One of our research interests is to determine the *in vivo* joint contact mechanics in the cat knee from accurate measurements of bone movements. For this purpose, we have tested a pulsed ultrasound technique to measure the movements of the tibia relative to the femur before and after transection of the anterior cruciate ligament. Our current system has a temporal and spatial resolution of about 1 ms and 16 μm, respectively. These resolutions can still be improved with a better data acquisition computer.

The pulsed ultrasound, or sono-micrometry system, is based on the transmission of ultrasound pulses from 1 mm-diameter, piezo-electric crystals. By measuring the time elapsed from the transmission of an ultrasound pulse in one crystal to the detection of the pulse in another crystal, the exact distance from one crystal to the next can be calculated based on the known average speed of ultrasound in the appropriate tissues.

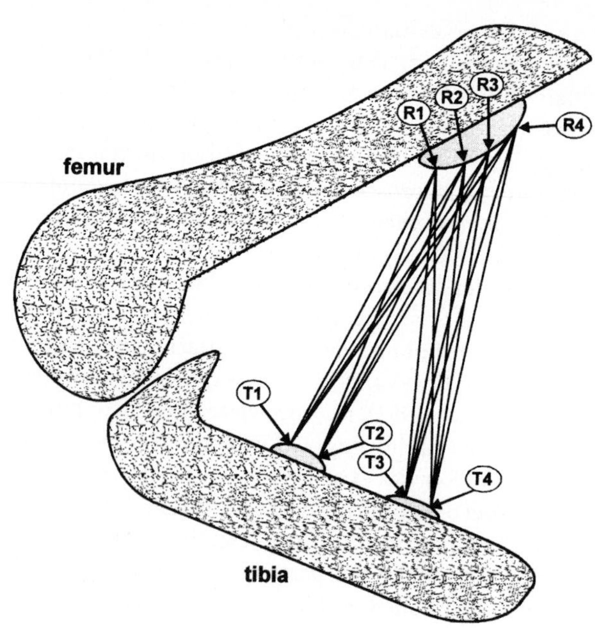

Fig. 13. Schematic drawing of the surgical placement of the four piezoelectric crystals for ultrasound transmission ($T_{1,2,3,4}$) and ultrasound receiving ($R_{1,2,3,4}$) on the cat femur and tibia. For additional details, please be referred to the text.

In our experiments, four piezo-electric crystals were rigidly attached to the cat femur and tibia (see Fig. 13). A set of four crystals on one bone is designated as the ultrasound transmitters, the other set as the corresponding ultrasound receivers. Having four transmitters and four receivers gives a total of 16 (4x4) intercrystal distances that can be calculated at any instant in time. From these known distances, and the known locations of the crystals on the bone (determined using Laser digitization), the location and orientation of one bone (e.g., tibia) relative to the other (e.g., femur) can be calculated for each instant in time. By three-dimensional reconstruction of the bones (from the Laser-digitized surface coordinates) and determining the relative location of the bones as a function of time, bone movements can be displayed on a graphics computer monitor, and can be analyzed using mathematical techniques similar to those used in video- and film-based movement analyses systems. Figure 14 shows an example of the raw data distance records between two transducers during cat locomotion before and after anterior cruciate ligament transection. One of the crystals was attached to the tibia, the other to the femur. The animal was walking at a speed of about 0.4 m/s in both instances. Comparing the two records, it is obvious that the distance between the two crystals is smaller after compared to before anterior cruciate ligament transection, indicating that the animal's knee became more flexed after transection. Also, after anterior cruciate ligament transection, there were distinct "wiggles" in the signal (Fig. 14, arrows), indicating rapid slipping of the tibia relative to the femur, which was not observed when the knee was intact. These rapid slips of the tibia relative to the femur following anterior cruciate ligament transection led to speculations that micro-instabilities might be responsible for the degenerative processes in the cat knee.

Theoretical Determination of Internal Forces

Internal, *in vivo* forces acting on the human and animal musculoskeletal system are hard to measure. Therefore, researchers in biomechanics have developed a number of theoretical models that allow for the calculation of the internal forces of interest. When talking about such models on the next few pages, it must be kept in mind that the theo-

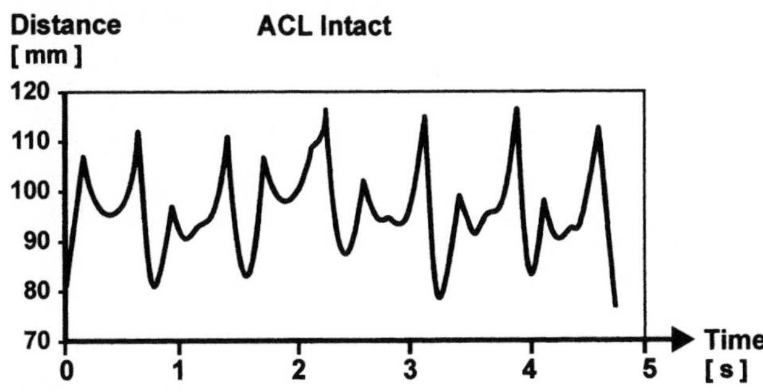

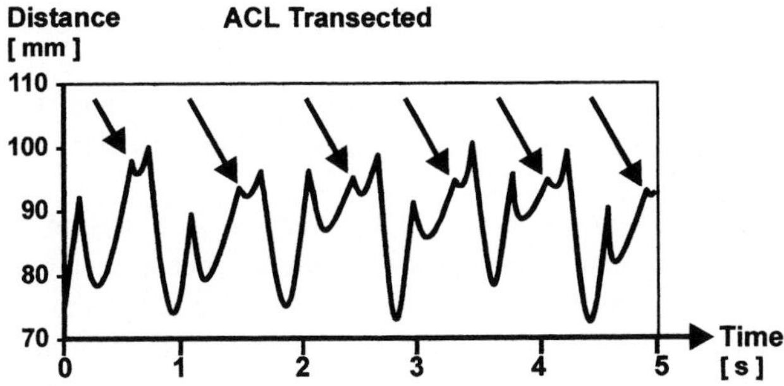

Fig. 14. Raw data records of the distances recorded between a pair of crystals located on the cat tibia and femur. Measurements were obtained during slow walking. Top: knee intact; bottom: anterior cruciate ligament-transected knee (about 4 weeks post-transection).

retical calculations of internal musculoskeletal forces are estimates of the actual forces. Very few biomechanical models aimed at determining internal forces have been validated rigorously (because of the experimental difficulties to do so), and none has been proven to give accurate internal force predictions for general movement conditions. Therefore, although not useful for making accurate quantitative force predictions, theoretical models of musculoskeletal forces are suited to give trends in the changes of internal forces with changing movement conditions. Many theoretical models are aimed at predicting the forces in bones, ligaments and articular cartilage. However, as discussed previously, all these forces are largely dependent on the forces exerted by muscles. Therefore, the following theoretical discussion will focus primarily on approaches aimed at predicting individual muscle forces during animal and human movement.

By far the most frequently used approach in biomechanics to predict individual muscle forces is mathematical optimization. In the following, this approach will be discussed. In order to use this approach and calculate the forces of individual muscles during a given movement task, the intersegmental forces and moments must be calculated in a first step. Once the intersegmental forces and moments are known, the so-called distribution problem is solved using mathematical optimization to obtain the corresponding muscle forces.

The General Distribution Problem

In order to calculate the forces exerted by individual muscles and, thus, to determine the force-sharing among muscles, the so-called distribution problem must be solved first (Crowninshield and Brand, 1981a). The distribution problem relates the forces and moments produced by structures in and around a joint to the intersegmental resultant force and moment. This approach is based on the idea that any distributed force/moment system can be replaced by a single force and moment that are equipollent to the distributed force/moment system. The joint equipollence relationship may be expressed as follows (please note: bold characters designate vectors) (Crowninshield and Brand, 1981a):

$$\mathbf{F}^0 = \sum_{i=1}^{m}\left(\mathbf{f}_i^m\right) + \sum_{j=1}^{l}\left(\mathbf{f}_j^l\right) + \sum_{k=1}^{c}\left(\mathbf{f}_k^c\right) \tag{1}$$

$$\mathbf{M}^0 = \sum_{i=1}^{m}\left(\mathbf{r}_i^m \times \mathbf{f}_i^m\right) + \sum_{j=1}^{l}\left(\mathbf{r}_j^l \times \mathbf{f}_j^l\right) + \sum_{k=1}^{c}\left(\mathbf{r}_k^c \times \mathbf{f}_k^c\right) \tag{2}$$

where **F** and **M** are the intersegmental resultant force and moment, respectively, and the superscript "0" designates the joint center 0 (which, from a mathematical point of view, may be defined arbitrarily within the joint); $\mathbf{f}_i^m$, $\mathbf{f}_j^l$ and $\mathbf{f}_k^c$ are the forces in the i-th muscle, the j-th ligament and k-th bony contact, respectively; $\mathbf{r}_i^m$, $\mathbf{r}_j^l$, $\mathbf{r}_k^c$, are location vectors from the joint center to any point on the line of action of the corresponding force; the "times sign" (×) denotes the vector (cross) product; and m, l, and c designate the number of muscles and ligaments crossing the joints, and the individual articular contact areas within the joint, respectively.

When solving the distribution problem (i.e., Eqs. 1,2), the purpose is to determine the ligamentous, muscular, and bony contact forces from the intersegmental resultant moment and force. The intersegmental resultants may be determined by using the inverse dynamics approach (Andrews, 1974), when the kinematics (obtained, for example, by filming the movement of interest, using high-speed video or film) and the external forces acting on the system of interest (except for the intersegmental resultants, which are to be determined) are known. It is important to note here that the intersegmental resultant force and moment are conceptual kinetic quantities; they cannot be measured, and they are not present in an identifiable structure.

Equations 1 and 2 represent the general distribution problem. When solving these two equations in practical situations, simplifying assumptions are typically made. Particularly when calculating the forces in muscles for the normal range of everyday movements, it is commonly assumed that ligaments do not transmit appreciable forces. Also, the joint center is usually assumed to fall on the line of action of the resultant joint contact force; therefore, the distributed joint contact forces do not contribute to the resultant intersegmental moments. When these two assumptions are made, Eqs. 1 and 2 are reduced to:

$$\mathbf{F}^0 = \sum_{i=1}^{m}\left(\mathbf{f}_i^m\right) + \sum_{k=1}^{c}\left(\mathbf{f}_k^c\right) \tag{3}$$

$$\mathbf{M}^0 = \sum_{i=1}^{m}\left(\mathbf{r}_i^m \times \mathbf{f}_i^m\right) \tag{4}$$

Equations 3 and 4 are two-vector equations in three-dimensional space; they give six independent scalar equations. For calculating individual muscle forces, it is preferable to use Eq. 4 only, because Eq. 3, although adding one vector (or three scalar) equation(s) to the description of the system of interest, also introduces at least one additional vector (or three scalar) unknown(s) in the form of the resultant(s) of the distributed forces of the joint contact(s), $\mathbf{f}_k^c$. Assuming that the resultant intersegmental moment, $\mathbf{M}^0$, has been determined using the inverse dynamics approach, Eq. 4 contains 2 · m vector unknowns, m unknown location vectors, $\mathbf{r}_i^m$, and m unknown muscle forces, $\mathbf{f}_i^m$. The location vectors, $\mathbf{r}_i^m$, (or the corresponding scalar moment arms) can be determined experimentally using cadaver specimens (e.g., Grieve et al., 1978; Herzog and Read, 1993; Spoor et al., 1990) or imaging techniques (e.g., Rugg et al., 1990; Spoor et al., 1990). Although the experimental determination of moment arms is by no means trivial (nor is it a problem that has been solved conclusively), it is usually assumed that moment arms are known. Also, the lines of action of the muscles are often determined experimentally; therefore, the only remaining unknowns in Eq. 4 are the magnitudes of the muscle forces, f_i^m. Since Eq. 4 gives three independent scalar equations, but the number of muscles crossing a given joint in humans (or in animals) typically exceeds three, Eq. 4 represents a mathematically indeterminate system (i.e., a system that contains more unknowns than equations).

Mathematically indeterminate systems generally have an infinite number of possible solutions. For example, the equation x+y=12 (which contains two unknowns, x and y) has an infinite number of possible solutions (e.g., x=6, y=6; x=1, y=11; x=–17, y=29, etc.). An indeterminate system may be made determinate, and so give a unique solution to a problem, by (a) adding equations to the system, or (b) eliminating unknowns until the number of unknowns and the number of equations match. For example, adding the equation x –y=4 to equation x+y=12 will result in a unique solution for x and y (i.e., x= 8, y=4). Unknowns may be eliminated by grouping them or by relating them in some fashion. For example, if we assume that x is the same as y, equation x+y=12 becomes 2x=12, and we have a unique solution for x (x=6).

Similarly, in biomechanics, the prediction of forces in individual muscles crossing a joint (often an indeterminate mathematical problem) has been solved by adding system equations or by eliminating unknowns. System equations may be added not only by considering the mechanical relation between muscular forces and joint moments but also by incorporating other known mechanical or neurophysiological relations between muscular forces and joint moments (e.g., Pierrynowski, 1982; Pierrynowski and Morrison, 1985). Reducing the number of unknown muscular forces when solving the distribution problem has been done by grouping individual muscles into functional units (e.g., Morrison, 1970; Paul, 1965). Neither the addition of system equations nor the reduction of system unknowns is a satisfactory solution because additional system equations are normally based on assumptions about the system behavior, and grouping muscles into functional units introduces a series of new problems (e.g., what is the line of action of a group of muscles or the moment arm about a given joint?), and it does not solve the problem of interest (i.e., what are the forces in individual muscles during movements?).

The most common approach to solving for individual muscle forces during movement has been mathematical optimization. This elegant way of solving the distribution problem is simple enough for analytic solutions to be obtained for many realistic musculoskeletal models, and the idea that human (or animal) movements obey some law of optimal control has both a strong appeal and a long history (Weber and Weber, 1836).

Although there are dozens of published papers that attempt to use optimization mathematics to predict individual muscle forces, only a few selected examples are mentioned here. These examples were chosen because they either contain a physiologically derived optimization model (rather than a purely mathematical model), or they show validations of the theoretical predictions using experimentally measured muscle forces.

Dul et al. (1984b) evaluated a series of published optimization algorithms and proposed an objective function for individual muscle force predictions, which maximizes the time a movement (or isometric contraction) can be maintained (Dul et al., 1984a). The objective function of Dul et al. (1984a) was the first to require input about the fiber-type distribution of the target muscles. Dul et al. (1984a) evaluated their optimization algorithm by first comparing the predicted contraction times to contraction times measured experimentally in lower limb muscles of humans during isometric tasks. They then attempted to validate their proposed optimization scheme by theoretically deriving the force-sharing relationship for cat soleus and gastrocnemius muscles from

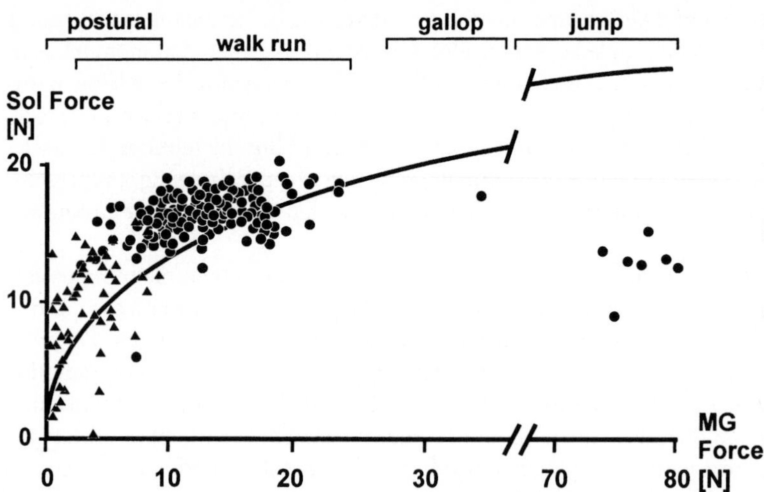

Fig. 15. Theoretical (line) and experimental (dots) force-sharing between cat soleus (S) and medical gastrocnemius (MG) for a variety of locomotor activities. (Adapted from Dul et al., 1984a)

the proposed model, and by comparing the predicted force-sharing between these two muscles to the maximal forces measured in cat soleus and gastrocnemius for standing, walking, trotting, galloping, and jumping. The theoretically predicted results were said to fit the experimental results well (and clearly fitted the experimental results better than any of the predictions from previously published works). Nevertheless, the comparisons made by Dul et al. (1984a) had several limitations: (a) The musculoskeletal input required for the theoretical model and the muscle force measurements were obtained from different animals; (b) The maximal forces in soleus and gastrocnemius muscles during cat locomotion do not occur at the same instant in time (Herzog and Leonard, 1991; Herzog et al., 1993), whereas the force-sharing equation requires that comparisons of force values between muscles are made for the same instants in time; (c) Rather than the force-time histories throughout the movements, only maximal forces were used for comparisons.

Furthermore, the force-sharing predictions of Dul et al. (1984a) had the following conceptual shortcomings: (a) Force-sharing between muscles was dependent only on the maximal isometric force and the fiber-type distribution (two constants) of the muscles; therefore, all force-sharing relations derived from their algorithm had a unique functional relationship, which means that a given force in muscle 1 was always associated with a given force in muscle 2 (see Fig. 15). Direct muscle force measurements at a variety of speeds of locomotion in the cat, however, demonstrated that a given force in the soleus may be associated with a large range of forces in a synergistic muscle (i.e., the gastrocnemius, see Fig. 16). (b) The force-sharing relations derived from the model by Dul et al. (1984a) increase steadily; that is, if the force in muscle 1 increases, the force in muscle 2 has to increase as well. Again, experimental data from cat soleus and gastrocnemius demonstrate convincingly that gastrocnemius forces increase steadily from standing to walking to running to jumping; whereas soleus forces tend to increase slightly from standing to walking and remain constant for a large range of speeds of locomotion, and they tend to decrease for jumping (see Fig. 15).

Herzog (1987a; 1987b) presented an optimization algorithm for the solution of individual muscle forces which had an objective function that contained explicitly the instantaneous contractile conditions (length, and rate of change in length) and the force-length-velocity properties of the muscles of interest. This algorithm allowed for the pre-

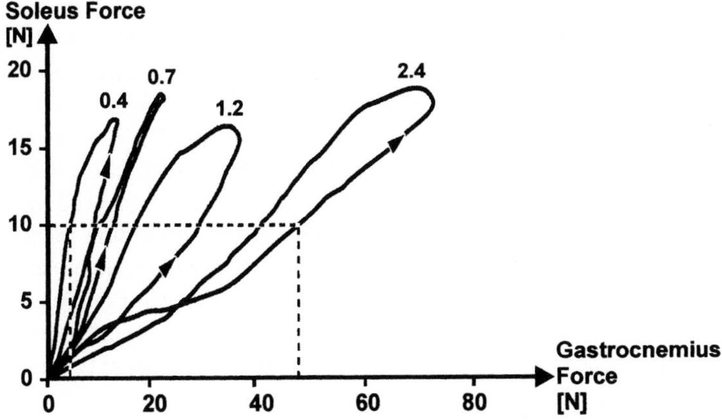

Fig. 16. Experimental force-sharing obtained from cat soleus and gastrocnemius for walking at nominal speeds of 0.4, 0.7, and 1.2 m/s, and for trotting at a nominal speed of 2.4 m/s. All curves shown represent an average curve obtained from one animal over at least 10 consecutive step cycles.

diction of a wide range of muscle forces for one muscle while the other muscle's force remained constant. Therefore, the force-sharing "loops" measured experimentally between two muscles (see Fig. 16) could, at least conceptually, be accommodated. Furthermore, the algorithm predicted smooth force transitions in muscles, whereas previous (strictly static) algorithms predicted large instantaneous force changes. The drawbacks of the algorithm proposed by Herzog (1987a; 1987b) were associated with its difficult implementation. For example, the force-length-velocity properties of most muscles are unknown and their behavior under submaximal levels of contraction can at best be approximated.

Davy and Audu (1987) introduced dynamic optimization for predicting the forces of nine lower limb muscle groups during the swing phase of walking. The dynamic algorithm was motivated by the fact that force-time histories of muscles using static optimization often showed unrealistic discontinuities. The results of the dynamic optimization proposed by Davy and Audu (1987) were compared to the static optimization algorithm of Crowninshield and Brand (1981b) and to the EMG envelopes during the swing phase of walking (Pedotti et al., 1978). Lacking a rigorous validation, it is difficult to assess the appropriateness of the dynamic algorithm proposed by Davy and Audu (1987). Furthermore, their results of the comparison of the static and dynamic algorithms must be considered with caution because the data shown for the static algorithm contained co-contraction of two single-joint antagonistic muscles in a planar model, an impossible solution (Herzog and Binding, 1992) for the algorithm of Crowninshield and Brand (1981b).

The first validation of optimization algorithms predicting the force-sharing among synergistic muscles was performed by Herzog and Leonard (1991). In this study, the force-sharing among the cat soleus, gastrocnemius, and plantaris muscles was predicted theoretically based on the most common optimization algorithms proposed up to that time. The required input parameters for the optimization algorithms were largely determined directly from the experimental animals. The predicted force-sharing behavior was then compared to the actual force measurements obtained from the target muscles for a variety of locomotor conditions. It was found that the theoretical algorithms were not able to predict actual force-sharing behavior among the muscles of a synergistic group. In the study by Herzog and Leonard (1991), static linear and non-linear optimization algorithms were evaluated, while dynamic algorithms or algorithms requiring instantaneous contractile conditions were ignored. An evaluation of the algorithm proposed by Herzog (1987a) containing the instantaneous contractile conditions of the muscles was performed in the form of a pilot study (Herzog et al., 1988), and although the initial comparisons between the theoretical and experimental force-sharing behavior of cat soleus and gastrocnemius were encouraging, a thorough analysis revealed that the predictions based on Herzog's (1987a) work did not predict experimental findings well (unpublished results). The question as to whether optimization-based theoretical algorithms are useful in predicting muscle forces remains unresolved; further research aimed at developing good theoretical predictions of individual muscle forces is required. Experimental data on the force-sharing among synergistic muscles are now available for comparison from a variety of laboratories, which was not the case 20 years ago. It is hoped that researchers will take advantage of this situation in the coming decade, and start to use these experimental data for rigorous validation of theoretical models aimed at predicting individual muscle forces.

Future Considerations

The analysis of human and animal movement using quantitative biomechanical techniques is relatively new. However, in the past 40 years, it has undergone a dramatic change from an emphasis on describing the external mechanics to a focus on explaining the control and mechanics of movements using analyses of the internal kinematics and kinetics.

I am predicting that we are just on the verge of another major shift in the biomechanical analysis of human and animal movement. This shift changes the focus to cellular and molecular mechanics. The reasons for this shift are twofold: first, an interest in the biological effects of the forces acting on the musculoskeletal system, and second, a curiosity about the basic mechanisms underlying the mechanical properties of tissues.

Let me start with the second point and take as an example the mechanical properties of skeletal muscle. Although the basic mechanical properties of muscle, the force-length, force-velocity, and force-time property of maximally activated muscle are well described (Abbott and Aubert, 1952; Gordon et al., 1966; Hill, 1938), the corresponding mechanisms of muscle contraction and force production are not. For example, the energetic cost of muscular contraction (i.e., the biochemical steps in ATP hydrolysis and how these biochemical steps relate to the corresponding molecular events of force production) are not well understood. Furthermore, on the whole muscle level, basic properties such as force depression following muscle shortening, or force-velocity properties of elongating muscles for submaximal activation, are ill characterized and even less understood.

Research on the muscle fiber (cell) level, and particularly research on the molecular level, promises new insights into how skeletal muscles contract. Several techniques have been developed that allow one to measure the forces of single cross-bridges attached to actin and to quantify the displacement of actin during a single cross-bridge cycle. For example, using the so-called laser trap technique, the ends of an actin filament can be fixed in laser beams and the actin filament fixed in this way is brought into contact with a single (S1) cross-bridge head. A similar technique requires the attachment of a single cross-bridge head to a fine glass needle from which the cross-bridge can form an attachment with a single actin filament (Fig. 17). The pico-Newton forces and nanometer displacements caused by the cross-bridge can be measured using this technique. From these types of experiments, insight can be gained into the basic step of force production and into the energetics in skeletal muscle (Finer et al., 1994; Ishijima et al., 1991, 1994; Kitamura et al., 1999).

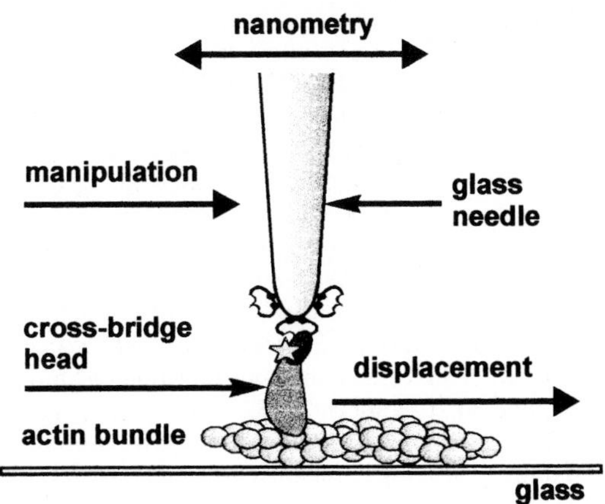

Fig. 17. Schematic drawing of the setup for interaction of a single cross-bridge head with an actin filament. With this setup, piconewton forces produced by the cross-bridge and nanometer displacements caused by cross-bridge head rotation can be quantified.

Coming back to the first point listed above, it is my opinion that researchers in biomechanics are becoming increasingly more interested in the effects of forces on musculoskeletal structures. In probably 99% of all biomechanical research done to date, musculoskeletal structures have been treated as mechanical rather than biological systems. The basic difference between a mechanical and biological element is that the mechanical structure cannot adapt, whereas the biological structure can and will adapt. For example, it has been known for a long time that skeletal muscle hypertrophies when loaded adequately and atrophies when not used. However, the detailed molecular mechanisms of skeletal muscle adaptation following exercise are not known.

Similarly, articular cartilage chondrocytes are known to increase proteoglycan synthesis when dynamically loaded at a range of frequencies and loads, and are known to decrease proteoglycan synthesis when statically loaded (Burton-Wurster et al., 1993; Sah et al., 1989; Torzilli et al., 1997; Wong et al., 1997). However, it is not known how mechanical loading of articular cartilage is translated into the up- or down-regulation of proteoglycan synthesis. But it is known that the material properties, the functional properties, and the load transmission properties change as articular cartilage adapts, for example, in degenerative diseases such as osteoarthritis (Herzog et al., 1998). One of the big challenges of future biomechanical research in the area of animal and human movement will be to elucidate the effects of movement and loading on the adaptation of tissues comprising the musculoskeletal system. Such adaptive behavior is of particular interest in musculoskeletal diseases that appear to be directly related to the mechanical loading of musculoskeletal tissues; for example, osteoarthritis, osteoporosis, tendonitis, muscular atrophy, joint stiffness and pain, and many more.

References

Abbott BC, Aubert XM (1952) The force exerted by active striated muscle during and after change of length. J Physiol 117:77–86

Abraham LD, Loeb GE (1985) The distal hindlimb musculature of the cat. Exp Brain Res 58: 580–593

Andrews JG (1974) Biomechanical analysis of human motion. Kinesiol 4:32–42

Arms SW, Johnson RJ, Johnson RJ, Pope MH (1983) Strain measurements in the medial collateral ligament of the human knee: an autopsy study. J Biomech 16:491–496

Basmajian JV, De Luca CJ (1985) Muscles alive: their functions revealed by electromyography. Los Angeles: Williams and Wilkins

Bigland-Ritchie B, Donovan EF Roussos CS (1981) Conduction velocity and EMG power spectrum changes in fatigue of sustained maximal efforts. J Appl Physiol 51:1300–1305

Bigland B, Lippold OCJ (1954) The relation between force velocity and integrated electrical activity in human muscles. J Physiol 123:214–224

Bouisset S, Goubel F (1971) Interdependence of relations between integrated EMG and diverse biomechanical quantities in normal voluntary movements. Activitas Nervosa Superior 13:23–31
Brown TD, Sigal L, Njus GO, Njus NM, Singerman R, Brand RA (1986) Dynamic performance characteristics of the liquid metal strain gage. J Biomech 19:165–173
Burton-Wurster N, Vernier-Singer M, Farquhar T, Lust G (1993) Effect of compressive loading and unloading on the synthesis of total protein proteoglycan and fibronectin by canine cartilage explants. J Orthop Res 11:717–729
Crowninshield RD, Brand RA (1981b) A physiologically based criterion of muscle force prediction in locomotion. J Biomech 14:793–801
Crowninshield RD, Brand RA (1981a) The prediction of forces in joint structures: Distribution of intersegmental resultants. In: Exercise and Sport Sciences Reviews, Doris I Miller, pp 159–181
Davy DT, Audu ML (1987) A dynamic optimization technique for predicting muscle forces in the swing phase of gait. J Biomech 20:187–201
Dul J, Johnson GE, Shiavi R, Townsend MA (1984a) Muscular synergism. II. A minimum-fatigue criterion for load sharing between synergistic muscles. J Biomech 17:675–684
Dul J, Townsend MA, Shiavi R, Johnson GE (1984b) Muscular synergism. I. On criteria for load sharing between synergistic muscles. J Biomech 17:663–673
Elftman H (1938) The force exerted by the ground in walking. Arbeitsphysiologie 10:485–491
Finer JT, Simmons RM, Spudich JA (1994) Single myosin molecule mechanics: piconewton forces and nanometre steps. Nature 368:113–119
Fukubayashi T, Kurosawa H (1980) The contact area and pressure distribution of the knee. Acta Orthop Scand 51:871–879
Gordon AM, Huxley AF, Julian FJ (1966) The variation in isometric tension with sarcomere length in vertebrate muscle fibres. J Physiol 184:170–192
Grieve DW, Pheasant S, Cavanagh PR (1978) Prediction of gastrocnemius length from knee and ankle joint posture. In: Asmussen E, Jorgensen K (eds) Biomechanics VI-A. Baltimore: University Park Press, pp 405–412
Guimaraes AC, Herzog W, Hulliger M, Zhang YT, Day S (1994a) Effects of muscle length on the EMG-force relation of the cat soleus muscle using non-periodic stimulation of ventral root filaments. J Exp Biol 193:49–64
Guimaraes AC, Herzog W, Hulliger M, Zhang YT, Day S (1994b) EMG-force relation of the cat soleus muscle: Experimental simulation of recruitment and rate modulation using stimulation of ventral root filaments. J Exp Biol 186:75–93
Hagberg JM, Mullin JP, Giese MD, Spitznagel E (1981) Effect of pedaling rate on submaximal exercise responses of competitive cyclists. J Appl Physiol 51:447–451
Hale JE, Brown TD (1992) Contact stress gradient detection limits of Pressensor film. J Biomech Eng 114:353–357
Haut RC, Ide TM, De Camp CE (1995) Mechanical responses of the rabbit patello-femoral joint to blunt impact. J Biomech Eng 117:402–408
Herzog W (1987b) Considerations for predicting individual muscle forces in athletic movements. Int J Sport Biomech 3:128–141
Herzog W (1987a) Individual muscle force estimations using a non-linear optimal design. J Neurosci Methods 21:167–179
Herzog W, Adams ME, Matyas JR, Brooks JG (1993b) A preliminary study of hindlimb loading morphology and biochemistry of articular cartilage in the ACL-deficient cat knee. Osteoarth Cart 1:243–251
Herzog W, Archambault JM, Leonard TR, Nguyen HK (1996a) Evaluation of the implantable force transducer for chronic tendon-force recordings. J Biomech 29:103–109
Herzog W, Binding P (1992) Predictions of antagonistic muscular activity using nonlinear optimization. Math Biosci 111:217–229
Herzog W, Diet S, Suter E, Mayzus P, Leonard TR, Muller C, Wu JZ, Epstein M (1998) Material and functional properties of articular cartilage and patellofemoral contact mechanics in an experimental model of osteoarthritis. J Biomech 31:1137–1145
Herzog W, Hasler EM, Leonard TR (1996b) In situ calibration of the implantable force transducer. J Biomech 29:1649–1652
Herzog W, Hoffer JA, Abrahamse SK (1988) Synergistic load sharing in cat skeletal muscles. Proceedings of the 5th Biennial Conference of the CSB, Ottawa, pp 78–79
Herzog W, Leonard TR (1991) Validation of optimization models that estimate the forces exerted by synergistic muscles. J Biomech 24:31–39
Herzog W, Leonard TR, Guimaraes ACS (1993a) Forces in gastrocnemius soleus and plantaris tendons of the freely moving cat. J Biomech 26:945–953

Herzog W, Leonard TR, Stano A (1995) A system for studying the mechanical properties of muscles and the sensorimotor control of muscle forces during unrestrained locomotion in the cat. J Biomech 28:211–218

Herzog W, Read LJ (1993) Lines of action and moment arms of the major force-carrying structures crossing the human knee joint. J Anat 182:213–230

Hill AV (1938) The heat of shortening and the dynamic constants of muscle. In: Proceedings of the Royal Society of London, pp 136–195

Hof AL, Van Den Berg J (1981a) EMG to force processing. I. An electrical analogue of the Hill muscle model. J Biomech 14:747–758

Hof AL, Van Den Berg J(1981b) EMG to force processing. II. Estimation of parameters of the Hill muscle model for the human triceps surae by means of a calf ergometer. J Biomech 14:759–770

Hof AL, Van Den Berg J (1981c) EMG to force processing. III. Estimation of model parameters for the human triceps surae muscle and assessment of the accuracy by means of a torque plate. J Biomech 14:771–785

Hof AL, Van Den Berg J (1981d) EMG to force processing. IV. Eccentric–concentric contractions on a spring-flywheel set up. J Biomech 14:787–792

Hoffer JA, Sugano N, Loeb GE, Marks WB, O'Donovan MJ, Pratt CA (1987) Cat hindlimb motoneurons during locomotion. II. Normal activity patterns. J Neurophysiol 57:530–553

Ishijima A, Doi T, Sakurada K, Yanagida T (1991) Sub-piconewton force fluctuations of actomyosin in vitro. Nature 352:301–306

Ishijima A et al (1994) Single-molecule analysis of the actomyosing motor using nano-manipulation. Biochem Biophys Res Commun 199:1057–1063

Kitamura K, Tokunaga M, Hikikoshi I, Yanagida T (1999) A single myosin head moves along an actin filament with regular steps of 5.3 nanometers. Nature 397:129–134

Komi PV, Tesch P (1979) EMG frequency spectrum muscle structure and fatigue during dynamic contractors in man. Eur J Appl Physiol 42:41–50

Korvick DL, Pijanowski GJ, Schaeffer DJ (1994) Three-dimensional kinematics of the intact and cranial cruciate ligament-deficient stifle of dogs. J Biomech 27:77–87

Lafortune MA, Cavanagh PR, Sommer HJ III (1992) Three-dimensional kinematics of the human knee during walking. J Biomech 25:347–357

Liggins AB, Hardie WR, Finlay JB (1995) Spatial and pressure resolution of Fuji pressure-sensitive film. Exp Mech 35:66–173

Lindström L, Kadefors R, Petersén I (1977) An electromyographic index for localized muscle fatigue. J Appl Physiol 43:750–754

Lippold OCJ (1952) The relation between integrated action potential in human muscle and its isometric tension. J Physiol 117:492–499

Loeb GE, Gans C (1986) Electromyography for experimentalists Chicago: The University of Chicago Press

Marey M (1873) De la Locomotion Terrestre chez les Bipedes et les Quadrupedes. J de l'Anat et de la Physiol 9:42

Milner-Brown HS, Stein RB (1975) The relation between the surface electromyogram and muscular force. J Physiol 246:549–569

Moritani T, Devries HA (1978) Re-examination of the relationship between the surface integrated electromyogram and force of isometric contraction. Am J Phys Med 57:263–277

Morrison JB (1970) The mechanics of muscle function in locomotion. J Biomech 3:431–451

Paul JP (1965) Bioengineering studies of the forces transmitted by joints. II. Engineering analysis. In: Kenedi RM (ed) Biomechanics and related bioengineering topics. London: Pergamon Press, pp 369–380

Pedotti A, Krishnan VV, Stark L (1978) Optimization of muscle-force sequencing in human locomotion. Math Biosci 38:57–76

Petrofsky JS, Glaser RM, Phillips CA (1982) Evaluation of the amplitude and frequency components of the surface EMG as an index of muscle fatigue. Ergonomics 25:213–223

Pierrynowski MR (1982) A physiological model for the solution of individual muscle forces during normal human walking. PhD Thesis SFU

Pierrynowski MR, Morrison JB (1985) A physiological model for the evaluation of muscular forces in human locomotion: theoretical aspects. Math Biosci 75:69–101

Reinschmidt C, Van Den Bogert AJ, Lundberg A, Nigg BM, Murphy N, Stacoff A, Stano A (1997) Tibiofemoral and tibiocalcaneal motion during walking: external versus skeletal markers. Gait and Posture 6:98–109

Rugg SG, Gregor RJ, Mandelbaum BR, Chin L (1990) In vivo moment arm calculations at the ankle using magnetic resonance imaging (MRI). J Biomech 23:495–501

Sah RL, Kim YL, Doong J-YH, Grodzinsky AJ, Plaas AHK, Sandy JD (1989) Biosynthetic response of cartilage explants to dynamic compression. J Orthop Res 7:619–636
Savelberg HCM, Herzog W (1997) Prediction of dynamic tendon forces from electromyographic signals: An artificial neural network approach. J Neurosci Methods 78:65–74
Sherif MH, Gregor RJ, Liu LM, Roy RR, Hager CL (1983) Correlation of myoelectric activity and muscle force during selected cat treadmill locomotion. J Biomech 16:691–701
Singerman RJ, Petersen DR, Brown TD (1987) Quantitation of pressure sensitive film using digital image scanning. Experimental Mechanics 27:99–105
Spoor CW, Van Leeuwen JL, Meskers CGM, Titulaer AF, Huson A (1990) Estimation of instantaneous moment arms of lower-leg muscles. J Biomech 23:1247–1259
Tashman S, Dupré K, Goitz H, Lock T, Kolowich P, Flynn M (1995) A digital radiographic system for determining 3D joint kinematics during movement. ASB 249–250 (Abstr)
Tashman S, Kolowich PA, Lock TR, Goitz HT, Radin EL (1997) Dynamic knee instability following ACL reconstruction in dogs. Proceedings of the Orthopaedic Research Society 43:97
Torzilli PA, Arduino JM, Gregory JD, Bansal M (1997) Effect of proteoglycan removal on solute mobility in articular cartilage. J Biomech 30:895–902
Van Den Bogert AJ, Hartman W, Schamhardt HC, Sauren AAHJ (1988) In vivo relationship between force EMG and length change in deep digital flexor muscle of the horse. In: Hollander AP, Huijing PA, Van Ingen Schenau GJ (eds) Biomechanics XI-A. Amsterdam: Free University, pp 68–74
Van Ruijven LJ,Weijs WA (1990) A new model for calculating muscle forces from electromyograms. Eur J Appl Physiol 61:479–485
Walmsley B, Hodgson JA, Burke RE (1978) Forces produced by medial gastrocnemius and soleus muscles during locomotion in freely moving cats. J Neurophysiol 41:1203–1215
Weber W, Weber E (1836) Mechanik der menschlichen Gehwerkzeuge. Göttingen: W Fischer-Verlag
Wilkie DR (1968) Studies in biology. No 11: Muscle. London: Edward Arnold
Wong M, Wuethrich P, Buschmann MD, Eggli P Hunziker E (1997) Chondrocyte biosynthesis correlates with local tissue strain in statically compressed adult articular cartilage. J Orthop Res 15:189–196
Wu JZ, Herzog W, Epstein M (1998) Inserting a Pressensor film into an articular joint changes the contact mechanics of the joint. J Biomech Eng 120:655–659
Xu WS, Butler DL, Stouffer DC, Grood ES, Glos DL (1992) Theoretical analysis of an implantable force transducer for tendon and ligament structures. J Biomech Eng 114:170–177

Detection and classification of synergies in multijoint movement with applications to gait analysis

Christopher D. Mah*

Introduction

Multi-joint movements such as kicking a ball, locomotion, or reaching with the arm force the central nervous system (CNS) to deal with many degrees of freedom, as described by Bernstein (1967). Surplus degrees of freedom arise not just because of flexible task requirements, but also because of redundant musculature, (Kuo 1994; Zajac & Gordon 1989) and redundant means to generate similar or identical motion with different force profiles (Winter 1987). The richness of the possibilities available to the CNS presents a potential difficulty for the formulation of motor control strategies by the CNS, and is a puzzle for motor control theorists who wish to understand and exploit these strategies. In this chapter we explore the practical and theoretical sides of several methods which have been used to simplify, and sometimes understand, complex multi-joint motor behavior.

A central principle in several approaches to multi-joint movement is reduction in the number of variables to be controlled by the CNS. Through learned habits, engineering principles (Bizzi et al. 1991) optimization principles (Englebrecht & Fernandez, 1997) neurophysiologically determined control mechanisms (Mussa-Ivaldi et al. 1994) or through the non-linear properties of muscle and reflexes (Gribble et al. 1998) it is assumed that the CNS is able to reduce the number of variables which must be controlled centrally, from many to few. Such reductions in dimensionality can be detectable in the structure of the data if the CNS command is formulated in reduced variables, and if the relationship between the reduced and full set of motor control variables is known. For example, if hip and knee angles always preserve the same ratio of angular velocities during a part of the stance phase of gait, then a single central command signal could do double duty in controlling both angles during this time. Similarly, if we were to accept that the human arm achieves a desired posture by selecting the equilibrium lengths of muscles, then this simple control signal would also determine the properties of the trajectory from initial to final posture, and thus simplify trajectory control. The idea of a control strategy formulated in reduced variables is conveyed by the term *synergy*.

In this chapter, the term *synergy* refers to the relationship between a reduced control variable and its full-blown consequences in multijoint movement. This terminology is consistent with Bernstein's (1967) formulation of the problem of degrees of freedom. Generally the underlying control variable is elusive, and difficult to measure independently. But whatever their origins, synergies will produce stereotyped behaviors which can be detected under the right conditions. We shall not say much here about the theories which point to the primacy of one control variable or another. We also do not discuss in detail the practical side of kinematic data collection, which is thoroughly covered in other works (Winter 1987; Whittle 1991; Cavanagh et al. 1999). Briefly, the data sets of interest here are usually collected by recording, during movement, the positions of markers fixed to body

*Dept. of Physical Medicine and Rehabilitation, Northwestern University, 345 E. Superior, #1406, Chicago, IL 60611, USA, phone: (312) 908-4109, fax: (312) 908-2208, email: c-mah@nwu.edu

segments and joints and converting these marker positions into joint and segmental angles using standard geometric procedures. Rather, we concentrate on techniques which allow one to classify data sets which are consistent with synergetic forms of control, and distinguish between those exhibiting different patterns of control.

Dimensionality and Data Reduction

The techniques we will use for kinematic analysis reduce the number of variables without loss (or minimal loss) of information (Johnson & Wichern 1988; Harmon 1967; Krzanowski 1988). When this is possible, it is possible because the important part of the data set never really had as many variables as it appeared to have. For example, a fingerprint is a record of the 3-dimensional structure of skin on the fingertip, transferred to a 2-dimensional image. The level of oil on a dipstick is a one-dimensional record of the 3-dimensional shape occupied by the liquid oil in a crankcase. And as we see below, the motions of joint angles during gait is a rather small subset of the possible motions of the lower limb (Mah et al. 1994). In these examples, special features of the dataset make the extra dimensions uninteresting, usually because variation in these directions is small. Fortunately, this is approximately true for many multivariate data sets and the process of finding the important variables can be informative.

The number of variables needed to describe a data set accurately is, for practical purposes, the dimensionality of that data set. Even though this definition seems to depend on specific variables, it does not. For example, a plane lying in 3-dimensional space, like a sheet of paper, might have many different (X, Y) coordinate systems drawn on it, and still remain the same plane. Since the choice of coordinates is arbitrary, facts which depend on a particular choice of coordinate system do not reflect intrinsic properties of the data. If a 3-dimensional data set lies symmetrically on a plane, then it is mainly interesting to know what plane it lies in. The choice of variables on the plane of the data set is not critical, though of course, a sensible choice can make things easier to understand (Harmon 1967).

Principal Component Analysis Made Simple

Principal component analysis (PCA) is a technique which attempts to find the best variables to describe a data set when the objective is to use as few variables as possible. There are many software packages (e.g. SYSTAT, MINITAB) available which can perform PCA, and the computation involved for most analyses of movement will not take much computer time. The correct interpretation of the results of these analyses is not always obvious, however. Since this section is tutorial in nature, we do not give the most general computational recipes, which are available elsewhere (Johnson & Wichern, 1988). Instead we concentrate on developing clear interpretation of results through specialized examples. Figure 17 illustrates a geometrical interpretation of the calculations we will describe. They will also be sketched in algebraic form so that the interested reader can perform them in a scientific calculation package such as MATLAB with the data set of his or her choice.

There are a number of data analytic techniques which attempt to find a set of reduced variables just sufficient to describe a data set. Principal component analysis is the simplest of these methods. While PCA can be applied in many different ways, it always works basically as follows.

A multivariate data set consists of n measurements y_i, $i = 1, n$ repeated N times. For example, y_1 might be the height of a father, and y_2 the adult height of his first son repeated for a number of father and son pairs. In Figure 17, $n = 2$ and $N = 17$ is the number of points in the ellipse. Given a multivariate data set $Y_k = (y_1, y_2, \ldots, y_n)_k$ observed at many points, $k = 1, N$, we try to find a linear combination of the y_i, say $P_1 = \sum_i a_{1i} y_i$ in which

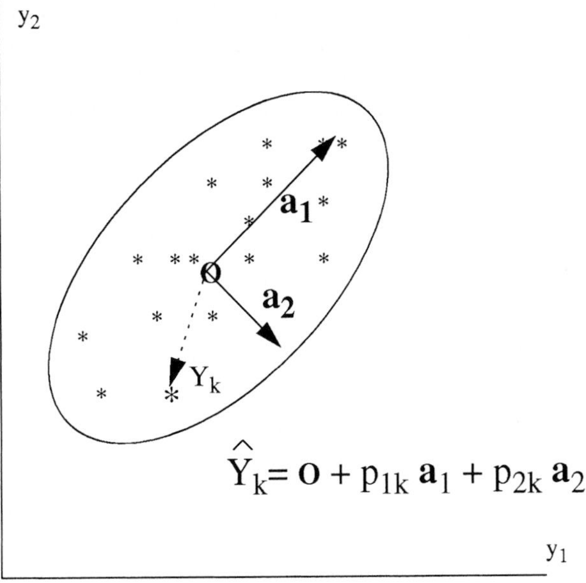

Fig. 17. The geometrical aspects of principal component analysis. The asterisks indicate observations (y_1, y_2), and the solid arrows indicate the principal component directions, $\boldsymbol{a}_1$ and $\boldsymbol{a}_2$, which correspond to the orthogonal major and minor axes of an ellipse. The displayed equation illustrates how each data point Y_k can be approximated by a combination of the principal components where the ^ symbol indicates that the expression $\hat{Y}_k$ is an approximation for Y_k.

the numbers a_{1i} are chosen to make the manufactured variable P_1 have maximum variance. With this choice, P_1, and the associated a_{1i} will define the first principal component of the data. The capital letter P in P_1 indicates a variable which might take any of the values allowed by its definition, while a lower case p will refer to a numerical instance of P. The idea is to find the best single direction vector $\boldsymbol{a} = (a_1, a_2, \ldots, a_n)$ in n-space to summarize all the variation in the data. For simplicity we write a_i instead of a_{1i} for now and remove the k labels from the data points. The squares of the a_i are constrained to add up to 1.0 because otherwise the a_i could be made arbitrarily large and there would be no definite maximum of the variance of P_1. Thus, we want to maximize the expectation:

$$\text{Var}(P_1) = E\left(\left(\sum_i a_i y_i - \bar{P}_1\right)\left(\sum_j a_j y_j - \bar{P}_1\right)\right) = \sum_{ij} a_i a_j \sigma_{ij}, \tag{32.259}$$

with the definition $\sigma_{ij} = E\big((y_i - \bar{y}_i)(y_j - \bar{y}_j)\big)$. The last equals sign in equation (32.259) uses the fact that $\bar{P}_1 = \sum_i a_i \bar{y}_i$ where, for neatness, we have written $\bar{P}_1$ instead of $E(P_1)$ and $\bar{y}_i$ instead of $E(y_i)$. The summation sign $\sum_i$ is a short way to indicate that terms with all possible values of i are to be added up. The reader unfamiliar with expectations may think of $E()$ as the average over the data. For instance if $n = 2$, $\text{Var}(P_1) = a_1^2\sigma_{11} + a_2^2\sigma_{22} + 2a_1a_2\sigma_{12}$. Thus, the function to be maximized is a quadratic function in the unknowns, $a_i, i = 1, n$.

In matrix notation, this is the same as maximizing the quadratic function: $\boldsymbol{a}^{\mathrm{T}}\boldsymbol{R}\boldsymbol{a}$ (where $\boldsymbol{R} = [\sigma_{ij}]$ is the covariance matrix of the data) by choice of $\boldsymbol{a}$, where the transpose operation ^{T} changes a row to a column vector and vice versa. This optimization is a well-studied problem in linear algebra, which is solved in one step by publicly available software programs such as MATLAB. The quadratic function has n local maxima, so that there are n different linear combinations $P_m, m = 1, n$ which achieve these maxima. The n different vectors $\boldsymbol{a}$ are the eigenvectors of $\boldsymbol{R}$. These P_m are the principal components of the data, and the associated $\boldsymbol{a}$ can be used to approximate the data efficiently. In the worst case, all n approximating directions might have to be used. However, each of the P_m, $m = 1, n$ are candidate explanatory variables for the data, and those with smallest variance will be thrown away in order to achieve dimensionality reduction. Conventionally the P_m are ordered by their variances, which are equal to the eigenvalues of $\boldsymbol{R}$. Provided that the variables in Y_k, $k = 1, N$ have the same units of measurement, the details of this ordering are of limited significance, and can easily vary from sample to sample.

Explaining Data

The P_m are explanatory because each data point Y_k in the data set can be reconstructed using one or more of the P_m. In Figure 17 the data point Y_k is approximated using all (both) of the $\boldsymbol{a}_m$. In this case the approximation is exact and $\hat{Y}_k = Y_k$. To achieve dimensionality reduction, just one P_m would be used, however. If we approximate or reconstruct each data point with one P_m, it is done using the vector equation:

$$\hat{Y}_k = p_{mk}\boldsymbol{a}_m, \qquad k = 1, N, \tag{32.260}$$

where the boldface letter $\boldsymbol{a}_m$ indicates the mth set $(a_1, a_2, \ldots, a_n)_m$ and $p_{1k} = \sum_i a_{1i}y_{ik}$ as in the definition of P_1. Thus, in equation (32.260) we have put the m labels back on the $m = 1, n$ sets of a_i, and p_{mk} and the k labels are restored to the data points.

The best single P_m to use is P_1, $(m = 1)$, which has largest variance. With this approximation (32.260) the expectation of the error $E\left(\sum_i (y_{ik} - p_{1k}a_{1i})^2\right)$ is a minimum. This happens because of a geometrical fact which holds true in any number of dimensions, namely, that the shortest distance from a point to a line or plane is a perpendicular to the line or plane. Suppose that the line corresponds to the direction $\boldsymbol{a}$, and the perpendicular corresponds to the error. Once $\boldsymbol{a}$ is selected, the available variation $v_{\text{tot}}^2 = E\left(\sum_i (y_{ik} - \bar{y}_i)^2\right)$ in the data set is divided into the parts along $\boldsymbol{a}$, v_{a}^2; and the error variation v_{e}^2. Because error and approximation vectors are perpendicular, the pythagorean theorem gives:

$$v_{\text{tot}}^2 = v_{\text{a}}^2 + v_{\text{e}}^2. \tag{32.261}$$

Since v_{tot}^2 is fixed in (32.261) the minimum of v_{e}^2 and the maximum of v_{a}^2 occur for the same choice of $\boldsymbol{a}$. Thus, the equation (32.261) shows that by maximizing the *variance explained*, v_{a}^2 also minimizes the error of approximation. Something similar works for approximations involving more of the P_m's, by constructing the approximation to add up two or more expressions like the right-hand side of (32.260).

For further descriptions of PCA and associated computational algorithms, the reader is referred to the references, especially Johnson and Wichern, (1988). Variations of PCA arise because of different ways to compute the covariance matrix, especially when data are collected over time, or the Y_i are functions defined over space. Further extensions of this technique are described in the next section, but first 3 examples will be given to illustrate some of the potential difficulties in interpreting PCA.

Principal Component Analysis Examples

Like other statistical procedures which summarize data, PCA can be misleading if we do not also look critically at the raw data. For example, it is not helpful to report the mean income of a group of people if the data set contains one or more unrepresentative values. Three examples may help to illustrate some similar difficulties in multivariate analyses.

- **Example 1. Sometimes different numbers represent the same thing.** Since the ordering of the principal components is only determined by variance explained, no definite ordering is possible when two eigenvalues have a similar size. In Figure 17, for example, the ellipse has a definite orientation because it is longer in one direction than the other. If the ellipse degenerates to a circle, then there will be no definite way to choose a_1 and $\boldsymbol{a}_2$. When this happens, random fluctuations can give numbers which fluctuate greatly from sample to sample. For instance, consider a multivariate data set with 3 variables (x, y, z) (not shown in Figure 17) in which 2 of the variables x and z have a circular distribution, and there is little y variation. The principal component analysis might give $\boldsymbol{a}_1 = (1, 0, 1)$ and $\boldsymbol{a}_2 = (-1, 0, 1)$ on one sample and $\boldsymbol{a}_1 = (1, 0, 0)$ and $\boldsymbol{a}_2 = (0, 0, 1)$ on the next sample. Both results are correct, because they both are acceptable coordinate systems for the plane of the data. The reason that the first principal component looks unstable is that the first principal component is not sharply defined and as a result its direction is nearly random.

• **Example 2. Definitions of variables can create artificial relationships**. Suppose Q is the dollar amount in a person's chequing account in thousands of dollars, and H is the last 4 digits of their phone number divided by 1000. We can reasonably suppose that these are uncorrelated. So taking these measures for 100 randomly chosen people, the covariance matrix for Q and H might look like:

$$R_{QH} = \begin{bmatrix} 8 & 0 \\ 0 & 16 \end{bmatrix}. \tag{32.262}$$

Since the off-diagonal coefficients are zero, this matrix correctly represents the lack of correlation between the two variables. The set of pairs (Q, H) lies roughly in an ellipse with axes oriented along x and y axes. If we set up two new variables, $V = Q$ and $W = Q + H$, which contain the same information, then using the usual rules of covariance, $(\mathrm{Var}(X) = \mathrm{Cov}(X, X)$ and $\mathrm{Cov}(Q, H) = 0)$, we find a covariance matrix:

$$R_{VW} = \begin{bmatrix} 8 & 8 \\ 8 & 24 \end{bmatrix} \tag{32.263}$$

which has a different structure. In fact, by doing a principal component analysis on these data, it can be calculated that a special combination of V and W, namely, $P_1 = 0.289V + 0.957W$, is the linear combination of V and W with maximum variance ($= 85\,\%$ of the total). Another way to say this is that 85 % of the variation in the data set is along the direction $(0.289, 0.957)$ in (v, w) space. This implies a strong covariation between V and W, which suggests a synergy like the angular velocity covariation described in the Introduction. However, because of the way these data were constructed, we know the relationship is only an artifactual result of how the variables V and W were defined.

• **Example 3. The tent effect: a thought experiment**. Consider a fictitious experiment in which 3 subjects each perform a tracking task in which they track a target on a flat surface by pointing with a finger. Wrist and finger motion is not permitted. Shoulder and elbow angles are recorded, including elbow flexion-extension, shoulder flexion-extension and shoulder adduction-abduction, for a total of 3 variables. In this situation there is a non-unique relationship between hand position and arm position. Therefore, the subject's arm motion is not completely prescribed by the tracking task. When subjects perform the task, suppose it is found that each uses only 1 of the possible degrees of freedom in joint angle space but that each subject makes a different choice of posture or strategy. The overall data set has all three dimensions, but for purposes of this example, each subject uses only one. Principal component analysis with all subject data pooled would give 3 principal components (PCs) and little or no data reduction, while the corresponding analysis on each subject could give one PC and considerable data reduction. With three one-dimensional data objects (the data for each subject), the overall data set can acquire three dimensions, just as tent poles allow a tent to enclose a 3-dimensional volume.

In practice, it would not be so easy to identify different parts of the variability. Principal component analysis is more useful for data reduction than for answering theoretical questions about dimensionality.

Application to Gait Analysis

In work with colleagues at the Neuroscience Research Group at the University of Calgary (Mah et al., 1994, 1996, 1999) we have applied a technique derived from principal component analysis of kinematic data to the problem of documenting and describing changes in gait over time and across movement conditions. The aims of this work were largely practical. That is, we sought to detect distributed changes in gait in multivariate recordings and document them over time. Neurophysiological and biomechanical explanation was not the aim of these analyses. However, as will be seen, the simplicity afforded by this form of analysis can be helpful in forming explanatory hypotheses, because the analysis can reveal patterns in the multivariate data which are not obvious from other approaches.

This form of PCA can be applied to any situation in which multiple variables are collected as a function of time. However, because it makes certain assumptions about the data, it is most appropriate when synergetic forms of multijoint control can reasonably be expected. Since the approach has been documented in detail in several publications, the presentation will be concise, and we concentrate on interpretation of results of the method.

A data series $Y_k(t_j) = (y_1(t_j), y_2(t_j), \ldots, y_n(t_j))_k$ is observed at time points $t_j, j = 1, N$ and where $0 < t_j < T$ within each gait cycle, and the $y_i(t)$ are joint and segmental angle traces recorded over $k = 1, M$ isolated gait cycles. For example, in the gait data to which the method was first applied, the input data were trunk, hip, knee and foot angles each projected into sagittal and frontal planes, for a total of $n = 8$ variables. The covariance matrix of the $y_i(t)$ is computed as $\left[\sigma_{ij}\right]$ as in (32.259). Typically when the objective is to study individual differences by comparison with a standard, the input to the PCA is a data set consisting of multiple cycles for a single subject collected under uniform conditions. However, this computation is also possible over a single cycle, multiple cycles for the same subject, or indeed, over a pool of gait cycles for several subjects. When studying a more diverse collection of gait cycles, for example, gait cycles collected over a long period of time, the original input to the PCA consists of all the gait cycles collected. Accuracy of approximation with a fixed number of principal components will diminish with the diversity of the input data, because the approximation of more complex data is more demanding.

It is assumed that the data series consists of a low-dimensional sum of synergies $\sum_i q_i(t)W_i$ where the W_i are L fixed vectors, and the $q_i(t)$, $i = 1, L$ are arbitrary scalar functions of time. This assumes that movement can be decomposed into hypothetical synergies in which angular velocities have fixed ratios. Thus, for example, if the data are hip, knee and ankle angle, one possible choice of $W_1 = (1.0, 2.0, 3.0)^{\mathrm{T}}$ implies knee angular velocity is twice that of the hip, and the ankle angular velocity is 3 times that of the hip. For the analysis to work, a few of the W_i directions must dominate the data. We do not give a detailed justification of this assumption here (see Mah et al. 1994 for additional references). However, our assumption is often satisfied, as illustrated in example 4 below, and when it is wrong this quickly becomes evident from the data.

The PCA tries to approximate the sum $\sum_i q_i(t)W_i$ as $\sum_m p_m(t)V_m$, where the V_m are the principal component vectors corresponding to the $\boldsymbol{a}$ in the previous section. We use a new letter to emphasize that the V_m now represent known synergies used to reconstruct movement. In the approximation, the V_m are fixed and the profiles $p_m(t)$ (called principal component scores) represent the time evolution of the movement. As the $p_m(t)$ vary, the joint and segmental angles associated with each component of V_m maintain constant ratios of angular velocities. Since a small number of the terms $L < n$, are typically sufficient to approximate gait data with small error, this gives a simplified description of gait, and possibly other kinds of movement data. The reason that the V_m and W_i are not the same is that PCA can discover the plane of the data, but not the original coordinate system W_i on this plane.

• **Example 4. Computation of movement PCs.** Table 1 gives angular data for an average of 20 gait cycles of a single 70-year-old male subject walking at a self-selected speed. The columns are trunk angle with the vertical, hip and knee angles, and foot angles with horizontal, projected into sagittal (even-numbered columns) and frontal (odd-numbered columns) planes. For compactness in this example, the measurements are given at only 21 equally spaced time points during the gait cycle. The reader may verify that the 8 eigenvalues (in descending order) of the resulting covariance matrix are approximately: $(1.2325, 0.2426, 0.0111, 0.0022, 0.0007, 0.0002, 0.0002, 0.0) \times 1000$, and the principal component vectors V_m are as given in Table 2. Since eigenvalues correspond to variances explained, it can be calculated that the first two PCs explain more than 99 % of the variance in this data set.

Table 32.6. Sample angle data.

$y_1(t)$	$y_2(t)$	$y_3(t)$	$y_4(t)$	$y_5(t)$	$y_6(t)$	$y_7(t)$	$y_8(t)$
1.35	3.33	31.01	6.89	11.62	6.97	4.19	11.52
0.89	3.49	31.22	7.33	16.57	6.50	9.56	12.91
0.25	3.72	31.37	7.82	26.17	7.08	21.25	12.22
0.18	3.87	29.38	7.59	32.41	6.47	26.30	12.83
0.76	3.83	24.74	7.21	32.33	5.26	27.09	14.00
1.89	3.48	19.32	7.58	29.00	5.12	27.88	14.07
2.91	3.12	13.21	8.25	23.82	5.55	28.82	14.20
3.68	3.27	6.97	8.21	18.87	5.86	29.80	15.01
4.44	3.34	1.06	8.22	15.14	5.99	31.24	15.83
5.34	3.36	−3.23	8.04	14.22	5.89	34.10	15.85
6.01	3.38	−5.80	7.44	17.40	5.55	40.02	15.10
6.08	3.30	−6.42	6.54	24.60	4.79	50.86	12.81
5.32	3.36	−4.27	4.20	37.28	2.72	73.35	7.74
4.23	3.65	2.25	1.40	53.87	0.57	98.97	5.16
2.95	3.97	10.52	−0.13	67.69	−0.05	104.63	6.68
1.56	4.30	18.45	−0.53	72.57	0.36	94.12	8.81
0.50	4.57	24.56	−0.23	68.15	1.59	75.38	9.80
−0.22	4.21	28.34	1.65	56.83	3.73	54.01	9.66
−0.55	3.55	29.49	4.29	39.70	6.01	32.99	10.94
−0.54	3.59	28.47	4.85	20.07	6.75	14.17	10.60
−0.43	4.04	27.94	4.01	7.49	5.19	1.48	10.84

Distortion Analysis

In a second step, called *distortion analysis*, the multivariate profile $P(t) = (p_1(t), p_2(t), \ldots, p_L(t))$, where $0 < t < T$ is studied for each trial or condition. The meaning of the capital letter P is expanded here to cover the multivariate version of the linear combinations defining the principal component calculation. As noted above, each individual trial or condition can be approximated using the truncated vector sum:

$$\hat{Y}(t) = \sum_{m}^{L} p_m(t) V_m \tag{32.264}$$

Table 32.7. Principal components V_m.

V_1	V_2	V_3
−0.0159	−0.1357	0.0470
−0.0051	0.0162	0.0046
0.1319	0.8252	−0.3953
0.0689	−0.0755	0.2179
−0.5027	0.5081	0.6175
0.0597	−0.0049	0.1017
−0.8474	−0.1811	−0.3685
0.0579	−0.0601	0.5163

When the technique works well, $L \leqslant 3$ is small and this profile can be displayed graphically, as a kind of movement signature as shown in Figure 18 below (Mah et al. 1999). This plot is similar to plotting angles against angles but because the $p_m(t)$ are chosen to optimally summarize the data, it is a more complete record of the movement.

Plotting the $p_m(t)$ in pairs against each other is a useful exploratory technique, and can often throw subtle timing changes into relief (Mah et al. 1999). This technique remains subjective, however, and we have supplemented it with a quantitative analysis. Distortion analysis is similar to Generalized Procrustes Analysis (Gower 1975; Haggard et al. 1995). It is a further compression of the data obtained by referring each unit of data (gait cycle) to a standard one, by means of a linear transformation. The premise of this analysis is similar to that of computer programs which can graphically approximate faces. While all faces are different, most have eyes, noses, cheeks, lips and ears; and a reasonable approximation of any face can be achieved by taking a standard face, and slightly rescaling the sizes and shapes of the standard features. The final description then consists of the standard face together with the numerical scale factors applied to different features. While there are some inaccuracies which result from this approximation, the procedure also yields a highly effective compression of the data. Given 100 faces, we need only store the single standard face, and the scale factors for each unit; not the full description of each face. An analogous procedure can be applied to the multivariate profiles of the $p_m(t)$ in our distortion analysis of gait data to obtain a drastically simplified description of each gait cycle.

Equation (32.265) gives the relationship between the standard profile P_{standard}, and the approximating profile $\hat{P}_k(t)$ computed using the scale factors in the distortion matrix $\boldsymbol{D}_k$. The approximation is given by the equation:

$$\hat{P}_k(t) = \boldsymbol{D}_k P_{\text{standard}}(t), \qquad k = 1, M; \quad 0 < t < T \tag{32.265}$$

where $P_{\text{standard}}(t)$ is a standard profile, and $\boldsymbol{D}_k$ is a matrix which depends only on the kth sample unit (gait cycle), but not time. The original profile $P_k(t)$ has components $p_{mk}(t) = Y_k(t) \cdot V_m$, where $\cdot$ denotes a dot product, which is a short way to write the summation from the definition of p_{mk} in equation (32.260). The matrix $\boldsymbol{D}_k$ is enough information to approximate $P_k(t)$, but it is much more compact. Each row of the distortion matrix $\boldsymbol{D}_k$ is obtained by a multiple linear regression analysis. In our applications $L = 3$, so that $\boldsymbol{D}$ is 3×3, and it is sufficient for most purposes to consider only 4 of these coefficients. This reduces the description of each gait cycle, which might originally consist of 8 angles sampled at 100 time points, to an approximation involving 4 to 9 coefficients. The approximation (32.265) is more lossy than the first step, (32.264) and can lead to higher levels of error. However, the success or failure of the procedure may be easily and directly verified by examining the percentage of variance explained in each case. This compression is important, not because of storage considerations, but because it is a constructive way to show that the control of gait is much less complex than it appears to be. The reader may refer to Figures 18 and 19 for an illustration of the technique.

Figure 18 shows the association between the multivariate profiles generated by the operation of equation (32.265), the joint and segmental angles, shown as stick figures and the distortion matrix $\boldsymbol{D}$. The data used to generate the principal components and P_{standard} used in Figure 18 are the same as those of example 4 (but with 101 instead of 21 time points in the sample). The loops pictured represent $(p_1(t), p_2(t))$ from a multivariate profile involving 3 components $(p_1(t), p_2(t), p_3(t))$. When $\boldsymbol{D} = \mathbf{diag}(1, 1, 1)$ (a matrix with ones on the diagonal and zero elsewhere), equation (32.265) generates exactly the profile P_{standard}. The standard profile is shown as a thin dotted line in each plot.

To generate a different version of the gait cycle for each quadrant, the first two entries of the first column of a 3×3 distortion matrix $\boldsymbol{D}$ have been altered. For example at point A, the first column of $\boldsymbol{D}$ is altered from $(1,\ 0,\ 0)^{\mathrm{T}}$ to $(0.8,\ 0.2,\ 0.0)^{\mathrm{T}}$. This causes the plot of the profile (shown as a heavy line) to be rotated counter-clockwise, slightly shortened, and gives the stick figure sequence at the upper left. Similarly, the point B corresponds to

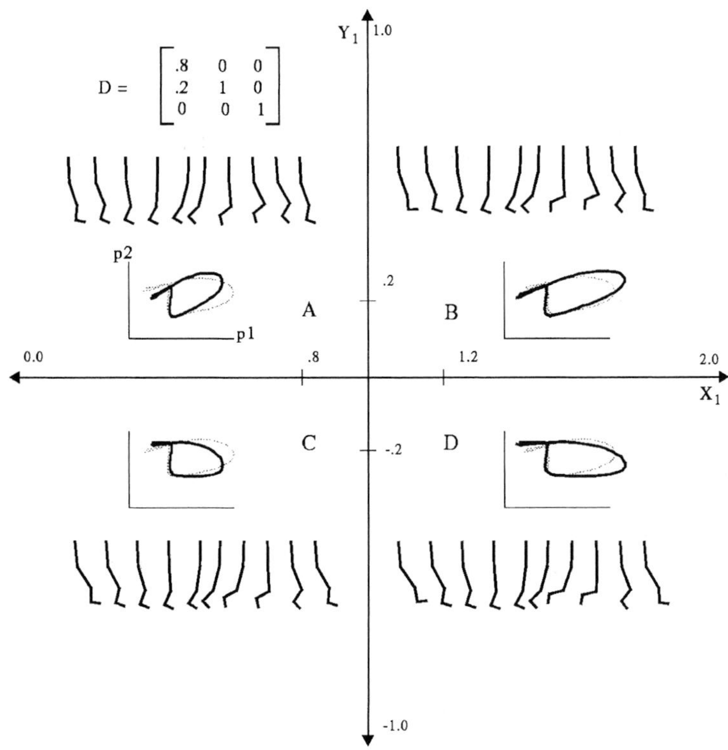

Fig. 18. The four insets show the interpretation of four points (A, B, C, D) in the plane formed by combinations of two entries in the first column of the distortion matrix $\boldsymbol{D}$. Each inset includes a plot of the principal component score $p2$ against the score $p1$ and a stick figure sequence which correspond to the selected value (A, B, C, or D) of $\boldsymbol{D}$. The complete $\boldsymbol{D}$ matrix corresponding to the point A is shown for the inset at upper left. Axis labels are obtained by labelling the rows of $\boldsymbol{D}$ as X, Y, Z, and the columns as $_1$, $_2$ and $_3$. The standard profile is shown as a thin line in each of the plots of $p2$ against $p1$. The data subjected to PCA and used to generate the standard are those of the single male subject described further in the text. Note that stretching and compression of the plot of $p2$ against $p1$ is predominantly in the horizontal direction because the first column of D is associated with a transformation of $p1$.

an alteration of the first column to $(1.2,\ 0.2,\ 0.0)^{\mathrm{T}}$, with a horizontal lengthening of the loop, and the corresponding stick figure shows greater knee flexion during swing phase than the sequence at point A. While these sequences do not approximate any actual data points, this cartoon shows how subtly differing gait cycles might be systematically classified by variation in the coefficients of $\boldsymbol{D}$.

Figure 19 shows a map of gait cycle parameters analogous to that of Figure 18, but the second column of the matrix has been altered from $(0,\ 1,\ 0)^{\mathrm{T}}$, which was its value in $diag(1, 1, 1)$, to the values depicted at the points I, J, K and L. The first column is unchanged. These alterations are twice as large as those in Figure 18 but the resulting changes in the stick figure sequences are roughly comparable in magnitude to those induced by the alterations of $\boldsymbol{D}$ of Figure 18. This is because the profile $p_2(t)$ associated with the non-zero entry on the diagonal in $\mathbf{diag}(0,\ 1,\ 0)$ represents the time evolution of the second principal component. Because the second principal component explains less variance than the first, it needs a larger scale factor to produce an effect of similar magnitude. In other words, separation between points on the scatterplot of the first column of $\boldsymbol{D}$ reveals larger and more obvious changes in gait pattern, while the same differences on the scatterplot of the second column reflect finer changes.

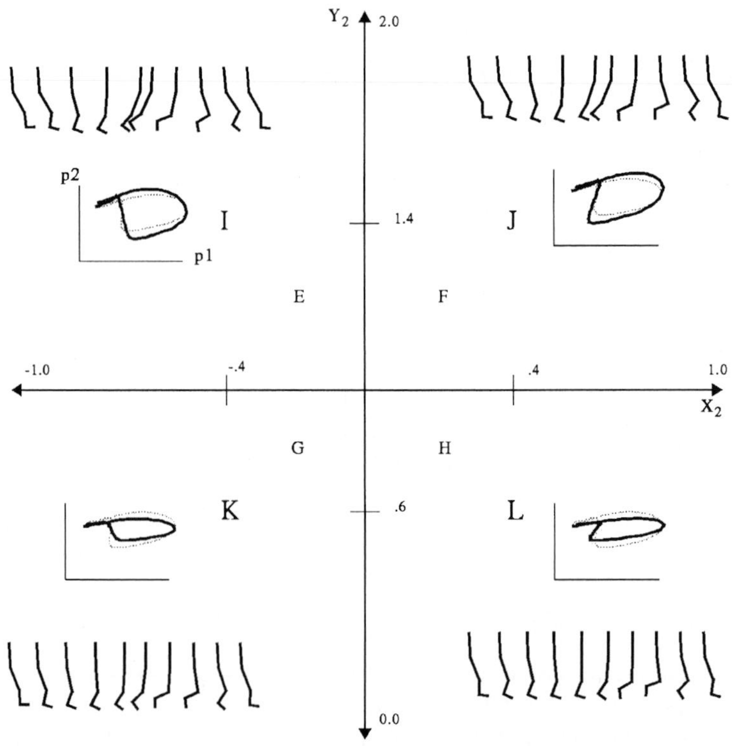

Fig. 19. The four insets show the interpretation of four points (I, J, K, L) in the plane formed by combinations of two entries in the second column of the distortion matrix $\boldsymbol{D}$. The insets follow the format of Figure 18, and the data used to generate these plots come from the same male subject as in Figure 18. Axis labels are obtained by labelling the rows of $\boldsymbol{D}$ as X, Y, Z, and the columns as $_1$, $_2$ and $_3$. Note that stretching and compression of the plot of $p2$ against $p1$ is predominantly in the vertical direction because the second column of $\boldsymbol{D}$ is associated with a transformation of $p2$.

Standards Should Match What They Will Represent

The standard profile $P_{standard}$ in equation (32.265) is usually computed from the average of the same gait cycles used in the PCA, but other choices are possible. It is necessary to make a judicious choice of standard for each application (Mah et al. 1996). For example, in a hypothetical study of facial expression which might parallel a gait study, we might ask whether the standard should be an average face (which represents no specific real face), a designated gender- and age-matched face, or should there be separate standards for smiling and frowning faces? The answer to this question depends on the research question being asked. More accurate representation will be obtained when the standard is close to the units (faces, gait cycles) to be represented. Thus, if the purpose is to document changes in individual behavior, the best standard for each individual is an average of a single condition for that individual. For example, a single condition standard might consist of all gait cycles in a single session conducted a month before a surgical procedure. If the purpose is to compare behavior across subjects, the average of all the individuals or a gender- and age-matched standard might be more useful.

A standard is not an experimental control, because no scientific conclusion will be drawn about the difference between the standard and the units represented using the standard. A good standard is one which gives a good approximation of the data in (32.265) and it does not matter whether the standard was produced under conditions matching the experimental ones. In fact, data from a single subject is likely to work better as a standard than standards which are averages over a heterogeneous collection of gait cycles. While such an average may be more representative, the average is different from any of the units used to compute it.

Simplicity of representation has a cost. There can be a significant loss of accuracy in the distortion approximation. This does not impair the functionality of the distortion method as an exploratory technique, because it remains possible to return to the original data with the insights gained. And like the approximation in PCA, the quality of each distortion approximation may be directly assessed from the data. In a diverse data set no single standard can guarantee accurate representation of all the data, but accurate discrimination often remains possible even when approximation accuracy declines, because the essential features of different conditions continue to be reflected in the $\boldsymbol{D}_k$.

Classification of Gaits

Further refinements to these techniques will simplify interpretation of the results of PCA and distortion analyses; however, what we have described so far has an important practical application to the classification and documentation of gait patterns. Scatterplots of the columns of $\boldsymbol{D}$ can be used without accompanying distortion plots or stick figure sequences as a simple way to document complex differences in gait pattern between conditions. Specifically, if two conditions produce two distinct clusters of distortion coefficient points, it can be concluded that different gait patterns have been induced in the two conditions. This kind of gait classification is illustrated in Figure 20. The scatterplots shown use the same standard as Figures 18 and 19, but the entries in $\boldsymbol{D}$ for individual trials are from a normal subject asked to simulate 3 different pathological gaits (Mah, Chaudry, Liew, Hulliger & O'Callaghan, unpublished data). There were 7 conditions (including normal walking) since the gaits were viewed from both right and left sides, to make a total of 6 pathological conditions. The gaits simulated were as follows (Hoppenfeld 1976; Seidel et al. 1991):

- **Scissoring**. This gait is like walking with an orange between the thighs. The knees are in contact and the thighs cross forward on each other with each step. The steps are short with the ball of the foot dragging across the floor. As a consequence of the thighs crossing each other, the advancing foot moves toward and crosses the midline of the body. The gait is stiff and each leg is advanced slowly.

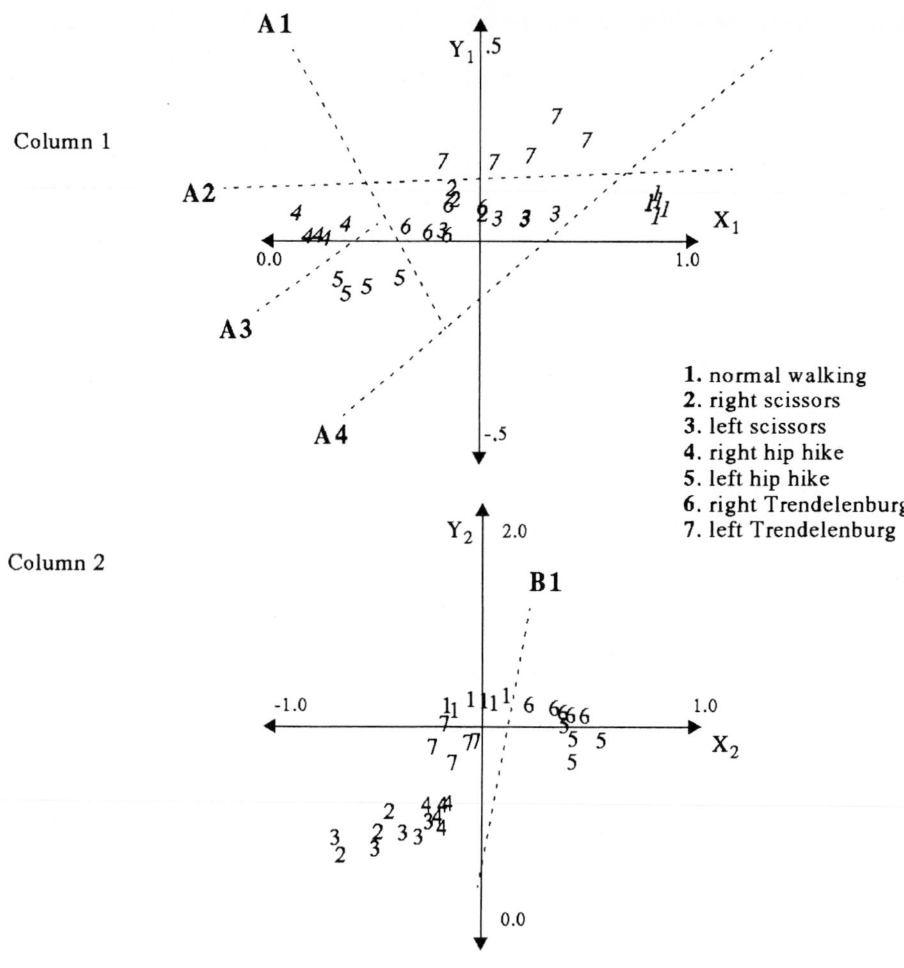

Fig. 20. The scatterplots of (X_1, Y_1) and (X_2, Y_2) from the first and second columns of $\boldsymbol{D}$, respectively, for each trial of a session in which the subject was asked to simulate 3 different pathological gaits. The lines $A1$, $A2$, $A3$, $A4$ and $B1$ were drawn by eye to separate the different gait conditions. The standard used to generate the coefficients of $\boldsymbol{D}$ was that same as the one used in Figures 18 and 19. Axis labels are obtained by labelling the rows of $\boldsymbol{D}$ as X, Y, Z, and the columns as $_1$, $_2$ and $_3$. Samples of gait from each gait simulation condition were viewed from right and left on separate trials.

- **Hip Hiking.** Hip hiking is an abnormal pattern which only occurs on one side of the body (in this case, the right side). Hip hiking occurs on the side of the advancing leg. As the name implies, the hip is lifted as if one were trying to clear an obstacle without bending the leg. The thigh and lower leg are moved forward with the hip while using the muscles of the leg as little as possible. The advancing thigh and lower leg are kept straight. The advancing foot is kept level with the ground and may tilt slightly downward but not upward (as it would normally). As a result of this position, the foot does not land heel first. Instead the front of the foot (toes and the ball of the foot) makes the initial contact as the foot flattens onto the floor.
- **Trendelenburg gait.** This pattern is unilateral (right side) and involves tilting of the hip. The gait is like a "chorus girl swing", involving an obvious side to side movement. When the right leg is supporting the weight of the body, the right hip elevates and the trunk will naturally bend laterally to the right and "fall" over the hip. When this is occurring, the supporting thigh and lower leg are straight.

Figure 20 shows that the coefficients of $\boldsymbol{D}_k$ for each individual trial allow perfect discrimination between all of the above gait conditions. To obtain this discrimination, information from scatterplots of both the first and second columns of the matrix must be combined. Specifically, the lines **A1**, **A2**, **A3** and **A4** separate all conditions except right Trendelenburg from the left and right scissoring conditions. Right and left scissoring cannot be separated, presumably because this subject followed the instruction to produce a symmetric gait. Finally, however, the line **B1** on the second scatterplot separates the right Trendelenburg gait from the scissoring condition. Since discrimination is perfect, no statistics are required here.

While gait conditions are often separated by simple factors, such as angular excursion size, distortion coefficients are also sensitive to relatively subtle changes in timing and intersegmental coordination. Other examples of gait pattern classification, including documentation of recovery of gait function after stroke, may be found in Mah et al. (1994) and Mah et al. (1999).

Mapping Gait Parameter Space

In the previous example, predefined verbal descriptions of the gaits observed were available, and given to the subject as instructions. In a more realistic application, we do not know in advance what is happening during the observed gait and it is helpful if the distortion coefficient variations can be displayed in a standard form with a simple interpretation. For this purpose, we have used *bunch displays* to illustrate the changes in control associated with the transition from one set of coefficients in $\boldsymbol{D}$ to another. These are superimposed stick figure displays which illustrate what happens as the matrix $\boldsymbol{D}$ changes. An example of two such displays is shown in Figure 21. The open-headed arrows respectively show linear changes from point A to D and from point C to B in the plot of the coefficients from the first column of $\boldsymbol{D}$. Each triple sequence of stick figures corresponds to one of the arrows, with the head of the arrow corresponding to the lightly shaded member of the stick figure sequence, the more extreme solid stick sequence to the tail, and the intermediate one to the middle point of the arrow. In the first half of the cycle, stick figures are aligned at the toe, because in stance phase the foot is on the floor, while in the second half of the cycle, stick figures are aligned at the shoulder in order to depict pendular motion of the lower body. This may result in a discontinuity between the 5th and 6th figures. The exact time of heel strike and toe-off may not correspond exactly to these divisions but experience suggests that this choice makes the displays easier to read.

The set of all possible coefficients of $\boldsymbol{D}$ will be referred to as *parameter space*. When variation in parameter space is due to a control mechanism which causes covariation across gait cycles, a bunch display can help to clarify its mechanism. One example is illustrated by the bunch display for the parameter space displacement CB in Figure 21. In this sequence, the head of the arrow (light sticks) undergoes less rotation over the ankle during stance, while the tail (extreme solid sticks) undergoes more rotation. During swing, the solid sticks start out behind the light sticks and end up ahead of the light sticks at the end of swing, indicating a swing phase with larger angular excursions. Such a covariance might occur because the base of support must be moved forward more to remain under the body's center of gravity and avoid a fall when there is a greater rate of forward rotation over the ankle during stance. This pattern of control would not be observed on any single gait cycle, since only one particular amount of forward rotation during stance occurs on a single cycle. When a number of different gait cycles are examined, however, the covariance of these quantities can demonstrate the control mechanism. It is reasonable to expect that this variation is correlated with walking speed.

The stick figures in this display are not a literal representation of the gait cycle. Body segment lengths are standardized, time is normalized, and the mean values of the joint

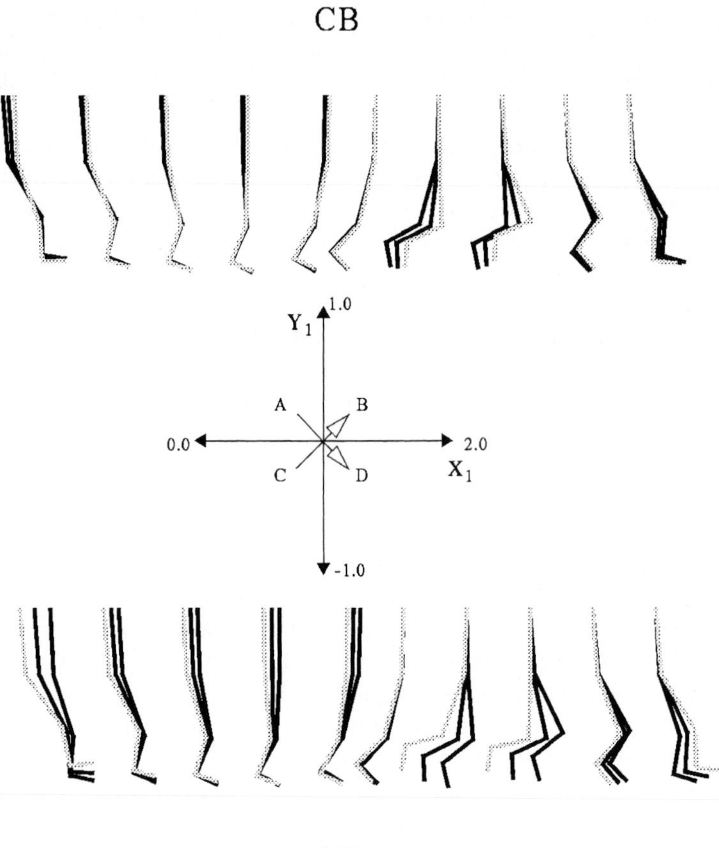

Fig. 21. Two orthogonal bunch displays. The top set of superimposed stick figure sequences corresponds to the arrow *CB* and the bottom set corresponds to the arrow *AD* in the central plot. The central plot of X_1 versus Y_1 is a plot of the first column of the $\boldsymbol{D}$ matrix, with axis labels defined as in Figures 18–20. The light-colored stick figure sequence corresponds to the head of the arrow, the extremely dark-colored sequence to the tail, and the intermediate member to the mid-point of the arrow. The first 5 sticks of each sequence are aligned so that the toes align vertically, as might be expected at heel strike and during stance, and the second 5 sticks are aligned at the shoulder to indicate pendular motion of the body during swing. These displays facilitate the interpretation of gait cycle parameter variation along the directions depicted by the arrows.

angles are forced to match across the different cycles. Not all variation in parameter space can be interpreted because not all bunch displays are physically realizable. Inferences about mechanism come with no guarantee, and need to be verified by other means. However, the methodology described here can document the parametric variation associated with intercycle covariance and can make it easier to guess the reason for it. We have written software with a point and click interface to map (gait) parameter space using bunched stick displays. This software is written in *C* and runs under Linux and Openwindows. While it is not yet publicly available, the reader with some programming skills will find it straightforward to emulate these functions with the information given here.

A second important example of a covariance mechanism is that permissible variation across a large number of diverse gait samples tends to enforce constant length control. That is, variation during stance involves configurations of joint angles preserving the distance between hip and floor contact point, while variation during swing involves configurations of joint angles preserving distance between toe and floor. This presumably arises because it is energy-conserving to keep the center of mass of the body the same distance from the floor during stance, and the toe the same small distance from the floor during swing. This type of variation is illustrated in Figure 22.

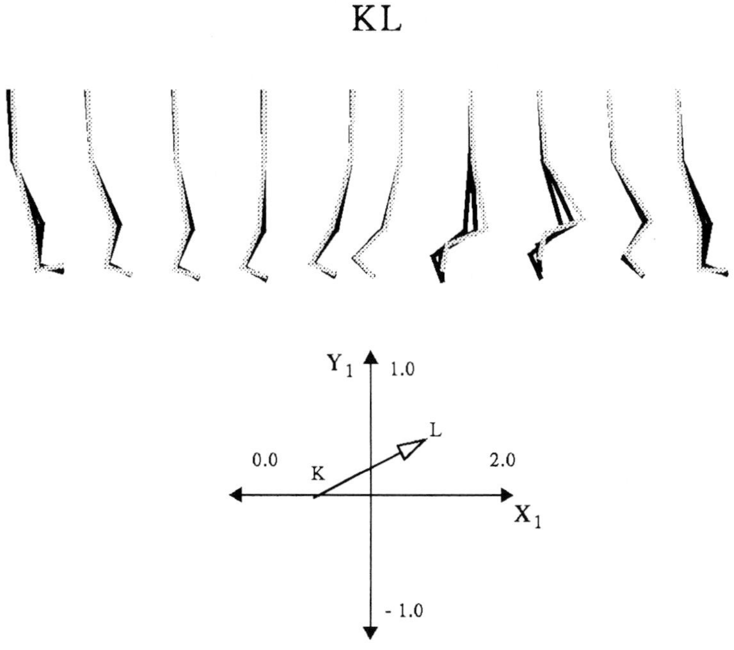

Fig. 22. The parametric variation in the $X_1 - Y_1$ plane corresponding to a control strategy which attempts to hold the distance from hip to toe roughly constant during swing. In addition the distance from hip to point of contact with the floor (toe) is held roughly constant during stance. Length control is accomplished by intercompensation between hip, knee and foot angles. Alignment and labelling conventions follow those used in Figure 20 and Figure 21.

Comparison and Contrast

A practical application of bunch displays is to display the meaning of the separation between clusters of points in parameter space. If there are two discrete clouds of points, this creates two natural directions, namely, the direction defined by the line drawn between the centers of the two clouds and the direction perpendicular to that line. Since the center-to-center direction separates the two conditions, this line represents the *contrast* between the two conditions. The perpendicular direction is variation which plays no role in separation of the two clusters. The perpendicular line is the direction of *comparison*, because this parametric variability represents variability which both gait conditions share. Unlike the original principal components, these directions provide a coordinate system for the parameter space which is defined in functional terms.

Figure 23 demonstrates the contrast and comparison between the left (contralateral) hip hike condition and the right side (ipsilateral) Trendelenburg gait condition. It can be seen that the contrast between the conditions suggests that the hip is somewhat less flexed during swing in the hip hiking condition, but the overall limb configuration variability has the general properties of constant length control. The comparison line, which is the direction both gaits might vary along while retaining their identities, looks like a speed-mediated covariance as described above, but with less participation at the knee than in the hypothetical mechanism of Figure 21.

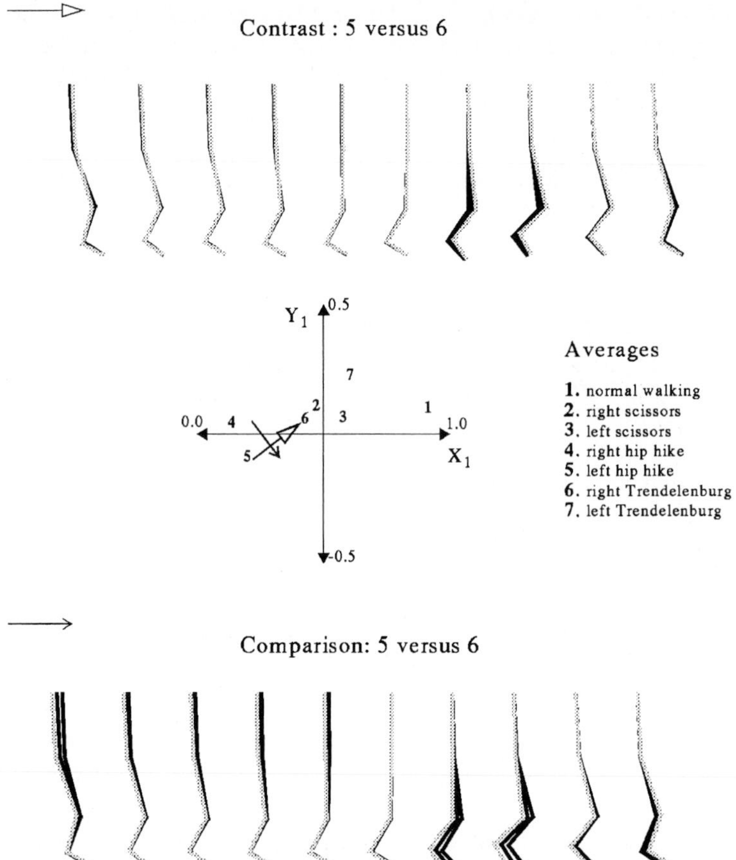

Fig. 23. A comparison line and a contrast line for the gait conditions **5**: left hip hiking and **6**: right Trendelenburg. The contrast arrow (unfilled closed head) is obtained by drawing an arrow between the mean values of the two point clusters (compare Figure 20), and the comparison arrow (open head) is drawn perpendicular to the contrast arrow. It is seen that the bunch display for the contrast line suggests that length control plays a role in separating the conditions. The comparison line display is similar to a putative speed-based covariation mechanism discussed in the text, suggesting that speed variation is not likely to be relevant to the contrast.

When gait data are collected over time, or across the levels of some other experimental variable, it is common to see a trend in the variation of one or more of the gait parameters. Even when the reason for the trend is unknown, the existence of a one-to-many mapping, or candidate synergy is often clear from the data. While displaying the meaning of the trend direction as a bunch display often provides rather clear hypotheses about its mechanism, it remains important to verify any such hypotheses on the original data and by other experimental means (Mah et al. 1999).

Force Fields and the Problem of Degrees of Freedom

Fixed synergies provide an explanation for the fact that stereotyped patterns of multijoint movement which acheive stereotyped results can be organized by the CNS. The most primitive forms of many multijoint movement patterns, such as simple locomotion (Ostry et al., 1991) or even reaching with the arm (Bizzi et al. 1991), can be acheived with relatively simple control strategies. It is more difficult to account for the fact that the motor system not only controls stereotyped actions in complex multijoint structures but is also able to effortlessly modulate and fine-tune this control according to the requirements of a particular situation. Thus the fixed synergies discussed so far are only a partial solution to the full problem of degrees of freedom. The best way to organize movement may shift from situation to situation, and thus the parameters being controlled may also shift. When fixed synergies are shown to account for a particular movement, this gives insight into how control has been simplified, but it does not account for the control itself. To do this, it is helpful to consider force fields.

Bizzi and co-workers, (Bizzi et al. 1984) have observed that the terminal positions of single joint reaching movements in deafferented monkeys were resistant to initial displacements of arm. Since sensory input presumably was not used to correct these displacements caused by the experimenter, it was concluded that feedforward movement control strategy must give appropriate forces for all positions near the terminal position. That is, the basic elements of feedforward motor control are not forces, but force fields, defined over space, and possibly other state variables. The ability of human subjects to compensate for externally imposed force fields (Conditt et al. 1997; Gandolfo et al. 1996; Shadmehr et al. 1994) after a period of learning suggests that these force fields are not hard-wired, but plastic, and modulated by sensory input.

The motor apparatus is a dynamical system which must produce behaviors of many kinds using the forces produced by muscle. Since force fields can produce both synergies, as described in the Introduction, and also behavior which is essentially arbitrary, they represent an important approach to the full problem of degrees of freedom.

Force Fields and Movement

Independent multijoint movement is the result of muscle activations which produce forces. It is natural to ask whether it makes more sense to represent synergies in terms of forces, rather than in terms of movement, or whether the two representations are in some sense equivalent. However, the measurement of forces is more technically difficult than the measurement of movement. Moreover, the non-linear properties of muscle and the recruitment properties of motor units limit the generality of observations on force to situations in which muscle lengths, velocities and force levels are similar to those of the original measurement (Zajac 1989).

Two approaches to force measurements can be taken. First they can be regarded as a supplement to observations of movement. Often, the information contained in force measurements is equivalent to the information in movement, because if movement is a consequence of a dynamical equation

$$f(t) = \varrho(\ddot{q}, \dot{q}, q), \tag{32.266}$$

forces $f(t)$ may always be obtained by algebraically substituting the joint angles q and their derivatives into the equation of motion. There are some situations where this is problematic, such as when the dynamics are unknown, or when a closed chain system creates ambiguity, as in the double support phase of gait.

Force Fields as Control Laws

A second approach considers the forces as the expression of a control law. In this case the forces (torques) $f(q, \dot{q}, t)$ in (32.266) are definite, unknown functions of $q(t)$, $\dot{q}(t)$ and t (and possibly also delayed versions of q and $\dot{q}$). We refer to such functions, which specify the control forces, as *force fields*. This approach is especially interesting for neural control of movement since it can address questions of dynamic stability of movement, and because such fields may have a physiological basis (D'Avella & Bizzi 1998). However, it imposes a daunting task on the experimenter, who must now determine not only the forces during movement, but also the functional dependence of these forces on unknown variables.

An example may help to make this clear. Consider a pendulum which consists of a mass m attached to a massless beam of length l with forcing function $F(x,t)$, where x is the lateral deviation of the pendulum from the vertical equilibrium position. Then the linearized equation of this pendulum is:

$$m\ddot{x} + \alpha x = F(x,t) \tag{32.267}$$

with $\alpha = mg/l$ where g is the acceleration due to gravity, $\ddot{x}$ is the lateral acceleration of the mass and $F(x,t)$ is a force input which may depend on space or time. Suppose there is no time dependence in F, and set it equal to $F(x) = \alpha x + 2m\sqrt{1-x^2} - 4mx\,\mathrm{Arcsin}(x)$. This $F(x)$ is a special force field which depends on the space coordinate x, chosen to give an exact solution for this example. With this $F(x)$, direct substitution shows that one possible solution of (32.267) is $x(t) = \sin(t^2)$ (to verify this, differentiate the latter twice and substitute back into the equation with the given $F(x)$ to verify that it is true). Suppose we observed the motion of the pendulum for this F with initial conditions giving this solution. The motion $x(t) = \sin(t^2)$ can be recorded but this is not sufficient to identify the force field acting, because another choice of F, in particular $F(x,t) = \alpha x + 2m\cos(t^2) - 4mt^2\sin(t^2)$, implies exactly the same motion, as do $F(x,t) = \alpha\sin(t^2) + 2m\cos(t^2) - 4mt^2x$ and many other functions which give the same force as a function of time as this solution evolves. A force field which is a function of time cannot be distinguished from a function of other variables by observing a single motion history.

To identify force fields, it is necessary to observe a big enough sample of the dynamics to distinguish between the possible functional forms of F. Usually this means introducing experimental perturbations which explore different combinations of time, velocity, and space dependence in F as the system evolves, and assuming that the perturbations do not actually change the form of F (Gomi & Kawato, 1996; Tsuji et al. 1995).

The contribution of force fields to observed kinematic synergies and their role in the organization of movement continue to be the subject of current research (Gandolfo et al. 1996; Gomi & Kawato, 1996; Mussa-Ivaldi & Mah, 1998).

Acknowledgements

Preparation of this manuscript was supported in part by a grant from the Canadian Networks of Centers of Excellence. I also gratefully acknowledge the support of Dr. F.A. Mussa-Ivaldi, and the contributions of Jennifer Stevens, Jonathan Dingwell and James Patton, who critically reviewed earlier versions of the manuscript.

References

Bernstein N (1967) The coordination and regulation of movement. Pergamon Press, London

Bizzi E, Accornero N, Chapple W, Hogan N (1984) Posture control and trajectory formation during arm movement. J Neurosci 4: 2738–2744

Bizzi E, Mussa-Ivaldi FA, Giszter SF (1991) Computations underlying the execution of movement: A biological perspective. Science 253: 287–291

Cavanagh PR, Dingwell JB (1999) Gait analysis and foot pressure studies. In: Myerson MS (ed) WB Saunders Co., Philadelphia, In Press.

Conditt MA, Gandolfo F, Mussa-Ivaldi FA (1997) The motor system does not learn the dynamics of the arm by rote memorization of past experience. J Neurophysiol 78: 554–560

D'Avella A, Bizzi E (1998) Low dimensionality of supraspinally induced force fields. Proc Natl Acad Sci USA 95: 7711–7714

Englebrecht SE, Fernandez JP (1997) Invariant characteristics of horizontal-plane minimum-torque-change movements with one mechanical degree of freedom. Biol Cyb 76: 321–329

Gandolfo F, Mussa-Ivaldi FA, Bizzi E (1996) Motor learning by field approximation. Proc Natl Acad Sci USA 93: 3843–3846

Gomi H, Kawato M (1996) Equilibrium-point control hypothesis examined by measured arm stiffness during multijoint movement. Science 272: 117–120

Gower JC (1975) Generalized procrustes analysis. Psychometrika 40: 33–51

Gribble PL, Ostry DJ, Sanguinetti V, Laboissiere R (1998) Are complex control signals required for human arm movement? J Neurophysiol 79: 1409–1424

Harmon HH (1967) Modern factor analysis. The University of Chicago Press, Chicago

Haggard P, Hutchinson K, Stein J (1995) Patterns of coordinated multi-joint movement. Exp Bn Res 107: 254–266

Hoppenfeld S (1976) Physical Examination of the spine and extremities. Appleton-Century-Crofts, Norwalk, Connecticut

Johnson RA, Wichern DW (1988) Applied multivariate statistical analysis. Prentice-Hall, Englewood Cliffs, NJ

Krzanowski WJ (1988) Principles of multivariate analysis: a user's perspective. Oxford University Press, Oxford

Kuo AD (1994) A mechanical analysis of force distribution between redundant, multiple degree-of-freedom actuators in the human: Implications for the central nervous system. Human Movement Science 13: 635–663

Mah CD, Mussa-Ivaldi FA (1998) Do delayed velocity-dependent forces contribute to voluntary movement? Soc Neurosci Abs, Proc 28th annual meeting. Los Angeles

Mah CD, Hulliger M, Lee RG, O'Callaghan I (1994) Quantitative analysis of human movement synergies: constructive pattern analysis for gait. J Motor Beh 26: 83–102

Mah CD, Hulliger M, Lee RG, O'Callaghan I (1996) Quantitative analysis techniques for human movement: finding multivariate patterns in large data sets. In: Witten M, Vincent DJ (eds). Computational Medicine, Public Health, and Biotechnology: Building a Man in the Machine, Part II, World Scientific Press pp 1056–1069.

Mah CD, Hulliger M, Lee RG, O'Callaghan I (1999) Quantitative kinematics of gait patterns during the recovery period following stroke. In Press.

Mussa-Ivaldi FA, Giszter SF, Bizzi E (1994) Linear combinations of primitives in vertebrate motor control. Proc Natl Acad Sci USA 91: 7534–7538

Ostry DJ, Feldman AG, Flanagan JR (1991) Kinematics and control of frog hindlimb movements. J Neurophysiol 65: 547–562

Seidel HM, Ball JW, Dains JE, Benedict GW (1991) Mosby's Guide to Physical Examination, 2nd Edition. Mosby Year Book, Toronto

Shadmehr R, Mussa-Ivaldi FA, Bizzi E (1994) Adaptive representation of dynamics during learning of a motor task. J Neurosci 14: 3208–3224

Tsuji T, Morasso PG, Goto K, Ito K (1995) Human hand impedance characteristics during maintained posture. Biol Cyb 72: 475–485

Winter DA (1987) Biomechanics and motor control of human gait. University of Waterloo Press, Waterloo Ontario

Whittle M (1991) Gait analysis: an introduction. Butterworth-Heinemann, Oxford

Zajac FE (1989) Muscle and tendon: Properties, models, scaling, and application to biomechanics and motor control. Crit Rev Biomed Eng 17: 359–411

Zajac FE, Gordon ME (1989) Determining muscle's force and action in multi-articular movement. Exercise and Sport Science Reviews 17: 187–231

Chapter 33

Magnetic Stimulation of the Nervous System

PETER H. ELLAWAY, NICHOLAS J. DAVEY and MILOS LJUBISAVLJEVIC

Introduction

In 1985 Barker and his colleagues (Barker et al., 1985a) produced the first practical magnetic stimulator which, when placed over the skull, was capable of exciting neurons within the human brain. The advance for the study of neural function in intact man was enormous in practical terms because the technique was non-invasive, painless and well tolerated by subjects, even children. In contrast, electrical stimulation via an electrode placed on the skull is quite painful and not well tolerated even by highly motivated subjects such as members of a research team. The application of electrical current to the scalp for stimulation of brain cells is limited by the fact that current flow is attenuated by skin and bone. The intensity of surface current needed to achieve excitation of nerve cells at a depth within the brain needs to be so high that it excites small myelinated axons and unmyelinated axons of free nerve endings in the skin of the scalp and the meninges – hence the pain. The rapidly changing magnetic field produced by a brief current pulse in a wire coil placed over the head is not attenuated by tissues of the head. Although there is a rapid decrease in intensity of the magnetic field with distance from the coil, the magnetic stimulators currently available commercially are able to stimulate neurons within the grey matter of the cerebral cortex but do not appear capable of exciting axons deep within the white matter or nuclei below the cortex.

Originally, the design of the magnetic stimulator was intended as a practical alternative to electrical stimulation of peripheral nerves. Present-day magnetic stimulators are quite suitable in this respect and have the distinct advantage that they can be used through clothing. However, it was undoubtedly the discovery that electromagnetic stimulation could excite neurons in the motor cortex (Barker et al., 1985b) that has led to the widespread use of magnetic stimulators in the study of motor control.

Peter H. Ellaway, Charing Cross Hospital, Imperial College School of Medicine, Department of Sensorimotor Systems (Rm 10L09), Division of Neuroscience & Psychological Medicine, Fulham Palace Road, London, W6 8RF, UK (phone: + 44-181-846-7293; fax: + 44-181-846-7338; e-mail: p.ellaway@ic.ac.uk)
Nicholas J. Davey, Charing Cross Hospital, Imperial College School of Medicine, Department of Sensorimotor Systems (Rm 10L09), Division of Neuroscience & Psychological Medicine, Fulham Palace Road, London, W6 8RF, UK (phone: + 44-181-846-7284; fax: + 44-181-846-7338; e-mail: n.davey@ic.ac.uk)
Milos Ljubisavljevic, Institute for Medical Research, Dr Subotica 4, PO Box 721, Belgrade, 11001, Yugoslavia (phone: +381-11-685-788; fax: +381-11-643-691; e-mail: milos@imi.bg.ac.yu)

Subprotocol 1
Apparatus and Mechanisms

Procedure

The general principle of operation of a magnetic stimulator is that a high-intensity, transient electrical current passing through a coil of insulated wire creates a rapidly changing magnetic field in the vicinity of the coil. That magnetic field will induce current flow in any adjacent biological tissue and will depolarize excitable cells such as nerve axons. Commercial magnetic stimulators consist of a device that creates an intense current (several thousand amperes) for a brief period (100 µs – 1 ms) in a wire coil which, in turn, generates a magnetic field strength of a few Tesla. In practice, electrical circuitry charges storage capacitors to a level set by a dial on a front panel. A hand-operated switch, or device-generated pulse, is used to trigger the discharge of the capacitors into the stimulating coil. Coils are encapsulated in a tough plastic that can withstand the recoil of the wire coils as well as insulate the device from the subject.

Standard Magnetic Stimulator and Round Coil

Conventional magnetic stimulators are connected to flat coils of wire, several centimeters in diameter, with an open center. A thyristor may be used as a switch to discharge the storage capacitors and generate a large current in the coil. A thyristor will only conduct in one direction, so the resulting current pulse is monophasic and there is no reversal of the current. Assuming that the initial current flow in the coil is in one direction when the coil is viewed from above (say clockwise), then the current induced in tissues below the coil will flow in the same plane but in an anti-clockwise direction (Fig. 1). This has important practical implications. If such a coil is placed with its center over the vertex of the cranium, an induced current flowing anti-clockwise in the cortex will tend to stimulate neurons in the right motor cortex, typically resulting in muscle twitches of the left hand and arm (Day et al., 1990). If the coil is turned over so that the initial current flow in the coil is anti-clockwise, then the induced current is clockwise and muscles of the right upper limb are stimulated. Muscles of the upper rather than the lower limb are stimulated because the wires of the coil lie over that part of the motor cortex devoted to the arm and hand musculature.

Types of Stimulator

There are several companies manufacturing magnetic stimulators (Cadwell, USA; Dantec, Denmark; Digitimer, UK; The MagStim Co. Ltd, UK; Nihon-Kohden, Japan). They have similar design features and all operate basically in the way described above for the MagStim stimulator. One difference is that a stimulator (e.g. Cadwell) may produce a biphasic rather than a monophasic current pulse. Since the induced current is proportional to the rate of change of the magnetic field, and this change occurs within an order of 100–200 ms for all models, the latency of excitation is unlikely to be substantially different for the different models. Several papers have reviewed operation of the different models (Cohen et al., 1990; Maccabee et al., 1990; Brasil-Neto et al., 1992a). The general conclusion from these studies is that the location and extent of the cortex excited by the device depends more upon the shape and orientation (see below) of the coil than on the type of stimulator.

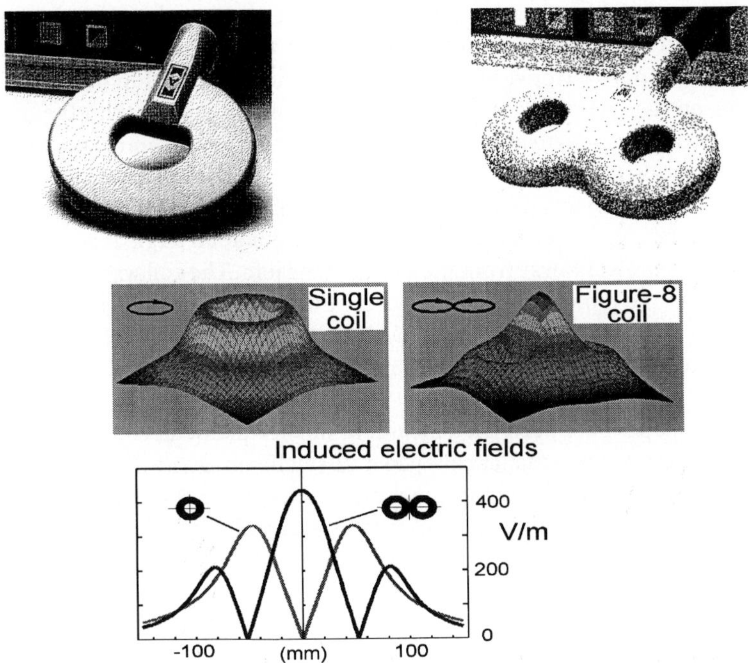

Fig. 1. Magnetic stimulator coils. Circular (*left*) and double circular (*right*) magnetic stimulator coils and representations of their electric field profiles. The coils may be used for transcranial stimulation of the brain and transcutaneous stimulation of peripheral nerves. (After Jalinous 1998. With permission).

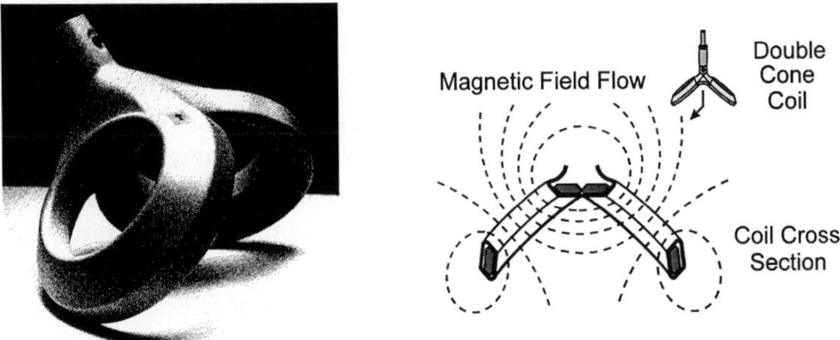

Fig. 2. The double-cone magnetic stimulator coil and a representation of its magnetic field. This coil is particularly useful for stimulating the motor cortical output to lower limb muscles. (After Jalinous 1998. With permission).

Types of Coil

A range of differently shaped coils (Figs. 1 and 2) are now available commercially. The double figure-of-eight coils consist of two windings, with an average diameter of 7 cm each, positioned side by side. The coils are flat, lie in the same plane, and have a handle projecting away from the mid-point, again in the same plane. Current flows in opposite directions in the two coils but effectively in the same direction, and towards the handle in the center. The electric field profile is therefore greater at the center where the coils join than elsewhere under the rim of either coil (Fig. 1). It follows that the induced current is at a maximum under the center join and in a direction away from the handle. The geometry of the figure-of-eight coil makes it particularly useful. Changing the orientation and position of the coil over the cranium, and adjusting the intensity of the output, allows the operator to be more selective with regard to specific targets within the cortex. For further reading on the selectivity of stimulation that can be achieved with the figure-of-eight coil consult Brasil-Neto et al., (1992a), Davey et al., (1994) and Nakamura et al., (1995).

The double-cone coil (Fig. 2) was developed to improve stimulation of muscles in the legs. It also has two windings butted up against each other but at an angle typically of 100 degrees. The handle projects away from the center of the join. The coil was designed specifically to stimulate the motor cortex at the level of the central sulcus. The handle allows the coils to be positioned over the head such that the highest-intensity field is generated in close proximity to the motor cortex representing the muscles of the lower limb.

A number of other shapes have been tried including smaller-diameter coils intended for pediatric purposes. In general, smaller coils generate smaller magnetic fields and are more limited in the range of neural structures that can be stimulated. Small coils do not appear to improve the ability to focus on a target such as part of the motor cortex innervating a particular muscle.

Paired Magnetic Stimulation

A frequent requirement in neurophysiology is the facility to deliver a pair of stimulating pulses to the same location: the condition-test regime. The BiStim module (The MagStim Co. Ltd.) connects two stimulators to the same coil. It allows separate control over the output and timing of the two stimulators. Since only the one coil is used to deliver the pair of stimuli, the orientation of the induced current can be maintained for the two stimuli. The ability to deliver pairs of stimuli to the same location has greatly facilitated the study of interacting facilitatory and inhibitory circuits within the cerebral cortex and the spinal cord (see later). It has also proved useful in the monitoring of cortical motor evoked potentials during spinal cord surgery when the response to a single pulse may be weak or absent and voluntary facilitation is precluded by anesthesia.

Tip: There may be situations when two stimulating coils are required to deliver the condition and test stimuli to different locations over the cranium. Provided there is adequate space to position the coils at the desired locations, and in the preferred orientations, then the coils can be powered by two separate magnetic stimulators triggered by an appropriate external timing device. However, a problem arises if the condition-test interval is reduced to zero, i.e. they are discharged simultaneously. The magnetic fields may interact if the coils are physically close to each other and the power to stimulate cerebral circuits is greatly reduced, giving the erroneous impression of neurophysiological inhibition or disfacilitation. It is adequate to separate the discharges by a minimum of 1 ms to avoid this situation (Ellaway et al., 1998).

Repetitive Stimulators

The rate at which a conventional magnetic stimulator can produce a repeated output depends upon the magnitude of the output. Typically, the output can be set at increments of 1% of the maximum (100%) output. At low settings the minimum interval between consecutive stimuli is of the order of a second or two. The limitation is caused by the time needed to recharge the capacitors after each output. In practice the maximum repetition rate at an output level sufficient to excite the motor cortex is one stimulus every two to three seconds. At the highest levels of output, approaching the maximum, it may be possible to stimulate only every 5–6 seconds.

Clearly, much of brain function is based on the repetitive discharges of neurons. Although the corticospinal output generated by a single magnetic stimulus may consist of an initial direct volley (D wave) followed by multiple indirect components (I waves), these are short-lived (Edgley et al., 1990), lasting less than 10 ms. There is no evidence that the single stimulator produces longer repetitive firing within the nervous system. The requirements of research into CNS function prompted the need for a rapid-rate stimulator. The current rapid-rate stimulators can generate stimuli at rates between 10 and 100 Hz but only for a few seconds. The immediate applications of rapid-rate magnetic stimulation have been in the areas of clinical, cognitive and behavioral research such as investigations into schizophrenia, obsessive compulsive disorder and speech.

Safety Considerations

Magnetic stimulation is non-invasive and has the advantage over electrical stimulation in that no electrodes need be attached to the skin. Thus, there is no possibility of burns or harmful shock from direct flow of electric current from device to human subject. There are, however, two types of safety issue to be considered with magnetic stimulation.

First are the possible dangers during operation of the stimulator and the precautions that can be taken to avoid them. Stimulation of the motor cortex is, of course, likely to produce muscle twitches which may be unexpected by the subject. The subject should therefore be seated or supported. Care should also be taken to ensure that hands or limbs are not likely to be damaged by the movements. Any movement of ferromagnetic items exposed to the transient magnetic field could damage tissue. Thus, all metallic apparel and jewelry etc. should be removed, as should credit and charge cards! Subjects and patients who may have received permanent metal implants or surgical clips in clinical operations should not be subjected to magnetic stimulation. It is not advisable to place the head of the subject in any form of restraint. This might be thought advantageous in order to keep the head still and to avoid any movement relative to the coil. The danger arises if the subject faints. A faint with head restraint could expose the subject to prolonged cerebral anoxia. Finally, the power of commercially available stimulators is insufficient to evoke an ectopic beat of the heart so that precautions do not need to be taken if the stimulator is applied to the chest or abdomen.

Second are the possible long-term effects of magnetic stimulation. There are several studies that have sought to reveal any consequences of exposure to magnetic fields such as may be caused by overhead power lines. Findings, in general, have been equivocal. Calculations by Jalinous (1994) show that the maximum electric field and current density induced by magnetic stimulation are equivalent to those produced by surface electrical stimulation that is not known to produce deleterious effects. The UK guideline for static magnetic fields used in magnetic resonance imaging is 2.5 Tesla. This is comparable to the maximum field strength of a magnetic stimulator (order of 2.0 Tesla), which is thus thought unlikely to be harmful, particularly since it is transient. The sound emitted during the discharge of the stimulator, a sharp click, is due to the expansion of the wire coils within the plastic case. Work in animals has suggested that this could cause hearing loss after repetitive use close to the ear. Barker and Stevens (1991) measured the sound output of a Magstim 200, using full power to a 9 cm round coil, and found it to be 117 dB at a distance of 50 mm from the coil. This is within the guidelines of the U.K. Noise at Work Regulations (1989) provided that fewer than 4000 stimuli are experienced in any one day. In practice, most applications will be at considerably lower power output and will involve far fewer stimuli.

Epilepsy deserves a special comment since transcranial magnetic stimulation would appear to synchronize neural activity at a particular location and might be the source of an epileptic focus. It is now 14 years since the introduction of transcranial magnetic stimulation and, after hundreds of studies with thousands of subjects, there have been no reports of seizures. There does remain concern that subjects prone to epileptic attacks might be susceptible to magnetic stimulation (Homberg and Netz, 1989). Certainly, the risk of rapid-rate stimulation has not been fully evaluated in this respect. The researcher is advised to consult current opinion in the literature if intending to use magnetic stimulation in epileptic subjects.

Subprotocol 2
EMG Recording and Analysis Protocol

Procedure

The motor responses evoked by magnetic stimulation are routinely recorded in a way that does not differ significantly from conventional EMG recording (cf. Chapter 26, 27, 31). Nevertheless, certain issues deserve specific attention.

Motor evoked potential (MEP) responses to magnetic stimulation of the motor cortex may be recorded with surface electrodes in a belly-tendon montage. Electrode-skin impedance should previously be lowered by gently abrading the skin and cleansing with cotton pad and alcohol. Depending on the type of electrodes, conductive gel may be applied. Coaxial needle electrodes should be used when selective recordings from deeply placed muscles are necessary or when recording from hypotrophic muscles in patients. The main advantage is to prevent or minimize cross-talk contamination from adjacent muscles. Needle electrodes are also used when single motor unit recordings are required. For most recordings a standard EMG amplifying-filtering set-up may be used. Ideally, for human use, recording leads should be connected to an EMG amplifier through optical isolators for safety reasons. Electronic filtering may be kept relatively open (i.e. 1–3000 Hz) or narrower (100–2000 Hz) to minimize mains hum (50 or 60 Hz), and to reduce both the size of the magnetic stimulus artifact and any non-biological high-frequency interference. In order to avoid induction of large voltages in the recording leads during stimulation, they should be either twisted together or screened. Stimulus-induced voltages may drive the recording amplifier into saturation, from which it will take tens of milliseconds to recover, thus impairing the recording of the response. Responses may be either recorded on tape, usually video or digital, and analyzed later, in which case an on-line oscilloscope feedback has to be available, or they can be sampled and stored directly on the computer. The sampling rate should be high enough to avoid aliasing (for more details see Chapter 45), usually between 4 and 5 kHz. The responses are best displayed as peri-stimulus epochs so as to identify any pre-stimulus EMG activity. The duration of the post-stimulus analysis time is usually around 50 ms for upper limb and 100 ms for lower limb MEP recording. The duration of pre-stimulus recording time can vary from 50 to 200 ms or even more depending on the paradigm. When a silent period (SP) is to be analyzed, longer post-stimulus epochs should be used, from 500 ms up to 1 s.

From the evoked response several parameters may be measured. The MEP, if of a simple diphasic form, could be expressed as peak-to-peak amplitude between the two peaks of opposite polarity, as shown in the top trace of Fig. 3. Alternatively, the MEP area under the rectified curve can be measured, as shown in the middle trace of Fig. 3. When

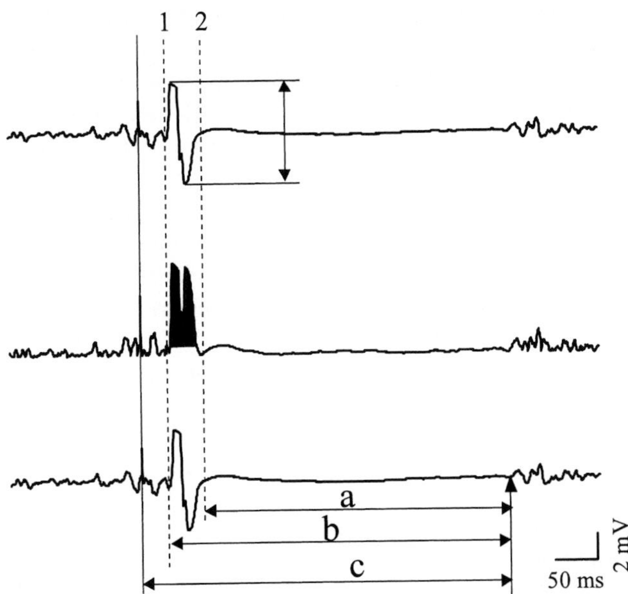

Fig. 3. Electromyographic responses to cortical stimulation. Top and bottom: Unrectified traces. Motor evoked response (MEP) recorded from the first dorsal interosseus muscle during weak voluntary activation. Middle: Full-wave rectified trace. Vertical continuous line at left: time of magnetic stimulus; vertical dashed lines 1 and 2: beginning and end point of MEP, respectively; single vertical arrow below bottom trace: reoccurrence of EMG activity after the silent period (SP). The time from the magnetic stimulus to vertical line 1 corresponds to the latency of the MEP. The time between two vertical dashed lines corresponds to the MEP duration. The vertical double-arrow in the top trace indicates the peak-to-peak amplitude of the MEP. The shaded area under the rectified curve in the middle trace corresponds to the MEP area. Horizontal arrows a, b and c correspond to three different approaches for measuring the time to resumption of the EMG signal following an SP: a, from the end of the MEP; b, from the beginning of the MEP; c, from the magnetic stimulus.

calculating the area, it should be noted that with increasing stimulus strength, the MEP might become polyphasic and prolonged. Such an MEP may contain repeated discharges of some motor units, and this potential complication needs to be recognized. More sophisticated measurements might include the calculation of the ratio between areas of background EMG and MEPs obtained during maximal voluntary contractions. Another approach could be to express the response as a ratio between MEPs obtained with cortical stimulation and evoked responses obtained with supramaximal electrical stimulation of peripheral nerves (the M-wave; cf. Chapter 26). The latency of an MEP is measured as the time from the stimulus onset to the beginning of the motor evoked response. The duration of any silent period (SP) following an MEP may be measured either from the stimulus artifact, from the onset latency of the MEP or from the end point of the MEP, to the restoration of the EMG activity after the MEP, as indicated by the three horizontal arrows below the bottom trace of Fig. 3. None of these three approaches reveals the actual start of any inhibitory process or disfacilitation contributing to the SP, which start may be concurrent with components of the MEP. The choice of measurement for the SP duration is therefore to some extent arbitrary.

The signals are usually evaluated by positioning screen cursors on the depicted points on the recorded traces corresponding to the particular parameter. This kind of measurement is dependent on the experience of the experimenter and any adopted criteria. What is more, the precise delineation of the onset of the response, and therefore the measurement of the latency, might be extremely difficult and arbitrary with concomitant strong voluntary contraction due to excessive amount of background EMG. In this case superimposition of two or more tracings of reproducible morphology could help in correctly determining the exact take-off point of the parameters of interest. Another way to overcome this problem is to average several responses. This may also overcome the inherent variability of MEPs evident with magnetic stimulation (Ellaway et al., 1998).

Discharge properties of single motor units (α-motoneurons) upon magnetic stimulation can be examined in several ways, construction of the post-stimulus time histograms (PSTH) being a widely used method (cf. Chapter 18). In brief, if during activation of a motor unit by gentle voluntary contraction the magnetic stimulus is repeated several times, a cross-correlation between the stimuli and the motor unit action potentials can be performed. Such a cross-correlogram yields information about changes of motoneuronal firing probability related to the stimulus and is therefore regarded as a tool to detect stimulus-induced motoneuronal excitation and inhibition (see Stephens et al. 1976). Any excitation of the motor unit under investigation evokes a peak within a PSTH, whereas an inhibition evokes a trough (Gerstein and Kiang, 1960). Nevertheless, a straightforward association of all peaks with excitation and all troughs with inhibition is not justified (for more information see Kirkwood 1979). To complement the analysis of the "raw" PSTH, an efficient method is provided by the calculation of cumulative sums (CUSUM) derived from the PSTH (Ellaway 1978). The CUSUM is more sensitive for the detection and statistical verification of neuronal excitation and inhibition (Davey et al. 1986). Moreover, the CUSUM is useful for the quantitative comparison of two PSTH waveforms (Miles et al. 1989).

As already described, the responses to magnetic stimulation of the motor cortex can be expressed in several ways. Nevertheless, the selection of the control response or control period for any of the parameters of interest, although dependent on the working hypothesis, should be conducted carefully. Several issues should be considered: the influence of any existing background motor activity on the observed parameter, the operating point on the input-output relation between stimulus intensity and motor response (for more details see Devanne et al. 1997), the detection of cross-talk artifacts in EMG recordings etc. These issues relate to two important questions: What is activated by the brain stimulus, and how and where are those brain processes activated?

Applications

Peripheral Nerve Stimulation

The first human compound muscle action potentials (CMAP) evoked using a magnetic stimulus applied to distal forearm came in 1982 (Polson et al. 1982). Since then magnetic stimulation has been used increasingly for research and clinical investigation of peripheral nerves. Several of its features appear advantageous, challenging the routine electrical stimulation technique. First, it is easy and quick to apply, to the extent that it can be applied through clothing without excessive inconvenience for the subjects and patients. More importantly, the magnetic technique is less painful as compared with electrical stimulation and allows for the depolarization of deep-lying nerves (Jalinous, 1991). Nevertheless, judgments on the practical applicability and reliability of the results vary considerably among researchers and clinicians. Moreover, due to physiological, anatomical and technical factors, several difficulties and restrictions are associated with the application of magnetic stimulation, and these are closely related to each other.

Focality

Direct stimulation of muscles does not occur (Ellaway et al, 1997), and accurate prediction of the site of nerve activation is uncertain (Nilsson et al, 1992). A number of studies have shown that the actual stimulation point for a given coil varies for different subjects, nerves and stimulation sites. The focality of magnetic stimulation is poorer than for electrical stimulation due to the physical properties of magnetic stimulation. A nerve is stimulated preferentially at low-threshold points along its course (Maccabee et al. 1993). Stimulation most likely occurs at locations corresponding to the first spatial derivative of the electric field, i.e. where the electrical field changes most steeply over distance. Since the human body is not a homogeneous conducting volume, excitation takes place where the nerve bends or where it is near regions with decreased field homogeneity. A practical example is the sudden change of the muscle response when the coil is shifted longitudinally while stimulating the median nerve at elbow. Therefore, in order to achieve maximum focality and minimal stimulus strength, the orientation of the coil with regard to targeted excitable tissue must be manipulated. Although the optimum orientation can vary between subjects and different coil architectures, there are some general guidelines. Peripheral nerves are most easily excited distally with a figure-8 coil, when the anterior divergence of the junction of the coil is directed distally. Using a round magnetic coil, peripheral nerves are most easily excited by its tangential-edge, in which case the current is induced parallel to the long axis of the nerve (Maccabee et al., 1991). In practice, the site of stimulation has been determined for a figure-of-eight coil to be 3 cm from its center toward the coil handle. Note, however, that peripheral nerves may be excited at two separate loci, corresponding to the cathode- and anode-like behavior of electrical stimulation, separated from each other by 7–8 cm when stimulating with a large 90 mm coil, and 3–4 cm when stimulating with a small 70 mm coil.

Maximal CMAPs

To evoke a maximum CMAP, the induced electrical current must be directed from proximal to distal, i.e. along the course of the nerve. Even then a maximum CMAP equal to the electrically evoked maximum cannot be elicited at every site with magnetic pulses. Maximum CMAPs can usually be elicited at proximal stimulation sites of the extremities including the radial and femoral nerves as well as from other nerves at the knee. At more distal stimulation points in the upper and lower extremities, magnetic stimulation is generally more time-consuming, mainly because many stimuli are necessary to optimize the coil position, and less effective due to persistent artifacts. For the stimulation of the deep nerves no maximum CMAP can be evoked regardless of the coil used. Another problem in achieving maximal CMAPs is that the working range of the magnetic stimulator is much smaller than for electrical stimulators. The limitation on stimulus intensity may prevent adequate investigation of pathologically altered nerves with magnetic stimulation.

Geometry of Commercially Available Magnetic Coils

Geometry is another limitation of magnetic stimulation when used for peripheral nerve stimulation. From practical and theoretical points of view, a coil should be as small as possible. If the outer diameter of the coil is too large it becomes difficult to keep the distance between the coil and the nerve small enough for effective stimulation. In practice, it is also difficult to achieve flat placement of the coil on the surface of the body, say at Erb's point or at the ankle. Another problem is that spread of the magnetic field can excite other nerves not under investigation. This can be particularly disadvantageous if it induces undesired local muscle contractions and movements of extremities. The situation can be partly improved by reducing the contact area between the coil and the body. For circular coils that means applying only the outer part of the coil. The body of the coil should be angled away from the body surface.

In conclusion, electrical stimulation is generally superior to magnetic for peripheral nerve conduction studies. This is particularly so when detecting focal changes in nerve conduction velocity at common entrapment sites such as in the cross-elbow segment of the ulnar nerve or when testing pathologically affected nerves.

Phrenic Nerve Stimulation

Magnetic stimulation can be successfully employed for bilateral or unilateral stimulation of phrenic nerves. In this case it does have several advantages over electrical stimulation. It is easier to apply and is tolerated better by patients. In brief, for bilateral phrenic nerve activation, a standard, round 90 mm stimulating coil should be positioned and discharged over C6/7 while diaphragmatic muscle action potentials are recorded from 7th and 8th intercostal space. It may be necessary to move the coil up or down the vertebral column in order to achieve the optimal response. For unilateral activation, the magnetic coil should be placed at the side of the neck.

Stimulation of Spinal Cord?

It has not proven possible to stimulate the spinal cord with magnetic pulses. It is thought that the distancing effect of bone and cartilage make it impossible to achieve sufficiently large induced currents to stimulate the spinal cord.

The motor evoked potentials obtained when stimulating over the spinal cord occur due to activation of spinal cord roots. The site of stimulation probably corresponds to the vicinity of the neuro-foramina where the induced electric field is most intense.

Recently a new method of measuring conduction time using magnetic stimulation within the cauda equina has been presented (Maccabee et al., 1996). In brief, ipsilateral CMAPs may be elicited in lower limb muscles, and striated sphincter muscle, by a magnetic coil placed over either the proximal or distal cauda equina. For activation of the proximal part, the junction of a figure-of-eight coil has to be oriented longitudinally so as to induce a cranially directed current. To activate the distal cauda equina, lumbar roots are optimally excited by a horizontally oriented magnetic coil junction. Sacral roots are best excited by a longitudinally oriented magnetic coil junction (see Maccabee et al., 1996). With this method, improved detection and classification of peripheral neuropathies affecting the lower limbs and striated sphincters may be achieved.

In conclusion, magnetic stimulation may be used for stimulating peripheral nerve, but with appropriate caution and a clear understanding of its limitations.

Transcranial Stimulation of the Primary Motor Cortex

Choice of Stimulating Coil and the Motor Homunculus

The representation of muscles within the primary motor cortex has been well documented by Penfield (Penfield and Rasmussen, 1950), and the motor homunculus (see Mapping Studies) can act as a useful guide when attempting to activate specific muscle groups. The choice of stimulating coil, its precise location over the cranium and the orientation of the induced current in the brain are all important factors when trying to maximize the corticospinal volley elicited by magnetic stimulation.

The Motor Evoked Potential – MEP

The most obvious component of the electromyographic muscle response to magnetic stimulation of the motor cortex is the short-latency motor evoked potential (MEP) accompanying an overt muscle twitch (Fig. 4A). The MEP is the result of relatively synchronized descending volleys in a number of pyramidal tract neurons (PTNs) that project to the motoneuron pool of the target muscle. The absolute size of the MEP response will be related both to the size of the descending volleys and the level of excitability of the motoneuron pool affected by other descending and segmental inputs.

The precise site at which a transcranial magnetic pulse activates pyramidal tract neurons (PTNs) is still controversial. Current opinion suggests that PTNs are excited either presynaptically via interneurons or at the axon hillock. Whichever is the case, the size of the descending volley will be related to the level of synaptic input impinging on the

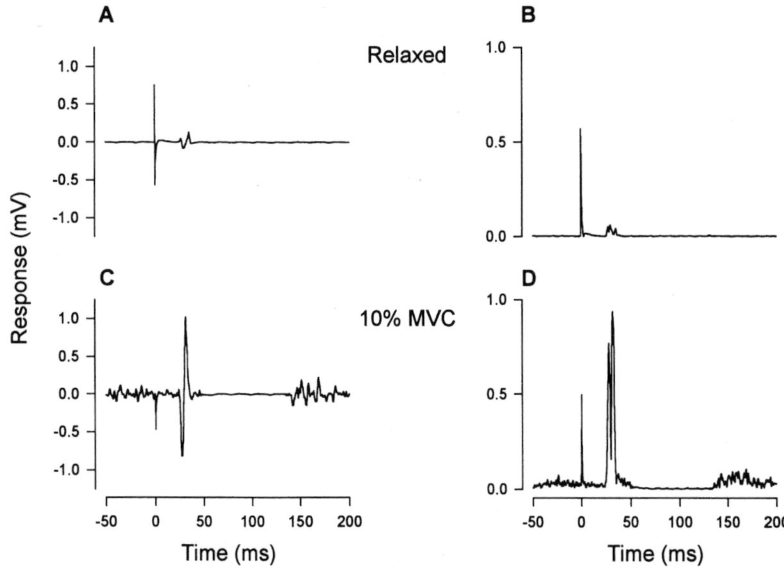

Fig. 4. MEP and facilitation. Electromyographic surface recordings from right thenar muscles of responses to TMS at a strength 60% of maximum stimulator output. A circular coil (9 cm average diameter) was placed over the vertex with an orientation (A-side up) appropriate for stimulation of the left motor cortex. A, C: Responses to single stimuli. B, D: Averages of full-wave rectified responses to six stimuli. The MEP at rest (A, B) is facilitated during a voluntary contraction (C, D). The MEP at a latency of approximately 25 ms is followed by a silent period lasting, on average, about 85 ms.

PTNs. Note that TMS elicits an MEP that is 1–2 ms longer in latency to the stimulus than is the response to transcranial, anodal electrical stimulation. This in itself suggests that TMS might activate pyramidal neurons presynaptically. When TMS is applied at high intensities (twice threshold or above) the MEP response can be seen to jump to the shorter latency seen with electrical stimulation that is thought to excite corticospinal axons directly, deeper in the white matter of the cortex.

Determination of Threshold for the MEP Response

MEP response amplitudes to TMS vary extremely from response to response (Fig. 5). This variability is probably a result of continuously changing excitability of both corticospinal neurons and motoneurons. When estimating the threshold stimulus intensity for evoking an MEP response, the variability presents a problem. Threshold may be determined as the stimulus strength that evokes an MEP response on more than 50% of stimulus presentations (Rossini et al., 1994). Threshold is usually expressed as percentage of the maximal output of the stimulator. It is therefore important to quote the type of stimulator and coil.

In some instances it may be difficult to state categorically whether an MEP response is present in a particular trial. This is especially so if background activity is present in the muscle, which can obscure small stimulus-evoked responses. In these cases the following procedure may be employed. The responses to a number (preferably n (20) of trials at a stimulus strength well above threshold are averaged. The latency and duration of the MEP are determined from the averaged response (see Fig. 5). Returning to lower stimulus strengths, the integrated EMG is measured in each record during the period where the response has been identified at higher stimulus strengths. These integrals can then be compared, using appropriate statistics, with pre-stimulus periods of the same duration. If the mean EMG integral in the expected period is greater than that seen in the pre-stimulus period, we can safely say that the stimulus was above threshold. Note

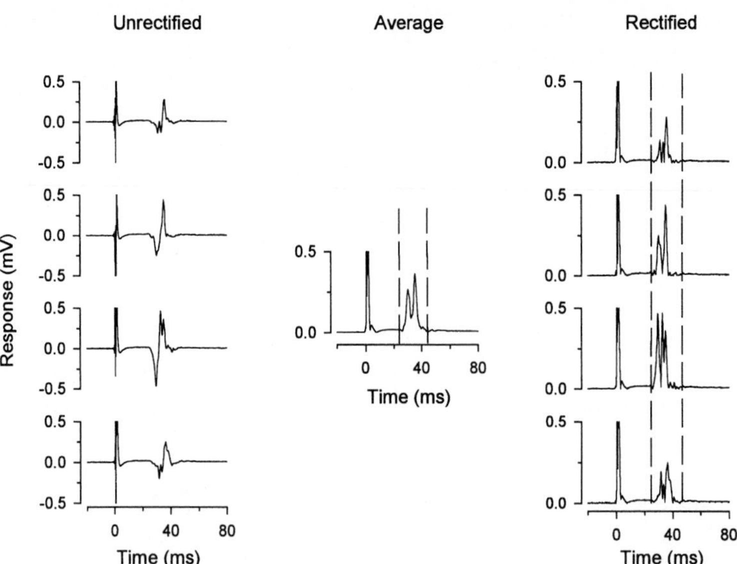

Fig. 5. Variability of MEP amplitude. Protocol for measuring the amplitude of an MEP. Left: Four successive responses of thenar muscles to TMS of the motor cortex at 70% MSO. Middle: Average of fifty full-wave rectified responses. Vertical dashed lines delimit the MEP. Right: The same four responses shown at left but now rectified. Dashed lines as defined in the averaged trace (middle). The size of an individual MEP may be taken either as the mean amplitude or as the area under the "curve" between the limits set by the cursors. (With permission modified from Ellaway et al, 1998.)

that this technique may yield a threshold lower than that defined as eliciting a response on 50 % of trials.

The threshold is influenced by the excitability level prevailing in the corticospinal pathway under test. The principal determinant of excitability in the corticospinal pathway is voluntary drive. When voluntary activity is present in the muscle, the threshold drops, reflecting the increased synaptic excitability. It is important to take this into consideration when assessing threshold, and to ensure that the muscle under study is completely relaxed. This process can be made easier by giving the subject feedback of EMG signals through a loudspeaker.

Determining the Latency of the MEP Response

The latency of the MEP response to magnetic stimulation of the motor cortex is determined by a number of factors. Principal components of the latent period are the conduction delays in the PTNs and the motoneurons. Additional delays will occur in the neuromuscular junction and as action potentials are conducted along muscle fibers to the recording electrode. These factors are all reasonably constant within an individual. Synaptic delay in the cortex and in the motoneuron pool can vary slightly according to the degree of synaptic facilitation present. For this reason it is important to state whether latency has been determined during voluntary contraction or at rest. In order to eliminate small variations in latency from stimulus to stimulus, it is normal practice to average the rectified EMG responses to several stimuli (Fig. 4B,D).

Facilitation as a Useful Tool

If it is necessary to isolate the response of a single muscle, the process of facilitation with voluntary effort can be very useful. The operator asks the subject to make a weak contraction of the muscle under study, so specifically facilitating its response to the TMS compared with other neighboring muscles.

During facilitation the amplitude of the response follows a particular pattern of increase as the level of voluntary effort increases. In hand muscles with a high degree of corticospinal input to their motoneuron pool, most of the facilitation occurs at low forces and is more or less maximum at about 10–15% MVC (Fig. 4). In leg muscles, the pattern of increase is much more gradual, peaking closer to 50% MVC.

Facilitation by voluntary contraction may occur in the motor cortex and/or at the level of the spinal motoneuron. The excitability of the motoneuron pool can be tested in isolation using a number of techniques. It is possible to evoke an H-reflex response (cf. Chapter 26) in some muscles, which may be modulated in amplitude as spinal motoneuron excitability changes. Electrical stimulation of a muscle nerve activates Ia afferent axons projecting from muscle spindles to the motoneuron pool of the same muscle. Differences in the amount of facilitation of the MEP and H-reflex to a particular maneuver can then be assessed. There are a number of drawbacks in assessing what component of any facilitatory input may be attributed to the corticospinal neurons or to spinal motoneurons. Firstly the population of motoneurons involved might be different in the H-reflex from the cortically evoked response. Secondly, any facilitation of the H-reflex might be due to subliminal facilitation of motoneurons by activity in the corticospinal pathway itself. Finally, a limitation of the technique is that H-reflexes are not easily evoked in the muscles of the hand at rest.

As an alternative, it is possible to stimulate corticospinal axons directly, using anodal electrical stimulation of the cortex. Any changes in excitability, e.g. during voluntary activation, will therefore be likely to occur in the spinal motoneurons. Similar problems apply to this method as with the H-reflex in that it is not certain that the same population of PTNs are activated with TMS and electrical brain stimulation. Furthermore, as mentioned above, electrical stimulation of the brain is extremely painful due to activation of cutaneous afferents in the scalp and in the meninges. This alone makes it unsuitable for routine use.

Direct stimulation of descending axons in the spinal cord can also be achieved using electrical stimulation over the cervical vertebrae. Since the target axons are deep within the spine itself, high voltage stimulation is required in order to excite them. As with electrical stimulation of the motor cortex, this procedure is quite painful and not recommended for routine use. Furthermore, the stimulus could be activating axons in any number of different descending tracts (e.g. vestibulospinal, reticulospinal) in addition to the corticospinal pathway.

Triple Stimulation Technique

One problem in interpreting the MEP response to TMS is caused by the variability in latency of response of individual motor units. The individual motor unit action potentials contributing to the MEP are not so tightly synchronized as is the response to peripheral nerve stimulation, the M-wave. Phase cancellation (cf. Chapter 26) occurs, and the MEP is thus smaller than the equivalent M-wave. An ingenious but laborious method has been proposed to enable a more accurate assessment of the response to TMS (Magistris et al., 1998). This triple stimulation technique involves TMS of the cortex and electrical stimulation of the peripheral nerve at two sites. Using a hand muscle as example, TMS applied to the scalp is followed by supramaximal electrical stimulation at the wrist. The delay is set to be just less than the latency of the MEP minus the latency of the M-wave. The action potentials excited by TMS collide in the peripheral nerve with the "antidromic" motor volley. Only those motor axons NOT contributing to MEP continue up towards the spinal cord. A second supramaximal electrical stimulus is applied at Erb's point in the neck. The delay for this third stimulus needs to be calculated according to the stature of the subject. This third "orthodromic" volley collides with the antidromic volley. Now it is only impulses in the motor axons that DID participate in the original MEP and continue on to the muscle. The motor potential evoked in the muscle is equivalent to the original MEP but is now synchronized and can be compared directly with the maximal M-wave. Using this technique it is possible, for example, to show that TMS can excite virtually all motoneurons supplying a target muscle.

The Silent Period and Inhibition

In an active muscle, the MEP is followed by a period of relative or absolute suppression of EMG, the so-called silent period (Fig. 4C,D). Several factors contribute to the silent period. It probably represents elements of inhibition both at the level of the cortex and the motoneuron, but refractoriness of motoneurons and peripheral reflex inputs also contribute.

The EMG of an active muscle can also be suppressed by using weak TMS, at a strength below threshold for the MEP (Fig. 6). In order to identify inhibition of drive to a muscle, the EMG record should be full-wave rectified and then averaged with respect to the magnetic stimulus. An unrectified EMG recording contains both positive and negative signals, which will cancel each other during averaging. After rectification, all components of the signal are positive and will "add up", rather than cancel out, when averaged. Inhibition can be examined at a steady level of contraction by asking the subject to contract to a specific percentage of MVC. Visual feedback of force from a transducer or of integrated EMG makes it easy for the subject to achieve the desired constant level of force. The locus for such inhibition (or disfacilitation) is the same as for the MEP (Davey et al., 1994). The inhibition has a latency several ms longer than that of the MEP and is of variable duration. It is likely to be caused by suppression of cortical output rather than inhibition of motoneurons. Suppression of EMG can also be achieved using TMS applied to the opposite motor cortex, i.e. the side ipsilateral to the target muscle. The mechanism is known to involve transmission through the corpus callosum to the contralateral cortex.

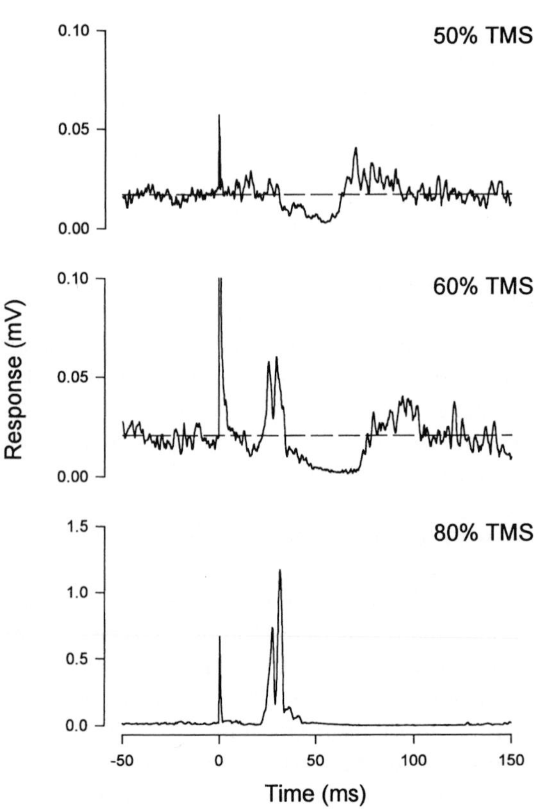

Fig. 6. Inhibition in response to cortical stimulation. Averages of full-wave rectified responses of the right thenar muscles to TMS of the left cortex (as in Fig. 4) during voluntary contraction at a strength of 10% MVC. Top: TMS (50% MSO) below threshold for an MEP suppresses the EMG for a period of 32 ms with a latency of 31 ms. Middle: TMS (60% MSO) close to threshold for an MEP. Bottom: TMS (80% MSO) well above threshold for an MEP (latency 22 ms). There is an increase in duration of the period of suppressed EMG with increasing strength of TMS. Averages of 30 (top), 20 (middle) and 6 (bottom) traces. Note the dissimilar voltage scales in the records.

Late Excitation from TMS

The silent period is often followed by a burst of EMG that is of greater intensity than the level of the background EMG signal. As the relative strength of TMS is altered, the duration of the silent period and the latency of the late EMG burst can vary. One source of the late burst is simply the relative synchronization of motor unit activity as a number of motoneurons resume firing. However, there may be other contributing factors. In certain pathological conditions, such as spinal cord injury, a late burst may be observed in isolation, without a preceding MEP or silent period. It may represent activity in more slowly conducting corticospinal axons or be caused by afferent feedback following activation of other muscles.

Test-Conditioning Using Paired Magnetic Stimulation

In a paired conditioning-test paradigm, a second magnetic stimulus may be used to measure the change in excitability of the corticospinal system at different times after activation with a conditioning stimulus. The conditioning and test stimuli may be delivered through the same stimulating coil using a module (for example BiStim; MagStim Company) to combine the outputs from two magnetic stimulators. The intensity of the test stimulus should be supra-threshold but not maximal to allow identification of both facilitation and inhibition of the test MEP by the conditioning stimulus. TMS at 1.2x threshold for an MEP in the relaxed muscle is appropriate, for example, when investigating responses in hand muscles. The strength of the conditioning stimulus can be altered depending on the characteristics of the response being investigated. The interval between the test and conditioning stimuli is varied and the size of the MEP response (to the test stimulus) is measured at each interval. In this way, the profile of facilitation or inhibition can be determined by plotting MEP amplitude against test-conditioning stimulus interval.

The paired stimulation technique can be used purely as a tool to facilitate MEP responses in muscles that have a weak corticospinal innervation and thus have a high threshold to single pulse stimulation. Axial muscles such as those of the lower back, abdominal area and the diaphragm can conveniently be studied using paired stimulation with a brief (~2 ms) interval between stimuli. The conditioning stimulus is used to increase the excitability of corticospinal neurons supplying the target muscle in order that a test stimulus of lower (and less traumatic) stimulus intensity can be used.

It is, of course, also possible to investigate connections between the motor cortex and other cortical areas by using two independent stimulation coils. For example, trans-callosal inhibition between the motor cortices of the two hemispheres has been investigated by delivering a conditioning stimulus over one hemisphere while testing the excitability of the opposite cortex.

Ipsilateral Corticospinal Innervation and Axial Muscles

Approximately 80% of the PTNs cross side in the medullary pyramids and descend to innervate muscles contralateral to the hemisphere of origin. The other 20% do not cross and descend ipsilaterally. Motoneurons that innervate more proximal limb muscles and axial muscles appear to have some ipsilateral as well as contralateral corticospinal innervation. Abdominal, paraspinal and intercostal muscles and the diaphragm operate bilaterally in executing many of their tasks. These muscles have a relatively small and medially placed corticospinal representation in Penfield's motor homunculus. This location makes their cortical representation difficult to test as the field generated by the magnetic coil may affect both cortices. Targeting one cortex, rather than stimulating both, may be improved by using a figure-of-eight stimulating coil. Stimulation should begin with coil placement a few cm away from the midline. The coil may then be moved progressively towards the mid-line and on over to the opposite hemisphere. Inadvertent stimulation of the opposite cortex, as distinct from genuine ipsilateral responses, can be distinguished as follows. A genuine ipsilateral response will have a higher threshold, a smaller amplitude and a latency 1–2 ms longer than the contralateral response. Any bilateral response due to inadvertent, coincident stimulation of both cortices will be evident as the coil is moved across the cranium. One response will grow as the other diminishes in size.

Mapping Cortical Representation with TMS

Recent advances in a wide range of technologies have made the concept of mapping the human brain, at least at the macroscopic level, a reality. Several rapidly developing methods have become available over the past several years, among them TMS. No matter which method is employed, when assessing brain mapping techniques it is useful to keep in mind the following variables: spatial and temporal resolution, the volume of tissue surveyed for a given measurement, repeatability and the minimum interval between studies. It is also important to assess whether the technique is invasive or not.

As previously mentioned, TMS provides a non-invasive means of stimulating the cortex. With experience it can provide good spatial and temporal resolution. With improved, focal coils, the experimental resolution can be brought down to less than a square centimeter and the rapid rise time and short duration of the magnetic pulse offer millisecond precision. TMS has been used to map the motor cortex, but recent studies have used TMS also to map other cortical regions, including Broca's area, sensory and visual cortices and the supplementary motor area (SMA).

How to Map? The relative simplicity of recording electromyographic muscle responses facilitates mapping of the motor cortex. Two general approaches can be employed for mapping the cortical representation of a muscle (Hallett 1996). The first is to use a pre-assessed standard stimulus magnitude and then move the stimulating coil systematically over the scalp and, at each site, measure the MEP from a muscle of particular interest, as shown in Fig. 7. This will produce a map of MEPs with variable amplitudes, with the highest amplitudes clustered at the center of the map, gradually decreasing towards the edges. In this case, the site corresponding to the maximal amplitude can be called the "optimal position". A second approach involves searching the stimulus intensity at the corresponding sites across the scalp that will produce a threshold response in the muscle. This approach will result in a map of threshold intensities, and in this case the smallest intensity necessary to induce a threshold response will be centered in the middle of the map, increasing towards the edges. Again, an optimal position can be described, corresponding to the site of the lowest intensity. At the moment, there are no specific recommendations as to which method to use, one of the reasons being the lack of systematic comparisons of these two mapping methods. The first method is certainly easier, and the choice could depend on the experimental paradigm.

Whichever method is used, an important task while mapping is to ensure constant and comparable positions of the stimulation points and stimulation coil on the scalp throughout the recording session and between the subjects. The direct approach would be to mark the stimulation spots directly on the skin of the skull with a relatively durable pen usually making a grid of several rows of stimulation spots equally distributed (Fig. 7). Some subject might not accept this for practical reasons, while in others it might not be visible enough. Another approach is to use a snugly fitting elastic cap that should be stretched over the subject's scalp in such a way as to ensure minimal or preferably no movement of the cap and to maximize the comfort of the subject. A square grid could be pre-marked on the cap. Instead of a cap, a flexible, latex-coated nylon grid could be used. It should be applied to the skull and attached at appropriate areas by collodion. Recently, a new method for reproducible coil positioning in TMS mapping was presented (Miranda et al., 1997). It employs a specially designed 3D digitizer that consists of a

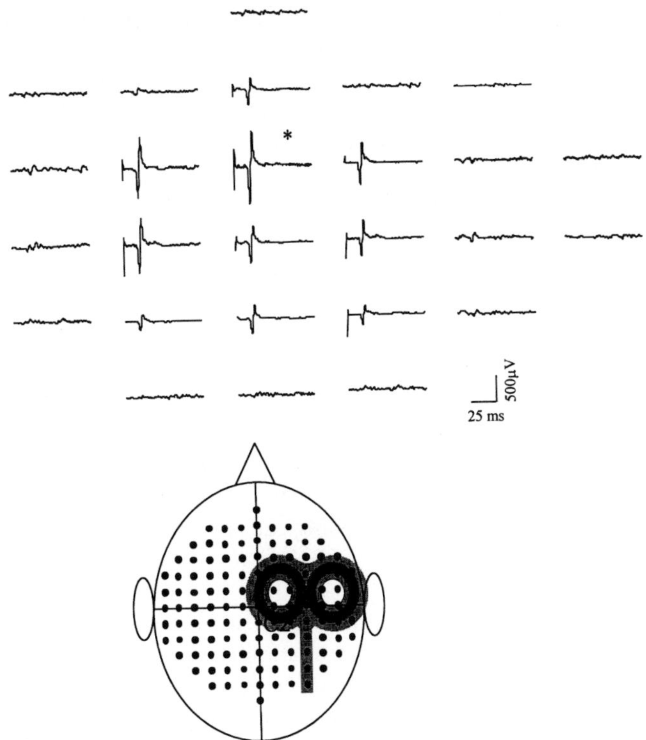

Fig. 7. Cortical mapping. Top: Map of responses recorded from abductor pollicis brevis, during weak voluntary activation, at sites 1 cm apart on the scalp. The response marked with * corresponds to the position 3 cm lateral to the vertex and 1 cm anterior to the inter-auricular line. Each trace is the average of 10 trials of stimulation. Bottom: Schematic representation of the grid on the scalp with a middle position of the wings of figure-of-eight double coil placed at the spot corresponding to the optimal position.

source of low frequency electromagnetic radiation, placed on the cervical collar, and a stylus-shaped sensor glued to the stimulating coil. They are connected to the computer that computes the orientation angles of the stylus and the co-ordinates relative to the default reference frame, enabling reliable and reproducible positioning of the stimulating coil on the scalp.

The distribution and density of the stimulating points on the scalp should be chosen relative to the representational areas of the cortex examined. They should be oriented to the vertex (Cz) as the origin of the grid, or they could be given following frontal and parietal midline (Fz, Pz) and pre-auricular reference locations, according to the international 10–20 system of electrode placement (cf. Chap. 35). Alternatively, a Cartesian reference frame defined by the two tragi, the inion and the nasion, could be used. The density of the stimulating spots is usually 0.5, 1 or 2 cm in sagittal and coronal directions. However, in order to achieve a reliable map in a reasonable period of time, 1-cm spacing is necessary (Brasil-Neto et al. 1992b). At the same time, available coil architecture limits the precision of finding the "optimal position" to less than 0.5 cm. Also, the distance between the spots could be expressed as a percentage of the total distance between the reference points (tragi, inion, nasion) defining the vertex (Cz). In fact this approach could offer higher accuracy and facilitate the comparison of the results obtained with magnetic stimulation with the findings of other mapping techniques (see Chaps. 34–39) as well as between the studies.

Available results indicate that TMS appears to be capable of distinguishing the representations of different upper extremity parts on the scalp in one dimension (Cohen et al. 1990), and when combined with appropriate statistical methods, in two dimensions (see later, Wassermann et al. 1992). Most recent results combining TMS and PET (cf. Chapter 39) have succeeded in relating the scalp map to the underlying cortical topography, showing high accuracy of TMS for locating the primary motor area, at a resolution of around 5 mm (Wassermann et al. 1996). Size, area and volume of TMS maps reflecting muscles have also proven very reproducible over several weeks (Wang et al. 1994; Mortifee et al. 1994). The results obtained show overlapping representations of different muscles, which is consistent with the results of direct cortical stimulation studies in humans and animals.

Apart from excitatory actions, TMS of the cerebral cortex also has inhibitory effects on muscle activity. It has been suggested that the inhibition of ongoing muscle activity arises mainly from the action of the stimulus on different neural structures. Mapping of these structures is performed in a similar way as of excitatory loci. Available data indicate similar representations of excitatory and inhibitory cortical circuits (Davey et al, 1994) or inhibitory areas being slightly larger, encompassing and surrounding the MEP areas (Wilson et al., 1993; Wassermann et al. 1993).

Which Coil to Use and How to Stimulate?

To achieve a stimulation as focal as possible, the design, position and orientation of the stimulating coil should be considered. Two main types of stimulating coils could be used for mapping, the classical circular coil and a figure-of-eight shaped coil (see above). When applied over the vertex, the round coil induces a widespread activation of several muscles and is not nearly as focal as a double figure-of-eight coil. Nevertheless, some caution is necessary with the double coil because there is also peripheral stimulation on either side of the coil winding at approximately half the intensity of that under the center of the coil (Jalinous 1992). On the other hand, for some types of round stimulating coils, the highest density of magnetic field is anterior (Dantec), and when only this part of the coil is kept in contact with the scalp and the stimulus intensity is kept low, it becomes acceptably focal for brain mapping. Another important feature is the type of the current pulses flowing in the magnetic coil. As previously described they can have one or more phases depending on the manufacturer. The current induced in the tissue is proportional to the rate of change of the magnetic field, and therefore always has two or more phases directly depending on the relative duration of the phases. The optimal direction of currents induced in the brain tissue further depends on the orientation of the coil (applicable to coils with monopolar current pulses). For excitation and inhibition of hand and arm muscles, the optimal direction of *induced* current flow is at approximately 45 degrees towards the central sulcus, flowing diagonally from back to front, i.e. antero-medially (see Davey et al., 1994). Careful manipulation of the orientation of the stimulating coil can further improve spatial resolution of cortical mapping and may reveal differences in orientation of spatially overlapping, but functionally different neural assemblies.

Several authors have noticed increased variability in MEP amplitude with magnetic stimulation of sub-optimal scalp positions, e.g. in the map perimetry (Brasil-Neto et al. 1992b; Mortifee et al. 1994). This is less likely in the optimal positions but it raises caution regarding the number of stimuli as well as the intensity sufficient to construct a reliable map. Also, it has been noticed that muscles with higher thresholds, usually more proximal ones, have the largest variability in MEP amplitude (Brasil-Neto et al. 1992b). Therefore, maps of cortical representation areas for proximal muscles as well as those in which the percentage of the peripheral motoneuron pool recruited with TMS is low should be defined by using more stimuli than for muscles with a high percentage of peripheral motoneuron pool recruited, e.g. distal muscles. The number of stimuli applied at each site, albeit so far undetermined, should be as large as possible, but at the same time has to be limited to a reasonable number in order not to make the whole procedure impractical. Ten to twenty stimuli are usually adequate.

Post-Processing – What to Measure?

As already described, when mapping muscle representation, stimulation should begin at a certain position and continue in various directions until the area where MEPs are produced is surrounded by inactive positions. The criterion for positive locations should be the appearance of an MEP $\geq 100\,\mu V$ in three out of six successive trials. One method of proceeding is to measure the peak-to-peak amplitude of all MEPs elicited at each scalp position for a particular muscle, average them and display the averages as a conventional 3D or contour map plot. Maps could also be constructed by expressing MEPs as a percentage of the M-wave or CMAP, elicited by supra-maximal stimulation of the nerve for a particular muscle. This way of examining cortical maps is especially important when performing mapping in a disease, such as ALS (Amyotrophic Lateral Sclerosis), in which the CMAP changes dramatically as the disease progresses. When the percentage of CMAP (% CMAP) is calculated, each muscle can be described by a 2-dimensional map matrix in which responses are represented as % CMAP (Wassermann et al. 1992). Each map is then characterized by the maximal amplitude (highest % CMAP recorded), the total volume (the sum of the % CMAPs at all scalp positions), the area (the number of excitable scalp positions), the center of gravity (a map position representing the amplitude-weighted center of the excitable area; for more details see Wassermann et al. 1992), the optimal position (the cortical spot where the maximal MEP amplitude is evoked), and the threshold as defined previously. Apart from MEP amplitude, the latency or threshold for eliciting a MEP could also be measured and cortical maps constructed.

Mapping of Other Cortical Areas

While mapping of the motor cortex is performed by monitoring muscle responses to cortical stimulation, mapping of other cortical areas is mostly achieved by coupling the magnetic stimulation with a functional task related to that part of the brain. TMS applied over the prefrontal cortex can interrupt or prolong vocalization (Amassian et al. 1991), while TMS of the sensorimotor cortex can occasionally trigger somatotopically organized paresthesias (Amassian et al. 1991), attenuate the perception of somatosensory stimuli or even a sensation of movement (Pascual-Leone et al. 1994). TMS has also been successfully used to investigate the long- and short-term reorganization of sensorimotor maps in people suffering either from the overuse or from a lack of sensory input from a particular sense organ or body part (Cohen et al., 1991a; Pascual-Leone et al., 1993). Learning involves cortical reorganization, and the application of TMS has shown that some visual areas are important in learning complex visual detection tasks, but are no longer required once the task is learned (Walsh et al., 1998). Most studies relating TMS with memory functions were performed in order to demonstrate that the technique does not produce permanent memory deficits (Ferbert et al. 1991). TMS applied over the occipital cortex impairs some aspects of verbal learning, serial processing of visual input and memory scanning rate (Beckers and Homberg, 1991). TMS was also applied to map cortical sites involved in inducing delay of voluntary movements (Taylor et al. 1995), and to determine the function of the dorsolateral frontal cortex in a delayed response task (Pascual-Leone and Hallett 1994).

In summary, TMS is a functional, noninvasive, simple and relatively inexpensive technique that can provide accurate localization of different cortical structures. Combined with other techniques, such as PET, it can provide great advancements in understanding of maturational, adaptive, pathological and other processes in the brain as well as serve as a useful clinical tool for planning various interventions.

Stimulation of Other Parts of the Brain

TMS may be used to study the function of other (non-motor) cortical areas of the brain. It is a reasonable hypothesis that TMS might disrupt briefly the normal function of a brain area and so lead to a detectable change in that same process. Stimulation of non-motor cortices rarely evokes an immediate, recognizable response. Even when the primary sensory cortex is the target, subjects do not report obvious sensory experiences (Cohen et al, 1991b). However, TMS has been shown to affect complex neural processing. For example, TMS over the supplementary motor area and dorsolateral prefrontal cortex can affect spatial working memory, as evidenced from disruption of saccadic eye movements to remembered targets, but it does not evoke overt time-locked eye movements (Muri et al, 1994; Muri et al, 1996). Also, stimulation of the premotor cortex or supplementary motor area can alter reaction times, without necessarily producing direct evoked responses (Davey et al, 1998, Stedman et al, 1998; see also Pascuel-Leone et al, 1992; Masur et al., 1996).

This type of investigation has a number of associated problems not met in studies in which TMS elicits evoked responses. The first concerns analysis. How does one test for changed reaction times or memory-guided saccades? These processes are variable in themselves, which makes their disruption hard to detect. A second problem is that, in the absence of an evoked response, it is not clear what strength of TMS to use or how to determine the threshold for an effect on the system. Previous workers have simply used the threshold for MEPs in response to stimulation of the motor cortex as a guide. However, differences in brain anatomy and nerve cell biophysics imply that the excitability of a particular brain area may not be related to that of the motor cortex. In addition to establishing a suitable stimulation strength, an empirical approach to the exact coil location and orientation will also be needed in novel investigations of non-motor areas.

There have been conflicting reports on whether it is possible to stimulate the cerebellum through the intact scalp using TMS (Werhahn et al, 1996; Ugawa et al, 1997; Cruz-Martinez & Arpa, 1997). Activation of cerebello-cerebral connections might be expected to alter the excitability of the motor cortex. However, due to the position of the coil, placed at the back of the skull, it is possible that any changes in motor cortex excitability might be the result of peripheral nerve activation in the neck.

Repetitive TMS

The maximum repeat rate of single pulse magnetic stimulators varies between once every 5 seconds to once per second depending on the strength of the output stimulus. The interval between stimuli is determined by the time needed to recharge the capacitors after each magnetic shock. Recently, new power-supply technology has enabled the development of machines able to deliver a short train of stimuli (typically 20 Hz for 0.5 s), so-called repetitive (or rapid-rate) transcranial magnetic stimulation (rTMS). It is expected to be more effective than single TMS pulses both in disrupting brain function in focal areas of the brain and in facilitating responses (see earlier). This has opened up new research possibilities and has made rTMS a therapeutic tool (George et al, 1997; Greenberg et al, 1997) and a functional stimulator (Sheriff et al, 1996). So far, rTMS has proven safe and well tolerated by patients. rTMS would appear to have the great advantage, over electro-convulsive therapy (ECT) for example, that it does not cause the patient to convulse and can be administered in conscious individuals. As a research tool, however, rTMS is clearly in its infancy (e.g. see Grafman et al, 1994; Brandt et al, 1998).

References

Amassian VE, Cracco RQ, Maccabee PJ, Bigland-Ritchie B, Cracco, JB. (1991) Matching focal and non-focal magnetic coil stimulation to properties of human nervous system: mapping motor unit fields in motor cortex contrasted with altering sequential digit movements by premotor-SMA stimulation. EEG clin. Neurophysiol. Suppl. 43: 3–28

Barker AT, Jalinous R, Freeston IL (1985a) Non-invasive stimulation of the human motor cortex. Lancet May 11 (8437): 1106–1107

Barker AT, Freeston IL, Jalinous R, Merton PA, Morton HB (1985b) Magnetic stimulation of the human brain. J Physiol 369, 3P.

Barker AT, Stevens JC. (1991) Measurement of the acoustic output from two magnetic nerve stimulator coils. J. Physiol. 438, 301P

Brandt SA, Ploner CJ, Meyer BU, Leistner S, Villringer A (1998) Effects of repetitive transcranial magnetic stimulation over dorsolateral prefrontal cortex and posterior parietal cortex on memory-guided saccades. Exp Brain Res 118, 197–204

Brasil-Neto JP, Cohen LG, Panizza M, Nilsson J, Hallett M. (1992a) Optimal focal transcranial magnetic activation of the human motor cortex: effects of coil orientation, shape of the induced current pulse, and stimulus intensity. J Clin Neurophysiol 9:132–136.

Brasil-Neto JP, McShane LM, Fuhr P, Hallett M, Cohen LG. (1992b) Topographic mapping of the human motor cortex with magnetic stimulation: factors affecting accuracy and reproducibility. EEG Clin Neurophysiol 85:9–16.

Beckers B, Homberg V (1991) Impairment of visual perception and visual short-term memory scanning by transcranial magnetic stimulation of occipital cortex. Exp Brain Res, 87:421–432.

Cohen LG, Bandinelli S, Findley T, Hallet M (1991a) Motor reorganization after upper limb amputation in man. Brain 114:614–627.

Cohen LG, Topka H, Cole RA, Hallett M (1991b) Leg parasthesias induced by magnetic brain stimulation in patients with thoracic spinal cord injury. Neurology 41, 1283–1288

Cohen LG, Roth BJ, Nilsson J, Dang N, Panizza M, Bandinelli S, Friauf W, Hallett M. (1990) Effects of coil design on delivery of focal magnetic stimulation. Technical considerations. EEG clin. Neurophysiol. 75, 350–357.

Cruz-Martinez A, Arpa J (1997) Transcranial magnetic stimulation in patients with cerebellar stroke. Eur Neurol. 38, 82–87

Davey NJ, Ellaway PH, Stein RB (1986) Statistical limits for detecting change in the cumulative sum derivative of the peristimulus time histogram. J Neurosci Methods 17: 153–166.

Davey NJ, Rawlinson SR, Maskill DW, Ellaway PH (1998) Facilitation of a hand muscle response to stimulation of the motor cortex preceding a simple reaction task. Motor Control 2, 241–250

Davey, NJ, Romaiguère P, Maskill DW & Ellaway PH. (1994) Suppression of voluntary motor activity revealed using transcranial magnetic stimulation of the motor cortex in man. J. Physiol. 477, 223–235.

Day BL, Dressler D, Hess CW, Maertens de Noordhout A, Marsden CD, Mills K, Murray NMF, Nakashima K, Rothwell JC, Thompson P. (1990) Direction of current in magnetic stimulating coils used for percutaneous activation of brain, spinal cord and peripheral nerve. J. Physiol. 430, 617

Devanne H, Lavoie BA, Capaday C (1997) Input-output properties and gain changes in the human corticospinal pathway. Exp Brain Res 114: 329–338.

Edgley SA, Eyre JA, Lemon RN, Miller S. (1990) Excitation of the corticospinal tract by electromagnetic and electrical stimulation of the scalp in the macaque monkey. J. Physiol. 425, 301–320.

Ellaway PH (1978) Cumulative sum technique and its application to the analysis of peristimulus time histograms. EEG Clin Neurophysiol 45: 302–304.

Ellaway PH, Davey NJ, Maskill DW, Rawlinson SR, Lewis HS, Anissimova NP. (1998) Variability in the amplitude of skeletal muscle responses to magnetic stimulation of the motor cortex in man. EEG. clin. Neurophysiol. 109, 104–113.

Ellaway PH, Rawlinson SR, Lewis HS, Davey NJ, Maskill DW. (1997) Magnetic stimulation excites skeletal muscle via motor nerve axons in the cat. Muscle Nerve. 20, 1108–1114.

Ferbert A, Musmann N, Menne A (1991) Short-term memory performance with magnetic stimulation of the motor cortex. Eur Arch Psychiat Clin Neuros 241:135–138.

George MS, Wassermann EM, Kimbrell TA, Little JT, Williams WE, Danielson AL, Greenberg BD, Hallett M, Post RM (1997) Mood improvement following daily left prefrontal repetitive transcranial magnetic stimulation in patients with depression: a placebo-controlled crossover trial. Am J Psychiatry 154, 1752–1756

Gerstein GL, Kiang NY-S (1960) An approach to quantitative analysis of electrophysiological data from single neurons. Biophys J 1: 15–28

Grafman J, Pascual-Leone A, Alway D, Nichelli P, Gomez-Tortosa E, Hallett M (1994) Induction of a recall deficit by rapid-rate transcranial magnetic stimulation. Neuroreport 5, 1157–1160

Greenberg BD, George MS, Martin JD, Benjamin J, Schlaeper TE, Altemus M, Wassermann EM, Post RM, Murphy DL (1997) Effect of prefrontal repetitive transcranial magnetic stimulation in obsessive-compulsive disorder: a preliminary study. Am J Psychiatry 154, 867–869

Hallett M. (1996) Transcranial magnetic stimulation: a tool for mapping the central nervous system. EEG clin Neurophysiol Suppl 46: 43–51.

Homberg V, Netz J. (1989) Generalized seizures induced by transcranial magnetic stimulation of motor cortex. Lancet Nov 18; 2(8673): 1223.

Jalinous R. (1991) Technical and practical aspects of magnetic nerve stimulation. J Clin Neurophysiol, 8, 10–25.

Jalinous R. (1992) Fundamental aspects of magnetic stimulation. In: M.A. Lissens (Ed.), Clinical Application of Magnetic Transcranial Stimulation, Peeters Press, Louvain, 1992: 1–20.

Jalinous R. (1994) Guide to magnetic stimulation. The MagStim Company Ltd.

Kirkwood PA (1979) On the use and interpretation of cross-correlation measurements in the mammalian central nervous system. J Neurosci Methods 1: 107–132.

Maccabee PJ, Amassian VE, Cracco RQ, Eberle LP, Rudell AP (1991) Mechanisms of peripheral nervous system stimulation using the magnetic coil. EEG Clin Neurophysiol, 43, 344–361.

Maccabee PJ, Amassian VE, Eberle LP, Cracco RQ. (1993) Magnetic coil stimulation of straight and bent amphibian and mammalian peripheral nerves in vitro: locus of excitation. J Physiol, 460, 201–219.

Maccabee PJ, Eberle L, Amassian VE, Cracco RQ, Rudell A, Jayachandra N. (1990) Spatial distribution of the electrical field induced by round and figure "8" magnetic coils: relevance to activation of sensory nerve fibers. EEG clin. Neurophysiol. 76, 131–141.

Maccabee PJ, Lipitz ME, Desudicht T, Golub RW, Nitti VW, Bania JP, Willer JA, Cracco RQ, Cadwell J, Hotston GC, Eberle LP, Amassian VE. (1996) A new method using neuromagnetic stimulation to measure conduction time within cauda equina. EEG Clin Neurophysiol, 101, 153–166.

Magistris MR, Rosler KM, Truffert A, Myers JP. (1998) Transcranial stimulation excites virtually all motor neurons supplying the target muscle. A demonstration and method improving the study of motor evoked potentials. Brain 121, 437–450.

Masur H, Schneider U, Papke K, Oberwittler C (1996) Variation of reaction time can be reduced by the time locked application of magnetic stimulation of the motor cortex. Electromyogr clin Neurophysiol 36, 495–501

Miles TS, Türker KS, Le TH (1989) Ia reflexes and EPSPs in human soleus motor neurons. Exp Brain Res 77: 628–636

Miranda PC, de-Carvalho M, Conceicao I, Luis ML, Ducla-Soares E (1997) A new method for reproducible coil positioning in transcranial magnetic stimulation mapping. EEG clin Neurophysiol 105: 116–123

Mortifee P, Stewart H, Schulzer M, Eisen, A (1994) Reliability of transcranial magnetic stimulation for mapping the human motor cortex. EEG clin Neurophysiol 93: 131–137

Muri RM, Rosler KM, Hess CW (1994) Influence of transcranial magnetic stimulation on the execution of memorized sequences of saccades in man. Exp Brain Res 101, 521–524

Muri RM, Vermersch AI, Rivaud S, Gaymard B, Pierrot-Deseilligny C (1996) Effects of single pulse transcranial magnetic stimulation over the prefrontal and posterior parietal cortices during memory-guided saccades in humans. J Neurophysiol. 76, 2102–2106

Nakamura H, Kitagawa H, Kawaguchi Y, Tsuji H, Takano H, Nakatoh S. (1995) Intracortical facilitation and inhibition after paired magnetic stimulation in humans under anaesthesia. Neurosci Letts. 199, 155–157.

Nilsson J, Panizza M, Roth BJ, Basser PJ, Cohen LG, Caruso G, Hallett M (1992) Determining the site of stimulation during magnetic stimulation of a peripheral nerve. EEG clin Neurophysiol 85: 253–264

Pascual-Leone A, Brasil-Neto JP, Valls-Sole J, Cohen LG, Hallett M (1992) Simple reaction time to focal transcranial magnetic stimulation. Comparison with reaction time to acoustic, visual and somatosensory stimuli. Brain 115, 109–122

Pascual-Leone A, Cammarota A, Wasserman EM, Brasil-Neto J, Cohen LG, Hallet M (1993) Modulation of motor cortical outputs to the reading hand of Braille readers. Ann Neurol 34:33–37.

Pascual-Leone A, Cohen LG, Brasil-Neto J, Valls-Sole J, Hallet M (1994) Differentiation of sensorimotor neuronal structures responsible for induction of motor evoked potentials, attenuation in detection of somatosensory stimuli, and induction of sensation of movement by mapping of optimal current directions. EEG Clin. Neurophysiol. 93:230–236.

Pascual-Leone A, Gates JR, Dhuna A (1991) Induction of speech arrest and counting errors with rapid-rate transcranial magnetic stimulation. Neurology 41:697–702.

Pascual-Leone A, Hallett M (1994) Induction of errors in a delayed response task by repetitive transcranial magnetic stimulation of the dorsolateral prefrontal cortex. NeuroReport 5:2517–2520.

Penfield W, Rasmussen AT. (1950) "Cerebral cortex of man. A clinical study of localization of function." Macmillan, New York.

Polson MJR, Barker AT, Freeston IL. (1982) Stimulation of nerve trunks with time-varying magnetic fields. Medical and Biological Engineering and Computing 20, 243–244.

Rossini PM, Barker AT, Berardelli A, Caramia MD, Caruso G, Cracco RQ, Dimitrijevic MR, Hallet M, Katayama Y, Lucking CH, Maertens-de Noordhout AL, Marsden CD, Murray NMF, Rothwell JC, Swash M, Tomberg C. (1994) Non-invasive electrical and magnetic stimulation of the brain, spinal cord and roots; basic principles and procedures for routine clinical application. EEG clin. Neurophysiol. 91, 79–92

Sheriff MKM, Shar PJR, Fowler C, Mundy AR, Craggs MD (1996) Neuromodulation of detrusor hyper-reflexia by functional magnetic stimulation of the sacral roots. Brit J Urolog 78, 39–46

Stedman A, Davey NJ, Ellaway PH (1998) Facilitation of human first dorsal interosseus muscle responses to transcranial magnetic stimulation during voluntary contraction of the contralateral homonymous muscle. Muscle Nerve 21, 1033–1039

Stephens JA, Usherwood TP, Garnett R (1976) Technique for studying synaptic connections of single motoneurons in man. Nature 263: 343–344.

Taylor JL, Wagener DS, Colebatch JG (1995) Mapping of cortical sites where transcranial magnetic stimulation results in delay of voluntary movement. EEG Clin Neurophysiol 97:341–348

Ugawa Y, Terao Y, Hanajima R, Sakai K, Furubayashi T, Machii K, Kanazawa I (1997) Magnetic stimulation over the cerebellum in patients with ataxia. EEG clin Neurophysiol 104, 453–458

Walsh V, Ashbridge E, Cowey A (1998) Cortical plasticity in perceptual learning demonstrated by transcranial magnetic stimulation. Neuropsychologia 36:363–367

Wang B, Toro C, Wassermann EM, Zeffiro T, Thatcher RW, Hallett, M. (1994) Multimodal integration of electrophysiological data and brain images: EEG, MEG, TMS, MRI and PET. Thatcher RW, Hallet M, Zeffiro, T., John ER, and Huerta M. 251–257. Academic Press, San Diego. Functional Neuroimaging: Technical Foundations.

Wassermann EM, McShane LM, Hallett M, Cohen LG. (1992) Noninvasive mapping of muscle representations in human motor cortex. EEG Clin Neurophysiol 85:1–8.

Wassermann EM, Pascual-Leone A, Valls-Sole J, Toro C, Cohen LG, Hallet M. (1993) Topography of the inhibitory and excitatory responses to transcranial magnetic stimulation in a hand muscle. EEG Clin. Neurophysiol. 89:424–433.

Wassermann EM, Wang B, Zeffiro T, Sadato N, Pascual-Leone A, Toro C, Hallet M (1996) Locating the Motor Cortex on the MRI with Transcranial Magnetic Stimulation and PET. Neuroimage 3:1–9.

Werhahn KJ, Taylor J, Ridding M, Meyer BU, Rothwell JC (1996) Effect of transcranial magnetic stimulation over the cerebellum on the excitability of the human motor cortex. EEG clin Neurophysiol 101, 58–66

Wilson SA, Thickbroom G, Mastaglia F (1993) Topography of excitatory and inhibitory muscle responses evoked by transcranial magnetic stimulation in the human motor cortex. Neurosci Letters 154:52–56.

Chapter 34

In-vivo Optical Imaging of Cortical Architecture and Dynamics

Amiram Grinvald, D. Shoham, A. Shmuel, D. Glaser, I. Vanzetta, E. Shtoyerman, H. Slovin, C.Wijnbergen, R. Hildesheim and A. Arieli

General Introduction

A number of new imaging techniques are available to scientists to visualize the functioning brain directly, revealing unprecedented details. These imaging techniques have provided a new level of understanding of the principles underlying cortical development, organization and function. In this chapter we will focus on optical imaging in the living mammalian brain, using two complementary imaging techniques. The first technique is based on intrinsic signals. The second technique is based on voltage-sensitive dyes. Currently, these two optical imaging techniques offer the best spatial and temporal resolution, but also have inherent limitations. We shall provide a few examples of new findings obtained mostly in work done in our laboratory. The focus will be upon the understanding of methodological aspects which in turn should contribute to optimal use of these imaging techniques. General reviews describing earlier work done on simpler preparations have been published elsewhere (Cohen, 1973; Tasaki and Warashina, 1976; Waggoner and Grinvald, 1977; Waggoner, 1979; Salzberg, 1983; Grinvald, 1984; Grinvald et al., 1985; De Weer and Salzberg, 1986; Cohen and Lesher, 1986; Salzberg et al., 1986; Loew, 1987; Orbach, 1987; Blasdel, 1988, 1989; Grinvald et al., 1988; Kamino, 1991; Cinelli and Kauer, 1992; Frostig, 1994).

Advantages of Optical Imaging of Cortical Activity

The processing of sensory information, coordination of movement or more complex cognitive brain functions is carried out by millions of neurons, forming elaborate networks. Individual neurons are synaptically connected to hundreds or thousands of other neurons which shape their response properties. These connections may be local, thereby spanning a short distance, or long-range, within the same cortical area or between different cortical areas. The manner in which these neurons, and their intricate connections, endow the brain with its remarkable performance is a central question in brain research.

In the mammalian brain, cells which perform a given function, or share common functional properties, are often grouped together (Mountcastle, 1957; Hubel and Wiesel, 1965). It is unlikely that we shall be able to discover the principles underlying the neural code and its implementation without knowing what is the functional processing performed by a given ensamble of neurons. Therefore, attaining an understanding of the three-dimensional functional organization of a given cortical area is a key step towards revealing the mechanisms of information processing there. Thus, experimental methods

Correspondence to: Amiram Grinvald, The Weizmann Institute of Science, Department of Neurobiology, The Grodetsky Center for Research of Higher Brain Functions, Rehovot, 76100, Israel (phone: + 972-8-9343833; fax: + 972-8-9343833; e-mail: Bngrinva@weizmann.weizmann.ac.il)

that allow the visualization of the functional organization of the cortical columns in a given cortical region are of special importance, particularly those methods providing high spatial and temporal resolution. Hubel and Wiesel were perhaps the first to realize the need for a functional brain imaging technique and used any new imaging method that became available to address questions regarding the cortical functional architecture; questions which could not be resolved with single unit recordings. Several imaging techniques have been developed that yield information about the spatial distribution of active neurons in the brain, with each technique having significant advantages as well as limitations. An example of this is the 2-deoxyglucose method (2-DG) (cf. Chapter 15) which permits postmortem visualization of active brain areas or even single cells, while its time resolution is minutes or hours rather than milliseconds. Furthermore, 2-DG is a one-time approach: Only a single stimulus condition in a single animal can be assayed (although the two-isotope 2-DG method permits the mapping of activity resulting from two stimulus conditions). Positron-Emission Tomography (PET) (cf. Chapter 39) and functional Magnetic Resonance Imaging (f-MRI) (cf. Chapter 38) offer spectacular three-dimensional localization of active regions in the functioning human brain, but currently offer only low temporal and spatial resolution. Other *in vivo* imaging techniques have also been applied with success, but still suffer from either a limited spatial resolution, limited temporal resolution, or a combination thereof. These methods include radioactive imaging of changes in blood flow, electroencephalography (cf. Chapter 35), magnetoencephalography (cf. Chapter 37), and thermal imaging (e.g., Shevlev 1998). In this chapter we shall start with optical imaging, based on intrinsic signals. Although this technique offers the highest spatial resolution thus far obtained in-vivo, its temporal resolution is inadequate in studying the dynamics of cortical processing.

Indeed, the visualization of cortical functional organization does not require high temporal resolution. In contrast, a more complete understanding of the mechanisms of cortical function, at the level of neuronal assemblies, requires methods that can monitor cortical dynamics, that is to say the flow of neuronal signals, from one group of neurons to the next, on a millisecond time basis. To date, single- or multi-unit recording techniques have provided the best tools in studying the functional response properties of single cortical neurons. These methods, however, are not optimal for a detailed study of neuronal networks or neuronal assemblies. Due to the tremendous effort required for the careful analysis of neuronal networks of simple invertebrate ganglia, it is apparent that new approaches must be utilized. Notwithstanding the fact that multi-electrode techniques offer promise, the size and placement of these electrode arrays pose severe problems. In addition, multiple recordings are only practical extracellularly, thus obscuring essential information contained in the dynamics of subthreshold synaptic potentials. Two imaging techniques, electroencephalography and magnetoencephalography, have been developed to study the dynamics of cortical processing in the intact human brain. However, these two methods do not currently have the adequate spatial resolution to resolve individual cortical columns. In the second part of this chapter, we shall review real-time optical imaging, which currently provides excellent temporal and spatial resolution, whenever applicable. It is particularly useful for cortical studies in animal models.

Optical Imaging Based on Intrinsic Signals

At present, the easiest and most effective strategy to image the functional architecture is based on slow intrinsic changes in the optical properties of active brain tissue, thus permitting visualization of active cortical regions at a spatial resolution greater than 50 μm. This can be accomplished without some of the problems associated with utilizing extrinsic probes. The sources for these activity-dependent intrinsic signals include either changes in physical properties of the tissue itself, which affect light scattering (for review see Cohen, 1973), and/or changes in the absorption, fluorescence or other optical properties of intrinsic molecules having significant absorption or fluorescence. The existence of small intrinsic optical changes associated with metabolic activity in many tissues has been recognized since the pioneering experiments of Kelin and Millikan on the absorption of cytochromes (Kelin, 1925) and hemoglobin (Millikan, 1937). The first optical recording of neuronal activity was made fifty years ago by Hill and Keynes (1949), who detected light-scattering changes in active nerves. Changes in absorption or fluorescence of intrinsic chromophores were extensively investigated by Chance and his colleagues (Chance et al., 1962), and Jobsis and his colleagues (Jobsis et al., 1977; Mayevsky and Chance, 1982). However, the intrinsic optical signals are usually very small or very noisy. Only recently has it become possible to use optical detection of intrinsic signals for the imaging of the functional architecture of the cortex (Grinvald et al., 1986). For a more extensive review of the methodology see Bonhoeffer and Grinvald (1996), parts of which are summarized here.

Real-Time Optical Imaging of Neuronal Activity Based on Voltage-Sensitive Dyes

To explore cortical dynamics and to accomplish real-time visualization of neuronal activity, the imaging based on intrinsic signals is not useful. Since most of the intrinsic signals are slow, the alternative is to utilize fast extrinsic probes. In such experiments, the preparation under study is first stained with a suitable voltage-sensitive dye (cf. also Chaps. 4, 16). The dye molecules bind to the external surface of excitable membranes and act as molecular transducers that transform changes in membrane potential into optical signals. The resulting changes in the absorption or the emitted fluorescence occur in microseconds and are linearly correlated with the membrane potential changes of the stained cells. These changes are then monitored with light-measuring devices. By using an array of photo-detectors positioned in the microscope image plane, the electrical activity of many targets can be detected simultaneously (Grinvald et al. 1981). The development of suitable voltage-sensitive dyes has been the key to the successful application of optical recording, because different preparations often require dyes with different properties (Ross and Reichardt 1979: Cohen and Lesher 1986; Grinvald et al., 1988). Optical imaging with voltage-sensitive dyes permits the visualization of cortical activity with a sub-millisecond time resolution and a spatial resolution of 50–100 microns. The instrumentation to record these fast optical signals with a higher spatial resolution over a large area requires fast detectors with many more pixels which are currently being developed. It is important to note that optical signals recorded from the cortex are different from those recorded from single cells or their individual processes and thus should be interpreted with care: In simpler preparations where single cells are distinctly visible, the optical signal appears just like an intracellular electrical recording (Salzberg et al., 1973, 1977; Grinvald et al., 1977, 1981, 1982). However, in optical recordings from cortical tissue single-cell activity is not resolved and the optical signal represents the sum of membrane-potential changes, in both pre- and post-synaptic intermingled neuronal elements, as well as a possible contribution from the depolarization of neighboring glial cells. Since the optical signals measure the integral of the membrane potential changes, slow sub-threshold synaptic potentials in the extensive dendritic arborization are easily detected by optical recording. Thus, optical signals, when properly dissected, can provide information concerning elements of neuronal processing, which is usually not available from single-unit recordings. Real-time optical imaging of cortical activity is a particularly attractive technique for providing new insights into the temporal aspects of the function of the mammalian brain. Among its advantages over other methodologies are:

- Direct recording of the summed intracellular activity of neuronal populations, including fine dendritic and axonal processes;
- The possibility of repeated measurements from the same cortical region with different experimental and/or stimulus conditions over an extended time;
- Imaging of spatio-temporal patterns of activity of neuronal populations with sub-millisecond time resolution; and
- Selective visualization of neuronal assemblies.

Below we first summarize the basics of optical imaging based on intrinsic signals (Part 1). Next, we discuss real-time optical imaging based on voltage-sensitive dyes in the neocortex (Part 2). Although the methodologies for these two optical imaging techniques have a lot in common, the large differences between the two justify a separate discussion. At the end of this chapter there is a discussion of the powerful and intimate combination of these two imaging techniques with other neurophysiological approaches (Part 3) and a comparison of their merits and limitations (Part 4).

Part 1: Optical Imaging Based on Intrinsic Signals

Introduction

The basic experimental setup for optical imaging experiments is shown in Fig. 1. The animal head is rigidly attached (not shown). The exposed brain is illuminated with flexible light guides, and digital pictures are acquired by the camera which views the cortex

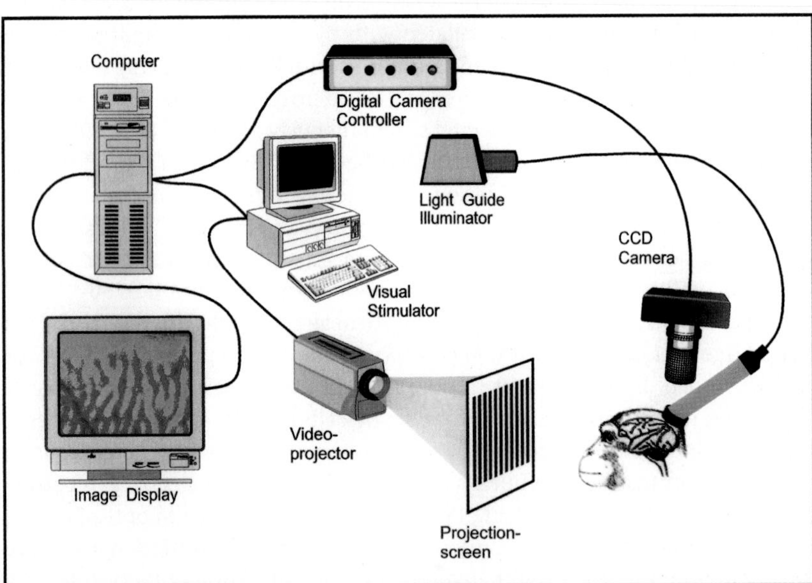

Fig. 1. Setup for optical imaging of functional maps *in vivo*. Images are taken of the animal's exposed cortex which is sealed in an oil-filled chamber. The cortex is illuminated with light of 605 nm wavelength. The images are acquired with the camera, during which time the animal is visually stimulated with moving gratings which are projected onto a frosted glass screen by a video projector. The acquired images are digitized by a computer controlling the entire experiment. The signal-to-noise ratio of the functional maps is improved by averaging several stimulus sessions. Functional maps are subsequently analyzed, and are displayed on a color monitor. A color-coded ocular dominance map is shown here. To determine the quality of the maps during the imaging sessions, the data can be sent to a second computer for detailed, quasi on-line analysis. (Modified from Ts'o et al., 1990)

through a cranial window. The data are analyzed either on the computer controlling the experiment or on a separate analysis computer (not shown), and the resulting functional maps are displayed on a color video monitor.

The initial optical imaging studies investigated the well-known structural elements of the functional architecture, such as ocular dominance in the primary visual cortex and the "stripes" in V2 (Ts'o et al., 1990), or the pinwheel-like organization of orientation preference (Bonhoeffer and Grinvald, 1991; Bonhoeffer and Grinvald, 1993; Bonhoeffer et al., 1995; Das and Gilbert, 1995, 1997). Figure2 illustrates some maps of orientation and ocular dominance columns. Subsequently, methodological improvements made it possible to investigate more subtle features of cortical organization, such as di-

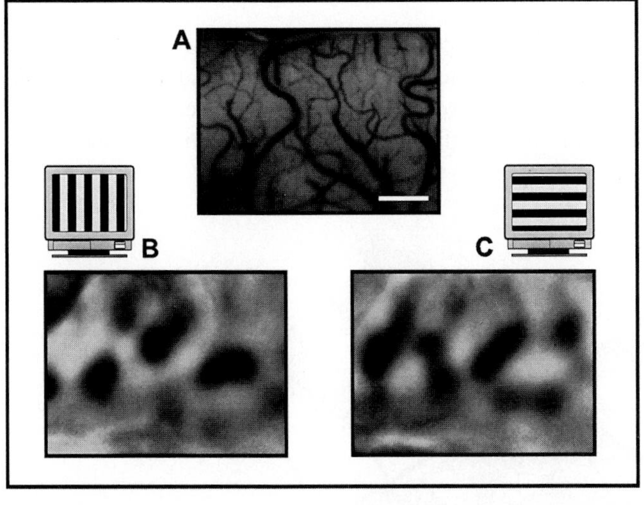

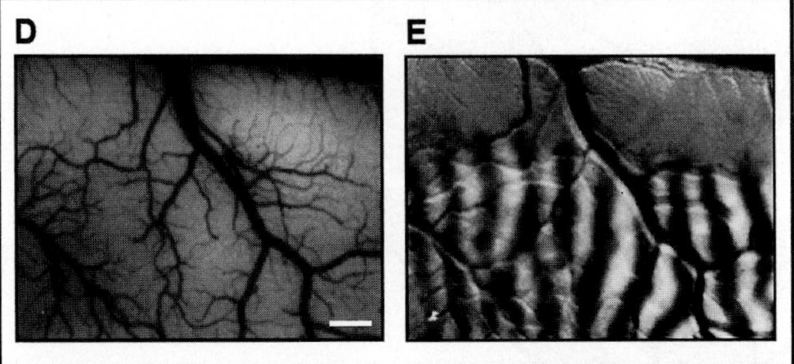

Fig. 2. Imaging the functional architecture. Top: Orientation mapping: An activity map for one orientation is obtained straightforwardly as follows. The cortical image captured when the animal viewed lines of this orientation is divided by (or subtracted from) the average of the images captured when the animal viewed all of the orientations. A: Image of the cortical surface illuminated with green light to emphasize the vasculature. B, C: Activity maps evoked by visual stimulation with horizontal and vertical gratings. Black patches denote the cortical functional domains which were activated by each stimulus. The amplitude of the functional domain, shown in panels B and C, is about 1000 times smaller than the light intensity of the recorded cortical image shown in A. (Figure modified from Bonhoeffer and Grinvald, 1991.) Bottom: Ocular dominance map in V1 of a monkey. D: The imaged cortical area under green illumination to emphasize the vascular pattern. E: Ocular dominance map (OD) as obtained when one eye is stimulated and the recorded cortical image is subtracted from a cortical image obtained when the other eye is stimulated. The ocular dominance pattern shows a clear demarcation between V1 and V2 where no OD pattern can be observed. A general feature of the ocular dominance pattern is that the OD bands usually terminate perpendicular to the V1/V2 border. (Grinvald and Bonhoeffer, unpublished results, see Ts'o et al., 1990).

rection-selective columns or spatial-frequency columns (Malonek et al., 1994; Shmuel and Grinvald, 1996; Weliky et al., 1996; Shoham et al., 1997). Similar progress has been obtained in exploring other visual areas. Ts'o and collaborators (1991, 1993) and Malach and his colleagues (1994) succeeded in imaging the separate pathways in thin, thick and pale stripes in monkey V2. It has even become possible to demonstrate functional columns in visual areas further up the processing stream, in areas V4 (Ghose et al., 1994) and MT (Malonek et al., 1994; Malach et al., 1997). Recently, Tanaka and his colleagues have used this method to image the functional organization in the infero-temporal area, one of the final stages of the visual pathway critical for object recognition (Wang et al., 1994). They showed that presentation of certain visual stimuli activated patchy regions, around 500 μm in diameter. A most striking report from these investigators suggested that the same face shown from different angles activated adjacent and partially overlapping clusters of neurons.

One outstanding question which has recently been resolved by optical imaging is how the various columnar subsystems in the primary visual cortex are organized with respect to one another. Figure 3 shows the schematic relationship between orientation columns, ocular dominance columns and the blobs in monkey primary visual cortex

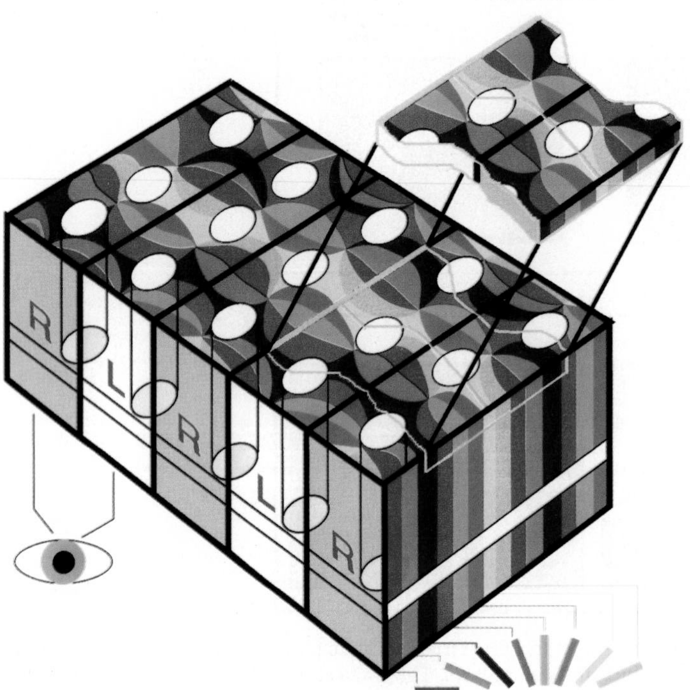

Fig. 3. 3-D schematic map showing the relationships between ocular dominance and orientation preference maps and the cytochrome oxidase blobs. Black lines mark the borders between columns of neurons that receive signals from different eyes. This segregation is partially responsible for depth perception. White ovals represent groups of neurons responsible for color perception (blobs). The "pinwheels" are formed by neurons involved in the perception of shape, with each color marking a column of neurons responding selectively to a particular orientation in space. Note that both the blobs and the centers of the pinwheels lie at the center of the R or L columns. The iso-orientation lines (appearing as a border between two colors) tend to cross borders of ocular dominance columns (black lines) at right angles. The top "slice" above the "ice cube" model depicts two adjacent fundamental modules (400 μm (800 μm). Each module contains a complete set of about 60,000 neurons, processing all three features of orientation, depth and color. This scheme is simplified in that clockwise and counterclockwise pinwheels are perfectly interconnected. In reality this relationship does not exist (Modified from Bartfeld and Grinvald, 1992).

(Bartfeld and Grinvald 1992). These three subsystems are responsible for the perception of shape, depth and color, respectively. The following relationships have been found: 1) orientation-preference is organized mostly radially, in a pinwheel-like fashion; 2) orientation domains are continuous and have fuzzy boundaries; 3) iso-orientation lines tend to cross ocular dominance borders at 90°; 4) orientation pinwheels are centered on ocular dominance columns; 5) blobs are centered on ocular dominance columns; 6) the centers of the blobs and the centers of the orientation pinwheels are segregated; 7) there is a regular mosaic-like organization for each type of functional domain, without there being an overall pattern of repeating hypercolumns. This latter finding is probably related to the existence of short-range (<1 mm) rather than long-range interactions during development. Some similar results in the monkey have been independently reported by Blasdel (1992a,b) and by Obermayer and Blasdel (1993), and for the cat by Hubener and his colleagues (1997).

Another set of significant questions which optical imaging is attempting to resolve is whether there is a relationship between the retinotopic map and other functional maps such as orientation (Das and Gilbert 1995, 1997) or the spatial arrangement of long-range horizontal connections. In Das and Gilbert's experiments, the retinotopic mapping was done with dense single-unit recording, thus it may have been hampered by the well-known receptive field scatter. Blasdel and Salama (1986) already showed that retinotopy can be directly mapped with optical imaging. This approach has been extended by Fitzpatrick and his colleagues (Bosking et al. 1997) to obtain a more complete retinotopic map. They used it to explore the relationship between long-range horizontal connections (related to a given orientation) and the map of visual space in the tree shrew. An example of high-resolution retinotopic maps in area V1 of owl monkey from the work of Shmuel and his colleagues is illustrated in Fig. 4.

Although to date most optical imaging studies have been done in the visual cortex, this is by no means the only sensory system that can be studied using this method. Indeed, this methodology has also proven useful for investigating functional architecture in the somatosensory cortex of the rat (Grinvald et al., 1986; Gochin et al., 1992; Frostig et al., 1994) and monkey (Shoham and Grinvald., 1994) and in the auditory cortex of the guinea pig (Bakin et al., 1993), gerbil (Hess and Scheich, 1994) and chinchilla (Harrison et al., 1998).

Certain outstanding questions cannot be explored by performing acute experiments but require long-term chronic recordings. Particularly important with regard to the feasibility of chronic optical imaging was the finding that cortical maps can be obtained through the intact or a thinned dura and even through a thinned bone (Frostig et al., 1990; Masino et al., 1993; Bosking et al., 1997). These results were achieved using near infrared light, which penetrates the tissue considerably better than light of a shorter wavelength.

Such studies of anesthetized preparations do not indicate whether and how the functional organization of a given cortical area is influenced by the behavior of the animal. Therefore, explorations of behaving animals are of great interest. It has been demonstrated that optical imaging based on intrinsic signals can be used to investigate the functional architecture of the cortex in the awake behaving monkey (Grinvald et al., 1991; Vnek et al., 1998; Shtoyerman et al., 1998), the awake cat (Tanifuji et al., unpublished results) and even the freely moving cat (Rector et al., 1997). Recently, ocular dominance and orientation columns were repeatedly imaged in the behaving macaque for a period of nine months. This recent progress encourages us to believe that optical imaging techniques can be successfully implemented for studying higher brain functions in behaving primates and other species.

An area of investigation where chronic optical imaging has been applied fruitfully is the study of postnatal experience-dependent plasticity and development in the neocor-

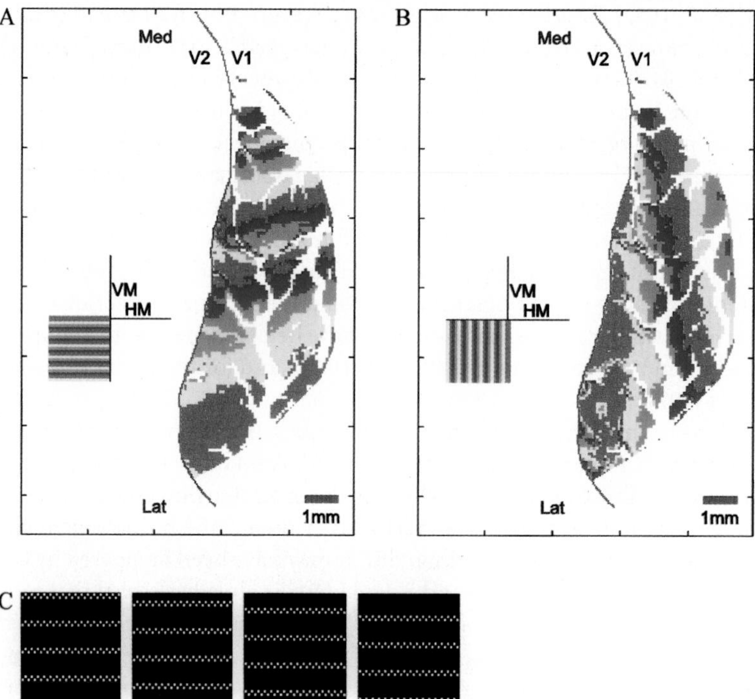

Fig. 4. Topography of area V1 in the owl monkey. A: Horizontal strips in the visual space, as mapped to V1. The horizontal grating patterns were composed of flashing checkers. The cortical images corresponding to the different stimuli were combined by pixel-wise vector summation. Each pixel in the cortical images was assigned to the particular phase of the corresponding stimulus, and the magnitude correlated to the response recorded at the specific pixels. The vectors were summed to obtain the topographic map presented in A. Each colored strip in the icon stands for a strip in the visual space whose width is 0.5°. The V1/V2 border was determined according to cytochrome oxidase histology. B: Vertical strips in the visual space, as mapped to V1. The format of presentation is identical to the one used in A. Vertical gratings were used. C: The set of horizontal gratings used for the retinotopic mapping depicted in A. (Shmuel and Grinvald, unpublished results).

tex (Kim and Bonhoeffer, 1994; Chapman and Bonhoeffer, 1994; 1998; Bonhoeffer and Goedecke, 1994; Crair et al 1997a,b, 1998). Such studies require longitudinal experiments, to determine changes in the cortical functional architecture over long periods of time. The technique of optical imaging is particularly beneficial in these studies, since it offers both the required spatial resolution and the ability to perform prolonged, comparative studies.

Another important application of optical imaging is a clinical one, the mapping of functional borders during neurosurgery. It has been reported (MacVicar et al., 1990; Haglund et al., 1992) that optical imaging can be used to visualize activation of the human cortex, in response to bipolar stimulation and during speech. Therefore, we face the exciting prospect that optical imaging may assist neurosurgeons in precisely locating the foci of epileptic events, or the borders of functional areas close to the site of surgical procedures, as well as in obtaining high resolution functional maps from that region.

Finally, is it science fiction or can one hope to image human brain function non-invasively using light, through the intact skull? The pioneering experiments by Jobsis (1977) on the cat, using trans-illumination with near infrared light, and subsequent studies on human infants (Wyatt et al., 1986, 1990) have suggested that progress can be made in this direction. Furthermore, Chance et al. (1993a,b), in a very innovative experiment, showed that although light reflected from the cortex is drastically attenuated by the thick skull, it can nevertheless be detected using photo-multipliers. Exciting progress along this line will be discussed in Sect 1.6.1.

Methods

1.1 Sources of Intrinsic Signals

In order to optimally image functional maps in the neocortex and to interpret these maps properly, it is crucial to understand the mechanisms underlying the intrinsic signals, and particularly their relation to the electrical activity of neurons. The idea that electrical activity is related to microvascular changes is not a new one. More than a century ago, Roy and Sherrington (1890) postulated that "the brain possesses an intrinsic mechanism by which its vascular supply can be varied locally in correspondence with local variations of functional activity". Modern imaging techniques have indeed demonstrated that there is a strong coupling between neuronal activity, local metabolic activity and blood flow (Kety et al., 1955; Lassen and Ingvar, 1961; Sokoloff, 1977; Raichle et al., 1983; Fox et al., 1986).

1.1.1 Signal Sources

Figure 5A illustrates the time course of intrinsic optical signals in several preparations, measured at different wavelengths. It is apparent that the signal is slow and has a time course very different from that of the evoked electrical activity. Furthermore, *in vivo* the time course strongly depends on the wavelength used. Figure 5B illustrates the time course of the orientation maps obtained from intrinsic signals imaged at a wavelength of 605 nm. Again, it is apparent that the functional maps are strongest long after the peak of evoked electrical activity has occurred (not shown).

Although the intrinsic signal has different components originating from different sources, it has been shown that functional maps obtained at different wavelengths are very similar. Therefore, it appears that all of these components can be used for functional mapping (Frostig et al., 1990), albeit with a different signal-to-noise ratio and different spatial resolution.

The main conclusion regarding the origin of the intrinsic signal was that, following sensory stimulation, there is an initial increase in the concentration of deoxyhemoglobin, due to increased oxygen consumption. This increase is referred to as "the initial dip" by the f-MRI community. It is followed by a larger decrease, due to large but delayed changes in blood flow, supplying highly oxygenated blood to the activated cortical area. The interpretation of our previous optical measurements (Grinvald et al., 1986, Frostig et al., 1990) is thus apparently in contrast with the interpretation of the PET results of Fox and Raichle (1988). Furthermore, our result was also inconsistent with the inability of most f-MRI measurements to confirm the initial dip. This issue seems important, not only because it posed the biological question of whether the first event that followed a sensory stimulation was an increase in oxygen consumption, but also because the interpretation of results from the three related functional imaging approaches often depends on the understanding of what exactly is measured by each technique. This consideration motivated us to continue exploring the mechanisms using the following two new approaches.

In the original studies on the physiological events underlying the intrinsic signals, changes in reflection were not measured simultaneously at different wavelengths. To overcome this problem, Malonek and Grinvald (1996) used *Optical Imaging Spectrosco-*

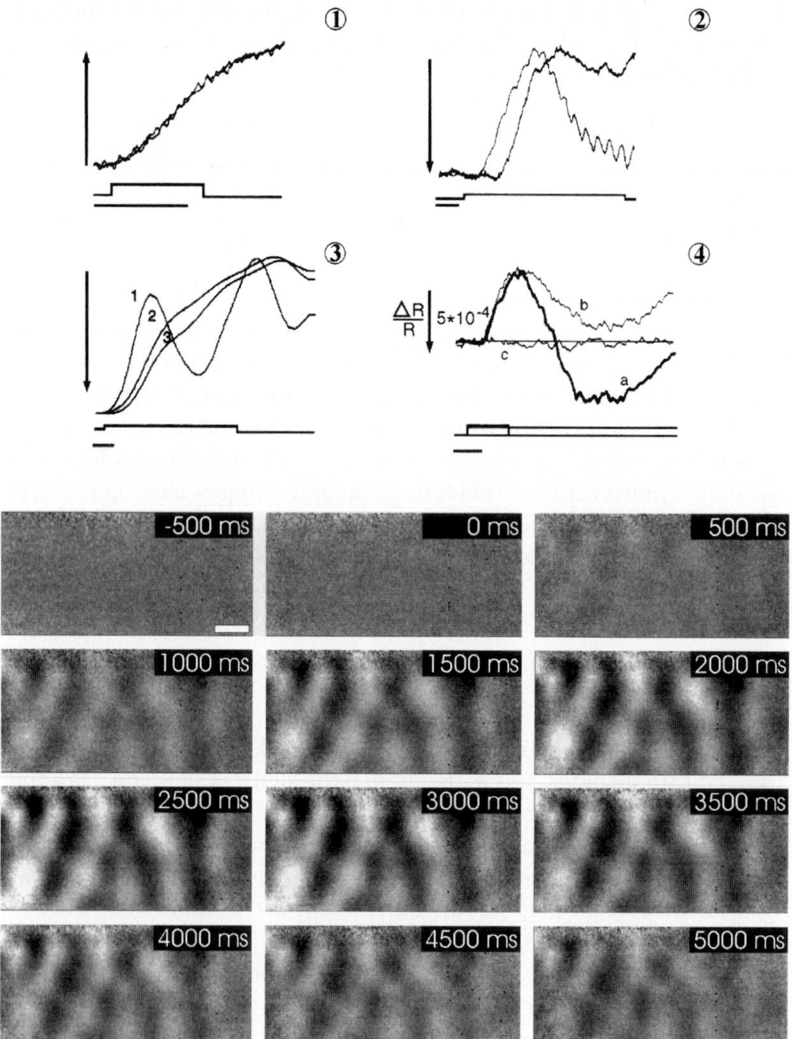

Fig. 5. Time course of various components of the intrinsic signal. Upper panel: (1) Light-scattering signals of an identical time course are measured at 540nm (vertical arrow 1×10^{-2}) and 850nm (vertical arrow 7×10^{-3}) in transmission experiments in a blood-free preparation (hippocampal slice). Electrical stimulation frequency: 40 Hz. (2) The time course of reflection signals measured in cat cortex at 600nm (thin trace; vertical arrow: 3×10^{-3}) and 930nm (thick trace; vertical arrow: 1.4×10^{-3}), in response to the onset of a visual stimulus. Stimulus duration: 8 s. The amplitude of the signal observed with light of 600nm declined even though the cortex was still electrically active. (3) Reflection signals from monkey striate cortex in response to a visual stimulus. Trace 1: measurement at 600nm (vertical arrow 2.5×10^{-3}). Trace 2: measurement at 570nm (vertical arrow 2×10^{-2}). Trace 3: measurement at 840nm (vertical arrow 1×10^{-3}). (4) The time course of intrinsic signals observed at 600nm when the cortex is activated with a 2 s stimulus (a), an 8 s stimulus (b) or when it is not activated at all (c). A large undershoot of the signal is observed after ~5 s even if stimulation of the cortex has not ceased (b). Horizontal bars: 1 s. – Lower panel: Twelve consecutive frames showing functional maps recorded at 605 nm. Each frame lasted 500ms. The labels indicate the times at which the frames were completed with respect to the onset of the stimulus. The first frame was taken before the onset of the visual stimulus. No sign of a functional map is apparent. Approximately one second after the onset of the stimulus the first indications of functional maps occur. The amplitude of the functional map is maximal after 3 s and its strength does not further increase with time. The stimulus duration in this case was 2.4 s. Note that this time course is only an example and that it can substantially differ from experiment to experiment.
(Modified from Grinvald et al., 1986 and Frostig et al., 1990. Lower panel of the figure courtesy of Amir Shmuel)

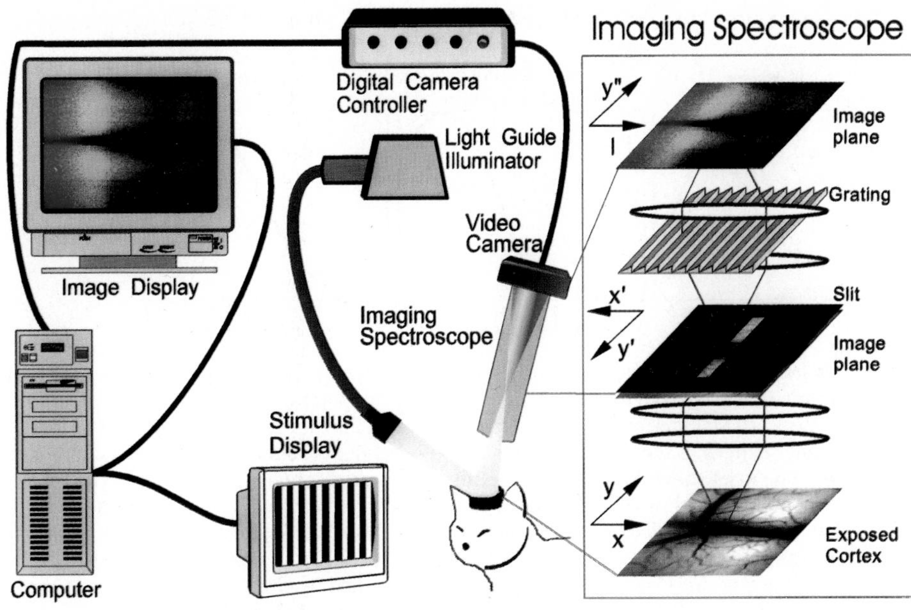

Fig. 6. Scheme of the imaging spectroscope. Left: General setup for intrinsic imaging, as in Fig. 1, except that the macroscope was replaced by an imaging spectroscope shown at the right. Right: Components of the imaging spectroscope. It contains 2 tandem lens macroscopes, diffraction gratings and an opaque disk with a transparent slit. The cortical surface (right bottom image) is illuminated with white light ((=3D 500–700 nm) and imaged through the first macroscope onto the first image plane where an opaque disk with a transparent slit (200 (m width, 15 mm length) is positioned. The light thus isolated from the "slit-like cortical image" is collimated, passed through a diffraction grating whose light dispersion is perpendicular to the slit, and then focused on the camera target positioned at the second image plane (right top color image). The axes on the right show the different transformations that the cortical image undergoes in the optical system. Temporal resolution is up to 100 ms. Spectral resolution is 1–4 nm. (Modified from Malonek and Grinvald, 1986)

py, a new technique providing simultaneous spectral information from many cortical locations in the form of a spatio-spectral image. The images obtained with imaging spectroscopy show the spectral changes at many wavelengths for each cortical point (y vs. λ), as a function of time relative to chosen stimuli (see Fig. 6). Using this technique, Malonek and Grinvald measured the spatial, temporal and spectral characteristics of light reflected from the surface of the visual cortex following natural stimulation. Obtaining cortical spectra in this manner helps in:

- Identifying signal sources by curve-fitting to known spectra;
- Determining the spatial precision of the signals;
- Evaluating the dynamics of the signals;
- Exploring the dynamics in different vascular compartments.

Additional technical details have been published elsewhere (Malonek and Grinvald,1996; Malonek et al., 1997).

As pointed out recently by Mayhew and his colleagues (Mayhew et al., 1998), the linear curve-fitting procedure employed by Malonek and Grinvald was oversimplified because it neglected the wavelength dependency of the path-length of the illumination and of the reflected light. This justified criticism raised the question of whether the results of our simplified analysis were indeed correct. Note, however, that even when using their more precise and sophisticated non-linear model, Mayhew and his colleagues have found that the initial dip also exists in the rat whisker barrel system (personal commu-

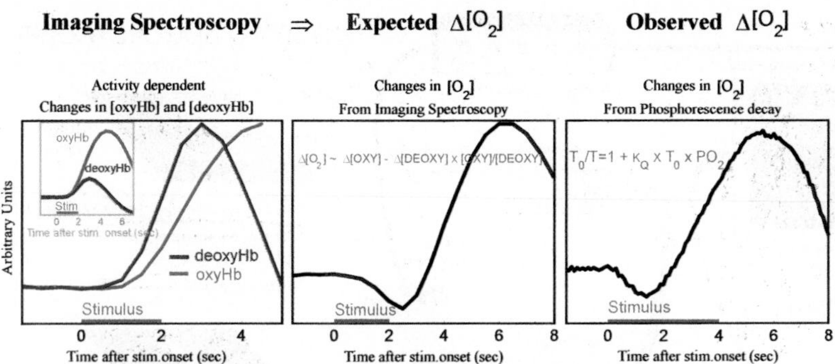

Fig. 7. Comparing imaging spectroscopy results with direct [oxygen] measurements. The left panel shows the time course derived from analysis of the imaging spectroscopy results using the simplified linear model of Malonek and Grinvald (1986). To show the different time courses of the activity-dependent changes in the concentrations of oxy- and deoxyhemoglobin, the two curves are normalized. The inset shows the relative amplitude of these components on a longer time scale. The middle panel shows the expected concentration change in oxygen itself based on the concentration change in oxy- and deoxyhemoglobin. The right panel shows the changes in oxygen concentration within the vascular bed, measured directly by measuring the phosphorescence decay time of the oxygen probe. (Vanzeta and Grinvald, unpublished results)

nication from J. Mayhew). Because the selected model is indeed critical and no good model currently exists, we decided to bypass imaging spectroscopy and measure the oxygen concentration directly. Before trying a new technique we first obtained an estimate of the kinetics of oxygen concentration changes relying on the imaging spectroscopy results? The familiar imaging spectroscopy results from the anesthetized cat are shown in Fig. 7 (left). From this data one can predict that the kinetics of oxygen concentration changed within the micro-vascular system by using a simple equation, predicting free oxygen kinetics from the concentration change in oxyhemoglobin and deoxyhemoglobin (Fig. 7, middle).

To measure the stimulus-dependent oxygen concentration change directly, we simply used measurements of the phosphorescence decay of Oxyfor 3 injected into the microcirculation. This method was invented by Wilson and his colleagues (Rumsey et al, 1988), and it is based on the fact that phosphorescence lifetime depends on the oxygen concentration in the immediate environment of the phosphorescent molecule. Thus, by measuring phosphorescence lifetime before, during and after cortical activation, the concentration of free oxygen mostly in the capillaries can be directly calculated. Although this method appears quite powerful, it has never been used to study oxygen dynamics in the sensory-stimulated cortex. Vanzeta and his colleagues have demonstrated that this is indeed feasible, and were able to directly measure changes in oxygen concentration in response to visual stimulation (Shtoyerman et al., 1998; Vanzetta and Grinvald, 1998).

Figure 7 (right) depicts the detected oxygen kinetics obtained by using the Stern-Volmer equation and the decay time of the phosphorescence, which was measured using a photo-multiplier. Evidently, this kinetic is very similar to that expected from the imaging spectroscopy data (Fig. 7, middle), thus confirming the previous interpretation of Frostig et al. and Malonek and Grinvald.

These results also suggest that Wilson's method (Rumsey et al, 1988) can probably provide the missing quantitative information on the kinetics of oxygen concentrations, under different physiological conditions, within the different vascular compartments.

From all of the above studies the following picture emerges concerning the sources of the intrinsic signals. One component of the intrinsic signal originates from activity-dependent changes in the oxygen saturation of hemoglobin. This change in oxygenation itself contains two different components. The first component is an early one: an increase in the deoxyhemoglobin concentration, resulting from elevated oxygen consumption of the neurons due to their metabolic activity. This causes a darkening of the cortex. The second component is a delayed one: an activity-related increase in blood flow, causing a decrease in the deoxyhemoglobin concentration. This is because the blood rushing into the activated tissue contains higher levels of oxyhemoglobin. Another signal component originates from changes in blood volume. These are probably due to local capillary recruitment or a rapid filling of capillaries and dilation of venules in an area containing electrically active neurons. These blood-related components dominate the signal at wavelengths between 400 to ~600 nm. The last significant component of the intrinsic signal arises from changes in light scattering that accompany cortical activation (Tasaki et al., 1968 Cohen et al., 1968). These are caused by ion and water movement, expansion and contraction of extracellular spaces, capillary expansion, or neurotransmitter release (see review by Cohen, 1973). The light-scattering component becomes a significant source of intrinsic signals above 630 nm, and dominates the intrinsic signals in the near infrared region above 800 nm.

The intrinsic signals that can be measured from the living brain are small. In optimal cases, the change in light intensity due to neuronal activity is about 0.1 to 0.2% (at 605 nm) or up to 6% (at 540 nm) of the total intensity of the reflected light. This means that intrinsic signals cannot be seen with the naked eye, and that they have to be extracted from the cortical images with appropriate data acquisition and analysis procedures. A major problem is that the biological noise associated with these measurements is, in many cases, larger than the signals themselves. Therefore, it is crucial to employ the proper procedures to extract the small signal of interest from the raw data. Such procedures have been developed, yielding high-resolution functional maps. To demonstrate the reliability of the data, the reproducibility of the optical maps obtained from the same area of cortex must be verified. The high degree of reproducibility which has been observed (e.g., Bonhoeffer and Grinvald, 1993) gives confidence in the precision and reliability of the cortical maps obtained in optical imaging. Note, however, that reproducibility of the optical maps is a necessary but not sufficient criterion. There exist signals from vessels which are also reproducible, but do not co-localize with electrical activity. Therefore, electrophysiological or histological confirmations are also necessary, whenever applicable, as discussed later.

1.1.2 Does the Intrinsic Signal Measure Spiking Activity?

It is clear from the above discussion that the answer to the question as to what the intrinsic signal measures depends on the wavelength used for that measurement. At the peak of the oxymetry signal it appears to measure primarily oxygen consumption. Clearly, action potentials create metabolic demands. However, in view of the large concentration of mitochondria within dendrites, one may wonder whether the sub-threshold synaptic activity or calcium action potentials gives rise to large oxygen consumption by the dendrites. Since this issue has not been fully resolved, it is not clear yet if the oxymetry signal reflects mostly sub-threshold synaptic potential as suggested by Das and Gilbert (1995, 1997), or a larger contribution originating from spiking activity as suggested by Sur and his colleagues (Tooth et al., 1996). The appearance of functional maps obtained with intrinsic imaging and the fact that the spread beyond the retinotopic border is larger with voltage-sensitive dyes than with intrinsic signals (Glaser Shoham and Grinvald, unpublished results) suggests that spiking neurons contribute more significantly to the intrinsic signals than sub-threshold activation. Either way, it has been shown that the amplitude of the differential intrinsic signals is well correlated with spike rates in cat area 18 (Shmuel et al and Grinvald, 1996; Shmuel et al., unpublished results). A similar conclusion has been reached by Frostig and his colleagues in the rat whisker barrel system (Frostig et al., 1994).

Similarly, at longer wavelengths where light scattering is dominant, both sub-threshold synaptic potentials and action potentials are likely to contribute because both are known to produce light-scattering signals (Cohen 1973; Salzberg et al 1983). It seems that the quantitative answer to this important question might have to await further experiments.

1.2 Animal Preparation for Optical Imaging

The preparation of animals for optical imaging experiments is very similar to the conventional preparation for *in vivo* electrophysiological experiments. However, there are a few procedural aspects requiring careful attention.

Initial anesthesia is conventionally done with a mixture of ketamine and xylazine, after which a venal catheter is inserted and intubation or tracheotomy is performed. It is well known that anesthesia has a strong effect on the coupling between cerebral blood flow and neuronal activity (e.g., Buchweitz and Weiss, 1986), and therefore the level of anesthesia has to be carefully monitored and the anesthetic agent used for the experiments has to be chosen with great care. It is known that both barbiturates and gas anesthetics (halothane, isofluorane) work well for optical imaging; other anesthetics may be suitable as well. Nevertheless, it is important to emphasize that it cannot be taken for granted that an experiment using a different anesthetic will work. Moreover, it should be kept in mind that the effects of anesthetic agents may be species-dependent.

Since imaging of intrinsic signals at 590–605 nm measures the hemoglobin saturation, it is very important to ensure proper ventilation of the animal. It is therefore advisable to carefully monitor the value of expired CO_2. Additionally, non-invasive monitoring of blood oxygen saturation has proven very useful to control this parameter, which naturally influences the quality of the signals to a great extent. Moreover, the oxygenation of the tissue can be estimated by eye, and with some experience, it is possible to predict the quality of the functional maps from the visual appearance of the cortical tissue, judged by the difference in color between the pial arteries and veins.

Since the craniotomy for optical imaging experiments and the openings in the dura are relatively large (up to ~600 mm^2), pulsations of the brain due to respiration and heartbeat are a major problem. Therefore, an elaborate chamber system (a "cranial window") has to be used in order to stabilize the brain. This chamber, described in detail in the next section, has to be mounted on the skull with dental cement before the skull is opened. The normal procedure then is to make the trepanation of the skull, to mount the chamber with dental cement before taking the piece of bone out of the skull, and only after mounting the chamber, to remove the bone and open the dura. Great care has to be taken during trepanation or drilling not to damage the cortex. If the drilling is carried out with a high-speed drill, the production of excessive heat is particularly dangerous.

Opening of the dura can also be more problematic than in conventional electrophysiological experiments, since a large piece of the dura has to be resected and therefore large blood vessels can sometimes not be circumvented. These dural blood vessels can often be shut off by simply clamping them with a thread or with forceps. Alternatively, to avoid contact between blood and the exposed cortical surface, particularly problematic in the primate, one can cut the superficial dural vessels prior to the full resection of lower layers of the dura.

If, due to hypoventilation or for other reasons, a cortical edema develops, the large opening in the skull causes this condition to be much more traumatic to the cortex (and the experimenter). There are several ways to deal with this problem, and they include injection of high molecular weight sugars (e.g., mannitol), lowering the position of the body, hyperventilation, applying 10–20 cm of hydrostatic pressure in a closed chamber for a limited period of time, and puncturing the *cisterna magna* (call 1-800-cisterna magna).

1.3 The Setup

1.3.1 The Chamber

Optical imaging of intrinsic signals can, under favorable circumstances, provide activity maps with a spatial resolution better than 50 μm. In order to achieve this resolution, it is important that movement of the brain which normally occurs due to blood-pressure pulsations and respiration is minimized as far as possible. This can be achieved in several ways as follows.

First and foremost, it is important that an optimally designed chamber is used for the optical imaging experiment. Such a chamber is shown in Fig. 8. It can be made of stainless steel, with an inlet and an outlet to which tubing can be attached, and it can be

A

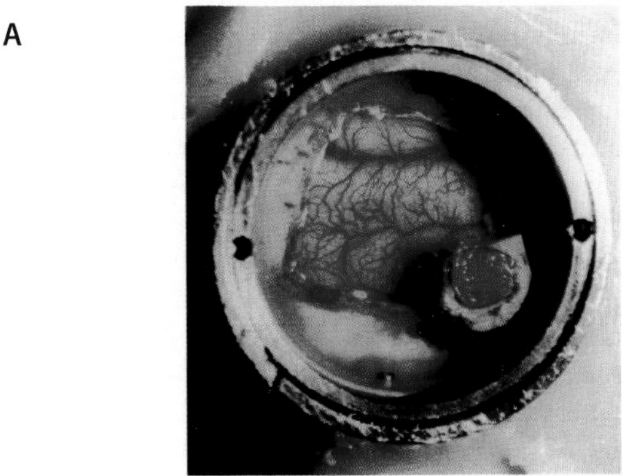

B

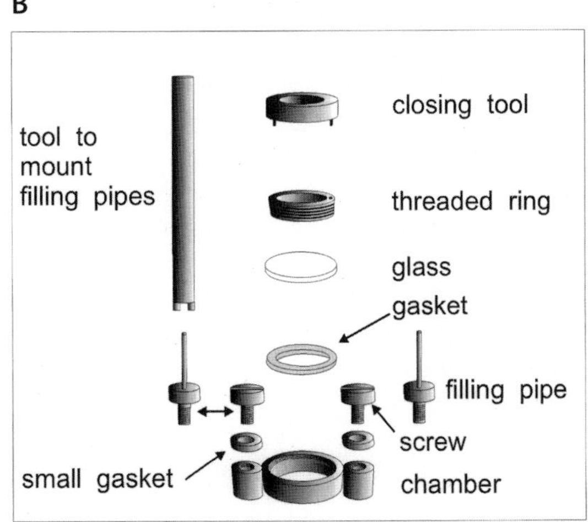

Fig. 8. Chamber for optical imaging. A: Photograph of an optical recording chamber mounted onto the skull of a macaque monkey. The exposed area is mostly V1 with a small strip of V2. The large vessel in the upper part of the cranial window marks the lunate sulcus. B: Schematic diagram of a chamber as it is used for chronic imaging. Chambers are made of stainless steel for acute experiments and of titanium for chronic experiments. The chamber is filled through tubing attached to the metal filling pipes. It is then closed with a round cover glass being pressed onto a gasket by a threaded ring that is screwed into the chamber. In the case of acute experiments the tubing can stay attached. For chronic experiments the tubing is exchanged for screws. (Modified from Bonhoeffer and Grinvald et al., 1996)

sealed with a round coverslip pressed onto a silicone gasket with a threaded ring. This chamber is mounted onto the skull with dental cement. To achieve a perfect seal, the inside of the chamber is treated with dental wax which is melted into the remaining gaps between the chamber and the skull. After the chamber is mounted, the remaining cerebrospinal fluid or saline is removed from the cortex. This can be very conveniently done with small triangles made from cellulose fibers (Sugi, Kettenbach, Eschenburg, Germany). The chamber is then filled with silicone oil (e.g., Dow Corning 200, 50°cSt) such that no pressure is applied to the cortex. This is best achieved by having the silicone oil flow into the chamber from an upright syringe without the piston. Adjusting the level of the syringe with respect to the chamber allows precise regulation of the pressure of the oil in the chamber (5–10cm above the cortical surface appears optimal). If the chamber is filled perfectly, that is to say without any air bubbles or cerebrospinal fluid droplets, this arrangement provides an ideal optical interface and, at the same time, stabilizes the brain perfectly.

In long-term experiments, it is essential to modify several features of the stainless steel chamber as described above. For such chronic recordings one has to be able to close the inlets of the chamber with screws as shown in Fig. 8B. Moreover, if the window of the chamber is large, it is important to have a metal lid that can be screwed onto the chamber instead of the breakable cover glass. Finally, it is advisable to construct the chamber not from stainless steel, but from titanium, because titanium – although difficult to machine – is very strong and at the same time very light and, above all, highly inert to bodily fluids. Even many months after implantation we have never observed any difficulties with a chamber made from this material. In long-term experiments, it is also of great importance that the chamber be mounted on the skull such as to prevent the chamber from detaching even after long survival times. This is particularly problematic in young animals in which the bone is often still relatively soft. It has proven useful to clean and degrease the skull with ether and in addition to place screws in the bone next to the chamber. These screws are then also covered with the dental cement for mounting the chamber and thus help to anchor the chamber firmly onto the skull. When taking these precautions, we have never had problems with chambers loosening from the skull.

Under some experimental circumstances it is essential to obtain electrical recordings simultaneously with the optical imaging data. To achieve this, we use a rubber gasket, which can be glued into a ~3 mm diameter hole made in the round cover glass (see the red rubber gasket in Fig. 8A). The sealed chamber is a necessity for successful optical imaging, but also represents a significant improvement for electrical recordings *per se*, which under these conditions are exceptionally stable over long periods of time (see Sect 3.2 for a detailed description of a more advanced chamber).

Another measure can be taken to further reduce the influence of heartbeat and respiration on the stability of the acquired images. The data acquisition can be triggered by the heartbeat synchronized with respiration. This is achieved by stopping respiration after exhalation for a short time (less than one second). Respiration is then started again when the next heartbeat is detected, e.g., by an appropriate Schmidt-trigger circuit. If the data acquisition is also started at that time, the images will always be collected in the same phase of heartbeat and respiration, and therefore the procedures of data analysis described below (division of the images) cause the noise originating from heartbeat and respiration to cancel out. Synchronization between heartbeat, respiration and data acquisition reduces the noise, in a well-sealed chamber, by a factor of approximately 1.5 (Grinvald et al., 1991).

1.3.2 The Macroscope

After an optimally stabilized cortex has been obtained, its image has to be projected onto the camera with the help of some type of lens. Initially photographic macrolenses were used to do this, but one problem with these lenses is that they have a large depth of field, causing the appearance of very large blood-vessel artifacts in the functional maps. These artifacts often hampered the observation of subtle features in the recorded maps. To alleviate this problem, Ratzlaff and Grinvald (1991) constructed a "macroscope": a tandem-lens arrangement with an exceedingly shallow depth of field.

This device is essentially a microscope, with a low magnification (around 0.5–10), composed of two front-to-front, high-numerical-aperture photographic lenses. The macroscope provides an unusually high numerical aperture compared to a commercial, low-magnification microscope objective. Consequently, this optical system has a very shallow depth of field (e.g., 50 μm, nominal for two coupled 50 mm lenses having an f number of 1.2). Therefore, when focused 300–500 μm below the cortical surface, the surface vasculature is sufficiently blurred and artifacts from the surface vasculature virtually disappear because they spread over a much larger area (Malonek et al., 1990; Ratzlaff and Grinvald 1991).

1.3.3 Lenses

The macroscope can easily be built by connecting two camera lenses "front-to-front". The magnification of this tandem-lens combination is given by f_1/f_2, where f_1 is the focal length of the lens close to the camera and f_2 is the focal length of the lens close to the cortex. To improvise a macroscope from conventional 35-mm camera lenses, one needs items 1 and 2 listed below, plus one or more of the combinations of lenses that are listed:

- C-mount to camera adapter (e.g., "Pentax" if one uses Pentax lenses).
- Adapter for the tandem-lens arrangement. A solid ring with proper threads on each side to connect the front part of each of the camera lenses. (This lens thread is usually used for standard camera filters.) To minimize vibration and to protect the lenses, it is advantageous to add some kind of rod to this adapter. This rod can then be used to attach the tandem lens to other solid parts of the camera manipulator.
- For a magnification of 1 (covering approx. 9×6 mm^2), use two 50 mm Pentax lenses with 1/f of at least 1.2. Alternatively, one can use a video lens with a shorter working distance (3 cm) but even higher numerical aperture (0.9).
- For a magnification of 2.7 (covering approx. 3.3×2.2 mm^2), use one 50 mm and one 135 mm lens. Pentax offers a 135 mm lens with numerical aperture of 1.8. If the 50 mm and 135 mm lenses are installed in the reverse order, then the tandem-lens will cover a very large portion of the cortex (approx. 22×14 mm^2).
- A 2× standard camera extender provides flexibility for additional magnification or demagnification.
- A zoom lens covering the range of 25–180 or 16–160 mm can be used as the top lens. A zoom lens has the advantage of allowing adjustment of the magnification without replacement of lenses. This flexibility is however achieved at the cost of a lower aperture that causes a larger depth of field.
- For imaging human cortex during neurosurgery, a zoom lens with a larger working distance starting at approximately 10 cm is preferable to the tandem-lens combination.

In the tandem-lens combination, commercial home-video CCD lenses may also be used as the lens next to the camera. However, the use of such lenses next to the cortex may be problematic whenever the working distance is important. The advantage of using the home-video lenses is that the numerical aperture of home-video CCD lenses is often larger than that of a 35mm camera lens.

The camera can also be mounted on a conventional microscope or an operating microscope, preferably one that offers a high numerical aperture and, consequently, a short working distance (5–7cm). Numerical aperture, illumination, working distance and exquisite mechanical stability should all be considered in the final design. The macroscope offers additional advantages in fluorescence imaging that will be discussed in Sect 2.7.2.

1.3.4 Camera Mount

The video camera should be rigidly mounted to a vibration-free support. The best arrangement is to mount the camera to an xyz-translator. The z-translator is used for focusing the camera. Preferably, it should have both a coarse, large-travel distance control as well as a fine focus control. Furthermore, it is advantageous to construct the camera holder to permit rotation of the camera around its optical axis, as well as its tilting in any desirable angle.

1.3.5 The Camera

Shot-noise

When there is a requirement to measure signals as small as one part in a thousand, the quantal nature of light has to be considered. The emission of light is a stochastic process in which the time intervals at which light quanta are emitted fluctuate randomly. Therefore, if one wants to measure a small change of one part in a thousand, one has to ensure that the additional photons measured are due to this small signal, and that they are not caused by the statistical fluctuations of the light emitting process. The number of photons that can be attributed to statistical fluctuations equals the square root of the total number of photons emitted. Consequently, the number of photons needed to detect a signal change of 0.001 with a signal-to-noise ratio of 10 is 100,000,000. Thus, since intrinsic signals are in the range of 0.001 fractional change of the absolute reflected light, the light intensity (and also the well capacity, see below) has to be chosen such that this number of photons will be accumulated during the recording time. It should be noted, however, that it is not necessary to accumulate this number of photons for every image frame, since later averaging by frame accumulation can also help to overcome the shot-noise limit. This topic is discussed in more detail below in the methodological section describing real-time optical imaging.

Video Cameras

Some 20 years ago, Schuette and collaborators were the first to attempt the use of video cameras to image cortical activity (Schuette et al., 1974, Vern et al., 1975). A decade later Gross et al. (Gross and Webb, 1984; Gross et al., 1985) took advantage of more modern video technology and used a video acquisition system with frame grabbers to measure voltage changes across neuronal membranes with voltage-sensitive dyes. Blasdel and Salama (1986) then used a similar technology and obtained spectacular images of the functional architecture of macaque visual cortex *in vivo*. It is clear that compared to photodiode arrays, the increased spatial resolution is achieved at the expense of temporal resolution: Video systems usually have a temporal resolution of at most 16.6ms (Kauer, 1988). However, for intrinsic signal imaging, time resolution is not a critical parameter. A more important underlying problem is the limited signal-to-noise ratio of standard video cameras of approximately 200:1. However, some modern cameras have overcome this problem and offer a signal-to-noise ratio close to 1000:1.

Slow-Scan CCD Cameras

Slow-scan digital CCD cameras offer a very good signal-to-noise ratio while retaining the advantages of higher spatial resolution and moderate cost and complexity. The disadvantage of long, relatively low readout speed is of little importance for the slow intrinsic signals and the cameras are therefore well suited for such signal experiments. Such cameras were first introduced into biology by Connor (1986) to study the distribution of calcium ions in single cells. Ts'o and co-workers (1990) then used these cameras to image intrinsic signals and the functional architecture of the visual cortex in the living brain.

Several parameters of slow-scan CCD cameras influence the quality of the functional maps. Due to their importance we discuss some of these aspects in detail.

Well Capacity

The well capacity denotes the number of electrons that can be accumulated by one pixel of the CCD chip before there is an overflow of charge. Therefore, it is important that CCD cameras have well capacities as big as possible. The capacity is normally directly related to the area of a single pixel on the silicon wafer, which constitutes the light-sensitive area. Good well capacities are in the range of 700,000, but cameras with somewhat smaller well capacities can also be used. To our knowledge the existing CCD chips have limited well capacities not because of fundamental engineering problems, but simply due to the fact that most other applications do not require a large well capacity. One way to increase the effective well capacity in existing chips is to use "on-chip binning" whereby the charge for several adjacent pixels is combined. However, in the latter case, there is an additional practical limitation, comparable to the well capacity, which is the capacity of the readout register. Normally, on-chip binning is limited to 2x2 or at most 3x3, at the maximally permitted light level.

Frame Transfer

Due to the functional design of the CCD camera, illumination of the area containing the image information should be avoided by all means during read-out of the CCD chip. Therefore, in many cameras a mechanical shutter is closed during the read-out time. For optical imaging purposes, this approach is problematic since read-out times for 12- or 16-bit digitization, even with relatively low spatial resolution, are in the order of 50 ms. If the shutter is closed during this read-out time, successive frames would not really adjoin temporally. Additionally, due to the large number of exposures in a single experiment and the limited lifetime of a mechanical shutter, this mode of operation is not practical. One possible solution is to use a camera providing the so-called frame-transfer mode. Utilizing this mode, half of the light-sensitive area of the CCD chip is covered with an opaque mask. After one exposure the accumulated charges from this illuminated area are shifted to the light-insensitive area within a short time of approximately one millisecond. The new exposure can then take place immediately while the information of the previous frame can be read out from the "light protected" area. Since using this mode of operation enables an optical imaging experiment to be run with minimal shutter actions and additionally sequences of frames to be recorded which are truly "back to back", it is clearly the preferred mode of operation for a CCD camera in such experiments.

1.3.6 Differential Video Imaging

Several modern but economic video cameras are based on CCD-type sensors rather than the old vidicon targets, and offer much better signal-to-noise ratios of close to 1000:1. Eight-bit frame-grabbers are now an industry standard offering a cost-effective method to digitize the video signal at video rates. However, when a high-quality camera output is digitized using only 8 bits, one loses the advantage of the low-noise video camera. In recent years, several image enhancement approaches have been developed. However, most of these techniques use procedures that enhance the image only after its initial 8-bit digitization. The disadvantage of such approaches is that any changes in intensity that are smaller than 1 part in 256 levels (8 bits) are lost because the image is digitally recorded with a precision of only 8 bits. Another common approach used in image enhancement is to subtract a DC level from the data and amplify the resulting signal prior to its digitization. This approach, however, is applicable only to flat images with very low contrast.

An alternative approach applicable to image enhancement is the use of analog differential subtraction of a stored "reference image", from the incoming video images, in order to "flatten" the images. An apparatus based on this principle is commercially available under the name Imager 2001 (Optical Imaging, Germantown, N.Y., USA, http://www.opt-imaging.com). This apparatus offers optimal enhancement of both low-contrast and high-contrast stationary images, thus producing images that are superior to those obtained by alternative image enhancement methods. It uses analog circuitry to subtract a "selected" reference image from the incoming camera images, and then performs a preset analog amplification (4–20x) of the differential video signal. Only then an 8-bit image processor digitizes this "enhanced" differential signal. With this approach, the accuracy of the acquired images is limited only by the signal-to-noise ratio of the camera used. This noise can be further reduced to a desired level by "on-the-fly" averaging, trial averaging and off-line frame averaging. The digital image of a given reference image, and the corresponding enhanced sequence of images, can be combined later. The resulting accuracy is comparable to 10–13 bit digitization and therefore better than the signal-to-noise ratio of the camera.

Another significant advantage to this approach is that the enhanced differential image is displayed on the monitor in real time, at a video rate, thus providing important on-line feedback for the experimenter. This feedback enables immediate problem detection and resolution at the very first stage of the experiment, rather than waiting until the data have been analyzed (which in many cases is only after an hour, in a conventional CCD system). Due to the strongly amplified picture, one can immediately notice minor optical changes which could later result in large artifacts. These include moving bubbles of air or cerebrospinal fluid in the closed chamber, excessive noise due to imperfect stabilization of the cortex, minute bleeding, excessive vascular noise, etc.

Performance Comparison of Video Imaging and Slow Scan CCD Cameras
Cooled CCD digital cameras excel in providing high-quality images at low light levels. However, at moderate light levels, when the detector noise is not the limiting factor, high-quality video cameras can provide better images because of their higher frame rate. Video systems can digitize data of up to 768 by 576 pixels at video rate at an accuracy of 10–13 bits. Therefore, under these circumstances video systems can provide even better images than high-grade digital CCD cameras. This can be demonstrated by challenging both cameras with a bright image containing a modulated signal of only 1 part in 1000. If both systems are operated under the optimal conditions, the digital CCD images have a signal-to-noise ratio approximately threefold worse than images acquired by the differential video system. Furthermore, in this test the binned CCD image has only 192x144 pixels, whereas the video image has a resolution of only 768x576 pixels. Further binning of the video image could therefore be used to achieve a tenfold advantage in signal-to-noise. This advantage, however, would not be fully realized during a functional imaging experiment because the "biological noise" is the limitting factor.

1.3.7 Biological Sources of Noise

To assess the noise level associated with functional imaging, the most appropriate procedure is to test the reproducibility of activity maps. This test shows that the noise is usually composed of high spatial-frequency components, which are dominated by shot-noise rather than biological noise, and low spatial-frequency components, for which the reverse is true. Since normally the spatial frequency of the biological noise is similar to or smaller than the periodicity of the functional domains of interest, it appears that the biological noise limits the reproducibility of the functional maps.

The main source of this biological noise is presumably very slow changes in the overall saturation level of blood in the vascular bed. A slow change of only 1% in the oxygen saturation level of hemoglobin will not significantly affect the physiological state of the cortex, but will introduce an optical change, which is much larger than the small mapping signals. Such changes then cause both large blood vessel artifacts and intensity changes over a cortical area larger than the size of a functional domain, originating from the capillary bed. Another, related, phenomenon is the regular slow oscillations in the saturation level of the blood occurring at frequencies of 0.08–0.18 Hz. These oscillations can be directly visualized using the real-time differential video enhancement system. They appear as slow waves of darkening which scan the cortex. This darkening may be much larger than the size of the mapping signal, and since it is not synchronized to the heartbeat or respiration, it introduces large slow noise. Although biological noise cannot be totally eliminated, procedures have recently been found which help to minimize its effect on the functional maps. Such procedures will be discussed in Sect 1.5. Additional information on optimization of the signal-to-noise ratio is provided in the methodology section for voltage-sensitive dyes.

1.3.8 The Illumination

The choice of wavelengths for the illuminating light depends on the sources of intrinsic signals that one is attempting to utilize. Moreover, it is important that the wavelengths used provide sufficient penetration into the tissue, otherwise only signals originating from very superficial cortical layers will be measured. In situations where the oxymetry component dominates the intrinsic signal it is advantageous to use a filter of 595–605 nm wavelength. However, in many cases, there is also a strong light-scattering component that is useful for functional mapping. In these cases considerably longer wavelengths (up to 750 nm) are preferable, because they provide deeper penetration into the tissue (up to 2 mm). Limiting the wavelength to750 nm is done for strictly practical reasons, since this wavelength is still visible to the human eye, and it is thus much easier for the experimenter to adjust the illumination to achieve an evenly illuminated brain. However, functional maps have also been obtained at wavelengths of 900 nm and above.

In order to relate the obtained activity maps to anatomical landmarks, it is also useful to record pictures of the blood-vessel pattern. This pattern can be seen particularly well if the brain is illuminated with green light. A standard filter of 546 nm wavelength is therefore very useful in order to obtain high contrast pictures of the vascular patterns. In addition, it is often useful to reduce the lens aperture to gain a high depth of field and thereby eliminate blurring of cortical vasculature due to a "curved" cortical surface.

1.3.9 How to Choose the Wavelength for Imaging

Based on experience accumulated in our lab, the best signal-to-noise ratio of the functional maps is obtained at the peak of the difference spectra between oxy- and deoxyhemoglobin, using a filter with peak transmission at 595–605 nm (orange color). This result has been repeatedly observed in cats and monkeys, when imaging was performed at multiple wavelengths. However, in preparations that are not at the optimal conditions where the coupling between electrical activity and the micro-circulation is impaired, it is better to use near infrared light for functional mapping. Thus, we recommend to start the imaging at orange wavelength and only if the intrinsic signals are slower than normal or the vascular noise is large (e.g., 0.1 Hz oscillations) to proceed with near infrared imaging. Recently it has been argued that the blood-vessel artifacts are more prominent when using orange light relative to those obtained using longer wavelengths (McLoughlin et al. 1998). Our results, obtained in the "best" preparations, are not in line with these results.

The optimal wavelength for functional imaging may depend on the cortical area being imaged. It has been repeatedly observed in several species that the use of green light for imaging the auditory cortex provides better tonotopic maps than the use of orange light (e.g., Harrison et al. 1998). This topic warrants further exploration. We speculate that this difference between the auditory cortex and the visual cortex is related to differences the relative amplitudes of evoked and spontaneous activity in these two sensory modalities.

1.3.10 Light Guides / Modes of Illumination

Although theoretically epi-illumination through the lens used to obtain the images should be ideal to achieve a uniform illumination, practical experience shows that in general this way of illuminating the brain is not advantageous. Since the brain is not a flat structure and since some parts of the brain absorb light more strongly than others, epi-illumination will necessarily cause an uneven image which is difficult to correct. Therefore, for *in vivo* imaging studies, two or three adjustable light guides attached to the camera or the stereotaxic frame have proven by far the most useful system providing relatively evenly illuminated images of the cortex.

1.3.11 Lamp Power Supply

A high-quality regulated power supply is absolutely essential in order to achieve a strong and stable light source. It should provide an adjustable DC output of up to 15 volts and 10 amps. Ripple and slow fluctuations should be smaller than 1:1000. An excessive ripple can usually be reduced by adding very large capacitors in parallel with the output.

1.3.12 Lamp Housing

A standard 100 W, tungsten halogen lamp housing (e.g., Newport, Oriel, and all microscope companies) with a focusing lens is suitable for the illumination. It should have an adapter that is connected to the lamp on one side, with room for at least two filters. The other side of the adapter should connect to the back of a dual- or triple-port light guide. It is preferable to use a liquid light guide rather than a fiber optic one, since the former provides more uniform illumination. The front portion of the light guide should be attached to adjustable lenses, permitting proper output focusing of the light guide on the cortex. Schott (Mainz, Germany) offers suitable light guides and small, adjustable lenses that attach next to the cranial window.

1.3.13 Filters and Attenuators

The considerations determining the choice of wavelengths used in optical imaging experiments have been discussed above in Sect 1.3.9. A "starter kit" of filters would include the following interference filters:

- Green filter 546 nm (30 nm wide)
- Orange filter 600 nm (5–15 nm wide)
- Red filter 630 nm (30 nm wide)
- Near infrared filters at 730, 750 or 850 nm (30 nm wide)
- Heat filter KG2
- Long wavelength heat filter above 720 nm: RG9

A 3 OD attenuator ([a]1000 attenuation) is also often useful in order to artificially produce a signal of 1 in 1000 to test the apparatus (see Sect 1.4.5).

1.3.14 Shutter

It is advantageous, in particular when experimenting with higher light intensities, to be able to control the illumination so that the cortex is illuminated only during data acquisition. To achieve this, the data acquisition program controls an electro-mechanical shutter, which is mounted between the lamp and the light guides. It is important to ensure that this shutter introduces only minimal vibrations into the system. Good shutters can be obtained from a variety of sources (e.g., Uniblitz, Prontor).

1.4 Data Acquisition

1.4.1 The Basic Experimental Setup

The basic experimental setup for optical imaging experiments was shown in Fig. 1 above. The animal head is held rigidly in a stereotaxic frame in the case of the anesthetized animal or a head holder for the awake animals. Because the signals are small, vibrational noise poses a serious problem. Therefore, it is recommended to use a vibration isolation table, particularly if the experimentation room is on a high floor of a building, close to the subway, etc. Microphonic noise should also be avoided. It is highly recommended to eliminate any relative vibrations among all the components related to the optics and the preparation. For example, in the awake, behaving monkey, after proper focusing, the camera lens is directly locked onto the skull itself.

1.4.2 Timing and Duration of a Single Data-Acquisition Trial

Since the time course of the intrinsic signals is of considerable importance for the proper evaluation and analysis of the data, it is advisable to perform optical imaging experiments such that the time course can be reconstructed from the data. A practical approach is to divide the data acquisition time into 5–10 "frames" where the data of each frame is stored separately on the computer's storage device. These frames can later be analyzed separately or averaged together. If, for instance, the data acquisition time is 3 s and 10 frames are stored, the duration for every frame amounts to 300 ms. Such a 300 ms frame can be a single exposure on a slow scan CCD camera or the sum of 10 genuine video frames. Hereafter, when we use the term frame, we refer to an image produced in the above manner and not to a genuine video frame.

For obtaining functional maps, the optimal timing and duration of the data acquisition relative to the stimulus onset have been clarified by studies of the multiple sources of the intrinsic signal. First, as discussed above, the various components of the intrinsic signal have different time courses. Second, the individual signals are physiologically regulated with different spatial precision. Two of the components, the oxygen delivery and the light-scattering signal, offer the best spatial resolution. Furthermore, they both show a faster rise time relative to the blood flow component. Therefore, one should aim at recording signals a few hundred milliseconds after the stimulus onset. However, since it is essential in some methods of data analysis to have one "baseline frame" (see Sect 1.5.7), one should actually start to acquire cortical images at least one frame duration prior to the onset of the signal.

How long should the duration of a single trial be? Because the blood-flow component peaks after 3–5 s and because it offers lower spatial resolution, it is preferable to stop data collection after approximately 3–4 s. Another reason for limiting the trial duration is related to the hyperemia induced in the cortex by prolonged stimuli, resulting in large vascular noise. Therefore, usually the best signal-to-noise ratio is obtained using a duration of 2 s for the stimulus and a period of 3 seconds for data acquisition. It should be noted, however, that the studies on the mechanisms of intrinsic signals were conducted mostly in anesthetized cats and monkeys, and it is not clear whether these results will also hold for other species or in awake animals. Furthermore, it has been found in several species that the optimal parameters such as wavelength and stimulus duration depend on the sensory modality that a given cortical area represents. For example, as already mentioned, several groups have concluded that in auditory cortex better maps are obtained at 540 nm illumination, thus imaging mostly blood volume changes, rather than at ~600 nm which emphasizes oxymetric changes.

1.4.3 Inter-Stimulus Intervals

A closely related issue is the choice of inter-stimulus intervals. The time course of intrinsic signals shows that the signals themselves as well as the functional maps decay back to baseline in 12–15 s for a stimulus lasting 2 s. Under such conditions the stimulus interval should not be too short, in order to avoid systematic errors in the resulting functional maps. However, in practice, the choice of excessively long stimulus intervals also results in maps of lower quality, since fewer images can be averaged in the same amount of time. Moreover, systematic errors can at least partly be avoided by randomizing the sequence of stimuli whenever possible. An inter-stimulus interval of 8–12 s has proven a good compromise.

1.4.4 Compression Data

High-resolution optical imaging can produce vast amounts of data. Ideally, to maximize the quality of the map during off-line analysis, one would greatly benefit from storing every single frame acquired. Normally, this is impractical. The amount of storage space needed for a typical experiment lasting one hour, with a data acquisition time of 3 s and an inter-stimulus interval of 10 s, would already amount to 30 Gigabytes (3 s × 30 video frames/s × (768×576) pixels × 360 trials × 2 bytes/pixel (30 Gigabytes). Twenty hours of data collection (not an exceptional amount) would then already require 0.5 Terabytes. Beyond the problem of storage space, enormous amounts of time would obviously be required for data transfer and image analysis. These considerations then call for a massive reduction in the amount of data. The first reduction in data occurs when genuine video frames are accumulated into the data frames described above. This usually reduces the amount of data by a factor of 20. Moreover, data accumulated under identical stimulus conditions are normally averaged 8–32 times, which again reduces the amount of data by this number. Thus, the first two procedures reduce the amount of data by a factor of ~400, so that the resulting data can be stored more easily. Nevertheless, these experiments still result in large amounts of data, and it is therefore advisable to have a set of programs that quickly further reduce the amount of data, for instance, by 2×2 or 3×3 binning and by additional averaging over time (summing the single frames together and/or adding the different "blocks" of data). Thus, further data size reduction substantially decreases the time needed for the initial data analysis. Later, of course, more sophisticated analysis can be performed on the original more complete data set.

There may be exceptional conditions, however, such as the imaging of the human brain during neurosurgery, in which it is advisable to store every single frame. However, under most other conditions it is advantageous to compress the data as discussed above, at least for the initial analysis.

1.4.5 Testing LED

Since optical imaging experiments are relatively complex, it is essential that the apparatus be thoroughly tested before an experiment is performed. Two issues are particularly important: 1) Test the signal-to-noise ratio of the system, and 2) Test whether data acquisition and data analysis match properly.

1. The following engineering procedure may be used to generate a test signal comparable to a typical intrinsic signal recorded from a living brain. An LED display of the number "8" (with 7 individual LED segments) is connected to the data acquisition system such that for each of the stimulus conditions, the different segments of the LED display are switched on. The brightness of this LED is regulated such that it uses the full dynamic range of the camera. This brightness is then attenuated by a factor of 1000 with an optical attenuator of 3 OD. If this whole arrangement is then illuminated with red light, so that its brightness is again almost at the saturation level of the camera, one has a device that produces modulations of 1 in 1000 at a relatively high absolute light intensity. The optical imaging apparatus can then be tested by acquiring data under these conditions, and seeing whether this very weak modulation can be picked up by the system with an appropriate signal-to-noise ratio.
2. In order to test the consistency of data acquisition, data analysis and also stimulation, it is very useful to have a visual stimulator produce patterns on the screen, which can easily be distinguished from each other. If this visual stimulator is then controlled by the data acquisition program, in exactly the same manner in which the real stimulus would be controlled, and if the camera is pointed directly onto the screen, one has a very simple testing procedure. Data analysis of the pattern that the camera imaged from the screen will immediately reveal any inconsistencies in the stimulation, data-acquisition and data-analysis procedures.

1.5 Data Analysis for Mapping the Functional Architecture

1.5.1 Signal Size

Intrinsic signals as can be measured from the living brain are small. In optimal cases, the changes in light intensity due to neuronal activity are no more than 0.1–6% of the total intensity of the reflected light. One difficulty in extracting the activity-related functional maps from the images is that the biological tissues to be investigated can never be illuminated in a perfectly even manner. The evenness of the illumination is always one to two orders of magnitude worse than the signal to be detected. An additional problem is that the biological noise associated with these measurements is in many cases larger than the signals themselves. Therefore, proper analysis procedures must be applied to extract the small signal of interest from the raw data.

1.5.2 Mapping Signal

In order to fully understand the data analysis performed in optical imaging experiments, it is important to distinguish between the *global signal* and the *mapping signal.*

The definition of the mapping signal is conveniently explained following the experimental results shown in Fig. 9. In this experiment the cortical surface was mapped using both single unit recordings and optical imaging. Panel A shows the change in intensity of the reflected light measured at a wavelength of 605 nm. This measurement was carried out from site "a" of the cortical map shown in panel C (**?????**). At this site, units responded vigorously to a horizontal moving grating and did not respond to a vertical stimulus. Note, however, that at this cortical site intrinsic signals were also observed for the ineffective stimulus. In fact, the amplitude of the reflected light signal is reduced by only 30% relative to the optimal stimulus. These changes in reflected light relative to baseline are called the *global signals.* Evidently even at a wavelength of 605 nm, the signals are not limited to the cortical sites showing spiking activity. The difference in reflected light for the two signals is shown in the bottom trace of panel A. The opposite situation is shown in panel B, where the optical signals were taken from a cortical site in which single-unit responses were observed only for the vertical stimulus (site "b" in panel C). The difference signal between activation with a vertical grating and that with a horizontal grating is referred to as the *mapping signal.* It is defined as that component of the global signal whose amplitude and spatial pattern correspond to the pattern of the supra-threshold electrical activation of the cortex.

It is important to keep in mind not only that the mapping signal is usually much smaller than the global signal, but also that the time course of the mapping signal can be substantially different from the time course of the global signals.

1.5.3 Cocktail Blank vs. Blank

In order to obtain activity maps from the cortex, one has to acquire images while the cortex is stimulated. These images must then be compared to a baseline image to correct for the uneven illumination. The easiest way to accomplish these two goals in one step

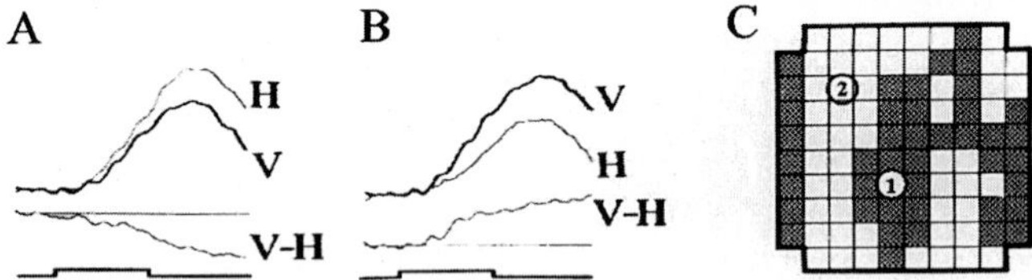

Fig. 9. The global and the mapping component of intrinsic signals. Demonstration of the spread of the intrinsic signal beyond the site of cortical spiking activity. A: Amplitude and time course of the reflected light signals evoked by drafting gratings. Trace H shows the response to a grating of horizontal lines whereas trace V shows the response to gratings of vertical orientation. These two traces were taken from a cortical site labeled a in panel C, where single-unit responses were detected *(only)* in response to gratings of horizontal lines. The difference in the signals is the trace labeled V-H. Trial duration 2.6 s. B: Similar to A except that the optical signals were recorded from cortical site b where units responded exclusively to the gratings with vertical lines. Note that the time course of the mapping component (traces labeled V-H) is very different from that of the global signals. C: A two-state map of orientation columns. The shaded area shows cortical regions where the reflected light signal was larger for the H grating than for the V grating. (Figure modified from Grinvald et al., 1986.)

is to divide the respective activity maps by a cortical image which is independent of the stimulus used. This can be done in two ways. The first is to use an image of the *unstimulated cortex* and take it as the baseline image for all subsequent analyses of activity maps; the images obtained under the different stimulus conditions are then divided by this so-called blank image. The second possibility is to try and obtain an image of the *uniformly activated* cortex, the so-called cocktail blank. The cocktail blank can be obtained by presenting a set of stimuli which, between them, supposedly activate the cortex in a relatively uniform manner. The individual responses to each of these different stimuli are then summed up to produce the cocktail blank. This resulting image is then taken as the baseline image and all activity maps are divided by this "cocktail blank."

Both procedures have their advantages and disadvantages. The advantage of taking a blank picture (i.e., the inactivated cortex) is that no assumption is made about the complete set of stimuli that are required to activate the cortex uniformly. Therefore, it is in a way the "purest" way to treat the data. The disadvantage is, however, that the blank picture is obtained from an unstimulated cortex. This can be problematic, for example, when analyzing iso-orientation maps. In primary visual cortex the respective activity maps (obtained with gratings of different orientations) have to be corrected for uneven illumination. If this is done with a genuine "blank", a stimulated cortex is compared with unstimulated cortex. This often causes very strong blood vessel artifacts in the maps, and therefore some structures which can be seen when using the cocktail blank may be grossly distorted in these maps.

Therefore, in many instances the cocktail blank is the better choice for obtaining good orientation-preference maps in the visual cortex, given the only difference between the single-activity map and the baseline image is the orientation of the stimulus, and the picture is not confounded with an additional difference in overall activity. The disadvantage, however, is the requirement to make assumptions about the functional architecture of the respective cortical area. For example, when recording iso-orientation maps from the primary visual cortex of the cat, it is normally assumed that a complete set of all orientations activates the cortex evenly, and this stimulus set is thus used to calculate the cocktail blank. Figure 19 in Bonhoeffer and Grinvald (1993) shows that this assumption may not always be true. In this case, the cocktail blank obtained for all different orientations was divided by the blank. It is apparent that a patchy structure remains in the resulting map, suggesting that the combination of all different orientations (at the one spatial frequency used) does not evenly activate the cortex. In fact, it was speculated by Bonhoeffer and Grinvald (1993) that this clear pattern obtained when the cocktail blank was divided by the genuine blank might indicate that spatial-frequency maps may exist in cat area 18. Indeed, such spatial-frequency maps have recently been demonstrated using optical imaging (Shoham et al., 1997). This example shows that under some circumstances a cocktail blank can be inadequate, since it imposes a spatial pattern onto the activity maps. It is therefore important to perform such tests, and critically inspect the resulting map, if any. If a map does emerge from this procedure, it is important to determine the amplitude of this map and compare it to the amplitude of the functional maps under consideration. This comparison then provides an estimate of the extent of distortion of these functional maps using a specific cocktail blank. Furthermore, this procedure can also give hints to the existence of additional stimulus attributes that could be represented on the cortex in a clustered fashion.

When analyzing optical imaging data, the above problems have to be treated cautiously. Note that in some cortical areas like the somatosensory cortex, the cocktail-blank approach is not relevant. It should always be kept in mind that the assumptions made for the analysis can influence the appearance of the maps. Often neurophysiologists are interested in maps that correspond to electrical activity rather than to the spread of irrelevant global signals beyond the electrically active region.

1.5.4 Single-Condition Maps

For standard analysis, all the stimulus conditions acquired with one particular stimulus are summed, and this picture is then divided by a "cocktail blank". Figure 10 shows four orientation maps from the visual cortex obtained in such a way. Cortical images were acquired while a moving grating of one orientation stimulated the cat's visual system. These images were then divided by the cocktail blank obtained by summing the images acquired for all different orientations. In the resulting maps, shown in Fig. 10, very clear activity patches can be seen. As expected from standard electrophysiology, the pattern for orientations which are 90° apart are roughly complementary.

In some previous publications there has been some confusion as to how the data analysis was performed. Thus, it is important to distinguish between "differential maps" and "single-condition maps". Single-condition maps are calculated by taking the activity map obtained with one particular stimulus and dividing this image by the blank image (either cocktail-blank or blank). The resulting map then shows the activation pattern that this particular stimulus causes. In differential maps further assumptions are made about the underlying cortical architecture. To maximize the contrast, one activity map is divided by (or subtracted from) the activity map that is likely to give the complementary activation pattern. As noted, specific assumptions (which may or may not be true) underlie this analysis, and therefore these data (despite their improved signal-to-noise ratio) must be treated with caution. One example for the difficulties in interpreting differential maps is related to the question of what the gray regions in such a map represent (e.g., pixel value of ~128). These regions can correspond either to cortical areas not activated by the two stimuli used, or alternatively to cortical regions strongly activated by the two stimuli but with equal magnitude. Single-condition maps using the genuine blank are free from these problems. In fact, when this approach was first introduced by Ts'o and Grinvald it led to the first optical imaging of stripes in area V2 in the macaque monkey. These stripes can be revealed by subtraction of an image obtained

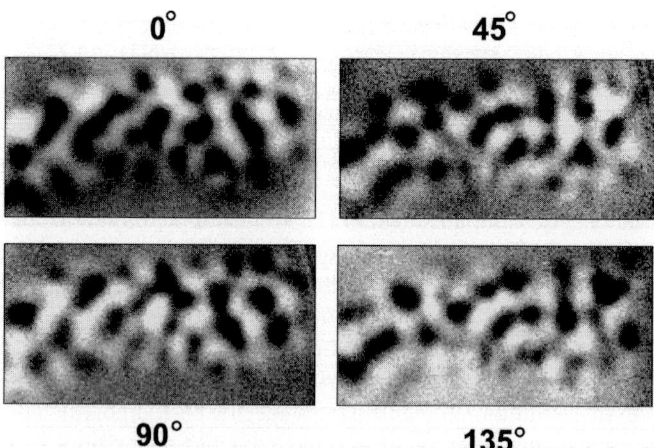

Fig. 10. Single-condition activity maps of cat visual cortex for four different orientations. Four images acquired during stimulation of a kitten with moving gratings of different orientations. To correct the effect of uneven illumination of the cortex and non-specific vascular responses, all images were divided by the sum of the images obtained for all the different orientations (cocktail blank). Dark areas in the image are regions of stronger light absorption and hence regions of strongest activity. Note that the signal-to-noise ratio in this experiment is extremely good. Bonhoeffer and his colleagues have found that, by and large, the signal-to-noise ratio in young kittens is by far better than in adult animals. (Modified from Bonhoeffer et al., 1995.)

during monocular stimulation from the "blank" image. They cannot be detected by differential analysis using data obtained with right and left eye stimulation.

While discussing the relative merits of using differential maps, cocktail blank and the genuine blank, it is also important to remember that the intrinsic signals used for mapping are not directly related to electrical activity. Most importantly, the intrinsic signal spreads beyond the region of spiking activity in a patch of cortex. The amount of such spread depends on the wavelength used. Therefore, if a genuine blank is used, the resulting map is in fact a map of the global signal and does not exactly correspond to a map of the electrically active cortex. This problem is largely minimized when a proper cocktail blank is used. Similarly, with differential maps much of the global component may be canceled out, especially if the spatial frequency of the activated domains is relatively high.

If differential maps must be calculated because single-condition maps do not have an acceptable signal-to-noise ratio to reveal the desired activity pattern, two different methods can be used to calculate these maps. One possibility is to calculate the ratio between the maps A and B and thereby produce the differential map. The other possibility is to subtract the two maps and then divide the result by a general illumination function like the cocktail blank: (A-B)/cocktail°blank. The rationale for the first calculation is to see the ratio between the activation patterns A and B. The second calculation is somewhat more intuitive: It assumes that the difference between the two maps is the important entity. This difference then has to be corrected for the uneven illumination. Although, at first glimpse, these two calculations seem very different, it can be demonstrated that they are equivalent under the assumption that the amplitude of the maps is small compared to the absolute intensity of the image (Bonhoeffer et al., 1995).

1.5.5 Color Coding of Functional Maps

In some cases the full information contained in the data can best be displayed with a color code (Blasdel and Salama, 1986; Ts'o et al., 1990; Bonhoeffer and Grinvald, 1991; Blasdel, 1992a,b). For instance, for a more comprehensive analysis of the organization of iso-orientation domains as displayed in Fig. 10, a color-coded display is advantageous. The responses for the four different gratings in such an experiment can be summed vectorially on a pixel-by-pixel basis. For every point in the cortex, four vectors are summed, their lengths being the magnitude of the "single-condition responses" and their angles corresponding to the orientation of the gratings that produced the responses. (To map the 180° onto a full 360° circle, the angles of the different orientations are first multiplied by two). There are several ways to display the results of the vector analysis, each of which emphasizes a particular aspect of the organization of the iso-orientation domains. One approach introduced by Blasdel and Salama (1986) is to display only the angle of the resulting vector (having divided it by two) to produce an "angle map". Thus, colors from yellow through green, blue, red and back to yellow are codes for the angle of the preferred grating for this piece of cortex. An "angle map" from a similar experiment in a large cortical area of the primary visual cortex of the macaque monkey is shown in Fig. 11. Additional information may be provided by also displaying the magnitude of the resulting vector as the brightness (i.e., intensity) of the color. The resulting "polar" map, first introduced by Ts'o et al. (1990), then shows the preferred orientation (hue of the color) and the magnitude of the vector (intensity of the color) at the same time. Note, however, that a vector with a low magnitude can either be the result of various stimulus orientations evoking the same strong response, or simply of a weak response to all orientations. HLS maps were designed to overcome this ambiguity. For details see Bonhoeffer and Grinvald (1996).

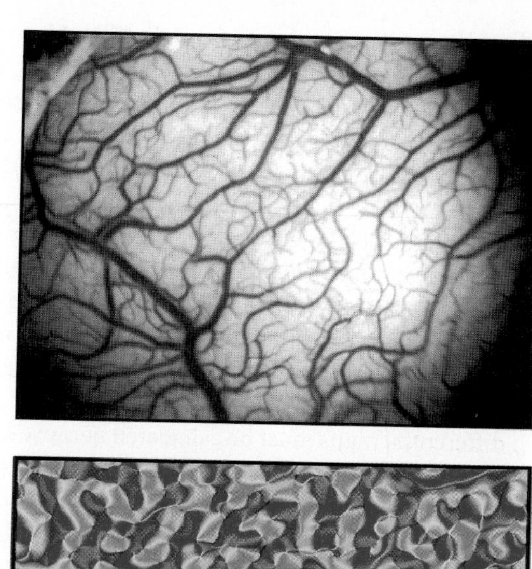

Fig. 11. Orientation preference map obtained by vector addition: Angle map showing the orientation preference for every region of the imaged cortex. In computing the local orientation preference, the activity maps obtained with different orientations were added vectorially on a pixel-by-pixel basis. The angle of the resulting vector is then color-coded according to the scheme at the bottom of the figure. Yellow stands for sites responding best to moving gratings of horizontal orientation, regions preferring moving gratings of vertical orientation are coded in blue, etc. This map very clearly shows pinwheel-like structures around orientation centers.

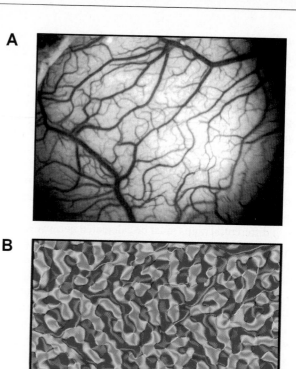

Whereas color-coding provides the reader with useful information, it may also be used to paint weak data in ways that disenables the skeptical reader to evaluate the data independently. We feel that it is important to show the grayscale raw data alongside color-coded data resulting from the additional processing.

1.5.6 Reproducibility of Optical Maps

To demonstrate the reliability of the data, reproducibility tests should be performed and described whenever possible. An example for reproducibility of angle maps is depicted in Fig. 5 in Bonhoeffer and Grinvald (1993). The reproducibility of the maps was quantitatively determined by calculating the RMS for the differences between the optimal angles detected at each pixel. This amounted to only 9.8°. Furthermore, for 88% of the pixels the deviation was less than 10°.

The high degree of reproducibility shown there validates the precision and reliability of the cortical maps obtained. The reproducibility in the location of the pinwheel centers suggests that the resolution of differential optical imaging may be better than 50 μm. Reproducibility studies are required for every optical imaging experiment. This is essential, in particular when the functional maps are weak, thus providing a very effective method for distinguishing artifacts from solid, reproducible maps of the cortical functional architecture.

1.5.7 First Frame Analysis

In Sect 1.3.7, we already mentioned that slow noise of biological origin is often the limiting factor in producing high-quality functional maps. One approach used to remove slow noise with a frequency lower than 0.3 Hz is to use the so-called first frame analysis introduced by Shoham. Whenever the noise is significantly slower than the duration of a trial, it is manifested as a fixed pattern in all the frames acquired during that trial. If the first frame is taken prior to any evoked response, it will only contain the slow noise but no signal that depends on stimulus-evoked activity. Thus, to minimize the slow noise, the first frame is subtracted from all subsequent frames before any additional analysis is done. Figure 17 in Bonhoeffer and Grinvald (1996) illustrates the remarkable improvement this approach often yields in the functional maps. This figure depicts an ocular dominance map using the standard differential analysis for activity maps evoked by a small stimulus in the macaque primary visual cortex. The ocular dominance bands and the retinotopic border of the stimulated area are hardly detectable. In contrast, the functional maps obtained after the first frame was subtracted from all subsequent frames clearly reveal the typical pattern of ocular dominance columns as well as the retinotopic border of the stimulus. Similar results for orientation maps are also shown in that figure. These data further underline the importance of acquiring at least one frame prior to the onset of any evoked response. Experience accumulated in our lab indicates that this approach can often salvage experiments that otherwise would have been useless due to their large vascular artifacts. Despite the significant improvements achieved with first frame analysis, this procedure also has its disadvantage: the introduction of high-frequency noise into the maps. However, this problem can be solved by obtaining a few frames prior to the onset of the evoked response and averaging them out to minimize this high-frequency noise.

1.5.8 More Sophisticated Image Analysis

The image analysis described above is rather simple relative to the image analysis used to process data obtained by other imaging techniques such as functional MRI or PET imaging. To avoid the need for further image processing relying on statistical analysis in optical imaging, it is desirable to obtain a good signal-to-noise ratio. Since photons are cheap and signal averaging is possible, this is usually feasible. Therefore, this avenue is recommended. However, in some cases even the best data still contains biological noise that must be removed before the proper functional maps can be obtained.

Kaplan and his colleagues applied Principle Component Analysis (PCA) (cf. Chapters 17, 32) to remove such noise from the images they obtained (Sirovich et al., 1995; Everson et al., 1998). They describe a particularly promising method to eliminate components whose time course does not correlate with the timing of the stimulus presentation. More recently, Obermayer and his colleagues have compared this method with ICA analysis and concluded that the latter technique is advantageous in several cases, particularly for obtaining cleaner single-condition maps (Stetter et al., 1998). It seems clear that additional development of such sophisticated image analysis and noise reduction techniques is desirable. However, as previously mentioned, great caution should be taken in interpreting such processed data, particularly if the raw data cannot be evaluated independently.

1.6 Chronic Optical Imaging

One of the foremost strengths of optical imaging using intrinsic signals is that it is a relatively non-invasive technique and, therefore, allows the recording of multiple activity maps from one cortical area within a single experiment. Furthermore, it is even possible to repeatedly image activity maps in single animals and therefore observe the functional architecture over a period of many weeks or months.

1.6.1 Infrared Imaging Through the Intact Dura or the Thinned Skull

A particularly promising result with regard to the feasibility of chronic optical imaging was the finding that cortical maps can be obtained through the intact dura or even through a thinned but closed skull (Frostig et al., 1990; Masino et al., 1993). This advance was achieved by using infra-red light which penetrates the tissue considerably better than light of a shorter wavelength. In the experiments seminal for many future chronic imaging studies, in particular those on young animals, Frostig et al. (1990) imaged orientation columns of the adult cat through the intact rather opaque dura. The dura was later removed, and it was shown that the columns imaged through the dura were identical to the ones obtained directly from the exposed cortex. Recently, spectacular orientation columns were imaged through the thin skull by Fitzpatrick and his colleagues (William et al., 1997).

1.6.2 Chronic Optical Imaging in the Awake Monkey

Experiments in behaving monkeys offer many advantages in the study of higher cognitive functions. Since such studies require long periods and extensive efforts devoted to training the animal, it is essential that the imaging should not be restricted to a single experiment, thus requiring chronic recordings.

A number of issues have to be resolved in order to be able to carry out optical imaging in behaving monkeys. Most of these problems do not exist in experiments on anesthetized and paralyzed animals. The first issue concerns the fact that behaving animals – unlike anesthetized animals – can of course not be immobilized with a paralytic agent. Therefore, there is a risk that the movements of the awake animal might cause the cortical surface to move relative to the camera's field of view. Since the mapping signals associated with evoked neuronal activity are often 10,000-fold smaller than the reflected light intensity, a motion of some tens of microns might be sufficient to ruin an experiment. The second issue is the large heartbeat and respiratory noise. In experiments on anesthetized animals, these periodic noises (often larger than the visually evoked signals) were removed by synchronizing the respiration to the heartbeat (see Sect 1.3.1) and triggering the stimulus and data acquisition on the electrocardiogram (ECG). Thus, the non-visually evoked signals could be almost completely eliminated by subtraction of two sets of images, both triggered on the ECG. In an animal that is not ventilated, this synchronization is of course unlikely to occur spontaneously, and it is therefore possible that the heartbeat and the respiration signals are so large as to obscure the mapping signal. The third issue is whether the intrinsic optical signal useful for imaging depends on the level of anesthesia.

Grinvald and his colleagues (1991) showed that all these problems can be overcome and that imaging based on intrinsic signals is useful for exploring the cortical functional architecture of behaving primates. A chronic sealed chamber as in Sect 1.3.1 has to be mounted on the monkey skull over the primary visual cortex. Restriction of head position by a solid head holder was already sufficient to eliminate the movement noise in the awake behaving monkey. Under some circumstances, the quality of maps obtained in the awake behaving monkey was equal to or even better than the functional maps obtained from anesthetized preparations. Furthermore, the wavelength dependency and time course of the intrinsic signals were similar in anesthetized and awake monkeys, suggesting that the signal sources were similar. However, noise was detected when the monkey moved his arms to press a lever or when he swallowed juice. Thus, the data acquisition had to be terminated at this point. Therefore, the standard behavioral paradigms must be carefully adapted to the additional requirement of optical imaging. These additional requirements must include noise considerations as well as timing considerations related to the slow time course of the intrinsic signals, the optimal duration of the stimulus, the optical data acquisition and the relatively long inter-stimulus interval.

1.6.3 Maintenance of Cortical Tissue over Long Periods of Time

As mentioned above, experiments with awake behaving monkeys require chronic recordings. The foremost problem with such experiments is to maintain the cortical tissue in good optical condition for long periods of time. To achieve this and at the same time provide a good optical access to the brain, Shtoyerman, Arieli and Grinvald examined the feasibility of implanting a transparent artificial dura made of silicon. Using this artificial dura, it has been possible to image both the orientation and ocular dominance columns repeatedly over a long period of time. Figure 12 shows an example of chronic imaging of ocular dominance columns. The cortex covered with this artificial dura appeared in perfect condition over periods as long as 36 weeks. Thus, with this approach it is now feasible to study higher brain functions with intrinsic optical imaging.

1.6.4 Developmental Studies Using Chronic Recordings in Anesthetized Preparations

As mentioned before, one of the key advantages of optical imaging when used in studying the development of the brain is that it is a relatively non-invasive procedure. Although the skull has to be opened in order to obtain optical access to the brain, the brain itself is not touched during the recording procedure. Particularly in young animals, it is even possible to obtain very good maps through the intact dura. This minimizes the risk

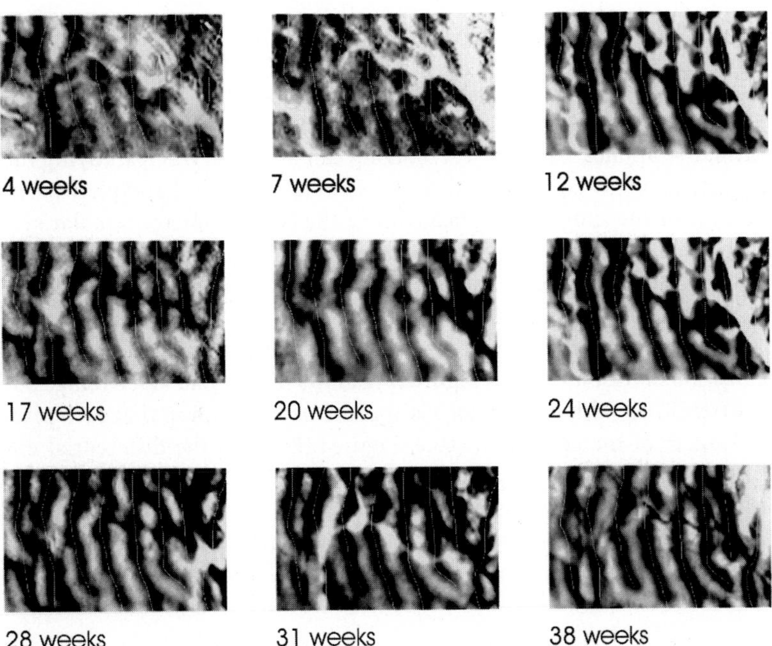

Fig. 12. Stability of ocular dominance maps in the behaving monkey cortex, protected by an artificial dura: To protect the exposed cortex, an artificial dura was implanted over macaque monkey primary visual cortex. Nine ocular dominance maps are shown which were obtained over a period of more than nine months. To assist the reader in inspecting the stability of these cortical maps, the yellow lines are plotted at the same cortical locations in all the figures relative to some large blood vessels. (Shtoyerman, Arieli and Grinvald, unpublished result.)

of infections, and most importantly, it leaves the brain in its "natural environment" and therefore in the optimal possible condition. Using this procedure, one can obtain chronic recordings over many weeks or even months. The longest chronic experiment that has been performed on young ferrets lasted four months (Chapman and Bonhoeffer 1998) and in this case a much longer survival time would have been possible without any complications.

In such experiments, as in any chronic experiment, sterile techniques have to be observed. It is furthermore crucial to give the recovering (young) animal the utmost care so that it can quickly overcome the stress of the initial chamber implantation. Once the chamber is implanted, the surgical stress the animal undergoes before every recording is minimal. The chamber has to be opened and cleaned (a procedure that takes approximately 20 minutes), it has to be refilled and the imaging session can begin. In many cases, anesthesia time when recording from pre-implanted animals can be held below 3 hours. These short anesthesia times are an important advantage, in particular when working on very young animals.

For long-term chronic experiments designed to last longer than two weeks, it is advisable to use the above procedures and to obtain the activity maps through the intact dura. If the dura needs to be resected in order to get activity maps of sufficient quality, the following procedure has proven very useful. After every recording session, before letting the animal wake up from its anesthesia, cover the cortex with agar containing a few drops of antibiotic (chloramphenicol). This has a twofold beneficial effect: First, it minimizes the risk of infections of the cortex and the surrounding tissue; secondly, the agar prevents proliferating cells of the surrounding tissue from invading the cortex and forming a connective tissue spreading over the cortical surface. When the next experiment is performed on the animal 3 to 7 days later, the layer of agar can simply be pulled off. With these precautions, the exposed cortical tissue can be kept healthy for up to 3 weeks.

1.7 On-Line Visualization of Changes in Cerebral Blood Volume and Flow

In addition to the above applications, optical imaging can be useful when studying the responses of the cortical micro-circulation to a variety of physiological and pharmacological manipulations or the spontaneous behavior of the cortical micro-vascular system as a function of time. The sensitivity of the video differential image to small optical changes is so large that slow 0.1 Hz oscillations in oxygen saturation level can be directly observed on the monitor screen.

Figure 13 demonstrates the sensitivity provided by a differential enhanced image allowing the visualization of the blood flow. Figure 13A shows the normal video image obtained from cat cortex. The width of the image is approximately 1 mm so that each pixel is viewing $\sim 1.5 \times 1.5\,\mu m^2$ of the cortical surface. Figure 13B shows the differential enhanced image that appears flat, and only the contours of the blood vessels are seen. The upper and lower rows of enlarged images are taken from site (1) and site (2), respectively (in Fig. 13A). In these pictures the movement of a small "black" particle can be seen. This particle is a small cluster of red blood cells within a capillary.

1.8 Three-Dimensional Optical Imaging

It is commonly assumed that the neocortex is organized in a columnar fashion, that is to say, neurons lying below each other have very similar functional properties. Although generally true, the concept of cortical columns was of course never meant to imply that

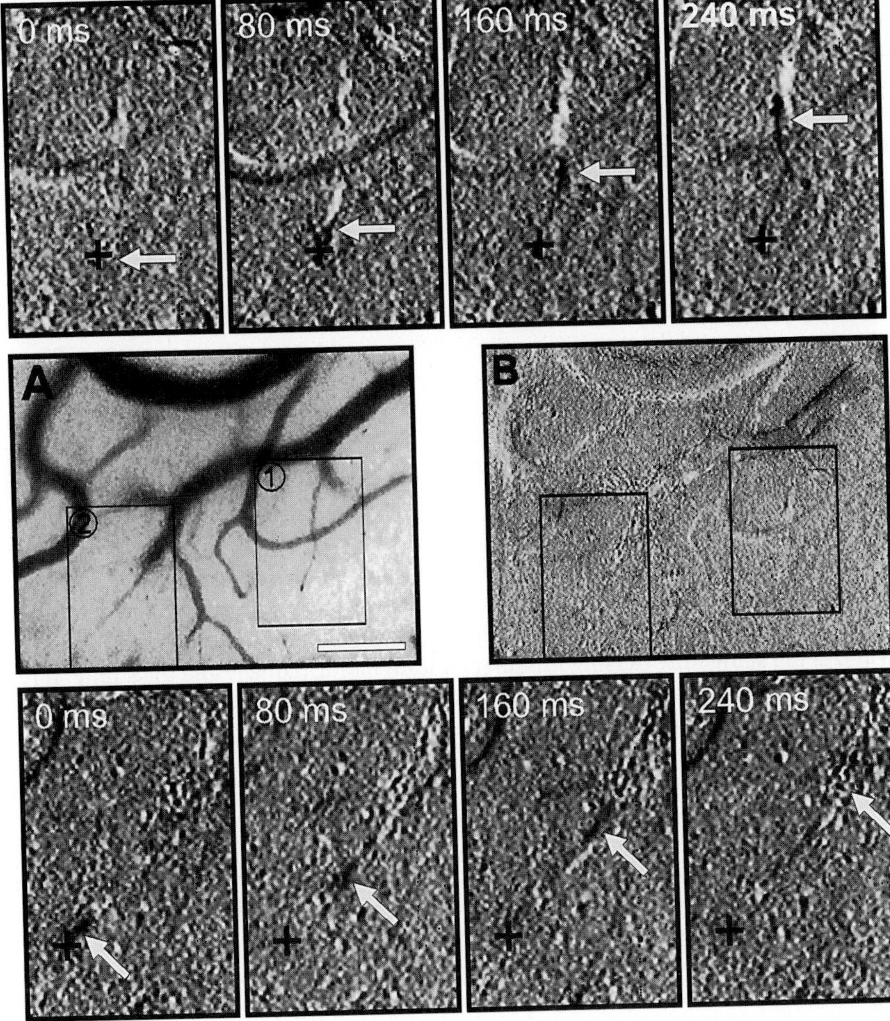

Fig. 13. Sensitivity of the differential enhanced image. Left column: Regular video image obtained from cat cortex. Each pixel has a size of approximately 1.5×1.5 μm². Middle column: Differential enhanced image corresponding to the normal images shown in the left column. It appears flat, and only the contours of the blood vessels are seen. The upper and lower rows of enlarged images are taken from site (1) and site (2), respectively (labeled rectangles in left column). Time resolution of individual frames: 80 ms. Movements of red blood cells within small capillaries can be detected. Scale bar is 200 μm. (Grinvald, unpublished results).

neurons positioned below each other have absolutely identical response properties. Therefore, in order to better understand the full functional architecture of the neocortex, it would be of great value to be able to obtain three-dimensional optical imaging data from the cortex. It is clear that light penetration is an important limitation, but since near infrared light penetrates the cortical tissue considerably better than visible light, near infrared imaging should facilitate the visualization of deeper cortical structures.

Optical sectioning as first described by Agard and Sedat (1983) is one approach to obtain three-dimensional information. Using this method, optical images are taken at different depths. Subsequently, the out-of-focus contribution is removed by mathematical deconvolution procedures. *In vivo* optical sectioning studies carried out by Malonek several years ago showed that focusing 800 μm below the cortical surface with a wavelength of 750 nm yields only a contribution 25% from this depth, which corresponds roughly to upper layer 4 in the monkey (Malonek et al., 1990). This study suggested that optical sectioning of the functional organization would only be feasible for cortical layers I, II and III. However, more recent experiments have failed to accomplish this goal in spite of the improvements in signal-to-noise ratio (Vanzeta and Grinvald, unpublished results).

Confocal microscopes (Egger and Petran, 1967; Boyde et al., 1983; Blouke et al., 1983; Lewin, 1985; Wijnaendts et al., 1985) can dramatically improve the three-dimensional resolution and reduce the light-scattering perturbation of a clear image. Using such a microscope, Egger and Petran (1967) were able to visually resolve single neurons 500 μm below the surface of the frog optic tectum. Although it is technically very demanding to achieve such a resolution *in vivo*, it should be kept in mind that a coarse resolution of ~50 μm would already be a major improvement for functional imaging. Thus, although technically difficult, the use of confocal microscopes is a promising approach to achieve optical sectioning.

Another possibility to accomplish three-dimensional imaging is based on the two-photon absorption technique developed by Webb and his colleagues (Denk et al., 1990). In several preparations this new technique has already proven more effective than conventional confocal microscopy in accomplishing an excellent three-dimensional resolution and at the same time causing minimal photodynamic damage and bleaching. It remains to be tested which of the three approaches, if any, will indeed provide adequate three-dimensional resolution for optical imaging.

1.9 Optical Imaging of the Human Neocortex

The spatial resolution of functional maps in the anesthetized and awake primate brain indicates that this approach will also be useful as a mapping tool in human neurosurgery. In patients undergoing surgical removal of tumors, it should be possible to precisely map functional borders on the cortical surface during the surgical procedure intra-operatively. This would allow the neurosurgeon to select the best resection strategy, minimizing potential damage to the patient's brain. Another potential application of intrinsic imaging in human neurosurgery is in the visualization of epileptic foci, whenever they are on the surface of the brain, with a precision much better than that currently achieved with electrical recordings (approximately 1 cm). This is of course also beneficial to the patients, since it may allow much smaller resection of the pathological tissue. Initial attempts to use optical imaging in human neurosurgery have been made by at least five groups.

1.9.1 Imaging during Neurosurgery

The first step in this direction was taken by MacVicar and his colleagues (1990) and by Haglund, Ojeman and Hochman (1992). Haglund and coworkers obtained maps from human cerebral cortex during electrical stimulation, epileptiform after-discharges and cognitively evoked functional activity in awake patients. They also reported that surrounding the after-discharge activity, optical changes were of opposite sign, possibly representing an inhibitory surround. Large optical signals were found in the sensory cortex during tongue movement and in Broca's and Wernicke's language areas during naming exercises. The large amplitude of the signals was surprising, and some of the time-course traces shown in this report were very different from those normally observed in experimental animals. It is clear that activity-dependent signals were indeed observed from the human cerebral cortex. However, the possibility that some of the maps and signals in this study were contaminated by noise and signals from the micro-vascular system could not be completely ruled out.

More recently, Shoham and Grinvald performed optical imaging studies on humans, to delineate borders of functional areas during neurosurgery. In a preliminary report (Shoham and Grinvald, 1994), they described the mapping of the human hand representation in the somatosensory cortex, which was then confirmed with differential EEG recording from a matrix of 16 surface electrodes (Shoham and Grinvald, 1994). Furthermore, Goedecke et al. (unpublished results) applied the optical imaging methodology to map neocortical epileptic foci. In both these studies it was found that the noise associated with the optical imaging of the human cortex was much larger than in the animal experiments using the cranial window technique. Both groups also observed very large activity-independent vascular noise. Despite these current technical difficulties, it appears that optical imaging of functional borders in the human cerebral cortex is in some cases feasible (Cannestra et al., 1998).

1.9.2 Optical Imaging through the Intact Human Skull

Is it science fiction or is it feasible to image human brain function using light, non-invasively, through the intact skull? Although the pioneering studies (Jobsys 1977; Wyatt et al., 1986, 1990; Chance et al. 1993a,b) did not attempt to produce images of brain activity, a related technique can also be used for imaging purposes. In a highly scattering medium such as the brain, photons essentially follow a random path. Some of the migrating photons reach the surface, exit the medium and do not reenter. If the relative positions of the light source and detector on the surface of the skull are known (usually 1–2 cm apart), the photon density in space can be calculated. Therefore, changes in either activity-dependent scattering or absorption will affect the photon flux reaching the detector and can be localized. Kato and coworkers (1993) employed an array of such illuminator-detector pairs to obtain low-resolution optical images in the human cerebral cortex. Thus, using near infrared light, the evoked responses in the auditory and motor cortex could be detected optically and confirmed with electroencephalography. This report and others suggest that relatively inexpensive optical imagers can be designed to explore cortical functional organization in human subjects, offering a spatial resolution of a few millimeters (Hoshi and Tamura, 1993; Gratton et al., 1994).

The changes in light intensity reaching a detector on the skull may originate from either changes in light absorption or changes in light scattering. From the animal experiments, it appears that a prominent component of the light absorption change originates from hemodynamic changes. Such changes are known to be slow, exhibiting a rise time of two to six seconds. Therefore, they could be utilized to provide imaging information related to the question *where* (position information). However, these hemodynamic signals are inadequate to provide answers to the question *when* (timing information) in the relevant millisecond time domain. As already mentioned, it is well known that light-scattering signals have a component that follows neuronal activity with a millisecond precision (Hill and Keynes, 1949; Cohen et al., 1968; Tasaki et al., 1968; Grinvald et al., 1982; Salzberg et al., 1983). Since it is known from animal experimentation that the fast light-scattering component is rather small relative to the slow light-scattering component and hemodynamic changes, it is a question of prime importance whether light-scattering signals can be separated from the absorption changes related to the micro-vascular responses. Furthermore, one wonders whether it would be possible to resolve the small but fast light-scattering component.

Since absorption and scattering objects produce different effects on the travel time of photons through the media, it should be possible to separate the absorption and scattering components using either time-resolved spectroscopy (Bonner et al., 1987; Sevick et al., 1991; Benaron and Stevenson, 1993) or frequency-domain optical techniques (Gratton et al., 1990; Maier and Gratton, 1993; Mantulin et al., 1993). Preliminary reports by Gabriele Gratton and his colleagues (1995) suggest that the small but fast light-scattering component can be resolved in human subjects performing tapping tasks or visual information processing. Thus, it seems possible that this recent mode of optical imaging will offer not only a spatial resolution comparable to PET and f-MRI, but would also provide a millisecond time resolution like electroencephalography or magnetoencephalography. The relative cost and simplicity of such optical devices justifies extensive investigation in this area. If this approach proves successful, it could possibly soon be used to explore human cognitive function at the neurophysiological level at a lower cost relative to the alternative methodologies.

Part 2: Voltage-sensitive Dye Imaging in the Neocortex

Introduction

A primary question in brain research is how the dynamic properties of single neurons and their intricate synaptic connections are associated to form brain networks capable of such remarkable performance. Research into this question has revealed properties of neocortical networks that are not reflected in the electrical activity of single neurons. Such properties can be revealed and subsequently fully understood only by studying the activity in neuronal populations, as opposed to the activity of single neurons alone. In this overview section we shall provide examples of some results which have been obtained based on voltage-sensitive dye imaging, providing temporal resolution in the millisecond time domain. Indeed, the imaging based on intrinsic signals discussed above is an excellent tool for functional mapping. However, because of the slow time course of the intrinsic signals, this type of imaging can only be used to address the WHERE question, and cannot address the WHEN question. The slow time course of a signal does not necessarily mean that it cannot be used to determine the relative timing of the much faster electrical events, in the same indirect way recently utilized during f-MRI measurements by Menon and his colleagues (Menon et al. 1998). However, a direct measurement would always be more advantageous. Therefore, we predict that the imaging of cortical dynamics based on voltage-sensitive probes will lead to more significant discoveries than the imaging based on the slow intrinsic signals.

Recordings of optical signals using voltage-sensitive dyes were first made by Tasaki et al. (1968) and by Cohen and his colleagues in the squid giant axon and in individual leach neurons (Salzberg et al., 1973). To perform optical imaging of electrical activity, the preparation under study is first stained with a suitable voltage-sensitive dye. The dye molecules bind to the external surface of excitable membranes and act as molecular transducers that transform changes in membrane potential *per se* into optical signals. These optical signals originate from electrical activity-dependent changes in the absorption or the emitted fluorescence, occurring in microseconds, and are linearly correlated with both the membrane potential changes and the membrane area of the stained neuronal elements.

Figure 14 shows the structure of a voltage-sensitive dye and depicts a possible mechanism explaining the dye response to a voltage change. A typical dye is a long conjugated molecule with a large dipole moment consisting of a hydrophobic tail at one end and a fixed charge at the other hydrophilic end. The hydrophobic tail anchors the dye in the

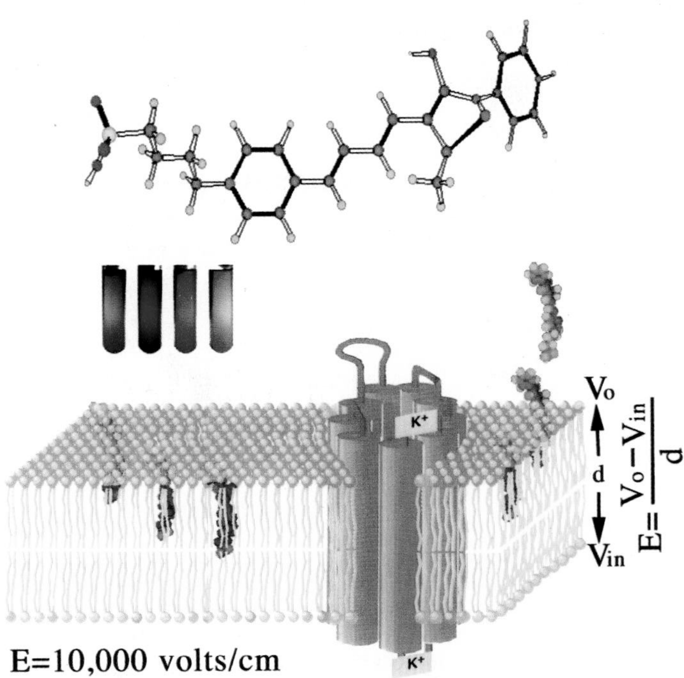

Fig. 14. Voltage-sensitive dyes. Top: Chemical structure of the voltage-sensitive dye RH 795. The four test tubes below the dye structure contain another voltage-sensitive dye dissolved in four different solvents of very different polarity. The significant changes in its color indicate that this dye is sensitive to its micro-environment. Bottom: Schematic illustration of the dye's interactions with the lipid bilayer, depicting one out of many possible mechanisms responsible for the dye's ability to transduce the membrane-potential change into an optical signal. Dye molecules not bound to the lipid bilayer are not fluorescent (depicted in blue). Once the dye binds it becomes fluorescent (red). The intensity of the fluorescence depends on the extent that the dye hydrophobic portion interacts with the hydrophobic portion of the bilayer. Since the dye is both charged and polar, it may change its interaction with the bilayer, depending on the electric field across the bilayer. An action potential gives rise to a change in electric field of ~ 20,000 Volts/cm across the bilayer. Such a large electric field change can also induce an electro-chromic effect even if the dye does not move during an action potential.

lipid bilayer, while the fixed charge tends to prevent the dye from crossing the neuronal membranes freely. In addition, the large dipole makes the dye sensitive to the micro-environment or to the changes in the electric field across the neuronal membrane during activity. The dye's sensitivity to voltage changes can be explained by several possible mechanisms: a direct electro-chromic effect or the motion of the probe in and out of the membrane as a function of the changing electric field across the membrane, affecting its optical properties such as the color or fluorescence quantum yield.

These optical changes are monitored with light imaging devices, positioned in a microscope image plane. The apparatus required for such imaging *in vivo* is similar to that required for imaging based on an intrinsic signal, with a few important modifications:

- Fast camera (2000 Hz to 300 Hz),
- Macroscope with epi-illumination option and
- Elaborate software for data acquisition, display and analysis.

A scheme of the apparatus is shown in Fig. 15.

In vivo imaging based on voltage-sensitive dyes began in 1984 (Grinvald et al, 1984, Orbach et al., 1985), but has matured only in recent years. (Shoham et al., 1998; Glaser et al, 1998; Sterkin et al., 1998). Therefore, the old notion that dye imaging is just too dif-

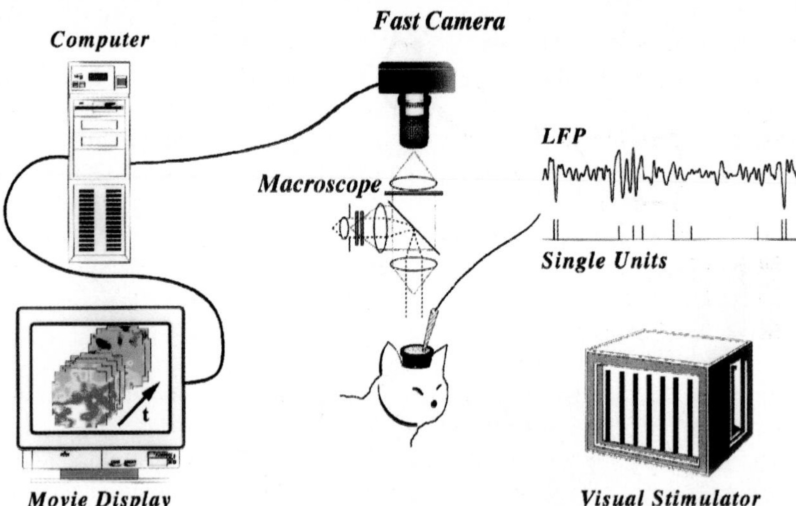

Fig. 15. Real-time optical imaging system. The exposed cortex is stained for 2 hours by topical application of a voltage-sensitive dye. An image of a 1 to 7 mm large square area of visual cortex is projected onto a fast camera with the aid of a macroscope. Computer-controlled visual stimuli are presented on a video monitor. Images of the cortical fluorescence are taken at 1000 Hz. The output of the fast camera is displayed in slow motion on an RGB monitor using either gray scale, color-coded or surface plot images. Local field potentials (LFP), multi- or single-unit activity or intracellular recording are performed simultaneously. (Glaser et al., 1998; Shoham et al., 1998)

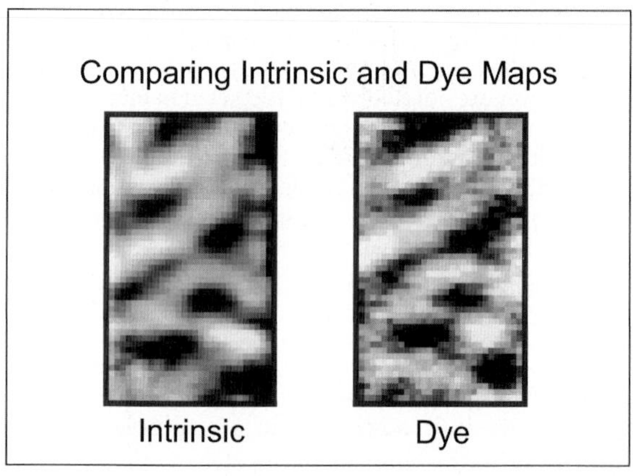

Fig. 16. Comparison of high-resolution functional maps obtained with the two techniques. The right panel shows the familiar pattern of orientation domains obtained with the fast high-resolution Fuji camera and the new voltage-sensitive dye RH-1692. One frame from the movie is shown. The left panel shows the map obtained by intrinsic imaging prior to the staining . For the intrinsic imaging map all the frames for a 3 s duration were integrated (Modified from Glaser et al., 1998).

ficult and problematic is no longer valid. Furthermore, it was generally believed that one must sacrifice spatial resolution for high temporal resolution. Recent technical advances in both fast cameras and design of voltage-sensitive dyes now facilitate the high-resolution imaging of functional domains with millisecond time resolution. Figure 16 compares the orientation domains imaged with a voltage-sensitive dye to those imaged in the same cortical area with intrinsic signals. The two high-resolution maps appear identical and the signal-to-noise ratio of the dye recording is even better. Below, we shall briefly discuss some general aspects of imaging based on voltage-sensitive probes.

The first voltage-sensitive dye imaging, replacing optical recording studies, used a 12x12 "diode array camera" for imaging (Grinvald et al., 1981). Higher resolution was subsequently achieved, mostly due to heroic efforts made by two groups in Japan, Kamino and his colleagues (Hirota et al., 1995) and Matsumoto and his colleagues (Iijima et al., 1992). Further improvement in spatial resolution was achieved by Toyama and his colleagues using a stroboscopic light (Toyama and Tanifuji, 1991; Tanifuji et al., 1993).

Ultimately, however, the fast camera specifications are not the only factors limiting the spatial resolution achieved with voltage-sensitive dye imaging. Rather, the limiting factors are the properties of the dyes currently in use: the signal-to-noise ratio that can be obtained with them, and the photodynamic damage or pharmacological side effects that an extrinsic probe may cause.

The development of suitable voltage-sensitive dyes is key to the successful application of optical recording for several reasons. First, different preparations often require dyes with different properties (Ross and Reichardt, 1979; Cohen and Lesher, 1986; Grinvald et al., 1988). Second, the use of dyes is associated with several difficulties that still need to be overcome. Under prolonged or intense illumination, the use of dyes causes photodynamic damage. Additional difficulties are bleaching, the limited depth of penetration into the cortex, and possible pharmacological side effects.

In simple preparations, such as tissue cultured neurons or invertebrate ganglia, where single cells are distinctly visible by a single pixel, the dye signal looks just like an intracellular electrical recording (Salzberg et al., 1973, 1977; Grinvald et al., 1977, 1981, 1982). However, it is important to note that the dye signals recorded or imaged from the neocortex are different from those recorded from single cells or their individual processes in simpler nervous systems. In optical recordings from cortical tissue, the optical signal does not have single-cell resolution. Rather, it represents the sum of membrane potential changes in both pre- and post-synaptic neuronal elements, as well as a possible contribution from the depolarization of neighboring glial cells (Konnerth et al., 1986; LevRam and Grinvald 1986). Since the optical signals measure the integral of the membrane potential changes over membrane area, optical recording can easily detect slow sub-threshold synaptic potentials in the extensive dendritic arborization. Thus, optical signals, when properly analyzed, can provide information about aspects of cortical processing by neuronal populations which usually cannot be obtained from single unit or intracellular recordings.

Studying the activity in neuronal populations is performed by measuring the sum of the optically detected electrical activity of all the neuronal elements at a given cortical site (cells bodies, axons and dendrites). If one pixel is viewing an area of 50 by 50 microns, then the recording is composed of activity contribution from 250–500 neurons and their processes. For some questions it is not clear how meaningful this type of information is, because neurons at a given cortical site may perform different computations and belong to different *neuronal assemblies.* What one really needs is to be able to image the dynamics of individual *neuronal assemblies*, rather than activity originating from functionally intermixed populations. Upon combining the imaging of population activity with single unit recordings, spatio-temporal patterns of the coherent activity in neuronal assemblies can be imaged, as discussed below. Thus, imaging based on voltage-sensitive dyes *in vivo* is turning into a powerful tool, currently the only one able to do this (Arieli et al., 1995; Kenet et al., 1998).

In summary, real-time optical imaging of cortical activity using voltage-sensitive dyes is a particularly attractive technique for providing new insights into temporal aspects of mammalian brain function. Among its advantages over other methods are:

- Direct recording of the summed intracellular membrane potential changes of neuronal populations, including fine dendritic and axonal processes;
- Ability to measure repeatedly from the same cortical region over an extended period of time, using different experimental or stimulus conditions;
- Imaging spatio-temporal patterns of activity of neuronal populations with a sub-millisecond temporal resolution; and
- Selective visualization of neuronal assemblies to be discussed below.

Several related reviews have been published elsewhere (Tasaki and Warashina, 1976; Cohen et al., 1978; Waggoner, 1979; Waggoner and Grinvald, 1977; Salzberg, 1983; Grinvald, 1984; Grinvald et al;., 1985; De Weer and Salzberg, 1986; Cohen and Lesher 1986; Salzberg et al., 1986; Loew, 1987; Orbach, 1987; Grinvald et al., 1988; Kamino 1991; Cinelli and Kauer, 1992; Frostig et al., 1994). Below we shall briefly review several examples from our work illustrating some of the issues which can be resolved using this approach.

Imaging of Population Activity

Brief History

The original results obtained from voltage-sensitive dye imaging in mammalian brain slices or isolated but intact brains suggested that optical imaging could also be a useful tool for the study of the mammalian brain *in vivo* (Grinvald et al., 1982a; Orbach and Cohen, 1983). However, accomplishing it required some efforts. Preliminary *in vivo* experiments in rat visual cortex in 1982 revealed serious complications associated with *in vivo* imaging. One of these complications was the appearance of large amounts of noise due to respiratory and blood pressure pulsation. In addition, the relative opacity and the packing density of the cortex limited the penetration of the excitation light and the ability of the available dyes to stain deep layers of the neocortex. Subsequently, other dyes (e.g., RH-414; Grinvald et al., 1982b) were developed by Rina Hildesheim and were proven better in extensive dye-screening experiments on rat cortex. In addition, an effective remedy for the large heartbeat noise was found: synchronizing the data acquisition with the ECG and subtracting a no-stimulus trial. These improvements facilitated the imaging of the retinotopic responses in the frog optic tectum (Grinvald et al., 1984), *in vivo* real-time imaging of the whisker barrels in rat somatosensory cortex (Orbach et al., 1985), and experiments on the salamander olfactory bulb (Kauer et al., 1987; Kauer, 1988; Cinelli and Kauer, 1995a,b). The development of more hydrophilic dyes improved the quality of the results obtained in the cat and the monkey visual cortices (e.g., RH-704 and RH-795; Grinvald et al., 1986, 1994). Finally, the design of dye-like RH 1692, whose fluorescence is excited outside the absorption band of hemoglobin, led to a tenfold reduction in the hemodynamic noise associated with *in vivo* imaging (Shoham et al., 1998;Glaser et al., 1998). Below we review a few examples of *in vivo* imaging studies.

What is the Cortical Point Spread Function?

For decades neurophysiologists have been characterizing the receptive field of single cortical neurons, but had difficulties in answering the complementary question of what the cortical point spread function is, that is to say, what the cortical area is that is activated by a point stimulus. One outstanding question that has been resolved by real-time optical imaging is how far direct activation by a sensory point stimulus spreads across the cortical surface via local cortical circuits. Because dendrites cover a much larger area relative to that covered by cell somata (about 1000-fold), the voltage-sensitive dye signal in cortical tissue reflects mostly postsynaptic potentials in the fine dendrites of cortical cells rather than action potentials in cell somata. Thus, it is very different from single-unit recording techniques, which emphasize spike activity next to the cell bodies. We took advantage of this property of voltage-sensitive dye imaging to try to resolve the questions posed above.

The frog retinotectal connections offer a system that is topographically well organized. Each spot of light on the retina activates a small region in the optic tectum. The first optical imaging study investigating the spread of activation focused on visualizing the topographic distribution of sensory responses in the frog. The optical signals obtained from the tectum in response to discrete visual stimuli were found to correspond well to the known retinotopic map of the tectum. However, in addition to a focus of excitation, the spatial distribution of the signals showed smaller, delayed activity (3–20 ms) covering a much larger area than expected on the basis of classical single-unit mapping (Grinvald et al., 1984). Similar mapping experiments were performed in the rat somatosensory cortex, where the simple somatotopic organization of the whisker barrels offered a convenient preparation to explore the issue of activation spread in the mammalian brain. When the tip of a whisker was gently moved, optical signals were observed in the corresponding cortical barrel field. However, a discrepancy was noted between the size of an individual barrel as recorded optically (a diameter of 1300 μm) and the histologically defined barrel (a diameter of only 300–600 μm) in layer IV of the cortex (showing neuronal somata rather than processes). The possible cause for this difference is that most of the optical signal originates from the superficial cortical layers in which neurons extend long processes to neighboring barrels (Orbach et al., 1985). Thus, pre- and postsynaptic activity in these processes probably accounted for the detected spread.

More recent retinotopic imaging experiments in monkey striate cortex also showed activity over a cortical area much larger than that predicted on the basis of standard retinotopic measurements in layer IV (Grinvald et al., 1994), but this was consistent with the anatomical finding of long-range horizontal connections in the visual cortex (Gilbert and Wiesel, 1983). The results of these experiments were used to calculate the cortical point spread function, which reflects the extent of cortical activation by retinal point stimuli. Figure 17 illustrates the cortical point spread function in the macaque primary visual cortex. To show the relationship between the observed spread and individual cortical modules, the experimentally determined point spread function was projected on a histological section of cytochrome oxidase blobs (Figure 14B). The stimulus used here activated only neurons residing in the marked small square, which contained only four blobs. However, more than 200 blobs had access to the information carried by the signal spread, albeit at lower amplitude. The apparent "space constant" for the spread was 1.5 mm along one cortical axis parallel to the ocular dominance columns and 3 mm along the other axis, perpendicular to the ocular dominance columns. The spread velocity was 0.1–0.2 m/s. The extensive lateral spread beyond the retinotopic border raises the possibility that the degree of distributed processing in the primary visual cortex is much larger than previously estimated – certainly a non-trivial challenge for theoreticians studying cortical networks.

These previous imaging experiments were performed with a low-resolution diode array. Because of the low spatial resolution, these experiments where not able to resolve the question of whether the observed spread was uniform or was restricted to well-defined cortical domains as expected from the known spatial clustering of the long-range horizontal connections. Recently, we repeated such experiments on the cat visual cortex, using small stimuli, and imaged the evoked activity with the high spatial resolution Fuji camera. We observed areas of direct retinotopic activation as well as spread and found that indeed the spread was more pronounced as well as faster in the patches that corresponded to the same orientation and were activated directly by the retinotopic stimulus (Glaser, Shoham and Grinvald, unpublished results).

Dynamics of Shape Perception

Using the recent advances in real-time optical imaging, we have investigated the dynamics of orientation selectivity in the millisecond time domain in cat area 18. The first issue we examined was whether the initial cortical response to an oriented stimulus is

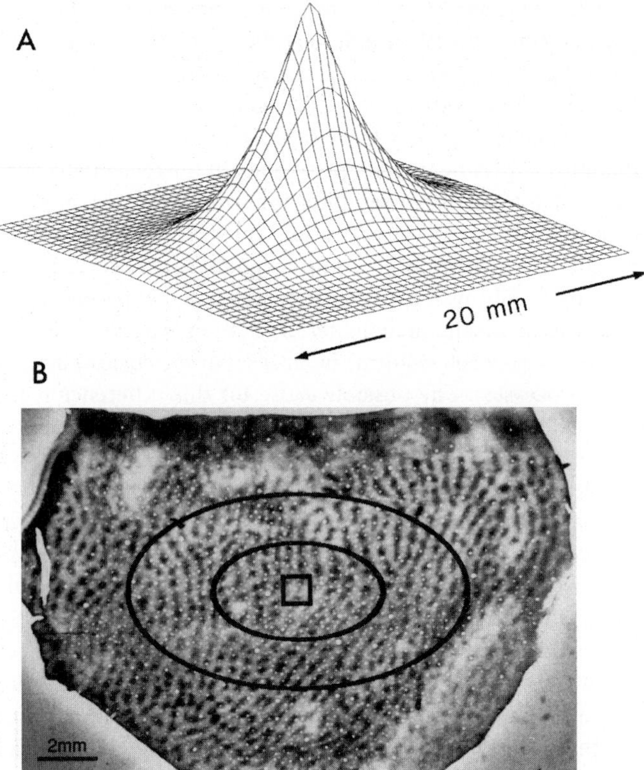

Fig. 17. Number of functional domains which may be involved in the processing of a small retinal image. A: Calculation of the activity spread from a small patch in layer 4 (the 1x1 mm square in B) within the upper cortical layers. At an eccentricity of ~6°, close to the V1/V2 border, such cortical activation would be produced by a retinal image of approximately 0.5x0.25° presented to both eyes. The "space constants" for the exponential spread were assumed to be 1.5 mm and 2.9 mm perpendicular and parallel to the vertical meridian, respectively. B: Mosaics of cytochrome oxidase blobs, close to the V1/V2 border. The thin and thick stripes of V2 are also evident in the upper part of the histological section. The center "ellipse" shows the contour where the amplitude of cortical activity drops to 1/e (37%) of its peak. The larger "ellipse" shows the contour where the spread amplitude drops to $1/e^2$ (14%). More than 10,000,000 neurons are included in the cortical area, bound by the large ellipse containing a regular mosaic of about 250 blobs. They can all sense the point stimulus from far away. (Modified from Grinvald et al., 1994)

sharply tuned, and this was found to be true. During the 100–200 ms following the earliest cortical response, the tuning width did not significantly change as a function of time, but the amplitude of the response increased approximately fivefold.

The second question we addressed explored the origin of the well-known effect referred to as "masking by light". Centuries ago it was shown that a sudden luminance change interferes with shape perception. Previous studies have shown that a luminance change affects shape processing at retinal, geniculate and cortical levels. We examined this by checking the dynamics of orientation tuning as a function of the previous luminance change. We found the cortical correlates of this effect. In one condition, the gray level of the screen prior to the onset of a high-contrast drifting grating was adjusted to the mean luminance of that grating. In another set, a dark or bright screen preceded the grating. Utilizing this latter set, the onset of the oriented grating also produced a sudden change in global luminance.

In the case where the grating onset was not accompanied by a luminance change, differential orientation maps were evident in the first response, 55 ms after stimulus onset.

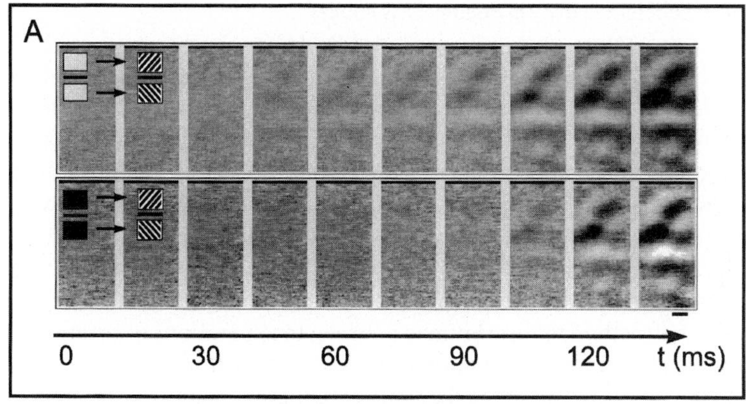

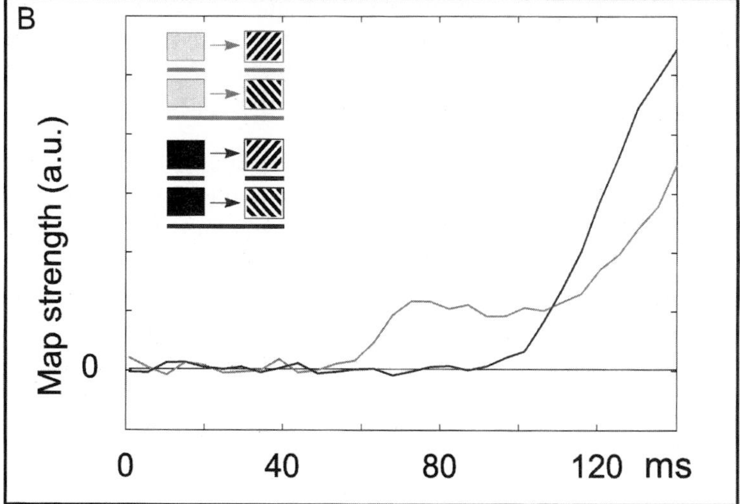

Fig. 18. "Masking by light": A sudden luminance change delays the onset of cortical orientation maps. A: Frames from a movie showing the development of orientation maps as a function of time in the millisecond time domain. The top row shows the development in the isoluminance case when a gray screen preceded the gratings. The lower row of frames shows the same except that a black screen preceded the onset of the grating, thus creating a sudden luminance change. Comparison of the two movies indicates that the onset of the orientation maps was delayed in the sudden luminance case. B: Time course of the development of orientation preference obtained by estimating the maps strength in each of the frames shown in A. When we inspected the time course of the response to a single orientation, the signal was faster in the case of sudden luminance change. However, initially it was independent of the stimulus orientation and therefore no orientation maps were seen (not shown). (Glaser and Grinvald unpublished results)

They peaked more than 100 ms later. However, when a sudden luminance change coincided with the grating onset, a new untuned response appeared at an earlier latency of 35 ms from stimulus onset. Furthermore, the orientation maps were delayed by an additional 45 ms. This result is illustrated in Fig. 18 in the form of frames from two movies depicting the development of orientation tuning with and without a luminance change.

In addition, we conducted a series of dichoptic experiments. Here one eye saw the grating stimulus as above, and the other eye saw either nothing or a sudden light flash simultaneously. Dichoptic interaction also produced the same effect, with the flash of light to one eye "delaying" the response to the oriented stimulus in the other eye (Glaser et al. 1998). This result suggests that the masking by light effect is not of retinal origin alone.

Selective Visualization of Coherent Activity in Neuronal Assemblies

For decades starting from the seminal work of Hebb, neurophysiologists have aspired to accomplish visualization of neuronal assemblies in action, but without success. A neuronal assembly may be defined as a group of neurons that cooperate to perform a specific computation required for a specific task. The activity of cells in an assembly is time-locked (coherent). However, the cells that comprise an assembly may be spatially intermixed with cells in other neuronal assemblies that are performing different computational tasks. Therefore, techniques that can visualize only the average population activity in a given cortical region are not adequate for the study of neuronal assemblies. What is needed, then, is a method to discriminate between the operations of several co-localized assemblies. A significant contribution of real-time optical imaging to neurophysiology has been the visualization of the dynamics of coherent neuronal assemblies. This goal has been accomplished by making use of the fact that activity of the neurons in an assembly is time-locked. The firing of a single neuron serves as a time reference to selectively visualize only the population activity that it is synchronized with, i.e., only the activity in the assembly it belongs to (Arieli et al., 1995). This approach is schematically illustrated in Figure 19.

To study the spatio-temporal organization of neuronal assemblies, we combined single-unit recordings and subsequent spike-triggered averaging of the optical recordings. The visual cortex (area 18) of an anesthetized cat was stained with the voltage-sensitive dye RH-795, and either on-going (spontaneous) or evoked activity was recorded continuously for 70 s. We simultaneously recorded optical signals from 124 sites, together with electrical recordings of local field potentials (LFP) and single unit recordings (1–3 isolated units recorded with the same electrode). Indeed, with sufficient averaging, the activity of neuronal assemblies not time-locked to the reference neuron was averaged out, enabling the selective visualization of the reference neuron's assembly (cf. Fig. 20). The spike-triggered averaging analysis showed that the averaged optical signal at the electrode site had a peak that also coincided with the occurrence of a peak in the LFP. The optical signals recorded from the dye were similar to the local field potential recorded from the same site. This result indicates that many neurons next to the electrode site had coherent firing patterns. Surprisingly, however, the fast components of the optically observed signals were heterogeneous in the field of view of 2x2 mm of cortex, indicating that optical recording provides a better spatial resolution relative to field potential recordings. The slow components of the coherent activity were distributed much more uniformly across large cortical areas.

Ongoing Activity Plays an Important Role in Cortical Processing of Evoked Activity

Is ongoing activity randomly distributed in space and time? Is it noise, or is it an expression of an intrinsic cortical mechanism, which is useful for cortical processing? How large is it?

We used the approach described in the previous section to image coherent activity during ongoing activity and during sensory evoked activity. We found that in 88% of the neurons recorded during spontaneous activity (eye closed), a significant correlation was found between the occurrence of a spike and the optical signal recorded in a large cortical region surrounding the recording site. This result indicates that spontaneous activity of single neurons is not an independent process, but is time-locked to the firing or to the synaptic inputs from numerous neurons, all activated in a coherent fashion even without a sensory input.

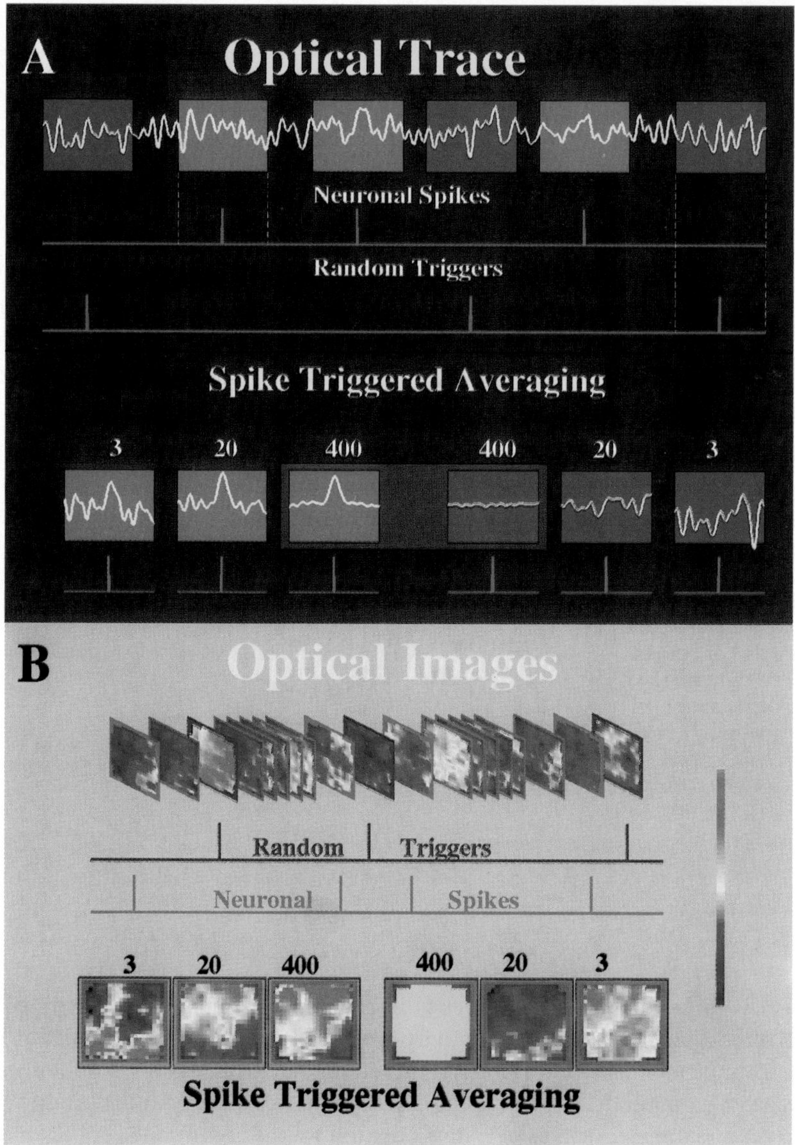

Fig. 19. Procedure for selective visualization of the dynamics of coherent neuronal assemblies. A: Time course of the optical signal obtained by spike-triggered averaging (STA). The top yellow trace shows the amplitude of the optical signal reflecting compound electrical activity from a given cortical site, measured for 8 s. The red trace below shows the simultaneously recorded action potentials from the reference neuron. The long recording session was subdivided into 1 s time segments (red windows on the top trace) each centered on the timing of the action potential. The blue trace below shows random virtual spikes that are used as a control for the procedure (blue windows). The bottom traces in the red windows show the time course of the spike-triggered averaged signal after averaging 3, 20 and 400 time segments, during which the action potential occurred. The traces in blue windows show the results obtained from averaging the control virtual spikes. A clear coherent activity is detected already after averaging 20 events. B: Spatial patterns of movies obtained by spike-triggered averaging. The top shows a series of images in the form of a movie instead of showing the activity at a single cortical site depicted in panel A above. The two traces below show the timing of simultaneously recorded action potential and virtual action potential that served as a control. The bottom frames show the spatial pattern at a given time, resulting from spike-triggered averaging. Note that the control patterns are rather flat already after averaging twenty random events without real action potentials, see the three blue windows at the bottom right. (Sterkin et al., 1998)

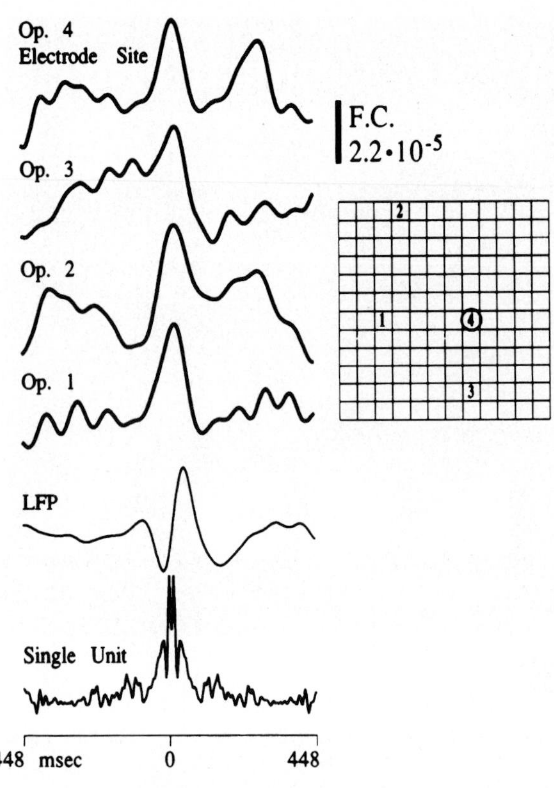

Fig. 20. Spike-triggered averages (STA) of spontaneous optical and electrical activity. 264 spikes of a single neuron were used for the STA. The wide auto-correlation density for the single unit activity of the triggering neuron shows that the neuron had a tendency to fire in bursts of spikes (bottom trace). At the same time, the average LFP exhibited a negative wave followed by a positive one. Four optical traces from different cortical loci, shown at the top, had positive peaks at the time of occurrence of the peak of the auto-correlation of the spikes. Note the second peak evident in site 4 and absent in sites 1 and 3. Time zero on the bottom scale indicates the firing time of the neuron. The location of the four diodes is shown in the array on the right. Each diode sampled an area of 200x200 µm. The trace marked by "electrode site" is above the recording site labeled by a circle. Raw data filtered 0–30 Hz, (= 3.5 ms and corrected for the heartbeat signal. (Modified from Arieli et al., 1995)

Surprisingly, we found that the amplitude of this coherent ongoing activity, recorded optically, was often almost as large as the activity evoked by optimal visual stimulation. One extreme example is illustrated in Fig. 21. Inspecting our entire data set we found that, on average, the amplitude of the ongoing activity that was directly and reproducibly related to the spontaneous spikes of a single neuron was as high as 54% of the amplitude of the visually evoked response by optimal sensory stimulation, recorded optically. Furthermore, coherent activity was detected even at distant cortical sites up to 6 mm apart.

The spontaneous activity of two adjacent neurons, isolated by the same electrode and sharing the same orientation preference, was often correlated with two different spatio-temporal patterns of coherent activity, suggesting that adjacent neurons in the same orientation column can belong to different neuronal assemblies.

Another important finding has been the discovery that the coherent spatial pattern for ongoing activity and evoked activity were often similar (Shoham et al., 1991; Kenet et al., 1998). These results suggest that intrinsic ongoing activity in neuronal assemblies may play an important role in shaping spatio-temporal patterns evoked by sensory stimuli. It may provide the neuronal substrate for the dependence of sensory information processing on context, behavioral and conscious states, memory retrieval, top-down or bottom-up activity streams, and other aspects of cognitive function. Therefore, it is important to be able to study the dynamics of ongoing and evoked activity without signal averaging.

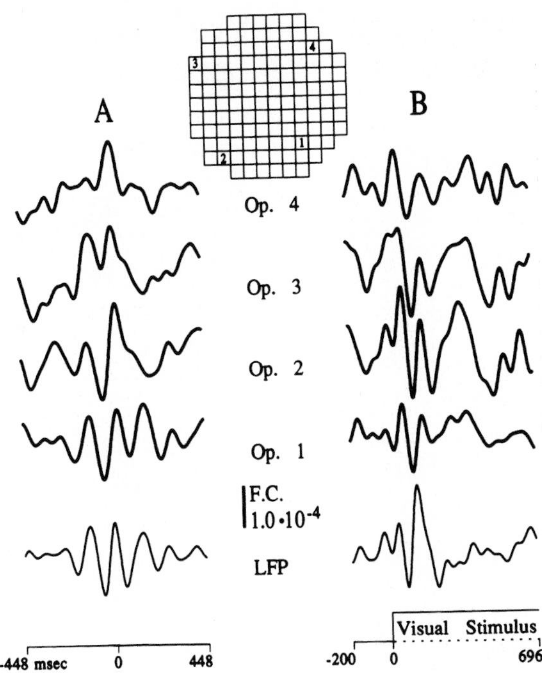

Fig. 21. The amplitude of coherent ongoing activity is comparable to that of evoked activity. A: STAs of spontaneous optical and electrical signals: 35 spikes of a single neuron were used for the STA. The first optical trace above the LFP (Op. 1) shows a similar wave as the LFP but with a phase shift. The three other traces (Op. 2–4) from different cortical loci had a different time course with peaks that coincided with various peaks of the LFP. This comparison suggests that different components of the LFP originated at different sites across the cortex. B: The visually evoked response (VER) of the LFP and the optical signal were obtained in a subsequent recording session, from exactly the same area. The signals were averaged on the onset of 35 grating stimuli delivered in the preferred orientation of the unit activity. The averaged LFP and optical signal (Op. 1–4) show a significant evoked response. Note that the amplitudes of the STA for spontaneous activity and of the VER were similar. The scale at the bottom shows the timing of the visual stimuli. Raw data filtered 2–14Hz, $\sigma = 7$ms in A and B. (Modified from Arieli et al., 1995)

Dynamics of Ongoing Activity; Imaging without Signal Averaging

In spite of the large biological noise originating from respiration and blood pressure pulsations, reliable real-time optical imaging was accomplished even without signal averaging, using off-line correction procedures instead. This allowed us to explore for the first time the dynamics of spontaneous ongoing population activity and its interaction with evoked activity. (Arieli et al., 1995, 1996; Sterkin et al., 1998).

In the mammalian visual cortex, evoked responses to repeated presentation of the same stimulus exhibit a large variability. It has been found that this variability in the spatio-temporal patterns of evoked activity results from the ongoing activity, reflecting the dynamic state of the cortical network. In spite of the large variability, the evoked responses in single trials can be predicted by taking account of the preceding ongoing activity (see Fig. 22). This prediction is valid as long as the ongoing activity pattern that presumably continues to change during the evoked response is still similar to the initial state (e.g., 50–100 ms; see Arieli et al., 1996).

These findings indicate that old notions of what is “noise” in brain activity should be revised. Since often the ongoing activity is very large, we do expect it to play a major role in cortical function.

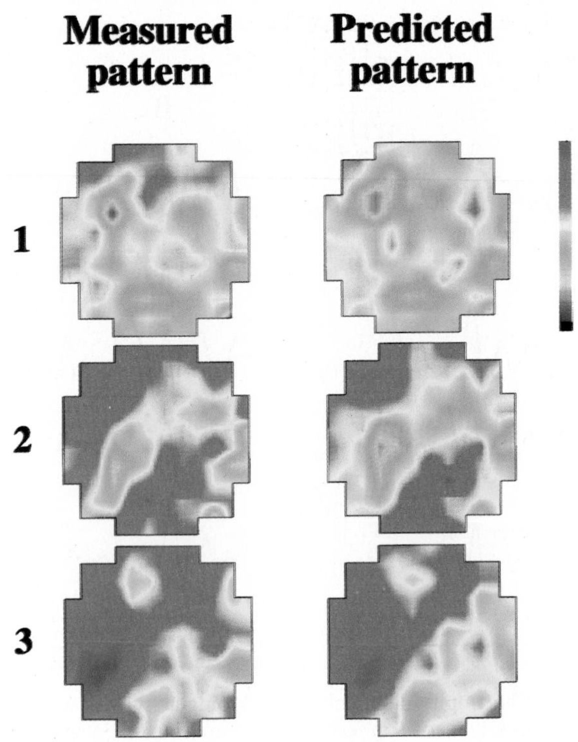

Fig. 22. Impact of ongoing activity, predicting cortical evoked responses in spite of their large variability. Three examples of comparing predicted and measured responses are shown. A single-trial response to a stimulus was predicted by summing the reproducible response component (estimated by averaging many trials of evoked responses) and the ongoing network dynamics (approximated by the initial state, i.e., the ongoing pattern just prior to the onset of the evoked response). Left columns: Measured responses. Right columns: Predicted responses, obtained by adding the initial state in each case. (Modified from Arieli et al., 1996)

Methods

As can be seen from the introductory section, optical imaging based on voltage-sensitive dyes has improved substantially, and the notion that dye imaging is not really practical *in vivo* is no longer justified. However, dye imaging is by no means simple. The many technical considerations discussed below are aimed at allowing optimization of experiments, and performance of proper controls and proper analysis of the data. They should also assist in reaching the proper interpretations of the results. Gaining of an in-depth understanding of the subtle technical aspects is fundamental for the many different types of scientific explorations of cortical dynamics that will undoubtedly lead to significant discoveries, by the many groups beginning to use this approach.

2.1 Optimization of the Measurement

The purpose of the following section is to provide a simplified method to predict the expected signal-to-noise ratio in a given experimental situation. This parameter is the single most important determinant of both the ease and the eventual success of an experiment employing optical methods. Somewhat counter-intuitively, signal size, no matter how small it is, is not very important, what counts is the signal-to-noise ratio.

2.2 Signal Size and Signal-to-Noise Ratio

The size of the activity-dependent dye signals in the *in vivo* neocortex is often small (only 10^{-4} to 5×10^{-3} of the light intensity reaching the detector), and thus obtaining a good signal-to-noise ratio is not always an easy task. Therefore, whenever the use of an alternative methodology is easy, it is preferable. As a rule of thumb, the size of the expected signal depends on the size of the membrane area of the target and on the size of the membrane potential change. The exact consequences of these relationships often contradict the intuitive feeling about the feasibility of an experiment. For example, in electrophysiological experiments in cell culture, the recording of a 100 mV action potential from a 100 μm neuron is not much easier than the recording of only 1 mV synaptic potential from a 10 μm neuron. In contrast, in fluorescence experiments, the signal-to-noise ratio for the latter experiment would be 1000-fold smaller (100-fold smaller voltage change and 100-fold smaller membrane area but 10 times smaller shot-noise; see Eq. 2 below). This means that, in order to get the same signal-to-noise-ratio for the action potential in the large neuron as for the synaptic potential in the small neuron, using dim light, the number of trials to be averaged is 1,000,000 larger. On the other hand, the recording of small sub-threshold changes in membrane potential from a dendritic population *in vivo* is easy because of its large membrane area.

2.3 Sources of Noise in Optical Measurements

When several sources of noise exist, the total noise level is roughly equal to the square root of the sum of the squares of each noise level. Therefore, practically, if one type of noise is twice as large as the rest, the contribution of the rest can be neglected. At least five independent sources of noise exist in the optical experiments:

- Light shot-noise due to random fluctuations in the photo-current originating from the quantal nature of light and electricity. The shot-noise is proportional to the square root of the detected light intensity, therefore an increase of the illumination intensity will reduce the relative noise level.
- Dark noise of the detectors: This noise is fairly constant. It is exceptionally low in photo-multipliers and in cooled CCD arrays but in the photodiode-amplifier combination it is much higher, equivalent to the shot-noise originating from 10^7–10^8 photons per millisecond. Therefore, for the detection of lower light-level fluorescence or phosphorescence signals (e.g., from segments of neuronal processes), photo-multipliers or CCDs are preferable. Inexpensive detector arrays of this type are not yet available. The dark noise in the fast, high-resolution cameras made by Adaptive Optics Association or Fuji Camera is currently far from optimal.
- Fluctuations in the stability of the illumination/excitation light. They depend on the nature of the light source, its power supply and the connections between the power supply and the bulb. The stability of a tungsten-halogen lamp is excellent up to of 10^{-6} (10 Hz – 1 kHz) when optimally operated. Mercury arc lamps can be stabilized to 5.10^{-5} to 10^{-4} with a negative feedback circuit (J. Pine, unpublished results). Xenon arc lamps are somewhat more stable than the mercury lamp and can replace it whenever the use of the narrow mercury lines is unimportant. Wide-band filters with the xenon lamp may provide nearly equal intensity. When one turns on an arc lamp for the first time, it should be left running continuously for 10–15 hours, to minimize arc wandering in later operations. Note that arc-wandering noise is not trivially eliminated by a stabilization circuit and that it is quite large and may be frequent. Therefore, the easiest approach is to implement a rejection procedure for trials containing arc wandering.
- Vibration noise originating from relative movements of any of the components along the optical light path including the preparation. It is especially large if the preparation is not uniform optically (e.g., the pigmented optic tectum). Sufficient care in the setup design and the use of a vibration isolation table can minimize this source of noise.

- Noise originating from movements within the preparation. In the vertebrate brain the predominant noise originates from movement due to blood circulation, heartbeat or respiration. This noise could not be sufficiently reduced by sealed chambers ($DI/I = 10^{-2} - 10^{-4}$, depending on the wavelength used, i.e., how is it related to the absorption spectra of intrinsic chromophores such as hemoglobin). For example, in the pioneering *in vivo* experiments (Grinvald et al., 1984; Orbach et al. 1985), this noise was 5–20 fold greater than the evoked optical signals!

Because the optical noise resulting from heartbeat is synchronized with the heartbeat and the electrocardiogram (ECG), it was relatively easy to reduce it by subtracting the result of a trial with a stimulus from a subsequent trial without a stimulus, with both trials triggered by the peak of the electrocardiogram. This procedure, together with resynchronization of the respirator with the ECG (by briefly stopping it and restarting with the next ECK peak), reduced the noise by a factor of ~10 depending on the heartbeat reproducibility. In signal-averaging experiments, further improvement was achieved by a computer program that rejected exceptionally noisy trials from the accumulated average. If the heartbeat noise becomes a limiting factor, the reproducibility of the heartbeat optical signals could be improved by pharmacological stabilization. This has been accomplished by i.m. injections of sympathetic and parasympathetic blockers such as hexamethonium.

To minimize the overall noise level associated with a given experiment, three important *rules* must be followed:

- If the signals are small, it is very important to minimize the contribution of all the dominant noise sources. If the light shot-noise is the limiting factor, then the light intensity should be increased as much as possible, up to the level at which the light shot-noise is equal to the light source instability fluctuations. However, one should not increase the illumination intensity beyond that level, or beyond the level where photo-dynamic damage or bleaching becomes the limiting factor. For signals exhibiting a small fractional change, the use of CCDs and video cameras is not optimal because of their small well capacity (see below), which means that these devices will saturate before the appropriate light level required to resolve small signals can be reached.
- If other types of noise are predominant, the light level should be reduced to the point where the shot-noise is equal to the predominant noise. Lowering the light level would minimize photo-dynamic damage and bleaching, and would thus permit more extensive signal averaging to improve the signal-to-noise ratio. Alternatively, the dye concentration may be reduced to minimize pharmacological side effects and/or photo-dynamic damage.
- If constant periodical noise is predominant (e.g., from heartbeat, respiration), it is relatively easy to correct it by proper synchronization and subtraction procedures. However, whereas this subtraction procedure may reduce the noise by a factor of 5 to 10, subtraction cannot eliminate the noise because of the imperfect reproducibility of the optical signal caused by the heartbeat pulsation or respiration. One possible way to remove this noise is to put the preparation on a heart-lung machine. For example, the frog optic tectum is covered by dark pigments overlying the blood vessels. Because of the normal blood flow, these pigments move. Therefore, signals from detectors covering blood vessels may be 10 times more noisy than those which cover a pigment-free area of the tectum. This noise and the heartbeat noise were totally eliminated when the frog was perfused with a laminar flow of saline through the aorta (Kamino and Grinvald, unpublished results). Another possible way to remove the noise is to use dual-wavelength recordings. A measurement at one wavelength where the optical signals are sensitive to the membrane potential change can be divided on-

line or off-line by another measurement at a wavelength range where the signal does not depend on membrane potential. In pursuing that approach, one should bear in mind that optical signals from the cortex are wavelength-dependent. Therefore, it would be advantageous to use a single excitation wavelength and dual emission, similar to the measurements with Indo-2. Alternatively, to achieve this goal, one can combine voltage-sensitive dye with a second dye that does not respond but has an emission spectrum which does not overlap at least a portion of the emission spectrum of the voltage dye. Occasionally, this problem is moot in some preparations where imaging has recently been accomplished without signal averaging (Sterkin et al., 1998).

2.4 Estimation of the Signal-to-Noise Ratio

This section is intended to provide a recipe that should help make the decision as to which measurement mode is preferable in a given situation (i.e., absorption, fluorescence, reflection). In general, it has been shown, on theoretical grounds, that recording of changes in fluorescence rather than transmission is the method of choice whenever the number of probe molecules is small (Rigler et al., 1974). Therefore, for obtaining large optical signals from small processes of single nerve cells, fluorescence changes give better signal-to-noise ratio than transmission changes. However, in other preparations, other considerations may become important, as outlined below.

The signal-to-noise ratio in transmission experiments (shot-noise limited) is given (Braddick, 1960) by:

$$(S/N)_T = (\Delta T/T)(2q\tau)^{1/2}\left(T^{1/2}\right), \qquad (1)$$

where τ is the rise time of the detector circuit ($\tau = 1/4\Delta f$) where Δf is the power bandwidth; T is the transmitted light intensity reaching the detector; q is the quantum efficiency of the detector; and $\Delta T/T$ is the fractional change in transmission. In fluorescence experiments the signal-to-noise ratio (shot-noise limited) is given by:

$$(S/N)_T = (\Delta F/F)(2q\tau)^{1/2}(gF)^{1/2} \qquad (2)$$

where: F is the fluorescence intensity originating from the preparation (it is linearly proportional to the illumination intensity); g is a geometrical factor related to the collection efficiency of the fluorescence detector; $\Delta F/F$ is the fractional change in fluorescence.

T, the transmitted light intensity, is usually larger than F by 3–4 orders of magnitude, for equal illumination intensity. However, from the above equations, it is evident that if $\Delta F/F$ is much larger than $\Delta T/T$, fluorescence measurements can give a better signal-to-noise ratio. Indeed, the largest observed $\Delta F/F$ from a single neuron was 2.5×10^{-1} in a fluorescence measurement (Grinvald et al., 1983), but the largest $\Delta T/T$ was only $5\times10^{-4}-10^{-3}$ in a transmission measurement, both for a 100 mV potential change (Grinvald et al., 1977; Ross and Reichardt, 1979).

The situation is different for recording population activity in multi-layer preparations. If $\Delta T/T$ and $\Delta F/F$ are, respectively, the corresponding fractional changes in transmission and fluorescence measurements in isolated neurons, then, in multi-layer preparations having n layers of neurons, to a first approximation $(\Delta T/T)_n = (n\,\Delta T/T)$. As long as the total dye absorption is a few percent, T is relatively insensitive to n, thus the signal-to-noise ratio will increase linearly with n (Eq. 1).

On the other hand, in fluorescence experiments, $(\Delta F/F)_n = (n\Delta F)/(nF) = \Delta F/F$. Because $\Delta F/F$ is relatively independent of n, in multi-layer preparations, the signal-to-noise ratio will increase only as $n^{1/2}$ (because only F will increase) (see Eq. 2). Therefore, for the detection of activity of a population of neurons in multi-layer preparations (large n), transmission measurements may be more suitable than fluorescence (whenever applicable) and the signals may be huge (e.g., Salzberg at el., 1983).

Transmission measurements are also much less sensitive to nonspecific binding of the dye to the non-active membranes or any other binding sites; such binding does not affect ΔT and affects T only moderately. In fluorescence experiments, non-specific dye fluorescence does not affect ΔF, but increases F linearly according to the amount of non-specific binding sites. Therefore, the signal-to-noise ratio will deteriorate as the square root of the non-specific binding. For example, in a hypothetical situation in which 25 neurons, on top of each other, are viewed by a photo-detector, but only one is active, the signal-to-noise ratio in transmission will hardly deteriorate (relative to the situation that only one active neuron is present there). In contrast, the signal-to-noise ratio in fluorescence will deteriorate by a factor of 5. On the other hand, in transmission measurements, it is important that the detector will have the same shape and size as the target image, otherwise the signal-to-noise ratio will deteriorate as the square root of the extra area viewed by the detector. This does not happen in fluorescence if only the target is fluorescing (e.g., processes of stained single neurons). A gain in the signal-to-noise ratio can be achieved by masking the irrelevant part of the image to prevent that light from reaching the detector.

A useful equation for estimating the expected signal-to-noise ratio in fast measurements is given by:

$$(S/N) = \Delta I_p / I_p^{1/2} = (\Delta I / I)(I_p)^{1/2} \tag{3}$$

where I_P is the photo-current from the detector in electrons/ms and $\Delta I/I$ is the actual fractional size of the signal, corrected for the non-specific binding, the fraction of active neuronal elements and the target area relative to the detector area. Thus, from the above discussion, one should be able to predict the expected fractional changes and signal-to-noise ratios for absorption or fluorescence measurements. Using the reasonable assumption that proper probes for the given preparations can be found, the fractional change for a single neuron is $0.5\text{–}2 \times 10^{-1}$ and $1\text{–}10 \times 10^{-4}$, for fluorescence and transmission, respectively. An estimate of the intensity reaching the photo-detector is also required; it can be made from a preparation properly stained with a voltage-sensitive dye without measuring any activity-dependent signals, even if the dye is not optimal for the preparation. The simple current-to-voltage amplifier required for such measurements has been described (Cohen et al., 1974). The output voltage of the amplifier V can be used to estimate the photo-current I_P, such that $I_P = V/R$ where R is the feedback resistor of the amplifier.

Transmission measurements are applicable for dyes exhibiting a voltage-dependent change in absorption rather than fluorescence. Because trans-illumination is often not practical for *in vivo* imaging of the neocortex, the absorption signal from a given probe can be picked up by a reflection measurement. To determine the expected signal-to-noise ratio in reflection measurement, one must determine $\Delta R/R$ and R, where R denotes the reflected light intensity and $\Delta R/R$ is the fractional change in reflected light. $\Delta R/R$ usually is similar to $\Delta T/T$. The relationship between the intensity of illumination and R is complex. This factor is species-specific and wavelength-dependent, and we do not know how to quantify it for a general case. However, it appears that in the neocortex, reflection measurements with absorption probe may provide a good signal-to-noise ratio.

2.5 Fast and Slow Measurements

The above equations also explain why it is much easier to obtain a good signal-to-noise ratio when the signals are slow, for example, in imaging based on slow intrinsic signals. The measurement of small slow signals with a time constant of a second rather than a millisecond provides 33-fold improvement in the signal-to-noise ratio (i.e., the number of sweeps averaged could be reduced 1000-fold) if shot noise was the limitting factor. In addition, one can reduce the light level and dye concentration, thus minimizing pharmacological side effects, photo-toxic effects and bleaching whenever a slow measurement is applicable.

2.6 Reflection Measurements

Often when transmission experiments are not feasible in preparations stained with absorption dyes, reflection measurements of neuronal activity can be performed. The reflected light signal from a squid giant axon stained with an absorption dye was 100–200 fold larger than the light-scattering signal from the unstained axons (Ross et al., 1977). The wavelength dependence of the reflection signal was similar to the action spectrum of the absorption signal. This indicated that the reflection signal was dye-related. The extrinsic reflection signal is expected whenever there is a change in absorption.The amount of reflected light depends on the intensity of the incident light. A change in absorption due to the presence of the dye affects the intensity of the incident light and therefore also that of the reflected light. When the signal in the reflected light originated from such an extrinsic absorption signal, the fractional change of the reflection signal was nearly equal to the fractional change in transmission. The measurement of extrinsic absorption signals via the reflection mode has the advantage that it should facilitate the use of absorption dyes in opaque preparations, where transmission measurements are not feasible, or from thick preparations (e.g., thick slices, cortex). Indeed such measurements proved useful in heart tissue (Salama et al., 1987; Salama and Morad, 1976). However, the amount of reflected light is usually much smaller than the amount of transmitted light, and therefore, whenever transmission measurements are feasible, they provide better signal-to-noise ratio than reflection measurements when considering photon noise.

The separation of light-scattering signals from intrinsic signals that originate from an intrinsic absorption or fluorescence changes proved to be a rather difficult task. Dual- or triple-wavelength measurements were used (Jobsis at el., 1977 ; Lamanna et al., 1985; Pikarsky et al., 1985). Similarly, slow extrinsic signals from *in vivo* preparations stained with dyes may be contaminated with intrinsic signals of several origins. Unless the dye signals are much larger than the intrinsic signals, the recording will be largely distorted by a contribution from all of the components of the intrinsic signal. This currently occurs with voltage-sensitive dyes. In particular, the delayed signal after a few hundred milliseconds is contaminated with the intrinsic signal that peaks in 2–4 seconds. When making functional maps, this can introduce distortions. These distortions can be minimized with appropriate analysis, subtraction procedures and by obtaining images at several wavelengths. The ultimate accuracy of such procedures remains to be demonstrated.

2.7 Apparatus for Real-Time Optical Imaging

2.7.1 The Imaging System

Figure 15 above depicts the computer-based fast imaging apparatus. The apparatus is very similar to the imager based on intrinsic signals. A macroscope is rigidly mounted on a vibration-isolation table and the preparation is illuminated for epi-fluorescence by a 12 V/100W or 15V/150W tungsten/halogen lamp. In similar fluorescence experiments, brighter light sources, such as an He-Ne laser, a mercury or a xenon lamp are employed. Changes in fluorescence are detected in the macroscope image plane by a high-resolution fast camera such as the Fuji HR-Deltaron (128x128) or a 10x10, 12x12, 16x16, 24x24 photodiode array (Centronics, Hamamatsu, RedShirtImaging or Sci-Media Ltd.). Each photodiode receives light from a small area of the preparation, depending on the macroscope magnification used. In some low-light-level fluorescence or phosphorescence experiments, a single photodiode or a photomultiplier is used instead of the array detector.

2.7.2 Which Microscope to Use?

Conventional microscopes have been used for *in vivo* imaging. However, the macroscope with its large numerical aperture for a low magnification and the large working distance offers the following considerable advantages:

- It is easier to use microelectrodes for intracellular or extracellular recordings (see Sect 3.2).
- The signal-to-noise ratio is better because of the macroscope's high numerical aperture.

For many applications, maximizing the light intensity provides a better signal-to-noise ratio. In fluorescence experiments, using epi-illumination, the signal-to-noise ratio is related to the square power of the objective numerical aperture. In many of the *in vivo* applications, the sub-micron spatial resolution of objectives and condensers far exceeds the requirements for optical imaging of neuronal activity, and the macroscope is more than adequate. Under some conditions the total gain in light intensity may be more than 100-fold that of a standard objective for low magnification.

2.7.3 Fast Cameras

Parallel read-out of a diode array yields a much better signal-to-noise ratio and higher speed than serial readout in modern high-resolution fast cameras. In addition, the dark noise in modern fast cameras is rather high, particularly with the Fuji camera. It seems that an ideal detector for the current voltage-sensitive dyes remains to be developed. At least two companies are currently involved in such developments (see http://www.RedShirtImagin.com and brainvis@edonagasaki.co.jp). The optimal requirements are:

- *Detector target size:* about 20x20mm or 30x30mm is optimal for the macroscope;
- *Number of pixels:* 64x64 or 128x128 seems optimal.
- *Speed:* up to 2000Hz, but 300Hz may be sufficient for many *in vivo* experiments,
- *Saturation capacity:* (photo-electrons) 10^8
- *S/N:* at least 1:5000 with low dark noise.
- *Number of readout ports:* 8–12 appears to be practical
- *Digitizers:* 12–14 bits seem optimal. 8 bits are also possible if differential imaging is done by subtracting a reference image from the subsequent incoming frames.
- *Fill factor:* close to 100%
- *Quantum efficiency:* maximal

Ichikawa and his colleagues (1998) are currently trying to utilize existing detectors employed in other video applications and perform extensive binning and faster readout. The new CCD-based system proved successful in slice experiments (Ichikawa et al. 1998). It remains to be seen if this direction would prove useful also in the neocortex.

Bullen and Saggau (1998) (cf. also Chapter 4) have decided to try an alternative approach avoiding the camera altogether. Instead, they successfully developed a fast laser scanning technique. It remains to be seen how suitable such a system is for exploring the neocortex.

2.7.4 Visualization of the Electrical Activity

Inspection of the large amount of optical data is time-consuming, but feedback during the experiment is important. Thus, it is important to display the data as a slow motion movie. For detecting patterns black-and-white or surface plot display seems more advantageous than color display, if the investigator wishes to fully use the processing power of his own visual system.

2.7.5 Computer Programs

Much effort must be put into the choice and development of the software required to perform the data acquisition, smooth interfacing to standard physiological experiments, automatic control of the experiments, data analysis and display. It is recommended to rely on existing software rather than to re-invent the wheel. The available software is camera-dependent, which further complicates the situation. It seems likely that a suitable commercial package for both the hardware and the software will become commercially available soon.

2.8 Spatial Resolution of Optical Recording

2.8.1 Microscope Resolution

The spatial resolution of optical imaging of neuronal activity can approach the spatial resolution of the microscope used. The spatial resolution of the microscope for flat two-dimensional preparations is excellent (<1 μm). However, the microscope's spatial resolution in a three-dimensional preparation is relatively poor, and therefore the spatial resolution for optical imaging *in vivo* is hampered. For example, in the salamander olfactory bulb, it was estimated to be 300 μm for a 10x objective (Orbach and Cohen., 1983). In the frog optic tectum experiments (Grinvald et al., 1984), it was estimated to be 200 μm with a 10x objective and about 80 μm with a 40x objective. Several reviews have suggested three ways to improve the spatial resolution (Cohen and Lesher.,1986; Grinvald., 1984, 1985; Grinvald and Segal., 1983; Grinvald et al., 1986; Orbach., 1988):

- Design of custom-made long-working distance objectives with high numerical aperture for optical sectioning measurements;
- Mathematical de-convolution of results obtained from measurements at different focal planes and use of the mathematical equation for the point spread function (i.e., the de-focus blurring function) of a given objective;
- Use of a confocal detection system and focal laser micro-beam illumination and scanning in three dimensions, instead of continuous illumination of the whole field under investigation.

The confocal microscope can dramatically improve the three-dimensional resolution and reduce the light-scattering perturbation of a clear image. Only a small spot in the preparation is illuminated at a given time and coincident detection is employed at the image plane, only from the precise spot where the unscattered image should appear. In this way both out-of-focus contributions to an image and the effects of light scattering are considerably diminished. However, at present, attaining the signal-to-noise ratio required for voltage-sensitive dye measurements with a confocal system *in vivo* has not been documented. The two photon microscope offers additional advantages. It will be worthwhile to develope it for imaging based on voltage-sensitive dyes. A line scan would already provide a significant advance.

Another solution to the three-dimensional resolution problem is to stain only a very thin layer in the preparation by iontophoretic application or by pressure injection of the dye. Specific staining restricted to a deep layer below the surface (e.g., from the ventricles) would also increase the depth of the loci accessible to optical measurements. Such approaches remain to be tested.

2.8.2 Effect of Light Scattering on the Spatial Resolution

Light scattering from cellular elements, especially in preparations which are relatively opaque, leads to the deterioration of images resolved by conventional microscope optics. Thus, light scattering both blurs the images of individual targets and causes an expansion of the apparent area of detected activity. These effects of light scattering on optical recordings were investigated in the vertebrate and mammalian brain. To quantify this problem, one can image fluorescent beads at the tip of a micro-electrode at different cortical depths. In the cat cortex, the light scattering appeared to be much higher than that found by other investigators using phantoms such as milk (Vanzeta, Kam and Grinvald, unpublished results). Thus, it appears to us that the light scattering is the limiting factor in improving the three-dimensional resolution of *in vivo* optical imaging.

2.9 Interpretation and Analysis of Optical Signals

2.9.1 What Does the Optical Signal Measure in the Neocortex?

As already mentioned, fast dye signals and intracellular electrical recordings follow an identical time course. Therefore, optical recordings of fast signals from well-defined cellular elements can replace intracellular recordings in several situations when the use of electrodes is difficult. However, this replacement is not always advantageous since with intracellular electrodes, the resting potential can readily be measured or manipulated (cf. Chapter 5), in contrast to dyes and optical imaging. The inability to evaluate reversal potentials for various EPSP's and ISPP's limits the interpretation of some observed scattering by the tissue. Therefore, in several preparations, the identification of signal sources or the interpretation of signal amplitudes present problems. Nevertheless, optical recording of population activity has advantages over electrical field potential recordings because the optical signal is restricted to its site of origin and reflects membrane-potential optical responses. Furthermore, recording of the intracellular population activity provides information about the weighted sum of the membrane potential changes in all the cellular elements imaged on a given pixel. The weighting factors for each cellular site depend on the membrane area, the density of dye binding, its sensitivity, the illumination intensity and the collection efficiency of the imaging optics. The latter depends on the objective, the object depth, and the extent of light scattering by the tissue. Therefore, in several preparations, the identification of signal sources or the interpretation of the signal amplitudes present a series of problems. Nevertheless, optical recording of population activity has advantages over electrical field-potential recordings, because the optical signal is restricted to its site of origin and reflects intracellular membrane-potential changes, rather than extracellular voltage changes (Grinvald et al., 1984; Grinvald et al., 1982; Orbach and Cohen., 1983; Orbach et al., 1985).

Several reservations have been raised regarding the origin of the dye signals *in vivo*, particularly concerning the contribution from glial cells or extracellular currents. Recently, Sterkin et al. have shown that with RH 1692 the dye signal does indeed reflect the membrane potential change of neurons primarily. This result is illustrated in Fig. 23. It was obtained by combining dye imaging *in vivo* with intracellular recording from a single neuron in a deeply anesthetized preparation.

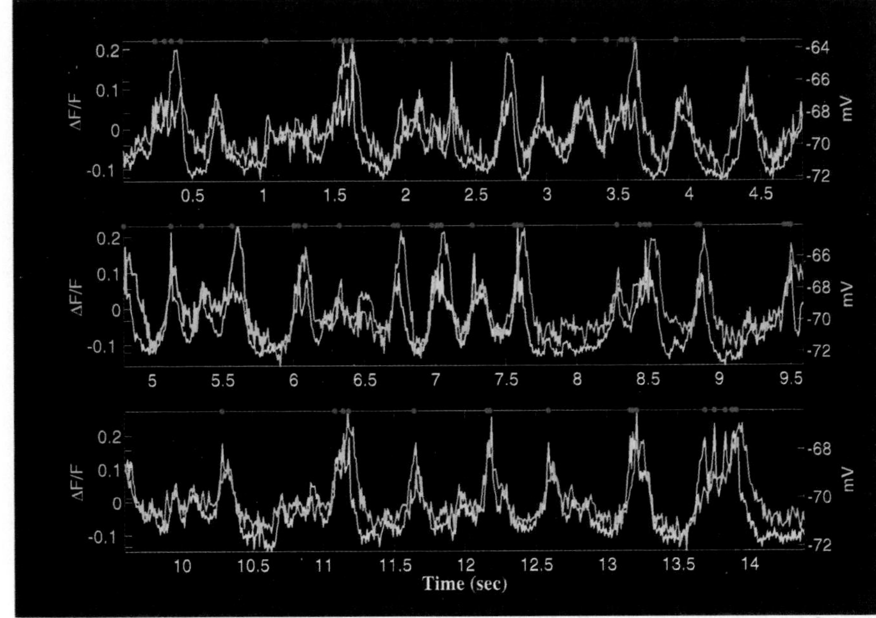

Fig. 23. Similarity between the cortical dye signal from a small population of neurons and the intracellular recording. Two traces showing simultaneous intracellular and optical recordings for 15 s, performed in a deeply anesthetized cat, a condition in which spontaneous changes in membrane potential are highly synchronized in a large population of neurons. The intracellular recording is depicted by the yellow traces. The action potentials were truncated and occurred at times marked by the red dots. The optical signal from the population next to the electrode is depicted by the blue traces. (Modified from Sterkin et al., 1998)

2.9.2 Which Cortical Layers Are Being Imaged?

Optical imaging in the neocortex is performed by taking images of the stained cortex with a camera oriented perpendicular to the cortical surface. It is known that the cortex is stained up to a depth of 1.5 mm, but the upper cortical layers are stained much better than the lower ones. It is also known that when utilizing conventional optics, the image contained mostly neuronal elements in the upper 0 to 400–800 μm. Thus, the question arises as to which cortical layers contribute to the detected optical signals. Intuitively one would guess that most of the activity originates from layer I and perhaps some from layers II and III. This is not the case considering also the exact morphology of cortical neurons because the apical dendrites of layer V and VI neurons do reach the upper cortical layers. Therefore, activity in the dendrites of neurons in deep cortical layers also contributes to the activity detected when the camera is focused on the upper layers.

The exact contribution of each cortical layer remains to be estimated and explored. It is clear that cross-correlation of the activity of identified neurons in any cortical layer with the population activity, detected optically, may help resolve activity that exists only in that cortical layer. Such experiments were not yet systematically performed.

2.9.3 Amplitude Calibration

The amplitude of the optical signals *in vivo* cannot be calibrated in millivolts of membrane potential change, because the size of optical signals is related also to the membrane area, the extent of binding at each site and the sensitivity of the dye for a given membrane. Therefore, great caution should be exercised when interpreting the absolute amplitude of the optical signals. Usually, only a comparison of the relative amplitudes observed under different experimental situations is meaningful. To get a rough estimate, one can compare the naturally evoked signal to the one evoked when all the neurons are active, for example, during inter-ictal events in the presence of GABA blockers.

2.9.4 Dissection of "Intracellular Population Activity" into Its Multiple Components

In the three-dimensional neocortex, with its multiple cortical layers and heterogeneous neuronal elements viewed by each single pixel, the identification of the signal sources presents a serious challenge. Utilizing real-time optical imaging of the active cortex, it is possible to detect spatial activity patterns, but it is not easy to determine whether these signals mostly originate from postsynaptic changes in membrane potential in dendrites, from presynaptic action potentials, from dendritic back-propagation, from dendritic calcium action potentials or from cell somatic potential changes and a combination thereof. Even the site of activity initiation may be difficult to detect if the activity there is small relative to reverberating activity, and if the latency is not detectably shorter than that in other monitored cortical sites. Furthermore, in the neocortex, changes in membrane potential in glial cells may significantly contribute to the optical signal.

These difficulties should not be underestimated. Proper identification of the origin of the signal may be important when novel findings are suggested by the optical imaging data, yet it may not be trivial to confirm them in alternative ways. To separate the various components of the signal, proper manipulation of stimulus parameters, such as intensity, location or frequency and/or pharmacological manipulations, or varying the ionic composition, etc., may be helpful (see examples in earlier work in mammalian brain slices; Grinvald et al., 1982). Thus, great caution should be exercised in the interpretation of the origin of an optical signal, especially if it is slow. In general, slow signals may also originate from a change in surface charge (e.g., Eisenbach et al., 1984) due to the dye's interactions with a changing ionic environment in restricted volumes (e.g., Beeler et al., 1981), probably attributed to the large change in [K^+] or other slow physical changes (Cohen, 1973).

To facilitate the interpretation of population activity, additional approaches can be used:

- Probes that are specific to a given cell type may be useful in such an analysis, yet it seems unlikely that such probes would be universal and perform equally well in all preparations. Advances in new genetic engineering approaches recently applied to the design of voltage probes may solve this problem. Miyawaki et al., 1997; Siegel and Isacoff 1977.
- Optical studies of single neurons in complex multi-cellular preparations by iontophoretic injection of a suitable fluorescent dye into single neurons can serve to elucidate the origin of the signal (e.g., Grinvald et al., 1987).
- Specific retrograde or anterograde labeling of a given neuronal population by remote extracellular injection of appropriate dyes at the proper site can selectively identify the activity of a given population of neurons (e.g., Wenner et al., 1996).
- Combination of conventional electrophysiological approaches with the optical measurements should also aid in the interpretation of the optical data. Particularly useful would be electrical recording from each of the cortical layers with a single electrode having multiple recording sites at different cortical depths.
- Elimination of a specific population of neurons can also help the analysis. One way to eliminate a specific type of cells may be the use of complement (killers) to specific antibodies or genetically engineered mutants. Yet another way is genetic manipulation to eliminate a particular receptor type etc.

2.10 Present Limitations

Although the state of the art in voltage-sensitive dye imaging is already well advanced, it is important to mention the limitations associated with the use of this optical imaging technique; since an intimate understanding is imperative for successful and optimal use.

2.10.1 Signal-to-Noise Ratio

In several preparations, the signals are presently small and the activity of single neurons could not yet be monitored. In other types of experiments, the signals are very large. Clearly, it is advantageous to use the technique for investigations and preparations where signals are large. A struggle with small signals can be justified only if an important question cannot be resolved by alternative methods. Maximal signal averaging, whenever applicable, should be used to improve the signal-to-noise ratio. Signal averaging will also reduce the amplitude of the large ongoing activity not related to the stimulus.

2.10.2 Contamination by Light Scattering or Other Intrinsic Signals

If the light absorption or fluorescence signals is small, the activity-dependent (dye-unrelated) light-scattering or slow hemodynamic intrinsic signals from the preparation may distort the voltage-sensitive optical signals. A solution to this problem is to try and subtract the light-scattering signals from the total optical response by measuring it independently in the proper way. This task may not be easy. If the dye signals are measured for a duration longer than a few hundred milliseconds, they would be "contaminated" by the slow intrinsic signal. How serious is the distortion of the dye signal by the intrinsic signal? Evidently, the magnitude of the distortion of the "voltage signals" depends on the relative size of the intrinsic signal and the dye signal at the wavelength used for the imaging.

Figure 24 illustrates the time course of the voltage-sensitive dye signal and the intrinsic signal measured at the same wavelength. Using an expanded time course for these two measurements, one can see that the intrinsic signal is delayed, therefore at an early time interval following the stimulus, the voltage signal is not significantly distorted by the intrinsic signal. The inset shows the same recording for a much longer time interval. Here one can see that the slow intrinsic signal becomes so large as to distort the voltage signal dramatically. The delayed "hyperpolarization" is simply an artifact of the intrinsic signal. This example emphasizes the importance of this issue whenever slow measurements are carried out.

A comparison of the time courses in Fig. 24 with those depicted in Fig. 1 in Blasdel and Salama (1986) indicates that no voltage signal was present in the latter paper, the detected signal being merely the intrinsic "light-scattering artifact", as mentioned above. Such slow and fast artifacts have been described in a previous publication (Grinvald et al., 1982). This example shows once again that a certain artifact associated with one technique, i.e., light scattering artifacts in dye recording in the neocortex, can be turned into a useful tool when used for another technique, i.e., optical imaging based on intrinsic signals.

2.10.3 Pharmacological Side Effects

Light-independent pharmacological side-effects can be expected whenever extrinsic probe molecules are bound to neuronal membranes, especially if high concentrations of dye are used. Excessive dye binding may change the threshold, specific ionic conduct-

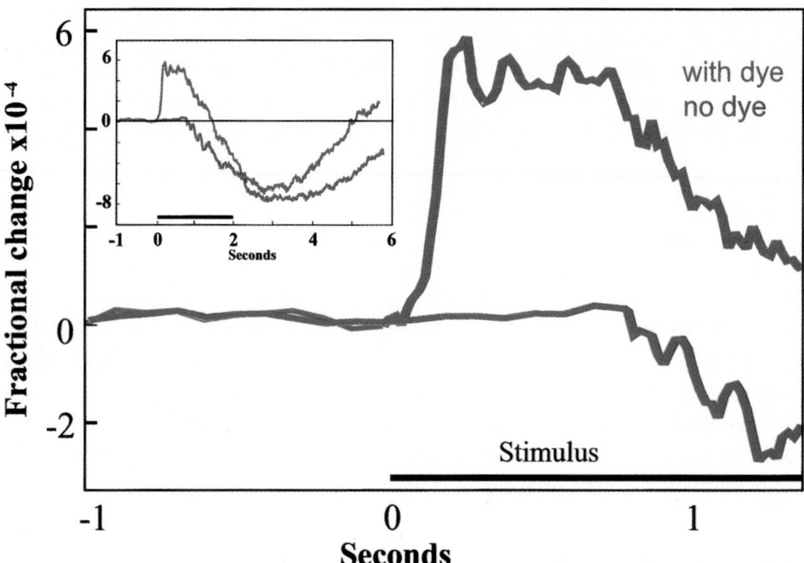

Fig. 24. Comparison of the time courses of the dye and the intrinsic signal, and intrinsic signal artifact. The time course of the voltage-sensitive dye signal is depicted by the red trace. The time course of the intrinsic signal recorded prior to the staining at the same wavelength is shown by the green trace. The inset shows the same signals recorded for a period of 6 s. The large undershoot in the dye signal is an artifact of the intrinsic signal rather than a net hyperpolarization.

ances, synaptic transmission, membrane resistance, pump activity etc. The optimal dye concentration depends on the binding constant of the dye to the membrane. With topical application of the dye for two hours, a gradient of staining is established where deep layers are much less stained because dye diffusion is restricted to narrow extracellular spaces. An important rule, therefore, is to use the minimal concentration that can still provide useful signals.

Following extensive screening for an optimal dye, it was observed that most of the selected dyes did not cause significant pharmacological side effects. This conclusion is based on the response properties of single units from stained cortex. However, the need for careful controls verifying the absence of pharmacological effects of the dyes cannot be overemphasized. Again, it is not always easy to provide such controls, because in several kinds of optical experiments it is hard to obtain the same information with alternative methods. One effective way to check if there are pharmacological side effects is to compare the intrinsic signals or the field potentials before and after staining of the neocortex. Such controls are essential to prevent misinterpretation of novel information regarding complex brain function. Without such controls, an interpretation of the results may be relevant only to a dye-modified nervous system rather than the normal system being explored.

2.10.4 Photo-Dynamic Damage

In the presence of intense illumination, the dye molecules sensitize the formation of reactive singlet oxygen. These reactive radicals attack membrane components and damage the neurons. Such photo-dynamic damage limits the duration of some experiments. With the significantly improved present dyes, the imaging session is limited to approximately 20–60 minutes of continuous illumination before significant damage occurs. If each trial lasts 500 ms, this means that 2400–7200 trials can be averaged. Thus, it appears that this long-lasting problem has been solved, whenever signal averaging is feasible.

Cells can repair damage to their membrane, provided that the damage is not too large. A preliminary observation suggested that brief exposure to bright light, followed by a long inter-stimulus interval (e.g., 30 s) reduces the accumulated damage in signal-averaging experiments to a level that is much lower than that observed for continuous illumination (T. Bonhoeffer, personal communication).

Repeated imaging sessions in the behaving monkey would facilitate studies of cortical dynamics related to higher brain functions. Therefore, an important question is whether repeated optical measurements, based on the present dyes can be done from the same cortex over a period of a few weeks or months. Preliminary experiments studying the development of synaptic connections in culture showed that cultured neurons survive both the staining and the optical recordings, provided each optical measurement is done only for short times before significant irreversible damage occurs (J. Pine, personal communication). Furthermore, repeated voltage-sensitive dye measurements have already been reported for the awake monkey (Inase et al., 1998; Slovin et al., 1999).

2.10.5 Dye Bleaching

Dye bleaching may affect the time course of the optical signals, especially in fluorescence measurements if an intense light source was used. An offline correction procedure has been described (Grinvald et al., 1987). Subtraction procedures of the type used to remove the heartbeat noise are even more effective (Grinvald et al., 1984, 1994). Bleaching during the optical measurement also limits the duration of the measurement. Thus, to minimize bleaching, the exposure time of the preparation to light should be reduced to a minimum. If the bleaching reduces the signal size prior to the onset of photodynamic damage, then the preparation can be restained and the original signal size restored.

2.11 Design, Synthesis and Evaluation of Improved Optical Probes

To overcome the limitations described above, much effort has been directed at the design and/or testing of adequate probes to enable optical monitoring of neuronal activity. Of more than 3,000 dyes already tested (Cohen et al., 1974; Loew et al., 1979,1985; Ross et al., 1977; Gupta et al., 1981; Grinvald et al., 1982; Grinvald et al., 1983; Grinvald et al., 1987; Lieke et al., 1988; Grinvald et al., 1994, Glaser et al., 1998), about two hundred have proven to be sensitive indicators of membrane potential while causing minimal pharmacological side effects or light-induced photo-chemical damage. Most of the initial screening experiments on voltage-sensitive dyes were done with the squid giant axon (Tasaki et al., 1968; Davila et al., 1974; Tasaki and Warashina., 1976). At present screening experiments are performed in the rat somato sensory cortex.

Much progress has been made. Today, the best fluorescence dye, RH 421, yields a change in fluorescence intensity of 25%/100 mV of membrane potential change (in neuroblastoma cells; Grinvald et al., 1983). This value is 120 times greater than the signal size that was obtained with leech neurons in the pioneering experiments of Salzberg and his colleagues (Salzberg et al., 1973). This improvement in signal size is attributable to some improvement in the sensitivity of the new dye itself and also to the fact that, in fluorescence measurements, signal size for isolated neurons maintained in culture is expected to be much larger relative to that for *in vivo* cortical imaging, where only a fraction of the neurons are active and a large background from glial-bound dye and the extracellalar space is present. In addition, there has recently been a considerable improvement in the quality of fluorescent probes, especially styryl dyes like RH 795 or the new oxonol dye RH 1692, designed for *in vivo* imaging of mammalian cortex. Because the development of high-quality voltage-sensitive probes is essential for the widespread application of this technology, it is unfortunate that only five laboratories have actually been involved in voltage-sensitive dye synthesis. The laboratory of Dr. A. Waggoner produced about 500 dyes until 1978 in collaboration with Cohen. This pioneering work was then continued by four other groups: Loew and his colleagues, Fromherz and his colleagues, Tsien and his colleagues, and Hildesheim in our group. In retrospect, it appears that faster progress could have been achieved if:

- More laboratories would synthesize extrinsic probes.
- Synthesis of dyes having alternative structures would also be attempted (other than cyanines, merocyanines and oxonols). There are more than 200,000 possible derivatives of these dyes, which have been mostly synthesized by the color photographic film industry.
- Dye screening would be done on the relevant preparations rather than primarily on squid giant axons or rats.
- Synthesis and testing would be done simultaneously for quick feedback.
- Theoretical approaches for the design of "ideal electrochromic" probes of the type introduced by Loew and his colleagues (Loew et al., 1978) would be incorporated.
- More theoretical work on the mechanisms underlying the probe signal and their relation to signal size would be done (Zhang et al., 1998).
- More investigations of the biophysical mechanisms underlying the probe sensitivity would be carried out. Thus, experiments on simple model systems, like lipid bilayers, spherical bilayers and vesicles, could contribute to better dye design (Loew, 1987: Roker et al., 1996).
- Better understanding of the relationship between dye structure and spectral properties would be achieved (extinction coefficient, quantum yield in different environments, photo-damage and bleaching).

2.11.1 Dye Design

Spectroscopic parameters which are a prerequisite to achieve proper dye sensitivity have been discussed elsewhere (Waggoner and Grinvald.1977). The structural requirements for the dyes to achieve proper interaction with a changing electric field across the lipid bilayer were also discussed (Loew et al., 1978; Loew and Simpson 1981), but several useful dyes do not seem to obey the suggested ideal structure. In fact, most of the useful dye families were discovered in persistent screening experiments rather than by chemical engineering. This situation resembles that of the pharmaceutical industry where despite the major efforts for rational design of suitable drugs, only one out of about 100,000 finds its way to the patients. With dye synthesis, the level of frustration has been similar, even though the rate of success was higher.

In contrast, the tuning of the dye analogs to achieve optimal performance in a given preparation greatly benefited from the accumulated experience in this field and rational design. For example, the large improvement in recent *in vivo* imaging is primarily due to the decision to use new dyes requiring a longer excitation wavelength >620 nm, where the hemodynamic activity-dependent signals and the heartbeat and respiratory pulsations are 5–12 fold smaller relative to those observed at 540 nm, using RH 795 (Sterkin et al., 1998; Glaser et al., 1998; Shoham et al., 1998).

2.11.2 Dye Synthesis

Initially most of the dyes were tested on the squid giant axon in heroic screening efforts made by Cohen and his colleagues (Cohen et al., 1974; Gupta et al., 1981; Ross et al., 1977). In retrospect, it is clear that the dyes should be tested directly on the relevant preparation under investigation. It is crucial to screen many dyes (from 6 to 100) when applying the technique on a new preparation or on a new system. For example, in the application to the mammalian brain, about 40 dyes were tested on the rat cortex before RH-414 was selected as a useful dye. However, the initial experiments with RH-414on cat visual cortex were not successful (Orbach et al, 1985), either because the dye did not penetrate the upper layer of the cortex, generating improper staining, or because the dye was not voltage-sensitive in cat visual cortex. In addition, it was found that RH-414 had pharmacological side effects. Staining led to arterial constriction, which significantly reduced the blood flow (Grinvald et al., 1986). Furthermore, it often modified the time course of electrically recorded evoked potentials. Therefore, many dyes were tested directly in the cat and the monkey visual cortex (Grinvald et al., 1994). Thus, it is preferable to screen the dyes first on the rat whisker barrel system and only then on the relevant preparation.

2.11.3 Other Improvements May Provide Larger Signals

Thus far, the direct measurement of changes in absorption or fluorescence of extrinsic probes has proved useful in the neocortex. Yet it is important to test whether the monitoring of other spectroscopic properties can provide better performance. For example, Ehrenberg and Berezin (Ehrenberg and Berezin, 1984) have used resonance Raman spectroscopy to study surface potential, but so far this approach has not provided better performance for measuring voltage transients. Similarly, it still remains to be seen whether monitoring of other spectral parameters, such as fluorescence polarization, circular dichroism, energy transfer between two chromophores (Gonzalez and Tsien 1995, 1997; Cacciatore, et al., 1998), delayed emission, infra-red absorption, etc., may provide better performance.

We also expect additional development of new types of voltage-sensitive probes along completely different lines: genetic engineering of suitable *in vivo* probes could make the experiments much easier and, therefore, of wide-spread use for numerous applications. Pioneering efforts in this direction appear fruitful (e.g., Miyawaki et al., 1997; Siegel and Isacoff 1997). At present it seems that the use of such probes would become practical in transgenic mice in the near future. It remains to be seen how long the road will be for making use of this approach also in other species, particularly monkeys.

Part 3: Combining Optical Imaging with Other Techniques

3.1 Targeted Injection of Tracers into Pre-Defined Functional Domains

It is clear that the study of cortical organization and function can greatly benefit from a combination of optical imaging with other techniques, such as tracer injections, electrical recording and micro-stimulation. Since optical imaging of the functional architecture can quickly and easily provide a picture of how certain functional parameters are represented on the cortical surface, it is an ideal tool to guide targeted electrophysiological recordings or tracer injections. Furthermore, using such an integration, morphological data such as the dendritic and axonal branching of single cells can be directly

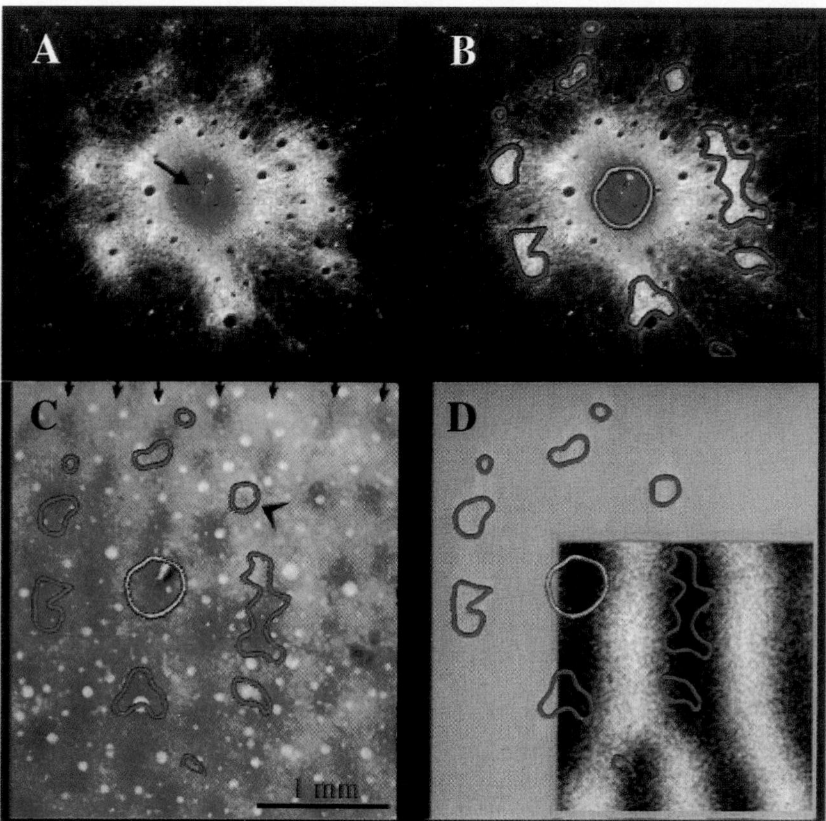

Fig. 25. Optical imaging guiding anatomical investigations. A-C: Biocytin injection targeted at a monocular site in macaque monkey area V1. A: Dark-field photo-micrograph of a tangential cortical section showing a biocytin injection (arrow) that was placed in the center of an ocular dominance column. Note the extensive local halo around the injection site followed by clear axonal patches further away. B: Same micro-graph as in A, but after the main patches were delineated (red contours). The effective tracer-uptake zone, defined under high magnification viewing (not shown), is depicted by the yellow contour at the center. C: Same injection as in A and B, but superimposed on an optically imaged map of ocular dominance columns. The injection was targeted at a contralateral eye column (coded black) at the top left corner of the optically imaged area. Note the tendency of the biocytin patches to skip over the ipsilateral eye column (coded white). Scale bar shown in C: 1 mm. (Modified from Malach et al., 1993)

correlated with the functional organization in the very same piece of tissue (Malach et al., 1993,1994,1997; Kisvarday et al., 1994; Bosking et al., 1997). An example of this application is illustrated in Fig. 25.

3.2 Electrical Recordings from Pre-Defined Functional Domains

In many cases it is important to use metal or glass electrodes (cf. Chapter 5) simultaneously with optical imaging. For this purpose, it is useful to have a sealed cranial window coupled to a manipulable electrode. Arieli and Grinvald have recently designed such a device illustrated in Fig. 26. This device consists of a chamber with a square cover glass that is much larger than the chamber diameter. The cover glass can be moved relative to the base of the cranial window. A hole in the glass covered with a rubber gasket allows the insertion of the electrode into the sealed chamber. This device has proved very use-

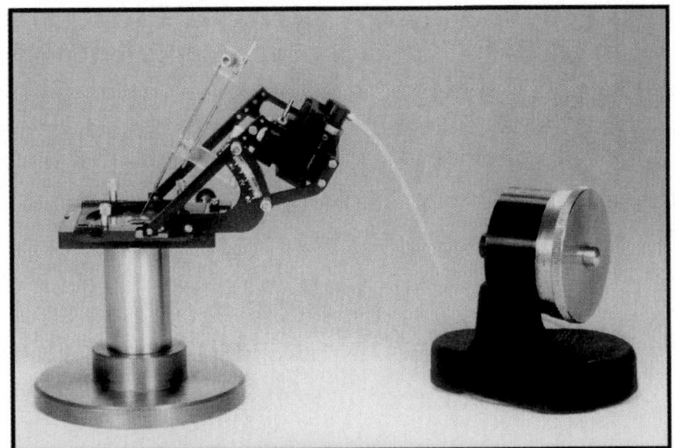

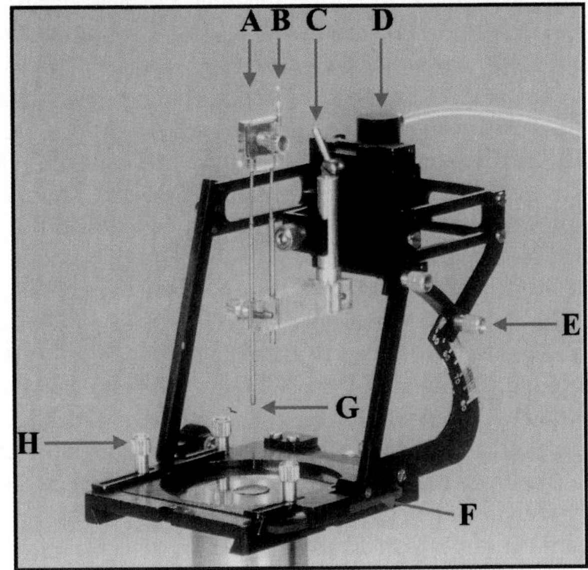

Fig. 26. Cranial window combined with an electrode manipulator. Top panel: The X-Y and axial microdrive on its stand at an angle of ~60(. The microdrive can work in the range from 30(to 90(. The hydraulic microdrive itself is a Narishige hydraulic microdrive MO-11N that can also be coarsely positioned with a manual manipulator. Bottom panel: **A:** tip of a tungsten microelectrode. **B:** pin for electrical connection with the electrode. **C:** bolt to lock the coarse positioning of the manual manipulator. It enables to manually advance the electrode guide (the small metal tubing plus plastic part plus the electrode) in order to penetrate the rubber gasket. Only then the electrode is pushed forward out of the protecting penetration tube and fine movement is achieved using the hydraulic microdrive. **D:** hydraulic microdrive connected to the black tiltable frame (which has two holes in its connecting piece in order to attach the Narishige to it). **E:** Two screws lock the microdrive at various angles (range of 30(to 90(). **F:** Sliding glass. **G:** needle that holds and protects the microelectrode, to penetrate the rubber gasket. **H:** Four screws lock the upper part of the microdrive to its X-Y platform (from Arieli and Grinvald, unpublished results).

ful for electrical confirmation of optically obtained functional maps, as well as for targeted microelectrode recordings. Shmuel et al. (1996) and Shoham et al. (1997) have used it to perform both perpendicular and nearly tangential penetrations, studying the relationships between unit responses and optically imaged functional domains for orientation and direction and for spatial frequency, respectively. Lampl, Ferster and their colleague have used it for simultaneous optical imaging and intracellular recordings in-vivo (Sterkin et al. 1998).

3.3 Combining Micro-Stimulation and Optical Imaging

The same device has proved equally advantageous for micro-stimulation during optical imaging experiments. We made use of it for evaluating the extent of cortical activation in response to micro-stimulation of neuronal clusters at identified functional domains (Glaser, Arieli, Seidman and Grinvald, unpublished results). The minimal diameter of the activated area (as judged by optical imaging) was approximately 500 µm. In view of recent micro-stimulation studies in the behaving monkey (Newsome et al., 1989; Salzman et al., 1990, 1992), such studies are of great importance as they should lead to an estimate of the number of neurons which are affected by such electrical micro-stimulations *in vivo*. The combination of micro-stimulation with optical imaging can also be used to dissect the nature of the optical signal by adjusting the various parameters of the electrical stimulus. Furthermore, it can be used to explore functional connectivity between multiple cortical sites.

Part 4: Comparison of Intrinsic and Voltage-sensitive Dyes

Optical Imaging

The two methods of optical imaging, intrinsic signals imaging and voltage-sensitive dyes imaging, each have their advantages and disadvantages. A central difference between the two methods is in the temporal resolution they provide. The principal shortcoming of intrinsic signal optical imaging is its limited time resolution, whereas voltage-sensitive dyes imaging offers a time resolution of sub-milliseconds. Therefore, it is clear that intrinsic signals imaging cannot replace optical imaging based on voltage-sensitive dyes whenever the temporal aspect of neural coding is at issue. For example, only voltage-sensitive dyes imaging could be used to study the ongoing activity in the absence of stimulation (Arieli et al., 1996).

On the other hand, the imaging based on intrinsic signals is by far easier than the imaging based on extrinsic probes. The fine temporal resolution of the voltage-sensitive dye signals, comes at a price. Due to the square-root relationship between the number of photons sampled and the signal-to-noise ratio, it is easier to obtain a good signal-to-noise ratio when signals are slow. An additional advantage of intrinsic signal optical imaging is that, since it makes no use of dyes, it does not suffer from the problems of photo-dynamic damage and pharmacological side effects. Furthermore, since optical imaging with intrinsic signals is less invasive, potential long term chronic over many months from the same cortical area is easier. Furthermore, imaging based on intrinsic signals can be done through the intact dura or even a thinned bone, whereas methods to do the same with imaging based on voltage-sensitive dyes have not yet been reported.

Another disadvantage that has been attributed to voltage-sensitive dyes imaging is that it sacrifices spatial resolution to gain better temporal resolution. Actually, however, no such trade-off exists. The recent improvements in dyes and in the spatial resolution of fast cameras have made it possible to obtain high-resolution functional maps of orientation columns, "lighting up" in milliseconds (Shoham et al, 1993), with a signal-to-noise ratio, which is in some cases even better than that obtained with the slow intrinsic signals (Glaser et al., 1998).

Conclusions and Outlook

Optical imaging based on intrinsic signals is a method which allows investigators to map the spatial distribution of functional domains, offering unique advantages. Currently, no alternative imaging technique for the visualization of functional organization in the living brain provides a comparable spatial resolution. It is this level of resolution which allows to reveal *where* processing is performed – a necessary step for the understanding of the neural code at the population level. A key advantage of intrinsic imaging is that the signals can be obtained in a relatively non-invasive manner and over long periods of time. This is particularly important for chronic recordings, which can last up to many months. Optical imaging during chronic experiments allows the study of cortical development and plasticity and of higher brain functions in behaving monkeys. A major challenge is to apply intrinsic signal imaging also to the human brain. So far, the quality of the results obtained in humans is lower than that achieved in animal experimentation. Nevertheless, it seems likely that optical imaging will prove to have clinical benefits. Thus, during clinical use, the functional architecture of various sensory cortical areas could be mapped at unprecedented resolution, an order or two better than that currently accomplished by PET or f-MRI. Finally, completely non-invasive optical imaging through the intact human skull may provide an imaging tool offering both the spatial and the temporal resolutions required to expand our knowledge of the principles underlying the remarkable performance of the human cerebral cortex.

Real-time optical imaging in the neocortex is based on the use of voltage-sensitive dyes. Extensive efforts have been successfully made to overcome its limitations. The results discussed here indicate that the technique has matured, enabling explorations of the neocortex in ways never feasible before. We expect additional development of new voltage-sensitive dyes and/or future genetic engineering of suitable *in vivo* probes, which could make the experiments much easier and, therefore, of wide-spread use for numerous applications.

We predict that a multi-purpose imaging system for either slow intrinsic imaging or fast voltage-sensitive dyes imaging will become available in the near future. Such a system should allow each laboratory to use each one of these imaging techniques alone, or both of them combined. This system would allow to use the advantages of each of these approaches and to minimize their limitations, all according to the specific questions being explored.

Real-time optical imaging based on voltage-sensitive dyes allows investigators to obtain information about both the spatial and the temporal aspects of neuronal activity, with a resolution good enough to see the fine structure of individual functional domains as well as coherent neuronal assemblies within the neocortex. No alternative imaging technique for visualizing functional organization in the living brain provides a comparable spatial or temporal resolution. It is this level of resolution which allows voltage-sensitive dyes imaging to address questions both of *where* and of *when* processing is performed. To increase the dimensionality of neurophysiological data obtained from the same patch of cortex, real-time optical imaging can be combined with targeted tracer injections, micro-stimulation or intracellular and extracellular recordings. These combined approaches allow to address also the question of how, thus promising that this technique will play a prominent role in the study of neural coding at the population level.

References

Arieli, A., Shoham, D., Hildesheim, R. and Grinvald., A. (1995). Coherent spatio-temporal pattern of on-going activity revealed by real time optical imaging coupled with single unit recording in the cat visual cortex. J. Neurophysiol., 73, 2072–2093

Arieli, A., Sterkin, A., Grinvald, A., and Aertsen, A. (1996). Dynamics of on-going activity: Explanation of the large in variability in evoked cortical responses. Science, 273, 1868–1871.

Arieli, A., et al., (1996). The impact of on going cortical activity on evoked potential and behavioral responses in the awake behaving monkey. NS abstract.

Bakin, J.S., Kwon, M.C., Masino, S.A., Weinberger, N.M., Frostig, R.D. (1993). Tonotopic organization of guinea pig auditory cortex demonstrated by intrinsic signal optical imaging through the skull, Neurosci. Abstr., 582(11).

Bartfeld, E., and Grinvald, A. (1992). Relationships between orientation preference pinwheels, cytochrome oxidase blobs and ocular dominance columns in primate striate cortex. Proc. Natl. Acad. Sci. USA 89, 11905–11909.

Beeler, T.J., Farmen, R.H., and Martonosi, A.N. (1981). The mechanism of voltage-sensitive dye responses on saroplasmic reticulum. J. Member. Biol. 62: 113–137.
Blasdel GG (1989). Visualization of neuronal activity in monkey striate cortex. Annu Rev Physiol 51: 561–581.
Blasdel GG., Salama G. (1986) Voltage-sensitive dyes reveal a modular organization in monkey striate cortex. Nature 321:579–585.
Blasdel, G. G. (1988). in: "Sensory Processing in the mammalian brain: Neural substrates & Experimental strategies". Ed., Lund, J.S. Oxford Univ. Press pp 242–268.
Blasdel, G.G. (1989). Topography of visual function as shown with voltage sensitive dyes. In: Sensory systems in the mammalian brain, J.S. Lund, ed., pp. 242–268, Oxford University Press, New York.
Blasdel, G.G. (1992a). Differential imaging of ocular dominance and orientation selectivity in monkey striate cortex, J. Neurosci., 12, 3115–3138.
Blasdel, G.G. (1992b). Orientation selectivity, preference, and continuity in monkey striate cortex, J. Neurosci., 12, 3139–3161.
Bonhoeffer, T., and Grinvald, A. (1991). Iso-orientation domains in cat visual cortex are arranged in pinwheel-like patterns, Nature, 353, 429–431.
Bonhoeffer, T., and Grinvald, A. (1993). The layout of Iso-orientation domains in area 18 of cat visual cortex: Optical Imaging reveals pinwheel-like organization, J. Neurosci., 13, 4157–4180.
Bonhoeffer, T., and Grinvald, A. (1996). Optical Imaging based on Intrinsic Signals. The Methodology. in Brain Mapping. The Methods. A. Toga and J. Mazziotta (eds). Academic Press.
Bonhoeffer, T., Goedecke, I. (1994). Kittens with alternating monocular experience from birth develop identical cortical orientation preference maps for left and right eye, Soc. Neurosci. Abstr., 20, 98(3), 215.
Bonhoeffer, T., Kim, A., Malonek, D., Shoham, D., and Grinvald, A. (1995). The functional architecture of cat area 17. Eur. J. Neurosci., 7, 1973–1988.
Bosking, W.H., Zhang, Y., Schofield, B., Fitzpatrick, D. (1997). Orientation selectivity and the arrangement of horizontal connections in tree shrew striate cortex. J-Neurosci. Mar 15, 17(6), 2112–27.
Bullen, A.,and Saggau, P. (1998). Indicators and optical configuration for simultaneous high resolution recording of membrane potential and intracellular calcium using laser scanning microscopy. Pflugers archiv european journal of physiology. 436, 788–796.
Cannestra, AF., Black, KL., Martin. NA., Cloughesy, T., Burton, JS., Rubinstein, E., woods, RP.,and Toga, AW. (1998). Topographical and temporal specificity of humann intraoperative optical intrinsic signals. Neurosci., 9, 2557–2563.
Cacciatore, TW., Brodfuehrer, PD.,, Gozalas JE Tsien, RY, Kristan, WB and Kleinfeld D. (1998). Neurons that are active in phase with swimming in leech, and their connectivity are revealed by optical techniques. Neurosci. Abs. 24, 1890.
Chance, B., Cohen, P., Jobsis, F., Schoener, B. (1962). Intracellular oxidation-reduction states *in vivo*. Science, 137, 499–508.
Chance, B., Kang, K., He, L., Weng, J., Sevick, E. (1993b). Highly sensitive object location in tissue models with linear in-phase and anti-phase multi-element optical arrays in one and two dimensions, PNAS 90(8), 3423–3427.
Chance, B., Zhuang, L., Unah, C., Alter, C., Lipton, L. (1993a). Cognition-activated low-frequency modulation of light absorption in human brain, PNAS 90(8), 3770–3774.
Chapman, B., and Bonhoeffer, T. (1998). Overrepresentation of horizontal and vertical orientation preference in developing ferret area-17. Proceedings of the national academy of science of the United States of America, 95, 2609–2614.
Chapman, B., Bonhoeffer, T. (1994). Chronic optical imaging of the development of orientation domains in ferret area 17, Soc. Neurosci. Abstr., 20, 98(2), 214.
Cinelli, A.R., and Kauer, J.S. (1992). Voltage sensitive dyes and functional-activity in the olfactory pathway. Annu. Rev. Neurosci., 15, 321–352.
Cinelli, A.R., and Kauer, J.S. (1995). Salamender olfactory-bulb neuronal-activity observed by video-rate, voltage sensitive dyes imaging. 2. Spatial and temporal properties of responses evoked by electrical stimulation. J. Neurophys., 73, 2033–2052.
Cinelli, A.R., Neff, S.R., and Kauer, J.S. (1995). Salamender olfactory-bulb neuronal-activity observed by video-rate, voltage sensitive dyes imaging. 1. Caracterization of the recording system. J. Neurophys., 73, 2017–2032.
Cohen, L. B., Salzberg, B. M., Davila, H. V., Ross, W. N., Landowne, D., Waggoner, A. S., and Wang, C. H. (1974). Changes in axon fluorescenceduring activity: Molecular probes of membrane potential. J. Membrane Biol. 19:1–36.

Cohen, L.B. (1973). Changes in neuron structure during action potential propagation and synaptic transmission, Physiol. Rev., 53, 373–418.

Cohen, L.B., and Lesher, S. (1986). Optical monitoring of membrane potential: methods of multisite optical measurement. Soc. Gen. Physiol., Ser. 40:71–99

Cohen, L.B., and Orbach, H.S. (1983). Simultaneous monitoring of activity of many neurons in buccal ganglia of pleurobranchaea and aplysia. Soc. Neurosci.Abstr. 9: 913.

Cohen, L.B., Keynes, R.D., Hille, B. (1968). Light scattering and birefringence changes during nerve activity, Nature, 218, 438–441.

Cohen, L.B., Landowne, D., Shrivastav, B.B., and Ritchie, J.M., (1970). Changes in fiuorescence of squid axons during activity. Biol. Bull. Woods Hole 139: 418–419.

Cohen, L.B., Slazberg, B.M., Grinvald, A. (1978) Optical methods for monitoring neurons activity. Ann. Rev. of Neurosci., 1, 171–182.

Coppola, DM., White,LE., Fitzpatrick, D., and Purves, D.(1998). Unequal representation of cardinal and oblique in ferret cortex. Proceedings of the national academy of science of the United States of America, 95, 2621–2623.

Crair, M.C., Gillespie, D.G., and Stryker, M.P. (1998) The role of visual experience in the development of columns in cat visual cortex. Science, 279, 566–570

Crair, M.C., Ruthazer, E.S., Gillespie, D.C. (1997). Stryker-MP Ocular dominance peaks at pinwheel center singularities of the orientation map in cat visual cortex. J-Neurophysiol., Jun, 77(6)

Crair, M.C., Ruthazer, E.S., Gillespie, D.C. (1997). Stryker-MP Relationship between the ocular dominance and orientation maps in visual cortex of monocularly deprived cats. Neuron. Aug, 19(2), 307–18

Das, A., and Gilbert, C.D. (1995). Long-range horizontal connections and their role in cortical reorganization revealed by optical recording of cat primary visual cortex Nature, 375(6534), 780–4.

Das, A., and Gilbert, C.D. (1997). Distortions of visuotopic map match orientation singularities in primary visual cortex. Nature, 387, 594–8

Davila, H.V., Cohen, L.B., Salzberg, B.M., and Shrivastav, B.B., (1974). Changes in ANS and TNS fluorescence in giant axons from loligo. J. Member. Biol.15: 29–46.

De Weer, P. and B.M. Salzberg (eds) (1986). Optical methods in cell physiology., Soc. Gen. Physiol. Ser. Vol. 40 John Wiley and Sons inc. New York.

Denk, W., Strickler, J.H., and Webb, W.W. (1990) Two-photon laser scanning fluorescence microscopy. Science, 248:73–76

Egger, M.D., and M. Petran. (1967). New reflected light microscope for viewing unstained brain and ganglion cells. Science, 157:305–307.

Ehrenberg, B., and Berezin, Y. (1984). Surface potential on purple membranes and its sidedness studied by resonance ramam dye prob. Biophys. J. 45: 663–670.

Eisenbach, M., Margolon, Y., Ciobotariu, A., and Rottenberg, H. (1984). Distinction between changes in membrane potential and surface charge upon chemotactic stimulation of escherichia coli. Biophys. J. 45: 463–467.

Everson, R.M., Prashanth, A.K., Gabbay, M., Knight, B.W., Sirovich, L., Kaplan, E. (1998). Representation of spatial frequency and orientation in the visual cortex. PNAS 95(14), 8334–8338.

Fox, P.T., Mintun, M.A., Raichle, M.E., Miezin, F.M., Allman, J.M., and van Essen, D.C. (1986). Mapping human visual cortex with positron emission tomography. Nature, 323, 806–809

Frostig, R.D. (1994). What does in vivi optical imaging tell us about the primary visual cortex in primates. In Cerebral Cortex, 10, 331–358, Eds. Peters A, and Rockland K.

Frostig, R.D., Lieke, E.E., Ts'o, D.Y., and Grinvald, A. (1990). Cortical functional architecture and local coupling between neuronal activity and the microcirculation revealed by *in vivo* high-resolution optical imaging of intrinsic signals. Proc. Natl. Acad. Sci. USA 87, 6082–6086.

Frostig, R.D., Masino, S.A., Kwon, M.C. (1994). Characterization of functional whisker representation in rat barrel cortex: Optical imaging of intrinsic signals vs. single-unit recordings, Neurosci. Abstr., 566(8).

Ghose, G.M., Roe, A.W., Ts'o, D.Y. (1994). Features of functional organization within primate V4, Neurosci. Abstr., 350(10).

Gilbert, C.D., Wiesel, T.N. (1983) Clustered intrinsic connections in cat visual cortex.J Neurosci 3:1116–1133.

Glaser, D.E., Hildesheim, R., Shoham, D., and Grinvald, A. (1988). Optical imaging with new voltage sensitive dues reveals that sudden luminance changes delay the onset of orientation tuning in cat visual cortex. Neurosci. Abs., 24:10.3

Glaser,D.E., Shoham,D.,and Grinvald,A. (1998). Sudden luminance changes delay the onset of cortical shape processing. Neurosci.Lett.,Suppl 51(S15).

Gochin, P.M., Bedenbaugh, P., Gelfand, J.J., Gross, C.G., Gerstein, G.L. (1992). Intrinsic signal optical imaging in the forepaw area of rat somatosensory cortex, PNAS 89(17), 8381–8383.

Gonzalez, JE. and Tsien, RY (1995). Voltage Sensing by fluorescence resonance energy transfer in single cells. Biophys. J. 69: 1272–1280.

Gonzalez, JE. and Tsien, RY (1997). Improved indicators of cell membrane potential that use fluorescence resonance energy transfer. Chem. Biol. 4: 269–277.

Gratton, G. (1997). Attention and probability effects in the human occipital cortex: an optical imaging study. Neuroreport, 8(7), 1749–53

Gratton, G., Corballis, P.M., Cho, E., Fabiani, M., Hood, D.C. (1995). Shades of gray matter: noninvasive optical images of human brain responses during visual stimulation. Psychophysiology, 32(5), 505–9

Gratton, G., Fabiani, M., Corballis, PM., Gratton, E. (1997). Noninvasive detection of fast signals from the cortex using frequency-domain optical methods. Ann-N-Y-Acad-Sci., 820, 286–98

Grinvald ,A, Fine, A., Farber, I.C., Hildesheim, R. (1983) Fluorescence monitoring of electrical responses from small neurons and their processes. Biophys. J. 42:195–198.

Grinvald, A. (1984). Real time optical imaging of neuronal activity: from single growth cones to the intact brain. Trends in Neurosci., 7, 143–150.

Grinvald, A. (1985). Real-time optical mapping of neuronal activity: from single growcones to the intact mammalian brain. Annu. Rev. Neurosci., 8, 263–305.

Grinvald, A., and Segal, M., (1983). Optical monitoring of electrical activity; detection of spatiotemporal patterns of activity in hippocampal slices by voltage-sensitive probes. In Brain Slices, ed. R. Dingledine. New York: Plenum Press, pp. 227–261.

Grinvald, A., Anglister, L., Freeman, J.A., Hildesheim, R., and Manker, A. (1984). Real time optical imaging of naturally evoked electrical activity in the intact frog brain. Nature, 308, 848–850

Grinvald, A., C.D. Gilbert, R. Hildesheim, E. Lieke., and T.N. Wiesel. (1985). Real time optical mapping of neuronal activity in the mammalian visual cortex *in vitro* and *in vivo*. Soc. Neurosci. Abstr., 11:8.

Grinvald, A., Cohen, L.B., Lesher, S., and Boyle, M.B., (1981). Simultaneous optical monitoring of activity of many neurons in invertebrate ganglia, using a 124 element "Photodiode" array. J. Neurophysiol., 45, 829–840

Grinvald, A., Frostig, R.D., Lieke, E.E., Hildesheim, R. (1988). Optical imaging of neuronal activity. Physiol Rev., 68, 1285–1366.

Grinvald, A., Frostig, R.D., Siegel, R.M., and Bartfeld, E. (1991). High resolution optical imaging of neuronal activity in awake monkey, PNAS, 88, 11559–11563.

Grinvald, A., Hildesheim, R., Farber, I.C., and Anglister, L. (1982b). Improved fluorescent probes for the measurement of rapid changes in membrane potential. Biophys. J., 39, 301–308

Grinvald, A., Lieke, E., Frostig, R.D., Gilbert, C.D., and Wiesel, T.N. (1986a). Functional architecture of cortex revealed by optical imaging of intrinsic signals, Nature, 324, 361–364.

Grinvald, A., Lieke, E.E., Frostig, R.D., Hildesheim, R. (1994), Cortical point-spread function and long-range lateral interactions revealed by real-time optical imaging of macaque monkey primary visual cortex, J. Neurosci., 14(5), 2545–2568.

Grinvald, A., Manker, A., and Segal, M. (1982a). Visualization of the spread of electrical activity in rat hippocampal slices by voltage sensitive optical probes. J. Physiol., 333, 269–291

Grinvald, A., Salzberg, B.M., and Cohen, L.B. (1977). Simultaneous recordings from several neurons in an invertebrate central nervous system. Nature, 268, 140–142

Grinvald, A., Salzberg, B.M., Lev-Ram, V., and Hildesheim, R. (1987). Optical recording of synaptic potentials from processes of single neurons using intrcellular potentiometric dys. Biophys. J. 51:643–651.

Grinvald, A., Segal, M., kuhnt, U., Hildesheim, R., Manker, A., Anglister, L., and Freeman, J.A. (1986) Real-time optical mapping of neuronal activity in vertebrate CNS in vitro and in vivo. Soc. Gen. Physiol. Ser. 40: 165–197.

Gupta, R.G., Salzberg, B.M., Grinvald, A., Cohen, L.B., Kamino, K., Boyle, M.B., Waggoner, A.S., Wang, C.H. (1981). Improvements in optical methods for measuring rapid changes in membrane potential. J. Mem. Biol. 58, 123–137.

Haglund, M.M., Ojemann, G.A., and Hochman, D.W. (1992). Optical imaging of epileptiform and functional activity in human cerebral cortex, Nature, 358, 668–671.

Harrison, VH., Harel, n., Kakigi, A., Raveh, E., and Mount, RJ. (1998). Optical imaging of intrinsic signals in chinchilla auditory cortex. Audiol Neurootal, 3, 214–223.

Hess, A., Scheich, H. (1994). Tonotopic organization of auditory cortical fields of the Mongolian Gerbil 2DG labeling and optical recording of intrinsic signals, Neurosci. Abstr., 141 (5).

Hill, D.K., and Keynes, R.D. (1949). Opacity changes in stimulated nerve, J. Physiol., 108, 278–281.

Hirota, A., Sato, K., Momosesato, Y., Sakai, T., and Kamino, K. (1995). A new simultaneous 1020 site optical recording system for monitoring neuronal activity using voltage sensitive dyes. J. Neurosci. Methods. 56, 187–194.

Horikawa, J., Hosokawa, Y., Nasu, M., and Taniguchi, I., (1997). Optical study of spatiotemporal inhibition evoked by 2 tone sequence in the guinea pig auditory cortex. J. of comparative physio.A sensory neural and behavioral physiology. 181, 677–684.

Hoshi, Y., Tamura, M. (1993). Dynamic multichannel near-infrared optical imaging of human brain activity, J. Appl. Physiol., 75, 1842–1846.

Hosokawa,Y., Horikawa, J., Nasu, M., and Taniguchi, I. (1997). Real time imaging of neural activity during binaural interaction in the guinea pig auditory cortex. J. of comparative physio.A sensory neural and behavioral physiology. 181, 607–614.

Hubel, D.H. and T.N. Wiesel. (1965). Receptive fields and functional architecture in two non-striate visual areas (18 and 19) of the cat. J. Neurophysiol. 28:229–289.

Hubel, D.H., and Wiesel, T.N. (1962). Receptive fields, binocular interactions and functional architecture in the cat's visual cortex. J. Physiol., 160, 106–154.

Hubener, M., Shoham, D., Grinvald, A., and Bonhoeffer, T. (1997). Spatial relationships among three columnar systems in cat area 17. J. Neurosci., 17, 9270–9284.

Ichikawa, M., Tominaga, T., Tominaga, Y., yamada, H., Yamamato, Y. and Matsomoto, G. (1998). Imaging of synaptic excitation at high special and temporal resolution at high temporal resolution using va novel CCD system in rat brain slices. Neurosci. Abs. 24, 1812.

Iijima, T., Matosomoto, G., Kisokoro, Y. (1992). Synaptic activation of rat adrenal-medula examined with a lrage photodiode array in combination with voltage sensitive dyes. Neurocscience 51, 211–219.

Inase, M., Iijima, T., Takashima, I., Takahashi, M., Shinoda, H., Hirose, H., Niisato, K.,Tsukada, K. (1998). Optical recording of the motor cortical activity during reaching movement in the behaving monkey. Soc. Neurosci.Abstr., Vol. 24, (404).

Jobsis, F.F. (1977). Noninvasive, infrared monitoring of cerebral and myocardial oxygen sufficiency and circulatory parameters, Science, 198, 1264–1266.

Jobsis, F.F., Keizer, J.H., LaManna, J.C., and Rosental, M.J. (1977). Reflectance spectrophotometry of cytochrome aa_3 *in vivo*. J. Appl. Physiol.: Respirat. Environ. Exercise Physiol., 43:858–872.

Jobsis, F.F., Keizer, J.H., LaManna, J.C., Rosental, M.J. (1977). Reflectance spectrophotometry of cytochrome aa_3 *in vivo*. J. Appl. Physiol.: Respirat. Environ. Exercise Physiol., 43, 858–872.

Kamino, K. (1991). Optical approaches to ontogeny of electrical activity and related functional-organization during early heart developnment. Phys.Rev., 71, 53–91.

Kato, T., Kamei, A., Takashima, S., Ozaki, T. (1993). Human visual cortical function during photic stimulation monitoring by means of near-infrared spectroscopy. J. Cereb. Blood Flow & Metab., 13, 516–520.

Kauer, J.S. (1988). Real-time imaging of evoked activity in local circuits of the salamander olfactory bulb. Nature, 331, 166–168.

Kauer,J.S., Senseman, D.M., Cohen, M.A. (1987). Odor-elicited activity monitored simultaneously from 124 regions of the salamander olfactory bulb using a voltage-sensitive dye. Brain Res ,25:255–61.

Kelin, D. (1925). On cytochrome, a respiratory pigment, common to animals, yeast, and higher plants. Proc. R. Soc. B, 98, 312–339

Keller,A., Yagodin, S., Aroniadouanderjaska, V., Zimmer, LA., Ennis, M., Sheppard,NF., and Shipley MT. Functional organization of rat olfactory bulb glomeruli revealed by optical imaging. J. Neurosci., 18, 2602–2612.

Kenet, T., Arieli, A., Grinvald, A., and Tsodyks, M. (1997). Cortical population activity predicts both spontaneous and evoked single neuron firing rates. Neurosci. Lett. S48, p27

Kenet, T.,Arieli, A., Grinvald,A., Shoham, D., Pawelzik, K., and Tsodyks, M. (1998). Spontaneous and evoked firing of single cortical neurons are predicted by population activity. Soc. Neurosci.Abstr., Vol. 24, (1138).

Kety, S.S., Landau, W.M., Freygang, W.H., Rowland, L.P., Sokoloff, L. (1955). Estimation of regional circulation in the brain by uptake of an inert gas, American Physiological Society Abstracts: 85.

Kim, D.S., Bonhoeffer, T. (1994). Reverse occlusion leads to a precise restoration of orientation preference maps in visual cortex, Nature, 370, 370–372.

Kisvarday, Z.F., Kim, D.S., Eysel, U.T., Bonhoeffer, T. (1994). Relationship between lateral inhibitory connections and the topography of the orientation map in cat visual cortex, Europ. J. Neurosci., 6, 1619–1632.

Konnerth, A., and Orkand, R.K. (1986). Voltage sensitive dyes measurpotential changes in axons and glia of frog optic nerve. Neuroscience Lett., 66, 49–54.

Lamanna, J.C., Pikarsky, S.N., Sick, T.J., and Rosenthhal, M., (1985) A rapid-scanning spectrophotometer designed for biological tissues in vitro or in vivo. Anal. Biochem. 144: 483–493.

Lev-Ram R. and A. Grinvald. K^+ and Ca^{2+} dependent communication between myelinated axons and oligodendrocytes revealed by voltage-sensitive dyes. Proc. Natl. Acad. Sci. USA, 83, 6651–6655, 1986

Lassen, N.A., Ingvar, D.H. (1961). The blood flow of the cerebral cortex determined by radioactive krypton, Experimentia Basel, 17, 42–43

Lieke, E.E., Frostig, R.D., Arieli, A., Ts'o, D.Y., Hildesheim, R., Grinvald, A. (1989). Optical imaging of cortical activity; Real-time imaging using extrinsic dye signals and high resolution imaging based on slow intrinsic signals. Annu Rev of Physiol 51:543–559.

Lieke, E.E., Frostig, R.D., Ratzlaff, E.H., Grinvald, A. (1988). Center/surround inhibitory interaction in macaque V1 revealed by real-time optical imaging. Soc Neurosci Abstr 14:1122.

Loew, L. M., Cohen, L. B., Salzberg, B. M., Obaid, A. L., and Bezanilla, F. (1985). Charge shift probes of membrane potential. Characterization of aminostyrylpyridinum dyes on the squid giant axon. Biophys J. 47:71–77.

Loew, L.M. (1987). Optical measurement of electrical activity., CRC Press Inc., Boca Raton

Loew, L.M., and Simpson, L.L. (1981). Charge shift probes of membrane potential. Biophys. J., 34:353–363.

Loew, L.M., Bonneville, G.W., and Surow, J. (1978). Charge shift probes of membrane potential theory. Biol. Chemistry. 17: 4065–4071.

Loew, L.M., Scully, S., Simpson, L., and Waggoner, A.S., (1979). Evidence for a charge shift electrochromic mechanism in a probe of membrane potential. Nature Lond. 281: 497–499.

MacVicar, B.A., Hochman, D. (1991). Imaging of synaptically evoked intrinsic optical signals in hippocampal slices. J. Neurosci., 11, 1458–1469.

MacVicar, B.A., Hochman, D., LeBlanc, F.E., Watson, T.W. (1990). Stimulation evoked changes in intrinsic optical signals in the human brain. Soc. Neurosci. Abstr., 16, 309.

Malach, R., Amir, Y., Harel, M., and Grinvald, A. (1993). Novel aspects of columnar organization are revealed by optical imaging and in vivo targeted biocytin injections in primate striate cortex. Proc. Natl. Acad. Sci. USA 90, 10469–10473.

Malach, R., Amir, Y., Harel, M., Grinvald, A. (1993). Relationship between intrinsic connections and functional architecture revealed by optical imaging and *in vivo* targeted biocytin injections in primate striate cortex. PNAS 90, 10469–10473.

Malach, R., Schirman, T.D., Harel, M., Tootell, R.B.H., and Malonek, D. (1997) Organization of intrinsic connections in owl monkey area MT. Cerebral Cortex, 7, 386–393.

Malach, R., Tootell, R.B.H., and Malonek, D. (1994) Relationship between orientation Domains, Cytochrome Oxidase Stipes and intrinsic horizontal connections in Squirrel monkey area V2. Cerebral Cortex, 4, 151–165.

Malach, R., Tootell, R.B.H., and Malonek, D. (1994). Relationship between orientation Domains, cytochrome oxidase stripes and intrinsic horizontal connections in squirrel monkey area V2.

Malonek, D., Dirnagl, U., Lindauer, U., Yamada, K., Kanno, I., and Grinvald, A. (1997).Vascular imprints of neuronal activity. Relationships between dynamics of cortical blood flow, oxygenation and volume changes following sensory stimulation., Proc. Natl Acad. Sci. USA, 94, 4826–14831,

Malonek, D., Grinvald, A. (1996). The imaging spectroscope reveals the interaction between electrical activity and cortical microcirculation; implication for functional brain imaging. Science, 272, 551–554.

Malonek, D., Shoham, D., Ratzlaff, E., and A. Grinvald (1990) *In vivo* three dimensional optical imaging of functional architecture in primate visual cortex. Neurosci. Abstr. 16, 292.

Malonek, D., Tootell, R.B.H., Grinvald, A. (1994). Optical imaging reveals the functional architecture of neurons processing shape and motion in owl monkey area MT. Proc. R. Soc. Lond. B, 258, 109–119.

Masino, S.A., Kwon, M.C., Dory, Y., Frostig, R.D. (1993). Structure-function relationships examined in rat barrel cortex using intrinsic signal optical imaging through the skull, Neurosci. Abstr., 702(6).

Mayevsky, A., and Chance, B. (1982). Intracellular oxidation-reduction state measured in situ by a multichannel fiber-optic surface fluorometer. Science, 217, 537–540

Mayhew, J., Hu, DW., Zheng, Y., Askew, S., Hou, YQ.,Berwick,J., Coffey, PJ., and Brown, N. (1998). An evaluation of linear model analysis techniques for processing images of microcirculation activity. Neuroimage, 7, 49–71.

Mc-Loughlin, N.P., and Blasdel, G.G. (1998). Wavelength dependent differences between optically determined functional maps from macaqe striate cortex. Neuroimage. 7, 326–36.

Menon RS., Luknowsky, DC. And Gati, JS (1998). Mental chronometry using latency resolved functional MRI. Proc. Natl./Acad. Sci. USA. 95,10902–10907.

Millikan, G.A. (1937). Experiments on muscle hemoglobin *in vivo*; the instantaneous measurement of muscle metabolism. Proc. R. Soc. B, 123, 218–241

Miyawaki-A; Llopis-J; Heim-R; McCaffery-JM; Adams-JA; Ikura-M; Tsien-RY (1997) Fluorescent indicators for Ca2+ based on green fluorescent proteins and calmodulin. Nature, 388, 834–5 Mountcastle, V.B. (1957). Modality and topographic properties of single neurons of cat's somatic sensory cortex. J. Neurophysiol., 20, 408–434.

Newsome, W.T., Britten, K.H., Movshon, J.A. (1989). Neuronal correlates of a perceptual decision, Nature, 341, 52–54.

Obermayer, K., and Blasdel, G.G. (1993) Geometry of orientation and ocular dimonance columns in primate striate cortex. J. Neuroscie., 13, 4114–4129.

Ogawa, S., Lee, T.M., Kay, A.R., (1990b). Brain magnetic resonance imaging with contrast dependent on blood oxygenation, PNAS 87, 9868–9872.

Ogawa, S., Lee, T.M., Nayak, AS. (1990a). Oxygenation-sensitive contrast in magnetic resonance image of rodent brain at high magnetic fields, Magn. Reson. Med., 14, 68–78.

Orbach, H.S. (1987). Monitoring electrical activity in rat cerebral cortex. In Optical measurement of electrical activity,. ed. L.M. Loew, CRC Press, Boca Raton

Orbach, H.S. (1988). Monitoring electrical activity in rat cerebal cortex. In: Spectroscopic Membrane Probes, edited by L. M. Loew. Boca Raton, FL: CRC, Vol III, p. 115–136.

Orbach, H.S., and Cohen, L.B. (1983). .Optical monitoring of activity from many areas of the in vitro and in vivo salamader olfactory bulb:a new method for studying functional organization in the vertebrate central nervous system. J. Neurosci., 3, 2251–2262

Orbach, H.S., Cohen, L.B., and Grinvald, A. (1985). Optical mapping of electrical activity in rat somatosensory and visual cortex. J. Neurosci., 5, 1886–1895

Pikarsky, S.M., Lamanna, J.C., Sick, T.J., and Rosenthal, M. (1985). A computer-assisted rapid-scannig spectrophotometer with applications to tissues in vitro and in vivo. Comput. Biomed. Res. 18: 408–421.

Raichle, M.E., Martin, W.R.W., Herscovitz, P., Minton, M.A., Markham, J.J. (1983). Brain blood flow measured with intravenous H2(15)0. II. Implementation and validation, J. Nucl. Med., 24(9), 790–798.

Ratzlaff, E.H., and Grinvald, A. (1991). A tandem-lens epifluorescence macroscope: hundred-fold brightness advantage for wide-field imaging. J. Neurosci. Methods 36: 127–137.

Rector, DM., Poe, GR., Redgrave, P., and Harper, RM. (1997). CCD video camera for high sensitivity light measurements in freely behaving animals full source. J. of Neuro. Metho., 78, 85–91.

Rigler, R., Rable, C.R., and Jovin, T.M. (1974). A temperature jump apparatus for fiuorescence measurements Rev. Sci. Instrum. 45: 581–587.

Roker, C, Heilemann, A, Fromherz, P. (1996) Time-resolved fluorescence of a hemicyanine dye: Dynamics of rotamerism and resolvation. J Phys. Chem. USA. 100: 12172–12177.

Ross, W.N., and Reichardt, L.F. (1979). Species-specific effects on the optical signals of voltage sensitive dyes. J. Membr. Biol., 48, 343–356

Ross, W.N., Salzberg, B.N., Cohen, L.B., Grinvald, A., Davila, H.V., Waggoner, A.S., Chang, C.H (1977) Changes in absorption, fluorescence, dichroism and birefringence in stained a: optical measurement of membrane potential. J Mem Biol 33:141–183.

Ross, W.N., Salzberg. BN., Cohen, L.B., Grinvald, A., Davila, H.V., Waggoner, A.S, Chang, C.H. (1977) Changes in absorption, fluorescence, dichroism and birefringence in stained axons: optical measurement of membrane potential. J Mem Biol 33:141–183.

Roy, C., Sherrington, C. (1890). On the regulation of the blood supply of the brain, J. Physiol., 11, 85–108.

Rumsey, W.L., Vanderkooi, J.M., Wilson, D.F. (1988). Imaging of phosphorescence; A novel method for measuring oxygen distribution in perfused tissue. Science, 241: 1649

Salama, G., and Morad, M. (1976). Merocyanine 540 as an optical prob of transmembrane electrical activity in the heart. Science Wash. DC. 191: 485–487.

Salama, G., Lombardi, R., and Elson, J. (1987). Maps of optical action potentials and NADH Fluorescence in intact working hearts. AM.J.Physiol. 252 (Heart Circ. Physiol.21): H384-H394.

Salzberg, B.M., (1983). Optical recording of electrical activity in neurons using molecular probes. In Current Methods in Cellular Neurobiology,, eds. J. Barber, and J. McKelvy. New York: John Wiley & Sons, p. 139–187

Salzberg, B.M., Davila, H.V., and Cohen, L.B. (1973) Optical; recording of impulses in individual neurons of an invertebrate central nervous system . Natrure, 246, 508–509.

Salzberg, B.M., Grinvald, A., Cohen, L.B., Davila, H.V., and Ross, W.N. (1977). Optical recording of neuronal activity in an invertebrate central nervous system; simultaneous recording from several neurons. J. Neurophys., 40, 1281–1291.

Salzberg, B.M., Obaid, A.L., and Gainer, H. (1986). Optical studies of excitation secretion at the vertebrate nerve terminal Soc. Gen Physiol., 40, 133–164.

Salzberg, B.M., Obaid, A.L., and Gainer, H. (1986). Optical studies of excitation secretion at the vertebrate nerve terminal Soc. Gen Physiol, 40, 133–164.

Salzberg, B.M., Obaid, A.L., Senseman, D.M., and Gainer, H. (1983). Optical recording of action potentials from vertebrate nerve terminals using potentiometric probs provide evidence for sodium and calcium components.Nature Lond. 306: 36–39

Salzman, C.D., Britten, K.H., Newsome, W.T. (1990). Cortical microstimulation influences perceptual judgements of motion direction, Nature, 346, 174–177.

Salzman, C.D., Murasugi, C.M., Britten, K.H., Newsome, W.T. (1992). Microstimulation in visual area MT: Effects on direction discrimination performance, J. Neurosci., 12, 2331–2355.

Shevelev, IA. (1998). Functional imaging of the brain by infrared radiation (thermoencephaloscopy) Progress in Neurobiology. 56, 269–305.

Shmuel, A., and Grinvald, A. (1996). Functional organization for direction of motion and its relationship to orientation maps in cat area 18. J. Neurosci., 16, 6945–6964.

Shoham ,D., Glaser, D., Arieli, A., Hildesheim, R., and Grinvald, A. (1998). Imaging cortical architecture and dynamics at high spatial and temporal resolution with new voltage-sensitive dyes. Neurosci. Lett., Suppl 51(S38).

Shoham, D., Hubener, M., Grinvald, A., and Bonhoeffer, T. (1997). Spatio-temporal frequency domains and their relation to cytochrome oxidase staining in cat visual cortex. Nature, 385, 529–533.

Shoham, D., Gottesfeld, Z., and Grinvald, A. (1993). Comparing maps of functional architecture obtained by optical imaging of intrinsic signals to maps and dynamics of cortical activity recorded with voltage sensitive dyes. Neurosci. Abs., 19:618.6

Shoham, D., Grinvald, A. (1994). Visualizing the cortical representation of single fingers in primate area S1 using intrinsic signal optical imaging, Abstracts of the Israel Society for Neuroscience 3, 26.

Shoham, D., Ullman, S., and Grinvald, A. (1991). Characterization of dynamic patterns of cortical activity by a small number of principle components. Neurosci. Abs., 17:431.8

Shtoyerman, E., Vanzetta, I., Barabash, S., Grinvald, A. (1998). Spatio-temporal characteristics of oxy and deoxy hemoglobin concentration changes in response to visual stimulation in the awake monkey. Soc. Neurosci.Abstr., Vol. 24, (10).

Siegel, M.S., Isacoff, E.Y. (1997) A genetically encoded optical probe of membrane voltage.Neuron., 19, 735–41

Sirovich, L., Everson, R., Kaplan, E., Knight, B.W., Obrien, E., Orbach,D. (1996). Modeling the functional-organization of the visual cortex. Physica D, 96 355–366.

Slovin, H., Arieli, A. and Grinvald, A. (1999) Voltage-sensitive dye imaging in the behaving monkey. Fifth IBRO Congress. Abstr. pp 129.

Sokoloff, L. (1977). Relation between physiological function and energy metabolism in the central nervous system. J. Neurochem., 19, 13–26.

Sterkin, A., Arieli, A., Ferster, D., Glaser, D.E., Grinvald, A., and Lampl, I. (1998). Real-time optical imaging in cal visual cortex exhibits high similarity to intracllular. Neurosci.Lett.,Suppl 51(S41)

Stetter, M., Otto, T., Sengpiel, F., Hubener, M.,Bonhoeffer, T., Obermayer, K. (1998).Signal extraction from optical imaging data from cat area 17 by blind separation of sources. Soc.Neurosci.Abstr., Vol. 24, (9).

Swindale, N.V., Matsubara, J.A., and Cynader, M.S. (1987). Surface organization of orientation and direction selectivity in cat area 18. J. Neurosci., 7, 1414–1427.

Tanifuji, M., Yamanaka, A., Sunaba, R., and Toyama, K. (1993). Propagation of excitation in the visual cortex studies by the optical recording. Japanese J. Physiol., 43, 57–59.

Taniguchi, I., Hrikawa, J., Hosokawa, Y., and Nasu, M. (1997). Optical Imaging of cortical activity induced by intracochlear stimulation. Biomed. Resea.Tokyo. 18, 115–124

Tasaki, I., and A. Warashina. (1976). Dye membrane interaction and its changes during nerve excitation. Photochem. Photobiol., 24, 191–207.

Tasaki, I., Watanabe, A., Sandlin, R., Carnay, L. (1968). Changes in fluorescence, turbidity and birefringence associated with nerve excitation, PNAS 61, 883–888.

Toth, T.J., Rao, S.C., Kim, D.S., Somers, D., Sur, M. (1996). Subthreshold facilitation and suppression in primary visual cortex revealed by intrinsic signal imaging. Proc.Natl.Acad.Sci.U.S.A, 91:9869–74

Toyama K. and Tanifuji M (1991). Seeing ecxcietation propagation in visual cortical slices Biomed. Res., 12, 145–147.Ts'o, D.Y, Roe, A.W., Shey, J. (1993). Functional connectivity within V1 and V2: Patterns and dynamics, Neurosci. Abstr., 618(3).

Ts'o, D.Y., Frostig, R.D., Lieke, E., and Grinvald, A. (1990). Functional organization of primate visual cortex revealed by high resolution optical imaging, Science, 249, 417–420.

Ts'o, D.Y., Gilbert, C.D., Wiesel, T.N. (1991). Orientation selectivity of and interactions between color and disparity subcompartments in area V2 of Macaque monkey, Neurosci. Abstr., 431(7).

Vanzetta, I., Grinvald, A. (1998). Phosphorescence decay measurements in cat visual cortex show early blood oxygenation level decrease in response to visual stimulation. Neurosci.Lett.,Suppl 51(S42).

Vnek, N., Ramsden, B.M., Hung, C.P., Goldman-Rakic, P.S., Roe, A.W. (1998). Optical imaging of functional domains in the awake behaving monkey. Soc. Neurosci.Abstr., Vol. 24, (1137).

Vranesic, I., Iijima, T., Ichikawa, M., Matsumoto, G., Knopfell, T. (1994). Signal transmission in the parallel fiber Purkenje-cell system visualized by high resolution imaging. Proc. Natl. Sacad. Sci. USA. 91, 13014–134017.

Waggoner, A.S. (1979). Dye indicators of membrane potential. Ann. Rev. Biophys. Bioener., 8,47–63

Waggoner, A.S., and Grinvald, A. (1977). Mechanisms of rapid optical changes of potential sensitive dyes. Ann. N.Y. Acad. Sci., 303, 217–242.

Wang,G., Tanaka, K., Tanifuji, M. (1994). Optical imaging of functional organization in Macaque inferotemporal cortex, Neuroscience Abstracts 138(10), 316.

Wenner, P, Tsau, Y, Cohen, LB, O'donovan MJ and Dan, Y. (1996) Voltage-sensitive dye recording using retrogradely transported dye in the chicken spinal cord: Staining and signal characteristics. J. Neurosci. Methods 70, 111–120.

William, H.B., Zhang,Y., Schofield, B., Fitzpatrick, D. (1997). Orientation Selectivity and the arrangement of horizontal connections in tree shrew striate cortex. J. Neurosci., 15, 2112–2127.

Wu, LY., Lam, YW., Falk, CX.,Cohen, LB., Fang, J., L, L., Prechtl, JC., Kleinfeld, D., and Tsau, Y. (1998). Voltage sensitive dyes for monitoring multineuronal activity in the intact central nervous system. Histoch. J. 30, 169–187

Wyatt, J.S., Cope, D., Deply, D.T., Richardson, C.E., Edwards, A.D., Wray, S. (1990). Reynolds EOR. Quantitation of cerebral blood volume in human infants by near-infrared spectroscopy, J. Appl. Physiol., 68, 1086–1091.

Wyatt, J.S., Cope, M., Deply, D.T., Wray, S., Reynolds, E.O.R. (1986). Quantitation of cerebral oxygenation and haemodynamics in sick newborn infants by near-infrared spectrophotometry, Lancet, 2, 1063–1066.

Zhang, J., Davidson, RM., Wei, MD., and Loew, LM. (1998). Membrane electric properties by combined patch-clamp and flurescence ratio imaging in single neurons. Biophys. J. 74, 48–53.

Chapter 35

Electroencephalography

Alexey M. Ivanitsky, Andrey R. Nikolaev and George A. Ivanitsky

Introduction

The term "electroencephalography", i.e., recording electrical brain activity from human (or animal) scalp, designates two close but not identical meanings: a branch of neuroscience and a clinical diagnostic technique. This chapter deals predominantly with the electroencephalography in its first meaning, i.e., with its scientific aspects. Clinical electroencephalography is beyond the scope of this chapter, since it is mainly based on phenomenological descriptions of brain electrical potential details in various diseases and is a subject of specialized handbooks. Nevertheless, everything written below is also correct and applicable for clinical electroencephalogram (EEG). The term "electrocorticogram" is applied to brain potentials when recorded directly from the cortex, and "electrosubcorticogram" when recorded from subcortical structures (usually via implanted electrodes).

EEG Origin

Generators of electric fields which can be registered with scalp electrodes are groups of neurons with uniformly oriented dendrites. The neurons incessantly receive impulses from other neurons. These signals affect dendritic synapses, inducing excitatory and inhibitory postsynaptic potentials. Currents derived from synapses move through the dendrites and cell body to a trigger zone in the axon base and pass through the membrane to the extracellular space along the way. The EEG is a result of the summation of potentials derived from the mixture of extracellular currents generated by populations of neurons. Hereby the EEG depends on the cytoarchitectures of the neuronal populations, their connectivity, including feedback loops, and the geometries of their extracellular fields (Freeman 1992). The main physical sources of the scalp potentials are the pyramidal cells of cortical layers III and V (Mitzdorf 1987).

The appearance of EEG rhythmic activity in scalp recordings is only possible as a result of the synchronized activation of massifs of neurons, the summed synaptic events of which become sufficiently large (Steriade et al. 1990). The rhythmic activity may be generated by both pacemaker neurons having the inherent capability of rhythmic oscillations, and neurons which cannot generate a rhythm on their own but can synchronize

Correspondence to: Alexey M. Ivanitsky, Russian Academy of Sciences, Institute of Higher Nervous Activity and Neurophysiology, 5a Butlerov str., Moscow GSP-7, 117865, Russia (phone: +07-095-334-7809; fax: +07-095-338-8500; e-mail: alivanit@aha.ru; home page: www.psi.med.ru)
Andrey R. Nikolaev, Russian Academy of Sciences, Institute of Higher Nervous Activity and Neurophysiology, 5a Butlerov str., Moscow GSP-7, 117865, Russia
George A. Ivanitsky, Russian Academy of Sciences, Institute of Higher Nervous Activity and Neurophysiology, 5a Butlerov str., Moscow GSP-7, 117865, Russia

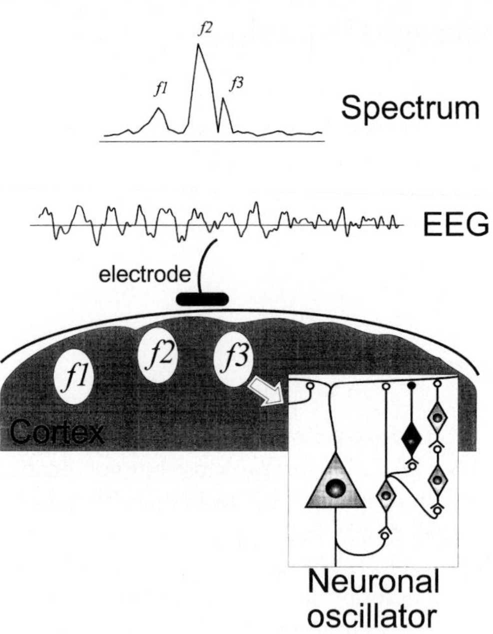

Fig. 1. Neuronal oscillators inside the cortex, discharging at their intrinsic frequencies (*f1*, *f2*, *f3*), produce extracellular currents summed on the scalp surface as EEG signal. Spectral analysis decodes these oscillators´ activity out of EEG records. In the rectangular window, a hypothetical scheme of neuronal oscillators is depicted. The axonal collateral of the basic neuron activates the circuits with excitatory and inhibitory interneurons. The inhibitory neuron of the scheme is depicted in black.

their activity through excitatory and inhibitory connections in such a manner that constitute a network with pacemaker properties. The latter may be designated as neuronal oscillators (Madler et al. 1991; Kasanovich and Borisyuk 1994; Abarbanel et al. 1996). The oscillators have their own discharge frequency, varying among different oscillators and dependent on their internal connectivity (Figure 1), in spite of close intrinsic electrophysiological properties of single neurons which constitute different oscillators. The neuronal oscillators start to act in synchrony after application of external sensory stimulation (Lopes da Silva 1991; Basar 1992) or hidden signals from internal sources, for example, as a result of cognitive loading (Basar et al. 1989).

The detailed circuitry of the neuronal oscillators underlying EEG rhythms was discussed in the Report of International Federation on Clinical Neurophysiology (IFCN) Committee on Basic Mechanisms (Steriade et al. 1990).

EEG Rhythms

The usual classification of the main EEG rhythms based on their frequency ranges is as follows: delta 2–4 Hz, theta 4–8 Hz, alpha 8–13 Hz, beta 13–30 Hz, gamma higher than 30 Hz. However, this classification only partially reflects the functional variation of rhythmic activities. For example, EEG rhythms within the alpha range may be distinguished by their dynamics, place of generation and relation to certain behavioral acts (Niedermayer 1997; Lutzenberger 1997; Pfurtscheller et al. 1997).

Since the pioneering work of H. Berger (1929), the main EEG rhythm – the *alpha rhythm* (Berger's rhythm) has been known. This rhythm is typical of the resting condition and disappears when the subject perceives a sensory signal or when he/she makes mental efforts. It was shown that the alpha rhythm is generated by reverberating propagation of nerve impulses between cortical neuronal groups and some thalamic nuclei, interconnected by a system of excitatory and inhibitory connections and resulting in rhythmic discharges of large populations of cortical neurons (Llinas 1988; Lopes da Silva 1991). In the visual cortex, however, the alpha rhythm can be generated by intracortical networks involving layer V pyramidal neurons, the latter being the main potential sources (Lopes da Silva and Storm van Leeuwen 1977; Steriade et al. 1990).

The *theta rhythm* originates as a result of interactions between cortical and hippocampal neuronal groups (Miller 1991). The neuronal oscillators, which generate the *beta rhythm*, presumably are located inside the cortex (Lopes da Silva 1991). The basis for *gamma oscillations* is interneuronal feedback with quarter-cycle phase lags between neurons situated close to each other in local areas of the cortex (Freeman 1992).

Most of the rhythms are rather widespread in brain structures. Induced gamma, theta and alpha rhythms were found in cortex, hippocampus, thalamus, and brain stem (Basar 1992). Freeman (1988) used the expression "common modes" for the existence of similar frequencies in various networks of the brain. This may play a role in the integration of activities of neuronal oscillators distributed over various brain structures. The candidate mechanism for such integration is synchronization of the distant neuronal oscillators' activity on a fine temporal scale (this *synchronization* of spatially separated oscillators should be distinguished from a term *synchronization* usually applied to the enhancement of EEG rhythm amplitude due to the synchronized activity of large neuronal populations under one electrode).

The idea of brain potential synchrony as a leading mechanism for neuronal communications descends from some basic ideas of the Russian classic neurophysiological school of N.E. Vvedensky and A.A. Ukhtomsky (see Rusinov 1973). At the beginning of the century they postulated that the number of excitation cycles per time unit, i.e., discharge frequency, is a fundamental parameter, characterizing the neural structure's functional state (the "functional lability" parameter). A.A. Ukhtomsky proposed that the coincidence of the functional lability of two structures promotes their functional connections. Developing these ideas, M.N. Livanov (1977) and V.S. Rusinov (1973) suggested that EEG rhythms reflect the parameter of functional lability. Subsequently, the EEG synchronization can promote, and reflect, the functional connectivity between two or more cortical areas. The reason is that in this case signals from one neuronal oscillator repeatedly reach the other oscillator in one and the same phase of its excitation cycle. When this phase is the exaltatory one, the excitation threshold of the second oscillator is lowered, facilitating its neurons' response and their recruitment in a concerted activity with the first oscillator neurons. On the contrary, when the phase of the second oscillator is the refractory one, the message cannot be received and this connection becomes silent. Thus both the frequency coincidence and appropriate phase relationship favor the neural communications. In this process the phase relationship controls the switch of connections from the active to the inactive state as well as its direction (both oscillators act as if in a dialogue and each can be sender or recipient of the message). The idea that synchrony of potentials promotes the neural connectivity was proven in a crucial experiment carried out by M.N. Livanov (1977). In this experiment a computer pursued the correlation coefficient between the EEGs in visual and motor cortical areas of the rabbit. It appeared that, if the correlation coefficient exceeded a certain level, the visual signal triggered paw movements, and if this coefficient was low, no motor reaction occurred.

The concept of EEG rhythm synchronization as a basic mechanism and a marker for cortical connections was later confirmed in a number of studies, including mathematical modeling of neural processes (Malsburg 1981; Abarbanel et al. 1996). The study of EEG synchrony is now used as one of the main tools in the study of circuitry of neural communications both in animal experiments and in cognitive neuroscience (French and Beaumont 1984; Sviderskaya 1987; Gevins and Bressler 1988; Gray and Singer 1989; Ivanitsky 1990, 1993; Petsche et al. 1992; Petsche 1996; Bressler et al. 1993; Ivanitsky 1993; Andrew and Pfurtscheller 1996).

Subprotocol 1
EEG Recording

Procedure

As mentioned at the beginning of the chapter, EEG recording is a routine procedure, particularly in clinics. Therefore the equipment for EEG is manufactured in almost all developed countries and its advertising and specification is presented in the journals of appropriate profile. All this equipment is supplied with detailed instructions for its use. Nevertheless, it is worthwhile to present below some details of EEG recording procedure useful for researchers inexperienced in this field. EEG recording usually includes the following steps:

1. A subject is seated in a comfortable chair in a dimly illuminated room;
2. Electrodes are placed on his head according to a certain scheme;
3. The reference electrodes are chosen;
4. Parameters of the electroencephalograph and software for EEG acquisition and storage are established;
5. Calibration of the electroencephalograph and data acquisition software is executed;
6. EEG is recorded;
7. Artifacts are removed.

EEG cabin

1. The EEG recordings are usually performed in a room shielded from outer electrical and magnetic fields. Howener, modern amplifiers can reject these effects. During the recording procedure the subject should avoid movements, which can cause artifacts in a record.

Electrodes and their placement schemes

2. The most appropriate electrodes for the EEG scalp recording are Ag-AgCl which avoid potential shifts due to electrode polarization. To get a good contact (i.e., with an impedance below 5 Kilo-Ohms) between the electrode and skin surfaces, the skin has to be cleaned with ether or alcohol for fat or dirt removal. Some abrasives were used in earlier practice to lower the impedance, but they are unacceptable today due to risk of bacterial, HIV and prion infection. An electrode gel or salt solutions are used to improve potential conduction between the skin and electrode surfaces.The most popular scheme for electrode placement is the so-called 10/20 scheme (Jasper 1958), which is shown in Fig. 2. Additional electrodes may be placed between the basic ones. According to "IFCN Standards for digital recording of clinical EEG" (Nuwer et al. 1998), amplification and channel acquisition must be available for at least 24 EEG channels. For artifact removal electrooculogram records are taken. Now the most common way to place the electrode array on the scalp is the use of a cap with the electrodes fixed on it. These caps (or helmets) are available with different numbers of electrodes (19, 32, 64, up to 256 electrodes) and in several sizes, including one for children (see, for example, the catalogues of "Electro Cap", "Geodesic Sensor Net" and "NeuroScan"). Such devices can be placed and removed rapidly and cause a minimal unpleasant feeling. The latter is especially important for psychophysiological experiments, when a rather long recording is required. These caps automatically provide the electrode placement with appropriate, usually equal, interelectrode distance.

Reference electrodes

3. One of the important questions in EEG recording is the placement of reference electrodes, relative to which the electric brain potentials at all other electrodes are meas-

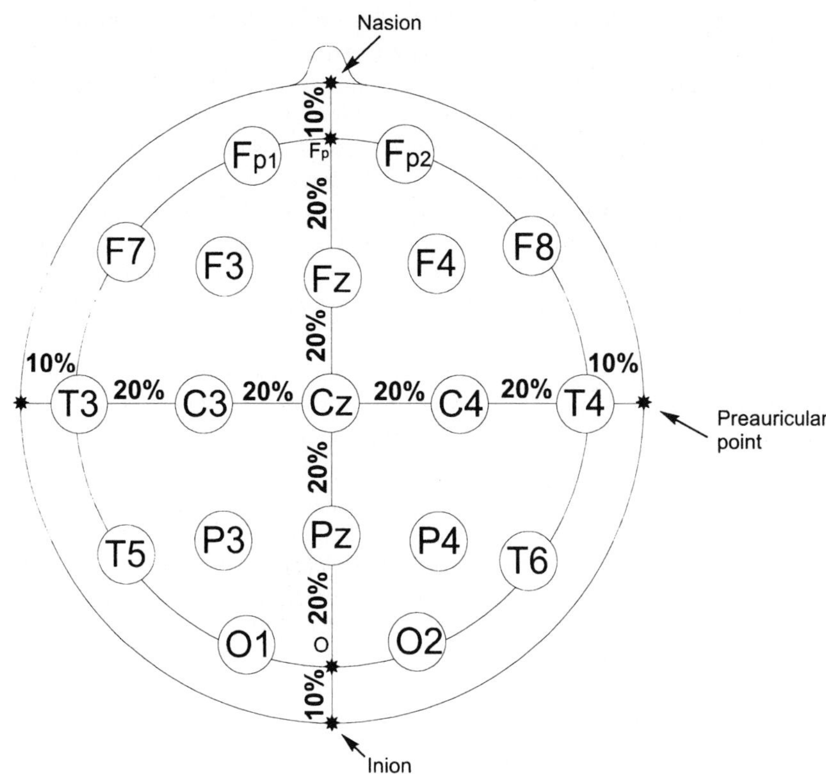

Fig. 2. 10–20 electrode placement scheme. According to this scheme three distances are measured: that between two preauricular points, that between the nasion (nose bridge) and inion (the occipital bone mount), both across vertex, and the circumference between the last two points of the skull. These distances are divided in proportion of 10-20-20-20-20-10% in both orthogonal axes and in circumference, and a net of imaging quadrates is built on the head surface. The electrodes are placed in quadrate angles.

ured. The reference electrodes should be placed on a presumed "inactive" zone. Frequently, this is the left or right earlobe or both of them. If one earlobe electrode is used as a reference, the topography of EEG rhythms is rather close to true, but there is the systematic decrease of EEG amplitudes at the electrodes which are closer to the reference side. If «linked» earlobes are used, this kind of asymmetry is avoided, but this linking distorts the EEG picture since the electric current flows inside the linking wire. This affects the intracranial currents that form the EEG potentials. In addition, low-amplitude EEG is observed in both temporal areas. Alternatively, the EEG may be recorded with any scalp electrode as a reference, and then the average reference is computed as a mean of all electrodes. It avoids all kinds of asymmetry and makes the EEG recorded in various laboratories comparable. In some cases, however, using the common reference may reveal rhythms not at their actual location. Sometimes the so-called bipolar recording is used when the potential is measured between two active electrodes. This scheme is good for exact localization of some local potential changes, i.e., a focus of pathological activity. A comprehensive review of the reference problem may be found in Lehmann (1987).

Parameters for computerized EEG acquisition and storage

4. For acquisition and storage of EEG data, "IFCN Standards" recommend a minimum sampling rate of the analog-to-digital conversion (*ADC*) of 200 samples/second (Nuwer et al. 1998). This rate allows one to analyze frequencies up to 100 Hz, as the maximal allowed frequency of the input signal (the *Nyquist frequency*) should be half the sampling rate (see Chap. 45). If the signal is sampled at too low a rate, *aliasing* (falsification of the signal) may occur with unpredictable errors in the digital waveform compared to the original one. Prior to sampling, an anti-aliasing low-pass filter must be used. ADC should be done at a resolution of at least 12 bits in order for the EEG to be resolved down to 0.5 μV. Whenever possible, the low-pass filter should be set to 0.16 Hz or less for recording. Routine use of higher settings of this frequency for recording are discouraged, as they should be reserved for specific or difficult clinical recordings only. A 50–60 Hz notch filter should be available, but not routinely used. Interchannel cross-talk must be less than 1%, i.e., 40 dB down or better.

Calibration

5. The calibration is needed to determine the exact amplitude of EEG signals and to evaluate the amplifier noise and other possible artifacts produced by it as well as by connection wires. Usually sine, triangle, and rectangle impulses of known amplitude are generated for this purpose by a special circuitry on the input of the main amplifier of the electroencephalograph. The calibration signal thus passes through the large part of the EEG signal's path in the recording system. The calibration impulses should be recorded and then used to measure the true EEG amplitude and to evaluate the equipment noise in quantitative EEG analysis. Modern software usually includes automatic comparison of EEG and calibration signals showing the actual brain electrical potential values.

Artifacts

6. EEG artifacts appear due to external electrical or magnetic fields and the subject's movements during the recording procedure. The latter are caused both by muscle electrical potential fields and electrode displacement. Visual and automatic search of high amplitude artifacts is usually not difficult. For example, eye movement artifacts can be eliminated via special algorithms (Gratton et al. 1983). Search and rejection of low amplitude artifacts is possible only by collation of results of frequency analysis, topographical mapping, and original EEG records. The topographical distribution of the main artifacts is discussed by Lee and Buchsbaum (1987). Eye movements are mainly reflected at frontal sites. Muscle activity is of high frequency and has lateral topography. Artifacts due to bad electrode placement have simple forms and are restricted by given EEG derivation.

Subprotocol 2
EEG Signal Analysis

Procedure

The usage of computers for EEG analysis has revolutionized electroencephalography. Still, the existence of a plethora of analysis methods indicates that no single one is able to reflect the vast variety of rhythm changes in numerous brain states and experimental conditions. The most widespread points of interest in EEG analysis are as follows:
- Rhythm power and frequency,
- EEG segmentation,
- EEG synchrony in distant brain areas,
- EEG nonlinear dynamics,
- Dipole Source Localization.

Below are given summaries of the main methods of EEG analysis in relation to various states of the normal and pathological brain. The appropriate software packages can be divided into routine ones offered by companies manufacturing EEG equipment and those designed for advanced analysis and created in EEG research laboratories.

Rhythm Power and Frequency

Rhythmic properties of EEG signals are of great importance for EEG analysis as they reflect functional states of neural structures and encode informational processing in the brain.

Fourier Transform

The Fourier transform (FT) is a method to uncover the rhythmic structure of EEG signals. It is based on the mathematical fact that any signal defined in a given time interval can be decomposed into a sum of sinusoidal waves of different frequencies, amplitudes and phases.

The FT is a method to estimate a signal's frequency spectrum (see also Chapter 18). Mathematically, any signal in a given time interval can be decomposed into a sum of mutually orthogonal sinusoidal waves of different frequencies, amplitudes and phases. The FT is a complex function of frequency that describes these components' amplitudes and phases. The FT is thus a way to obtain a *spectrum* of a signal. If some rhythms are present in an analyzed segment of EEG, they can be recognized as peaks in a spectrum obtained via FT.

There are certain subtle points with the FT application. The first one is that FT is a complex function that cannot be interpreted immediately since it has no physical meaning of its own. There are, however, certain functions, derived from FT, that do have such a meaning. The squared module of FT is a real function that describes the frequency components' power and is hence called a power spectrum. The square root of the power spectrum, i.e., FT's module, describes the components' amplitudes, whereas the *arctangent* of the ratio of FT's imaginary and real parts defines the components' phases. The other subtle point is actually a serious problem. Provided that the EEG is a signal with a high degree of stochasticity (randomness), the FT obtained from a single EEG trial cannot serve as a reliable estimator of its spectrum. For a completely random signal (gaussian noise), the variance of each frequency component's amplitude, obtained via FT from one trial, is equal to the value of this amplitude itself. For an EEG that is not completely random, this variance may be smaller but still too large. The problem is overcome by averaging spectra computed from single trials, the spectrum estimation thus becoming an accumulative procedure. In most cases, the power spectra of the single trials are used for averaging, resulting in a reliable power spectrum estimation.

The other problem, which is closely related to spectra averaging and has been much discussed in the literature during the recent years, is the non-stationarity of EEG signals. Stationarity means that the frequency content of a signal does not change with time. The real EEG signal might be expected to be stationary (still not completely) only in some very special conditions, such as a subject's resting state with his/her eyes closed. At the same time it is often believed that, in order to get true spectrum estimation, the FT should be applied only to stationary segments of EEG. Indeed, according to stochastic signal theory, the estimation of a random signal's spectrum by averaging converges to its "true" spectrum only if a signal is stationary along the whole epoch of analysis. However, as mentioned earlier, the EEG is not quite a stochastic signal since many of its components have definite neurophysiological mechanisms at their background, although many of these underlying mechanisms may remain unknown to the investigator. The EEG thus has properties of *both* a stochastic and a deterministic signal. The requirement of stationarity and the necessity of averaging emerge from the assumption that the analyzed signal be stochastic, which may not hold true for the real EEG.

Another consideration in favor of the possibility to apply the FT to non-stationary segments of EEG is as follows. Practically, if a certain rhythm appears only in part of a (non-stationary) segment of EEG, one can still see it as a peak in the relevant spectrum. However, from this spectrum obtained via FT, it is absolutely impossible to know when, i.e., in what time interval of an analyzed segment, this rhythm appeared. In general, the FT provides *all* frequency information on the transformed signal but *completely* loses the time information on it. For stationary signals, this does not cause any problem since its frequency content does not vary with time. For non-stationary signals, such as EEG segments with transient rhythms, the problem is canceled if we are not interested in the fine time-frequency structure of a signal but want to know only whether a certain rhythm appeared anywhere in the course of the analyzed segment. In such a case we can ignore non-stationarity.

An application of FT to non-stationary segments of EEG is presented in a paper by G. Ivanitsky et al. (1997). Besides applying FT to non-stationary EEG segments, the authors did not use explicit averaging of spectra and described a way to recognize specific features of single EEG trials by their non-averaged spectra. This becomes possible by applying a special analysis tool – namely, an artificial neural network that performs some sort of implicit averaging of data at some stage of learning (see the details later in this chapter).

There is a practically and theoretically significant property of the FT: its *frequency resolution* is the inverse of the width of the analyzed time window: $\Delta f = 1/T$. This basic fact is often called the uncertainty principle since it implies that it is impossible to receive precise time and frequency information on a signal simultaneously. Indeed, large time windows provide good frequency resolution but poor time resolution, and *vice versa*. To estimate the time course of an EEG spectrum, the so-called *windowed FT* is applied. In this method, the FT is calculated in a window of constant duration that moves along the EEG record (Makeig 1993; Nikolaev et al. 1996). To increase the reliability of the spectral parameters in this technique, it is recommended to average the spectral samples obtained in the same time interval in a number of tasks in one subject and across subjects.

For practical purposes, the discrete fast Fourier transform (FFT) is suited best. This method significantly economizes computational time at the cost of the following constraint: the length of the time window for FT calculation must be a power of two of the number of discrete time samples. A large number of EEG features can be obtained from the FT analysis, for example, the absolute power within a frequency band, the relation of power in different bands, spatial asymmetry of band power, peak frequencies, and peak asymmetries (Davidson et al. 1990; Sabourin et al. 1990; Lutzenberger et al. 1994; Wilson and Fisher 1995). The procedures and precautions for EEG spectral analysis were reviewed in the Handbook of Electroencephalography and Clinical Neurophysiology (1987) and by Jervis et al. (1989).

The FT is the most widely used method of rhythm analysis and should be considered as a principal pilot method in EEG studies. It is included in almost all commercially available EEG processing software packages.

Wavelet Transform

The wavelet transform (WT) is a method to search for transient trains of rhythms in non-stationary EEG records. The WT provides a compromise between time-domain and frequency-domain localization of a train. As FT, the WT is a decomposition of EEG into a sum of orthogonal signals, these special signals now being not sine waves, but a family of short oscillatory trains of various duration and frequency content, the so-called wavelets.

As stated earlier in the section describing the FT, any continuous signal can be decomposed into a sum of mutually orthogonal signals. For classical FT these signals are sine waves of different frequencies, amplitudes and phases. It is possible, however, to use other sets of mutually orthogonal signals, since this class of signals is wide. What particular kind of signals to choose as a basis for decomposition depends on the kind of elementary components that are expected to be the most interesting part of the analyzed EEG. If the components of interest are sustained rhythms that exist over comparatively long time periods (comparable with the analysis epoch) or if the time of appearance of transient rhythms inside the analyzed epoch is not the point of interest, then the choice of sine waves (and hence classical FT) is reasonable. If the EEG signal is supposed to be essentially non-stationary and composed of many transient rhythmic trains, and if additionally we are interested not only in the frequency of these transient oscillations but also in the temporal position of rhythmic trains, then the family of wavelets is a choice.

The FT provides detailed frequency information on a signal in a given time interval, but completely loses information on how the frequency content changes with time *inside* the interval. In contrast, the WT finds a compromise between time-domain and frequency-domain localization of rhythmic trains and hence provides knowledge of both the time and frequency structure of a signal inside the analyzed window, although with some limited resolution. A series of FT windows (the windowed FT) may also provide this dual information, but in this case the resolution in both the time and frequency domain is fixed, does not depend on the analyzed frequency, but depends entirely on the width of the window. One of the advantages of WT over windowed FT is that the WT's time resolution is variable and depends on the frequency of a component – for high frequencies the time resolution is better than for low frequencies.

A full set of wavelet functions, used for decomposition, consists of a so-called *mother* wavelet and a set of its time-shifted and compressed or dilated copies. A mother wavelet is some zero mean function localized both in the frequency and time domains, e.g., the oscillatory signal amplitude is modulated by a bell function. Some particular examples of mother wavelets are the *Morlet wavelet* and the *Mexican hat function* as shown in Fig. 3 (Torrence and Compo 1998). Any compressed or dilated wavelet is characterized by a scale parameter (s) which is actually a dilation coefficient and is inversely related to frequency (larger scales correspond to dilated signals and smaller scales correspond to compressed signals). As mentioned earlier, certain sets of such functions may serve as full orthogonal bases for decomposition of arbitrary signals defined in a given time interval (this can be strictly proved mathematically; Daubechies 1992).

The computation of a wavelet transform proceeds according to the following steps:

1. The mother wavelet is chosen to serve as a prototype for all other wavelets in the process.
2. A definite set of scale parameters s is selected.
3. For every s, a definite set of time shifts τ is selected.

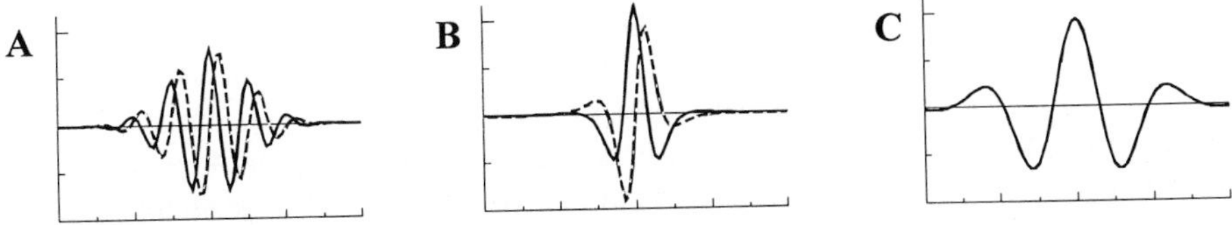

Fig. 3. Three different wavelet bases, (a) Morlet, (b) Paul, (c) Mexican hat. The graphs plot, on the ordinate, the real part (solid) and the imaginary part (dashed) in unspecified units as functions of time (abscissa) in unspecified units. All other wavelets in a set are compressed/dialted and translated copies of a mother wavelet.

4. For every given s and for every given τ, the scalar product of the corresponding wavelet with the given realization of the EEG is calculated (a scalar product is the sum of count-by-count products of these functions).
5. The procedure is repeated for every s and for every τ selected. The result is a discrete function of two variables – s and τ – defined on the $s\tau$ plane. Since s corresponds to the inverse frequency and τ corresponds to the time delay, this function describes the time-frequency characteristics of the original EEG trial. A high value of this function at some s and τ indicates that a rhythmic train appeared at *approximately* this frequency (defined by s) and at *approximately* this time.

Note: It should be noted that
- at any scale s, the wavelet function has not one frequency, but a band of frequencies, and the bandwidth is inversely proportional to s. This means that the finer the resolution in time, the smaller is the resolution in frequency, and *vice versa*;
- s^{-1} is not exactly the frequency at which the spectrum of the wavelet function reaches its maximum value, and the relation between these two values depends on the type of the mother wavelet;
- different mother wavelets result in different values of the wavelet transform, but qualitatively the results are similar.

In EEG studies the WT makes it possible to describe time-frequency characteristics of both transients in spontaneous EEG and time-varying rhythms in event-related brain activity (Schiff et al. 1994). The increase in time resolution of the method with frequency makes it especially useful for the analysis of high-frequency (gamma) EEG bands (Tallon-Baudry et al. 1996).

Autoregressive Models

The idea of the autoregressive (AR) method is based on the assumption that the real EEG can be approximated by a so-called AR process. With this assumption settled, the order and parameters of the appropriate AR model are chosen in a way to fit the measured EEG as closely as possible. In turn, for every particular AR model, the power spectrum of the corresponding AR process can be found analytically. Thus, the AR method provides an alternative way of estimating EEG spectral properties.

The method describes any new value X_t of the EEG observed at a discrete time point t as a weighted sum of the values at a fixed number of immediately preceding points (its "history") plus a value equal to the difference between the prediction derived from the "history" and the actual value at this point:

$$X_t = a_1 X_{t-1} + \ldots + a_p X_{t-p} + Z_t$$

where p is the model's order, $a_{1\ldots}$ a_p are regression coefficients, and Z_t is a prediction error. If the time series of prediction error values turns out to be white noise, the analyzed process is a true *AR* signal. If it turns out not to be white noise, the measured process is not an AR process, but it is still possible to find regression coefficients that entirely determine the AR process in a way to give the best possible prediction (i.e., minimize a prediction error) of measured X_t, and thus approximate the real EEG by some AR process as closely as possible. The calculation of a set of such AR coefficients is tantamount to determining what the system is. The approximation of the real EEG by an AR process is done within the selected order of the AR model, so choosing the optimal order is an important task. The high-order models, being more detailed, can in principle provide better prediction than low-order models, but, at the same time, the reliability of the AR coefficient's estimation decreases, which implicitly decreases the prediction accuracy. So, selecting the model order is a search for a compromise between minimizing mathematical artifacts and building a more precise model. The optimal choice is provided by statistics like Akaike's Information Criterion (AIC) or the Final Prediction Error (FPE) (see Priestley 1981, for a review). Practically, in EEG analysis the order rarely exceeds 11.

The AR approach to EEG studies is used

- in obtaining the spectral characteristics of EEG signals (reviewed by Madhavan et al. 1991),
- in the discrimination of stationary parts of the EEG (i.e., segmentation) (Pretorius et al. 1977; Gath et al. 1992),
- in artifact rejection (replacement of artifactual EEG pieces with pieces predicted by the AR model), and also
- in testing specific hypotheses of EEG genesis.

For every particular AR process, the power spectrum can be estimated analytically. Provided that the EEG is approximated by some AR process, the latter´s estimated spectrum can be considered an estimate of the EEG spectrum. Spectral analysis based on the AR model is hence an alternative to the conventional Fourier-based techniques and is particularly advantageous when short data segments are analyzed, since from a formal point of view the frequency resolution of an analytically derived AR spectrum is infinite and does not depend on the length of the analyzed data. One should realize, however, that the AR-based estimation of the EEG spectrum is valid only so far as the main assumption of the method is valid, i.e., that the EEG can be approximated well enough by an AR process.

The AR models are only applicable to stationary signals. As this assumption rarely holds for EEG, the concept of local stationarity was proposed: within small intervals of time the EEG signal departs only slightly from stationarity (Florian and Pfurtscheller 1995). In this study, in view of the property of local stationarity, the EEG trials were divided into segments, such that, within each segment, the data were stationary. The AR spectral estimation was then applied to locally stationary segments of EEG.

Event-Related Desynchronization

Event-related desynchronization (ERD) is a short-lasting, topographically localized attenuation of rhythms within the alpha (beta) band. Desynchronization is considered as a sign of activation of cortical areas while synchronization is characteristic of cortical areas at relative rest.

The ERD procedure is usually applied to sequentially recorded event-related EEG trials. It is based on averaging techniques and consists of several steps:

1. Some primary information about which frequencies are indeed affected (i.e., desynchronized) is obtained by comparing spectra before and after an investigated event;
2. The frequencies changed most are bandpass-filtered from the original trials;
3. The filtered trials related to task performance intervals are averaged (A);
4. All trials in the reference interval are also averaged (R);
5. The ERD is defined as:

$$ERD\% = \frac{R - A}{R} * 100\%$$

A positive ERD value indicates a decrease in band power (desynchronization), whereas a negative value indicates an increase in band power (synchronization) with respect to the reference interval. Desynchronization and synchronization of alpha frequency components can be observed within the same time interval at different locations on the scalp (Pfurtscheller and Klimesch 1992).

Pfurtscheller's group showed that the ERD may be a precise index of various brain operations: movement (Pfurtscheller et al. 1997), visual processing (Pfurtscheller et al. 1994), reading and recognition (Pfurtscheller and Klimesch 1992), and memory processes (Klimesch et al. 1994).

EEG Segmentation

One of the difficult and intriguing problems of EEG analysis is the discrimination of EEG portions that are probably related to some behavioral, mental or pathological state. Accordingly, the EEG might be considered as composed of stationary segments of variable lengths.

The procedure of segmentation aims at finding boundaries between stationary parts of an EEG signal.

Single-Channel-Based Segmentation

A procedure for *adaptive segmentation*, where the duration of the segments is determined by the particular EEG itself, is based on spectral changes in one channel; it was proposed by Pretorius et al. (1977) and has since been slightly modified by Creutzfeldt et al. (1985):

1. At the beginning of the EEG recording, the *autocorrelation function* (see Chap. 18) of an initial reference EEG of nominally 1 s duration is computed;
2. Then the autocorrelation function of the test EEG is computed within a "window" of 1 s duration in the first EEG interval;
3. This autocorrelation function is compared with the autocorrelation function of the reference EEG;
4. The difference is collated with a predetermined threshold established empirically;
5. The window progressively moves along the EEG. When the difference exceeds the threshold for more than a specified minimum time, a segmentation line is placed;
6. Once a segment boundary has been placed, the process begins anew with a new reference EEG portion taken just after the segment boundary.

The obtained segments need further sorting which is provided, for example, by a clustering algorithm; that is, similar segments are grouped in a feature space spanned by mean frequency and power of the segment (Creutzfeldt et al. 1985).

Other methods of both adaptive and non-adaptive models are reviewed elsewhere (Pardey et al. 1996). Segmentation methods find their widest application in the *detection of epileptic seizures* (Creutzfeldt et al. 1985; Inouye et al. 1990; Gath et al. 1992; Pietila et al. 1994).

Segmentation Based on Changes in Scalp Potential Topography

The identification of short EEG segments separated by changes in EEG topography was first proposed by Viana Di Prisco and Freeman (1985). Lehmann at al. (1987) demonstrated that monitoring changes in electric field distribution over the scalp can be used for breaking a series of momentary voltage maps into segments, within which the maps preserve some principal spatial characteristics. The procedure is then as follows:

1. A "global field power" (i.e., the standard deviation of the channels' voltages) is computed for each momentary map, and then only maps with maximum "global field power" are considered further;
2. For each of these, the locations of the potential maximum and minimum are determined. A segment is accepted to continue as long as the extrema of successive maps remain in spatial windows that are defined by the location of the extrema in the first map of the segment (see Fig. 5);
3. The obtained segments are classified. It has been suggested that different segment classes manifest different functional brain states (microstates).

The mean duration of the microstates was found to be consistent with the duration of elementary steps of cognitive processes, and different topographies of the microstates were associated with different cognitive modalities (Wackermann et al. 1993; Koenig and Lehmann 1996). Space-oriented segmentation may also reveal specific map changes in patients with Alzheimer dementia (Ihl at al. 1993), in schizophrenics (Merrin at al. 1990; Stevens at al. 1997), and in depressive patients (Strik et al. 1995).

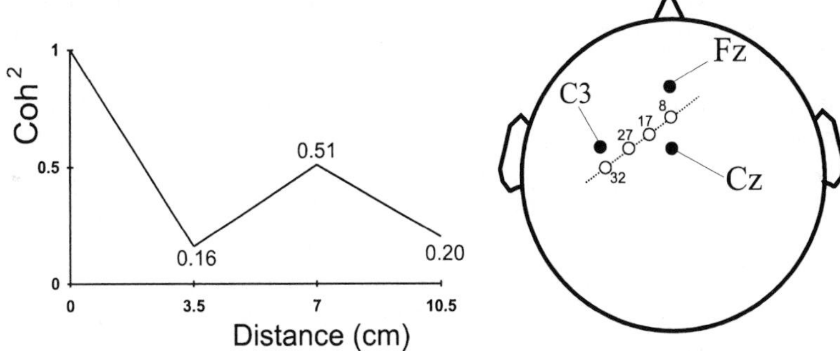

Fig. 4. Coherence (10–12 Hz) calculated between electrodes located on a line connecting the electrode 2.5 cm behind C3 and the electrode 2.5 cm behind Fz. It can be seen that the coherence does not decrease continuously with increasing distance. (Adapted with permission from Pfurtscheller et al. 1997)

EEG Synchrony in Distant Brain Areas

Coherence

Coherence computation (cf. Chapter 18) is a method to find out whether the same frequency components present in two signals (e.g., two EEG channels) preserve their phase shift from trial to trial. The phase shift stability observed at a certain frequency may indicate that the corresponding rhythms in two EEG channels are of the same origin or interact with each other.

The coherence method is based on the FT and is designed to evaluate the stability of phase shifts between the same frequency components of two simultaneously recorded signals (e.g., two different channels of EEG) regardless of these components' amplitudes. The coherence is a function of frequency. If this function has a peak at a certain frequency, the phase shift between corresponding oscillatory components of the two signals is nearly the same in the majority of the analyzed trials.

The calculation of the coherence is an accumulative procedure and proceeds as follows. First a set of time windows involved in processing is selected. The choice of these windows depends on the experimental paradigm and research conditions. They may be sequential time intervals of equal duration cut from continuous EEG, or time pieces corresponding to a subject's condition, or time intervals preceding and/or following external events (such as in the case of the ERP), etc. Then the FTs for both signals are calculated in every time window and one of them is multiplied with the complex conjugate of the other. Then these products, obtained for every time window, are summated. The whole procedure of obtaining the coherence function can be described by the following formula:

$$Coh(f) = \frac{\left| \sum_{i=1}^{N} F_1(f) \cdot F_2^*(f) \right|^2}{\sum_{i=1}^{N} |F_1(f)|^2 \cdot \sum_{i=1}^{N} |F_2(f)|^2}$$

where $Coh(f)$ is the coherence function, f is the frequency, N is the number of EEG realizations involved in averaging, $F_1(f)$ and $F_2(f)$ are the Fourier transforms of the EEG signals in two different channels, and * symbol denotes complex conjugation.

If, for a given frequency, the phase shift between frequency components stays constant from realization to realization, the accumulation in the numerator of the formula proceeds well. If, in contrast, the phase shift jumps randomly, the accumulation proceeds badly. After the accumulation is finished, the resultant complex product function is module-squared and normalized by the product of the single signals′ power spectra sums. The goal of the normalization is to fit the coherence function into the range from 0 to 1. "Zero" means no coherence exists between two signals at a given frequency, and "one" means that the phase shift is absolutely still, such that the coherence function depicts only the phase shift stability while not depending on the components' amplitudes.

The coherence values are interpreted in terms of connectivity between brain structures, as mentioned in the section on "EEG rhythms". A lot of studies have been performed to find relations between coherence topography and various brain states (Livanov 1977; French and Beaumont 1984; Gray and Singer 1989; Petsche et al. 1992; Petsche 1996; Weiss and Rappelsberger 1996). The coherence is usually high between EEG recordings from adjacent electrodes and falls dramatically with increasing interelectorde distance, which may be explained by volume conduction. But if rhythmic activities dominate the EEG, the degree of cooperativity increases very significantly, such that coherent activity can occur over large extents of the cortical surface (Lopes da Silva 1991). In some cases the dependence of coherence on distance may not be gradual: High coherence may be registered between spatially separate cortical areas although their coherence with intermediate sites is low (Bressler et al. 1993; Andrew and Pfurtscheller 1996). This then indicates that the synchronization of EEG rhythms does not simply result from direct volume conduction but is related to interactions between distant cortical areas that participate in mutual functioning (cf. Fig. 4).

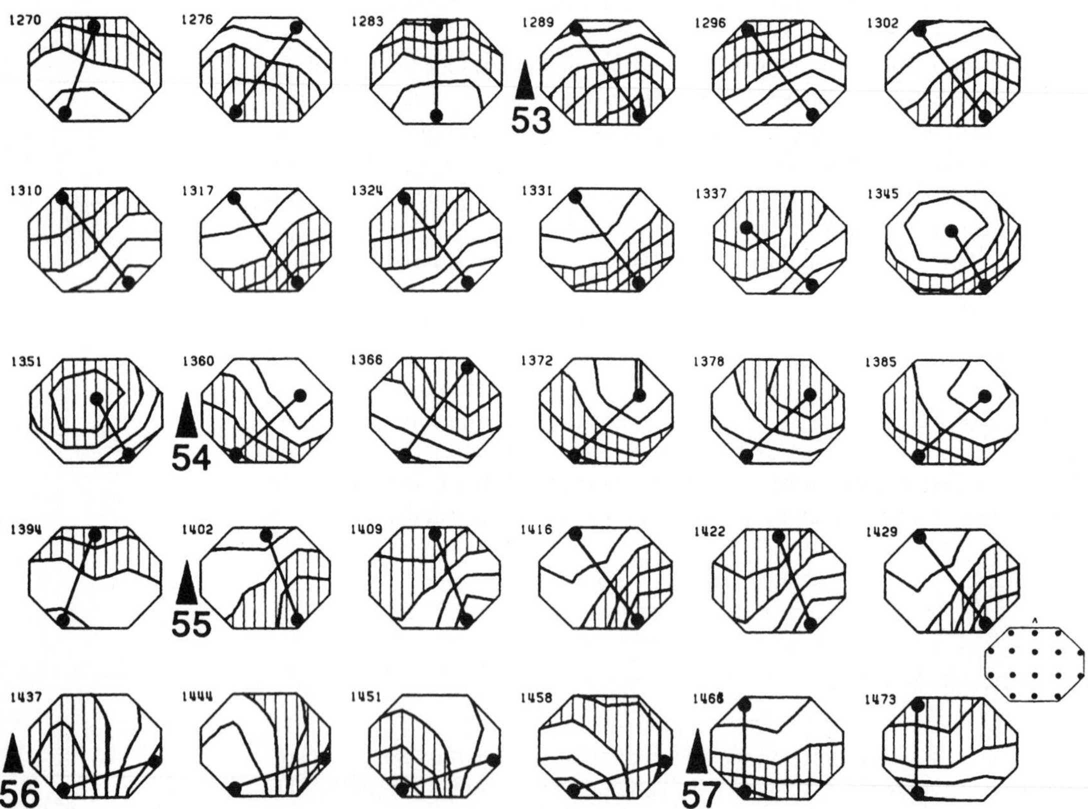

Fig. 5. Adaptive segmentation of a series of momentary scalp field maps covering 1585 ms of alpha activity recorded from 16 electrodes. The maps at successive times of maximal global field power are used. Average time between maps is 54.7 ms; numbers identify original sampling time points (128/s); white positive, hatched negative relative to average reference; isopotential contour lines in steps of 10 μV. In each map, the locations of the maximal and minimum potentials are marked by dots and connected by a line. Segment terminations are marked by vertical arrows. (Adapted with permission from Lehmann et al. 1987)

Intracortical Interaction Mapping (IIM)

The basic idea of IIM is that the EEG spectral components reflect the activity of the main cortical oscillators and that the coincidence in their frequency properties reports on communication between cortical areas. It is proposed that IIM reveals cortical connections that are invariant to transient phase shifts between the interacting cortical oscillators.

One drawback of coherence analysis is the reverse side of its advantage: The coherence is strongly influenced by phase-shift changes between two interacting oscillators during the analysis epoch. It was said above that phase shifts control both the neural connection mode (active vs. silent) and its direction. But changing these variables presumably is an important element of brain processes underlying mental functions, and the temporal scale of these transformations can be considerably faster than any analysis epoch. Thus, some important connections can be missed by coherence methods. To overcome this difficulty, a technique called intracortical interaction mapping (IIM) has been proposed (Ivanitsky 1990; 1993). The method is based on three premises:

- Cortical neuronal groups are specialized in their function, and complex brain functions are based on these groups' cooperation;
- The neurons inside a group are organized such that they acquire the features of neuronal oscillators, which can be revealed by EEG spectral analysis;
- The coincidence of frequency properties of two or more oscillators (i.e., EEG spectral components in two or more recordings) can provide evidence for their functional connectivity.

The data obtained with IIM are usually presented in the form of a brain map whose construction includes the following operations:

1. Calculation of fast Fourier transforms (FFTs) from segments of EEG or event-related potential (ERP) recordings;
2. Spectral smoothing if needed;
3. Selection of the major spectral peaks from each of the major EEG bands. The criteria for selection are determined by the investigator based on a comparison of the peak with the mean power level of the spectra.
4. Search for coincidences of peaks (with a defined level of precision, usually one spectral quantum) with peaks in the spectra of all the other sites;
5. Calculation of the number of coinciding peaks for each site and EEG band. This number is normalized with respect to the number of sites minus one;
6. Construction of brain maps. Two versions of the maps are proposed: interpolation map and "arrows" evaluating the actual connections between cortical areas. The step 5 is excluded in the "arrows" version. (see section "EEG mapping" and Fig. 6 for details).

The method was tested in a fairly simple experiment in which subjects repeatedly bent their left- or right-hand fingers. The system of connections between contralateral central and frontal cortical areas was mapped by using only IIM, not the coherence method. IIM appeared to be effective in studying brain mechanisms of thinking and emotions (Ivanitsky 1993; Nikolaev et al. 1996) as well as in studies of mental pathology (Strelets and Alyeshina 1997). Being invariant to phase shifts during the analysis epochs, the method thus reveals some labile functional interactions which can be missed by the coherence method.

EEG Non-Linear Dynamics

Neuronal networks that generate EEG signals are complex systems with non-linear dynamics. Therefore, methods of analysis of non-linear dynamic systems (cf. Chaps. 22–24) have been applied in search of new ways to uncover additional characteristics of brain functional processing. The basic idea, inspired by recent advances in the area of non-linear dynamics and chaos theory (cf. Chapter 24), is to view an EEG signal as the output of a deterministic system of relatively simple complexity, but containing non-linearities. A main fact here is that EEG signals are irregular oscillations, which can be found in complex systems with a chaotic attractor (Babloyantz 1985). This suggests that studying the geometric dynamics of EEGs may lead to a more appropriate analysis than the classical, stochastic methods. The techniques applied to EEG signals involve the computation of the

- *correlation dimension* (Grassberger and Procaccia 1983),
- *Lyapunov exponent* (Wolf et al. 1985), and
- *Kolmogorov entropy.*

Combinations of spectral and nonlinear measures yield a better overall discrimination between various brain states (Fell et al. 1996; Stam et al. 1996). This vast area cannot be reviewed here in detail. Reviews of the fundamentals of chaos theory including the derived analytical techniques are provided by Elbert et al. (1994) and Jansen (1996).

In general, the dimensionality reflects the complexity of the dynamics of neuronal networks, which generate the EEG, and its increase implies a growth in the number of such networks. Therefore, it is believed that a higher dimensionality reflects a higher mental activity (Babloyantz 1989; Lutzenberger et al. 1992; Pritchard and Duke 1992; Elbert et al. 1994; Schupp et al. 1994).

Recent re-examination of the evidence for the non-linear structure of the EEG has shown that it may be more appropriately modeled by linearly filtered noise (Theiler and Rapp 1996). However, non-linear components were detected in the EEG inconsistent with the hypothesis of low-dimensional chaos (Palus 1996; Pritchard et al. 1996).

Dipole Source Localization

An electric current dipole is the elementary physical source of an electric brain potential. At any given moment, a tremendous amount of dipoles are active inside the brain. Under certain conditions, however, it is possible to determine the location and strength of a few principal dipoles, i.e., the dipoles that make a major contribution to the generation of an electric field. In another approach, the entire brain activity is modelled as the activity of a single dipole that produces a superficial potential close to the observed one. In this case the dipole is called an equivalent dipole.

As mentioned earlier in this chapter, the pyramidal cells of the cortex (namely, their dendrites) are the principal physical sources of the brain electrical activity. When a neuron´s postsynaptic potential builds up under the influence of synaptic inputs, an electric charge sink appears on its dendrite and an electric charge source appears at its axon's trigger zone (Plonsey 1982). This pair of charges (source and sink) constitutes an electric current dipole which can be thought of as being produced by a microscopic electromotive force that acts inside the dendrite. The electric charge separation produced by this force is compensated by the distributed electric currents flowing over the whole volume of the head acting as a conductor. These currents (induced by many dipoles) flow mainly inside the brain tissue, but also penetrate into the skull and the scalp. The potential differences between points on the scalp (which underlie the superficially measured EEG) are merely the voltage drops (according to Ohm's law) produced by the above-mentioned electric currents. Thus, for any given dipole position in the brain, there is a unique distribution of scalp potentials. The question then arises whether it is possible to determine the in-brain active dipole configuration based on the superficially measured potential values? In other words, whether it is possible to solve *the inverse problem of the EEG*? The general answer to this question is "no", at least in the sense that no unique solution exists, because the given potential distribution may be produced by a multiple of dipole configurations. However, the solution may become more confined if additional constraints are incorporated (Ilmoniemi 1993; Scherg and Berg 1991). Today, the following constraints (or their combinations) are the most popular:

- A fixed small number of (principal) dipoles are assumed to exist (usually one or two);
- Dipoles are assumed to reside on a predetermined surface, i.e., in the cortex, and they are allowed to have only definite orientation (dependent on their location);
- The solution has a minimal norm among all solutions that fit the measured superficial potential, or it is the smoothest possible solution (Pascual-Marqui et al. 1994).

Which constraints should be taken into account depends on the physiological model under consideration.

The solution of the *inverse problem* relies on the solution of the *forward problem*, which is the mathematical calculation of the superficial potential for any possible dipole configuration. Practically, the solution of the forward problem is often implemented as a computer simulation. This can be based on a realistic head model that takes into account the real shape of anatomical structures (brain, skull, scalp, ventriculi) and the conductivity differences between them, or on a (to some extent) simplified model, the simplest one being just a spherically bounded uniform volume conductor (Cuffin 1996; Yvert et al. 1995). The optimum head model is chosen according to the physiological or clinical task to be solved. Then, while solving the inverse problem, that dipole configuration is searched for that meets the constraints and at the same time yields the smallest least-square error between the mathematically calculated and the observed potentials. Appropriate mathematical methods have been elaborated (for review see van Oosterom 1991).

At present, there are several fields of application of the dipole-source analysis.

- *Localization of epileptiform activity sources.* In some cases, distinct large-amplitude spikes are likely to be generated by a single source; and the goal is to localize it with high accuracy. In this case, a detailed realistic head model with a single dipole constraint is appropriate (Roth et al. 1997). In other cases, however, the sources are likely to be distributed more diffusely, so models that can evaluate multiple dipoles (at the cost of accuracy) are preferable (Lantz et al. 1997; Scherg and Ebersole 1994).
- *Neuroscience research.* Here, the method is often applied to evoked potentials (EP) and event-related potentials (ERP) (e.g., Plendl et al. 1993; see also Chapter 36). In many research applications (e.g., in studies of cognition mechanisms), multiple principal dipoles are active inside the brain at a given time. Due to limitations of the method, it is hardly possible to determine their exact number and location. At the same time, evaluation of the dipoles' approximate number and location may still provide valuable scientific information, especially if the dipole localization method is combined with other methods, such *as positron emission tomography* (PET; see Chapter 39) or *functional magnetic resonance imaging* (fMRI; see Chapter 38), since the EEG-based method can provide information that other methods do not supply: the fine time course of brain processes. An example of such a fruitful combination of methods is the study performed by Abdullaev and Posner (1998).
- *Information reduction.* In many cases, the dipole-source localization can be seen as a way for information reduction. For example, the simplest head model (uniform sphere) with one dipole is often used in a situation where the source is evidently spread and/or multiple (e.g., ERP). The single dipole is then considered to be an equivalent dipole, which is, of course, an abstraction. However, its location may represent the location of active brain structures; this may be especially illustrative in comparative studies concerned with functional lateralization.

Surface Laplacian

The surface Laplacian provides a way to improve the spatial resolution of scalp-recorded EEG.

EEG transmission through the low-conductance skull results in spatial low-pass filtering which causes the potentials to blur. The so-called surface Laplacian appears to provide improved spatial resolution of scalp-recorded EEG by computing the two-dimensional second derivative in space of the potential field at each electrode. This converts the potential into a quantity proportional to the current entering and exiting on the scalp at each electrode site, and eliminates the effect of the reference electrode. Comparison of ordinary potential-based methods with surface Laplacian methods indicates that the latter provides a better spatial discrimination among electrode sites (Law et al. 1993). It is especially important for multi-electrode EEG systems such as the "High resolution EEG" with 124 electrodes (Gevins 1996).

Subprotocol 3
Secondary EEG Analysis

Procedure

The results of primary EEG analysis should be subjected to secondary analysis. The most common goals are a search for differences between obtained EEG parameters related to specific groups, brain states, mental tasks, and so on. This analysis is performed by comparative statistics such as t-test, ANOVA, and non-parametric methods. However, the complex nature of the EEG signal often makes it impossible to obtain reliable estimates based on one parameter only. To take into account simultaneous dynamic changes of many EEG parameters in many brain areas, it is necessary to use systems of multi-parametric evaluation which can provide principally new results. Below is an example of such a system.

EEG Classification Using Artificial Neural Networks

Artificial neural networks (ANNs) are a convenient instrument to reveal EEG features pertinent to certain mental functions, and to perform classification of the EEG based on these features.

Each EEG component is caused by some brain process, whose nature, however, is often unknown. This idea is linked to the problem of a normal subject's EEG classification. Since the EEG is a complicated and multi-component process, it is usually difficult to decide which parameters should be chosen as a base for such classification. It seems reasonable to use EEG features pertinent to certain psychic functions, which are considered to be the point of interest in a given research. If so, one must have some means of relating EEG features to psychic processes, provided that no a priori knowledge on this item is available.

This can be achieved by means of artificial intelligence systems, and in particular, artificial neural networks (ANNs; cf. Chapter 25). This technology, which is actually some sort of multi-parametric analysis, proves to be a powerful and convenient instrument for feature extraction and data classification. There are at present a variety of different ANN paradigms, some of which have been successfully utilized for EEG classification

(Bankman et al. 1992; Gabor and Seyal 1992; Peltoranta and Pfurtscheller 1994). Below we describe one possible way to use a simple two-layered learned ANN for the purpose of classifying a normal subject's EEG (Ivanitsky et al. 1997).

First the network is taught to distinguish between patterns of EEG corresponding to different mental states and to find state-dependent features. At this stage called "learning", a set of EEG patterns is formed, for each of which the class defined by the mental operation performed by the subject is known. Then this learned set of patterns is presented to the network in a random order many times, until the network is able to recognize the patterns correctly. It achieves this by gradually changing its internal weights. After the learning stage, the weights are fixed and can no longer be changed. The network is then used to recognize, and thus classify, new data that were not included in the learning set of patterns and whose class is not yet known.

Analysis of the final ANN's internal weights (that were settled during learning) makes it possible to find out what EEG features are mainly used by the ANN for classification, or, in other words, what features are most characteristic for psychic processes of interest.

The described technology may thus be useful in two ways. Firstly, it can be used in practice for on-line monitoring of brain activity and detection of certain mental states. Secondly, it can be used as an instrument to investigate EEG features that correspond to certain psychic functions or, in other words, it may be useful as a research tool. For example, in our study the new EEG data corresponding to two thinking operations (verbal vs. spatial) of the same subject were discriminated at a mean accuracy of 87%.

Subprotocol 4
Presentation of Results

Procedure

In contrast to the common EEG record in form of a multi-channel time series, EEG data may be presented in more sophisticated and advanced ways such as EEG mapping.

EEG Mapping

EEG mapping is a presentation of EEG parameters on a schematic head surface obtained by interpolating the data recorded at each single electrode site into interelectrode space. This provides the researcher with a visual image presenting multi-channel EEG in the most integrative and illustrative form.

The main feature and advantage of the map are that it presents the EEG in the form of an image that displays data sets as a whole. Visual map analysis therefore involves both the mechanisms of imaginative and abstract thinking, making the process more efficient.

Mapping became widespread with the introduction of computers in EEG studies (Harner and Ostergren 1978; Duffy 1981), although the topographical presentation of EEG was realized earlier in a number of studies. In the fifties, Livanov (1977) constructed the 50-channel "Electroencephaloscope" for on-line visualization of EEG amplitude dynamics. The maps in this device were formed by 50 flickering lightening points on the head contour and 50 columns below presented on a cathode-ray tube. Both the brightness of the flickering points and the columns´ altitude indicated the momentary EEG amplitude under each of 50 electrodes, forming a rather intriguing and vivid picture.

Now the map is usually built by linear interpolation of the potential values of three or four neighboring electrodes, which are summarized as a mathematical average, inversely proportional to the distances from each of these electrodes. More complicated is *surface spline interpolation*, which can exhibit maxima and minima between electrodes (this is not possible in linear interpolation) and produces smoother maps (Ashida et al. 1984).

An "Atlas of brain maps" especially dedicated to the mapping of brain electrical activity was published by Maurer and Dierks (1991). In the clinic, the mapping helps reveal the exact site of pathological activity, for example, slow waves or epileptic spikes. For research purposes the mapping is useful in the comparison of activated and inhibited cortical areas. Now a number of companies offer special devices for EEG mapping. These devices include electrodes, multi-channel amplifiers, and computers with a number of programs.

Almost every study aimed at elucidating the topography of brain rhythms (see previous sections) can be presented in the form of a brain map. For example, it may be potential and spectral power maps as well as more advanced maps displaying the connectivity between cortical areas. The latter maps are based on the synchronization of brain rhythms. As mentioned above, the coherence differences between the rest state and some mental operations have been evaluated as probability maps (Petsche 1996). Intracortical interaction mapping has been used to construct connectivity maps of two types: an interpolation map identifying connection centers, and an "arrows" map showing connections between particular cortical regions (Figure6).

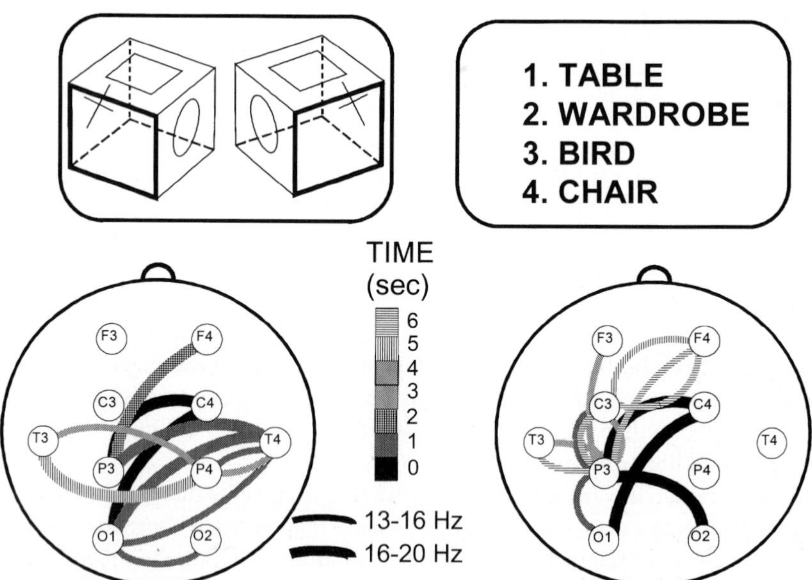

Fig. 6. Cortical connections ("arrows") in two thinking tasks in beta-band frequencies (13–20 Hz). The spatial task was a comparison of geometrical figures, the verbal task included the search for one in four words related to another semantic category. Only statistically significant connections in comparison to visual motor control in a group of 43 subjects are shown. The hatching indicates the time of connection appearance in the solving process (see the scale). The thickness of lines designates the beta-subbands in which the connections are formed (Nikolaev et al. 1996).

Advantages of the EEG in Comparison with High-Technology Brain Imaging Methods

At present electroencephalography meets a real challenge from new high-technology methods for living brain imaging, such as PET and fMRI. These techniques provide a very detailed and exact picture of brain structures involved in normal functioning or when damaged by pathological processes.

What are the *advantages of EEG* studies? Some of the advantages are evident. The EEG

- has a *high-time resolution*,
- is rather *simple in use* and *cheap*,
- almost *does not disturb* a subject,
- can be *recorded near the patient's bed*,
- can be used for *long-term monitoring* of sleep stages or epilepsy,
- is also a *convenient tool for psychophysiological research* when the subject has to perform some behavioral tasks or is out of laboratory.

There is nevertheless one more, not so evident, but very valuable advantage of EEG studies. In fact, PET and fMRI (see Chaps. 38, 39) measure secondary metabolic changes in brain tissue, but not primary, i.e., electrical, effects of neural excitation. This is the case in EEG recording. The EEG can thus reveal a major parameter of neural activity – its rhythmic property, which reflects the essence of neural excitation. Therefore, while recording electrical (as well as magnetic) field patterns, the physiologist has access to the actual mechanisms of the brain information processing.

Primary effect studies can discover the circuitry of brain processes, revealing not only "where" but also "how" information is processed in the brain. This approach is essential for the solution of the mind-brain problem.

As the EEG spectrum may reflect particular cognitive conditions, EEG recording can be used for on-line diagnostics of cognitive operations with a possible detection of thinking errors. EEG-based biofeedback may be applied to the correction of brain malfunction, including the signaling of wrong mental operations. It is essential in high-technological processes where the error value is high.

Thus, the EEG which previously was treated as the result of a simple summation of activities of huge masses of neurons, is now an effective tool in exploring the intimate mechanisms of informational processing in the human brain (Mountcastle 1992). One may expect, therefore, that innovative EEG methods have a good perspective in future neuroscience.

Acknowledgement: The authors would like to thank Prof. A.A. Frolov for very valuable suggestions and also Dr. N. Polikarpova and Dr. E. Cheremushkin for useful advice.

References

Abarbanel GDJ, Rabinovich MI, Selverston A, Bazhenov MV, Huerta P, Sustchik MM, Rubchinsky LL (1996) Synchronization in neuronal ensembles. Usp fizich nauk 166: 363–390 (in Russian).

Abdullaev Ya, Posner M (1998) Event-related brain potential imaging of semantic encoding during processing single words. NeuroImage 7: 1–13

Andrew C, Pfurtscheller G (1996) Dependence of coherence measurements on EEG derivation type. Med Biol Eng Comput 34: 232–238

Ashida H, Tatsuno J, Okamoto J, Maru E (1984) Field mapping of EEG by unbiased polynomial interpolation. Comput Biomed Res 17: 267–276

Babloyantz A (1985) Evidence of chaotic dynamics of brain activity during the sleep cycle. Phys Lett 111: 152–156.

Babloyantz A (1989) Estimation of correlation dimensions from single and multichannel recordings: a critical view. In: Basar E, Bullock TH (eds) Brain dynamics. Springer Verlag, Berlin Heidelberg New York, pp 122–131

Bankman IN, Sigillito VG, Wise RA, Smith PL (1992) Feature-based detection of the K-complex wave in the human electroencephalogram using neural networks. IEEE Trans Biomed Eng 39: 1305–1310

Basar E, Basar-Eroglu C, Roschke J, Schutt A (1989) The EEG is a quasi-deterministic signal anticipating sensory-cognitive task. In: Basar E, Bullock TH (eds) Brain dynamics. Springer Verlag, Berlin Heidelberg New York, pp 43–71

Basar E (1992) Brain natural frequencies are causal factor for resonances and induced rhythms. In: Basar E, Bullock TH (eds) Induced rhythms in the brain, Birkhauser, Boston, pp 425–467

Berger H (1929) Ueber das Elektrenkephalogramm des Menschen. Arch Psichiatr Nervenkr 87: 527–570

Bressler SL, Coppola R, Nakamura R (1993) Episodic multiregional cortical coherence at multiple frequencies during visual task performance. Nature 366: 153–156

Creutzfeldt OD, Bodenstein G, Barlow JS (1985) Computerized EEG pattern classification by adaptive segmentation and probability density function classification. Clinical Evaluation. Electroencephalogr Clin Neurophysiol 60: 373–393

Cuffin BN (1996) EEG localization accuracy improvements using realistically shaped head models. IEEE Trans Biomed Eng 43:299–303

Daubechies I (1992) Wavelets. Philadelphia S.I.A.M.

Davidson RJ, Chapman JP, Chapman LJ, Henriques JB (1990) Asymmetrical brain electrical activity discriminates between psychometrically-matched verbal and spatial cognitive tasks. Psychophysiology 27: 528–543

Duffy FN (1981) Brain electrical activity mapping (BEAM): computerized access to complex brain function. Int J Neurosci 13: 55–65

Elbert T, Ray WJ, Kowalik ZJ, Skinner JE, Grag KE, Birbaumer N (1994) Chaos and physiology: deterministic chaos in excitable cell assemblies. Physiol Rev 74: 1–47

Fell J, Roschke J, Mann K, Schaffner C (1996) Discrimination of sleep stages: a comparison between spectral and nonlinear EEG measures. Electroencephalogr Clin Neurophysiol 98: 401–410

Florian G, Pfurtscheller G (1995) Dynamic spectral analysis of event-related EEG data. Electroencephalogr Clin Neurophysiol 95: 393–396

Freeman WJ (1988) Nonlinear neural dynamics in olfaction as a model for cognition. In: Basar E (ed) Dynamics of sensory and cognitive processing by the brain. Springer Verlag, Berlin Heidelberg New York, pp 19–28

Freeman WJ (1992) Predictions on neocortical dynamics derived from studies in paleocortex. In: Basar E, Bullock TH (eds) Induced rhythms in the brain, Birkhauser, Boston, pp 183–199

French CC, Beaumont JG (1984) A critical review of EEG coherence studies of hemisphere function. Int J Psychophysiol 1: 241–254

Gabor AJ, Seyal M (1992) Automated interictal EEG spike detection using artificial neural networks. Electroencephalogr Clin Neurophysiol 83:271–280

Gath I, Feuerstein C, Pham DT, Rondouin G (1992) On the tracking of rapid dynamic changes in seizure EEG. IEEE Trans Biomed Eng 39: 952–958

Gevins AS, Bressler SL (1988) Functional topography of the human brain. In: Pfurtscheller G (ed) Functional brain imaging. Hans Huber, Bern pp 99–116

Gevins AS (1996) High resolution evoked potentials of cognition. Brain Topogr 8: 189–99

Grassberger P, Procaccia I (1983) Measuring the strangeness of strange attractors. Physica 9D: 189–208.

Gratton G, Coles MGH, Donchin E (1983) A new method for off-line removal of ocular artifact. Electroencephalogr Clin Neuroph 55: 468–484.

Gray C, Singer W (1989) Stimulus-specific neuronal oscillations in orientation columns of cat visual cortex. Proc Natl Acad Sci USA 86: 1698–1702

Handbook of Electorencephalography and Clinical Neurophysiology (1987) Methods of analysis of brain electrical and magnetic signals. Gevins AS, Remond A (eds.) Vol 1. Amsterdam, Elsevier

Harner RN, Ostergren KA (1978) Computed EEG topography. In: Contemporary Clin Neurophysiol (EEG Suppl 34)

Ihl R, Dierks T, Froelich L, Martin EM, Maurer K (1993) Segmentation of the spontaneous EEG in dementia of the Alzheimer type. Neuropsychobiology 27: 231–236

Inouye T, Sakamoto H, Shinosaki K, Toi S, Ukai S (1990) Analysis of rapidly changing EEGs before generalized spike and wave complexes. Electroencephalogr Clin Neurophysiol 76: 205–221

Ilmoniemi RJ (1993) Models of source currents in the brain. Brain Topography Summer 5:331–336

Ivanitsky AM (1990) The consciousness and reflex. Zhurn Vyssh Nerv Deyat 40: 1058–1062 (in Russian).
Ivanitsky AM (1993) Consciousness: criteria and possible mechanisms. Int J Psychophysiol 14: 179–187
Ivanitsky GA, Nikolaev AR, Ivanitsky AM (1997) The application of artificial neural networks for thinking operation type recognition with EEG. Aerosp Envir Medic 31: 23–28 (In Russian)
Jansen BH (1996) Nonlinear dynamics and quantitative EEG analysis. Electroencephalogr Clin Neurophysiol Suppl 45: 39–56
Jasper H (1958) Report on committee on methods of clinical exam in EEG. Electroencephalogr Clin Neurophysiol 7: 370–375
Jervis BW, Coelho M, Morgan GW (1989) Spectral analysis of EEG responses. Med Biol Eng Comput 27: 230–238
Kasanovich YaB, Borisyuk RM (1994) The synchronization in neuronal network of the phase oscillators with the central element. Mathem Model 6: 45–60 (in Russian).
Klimesch W, Schimke H, Schwaiger J (1994) Episodic and semantic memory: an analysis in the EEG theta and alpha band. Electroencephalogr Clin Neurophysiol 91: 428–441
Koenig T, Lehmann D (1996) Microstates in language-related brain potential maps show noun-verb differences. Brain Lang 53: 169–182
Lantz G, Michel CM, Pascual-Marqui RD, Spinelli L, Seeck M, Seri S, Landis T, Rosen I (1997) Extracranial localization of intracranial interictal epileptiform activity using LORETA (low resolution electromagnetic tomography). Electroencephalogr Clin Neurophysiol 102:414–22
Law SK, Nunez PL, Wijesinghe RS (1993) High-resolution EEG using spline generated surface Laplacians on spherical and ellipsoidal surfaces. IEEE Trans Biomed Eng 40: 145–153
Lee S, Buchsbaum MS (1987) Topographic mapping of EEG artifacts. Clin Electroencephalogr 18: 61–67
Lehmann D (1987) Principles of spatial analysis. In: Gevins AS, Remond A (eds.) Handbook of Electorencephalography and Clinical Neurophysiology Methods of analysis of brain electrical and magnetic signals. Vol 1. Amsterdam, Elsevier
Lehmann D, Ozaki H, Pal I (1987) EEG alpha map series: brain micro-states by space-oriented adaptive segmentation. Electroencephalogr Clin Neurophysiol 67: 271–288
Livanov MN (1977) Spatial Organization of Cerebral Processes. New York: Wiley and Sons.
Llinas RR (1988) The intrinsic electrophysiological properties of mammalian neurons: insights into central nervous system function. Science 242: 1654–1664
Lopes da Silva FH, Storm van Leeuwen W (1977) The cortical source of the alpha rhythm. Neurosci Lett 6: 237–241
Lopes da Silva FH (1991) Neural mechanisms underlying brain waves: from neural membranes to networks. Electroencephalogr Clin Neurophysiol, 79: 81–93
Lutzenberger W, Elbert T, Birbaumer N, Ray WJ, Schupp H (1992) The scalp distribution of the fractal dimension of the EEG and its variation with mental tasks. Brain Topogr 5: 27–34
Lutzenberger W, Pulvermuller F, Birbaumer N (1994) Words and pseudowords elicit distinct patterns of 30-Hz EEG responses in humans. Neurosci Lett 176: 115–118
Lutzenberger W (1997) EEG alpha dynamics as viewed from EEG dimension dynamics. Int J Psychophysiol 26: 273–285
Madhavan PG, Stephens BE, Klingberg D, Morzorati S (1991) Analysis of rat EEG using autoregressive power spectra. J Neurosci Methods 40: 91–100
Madler C, Schwender D, Poeppel E (1991) Neuronal oscillators in auditory evoked potentials. Int J Psychophysiol 11: 55
Makeig S (1993) Auditory event-related dynamics of the EEG spectrum and effects of exposure to tones. Electroencephalogr Clin Neurophysiol 86: 283–293
Malsburg C vd (1981) The correlation theory of brain function. Intern Report 81–2. Department of Neurobiology Max Plank Institute for Biophysical Chemistry
Maurer K, Dierks T (1991) Atlas of brain mapping. Topographic mapping of EEG and evoked potentials. Springer Verlag, Berlin Heidelberg New York
Merrin EL, Meek P, Floyd TC, Callaway E (1990) 3D topographic segmentation of waking EEG in medication-free schizophrenic patients. Int J Psychophysiol 9: 231–236
Miller R (1991) Cortico-hippocampal interplay and the representation of contexts of the brain. Springer Verlag, Berlin Heidelberg New York
Mitzdorf U (1987) Properties of the evoked potential generators: current source-density analysis of evoked potential in cat cortex. Int J Neurosci 33: 33–59
Mountcastle V (1992) Preface. In: Basar E, Bullock TH (eds) Induced rhythms in the brain, Birkhauser, Boston, pp xvii-xix

Niedermayer E (1997) Alpha rhythms as physiological and abnormal phenomena. Int J Psychphysiol 26: 31–50

Nikolaev AR, Anokhin AP, Ivanitsky GA, Kashevarova OD, Ivanitsky AM (1996) The spectral EEG reconstructions and the cortical connections organization in spatial and verbal thinking. Zhurn Vyssh Nerv Deyat 46: 831–848 (in Russian).

Nuwer MR, Comi G, Emerson R, Fuglsang-Frederiksen A, Guérit JM, Hinrichs H, Ikeda A, Luccas FJC, Rappelsberger P (1998) IFCN Standards for digital recording of clinical EEG. Electroencephalogr Clin Neurophysiol 106: 259–261

Oosterom van A (1991) History and evolution of methods for solving the inverse problem. J Clin Neurophysiology 8:371–380

Palus M (1996) Nonlinearity in normal human EEG: cycles, temporal asymmetry, nonstationarity and randomness, not chaos. Biol Cybern 75: 389–396

Pascual-Marqui RD, Michel CM, Lehmann D (1994) Low resolution electromagnetic tomography: a new method for localizing electrical activity in the brain. Int J Psychophysiol 18:49–65

Pardey J, Roberts S, Tarassenko L (1996) A review of parametric modeling techniques for EEG analysis. Med Eng Phys 18: 2–11

Peltoranta M, Pfurtscheller G (1994) Neural network based classification of non-averaged event-related EEG responses. Med Biol Eng Comput 32:189–196.

Petsche H, Lacroix D, Lindner K, Rappelsberger P, Schmidt-Henrich E (1992) Thinking with images or thinking with language: a pilot EEG probability mapping study. Int J Psychophysiol 12: 31–39

Petsche H (1996) Approaches to verbal, visual and musical creativity by EEG coherence analysis. Int J Psychophysiol 24: 145–159

Pfurtscheller G, Klimesch W (1992) Event-related synchronization and desynchronization of alpha and beta waves in a cognitive task. In: Basar E, Bullock TH (eds) Induced rhythms in the brain. Birkhauser, Boston, pp 117–128

Pfurtscheller G, Neuper C, Mohl W (1994) Event-related desynchronization (ERD) during visual processing. Int J Psychophysiol 16: 147–153

Pfurtscheller G, Neuper C, Andrew C, Edlinger G (1997) Foot and hand area mu rhythms. Int J Psychophysiol 26: 121–135

Pietila T, Vapaakoski S, Nousiainen U, Varri A, Frey H, Hakkinen V, Neuvo Y (1994) Evaluation of a computerized system for recognition of epileptic activity during long-term EEG recording. Electroencephalogr Clin Neurophysiol 90: 438–443

Plendl H, Paulus W, Roberts IG, Botzel K, Towell A, Pitman JR, Scherg M, Halliday AM (1993) The time course and location of cerebral evoked activity associated with the processing of colour stimuli in man. Neurosci Lett 150: 9–12

Plonsey R (1982) The nature of sources of bioelectric and biomagnetic fields. Biophys J 39: 309–312

Pretorius HM, Bodenstein G, Creutzfeldt OD (1977) Adaptive segmentation of EEG records: a new approach to automatic EEG analysis. Electroencephalogr Clin Neurophysiol 42: 84–94

Priestley MB (1981) Spectral analysis and time series. Academic Press, London

Pritchard WS, Duke DW (1992) Dimensional analysis of no-task human EEG using the Grassberger-Procaccia method. Psychophysiology 29: 182–192

Pritchard WS, Krieble KK, Duke DW (1996) Application of dimension estimation and surrogate data to the time evolution of EEG topographic variables. Int J Psychophysiol 24: 189–195

Roth BJ, Ko D, von Albertini-Carletti IR, Scaffidi D, Sato S (1997) Dipole localization in patients with epilepsy using the realistically shaped head model. Electroencephalogr Clin Neurophysiol 102:159–66

Rusinov VS (1973) The dominant focus. Electrophysiological investigation. Consultants Bureau, New York London

Sabourin ME, Cutcomb SD, Crawford HJ, Pribram K (1990) EEG correlates of hypnotic susceptibility and hypnotic trance: spectral analysis and coherence. Int J Psychophysiol 10: 125–142

Scherg M, Berg P (1991) Use of prior knowledge in brain electromagnetic source analysis. Brain Topography 4:143–150

Scherg M, Ebersole JS (1994) Brain source imaging of focal and multifocal epileptiform EEG activity. Neurophysiol Clin 24: 51–60

Schiff SJ, Aldroubi A, Unser M, Sato S (1994) Fast wavelet transformation of EEG. Electroencephalogr Clin Neurophysiol 91: 442–455

Schupp HT, Lutzenberger W, Birbaumer N, Miltner W, Braun C (1994) Neurophysiological differences between perception and imagery. Cogn Brain Res 2: 77–86

Stam CJ, Jelles B, Achtereekte HA, van Birgelen JH, Slaets JP (1996) Diagnostic usefulness of linear and nonlinear quantitative EEG analysis in Alzheimer's disease. Clin Electroencephalogr 27: 69–77

Steriade M, Gloor P, Llinas RR, Lopes de Silva FH, Mesulem MM (1990) Report of IFCN Committee on Basic Mechanisms. Basic mechanisms of cortical rhythmic activity. Electroencephalogr Clin Neurophysiol 76: 481–508

Stevens A, Lutzenberger W, Bartels DM, Strik W, Lindner K (1997) Increased duration and altered topography of EEG microstates during cognitive tasks in chronic schizophrenia. Psychiatry Res 66: 45–57

Strelets VB, Alyeshina TD (1997) EEG rhythms disturbances and function impairments in different types of mental pathology. In: Third International Hans Berger Congress. Quantitative and topological EEG and MEG analysis. Friedrich Schiller University, Jena, pp. 161–165

Strik WK, Dierks T, Becker T, Lehmann D (1995) Larger topographical variance and decreased duration of brain electric microstates in depression. J Neural Transm Gen Sect 99: 213–222

Sviderskaya NE (1987) The synchronous brain electrical activity and mental processes. Nauka, Moscow (In Russian)

Tallon-Baudry C, Bertrand O, Delpuech C, Pernier J (1996) Stimulus specificity of phase-locked and non-phase-locked 40 Hz visual responses in human. J Neurosci 16: 4240–4249

Theiler J, Rapp PE (1996) Re-examination of the evidence for low-dimensional, nonlinear structure in the human electroencephalogram. Electroencephalogr Clin Neurophysiol 98: 213–222

Torrence C, Compo G (1998) A practical guide to wavelet analysis. Bul Am Meteorol Soc 79: 61–78

Viana Di Prisco G, Freeman WJ (1985) Odor-related bulbar EEG spatial pattern analysis during appetitive conditioning in rabbits. Behav Neurosci 99: 946–978

Wackermann J, Lehmann D, Michel CM, Strik WK (1993) Adaptive segmentation of spontaneous EEG map series into spatially defined microstates. Int J Psychophysiol 14: 269–283

Weiss S, Rappelsberger P (1996) EEG coherence within the 13–18 Hz band as a correlate of a distinct lexical organization of concrete and abstract nouns in humans. Neurosci Lett 209: 17–20

Wilson GF, Fisher F (1995) Cognitive task classification based upon topographic EEG data. Biol Psychol 40: 239–250

Wolf A, Swift JB, Swinney HL, Vastano JA (1985) Determining Lyapunov exponents from a time series. Physica D 16: 285.

Yvert B, Bertrand O, Echallier JF, Pernier J (1995) Improved forward EEG calculations using local mesh refinement of realistic head geometries. Electroencephalogr Clin Neurophysiol 95:381–92

Chapter 36

Modern Techniques in ERP Research

Daniel H. Lange and Gideon F. Inbar

General Introduction

Evoked Potentials (EP) are defined as averaged electric responses of the nervous system to sensory stimulation (Gevins 1984). They consist of a sequence of transient waveforms, each with its own morphology, latency and amplitude. In clinical settings, EPs are elicited by visual or auditory stimulation, or by electric stimulation of sensory nerves (Chiappa 1983). These EPs are usually recorded from the scalp, although in special cases like during brain surgery, electrodes may be placed on the surface of the brain or even deep in brain tissue. The term Event Related Potential (ERP) is now commonly used to denote both EP as well as other brain responses that are the result of cognitive processes accompanying and following external stimuli, or of preparatory mechanisms preceding motor action. However, due to historical reasons and to avoid confusion, *we shall generally use the term EP for all types of brain responses including ERP.*

The shape, size, and timing of an EP recorded from the scalp may depend on many factors, including the duration of the potential scalp responses to single stimuli. These responses are usually of a very low amplitude, with characteristic Signal to Noise Ratio (SNR) in the order of 0 dB down to –40 dB (e.g. Visual EPs ~ 0 dB, Cognitive EPs ~ –5 dB, Movement Related EPs ~ –20 dB, and Brainstem EPs ~ –40 dB), and thus may be partly or totally obscured by the ongoing background EEG activity. The conventional method of EP extraction is synchronous averaging of repeatedly elicited responses, where the uncorrelated EEG contribution averages out, and thus enhancing the neural activity that is time-locked to the stimulus (see Fig. 1). This is a valid procedure, provided that the time-locked responses remain identical throughout the session (Aunon et. al. 1981; Rompelman and Ross 1986a,b). In practice, however, responses are never identical and trial to trial variability may in fact be quite substantial (Popivanov and Krekule 1983). Signal variability may be encountered due to a variety of reasons, such as variability due to different behavioral outcome with identical stimuli, or progressive changes in the evoked potential morphology due to factors like sensory adaptation or variable performance (Rompelman and Ross 1986a,b). Therefore, the basic assumption underlying signal averaging is usually violated, though its effect may be somewhat reduced with moving-average techniques. This prevents tracking and analysis of dynamic brain processes, which calls for improved methods that would enable analysis of evoked potential waveforms on a *single-trial* basis.

It should be realized that evoked brain responses generally constitute complex waveforms which may include several, possibly overlapping, signal components. Two basic approaches are used for defining EP components. The first is based on peak analysis, where a peak is defined as the most positive or negative voltage within a specified time

Daniel H. Lange, Hewlett Packard Laboratories-Israel, Haifa, 32000, Israel (e-mail: lange@hpl.hp.com)
Gideon F. Inbar, Technion – IIT, Department of Electrical Engineering, Haifa, 32000, Israel (phone: +972-4-82 94 718; fax: +972-4–8323041; e-mail: inbar@ee.technion.ac.il)

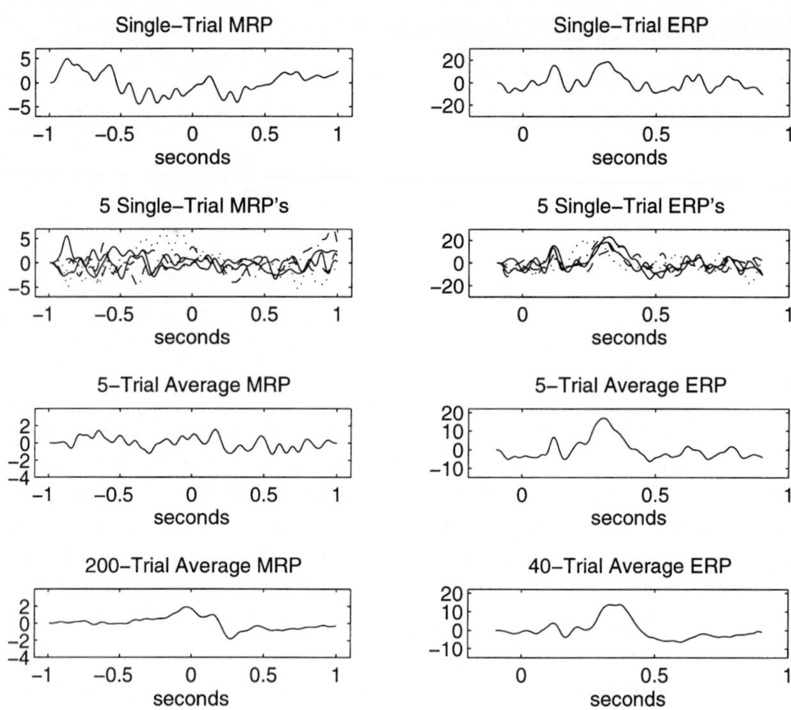

Fig. 1. Examples of MRP (left) and ERP (right) signals. The MRPs and ERPs are characterized by low SNR in the order of –15dB and –5dB, respectively. First row: Raw measurements. Second row: Five raw measurements superimposed. Third row: 5-trial averages. Last row: Ensemble averages.

interval. The second approach, adopted throughout this chapter, attempts to relate the concept of a component to neuronal populations of specific localization and electric orientation which become active during the performance of specific processing. The decomposition problem of the global response into its constituent components cannot be solved without introducing constraints to reduce the dimensionality of this ill-posed inverse problem. This issue becomes essential if variability analysis of EPs is required (Donchin 1966).

The first part of this chapter presents a review of recent developments in the field of EP processing, with emphasis on single-trial estimation methods. The review is not technically oriented, and practical issues regarding implementability should be drawn from the cited literature. The second part of the chapter presents a novel approach to single-trial EP processing, developed recently by the authors of the chapter. The proposed approach is discussed in detail, aiming at laying a solid basis for implementation by the reader. However, additional information may be found in other recent publications (Lange et al. 1997; Lange et al. 1998).

Part I: Review of EP Processing Methods

Introduction

Evoked Potential estimation and classification have been sporadically dealt with in the Biomedical Engineering literature. Most publications have addressed the issue of improving the Signal to Noise Ratio (SNR) of the *averaged* response rather than the extraction of single brain responses, probably due to the objective difficulty of analyzing the faint evoked responses embedded within the spontaneous cerebral activity. We shall present a brief review of the historical development of EP processing methods, from the conventional averaging via spectral filtering to the modern single-trial techniques.

Processing Methods

Averaging Methods

Ensemble averaging has been the traditional and most common method for EP analysis (Rompelman and Ross 1986a). In spite of its well-known limitations, including but not limited to its restricted ability to adequately estimate trial-varying responses (Rompelman and Ross 1986b), quite surprisingly averaging is still the most popular EP analysis method. The reasons for this apparent conservatism may vary – from the simplicity of averaging and the consistent results it produces, via limited technical resources of neurophysiologists to apply and assess the highly complicated methods, to the unclear cost-benefit value in using the more advanced methods.

The first attempt to relax the invariant time-locked signal constraint was based on *cross-correlation averaging* to compensate for latency variation jitter of the evoked responses (Woody 1967). The single responses are re-aligned according to cross-correlation measures and improved templates can be obtained. However, cross-correlation methods are highly sensitive to the fall of SNR, breaking down for SNR levels lower than around 5 dB. In addition, this method does not compensate for changes in response morphology, which limits the analysis of variable responses. Cross-correlation averaging was later expanded to allow shifts of single components; however, this made the analysis even more sensitive to low SNR and produced irregularities due to the artificial breaking of the signal into components. Consequently, several *latency correction procedures* have been proposed, among the more recent ones those described in Gupta et al. (1996), Kaipio and Karjalainen (1997), Kong and Thakor (1996), Meste and Rix (1996), Nakamura (1993), and Rodriguez et al. (1981).

Filtering Methods

The first SNR improvement technique that was proposed for EP enhancement applied *Wiener filtering* to averaged EPs, by calculating the filter coefficients from the averaged spectra and the spectrum of the averaged response (Walter 1969; Doyle 1975):

$$H_N(w)=\frac{\phi_{ss}(w)}{\phi_{ss}(w)+\frac{1}{N}\cdot\phi_{xx}(w)}$$

where ϕ_{ss}, ϕ_{xx} are the spectral density functions of the signal (EP) and noise (EEG), respectively, and N accounts for the number of trials used in the averaging process. The motivation to use *a posteriori* filtering is based on the multi-dimensional nature of EP estimation. Rather than estimating a noisy scalar signal, where the average may well be the maximum likelihood estimate (e.g., with Gaussian noise), it may be possible to improve beyond averaging due to the coherence of the estimated waveform. Yet the success has been controversial mainly due to the inherent difficulty of spectral estimation of the transient responses (Carlton and Katz 1987). A time-varying *a posteriori* Wiener filtering structure was also suggested and its performance shown to be superior to time-invariant filtering (Yu and McGillem 1983). The time-varying Wiener filter was enhanced by using constant relative bandwidth filters, followed by time-varying attenuators and a summing network controlled by a time-varying SNR estimator in the corresponding frequency bands (de Weerd 1981). Other *a posteriori* methods have also been developed (e.g., Furst and Blau 1991; Lange et al. 1995), yielding similar improvements to those obtained with Wiener filtering.

Some improvements in the capability of tracking steady-state EPs have been obtained by *adaptive filtering* methods (Thakor 1987). Several adaptive algorithms have evolved from the latter approach, concerned mainly with outperforming averaging in tracking of steady-state EPs (e.g., Laguna et al. 1992; Svensson 1993; Vaz and Thakor 1989).

Parametric Methods

The first attempt to extract transient evoked potentials on a single-trial basis utilized an *ARX* (AutoRegressive with eXogeneous input) model with the average response driving the exogenous input (Cerutti et al. 1987). The basic idea was using the averaged response as a model for the single response, extracted via identification of the model parameters. Elimination of eye-movement artifacts was also incorporated using a similar model (Cerutti et al. 1988), which was used later for topographic mapping of multichannel single-trial brain responses (Liberati et al. 1992). An improvement to the *ARX* estimator was recently proposed, enhancing the conditions for the EP filter identification and increasing robustness to the strong ongoing activity (Lange and Inbar 1996a).

A different parametric model for single-trial EP extraction was proposed at almost the same time (Spreckelsen and Bromm 1988), which used autoregressive (AR) modeling for the EEG and an impulse response model of a linear system for the EP. The average response was used as the filter impulse response, allowing for amplitude and latency variations of each single-trial response with respect to the averaged response.

Evoked potential reconstruction by means of a *Wavelet transform* was recently proposed (Bartnik et al. 1992). The approach was to compare the wavelet representation of pre- and post-stimulus interval signals, and use the differentiating coefficients for reconstruction of the additive EP contribution. The principle underlying this approach is not new and was already investigated previously (e.g., Madhaven 1992). However, the proposed implementation is novel, based on decomposing the noisy signals via decreasing scale. The process is comparable to taking a picture from increasing distances by a factor of α. On the *j*-th step the resolution decreases by a factor of α^j, where the picture of α^j resolution can be restored from the higher detailed picture of α^{j-1} resolution. The algorithm projects the background activity of the pre- and post-stimulus signals onto an orthonormal basis derived from a Wavelet function, and finds the principal basis components which best differentiate the two projections, from which it constructs the single response. A major problem lies in identifying the optimal scaling function $\Phi(x)= \alpha^j\Phi(\alpha^j x)$ to fit the signals at hand, and a direct use of the average response is necessary to determine $\Phi(x)$. A theoretical drawback may be the critical assumption that any irregularities during the epoch must be related to the specific time-locked task, an assumption which cannot be validated.

Single-trial extraction using statistical outlier information has also been attempted (Birch et al. 1993), generalized later to extraction of low SNR events (Mason et al. 1994). The *outlier processing method* is based on building an AR model whose parameters represent the underlying ongoing EEG process. The model was identified by using a robust parameter estimation method, suppressing non-conforming activity during the model building process. Then the model was used to reconstruct the ongoing EEG, where its difference from the measured process was considered to be the additive outlier content.

The method's advantage lies in its minimal assumption about the signal, requiring only that it last no longer than about 20% of the entire epoch. As the percentage increases, so do the inaccuracies due to the AR model builder sensitivity to outlier contamination of the EEG. This method suffers from the same theoretical drawback as the previous approach, due to the inherent variability of the EEG signal.

Current State of Modern EP Analysis

Current single-trial extraction methods may generally be divided into two main categories. The first includes template-based Linear Time Invariant (LTI) models, displaying some success with responses characterized by small trial-to-trial variabilities (e.g., brainstem signals), as well as with variability of global magnitude and/or phase of the evoked responses. The second category includes methods which attempt to reconstruct the EP by means of identification of statistical changes of the EEG signal over the transition from pre-stimulus to post-stimulus interval; these methods generally suffer from high estimation variances, probably due to physiological signal contamination of the background EEG.

Generally a tendency towards parametric techniques is observed, as spectral methods are inadequate to deal with the spectral overlap of the EP and EEG at the unfavorable SNR characteristic of EP signals. In the following, we shall describe a novel technique recently developed by our team which enables tracking of evoked potential component variations on a trial-to-trial basis with SNRs of down to −20 dB. Our method presents a pioneering approach to the single-trial estimation problem, allowing for the first time to track single-trial EP components.

Part II: Extraction of Trial-Varying EPS

Introduction

When using the term *evoked potential*, one usually refers to the *average* evoked response, obtained by averaging many synchronized brain responses time-locked to a repeating stimulus or event. The single evoked response has not been useful since it is embedded deep within massive background brain activity. Conventional EP processing relies on several major assumptions, whose validity may depend on factors related to the experimental paradigm as well as to the subject's state of mind:

- A repetitive, invariable time-locked signal (EP)
- A stochastic, stationary noise (EEG)
- Additivity of signal and noise
- Uncorrelated signal and noise

The first assumption is generally not valid, except in some special cases, and is often the source of error in many template-based processing methods which rely heavily on the average response as a template for analysis. Examples of EP variability are numerous – the responses are known to be influenced by the mental state of the subject, fatigue, habituation, level of attention, quality of performance, and so on. This assumption, which is considered to be the main drawback of most EP processing methods, is relaxed in the method presented in the following.

The second assumption means that the spontaneous EEG is uncorrelated over long time periods. It is easily verifiable by EEG correlograms demonstrating the non-correlation for periods in the order of 1–2 seconds, and is also adopted here.

The third assumption, which is more of a convention, states that the evoked response is added to the background activity without affecting it, so that any part of the response that is influenced by the stimulus is incorporated in the EP. This is a common assumption underlying almost all EP analysis frameworks, and is adopted in this work.

The fourth assumption is essential for methods based on correlation, as phase matches of EEG and EP would give rise to errors in estimating single EP waveforms. However, it should be noted that, by definition, the sum of all EEG components which are correlated with the stimulus are regarded as the EP, making it more of a definition than an assumption. This assumption is also adopted in this work. In the following, we propose a radically new approach to the single-trial evoked potential estimation problem; relaxing substantially the constraint of the above assumption regarding EP invariability, a novel framework for dynamic EP extraction on a single-trial basis is presented.

Proposed Framework

The proposed framework consists of three data-driven serial layers (Fig. 2): (1) a pattern identification layer, consisting of an unsupervised identification mechanism, (2) a statistical decomposition layer, consisting of a linear decomposition unit, and (3) a parametric synthesis unit based on the additivity of the EEG and EP contributions. Extensive

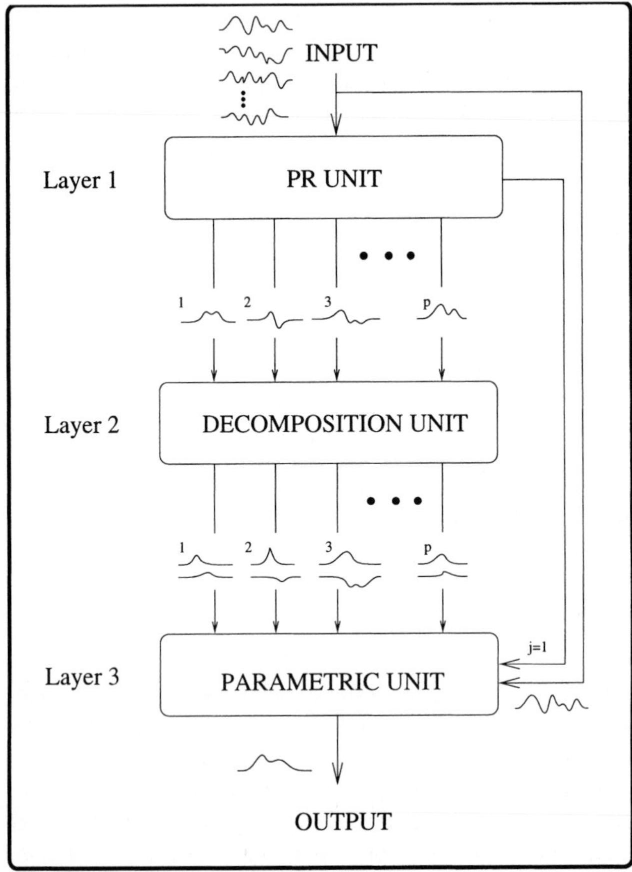

Fig. 2. A signal flowchart of the proposed processing framework. The ensembled data are identified and classified by the pattern recognition (PR) unit, the resulting library response items are decomposed by the decomposition unit, and finally the single sweep recording is modeled using its respective decomposed library response item yielding the single-trial EP estimate.

analysis and performance evaluation studies can be found in our recent publications (Lange and Inbar 1996b; Lange et al. 1997). The functionality of the layers can be summarized as follows:

- The first layer consists of an unsupervised learning structure, in the form of a competitive artificial neural network (ANN), which is used to dynamically identify the various response types embedded within the recorded ensemble. The constraint of this layer lies in the assumption that the single brain responses are not randomly distributed, but rather belong to a relatively small family of brain responses. Identifying the waveform family, the network spontaneously builds a library of the embedded data types.
- The second layer consists of a linear statistical decomposition scheme, responsible for separation of the identified library items into constituent components. The purpose of decomposition is to allow certain variability of each response with relation to its respective library item, and thus to enable tracking of trial-to-trial variability of the evoked responses. An additional assumption employed has to do with the statistical nature of cortical neural activity. Gaussian distributions are used to model the firing instants of a synchronously activated ensemble of cortical neurons. As the true neural activation characteristics are obviously unknown, and any assumption regarding the statistics of cortical neural activation might be severely violated, a robust method which does not inflict significant distortions on the separated signal components is used.
- The third and final layer consists of a parametric synthesis model, emulating the single sweep generation mechanism. Dealing with low SNR signals, the model relies on preliminary information extracted via the previous processing layers, utilizing the identified EP waveform types. Once its parameters are identified, the model generates the single sweep recording from its assumed contributions – the EEG and EP, where the identified EP is the desired output of the integrated system.

Layer 1 – Unsupervised Learning Structure

Introduction

Machine learning can be implemented by means of a dynamic neural structure, which has an ability to learn from its environment, and through learning to improve its own performance (cf. Chapter 25). Unsupervised learning, in the form of a self-organizing neural structure, is used to discover significant patterns or features in the input data through a spontaneous learning paradigm. To achieve such spontaneous learning, the algorithm is equipped with a set of rules of a local nature, which enable it to learn its environment via some sort of mapping with specific desirable properties (Duda and Hart 1976).

The learning process is based on positive feedback or self re-enforcement, stabilized by means of a fundamental rule: An increase in strength of some synapses in the network must be compensated for by a decrease in other synapses. In other words, a competition takes place for some limited resources, preventing the system from exploding due to the positive feedback-based learning process. We have chosen to use Competitive Learning, a well-established branch of the general theme of unsupervised learning, implemented so that the network weights actually converge to the embedded signal types (Lange et al. 1998). The elementary principles of competitive learning are (Rumelhart and Zipser 1985):

- Start with a set of units that are all the same except for some randomly distributed parameter which makes each of them respond slightly differently to a set of input patterns.

- Limit the "strength" of each unit.
- Allow the units to compete in some way for the right to respond to a given subset of inputs.

Applying these three principles yields a learning paradigm in which individual units learn to specialize on sets of similar patterns and thus become "feature detectors". Competitive learning is a mechanism well-suited for regularity detection (Haykin 1994), where there is a population of stimulus patterns, each of which is presented with some probability. The detector is supposed to discover statistically salient features of the input population, without requiring an *a priori* set of categories into which the patterns should be classified. Thus the detector needs to develop its own featural representation of the population of input patterns capturing its most salient features.

Theory

A typical architecture of a competitive learning system appears in Fig 3. The system consists of a set of hierarchically layered neurons in which each layer is connected via excitatory connections to the following layer. Within a layer, the neurons are divided into sets of inhibitory clusters in which all neurons within a cluster inhibit all other neurons in the cluster, resulting in a competition among the neurons to respond to the pattern appearing on the previous layer. The stronger a neuron responds to an input pattern, the more it inhibits the other neurons of its cluster.

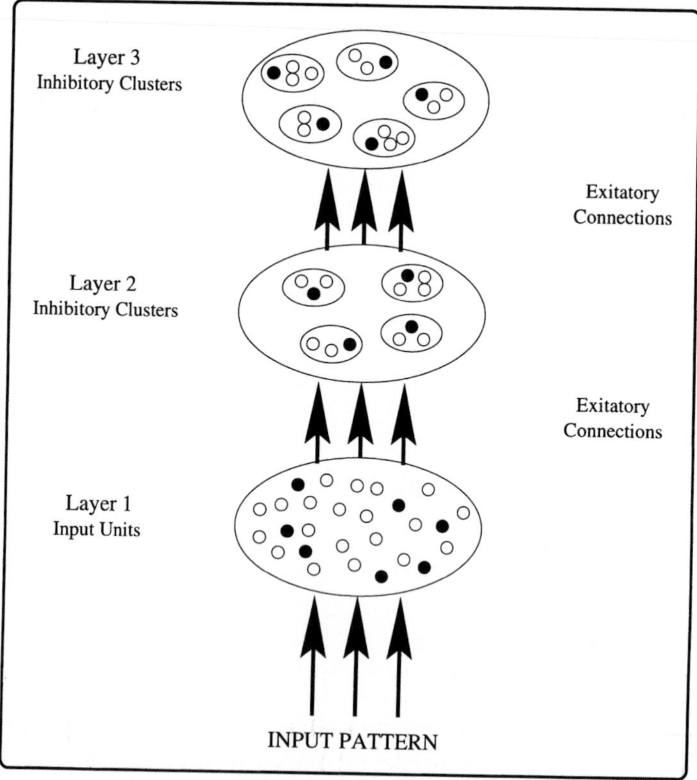

Fig. 3. The architecture of a competitive learning structure. Competitive learning takes place in a context of hierarchically layered units, which are presented as filled (active) and empty (inactive) dots. The winning neurons suppress the activity of neighboring neurons while exciting the following layers.

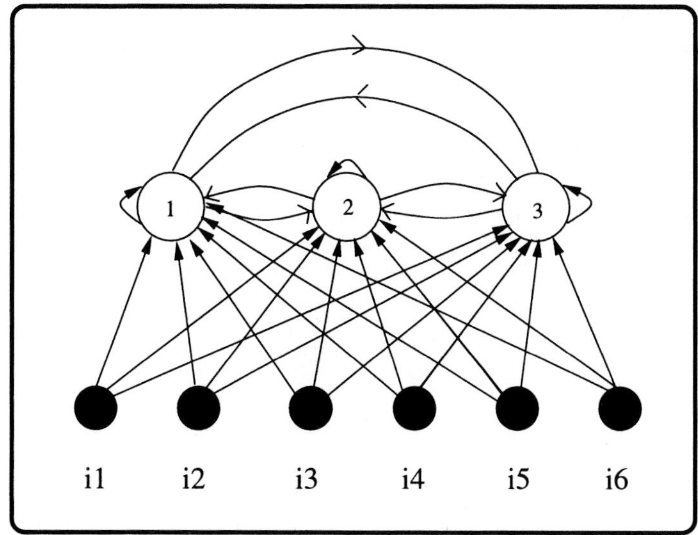

Fig. 4. A single-layer competitive learning ANN (6-input nodes, 3-output neurons). Open arrows represent inhibitory connections and dark arrows represent excitatory connections. Lateral inhibition is implemented to suppress the activity of neighboring neurons.

There are many variations of the competitive learning scheme. We have selected a single-layer structure, where the output neurons are fully connected to the input nodes and the non-linearity is implemented in the learning phase only. The advantage of using this structure lies in enhanced analysis capabilities of the converged network, as the weights actually converge to the embedded signal patterns and thus form a Pattern Identification network. The general network structure is depicted in Fig. 4.

For neuron j to be the winning neuron, its net internal activity level v_j for a specified input pattern x_i must be the largest among all neurons in the network. The output signal y_j of a winning neuron j is set equal to one, and all other neuron outputs that lose the competition are set equal to zero.

Let w_{ji} denote the synaptic weight connecting input node i to neuron j. Each neuron is given a fixed amount of synaptic energy, which is distributed among its input nodes as follows:

$$\sum_i w_{ji}^2 = 1 \quad , \quad \text{for all } j$$

A neuron learns by shifting synaptic weights from its inactive to active input nodes. If a neuron does not respond to some input pattern, no learning occurs in that neuron. When a single neuron wins the competition, each of its input nodes gives up some proportion of its synaptic weight, which is distributed equally among the active input nodes. According to the standard competitive learning rule for a winning neuron, the change Δw_{ji} applied to synaptic weight w_{ji} is defined by:

$$\Delta w_{ji} = \eta\left(x_i - w_{ji}\right)$$

where η is the learning-rate coefficient. The effect of this rule is that the synaptic weight of a winning neuron is shifted towards the input pattern. In our case, where the signals are assumed to be embedded within an additive Gaussian noise, the network weight structure converges to a matched filter bank, operating as a pattern identification network as well as an optimal signal classifier.

Finally, a common problem with random initialized competitive networks is the phenomenon of *stuck vectors*. The training process may result, in extreme cases, in all weight vectors but one becoming stuck, sometimes also referred to as dead neurons. A single weight vector may always win and the network would not learn to distinguish between any of the classes. This happens because of two reasons (Freeman and Skapura 1992): First, in a high-dimensional space, random vectors are all nearly orthogonal, and second, all input vectors may be clustered within a single region of the space. One solution of this problem is to include a variable bias, in the form of a time-constant, giving advantage to neurons which rarely or never win over neurons which always win. The bias of a dying neuron is increased proportionally to the number of winnings of the other neurons, and decreased after it gains victory over the other neurons to allow fair competition. This bias is used during training only and discarded thereafter.

Statistical Evaluation

Identification Property

The essential identification feature of the proposed network is, ideally, an inherent convergence of the network weights to the embedded EP waveforms, thus operating as a Pattern Identification network. In an optimal scenario, in which the embedded EP patterns and background EEG are uncorrelated, each of the competing neurons tends to fixate on a different signal type by mapping itself to a specific signal waveform. It is important to stress here that the following mathematical formulations are limited to a single-neuron system detecting a deterministic embedded signal, or equivalently, to an optimal multi-neuron system where each neuron responds only to its matching input. Despite this over-simplification, the following analysis provides general assessment of the system performance.

Each single measurement can be represented as follows:

$$x_i(t) = \sqrt{E_i} s_i(t) + e_i(t), \quad i = 1,2,\ldots,N$$

where x_i, E_i and e_i represent the recorded i-th single sweep, the energy of the i-th EP, the normalized EP waveshape and the embedding background EEG, respectively. P and N refer to the number of signal categories and number of trials. We assume $P<N$ correctly identified signal categories, where

$$s_i \in \{S_j\}_{j=1}^{P},$$

and use normalized inputs and weights as we are interested only in waveshape modifications. Using a Gaussian model for the background EEG, it can be shown that in each iteration the winning neuron shifts its weights towards the respective EP pattern. First we calculate the neural outputs:

$$o^k = \langle x_i, \omega^k \rangle, \quad k = 1\ldots P$$

Then, selecting the winning neuron $l = \arg\max\{o^k\}$, we update the weights of the winning neuron only:

$$\omega_n^l = \omega_{n-1}^l + \eta \cdot \left(x_i - \omega_{n-1}^l\right)$$

The winning neuron's output to a matching single-trial measurement increases monotonically (note: $|o^l(\cdot)| \leq 1$):

$$
\begin{aligned}
o_n^l &= \left\langle x_i, \omega_{n-1}^l + \eta \cdot \left(x_i - \omega_{n-1}^l\right)\right\rangle \\
&= o_{n-1}^l + \left\langle x_i, \eta \cdot \left(x_i - \omega_{n-1}^l\right)\right\rangle \\
&= o_{n-1}^l + \left\langle x_i, \eta \cdot \left\langle x_i, x_i\right\rangle - \eta \cdot \left\langle x_i, \omega_{n-1}^l\right\rangle\right\rangle \\
&= o_{n-1}^l + \left\langle x_i, \eta \cdot \left(1 - o_{n-1}^l\right)\right\rangle
\end{aligned}
$$

and thus:

$$o_n^l \geq o_{n-1}^l$$

It should stressed that this result relies on an oversimplified model which assumes strictly correct input clustering. In real-life conditions this assumption may be violated, and monotonic behavior can only be achieved after an initial transient learning phase, its length depending on the SNR. The clustering performance depends also on intra- versus intercluster variability, its principles discussed in detail in Duda and Hart, 1976.

Hence it was shown that matching neurons approach the embedded patterns monotonically, and thus each neuron will finally converge to a respective embedded ERP pattern; the signal identification process yields a matched filter bank, whose classification performance can be analyzed from an information-theoretic perspective. The pattern identification procedure is unbiased and the variance could be made as small as desired by decreasing the learning rate coefficient, as presented in the following sections.

Identification Bias

The competing neurons are mapped to the input space, and the correlation of the neurons' weights with their matching input patterns is ever increasing, i.e., the learning processes of the competing neurons are assumed to be independent. Again, it should be stressed that this is a best-case analysis, valid only under the strict assumption of independence. It is thus sufficient to evaluate a single neuron system detecting a constant signal pattern embedded within random noise realizations, although in real life a multidimensional system will probably suffer from some degree of cross effects biasing the solution.

Recalling the competitive learning rule, applied to the winning neuron, we have:

$$\omega_n = \omega_{n-1} + \eta \cdot \left(x_i - \omega_{n-1}\right); \quad 0 < \eta \ll 1$$

where x_i is an arbitrary input vector. Rearranging and using the additive signal and noise model yields:

$$\omega_n = \omega_{n-1} \cdot (1-\eta) + \eta \cdot \left(s + e_i\right),$$

where s and e_i represent the embedded signal pattern and the embedding noise realizations, respectively. This difference equation can be solved as follows:

$$
\begin{aligned}
\omega_n &= (1-\eta)^n \cdot \omega_0 + \begin{bmatrix} (1-\eta)^{n-1} \cdot \eta \cdot (s+e_i) + (1-\eta)^{n-2} \cdot \eta \cdot (s+e_2) \\ + \ldots + (1-\eta) \cdot \eta \cdot (s+e_{n-1}) + \eta \cdot (s+e_n) \end{bmatrix} \\
&= (1-\eta)^n \cdot \omega_0 + \eta \cdot \sum_{i=0}^{n-1} (1-\eta)^i \cdot (s+e_{n-i}) \\
&= (1-\eta)^n \cdot \omega_0 + \eta \cdot s \cdot \sum_{i=0}^{n-1} (1-\eta)^i + \eta \cdot \sum_{i=0}^{n-1} (1-\eta)^i \cdot e_{n-i} \\
&= (1-\eta)^n \cdot \omega_0 + s \cdot \sum_{i=0}^{n-1} \left(1 - (1-\eta)^n\right) + \eta \cdot \sum_{i=0}^{n-1} (1-\eta)^i \cdot e_{n-i}
\end{aligned}
$$

Taking the limit as n approaches infinity, where $0 < \eta < 1$, $\tilde{e}_i = e_{n-i}$, yields:

$$\omega_\infty = s + \eta \cdot \sum_{i=0}^{\infty} (1-\eta)^i \cdot \tilde{e}_i$$

and calculating the expected mean value provides the unbiased result:

$$E[\omega_\infty] = s + \eta \cdot \sum_{i=0}^{n-1} (1-\eta)^i \cdot E[\tilde{e}_i]$$

$$E[\omega_\infty] = s$$

Identification Variance

Assuming zero-mean Gaussian EEG with σ^2 variance, and recalling the solution of the learning rule equation:

$$\omega_n = s + \eta \cdot \sum_{i=0}^{n-1} (1-\eta)^i \cdot \tilde{e}_i,$$

we can calculate the identification variance (I denotes the identity matrix):

$$\begin{aligned} E(\omega_n - s)(\omega_n - s)^T &= E\left[\left(\eta \cdot \sum_{i=0}^{n-1} (1-\eta)^i \cdot \tilde{e}_i\right)\left(\eta \cdot \sum_{i=0}^{n-1} (1-\eta)^i \cdot \tilde{e}_i^T\right)\right] \\ &= \eta^2 \cdot \sum_{i=0}^{n-1} \sum_{j=0}^{n-1} (1-\eta)^i (1-\eta)^j \cdot E\left[\tilde{e}_i \cdot \tilde{e}_i^T\right] \\ &= \eta^2 \cdot \sum_{i=0}^{n-1} (1-\eta)^{2i} \cdot \sigma^2 \cdot I \\ &= \eta^2 \cdot \frac{1-(1-\eta)^{2n}}{1-(1-\eta)^2} \cdot \sigma^2 \cdot I \end{aligned}$$

Taking the limit as n approaches infinity yields the asymptotic identification variance:

$$C_{\omega\omega} = E(\omega_n - s)(\omega_n - s)^T = \frac{\eta}{2-\eta} \cdot \sigma^2 \cdot I$$

Bounding the Learning Rate

Decreasing the learning rate during training is a common method to ensure convergence. However, this fine tuning might obscure the fact that there may well be several local minima and the obtained solution might not be the optimal one. In this work we have chosen to quantify the problem as follows: the learning rate coefficient directly influences the amount of weight fluctuations in steady state due to the variance of estimation, and thus should be kept small. We use it to limit the power of noise affecting the steady-state solution by providing an upper bound on the learning rate which ensures high-quality estimations. Requiring $C_{\omega\omega} \le \alpha \cdot E \cdot I$ where $C_{\omega\omega}$, α and E are the estimation covariance matrix, a distortion coefficient, and the energy of the signal yields:

$$\frac{\eta}{2-\eta} \cdot \sigma^2 \le \alpha \cdot E$$

and solving for η provides the bound:

$$\eta \le \frac{2\alpha \dfrac{E}{\sigma^2}}{1 + \alpha \dfrac{E}{\sigma^2}}$$

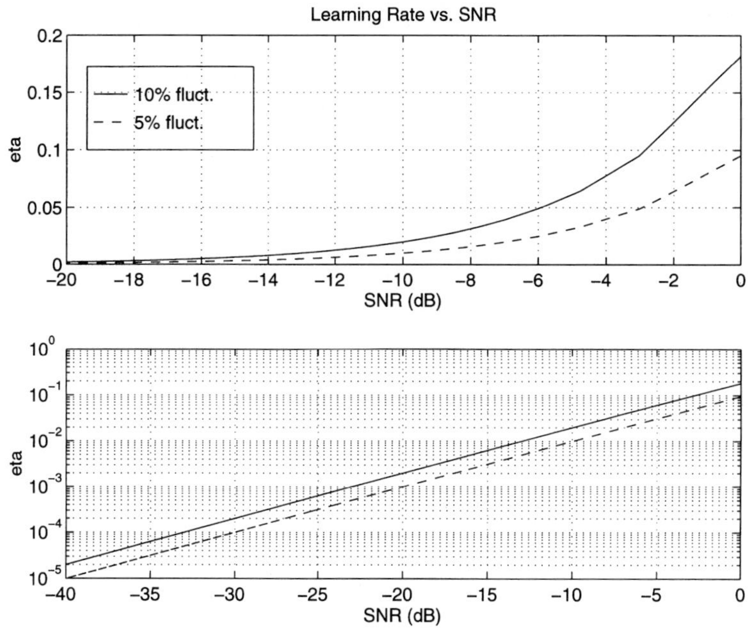

Fig. 5. Maximum Learning Rate as a function of the SNR, at noise fluctuation levels of 10% (solid) and 5% (dashed) of the signal energy (top: linear scale, bottom: logarithmic scale). The noise fluctuations decrease with lowering of the learning rate, at the expense of an extended learning cycle.

e.g., for SNRs of 0dB,–20dB and –40dB, and for a distortion coefficient of α=0.1, the learning rate coefficient should not exceed 0.18, 0.0198 and 0.002, respectively (see Fig. 5).

Learning Cycle Duration

The learning process is a non-linear one, and the convergence time depends on several unknown factors like the degree of correlation among the various signal patterns. However, it is possible to provide a "thumb rule" for the required learning cycle duration, which may prove useful at least for initial assessment of the data at hand.

Referring to the estimation variance, an exponential envelope of time constant τ_n can be fitted to the geometric term by assuming the unit of time to be the duration of one iteration cycle, and by choosing a time constant τ_n such that:

$$(1-\eta) = \exp\left(-\frac{1}{\tau_n}\right)$$

thus τ_n can be expressed in terms of the learning rate coefficient η:

$$\tau_n = -\frac{1}{\ln(1-\eta)}$$

The time constant τ_n defines the time required for a decay of the noise contribution to $1/e$ of its initial value for a single neuron. For multiple neurons, the required learning cycle should be multiplied by the number of competing neurons.

With low SNR EPs dictating slow learning rates ($\eta \ll 1$), the time constant τ_n may be approximated by (P denotes the number of competing neurons):

$$\tau_n \approx \frac{P}{\eta}$$

Example 1

An important task in EP research is the identification of effects related to cognitive processes triggered by meaningful versus non-relevant stimuli (e.g., Pratt et al. 1989). A common procedure to study these effects is the classic *odd-ball paradigm*, where the subject is exposed to a random sequence of stimuli and is instructed to respond only to the task-relevant stimuli (also referred to as Target stimuli). Typically, the brain responses are extracted via selective averaging of the recorded data, and ensembled according to stimulus context. This method of analysis assumes that the brain responds equally to the members of each type of stimulus. However, the validity of this assumption is unknown in the above case where cognition itself is studied. Using the proposed approach, *a priori* grouping of the recorded data is not required, thus overcoming the above severe assumption on cognitive brain function. The experimental paradigm and the identification results of applying the proposed method are described in the following.

Cognitive event-related potential data were acquired during an odd ball-type paradigm from electrode Pz referenced to the mid-lower jaw (Jasper 1958; see Chapter 35), with a sample frequency of 250 Hz. The subject was exposed to repeated visual stimuli, consisting of the digits "3" and "5", appearing on a PC screen. The subject was instructed to press a push-button upon the appearance of "5" – the Target stimulus – and to ignore the appearances of the digit "3" (Lange et al. 1995).

With odd ball-type paradigms, the Target stimulus is known to elicit a prominent positive component in the ongoing brain activity, related to the identification of a mean-

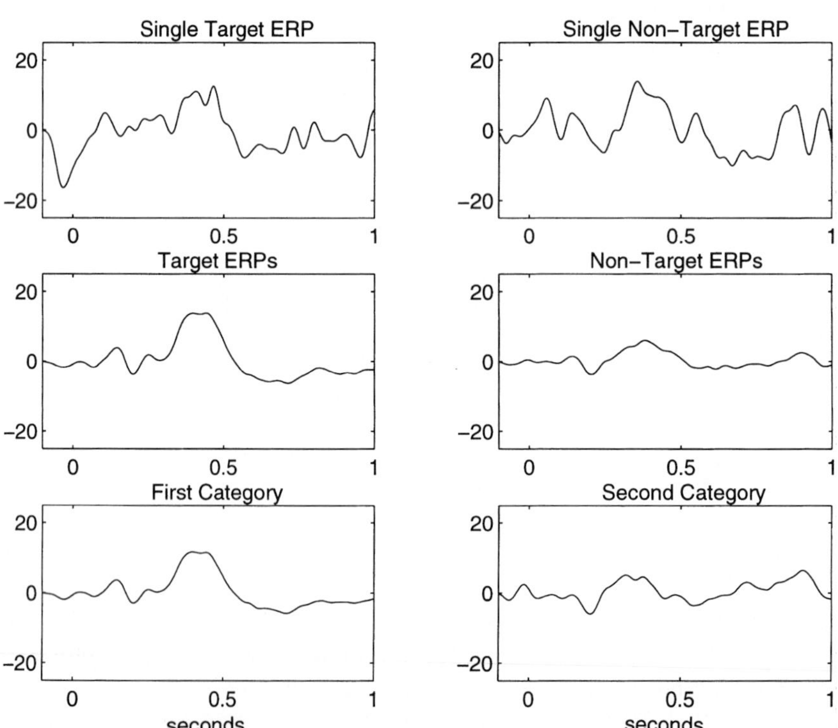

Fig. 6. Stimulus-related selective averaging versus spontaneous categorization. Top row: Sample raw Target and Non-Target sweeps. Middle row: Target and Non-Target ERP templates. Bottom row: The ANN categorized patterns. The spontaneously categorized EPs appear similar to the stimulus-related averages.

ingful stimulus. This component has been labeled *P300*, indicating its polarity (positive) and timing of appearance (300 ms after stimulus presentation). The parameters of the *P300* component (latency and amplitude) are used by neurophysiologists to assess, among other things, effects related to the relevance of stimulus and level of attention (e.g., Lange et al. 1995; Picton and Hillyard 1988; Schwent and Hillyard 1975).

The competitive ANN was trained with 80 input vectors, half of which were Target EPs and the other half were Non-Target. The network converged after approximately 300 iterations (per neuron). A sample of two single-trial post-stimulus sweeps (upper row), of the Target and Non-Target averaged EP templates (middle row), and of the NN identified signal categories (lower row), are presented in Fig. 6. The automatic identification procedure has provided two signal categories, similar to the stimulus-related selectively averaged signals, but requiring further examination due to the slight differences between the selectively averaged waveforms and the categorization obtained by the ANN. The categorization process was consequently repeated, this time using Target and Non-Target data separately; the results are presented in Fig. 7. The categorization of Target data yielded 3 EP patterns, increasing in latency and corresponding to previous findings of increased latency with prolonged reaction times (Lange et al. 1997). Non-Target epoch analysis yielded a Target-like *P300* waveform meaning that, at least occasionally, Target-like *P300* appears even with Non-Target stimuli. This accounts for the above differences and obviously requires further investigation as to the credibility of selective event-related data averaging when applied in the context of analysis of cognitive brain function.

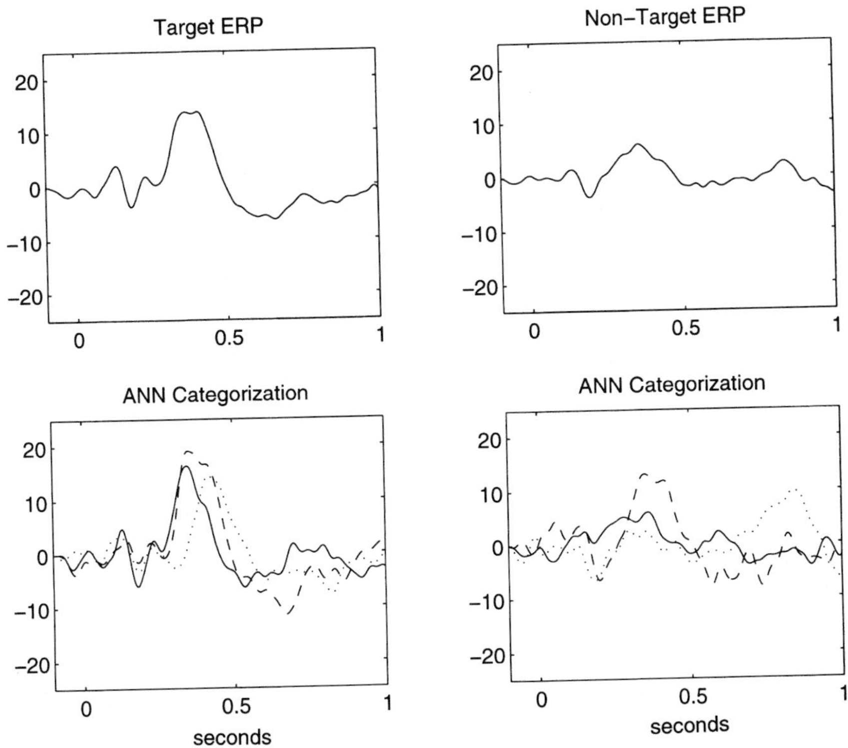

Fig. 7. Spontaneous categorization of separated Target and Non-Target ERPs. Top: Target and Non-Target ERPs. Bottom: The ANN categorizations. The categorized Non-Target patterns include a P_{300}-like waveform (dashed) indicating that some of the Non-Target trials may include a Target-like P_{300} contribution.

Layer 2: Decomposition of EP Waveform

Introduction

Transient evoked brain potentials are elicited in response to various external stimuli, having different time-courses with varying characteristics depending on various parameters among which are the nature and complexity of the applied stimulus, its associated task, and the subject's state of mind. It is well established that different components of the evoked potential complex may originate from different functional brain sites, and can sometimes be distinguished by their respective latencies and amplitudes (Donchin 1966).

An important issue in evoked potential research is assessment of the extent to which an EP complex varies with the manipulation of experimental parameters, in an attempt to utilize the identified variations as insight into brain and central nervous system function. Physiological interpretations are often given to local variations of EP peaks by analyzing the entire signal complex, while ignoring cross-peak effects caused by the close proximity of the peaks, where a slight change of amplitude or latency of a single component may significantly alter the appearance of adjacent component peaks.

The issue of EP decomposition is thus of major significance, as component analysis is at the core of EP variability analysis. Changes in component parameters (latency and amplitude) can have profound implications in EP-based diagnostic evaluations. Decomposition methods vary from simple peak analysis methods, through analytical modeling attempts to approximate EP components, and to purely mathematical methods which often do not consider the physiological origin of the signals at hand (Roterdam 1970).

Evoked Potentials have been decomposed via three common methods: (1) Peak Analysis, treating peaks of the EP complex as independent components, (2) Inverse Filtering, where the EP complex is inversely filtered to extract the component parameters, and (3) Principal Component Analysis, which is equivalent to Karhunen-Loeve' decomposition of the data ensemble.

We present a statistical method for robust decomposition of an EP complex into a set of distinctive components (Lange and Inbar 1986b). The model assumes linear superposition of the EP constituent components, and Gaussian-distributed firing instants of the neuronal population associated with each component. The decomposition provides a loss-less description of the EP complex, which is demonstrated via computer simulations as robust to violations of the model assumptions (Lange et al. 1998). The decomposition method is applied to experimental data, demonstrating its separation performance of severely overlapping EP components.

Model and Assumptions

We adopted Donchin's definition of evoked potential components, suggesting that an evoked potential complex represents a sequence of events, triggered successively in different cortical structures by the stimulus (Donchin 1966). The contribution of one such event to the EP complex was considered as an EP component. In addition, it was assumed that the EP waveform is formed by a superposition of the components, where trial-to-trial variability may be accounted for by adapting the respective amplitudes and latencies of the individual components. In devising an appropriate decomposition rule, the following two points were considered:

- The decomposed waveforms should look “natural”, that is, they should not appear as synthetic to the neurophysiologist. This implied using a loss-less procedure to ensure that fine detail, perhaps insignificant in terms of mean-square error (MSE) but with potentially substantial diagnostic value, would not be lost.

- Dealing with an inverse problem task of inherently infinite solutions, the method should be of a data-driven type rather than a mathematically constrained solution. This has led to using a statistical modeling approach of mass neural activity, as described in the following.

Decomposition Rule

The EP components are generated by vast neural populations which fire synchronously. It is therefore only logical to use a decomposition method based on the nature of neuronal activity. In the general case, if neural activity has to pass through N synaptic stations each producing a delay of D with a variance σ^2, the total delay would be (Abeles et al. 1993):

$$DELAY = N \cdot D \pm \sigma \cdot \sqrt{N}$$

Thus the *firing instants* of a mass population of synchronized neurons may be assumed to be governed by Gaussian probability distributions. Moreover, due to the relatively short duration of the action potential (a few ms) compared to the EP component duration (hundred ms), the probability distributions of the component neuronal sources may be approximated from the components of the measured signal complex (Lange et al. 1997). An initial approximation of the mean value is extracted from the points of maximum activity (peak amplitude), and the approximate variance is implied by peak width. The initial approximation serves as a starting point for a *standard gradient search* of the optimal parameters (Haykin 1986).

The EP waveform is assumed to consist of a superposition of P components:

$$s(t) = \sum_{i=1}^{P} K_i \cdot \upsilon_i(t - \tau_i)$$

where $\upsilon_i(t)$ represents the basic shape of the i-th component, τ_i indicates the component latency, and K_i refers to the component amplitude. Let us denote by A_t the set of neurons $\{a_t\}$ firing at time instant t, with A_t^j being a subset of A_t, which represents the neural batch responsible for the j-th component. Let $\{d_i,\ i=1,2,\ldots,p\}$ denote the set of probability distributions of neuronal firing instants of the p components.

The decomposition is calculated by dividing each data point among the component sources according to their relative contributions, which can be estimated as the probability of a neuron from batch j to fire at instant t normalized by the sum of probabilities of neurons from all neural batches to fire at time instant t.

$$\begin{aligned} \upsilon_j(t) &= s(t) \cdot \frac{Pr\{a_i^j\}}{\sum_{i=1}^{P} Pr\{a_t^i\}} \\ &= s(t) \cdot \frac{\lim_{\Delta \to 0} \int_{t-\Delta}^{t+\Delta} d_j(s)ds}{\sum_i \lim_{\Delta \to 0} \int_{t-\Delta}^{t+\Delta} d_i(s)ds} \\ &= s(t) \cdot \frac{2\Delta \cdot d_j(t)}{\sum_i 2\Delta \cdot d_i(t)} \end{aligned}$$

yielding:

$$\upsilon_j(t) = s(t) \cdot \frac{d_j(t)}{\sum_{i=1}^{P} d_i(t)} \qquad ; \qquad j = 1,2,\ldots,P.$$

Applying the assumption of Gaussian-distributed firing instants yields the decomposition rule:

$$\upsilon_j(t) = s(t) \cdot \frac{\sigma_j^{-1} \exp\left(-\frac{(t-\tau_j)^2}{2\sigma_j^2} \right)}{\sum_{i=1}^{p} \left[\sigma_i^{-1} \exp\left(-\frac{(t-\tau_i)^2}{2\sigma_i^2} \right) \right]}$$

As noted above, initial estimates of τ_i and σ_i are extracted from peak attributes of the signal complex and are used as a starting point for a gradient search algorithm based on a least-squares fitting criterion. It is worth mentioning here that the assumed Gaussian distributions used for the decomposition process are not imposed on the resulting decomposed waveforms, which is the case with some other methods (e.g., Geva et al. 1996). The decomposition would perfectly describe the underlying components if the assumed distributions are correct, nevertheless moderate deviations would not significantly distort peak amplitudes and latencies but might affect the overlapping component tails. This is not crucial, however, since neurophysiologists are mostly concerned with peak parameters (latency and amplitude), rather than with waveform morphology.

Example 2

Evoked Potential data which included a substantial temporal overlap of a stimulus-related (*P200*) component and a cognitive (*P300*) component were investigated, where the degree of temporal overlap depended on the physical magnitude of the stimulus. The experiment included an odd ball-type task, where upon recognition of an auditory rare stimulus the subject was instructed to hit a push-button. The paradigm was performed in two sessions, with a 20 dB power decrease of the Target stimulus intensity in the second session. Fig. 8 shows the EPs associated with the loud and soft Targets, and their respective decompositions. Obviously, the resulting components do not appear Gaussian. However, near-peak characteristics can be approximated with a Gaussian waveform, justifying the decomposition technique; alternatively, the left components of the soft Target stimulus could be modeled as a sum of two Gaussian waves and decomposed accordingly. The results present means for quantitative analysis of the overlapping signals, which could not be performed with the raw averaged waveform.

Layer 3: EP Reconstruction

Using the pattern identification and statistical decomposition units described in the previous sections as preprocessing stages, a comprehensive parametric signal generation model is constructed. The EEG is modeled via AR analysis and the EP is modeled as a sum of finite impulse response (FIR) filtered EP components (Box and Jenkins 1976). The model is designed to account for small latency and amplitude variations of the signal components from trial to trial, assuming that large variations will have been taken care of by the pattern identification structure described earlier. A block diagram of the parametric signal generation mechanism is presented in Fig. 9. The model assumptions are as follows:

1. The EP and EEG are additively superpositioned in each recorded sweep.
2. The signal and noise in each sweep are uncorrelated.
3. The post-stimulus EEG in each sweep can be adequately modeled with a pre-stimulus adapted AR model.
4. The single-trial EP can be modeled as a superposition of latency- and amplitude-corrected components of a specific library waveform.

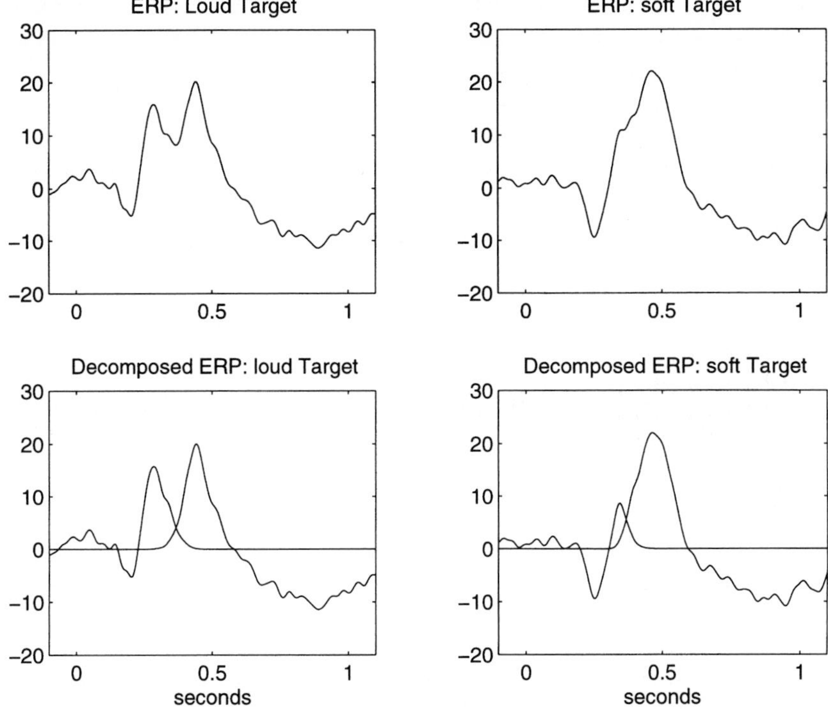

Fig. 8. Experimental results of an auditory odd-ball paradigm with alternating stimulus intensities. The temporal overlap of P_{200} and P_{300} is increased with low stimulus intensities. Left: Loud target ERP and its reconstructed components. Right: Soft target highly overlapping ERP and its reconstructed components.

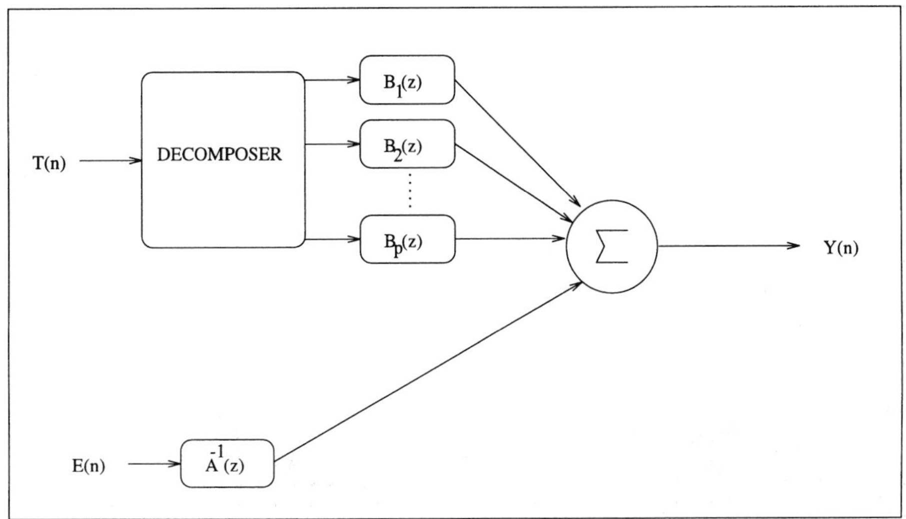

Fig. 9. A block diagram of the signal synthesis model. *T*, *E* and *Y* represent the library matching item (template), a white Gaussian noise series and the recorded single sweep, respectively. $B_i(z)$ are the component latency- and amplitude-correcting filters. Once optimized, the model provides the two assumed signal contributions, namely the ongoing brain activity and the single-trial evoked brain response.

The first two assumptions are conventional assumptions in EP analysis. The third assumption is borrowed from the field of EEG analysis, where it has gained popularity almost as equal to averaging in EP analysis. The fourth assumption reflects the essence of variability analysis capabilities of the proposed model.

As aforementioned, this last processing layer is the most restricting one, its validity resulting from the flexibility and robustness of the previous layers. The first layer identifies a family of responses creating the library of EP template responses, the second layer decomposes the obtained library items, and thus the third layer can rely on the above fourth assumption with much confidence.

The Parametric Model

The notations used throughout the formulations are those of Fig. 9. The model equation in the Z-transform domain (cf. Chapter 21) is given by:

$$Y(z) = \sum_{i=1}^{p} B_i(z) \cdot T_i(z) + \frac{1}{A(z)} \cdot E(z),$$

where $Y(z)$, $T(z)$ and $E(z)$ represent the measured process, a library item (EP template), and a Gaussian white noise series, respectively. Assuming stationarity of the background EEG, which is verified in the experimental analysis section, $A(z)$ may be identified from pre-stimulus data via autoregressive modeling, and then used for post-stimulus analysis. The template and measured EEG are filtered through the identified $A(z)$ to whiten the background EEG signal and thus facilitate a closed-form least-square solution of the model. Having to estimate only $B_i(z)$ from the post-stimulus data simplifies the solution process, avoiding the need for iterative identification of the model parameters. The model can be described by the following regression-type formula, where the apostrophe denotes whitened signals:

$$y'(n) = \sum_{i=1}^{p} \sum_{j=-d}^{d} b_{i,j} \cdot T_i' \ (n-j) + e(n)$$

Matrix notation is used to solve the model. Let y'^T be the whitened measurements vector, let A^T be the input matrix, and let b^T be the filter coefficients vector, as defined below:

$$y'^T = \left[y'(d+1), y'(d+2), \ldots, y'(N-d) \right]$$

$$A^T = \begin{pmatrix} T_1'(2d+1) & T_1'(2d+2) & \cdots & T_1'(N) \\ T_1'(2d) & T_1'(2d+1) & \cdots & T_1'(N-1) \\ \vdots & \vdots & \vdots & \vdots \\ T_1'(1) & T_1'(2) & \cdots & T_1'(N-2d) \\ T_2'(2d+1) & T_2'(2d+2) & \cdots & T_2'(N) \\ \vdots & \vdots & \vdots & \vdots \\ T_2'(1) & T_2'(2) & \cdots & T_2'(N-2d) \\ \vdots & \vdots & & \vdots \\ \vdots & \vdots & & \vdots \\ T_p'(2d+1) & T_p'(2d+2) & \cdots & T_p'(N) \\ \vdots & \vdots & & \vdots \\ T_p'(1) & T_p'(2) & \cdots & T_p'(N-2d) \end{pmatrix}$$

$$b^T = \left[b_{1,-d}, b_{1,-d+1}, \ldots, b_{1,d}, b_{2,-d}, \ldots, b_{p,d} \right]$$

Using the defined notations, the model can be expressed as follows:

$$y' = A \cdot b + \varepsilon,$$

where ε is the vector of prediction errors:

$$\varepsilon^T = \left[e(2d+1), e(2d+2), .., e(N)\right].$$

The sum of square errors can be written as:

$$\xi(b) = \varepsilon^T \cdot \varepsilon,$$

where

$$\varepsilon^T = y'^T - b^T \cdot A^T,$$

and thus the sum of square errors may be specified as follows:

$$\xi(b) = y'^T y' - y'^T Ab - b^T A^T y' + b^T A^T Ab.$$

Deriving ξ(b) with respect to *b* and equating to zero yields the desired solution:

$$\frac{\partial \xi}{\partial b} = -2A^T y' + 2A^T Ab.$$

$$A^T y' = A^T Ab.$$

Therefore, the optimal vector of parameters in the least-square sense is (Lange et al. 1997):

$$\hat{b} = \left(A^T \cdot A\right)^{-1} \cdot A^T \cdot y'.$$

Examples 3 and 4

In the following we shall demonstrate the estimator's performance by two applications. The first deals with motor potentials accompanying free finger movements versus suddenly loaded movements, uncovering a significant component related to the sudden loading. The second application demonstrates the outcome of an odd-ball paradigm, revealing dynamic features of the single trial evoked potentials correlated with the reaction time of the subject.

Statistical evaluation can be shown to lead to estimation bias and variance proportional to the mean and variance of the whitened signal (Lange et al. 1998). Stationarity of the mean is guaranteed since the signals are band-passed prior to sampling.

Movement-Related Potentials (MRPs)

Averaging methods reveal relationships between specific components of MRPs and actual movement parameters. We wished to test whether such effects could be described on a single-trial basis, which would thus improve the existing analysis tools by enabling dynamic tracking of time-varying features. We used evoked potential data recorded during a self-paced finger flexion experiment, recorded differentially between *C3* and *C4* according to the conventional 10–20 electrode placement system (Jasper 1958; cf. Chapter 35), sampled at *250 Hz* and decimated by a factor of *3* prior to processing. The self-paced movements were occasionally disturbed by a mechanical device without pri-

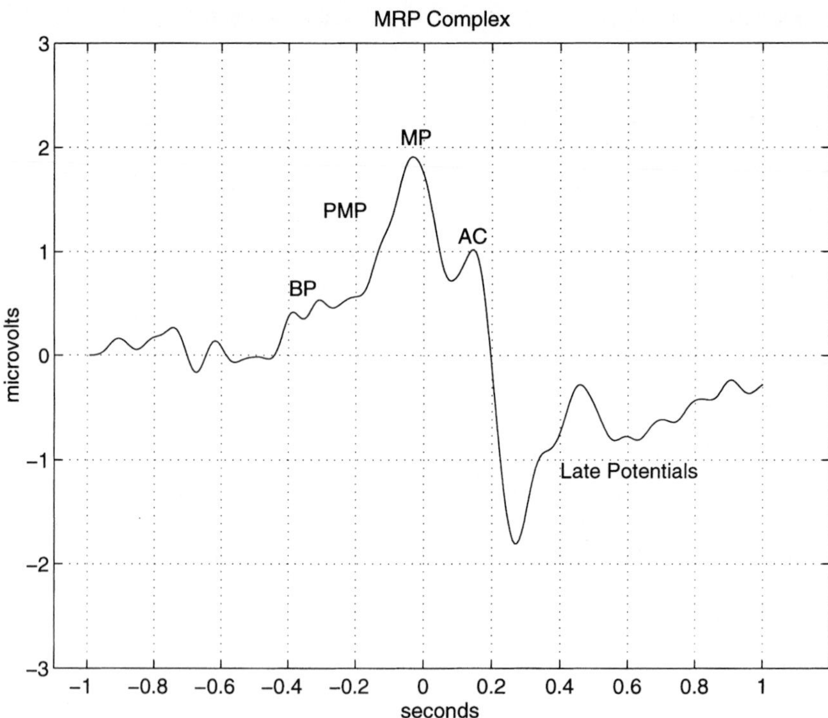

Fig. 10. Grand average of 200 single trials recorded during a typical finger tapping experiment (*t= 0* corresponds to movement onset). Activity begins with the Bereitschafts-Potential (BP), reflecting preparatory mechanisms, followed by a Pre-Movement Potential (PMP), representing the update of motor strategy; then appears the Motor Potential (MP) related to the descending motor command, followed by an Afferent Component (AC) representing the sensory feedback from the moving limb, and concluded by a late activity related to performance assessment.

or knowledge of the subject, who was blindfolded during the experiment. The average response is presented in Fig. 10.

In order to study the effect of the sudden loading, we applied the estimator to two classes of responses – those measured during free (unloaded) movements vs. disturbed (loaded) movements. Comparing the estimation results of the EPs during free and disturbed movements revealed a unique peak appearing at about 150 ms after movement onset with the disturbed movements only (see Fig. 11). This unique peak seems to be in response to the afferent proprioceptive feedback informing the brain of the change of load. This result, which resembles results of similar tests carried out with averaging techniques (Kristeva et al. 1979), motivated us to further pursue this line of investigation and try to track dynamic signal variations under a well-known experimental paradigm. Since cognitive evoked potentials have been studied extensively for a long time, and recordings are easily accessible, cognitive potentials were chosen for the next test.

Cognitive Event-Related Potentials (ERPs)

The main motivation for the development of the EP estimation system was to facilitate trial-by-trial tracking of evoked potentials. To demonstrate tracking performance, we applied the estimator to cognitive evoked potential data recorded from Pz referenced to the mid-lower jaw during a typical odd ball-type paradigm. A detailed description of the experiment can be found in Lange et al. (1995). In addition to the evoked potential

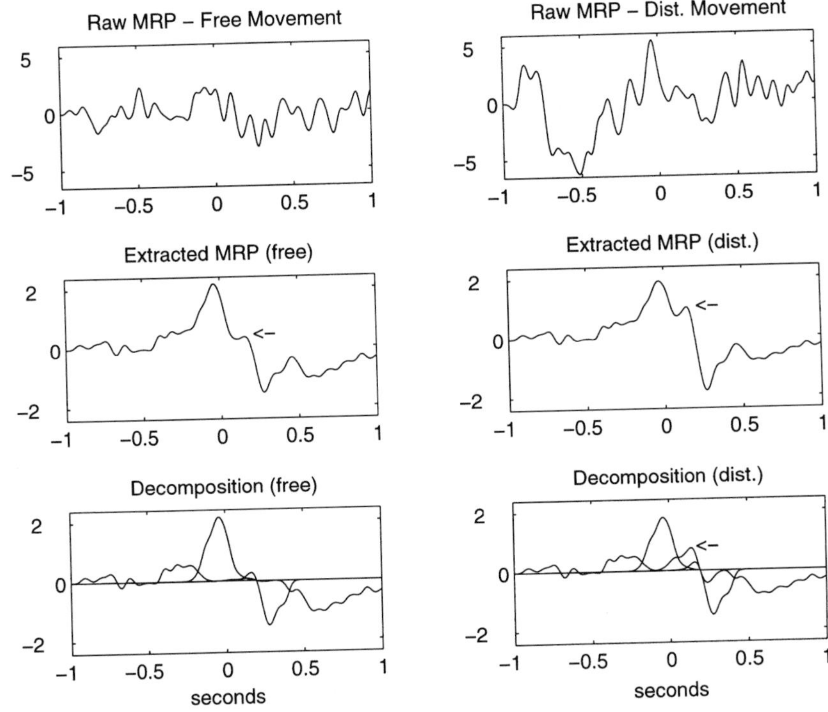

Fig. 11. Experimental results. Top row: Raw EEG signals recorded during free and disturbed movements. Middle row: Extracted evoked potentials. Note the unique peek (arrow) which is significantly larger with the disturbed movement. Bottom row: Evoked potential decomposition; the afferent component appears during the disturbed movement only.

data, reaction times to target stimuli were noted in an attempt to establish a relation between behavioral performance measures and the *P300* complex. In a previous analysis which was reported in Lange et al. (1995), we found that the latency of the *P300* component increased with the increase in reaction time. This provides one possible explanation for the appearance of two peaks in the averaged response as demonstrated in Fig. 12. Similar results were commonly reported in ERP literature. However, applying the current estimation procedure, while identifying *P300a* and *P300b* to be two distinctive components, yields a different result as can be seen in Fig. 13. In this case, it seems that rather than a change in the latency of *P300*, we obtain a reciprocal change of magnitude of each sub *P300* component: with the increase in reaction time, *P300a* decreases and *P300b* increases. This result may be due to the relation between attention allocation and prompt performance; with the drop in attention reflected by decreased amplitude *P300a* and resulting in prolonged reaction times, computation, reflected by *P300b*, is delayed but increased in amplitude. This effect may indicate a compensation for decreased attention by increased computational effort. This result is also in accordance with limited findings in the ERP literature, which differentiate the positive P*300* component into two constituents, based on averaged EPs recorded during cognitive experimentation (Verleger and Wascher 1995). Interestingly, the estimator has identified changes of respective component amplitudes without a change of latencies, even though it can compensate for latency variations of single components.

We have thus shown that dynamic variations of evoked potential components which may not be detectable from the averaged data can be extracted from the raw data using the suggested estimation method, overcoming the severe problems of a low SNR and temporally overlapping signal components.

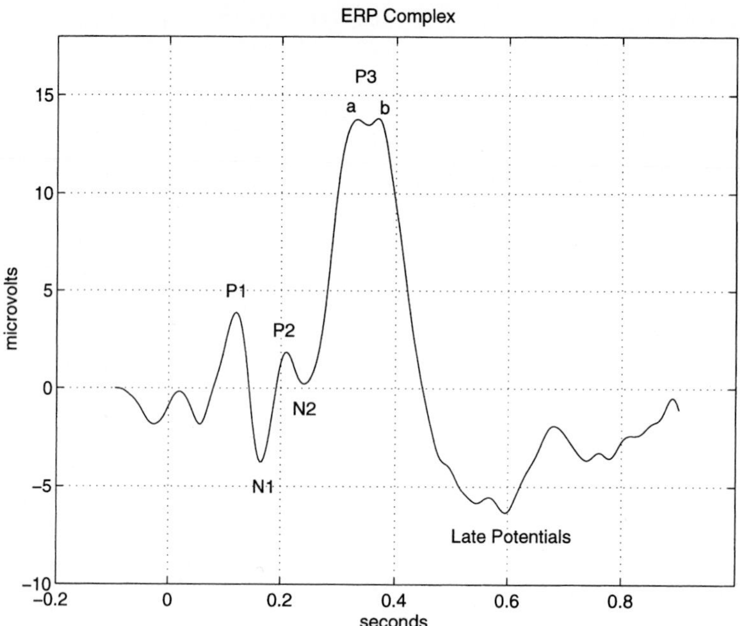

Fig. 12. Grand average of 40 ERP target trials recorded during a typical odd ball-type paradigm. P_1 is stimulus related, N_1 is attention related, and P_3 (P_{300}) is a cognitive, memory access-related component. The EP is concluded with a late activity which may reflect performance assessment. The P_{300} component seems to be comprised of two contributions, P_{300a} and P_{300b}, which are believed to co-vary with relation to performance indices.

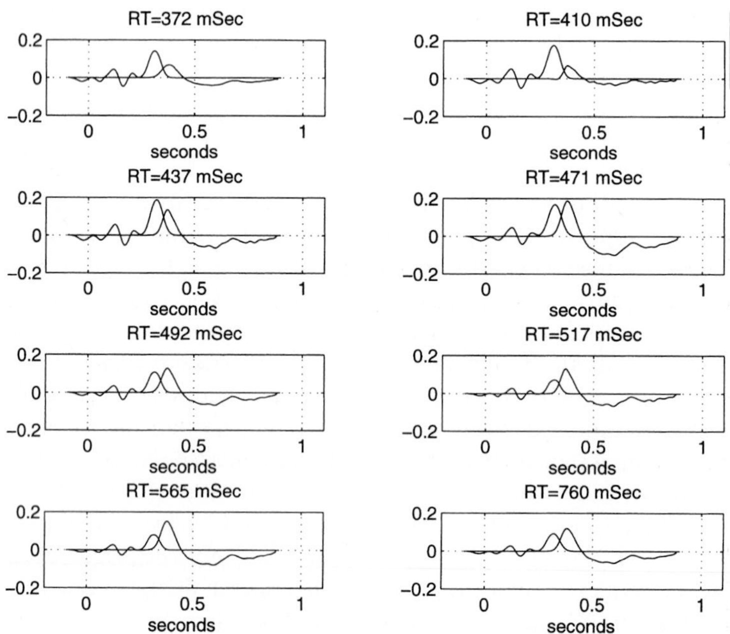

Fig. 13. Results of applying the proposed estimator to the ERP data: P_{300a} decreases and P_{300b} increases with the increase in reaction time. This phenomenon can be explained by relating the P_{300a} to the invested effort in stimulus detection, where P_{300b} can be related to the consequent effort required for the decision-making.

Discussion

The processing begins with spontaneous identification of the EP response types, embedded within the ensembled data. The identification process can be implemented online in real time, with a continuous dynamic update of the response types. The current implementation is based on a fixed network structure, requiring *a priori* selection of the network size which corresponds to the anticipated size of the response library. Preliminary studies have shown that the network can adapt its own structure by increasing or decreasing its dimension. A possible criterion for such dynamic behavior is the degree of correlation among the various library items, that is, the network may increase until no significant changes of the obtained items can be observed.

The single evoked responses are modeled as a sum of several components derived from the library of parent responses. Responses associated with the same parent waveform may differ one from another due to amplitude and latency variations of the constituent components. The set of components is related to both sequential and parallel neuronal processing stages, distinguished by manipulating the experimental paradigm to produce independent variations of the different components. The separation of the signal into distinct components is based on a statistical approach, assuming Gaussian-distributed neural activity yielding a set of components which sum up exactly to the original waveform. The separation process is robust to major violations of the assumptions, e.g., non-Gaussian or non-symmetrical components. It should be noted that different assumptions regarding neural activity, if found more appropriate, may easily be incorporated in the model by substitution of the statistical neural activation hypothesis with an alternative hypothesis.Component smearing due to latency variations was not considered; however, it may be considerably corrected for by component realignment throughout the ensemble using the identified lag of each component, after which an improved template may be obtained by processing the realigned data, similar to common realignment procedures (e.g., Spreckelsen and Bromm 1988). In addition, artifacts due to eye movements may be eliminated by adding an appropriate input channel whose contribution should be adaptively subtracted from the estimated response (e.g., Cerutti et al. 1987).

The system was applied to two major types of EPs: Movement-Related Potentials and Cognitive Event-Related Potentials. In both cases the system was able to extract the single-trial brain responses, the analysis of which revealed dynamic behavior of the responses, like the afferent peak appearing with loaded movements in MRPs, and a reciprocal change of cognitive components correlated with performance indices (reaction time) in ERPs. Moreover, it has been shown that selective averaging may not be adequate, as demonstrated in the case of ERPs, where some of the Non-Target responses presented similar characteristics to the Target responses. Such effects could not be detected with averaging techniques, nor with single-trial methods relying on the average response as a reference signal. The ability to objectively track EP variations on a single-trial basis, as presented in the experimental results throughout this work, emphasizes the added value of using the processing framework described herein.

Conclusion

Current single-trial EP processing methods are usually restricted to average-like single-trial estimates, or use EEG non-stationarities as indices of EP manifestations. While the first approach is not flexible in the sense that single-trial EPs which do not resemble the template will not be detected, the second approach is over-sensitive as it might associate "non-conforming" EEGs with "contaminating" EPs. The processing framework proposed herein is to some extent a compromise between the two above approaches, as it is not limited to average-like responses on the one hand, yet is not sensitive to EEG non-stationarities on the other hand. The multi-layer processing approach provides a flexible yet robust estimation performance, enabling extraction of the low SNR, trial-varying evoked brain responses and tracking of their constituent components.

Acknowledgement: The authors would like to thank Prof. Hillel Pratt for providing the ERP data and for many fruitful discussions.

References

Abeles M, Prut Y, Vaadia E, and Aertsen A (1993) Integration, Synchronicity and Periodicity. In: Aertsen A (ed.) Brain Theory: Spatio-Temporal Aspects of Brain Function. Elsevier Science Publishers B.V.. 149–181.

Aunon JI, McGillem CD, Childers DG (1981) Signal Processing in Evoked Potential Research: Averaging and Modeling. CRC Crit. Rev. Bioeng 5:323–367.

Bartnik EA, Blinowska, KJ, Durka, PJ (1992) Single Evoked Potential Reconstruction by Means of Wavelet Transform. Biol. Cybern. 67:175–181.

Birch GE, Lawrence PD, Hare RD (1993) Single-Trial Processing of Event Related Potentials using Outlier Information. IEEE Trans. Biomed. Eng. 40:59–73.

Box GEP, Jenkins GM (1976) Time Series Analysis: Forecasting and Control. Holden-Day.

Carlton EH, Katz S (1987) Is Wiener Filtering an Effective Method of Improving Evoked Potential Estimation? IEEE Trans. Biomed. Eng. 34(1).

Cerutti S, Baselli G, Liberati D, Pavesi G (1987) Single Sweep Analysis of Visual Evoked Potentials through a Model of Parametric Identification. Biol. Cybern. 56:111–120.

Cerutti S, Chiarenza G, Liberati D, Mascellani P, Pavesi G (1988) A Parametric Method of Identification of Single Trial Event Related Potentials in the Brain. IEEE Trans. Biomed. Eng. 35:701–711.

Chiappa KH (1983) Evoked Potentials in Clinical Medicine. New York: Raven.

de Weerd JPC (1981) A Posteriori Time-Varying Filtering of Averaged Evoked Potentials. I. Introduction and Conceptual Basis. Biol. Cybern. 41: 211–222.

Donchin E (1966) A Multivariate Approach to the Analysis of Average Evoked Potentials. IEEE. Trans. Biomed. Eng 13.

Doyle DJ (1975) Some Comments on the Use of Wiener Filtering in the Estimation of Evoked Potentials. Electroencephalogr. Clin. Neurophysiol. 28: 533–534.

Duda RO, Hart PE (1976) Pattern Classification and Scene Analysis. Wiley: New-York.

Freeman JA, Skapura DM (1992) Neural Networks: Algorithms, Applications, and Programming Techniques. Addison-Wesley Publishing Company, USA.

Furst M, Blau A (1991) Optimal A-posteriori Time Domain Filter for Average Evoked Potentials. IEEE Trans. Biomed. Eng. 38:827–833

Geva AB, Pratt H, Zeevi YY (1996) Spatio-Temporal Source Estimation of Evoked Potentials by Wavelet Type Decomposition: Wavelet-Type Source Estimation of EPs. In: I Gath, GF Inbar (eds.) Advances in Processing and Pattern Analysis of Biological Signals. Plenum Press: New York, 103–122

Gevins AS (1984) Analysis of the Electromagnetic Signals of the Human Brain: Milestones, Obstacles and Goals. IEEE Trans. Biomed. Eng. 31: 833–850.

Gupta L, Molfese DL, Tammana R, Simos PG (1996) Nonlinear Alignment and Averaging for Estimating the Evoked Potential. IEEE Trans. Biomed. Eng. 43:348–356.

Haykin S (1986) Adaptive Filter Theory. Prentice-Hall: N.J.

Haykin S (1994) Neural Networks: a Comprehensive Foundation. Macmillan College Publishing Company, Inc., USA.

Jasper HH (1958) The Ten-Twenty Electrode System of the International Federation. Electroencephalogr. Clin. Neurophysiol. 10: 371–375

Kaipio JP, Karjalainen PA (1997) Estimation of Event Related Synchronization Changes by a New Tvar Method. IEEE Trans. Biomed. Eng. 44:649–656.

Kong X, Thakor NV (1996) Adaptive Estimation of Latency Changes in Evoked Potentials. IEEE Trans. Biomed. Eng 43:189–197.

Kristeva R, Cheyne D, Lang W, Lindinger G, Deecke L (1979) Movement Related Potentials Accompanying Unilateral and Bilateral Finger Movements with Different Inertial Loads. Electroencephalogr. Clin. Neurophysiol. 75: 410–418.

Laguna P, Jane R, Meste O, Poon PW, Caminal P, Rix H, and Thakor NV (1992) Adaptive Filter for Event-Related Bioelectric Signals Using an Impulse Correlated Reference Input: Comparison with Signal Averaging Techniques. IEEE Trans. Biomed. Eng. 39: 1032–1043.
Lange DH, Pratt H, Inbar GF (1995) Segmented Matched Filtering of Single Event Related Evoked Potentials. IEEE Trans. Biomed. Eng. 42: 317–321.
Lange DH, Inbar GF (1996a) A Robust Parametric Estimator for Single-Trial Movement Related Brain Potentials. IEEE Trans. Biomed. Eng. 43: 341–347.
Lange DH, Inbar GF (1996b) Parametric Modeling and Estimation of Amplitude and Time Shifts in Single Evoked Potential Components. In: I Gath, GF Inbar (eds.) Advances in Processing and Pattern Analysis of Biological Signals. Plenum Press.
Lange DH, Pratt H, Inbar GF (1997) Modeling and Estimation of Single Evoked Brain Potential Components. IEEE Trans. Biomed. Eng. 44: 791–799.
Lange DH, Siegelman HT, Pratt H, Inbar GF (1998) A Generic Approach for Identification of Event Related Brain Potentials via a Competitive Neural Network Structure. Proc. NIPS*97 - Neural Information and Processing Systems: Natural & Synthetic, Denver, 1998.
Liberati D, DiCorrado S, Mandelli S (1992) Topographic Mapping of Single Sweep Evoked Potentials in the Brain (1992). IEEE Trans. Biomed. Eng. 39: 943–951.
Madhaven PG (1992) Minimal Repetition Evoked Potential by Modified Adaptive Line Enhancement. IEEE Trans. Biomed. Eng. 39: 760–764.
Makhoul J (1975) Linear Prediction: A Tutorial Review. Proc. IEEE. 63.
Mason SG, Birch GE, Ito MR (1994) Improved Single-Trial Signal Extraction of Low SNR Events. IEEE Trans. Sig. Proc. 42: 423–426.
Meste O, Rix H (1996) Jitter Statistics Estimation in Alignment Processes. Signal Processing 51:41–53.
Nakamura M (1993) Waveform Estimation From Noisy Signals with Variable Signal Delay using Bispectrum Averaging. IEEE Trans. Biomed. Eng. 40:118–127.
Picton TW, Hillyard SA (1988) Endogenous Event Related Potentials. In: TW Picton (ed.) Handbook of Electroencephalogr. Clin. Neurophysiol. Amsterdam: Elsevier, .3: 361–426.
Popivanov P, Krekule I (1983) Estimation of Homogenity of a Set of Evoked Potentials with Respect to its Dispersion. Electroencephalogr. Clin. Neurophysiol. 55: 606–608.
Pratt H, Michalewski HJ, Barrett G, Starr A (1989) Brain Potentials in Memory-Scanning Task: Modality and Task Effects on Potentials to the Probes. Electroencephalogr. Clin. Neurophysiol. 72:407–421.
Rodriguez MA, Williams RH, Carlow TJ (1981) Signal Delay and Waveform Estimation using Unwarped Phase Averaging. IEEE Trans. Acoust. Sp. & Sig. Proc. 29: 508–513.
Rompelman O, Ros HH (1986a) Coherent Averaging Technique: A Tutorial Review. Part 1: Noise Reduction and the Equivalent Filter. J. Biomed. Eng. 8: 24–29.
Rompelman O, Ros HH (1986b) Coherent Averaging Technique: A Tutorial Review. Part 2: Trigger Jitter Overlapping Responses and Non-Periodic Stimulation. J. Biomed. Eng. 8:30–35.
Roterdam AV (1970) Limitations and Difficulties in Signal Processing by Means of the Principal Component Analysis. IEEE Trans. Biomed. Eng. 17: 268–269.
Rumelhart DE, Zipser D (1985) Feature Discovery by Competitive Learning. Cognitive Science, 9: 75–112.
Schwent VL, Hillyard SA (1975) Evoked Potential Correlates of Selective Attention with Multi-Channel Auditory Inputs. Electroencephalogr. Clin. Neurophysiol. 38: 131–138.
Spreckelsen MV, Bromm B (1988) Estimation of Single-Evoked Cerebral Potentials by Means of Parametric Modeling and Kalman Filtering. IEEE Trans. Biomed. Eng. 35: 691–700.
Svensson O (1993) Tracking of Changes in Latency and Amplitude of the Evoked Potential by Using Adaptive LMS Filters and Exponential Averagers. IEEE Trans. Biomed. Eng. 40: 1074–1079.
Thakor NV (1987) Adaptive Filtering of Evoked Potentials. IEEE Trans. Biomed. Eng. 34: 6–12.
Vaz CA, Thakor NV (1989) Adaptive Fourier Estimation of Time-Varying Evoked Potentials. IEEE Trans. Biomed. Eng. 4: 448–455.
Verleger R, Wascher E (1995) Fitting Ex-Gauss Functions to P3 Waveshapes: an Attempt at Distinguishing between Real and Apparent Changes of P3 Latency. Journal of Psychophysiology, 9: 146–158.
Walter DO (1969) A Posteriori Wiener Filtering of Average Evoked Responses. Electroencephalogr. Clin. Neurophysiol., suppl. 27: 61–70.
Woody CD (1967) Characterization of an Adaptive Filter for the Analysis of Variable Latency Neuroelectric Signals. Med. & Biol. Eng. 5: 539–553.
Yu K, McGillem CD (1983) Optimum Filters for Estimating Evoked Potential Waveforms. IEEE Trans. Biomed. Eng. 30: 730–737.

Chapter 37

Magnetoencephalography

Volker Diekmann, Sergio N. Erné and Wolfgang Becker

Introduction

Magnetic Brain and Nerve Activity

As known from physics, each current is accompanied by a magnetic field. So are the ionic currents within the brain and the nerves. Using highly sensitive sensors, so-called SQUIDs ("superconducting quantum interference device") developed during the last three decades, magnetoencephalography (MEG) measures these extremely weak fields outside the head. MEG can pick up the fields associated with the concerted action of a few thousand neurons and thus non-invasively monitor brain activity. With good approximation these fields reflect only intracellular (mostly intradendritic) currents and are insensitive to the extracellular current distribution, in contrast to the scalp potentials measured with EEG (cf. Chapter 35).

A useful concept for understanding biomagnetic fields is the current dipole. An exact definition of the dipole notion can be gleaned from textbooks of physics. However, for all practical purposes it can be envisioned as a current confined for a short way to a linear isolated conductor immersed in a volume conductor into which it discharges at one end of the linear conductor and from where it is collected at the other end. In physical terms the dipole is a vector characterized by a *position* and a moment; the *moment* indicates its strength (=current amplitude x length of current path, dimension [Am]) and orientation. The intradendritic currents caused by postsynaptic potentials or the currents associated with a propagating spike in an axon or a peripheral nerve are biological examples of current dipoles. In an irregular, non-homogeneous medium like the head or body, both the dipole current proper (=intracellular current) and the extracellular volume currents contribute to the magnetic field. However, if the conducting medium can be approximated by rotationally symmetric compartments (e.g., a sphere in the case of the head), the magnetic field is essentially determined by the intracellular currents because the fields of the extracellular volume currents cancel each other. Moreover, only current dipoles which are oriented tangentially with respect to the spherical surface contribute to the field whereas normally oriented ones are silent (see Williamson and Kaufmann (1990) and Hämäläinen et al. (1993) for a detailed description of these neuroelectric and neuromagnetic fundamentals).

Volker Diekmann, Universität Ulm, Sektion Neurophysiologie, Albert-Einstein-Allee 47, D-89081 Ulm, 89081, Germany (e-mail: volker.diekmann@medizin.uni-ulm.de)
Sergio N. Erné, Universität Ulm, Zentralinstitut für Biomedizinische Technik, Albert-Einstein-Allee 47, D-89081 Ulm, 89081, Germany (e-mail: sergio-nicola.erne@zibmt.uni-ulm.de)
Correspondence to: Wolfgang Becker, Universität Ulm, Sektion Neurophysiologie, Albert-Einstein-Allee 47, D-89081 Ulm, Germany (phone: +49-731-50-25500; fax: +49-731-50-25501; e-mail: wolfgang.becker@medizin.uni-ulm.de)

Origin of MEG

It is commonly accepted that the magnetic fields of the brain, like the electric potentials picked up at the scalp (EEG, cf. Chapter 35), mostly originate from the dendritic currents set up by postsynaptic potentials (PSP) rather than from the axonal currents associated with action potentials. Indeed, the depolarization and repolarization zones of an action potential travelling along an axon (cf. Chapter 5) can be approximated by two closely spaced but oppositely orientated current dipoles forming a quadrupole whose field strength decreases as $1/r^3$ (r, distance from current source). Dendritic currents, on the other hand, can be approximated by single current dipoles whose fields only decrease as $1/r^2$.

Assuming a typical cortical dendrite of 100 –200 μm length and a diameter of 1 μm, the current dipole resulting from a single PSP at one synapse has an estimated strength of $20 \cdot 10^{-15}$ Am (Hämäläinen et al., 1993). At a distance of 4 cm (typical distance between cortical activity and MEG sensors) it would produce a field of only 10^{-19} T. On the other hand, auditory magnetic fields (a preferred topic of MEG studies) reach intensities of about 10^{-13} T (100 fT). This implies that the coherent activity of at least 1 million synapses contributes, provided the involved dendrites all are parallel to each other and tangential to the head surface. In reality, probably a much larger number of synapses is involved because the individual dendrites do not all have similar orientation, hence their fields partially cancel.

Equivalent Current Dipoles

The analysis of neuromagnetic (and/or neuroelectric) fields often is aimed at determining the intracranial 3-D localization of one or several aggregates of neurons that are active in relation to a brain state, a sensory event, a motor action, or the like. The dendritic currents set up within such an aggregate form a spatial distribution of elementary current dipoles. Viewed from some distance, such a distribution can be approximated by a single current dipole, the "equivalent current dipole" (ECD) which is a weighted vector sum of the elementary dipoles. The ECD is the most frequently used physical model for the localization of brain activity. The usage of this model is justified as long as the aggregate of active neurons is small in relation to its distance from the sensors measuring the magnetic field (or the electric potential). More realistic models of the biological current sources are needed if this condition is not met. Often a combination of a few ECDs already provides a good first approximation; sometimes also a multipole expansion may be useful (see a textbook of physics for details on multipole expansions, e.g. Jackson (1998)).

In many cases, the extracranial magnetic field of an ECD can be successfully determined by modelling the head as a homogeneous conducting sphere fitted to the local curvature of the brain. Similarly, the electric potential generated by an ECD at the head surface (electroencephalogram, EEG; cf. Chapter 35) is obtained by modelling the head as a four-shell concentric sphere consisting of the brain, cerebrospinal fluid, skull and scalp compartments, all with different conductivities. Figure 1 shows the patterns of the magnetic field and the electrical potential of a tangential current dipole that are obtained with these idealized models. The magnetic isodensity lines in Fig. 1a refer to the normally oriented field component (with respect to head surface); this component vanishes along the central projection of the dipole upon the head surface and reaches extrema to the left and right of it. The distance between these extrema increases with the ECD's depth below head surface. Fig. 1b shows the pattern of the electric potential of the ECD on the surface of the outermost spherical shell. The pattern is similar to the magnetic one except for a rotation by 90°.

Comparison between Brain Magnetic Fields and Ambient Fields

Typical physiological neuromagnetic fields outside the head or body are of the order of 1 pT for spontaneous brain activity (e.g. alpha activity), 0.1 pT for evoked brain activity or peripheral nerve compound action fields, and 0.01 pT for high-frequency evoked activity (somatosensory evoked burst brain activity around 600 Hz). These biological

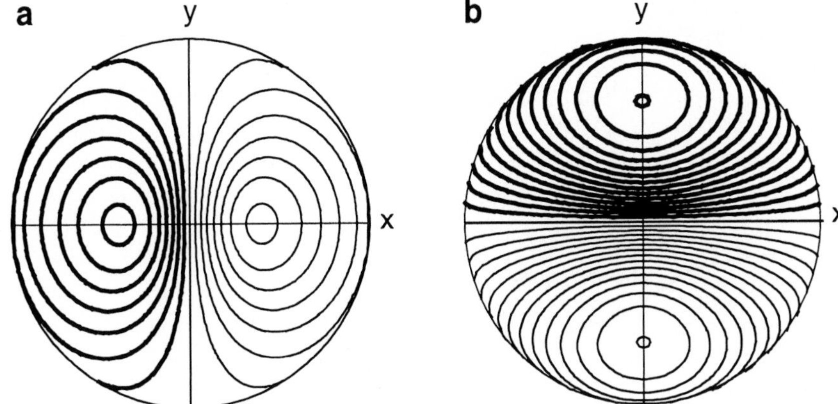

Fig. 1. Radial component of magnetic field (a) and electric potential (b) generated by a current dipole inside a spherical head model. The model consists of four concentric spherical shells with radii of 7.6, 7.8, 8.4 and 9 cm, enclosing compartments with conductances of 0.0033, 0.000042, 0.01 and 0.0033 S/m which represent the brain (innermost sphere), the cerebrospinal fluid, the skull, and the scalp, respectively. The dipole is positioned at (x, y, z) = (0, 0, 5.6) cm, has a moment of 10 nAm, and points into the y-direction (in other words, the dipole is 2 cm below the brain surface and tangential to it). The iso-density lines in (a) show the field that would be seen by magnetic sensors placed at a distance of 1 cm from the volume surface (line spacing, 20 fT; bold lines, outflux). The isopotential lines in (b) correspond to the potential which would be picked up by electrodes at the surface of the volume (line spacing, 0.05 μV; bold lines, positivity). Note that the patterns in (a) and (b) are qualitatively similar except for a rotation by 90.

fields are eight to ten orders of magnitude smaller than the earth magnetic field, and also several orders of magnitude weaker than the variations of the ambient magnetic field occurring in a normal laboratory or clinical environment. Electrical motors, power lines, and deformations of the earth magnetic field by moving ferromagnetic material (elevators, cars) all contribute to the magnetic noise at such a site (Fig. 2), and so do laboratory instruments such as computers, watches, stimulus generators, and the like. In the frequency range below 1 Hz geomagnetic fluctuations may also interfere with biomagnetic measurements. Therefore, brain magnetic fields cannot be recorded without sophisticated shielding measures (see Sect.: *Equipment*) and other noise suppression strategies (e.g., gradiometers, see below).

Measuring Principle

SQUIDs

At the heart of all contemporary sensors used to measure fields of biological origin are DC-SQUIDs which convert magnetic flux into voltage. They consist of a super-conducting loop containing two so-called Josephson junctions formed by very thin insulating barriers. A bias current forced across these junctions creates a voltage which, owing to quantum mechanical phenomena, is related to the magnetic flux threading the superconducting loop (flux = field normal to loop × loop area; for details on DC-SQUIDs see Clarke et al. (1975) and (1976), Erné (1983) and Weinstock (1996); for details on the Josephson effect see a textbook on superconductivity, e.g., Buckel (1990)).

Gradiometers

Many MEG systems do not couple the flux directly into the SQUID but use a superconducting flux transformer (Fig3A) consisting of one (or several) pickup coil(s) L_p and coil L_s which couples the magnetic flux into the SQUID. A single pickup coil coupled to a SQUID forms a magnetometer. Frequently, however, two or more pickup coils are

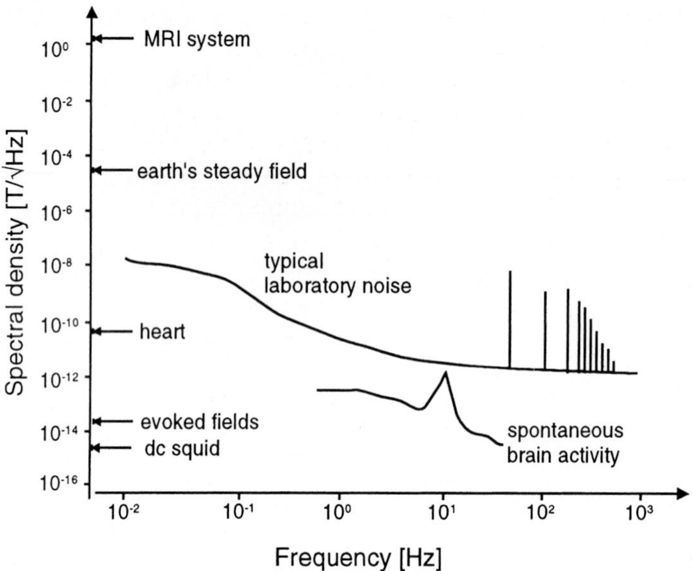

Fig. 2. Magnitude of various natural, technical, and physiological magnetic fields as function of frequency (spectral density). Laboratory noise is for an unshielded environment and contains spectral lines corresponding to the main frequency and its harmonics. Peak of spontaneous activity at 10 Hz corresponds to alpha activity.

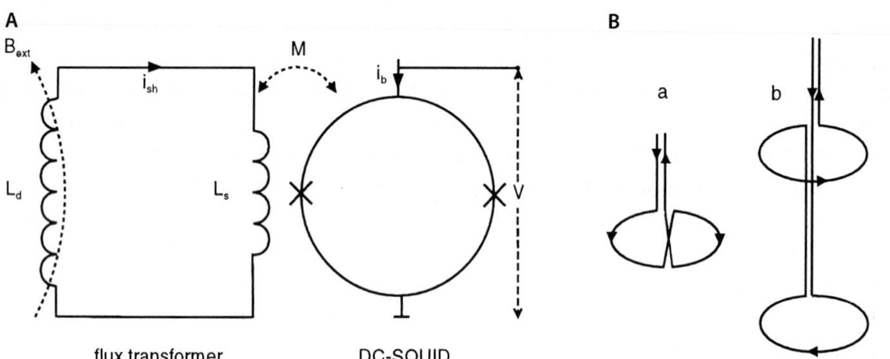

Fig. 3. (A) Principle of field sensing by SQUIDs. A bias current i_b flows through both halves of the superconducting loop of a DC-SQUID and produces the voltage V across the two Josephson junctions. The external magnetic field B_{ext} threading the detection coil (pick-up coil) L_d excites a shielding current i_{sh} and, hence, a flux in the primary coil L_s. This flux is coupled, via the mutual inductance M, into the SQUID loop where it causes a change of the junction voltage V. Note: some recent instruments do without separate pick-up and transformer coils and use the SQUID loop itself to pick up the field. (B) Two types of first order gradiometers: (a) planar gradiometers consist of 2 coils of opposite sense which pick up the flux difference between adjacent areas in the same plane; (b) axial gradiometers sense the flux difference between 2 axially displaced areas.

linked, forming a gradiometer (Fig. 3B). A simple gradiometer consists of two either coplanar (a) or collinear (b) superconducting loops of identical area wound in opposite sense. Its net-induced flux is proportional to the difference in field intensity at the two loop sites, i.e., to the local gradient of the field. Because the field of faraway sources has a lower gradient than that of nearby sources, gradiometers selectively reduce the influence of distant noise sources in comparison to that of nearby biological sources.

Hämäläinen et al. (1993) give a survey of the different types of gradiometers used in MEG. Instead of physically coupling the two coils of a gradiometer (Fig3B), they can also be coupled by software. In this scheme, the two coil fluxes are independently processed by separate SQUIDs whose outputs are digitized and then subtracted from each other. There are several advantages to this scheme: (1) The subtraction can be easily fine-tuned so as to offset the unavoidable differences between the two coils (note that a difference in area of, say, 0.1 % limits the reduction of noise to a factor of 10^3, whereas ambient noise frequently exceeds the signal of interest by a factor of 10^4 – 10^5). (2) The number of coils can be reduced. It is sufficient to measure the ambient noise at some distance from the biological object with a few dedicated compensation coils linked to separate SQUID channels. Using appropriately determined weights, these noise measurements can be subtracted from the signals picked up by any of the coils close to the investigated object (Becker et al.,1993; Diekmann et al., 1996). Most of the modern large-scale helmet-type MEG systems (and also their multi-channel analogues in magnetocardiography (MCG)) are based on software gradiometers or a combination of hardware and software gradiometer designs.

Principles of Data Analysis

Measuring the spatial distribution of the magnetic field around the head by means of an MEG machine is only a first step in determining the 3-D location of sources. The problem of determining the ECD(s) underlying a given field distribution has come to be known as the inverse problem. Without additional, physiological information there is no unique solution to this problem (Helmholtz, 1853). In order to reduce the number of possible solutions, one has to proceed from initial assumptions which should reflect as much *a priori* knowledge as possible. For example, in a sensory stimulation experiment it may be reasonable to assume that the field measured at a certain latency originates from a single ECD located in an approximately known part of the cortex. The field predicted by this assumed configuration can be determined ("forward problem") and compared to the measured one. The difference between the two fields can be used to correct the initial assumption. This process can iteratively be repeated until a desired error limit is reached. However, often the iteration turns out to be unable to reduce the error between the assumed and actual fields below the desired limit; one then must try and improve the initial assumptions.

Fields of Application, Advantages and Limits of MEG

Because MEG is based on the same electrophysiological principles as EEG, also its fields of application are basically similar to those of EEG. In comparison to EEG there are several advantages to MEG, but also drawbacks:

- Source localization is easier because MEG is less affected by variations in conductivity across compartment borders (brain tissue, skull, scalp) than is EEG (Haueisen et al., 1997b). Therefore, MEG is an appropriate choice when the objective of the measurement is to infer the 3-D locus of the source(s) of a given brain activity rather than to merely describe activity in terms of its temporal pattern. Such brain activities may be spontaneous rhythms (e.g., theta-rhythm), pathological events (e.g., epileptiform activity), sensory evoked activity (e.g., auditory evoked responses), cognition-related changes of activity (e.g., magnetic equivalent of P300). However, one must keep in mind that, unlike EEG, MEG does not well reflect radially oriented sources (which are likely to occur in areas where the cortex is tangential to the skull (gyri)).

- Unlike with EEG, subjects need not be equipped with electrodes for MEG measurements. With many channels and a large number of subjects to be screened, this can result in significant time savings. Also the question as to the choice of the appropriate "indifferent" electrode does not arise. However, in thorough clinical examinations MEG is not a replacement of EEG, but should be used as a complementary method. Because subjects' heads must be immobilized in close contact to the sensor-carrying cryostat, MEG is difficult to record during sleep and cannot be used in mobile situations and in restless subjects. Also, MEG cannot be applied in subjects with ferromagnetic contaminations (e.g., magnetic fragment inclusions in tissue after injury, metal dust, certain dental implants) which can generate large signals in relation to minute, pulsation-evoked movements.
- MEG can detect the quasistatic fields associated with spreading depressions (for a review see Templey (1992)). Used in conjunction with specialized equipment, MEG can also record the early DC-fields occurring after a nerve injury (magnetoneurography; Curio, 1995), a technique which is still under development.
- In comparison to functional imaging methods such as positron emission tomography (PET) or functional MRI (fMRI), MEG (and EEG) offers a much better time resolution ($\approx$ 1 ms); PET's resolution is of the order of many seconds to minutes, and recent fMRI techniques achieve a resolution of 100 ms if single slices are considered. However, with regard to spatial resolution, MEG is mostly inferior to these imaging techniques. Under favorable conditions (superficial source, large signal-to-noise ratio (SNR), large array of sensors), MEG localization accuracy is comparable or superior to the relative precision of modern PET instruments (4–6 mm) but probably worse than that of fMRI (1 mm); note however that for deeper sources, MEG accuracy deteriorates considerably (several cm), whereas PET and MRI accuracy remains about constant also for deeper sources.

Materials

The main parts of the measurement equipment are a cryostat (dewar) including the cooled SQUID sensors, a non-magnetic patient table or chair inside a magnetically shielded room (MSR), and high performance electronics for simultaneously recording a large number of channels.

Sensors

As already mentioned in the Introduction, very sensitive magnetic sensors are needed to record biomagnetic activity. Good quality sensors should contribute in only a minor way to the total signal noise. Only DC-SQUIDS cooled by liquid helium fulfill these requirements. Typical noise values for multi-channel devices designed for investigations in humans are below 10 fT/Hz at 1 Hz. Today's most sensitive multi-channel biomagnetic measurement system in use has been reported by the Physikalisch-Technische Bundesanstalt in Germany (Drung, 1995) with a typical white noise level of 2.7 fT/Hz.

Cryo-cooling

In order to become superconductive, the sensors of any MEG system (SQUIDs) must be immersed in a cooling medium. For low-temperature superconductivity (used in all commercial systems) liquid helium must be used as a cryogenic liquid. Fiberglas and similar plastics have proven to be ideal material for building non-magnetic containers of liquid helium (dewars) inside of which SQUIDs can be operated. With such materials planar as well as helmet-shaped, dewars and corresponding sensor geometries can be produced. If properly designed, such dewars will allow maintenance intervals of at least 7 days before the loss of helium by vaporization requires a refill.

An alternative method to maintain SQUIDs at superconducting temperatures are cryo-coolers. In principle, these allow the continuous operation of an MEG system by completely avoiding the handling of liquid helium (Fujimoto et al., 1993). In such a system the superconductors are cooled by gaseous helium having a temperature of about 5 K. The cooled helium gas is continuously produced by a compression and expansion process. Because this process requires the movement of mechanical components, early cryo-coolers produced large magnetic and mechanical noise. Newest developments using a so-called 2-stage Gifford-McMahon/Joule-Thomson cryo-cooler reach noise levels comparable to that of the SQUID sensors (white noise level 10 fT /√Hz (Sata et al., 1998)).

Magnetic and Electric Shielding

Historically the development of biomagnetism is strongly connected with that of magnetically shielded rooms (MSRs). As a result of this development, currently available MSRs allow biomagnetic measurements to be conducted in a normal clinical environment, i.e., in buildings with a relatively high level of electromagnetic interference; for a list of commercial suppliers of MSRs see Sect. *Suppliers*.

The most straightforward and reliable way of reducing the effect of external magnetic disturbances is to perform the measurements in a magnetically and electrically shielded room (MSR). There exist several methods of active and passive shielding: ferromagnetic shielding, eddy-current shielding, active compensation and high T_c-superconducting shielding.

Passive Shielding

Passive shielding is based on diverting the flux by means of layers of high permeability metal (magnetic shielding) and/or by attenuation of interfering AC fields with conductive shells made of copper or aluminum (eddy-current shielding). Modern MSRs consist of several shells of high permeability metal and additional shells of copper or aluminum. The development of MSRs is documented by a number of publications in the last 28 years: MIT-MSR (Cohen, 1970) (three shells), Berlin MSR (Mager, 1981) (six shells), Low Temperature Laboratory, Technical University of Helsinki MSR (Kelhä et al., 1982) (three shells), COSMOS MSR (Harakawa et al., 1996) (four shells) and the new Ulm MSR (Pasquarelli et al. 1998a) (three shells). For detailed information see Andrä and Nowak (1998). In Fig. 4 A the shielding performance of a typical modern MSR is shown.

Active Shielding

Active shielding is gaining increasing interest because it may help in achieving high performance shielding at affordable prices especially at low frequencies (below 1 Hz) where only the magnetic shielding is available. In particular, it can be used in addition to a standard MSR product to achieve performances which would require an expensive and heavy custom design if it were to be obtained with a purely passive MSR. Generally speaking, active shielding consists of a reference sensor triplet measuring the field inside the MSR, a signal-conditioning network and a power stage producing compensatory fields by driving three orthogonal Helmholtz-like coil pairs around the MSR. At the University of Ulm an additional active shielding system was implemented using a fully analogue technique (Pasquarelli et al. 1998b). The results (Fig. 4 B) show that the residual noise essentially equals the background noise of the SQUID sensor used to detect the

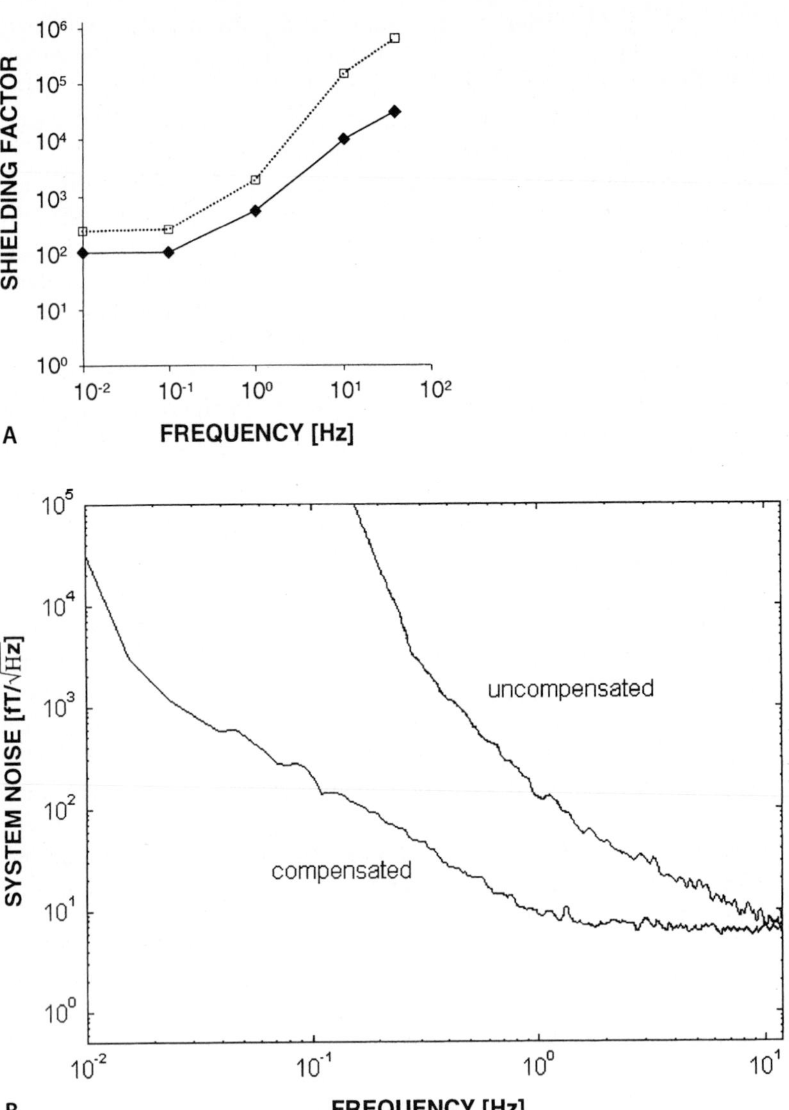

Fig. 4. (A) Shielding performance of a typical modern magnetic shielding room as function of the frequency (black diamonds, continuous lines) and of the Ulm magnetic shielded room (open squares, dotted lines; from Pasquarelli et al., 1998a). (B) Reduction of system noise by an active compensation system (from Pasquarelli et al. 1998b).

noise. The slew-rate of the compensation system does not play a significant role, because (i) the operation of the active shielding is limited to the frequency range DC-10 Hz and (ii) at higher frequencies the disturbances are strongly rejected by the eddy-current shielding of the MSR. Other interesting results were obtained in the Netherlands (ter Brake et al. 1991), at the University of Jena (Platzek and Nowak,1998) and by a few commercial companies.

Superconductive Shielding A perfect shielding down to DC frequency can, in principle, be achieved by using a superconducting material. This approach has been taken by Matsuba et al. (1995) whose SCUTUM shielding system uses the new high-temperature (high T_c) superconductors operating at the temperature of liquid nitrogen as shielding material.

Commercial Biomagnetic Measurement Systems

Several companies offer commercial biomagnetic instrumentation for brain investigations. The various commercial multi-channel magneto(gradio)meters differ in the number of sensors, in the sensor geometry (planar or helmet) and in the average noise level of their SQUIDs . For MEG, helmet-shaped dewars and sensor configurations give the best results ("whole head systems"). As of October 1998 the following MEG systems are on the market (to identify suppliers see list in Sect.: *Suppliers*):

- AtB Argos 150 system:
  Whole head system using integrated magnetometers at 153 measurement sites and three triplets for on-line noise cancellation (software gradiometer principle). The acquisition system allows the simultaneous recording of 64 EEG channels, the maximum sampling rate being 10 kHz. The shape of the helmet has been optimized to fit 97% of the Caucasian population.
- BTi system 2500 WH:
  Whole head system featuring a helmet-style dewar with 148 sensors and a 256-channel data acquisition system that accommodates up to 92 channels of simultaneously recorded EEG. Designed to operate with seated as well as with lying patients.
- CTF system:
  Whole head system with two helmet versions, optimized for North American and Asian populations, respectively. The acquisition system provides for the simultaneous recording of up to 64 EEG-channels.
- Daikin System:
  Daikin offers a 150 channel system with helmet-shaped dewar. It is the only system working with a cryo-cooler instead of using liquid helium.
- Neuromag system:
  The Neuromag 122 MEG system was the first whole head MEG device to be developed. It consists of 122 planar gradiometers located at 61 measurement points. The helmet-shaped dewar fits 95 % of the population. The Neuromag company's latest development is Neuromag Vectorview for combined MEG/EEG recordings. The system consists of 306 MEG channels with 102 triple sensor elements. These sensor elements comprise two orthogonal planar gradiometers and one magnetometer coupled to a multi-SQUID chip. A fully integrated EEG amplifier system with up to 128 channels is provided as part of the system.
- Shimadzu system:
  The Shimadzu MEG system currently is the only one that can be used as multi-channel vector magnetometer or gradiometer.
- Yokogawa system:
  Multi-channel system with characteristics similar to those of the systems listed above, except that it is the first whole head system that was designed to operate exclusively on lying patients.

Requirements to be Met by an MEG System

Taking currently available technology (see above list) as a criterion, the following minimum requirements should be observed when selecting the equipment for an MEG system:

- The MSR should be a combination of active and passive shielding with a shielding performance approximately equal to that shown in Fig. 4 A,B. Specifically, at 0.01 Hz the shielding factor should be about 40 dB, and at 30 Hz it should be better than 80 dB.
- The cryostat should be helmet-shaped with a maintenance interval of at least 7 days for liquid helium refilling. It should fit the head in at least 95% of the adult popula-

tion, and for a sufficiently dense sampling of the head surface at least 80 magnetometers or gradiometers should be evenly distributed over its inner surface.
- SQUID sensitivity should be close to 5 fT/Hz at 1 kHz.
- There should be the possibility to record at least 64 EEG signals in addition to MEG. Furthermore, at least 8 auxiliary monitoring channels (breathing, etc.) should be available.
- Ideally, a lying patient's/subject's position comparable to MR tomographs should be possible.
- The patient/subject handling setup must consist of non-magnetic material and must also be absolutely free of even minute ferromagnetic contaminations.
- The hard- and software used for data acquisition should allow fast pre-processing (e.g., band-pass filtering, on-line noise compensation, on-line averaging with artifact rejection based on statistical criteria) and provide for large storage capabilities.

Procedure

Maintenance Procedures

Two types of maintenance procedures can be distinguished: routine maintenance and special maintenance in case of faulty or degraded performance of the MEG system.

Routine Maintenance

In most modern MEG systems, the routine maintenance consists only of the weekly filling of the cryostat(s) with liquid helium and conventional computer procedures (data backup).

Special Maintenance

Degradation of the shielding quality of the MSR is an example where special maintenance is required. The effectiveness of the high permeability shell of an MSR can suffer from magnetic or mechanical stress (e.g., switching of strong electrical machines in close proximity, deformation by mechanical impact). In such a case the MSR must be de-magnetized in order to move the working point of the magnetic material back to its optimum. This can be achieved by winding a coil around the walls of the MSR (so as to make it part of a choke coil) and energizing it with an alternating current (e.g., from the 50 or 60 Hz mains) that is slowly increased up to a peak value of 400–800 Ampereturns and then slowly returned to zero.

Subject Preparation

Magnetic Decontamination

Subjects must be free of any magnetic contamination which can cause all sorts of movement (breathing, cardiac pulsation, etc.) related artifacts. Search subjects for
- Clothes with metallic parts like zip-fasteners, buttons, clasps, etc.
- Metallic fittings in bras and corsets (a common source of breathing-related artifacts in women).
- Purses, keys, jewelry from non-precious metals
- Dentures or dental implants which sometimes contain ferromagnetic material.
- Ferromagnetic deposits in the lungs or encapsulated ferromagnetic fragments in the body (e.g., after an injury). The only chance to reduce the artifacts from these sources is to demagnetize the subject by a degaussing procedure. To this end the subject must be exposed to an alternating magnetic field which is slowly reduced to zero. In our experience, commercially available coils for the de-magnetization of TV tubes can be used for this purpose (e.g., type Bernstein 2–505, Werkzeugfabrik Steinrücke; see

Sect.: *Suppliers*). The coil is eccentrically rotated in a frontal plane close to the subject at a frequency of about 1 Hz. After about 5–8 rotations the coil is slowly withdrawn from the patient by screw-like rotations until a distance of about 2 m has been reached and a total of 20 rotations have been performed; the coil then is tilted by 90° and switched off.

Determination of Head Coordinate System – Marker Placing

In most cases the aim of an MEG investigation will be to find the location of a biological source with respect to head anatomy. Therefore, a link must be constructed between the coordinate system defining the position of the MEG sensors and a head coordinate system (based on anatomical landmarks) and, ultimately, the coordinate system of an imaging system (e.g., MRI). There are mainly 2 procedures in use:

- One is to scan the head surface or selected fiducial points by means of a 3D-scanner (e.g., Isotrak II, Polhemus, cf. Sect.: *Suppliers*) referenced to the MEG system. Using a fitting procedure, the scanned surface (or fiducial points) can be matched to the anatomical contours of the head (or to the fiducial points) as identified by MR images.
- Another solution uses markers placed on the subject's head. During MEG measurements markers carrying small coils are used. When energized by a current, these coils form magnetic dipoles which can be localized by the MEG system. During MR or PET imaging the same markers are filled with an appropriate contrast agent. In the case of PET, which does not allow the direct identification of anatomical structures with millimeter precision, this procedure provides the only reliable link to the MEG coordinates.

  A proven procedure is as follows:

  Three markers are placed on the nasion and the two preauricular points, respectively, using double-sided adhesive tape. The markers consist of three coils each with exactly known geometry (printed circuits; Becker at al., 1992) mounted on the 3 faces of a small, blunt prism (Diekmann et al., 1995). Additional markers consisting of single flat coils may be placed on the midline of the head (e.g., vertex, occiput, etc.; coils may be integrated into EEG electrodes, cf. below) in order to computationally stabilize the coordinate transformation. The marker coils are either sequentially energized with a sinusoidal low frequency (e.g., 8 Hz) current, or simultaneously using currents of different frequencies. Current amplitude should be precisely stabilized but adjustable. By starting with small values (e.g., 0.1 mA) and increasing coil current stepwise, first the sensors close to the markers and later the more distant ones are exposed to an optimal signal (high signal-to-noise ratio without exceeding SQUIDs' maximum slew rate). From these measurements, the coil positions are determined by an iteration procedure similar to that used to localize biological sources (see Sect.: *Data analysis: Inversion Algorithms*) except that very precise initial assumptions can be made with regard to the sources (magnetic dipoles defined by exactly known currents and coil geometries), their number, and their relative orientation (defined by the prism geometry).

  For MR imaging, a cylindrical hole (5 mm depth, 6 mm diameter) in the center of the marker is filled with, e.g., an aqueous solution of 10 mmol/l $CuSO_4$ for contrast. For PET imaging the hole is filled with a positron-emitting radio-nuclear solution.

Preparation for Simultaneous EEG Recording – EEG Electrodes

As already pointed out, in many cases it may be attractive to combine MEG with EEG, and for some clinical purposes this combination is almost mandatory. The EEG electrodes must be free of ferromagnetic material or contamination and should be flat in order to keep the distance between head and cryostat small.

The number and placement of electrodes depend on the clinical or scientific question. If superficial sources are to be localized with simultaneous MEG and EEG measurements using a unified physical model (see Sect.: *Data analysis*: *Inversion Algo-*

rithms), the head area under investigation must be sampled in a sufficiently dense way. This may require replacing the standard ten-twenty system by a grid of 2–3 cm width.

Ideally, EEG electrodes for the combination with MEG should also carry a coil (Becker et al. 1992) by means of which electrode position on the head can be determined in much the same way as the position of the markers (cf. above). Knowing the positions of the electrodes, one can determine the shape of the head and match it to its MR image, and thus obtain the transform linking the MEG and MRI coordinate systems (as with 3-D scanners, cf. above).

Subject Positioning

Every movement of the subject produces changes of the magnetic field picked up by the MEG sensors. Therefore, the head of the subject must be immobilized with respect to the MEG system, e.g., by vacuum pads or inflatable balloons (for helmet-shaped cryostats). Subjects should be lying or sitting on a comfortable, mechanically stable, nonmagnetic bed or chair. Every effort should be made to give the subject a comfortable position as this will minimize his attempts to move. Depending on the subject's task and position during measurements, vacuum pads can be used to support and stabilize the trunk and the extremities.

Connection to Stimulation and Monitoring Equipment

Equipment for stimulation and monitoring inside the shielded room must produce no fields of its own, and connections to outside equipment must not interfere with the functioning of the SQUID electronics. In particular, cables into the shielded room must be thoroughly filtered to suppress radio frequencies, and ground loops must be avoided.

To understand the importance of these problems, note that electromagnetic interference (EMI) from radio frequencies can seriously degrade the signal-to-noise ratio (SNR) of the SQUIDs or even prevent them from working. Indeed, high-frequency fields can have two adverse effects upon SQUID operation: (1) The microscopic quantum effect upon which the working principle of SQUIDs is based (Josephson effect) becomes seriously disturbed, causing an instability of the working point or even a complete inhibition of operation. (2) The limited bandwidth of readout electronics is unable to follow the very rapid variation of high-frequency fields, causing the system to unlock. To avoid EMI, every low-frequency electric line (power or signal) penetrating the MSR wall must be conducted via a feed-through filter with proper bandwidth and adequate current/voltage rating. High-speed signals like triggers and data lines should be transmitted by means of fiber optics to achieve both a large bandwidth and EMI safety.

Ground loops (multiple ground connections) can add severe line interferences to the measured signal. The amplitude of these interferences may exceed the magnetic brain signals by two orders of magnitude. Therefore, as in every complex electronic system, the grounding scheme should be star-like and not a polygon. This may require insulators to avoid a direct grounding of appliances by electrical contacts between their cases and the MSR shell. For devices working above the line frequency (intercom, TV set) the ground loops can be broken by decoupling the signal connections either inductively with transformers or capacitively closing the return path with a capacitor.

Specific Equipment

- Visual stimuli must be projected into the magnetic shielding room through holes in the cabin wall, or must be viewed by the subject on an outside monitor through such a hole, or can be displayed via optical fiber systems.
- Auditory stimuli must be conducted via tubes to the subject's ears; to this end, a pair of 10 mm diameter PVC tubes can be installed which, inside the cabin, would end in flexible hoses leading to the subject's ears; outside they would be coupled to small electro-acoustic transducers. Conventional headphones or loudspeakers cannot be used as they produce particularly large magnetic artifacts. For simple signals, also

piezoelectric buzzers are feasible if the energizing cable is run far enough from the sensors.

- For tactile stimulation, pneumatic, hydraulic or piezoelectric stimulators can be used if the actuators attached to the subject do not contain ferromagnetic material. To deliver painful stimuli, light pulses can be fed through a glass fiber cable from a laser outside the shielded room. It is also feasible to electrically stimulate somatosensory nerves by means of conventional stimulation electrodes, provided the cables carrying the stimulation current are very carefully twisted.
- For voice intercoms capacitor microphones and piezoelectric loudspeakers should be used inside the shielded room, again with the caveat that connecting cables should run at sufficient distance from the sensors.
- TV cameras for monitoring subject behavior (a highly desirable feature) should be encapsulated, except for the objective, in an electrically and magnetically shielding box.
- Equipment needed with patients such as monitoring (e.g., CO_2) and intervention batteries should be installed outside the shielded room and connected via cables (same caveats as above) and plastic tubes to the patient.

Conduction of Measurement

Choice of Acquisition Parameters

Before starting an MEG recording, the parameters of the data acquisition program must be chosen. With regard to filtering and sampling frequency, similar general considerations apply as when recording other biological time functions:

- Upper and lower corner frequencies will be set corresponding to the expected frequency content of the signal. Both high- and low-pass filtering contribute to improving the signal-to-noise ratio; high-pass filtering also limits the required sampling frequency and, hence, the amount of data to be handled. Magnetic brain signals mostly contain frequency components of less than 1 kHz. Normal spontaneous magnetic brain activity, middle and long latency evoked fields and also epileptiform magnetic activity range from 0 – ~70 Hz. Short latency evoked fields and nervous compound action fields may contain frequency components up to several hundred Hz.
- Sampling frequency basically must satisfy the sampling theorem and avoid aliasing phenomena (see Chaps. 35, 45). The choice of the sampling frequency therefore will mainly depend on the upper corner frequency but also on filter steepness above that frequency. Typical sampling rates for spontaneous brain activity range from 200 to 500 Hz, while compound action fields (peripheral nerves) may require up to a few kHz.
- Data can be stored continuously or in segments containing only the events of interest. Continuous storage preserves the possibility to analyze the data off-line with different epoch lengths but may be demanding in terms of storage capacity. Segmented storage needs less capacity but limits the possibilities of later analysis. The experimenter also has to decide whether he wants to store raw or preprocessed data (e.g., magnetometer vs. gradiometer signals) or even only averages produced by on-line algorithms. For clinical purposes the latter strategy may in some cases be sufficient, whereas for most scientific purposes single sweeps are preferable. Needless to say that the above choices may also depend on the software available to users of commercial systems who do not want to write their own programs.

Calibrations

Each session of MEG recording is accompanied by a variable number of technical measurements needed for calibration, noise suppression, etc. Once the subject is positioned and prior to acquiring any physiological data, calibration routines have to be performed. Depending on the particular MEG system being used, these may include the

- Determination of software gradiometer coefficients.
- Calibration of active shielding equipment.
- Measurement of head position with respect to the sensor coordinates. This is a mandatory routine with all systems whether using 3-D scanners or magnetic markers (see above *Determination of Head Coordinate System – Marker Placing*). It must be carried out very carefully because all inferences regarding 3-D source locations within the brain depend critically on it. From the beginning of this routine and throughout the subsequent biological measurements subjects must not move their heads. If head movements are nonetheless detected while measurements of biological activity are in progress, it is strongly recommended to interrupt the session and to repeat the calibration of head position. If the experimental design does not tolerate an interruption, head position should be measured at the conclusion of the experiment and be compared to its pre-session values to make sure the movements were below a predefined tolerance threshold. Even without a suspected head movement it is good practice to check in this way at the conclusion of a measurement whether the head was indeed stable.

Biological Measurement Proper

There are few generally valid recommendations that can be given for the conduction of the biological measurement itself which will largely depend on the experimental or clinical context of the measurement. As far as possible the technical quality of the acquired data should be continuously checked on-line by visual inspection. As far as the experimenter has control over the environment of the MEG site, he will make sure that no ferromagnetic masses (e.g., elevators) are being moved during his measurements. In order to identify the cause of those artifacts which possibly are related to the subject's behavior, it is advantageous if the experimenter can view the subject by means of a TV system or can leave an assistant with the subject inside the shielded room. In clinical studies it is good practice anyway to have a person sitting by the patient, noticeably in cases of epilepsy and claustrophobia.

Data Analysis

Preprocessing

An important first step in the analysis of MEG data is preprocessing. Preprocessing is aimed at optimizing the signal-to-noise ratio (SNR). The MEG data sampled during a measurement constitute a time series. Therefore, all techniques developed to improve the SNR of time series, and noticeably those in use for EEG, can also be applied to MEG data.

Digital filtering will be used to restrict the bandpass of the data to a specific "band of interest" or to eliminate specific noise bands like low-frequency drift or power line components.

If repeatable situations are being investigated (e.g., sensory evoked activity), the signal-to-noise ratio can be improved by averaging. Under the assumption that the data measured during any epoch, $g(t)$, results from the superposition of always the same biological signal, $s(t)$, and an uncorrelated Gaussian noise component, $n(t)$, that is,

$$g(t) = s(t) + n(t),$$

averaging of N epochs will improve the SNR by $1/\sqrt{N}$ (see Chapter 45). In practice, however, the improvement is smaller because the response $s(t)$ of a biological system may change over time (e.g., by fatigue or adaptation processes), and because the noise, being itself mostly of biological origin (spontaneous brain activity, heart activity, etc.), may have non-Gaussian character and may exhibit temporal and spatial correlations.

- To cope with biological noise, procedures aimed at eliminating spatially coherent components (e.g., the artifacts produced by eye movements) often are used. One way to realize such "filters" (in a broader sense) is to record the noise close to its sources (e.g., the eye movements) in separate channels and to identify and subtract correlated activity in the signal channels (for a suitable algorithm see, e.g., Widrow (1985) or Abraham-Fuchs et al., 1993). A method to eliminate or reduce biological noise sources without recording the "noise" in a separate channel was proposed by Huotilainen et al. (1995) and is based on the signal space projection method. The idea is as follows: The noise (e.g., an eye blink) is identified in the signal channels by marking a time instant with maximal noise interference. This time instant represents a multidimensional "noise" vector with a certain amplitude and direction in the signal space. Then, in principle, the noise component in the signal during instants of less than maximal interference can be reduced by eliminating that part of the signal which has the same direction as the "noise" vector.
- Time jitter of the signal $s(t)$ (or of the trigger signal) with respect to the provoking stimulus may also be a reason for a decrease in SNR. In this case SNR can be improved by estimating the stimulus-response delay prior to the averaging, e.g., by a matched-filter technique (e.g., Whalen 1971).

More sophisticated procedures to overcome the limitations of straightforward averaging have been described by de Werd (1981) and Bertrand et al. (1990). Remember, however, that prevention of noise during recording should be the preferred method of optimizing SNR.

Orienting Overview, Field Maps

After preprocessing, the magnetic field distribution can be determined for each sampled time instant and can be visualized in form of maps (mapping programs are part of almost every commercial system). Typically, these maps show contours of constant field intensity on the surface of the head, or on a sphere approximating the head, or on a plane tangential to some point of the head. They convey an orienting overview of the results of a measurement and are useful in generating first ideas about the underlying sources and their temporal behavior.
For example, Fig. 5A shows the field evoked by electrical shocks to the left N. medianus sampled at 9 different stimulus latencies by an array of 22 sensors over the central right hemisphere The temporal sequence of maps in Fig. 5A reveals two roughly dipolar field patterns, one fully developed around 18 ms, the second one around 28 ms; a first guess would assume two ECDs located in roughly similar areas, one (18 ms) approximately pointing into the positive y-direction, the second one (28 ms) into the opposite, negative y-direction.

Note, however, that with increasing complexity of the field pattern (Fig. 5B) it becomes increasingly difficult if not impossible to make sensible guesses on the basis of visual inspection alone.

Solving the Inverse Problem

1. Choice of Source Model

The choice of the source model is as critical a step for the localization procedure as the choice of the head or trunk model. In addition to the field maps, physiological constraints must also be taken into account. If, for example, the field pattern of a spike discharge in a patient with focal epilepsy has dipolar character at a certain instant of interest, it is reasonable to start with a single ECD. Similarly, when studying the response to stimuli which are known to evoke bilateral activity, one obviously has to consider at least two ECDs. However, besides physiology the choice of the source model must also take into account practical constraints such as speed, computing resources, availability of specific software packages, and also the goal of an investigation.

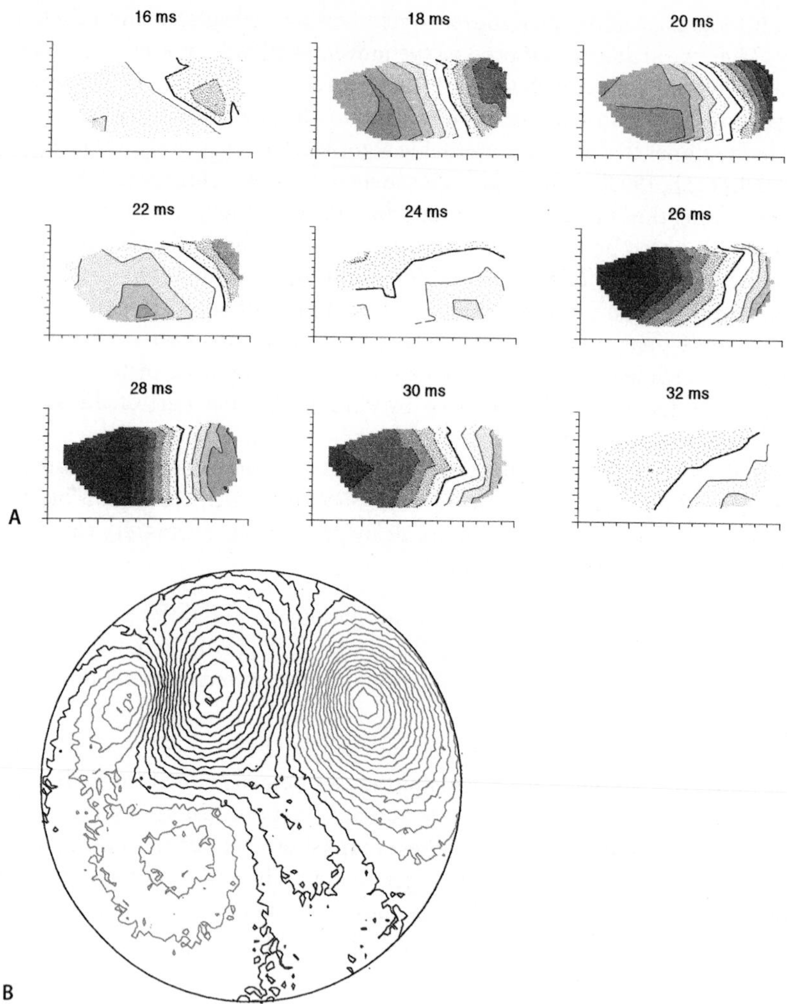

Fig. 5. (A) Iso-density maps of the radial field component over the somatosensory cortex area sampled at various latencies after an electrical shock to the N. medianus; average (N=274) from one subject. Magnetic flux out of head marked red, flux into head blue; bold black line marks loci of zero flux. Line distance (delta field increment) 20 fT. Note dipolar character and reversed direction of fields at 18 and 28 ms. (B) Complex field pattern with several field maxima and minima exemplifying a case where it is almost impossible to determine the number of active sources from the field map. The pattern was obtained by simulating five simultaneously active dipoles each of 10 nA·m field strength in a sphere of 9 cm radius with added Gaussian noise of 9 nA·m peak value (delta field increment 18 fT).

If only one particular instant in time is of interest (e.g., the instant of peak amplitude of an epileptic spike), the choice merely involves the decision whether and how many discrete ECDs should be considered (as in the examples above) or whether the complexity of the field pattern warrants a distributed sources approach (described further below).

If the whole time course of an activity is to be analyzed, various models have to be taken into consideration; here we will restrict ourselves to three classes:

- *Single Moving Dipole Model:* If field maps change with time, yet remain essentially dipolar, and unless physiology suggests a different scenario, the biological current sources may be summarized by a single ECD whose parameters (position, moment) are determined for each time instant by a separate fit of the field existing at this instant. Thus, the position of the dipole which in a sense represents the 'center of gravity' of the sources, is allowed to move as time progresses. This model is widely used and has proven its applicability in many scientific and also some clinical studies (see, e.g., Aine et al., 1998). Note that the fit of only a particular single instant described above often is also referred to as a moving dipole model because the dipole position is moving with each iteration of the fitting procedure.

- *Fixed and Rotating Dipole Models:* If the field maps do not have simple dipolar character and/or if physiological constraints suggest that more than one distinct source is active at a time, one often makes the assumption that the biological sources have fixed locations and orientations, with only their strength varying over time. Alternatively, one also may allow the current orientations to change with time. These assumptions lead to the so-called *fixed dipole* and *rotating dipole* models, respectively. In addition to the location of their dipoles, these models also specify time courses of the dipoles' strength and, in the case of rotating dipoles, their orientations. An essential prerequisite for the application of these models is a high SNR. Moreover, it is often difficult to determine the appropriate number of dipole sources. Finally, fixed dipole models fail if the distances between the sources are too small for a given sensor configuration. With the typical sensor-to-source distance of current instruments (about 4 cm), sources spaced 2 cm or closer are difficult to separate. Fixed and rotating dipole models originally were introduced for the analysis of electric brain data (EEG) by Scherg in 1984 (for a review see Scherg (1990) and Mosher et al. (1992)) and are now being successfully applied especially to evoked electric and magnetic activity.
- *Current Distribution Models:* Rather than localizing a few discrete ECDs, these models seek to determine a current density distribution which would explain the measured data. As already pointed out, no unique solution exists for the inverse problem. Therefore, physical and/or mathematical constraints must be imposed to overcome the problem. Depending on these constraints, several classes of distributed-sources solutions of the inversion problem can be distinguished. The method first used is the conventional minimum norm estimate (for an early review see Hämäläinen et al., 1993), which requires the Euclidean or L_2-norm of the current sources to be minimal. Later refinements based on the introduction of additional constraints led to several variants (LORETA: Pascual-Marqui et al., 1994; MFT: Ioannides et al., 1995; CCI: Fuchs et al., 1995; FOCUSS: Gorodnitsky and Rao, 1997; VARETA: Valdes-Sosa et al., 1998). A method to estimate the current distribution by spatial filtering was introduced by Robinson and Rose (1992).

All these methods should be used cautiously. They are not general-purpose tools for solving the inverse problem! They may produce ghost images, and most share the property of producing blurred images from point sources. Also, the conventional minimum-norm estimate favors sources near the sensors. Minimum-norm solutions may be used (1) if no or only very limited additional information is available and cannot be supplemented by reasonable assumptions, or (2) if it is sufficient for the purpose of the investigation to estimate only regions with higher current densities. Minimum-norm estimates have been applied to the late components of evoked activity and to the localization of epileptic and pathological brain rhythms (e.g., Grummich et al., 1992; Bamidis et al., 1995; Ioannidis et al., 1995).

2. Choice of Volume Model

The choice of the volume (head, extremities, trunk) model can critically influence the localization precision, especially when combined MEG and EEG data (or pure EEG data) are modeled. With MEG data from superficial cortical sources a simple *spherical model* often yields an acceptable precision, provided the skull over the area of interest can be reasonably approximated by a sphere. Realistically shaped models increase precision noticeably only when the skull and/or the cortex surfaces significantly deviate from sphericity, and this at the cost of significantly more programming effort, computing power, and time. The last point holds especially for *finite element models* which have the advantage that they can also account for the anisotropy of the brain's conductivity (the conductivity along a fiber bundle is larger than across the bundle). Thus, the choice of the volume model must take into consideration (1) the presumed area of origin of the data, (2) the desired precision of the localization, and (3) the available computing facilities. Three classes of volume models can be distinguished:

- *Simple Geometrical Models*: Sphere, spherical shells, infinite half space. These models have the advantage that there exist analytical solutions to the forward problem, which can be computed very fast. They should be used (1) if only centimeter precision is required, or (2) if the source space lies in an area where the head's surface can be well approximated by a sphere (e.g., the central area with the primary motor and somatosensory cortices, or the visual cortex in the occipital area). Simple geometric models may also be used if computing speed is more import than precision, or if large amounts of data must be analyzed with limited computing facilities. Which geometric model is appropriate depends on the type of data that is being considered:
  - *MEG Data*: A simple sphere can be chosen which should be fitted to the local curvature of the brain region under investigation. The boundary of the brain is best derived from MR images of the subject under examination. If MR images are not available, one can fit a sphere to the head contours recorded with a 3D scanner or determined by magnetically localizing EEG electrodes (see Sect.: *Subject Preparation*); the model sphere then should be chosen to be about 1.5 cm smaller in radius to account for the thickness of the scalp and the bone.
  - *Combined MEG/EEG (or Pure EEG Data)*: A 4-shell spherical model must be fitted to the brain (innermost sphere), cerebrospinal fluid, skull and scalp compartments of the head, respectively. In our experience it is sufficient to use concentric shell models. Compartment boundaries must again be derived from MR images of the individual or, if not available, must be estimated from his head contours using a thickness of 6–7 mm for each of the scalp and skull compartments, and of 2 mm for the cerebrospinal fluid compartment.
  - *Magnetoneurography:* The trunk or the extremities can be modeled by an infinite half-space with its surface tangential to the border between air and body.

- *Realistically Shaped Boundary Element Models (BEM)*: These models dissect the brain (the trunk, the extremities) into piecewise homogeneous, isotropic compartments. The (closed) surfaces of these compartments, which separate areas of different conductivity, must be determined from MR or CT scans. A first step in calculating the magnetic field is to determine the electric potentials at these surfaces (Barnard et al., 1967; Meijs et al., 1987). To this end, the surfaces are generally decomposed into triangles. The number of triangles depends on the details of the surface that is being modeled and on the spatial frequency of the source distribution in the vicinity of the surface; to handle sources near a boundary (high spatial frequencies), the number of triangles must not be too small in order to estimate the potential distribution on the surface with sufficient precision. To ease the computing burden, which increases with the number of triangles, one often may wish to find a compromise, e.g., by using a mesh of triangles with local refinements at critical locations. Simulations show that about 2000–3000 triangles are needed to model the brain, the skull or the scalp surfaces when a millimeter localization precision is to be achieved (Yvert et al., 1996). Because numerical methods must be used, the computing burden of the boundary element method is by several orders of magnitude larger than with simple geometric models. However, with the advent of cheap computers of ever increasing capacity, BEM is likely to become the state of the art method when MRI scans are available and localization precision needs to be better than can be achieved with spherical models; this holds especially true if also EEG data must be accounted for. Examples of the use of BEM can be found in Meijs et al. (1987), Hämäläinen and Sarvas (1989), Pruis et al. (1993), Cuffin (1995), and Haueisen et al. (1997a).

- *Finite Element Models (FEM):* These models are able to also account for anisotropies of the volume (head, extremities, trunk) conductivity, for regions not covered by the scalp (eyes), and for fluid-filled volumes (ventricles, blood vessels, cysts etc.). To obtain an FEM, the 3D volume of the head must be divided into many small, piecewise homogeneous volume elements. As numerical methods have to be used in 3D space instead of on 2D surfaces, the computing burden increases dramatically in comparison to BEM. If anisotropies are to be taken into account, this burden becomes even more inflated. Therefore, FEM should be restricted to cases where only small amounts of data have to be analyzed and today's best-known precision is required (e.g., when using FEM as a gold standard for the comparison of different methods), or when computing time plays only a minor rule. An as yet not fully solved problem, which prevents FEM from reaching its full precision, is our inaccurate knowledge of the conductivity distribution within the living organism. Up to now no measurement procedure is known that would determine the conductivity parameters of the head's different compartments in the frequency band of interest ($< \sim 10$ kHz) with sufficient precision. Examples for the application of FEM can be found in Yan et al. (1991), Pruis et al. (1993), Haueisen et al. (1995), Gevins at al. (1996, 1997), Buchner et al. (1997), and Haueisen et al. (1997b).

3. Inversion Algorithms

When modeling sources by ECDs, the "inverse problem" consists of finding the dipole parameters that could explain the observed field values. The problem is solved by a fitting procedure which alternates between two computational steps: A forward step first calculates the field the currently assumed source(s) would generate, using the chosen volume model. Thereafter a correction step compares the calculated and measured field distributions; from their difference (error) it derives corrective terms to adjust the currently assumed ECD parameters in such a way as to reduce the error upon the next forward step. These steps are iterated until the error falls below a predefined minimum. When computing the correction terms one can take advantage of the fact that the field depends linearly on the dipole moments and nonlinearly on the dipoles' locations. Only the non-linear part of the problem needs to be treated by a numerical optimization algorithm to obtain a correction of position. Thereafter, the corrected dipole moment can easily be computed by solving a linear equation system. The optimization algorithm should be fast, yet not lead to local error minima upon iteration. If prior knowledge exists about the location of a source (e.g., the position of a marker coil on the head), fast gradient methods like the Levinson-Marquart algorithm (Marquart, 1963) may be used. In all other cases, methods which contain a random search component are preferable. In our experience, Nelder and Mead's simplex method (Press et al., 1992) is one which rarely ends in local minima, but it is also a fairly slow one. At any rate, it is always good practice to repeat the inversion procedure with different starting values (=assumptions about ECD position and orientation) in order to ascertain that a solution corresponds to a global, and not a local, error minimum.

When MEG and EEG data are to be combined, they must first be normalized in some way because they represent different physical units (MEG: Tesla; EEG: Volt). A proved method (Diekmann et al., 1998) is to normalize the MEG data to σ_m where σ_m^2 is the residual variance of a separately fitted pure MEG model, and the EEG data correspondingly to σ_e. As a beneficial side effect, this normalization method gives greater weight to the modality (MEG or EEG) that produces the better explanation of the corresponding data (i.e., the model of which has lower residual variance).

4. Source Parameter Confidence Limits

Inevitably, each biomagnetic measurement contains noise. Therefore, even if the assumed source and volume models were absolutely correct, the source parameters produced by the inversion procedure can only be estimations of the real parameters and need to be qualified by confidence limits. In case of uncorrelated noise the confidence intervals may be easily deduced from the least square errors of the inversion iteration process. However, there is not only instrument noise but also "biological noise". In MEG, the spontaneous brain activity as a whole constitutes a background noise for the signal of interest (a particular rhythm, an epileptic spike, an evoked activity); this background is highly correlated over time and space. Conventionally calculated confidence regions become unwarrantably small when this correlation is ignored. A better measure is the covariance in time and space of the background activity (see van Dijk et al., 1993). Most reported methods for computing confidence limits are either based on a linearization of the confidence region in the neighborhood of the solution vector (for a review see Hämäläinen et al., 1993) or are based on computer simulations (e.g., Kuriki et al., 1989; Mosher et al., 1993; Radich and Buckley, 1995)). Mosher et al. (1993) suggested the use of Cramer-Rao lower bounds to characterize the precision of localizations. Cramer-Rao lower bounds do not only account for the noise but also for the source-sensor geometry (e.g., distance and relative orientation of sensors with respect to source) which also influences the precision of localizations. However, as their name implies, Cramer-Rao lower bounds provide only lower limits for the errors. They are not confidence limits in the conventional sense. Thus, they can only be used as relative measures, for example when different solutions are compared, or when summarizing measures are computed (see Sect.: *Summarizing Methods*).

5. Physiological Constraints

Because no unique solution exits for the biomagnetic inverse problem, additional physiological information should be used to reduce the uncertainty of solutions. For example, it is reasonable to set an upper limit to the strength of a single source. A physiologically meaningful value for this limit can be deduced if the surface of activated brain area can be estimated. According to Lopez da Silva et al. (1991) an activated area of 3–4 cm^2 produces a dipole moment of 100 nAm. Other physiological information may concern the number of likely sources. For example, acoustically evoked cortical activity is likely to stem from at least two simultaneously active sources, one in each hemisphere. In clinical investigations, images (MR or CT) of a patient's brain may reveal cortical lesions and therefore exclude these areas as possible loci of activity.

6. Integration into MRI

In most cases one will wish to transfer the results of the inversion calculus to MR images of the brain for visualization. This is easily accomplished provided a head-based reference system for MEG data and a transform linking it to MRI (or PET) has been determined as described in Sect.: *Subject Preparation.* Using the transform, any source localization can be projected into MR (or PET) images (Fig. 6).

7. Summarizing Methods

When repeated single events are being localized (e.g., epileptic spikes), the resulting ECDs generally are scattered over a certain area. In general, this scatter can only partially be attributed to the interference of biological and technical noise. Rather, it reflects the individuality of the observed events which may arise anywhere within a certain territory (e.g., epileptogenic zone) and may have different time courses . It is therefore not

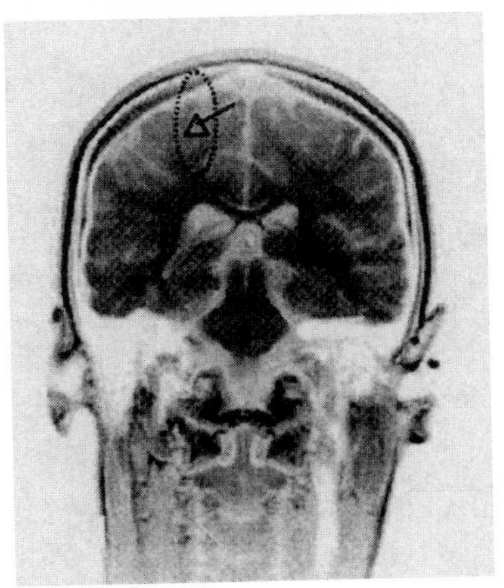

Fig. 6. Example of the localization of an equivalent current dipole representing activity during the peak of an epileptic spike (radiological view). Black triangle in right cortex indicates position, black line orientation of ECD projected onto a coronal magnetic resonance (MR) image of the patient; dotted ellipse shows Cramer-Rao lower bound of ECD location. The MR image represents the slice closest to the calculated dipole location. Black dot next to the left ear shows the image of one of the markers used for matching the MR and MEG coordinate systems.

correct to first average these events and then search for a corresponding "average ECD". Other ways to summarize the localizations (i.e., to give a measure of their central tendency) are required. A recommendable method is the dipole density distribution introduced by Vieth et al. (1992), which associates a continuous 3D-distribution (dimension: dipoles per volume) to each localization, with parameters reflecting localization uncertainty. For example, 3D-Gaussians are centered on the location of each individual ECD with variances equal to the Cramer-Rao lower bounds and their volume integrals equaling unity (or more accurately: 1 dipole). By summing the individual distributions a global dipole density distribution is obtained whose mean, median, centers of gravity etc. may be computed to describe the ensemble of localizations (Diekmann et al., 1998).

Results

Epileptic Activity

Figure 7 shows an epoch of magnetic brain activity with an epileptiform spike event recorded in a patient suffering from focal epilepsy. Figure 8 shows the equivalent dipole localizations (marked by the triangles) of 15 such events projected onto an MR image of the patient. The dipoles were fitted to the time instant of the spike's first amplitude peak (dotted line in Fig. 7). Figure 9 shows a plot of the dipole density distribution of the ECDs of Fig. 8. The plot is given in the form of isodensity contours projected onto the same MR slice as shown in Fig. 8. The distribution's center of gravity is shown as a white diamond; in the present example it does not coincide with an area of peak density because there were several local density peaks (yet, statistically, these did not warrant the assumption of two or more *separate* epileptogenic areas).

Evoked Brain Activity

As an example of evoked brain activity, Fig. 10 shows the averaged time course of the magnetic brain activity recorded by two arrays of pick-up coils in a somatosensory stimulation experiment. A 0.3 ms supra-threshold electric stimulation of the right medianus nerve at the wrist produced a biphasic temporal pattern in most channels, with a reversal of field direction about 26 ms after the stimulus (vertical lines) and 2 peaks at

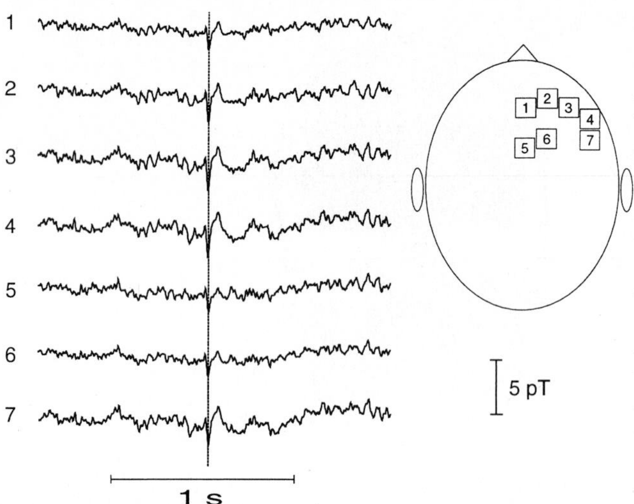

Fig. 7. Time course of magnetic brain activity in a patient suffering from focal epilepsy. The recording was made from the right fronto-central area of the brain and shows a single spike event (dotted vertical line) with maximum peak amplitude around the position of pick-up coil #3 (see right inset for the positions of the rectangular pick-up coils used in this measurement). Data low-pass filtered 0–64 Hz.

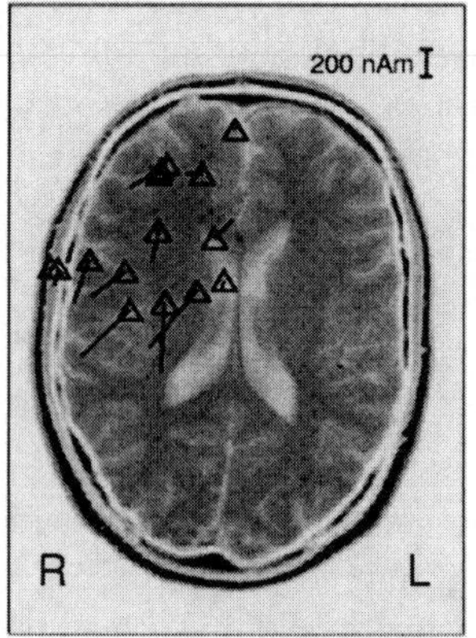

Fig. 8. Localizations (black triangles) and moments (black lines) of equivalent current dipoles fitted to the time instants of the peak amplitude of 15 epileptogenic spike events similar to that shown in Fig. 7. The data have been projected onto the patient's transversal MRI slice with the smallest distance from the median location of the 15 dipoles. R right, L left. Note right fronto-central distribution of ECDs in this patient.

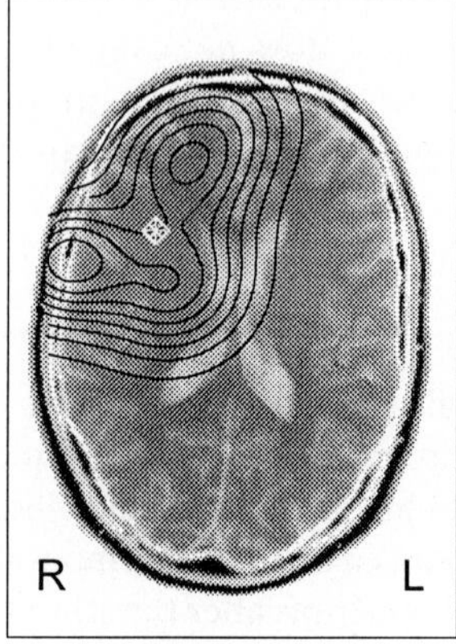

Fig. 9. Iso-density plot of the dipole density distribution (DDD) computed for the 15 dipoles whose locations are shown in Fig. 8. The iso-density lines (spacing: 0.05 dipoles/cm^3) have been projected onto the patient's transversal MRI slice closest to the DDD's center of gravity (white crossed diamond). R right, L left. The two peaks of the distribution correspond approximately to a density of 1 dipole in a cube of 1.4 cm edge length.

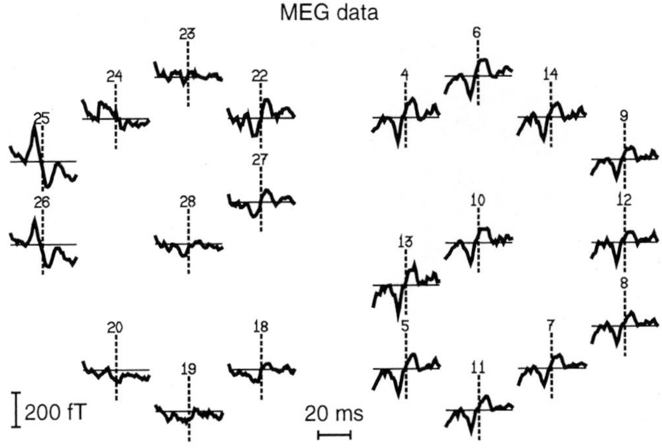

Fig. 10. Topographical representation of averaged (N=274) magnetic activity evoked by suprathreshold electrical stimulation of the right medianus nerve at the wrist of a subject. The sensor array is centered approximately at the C3 EEG-electrode position of the international 10–20-system (cf. Chapter 35 on electroencephalograpy) with the nose on top. Note that many sites show a biphasic temporal pattern of the magnetic field with a reversal of direction about 26 ms after stimulation (vertical dashed line).

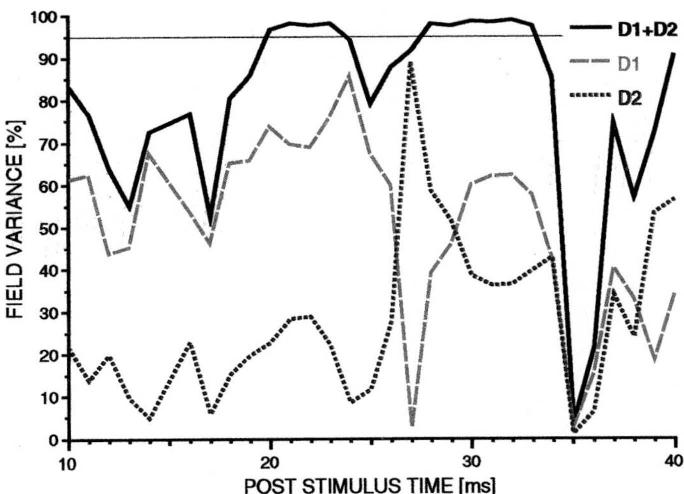

Fig. 11. Time course of the spatial field variance accounted for by the ECDs of a fixed, two-dipole model applied to the somatosensory evoked activity shown in Fig. 10. The variance is given as percentage of the total spatial variance of the averaged field. The complete model (D1+D2; black solid line) explains more than 95% of the measured field variance (horizontal hair line) in the time epochs 20–23 ms and 27–34 ms after stimulation. During the first epoch, which corresponds to the first peak of magnetic activity in Fig. 10, the dipole D1 (dashed magenta line) accounts for most of the variance. During the second epoch, which corresponds to the peak following the reversal of field direction in Fig. 10, the contribution of dipole D2 (dotted blue line) abruptly increases exceeding that of D1 for about 2 ms, and then settles at about 40% of the total field variance for the rest of the epoch.

approximately 22 and 30 ms. The resulting spatiotemporal pattern was successfully fitted by a fixed dipole model. An analysis of the fit shows that two dipoles can describe approximately 98% of the field variance during the poststimulus epochs 20–23 ms and 28–34 ms, i.e., during the occurrence of the two temporal field maxima (Fig. 11). Both dipoles have been localized to the contralateral sensory-motor brain areas of the subject

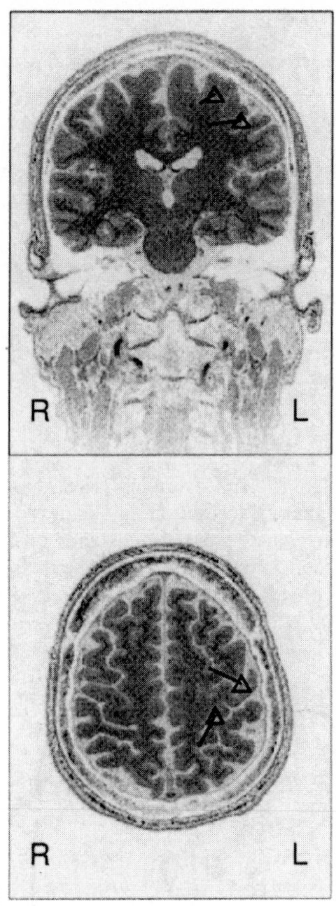

Fig. 12. Localization of the two ECDs obtained with the fixed, two-dipole model fitted to the evoked magnetic field distribution shown in Fig. 10. The fit covers the time epoch from 10 to 40 ms after stimulation. The time course of its "explanatory power" (percent variance) was shown in Fig. 11. Dipole D1 (lateral black triangle) is mainly orientated in medial direction (black line), whereas dipole D2 (medial black triangle) has a mainly posterior direction. Dipoles are projected onto those coronal (top) and transversal (bottom) slices of the subject's MRI which had the smallest distance from the median location of the two dipole locations. Both locations fall into the subject's contralateral sensorimotor area.

(Fig. 12). According to the fixed dipole model, the more laterally located ECD in Fig. 12 is the dominant component during the first peak (epoch 20–23 ms) of the evoked field; as shown in Fig11 (curve D1), this dipole explains most of the variance during the first peak, whereas both dipoles contribute about equally to the field during the second peak (epoch 27–34 ms; compare curved D1 and D2).

Troubleshoot

Technical Problems

Flux Trapping The performance of some SQUIDs and their electronics may degrade to unacceptable levels if the changes of the magnetic field exceed a critical velocity. The degradation often becomes visible in the form of an increased high-frequency noise. It probably owes to the SQUID electronics being not fast enough to fend off surges of the field so that large amounts of flux become trapped in the SQUIDs. Field surges may occur, for instance, if a person carrying ferromagnetic objects (a key, a screwdriver, etc.) moves close to the sensors, or if large currents are switched nearby. To eliminate a trapped flux, the SQUID must be warmed up above the critical temperature of the superconductor. Usually this can only be achieved by warming up the whole MEG systhem, a procedure that may be expensive and time-consuming. Therefore, steep field changes should be avoided in the neighborhood of such SQUID sensors.

Magnetic Contaminations, Stimulus Artifacts

As already mentioned (cf. *Subject Preparation*) magnetic contaminations of the subject, but also of objects attached to the subject for experimental purposes (electrodes, stimulators, etc.), may produce large artifacts during magnetic registrations. In case of objects attached to the subject, ferromagnetic soiling may result from the fabrication process (e.g., when tools give off ferromagnetic dust or fines) and can possibly be eliminated by carefully cleaning with a diluted hydrochloric acid solution and subsequent washing in water. Magnetic contamination should be suspected if artifacts are correlated with movements of the subject or with his heartbeat or respiration. Artifacts produced by stimulation equipment often can be identified, and then eliminated or minimized, by simulating a recording session without the subject in place but with the stimulating equipment in approximately the same position as in a real experiment. Systematically varying the position of the current-conducting wires in the mock setup may give hints at how to eliminate or minimize the stimulation artifacts. If the artifacts fail to manifest in the mock situation, it may be necessary to use phantoms (e.g., resistor networks or spheres filled with saline solutions) in order to imitate experimental conditions with the subject in place.

Vibration Artifacts

Vibrations can also ruin magnetic recordings. They should be suspected if there are very regular signals with waxing and waning amplitudes. Discrete low frequency (< 30 Hz) noise often is of mechanical origin and can be traced back to vibrations of the dewars, the gantry, or of large metallic constructions near the recording place. A striking example from the authors' laboratory are vibrations of the walls of the shielded room excited by a rescue helicopter landing at a distance of 150 m. Generally, however, the identification of these generators may be difficult, especially if the artifacts occur only sporadically and/or in the same frequency band as the magnetic brain activity. The frequency of the vibration may hint at the involved masses (the lower the frequency, the higher the masses). The only way to eliminate or attenuate vibration-induced artifacts is to change the mechanical design of the vibrating objects and/or their support.

Low-Frequency Noise

Irregular low-frequency noise may be generated by transient distortions of the earth magnetic field due to the movement of large ferromagnetic objects like cars, elevators, doors, etc. This kind of noise must be minimized by appropriately choosing the recording site and using a shielded room with sufficient field attenuation at low frequencies. If these conditions cannot be met, various techniques may be used during the preprocessing step (e.g., software gradiometers removing coherent low-frequency noise, or appropriate high-pass filters if this is compatible with the purpose of the measurement). Alternatively, an active shielding system may be installed (see Sect.: *Materials*).

When low-frequency noise seems to increase with time, the permeability of the MSR may have suffered. It can be restored by de-magnetization (see Sect.: *Maintenance Procedures*).

Power-Line Noise

Power-line artifacts (50 or 60 Hz and its harmonics) also may sometimes present a serious problem. They may stem from nearby power lines carrying large currents, or may be due to a badly chosen grounding point of the magnetically shielded room or to ground loops. The first possibility should already be addressed during the planning of the recording site. With regard to the problem of grounding we note that the conductive shell of a shielded room can be imagined as a capacitive antenna which picks up the electromagnetic field of ambient power lines and capacitively couples a certain amount of it into the electronic equipment inside the MSR. Ideally, grounding prevents this coupling by keeping the shell at the ground iso-potential state. However, because it is almost impossible to determine the actual path of the currents flowing to the ground, the best connection must be determined by successive trials at different points. Also, it has to be taken into account that the quality of the grounding contacts into the soil may significantly change over time (seasonal variation of soil humidity, oxidation of the conductor, etc.). Last but not least, the MSR and SQUID electronics should use "private" ground conductors and should not share them with other facilities in order to avoid a cross-coupling of the ground currents (common mode noise). Finally, ground loops must be avoided as discussed in Sect.: *Subject Preparation – Connection to Stimulation and Monitoring Equipment.*

A slow degradation with time of the shielding room's ability to suppress power-line noise sometimes hints at an interruption of the eddy-current shield at the door where the electrical contact between the door and the other parts of the conductive shell enclosing the room may deteriorate.

High-Frequency Interference

As already mentioned in the subject preparation paragraph, many SQUID systems are very sensitive to high-frequency (hf) irradiation. Cables led into the shielded room without HF rejection filters may transmit signals (e.g., radio or TV signals) from outside the room. The identification of HF as a cause of malfunction may be difficult because its effects upon SQUID output are hard to predict.

Biological Problems

The subject itself may be a source of artifacts. Movements of the head (or of the body in the case of magnetoneurography) cause artifacts by moving the field past the sensors. Movements of the eyes and the activity of the heart generate currents, and by the same token, magnetic fields (magnetooculogram, magnetocardiogram) which may interfere with the intended measurements. Countermeasures have been described in Sect.: *Subject Preparation* and *Preprocessing.*

Applications

Clinical

Neuromagnetic measurements are currently being applied, or will probably be applied, to the following problems
- localization of epileptic foci
- localization of pathological rhythmic activity in stroke and after transient ischemic attacks (TIA)
- presurgical mapping of eloquent areas
- follow-up of rehabilitation
- replacement of EEG in mass screening (saves time, but see caveat in Sect.: *Fields of Application, Advantages and Limits of MEG*)
- localization of peripheral nerve dysfunction (conduction block, rupture).

Obviously, besides these topics of potentially practical diagnostic value, most research into clinical problems that hitherto has been carried out with EEG can also make use of MEG. To mention but one area of research, MEG is being applied in the search for biological markers of mental diseases, e.g., deviations in functional brain anatomy.

Research

Neuromagnetic methods are being successfully used in studies of
- sensory evoked brain activity (acoustic, somatosensory, visual, pain),
- cognitive event-related activity (pattern recognition, learning, memory, processing activity, etc.)

- cortical motor activity
- functional mapping
- high-frequency cortical oscillations (e.g., test of binding hypothesis)
- cortical plasticity (remapping of brain areas)
- processing in deep structures (thalamus, hippocampus)

References

Abraham-Fuchs K, Strobach P, Härer W, Schneider S (1993) Improvement of neuromagnetic localization by MCG artifact correction in MEG recordings. In : Baumgartner C, Deecke L, Stroink G, Williamson SJ (eds) Biomagnetism: Fundamental Research and Clinical Applications. Elsevier Science, IOS Press, Amsterdam, Oxford, Burke, Tokyo, pp.787–791

Aine C, Okada Y, Stroink G, Swithenby S, Wood CC (1999, in press) Advances in Biomagnetism Research: Biomag96, Springer, New York

Andrä W, Nowak H (1998) Magnetism in Medicine – A Handbook. Wiley VCH, Weinheim

Bamidis PD, Hellstrand E, Lidholm H, Abraham-Fuchs K, Ioannides AA (1995) MFT in complex partial epilepsy: spatio-temporal estimates of ictal activity. NeuroReport 7: 17–23

Barnard CL, Duck IM, Lynn MS (1967) The application of electromagnetic theory to electrocardiology. I. Derivation of the integral equations. Biophys. Journal 7: 443–462

Becker W, Diekmann V, Jürgens R (1992) Magnetic localization of EEG electrodes for simultaneous EEG and MEG measurements. In: Dittmer A, Froment JC (eds) Proceedings of the Satellite Symposium on Neuroscience and Technology, 14th Annual International Conference of the IEEE Engineering in Medicine and Biology Society. Lyon, pp 34–36

Becker W, Diekmann V, Jürgens R, Kornhuber C (1993) First experiences with a multichannel software gradiometer recording normal and tangential components of MEG. Physiol Meas 14: A45–50

Bertrand O, Bohorquez J, Pernier J (1990) Technical requirements for evoked potential monitoring in the intensive care unit. In Rossini PM, Mauguière F (eds) New Trends and Advanced Techniques in Clinical Neurophysiology. Elsevier, Amsterdam, Vol EEG suppl 41: 51–70

Buchner H, Knoll G, Fuchs M, Rienacker A, Beckmann R, Wagner M, Silney J, Pesch J (1997) Inverse localization of electric dipole current sources in finite element models of the human head. EEG clin Neurophysiol 102: 267–278

Buckel W, Superconductivity. Fundamentals and Applications (1990). Wiley-VCH, Weinheim

Clarke J, Goubau WM, Ketchen MB (1975) Thin-film dc SQUID with low noise and drift. Appl Phys Lett 27: 155–156

Clarke J, Goubau WM, Ketchen MB (1976) Tunnel junction dc SQUID fabrication, operation, and performance. J Low Temp Phys 25: 99–144

Cohen D (1970) Large-volume conventional magnetic shields. Rev de Physique Appl 5 : 53–58

Cuffin BN (1995) A method for localizing EEG sources in realistic head models. IEEE Trans Biomed Eng 42: 68–71

Curio G, Neuromagnetic recordings of evoked and injury related activity in the peripheral nervous system (1995). In : Baumgartner C, Deecke L, Stroink G, Williamson SJ (eds) Biomagnetism: Fundamental Research and Clinical Applications. Elsevier Science, IOS Press, Amsterdam, Oxford, Burke, Tokyo, pp. 709–714

de Weerd JPC (1981) A posteriori time-varying filtering of averaged potentials. I. Introduction and conceptual basis. Biol. Cybern 41: 211–222

Diekmann V, Jürgens R, Becker W (1995) Magnetische Lokalisation von Markern. Biomed Tech 40, suppl 1: 211–212

Diekmann V, Jürgens R, Becker W, Elias H, Ludwig W, Vodel W (1996) RF-SQUID to DC-SQUID upgrade of a 28-channel magnetencephalography (MEG) system. Meas Sci Technol 7: 844–852

Diekmann V, Becker W, Jürgens R, Grözinger B, Kleiser B, Richter HP, Wollinsky KH (1998) Localisation of epileptic foci with electric, magnetic and combined electromagnetic models. EEG clin Neurophysiol 106: 297–313

Drung D (1995) The PTB-SQUID system for biomagnetic applications in a clinic. IEEE Trans Appl Supercon 5: 2112–2117

Erné SN (1983) Squid sensors. In : Williamson SJ, Romani GL, Kaufmann L, Modena I (eds) Biomagnetism: An Interdisciplinary Approach. Plenum Press, New York and London, pp. 69–84

Fuchs M, Wagner M, Wischmann HA, Dössel O (1995) Cortical current imaging by morphologically constrained reconstructions. In : Baumgartner C, Deecke L, Stroink G, Williamson SJ (eds) Biomagnetism: Fundamental Research and Clinical Applications. Elsevier Science, IOS Press, Amsterdam, Oxford, Burke, Tokyo, pp. 320–325

Fujimoto S, Ogata H, Kado S (1993) Magnetic Noise produced by GM-Cryocoolers. Cryocoolers 7: 560–568

Gevins A, Smith ME, Le J, Leong H, Bennett J, Martin N, McEvoy L, Du R, Whitfield S (1996) High resolution evoked potential imaging of the cortical dynamic of human working memory. EEG clin Neurophysiol 98: 327–348

Gevins A, Smith ME, McEvoy L, Yu D (1997) High-resolution EEG mapping of cortical activation related to working memory: effects of task difficulty, type of processing, and practice. Cereb Cortex 7 : 374–385

Gorodnitsky IF, Rao BD (1997) Sparse signal reconstruction from limited data using FOCUSS: a re-weighted minimum norm algorithm. IEEE Trans Sig Proc 45: 600–616

Grummich P, Kober H, Vieth J (1992) Localization of the underlying currents of magnetic brain activity using spatial filtering. Biomed. Eng. 37 (suppl 2): 158–159

Hämäläinen MS, Sarvas J (1989) Realistic geometry model of the human head for the interpretation of neuromagnetic data. IEEE Trans Biomed Eng 36: 165–171

Hämäläinen MS, Hari R, Ilomoniemi RJ, Knuutila J, Lounasmaa OV (1993) Magnetencephalography – theory, instrumentation, and applications to noninvasive studies of the working human brain. Rev Mod Phys 65: 413–497

Harakawa K, Kajiwara G, Kazami K, Ogata H, Kado H (1996) Evaluation of High-Performance Magnetically Shielded Room for Biomagnetic Measurement. IEEE Trans Magn 32: 5256–5260

Haueisen J, Ramon C, Czapski P, Eiselt M (1995) On the influence of volume currents and extended sources on neuromagnetic fields: a simulation study. Ann Biomed Eng 23: 728–739

Haueisen J, Bottner A, Funke M, Brauer H, Novak H (1997a) Effect of boundary element discretization on forward calculation and the inverse problem in electroencephalography and magnetoencephalography. Biomed Technik, 42: 240–248

Haueisen J, Ramon C, Eiselt M, Brauer H, Novak H (1997b) Influence of tissue resistivities on neuromagnetic fields and electric potentials studied with a finite element model of the head. IEEE Trans Biomed Eng 44: 727–735

Helmholtz H (1853) Ueber einige Gesetze der Vertheilung elektrischer Ströme in körperlichen Leitern mit Anwendung auf die thierisch-elektrischen Versuche. Ann Phys Chem 89: 211 –233; 353 –377

Huotilainen M, Ilimoniemi RJ, Tiitinen H, Lavikainen J, Alho K, Kajola M, Näätänen R (1995) The projection method in removing eye-blink artefakts from multichannel MEG measurements. In: Baumgartner C, Deecke L, Stroink G, Williamson SJ (eds) Biomagnetism: Fundamental Research and Clinical Applications, Elsevier Science. IOS Press, Amsterdam, Oxford, Burke, Tokyo, pp 363–367

Ioannides AA, Liu MJ, Liu LC, Bamidis PD, Hellstrand E, Stefan KM (1995) Magnetic field tomography of cortical and deep processes: examples of "real-time mapping" of averaged and single trial MEG signals. Int J Psychophysiology 20: 161–175

Jackson JD (1998) Classical Electrodynamics, 3rd Ed., Wiley, New York

Kelhä VO, Pukki JM, Peltonen RS, Penttinen AJ, Ilmoniemi RJ, Heino JJ (1982) Design, construction and performance of a large-volume magnetic shield. IEEE Trans Magn 18: 260–270

Kuriki S, Murase M, Takeeuchi F (1989) Locating accuracy of a current source of neuromagnetic responses: simulation study for a single current dipole in a spherical conductor. EEG clin Neurophysiol 73: 499–506

Lopez da Silva FH, Wieringa HJ, Peters MJ (1991) Source localization of EEG versus MEG: empirical comparison using visually evoked responses and theoretical considerations. Brain Topography, 4: 133–142

Mager A (1981) Berlin magnetically shielded room. In Erné SN, Hahlbohm HD, Lübbig H (eds.), Biomagnetism. WdG, Berlin, pp 51–78

Marquart DM (1963) An algorithm for least squares-estimation of nonlinear parameters. J Soc Indust Appl Math 11: 431–441

Matsuba H, Shintomi K, Yahara A, Irisawa D, Imai K, Yoshida, H, Seike S (1995) Superconducting shield enclosing a human body for biomagnetism measurement. In Baumgartner C, Deecke L, Stroink G, Williamson SJ (eds.) Biomagnetism: Fundamental Research and Clinical Applications. Elsevier Science, IOS Press, Amsterdam, pp 483–489

Meijs JWH, Bosch FGC, Peters MJ, Lopes da Silva FH (1987) On the magnetic filed distribution generated by a dipolar current source situated in a realistically shaped compartment model. EEG clin Neurophysiol 66: 286–298

Mosher JC, Lewis PS, Leahy RM (1992) Multiple dipole modeling and localization from spatio-temporal MEG data. IEEE trans Biomed Eng 39: 541–557

Mosher JC, Spencer ME, Leahy RM, Lewis PS (1993) Error bounds for EEG and MEG dipole localization. EEG clin Neurophysiol 86: 303–321

Pascual-Marqui RD, Michel CM, Lehmann D (1994) Low resolution electromagnetic tomography: a new method for localizing electrical activity in the brain. Int J Psycho-Physiol 18: 49–65

Pasquarelli A, Kammrath H, Tenner U and Erné SN (1998a) The new Ulm Magnetic Shielded Room. Book of abstracts BIOMAG98, Sendai, Japan, p 63

Pasquarelli A, Tenner U and Erné SN (1998b) Use of an Additional Active Shielding System to enhance the low-frequency performances of a Magnetic Shielded Room. Proceedings of IWK98, Ilmenau, Germany

Platzek D, Nowak H (1998) Active Shielding and its Application on MEG-DC Measurements, Book of abstracts BIOMAG98, Sendai, Japan, p 32

Press WH, Teulosky SA, Vetteling WT Fannery BP (1992) Numerical recipes in FORTRAN: the art of scientific computing, 2nd ed, Cambridge University Press, Cambridge, pp. 402–406

Pruis GW, Gilding BH, Peters MJ (1993) A comparison of different numerical methods for solving the forward problem in EEG and MEG. Physiol Meas 14 Suppl 4A: A1-A9

Radich BM, Buckley KM (1995) EEG dipole localization bounds and MAP algorithms for head models with parameter uncertainties. IEEE Trans Biomed Eng 42: 233–241

Robinson SE, Rose DF (1992) Current source image estimation by spatially filtered MEG. In: Hoke M, Erné SN, Okada YC, Romani GL (eds) Biomagnetism: Clinical Aspects. Elsevier Science, Amsterdam, New York, pp. 761–765

Sata K, Fujimoto S, Yoshida T, Miyahara S, Yoshii K, Kang YM (1998) A helmet-shaped MEG measurement system cooled by a GM/JT Cryocooler. Abstract proceedings BIOMAG98, Sendai, Japan, p 65

Scherg M (1990) Fundamentals of dipole source potential analysis. In: Grandori F, Hoke M, Romani GL (eds) Auditory evoked magnetic fields and electric potentials. Advances in Audiology, vol 6, Karger, Basel, pp 40–69

Templey N (1992) Spreading depression and related DC phenomena. In : Hoke M, Erné SN, Okada YC, Romani GL (eds) Biomagnetism: Clinical Aspects. Elsevier Science, Amsterdam, New York, pp. 329–335

ter Brake HJM, Flokstra J, Jaszczuk W, Stammis R, van Ancum GK, Martinez A, Rogalla H (1991) The UT 19-channel DC SQUID based neuromagnetometer. Clin Phys Physiol Meas 12: 45–50

Valdes-Sosa P, Marti F, Gracia F, Casanova R (1999 in press) Variable resolution electric-magnetic tomography. In: Aine C, Okada Y, Stroink G, Swithenby S, Wood CC (eds) Advances in Biomagnetism Research: Biomag96. Springer, New York

Van Dijk BW, Spekreijse H, Yamazaki T (1993) Equivalent dipole source localization of EEG and evoked potentials: sources of errors or sources with confidence? Brain Topography 5: 355–359

Vieth J, Kober H, Weise E, Daun A, Moegner A, Friedrich S, Pongratz H (1992) Functional 3D localization of cerebrovascular accidents by magnetoencephalography. Neurol Res, Suppl. 14: 132–134

Weinstock H (ed) (1996) SQUID Sensors: Fundamentals, Fabrication and Applications NATO ASI, Kluwer Academic Pub., Dordrecht, Boston, London, 703 pages

Whalen AD, Detection of signal in noise (1971), Academic Press, New York, London

Widrow B (1985) Adaptive signal processing. Prentice Hall US

Williamson SJ, Kaufmann L (1990) Theory of neuroelectric and neuromagnetic fields. In: Grandori F, Hoke M, Romani GL (eds) Auditory Evoked Magnetic Fields and Electric Potentials. Advances in Audiology, vol 6, Karger, Basel, pp 1–39

Yan Y, Nunez PL, Hart RT (1991) Finite element model of the human head: scalp potentials due to dipole sources. Med Biol Eng Comput 29: 475–481

Yvert B, Bertrand O, Echallier JF, Pernier J (1996) Improved dipole localization using local mesh refinement of realistic head geometries: an EEG simulations study. EEG clin Neurophysiol 99: 79–89

Suppliers

Biomagnetic instrumentation for brain investigations (as of October 1998):

- Advanced Technologies Biomagnetics, Chieti, Italy
- Biomagnetic Technologies Inc., San Diego, CA 92121–3719, USA
- CTF Systems Inc., V3C 1M9 Port Coquitlam, B.C., Canada
- Neuromag Ltd., P.O. Box 68, 00511 Helsinki, Finland
- Daikin Industries Ltd., Tsukuba, Ibaraki, 305–0841, Japan
- Shimadzu Scientific Instruments Ltd., Kyoto, Japan
- Yokogawa Electric Corporation, Japan.

Magnetically shielded rooms (MSRs) (as of October 1998):

- Ammuneal Co., Philadelphia, USA
- Euroshield Oy, Eura, Finland
- Imedco AG, CH-4616 Hagendorf, Switzerland
- Vacuumschmelze, Hanau, Germany

3-D digitizing systems:

- Polhemus Incorporated, Colchester, VT 05446 (USA)

Demagnetizing coils:

- Werzeugfabrik Steinrücke GmbH, D-42897 Remscheid, Germany

Chapter 38

Magnetic Resonance Imaging of Human Brain Function

Jens Frahm, Peter Fransson and Gunnar Krüger

Introduction

Noninvasive methods for studying metabolic and functional properties of the central nervous system provide new tools for understanding the human brain at the system level. As a bridging technology between basic neurobiologic research in (transgenic) animals, system-oriented studies in humans, and medical applications to patients with neurologic disease, magnetic resonance is expected to gain further importance in linking advances in molecular neurobiology and neurogenetics to cerebral metabolism and physiology and even beyond to human brain function.

MRI

Over the past two decades magnetic resonance imaging (MRI) has evolved into the premier modality for mapping structural anatomy at high spatial resolution and with exquisite soft tissue contrast (Stark and Bradley 1998). Parallel advances extended the range of evaluations to cellular metabolism by localized MR spectroscopy (cf. Chapter 40), adding a biochemical dimension to anatomic imaging (Bachelard 1997). Functional aspects became available through magnetic resonance angiography, perfusion studies, diffusion contrast, and magnetization transfer techniques. For neuroscientists, the most fascinating development is the discovery that suitable MRI techniques are able to map the functional anatomy of the human brain – visualizing the processes of thinking and feeling.

Neuroimaging

The use of MRI for functional neuroimaging is attractive because it
- is noninvasive,
- offers repeated studies of individual subjects and patients,
- provides access to submillimeter spatial resolution,
- yields an efficient temporal resolution of about 1s, and
- allows for flexibility in the design of cognitive paradigms.

Moreover, high-field clinical MRI units are much more widely distributed and less costly than positron emission tomography (PET) systems (cf. Chapter 39) as the major alternative for functional mapping apart from single photon emission computed tomography (SPECT) as well as electroencephalographic (EEG; cf. Chapter 35) and magnetoencephalographic (MEG; cf. Chapter 37) recordings.

Correspondence to: Jens Frahm, Biomedizinische NMR Forschungs GmbH, Max-Planck-Institut für biophysikalische Chemie, Göttingen, D-37070, Germany (phone: +49-551-201-1721; fax: +49-551-201-1307; e-mail: jfrahm@gwdg.de)
Peter Fransson, Karolinska Institutet, MR Research Center, Department of Clinical Neuroscience, Stockholm, Sweden
Gunnar Krüger, Stanford University, Department of Radiology, Palo Alto, CA, 94305, USA

Oxygenation Experimental work in animals first demonstrated that the level of cerebral blood oxygenation influences the signal intensity of T2*-weighted gradient-echo MR images (Ogawa et al 1990, Turner et al 1991). Whereas oxyhemoglobin is diamagnetic, deoxygenated hemoglobin serves as an endogenous paramagnetic contrast agent that dephases the nuclear spins of water protons in its direct vicinity, i.e., primarily within the venous microvasculature and surrounding tissue.

Contrast The physical effect is a signal loss in MR images which integrate pertinent magnetizations over an image pixel during the acquisition process. The degree of signal loss depends on the absolute concentration of deoxyhemoglobin per voxel and scales with voxel size and gradient-echo time, i.e., the differential precession of spin moments between excitation and detection. For example, if an increase in cerebral blood flow and oxygenation decreases the deoxyhemoglobin concentration, then both the effective spin-spin relaxation time T2* and the corresponding signal intensity of a gradient-echo image show a concomitant increase. This blood oxygenation level dependent (BOLD) contrast may be enhanced by acquiring the signal at prolonged echo times of, for example, 30 to 50 ms depending on field strength and MRI technique.

Function The demonstration of oxygenation sensitivity in animal imaging was rapidly followed by the observation that BOLD contrast may be exploited for functional mapping of the human brain (Kwong et al 1992, Ogawa et al 1992, Bandettini et al 1992, Frahm et al 1992, Blamire et al 1992). The driving force is the observation that a change in neuronal activity – often termed activation – causes a rise in cerebral blood flow that at least transiently "uncouples" from oxygen consumption (Fox and Raichle 1986, Frahm et al 1996) and therefore results in a venous hyperoxygenation, i.e., decreased deoxyhemoglobin. Although the full details of the hemodynamic and metabolic correlates associated with brain activation are not yet fully understood, the corresponding MRI signal intensity rise after stimulus onset is easily detectable and well documented for visual, auditory, and sensorimotor cortical processing.

Mechanisms Similar to neuroimaging by PET, MRI studies of brain function rely on `secondary signals' that link changes in neuronal activity to correlated alterations of
- metabolism (e.g., glucose and oxygen consumption),
- perfusion (blood flow and blood volume), and
- blood oxygenation (deoxyhemoglobin).

The mainstream of techniques focuses on perfusion-related changes of the intravascular concentration of paramagnetic deoxyhemoglobin. Figure 1 conceptionally sketches the transformation of a change in focal neuronal activity into an MRI-derived activation map. The scheme indicates putative interferences at several levels ranging from data acquisition and signal processing to physiologic mechanisms and details of the paradigm design.

Key Message Thus, a key to understanding MR functional neuroimaging is the fact that it does not measure neuronal activity directly. An important question therefore is: how can we reliably detect a change in neuronal activity at the physiologic level? To answer this question we need to examine the neurovascular and neurometabolic coupling under various experimental and clinical conditions, e.g., with respect to confounding modulations caused by medication, brain disease, or even paradigm timings, and to carefully optimize the experimental strategies.

Outline The purpose of this chapter is to review the mainstream of MRI techniques proposed for successfully mapping brain activation. It introduces into the methodology and raises

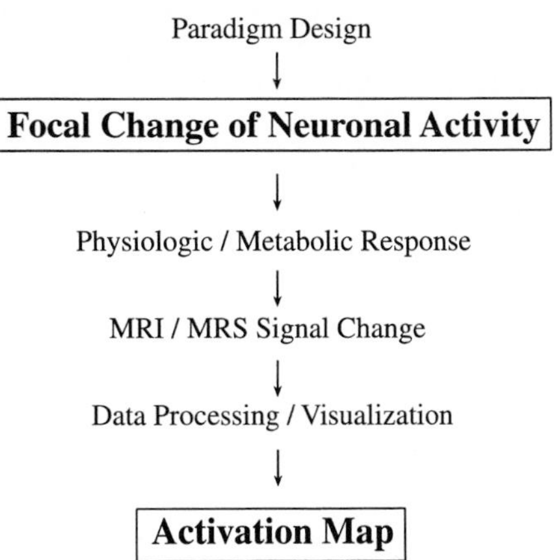

Fig. 1. Schematic diagram demonstrating the transformation of a focal change of neuronal activity into an activation map. MRI responses to brain activation (i) reflect differences in the experimental paradigm and its particular design, (ii) are sensitive to modulations of the neurovascular and/or neurometabolic coupling, and (iii) may be influenced by technical aspects of data acquisition and data processing including visualization and statistical treatment.

the awareness of potential pitfalls such as physiologic interferences. To accomplish these goals, the chapter will address

- technical aspects of MRI data acquisition,
- data analysis from dynamic MRI signals to activation maps
- physiologic aspects underlying MRI signals, and
- the design of suitable activation paradigms.

Technical Aspects of MRI Data Acquisition

Sequences for Brain Mapping

Principles

MRI comprises a large variety of radiofrequency (RF) and magnetic field gradient sequences for the acquisition of image data. Whereas RF pulses accomplish signal excitation, gradient pulses are required for a spatial discrimination of the excited MRI signals within a three-dimensional object.

In particular, RF excitation of a two-dimensional cross-section is based on the simultaneous application of an RF pulse and a slice selection gradient. Spatial information about the selected image section is encoded into the phase and frequency of its MRI signal by perpendicular gradients applied before and during signal acquisition. Image reconstruction is usually performed by two-dimensional Fourier transformation. Three-dimensional extensions are possible by adding an independent phase-encoding gradient along the direction of the slice selection gradient. Basic MRI principles, details of imaging sequences, and medical applications are summarized in Stark and Bradley (1998).

Fast Imaging

MRI sequences that are best suited for functional neuroimaging should be both fast and sensitive to changes in the deoxyhemoglobin concentration. These conditions are fulfilled for gradient-echo imaging techniques with prolonged echo times. Basic RF and gradient schemes for the most commonly used high-speed EPI (echo-planar imaging) and rapid FLASH (fast low angle shot) technique are shown in Figure 2. The diagrams illustrate the aforementioned combinations of RF pulses and gradients for cross-sectional imaging.

EPI

The main characteristic of single-shot EPI is that the technique acquires all gradient echoes needed for image reconstruction after excitation by a single slice-selective RF pulse. The individual echoes are generated by multiply alternating the sign of the fre-

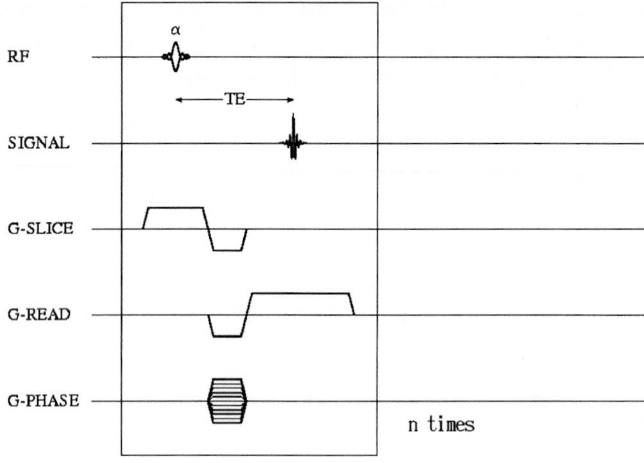

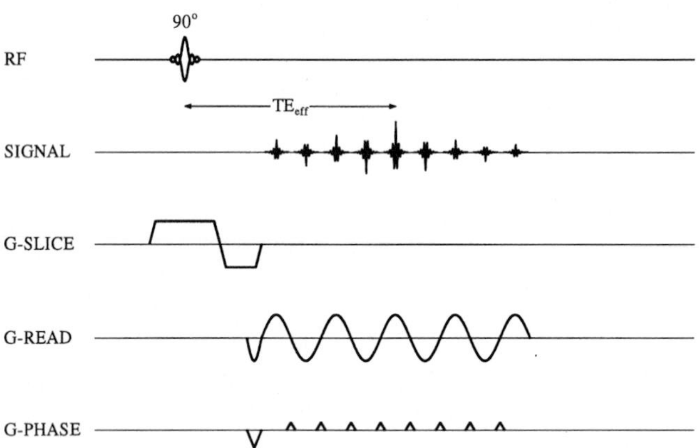

Fig. 2. Schematic radiofrequency (RF) and magnetic field gradient (G-SLICE, G-READ, G-PHASE) sequences for (top) FLASH (fast low angle shot) MRI and (bottom) single-shot gradient-echo EPI (echo-planar imaging) with blipped phase-encoding and symmetric coverage of *k* space. FLASH sequences employ multiple RF pulses with low flip angles $\alpha < 90°$ and generate only a single gradient echo (SIGNAL) per repetition interval by inversion of the frequency-encoding read gradient. EPI sequences acquire all differently phase-encoded gradient echoes after the application of a single RF excitation pulse.

quency-encoding "read" gradient (corresponding to 180° rotations of its direction). They represent individual lines of data points (Fourier lines) in the two-dimensional data space (*k* space). Phase encoding of shifted lines is accomplished by a "blipped" gradient between successive gradient echoes.

Data acquisition of EPI sequences is done during the decay of the excited MRI signal which is characterized by the effective spin-spin relaxation time T2*. The T2* sensitivity of the resulting image is determined by the gradient echo time TE, i.e., the period between RF excitation and signal acquisition. Because the individual echoes in EPI span a large range of different echo times, the effective TE value is given by the Fourier line representing the lowest spatial frequency. This Fourier line corresponds to the gradient echo with zero phase encoding and dominates the overall image contrast.

Typical imaging times are on the order of 100 ms, matrix sizes are usually 64×64 or 64×128 data points.

An alternative high-speed MRI technique that may be used for functional neuroimaging is spiral scanning where coverage of the data space starts in the center and follows an outward spiral. Pertinent gradient schemes and image reconstruction schemes are slightly more complex than Fourier imaging of a rectangular grid of data points commonly employed for EPI and FLASH. Spirals

The FLASH technique resembles a conventional MRI sequence in using multiple RF excitations and repetition intervals that each generate only one phase-encoded MRI signal. FLASH

In order to keep the overall imaging time short, the technique relies on a combination of RF pulse excitations with flip angles less than 90° and the acquisition of gradient echoes. Because the choice of a large TE value for obtaining sufficient T2* contrast directly prolongs the repetition period, typical imaging times are in the range of several seconds. The trade-off between temporal and spatial resolution is freely controllable via the number of phase-encoded gradient echoes, i.e., the number of repetition intervals.

The FLASH option is best exploited for mapping brain activation at high spatial resolution.

Image Contrast

The basic MRI contrast parameters are given by the proton spin density (e.g., water concentration), the proton T1 relaxation time denoting the exponential recovery of the longitudinal magnetization after excitation, and the proton T2 relaxation time describing the decay of the excited and detected transverse magnetization. In addition, gradient-echo signals such as those used in EPI and FLASH are subject to an even faster decay process phenomenologically characterized by the effective relaxation time T2*. It combines true T2 relaxation with the effects of magnetic field inhomogeneities including distortions that are caused by paramagnetic molecules such as deoxyhemoglobin. Relaxation

In order to reduce the MRI sensitivity to signal fluctuations caused by vascular flow, brain tissue pulsations, and head motions, it is recommended to avoid T1 contrast when acquiring raw images for functional brain mapping. Instead, the images should be spin-density weighted in combination with T2* sensitivity, i.e., prolonged TE. Spin Density

Spin density contrast is achieved under fully relaxed conditions using sufficiently long recovery periods before re-exciting the longitudinal magnetization, or by lowering the flip angle well below 90° for acquisitions with repetition times TR < 3 T1.

At 1.5–2.0 T field strength suitable combinations of TR and flip angle are 6000 ms/90°, 2000 ms/70°, 1000 ms/50°, 500 ms/30°, 200 ms/20°, 100 ms/15°, and 50 ms/10°. These parameters apply to both serial EPI and FLASH sequences.

Examples of spin-density weighted EPI and FLASH images with T2* sensitivity are shown in Figure 3. For a fixed field-of-view of 200 mm the images illustrate the achievable image quality and contrast for pixel matrices of 64 × 64 and 128 × 128 (EPI and FLASH) as well as 256 × 256 (FLASH) in comparison to a flow-sensitized anatomic FLASH image of the same section. Images

Macroscopic Susceptibility Artifacts

The magnetic field pertubations induced by paramagnetic deoxyhemoglobin in and around the cerebral microvasculature are microscopic in nature (i.e., extending over distances of less than 50 μm) and well below the linear dimension of a typical image voxel (e.g., 1 mm).

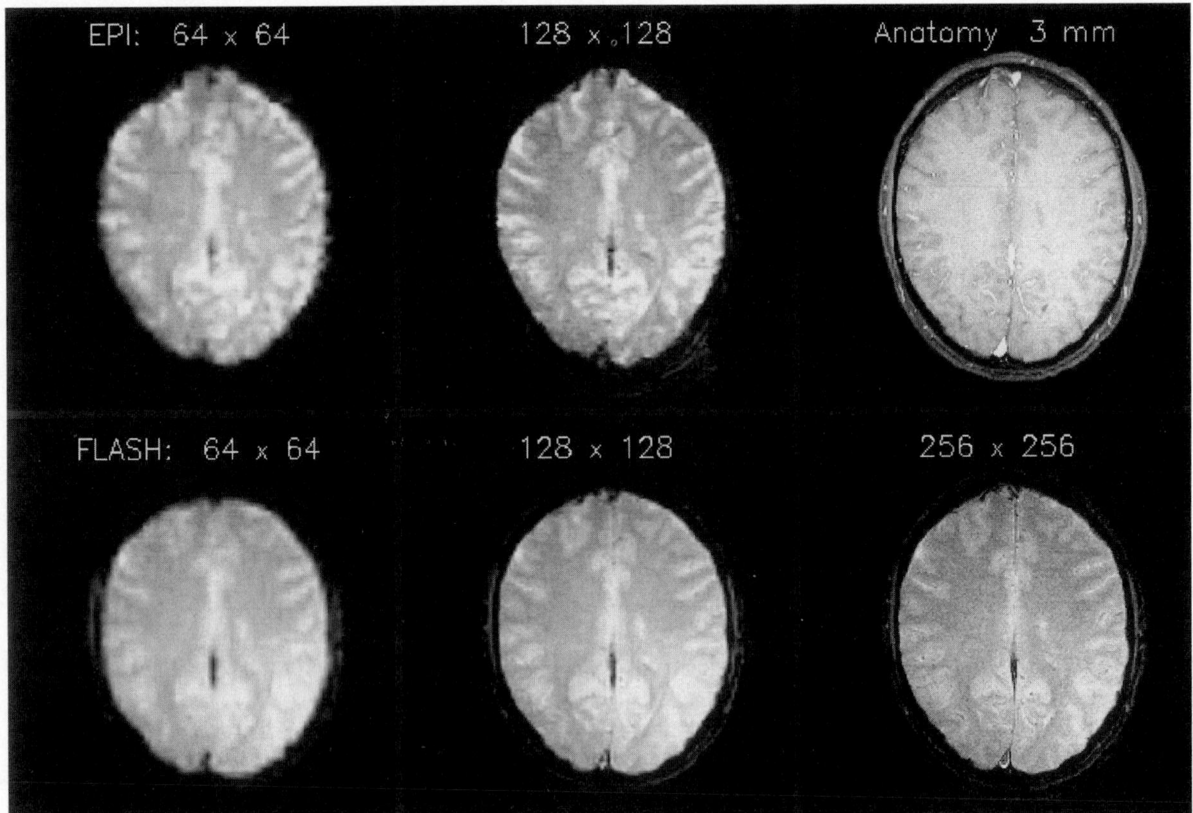

Fig. 3. Contrast and resolution of (top) EPI and (bottom) FLASH images commonly employed for functional neuroimaging (2.0 T, 200 mm field-of-view, 3 mm slice thickness) in comparison to a flow-sensitized T1-weighted FLASH image of the section anatomy (top right). Imaging parameters were adjusted such as to avoid T1 weighting and to ensure spin-density contrast (i.e., TR = 6000 ms and flip angle 90° for EPI as well as TR = 62.5 ms and flip angle 10° for FLASH) in combination with T2* sensitivity (i.e., an effective TE = 54 ms for EPI and TE = 30 ms for FLASH).

In addition, however, EPI and FLASH images with prolonged gradient echo times exhibit unwanted sensitivities to macroscopic magnetic field inhomogeneities from structural effects that usually extend over several voxels. They mainly originate from strong magnetic susceptibility differences at air-tissue interfaces, e.g., near the acoustic canals and the cavities at the cranial base. The resulting images are affected by signal loss and geometric distortions.

The brain section chosen for the images shown in Figure 3 suffers from only few macroscopic inhomogeneities. Nevertheless, the presence of small susceptibility differences, e.g., at the borderline of the frontal brain, leads to visible image degradation in the low-resolution 64 × 64 EPI image.

Signal Loss Inhomogeneity problems become dramatic when moving to a section through the lower part of the brain as shown in Figure 4. The images clearly demonstrate that the sensitivity to macroscopic inhomogeneities poses a key problem for MR functional neuroimaging. For example, psychiatric interest in studying functional encoding in (anterior) hippocampal structures may be severely hampered by this problem.

Voxel Size Susceptibility artifacts are less pronounced in FLASH images because of their conventional way of covering *k* space and their use of a fixed time for all gradient echoes. They

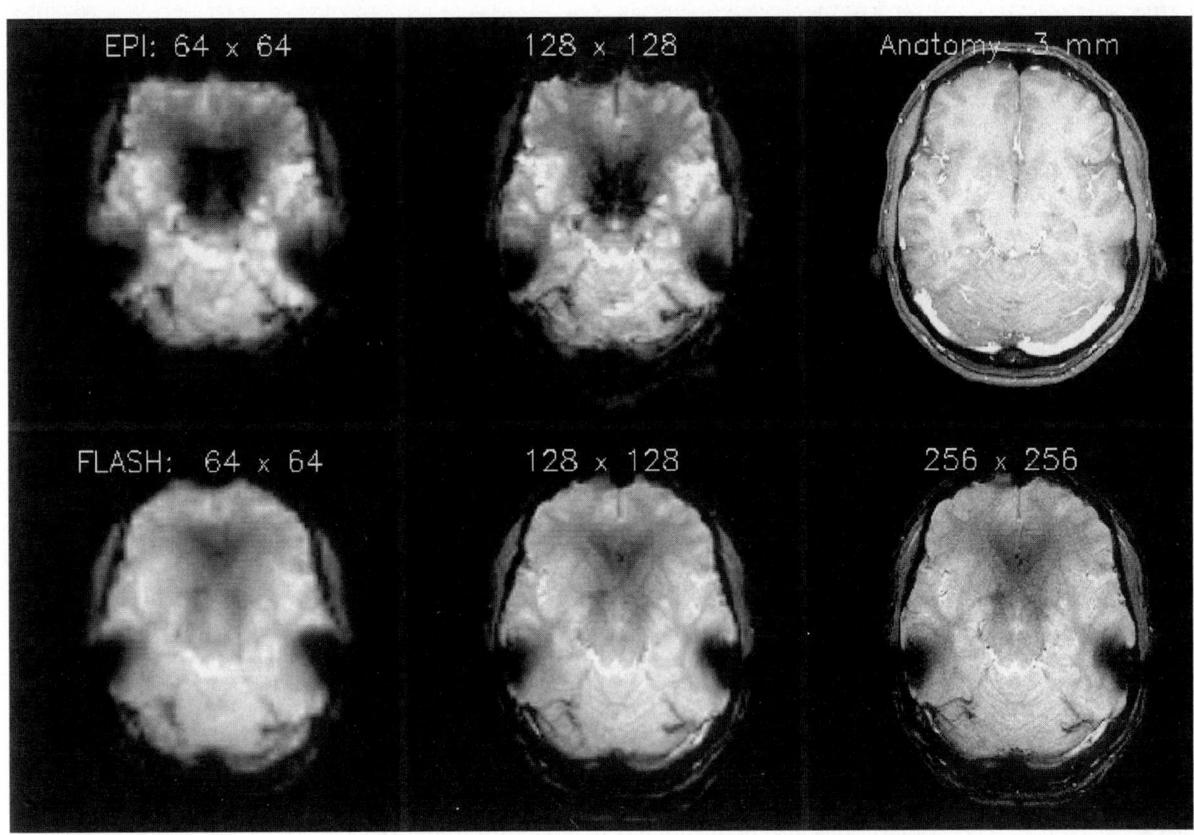

Fig. 4. Macroscopic susceptibility artifacts caused by air-tissue interfaces in (top) EPI and (bottom) FLASH images commonly employed for functional neuroimaging in comparison to a flow-sensitized anatomic FLASH image of the same section (top right). The data originate from the same subject as in Figure 3 using identical imaging parameters except for choosing a section through the lower part of the brain. The artifacts represent signal losses (EPI, FLASH) and geometric distortions (EPI) that increase with pixel size, i.e., reduced spatial resolution.

can be further reduced by using very small voxel sizes, i.e., very high spatial resolution and thin sections, which decrease the influence of macroscopic field gradients but retain the sensitivity to deoxyhemoglobin-induced intravoxel effects. These strategies are at the expense of temporal resolution and volume coverage.

Flow and Motion Sensitivity

Confounding signal fluctuations in serial MRI acquisitions may be induced by vascular flow, breathing, and involuntary subject movements. For both FLASH and EPI, it is therefore recommended to adjust flip angles and repetition times such as to avoid T1 weighting and thereby minimize the sensitivity of the steady-state longitudinal magnetization to amplitude fluctuations.

Intrascan phase effects such as ghosting may be minimized by optimized gradient waveforms and/or the use of navigator echoes. Moreover, the maximum imaging times should be kept to below 10 s to avoid complications from respiratory motions with time periods on the order of 5–6 s.

Head Motion

A most relevant interscan motion problem may result from stimulus-correlated head or body movements during the applied paradigm. For example, a typical observation is a slight nodding movement of the subject's head as well as a continuous drift or rotation during an experiment of several minutes. Whereas the latter effects may be alleviated with the help of suitable post-acquisition motion correction algorithms, the former influence may even be difficult to detect unless major correlated motions lead to a spatially unspecific emphasis of all contrast borders in the resulting activation map.

Control

In general, it is recommended to control the MRI data quality by a rapid cine display of the serial images. Often it might be preferable to repeat an experiment rather than to rely on a proper correction of artifacts. Motion problems may become dramatic when studying patients instead of motivated healthy volunteers. In special cases it may be helpful to use fixation devices such as bite bars.

Temporal and Spatial Resolution

Depending on study purpose and the underlying scientific or clinical question, the necessary choice of a compromise between
- temporal resolution,
- volume coverage, and
- spatial (in-plane) resolution

should be based on single-shot EPI (spirals) or FLASH, respectively.

EPI

EPI emphasis is on high speed yielding imaging times on the order of 100ms which translate into a maximum temporal resolution of 100ms for single slice acquisitions. Alternatively, EPI offers excellent volume coverage by multi-slice imaging as up to 20sections may be scanned with a repetition time of 2s.

In general, the in-plane resolution of EPI sequences is limited because of the limited number of gradient echoes that can be acquired during the T2* signal decay after RF excitation. A typical in-plane resolution for EPI is 3×3 mm^2, but 2×2 mm^2 are possible. Slice thicknesses are generally 3–4mm, but some groups still use 5–10mm sections.

FLASH

FLASH applications are complementary to the more commonly used EPI technique. They facilitate access to high spatial resolution at the expense of temporal resolution and volume coverage. Attempts to maximize the in-plane resolution may even be limited to single sections.

Typically, FLASH images with a 96×256 data matrix, a rectangular field-of-view of 150×200 mm^2, and zero-filling by a factor of 2 during Fourier transformation result in an in-plane resolution of 0.78×0.78 mm^2 and a temporal resolution of a few seconds. Pertinent strategies may be further optimized to allow for high-resolution mapping of the columnar organization of the human cortex. Section thicknesses are 3–4mm but may be reduced to 1mm.

Magnetic Field Strength

1.5–3.0 T

The choice of the "optimum" magnetic field strength for MR functional neuroimaging is an unresolved issue. The results presented here are obtained at 2.0T (Siemens Vision, Erlangen, Germany), while most clinical systems are likely to continue operating at 1.5T (or below). More recently, a few dedicated head systems became available at 3.0T.

T2* etc.

In general, the effects of magnetic field inhomogeneities and susceptibility differences – or in other words the 1/T2* relaxation rate – increase with field strength. Thus, because the T2* relaxation time becomes shorter, the useful length of the EPI data acquisition period is further reduced. On the other hand, the possibility of selecting shorter gradient echo times at higher fields may be an advantage for FLASH sequences. Other considerations refer to the fact that not only the signal-to-noise ratio but also the unwanted sensitivity to motion and macroscopic inhomogeneities increase with field strength.

Contrast

It is therefore not yet clarified whether – or under which circumstances – putative improvements in "functional contrast-to-noise" as the crucial criterion justify the investment of a 3.0 T system. In any case, it is recommended to rely on commercially available equipment, i.e., complete MRI systems from the leading manufacturers in diagnostic imaging.

Data Evaluation and Visualization

A Simple Experiment

Functional information about the human brain is extracted from serial MR images acquired over a period of several minutes. During this period the subject reacts in response to a paradigm that involves at least two different conditions, i.e., functional states exhibiting different degrees of neuronal activity.

Anatomy

Figure 5 summarizes some of the elements of a simple visual mapping experiment. It compares the neuronal representations of processing flickerlight and darkness in a single oblique MRI section along the calcarine fissure. Typically, the experiment starts with the acquisition of the individual 3D anatomy for retrospective referencing, the definition of the desired section(s), and a visualization of the local anatomy and (macro-) vasculature (top panels).

BOLD Images

These preparatory steps are followed by application of the paradigm with simultaneous dynamic acquisitions of the functional images. The middle panels of Figure 5 depict spin-density weighted FLASH images with T2* sensitivity at a temporal resolution of 6 s and a spatial resolution of 0.78×0.78 mm^2. Relative to images acquired during darkness (middle left) flickerlight stimulation (middle right) reveals a focal increase in MRI signal intensity, i.e., a decrease in deoxyhemoglobin, in selected areas of visual cortex. Visual stimulation was accomplished by a projection setup covering $40° \times 30°$ of the subject's visual field (Schäfter & Kirchhoff, Hamburg, Germany).

Raw Data

The direct experimental result is a series of consecutive images yielding MRI signal intensity time series for each image pixel. Using multi-slice acquisitions (e.g., 20 sections), high temporal resolution (e.g., 2 s), moderate spatial resolution (e.g., 128×128 data matrix), and a 5 min activation protocol, the raw data of a single experiment already amounts to 3000 images or 96 Mbyte. With a cooperative volunteer and a measuring time of 1.5 hours at least 10 successive experiments are possible yielding about 1 Gbyte per session. Thus, also data storage and on-line computational times of sequential experiments have to be accounted for in planning a study.

Difference Map

The desired result is a single "activation" map per section that transforms any paradigm-specific MRI signal change into a color-coded quantity. Pertinent pixel values are summarized in a map which is commonly overlayed onto an anatomic image or functional raw image.

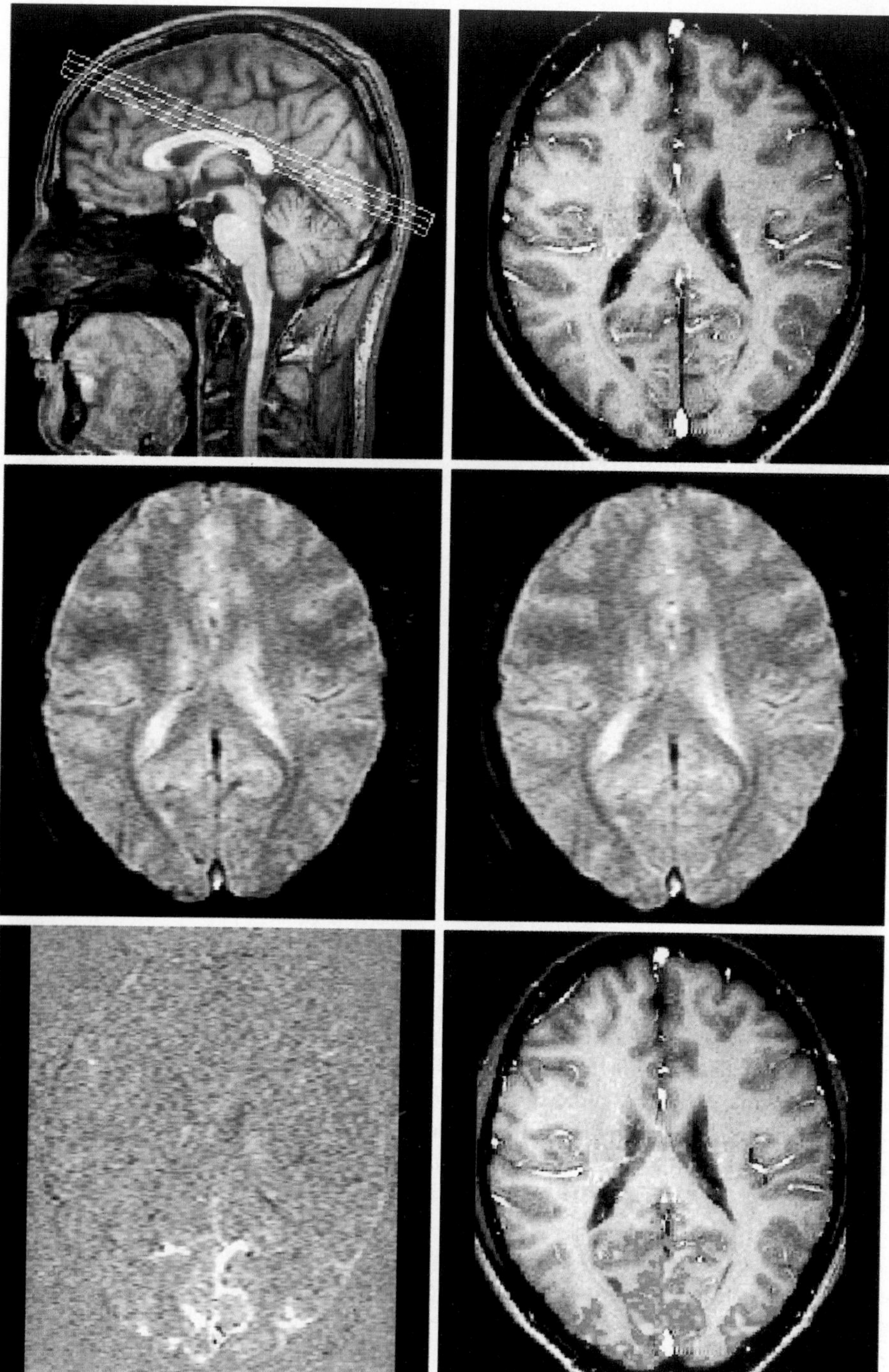

Fig. 5. Functional mapping of visual activation. (Top left) Definition of section orientation (2.0T, 3D FLASH, TR/TE = 15/6ms, flip angle 20°) and (top right) section anatomy and vasculature (FLASH, TR/TE = 70/6ms, flip angle 50°). In going from darkness (middle left) to flickerlight stimulation (middle right), dynamic acquisitions of spin-density weighted images with T2* sensitivity (FLASH, 96 × 256 matrix, rectangular FOV 150 × 200 mm^2, slice thickness 4mm, TR/TE = 62.5/30ms, flip angle 10°, temporal rsolution 6s) reveal a focal increase in MRI signal intensity, i.e., a decrease in deoxyhemoglobin, in visual cortex. (Bottom left) Difference map and (bottom right) color-coded activation map obtained by a pixel-by-pixel cross-correlation of signal intensity time courses with a reference function representing the stimulus protocol, compare Figure6. (Modified from Kleinschmidt et al 1995a).

As shown in the bottom left part of Figure 5 the most simple approach is time-locked averaging of images representing the same functional state and subsequent subtraction, i.e., summation of images that are acquired during one condition (e.g., darkness) and subtraction of the result from that obtained for a different condition (e.g., flickerlight). The resulting difference map highlights cortical areas that are involved in processing flickerlight relative to darkness.

Correlation Map

A much more robust and also more sensitive approach than image subtraction stems from a correlation analysis. The technique exploits the temporal structure of the applied stimulation protocol and compares it to the oxygenation-sensitive MRI signal intensity time courses on a pixel-by-pixel basis (Bandettini et al 1993). Such strategies eliminate any statistical signal fluctuations as a source of contrast and are most helpful when recording multiple stimulation cycles that alternate between conditions. The bottom right part of Figure 5 depicts a map of correlation coefficients. Small activated areas of lower significance such as the frontal eye fields are more clearly distinguished than in the corresponding difference map.

Activity?

Whether the oxygenation-sensitive MRI responses reflect the enhanced firing or synchronization of neurons or indicate the recruitment of new though overlapping populations of specialized neurons within visual cortex cannot be decided from the MRI experiment. We need to keep in mind that even high-resolution studies in general are too coarse to resolve the columnar organization of the human cortex.

MRI Signal Intensity Time Courses

Figure 6 depicts selected time courses from regions-of-interest (ROI) in the primary visual cortex, the frontal eye field, and non-activated frontal gray matter for the experiment shown in Figure 5. The time courses reveal stimulus-induced responses to visual activation that tightly follow the sixfold application of a stimulation cycle comprising 18 s of flickerlight (shaded boxes) and 36 s of darkness.

The bottom trace in Figure 6 refers to a reference vector that mimics the temporal structure of the protocol shifted by one image (6 s) to account for hemodynamic latencies of the flow-induced change in blood oxygenation. This boxcar function was employed to calculate the color-coded map of correlation coefficients shown in the bottom right panel of Figure 5.

Calibration?

Mapping of brain activation relies on suitable physiologic models and mathematical methods that identify stimulus-related changes in MRI signal intensity time courses. Because the dynamic MRI signal may be characterized by its amplitudes, frequencies, and phases (relative to the applied paradigm), there are manifold ways of extracting functional information.

Although amplitude-based methods such as differences, variances, or statistical treatments (e.g., t or χ^2 tests) are possible, a reliable quantification and calibration of MRI signal strengths in terms of "oxygenation" or degree of "activity" is not possible. Attempts to calibrate T2* relaxation rates with respect to the arterial oxygen saturation are time-consuming and mainly performed in physiologically well-controlled animal experiments.

Mathematics

It is therefore well accepted to rely on relative amplitudes and to analyze the temporal evolution of signal changes. Pertinent techniques attempt to classify temporal properties by *correlation analyses* or equivalent frequencies by *Fourier analyses*. Further ap-

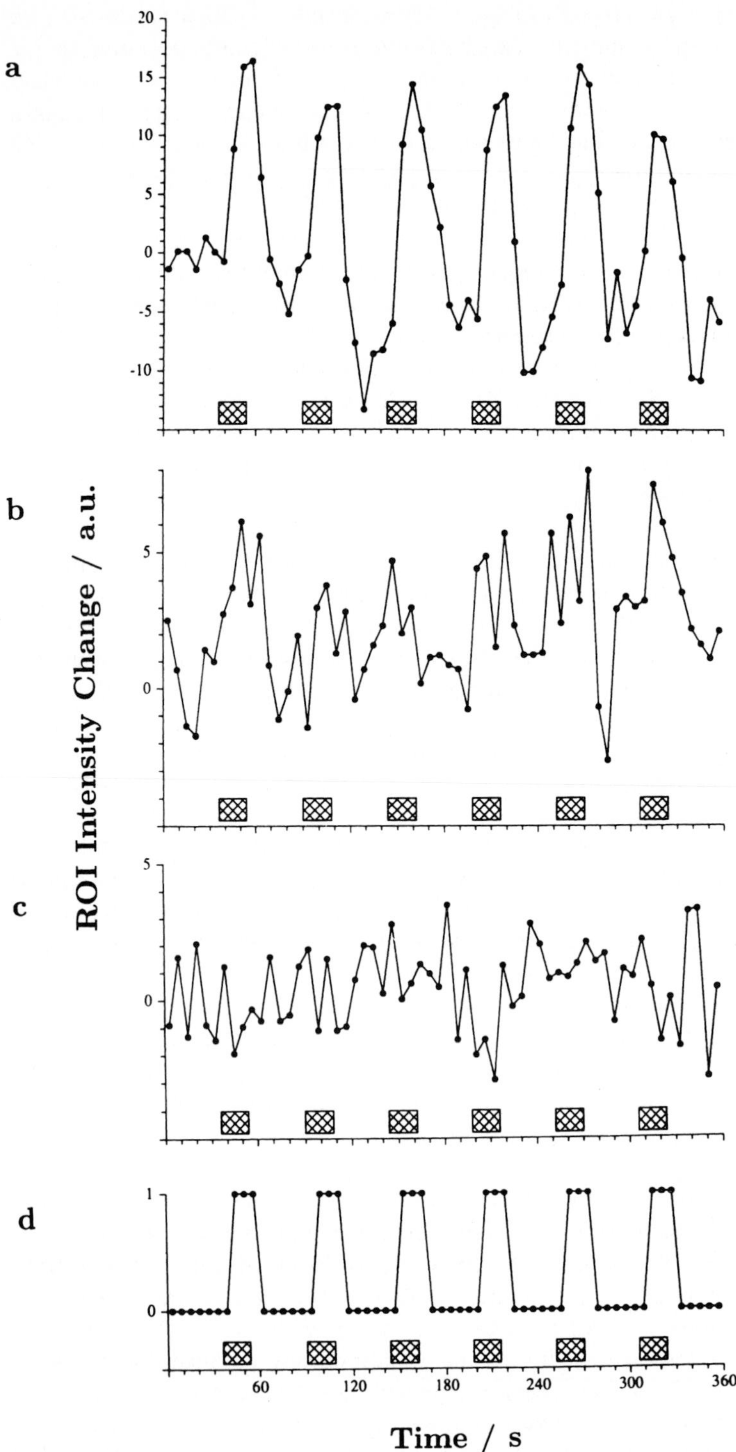

Fig. 6. Time courses of oxygenation-sensitive MRI signal intensities (2.0T, FLASH, TR/TE = 62.5/30ms, flip angle 10°, temporal resolution 6s) from regions-of-interest in (a) primary visual cortex, (b) frontal eye field, and (c) non-activated frontal gray matter (control) of a single subject in response to visual activation (18s of flickerlight vs. 36s of darkness). (d) Reference function that mimics the temporal structure of the stimulation protocol shifted by one image (6s) to account for hemodynamic latencies. This boxcar function was employed to calculate the thresholded map of correlation coefficients shown in Figure 5.

proaches are *principal component analyses* and *fuzzy clustering algorithms.* However, neither is this list complete nor is a comprehensive description within the scope of this article.

Mathematical strategies for data analysis also need to take into account the noise characteristics of the experimental data. In optimized MRI systems the noise is of a physiologic rather than instrumental origin. Of course, the specifics also depend on the spatial integration (resolution) of vascular and/or tissue components and the temporal resolution relative to the hemodynamic responses and other modulatory effects such as breathing. Moreover, any post-acquisition data manipulation, the use of temporal and/or spatial filters or even artificial "baseline" or "drift" corrections may have strong influences on the resulting activation maps.

Statistical Maps and Beyond

Correlations

A preferred and robust method for data analysis is a cross-correlation of pixel intensity time courses with a reference function (Bandettini et al 1993). The method is robust against uncorrelated signal fluctuations, more sensitive to small or weakly activated areas than amplitude-based techniques, and flexible with regard to the applied paradigm and its temporal structure.

The choice of a reference function for cross-correlation may be a measured pixel intensity time course, a mathematical representation of the stimulation protocol, a real or hypothesis-driven model function, or even an EEG signal recorded in parallel to the MRI acquisition. The activation paradigm does not need to be periodic when taking the value of the cross-correlation function at zero time lag, i.e., the well-known correlation coefficient, which also results in fast calculations.

Histograms

An unprocessed map of correlation coefficients is a neutral representation of values between −1 (antiphase correlation), zero (no correlation), and +1 (perfect correlation) assigned to each pixel. To identify specific areas that are activated in association with an applied paradigm, pertinent maps need to be thresholded based on the statistical significance of the underlying MRI signal alterations.

Figure 7 shows two histograms of correlation coefficients obtained for dynamic MRI acquisitions in the absence and presence of visual stimulation, respectively. Without activation the "noise" distribution of correlation coefficients is symmetric and centered at zero. Brain activation causes a deviation from the noise distribution at high positive (or negative) correlation coefficients.

Thresholds

Because differences in systemic hemodynamic responsiveness and motion stability cause intertrial variabilities that affect the center and width of the underlying noise distribution, it is not advisable to use absolute correlation coefficient thresholds for defining activation. Moreover, a single fixed threshold would result in a trade-off between specificity (e.g., by emphasizing only a few highly significantly activated foci at the expense of adequate area delineation) and sensitivity (e.g., by using a low threshold that allows for extended areas at the expense of including considerable noise).

Invariance against intertrial differences as well as adequate visualization of activated areas may be achieved by the following procedure (Kleinschmidt et al 1995a) schematically indicated in the right-hand part of Figure 7.

Noise

A symmetrized noise distribution is reconstructed from the actual activation map by fitting the central part of the distribution to a symmetric function after identifying the center position and the full width at half maximum. This procedure allows rescaling of

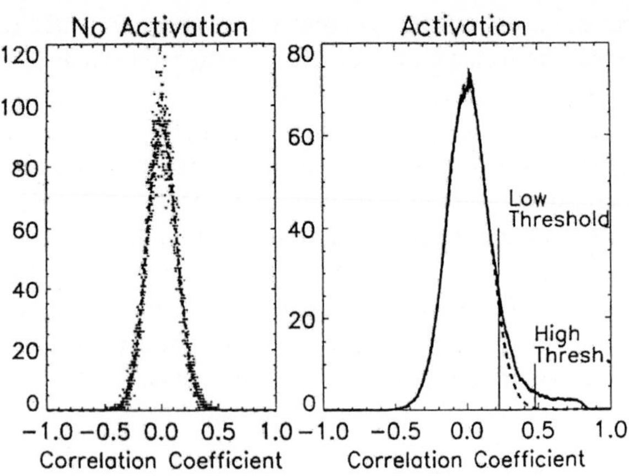

Fig. 7. Distribution of correlation coefficients for an experiment (left) without and (right) with visual activation using darkness as control state. In the right histogram the presence of activation is indicated by a deviation of the actual distribution (solid line) from the symmetrized noise floor (dashed line). The noise distribution is reconstructed from the central part of the actual distribution. The area between both distributions refers to the elevated number of pixels with high positive correlation coefficients.

correlation coefficients into percentile ranks with respect to the integral of the noise distribution. The percentile rank describes the probability of pixels at or above a given correlation coefficient to be noise, i.e., the error probability when assuming that such a pixel is "activated".

Specificity

While the procedure is still based on correlation coefficients, it generates robustness against variability of correlational noise. A high percentile rank (or correspondingly low error probability) is used as a threshold to define primary sites of activation. For the example shown in Figure 5 the activation map (bottom right) represents a thresholded map of correlation coefficients at the 99.99 % percentile for identifying significant activation with high specificity.

Sensitivity

The full spatial extent of an activated brain area may be mapped by adding directly neighboring pixels to highly significant foci provided their lower correlation coefficients are high enough to contribute to the deviation from the noise distribution. The rationale behind this is that inspection of correlation coefficient maps without any threshold reveals the activation foci to be embedded within clusters of elevated correlation coefficients. This tight topographic relationship as the "peak of the mountain" may be exploited to improve sensitivity without losing specificity: while identifying significant activation by a high threshold, adequate area delineation may be achieved by iteratively incorporating nearest neighbors that exceed a lower threshold derived from the difference between the actual and the noise distribution.

SPM etc.

Most research teams have developed their own analysis programs that may or may not be available on request. A popular, semi-automated, and more standardized program is SPM (Friston 1995). The package was originally developed for Statistical Parametric Mapping using low-resolution (static) PET images, but now contains tools suitable for the analysis of MRI data sets.

Regardless of which program is selected, both the methodologic novice and the cognitive neuroscientist who only wants to "use" the technique as a problem solver are strongly advised to follow a data-driven approach rather than to blindly rely on the performance of a black box evaluation program. In particular, it is recommended to frequently check the quality of the acquired raw images and to visualize unprocessed MRI signal intensity time courses.

Visualization

Once data acquisition and evaluation have led to an activation map, additional processing may be necessary to accomplish further scientific goals or clinical needs. In most cases pertinent tasks aim at a more advanced visualization of cross-sectional maps. They include image integration by matching brain function and co-registered individual 3D anatomy as well as surface rendering of combined data sets (e.g., for presurgical planning).

In neuroscience, transformation of individual activation maps into standardized brain atlases may serve the purpose of intersubject averaging to enhance sensitivity or to separate common and individualized cognitive strategies. In addition, the procedure allows for a comparison of the functional brain anatomy with complementary information on the cerebral microvasculature or cytoarchitecture. Standardized maps are also a prerequisite for the combination of data across modalities. Many of the aforementioned programs are still under development.

Physiologic Aspects of Brain Activation

Coupling

The critical link between a focal change in neuronal activity and MRI-detectable observables is the neurovascular coupling (e.g., see Villringer and Dirnagl 1995). It is therefore mandatory to control the influence of hemodynamic and metabolic parameters on the mapping process.

Hemoglobin

Provided that proper acquisition parameters are chosen, dynamic MRI of brain activation relies on BOLD contrast reflecting changes in the absolute concentration of deoxyhemoglobin. This understanding is not only supported by animal experiments but directly confirmed by simultaneous recordings using oxygenation-sensitive MRI and near-infrared optical spectroscopy of human motor activation (Kleinschmidt et al 1996).

This section introduces some of the mechanisms that determine the net cerebral blood oxygenation and putatively interfere with activation-induced responses to functional challenge.

Modulations of the Hemodynamic Response

Apart from the desired task-related adaptation in neuronal activity, modulations of the hemodynamically driven BOLD MRI contrast may arise from intrasubject and intersubject differences in other cognitive or emotional processes (e.g., attention and anxiety) or even systemic adjustments (e.g., heart rate and blood pressure). In addition, amplitude and temporal characteristics may be subject to changes in neurovascular responsiveness in relation to age (e.g., in early infancy and in the elder population), nutrition and medication (e.g., vasoactive and psychotropic drugs), or neuropathology (e.g., cerebrovascular disease). It is also likely that BOLD responses reveal a topographic dependence on brain systems as regional differences in neuroanatomy, microvasculature, and neurotransmitter systems are well established.

Neuroactive Drugs

Obvious examples of an interference with the cerebral blood oxygenation are direct pharmacologic interventions.

Figure 8 shows BOLD MRI responses to the administration of psychotropic drugs such as diazepam and metamphetamine. The effects of depression and stimulation of cerebral activity are significantly different for the two drugs applied. Relative to signal strength during injection, metamphetamine elicits a 1.0 % signal increase 3–4 min after the end of injection, whereas diazepam abolishes this effect.

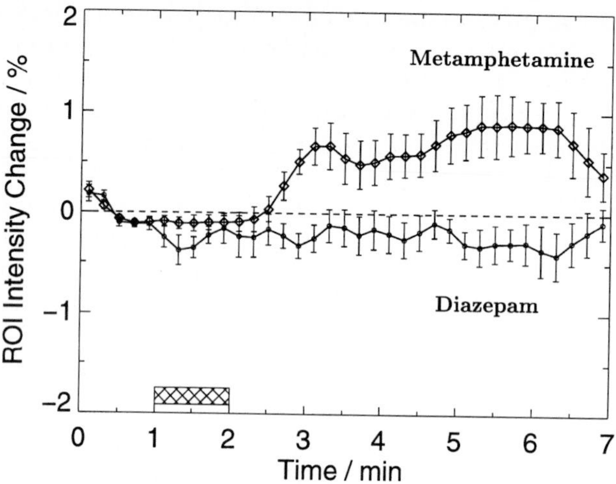

Fig. 8. Time courses of oxygenation-sensitive MRI signal intensities (2.0T, FLASH, TR/TE = 62.5/30ms, flip angle 10°, temporal resolution 12s) in response to pharmacologic stimulation and depression of cerebral activity by the intravenous administration (cross-hatched bar) of 15mg metamphetamine (upper curve) and 10mg diazepam (lower curve). The data originate from a whole brain section and represent a grand average across healthy young subjects ($n = 7$). Drug-specific responses are a signal increase after metamphetamine and a steady or mildly decreased signal intensity after diazepam, respectively. Vertical lines indicate the standard error of the mean. (Modified from Kleinschmidt et al 1999).

The metamphetamine-induced signal increase is more pronounced in subcortical gray matter structures and cerebellum than in frontotemporal cortical gray matter. These differences support the hypothesis that pertinent responses not only reflect global cerebral hemodynamic adjustments but also localized perfusion changes coupled to alterations in synaptic activity. The occurrence of a placebo response is best explained by expectancy and may provide a confounding factor in the design of functional activation experiments.

Vasoactive Drugs

The application of vasoactive drugs such as acetazolamide (vasodilation) and aminophylline (vasoconstriction) provides an even more direct interaction with cerebral blood flow and its effect on the deoxyhemoglobin concentration. Figure 9 shows pertinent responses as well as their modulation of activation-induced signal changes in visual cortex.

Vasodilation

Administration of acetazolamide yields a generalized increase of the oxygenation-sensitive MRI signal in cortical (and subcortical) gray matter. As shown in the upper part of Figure 9 the process starts about 1 min after drug application and reaches a plateau phase after 5–10 min. Assuming unchanged oxidative metabolism, it reflects a venous hyperoxygenation, i.e. decreased deoxyhemoglobin, due to elevated cerebral perfusion. This is a consequence of hypercapnia caused by inhibition of carbonic anhydrase.

The marked reduction of activation-induced MRI signal changes after vasodilation indicates an attenuation of vasomotor activity. Taking intersubject variability into account, the data reveal an individually modulated autoregulatory responsiveness to functional challenge. In cases of a truly "exhausted" reserve capacity, autoregulation may fail to generate any MRI signal change which effectively eliminates the detectability of brain activation at the physiologic level.

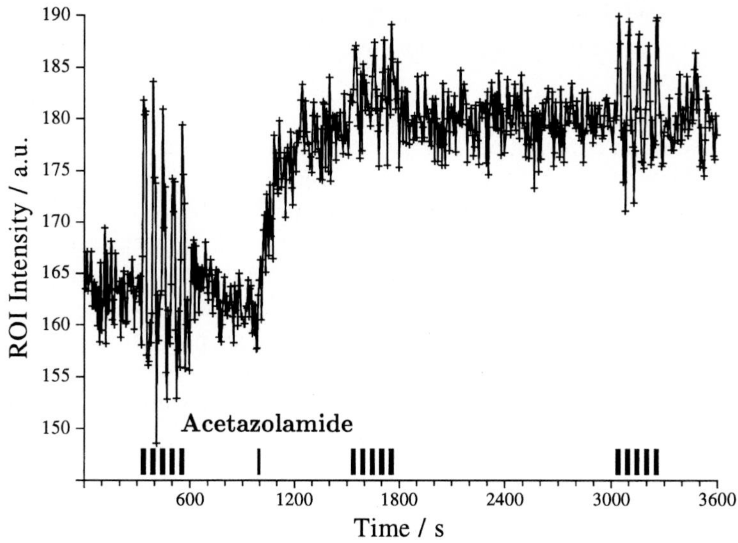

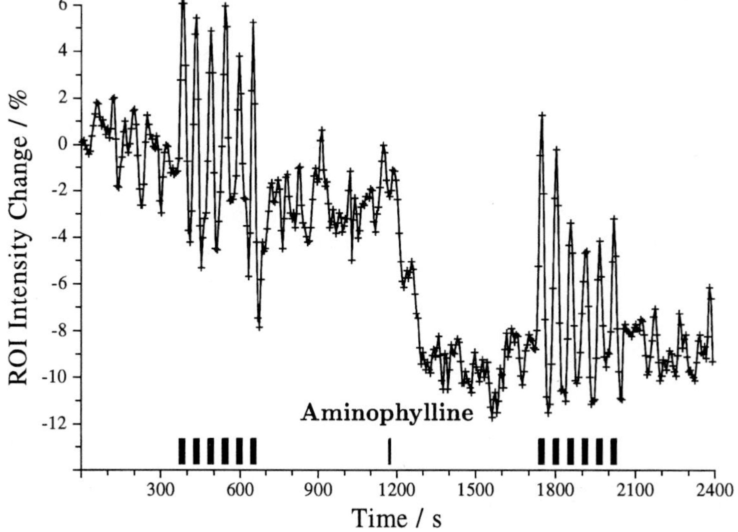

Fig. 9. Time courses of oxygenation-sensitive MRI signal intensities (2.0T, FLASH, TR/TE = 62.5/30ms, flip angle 10°, temporal resolution 6s) in response to visual activation (5 cycles, 18s flickerlight vs. 36s of darkness, solid bars) before and after the intravenous administration of 1.0g acetazolamide (top) and 0.2g aminophylline (bottom). The data originate from a region-of-interest in the visual cortex of two different subjects consecutively imaged over a period of 60min (top) and 40min (bottom), respectively. Whereas vasodilation results in MRI signal increases (i.e., decreased deoxyhemoglobin) and significant attenuation of functional responses, vasoconstriction yields a corresponding signal decrease (i.e., increased deoxyhemoglobin) but no change of the response strength to visual activation. (Modified from Bruhn et al 1994 and Kleinschmidt et al 1995b).

Vasoconstriction

The administration of aminophylline results in a fast decrease of the MRI signal intensity as shown in the lower part of Figure 9. The respective increase of deoxyhemoglobin may be explained by the combination of reduced blood flow and unchanged or even slightly enhanced oxygen consumption as aminophylline includes a neural excitatory action in addition to its vasoactivity. The fact that the post-injection response strength to visual activation is neither diminished nor enhanced suggests the involvement of different mediator systems for hemodynamic adjustments of pharmacologic and functional challenges, respectively.

Pathology Special care is required when applying physiologic mapping of neuronal activation to patients. Apart from direct effects of space-occupying lesions, neurodegenerative and neurometabolic disorders may have consequences affecting cerebral autoregulation or other aspects of the neurovascular (neurometabolic) coupling. Cerebrovascular disease that leads to hemodynamic compromise has already been demonstrated to completely disrupt the expected MRI response in individual patients. Further examples are patients receiving medication either related or unrelated to their neurologic problem (e.g., psychiatric patients in the elder population) or studies under anesthesia. In the latter case the affected regulation of cerebral blood flow most likely precludes any meaningful applications.

Response Functions and Protocol Timings

There is still limited knowledge about the temporal evolution and interplay of activation-induced adjustments of cerebral blood flow, blood volume, and oxidative metabolism in humans and its effect on the "functional contrast". In general, the development of the real-time BOLD MRI signal will be the result of multiple regulative processes with different time constants. Such contributions may critically affect the interpretation of MRI responses, in particular when these are acquired with use of repetitive stimulation protocols.

The question arises whether focal changes in neuronal activity result in a characteristic physiologic MRI response function. Figure 10 demonstrates typical signal patterns for visual activation protocols ranging from sustained stimulation to the presentation of single brief stimuli. The most important finding from these studies is the identification of a slow physiologic process that complements the initial signal rise after stimulation onset.

Fast Process The fast positive BOLD response is usually ascribed to a hyperoxygenation caused by a rapidly increased blood flow (i.e., oxygen delivery) that is not matched by an immediately effective upregulation of oxidative metabolism (i.e., oxygen consumption).

Slow Process In contrast, the slow signal changes refer to an attenuation of the hemodynamic response curve during ongoing stimulation (i.e., a relative deoxygenation) and the recovery of a pronounced signal undershoot after the end of stimulation to pre-stimulation baseline, e.g., seen for sustained activation in panel A of Figure 10.

Whereas checkerboard responses remain almost unaffected during ongoing activation (less than 20 % signal attenuation), all other stimuli including flashing diffuse light and real movies result in more than a 50 % decrease of the initial signal response after 6 min of stimulation. Potential factors that might explain this discrepancy include stimulus-dependent regulations of blood flow, blood volume, and oxygen consumption or differential degrees of habituation and adaptation.

Mechanisms The most likely candidates for the slow process are adjustments of oxidative glucose consumption (Frahm et al 1996) and a modulation of the venous blood volume (Buxton et al 1998). These mechanisms may require a certain delay to become quantitatively detectable against the dominating flow-related MRI signal changes both after the onset

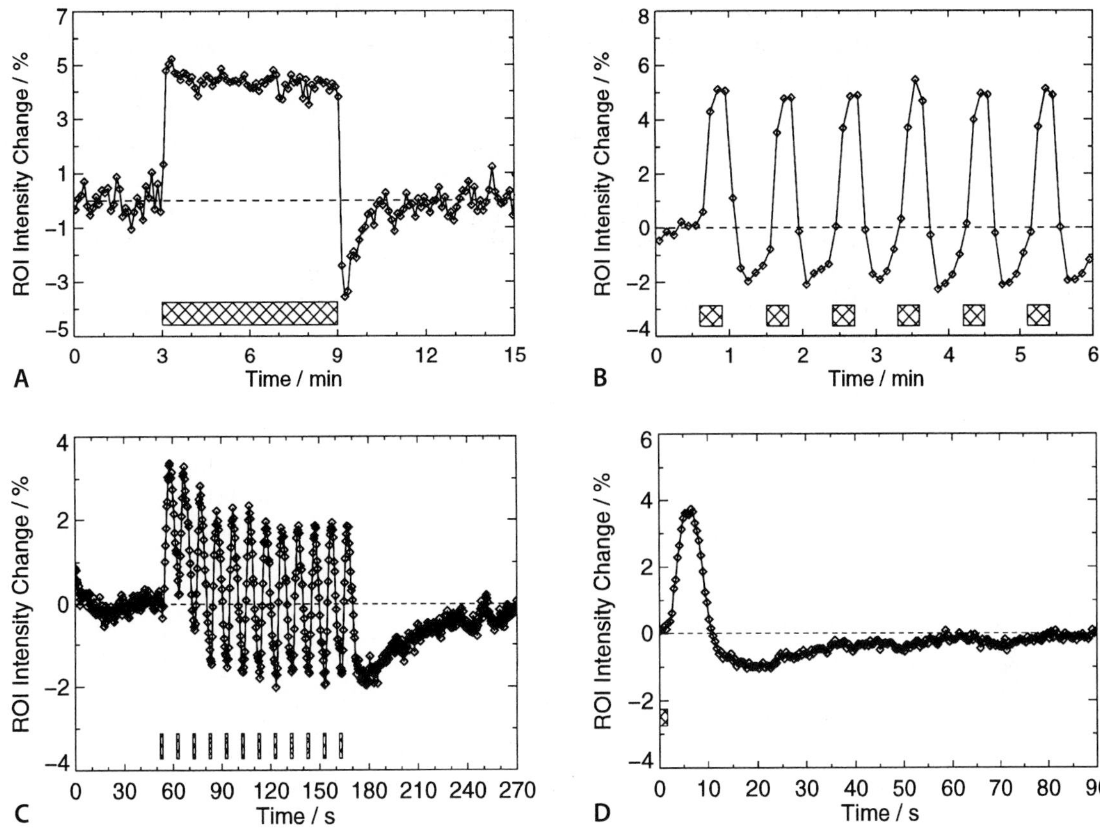

Fig. 10. Time courses of oxygenation-sensitive MRI signal intensities (2.0T) for various protocols of visual activation (checkerboard vs. darkness, cross-hatched bars). (a) Responses to sustained activation ($n = 7$) for a period of 6min and (b) to repetitive activation ($n = 7$) by a 18 s/36s protocol (FLASH, TR/TE = 62.5/30ms, flip angle 10°, temporal resolution 6s). (c) Responses to repetitive activation (n=8) by a 1.6 s/8.4s protocol and (d) to single events (n=5) by a 1.6 s/90s protocol (EPI, TR/TE = 400/54ms, flip angle 30°, temporal resolution 0.4s). The curves represent mean values of all pixels with statistically significant stimulus-related signal alterations averaged across subjects. (Modified from Krüger et al 1998 and Fransson et al 1998a).

and end of stimulation. However, independent of the underlying mechanism, the existence of two processes with different time constants is directly relevant for mapping studies that are commonly based on short protocol timings.

Repetitions

The time courses for repetitive visual activation in panels B and C of Figure 10 represent 6 cycles of a 18 s/36s protocol and 12 cycles of a 1.6 s/8.4s protocol. Both studies reveal a "baseline drift" during cyclic stimulation that reaches a new steady state after about 2min. This observation is in close agreement with the presence of a slow physiologic process and seems to represent the accumulation of undershoot contributions from preceding stimuli. The end of repetitive stimulation is characterized by a marked undershoot with an amplitude of up to –2 % below pre-stimulation baseline and a duration of 60–90s.

The relative peak-to-peak signal difference or "functional contrast" between successive stimuli remains almost unchanged throughout the experiments at about 7 % for the 18s stimulus and 4 % for the 1.6 s stimulus, respectively.

Single Trials

The qualitative agreement of responses for different stimulus durations motivated studies of single events as shown in panel D of Figure 10. A close inspection of the response profiles obtained for brief visual stimuli with long interstimulus intervals reveals a 1.5–2.0 s hemodynamic latency, a positive 4 % signal increase at 5–7 s after stimulus onset, and a signal undershoot of about –1 % at 15–20 s after stimulus onset. Return to prestimulation baseline is accomplished 60–90 s after the end of the stimulus. It indicates that rather long periods are required if activation-induced BOLD MRI signal alterations need to be physiologically 'decoupled' from each other.

Conversely, shorter interstimulus periods physiologically 'couple' repetitive stimuli in such a way that their BOLD MRI responses become contaminated by the pre-stimulation history.

Similar findings hold true for subsecond visual stimuli as well as for paradigms replacing darkness as the only control state.

Noise

Response functions such as in panel D of Figure 10 create an important link between brief cortical events and much slower physiologic fluctuations that modulate the dynamic BOLD MRI signal even in the absence of functional challenge. Excluding instrumental noise, an analysis of such low frequency "noise" in the primary sensorimotor and visual system has already been used to generate maps of functional connectivity (Biswal et al 1995).

Initial Dip

Data from near-infrared optical imaging of exposed visual cortex in animals have suggested a rise of deoxyhemoglobin within the first 2 s of activation due to an immediate increase in oxygen consumption at the site of neuronal activation. However, it is not entirely clear to which degree optical data from animals may be extrapolated to MRI studies of humans. For example, both methodologies focus on different tissue elements: whereas MRI detects intravoxel susceptibility changes from all vascular components that contribute to one image voxel, optical experiments commonly exclude all signals from vessels that are identifiable by visual inspection.

In this context, it should be emphasized that time-locked averaging of multiple activation cycles from repetitive protocols with insufficient recovery periods (e.g., traces B and C of Figure 10) may give rise to an artifactual "initial dip". The effect is caused by a "wrap around" of real-time signal contributions and particularly applies to the undershoot of preceding activations which folds back into the "silent" latency phase of a subsequent activation cycle. When ensuring complete decoupling of successive stimuli by using long interstimulus recovery periods (e.g., in trace D of Figure 10), the wrap around effect and the resulting dip vanish.

Event-Related Recordings

Rather than only reducing the stimulus duration, mapping protocols may be optimized by decreasing the interstimulus periods. Pertinent protocols attempt to identify individual "events" in a stream of rapidly presented stimuli and should facilitate the MRI adaptation of paradigms commonly applied in cognitive neuropsychology.

Figure 11 demonstrates real-time BOLD MRI responses to a series of 1 s visual stimuli when the interstimulus interval is reduced from 89 s to 6 s and 3 s, respectively.

Contrast

A key observation is a gradual decrease of the peak-to-peak difference between stimulus-related positive and negative signal deflections. The effect becomes dramatic for interstimulus durations of less than about 5–6 s which corresponds to the time-to-maximum response for brief visual stimuli.

The attenuation of the functional contrast reflects the insufficient time for the development of both the full positive BOLD MRI response and its effective reversal after the end

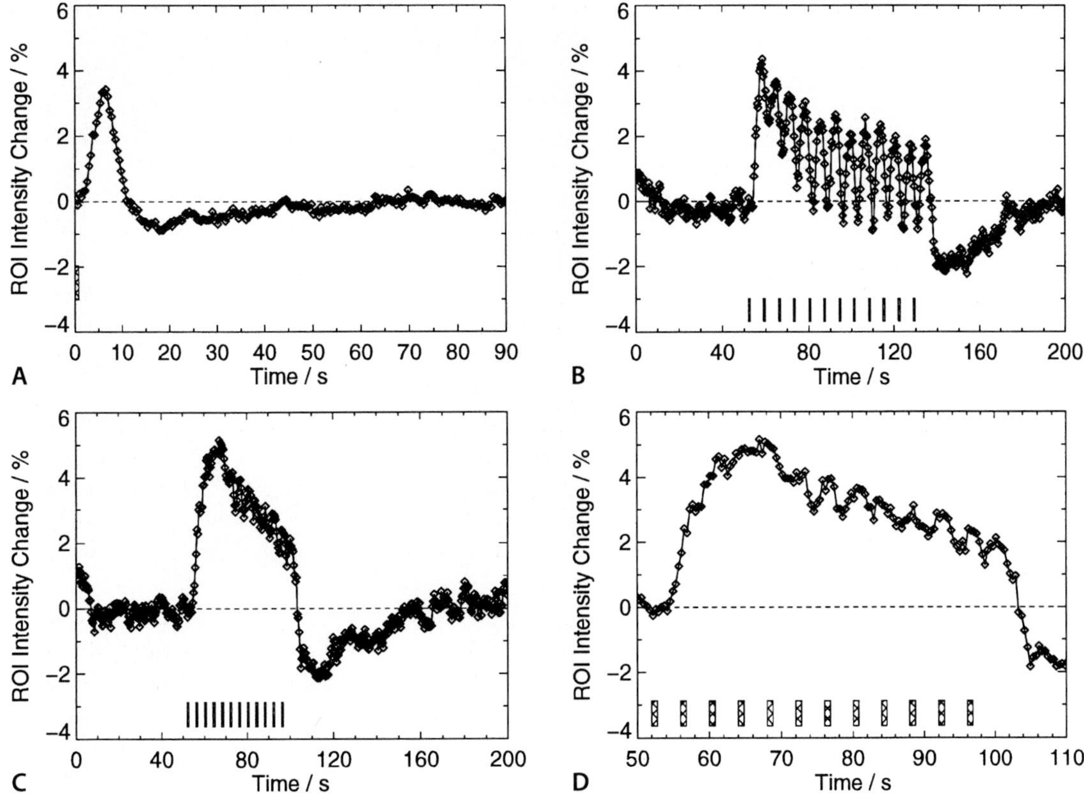

Fig. 11. Time courses of oxygenation-sensitive MRI signal intensities (2.0T, EPI, TR/TE = 400/54ms, flip angle 30°, temporal resolution 0.4s) for protocols comprising a 1s period of visual activation (checkerboard vs. darkness, cross-hatched bar) but different interstimulus periods. The real-time responses refer to (a) a 1 s/89s, (b) a 1 s/6s, and (c) a 1 s/3s protocol. (d) Enlarged view of the responses shown in (c). The curves represent mean values of all pixels with statistically significant signal alterations averaged across subjects (*n*=6) and stimulation cycles. (Modified from Fransson et al 1998b).

of the stimulus. The cumulative effect from several undershoots leads to a new equilibrium with an apparent "baseline" well below the pre-stimulation baseline and a pronounced undershoot after the end of the last stimulation cycle. It also affects the individual real-time MRI responses in such a way that the almost nonexistent functional contrast during the initial phase of the event-related paradigm enhances during the first 4 (Figure 11B) or 6 cycles (Figure 11C,D) for interstimulus intervals of 6s and 3s, respectively.

Mapping

The activation maps shown in Figure 12 reveal adequate spatial congruence of activated brain regions for the 1 s/89s, 1 s/6s, and 1 s/3s protocols. They are obtained by cross-correlation of pertinent time courses with suitable reference functions mimicking the protocol timings as well as hemodynamic latencies, rise and fall times.

Notwithstanding ongoing discussions about a linear or nonlinear system analysis of responses elicited by event-related paradigms, the results confirm the ability to map condition-specific populations of neurons as long as dynamic MRI data sets provide sufficient functional contrast at the physiologic level. More specifically, this finding only applies to processing of identical stimuli in the visual cortex and under conditions where the individual responses can be separated in time.

Deactivation

In general, complex paradigms may pose further complications because they tend to involve activity switches between arbitrary conditions that may yield both MRI signal in-

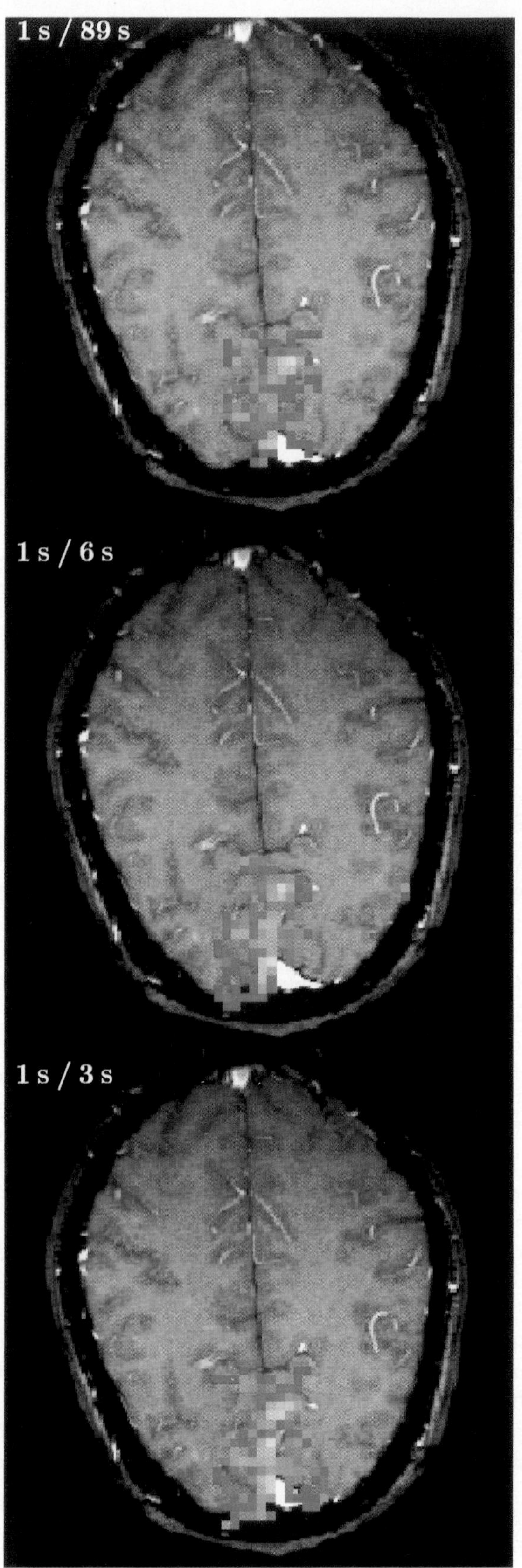

Fig. 12. Spatial congruence of visual cortical areas of a single subject activated by (top) a 1 s/89 s, (middle) a 1 s/6 s, and (bottom) a 1 s/3 s protocol using the same paradigm as in Figure 10. The maps are superimposed onto a reference image (rf-spoiled FLASH, TR/TE = 70/6 ms, flip angle 60°) delineating anatomy and macrovasculature. (Modified from Fransson et al 1998b).

creases and decreases. Future work will need to address "deactivation" at the physiologic level and develop evaluation strategies to analyze bidirectional switches between states of variable degrees of neuronal activity.

Paradigm Design

A most promising feature of MR functional neuroimaging is the flexibility in the design of experimental paradigms. A preliminary list of possibilities includes
- block structures comparing two functional states,
- multiplexed or interleaved block structures with multiple states,
- event-related paradigms,
- triggered acquisitions, and
- noise analyses without a functional challenge.

Some Basic Structures

Blocks

So far, many functional MRI examinations employ a design based on a block structure A vs. B. The paradigm compares the neuronal representations of two distinct conditions - often under the simplified assumption of an activated state A (e.g., motor activity, visual stimulation) and a control condition B (e.g., motor rest, darkness).

Even the simple block structure requires further consideration. For example, the statistical power of the data analysis depends on the number of images acquired per condition as well as on the number of repetitions (or activation cycles). In addition, the choice of a temporally symmetric or asymmetric presentation of A and B conditions as well as their absolute durations may influence the task performance, the data analysis, and the resulting map. An important aspect may be the order or direction of comparisons: if A vs. B corresponds to activation processes, B vs. A will indicate deactivation processes in either overlapping or different regions even when the data is taken from the same experiment.

Multiplexing

Without the doubtful assumption of a "resting" brain, more reasonable concepts adopt the view of qualitatively equivalent states of activity that only quantitatively differ in the degree of neuronal firing or synchronization or the extent of neuronal subpopulations involved. These ideas lead to the development of multiplexed or interleaved paradigm structures, where multiple conditions may be compared to each other or suitable combinations. A simple example demonstrating marked differences in the underlying question as well as in the results stems from the replacement of a task vs. control structure by a task 1 vs. task 2 design in a study of finger somatotopy in the human motor hand area (see below).

Multiplexed designs offer better chances to optimize cognitive paradigms with respect to maintaining attention and avoiding habituation and expectancy. They are also efficient in allowing for multiple differential comparisons within a single experimental run. In practice, the use of physiologically optimized durations of 8–10 s per condition as well as of statistically recommended 4–6 repetitions limits the number of different states to about 10 as measurement times should be kept to below 10 min with regard to subject compliance and motion problems.

A special variant are serial or incremental variations of stimulus features such as eccentricity or angle of the visual field or frequency of an acoustic tone. Suitable analysis strategies provide direct access to the retinotopic or tonotopic organization of primary visual and auditory cortices, respectively.

Events Event-related paradigms minimize the durations of stimulus presentations and interstimulus intervals in a multiplexed block paradigm and attempt to unravel the transient responses to and spatial representations of differential cortical events. A typical example is the "odd-ball" paradigm which tests the brain's capacity for identifying individual responses to novel events in a rapid stream of otherwise identical stimuli.

Nevertheless, the ability to separately analyze neuronal activity changes at the physiologic level depends on the ability to distinguish pertinent responses in time or space.

In general, we have to face activations in brain regions showing multi-functional encoding, i.e., overlap of subpopulations of neurons with different functional representations. Such observations already apply to visual and sensorimotor areas (see below), but are even more likely to occur for higher order cognitive processing. Accordingly, in order to distinguish responses with similar spatial encodings, we need to retain a sufficient "decoupling" of physiologic responses in time and thereby ensure residual functional contrast for spatial mapping.

It has also been proposed to temporally randomize stimulus presentations in event-related paradigms to alleviate the degradation of the functional contrast for short interstimulus periods.

EEG Another class of experiments are dynamic MRI acquisitions where the data analysis is performed in accordance with an (externally) recorded "trigger" signal. Although still hampered by technical difficulties, a prominent application is the correlation of oxygenation-sensitive MRI signal changes with specific patterns in the EEG. For example, it is conceivable to detect and map the physiologic correlates of epileptic activity in the interictal state.

Other possibilities are correlations with the application of repetitive transcranial magnetic stimulation (e.g., in treating depression) or involve the recording of psychophysical or behavioral parameters.

Selected Applications

Foreseeable clinical applications range from contributions to presurgical planning (e.g., sensorimotor activity, hemispheric language dominance) to mapping of disease-related dysfunction (e.g., blindsight) and cortical (re-) organization. The latter refers to both short-term plasticity (e.g., learning, memory) and long-term changes during early brain development or rehabilitation and recovery after brain injury. Still rather premature applications aim at the functional mapping of cognitive and emotional deficits of psychiatric patients.

Somatotopy A simple though illustrative example of the importance of a proper paradigm is the analysis of finger somatotopy in the human motor cortex. Fine-scale somatotopic encoding in brain areas devoted to sensorimotor processing may be questioned when using a design of self-paced isolated finger movements vs. motor rest. Pertinent experiments result in extensive overlap of spatial representations covering the entire hand area for almost all fingers. The design asks for all locations of finger representations and addresses somototopy only in the sense of true functional (spatial) segregation.

In contrast, Figure 13 demonstrates markedly different results when the paradigm directly switches between individual finger movements. This design asks for cortical response preference or "qualitative" somototopy rather than for mere neuronal activity. Using functional maps at high spatial resolution, the comparative paradigm reveals foci of differential activation which display an orderly medio-lateral progression in accordance with the classical cortical motor homunculus.

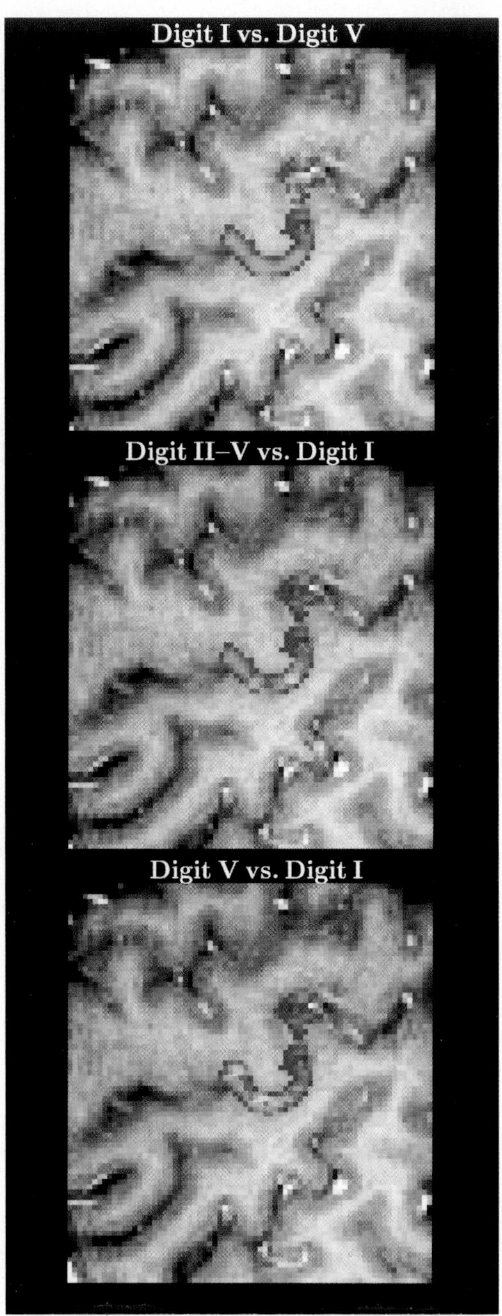

Fig. 13. Functional predominance of cortical representations of individual finger movements of the right hand (coded in red) in the left-hemispheric primary motor hand area (coded in blue) of a right-handed subject. The hand area was delineated by a sequential finger-to-thumb opposition task of the right hand vs. motor rest. Finger movements were studied by a comparative paradigm design based on 6 cycles of a 18 s/36 s protocol. Despite overlap of representations, functional predominance of (top) the thumb vs. little finger, (middle) index through little finger vs. thumb, and (bottom) little finger vs. thumb follows a somatotopic arrangement in covering the lateral, intermediate, and medial portion of the hand area, respectively (2.0 T, FLASH, TR/TE = 62.5/30 ms, flip angle 10°, measuring time 6 s, spatial resolution 0.78 × 1.56 mm^2 interpolated to 0.78 × 0.78 mm^2 during image reconstruction, section thickness 4 mm). (Modified from Kleinschmidt et al 1997).

From these design-specific findings it may be concluded that somatotopy within the hand area of primary motor cortex does not present as qualitative functional segregation but as quantitative predominance of certain digit representations embedded within a physiologically synergetic and anatomically interconnected joint hand area.

Language

Another example of paradigm-specific responses deals with the identification of an early visual area involved in language comprehension. Exploiting the MRI potential for testing a large number of differential paradigms within single sessions, the mapping process itself identified the relevant properties for separating "reading" from "seeing".

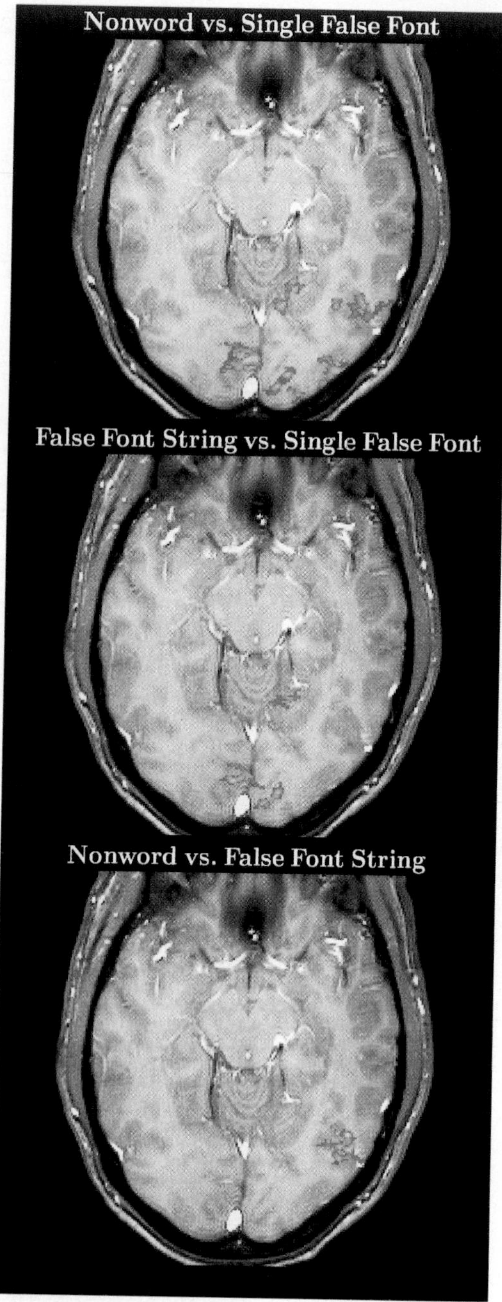

Fig. 14. Identification of an early visual area involved in language comprehension. (Top) Activations in ventral occipital cortex representative of a "global" paradigm comparing a nonword and a single false font with no lexical meaning. (Middle) Activations representative of horizontal length encoding ("seeing") obtained by comparing a string of false fonts vs. a single false font. (Bottom) Lateralized activations in a left-hemispheric region representative of processing lexical input ("reading") obtained by comparing a nonword vs. a length-matched string of false fonts. Experimental parameters were as in Figure 13. (In collaboration with Peter Indefrey and Colin Brown).

The top map in Figure 14 depicts responses in the ventral stream of the occipital (visual) cortex that are representative for viewing a pronounceable though meaningless "nonword" (e.g., debam) vs. a single false font with a complexity similar to that of a letter but no lexical character. Although the observed hemispheric lateralization suggests specific contributions representative for processing lexical input, the distributed areas may be further split into subcomponents by using paradigms with even more subtle cognitive differences than the leading experiment.

The middle part of Figure 14 indicates that the medial components in this subject correspond to simple length encoding along the horizontal median unraveled by a direct comparison of a string of false fonts vs. a single false font. Complementarily, the strictly lateralized left-hemispheric region in the bottom map is indeed suggestive of early visual processing of lexical input as the used paradigm compared presentations of a non-word vs. a length-matched string of false fonts.

Acknowledgement: We are indebted to a large number of colleagues and co-workers including Harald Bruhn, Wolfgang Hänicke, Andreas Kleinschmidt, and Klaus-Dietmar Merboldt for invaluable contributions over the past 5 years.

References

Bachelard HS (1997) Magnetic resonance spectroscopy and imaging in neurochemistry. In: Agranoff B, Suzuki K (eds) Advances in neurochemistry, Vol 8. Plenum, New York

Bandettini PA, Wong EC, Hinks RS, Tikofsky RS, Hyde JS (1992) Time course EPI of human brain function during task activation. Magn Reson Med 25:390–397

Bandettini PA, Jesmanowicz A, Wong EC, Hyde JS (1993) Processing strategies for time-course data sets in functional MRI of the human brain. Magn Reson Med 30:161–173

Biswal B, Yetkin FZ, Haughton VM, Hyde JS (1995) Functional connectivity in the motor cortex of resting human brain using echo-planar MRI. Magn Reson Med 34:537–541

Blamire AM, Ogawa S, Ugurbil K, Rothman D, McCarthy G, Ellermann JM, Hyder F, Rattner Z, Shulman RG (1992) Dynamic mapping of the human visual cortex by high-speed magnetic resonance imaging. Proc Natl Acad Sci USA 89:11069–11073

Bruhn H, Kleinschmidt A, Boecker H, Merboldt KD, Hänicke W, Frahm J (1994) The effect of acetazolamide on regional cerebral blood oxygenation at rest and under stimulation as assessed by MRI. J Cereb Blood Flow Metab 14:742–748

Buxton RB, Wong EC, Frank LR (1998) Dynamics of blood flow and oxygenation changes during brain activation: The balloon model. Magn Reson Med 39:855–864

Fox PT, Raichle ME (1986) Focal physiological uncoupling of blood flow and oxidative metabolism during somatosensory stimulation in human subjects. Proc Natl Acad Sci USA 83:1140–1144

Frahm J, Bruhn H, Merboldt KD, Hänicke W (1992) Dynamic MRI of human brain oxygenation during rest and photic stimulation. J Magn Reson Imaging 2:501–505

Frahm J, Krüger G, Merboldt KD, Kleinschmidt A (1996) Dynamic uncoupling and recoupling of perfusion and oxidative metabolism during focal brain activation in man, Magn Reson Med 35:143–148

Fransson P, Krüger G, Merboldt KD, Frahm J (1998a) Temporal characteristics of oxygenation-sensitive MRI responses to visual activation in humans. Magn Reson Med 39:912–919

Fransson P, Krüger G, Merboldt KD, Frahm J (1998b) Physiologic aspects of event related paradigms in magnetic resonance functional neuroimaging. NeuroReport 9:2001–2005

Friston KJ, Holmes AP, Worsley KP, Poline JB, Frith CD, Frackowiak RSJ (1995) Statistical parametric maps in functional imaging: a general linear approach. Hum Brain Map 2:189–210 [http://www.fil.ion.ucl.ac.uk/spm]

Kleinschmidt A, Requardt M, Merboldt KD, Frahm J (1995a) On the use of temporal correlation coefficients for magnetic resonance mapping of functional brain activation. Individualized thresholds and spatial response delineation. Intern J Imag Sys Technol 6:238–244

Kleinschmidt A, Bruhn H, Steinmetz H, Frahm J (1995b) Pharmacologic manipulation of vasomotor tone studied by magnetic resonance imaging of cerebral blood oxygenation. J Cerebr Blood Flow Metab 15:S527

Kleinschmidt A, Obrig H, Requardt M, Merboldt KD, Dirnagl U, Villringer A, Frahm J (1996) Simultaneous recording of cerebral blood oxygenation changes during human brain activation by magnetic resonance imaging and near-infrared spectroscopy. J Cereb Blood Flow Metab 16:817–826

Kleinschmidt A, Nitschke MF, Frahm J (1997) Somatotopy in the human motor cortex hand area. A high-resolution functional MRI study. Eur J Neurosci 9:2178–2186

Kleinschmidt A, Bruhn H, Krüger G, Merboldt KD, Stoppe G, Frahm J (1999) Effects of sedation, stimulation, and placebo on cerebral blood oxygenation. A magnetic resonance neuroimaging study of psychotropic drug action. NMR Biomed in press

Krüger G, Kleinschmidt A, Frahm J (1996) Dynamic MRI of cerebral blood oxygenation and flow during sustained activation of human visual cortex. Magn Reson Med 35:797–800

Krüger G, Kleinschmidt A, Frahm J (1998) Stimulus dependence of oxygenation-sensitive MRI responses to sustained visual activation. NMR Biomed 11:75–79

Kwong KK, Belliveau JW, Chesler DA, Goldberg IE, Weisskoff RM, Poncelet BP, Kennedy DN, Hoppel BE, Cohen MS, Turner R, Cheng HM, Brady TJ, Rosen BR (1992) Dynamic magnetic resonance imaging of human brain activity during primary sensory stimulation. Proc Natl Acad Sci USA 89:5675–5679

Ogawa S, Lee TM, Kay AR, Tank DW (1990) Brain magnetic resonance imaging with contrast dependent on blood oxygenation. Proc Natl Acad Sci USA 87:9868–9872

Ogawa S, Tank DW, Menon R, Ellermann JM, Kim SG, Merkle H, Ugurbil K (1992) Intrinsic signal changes accompanying sensory stimulation: functional brain mapping with magnetic resonance imaging. Proc Natl Acad Sci USA 89:5951–5955

Stark DD, Bradley WG (1998) Magnetic Resonance Imaging, 3rd Edition. Mosby, St Louis

Turner R, Le Bihan D, Moonen CTW, DesPres D, Frank J (1991) Echo-planar time course MRI of cat brain oxygenation changes. Magn Reson Med 22:159–166

Villringer A, Dirnagl U (1995) Coupling of brain activity and cerebral blood flow: Basis of functional neuroimaging. Cerebrovasc Brain Metab Rev 7:240–276

Chapter 39

Positron Emission Tomography of the Human Brain

Fabrice Crivello and Bernard Mazoyer

Introduction

Positron Emission Tomography (PET) is a functional *in vivo* neuroimaging technique that has the potential to provide three-dimensional quantitative maps of a variety of physiological parameters such as blood flow, glucose metabolic rate, protein synthesis, receptor density and affinity (Phelps et al. 1979; 1986). PET is a non-invasive technique that can be used in humans in various domains of neuroscience research both in normal volunteers and in patients. Typical examples of PET research in neuroscience are found in cognitive function mapping, and in pathophysiological and pharmacological studies of neurological or psychiatric disorders.

Basic Principles in PET Imaging

PET Tracer Selection

PET is based on the imaging of the distribution of a biological molecule labeled with a positron-emitting nucleus, after it has been introduced in a subject. The most critical issue in designing a PET experiment is the selection of the radiotracer to be used (Stöcklin and Pike, 1993). The key factor in tracer selection is the nature of the biological process (or parameter) of interest, for it will determine the appropriate kind of tracer: glucose analogs for studying glycolysis activity, amino acids for measuring protein synthesis rate, antagonists of neurotransmitters to obtain estimates of receptor density and affinity. PET tracers are produced by dedicated radiochemistry labs. PET radiochemistry is rendered especially difficult by the short half-life of many positron emitters (see Table 1 below). More often than not, dedicated syntheses have to be designed for each new tracer in order to obtain sufficient amounts of adequate specific radioactivity, a process that may span several years.

PET Physics

Positron (e^+) emitters are nuclei having an excess of protons as compared to their numbers of neutrons: as a consequence, such nuclei will undergo a nuclear transition consisting of the conversion of a proton into a neutron (or equivalently of a u quark into a

Positron Emission

Fabrice Crivello, Université de Caen, Groupe d'Imagerie Neurofonctionnelle, UPRES EA2127 & LRC CEA 13V, GIP Cyceron, BP 5229, Caen GIP, 14074, France
Correspondence to: Bernard Mazoyer, Université de Caen, Groupe d'Imagerie Neurofonctionnelle, UPRES EA2127 & LRC CEA 13V, GIP Cyceron, BP 5229, Caen GIP, 14074, France (phone: +33-231-470-271; fax: +33-231-470-222; e-mail: mazoyer@cyceron.fr)

d quark) with emission of a positron, a positively charged particle that is the antiparticle of the electron (see Fig. 1). For the oxygen 15 nucleus, the reaction would follow the equations:

$$p^{+}(uud) \rightarrow n(udd) + e^{+} + <n_e>$$

$$^{15}O(8p^{+},7n) \rightarrow {}^{15}B(7p^{+},8n) + e^{+} + <n_e>$$

Where $<n_e>$ denotes the electron neutrino. Note that the neutrino and the positron share the excess of energy in the reaction and are emitted with kinetic energies that depend upon the type of nucleus. An important characteristic of e^+ emitters is their short half-life (see Table 1).

Positron Annihilation

In matter, positrons lose their energy mainly by collisions in a way similar to that of electrons. Once it has lost its kinetic energy, a positron combines with a free electron of the medium: the electron-positron pair annihilates and produces a pair of anti-parallel (conservation of momentum) photons of 511 keV energy (conservation of electron/positron energy at rest) (see Fig. 1).

Note that positron lifetime in matter is very short (on the order of 10^{-23} s) but long enough to allow travelling of the positron and consequently to generate a significant difference between the location of the positron emission and that of its annihilation, an effect known as the *positron range*.

Positron Annihilation Detection

Positrons cannot escape the biological matter and cannot be detected as such. Rather, it is the detection of the product of their annihilation with electrons, namely, *pairs of 511 keV anti-parallel gamma rays*, that is at the origin of the PET signal. Detection of such events is achieved using circular arrays of classical gamma ray detectors (see Fig. 2) consisting of scintillators (usually made of bismuth germanate, BGO) optically coupled with photomultipliers (PMT). Pairs of detectors are electronically coupled so as to ensure that only coincident gamma rays are detected (collimation). The nature, number, size of crystals and PMTs determine the final resolution of the PET image, typically 5 mm in each space direction.

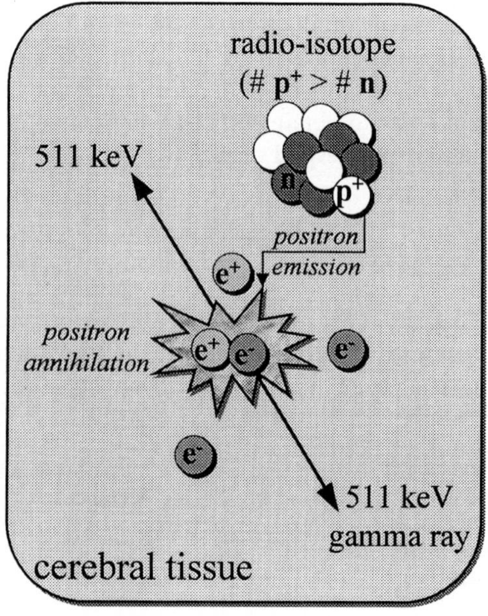

Fig. 1. Positron emission and annihilation. A positron is emitted from a labeled isotope, losing its energy by scattering from electrons of the surrounding medium before annihilating with an electron in order to produce a pair of anti-parallel 511-keV gamma rays.

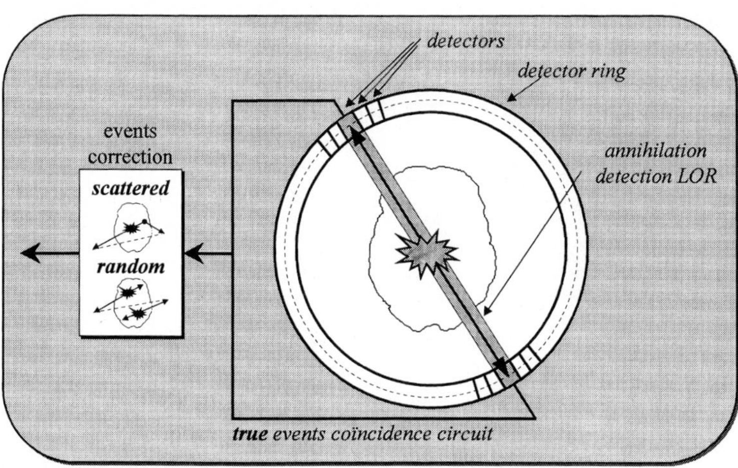

Fig. 2. Schematic diagram illustrating the acquisition geometry of a PET scanner. A PET scanner consists of a circular array of scintillation detectors that flash when gamma rays interact with them. Pairs of detectors are electronically coupled to form Lines Of Response (LOR) which look for two coincident events (named true events). The total amount of detected events is then corrected for random coincidences and scatter events in order to obtain a fully quantitative image.

PET Imaging

PET Image Formation

During a PET scan, the number of coincident events over a period of time is recorded for every pair of detectors crossing the field of view (also called Lines Of Response, LOR). From the total number of events recorded in each LOR, it is possible to reconstruct the three-dimensional distribution of the radioactive concentration averaged over the duration of event acquisition (Bendriem and Townsend 1997). Corrections are applied to account for random coincidences (coincident gamma rays originating from two different positrons), gamma ray attenuation by the head (using an external source, see below: transmission scan), and scattered coincidences (coincident gamma rays originating from the same positron but having experienced Compton diffusion while going through matter). The usual end product is a stack of contiguous brain slices in which each image element (voxel) contains the absolute value of the local radioactive concentration of positrons. When needed, time series of such volumes can be acquired.

PET Image Analysis

Positron concentration images are of limited interest. Rather, the ultimate goal of a PET experiment is to obtain local estimates of relevant biological parameters such as blood flow, glucose metabolism and receptor density. The art of PET image analysis is that of converting, through physiological modeling, a time series of radioactive concentration volumes into biochemical parametric images (Budinger et al. 1985). Using *a priori* knowledge on the behavior of the radiotracer and its metabolism in the tissue, a compartment model (see Fig. 3) is built and a set of differential equations derived that describe the variation of radioactivity over time.

The model shown in Fig. 3 is used for measuring regional cerebral blood flow (rCBF): For a given cerebral tissue volume V_t, irrigated by its arterial and venous capillaries, the regional cerebral blood flow (identified by the parameter F / V_t) can be easily obtained by resolving the following equation:

$$C_t(t) = \left(\frac{F}{V_t}\right).C_a(t) \otimes \exp^{-\left(\left(\frac{F.C_v(t)}{V_t.C_t(t)} - \lambda\right)t\right)}$$

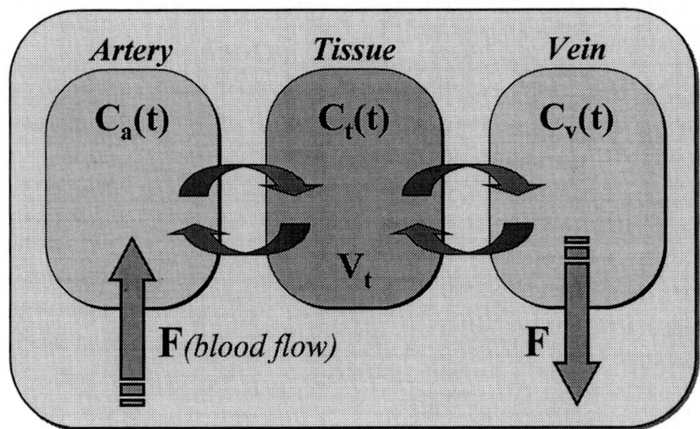

Fig. 3. Compartment model used for blood flow measurement with PET representing the kinetics of the radioisotope used through the blood-brain barrier. $C_t(t)$, $C_a(t)$, $C_v(t)$ are the radioactive concentrations of the cerebral tissue (with a volume equal to V_t), the arterial and the venous compartments, respectively.

Where λ is the radioactive decay constant of the isotope used, $C_t(t)$, $C_a(t)$, $C_v(t)$ are respectively the radioactive concentration of the cerebral tissue, the arterial and the venous compartments, and $\otimes$ is the convolution product (for convolution see Chapter 21).

Model parameters can be estimated by fitting the predicted values of the local radioactivity model to those measured in the PET volumes (Mazoyer et al. 1986). This process sometimes requires knowledge of the blood radioactivity, which can be obtained through counting of blood samples drawn during PET data acquisition through radial artery cannulation. Additional details on the state of the art of PET data acquisition and analysis can be found in Carson et al. (1998).

PET vs. Other Neuroimaging Modalities

In vivo human neuroimaging techniques can be subdivided into two major complementary categories: structural and functional. X-ray computerized tomography (CT) and magnetic resonance imaging (MRI) are of the former type, whereas PET, single-photon computerized tomography (SPECT), and functional MRI (FMRI) are of the latter type.

PET vs. FMRI

In the mid 1980s, while MR (see Chapter 38) was establishing itself as the reference technique for *in vivo* structural imaging, PET emerged as the functional imaging method of choice, thanks to the availability of short-lived tracers and to technological advances that allows one to obtain quantitative maps of physiological parameters of interest. For ten years, cognitive neuroimaging was thus preferentially, if not exclusively, performed using PET activation studies and has experienced a rapid development. Meanwhile, the sensitivity of the PET approach coupled to the synthesis of a variety of positron-emitting radiolabeled ligands also has allowed the *in vivo* investigation of neurotransmission, providing density maps for several major neurotransmitter systems (Frost and Wagner Jr. 1990).

The advent of functional MR in the mid 1990s has had a major impact on the respective roles of PET and MR in the neuroimaging domain, due to the superior spatial and temporal resolutions of BOLD FMRI, and the large availability of commercial MR scanners on which this technique can be implemented. Two characteristics, however, remain specific to PET, namely, its ability to provide absolute values of physiological parameters and its unique suitability, only shared in part with magnetic resonance spectroscopy, to investigate *in vivo* neurotransmission. Accordingly, provided that some technical challenges can be adequately addressed, it can be anticipated that PET and FMRI will be synergistically used to improve our knowledge on

1. the mechanisms of neurovascular coupling during cognitive activity,
2. the modulation of cortical networks by pharmacological agents,
3. the relationships between neurotransmission parameter distribution and cognitive functions,
4. the mechanism of endogenous transmitter release during cognitive activities.

PET vs. SPECT

PET has a number of *advantages* over SPECT. PET makes use of positron emitters, which are isotopes of natural components of biological molecules; therefore, tracers labeled with positron emitters will retain the same exact biochemical properties as their unlabeled equivalents (no isotopic effect). In contrast to PET, SPECT is based on single-photon emitters such as ^{99m}Tc or ^{123}I which are heavy atoms with relatively long half-lives (6 and 13.2 hours, respectively). SPECT ligands are thus usually less specific and more difficult to synthesize than their PET counterparts. Because of their shorter half-lives, positron emitters are also better suited in terms of the radioactive dose delivered to the subject. For example, activation studies with ^{133}Xe are limited to a few inhalations, whereas up to 12 radioactive water injections can be delivered to a normal volunteer in a PET protocol (see below). Finally, absolute local radioactive concentration maps can be obtained with PET, thanks to the detection of pairs of gamma rays by electronic collimating and to the ability one has to adequately correct for head attenuation using an external source. By contrast, physical collimating and the difficulty of attenuation correction in SPECT limit its capacity to provide quantitative data.

PET has, however, one *disadvantage* over SPECT, namely, its cost and the requirement for a nearby cyclotron. Newer SPECT/PET machines are now able to perform both 511 keV and lower energy gamma ray detection which, combined with the use of FDG delivery from a remote production site, may allow a wider dissemination of the technique in the near future.

PET Regulations and Safety

In most countries, PET is considered as a research tool and has not yet been approved as a standard medical imaging tool. There are two major safety issues in PET: dosimetry and pharmacology. Dosimetry is strictly regulated: In EU, for example, the total dose delivered to a subject who participates in a PET research protocol cannot exceed the dose the general population on average receives over one year (5 mSv). For some protocols (see PET activation studies below) this puts a stringent limit on the kind of experiment one can achieve with PET. Pharmacology is an even more stringent criterion that decides whether or not a PET experiment can be set up or not. Approval for each PET tracer must be obtained before using it in humans, and apart from very few specific cases, the use of PET tracers in humans remains of experimental nature. Using a new tracer will thus require conducting preliminary studies in animals and primates before being able to perform human studies.

Outline

PET is basically an *in vivo* autoradiographic technique in which molecules of biological interest are first labeled with cyclotron-produced short-lived positron-emitting nuclei before being intravenously injected into a human subject. A positron tomograph, surrounding the subject's head, dynamically records coincident 511 keV anti-parallel pho-

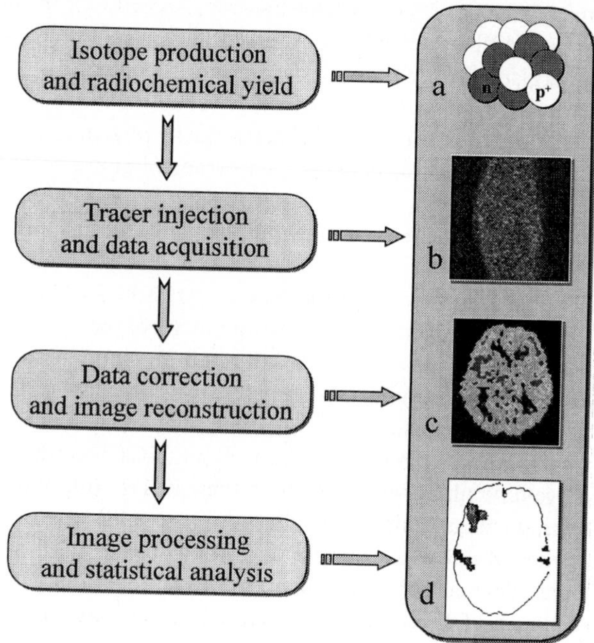

Fig. 4. Flowchart of a classical PET experiment. a) Production of radioactive tracer; b) sinogram for which each element represents the number of counts detected by a particular detector pair; c) raw PET image; d) final statistical map.

ton pairs that originates from positron emissions and local annihilations. A tomographic reconstruction algorithm applied to acquisition data provides, after correction for scatter and attenuation, the 3D distribution of the absolute concentration in the brain of the positron emitter. Mathematical modeling is eventually used to convert time series of radiotracer concentration maps into physiological parameter maps. Because of its intrinsic coarse spatial resolution and low signal-to-noise ratio, intersubject averaging of PET images is often required and achieved using atlas or individual MRI structural data. An overview of the entire PET procedure is given in Fig. 4.

Materials

Positron-Emitting Radionuclides

Since positrons are electron anti-particles, they belong to anti-matter and do not exist in free state. Consequently, e^+ radionuclides must be produced with the help of a cyclotron. In this kind of circular accelerator, a beam of heavy particles is speeded up to sub-relativistic energy and eventually collides with a gaseous target to produce the desired nucleus. Table 1 summarizes the characteristics of the production of the most widely used positron emitters in PET neuroscience, namely, Oxygen-15 (^{15}O), Carbon-11 (^{11}C) and Fluorine-18 (^{18}F), which have 2, 20 and 110 minute half-lives, respectively.

Medical Cyclotron

The relatively short half-life of positron emitters constitutes both an advantage and a limitation: It allows performing human studies at low subject radiation exposure but they require a cyclotron to be available in the vicinity of the experiment room. Small-sized cyclotrons have been designed to fulfill this requirement that can be installed in the dedicated and shielded room of a lab building and can be operated by technicians.

Table 1. Physical properties of the most frequent radioisotopes used for positron emission tomography of the human brain.

Radioisotope	Half-life (min)	Energy (max) (MeV)	Range in water (mm)
^{11}C	20	0.95	1.1
^{13}N	10	1.19	1.4
^{15}O	2.2	1.73	2.5
^{18}F	110	0.63	1.0
^{68}Ge	3.9105	1.90	1.7

Positron-Labeled Radiochemicals

Positron-emitting nuclei, such as Oxygen-15, Carbon-11 and Fluorine-18, are isotopes of major components of biological molecules. Accordingly, almost every molecule of biological relevance can potentially be labeled by a positron emitter, provided that a fast radiosynthesis can be designed. This requires the use of semi- or fully automated radiochemistry in dedicated hot cells.

Hundreds of e^+ labeled radiotracers have been successfully synthesized that can be roughly classified in three categories, depending on the biological pathway they trace (Stöcklin and Pike 1993, see Table 2):

- Hemodynamic tracers are used to map cerebral blood flow and cerebral blood volume. As shown in Table 2, Oxygen-15 labeled tracers are of very simple structures due to the very short half-life of ^{15}O. In addition, a dedicated delivery system, either for inhalation or for intravenous injection, must be acquired and placed close to the PET camera;
- Metabolic tracers are used to map pH, glucose or oxygen metabolism, amino-acid metabolism and protein synthesis rate;
- Neuropharmacological tracers are designed to trace the dynamics of pre- or post-synaptic sites of neurotransmitters and drugs, including enzymes involved in their metabolism.

Positron Camera

A state-of-the-art PET camera, such as the ECAT HR+ (Siemens, Erlangen, Germany) is made of a cylindrical array, 90 cm in diameter and over 15 cm long so as to cover the entire brain. The array is made of 4 rings of blocks of 64 (8x8) crystals, each block being optically coupled to 4 large-size photomultipliers (PMTs). Front end and coincidence electronics are placed within the camera gantry, which also accommodates a set of retractable tungsten septa and 3 germanium-68 rod sources. When septa are extended in the field of view, 511 keV gamma rays are physically collimated and coincident lines define axial planes data (2D mode). When septa are retracted, coincident lines can be acquired in all directions in space and define a volumetric acquisition (3D mode). Although it yields more scattered events and dead-time losses, 3D should be the preferred mode of acquisition because of its high sensitivity. The 2D mode should be reserved when too much activity is present in the field of view, such as when using inhalation of ^{15}O gases. A PET image volume typically consists of a stack of 63 contiguous transaxial brain images, each of 128x128 size, where each voxel has a 2x2x2-mm^3 size and represents 2 Mbytes of data.

The rod sources are used to obtain the head attenuation coefficient map: When extended in the field of view, a 2D acquisition is performed (transmission scan) that will

Table 2. Most frequent positron-labeled radiopharmaceuticals used for positron emission tomography of the human brain and the corresponding local cerebral physiological parameters that can be estimated.

Radio-tracer	Local physiological parameter
Hemodynamics	
$H_2{}^{15}O$, $C^{15}O_2$, ^{15}O-Butanol	Blood flow
$C^{15}O$	Blood volume
Energy metabolism	
$^{15}O_2$	Oxygen metabolic rate
^{18}F-Deoxyglucose	Glucose metabolic rate
^{11}C-1-Glucose	Glucose metabolic rate (aerobic)
Amino-acid metabolism	
^{11}C-Methionine, ^{18}F-Tyrosine	Protein synthesis rate
Fatty acid metabolism	
^{11}C-Palmitate	
Neurotransmitter metabolism	
^{18}F-DOPA	Dopamine
^{11}C-Tryptophan	Serotonin
^{11}C-Deprenyl	Monoamine-oxydase-B turnover
Neuroreceptor binding kinetics	
^{11}C-Flumazenil	Benzodiazepine receptor
^{11}C-SCH 23390	Dopamine D1 receptor
^{11}C-Raclopride	Dopamine D2 receptor
^{11}C-Nomifensine	Dopamine re-uptake
^{18}F-Altanserin	Serotonin 5HT2A receptor
^{18}F-Setoperone	Serotonin 5HT2A receptor
^{11}C-Scopolamine	Muscarinic cholinergic receptor
^{11}C-Doxepin	Histamine H1

provide measurements of attenuation coefficient along each coincident line. Given the ^{68}Ge half-life (268 days) and the required counting statistics, 3 mCi rod sources are usually necessary and replaced every 6 months.

Besides the PET camera itself, other devices may be necessary for conducting specific experiments. The most common are automatic water injectors and arterial blood samplers and counters. Water injectors are used for $H_2{}^{15}O$ infusion (see below): They consist of a small reactor in which $H_2{}^{15}O$ is continuously produced and of a computer controlled delivery system that automatically delivers to the patient the desired amount of radioactivity in a predefined volume and speed. Blood samplers and counters are used for measuring the blood radioactivity at the camera site: They are very useful devices when absolute quantification of blood flow or glucose metabolism is required. They are simply made of a peristaltic pump coupled to a 511 keV photon detection/counting device.

Image Processing

PET cameras are usually delivered with a dedicated graphic workstation that controls the PET camera acquisition and daily check, and the reconstruction process. Additional graphic workstations (SUN UltraSparc, or SGI O_2 with a minimum RAM of 128 Mbytes)

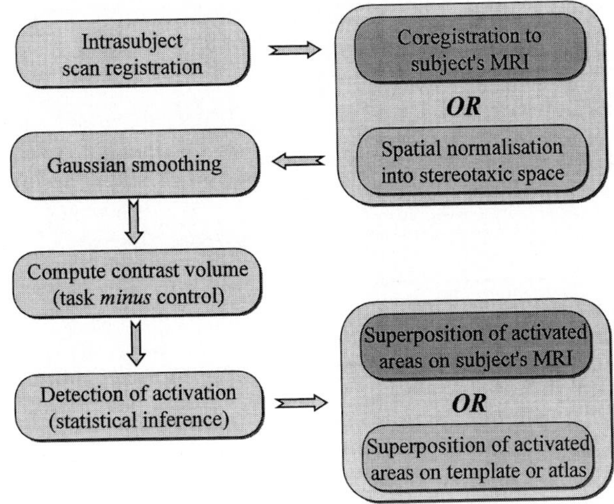

Fig. 5. Flowchart illustrating the different processing steps needed to process the dataset of a PET activation experiment.

are designed to further process the images (cf. Fig. 5). In addition to the manufacturer software, it is strongly recommended to purchase image and statistical toolbox licenses such as IDL and MATLAB.

These are required to run image-processing software packages that have become standards in the field of neuroimaging, such as SPM (Statistical Parametric Mapping, Friston et al., 1995a,b), MPItool (Multi Purpose Imaging Tool, Pietrzyk et al., 1994), or AIR (Automated Image Registration, Woods et al., 1997a,b). Given the large volume of data generated by the acquisition and the processing stages, it is mandatory to carefully design the storage and archiving system with the following rules in mind: 1) raw data (sinograms before any reconstruction) should always be archived on a secure medium (digital linear tape, CD-ROM) immediately following the completion of the experiment; 2) reconstructed and processed images should be kept on a fast access medium (disk) to allow easy access until the research protocol is completed; 3) double archiving on separate and distant media of everything should always be performed.

Procedure

PET procedures in human neuroscience mainly vary as to the radiotracer they use; however, they share a number of common features that will be illustrated using a widely used PET protocol, namely, cognitive function mapping with radioactive water.

Cognitive Neuroimaging with $H_2{}^{15}O$

Preparation

1. *Automatic injector preparation.* Insert new sterile tubing, valves and saline solution pockets. Start cyclotron and water production. Calibrate the radioactive water production system so that each injection will deliver 5 to 6 mCi. Save a sample in a sterile vial and send to laboratory for quality control.
2. *PET camera room preparation.* Install all necessary equipment for stimulus delivery (computer screen, headphones, etc.) and behavioral response recording (electrooculogram, electrocardiograms, mouse pad, etc.). Place black curtain around PET camera if necessary.
3. *Subject preparation.* Position helmet (head holder) on subject's head and install any additional equipment (mirror, headphones, etc.). Install subject in supine position

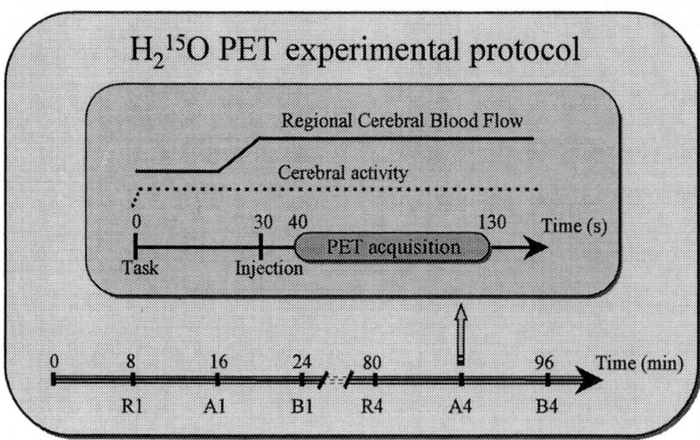

Fig. 6. Schematic diagram of a classical PET activation study showing the temporal resolution of this functional brain mapping technique.

on camera's bed. Fix helmet on head holder. Insert intravenous injection kit in subject's left arm; connect to water injector.

4. *Subject positioning.* Enter subject in PET camera and align desired anatomical landmarks on camera laser beams. Use ink pen to mark reference landmarks on subject's skin (to be used for head position control throughout PET experiment)

Scanning

5. *Start transmission scan.* A typical brain transmission scan on an ECAT HR+ will last about 10 minutes during which 130 M-events will be collected.
6. *Radioactive water production.* Start accumulation of radioactivity in water injector for 6 minutes. Prepare subject for cognitive task to be performed. One minute before water injection, turn laser beams on and adjust subject's head.
7. *Emission scanning* (see Fig. 6). Start water injection, stimulus delivery and PET scanning. In a typical setting water injection is started first and stimulus delivery 30 seconds later, once water is entering the subject's venous system. PET scanning starts when radioactivity reaches the brain, namely, when total event count-rate rises above 400% of background noise. Coincidences are acquired for 90 seconds in a single frame. Turn laser beams on and check head position upon scan completion. Record injected activity in lab book. Restart water production for next scan if needed. Debrief subject.
8. *Image reconstruction.* A set of 63 transaxial images are reconstructed on-line and checked for quality. Typical reconstruction parameters are: 128x128 matrix, 3D filter parameters: correction for attenuation, scatter and randoms. Store raw data on long-term archiving system and reconstructed images in experimenter's file system.
9. *Repeat steps 6 to 8* until protocol is completed. A typical brain activation protocol will consist of 8 to 12 repeated injections 8 minutes apart (see Fig. 6).
10. Upon protocol completion, debrief subject and check water injector calibration.

Image Processing

The different steps involved in image processing of PET datasets are illustrated in Fig. 5, and a non-exhaustive list of the free software packages available for each of these steps is given in Table 3.

11. *Visualize* each acquired volume to detect large artifacts. Convert images from manufacturer to standard (ANALYZE) format.
12. *Single case study.* Register each acquisition to the first one using the AIR package. When available, register 3D MRI acquisition to PET volumes. Define brain outer

Table 3. Free software packages available in the neuroimaging field for PET data processing.

Image processing step	Software package
PET-PET registration	Unsupervised : AIR*, SPM* Manual : MPItool*
PET-MRI registration	Unsupervised : AIR, SPM Manual : MPItool
Spatial normalization of PET images	Unsupervised : AIR, SPM
Smoothing	AIR, SPM, MPItool
Statistical analysis	Individual : Internal routines Multisubject : SPM
Compartment analysis of PET data	RFIT*
Display	SPM, MPItool

* See supplier

contour on MRI, compute brain average radioactivity for each acquisition and use it as a normalization factor. Compute contrast volumes (Task minus Control), threshold and superimpose onto MRI. Visualize results.

13. *Multisubject study.* Use SPM.

Results

In this section we will describe a full PET experiment (experimental design, acquisition, data analysis and results) obtained from a silent verb generation protocol performed with oxygen-15-labeled water (Crivello et al. 1995).

Experimental Design

Starting from the 260 Snodgrass Vanderwart Pictures, a list of unambiguous nouns was selected. One- or two-syllable nouns inducing at least the generation of three verbs within the 10 seconds following noun presentation, and having high associative strength for the first response, were retained for the lists to be presented during the PET experiment. Six right-handed young male volunteers (aged from 21 to 25 years) participated in this study. Using PET and oxygen-15-labeled water, normalized regional cerebral blood flow (NrCBF) was measured six times in each subject by replicating three times a series of two conditions: 1) silent rest control; 2) silent generation of verbs semantically associated to nouns presented at 0.1 Hz via earphones. All measurements were performed with eyes closed in total darkness.

Scanning Procedure

For each condition, 31 contiguous brain slices were acquired simultaneously on an ECAT 953B/31 PET camera giving an in-plane resolution of 5 mm. Emission data were acquired with septa extended (2D acquisition). Following intravenous bolus injection of 60 mCi of oxygen-15-labeled water, a single 80 second scan was reconstructed (including a correction for head attenuation using a measured transmission scan) starting at the arrival of the radioactivity in the brain. The between-scan interval was 15 min. In addition to the PET data, axial and sagittal series of 3-mm-thick T1-weighted high-resolution magnetic resonance images (MRI) covering the whole brain were acquired for each subject on an 0.5 T GE-MRMAX imager.

Data Analysis

As a pre-processing step, the AIR package was used to perform individual PET-PET and PET-MRI registration. The PET data analysis was then conducted in two different ways.

Single Subject Analysis

An individual detection algorithm was applied to the individual NrCBF difference images averaged across the three replicates without any spatial normalization of the original data. Each subject's difference images were composed of 31 slices with a 3.375 mm thickness. The significance level was set at 0.05 per plane after Bonferroni correction. An accurate localization of the objects detected with the individual algorithm was made based on a detailed anatomical analysis of each subject's brain anatomy. Using dedicated software (Voxtool), MRI axial slices were used to reconstruct a three-dimensional brain volume that was further segmented and allowed the display of both hemisphere's surfaces together with sections in three orthogonal directions. The principal sulci, in particular those limiting the inferior frontal gyrus, were identified in each subject's MRI. In this study, single-case analysis was not possible using SPM. Indeed, only six 2D acquisitions were available per subject, resulting in a degree of freedom too low (df = 4) to proceed the statistical analysis.

Intersubject Stereotactically Averaged Data

Scans from each subject were realigned and stereotactically normalized into the Talairach space with the image-processing transformation included in the SPM package. The resulting images were then smoothed with a 16 mm Gaussian kernel filter and finally processed using the statistical procedure implemented in SPM with a significance threshold of 0.001 uncorrected for multiple comparisons.

Results

Single-subject Analysis

Significant activation was detected in the left inferior frontal gyrus for subjects 2, 5 and 6 and in the right side in subject 4 (see upper part of Fig. 7). In subjects 2, 5 and 6 the activation lay in the left inferior frontal gyrus (Broca's area), in the *pars triangularis* for subject 6, and in the *pars opercularis* in subjects 2 and 5. Subject 4 is remarkable in that his activation was located in the *pars opercularis* of the right inferior frontal gyrus extending anteriorly to the *pars triangularis* (the right analog of Broca's areas), leading us to conclude a right hemisphere dominance for language in this strictly right-handed subject.

Stereotactically Averaged Intersubject Data

Results obtained with SPM running at a 0.001 uncorrected significance level are illustrated in the lower part of Fig. 7. The major activations were located in the supplementary motor area, the left inferior frontal gyrus, and a bilateral activation of the superior temporal gyrus. Note the activation of the primary visual cortex.

Comparison between Intrasubject and Intersubject Averaged Functional Anatomy

The averaged results partly reflect the regions that were found activated in a majority of subjects like the left inferior frontal gyrus (3 among the 6). However, intersubject averaged activation may also result from a strong activation present in one or two subjects as was observed for the superior temporal cortex or the visual cortex activation detected in subject 5 only. By contrast, some areas were found activated in only one subject and had no corresponding activations detected in the intersubject averaged image. These activations may be either true positives, thereby revealing significant individual differences in functional neuroanatomy, or false positives. The results presented here exhibit a strong functional inter-individual variability in a group of six strictly right-handed subjects. Individual functional patterns demonstrate a high level of variability in terms of

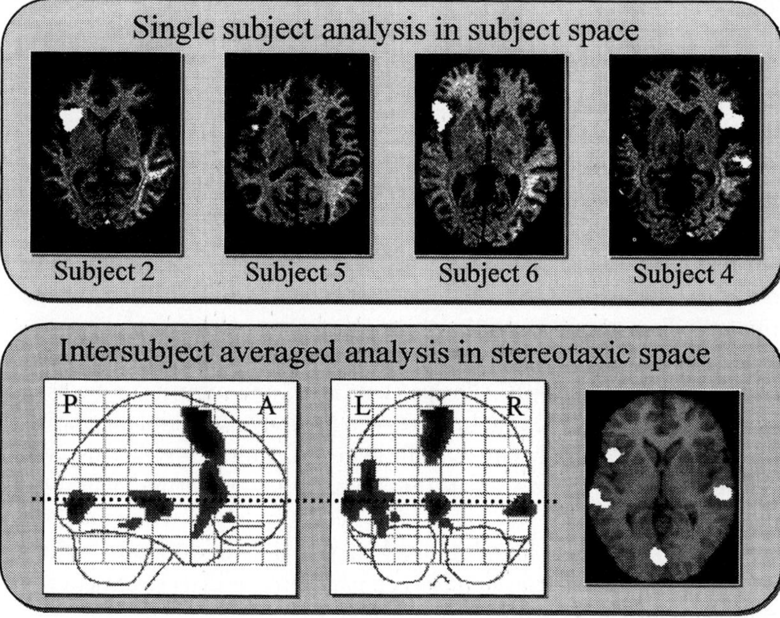

Fig. 7. *Upper part*: results of the single subject analysis of the verb generation versus control comparison. After realignment of the MRI and PET volumes in the same coordinate space, inferior frontal gyrus activations detected in four subjects are superimposed on their corresponding axial MRI slice. A detailed anatomical analysis allows to precisely localize these activations in the left inferior frontal gyrus for subjects 2, 5 and 6 and in the right inferior frontal gyrus for subject 4. *Lower part*: Intersubject averaged analysis performed with SPM for the same contrast. Note the absence of right inferior frontal gyrus activation in this mean activation map, although it was detected in subject 4.

volumes and intensity of the activated areas. This example is a perfect illustration of the necessity of individual analysis for PET activation study and its complement in comparison to the classical inter-individual averaged analysis. The conjoint use of a sensitive individual detection algorithm and a precise cerebral anatomical analysis nowadays allows the investigation of precise and quantitative relations between structure and function of the human brain.

Troubleshoot

Below are some of the most frequent problems encountered during a PET activation protocol and tips on how to remedy them.

- Count-rate is too low, total number of recorded events 25% less than usual. Check oxygen-15 production line (cyclotron current, target, gas supply). Check water injector line (saline solution kit +++, catheter entry point). If no problem is detected, the most likely reason is the arm position of the subject that slows down the water bolus. The solution should slightly move up or down the subject's arm.
- Patient head motion during acquisition. This is a frequent and complicated problem to solve. The strategy is to try first to register together all the reconstructed emission scans in the series acquired in this subject, using the same transmission data. If matching is not satisfactory, it will be necessary to use a special program that matches sinograms before reconstruction.

Applications

In human basic neuroscience, PET has been applied to mapping human normal cognition. In this field, despite the supremacy of FMRI, it will remain a valuable tool to study neurophysiological bases of cognitive processes, to conduct multimodal experiments (PET and FMRI or EEG), and to correlate cognition with neurotransmission (see above). A significant use of PET in the future could also be in the field of drug design and testing (Comar 1995).

In clinical neuroscience, PET has been and still is a particularly useful tool to investigate cerebrovascular diseases (Baron et al. 1989), movement disorders such as Parkinson's or Huntington's disease and dementia syndromes such as Alzheimer's disease. Alterations of cognitive functions can be mapped and their recovery under treatment followed up with activation studies. FDG (Phelps et al. 1979) and oxygen can be used to map low energy areas and study the coupling between hemodynamic and metabolic activity. Receptor ligands can be used to unravel the neurotransmitter systems involved in the diseases (Baron et al. 1991) or to evaluate the clinical efficacy of drugs, neural grafts or gene therapy.

Similar types of application are found in psychiatry, mainly in the study of depression and schizophrenia, with a special emphasis on psychotropic drug treatment evaluation and planning (Sedvall et al. 1996).

In neuro-oncology, PET is a tool of choice in three areas: pre-surgical mapping of functional areas using brain activation protocols (Herholz et al. 1997), tumor grading using FDG and methionine (Mazoyer et al. 1993), and evaluation of therapy (surgery, X-ray, anti-mitotic drugs).

Acknowledgement: The authors are deeply indebted to their colleagues from the Cyceron cyclotron staff and to Nathalie Mazoyer (GIN) for supplying details of the PET procedure and images for the illustrations.

References

Baron JC, Frackowiak RSJ, Herholz K, Jones T, Lammertsma AA, Mazoyer B, Wienhard K (1989) Use of PET methods for measurement of cerebral energy metabolism and hemodynamics in cerebrovascular disease. J Cereb Blood Flow Metab 9:723–742

Baron JC, Comar D, Farde L, Martinot JL, Mazoyer B (1991) Brain dopaminergic system imaging with positron emission Tomography. Kluwer Academic Publishers, Dordrecht

Bendriem B, Townsend D (1997) The art of 3D PET. Kluwer Academic Publishers, Dordrecht

Budinger TF, Huesman RH, Knittel B, Friedland RP, Derenzo SE (1985) Physiological modeling of dynamic measurements of metabolism using positron emission Tomography. In: Greitz T et al (eds) The metabolism of the human brain studied with positron emission Tomography. New York: Raven Press, pp 165–183

Carson RE, Daube-Witherspoon ME, Herscovitch P (1998) Quantitative functional brain imaging with positron emission Tomography. Academic Press, London

Comar D (1995) PET for drug development and evaluation. Kluwer Academic Publishers, Dordrecht

Crivello F, Tzourio N, Poline JB, Woods RP, Mazziotta JC, Mazoyer B (1995) Intersubject Variability in functional neuroanatomy of silent verb generation: assessment by a new activation detection algorithm based on amplitude and size information. Neuroimage 2:253–263

Friston KJ, Ashburner J, Frith CD, Poline JB, Heather JD, Frackowiak RSJ (1995a) Spatial registration and normalization of images. Human Brain Mapping 2:165–189

Friston KJ, Holmes AP, Worsley KJ, Poline JB, Frith CD, Frackowiak RSJ (1995b) Statistical parametric maps in functional imaging: A general approach. Human Brain Mapping 2:189–210

Frost JJ, Wagner Jr. HN (1990) Neuroreceptors, neurotransmitters and enzymes. Raven Press, New York

Herholz K, Reulen H-J, von Stockhausen H-M, Thiel A, Ilmberger J, Kessler J, Eisner W, Yousry TA, Heiss W-D (1997) Preoperative activation and intraoperative stimulation of language-related areas in patients with glioma. Neurosurg 41:1253–1262

Mazoyer BM, Huesman RH, Budinger TF, Knittel BL (1986) Dynamic PET data analysis J Comput Assist Tomogr 10:645–653

Mazoyer BM, Heiss WD, Comar D (1993) PET studies on amino acid metabolism and protein synthesis. Kluwer Academic Publishers, Dordrecht

Phelps ME, Huang SC, Hoffman EJ, Selin C, Sokoloff L, Kuhl DE (1979) Tomographic measurement of local cerebral glucose metabolic rate with (F-18)2-fluoro-2-deoxy-D-glucose: validation of the method. Ann Neurol 6:371–388

Phelps ME, Mazziotta JC, Schelbert H (1986) Positron emission tomography and autoradiography. Principles and applications for the brain and heart. Raven Press, New York

Pietrzyk U, Herholz K, Fink G, Jacobs A, Mielke R, Slansky I, Würker M, Heiss W-D (1994) An interactive technique for three-dimensional image registration: Validation for PET, SPECT, MRI and CT brain studies. J.Nucl.Med. 35:2011–2018

Sedvall CG, Brené S, Farde L, Karlsson P, Nordström A-L, Nyberg S, Pauli S, Halldin C (1996) Neuroreceptor imaging by PET: implications for clinical psychiatry and psychopharmacology. Int Acad Biomed Drug Res 11:198–207

Stöcklin G, Pike VW (1993) Radiopharmaceuticals for positron emission Tomography. Methodological aspects. Kluwer Academic Publishers, Dordrecht

Woods RP, Grafton ST, Holmes CJ, Cherry SR, Mazziotta JC (1997a) Automated Image Registartion: I. General methods and intrasubject validation. J.Comput.Assist.Tomogr. 22:139–152

Woods RP, Grafton ST, Watson JDG, Sicotte NL, Mazziotta JC (1997b) Automated Image Registration : II. Intersubject validation of linear and nonlinear models. J.Comput.Assist.Tomogr. 22:153–165

Suppliers

PET camera and medical cyclotrons

- Siemens- CTI : http://www.cti-pet.com
- General Electric : http://www.ge.com/medical/nuclear

Pharmaceutical company and suppliers of radioligand precursors

see Comar 1995, pp149–150

Image processing software

- **AIR** : http://bishopw.loni.ucla.edu/AIR3
- **SPM** : http://www.fil.ion.ucl.ac.uk/spm
- **MPItool** : http://www.uni-koeln.de/institute/mpifnf/mpitool
- **RFIT:** http://cfi.lbl.gov

Chapter 40

Magnetic Resonance Spectroscopy of the Human Brain

Stefan Blüml and Brian Ross

Introduction

Background

Twenty years ago, the optimum techniques available to address a metabolic question in the brain were considered to be – in order of reliability – brain slices *in vitro*, arterio-venous differences *in vivo* across the jugular bulb and carotid artery, isolated intact brain perfusion *in situ* and, the then newly emerging techniques of isolated brain-cell preparation (Ross 1979). Almost at the same time, however, there appeared a seminal paper describing the transfer, after 25 years, of the chemists' major investigational tool, nuclear Magnetic Resonance Spectroscopy (MRS), to the intact mammalian brain *in vivo* (Thulborn et al. 1981). Soon thereafter followed *in vivo* MRS of human muscle (Ross et al. 1981), and of neonatal (Hamilton et al. 1986) and adult human brain (Bottomley et al. 1983). Today, few university hospitals in the world are without a whole-body MR scanner capable of assaying metabolites non-invasively in the human brain, using robust MRS methods. Together with MRS, physiological MRI, fMRI and PET (Chaps. 38 and 39, respectively), the neuroscientist can now reverse the order of preference when considering a technique with which to address a metabolic question in the brain. It is almost certainly "easiest" to turn first to the intact human brain with *in vivo* magnetic resonance spectroscopy. How best to do this in practice is the subject of this chapter.

As a spectroscopy technique, the product of MRS is a print-out of peaks of different radio-frequency and intensity, recording molecules which possess the intrinsic property of the NMR technique, nuclear spins, unique resonance frequencies, spin-couplings and relaxation properties (all terms defined in Glossary). Many classical neurochemical events are readily documented in the human brain *in vivo* through MRS. However, the molecules which yield the optimum MR-signals are not always those of which the neuroscientist first thinks, or even wanted, in her or his pre-conceived experiment. As a result, in the first 20 years of *in vivo* brain MRS, some "new" neurometabolites have come to the fore.

Principles of Neurospectroscopy

We define neurospectroscopy as the field of study resulting from MRS examination of the human brain. Diseases and pathologies of the brain are commonly classified as:

- structural (including degenerative, tumor and embryogenic defects);

Stefan Bluml, Huntington Medical Research Institute, Clinical Magnetic Resonance Spectroscopy Unit, 660 South Fair Oaks Avenue, Pasadena, CA, 91105, USA (phone: +01-626-397-3271; fax: +01-626-397-3332; e-mail: soccss@hmri.org)

- physiological (essentially interruption of blood supply); and
- biochemical or genetic.

Of the latter, some are receptor- and neurotransmitter-related (e.g., dopamine in Parkinson's disease) but many are directly or indirectly related to disturbances of the pathways of oxidative, anabolic and catabolic intermediary metabolism, the tricarboxylic acid (TCA) or Krebs cycle, glutamine/glutamate turnover, glycolysis, ketogenesis or fatty acid metabolism. PET, and to a lesser extent SPECT, MRI, fMRI and "diffusion-imaging" address blood flow, glucose turnover and oxygen consumption, and PET and SPECT are uniquely able to "image" targeted receptor ligands. However, until the advent of NMR, no direct non-invasive assay of the products of gene expression, the cerebral metabolites, was available. There was no neuronal marker, no astrocyte marker and no technique to directly determine energy metabolism. These gaps are now filled by neurospectroscopy and, with increased clinical experience, a diagnostic "need" for magnetic resonance spectroscopy (MRS) of the brain emerges.

Methods for Human Neurospectroscopy

At least 20 methods are available for human neurospectroscopy. Table 1 outlines the currently available methods of MRS as applied to the brain. Localized ^{1}H MRS includes long echo time, short echo time, STEAM, PRESS, quantitative, chemical shift imaging (CSI), metabolite imaging, "fast" metabolite images, automated, functional measurements. ^{31}P MRS includes pulse-acquire, DRESS, ISIS, PRESS, proton decoupled ^{31}P MRS, nuclear Overhauser effect (NOE) enhanced, CSI, fast phosphocreatine metabolite imaging, magnetization transfer (flux) measurements. ^{13}C includes pulse-acquire, CSI, proton decoupled, NOE enhanced, polarization transfer, ^{13}C labeled glucose infusion flux studies. Each method has contributed to some aspect of neurochemical research or has found clinical application. All of the methods offer useful and specific information. An example would be the renewed interest in ^{31}P MRS with proton-decoupling to identify separately the components of the "choline" peak seen in routine ^{1}H MRS (see "Results: Cholines").

Milestones of Neurospectroscopy

The milestones of neurospectroscopy are summarized in Table 2. MRS in the brain began with ^{31}P spectroscopy in anesthetized rats and other small animals. Non-invasive assays of adenosine triphosphate (ATP) and phosphocreatine (PCr) (expressed as "metabolite ratios") and of intracellular pH gave exciting new insights. Direct metabolic rate determination *in vivo*, using ^{31}P magnetization transfer, was among the first biological applications of this now widespread technique. MRS confirmed the dependence of cerebral energetics upon oxidative metabolism and glycolysis. A practical future for MRS was demonstrated in the gerbil "stroke" model, when carotid ligation was clearly shown to produce ipsilateral changes of anaerobic metabolism: loss of PCr and ATP, increase of inorganic phosphate (Pi) and acidification of the affected hemisphere (Thulborn et al. 1981). A satisfying synthesis of some of the new neurophysiology with clinical management comes in studies of a stroke model. *Spreading depression* is the term applied to the depolarizing condition which is mimicked by high K^+ cell incubation (Badar-Goffer et al. ,1992, have explored this extensively in tissue slices), and which likely occurs with energy failure and hypoxia after stroke. Hossman (1994) and Gyngell et al. (1995) have brought diffusion weighted imaging (DWI), MRS and electrophysiology together to provide new insights into growth of infarcts in rats. In short, the depolarization, which *in vitro* accelerates glycolysis, is detected in regions where excess lactate appears in 'bursts' within penumbra or spreading depression. The authors conclude that this physiological evidence of deterioration can be monitored by MRS, and segregated from true infarcts by diffusion weighted imaging. Removal of lactate and recovery of the neuronal marker N-acetyl aspartate (NAA see below) are likely to be excellent end-points in the

Table 1. Current methods of magnetic resonance spectroscopy available for the brain. The first applications of MRS with surface coils demonstrated the potential of MRS for non-invasive insights into brain metabolism (Ackerman et al. 1980). The lack of a proper localization, using this easy and straightforward technique, has been partly overcome by improvements (Bottomley et al. 1984). However, in the clinical setting of neurospectroscopy, more advanced localization techniques are used. Currently, ISIS (Ordidge et al. 1986), STEAM (Frahm et al. 1987; Merboldt et al. 1990) or PRESS (Bottomley 1987) sequences are most frequently used as single-voxel and/or CSI techniques for localized MRS. Proton MRS performed with long or, better, short echo times allows the quantitation of important metabolites (Frahm et al. 1989; Ross et al. 1992; Kreis et al. 1990; Narayana et al. 1991; Hennig et al. 1992; Kreis et al. 1993a; Michaelis et al. 1993; Barker et al. 1993; Christiansen et al. 1993; Danielsen and Henriksen 1994). The automation of the measurement (shimming, water suppression, acquisition) (Webb et al. 1994) and of the data processing (phasing, fitting) have led to a fully automated exam. Additional information is obtained by spectral editing techniques which are currently used for the identification of low concentration metabolites or overlapping resonances (Provencher 1993).

Clinical method	Rf coil available	Implementation on clinical 1.5T
Localized ^{1}H MRS		
Long echo	*	*
Short echo	*	*
Quantitation	*	*
Phase-encoded imaging of metabolites, CSI	*	*
Fast metabolite imaging	*	*
Automation	*	*
Osmolality	*	*
Editing	*	*
Functional MRS	*	*
Localized ^{31}P MRS		
Pulse-acquire		*
Decoupled ^{1}H-^{31}P		*
Phase-encoded imaging, CSI	*	*
Fast phosphocreatine imaging	*	*
Magnetization transfer (flux)		*
Localized ^{13}C MRS		
Natural abundance	*	*
^{13}C enriched-flux measures	*	*
^{1}H-^{13}C heteronuclear methods	*	*
Localized ^{15}N MRS		
^{15}N enriched-flux measures		
^{1}H-^{15}N heteronuclear method(s)		
Localized ^{19}F MRS		
^{19}F drug detection		
^{19}F imaging and blood flow methods	*	*
(^{19}F probes for Ca2+ and Mg2= determination)†		*
^{7}Li MRS		
^{23}NaMRS		

† Toxic in vivo

newly emerging clinical trials of brain salvage post stroke (Gyngell et al. 1995). It is hard now to realize that prior to MRS, rapid-freezing of whole animals, brain-blowing and surgical biopsy were the only effective sources of knowledge of such events.

MRS in the human brain, which began with newborns, verified the predictions of animal studies that hypoxic-ischemic disease of the brain could be monitored by the changes in high-energy phosphates, Pi and pH (Hamilton et al. 1986; Cady et al. 1983; Hope et al. 1984). The predictive value of MRS has been demonstrated in several hun-

Table 2. Milestones in *in vivo* MRS. Twelve Most Prevalent Uses of Neurospectroscopy 1984–1999

1. Differential diagnosis of coma: A. Neurodiagnosis of symptomatic patients
2. Subclinical hepatic encephalopathy and pre-transplant evaluation
3. Differential diagnosis of dementia (rule out AD)
4. Therapeutic monitoring in cancer ± radiation
5. Neonatal hypoxia
6. "Work-up" of inborn errors of metabolism
7. "Added-value" in routine MRI
8. Differential diagnosis of white matter disease, especially MS, ALD and HIV
9. Prognosis in acute C.V.A. and stroke
10. Prognosis in head injury
11. Surgical planning in temporal lobe epilepsy
12. Muscle disorders

dred newborn infants. The outcome after severe hypoxic ischemic encephalopathy in newborn humans is determined by the intracerebral pH and Pi/ATP ratio.

Wide-bore high-field magnets permitted extension to adults and infants beyond a few weeks of age (Bottomley et al. 1983). Three areas of human neuropathology have as a result been extensively illuminated by ^{31}P MRS. Using brain tumor as a target of newly evolving localization techniques (the early studies in infants employed no localization beyond that conferred by the "surface coil"), Oberhaensli et al. (1986) began the slow process of overturning three decades of thought about brain tumors in particular, showing their intracellular pH to be generally alkaline, not acidic. A new generation of drugs in oncology will be designed to enter alkaline intracellular environments, rather than the acidic environment measured in the interstitial fluid.

^{31}P MRS in adult stroke exactly mirrored the findings in hypoxic-ischemic disease of newborns, and even appears to offer predictive value through intracellular pH and Pi/ATP (Welch 1992). This work in turn has provoked a large body of research in experimental models of stroke which now guides the human application of MRS.

The third area of work stimulated by the advent of ^{31}P MRS was that of neurodegenerative diseases, including Alzheimer disease (Pettegrew et al. 1994). Commencing with *in vitro* studies of tissue extracts, two hitherto unrecognized groups of compounds, seen in the ^{31}P spectrum as phosphomonoesters (PME) and phosphodiesters (PDE), were empirically shown to be altered. The metabolic significance of these "peaks" was incompletely understood. Nevertheless, a promising new area of neurochemistry was opened by ^{31}P MRS, and then extended to *in vivo* brain analysis. By providing non-invasive assays of less well-known metabolites and pathways, MRS has identified a "New Neurochemistry." A perfect example of this new knowledge emerged with the advent of water suppressed ^{1}H MRS of the brain *in vivo*.

N-acetylaspartate (NAA), a neuronal marker (Tallan 1957), was re-discovered in 1983 (Prichard et al. 1983). Of the many expected and new resonances now identified in human neuro-MRS, none has yielded more diagnostic information than NAA. Its identity, concentration and distribution are now well established. Early experiments in animals showed loss of NAA in stroke. Very large numbers of studies in man show NAA absent, or reduced in brain tumor (glioma), ischemia, degenerative disease, inborn errors and trauma, so that to a first approximation, the histochemical identification of NAA (and N-acetylaspartyl glutamate (NAAG)) with neurons and axons, and its absence from mature glial cells, is confirmed. The use of ^{1}H MRS as an assay of neuronal "number" appears well justified.

The first generation of human MRS studies was performed without image-guidance. While MRI is not essential to our understanding of neurochemistry, the combined use of these two powerful tools permitted the direct demonstration that there is often a dissociation in space, between anatomically obvious events in the brain and biochemical changes. Metabolite imaging has confirmed this important principle in stroke, tumors, multiple sclerosis and degenerative diseases.

A simplified method of localization permitted the routine use of MRS to assay neurochemistry in a single place, albeit rather large, in the cerebral cortex, cerebellum or mid-brain. This method, now generally known as "single-voxel MRS", is largely responsible for showing that biochemical disorders commonly underlie neurological disease (Prichard et al. 1983; Hanstock et al. 1988; Frahm et al. 1988, 1989, 1990; Kreis et al. 1990; Ross et al. 1992; Michaelis et al. 1991; Stockler et al. 1996). MRS is therefore well poised for "early" diagnosis. Reversible biochemical changes accompany several physiological events, and provide a biochemical basis for functional imaging (fMRI) (Merboldt et al. 1992). The inborn errors of metabolism and hereditary diseases, and several of the major neurological scourges of our time, reveal functional biochemical disturbance. Neonatal hypoxia, cerebral palsy, neuro-AIDS, dementias, stroke, epilepsies, neuro-infections and many encephalopathies are now seen to include a biochemical component.

Automation and quantitation are the final ingredients required to make MRS an indispensable tool in human neuroscience. Automation permits universal access, including urgent MRS in acute, reversible neurological diseases, and large-scale clinical trials. Quantitation, a long-overlooked area, gives the precision of measurement which will be required to conclusively demonstrate incremental metabolic responses to intervention or therapy.

Chemical Composition of the Brain

Although obviously an oversimplification, in MRS-terms brain may be biochemically defined as water plus dry-matter.

Brain-Water

The water is, as in other tissues, divided into intracellular and extracellular (circa 85% and 15%, respectively). Intracellular water, which is further divided into cytoplasmic and mitochondrial compartments, circa 75% and 25%, respectively, contains all of the important neurochemicals. These are either unique to intracellular water, such as lipids, proteins, amino acids, neurotransmitters and low molecular weight substances, or at least have a very different concentration from the extracellular and cerebrospinal fluid (CSF) compartments. Glucose is an exception, being found in proportions 5:3:1 in blood, CSF and brain-water, respectively. Amino acids are generally distributed 20:1, brain water:CSF or blood. Brain water and extra-cellular fluid (ECF) are distinct from the large CSF compartment, the volume of which depends greatly upon the location selected, and from the intravascular blood, which comprises up to 6% of brain water. MRS-assays of brain water represent the sum of icf and ecf (Ernst et al. 1993a).

Brain Dry-Matter

Seen through the MR image and the MRS assays of water, the 20% or so of brain which really "matters" is largely invisible! Hence the occasional use of the terms "missing" or "invisible" found in the MRS literature. Covered by these terms are all macromolecules (DNA, RNA, most proteins and phospholipids), as well as cell membranes, organelles, including the dry-matter of the mitochondria, the christae, and myelin. The term is probably equivalent to the biochemist's "dry-weight" and can be used as a more constant unit by which to determine the concentration of key neurochemicals. This is particularly relevant in pathologies in which brain water (or wet-weight/dry-weight) may alter, such as metabolic disorders, edema, tumors, inflammation, stroke or infarction. Metabolite concentrations may therefore more accurately be compared as mmol/g dry-weight than by the more usual mmol/g wet weight, or per ml of brain water.

Myelin and Myelination For the most part myelin is inaccessible to *in vivo* MRS (because of the manner in which myelin water contributes to the MR signal, the contrast between white and gray matter is striking in MRI). The composition of myelin is nevertheless of some interest to the *in vivo* spectroscopist because of the changes which may occur in demyelinating and many other diseases. While major components like phosphatidyl choline, phosphatidyl-ethanolamine, -serine and -inositol are probably entirely immobile and NMR-invisible, their putative breakdown products, such as phosphoryl choline, glycerophosphoryl choline, choline, and *myo*-inositol, are a normal feature of the ^{31}P or ^{1}H brain spectrum. These molecules will be frequently encountered in discussions of clinical spectroscopy, even though their precise relationship to myelin is far from clear. This is nowhere more important than in the detection of developmental changes in the brain, which are accompanied by dramatic changes in the MR spectrum (Blüml et al 1999).

Edema This all-important concept in neurophysiology and in clinical diagnosis by DWI and MRI has not yet been clearly defined in MRS assays of brain water. One possibility is that edema, as seen in MRI, represents less than 1% of total brain water, and falls within the limits of error of present methods of NMR water assay. These methods rely heavily upon differences in *T2* between water in various states. While it is *T2* which distinguishes edema in MRI, the differences are either too small or too local to be measured directly with MRS.

Metabolism Amino acids, carbohydrates, fatty acids and lipids, including triglycerides, form a complex network of biosynthetic and degradative pathways. The network is maintained by the thermodynamic equilibrium of hundreds of identified enzymes, and relative rates of flux through the various pathways are equally closely controlled. Hence, the concentrations of all but a few key molecules (messengers and neurotransmitters) are kept remarkably constant. It is for this reason that a thoroughly reproducible brain "spectrum" can be obtained with MRS. Conversely, predictable and reversible changes, such as increased lactate and glutamate, reduced ATP and increased ADP, due to the altered redox state of the pyridine nucleotide coenzymes of electron transport that does occur. This makes MRS the tool for short-term studies on the brain.

Compartmentation Mitochondrial energetics, the enzymes of which are controlled by non-Mendelian genetics, consists of the electron-transport chain and of oxidative phosphorylation which provides virtually all of the high-energy phosphate bonds to maintain ion pumps, neurotransmission, cell volume and active transport of nutrients. ATP, the essential "currency" of this process, is buffered in brain by another high-energy system, that of creatine-kinase, creatine (Cr), and phosphocreatine (PCr). These molecules are readily observed in MR spectra. Cytoplasmic enzymes control aerobic glycolysis and the formation of lactate, which supplements ATP synthesis. Glycolysis is massively activated by the Pasteur effect under hypoxic conditions which obviously limit mitochondrial energy production. Lactate and glutamate are both formed in excess when the mitochondrial redox state changes. It is possible that a similar activation of glycolysis accompanies "functional" changes (as in fMRI) and electrical activation (in seizures). It should be noted, however, that mitochondrial metabolite pools are to a variable extent NMR-invisible and may not contribute to the final brain spectrum.

Fuels of Oxidative Phosphorylation Glucose dominates the fuel supply for brain, and its supply via blood flow is strenuously protected. Vascular occlusion, because it brings with it glucose deprivation, oxygen lack, and CO_2 and H^+ accumulation, results in rather different neurochemical insults from that of pure hypoxia, such as is seen in respiratory failure or near-drowning. Thus, hypoxia and ischemia are different to the spectroscopist whereas the terms might not need to be distinguished for the purposes of MRI.

Under severe conditions of starvation, when glucose is not available, fatty acids and ketone bodies can sustain cerebral energy metabolism, and this may be the normal state of affairs for the milk-fed newborn. Unlike other tissues, the brain does not apparently require insulin to utilize glucose, so in diabetics the marked alterations in cerebral metabolism (and in the MR spectrum) are secondary to the systemic metabolic disorder.

Brain Metabolism: A Summary

Figure 1A-D depicts some of the neurochemical pathways which have become more relevant since the advent of neuro-MRS. The energetic interconversion of ATP, PCr and Pi, together with intracellular pH, is readily monitored by ^{31}P MRS. The major peaks of the

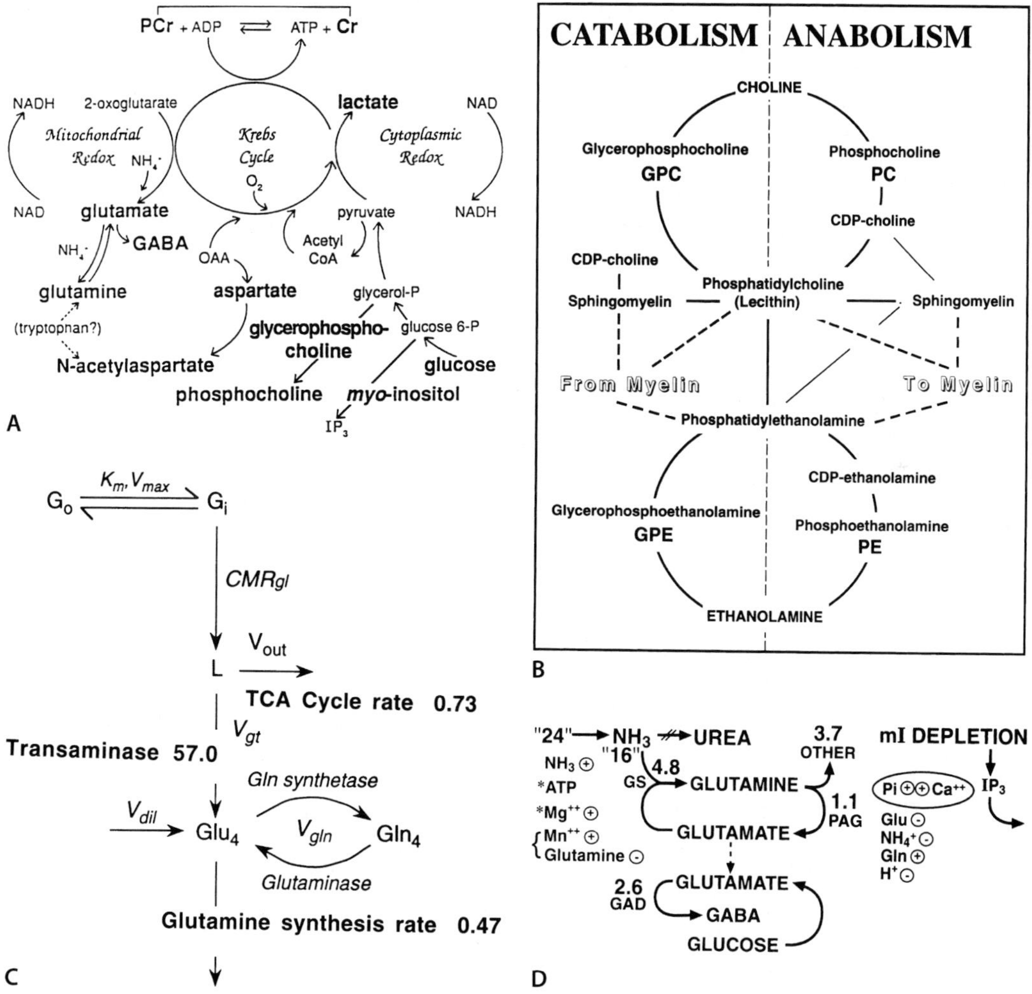

Fig. 1. Neurochemical Pathways: The New Neurochemistry. Reactions and metabolites now readily observed *in vivo* are shown diagramatically. 1A. Proton and phosphorus MRS. 1B. Choline and ethanolamine metabolites of "myelination" observed through proton-decoupled phosphorus MRS (figure courtesy of Prof. D. Leibfritz). 1C. Carbon fluxes detected with ^{13}C-glucose enrichment. 1D (figure courtesy of Dr. G. Mason). Nitrogen fluxes detected with ^{15}N-ammonia enrichment (figure courtesy of Dr. K. Kanamori). Note: Flux rates are indicated by numbers given in bold type C (micromoles/min/g brain) and D (micromole/hr/g brain).

^{1}H MR spectrum, N-acetylaspartate (NAA), total creatine (creatine plus phosphocreatine; Cr), total choline (as reflection of phosphoryl choline and glycerophosphoryl choline; Cho), *myo*-inositol (mI), and glutamate plus glutamine (Glx), were only infrequently encountered in neurochemical discussions of physiology or disease, before the advent of MRS. They now join glucose uptake and oxygen consumption as the most easily measured neurochemical events, and must become increasingly important in neurological discussion. The interconversion of phosphatidylethanolamine and phosphatidylcholine (by transmethylation) explains the close links between myelin products now quantifiable through proton-decoupled ^{31}P MRS (Fig. 1B).

Because of the concentration limit (of protons) at about 0.5–1.0 mM for NMR-detection, virtually all true neurotransmitters, including acetylcholine, norepinephrine, dopamine, serotonin (the exceptions are glutamate, glutamine and GABA), are currently beyond detection by conventional neuro-MRS. Similarly, the hormone messengers inositol-polyphosphates and cyclic AMP are not detected. This leaves important gaps in the New Neurochemistry.

Another evident shortcoming of NMR is the inaccessibility of most macromolecules because of their limited mobility. Accordingly, phospholipids, myelin, proteins, nucleosides and nucleotides, as well as RNA and DNA, are effectively "invisible" to this family of methods. Exceptions may be glycogen, a macromolecule in brain (Grütter et al. 1999), heart, skeletal muscle and liver, which is readily detectable in ^{13}C spectra, and the broad signals from phospholipids in the ^{31}P spectrum and from low molecular weight proteins in ^{1}H spectra in the brain.

Flux measurements using enriched stable isotopes of ^{13}C (Fig. 1C) or ^{15}N (Fig. 1D) extend the range of neurochemical events accessible to *in vivo* MRS. More than 50 metabolites can now readily be determined by combination of these techniques (Table 3).

Technical Requirements and Methods

Many years of experience with clinical MRI have resulted in a widespread and comfortable understanding of the MR process (cf. Chapter 38).

The patient (or subject) must be able to lie on a bed, which enters a confined space where electromagnetic fields are repeatedly switched on and off. The subject must tolerate the accompaniment of considerable noise. After an interval of one to several minutes, all of the radio signals generated by resonating protons are mathematically mapped to produce an image. The anatomical display in cross-section is now the norm for all who work in the brain.

MRS is produced in the same way, with three *additional steps*. Using the image just obtained, a volume of interest (VOI; voxel) is selected for MRS and the field within is further refined in a process called *shimming* (shim: old English, a wedge or ploughshare). Then, for ^{1}H MRS (but not for MRS of other nuclei), the protons of H_2O within the VOI are rendered silent by suppressing their particular frequency band (termed *water suppression*). Finally, using the same constellation of switched magnetic fields already familiar from MRI, a frequency profile or spectrum is acquired. Intensity at any given frequency is proportional to concentrations of protons. Frequency is a measure of chemical structure; thus the spectrum is a typical output of metabolite composition of the sample.

In MRI, where localization is "everything," only a single peak (^{1}H of water) is mapped. In MRS, localization must retain chemical shift information for the acquisition of metabolite profiles (Ordidge et al. 1985). Single-volume (or voxel) MRS uses methods which allow to measure MR signals originating from one region of interest and ensures that unwanted MR signals outside this area are excluded. Alternative strategies exist. *Se-*

Table 3. Cerebral metabolites measured *in vivo* by MRS. Current techniques of MRS on "clinical" MR scanners permit quantifiable assays of each of the following metabolites, fluxes or neurochemical events in examination times tolerated by the average volunteer or patient. The approximate precision of each assay is given. Not all tests are available on every commercial scanner.

Name of Metabolite	Technique	Precision
acetoacetate	^1H	±0.5mM
acetone	^1H	±0.3mM
adenosine-triphosphate (ATP)	^{31}P	±0.2mM
aspartate	^1H	±2mM
atrophy index	^1H	±1%
beta-hydroxybutyrate	^1H	±0.5mM
brain dry-matter	^1H	±2%
choline	^1H	±0.1mM
creatine	^1H	±1mM
CSF-peak aqueduct flow	^1H -Cine	±3ml/min
CSF-volume	^1H	±1%
ethanol	^1H	±1mM
fluoxetine	19F	±1μg/ml
(tri)fluoperazine	19F	±1μg/ml
GABA	^1H	±1mM
glucose	^1H, ^{13}C	±1.0mM
glucose transport rate (t1/2)	enriched^{13}C	±1.5min
glutamate	^{13}C	±2mM
glutamine	^1H	±1mM
glycerol	^{13}C	±5mM
glycerophosphorylcholine	$\{^1H\}$-^{31}P	±0.2mM
glycerophosphoethanolamine	$\{^1H\}$-^{31}P	±0.2mM
glycine	^1H	±1mM
glycolysis rate	enriched^{13}C	±0.37/min/g
guanosine-phosphate	^{31}P	±1mM
hydrogen-ion (pH)	^{31}P	±0.02pH units
inorganic phosphate	^{31}P	±0.2mM
inositol-1-phosphate	$\{^1H\}$-^{31}P	yes or no
isoleucine	^1H	±1mM
lactate	^1H	±0.5mM
leucine	^1H	±1mM
lipid	^1H; ^{13}C	±1mM
lithium	^7Li	±0.1mM
macromolecules	^1H; 1R	±1.0mM
magnesium (Mg^{++})	^{31}P	±200μM
mannitol	^1H	±2mM
myoinositol	^1H; ^{13}C	±1mM
NAA	^1H	±0.7mM
NAAG	^1H	±0.3mM
oxidized hemoglobin	fMRI (1H)	±0.2%
phenyl-alanine	^1H	±2mM
phospho-choline	dc^{31}P	±0.2mM
phosphocreatine	^{31}P	±0.2mM
phosphodiesters	^{31}P	±2mM
phosphoethanolamine	dc^{31}P	±0.1mm
phospholipid (membrane)	dc^{31}P	±30%
phosphomonoesters	^{31}P	±2mM
propane-diol	^1H	±1mM
pyridine nucleolide(s) (NAD, NADP)	dc^{31}P	±1mM
scylloinositol	^1H	±0.2mM
sodium	^{23}Na	±
taurine* (*see also scylloinositol)	^1H	±1mM
TCA-cycle rate	enriched^{13}C	± 0.1μmole/min/g
transaminase rate	enriched^{13}C	± 10μmole/min/g
triglyceride	^{13}C	±5mM
valine	^1H	±1mM
water content	^1H	±3%

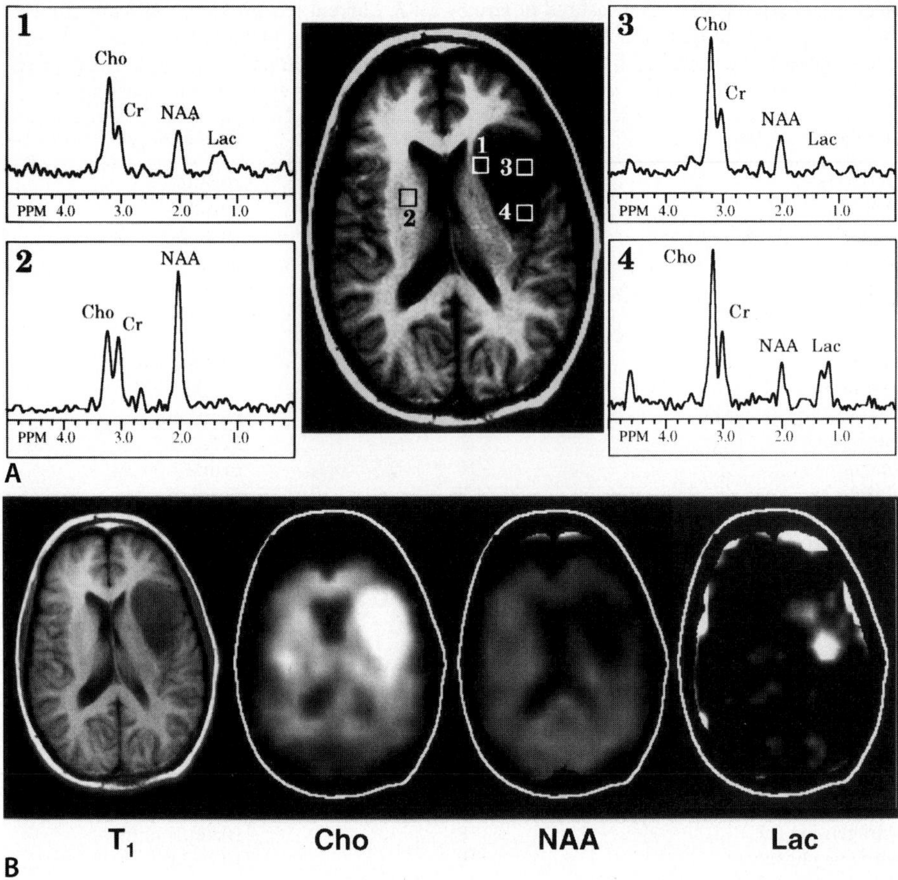

Fig. 2. CSI: Heterogeneous metabolism of brain tumor. (A) Local concentrations of each of the four principal metabolites to be recorded during a single MRS-examination. (B) Results are displayed as metabolite images (Courtesy of P.B. Barker, D.Phil. (Oxon)).

lective excitation is analogous to MRI, in which a single metabolite frequency is excited and an image is reconstructed. Mapping of cerebral phosphocreatine (PCr) has employed this technique (Ernst et al 1993b). *Chemical-shift-imaging* (CSI) acquires simultaneously multiple spectra from slices or volumes of the brain (Fig. 2A) and metabolite specific images are readily formed from the resulting peak-intesities (Fig. 2B). While theoretically the most time-efficient method of *in vivo* neurochemical analysis, in practice CSI brings with it many unwanted features such as loss of metabolic information, unexpected quantitative variability across the "slice", and inconveniently bulky data sets. At this point, *single-voxel MRS* dominates the field of *in vivo* brain MRS.

For single-voxel MRS, manufacturers provide one or more of the following capabilities:

- STEAM (stimulated echo acquisition mode);
- PRESS (point resolved spectroscopy) and
- ISIS (image selected in vivo spectroscopy).

Technical details, selection criteria and the necessary physics are extensively discussed in an earlier Springer-Verlag handbook: In vivo magnetic resonance spectroscopy; Berlin 1992. Eds. Diehl P, Fluck E, Gunter H, Kosfeld R and Seelig J.

Hardware Equipment

Relatively few research organizations have the capability to design and build human MR scanners, and stringent national and international regulations control their use. The MR equipment is large, heavy and expensive to site, in magnetically shielded rooms, usually remote from other equipment. Whole-body or slightly smaller "head-only" units are available. Clinical equipment is rarely more than 1.5 or 2.0 Tesla, but even in the clinic, 3 Tesla, 4 Tesla or 4.7 Tesla are becoming commonplace. Seven Tesla and 8.4 Tesla are in the exploratory stage but likely to become equally indispensable tools in the neuroscience research institute of the future.

A Word About Safety

Three different magnetic fields are applied in MRS:
- static magnetic field B_0;
- gradient fields for localization purposes, and
- rf fields to excite the magnetization.

These fields are generally remarkably safe, with no known biological hazards. Fast switching gradients have been considered as associated with risk, but never more than vaguely identified. While there exist "exotic" techniques, such as echo planar spectroscopic imaging (EPSI), the vast majority of MRS techniques switch gradients a magnitude slower than routinely applied in MR imaging. Prolonged irradiation of RF is identified as hazardous, to the extent that energy is "deposited" in the human head. A sensible government limit (SAR: see Glossary) has provided binding guidelines for nearly 20 years. Provided instruments are correctly calibrated and fitted with the necessary power monitor and automatic trip, no harm can come to subjects – voluntary or patients – during MRS studies. Isolated reports of burns from faulty electrical equipment, home-built RF coils, guide wires and electrodes are rarely serious but are a sure sign of sloppy science and cannot be tolerated in a first-class neuroscience institute.

A different class of safety deals with magnetic objects. Scissors, knives, scalpels etc. brought into the vicinity of the magnetic field become fast projectiles and can cause serious injuries. Also implanted metallic objects can cause serious harm to a patient as a consequence of electrical interaction, torque, or heating before or during an examination. Therefore physical safety around MR equipment is a matter of great concern. Every unit should have a safety officer, a clear code of conduct and probably metal detectors at all public entrances. Lax security results inevitably in "accidents" which could have been avoided with conscientious management. Safety lies in excellence of the design of the MR-suite and military-style discipline.

Purchasing Equipment

Almost all vendors of human MR equipment have first and foremost to take care of medical imaging (this is their natural market), and must then design engineering, NMR and physics solutions to simple MRS questions within a very unpropitious environment. First choice for neuroscience research is therefore a manufacturer with primarily scientists as customers. Investing in MR equipment is almost the most expensive decision a neuroscientist will make in his/her career. Only PET is more costly. Both qualify as "big science", hitherto the exclusive domain of physics. Careful, open discussion and clearly written specifications will guarantee early operation of a newly acquired MR system.

Coping with Upgrades

In fast moving fields, MR equipment can change faster than the neuroscientific question. Upgrades for MRI can be counterproductive for MRS. The investigator would be wise to build this into his/her research strategies by completing studies of patients and controls in a timely fashion, by re-calibrating equipment frequently (at least monthly) and by cross-correlating results for each software upgrade. Hardware upgrades, although less frequent, are costly and destructive of scientific effort if not carefully planned. Manufacturers will agree to a development plan acceptable to the purchaser – even if upgrades were not foreseen by either party! In practice, an MRS group does well not to accept upgrades early, for the reasons mentioned above. Upgrades aimed at enhancing MRI normally impact adversely on MRS capabilities. Benefits to MRS from the upgrades can lag by several years.

RF Coils and Gradients

MRI and ^{1}H MRS of brain is usually undertaken with a standard-volume head coil constructed like a helmet to fit over the entire head, and is provided by the manufacturer. For specialized purposes, and because in general they furnish the much-needed extra signal, surface coils or an assembly of surface coils (termed "phased-array") are used. Volume coils are more flexible and have a special advantage when quantitation of cerebral metabolites is the goal.

Since a ^{1}H head coil is standard equipment on all scanners, proton MRS is in principle available on all clinical scanners and by far the most widely used MRS technique. However, RF coils are frequency-specific so that MRS studies with other nuclei than proton demand a different (and costly) head coil. Often these coils are not provided directly by the manufacturer but need to be purchased from smaller suppliers. For proton-decoupled MRS, which has increased sensitivity and other advantages in MRS of nuclei other than proton (X-nucleus), combined RF coils, H + X, are the norm. Again, these types of coils are often not among the options offered by the manufacturer of the system but need to be purchased separately.

The rapid technical progress in gradient coils in recent years is mainly driven by MRI applications. In particular, functional MRI and diffusion weighted MRI, utilizing echo planar imaging (EPI), require fast switching gradients with fast rise times. While in general MRS also benefits from more powerful gradient systems, sometimes larger eddy currents from speed and power optimized gradient coils may affect spectral quality adversely on newer systems.

Hetero-Nuclear Detection and ^{1}H Decoupling

Even if RF coils tuned to nuclei other than 1H are available, not all MRI scanners can progress beyond 1H detection. Broadband amplifiers are essential if anything other than 1H MRS is to be undertaken. Similarly, RF receivers tuned to the resonance frequency of the nucleus to be observed must be provided along with several other hardware features which are sometimes difficult to retrofit.

Along with the capability to perform heteronuclear (other than^{1}H) MRS comes the need to enhance sensitivity and specificity of chemical analyses by proton-decoupling. This involves simultaneous excitation at a different frequency, and accordingly requires a second RF channel and amplifier.

Examples of spectra which illustrate ^{1}H MRS quantitation, signal advantage from a surface coil compared to a volume coil, broadband-heteronuclear MRS and finally, the effects of proton-decoupling, all from the same clinical MR scanner retrofitted with the above-mentioned components, are shown at the end of this chapter.

Pulse Sequences – Localization Methods

A necessary step *in vivo* is the segregation of extra-cerebral from intra-cerebral metabolites and of one intra-cerebral location from another. Localization sequences select cubes, rhomboid shapes, "slices" or multiple boxes, none of which confirm to recognizable structures of the human brain. The choice of the technique is often pre-ordained by the manufacturer and is less important than the need for absolute consistency within any research program. Although marked differences in spectral appearance result, data is interchangeable between two different sequences, with some loss of precision.

STEAM and PRESS

Two widespread techniques for single-voxel, water-suppressed proton spectroscopy are STEAM = *stimulated echo acquisition mode* and PRESS = *point-resolved spectroscopy.* Both techniques aim to excite the volume of interest with a minimum excitation of the rest of the sample. Three slice-selective pulses, along each of the spatial directions (x,y,z), are applied in a single-pulse sequence. Therefore signal from the overlap is generated in a "single shot" experiment. This proved to be of great value for optimizing sequence parameters and measurement conditions like shimming. The difference between these two techniques is that STEAM does the excitation with three 90° pulses while PRESS uses one 90° pulse and two 180° pulses (see Glossary). The principle of PRESS, a double spin-echo technique, is easier to understand. The first 90° pulse excites all the magnetization in one slice. The second pulse (180°) is applied in a slice perpendicular to this slice and generates a (first=SE12) spin echo stemming from a column or row of intersection between these slices. The third pulse (180°) perpendicular to both previous excited slices generates a (second=SE123) spin echo of only magnetization from the intersection of all three slices. These radiofrequency pulses are usually applied within 30–300 ms, and transverse magnetization generated for example by SE12, with a decay time of 500 ms, may interfere with the data acquisition of the double spin echo (SE123). For this reason, and also to dephase any FID from imperfect 180° pulses, additional so-called crusher gradients are applied between RF pulses and before the data readout.

STEAM, by utilizing 90° pulses, does not generate a spin echo, but a stimulated echo, which gives only half the possible signal from the volume of interest.

PRESS appears to be the more favorable since acquisition times can be cut down or spectra from smaller volumes can be acquired within the same acquisition time with comparable signal-to-noise (S/N) ratio. However, for reasons not discussed in detail here, STEAM can incorporate more gradient crusher power, which is crucial in preventing unwanted out-of-volume signal to interfere with the acquisition. Further, STEAM demands less RF power and VOI mislocation is a less severe problem. Note that the time between the second and third RF pulse (TM period) does not affect the echo times. Thus, short echo times can be achieved more easily with STEAM. Short echo times are important if the investigator is interested in the amino acids glutamate and glutamine which cannot be observed in long echo time modes due to their J-coupling. STEAM is therefore more robust and the appropriate choice for applications where the signal-to-noise and voxel size are not crucial: global brain diseases rather then focal diseases like brain tumors or MS.

Both STEAM and PRESS can be combined with chemical shift selective (CHESS) pulses to selectively saturate the water signal prior to the localization sequence. Since both techniques utilize echoes, they have the advantage over other methods that the window of data acquisition can be moved away from slice selection pulses and the switching of the corresponding gradients. This reduces spectral distortions due to eddy currents and it allows us to generate *T2* weighted spectra where the interpretation is simplified by the disappearance of coupled complex peaks.

ISIS

ISIS (*Image Selected In vivo Spectroscopy*) utilizes a slice-selective inversion pulse (180°) followed by a non-selective 90° excitation pulse which generates an FID. Localization is therefore achieved without the formation of an echo which makes ISIS well-suited for studies of nuclei with short *T2*-relaxation times such as ^{31}P. However, the FID is not fully localized until a series of eight FIDs with various on/off combinations of the selective inversion pulses applied along each of the spatial directions (x,y,z) are collected and summed. In contrast to STEAM and PRESS, ISIS is therefore not a "single shot" experiment and is more sensitive to motion artifacts. Another disadvantage is that localized shimming is not feasible because eight FIDs must be collected before localization is achieved. ISIS is mostly used in heteronuclear studies, where its advantage of avoiding T2-relaxation is valuable.

Chemical Shift Imaging (CSI)

CSI, or spectroscopic imaging (SI), is a method to collect spectra from multiple voxels covering a whole slice or volume simultaneously. Spatial localization is done by phase encoding in one (1D CSI), two (2D CSI), or three (3D CSI) directions. CSI therefore combines features of MRI and MRS. The potential *advantages* of CSI are three-fold:

- CSI is probably the most efficient way to acquire weak MR signals from the whole brain *in vivo*. STEAM, PRESS or ISIS acquire signals from the whole brain or from whole slices, canceling or discarding unwanted signals from regions outside the region of interest.
- CSI can be performed without pre-selection of the location(s) within the brain because CSI allows the adjustment of any voxel position after a measurement is finished (grid shifting).
- The heterogeneity of cerebral metabolism defined within a single experiment can be reconstructed as images and co-registered on MR images (Kumar et al. 1975; Brown et al. 1982) (Fig. 2).

However, a number of *practical problems* are associated with CSI, like achieving a good shim and uniform water suppression over a large volume. Lipid signals from the skull can interfere with metabolite signals as a result of incomplete localization. Other problems encountered include patient motion due to poor compliance with relatively long data acquisition times, loss of biochemical detail when more robust long TE sequences are employed, and errors sometimes encountered in individual spectra of the multi-voxel data sets (Ross et al. 1989).

Other Commonly Used Metabolite Imaging Strategies

- *Selective excitation* (Ernst et al. 1993b; Greenman et al. 1998). PCr images can be produced for skeletal muscle and for the brain.

- *Fractal* (Ernst et al. 1993b). The spatial separation imposed in low field images by the chemical shift differences between Pi, PCr, ATP, PME and PDE can be exploited to produce images of each metabolite.
- *Water-spectroscopy*. The intense water signal can be recorded for single voxels with ultra-short acquisition times. This technique is a useful addition to fMRI (Ernst and Hennig 1994).

Editing Techniques

Editing techniques exploit unique properties of molecules other than the chemical shift, most commonly homo- or heteronuclear J-coupling. Many editing sequences utilize the fact that in an echo sequence the phase of J-coupled spins is modulated during the echo delay. A series of spectra acquired with different echo times each may allow the separation and identification of overlapping signals from different molecules due to their different J-modulation. Metabolite editing confers some specificity on the process of peak identification in high-resolution NMR-techniques but has so far contributed little new information to *in vivo* human brain studies. Practical *in vivo* sequences have been proposed by Ryner et al. (1995) and Hurd et al. (1998) and tested in human subjects. While many creative editing sequences from high-resolution NMR are available in the literature, in practice, signal-to-noise limitations preclude their use *in vivo*. For example, zero-quantum filter for lactate editing is accomplished with a 2:1 signal loss; simple short-echo time sequences without metabolite-specific editing may work just as well. Recent examples of successful *in vivo* editing include GABA (Rothman et al. 1993; Gruetter et al. 1997; Hetherington et al. 1998) and beta-hydroxy butyrate (Shen et al. 1998).

Post-Processing and Quantitation

A weakness of early *in vivo* MRS studies of human brain has been the failure to provide quantitative output, for anything other than chemical shift. This approach was acceptable as a technical compromise but severely limits integration of MRS data with other neurobiological data and unnecessarily confuses the published literature in this field where variable spectral quality and variable methods of describing experimental findings abound. The objective of post-processing and quantitation is to reduce the somewhat less interpretable information of spectra (at least for any person who does not deal with MRS in a daily routine) to a limited number of parameters, which can then be interpreted and hopefully yield biochemically useful information. The "natural" parameters appear to be metabolite concentrations in moles per unit volume linking MRS with existing norms of biological chemistry. However, the commonest approach for the quantitation of spectra are peak ratios which can be obtained by using already installed automated software by the instrument manufacturers.

We recommend that MRS should adhere to existing norms for biological chemistry, describing identifiable metabolites of known molecular weight in SI units of moles/unit tissue mass or volume. Volume is preferable since MRI often defines the sample from which signal originates. In the following we outline briefly some quantitation strategies; for a more detailed description we refer to Kreis (1997).

General Approaches to Quantification of Cerebral MRS Signals *In Vivo*

NMR is intrinsically quantifiable since the signal is proportional to the number of nuclear spins. Care must be taken to account for all of the known variables in the *in vivo* technique used to acquire the signal. When this is achieved, key metabolite concentrations determined for human brain should be compared with the direct determinations from other *in vivo* techniques, or more commonly with reliable enzymatic assays of the metabolite in quick-frozen animal brain. For ^{1}H MRS, remarkable uniformity is achieved in assaying the key metabolites, with methodological variations < 5% between international reference laboratories (Kreis 1997).

Internal Reference Methods

For metabolites identified by MRS which have not been previously assayed *in vivo*, accurate quantification of one metabolite may provide a reliable internal concentration reference.

Note: This is NOT the same as expressing MRS results in terms of metabolite peak ratios.

Brain water is often the easiest internal reference (Thulborn and Ackerman 1983), and can be used with the clear understanding that this reference concentration may be influenced by pathology. An advantage is that the signal-to-noise ratio of the unsuppressed water signal (concentration ~50 Mol/l) is sufficient to not only provide an absolute standard but also to obtain T2 times and compartmentation information.

External Reference Methods

External reference methods utilize a vial containing a substance of known concentration mounted on the RF coil to obtain a reference signal. Cerebral metabolites peak areas can be compared with the reference signal and quantified. Since the reference signal is acquired from a different location, acquisition parameters may need to be re-adjusted to correct for B1 inhomogeneities. A method to avoid these adjustments is to use the external reference signal only to obtain the coil loading, without re-adjusting parameters. Signal intensities from the external reference and from *in vivo* metabolites can be compared with signal intensities from a simulation experiment, where MRS on a sphere containing known concentrations of metabolites and the external reference is carried out as *in vivo*. Note that in principle the additional simulation experiment needs to be done only once; in practice, however, a calibration measurement should be repeated frequently (once a month).

Brain Compartment Assays

A simplified approach which recognizes that brain metabolites are dissolved in intracellular fluid, separated by physicochemical boundaries from vascular, cerebro-spinal fluid space, and are (for the most part) not detected in brain solid materials, is a combination of internal and external reference methods. First, to obtain the "dry material" within the VOI, the fully relaxed, loading-corrected *in vivo* water signal is compared with that of pure water measured in the simulation experiment. The difference between the two signals is interpreted as the missing signal due to NMR invisible dry material *in vivo*. Second, a discrimination of brain water and CSF can be achieved based on their differences in T2 relaxation times by measuring the water signal at different echo times TE. Third, metabolite concentrations are calculated using the internal brain water signal corresponding to a now known brain water concentration as reference.

Automation: The Way of the Future

Purists in MR spectroscopy pride themselves on being able to acquire good quality spectra even under difficult conditions. For good neuroscience, we believe that automation is essential. This removes some of the subjective element, and above all, avoids human error in long sequences of decisions and computer-keyboard or screen-driven actions. "Complete" automation (after prescribing VOI) has been achieved for ^{1}H MRS. For other procedures, macros should be constructed for as many of the successive steps as possible.

Applied MRS – Single-Voxel ^{1}H MRS

When expertly performed, quantitative MRS brain studies provide results with a variance between 5 and 10% across large populations of normal subjects. The variance in single subjects may be as little as ±3%, although measurements of complex-coupled peaks do not achieve better than ±10–15%. The single greatest variable is not, we believe, true biological or diet-imposed variability but inaccuracy in positioning the patients' head and/or the VOI of localized MRS.

Below the reader can find a few suggestions on how to set up single-voxel proton MRS on a clinical scanner. These suggestions are heavily biased by the authors' experience with a General Electric 1.5 T Signa scanner. Nevertheless, the following section may be useful for users on any system since the underlying principles are the same on any scanner. Focusing for the moment on ^{1}H MRS, simple protocols with fixed echo time (TE) and repetitive time (TR) are best embedded in "macros," to prevent acquisition errors and to rigorously standardize techniques which may be required for longitudinal studies over many years. STEAM 20 ms (MPI Göttingen), STEAM 30 ms (HMRI, Pasadena) and PRESS 30 ms have been "safe" choices and can be recommended. STEAM (or PRESS) 135 ms and 270 ms (or more recently 144 ms and 288 ms) are satisfactory long-echo times. Short-echo times TE give the smallest T2 losses and therefore the best signal-to-noise (S/N) ratio and also the smallest susceptibility to T2 changes in pathology. However, background signals may cause considerable baseline distortions at extremely short TEs. To ensure reliability and quality we recommend performing a test as described below for single-voxel proton MRS based on axial localizers.

Identifing the Volume of Interest (VOI)

A few simple *preparations* should be done before starting the examination. We recommend to obtain a film with axial T1w images and to try to identify standard gray matter (GM) and white matter (WM) locations as used in the literature (see Fig. 3 and description below for HMRI definition). Mark the voxels with a pen on the film. Do the same on a set of T2w images. Accuracy in prescribing the voxel is of great importance! Remember, the slice thickness of an MRS voxel is typically 20 mm while the slice thickness of MRI is typically 5 mm. Therefore slices below and above the center slice for MRS need to be reviewed. Also check slices outside the range of the MRS voxel for being not too close to the skull. Confirm the voxel location with the following prescription:

Gray matter

Center across falx at level 1 cm above posterior commissures of corpus callosum. Start inferior one slice (5 mm) above the slice with the a) internal capsule, b) angular artery in sylvian fissure, c) occipitoparietal fissure, d) vein of Galen, e) internal cerebral vein, f) frontal horn of lateral ventricle. Note: If the head is tilted not all of these landmarks

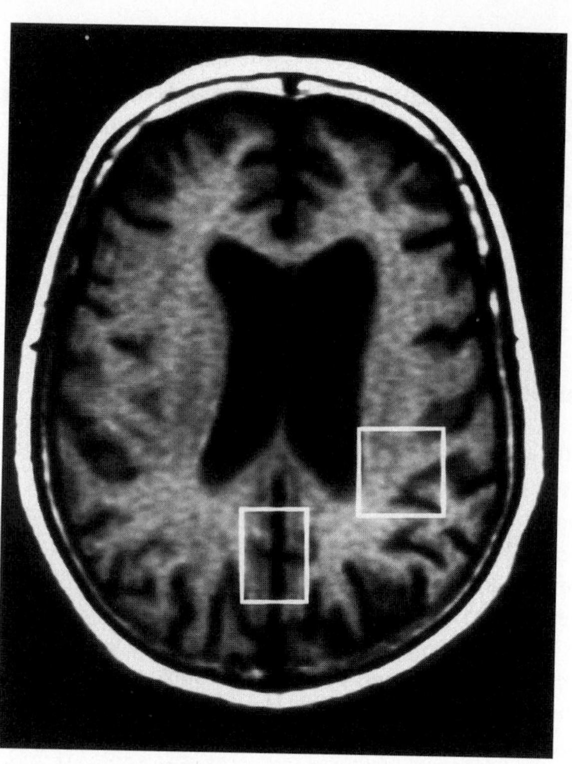

Fig. 3. MRI for voxel placement

may be visible. By choosing a voxel shape of 27 mm in A/P and 21 mm in R/L, most of the tissue within the voxel is gray matter. The recommended slice thickness (S/I dimension) is 20 mm. Review above and below center position at least two slices (for 5 mm MRI slice thickness and 20 mm voxel slice thickness).

White matter

Center left (or right) in parietal cortex. Stay in largely white matter, but allow up to 25% of gray matter. Landmarks are the center posterior rim of left (right) lateral ventricle ~1–1.5 cm above posterior commissures of corpus callosum. Review above two and below at least two slices (for 5 mm MRI slice thickness and 20 mm voxel slice thickness). The S/I center position should be not more than 0–5 mm (superior) off from the gray matter center.

There is an element of subjectivity and "in-house" training in voxel placement. Landmarks are defined on the subject's MRI. Voxel dimensions should be held constant for a given location. Here we encounter the first serious variable, which is the size of the head. Dimensions are tailored accordingly. Signal from outside the VOI is included in the final spectrum, because to the extent that pulse imperfections occur, the limits of the VOI noted on the screen are inaccurate. The skilled spectroscopist avoids the worst problem – excess lipid signal – by selecting a VOI 5–7 mm away from the outer borders of the brain.

Finally, the best results should not depend on endless compliance of the subject. Protocols which include both MRI and MRS would do well to complete MRS and scout MRI if possible before, rather than after the bulk of the MRI protocol.

Subject Positioning

Plan the first tests at a time when there is no pressure for fast work. The subject is always supine. The RF head coil center is always "landmarked" within 1 mm of the same posi-

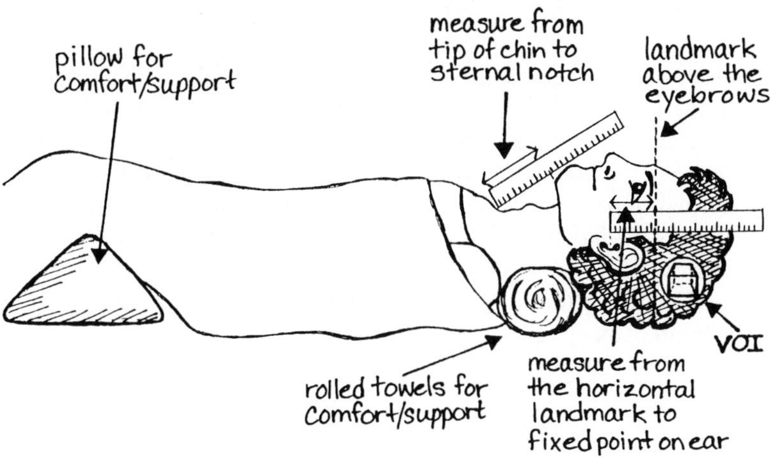

Fig. 4. Positioning subject for MRS of brain.

tion on the subjects' head. The head is accurately positioned in all three planes. *Sagittal*: bore center at the mid-point of brow, nose and chin. *Coronal* (the commonest source of error) is selected only after defining the angle at which the subject is lying. Use a ruler to define the distance of tip of chin to sternal notch and record this value for future MRS examinations. *Axial*: select a bony landmark, usually the supraorbital ridge, and drop a vertical through two other landmarks – say outer angle of the eye and tragus of the ear lobe (Fig. 4). For even the most trivial procedure, it is recommended that the head be fixed by Velcro straps. If an external reference is used for quantitation, it is important for automation and reproducibility that the vial is placed at the same position in all experiments.

Data Acquisition

Perform axial MRI and display image appropriately for center position. After the scout MRI, angulation errors should be noted. Unless these are severe, the MRS can proceed.

MRS is prescribed in different ways on different instruments. However, they all have in common the need to define coordinates orthogonal to the magnet bore (this explains the need to minimize angulation errors in initial positioning of the patient; they cannot be accurately corrected despite the availability of all the desired landmarks on MRI).

Start with either a standard gray matter or white matter location. Do not forget how important it is to check the location in all slices covering the volume of interest (VOI). The MRI slice thickness is usually 5 mm, while the MRS slice thickness is 20 mm. Write down voxel position and voxel size.

The next step is the shimming and the adjustment of the scan parameters. A well-shimmed VOI is a prerequisite for good MRS, as resolution and line shape have significant impact on the accuracy of the quantitation. Most scanners provide automated shimming which is generally faster than manual shimming and is the method of choice. Automated adjustment of RF transmitter gain, receiver gain and transmitter frequencies is standard on all modern scanners. Also, the adjustment of the water suppression is automated on most systems. On a GE scanner using the PROBE (=PROton Brain Exam), all of the above steps are fully automated and can be done by pushing one button. Acquire the spectrum and print the spectrum or document it on a film. Save the spectrum file for later off-line processing.

Processing and Quantitation

Determine peak ratios and compare the spectrum with literature spectra. Stability and reproducibility are the key to successful longitudinal studies in neurophysiology and neuropathology. Fig. 5 shows the results achieved in single subjects, volunteers and a patient with probable Alzheimer disease, illustrating the robustness of ^{1}H MRS over a 13-month period in normal and diseased human subjects. Dramatic neurochemical events, with a time constant of weeks or months, can also be readily observed (Ross and Michaelis 1994). Two examples are shown in Figs. 6 where ^{1}H MRS monitors the restoration of biochemical abnormalities after liver transplantation, and the appearance of ketone peak in the brain of an epileptic patient undergoing a ketogenic diet treatment (Seymour et al. 1999).

Where to Place the Voxel in Clinical Exams

The clinical question determines voxel location and size. In global diseases standard locations should be selected. The decision whether gray or white matter depends on the question asked. For example, to predict outcome in head trauma, it is most important

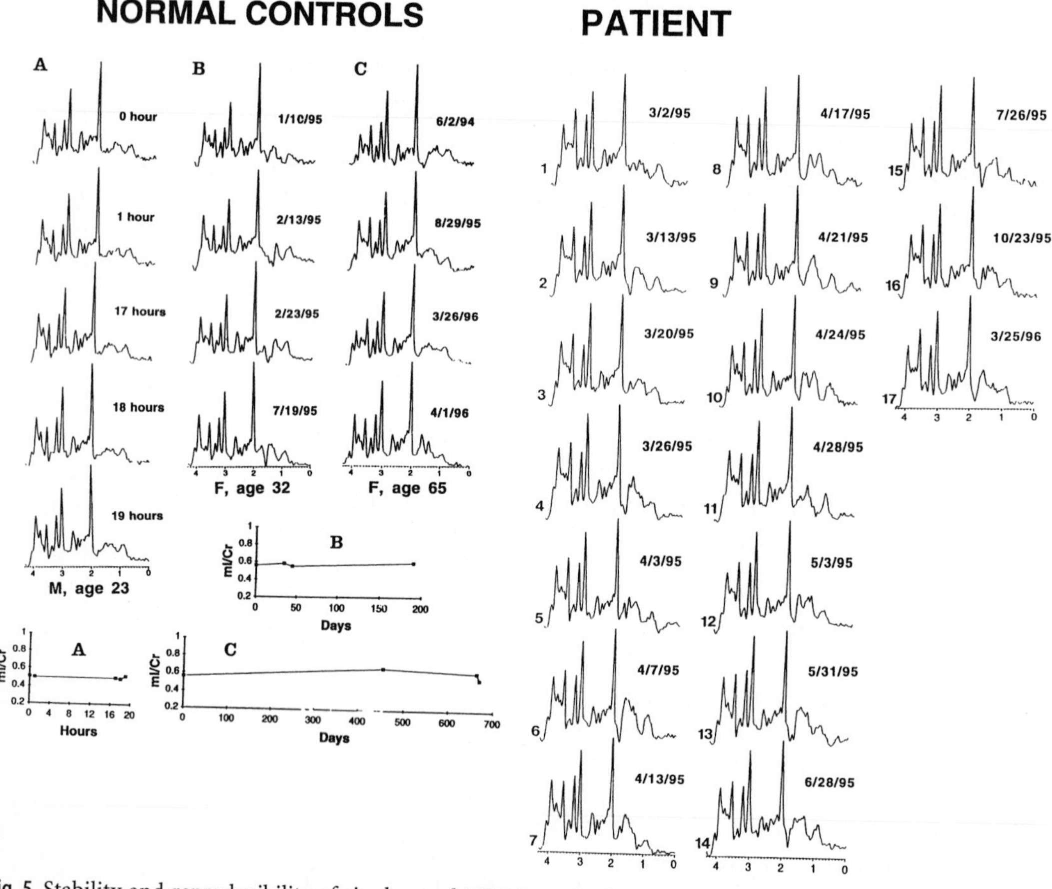

Fig. 5. Stability and reproducibility of single-voxel MRS in controls and patients. Repeated single-voxel MRS (STEAM, TE = 30 ms, TR = 1.5 s, TM = 13.7 ms) in healthy controls (*left*) and seventeen single-voxel MRS carried out in a patient with clinically suspected Alzheimer's disease (*right*) over a period of 13 months. All spectra can be distinguished from controls by elevated mI/Cr and reduced NAA/Cr, a feature of probable Alzheimer's disease.

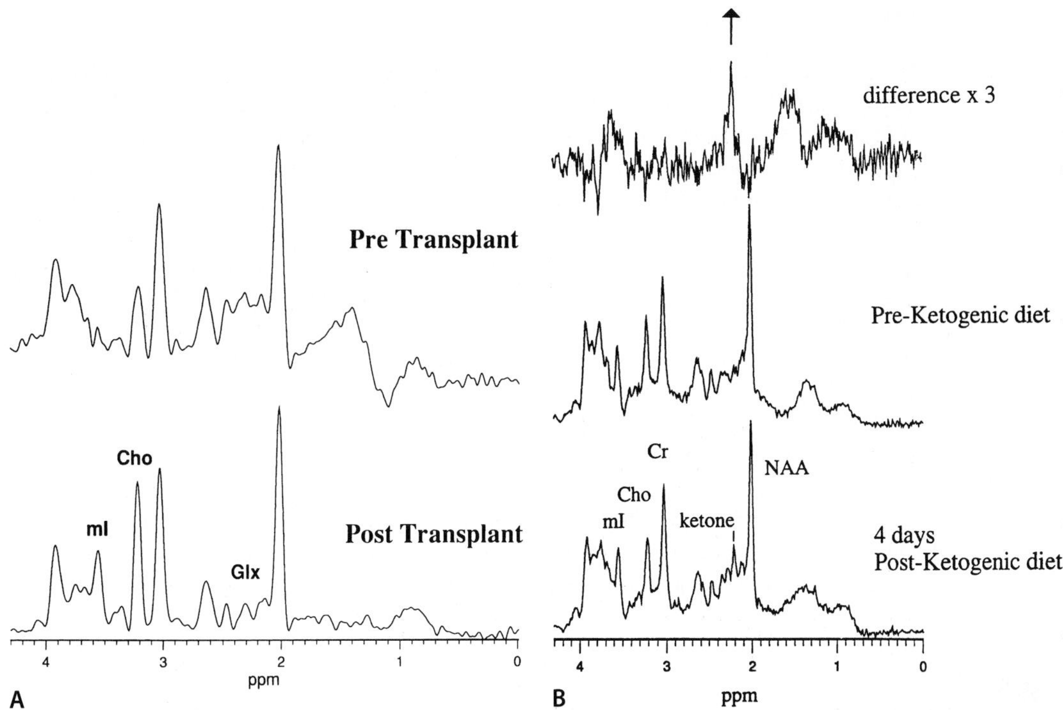

Fig. 6. The potential of MRS in diagnosis and treatment monitoring. (A) Restoration of biochemical abnormalities of the brain post-liver transplant. The patient is a 30-year-old man with acute-on-chronic hepatic encephalopathy (HE) secondary to hepatitis and subsequently successfully treated by liver transplantation. Spectra were acquired 6 months apart from the same parietal white matter location (15.0cc. stimulated-echo acquisition mode (STEAM) TR 1.5s, TR 30 ms; NEX 128) and scaled to the same creatine (Cr) intensity for comparison. The obvious abnormalities before liver transplantation; increased α,β,andγ-glutamine, and reduced choline (Cho)/Cr and myo-*inositol* (mI)/Cr (upper spectrum) were completely reversed 3 months after transplantation and Cho/Cr exceeded normal (lower spectrum). **(B)** Comparison of 1H MRS in pre- and post-ketogenic diet in the same child. A 1H MR spectrum from occipital gray matter of a patient before initiation of ketogenic diet reveals normal proton MRS (middle trace). Four days after starting the diet the 1H spectrum acquired at the same location shows the presence of a single peak at 2.2 ppm (lower trace). The difference spectrum is shown in the upper trace (Seymour et al. 1999).

to rule out global hypoxic injury. For this question the gray matter location is more sensitive and would be the first choice. In tumors the voxel should be placed in the center of the suspicious region, minimizing partial volume with apparently normal tissue. In case of a small lesion, two voxels at the same center position but with different volumes can be measured to estimate partial volume. There is some controversy about MRS after contrast agent. However, since, for short-echo time MRS in particular, there is no evidence for a significant impact of contrast agents on the spectral quality, the improved information about the region of interest after contrast agent may be favorable when MRI from a separate examination with contrast agent is not available. For suggestions where to place the voxel in a clinical situation see Table4.

Results: Neurospectroscopy

Proton MRS is by far the most widely used spectroscopy technique in the brain. This is due to the fact that standard MRI hardware components are used, making ^{1}H MRS available, that the concentrations of proton are relatively high in the brain, and that the MR sensitivity to protons is higher than the sensitivity to other nuclei.

How to "Read" a Proton Spectrum

The proton spectrum of the normal human brain is most readily understood by referring to Fig. 7. Each metabolite has a "signature" (Ross et al. 1992), which, when added to the other major metabolites, results in a complex spectrum of overlapping peaks. For all practical purposes, at the moment, due to ease and universal access, proton spectroscopy is synonymous with neurospectroscopy. Although such spectra are familiar to all, it is crucial to adopt a rigorous approach to acquiring and interpreting spectra.

Figure 8 is composed of 4 spectra from "gray matter" acquired by an automated procedure (PROBE = PROton Brain Exam), using a 1.5 T scanner, STEAM and short echo time (TE = 30 ms). The equivalent spectra acquired at long echo (TE = 135 or 270 ms) would look substantially different, but could be similarly interpreted by referring to a "normal" spectrum acquired under identical conditions.

The top spectrum (control) is taken from a healthy volunteer. Reading from right to left there are two broad resonances which are believed to be due to intrinsic cerebral proteins and/or lipids. The first and tallest sharp peak, resonating at 2.0 ppm, is assigned to the neuronal marker N-acetylaspartate (NAA). The next cluster of small peaks consists of the coupled resonances of β- and γ-glutamine plus glutamate (Glx). The tallest peak of this cluster at approximately 2.6 ppm is actually NAA which has three peaks,

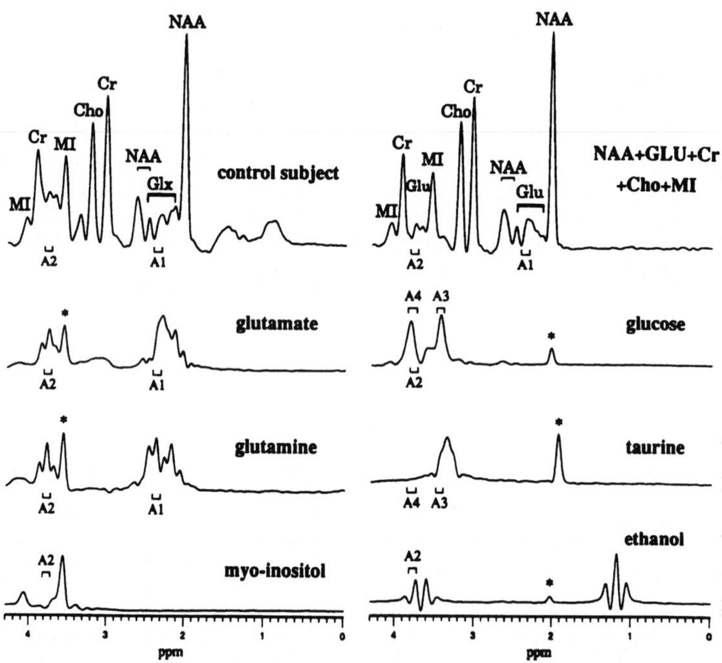

Fig. 7. Localized *in vivo* and *in vitro* ^{1}H MR spectra acquired with a stimulated echo sequence at 1.5T. At the top left is a normal brain spectrum (the sum of results from 10 age-matched control subjects). At the top right is a reference spectrum from an aqueous solution composed of 36.7 mmol/l N-acetylaspartate (NAA), 25.0 mmol/l Cr, 6.3 mmol/l choline chloride, 30.0 mmol/l glucose (Glu), and 22.5 mmol/l mI (adusted to a pH of 7.15 in a phosphate buffer). The remaining spectra were recorded from solutions of individual biochemicals. To simulate *in vivo* conditions, all spectra were subjected to a line-shape transformation yielding Gaussian peaks of approximately 4 Hz line width. The integration ranges used to detect changes in the cerebral levels of Glx (Glu or glutamine [Gln]) (A1 + A2) and Glu (A3 + A4) are indicated. The peaks labeled * originated from glycine, NAA or acetate, which were added to the various solutions as chemical shift references (the methyl peak of NAA was set to 2.02 ppm). All spectra were scaled individually and cannot be used for direct quantitation (modified from Kreis et al. 1992).

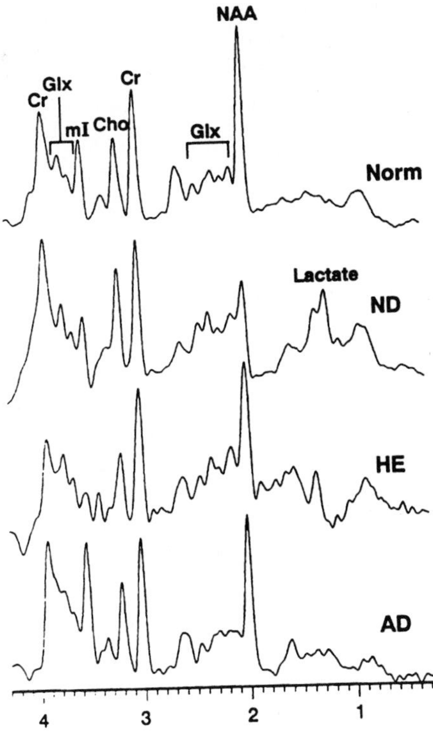

Fig. 8. [1]H MR spectra of gray matter acquired with PROBE. ^{1}H MRS of gray matter acquired with PROBE. The top spectrum is taken from a healthy volunteer. Spectra from global hypoxia due to near-drowning (ND) (spectrum 2), hepatic encephalopathy (HE), and probable Alzheimer's disease (AD) are shown. All spectra were acquired on a 1.5 T scanner using stimulated-echo acquisition mode (STEAM) and short echo time TE = 30 ms; repetition time TR = 1.5 s.

Table 4. Locations for single-voxel MRS in various diseases

Diagnosis	Preferred Location	Information needed
global hypoxia	gray matter	
Trauma	Far away from blood / lesions. Gray matter (1st choice) to rule out hypoxic injury, white matter is 2nd choice. If there is no suspicion of hypoxic injury WM is 1st choice.	Anticonvulsants?; conscious/unconscious Date of Injury?
Stroke (chronic)	center (1st choice) and rim (2nd choice)	Date of CVA
Liver diseases; hepatic encephalopathy	gray matter (1st choice), however, earliest changes in liver disease are in white matter (2nd choice, Cho reduction)	Lactulose?; Neomycin?
Dementia	gray matter	Clinical Diagnosis Symptoms?
MS	No evidence of lesions: White matter (1st choice). Lesions: Lesion (1st choice) and contra lateral side (2nd choice)	
HIV	Lesions: Lesion (1st choice), contralateral side (2nd choice). No lesions: White matter	AIDS +, CD-4 medication, clinical diagnosis of lesion
HIV (AIDS dementia)	Through lesion / if no lesion gray matter: as above dementia	
Tumors, rule out tumors	Center of lesion-suspicious region, sometimes smaller voxel necessary (adjust (increase) number of total scans); contralateral side as control	Type of tumor, chemo-, radiation therapy
unknown not focal disease	gray matter	

one of which overlaps the glutamine resonance. The second tallest resonance (at ~3.0 ppm) is creatine plus phosphocreatine (Cr), and adjacent to this is another prominent but smaller peak, assigned to "choline" (Cho). A small peak to the left of Cho is that of *scyllo*-inositol (sI). A prominent peak at 3.6 ppm is assigned to *myo*-inositol (mI). To the left of mI, two small peaks of the a-Glx triplet are clearly seen, and to the left is the second Cr peak. Variations in the degree of water suppression affect the peak intensities of metabolites closest to the water frequency at 4.7 ppm, i.e., the second Cr peak and its immediate neighbors. However, this effect of water suppression has no influence on the diagnostic value of spectra. The three major resonances (NAA, Cr and mI) provide a steep angle up from left to right in normal spectra acquired at short TE.

In the near-drowning spectrum (spectrum 2, "ND"), a lipid peak and overlying lactate doublet peak (at 1.3 ppm) replace the normally nearly "flat" baseline. NAA is almost completely depleted and there is a characteristic pattern of increased glutamine resonances (2.2–2.4 ppm). Cho/Cr peak ratio is apparently increased, compared with the normal above, but in this case the impression is created by reduction in Cr intensity (which can only be ascertained from a quantitative spectrum).

The patient with acute hepatic encephalopathy ("HE") shows peaks in the lipid/lactate region which cannot be reliably interpreted. NAA/Cr is clearly reduced, while the cluster of peaks designated Glx is obviously increased. Cho/Cr is if anything slightly less than normal, but the most striking change is the almost complete absence of myoinositol (mI). This also makes the increased a-Glx peaks to the left of mI more easily visible.

The lower spectrum "probable Alzheimer disease ("AD")" also has a characteristic appearance, with NAA much reduced (NAA/Cr close to 1). Glx is if anything reduced, while Cho/Cr is in this case slightly higher than the normal. The prominent mI peak is almost equal to Cr and NAA intensities, giving the spectrum its characteristic "flat" appearance.

In each case MRI was essentially normal, and gave little or no diagnostic information, while the spectrum is now well established as characteristic for the disease state described.

Normal Brain Development

The evolution of MRS changes in the newborn brain, from *in utero* (in a single near-term fetus) (Heerschap and van den Berg 1993) to post-partum 300 plus weeks of gestational age, is now well described (Figs. 9 and 10) (Kreis et al. 1993b). The findings add significantly to the information routinely obtained in MRI. Van der Knaap et al. (1993) have correlated the evolution of changes in the ^{31}P and ^{1}H MRS with the development of myelination. Normative curves for normal development now established for two cerebral locations (Fig. 10) confirm earlier published long-echo time and ^{31}P findings. *Myo*-inositol dominates the spectrum at birth (12 mmol/kg), while choline is responsible for the strongest peak in older infants (2.5 mmol/kg). Creatine (plus phosphocreatine) and N-acetyl groups (NA, of which the major component is NAA) are at significantly lower concentrations in the neonate than in the adult (Cr ~6 and NAA ~5 mmol/kg). NAA and Cr increase, while Cho and particularly mI decrease during the first few weeks of life (Table 5) (Kreis et al. 1993b). Increased NAA and Cr are determined by gestational age, whereas the falling concentration of mI correlates best with postnatal age. Absolute metabolite concentrations depend upon metabolite T1 and T2 relaxation. While T_1 values alter significantly with age for the metabolites NAA, Cr and mI, that for Cho is not altered. T_2 of NAA does appear to show important changes between newborns and adults whereas those of Cr, Cho and mI seem to be unimportant.

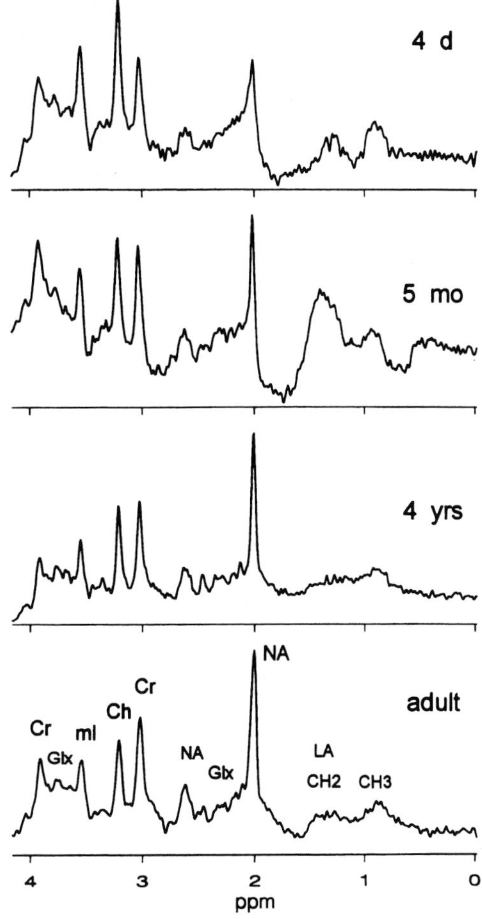

Fig. 9. Typical cerebral MR spectra for subjects of different ages. Typical cerebral proton MR spectra from subjects of different ages. Relative amplitude of the main peaks in STEAM spectra vary drastically with age. Spectra were obtained from a periventricular area in the parietal cortex. Acquisition parameters: echo time TE = 30ms, repetition time TR = 1.5 s, 144–256 averages, voxel sizes 8–10 cm^3 for children, 12–16 cm^3 for adults (modified from Kreis et al. 1993b).

Table 5. Absolute concentrations of different age groups in mmol/kg brain tissue (mean ± 1 SEM) and significance tests for differences found.

Group	n (ROIs)	GA (weeks)	pn age (weeks)	NAA	Cr	Cho	mI
< 42 GA	11	38.7 ± 0.7	3.9 ± 1.6	4.82 ± 0.54	6.33 ± 0.32	2.41 ± 0.11	10.0 ± 1.1
42–60 GA	8	50.2 ± 1.8	9.3 ± 1.9	7.03 ± 0.41	7.28 ± 0.28	2.23 ± 0.06	8.52 ± 0.92
< 2pn	6	40.0 ± 1.0	0.8 ± 0.2	5.52 ± 0.64	6.74 ± 0.43	2.53 ± 0.12	12.4 ± 1.4
2–10 pn	7	41.9 ± 1.5	3.8 ± 0.9	5.89 ± 0.21	6.56 ± 0.44	2.16 ± 0.09	7.69 ± 0.62
Adult	10		1,440 ± 68	8.89 ± 0.17	7.49 ± 0.12	1.32 ± 0.07	6.56 ± 0.43
p < 42 GA vs. adult				< 0.0001	< 0.0001	< 0.0001	0.001
p < 42 GA vs. 42–60 GA				0.0007	0.02	0.07	0.10
p 42–60 GA vs. adult				< 0.0001	0.16	< 0.0001	0.0003
p 42–60 2 pn vs. adult				< 0.0001	0.005	< 0.0001	< 0.0001
p 2pn vs. 2–10 pn				0.7	0.8	0.02	0.004
p 2–10 GA vs. adult				< 0.0001	0.001	< 0.0001	0.007

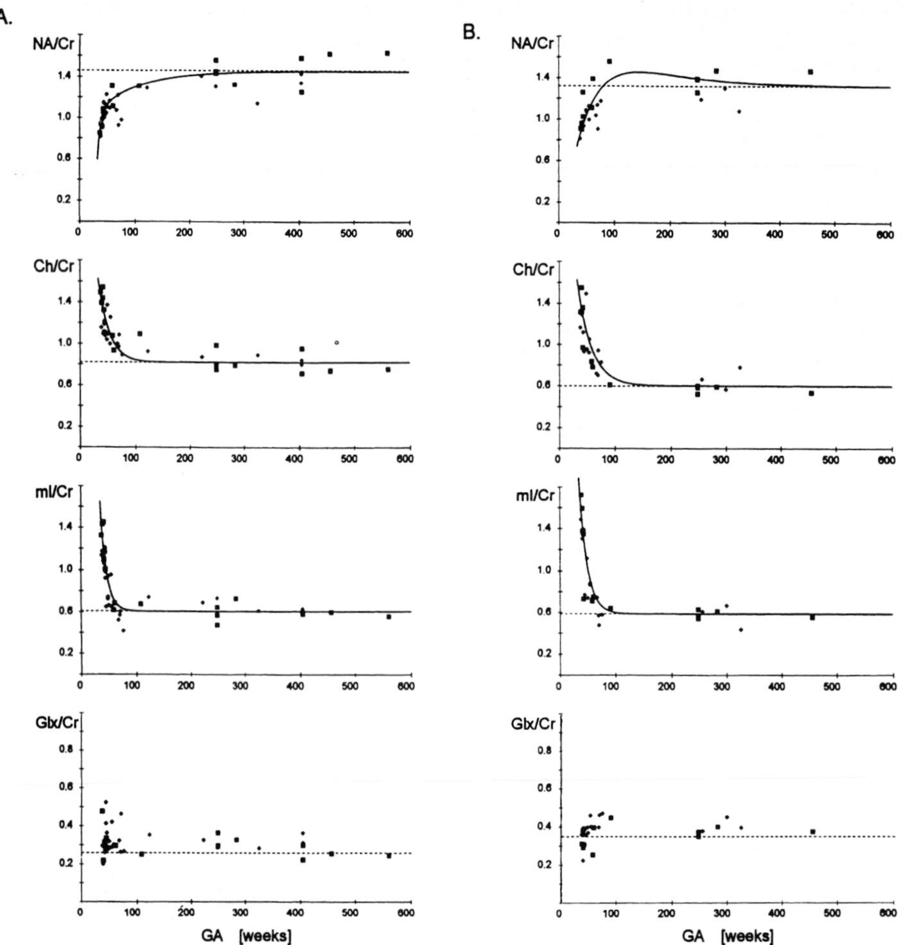

Fig. 10. Time courses of metabolite peak amplitude ratios versus gestational age of the subject. Time courses of metabolite peak amplitude ratios versus gestational age of the subject. (A) and (B) contain the normative curves for the parietal (mostly white matter) and occipital (predominantly gray matter) locations, respectively. The ratios were calculated as detailed in Kreis et al. (1991). Cho/Cr and mI/Cr were fitted to a mono-exponential, whereas NAA/Cr was fitted to a bi-exponential model function. No developmental curve was calculated for Glx/Cr data, because no clear trend was visible. The curves are well defined for the 1st year of life, where the most dramatic changes take place. Features in the later stages of development are less accurately described by the present data. The normative curves are specific for the acquisition parameters used. Open symbols represent data from parietal cortex, and filled symbols data from occipital cortex.

Quantitative ^{1}H MRS is expected to be of particular value in diagnosis and monitoring of pathology in infants (Van der Knaap et al. 1993), since metabolite ratios are often misleading. This may be particularly useful in that period before myelination is apparent in the developing brain.

Nonspecific cerebral damage in inborn errors of metabolism may consist of disturbance of brain maturation, demyelination, or neuronal degeneration. In demyelinating disorders, it is primarily the myelin sheath that is lost; secondarily, axonal damage and loss occur.

In demyelinating disorders, the rarefaction of white matter implies that the total amount of membrane phospholipids per volume of brain tissue decreases. Myelin sheaths consist of condensed membranes with a high lipid content.

Myelination in normal brain commences in the 6th month of fetal development and continues to adult years. Peak myelin production, however, occurs from 30 weeks gestation to 8 months postnatal development with young adult-like myelination observed at age 2 years (Brody et al. 1987). Phospholipids containing ethanolamine (E) and phosphoglycerides containing choline (Cho) are constituents of sphingomyelin and lecithin, respectively, both of which are components of the myelin sheath. ^{1}H decoupled ^{31}P MRS ($\{^{1}H\}$-dc^{31}P) is able to separate the complex phosphomonoester and diester peaks into their components of phosphoethanolamine (PE), glycerophospho-ethanolamine (GPE), phosphatidylcholine (PC) and glycerophosphatidylcholine (GPC). Quantitative $\{^{1}H\}$-dc^{31}P can be used to investigate and quantify age-related changes in the choline and ethanolamine constituents of normal, developing and diseased human brain *in vivo* (Bluml et al. 1998a, 1998b) (Fig. 11).

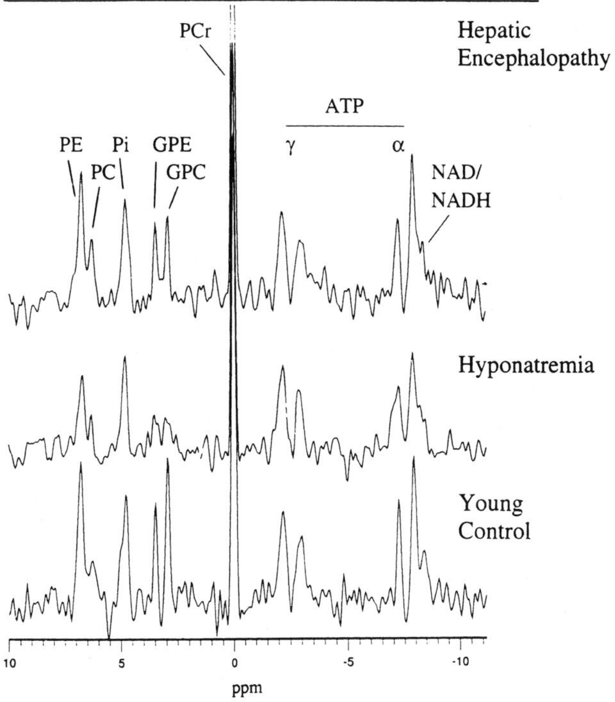

Fig. 11. Representative $\{^{1}H\}$-^{31}P MRS spectra of diseased brain in adults: HE a disorder of cerebral osmoregulation? Accumulation of ammonia in the blood causes a series of very specific changes in ^{1}H MRS of hepatic encephalopathy (HE). Expected and readily explained by the conversion of ammonia and glutamate into glutamine is an increase of cerebral glutamine. However, changes of other metabolites such as a depletion of myo-inositol and a reduction in choline are not completely understood. Proton decoupled ^{31}P MRS can significantly improve our understanding of the pathophysiology of HE by measuring the Cho constituents separately, quantifying brain phosphoethanolamines, and high energy metabolites. (**A**) In this HE patient a reduction in GPE and the recognized osmolyte GPC can readily be detected while PC is unchanged. When groups of patients and controls were compared, statistically significant reduction of PE, Pi, and ATP were observed. These findings support the hypothesis of a disturbed osmoregulation. The suggestion that HE is accompanied by cerebral energy failure is not supported by the data since the classical pattern of reduced PCr and elevated Pi was not observed. (**B**) Hyponatremia is recognized to cause a disorder of cerebral osmoregulation. The spectrum from a patient with hyponatremia of unknown cause (not HE) appears to have a similar pattern as in HE but with the abnormalities even more pronounced. (**C**) Typical spectrum from a young control for comparison (figure modified from Bluml et al. 1998b).

Using a modified PRESS (TE=12ms, TR=3s) sequence, and [PCr] obtained from each quantitative non-decoupled ^{31}P MRS as an internal reference, quantification of [PE], [PC], [GPE] and [GPC] shows the age-related changes of membrane metabolites *in vivo*. The {^{1}H}-dc^{31}P spectra from controls of different ages show clear differences related to age. The largest change is observed in [PE], which is initially high and decreases to adult levels by about age 10 years. [PC] is also high in the neonate and decreases to young adult concentrations by about age 4 years. Phosphocreatine (PCr) concentration increases and reaches adult levels at age 4 years. [GPE] and [GPC] show a slight generally increasing trend with age (Fig. 12.1 and Fig. 12.2).

Pathology Comparing MR spectra with those from relevant age-matched normal subjects, a number of examples of inappropriate development have been identified. *Canavan's disease* (Fig. 13 left) (aspartoacylase deficiency) is presented in proton MRS (see Fig. 22 for sequential proton MRS for treatment monitoring in this patient) with highly elevated NAA, elevated mI and reduced total Cr and Cho. Additional information about the pathophysiology of this rare inborn error is obtained by {^{1}H}-^{31}P MRS, the membrane metabolites PC, GPE and GPC, as well as the high energy metabolites PCr and ATP appearing to be reduced.

A patient (Fig. 13B) initially diagnosed with and treated for hydrocephalus did not do well clinically and showed abnormal myelination in follow-up exams. GPC was found to be reduced while PCr appears to be elevated.

A patient (Fig. 13C) with amino acid inborn error disease (*glutaric aciduria II*) showed elevated GPE, GPC and PCr.

In consolidated *late hypoxic injury*, increased GPE, GPC and PCr may reflect a partial volume effect of increased glial cell density (Fig. 13D).

In all patients information separately acquired with proton MRS is consistent with {^{1}H}-^{31}P MRS findings insofar as total Cr (=free Cr + PCr) and total Cho (GPC + PC + minor contributions from other metabolites) parallels changes in PCr or GPC and PC (see Fig. 13 right panel).

Rapid changes in [PE] and [PC], which are important precursors of phospholipids, may be related to the high rate of synthesis of membranes and myelin in the young developing brain.

What Have We Learned From Neurospectroscopy?

Table 6 summarizes abnormalities observed by ^{1}H MRS in diseases. When observing neuropathological events through MRS, there seems to be a rather limited range of metabolic changes in response to disease. Tumor, MS, stroke, inflammation and infections may produce very similar patterns of change. This is not surprising and should not be condemned as lack of specificity. Rather, it should teach us more about the brain's response to injury and its prevention and repair.

N-Acetylaspartate

Most observations with ^{1}H MRS strongly support the original formulation of NAA as a "neuronal marker". However, this simple conclusion must be modified in some particulars. In addition to neurons, there is evidence that NAA is found in a precursor cell of the oligodendrocyte. The time course of appearance of NAA in human embryology is still unknown, but the best estimate is that NAA biosynthesis may begin in the middle trimester, i.e., it is not dependent upon the existence of MRI-visible myelin, which is

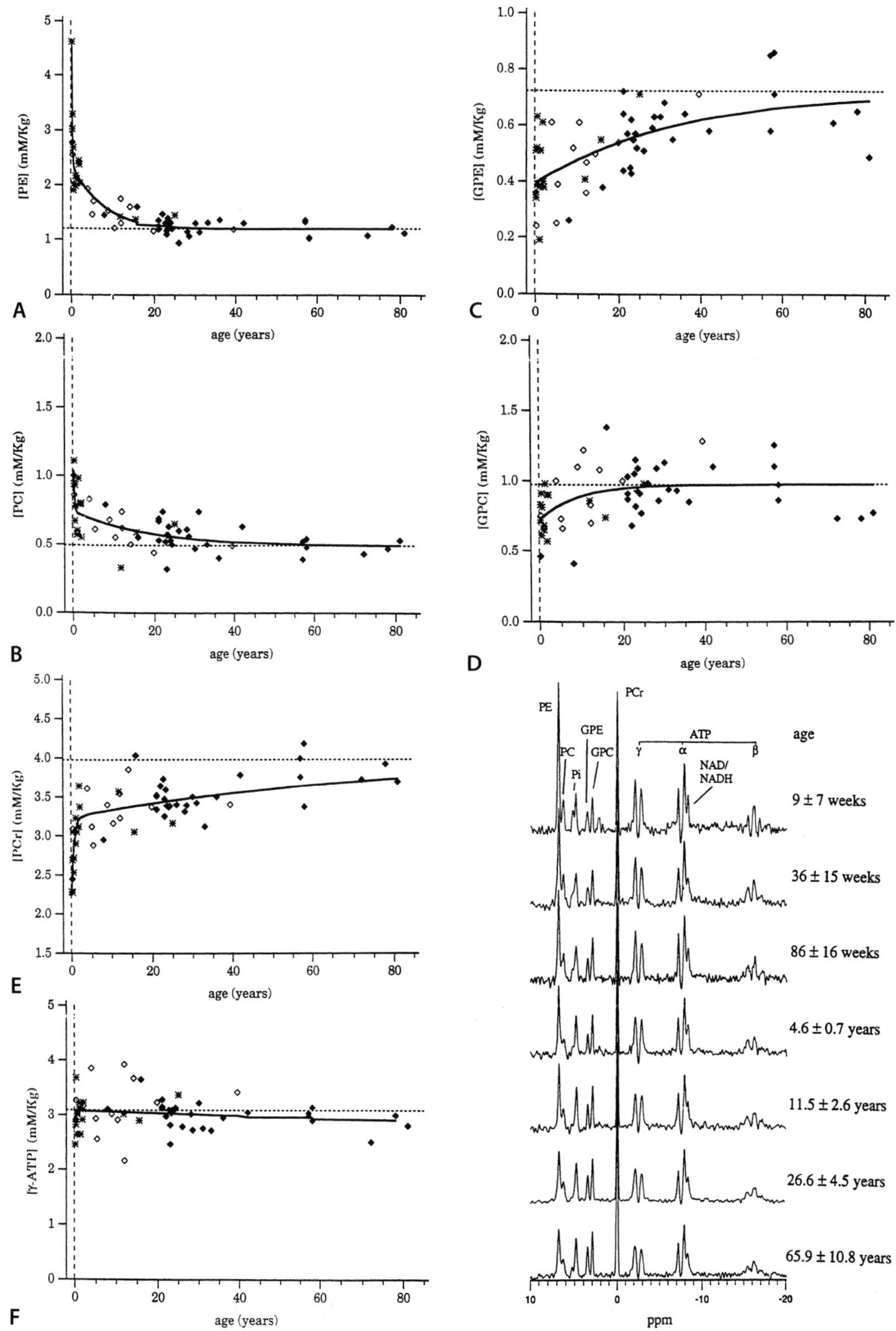

Fig. 12. A–F: Age-related changes of absolute concentrations of the membrane metabolites PE, PC (A,B), GPE, GPC (C,D) and high energy metabolites PCr and ATP (E,F). [PE] and to a lesser extent [PC] decrease with age. Adult levels are reached at age >12 years. [GPE] and [GPC] increase only slightly with age. [PCr] increases while [ATP] shows a mild decrease with age. Age was corrected for gestational age where applicable.

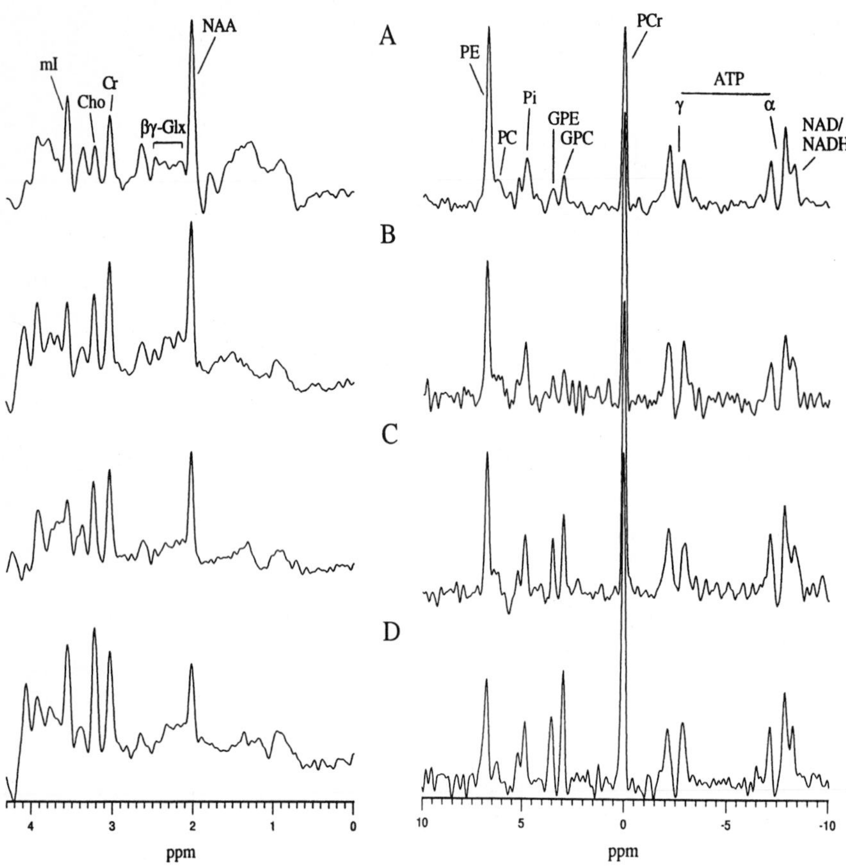

Fig. 13. Representative ^{1}H MR (*left*) and {^{1}H}-^{31}P (*right*) spectra of diseased brain in babies and children. (**A**) Canavan's disease (aspartoacylase deficiency) is presented in proton MRS (see Fig. 22B for sequential proton MRS for treatment monitoring in this patient) with highly elevated NAA, elevated mI and reduced total Cr and Cho. Additional information about the pathophysiology of this rare inborn error is obtained by {^{1}H}-^{31}P MRS, the membrane metabolites PC, GPE and GPC, as well as the high energy metabolites PCr and ATP appearing to be reduced. (**B**) This patient initially diagnosed with and treated for hydrocephalus did poorly clinically and showed abnormal myelination in follow-up exams. GPC was found to be reduced while PCr appears to be elevated. (**C**) A patient with amino acid inborn error disease (glutaric aciduria II) showed elevated GPE, GPC and PCr. (**D**) In consolidated late hypoxic injury, increased GPE, GPC and PCr may reflect a partial volume effect of increased glial cell density. In all patients, separately acquired proton MRS is consistent with {^{1}H}-^{31}P MRS findings insofar as total Cr (=free Cr + PCr) and total Cho (GPC + PC + minor contributions from other metabolites) parallel changes in PCr or GPC and PC (right panel).

←

Fig. 12.2. Normal age-related changes in the spectral appearance of {^{1}H}-^{31}P MRS. Shown are averaged spectra from control subjects each calculated from a group of subjects within the age range as indicated in the figure. In the absence of true normal control subjects below 21 years of age, neonates, babies, or children with apparently normal proton MRS and an absence of clinical symptoms in follow-up examinations served as controls. Most of those patients were diagnosed with slowly progressive hydrocephalus which can be successfully treated with shunt surgery. Different S/N in spectra is due to different numbers of subjects included in each age range. Most apparent is the striking reduction of PE, being the most prominent peak in a newborn, within the first weeks of life. This is contrasted by only a small reduction of the second phosphomonoester PC. Inorganic phosphate Pi peaks are observed at 4.8 and 5.1 ppm, representing an intercellular compartment and an extracellular or CSF compartment. The two peaks are separated due to the different pH values intra- and extracellularly, 7.0 vs. 7.2. The peak orginating from extracellular Pi can be readily observed in the baby spectra because of the increased ventricles in hydrocephalus. GPE and GPC show a slight generally increasing trend with age. A resonance at 2.2 ppm consistent with glycerophosphoryl serine and/or phosphoenolpyruvate was observed only in the neonate spectrum (age 9 ± 7 weeks). PCr increases with age while ATP slightly decreases. All spectra were processed and scaled identically to allow direct comparison (figure modified from Bluml et al 1999).

Table 6. Differential diagnostic uses of magnetic resonance spectroscopy

Metabolite (normal cerebral concentration)	Increased	Decreased
Lactate (Lac) (1 m*M*; not visible)	(often) Hypoxia, anoxia, near-drowning, ICH, stroke, hypoventilation (inborn errors of TCA, etc.), Canavan's, Alexander's, hydrocephalus	(unknown)
N-acetylaspartate (NAA) (5, 10 or 15 mM)	(rarely) Canavan's	(often) Developmental delay, infancy, hypoxia, anoxia, ischemia, ICH, herpes II, encephalitis, near-drowning, hydrocephalus, Alexander's, epilepsy, neoplasm, multiple sclerosis, stroke, NPH, diabetes mellitus, closed head trauma
Glutamate (Glu) and/or glutamine (Gln) (Glu = ? 10 mM; Gln = ? 5 mM)	Chronic hepatic encephalopathy (HE), acute HE, hypoxia, near-drowning, OTC deficiency	(unknown) Possibly Alzheimer's disease
Myo-inositol (mI) (5 mM)	Neonate, Alzheimer's disease, diabetes mellitus, recovered hypoxia, hyperosmolar states	Chronic HE, hypoxic encephalopathy, stroke, tumor
Creatine (Cr) + phosphocreatine (Pcr) (8 mM)	Trauma, hyperosmolar, increasing with age	Hypoxia, stroke, tumor, infant
Glucose (G) (~1 mM)	Diabetes mellitus, ? parental feeding (G), ? hypoxic encephalopathy	Not detectable
Choline (Cho) (1.5 mM)	Trauma, diabetes, "white" vs. "gray", neonates, post-liver transplant, tumor, chronic hypoxia, hyperosmolar, elderly normal, ? Alzheimer's disease	Asymptomatic liver disease, HE, stroke, nonspecific dementias
Acetoacetate; acetone; ethanol; aromatic amino acids; xenobiotics (propanediol; mannitol)	Detectable in specific settings diabetic coma; ketogenic diet (Seymour et al. 1998) etc.	

ICH= intracerebral hemorrhage; TCA= tricarboxylic acid cycle; NPH= normal pressure hydrocephalus; OTC= ornithine transcarbamylase.

only slowly added to the brain in the months after birth. Furthermore, the finding of approximately equal concentrations of NAA in white and gray matter of the human brain makes it inescapable that NAA is also a component of the axon or the axonal sheath in man. In addition to NAA, there is good evidence now for the existence of NAAG in human (as well as animal) brain, with the preponderance in white matter, and posterior and inferior regions of adult brain, especially the cerebellum.

Human pathobiology also supports the idea of NAA as a neuronal marker, loss of NAA being generally an accompaniment of diseases in which neuronal loss is documented. Glioma, stroke, the majority of dementias, and hypoxic encephalopathy all show loss of NAA.

That NAA is an "axonal marker", too, is supported by the loss of NAA in many white matter diseases (leukodystrophies of many kinds have been studied), in MS plaques and in white matter in hypoxic encephalopathy.

If NAA is a neuronal marker, can we ever expect to see recovery of NAA in practice? A clear example of neuronal recovery or regeneration is provided by the fetal neural transplant into adult human brain (Hoang et al. 1997, 1998). Convincing evidence of the presence of NAA in the grafted region of the putamen is provided by sequential examinations in such a patient, by means of localized ^{1}H MRS (Fig. 14). Most other examples proposed are perhaps best understood not as evidence of neuronal recovery (still to be viewed as unlikely), but as one of four possible alternatives:

1. Axonal recovery, after a less than lethal insult to the neuron, as for example in MS plaques, or in the rare MELAS syndrome.

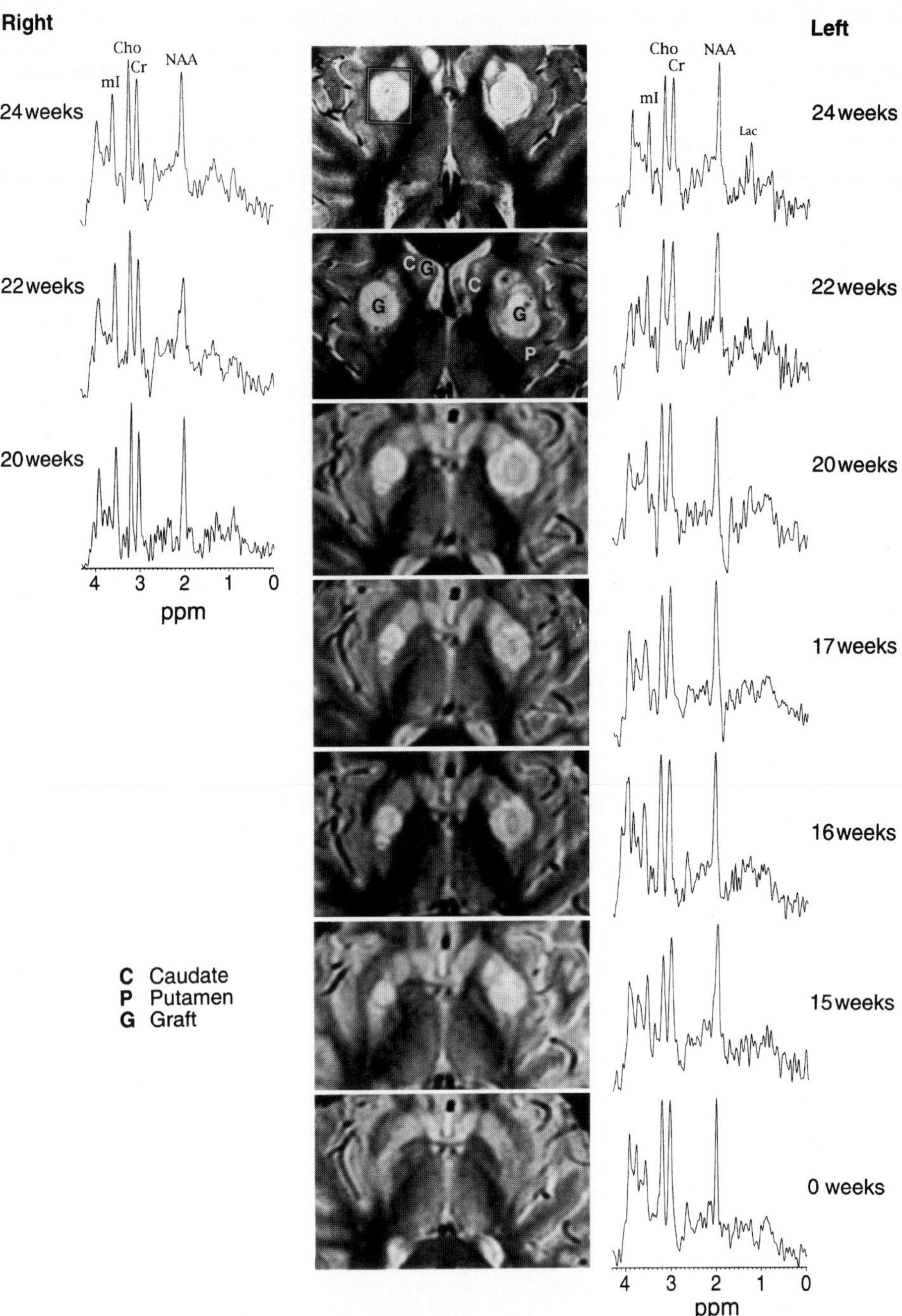

Fig. 14. Time course of development of bilateral fetal grafts in a patient with Huntington's disease. In ascending order, sequential MRI studies demonstrate the increasing volume of bilateral grafts placed in the putamen and caudate. In the latest examination, a cyst has developed on the left. Localized ^{1}H MR spectra are depicted for each examination of left- and right-sided grafts. Peaks identified as NAA, Cr, Cho and mI reflect a near-normal adult (rather than fetal) neuro-chemistry for those well-established neuro-transplants. Corresponding to the developing cyst, lactate is noted in the latest ^{1}H MR spectrum (top left), and may be an early indicator of graft "failure" (figure modified from Ross et al. 1999).

2. An "artifact" of cortical atrophy. Thus, as neuronal death occurs, the consolidation of surviving brain tissue is well documented by neuropathological studies, and by cortical atrophy on MRI. MRS which determines local ratios or even concentrations of NAA will record a real increase in the local concentration of this metabolite.
3. Survival (or recovery) of a peak at 2.01 ppm in the proton spectrum may not be due to NAA or NAAG, but to several other metabolites that contribute to this spectral region. A small decrease in the NAA peak in diabetes mellitus may be explained better in terms of another metabolite.
4. NAA appears also to be a reversible cerebral osmolyte, increasing or decreasing in response to hyper-osmolar states, and decreasing noticeably in hypo-osmolar states, such as sodium depletion and possibly hydrocephalus (Bluml et al. 1997).

Creatine and Phosphocreatine

We know that, in human brain at least, these two compounds, which are in rapid chemical, enzymatic exchange, represent a single T_2 species. MRS estimates 8 mM Cr + PCr in human gray matter, compared with published values of 8.6 mM for rapidly frozen rat brain. Cr concentration in human gray matter significantly exceeds that measured in white matter, in contrast to the results of tissue culture studies, in which Cr appears to be more related to astrocytes than to neurons (Flögel et al. 1995).

As with NAA, MRS studies have thrown very interesting light upon the factors which might control Cr + PCr in the human brain. In addition to the well-known regulation by enzyme equilibrium which permits a presumably crucial role of PCr in energetics of ATP synthesis, two new concepts have emerged. The first is that cerebral Cr is controlled by distant events, due to the complex biosynthetic pathway through liver and kidney enzymes. Before Cr can be available for transport to the brain it must be synthesized (Fig. 15). Absolute cerebral [Cr] falls in chronic liver disease, and recovers after liver transplantation. Even more striking is the recent discovery of a new human inborn error

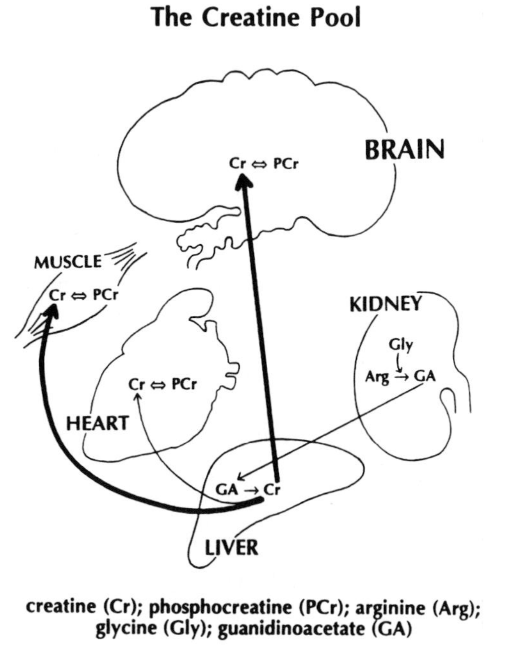

Fig. 15. The Cr pool. Creatine synthesis requires participation of kidney and liver. Tissues may express creatine kinase, in which case phosphocreatine (PCr) will be present. Other tissues lack PCr.

The Creatine Pool

BRAIN
Cr ⇔ PCr
MUSCLE
Cr ⇔ PCr
HEART
Cr ⇔ PCr
KIDNEY
Gly
Arg → GA
LIVER
GA → Cr

creatine (Cr); phosphocreatine (PCr); arginine (Arg); glycine (Gly); guanidinoacetate (GA)

of Cr biosynthesis which manifests as absence of cerebral Cr from the proton spectrum, which can be corrected by dietary administration of creatine (Hanefeld et al. 1993).

The third method of regulation of cerebral Cr content is surprising, in view of the crucial nature of cerebral energy conservation. This is the marked modification of cerebral Cr by osmotic (Donnan) forces, increased in hyperosmolar states and decreased in the very common setting of hypo-osmolar states due to sodium depletion. The explanation for this apparent over-riding of the all-important enzyme equilibrium (Gibbs) forces is probably the same as that recently discovered for the mammalian heart and for cancer cells. Namely, Gibbs equilibrium and Donnan equilibrium are very closely linked. When all equilibria are interdependent, then the total [Cr + PCr] may rise or fall to maintain the osmotic equilibrium. We presume, but cannot tell from ^{1}H MRS alone, that even under these circumstances, the ratio of PCr/Cr continues to comply with the over-riding requirements of the thermodynamic equilibrium between PCr and ATP (Fig. 16; modified from R.L. Veech, with permission).

Cr

1. Single T_2 species
2. Rapid exchange
3. Invisible pool?—not likely (8.0 vs 8.6 mM)
4. Neuron vs glial marker—(8.0 vs 6.3 mM; GM vs WM; unlikely)

3rd "force" controls [Cr]

(a) Enzyme equilibrium: early hypoxia, no change in [Cr + PCr]; Cr/PCr ↑

(b) Biosynthesis in liver and kidney: liver disease ⬇ [Cr]

(c) Gibbs-Donnan equilibrium [Cr] ⬆ or ⬇

Gibbs-Donnan Equilibrium
Control Cell Volume and Biochemistry in Brain

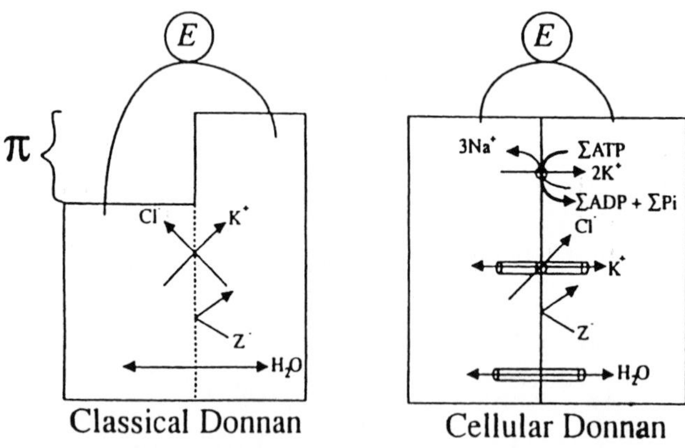

$$[\Sigma P_i] = \frac{[\Sigma 3PG]\,[\Sigma LACT]\,[\Sigma P\text{-}Cr]\,[H^+]}{[\Sigma DHAP]\,[\Sigma PYR]\,[\Sigma Creatine]} \times \frac{K_{LDH} \times K_{CK} \times K_{TPI}}{K_{G+G}}$$

1. Electrolytes, metabolites **and** osmolytes "interact."
2. Metabolites (even PCr) might be osmolytes.

ERGO: We will observe reversible changes in [Cr] **and** [NAA]

Fig. 16. Creatine: Gibb-Donnan Equilibrium.

An interesting example of this complexity is the observation that the concentration of [Cr] and of [PCr] in increased in the late-hypoxic-encephalopathy brain (Fig. 13D). This secondary effect is presumably a reflection of a new steady state, in which creatine kinase equilibrium is maintained, but the residual cell population ("gliosis") is defined by a higher total creatine content (Bluml et al. 1999).

Cholines

A number of new ideas concerning the choline resonance and its constituent metabolites have emerged from clinical studies. Although theoretically associated with myelin, the choline concentration in cerebral white matter is not much higher than that in gray matter, even though this is the impression one gains from constantly seeing ^{1}H spectra, in which Cho/Cr is much higher and nearer to 1.0 in short echo-time spectra of white matter. The explanation lies in the difference of [Cr] concentration between the two locations, the [Cr] concentration being approximately 20% higher in gray matter. Thus, the [Cho] is only a little higher, 1.6 mM in white matter and 1.4 mM in gray matter. The choline head groups of phosphatidyl choline contribute hardly at all to the proton spectrum of the human brain *in vivo* since the total of free choline, plus phosphoryl choline plus glycerophosphoryl choline determined by chemical means in human brain biopsies and post-mortem samples is very close to 1.5 mM.

As with Cr, osmotic events are among the many local and systemic events which alter its concentration in brain. The finding that many focal, inflammatory and hereditary diseases result in increased choline concentration has lead to the speculation that these metabolites represent breakdown products of myelin. Conversely, the finding that several systemic disease processes also modify cerebral choline indicates that biosynthesis and hormonal influences outside the brain, possibly in the liver, can markedly alter the composition and concentration of the choline peak. These remain to be elucidated. While proton spectroscopy offers little hope of distinguishing the different components (in vivo ^{1}H MRS at very high field (e.g., 7 Tesla) is showing some promise; Fig. 17), proton decoupled phosphorus spectroscopy undoubtedly can do so, giving the opportunity to use disease processes to further understand these interesting metabolites. Fig. 1B (modified from Prof. D. Leibfritz) integrates a number of these ideas concerning cerebral choline and ethanolamine metabolites, which are directly accessible through MRS.

Myo-inositol and *Scyllo*-inositol (sI)

Some remarkable facts have emerged concerning this simple sugar-alcohol which was rediscovered with the advent of short TE in *in vivo* human brain spectroscopy. Its con-

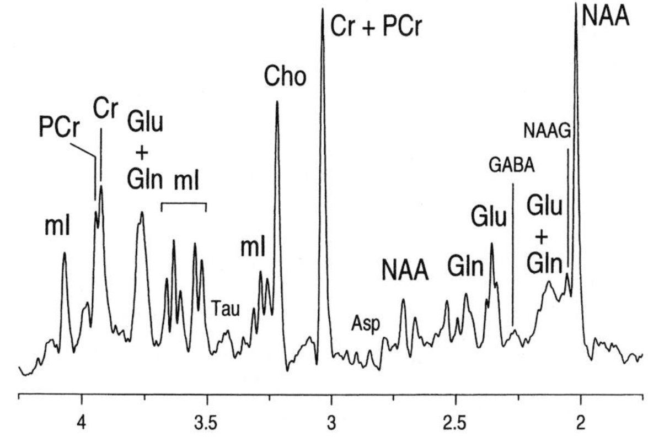

Fig. 17. Ultra high field resolves "choline" region *in vivo*. ^{1}H NMR spectrum acquired from a 1 ml volume lateral to the ventricle in dog brain at 9.4 Tesla. Processing consisted of zero-filling, 3 Hz Lorentz-to-Gauss line-shape conversion and FFT. Peaks were tentatively assigned based on dominant constituent and published chemical shifts. (Courtesy Dr. R. Grütter)

centration fluctuates more than any of the other major compounds detected in the proton spectrum – over 10-fold, from the three-times adult normal values in newborn infants and hypernatremic states, to almost zero, in hepatic encephalopathy. mI has been recognized as a cerebral osmolyte since 1990, and its cellular specificity is believed to be as an astrocyte "marker". Like Cho, mI has been labeled as a breakdown product of myelin (because it is seen at apparently increased concentration in MS plaque, HIV infection and metachromatic leukodystrophy). However, the evidence is particularly indirect on this point. Despite attempts to confine the role of mI to that of a chemically inert osmolyte or cell marker, it is important to remember that mI is at the center of a complex metabolic pathway which contains among other products the inositol-polyphosphate messengers, inositol-1-phosphate, phosphatidyl inositol, glucose-6-phosphate and glucuronic acid (Fig. 18). Any or all of these products may be involved in diseases which result in marked alterations in mI or sI concentration. The differentiation of inositol phosphate from mI which is difficult to achieve with ^{1}H MRS is likely to be achieved by a combination of proton decoupled ^{31}P MRS and natural abundance ^{13}C MRS (Ross et al. 1997).

Glutamine and Glutamate (Glx)

Provided care is taken, and the appropriate sequences applied, even at 1.5 T, the two amino acids which contribute to the spectral regions 2.2–2.4 and 3.6–3.8 ppm can be separated. Glutamine, particularly when present at elevated concentrations, can be determined with some precision. Even better separation is achieved at 2.0 T, when glutamate can be unequivocally identified and quantified. It is glutamine concentration, rather than that of glutamate, which appears to respond to disease. Increased cerebral glutamine concentration occurs in many settings, from Reyes syndrome and hepatic encephalopathy to hypoxic encephalopathy. It is the latter case which appears contrary to popular neurochemical theory.

However, the determination of *in vivo* cerebral glutamate and glutamine concentrations using ^{1}H MRS is compromised by the complex spectral appearance of glutamate/glutamine due to J-coupling. Further, other metabolites contributing to the signal at the chemical shift of glutamate/glutamine render their quantitation difficult. Studies demonstrated the potential of natural abundance *in vivo* ^{13}C for direct determination of cerebral metabolites at 2.1 and 4 T experimental systems (Fig. 19) (Gruetter et al. 1994, 1996). However, in a recent study it was shown that even on a 1.5 Tesla clinical scanner, glutamate and glutamine can be separated from each other, and natural abundance ^{13}C MRS provides enough S/N ratio for their *in vivo* quantitation (Fig. 20) (Bluml 1998).

A much closer look at glutamate turnover is achieved through the use of either ^{13}C or ^{15}N MRS (the former in human brain). Mason et al. (1995) and Gruetter et al. (1994) determined the rates of the TCA cycle, glucose consumption, glutamate formation from 2-oxoglutarate (Fig. 21) and finally the rate of glutamine synthesis (GS) *in vivo*. Their data is consistent with the long-held view of two glutamate compartments. While no explanation is yet available for the accumulation of glutamine rather than glutamate in hypoxic brain, the work of Kanamori and Ross (1997) with ^{15}N MRS offers some clues (Fig. 1D). Thus, the rate of PAG, the sole pathway of glutamine breakdown to glutamate, is under tight metabolic control in the (rat) brain. Since there is a cycle converting glutamate to glutamine and back, it may be that PAG holds the answer to the regulation of cerebral glutamate concentration in hypoxia. On the horizon are new insights through ^{13}C and ^{15}N; Figs. 1C and 1D illustrate the present state of our knowledge of in vivo flux rates, measured in the brain.

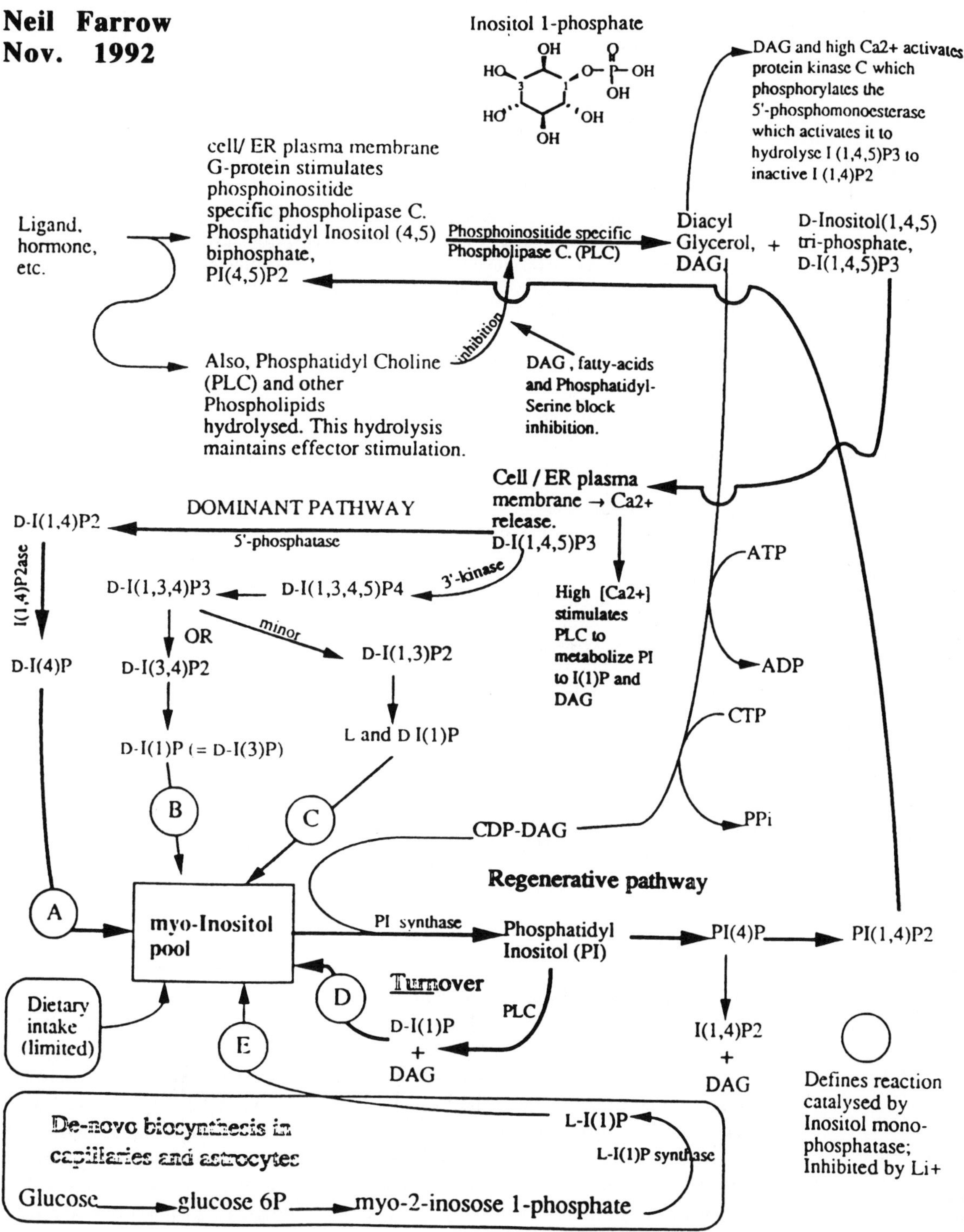

Fig. 18. Reactions involving myo-inositol (mI)

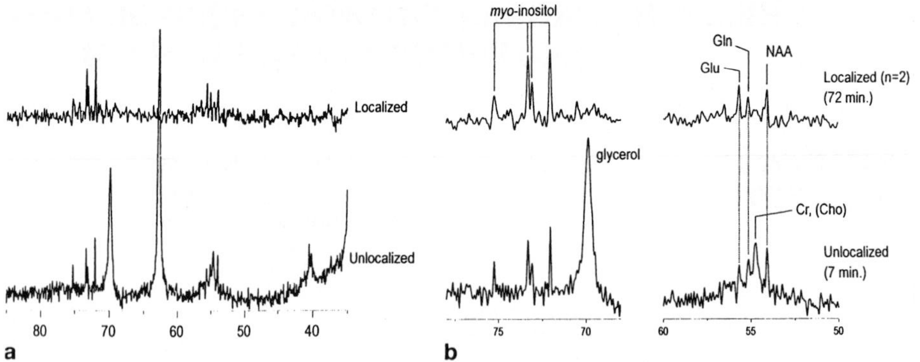

Fig. 19. Natural abundance proton decoupled ^{13}C MRS at 4T. Shown is a comparison of localized and unlocalized natural abundance 13C MRS at 4 Tesla. Peaks from myo-inositol at 72.0, 73.0, 73.3 and 75.1 ppm, as well as the C2 glutamate at 55.7 ppm, C2 glutamine at 55.1 ppm, and C2 NAA at 54.0 ppm can readily be detected. Peaks from creatine and choline overlap at 54.7 ppm and are eliminated from the localized spectrum by the polarization transfer technique used to acquire the localized spectrum (with permission from Gruetter et al. 1996).

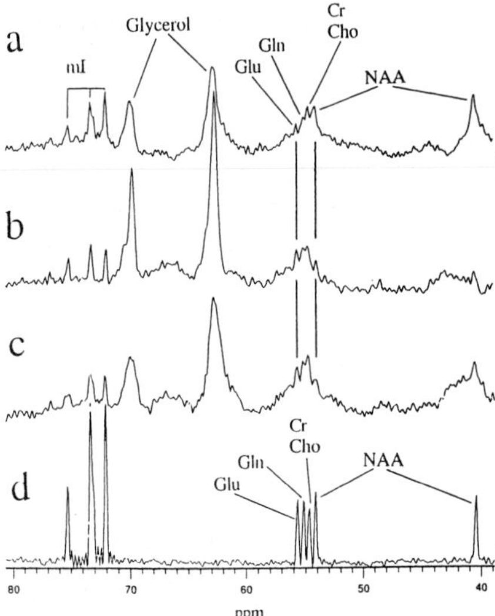

Fig. 20. Natural abundance proton-decoupled ^{13}C MRS in Canavan´s disease acquired on a clinical scanner at 1.5T. Details of *in vivo* {^{1}H}-^{13}C spectra from a child diagnosed with Canavan´s disease (**a**), 7-month and 3-years-old controls (children with unrelated diseases) (**b+c**) aligned with a spectrum from a model solution (**d**). mI at 72.1, 73.3 and 75.3 ppm and NAA at 40 and 54 ppm can be clearly identified. Peaks at 54.6 and 55.2 ppm are consistent with Cr/Cho and Gln. Note that Glu at 55.7 ppm is obviously depleted in Canavan´s disease when compared with control spectra. NAA and mI resonances appear to be elevated in Canavan´s disease, confirming ^{1}H MRS results. Glycerol peaks at 62.9 ppm and 69.9 ppm are well decoupled (figure courtesy of Bluml, 1998).

Multinuclear MRS

Solving a Problem: Multinuclear MRS of the Brain *In Vivo*

With the extraordinary family of MRS techniques now available to the human neuroscientist, we await an explosion of new knowledge. For any science to be dominated by methods of investigation, this "can result in concentration of effort on the possible rather than the real questions (Windhorst et al. editors, this volume)." Nevertheless, MRS is so rich in information, and conveniently correlated with MRI, fMRI and related procedures, that the concept of "one-stop-shopping" is close at hand. Within MRS, combining carefully quantified multi-nuclear studies can expand the utility of the equipment. This approach is illustrated by the example of *Canavan's disease* (Fig. 4), where a single patient underwent a series of non-invasive MRS examinations. Regions of different cell type composition (e.g., gray vs. white matter) were readily identified on MRI. Localized

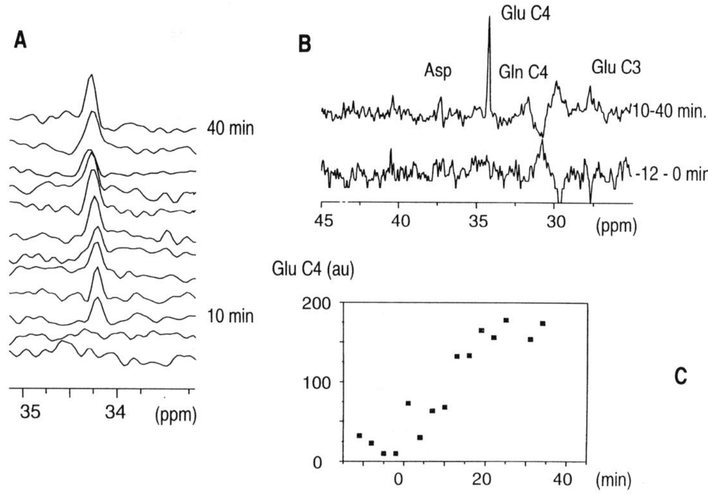

Fig. 21. *In vivo* proton-decoupled ^{13}C MRS with ^{13}C labeled glucose infusion at 4T. ^{13}C enriched glucose infusion studies allow the determination of glycolysis and TCA cycle flux rates *in vivo*. Direct ^{13}C NMR detection of label accumulation in a 22.5 ml volume in the visual cortex was measured by Greutter et al. (1996) 4T. (**A**) Stack plot with 3 min time resolution of the region containing the Glu C4 resonance at 34.2 ppm. (**B**) Pre-infusion spectra with data accumulation extended to 12 min and spectrum acquired within 30 min after start of infusion. (**C**) Time course of the C4 glutamate resonances after infusion of 1-^{13}C labeled glucose.

^1H MRS provided quantitation of brain water, and the neuronal marker NAA. Reduced aspartoacylase activity in this disease is expected to result in an elevation of NAA, readily observed and monitored by ^1H MRS. Additional information about the pathophysiology of this rare inborn error was obtained by proton-decoupled ^{31}P MRS insofar as the membrane metabolites PC, GPE and GPC appear to be reduced. Alterations in those putative membrane metabolism markers may reflect delayed or abnormal myelination. Total Cho from ^1H MRS can be correlated with myelin metabolite and osmolyte determinations from proton-decoupled ^{31}P MRS. A reduction in cerebral PCr confirms low total Cr (= free Cr + PCr) from ^1H MRS. The significance of the slightly decreased ATP is uncertain. Natural abundance $\{^1H\}$-^{13}C MRS confirmed elevated NAA and mI and detected a striking reduction of glutamate. This may be a result of the sequestration of aspartate in NAA and the reduction of free aspartate. After intravenous infusion of I–^{13}C Glucose, characteristic enrichment of glutamate, aspartate and glutamine are indicative of an intact tricarboxylic acid cycle in Canavan disease (Blüml unpublished).

Conclusions

Neuroscience has thrown up a myriad of biochemical questions, many of which are beyond the scope of *in vivo* MRS. The ideal question involves global (or at least "million-neuron") events and millimolar changes in concentration or flux, on time-scales of minutes rather than seconds. By applying these three filters at the outset (volume, concentration and time), the likelihood of a successful outcome for the early studies is increased.

As a new technology, MRS is perhaps less bound by rules of "hypothesis-driven" research, witness the unexpected inventions of fMRI, diffusion tensors, magnetization transfer, relaxation time and nuclear-Overhauser-effect determination. *In vivo* MRS brings great benefits to neuroscience, not least because studies previously only conceivable in experimental animals become extremely inviting in man!

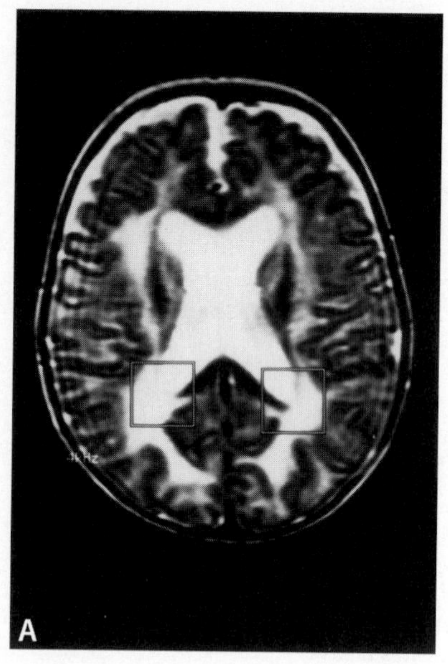
A

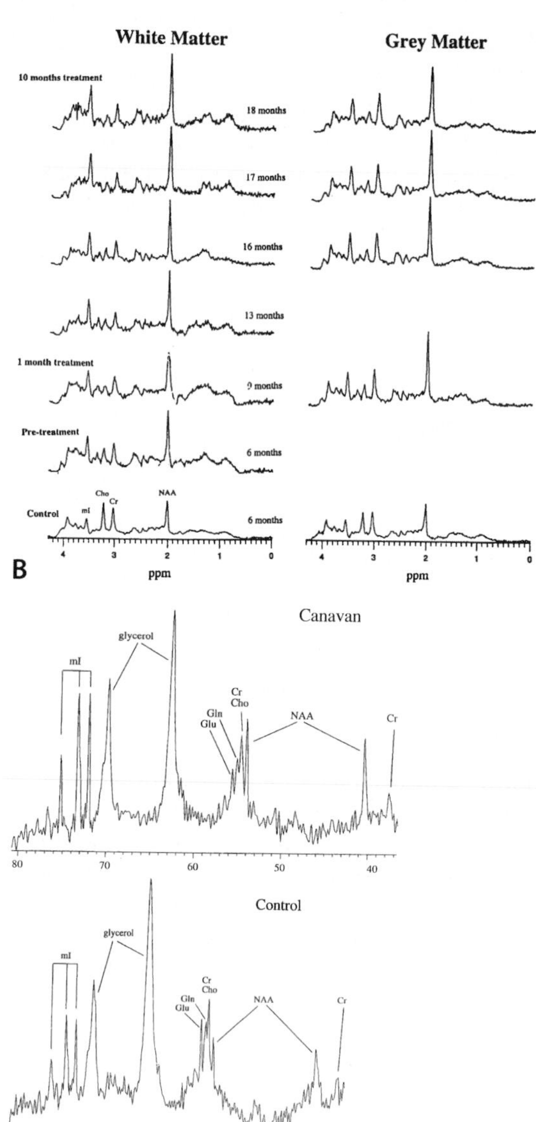
White Matter
Grey Matter
10 months treatment
18 months
17 months
16 months
13 months
1 month treatment
9 months
Pre-treatment
6 months
Control
Cho
Cr
mI
NAA
6 months
ppm
B
Canavan
glycerol
mI
Cr
Cho
Gln
Glu
NAA
Cr
Control
glycerol
mI
Cr
Cho
Gln
Glu
NAA
Cr
ppm
D

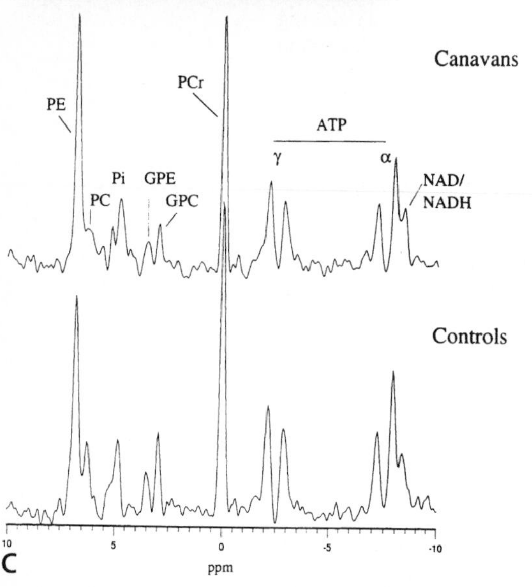
Canavans
PE
PCr
ATP
γ
α
Pi
GPE
PC
GPC
NAD/
NADH
Controls
ppm
C

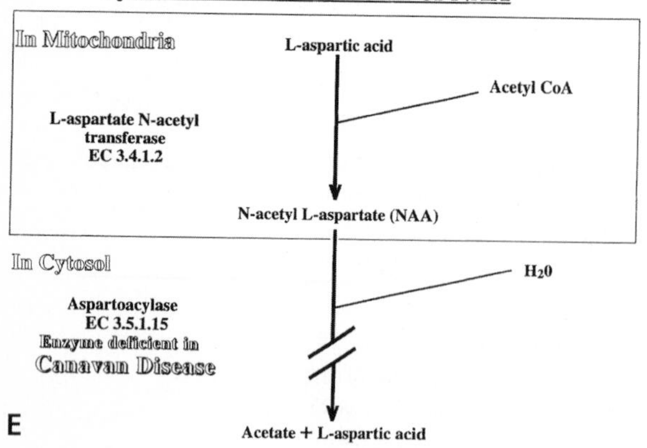
Synthesis and Breakdown of NAA
In Mitochondria
L-aspartic acid
Acetyl CoA
L-aspartate N-acetyl
transferase
EC 3.4.1.2
N-acetyl L-aspartate (NAA)
In Cytosol
H₂0
Aspartoacylase
EC 3.5.1.15
Enzyme deficient in
Canavan Disease
E
Acetate + L-aspartic acid

←

Fig. 22. Multi-nuclear MRS of the brain in vivo: Canavan's disease. In Canavan's disease the aspartoacylase deficiency results in hypomyelination with megalocephaly, blindness and spasticity and death within the first few years of life. The potential of multi-nuclear *in vivo* MR spectroscopy is illustrated on a single Canavan´s disease patient who underwent ^{1}H, {^{1}H}-^{31}P, and {^{1}H}-^{13}C MRS at a standard clinical 1.5T scanner equipped with a second rf channel. Quantitative information is transferrable from one assay to the next, greatly enhancing the study. **(A)** Standard MRI in order to detect anatomical abnormalities and to identify regions or volumes of interest (VOI). **(B)** Aspartoacylase deficiency prevents breakdown of NAA; the accumulation of NAA can readily be detected by ^{1}H MRS. Sequential ^{1}H MRS further offers the possibility of monitoring therapy. Other abnormalities such as low cerebral Cho, high mI, excess of scyllo inositol, and a subtle reduction of total Cr can be observed consistently. **(C)** Additional information about the pathophysiology of this rare inborn error is obtained by {^{1}H}-^{31}P MRS insofar as the membrane metabolites PC, GPE and GPC appear to be reduced. Alterations in those putative membrane metabolism markers may reflect delayed or abnormal myelination. A small reduction in cerebral PCr and ATP was further detected. **(D)** Natural abundance {^{1}H}-^{13}C MRS confirms elevated NAA and mI and detects a striking reduction of glutamate. This may be a result of the sequestration of aspartate in NAA and the reduction of free glutamate. Glutamate is a precursor of aspartate **(E)**. In summary, as demonstrated in this rare inborn error disease, *in vivo* MRS can be used to quantify and monitor metabolic abnormalities and may provide significant contributions to our understanding of human neuropathophysiologies. (Patients examined courtesy Rueben Matalon MD and Michel Philippert MD)

References

Ackerman JJH, Bore PJ, Wong GG, Gadian DG, Radda GK. (1980) Mapping of metabolites in whole animals by 31P NMR using surface coils. Nature 283:167–170.

Badar-Goffer RS, Ben-Yoseph O, Bachelard HS, Morris PG. (1992) Neuronal-glial metabolism under depolarizing conditions. Biochem. J 282:225–230.

Barker PB, Soher BJ, Blackband SJ, Chatham JC, Mathews VP, Bryan RN. (1993) Quantitation of proton NMR spectra of the human brain using tissue water as an internal concentration reference. NMR Biomed 6:89–94.

Bluml S. (1999) In vivo quantification of cerebral metabolite concentrations using natural abundance ^{13}C MRS at 1.5 Tesla. J. Magn. Reson 136:219–225.

Bluml S, McComb JG, Ross BD. (1997) Differentiation between cortical atrophy and hydrocephalus using 1H MRS. Magn. Reson. Med 37:395–403.

Bluml S, Seymour KJ, Ross BD. (1999) Developmental changes in choline- and ethanolamine- containing compounds measured with proton-decoupled ^{31}P MRS in human brain. Magn. Reson. Med \ (in press).

Bluml S, Zuckerman E, Tan J, Ross BD. (1998b) Proton-decoupled ^{31}P magnetic resonance spectroscopy reveals osmotic and metabolic disturbances in human hepatic encephalopathy. J. Neurochem 71:1564–1576.

Bottomley PA. (1987) Spatial localization in NMR spectroscopy *in vivo*. Ann. NY Acad. Sci 508:333–348.

Bottomley PA, Foster TB, Darrow RD. (1984) Depth-resolved surface spectroscopy (DRESS) for *in vivo* 1H, 31P and 13C NMR. J. Magn. Reson 59:338–342.

Bottomley PA, Hart HR, Edelstein WA, et al. (1983) NMR imaging/spectroscopy system to study both anatomy and metabolism. Lancet 1:273–274.

Brody BA, Kinney HC, Kloman AS, Giles FM. (1987) Sequence of central nervous system myelination in human infancy: An autopsy study of myelination J. Neuropathol. Exp. Neurol 46:283.

Brown TR, Kincaid BM, Ugurbil K. (1982) NMR chemical shift imaging in three dimensions. Proc Natl Acad Sci USA 79:3523–3526.

Cady EB, Dawson MJ, Hope PL, et al. (1983) Non-invasive investigation of cerebral metabolism in newborn infants by phosphorus nuclear magnetic resonance spectroscopy. Lancet 1:1059–1062.

Christiansen P, Henriksen O, Stubgaard M, Gideon P, Larsson HBW. (1993) *In vivo* quantification of brain metabolites by means of 1H MRS using water as an internal standard. Magn. Reson. Imaging 11:107–118.

Danielsen ER, Henriksen O. (1994) Quantitative proton NMR spectroscopy based on the amplitude of the local water suppression pulse. Quantification of brain water and metabolites. NMR Biomed 7, 311–318.

Ernst T, Hennig J. (1994) Observation of a fast response in functional MR. Magn. Reson. Med 32:146–149.
Ernst T, Kreis R, Ross BD. (1993a) Absolute quantitation of water and metabolites in the human brain. Part I: Compartments and water. J. Magn. Reson 102(1):1–8.
Ernst T, Lee JH, Ross BD. (1993b) Direct 31P imaging in human limb and brain. J. Comput. Assist. Tomogr 17(5):673–680.
Flögel U, Niendorf T, Serkowa N, Brand A, Henke J, Leibfritz D. (1995) Changes in organic solutes, volume, energy state, and metabolism associated with osmotic stress in a glial cell line: A multinuclear NMR study. Neurochem. Res 20(7):793–802.
Frahm J, Bruhn H, Gyngell ML, Merboldt K-D, Hänicke W, Sauter R. (1989) Localized proton NMR spectroscopy in different regions of the human brain *in vivo*. Relaxation times and concentrations of cerebral metabolites. Magn. Reson. Med 11:47–63.
Frahm J, Merboldt K-D, Hänicke W. (1987) Localized proton spectroscopy using stimulated echoes. J. Magn. Reson 72:502–508.
Frahm J, Michaelis T, Merboldt K-D, Bruhn H, Gyngell ML, Hänicke W. (1990) Improvements in localized proton NMR spectroscopy of human brain. Water suppression, short echo times, and 1 ml resolution. J. Magn. Reson 90:464–473.
Frahm J, Michaelis T, Merboldt K-D, et al. (1988) Localized NMR spectroscopy *in vivo*: progress and problems. NMR in Biomed 2(5/6):188–195.
Greenman RL, Axel L, Lenkinski RE. (1998) Direct imaging of phosphocreatine in the human myocardium using a RARE sequence at 4.0 T. Proceedings, 6th International Society for Magnetic Resonance in Medicine. Sydney, Australia: p. 1922.
Gruetter R, Adriany G, Merkle H, Anderson PM. (1996) Broadband decoupled, ^{1}H-localized ^{13}C MRS of the human brain at 4 Tesla. Magn. Reson. Med 36:659–664.
Gruetter R, Mescher M, Kirsch J, et al. (1997) ^{1}H MRS of neurotransmitter GABA in humans at 4 Tesla. Proceedings, 5th International Society for Magnetic Resonance in Medicine. Vancouver, Canada: , 1217.
Gruetter R, Novotny EJ, Boulware SD, et al. (1994) Localized ^{13}C NMR spectroscopy in the human brain of amino acid labeling from D-[1-^{13}C]glucose. J. Neurochem 63:1377–1385.
Gyngell ML, Busch E, Schmitz B, et al. (1995) Evolution of acute focal cerebral ischemia in rats observed by localized ^{1}H MRS, diffusion weighted MRI and electrophysiological monitoring. NMR in Biomed 8:206–214.
Hamilton PA, Hope PL, Cady EB, Delpy DT, Wyatt JS, Reynolds EOR. (1986) Impaired energy metabolism in brains of newborn infants with increased cerebral echodensities. Lancet 1242–1246.
Hanefeld F, Holzbach U, Kruse B, Wilichowski E, Christen HJ, Frahm J. (1993) Diffuse white matter disease in three children: an encephalophathy with unique features on magnetic resonance imaging and proton magnetic resonance spectroscopy. Neuropediatrics 24(5):244–248.
Hanstock CC, Rothman DL, Prichard JW, Jue R, Shulman R. (1988) Spatially localized 1H NMR spectra of metabolites in the human brain. Proc. Natl. Acad. Sci. USA 85:1821–1825.
Heerschap A, van den Berg P. (1993) Proton MR spectroscopy of the human fetus *in utero*. Proceedings, 12th Society of Magnetic Resonance in Medicine. New York: 318.
Hennig J, Pfister H, Ernst T, Ott D. (1992) Direct absolute quantification of metabolites in the human brain with *in vivo* localized proton spectroscopy. NMR Biomed 5:193–99.
Hetherington HP, Newcomer BR, Pan JW. (1998) Measurements of human cerebral GABA at 4.1 T using numerically optimized editing pulses. Magn. Reson. Med 39:6–10.
Hoang T, Dubowitz D, Bluml S, Kopyov OV, Jacques D, Ross BD. (1997) Quantitative ^{1}H MRS of neurotransplantation in patients with Parkinson's and Huntington's disease. 27th Annual Meeting, Society for Neuroscience. New Orleans, LA. Oct. 25–30: 658.9.
Hope PL, Cady EB, Tofts PS, et al. (1984) Cerebral energy metabolism studied with phosphorus NMR spectroscopy in normal and birth-asphyxiated infants. Lancet (Aug):366–370.
Hossman K-A. (1994) Viability thresholds and the penumbra of focal ischemia. Ann. Neurol 36:557–565.
Hurd RE, Gurr D, Sailasuta N. (1998) Proton spectroscopy without water suppression: The oversampled J-Resolved experiment. Magn. Res. Med. 40, 343 - 347.
Kanamori K, Ross BD. (1997) *In vivo* nitrogen MRS studies of rat brain metabolism. In: Bachelard H, ed. Advances in Neurochemistry. New York: Plenum Publishing 66–90. vol 8).
Kreis R. (1997) Quantitative localized ^{1}H MR spectroscopy for clinical use. Progress in Nuclear Magnetic Resonance Spectroscopy 31:155–195.
Kreis R, Ernst T, Ross BD. (1993a) Absolute quantitation of water and metabolites in the human brain. Part II: Metabolite concentrations. J. Magn. Reson 102(1):9–19.

Kreis R, Ernst T, Ross BD. (1993b) Development of the human brain: *In vivo* quantification of metabolite and water content with proton magnetic resonance spectroscopy. Magn. Reson. Med 30:1–14.

Kreis R, Farrow NA, Ross BD. (1990) Diagnosis of hepatic encephalopathy by proton magnetic resonance spectroscopy. Lancet 336:635–6.

Kreis R, Farrow NA, Ross BD. (1991) Localized 1H NMR spectroscopy in patients with chronic hepatic encephalopathy. Analysis of changes in cerebral glutamine, choline and inositols. NMR Biomed 4:109–16.

Kreis R, Ross BD, Farrow NA, Ackerman Z. (1992) Metabolic disorders of the brain in chronic hepatic encephalopathy detected with 1H MRS. Radiology 182:19–27

Kumar A, Welti D, Ernst RR. (1975) NMR Fourier zeugmatography. J Magn Reson 18:69–83.

Mason GF, Gruetter R, Rothman DL, Behar KL, Shulman RG, Novotny EJ. (1995) NMR determination of the TCA cycle rate and a-ketoglutarate/glutamate exchange rate in rat brain. J. Cereb. Blood Flow Metab 15:12–25.

Merboldt K-D, Bruhn H, Hanicke W, Michaelis T, Frahm J. (1992) Decrease of glucose in the human visual cortex during photic stimulation. Magn. Reson. Med 25:187–194.

Merboldt KD, Chien D, Hanicke W, Gyngell ML, Bruhn H, Frahm J. (1990) Localized 31P NMR spectroscopy of the adult human brain *in vivo* using stimulated-echo (STEAM) sequences. J. Magn. Reson 89:343–361.

Michaelis T, Merboldt K-D, Bruhn H, Hänicke W, Frahm J. (1993) Absolute concentrations of metabolites in the adult human brain in vivo: Quantification of localized proton MR spectra. Radiology 187:219–27.

Michaelis T, Merboldt K-D, Hänicke W, Gyngell ML, Bruhn H, Frahm J. (1991) On the identification of cerebral metabolites in localized 1H NMR spectra of human brain *in vivo*. NMR Biomed 4:90–8.

Narayana PA, Johnston D, Flamig DP. (1991) *In vivo* proton magnetic resonance spectroscopy studies of human brain. Magn. Reson. Imag 9:303–308.

Oberhaensli RD, Hilton Jones D, Bore PJ, Hands LJ, Rampling RP, Radda GK. (1986) Biochemical investigation of human tumours *in vivo* with phosphorus-31 magnetic resonance spectroscopy. Lancet 2:8–11.

Ordidge RJ, Bendall MR, Gordon RE, Connelly A. (1985) Volume selection for *in vivo* biological spectroscopy. In: Govil G, Khetrapal CL, Saran A, ed. Magnetic Resonance in Biology and Medicine. New Delhi: Tata McGraw-Hill, 387–397.

Ordidge RJ, Connelly A, Lohman JAB. (1986) Image-selected *in vivo* spectroscopy (ISIS). A new technique for spatially selective NMR spectroscopy. J. Magn. Reson 66:283–294.

Pettegrew JW, Minshew NJ, Cohen MM, Kopp SJ, Glonek T. (1984) P-31 NMR changes in Alzheimer and Huntington diseased brain. Neurology 34:281.

Prichard JW, Alger JR, Behar KL, Petroff OAC, Shulman RG. (1983) Cerebral metabolic studies *in vivo* by 31P NMR. Proc Natl Acad Sci USA 80:2748–2751.

Provencher SW. (1993) Estimation of metabolite concentrations from localized *in vivo* proton NMR spectra. Magn. Reson. Med 30:672–79.

Ross BD (1979) Techniques for investigation of tissue metabolism. In: Kornboj HC, Metcalf JC, Northcote D, Pogson CI, Tipton KF, (eds) Techniques in Metabolic Research, Part I. Elsevier,: 1–22. vol B203).

Ross BD, Bluml S, Cowan R, Danielsen ER, Farrow N, Gruetter R. (1997) In vivo magnetic resonance spectroscopy of human brain: the biophysical basis of dementia. Biophysical Chemistry 68:161–172.

Ross BD, Hoang T, Blüml S, Dubowitz D, Kopyov OV, Jacques DB, Alexander Lin, Kay Seymour and Jeannie Tan. In vivo magnetic resonance spectroscopy of human fetal neural transplants. NMR in Biomed 1999, 12:221–236.

Ross BD, Kreis R, Ernst T. (1992) Clinical tools for the 90's: magnetic resonance spectroscopy and metabolite imaging. Eur. J. Radiol 14:128–140.

Ross BD, Michaelis T. (1994) Clinical applications of magnetic resonance spectroscopy. Magn. Reson. Quarterly 10(4):191–247.

Ross BD, Narasimhan PT, Tropp J, Derby K. (1989) Amplification or obfuscation: Is localization improving our clinical understanding of phosphorus metabolism? NMR in Biomed 2(5):340–345.

Ross BD, Radda GK, Gadian DG, Rocker G, Esiri M, Falconer-Smith JC (1981) Examination of a case of suspected McArdle's syndrome by 31P nuclear magnetic resonance. N. Engl. J. Med 304(22):1338–42.

Rothman DL, Petroff OAC, Behar KL, Mattson RH. (1993) Localized ^{1}H NMR measurements of γ-aminobutyric acid in human brain in vivo. Proc. Natl. Acad. Sci 90:5662–5666.

Ryner LN, Sorenson JA, Thomas MA. (1995) Localized 2D J-Resolved ^{1}H MR spectroscopy: strong coupling effects in vitro and in vivo. Magn. Reson. Imag 13(6):853–869.

Seymour, K., Bluml S, Sutherling J, Sutherling W and Ross B.D. (1998) Identification of cerebral acetone by 1H MRS in patients with epilepsy, controlled by ketogenic diet 1999, MAGMA, 8:33–42

Shen J, Novotny EJ, Rothman DL. (1998) In vivo lactate and β-hydroxybutyrate editing using a pure phase refocusing pulse train. Proceedings, 6th International Society for Magnetic Resonance in Medicine. Sydney, Australia: 1883.

Stockler S, Hanefeld F, Frahm J. (1996) Creatine replacement therapy in guanidinoacetate methyltransferase deficiency, a novel inborn error of metabolism. Lancet 348:789–90.

Tallan HH. (1957) Studies on the distribution of N-acetyl-L-aspartic acid in brain. J. Biol. Chem 224:41–45.

Thulborn KR, Ackerman JJH. (1983) Absolute molar concentrations by NMR in inhomogeneous B1. A scheme for analysis of *in vivo* metabolites. J. Magn. Reson 55:357–371.

Thulborn KR, du Boulay GH, Radda GK (1981) Proceedings of the Xth Int. Symp. on Cerebral Blood Flow and Metabolism.

Van der Knaap MS, Ross BD, Valk J. (1993) Uses of MR in inborn errors of metabolism. In: Kucharcyzk J, Barkovich AJ, Moseley M, ed. Magnetic Resonance Neuroimaging. Boca Raton: CRC Press, Inc. 245–318.

Webb PG, Sailasuta N, Kohler S, Raidy T, Moats RA, Hurd RE. (1994) Automated single-voxel proton MRS: Technical development and multisite verification. Magn. Reson. Med 31(4):365–373.

Welch KMA. (1992) The imperatives of magnetic resonance for the acute stroke clinician (Plenary Lecture). 11th Society of Magnetic Resonance in Medicine. Berlin: 901.

Glossary

Much of the nomenclature of the chemical analytical technique known as in vivo magnetic resonance is relatively unfamiliar to the neuroscientist. For this reason we here include an extensive glossary of the most useful terms employed in MRS.

Absolute quantitation Aiming to measure concentrations of metabolites in mMol per kg brain tissue or volume. Absolute quantitation requires additional measurements and is therefore more time-consuming than quantifying spectra from peak ratios. Several methods for absolute quantitation have been suggested and described in detail in the literature. See also *quantitation.*

B_0 Strong magnetic field generated by the superconducting magnet. B_0 is meant to be constant in time and space within the limits of technical feasibility. A typical field strength for clinical scanners is 1.5 Tesla (~ 30,000 times the magnetic field of the earth). The resonance frequency of protons is ~ 64 MHz (10^6 Hertz), phosphorus ~ 26 MHz, carbon (^{13}C) 16 MHz at this field strength.

B_1 Radio frequency (RF) magnetic field generated by *radio frequency coils.* In MR examinations B_1 fields and *gradient* fields are varied, B_0 is constant.

Bandwidth A) Radio frequency pulse bandwidth. Every RF pulse of finite duration excites magnetization within a certain band of frequencies. The bandwidth needed for an MR experiment is determined by the *chemical shift* range of the nucleus investigated which should be well within the chosen RF bandwidth to allow homogeneous excitation of all metabolites. Failure to recognize this can result in *chemical shift artifacts.* B) Receiver bandwidth; not all of the MR signal is used to calculate the spectrum. Noise signal is statistically spread over all frequencies while the MR signal stems from a rather small frequency range. By allowing frequencies to pass through the receiver only within a certain bandwidth, the *S/N* of an MRS experiment is greatly increased. The receiver bandwidth has to be larger than the chemical shift range.

Chemical shift In addition to the external B_0 field, there are internal local fields which affect the resonance frequency of the *spins.* A common example is ethanol (CH_3CH_2OH) that has three proton resonances since it has three chemically distinct proton groups: i) three protons of the CH_3 group, ii) two protons from CH_2, and iii) one proton sited next to the oxygen. The different electron structure surrounding those protons causes the chemical shift. In contrast to *J-coupling,* the chemical shift is measured in parts per million (*ppm*) relative to a standard because it scales with the B_0 field strength. The chemical shift is the same in all human MR scanners.

Chemical shift artifact

The problem of chemical shift artifacts arises from the different frequencies of the resonances associated with various chemical structures. When a *gradient* is applied to a sample containing chemically shifted species, there will be a displacement of the sensitive volume for each of the different species. The *bandwidth* of the RF pulse, with respect to the chemical shift range of the chemical structures, sets the percentage of overlap one can expect. Increasing the bandwidth of the RF pulse reduces chemical shift artifacts. At clinical field strengths of 1.5 Tesla an RF bandwidth of > ~ 2000 Hz is sufficient, for ^{1}H MRS.

Chemical shift imaging (CSI)

Localization method. This technique encodes all spatial information into the phase of the magnetic resonance signal. In contrast to standard 2D MRI (cf. Chap 35) where one spatial dimension is phase encoded while the second dimension is frequency encoded, data acquisition is performed in the absence of a frequency encoding gradient so that the chemical shift information can be retained. Due to the phase encoding, many spectra from a slice or from a 3D volume can be acquired simultaneously, and CSI is an excellent technique to obtain metabolic maps. However, CSI is less suitable for short echo times, the quality of individual spectra is inferior to spectra acquired with single-voxel techniques, and acquisition times are longer.

Chemical shift selective excitation

While various degrees of selectivity can be required, the usual meaning of selective excitation is that a single resonance line needs to be excited while the adjacent lines are left unperturbed. In ^{1}H MRS, specially designed *RF pulses* with a narrow bandwidth of typically 50–75 Hz (at 1.5 Tesla) are used to selectively excite the water resonance prior to a localization sequence to achieve *water suppression*. To study creatine kinase with ^{31}P MRS, chemical shift selective excitation is used to saturate the γ-ATP resonance in a *magnetization transfer* experiment. To image PCr, independent of other ^{31}P-metabolites, chemical shift selective excitation is applied.

Coils

see *radio frequency coils*

Data processing

To improve interpretation and quantitation of the *Fourier transformed FID*, several post processing steps are applied. This includes *eddy current* corrections, elimination of residual water signal in proton spectroscopy, and phase and line-shape corrections. It is beyond the scope of this chapter to discuss strategies in detail.

CHESS (chemical shift selective)

CHESS *RF pulses* are often used for *water suppression* in ^{1}H MRS. They consist of one or more narrow *bandwidth* (typically 50–75 Hz at 1.5 T) RF pulses which selectively excite the water resonance but do not affect metabolite resonances (see also chemical shift selective excitation).

Decoupling

As a result of *J-coupling* (or scalar coupling) between spins, signals are commonly split into *multiplets*. Such multiplets can be collapsed into single lines by applying irradiation at the resonance frequency of the nuclei to which the spins of interest are coupled. Decoupling significantly simplifies the interpretation of spectra and increases the *S/N*. This can involve considerable irradiation power. However, a number of techniques are described in the literature that allow sufficient decoupling to be carried out without excessive power requirements. Additional S/N gain is achieved by the build-up of a nuclear Overhauser effect (*NOE*). Spin decoupling is most commonly used in ^{13}C MRS.

Dipolar coupling

A nuclear spin creates a local field of its own which can be felt by neighboring spins. This interaction is called dipole-dipole interaction or dipolar coupling. The strength of this interaction depends on the distance between the spins and relative orientation of the spins. In vivo where molecules can tumble rapidly and change their relative position fast, dipolar coupling cannot be observed in form of line splitting. However, dipolar coupling is a mechanism that causes *T1*- and *T2-relaxation*.

DRESS

Depth resolved surface coil spectroscopy. This localization method was originally designed for ^{31}P MRS using a *surface coil*. In a plane parallel to the plane of the coil a slice is excited by an RF pulse in the presence of a *gradient*.

Echo time (TE)

Time interval in which the magnetization is in the transverse plane and undergoes *T2 decay* in sequences using *spin echos*.

Eddy currents

Induced by the interaction between gradient coils and conducting elements within the magnetic structure. The effect of these eddy currents is a time-dependent variation of the main magnetic field B_0. Eddy currents cause distortions of the shape of resonances as well as an increase in the line width. Eddy current distortions are best reduced by using *shielded gradients*. Additionally, data processing methods for correcting the effects of eddy currents on the spectrum have been developed.

Editing techniques Often the interpretation of spectra is compromised by the superimposition of resonance lines from different molecules at the same or at similar chemical shifts. To overcome these problems, it is necessary to use spectral simplification or editing techniques. Editing techniques exploit unique properties of molecules other than the chemical shift, most commonly *homo-* or *hetero-nuclear J-coupling* (or scalar coupling). Many editing sequences utilize the fact that in an echo sequence the phase of J-coupled spins is modulated during the echo delay. A series of spectra acquired with different echo times each may allow the separation and identification of overlapping signals from different molecules due to their different J-modulation.

FID (free induction decay) is the response of magnetization to a radiofrequency pulse. The *spectrum* is calculated from the FID by a *Fourier transform* .

Flip angle see *RF pulse*

Fourier transform Mathematical operation (cf. Chapter 19). The observed MR signal (*FID*) from a sample in the time domain is a superimposition of individual MR signals originating from different species in the sample which resonate at different frequencies due to their different *chemical shifts*. The Fourier transform generates a frequency spectrum which allows easy identification of the individual components by their different frequencies.

Gradients, gradient coils Set of coils to generate a static but spatially varying magnetic field additional to the B_0 field. By doing this the resonance frequency of magnetization varies accordingly and spatial information is frequency encoded. In the presence of gradients, radio frequency pulses excite magnetization within a certain *bandwidth* of frequencies and therefore within a certain slice of the sample. Gradients are essential for localized spectroscopy such as *STEAM* and *PRESS*. See also *shielded gradients*.

Hetero-nuclear J-coupling *J-coupling* between different species of spins, e.g., proton and carbon.

Homo-nuclear J-coupling *J-coupling* between the same species of spins, e.g., proton and proton.

Inverse detection See *polarization transfer*.

ISIS Image-selected *in vivo* spectroscopy is based on a cycle of eight acquisitions which need to be added and subtracted in the right order to get a single volume. ISIS is considerably more susceptible to motion then *STEAM* or *PRESS* and is mostly used in heteronuclear studies, where its advantage of avoiding *T2-relaxation* is valuable.

J-coupling (or scalar coupling) Many resonances split into multiplet components. This is the result of an internal indirect interaction of two spins via the intervening electron structure of the molecule. The coupling strength is measured in Hertz (Hz) rather than *ppm* because it is independent of the external B_0 field strength.

Localization Acquisition of an MR signal from a certain region of the brain. See *STEAM, PRESS, ISIS, CSI, gradients, VOI*.

Long echo time MRS Term used for proton MRS carried out with *echo times* (*TE*) longer than ~135 ms. Long TE MRS has the advantage of offering simplified spectra which are easy to read, because lipid signal and signal from metabolites with a complex spectral pattern like the amino acids glutamate and glutamine has decayed due to *J-coupling* and *T2 relaxation* effects at the time of data acquisition. Also, *eddy current* problems are less severe in long TE MRS because the data acquisition has been moved away from slice selection gradient pulses. However, the information content of long TE MRS and the *S/N* is inferior to *short echo time MRS*. On newer MR systems *shielded gradients* provide adequate protection against eddy currents favoring short echo time MRS.

Mg (magnesium) measurement Free adenosine-triphosphate (ATP) and Mg.ATP (ATP magnesium complex) have a slightly different spectral pattern in solution. The chemical shift difference between the α and β ATP phosphorus nuclei resonance possibly provides a non-invasive determination of the *in vivo* magnesium concentration.

Magnetization The magnetization vector is viewed as being the result of summing the magnetization from a large number of individual spins. Being a vector, the magnetization has amplitude and direction. In equilibrium the magnetization vector points are parallel to the B_0-field which is usually assigned as the z-direction. The magnetization vector or parts of it can be flipped into the xy- or transverse plane (non-equilibrium condition) where it precesses at a frequency which depends on the B_0 field strength.

Magnetization transfer

When two spins are coupled through processes like *chemical exchange*, *J-coupling*, or *dipolar coupling*, then any of these interactions can be used to transfer magnetization from one spin system to another. A common example is the study of creatine kinase, PCr + ADP × Cr + ATP, where the γ-phosphate of adenosine-triphosphate (ATP) is in (slow) exchange with the phosphate of phosphocreatine (PCr). As a result, in a saturation experiment where the resonance of γ-ATP is eliminated by a *chemical shift selective excitation*, a decrease of the PCr resonance will be observed depending on the length of the saturation pulse.

Metabolite imaging

Special techniques such as *chemical shift imaging* allow the simultaneous acquisition of many spectra from a slice of the brain or from a whole 3D volume. When the intensity of only one metabolite is mapped into an image, a qualitative metabolite image displaying local variation of the metabolite concentration is observed. Another technique is the direct imaging of a metabolite using MRI techniques. The transmitter frequency is adjusted not on the water resonance but on the resonance of the metabolite. Metabolite imaging suffers from the intrinsic low concentration of metabolites and the correspondingly low *S/N* which limits the resolution of the metabolite maps. Nevertheless, it provides crucial information in focal diseases such as tumors where it guides the physician to the right spot for biopsies.

Multiplet

Due to *J-coupling* (or scalar coupling) to neighboring spins, a single resonance line from a nucleus splits up in several lines. In the case of *hetero-nuclear* J-coupling this splitting can be removed by irradiation of RF at the resonance frequency of one nucleus while acquiring the signal from the second (see *decoupling*).

NOE

Nuclear Overhauser effect, the magnetization of protons dipolar-coupled to ^{13}C nuclei can be used to enhance the ^{13}C signal. While the term NOE is mainly associated with ^{13}C MRS, NOE enhancement can also be observed in ^{31}P MRS and with other nuclei.

On/off resonance

Every nucleus with a spin resonates at a certain frequency in an external magnetic field. In order to excite these spins the frequency *bandwidth* of an *RF pulse* must contain this frequency. If the frequency range of the RF pulse is adjusted so that the frequency of a certain species of nuclei lies in the center of the frequency band, the on-resonance condition is achieved.

Peak ratios

The commonest approach for the *quantitation* of spectra. Either peak amplitudes or peak areas can be used. In proton MRS most commonly creatine (Cr) is used as an internal reference. When peak ratios are calculated from peak amplitudes, errors can be made due to the dependency of the amplitudes on the shim quality. Differences in *T1*- and *T2-relaxation* introduce further ambiguity, which is minimized by the use of short echo times (*TE*) and long repetition times (*TR*). In general, *absolute quantitation* is more reliable than quantitation with peak ratios.

Phasing, phase correction

A post-processing step. Due to hardware settings and sequence timing following *Fourier transform*, a mixture of absorption and dispersion signals is observed in the *spectrum*. Spectrum analysis and *quantitation* is performed on the pure absorption signal which needs to be extracted by a phase correction procedure.

pH measurement

^{31}P MRS allows accurate non-invasive monitoring of the intracellular pH by measuring the precise *chemical shift* between the phosphocreatine (PCr) and inorganic phosphate (Pi) resonance in the *spectrum*. The exact resonance position of the Pi peak changes as a function of pH because the relative concentrations of the phosphate containing molecules HPO_4 and H_2PO_4 changes.

Polarization transfer

Many interesting nuclei like ^{13}C suffer from low inherent sensitivity compared with proton MR. Techniques like DEPT (**d**istortionless **e**nhancement by **p**olarization **t**ransfer) and INEPT (**in**sensitive **n**uclei **e**nhanced by **p**olarization **t**ransfer) improve the ^{13}C sensitivity by transferring the higher polarization of coupled protons to the carbon nuclei. Special hardware with two RF channels is needed for polarization transfer experiments. A modification of DEPT and INEPT is reverse DEPT and inverse INEPT where the polarization is transferred back to utilize the higher sensitivity of the protons for observation (inverse detection).

Ppm

Parts per million, a measure of the *chemical shift* of substances. The CH_3 protons of N-acetyl-aspartate and of creatine resonate in a magnetic field of 1.5 Tesla at 64 Hertz or 1 ppm apart (resonance frequency ~64 MHz). In a magnetic field of 4.0 Tesla the difference is 171 Hz which is also 1 ppm of the B_0 field strength. Measuring the chemical shift in ppm is therefore favored as it is independent of the magnetic field strength.

PRESS Point-resolved spectroscopy, utilizes three 180° slice selective pulses along each of the spatial directions and generates signals from the overlap in form of a spin echo. At the same *echo time* (*TE*), PRESS has the advantage over *STEAM* that it recovers the full possible signal and is therefore the method of choice for applications where *S/N* is crucial.

PROBE proton brain exam, application package available on General Electric MR systems for fully automated single-voxel proton spectroscopy.

Quantitation Providing quantitative MRS data is a prerequisite in order to avoid misinterpretation and to allow global comparisons of data from classical measurement techniques. The use of one of the metabolites, most frequently creatine (Cr), as an internal reference and expressing the results as *peak ratios* is the most simple and common approach. The dependence of peak ratios on sequence parameters (*long vs. short echo time MRS*) makes this approach uncertain, and close attention to the methods applied has to be paid when data from different sites are compared. Absolute quantitation, aiming to measure concentrations of metabolites in mMol per kg brain tissue or volume, is superior and independent of the applied technique. However, absolute quantitation requires additional measurements and is therefore more time-consuming.

RF (radio frequency) coils Devices used to transmit radio frequency pulses and to receive the radio frequency response from the brain or sample. All manufacturers offer volume coils which are specifically designed for the imaging of the head. These coils have very uniform B_1 fields and are well suited for spectroscopy examinations. Surface coils are simple circular, rectangular or square loops of wire, foil or tubing. Most of the signal observed by a surface coil originates roughly from a hemisphere of the radius of the coil. Surface coils are appropriate to obtain high *S/N* ratio spectra from surface tissue.

RF (radio frequency) pulse An RF pulse modifies the magnetic field seen by the spins for a short time to cause the *magnetization* to be flipped by a certain angle out of its equilibrium position. The flip angle depends on the amplitude of the RF pulse and the duration.

S/N (signal-to-noise) The definition and the measurement of absolute S/N depend on acquisition parameters and steps involved in pre-processing of the data and are beyond the scope of this chapter. In general, S/N is the ratio between the amplitude of an observed resonance and the size of random noise signal received by the MR scanner. In practice, it is important to observe a few rules about S/N: i) To improve the S/N by a factor of two, four times the acquisition time is necessary. ii) S/N scales with the volume; half the volume gives half the S/N. iii) The area of a resonance line is constant. Therefore by improving the shim and narrowing the width and increasing the amplitude of a resonance line the S/N can be improved. iv) S/N decreases with shorter repetition times (*TR*) due to *T1-saturation* and decreases with longer echo times (*TE*) due to *T2-decay*.

SAR Specific absorption rate. Due to inductive and dielectric losses, energy from radio frequency pulses is absorbed by tissue and mainly transferred into rotational and translational movements of water molecules, which causes an increase in tissue temperature. Limits for the human brain are established by FDA guidelines as 4 Watts/kg brain tissue. The SAR needs to be observed in experiments with intense irradiation of radio frequency pulses like in proton-decoupled phosphorus or carbon MRS.

Scalar coupling See *J-coupling*

Selective excitation See *chemical shift selective excitation.*

Shielded gradients Shielded gradients eliminate the source of the *eddy currents* by minimizing the interaction of gradient coils with other conducting elements within the magnet structure. A second set of gradient coils placed outside the first set runs in opposition to the inner set, ensuring there is no magnetic field outside the cylinder of the gradient coil.

Shimming Optimizing the magnetic field homogeneity. A well-shimmed *VOI* is a prerequisite of MRS. With better shimming resonance lines get narrower, the overlap between neighboring lines is reduced, and the interpretation and quantitation of the spectra is easier. On many systems automated shimming procedures are available.

Short echo time (TE) MRS Term used for proton MRS carried out with *echo times* (*TE*) shorter than ~135ms. Short TE MRS has the advantage of providing more information in a single spectrum. In particular, lipid signals and signals from metabolites with a complex spectral pattern like the amino acids glutamate and glutamine can be observed and quantified. On older systems without *shielded gradients, eddy currents* may sometimes cause problems because the data acquisition starts after a short delay following a slice selection gradient pulse. However, the information content of short TE MRS and the *S/N* is superior to *long echo time MRS.*

Spectral resolution Separation of individual peaks of the spectrum. The spectral resolution depends on acquisition conditions such as the *shim*, acquisition parameters such as the *receiver bandwidth*, and the B_0 field strength.

Spectrum Result of an MRS experiment. The spectrum is the *Fourier transform* of the time domain signal (*FID*) measured. The biochemical fingerprint is stored in a two-dimensional form. The x-axis or *chemical shift* axis identifies metabolites by their unique chemical shifts, measured in *ppm* (= parts per million) relative to a standard. The y-axis reflects the number of spins, i.e., the metabolite concentration.

Spin The true definition of spin is quantum mechanical. Quantum mechanics can be ignored as long as the interaction of a single spin with an external field is not discussed. For the scope of this chapter we utilize a net magnetization vector which is the sum of the magnetization from a large number of individual spins.

Spin decoupling See *decoupling*

Spin echo A spin echo sequence in its simplest form consists of a 90° *RF pulse* followed after a delay time TE/2 by a 180° pulse. This results in the formation of an echo signal at the *echo time TE*. The echo signal amplitude will be diminished by a factor of exp(-TE/T2).

Spin-spin relaxation See *T2-relaxation*

STEAM Stimulated echo acquisition mode, localization method utilizing three 90° slice selective pulses, along each of the spatial directions. Signal, in form of a *stimulated echo*, from the overlap is generated in a "single shot" experiment. In contrast to *PRESS*, only half of the possible signal is recovered when the same *echo time* is used. However, STEAM allows shorter echo times than PRESS, partially compensating for lower *S/N*. Secondly, the RF bandwidth of 90° pulses is superior to the bandwidth of 180° pulses utilized by PRESS. STEAM is therefore the method of choice when short echo times, minimal *chemical shift artifacts* and robustness is required.

Stimulated echo Three 90° *RF pulses* are needed to generate a stimulated echo. The feature of a stimulated echo is that the magnetization is transiently stored by the second RF pulse along the z-axis and does not undergo *T2-relaxation*. This allows the design of a robust localization sequence with short *echo times* (*TE*). The third pulse flips the magnetization back into the transverse plane. See *STEAM*, PRESS.

Surface coil See *radio frequency coils*

T1-relaxation, T1-relaxation time After the *magnetization* vector has been flipped into the transverse plane, new magnetization builds up along the z-axis. The time after 63% (1–1/e) of the equilibrium magnetization has built up is called the T1-relaxation time. T1 and T2 relaxation is caused by time-dependent fluctuations of local magnetic fields arising mostly from the motion of molecules with electric or magnetic dipoles at the site of the spins. For accurate *absolute quantitation* the relaxation times of all metabolites must be known in order to correct peak intensities appropriately.

T1-saturation The repetition times *TR* are usually in the range of the *T1-relaxation* times. As a consequence of this, not all the *magnetization* has recovered, for example, when TR = T1, only 63% of the equilibrium magnetization can be used for each scan (with the exception of the first scan) when 90° flip angles are used for excitation. This effect is called T1-saturation. The extreme case of saturation occurs when several *RF pulses* are applied within a very short time followed by dephasing *gradients*. This technique is used in localized ^{1}H MRS to remove the dominant water signal (see *water suppression*).

T2-relaxation, T2-relaxation time The *magnetization* vector can be flipped into the transverse plane by using an *RF pulse*. The so generated transverse magnetization undergoes an exponential decay. The time after the magnetization has relaxed to 37% (1/e) of its amplitude is called the T2 or transverse or spin-spin relaxation time. See also *T1-relaxation*.

TE Echo time

TM Time delay between the second and the third 90° *RF pulse* in a *STEAM* sequence. The *echo time TE* and TM are independent parameters in STEAM.

TR (repetition time) In a typical ^{1}H MRS experiment, 128–256 single scans are averaged to improve *S/N*. The time between each initial excitation of the magnetization is called the repetition time (see *T1-saturation*).

Transverse relaxation See *T2-relaxation*

VOI Volume of interest. Localized spectroscopy allows the operator to acquire signals from different regions of the brain, typical volumes are 1–20 ml (1 x 1x1 cm – 2 x 3x3 cm) for proton MRS. In phosphorus MRS significantly larger VOIs are selected to compensate for the lower *S/N*. Localization greatly increases the specificity of MRS. (also known as ROI = region of interest)

Volume coils See *radio frequency coils*

Voxel See *VOI*

Water suppression The ^{1}H spectrum from the brain is dominated by the signal from water since the concentration of brain water is approximately 10,000 times higher (55 Mol/kg) than the concentrations of metabolites (1–10 mMol/kg). Water suppression is essential mainly to obtain a flat baseline because it is in general difficult to adequately remove the unsuppressed water signal from the spectrum in a post-processing step. The most common approach is to selectively saturate (see *T1-saturation*) the water resonance by one or more chemical shift selective (*CHESS*) *RF pulses* prior to the localization sequence. Water suppression is also important in inverse detection MRS (^{1}H-X) methods (see *polarization transfer*) which seek to harness the much greater signal intensity of ^{1}H vs. ^{13}C or ^{15}N, where the dynamic range of signal receivers must cope with the disparity between water and metabolite resonance intensities.

Chapter 41

Monitoring Chemistry of Brain Microenvironment: Biosensors, Microdialysis and Related Techniques

JAN KEHR

General Introduction

Invasive techniques for continuous *in vivo* monitoring of substances involved in *chemical neuronal signaling* and *cellular metabolism* have became an integral part of the experimental armamentarium within the fields of functional *neuroanatomy, neuropsychopharmacology* and *neuropathology* (Boulton et al 1988; Justice 1987; Marsden 1984;

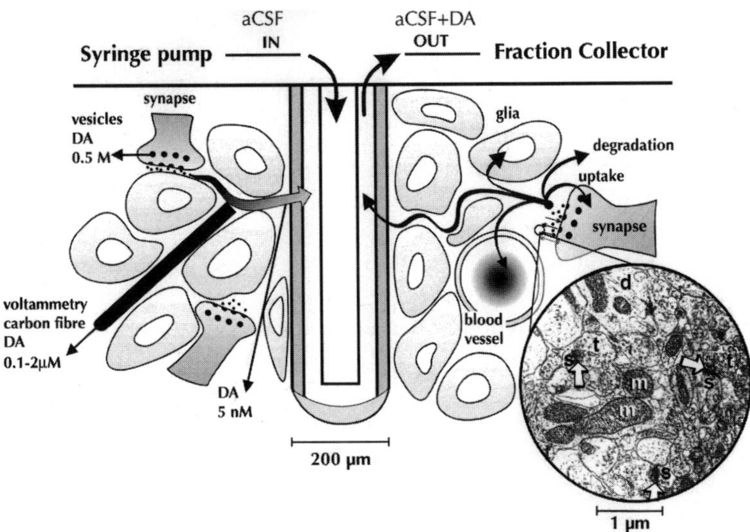

Fig. 1. Schematic drawing of brain microenvironment which serves as a target place for implantation of sensing devices. Molecules trafficking across the cellular interspace can be monitored either *in situ* by biosensors, such as voltammetric electrodes, or sampled by microdialysis probes at regular intervals and collected fractions analyzed by suitable analytical techniques. However, as shown in the figure, only a diminutive amount of vesicularly released transmitter, in this case DA, can manage to diffuse over larger distances from its release site (synaptic cleft) and reach the surface of the sensing/sampling device. The electron microscopic (EM) picture of the rat striatal neuropil illustrates the minimal anatomical size of the synaptic contacts (s – synapses) compared to other cellular structures (d – dendrites, t – axon terminals, m – mitochondria) and the ratio between the volume of the synaptic clefts (arrows) and the overall extracellular space. In reality, the extracellular volume fraction is larger than shown in the EM picture and it is about 20% of the total tissue volume. The concentrations of neurotransmitters in this space are at least 10^6 times lower than those originally released into the synaptic clefts. Thus, both biosensors and microdialysis probes provide only an approximate measure of actual changes in synaptic release and uptake mechanisms for classic neurotransmitters following various behavioral or pharmacological stimuli.

Jan Kehr, Karolinska Institute, Department of Neuroscience, Doktorsringen 12, Stockholm, 17177, Sweden (phone: +46-8-728-7084; fax: +46-8-302875; e-mail: Jan.Kehr@neuro.ki.se)

Robinson, Justice 1991). A target place for most of the *in vivo* sensing devices is the extracellular space filled with the fluid which under normal circumstances comprises about 20% of the total brain tissue volume. It is suggested that the ionic composition of the extracellular fluid (ECF) is the same as the concentration of ions in the cerebrospinal fluid, due to the absence of tight junctions between ependymal and pial cells. However, the ECF consists also of a number of long-chain *glycosaminoglycans*, *proteoglycans* and *glycoproteins* tethered to membranes, and generally, it contains higher concentrations of nutrients transported from the blood capillaries to the cells and metabolic molecules moving in the opposite direction. *Neurotransmitters* and *neuromodulators* released from the nerve terminals as well as *neurotrophic factors* and other *cytokines* all have to traverse the extracellular space on their way to the target receptors. Recently, the concept of synaptic transmission was broadened under the term *volume transmission* incorporating these inter-cellular communications over the long distances (Fuxe, Agnati 1991). It is believed that there is a slow movement of the ECF towards cortical subarachnoid space, ventricles and the perivascular Virchow-Robin space. For molecules such as neurotransmitters, the most dominant process is diffusion within the ECF compartment, due to a steep gradient of concentrations being up to 6–7 orders of magnitude higher at the release sites (a synaptic cleft) than in the outer interstitial space. For example, it was estimated that striatal dopamine (DA) vesicles with a radius of 25 nm contain 25 mM dopamine, while the concentration of dopamine in the synaptic cleft, which is about 15 nm wide and 300 nm long, was around 1.6 mM (Garris et al 1994). However, the actual extracellular dopamine concentrations measured by *in vivo voltammetry* using a thin (5 μm O.D.) carbon fiber electrode was not higher than 20 nM in pargyline-treated rats (Gonon, Buda 1985) or 0.1–2 μM during electrical stimulation (Kawagoe et al 1992). Using the devices of even larger diameter such as a 200 μm O.D. *microdialysis* probe, the calculated basal extracellular dopamine concentration was as low as 5 nM (Parsons, Justice 1992). Thus, all the implantable sensors described in this chapter can measure only an "echo" of the actual synaptic event, providing that the overflow of the detected substance is higher than its reuptake, sequestration, enzymatic or other inactivation mechanisms. This situation is schematically depicted in Fig. 1 illustrating the distance and the possible diffusion routes of released neurotransmitters through the ECF compartment and, to the surface of a sampling or sensing device. The spatial and temporal resolution of the detecting/sampling probes is dependent both on their geometry and the detection principle. The response time can be as low as 0.1–1 s for directly detecting voltammetric electrodes, while a relatively laborious and time-consuming analysis of neurotransmitters or neuropeptides sampled by microdialysis requires the fractions to be collected in 5–30 min intervals. Thus, microdialysis is most suitable for studies of those events where a typical time-course lasts for at least 10 min, the latter criterion being fulfilled for most profiles of the psychoactive drugs, neuropathological models of brain diseases and certain behaviors such as eating, drinking, motor activity etc.

Principally, the invasive techniques for *in vivo* monitoring of brain chemistry can be divided into two main groups: 1) intracorporeal biosensors and 2) continuous sampling devices. The sensing devices and related detection techniques within the first group are highly selective towards an analyte which is detected directly at the surface of the biosensor implanted into the brain tissue. In the simplest case, the selectivity for a detected endogenous substance is assured by its intrinsic chemical properties such as a relatively specific oxidation potential for dopamine. A more common way is to utilize the highly specific molecular interactions such as those between ions-ionophores, enzymes-substrates and antibodies-antigens, which all lead, either directly or via an intermediate product, to the changes in physical (electrical or optical) signals.

Implantable sampling devices allow continuous removal of samples (superfusates, dialysates) which mirror chemical composition of the ECF under physiological conditions in conscious freely moving animals. The sampling device, for example, a microdialysis probe, is continuously perfused at low flow-rate (0.1–2 µl/min) with physiological solution, typically a Ringer solution or artificial cerebrospinal fluid (aCSF). Samples are collected at regular intervals and analyzed by suitable analytical techniques. The implantation of the sensing device or its guide cannula occurs in anesthetized animals and requires a good knowledge of brain anatomy, stereotaxy and microsurgical techniques.

Part 1: General Methods

Stereotaxic Surgery on Small Rodents

Introduction

Implantation of microdialysis probes, electrodes, guide or injection cannulae into the brain requires their precise positioning into the brain loci by use of a stereotaxic instrument. Coordinates for a given location can be found in a number of brain stereotaxic atlases, depending on the species, for example, for rat : Paxinos and Watson (1982); mouse: Franklin and Paxinos (1997); guinea pig: Luparello (1967). Two suture crosses (lambda and bregma) of the sagittal and coronal sutures of the skull bones are most often taken as the main landmarks for the system of brain coordinates. A typical atlas for a rat (Paxinos and Watson 1982) or a mouse (Franklin and Paxinos 1997) brain uses bregma as a zero point and positioning of the brain in such a way that bregma and lambda will lie in the horizontal plane. The animal´s head is fixed into a frame of the stereotaxic instrument by placing the ear bars into the bony ear canal (the interaural line) and by attaching the incisor bar at the bottom edge of the upper jaw just behind the first incisors. The animals must be deeply anesthetized during the stereotaxic surgery. The following protocol describes the stereotaxic procedure for a rat.

Materials

Instrumentation
A surgical theater for a rat includes a

- Stereotaxic instrument equipped with one or two micropositioning arms (David Kopf),
- Gas anesthetic (halothane, enflurane or similar) with a mixing chamber, N_20 and O_2 cylinders with flowmeters,
- A CMA/150 Temperature Controller (CMA/Microdialysis),
- A drill and
- A stereomicroscope.

If the experiments are conducted acutely, on anesthetized animals, the sensors must be connected to the respective measuring apparatus, whereas

- CMA/11 or CMA/12 microdialysis probes must be continuously perfused by use of a
- CMA/100 Microinjection pump.

Optional devices, such as

- A liquid switch (CMA/110 or CMA/111) and
- A CMA/170 Refrigerated Fraction Collector

could be included within the system.

Perfusion fluid
Ringer solution (147 mM Na^+, 2.4 mM Ca^{2+}, 4 mM K^+, 155.6 mM Cl^-) is prepared by dissolving: 8.591 g NaCl, 298.2 mg KCl and 228.1 mg $CaCl_2.2H_2O$ in 1000 ml deionized water, final pH is around 6–6.5. In some experiments on awake rats, the use of artificial cere-

brospinal fluid (aCSF, for example: 148 mM NaCl, 1.4 mM $CaCl_2$, 4 mM KCl, 0.8 mM $MgCl_2$, 1.2 mM Na_2HPO_4, 0.3 mM NaH_2PO_4, final pH 7.4) is preferred.

Procedure and Results

1. Gently pick up the rat (270–300 g body weight) from its cage and place it into the plexi-glass tube connected to the anesthetic gas (5% enflurane in $N_2O:O_2$, flow rate 0.8 l/min each). Close the tube and wait for some 2–3 min until the animal has lost its consciousness.
2. Remove the rat from the tube and place it on the heating pad. Put the rectal thermometer under the rat and set the temperature controller to 37.5 °C. Reduce the concentration of enflurane to 3% and connect the gas to the tubing with a face mask. Place the rat´s upper jaw to the incisor bar which is not fastened to the frame and can move freely in the horizontal plane. Put the rat nose into the face mask.
3. Check the face mask is properly placed over the nose, open the animal's mouth and pull out the tongue with tweezers. Optionally, fix the face mask with a piece of plastic tape, but do not tighten. Inject a small amount of atropine (0.08–1.5 mg/kg) to suppress salivation and to reduce the possibility of the animal's aspirating saliva, especially during the long-lasting surgery. To protect the cornea of the eyes during anesthesia, place a drop of mineral oil in the eyes or apply ophthalmic ointment. Check the depth of anesthesia by pinching the rat's paw with a pair of tweezers.
4. Fasten the screw fixing the ear bar to the frame. Hold the animal's head with one hand from a side, while trying to find the bony ear canal by moving the head up and down. This step requires some training and experience. When you feel that the skull sits steadily, hold the head slightly pressed towards the ear bar, while trying to fit the opposite bar to the second ear canal as shown in Fig. 2A. Once it is in the proper position, the head is fixed horizontally with the interaural line and it can be moved a bit up and down around this axis.
5. Tighten the incisor bar which should be some 3.5 mm beneath the interaural line. Now, the animal's head should be fixed and should not move either vertically or horizontally. Position the nose clamp over the nose.
6. Clip the hair from the top of the head with electric clippers or scissors. Clean the exposed skin with 70% ethanol. Hold the skin tightly on the sides of the head and make a middle scalp incision of some 2–3 cm. Try to cut down the skin with the first stroke. Keep the flaps aside by clamping homeostatic forceps. Expose the bare skull by scraping the periosteum so you can see the skull suture lines. On the properly cleaned and dried skull the coronal and sagittal sutures, as well as bregma and lambda, should be easily identified.
7. Mount the sensor needle or the probe to the holder of the vertical arm. Using a stereomicroscope, place the tip of the needle just above bregma and read the coordinates giving a zero point for the horizontal (anteroposterior, AP; and lateral, L)

→

Fig. 2. Stereotaxic surgery on a rat. (A) The ear bars should be placed in the bony ear canals, and the incisor bar together with a nose clamp should fix the animal´s head in the stereotaxic frame. (B) The suture lines on the exposed skull serve as the landmarks for stereotaxic coordinates: lambda (lower suture T-cross) and bregma (upper suture cross) should be in the same horizontal line according to Paxinos and Watson´s atlas (1982). A hole in the skull is drilled with a trephine drill for implantation of the probe into the lateral striatum. (C) For probes to be implanted chronically two additional holes are drilled and small screws are fixed into the skull for better attachment of the dental cement. (D) An example of an acutely implanted microdialysis probe for experiments on anesthetized rats. (E) Implantation of a guide cannula with a dummy (blue plastic body). (F) Application of cold hardening dental cement on a dried surface of the skull and around the guide cannula. (G) Schematic drawing illustrating brain microdialysis experiment on anesthetized rat. Besides a stereotaxic apparatus, typical equipment necessary for this experiment consists of a precision syringe pump, liquid switch, fraction collector and a controller of rat body temperature.

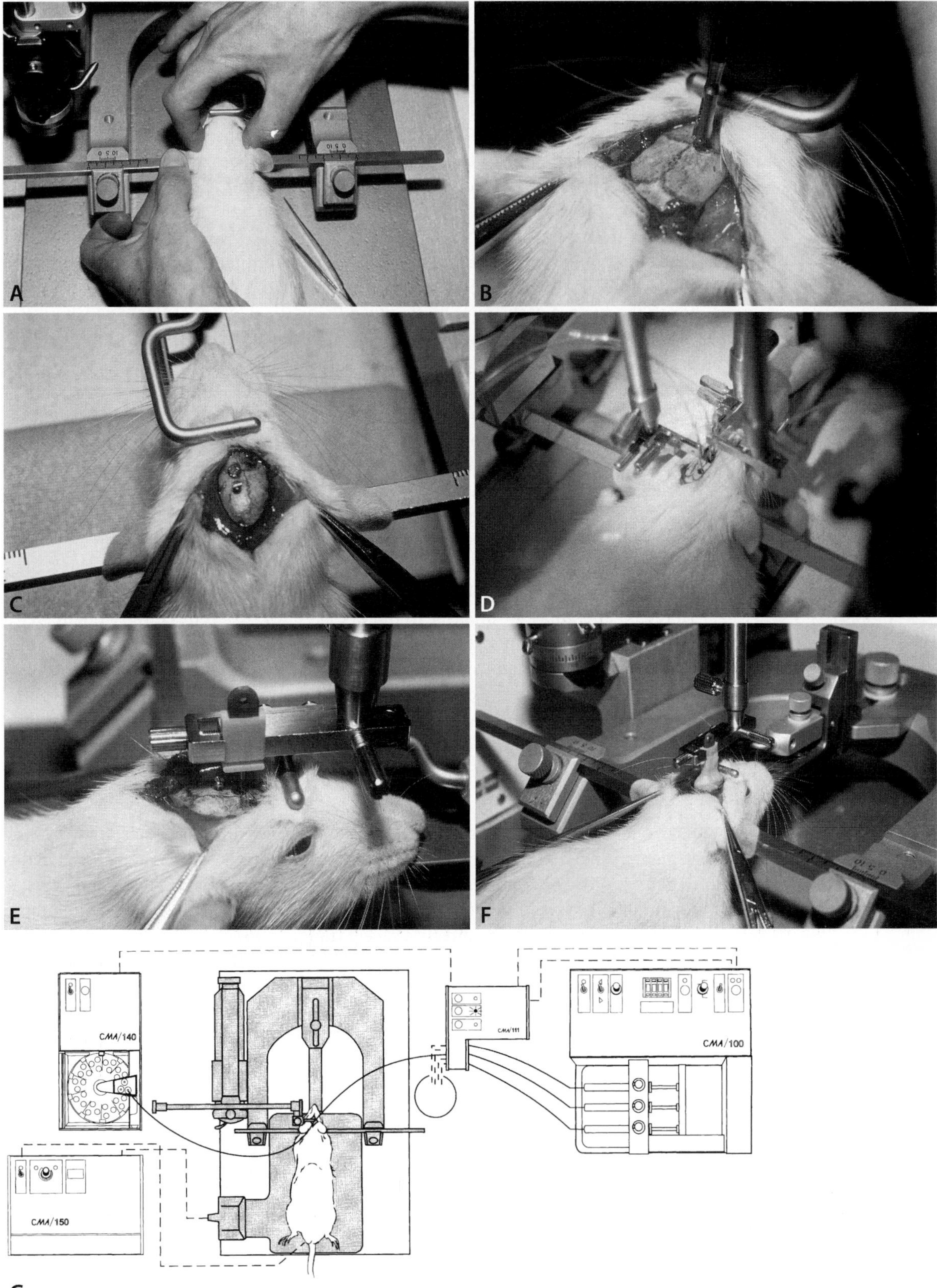
A
B
C
D
E
F
CMA/140
CMA/111
CMA/100
CMA/150
G

plane on the micrometric screws. Place the needle just above the lambda – the AP and L coordinates must be the same. If not, adjust the teeth bar until bregma and lambda are at the same horizontal level. Add the coordinates for a given brain structure and move the probe to this point, for example, the coordinates for lateral striatum are A-P +1.3mm and L 2.2mm. Mark this point and move the arm aside.

8. Drill a hole in the calculated position with a fine trephine drill (Fig.2B). For experiments on awake animals, the CMA/11 or CMA/12 probes or their respective guide cannulae must be cemented to the skull. Drill two additional small holes for the attachment screws some 2–3mm around the insertion hole. Fix the screws and clean the hole from any bone debris (Fig.2C).
9. Place the cannula/probe at the same A-P and L coordinates above the drilled hole; it should be in the middle of the opening. Move the probe vertically until it will touch the surface of the dura mater. Read the vertical (V) coordinate which gives a zero point. Pull up the arm so you can easily access the dura. Cut the dura with microscissors or with a needle. Dry the outflowing CSF with a cotton web.
10. Place the cannula/probe again to the A-P, L and zero V coordinates. Slowly insert the cannula/probe into the brain at a depth calculated from the brain atlas, for example, for lateral striatum it is –6.2mm. An example of implanted microdialysis probe into the striatum of the anesthetized rat is shown in Fig.2D. Experiments on such acute preparations can start within 2–3 hours. When implanting only a guide cannula, the depth must be shortened by the length of the microdialysis membrane or sensor, for example, by 2.5mm for CMA/12 probe with 2mm membrane + 0.5mm glue tip (Fig.2E).
11. Prepare a small amount of cold hardening acrylic dental cement (Dentalon Plus, Heraeus) and apply a thin layer to the dry skull. In two to three subsequent steps, apply the small amounts of the cement around the screws and the probe (Fig.2F).
12. Suture the skin with silk suture by simple interrupted sutures placed about 3mm apart. Place a soft plastic collar around the animal´s body just behind the forepaws. Remove the animal from the stereotaxic instrument; place it in the cage. Control the body temperature with an infrared lamp until the animal regains its consciousness. Give the animal 2ml saline s.c. and some drops in the eyes and the mouth.

A proper placement of the anesthetized animal in the stereotaxic instrument and all the necessary equipment for running a microdialysis experiment are shown in Fig.2G. Many experiments, especially those involving implantation of fragile voltammetric carbon fibers or glass capillary-based ion-selective electrodes, are run under these conditions. A limited number of simple microdialysis experiments can be performed on acute preparations as well. Anesthesia, but mostly a severe trauma caused by the implantation of relatively thick (thinnest membrane available is of 0.2mm O.D.) probes alter tissue homeostasis to such a degree that the technique may no longer reflect the "normal" extracellular concentrations of endogenous compounds. Thus, for microdialysis experiments the use of awake animals with chronically implanted guide cannulae is recommended.

Troubleshooting

- The most complicated part of the stereotaxic surgery is the proper fixation of the skull in the stereotaxic frame. It is advisable to start training on the prepared rat skull or on the dead animal where the skull bones can be easily exposed.
- Do not use the sharp ear bars and high pressure, especially when implanting devices for later work on awake animals. They can cause bleeding and damage to the bones.
- When using a trephine drill, always use a stereomicroscope and be careful the drill does not suddenly penetrate through the bone and damage the cortex.

- If there is much bleeding from the skin flaps, it could be due to the insufficient anesthesia. Increase enflurane to 5% for a while.

Comments

The proper position of any stereotaxically implanted device into the desired brain structure should be checked post mortem and compared to the coordinates of the brain atlas. After sacrificing the animals, the brains should be removed and stored frozen until the neuroanatomical examination. Usually, 10–50 μm thick brain slices should be cut using a cryotome. The implanted cannula leaves, after its removal, a track in the tissue, which is clearly visible even on frozen slices. For additional inspection the slices can be stained by some rapid procedure, for example, by thionin staining.

Applications

Stereotaxic surgery on small rodents is one of the essential techniques needed for studies of brain function in living animals. Besides the stereotaxic implantation of devices for monitoring (biosensors) and sampling (microdialysis probes), the technique is frequently used for *microinjections* of compounds which do not penetrate the blood-brain barrier. Microinjections of drugs, toxins, large molecules etc. can be done directly into the brain parenchyma (0.1–0.5 μl) or into the ventricles (1–10 μl).

Microdialysis Experiments on Awake Rats

Introduction

General anesthetics cause severe disturbances in chemical neurotransmission and cell metabolism. Thus, in many cases, it is desirable to conduct the experiments on conscious animals. For correlative studies of behavior and transmitter release, the post-operative recovery period should be at least one week. At that time the tissue reactions such as gliosis and activation of microglia may alter or completely block the diffusion of neuron-derived substances to the implanted sensors or probes. To avoid this problem, the experiments are often run within the first three days after the implantation or the probes are implanted on awake animals by use of the guide cannulae. Biosensors are used predominantly on anesthetized animals, whereas most of the recent papers on microdialysis report the data obtained on awake rats. The stereotaxic implantation of the guide cannula for a microdialysis probe is described in the previous protocol.

Materials

Instrumentation

- A CMA/120 system for microdialysis on awake animals consists of a round cage and a counter-balance arm with a dual-channel swivel (TSC-23, BASJ, Japan).
- A CMA/100 Microinjection pump and
- A CMA/170 Refrigerated Fraction Collector are used for constant flow delivery and automated collection of samples.

Additional microdialysis accessories are:

- 1 ml gas-tight syringes,
- A CMA/111 syringe selector,
- CMA/11 or 12 microdialysis probes,
- FEP tubing (0.1 mm I.D.) with tubing adapters and
- 300 μl vials with caps (Chromacol Ltd., UK) for collecting fractions.

Perfusion fluid

Ringer and aCSF solutions were prepared as described above.

Procedure and Results

1. A day before the experiment, place the rat into the cage which is provided with a water bottle and food pellets. Connect the metal wire from the swivel to the collar of the rat.
2. Fill the syringes with Ringer solution (for a composition, see previous protocol) and Ringer containing the test drug, and mount them to the CMA/100 Microinjection pump. Use the FEP tubing and PUR alcohol-soaked adapters to connect the syringes to the CMA/111 syringe selector, the swivel, the CMA/11 (or 12) microdialysis probe and to the CMA/170 Refrigerated Fraction Collector operating at +4 °C.
3. The probe should be sufficiently washed with Ringer solution and it should not contain any trapped air bubbles. Reduce the flow rate from initial 5–10 μl/min to 0.5 μl/min.
4. Pick up the animal from its cage and hold it firmly against your body (Fig. 3A). Remove the dummy from the guide cannula (Fig. 3B) and insert the microdialysis probe (Fig. 3C,D). Place the animal back into the cage. With some pieces of tape fix the inlet and outlet tubing to the metal wire. Be sure the animal cannot reach the tubing and damage the connections during its active period (Fig. 3E,F).
5. Perfuse the probe at low flow-rate of 0.5 μl/min overnight.
6. On the day of experiment, place the newly filled syringes into the pump and start perfusion at a higher flow-rate (typically between 1 or 2 μl/min).
7. Install the program of the fraction collector, activate cooling and select the number of fractions and sampling time (volume).
8. Following additional 60–90 min of stabilization, the experiment can be started by collecting some first 6 fractions (at 10 or 20 min intervals) for estimation of basal levels.
9. A very simple test to provoke transmitter release is to perfuse with a Ringer solution containing 100 mM K^+ at the same osmolarity, i.e., only 51 mM Na^+, 100 mM K^+, 2.4 mM Ca^{2+} and 155.6 mM Cl^-. After collecting the basal samples, switch the syringe to one containing Ringer solution at 100 mM K^+ for a duration of one fraction and then switch back to the original Ringer solution. Continue to collect samples for an additional 2 hours.
10. After the experiment is finished, reduce the flow to 0.5 μl/min and let the animal recover overnight. The following day, the experiment (points 6–9) can be repeated.

The procedure of implanting the microdialysis probe in awake animals is depicted in Fig. 2A-F. During the whole procedure the animal is held steadily towards the operator´s body while removing the dummy and inserting the microdialysis probe into the guide cannula. As seen in the photographs, in a properly operated and fully recovered animal, the insertion of the probe should not cause any stress or pain. Once the animal is connected to the swivel and placed into the round cage (Fig. 2E,F) a certain period of time is needed to reach the steady state of recovered substances. Usually, at about 12 hours, perfusion at low flow-rate is satisfactory to equilibrate the levels of amino acid neurotransmitters, 5-HT, catecholamines, adenosine and others.

Troubleshooting

- A most common problem with any chronically implanted brain device is that the whole assembly is not properly fixed to the skull, gets loose and falls apart. The reasons for this could be: a) lack of enough dry surface of the skull before the application of the dental cement, b) wrong type or quality of the dental cement, c) too loose fixation screws, too big holes drilled, d) wrong type of cage during the recovery period, where the cemented assembly on the animal´s head could fasten in the grid or sharp horns.
- The rats connected to the perfusion system usually sleep during the day or show minimal motor activity. However, during the night, they often manage to reach the tubing and damage it. If it does not help to fix the tubing to the metal wire with a tape, use a metal theter to protect the tubing and to connect the collar to the swivel.

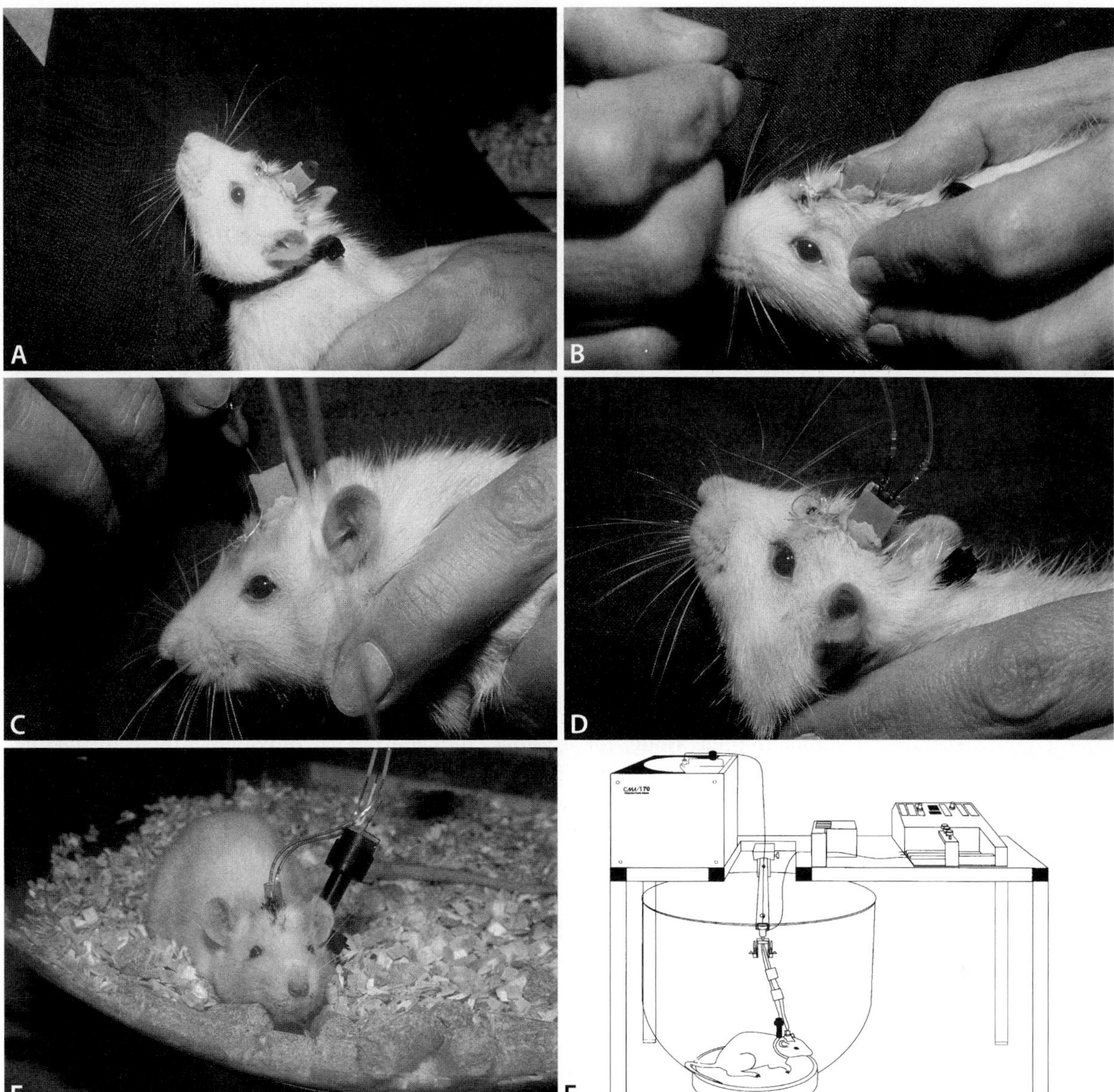

Fig. 3. Implanting the microdialysis probe through a guide cannula into an awake animal. (A) Pick up the animal; it should be used (trained) to handling in order to avoid stress. (B) Remove the dummy from the guide and (C) insert the microdialysis probe (CMA/12). (D) The probe is already connected to the pump and fraction collector and it is continuously perfused with aCSF or Ringer solution. (E) Place the rat into the cage. Connect a metal wire from the swivel to the neckband. (F) A typical setup for a microdialysis experiment on a freely moving animal: the inlet and outlet tubing of the probe is connected via the swivel in the balance arm to the pump and the switch on the inlet side and the refrigerated fraction collector on the probe outlet.

- Sometimes one may observe that the sampling volume is lower than theoretically calculated. The reason for this could be a leakage somewhere in the perfusion system. Most often it is a leaking syringe or a swivel. The piston seal of the syringe can be replaced; the swivel should be replaced with a new one.
- The membrane of the probe can get damaged if inserted improperly. Often the animal´s behavior is radically changed due to a developing lesion from the leaking probe. Stop the perfusion immediately and exclude the animal from the experiment.

Comments

Microdialysis on chronically implanted rats can be run for up to four days after the probe implantation as demonstrated for neurotransmitters dopamine (DA), GABA and acetylcholine (ACh) (Osborne et al 1991). Earlier, it was demonstrated that one week after the probe implantation, both basal and stimulated dopamine release were significantly reduced and showed slower kinetics as a consequence of tissue gliosis (Westerink, Tuinte 1986). Histochemical techniques revealed that severe gliosis around the implanted device takes place at about four days after the surgery (Hamberger et al 1985, Benveniste, Diemer 1987, Benveniste et al 1987) and the gliotic reaction is dependent on the size of the implanted probe (Zini et al 1990). However, a complete recovery of physiological functions occurs at the earliest at 5–7 days after the surgery as revealed by telemetry (Drijfhout et al 1995). Another interesting question is the possibility of running within-group experiments by repeated insertions of a microdialysis probe into the same animal. It was shown that this approach could be applied to repeated measurements of acetylcholine (Moore et al 1995) and dopamine release, but with no particular advantage over the chronically implanted probes (Robinson and Camp 1991).

Applications

In vivo microdialysis on awake animals offers a *combination of neurochemical (transmitter release) and functional (behavior) analysis* (Zetterström et al 1986; Young 1993; Ögren et al 1996). A very useful way to study functional neuroanatomy and the role of specific brain circuits in various behaviors is a so-called *dual-probe approach*. Here, one probe is implanted at the cell body level, whereas the second probe is implanted in the terminal area. The first probe is used for chemical stimulation by infusing the drugs while the second probe is used to measure the neurotransmitter release (for review, see Westerink et al 1998). Similarly, the release can be provoked electrically by implanting a stimulating electrode, for example, into the medial forebrain bundle to measure dopamine release in striatum (Imperato and DiChiara 1984) or into the neurons of substantia nigra reticulata to measure release of GABA in the ventromedial thalamus (Timmerman and Westerink 1997).

Today, *all known neurotransmitters and neuromodulators* can be recovered by microdialysis sampling. Further, a number of large molecules such as *peptides* and even *certain proteins and enzymes* can diffuse through some microdialysis membranes with higher molecular cut-off. *Distribution and pharmacokinetics of drugs* can be studied by applying various models of quantitative microdialysis in order to estimate absolute extracellular concentrations of the drugs and their metabolites (for review, see Kehr 1993; Elmquist and Sawchuk 1997).

Part 2: Implantable Sensors

Depending on the physical detection principle, the intracorporeal sensors can be divided into
- Amperometric,
- Potentiometric and
- Optical.

Another distinction can be made on the basis of chemistry involved in the process of sensing:
- Electrodes measuring directly the concentrations of endogenous ions or
- Electrodes measuring the concentrations of those endogenous ions produced by electrochemical reactions on the electrode surface,
- Optodes measuring the changes of light introduced externally (reflectance, absorbance, fluorescence) or

- Optodes measuring the changes of light introduced by (bio)chemical means (luminescence).

Biosensors are electrodes or optodes utilizing a certain selective biochemical (enzymatic) or immunochemical (antibodies) reaction.

Potentiometric Electrodes

A typical potentiometric sensor is a *pH (micro)electrode* selective to H^+ ions, followed by ion-selective microelectrodes (ISM) for ions such as K^+, Na^+, Ca^{2+}, Cl^-, CO_2, NO (for review, see Ammann 1986). The ISM capillary is filled with the electrolyte containing a liquid ion-exchanger (ionophore) forming a liquid-membrane permeable only for a selected ion. If the electrode is introduced into the solution or the nervous tissue, a concentration gradient is created for a selected ion across the membrane, generating a potential difference which is logarithmically proportional to the concentration of a given ion and can be measured against a suitable reference electrode. The final electromotive force (EMF) of such electrochemical cells measured in mV is described by the Nernst equation, modified by Nicolsky and Eisenman, 1967:

$$\mathrm{EMF} = \mathrm{E}_0 + \mathrm{s}\,\log\left[\mathrm{a}_\mathrm{i} + \Sigma \mathrm{K}_{\mathrm{ij}}^{\mathrm{pot}} \left(\mathrm{a}_\mathrm{j}\right)^{\mathrm{zi/zj}}\right]$$

where E_o is a constant potential difference which includes interfacial potentials, the slope s is defined as 2.303 (RT/z_i F) where R is the universal gas constant, T absolute temperature, F is Faraday´s constant, z_i the charge number and a_i the concentration (activity) of the ion being measured. The sum of interfering ions modulates the electrode response depending on their concentrations a_j and potentiometric selectivity coefficients $K_{ij}{}^{po}$. At 25 °C and with no interferences present, the electrode gives a potential change of 59.16 or 29.58 mV for each concentration step (order of magnitude) for monovalent or divalent ions, respectively.

Measuring Extracellular K^+ Ions by ISM

Introduction

A typical modern ISM consists of a double-barreled glass microelectrode with a liquid-membrane/ionophore in one tip while the second tip is filled with isotonic NaCl working as a conventional potential-sensing electrode. A third electrode is a nonpolarizable Ag/AgCl reference electrode. A number of ready-made ionophore cocktails are available from Fluka, Switzerland, for example, a *valinomycin-based ionophore I, cocktail B* (Fluka 60398) *or high-impedance cocktail A* (Fluka 60031) for measuring extracellular K^+ ions. Today an earlier type of liquid ion exchanger (Corning 477317) for potassium ions is also used to sense tetramethylammonium ions in diffusion studies (for review, see Nicholson and Syková 1998).

Materials

- A differential amplifier with high-input impedance and low bias current such as FD223 from World Precision Instruments (WPI) or GeneClamp 500 from Axon Instruments should be used for measurements of ion potentials.
- The signals are recorded on a chart recorder or using a computer program TIDA (Heka).
- The double-barreled glass capillaries and a programmable multi-barrel pipette puller PMP-100 are available from WPI.
- Ionophore cocktails, trimethylchlorsilane and all other chemicals are available from Fluka or Sigma.

Procedure

1. Clean the glass tubes in concentrated chromic acid, in an ultrasound bath or similar and wash them carefully with distilled water.
2. Prepare a double-barreled microelectrode by pulling a ready-made two-barrel borosilicate glass tubing in a micropipette puller. Double-barreled microelectrodes can also be prepared from two single glass capillaries, by heating them over a Bunsen burner and twisting them in the middle by 360° before the final pulling in the puller.
3. One barrel of the electrode, which should be filled with the ion-exchanger, should be silanized by vapors of a silanization reagent, for example, trimethylchlorsilane or dimethylaminotrimethylsilane. Place the electrode in the oven at 250 °C. Connect the one barrel tightly to a small reservoir containing the silane compound using a piece of plastic tubing. Connect a nitrogen gas cylinder to the reservoir and blow nitrogen at a pressure of 2 kg/cm^2. The nitrogen stream serves as a carrier of silane vapors to the glass capillary. Connect the second barrel (a reference electrode) to the nitrogen stream only.
4. Fill the tip of the silanized barrel with the ion-exchanger by back-filling technique. Using micromanipulators, a stereomicroscope and a thin capillary pulled from thermoplastic tubing, dispense a small droplet of the ion-exchanger to the tip of the glass barrel. The liquid exchanger will be drawn into the capillary and form a plug a few millimeters thick. Fill the resting volume of the barrel with an internal filling solution (for example, 0.5 M KCl). Fill the other barrel with 150 mM NaCl.
5. Inspect the barrel under the microscope and remove any air bubbles with a fused silica fiber or capillary.
6. Insert a chlorinated silver wire into each barrel and close both ends with bee wax or with a hot-melt adhesive, for example, PolyFil from WPI. Connect the electrodes to the differential amplifier.
7. Anesthetize the rat with pentobarbital (50 mg/kg i.p.) or even better, with halothane or enflurane. Fix the rat skull into the stereotaxic frame and drill the holes for the probe/electrode implantation as described in the protocol 1.1.
8. Insert the double-barreled ISM into the brain and place a reference Ag/AgCl electrode (Dri-Ref from WPI) on the surface of the cortex.
9. Following a stabilization period and after recording the basal EMF for K^+ ions, induce a complete ischemia by intraperitoneal injection of 1 ml saturated $MgCl_2$ which will result in cardiac arrest within 1–2 min.

Results

Figure 4 shows the simultaneous increase in the concentration of K^+ ions and anoxic depolarization (DC potential) following the cardiac arrest in rat pups at two postnatal periods (P10 and P22). The DC potentials were measured between the nonpolarizable reference electrode and a potential-sensing barrel electrode. As seen in the figure, the increase in extracellular potassium occurs in several phases. It was hypothesized that the first significant change of 6–12 mM K^+ ions, accompanied by a negative DC shift of 4–8 mV and a slight decrease in extracellular volume, is predominantly due to neuronal cell swelling. The subsequent dramatic increase of extracellular K^+ by 50–70 mM and DC shift by 20–25 mV is caused preferentially by depolarization of glial cells (Vorísek and Syková 1998). The average increase in extracellular K^+ following the cardiac arrest-induced brain ischemia was about 60 mM and was not significantly different between the two pup groups. However, the calculated time courses show faster kinetics of brain swelling in older rat brains, as revealed by measuring the diffusion kinetics of micro-injected tetraethylammonium ions and registering K^+ ions or DC potentials (Vorisek and Syková 1998).

Troubleshooting

- The stability and the low noise of the ISM are primarily dependent on the quality of the filling procedure with the ion-exchanger and the internal salt solution. A Corning 477413 liquid ion-exchanger has a lower resistance (about 10^9–10^{10} Ω) than valino-

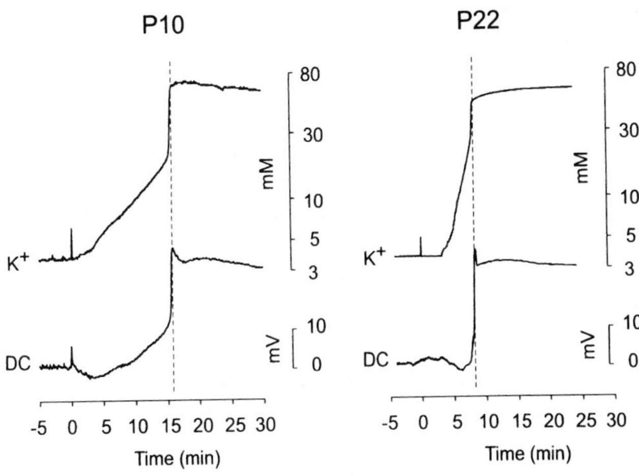

Fig. 4. Simultaneous recording of extracellular K^+ ions and anoxic depolarization (DC) in layer V of a rat at P10 and P22. Both K^+ and DC recordings are from one K^+ ion-selective microelectrode. Note that the peaks in K^+ and in DC curves are reached at the same time, the time course is shorter at P22 than at P10, and after the negative peak, there is a delayed slow DC outlasting the rise in K^+ ions.

mycin but interferes with other cations such as acetylcholine and other ammonium ions. The loss of electrode sensitivity is often caused by a leakage of the ion exchanger out of the micropipette. The reason can be the improperly pulled or silanized capillary.

- Measurement of the ISM's resistance is a very good indicator of the status of electrode filling, storage and lifetime. Basically, any ohmmeter can be used for this purpose; some special devices such as Ωmega-Tip-Z (from WPI) are constructed for resistance measuring of electrolyte-filled microelectrodes.
- Drift and noise of a typical ISM are in the range of 0.3 mV/h and 0.2 mV, respectively. The most common reasons for electrode failure are mechanical impurities, such as dust or air bubbles, therefore work in a dust-free environment is strongly recommended. All solutions should be filtered through a 0.2 μm membrane filter and air should be removed by vacuum or sonication.

Comments

Before the use in animals, each electrode type should be calibrated in solutions of aCSF (see protocol 1.1. for details) containing various concentrations of respective ions. The calibration curve for K^+ in the presence of Na^+ ions is not linear in the range of 1–5 mM K^+, which means that a correction factor should be estimated for calculations of basal extracellular K^+ concentrations. For example, the basal extracellular K^+ levels in the mammalian cortex were shown to vary between 2.8 and 3.4 mM (for review, see Syková 1983). The deviation from linearity at this concentration of K^+ ions in the presence of 150 mM NaCl was shown to be about 10 mV (Kriz et al 1975), which means an error of about 17 % for *in vivo* estimation of basal extracellular K^+ concentrations.

Applications

All changes in transmembrane ionic gradients are accompanied by changes in *potassium ion homeostasis*. Thus, measuring of K^+ efflux by ISMs provides an important measure of *neuronal and glial cell activity* under various physiological and pathological conditions (for review, see Syková 1983). Recently, the application of ISMs was extended to studies of the physiology of the extracellular space, namely, the *estimation of extracellular volume fraction and tortuosity factors* in white and gray matter, during development, aging and experimental brain damage (for review, see Syková 1992; Syková 1997; Nicholson and Syková 1998).

Amperometric Electrodes

The amperometric biosensors measuring the electrical current provide wider dynamic range, linear response to concentration changes and are applicable to larger number of analytes than their potentiometric counterparts. The original idea of using voltammetry for *in vivo* detection of biogenic amines was pioneered by Adams and coworkers in the early 1970s (Kissinger et al 1973; for review, see Adams 1990). Carbon fibers proved to be an excellent material for the construction of the voltammetric electrodes (Gonon et al 1978). Typically, the fiber is inserted and sealed into the pulled glass capillary. The fiber is either cut at the edge of the capillary (microdisc electrode) or it protrudes from the capillary by about 0.5 mm (cylindrical electrode), the latter approach being preferable due to reduced tissue damage and better reproducibility. Catechol and hydroxyindole compounds can be easily oxidized at low potentials on carbon electrodes (+0.15 to +0.25 V and +0.3 to +0.4 V vs. Ag/AgCl reference electrode) according to the following reactions:

HO, HO — R → O, O — R $+ 2e^{-} + 2H^{+}$

for dopamine (DA), noradrenaline (NA) and DOPAC,

HO — R, N, H → O — R, N $+ 2e^{-} + 2H^{+}$

for 5-hydroxy-tryptophan (serotonin: 5-HT), 5-HIAA and 5-hydroxy-tryptophol (5-HTP). Besides acidic metabolites DOPAC and 5-HIAA, there is only a small number of possible interfering molecules, of which the most critical are ascorbic and uric acids. Unfortunately, their concentrations in the ECF are usually several hundred times higher than that of monoamines DA (or NA) and 5-HT. To improve selectivity and to prevent the diffusion of these acidic substances to the electrode surface, the carbon fibers are electrochemically pretreated (Gonon et al 1980) and/or coated with a film of a liquid cation-exchanger Nafion (Gerhardt et al 1984a). For selective detection of neurotransmitters DA or NA and 5-HT the following voltammetric techniques have mainly been used:

- fast cyclic voltammetry (Armstrong James et al 1980),
- chronoamperometry (Hefti and Melamed 1981) and
- differential pulse voltammetry (Gonon et al 1980).

In all three methods, the applied voltage is not maintained at a constant value, but it sweeps over a given potential interval in a waveform. Thus, fast cyclic voltammetry is performed by repetitive scans at rates of 300 V/s within a potential window from –0.4 to +0.8/1.0 V (Ewing et al 1983; Millar, Barnett 1988). The first part of a scan (anodic curve) leads to oxidation of monoamines to respective quinones (as depicted in the reaction schemes above), resulting in a maximal current of about +0.6 V for DA. The second, negative (cathodic curve) sweeping scan reverses the reactions, reducing the quinones back to the original catechols or hydroxyindoles. The final cyclic voltammogram is a measure of specificity, while the current peak is a measure of the concentration of the compound in the surrounding medium.

In chronoamperometry the voltage within a potential window of anodic and cathodic peaks is increased for a fixed time of 50 ms to 1 s. The current rises in a sharp spike which decays to steady (residual current) level. This time-dependent current profile is directly proportional to the concentration of the electrolyzed analyte. However, a certain drawback of this technique is worsened selectivity, since any electroactive species can contribute to the current at a given potential pulse. Differential pulse voltammetry combines the features of two previous techniques: the linearly increasing ramp on which are superimposed small amplitude (about 50 mV) pulses. A typical potential range of the scans is from −0.2 V to +0.45 V at a relatively slow scan rate of 5 mV/s. Thus, each scan takes about 1–2 min and oxidizes some amount of analyte in the vicinity of the electrode. Additional time is required for diffusion of a new analyte to the surface of the electrode. This means the scans can be repeated only every 3–5 min. However, the advantage of the method is about ten times higher sensitivity, for example, for DA the limit of detection is about 5 nM.

Measurement of Dopamine by Chronoamperometry

Introduction

Chronoamperometry represents a simpler form of scanning voltammetric techniques where short constant voltage pulses are applied to the electrode system and the current induced is measured at a certain time point, typically at the end of each pulse. This allows elimination of the charging currents caused by the voltage scans/pulses, and measuring more specifically the faradaic currents which reflect the actual concentration of the oxidized substance. On the other hand, the constant voltage pulses, for example +0.55 V, will lead to the oxidation of not only dopamine (DA) but also all other monoamines, their metabolites and some other electroactive substances such as ascorbic and uric acids. Therefore, the pretreatment of the carbon fiber electrodes by electrical pre-conditioning and/or Nafion coating is necessary in order to achieve better selectivity towards the DA molecules.

Materials

- Chronoamperomeric measurements are performed by use of an IVEC-10 instrument and the
- microinjections of KCl were done by a micropressure system BH-2 (both devices available from Medical Systems Corp.).
- Single carbon fiber electrodes are constructed from carbon fibers of 30–35 μm in diameter (AVCO Specialty Materials).
- The glass capillaries (World Precision Instruments) are pulled to have an appropriate size of the tip for accommodation of the carbon fiber or, for the pressure KCl injections to the diameters of 5–10 μm.
- Nafion is available from C.G. Processing, Delaware, U.S.A.; all other chemicals are available from Sigma.

Procedure

1. Pull a glass capillary using a conventional puller in order to achieve a thin tip about 10 mm long.
2. Insert the carbon fiber into the glass capillary and push it as far as possible to the tip. Cut the tip of the capillary with fine scissors and push the fiber through the opening so it will protrude some 3–5 mm.
3. Cut the fiber at the opposite end of the glass capillary. Prepare a piece of Teflon-coated silver wire by scratching the isolation film from both ends of the wire. Seal one end of the wire to the carbon fiber at the proximal end of the capillary by applying a conductive glue and cover with epoxy resin to strengthen the sealing.
4. Seal the tip of the capillary using a low viscosity resin which will be easily drawn into the capillary by capillary forces. Take care not to coat the protruding fiber.
5. Cut the exposed, non-coated carbon fiber to about 0.5 mm length. Clean the fiber by chromic and sulphuric acids and wash it thoroughly in distilled water.
6. Pretreat the electrodes by applying 70 Hz triangular waves:
7. Immerse the electrode tip into a drop of 5% Nafion solution and connect it to the voltammeter. Place a Pt wire into the solution and connect it as reference and aux-

iliary electrode. Apply a potential of +3.7 V for 2 s several times. Wash the electrode and dry before the implantation.

8. Anesthetize the rat and place it into the stereotaxic frame as described above.
9. Implant the carbon fiber electrode into the striatum and the Ag/AgCl reference electrode and the Ag auxiliary electrode into the cortical surface. Fill a separately prepared micropipette with 120 mM KCl and 2.5 mM $CaCl_2$ solution and connect it to the BH-2 micropressure ejection system. Using the second stereotaxic manipulator, implant the pipette to the stiatum at a distance of 300 μm from the carbon fiber.
10. Apply square-wave voltage pulses of +0.55 V at a frequency of 5 Hz (0.2 s) with 0.1 s intervals at +0 V between the pulses. After establishing a baseline response inject KCl solution into the tissue and continue monitoring for additional 3 min.

Results

A representative potassium-evoked high-speed chronoamperogram from the dorsal striatum of the anesthetized rat is shown in Fig. 5. The Ox and Red curves show the oxidation and simultaneous reduction currents of DA following K^+ injection from the micropipette adjacent to the Nafion-coated recording electrode. The curves could be used to calculate the parameters reflecting the kinetics of DA release and uptake. In a control group of animals the average amplitude was 2.65 μM, the rise time was 38 s and the half-decay time was 74 s. The reduction/oxidation current ratio for DA was approximately 0.5, which corresponded to the ratio measured *in vitro* in 2 μM DA solution. The ratio for 2 μM 5-HT was less than 0.2 and zero for 250 μM ascorbic acid, which did not show any reduction current (insets in Fig. 5). Thus it may be concluded that in the rat striatum the measured Ox and Red curves correspond predominantly to extracellular DA. In addition, the release amplitude of DA curves was dramatically reduced from 2.65 μM in control rats to 0.35 μM in 6-hydroxydopamine denervated rats (Strömberg et al 1991).

Troubleshooting

- The major concern of all *in vivo* voltammetric techniques is the specificity of a recording electrode towards a given analyte, in this case DA. Thus, the performance of

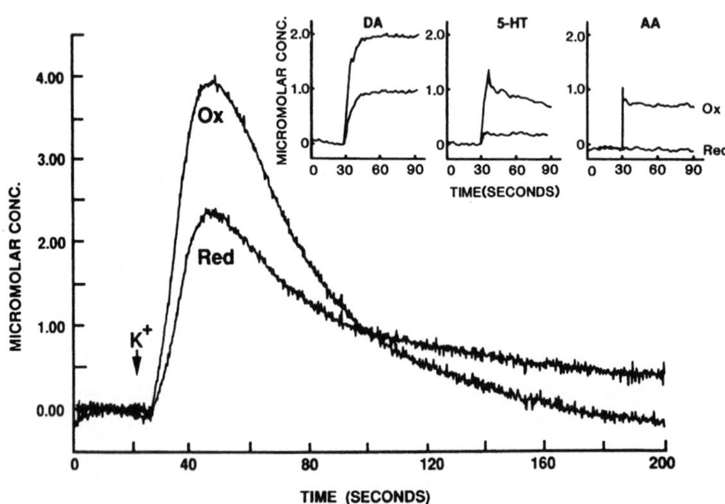

Fig. 5. High-speed *in vivo* electrochemical recording of potassium-evoked overflow of DA in dorsal striatum of the anesthetized rat. The Ox curve (upper trace) represents the oxidation current signal while the Red curve (lower trace) is the simultaneous reduction current signal from the Nafion-coated recording electrode. The inset shows representative oxidation and reduction current responses *in vitro* for a Nafion-coated electrode to additions of 2 μM DA, 2 μM 5-HT, and 250 μM ascorbic acid all prepared in 0.1 M phosphate-buffered saline.

all Nafion-coated electrodes should be tested at first *in vitro* in solutions of various electroactive substances, of which ascorbic acid is the most typical interference. Also, DA metabolites DOPAC and HVA and monoamines such as 5-HT should be tested.
- The poor electrode selectivity can be caused by insufficient Nafion coating. Dry the electrode at 80 °C for some 5–10 min before dipping it into the Nafion solution and dry again. Repeat the procedure (5–10 times) until the electrode exhibits the appropriate selectivity.
- The baseline stability should be checked as follows: Place the recording and the reference electrodes in the phosphate-buffered saline (PBS: 0.1 M, pH 7.2) and start recording. If the electrode drifts upwards, it may indicate the leak at the interface between the fiber and glass capillary or an adsorbed particle or air bubble on the tip of the recording electrode.
- Every electrode should be calibrated (PBS in the presence of 250 μM ascorbic acid) and it should show a linear response in the range of 0.1–10 μM DA.

Comments

Voltammetric methods provide higher temporal and spatial resolutions than any other chemical sensing techniques and in this respect are superior to, for example, microdialysis sampling. However, *in vivo* electrochemistry has a very limited use in preparations where DA and NE coexist, or in neuroanatomical areas where the measured amine concentration (release) is much lower than that of the other amine. Thus it is rather easy to measure DA release in the basal ganglia, NE release in the locus coeruleus or 5-HT release in the raphé nuclei, whereas monitoring of monoaminergic neurotransmission in other nuclei such as prefrontal and frontal cortices or hippocampus can be performed, at present, only by microdialysis. In addition, *in vivo* voltammetry data are predominantly collected under stimulated conditions, whereas microdialysis allows studying the effects of drugs or behavioral manipulations on basal, non-evoked extracellular levels of neurotransmitters.

Applications

In vivo voltammetric techniques have been extensively used in neuroscience to study *electrochemical signals associated with release-reuptake processes of monoamine neurotransmitter substances* (for review, see Boulton et al 1995). Here, the fast methods such as fast cyclic voltammetry, differential pulse voltammetry and chronoamperometry are of major importance. The voltammetric methods, in line with electrophysiological techniques, allow to measure rapid changes in neuronal communication, in particular the *release of DA, NE or 5-HT* following brief stimulation by electrical or chemical means. Several mathematical models have been proposed to calculate kinetic parameters of stimulated DA release, clearance and diffusion. This allowed *quantitative topographic analysis of heterogeneity of DA release in the rat striatum* by fast cyclic voltammetry as was already mentioned in the General Introduction and references to that section. Chronoamperometric techniques have been used mainly for *functional characterization of neuronal transplants in DA or NE denervated animals* (Gerhardt et al 1994b). Most of the applications are related to the studies in the living brain; however, recently a number of reports have described the feasibility of voltammetry for *in vitro* studies in brain slices, sympathetic nerve terminals, chromaffin cells and giant neuronal cells of invertebrates (for review, see Boulton et al 1995).

Biosensors

Today, probably the most typical device associated with the term biosensor is an electrochemical sensing strip which measures capillary blood glucose (for review, see Freitag 1993). However, the definition of a biosensor can be much broader, covering not only electrochemical devices which convert (bio)chemical reactions directly into an electrical response, but also all optical assays at a liquid/solid phase boundary where specific chemical interactions alter some physical property of light. Thus, any disposable diagnostic device using a simple visual positive/negative scale can be classified as a biosensor or immunosensor. The same chemical principle as used for these single-use sensors can be applied for continuous-use implantable sensing electrodes or optodes. There are probably hundreds of biosensors and detecting principles described for purposes of *in vitro* diagnostics, but of those, there is only a limited number of some 5 to 10 devices which can be adopted for *in vivo* conditions and can operate in real-time, with sufficient stability and specificity towards the target substrate. Clinically, the only implantable sensor used for neuromonitoring in severe human head injury has been a device measuring pH, partial tension of oxygen (pO_2) and carbon dioxide (pCO_2) (Zauner et al 1997). The electrochemical sensors used *in vivo* are constructed typically of carbon fibers or Teflon-insulated Pt wire, whereas the optodes use quartz optical fibers. Oxidoreductase or dehydrogenase catalyzed reactions can be coupled to the electrodes/optodes by various enzyme immobilization techniques (Barker 1987). The simplest example is the above-mentioned glucose sensor, where glucose oxidase converts glucose to gluconic acid and hydrogen peroxide; the latter molecule being easily detected electrochemically. Similarly, this approach can be applied to other substances such as lactate, glutamate, choline and amino acids, where the corresponding oxidase enzymes have been isolated and are commercially available. The enzymes may be immobilized by known immobilization methods such as glutaraldehyde or avidin-biotin complexes (Pantano and Kuhr 1993). The major problems associated with the application of biosensors *in situ* are the successive loss of sensitivity and fouling. These are typically due to the direct exposure of the sensing enzymes to cleavage by extracellular proteases or by inhibiting the catalytic activity of the immobilized enzymes by other toxic molecules, and the direct adsorption of proteins and lipids onto the electrode surface. Pretreatment of the electrodes by electropolymerization of diaminobenzenes or pyrroles helps to entrap the enzymes at the electrode surface (Malitesta et al 1990; Sasso et al 1990). The sensors based on electropolymerized films of *o*-phenylenediamine with immobilized enzymes show fast response time, are free of protein and lipid fouling and have minimal interference from endogenous electroactive species (Lowry et al 1998b). Another approach to protect the sensing electrode from poisoning and interferents is to cover its surface with a semipermeable membrane and ultimately create a separate measuring compartment between the transducer and the membrane (Barker 1987). Classical "chemical" representatives of such electrodes could be gas ion-selective electrodes, for example, an ammonia electrode. One recent modification of this type of design is the dialysis electrode (Walker et al 1995). For enzyme-based sensors, the electrodes are inserted into a tubular dialysis membrane which is closed (glued) at its distal tip. The immobilization step is not necessary, since this microcontainer can be simply filled with any cocktail of enzymes. The construction resembles that of a microdialysis probe; however, there is no perfusion through the probe during the experiment. The microdialysis biosensor offers faster temporal response than microdialysis probes; however, spatial resolution and tissue damage are about the same. For example, a "dialysis electrode" was used to measure extracellular glutamate and ascorbic acid release in rat hippocampus after electrical stimulation of the perforant path (Walker et al 1995) or glutamate release under acute ischemia (Asai et al 1996). The oxidase-based biosensors usually operate in a simple amperometric mode at a constant potential of +0.7 V vs. Ag/AgCl reference electrode. The most recent contribution to continuous monitoring of brain chemistry is a nitric oxide sensor utilizing sensor technology similar to that used with biosensors. The original electrode constructed by Shibuki (1990) was a membrane-based electrode utilizing the following chemical reaction:

$$2NO + 4OH^- \rightarrow 2HNO_3 + 2H^+ + 6e^-$$

The membrane was permeable only to free gases, and was thus able to separate the chemical solutions inside the membrane from the outside compartment. Further improvement was achieved by using carbon fibers coated with a film of semiconducting polymeric porphyrin and Nafion (Malinski and Taha 1992). The porphyrinic biosensor was used to detect NO in single cells, cerebral tissue, the cardiovascular system, and also in humans (for review, see Kiechle and Malinski 1996).

Biosensor for Glucose

Introduction

The glucose biosensor utilizes the high substrate specificity and turnover rate of glucose oxidase (GOx) isolated from *Aspergillus niger*, which converts D-glucose to electrochemically detectable hydrogen peroxide according to the following reactions:

$$\text{D-glucose} + \text{GOx/FAD} \rightarrow \text{D-glucono-}\delta\text{-lactone} + \text{GOx/FADH}_2$$

$$\text{GOx/FADH}_2 + O_2 \rightarrow \text{GOx/FAD} + H_2O_2$$

$$H_2O_2 \rightarrow O_2 + 2H^+ + 2e^-$$

where FAD is flavin adenine dinucleotide. The GOx enzyme is immobilized on the Pt wire via the electropolymerized *o*-phenylenediamine film (Lowry et al 1998b, c; Malitesta et al 1990; Sasso et al 1990).

Materials

- Any low-noise potentiostat such as a Biostat II (Electrochemical and Medical Systems) or a BAS 100 (Bioanalytical Systems), which can operate in the mode of constant-potential amperometry, can be used for the measurements.
- The data is recorded using either a strip-chart recorder or a computer installed with a suitable data acquisition board (available from National Instruments or World Precision Instruments).
- The PTFE-coated platinum/iridium (125 μm bare diameter) wire and the Ag (200 μm bare diameter) wire (reference and auxiliary electrodes, see below) are available from Advent Research Materials (Suffolk, UK).
- Glucose oxidase (EC 1.1.3.4, Grade I) may be obtained from Boehringer Mannheim, and all other chemicals from Sigma.

Procedure

1. Cut the PTFE insulation by about 5 mm from each end of a 5 cm length of Pt/Ir wire. Seal each tip of the insulation film to the Pt/Ir wire with epoxy glue (Devcon 5 Minute Fast Drying Epoxy) to produce about 4 mm active length of bare wire. Scratch the surface of the exposed tip carefully with a scalpel blade. Solder a gold electrical contact (Semat) to the opposite end.
2. Prepare a 300 mM solution of *o*-phenylenediamine (*o*-PD) in a phosphate-buffered saline solution (PBS: 150 mM NaCl, 40 mM NaH_2PO_4, 40 mM NaOH, pH 7.4) and remove oxygen by saturating the solution with N_2 (approx. 20 min). Place the electrodes in the *o*-PD solution for about 60 min.
3. Remove the electrodes and place them in a solution of GOx (850 U) in 5 μl *o*-PD. The enzyme will absorb within 5–10 min as indicated by the yellow color of the electrode surface.
4. Transfer the electrodes to about 10 ml of freshly prepared *o*-PD solution. Connect the Pt/Ir electrodes (working electrodes) against a reference (saturated calomel electrode) and an auxiliary (Pt wire) electrode and apply a potential of +650 mV for about 15 min.

5. Rinse the electropolymerized electrodes with PBS solution and store them in PBS at +4 °C.
6. For simultaneous measurements of brain tissue oxygen a carbon paste electrode (Lowry et al 1997) is used: prepare the carbon paste by thoroughly mixing 2.8g of carbon powder (UCP-1-M, Ultra Carbon Corp.) and 1ml of silicone oil (Sigma-Aldrich). Pack the paste into a 2mm cavity produced by sliding the insulation of a teflon-coated silver wire (125 μm bare diameter) over the end of the wire.
7. Prior to implantation, the biosensor electrodes should be preconditioned in 5mM glucose for 10 hours. *In vitro* glucose (0–100mM) and ascorbic acid (up to 1mM) calibrations should then be performed in order to ensure suitable sensitivity and selectivity in 0–100mM glucose solutions with or without ascorbic acid (up to 1mM).
8. Implant the calibrated glucose sensor and the carbon paste electrode into the rat brain in a similar manner to that described above. Follow the instructions of the protocol for the stereotaxic surgery and implantation of the electrodes: Place the glucose sensor in the left striatum and the carbon paste oxygen electrode in the right striatum. Place the reference electrode in the cortex and the auxiliary electrode between the skull and dura. Fix the electrodes to the skull with dental cement and screws as detailed previously.
9. The animals should recover from the surgery for at least five days. Normally, the animals are housed under a 12h light/dark cycle; food and water is available ad libitum.
10. On the day of experiment, connect the electrodes to the potentiostat. Set the potential of the oxygen electrode to –550mV (reductive mode) and that of the glucose sensor to +700mV vs. an implanted Ag reference electrode. Wait until the background currents for the electrodes have stabilized, typically 30–60min.
11. Administer the desired drug, for example, insulin (1ml, 15 U/kg), as shown below where insulin was injected intraperitoneally into the rats which were fasted for 24 hours before the experiment. The signals from the glucose and oxygen sensors were recorded simultaneously during an additional three hours.

Results

The effect of systemically injected insulin on the levels of brain extracellular glucose and brain tissue oxygen levels is shown in Fig.6. As seen in the figure, 40min following insulin injection, the extracellular glucose concentration was lowered by about 14% and this reduction lasted for about 20 min. In contrast, the oxygen signals measured at the carbon paste electrode increased during this period. Calculations using the post-*in vivo* calibrations allowed conversion of the change in amperometric signals (DI in nA) for oxygen to be converted into oxygen concentration (in μM). Thus, insulin administration caused an increase in tissue oxygen by about 70 μM, which is by more that 100% assuming a basal extracellular oxygen concentration of 50 μM (Zimmerman and Wightman 1991). These results demonstrate that the glucose sensor follows the changes in concentration of glucose in the extracellular fluid and that the accompanying changes in tissue oxygen levels do not interfere with this measurement.

Troubleshooting

- The quality of the *o*-PD film is critical for the stability of the electrode response at different temperatures. The electrodes prepared by the present procedure should be insensitive to temperature changes, which allows the use of *in vitro* calibration data obtained at room (22 °C) temperature for conversion of *in vivo* amperometric signals to glucose concentrations at physiological (37.6 °C) temperatures.
- Since the biosensors are being used in a complex biological matrix, it is important to perform pre-*in vivo* calibrations for glucose and the principal interferent, ascorbic acid. These should be performed immediately prior to implantation to ensure the integrity of the sensor.

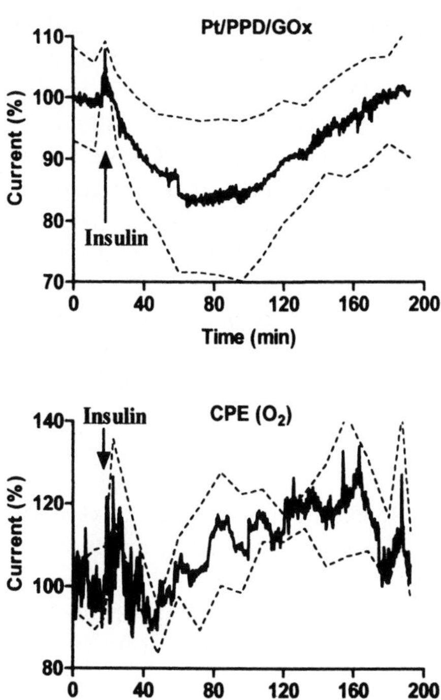

Fig. 6. The effect of intraperitoneal injection of insulin (1 ml, 15 U/kg) on the Pt/poly-PD/GOx (top) and carbon paste electrode (CPE) O_2 (bottom) signals recorded simultaneously with bilaterally implanted sensors in the striatum of freely moving rats. The tissue O_2 levels at the CPEs were monitored using both differential pulse amperometry and constant potential amperometry. Data (n=4) are normalized with the basal level before injection taken as 100%. The hashed lines represent the SEM, which is plotted at 12 min intervals for clarity.

- The common drawback of most of the directly sensing *in vivo* devices is a successive decrease in sensitivity due to surface fouling by proteins, lipids and other biomolecules. The sensitivity of the biosensor can drop by as much as 50% within several hours after implantation into the brain tissue. The procedure of preconditioning the sensor by continuous recording in a 5 mM glucose solution at +700 mV vs. the calomel reference electrode for at least 8 hours is strongly recommended.

Comments

It was demonstrated earlier that the glucose oxidase/poly(*o*-phenylenediamine) biosensor does not respond to changes in tissue oxygen or ascorbic acid which are the most typical interferent candidates for the amperometric sensors applied in monitoring of the brain microenvironment (Lowry et al 1998b). Thus, electropolymerization by *o*-PD provides a general means to prevent interferents and other electroactive substances from diffusing through the polymer film (about 10 nm thick) and reaching the surface of the electrode wire (Sasso et al 1990). Additionally, the *o*-PD polymer acts as an efficient trap to immobilize the oxidoreductase enzymes. The glucose biosensor removes or utilizes 41 times less extracellular glucose compared to microdialysis sampling at a typical flow rate of 2 µl/min (Lowry et al 1998b). This suggests that the biosensor-based measurement causes minimal perturbation of glucose homeostasis and thus may reflect true extracellular glucose concentrations.

Applications

Extracellular glucose levels have been measured in relation to changes in cerebral blood flow (Fellows and Boutelle 1993) and neuronal activity (Fellows et al 1992) using microdialysis combined with an on-line *ex vivo* biosensor assay system. Perfusions with a sodium channel blocker tetrodotoxin caused a significant increase, whereas depolarization of cell membranes by veratridine caused a rapid decrease of extracellular glucose levels (Fellows et al 1992). A recently developed intracorporeal glucose biosensor allows real-time glucose monitoring over periods of up to five days and estimation of the extracellular glucose concentration by combining the biosensor with a microdialysis probe (Lowry et al 1998b,c). Previously, the microdialysis perfusion method based on searching the point of zero net flux was used to calculate basal extracellular glucose concentrations in the striatum of awake rats. The estimated glucose levels were in the range of 0.35 μM to 0.47 μM (Fray et al 1997; Fellows et al 1992), which is in good agreement with the concentrations measured with glucose biosensors 0.35 and 0.49 μM (Lowry et al 1998b,c). This glucose biosensor-oxygen sensor combination has also been used in several other *in vivo* neurochemical studies: investigating *oxygen and glucose utilization during neuronal activation* (Lowry and Fillenz 1997), studying the *source of brain extracellular glucose* (Lowry and Fillenz 1998), and studying the *relation between cerebral blood flow and extracellular glucose during mild hypoxia and hyperoxia* (Lowry et al 1998a).

Optical Sensors

For *in vivo* measurements of optical signals on conscious animals, the optical fibers are stereotaxically implanted into the brain or attached to the brain surface. The light source is connected to a single fiber or to the fiber bundle, while the other fibers are linked to the photosensing/registration device (photomultiplier, CCD camera etc.). This approach can involve a variety of optical techniques, detecting:

- Brain oxygenation – near-infrared spectroscopy (Imamura et al 1997),
- Intrinsic signals – reflectance imaging (Grinvald et al 1986; Rector et al 1993),
- Native fluorescence of endogenous compounds such as NADH (Mayevsky, Chance 1973) and
- All methods where exogenous markers, dyes or reagents are used for activation of optical signals, for example, visualization of Ca^{2+} (Hirano et al 1996) or release of acetylcholinesterase (Clarencon et al 1993).

A large number of fluorescence and luminescence dyes/molecular probes are available for selective *in vitro* labeling of ions and molecules in brain slices or cell cultures. Obviously, the latter group can also cover all non-invasive methods of brain imaging, such as *positron emission tomography* (PET; cf. Chapter 39), single-photon emission tomography and *NIR spectroscopy* (cf. Chapter 40). Recently developed reporter-gene technology using transgenic animals expressing firefly luciferase or green fluorescent protein permits non-destructive monitoring of gene transformation or temporal changes in gene expression, such as c-fos (Geusz et al 1997; for review, see Welsh and Kay, 1997).

Part 3: Continuous Sampling Devices

Cortical Cup Technique

Techniques of extracting biologically active substances by superfusing various organs have been known since the 1920s. Today, *in vitro* studies of stimulated neurotransmitter release, firstly developed for brain slices (McIlwain 1955), can also be conducted on subcellular structures such as synaptosomes and vesicles or in cell cultures. The simplest *in vivo* alternative to brain slice superfusions is the so-called cortical cup technique (MacIntosh and Oborin 1953; for review, see Myers 1972; Moroni and Pepeu 1984). A small cylindric cup is fixed to the exposed cortical surface of an anesthetized animal. The cup is filled with the artificial cerebrospinal fluid (aCSF). Optionally, it can be connected to the superfusion apparatus which allows periodical refilling (Mitchell 1963) or continuous perfusion (Celesia and Jasper 1966) of the cup. An obvious limitation of the technique is that it can be used only for studies of neurotransmitter release from the cortical surface. Thus, only about 30 papers have been published on measuring *in vivo* release of acetylcholine, amino acids, nucleosides and free radicals by cortical cup technique. The majority of the studies was done by Phillis and coworkers, who have used the technique until recently (Phillis et al 1998).

Push-pull Cannula

Much larger popularity was gained by a push-pull technique, which is an "ultimate cup" perfusion within the size of an injection needle. The main advantage is that the needle can be implanted into the brain, thus making any brain structure accessible for *in vivo* sampling. The brain push-pull cannula was firstly described by Gaddum (1961) as a modification of earlier devices used for perfusions of ventricular space (Bhattacharya and Feldberg 1959). There are two principal constructions of the push-pull cannulae, where two (push and pull) capillaries are placed either side-by-side at the same distal level, or are concentric, where the thinner capillary extends the end of the outer capillary by some 0.5–1 mm. The concentric design allows chronic studies by use of guide cannulae permanently implanted into the awake animals (Myers 1977). Once the cannula is implanted into the tissue, the perfusion fluid (aCSF) is simultaneously delivered (pushed) and removed (pulled), using either a peristaltic pump or a dual syringe pump with reciprocally operating syringes. Both pumps provide about the same precision of liquid pumping (Privette and Myers 1989) at typical push-pull flow-rates of 25–75 μl/min; the latter mode requires a sampling valve to be installed in the closed loop system. Ideally, the exchange of substances occurs by free diffusion at the boundary between the perfusion fluid and the extracellular fluid. The tricky part is to keep the exact equilibrium between delivered and removed liquid during the entire experiment. The major obstacle is clogging of the pull cannula by tissue or blood clotting, causing immediate accumulation of infused fluid and severe lesions. In spite of many technical difficulties and limitations of the push-pull technique, a number of investigators have succeeded in applying this approach in some essential studies on *in vivo* release of classic neurotransmitters, peptides and proteins (for review, see Myers and Knott 1986; Philippu 1984). A capillary push-pull cannula has been described (Zhang et al 1990) which has a tip diameter of about 150–190 μm and is perfused at flow-rates of only 5–10 μl/min, i.e., at conditions very similar to microdialysis sampling.

Microdialysis

A very first step to apply dialysis principles for sampling of the brain microenvironment was taken by Bito and collaborators in 1966. Mongrel dogs were implanted with sterile dialysis sacs (8–12 mm long, flat width 9 mm) filled with 6% dextran in saline into the cortex and subcutaneously in the neck. Ten weeks later, the contents of the sacs were analyzed for amino acids and ions and compared to the concentrations in plasma and CSF. For most of the amino acids, they found a concentration gradient of the following order: blood plasma > ECF > CSF. They suggested that the brain sac fluid cannot be a dialysate of either blood or CSF, thus brain extracellular space must represent a third fluid compartment, supporting the idea of a carrier-mediated transport system for amino acids. However, this was a static approach, providing only one dialysis sample per animal. The possibility of continuous perfusion of dialysis bags in order to measure time-course of infused and recovered compounds was explored by Delgado and coworkers at the beginning of the 1970s. Delgado et al (1972) reported the construction of a "dialytrode" for monkeys, which is basically a push-pull cannula with a small (5x1 mm) polysulfone membrane bag glued on its tip. The authors described some conceptual experiments, derived from established push-pull protocols: A) infusion of compounds or labeled precursors into the brain and correlation with brain electrical activity or with newly synthesized labeled compounds and B) sampling of endogenous compounds such as amino acids and glycoproteins. However, the method failed to reproduce the data obtained earlier with push-pull "chemitrodes" (Roth et al 1969); for example, there was no detectable dopamine in the perfusates after prelabeling with ^{14}C L-DOPA. It was Ungerstedt and Pycock (1974) who finally succeeded to measure amphetamine-induced release of dopamine-like radioactivity after prelabeling with dopamine through a hollow fiber dialysis probe implanted in the rat striatum. A rapid advance in development of highly sensitive HPLC analytical techniques and technologies of manufacturing small-bore dialysis tubing during the following years accelerated the research on applications of single fiber dialysis – microdialysis – in neurobiology, pharmacology and physiology from animals (Ungerstedt et al 1982; Hamberger et al 1982; Ungerstedt 1984) to humans (Meyerson et al 1990; Hillered et al 1990; Ungerstedt 1991; During and Spencer 1993). Several books on *in vivo* monitoring include chapters on microdialysis, and recently a monograph "Microdialysis in the Neurosciences" was edited by Robinson and Justice in 1991.

Determination of Dopamine Release by Microcolumn Liquid Chromatography with Electrochemical Detection (LCEC)

Introduction

In vivo sampling of extracellular dopamine was one of the first successful applications of microdialysis (Ungersted and Pycock 1974, see previous section) and even today dopamine monitoring is described in more than 30% of all papers on microdialysis. The development of electrochemical detectors in liquid chromatography (Kissinger et al 1973) opened a new alternative to simplify and automatize the determination of monoamines as compared to already existing radioenzymatic assays. Following HPLC separation, the catecholamines dopamine (DA), adrenaline (A), noradrenaline (NA), and indoles like serotonin (5-HT) and their major metabolites are oxidized at the glassy carbon electrode operating at potentials between +550 to +750 mV (for details, see above). Separation of catecholamines was initially achieved on cation exchanger resins packed in large columns with low separation efficiency (Zetterström et al 1983) but soon these materials were replaced by reversed-phase silica-packed columns. Using the ion-pairing reagents, both basic (catecholamines) and acidic (metabolites) could be separated in the same run. Recent miniaturization of the dimensions of the separation columns down to and below 1 mm I.D. (microbore columns) favors the sensitivity of the electrochemical detector. Using these columns, the limits of detection of 1 fmol or even sub-fmol/sample are reported for dopamine and 5-HT in the microdialysis samples (Wages et al 1986).

Materials

Instrumentation

- A liquid chromatograph consists of a Micro LC Pump LC100 (ALS) which can provide a pulse-free delivery of the mobile phase at a constant flow of 70 µl/min,
- A CMA/260 Degasser (CMA/Microdialysis) which is installed between the mobile phase reservoir and the pump inlet and
- A CMA/200 Refrigerated Microsampler is used for automated injections of the microdialysis samples.

- The electrochemical detector LC4B (Bioanalytical Systems) is equipped with a radial flow cell (ALS).
- The potential of a 6 mm I.D. glassy carbon electrode is set to +700 mV vs. Ag/AgCl/3 M NaCl reference electrode.
- Data are recorded on an SP-4290 integrator (Spectra-Physics) or using an EZChrom data acquisition system (Scientific Software).
- A microbore column is a 150 x 1 mm I.D. stainless-steel capillary packed with C18 silica, 3 μm particle size (ALS).

Mobile phase and chemicals

Prepare a buffer containing 0.05 M citric acid, 0.2 mM Na_2EDTA, 0.65 mM sodium octylsulphonate and adjust pH to 3.2 with concentrated sodium hydroxide. Fill up to 1000 ml and filter the solution through a 0.45 μm filter. Mix 970 ml of the citrate buffer with 30 ml acetonitrile (ACN). This will give the final mobile phase containing 3% ACN. Standards are usually prepared as 10 mM stock solutions in 10 mM HCl in 1 ml aliquots and kept frozen at −70 °C. Working calibration standards are prepared daily by diluting stock solutions down to 10 nM DA and 5-HT and 100 nM DOPAC, HVA and 5-HIAA in Ringer solution. All chemicals are available from Sigma.

Procedure

1. Purge the pump with the mobile phase and install the microbore column at the outlet of the injection valve. Stabilize the flow rate at 70 μl/min.
2. Mount the new or newly polished working electrode in the electrochemical cell. Connect the cell to the column outlet. Wait until the cell internal volume is filled with the mobile phase and then fit the reference electrode into the cell.
3. Set the working potential of the electrochemical detector to +700 mV and operation in an oxidative mode. Set the active filter to 0.1 Hz.
4. Turn on the cell and wait until the basal signal is stabilized to some 1 to 3 nA. Reduce the background current to an appropriate baseline level by activating the offset function.
5. Implant the microdialysis probe (CMA/11 with a 2 mm membrane length) into the rat or mice striatum as described in protocol 1.1. Perfuse the probe with Ringer solution at a flow-rate of 2 μl/min. Following the initial 2–3 hour stabilization period, collect the fractions in 20 min intervals.
6. Place the microdialysis samples and the calibration standards in 300 μl vials into the CMA/200 autosampler. Program the autosampler to inject 10 μl and set the analysis time to 25 min.

Results

Chromatograms of a calibration standard and microdialysis samples from the striatum of an anesthetized mouse (C57 strain) are shown in Fig. 7A,B,C. A typical separation time for DA and metabolites (Fig. 7A) was about 15 min but can be significantly reduced by modifying the concentration of the ion-pairing reagent, organic modifier and pH of the mobile phase. At basal conditions (Fig. 7B), the concentration of DA was 0.53 nM (5.3 fmol/10 μl injected) whereas the levels of metabolites were three orders of magnitude higher: DOPAC 146.4 nM, HVA 374.4 nM and 5-HIAA 64.2 nM. Administration of amphetamine (5 mg/kg i.p.) (Fig. 7C) caused a rapid, 13-fold increase in extracellular DA to 7 nM within the first 20 min, while its metabolites DOPAC and HVA were reduced to 166 and 311 nM, respectively. The value of 5-HIAA was unchanged (63.9 nM). The high sensitivity of the assay makes it possible to measure DA outflow from various structures of the mouse brain, or from the rat microdialysates sampled in intervals as short as 1 min (Church and Justice 1987). The high spatial and temporal resolution is often necessary for studies of DA release under various behavioral tasks such exploratory activity, feeding or drinking, self-administration of drugs etc.

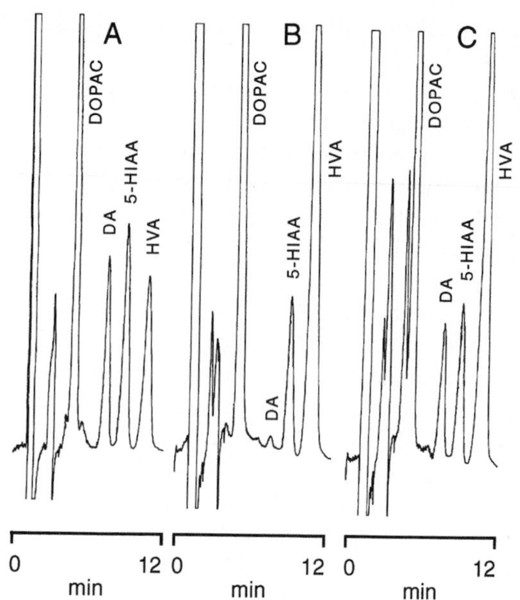

Fig. 7. Chromatograms of (A) a 10 µl standard mixture containing 10 nM DA and 100 nM each of DOPAC, 5-HIAA and HVA, (B) a 10 µl microdialysis sample collected at basal conditions from the striatum of the anesthetized mouse, containing 0.53 nM DA, 146.4 nM DOPAC, 374.4 nM HVA and (C) a 10 µl microdialysis sample from the same mouse 20 min after amphetamine injection (5 mg/kg i.p.). DA levels increased to 7 nM, whereas DOPAC and HVA concentrations were reduced to 166 and 311 nM, respectively. The sensitivity of the detector cell in all cases was 0.31 nA full scale for DOPAC and DA, and 1.25 nA F.S. for 5-HIAA and HVA.

Determination of Serotonin in Microdialysis Samples by LCEC

Materials

Instrumentation

The same apparatus as used for estimation of DA (see protocol 3.4) is used for determination of 5-HT and 5-HIAA.

Mobile phase and chemicals

Prepare 0.05 M citrate buffer in the same way as described in protocol 3.4 for determination of DA and metabolites. Mix 935 ml of the citrate buffer with 65 ml acetonitrile (ACN). This will give the final mobile phase containing 6.5% ACN. Standards are usually prepared as 10 mM stock solutions in 10 mM HCl in 1 ml aliquots and kept frozen at –70 °C. Working calibration standards are prepared daily by diluting stock solutions down to 10 nM 5-HT and 100 nM 5-HIAA in Ringer solution.

Procedure

Follow the same procedure as described for DA determination by LCEC in protocol 3.4. The microdialysis probe is implanted in the rat prefrontal cortex at the coordinates (mm) A-P +2.7, L +0.7 and V –3.5 according to the atlas of Paxinos and Watson, 1982.

Results

The chromatograms of (A) the standard solution containing 10 fmol 5-HT and 500 fmol 5-HIAA in 10 µl injected and (B) the microdialysis sample collected from the prefrontal cortex of the awake rat are shown in Fig. 8A,B. The estimated concentration of 5-HT in the cortical microdialysate was 6.2 fmol/10 µl. The limit of detection (defined as signal-to-noise ratio of 2) is about 1 fmol 5-HT, which allows determination of basal 5-HT release in most of the anatomically relevant areas without a need to use 5-HT uptake blockers, for example, 1 µM citalopram in the perfusion medium.

Troubleshooting

- The electrochemical detector, if not properly operated, can be a source of many troubles and frustrations. A high background noise, noise spikes, drifting baseline, low peak heights are some of the typical problems in LCEC. There are a number of possible reasons: a) dissolved oxygen or contaminants in the mobile phase, b) pulsating pump or leakage in the HPLC system, c) irreversibly contaminated or deteriorated

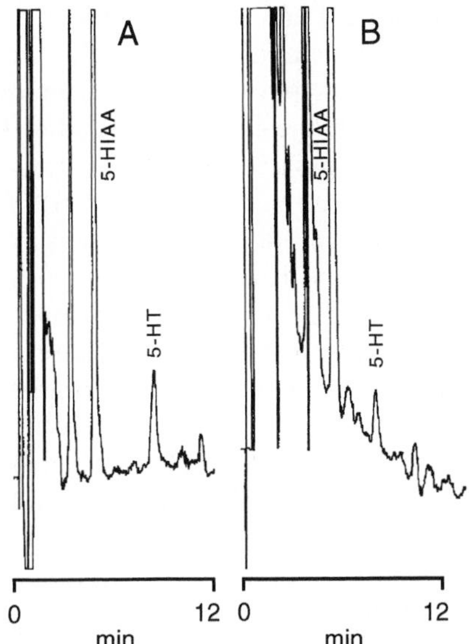

Fig. 8. Chromatograms of (A) a 10 μl standard mixture containing 1 nM 5-HT and 50 nM 5-HIAA and (B) a 10 μl microdialysis sample collected at basal conditions from the prefrontal cortex of the awake rat. Estimated 5-HT concentration was 6.2 fmol. The sensitivity of the detector cell was 0.62 nA full scale.

column, d) faulty operated detector cell due to a dirty surface of the working electrode or an old, unstable reference electrode, e) mobile phase dropping from the outlet tube at waste, f) electrical disturbances, not grounded instruments, g) temperature variations in the laboratory.

- Periodical disturbances such as those mentioned under b), e) and f) can be identified relatively easily by changing the flow-rate of the pump, switching on and off all the switches and relays in the room.
- The high background noise can be eliminated by preparing a new mobile phase (check the purity of distilled/deionized water) or by polishing the glassy carbon electrode.
- To polish the electrode use a dispersion of 0.5 μm aluminum particles in water and a smooth polishing pad. Drop one drop of aluminum solution on the wetted pad and polish the electrode by circling and applying low pressure towards the pad. Rinse the electrode properly with water, alternatively for 30 s in an ultrasound bath.
- When using the new columns, the retention times of DA and metabolites could be slightly changed, often prolonged. Increase concentration of acetonitrile by 1–2% to speed-up the separation. If DA peak elutes too close to DOPAC, increase concentration of ion-pairing reagent by 0.1–0.2 mM, which should preferentially prolong the retention time of DA without affecting retention of acidic metabolites.

Comments

The increasing number of reversed-phase materials with largely different separation properties can cause some variation in retention order and retention times of monoamines, their acidic and neutral metabolites. The preferred materials for separation of basic compounds are so-called end-capped reversed-phase silica, with low free silanol residues and moderate hydrophobicity (12–15% C, measured as % of bound carbon of the octyldecylsilane molecules). A typical elution order of monoamines and their metabolites separated in isocratic, ion-paired reversed-phase mode is: MHPG, NA, DOPAC, DA, 5-HIAA, HVA, 5-HT. Unfortunately, a complete separation of these compounds in microdialysis samples at a reasonably high sensitivity and specificity is very difficult. For these reasons, typically three separate systems based on three different mobile phases are used: 1) a mobile phase with low elution strength for determination of NA and MHPG (Abercrombie, Zigmond 1989), 2) intermediate-strength mobile phase for determination of DA and metabolites (Caliguri and Mefford 1984; Church and Justice 1987), and 3) determination of 5-HT and 5-HIAA (Sarre et al 1992) using a mobile phase with higher content of an organic modifier or a modifier with higher elution strength.

Column liquid chromatography with electrochemical detection of monoamines and their metabolites is certainly the most widespread method of analyzing trace levels of these substances in microdialysis and other biological samples. However, some alternative techniques such as fluorescence detection in HPLC or radioenzymatic assays have been used for determination of "difficult" analytes such as NA and 5-HT. Thus, 5-HT as a tryptophan derivative can be detected also by fluorescence detection, either in its native form (Kalen et al 1988) or by recently developed precolumn derivatization with benzylamine (Ishida et al 1998). The latter procedure enables sensitive determination of NA (Yamaguchi et al 1999) or simultaneous determination of 5-HT and NA. DA and NA can be simultaneously analyzed by precolumn derivatization with 1,2-diphenylethylenediamine (Kehr 1994). This method was also applied for on-line determination of NA in microdialysates from rat pineal gland (Drijfhout et al 1996).

Applications

LCEC has become a major methodology used for the *determination of DA, 5-HT, NA and their metabolites* in the microdialysis samples. *In vivo* voltammetry is probably the closest alternative to microdialysis combined with LCEC for measurements of extracellular monoamines. The main advantage of monitoring monoamines by microdialysis/LCEC is the relative simplicity of this procedure and the possibility to measure transmitter release *in vivo* in awake animals. This allows to search for *chemical correlates to behavior* and to study *in vivo extracellular chemistry and pharmacology in most of the behavioral models of mental and degenerative diseases*. Similarly, various chemical markers of *neuronal damage in animal models of brain trauma*, such as *stroke*, are easily measured by microdialysis sampling. DA, NA and 5-HT have been implicated in the pathophysiology of a number of diseases, such as *Parkinson's disease, schizophrenia, mood disorders, depression* and *anxiety, sleep disorders, stress* as well as *food intake or drug and alcohol reinforcement*. Today, the microdialysis technique combined with HPLC analysis of DA or 5-HT is a well-established research tool in studies of mechanisms underlying *drug abuse, psychosis and depression* and in the development of new drugs for therapies of these diseases.

Determination of Aspartate and Glutamate in Microdialysis Samples by HPLC with Fluorescence Detection

Introduction

Detection of amino acids at low concentrations requires their chemical modification (derivatization), resulting in highly fluorescent or electrochemically active compounds. Several fluorogenic reagents are available, among which *o*-phthaldialdehyde/2-mercaptoethanol (OPA/MCE) based reagent is the most common for analysis of physiological amino acids. In the alkaline medium (borate buffer, pH 9.5), the primary amino acids and some related amines react rapidly already at room temperature with OPA in the presence of nucleophile (MCE) to form various substituted isoindoles. These compounds, when excited at 340 nm, emit fluorescence light at 450 nm. However, the fluorescence isoindoles are rather unstable and subsequently degrade to non-fluorescent derivatives (Kehr 1993), as depicted in the reaction scheme in Fig. 9. Precolumn derivatization with OPA/MCE allows easy automation and a use of reversed-phase columns for gradient separations (Lindroth and Mopper 1979). Several instrumental systems were described for rapid separation of amino acid neurotransmitters aspartate (Asp) and glutamate (Glu) (Kehr 1998a) or γ-aminobutyric acid (GABA) (Ungerstedt and Kehr 1988; Kehr 1998b). Of those, a microbore chromatography at isocratic conditions incorporating column wash-out steps with acetonitrile (ACN) is the simplest and most cost-

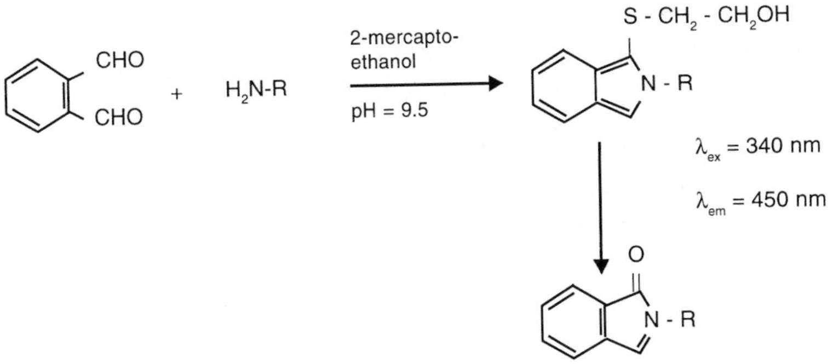

Fig. 9. Derivatization reaction of primary amino acids with *o*-phthaldialdehyde in the presence of a nucleophile (2-mercaptoethanol) and under alkaline conditions yields highly fluorescent isoindoles which gradually degrade to non-fluorescent derivatives.

effective HPLC system for automated analysis of aspartate (Asp) and glutamate (Glu) based on pre-column derivatization and fluorescence detection.

Materials

Instrumentation

- An HPLC analyzer consists of an autosampler (CMA/200 Refrigerated Microsampler) with a modified version of the program allowing in-between injections of acetonitrile in order to wash the microbore column,
- A CMA/260 degasser,
- A CMA/280 fluorescence detector equipped with 6µl cell (all purchased from CMA/Microdialysis).
- A Micro LC 100 Pump (ALS) delivers mobile phase at a constant flow-rate of 70µl/min.
- Data are recorded using an SP-4290 integrator (Spectra-Physics, San Jose, CA, U.S.A.) or an EZChrom data acquisition system (Scientific Software).
- A 100x1mm I.D. chromatographic column is packed with C18 silica, 5µm particle size (ALS).

Mobile phase

Prepare 0.05M solution of disodium hydrogenphosphate, adjusted to pH 6.9 with concentrated phosphoric acid. Use only high-purity deionized or distilled water. Prepare a mobile phase by mixing 875ml 0.05M phosphate buffer with 100ml methanol and 25ml tetrahydrofuran.

OPA/MCE and ACN reagents, amino acid standard

Pipette 100µl of concentrated MCE (Sigma) and 900ml methanol into a 1.5ml Eppendorf tube. Pipette 1000µl OPA solution (Sigma incomplete, Cat. No. P7914) into a 1.5ml glass vial and add 14µl MCE working solution. Place a cap with silicone rubber septum on the vial; use a crimper to seal the vial. Mix properly. Pipette 1.2ml ACN into a 1.5ml glass vial and seal it with a silicone rubber septum. Prepare the reagents fresh every day. Dilute amino acid standard solution (Sigma, AA-S-18) with deionized water or Ringer solution to a final concentration 0.1 µM for Asp and Glu.

Procedure

1. A day before the experiment, insert the microdialysis probe (CMA/11 with a 2mm membrane length) into the guide cannula implanted in the rat brain (in the case described below frontal cortex (Fr2) area: A-P +1.7, L + 2.7, V –0.5mm) as described above. Perfuse the probe with Ringer solution at a low flow-rate of 0.5µl/min overnight and then increase to 2 µl/min. Following the initial 2–3 hour stabilization period, collect the fractions in 10min intervals.

2. Purge the pump with the mobile phase and set the flow-rate to 70 μl/min. Wait 10–15 min and observe the pressure and its stabilization.
3. Install the program for the autosampler, set the following parameters: Injected volume: 10 μl; reagent volume: 3 μl; analysis time: 2 min 30 s; standard injections/calibration: 1; wait in loop: 0 min 45 s. This means that 2.5 min after each 10 μl sample injection the autosampler will flush the needle and then pipette 10 μl ACN. Following additional 45 s the injection of ACN will rapidly flush out the rest of amino acids still retained in the column.
4. The fluorescence detector requires some 30–40 min to stabilize. Set amplify to 10 and rise time to 1 s.
5. Place the sealed glass vials with OPA/MCE and ACN reagents in their respective positions in the autosampler. Pipette 10 μl of amino acid standard working solution and blanks (perfusion medium) to the vials and seal them with caps with PTFE septa. Pipette 10 μl volumes from the collected microdialysis fractions into the sampling vials, seal them and place into the cassettes of the sampler's carousel.
6. Calibrate the system regularly between every 6–8 samples.

Results

The chromatogram of a 10 μl standard mixture of amino acids (AA-S-18, Sigma) containing 2.5 pmol each of Asp and Glu and the chromatogram of a 10 μl microdialysate sampled from the ventral hippocampus of awake rat are shown in Fig. 10A,B. As seen, the separation of Asp and Glu occurs within the first 3 min, thereafter the retained amino acids were washed out by ACN injection (a large third peak). The total run-to-run analysis time is about 8 min, which allows analysis of fully loaded autosampler (60 samples) within one working day (8h). The use of a microbore column improves the sensitivity of the assay down to 20–30 fmol Asp and Glu/10 μl samples. This means that samples as low as 1–2 μl can be analyzed, which allows to achieve a temporal resolution of about 30–60 s for off-line mode of microdialysis monitoring (Kehr 1998a).

Determination of GABA in Microdialysis Samples by HPLC with Fluorescence (FL) and Electrochemical (EC) Detection

Introduction

Determination of GABA in microdialysis samples requires an ultrasensitive analytical method, since basal GABA levels in a typical microdialysis sample are often in the range

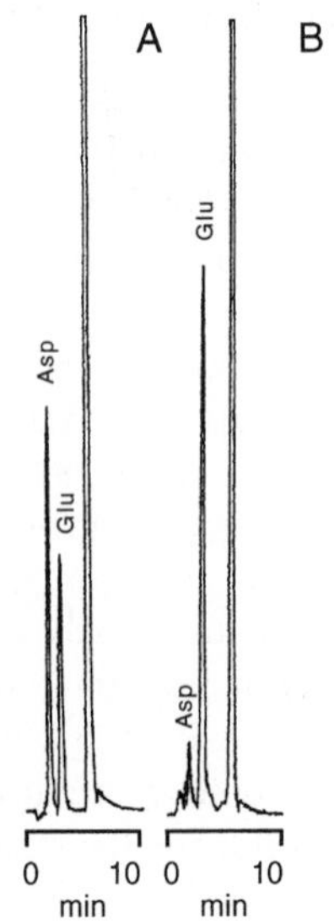

Fig. 10. Chromatograms of (A) a standard mixture of amino acids (AA-S-18). Asp and Glu concentrations were 2.5 pmol each in a 10 μl sample derivatized with 3 μl OPA/MCE reagent; a 10 μl volume was injected onto a column. (B) a 10 μl microdialysis sample from the ventral hippocampus of a conscious rat. Estimated Asp and Glu concentrations were 0.04 μM and 0.52 μM, respectively.

of 0.1–0.5 pmol (5–50 nM). Amino acids derivatized with *o*-phthaldialdehyde/thiol based reagents can be detected both by fluorescence (FL) and electrochemical (EC) detection (Joseph and Davies 1983; Allison et al 1984; Kehr 1998a,b). However, the EC detector is extremely sensitive to changes in composition of the mobile phase, which leads to a steep increase of the detector baseline during gradient elution. Therefore, for HPLC with EC detection several isocratic elution methods were developed for Asp and Glu (Kehr 1998a) or for GABA (Kehr and Ungerstedt 1988). The rapid method for GABA determination with sensitivity down to 50 fmol and a retention time of 3.5–4 min allows to characterize, by *in vivo* microdialysis, basal and evoked release of GABA in anesthetized (Kehr and Ungerstedt 1988; Drew et al 1989) and awake animals (Osborne et al 1991). Using FL detection of OPA/MCE derivatized amino acids, the limit of detection for GABA is only 0.5–1 pmol/sample for a typical chromatographic separation on a normal bore reversed-phase column with gradient elution (Westerberg et al 1988). To achieve detection levels of GABA below 100 fmol, it is thus necessary to "scale down" the chromatographic system by using a microbore column. The system can be further simplified by using isocratic separation with an automated washout step between the injections (Kehr 1998b).

Materials

Instrumentation

- Two liquid chromatographic analyzers are used for determination of GABA in the microdialysis samples:
  - HPLC with FL detection – the same instrument as that used for determination of Asp and Glu (see Materials in protocol 3.5 and
  - HPLC with EC detection – the same instrument and detector settings as described for electrochemical determination of DA (see above) except for the diameter of the glassy carbon electrode which is only 2 mm.
- The same type of microbore column is used for both systems: a 150 x 1 mm I.D. stainless-steel capillary, packed with C18 silica, 3 µm particle size (ALS).

Mobile phase

Prepare the mobile phase for OPA/MCE derivatization and FL detection of GABA consisting of 0.1 M sodium acetate buffer, pH 5.4, and 20% (v/v) ACN. Briefly, dissolve 13.61 g sodium acetate.$3H_2O$ in about 900 ml redistilled/deionized water and adjust pH to 5.4 with concentrated phosphoric acid. Dilute up to 1000 ml. Mix 800 ml acetate buffer with 200 ml ACN. The OPA/tBSH-derivatized GABA is separated by HPLC system with EC detection using the same mobile phase but the ACN content is increased to 50%. Flow-rates are 50 µl/min for both systems.

OPA/MCE and OPA/tBSH reagents, amino acid standard.

Pipette 100 µl of concentrated MCE (Sigma) or tBSH (2-methyl-2-propanethiol, Fluka) each into a 1.5 ml Eppendorf tube and add 900 ml methanol. Pipette 1000 µl OPA solution (Sigma incomplete, Cat. No. P7914) into a 1.5 ml glass vial and add 14 µl MCE working solution (OPA/MCE reagent) or 25 µl tBSH working solution (OPA/tBSH reagent). Place a cap with a silicone rubber septum on the vial; use a crimper to seal the vial. Mix properly. Pipette 1.2 ml ACN into a 1.5 ml glass vial and seal it with a silicone rubber septum. Prepare the reagents fresh every day. Dilute amino acid standard solution (Sigma, ANB) with deionized water, or Ringer solution to a final GABA concentration of 50 nM. Similarly, prepare a standard solution containing only 50 nM GABA and a solution of physiological amino acids without GABA (Sigma AN) at concentrations of 250 nM each.

Procedure

OPA/MCE derivatization and HPLC with FL detection.

1. Purge the pump with the mobile phase and set the flow rate to 50 µl/min. Wait 10–15 min and observe the pressure stabilization.

2. Install the program for the autosampler (the same as used for Asp/Glu assay, protocol 3.5), set the following parameters: injected volume: 10 µl; reagent volume: 3 µl (OPA/MCE); analysis time: 2 min 30 s; standard injections/calibration: 1; wait in loop: 13 min 30 s. This means that 2.5 min after each 10 µl sample injection the autosampler will flush the needle and then pipette 10 µl ACN. Following additional 13.5 min the injection of ACN will rapidly flush out the rest of amino acids still retained in the column.
3. The fluorescence detector requires some 30–40 min to stabilize. Set amplify to 100 and rise time to 1 s.
4. Place the sealed glass vials with OPA/MCE and ACN reagents in their respective positions in the autosampler. Pipette 10 µl of amino acid standard working solution, blanks (perfusion medium) and standard without GABA (Sigma AN) to the vials and seal them with caps with PTFE septa. Pipette 10 µl volumes from the collected microdialysis fractions into the sampling vials, seal them and place into the cassettes of the sampler´s carousel.
5. Calibrate the system regularly between every 6–8 samples.
6. Collect the microdialysis samples from the striatum of the awake rats (for details, see above).

OPA/tBSH derivatization and HPLC with EC detection.
1. Prepare the HPLC system as described for DA analysis in the Procedure section of protocol 3.4. Following the stabilization of the basal signal, the background current should be in the range of 10–15 nA.
2. Collect the microdialysis samples from the ventral hippocampus of the awake rat (for details, see above).

Results

GABA in the microdialysis samples can be determined by pre-column OPA derivatization and microbore HPLC coupled to either FL (Fig. 11) or EC (Fig. 12) detection. Figure 11A,B,C,D shows the chromatograms of (A) 10 µl standard mixture of amino acids (AN, Sigma), (B) the same standard spiked with GABA at a final concentration of 50 nM and the chromatograms illustrating the separation of typical microdialysis samples from the lateral striatum and substantia nigra of the awake rat are shown in Fig. 11C and D, respectively. Sample volumes of up to 20 µl can be injected onto the microbore column without a loss of baseline separation of GABA in these microdialysis samples (Kehr 1998b). The limit of quantification at a signal-to-noise ratio of two was 1.14 nM, or 23 fmol in a 20-µl sample. A similar detectability down to 10 fmol GABA was achieved using the EC detection of OPA/tBSH-derivatized GABA. Chromatograms in Fig. 12A,B show the separation of GABA in (A) a 10 µl mixture of amino acids (ANB standard) containing 500 fmol GABA and (B) a microdialysis sample collected from the ventral hippocampus of the awake rat. The estimated GABA concentration in this sample was 157 fmol. An excellent correlation (r=0.99, n=10) was found between GABA levels detected by FL and EC detection (Kehr 1998b).

Determination of Physiological Amino Acids in Microdialysis Samples by Microcolumn HPLC with Gradient Elution and Fluorescence (FL) Detection

Introduction

Determination of OPA-derivatized amino acids and related amino compounds in body fluids such as plasma, urine, CSF or lately in microdialysis samples requires the use of gradient elution HPLC in combination with fluorescence detection as originally described by Lindroth and Mopper (1979). The relative drawback of OPA is its specificity only towards primary amines and the instability of resulting fluorophores. Fluorescence

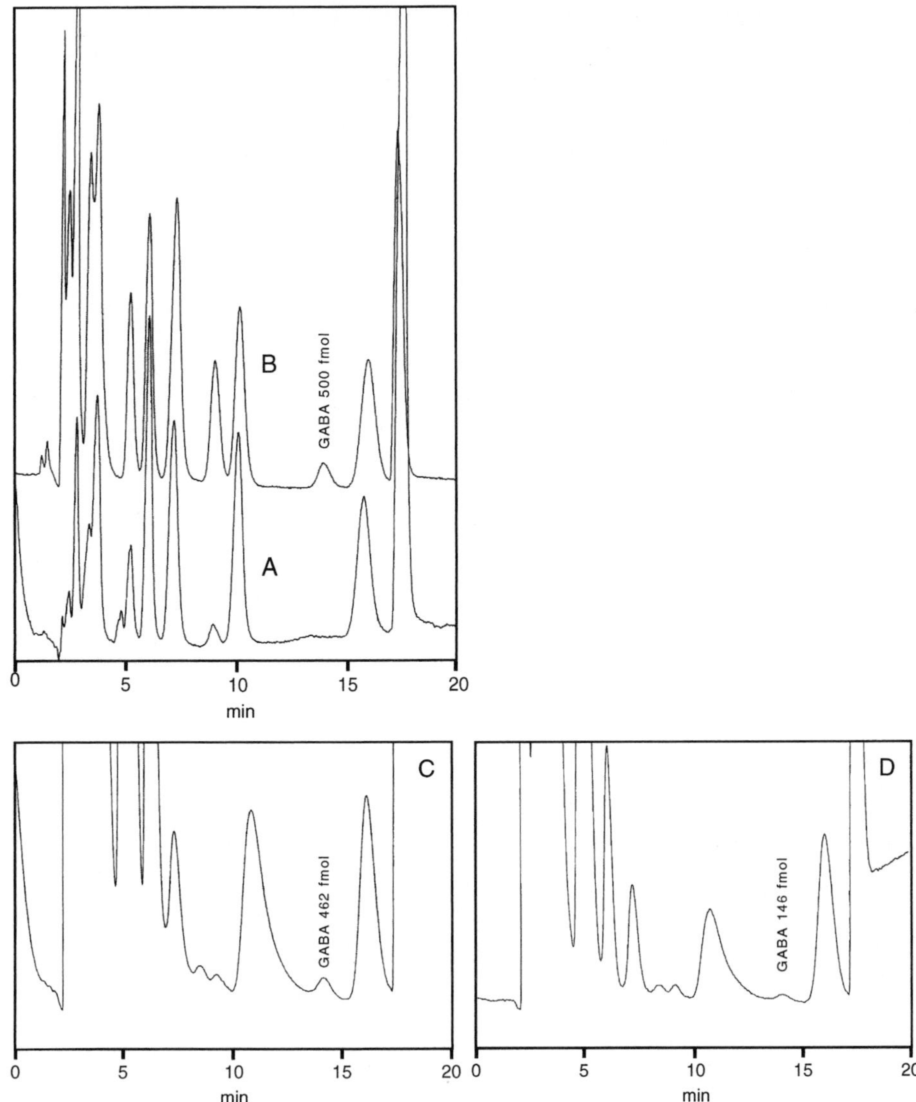

Fig. 11. Isocratic separations of GABA by microbore HPLC with fluorescence detection using automatic washout steps between injections: (A) a 10 μl amino acid standard mixture (AN, not containing GABA) at concentrations of 250 nM each, and (B) the same standard spiked with 500 fmol GABA. (C,D) GABA levels estimated in some typical microdialysis samples: 46.2 nM GABA in striatum (C) and 14.6 nM GABA in substantia nigra (D) of awake rats.

quantum yield of isoindole derivatives is dependent on each particular amine, excess of OPA and on the nature of the nucleophile (Kehr 1993). Besides thiols, other reducing agents such as cyanide or sulphite have also been reported. However, of all these nucleophiles, only 2-mercaptoethanol (MCE) and 3-mercaptopropionic acid (3-MPA) have gained broader use for FL determination of physiological amino acids. Mercaptans-based reagents offer feasible optimization of gradient separations of complex biological samples and high fluorescence yields. The first applications of OPA/MCE derivatization for analysis of amino acids in the microdialysis samples (Lehman et al 1983; Tossman et al 1986) were based on the use of normal-bore (4.6 mm I.D.) columns requiring relatively large sample volumes (20–40 μl), and high flow-rates (1–1.5 ml/min), resulting in

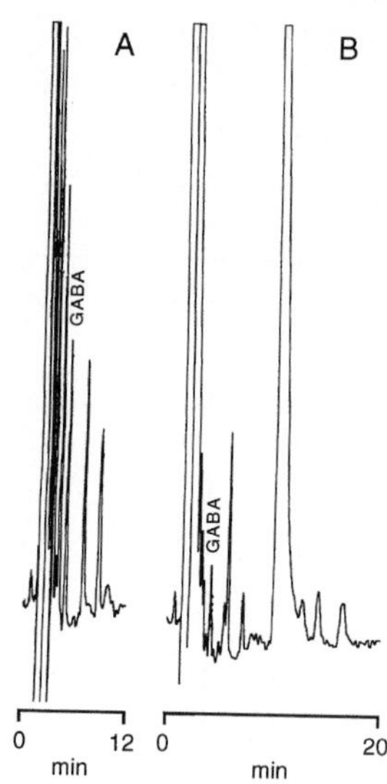

Fig. 12. Isocratic separations of GABA on microbore HPLC with electrochemical detection: (A) a 10 µl amino acid standard mixture (ANB) containing 50 nM GABA, and (B) a 10 µl microdialysis sample from ventral hippocampus of a conscious rat containing 15.7 nM GABA. The sensitivity of the detector cell was 0.62 nA full scale.

rather poor sensitivity and selectivity for neurotransmitter amino acids (Asp, Glu, Gly, GABA). Miniaturization of the LC system provides, besides improved sensitivity and reduced sample volumes down to 1–10 µl, also a number of practical advantages such as dramatically reduced consumption of the mobile phases and organic solvents.

Materials

Instrumentation

- The amino acid micro-LC analyzer consists of two CMA/250 LC Pumps, a CMA/252 Gradient Pump Controller, a CMA/200 Refrigerated Microsampler, a CMA/260 Degasser and a CMA/280 Fluorescence detector (CMA/Microdialysis).
- The gradients are mixed in a high-pressure mode, using a PEEK mixing tee connected to the outlets of the pumps and 25 cm of PEEK tubing (0.5 mm I.D.) connected to the injector valve and serving as a gradient mixer.
- The chromatograms are recorded and analyzed using an EZChrom data acquisition system (Scientific Software).
- The microbore column is a 150x1 mm I.D. stainless-steel capillary, packed with C18 silica, 5 µm particle size (ALS).

Mobile phase, gradient program, OPA/3-MPA reagent.
Prepare the mobile phases A and B as follows: A) dissolve 6.9 g $NaH_2PO_4.H_2O$ in 900 ml deionized water. Adjust pH to 6.95 with concentrated NaOH and dilute to 1000 ml. Filter the buffer through a 0.45 µm filter. Mix 985 ml of 50 mM phosphate buffer with 10 ml methanol (MeOH) and 5 ml tetrahydrofuran (THF). This will give mobile phase A: 50 mM phospate buffer, pH 6.95, 1 % MeOH and 0.5% THF. Mobile phase B is prepared by mixing 900 ml MeOH, 50 ml THF and 50 ml deionized water to give final 90% MeOH, 5% THF, 5% water. The gradient elution program is set as follows (time (min)-mobile phase %A/%B): 0–100/0; 30–80/20; 47–60/40; 55–40/60; 57–40/60; 60–100/0; the flowrate 50 µl/min. Pipette 100 µl of concentrated 3-MPA (Sigma) into a 1.5 ml Eppendorf tube and add 900 ml methanol. Pipette 1000 µl OPA solution (Sigma incomplete, Cat. No. P7914) into a 1.5 ml glass vial and add 25 µl 3-MPA working solution (OPA/3-MPA reagent). Place a cap with a silicone rubber septum on the vial; use a crimper to seal the vial. Mix properly. Prepare the reagent fresh every day.

Mix and dilute amino acid standard solution (Sigma, AA-S-18) with additional amino acids: Asn, Gln, Tau, GABA, Trp, ethanolamine, citruline and ornithine prepared separately in deionized water, or Ringer solution, to final concentrations of 1 µM.

Procedure

1. Purge the pumps with the mobile phases and set the flow rate to 50 µl/min. Run the gradient program without injecting any sample and observe pressure fluctuations. Wait 15–30 min for pressure stabilization.
2. Install the program for the autosampler. Set the amplification of the fluorescence detector to 10 and rise time (filter) to 1 s.
3. Place the OPA/3-MPA reagent to the autosampler. Pipette 10 µl of the amino acid standard solution and water or Ringer solution blanks into the vials, seal with the caps and place them into the autosampler. Set the derivatization parameters as follows: reagent volume 5 µl; mix 3 times 15 µl using liquid; reaction time 60 s; analysis time 65 min.
4. Calibrate the system with the amino acid standards. If there are any doubts about the identification of some peaks, run the standards of individual amino acids and compare the retention times.
5. For the case described below, collect the microdialysis samples from the striatum of awake rat, following the procedures described above. Pipette 10 µl microdialysis samples into the vials and analyze them at the same conditions as the standards.
6. Calibrate the system regularly between every 6–8 microdialysis samples.

Results

Separation of complex mixtures of amino acids in the form of their isoindole derivatives can be achieved by use of binary gradients and microbore HPLC, as shown in Fig. 13A. Derivatization of primary amino acids with OPA reagent using a thiol nucleophile containing a carboxylic group leads to derivatives which are more polar (hydrophilic) and thus less retained at the lipophilic surface of the reversed-phase material. It is mainly the amino acid moiety of each derivative which determines the strength of the hydrophobic interactions with the stationary phase. Thus, most polar amino acids such as Asp and Glu elute at the very beginning of the chromatogram, while the neutral and short-chain amino acids elute in the middle, and long-aliphatic chain amino acids elute at the end of the chromatographic run (Fig. 13A). A chromatogram of 10 µl microdialysis sample from rat striatum collected under basal conditions is shown in Fig. 13B. As seen, the peaks of amino acid neurotransmitters Asp, Glu and GABA are about 2–3 orders of magnitude smaller than those of the metabolic amino acids such as Gln, Ser, Thr, Ala, Tau, Met, Val, Leu, Iso etc. In fact, quantification of basal GABA outflow (0.12 µM) is at the limit of detection for this assay. The analysis time is about 60 min with additional 5-10 min re-equilibration period. The size of the detector cell is 6 µl, which is too large for micro-flow separations and may lead to a considerable peak broadening and reduction of separation efficiency. Therefore, a relatively long analysis time is needed for a full separation of amino acids. It is likely that using a micro-cell of 1–2 µl and higher flow-rates could improve the separation efficiency and speed up the analysis time to some 30–40 min.

Troubleshooting

(For all protocols in this section)

- The major difference between the high-sensitivity analysis of amino acids and, for example, monoamines is the problem of contaminants. It is rather easy to obtain false data, just because of using glassware, chemicals and other accessories contaminated with biological material. Always use the highest-purity water for preparation of HPLC buffers.
- Use sterile Ringer or aCSF solutions in microdialysis experiments.
- Always run a series of blank samples (water, Ringer, aCSF or even the mobile phase itself) to be sure the level of interference or background peaks is negligible when compared to the levels in your standards/microdialysis samples. For example, for Glu determined by the above-described methods a typical background peak of 10 µl derivatized water should be below 20 fmol.

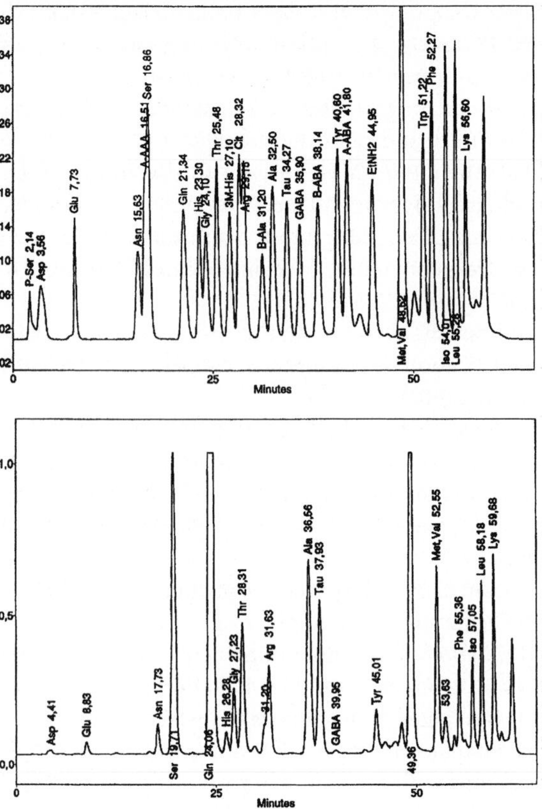

Fig. 13. A) A chromatogram of a 10 μl standard mixture containing 1–2.5 μM each of the following amino acids and related amines: P-Ser (Phospho-Ser), Asp, Glu, Asn, A-AAA (α-aminoadipic acid), Ser, Gln, His, Gly, Thr, 3M-His (3-methyl-His), Cit (citruline), Arg, B-Ala (β-Ala), Ala, Tau, GABA, B-ABA (β-aminobutyric acid), Tyr, A-ABA (α-aminobutyric acid), EtNH2 (ethanolamine), Met, Val, Trp, Phe, Iso, Leu and Lys. (B) A 10 μl microdialysis sample from the striatum of awake rat containing (in μM): Asp 0.125, Glu 0.413, Asn 0.713, Ser 7.724, Gln 11.524, His 0.82, Gly 1.65, Thr 3.041, Arg 1.958, Ala 4.654, Tau 4.095, GABA 0.125, Tyr 0.772, Met+Val 1.504, Phe 1.401, Iso 1.056, Leu 2.111 and Lys 4.5.

- The OPA reagent ages rather fast, therefore it is important to calibrate the system regularly and to prepare the fresh reagent every day. Similarly, many amino acids are unstable, especially basic amino acids Gln, Asn, as well as GABA, Tau, His, ethanolamine, Trp. Thus, using the old reagent and old standards can reduce or even broaden the peaks of some amino acids and at the same time, a number of unidentified new peaks can appear in the chromatogram.
- Special care must be taken when handling the organic thiols. The considerable odor of the 2-methyl-2-propanethiol (tBSH) reagent may cause some practical problems with the location of the HPLC apparatus, storage and handling of the reagent, etc. Always use diluted (stock) solutions of MSE and tBSH in as small volumes as possible. Keep the solutions in a desiccator; optionally use some odor absorbent, which you can place in the desiccator and in the refrigerated part of the autosampler.

Comments

It is convenient to prepare the OPA/MCE reagent from the ready-made solution, for example, OPA-incomplete from Sigma. This reagent contains 1 mg OPA/ml solution, which is a borate buffer (pH 10.4) containing Brij 35 (polyoxyethylene 23 lauryl ether) working as a stabilizer, and methanol. However, the OPA reagent can also be prepared in the laboratory as follows (Kehr 1998a): dissolve 13.4 mg OPA (specially purified grade) in 5 ml of a solution consisting of 50% 0.2 M borate buffer (prepared from boric acid and titrated with NaOH to pH 9.5) and 50% methanol. Then add 28 μl concentrated MCE. This stock solution can be used for about 14 days. The working reagent is prepared fresh daily by diluting the stock solution 4 times with borate-methanol buffer. This gives the final 5 mM OPA and 20 mM MCE solution. This reagent should quantitatively derivatize samples with a total amino acid content of up to 1 mM, which means at about 50 pmol/μl for an amino acid in a microdialysis sample.

Applications

Glu is considered to act as a major excitatory and GABA as a major inhibitory neurotransmitter in the brain (Fonnum 1984; Curtis et al 1970). The role of Asp is more controversial. It has been argued that Asp levels reflect only general metabolism (Orrego and Villanueva 1993), while several other studies suggest its excitatory role, at least in the cortico-striatal pathway (for review, see Herrera-Harschitz et al 1997). Microdialysis is frequently used for studies of Glu and GABA overflow following various pharmacological, but mostly neuropathological manipulations (for review, see Fillenz 1995; Timmerman and Westerink 1997). However, the applicability of microdialysis sampling to monitor neuronally released Glu and GABA has been hampered by the existence of large non-exocytotically released pools of these amino acids in the extracellular fluid. In fact, microdialysis failed to demonstrate the vesicular origin of Glu and GABA and the exocytotic mechanism of their release, since neither the local perfusions with Na-channel blocker tetrodotoxin (TTX) nor with calcium-free medium could reduce basal Glu or GABA levels. On the contrary, several authors reported increased levels of extracellular Glu (Herrera-Marschitz et al 1996) and GABA (Drew et al 1989) under these conditions. A number of attempts have been made to elucidate the neuronal pool of Glu and GABA (for review, see Timmerman and Westerink 1997): a) Inhibition of synthesis or metabolism, b) stimulation by high potassium, veratridine, ouabain and in combination with TTX, c) the use of reuptake inhibitors, d) chemical lesions, e) chemical stimulation using a dual-probe approach, f) electrical stimulation. In the latter case, it has been demonstrated for both Glu and GABA that the release is only short-lasting despite the prolonged period of stimulation. This makes it even more important to develop rapid and high-sensitive HPLC methods allowing to analyze microdialysis samples collected in 30–60 s intervals (Kehr 1998a). Further improvement of temporal resolution for detection of electrically stimulated Glu release can be achieved by applying the biosensor approach using a microdialysis electrode (Walker et al 1995) or by recently developed capillary electrophoresis with laser-induced fluorescence detection (Bert et al 1996; Lada et al 1998).

Determination of Acetylcholine in Microdialysis Samples by Microbore Liquid Chromatography/Electrochemistry on Peroxidase Redox Polymer Coated Electrodes

Introduction

Determination of acetylcholine (ACh) and choline (Ch) by liquid chromatography/electrochemistry (LCEC) was first reported by Potter and coworkers (1983). The original method using post-column enzymatic reactions to produce hydrogen peroxide was further developed by incorporating *immobilized enzyme reactors* (IMER) where acetylcholine esterase and choline oxidase were attached to the particles in the packed-bed reactor (Eva et al 1984). Conversion of separated cholines to H_2O_2 is easily detected electrochemically by oxidation of H_2O_2 on a platinum (Pt) electrode at +500 mV. The high sensitivity of the latter method allowed determination of extracellular ACh in brain tissue sampled by *in vivo* microdialysis (Damsma et al 1987; Tyrefors and Gillberg 1987; Ajima and Kato 1987), though in most cases only in the presence of cholinesterase inhibitors, such as physostigmine or neostigmine in the perfusion medium. Recently, the "wired" peroxidase redox polymer coated electrodes were introduced in LCEC of various oxidizable analytes including ACh detection in microdialysis samples (Huang et al 1995; Kato et al 1996; Kehr et al 1996). The advantages of these electrodes over the Pt-electrodes are their lower detection limits, lower background noise and particularly a much faster stabilization of the background current. The principle of electrochemical detection utilizing a "wired" horseradish peroxidase (HRP) was originally developed by Vreeke and coworkers (1992) for amperometric biosensors. Briefly, H_2O_2 generated by peroxidase enzymes is reduced by HRP trapped in the Os^{II} (2,2′-bipyridine)$_2$Cl -poly(4-vinylpyridine) polymer (HRP-Os(PVP)). Next, HRP oxidizes Os^{II} to the Os^{III} (PVP) complex, which is finally reduced to its original form by accepting one electron from the electrode surface operating at 0 mV. The limit of detection for the laboratory coated glassy carbon electrode is 10 fmol ACh/5 µl (Huang et al 1995) and around 5 fmol ACh/10 µl on pre-coated gold film electrodes (Kehr et al 1996).

Materials

Instrumentation

- An Intelligent Micro-LC pump, LC100 (ALS), a CMA/260 Degasser and a CMA/200 Refrigerated Microsampler (CMA/Microdialysis), equipped with a 10 µl loop and operating at +6 °C, a BAS LC4B Amperometric Detector (Bioanalytical Systems) equipped with a Radial-flow electrochemical cell (ALS) are used.
- The glassy carbon electrodes (6 mm I.D. from ALS) or gold film electrodes (ALS) are coated with a peroxidase redox polymer solution (BAS). The potential of working electrode vs. Ag/AgCl reference electrode is 0 mV.
- Cholines are separated on a 530x1 mm microbore column, packed with a polymeric strong cation-exchanger, 10 µm particle size (BAS) at a flow rate of 120 µl/min. The IMER cartridges (2x5 mm) are available from ALS.
- The chromatograms are recorded and integrated on a Chrom Jet integrator (Spectra Physics).

Chemicals

A mobile phase for choline separations consists of 50 mM disodium hydrogenphosphate, 0.5 mM disodium-EDTA (both from Merck) and 0.05 % v/v Kathon CG (Rohm and Haas), final pH 8.5. In the case described below, acetylethylhomocholine (AEHCh) was kindly provided by Dr. J. Ricny from Inst. of Physiology, Prague, Czech Republic. Acetylcholine (ACh), choline (Ch), butyrylcholine (BCh) and all other chemicals are available from Sigma (St.Louis, MO, USA). Stock solutions of 100 µM choline and physostigmine are prepared in distilled water and kept frozen at –70 °C, working solutions are prepared daily by diluting AEHCh and BCh down to 1 µM in Ringer solution and ACh and Ch to a final concentration of 30 nM each in Ringer containing 10 µM physostigmine.

Procedure

1. Polish the glassy carbon electrode on a soft pad with an aqueous dispersion of 0.5 µm alumina (ALS). After rinsing with distilled water and drying in air, pipette 2 µl of a redox polymer solution (BAS) onto the electrode's surface. A thin film should be created by a gentle circle movement of a pipette plastic tip over the electrode's surface.
2. If the polymer solution tends to associate in droplets due to a lipophilic surface, the electrode should be washed with water and re-polished again with alumina. After drying, pipette 10 µl of 1 M NaOH onto the electrode for 1–2 min, in order to improve hydrophilicity of the surface, followed by intensive washing before repeated coating.
3. The polymer electrode is left to harden in a dust-free box at room temperature for at least 12 hours before use. Gold electrodes can be purchased already pre-coated with the redox polymer from ALS-Japan.
4. Purge the HPLC pump with the mobile phase and install the microbore column into the HPLC system. Stabilize the flow-rate at 120 µl/min.
5. Mount the working electrode into the electrochemical cell. Connect the cell to the column outlet. Wait until the cell internal volume is filled with the mobile phase and then fit the Ag/AgCl/3 M NaCl reference electrode into the cell.
6. Set the working potential of the electrochemical detector to 0 mV and operate in a reductive mode. This will convert the negative electric current to the positive signals at the detector´s output. Set the active filter to 0.1 Hz.

7. Turn on the cell and wait until the basal signal is stabilized to some –5 to –10 nA. Reduce the background current to an appropriate baseline level by activating the offset function.
8. Place the microdialysis samples and the standards in 300 μl vials into the CMA/200 autosampler. Program the autosampler to inject 10 μl and set the analysis time to 30 min. For analyses using an internal standardization method, pipette a solution of AEHCh and/or BCh at concentration 1 μM each into a 1.5 ml Chromacol glass vial and place it in the reagent position of the CMA/200 autosampler. Program the autosampler to pipette 1 μl of internal standard solution to each sample, mix and thereafter inject 10 μl onto the column.

Results

The protocol describes the method for simultaneous determination of ACh, Ch and physostigmine in microdialysis samples by microbore LCEC. Applying the HRP-Os(PVP) coated electrodes technology together with the internal standardization with AEHCh or BCh, the sensitivity and precision of the assay is improved as compared to the direct detection of hydrogen peroxide on the Pt-electrode. The high sensitivity of the redox polymer-coated electrodes allows the detection of basal extracellular ACh levels at perfusions without esterase inhibitor, as illustrated in Fig. 14A,B. In this case, the mean basal ACh levels were 1.6 nM (ranging from 0.9 to 2.5 nM) in 10 μl samples. However, for many pharmacological studies where basal ACh levels are expected to be reduced, the addition of esterase inhibitor is still necessary. A typical chromatogram of a microdialysis sample from ventral hippocampus of the awake rat is shown in Fig. 14C. Estimated ACh and Ch levels in 10 μl samples were 50.2 nM and 0.31 μM, respectively. Physostigmine concentration was reduced from an initial 10 μM to 8.86 μM, indicating

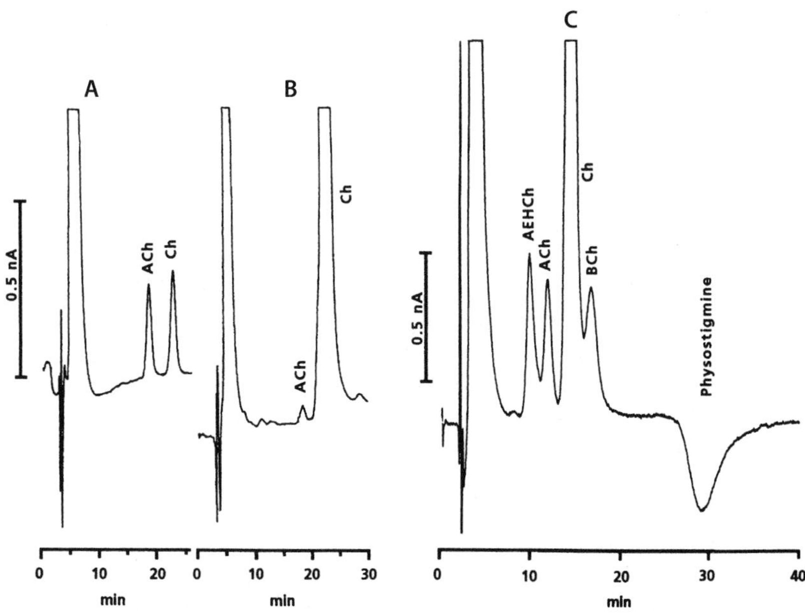

Fig. 14. Separation of (A) 3 μl standard containing 150 fmol ACh and Ch each on a gold redox polymer pre-coated electrode and (B) 10 μl of a typical microdialysis sample from ventral hippocampus of awake rat at basal levels. The CMA/12 microdialysis probe was perfused with aCSF without physostigmine at 1.25 μl/min. Estimated ACh concentration was 2.4 nM. (C) Separation of ACh, Ch and physostigmine in 10 μl of a microdialysis sample mixed with 1 μl of internal standards AEHCh and BCh, 1 μM each. The CMA/12 microdialysis probe was perfused with aCSF with 10 μM physostigmine at 1.25 μl/min. Estimated levels were 50.2 nM ACh, 306 nM Ch and 8.86 μM physostigmine.

11.4% *in vivo* delivery of physostigmine into the ventral hippocampus. The internal standardization method reduced the variation by about 1.5 % and for a given set of 36 microdialysis samples from 3 animals shortened the total analysis time by about 3 hours when compared to the external standardization at every 7th injection (Kehr et al 1998).

Troubleshooting

- The most common problem is the loss of sensitivity due to aging – reducing activity of IMER and the electrode film. It is strongly recommended to keep the whole assembly (column, IMER, electrode cell) in the refrigerator when not used.
- To prevent growth of microorganisms in the mobile phase and flushing medium (water), use Kathon CG (or an alternative biocide ProClin 300) which is a mixture of two isothiazolones with low toxicity and high antimicrobial activity.
- Stock solutions of Ch and ACh should be prepared in 10 mM HCl. Volumes of 1 ml aliquots can be stored at least 6 months in a freezer at –70 °C. Working standard solutions should be prepared fresh every day.
- The microbore column will successively increase the backpressure without any significant effects on peak shapes and retention times of ACh and Ch. Sometimes a reversal of the column helps to regain its permeability.

Comments

The HRP-Os(PVP)-coated electrodes for detection of H_2O_2 originating from enzymatically converted ACh, Ch and their derivatives provide a significant methodological improvement compared to the method of direct oxidation of H_2O_2 on a Pt-electrode. Faster signal stabilization, higher signal-to-noise ratio and lower limits of detection for ACh and Ch have been reported previously (Huang et al 1995; Kato et al 1996; Kehr et al 1996). However, a great variability in the stability of laboratory-coated electrodes has been observed; usually the electrode sensitivity diminishes by about 10%/day starting from the first injection (Kehr et al 1998). Thus, similarly as for the reported reduction of ACh peaks on pre-coated gold electrodes (Kehr et al 1996) and Pt-electrodes (Carter, Kehr, 1997), the application of internal standardization could improve the precision of ACh detection for both off-line and on-line methods.

Applications

The cholinergic system of the *basal forebrain* plays a crucial role in *learning and memory*. The cholinergic hypofunction is associated with progressive reduction of cognitive function such as in *senile dementia of Alzheimer type*. High-sensitive determination of extracellular ACh in microdialysis samples is an elegant scientific tool for *in vivo* tests of the efficacy of various cholinomimetics and other drugs as possible candidates in cholinergic replacement therapy. Physostigmine can be taken as a reference compound and its distribution and kinetics can be measured simultaneously with ACh outflow. The internal standardization method is especially suitable for long-term off-line or on-line monitoring. Additionally, other choline derivatives such as butyrylcholine can be detected by the present method.

Microdialysis in the Human Brain

Introduction

During the past few years, the microdialysis technology has been developed for use in humans, which has enabled its rapid penetration into the clinical fields, namely, *(neuro)intensive care monitoring* (Hildingsson et al 1996; Persson et al 1996; Ungerstedt 1997; Zauner 1998; for review, see Landolt and Langermann 1996), *diabetes* and *clinical pharmacology*. For the first time, it is possible to follow brain biochemistry in patients with *brain trauma*, *stroke* and *subarachnoid hemorrhage*. Monitoring changes in energy metabolism (glucose, lactate, pyruvate), excitotoxic release of glutamate as well as glycerol as an index of membrane decomposition provide valuable information about the patient's status. Samples as low as 18 µl are collected every hour and analyzed bedside by a spectophotometric analyzer. Simple enzymatic assays, implementing specific oxidases and colorimetric reaction of enzymatically yielded hydrogen peroxide with 4-aminoantipyrine, are used for all the analytes. The only exception is urea which is based on measurement of reduction of NADH absorbance in a two-step enzymatic reaction.

Materials

Instrumentation

- A system for bedside monitoring in neurointensive care includes a CMA 600 Microdialysis analyzer (CMA/Microdialysis), a CMA 106 Microdialysis pump and a CMA 70 Brain microdialysis catheter. The catheter is a class III product according to European Medical Device Directive and is available in various lengths, allowing its stereotaxic implantation into the human brain.
- The reagents and calibrators for glucose, lactate, glycerol, urea, pyruvate and glutamate are available from CMA/Microdialysis.

Procedure

1. Insert the CMA 70 catheter into the brain parenchyma during the neurosurgical procedure of implanting the devices for monitoring the intracranial pressure and CSF drainage.
2. Fill the syringe of the CMA 106 pump with perfusion fluid (sterile Ringer solution). Remove any air bubbles. Connect the Luer connection of the syringe to the inlet tubing of the catheter. Place a microvial into the holder.
3. Fit the syringe into the pump and close the cover. The pump will start automatically with flush cycle, which lasts for 16 min.
4. Place a new microvial into the holder and start sampling. The vials are replaced manually every 60 min.
5. Prepare the reagents for the CMA 600 analyzer as follows: Transfer the contents of buffer bottles for each reagent to its respective reagent bottles. Place the cups with the septa on the reagent bottles. Dissolve the contents completely, let the reagents equilibrate at room temperature for at least 30 min.
6. Prepare the calibrator bottle by removing the rubber stopper and replacing it with a septum containing a cap.
7. Place 4 reagents, the calibrator and the samples in microvials into the sample tray of the CMA 600 analyzer.
8. Start the analysis by pressing the start button at the instrument or from the computer software.

Results

The enzymatic assays are based on kinetic measurements of the reaction rate, which is proportional to the analyte concentration. The sample and the reagent are mixed in a flow stream during the aspiration to the cuvette. For every analyte 30 s absorbance readings are satisfactory to calculate the slope of the reaction, thus making it possible to complete the analysis of four substances in each sample in a few minutes. The limits of detection (in µM) are for glucose 100; lactate 100; glycerol 10; urea 500; pyruvate 10 and glutamate 1 for sample volumes between 0.2 to 2 µl and reagent volumes between 5 to 15 µl per test. Typical extracellular concentrations of these substances sampled from human cortex are at least ten times above the detection limits. The following example (Fig. 15) illustrates the concentration profiles of glucose (B), lactate (C), glycerol (D) and glutamate (E) sampled by microdialysis from a patient with severe head trauma. The accompanying measurement of intracranial pressure (A) in the injured brain is also shown. As seen, there is a good correlation between the increase of intracranial pressure as a result of growing brain edema and the loss of energy substrates (glucose) with accompanying acidosis (lactate) and a massive release of glutamate. Although the latter observations have been reported in a number of animal studies using microdialysis in

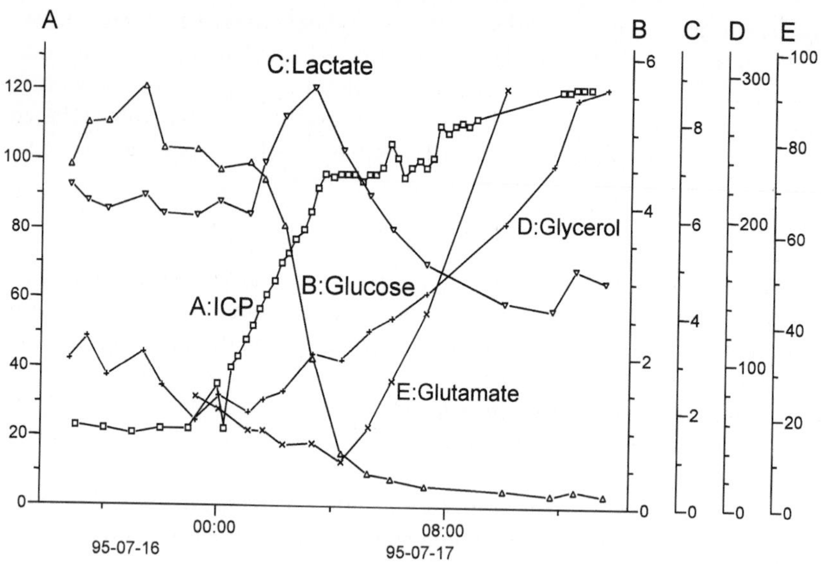

Fig. 15. Microdialysis monitoring of a patient with severe brain trauma. The following parameters were recorded simultaneously: (A) Intracranial pressure (ICP, mmHg), (B) glucose (mM), (C) lactate (mM), (D) glycerol (μM), (E) glutamate (μM). A critical period of 16 hours is shown during which the patient's status was severely impaired as demonstrated by a rapid increase in ICP followed by a loss of energy substrates (glucose, glutamate), acidosis (lactate) and consecutive damage of cell membrane structure (glycerol).

various models of brain injury (Benveniste et al 1984; Hillered et al 1989), the long-term monitoring of chemical markers of human brain damage have become available only recently.

Comments Clinical microdialysis in subcutaneous adipose tissue or resting skeletal muscle can be performed in an ambulatory environment and does not require special training. A CMA 60 microdialysis catheter with a special introducer should be used for this purpose. In the USA, the CMA 60 and CMA 70 microdialysis catheters are considered as investigational devices and should be used only in IRB approved studies.

Applications The enzymatic assays of glucose, lactate, pyruvate, glycerol, glutamate and urea are developed for microdialysis samples from humans, i.e., for sampling under the conditions of reaching almost 100% recovery of these substances from the exracellular fluid. However, the sensitivity of miniaturized spectrophotometric methods allows detecting most of the analytes even in samples from animals in experimental brain research.

Clinical microdialysis offers to the physician new information on *time-courses of substances involved in cell survival or death*. This in turn opens a unique opportunity for an early intervention to stop the onset of developing brain damage. Unfortunately, here a possible choice of treatments is still very limited. In practice, besides barbiturates and anticoagulants in case of stroke, no specific anti-ischemic drugs are available today. Several physiologically based treatments such as hypothermia and hypotension and partial hypoxia in the post-ischemic recovery period are currently under investigation. Brain microdialysis can help to find the most efficient therapy.

Acknowledgement:

General Introduction: We thank Dr. K. Drew, Inst. of Arctic Biology, Univ. Alaska, Fairbanks, for providing the EM photo used in Fig. 1.

Part 2, Section 2.2: We thank Dr. E. Syková, Inst. of Experimental Medicine, Prague, Czech Republic, for kind revision of this protocol.

Part 2, Section 2.3: We thank Dr. I. Strömberg, Dept. Neuroscience, Karolinska Institute, Stockholm, for kind revision of the protocol.

Part 2, Section 2.6: We thank Dr. J. P. Lowry, Biomedical Sensors Laboratory, Neurochemistry Research Unit, Dept. of Chemistry, National University of Ireland, Maynooth, Ireland, for kind revision of this protocol and helpful comments.

Part 3, Section 3.7: We thank Prof. U. Ungerstedt, Dept. Pharmacology and Physiology, Karolinska Institute, Stockholm, for providing the data and kind revision of the protocol.

References

Abercrombie ED, Zigmond MJ (1989) Partial injury to central noradrenergic neurons: reduction of tissue norepinephrine content is greater than reduction of extracellular norepinephrine measured by microdialysis. J Neurosci 9:4062–4067

Adams RN (1990) In vivo electrochemical measurements in the CNS. Prog. Neurobiol 35:297–311

Ajima A, Kato T (1987) Brain dialysis: detection of acetylcholine in the striatum of unrestrained and unanesthetized rats. Neurosci Lett 81:129–132

Amman D, (1986) Ion-selective microelectrodes. Principles, design and applications. Springer Verlag, Berlin, Heidelberg , New York, Tokyo

Armstrong-James M, Millar J, Kruk ZL (1980) Quantification of noradrenaline iontophoresis. Nature 288:181–183

Asai S, Iribe Y, Kohno T, Ishikawa K (1996) Real time monitoring of biphasic glutamate release using dialysis electrode in rat acute brain ischemia. Neuroreport 7:1092–1096

Barker SA (1987) Immobilization of the biological component of biosensors. In: Turner AF, Karube I, G. Wilson GS (eds) Biosensors: Fundamentals and applications, Oxford University Press, New York, pp 85–99

Benveniste H, Drejer J, Schousboe A, Diemer NH (1987) Regional cerebral glucose phosphorylation and blood flow after insertion of a microdialysis fiber through the dorsal hippocampus in the rat. J Neurochem 49:729–34

Benveniste H, Diemer NH (1987) Cellular reactions to implantation of a microdialysis tube in the rat hippocampus. Acta Neuropathol (Berl) 74:234–238

Benveniste H, Drejer J, Schousboe A, Diemer NH (1984) Elevation of the extracellular concentrations of glutamate and aspartate in rat hippocampus during transient cerebral ischemia monitored by intracerebral microdialysis. J Neurochem 43:1369–1374

Bert L, Robert F, Denoroy L, Stoppini L, Renaud B (1996) Enhanced temporal resolution for the microdialysis monitoring of catecholamines and excitatory amino acids using capillary electrophoresis with laser-induced fluorescence detection. Analytical developments and in vitro validations. J Chromatogr 755:99–111

Bhattacharya BK, Feldberg W (1959) Perfusion of cerebral ventricles: assay of pharmacologically active substances in the effluent from the cisterna and the aqueduct. Br J Pharmacol Chemother 13:163–174

Bito L, Davson H, Levin E, Murray M, Snider N (1966) The concentrations of free amino acids and other electrolytes in cerebrospinal fluid, in vivo dialysate of brain, and blood plasma of the dog. J Neurochem 13:1057–1067

Boulton AA, Baker GB, Walz W (eds) (1988) The neuronal microenvironment, Neuromethods Vol 9. Humana Press, Clifton, NJ

Boulton A, Baker GB, Adams RN (eds) (1995) Voltammetric methods in brain systems. Humana Press, Totowa, NJ

Caliguri EJ, Mefford IN (1984) Femtogram detection limits for biogenic amines using microbore HPLC with electrochemical detection. Brain Res 296:156–159

Carter A, Kehr J (1997) Microbore high-performance liquid chromatographic method for measuring acetylcholine in microdialysis samples: optimizing performance of platinum electrodes. J Chromatogr 692:207–212

Celesia GG, Jasper HH (1966) Acetylcholine released from cerebral cortex in relation to state of activation. Neurology 16:1053–1063

Church WH, Justice JB Jr (1987) Rapid sampling and determination of extracellular dopamine in vivo. Anal Chem 59:712–716

Clarencon D, Testylier G, Estrade M, Galonnier M, Viret J, Gourmelon P, Fatome M (1993) Stimulated release of acetylcholinesterase in rat striatum revealed by in vivo microspectrophotometry. Neuroscience 55:457–462

Delgado JM, DeFeudis FV, Roth RH, Ryugo DK, Mitruka BM (1972) Dialytrode for long term intracerebral perfusion in awake monkeys. Arch Int Pharmacodyn 198:9–21

Curtis DR, Felix D, McLennan H (1970) GABA and hippocampal inhibition. Br J Pharmacol 40:881–883.

Damsma G, Westerink BHC, de Vries JB, Van den Berg CJ, Horn AS (1987) Measurement of acetylcholine release in freely moving rats by means of automated intracerebral dialysis. J Neurochem 48:1523–1528

Drew KL, O´Connor WT, Kehr J, Ungerstedt U (1989) Characterization of extracellular gamma-aminobutyric acid and dopamine overflow following acute implantation of a microdialysis probe. Life Sci 45:1307–1317

Drijfhout WJ, Van der Linde AG, Kooi SE, Grol CJ, Westerink BHC (1996) Norepinephrine release in the rat pineal gland: the input from the biological clock measured by in vivo microdialysis. J Neurochem 66:748–755

Drijfhout WJ, Kemper RH, Meerlo P, Koolhaas JM, Grol CJ, Westerink BH (1995) A telemetry study on the chronic effects of microdialysis probe implantation on the activity pattern and temperature rhythm of the rat. J Neurosci Methods 61:191–196

During MJ, Spencer DD (1993) Extracellular hippocampal glutamate and spontaneous seizure in the conscious human brain. Lancet 341:1607–1610

Eisenman G (1967) Glass electrodes for hydrogen and other cations: principles and practice. M. Dekker, New York

Elmquist WF, Sawchuk RJ (1997) Application of microdialysis in pharmacokinetic studies. Pharmaceut Res 14:267–288

Eva C, Hadjiconstantinou M, Neff NH, Meek JL (1984) Acetylcholine measurement by high-performance liquid chromatography using an enzyme-loaded postcolumn reactor. Anal Biochem 143:320–324

Ewing AG, Bigelow JC, Wightman RM (1983) Direct in vivo monitoring of dopamine released from two striatal compartments. Science 221:169–171

Fellows LK, Boutelle MG (1993) Rapid changes in extracellular glucose levels and blood flow in the striatum of the freely moving rat. Brain Res 604:225–231

Fellows LK, Boutelle MG, Fillenz M (1992) Extracellular brain glucose levels reflect local neuronal activity:a microdialysis study in awake, freely moving rats. J Neurochem 59:2141–2147

Fillenz M (1995) Physiological release of excitatory amino acids. Behav Brain Res 71:51–67

Fonnum F (1984) Glutamate: A neurotransmitter in mammalian brain. J Neurochem 42:1–11

Franklin KBJ, Paxinos G (1997) The Mouse Brain in Stereotaxic Coordinates. Academic Press, San Diego

Fray AE, Boutelle M, Fillenz M (1997) Extracellular glucose turnover in the striatum of unanaesthetized rats measured by quantitative microdialysis. J Physiol 504:721–726

Freitag R (1993) Applied biosensors. Curr Opin Biotechnol 4:75–79

Fuxe K, Agnati L (1991) Volume transmission in the brain, novel mechanisms for neuronal transmission. In: Fuxe K, Agnati L (eds) Advances in neuroscience. Raven Press, New York, pp 1–11

Gaddum JH (1961) Push-pull cannulae. J Physiol 155:1P-2P

Garris PA, Ciolkowski EL, Pastore P, Wightman RM (1994) Efflux of dopamine from the synaptic cleft in the nucleus accumbens of the rat brain. J Neurosci 14:6084–6093

Gerhardt GA, Oke AF, Nagy G, Moghaddam B, Adams RN (1984a) Nafion-coated electrodes with high selectivity for CNS electrochemistry. Brain Res 290:390–395

Gerhardt GA, Palmer M, Seiger A, Adams RA, Olson L, Hoffer BJ (1984b) Adrenergic transmission in hippocampus-locus coeruleus double grafts in oculo: demonstration by in vivo electrochemical detection. Brain Res 306:319–325

Geusz ME, Fletcher C, Block GD, Straume M, Copeland NG, Jenkins NA, Kay SA, Day RN (1997) Long-term monitoring of circadian rhythms in c-fos gene expression from suprachiasmatic nucleus cultures. Current Biol 7: 758–766

Gonon F, Cespuglio R, Ponchon JL, Buda M, Jouvet M, Adams RN, Pujol JF (1978) Mesure électrochimique continue de la libZration de DA rZalisZ in vivo dans le nZostriatum du rat. C R Acad Sci 286:1203–1206

Gonon F, Buda M, Cespuglio R, Jouvet M, Pujol JF (1980) In vivo electrochemical detection of catechols in the neostriatum of anaesthetized rats: dopamine or DOPAC? Nature 286:902–904

Gonon F, Buda M (1985) Regulation of dopamine release by impulse flow and by autoreceptors as studied by in vivo voltammetry in the rat striatum. Neuroscience 14:765–774

Grinvald A, Lieke E, Frostig RD, Gilbert CD, Wiesel TN (1986) Functional architecture of cortex revealed by optical imaging of intrinsic signals. Nature 324:361–364

Hamberger A, Berthold C-H, Jacobson I, Karlsson B, Lehmann A, Nyström B, Sandberg M (1985) In vivo brain dialysis of extracellular neurotransmitter and putative transmitter amino acids. In: Bayon A, Drucker-Colin R (eds) In vivo perfusion and release of neuroactive substances. Alan R Liss, New York, pp 473–492

Hamberger A, Jacobson I, Molin S-O, Nyström B, Sandberg M, Ungerstedt U (1982) Metabolic and transmitter compartments for glutamate. In: Bradford H (ed) Neurotransmitter interaction and compartmentation. Plenum, New York, pp 359–378

Hefti F, Melamed E (1981) Dopamine release in rat striatum after administration of L-dope as studied with in vivo electrochemistry. Brain Res 225:333–346

Herrera-Marschitz M, You ZB, Goiny M, Meana JJ, Silveira R, Godukhin OV, Chen Y, Espinoza S, Pettersson E, Loidl CF, Lubec G, Andersson K, Nylander I,Terenius L, Ungerstedt U (1996) On the origin of extracellular glutamate levels monitored in the basal ganglia of the rat by in vivo microdialysis. J Neurochem 66:1726–1735

Herrera-Marschitz M, Goiny M, You ZB, Meana JJ, Pettersson E, Rodriguez-Puertas R, Xu ZQ, Terenius L, Hskfelt T, Ungerstedt U (1997) On the release of glutamate and aspartate in the basal ganglia of the rat: interactions with monoamines and neuropeptides. Neurosci & Biobehav Rev 21:489–495

Hildingsson U, Sellden H, Ungerstedt U, Marcus C (1996) Microdialysis for metabolic monitoring in neonates after surgery. Acta Paediatrica 85:589–594

Hillered L, Hallstrom A, Segersvard S, Persson L, Ungerstedt U (1989) Dynamics of extracellular metabolites in the striatum after middle cerebral artery occlusion in the rat monitored by intracerebral microdialysis. J Cerebral Blood Flow Metabol 9:607–616

Hillered L, Persson L, Ponten U, Ungerstedt U (1990) Neurometabolic monitoring of the ischaemic human brain using microdialysis. Acta Neurochirurgica 102:91–97

Hirano M, Yamashita Y, Miyakawa A (1996) In vivo visualization of hippocampal cells and dynamics of Ca2+ concentration during anoxia: feasibility of a fiber-optic plate microscope system for in vivo experiments. Brain Res 732:61–68

Huang T, Yang L, Gitzen J, Kissinger PT, Vreeke M, Heller A (1995) Detection of basal acetylcholine in rat brain microdialysate. J Chromatogr 670:323–327

Imamura K, Takahashi M, Okada H, Tsukada H, Shiomitsu T, Onoe H, Watanabe Y (1997) A novel near infra-red spectrophotometry system using microprobes: its evaluation and application for monitoring neuronal activity in the visual cortex. Neurosci Res 28:299–309

Imperato A, Di Chiara G (1984) Trans-striatal dialysis coupled to reverse phase high performance liquid chromatography with electrochemical detection: a new method for the study of the in vivo release of endogenous dopamine and metabolites. J Neurosci 4:966–977

Ishida J, Yoshitake T, Fujino K, Kawano K, Kehr J, Yamaguchi M (1998) Serotonin monitoring in microdialysates from rat brain by microbore liquid chromatography with fluorescence detection. Anal Chim Acta 365:227–232

Justice JB (1987) Voltammetry in the neurosciences, Humana Press, Clifton, NJ

Joseph MH, Davies P (1983) Electrochemical activity of o-phthalaldehyde-mercaptoethanol derivatives of amino acids. Application to high-performance liquid chromatographic determination of amino acids in plasma and other biological materials. J Chromatogr 277:125–36

Kalen P, Kokaia M, Lindvall O, Björklund A. (1988) Basic characteristics of noradrenaline release in the hippocampus of intact and 6-hydroxydopamine lesioned rats as studied by in vivo microdialysis. Brain Res 472:374–379

Kalen P, Strecker RE, Rosengren E, Björklund A (1988) Endogenous release of neuronal serotonin and 5-hydroxyindoleacetic acid in the caudate-putamen of the rat as revealed by intracerebral dialysis coupled to high-performance liquid chromatography with fluorimetric detection. J Neurochem 51:1422–1435

Kato T, Liu KJ, Yamamoto K, Osborne PG, Niwa O (1996) Detection of basal acetylcholine release in the microdialysis of rat frontal cortex by high-performance liquid chromatography using a horseradish peroxidase-osmium redox polymer electrode with pre-enzyme reactor. J Chromatogr 682:162–166

Kawagoe KT, Garris PA, Wiedemann DJ, Wightman RM (1992) Regulation of transient dopamine concentration gradients in the microenvironment surrounding nerve terminals in the rat striatum. Neuroscience 51:55–64

Kehr J, Dechent P, Kato T, Ögren SO (1998) Simultaneous determination of acetylcholine, choline and physostigmine in microdialysis samples from rat hippocampus by microbore liquid chromatography/electrochemistry on peroxidase redox polymer coated electrodes. J Neurosci Methods 83:143–150

Kehr J, Yamamoto K, Niwa O, Kato T, and Ögren SO (1996) Disposable "chip" electrodes for LCEC determinations of acetylcholine and GABA in microdialysis samples. In: González-Mora JL, Borges R, Mas M (eds) Monitoring molecules in neuroscience. University of La Laguna, Tenerife, pp. 27–28

Kehr J (1998b) Determination of g-aminobutyric acid in microdialysis samples by microbore column liquid chromatography and fluorescence detection. J. Chromatogr 708:49–54

Kehr J (1998a) Determination of glutamate and aspartate in microdialysis samples by reversed-phase column liquid chromatography with fluorescence and electrochemical detection. J Chromatogr 708:27–38

Kehr J (1993) A survey on quantitative microdialysis: theoretical models and practical implications. J Neurosci Methods 48:251–261

Kehr J, Ungerstedt U (1988) Fast HPLC estimation of gamma-aminobutyric acid in microdialysis perfusates: effect of nipecotic and 3-mercaptopropionic acids. J Neurochem 51:1308–1310

Kehr J (1993) Fluorescence detection of amino acids derivatized with o-phthalaldehyde (OPA) based reagents. Application note no 16, CMA/Microdialysis, Stockholm

Kehr J (1994) Determination of catecholamines by automated precolumn derivatization and reversed-phase column liquid chromatography with fluorescence detection. J Chromatogr 661:137–142

Kiechle FL, Malinski T (1996) Indirect detection of nitric oxide effects: a review. Annals Clin Lab Sci 26:501–511

Kissinger PT, Hart JB, Adams RN (1973) Voltammetry in brain tissue – a new neurophysiological measurement. Brain Res 55:209–213

Kissinger PT, Refshuage CJ, Dreiling R, Blank L, Freeman R, Adams RN (1973) An electrochemical detector for liquid chromatography with picogram sensitivity. Anal Lett 6:465–477

Kriz N, Syková E, Vyklicky L (1975) Extracellular potassium changes in the spinal cord of the cat and their relation to slow potentials, active transport and impulse transmission. J Physiol 249:167–182

Lada MW, Vickroy TW, Kennedy RT (1998) Evidence for neuronal origin and metabotropic receptor-mediated regulation of extracellular glutamate and aspartate in rat striatum in vivo following electrical stimulation of the prefrontal cortex. J Neurochem 70:617–625

Landolt H, Langemann H (1996) Cerebral microdialysis as a diagnostic tool in acute brain injury. Eur J Anaesth 13:269–278

Lehman A, Isacsson H, Hamberger A (1983) Effects of in vivo administration of kainic acid on extracellular amino acid pool in the rabbit hippocampus. J Neurochem 40:1314–1320

Lindroth P, Mopper K (1979) High performance liquid chromatographic determination of subpicomole amounts of amino acids by pre-column fluorescence derivatisation with o-phthalaldehyde. Anal Chem 51:1667–1674

Lowry JP, Fillenz M (1997) Evidence for uncoupling of oxygen and glucose utilisation during neuronal activation in rat striatum. J Physiol (London) 498:497–501

Lowry JP, Boutelle MG, Fillenz M (1997) Measurement of brain tissue oxygen at a carbon paste electrode can serve as an index of increases in regional cerebral blood flow. J Neurosci Methods 71:177–182

Lowry JP, O'Neill RD, Boutelle MG, Fillenz M (1998c) Continuous monitoring of extracellular glucose concentrations in the striatum of freely moving rats with an implanted glucose biosensor. J Neurochem 70:391–396

Lowry JP, Miele M, O'Neill RD, Boutelle MG, Fillenz M (1998b) An amperometric glucose-oxidase/poly(o-phenylenediamine) biosensor for monitoring brain extracellular glucose: in vivo characterisation in the striatum of freely-moving rats. J Neurosci Methods 79:65–74

Lowry JP, Demestre M, Fillenz M (1998a) Relation between cerebral blood flow and extracellular glucose in rat striatum during mild hypoxia and hyperoxia. Developmental Neurosci 20:52–58

Lowry JP, Fillenz M (1998) Studies of the source of glucose in the extracellular compartment of the rat brain. Developmental Neurosci 20:365–368

Luparello TJ (1967) Stereotaxic atlas of the forebrain of the guinea-pig. Wiliams and Wilkins, Baltimore

MacIntosh FC, Oborin PE (1953) Release of acetylcholine from intact cerebral cortex. In: Abstracts of communications, XIX International physiological congress, Montreal, pp 580–581

Malinski T, Taha Z (1992) Nitric oxide release from a single cell measured in situ by a porphyrinic-based microsensor. Nature 358:676–678

Malitesta C, Palmisano F, Torsi L, Zambonin PG (1990) Glucose fast-response amperometric sensor based on glucose oxidase immobilised in an electropolymerized poly(o -phenylenediamine) film. Anal Chem 62:2735–2740

Marsden CA (ed) (1984) Measurement of neurotransmitter release in vivo. Methods in neurosciences, vol 6. Wiley, New York

Mayevsky A, Chance B (1973) A new long-term method for the measurement of NADH fluorescence in intact rat brain with chronically implanted cannula. Adv Exp Med Biol 37A:239–244

McIlwain H (1955) Biochemistry and the central nervous system. Little, Brown, Boston

Meyerson BA, Linderoth B, Karlsson H, Ungerstedt U (1990) Microdialysis in the human brain: extracellular measurements in the thalamus of parkinsonian patients. Life Sci 46:301–308

Millar J, Barnett TG (1988) Basic instrumentation for fast cyclic voltammetry.

Mitchell JF (1963) The spontaneous and evoked release of acetylcholine from the cerebral cortex. J Physiol 165:98–116

Moore H, Stuckman S, Sarter M, Bruno JP (1995) Stimulation of cortical acetylcholine efflux by FG 7142 measured with repeated microdialysis sampling. Synapse 21:324–331

Moroni F, Pepeu G (1984) The cortical cup technique. In: Marsden CA (ed) Measurement of neurotransmitter release in vivo. Methods in neurosciences, vol 6. Wiley, New York, pp 63–79

Myers RD (1972) Methods for perfusing different structures of the brain. In: Myers RD (ed) Methods in psychobiology, vol 2, Academic Press, New York, pp 169–211

Myers RD, Knott PJ (eds) (1986) Neurochemical analysis of the concious brain: voltammetry and push-pull perfusion. Ann NY Acad Sci, vol 473. New York Academy of Sciences, New York

Myers RD (1977) An improved push-pull cannula system for perfusing an isolated region of the brain. Physiol Behav 5:243–246

Nicholson C, Syková E (1998) Extracellular space structure revealed by diffusion analysis. Trends Neurosci 21: 207–215

Ögren SO, Kehr J, Schött PA (1996) Effects of ventral hippocampal galanin on spatial learning and on in vivo acetylcholine release in the rat. Neuroscience 75: 1127–1140

Orrego F, Villanueva S (1993) The chemical nature of the main excitatory transmitter: a critical appraisal based upon release studies and synaptic vesicle localization. Neuroscience 56:539–555

Osborne PG, O'Connor WT, Kehr J, Ungerstedt U (1991)In vivo characterisation of extracellular dopamine, GABA and acetylcholine from the dorsolateral striatum of awake freely moving rats by chronic microdialysis. J Neurosci Methods 37:93–102

Osborne PG, O'Connor WT, Kehr J, Ungerstedt U (1991) In vivo characterisation of extracellular dopamine, GABA and acetylcholine from the dorsolateral striatum of awake freely moving rats by chronic microdialysis. J Neurosci Methods 37:93–102

Pantano P, Kuhr WG (1993) Dehydrogenase-modified carbon-fiber microelectrodes for the measurement of neurotransmitter dynamics. 2. Covalent modification utilizing avidin-biotin technology. Anal Chem 65:623–630

Parsons LH, Justice JB Jr (1992) Extracellular concentration and in vivo recovery of dopamine in the nucleus accumbens using microdialysis. J Neurochem 58:212–218

Paxinos G, Watson C (1982) The Rat Brain in Stereotaxic Coordinates. Academic Press, Sydney

Persson L, Valtysson J, Enblad P, Warme PE, Cesarini K, Lewen A, Hillered L (1996) Neurochemical monitoring using intracerebral microdialysis in patients with subarachnoid hemorrhage. J Neurosurg 84:606–616

Philippu A (1984) Use of push-pull cannulae to determine the release of endogenous neurotransmitters in distinct brain areas of anesthetized and freely moving animals. In: Marsden CA (ed) Measurement of neurotransmitter release in vivo. Methods in neurosciences, vol 6. Wiley, New York, pp 3–37

Phillis JW, Song D, O'Regan MH (1998) Tamoxifen, a chloride channel blocker, reduces glutamate and aspartate release from the ischemic cerebral cortex. Brain Res 780:352–355

Potter PE, Meek JL, Neff NH (1983) Acetylcholine and choline in neuronal tissue measured by HPLC with electrochemical detection. J Neurochem 41:188–194

Privette TH, Myers RD (1989) Peristaltic versus syringe pumps for push-pull perfusion: tissue pathology and dopamine recovery in rat neostriatum. J Neurosci Methods 26:195–202

Rector DM, Poe GR, Harper RM (1993) Imaging of hippocampal and neocortical neural activity following intravenous cocaine administration in freely behaving cats. Neuroscience 54:633–641

Robinson TE, Justice JB Jr (1991) Microdialysis in the neurosciences,Techniques in the behavioral and neural sciences Vol 7. Elsevier, Amsterdam London New York Tokyo

Robinson TE, Camp DM (1991) The feasibility of repeated microdialysis for within-subjects design experiments: studies on the mesostriatal dopamine system. In: Robinson TE, Justice JB Jr (eds) Microdialysis in the neurosciences,Techniques in the behavioral and neural sciences Vol 7. Elsevier, Amsterdam London New York Tokyo, pp 189–234

Roth RH, Allikmets L, Delgado JM (1969) Synthesis and release of noradrenaline and dopamine from discrete regions of monkey brain. Arch Int Pharmacodyn 181:273–282

Sarre S, Michotte Y, Marvin CA, Ebinger G (1992) Microbore liquid chromatography with dual electrochemical detection for the determination of serotonin and 5-hydroxyindoleacetic acid in rat brain dialysates. J Chromatogr 582:29–34

Sasso SV, Pierce RJ, Walla R, Yacynych AM (1990) Electropolymerized 1,2-diaminobenzene as a means to prevent interferences and fouling and to stabilise immobilised enzyme in electrochemical biosensors. Anal Chem 62:1111–1117

Shibuki K (1990) An electrochemical microprobe for detection of nitric oxide release in brain tissue. Neurosci Res 9:69–76

Strömberg I, Van Horne C, Bygdeman M, Weiner N, Gerhardt G (1991) Function of intraventricular human mesencephalic xenografts in immunosupressed rats: An electrophysiological and neurochemical analysis. Exp Neurol 112:140–152

Syková E (1997) The extracellular space in the CNS: Its regulation, volume and geometry in normal and pathological neuronal function. The Neuroscientist 3:28–41

Syková E (1992) Ionic and volume changes in the microenvironment of nerve and receptor cells. In: Ottoson D (ed) Progress in sensory physiology. Springer-Verlag, Heidelberg, pp 1–167

Syková E (1983) Extracellular K^+ accumulation in the central nervous system. Prog biophys molec biol 42:135–189

Timmerman W, Westerink BH (1997) Brain microdialysis of GABA and glutamate: what does it signify?.Synapse 27:242–261

Tossman U, Jonsson G, Ungerstedt U (1986) Regional distribution and extracellular levels of amino acids in rat central nervous system. Acta Physiol Scand 127:533–545

Tyrefors N, Gillberg PG (1987) Determination of acetylcholine and choline in microdialysates from spinal cord of rat using liquid chromatography with electrochemical detection. J Chromatogr 423:85–91

Ungerstedt U (1984) Measurement of neurotransmitter release by intracranial dialysis. In: Marsden CA (ed) Measurement of neurotransmitter release in vivo. Methods in neurosciences, vol 6. Wiley, New York, pp 81–105

Ungerstedt U (1991) Microdialysis–principles and applications for studies in animals and man. J Int Med 230:365–73

Ungerstedt U, Pycock C (1974) Functional correlates of dopamine neurotransmission. Bull Schweiz Akad Med Wis 30:44–55

Ungerstedt U (1997) Microdialysis–a new technique for monitoring local tissue events in the clinic. Acta Anaesth Scand Suppl 110:123

Ungerstedt U, Herrera-Marschitz M, Jungnelius U, Ståhle L, Tossman U, Zetterström T (1982) Dopamine synaptic mechanisms reflected in studies combining behavioural recordings and brain dialysis. In: Kotisaka M, Shomori T, Tsukada T, Woodruff GM (eds) Advances in dopamine research. Pergamon Press, New York, pp 219–231

Vreeke M, Maidan R, Heller A (1992) Hydrogen peroxide and ß-nicotinamide adenine dinucleotide sensing amperometric electrodes based on electrical connection of horseradish peroxidase redox centers to electrodes through a three-dimensional electron relaying polymer network. Anal Chem 64:3084–3090

Wages SA, Church WH, Justice JB Jr (1986) Sampling considerations for on-line microbore liquid chromatography of brain dialysate. Anal Chem 58:1649–1656

Walker MC, Galley PT, Errington ML, Shorvon SD, Jefferys JG (1995) Ascorbate and glutamate release in the rat hippocampus after perforant path stimulation: a "dialysis electrode" study. J Neurochem 65:725–731

Walker MC, Galley PT, Errington ML, Shorvon SD, Jefferys JG (1995) Ascorbate and glutamate release in the rat hippocampus after perforant path stimulation: a "dialysis electrode" study. J Neurochem 65:725–731

Welsh S, Kay SA (1997) Reporter gene expression for monitoring gene transfer.

Westerberg E, Kehr J, Ungerstedt U, Wieloch T (1988) The NMDA-antagonist MK-801 reduces extracellular amino acid levels during hypoglycemia and prevents striatal damage. Neurosci. Res. Commun. 3:151–158

Westerink BH, Tuinte MH (1986) Chronic use of intracerebral dialysis for the in vivo measurement of 3,4-dihydroxyphenylethylamine and its metabolite 3,4-dihydroxyphenylacetic acid. J Neurochem 46:181–185

Westerink BH, Drijfhout WJ, vanGalen M, Kawahara Y, Kawahara H (1998) The use of dual-probe microdialysis for the study of catecholamine release in the brain and pineal gland. Adv Pharmacol 42:136–140

Yamamguchi M, Yoshitake T, Fujino K, Kawano K, Kehr J, Ishida J (1999) Norepinephrine monitoring in microdialysates from rat brain by microbore-high-performance liquid chromatography with fluorescence detection. Anal Biochem Submitted

Young AM (1993) Intracerebral microdialysis in the study of physiology and behaviour. Rev Neurosci 4:373–395

Zauner A, Doppenberg EM, Woodward JJ, Choi SC, Young HF, Bullock R (1997) Continuous monitoring of cerebral substrate delivery and clearance: initial experience in 24 patients with severe acute brain injuries. Neurosurgery 41:1082–1091

Zauner A, Doppenberg E, Soukup J, Menzel M, Young HF, Bullock R (1998) Extended neuromonitoring: new therapeutic opportunities? Neurol Res 20 Suppl 1:S85–90

Zetterström T, Herrera-Marschitz M, Ungerstedt U (1986) Simultaneous measurement of dopamine release and rotational behaviour in 6-hydroxydopamine denervated rats using intracerebral dialysis. Brain Res 376:1–7

Zetterström T, Sharp T, Marsden CA, Ungerstedt U (1983) In vivo measurement of dopamine and its metabolites by intracerebral dialysis: changes after d-amphetamine. J Neurochem 41:1769–1773

Zhang X, Myers RD, Wooles WR (1990) New triple microbore cannula system for push-pull perfusion of brain nuclei of the rat. J Neurosci Methods 32:93–104

Zimmerman JB, Wightman RM (1991) Simultaneous electrochemical measurements of oxygen and dopamine in vivo. Anal Chem 63:24–28

Zini I, Zoli M, Grimaldi R, Pich EM, Biagini G, Fuxe K, Agnati LF (1990) Evidence for a role of neosynthetized putrescine in the increase of glial fibrillary acidic protein immunoreactivity induced by a mechanical lesion in the rat brain. Neurosci Lett 120:13–16

Suppliers

1. CMA/Microdialysis AB, Box 2, 171 18 Solna, Sweden
2. David Kopf Instruments, 7324 Elmo St., Tujunga, CA 91042, U.S.A.
3. Fine Science Tools, Inc., 373-G Vintage Park Drive, Foster City, CA 94404, U.S.A.
4. Heraeus, W.C. GmbH, Heraeusstrasse 12–14, D-63450 Hanau, Germany.
5. World Precision Instruments (WPI), Astonbury Farm Business Centre, Aston, Stevenage, Hertfordshire SG2 /EG, UK.
6. Axon Instruments, Inc., 1101 Chess Drive, Foster City, CA 94404–1102, U.S.A.
7. HEKA elektronik GmbH, Wiesenstrasse 71, 6734 Lambrecht/Pfalz, Germany.
8. Fluka Chemie AG, Industriestrasse 25, 9471 Buchs, Switzerland
9. Sigma, P.O. Box 14508, St. Louis, MO 63178-9916, U.S.A.
10. Medical System Corp., U.S.A.
11. AVCO Specialty Materials, U.S.A.
12. C.G. Processing, Delaware, U.S.A.
13. Electrochemical and Medical Systems, Newbury, UK.
14. Bioanalytical Systems Inc., 2701 Kent Avenue, West Lafayette, IN, U.S.A.
15. National Instruments, Austin, TX, U.S.A.
16. Advent Research Materials, Suffolk, UK.
17. Ultra Carbon Corporation, Bay City, MI, U.S.A.
18. Boehringer Mannheim, Sandhofer Str. 116, 68305 Mannheim, Germany.
19. Sigma-Aldrich, P.O. Box 14508, St. Louis, MO 63178-9916, U.S.A.
20. Devcon 5 Minute Fast Drying Epoxy, ITW Brands, IL, USA.

21. Semat Technical Ltd, Herts, UK.
22. ALS, Tokyo, Japan.
23. Spectra Physics, San Jose, CA, USA.
24. Scientific Software, San Ramon, CA, U.S.A.
25. Rohm and Haas, Philadelphia, PA, USA.

Chapter 42

Invasive Techniques in Humans: Microelectrode Recordings and Microstimulation

Jonathan Dostrovsky

Introduction

The use of microelectrodes for recording single and multiunit activity in the human brain was pioneered by Albe-Fessard and colleagues (Albe-Fessard et al., 1963) as an aid in localising targets within the brains of patients undergoing stereotactic surgery. Use of microelectrodes in human recordings has occurred since then only on a very small scale by a handful of groups (see Tasker et al., 1998). However, in the past few years there has been a resurgence of interest in functional stereotactic neurosurgery for the treatment of movement disorders, in particular Parkinsons's disease, and this has resulted in an increased interest and increased opportunity for human microelectrode recordings. Indeed there are now several companies that make microelectrodes and recording systems specifically designed for human use. The information obtained from these recordings is not only of clinical importance for localising the appropriate site for implanting a chronic stimulation electrode or making a radiofrequency lesion (Lozano et al., 1996; Tasker et al., 1987; Tasker et al., 1986) but can also provide interesting and unique information relevant to understanding brain function and pathophysiological mechanisms underlying the patient's disorder. Another related and more common technique for localising targets in stereotactic surgery is stimulation. This technique has usually utilised large electrodes and relatively high currents. Our group was the first to combine microelectrodes for both recordings and microstimulation (Lenz et al., 1988a; Dostrovsky et al., 1993a) and this has proved a very useful technique and is now being used by other groups.

Use of the techniques described in this chapter allows one to record the activity of neurons within the cortex and underlying structures such as thalamus, basal ganglia and rostral brainstem under local anesthesia. The patient can be asked to perform various motor and mental tasks during the recordings and to report on the effects of stimulation within the brain.

This chapter describes the techniques required to obtain recordings and perform microstimulation, and assumes a basic knowledge and some experience in extracellular recording methods in experimental animals (see Chapter 5). This paper does not attempt to describe the neurosurgical techniques involved in applying the methods described in this chapter. The methods described here are based on those which we have been using successfully for over 10 years at the Toronto Hospital, and are provided solely as a guide. Use of this invasive technique is associated with various risks which are described in the clinical literature and are the responsibility of those carrying out the procedure. The author accepts no liability for any complications and problems that might arise from the application of the methods described in this chapter. The reader may also find it helpful to refer to Chapter 5 in this volume.

Jonathan Dostrovsky, University of Toronto, Department of Physiology, Toronto, ON, M5S 1A8, Canada (phone: 416-978-5289; fax: 416-978-4940; e-mail: j.dostrovsky@utoronto.ca)

Outline

This chapter will describe the necessary equipment and procedures for obtaining microelectrode recordings and performing stimulation during stereotactic surgery. Many of the procedures are similar to those one would utilise in animal studies and will only be briefly described. Particular emphasis will be placed on those aspects that are unique or require particular attention in the human. Examples of the types of data that can be obtained using these techniques are shown.

Materials

- Microelectrodes
- Microelectrode holder and microdrive with attachment to stereotactic frame
- MR or CT compatible stereotactic frame
- High impedance amplifier, preferably with headstage close to the electrode
- Isolation amplifier (recommended)
- Signal conditioning equipment (filters, amplification) if not part of main amplifier
- Oscilloscope
- Audio monitor
- Rate meter, simple window discriminator (optional)
- Stimulator capable of producing 1–10 s trains of 100–300 Hz monophasic or biphasic pulses of 0.1–0.5 ms duration.
- Stimulus isolation unit capable of delivering 1–100 μA constant current pulses. A battery-powered optically isolated model for safety is recommended.
- Optional research-related equipment (e.g. data storage device).

Procedure

The following brief summary of the entire procedure is followed by a more detailed discussion of electrodes, equipment and in particular those issues that are especially relevant to human recordings such as safety, sterilisation, electrical noise, and stability.

Overview

Under local anesthesia a stereotactic frame is attached to the patient's head and the frame co-ordinates relative to the Anterior and Posterior commissures (AC, PC) and/or brain target are determined using MR or X-ray (CT) imaging. A small hole is then made in the skull under local anesthesia and a guide tube aimed at the target but terminating 10 mm or more above the suspected target is introduced. A microelectrode inside a smaller protective guide tube is then inserted into the main guide tube and is slowly driven down towards the target with the aid of a microdrive. The electrode is attached to a preamplifier. The signals from the electrode are further amplified and filtered and displayed on an oscilloscope and fed into an audio amplifier so that the signals can be heard via a speaker or headphones. Stimulation trains are delivered through the microelectrode at selected sites and intervals and any motor effects observed and the patient questioned for any stimulation-evoked sensations. Optional equipment, primarily for research purposes, is described below.

Electrodes

The microelectrodes used for human recordings are metal microelectrodes, commonly made of tungsten or platinum-iridium. These are either coated with glass or some other insulating coating (see Chapter 5). We currently use parylene-C coated tungsten microelectrodes (Microprobe, Inc., e.g. WE300325A). Since recordings in stereotactic surgery are generally obtained from structures underlying the cortex such as thalamus or basal ganglia, the electrode must be very long, much longer than those used in animal studies. Since it is difficult to etch and insulate the long electrodes, what is generally done is to insert a short microelectrode (such as used in animal experiments) into a stainless steel tube and then insulate the tube with polyimide (Kapton) tubing. The length of the exposed tip of the microelectrode can range from 15–40 µm or even smaller and initial impedance from 1–2 MΩ. The shank of the electrode is stripped of insulation, bent and inserted into a long thin stainless steel tubing (e.g. 25 gauge, Small Parts Inc.) and insulated with a covering of polyimide tubing (23 gauge for 25 gauge tube, Micro ML). With the aid of a dissecting microscope, the insulating tube is pushed down towards the electrode to overlap and cover the insulated shank of the electrode and epoxy resin glue is used to seal and insulate the junction. Although the electrode can be used without further treatment, we find that with our tungsten electrodes improved recordings can be obtained by plating the electrode tip with gold and platinum (Merrill and Ainsworth,1972). Plating generally reduces the impedance by a factor of 5 to 10 to a final impedance of a few hundred kΩ. To ensure that there are no breaks in the insulation, the impedance of the electrode is tested as the tip is dipped into saline and slowly immersed until the junction of the electrode and the polyimide tubing is below the surface. The impedance should not significantly change as the electrode is immersed deeper. In addition we usually apply a low DC voltage (3–5 V) to the electrode while it is immersed. This should result in bubble formation only at the tip. The completed electrodes are inserted into labelled protective carrier tubes. There are now several companies that sell complete long electrodes suitable for human recordings (e.g. Apollo Microsurgicals, ARS, Frederick Haer Inc., Radionics). We also sometimes use a larger tipped electrode, constructed by insulating the 25 gauge stainless steel tubing to within 1.5 mm of the tip with polyimide tubing and bevelling and polishing the tip to remove any sharp edges. This is used for macrostimulation and/or for microinjection of lidocaine, as has been previously described by us for the thalamus (Dostrovsky et al., 1993b).

Equipment

Any amplifier designed for microelectrode extracellular neuronal recordings can be used (cf. Chapter 5; see safety issues below). We use both a WPI DAM 80 model preamplifier (World Precision Instruments) or the Guideline System 3000 system (Axon Instruments). When using the WPI amplifier, the headstage is mounted on the arc car adapter, which also houses the slave cylinder of a hydraulic microdrive. In the case of the Axon Instruments system the leads from the headstage can be much longer and the headstage is located on a table close to the head of the patient. Many amplifiers for extracellular recordings include low- and high-pass filters and sometimes line frequency notch filters and sufficient amplification so that no further filtering or amplification is necessary. However, in some cases it may be necessary to connect the amplifier output to additional filters (e.g. Krohn-Hite, model 3700) and amplifiers (see also section on electrical safety below). The output of the amplifier is displayed on an oscilloscope and/or computer monitor of a computerised data acquisition system (e.g. CED 1401 with spike 2, DataWave, Guideline System 3000). The signal is also fed into an audio system, preferably with baseline noise suppression (e.g. Grass AM8). In some cases it is useful to have a window discriminator connected to the amplifier output so that single units can be discriminated and their firing rate displayed (e.g. Winston Electronics, Guideline System 3000). The output of the window discriminator, when it is fed into an audio monitor and/or on-line display of neuronal firing rate, is also sometimes useful for identifying single unit neuronal responses to active or passive movements. Radionics Inc. and ARS also market a system for microelectrode recordings in functional stereotactic surgery.

Microstimulation

In order to use the microelectrode for microstimulation, the output of the stimulator must be fed to the microelectrode. This can either be done by manually disconnecting the amplifier lead from the electrode and attaching the output lead from the stimulator (negative to electrode tip), or by using a specially designed amplifier or circuit that allows the stimulus to be applied without changing leads (e.g. Guideline System 3000, ARS system). The stimulator must be capable of delivering trains of monophasic or biphasic pulses of 1 to 100μA in the range 0.05 to 1.0 ms duration. One second trains of 300 Hz are the most commonly used. These stimulators are either integrated within the recording system as in the Guideline System 3000 or can be produced by separate components (e.g. WPI Anapulse modelA310) stimulus generator with a constant current stimulus isolation unit (A360).

Microinjection

In order to assess what the effects of a lesion might be, it is possible to inject a small volume of local anesthetic into the target site. Although we have observed effects of such injections on motor performance and tremor in movement disorder patients, it must be stressed that there has been no systematic study of the predictive value of this method. The microelectrode is replaced by a stainless steel tube of equal diameter (see section on microelectrodes above). Prior to insertion into the brain the tube is filled with 2% lidocaine without preservatives and connected via fine polyethylene tubing (PE50) to a 25μl Hamilton syringe. We generally inject 1 to 5μl initially and sometimes inject an additional amount if no effect is observed within a few minutes. See Dostrovsky et al.,1993b for further details.

Optional Equipment

Strobe Light For determining when the electrode is in the optic tract (for localising sites in globus pallidus) it is useful to use a strobe light. However, a flashlight or even turning the room lights on and off is quite adequate. If one wants to record visual evoked potentials (slow waves) from the optic tract, then a strobe light is essential.

EMG and/or Accelerometer Primarily for research purposes it may be useful to attach EMG electrodes and/or an accelerometer to the limbs. When assessing the effects of microstimulation or macrostimulation on movements (usually tremor) in order to localise the optimum target for deep brain stimulation or a lesion, it is sometimes helpful to assess the effects of stimulation by observing the EMG or accelerometer output. This requires on-line display of these records, and a computer terminal display is most convenient for this purpose (e.g. Spike2, CED) and appropriate amplifiers and filters for conditioning the EMG and accelerometer signals are required.

Storage of Recordings for Off-line Analysis For off-line analysis it is essential to store the unit recordings and any associated signals from transducers, EMG etc. This can be achieved by recording on video tape with a digital recorder (e.g. Instrutech Corp, VR-100-B) or high capacity digital storage medium such as CDs, ZIP disks, etc. It may also be useful to videotape the procedure using a standard videocamera. The output of the videocamera can be fed directly into a standard Hi-Fi VCR. One can then use one of the two audio channels for recording the neuronal recordings (the output of the microelectrode amplifier).

Special Considerations

Electrical Safety

There are two major technical issues which are unique to human studies and relate to safety. The first is safety from electrical shock to the patient. Although most equipment designed for use on animals is likely to be very safe and reliable and the possibility of hazardous electrical shock due to malfunction remote, most hospitals/countries have strict regulations regarding electrical isolation of any equipment connected to the patient. Equipment designed for use on animals generally does not meet these regulations or has not been tested and approved for use on humans. Thus if one is to use equipment that has not been specifically approved for use on humans, it is generally necessary to have the equipment tested and approved by the applicable local authorities. Increased safety, although not necessarily meeting local specifications, can generally be achieved by connecting all equipment to mains via an isolation transformer and/or ground fault interrupt circuits. In addition, it is advisable to use a battery-operated final amplifier and stimulus isolation unit. Ideally the amplifier, even if battery-powered, should be connected to the rest of equipment via a medically approved isolation amplifier. At the present time the only complete recording system known to the author which has been specifically designed for human use and which has been approved in the USA for use on humans is that by Axon Instruments, although several other manufacturers supply equipment for use on humans.

Sterility

Obviously the microelectrodes must be sterilised. Gas sterilisation is the method of choice. The electrodes are placed in a perforated box and sent for sterilisation prior to the operation and must be prepared beforehand.If more than one procedure is to be carried out within a day, then it is essential to have at least one additional box of microelectrodes, as once a box is opened for one case, it is not generally permissible to use other electrodes from that box on another patient. Depending on the construction of the electrode it may be possible to autoclave the electrodes prior to use. Immersion in a sterilising fluid (e.g. 2% glutaraldehyde) and then rinsing with sterile water is another option.

Everything that comes in contact with the stereotactic frame and electrode carrier needs to be sterile as well. We place the leads that connect the electrode and ground the frame, as well as the guide tube, in with the microelectrodes for gas sterilisation. The electrode/microdrive carrier which is all metallic is autoclaved immediately before the procedure. Sterilising the microdrive and preamplifier may be more difficult. In our case we use a hydraulic microdrive and the slave cylinder which attaches to the frame is sterilised in glutaraldehyde and rinsed in sterile water immediately before the procedure. If the preamplifier is well sealed, it may be possible to gas-sterilise it. If not, then a sterilisable sleeve that covers it and its cable should work. We used to use a small headstage (DAM80, WPI) that fits into a chamber in our electrode holder/carrier and thus covers most of the nonsterile headstage.The initial length of the output cable is covered with sterile adhesive plastic sheet. We now use a system that does not require the amplifier headstage to be close to the electrode and therefore only requires the lead to be sterilised by gas sterilisation.

Electrical Noise

Due to the high gains and relatively high electrode impedances used in extracellular single unit recordings, pickup of electrical noise is frequently a problem and sometimes a challenge to eliminate. In the operating room environment this can be an even greater problem due to the operation of various types of electrical equipment in the room or adjacent rooms. Probably the worst source of noise is monopolar cautery equipment in adjacent operating rooms and it may not be possible to completely eliminate it. A shielded operating room would be desirable but is usually not practical due to the high cost involved. Heart rate and other monitors attached to the patient sometimes cause interference and must be disconnected during the recording session. Fluorescent lights are frequently a source of interference and may need to be shielded with grounded wire mesh or else turned off.Regular incandescent lamps and surgical lights usually do not cause any problems. Noise resulting from mains (power line) oscillations (60 or 50 Hz) can be very effectively eliminated by using a Humbug filter (Quest Inc.).

Microphonics and Speech

Under some conditions there may be feedback oscillation resulting from coupling between the audio monitor loudspeaker and the frame/microelectrode. This appears to result most commonly from oscillation of the microelectrode within the guide tube but other loose junctions in the frame can also sometimes lead to this problem. Turning the volume down or using headphones will eliminate this problem. Changing electrodes will sometimes also resolve the problem. If the patient speaks during the recording, the speech will frequently cause interference in the recordings. Changing electrodes sometimes resolves the problem.

Stability

Since the frame is rigidly attached to the skull, there should be no problem with electrode movement relative to the skull. However, in some patients cardiac and/or respiratory induced movements of the brain relative to the skull can result in poor stability and fluctuating action potential amplitude. In addition, large passive or active movements of the legs, arm and/or shoulder will sometimes result in brain movement and loss of the unit under study. Coughing almost always results in brain movement and loss of the unit.

Stereotactic Coordinates

In order to set the approximate coordinates for the initial electrode trajectory, it is necessary to obtain these using MR or CT imaging (injection of contrast medium into the ventricles was used before the common use of CT and MR imaging and can also be used). The target can be determined directly from the image but since it is frequently difficult to visualise precisely this target, we always use the coordinates of the AC and

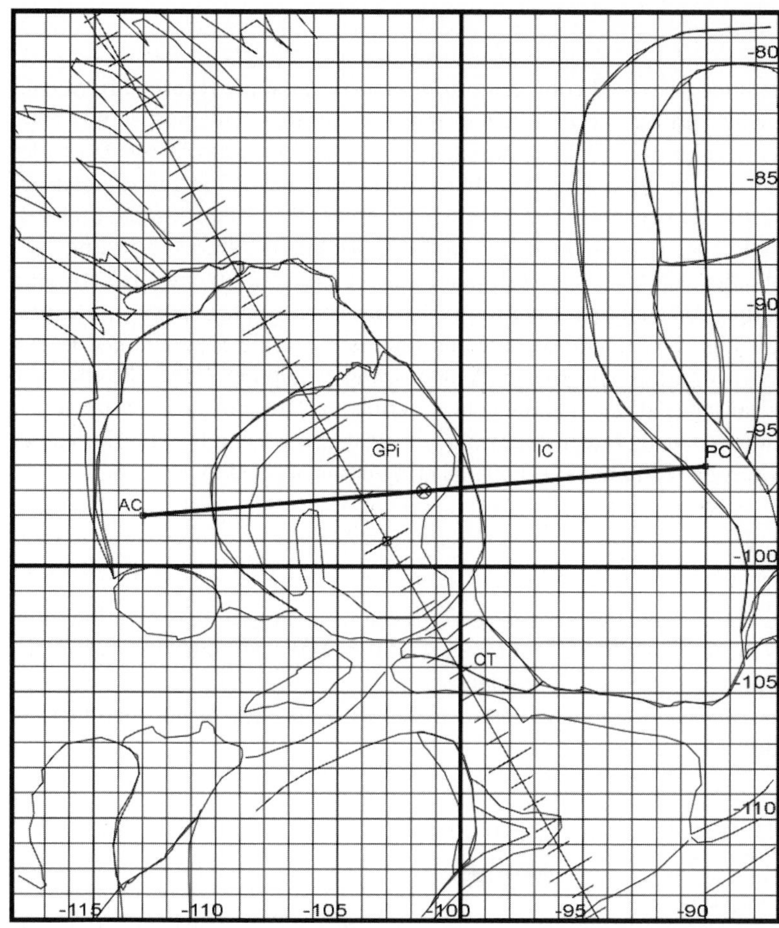

Fig. 1. An example ofa computer generated stereotactic map with superimposed frame coordinates and electrode track. The map is that of a 20 mm lateral sagittal plane and shows the locations of the anterior and posterior commissures (AC, PC) and the midcommissural point (circle with x). Also illustrated in this example is a typical electrode track passing through the globus pallidus and optic tract. The grid and electrode track are marked in mm and the numbers refer to the Lexell frame co-ordinates. Abbreviations: GPi – internal segment of globus pallidus; IC – internal capsule; OT – optic tract.

PC commissures relative to the frame coordinates and use a program to superimpose a standard stereotactic atlas map of the brain on the frame coordinates. The stereotactic coordinates of AC and PC are calculated using the MRI or CT scanner's computer software. The atlas map which is based on the Schaltenbrand and Wahren atlas (Schaltenbrand and Wahren,1977) is then stretched or shrunk by means of a computer program so as to match the patient's AC and PC coordinates with the locations of these points on the standard atlas maps and also draw the trajectories of the electrode tracks on the map (see Figure 1).

Brain Shift

One possible complication that one needs to be aware of is the possibility of brain shift having occurred between the time the scan was made and the electrode penetrations are made and also during the recording session. Loss of CSF from the hole made in the skull is probably the main reason for this occurrence.

Results

Using the techniques described above, our group has successfully recorded from over 500 cases involving localisation of targets in motor or sensory thalamus, subthalamic nucleus, periventricular gray, globus pallidus and anterior cingulate cortex. Examples of various types of data obtained using the techniques described in this chapter are shown. Figure 2 shows recordings from two single units in the ventrocaudal thalamic nucleus activated by tactile stimuli applied to the digits. Figure 3 shows recordings of a tremor cell in globus pallidus. Figure 4 shows the effect of increasing the microstimulation current on the projected field (sensation location) size and the intensity of the perceived sensation. Figure 5 shows recording and microstimulation data obtained from a microelectrode track through the thalamus of an amputee. Figure 6 shows the effects of thalamic electrical stimulation and lidocaine injection on tremor in a parkinsonian patient. Figure 7 shows a firing rate histogram of a neuron recorded in the anterior cingulate cortex that responded to noxious thermal stimulation applied to the arm.

Troubleshoot

The main types of problems that arise in carrying out recordings in the operating room are the same as those encountered in animal recordings in the laboratory and generally caused by electrode and grounding problems.

- *High noise levels* usually signify a loose connection to the electrode or that the electrode impedance has gone up.

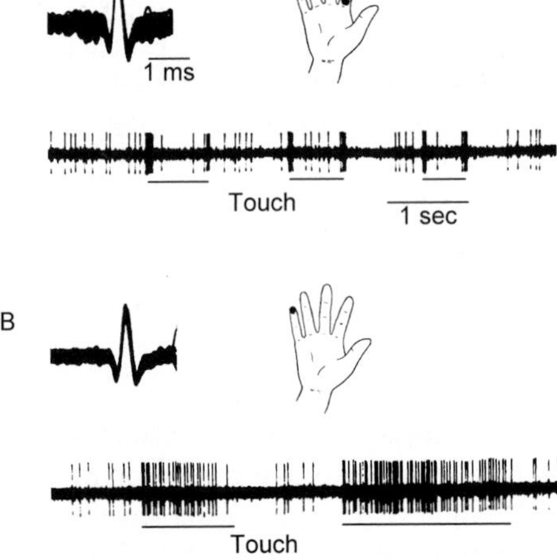

Fig. 2. Oscilloscope tracings of recordings from two single units in the ventrocaudal thalamic nucleus activated by tactile stimuli applied to the digits. The receptive field location of each unit is indicated.(Modified from Lenz et al.,1988b) with permission.

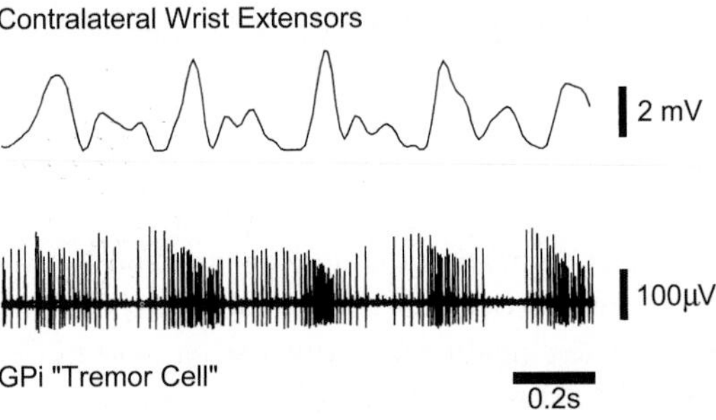

Fig. 3. Example showing firing pattern of a "tremor cell" recorded in globus pallidus, internal segment (GPi). The top trace shows the rectified and filtered EMG from contralateral wrist extensors. (Reproduced with permission from Tasker et al.,1998).

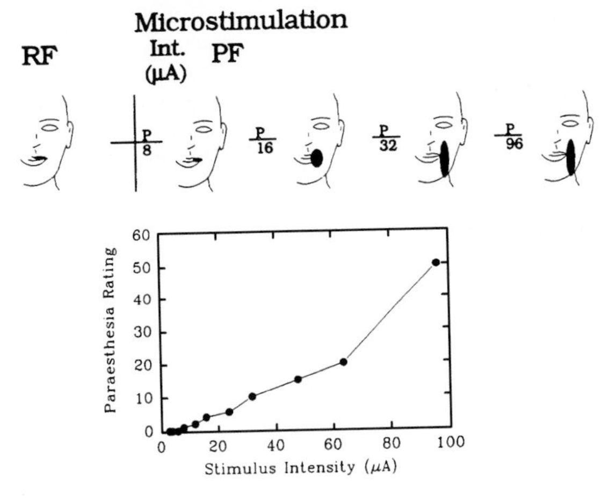

Fig. 4. Effects of microstimulation (0.2 ms pulse width, 300 Hz) in sensory thalamus (Vc). The receptive field at the stimulation site was on the upper lip and stimulation produced a tingling sensation (p – parasthesia) in the same location at threshold (8 μA) and in an increasing area with increased intensity of stimulation as shown. Also shown is the patient's rating on a scale of 0 to 100 of the perceived intensity of the parasthesia at different stimulation intensities. (Modified from Dostrovsky et al., 1993a with permission)

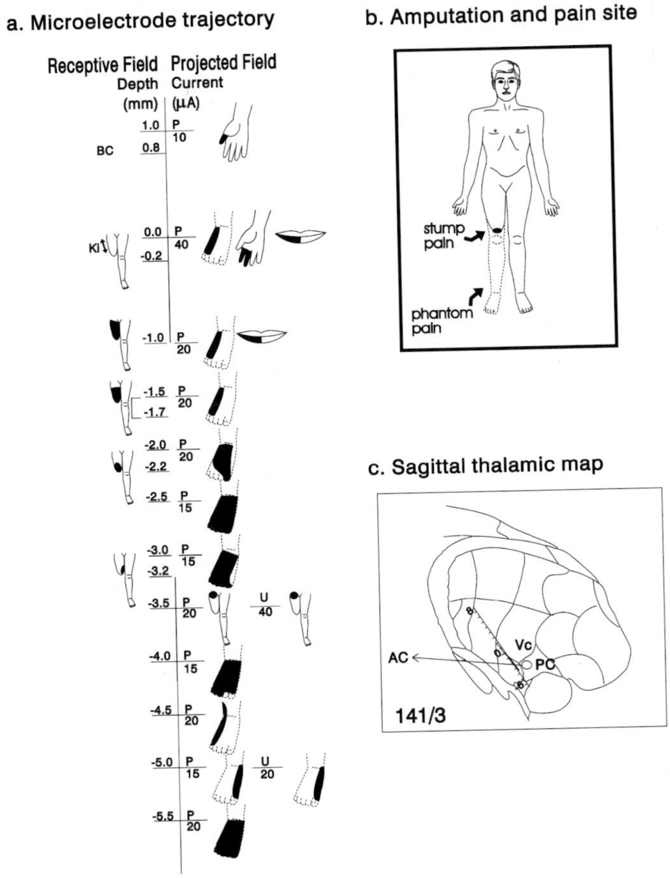

Fig. 5. Example of data obtained from one electrode track. On the left are shown the receptive fields and information regarding the recordings and on the right the effects of microstimulation. The insert b shows the site of amputation and locations of pain in this phantom limb patient, and inset c shows the location of the electrode track relative to the thalamic nuclei. Abbreviations: Ki – kinesthetic, BC – bursting cell, P – parasthesia, U – unpleasant parasthesia, Vc – ventrocaudal nucleus, AC – anterior commissure, PC – posterior commissure. (Reproduced with permission from Davis et al., 1998).

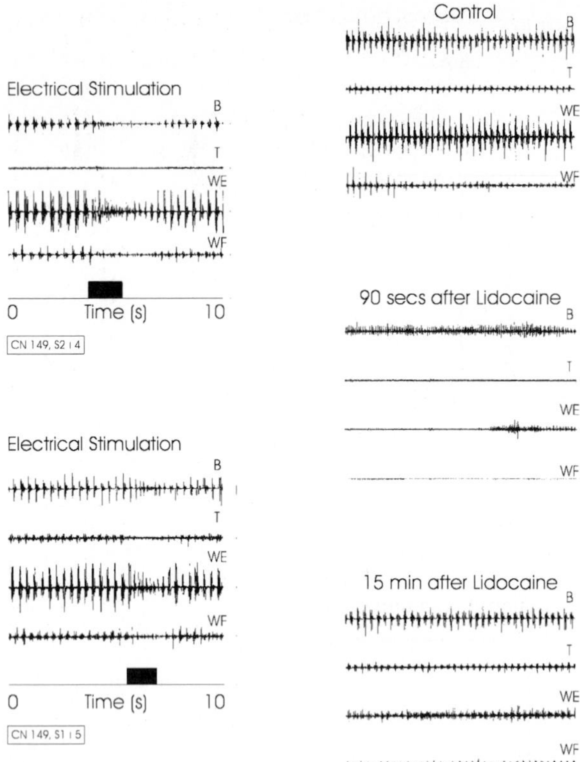

Fig. 6. Effects of microstimulation and of microinjection of lidocaine in motor thalamus on tremor in a parkinsonian patient. On the left are shown effects of stimulation (0.2 ms pulse width, 300 Hz, 100 μA train) at two different sites on arm tremor. Each trace shows EMG recordings from biceps (B), triceps (T), wrist extensors (WE) and wrist flexors (WF). The time of stimulation is indicated by the black bar. On the right are displayed the EMG traces as displayed on the left recorded just before, 90 s and 15 minutes after injection of 2 μl of 2% lidocaine at the site where microstimulation reduced tremor (shown at top left)

Fig. 7. A firing rate histogram of a neuron recorded in the anterior cingulate cortex and which responded to noxious thermal stimulation applied to the arm. The temperature of the thermode is shown on the top trace. (Modified from Dostrovsky et al., 1995 with permission)

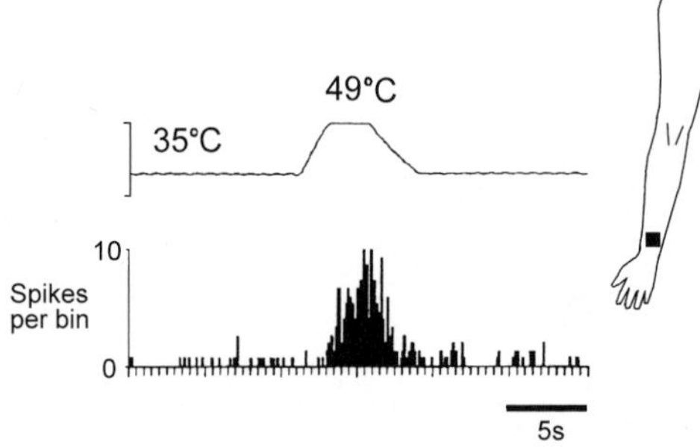

- *Line frequency noise* usually signifies a poor ground connection or a ground loop, or interference from equipment or fluorescent lights.
- *Lack of neuronal recordings* can signify an electrode problem (e.g. break in insulation, bent tip) or amplifier problems.
- *Lack of stimulation effects* (when one suspects the electrode is in a region where stimulation should produce an effect) might be the result of a short somewhere between the stimulator output and the electrode, or if the electrode impedance is high, insufficient driving voltage to deliver sufficient current.

Acknowledgement: The author wishes to acknowledge the contributions of Dr. Fred Lenz, who initially implemented microelectrode recordings in our group, and my present colleagues Drs. Karen Davis and Bill Hutchison, who are currently responsible for carrying out the recordings, and of Dr. Ron Tasker, who pioneered functional stereotactic surgery and microelectrode recordings in Toronto, and Dr. Andres Lozano, the other neurosurgeon involved in many of our current stereotactic cases.

References

Albe-Fessard D, Guiot G, and Hardy J (1963) Electrophysiological localization and identification of subcortical structures in man by recording spontaneous and evoked activities. EEG Clin Neurophysiol 15: 1052-1053

Davis KD, Kiss ZHT, Luo L, Tasker RR, Lozano AM, and Dostrovsky JO (1998) Phantom sensations generated by thalamic microstimulation. Nature 391: 385–387

Dostrovsky JO, Hutchison WD, Davis KD, Lozano AM (1995) Potential role of orbital and cingulate cortices in nociception. In: Besson JM, Guilbaud G, Ollat H (eds) Forebrain areas involved in pain processing. John Libbey Eurotext, Paris, pp 171–181

Dostrovsky JO, Davis KD, Lee L, Sher GD, Tasker RR (1993a) Electrical stimulation-induced effects in human thalamus. In: Devinsky O, Beric A, Dogali M (eds) Electrical and magnetic stimulation of the brain and spinal cord. Advances in Neurology, Volume 63. Raven Press, New York, pp 219–229

Dostrovsky JO, Sher GD, Davis KD, Parrent AG, Hutchison WD, and Tasker RR (1993b) Microinjection of lidocaine into human thalamus: A useful tool in stereotactic surgery. Stereotact Funct Neurosurg 60: 168–174

Lenz FA, Dostrovsky JO, Kwan HC, Tasker RR, Yamashiro K, and Murphy JT (1988a) Methods for microstimulation and recording of single neurons and evoked potentials in the human central nervous system. J Neurosurg 68: 630–634

Lenz FA, Dostrovsky JO, Tasker RR, Yamashiro K, Kwan HC, and Murphy JT (1988b) Single-unit analysis of the human ventral thalamic nuclear group: somatosensory responses. J Neurophysiol 59: 299–316

Lozano A, Hutchison W, Kiss Z, Tasker R, Davis K, and Dostrovsky J (1996) Methods for microelectrode-guided posteroventral pallidotomy. J Neurosurg 84: 194–202

Merrill E, and Ainsworth A (1972) Glass-coated platinum-plated tungsten microelectrodes. Med Biol Eng 10: 662–672

Schaltenbrand G, Wahren W (1977) Atlas for Stereotaxy of the Human Brain. Thieme, Stuttgart

Tasker RR, Davis KD, Hutchison WD, Dostrovsky JO (1998) Subcortical and thalamic mapping in functional neurosurgery. In: Gildenberg P, Tasker RR (eds) Textbook of Stereotactic and Functional Neurosurgery. McGraw-Hill, NY, pp 883–909

Tasker RR, Lenz FA, Yamashiro K, Gorecki J, Hirayama T, and Dostrovsky JO (1987) Microelectrode techniques in localization of stereotactic targets. Neurol Res 9: 105–112

Tasker RR, Yamashiro K, Lenz FA, Dostrovsky JO (1986) Microelectrode techniques in stereotactic surgery for Parkinson's disease. In: Lunsford (ed) Stereotactic Surgery.

Suppliers

Company: Apollo Microsurgicals, 114 Winding Woods Rd., London, Ontario, N6G 3G8, Canada (fax: 519-641-0452; home page: execulink.com/~apollom/)

Company: Atlantic Research Systems, Inc. (ARS), 2932 Ross Clarke Circle, No. 165, Dothan, Alabama, 36301, USA (phone: +01-404-321-1848)

Company: Axon Instruments, 1101 Chess Drive, Foster City, California, 94404, USA (phone: +01-415-571-9400; fax: +01-415-571-9500.)

Company: Cambridge Electronic Design Limited (CED), Science Park, Milton Rd., Cambridge, CB4 4FE, England (phone: 0223-420186; fax: 0223-420488)

Company: Datawave Technologies Corp., 30 Main St., Ste. 209, Longmont, CO, 80501, USA (phone: +01-303-776-8214; fax: +01-303-776-8531)

Company: Frederick Haer Inc. (FHC), 9 Main Street, Bowdoinham, Maine, 04008, USA (phone: +01-207-666-8190 (USA & Canada: 800-326-2905); fax: +01-+207-666-8292 (USA & Canada: 800-639-3313); e-mail: fhcinc@fh-co.com; home page: fh-co.com)

Company: Grass Instruments, (Astro-Med Inc.), 600E Greenwich Ave., W. Warick, RI, 02893, USA (phone: +01-401-828-4000; fax: +01-401-822-2430; home page: astro-med.com)

Company: Instrutech Corporation, 475 Northern Blvd., Suite 31, Great Neck, New York, 11021, USA (phone: +01-516-829-5942; fax: +01-516-829-0934; e-mail: instrutk@panix.com.)

Company: Krohn-Hite Corporation, 255 Bodwell Street, Avon, Massachusetts, 02322, USA (phone: +01-508-580-1660; fax: +01-508-580-1660; e-mail: ; home page:)

Company: Micro ML, 6-12 124 Street, College Point, New York, 11356-1134, USA (phone: +01-718-886-0769; fax: +01-718-886-9758)

Company: Micro Probe Inc., 11715 Tifton Drive, Potomoc, Maryland, 20871, USA (phone: +01-301-765-0600 or 800-290-8282)

Company: Radionics, Inc., 22 Terry Avenue, Burlington, Massachusetts, 01803, USA (phone: +01-781-272-1233; fax: +01-781-272-2428; home page: radionics.com)

Company: Small Parts Inc., P.O. Box 4650, Miami Lakes, Florida, 33014-9727, USA (phone: +01-305-558-1255 or 800-220-4242; fax: +01-1-800-423-9009; home page: smallparts.com)

Company: Quest Scientific, c/o Pacer Scientific Instruments, 5649 Valley Oak Drive, Los Angeles, California, 90068-2556, USA (phone: +01-213-462-0636; fax: +01-213-462-1430)

Company: Winston Electronics, P.O. Box 16156, San Francisco, California, 94116, USA (phone: +01-415-589-6900)

Company: World Precision Instruments, (WPI), 175 Sarasota Center Blvd, Sarasota, Florida, 34240-9258, USA (phone: +01-941-371-1003; fax: +01-941-377-5428)

Chapter 43

Psychophysical Methods

Walter H. Ehrenstein and Addie Ehrenstein

Introduction

When Fechner (1860/1966) introduced the new transdisciplinary research program of "*Psychophysik*", his goal was to present a scientific method of studying the relations between body and mind, or, to put it more precisely, between the physical and phenomenal worlds. The key idea underlying Fechner's psychophysics was that body and mind are just different reflections of the same reality. From an external, objective viewpoint we speak of processes in the brain (i.e., of bodily processes). Considering the same processes from an internalized, subjective viewpoint, we can speak of processes of the mind. In suggesting that processes of the brain are directly reflected in processes of the mind, Fechner anticipated one of the main goals of modern neuroscience, which is to establish correlations between neuronal (objective) and perceptual (subjective) events.

The goal of this chapter is to present Fechner's techniques and those extensions and modifications of psychophysical methods that may be helpful to the modern neuroscientist with the time-honored objective of discovering the properties of mind and their relation to the brain.

Inner and Outer Psychophysics

In Fechner's time there were no physiological methods that enabled the objective recording and study of sensory or neuronal functions. Sensory physiology at that time was essentially "subjective" in that it had to rely on subjective phenomena, that is, on percepts rather than on receptor potentials or neuronal activity. Nonetheless, Fechner referred to neuronal functions in his concept of *inner psychophysics*, or the relation of sensations to the neural activity underlying them (Scheerer 1992). This he distinguished from *outer psychophysics*, which deals with the relation between sensations and the corresponding physical properties and variations of the stimulus itself (see Fig. 1).

For much of the century following Fechner's publication of *Psychophysik* in 1860, inner psychophysics remained a theoretical concept, whereas the notion of outer psychophysics provided the basis for methods to study sensory and brain processes. The study of subjective phenomena with psychophysical techniques has shaped not only the development of experimental psychology, but also of sensory physiology. Psychophysical methods were used by pioneers in the field of sensory research, such as Aubert, Exner, Helmholtz, Hering, von Kries, Mach, Purkinje and Weber, and provided the basis for

Walter H. Ehrenstein, Institut für Arbeitsphysiologie an der Universität Dortmund, Ardeystr. 67, 44139 Dortmund, Germany (phone: +49-231-1084-274; fax: +49-231-1084-401; e-mail: ehrenst@arb-phys.uni-dortmund.de)
Addie Ehrenstein, Universiteit Utrecht, Department of Psychonomics, Heidelberglaan 2, 3584 CS Utrecht, The Netherlands (phone: +31-30-2534907; fax: +31-30-2534511; e-mail: A.Ehrenstein@fss.uu.nl)

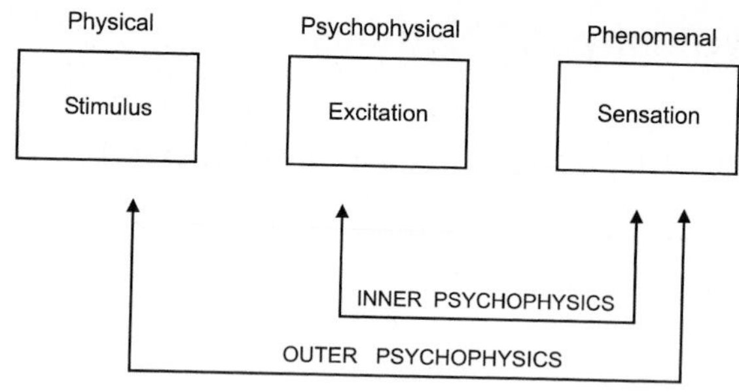

Fig. 1. Fechner's conception of psychophysics. Whereas outer psychophysics was assumed to be based on the methods of physics to describe and control the stimulus, inner psychophysics was a theoretical concept and relied on the methods of outer psychophysics to infer the rules of sensory and neuronal stimulus processing and transformation.

many fundamental insights into and understanding of sensory mechanisms. This psychophysical approach to sensory physiology has come to be referred to as *subjective sensory physiology* (see Jung 1984).

Correlational Research

With the development of various objective methods, such as electrophysiology (e.g., electroencephalography: EEG, Chapter 35 ; visually evoked potentials: VEP, Chapter 36; and single-unit recordings, Chapter 5), magnetoencephalography (MEG, Chapter 37), positron emission tomography (PET, Chapter 39) and functional magnetic resonance imaging (fMRI, Chapter 38), it has become possible to study sensory and brain processes and their locations directly. The relative ease of use and non-invasiveness of most of these techniques has made possible a new interplay between classic psychophysics and modern neuroscience (see Fig. 2). Psychophysical methods have, however, maintained their importance and are used in conjunction with the various objective methods to confirm and complement neurophysiological findings. The complementary research approach that concerns itself with subjective and objective correlates of sensory and neural processes has come to be called *correlational research* (Jung 1961a 1972). This approach, which compares psychophysical and neuronal data on a quantitative, descriptive level (neutral with respect to the question of a material or causal relationship between mental and brain processes), was first established in the study of vision by Jung and colleagues (Jung 1961b, Jung & Kornhuber 1961; Jung & Spillmann 1970; Grüsser and Grüsser-Cornehls 1973). The correlational research approach was soon followed in other sensory areas (see, e.g., Keidel and Spreng 1965; Werner and Mountcastle 1965; Ehrenberger et al. 1966; Borg et al. 1967; Hensel 1976) and has by now become an established venue of research in modern neuroscience (e.g., Spillmann and Werner 1990; Gazzaniga 1995; Spillmann and Ehrenstein 1996;).

As indicated in Fig. 2, the goals of inner psychophysics can be achieved now that the means to directly correlate phenomenal, subjective findings with objective evidence of

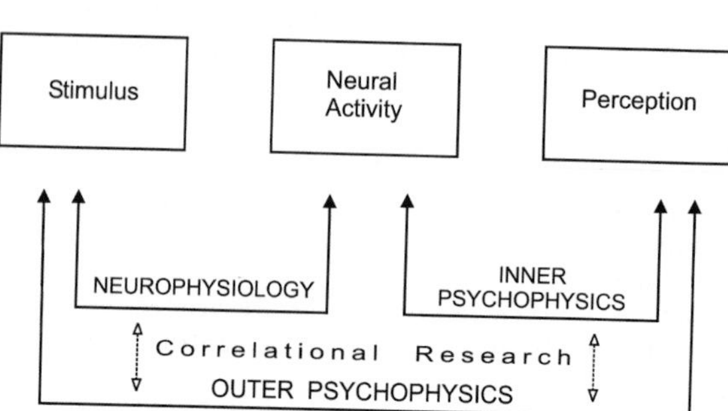

Fig. 2. Modern conception of psychophysics. Because of advanced neurophysiological methods, neural activity can be measured objectively, thus allowing for quantitative correlations between psychophysical and neural correlates of perception

sensory and neuronal activity are available. Thus, Fechner's conception of inner psychophysics is no longer dependent on the methodology of outer psychophysics alone. With further progress in correlational research, greater steps in inferring subjective events and perceptual performance by objective techniques are sure to come. For example, perceptual performance losses due to a brain lesion of a given size and location can be examined in great detail with psychophysical tasks. Moreover, in the context of the immensely increased knowledge of sensory and brain functions, inner psychophysics can be addressed much more specifically by choosing stimuli to selectively tap a given mechanism at a certain location. In turn, the hypothesized perceptual (behavioral) significance of a given mechanism or brain area can be determined by means of psychophysical testing (e.g., Wist et al. 1998).

How to Measure Perceptual Experience

Psychophysics starts out with a seeming paradox: It requires the objectification of subjective experience. No apparatus is necessary to obtain percepts; they are immediately present and available to each of us. Thus, the problem is not how to obtain perceptual experience, but how to describe and investigate individual percepts so that they can be communicated and shared by others.

Psychophysics tries to solve this problem by closely linking perceptual experience to physical stimuli. The basic principle is to use the physical stimuli as a reference system. Stimulus characteristics are carefully and systematically manipulated and observers are asked to report their perception of the stimuli. The art of psychophysics is to formulate a question that is precise and simple enough to obtain a convincing answer. An investigation might begin with a simple question such as, "Can you hear the tone?" That is, the task may be one of *detection*.

Sometimes we are not only interested in whether detection has occurred, but in determining which characteristics of the stimulus the observer can identify, e.g., sound characteristics or spatial location. Thus, the problem of sensing something, that of detection, may be followed by that of *identification*.

Detection and identification problems are solved quickly and almost simultaneously when they concern stimuli which are strong and clear. However, under conditions of weak and noisy signals we often experience a stage at which we first detect only that something is there, but fail to identify exactly what or where it is. In such a situation we try to filter out the consistent signal attributes, for instance, the sound of an approaching car, from inconsistent background noise. In such a case, the task is one of *discrimination* of the stimulus, or signal, from a noisy background, and the task is performed under uncertainty. As the car approaches and its sound becomes stronger, the probability of correct discrimination between signal and noise is enhanced. Even if we clearly perceive and identify an object, we may still be faced with the further problem of perceptual judgment, such as, "Is this car dangerously close?" or "Is the rattle under the hood louder than normal?" Questions such as these, concerning "How much x is there?", are part of another fundamental perceptual problem, that of *scaling*, or interpreting, the magnitude of the stimulus on a psychophysical scale.

Outline

In the following sections we will describe the principles of psychophysical methods and give three examples to illustrate their application. First, we present methods that are based on *threshold psychophysics*, starting with the classical procedures along with modern modifications of the classical procedures that allow for adaptive testing. Techniques for control of observer criteria and strategies are also discussed. Second, we describe the methods of *suprathreshold psychophysics*, including the use of reaction time, category scaling, magnitude estimation and cross-modality matching. A third section deals with *comparative psychophysics*, that is, with the special conditions and methods of psychophysical testing in animals.

The description of methods is followed by three specific *examples of psychophysical research*. These examples illustrate how to: (1) study basic mechanisms of adaptation in auditory motion perception, (2) assess impairment of visual function in neurological patients, and (3) measure perceptive fields in monkey and man.

Methods and Procedures

In the following, we will describe the psychophysical tasks and methods that have proven to be most useful in sensory research. Most of the principles are classic, with some having already been worked out by Fechner. The methods of stimulus presentation, response recording and data analysis, however, have been modernized, especially with regard to currently available computer-assisted procedures (see also Chapter 45).

Methods Based on Threshold Measurements

The most basic function of any sensory system is to detect energy or changes of energy in the environment. This energy can consist of chemical (as in taste or smell), electromagnetic (in vision), mechanical (in audition, proprioception and touch) or thermal stimulation. In order to be noticed, the stimulus has to contain a certain level of energy. This minimal or liminal amount of energy is called the *absolute threshold*, and is the stimulus intensity that, according to Fechner, "lifts its sensation over the threshold of consciousness." The absolute threshold is thus the intensity that an observer can just barely detect. Another threshold, known as the *difference threshold*, is based on stimulus intensities above the absolute threshold. It refers to the minimum intensity by which a variable *comparison* stimulus must deviate from a constant *standard* stimulus to produce a noticeable perceptual difference.

Method of Adjustment

The simplest and quickest way to determine absolute and difference thresholds is to let a subject adjust the stimulus intensity until it is just noticed or until it becomes just unnoticeable (in the case of measurements of the absolute threshold) or appears to be just noticeably different from, or to just match, some other standard stimulus (to measure a difference threshold). The observer is typically provided with a control of some sort that can be used to adjust the intensity, say of a sound, until it just becomes audible (or louder than a standard sound), and then the stimulus intensity is recorded to provide an estimate of the observer's threshold. Alternatively, the observer can adjust the sound from clearly audible to just barely inaudible (or to match the standard sound), providing another estimate of the threshold. Typically, the two kinds of measurement, that is, series in which the signal strength is increased (ascending series) and series of decreasing signal strengths (descending series) are alternated several times and the results are averaged to obtain the threshold estimate. For example, if a 500-Hz tone is first *heard* at 5 dB on one ascending trial and at 5.5 dB on another, and the tone is first *not* heard at 4 dB on one descending trial and at 4.5 dB on another, the resulting threshold estimate is 4.75 dB.

The following methods of threshold determination differ from the adjustment method in that they do not allow the observer to control the stimulus intensity directly. As they rely on the experimenter's rather than on the subject's control, they provide a more standardized method of measurement.

Method of Limits

In the method of limits, a single stimulus, say a single light, is changed in intensity in successive, discrete steps and the observer's response to each stimulus presentation is recorded. As in the previous method, the stimulus should initially be too weak to be detected, so that the answer is "not seen"; intensity is then increased in steps until the stimulus becomes visible (ascending series), or it is changed from a clearly visible intensity until it becomes invisible (descending series). The average of the intensity of the last "seen" and the first "not seen" stimuli in the ascending trials, or vice versa in the descending trials, is recorded as an estimate of the absolute threshold (for an example, see Table 1). Ascending and descending series often yield slight but systematic differences in thresholds. Therefore, the two types of series are usually used in alternation and the results are averaged to obtain the threshold estimate.

The determination of the difference threshold requires stimuli, such as two flashes of light, which may be presented simultaneously, one next to the other, or successively, one after the other. While the intensity of the standard stimulus is kept constant, the intensity of the comparison stimulus is changed in a series of steps. The comparison stimulus is either initially weaker (ascending series) or initially stronger (descending series) than the standard. A series terminates when the observer's response changes from "weaker" to "stronger" or vice versa. The difference threshold is then the intensity difference between the stimuli of the first trial on which the response differs from the previous one. As before, ascending and descending series are alternated and the results averaged to obtain the threshold estimate.

Method of Constant Stimuli

In the method of constant stimuli the experimenter chooses a number of stimulus values (usually from five to nine) which, on the basis of previous exploration (e.g., using the Method of Adjustment) are likely to encompass the threshold value. This fixed set of

Table 1. Method of Limits. Determination of Absolute Threshold.
Response (Stimulus Perceived): yes (Y), no (N).

Stimulus Intensity	Alternating Ascending and Descending Series					
0	N		N		N	
1	N		N		N	
2	N		N	N	N	
3	N	N	N	Y	N	N
4	N	Y	N	Y	N	Y
5	N	Y	Y	Y	Y	Y
6	Y	Y		Y		Y
7		Y		Y		Y
Transition Points	5.5	3.5	4.5	2.5	4.5	3.5

Threshold = Average Transition Points = (5.5.+3.5+4.5+2.5+4.5+3.5)/6 = 24/6 = 4

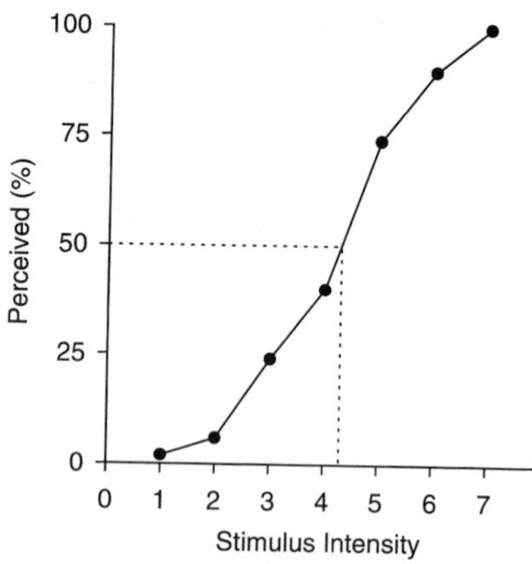

Fig. 3. Psychometric function which shows the relationship between the percentage of times that a stimulus is perceived and the corresponding stimulus intensity. The threshold is defined as the intensity at which the stimulus is detected 50 percent of the time.

Table 2. Method of Constant Stimuli (50 Presentations for each Stimulus Intensity)

Stimulus Intensity (arbitrary units)	1	2	3	4	5	6	7
Frequency of Perceived Stimuli	1	3	12	20	37	45	50
Percentage of Perceived Stimuli	2	6	24	40	74	90	100

stimuli is presented multiple times in a quasi-random order that ensures each will occur equally often. After each stimulus presentation, the observer reports whether or not the stimulus was detected (for the absolute threshold) or whether its intensity was stronger or weaker than that of a standard (for computing a difference threshold). Once each stimulus intensity has been presented multiple times (usually not less than 20), the proportion of "detected" and "not detected" (or, "stronger" and "weaker") responses is calculated for each stimulus level (for an example, see Table 2). The data are then plotted with stimulus intensity along the abscissa and percentage of perceived stimuli along the ordinate. The resulting graph represents the so-called *psychometric function* (see Fig. 3).

If there were a fixed threshold for detection, the psychometric function should show an abrupt transition from "not perceived" to "perceived." However, psychometric functions seldom conform to this all-or-none rule. What we usually obtain is a sigmoid (S-shaped) curve that reflects that lower stimulus intensities are detected occasionally and higher values more often, with intensities in the intermediate region being detected on some trials but not on others. There are various reasons why the psychometric function obeys an S-shaped rather than a sharp step function. A major source of variability are the continual fluctuations in sensitivity that are present in any biological sensory system (due to spontaneous activity or internal noise). Those inherent fluctuations mean that an observer must detect activity elicited by external stimulation against a background level of activity.

In any case, the threshold thus occurs with a certain *probability* and its intensity value must be defined statistically. By convention, the absolute threshold measured with the method of constant stimuli is defined as the intensity value that elicits "perceived" responses on 50% of the trials. Notice that in the example shown in Table 2 and Fig. 3, no stimulus level was detected on exactly 50% of the trials. However, level 4 was detected 40% of the time and level 5, 74% of the time. Consequently, the threshold value of 50% lies between these two points. If we assume that the percentage of trials in which the stimulus is detected increases linearly between these intensities (which is justified given that sigmoid functions are approximately linear in the middle range), we can determine the threshold intensity by linear interpolation as follows:

$$T = a + (b - a) \cdot \frac{50 - p_a}{p_b - p_a}$$

where T is the threshold, a and b are the intensity levels of the stimuli that bracket 50% detection (with a being the lower intensity stimulus), and p_a and p_b the respective percentages of detection. For the present case we obtain the following result:

$$T = 4 + (5 - 4) \cdot \frac{50 - 40}{74 - 40} = 4 + \frac{10}{34} = 4.29$$

Although the method of constant stimuli is assumed to provide the most reliable threshold estimates, its major drawback is that it is rather time-consuming and requires a patient, attentive observer because of the many trials required.

Adaptive Testing

Adaptive testing procedures are used to keep the test stimuli close to the threshold by adapting the sequence of stimulus presentations according to the observer's response. Since a smaller range of stimuli need be presented, adaptive methods are relatively efficient. An example of such an adaptive procedure is the *staircase method* first introduced by von Békésy (1947), who applied it to audiometry.

Staircase Method

The staircase method is a modification of the Method of Limits. A typical application of this method is shown in Fig. 4, where the stimulus series starts with a descending set of stimuli. Each time the observer says "yes" (I can detect the stimulus), the stimulus intensity is decreased by one step. This continues until the stimulus becomes too weak to be detected. At this point we do not, as in the method of limits, end the series, but rather reverse its direction by increasing the stimulus intensity by one step. This procedure continues with increasing the intensity if the observer's response is "no" and decreasing the intensity if it is "yes." In this way, the stimulus intensity flips back and forth around the threshold value. Usually six to nine such reversals in intensity are taken to estimate the threshold, which is defined as the average of all the stimulus intensities at which the observer's responses changed, i.e., the transition points as defined in the Methods of Limits (see Table 1).

In the staircase method, most of the stimulus values are concentrated in the threshold region, making it a more efficient method than the method of limits. A problem with this *simple staircase* procedure is that an observer may easily become aware of the

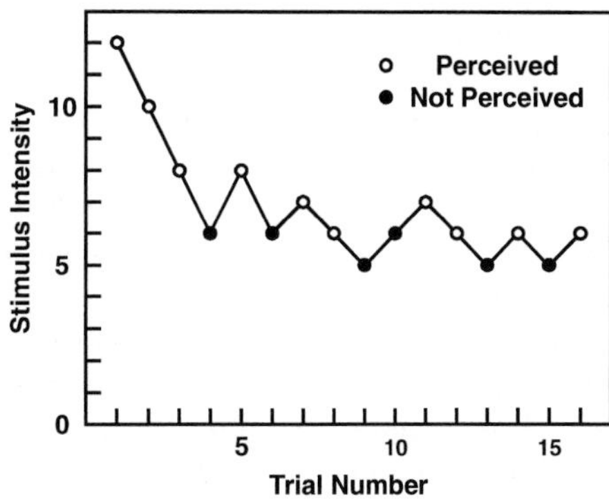

Fig. 4. Adaptive testing technique using a single staircase procedure. This example shows a descending staircase for which stimulus intensity is decreased when the stimulus is perceived and increased when it is not perceived.

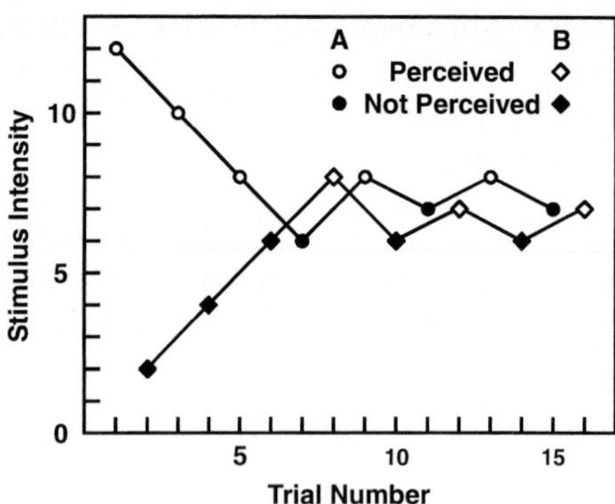

Fig. 5. Example of using two interleaved staircases. Stimuli from the respective descending and ascending staircases are presented on alternate trials.

scheme that governs stimulus presentation, which could lead him or her to anticipate the approach of threshold and change his or her response before the threshold is actually reached.

To overcome this problem, one may use two (or more) *interleaved staircases*, as shown in Figure 5. On trial 1, staircase A begins with a well above threshold intensity. On trial 2, staircase B starts with a well below threshold stimulus. On trial 3, the next stimulus of staircase A is presented, on trial 4, the next stimulus from staircase B, and so on. Over the course of the trials, both staircases converge at the threshold intensity. The two staircases may also be interleaved in a random rather than regular sequence to prevent the observer from figuring out which staircase to expect from trial to trial (Cornsweet 1962).

Best PEST Procedure

An even more efficient adaptive testing method is that of *Parameter Estimation by Sequential Testing* (PEST), which uses maximum-likelihood estimation to select the most efficient ("best") stimulus intensity for a given trial (Best PEST; Lieberman and Pentland 1982). This method, which can easily be implemented on a personal computer, is similar to, but faster and more accurate than, conventional staircase procedures. Basically, it adjusts the amount of change in the stimulus on the basis of information already gathered according to the observer's performance, thereby allowing the collection of more precise information about the threshold on subsequent trials. The Best PEST procedure usually assumes that the underlying psychometric function has a standard sigmoid form; it can, however, also accommodate differently shaped functions as specified by the researcher (Lieberman and Pentland 1982).

Adaptive Variants of the Method of Constant Stimuli

A problem with using constant stimuli is that, because only stimuli near the threshold provide relevant information, many of the stimuli presented are too far away from the threshold to be of use. This inefficiency can be avoided by pre-testing (e.g., using the Method of Adjustment) in order to determine the exact range of critical stimuli to use, tailored to the sensitivity of the observer. Alternatively, one may adapt the stimulus set while the experiment itself is in progress (Farell and Pelli 1999). This approach of *sequential estimation* is preferable over pre-testing since a stimulus set optimized during a short pre-testing period might not be optimal for the entire experiment due to fluctuations in the observer's sensitivity. The most sophisticated and efficient versions of these adaptive strategies combine the experimenter's prior knowledge of the appropriate stimulus range and the observer's response on past trials to choose the signal strength for the next trial using a Bayesian adaptive psychometric method (Watson and Pelli 1983; King-Smith et al. 1994). In the Bayesian adaptive method the threshold is treated as a normally distributed random variable. After each response, the threshold's Gaussian probability density function is updated using Bayes's rule to integrate the prior probability-of-detection information. Each trial is placed at the current maximum-likelihood estimate of threshold, i.e., the mode. The final threshold estimate is also the mode (Farell and Pelli 1999).

Forced-Choice Methods

All the psychophysical methods discussed so far rely on the observer's subjective report of what he or she has perceived. These methods may be termed "subjective," because the experimenter cannot control whether the observer's report is correct or not. In such subjective experiments, results may depend on the criterion that the observer uses for judging whether or not a stimulus was perceived. The *forced-choice method* provides a more objective approach. In this method, the observer is required to make a positive response on every trial – regardless of whether he or she "saw" (or heard, etc.) the stimulus. For example, the stimulus (e.g., a light) may be presented above or below a fixation point, and the observer may be required to indicate – on every trial – which position was occupied by the stimulus. The forced-choice method was first devised by Bergmann (1858; see also Fechner 1860/1966, p. 242). In order to measure visual acuity, Bergmann varied the orientation of a test grating, and instead of asking whether a particular grating was visible, he forced the observer to identify the grating's orientation. A century later, this approach was "re-invented" and achieved an established position in psychophysics (Blackwell 1952).

The use of forced-choice methods reveals that many observers can discern lights so dim or sounds so weak that they claim they cannot see or hear them. For example, if one first measures the absolute threshold for a light by the method of adjustment and then uses this threshold intensity in a forced-choice experiment in which the dim light is flashed either above or below a fixation point and the task is to indicate its location, the performance is often found to be correct throughout. This accomplishment stands in contrast to the observer's prevailing impression that her or his responses were mere guesses and that nothing was actually visible. If the light is then presented again, now at an intensity somewhat below the previously determined threshold, one still can obtain 70 to 75 percent correct choices, which is well above chance level. Typically, forced-choice testing confirms that stimulus intensities can be discerned below the absolute thresholds defined by an unforced, more subjective procedure (Sekuler and Blake 1994).

It appears that the amount of stimulus information necessary to support a decision is greater in an unforced-choice than in a forced-choice situation. A comparison between unforced and forced-choice testing is also useful to factor out possible *criterion* differences among observers. The criterion can be defined as an implicit rule that an observer obeys in converting sensory information into overt responses. It has received much consideration within the theory of signal detection, to be discussed in the next section.

Signal Detection Approach

This psychophysical approach that concentrates on sensory decision processes was derived from *Signal Detection Theory* (SDT; Green and Swets 1966). Its precursors are found in Fechner's *Theory of Discrimination* (Fechner 1860/1966, pp. 85–89; see also Link 1992) and in Thurstone's (1927) *Law of Comparative Judgment*. SDT provides the basis for a set of methods used to measure both the *sensitivity* of the observer in performing some perceptual task and any *response bias* that the observer might have. According to SDT, the sensory evidence that indicates the presence of a stimulus (the "sig-

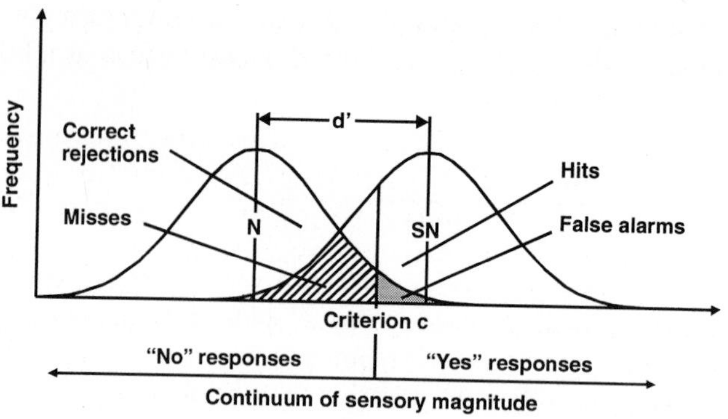

Fig. 6. Assumed psychophysical distributions in a signal detection task. Sensory magnitude (from weak to strong) is plotted from left to right along the horizontal axis and sensory excitation is plotted on the vertical axis. The distribution of noise alone (N) is shown on the left, that of signal plus noise (SN) on the right. The presence of signals is assumed to shift the sensory magnitude, but leaves the shape of distribution unchanged. The index *d'* refers to observer *sensitivity* and *c* to the observer's *criterion.*

nal") can be represented on a continuum (the continuum of sensory evidence). The strength of the signal, or, to put it in SDT terms, the evidence that a signal is indeed present, is assumed to vary from trial to trial. That is, the signal is assumed to be characterized best as a distribution of values on a continuum of sensory evidence rather than as a single value. Also, on any given trial there is some "noise" present in addition to the signal. Therefore, trials on which a signal is present are typically called signal-plus-noise trials.

Even on trials where no stimulus is present, there is assumed to be some evidence suggesting that a stimulus might be present. For instance, there may be some background noise or variability in the sensory registration process that is interpreted as evidence of the to-be-detected stimulus. Thus, a distribution of values of "noise" strength is also assumed.

Signal detection methods can be applied whenever there is some overlap of the signal-plus-noise and noise distributions. That is, whenever there is some range of values on the sensory evidence continuum for which the observer is unsure whether a signal was presented or not. If the overlap of the distributions is minimal, the signal-plus-noise and noise trials are relatively easy to tell apart and the observer will appear to be very sensitive. If the distributions overlap more, so that the means of the distributions are relatively close together, observer sensitivity in detecting the signal will be relatively low. Thus, the distance between the signal-plus-noise and noise distributions can be taken as a measure of sensitivity (see Figure6).

Another assumption of SDT is that, for a given session, the observer sets some response "criterion." If the stimulus energy on a given trial exceeds the criterion, the observer responds, "Yes, a signal was present"; if the criterion is not exceeded, the response is "No, a signal was not present". Because the observer does not know whether the signal was presented or not on a given trial, the criterion is assumed to be the same for both signal-plus-noise and noise trials.

Signal Detection Procedures

Observer sensitivity and response bias are measured by examining performance on both signal-plus-noise and noise trials. First, the "hit" and "false alarm" rates are computed. The hit rate is the probability that the observer said "Yes" when a signal was in fact present. The false alarm rate is the probability that the observer responded "Yes" when the signal was not present. Miss (saying "No" when the signal was present) and correct rejection (saying "No" when no signal was present) rates can easily be computed from the hit and false alarm rates, respectively (e.g., 1 – Hit Rate = Miss Rate), but these measures are not needed for further computations.

As already mentioned, the hit rate is the probability of correctly identifying a signal. As such, it can be defined as the area of the signal-plus-noise distribution that lies to the right of the response criterion (see Figure6). Similarly, the false alarm rate is the area of the noise distribution to the right of the criterion. It is often assumed that both the signal-plus-noise and noise distributions are normal and have equal variance. Under this assumption, a table of areas under the normal curve can be used to convert the hit and false alarm rates to measures of distance along the sensory evidence continuum, and, thus, to compute a common *measure of sensitivity, d'*. The formula for d′ and an example are given later in the chapter.

An observer's performance will depend not only on how much overlap there is between the signal and noise distributions, but also on where the criterion is located. If it is set at a relatively low level of sensory evidence, the observer will detect most of the signals (i.e., the signal energy will most often exceed the criterion value), but many false alarms will be made as well. Such a lenient observer is called "liberal." If the criterion is set to a higher level of evidence, fewer false alarms will be made, but the hit rate will also decrease. High criterion settings characterize a stringent or "conservative" observer. High hit rates combined with low false alarm rates are representative of good performance (i.e., high sensitivity).

Isosensitivity Functions

Receiver Operating Characteristic (ROC) curves can be created for an observer by plotting the hit rate against false alarm rate. All of the points along a single ROC curve reflect the same sensitivity or d'. In other words, ROC curves are *isosensitivity functions* (see Fig.7). Since the degree to which the ROC curve approaches the upper left-hand corner of the ROC plot depends on the sensitivity of the observer, different curves are generated by manipulating the discriminability of the stimulus or the sensitivity of the observer.

Different points along a single ROC curve represent different levels of bias. Technically, bias is defined as the slope of the ROC curve at a given point. Observers may be naturally biased to respond either liberally or conservatively, but bias can also be manipulated by the experimenter. For instance, the probability of a signal trial may be changed (a lower percentage of signal trials will tend to lead to more conservative performance) or observers may be given different incentives to make relatively few false alarms or, alternatively, relatively few misses.

Computing Observer Sensitivity

As mentioned above, a popular measure of observer sensitivity is d'. It is defined as the distance between the means of the signal-plus-noise and noise distributions (see Fig.6) and is calculated from the observer's hit and false alarm rates. The z-score is a measure

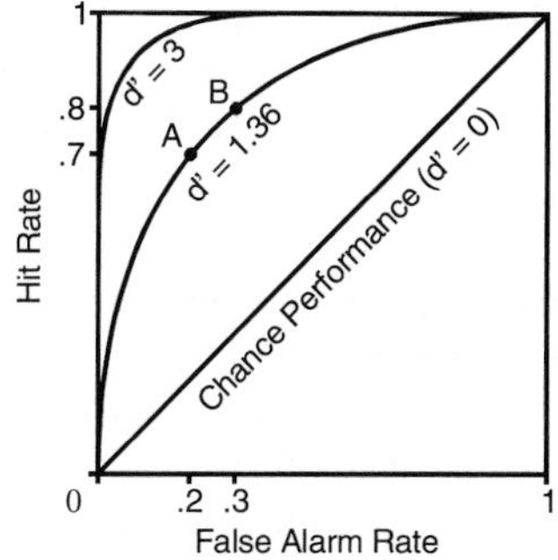

Fig. 7. Receiver operating characteristic curves for three sensitivities (d'). Note that the same sensitivity can result from different hit and false alarm rates: For a given d', a more conservative criterion is indicated by both lower hit and false alarm rates (A). A more liberal criterion is reflected in higher hit and false alarm rates (B).

of the distance of a score from the mean of a distribution in standard deviation units. Thus, z-scores can be used to measure sensitivity. The z-transformation (z is the inverse of the normal distribution function) converts the hit and false alarm rates to z-scores. The difference between z(hit rate) and z(false alarm rate) tells us how far apart in standardized units the means of the signal-plus-noise and noise distributions are. The formula for d' thus is:

$$d' = z(\text{hit rate}) - z(\text{false alarm rate})$$

A d' of 0 describes chance-level discrimination, i.e., a complete overlap of the signal-plus-noise and signal distributions and a complete lack of discrimination. A d' of 1 is considered moderate performance, a d' of 4.65 (when the hit rate reaches 0.99 with a false alarm rate of 0.01) is considered a ceiling value or optimal performance (see Fig. 7).

Example

As an example of applying signal detection methods, consider the case of evaluating the effects of alcohol on performance in a visual detection task. Suppose that 100 trials were presented to an observer, 50 noise trials and 50 signal plus noise trials, on two different occasions. One set of 100 trials was performed without ingesting alcohol, and the other after, say, two drinks.

For each set of trials, we compute the hit and false alarm rates. Suppose that without alcohol, 35 of the signals were correctly detected, giving a hit rate of 35/50 = 0.70, and suppose that 10 of the noise trials were erroneously classified as signal trials, giving a false alarm rate of 10/50 = 0.20. Suppose that the hit rate and false alarm rate after ingesting alcohol were 0.80 and 0.30, respectively. Looking at the hit rates, it appears that performance actually improved after ingesting alcohol. To test whether this is really the case, we need to compute d'. To compute d', we simply look up the hit and false alarm rates in the body of a table of areas under the normal curve, and take the corresponding z-scores (e.g., Macmillan and Creelman 1991). We compute d' = z(hit rate) – z(false alarm rate) for the two sets of observations. Without alcohol, this result is 0.542 – (–0.842) = 1.366. With alcohol, the result is 0.842 – (–0.542) = 1.366.

Naturally, such perfect agreement is seldom seen in the real world (and a statistical test of differences in d' may be required), but the example is clear: *Sensitivity*, as measured by **d'**, was not affected by alcohol. What did change is the *bias* of the observer. One measure of bias, **c**, can be computed as **c** = –0.5[z(hit rate) + z(false alarm rate)]. Accordingly, we find a bias of 0.159 for the alcohol-free performance and a bias of –0.159 for performance with alcohol. Positive values of **c** reflect more conservative performance, and negative values of **c** more liberal performance. Thus we can see, in our example, that alcohol had the effect of making our observer more liberal. These data are illustrated in Fig. 7; point A on the ROC curve refers to alcohol-free performance, point B to performance with alcohol.

Suprathreshold Methods

So far we have dealt with methods to determine the observer's threshold or ability to detect ambiguous signals. Since threshold stimuli are per definition difficult to detect or discriminate, these methods cannot be used in situations in which all stimuli are easily perceived. Even when all stimuli are above threshold, not all stimuli are equally easy to detect. For example, some stimuli are more conspicuous than others; they "pop out" and are rapidly discerned. For instance, a red object stands out more against a green background than does a blue one of similar intensity. Methods for comparing performance with above-threshold stimuli are discussed in this section.

Chronometric Methods

To measure suprathreshold stimulus differences, Münsterberg (1894) advocated the use of response or reaction times as the proper tool in psychophysics. Reaction time (RT) is defined as the time between the onset of a stimulus and the overt motor response elicited by the stimulus. In the case where two stimuli are presented on each trial and the observer is to indicate whether they are the same or different, RT can be seen as a measure of the ease with which we can discriminate between the two stimuli, or, according to Münsterberg, as "a measure of their subjective difference" (see also Petrusic 1993).

In a typical experiment, on any given trial the observer presses a key to indicate whether the two stimuli were the same or different on a certain dimension, e.g., color. Typically, RT increases as the two stimuli become more similar, or in other words, it increases as the difference between the stimuli decreases. An example with results from two observers in a brightness discrimination task is given in Fig. 8. The stimuli were pairs of squares with luminances ranging from 0.1 to 10 candela/m^2 in ten logarithmic steps, resulting in 9 perceptually equal stimulus separations (Ehrenstein et al. 1992).

We can distinguish between two varieties of RT: simple and choice. In *simple RT* tasks, a response is made upon detection of any stimulus event. In *choice RT* tasks, the response to be made depends on the identity of the stimulus. Thus, simple RT can be said to require only the detection of a stimulus event, whereas choice RT requires sensory discrimination and further processes, such as stimulus identification and response selection (Sanders 1998).

Generally, simple RTs are used to study sensory performance. When the stimulus intensity is low and near threshold (although still strong enough to be readily distinguishable), the response is slower than when the stimulus is well above threshold. Typically, both simple and choice RT decrease as stimulus intensity and separation increase up to a certain stimulus strength, with little change in RT thereafter (see Fig. 8). Thus, RT differentiates better between sensory performance in the lower than upper intensity range of stimuli.

Simple RT reflects not only variations of stimulus intensity; it can also be affected by other stimulus attributes such as spatial frequency. For instance, with sine-wave gratings of equal mean luminance (i.e., without differences in average intensity) one finds shorter RTs for low spatial frequencies (coarse gratings) than for high spatial frequencies (fine gratings). Breitmeyer (1975), who first showed this spatial-frequency effect on simple RT, concluded that low spatial frequencies preferentially stimulate *transient* visual mechanisms whereas the high spatial frequencies trigger *sustained* visual processing. Breitmeyer's study is a good example of using an RT task to test sensory mecha-

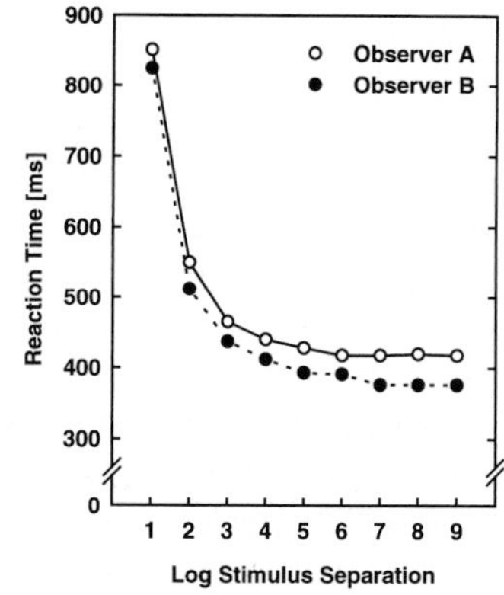

Fig. 8. Reaction times for same-different judgments as a function of perceptually equal steps of brightness contrast (Ehrenstein et al., 1992). In both observers (A and B), reaction time decreases with increasing contrast in a hyperbolic fashion.

nisms (and their preferred sensitivities) rather than, as done in conventional psychophysics, relating changes in stimulus intensity to that of RT. The distinction between transient and sustained mechanisms has meanwhile received further support from functional anatomic studies (Livingstone and Hubel 1988), resulting in the distinction of two visual pathways, the magnocellular system with good temporal but poor spatial resolution and the parvocellar system with good spatial but poor temporal resolution.

Choice RT is often used to study more complex sensorimotor or cognitive behavior as it involves stages, in addition to detection, such as discrimination, identification and response selection (Proctor and van Zandt 1994; Sanders 1998). In sensory research, however, one has to be careful in ruling out or accounting for these additional, potentially complicating factors. Typically, the observer is provided with two response keys, one for "same" and one for "different" responses. Pairs of stimuli that vary in similarity from identical to very different are presented, and RTs are collected. The observer has to respond as rapidly as possible, but to keep incorrect responses to a minimum (incorrect responses are often identified to the observer, e.g., by sounding a buzzer following an incorrect response, and excluded from the final analysis). Identical stimuli are shown on some trials so that the observer has to decide whether the stimuli are in fact different. However, the subsequent data analysis usually uses only responses to dissimilar pairs of stimuli. RTs for multiple trials (typically 15 to 30) with a given stimulus pair are averaged, and the resulting means or medians are plotted as a function of stimulus separation (as in Fig. 8).

Rank-orderings of RTs can also be further analyzed by using *Multidimensional Scaling* (MDS) methods. MDS refers to various computational procedures by which the degree of perceived similarity between stimuli is mapped to spatial distances. For example, by means of MDS, it is possible to construct the color space of a color-normal observer in comparison to that of a color-deficient observer (Müller et al. 1992).

Chronometric methods are also employed to study behavior that involves cognitive (e.g., attention, short-term memory) processes in addition to perceptual functions, e.g., in order to infer the amount of information processed in the working memory or the complexity of stimulus-response relations (see Ehrenstein et al. 1997; Sanders 1998). Although these studies involve various aspects of perception, they clearly transcend the reach of sensory psychophysics.

Scaling Methods

In addition to the use of RTs, there are other psychophysical approaches that apply to suprathreshold stimulus intensities. Most notable is that of *scaling*. A scale is a rule by which we assign numbers to objects or events. Psychophysical scaling has the goal of assigning numbers to perceptual events.

Category Scaling

First attempts to use scaling as a psychophysical method go back to Sanford in 1898 (Coren et al. 1994, p. 49). Sanford had observers judge 109 envelopes, each of which contained different weights (ranging between 5 and 100 grams), by sorting the envelopes into 5 categories (with category 1 for the lightest weights, category 5 for the heaviest, and the remaining weights distributed in the other categories). The average of all the weights placed in each category was plotted on a logarithmic ordinate, against linear units (category numbers) on the abscissa to obtain a measure of subjective magnitude. This method is called *category scaling*, and since category scaling uses judgment of single stimuli rather than relative comparison between stimuli, it is also referred to as "absolute judgment" (e.g., Haubensak 1992; Sokolov and Ehrenstein 1996).

Category scaling requires more stimulus levels than categories and, as we cannot know in advance what a correct category assignment might be, there are no right or wrong responses. Sensation scales based on category judgments turn out to be relatively stable over variations such as the number of categories used and the labels applied to the categories (number versus words). They depend, however, on the stimulus spacing, that is, on whether the stimuli vary in steps that are equal in absolute or relative (logarithmic) terms (Marks 1974). Linear spacing is useful if a smaller stimulus range (1–2 log units) is to be scaled; logarithmic spacing is more appropriate when scaling larger stimulus ranges. Category scaling has the advantage of allowing direct responses of the observer to variations in the stimulus. It is, however, limited in that responses are necessarily restricted to relatively few categories. Thus, similar stimuli that may give rise to different sensations may nonetheless be grouped into the same category. Another scaling technique, magnitude estimation, avoids these problems.

Magnitude Estimation

In its basic formulation, magnitude estimation is relatively simple. The observer is presented with a series of stimuli in irregular order and instructed to tell how intense they appear by assigning numbers to them. For example, the observer may be asked to assign to the first stimulus, say a sound signal, any number that seems appropriate and then to assign successive numbers in such a way that they reflect subjective differences in perceived intensity. In order to make each number match the perceived intensity, one may use whole numbers or decimals. That is, if one sound appears slightly louder than another, the louder sound may be assigned 1.1, 11 or 110 with the softer sound being 1.0, 10 or 100, respectively. Usually there is no limit to the range of numbers that an observer may use, although sometimes the experimenter may arbitrarily assign a given number to one particular stimulus level. In this case the subject might be told to let a rating of 100 correspond to the loudness of a 65-dB sound, but from there on the ratings would be entirely up to the subject. Given this latitude in assigning numbers to perceived intensities, one might expect the outcome to be a jumble of random numbers, but in fact this does not happen. Whereas the particular numbers assigned to particular stimulus intensities will vary from observer to observer, the order and spacing between numbers show a high degree of regularity among individuals.

When applying this *direct ratio-scaling* method, the experimenter should use as large a range of stimulus values as possible or practical. This is to assure that the sensory magnitudes will be well distributed over the perceptual range. However, the sessions should be made a reasonable length so as not to overtax the subjects. It is preferable to use multiple sessions rather than one extremely long session.

Instructions in a magnitude estimation task should reflect that magnitude estimation is an estimating rather than a matching approach. That is, it should be made clear that there is no fixed standard stimulus and that there is no modulus. Early experiments typically used both a standard and a modulus. For example, a stimulus might be selected from the middle of the range, assigned the number "10", and then be presented at a given interval or per subject request. However, the choice of standard has been shown to influence the shape of the obtained psychophysical function and has not proved to have any clear benefits (see Marks 1974).

If the underlying relations are expected to be similar for all individuals tested, the data may be averaged. To average magnitude estimates, the *geometric mean*:

$$M_g = \sqrt[n]{x_1 \cdot x_2 \cdot \cdots \cdot x_n} = \left(\prod_{i=1}^{n} x_i \right)^{\frac{1}{n}}$$

is preferred over the *arithmetic mean*:

$$M_a = \frac{x_1 + x_2 + \cdots + x_n}{n} = \frac{1}{n} \sum_{i=1}^{n} x_i$$

since it is less susceptible to the extreme values that are likely to occur in experiments where subjects pick their own numbers. The distribution of magnitude estimates given to a single stimulus level often approximates log-normality (i.e., is approximately normal after a log transformation – this is another way of saying the distribution has a long tail at the high end of the distribution), and the geometric mean gives an unbiased estimate of the expected value of the logarithms of the magnitude estimates. However, since the geometric mean will be zero when any of the observations are zero, the median (which is generally not preferred because it "wastes" information about the distribution of values) may be preferred when there are zero values (as will occur if some subjects cannot detect some of the stimuli). As Marks (1974) recommends: "Use the geometric mean whenever possible (even if it means dropping a few stimuli that were given zero values), but use the median in a pinch."

Once an average is decided on, plot the data. Magnitude estimates often plot as lines on double-logarithmic coordinates. Note that a straight line on a double log plot indicates a simple power function between stimulus and sensation.

Determining Psychophysical Functions

It is often the case that the lines representing the relation of a sensation to stimulus values are not quite straight when plotted on log-log paper. Thus, simply fitting a power function to the data may not adequately describe the data. Simply connecting the data points with line segments can serve to show the basic trends in the data, and drawing free-hand curves can sometimes serve to display the trends more clearly, but at the cost of precision. It can be of advantage to fit some simple equation to the data to characterize the psychophysical functions. This can serve to summarize a good deal of data with just a few parameters.

How should one deal with departures from linearity in log-log plots of psychophysical functions? The typical finding is that the functions become steeper as the stimulus values approach zero. More correctly, one could say that the functions become steeper as the threshold of sensation is approached. Stevens (1975) and others have suggested that a threshold parameter should be subtracted from the value of each stimulus. This suggestion follows from the reasoning that, only after the threshold is passed, can we expect a power relation.

Thus, one equation one might fit to the data is:

$$\Psi = \mathrm{k}\,(\phi - \phi_0)^{\beta},$$

where Ψ is estimated sensory magnitude, ϕ is stimulus intensity, ϕ_0 is the threshold parameter as defined above, and k and β are constants.

Cross-Modality Matching

Another popular procedure for scaling sensory experience relies exclusively on sensory magnitudes, avoiding the assignment of numbers altogether. In this method, called *cross-modality matching,* an observer equates perceived intensities arising from the stimulation of two different sensory modalities (Stevens 1975). That is, an observer adjusts the intensity of one stimulus (e.g., a light) until it appears to be as intense as a stimulus presented in another modality (e.g., a tone). Cross-modality matching can be regarded as a combination of the methods of scaling with that of adjustment. The adjusted stimulus value, such as the luminance of a light, is recorded by a photometer and taken as an estimate of perceived sound intensity. These data (averaged across repeated trials) are plotted on log-log axes as for magnitude estimation data. Despite the fact that the observers no longer make numerical estimates, the data typically still obey the power function (see above).

Since the subject needs to adjust one of the sensory variables to give a response, rather than simply report a number or assign a category label, cross-modality matching is often more difficult to use than the other scaling methods. Nevertheless, because of its conceptual simplicity (with respect to measurement theory), in that the observer relies on sensory rather than "cognitive" magnitudes (numbers or verbal categories), cross-modality matching can be regarded as an optimal method of suprathreshold sensory psychophysics.

Comparative Psychophysics

Although psychophysical methods were originally developed to study human perception, they can also be used in the study of sensory performance of non-human species. This so-called *comparative psychophysics* has been applied to a wide range of species, including mammals, birds, fish and insects (e.g., Berkley and Stebbins 1990; Blake 1999). Comparative psychophysics is essentially the same as human psychophysics in that it uses highly restricted sorts of stimuli and even more restricted simple responses. Although it is much easier to work with human subjects, the advantages of animal work are obvious. It affords the comparison of anatomical, physiological and behavioral information within a species as well as comparison across species with differently organized neuronal and sensory systems.

Because verbal instructions are of no use in comparative psychophysics, animals need to be trained non-verbally to perform in a sensorimotor task (see also Chapter 44). This training essentially relies on the two main forms of behavioral control or reinforcement, appetitive or aversive. Appetitive control is accomplished by giving food or water, respectively, to a hungry or thirsty animal for correct performance; aversive control relies on presentation of some undesirable stimulus (such as a loud tone or a brief electric shock) in association with a sensory event or in consequence of an incorrect response. Aversive control should evoke some brief, mild degree of discomfort rather than pain.

Behavioral techniques for establishing stimulus control and assessing sensory performance include reflex responses, stimulus-associated (classical) conditioning and operant, behavior-linked conditioning.

Reflex Methods

Reflex responses, such as optokinetic nystagm, the looming reflex and orienting reactions, require no training because the animal's behavior occurs automatically and stereotypically in response to the adequate (releasing) stimulus. For instance, orienting reactions to light refer to the fact that some animals are naturally attracted to light (phototaxis) while others are repelled by light (photophobia). Both stereotypes of behavior can be used in determining a psychophysical threshold for seeing, indicated by the intensity of an optic stimulus necessary to just elicit an orienting reflex. Although reflex methods are attractive to the investigator in that no training is required, their use is limited to reflexes elicited by just a few stimuli. These methods are used when they happen to be so appropriate that conditioning is unnecessary, or when it is not clear that the animal can be conditioned. If a training method seems called for, one may decide on stimulus-associated or operant conditioning methods.

Stimulus-Associated Conditioning

Procedures that use stimulus-associated (also called "classical") conditioning have proven particularly popular in comparative psychophysics because initial training is simple and relatively rapid. With these procedures, a sensory stimulus (the conditioned stimulus) is repeatedly paired with an unconditioned stimulus that itself reliably elicits a reflexive (unconditioned) response. Following relatively few such pairings, the sensory stimulus elicits the response on its own (i.e., the conditioned response). For example, an electric shock delivered across the animal's body produces heart-rate acceleration (unconditioned response) which can be registered by implanted electrodes connected to an electrocardiograph. Initially neutral sensory stimuli are paired with the shock stimulus several times until they elicit heart-rate acceleration in the absence of electric shock.

In contrast to the finding that "classical conditioning does not lend itself to the measurement of discrimination thresholds" (Blake 1999, p. 145) in the visual modality, various examples of successful application of classical conditioning in measuring *discrimination thresholds* exist for the auditory modality (Delius and Emmerton 1978; Grunwald et al. 1986; Lewald 1987a,b; Klump et al. 1995).

For example, studies of sound localization in pigeons (Lewald 1987a,b) have used a conditioned (acoustic) stimulus paired with an unconditioned stimulus (a weak electric shock causing an increase in heart rate). In these studies, each experimental series started with a random number (between 5 to 11) of identical presentations of a *reference tone* from one loudspeaker. Subsequently, the conditioned or *test tone* was presented (with the same frequency, intensity and duration as the reference tone, but from another loudspeaker location) and followed by an electric shock. The pigeon was repeatedly exposed to these stimulus conditions until its heart rate accelerated selectively at the onset of the test tone. Starting with a simple task in which the sound locations were well separated, the pigeon was then presented with increasingly more difficult discriminations (smaller angles between the sound locations) until its heart rate failed to differentiate between the test and the reference stimuli. An easier task was then presented to enable the bird to relearn its response before the limits of discrimination, which define the difference threshold, were tested again.

In visual psychophysics, the response readily generalizes to stimuli other than the conditioned stimulus so that it is difficult to test conditions in which an animal is required to respond to one sensory stimulus but not to another. For this reason, stimulus-associated conditioning in vision works primarily for the measurement of *absolute thresholds* where the animal simply detects the presence of a sensory stimulus.

In addition to the limitation that classical conditioning applies unequally well to auditory and visual modalities (Delius and Emmerton 1978; Blake 1999), a general disadvantage of this method is the use of aversive stimuli which may put the welfare of the animal at risk.

Operant Conditioning

The technique of operant or instrumental conditioning refers to procedures where the animal's behavior determines the outcome of a given trial. For comparative psychophysics, instrumental behavior is controlled by a sensory stimulus. For example, animals may learn that a reward is contingent on the presence of one stimulus, but not on another, i.e., the type of stimulus cues the animal about which response produces reward. If an animal is trained to respond to one of two visual stimuli presented simultaneously, it must select which one of the stimuli is reliably connected with reward delivery.

Absolute thresholds are determined by using one stimulus which is always sufficiently weak so as to be invisible (e.g., a zero-contrast grating) paired with an initially visible test grating whose contrast is varied systematically over trials (e.g., by a staircase procedure), to find the luminance contrast at which it becomes barely distinguishable from the blank comparison stimulus.

Difference thresholds are determined with both stimuli visible, but differing in some dimension (e.g., grating contrast, spatial frequency). The animal is initially trained to respond to one of the two easily discriminable stimulus patterns; then the difference between the stimuli is diminished until the value at which the two are barely discriminable is reached. The trial is terminated when the animal fails to respond correctly. As in human psychophysics, the intensity or value of the stimulus can be varied according to the method of constant stimuli or adaptive staircase procedures.

Animal Studies and Inner Psychophysics

If we argued that we cannot know what an animal actually perceives, we should be aware that this problem generalizes to all other individuals, animal and human. Thus the challenge in both animal and human psychophysics is to design and control stimulus conditions that rule out all possible extraneous cues that might support successful detection or discrimination. The study of animal perception has been much advanced by the development and refinement of psychophysical paradigms that can be used to train animals to perform tasks involving sensory detection and discrimination. Some of these methods have also proved useful for the study of human abilities, especially in infants (Atkinson and Braddick 1999).

Comparative psychophysics can be regarded as coming closest to Fechner's *inner psychophysics* in that it allows for the most direct linking of neural activity to corresponding perceptual performance. For example, by recording single cell activity from alert behaving animals, variations in neural responsiveness can be simultaneously compared with perceptual performance under the same task demands and stimulus conditions within the same animal. Such experiments have shown, e.g., that when the stimulus displays are matched for preferred size, speed and direction of motion, *neurometric* and *psychometric* functions (based on neuronal firing rates and perceptual response rates, respectively) become statistically indistinguishable (see Britten, et al. 1992).

Experimental Examples

A: Auditory Motion Aftereffect

(Source: Ehrenstein 1994, in which further references are given).

Problem

After being exposed to visual motion for some time, and then shifting our gaze to a nonmoving stimulus, we perceive apparent motion in the direction opposite to that of the previously observed motion. For example, when we shift our gaze from a running waterfall to adjacent rocks, the rocks appear to be moving upward (this is called the *waterfall illusion*). Motion aftereffects are quite compelling, both when induced experimentally (e.g., by gazing at a rotating spiral) or naturally (e.g., by looking at a waterfall). Whereas motion aftereffects are well known and well investigated in vision, little is known about possible analogs of these effects in the auditory modality. Auditory motion aftereffects may be studied by first allowing observers to adapt to auditory motion in a given direction, and then asking them to use a fast psychophysical method, e.g., that of adjustment to indicate the location of a static sound. Biases in the localization of the sound would indicate that observers were influenced by motion aftereffects.

Methods

Apparatus To protect the observer (and the results) from the influence of ambient noise, the experiment was conducted in an acoustically shielded, sound-attenuating room. The sound signal, a narrow-band (0.5 octave band-pass) signal with 1 kHz mean frequency and 60 dB sound pressure level, was generated by a noise generator (Rohde and Schwarz, SUF) linked to a function generator (WAVETEK VCG/VCA, Model 136). Sound pressure level was calibrated using an artificial ear (Brüel and Kjaer 4152) in combination with a sound-level meter (Brüel and Kjaer 2209). By means of a ramp generator and multiplier the signal was presented via headphones (KOSS, Type PRO14AA) with reciprocally increasing intensity (in one ear) and decreasing intensity (in the other ear) to obtain simulated auditory motion. That is, the tone was perceived as moving within the head along a line between the two ears. For instance, with a ramp timing of 3.11 to 0.11 s and a rising ramp presented to the right ear (with the left ear receiving the inverted signal), the sound is heard to move at a moderate speed from left to right, and then to suddenly (within 0.11 s) return from the right ear to its starting position at the left ear. If the ramp inputs to both ears are inverted, motion from right to left is perceived.

When the signal is not ramped, the same narrow-band stimulus can be presented at various interaural intensities by means of a two-channel logarithmic attenuator. The result of this is that the signal is perceived at different intracranial locations. The perceived location of the sound could be shifted by the observer by adjusting the knob of a ten-turn potentiometer connected to the attenuator, with each rotation resulting in a constant variation of the interaural intensity (ΔI), and with clockwise rotation shifting the perceived location of the sound to the right and counter-clockwise rotation shifting the sound to the left. The actual potentiometer settings were transformed into the respective ΔI-values (dB) and recorded for later analysis.

Selection of Subjects Before testing, subjects were selected according to normal hearing ability, assessed by a standard hearing test (e.g., PHONAK Selector A).

Procedure

1. Subjects were instructed to adjust the ΔI of the sound by turning a knob of the potentiometer so that the sound was perceived as being exactly on the midline between the two ears.
2. Initial practice trials were given to acquaint the subject with the task and, especially, with the psychophysical adjustment procedure. Since the dichotic midpoint has a very salient perceptual quality, most subjects learn the task quite easily.
3. Before the trials on which the subject adapted to simulated auditory motion, each subject made a total of 10 settings of the intracranial auditory midline. Beginning with $\Delta I = 0$, four randomly presented initial starting points ($\Delta I = 3, 6, -3, -6$ dB) were presented and the sequence was repeated once. These settings were used as control measurements.
4. Following the control trials, 10 trials of sound localization after motion adaptation were presented. Simulated motion was presented as an adapting stimulus for approximately 90 s (28 ramp cycles). Immediately following this adaptation, the same narrow-band stimulus was presented without ramping at various interaural intensities (resembling different intracranial locations) and the observer adjusted the location of the sound to correspond to the auditory midline in the same way as in the control measurements. These settings were made over approximately 4 minutes, and the times of the settings were marked for later analysis to plot the time course of any mo-

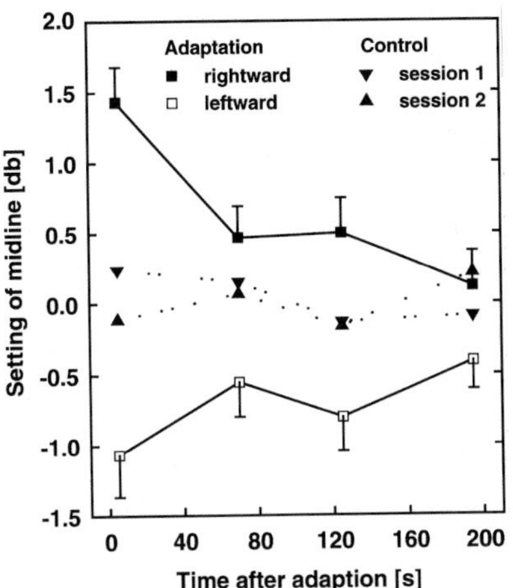

Fig. 9. Inter-aural midline settings as a function of time after adapting to leftward or rightward dichotic motion (Ehrenstein 1994). Settings are maximally displaced (in opposite direction to that of adapting motion) immediately after adaptation and soon return to control-measurement level.

tion aftereffect. To control for possible effects of different initial starting points, the measurements at each of the respective four non-zero starting points were pooled together.

5. In order to avoid possible confounding effects of motion adaptation in one direction with those caused by adaptation in the other direction, the experiment was performed in two different sessions on different days, and each subject was tested for only one direction of motion within a given session.

Results

Following the exposure to simulated sound motion no counter-motion is heard. That is, the test stimulus appears to be stationary. However, the observer's settings of the inter-aural midline of the sounds were displaced in the direction opposite to the direction of adaptation (on average by 1.2 dB). This displacement effect was reduced as a function of time after adaptation (see Fig. 9).

Conclusion

Having established that an aftereffect of auditory motion occurs and having determined the strength of the effect and its time course by the adjustment method, it would be appropriate to investigate it further with a more sophisticated psychophysical method, for example, with a computer-assisted adaptive staircase procedure or method of constant stimuli.

The observed auditory aftereffect may be analogous to visual motion aftereffects since it is direction-specific; however, because it does not induce apparent motion and occurs with dichotic stimulation, it might also resemble disparity-specific stereoscopic aftereffects.

B: Interocular Latency Differences in Neurological Patients

(Source: Ehrenstein, Manny and Oepen 1985; with further references)

Problem

The optic nerves are among the earliest and most frequently involved sites of demyelinating plaques in multiple sclerosis (with fibers of the foveal ganglion cells being particularly affected). In the early stages of multiple sclerosis, the conduction time for visual stimuli is often prolonged for one eye. Typically, the change in conduction time is revealed by a prolonged latency in the visually evoked cortical potential (VEP). In this example, we look at a psychophysical procedure for testing foveal differences between conduction times of the two eyes.

Methods

Rationale

Assume that due to demyelination the conduction time of the right visual pathway is prolonged by 20 ms. Then a stimulus presented to the right eye has to be delivered 20 ms before a stimulus to the left eye in order for the two to appear simultaneous.

Subjects

Subjects were patients clinically suspected of multiple sclerosis (based on different groups of definitive, probable, or possible neurological symptoms) and healthy control subjects, matched for age and sex (incidence of multiple sclerosis is much higher in females than in males). All subjects had normal or corrected-to-normal visual acuity.

Apparatus

The stimulus consisted of a small cross formed by four rectangular light-emitting diodes (LEDs). The LEDs were covered by polarizing filters so that an observer wearing polarizing glasses could see the horizontal bar only with her or his left eye and the vertical bar only with the right eye. A circular, non-polarized LED in the center of the cross served as a fixation cross, visible to both eyes. The horizontal and vertical bars of the LED cross were presented for 80 ms with the onset asynchrony (Δt) of the two bars varied by a 4-channel tachistoscope timer (Scientific Prototype) controlled by a computer in steps of either 30 or 15 ms for the patients, and 15 ms for normal controls (see Fig. 10).

Procedure

A chin-and-forehead rest supported the head of the subject. A warning click preceded each trial by 500 ms. The observer's task was to press one of two keys in order to indicate which bar, horizontal or vertical, appeared first. Observers were allowed to withhold a response whenever both bars appeared to occur simultaneously, or when the order of succession was uncertain.

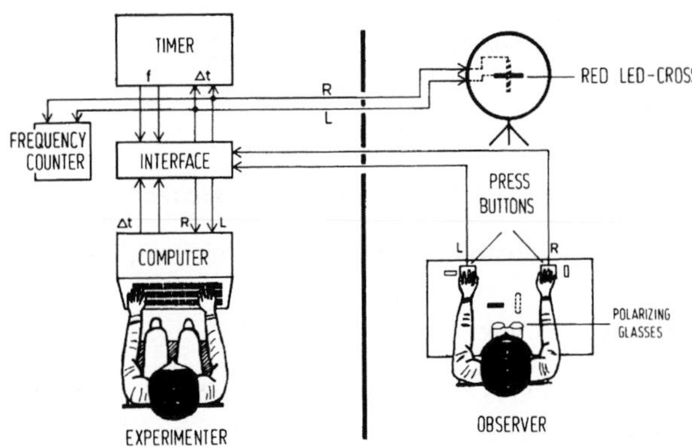

Fig. 10. Schematic representation of the experimental set-up to measure interocular time thresholds (Ehrenstein et. al., 1985).

The threshold for stimulus onset asynchrony was determined by a modified method of constant stimuli. A series of trials was presented that started with one bar (e.g., the horizontal bar, left-eye stimulus) preceding the other bar by 150 ms, with the interval being reduced by 15-ms steps until the vertical bar (right eye stimulus) preceded the horizontal bar by 150 ms; the order of presentation was then reversed, going from vertical bar (right eye) first to horizontal bar (left eye) first. So as not to overtax the patients, only a total of 10 such sequences of trials were given. Thresholds were determined as 50 % of the responses left before right, or right before left, respectively. The interocular latency difference (the arithmetic mean of the two time thresholds "right eye before left" and "left eye before right") was computed for each subject. For example, if an observer needs a delay of –67.5 ms to just see "left before right" and a delay of +45 ms to just see "right before left", the latency difference is (–67.5 + 45) / 2 = –11.25 ms.

Initial practice trials were conducted to acquaint the observers with the task, but also to check whether the patients' time discrimination could be tested within the range of 150 ms, or whether a larger Δt (i.e., 300 ms with a step size of 30 ms) would be needed. Catch trials with zero or maximal temporal delays were inserted at random intervals to check on the observer's attention and reliability. Non-zero delay trials on which the subject withheld the response were repeated. Patients with motor problems, who found pressing the response key difficult, were allowed to give their responses verbally.

VEPs (P2-latencies) of the patients, recorded under clinical routine conditions in response to foveal stimulation (contrast reversal of a small rectangle), served as an objective reference for comparison with the psychophysical data.

Results

Patients suspected of multiple sclerosis exhibited much higher interocular latency differences (up to 29 ms) than normal controls (up to 12 ms). This indicates unilateral or asymmetrical impairment of the visual pathways in these patients. VEP and psychophysical latency differences showed a highly significant positive correlation ($r = 0.59$; $p < 0.01$, see correlation diagram in Fig. 11). Psychophysical latencies were better measures of performance than VEP latencies, since they extended over a slightly larger time range.

The psychophysical data also yielded a higher diagnostic validity than the VEP data (higher hit rate in indicating a neurologically confirmed history of optic neuritis). Thus, the diagnostic rate can be significantly raised: from approximately 60% (based on VEP evidence alone) to 90% (VEP and psychophysical testing combined).

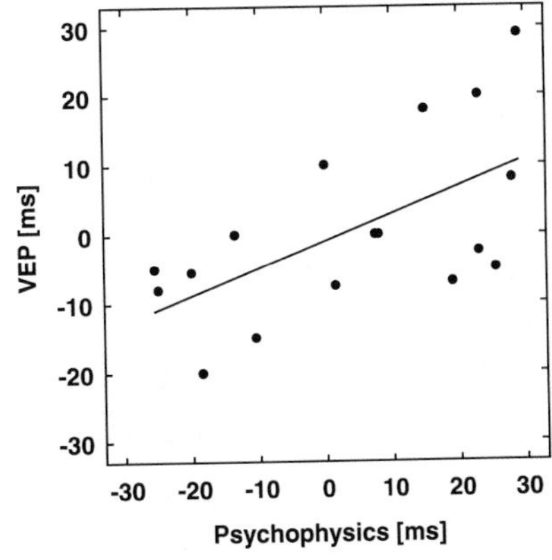

Fig. 11. Correlation diagram of the interocular latency differences obtained by VEP recordings and psychophysically in 17 patients suspected of multiple sclerosis (redrawn from Ehrenstein et al., 1985).

Conclusion

It seems to be useful to combine VEP and psychophysical latency measures to diagnose multiple sclerosis. As a clinical routine, the established VEP examination might be preferred, since the recording process is simple and less dependent on the patient's cooperation than the psychophysical procedure. However, psychophysical examination of latency differences is indicated in cases where the VEP provides unclear or conflicting evidence (e.g., in patients with a confirmed history of optic neuritis, but normal VEP latencies).

C: Perceptive Fields in Monkey and Man

(Source: Oehler 1985; see also Spillmann, Ransom-Hogg and Oehler 1987 with further references).

Problem

A key concept in sensory physiology is that of the receptive field. A receptive field is defined as the area on the receptor surface (e.g., retina) within which a change in stimulation leads to a change in the discharge rate of a corresponding single neuron. Jung and Spillmann (1970) introduced the term *perceptive field* to denote the psychophysical correlate of a receptive field, as in the perception of illusory changes in brightness in Hermann grids. Westheimer (1965) introduced a psychophysical paradigm that affords precise measurement of perceptive fields. In the present study perceptive fields were measured by using the same Westheimer paradigm in monkey and human observers.

Methods

Rationale The Westheimer paradigm requires measuring the threshold for a small test spot as a function of background size (Fig. 12). To reduce the effects of stray light, both the test spot and background are centered on a large ambient field. The test spot is varied in luminance according to one of the methods to determine the absolute threshold, such as the method of constant stimuli.

Typical results obtained with this method are shown schematically in Fig.13. The threshold for detecting the test spot first increases with the diameter of the background, reaching a peak and then decreases and finally levels off with further increases in the background area.

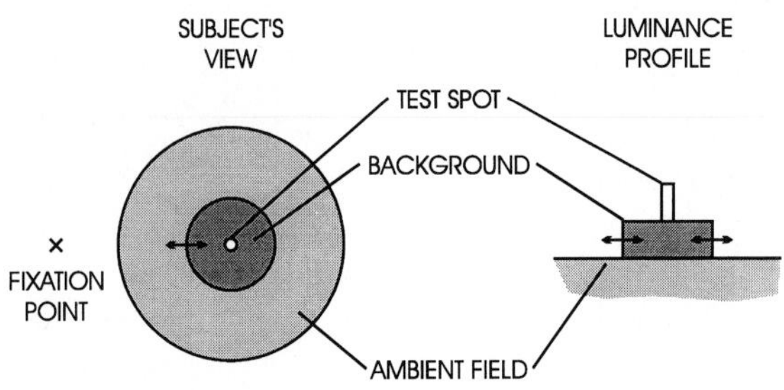

Fig. 12. Stimulus configuration used to measure visual perceptive fields (Westheimer paradigm); redrawn from Spillmann et al. (1987).

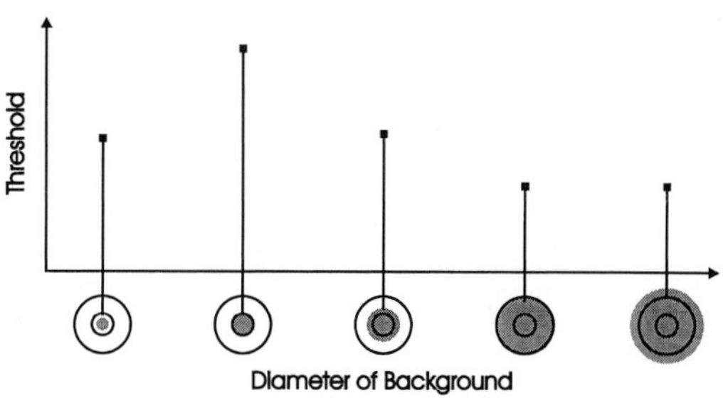

Fig. 13. Typical threshold data as a function of background size and their presumed relation to receptive field organization of an on-center neuron. The threshold increases until the background reaches center size. The threshold decreases once the background encroaches on the inhibitory surround region until the total field size is reached and finally levels off; a further increase in background size does not change the adapting level (redrawn from Spillmann et al., 1987).

The neurophysiological interpretation of this function, relating change in background area to threshold for detection, is depicted (for an on-center neuron) below the abscissa in Fig. 13. As the size of the background increases, the neuronal adapting level is increased, which causes desensitization due to spatial summation of receptor input. As a consequence, the threshold value continues to increase until the background covers the entire receptive field center. As the background encroaches upon the antagonistic surround, the adaptation level is lowered, which leads to sensitization caused by lateral inhibition, hence the threshold decreases. This decrease continues until the background covers the entire receptive field. A further increase in background size does not change the adapting level; hence the threshold remains constant.

According to this interpretation, two measures of perceptive fields can be derived: the center size, indicated by the diameter at which the threshold curve peaks, and the total field size, corresponding to the background diameter at which the threshold curve levels off.

Subjects

Two rhesus monkeys (Macaca mulatta) were tested; one was an adult female of 3.5 kg body weight, the other an adolescent 6 kg male; both were emmetropic as determined by skiascopy (assessing refraction by light reflection through the pupil of the eye). Threshold measurements were repeated with two emmetropic human observers, a 30-year-old woman and a 21-year-old man.

Apparatus

The animal, a rhesus monkey, was placed on a "primate chair" with her or his head fixed to it by means of a permanently implanted metal headpiece. Attached to the chair were a response lever and an infrared eye movement monitor. One eye of the monkey was occluded; the monkey used the other eye to fixate the center of a semi-cylindrical screen (see Fig. 14). The stimulus configuration (as shown in Fig. 12) was back-projected onto a translucent screen.

The optical system for presenting the stimuli consisted of two channels, one producing the test spot, the other the background stimulus. The test spot's luminance (produced by a halogen bulb) was varied by a neutral density wedge, and its duration (100 ms) was controlled by an electromagnetic shutter. The background diameter was varied by a series of apertures. The background was centered relative to the test spot by means of an adjusting mirror (*M* in Fig. 14). Background and test spot beams were united by means of a pellicle (*P* in Fig. 14). The ambient field was superimposed onto the luminance configuration using another projector, which projected onto the front side of the screen.

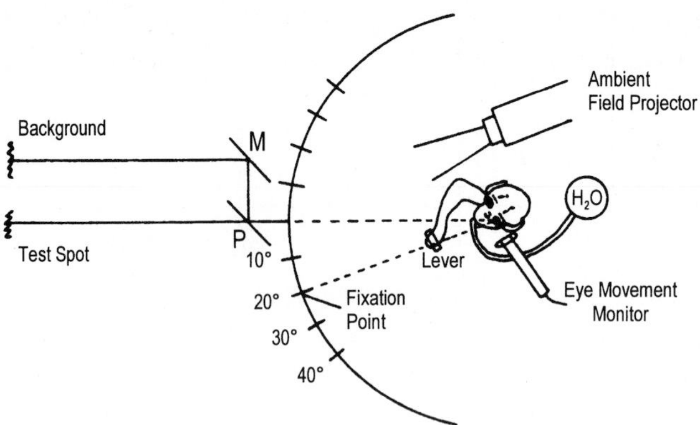

Fig. 14. Schematic apparatus used to measure perceptive fields in monkey and man (redrawn from Oehler 1985).

Training A reaction-time paradigm was used to train the animal in two steps. First, the animal was trained to fixate on the stimulus location using a dimming task in which the monkey had to pull a lever at the onset of a red fixation light, hold the lever for a random duration (0.4 to 4.9 s), and release it within an allowed RT of 0.8 s following the dimming of the fixation point. Correct responses were rewarded by a drop of water, whereas responses occurring too early or too late were indicated by a high or low tone and punished by a prolonged "time-out." In the second step, the dimming of the fixation point (which was now held at a constant intensity) was replaced by the onset of the test spot as signal for the monkey as when to release the lever.

As the threshold measurements were to be made at defined eccentricities up to 40 degrees in the periphery, the monkey was required to dissociate between the locus of fixation and the focus of visual attention. To control for accurate fixation, eye movements were continuously recorded and trials automatically interrupted when the animal departed from proper fixation.

At the beginning of the training sessions, every correct response was rewarded. After a few days of training, subthreshold stimuli were interspersed with clearly visible stimuli (in an intermittent reinforcement schedule). Before each experimental session, the animal was water-deprived for 22 hours; during each session the monkey obtained 120–200 ml of water.

Threshold Measurement The method of constant stimuli was used. Releasing the lever within the given time was regarded as a correct response ("stimulus seen"), whereas releasing too late was regarded as "not seen." (Releasing the lever too early occurred rarely, presumably because it was punished by a long time-out.) Each stimulus intensity was presented 10 times in quasi-random order and the rate of correct responses was plotted as a function of test spot luminance with "50% seen" as the threshold value.

Although human observers do not require a special training session to learn the task, threshold measurements were made for comparison under essentially the same conditions as for the monkey, i.e., using the reaction-time paradigm and using a lever to make responses.

Results

With increasing retinal eccentricity, the threshold curves peak at increasingly larger background sizes, indicating an increase of perceptive center size from 0.25° (at 5° eccentricity) to 1.5° (at 40° eccentricity). Total perceptive field size, indicated by the points where the threshold curves level off, increases from about 1° to 3° (from 0.5° to 40° eccentricity). Rhesus monkey and human data are fairly similar with respect to center size, whereas total perceptive fields are larger for humans.

Conclusion

Psychophysical methods can be used to study neuronal organization underlying visual perception in human and non-human primates. Human perceptive field sizes can be compared to those obtained in trained monkeys; these in turn can be related to receptive field sizes and their morphological substrate, dendritic fields, in monkeys. This allows for correlations among psychophysical, neurophysiological and histological measurements.

Concluding Remarks

The objective of this chapter was to outline some of the main methodological issues of psychophysics in their relation to neuroscientific approaches. As illustrated by the three experimental examples, psychophysical methods can be applied to various aspects of sensory and perceptual problems, from basic and comparative research to clinical diagnostics. Due to the current progress in the cognitive and brain sciences, along with the cumulative progress of psychophysics itself, psychophysics is beginning to emerge as a discipline in its own right. Its methods apply not only to questions concerning the sensory environment and perceptual performance afforded by it (outer psychophysics), but more and more to the functional states and processes of sensory and neural systems (inner psychophysics).

The limitations in the use of psychophysical methods reside in that they require a conscious and cooperating subject, an understanding of the task and a way to reliably report the sensed events. In clinical applications, there is cause for some concern that observers may purposely "cheat" in order to simulate or exaggerate a performance loss. To some extent, methods such as those of the signal detection approach allow the experimenter to check on the observer's response bias and sensitivity, but these techniques cannot be used to correct for purposely biased reports. Here we reach a principal limitation of subjective testing: Cooperation on the part of the subject is required. This limitation of psychophysical methods can be overcome by combining subjective with objective methods in the examination of sensory function.

We have made a distinction between methods that rely on threshold stimuli (i.e., methods that force the subject to reach a performance limit, i.e., to fail) and suprathreshold conditions which more realistically cover the daily-life range of stimulus intensities. Both approaches of measurement have advantages. Threshold measures allow for an accurate analysis of sensitivity of a detailed sensory function, whereas suprathreshold measures afford the study of performance and integrative functioning of sensory systems over their entire operational range. It is difficult, however, to prescribe in advance the appropriate technique for a particular application (some useful sources of information about current methods and software developments as well as suppliers of experimental equipment and computer software are given in the Chapter Appendix).

As a general rule, one should look for the easiest and most convincing method to answer the experimental question. More sophisticated procedures are not *per se* better than simpler methods, particularly if they introduce difficult technical problems. For example, if accurate timing over very short time intervals is necessary, the use of easily controllable (and inexpensive) LEDs might be preferable to using a computer monitor that relies on a cathode-ray tube display with complex problems of timing accuracy (depending, e.g., on the sampling rate and on rise and decay functions of the phosphor; see Robson 1999). Sometimes, new methods evolve out of special experimental questions. The problem of alignment of auditory and visual spatial coordinates, for example, led to the development of a new psychophysical technique, i.e., that of using laser pointing to acoustic targets (Lewald and Ehrenstein 1998).

The methods presented here are recipes for psychophysical research, but they need to be complemented by detailed information from the respective field of research application and should be followed flexibly, to stimulate rather than hinder creative modification and development of new experimental paradigms.

References

Atkinson J, Braddick O (1999) Research methods in infant vision. In: Carpenter RHS, Robson JG, eds., Vision research. A practical guide to laboratory methods. Oxford University Press, Oxford, pp. 161–186

Békésy G von (1947) A new audiometer. Acta Otolaryngol 35: 411–422

Bergmann C (1858) Anatomisches und Physiologisches über die Netzhaut des Auges. In: Henle J, Pfeufer C von (eds) Zeitschrift für rationelle Medicin, Dritte Reihe, II. Band. Winter, Leipzig & Heidelberg, pp 83–108

Berkley MA, Stebbins WC, eds. (1990) Comparative perception. Wiley, New York.

Blackwell HR (1952) Studies of psychophysical methods for measuring visual thresholds. J Opt Soc Amer 42: 624–643

Blake R (1999) The behavioural analysis of animal vision. In: Carpenter RHS, Robson JG, eds., Vision research. A practical guide to laboratory methods. Oxford Univ. Press, Oxford, pp. 137–160.

Borg G, Diamant H, Ström L, Zotterman Y (1967) The relation between neural and perceptual intensity: A comparative study on the neural and psychophysical response to taste stimuli. J Physiol 192: 13–20

Breitmeyer BG (1975) Simple reaction time as a measure of the temporal response properties of transient and sustained channels. Vision Res 15: 1411–1412

Britten KH, Shadlen MN, Newsome WT, Movshon JA (1992) The analysis of visual motion: A comparison of neuronal and psychophysical performance. J Neurosci 12: 4745–4765

Coren S, Ward LM, Enns JT (1994) Sensation and perception. Harcourt Brace & Co, Forth Worth TX , 4th ed

Cornsweet TN (1962) The staircase-method in psychophysics. Amer J Psychol 75: 485–491.

Delius JD, Emmerton J (1978) Stimulus dependent asymmetry in classical and instrumental discrimination learning by pigeons. Psychol Rec 28: 425–434

Ehrenberger K, Finkenzeller P, Keidel WD, Plattig KH (1966) Elektrophysiologische Korrelation der Stevensschen Potenzfunktion and objektive Schwellenmessung am Vibrationssinn des Menschen. Pflügers Arch 290: 114–123

Ehrenstein A, Schweickert R, Choi S, Proctor RW (1997) Scheduling processes in working memory: Instructions control the order of memory search and mental arithmetic. Q J Exp Psychol 50A: 766–802

Ehrenstein WH (1994) Auditory aftereffects following simulated motion produced by varying interaural intensity or time. Perception 23: 1249–1255

Ehrenstein WH, Manny K, Oepen G (1985) Foveal interocular time thresholds and latency differences in multiple sclerosis. J Neurol 231: 313–318

Ehrenstein WH, Hamada J, Müller M, Cavonius CR (1992) Psychophysics of suprathreshold brightness differences: a comparison of reaction time and rating methods. Perception 21, suppl 2: 82

Farell B, Pelli DG (1999) Psychophysical methods, or how to measure a threshold, and why. Carpenter RHS, Robson JG, eds., Vision research. A practical guide to laboratory methods. Oxford University Press, Oxford, pp. 129–136

Fechner GT (1860/1966) Elemente der Psychophysik. Breitkopf & Härtel, Leipzig (reprinted in 1964 by Bonset, Amsterdam); English translation by HE Adler (1966): Elements of psychophysics. Holt, Rinehart & Winston, New York

Gazzaniga MS, ed (1995) The cognitive neurosciences. MIT Press, Cambridge, MA

Green DM, Swets JA (1966) Signal detection theory and psychophysics. Wiley, New York

Grüsser OJ, Grüsser-Cornehls U (1973) Neuronal mechanisms of visual movement perception and some psychophysical and behavioral correlations. In: Jung R (ed) Handbook of sensory physiology, vol. VII/3A, Springer, Berlin, pp. 333–429

Grunwald E, Bräucker R, Schwartzkopff J (1986) Auditory intensity discrimination in the pigeon (Columba livia) as measured by heart-rate conditioning. Naturwiss 73: 41

Haubensak G (1992) The consistency model: A process model for absolute judgments. J Exp Psychol: Hum Perc Perf 18: 303–309

Hensel H (1976) Correlations of neural activity and thermal sensation in man.In: Zotterman Y (ed) Sensory functions of the skin in primates. Pergamon Press, Oxford, pp 331–353

Jung R (1961 a) Korrelationen von Neuronentätigkeit und Sehen. In: Jung R, Kornhuber HH (eds.) Neurophysiologie und Psychophysik des visuellen Systems. Springer, Berlin, pp. 410–435

Jung R (1961 b) Neuronal integration in the visual cortex and its significance for visual information. In: Rosenblith W (ed.) Sensory communication. M.I.T. Press, New York, pp. 629–674.

Jung R (1972) Neurophysiological and psychophysical correlates in vision research. In: Karczmar AG, Eccles JC (eds.) Brain and Human Behavior. Springer, Berlin, pp 209–258

Jung R (1984) Sensory research in historical perspective: Some philosophical foundations of perception. In: Brookhart JM, Mountcastle VB (eds) Handbook of Physiology, vol. III. American Physiological Society. Washington DC, pp 1–74

Jung R, Kornhuber H , eds (1961) Neurophysiologie und Psychophysik des visuellen Systems. Springer, Berlin

Jung R, Spillmann L (1970) Receptive-field estimation and perceptual integration in human vision. In: Young FA, Lindsley DB (eds), Early experience and visual information processing in perceptual and reading disorders. National Academy Press, Washington, DC, pp. 181–197

Keidel WD, Spreng M (1965) Neurophysiological evidence for Stevens' power function in man. J Acoust Soc Amer 38: 191–195

King-Smith PE, Grigsby SS, Vingrys AJ, Benes SC, Supowit A (1994) Efficient and unbiased modifications of the QUEST threshold method: Theory, simulations, experimental evaluation and practical implementation. Vision Res 34: 885–912

Klump GM, Dooling RJ, Fay RR, Stebbins WC, eds. (1995) Methods in comparative psychoacoustics. Birkhäuser, Basel

Lewald J (1987 a) The acuity of sound localization in the pigeon (Columba livia). Naturwiss 74: 296–297

Lewald J (1987 b) Interaural time and intensity difference thresholds of the pigeon (Columba livia). Naturwiss 74: 449–451

Lewald J, Ehrenstein WH (1998) Auditory-visual spatial integration: A new psychophysical approach using laser pointing to acoustic targets. J Acoust Soc Am 104: 1586–1597.

Lieberman HR, Pentland AP (1982) Microcomputer-based estimation of psychophysical thresholds: The Best PEST. Beh Res Meth Instr 14: 21–25

Livingstone M, Hubel D (1988) Segregation of form, color, movement and depth: Anatomy, physiology, and perception. Science 240: 740–750.

Link SW (1992) The wave theory of difference and similarity. Erlbaum, Hillsdale, NJ

Macmillan NA, Creelman CD (1991) Detection theory. A user's guide. Cambridge Univ. Press, Cambridge

Marks LE (1974) Sensory processes. The new psychophysics. Academic Press, New York

Müller M, Cavonius CR, Mollon JD (1992) Constructing the color space of the deuteranomalous observer. In: Drum B., Moreland JD, Serra J (eds.) Colour vision deficiencies X. Kluwer, Dordrecht, pp. 377–387

Münsterberg H (1894) Studies from the Harvard psychological laboratory: A psychometric investigation of the psycho-physic law. Psychol Rev 1: 45–51

Oehler R (1985) Spatial interactions in the rhesus monkey retina: a behavioural study using the Westheimer paradigm. Exp Brain Res 59: 217–225

Petrusic WM (1993) Response time based psychophysics. Beh. Brain Sci 16: 158–159

Proctor RW, Van Zandt T (1994) Human factors in simple and complex systems. Allyn & Bacon, Boston

Robson T (1999) Topics in computerized visual-stimulus generation. In: Carpenter RHS, Robson JG (eds) Vision research. A practical guide to laboratory methods. Oxford University Press, Oxford, pp 81–105

Sanders AF (1998) Elements of human performance: Reaction processes and attention in human skill. Erlbaum, Mahwah, NJ

Scheerer E (1992) Fechner's inner psychophysics: Its historical fate and present status. In: Geissler HG, Link SW, Townsend JT (eds) Cognition, information processing, and psychophysics. Erlbaum, Hillsdale NJ, pp 3–21

Sekuler R, Blake R (1994) Perception, 3rd ed. McGraw-Hill, New York

Sokolov AN, Ehrenstein WH (1996) Absolute judgments of visual velocity. In: Masin S (ed) Fechner Day 96, CLEUP, Padua, pp. 57–62.

Spillmann L, Ehrenstein WH (1996) From neuron to Gestalt: Mechanisms of visual perception. In: Greger R, Windhorst U (eds) Comprehensive human physiology, vol. 1. Springer, Berlin, pp 861–893

Spillmann L, Ransom-Hogg A, Oehler R (1987) A comparison of perceptive and receptive fields in man and monkey. Hum Neurobiol 6: 51–62

Spillmann L, Werner JS, eds (1990) Visual perception: The neurophysiological foundations. Academic Press, San Diego.

Stevens SS (1975) Psychophysics: Introduction to its perceptual, neural, and social prospects. Wiley, New York

Thurstone LL (1927) A law of comparative judgment. Psychol Rev 34:273–286.

Watson AB, Pelli DG (1983) QUEST: a Bayesian adaptive psychometric method. Perc Psychophys 33: 113–120.

Werner G, Mountcastle VB (1965) Neural activity in mechanoreceptive cutaneous afferents: Stimulus-response relations, Weber functions, and information transmission. J Neurophysiol 28: 359–397

Westheimer G (1965) Spatial interaction in the human retina during scotopic vision. J Physiol 181: 881–894

Wist ER, Ehrenstein WH, Schrauf M (1998) A computer-assisted test for the electrophysiological and psychophysical measurement of dynamic visual function based on motion contrast. J Neurosci Meth 80: 41–47.

Further Useful Sources of Information

There are two special issues on the use of cathode-ray tube displays in visual psychophysics; they contain papers both on techniques and a series of hardware and software notes: *Spatial Vision*, vol. 10/ 4 (1997) and vol. 11/1 (1997) VSB International Science Publishers, Zeist, NL, http://www.vsppub.com.

Current information about psychophysical methods and computer software can also be found in the following journals:

- **Behavior Research Methods, Instruments & Computers** Psychonomic Society, http://www.sig.net/~psysoc/home.htm.
- **Journal of the Acoustical Society of America** ASA, http://asa.aip.org.
- **Journal of Neuroscience Methods** Elsevier, http://elsevier.nl.
- **Journal of the Optical Society of America** OSA, http://www.osa.org.
- **Perception** Pion Ltd, http://www.perceptionweb.com.
- **Perception & Psychophysics** Psychonomic Society, http://www.sig.net/~psysoc/home.htm.
- **Vision Research** Pergamon/ Elsevier, http://elsevier.nl.

Suppliers

Most up-to-date information is now available on the internet. A list of all the interesting sites would soon be obsolete, but some starting points are http://asa.aip.org (for auditory psychophysics) and http://www.visionscience.com (for visual psychophysics).

Some Suppliers of Experimental Equipment

- Brüel & Kjaer Sound & Vibration, http://bk.dk
- Displaytech, http://www.displaytech.com
- Hewlett-Packard, http://www.hp.com

- ISCAN Inc., http://www.iscaninc.com
- Ledtronics, http://www.letronics.com
- Oriel Instruments, http://www.oriel.com
- Permobil Meditech, http://www.tiac.net/users/permobil
- SensoMotoric Instruments (SMI), http://www.smi.de
- Skalar Medical, http://www.wirehub.ul/~skalar/eye.htm
- Texas Instruments, http://www.ti.com

Some Suppliers of Computer Software

Eye Lines 2.5 (1997) by W. K. Beagley, http://www.alma.edu/EL.html
This software is for generating and running experiments in visual perception and sensorimotor coordination (on any Macintosh or IBM PC). Paradigms include the Method of Adjustment with a set of open-ended tools allowing for real-time adjustments of the size, orientation, position or brightness of stimulus elements on multiple stimulus fields. Data can be sorted by subject, trial or condition.
Reference: Beagley WK (1993) Eye Lines: Generating data through image manipulation, issues in interface design, and teaching of experimental thinking. *Behavior Research Methods, Instruments, & Computers* 25: 333–336.

VideoToolbox (1995) by D. Pelli, ftp.stolaf.edu/pub/macpsych
This is a collection of 200 subroutines written in C and several demonstration and utility programs for visual psychophysics with Macintosh computers. Some programs, like the threshold estimation program, are in standard C and can also be used on other computers.
Reference: Pelli DG, Zhand L (1991) Accurate control of contrast on microcomputer displays. *Vision Research* 31: 1337–1350.

Auditory Perception (1994) by Cool Spring Software, http://users.aol.com/Coolspring
This program for the Macintosh (with stereo sound output) provides tools to examine auditory acuity (for amplitude and pitch using pure tones and speech), neglect (auditory agnosia, using environmental sounds), and cerebral dominance (dichotic listening). Users may also enter their own auditory stimuli by means of an inexpensive sound digitizer.
Reference: Psychology Software News 5: 95 (April 1995)

Chapter 44

Analysis of Behavior in Laboratory Rodents

Ian Q. Whishaw, Forrest Haun and Bryan Kolb

Introduction

To see the world in a grain of sand
And a heaven in a wildflower
Hold infinity in the palm of your hand
And eternity in an hour
John Donne (1)

The nervous system is designed to produce behavior, and so behavioral analysis is the ultimate assay of neural function. In this chapter we provide an overview of the behavior of rodents. We also provide references for testing details. Most of the behavioral methodology comes from research on rats, but the ethograms of rodents are similar enough to allow for generalization of the methods, if not many aspects of behavior, to other species. The testing method can be conceive of as having a number of stages, sequentially involving the description of: (I) general appearance, (II) sensorimotor behavior, (III) immobility and its reflexes, (IV) locomotion, (V) skilled movement, (VI) species-specific behaviors, and (VII) learning. For convenience, tables summarizing each class of behaviors are given in the relevant sections that follow.

The thoughts expressed in the opening line of John Donne's poem (above) provide advice for behavioral neuroscientists, in both the surface and deep meaning of the words. The surface meaning is that the observation of details can provide insights into the larger structure of behavior. The behaviorist in the neuroscience laboratory who is attempting to diagnose the effects of a drug, a neurotoxin, or a genetic manipulation can heed the advice that it is often subtle cues that provide the insights into the effects of the treatment (Hutt and Hutt 1970). The deeper meaning is quite simply that one should believe what one sees and not be biased by theory to the extent that observed behavior is ignored, even when the particular behaviors seem at odds with theory.

To make this point in another way, the beginning student and even the seasoned worker may have been taught that the proper way to do science is to state a theory consisting of a number of postulates, logically deduce predictions about behavioral outcomes from the theory, and then compare the predictions with the obtained results of carefully controlled experiments, which leads to a revision or a confirmation of the theory. This way of doing science is potent, but unfortunately, this has not been an especially productive way of conducting behavioral neuroscience. This is because our current understanding of how the brain produces behavior is not sufficiently advanced to per-

Correspondence to: Ian Q. Whishaw, University of Lethbridge, Department of Psychology, Lethbridge, Alberta, T1K 3M4, Canada (phone: +403-329-2402; fax: +403-329-2555; e-mail: whishaw@uleth.ca)
Forrest Haun, NeuroDetective Inc., Quakertown, Pennsylvania, 18951, USA
Bryan Kolb, University of Lethbridge, Department of Psychology and Neuroscience and NeuroDetective Inc., Lethbridge, Alberta, T1K 3M4, Canada

mit the generation of non-trivial and readily testable theories. Put another way, there is no one-to-one congruence between behavioral effects and brain function (Vanderwolf and Cain 1994). Consider the following example.

Let us suppose that Professor Alpha believes that he has discovered a gene for learning. He predicts that if the gene is knocked out in an experimental animal, the animal, although otherwise normal, will no longer be able to learn. He develops a "knock-out" mouse that does not have the gene and then examines the learning ability of the mouse in an apparatus that is widely used for testing learning. Sure enough, the mouse is unable to solve the task and Professor Alpha publishes to much acclaim. Some time later Professor Alpha's knock-out mouse is examined in another laboratory where it is discovered that it has a defect in its retina rendering it functionally blind. Of course, the reader may argue that Professor Alpha is unlikely to be so naive, but in actuality errors of this sort are common (see Huerta et al. 1996 for an example of avoiding such an error). Even when the more obvious sensorimotor functions that could affect learning are examined, there are potentially dozens of other subtle problems that might keep the animal from learning.

A different way of proceeding in behavioral neuroscience is to use an empirical and inductive approach (Whishaw et al. 1983). Empirical means that an animal's behavior is carefully assessed, without regard to theories, in order to describe its condition. Inductive means that from the description, generalizations and conclusions are drawn about the effects of the treatment. Inductive science has been criticized, because, it has been argued, there is no way to tell which conclusions are correct and which are incorrect. We argue, however, that for behavioral research in general and behavioral neuroscience in particular, conclusions arrived at through induction can be subject to a rigorous evaluation using the theoretical method. The inductive technique is widely used as a first analytical step in clinical medicine (Denny-Brown et al. 1982) and in neuropsychology (Kolb and Whishaw 1996). For example, when a patient goes to see a physician with a specific complaint, the careful physician will administer a physical examination in which sensory processes, motor status, circulatory function etc. are examined. Only after such an examination does the physician venture a conclusion about the cause of the patient's symptoms. In neuropsychology, a wide-ranging battery of behavioral and cognitive tests are given to a patient and then the outcome of the tests is compared to the results obtained from patients who have received known brain damage. Similar clinical tests have been developed for rodents (Whishaw et al. 1983). Had Professor Alpha administered a physical and neuropsychology examination to his knock-out mouse, he may have noticed that the mouse was blind and therefore tested it in conditions in which vision would not be essential for performance. The testing protocol given here is designed to provide an "ethogram" that becomes the foundation for subsequent detailed testing. For studies of genetically-manipulated rodents in particular, this comprehensive behavioral evaluation is intended to cast as wide a net as possible, to capture multiple brain functions that may have been altered by even a single gene manipulation.

Methods

The three main ways of evaluating behavior are:
- End-point measures,
- Kinematics, and
- Movement description.

End-point measures are measures of the consequences of actions, e.g., a bar was pressed, an arm of the maze was entered, or a photobeam was broken (Ossenkopp et al. 1996). Kinematics provide Cartesian representations of action, including measures of distance, velocity and trajectories (Fentress and Bolivar 1996; Fish 1996; Whishaw and Miklyaeva 1996). Movements can be described using formal languages that have been adapted to the study of behavior, such as Eshkol Wachman Movement Notation (Eshkol and Wachmann 1958). This system has been used for describing behaviors as different as social behavior (Golani 1976), solitary play (Pellis 1983), skilled forelimb use in reaching (Whishaw and Pellis 1990), walking (Ganor and Golani 1980) and recovery from brain injury (Golani et al. 1979; Whishaw et al. 1993). For a comprehensive description of behavior, all three methods are recommended (see example below). Endpoint measures provide an excellent way of quantifying behavior, but animals are extremely versatile and can display compensatory behavior after almost any treatment. There are many ways that they can press a bar, enter an alley or intersect a photobeam. Kinematics provide excellent quantification of movement, but unless every body segment is described, ambiguity can exist about which body part produced a movement. Movement notation provides an excellent description of behavior but quantification is difficult.

Video Recording

As a prelude to behavioral analysis, we recommend that behavior be video-recorded. Regardless of the type of experiment, equipment for video recording is relatively inexpensive, as off-the-shelf camcorders and VCRs are suitable for most behavioral studies (Whishaw and Miklyaeva 1996). Equipment required is a hand-held video recorder that has an adjustable shutter speed. Most of the filming of human movements uses a shutter speed of about 1/100 of a second, but in order to capture blur-free pictures of the rapid movements made by rodents, who have a respiratory cycle, lick rate and whisker brushing rate of about seven times per second, shutter speeds of 1/1,000 and higher are required. Fast shutter speeds require fairly bright lights, but with a little habituation, rodents do not appear distressed by lights.

Video Analysis

To analyze the film, a videocassette recorder that has a frame-by-frame video advance option is necessary. A VHS model cassette recorder can be used and the record acquired on the recorder cassette can be copied over to the VHS tape. Optional, but not essential, equipment includes a computer with a frame grabber that allows individual frames of behavior to be captured for computer manipulation. Once a video tape of behavior is made, the behavior can be subject to frame-by-frame analysis. Each video frame provides a 1/30 s snapshot of behavior. If rat licking is being studied, a single lick cycle would be represented on three successive video frames, a resolution that is adequate for most purposes. One important feature of video frames, however, is that they are actually made up of two fields that are superimposed. Because computer-based frame grabbers can capture each field, a computer image of the video record increases resolution to 1/60 of a second. The analysis of some behaviors may require still greater resolution and higher speed cameras are available, but they are presently very expensive. For most studies, it is helpful to use a mirror to film the animal from below (Pinel et al. 1992) or so that the surface on the animal's body that faces away from the camera can also be seen (Golani et al. 1979). A typical recording set-up is illustrated in Fig. 1.

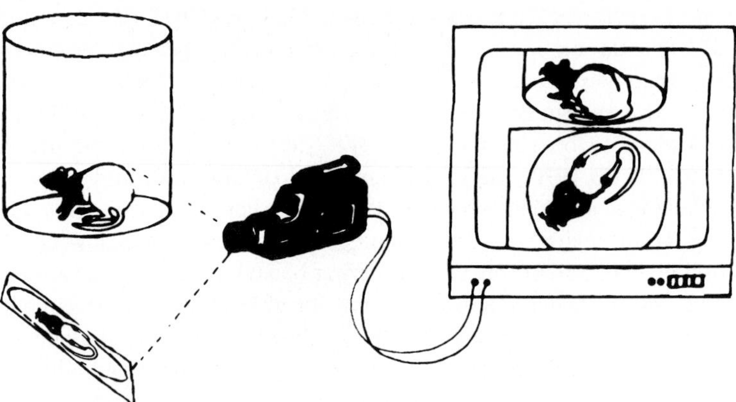

Fig. 1. Video-recording method. The video camcorder is placed so that it simultaneously records the rat from a lateral view and, through an inclined mirror, from a ventral view. Thus the rat can be seen from two perspectives on the monitor (After Pinel et al. 1992).

Example 1: Video-based Behavioral Analysis

The following example illustrates the complementary role of the different types of video-based behavioral analysis. The reaching for food by rats with unilateral motor cortex injury was studied using an end-point, videorecording, and movement notation analysis (Whishaw et al. 1991). The study began with an end-point measure. The animal was allowed to obtain a piece of food on a tray by reaching through a slot in its cage. To force the rat to use its non-preferred limb, a light bracelet was placed on the normal limb, thus preventing it from going between the bars. The end-point measure of behavior was the success in reaching for food with the limb contralateral to the lesion. The end-point measure revealed that motor cortex lesions impaired the grasping of food. Normal control animals have a success rate of about 70% and rats with motor cortex lesions had success rates that varied from near 20% to about 50%, the extent of impairment varying directly with the lesion size. Thus, this type of analysis shows that the forelimb representation of the motor cortex of the rat plays a role in skilled reaching movements. Note, however, that although this analysis identifies that the motor cortex lesion interferes with reaching for food, it does not identify the reason for the poor reaching. This question was addressed by the movement notation and kinematic analyses (Fig. 2). Analysis with the Eshkol-Wachman movement notation system, which is designed to express relations and changes of relations between parts of the body, revealed that the motor impairments were attributable to an inability to pronate the paw over the food in order to grasp it, as well as an inability to supinate the paw at the wrist to bring the food to the mouth. Finally, once a movement description was established with movement notation, a number of other aspects of the movement were measured and documented using a Cartesian coordinate system, with initial and terminating components of the movements serving as reference points. For this analysis, points on the body were digitized and the trajectory of the movements of the limb reconstructed. This analysis revealed that in addition to the inability to supinate and pronate properly, the animals with motor cortex injury also had an abnormal movement trajectory of the limb so that the aiming of the limb to the food was impaired.

Taken together, these three analyses provide a description of the various motor components that are affected by the motor cortex injury, as well as the subsequent effect that the impairments have on the behavior. Such an analysis is necessary not only to under-

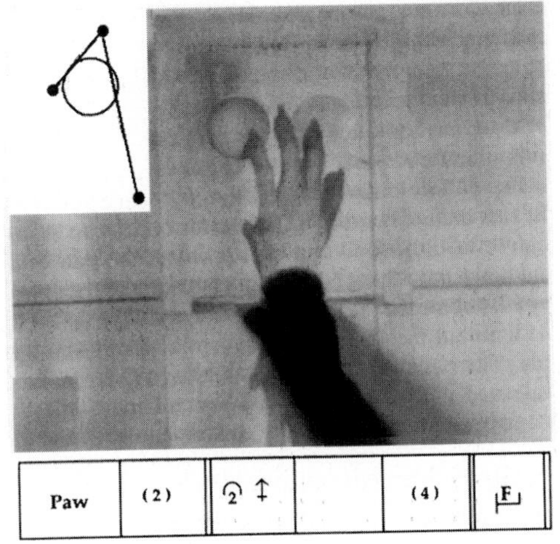

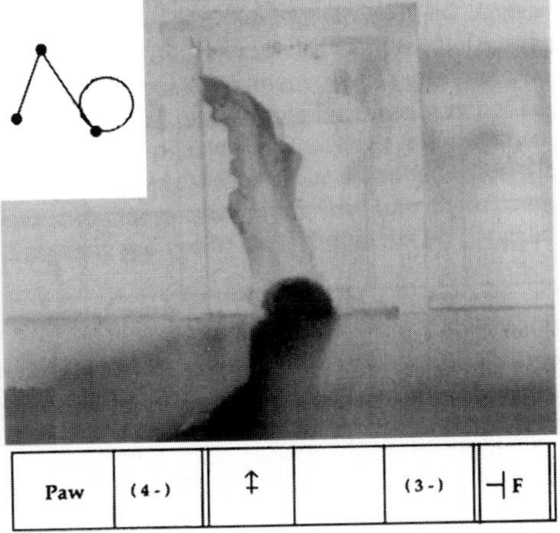

Fig. 2. Three methods of describing behavior. (1) End-point measure. The photographs are of a rat reaching through a slot in order to obtain a food pellet located on the shelf. In the top figure a control rat is about to grasp the food while in the bottom photo a rat with a motor cortex lesion has knocked the food pellet off the shelf and lost it. (2) Movement notation. On the bottom of each figure a movement notation score describes the movement. The first box indicates that it is the paw that is being described, the second box indicates the starting position of the paw and the last box indicates the end position of the paw. The notation in the three middle boxes indicates movement in three video frames (1/30 sec). Top: The paw advances and pronates and grasps the food pellet (F). Bottom: The paw advances and turns sideways without pronation to swipe the food away. (3). Cartesian reconstruction. The insert shows the trajectory of the tip of the third digit of the paw relative to the food pellet on the same three video frames. The three forms of analysis are necessary for a complete description of the movement and its result (After Whishaw et al. 1991).

stand the role of a neural structure in the control of movement but also to investigate the effects of different treatments on the observed impairments. Thus, it might be shown that a particular drug influenced "recovery of function", which could be documented by an improvement in an end-point measure. The subsequent movement notation and kinematic analyses would be required, however, to determine which aspects of the movement had been affected. For example, it has been shown that rats with motor cortex lesions show some recovery of skilled limb reaching but this is largely due to changes in a variety of whole body movements that compensate for the deficits in aiming (Whishaw et al. 1991). The impairments in pronation and supination remain.

The Neurobehavioral Examination

Many tests can be administered quite quickly, while an animal is in its home cage, whereas others are given when the animal is removed from its cage. Most tests require no special equipment and are intended to be simple, rapid, and inexpensive ways of evaluating an animal's condition. In administering the clinical examination, the stand-

Table 1. Examination of Appearance and Responsiveness

Appearance	Inspect body shape, eyes, vibrissae, limbs, fur and tail, and coloring.
Cage examination	Examine the animal's cage, including bedding material, nest, food storage, and droppings.
Handling	Remove the animal from its cage and evaluate its response to handling, including movements and body tone, and vocalization. Lift the lips to examine teeth, especially the incisors, and inspect digits and toenails. Inspect the genitals and rectum.
Body Measurements	Weigh the animal and measure its body proportions, e.g., head, trunk, limbs, and tail. Measure core temperature with a rectal or aural thermometer.

ard is that a healthy laboratory animal is clean, lively, inquisitive, but not aggressive. The tests that we describe may require innovative and liberal use paraphernalia found around the laboratory. For all of the tests described, it is assumed that control animals are also examined to provide the standard against which an experimental group is compared. Often in studies of genetically altered animals, several classes of control animals are needed in order to be able to attribute behaviors to particular genetic alterations (Crawley et al. 1997; Upchurch and Wehner 1989).

I. Appearance

The main features of the physical examination are given in Table 1. Animals should be examined in the home cage and also removed for individual inspection.

1. The appearance of the fur and its color should be noted.
2. The proportions of body parts should be examined, including the length of the snout, head and body, limb, and tail.
3. An examination of the eyes includes using a small flashlight to test pupillary responses. The pupils should constrict to light and dilate when the light is removed, which indicates intact midbrain function. Rodents secrete a reddish fluid, called Hardarian fluid, from the eye glands. This fluid is collected on their paws as the paws are rubbed across the eyes during grooming. It is then mixed with saliva when the animal licks its paws, and the mixture is spread onto the fur during subsequent grooming. This material maintains the condition of the fur and aids thermoregulation. Reddening of the fur around the eyes due to the accumulation of Hardarian material, usually more obvious in albino rats, indicates that the animal is not grooming. Rodents are fastidious groomers and a rich rough appearance to the fur also indicates normal grooming.
4. An animal's teeth should be inspected by gently retracting the lips. The incisor teeth of rodents grow continuously and they are kept at an appropriate length by chewing. Excessively long or crooked teeth indicate an absence of chewing or a jaw malformation. Long teeth can be safely cut short with a pair of cutters. If the teeth are inadequate for chewing, an animal can be fed a liquid diet.
5. During grooming, rats cut their toenails, especially the toenails of the hind feet, and so these should be short with rough tips, indicative of daily care. Long unkept toenails may indicate poor grooming habits, or problems with teeth and mouth, which are used for making the fine nail cutting movements. (We must note, however, that many inbred strains of animals display less than propitious toenail care).
6. During the examination, the genitals and rectum should be examined to ensure that they are clean, indicating an absence of internal infection or illness.

Example 2: Analysis of Appearance

The power of a simple analysis of appearance is illustrated in Fig. 3. In the course of studying the embryological development of neurons in the rat cortex we noticed a dramatic effect of our treatments on coat coloration pattern. Pregnant rats were injected with a standard dose of bromodeoxyuridine (BrdU, 60 mg/kg) on different embryonic days ranging from E11 to E21. In the rat, neocortical neurons are generated during this period and we were interested in the effects of different postnatal treatments on the numbers of neurons generated at different ages. The BrdU labels cells that are mitotic during the two or so hours following the injection. Labeled cells can later can be identified using immunohistological techniques. It turned out, however, that Long-Evans hooded rats with BrdU injections on days E11 to about E15 had a dramatically altered pattern of black and white coloration in their fur coats and this pattern varied with the precise age of BrdU treatment. The injections produced spots, much like those on a Dalmatian dog, the size of the spots varying with injection age (see Fig. 3). Because we knew

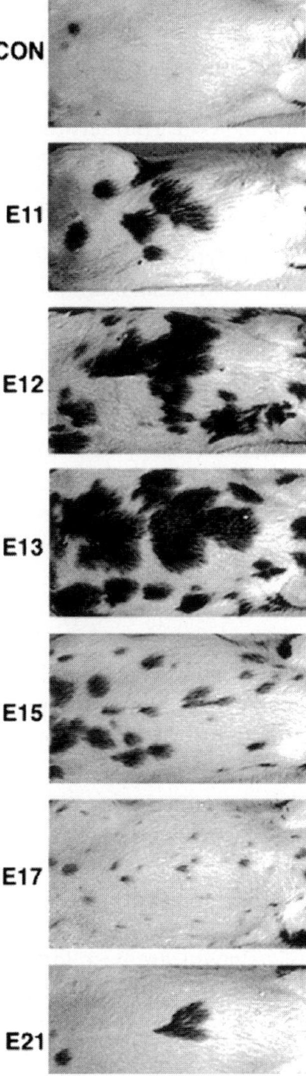

Fig. 3. Coat color on the ventrum of Long-Evans rats whose mother had received bromodeoxyuridine at different times of gestation. The changes in coat color are paralleled by changes in brain and behavior (After Kolb et al. 1997).

that the skin cells (specifically the melanocytes in this case) and the brain cells were derived from the same embryonic source, we immediately became suspicious that the BrdU was not only altering skin cells but also the pattern of migration of neurons (Kolb et al. 1997). This led us to analyze the behavior of the animals in more detail and it turned out that the BrdU treatment was producing marked changes in many behaviors. This discovery was only possible because we had observed the general appearance of the animals.

Body Weight

Animal suppliers provide weight curves for the strains of animals that they sell, so it is a simple matter to weigh an animal and compare its weight to a standard weight curve. Growth in rodents is extremely sensitive to nutrition, being accelerated or retarded by changes in nutritional status at any age. It is also influenced by the behavior of conspecifics, as subordinate animals are typically smaller than dominant conspecifics. Rats, especially males, continue to increase in size and weight throughout life, but size and rate of growth vary appreciably across strains. Differences in actual weight from expected weight may signal malnutrition, overfeeding, developmental disorders, or any of a variety of peripheral and central problems.

Body Temperature

At the time of weighing the animal's temperature can be recorded with a rectal or aural thermometer. Rodent core temperature is quite variable and can fall to about 35 °C when an animal is resting in its home cage and can increase to 41 °C when the animal is aroused upon removal from its home cage. Temperatures lower or higher than these ranges indicate hypothermia or fever. Animals display a variety of postures, reflexes, and complex behaviors in order to maintain temperature and these are described in detail by Satinoff (1983).

Response to Handling

During handling, an animal may typically make soft vocalizations. Excessive squeaking may indicate distress or sickness. Rodents maintained in group housing are usually unaggressive when handled by an experimenter. Animals raised in isolation, e.g., housed individually, may be very sensitive to handling and squeak and struggle or even display rage responses. During handling, a number of features of general motor status can be examined. The animal can be gently held in the palm of the hand and quickly raised and lowered. Limb muscles should tense and relax as the rat adjusts itself to the movements of the hand. Absence of muscle tone or excessive rigidity are both indicative of problems with motor status, e.g., drugs that stimulate dopamine function produce flaccid muscle tone whereas drugs that block dopamine function produce rigidity.

II. Sensorimotor Behavior

The objective of sensorimotor tests is to evaluate the sensory and motor abilities of animals (see Table 2). The tests evaluate the ability of animals to orient to objects in the environment in each sensory modality. The term *sensorimotor* derives from the recognition that it is ordinarily very difficult to determine whether the absence of a response is related to an inability to detect a stimulus or to an inability to respond to a detected stimulus. For the purposes of the present overview, such distinctions are not necessary, but it is worthwhile pointing out that some theoretical positions suggest that such distinctions are not possible (Teitelbaum et al. 1983).

Home Cage

Tests of sensory and motor behavior are best administered to an animal in its home cage, preferably a hanging wire mesh cage because the holes in the mesh allow easy access to the animal. (Sensorimotor behavior of animals is radically different if they are assessed in an open area where even neurologically intact animals will act as though

Table 2. Sensory and Sensorimotor Behavior

Home Cage	Response to auditory, olfactory, somatosensory, taste, vestibular, and visual stimuli. The home cage should provide easy viewing of an animal. Holes in the sides and bottom of the cage provide entries for probes to touch the animal or to present objects to the animal or to present food items. Animals are extremely responsive to inserted objects and treat capturing the objects as a "game". Slightly opening an animal's cage can attenuate its responses to introduced stimuli, showing that it notices the change.
Open Field	Response to auditory, olfactory, somatosensory, taste, vestibular, and visual stimuli. The same tests are administered. Generally animals taken out of their home cage are more interested in exploring and so ignore objects that they responded to when in the home cage.

they are neglecting sensory information, see below.) For the cage examination, put food pellets into the cage and place some paper towels beneath the home cage to catch residue. If the tray is slid out from beneath the animal's cage a day to two later, residue can be examined. Rodents are fastidious in their eating and toilet habits and so feces and urine will be found in one location on the tray and food droppings will be found in another location, an indication that the animal has compartmentalized its home spatially. Residue from food should be quite fine, indicating that the animal is chewing its food. Many rodents are central place foragers; they carry food to their home territory and store it for subsequent use. An examination of the inside of the cage should indicate that the food is piled in one corner of the cage.

Orienting

While the rat is in its home cage, its sensory responsiveness can be examined. In an analysis of animals recovering from lateral hypothalamic damage, Marshall et al. (1971) observed a rostrocaudal recovery of sensory responsiveness. Generally, normal animals are much more responsive to rostrally than to caudally applied stimulation. Schallert and Whishaw (1978) reported that a syndrome of hyperresponsiveness and hyporesponsiveness can occur after hypothalamic damage.

1. Take a cotton-tipped applicator, of the type used in surgery, and insert into the animal's cage, gently touching different parts of the animal's body, including its vibrissae, body, paws, and tail. The animal should perceive this as a "game" and vigorously pursue and bite the applicator, thus allowing assessment of the sensitivity of different body parts. Rubbing the applicator gently against the cage can be used to test the rat's auditory acuity, as it will orient to the sound. Placing objects on the rat provides an additional assessment of its sensory responsiveness, as a rat will quickly remove the object (Fig. 4). Pieces of sticky tape of various sizes placed on the ulnar surface of the forearm or bracelets tied with single or double knots provide tests of detection, obscuration, or neglect (Schallert and Whishaw 1984).
2. An animal's responsiveness to odors can be tested by placing a small drop of an odorous substance on the tip of the applicator. Animals will investigate food smells, recoil from noxious odors such as ammonia, and recoil from the odors of predators such as stoats and foxes (Heale et al. 1996).
3. Simple ingestive responses can be investigated by placing a drop of food substances on the blade of a spatula. In their home cage, rodents, unless water deprived, are usually indifferent to water, but they enthusiastically ingest sweet foods such as sugar water, milk, or a mash of sweet chocolate flavored cookie. Lip licking indicates that the animal's sensitivity to sweet food is normal. If the spatula is held adjacent to the cage, the animal will stick its tongue out to lick up the food, providing an indication of the motor status of the tongue. Bitter tasting food, such as quinine, elicits a series of rejection responses, including wiping its snout with its paws, wiping its chin on the floor of the cage, and tongue protrusion to remove the food. Grill and Norgren (1978)

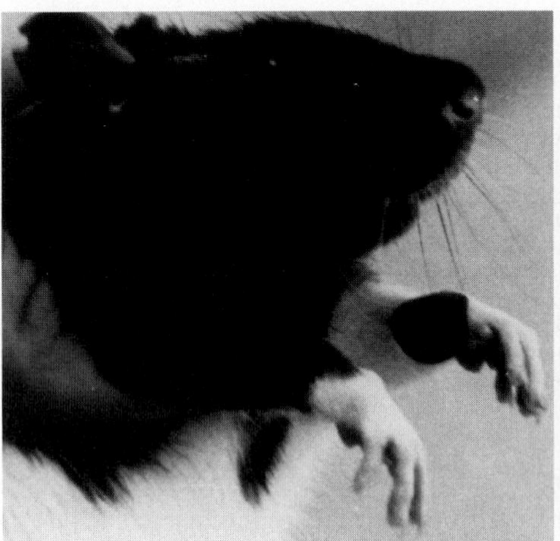

Fig. 4. Left: A sticky dot placed on the ulnar pad of the forearm provides a simple test for orienting. Remove the tape, and sensitivity can be tested by reducing the size of the dot. Right: Competition between stimuli can be tested with bracelets tied with one or two knots. If a stimulus on the good side obscures the stimulus on the bad side, a rat will persist in attempting to untie the difficult right knot while ignoring the easy left knot (After Schallert and Whishaw, 1984).

have described taste-responsive reflexes in rats that have subsequently been widely used to assess gustatory responses.

An animal's ability to eat and chew may be further examined by giving the animal a food pellet, a piece of cheese, or some other food substance of a standard size. Rodents sniff the food, grasp it in their incisors, sit back on their haunches while transferring the food to the paws, and eat the food from the paws while in a sitting position. Observing the processes of food identification, food handling, and eating speed all provide insights into the function of the front end of the animal's body. More detailed analysis of rat eating speed shows that the animals eat more quickly when exposed than when in a secure environment and eat more quickly at normal meal times than at other times (Whishaw et al. 1992a).

Open Field Behavior

Sensory tests may also be given to an animal that is removed from its home cage, but here the meaning of the responses changes. Normally an animal in a novel environment ignores food in favor of making exploratory movements. Ingestion of food outside the home means it has habituated to that location or is insensitive to novelty. Generally, it may take a number of days or weeks to habituate an animal to an environment that provides no hiding place, as is the case with most mazes. Animals also display a number of defensive responses when they find food, and will turn or dodge away from other animals with the food, or run or "hoard" the food to a secure location for eating. These behaviors can be used as "natural" tests of orienting and defensive behavior (Whishaw 1988).

III. Immobility and Its Reflexes

Posture and locomotion are supported by independent neural subsystems (see Table 3). A condition of immobility in which posture is supported against gravity is the objective of a large number of local and whole body reflexes. Thus, immobility should be viewed as a behavior with complex allied reflexes. Even animals that are catatonic and appear

Table 3. Posture and Immobility

Immobility and Movement with Posture	Animals usually have postural support when they move about and they maintain posture when they stand still and remain still while rearing. Posture and movement can be dissociated: in states of catalepsy postural support is retained while movement is lost.
Immobility and Movement without Posture	An animal has posture only with limb movement. When a limb is still, the animal collapses, unable to maintain posture when still. When still, the animal remains alert but has no posture, a condition termed cataplexy.
Movement and Immobility of Body Parts	Mobility and immobility of body parts can be examined by placing a limb in an awkward posture or placing it on an object such as a bottle stopper and timing how long it takes an animal to move it.
Restraint-Induced Immobility	Restraint-induced immobility, also called tonic immobility or hypnosis, is induced by placing an animal in an awkward position, e.g., on its back. The time it remains in such a position is typically measured. Animals will maintain awkward positions while maintaining body tone or when body tone is absent. During tonic immobility animals are usually awake.
Righting Responses	Supporting, righting, placing, hopping reactions are used to maintain a quadrupetal posture. When placed on side or back or dropped in a supine or prone position, adjustments are made to regain a quadrupetal position. Righting responses are mediated by tactile, proprioceptive, vestibular, and visual reflexes.
Environmental Influences on Immobility	Feeding fatigue potentiates immobility. Warming induces heat loss postures, e.g., sprawling, and thus potentiates immobility without tone. Cooling induces heat gain posture with shivering and thus potentiates immobility with muscle tone.

completely unresponsive may move quickly to regain postural support if they are placed in a condition of unstable equilibrium (Fig. 5). Postural and righting reflexes are mediated by the visual system, the vestibular system, surface body senses, and proprioceptive senses. Although responses mediated by each system are allied they are frequently independent (Pellis 1996).

Postural Reflexes

If the animal is placed on a flat surface and gently lifted by the tail, it should display a number of postural reflexes. The head should be raised and the forelimbs and hindlimbs extended outward, while the forequarters are twisted from side to side. When mice are raised and then quickly lowered, their digits should extend, a response not seen in rats. The posture and movements are typical of an animal searching for a surface on which to obtain support.

Asymmetries in movement are used as tests of brain asymmetries that might be produced by unilateral injury (Kolb and Whishaw 1985). For example, when suspended by the tail, adult animals with unilateral cortical lesions usually turn contralateral to the lesion whereas animals with unilateral dopamine depletions turn ipsilateral to the lesion. The posture of the limbs provides a sensitive measure of central motor status. Flexion of the forelimbs toward the body, including grasping of the fur of the ventrum, and flexure of the hindlimbs toward each other, including grasping of each other, can indicate abnormalities in descending pyramidal or extrapyramidal systems (Whishaw et al. 1981b). On a flat surface, animals typically rotate toward the side of injury, and when placed on irregular surfaces, they may favor a limb contralateral to an injury.

Postural Support

Placing an animal on a flat surface allows inspection of its postural support. An animal may maintain a condition of immobility in which it has posture or it may adopt a position of immobility in which it lacks posture. The two types of immobility are independent. An animal without posture while immobile may achieve posture as a part of locomotion. DeRyck et al. (1980) have demonstrated that morphine, an opioid agonist, pro-

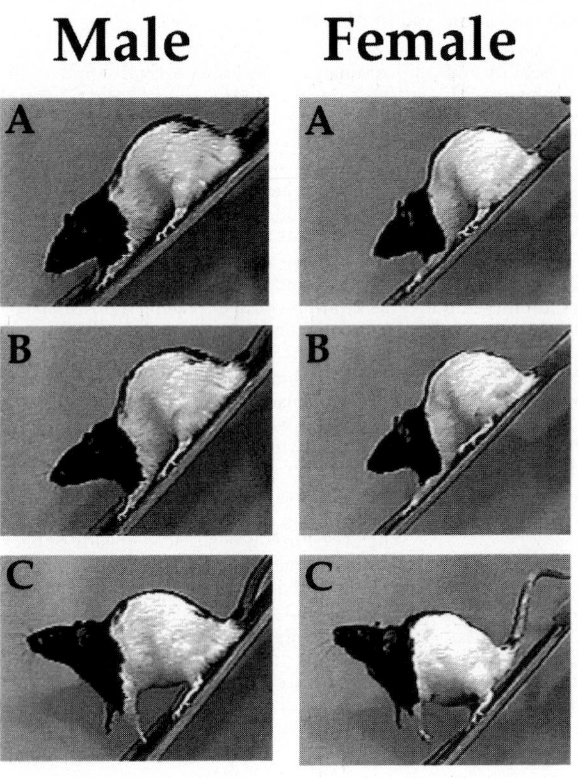

Fig. 5. Animals defend immobility when placed in a condition of instability. Rats treated with haloperidol (5 mg/kg) display immobility with postural support. When made unstable by tilting the substrate, they first brace (A) but eventually jump to regain a new supporting position when postural instability becomes too great (B, C). The tactics for maintaining postural stability prior to jumping are sexually dimorphic (After Field, Pellis and Whishaw, unpublished).

duces a condition of immobility without postural support whereas haloperidol, a dopamine antagonist, produces immobility with postural support.

The two types of immobility are part of normal behavior, as an animal that is cold and shivering has postural support while otherwise remaining immobile while an animal that is hot will sprawl without postural support as part of a heat loss strategy. Immobility with postural support is characteristic of an animal that pauses during a bout of exploratory behavior or an animal that rears and stands against a wall. Immobility without postural support is typical of an animal that is resting or sleeping. Although difficult to achieve in normal rodents, if an animal is gently restrained in almost any position, it may remain in that position when released. This form of immobility, sometimes called animal hypnosis or tonic immobility, is usually accompanied by a good deal of body tone as the latter term implies (Gallup and Maser 1977). If an animal is frightened, it may "freeze" in place in a condition of immobility with postural support. When hiding or attempting to escape it frequently crouches into immobility without postural support. (The suggestion that animal hypnosis can be used as a condition of anesthesia is not supported by research, which indicates on the contrary that hypnosis is but one of many forms of adaptive immobility.)

Placing Responses

Placing responses are movements of the head, body or limbs that are directed toward regaining a quadrupedal posture. If an animal is lifted by the tail, then as the animal is lowered toward a surface, contact of its long whiskers with the surface will trigger a placing reaction of extending the forelimbs to contact the surface. Placing reactions of each of the limbs can be tested by holding the rat in both hands while touching the dorsal surface of each paw against the edge of a table. Upon contact, the paw should be lifted and placed on the substrate (Wolgin and Bonner 1985). Placing responses are sensitive to damage to corticospinal systems.

Bracing Responses

If an immobile animal is gently pushed, it will often push back against displacement to maintain static equilibrium, a behavior termed a bracing response. If the push begins to make the animal unstable, it will step or turn away to relieve pressure. Animals that have been rendered cataleptic by a treatment, that is they are immobile with postural support, may be unable to step and thus are reduced to bracing to maintain stability (Schallert et al. 1979). Bracing can be examined in a single limb. Animals that are made hemi-Parkinsonian with a unilateral injection of 6-hydroxydopamine, a dopamine-depleting neurotoxin, can be held so that they are standing on a single forelimb. When gently pushed forward, they will step to gain postural support with their good forelimb while being reduced to bracing with their bad forelimb (Olsson et al. 1995; Schallert et al. 1992).

Righting Response

When placed in a condition of unstable equilibrium, animals will attempt to regain an upright posture in relation to gravity, a behavior termed righting. Tests of righting examine visual, vestibular, tactile, and proprioceptive reflexes. If an animal is dropped from the height of less than a meter, onto a cushion, it will adjust its posture so that it lands "on its feet". If released with feet facing down, it arches its back, and extends its limbs to parachute to the surface. If it is released in a prone position, it will right itself by first turning the forequarters of the body and then the hindquarters, a response mediated by vestibular receptors. Video recordings of righting responses show that righting is also visually modulated. Sighted animals right proximal to the landing surface, whereas in the absence of visual cues, animals initiate righting immediately upon release. If a rat is placed on its side or back, on the surface of a table, it will right itself so that it returns to its feet. The righting responses of parts of the body can be tested by holding the head, forequarters, or hindquarters. Details of righting reflexes and their sensory control are provided by Pellis (1996).

IV. Locomotion

Locomotor behavior includes all of the acts in which an animal moves from one place to another (see Table 4). It includes the acts of initiating movement, which is often referred to as warm-up, turning behavior, exploratory behavior, and a variety of movement patterns on dry land, water, or vertical substrates.

Table 4. Locomotion

General Activity	Video-recording, movement sensors, activity wheels, open field tests.
Movement Initiation	The warm-up effect: Movements are initiated in a rostral-caudal sequence, small movements precede large movements, and lateral movements precede forward movements, which precede vertical movements.
Turning and Climbing	Components of movements can be captured by placing animals in cages, alleys, tunnels, etc.
Walking and Swimming	Rodents have distinctive walking and swimming patterns. Rats and mice walk by moving limbs in diagonal couplets with a forelimb leading a contralateral hindlimb. They swim using the hindlimbs with the forelimbs held beneath the chin to assist in steering.
Exploratory Activity	Rodents select a home base as their center of exploration, where they turn and groom, and make excursions of increasing distance from the home base. Outward trips are slow and involve numerous pauses and rears while return trips are more rapid.
Circadian Activity	Most rodents are nocturnal and are more active in the night portion of their cycle. Peak activity typically occurs at the beginning and end of the night portion of the cycle. Embedded within the circadian cycle are more rapid cycles of eating and drinking, especially during the night portion of the circadian cycle.

Warm-up Movement is conceived of as being organized along three dimensions as is illustrated by warm-up. The initiation of locomotion, or warm-up, may be observed in an animal gently placed in the center of an open field (Golani et al. 1979). There are four principles of warm-up.

1. Lateral, forward, and vertical movements are independent dimensions of movement. During warm-up, an animal can be observed to alternate between lateral, forward, and vertical movements.
2. Small movements recruit larger movements. For example, a small lateral head turn will be shortly followed by a larger head turn and so forth until the animal turns in a complete circle.
3. Rostral movements precede caudal movements. That is, a head turn will precede movements of the front limbs, which will precede movements of the hind limbs.
4. The relationship between the movements is such that lateral movements precede forward movements which precede vertical movements. In a novel environment, warm-up may be lengthy while in a familiar environment it may precede quickly, e.g., an animal may simply turn and walk away. Almost any nervous system treatment or pharmacological treatment may affect warm-up. Conceptually, warm-up is thought to be a reflection of evolutionary and ontogenetic processes, and brain structural organization (Golani 1992). Functionally, warm-up allows an animal to systematically examine an environment into which it is moving.

Turning Rodents have a variety of strategies for turning. Turning may be incorporated into patterns of forward locomotion, in which case an animal turns its head and then "follows" using a normal walking pattern. Incorporated into this pattern, or used independently, it may make most of the turn with the hindquarters, thus pivoting in part or in whole around its hindquarters, or most of the turn with its forequarters, thus pivoting around its forequarters. These two patterns of turning are incorporated into a variety of other behaviors including locomotion, aggression, play, and sexual behavior. In rats, there is sexual dimorphism in the extent to which the patterns are used (Field et al. 1997). Females make greater use of forequarter turning whereas males make greater use of hindquarter turning. Dimorphism, in turn, may be related to the way the animals turn during sex and aggression, respectively. Animals may also turn by first rearing and then using the potential energy of the rear to pivot and fall in one direction or the other.

The incidence of turning as well as the form of turning is widely used as an index of asymmetrical brain function (Miklyaeva et al. 1995). For example, animals with unilateral dopamine depletions turn ipsilateral to their lesion when given amphetamine and contralateral to their lesion when given apomorphine. Some papers have suggested that direction of rotation can be used as an index of recovery after therapeutic treatments (Freed 1983). Miklyaeva and colleagues' analysis shows, however, that the depleted rats tend not to exert force with the limbs contralateral to their lesion and the impairment persists irrespective of turning direction or drug treatment (Fig. 6). Thus it is more appropriate to use analyses of limb use rather than turning direction in assessing functional recovery (Olsson et al. 1995; Schallert et al. 1992). The causes of turning direction induced by drugs remain enigmatic.

Walking and Running Although this may be a little difficult for the novice to observe, a rodent's major source of propulsion comes from its hindlimbs. During slow locomotion, the forelimbs are used for contacting and exploring the substrate and walls (Clarke 1992). Limb contact with irregular surfaces or the wall of a cage can be used as a test of normal forelimb function. At its simplest, the number of times a limb contacts a wall when an animal rears can be a sensitive measure of forelimb function (Kozlowski et al. 1996).

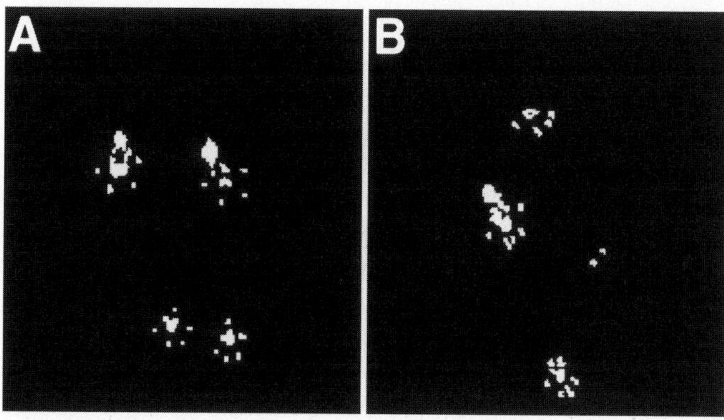

Fig. 6. Limb use can be examined using the reflectance technique, in which a light shone through the edge of a glass table top reflects from the paws of an animal standing on the table surface. The technique illustrates that a control animal distributes its weight evenly when standing while a rat with unilateral dopamine depletion rests its weight on its good (ipsilateral to lesion) limbs (After Miklyaeva et al, 1995).

When a rodent does walk, it moves with diagonal limb couplets. One forelimb and the contralateral hindlimb move together followed by the other forelimb and hindlimb. Rodents also have species typical movement patterns. For example, rats seldom walk. They move either hesitantly with turns and pauses or they trot. Patterns of locomotion are difficult to analyze by eye unless they are grossly abnormal, but a number of simple video-recording techniques have been used for detailed locomotor analysis (Clarke 1992; Miklyaeva et al. 1995). The structure of walking movements has been described by Ganor and Golani (1980).

Exploration

The movements of an animal can be observed by removing it from its cage and placing it in a small open environment or field. Golani and coworkers have described some of the geometric aspects of rodent exploratory behavior (Eliam and Golani 1989; Golani et al. 1993; Tchernichovski and Golani 1995). An animal will usually treat the place at which it is first placed as a "home base". It will pause, rear, turn, and groom at this location, often before exploring the rest of the open field. When it does rear, it will support itself by touching the wall of the field with its forepaws. As it begins to explore, it extends its forequarters and head and examines the area surrounding its home base. It will eventually begin to make trips away from its home base, usually along the edge of the walls of the enclosure. Exploration will proceed with brief and slow outward excursions followed by more rapid returns to the home base. The outward excursions become gradually longer until the surrounding area is explored. During the course of its exploration, the animal may choose another location as its home base and this can be identified because it circles and grooms at this location and uses the location as the home base for its outbound trips. Typically, return trips are more rapid than outbound trips. A ten-minute exploratory test can provide a wealth of behavior to analyze, including number of home bases, number of trips, kinematics of excursions and returns, number of stops, number of rears, incidence of grooming, duration of trips, etc. (Golani et al. 1993; Whishaw et al. 1994).

Another feature of open field behavior is habituation. Over time, animals normally will show a reduction in open field activity. Furthermore, they will show a shift in behavior and may spend more time grooming or sitting immobile once they are familiar with the environment. Animals with various forebrain injuries, such as frontal cortex or hippocampal lesions, may display slower habituation, even with extended exposure to the open field (e.g. Kolb 1974).

Swimming If a swimming pool is available, movements of swimming can be observed. Rats are semiaquatic, as their natural environments are usually along the margins of streams and rivers. Rats are proficient swimmers and propulsion comes entirely from the hindlimbs. Other rodents may be less proficient in water than rats, and some may use quite idiosyncratic swimming strategies. Hamsters, for example, inflate their cheek pouches and use them as "water-wings". In typical rodent swimming, the forelimbs are tucked up under the chin and the open palms of the paws are used for steering (Fish 1996; Salis 1972). Changes in the way that animals swim may occur with development and aging, under the effects of drugs and brain damage, but swimming itself is quite resistant to central nervous system damage (Whishaw et al. 1981b).

Circadian Activity Tests of circadian activity involve recording the general activity of animals across a day-night cycle. Usually the activity cycle is entrained by having lights come on at 0800 hrs and go off at 2000 hrs. Rodents are typically active during the dark portion of the cycle. The test requires a dedicated room in which lighting can be regulated with a timer. A test apparatus consists of a cage that has a photocell at each end. The photocells are connected to a microcomputer which records instances of beam breaks. The computer is programmed so that it records beam breaks at each photocell as well as instances in which beam breaks occur at successive photocells. Beam breaks at a single photocell provide a measure of stereotyped movements, such as head bobbing, grooming, circling etc. Successive beam breaks provide a measure of locomotion, i.e., walking from one end of the cage to the other. If an animal is placed in the activity cages at 1200, it is initially very active in exploring the apparatus but habituation occurs across the first hour or two. At 2000 hrs, when the light turns off, there is a burst of activity followed by bouts of increased activity across the dark cycle. Just prior to light on at 0800 hrs animals show another burst of activity. During subsequent lights-on periods, animals are typically inactive. These features of circadian activity can be recorded in a single twenty-four recording period and one recording session can frequently reveal distinctive differences between control and experimental groups. More detailed analysis of circadian activity can be examined in which the effects of light, sound, feeding and so forth are assessed (Mistleburger and Mumby 1992). Figure 7 illustrates circadian activity in eating speed in rats on different levels of food deprivation. Rats eat more quickly at their usual feeding times at the onset and offset of the lights-off period.

V. Skilled Movement

The term "skilled movements" is somewhat arbitrary in that it refers to movements in which the mouth or paws are used to manipulate objects. The term may also be used to include movements used to traverse difficult terrain, such as walking on a narrow beam or climbing a rope, and swimming (see Table 5). The cohesive feature of the movements is that they seem much more disrupted by cortical lesions than are species-typical movements or movements of locomotion on a flat surface. The distinguishing feature of the movements is that they require rotatory movements, irregular patterns movement, selective movements of a limb, and movements that break up the patterns of normal antigravity support (Fig. 8). Skilled movements in rodents and primates are quite comparable, which makes rodent models quite generalizable to humans (Whishaw et al. 1992b). Two commonly used tests of skilled movement are beam walking and skilled reaching.

Beam Walking When a normal rat walks a narrow beam, it has the surprising ability to move along rapidly with its feet placed on the dorsal surface of the beam. A sign of motor incoordination is that it grasps the edge of the beam with its digits as it walks. Following unilateral motor system damage, only the contralateral paws are likely to be used for grasping.

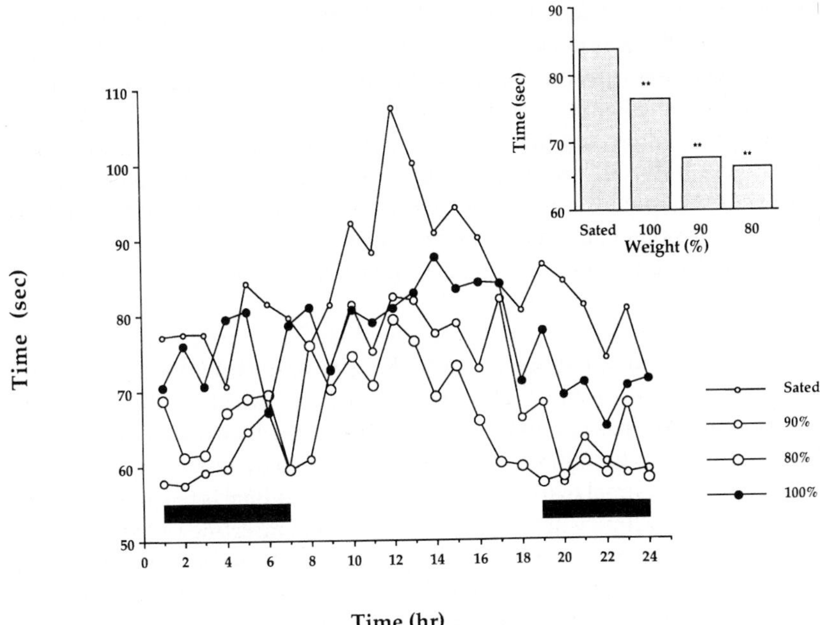

Fig. 7. Time taken to eat a one-gram food pellet as a function of time of day and food deprivation. Sated rats were never food-deprived, 100% represents rats previously food-deprived to 90 and 80% and returned to free feeding. Note that rats eat more quickly after having been subject to food deprivation, when hungry, and at usual feeding times (After Whishaw et al, 1992).

Table 5. Skilled Movements

Limb Movements	Bar-pressing, reaching and retrieving food through a slot, spontaneous food handling of objects or nesting material and limb movements used in fur grooming and social behavior. Rodents use limb movements that are order-typical and species-typical.
Climbing and Jumping	Movements of climbing up a screen, rope, ladder, etc. and jumping from one base of support to another.
Oral Movements	Mouth and tongue movements in acceptance or rejection of food such as spitting food out or grasping and ingesting food. Movements used in grooming, cleaning pups, nest building, teeth cutting.

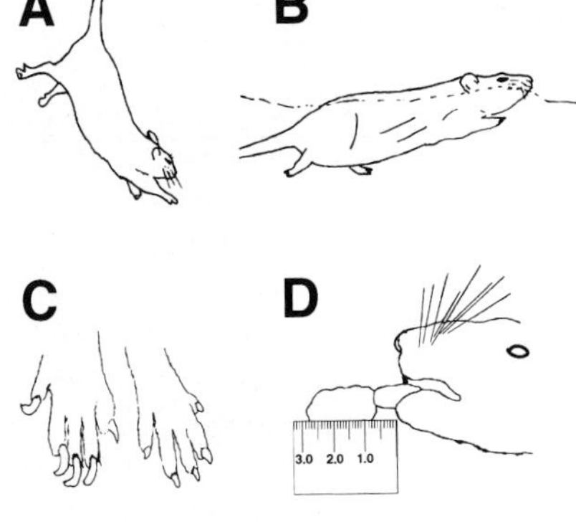

Fig. 8. Examples of skilled movements. A: When lifted in the air, a rat reaches with its forelimbs to regain postural support. B: When swimming, a rat tucks the forelimbs under the chin and tilts the paws to steer. C: Toenails are regularly clipped (right) and if skilled movements of chewing are impaired, they grow long (left). D: Skilled movements of the tongue are used to reach for food as illustrated by a rat licking through the bars of a cage to obtain food from a ruler. Maximum tongue extension is about 11 mm.

Grasping can be measured by video-recording the placement of the paws on the beam or by painting the feet of the animals so that paw placement can be visualized (Becker et al. 1987). The overall body posture of the animal also may be abnormal on a narrow beam. This can be quantified by measuring the angle of the back and head relative to the beam (e.g. Gentile et al. 1978). Another formal test related to beam walking is balancing on a rotating rod, the "rotorod" test. The test consists of a rotating rod upon which the animal balances. As the animals learn to balance, the rod is turned increasingly faster. The measure of motor skill is the time the animal spends on the rotorod as a function of speed with which the rod rotates (Le Marec et al. 1997).

Skilled Forelimb Movements

Rodents use their paws and digits to reach for, hold, and manipulate food. Tests of skilled forelimb use have an animal reach through a slot to obtain a food pellet. One form of the test has animals reach into a tray to retrieve food (Whishaw and Miklyaeva 1996). Limb preference and success (number of pellets per reach) are used as performance measures. Limb use by an animal can also be controlled by restricting the use of one limb, thus forcing the animal to use the other limb. If the animal is reaching through a slot, a bracelet placed on a limb will prevent the animal from inserting its limb through the slot while otherwise leaving use of the limb unimpaired. A second form of the test has an animal reach for a single food pellet while its movements are filmed (Whishaw and Pellis 1990). Video analysis of skilled reaching shows that rodents use a variety of whole body preparatory movements and the reach itself consists of a number of subcomponents such as limb lifting, aiming, pronating, grasping, and supination upon withdrawal. Animals with central nervous system damage may learn to compensate for their impairments and regain presurgical performance as judged by success measures, but movement analysis will indicate that the movements used in reaching are permanently changed (see Fig. 2). Use of the forelimbs can also be evaluated by watching spontaneous food retrieval and manipulation. With the exception of guinea pigs, most rodent species have five "order-typical" movements in spontaneous eating (Whishaw et al. 1998). They (1) identify food by sniffing, (2) grasp food in the incisors, (3) sit back on their haunches to eat, (4) take the food from the mouth with an inward movement of the forelimbs, and (5) grasp and handle the food with the digits. Each of these movements has its characteristic features that can be subject to further analysis and there are species-typical features of the movements in different rodents (Fig. 9).

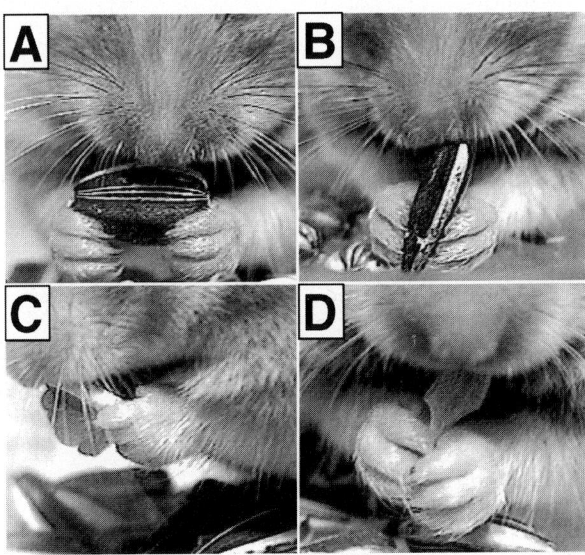

Fig. 9. Skilled paw and digit movements in a hamster. A,C: Food is held between digit 1 (thumb) and digit 2. B: Food is held with all digits. D: Food is held bilaterally with digit 1.

VI. Species-Typical Behaviors

Most movements performed by rodents are sufficiently stereotyped that they are recognizable from occurrence to occurrence and from animal to animal within a species. A variety of complex actions, however, are referred to as species-typical movements (see Table 6). Examples of species-typical behaviors include grooming, nest building, play, sexual behavior, social behavior, and care of young. With the resolution provided by a video record, it is possible to document successive behavioral acts in species-typical behavior in order to produce a description of their order or syntax.

Table 6. Species-Specific Behaviors

Grooming	Grooming movements are species-distinctive and are used for cleaning and temperature regulation. Begin with movements of paw cleaning and proceed through face washing, body cleaning, and limb and tail cleaning.
Food Foraging/ Hoarding	Food carrying movements are species-distinctive and used for transporting food to shelter for eating, scattering food throughout a territory, or storing food in depots. Size of food, time required to eat, difficulty of terrain, and presence of predators influence carrying behavior. Both mouth-carrying or cheek-pouching are used by different species. Rodents also engage in food wrenching, in which food is stolen from a conspecific, and dodging, in which the victim protects food by evading the robber.
Eating	Incisors are used for grasping and biting, rear teeth are used for chewing, tongue is used for food manipulation and drinking.
Exploration/ Neophobia	Species vary in responses to novel territories and objects. Objects are explored visually or with olfaction, avoided, or buried. Spaces are explored by slow excursions into space and quick returns to a starting point. Spaces are subdivided into home bases, familiar territories, and boundaries.
Foraging and Diet Selection	Food preferences are based on size and eating time of food, nutritive value, taste, and familiarity. For colony species the colony is an information source with acceptable foods identified by smelling and licking snout of conspecifics.
Sleep	Rodents display all typical aspects of sleep including resting, napping, quiet sleep and rapid eye movement sleep. Most rodents are nocturnal, thus sleeping during the day with major activity periods occurring at sun up and sun down. Cycles in natural habitats vary widely with seasons.
Nest-building	Different species are nest builders, tunnel builders, and build nests for small family groups or large colonies. All kinds of objects are carried, manipulated, and shredded for nesting material.
Maternal Behavior	Laboratory rodents typically have large litters that are immature when born. Pups are fed for the first two to three weeks of life and thereafter become independent.
Social Behavior	Colony or family rodents have rich social relations including territorial defense, social hierarchies, family groupings, and greeting behaviors. Solitary rodents may have simplified social patterns. Defensive and attack behavior in males and females is distinctive.
Sexual Behavior	Characteristic sexual behavior displayed by males and females. Males display territorial control or territory invasion, and engage in courtship and often group sexual behavior. Sexual behavior is often long-lasting with many bouts of chasing, mounting and intromission, and incidents of ejaculation. Mounting is followed by genital grooming and intromission is followed by immobility and high frequency vocalizations. Females engage in soliciting including approaches and darting, pauses and ear wiggling, and dodging and lordosis to facilitate male mounting.
Play Behavior	Many rodents have rich play behavior with the highest incidence in the juvenile period. Play typically consists of attack in which snout-to-neck contact is the objective and defense in which the neck is protected.

Grooming

Berridge (1990) provides a comparative description of the grooming behavior of a number of species of rodents, including most rodents commonly used in laboratories. His method involved filming the animal through a mirror placed beneath the animal's holding cage. Grooming was elicited by spraying a little water onto an animal's fur. A typical grooming sequence consists of an animal walking forward and making a few body shakes to remove the water from its fur. Then it sits back onto its haunches, in which posture it performs a number of grooming acts in a relatively fixed sequence. The animal first licks its paws and then wipes its nose with rotatory movements of the paws. This is followed by face washing, which consists of wiping the paws down across the face, with the successive wiping movements becoming larger until the paws reach behind the ears and then wash downward across the face. Once an animal has finished a sequence of head grooming, it turns to one side, grasps its fur with a paw and then proceeds to groom its body (Fig. 10). A single grooming bout thus begins at the snout and moves caudally down the body and may consist of more than one hundred individual grooming acts.

Berridge has examined the internal consistency of grooming, that is the extent to which one grooming act predicts another, to derive a grooming syntax. The syntax, in turn, provides the baseline against which central nervous system manipulations are contrasted. The grooming syntax of rats and mice, which are each other's closest relatives, is slightly different in that mice make fewer asymmetrical limb movements when face washing. The grooming of other less closely related rodents is different still. For instance, an animal may use only a single limb to wipe the face. Grooming syntax in turn becomes a powerful tool for the analysis of neural control of action patterns. For example, to answer the question of whether grooming is produced by a grooming center or has its control distributed across a number of neural systems, Berridge and coworkers have sectioned the brain at various rostrocaudal levels to find that different features of grooming control are represented at many different nervous system levels (Berridge 1989).

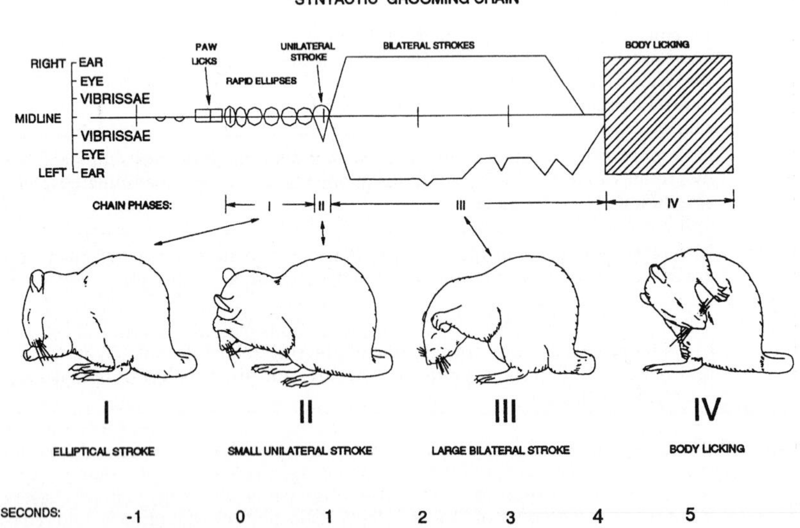

Fig. 10. Choreographic transcription of an idealized syntactic grooming chain. Time proceeds from left to right. The horizontal axis represents the center of the nose. The line above the horizontal axis denotes movement of the left forepaw. Small rectangles denote paw licks. Large rectangle denotes body licks. Chain phases are: (I) 5 to 8 rapid elliptical strokes around the nose (6.5 Hz); (II) unilateral strokes of small amplitude; (III) symmetrically bilateral strokes of large amplitude; (IV) licking the ventrolateral torso (After Berridge and Whishaw, 1992).

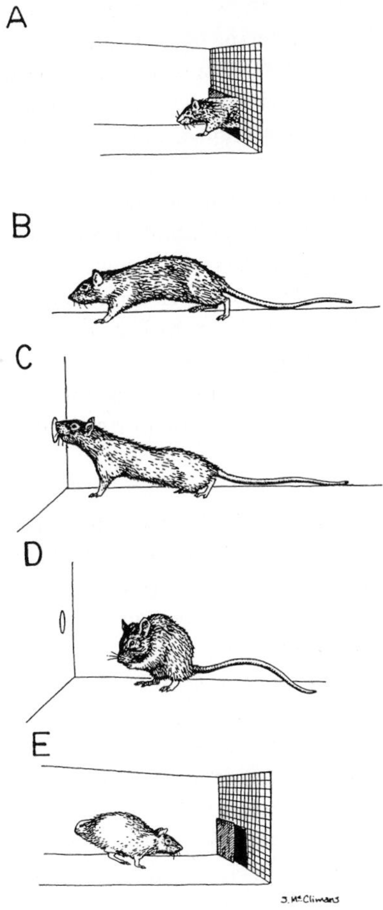

Fig. 11. Foraging in the rat for food items that are passed to it through a hole in the test apparatus. A: The animal "stops, looks and listens" before leaving its home base. B: The food location is approached cautiously. C: Food items that can be eaten quickly are swallowed. D: Larger items are eaten from a sitting position. E: Very large items are carried "home", usually at a gallop. (After Whishaw et al, 1990).

Food Hoarding

Rodents have a number of food handling patterns depending upon the time required to eat food (Fig. 11). Rodents are usually cautious in leaving the home base and cautious in approaching a food source. Small pieces of food that take little time to eat are consumed in situ, items that take a little longer to eat are consumed from a sitting posture, while items that take a long time to eat are carried to a secure location or to the home base. Rodents that cheek-pouch carry items of all sizes whereas animals that do not cheek-pouch eat smaller items at the location at which they are found and carry larger objects. The decision on whether to carry an object is based on an estimate of eating time. Food carrying animals may store food in a single central location, such as a nesting area, hide it at a number of locations, or hide individual items throughout a territory. Colony living rodents may carry food to a home area but here it is likely to be stolen and eaten quickly by other colony members (Whishaw and Whishaw 1996). Tests of food carrying are used to evaluate exploration, spatial abilities, time estimates, and social competition for food (Whishaw et al. 1990).

Foraging and Diet Selection

A rodent colony is frequently an information center, especially regarding novel foods (Galef 1993). Rodents, especially those that have been subjected to attempts at eradication, are extremely wary of novel objects and foods, a trait called neophobia. Thus, rodents in a colony may share information about types of foods, food safety, and food locations. An animal will sniff and lick the snout of a conspecific and so gain information about food types and sources. Sulfur dioxide on the breath of the conspecific indicates to the inquirer that the food is palpable. Tests of neophobia and information sharing can be used to study both learning and social behavior in animals (Galef et al. 1997). Pairing food items with a sickness-inducing agent is commonly used to study food aversion responses (Perks and Clifton 1997). Conditioned aversion is considered a special form of learning because the ingestion of food and subsequent illness may be separated by hours or days but will still be strongly acquired.

Nest Building

If rodents are provided with appropriate material, they will build nests (Kinder 1927). The nests contribute to thermoregulation and provide a place to gather young. The quality of a nest built by female rats will also wax and wane over the four-day esterous cycle. If an animal is given strips of paper about 2 cm wide and 20 cm long, each behavior involved in nest building can be recorded as it occurs, e.g., pick up material, carry, push, chew, etc. The quality of the nest constructed can be rated on a four-point scale over three to four days, the time required to fashion an optimal nest. Limbic system and medial frontal cortex lesions have been reported to disrupt nest building (Shipley and Kolb 1977).

Social Behavior

Social behavior can be defined as all behavior that influences, or is influenced by, other members of the same species. The term thus covers all sexual and reproductive activities and all behavior that tends to bring individuals together as well as all forms of aggressive behavior (e.g. Grant 1963). It is traditional to describe sexual behavior separately (see below), and in recent years aggressive behavior has also come to be seen as a separate form of social behavior as well (see below).

It is generally recognized that social behavior is not a unitary behavior with a unitary neurological basis. Rather, different aspects of social behavior have different neural and endocrine bases (e.g. Moyer 1968). It is therefore necessary to examine social behavior in a number of different situations before concluding that a particular treatment has produced a general change in social behavior; behavioral changes may be situation-specific. Most studies of social behavior in rodents usually utilize some type of test of free interaction in which a group of subjects are housed in a large group cage, often with several interconnected chambers (e.g. Lubar et al. 1973). A less natural version has animals paired in a novel situation, often repeatedly over days (e.g. Latane 1970). In both situations the behavior is videotaped and can either be analyzed by calculating specific behaviors, such as time in contact, or by writing a detailed description of the behaviors such as described by Grant and Mackintosh (1963). Other behaviors such as vocalizations (e.g. Francis 1977) or urine marking (e.g. Brown 1975) can also be recorded.

Aggression

Aggressive behavior is used to establish social hierarchies and to defend territories. As rodents vary widely in their tendency to live in colonies or in relative isolation, their tendency to engage in aggressive behavior also varies. Patterns of aggressive behavior are also usually distinctive in male and female animals. The targets of aggressive contact, e.g. the location on the body on which an animal attempts to inflicts bites, are usually quite distinctive from the targets of play contact (Pellis 1997). In rats, bites are typically directed to the back and rear. Fellow residents and strangers are usually identified by their odor. Aggressive behavior is widely used as an animal model of human aggression (Blanchard et al. 1989).

Sexual Behavior

Sexual behavior requires the integrity of hormonal systems and neural systems, requires developmental experiences, learning, context and an appropriate partner. Sexual behavior consists of at least two phases, courtship and consummation, and both are extremely complex requiring complex independent and interdependent actions by both the male and female. Dewsbury (1973) has described the social behaviors associated with sexual activity, including exploration, sniffing, grooming, soliciting, ear wiggling, hopping and darting in females, genital and nongenital grooming, as well as mounting, pelvic thrusting, ejaculation, lordosis, immobility, ultrasonic songs, etc. The patterning of movements of sexual behavior has been described by Sachs and Barfield (1976) for male rats and Carter et al. (1982) for female rats. Paradigms in which the female is given the opportunity to pace sexual activity are described by Mermelstein and Becker 1995. Paradigms in which interest in access to sexual partners and interest in sexual intercourse are described are presented by Everitt (1990). Michal (1973) provided a detailed discussion of the behavioral patterns of rats with limbic lesions.

Care of Young

Rodents are immature when born and are hairless with immature sensory and motor systems and receive extensive parental care (Grota and Ader 1969). Pups are cared for mainly by the mother with elaborate patterns of pup cleaning and feeding under hormonal and thermoregulatory control (Leon et al. 1978). Thermoregulation influences much of the pups' social behavior as they have elaborate huddling strategies (Alberts 1978).

Play

Rodents may display play behavior at any stage of development, but play is particularly prominent during the juvenile stage of life. Play is highly ritualized but incorporates many of the movements used by animals in other aspects of life, including sex, aggression, and skilled manipulation. In rodent play, the rich array of movements used by the participants appears to be orchestrated around the attempt of an initiator to thrust the tip of its snout into the neck of the recipient while the recipient attempts to avoid the contact (Fig. 12). The patterns of thrust and parry are distinctive in different rodent species (Pellis et al. 1996).

VII. Learning

Research on the neural basis of learning suggests that there are a number of at least partially independent learning and memory systems. These include short-term memory (thought to be a frontal lobe function), object memory (thought to be a function of rhi-

Fig. 12. A sequence of play fighting between two 30-day-old male rats. Note the repeated attack and defense of the **neck** (After Pellis et al. 1996).

Table 7. Learning

Classical Conditioning or Conditional Learning	Unconditioned stimuli are paired with conditioned stimuli and the strength of an unconditioned response to the conditioned stimuli is measured. Almost any arrangement of stimuli, environments, treatments, or behavior can be used.
Instrumental Conditioning	Animals are reinforced for performing motor acts such as running, jumping, sitting still, lever pressing, or opening puzzle latches.
Avoidance learning	Passive responses including avoiding preferred places or objects which have been associated with noxious stimuli such as electric shock. Active responses including moving away from noxious items or burying noxious items.
Object Recognition	Including simple recognition of one or more objects, matching to sample, and nonmatching to sample in any sensory modality. Tasks are formal in which an animal makes an instrumental response or inferential in which recognition is inferred from exploratory behavior.
Spatial Learning	Dry land- and water-based tasks are used. Spatial tasks can be solved using *allothetic cues,* which are external and relatively independent to movements, or *idiothetic cues,* which include cues from vestibular or proprioceptive systems, reafferance from movement commands, or sensory flow produced by movements themselves. Animals are required to move to/away from locations. *Cue tasks* require responding to a detectable cue. *Place tasks* require moving using the relationships between a number of cues, no one of which is essential. *Matching tasks* require learning a response based on a single information trial.
Memory	Memory includes *procedural memory* in which response and cues remain constant from trial to trial and *working memory* in which response or cues change from trial to trial. Tasks are constructed to measure one or both types of learning. Memory is typically divided into object, emotional, and spatial and each category can be further subdivided into sensory and motor memory.

nal cortex), emotional memory (thought to be functions of the amygdala and related circuits), and implicit and explicit spatial memory (thought to be a function of the neocortex and hippocampal formation, respectively). A quick overview of an animal's learning ability therefore requires a number of tests, in order to tap into all of these systems (see Table7). Widely used tests for rodents include

- Passive avoidance,
- Defensive burying,
- Conditioned place preference,
- Conditioned emotional response,
- Object recognition task,
- Swimming pool place and matching-to-place tasks.

Although these tests provide a rapid way of screening for learning and memory deficits, all of the tests are sensitive to quite a number of different functions and brain regions. At this point, we emphasize that the enormous number of tests in all their variations and the diversity of opinion concerning what they actually measure requires consultation with more comprehensive sources than provided here (see Olton et al. 1985; Vanderwolf and Cain 1994).

Memory

Memory is also often described as being either procedural or working memory. *Procedural memory* is memory for the rules of task solution. For example, the rule may be that food is found at the end of an alley or that an escape platform is located somewhere in the swimming pool. *Working memory* is trial unique memory; that is, "on the last trial I found food here". It is thought that each sensory or motor system may be involved in system-specific procedural or working memory. For example, the visual pathway underlying the perception of objects is likely to be involved in the storage of object information. Memory is also described as being either *short-term*, to be used only for the moment, or *long-term*, to be used for long durations. That is, for each of the kinds of memory described above, there are procedural and working memory, short-term, and long-term memory. Detailed discussion concerning terminology, tests, and their significance can be found in comprehensive sources (Dudai 1989; Martinez and Kesner 1991).

Passive Avoidance

The passive avoidance apparatus consists of a box with two compartments and a connecting door. One of the boxes is white and the other is black, and the floor of both boxes is constructed of grids which can pass a small electric shock. An animal is placed in the white compartment each day for three days and is removed after it has crossed over into the dark compartment. As most rodents display a strong preference for the dark, by the third day the animal's passage between the boxes is quite quick. Now, however, the animal is given a brief electric shock when it enters the dark compartment. After one hour to twenty-four hours, the animal is once again placed in the white compartment of the box and the measure of learning is the time taken to again enter the now noxious dark compartment. Usually, a five-minute cutoff for entry is used. Passive avoidance has been found to be a very sensitive measure of the effects of drugs that affect memory, such as the muscarinic blocker atropine (or scopolamine). Certain kinds of brain damage, including damage to the limbic system and globus pallidus and their transmitters, are similarly sensitive to the passive avoidance task (Bammer 1982; Slagen et al. 1990).

Defensive Burying

The strength of the defensive burying test is that it provides a natural test of an animal's response to a threatening or noxious object. A large number of modifications of the test and their uses have been described by Pinel and Treit (1983). Traditionally, animals have been thought to have two primary defensive strategies, flight or fight. The defensive burying paradigm reveals that the responses of rats and mice (but not gerbils and hamsters) to threat are much more complex than has been thought and include investigating the object, removing it, or burying it so that it is no longer threatening. At its simplest, an animal is placed in and briefly habituated to a box that contains sawdust on the floor. After habituation, a probe that can deliver a shock is inserted into the box through a hole in the wall. When the animal investigates the object, it receives a brief electric shock from the probe. The response of the animal is to first withdraw from the object, then investigate it cautiously, and finally to use the forelimbs to cover the object with sawdust. Measures of the strength of learning about the probe include the number of times the animal investigates the object, the length of time that it spends burying the object, and the depth of the sawdust that eventually covers the object. A variety of variations of the task have been used, including burying objects that deliver noxious sounds or odors. Animals will also bury other objects that are proximal to the offending object, indicating that the burying response can be secondarily conditioned to other objects. Defensive burying has been used to examine the effects of aging on behavior and also to examine the effects of potential anti-anxiety agents (Pinel and Treit 1983).

Object Recognition

Object recognition can be tested in a three-compartment box, also called the Mumby box (Mumby and Pinel 1994; Mumby et al. 1989), or other similar test situations (Ennaceur and Aggleton 1997). The central compartment of the box is a waiting area and is connected to two side boxes, choice box one and choice box two, by sliding doors. A door is opened and the animal is allowed to enter choice box one, where it finds two food wells, one of which is covered by a "sample" object. When it displaces the sample, it receives a food reward. The rat then shuttles into the waiting area until it is allowed access to choice box two. Here it again finds two food wells, one covered by an object identical to the previously encountered sample in choice box one, and a novel object. To obtain food reward, it must displace the novel object. New sample and novel trials are given with new sample and novel objects. The measure of success is the animal's memory that sample objects will not provide reinforcement on two successive trials. The rodent object recognition test is similar to nonmatching-to-sample tests previously developed for primates. Both short-term and long-term memory for objects can be measured by introducing a delay of variable duration after the sample trial. Object recognition in more natural environments uses very similar methodology to that described above, but the time spent sniffing and examining objects or animals placed in the animal's home environment is used as the measure of recognition.

Conditioned Place Preferences

The conditioned place preference task takes advantage of the fact that objects, events, or substances that an animal finds pleasant or noxious become conditioned to the location in which the object, event, or substance is encountered (e.g., Cabib et al. 1996; Schechter and Calcagnetti 1993). Typically, a two-compartment box is used and the measure of behavior is the time spent in the compartments. For example, if the experiment wishes to determine whether a drug treatment is perceived as being pleasant, the animal is exposed to one of the compartments of the box while under the effects of the drug. At some later date, while undrugged, the animal is given access to the original box or to a different box. If the animal spends more time in the original "conditioned" box, that can be taken as evidence that the animal perceived the treatment as positively rewarding, while if it shows a preference for the other box, it perceived the treatment as negatively rewarding. Conditioned place preferences can be modified to measure the strength of memories and their duration by varying intervals between sample and test trials.

Spatial Navigation

A large number of maze tests has been used to measure spatial navigation (Fig. 13). The central idea for all of the tests consists of having an animal: (1) learn to find food at one or more locations, or (2) escape to a refuge from different locations. Most of the tests are administered on dry land, but because of the excellent swimming ability of the rat, tests have been developed in a swimming pool. The two most widely used tests are the radial arm maze and the swimming pool place task.

The radial arm maze consists of a central box or platform from which protrude a number of arms (Jarrard 1983; Olton et al. 1979). The location of the arms is either fixed or marked by a cue on the arm (e.g., roughness of the surface of the arm, the color of the arm, a light at its end, etc.). Food is located at the end of one or more arms. The task of the animal is to learn the location of the food over a number of test days. This evaluates its ability to form a procedural memory for the task. The animal's performance can be interrupted to see if it can "pick up where it left off" in order to evaluate its working memory.

The swimming pool task has become extremely popular, mainly because the animals do not have to be food- or water-deprived to motivate them to perform (McNamara and Skelton 1993; Morris 1984; Sutherland and Dyck 1984; Whishaw 1985). Although the task is excellent for the rat, which is semiaquatic, it may be less useful for other species (Whishaw 1995). The apparatus consists of a circular round pool about 1.5 m in diameter, filled with tepid water made opaque with powdered milk, paint, sawdust, or floating beads. A platform about 10 cm sq is placed in the pool with its surface either visible or hidden about 1 cm beneath the surface of the water. The animal is placed into the water facing the wall of the pool, and in order to escape from the water, must reach the platform. On successive trials, the animal is placed into the water at new locations, and its response time decreases until it escapes by swimming directly to the hidden platform. Its ability to learn to escape to a platform hidden at a fixed location is thought to be a measure of spatial procedural memory. If the platform is moved repeatedly to new locations, the task becomes mainly a test of spatial working memory (Whishaw 1985). That

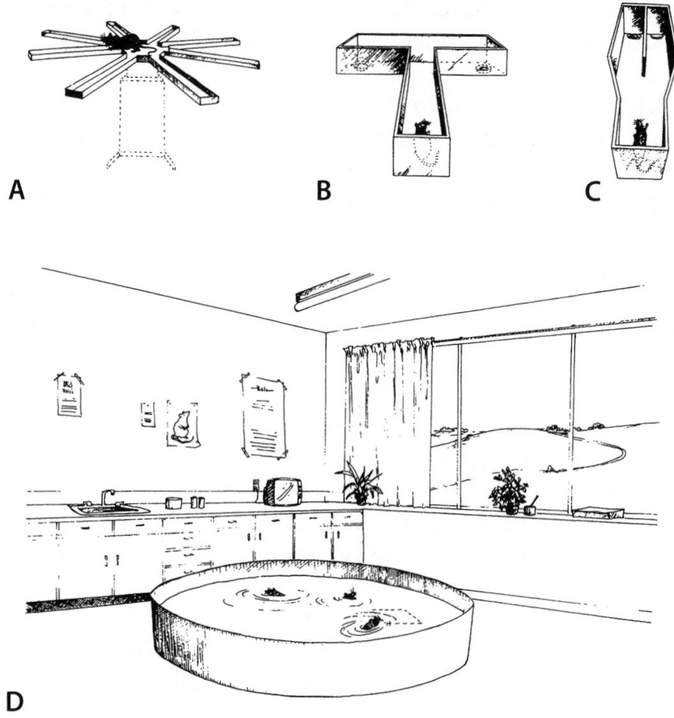

Fig. 13. Spatial tasks: A: radial arm maze in which food is located only at the end of some arms, B: T-maze in which food is located in one arm, C: Grice box in which food is located on one side, D: swimming pool task, in which an animal searches for a platform hidden just beneath the surface of the water. The tests are usually conducted in a open room that provides many spatial cues.

is, the animal has to match on its second trial the location at which the platform was found on the first trial.

At their simplest, spatial tasks attempt to measure three aspects of spatial behavior.

- Place tasks measure whether an animal can find a food item or an escape platform using the relational properties of ambient cues, usually visual cues.
- Cue tasks measure whether an animal can find a food item or an escape platform using a visible cue marking the location of the target.
- Response tasks measure whether an animal can use body cues, e.g., turn left or right, to locate an object.

It is widely assumed that these different types of spatial learning are mediated by different neural systems. Thus administration of more than one task can be used to dissociate spatial functions.

Comments: Generalizing from Behavioral Analysis

Because the ultimate goal of studies on rodents is to understand the brain-behavior relationships in humans, it is reasonable to ask to what extent the behavior of rodents is useful in understanding human brain-behavior relationships. Indeed, one difficulty with choosing any mammal or mammalian order to use as a model of brain function is that each species has a unique behavioral repertoire that permits the animal to survive in its particular environmental niche. There is, therefore, the danger that neural organization is uniquely patterned in different species in a way that reflects the unique behavioral adaptations of the different species. Stated differently, it is possible that the brain-behavior relationships of rodents are not representative of other mammalian species, especially primates.

We have emphasized elsewhere that although the details of behavior may differ somewhat, mammals share many similar behavioral traits and capacities (Kolb and Whishaw 1983). For example, all mammals must detect and interpret sensory stimuli, relate this information to past experience, and act appropriately. Similarly, all mammals appear to be capable of learning complex tasks under various conditions of reinforcement (Warren 1977), and all mammals are mobile and have developed mechanisms for navigating in space. The details and complexity of these behaviors clearly vary, but the general capacities are common to all mammals. Warren and Kolb (1978) proposed that behaviors and behavioral capacities demonstrable in all mammals could be designated as *class-common* behaviors. In contrast, behaviors that are unique to a species and that have been selected to promote survival in a particular niche are designated as *species-typical* behaviors. The distinction between these two types of behavior can be illustrated by the manner in which different mammals use their forelimbs to manipulate food objects. Monkeys will grasp objects with a single forepaw and often sit upright, holding the food item to consume the food. Rats too will grasp objects with one forepaw and then typically transfer the food to their mouth, assume a sitting posture, transfer the food back to both forepaws, and then eat. So many mammals use the forepaws to manipulate food (or other items) that it can be considered a class-common behavior. Nonetheless, the details vary from species to species. Some species-typical differences are large indeed, such as the use of the forelimbs in bats or carnivores versus rodents or primates. It would seem foolish indeed to use dogs as a model for studying the details of the neural control of object manipulation by humans, as their limb use is relatively rudimentary. But what about rodents? There are clearly species-typical differences among rodent species and between rodents and primates. The question is whether these species-typical differences necessarily reflect fundamental differences in the neural control of skilled forelimb use. One way to address this question is to examine the order-specific characteristics of forelimb use. That is, one can study the similarity in forelimb use in different species within an order. This type of analysis would allow us to determine the commonalities in behavior across species of an order, which would give us a better basis for comparing across orders (Whishaw et al. 1992b).

L'Envoie

The following example from our laboratory illustrates how behavioral analysis can provide insights into behavior. Aged Fischer 344 rats are widely used to examine the effects of aging on memory (Lindner 1997). We were interested in using these animals for studies that examine the ameliorative effects of exogenously supplied compounds thought to have neurotrophic properties. Since previous studies had used swimming pool spatial tasks, we first sought to confirm that 24-month-old Fischer rats were impaired relative to 6-month-old rats. We found that when the rats were given 8 trials a day on a task that required them to find a hidden platform at a fixed location in a swimming pool, the 24-month-old rats were indeed severely impaired relative to the 6-month-old rats. When given just one trial a day, however, they learned the task much more quickly, reaching the hidden platform as rapidly as the young rats after only 14 to 16 trials, indicating that learning per se was not impaired. Finally, when given a matching-to-place task in which they were given two trials a day with the platform in a new location each day, the aged rats were severely impaired and showed no improvement from the first (test) trial to the second (matching) trial, whereas the young rats showed a marked reduction in latency. Results from these three tests seemed to support the idea that the animals had a selective spatial deficit, since a very similar pattern of results is obtained from rats that have selective hippocampal lesions.

This conclusion was severely compromised by the results of further tests. In open field tests, the old rats were less active than the young rats as they walked, and they reared less. When given locomotion tests, they were slower swimmers and walked more slowly for food in a straight alley. When required to climb out of a 9-inch-deep cage to obtain food, they were extremely impaired. Further tests showed that the motor impairments of old rats were quite selective. The aged animals could protrude their tongue normally, eat one gram food pellets as quickly as the young rats, and they reached for food in a skilled reaching task as well as did the young rats. When given tests of righting, they were impaired relative to the young rats but when their forequarters and hindquarters were tested separately, forequarter righting was unimpaired whereas hindquarter righting was impaired. The results of these neurologic tests suggested that the old rats were selectively impaired in using their hindlimbs and this result was confirmed by kinematic analysis of hindlimb movements used in swimming and walking. Thus, it is unclear whether their "spatial" deficit was due to a learning impairment or related to impaired use of the hindlimbs.

These results are relevant to the discussion of methodology given in the opening of this paper. There is a general expectation that aged rats will be impaired in spatial memory tasks. Our tests confirmed this expectation. The comprehensive follow-up analysis showed, however, that the animals had a selective motor impairment in use of the hindlimbs to move themselves. The selectivity of the deficit was completely unexpected. Since commonly used spatial tasks require the animals to use their hindlimbs to move themselves, the results of the spatial tests are confounded by the rats' motor impairment.

The particular result from this study suggests two new hypotheses. Aged Fischer 344 rats may have only a motor deficit that impairs their swimming performance, or the animals may have both a learning deficit and a motor deficit. Subsequent testing, e.g., spatial tests that do not require movement, could be used to assess these hypotheses. Fortuitously, however, the finding of a selective motor deficit provides a very good model for studying motor impairments associated with aging.

The lesson from this example is, therefore, that a careful examination of behavior can provide insights into the specific impairments of an animal, can provide new models for behavioral analysis, and finally can assist in evaluating the specificity of animal models of functional disorders.

References

Alberts JR (1978) Huddling by rat pups: Multisensory control of contact behavior. J C Physiol Psychol 92: 220–230.

Bammer G (1982) Pharmacological investigations of neurotransmitter involvement in passive avoidance responding: a review and some new results. Neurosci Biobehav Rev 6: 247–296.

Becker JB, Snyder PJ, Miller MM, Westgate SA, Jenuwine MJ (1987) The influence of esterous cycle and intrastriatal estradiol on sensorimotor performance in the female rat. Pharm Biochem Behav 27: 53–59

Berridge KC (1989) Progressive degradation of serial grooming chains by descending decerebration. Behav Brain Res 33: 241–253.

Berridge KC (1990) Comparative fine structure of action: rules of form and sequence in the grooming patterns of six rodent species. Behavior, 113: 21–56.

Barridge KC, Whishaw IQ (1992) Cortex, striatum and cerebellum: control of a syntactic grooming sequence. Exp Brain Res 90:275–290.

Blanchard BJ, Blanchard DC, Hori K (1989) An ethoexperimental approach to the study of defense. In RJ Blanchard, PF Brain, DC Blanchard, S Parmigiani (eds) Ethoexperimental approaches to the study of behavior. London: Kluwer Academic Publishers, pp 114–137.
Brown RE (1975) Object-directed urine marking by male rats (Rattus norvegicus). Behavioral Biology 15: 251–254.
Cabib S, Puglisi-Allegra S, Genua C, Simon H, Le Moal M, Pizza PV (1996) Dose-dependent aversive and rewarding effects of amphetamine as revealed by a new place conditioning apparatus. Psychopharmacology 125: 92–96.
Carter CS, Witt DM, Kolb B, Whishaw IQ (1982) Neonatal decortication and adult female sexual behavior. Physiol Behav 29: 763–766.
Clarke KA (1992) A technique for the study of spatiotemporal aspects of paw contact patterns applied to rats treated with a TRH analogue. Behav Res Meth Instrum Comput 24: 407–411.
Crawley JN, Belknap JK, Collins A, Crabbe JC, Frankel W, Henderson N, Hitzemann RJ, Maxson SC, Miner LL, Silva AJ, Wehner JM, Wynshaw-Boris A, Paylor R (1997) Behavioral phenotypes of inbred mouse strains: implications and recommendations for molecular studies. Psychopharmacology 132: 107–124.
Denny-Brown, D, Dawson DM, Tyler HR (1982) Handbook of neurological examination and case recording. Harvard University Press: Cambridge Mass.
DeRyck M, Schallert T, Teitelbaum P (1980) Morphine versus Haloperidol catalepsy in the rat: A behavioral analysis of postural support mechanisms. Brain Res 201: 143–172.
Dewsbury DA (1973) A quantitative description of the behavior of rats during copulation. Behaviour 29: 154–178.
Donne, John (1572–1631) (1985) The complete English poems of John Donne, C.A. Patrides (ed). London, Dent, 1985.
Dudai Y (1989) The neurobiology of memory. Oxford: Oxford University Press.
Eliam D, Golani I (1989) Home base behavior of rats (Rattus norvegicus) exploring a novel environment. Behav Brain Res 34: 199–211.
Ennaceur A, Aggleton JP (1997) The effects of neurotoxic lesions of the perirhinal cortex compared to fornix transection on object recognition memory in the rat. Behav Brain Res 88: 181–193.
Eshkol N, Wachmann A (1958) Movement notation. Weidenfeld and Nicholson, London.
Everitt BJ (1990) Sexual motivation: A neural and behavioral analysis of the mechanisms underlying appetitive and copulatory responses of male rats. Neuroscience and Biobehavioral Reviews 14: 217–232.
Fentress JC, Bolivar VJ (1996) Measurement of swimming kinematics in small terrestrial mammals. In K-P Ossenkopp, M Kavaliers, PR Sanberg (Eds) Measuring movement and locomotion: From invertebrates to humans. RG Landes, Austin, Texas, pp 171–184.
Field EF, Whishaw IQ, Pellis SM (1997) A kinematic analysis of sex-typical movement patterns used during evasive dodging to protect a food item: the role of testicular hormones. Behav Neurosci 111: 808–815.
Fish FF (1996) Measurement of swimming kinematics in small terrestrial mammals. In K-P Ossenkopp, M Kavaliers, PR Sanberg (Eds) Measuring movement and locomotion: From invertebrates to humans. RG Landes, Austin, Texas, pp 135–164.
Francis RL (1977) 22-kHz calls by isolated rats. Nature 265: 236–238.
Freed W (1983) Functional brain tissue transplantation: reversal of lesion-induced rotation by intraventricular substantia nigra and adrenal medulla grafts, with a note on intracranial retinal grafts. Biol Psychiat 18: 1205–1267.
Galef BG (1993) functions of social learning about food: A causal analysis of effects of diet novelty on preference transmission. Animal Behav 46: 257–265.
Galef BG, Whiskin EE, Bielavska E (1997) Interaction with demonstrator rats changes observer rats' affective responses to flavors. J Comp Psychol 111: 393–398.
Gallup GG, Maser JD (1977) Tonic immobility: Evolutionary underpinnings of human catalepsy and catatonia. In JD Maser and MEP Seligman (eds) Psychopathology: Experimental models, WH Freeman, San Francisco, 334–462.
Ganor I, Golani I (1980) Coordination and integration in the hindleg step cycle of the rat: Kinematic synergies. Brain Res 195: 57–67.
Gentile AM, Green S, Nieburgs A, Schmelzer W, Stein DG (1978) Behav Biol 22, 417–455.
Golani I (1992) A mobility gradient in the organization of vertebrate movement (The perception of movement through symbolic language). Behavioral and Brain Sciences 15:249–266.
Golani I (1976) Homeostatic motor processes in mammalian interactions: a choreography of display. In PG Bateson and PH Klopfer (Eds\), Prospectives in Ethology, Vol 2. Plenum Press, New York, pp 237–134.

Golani I, Benjamini Y, Eilam D (1993) Stopping behavior: constraints on exploration in rats (Rattus norvegicus). Behav Brain Res 26: 21–33.

Golani I, Wolgin DL, Teitelbaum, P (1979) A proposed natural geometry of recovery from akinesia in the lateral hypothalmic rat. Brain Res 164: 237–267.

Grant EC (1963) An analysis of the social behaviour of the male laboratory rat. Behaviour 21: 260–281.

Grant EC, Mackintosh JH (1963) A comparison of the social postures of some common laboratory rodents. Behaviour 21: 246–259.

Grill HJ, Norgren R (1978) The taste reactivity test. I. Mimetic responses to gustatory stimuli in neurologically normal rats. Brain Res 143: 263–279.

Grota LJ, Ader R (1969) Continuous recording of maternal behavior in Rattus norvegicus. Anim Behav 21: 78–82.

Heale VR, Petersen K, Vanderwolf CH (1996) Effect of colchicine-induced cell loss in the dentate gyrus and Ammon's horn on the olfactory control of feeding in rats. Brain Res 712: 213–220.

Huerta PT, Scearce KA, Farris SM, Empson RM, Prusky GT (1996) Preservation of spatial learning in fyn tyrosine kinase knockout mice. Neuroreport 10: 1685–1689.

Hutt SJ, Hutt C (1970) Direct observation and measurement of behavior. Springfield IL, Charles C Thomas.

Jarrard LE (1983) Selective hippocampal lesions and behavior: effects of kainic acid lesions on performance of place and cue tasks. Behav Neurosci 97: 873–889.

Kinder EF (1927) A study of the nest-building activity of the albino rat. J Comp Physiol Psychol 47: 117–161.

Kolb, B (1974). Some tests of response habituation in rats with prefrontal lesions. Can. J. Psychol. 28, 260–267

Kolb, B., Gibb, R., Pedersen, B., & Whishaw, I.Q. (1997). Embryonic injection of BrdU blocks later cerebral plasticity. Society for Neuroscience Abstracts, 23: 677.16.

Kolb, B, Whishaw, IQ (1983) Problems and principles underlying interspecies comparisons. In TE Robinson (Ed.) Behavioral approaches to brain research. Oxford U Press: Oxford, pp. 237–265.

Kolb B, Whishaw IQ (1985) An observer's view of locomotor asymmetry in the rat. Neurobehav Toxicol Teratolog 7: 71–78.

Kolb B, Whishaw IQ (1996) Fundamentals of human neuropsychology. Freeman and Co, New York.

Kozlowski DA, James DC, Schallert TJ (1996) Use-dependent exaggeration of neuronal injury after unilateral sensorimotor cortex lesions. Neurosci 16:4776–4786.

Lassek, A.M. (1954) The pyramidal tract. Springfield, Ill: Charles C. Thomas.

Latane B. (1970) Gregariousness and fear in laboratory rats. J Exp Soc Psychol 5: 61–69.

Le Marec N, Stelz T, Delhaye-Bouchaud N, Mariani J, Caston J (1997) Effect of cerebellar granule cell depletion on learning of the equilibrium behaviour: study in postnatally X-irradiated rats. Eur J Neurosci 9: 2472–2478.

Leon M, Croskerry PG, Smith GK (1978) Thermal control of mother-young contact in rats. Physiol Behav 21: 793–811.

Lindner MD (1997) Reliability, distribution, and validity of age-related cognitive deficits in the Morris water maze, Neurobiol Learn Mem 68: 203–220.

Lubar JF, Herrmann TF, Moore DR, Shouse MN (1973) Effect of septal and frontal ablations on species-typical behavior in the rat. J Comp Physiol Psychol 83: 260–270.

Marshall J, Turner BH, Teitelbaum P (1971) Further analysis of sensory inattention following lateral hypothalamic damage in rats. J Comp Physiol Psychol 86: 808–830.

Martinez, J.L. Kesner, R.P. (1991). Learning and memory: A biological view Second Ed. New York: Academic Press

McNamara RK, Skelton RW (1993) The neuropharmacological and neurochemical basis of place learning in the Morris water maze. Brain Res Rev 18: 33–49.

Mermelstein PG, Becker JB (1995) Increased extracellular dopamine in the nucleus accumbens and striatum of the female rat during paced copulatory behavior. Behav Neurosci 109: 354–365.

Michal EK (1973) Effects of limbic lesions on behavior sequences and courtship behavior of male rats (Rattus norvegicus). Anim Behav 244: 264–285

Miklyaeva EI, Martens DJ, Whishaw IQ (1995) Impairments and compensatory adjustments in spontaneous movement after unilateral dopamine depletion in rats, Brain Res 681: 23–40.

Mistleburger RE, Mumby DG (1992) The limbic system and food-anticipatory circadian rhythms in the rat: ablation and dopamine blocking studies, Behav Brain Res 47: 159–168.

Morris R (1984) Developments of a water-maze procedure for studying spatial learning in the rat. Neurosci Meth 11: 47–60.

Moyer KE (1968) Kinds of aggression and their physiological basis. Com Behav Biol 2: 65–87.

Mumby DG, Pinel JPJ, Wood, ER (1989) Nonrecurring items delayed nonmatching-to-sample in rats: A new paradigm for testing nonspatial working memory. Psychobiology 18: 321–326.
Mumby DG, Pinel JPJ (1994) Rhinal cortex lesions impair object recognition in rats. Behav Neurosci 108: 11–18.
Olsson M, Nikkhah G, Bentlage C, Bjorklund A (1995) Forelimb akinesia in the rat Parkinson model: differential effects of dopamine. Neurosci 15: 3863–3875.
Olton DS, Becker JT, Handlemann GE (1979) Hippocampus, space and memory. Behav Brain Sci 2: 313–365.
Olton DS, Gamzu E, Corkin S (1985) Memory dysfunctions: An integration of animal and human research from preclinical and clinical perspectives. Ann NY Acad Sci 444.
Ossenkopp K-P, Kavaliers M, Sanberg PR (1996) Measuring movement and locomotion: From invertebrates to humans. RG Landes, Austin, Texas.
Pellis SM (1983) Development of head and food coordination in the Australian magpie *Gymnorhina tibicen*, and the function of play. Bird Behav 4: 57–62.
Pellis SM (1996) Righting and the modular organization of motor programs. In K-P Ossenkopp, M Kavaliers, PR Sanberg (Eds) Measuring movement and locomotion: From invertebrates to humans. RG Landes, Austin Texas, pp 115–133.
Pellis SM (1997) Targets and tactics: The analysis of moment-to-moment decision making in animal combat. Agg Behav 23: 107–129.
Pellis SM, Field EF, Smith LK, Pellis V (1996) Multiple differences in the play fighting of male and female rats. Implications for the causes and functions of play. Neurosci Biobehav Rev 21: 105–120.
Perks SM, Clifton PG (1997) Reinforcer revaluation and conditioned place preference. Physiol Behav 61: 1–5.
Pinel JPJ, Hones CH, Whishaw IQ (1992) Behavior from the ground up: rat behavior from the ventral perspective. Psychobiology 20: 185–188.
Pinel JPJ, Treit D (1983) The conditioned defensive burying paradigm and behavioral neuroscience, In TE Robinson (ed) Behavioral contributions to brain research. Oxford University Press, Oxford, pp 212–234.
Sachs BD, Barfield RJ (1976) Functional analysis of masculine copulatory behavior in the rat. In JS Rosenblatt, RA Hinde, E Shaw, C Beer (Eds) Advances in the study of behavior. Vol 7, Academic Press: New York, 1976.
Salis M (1972) Effects of early malnutrition on the development of swimming ability in the rat. Physiol Behav 8: 119–122.
Satinoff E (1983) A reevaluation of the concept of the homeostatic organization of temperature regulation. In E Satinoff and P Teitelbaum Handbook of behavioral neurobiology Vol 6: Motivation. Plenum, New York, pp 443–467.
Schallert, T, Norton D, Jones TA (1992) A clinically relevant unilateral rat model of Parkinsonian akinesia. J Neural Transpl Plast 3: 332–333.
Schallert T, DeRyck M, Whishaw IQ, Ramirez VD, Teitelbaum P (1979) Excessive bracing reactions and their control by atropine and l-dopa in an animal analog of Parkinsonism. Exp Neurol 64: 33–43.
Schallert T, Whishaw IQ (1978) Two types of aphagia and two types of sensorimotor impairment after lateral hypothalamic lesions: Observations in normal weight, dieted, and fattened rats. J Comp Physiol Psychol, 92: 720–741.
Schallert T, Whishaw I Q (l984) Bilateral cutaneous stimulation of the somatosensory system in hemidecorticate rats. Behav Neurosci 98: 375–393.
Schechter MD, Calcagnetti DJ (1993) Trends in place preference conditioning with a cross-indexed bibliography; 1957–1991. Neurosci Biobehav Rev 17: 21–41.
Shipley J, Kolb B (1977) Neural correlates of species-typical behavior in the Syrian golden hamster. J Comp Physiol Psychol 91: 1056–1073.
Slagen JL, Earley B, Jaffard R, Richelle M, Olton DS (1990) Behavioral models of memory and amnesia. Pharmacopsychiatry Suppl 2: 81–83.
Sutherland RJ, Dyck RH (1984) Place navigation by rats in a swimming pool. Can J Psychol 38: 322–347.
Tchernichovski O, Golani I (1995) A phase plane representation of rat exploratory behavior. J Neurosci Method 62: 21–27.
Teitelbaum P, Schallert T, and Whishaw IQ (1983) Sources of spontaneity in motivated behavior, In E Satinoff and P Teitelbaum, Handbook of Behavioral Neurobiology: 6 Motivation, Plenum Press, New York, 23–61.
Vanderwolf CH, Cain DP (1994) The behavioral neurobiology of learning and memory: a conceptual reorientation. Brain Res 19: 264–297.

Upchurch M, Wehner JM (1989) Inheritance of spatial learning ability in inbred mice: A classical genetic analysis. Behav Neurosci 103: 1251–1258.
Warren JM (1977) A phylogenetic approach to learning and intelligence. In A Oliverio (Ed), Genetics, environment and intelligence. Amsterdam: Elsevier, pp. 37–56.
Warren JM, Kolb B (1978) Generalizations in neuropsychology. In S Finger (Ed) Recovery from brain damage. New York: Plenum Press, pp.36–49.
Whishaw I Q (1985) Formation of a place learning-set in the rat: A new procedure for neurobehavioural studies. Physiol Behav 35:139–143.
Whishaw IQ (1988) Food wrenching and dodging: use of action patterns for the analysis of sensorimotor and social behavior in the rat. J Neurosci Method 24: 169–178
Whishaw IQ (1995) A comparison of rats and mice in a swimming pool place task and matching to place task: Some surprising differences. Physiol Behav 58: 687–693.
Whishaw IQ, Cassel J-C, Majchrazak M, Cassel S, Will B (1994) "Short-stops" in rats with fimbria-fornix lesions: Evidence for change in the mobility gradient, Hippocampus 5: 577–582.
Whishaw IQ, Dringenberg HC, Comery TA (1992a) Rats (Rattus norvegicus) modulate eating speed and vigilance to optimize food consumption: effects of cover, circadian rhythm, food deprivation, and individual differences. J Comp Psychol 106: 411–419.
Whishaw IQ, Kolb B, Sutherland RJ (1983) The analysis of behavior in the laboratory rat. In: TE Robinson (Ed) Behavioral approaches to brain research. Oxford University Press, Oxford, pp 237–264.
Whishaw IQ, Miklyaeva E (1996) A rat's reach should exceed its grasp: Analysis of independent limb and digit use in the laboratory rat. In K-P Ossenkopp, M Kavaliers, PR Sanberg (Eds) Measuring movement and locomotion: From invertebrates to humans. RG Landes, Austin Texas, pp 135–164.
Whishaw IQ, Nonneman AJ, Kolb B (1981a) Environmental constraints on motor abilities used in grooming, swimming, and eating by decorticate rats. J Comp Physiol Psychol 95: 792–804.
Whishaw IQ, Oddie, SD, McNamara RK, Harris TW, Perry B (1990) Psychophysical methods for study of sensory-motor behavior using a food-carrying (hoarding) task in rodents. J Neurosci Method 32: 123–133.
Whishaw IQ, Pellis SM (1990) The structure of skilled forelimb reaching in the rat: a proximally driven movement with a single distal rotatory component. Behav Brain Res 41: 49–59.
Whishaw IQ, Pellis SM, Gorny BP (1992b). Skilled reaching in rats and humans: Parallel development of homology, Behav Brain Res 47: 59–70.
Whishaw IQ, Pellis SM, Gorny BP, Pellis VC (1991) The impairments in reaching and the movements of compensation in rats with motor cortex lesions: an endpoint, videorecording, and movement notation analysis. Behav Brain Res 42: 77–91.
Whishaw IQ, Pellis SM, Gorny B, Kolb B, Tetzlaff W (1993) Proximal and distal impairments in rat forelimb use in reaching follow unilateral pyramidal tract lesions, Behav Brain Res 56: 59–76.
Whishaw IQ, Sarna JR, Pellis SM (1998) Rodent-typical and species-specific limb use in eating: evidence for specialized paw use from a comparative analysis of ten species, Behav Brain Res, in press.
Whishaw, IQ, Schallert, T, Kolb, B (1981b) An analysis of feeding and sensorimotor abilities of rats after decortication. J Comp Physiol Psychol, 95: 85–103.
Whishaw IQ, Whishaw GE (1996) Conspecific aggression influences food carrying: Studies on a wild population of Rattus norvegicus. Aggres Behav 22: 47–66.
Wolgin DL, Bonner R (1985) A simple, computer-assisted system for measuring latencies of contact placing. Physiol Behav 34: 315–317.

Chapter 45

Data Acquisition, Processing and Storage

M. LJUBISAVLJEVIC and M.B. POPOVIC

Introduction

Neurophysiological *signals* are mostly recorded as potentials, voltages, currents and electromagnetic field strengths generated by nerves and muscles. They carry information needed to understand the complex mechanisms underlying the behavior of the living system. Nevertheless, such information is rarely available directly, but has to be extracted from the raw signal(s). The whole process of extraction, from a sensor to the relevant information sought, can be considered as a state-of-the-art fermentation and distillation process aiming to extract the desired properties while preserving their unique characteristics. Usually, this line of action encompasses several stages which are generally defined as signal *acquisition* and *processing*. As presented in this book, modern neuroscience heavily relies on a number of different methods extracted from other sciences.

The overall outcome may be a single number, e.g., temperature, or it can be a more complicated result, e.g., the electromyogram (EMG) from contracting muscle(s) (cf. Chapters 26–28, 31). Anyway, in most cases, the result of the analysis contains only part of the information necessary to reconstruct the complete input signal. In this sense, the complete process can be thought of as a nonreversible signal transformation where the output signal is the desired result uniquely derived from the first. In order to succeed, knowledge of the specific properties of the signals as well as adequate signal processing and system engineering knowledge are critical for all phases of the process. It is rather unfeasible to recommend a single method of acquiring and processing biological signals. There are even several standard types of procedures for acquisition and processing of the same signal type. However, although each researcher could select his own approach for data recording and processing, some general guidelines should be followed.

Therefore, the general aim of this chapter is not to cover the vast variety of acquisition and processing approaches used in modern neuroscience research. Rather, it is to stress some of the general aspects associated with data acquisition and processing of signals. Theoretical aspects will be dealt with whenever needed to better explain practical issues, referring the reader to the cited literature when a deeper insight into the various subjects is sought. In order to assist a step-by-step implementation of particular neuroscience techniques presented in this book, we have organized this chapter as independent sections, giving each researcher the freedom to roam through it according to the actual goals of his study and his imagination.

Milo Ljubisavljevi , Dr Suboti a 4, Institute for Medical Research, PO Box 102, Belgrade, 11129, Yugoslavia (phone: +381-11-685-788; fax: +381-11-643-691; e-mail: milos@imi.bg.ac.yu), National Institute for Working Life, Department of Musculoskeletal Research, BOX 7654, Umea, Sweden
Mirjana. B. Popovi , Dr Suboti a 4, Institute for Medical Research, PO Box 102, Belgrade, 11129, Yugoslavia (phone: +381-11-685-788; fax: +381-11-643-691; e-mail: mira@imi.bg.ac.yu)

Outline

The conventional path of digital signal processing from its source to the final presentation is depicted in Fig. 1.

Most measurements begin with the *transducer*, a device that converts a measurable physical quantity, such as pressure, temperature or joint rotation, into an electrical signal. Transducers are available for a wide range of measurements and different quantities, and thus come in a variety of shapes, sizes and specifications.

Signal conditioning transforms a transducer's output signal so that an analog-to-digital converter can sample the signal. On the hardware level, signal conditioning incorporates amplification, filtering, differential applications, isolations, sample and hold, current-to-voltage conversion, linearization and more. In this chapter, amplification and filtering will be dealt with in some detail.

The output of the signal-conditioning device is connected to an *analog-to-digital converter* (ADC) input. The ADC converts the analog voltage to a digital signal that is transferred to the computer for processing, graphing and storage.

The generalized instrumentation system usually includes additional *signal processing*. Traditionally, this additional processing was handled either by using relatively simple digital-electronic circuits or, if a significant amount of processing was required, by connecting the instrument to the computer. The use of microcomputers generally results in fewer integrated-circuit packages. The most useful application of microcomputers for bio-medical instrumentation involves controller functions such as the capability for self-calibration and error detection, and automatic sequencing of events. All these functions enhance the reliability of the computerized biomedical instrument.

As mentioned above, the electronic devices used for acquiring a biological signal perform some initial processing such as filtering, or perform transformations of the signal (i.e., Fourier transformation) in order to estimate various signal parameters. When processing results are not required immediately following signal acquisition, *off-line* processing or *post-processing* methods may be used. By contrast, when results are needed immediately after signal acquisition, *real-time* or *on-line* processing methods must be applied. Depending on the signal-frequency bandwidth and application, the required digital sampling rate determines the type of hardware that can be applied for digital signal processing. In real-time processing applications, the computations must be performed on a continuous basis to keep pace with the sampled input signal. In post-processing applications, the input signal is collected and stored ahead of time, and the computational rate is driven primarily by the desire to get results quickly. In both cases, computational speed is desirable. However, in the real-time case it is absolutely mandatory to accomplish the task at all. In conclusion, off-line processing can be performed on general-purpose computers and real-time processing requires special dedicated machines or processors.

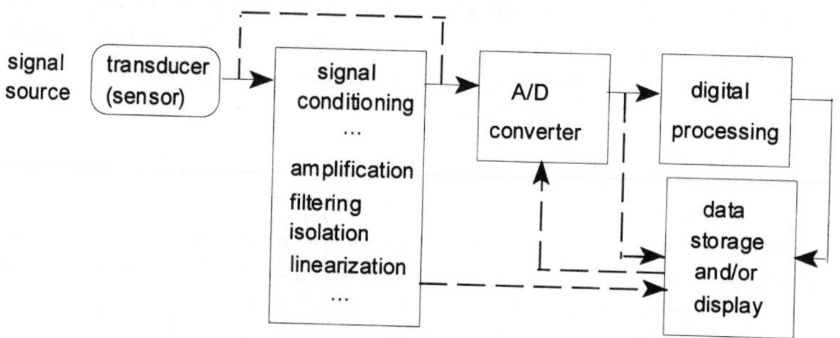

Fig. 1. General signal processing chart from the signal source to the final presentation.

In order to reveal important attributes of a signal in a more immediately interactive manner, results from the signal processing operations could be *displayed*. One or more displays allow the user to actively participate in the measurement itself. Display devices come in a wide variety depending on the use of the display. No matter what kind of display is used, its purpose is to convey information in a timely and non-permanent way with sufficiently high quality of presentation, so that the user can extract the information needed efficiently and accurately.

Frequently, there is a requirement for archiving either experimental or processed data. This can be done by various techniques and devices and is called *data storage*, or *backup* if it is a more permanent archive.

Part 1: Signal and Noise

What is a Signal?

As depicted in Fig. 1, the acquisition process starts with a signal. In general, signals are physicochemical phenomena that convey information, or they can be described as quantities that reveal the behavior of a system. As such they possess certain characteristic properties that require appropriate processing methods.

Two main types of signals are distinguished: *continuous (analog)* or *discrete*. Continuous signals are defined over a continuous range of a particular variable (usually time), while discrete signals are defined at discrete instants. Most of the signals of interest in neuroscience research are continuous, but some are discrete. In processing, continuous signals are represented as x (t), where t is time (in units of seconds), as shown in Fig. 2a. The units of x depend on what is being described; examples would be volts, amperes, or – often in signal processing – unspecified units. Discrete signals are designated by series of discrete numbers: x (k), where $1 \leq k \leq n$, as shown in Fig. 2b.

Another way to classify signals is as *deterministic* or *random* signals. Deterministic signals are those that can be described by explicit mathematical relationships. In contrast, random signals cannot be exactly expressed in that way, which is inherent in their nature. Although it might be possible to determine a mathematical relationship, we may not have all the information to describe it by an explicit equation. Random signals can be described only in terms of probabilities and statistical measures. Neurobiological signals are usually extracted from living organisms and contain various degrees of randomness. Randomness appears in neurobiological signals in two major ways: the source itself may be stochastic, or the measurement system introduces external, additive or multiplicative, noise to the signal, often because the measurement device has to be designed so as not to damage the biological system.

Measurements of any kind of signal can be classified as *static* or *dynamic*. Static measurements assume that the input is a fixed value, not changing in time: x(t)=const. Dynamic measurements assume that the value of the input fluctuates with time, so that

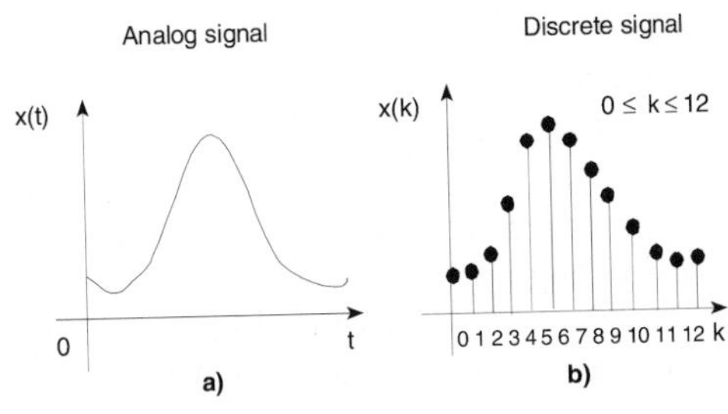

Fig. 2. Examples of a) a continuous (analog) and b) a discrete signal.

the measurement depends on the exact time at which it is made, or the input can be represented mathematically as a function of time x(t). The physical value represented by the function can be a scalar, such as pressure or temperature, or a vector such as force or velocity.

Noise

Noise is present in all signal sources and in all measuring systems. It is unavoidable in any electrical signal and affects the useful information that can be derived. Noise is crucially important when processing the low-amplitude signals which neuro-biological signals in fact are. Minimizing degradation of the desired signal by noise is of main concern in signal processing. There are no criteria for what constitutes acceptable signal amplitude or an acceptable noise level. The quality of a signal is determined by the simple ratio (*S/N*) of the amplitude of the desired signal, *S*, to the amplitude of the added noise, *N*. At what level will the *signal-to-noise ratio* begin to interfere with the analysis of the results? The answer to this question depends on both the nature of the interfering noise and the type of analysis to be performed.

A distinction is commonly made between noise of a random nature arising from basic physical processes and noise caused by interference which may or may not be correlated with the signal being measured. For example, one fundamental source of noise is the statistical variation in the electron density in a conductor and is present in all resistive elements. A frequent cause of interference noise is the electromagnetic or electrostatic interference arising from the presence of 50 or 60 Hz line current. Some signals are inherently of low amplitude and some environments are unavoidably polluted by noise sources of large amplitude. Periodic noise or pulse-like events with a pattern similar to that of the desired signal may prove confusing even when the signal being recorded tends to be 10 times greater than the amplitude of the noise. Noise that is time-locked to events under study may seriously degrade results even when it is invisible on the raw records. It interferes particularly with signal analysis by averaging (see below) or other statistical methods that are also time-locked to the event. Whenever the noise sources interact randomly with each other and with the recorded signal, the probability that the record will include an event large enough to be confused with a real signal increases with the time of recording.

Frequency Content of Noise

Generally, the noise content of a signal increases as the signal bandwidth increases, so that for dynamic signals requiring large bandwidth for adequate resolution, special care is needed in the design to insure that the measurement system has an optimum noise performance.

The frequency content of the noise can be determined by spectral analysis (see below), resulting in a spectral density graph as shown in Fig. 3. The graph is a typical example of many practical measurement systems in that it displays three distinct regions: a low-frequency region in which the mean square noise varies as 1/f, an intermediate-frequency region in which the noise spectrum is essentially flat, and a high-frequency region displaying an increasing noise spectrum. It is generally found that most components and systems display a low frequency region in which the noise varies as $1/f^n$ with n being approximately unity. Typically, an amplifier may exhibit a noise spectrum in which the 1/f noise is dominant for frequencies less than a few kHz. At intermediate frequencies the noise may be governed by thermal fluctuations in the electron density, giving rise to Johnson noise, which has a flat spectrum; noise of this type is commonly referred to as *white noise*. As the frequency approaches the cut-off frequency of the amplifier, other processes come into play and the noise amplitude generally increases.

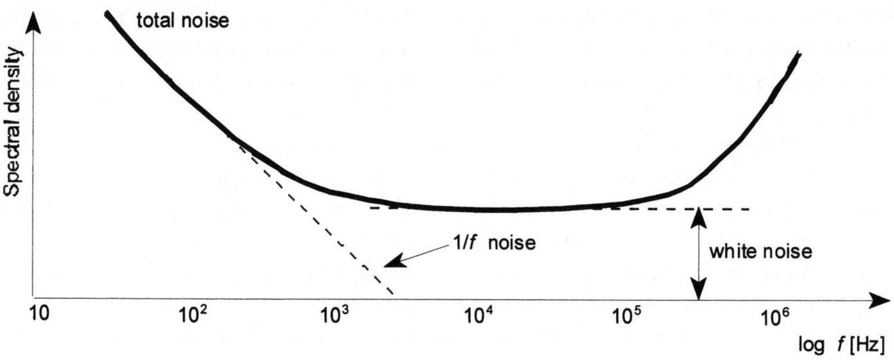

Fig. 3. Spectral density typical of a wide-band signal amplifier. Flat part comes from white noise.

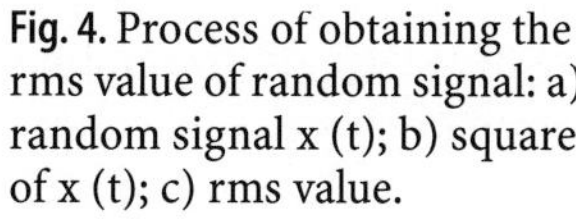
Fig. 4. Process of obtaining the rms value of random signal: a) random signal x (t); b) square of x (t); c) rms value.

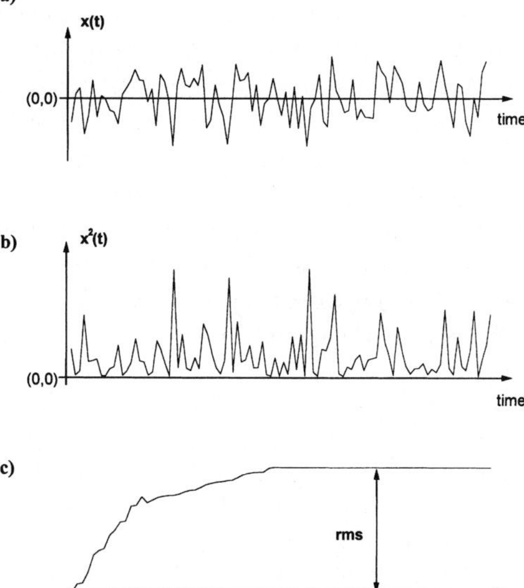

Of particular importance for the amplification of low-frequency signals or signals in which the direct-current (DC) component must be preserved is the 1/f noise. By modulating the input signal so as to convert it to a higher frequency, the amplifier can be used in a region where the 1/f component is less important. Once the signal has been amplified to such an extent that the noise generated by subsequent conditioning elements is no longer significant, the signal can be demodulated and the original, but amplified, signal can be recovered.

Rms

A consistent measure of the noise amplitude is the square root of the time-averaged (mean), squared value of the amplitude (*root mean square* or rms value). For a given waveform x (t), the rms value is defined as (Fig. 4c):

$$X_{rms} = \left(\frac{1}{T} \sum_{0}^{T} x(t)^2 \, dt \right)^{1/2}$$

By squaring the amplitude as in Fig. 4a, all negative amplitudes become positive, so that the mean value of the amplitude squared is a nonzero quantity as seen in Fig. 4b. In statistical terms, the rms is the standard deviation of the noise as seen in Fig. 4c. The rms value is the only amplitude characteristic of a waveform that does not depend on its shape. Therefore, the rms value is the most useful means to quantify signal amplitudes in alternate-current (AC) measurements. Although the rms value is a measure of the noise amplitude, it says nothing about the frequency content.

The rms value can also be computed precisely using data sampled from the original analog waveform. In this case, samples must be acquired at a rate greater than twice the highest frequency of the signal (see Sampling Theorem below). The samples are squared, the squared values are summed over some averaging interval T, the square root is taken of the sum of squared values, and this value is divided by the number of samples within T. These operations can be performed either directly on a computer or in digital signal processing (DSP) hardware. Many instruments use this sampling technique. The signal-frequency range of this technique is theoretically limited solely by available sampler and ADC rates (see below). Rms meters and measurement devices are available from several manufacturers.

Quantizing Noise In any process of analog-to-digital conversion (see below), the next step after sampling is to encode, or quantize, each sample value into a finite number of binary bits. The most common technique linearly maps a range of possible sample values $V_{p\text{-}p}$ into a fixed-size binary word of n bits. This coding requires that each sample be approximated by the nearest of 2^n possible values. The error introduced by this coding technique is a saw-tooth function. It is common practice to assume that the signal samples excide this error function so that individual sample errors can be modeled as random noise, called quantizing noise, with a uniform amplitude probability distribution. Using this assumption the rms value of the quantizing noise is $V_{2ms}^{2n} = V_{p\text{-}p} / 2^n \sqrt{12}$, where $V_{p\text{-}p}$ is the peak-to-peak full-scale range of the quantizer.

Eliminating Noise The quantification of noise and signal-to-noise ratio are a science in itself; the experimentalist primarily needs methods for recognizing the types of noise often encountered during specific recording and for minimizing their amplitude and effect. Obviously, the identification of the noise source will dictate the range of measures that will be effective and feasible. Here we summarize the most common reasons of low signal-to-noise ratios: electrode design, the first stage of the preamplifier, ground circuits and shielding. Unnecessary equipment that is not switched off, as well as AC power, are also noise sources. Interference can usually be reduced and often virtually eliminated by careful electrostatic and magnetic shielding; however, noise arising from basic physical phenomena usually sets a fundamental limit to the precision with which a given measurement can be performed.

Filtering Whereas filtering (for more details see below) is an excellent way to minimize noise and enhance the biological component of a noisy signal, it can also distort noise to the point that it looks entirely physiological. For that reason the raw, unfiltered signal coming from a preamplifier should be inspected regularly with a high-accuracy display, such as an oscilloscope with a fast sweep speed. Inspection will reveal whether unphysiologically fast, large noise spikes have been reduced and smoothed into recorded signal. Almost all noise sources contain a wide range of frequencies in their spectra, which means that any filter distorts the shape as well as attenuates the amplitude.

Averaging If the noise is uncorrelated with the signal, very significant reductions in the noise content of repetitive signals can be achieved by using an averaging process. The conditions and details are dealt with below.

Blanking In certain applications, especially those employing electrical stimulation with a high-voltage stimulator, relatively huge transients may be superimposed on the signal, causing artifacts that often drive amplifiers into saturation. The recorded signal may be corrupted and the valuable data lost. Typical examples would be extracellular recordings of nerve impulses (cf. Chapter 5) evoked by high-voltage stimulators, stimulus artifacts in EMG recordings (cf. Chapters 26–28, 31) due to high-voltage stimulation during functional electrical stimulation or currents saturating EMG amplifiers due to magnetic stimulation. The solution to overcome these problems is either to prevent the stimulus coupling or to suppress the artifact before it feeds into the AC-coupled amplifier. This is usually referred to as *blanking*. Blanking could be realized in various ways, one of the most often used is to provide a sample-and-hold function early in the signal pathway. In this case, the input of the amplifier is going to be forced into the halt mode, causing it to ignore the incoming signals during the period of the transient. The output of the amplifier will at the same time continue to supply the signal equal to the that just preceding the transient.

Random Number Generators

In some cases, random number generators are used to simulate the effect of noise-like signals and other random phenomena encountered in the physical world of signals. Such noise is present in electronic equipment and systems under measurement. Its presence usually limits our ability to communicate over large distances and to detect relatively weak signals. By generating such noise on a computer, we are able to study its effects through simulation of communication systems, and to assess the performance of such systems in the presence of noise.

Most computer software libraries include a uniform random number generator. The output of the random number generator is a random variable, and is in the range [0,1] with equal probability. For all practical purposes, the number of outputs is sufficiently large, so that it can justify the assumption that any value in the interval is a possible output from the generator. Noise encountered in physical systems is often characterized by the normal or Gaussian probability distributions. Again, different mathematical methods could be employed on the computer in order to obtain such probability distributions.

Part 2: Signal Conditioning

Amplification and Amplifiers

The main reason for neurophysiological signals to be amplified is the need to make them suitable for signal-processing hardware whose proper functioning depends on the signal being within a certain amplitude range. As already mentioned, many measurements of neurophysiological signals involve voltages at very low levels, typically ranging between 1 μV and 100 mV, superimposed with noise and interference from different sources. Amplifiers are commonly used to couple these low-level biopotentials from high impedance sources to make them compatible with devices such as recorders, displays and A/D converters for computerized equipment.

Amplifiers adequate for measuring neurophysiological signals have to satisfy several very specific requirements. They have to provide amplification specific to the signal, reject superimposed noise and interference signals, and guarantee protection against damage from voltage and current surges for patient and animal and electronic equipment. Amplifiers featuring these specifications are known as biopotential amplifiers.

Differential Amplifiers

The input signal to the amplifier consists of several components: the desired biopotential, undesired biopotentials, a power-line interference of 50 (60) Hz and its harmonics,

Fig. 5. Typical configuration for the measurement of biopotentials. Z_1 and Z_2 are source impedances from measured biological signal V_{biol}. V_c provides reference potential for the amplifier.

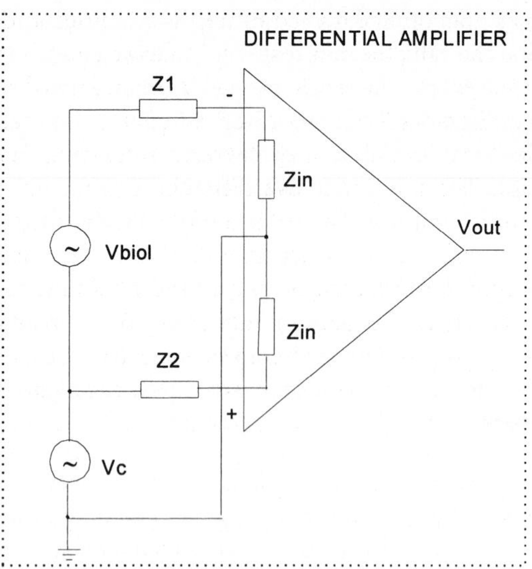

interference signals generated by the transducer (i.e., tissue/electrode interface), and noise. The differential measuring technique is used in biopotential amplifiers to minimize interference and noise, which are usually present in these low-level signals. Proper design of the amplifier provides rejection of a large portion of interference.

A typical configuration for the measurement of biopotentials is shown in Fig. 5. Three electrodes, two of them collecting the biological signal V_{biol} and the third providing the reference potential V_c, connect the subject to the amplifier. The output of the differential amplifier is, therefore, the difference between the two input signals times a certain gain factor. The desired biopotential appears as a voltage between the two input terminals of the differential amplifier and is referred to as the *differential signal.* The signal that appears between inputs and ground is called the *common-mode signal.* Thus, any common signal applied to both inputs (i.e., any common-mode signal) should result in zero output. In practice, the gains of the two signal paths are slightly different, resulting in a small output voltage even when identical voltages are applied to the two inputs. The line frequency interference signal shows only very small differences in amplitude and phase between the two measuring electrodes, causing approximately the same potential at both inputs. Differential voltage measurement eliminates common-mode noise, thus reducing the noise in analog input signals.

The ratio of the output voltage to the common input voltage is the *common-mode gain (G_{CM})*, which is usually much less than one. The ratio of the output voltage to the applied differential input voltage is the *differential gain (G_D)* of the bioamplifier and is usually much larger than one.

An index of how closely the bioamplifier approaches the ideal differential amplifier is given by the *common-mode rejection ratio (CMRR),* which is the ratio of the amplifier's differential gain to the common-mode gain. This value is expressed in decibels (dB), and is a function of frequency and source-impedance unbalance. Strong rejection of the common-mode signal is one of the most important characteristics of a good biopotential amplifier. It should be in order of at least 100 dB. Rejection of the common-mode signal is a function of both the amplifier CMRR and the source impedances Z_1 and Z_2. For the ideal amplifier CMRR is infinite and Z_1=Z_2 (Fig. 5), leaving the output voltage as the pure biological signal amplified by the differential mode gain. With CMRR finite or even slightly different Z_1 and Z_2, the common-mode signal is not completely eliminated, adding the interference term. The common-mode signal causes currents to flow through Z_1 and Z_2. The related voltage is amplified and thus not rejected by the amplifier. The output of a real amplifier will always consist of the desired output component resulting from the differential biosignal, an undesired component due to incomplete rejection of common-mode interference signals as a function of CMRR, and an undesired component due to source impedance imbalance allowing a small proportion of common-mode signal to appear as a differential signal to the amplifier.

In order to achieve optimum signal quality, the biopotential amplifier has to be adapted to the specific application. Obviously, each particular application for a biopotential amplifier will have a unique solution. Some of the amplifiers need to be extremely fast (e.g., when measuring action potentials from nerve cells; cf. Chapter 5), or have extremely high gain (e.g., when measuring EEG; cf. Chapter 35), or be extremely quiet (e.g., when measuring random noise in biological processes), etc. Based on the signal parameters, both appropriate bandwidths and gain factors are chosen. A final requirement for biopotential amplifiers is calibration. Since the amplitude of the biopotential often has to be determined very accurately, there must be a way to easily determine the gain or the amplitude range referenced to the input of the amplifier. For this purpose, the gain of the amplifier must be well calibrated. In order to prevent difficulties with calibrations, some amplifiers that need to have adjustable gains use a number of fixed gain settings rather than providing a continuous gain control. Some amplifiers have a standard, built-in signal source of known amplitude that can be momentarily connected to the input by the push of a button to check the calibration at the output of the biopotential amplifier.

Instrumentation Amplifiers (IA)

As mentioned, a very important stage in analog processing hardware is the amplification block. This stage is called an instrumentation amplifier and has several important functions: signal voltage amplification, rejection of the common-mode signal and proper driving of A/D converter input. Crucial to the performance of the preamplifier is the input impedance, which should be as high as possible. The input stage of an instrumentation amplifier usually consists of two voltage followers, which have the highest input impedance of any common amplifier configuration. A standard single operational amplifier (op-amp) design does not provide the necessary high input impedance, but two input op-amps provide high differential gain and unity common-mode gain without the requirement of close resistor matching (Fig. 6). The differential output of the first stage represents a signal with substantial relative reduction of the common-mode signal and is used to drive a standard differential amplifier, which further reduces the common-mode signal. Complete instrumentation amplifier-integrated circuits based on this standard instrumentation amplifier configuration are available from several manufacturers. All components, except the resistor that determines the gain of the amplifier, are contained on the integrated chip.

The output of the instrumentation amplifier has low impedance, which is ideal for driving the A/D converter input. The typical A/D converter does not have high or constant input impedance. It is important for the preceding stage to provide a signal with the lowest impedance practical. Instrumentation amplifiers have some *limitations*, including offset voltage, gain error, limited bandwidth and settling time. The offset voltage and gain error can be calibrated out as part of the measurement, but the bandwidth and settling time are parameters that limit the frequencies of amplified signals.

There is a special class of instrumentation amplifiers with programmable gain which can switch between fixed gain levels at sufficiently high speeds to allow different gains for different input signals delivered by the acquisition input system. These amplifiers are called *programmable gain instrumentation amplifiers* (PGIA).

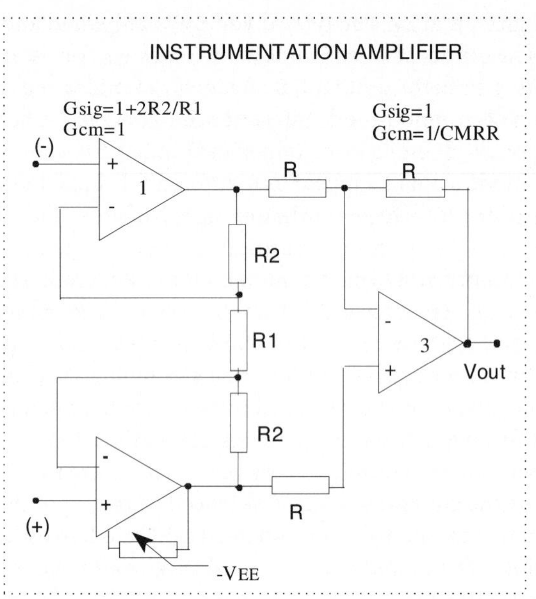

Fig. 6. Typical configuration of an instrumentation amplifier for biological application; an op-amp version.

Isolation Amplifiers

Biopotential amplifiers have to provide sufficient protection against electric shock to the user, patient and animal or other preparation. Electric safety codes and standards specify the minimum safety requirements for the equipment. To that end, isolation amplifiers may be used to break ground loops, to eliminate source-ground connections, and to provide isolation of patient and electronic equipment. They also contribute to preventing line frequency interferences. Isolation amplifiers are realized in three different technologies: transformer isolation, capacitor isolation and opto-isolation. An isolation barrier provides a complete galvanic separation of the input side, i.e., patient and preamplifier, from all equipment on the output side. Ideally, there should be no flow of electric current across the barrier. An index is the *isolation-mode voltage*, which is the voltage appearing across the isolation barrier, i.e., between input common and output common. The *isolation mode rejection ratio (IMRR)* is the ratio between the isolation voltage and the amplitude of the isolation signal appearing at the output of the isolation amplifier. Since the isolation mode rejection ratio is not infinite, there is always some leakage across the isolation barrier. Two isolation voltages are specified for commercial isolation amplifiers: the continuous rating and the test voltage. To eliminate the need for lengthy testing, the device is tested at about two times the rated continuous voltage.

Dynamic Range

Practical amplifiers introduce errors into the signal. Random noise is added by mechanisms such as thermal noise in resistors or shot noise in transistor junctions. Another problem is nonlinearity introduced into transistor junctions. To minimize the effect of added noise, it is desirable to keep the analog signal levels large compared with the noise sources. On the other hand, to minimize the distortion effects of nonlinearities, it is desirable to operate with low signal levels. The range of signal levels that can be processed between the lower limit imposed by the noise and the upper limit imposed by distortion is called the dynamic range of an amplifier. Expressed in decibels, a dynamic range up to 200 dB is achievable.

Additive noise introduced in amplifiers tends to be "white", meaning that the noise power is distributed evenly over a wide frequency range. Thus the noise density per Hz is constant over the bandwidth of the amplifier. A common basis of amplifier comparison is to specify the noise density in decibels relative to one milliwatt (dBm) per Hz.

Single-Ended and Differential Measurements

Data acquisition systems have provisions for single-ended and differential input connections. The essential difference between the two is the choice of the analog common connection. Single-ended multi-channel measurements require that all voltages be referenced to the same common node, which will result in measurement errors unless the common point is very carefully chosen; sometimes there is no acceptable common point. Differential connections cancel common-mode voltages and allow measurement of the difference between the two connected points. When given the choice, a differential measurement is always better. The rejected common-mode voltages can be steady DC levels or noise spikes. The best reason for choosing single-ended measurement will be for a higher channel count that is available in some devices. Most data acquisition products allow doubling the number of channels in a differential system by selecting single-ended operation.

Fundamentals of Filtering and Filters

Filtering is a signal processing operation that alters the frequency content of a signal. A *filter* is a frequency-selective device or a computer program that passes signals in one band of frequencies and rejects (or attenuates) signals in other bands. In signal processing, spectral components containing information on the desired signal are of foremost interest, and filters are designed so as to pass those spectral components while rejecting or attenuating components consisting mostly of noise. To design a satisfactory filtering device or program, it is necessary to have some knowledge of the structure of both the signal and the noise. Filtering is mostly performed to enhance the signal-to-noise ratio (S/R), eliminate certain types of noise or smooth the signal.

Frequency Response

The basic feature that determines filter behavior is its frequency response. The frequency response is a complex function which includes both *gain* and *phase* information. The gain and phase responses show how the filter alters the amplitude and phase of a sinusoidal input signal to produce a sinusoidal output signal. Since these two characteristics depend on the frequency content of the input signal, they can be used to describe the frequency response of the filter. The gain (dimensionless) and the phase (in degrees or radians) are mostly plotted versus frequency yielding frequency-response plots as exemplified for low-pass filters in Fig. 7.

Since the frequency ranges of interest often span several orders of magnitude, a logarithmic frequency scale compresses the data range, highlighting important features in the gain and phase responses. Special terminology involved with the use of logarithmic frequency intervals is *octave* and *decade*. The octave is the frequency range whose end points have a ratio of 2:1, while the one with a 10:1 ratio is called decade.

In signal processing, most often variables are not an issue, but the relation of two of the same kind. Furthermore, to simplify the mathematics that follows processing, it is

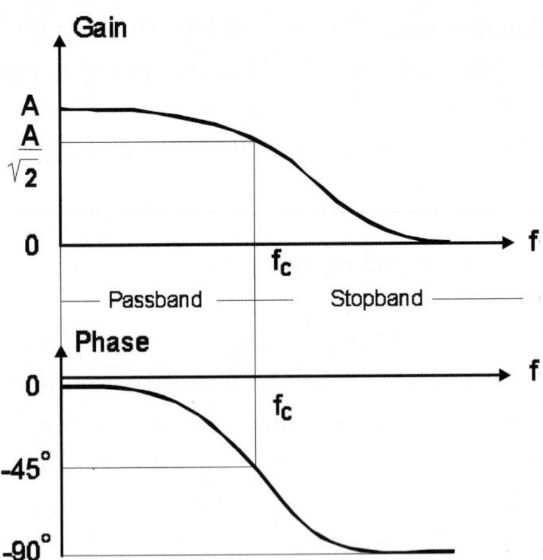

Fig. 7. Frequency response plots: a) gain; and b) phase.

useful to take logarithmic measure of the relation. In practice, variables are expressed through the measure's power or amplitudes. A relative (dimensionless) logarithmic measure of signal amplitudes is the *decibel (dB) scale*. It relates two signal powers, P_1 and P_2 or two corresponding signal amplitudes, A_1 and A_2. Since power is proportional to amplitude squared, the definition is:

$$10\log_{10}(P_1/P_2) = 10\log_{10}(A_1/A_2)^2 = 20\log_{10}(A_1/A_2)\ [\text{dB}].$$

When describing the amplitude response of filters, the term *attenuation* is often preferred to gain because many filters have a maximum gain of unity. Attenuation is the reciprocal of the gain. For example, if the gain is 0.1, then the attenuation is 10.

Passband and Stopband

The frequency range over which there is a little attenuation is called the passband. The range of frequencies over which the output is significantly attenuated is called the stopband. For example, at high frequencies the gain in Fig. 7 falls off, so that output signals at these frequencies are reduced in amplitude. At low frequencies the gain is essentially constant and there is relatively little attenuation.

Ideally, all the power would pass through the filter in the passband and no power would pass through in the stopband. In practical filters, however, between 80 and 95% of the input power shows up at the output in the passband and some small amount appears at output in the stopband. A real filter must have a *transition* zone between the stopband and passband (Fig. 8). It is defined by the cut-off frequency f_c and stopband frequency f_s. For the high-pass filter in Fig. 8. everything above f_c is passed; everything below f_s is stopped.

Cut-off Frequency

The frequency where the response starts to fall off significantly is called cut-off frequency (f_c). The term is closely associated with the boundary between a passband and adjacent stopband. It is defined as that frequency at which the signal voltage at the output of the filter falls to $1/\sqrt{2}$ of the amplitude of the input signal. Since the fall of $1/\sqrt{2}$ corresponds to -3 dB, this frequency is often called -3 dB frequency, or f_{-3}. Equivalently, the cut-off frequency (f_c) is the frequency at which the signal power at the output of the filter falls to half of the power of the input signal, which yields another term for cut-off frequency: half-power frequency.

Passband Ripple (Passband Flattens)

Real filters cannot be perfectly flat in the passband and stopband, though depending on the type of filter used, the amount of unevenness can vary. The amplitude variation

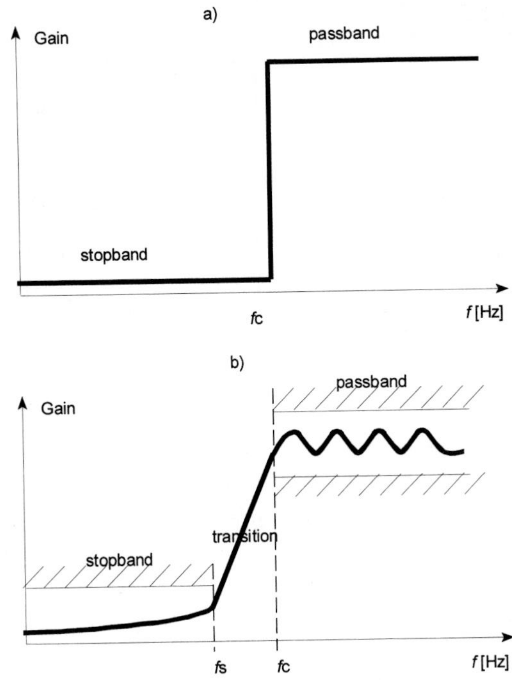

Fig. 8. Ideal *vs.* real high-pass filter magnitude response; a) and b), respectively.

within the passband is called the passband ripple (usually expressed in decibels) (Fig. 8). Some filters also have a predefined stopband ripple.

Phase Shift

The filter shifts the phase of sinusoidal components of the input signal as a function of frequency. If the phase shift in the passband is linearly dependent on the frequency of the sinusoidal components, the distortion of the signal waveform is minimal. When the phase shift in the passband is not linearly dependent on the frequency of the sinusoidal component, the filtered signal generally exhibits *overshoot*. That is, the initial response to a step input transiently exceeds the final value.

10–90% Rise Time

This term refers to the time it takes for a signal to rise from 10% of its initial value to 90% of its final value. As a general rule, when a signal with $t_{10-90} = t_s$ is passed through a filter with $t_{10-90} = t_f$, the rise time of the filtered signal is approximately $\sqrt{t_s^2 + t_f^2}$.

Filter Types

Four common types of filters can be distinguished in respect to the bandwidth: low-pass, high-pass, band-pass, and band-stop or band-reject. An illustration of how the amplitude of an input signal is altered by each of the four filter responses is shown in Fig. 9. For simplicity, consider an input signal consisting of three separate frequency components: f_1, f_2 and f_3. The cut-off frequency is designated by f_c for low-pass and high-pass filters, and by f_{c1} and f_{c2} for band-pass and band-stop filters. The low-pass filter passes frequency f_1 below its cut-off frequency f_c, and attenuates the frequencies above it: f_2 and f_3. A high-pass filter attenuates frequency f_1 below the cut-off frequency f_c and passes the frequencies above it: f_2 and f_3. A band-pass filter attenuates frequency components f_2 and f_3 outside the bandwidth determined by cut-off frequencies f_{c1} and f_{c2} and passes frequency component f_2 inside the same bandwidth. A band-reject (notch) filter does the opposite. Band-pass and band-reject filters can simply be thought

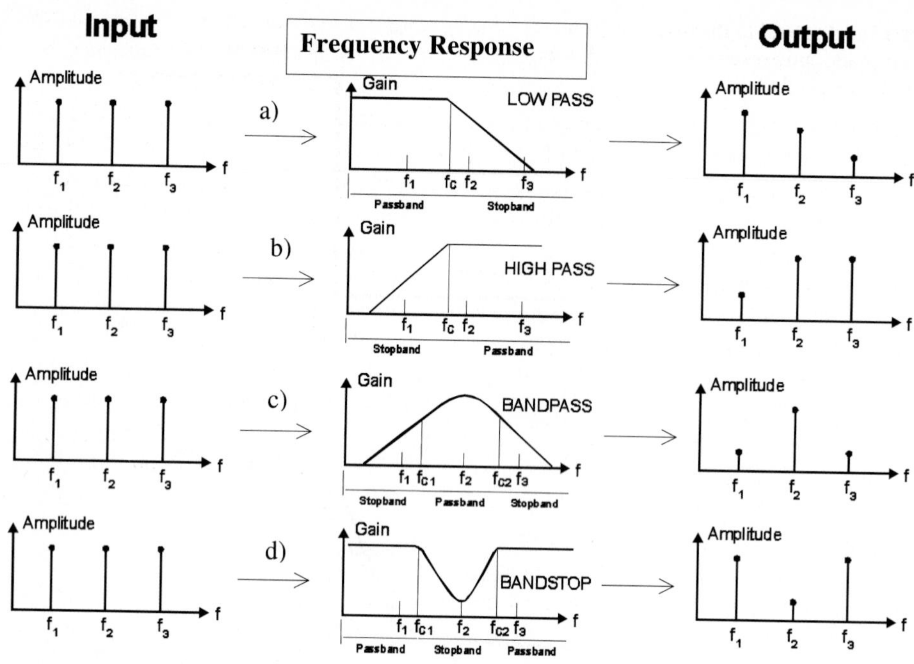

Fig. 9. Frequency responses for a) low pass; b) high pass; band pass; and band stop filters. (See text for details).

of as a series cascade of high-pass and low-pass filters, or as a parallel combination of high-pass and low-pass filters, respectively.

Order The order of a filter controls the intensity of its falloff with frequency. The higher the order of a filter, the better its performance, meaning the less the attenuation in the passband and the more complete the rejection of out-of-band signal components. The order is often described as the slope of the attenuation in the stopband, well above cut-off frequency, so that the slope of the attenuation has approached its asymptotic value. For instance, a first-order low-pass filter falls off near-linearly with frequency for high frequencies, at a rate of 6 dB per octave. A first-order filter can be constructed from one resistor and one capacitor. It is also known as a simple filter or a single-pole filter (see Analog Filter in Glossary, n=1). The order of the filter relates to the number of poles and the slope in the following way: 1 pole = 1st order = 6 dB/octave (=20 dB/decade); 2 poles = 2nd order = 12 dB/octave (= 40 dB/decade). In other words, voltage attenuation increases by a factor of 2 for each doubling of the frequency (octave) or by 10 for each tenfold increase in frequency (decade).

Filtering in Time and Frequency Domains

Neuropysiological signals can be processed either in the time domain or the frequency domain. Time-domain analysis refers to the analysis of a signal that is represented the same way it would appear on a conventional oscilloscope. Frequency-domain analysis refers to the analysis of signals that are transformed into the frequency domain (see below). To achieve optimal analysis, specific filtering should be applied.

For time-domain analysis, the filter should optimally minimally distort the time course of the signal. For example, it would not be very helpful to implement a very effective filter for elimination of high-frequency noise if it causes 15% overshoot. In general, the best filters to use for time-domain analysis are Bessel filters. They add less than 1% overshoot to pulses and have the linear change of phase with changing frequency. Because they do not alter the phase of the sinusoidal component of the signal, Bessel filters are sometimes called "linear-phase" or "constant-delay" filters. Unlike Bessel filters, Butterworth filters add considerable overshoot. In many experiments in neurophysiology, the signal noise increases rapidly with bandwidth. Therefore, a single-pole filter is inadequate. The four-pole Bessel filter is usually sufficient, but high-order filters are preferred. In experiments where the noise-power spectral density is constant with bandwidth, a single-pole filter is sometimes considered adequate. In the time domain, notch filters must be used with caution because signals that include sinusoidal components at the notch-filter frequency will be grossly distorted. On the other hand, if the notch filter is in series with low-pass or high-pass filters that exclude the notch frequency, distortion will be prevented. For example, notch filters are often used in electromyogram recordings, in which the line-frequency pickup is sometimes much larger than the signal. The 50 (60) Hz notch filter is typically followed by a 300 Hz high-pass filter.

Frequency-domain analysis is typically achieved using a fast Fourier transform (FFT). For this type of analysis the most important requirement is to have a sharp filter cut-off so that the noise above –3 dB frequency does not get folded back into the frequency of interest by the aliasing phenomenon (see below). The simplest and most commonly used filter for frequency-domain analysis in biological applications is the Butterworth filter. This filter type has the attenuation in the passband as flat as possible without having passband ripple. This means that the frequency spectrum is minimally distorted. Notch filters can be safely used in conjunction with frequency-domain analysis since they simply remove a narrow section of the power spectrum. Another approach could be to store "raw data" and then digitally remove the disturbing frequency components from the power spectrum.

Implementation

Basically, two filtering methods are available: special-purpose hardware designed for each filter structure or a general-purpose computer system with special-purpose software.

Hardware Filters

Filtering via hardware is performed by electrical circuits usually comprising capacitors, resistors, inductors and operational amplifiers which will allow one range of frequencies to pass through it and will block another range. The main drawback of using hardware for filtering is that each filter requires its own specific components, and a filter designed for one particular application cannot be used for another. Analog filters are designed as *active* or *passive*. Active filters arose from the need for filters that are compatible with modern integrated circuitry (IC) technology. They have become very attractive not only because passive filters are much more difficult to implement in IC technology, but also because active filters have other advantages over passive filters. Active filters offer higher sensitivity and flexibility, the ease of tuning and power gain. They also outperform passive filters in costs. There are many possible frequency responses that can be implemented by active filters. The most common filters are: Elliptic, Cauer, Chebyshev, Bessel and Butterworth.

Digital Filters

The alternative approach is to use general-purpose digital hardware (i.e., a computer) and implement the filter algorithm in the form of software. Software design can be accomplished via "high- or low-level" languages. This has the advantage of flexibility:

software designed for one application can easily be adapted to another, but has the disadvantage of being slower than special-purpose hardware. Digital filters are linear discrete systems governed by differential equations implemented in software. They consist of a series of mathematical calculations that process digitized data. Digital filter algorithms can implement all of the already mentioned filter frequency responses and more. Digital filtering is accomplished in three steps. First, the signal must be Fourier transformed (see below). Then, the signal's amplitude in the frequency domain must be multiplied by the desired frequency response. Finally, the transferred signal must be inverse-Fourier transformed back into the time domain. There are two types of discrete-time filters: *FIR* (Finite Impulse Response) and *IIR* (Infinite Impulse Response). Each filter type has its own set of advantages. The choice between FIR and IIR filters depends on the importance of these advantages to the design problem. If phase considerations are put aside, it is generally true that a given magnitude response specification can be met most efficiently with an IIR filter. In contrast, FIR filters can have precisely linear phase.

Finite Impulse Response Filters

FIR filters are almost entirely restricted to discrete implementations. Design techniques for FIR filters are based on directly approximating the desired frequency response of the discrete-time system. Most techniques for approximating the magnitude response of an FIR system assume a linear phase constraint. These filters have the advantage of not altering the phase of the signal. FIR filters are also known as nonrecursive filters. The output of a nonrecursive filter depends only on the input data. There is no dependence on the history of previous outputs. An example is the smoothing filter. Another example of a nonrecursive digital filter is the Gaussian filter. It is similar to a smoothing filter, except that the magnitudes of coefficients stay on the bell-shaped Gaussian curve.

Infinite Impulse Response Filters

IIR filters, unlike FIR filters, do not exhibit linear phase. However, the IIR design often results in filters with fewer coefficients than an equivalent FIR design. These are also known as recursive filters. The output of a recursive filter depends not only on the inputs, but on the previous outputs as well. That is, the filter has some time-dependent "memory". Digital filter implementations of analog filters such as Bessel, Butterworth, Elliptic and Chebyshev filters are recursive.

Digital over Analog Filtering?

Digital filtering is advantageous because the filter itself can be tailored to any frequency response without introducing the phase error. In contrast, analog filters are only available with a few frequency response curves, and all introduce some element of phase error. The delay introduced by analog filters necessarily makes recorded events occur later than they actually occurred. If it is not accounted for, this added delay can introduce an error in subsequent data analysis. A drawback of digital filtering is that it cannot be used for anti-aliasing because it occurs after sampling. Another problem with digital filters is that values near the beginning and end of the data cannot be properly computed. This is only a problem for a short data sequence. Adding values outside the sequence of data is arbitrary and can lead to misleading results.

Optimal and Adaptive Filtering

When the signal and noise are stationary and their characteristics are approximately known or can be assumed, an optimal filter can be designed *a priori*. Wiener and matched filters belong to this group. When no *a priori* information on the signal or noise is available, or when the signal or noise is non-stationary, *a priori* optimal filter design is not possible. Adaptive optimal filters are filters that can automatically adjust their own parameters, based on the incoming signal. The adaptation process is conducted so that a given performance index is optimized. Adaptive filters thus require little or no *a priori* knowledge of the signal and noise. Least-mean-square (LMS) filters belong to this group. The adaptive filter is required to perform calculations to satisfy the performance index and must have provision for changing its own parameters. Digital techniques, with or without a computing device, have clear advantages here over analog techniques. It is mainly for this reason that most adaptive filter implementations are performed by discrete systems.

Decimation Filtering

Often the information of interest in a signal is contained only in the low-frequency range, and the upper frequencies are not of interest. By filtering out the unwanted high frequencies and thereby narrowing the signal bandwidth, it is possible to meet the Nyquist sampling criterion with a lower sampling rate. If the bandwidth is reduced by a factor of k, the remaining signal can be fully described by saving every k-th sample and discarding the samples in between. This is called decimation by factor k, and the resulting output sample rate is f_s/k.

Part 3: Analog-to-Digital Conversion (Digitization)

Digital or Analog Processing?

Modern digital technology, both in terms of hardware and software, makes digital processing in many cases advantageous over analog processing. It may therefore be worthwhile to convert the analog signal to a discrete one so that digital processing can be applied. The conversion is done by *analog-to-digital (A/D) conversion systems* that sample and quantitize the signal at discrete times (see below). The factors that determine whether signals are processed digitally or in analog form include signal bandwidth, flexibility, accuracy and cost.

Cost

When a signal has a broad bandwidth, analog processing is more attractive because of the cost associated with high-speed digital signal-processing hardware. Most signal processing above 100 MHz is analog, while much of the signal processing up to 10 MHz is digital.

Accuracy

Accuracy is always an issue. The simplest analog-signal processing function is to change the signal amplitude with an attenuator or amplifier. The main reason for doing this is that the proper operation of the analog-signal processing hardware depends on the signal being within a certain amplitude range. Although the ideal result of this function would be to multiply the signal by a fixed gain, amplifiers and attenuators also introduce errors into the signal.

A digital signal, in principle, assumes only one of two states or levels, either "high" (logic 1), or "low" (logic 0). These states are represented by a voltage signal that is, according to the current standards, nominally defined as either 5 or 0 Volts. Actual digital signals fluctuate over a small range near their nominal values. The acceptable level of fluctuation depends on the technology in use. In addition to the amplitude of the signal, the time behavior is important, specifically the time required for the signal to change from one state to the other. Typically, this is in the order of milliseconds to nanoseconds in today's technology and depends on the slope of rise or fall expressed in volts per second.

Multiple Functions and Flexibility

One of the primary benefits of digital signal processing is the wide range of processing functions that can be implemented. Virtually any function that can be expressed mathematically can be performed with digital processing. Analog functions, on the other hand, are limited by the available components. Some functions may be theoretically implemented in analog form, but the inability to maintain sufficient accuracy may make the function impractical. Also, random noise and component nonlinearities limit the dynamic range of analog signal processing. By contrast, digital processing can be done with arbitrary precision by representing signals with high-precision numeric data types. Also, digital processing is exactly reproducible and is stable over time, temperature changes, and other environmental conditions. Calibration procedures are not mandatory for manufacturing and maintaining digital signal processing circuits. As mentioned above, the costs of digital signal processing strongly depend on the signal bandwidth. However, the costs of programmable digital signal processing (DSP) integrated circuits are substantially less than of those of analog circuitry that can perform the same function.

A/D Conversion

Neurophysiological signals are mainly analog. To process analog signals by digital means, it is first necessary to convert them into digital form, that is, to convert them to a sequence of numbers having finite precision. This procedure is called analog-to-digital (A/D) conversion, and the corresponding devices are called A/D converters (ADCs). The analog-to-digital converter stage is the last link in the chain between the analog domain and the digitized signal path. Analog-to-digital converters have played an increasingly important role in instrumentation in recent years. The expansion of ADCs has been driven by the development of high-performance integrated circuit (IC) technology. Advanced IC technology has led to the microprocessor and fast digital signal processing capability, which are essential in providing a low-cost transformation from the raw data generated by the ADC to the measurement results sought by the user.

Sampling The first step in A/D conversion is sampling and involves time discretization of the continuous signal x_a (t) into a series of n discrete numbers $\{x_a (k), 1 \le k \le n\}$. Here we consider a band-limited analog signal x_a (t) with maximal frequency f_{max}, which is also bounded in amplitude. Also, we assume uniform sampling with a constant sampling frequency f_s, so that the signal x_a (t) is transformed into a sequence of sampled data $\{x_a(kT_s), 1 \le k \le n\}$, where $T_s = 1/f_s$ is the sampling period or sampling interval, as illustrated in Fig. 10.

Sampling Theorem In many cases of interest it is desirable to be able to reconvert the processed digital signals back into analog form, that is, to apply D/A conversion. This puts certain demands on the sampling of the original analog signal. If this signal is sampled at greater than twice the frequency of the highest-frequency component in the signal, then the original

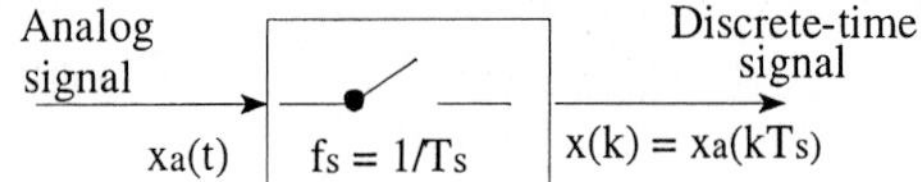

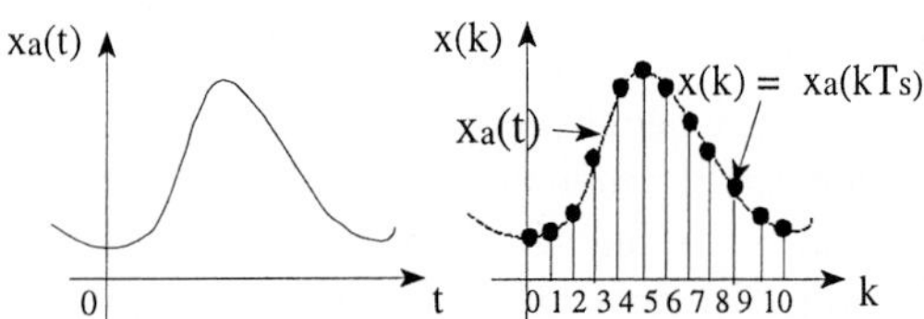

Fig. 10. Sampling of an analog signal $x_a(t)$ with sampling frequency $f_s = 1/T_s$.

signal can be reconstructed exactly from the samples. In other words, x_a (t) can be reconstructed from $\{x_a(kT_s), 1\leq k\leq n\}$ if $f_s>2 f_{max}$. This is known as Nyquist's *sampling theorem*, and $f_s/2$ is known as the *Nyquist frequency*. If the sampling frequency does not satisfy the sampling theorem, i.e., $f_s<2 f_{max}$, time discretization results in a phenomenon called *aliasing*. Any frequency above $f_s/2$ results in samples that are identical with a corresponding frequency in the range $-f_s/2 < f < f_s/2$. To avoid the ambiguities resulting from aliasing, we must select the sampling rate to be sufficiently high, i.e., $f_s>2 f_{max}$.

Aliasing

Consider the effect of sampling a signal that is not band limited to f_{max} in a case where the sampling frequency is $f_s = 2 f_{max}$. Any signal component having a frequency higher than f_{max} is folded back or falsely translated to another frequency somewhere between 0 and f_{max} by the act of sampling. Figure 11 shows an example of two cases for f_s. In Fig. 11b, the sampling frequency obeys $f_s > 2 f_{max}$, where f_{max} is the largest frequency of the signal $x_a(t)$. Note that the sampled signal in the frequency domain consists of non-overlapping functions. Consider the effect of a low-pass filter that passes all frequencies in the range $0<f<f_{max}$ undistorted, while zeroing all frequencies outside this range. The Fourier transform of the signal at the output of such a filter equals that of $x_a(t)$. Since the Fourier transform is unique, we can restore the original signal from its samples by such a low-pass filtering operation, provided the sampling frequency obeys: $f_s>2f_{max}$. In Fig. 11c, the sampling frequency is lower than twice the f_{max}: $f_s<2f_{max}$. All signal frequencies above $f_s/2$ show up as aliases or spurious lower frequency errors that cannot be distinguished from valid sampled data (black area).

Since knowledge of Fourier transform (X) implies (by Inverse Fourier transform) a knowledge of $x_a(t)$, it follows from the sampling theorem that if we sample fast enough, then we can reconstruct the original signal from its Fourier transformation.

How to Avoid Aliasing

The apparently straightforward solution to the aliasing problem is low-pass filtering of the analog signal before sampling. The filter's cut-off frequency must then be set at one-half the sampling rate or lower. Note, however, that slight fluctuations in the measured environment or the measured signal can cause alias signals to move, leaving errors in different locations throughout the data each time an A/D converter is in use. To circum-

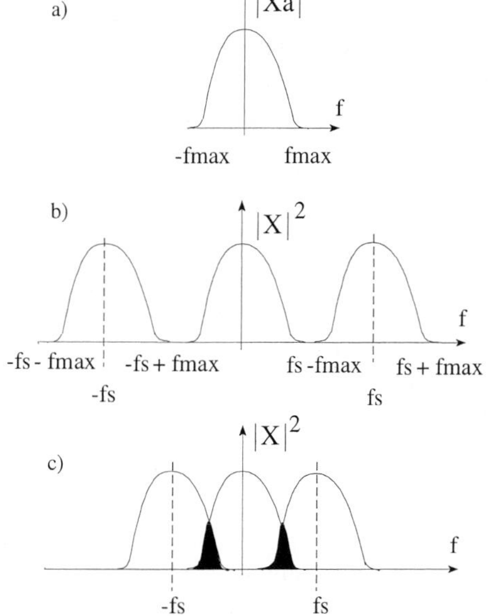

Fig. 11. Sampled band-limited signal in the frequency domain: a) Spectrum of the band-limited signal; b) Spectrum of the sampled signal when $f_s > 2f_{max}$; and c) Spectrum of the sampled signal when $f_s < 2f_{max}$. Aliasing results in overlapping spectrum regions (black area).

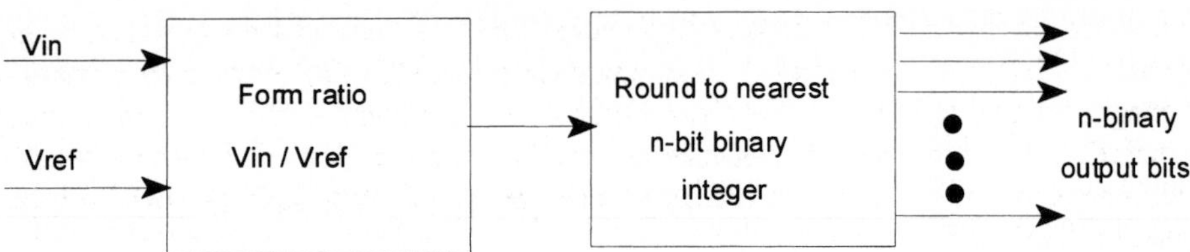

Fig. 12. Concept of the ADC: forming the ratio between V_{in} and V_{ref}, and rounding it to the nearest n-bit binary integer.

vent this problem, one could sample the analog signal at a very high rate far beyond the Nyquist frequency and then filter out the high frequencies with digital techniques. But such oversampling increases system costs by requiring faster A/D conversion for digital processing, more memory and higher bandwidth buses. It also leads to higher analysis costs by creating more data to process and interpret. Nevertheless, a practical alternative is to limit bandwidth of the signal below one-half the sample rate with a low-pass or *anti-alias filter*, which can be implemented on each channel in front of the A/D converter. Low-pass filtering must be done before the signal is sampled or multiplexed since there is no way to retrieve the original signal once it has been digitized and aliased signals have been created. The perfect reconstruction of a sampled signal would require an ideal rectangular low-pass filter, which is impossible to implement (see above). The need to use realizable filters instead leaves no choice but to sample at frequencies higher than the Nyquist rate. Sampling frequencies of 2.5 to 10 times f_{max} are often used.

Quantization Error A/D conversion is achieved in three steps: sampling, quantization and coding. In practice, A/D conversion is performed by a single device that takes an analog signal and produces a binary-coded number. The digital number represents the input voltage in discrete steps with finite resolution. A/D conversion can be viewed as forming a ratio between the input signal V_{in} and a known reference voltage V_{ref}, and then rounding the result to the nearest n-bit binary integer. The reference voltage is typically a precise value generated internally by commercial converters and sets the full-scale input range of the converter (Fig. 12). ADC *resolution* is determined by the number of bits that represent the digital number. An n-bit ADC has a resolution of 1 part in 2^n. For example, a 12-bit ADC has a resolution of 1 part in $2^{12} = 4096$. Twelve-bit ADC resolution corresponds to 2.44 mV for a 10 V range. Similarly, a 16-bit ADC resolution is 1 part in 65 536, which corresponds to 0.153 mV for a 10 V range. The rounding error is often called the *quantization error.*

Acquisition Methods Digital signal acquisition is performed in three distinct modes depending on the intended application. The three methods are: real-time sampling, sequential repetitive sampling and random repetitive sampling. The most straightforward application of digital capture technology is real-time sampling. In this method, a complete record of n samples is simultaneously captured on each and every channel in response to a single trigger event. Each waveform plotted on the display is derived entirely from the samples recorded in a single acquisition cycle and might represent the capture of a single non-repeating transient.

Implementations

Data acquisition ADCs typically run at speeds ranging from 20 kHz to 1 MHz. Many data acquisition systems have the capability of reading bipolar or unipolar voltages to the full resolution of the ADC. The unipolar-type range typically runs between 0 volts and some positive or negative voltage as V. The bipolar-type range typically runs from a negative voltage to a positive voltage of the same magnitude. Different ADC types offer varying resolution, accuracy, and speed specifications. The most popular ADC types are the parallel converter, integrating converter, voltage-to-frequency ADC and successive approximation ADC.

Parallel or Flash Converter

The parallel or flash converter is the simplest ADC implementation. It uses a reference voltage at the full scale of the input range and a voltage divider composed of 2^n+1 resistors in series, where n is the ADC resolution in bits. The value of the input voltage is determined by using a comparator at each of the 2^n reference voltages created in the voltage divider. Parallel ADCs are used in applications where very high bandwidth is required, but moderate resolution is acceptable. These applications essentially require instantaneous sampling of the input signal and high sample rates to achieve their broad bandwidth. Parallel converters are very fast because the bits are determined in parallel. Sample rates of 1 GHz have been achieved with parallel converters.

Integrating Converter

Integrating ADCs operate by integrating (averaging) the input signal over a fixed time, in order to reduce noise and eliminate interfering signals (integration corresponds to low-pass filtering with infinite time constant). To determine input voltage, integrating ADCs use a current proportional to the input voltage and measure the time it takes to charge or discharge a capacitor. This makes integrating ADCs the most suitable for digitizing signals that do not change very rapidly. The integration time is typically set to one or more periods of the local AC power line in order to eliminate noise from that source. With 50 Hz power, as in Europe, this would mean an integration time that is a multiple of 20 ms. In general, integrating converters are chosen for applications where high resolution and accuracy are important but where extraordinarily high sample rates are not. Resolution can exceed 28 bits at a few samples/s, and 16 bits at 100 ksamples/s. The disadvantage is a relatively slow conversion rate.

Voltage-to-Frequency Converter

Voltage-to-frequency ADCs convert an input voltage to an output pulse train with a frequency proportional to the input voltage. Output frequency is determined by counting pulses over a fixed time interval, and the voltage is inferred from the known relationship. Voltage-to-frequency conversion provides a high degree of noise rejection, because the input signal is effectively integrated over the counting interval. Voltage-to-frequency conversion is commonly used to convert slow and often noisy signals.

Successive Approximation Converter

Successive approximation ADCs employ a digital-to-analog converter (DAC) and a signal comparator. The converter effectively makes a bisection or binomial search by beginning with an output of zero. It provisionally sets each bit of the DAC, beginning with the most significant bit. The search compares the output of the DAC to the voltage being measured. If setting a bit to one causes the DAC output to rise above the input voltage, that bit is set to zero. Successive approximation is slower than parallel because the comparisons must be performed in a series and the ADC must pause at each step to set the DAC and wait for it to settle. Nonetheless, conversion rates over 200 kHz are common. Successive approximation is relatively inexpensive to implement for 12- and 16-bit resolution. Consequently, they are the most commonly used ADCs, and can be found in many PC-based data acquisition products.

Digital-to-Analog Converters

Digital-to-analog converters (D/A) convert a digital signal into an analog signal. The main function of D/A converters is to interpolate between discrete sample values. From a practical viewpoint, the simplest D/A converter is the zero-order hold, which simply holds constant the value of one sample until the next one is received. Additional improvement can be obtained by using linear interpolation to connect successive samples with straight-line segments. Even better interpolation can be achieved by using more sophisticated higher-order interpolation techniques. In general, suboptimum interpolation techniques result in passing frequencies above the folding frequency. Such frequency components are undesirable and are usually removed by passing the output of the interpolator through a proper analog filter, which is called a postfilter or smoothing filter. Thus D/A conversion usually involves a suboptimum interpolator followed by a postfilter.

Part 4: Data Processing and Display

Data Processing

Data processing involves a huge number of diverse techniques specifically tailored to a customer's demands. This plethora cannot be dealt with here. Instead a few issues of more general concern are briefly touched upon.

Signal Averaging

Averaging is a processing technique to increase the signal-to-noise ratio (S/N) on the basis of different statistical properties of signal and noise in those cases where the frequency content of signal and noise overlap (see above). In these cases, traditional filtering would reject signal and noise in parallel. Averaging is applicable only if signal and noise are characterized by the following properties:

- The data consist of a sequence of repetitive signals plus noise tied to a sequence of identifiable time flags.
- These signal sequences contain a consistent component x (n) that does not vary for all sequences (repetitive component of the signal).
- The superimposed noise w (n) is a broadband stationary process with zero mean.
- Signal x (n) and noise w (n) are uncorrelated, so that the recorded signal y_i (n) in the i-th signal sequence can be expressed as $y_i(n) = x_i(n) + w_i(n)$. The averaging process yields y as:

$$y(n) = \frac{1}{M}\sum_{i=1}^{M} y_i = x(n) + \sum_{i=1}^{M} w_i(n)$$

where M is the number of repetitions in the signal sequence.

If the desired signal is characterized by the above properties, then the averaging technique can satisfactorily solve the problem of separating signal from noise. Averaging is then performed in two steps: all recorded repetitions of signal + noise in a sequence are first superimposed, such that they are synchronized to the time flags, and then divided by M. Because the noise in each sequence is uncorrelated to the noise in any other sequence, the amplitude of the noise in the accumulated signal only increases by $\sqrt{M}$. After the division, the signal has a magnitude of unity compared to the noise having a magnitude of $1/\sqrt{M}$. Signal averaging thus improves the signal-to-noise ratio by a factor $\sqrt{M}$.

Although averaging is an effective technique, it suffers from several drawbacks. Noise present in measurements only decreases as the square root of the number of recorded repetitions. Therefore, a significant noise reduction requires averaging many repetitions. Also, averaging only eliminates random noise; it does not necessarily eliminate many types of system noise, such as periodic noise from switching power supplies. It is also important to remember that averaging is based on the hypothesis of a broadband distribution of the noise frequencies and the lack of correlation between signal and noise. Unfortunately, these assumptions are not always warranted for neurobiological signals. In addition, much attention must be paid to the alignment of the repetitions; slight misalignments may have a low-pass filtering effect on the final result. Still, with the easy access to A/D converters and digital computers, signal averaging is easily performed.

Fitting

Fitting a function to a set of data points may be done for any of the following reasons:

- A function may be fitted to a data set in order to describe its shape or behavior, without ascribing any biophysical meaning to the function or its parameters. This is done when a smooth curve is useful to guide the eye through the data or when a function is required to find the behavior of some data in the presence of noise.
- A theoretical function may be known to describe the data, such as a probability density function consisting of an exponential, and the fit is made only to extract the parameters. Estimates of the confidence limits on the derived parameters may be needed in order to compare data sets.
- One or more hypothetical functions might be tested with respect to the data, e.g., to decide how well the data are described by the best-fit function.

The fitting procedure begins by choosing a suitable function to describe the data. This function has a number of free parameters whose values are chosen so as to optimize the fit between the function and the data points. The set of parameters that gives the best fit is said to describe the data as long as the final fit function adequately describes the behavior of the data. Fitting is best performed by software programs. The software follows an iterative procedure to successively refine the parameter estimates according to a selected optimization criterion until no further improvement is found when the procedure is terminated. Feedback about the quality of the fit allows the model or initial parameter estimates to be adjusted manually before restarting the iterative procedure. Two aspects of fitting can be discussed: statistical and optimization.

Statistical Methods

Statistical aspects of fitting concern how good the fit is and how confident the knowledge of the fitting parameters is. They are thus concerned with the probability of occurrence of events. There are two common ways in which this word is used: direct and inverse probability. The direct probability is often expressed by the probability density function (pdf) in algebraic form. After a best fit has been obtained, the user may want to find out if the fit is good (the goodness of fit) and obtain an estimate of the confidence limits for each of the parameters.

Optimization Methods

Optimization methods are concerned with finding the minimum of an evaluation function (such as the sum of squared deviations between data values and values of the fitted function) by adjusting the parameters. A global, i.e., the absolute minimum, is clearly preferred. Since it is often difficult to know whether one has the absolute minimum, most methods settle for a local minimum, i.e., the minimum within a neighborhood of parameter values.

An Example: Linear Regression

Linear regression is the simplest fitting procedure. It determines the best linear fit to the data. Additionally, the following parameters are noted as parameter descriptions for linear regression: intercept value and its standard error, slope value and its standard error, correlation coefficient, p-value, number of data points and standard deviation of the fit. More information on fitting procedures can be found in statistical textbooks.

Frequency-Domain Analysis

Signals are most frequently given as a function of time. For many applications, it is advantageous, or even imperative, to transform the signal to an alternative, *frequency-domain* form in which the distribution of amplitudes and phase are given as a function of frequency. The design of digital signal processing algorithms and systems often starts with a frequency domain specification. In other words, it specifies which frequency ranges in an input signal are to be enhanced, and which suppressed. The low-pass, high-pass, band-pass and band-stop filters (see above) are good examples. The Fourier transform (FT) provides the mathematical basis for frequency-domain analysis. The Fourier transform is reversible, since the original signal as a function of time can be recovered from its Fourier transform. The two representations are thus related via the Fourier Transform (FT) and Inverse Fourier Transform (IFT). Not only is the Fourier transform useful for analyzing the frequency content of a signal, but it also has some properties that make it a useful intermediate step in a wide range of signal processing algorithms.

There are several major reasons for a frequency-domain approach. Sinusoidal and exponential signals take place in the natural world and in technology. Even when a signal is not of this type, it can be decomposed into component frequencies. The Fourier transformation (FT) has therefore become a basic tool in the analysis of many biological signals. The FT is also fundamental to linear systems theory in which, via the convolution theorem, the spectrum of the output is simply the product of the spectrum of the input and the frequency response function of the system under study (see above). Indeed, the first line of investigation of a biological system is often to model it as a linear system. Just as a signal can be described in the frequency domain by its spectrum, so a time-invariant system can be described by its frequency response. This indicates how each sinusoidal (or exponential) component of an input signal is modified in amplitude and phase as it passes through the system. In modeling, the response of a linear, time-invariant (LTI) processor to each such component is quite simple: it can only alter the amplitude and phase, not the frequency of that component. The overall output signal can then be found by superposition of the component responses. The product of frequency response and input signal spectrum gives the spectrum of the output signal. This process is generally simpler to perform, and to visualize, than the equivalent time-domain convolution.

The main features of the frequency-domain analysis are: a signal may always be decomposed into, or synthesized from, a set of *sine* and *cosine* components with appropriate amplitudes and frequencies; Fourier transformation of a signal provides its spectrum. A complementary process, Inverse Fourier Transformation (IFT), allows us to regenerate the original signal in the time domain. If the signal is an even function (symmetrical about the time origin), it contains only cosines. If it is an odd function (anti-symmetrical about the time origin), it contains only sines; If the signal is strictly periodic, its frequency components are related harmonically. The spectrum then has a finite number of discrete spectral lines and is called a line spectrum. It is described mathematically by a Fourier series. The trigonometric form of the Fourier series may be converted into an exponential form, by expressing each sine and cosine as a pair of imaginary exponentials. When a signal is aperiodic, it can be expressed as the infinite sum (integral) of sinusoids or exponentials, which are not related harmonically. The corresponding spectrum is continuous and is described mathematically by the Fourier transform. Approximation of the signal by a limited number of frequency components provides a best fit in the least-squares sense.

Fourier analysis is intuitively appealing in the case of long periodic signals, where there are many repetitions or cycles of some temporal pattern. However, measured biological signals, such as fast muscular movements or the underlying neuromuscular signals that drive such movements, may be single events in time, meaning that they change their behavior in a certain relatively brief interval. In general, biological data are always finite in time, having defined start and end transients. In such cases, Fourier analysis can be physiologically informative, but it is not the natural approach, and it is neither intuitive nor trivial. The Fourier spectrum of a transient depends strongly on the temporal separation and type of the edge discontinuities, and may be completely dominated by them rather than the signal during the transient.

Power Spectrum

In many applications we consider the distribution of the energy of the signal in the frequency domain, rather than the distributions of amplitude and phase. The power is proportional to the squared amplitude. Thus, when dealing with energy and power distribution, we lose information concerning the phase of the signal.

In the case when the signal is very long in duration, it is not feasible to measure the true spectrum because of the requirement to integrate over the entire signal length. It is common to approximate the spectrum in one of two ways: the short Fourier transform and the swept-spectrum measurement. In the short Fourier transform method, a segment of the signal is captured and weighted with a finite-length window function. The Fourier transform of this weighted segment is computed as an approximation of the actual spectrum. In the case of transient signals, it is sometimes possible to capture the entire signal in the short segment. By using a uniform window function in this case, the resulting spectrum is not an approximation but is the actual spectrum of the signal. The amplitude of the Fourier transform for transient signals is in units of energy per Hertz and is therefore called an *energy spectral-density function*. The integral of this energy density function over all frequencies will yield the total energy in the transient signal. An alternative analog signal-processing technique for estimating the power spectrum of a stationary signal is to filter the signal with a narrow-bandwidth filter and measure the amplitude of the filter output. By sweeping this filter across a range of frequencies, a measurement of the signal power versus frequency can be obtained. The rate of sweeping is limited by the bandwidth of the narrow-band filter. A good estimate of the maximum sweep rate is $B^2/2$, where B is the frequency bandwidth of the filter. With this sweep-rate limit, the measurement time required to produce a power spectrum is much longer than when using the FFT-based short Fourier transform technique.

Fourier Transform of Digital Signals

Fourier analysis has been applied to analog signals for almost two hundred years. Recent developments in digital processing have resulted in corresponding discrete-time (digital) techniques for analyzing the frequency components of signals, and the frequency-domain performance of systems. That is, the two Fourier representations, Fourier series and Fourier transformations, can be applied to both analog and digital signals. The Fourier transforms defined for analog signals are modified for finite-duration sampled signals. There are many similarities, as well as a few important differences, between discrete-time and continuous versions of Fourier representations. There is a third type of Fourier representation known as the *Discrete Fourier Transform (DFT)*, which is of key significance for the computer analysis of digital signals. The DFT is an important tool for discrete signal processing for the same reasons that the FT is important for continuous signal processing. The direct computation of the DFT requires approximately n^2 (n is a number of samples) complex multiplication and addition operations. Another, more efficient, method requiring only $n\log_2 n$ operations is known as *Fast Fourier Transform (FFT)*. The DFT is widely implemented using FFT algorithms. Many different FFT algorithms have been developed for software and hardware implementations. Two commonly used algorithms are known as the *decimation in time and decimation in frequency algorithms*. The popularity of the FT has grown because of the increasing availability of computer software packages that can generate DFTs at the press of a mouse button.

Digital Filters The availability of low-cost and efficient computers and dedicated processing circuits has made the implementation of digital means of filtering very attractive. Even when dealing with analog environments, where both input and output signals are continuous, it is often worthwhile to apply analog-to-digital conversion, perform the required filtering digitally, and convert the discrete filtered output back into a continuous signal.

Windowing Computing the Fourier transform of a signal involves integration over the entire duration of the non-zero portion of the signal. For signals of long duration, this can be impractical if not impossible. An alternative is to compute the transform of a finite-length segment of the signal multiplied by a "weighting" or "windowing" function. Since the Fourier transform of the product of two signals is the convolution of their individual transforms, the result is the Fourier transform of the original signal convoluted with the Fourier transform of the finite-length windowing function. By choosing a long, smooth time-domain window, its width in the frequency domain will be narrow, and little smearing will result from the convolution. Different functions produce several windows, such as Hanning, Hamming, Blackman, Bartlet, Kaiser and Tukey.

Examples of FT Applications

Example 1 A common use of Fourier transforms is to find the frequency components of a signal buried in a noisy time-domain signal. For illustration consider two frequencies of 50 Hz and 5 Hz, which are sampled at 1000 Hz, as shown in Fig. 13, upper two traces. In the middle of Fig. 13, zero-mean random signal is created by a random number generator. Two frequency components at 50 Hz and 5 Hz are then corrupted with the random signal forming the noisy signal, as shown in the second trace from the bottom. It is hard or even impossible to recognize the 50 and 5 Hz components in the noisy signal. By contrast, the power spectral density as seen at the bottom reveals strong peaks at 5 Hz and 50 Hz. The frequency content of the noisy signal is presented in the range from DC up to and including the Nyquist frequency (500 Hz).

Example 2 Most practical digital signals are aperiodic – that is, they are not strictly repetitive. For illustration consider two signals of predominantly low frequency content (Fig. 14a,b upper traces). A relevant technique to apply Fourier analysis on digital signals is the Fourier transform. There are several ways of developing the FT for a digital sequence. A common approach is via the continuous-time FT, as used in analog signal analysis. However, a digital approach is also common. The spectrum of a digital signal is always repetitive, unlike that of an analog signal. This is an inevitable consequence of sampling, and reflects the ambiguity of digital signals. It is informative enough to show one period of that repetition, as it is in the lower traces in Fig. 14 for digital signals a) and b).

Data Display

Using results from the signal processing operations, it is possible to create displays that reveal important attributes of a signal. The generation of one or more of these generic displays is often the end objective of measurement instrumentation. Display devices on

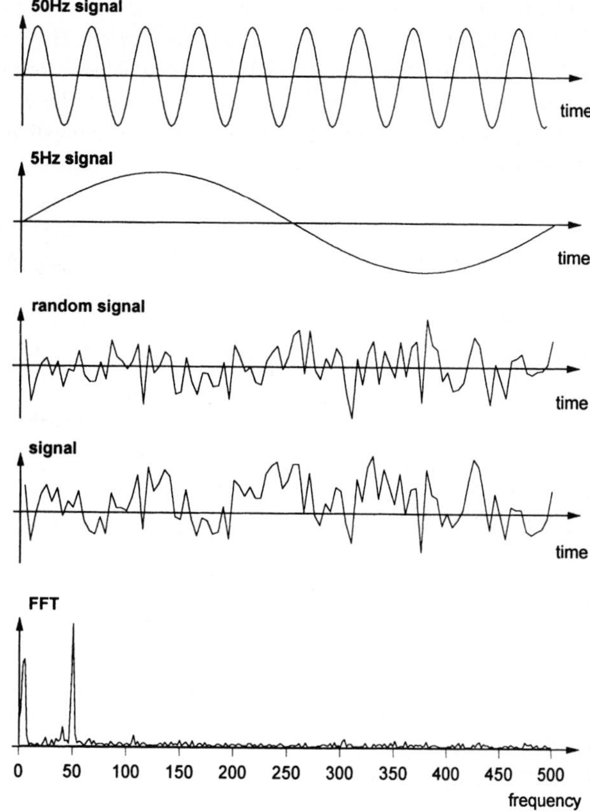

Fig. 13. An example of the use of FFT. From the top to the bottom: 50 Hz signal, 5 Hz signal, random signal, noisy signal, all presented in the time domain. At the bottom, the power spectral density clearly shows peaks at 50 and 5 Hz.

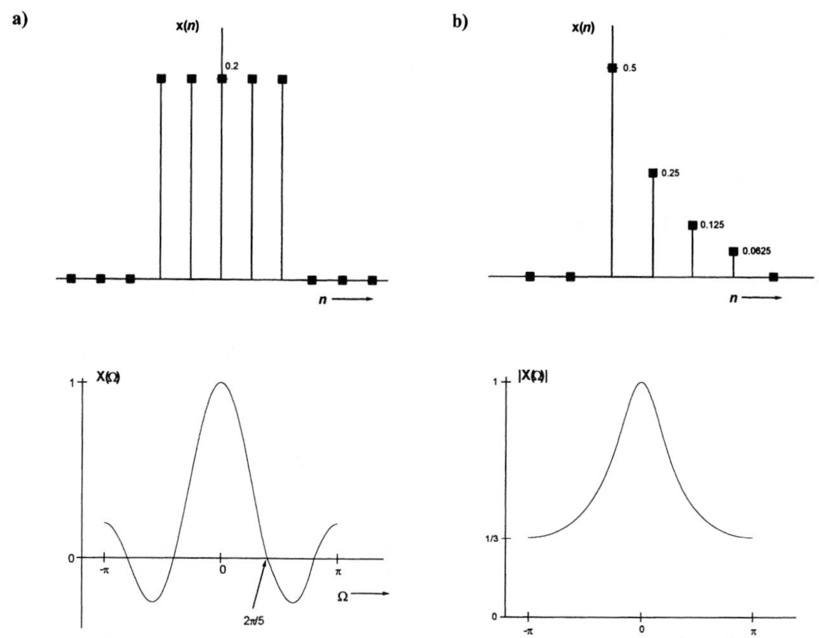

Fig. 14. Fourier transforms of the aperiodic digital signals: (a) signal defined as $x(n)=0.2\ \{\delta[n-2]+\delta[n-1]+\delta[n]+\delta[n+1]+\delta[n+2]\}$; (b) signal defined as $x(n)=0.5^{n+1}$ for $n \geq 0$ and $x(n)=0$ for $n<0$. n is the number of samples and δ is the delta function.

instruments come in a wide range and variety depending on the use of the display, but they can be grouped into three basic categories. The simple, single-purpose *indicator light* is used for conveying information that has a binary or threshold value, such as warnings, alerts and go-no-go messages. *Small alphanumeric* displays are quite common on instrument front panels. They are useful for displaying information that has relatively short messages composed of text only. The third category of display handles both *graphics and text* and is found on instruments whose measurement results are easier to interpret if the information is in a graphic form, such as an oscilloscope. Computer displays also fall in this third category, since they use graphic representations to signify instrument controls, as well as displaying results.

No matter what kind of display is used, its purpose is to convey information in a timely and non-permanent manner with sufficient quality of presentation so that the user can extract the information efficiently and accurately. In instrumentation, a display allows the user to observe measurement data in a more immediately interactive way and actively participate in the measurement itself.

A display must have sufficient quality for its purpose to allow the user to extract the information presented there efficiently and accurately. Factors affecting the image quality of a display include the following: amount of glare, resolution, design of characters, stability of the screen image, contrast in the image, color selection, and image refresh rate. The basic unit of a display is the smallest area that can be illuminated independently, called a *pixel*. The shape and number of pixels in a display are factors determining the resolution that the display can achieve. Current display technologies used in instrumentation are *cathode-ray tubes* (CRTs), *light-emitting diodes* (LEDs) and *liquid-crystal displays* (LCDs).

Cathode-ray tube display technology was the first developed. While it has some drawbacks, such as large size, weight and power consumption, in some applications it is still by far superior to the other technologies. In instruments, CRT display technology was first used in oscilloscopes, and it continues to be the best choice of display technologies for displaying information rapidly with a high graphic waveform content requiring high visibility. There are several types of CRT hardware displays in the PC world: monochrome display adapter (MDA), color graphics adapter (CGA), enhanced graphics adapter (EGA) and video graphics adapter (VGA). The different hardware implementations differ in their resolution and color capabilities, with MDA capable of displaying only monochrome text, the CGA capable of displaying graphics and text in 16 colors but at low resolution, and the EGA capable of displaying 16 colors but at higher resolution than CGA. VGA and super VGA use an analog display signal instead of the digital signal that MDA, CGA and EGA use, requiring a monitor designed specifically for use with VGA. VGA and super VGA are capable of high resolution and use 256 simultaneous colors.

Light-emitting diode display technology uses light-emitting diodes that illuminate pixels by converting electrical energy into electromagnetic radiation ranging from green to near infrared (550 to 1300 μm). LEDs in instruments are used most commonly as indicator lights and small alphanumeric displays. Relative to CRTs, LEDs are smaller, more rugged, have a longer life, and operate at a lower temperature.

Liquid-crystal display technology channels light from an outside source, such as a fluorescent lamp behind the display, through a polarizer and then through liquid crystals aligned by an electric field. The aligned crystals twist the light and pass it through the surface of the display. Those crystals which have not been aligned do not pass the light through. This creates the pattern of on/off pixels that produces the image the user sees. The advantages of LCD technology are low power consumption, physical thinness of the display, lightweight, raggedness, and good performance in bright ambient light conditions. LCD outperforms both LED and CRT technology for readability in bright light. LCDs are used for both small alphanumeric displays and larger graphics and text displays.

Part 5: Storage and Backup

Data storage refers to various techniques and devices for archiving experimental data, while backup might be regarded as a more permanent archive of experimental data by making copies on a second medium. In general two distinct approaches to archival storage can be undertaken. The first is to record data on a laboratory tape recorder (magnetic, VCR or DAT) while the second includes direct digitization and storage in the computer. Which approach to apply depends on several aspects like the type of data to be stored, the total amount of data to be stored, the necessity for on-line processing and the personal preferences of the researcher. It is also very common to combine these two methods, i.e., to record data simultaneously to the laboratory recorder and the computer. While the data recorded on the tape recorder are in a way permanently stored, data recorded on the computer hard disk are only temporarily stored and additional means of backup and archival storage must be secured. This is done as a precaution in the case the first medium fails. One of the cardinal rules in using computers is "Back up your data regularly". Since the data are already kept in binary format, it is most common to archive the data on conventional computer backup media. Choosing an appropriate device for a particular application requires understanding the ways that devices differ and weighing the trade-offs involved in using various devices. Performance and speed of backup, capacity and price per storage unit of the backup device, reliability, volatility, writable medium as well as random access should be carefully considered before making a choice.

Performance refers to the speed of a storage device and can be expressed as through-output and latency. Through-output is the rate at which a device can accomplish work, i.e., storing and retrieving data. Latency is the time it takes to do a portion of work. The ideal storage device has high through-output and low latency.

Reliability refers to the rate at which the storage device fails; this can also be inverted to express the expected time before failure. For example, the device failure rate is given as 1 error every 100 trillion accesses or 1 failure every 10 years.

Capacity refers to the amount of data that a device can store. As already mentioned, closely connected to this concept is the cost since it is usually possible to buy more devices to increase capacity.

Volatility refers to whether or not a device can retain information after power is turned off. Volatility can also be viewed as a component of reliability, since a power failure is one of the reasons for volatile devices to fail. In general, mass storage devices are non-volatile, i.e., they do not need power to store information.

Rewritable are those devices that can store new information. Almost all applications require the ability to read a storage device, but some do not require the ability to write new information to the device.

Recording and storing data on the laboratory recorder is commonly employed when dealing with continuous data such as spike train recordings or patch-clamp data. The tape recorder could be an FM recorder, a VCR recorder or a DAT recorder. Their main advantage is huge capacity, while their biggest disadvantage is that they are sequential-access devices. This means that to read any particular block of data, the tape has to be navigated to a certain position either through reading all preceding blocks of data or tape winding. This makes them relatively slow for general-purpose storage operations. Once the experiment has been completed, the experimental data stored on the tape can be transferred onto computer either directly, if previously stored in digital format on the VCR or DAT tapes, or they can be digitized through ADC conversion. This is a very flexible solution that enables additional signal conditioning.

Direct digitization and storage of experimental data is commonly employed for non-continuous, episodic data, for which storage demands are not as large as for continuous data. Nevertheless, increasing availability of DAT recorders on the market reduces certain advantages of direct digitization approaches such as higher acquisition rates.

In general, the optimal solution for data storage is arbitrary and depends on the type of research and data, amount of data, necessity for speed and on-line processing, and again on available solutions versus costs.

We will address in more detail various types of available computer mass storage mediums. Modern mass storage devices include all types of disk drives and tape drives. Mass storage is distinct from memory, which refers to temporary storage areas within the computer. Unlike main memory, mass storage devices retain data even when the computer is turned off. Mass storage is measured in kilobytes (1,024 bytes), megabytes (1,024 kilobytes), gigabytes (1,024 megabytes) and terabytes (1,024 gigabytes). It is also sometimes called *auxiliary storage*.

Backup could be archival, in which case all data are copied to a backup storage device. Archival backups are also called *full backup*. Incremental backup implies backup in which only those files that have been modified since the previous backup are copied.

The main types of mass backup media include:

- *Floppy disks*: They are relatively slow and have a small capacity, but they are portable, inexpensive and universal. Lately they are being replaced by devices with much higher capacity such as zip or jazz drives. These mediums are also portable and are relatively inexpensive. Their speed is still much lower then that of hard disks.
- *Hard disks*: Very fast and with more capacity than floppy disks, but also more expensive. Some hard disk systems are portable (removable cartridges), but most are not.
- *Optical disks*: Unlike floppy and hard disks, which use electromagnetism to encode data, optical disk systems use a laser to read and write data. Optical disks have very large storage capacity, but they are not as fast as hard disks. In addition, the inexpensive optical disk drives are read-only. Read/write varieties are more expensive.
- *Tape drives*: They are relatively inexpensive and can have very large storage capacities, but they do not permit random access of data. Their transfer speeds also vary considerably. Fast tape drives can transfer as much as 20 MB (megabytes) per minute. Tapes are usually called streamers or streaming tapes.
- *CD-ROMs*: abbreviation of Compact Disc Read Only Memory, also called a CD-ROM drive, a device that can read information from a CD-ROM. It is a type of optical disk capable of storing large amounts of data – up to 1 GB, although the most common size is 650 MB (megabytes). A single CD-ROM has the storage capacity of 700 floppy disks, enough memory to store about 300,000 text. All CD-ROMs conform to a standard size and format, so any type of CD-ROM can be loaded into any CD-ROM player. In addition, CD-ROM players are capable of playing audio CDs, which share the same technology. CD-ROMs are particularly well suited for information requiring large storage capacity.

  CD-ROM players can be either internal, in which case they fit in a bay, or external, in which case they generally connect to the computer's SCSI (Small Computer System Interface) or parallel port. Parallel CD-ROM players are easier to install, but they have several disadvantages: they are somewhat more expensive than internal players, they use the parallel port which means that another device such as a printer cannot use that port, and the parallel port itself may not be fast enough to handle all the data pouring through it.

  There are a number of features that distinguish CD-ROM players, the most important of which is probably their speed. CD-ROM players are generally classified as single-speed or some multiple of single-speed. For example, a 4x player accesses data at four times the speed of a single-speed player. Within these groups, however, there is some variation. Also, one should be aware of whether the CD-ROM uses the CLV (Constant Linear Velocity) or CAV (Constant Angular Velocity) technology. The reported speeds of players that use CAV are generally not accurate because they refer only to the access speed for outer tracks. Inner tracks are accessed more slowly. Two more precise measurements are the drive's access time and data transfer rate. The access time measures how long, on average, it takes the drive to access a particular piece of information. The data transfer rate measures how much data can be read and sent to the computer in a second. Finally, how the player connects to the computer should also be considered. Many CD-ROMs connect via an SCSI bus. If the computer doesn't already contain such an interface, it needs to be installed. Other CD-ROMs connect to an IDE (Integrated Device Electronics) or Enhanced IDE interface, which is the one used by the hard disk drive.

- *CD-R drive*: Stands for Compact Disk-Recordable drive, a type of disk drive that can create CD-ROMs and audio CDs. This allows users to "master" a CD-ROM with selected data. Until recently, CD-R drives were quite expensive, but prices have dropped dramatically. A particularly useful feature of many CD-R drives, called multisession recording, enables sequential adding of data to a CD-ROM over time. This is extremely important if you want to use the CD-R drive to create backup CD-ROMs. In order to create data archives, a CD-ROM-appropriate CD-R software package is also needed, and it is often the software package, not the drive itself, that determines how easy or difficult it is to create CD-ROMs. CD-RW (rewritable) disks are a type of CD disk that enable multiple writing sessions unlike CD-R disks that only enable sequential adding of data up to the maximum storage capacity of the disk. Therefore, CD-RW drives and disks can be treated just like a floppy or hard disk, writing data onto it multiple times. They became available in mid-1997 while their price has dropped dramatically since then.
- *DVD:* Short for Digital Versatile Disc or Digital Video Disc, a new type of CD-ROM that supports disks with capacities ranging from 4.7 GB to 17 GB and access rates from 600 KBps to 1.3 MBps. One of the best features of DVD drives is that they are backward-compatible with CD-ROMs. This means that DVD players can play old CD-ROMs, and video CDs, as well as new DVD-ROMs. Newer DVD players, called second-generation or DVD-2 drives, can also read CD-R and CD-RW disks.
- *DAT*: Acronym for digital audio tape, a type of magnetic tape that uses a scheme called helical scan to record data. A DAT cartridge contains a magnetic tape that can hold from 2 to 24 GBs of data. It can support data transfer rates of about 2 MB per second. Like other types of tapes, DATs are sequential-access media. The most common format for DAT cartridges is DDS (digital data storage).

Concluding Remarks

The rapid development of computer technology and data processing has made these techniques widely accessible and used. Furthermore, the ease of implementation of various data acquisition and processing techniques and tools as well as the comfort of generation and implementation of complex transformations of the signal data, with the excellent graphical presentations with just a couple of finger strokes on the computer keyboard, has tempted a number of researchers to slip into this clicking environment. Nevertheless, the essential question remaining is what do such manipulations and operational condensations of the signal and data signify in the context of a particular paradigm and set of observations. Are the signal processing and analyses undertaken the most appropriate ones, and if so, do they implement the optimal methods and what kind of errors, inaccuracies and ambiguities can result from them?

This has recently been recognized by several authors, leading to a number of publications and specialized issues aiming to critique modern signal processing and analysis approaches and at the same time provide introductory texts with a number of practical examples to which the techniques they discuss may or may not be applied.

As a bottom line, researchers should be aware of their own attitudes towards the phenomena under investigation and the tactical approaches they deploy in their experiments. Before jumping hastily into an experiment, one should stop and ask what is really required of the data to be collected, preferably before its accumulation. This requires a clear concept of the purpose of the experiment with respect to the type of observation to be made, measurements used and processing and analysis necessary for making final comparisons and conclusions as to the mechanisms investigated. It also requires a clear hypothesis about the nature and composition of the signals. It is only within such a conceptual and operational framework that sense can be made of the results acquired and processed with tricky techniques.

Finally, a simple and essential piece of advice to inexperienced researchers, particularly those with no mathematical or engineering background, is to stop and ask oneself what particular operation is to be done and why. The raw signal recorded contains all information available. Conditioning and processing may reveal hidden features, but can also hide features and frequently remove information that may be of importance. It is essential to remember that these procedures never add information to a signal, so at times it may be better not to process it from the beginning.

Acknowledgement: This work was supported by a Serbian Ministry of Science and Technology Grant.

References

Banks SP (1990) Signal processing in: Signal processing, image processing and pattern recognition. Prentice Hall, New York London Toronto Sydney Tokyo Singapore.

Bronzino JD (1995) The biomedical engineering handbook. CRC Press, IEEE Press.

Clyde FC Jr (1995) Electronic instrument handbook. McGraw-Hill, New York San Francisco Washington DC Auckland Bogota Caracas Lisbon London Madrid Mexico City Milan Montreal New Delhi San Juan Singapore Sydney Tokyo Toronto.

Cobbold RSC (1974) Transducers for biomedical measurements: principles and applications. John Wiley & Sons, New York Chichester Brisbane Toronto Singapore.

Glaser EM, Ruchkin DS (1976) Principles of neurobiological signal analysis. Academic Press, New York San Francisco London.

Loeb GE, Gans C (1986) Electromyography for experimentalists. The University of Chicago Press, Chicago and London.

Lynn PA, Fuerst W (1994) Digital signal processing with computer applications. John Wiley&Sons, New York Chichester Brisbane Toronto Singapore.

Normann RA (1985) Experiments in bioinstrumentation. Department of Bioingeneering University of Utah, Salt Lake City.

Sherman-Gold R (Ed) (1993) The axon guide for electrophysiology & biophysics, Laboratory techniques. Axon Instruments, Foster City, USA.

Proakis JG, Manolakis DG (1996) Digital signal processing, principles, algorithms, and applications. Prentice Hall, New Jersey.

Webster JG (1992) Medical instrumentation, application and design. Houghton Mifflin Co, Boston Toronto.

Glossary

Aliasing Aliasing or frequency folding is interference due to an insufficiently high sampling rate in A/D conversion when an input signal has frequency components at or higher than half the sampling rate. If the signal is not correctly band-limited before sampling to eliminate these frequencies, they will show up as aliases or spurious lower frequency errors that cannot be distinguished from valid sampled data.

Analog A term applied to any device, usually electronic, that represents values by a continuously variable physical property, such as voltage in an electronic circuit.

Analog Signal An analog signal consists of a voltage or current that varies continuously within a range of values.

Analog-to-Digital (A/D) Converter An analog-to-digital converter is a device that translates analog signals to digital signals suitable for input to the computer. It periodically measures (samples) the analog signals and converts each measurement into the corresponding digital value.

Analog Filter An analog filter is defined by a rational function of the form:

$$G(s) = \frac{N(s)}{D(S)} = \frac{\prod_{i=1}^{m}(s - z_i)}{\prod_{i=1}^{n}(s - p_i)}$$

where s is the Laplace variable (a complex number) and the complex numbers z_i, p_i are, respectively, the zeros and poles of the transfer function G. We define the loss function (or attenuation) of G (in dB) by

$$A(\omega) = 20\log_{10}\frac{1}{|G(\omega)|}.$$

Anti-alias Filter An anti-alias filter is always required to band-limit the signal before sampling and to avoid aliasing errors. Such filters are specified according to the sampling rate of the system, and there must be one filter per input signal. In practice, such filtering is commonly used prior to sampling.

Attenuation The weakening of a transmitted signal as it travels farther from its point of origin. Usually attenuation is measured in decibels (dB). The opposite of attenuation is amplification or gain.

Averaging Filtering technique based on the summation of M time-locked, stationary waveforms buried in real-world broadband noise. Averaging improves S/N by a factor of $\sqrt{M}$.

Backup As a noun, a duplicate copy of a program, a disk, or data, made either for archiving purposes or for safe-guarding files from loss if the active copy is damaged or destroyed. A backup is an "insurance" copy. As a verb, back up means to make a backup copy.

Bandwidth The difference between the highest and lowest frequencies of a certain signal or of the frequency range of an electronic device, within which it transmits, processes or stores a signal.

Blanking Brief suppression of a signal.

Butterworth Filter Filter built around the Butterworth polynomial; it is a mathematical approximation to an ideal filter, in which the magnitude of the transfer function in the frequency domain is maximally flat. Butterworth filters are optimal in the sense that they provide the least weakening without overshoot in the magnitude response.

Chebyshev Filter Variant of the Butterworth filter, in which the magnitude of the transfer function has a series of ripples in the passband that are of equal amplitude. Chebyshev filters are optimal in the sense that they provide the sharpest transition band for a given filter order.

Common-Mode (CM) Voltage Common-mode voltage is the voltage common to both input voltages. Ideally the instrumentation amplifier ignores the common-mode voltage and amplifies the difference between the two inputs.

Common-Mode Rejection Ratio (CMRR) CMRR is the degree to which the amplifier rejects common-mode voltages and is usually expressed in decibels: CMRR = 20 log (G_D/G_{CM})[dB], where G_D is differential gain and G_{CM} is common-mode gain.

Conditioning The use of special equipment to improve the ability of the signal processing line to transmit data.

Data Plural of the Latin *datum* (that which is given), meaning an item of information. Following classical usage, one item of information should be called a datum, and more than one item should be called data: "The datum is", but "the data are". In practice, however, data is frequently used for the singular as well as the plural form of the noun.

Data Acquisition The process of obtaining data from another source, typically one outside the system, such as by electronic sensing.

DC Offset The shift in the DC level of the signal.

Decibel (dB) One-tenth of a *bel* (derived from Alexander Graham Bell), a dimensionless unit of relative measurement commonly used in signal processing. Measurements in decibels fall on a logarithmic scale, and they compare the measured quantity against a known reference, or against another measured quantity of the same kind.

Differential Inputs Reduce noise picked up by the signal leads. For each input signal there are two signal wires. A third connector allows the signals to be referenced to the ground. The measurement is the difference in voltage between the two wires: any voltage common to both wires is removed.

Digital Digital representation that maps values onto discrete numbers, limiting the possible range of values to the resolution of the digital device.

Digital Filter Digital filter is an algorithm implemented in computer software that transforms digital input signal. Digital filters are usually specified in the frequency domain in terms of the frequency ranges that they leave unaffected and those that are removed from any input signal.

Digital Signal A signal in which information consists of discrete numeric values represented by binary patterns of 0s and 1s, or physically by "low" and "high" voltages.

Digital Signal Processor Abbreviated DSP. An integrated circuit designed for high-speed data manipulations, used in data acquisition applications, for example.

Digital-to-Analog (D/A) Converter A digital-to-analog converter is a device that transforms series of samples back into an analog signal. A D/A converter takes a sequence of discrete digital values as input and creates an analog signal whose amplitude corresponds, moment by moment, to each digital value.

Disc and Disk It is now standard practice to use the spelling disc for optical discs and the spelling disk in all other computer contexts, such as floppy disks, hard disk, and so on.

Discrete vs. Digital Signal A discrete signal may be specified *a priori*, without any reference to a continuous-time system, or it may be obtained by sampling a continuous-time signal. If x (t) is a continuous-time signal and we sample at intervals of length T_s, than we obtain the sequence $x(nT_s)$, with n=1…M, and M the maximal number of samples. When we wish to process a sampled signal by computer, then we must digitize each sample. Rounding $x(nT_s)$ to its nearest level results in a quantized signal value $x_q(nT_s)$. The quantized signal x_q is called a digital signal to distinguish from discrete signal.

Discrete Time System A discrete system is an algorithm that operates on an input sequence x and produces an output sequence y. Obvious properties of this kind are linearity and time invariance. The response to any input of linear, time-invariant (LTI) system can be found by convolving the input with the response of the system to the unit impulse. This implies that an LTI system is completely characterized by its impulse response.

Display Displays are used to reveal important attributes of a signal. In a computer environment, a display is the visual output of a computer, which is commonly CRT-based video display. With notebook computers, the display is usually LSD-based.

Fast Fourier Transform Abbreviated FFT. A set of algorithms used to compute the discrete Fourier transform of a function, which in turn is used for solving series of equations, performing spectral analysis, and carrying out other signal-processing and signal-generation tasks.

Finite Impulse Response (FIR) Filters Filters whose response to a single input impulse remains only as long as the next sample arrives to be included in the calculating formula.

Fitting The calculation of a curve or other line that most closely approximates a set of data points or measurements.

Fourier Transform
Transforms data from the time into the frequency domain and is a representation that is often easier to work with.

Frequency Response
Also called a spectral response. It embraces magnitude and phase-frequency characteristics.

Infinite Impulse Response (IIR) Filters
Their response to a single impulse extends indefinitely into the future. The output of IIR filters depends not only on the inputs, but on previous outputs as well.

Quantization Effects
Quantizing effects are introduced by analog-to-digital conversion and are due to coding technique.

Random Noise
A noise in which there is no relationship between amplitude and time and in which many frequencies occur without pattern or predictability.

Random Number Generator
Generator that creates a number or sequence of numbers characterized by unpredictability so that no number is any more likely to occur at a given time or place in the sequence than any other is. Because a truly random number generator is generally viewed as impossible, the process would be more properly called "pseudo-random number generator".

Resolution
The resolution of an A/D or D/A converter is the number of steps the range of the converter is divided into. The resolution is usually expressed as bits (n) and the number of steps is 2^n, so a converter with a 12-bit resolution divides its range into 2^{12} or 4096 steps. In this case a [0–10] volt range will be broken up to 0.25 millivolts.

Rms - Root Mean Square
The square root of the sum of the squares of a set of quantities divided by the total number of quantities. Used in monitoring and measuring AC signals.

Signal-to-Noise Ratio
Abbreviated as S/N or SNR. The amount of power by which a signal exceeds the amount of noise at the same point in transmission or processing.

Signal Processing
An umbrella term for the work performed by mostly electronic devices, more specifically the systematic manipulation of signals to transform it in some way in order to achieve a desired goal.

Spectrum
The range of frequencies that a signal contains in the frequency domain.

Settling Time
Settling time of the amplifier is the time necessary for the output to reach final amplitude to within small error (often 0.01%) after the signal is applied to the input.

Signal
A variation in the amplitude and polarity of an observed physical quantity produced by a mechanism we desire to understand by experimental investigation.

Smoothing
Techniques to smooth rugged variations in digitized data with a tendency to preserve features of the data such as peak height and width.

Storage
In computer terms, any physical device in which computer information can be kept.

Index

F

R

T

Installation and System Requirements

Contents The contents of the entire book are stored on the CD-ROM. You can search for words in the text via a full text index and browse through those pages which contain the word you want to look up. To do this, you need to install Adobe®Acrobat® Reader 4.0 with search function, which is included on this CD-ROM for Windows 95, Windows 98, Windows NT, and Macintosh.

If there is no Adobe®Acrobat® Reader version for your operating system included on this CD-ROM, you can download the most recent version free of charge from the internet via "http://www.adobe.de" or "http://www.adobe.com". The program is also available there for various Unix versions.

Installation Installation of Adobe, Acrobat, Reader 4.0 with search function:

Windows: Click on "Install.exe" to install the program for Windows 95 or Windows 98

Macintosh: To install Adobe® Acrobat 4.0 with Search
Start the Acrobat_Reader_Installer
in the directory /mac/reader
and follow the instructions
(if you don't see the umlauts -e.g. ä,ü,ö inthe editors names on the startscreen, please install Adobe Type Manager)

UNIX: Unix-versions of the Adobe®Acrobat® Reader 4.0 with Search can be found in the directory /unix/rdr_srch

Start After having installed the Adobe® Acrobat® Reader 4.0 with search function, click on"Start.pdf" on your CD-ROM to start the program.

Information You can get further information and the newest Adobe® Acrobat®Reader at http://www.adobe.com.

Recommondation We strongly advise to use Adobe®Acrobat® 4.0 to guarantee that equations are displayed properly.

Return or exchange
only possible when packaging unopened

System requirements

Windows

- i486™ Pentium® or Pentium Pro-Processor-based personal computer
- Microsoft® Windows® 95, 98 oder Windows NT® 4.0
- 10 MB RAM, recommended 16 MB RAM (Windows)
- 16 MB RAM, recommended 24 MB (Windows NT)
- 10 MB available hard-disk space

Macintosh

- Apple Power Mactintosh or compatible computer
- MacOS software version 7.1.2 or later
- 4.5 MB of available RAM (6.5 MB recommended)
- 12.5 MB of available hard-disk space
- 8 MB RAM available RAM

UNIX

IBM AIX: IBM AIX® 4.1. or later, Common Desktop Enviroment (CDE) 1.0 or Motif®
DEC OSF/1: DEC OSF/1 version 4.0 or later
SGI Irix: Silicon Graphics® IRIX™, 5.3 or later
HP-UX: HP 9000 Series Workstations model 700 or higher
Hewlett-Packard HP-UX version 9.0.3 or later
X Window System X11R5 with HP-VUE or Common Desktop Environment (CDE) 1.0 or later
32 MB of available RAM
12 MB of available hard-disk space